Multivariable Calculus, Linear Algebra, and Differential Equations

THIRD EDITION

MULTIVARIABLE CALCULUS, LINEAR ALGEBRA, AND DIFFERENTIAL EQUATIONS

STANLEY I. GROSSMAN
University of Montana and University College London

SAUNDERS COLLEGE PUBLISHING
HARCOURT BRACE COLLEGE PUBLISHERS
Fort Worth Philadelphia San Diego New York Orlando Austin
San Antonio Toronto Montreal London Sydney Tokyo

Text Typeface: Times Roman
Compositor: York Graphic Services, Inc.
Acquisitions Editor: Jay Ricci
Developmental Editor: Beth Sweet
Managing Editor: Carol Field
Production Management: York Production Services
Manager of Art and Design: Carol Bleistine
Art Director: Jennifer Dunn
Text Designer: York Production Services
Cover Designer: Jennifer Dunn
Director of EDP: Tim Frelick
Production Manager: Carol Florence
Marketing Manager: Monica Wilson
Cover art: © J.A.K. Graphics, Ltd.

Printed in the United States of America

MULTIVARIABLE CALCULUS, LINEAR ALGEBRA, AND DIFFERENTIAL EQUATIONS, 3rd ed.

0-03-003038-2

Library of Congress Catalog Card Number: 94-21766

3456 048 987654321

This book was printed on acid-free recycled content paper, containing
MORE THAN
10% POSTCONSUMER WASTE

To Kerstin, Aaron, and Erik

PREFACE

In 1977 the first edition of my book *Calculus* was published. It, like the four editions that followed, contained a comprehensive introduction to the calculus of one and several variables. Many instructors suggested that since a large number of students stopped after studying one-variable calculus, a shorter version of my text should be available.

To meet that need, Saunders published *Calculus of One Variable* in 1981, 1986, and 1993. This one-variable calculus text contains the first ten chapters of *Calculus, Fifth Edition.* The original plan was to publish, simultaneously, a second ''short book'' containing the last six chapters of the main text.

Most first-year calculus courses cover similar material. However, I soon found that this was not the case for the second-year course. Some schools cover only multivariable calculus in the second year. Others include some linear algebra, some differential equations, or both. Moreover, some universities include more advanced calculus material in the second year: topics such as Taylor's theorem in n variables and mappings from $\mathbb{R}^n$ to $\mathbb{R}^m$.

Thus, we made the decision to write a book that would be usable in a wide variety of courses. My goal has been to retain the flavor of the original calculus book while making a large number of traditionally ''post-calculus'' topics accessible to sophomores.

This book is the result. The third edition is divided into five parts:

I. Multivariable Calculus (Chapters 1–5)
II. Linear Algebra (Chapters 6–8)
III. Introduction to Intermediate Calculus (Chapter 9)
IV. Differential Equations (Chapters 10 and 11)
V. Review of Taylor Polynomials, Sequences, and Series (Chapter 12)

The last chapter is indeed a ''review.'' It is placed at the end so that it can be ignored by those who have seen the material before. However, for those students who have not previously studied infinite series, it should be covered first. Except for the last section, Chapter 12 does not contain references to anything in Chapters 1 through 11 of this book.

PREREQUISITES

This text covers a wide range of topics, but the only prerequisite is a course in one-variable calculus. Principally, I expect that a student using this book will know the following:

- □ how to compute limits
- □ how to differentiate any elementary function
- □ the basic techniques of integration including, especially, integration by parts and integration by a variety of substitutions
- □ the basic applications of differentiation and integration including curve sketching, computing areas, and computing volumes

The student who comes to the course with these skills will do well.

FEATURES

EXAMPLES

The book contains approximately 750 examples. Each example includes *all* the algebraic steps needed to complete the solution. In many instances, explanations are highlighted with a marginal note to make a step easier to follow.

EXERCISES The text contains over 5500 exercises. As in all mathematics books, these are the most important learning tool in the text. Problems are graded in order of increasing difficulty, and there is a balance between technique and proof. The more difficult problems are marked with * and a few exceptionally difficult ones with **.

SELF-QUIZ PROBLEMS Almost every problem set begins with multiple-choice and true-false questions that require little or no computation. Answers to these problems appear at the end of the problem set (not at the back of the book). These problems are designed to test whether the student understands the basic ideas in the section, and they should be done before tackling the more standard problems that follow. The Self-Quiz Problems are new to this edition.

CHAPTER REVIEW EXERCISES

At the end of each chapter, I have provided a collection of review exercises. Any student who can do these exercises can feel confident that he or she understands the material in the chapter.

CHAPTER SUMMARIES A detailed review of the important results of each chapter appears at the end of that chapter.

EXAMPLE AND FIGURE TITLES

Every example and figure in the book is titled so that a student can see both more quickly and more clearly the point of that example or figure.

APPLICATIONS The topics in this book comprise the heart of what is known as *applied* mathematics. Consequently, applications are an important part of the courses for which this book is intended. I won't list here the many applications in the text. However, I'd like to point to the many section-long applications that are listed in the table of contents. As the majority of the students taking this course will be students of engineering and the physical sciences, the majority of the applications are in these areas. However, there are a number of other fields covered. See, for example, the applications to population growth in Section 8.14 and the discussion of models of epidemics in Section 11.6.

SOME OTHER FEATURES

There are many other features which, I hope, will make this book more interesting. Here, is a list of a few of my favorite unusual items:

- A description of how to draw a plane (page 56)
- A discussion of Newton's method for two variables (Section 3.12)
- The use of parametric integration to compute $\int_0^\infty \frac{\sin x}{x}\,dx$ (page 261)
- The frequent appearance of the Summing Up Theorem which ties together seemingly disparate topics in the study of linear algebra. The theorem is first encountered in Section 6.2 (page 374). Successively more complete versions of this theorem are found on pages 427, 440, 478, 509, 533, 583, and 597.

- An answer to the question "When is a differential equation separable?" (on page 708)
- Using the Cayley-Hamilton theorem to find the principal matrix solution (e^{At}) to a linear, homogeneous system of differential equations with constant coefficients (Section 11.12)
- A fixed point convergence theorem (page 925)

HISTORICAL NOTES

Mathematics becomes more interesting if one knows something about the people who helped to develop it. Whenever someone's name is mentioned, I identify that person and try to say a bit about his discovery. In addition, I have added seventeen longer essays that describe a person or the development of an important idea in mathematics. They include:

- Sir William Rowan Hamilton (page 12)
- Josiah William Gibbs and the Origins of Vector Analysis (page 49)
- René Descartes (page 88)
- Distinguishing Between Ordinary and Partial Derivatives (page 175)
- Karl Weierstrass (page 220)
- Joseph Louis Lagrange (page 228)
- Leonhard Euler (page 251)
- Pappus of Alexandria (page 268)
- Isaac Newton (page 294)
- Alternative biography of Isaac Newton, written by Stephen Hawking (page 295)
- Carl Friedrich Gauss (page 353)
- Carl Gustav Jacobi (page 363)
- Arthur Cayley and the Algebra of Matrices (page 406)
- A Short History of Determinants (page 470)
- Compound Interest in the Seventeenth Century (page 696)
- The Struggle to Understand Infinite Sums (page 940)
- Colin Maclaurin (page 992)

CALCULATORS AND COMPUTERS

Many examples and exercises in this book require the use of a calculator. Such examples are marked with the symbol 🖩. Computer software is not used in this text. Software and materials explaining their use in calculus, linear algebra, and differential equations are available from the publisher. However, in five chapters of the book (Chapters 2, 3, 4, 10, and 12) I have added a few exercises that require the use of an appropriate software package.

I expect that most students taking this course will have access to a symbolic software manipulator such as DERIVE, MATHEMATICA, MAPLE, MATLAB, or MATHCAD. These tools greatly simplify many of the calculations carried out in this text. However, to try to incorporate these and other devices into the book would have made the text unwieldly. It is best, I believe, to leave the decisions whether to incorporate software into the text, what software to use, and how to use it up to the instructor.

ANSWERS

The answers to most odd-numbered exercises appear at the back of the book. In addition, a *Student's Solution Manual* containing detailed solutions to all odd-numbered problems is available, as is an *Instructor's Solutions Manual* containing detailed solutions to the even-numbered problems. Both manuals were prepared by Leon Gerber at St. John's University in New York.

NUMBERING IN THE TEXT

Numbering in the book is fairly standard. Examples, problems, theorems, and equations are numbered consecutively within each section, starting with 1. Reference to an example, problem, theorem, or equation outside the section in which it appears is by chapter, section, and number. Thus, what is labeled simply as Example 4 in Section 2.3 is referred to as Example 2.3.4 outside the section. In addition, in many cases cross-referenced page numbers are included to make it easy to find an important reference. Finally, ends of proofs are denoted by the symbol ■.

REFERENCES TO *Calculus* AND *Calculus of One Variable*

In a number of places in the book—especially in Chapters 1–5—I make use of results from one-variable calculus. These results can be found in most basic calculus texts. However, to make these references precise, I give page numbers on which they can be found in my books *Calculus* or *Calculus of One Variable.* These refer to *Calculus, Fifth Edition,* published in 1993 by Saunders and its one-variable counterpart.

ORGANIZATION

As mentioned earlier, the book is divided into five parts. The first part consists of Chapters 1–5. These are similar to Chapters 11–15 in *Calculus, Fifth Edition,* and include basic multivariable calculus material. The basic difference is that I introduce the space $\mathbb{R}^n$ in Section 1.10 and then, in later sections, I generalize to $\mathbb{R}^n$ basic topics in $\mathbb{R}^2$ and $\mathbb{R}^3$. Vectors in $\mathbb{R}^2$, $\mathbb{R}^3$, and $\mathbb{R}^n$ are discussed in Chapter 1 with vector functions in Chapter 2.

Chapter 3 contains an introduction to the calculus of two or more variables. The gradient is introduced in Section 3.5 as the natural extension of the ordinary derivative. Chapter 4 provides an introduction to multiple integration with an emphasis on applications. Chapter 5 contains a detailed introduction to vector analysis including a discussion, with proofs and applications, of Green's, Stokes's and the divergence theorems.

The second part of the book is an introduction to linear algebra in Chapters 6, 7, and 8. This material requires no multivariable calculus except a familiarity with vectors in $\mathbb{R}^2$, $\mathbb{R}^3$ and $\mathbb{R}^n$.

Chapters 6 and 7 contain introductions to matrices, determinants, and the Gauss–Jordan technique for solving systems of equations. Chapter 8 includes more advanced material on vector spaces, linear transformations, and eigenvalues and eigenvectors.

The third part of the book consists of the single Chapter 9. The chapter combines techniques from calculus and linear algebra and contains discussions of some of the most elegant results in the calculus including Taylor's theorem in n variables, the multivariable mean value theorem, and the implicit function theorem. None of the results here are found in standard calculus texts.

Chapters 10 and 11 comprise the fourth part of the book and provide a one-quarter or semester introduction to ordinary differential equations. Chapter 10 is independent of Chapters 1–9 and can be covered at any time. It contains detailed discussions of first-order and linear second-order equations. Also included are optional discussions of electric circuits and vibratory motion.

Chapter 11, on systems of differential equations, begins with three sections that require no matrix theory. The remainder of the chapter combines matrix theory and linear systems. The diagonalization technique is used in Section 11.9 to compute e^{At}, the principal matrix solution of a linear homogeneous system of differential equations. A method for computing e^{At}, even when A cannot be diagonalized, is presented in Section 11.12.

As discussed earlier, the final chapter contains a review of Taylor polynomials, sequences, and series. The first twelve sections can be covered at any time. The only prerequisite for them is one-variable calculus. Section 12.13 describes a method for solving linear differential equations using power series and uses some of the material in Chapter 10.

There are six appendices. Appendix 1 contains a review of mathematical induction. The very useful binomial theorem is described and proved in Appendix 2. In discussing eigenvalues and eigenvectors, and in describing solutions to certain second-order differential equations, it is necessary to know something about complex numbers. For that reason I have provided a discussion of complex numbers in Appendix 3.

Basic properties about determinants can be proved once we know that the determinant can be obtained by expanding by cofactors in any row or column. This central result is proved in Appendix 4.

A basic existence-uniqueness result for solutions of first-order initial-value problems is proved in Appendix 5. Finally, in Appendix 6, I prove that every vector space has a basis.

ACKNOWLEDGMENTS

I am grateful to many individuals who helped in the preparation of this text. Many of the reviewers of *Calculus, Fifth Edition,* provided useful criticism that improved the material in Chapters 1–5. Five reviewers painstakingly worked their way through the entire text of the first edition of this book and provided hundreds of detailed, insightful suggestions. I am particularly grateful to Professor Geoge Cain of the Georgia Institute of Technology, Professor Art Copeland at the University of New Hampshire, Professor Carl Cowen at Purdue University, Professor Charles Denlinger at Millersville State College, and Professor Keith Yale at the University of Montana.

I wish to thank the following individuals for their invaluable help in preparing the second edition: Alfred D. Andrew, Georgia Institute of Technology; James M. Edmondson, Santa Barbara Community College; Nathaniel Grossman, UCLA; Daniel S. Kahn, Northwestern University; T. J. Ransford, University of Leeds; John Venables, Lakewood Community College, White Bear Lake; and Paul Yearout, Brigham Young University.

The following individuals worked through the draft of this third edition. I am grateful to them for there many helpful suggestions:

George Cain, The Georgia Institute of Technology
Jen Chayes, University of California at Los Angeles
Robert Foote, Wabash College
Chris Freling, California State University, San Bernadino
James Herod, the Georgia Institute of Technology
Daniel S. Kahn, Northwestern University

Professor Leon Gerber, who prepared the Student's and Instructor's manuals, made many useful suggestions for the improvement of the problem sets. This book is a better teaching tool because of him.

I am grateful to V. C. Varadachari at Lakewood Community College in Minnesota and Mary Beth Young at Georgia Institute of Technology who worked all the odd-numbered problems in the book. Because of their accurate work, I can say with confidence that the answers at the back of the book are correct.

Some of the material in Chapters 10 and 11 will appear in my book (with W. R. Derrick) *Elementary Differential Equations with Boundary Value Problems, Fourth Edition,* to be published by HarperCollins in 1996. I wish to thank HarperCollins for permission to use this material here.

Some of the material here first appeared in *Mathematics for the Biological Sciences* (New York: Macmillan, 1974) written by James E. Turner and myself. I am grateful to Professor Turner for permitting its use.

A great deal of this book was written while I was a research associate at University College London. I wish to thank the Mathematics Department of UCL for providing office facilities, mathematical suggestions, and, especially, friendship during my annual visits there.

Special thanks go to the editorial and production staffs at Saunders College Publishing for the care and skill they brought to this product. My editor, Jay Ricci, made many helpful suggestions. Finally, I am grateful to Kirsten Kauffman at York Production Services for her great skill in turning my manuscript into a bound book.

Stanley I. Grossman
August 1994

TABLE OF CONTENTS

PART I

MULTIVARIABLE CALCULUS

PART II

LINEAR ALGEBRA

PART III

INTRODUCTION TO INTERMEDIATE CALCULUS

PART IV

DIFFERENTIAL EQUATIONS

PART V

SEQUENCES AND SERIES

Multivariable Calculus, Linear Algebra, and Differential Equations

PART I

MULTIVARIABLE CALCULUS

CHAPTERS INCLUDE:

CHAPTER 1

VECTORS IN THE PLANE AND IN SPACE

Before we can discuss calculus in two or more dimensions, we need to know something about the geometry of two and three dimensional space. This geometry is discussed in the language of vectors and vector functions.

The modern study of vectors began essentially with the work of the great Irish mathematician Sir William Rowan Hamilton (1805–1865) who worked with what he called quaternions.[†] After Hamilton's death, his work on quaternions was supplanted by the more adaptable work on vector analysis by the American mathematician and physicist Josiah Willard Gibbs (1839–1903) and the general treatment of ordered n-tuples by the German mathematician Hermann Grassmann (1809–1877).

Throughout Hamilton's life and for the remainder of the nineteenth century, there was considerable debate over the usefulness of quaternions and vectors. At the end of the century, the great British physicist Lord Kelvin wrote that quaternions, "although beautifully ingenious, have been an unmixed evil to those who have touched them in any way. . . . Vectors . . . have never been of the slightest use to any creature."

But Kelvin was wrong. Today nearly all branches of classical and modern physics are represented using the language of vectors. Vectors are also used with increasing frequency in the social and biological sciences. Quaternions, too, have recently been used in physics—in particle theory and other areas.

In this chapter, we will explore properties of vectors in the plane and in space. When going through this material, keep in mind that, like most important discoveries, vectors have been a source of great controversy—a controversy that was not resolved until well into the twentieth century.[‡]

[†] See the biographical sketch of Hamilton on p. 12.

[‡] For interesting discussions of the development of modern vector analysis, consult the book by M. J. Crowe, *A History of Vector Analysis* (Notre Dame: University of Notre Dame Press, 1967), or Morris Kline's excellent book *Mathematical Thought from Ancient to Modern Times* (New York: Oxford University Press, 1972), Chapter 32.

1.1 VECTORS AND VECTOR OPERATIONS

In many applications of mathematics to the physical and biological sciences and engineering, scientists are concerned with entities that have both magnitude (length) and direction. Examples include the notions of force, velocity, acceleration, and momentum. It is frequently useful to express these quantities both geometrically and algebraically.

Let P and Q be two different points in the plane. Then the **directed line segment** from P to Q, denoted by $\overrightarrow{PQ}$, is the straight-line segment extending from P to Q [see Figure 1(a)]. Note that the directed line segments $\overrightarrow{PQ}$ and $\overrightarrow{QP}$ are different since they point in opposite directions [Figure 1(b)].[†]

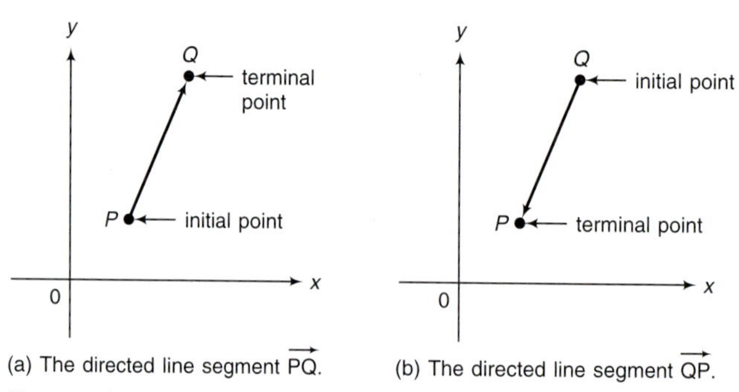

(a) The directed line segment $\overrightarrow{PQ}$. (b) The directed line segment $\overrightarrow{QP}$.

FIGURE 1
The directed line segments $\overrightarrow{PQ}$ and $\overrightarrow{QP}$ point in opposite directions.

The point P in the directed line segment $\overrightarrow{PQ}$ is called the **initial point** of the segment and the point Q is called the **terminal point**. The two important properties of a directed line segment are its magnitude (length) and its direction. If two directed line segments $\overrightarrow{PQ}$ and $\overrightarrow{RS}$ have the same magnitude and direction, we say that they are **equivalent** no matter where they are located with respect to the origin. The directed line segments in Figure 2 are all equivalent.

DEFINITION GEOMETRIC DEFINITION OF A VECTOR

The set of all directed line segments equivalent to a given directed line segment is called a **vector**. Any directed line segment in that set is called a **representation** of the vector. ■

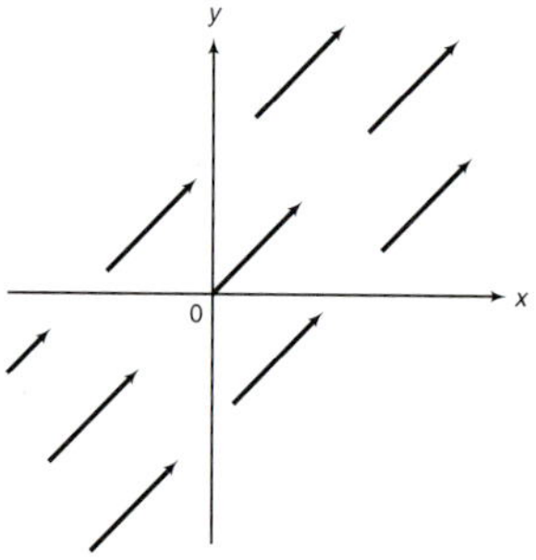

FIGURE 2
Equivalent directed line segments

REMARK: The directed line segments in Figure 2 are all representations of the same vector.

NOTATION: We will denote vectors by lowercase boldface letters such as **v**, **w**, **a**, **b**.

From the definition, we see that a given vector **v** can be represented in many different ways. In fact, let $\overrightarrow{PQ}$ be a representation of **v**. Then without changing magnitude or direction, we can move $\overrightarrow{PQ}$ in a parallel way so that its initial point is shifted to the origin. We then obtain the directed line segment $\overrightarrow{0R}$, which is another representation of the

[†] We give a formal definition of the direction of a vector on page 6. Here we rely on your intuition.

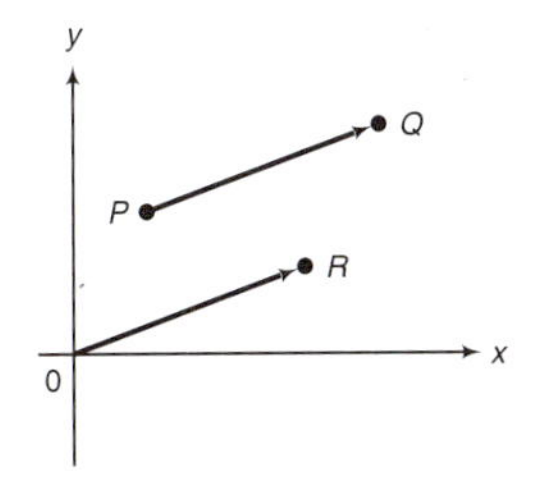

FIGURE 3
$\overrightarrow{0R}$ is the standard representation of the directed line segment $\overrightarrow{PQ}$.

We can move $\overrightarrow{PQ}$ to obtain an equivalent directed line segment with its initial point at the origin. Note that $\overrightarrow{OR}$ and $\overrightarrow{PQ}$ are parallel and have the same length.

vector **v** (see Figure 3). It is called the **standard representation** of the vector. Now suppose that R has the Cartesian coordinates (a, b). Then we can describe the directed line segment $\overrightarrow{0R}$ by the coordinates (a, b). That is, $\overrightarrow{0R}$ is the directed line segment with initial point $(0, 0)$ and terminal point (a, b). Since one representation of a vector is as good as another, we can write the vector **v** as (a, b). In sum, we see that a vector can be thought of as a point in the xy-plane.

REMARK: In a number of places in this text, we shall use phrases such as "a vector lies in a plane." This is common shorthand usage. The precise statement is "a vector has a representation as a directed line segment all of whose points lie in the plane." In the remainder of this book, we shall not worry about this distinction.

DEFINITION ALGEBRAIC DEFINITION OF A VECTOR

A **vector v** in the xy-plane is an ordered pair of real numbers (a, b). The numbers a and b are called the **components** of the vector **v**. The **zero vector** is the vector $(0, 0)$ and is denoted by **0**. Two vectors are **equal** if their corresponding components are equal. That is, $(a, b) = (c, d)$ if $a = c$ and $b = d$. ■

DEFINITION SCALAR

Since we will often have to distinguish between real numbers and vectors (which are pairs of real numbers), we will use the term **scalar**† to denote a real number. ■

DEFINITION MAGNITUDE OF A VECTOR

Since a vector is really a set of equivalent directed line segments, we define the **magnitude** or **length** of a vector as the length of any one of its representations.

Using the representation $\overrightarrow{0R}$, and writing the vector $\mathbf{v} = (a, b)$, we find that

$$|\mathbf{v}| = \text{magnitude of } \mathbf{v} = \sqrt{a^2 + b^2}. \tag{1}$$

■

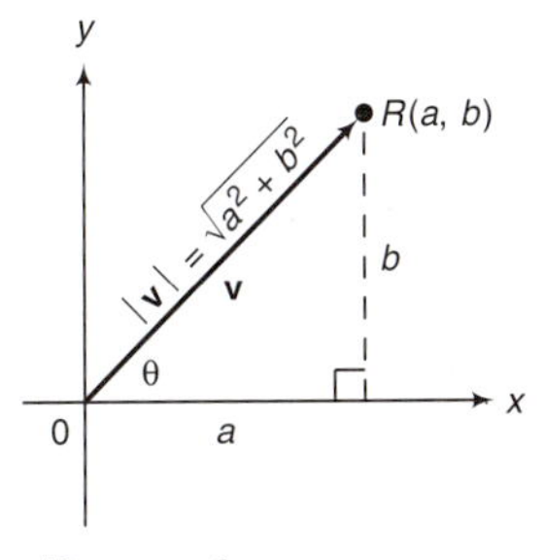

FIGURE 4
The magnitude of a vector

This follows from the Pythagorean theorem (see Figure 4). We have used the notation $|\mathbf{v}|$ to denote the magnitude of **v**. Note that $|\mathbf{v}|$ is a *scalar*.

---

†The term "scalar" originated with Hamilton. His definition of the quaternion included what he called a *real part* and an *imaginary part.* In his paper, "On Quaternions, or on a New System of Imaginaries in Algebra," in *Philosophical Magazine,* 3rd Ser., 25 (1844): 26–227, he wrote

> The algebraically *real* part may receive . . . all values contained on the one *scale* of progression of numbers from negative to positive infinity; we shall call it therefore the *scalar part,* or simply the *scalar* of the quaternion.

Moreover, in the same paper, Hamilton went on to define the imaginary part of his quaternion as the *vector* part. Although this was not the first usage of the word *vector,* it was the first time it was used in the context of the definition of a vector given above. In fact, it is fair to say that the paper from which the quotation above was taken marks the beginning of modern vector analysis.

EXAMPLE 1

CALCULATING THE MAGNITUDES OF SIX VECTORS

Calculate the magnitudes of the vectors **a.** $(2, 2)$; **b.** $(2, 2\sqrt{3})$; **c.** $(-2\sqrt{3}, 2)$; **d.** $(-3, -3)$; **e.** $(6, -6)$; **f.** $(0, 3)$.

SOLUTION:

a. $|\mathbf{v}| = \sqrt{2^2 + 2^2} = \sqrt{8} = 2\sqrt{2}$
b. $|\mathbf{v}| = \sqrt{2^2 + (2\sqrt{3})^2} = \sqrt{16} = 4$
c. $|\mathbf{v}| = \sqrt{(-2\sqrt{3})^2 + 2^2} = 4$
d. $|\mathbf{v}| = \sqrt{(-3)^2 + (-3)^2} = \sqrt{18} = 3\sqrt{2}$
e. $|\mathbf{v}| = \sqrt{6^2 + (-6)^2} = \sqrt{72} = 6\sqrt{2}$
f. $|\mathbf{v}| = \sqrt{0^2 + 3^2} = \sqrt{9} = 3$

DEFINITION DIRECTION OF A VECTOR

We now define the **direction** of the nonzero vector $\mathbf{v} = (a, b)$ to be the angle θ, measured in radians, that the standard representation of the vector makes with the positive x-axis. By convention, we choose θ such that $0 \leq \theta < 2\pi$.

It follows from Figure 4 that if $a \neq 0$, then

$$\tan \theta = \frac{b}{a}. \tag{2}$$

■

Tan θ is periodic of period π, so for $a \neq 0$ there are always *two* numbers in $[0, 2\pi)$ such that $\tan \theta = b/a$. For example, $\tan \pi/4 = \tan 5\pi/4 = 1$. In order to determine θ uniquely, it is necessary to determine the quadrant of $\mathbf{v}$ as we will see in the next example.

REMARK 1: The zero vector has a magnitude of 0. Since the initial and terminal points coincide, we say that *the zero vector has no direction.*

REMARK 2: It follows from the definition that *parallel vectors have the same direction.* This is because two parallel vectors of the same magnitude have the same standard representation.

EXAMPLE 2

CALCULATING THE DIRECTIONS OF SIX VECTORS

Calculate the directions of the vectors in Example 1.

SOLUTION: We depict these six vectors in Figure 5.

a. Here $\mathbf{v}$ is in the first quadrant, and since $\tan \theta = 2/2 = 1$, $\theta = \pi/4$.
b. Here $\theta = \tan^{-1} 2\sqrt{3}/2 = \tan^{-1} \sqrt{3} = \pi/3$.
c. We see that $\mathbf{v}$ is in the second quadrant, and since $\tan^{-1} 2/(2\sqrt{3}) = \tan^{-1} 1/\sqrt{3} = \pi/6$, we see from the figure that $\theta = \pi - (\pi/6) = 5\pi/6$.
d. Here $\mathbf{v}$ is in the third quadrant, and since $\tan^{-1} 1 = \pi/4$, $\theta = \pi + (\pi/4) = 5\pi/4$.

e. Since $\mathbf{v}$ is in the fourth quadrant, and since $\tan^{-1}(-1) = -\pi/4$, $\theta = 2\pi - (\pi/4) = 7\pi/4$.

f. We cannot use equation (2) because b/a is undefined. However, we see in Figure 5(f) that $\theta = \pi/2$.

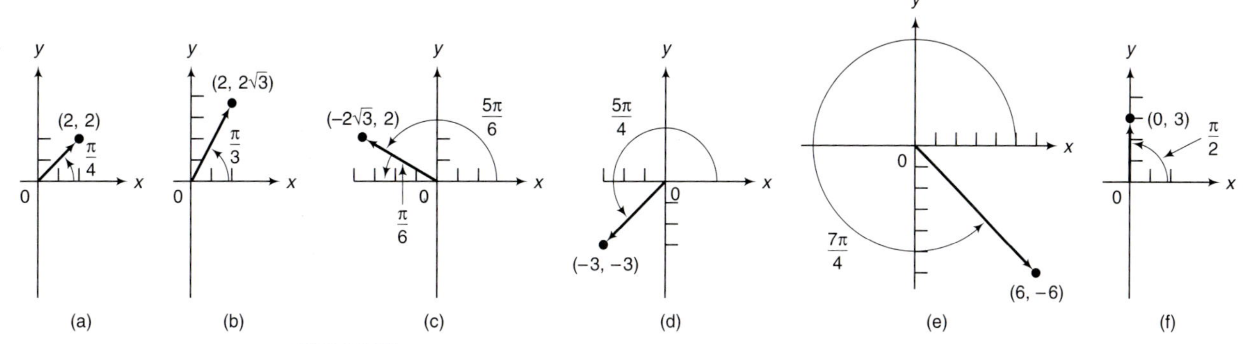

FIGURE 5
The directions of six vectors

In general, if $b > 0$, the direction of $(0, b) = \dfrac{\pi}{2}$ and the direction of $(0, -b) = \dfrac{3\pi}{2}$.

WARNING: The definition of the function $\tan^{-1} x$ has $-\pi/2 < \tan^{-1} x < \pi/2$. But $0 \le$ direction of $\mathbf{v} < 2\pi$. Thus,

$$\theta = \text{direction of } \mathbf{v} \text{ is not necessarily equal to } \tan^{-1}\frac{b}{a}.$$

In fact, $\theta = \tan^{-1} b/a$ only if θ is in the first quadrant. In the other cases, we compute θ as in Example 2.

Let $\mathbf{v} = (a, b)$. Then as we have seen, $\mathbf{v}$ can be represented in many different ways. For example, the representation of $\mathbf{v}$ with the initial point (c, d) has the terminal point $(c + a, d + b)$. This is depicted in Figure 6. It is not difficult to show that the directed line segment $\overrightarrow{PQ}$ in Figure 6 has the same magnitude and direction as the vector $\mathbf{v}$ (see Problems 44 and 57).

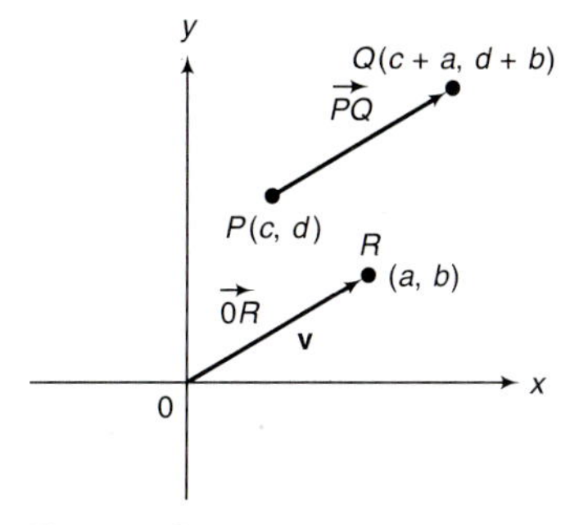

FIGURE 6
The directed line segments $\overrightarrow{PQ}$ and $\overrightarrow{0R}$ are equivalent.

EXAMPLE 3 FINDING A REPRESENTATION OF A VECTOR WHOSE INITIAL POINT IS NOT AT THE ORIGIN

Find a representation of the vector $(2, -1)$ whose initial point is the point $P = (5, -4)$.

SOLUTION: Let $Q = (5 + 2, -4 - 1) = (7, -5)$. Then $\overrightarrow{PQ}$ is a representation of the vector $(2, -1)$. This is illustrated in Figure 7.

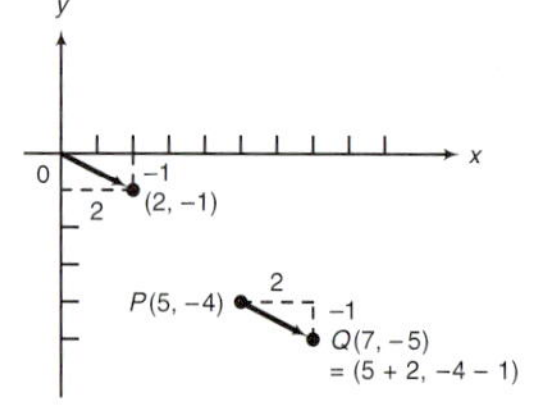

FIGURE 7
Equivalent directed line segments

We now turn to the question of adding vectors and multiplying them by scalars.

DEFINITION ADDITION AND SCALAR MULTIPLICATION OF A VECTOR

Let $\mathbf{u} = (a_1, b_1)$ and $\mathbf{v} = (a_2, b_2)$ be two vectors in the plane and let α be a scalar. Then we define

i. $\mathbf{u} + \mathbf{v} = (a_1 + a_2, b_1 + b_2)$
ii. $\alpha\mathbf{u} = (\alpha a_1, \alpha b_1)$
iii. $-\mathbf{v} = (-1)\mathbf{v} = (-a_2, -b_2)$
iv. $\mathbf{u} - \mathbf{v} = \mathbf{u} + (-\mathbf{v}) = (a_1 - a_2, b_1 - b_2)$. ■

We have:

> VECTOR ADDITION AND SCALAR MULTIPLICATION
>
> To add two vectors, we add their corresponding components. To multiply a vector by a scalar, we multiply each of its components by that scalar.

EXAMPLE 4 FINDING SUMS AND SCALAR MULTIPLES OF VECTORS

Let $\mathbf{u} = (1, 3)$ and $\mathbf{v} = (-2, 4)$. Calculate **a.** $\mathbf{u} + \mathbf{v}$; **b.** $3\mathbf{u}$; **c.** $-\mathbf{v}$; **d.** $\mathbf{u} - \mathbf{v}$; and **e.** $-3\mathbf{u} + 5\mathbf{v}$.

SOLUTION:

a. $\mathbf{u} + \mathbf{v} = (1 + (-2), 3 + 4) = (-1, 7)$
b. $3\mathbf{u} = 3(1, 3) = (3, 9)$
c. $-\mathbf{v} = (-1)(-2, 4) = (2, -4)$
d. $\mathbf{u} - \mathbf{v} = \mathbf{u} + (-\mathbf{v}) = (1 + 2, 3 - 4) = (3, -1)$
e. $-3\mathbf{u} + 5\mathbf{v} = (-3, -9) + (-10, 20) = (-13, 11)$

There are interesting geometric interpretations of vector addition and scalar multiplication. First, let $\mathbf{v} = (a, b)$ and let α be any scalar. Then,

$$|\alpha\mathbf{v}| = |(\alpha a, \alpha b)| = \sqrt{\alpha^2 a^2 + \alpha^2 b^2} = |\alpha|\sqrt{a^2 + b^2} = |\alpha||\mathbf{v}|.$$

> That is, multiplying a vector by a nonzero scalar has the effect of multiplying the length of the vector by the absolute value of that scalar.

Moreover, if $\alpha > 0$, then $\alpha\mathbf{v}$ is in the same quadrant as $\mathbf{v}$ and, since $\tan^{-1}(\alpha b/\alpha a) = \tan^{-1}(b/a)$, the direction of $\alpha\mathbf{v}$ is the same as the direction of $\mathbf{v}$. If $\alpha < 0$, then the direction of $\alpha\mathbf{v}$ is equal to the direction of $\mathbf{v}$ plus π (which is the direction of $-\mathbf{v}$). Thus:

> Scalar multiples of $\mathbf{v}$ have the same direction as $\mathbf{v}$ or $-\mathbf{v}$.

Example 5

Geometric Interpretation of the Scalar Multiple of a Vector

Let $\mathbf{v} = (1, 1)$. Then $|\mathbf{v}| = \sqrt{1 + 1} = \sqrt{2}$ and $|2\mathbf{v}| = |(2, 2)| = \sqrt{2^2 + 2^2} = \sqrt{8} = 2\sqrt{2} = 2|\mathbf{v}|$. Also, $|-2\mathbf{v}| = \sqrt{(-2)^2 + (-2)^2} = 2\sqrt{2} = 2|\mathbf{v}|$. Moreover, the direction of $2\mathbf{v}$ is $\pi/4$, while the direction of $-2\mathbf{v}$ is $5\pi/4$.

Now suppose we add the vectors $\mathbf{u} = (a_1, b_1)$ and $\mathbf{v} = (a_2, b_2)$, as in Figure 8. From the figure we see that the vector $\mathbf{u} + \mathbf{v} = (a_1 + a_2, b_1 + b_2)$ can be obtained by shifting the representation of the vector $\mathbf{v}$ so that its initial point coincides with the terminal point (a_1, b_1) of the vector $\mathbf{u}$. We can therefore obtain the vector $\mathbf{u} + \mathbf{v}$ by drawing a parallelogram with one vertex at the origin and sides $\mathbf{u}$ and $\mathbf{v}$. Then $\mathbf{u} + \mathbf{v}$ is the vector that points from the origin along the diagonal of the parallelogram.

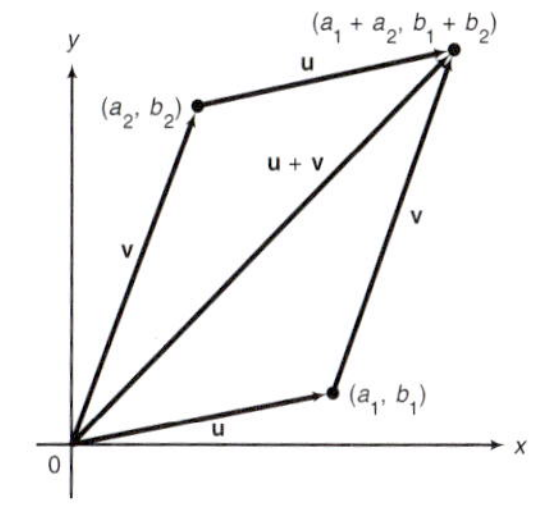

Figure 8
Sum of two vectors

Note: Since a straight line is the shortest distance between two points, it immediately follows from Figure 8 that

> **Triangle Inequality**
>
> $|\mathbf{u} + \mathbf{v}| \le |\mathbf{u}| + |\mathbf{v}|$.

For obvious reasons this inequality is called the **triangle inequality**.

We can also obtain a geometric representation of the vector $\mathbf{u} - \mathbf{v}$. Since $\mathbf{u} = \mathbf{u} - \mathbf{v} + \mathbf{v}$, the vector $\mathbf{u} - \mathbf{v}$ is the vector that must be added to $\mathbf{v}$ to obtain $\mathbf{u}$. This is illustrated in Figure 9.

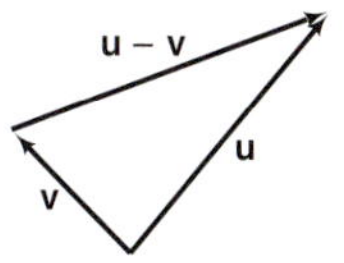

Figure 9
The vector $\mathbf{u} - \mathbf{v}$

The following theorem lists several properties that hold for any vectors $\mathbf{u}$, $\mathbf{v}$, and $\mathbf{w}$ and any scalars α and β. Since the proof is easy, we leave it as an exercise (see Problem 58). Some parts of this theorem have already been proven.

Theorem 1 Properties of Vectors

Let $\mathbf{u}$, $\mathbf{v}$, and $\mathbf{w}$ be any three vectors in the plane, let α and β be scalars, and let $\mathbf{0}$ denote the zero vector.

i. $\mathbf{u} + \mathbf{v} = \mathbf{v} + \mathbf{u}$

ii. $\mathbf{u} + (\mathbf{v} + \mathbf{w}) = (\mathbf{u} + \mathbf{v}) + \mathbf{w}$

iii. $\mathbf{v} + \mathbf{0} = \mathbf{v}$

iv. $0\mathbf{v} = \mathbf{0}$ (here the 0 on the left is the scalar zero)

v. $\alpha\mathbf{0} = \mathbf{0}$

vi. $(\alpha\beta)\mathbf{v} = \alpha(\beta\mathbf{v})$

vii. $\mathbf{v} + (-\mathbf{v}) = \mathbf{0}$

viii. $1\mathbf{v} = \mathbf{v}$

ix. $(\alpha + \beta)\mathbf{v} = \alpha\mathbf{v} + \beta\mathbf{v}$

x. $\alpha(\mathbf{u} + \mathbf{v}) = \alpha\mathbf{u} + \alpha\mathbf{v}$

xi. $|\alpha\mathbf{v}| = |\alpha||\mathbf{v}|$

xii. $|\mathbf{u} + \mathbf{v}| \le |\mathbf{u}| + |\mathbf{v}|$ ■

When a set of vectors together with a set of scalars and the operations of addition and scalar multiplication have the properties given in Theorem 1(i)–(x), we say that the vectors form a **vector space**. The set of vectors of the form (a, b), where a and b are real numbers, is denoted by $\mathbb{R}^2$. We will discuss properties of abstract vector spaces in Chapter 8.

$\mathbb{R}^2$

There are two special vectors in $\mathbb{R}^2$ that allow us to represent other vectors in $\mathbb{R}^2$ in a convenient way. We will denote the vector $(1, 0)$ by the vector symbol $\mathbf{i}$ and the vector $(0, 1)$ by the vector symbol $\mathbf{j}$ (see Figure 10). If (a, b) denotes any vector in $\mathbb{R}^2$, then since $(a, b) = a(1, 0) + b(0, 1)$, we may write

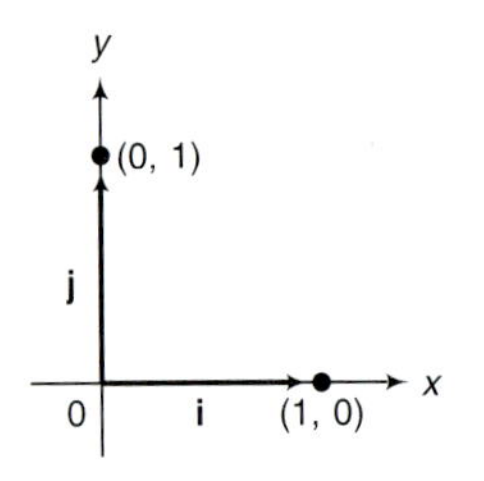

FIGURE 10
The vectors $\mathbf{i}$ and $\mathbf{j}$

WRITING A VECTOR IN TERMS OF I AND J

$$\mathbf{v} = (a, b) = a\mathbf{i} + b\mathbf{j}. \tag{3}$$

Moreover, any vector in $\mathbb{R}^2$ can be represented in a unique way in the form $a\mathbf{i} + b\mathbf{j}$ since the representation of (a, b) as a point in the plane is unique. (Put another way, a point in the xy-plane has one and only one x-coordinate and one and only one y-coordinate.) Thus Theorem 1 holds with this new representation as well.

HISTORICAL NOTE: The symbols $\mathbf{i}$ and $\mathbf{j}$ were first used by Hamilton. He defined his quaternion as a quantity of the form $a + b\mathbf{i} + c\mathbf{j} + d\mathbf{k}$, where a was the "scalar part" and $b\mathbf{i} + c\mathbf{j} + d\mathbf{k}$ the "vector part." In Section 1.4 we will write vectors in space in the form $b\mathbf{i} + c\mathbf{j} + d\mathbf{k}$.

When the vector $\mathbf{v}$ is written in the form $\mathbf{v} = a\mathbf{i} + b\mathbf{j}$, we say that $\mathbf{v}$ is *resolved into its horizontal and vertical components,* since a is the horizontal component of $\mathbf{v}$ while b is its vertical component. The vectors $\mathbf{i}$ and $\mathbf{j}$ are called **basis vectors** for the vector space $\mathbb{R}^2$.

Now suppose that a vector $\mathbf{v}$ can be represented by the directed line segment $\overrightarrow{PQ}$, where $P = (a_1, b_1)$ and $Q = (a_2, b_2)$. (See Figure 11.) If we label the point (a_2, b_1) as R,

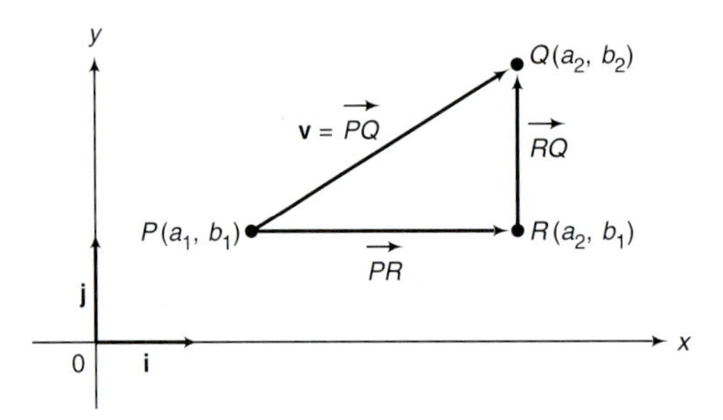

FIGURE 11
$\overrightarrow{PQ} = \overrightarrow{PR} + \overrightarrow{RQ} = (a_2 - a_1)\mathbf{i} + (b_2 - b_1)\mathbf{j}$

then we immediately see that

$$\mathbf{v} = \overrightarrow{PQ} = \overrightarrow{PR} + \overrightarrow{RQ}. \tag{4}$$

If $a_2 \geq a_1$, then the length of $\overrightarrow{PR}$ is $a_2 - a_1$, and since $\overrightarrow{PR}$ has the same direction as $\mathbf{i}$ (they are parallel), we can write

$$\overrightarrow{PR} = (a_2 - a_1)\mathbf{i}. \tag{5}$$

If $a_2 < a_1$, then the length of $\overrightarrow{PR}$ is $a_1 - a_2$, but then $\overrightarrow{PR}$ has the same direction as $-\mathbf{i}$ so $\overrightarrow{PR} = (a_1 - a_2)(-\mathbf{i}) = (a_2 - a_1)\mathbf{i}$ again. Similarly,

$$\overrightarrow{RQ} = (b_2 - b_1)\mathbf{j}, \tag{6}$$

and we may write [using (4), (5), and (6)]

$$\mathbf{v} = (a_2 - a_1)\mathbf{i} + (b_2 - b_1)\mathbf{j}. \tag{7}$$

This is the **resolution of v into its horizontal and vertical components**.

EXAMPLE 6

RESOLVING A VECTOR INTO ITS HORIZONTAL AND VERTICAL COMPONENTS

Resolve the vector represented by the directed line segment from $(-2, 3)$ to $(1, 5)$ into its horizontal and vertical components.

SOLUTION: Using (7), we have

$$\mathbf{v} = (a_2 - a_1)\mathbf{i} + (b_2 - b_1)\mathbf{j} = [1 - (-2)]\mathbf{i} + (5 - 3)\mathbf{j} = 3\mathbf{i} + 2\mathbf{j}.$$

We conclude this section by defining a kind of vector that is very useful in certain types of applications.

DEFINITION UNIT VECTOR

A **unit vector u** is a vector that has length 1. ■

EXAMPLE 7

A UNIT VECTOR

The vector $\mathbf{u} = (\frac{1}{2})\mathbf{i} + (\sqrt{3}/2)\mathbf{j}$ is a unit vector since

$$|\mathbf{u}| = \sqrt{\left(\frac{1}{2}\right)^2 + \left(\frac{\sqrt{3}}{2}\right)^2} = \sqrt{\frac{1}{4} + \frac{3}{4}} = 1.$$

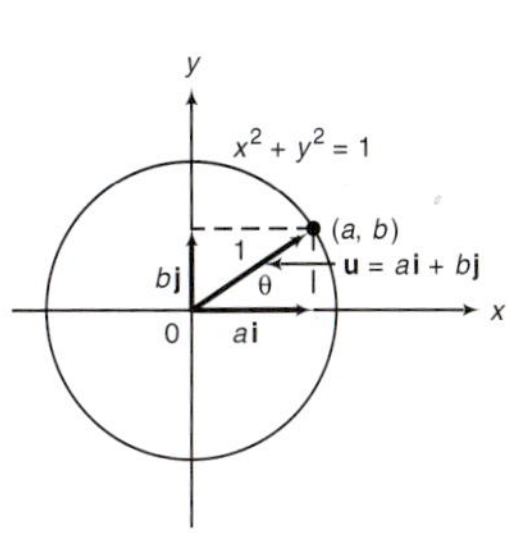

FIGURE 12
Representation of a unit vector

Let $\mathbf{u} = a\mathbf{i} + b\mathbf{j}$ be a unit vector. Then $|\mathbf{u}| = \sqrt{a^2 + b^2} = 1$, so $a^2 + b^2 = 1$ and **u** is a point on the unit circle (see Figure 12). If θ is the direction of **u**, then we immediately see that $a = \cos\theta$ and $b = \sin\theta$. Thus any unit vector **u** can be written in the form

$$\mathbf{u} = (\cos\theta)\mathbf{i} + (\sin\theta)\mathbf{j} \tag{8}$$

where θ is the direction of **u**.

EXAMPLE 8

WRITING A UNIT VECTOR AS $\cos\theta\mathbf{i} + \sin\theta\mathbf{j}$

The unit vector $\mathbf{u} = (\frac{1}{2})\mathbf{i} + (\sqrt{3}/2)\mathbf{j}$ of Example 7 can be written in the form (8) with $\theta = \cos^{-1}(\frac{1}{2}) = \pi/3$. Note that since $\cos\theta = \frac{1}{2}$ and $\sin\theta = \sqrt{3}/2$, θ is in the first quadrant. We need this fact to conclude that $\theta = \pi/3$. It is also true that $\cos 5\pi/3 = \frac{1}{2}$, but $5\pi/3$ is in the fourth quadrant.

Finally (see Problem 55):

> Let **v** be any nonzero vector. Then $\mathbf{u} = \mathbf{v}/|\mathbf{v}|$ is the unit vector having the same direction as **v**.

EXAMPLE 9 FINDING A UNIT VECTOR THAT HAS THE SAME DIRECTION AS A GIVEN VECTOR

Find the unit vector having the same direction as $\mathbf{v} = 2\mathbf{i} - 3\mathbf{j}$.

SOLUTION: Here $|\mathbf{v}| = \sqrt{4 + 9} = \sqrt{13}$, so

$\mathbf{u} = \mathbf{v}/|\mathbf{v}| = (2/\sqrt{13})\mathbf{i} - (3/\sqrt{13})\mathbf{j}$ is the required unit vector.

OF HISTORICAL INTEREST

SIR WILLIAM ROWAN HAMILTON, 1805–1865

Born in Dublin in 1805, where he spent most of his life, William Rowan Hamilton was without question Ireland's greatest mathematician. Hamilton's father (an attorney) and mother died when he was a small boy. His uncle, a linguist, took over the boy's education. By his fifth birthday, Hamilton could read English, Hebrew, Latin, and Greek. By his 13th birthday he had mastered not only the languages of continental Europe, but also Sanscrit, Chinese, Persian, Arabic, Malay, Hindi, Bengali, and several others as well. Hamilton liked to write poetry, both as a child and as an adult, and his friends included the great English poets Samuel Taylor Coleridge and William Wordsworth. Hamilton's poetry was considered so bad, however, that it is fortunate that he developed other interests—especially in mathematics.

Although he enjoyed mathematics as a young boy, Hamilton's interest was greatly enhanced by a chance meeting at the age of 15 with Zerah Colburn, the American lightning calculator. Shortly afterwards, Hamilton began to read important mathematical books of the time. In 1823, at the age of 18, he discovered an error in Simon Laplace's *Mécanique céleste* and wrote an impressive paper on the subject. A year later he entered Trinity College in Dublin.

Hamilton's university career was astonishing. At the age of 21, while still an undergraduate, he had so impressed the faculty that he was appointed Royal Astronomer of Ireland and Professor of Astronomy at the University. Shortly thereafter, he wrote what is now considered a classical work on optics. Using only mathematical theory, he predicted conical refraction in certain types of crystals. Later, this theory was confirmed by physicists. Largely because of this work, Hamilton was knighted in 1835.

Hamilton's first great purely mathematical paper appeared in 1833. In this work he described an algebraic way to manipulate pairs of real numbers. This work gives rules that are used today to add, subtract, multiply, and divide complex numbers. At first, however, Hamilton was unable to devise a multiplication for triples or n-tuples of numbers for $n > 2$. For 10 years he pondered this problem, and it is said that he solved it in an inspiration while walking on the Brougham Bridge in Dublin in 1843. The key was to discard the familiar commutative property of multiplication. The new objects he created were called *quaternions,* which were the precursors of what we now call vectors.

For the rest of his life, Hamilton spent most of his time developing the algebra of quaternions. He felt that they would have revolutionary significance in mathematical physics. His monumen-

(CONTINUED)

OF HISTORICAL INTEREST (CONCLUDED)

tal work on this subject, *Treatise on Quaternions,* was published in 1853. Thereafter, he worked on an enlarged work, *Elements of Quaternions.* Although Hamilton died in 1865 before his *Elements* was completed, the work was published by his son in 1866.

Students of mathematics and physics know Hamilton in a variety of other contexts. In mathematical physics, for example, one encounters the Hamiltonian function, which often represents the total energy in a system and the Hamilton-Jacobi differential equations of dynamics. In matrix theory, the Cayley-Hamilton theorem states that every matrix satisfies its own characteristic equation.

Despite the great work he was doing, Hamilton's final years were a torment to him. His wife was a semi-invalid and he was plagued by alcoholism. It is therefore gratifying to point out that during these last years, the newly formed American National Academy of Sciences elected Sir William Rowan Hamilton to be its first foreign associate.

For a fascinating account of Hamilton's discoveries, read the article ''Hamilton, Rodrigues, and the Quaternion Scandal—what went wrong with one of the major mathematical discoveries of the nineteenth century'' in *Mathematics Magazine* 62(5) December, 1989, 291–308.

PROBLEMS 1.1

SELF-QUIZ

I. A *vector* is ________.
a. two points in the xy-plane
b. a line segment between two points
c. a directed line segment from one point to another
d. a collection of equivalent directed line segments

II. If $P = (3, -4)$ and $Q = (8, 6)$, the vector $\overrightarrow{PQ}$ has length ________.
a. $|3| + |-4|$
b. $(3)^2 + (-4)^2$
c. $(3 - 8)^2 + (-4 - 6)^2$
d. $\sqrt{(8 - 3)^2 + (6 - (-4))^2}$

III. The *direction* of the vector (4, 8) is ________.
a. π **b.** $\tan^{-1}(8 - 4)$
c. $(\frac{8}{4})\pi$ **d.** $\tan^{-1}(\frac{8}{4})$

IV. If $\mathbf{u} = (3, 4)$ and $\mathbf{v} = (5, 8)$, then $\mathbf{u} + \mathbf{v} =$ ________.
a. (7, 13) **b.** (8, 12)
c. (2, 4) **d.** (15, 32)

V. If $\mathbf{u} = (4, 3)$ then the unit vector with the same direction as $\mathbf{u}$ is ________.
a. (0.4, 0.3) **b.** (0.8, 0.6)
c. $(\frac{4}{5}, \frac{3}{5})$ **d.** $(\frac{4}{7}, \frac{3}{7})$

In Problems 1–6, a vector $\mathbf{v}$ and a point P are given. Find a point Q such that the directed line segment $\overrightarrow{PQ}$ is a representation of $\mathbf{v}$. Sketch $\mathbf{v}$ and $\overrightarrow{PQ}$.

1. $\mathbf{v} = (2, 5)$; $P = (1, -2)$
2. $\mathbf{v} = (5, 8)$; $P = (3, 8)$
3. $\mathbf{v} = (-3, 7)$; $P = (7, -3)$
4. $\mathbf{v} = -\mathbf{i} - 7\mathbf{j}$; $P = (0, 1)$
5. $\mathbf{v} = 5\mathbf{i} - 3\mathbf{j}$; $P = (-7, -2)$
6. $\mathbf{v} = e\mathbf{i} + \pi\mathbf{j}$; $P = (\pi, \sqrt{2})$

In Problems 7–18, find the magnitude and direction of the given vector $\mathbf{v}$.

7. $\mathbf{v} = (4, 4)$ **8.** $\mathbf{v} = (-4, 4)$
9. $\mathbf{v} = (4, -4)$ **10.** $\mathbf{v} = (-4, -4)$
11. $\mathbf{v} = (\sqrt{3}, 1)$ **12.** $\mathbf{v} = (1, \sqrt{3})$
13. $\mathbf{v} = (-1, \sqrt{3})$ **14.** $\mathbf{v} = (1, -\sqrt{3})$
15. $\mathbf{v} = (-1, -\sqrt{3})$ **16.** $\mathbf{v} = (1, 2)$
17. $\mathbf{v} = (-5, 8)$ **18.** $\mathbf{v} = (11, -14)$

In Problems 19–26, the vector $\mathbf{v}$ is represented by $\overrightarrow{PQ}$ where P and Q are given. Write $\mathbf{v}$ in the form $a\mathbf{i} + b\mathbf{j}$. Sketch $\overrightarrow{PQ}$ and $\mathbf{v}$.

19. $P = (1, 2)$; $Q = (1, 3)$
20. $P = (2, 4)$; $Q = (-7, 4)$
21. $P = (5, 2)$; $Q = (-1, 3)$
22. $P = (8, -2)$; $Q = (-3, -3)$
23. $P = (7, -1)$; $Q = (-2, 4)$
24. $P = (3, -6)$; $Q = (8, 0)$
25. $P = (-3, -8)$; $Q = (-8, -3)$
26. $P = (2, 4)$; $Q = (-4, -2)$

27. Let $\mathbf{u} = (2, 3)$ and $\mathbf{v} = (-5, 4)$. Compute and sketch the following vectors:

a. $3\mathbf{u}$ **b.** $\mathbf{u} + \mathbf{v}$
c. $\mathbf{v} - \mathbf{u}$ **d.** $2\mathbf{u} - 7\mathbf{v}$

28. Let $\mathbf{u} = 2\mathbf{i} - 3\mathbf{j}$ and $\mathbf{v} = -4\mathbf{i} + 6\mathbf{j}$. Compute and sketch the following vectors:

a. $\mathbf{u} + \mathbf{v}$ **b.** $\mathbf{u} - \mathbf{v}$ **c.** $3\mathbf{u}$
d. $-7\mathbf{v}$ **e.** $8\mathbf{u} - 3\mathbf{v}$ **f.** $4\mathbf{v} - 6\mathbf{u}$

In Problems 29–34, find a unit vector having the same direction as the given vector $\mathbf{v}$.

29. $\mathbf{v} = 2\mathbf{i} + 3\mathbf{j}$ **30.** $\mathbf{v} = \mathbf{i} - \mathbf{j}$
31. $\mathbf{v} = (3, 4)$ **32.** $\mathbf{v} = (3, -4)$
33. $\mathbf{v} = -3\mathbf{i} + 4\mathbf{j}$ **34.** $\mathbf{v} = (a, a)$ $a \neq 0$

35. For $\mathbf{v} = 2\mathbf{i} - 3\mathbf{j}$, find $\sin\theta$ and $\cos\theta$.
36. For $\mathbf{v} = -3\mathbf{i} + 8\mathbf{j}$, find $\sin\theta$ and $\cos\theta$.

A vector $\mathbf{u}$ has a direction **opposite** to that of a vector $\mathbf{v}$ if and only if $|\text{direction } \mathbf{v} - \text{direction } \mathbf{u}| = \pi$. In Problems 37–42, find a unit vector $\mathbf{u}$ that has direction opposite the direction of the given vector $\mathbf{v}$.

37. $\mathbf{v} = \mathbf{i} + \mathbf{j}$ **38.** $\mathbf{v} = 2\mathbf{i} - 3\mathbf{j}$
39. $\mathbf{v} = (-3, 4)$ **40.** $\mathbf{v} = (-2, 3)$
41. $\mathbf{v} = -3\mathbf{i} - 4\mathbf{j}$ **42.** $\mathbf{v} = (8, -3)$

43. Let $\mathbf{u} = 2\mathbf{i} - 3\mathbf{j}$ and $\mathbf{v} = -\mathbf{i} + 2\mathbf{j}$. For each of the following, find a unit vector having the same direction.

a. $\mathbf{u} + \mathbf{v}$ **b.** $\mathbf{u} - \mathbf{v}$
c. $2\mathbf{u} - 3\mathbf{v}$ **d.** $3\mathbf{u} + 8\mathbf{v}$

44. Let $P = (c, d)$ and $Q = (c + a, d + b)$. Verify that the magnitude of $\overrightarrow{PQ}$ is $\sqrt{a^2 + b^2}$.

In Problems 45–52, find a vector $\mathbf{v}$ having the given magnitude and direction.

45. $|\mathbf{v}| = 3;\ \theta = \pi/6$ **46.** $|\mathbf{v}| = 8;\ \theta = \pi/3$
47. $|\mathbf{v}| = 7;\ \theta = \pi$ **48.** $|\mathbf{v}| = 4;\ \theta = \pi/2$
49. $|\mathbf{v}| = 1;\ \theta = \pi/4$ **50.** $|\mathbf{v}| = 6;\ \theta = 2\pi/3$
51. $|\mathbf{v}| = 8;\ \theta = 3\pi/2$ **52.** $|\mathbf{v}| = 6;\ \theta = 11\pi/6$

53. Show that the vectors $\mathbf{i}$ and $\mathbf{j}$ are unit vectors.

54. Show that the vector $\dfrac{1}{\sqrt{2}}\mathbf{i} - \dfrac{1}{\sqrt{2}}\mathbf{j}$ is a unit vector.

55. Show that if $\mathbf{v} = a\mathbf{i} + b\mathbf{j} \neq \mathbf{0}$, then

$$\mathbf{u} = \frac{a}{\sqrt{a^2 + b^2}}\mathbf{i} + \frac{b}{\sqrt{a^2 + b^2}}\mathbf{j}$$

is a unit vector having the same direction as $\mathbf{v}$.

56. Show that if $\mathbf{v} = a\mathbf{i} + b\mathbf{j} \neq \mathbf{0}$, then $a/\sqrt{a^2 + b^2} = \cos\theta$ and $b/\sqrt{a^2 + b^2} = \sin\theta$, where θ is the direction of $\mathbf{v}$.

57. Show that the direction of $\overrightarrow{PQ}$ in Problem 44 is the same as the direction of the vector (a, b). [*Hint:* If $R = (a, b)$, show that the line passing through the points P and Q is parallel to the line passing through the points 0 and R.]

58. Prove Theorem 1. [*Hint:* Use the definitions of addition and scalar multiplication of vectors.]

59. Show algebraically (i.e., strictly from the definitions of vector addition and magnitude) that for any two vectors $\mathbf{u}$ and $\mathbf{v}$,

$$|\mathbf{u} + \mathbf{v}| \leq |\mathbf{u}| + |\mathbf{v}|.$$

60. Show that if neither $\mathbf{u}$ nor $\mathbf{v}$ is the zero vector, then $|\mathbf{u} + \mathbf{v}| = |\mathbf{u}| + |\mathbf{v}|$ if and only if $\mathbf{u}$ is a positive scalar multiple of $\mathbf{v}$.

ANSWERS TO SELF-QUIZ

I. d **II.** d **III.** d **IV.** b **V.** b = c

1.2 THE DOT PRODUCT

In Section 1.1, we showed how a vector could be multiplied by a scalar but not how two vectors could be multiplied. Actually, there are several ways to define the product of two vectors, and in this section we will discuss one of them. We will discuss a second product operation in Section 1.6.

DEFINITION DOT PRODUCT

Let $\mathbf{u} = (u_1, u_2) = u_1\mathbf{i} + u_2\mathbf{j}$ and $\mathbf{v} = (v_1, v_2) = v_1\mathbf{i} + v_2\mathbf{j}$. Then the **dot product** of $\mathbf{u}$ and $\mathbf{v}$, denoted by $\mathbf{u} \cdot \mathbf{v}$, is defined by

$$\mathbf{u} \cdot \mathbf{v} = u_1v_1 + u_2v_2. \tag{1}$$

■

REMARK: The dot product of two vectors is a *scalar*. For this reason, the dot product is often called the **scalar product**. It is also called the **inner product**.

EXAMPLE 1

COMPUTING A DOT PRODUCT

If $\mathbf{u} = (1, 3)$ and $\mathbf{v} = (4, -7)$, then

$$\mathbf{u} \cdot \mathbf{v} = 1(4) + 3(-7) = 4 - 21 = -17.$$

THEOREM 1 PROPERTIES OF THE DOT PRODUCT

For any vectors $\mathbf{u}$, $\mathbf{v}$, $\mathbf{w}$, and scalar α,

i. $\mathbf{u} \cdot \mathbf{v} = \mathbf{v} \cdot \mathbf{u}$
ii. $(\mathbf{u} + \mathbf{v}) \cdot \mathbf{w} = \mathbf{u} \cdot \mathbf{w} + \mathbf{v} \cdot \mathbf{w}$
iii. $(\alpha\mathbf{u}) \cdot \mathbf{v} = \alpha(\mathbf{u} \cdot \mathbf{v})$
iv. $\mathbf{u} \cdot \mathbf{u} \geq 0$ and $\mathbf{u} \cdot \mathbf{u} = 0$ if and only if $\mathbf{u} = \mathbf{0}$
v. $|\mathbf{u}|^2 = \mathbf{u} \cdot \mathbf{u}$

PROOF: Let $\mathbf{u} = (u_1, u_2)$, $\mathbf{v} = (v_1, v_2)$, and $\mathbf{w} = (w_1, w_2)$.

i. $\mathbf{u} \cdot \mathbf{v} = u_1v_1 + u_2v_2 = v_1u_1 + v_2u_2 = \mathbf{v} \cdot \mathbf{u}$

ii.
$$\begin{aligned}(\mathbf{u} + \mathbf{v}) \cdot \mathbf{w} &= (u_1 + v_1, u_2 + v_2) \cdot (w_1, w_2) \\ &= (u_1 + v_1)w_1 + (u_2 + v_2)w_2 \\ &= u_1w_1 + u_2w_2 + v_1w_1 + v_2w_2 = \mathbf{u} \cdot \mathbf{w} + \mathbf{v} \cdot \mathbf{w}\end{aligned}$$

iii.
$$\begin{aligned}(\alpha\mathbf{u}) \cdot \mathbf{v} &= (\alpha u_1, \alpha u_2) \cdot (v_1, v_2) = \alpha u_1v_1 + \alpha u_2v_2 = \alpha(u_1v_1 + u_2v_2) \\ &= \alpha(\mathbf{u} \cdot \mathbf{v})\end{aligned}$$

iv. $\mathbf{u} \cdot \mathbf{u} = u_1{}^2 + u_2{}^2 \geq 0$ and $\mathbf{u} \cdot \mathbf{u} = 0$ if and only if $u_1 = u_2 = 0$.

v. $\mathbf{u} \cdot \mathbf{u} = (u_1, u_2) \cdot (u_1, u_2) = u_1{}^2 + u_2{}^2 = |\mathbf{u}|^2$ ■

The dot product is useful in a wide variety of applications. An interesting one follows.

DEFINITION ANGLE BETWEEN TWO VECTORS

Let $\mathbf{u}$ and $\mathbf{v}$ be two nonzero vectors. If $\mathbf{u}$ and $\mathbf{v}$ are not parallel, then the **angle** φ between $\mathbf{u}$ and $\mathbf{v}$ is defined to be the smallest angle between the representations of $\mathbf{u}$ and $\mathbf{v}$ that have the origin as their initial points. If $\mathbf{u} = \alpha\mathbf{v}$ for some scalar α, then we define $\varphi = 0$ if $\alpha > 0$ and $\varphi = \pi$ if $\alpha < 0$. By this definition

$$0 \leq \varphi \leq \pi.$$ ■

In Figure 1 we illustrate the angles between five pairs of vectors.

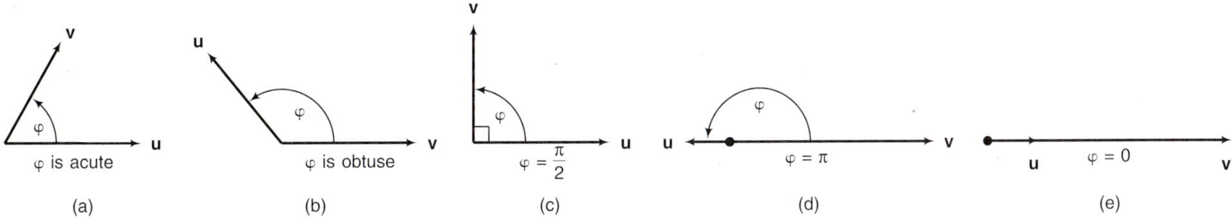

FIGURE 1
The angles between five sets of vectors

THEOREM 2

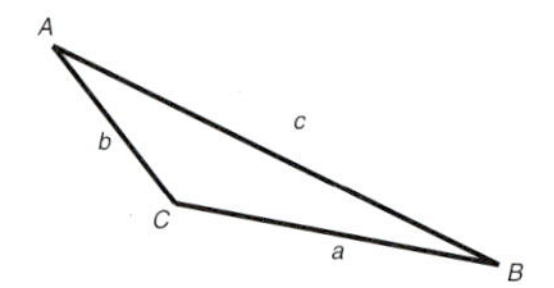

FIGURE 2
From the law of cosines, $c^2 = a^2 + b^2 - 2ab\cos C$

Let **u** and **v** be two nonzero vectors. Then if φ is the angle between them,

$$\cos\varphi = \frac{\mathbf{u}\cdot\mathbf{v}}{|\mathbf{u}||\mathbf{v}|} \quad \text{and} \quad \varphi = \cos^{-1}\left(\frac{\mathbf{u}\cdot\mathbf{v}}{|\mathbf{u}||\mathbf{v}|}\right). \tag{2}$$

PROOF: The law of cosines states that in the triangle of Figure 2,

$$c^2 = a^2 + b^2 - 2ab\cos C. \tag{3}$$

We now place the representations of **u** and **v** with initial points at the origin so that $\mathbf{u} = (a_1, b_1)$ and $\mathbf{v} = (a_2, b_2)$ (see Figure 3). Then from the law of cosines,

$$|\mathbf{v}-\mathbf{u}|^2 = |\mathbf{v}|^2 + |\mathbf{u}|^2 - 2|\mathbf{u}||\mathbf{v}|\cos\varphi.$$

But using Theorem 1 several times, we have

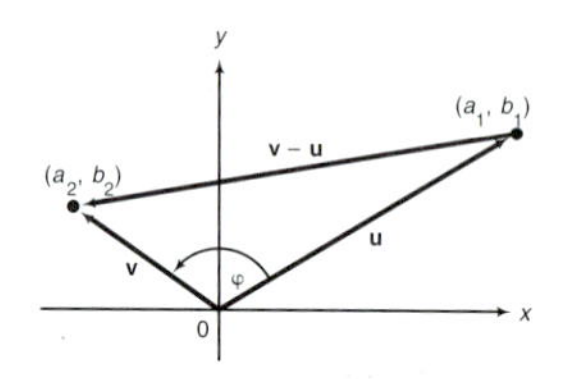

FIGURE 3
$|\mathbf{v}-\mathbf{u}|^2 = |\mathbf{u}|^2 + |\mathbf{v}|^2 - 2|\mathbf{u}||\mathbf{v}|\cos\varphi$

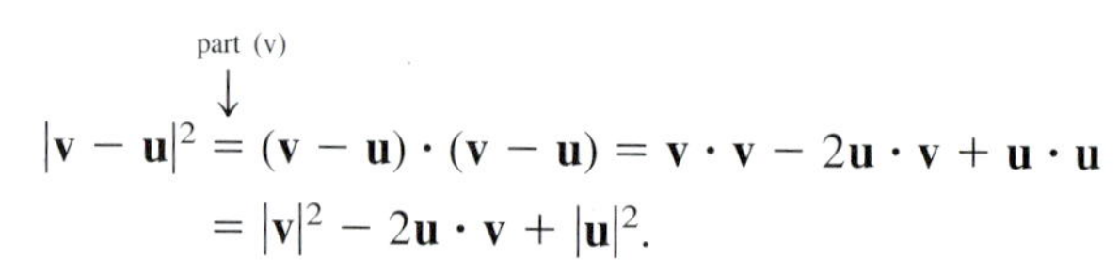

$$|\mathbf{v}-\mathbf{u}|^2 \overset{\text{part (v)}}{=} (\mathbf{v}-\mathbf{u})\cdot(\mathbf{v}-\mathbf{u}) = \mathbf{v}\cdot\mathbf{v} - 2\mathbf{u}\cdot\mathbf{v} + \mathbf{u}\cdot\mathbf{u}$$
$$= |\mathbf{v}|^2 - 2\mathbf{u}\cdot\mathbf{v} + |\mathbf{u}|^2.$$

Thus after simplification, we obtain

$$-2\mathbf{u}\cdot\mathbf{v} = -2|\mathbf{u}||\mathbf{v}|\cos\varphi,$$

from which the theorem follows. ■

REMARK: Using Theorem 2, we could define the dot product $\mathbf{u}\cdot\mathbf{v}$ by

$$\mathbf{u}\cdot\mathbf{v} = |\mathbf{u}||\mathbf{v}|\cos\varphi \tag{4}$$

if neither **u** nor $\mathbf{v} = \mathbf{0}$ and $\mathbf{u}\cdot\mathbf{v} = 0$ if **u** or **v** is the zero vector.

EXAMPLE 2

COMPUTING THE ANGLE BETWEEN TWO VECTORS

Find the cosine of the angle between the vectors $\mathbf{u} = 2\mathbf{i} + 3\mathbf{j}$ and $\mathbf{v} = -7\mathbf{i} + \mathbf{j}$.

SOLUTION: $\mathbf{u}\cdot\mathbf{v} = -14 + 3 = -11$, $|\mathbf{u}| = \sqrt{2^2 + 3^2} = \sqrt{13}$, and $|\mathbf{v}| = \sqrt{(-7)^2 + 1^2} = \sqrt{50}$, so

$$\cos\varphi = \frac{\mathbf{u}\cdot\mathbf{v}}{|\mathbf{u}||\mathbf{v}|} = \frac{-11}{\sqrt{13}\sqrt{50}} = \frac{-11}{\sqrt{650}} \approx -0.431455497,$$

so $\varphi = \cos^{-1}(-0.431455497) \approx 2.0169^{\dagger}$ ($\approx 115.6°$).

DEFINITION PARALLEL VECTORS

Two nonzero vectors **u** and **v** are **parallel** if the angle between them is 0 or π. ■

† When doing this computation yourself, make certain that the calculator is set to radian mode.

EXAMPLE 3 TWO PARALLEL VECTORS

Show that the vectors $\mathbf{u} = (2, -3)$ and $\mathbf{v} = (-4, 6)$ are parallel.

SOLUTION:

$$\cos\varphi = \frac{\mathbf{u}\cdot\mathbf{v}}{|\mathbf{u}||\mathbf{v}|} = \frac{-8-18}{\sqrt{13}\sqrt{52}} = \frac{-26}{\sqrt{13}(2\sqrt{13})} = \frac{-26}{2(13)} = -1,$$

so $\varphi = \pi$ (so that $\mathbf{u}$ and $\mathbf{v}$ have opposite directions).

THEOREM 3

If $\mathbf{u} \neq \mathbf{0}$, then $\mathbf{v} = \alpha\mathbf{u}$ for some nonzero constant α if and only if $\mathbf{u}$ and $\mathbf{v}$ are parallel.

PROOF: This follows from the last part of the definition of an angle between two vectors (see also Problem 69). ■

DEFINITION ORTHOGONAL VECTORS

The nonzero vectors $\mathbf{u}$ and $\mathbf{v}$ are called **orthogonal** (or **perpendicular**) if the angle between them is $\pi/2$. ■

EXAMPLE 4 TWO ORTHOGONAL VECTORS

Show that the vectors $\mathbf{u} = 3\mathbf{i} - 4\mathbf{j}$ and $\mathbf{v} = 4\mathbf{i} + 3\mathbf{j}$ are orthogonal.

SOLUTION: $\mathbf{u}\cdot\mathbf{v} = 3\cdot 4 - 4\cdot 3 = 0$. This implies that $\cos\varphi = (\mathbf{u}\cdot\mathbf{v})/(|\mathbf{u}||\mathbf{v}|) = 0$. Since φ is in the interval $[0, \pi]$, $\varphi = \pi/2$.

THEOREM 4

The nonzero vectors $\mathbf{u}$ and $\mathbf{v}$ are orthogonal if and only if $\mathbf{u}\cdot\mathbf{v} = 0$.

PROOF: This proof is not difficult and is left as an exercise (see Problem 70). ■

REMARK: The condition $\mathbf{u}\cdot\mathbf{v} = 0$ is sometimes given as the *definition* of orthogonality.

A number of interesting problems involve the notion of the *projection* of one vector onto another. Before defining this term, we prove the following theorem.

THEOREM 5

Let $\mathbf{v}$ be a nonzero vector. Then for any other nonzero vector $\mathbf{u}$, the vector

$$\mathbf{w} = \mathbf{u} - [(\mathbf{u}\cdot\mathbf{v})/|\mathbf{v}|^2]\mathbf{v} \text{ is orthogonal to } \mathbf{v}.$$

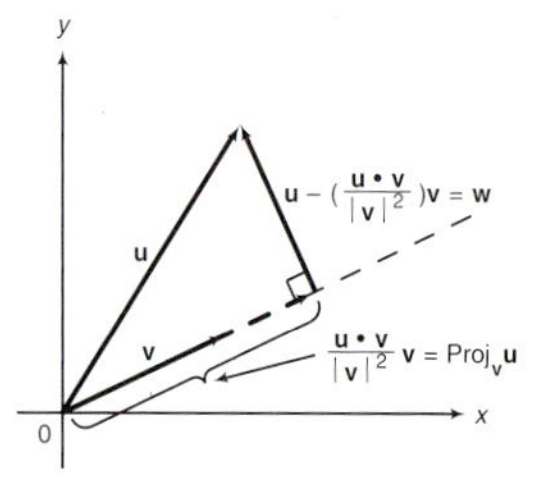

FIGURE 4
Projection of **u** onto **v**

PROOF:

$$\mathbf{w} \cdot \mathbf{v} = \left(\mathbf{u} - \frac{(\mathbf{u} \cdot \mathbf{v})\mathbf{v}}{|\mathbf{v}|^2}\right) \cdot \mathbf{v} = \mathbf{u} \cdot \mathbf{v} - \frac{(\mathbf{u} \cdot \mathbf{v})(\mathbf{v} \cdot \mathbf{v})}{|\mathbf{v}|^2}$$

$$= \mathbf{u} \cdot \mathbf{v} - \frac{(\mathbf{u} \cdot \mathbf{v})|\mathbf{v}|^2}{|\mathbf{v}|^2} = \mathbf{u} \cdot \mathbf{v} - \mathbf{u} \cdot \mathbf{v} = 0$$

The vectors **u**, **v**, and **w** are illustrated in Figure 4. ■

DEFINITION PROJECTION AND COMPONENT

Let **u** and **v** be nonzero vectors. Then the **projection of u onto v** is a vector, denoted by $\text{Proj}_{\mathbf{v}}\, \mathbf{u}$, which is defined by

$$\text{Proj}_{\mathbf{v}}\, \mathbf{u} = \frac{\mathbf{u} \cdot \mathbf{v}}{\mathbf{v} \cdot \mathbf{v}}\mathbf{v} = \frac{\mathbf{u} \cdot \mathbf{v}}{|\mathbf{v}|^2}\mathbf{v} = \left(\frac{\mathbf{u} \cdot \mathbf{v}}{|\mathbf{v}|}\right)\frac{\mathbf{v}}{|\mathbf{v}|}. \tag{5}$$

The **component** of **u** in the direction **v** is $\mathbf{u} \cdot \mathbf{v}/|\mathbf{v}|$. (6)

■

Note that $\mathbf{v}/|\mathbf{v}|$ is a unit vector in the direction of **v**.

REMARK 1: From Figure 4 and the fact that $\cos \varphi = (\mathbf{u} \cdot \mathbf{v})/(|\mathbf{u}||\mathbf{v}|)$, we find that **v** and $\text{Proj}_{\mathbf{v}}\, \mathbf{u}$ have

i. The same direction if $\mathbf{u} \cdot \mathbf{v} > 0$ and
ii. opposite directions if $\mathbf{u} \cdot \mathbf{v} < 0$.

REMARK 2: $\text{Proj}_{\mathbf{v}}\, \mathbf{u}$ can be thought of as the "**v**-component" of the vector **u**.

REMARK 3: If **u** and **v** are orthogonal, then $\mathbf{u} \cdot \mathbf{v} = 0$, so that $\text{Proj}_{\mathbf{v}}\, \mathbf{u} = \mathbf{0}$.

REMARK 4: An alternative definition of projection is: If **u** and **v** are nonzero vectors, then $\text{Proj}_{\mathbf{v}}\, \mathbf{u}$ is the unique vector having the properties

i. $\text{Proj}_{\mathbf{v}}\, \mathbf{u}$ is parallel to **v** and
ii. $\mathbf{u} - \text{Proj}_{\mathbf{v}}\, \mathbf{u}$ is orthogonal to **v**.

EXAMPLE 5 CALCULATING A PROJECTION

Let $\mathbf{u} = 2\mathbf{i} - 3\mathbf{j}$ and $\mathbf{v} = \mathbf{i} + \mathbf{j}$. Find $\text{Proj}_{\mathbf{v}}\, \mathbf{u}$.

SOLUTION: Here $(\mathbf{u} \cdot \mathbf{v})/|\mathbf{v}|^2 = -\frac{1}{2}$, so that $\text{Proj}_{\mathbf{v}}\, \mathbf{u} = -\frac{1}{2}\mathbf{i} - \frac{1}{2}\mathbf{j} = -\frac{1}{2}(\mathbf{i} + \mathbf{j})$. This is illustrated in Figure 5.

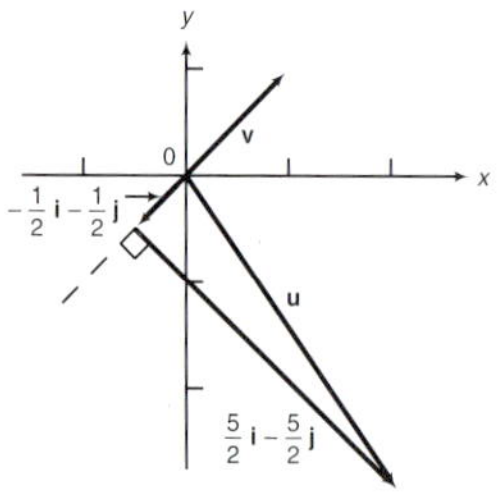

FIGURE 5
The projection of **u** onto **v** is the vector $-\frac{1}{2}\mathbf{i} - \frac{1}{2}\mathbf{j}$.

FORCE AND WORK

We will be dealing with mass and force in many problems in this text, so we take a moment here to explain units of weight, mass, and force in both the English and metric systems.

In the English system, we start with the unit of force, which is the *pound.* (It is also the standard unit of weight.) Since $F = mg$ by Newton's second law of motion, we obtain $m = F/g$. In the English system, the unit of mass is the *slug,* and a 1-lb weight (force) has a mass of $(1/32.2)\ \text{lb}/(\text{ft}/\text{sec}^2) = 1$ slug. Or we can equivalently define a slug as the mass of an object whose acceleration is 1 ft/sec^2 when it is subjected to a force of 1 lb.

In the metric system, we start with the standard unit of mass, which is the *kilogram.* In this case, to obtain the force, we must multiply the mass by the acceleration:

$$F\ (\text{in newtons}) = [\text{mass (in kilograms)}] \times g(=9.81\ \text{m/sec}^2).$$

In sum, we have the following:

> The pound is a force, but not a mass.
>
> The kilogram is a mass but not a force.

In one variable calculus we solved problems involving the notion of force. We started such problems with a statement like "a force of x newtons is applied to . . . ." Implicit in such a problem is the idea that a force with a certain magnitude (measured in pounds or newtons) is exerted in a certain direction. In this context force can (and should) be thought of as a vector.

If more than one force is applied to an object at the same point, then we define the **resultant** of the forces applied to the object to be the *vector sum* of these forces. We can think of the resultant as the *net* applied force.

EXAMPLE 6

COMPUTING THE RESULTANT OF THREE FORCES

A force of 3 N is applied from the left side of an object, one of 4 N is applied from the bottom, and a force of 7 N is applied from an angle of $\pi/4$ to the horizontal. What is the resultant of forces applied to the object?

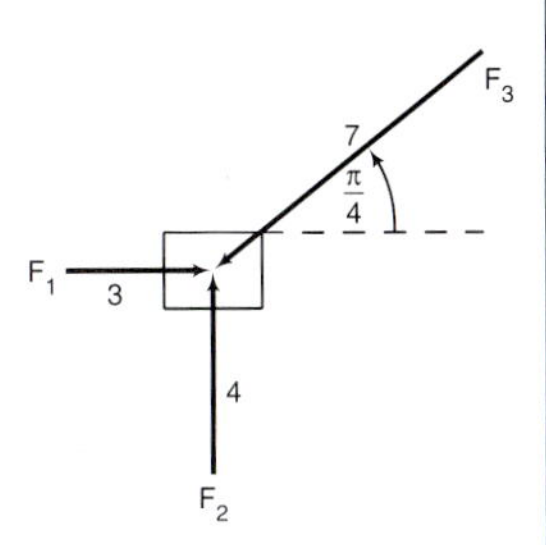

FIGURE 6
Three forces act on the same object

SOLUTION: The forces are indicated in Figure 6. We write each force as a magnitude times a unit vector in the indicated direction. For convenience we can think of the center of the object as being at the origin and that all three forces are applied at the center. Then $\mathbf{F}_1 = 3\mathbf{i}$; $\mathbf{F}_2 = 4\mathbf{j}$; and $\mathbf{F}_3 = -(7/\sqrt{2})(\mathbf{i} + \mathbf{j})$. This last vector follows from the fact that the vector $-(1/\sqrt{2})(\mathbf{i} + \mathbf{j})$ is a unit vector pointing toward the origin making an angle of $\pi/4$ with the x-axis. Then the resultant is given by

$$\mathbf{F} = \mathbf{F}_1 + \mathbf{F}_2 + \mathbf{F}_3 = \left(3 - \frac{7}{\sqrt{2}}\right)\mathbf{i} + \left(4 - \frac{7}{\sqrt{2}}\right)\mathbf{j}.$$

The magnitude of $\mathbf{F}$ is

$$|\mathbf{F}| = \sqrt{\left(3 - \frac{7}{\sqrt{2}}\right)^2 + \left(4 - \frac{7}{\sqrt{2}}\right)^2} = \sqrt{74 - \frac{98}{\sqrt{2}}} \approx 2.17\ \text{N}.$$

The direction θ can be calculated by first finding the unit vector in the direction of $\mathbf{F}$:

$$\frac{\mathbf{F}}{|\mathbf{F}|} = \frac{3 - (7/\sqrt{2})}{\sqrt{74 - (98/\sqrt{2})}}\mathbf{i} + \frac{4 - (7/\sqrt{2})}{\sqrt{74 - (98/\sqrt{2})}}\mathbf{j}$$
$$= (\cos\theta)\mathbf{i} + (\sin\theta)\mathbf{j}.$$

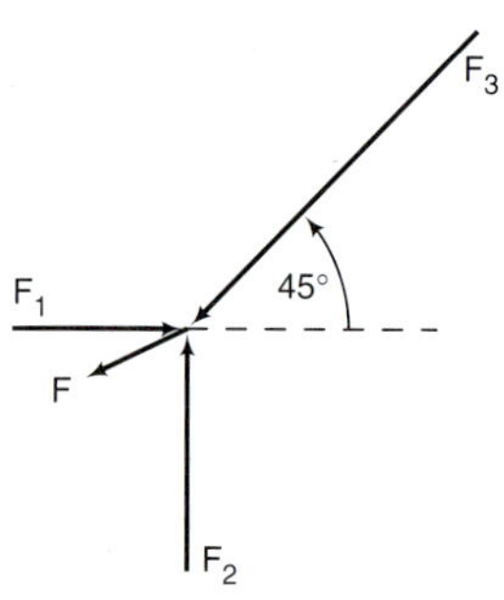

FIGURE 7
F is the resolvent of $\mathbf{F}_1$, $\mathbf{F}_2$, and $\mathbf{F}_3$.

Then

$$\cos\theta = \frac{3 - (7/\sqrt{2})}{\sqrt{74 - (98/\sqrt{2})}} \approx -0.8990 \qquad \text{and}$$

$$\sin\theta = \frac{4 - (7/\sqrt{2})}{\sqrt{74 - (98/\sqrt{2})}} \approx -0.4379.$$

This means that θ is in the third quadrant, and $\theta \approx 3.5949 \approx 206°$ (or $-154°$). This is illustrated in Figure 7.

We now turn to the notion of **work**. Consider a particle acted on by a force. In the simplest case, the force is constant, and the particle moves in a straight line in the direction of the force (see Figure 8). In this situation, we define the *work W done by the force on the particle* as the product of the magnitude of the force F and the distance s through which the particle travels:

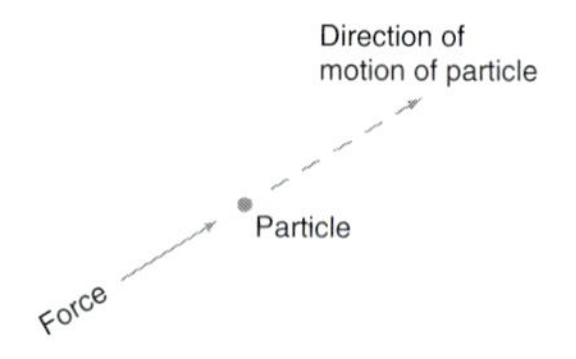

FIGURE 8
The direction of a force

> THE WORK DONE BY A CONSTANT FORCE OVER A DISTANCE *s*
>
> $W = Fs.$ **(7)**

One unit of work is the work done by a unit force in moving a body a unit distance in the direction of the force. In the metric system, the unit of work is 1 newton† meter ($N \cdot m$), called 1 *joule* (J). In the British system, the unit of work is the foot pound‡ ($ft \cdot lb$).

EXAMPLE 7 THE WORK DONE BY A CONSTANT FORCE (IN ENGLISH UNITS)

How much work is done in lifting a 25-lb weight 5 ft off the ground?

SOLUTION: $W = Fs = 25 \text{ lb} \times 5 \text{ ft} = 125 \text{ ft} \cdot \text{lb}$.

EXAMPLE 8 THE WORK DONE BY A CONSTANT FORCE (IN METRIC UNITS)

A block of mass 10.0 kg is raised 5 m off the ground. How much work is done?

SOLUTION: From Newton's second law, force = (mass) × (acceleration). The acceleration here is opposing acceleration due to gravity and is therefore equal to 9.81 m/sec^2. We have $F = ma = (10 \text{ kg})(9.81 \text{ m/sec}^2) = 98.1$ N. Therefore,

$$W = Fs = (98.1) \times 5 \text{ m} = 490.5 \text{ J}.$$

In formula (7) it is assumed that the force is applied in the same direction as the direction of motion. However, this is not always the case. In general, we may define

$$W = (\text{component of } \mathbf{F} \text{ in the direction of motion}) \times (\text{distance moved}). \qquad \textbf{(8)}$$

† 1 newton (N) is the force that will accelerate a 1-kg mass at the rate of 1 m/sec^2; 1 N = 0.2248 lb.
‡ 1 joule (J) = 0.7376 ft · lb, or 1 ft · lb = 1.356 J.

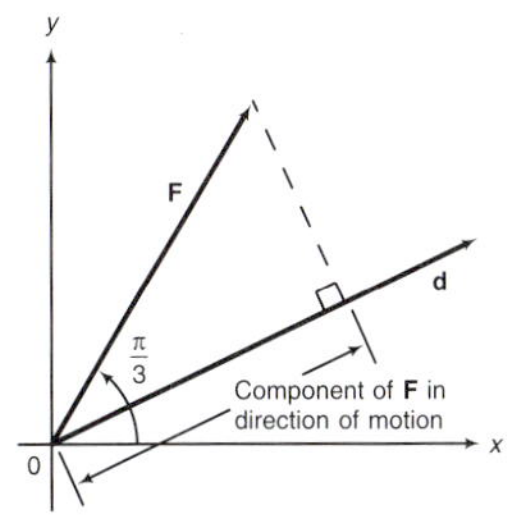

FIGURE 9
$\frac{\mathbf{F}\cdot\mathbf{d}}{|\mathbf{d}|}$ is the component of the force vector in the direction of the displacement vector **d**.

If the object moves from P to Q, then the distance moved is $|\overrightarrow{PQ}|$. The vector **d**, one of whose representations is $\overrightarrow{PQ}$, is called a **displacement vector**. Then from equation (6),

$$\text{component of } \mathbf{F} \text{ in direction of motion} = \frac{\mathbf{F}\cdot\mathbf{d}}{|\mathbf{d}|}. \quad (9)$$

Finally, combining (8) and (9), we obtain

$$W = \frac{\mathbf{F}\cdot\mathbf{d}}{|\mathbf{d}|}|\mathbf{d}| = \mathbf{F}\cdot\mathbf{d}. \quad (10)$$

That is, *the work done is the dot product of the force* **F** *and the displacement vector* **d**. Note that if **F** acts in the direction **d** and if φ denotes the angle (which is zero) between **F** and **d**, then $\mathbf{F}\cdot\mathbf{d} = |\mathbf{F}||\mathbf{d}|\cos\varphi = |\mathbf{F}||\mathbf{d}|\cos 0 = |\mathbf{F}||\mathbf{d}|$, which is formula (7).

EXAMPLE 9

COMPUTING WORK WHEN THE FORCE DOES NOT ACT IN THE DIRECTION OF MOTION

A force of 4 N has the direction $\pi/3$. What is the work done in moving an object from the point (1, 2) to the point (5, 4), where distances are measured in meters?

SOLUTION: A unit vector with direction $\pi/3$ is given by $\mathbf{u} = (\cos\pi/3)\mathbf{i} + (\sin\pi/3)\mathbf{j} = \frac{1}{2}\mathbf{i} + (\sqrt{3}/2)\mathbf{j}$. Thus $\mathbf{F} = 4\mathbf{u} = 2\mathbf{i} + 2\sqrt{3}\mathbf{j}$. The displacement vector **d** is given by $(5-1)\mathbf{i} + (4-2)\mathbf{j} = 4\mathbf{i} + 2\mathbf{j}$. Thus

$$W = \mathbf{F}\cdot\mathbf{d} = (2\mathbf{i} + 2\sqrt{3}\mathbf{j})\cdot(4\mathbf{i} + 2\mathbf{j})$$
$$= (8 + 4\sqrt{3}) \approx 14.93 \text{ N}\cdot\text{m}.$$

The component of **F** in the direction of motion is sketched in Figure 9.

PROBLEMS 1.2

SELF-QUIZ

I. $\mathbf{i}\cdot\mathbf{j} =$ ________.
a. 1 **b.** $\sqrt{(0-1)^2 + (1-0)^2}$
c. 0 **d.** $\mathbf{i} + \mathbf{j}$

II. $(3, 4)\cdot(3, 2) =$ ________.
a. $(3+3)(4+2) = 36$
b. $(3)(3) + (4)(2) = 17$
c. $(3-3)(2-4) = 0$
d. $(3)(3) - (4)(2) = 1$

III. The cosine of the angle between $\mathbf{i} + \mathbf{j}$ and $\mathbf{i} - \mathbf{j}$ is ________.
a. $0\mathbf{i} + 0\mathbf{j}$ **b.** 0
c. $\sqrt{2}$ **d.** $1/\sqrt{2+0}$

IV. The vectors $2\mathbf{i} - 12\mathbf{j}$ and $3\mathbf{i} + (1/2)\mathbf{j}$ are ________.
a. neither parallel nor orthogonal
b. parallel
c. orthogonal
d. identical

V. $\text{Proj}_{\mathbf{w}}\,\mathbf{u} =$ ________.
a. $\frac{\mathbf{u}\cdot\mathbf{w}}{|\mathbf{w}|}$ **b.** $\frac{\mathbf{w}}{|\mathbf{w}|}$
c. $\frac{\mathbf{u}\cdot\mathbf{w}}{|\mathbf{w}|}\frac{\mathbf{w}}{|\mathbf{w}|}$ **d.** $\frac{\mathbf{u}\cdot\mathbf{w}}{|\mathbf{u}|}\frac{\mathbf{u}}{|\mathbf{u}|}$

In Problems 1–10, two vectors are given. Calculate both the dot product of the two vectors and the cosine of the angle between them.

1. $\mathbf{u} = \mathbf{i} + \mathbf{j};\ \mathbf{v} = \mathbf{i} - \mathbf{j}$
2. $\mathbf{u} = 3\mathbf{i};\ \mathbf{v} = -7\mathbf{j}$
3. $\mathbf{u} = -5\mathbf{i};\ \mathbf{v} = 18\mathbf{j}$
4. $\mathbf{u} = \alpha\mathbf{i};\ \mathbf{v} = \beta\mathbf{j}$
5. $\mathbf{u} = 2\mathbf{i} + 5\mathbf{j};\ \mathbf{v} = 5\mathbf{i} + 2\mathbf{j}$

6. $\mathbf{u} = 2\mathbf{i} + 5\mathbf{j}$; $\mathbf{v} = 5\mathbf{i} - 3\mathbf{j}$
7. $\mathbf{u} = -3\mathbf{i} + 4\mathbf{j}$; $\mathbf{v} = -2\mathbf{i} - 7\mathbf{j}$
8. $\mathbf{u} = 4\mathbf{i} + 5\mathbf{j}$; $\mathbf{v} = 7\mathbf{i} - 4\mathbf{j}$
9. $\mathbf{u} = 11\mathbf{i} - 8\mathbf{j}$; $\mathbf{v} = 4\mathbf{i} - 7\mathbf{j}$
10. $\mathbf{u} = -13\mathbf{i} + 8\mathbf{j}$; $\mathbf{v} = 2\mathbf{i} + 11\mathbf{j}$

In Problems 11–18, determine whether the given vectors are orthogonal, parallel, or neither. Then sketch each pair of vectors.

11. $\mathbf{u} = 3\mathbf{i} + 5\mathbf{j}$; $\mathbf{v} = -6\mathbf{i} - 10\mathbf{j}$
12. $\mathbf{u} = 2\mathbf{i} + 3\mathbf{j}$; $\mathbf{v} = 6\mathbf{i} - 4\mathbf{j}$
13. $\mathbf{u} = 2\mathbf{i} + 3\mathbf{j}$; $\mathbf{v} = 6\mathbf{i} + 4\mathbf{j}$
14. $\mathbf{u} = 2\mathbf{i} + 3\mathbf{j}$; $\mathbf{v} = -6\mathbf{i} + 4\mathbf{j}$
15. $\mathbf{u} = 7\mathbf{i}$; $\mathbf{v} = -23\mathbf{j}$
16. $\mathbf{u} = 2\mathbf{i} - 6\mathbf{j}$; $\mathbf{v} = -\mathbf{i} + 3\mathbf{j}$
17. $\mathbf{u} = \mathbf{i} + \mathbf{j}$; $\mathbf{v} = \alpha\mathbf{i} + \alpha\mathbf{j}$
18. $\mathbf{u} = -2\mathbf{i} + 3\mathbf{j}$; $\mathbf{v} = -\mathbf{i} + 2\mathbf{j}$

In Problems 19–32, calculate $\text{Proj}_{\mathbf{v}}\,\mathbf{u}$.

19. $\mathbf{u} = 3\mathbf{i}$; $\mathbf{v} = \mathbf{i} + \mathbf{j}$
20. $\mathbf{u} = -5\mathbf{j}$; $\mathbf{v} = \mathbf{i} + \mathbf{j}$
21. $\mathbf{u} = 2\mathbf{i} + \mathbf{j}$; $\mathbf{v} = \mathbf{i} - 2\mathbf{j}$
22. $\mathbf{u} = 2\mathbf{i} + 3\mathbf{j}$; $\mathbf{v} = 4\mathbf{i} + \mathbf{j}$
23. $\mathbf{u} = \mathbf{i} + \mathbf{j}$; $\mathbf{v} = 2\mathbf{i} - 3\mathbf{j}$
24. $\mathbf{u} = \mathbf{i} + \mathbf{j}$; $\mathbf{v} = 2\mathbf{i} + 3\mathbf{j}$
25. $\mathbf{u} = 4\mathbf{i} + 5\mathbf{j}$; $\mathbf{v} = 2\mathbf{i} + 4\mathbf{j}$
26. $\mathbf{u} = 4\mathbf{i} + 5\mathbf{j}$; $\mathbf{v} = 2\mathbf{i} - 4\mathbf{j}$
27. $\mathbf{u} = -4\mathbf{i} + 5\mathbf{j}$; $\mathbf{v} = 2\mathbf{i} - 4\mathbf{j}$
28. $\mathbf{u} = -4\mathbf{i} - 5\mathbf{j}$; $\mathbf{v} = -2\mathbf{i} - 4\mathbf{j}$
29. $\mathbf{u} = \alpha\mathbf{i} + \beta\mathbf{j}$; $\mathbf{v} = \mathbf{i} + \mathbf{j}$
30. $\mathbf{u} = \mathbf{i} + \mathbf{j}$; $\mathbf{v} = \alpha\mathbf{i} + \beta\mathbf{j}$
31. $\mathbf{u} = \alpha\mathbf{i} - \beta\mathbf{j}$; $\mathbf{v} = \mathbf{i} + \mathbf{j}$
32. $\mathbf{u} = \alpha\mathbf{i} - \beta\mathbf{j}$; $\mathbf{v} = -\mathbf{i} + \mathbf{j}$
33. Let $\mathbf{u} = 3\mathbf{i} + 4\mathbf{j}$ and $\mathbf{v} = \mathbf{i} + \alpha\mathbf{j}$.
  a. Determine α such that $\mathbf{u}$ and $\mathbf{v}$ are orthogonal.
  b. Determine α such that $\mathbf{u}$ and $\mathbf{v}$ are parallel.
  c. Determine α such that the angle between $\mathbf{u}$ and $\mathbf{v}$ is $2\pi/3$.
  d. Determine α such that the angle between $\mathbf{u}$ and $\mathbf{v}$ is $\pi/3$.
34. Let $\mathbf{u} = -2\mathbf{i} + 5\mathbf{j}$ and $\mathbf{v} = \beta\mathbf{i} - 2\mathbf{j}$.
  a. Determine β such that $\mathbf{u}$ and $\mathbf{v}$ are orthogonal.
  b. Determine β such that $\mathbf{u}$ and $\mathbf{v}$ are parallel.
  c. Determine β such that the angle between $\mathbf{u}$ and $\mathbf{v}$ is $2\pi/3$.
  d. Determine β such that the angle between $\mathbf{u}$ and $\mathbf{v}$ is $\pi/3$.
35. A triangle has vertices $(1, 3)$, $(4, -2)$, and $(-3, 6)$. Find the cosine of each of its angles.
36. A triangle has vertices (a_1, b_1), (a_2, b_2), and (a_3, b_3). Find a formula for the cosine of each of its angles.
37. Let $P = (2, 3)$, $Q = (5, 5)$, $R = (2, -3)$, and $S = (1, 2)$. Calculate $\text{Proj}_{\overrightarrow{PQ}}\,\overrightarrow{RS}$ and $\text{Proj}_{\overrightarrow{RS}}\,\overrightarrow{PQ}$.
38. Let $P = (-1, 3)$, $Q = (2, 4)$, $R = (-6, -2)$, and $S = (3, 0)$. Calculate $\text{Proj}_{\overrightarrow{PQ}}\,\overrightarrow{RS}$ and $\text{Proj}_{\overrightarrow{RS}}\,\overrightarrow{PQ}$.
39. Find the distance between $P = (2, 3)$ and the line through the points $Q = (-1, 7)$ and $R = (3, 5)$. [*Hint:* Draw a picture and use the Pythagorean theorem.]
40. Find the distance between the point $(3, 7)$ and the line along the vector $\mathbf{v} = b\mathbf{i} - a\mathbf{j}$ which passes through the origin.

In Problems 41–49, find the resultant of the forces acting at the center of an object. Then find the force vector that must be applied so that the object will remain at rest.

41. 2 N (from right), 5 N (from above)
42. 2 N (from left), 5 N (from below)
43. 3 N (from left), 5 N (from right), 3 N (from above)
44. 10 lb (from right), 8 lb (from below)
45. 5 lb (from above), 4 lb (from direction $\pi/6$)
46. 6 N (from left), 4 N (from direction $\pi/4$), 2 N (from direction $\pi/3$)
47. 2 N (from above), 3 N (from direction $3\pi/4$)
48. 5 N (from direction $\pi/3$), 5 N (from direction $2\pi/3$)
49. 7 N (from direction $\pi/6$), 7 N (from direction $\pi/3$), 14 N (from direction $5\pi/4$)

In Problems 50–58, find the work done when the force with the given magnitude and direction moves an object from P to Q. All distances are measured in meters. (Note that work can be negative.)

50. $|\mathbf{F}| = 3$ N; $\theta = 0$; $P = (2, 3)$; $Q = (1, 7)$
51. $|\mathbf{F}| = 2$ N; $\theta = \pi/2$; $P = (5, 7)$; $Q = (1, 1)$
52. $|\mathbf{F}| = 6$ N; $\theta = \pi/4$; $P = (2, 3)$; $Q = (-1, 4)$
53. $|\mathbf{F}| = 4$ N; $\theta = \pi/6$; $P = (-1, 2)$; $Q = (3, 4)$
54. $|\mathbf{F}| = 7$ N; $\theta = 2\pi/3$; $P = (4, -3)$; $Q = (1, 0)$
55. $|\mathbf{F}| = 3$ N; $\theta = 3\pi/4$; $P = (2, 1)$; $Q = (1, 2)$
56. $|\mathbf{F}| = 6$ N; $\theta = \pi$; $P = (3, -8)$; $Q = (5, 10)$
57. $|\mathbf{F}| = 4$ N; θ is the direction of $2\mathbf{i} + 3\mathbf{j}$; $P = (2, 0)$; $Q = (-1, 3)$
58. $|\mathbf{F}| = 5$ N; θ is the direction of $-3\mathbf{i} + 2\mathbf{j}$; $P = (1, 3)$; $Q = (4, -6)$

†59. Two tugboats are towing a barge (see Figure 10). Tugboat 1 pulls with a force of 500 N at an angle of 20° with the horizontal. Tugboat 2 pulls with a force of x newtons at an angle of 30°. The barge moves horizontally (i.e., $\theta = 0$). Find x.

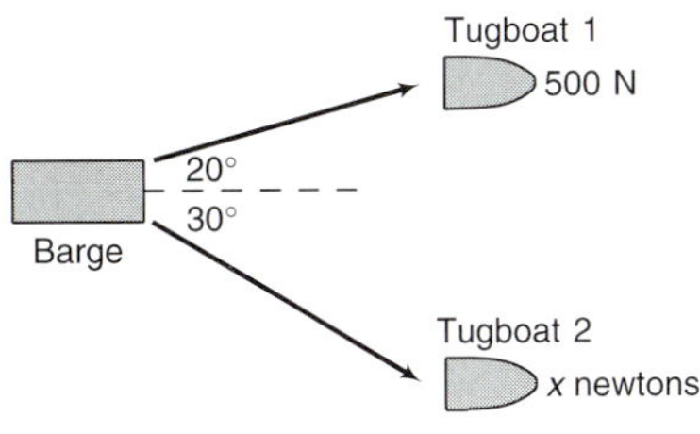

FIGURE 10

60. Answer the question of Problem 59 if the angles are 50° and 75°, respectively, and all other data remain the same.

61. In Problem 59, how much work is done by each tugboat in moving the barge a distance of 750 m?

62. In Problem 60, how much work is done by each tugboat in moving the barge a distance of 2 km?

63. Prove or Disprove: For any nonzero real numbers α and β, the vectors $\mathbf{u} = \alpha\mathbf{i} + \beta\mathbf{j}$ and $\mathbf{v} = \beta\mathbf{i} - \alpha\mathbf{j}$ are orthogonal.

64. Let **u**, **v**, and **w** be three arbitrary vectors. Explain why the triple dot-product $\mathbf{u} \cdot \mathbf{v} \cdot \mathbf{w}$ is *not defined.*

65. Prove or Disprove: There is at least one value of α for which $3\mathbf{i} + 4\mathbf{j}$ and $\mathbf{i} + \alpha\mathbf{j}$ have opposite directions.

66. Prove or Disprove: There is a unique value of β for which $-2\mathbf{i} + 5\mathbf{j}$ and $\mathbf{v} = \beta\mathbf{i} - 2\mathbf{j}$ have the same direction.

67. Let $\mathbf{u} = a_1\mathbf{i} + b_1\mathbf{j}$ and $\mathbf{v} = a_2\mathbf{i} + b_2\mathbf{j}$. Give a condition on a_1, b_1, a_2, and b_2 which will ensure that **v** and $\text{Proj}_{\mathbf{v}}\,\mathbf{u}$ have the same direction.

68. In the preceding problem, give a condition which will ensure that **v** and $\text{Proj}_{\mathbf{v}}\,\mathbf{u}$ have opposite directions.

69. Prove that the nonzero vectors **u** and **v** are parallel if and only if $\mathbf{v} = \alpha\mathbf{u}$ for some nonzero scalar α. [*Hint:* Show that $\cos\varphi = \pm 1$ if and only if $\mathbf{v} = \alpha\mathbf{u}$.]

70. Prove that the nonzero vectors **u** and **v** are orthogonal if and only if $\mathbf{u} \cdot \mathbf{v} = 0$.

71. Show that the vector $\mathbf{v} = a\mathbf{i} + b\mathbf{j}$ is orthogonal to the line $ax + by + c = 0$.

72. Show that the vector $\mathbf{v} = b\mathbf{i} - a\mathbf{j}$ is parallel to the line $ax + by + c = 0$.

73. **Parallelogram law**: Prove that for any vectors **a** and **b**,

$$|\mathbf{a} + \mathbf{b}|^2 + |\mathbf{a} - \mathbf{b}|^2 = 2|\mathbf{a}|^2 + 2|\mathbf{b}|^2.$$

[*Hint:* $|\mathbf{u}|^2 = \mathbf{u} \cdot \mathbf{u}$ for any vector **u**.]

74. **Polarization identity**: Prove that for any vectors **a** and **b**,

$$\mathbf{a} \cdot \mathbf{b} = \frac{1}{4}(|\mathbf{a} + \mathbf{b}|^2 - |\mathbf{a} - \mathbf{b}|^2).$$

*75. One form of the **Cauchy-Schwarz inequality** states that for any real numbers a_1, b_2, b_1, and b_2,

$$\left|\sum_{i=1}^{2} a_i b_i\right| \le \left(\sum_{i=1}^{2} a_i^2\right)^{1/2}\left(\sum_{i=1}^{2} b_i^2\right)^{1/2}.$$

Use the dot product to prove this formula. Under what circumstances can the inequality be replaced by an equality? [*Hint:* Show that $|\mathbf{a} \cdot \mathbf{b}| \le |\mathbf{a}||\mathbf{b}|$.]

76. Use the dot product and the Cauchy-Schwarz inequality to prove the **triangle inequality**:

$$|\mathbf{a} + \mathbf{b}| \le |\mathbf{a}| + |\mathbf{b}| \qquad \text{for all vectors } \mathbf{a}, \mathbf{b}.$$

77. Prove that the shortest distance between a point and a line is measured along a line through the point and perpendicular to the line.

78. Prove that the distance from the point (x_0, y_0) to the line with standard equation $Ax + By + C = 0$ is

$$\frac{|Ax_0 + By_0 + C|}{\sqrt{A^2 + B^2}}.$$

[*Hint:* Let **u** be a vector from (x_0, y_0) to some point on the line and let **v** be a vector with the same direction as the line. Remember Theorem 5 and the result of the preceding problem, then consider the length of $\mathbf{u} - \text{Proj}_{\mathbf{v}}\,\mathbf{u}$.]

† This symbol indicates that a calculator is needed to solve the problem.

ANSWERS TO SELF-QUIZ

I. c **II.** b **III.** b **IV.** c **V.** c

1.3 THE RECTANGULAR COORDINATE SYSTEM IN SPACE

In your first calculus course, you saw how any point in a plane could be represented as an ordered pair of real numbers. It is not surprising, then, that any point in space can be represented by an **ordered triple** of real numbers

$$(a, b, c), \tag{1}$$

where a, b, and c are real numbers.

DEFINITION THREE-DIMENSIONAL SPACE $\mathbb{R}^3$

The set of ordered triples of the form (1) is called **real three-dimensional space** and is denoted by $\mathbb{R}^3$. ■

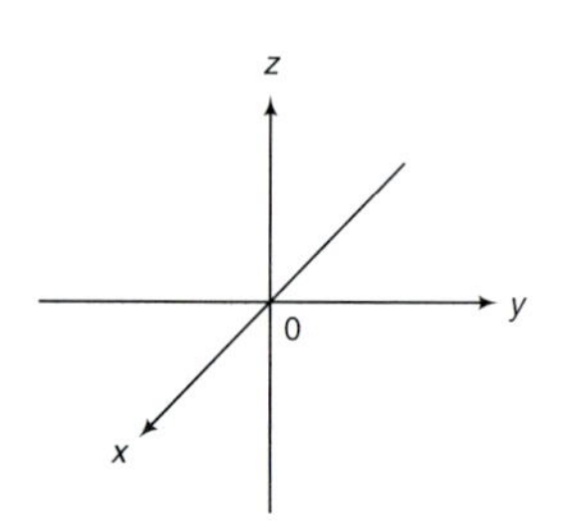

FIGURE 1
A right-handed system

There are many ways to represent a point in $\mathbb{R}^3$. Two others will be given in Section 1.9. However, the most common representation (given in the definition above) is very similar to the representation of a point in the plane by its x- and y-coordinates. We begin, as before, by choosing a point in $\mathbb{R}^3$ and calling it the **origin**, denoted by 0. Then we draw three mutually perpendicular axes, called the **coordinate axes**, which we label the ***x*-axis**, the ***y*-axis**, and the ***z*-axis**. These axes can be selected in a variety of ways, but the most common selection has the x- and y-axes drawn horizontally with the z-axis vertical. On each axis we choose a positive direction and measure distance along each axis as the number of units in this positive direction measured from the origin.

A standard way of drawing these axes is depicted in Figure 1. The system is called a **right-handed system**. In the figure the arrows indicate the positive directions on the axes. The reason for this choice of terms is as follows: In a right-handed system, if you place your right hand so that your index finger points in the positive direction of the x-axis while your middle finger points in the positive direction of the y-axis, then your thumb will point in the positive direction of the z-axis. This is illustrated in Figure 2. For the remainder of this text we will follow common practice and depict the coordinate axes using a right-handed system.

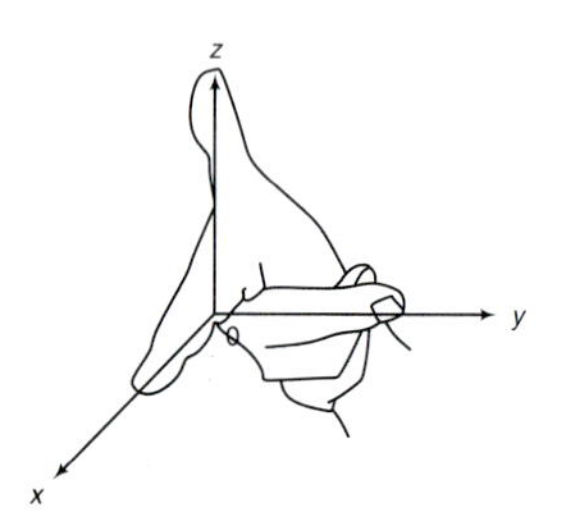

FIGURE 2
A right-handed system

If you have trouble visualizing the placement of these axes, do the following. Face any uncluttered corner (on the floor) of the room in which you are sitting. Call the corner the origin. Then the x-axis lies along the floor, along the wall, and to your left; the y-axis lies along the floor, along the wall, and to your right; and the z-axis lies along the vertical intersection of the two perpendicular walls. This is illustrated in Figure 3.

The three axes in our system determine three **coordinate planes** that we will call the ***xy*-plane**, the ***xz*-plane**, and the ***yz*-plane**. The xy-plane contains the x- and y-axes and is simply the plane with which we have been dealing to this point in most of this book. The xz- and yz-planes can be thought of in a similar way.

Having built our structure of coordinate axes and planes, we can describe any point P in space in a unique way:

$$P = (x, y, z), \tag{2}$$

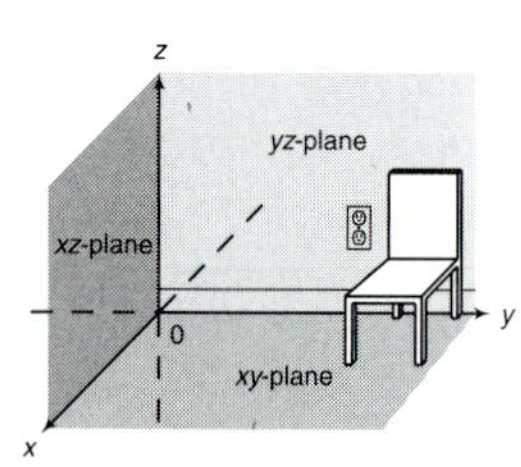

FIGURE 3
The three coordinate planes

where the first coordinate x is the directed distance from the yz-plane to P (measured in the positive direction of the x-axis), the second coordinate y is the directed distance from the xz-plane to P (measured in the positive direction of the y-axis), and the third coordinate z is the directed distance from the xy-plane to P (measured in the positive direction of the z-axis). Thus, for example, any point in the xy-plane has z-coordinate 0; any point in the xz-plane has y-coordinate 0; and any point in the yz-plane has x-coordinate 0. Some representative points are sketched in Figure 4.

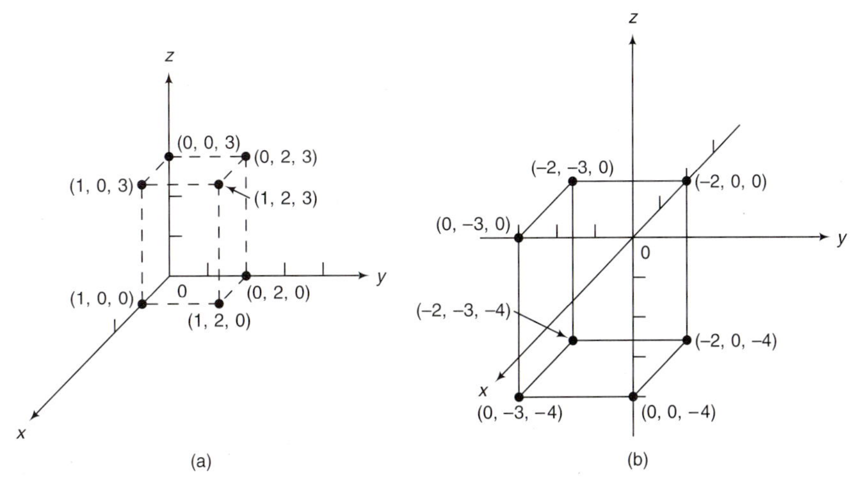

FIGURE 4
Typical points in $\mathbb{R}^3$

In this system, the three coordinate planes divide $\mathbb{R}^3$ into eight **octants**, just as in $\mathbb{R}^2$ the two coordinate axes divide the plane into four quadrants. The first octant is always chosen to be the one in which the three coordinates are positive (the other seven octants are usually not numbered).

The coordinate system we have just established is often referred to as the **rectangular coordinate system**, or the **Cartesian coordinate system**. Once we are comfortable with the notion of depicting a point in this system, then we can generalize many of our ideas from the plane.

THEOREM 1

Let $P = (x_1, y_1, z_1)$ and $Q = (x_2, y_2, z_2)$ be two points in space. Then the distance $\overline{PQ}$ between P and Q is given by

$$\overline{PQ} = \sqrt{(x_1 - x_2)^2 + (y_1 - y_2)^2 + (z_1 - z_2)^2}. \quad (3)$$

■

The proof of this theorem is left as an exercise (see Problem 40).

EXAMPLE 1

CALCULATING THE DISTANCE BETWEEN TWO POINTS IN $\mathbb{R}^3$

Calculate the distance between the points $(3, -1, 6)$ and $(-2, 3, 5)$.

SOLUTION: $\overline{PQ} = \sqrt{[3 - (-2)]^2 + (-1 - 3)^2 + (6 - 5)^2} = \sqrt{42}.$

DEFINITION GRAPH IN $\mathbb{R}^3$

The **graph** of an equation in $\mathbb{R}^3$ is the set of all points in $\mathbb{R}^3$ whose coordinates satisfy the equation. ■

One of our first examples of a graph in $\mathbb{R}^2$ was the graph of the unit circle $x^2 + y^2 = 1$. This example can easily be generalized.

DEFINITION SPHERE

A **sphere** is the set of points in space at a given distance from a given point. The given point is called the **center** of the sphere, and the given distance is called the **radius** of the sphere. ■

EXAMPLE 2 THE EQUATION OF THE UNIT SPHERE

Suppose that the center of a sphere is the origin (0, 0, 0) and the radius of the sphere is 1. Let (x, y, z) be a point on the sphere. Then from (3)

$$1 = \sqrt{(x-0)^2 + (y-0)^2 + (z-0)^2}.$$

Simplifying and squaring, we obtain

$$x^2 + y^2 + z^2 = 1,$$

which is the equation of the **unit sphere**.

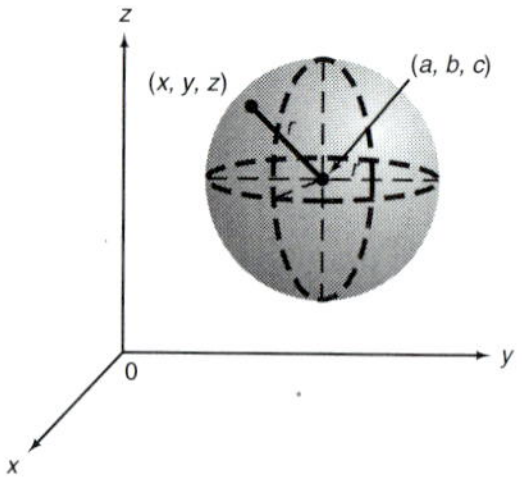

FIGURE 5
Sphere of radius r centered at (a, b, c)

In general, if the center of a sphere is (a, b, c), the radius is r, and if (x, y, z) is a point on the sphere, we obtain $r = \sqrt{(x-a)^2 + (y-b)^2 + (z-c)^2}$, or

EQUATION OF A SPHERE CENTERED AT (a, b, c) WITH RADIUS r

$$(x-a)^2 + (y-b)^2 + (z-c)^2 = r^2. \tag{4}$$

This is sketched in Figure 5.

EXAMPLE 3 FINDING THE EQUATION OF A SPHERE

Find the equation of the sphere with center at $(1, -3, 2)$ and radius 5.

SOLUTION: From (4) we obtain

$$(x-1)^2 + (y+3)^2 + (z-2)^2 = 25.$$

EXAMPLE 4 SHOWING THAT A SECOND-DEGREE EQUATION IS THE EQUATION OF A SPHERE

Show that

$$x^2 - 6x + y^2 + 2y + z^2 + 10z + 5 = 0 \tag{5}$$

is the equation of a sphere, and find its center and radius.

SOLUTION: We complete the square three times:

$$\begin{aligned} 0 &= x^2 - 6x + y^2 + 2y + z^2 + 10z + 5 \\ &= x^2 - 6x + 9 - 9 + y^2 + 2y + 1 - 1 + z^2 + 10z + 25 - 25 + 5 \\ &= (x-3)^2 - 9 + (y+1)^2 - 1 + (z+5)^2 - 25 + 5 \end{aligned}$$

or

$$(x-3)^2+(y+1)^2+(z+5)^2=30,$$

which is the equation of a sphere with center $(3, -1, -5)$ and radius $\sqrt{30}$.

WARNING: Not every second-degree equation in a form similar to (5) is the equation of a sphere. For example, if the number 5 in (5) is replaced by 40, we obtain

$$(x-3)^2+(y+1)^2+(z+5)^2=-5.$$

Clearly, the sum of squares cannot be negative, so there are *no* points in $\mathbb{R}^3$ that satisfy this equation. On the other hand, if we replaced the 5 by 35, we would obtain

$$(x-3)^2+(y+1)^2+(z+5)^2=0.$$

This equation can hold only when $x=3$, $y=-1$, and $z=-5$. In this case, the graph of the equation consists of the single point $(3, -1, -5)$.

PROBLEMS 1.3

SELF-QUIZ

I. True–False: The common practice, followed in this text, is to display the xyz-axes for $\mathbb{R}^3$ as a right-handed system.

II. The distance between the points $(1, 2, 3)$ and $(3, 5, -1)$ is ________.
a. $\sqrt{(1+2+3)^2+(3+5-1)^2}$
b. $\sqrt{2^2+3^2+2^2}$
c. $\sqrt{2^2+3^2+4^2}$
d. $\sqrt{4^2+7^2+2^2}$

III. The point $(0.3, 0.5, 0.2)$ is ________ the unit sphere.
a. tangent to **b.** on
c. inside **d.** outside

IV. $(x-3)^2+(y+5)^2+z^2=81$ is the equation of the sphere with ________.
a. center 81 and radius $(-3, 5, 0)$
b. radius 81 and center $(-3, 5, 0)$
c. radius -9 and center $(3, -5, 0)$
d. radius 9 and center $(3, -5, 0)$

In Problems 1–16, sketch the given point in $\mathbb{R}^3$.

1. $(1, 4, 2)$
2. $(3, -2, 1)$
3. $(-1, 5, 7)$
4. $(8, -2, 3)$
5. $(-2, 1, -2)$
6. $(1, -2, 1)$
7. $(3, 2, -5)$
8. $(-2, -3, -8)$
9. $(2, 0, 4)$
10. $(-3, -8, 0)$
11. $(0, 4, 7)$
12. $(1, 3, 0)$
13. $(3, 0, 0)$
14. $(0, 8, 0)$
15. $(0, 0, -7)$
16. $(5, 5, 5)$

In Problems 17–26, calculate the distance between the two given points.

17. $(8, 1, 6)$; $(8, 1, 4)$
18. $(3, -4, 3)$; $(3, 2, 5)$
19. $(3, -4, 7)$; $(3, -4, 9)$
20. $(-2, 1, 3)$; $(4, 1, 3)$
21. $(2, -7, 5)$; $(8, -7, -1)$
22. $(1, 3, -2)$; $(4, 7, -2)$
23. $(3, 1, 2)$; $(1, 2, 3)$
24. $(5, -6, 4)$; $(3, 11, -2)$
25. $(-1, -7, -2)$; $(-4, 3, -5)$
26. $(8, -2, -3)$; $(-7, -5, 1)$

27. Find the equation of the sphere with center $(2, -1, 4)$ and radius 2.

28. Find the equation of the sphere with center $(-1, 8, -3)$ and radius $\sqrt{5}$.

29. Use the result stated in Problem 38 to verify that the points $(3, 0, 1)$, $(0, -4, 0)$ and $(6, 4, 2)$ are collinear.

30. Use the result stated in Problem 39 to find the midpoint of the line joining the points $(2, -1, 4)$ and $(5, 7, -3)$.

31. Verify that $x^2+y^2+z^2-4x-4y+8z+8=0$ is the equation of a sphere. Find the center and radius of that sphere.

32. Verify that $x^2 + 3x + y^2 - y + z^2 + 2z - 1 = 0$ is the equation of a sphere. Find the center and radius of that sphere.

33. One sphere is said to be **concentric** to a second sphere if it has a different radius and the same center as the second sphere. Find the equation of a sphere of radius 1 concentric to the sphere given by $x^2 - 2x + y^2 - 4y + z^2 + z - 2 = 0$.

34. Find the equation of the sphere that has a diameter with endpoints $(3, 1, -2)$ and $(4, 1, 5)$. [*Hint:* First find the center and radius of the sphere.]

35. Find a number α such that the equation $x^2 - 2x + y^2 + 8y + z^2 - 5z + \alpha = 0$ has exactly one solution (x, y, z).

36. For the equation in the preceding problem, give a condition on α such that the resulting equation has no solution.

37. For fixed constants a, b, c, and d the equation

$$x^2 + ax + y^2 + by + z^2 + cz + d = 0 \qquad (6)$$

is a second-degree equation in three variables x, y, and z. Let

$$E = d - (a/2)^2 - (b/2)^2 - (c/2)^2.$$

Show that equation (6)

a. is the equation of a sphere if $E < 0$,

b. has exactly one solution if $E = 0$,

c. has no solutions if $E > 0$.

38. Three points P, Q, and R are **collinear** if and only if they lie on the same straight line. Show that, in $\mathbb{R}^2$, points P, Q, and R are collinear if and only if one of the following conditions holds:

- $\overline{PR} = \overline{PQ} + \overline{QR}$ (i.e., Q is between the others)
- $\overline{PQ} = \overline{PR} + \overline{RQ}$ (i.e., R is between the others)
- $\overline{QR} = \overline{QP} + \overline{PR}$ (i.e., P is between the others)

[Here $\overline{PQ}$ denotes the length of the directed line segment $\overrightarrow{PQ}$; $\overline{PQ} = |\overrightarrow{PQ}|$.] Apply the same idea in $\mathbb{R}^3$ to show that the points $(-1, -1, -1)$, $(5, 8, 2)$ and $(-3, -4, -2)$ are collinear.

***39.** Let $P = (x_1, y_1, z_1)$ and $Q = (x_2, y_2, z_2)$. Show that the midpoint of PQ is the point

$$R = \left(\frac{x_1 + x_2}{2}, \frac{y_1 + y_2}{2}, \frac{z_1 + z_2}{2}\right).$$

[*Hint:* Show that P, Q, and R are collinear and that $\overline{PR} = \overline{RQ}$.]

40. Prove Theorem 1. [*Hint:* Use the Pythagorean theorem twice. Show, with points labeled as in Figure 6, that

a. $\overline{PQ}^2 = \overline{PR}^2 + \overline{RQ}^2$,

b. $\overline{PR}^2 = \overline{PS}^2 + \overline{SR}^2$,

c. $\overline{PQ}^2 = \overline{PS}^2 + \overline{SR}^2 + \overline{RQ}^2$.]

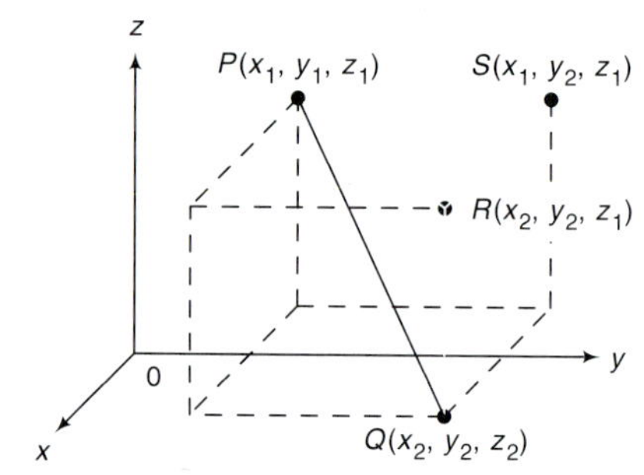

FIGURE 6

ANSWERS TO SELF-QUIZ

I. True **II.** c **III.** c **IV.** d

1.4 VECTORS IN $\mathbb{R}^3$

In Sections 1.1 and 1.2, we developed properties of vectors in the plane $\mathbb{R}^2$. Given the similarity between the coordinate systems in $\mathbb{R}^2$ and $\mathbb{R}^3$, it should come as no surprise to learn that vectors in $\mathbb{R}^2$ and $\mathbb{R}^3$ have very similar structures. In this section we will develop the notion of a vector in space. The development will closely follow the development in Sections 1.1 and 1.2 and, therefore, some of the details will be omitted.

Let P and Q be two distinct points in $\mathbb{R}^3$. Then the **directed line segment** $\overrightarrow{PQ}$ is the straight line segment that extends from P to Q. Two directed line segments are **equivalent** if they have the same magnitude and direction. A **vector** in $\mathbb{R}^3$ is the set of all directed line segments equivalent to a given line segment, and any directed line segment $\overrightarrow{PQ}$ in that set is called a **representation** of the vector.

So far, our definitions are identical. For convenience we will choose P to be the origin and label the endpoint of the vector R, so that the vector $\mathbf{v} = \overrightarrow{0R}$ can be described by the coordinates (x, y, z) of the point R. The numbers x, y, and z are called the **components** of $\mathbf{v}$. As in $\mathbb{R}^2$, two vectors are **equal** if their corresponding components are equal. Then the **magnitude** of $\mathbf{v} = |\mathbf{v}| = \sqrt{x^2 + y^2 + z^2}$ (from Theorem 1.3.1).

EXAMPLE 1

CALCULATING THE MAGNITUDE OF A VECTOR IN $\mathbb{R}^3$

Let $\mathbf{v} = (1, 3, -2)$. Find $|\mathbf{v}|$.

SOLUTION: $|\mathbf{v}| = \sqrt{1^2 + 3^2 + (-2)^2} = \sqrt{14}$.

Let $\mathbf{u} = (x_1, y_1, z_1)$ and $\mathbf{v} = (x_2, y_2, z_2)$ be two vectors and let α be a real number (scalar). Then we define

SUM AND SCALAR MULTIPLE OF VECTORS IN $\mathbb{R}^3$

$$\mathbf{u} + \mathbf{v} = (x_1 + x_2, y_1 + y_2, z_1 + z_2)$$

and

$$\alpha\mathbf{u} = (\alpha x_1, \alpha y_1, \alpha z_1).$$

This is the same definition of vector addition and scalar multiplication we had before and is illustrated in Figure 1.

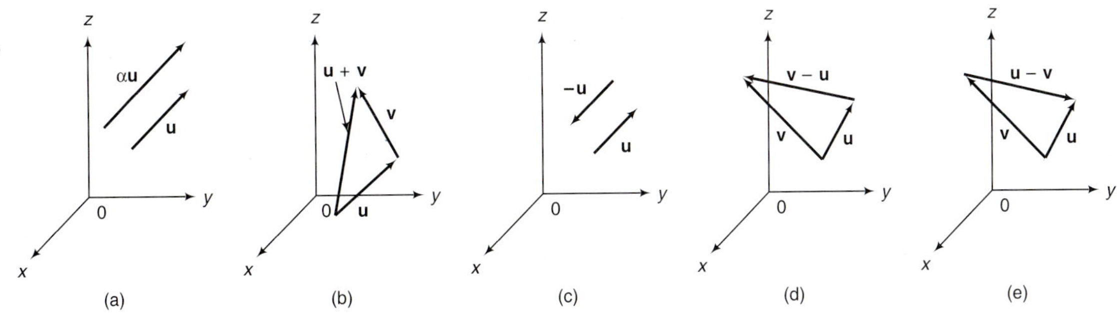

FIGURE 1
Illustrating the vector operations $\alpha\mathbf{u}$, $\mathbf{u} + \mathbf{v}$, $-\mathbf{u}$, $\mathbf{v} - \mathbf{u}$, and $\mathbf{u} - \mathbf{v}$

EXAMPLE 2

FINDING SUMS AND SCALAR MULTIPLES OF VECTORS

Let $\mathbf{u} = (2, 3, -1)$ and $\mathbf{v} = (-6, 2, 4)$. Find **a.** $\mathbf{u} + \mathbf{v}$; **b.** $3\mathbf{v}$; **c.** $-\mathbf{u}$; and **d.** $4\mathbf{u} - 3\mathbf{v}$.

SOLUTION:

a. $\mathbf{u} + \mathbf{v} = (2 - 6, 3 + 2, -1 + 4) = (-4, 5, 3)$
b. $3\mathbf{v} = (-18, 6, 12)$
c. $-\mathbf{u} = (-2, -3, 1)$
d. $4\mathbf{u} - 3\mathbf{v} = (8, 12, -4) - (-18, 6, 12) = (26, 6, -16)$

The following theorem extends to three dimensions the results of Theorem 1.1.1. Its proof is easy and is left as an exercise (Problem 56).

THEOREM 1 PROPERTIES OF VECTORS IN $\mathbb{R}^3$

Let **u**, **v**, and **w** be any three vectors in space, let α and β be scalars, and let **0** denote the zero vector (0, 0, 0).

i. $\mathbf{u} + \mathbf{v} = \mathbf{v} + \mathbf{u}$ **ii.** $\mathbf{u} + (\mathbf{v} + \mathbf{w}) = (\mathbf{u} + \mathbf{v}) + \mathbf{w}$
iii. $\mathbf{v} + \mathbf{0} = \mathbf{v}$ **iv.** $0\mathbf{v} = \mathbf{0}$
v. $\alpha\mathbf{0} = \mathbf{0}$ **vi.** $(\alpha\beta)\mathbf{v} = \alpha(\beta\mathbf{v})$
vii. $\mathbf{v} + (-\mathbf{v}) = \mathbf{0}$ **viii.** $(1)\mathbf{v} = \mathbf{v}$
ix. $(\alpha + \beta)\mathbf{v} = \alpha\mathbf{v} + \beta\mathbf{v}$ **x.** $\alpha(\mathbf{u} + \mathbf{v}) = \alpha\mathbf{u} + \alpha\mathbf{v}$
xi. $|\alpha\mathbf{v}| = |\alpha|\,|\mathbf{v}|$ **xii.** $|\mathbf{u} + \mathbf{v}| \leq |\mathbf{u}| + |\mathbf{v}|$
triangle inequality ■

DEFINITION UNIT VECTOR

A **unit vector u** is a vector with magnitude 1. If **v** is any nonzero vector, then $\mathbf{u} = \mathbf{v}/|\mathbf{v}|$ is a unit vector having the same direction as **v**. ■

EXAMPLE 3 FINDING A UNIT VECTOR THAT HAS THE SAME DIRECTION AS A GIVEN VECTOR

Find a unit vector having the same direction as $\mathbf{v} = (2, 4, -3)$.

SOLUTION: Since $|\mathbf{v}| = \sqrt{2^2 + 4^2 + (-3)^2} = \sqrt{29}$, we have

$$\mathbf{u} = \left(\frac{2}{\sqrt{29}}, \frac{4}{\sqrt{29}}, \frac{-3}{\sqrt{29}}\right).$$

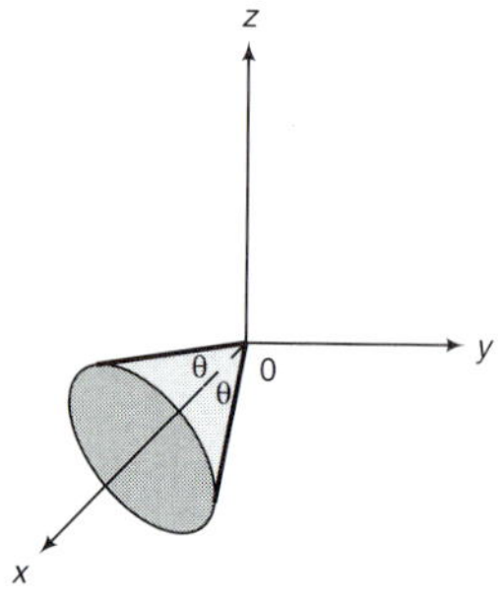

FIGURE 2
Every vector in the cone makes the angle θ with the positive x-axis

We can now formally define the direction of a vector in $\mathbb{R}^3$. We cannot define it to be the angle θ the vector makes with the positive x-axis, since, for example, if $0 < \theta < \pi/2$, then there are an *infinite number* of unit vectors making the angle θ with the positive x-axis, and these together form a cone (see Figure 2).

DEFINITION DIRECTION OF A VECTOR

The **direction** of a nonzero vector **v** in $\mathbb{R}^3$ is defined to be the unit vector $\mathbf{u} = \mathbf{v}/|\mathbf{v}|$. ■

REMARK: We could have defined the direction of a vector $\mathbf{v}$ in $\mathbb{R}^2$ in this way. For if $\mathbf{u} = \mathbf{v}/|\mathbf{v}|$, then $\mathbf{u} = (\cos\theta, \sin\theta)$, where θ is the direction of $\mathbf{v}$ (according to the $\mathbb{R}^2$ definition).

DIRECTION COSINES

It would still be useful to define the direction of a vector in terms of some angles. Let $\mathbf{v} = (x_0, y_0, z_0)$ be the vector $\overrightarrow{OP}$ depicted in Figure 3. We define α to be the angle between $\mathbf{v}$ and the positive x-axis, β the angle between $\mathbf{v}$ and the positive y-axis, and γ the angle between $\mathbf{v}$ and the positive z-axis. The angles α, β, and γ are called the **direction angles** of the vector $\mathbf{v}$. Then, from Figure 3,

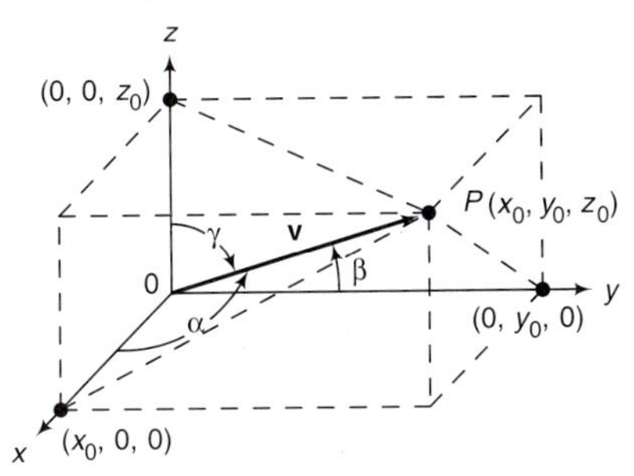

FIGURE 3
Direction angles

THREE DIRECTION ANGLES

$$\cos\alpha = \frac{x_0}{|\mathbf{v}|}, \qquad \cos\beta = \frac{y_0}{|\mathbf{v}|}, \qquad \cos\gamma = \frac{z_0}{|\mathbf{v}|}. \tag{1}$$

If $\mathbf{v}$ is a unit vector, then $|\mathbf{v}| = 1$ and

$$\cos\alpha = x_0, \qquad \cos\beta = y_0, \qquad \cos\gamma = z_0. \tag{2}$$

By definition, each of these three angles lies between 0 and π. The cosines of these angles are called the **direction cosines** of the vector $\mathbf{v}$.

Note from (1) that

$$\cos^2\alpha + \cos^2\beta + \cos^2\gamma = \frac{x_0^2 + y_0^2 + z_0^2}{|\mathbf{v}|^2} = \frac{x_0^2 + y_0^2 + z_0^2}{x_0^2 + y_0^2 + z_0^2} = 1. \tag{3}$$

If α, β, and γ are any three numbers between 0 and π such that condition (3) is satisfied, then they uniquely determine a unit vector given by $\mathbf{u} = (\cos\alpha, \cos\beta, \cos\gamma)$.

REMARK: If $\mathbf{v} = (a, b, c)$ and $|\mathbf{v}| \neq 0$, then the numbers a, b, and c are called **direction numbers** of the vector $\mathbf{v}$.

EXAMPLE 4 FINDING THE DIRECTION COSINES OF A VECTOR IN $\mathbb{R}^3$

Find the direction cosines of the vector $\mathbf{v} = (4, -1, 6)$.

SOLUTION: The direction of $\mathbf{v}$ is $\mathbf{v}/|\mathbf{v}| = \mathbf{v}/\sqrt{53} = (4/\sqrt{53}, -1/\sqrt{53}, 6/\sqrt{53})$. Then $\cos\alpha = 4/\sqrt{53} \approx 0.5494$, $\cos\beta = -1/\sqrt{53} \approx -0.1374$, and $\cos\gamma = 6/\sqrt{53} \approx 0.8242$. From these we use a calculator to obtain $\alpha \approx 56.7° \approx 0.989$ radian, $\beta \approx 97.9° \approx 1.71$ radians, and $\gamma = 34.5° \approx 0.602$ radian.

EXAMPLE 5

FINDING A VECTOR IN $\mathbb{R}^3$ GIVEN ITS MAGNITUDE AND DIRECTION COSINES

Find a vector $\mathbf{v}$ of magnitude 7 whose direction cosines are $1/\sqrt{6}$, $1/\sqrt{3}$, and $1/\sqrt{2}$.

NOTE: We can solve this problem because $(1/\sqrt{6})^2 + (1/\sqrt{3})^2 + (1/\sqrt{2})^2 = 1$.

SOLUTION: Let $\mathbf{u} = (1/\sqrt{6}, 1/\sqrt{3}, 1/\sqrt{2})$. Then $\mathbf{u}$ is a unit vector since $|\mathbf{u}| = 1$. Thus, the direction of $\mathbf{v}$ is given by $\mathbf{u}$, and so

$$\mathbf{v} = |\mathbf{v}|\mathbf{u} = 7\mathbf{u} = \left(\frac{7}{\sqrt{6}}, \frac{7}{\sqrt{3}}, \frac{7}{\sqrt{2}}\right).$$

In Section 1.1, we showed how any vector in the plane can be written in terms of the basis vectors $\mathbf{i}$ and $\mathbf{j}$. To extend this idea to $\mathbb{R}^3$, we define

THE VECTORS i, j, AND k IN $\mathbb{R}^3$

$$\mathbf{i} = (1, 0, 0), \qquad \mathbf{j} = (0, 1, 0), \qquad \mathbf{k} = (0, 0, 1). \tag{4}$$

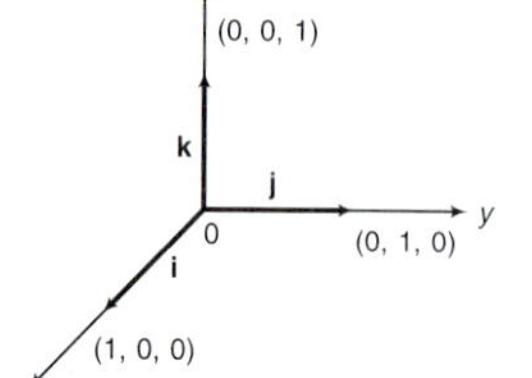

FIGURE 4
The vectors **i**, **j**, and **k**

Here $\mathbf{i}$, $\mathbf{j}$, and $\mathbf{k}$ are unit vectors. The vector $\mathbf{i}$ lies along the x-axis, $\mathbf{j}$ along the y-axis, and $\mathbf{k}$ along the z-axis. These vectors are sketched in Figure 4. If $\mathbf{v} = (x, y, z)$ is any vector in $\mathbb{R}^3$, then

$$\mathbf{v} = (x, y, z) = (x, 0, 0) + (0, y, 0) + (0, 0, z) = x\mathbf{i} + y\mathbf{j} + z\mathbf{k}. \tag{5}$$

That is, *any vector* $\mathbf{v}$ *in* $\mathbb{R}^3$ *can be written in a unique way in terms of the vectors* $\mathbf{i}$, $\mathbf{j}$, *and* $\mathbf{k}$.

Let $P = (a_1, b_1, c_1)$ and $Q = (a_2, b_2, c_2)$. Then as in Section 1.1, the vector $\mathbf{v} = \overrightarrow{PQ}$ can be written (see Problem 57)

$$\mathbf{v} = (a_2 - a_1)\mathbf{i} + (b_2 - b_1)\mathbf{j} + (c_2 - c_1)\mathbf{k}. \tag{6}$$

EXAMPLE 6

WRITING A VECTOR IN TERMS OF THE BASIS VECTORS I, J, AND K

Find a vector in space that can be represented by the directed line segment from $(2, -1, 4)$ to $(5, 1, -3)$.

$$\mathbf{v} = (5 - 2)\mathbf{i} + [1 - (-1)]\mathbf{j} + (-3 - 4)\mathbf{k} = 3\mathbf{i} + 2\mathbf{j} - 7\mathbf{k}.$$

We now turn to the notion of dot product (or scalar product) in $\mathbb{R}^3$.

DEFINITION DOT PRODUCT IN $\mathbb{R}^3$

If $\mathbf{u} = x_1\mathbf{i} + y_1\mathbf{j} + z_1\mathbf{k}$ and $\mathbf{v} = x_2\mathbf{i} + y_2\mathbf{j} + z_2\mathbf{k}$, then we define the **dot product** (or **scalar product** or **inner product**) by

$$\mathbf{u} \cdot \mathbf{v} = x_1x_2 + y_1y_2 + z_1z_2. \qquad (7)$$

■

As before, the dot product of two vectors is a *scalar*. Note that $\mathbf{i} \cdot \mathbf{i} = 1$, $\mathbf{j} \cdot \mathbf{j} = 1$, $\mathbf{k} \cdot \mathbf{k} = 1$, $\mathbf{i} \cdot \mathbf{j} = 0$, $\mathbf{j} \cdot \mathbf{k} = 0$, and $\mathbf{i} \cdot \mathbf{k} = 0$.

EXAMPLE 7 COMPUTING A DOT PRODUCT IN $\mathbb{R}^3$

For $\mathbf{u} = 2\mathbf{i} - 3\mathbf{j} - 4\mathbf{k}$ and $\mathbf{v} = -3\mathbf{i} + \mathbf{j} - 2\mathbf{k}$, calculate $\mathbf{u} \cdot \mathbf{v}$.

SOLUTION: $\mathbf{u} \cdot \mathbf{v} = 2(-3) + (-3)(1) + (-4)(-2) = -1$.

THEOREM 2 PROPERTIES OF THE DOT PRODUCT IN $\mathbb{R}^3$

For any vectors $\mathbf{u}$, $\mathbf{v}$, and $\mathbf{w}$ in space, and for any scalar α, we have

i. $\mathbf{u} \cdot \mathbf{v} = \mathbf{v} \cdot \mathbf{u}$
ii. $(\mathbf{u} + \mathbf{v}) \cdot \mathbf{w} = \mathbf{u} \cdot \mathbf{w} + \mathbf{v} \cdot \mathbf{w}$
iii. $(\alpha\mathbf{u}) \cdot \mathbf{v} = \alpha(\mathbf{u} \cdot \mathbf{v})$
iv. $\mathbf{u} \cdot \mathbf{u} \geq 0$, and $\mathbf{u} \cdot \mathbf{u} = 0$ if and only if $\mathbf{u} = \mathbf{0}$
v. $|\mathbf{u}| = \sqrt{\mathbf{u} \cdot \mathbf{u}}$.

PROOF: The proof is almost identical to the proof of Theorem 1.2.1 and is left as an exercise (see Problem 58). ■

THEOREM 3

If φ denotes the angle between two nonzero vectors $\mathbf{u}$ and $\mathbf{v}$, we have

$$\cos\varphi = \frac{\mathbf{u} \cdot \mathbf{v}}{|\mathbf{u}|\,|\mathbf{v}|}. \qquad (8)$$

PROOF: The proof is almost identical to the proof of Theorem 1.2.2 and is left as an exercise (Problem 59). For an interesting corollary to this theorem, see Problem 55. ■

EXAMPLE 8 CALCULATING THE COSINE OF THE ANGLE BETWEEN TWO VECTORS IN $\mathbb{R}^3$

Calculate the cosine of the angle between $\mathbf{u} = 3\mathbf{i} - \mathbf{j} + 2\mathbf{k}$ and $\mathbf{v} = 4\mathbf{i} + 3\mathbf{j} - \mathbf{k}$.

SOLUTION: $\mathbf{u} \cdot \mathbf{v} = 7$, $|\mathbf{u}| = \sqrt{14}$, and $|\mathbf{v}| = \sqrt{26}$, so that $\cos\varphi = 7/\sqrt{(14)(26)} = 7/\sqrt{364} \approx 0.3669$ and $\varphi = 68.5° \approx 1.2$ radians.

DEFINITION PARALLEL AND ORTHOGONAL VECTORS

i. Two nonzero vectors $\mathbf{u}$ and $\mathbf{v}$ are **parallel** if the angle between them is 0 or π.

ii. Two nonzero vectors $\mathbf{u}$ and $\mathbf{v}$ are **orthogonal** (or **perpendicular**) if the angle between them is $\pi/2$. ■

THEOREM 4

i. If $\mathbf{u} \neq \mathbf{0}$, then $\mathbf{u}$ and $\mathbf{v}$ are parallel if and only if $\mathbf{v} = \alpha\mathbf{u}$ for some constant α.

ii. If $\mathbf{u}$ and $\mathbf{v}$ are nonzero, then $\mathbf{u}$ and $\mathbf{v}$ are orthogonal if and only if $\mathbf{u} \cdot \mathbf{v} = 0$.

PROOF: Again the proof is not difficult and is left as an exercise (see Problem 60). ■

We now turn to the definition of the projection of one vector on another. First, we state the theorem that is the analog of Theorem 1.2.5 (and has an identical proof).

THEOREM 5

Let $\mathbf{v}$ be a nonzero vector. Then for any other vector $\mathbf{u}$,

$$\mathbf{w} = \mathbf{u} - \frac{\mathbf{u} \cdot \mathbf{v}}{|\mathbf{v}|^2}\mathbf{v}$$

is orthogonal to $\mathbf{v}$. ■

DEFINITION PROJECTION AND COMPONENT

Let $\mathbf{u}$ and $\mathbf{v}$ be nonzero vectors. Then the **projection**† **of u onto v**, denoted by $\text{Proj}_{\mathbf{v}}\,\mathbf{u}$, is defined by

$$\text{Proj}_{\mathbf{v}}\,\mathbf{u} = \frac{\mathbf{u} \cdot \mathbf{v}}{\mathbf{v} \cdot \mathbf{v}}\mathbf{v} = \frac{\mathbf{u} \cdot \mathbf{v}}{|\mathbf{v}|^2}\mathbf{v} = \left(\frac{\mathbf{u} \cdot \mathbf{v}}{|\mathbf{v}|}\right)\frac{\mathbf{v}}{|\mathbf{v}|}. \tag{9}$$

The **component** of $\mathbf{u}$ in the direction $\mathbf{v}$ is given by

$$(\mathbf{u} \cdot \mathbf{v})/|\mathbf{v}|. \tag{10}$$

■

EXAMPLE 9 CALCULATING A PROJECTION IN $\mathbb{R}^3$

Let $\mathbf{u} = 2\mathbf{i} + 3\mathbf{j} + \mathbf{k}$ and $\mathbf{v} = \mathbf{i} + 2\mathbf{j} - 6\mathbf{k}$. Find $\text{Proj}_{\mathbf{v}}\,\mathbf{u}$.

† The projection vector in $\mathbb{R}^2$ is sketched in Figure 4 of Section 1.2. The derivation and a sketch of a projection vector in $\mathbb{R}^3$ are virtually the same.

SOLUTION: Here $(\mathbf{u} \cdot \mathbf{v})/|\mathbf{v}|^2 = \frac{2}{41}$, so

$$\text{Proj}_{\mathbf{v}}\, \mathbf{u} = \frac{2}{41}\mathbf{i} + \frac{4}{41}\mathbf{j} - \frac{12}{41}\mathbf{k}.$$

The component of $\mathbf{u}$ in the direction $\mathbf{v}$ is $(\mathbf{u} \cdot \mathbf{v})/|\mathbf{v}| = 2/\sqrt{41}$.

Note that, as in the planar case, $\text{Proj}_{\mathbf{v}}\, \mathbf{u}$ is a vector that has the same direction as $\mathbf{v}$ if $\mathbf{u} \cdot \mathbf{v} > 0$ and the direction opposite to that of $\mathbf{v}$ if $\mathbf{u} \cdot \mathbf{v} < 0$.

PROBLEMS 1.4

SELF-QUIZ

I. $\mathbf{j} - (4\mathbf{k} - 3\mathbf{i}) =$ ________.
   a. $(1, -4, -3)$ **b.** $(1, -4, 3)$
   c. $(-3, 1, -4)$ **d.** $(3, 1, -4)$

II. $(\mathbf{i} + 3\mathbf{k} - \mathbf{j}) \cdot (\mathbf{k} - 4\mathbf{j} + 2\mathbf{i}) =$ ________.
   a. $2 + 4 + 3 = 9$
   b. $(1 + 3 - 1)(1 - 4 + 2) = -3$
   c. $1 - 12 - 2 = -13$
   d. $2 - 4 - 3 = -5$

III. The unit vector in the same direction as $2\mathbf{i} - 2\mathbf{j} + \mathbf{k}$ is ________.
   a. $\mathbf{i} - \mathbf{j} + \mathbf{k}$ **b.** $\frac{1}{5}(2\mathbf{i} - 2\mathbf{j} + \mathbf{k})$
   c. $\frac{1}{3}(2\mathbf{i} - 2\mathbf{j} + \mathbf{k})$ **d.** $\frac{1}{3}(2\mathbf{i} + 2\mathbf{j} + \mathbf{k})$

IV. The component of $\mathbf{u}$ in the direction $\mathbf{w}$ is ________.
   a. $\frac{\mathbf{u} \cdot \mathbf{w}}{|\mathbf{w}|}$ **b.** $\frac{\mathbf{w}}{|\mathbf{w}|}$
   c. $\frac{\mathbf{u} \cdot \mathbf{w}}{|\mathbf{w}|}\frac{\mathbf{w}}{|\mathbf{w}|}$ **d.** $\frac{\mathbf{u} \cdot \mathbf{w}}{|\mathbf{w}|}\frac{\mathbf{u}}{|\mathbf{u}|}$

In Problems 1–20, find the magnitude and the direction cosines of the given vector $\mathbf{v}$.

1. $\mathbf{v} = 3\mathbf{j}$ **2.** $\mathbf{v} = -3\mathbf{i}$
3. $\mathbf{v} = 14\mathbf{k}$ **4.** $\mathbf{v} = -8\mathbf{j}$
5. $\mathbf{v} = 4\mathbf{i} - \mathbf{j}$ **6.** $\mathbf{v} = \mathbf{i} + 2\mathbf{k}$
7. $\mathbf{v} = -2\mathbf{i} + 3\mathbf{j}$ **8.** $\mathbf{v} = \mathbf{i} + \mathbf{j} + \mathbf{k}$
9. $\mathbf{v} = \mathbf{i} - \mathbf{j} + \mathbf{k}$ **10.** $\mathbf{v} = \mathbf{i} + \mathbf{j} - \mathbf{k}$
11. $\mathbf{v} = -\mathbf{i} + \mathbf{j} + \mathbf{k}$ **12.** $\mathbf{v} = \mathbf{i} - \mathbf{j} - \mathbf{k}$
13. $\mathbf{v} = -\mathbf{i} + \mathbf{j} - \mathbf{k}$ **14.** $\mathbf{v} = -\mathbf{i} - \mathbf{j} + \mathbf{k}$
15. $\mathbf{v} = -\mathbf{i} - \mathbf{j} - \mathbf{k}$ **16.** $\mathbf{v} = 2\mathbf{i} + 5\mathbf{j} - 7\mathbf{k}$
17. $\mathbf{v} = -7\mathbf{i} + 2\mathbf{j} - 13\mathbf{k}$
18. $\mathbf{v} = \mathbf{i} + 7\mathbf{j} - 7\mathbf{k}$
19. $\mathbf{v} = -3\mathbf{i} - 3\mathbf{j} + 8\mathbf{k}$
20. $\mathbf{v} = -2\mathbf{i} - 3\mathbf{j} - 4\mathbf{k}$

In Problems 21–40, let $\mathbf{u} = 2\mathbf{i} - 3\mathbf{j} + 4\mathbf{k}$, $\mathbf{v} = -2\mathbf{i} - 3\mathbf{j} + 5\mathbf{k}$, $\mathbf{w} = \mathbf{i} - 7\mathbf{j} + 3\mathbf{k}$, and $\mathbf{t} = 3\mathbf{i} + 4\mathbf{j} + 5\mathbf{k}$; calculate the value of the specified expression.

21. $\mathbf{u} + \mathbf{v}$ **22.** $2\mathbf{u} - 3\mathbf{v}$
23. $-18\mathbf{u}$ **24.** $\mathbf{w} - \mathbf{u} - \mathbf{v}$
25. $\mathbf{t} + 3\mathbf{w} - \mathbf{v}$ **26.** $2\mathbf{u} - 7\mathbf{w} + 5\mathbf{v}$
27. $2\mathbf{v} + 7\mathbf{t} - \mathbf{w}$ **28.** $\mathbf{u} \cdot \mathbf{v}$
29. $|\mathbf{w}|$ **30.** $\mathbf{u} \cdot \mathbf{w} - \mathbf{w} \cdot \mathbf{t}$
31. angle between $\mathbf{u}$ and $\mathbf{w}$
32. angle between $\mathbf{t}$ and $\mathbf{w}$
33. angle between $\mathbf{v}$ and $\mathbf{t}$
34. angle between $\mathbf{v} - \mathbf{t}$ and $\mathbf{v} + \mathbf{t}$
35. $\text{Proj}_{\mathbf{v}}\, \mathbf{u}$ **36.** $\text{Proj}_{\mathbf{u}}\, \mathbf{v}$
37. $\text{Proj}_{\mathbf{t}}\, \mathbf{w}$ **38.** $\text{Proj}_{\mathbf{w}}\, \mathbf{t}$
39. $\text{Proj}_{\mathbf{w}}\, \mathbf{u}$ **40.** $\text{Proj}_{\mathbf{w}}(\mathbf{u} - \mathbf{t})$

41. The three direction angles of a certain unit vector are equal and are between 0 and $\pi/2$. What is the vector?

42. Find a vector of magnitude 12 that has the same direction as the unit vector of the preceding problem.

43. Let $P = (2, 1, 4)$ and $Q = (3, -2, 8)$. Find a unit vector in the direction of $\overrightarrow{PQ}$.

44. Let $P = (-3, 1, 7)$ and $Q = (8, -1, -7)$. Find a unit vector whose direction is opposite that of $\overrightarrow{PQ}$.

45. Verify that there is no vector whose direction angles are $\pi/6$, $\pi/3$, and $\pi/4$.

***46.** Find the possible angles between the diagonal of a cube and the diagonal of one of its faces.

47. Let $P = (-3, 1, 7)$ and $Q = (8, 1, 7)$. Find all points R such that $\overrightarrow{PR} \perp \overrightarrow{PQ}$.

48. Let $P = (-3, 2, -5)$ and $Q = (8, 2, -5)$. Verify that the set of all points R such that $\overrightarrow{PR} \perp \overrightarrow{PQ}$ and $|\overrightarrow{PR}| = 1$ is a circle. Then find its center and radius.

49. Find the distance between the point $P = (2, 1, 3)$ and the line passing through the points $Q = (-1, 1, 2)$ and $R = (6, 0, 1)$. [*Hint:* See Problem 1.2.77 and the hint for Problem 1.2.78.]

50. Find the distance between the point $P = (1, 0, 1)$ and the line passing through the points $Q = (2, 3, -1)$ and $R = (6, 1, -3)$.

51. Verify that the points $P = (3, 5, 6)$, $Q = (1, 2, 7)$, and $R = (6, 1, 0)$ are vertices of a right triangle.

52. Verify that the points $P = (3, 2, -1)$, $Q = (4, 1, 6)$, $R = (7, -2, 3)$, and $S = (8, -3, 10)$ are vertices of a parallelogram.

53. Use the dot product to find two unit vectors orthogonal to the vectors $(1, 2, 3)$ and $(-4, 1, 5)$.

54. Find two unit vectors perpendicular to both of the vectors $(-2, 0, 4)$ and $(3, -2, -1)$.

55. a. Use Theorem 3 to prove one form of the **Cauchy-Schwarz inequality**:

$$\left(\sum_{i=1}^{3} u_i v_i\right)^2 \leq \left(\sum_{i=1}^{3} u_i^2\right)\left(\sum_{i=1}^{3} v_i^2\right) \tag{10}$$

where the u_i's and v_i's are real numbers. (See Problem 75 in Section 1.2 for another form.)

b. Show that equality holds in (10) if and only if at least one of the vectors $\mathbf{u} = (u_1, u_2, u_3)$ and $\mathbf{v} = (v_1, v_2, v_3)$ is a scalar multiple of the other.

56. Prove Theorem 1.

57. Prove that formula (6) is correct. [*Hint:* Follow the steps leading to formula (1.1.7).]

58. Prove Theorem 2.

59. Prove Theorem 3.

60. Prove Theorem 4.

ANSWERS TO SELF-QUIZ

I. d **II.** a **III.** c **IV.** a

1.5 LINES IN $\mathbb{R}^3$

In your first calculus course you derived the equation of a line in the plane. There you saw that the equation of the line could be determined if you knew either (i) two points on the line or (ii) one point on the line and the direction (slope) of the line. In $\mathbb{R}^3$ our intuition tells us that the basic ideas are the same. Since two points determine a line, we should be able to calculate the equation of a line in space if we know two points on it. Alternatively, if we know one point and the direction of a vector parallel to the line, we should also be able to find its equation.

We begin with two points $P = (x_1, y_1, z_1)$ and $Q = (x_2, y_2, z_2)$ on a line L. A vector parallel to L is a vector with representation $\mathbf{v} = \overrightarrow{PQ}$, or [from formula (6) in Section 1.4],

$$\mathbf{v} = (x_2 - x_1)\mathbf{i} + (y_2 - y_1)\mathbf{j} + (z_2 - z_1)\mathbf{k}. \tag{1}$$

Now let $R = (x, y, z)$ be another point on the line. Then $\overrightarrow{PR}$ is parallel to $\overrightarrow{PQ}$, which is parallel to $\mathbf{v}$, so that by Theorem 4(i) in Section 1.4,

$$\overrightarrow{PR} = t\mathbf{v} \tag{2}$$

for some real number t. Now look at Figure 1. From the figure we have (in each of the three possible cases)

$$\overrightarrow{0R} = \overrightarrow{0P} + \overrightarrow{PR}, \tag{3}$$

and combining (2) and (3),

$$\overrightarrow{PR} = \overrightarrow{0R} - \overrightarrow{0P} = t\mathbf{v}.$$

Thus,

VECTOR EQUATION OF A LINE

$$\overrightarrow{0R} = \overrightarrow{0P} + t\mathbf{v} \tag{4}$$

or

$$(x, y, z) = (x_1, y_1, z_1) + t(x_2 - x_1, y_2 - y_1, z_2 - z_1). \tag{4'}$$

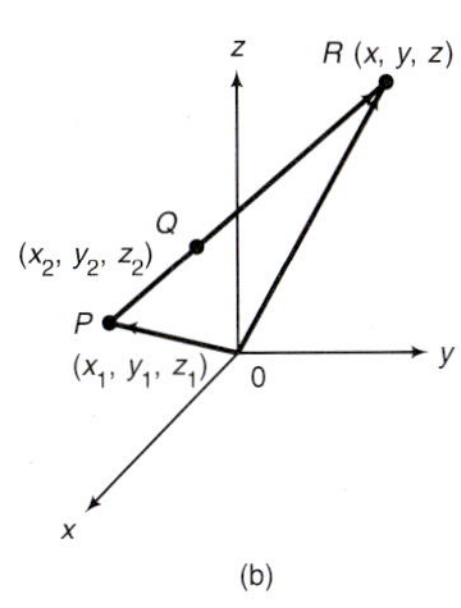

Equation (4) or (4′) is called the **vector equation** of the line L. For if R is on L, then (4) is satisfied for some real number t. Conversely, if (4) is satisfied, then reversing our steps, we see that $\overrightarrow{PR}$ is parallel to $\mathbf{v}$, which means that R is on L.

NOTE: Since $\mathbf{v} = \overrightarrow{PQ} = \overrightarrow{0Q} - \overrightarrow{0P}$, (4) can be rewritten as

$$\overrightarrow{0R} = \overrightarrow{0P} + t(\overrightarrow{0Q} - \overrightarrow{0P}),$$

or

$$\overrightarrow{0R} = (1 - t)\overrightarrow{0P} + t\overrightarrow{0Q}. \tag{5}$$

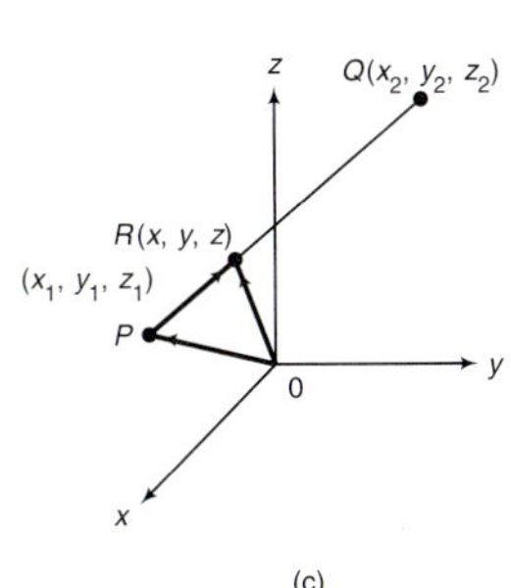

FIGURE 1
In all three cases
$\overrightarrow{0R} = \overrightarrow{0P} + \overrightarrow{PR}$

The vector equation (5) is sometimes very useful.

If we write out the components of equation (4′), we obtain

$$x\mathbf{i} + y\mathbf{j} + z\mathbf{k} = x_1\mathbf{i} + y_1\mathbf{j} + z_1\mathbf{k} + t(x_2 - x_1)\mathbf{i} + t(y_2 - y_1)\mathbf{j} + t(z_2 - z_1)\mathbf{k};$$

or

PARAMETRIC EQUATIONS OF A LINE

$$x = x_1 + t(x_2 - x_1),$$
$$y = y_1 + t(y_2 - y_1),$$
$$z = z_1 + t(z_2 - z_1). \tag{6}$$

The equations (6) are called the **parametric equations** of a line.

Finally, solving for t in (6), and defining $x_2 - x_1 = a$, $y_2 - y_1 = b$, and $z_2 - z_1 = c$, we find that

SYMMETRIC EQUATIONS OF A LINE

$$\frac{x - x_1}{a} = \frac{y - y_1}{b} = \frac{z - z_1}{c}, \quad \text{where } a, b, c \neq 0. \tag{7}$$

The equations (7) are called the **symmetric equations** of the line.

Here, a, b, and c are direction numbers of the vector $\mathbf{v}$. In Examples 3 and 4, we illustrate what happens if one or more of these numbers is zero.

EXAMPLE 1

DETERMINING EQUATIONS OF A LINE

Find a vector equation, parametric equations, and symmetric equations of the line L passing through the points $P = (2, -1, 6)$ and $Q = (3, 1, -2)$.

SOLUTION: First, we calculate

$$\mathbf{v} = (3 - 2)\mathbf{i} + [1 - (-1)]\mathbf{j} + (-2 - 6)\mathbf{k} = \mathbf{i} + 2\mathbf{j} - 8\mathbf{k}.$$

Then from (4), if $R = (x, y, z)$ is on the line.

$$\overrightarrow{0R} = x\mathbf{i} + y\mathbf{j} + z\mathbf{k} = \overrightarrow{0P} + t\mathbf{v} = 2\mathbf{i} - \mathbf{j} + 6\mathbf{k} + t(\mathbf{i} + 2\mathbf{j} - 8\mathbf{k}),$$

or

$$x = 2 + t, \qquad y = -1 + 2t, \qquad z = 6 - 8t. \qquad \text{Parametric equations}$$

Finally, since $a = 1$, $b = 2$, and $c = -8$, we find the symmetric equations

$$\frac{x - 2}{1} = \frac{y + 1}{2} = \frac{z - 6}{-8}.$$

REMARK: If in Example 1 we write $P = (3, 1, -2)$ and $Q = (2, -1, 6)$, then we find that $\mathbf{v} = (2 - 3)\mathbf{i} + (-1 - 1)\mathbf{j} + (6 - (-2))\mathbf{k} = -\mathbf{i} - 2\mathbf{j} + 8\mathbf{k}$ and the direction numbers are now $a = -1$, $b = -2$, and $c = 8$. A new set of parametric equations is

$$x = 3 - t, \qquad y = 1 - 2t, \qquad z = -2 + 8t$$

and the corresponding symmetric equations are

$$\frac{x - 3}{-1} = \frac{y - 1}{-2} = \frac{z + 2}{8}.$$

Note that $(2, -1, 6)$ is on this line because

$$\frac{2 - 3}{-1} = \frac{-1 - 1}{-2} = \frac{6 + 2}{8} = 1.$$

The point here is that for a line, there are an infinite number of sets of parametric or symmetric equations (two for every point on the line) and two sets of direction numbers.

EXAMPLE 2

FINDING SYMMETRIC EQUATIONS OF A LINE

Find symmetric equations of the line passing through the point $(1, -2, 4)$ and parallel to the vector $\mathbf{v} = \mathbf{i} + \mathbf{j} - \mathbf{k}$.

SOLUTION: We simply choose $\overrightarrow{0P} = \mathbf{i} - 2\mathbf{j} + 4\mathbf{k}$ and $\mathbf{v}$ as above. Then $a = 1$, $b = 1$, $c = -1$, and we obtain

$$\frac{x - 1}{1} = \frac{y + 2}{1} = \frac{z - 4}{-1}.$$

What happens if one of the direction numbers a, b, or c is zero?

EXAMPLE 3

FINDING SYMMETRIC EQUATIONS OF A LINE WHEN ONE OF THE DIRECTION NUMBERS IS ZERO

Find symmetric equations of the line containing the points $P = (3, 4, -1)$ and $Q = (-2, 4, 6)$.

SOLUTION: Here $\mathbf{v} = -5\mathbf{i} + 7\mathbf{k}$, and $a = -5$, $b = 0$, $c = 7$. Then a parametric representation of the line is

$$x = 3 - 5t, \qquad y = 4 \qquad z = -1 + 7t.$$

Solving for t, we find that

$$\frac{x - 3}{-5} = \frac{z + 1}{7} \qquad \text{and} \qquad y = 4.$$

The equation $y = 4$ is the equation of a plane parallel to the xz-plane, so we have obtained an equation of a line in that plane.

Now, what happens if two of the direction numbers are zero?

EXAMPLE 4

FINDING SYMMETRIC EQUATIONS OF A LINE WHEN TWO OF THE DIRECTION NUMBERS ARE ZERO

Find the symmetric equations of the line passing through the points $P = (2, 3, -2)$ and $Q = (2, -1, -2)$.

SOLUTION: Here $\mathbf{v} = -4\mathbf{j}$, so that $a = 0$, $b = -4$, and $c = 0$. A parametric representation of the line is, by equations (6), given by

$$x = 2, \qquad y = 3 - 4t, \qquad z = -2.$$

Now $x = 2$ is the equation of a plane parallel to the yz-plane, while $z = -2$ is the equation of a plane parallel to the xy-plane. Their intersection is the line $x = 2$, $z = -2$, which is parallel to the y-axis. In fact, the equation $y = 3 - 4t$ says, essentially, that y can take on any value (while x and z remain fixed).

The results of Examples 1, 3, and 4 are summarized in Theorem 1.

THEOREM 1

Let L be a line passing through the point (x_1, y_1, z_1) and parallel to the vector $\mathbf{v} = a\mathbf{i} + b\mathbf{j} + c\mathbf{k}$. Then symmetric equations of the line are as follows:

i. $\dfrac{x - x_1}{a} = \dfrac{y - y_1}{b} = \dfrac{z - z_1}{c}$, if a, b, and c are all nonzero.

ii. $x = x_1$, $\dfrac{y - y_1}{b} = \dfrac{z - z_1}{c}$,
if $a = 0$. Then the line is parallel to the yz-plane. If either b or $c = 0$, but $a \neq 0$, similar results hold.

iii. $x = x_1$, $y = y_1$, $z = z_1 + ct$,
if a and b are 0. Then the line is parallel to the z-axis. If a and c or b and c are 0, similar results hold. ■

WARNING: We repeat that the parametric or symmetric equations of a line are *not* unique. To see this, simply choose two other points on the line.

EXAMPLE 5 ILLUSTRATING THE FACT THAT SYMMETRIC EQUATIONS ARE NOT UNIQUE

In Example 1, the line contains the point $(5, 5, -18)$; set $t = 3$. Choose $P = (5, 5, -18)$ and $Q = (3, 1, -2)$. We find that $\mathbf{v} = -2\mathbf{i} - 4\mathbf{j} + 16\mathbf{k}$, so that

$$x = 5 - 2t, \qquad y = 5 - 4t, \qquad z = -18 + 16t.$$

[Note that if $t = \frac{3}{2}$, we obtain $(x, y, z) = (2, -1, 6)$.] The symmetric equations are now

$$\frac{x-5}{-2} = \frac{y-5}{-4} = \frac{z+18}{16}.$$

PROBLEMS 1.5

SELF-QUIZ

I. The line through the points $(1, 2, 4)$ and $(5, 10, 15)$ satisfies the equation ________.

a. $(x, y, z) = (1, 2, 4) + t(4, 8, 11)$

b. $\frac{x-1}{4} = \frac{y-2}{8} = \frac{z-4}{11}$

c. $(x, y, z) = (5, 10, 15) + s(4, 8, 11)$

d. $\frac{x-5}{4} = \frac{y-10}{8} = \frac{z-15}{11}$

II. The line through the point $(7, 3, -4)$ and parallel to the vector $\mathbf{i} + 5\mathbf{j} + 2\mathbf{k}$ satisfies the equation ________.

a. $\frac{x-7}{1} = \frac{y-3}{5} = \frac{z+4}{2}$

b. $(x, y, z) = (1, 5, 2) + t(7, 3, -4)$

c. $\frac{x-7}{8} = \frac{y-3}{8} = \frac{z+4}{-2}$

d. $(x, y, z) = (7, 3, -4) + s(8, 8, -2)$

III. The vector equation $(x, y, z) - (3, 5, -7) = t(-1, 4, 8)$ describes ________.

a. the line through $(-1, 4, 8)$ and parallel to $3\mathbf{i} + 5\mathbf{j} - 7\mathbf{k}$

b. the line through $(-3, -5, 7)$ and parallel to $-\mathbf{i} + 4\mathbf{j} + 8\mathbf{k}$

c. the line through $(3, 5, -7)$ and perpendicular to $-\mathbf{i} + 4\mathbf{j} + 8\mathbf{k}$

d. the line through $(3, 5, -7)$ and parallel to $-\mathbf{i} + 4\mathbf{j} + 8\mathbf{k}$

In Problems 1–22, find a vector equation, parametric equations, and symmetric equations for the specified line.

1. the line passing through $(2, 1, 3)$ and $(1, 2, -1)$

2. the line passing through $(1, -1, 1)$ and $(-1, 1, -1)$

3. the line passing through $(1, 3, 2)$ and $(2, 4, -2)$

4. the line passing through $(-2, 4, 5)$ and $(3, 7, 2)$

5. the line passing through $(-4, 1, 3)$ and $(-4, 0, 1)$

6. the line passing through $(2, 3, -4)$ and $(2, 0, -4)$

7. the line passing through $(1, 2, 3)$ and $(3, 2, 1)$

8. the line passing through $(7, 1, 3)$ and $(-1, -2, 3)$

9. the line passing through $(1, 2, 4)$ and $(1, 2, 7)$

10. the line passing through $(-3, -1, -6)$ and $(-3, 1, 6)$

11. the line passing through $(2, 2, 1)$ and parallel to $2\mathbf{i} - \mathbf{j} - \mathbf{k}$

12. the line passing through $(-1, -6, 2)$ and parallel to $4\mathbf{i} + \mathbf{j} - 3\mathbf{k}$

13. the line passing through $(1, 0, 3)$ and parallel to $\mathbf{i} - \mathbf{j}$

14. the line passing through $(2, 1, -4)$ and parallel to $\mathbf{i} + 4\mathbf{k}$

15. the line passing through $(-1, -2, 5)$ and parallel to $-3\mathbf{j} + 4\mathbf{k}$

16. the line passing through $(-2, 3, -2)$ and parallel to $4\mathbf{k}$

17. the line passing through $(-1, -3, 1)$ and parallel to $-7\mathbf{j}$

18. the line passing through $(2, 1, 5)$ and parallel to $3\mathbf{i}$

19. the line passing through (a, b, c) and parallel to $d\mathbf{i} + e\mathbf{j}$, $d, e \neq 0$

20. the line passing through (a, b, c) and parallel to $d\mathbf{k}$, $d \neq 0$

21. the line passing through $(4, 1, -6)$ and parallel to

$$\frac{x-2}{3} = \frac{y+1}{6} = \frac{z-5}{2}$$

22. the line passing through $(3, 1, -2)$ and parallel to

$$\frac{x+1}{3} = \frac{y+3}{2} = \frac{z-2}{-4}$$

In the plane, two lines that are not parallel have exactly one point of intersection. In $\mathbb{R}^3$, this is not the case. For example the lines L_1: $x = 2$, $y = 3$ (parallel to the z-axis) and L_2: $x = 1$, $z = 3$ (parallel to the y-axis) are not parallel and have no points in common. It takes a bit of work to determine whether two lines in $\mathbb{R}^3$ do have a point in common (they usually do not).

23. Determine whether the lines

$$L_1: x = 1 + t,\ y = -3 + 2t;\ z = -2 - t$$

and

$$L_2: x = 17 + 3s,\ y = 4 + s,\ z = -8 - s$$

have a point of intersection. [*Hint:* If (x, y, z) is a point common to both lines, then $x = 1 + t = 17 + 3s$, $y = -3 + 2t = 4 + s$, $z = -2 - t = -8 - s$. Find numbers s and t that satisfy all three of these equations or show that no such numbers s and t can exist.]

24. Determine whether the lines

$$L_1: x = 2 - t,\ y = 1 + t,\ z = -2 - t$$

and

$$L_2: x = 1 + s,\ y = -2s,\ z = 3 + 2s$$

have a point in common.

In Problems 25–30, determine whether the given pair of lines has a point of intersection. If so, find it.

25. L_1: $x = 2 + t$, $y = -1 + 2t$, $z = 3 + 4t$;
L_2: $x = 9 + s$, $y = -2 - s$, $z = 1 - 2s$

26. L_1: $x = 3 + 2t$, $y = 2 - t$, $z = 1 + t$;
L_2: $x = 4 - s$, $y = -2 + 3s$, $z = 2 + 2s$

27. $L_1: \dfrac{x-4}{-3} = \dfrac{y-1}{7} = \dfrac{z+2}{-8}$;
$L_2: \dfrac{x-5}{1} = \dfrac{y-3}{-1} = \dfrac{z-1}{2}$

28. $L_1: \dfrac{x-2}{-5} = \dfrac{y-1}{1} = \dfrac{z-3}{4}$;
$L_2: \dfrac{x+3}{4} = \dfrac{y-2}{-1} = \dfrac{z-7}{6}$

29. L_1: $x = 4 - t$, $y = 7 + 5t$, $z = 2 - 3t$;
L_2: $x = 1 + 2s$, $y = 6 - 2s$, $z = 10 + 3s$

30. L_1: $x = 1 + t$, $y = 2 - t$, $z = 3t$;
L_2: $x = 3s$, $y = 2 - s$, $z = 2 + s$

31. Verify that

$$L_1: \frac{x-1}{1} = \frac{y+3}{2} = \frac{z+3}{3}$$

and

$$L_2: \frac{x-3}{3} = \frac{y-1}{6} = \frac{z-3}{9}$$

are equations of the same straight line.

32. Verify that L_1: $x = 1 - 2t$, $y = -3 - 6t$, $z = 5 + 10t$ and L_2: $x = -5 + s$, $y = -21 + 3s$, $z = 35 - 5s$ are equations for the same line.

33. Let the line L be given in its vector form $\overrightarrow{0R} = \overrightarrow{0P} + t\mathbf{v}$. Find a number t such that $\overrightarrow{0R}$ is perpendicular to $\mathbf{v}$.

34. Apply your solution of the preceding problem to find the distance between the origin 0 and the line L which passes through the given P and is parallel to the given $\mathbf{v}$.

a. $P = (2, 1, -4)$; $\mathbf{v} = \mathbf{i} + \mathbf{j} + \mathbf{k}$
b. $P = (1, 2, -3)$; $\mathbf{v} = 3\mathbf{i} - \mathbf{j} - \mathbf{k}$
c. $P = (-1, -4, 2)$; $\mathbf{v} = -\mathbf{i} + \mathbf{j} + 2\mathbf{k}$

35. Find two different lines that pass through the point $(2, -3, 1)$ and are also perpendicular to the line

$$\frac{x+2}{4} = \frac{y-1}{-4} = \frac{z+2}{3}.$$

36. Prove or Disprove: There are at least two different lines which pass through the point $(2, -3, 1)$ and are also perpendicular to the line

$$\frac{x+2}{4} = \frac{y-1}{2} = \frac{z+2}{-1}.$$

37. Show that direction vectors of the lines

$$L_1: \frac{x-3}{2} = \frac{y+1}{4} = \frac{z-2}{-1}$$

and

$$L_2: \frac{x-3}{5} = \frac{y+1}{-2} = \frac{z-3}{2}$$

are orthogonal.

38. Let L_1 be given by

$$\frac{x-x_1}{a_1} = \frac{y-y_1}{b_1} = \frac{z-z_1}{c_1}, \qquad a_1, b_1, c_1 \neq 0,$$

and L_2 be given by

$$\frac{x-x_2}{a_2} = \frac{y-y_2}{b_2} = \frac{z-z_2}{c_2}, \qquad a_2, b_2, c_2 \neq 0.$$

Show that the direction vector of L_1 is orthogonal to the direction vector of L_2 if and only if $a_1a_2 + b_1b_2 + c_1c_2 = 0$.

39. a. Find symmetric equations for the line in the xy-plane that passes through the distinct points $(x_1, y_1, 0)$ and $(x_2, y_2, 0)$.

b. Show those equations can be rewritten in the form

$$y = mx + b, \qquad z = 0.$$

This shows the symmetric equations of a line in $\mathbb{R}^3$ generalize the slope-intercept equation of a line in $\mathbb{R}^2$.

***40.** Show that the lines L_1: $x = x_1 + a_1t$, $y = y_1 + b_1t$, $z = z_1 + c_1t$ and L_2: $x = x_2 + a_2s$, $y = y_2 + b_2s$, $z = z_2 + c_2s$ have at least one point in common or they are parallel if and only if

$$\begin{vmatrix} a_1 & a_2 & x_1 - x_2 \\ b_1 & b_2 & y_1 - y_2 \\ c_1 & c_2 & z_1 - z_2 \end{vmatrix} = 0.$$

[*Note:* See Section 7.1 if you need to learn or review how to evaluate a 3×3 determinant.]

ANSWERS TO SELF-QUIZ

I. a, b, c, d **II.** a **III.** d

1.6 THE CROSS PRODUCT OF TWO VECTORS

To this point, the only product of vectors we have considered has been the dot or scalar product. We now define a new product, called the *cross product* (or *vector product*), which is defined only in $\mathbb{R}^3$.

There are several ways to define the product of two vectors. In many applications, it is useful to find a vector that is orthogonal to two given vectors. The cross product of two vectors **u** and **v** is indeed a vector that is orthogonal to both **u** and **v**. How do we find such a vector?

Suppose that $\mathbf{u} = a_1\mathbf{i} + b_1\mathbf{j} + c_1\mathbf{k}$ and $\mathbf{v} = a_2\mathbf{i} + b_2\mathbf{j} + c_2\mathbf{k}$. Let $\mathbf{w} = a_3\mathbf{i} + b_3\mathbf{j} + c_3\mathbf{k}$ be a vector orthogonal to both **u** and **v**. Then

$$\mathbf{u} \cdot \mathbf{w} = a_1a_3 + b_1b_3 + c_1c_3 = 0 \tag{1}$$

and

$$\mathbf{v} \cdot \mathbf{w} = a_2a_3 + b_2b_3 + c_2c_3 = 0. \tag{2}$$

Multiply equation (1) by c_2 and equation (2) by c_1 and subtract:

$$\begin{array}{l} a_1c_2a_3 + b_1c_2b_3 + c_1c_2c_3 = 0 \\ a_2c_1a_3 + b_2c_1b_3 + c_1c_2c_3 = 0 \\ \hline (a_1c_2 - a_2c_1)a_3 + (b_1c_2 - b_2c_1)b_3 = 0 \end{array} \tag{3}$$

Equation (3) has the form

$$\alpha a_3 + \beta b_3 = 0 \qquad \text{where } \alpha = a_1c_2 - a_2c_1 \text{ and } \beta = b_1c_2 - b_2c_1.$$

One solution is $a_3 = \beta$ and $b_3 = -\alpha$. Thus one solution to (3) is

$$a_3 = b_1c_2 - b_2c_1 \qquad \text{and} \qquad b_3 = -(a_1c_2 - a_2c_1) = a_2c_1 - a_1c_2.$$

Inserting these values into equation (1) yields $c_3 = a_1b_2 - b_1a_2$ (check this). Thus the vector

$$\mathbf{w} = (b_1c_2 - b_2c_1)\mathbf{i} + (c_1a_2 - a_1c_2)\mathbf{j} + (a_1b_2 - b_1a_2)\mathbf{k}$$

is a vector orthogonal to both **u** and **v**.

CHECK:

$$\mathbf{u} \cdot \mathbf{w} = a_1b_1c_2 - a_1b_2c_1 + b_1c_1a_2 - b_1a_1c_2 + c_1a_1b_2 - c_1b_1a_2 = 0.$$

$$\mathbf{v} \cdot \mathbf{w} = a_2b_1c_2 - a_2b_2c_1 + b_2c_1a_2 - b_2a_1c_2 + c_2a_1b_2 - c_2b_1a_2 = 0.$$

This informal derivation motivates the following definition:

DEFINITION CROSS PRODUCT

Let $\mathbf{u} = a_1\mathbf{i} + b_1\mathbf{j} + c_1\mathbf{k}$ and $\mathbf{v} = a_2\mathbf{i} + b_2\mathbf{j} + c_2\mathbf{k}$. Then the **cross product (vector product)** of **u** and **v**, denoted by **u** × **v**, is a new vector defined by

$$\mathbf{u} \times \mathbf{v} = (b_1c_2 - c_1b_2)\mathbf{i} + (c_1a_2 - a_1c_2)\mathbf{j} + (a_1b_2 - b_1a_2)\mathbf{k}. \tag{4}$$

■

Note that the result of the *cross product* is a *vector,* while the result of the *dot product* is a scalar.

HISTORICAL NOTE: The cross product was defined by Hamilton in one of a series of papers discussing his quaternions, which were published in *Philosophical Magazine* between the years 1844 and 1850.

EXAMPLE 1

CALCULATING THE CROSS PRODUCT OF TWO VECTORS

Let $\mathbf{u} = \mathbf{i} - \mathbf{j} + 2\mathbf{k}$ and $\mathbf{v} = 2\mathbf{i} + 3\mathbf{j} - 4\mathbf{k}$. Calculate $\mathbf{w} = \mathbf{u} \times \mathbf{v}$.

SOLUTION: Using formula (1), we have

$$\begin{aligned}\mathbf{w} &= [(-1)(-4) - (2)(3)]\mathbf{i} + [(2)(2) - (1)(-4)]\mathbf{j} + [(1)(3) - (-1)(2)]\mathbf{k} \\ &= -2\mathbf{i} + 8\mathbf{j} + 5\mathbf{k}.\end{aligned}$$

NOTE: In this example $\mathbf{u} \cdot \mathbf{w} = \mathbf{v} \cdot \mathbf{w} = 0$. That is, **u** × **v** is orthogonal to both **u** and **v**.

Before continuing our discussion of the uses of the cross product, there is an easy way to remember how to calculate **u** × **v** if you are familiar with the elementary properties of 3 × 3 determinants. If you are not, we suggest that you turn to Sections 7.1 and 7.2, where these properties are discussed.

THEOREM 1

$$\mathbf{u} \times \mathbf{v} = \begin{vmatrix} \mathbf{i} & \mathbf{j} & \mathbf{k} \\ a_1 & b_1 & c_1 \\ a_2 & b_2 & c_2 \end{vmatrix}^{\dagger} \tag{5}$$

PROOF:

$$\begin{vmatrix} \mathbf{i} & \mathbf{j} & \mathbf{k} \\ a_1 & b_1 & c_1 \\ a_2 & b_2 & c_2 \end{vmatrix} = \mathbf{i}\begin{vmatrix} b_1 & c_1 \\ b_2 & c_2 \end{vmatrix} - \mathbf{j}\begin{vmatrix} a_1 & c_1 \\ a_2 & c_2 \end{vmatrix} + \mathbf{k}\begin{vmatrix} a_1 & b_1 \\ a_2 & b_2 \end{vmatrix}$$

$$= (b_1c_2 - c_1b_2)\mathbf{i} + (a_2c_1 - a_1c_2)\mathbf{j} + (a_1b_2 - b_1a_2)\mathbf{k}$$

which is equal to $\mathbf{u} \times \mathbf{v}$ according to the definition of the cross product. ■

EXAMPLE 2 USING A DETERMINANT TO CALCULATE A CROSS PRODUCT

Calculate $\mathbf{u} \times \mathbf{v}$, where $\mathbf{u} = 2\mathbf{i} + 4\mathbf{j} - 5\mathbf{k}$ and $\mathbf{v} = -3\mathbf{i} - 2\mathbf{j} + \mathbf{k}$.

SOLUTION:

$$\mathbf{u} \times \mathbf{v} = \begin{vmatrix} \mathbf{i} & \mathbf{j} & \mathbf{k} \\ 2 & 4 & -5 \\ -3 & -2 & 1 \end{vmatrix} = (4 - 10)\mathbf{i} - (2 - 15)\mathbf{j} + (-4 + 12)\mathbf{k}$$

$$= -6\mathbf{i} + 13\mathbf{j} + 8\mathbf{k}.$$

The following theorem summarizes some properties of the cross product.

THEOREM 2 PROPERTIES OF THE CROSS PRODUCT

Let $\mathbf{u}$, $\mathbf{v}$, and $\mathbf{w}$ be vectors in $\mathbb{R}^3$, and let α be a scalar.

i. $\mathbf{u} \times \mathbf{0} = \mathbf{0} = \mathbf{0} \times \mathbf{u}$.

ii. $\mathbf{u} \times \mathbf{v} = -(\mathbf{v} \times \mathbf{u})$.

iii. $(\alpha\mathbf{u} \times \mathbf{v}) = \alpha(\mathbf{u} \times \mathbf{v})$.

iv. $\mathbf{u} \times (\mathbf{v} + \mathbf{w}) = (\mathbf{u} \times \mathbf{v}) + (\mathbf{u} \times \mathbf{w})$.

v. $(\mathbf{u} \times \mathbf{v}) \cdot \mathbf{w} = \mathbf{u} \cdot (\mathbf{v} \times \mathbf{w})$. (This product is called the **scalar triple product** of $\mathbf{u}$, $\mathbf{v}$, and $\mathbf{w}$.)

vi. $\mathbf{u} \cdot (\mathbf{u} \times \mathbf{v}) = \mathbf{v} \cdot (\mathbf{u} \times \mathbf{v}) = 0$. (That is, $\mathbf{u} \times \mathbf{v}$ is orthogonal to both $\mathbf{u}$ and $\mathbf{v}$.)

vii. If $\mathbf{u}$ and $\mathbf{v}$ are parallel, then $\mathbf{u} \times \mathbf{v} = \mathbf{0}$.

---

† The determinant is defined as a real number, not a vector. This use of the determinant notation is simply a convenient way to denote the cross product. In this notation we always expand the determinant along the first row.

Proof:

i. Let $\mathbf{u} = a_1\mathbf{i} + b_1\mathbf{j} + c_1\mathbf{k}$. Then,

$$\mathbf{u} \times \mathbf{0} = \begin{vmatrix} \mathbf{i} & \mathbf{j} & \mathbf{k} \\ a_1 & b_1 & c_1 \\ 0 & 0 & 0 \end{vmatrix} = 0\mathbf{i} + 0\mathbf{j} + 0\mathbf{k} = \mathbf{0}.$$

Similarly, $\mathbf{0} \times \mathbf{u} = \mathbf{0}$.

ii. Let $\mathbf{v} = a_2\mathbf{i} + b_2\mathbf{j} + c_2\mathbf{k}$. Then,

$$\mathbf{u} \times \mathbf{v} = \begin{vmatrix} \mathbf{i} & \mathbf{j} & \mathbf{k} \\ a_1 & b_1 & c_1 \\ a_2 & b_2 & c_2 \end{vmatrix} = -\begin{vmatrix} \mathbf{i} & \mathbf{j} & \mathbf{k} \\ a_2 & b_2 & c_2 \\ a_1 & b_1 & c_1 \end{vmatrix} = -(\mathbf{v} \times \mathbf{u}),$$

since interchanging the rows of a determinant has the effect of multiplying the determinant by -1 [see Property 4 on page 460].

iii.
$$(\alpha\mathbf{u}) \times \mathbf{v} = \begin{vmatrix} \mathbf{i} & \mathbf{j} & \mathbf{k} \\ \alpha a_1 & \alpha b_1 & \alpha c_1 \\ a_2 & b_2 & c_2 \end{vmatrix} = \alpha\begin{vmatrix} \mathbf{i} & \mathbf{j} & \mathbf{k} \\ a_1 & b_1 & c_1 \\ a_2 & b_2 & c_2 \end{vmatrix} = \alpha(\mathbf{u} \times \mathbf{v})$$

The second equality follows from Property 2 on page 458.

iv. Let $\mathbf{w} = a_3\mathbf{i} + b_3\mathbf{j} + c_3\mathbf{k}$. Then

$$\begin{aligned} \mathbf{u} \times (\mathbf{v} + \mathbf{w}) &= \begin{vmatrix} \mathbf{i} & \mathbf{j} & \mathbf{k} \\ a_1 & b_1 & c_1 \\ a_2 + a_3 & b_2 + b_3 & c_2 + c_3 \end{vmatrix} \\ &= \begin{vmatrix} \mathbf{i} & \mathbf{j} & \mathbf{k} \\ a_1 & b_1 & c_1 \\ a_2 & b_2 & c_2 \end{vmatrix} + \begin{vmatrix} \mathbf{i} & \mathbf{j} & \mathbf{k} \\ a_1 & b_1 & c_1 \\ a_3 & b_3 & c_3 \end{vmatrix} \\ &= (\mathbf{u} \times \mathbf{v}) + (\mathbf{u} \times \mathbf{w}). \end{aligned}$$

The second equality is easily verified by direct calculation.

v.
$$\begin{aligned} (\mathbf{u} \times \mathbf{v}) \cdot \mathbf{w} &= [(b_1c_2 - c_1b_2)\mathbf{i} + (c_1a_2 - a_1c_2)\mathbf{j} + (a_1b_2 - b_1a_2)\mathbf{k}] \\ &\quad \cdot (a_3\mathbf{i} + b_3\mathbf{j} + c_3\mathbf{k}) \\ &= b_1c_2a_3 - c_1b_2a_3 + c_1a_2b_3 - a_1c_2b_3 \\ &\quad + a_1b_2c_3 - b_1a_2c_3 \\ &= \begin{vmatrix} a_1 & a_2 & a_3 \\ b_1 & b_2 & b_3 \\ c_1 & c_2 & c_3 \end{vmatrix} \end{aligned}$$

You should show that $\mathbf{u} \cdot (\mathbf{v} \times \mathbf{w})$ is equal to the same expression (see Problem 47).

Note: We provide an interesting geometric interpretation of the scalar triple product on page 48. See also Problem 50.

vi. We have already shown this. Here is an alternative proof. We know that $\mathbf{u} \cdot (\mathbf{u} \times \mathbf{v}) = (\mathbf{u} \times \mathbf{v}) \cdot \mathbf{u}$ [since the dot product is commutative—see Theorem 2(i) of Section 1.4]. But from parts (ii) and (v) of this theorem,

$$(\mathbf{u} \times \mathbf{v}) \cdot \mathbf{u} = \mathbf{u} \cdot (\mathbf{v} \times \mathbf{u}) = \mathbf{u} \cdot (-\mathbf{u} \times \mathbf{v}) = -\mathbf{u} \cdot (\mathbf{u} \times \mathbf{v}).$$

Thus $\mathbf{u} \cdot (\mathbf{u} \times \mathbf{v}) = -\mathbf{u} \cdot (\mathbf{u} \times \mathbf{v})$, which can only occur if $\mathbf{u} \cdot (\mathbf{u} \times \mathbf{v}) = 0$. A similar computation shows that $\mathbf{v} \cdot (\mathbf{u} \times \mathbf{v}) = 0$.

vii. If $\mathbf{u}$ and $\mathbf{v}$ are parallel, then $\mathbf{v} = \alpha\mathbf{u}$ for some scalar α [from Theorem 4(i) of Section 1.4], so that

$$\mathbf{u} \times \mathbf{v} = \begin{vmatrix} \mathbf{i} & \mathbf{j} & \mathbf{k} \\ a_1 & b_1 & c_1 \\ \alpha a_1 & \alpha b_1 & \alpha c_1 \end{vmatrix} = 0$$

since the third row is a multiple of the second row [see Property 5 on page 461]. ■

NOTE: We could have proved this theorem without using determinants, but the proof would have involved many more computations.

Part (vi), which motivated our definition of the cross product, is the most commonly used part of this theorem. We restate it below:

> The cross product $\mathbf{u} \times \mathbf{v}$ is orthogonal to both $\mathbf{u}$ and $\mathbf{v}$.[†]

What happens when we take cross products of the basis vectors $\mathbf{i}$, $\mathbf{j}$, $\mathbf{k}$? It is not difficult to verify the following:

$$\mathbf{i} \times \mathbf{i} = \mathbf{j} \times \mathbf{j} = \mathbf{k} \times \mathbf{k} = \mathbf{0},$$
$$\mathbf{i} \times \mathbf{j} = \mathbf{k}, \qquad \mathbf{k} \times \mathbf{i} = \mathbf{j}, \qquad \mathbf{j} \times \mathbf{k} = \mathbf{i},$$
$$\mathbf{j} \times \mathbf{i} = -\mathbf{k}, \qquad \mathbf{i} \times \mathbf{k} = -\mathbf{j}, \qquad \mathbf{k} \times \mathbf{j} = -\mathbf{i}.$$

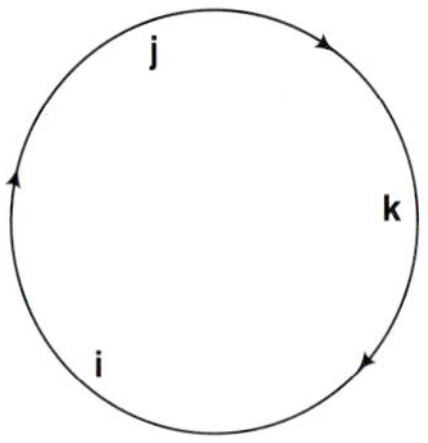

To remember these results, consider the circle in Figure 1. The cross product of two consecutive vectors in the clockwise direction is positive, while the cross product of two consecutive vectors in the counterclockwise direction is negative. Note that the formulas above show that the cross product is *not* associative, since, for example, $\mathbf{i} \times (\mathbf{i} \times \mathbf{j}) = \mathbf{i} \times \mathbf{k} = -\mathbf{j}$ while $(\mathbf{i} \times \mathbf{i}) \times \mathbf{j} = \mathbf{0} \times \mathbf{j} = \mathbf{0}$, so that

$$\mathbf{i} \times (\mathbf{i} \times \mathbf{j}) \neq (\mathbf{i} \times \mathbf{i}) \times \mathbf{j}.$$

j
k
i

FIGURE 1
Cross products of basis vectors

The cross product of two consecutive vectors in the clockwise direction is positive, while the cross product of two consecutive vectors in the counterclockwise directions is negative.

In general,

$$\mathbf{u} \times (\mathbf{v} \times \mathbf{w}) \neq (\mathbf{u} \times \mathbf{v}) \times \mathbf{w}.$$

This means that, in most cases, $\mathbf{u} \times \mathbf{v} \times \mathbf{w}$ is undefined because we get different answers depending on whether we multiply $\mathbf{u} \times \mathbf{v}$ or $\mathbf{v} \times \mathbf{w}$ first.

[†] Technically, this statement is correct only when $\mathbf{u} \times \mathbf{v} \neq \mathbf{0}$ because orthogonality is only defined for nonzero vectors.

Example 3 **Finding a Line Whose Direction Vector Is Orthogonal to Two Given Lines**

Find a line whose direction vector is orthogonal to the direction vectors of the lines $(x-1)/3 = (y+6)/4 = (z-2)/-2$ and $(x+2)/-3 = (y-3)/4 = (z+1)/1$ and that passes through the point $(2, -1, 1)$.

Solution: The directions of these lines are

$$\mathbf{v}_1 = 3\mathbf{i} + 4\mathbf{j} - 2\mathbf{k} \qquad \text{and} \qquad \mathbf{v}_2 = -3\mathbf{i} + 4\mathbf{j} + \mathbf{k}.$$

A vector orthogonal to these vectors is

$$\mathbf{w} = \mathbf{v}_1 \times \mathbf{v}_2 = \begin{vmatrix} \mathbf{i} & \mathbf{j} & \mathbf{k} \\ 3 & 4 & -2 \\ -3 & 4 & 1 \end{vmatrix} = 12\mathbf{i} + 3\mathbf{j} + 24\mathbf{k}.$$

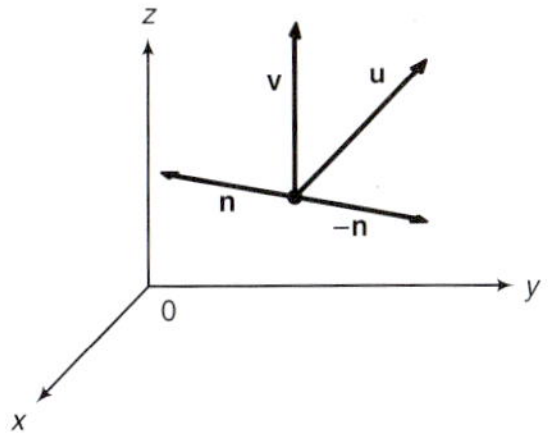

Then symmetric equations of a line satisfying the requested conditions are given by

$$L_1\colon \frac{x-2}{12} = \frac{y+1}{3} = \frac{z-1}{24}.$$

Note: $\mathbf{w}_1 = \mathbf{v}_2 \times \mathbf{v}_1 = -(\mathbf{v}_1 \times \mathbf{v}_2)$ is also orthogonal to $\mathbf{v}_1$ and $\mathbf{v}_2$, so symmetric equations of another line are given by

$$L_2\colon \frac{x-2}{-12} = \frac{y+1}{-3} = \frac{z-1}{-24}.$$

However, L_1 and L_2 are really the same line. (Explain why.)

Figure 2
Both $\mathbf{n}$ and $-\mathbf{n}$ are orthogonal to $\mathbf{u}$ and $\mathbf{v}$.

The preceding example leads to a basic question. We know that $\mathbf{u} \times \mathbf{v}$ is a vector orthogonal to $\mathbf{u}$ and $\mathbf{v}$. But there are always *two* unit vectors orthogonal to $\mathbf{u}$ and $\mathbf{v}$ (see Figure 2). The vectors $\mathbf{n}$ and $-\mathbf{n}$ ($\mathbf{n}$ stands for *normal*) are both orthogonal to $\mathbf{u}$ and $\mathbf{v}$. Which one is in the direction of $\mathbf{u} \times \mathbf{v}$? The answer is given by the **right-hand rule**. If the right hand is placed so that the index finger points in the direction of $\mathbf{u}$ while the middle finger points in the direction of $\mathbf{v}$, then the thumb points in the direction of $\mathbf{u} \times \mathbf{v}$ (see Figure 3).

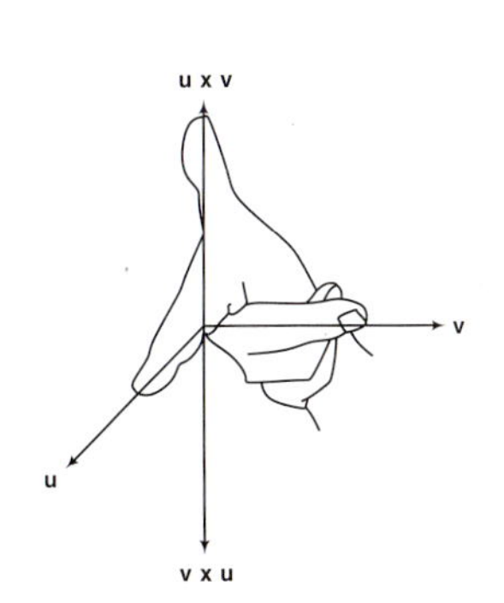

Figure 3
Right-hand rule for the cross product

We will not prove the right-hand rule here. However, we note that the right-hand rule for cross products holds if the three coordinate axes constitute a right-handed system as defined on page 24. If we instead had a left-handed system (obtained by interchanging the x- and y-axes), then the direction of $\mathbf{u} \times \mathbf{v}$ could be found by using a left-hand rule.

Having discussed the direction of the vector $\mathbf{u} \times \mathbf{v}$, we now turn to a discussion of its magnitude.

Theorem 3

If φ is the angle between $\mathbf{u}$ and $\mathbf{v}$, then

$$|\mathbf{u} \times \mathbf{v}| = |\mathbf{u}|\,|\mathbf{v}| \sin \varphi. \tag{6}$$

Proof: It is not difficult to show (by comparing components) that

$$|\mathbf{u} \times \mathbf{v}|^2 = |\mathbf{u}|^2|\mathbf{v}|^2 - (\mathbf{u} \cdot \mathbf{v})^2 \tag{7}$$

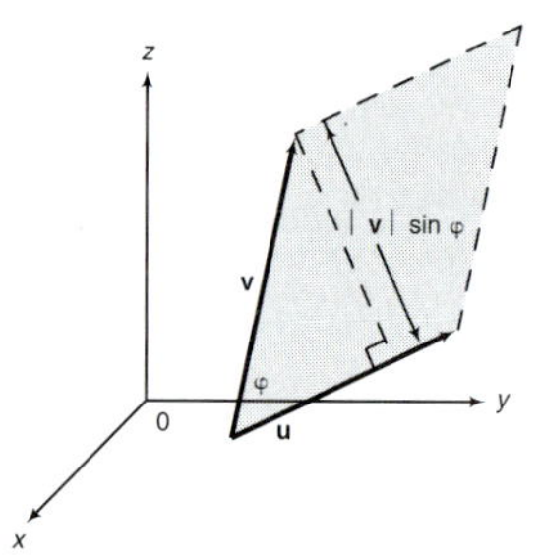

FIGURE 4
Area of a parallelogram

(see Problem 48). Then since $(\mathbf{u} \cdot \mathbf{v})^2 = |\mathbf{u}|^2|\mathbf{v}|^2 \cos^2 \varphi$ (from Theorem 3 in Section 1.4),

$$|\mathbf{u} \times \mathbf{v}|^2 = |\mathbf{u}|^2|\mathbf{v}|^2 - |\mathbf{u}|^2|\mathbf{v}|^2 \cos^2 \varphi = |\mathbf{u}|^2|\mathbf{v}|^2(1 - \cos^2 \varphi)$$
$$= |\mathbf{u}|^2|\mathbf{v}|^2 \sin^2 \varphi,$$

and the theorem follows after taking square roots of both sides. ■

There is an interesting geometric interpretation of Theorem 3. The vectors **u** and **v** are sketched in Figure 4 and can be thought of as two adjacent sides of a parallelogram. Then from elementary geometry we see that

$$\text{area of the parallelogram} = |\mathbf{u}||\mathbf{v}| \sin \varphi = |\mathbf{u} \times \mathbf{v}|. \tag{8}$$

EXAMPLE 4

FINDING THE AREA OF A PARALLELOGRAM

Find the area of a parallelogram with vertices at $P = (1, 3, -2)$, $Q = (2, 1, 4)$, and $R = (-3, 1, 6)$.

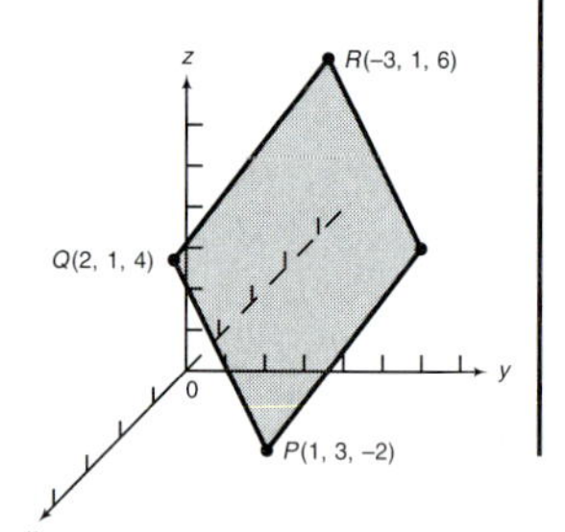

FIGURE 5
The parallelogram with vertices at $(1, 3, -2)$, $(2, 1, 4)$ and $(-3, 1, 6)$

SOLUTION: One such parallelogram is sketched in Figure 5 (there are two others). We have

$$\text{area} = |\overrightarrow{PQ} \times \overrightarrow{QR}| = |(\mathbf{i} - 2\mathbf{j} + 6\mathbf{k}) \times (-5\mathbf{i} + 2\mathbf{k})|$$
$$= \left\| \begin{vmatrix} \mathbf{i} & \mathbf{j} & \mathbf{k} \\ 1 & -2 & 6 \\ -5 & 0 & 2 \end{vmatrix} \right\| = |-4\mathbf{i} - 32\mathbf{j} - 10\mathbf{k}|$$
$$= \sqrt{1140} \text{ square units.}$$

GEOMETRIC INTERPRETATION OF THE SCALAR TRIPLE PRODUCT

Let **u**, **v**, and **w** be three vectors that are not in the same plane. Then they form the sides of a **parallelepiped** in space (see Figure 6). Let us compute its volume. The base of the parallelepiped is a parallelogram. Its area, from (6), is equal to $|\mathbf{u} \times \mathbf{v}|$.

The vector $\mathbf{u} \times \mathbf{v}$ is orthogonal to both **u** and **v** and is therefore orthogonal to the parallelogram determined by **u** and **v**. The height of the parallelepiped, h, is measured along a vector orthogonal to the parallelogram.

From our discussion of projections on page 18, we see that h is the absolute value of the component of **w** in the (orthogonal) direction $\mathbf{u} \times \mathbf{v}$. Thus, from equation (10) on page 21,

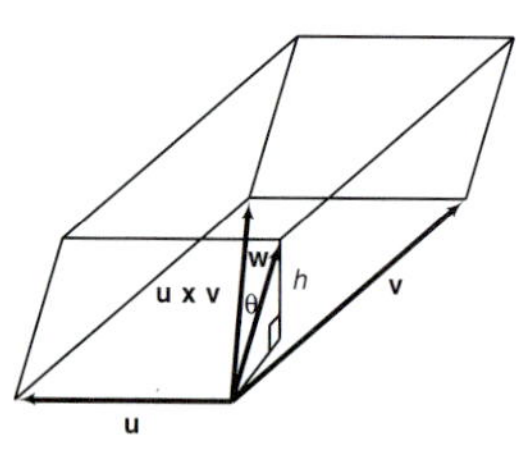

FIGURE 6
The parallelepiped with sides **u**, **v**, and **w**

$$h = |\text{component of } \mathbf{w} \text{ in the direction } \mathbf{u} \times \mathbf{v}| = \left| \frac{\mathbf{w} \cdot (\mathbf{u} \times \mathbf{v})}{|\mathbf{u} \times \mathbf{v}|} \right|.$$

Thus

$$\text{volume of parallelepiped} = \text{area of base} \times \text{height}$$
$$= |\mathbf{u} \times \mathbf{v}| \left[\frac{|\mathbf{w} \cdot (\mathbf{u} \times \mathbf{v})|}{|\mathbf{u} \times \mathbf{v}|} \right] = |\mathbf{w} \cdot (\mathbf{u} \times \mathbf{v})|.$$

That is:

> **VOLUME OF A PARALLELEPIPED**
>
> The volume of the parallelepiped determined by the three vectors **u**, **v**, and **w** is equal to $|(\mathbf{u} \times \mathbf{v}) \cdot \mathbf{w}|$ = the absolute value of the scalar triple product of **u**, **v**, and **w**.

We can derive another interesting and useful fact about the scalar triple product. Suppose that neither **w** nor $\mathbf{u} \times \mathbf{v}$ is the zero vector. If **w** is in the plane of **u** and **v**, then **w** is perpendicular to $\mathbf{u} \times \mathbf{v}$, which means that $\mathbf{w} \cdot (\mathbf{u} \times \mathbf{v}) = 0$. Conversely, if $(\mathbf{u} \times \mathbf{v}) \cdot \mathbf{w} = 0$, then **w** is perpendicular to $(\mathbf{u} \times \mathbf{v})$, so **w** is in the plane determined by **u** and **v**. We conclude that

> three vectors **u**, **v**, and **w** are coplanar if and only if their scalar triple product is zero.

OF HISTORICAL INTEREST

JOSIAH WILLARD GIBBS AND THE ORIGINS OF VECTOR ANALYSIS

JOSIAH WILLARD GIBBS, 1839–1903

As we have already noted, the study of vectors originated with Hamilton's invention of quaternions. Quaternions were developed by Hamilton and others as mathematical tools for the exploration of physical space. But the results were disappointing because quaternions proved to be too complicated for quick mastery and easy application. Fortunately, there was a solution. Quaternions contained a scalar part and a vector part and difficulties arose when these parts were treated simultaneously. Scientists soon learned that many problems could be dealt with by considering the vector part separately, and the study of vector analysis began.

This work was due principally to the American physicist Josiah Willard Gibbs (1839–1903). As a native of New Haven, Connecticut, Gibbs studied mathematics and physics at Yale University, receiving a doctorate in physics in 1863. He then studied mathematics and physics further at Paris, Berlin, and Heidelberg. In 1871 he was appointed professor of mathematical physics at Yale. He was a highly original physicist who published widely in mathematical physics. Gibbs's book *Vector Analysis* appeared in 1881 and again in 1884. In 1902 he published his *Elementary Principles of Statistical Mechanics.* Students of applied mathematics encounter the curious **Gibbs's phenomenon** of Fourier series.

Gibbs's pioneering book *Vector Analysis* was actually a small pamphlet based on Gibbs's lectures and printed for private distribution—primarily for the use of his students. Nevertheless, it created great excitement among those who were looking for an alternative to quaternions and the book soon became widely known. The material was finally turned into a standard book by E. B. Wilson. The book, *Vector Analysis* by Gibbs and Wilson, was published in 1901.

Gibbs's work is encountered by every student of elementary physics. In introductory physics, a vector in space is regarded as a directed line segment, or arrow. Gibbs gave definitions of equality, addition, and multiplication of vectors; these are essentially the definitions given in this

(CONTINUED)

OF HISTORICAL INTEREST

(CONCLUDED)

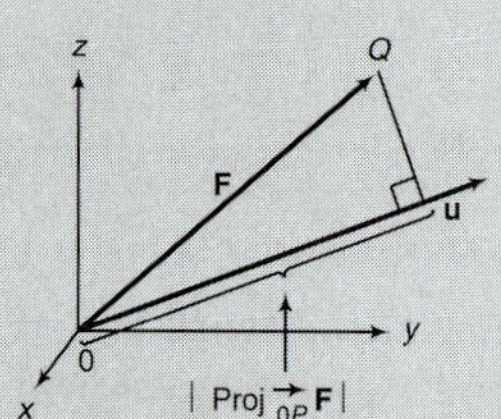

FIGURE 7
The effectiveness of **F** in the direction of $\overrightarrow{0P}$ is the component of **F** in the direction of $\overrightarrow{0P}$.

chapter. In particular, the vector part of a quaternion was written as $a\mathbf{i} + b\mathbf{j} + c\mathbf{k}$ and this is one way we now depict vectors in $\mathbb{R}^3$.

Gibbs defined the scalar product initially only for the vectors **i**, **j**, **k**:

$$\mathbf{i}\cdot\mathbf{i} = \mathbf{j}\cdot\mathbf{j} = \mathbf{k}\cdot\mathbf{k} = 1$$

$$\mathbf{i}\cdot\mathbf{j} = \mathbf{j}\cdot\mathbf{i} = \mathbf{i}\cdot\mathbf{k} = \mathbf{k}\cdot\mathbf{i} = \mathbf{j}\cdot\mathbf{k} = \mathbf{k}\cdot\mathbf{j} = 0.$$

The more general definition followed soon thereafter. Gibbs applied the scalar product in a problem involving force [remember, he was first a physicist]: If **F** is a force vector of magnitude $|\mathbf{F}|$ acting in the direction of the segment $\overrightarrow{0Q}$ (see Figure 7), then the effectiveness of this force in pushing an object along the segment $\overrightarrow{0P}$ (i.e., along the vector **u**) is given by $\mathbf{F}\cdot\mathbf{u}$. If $|\mathbf{u}| = 1$, then $\mathbf{F}\cdot\mathbf{u}$ is the component of **F** in the direction **u**. The cross product, too, has physical significance. Suppose that a force vector **F** acts at a point P in space in the direction $\overrightarrow{PQ}$ (see Figure 8). If **u** denotes the vector represented by $\overrightarrow{0P}$, then the moment of force exerted by **F** around the origin is the vector $\mathbf{u}\times\mathbf{F}$.

Both the scalar and cross products of vectors appear prominently in physical applications involving multivariable calculus. These include the famous Maxwell equations of electromagnetism.

Despite the phenomenal success of Gibbs's work, followers of Hamilton clung stubbornly to quaternions and severely criticized those who favored the much more applicable vector analysis. In 1890, the Scottish mathematician Peter Guthrie Tait (1831–1901) wrote:

> Even Prof. Willard Gibbs must be ranked as one of the retarders of quaternion progress, in virtue of his pamphlet on *Vector Analysis,* a sort of hermaphrodite monster, compounded on the notations of Hamilton and Grassman.

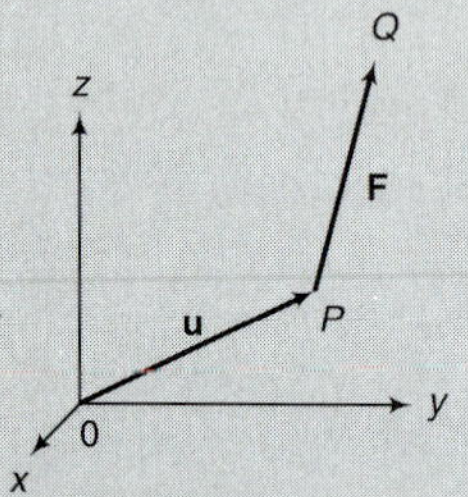

FIGURE 8
The force **F** acts at a point P in space. The vector $\mathbf{u}\times\mathbf{F}$ is the moment of force around the origin.

But vectors were not monstrous and Tait's view soon became a minority one.

In studying mathematics at the end of the twentieth century, it is important that we not lose sight of the fact that much of modern mathematics was developed to solve real-world problems. Vectors were developed by Gibbs and others to make it easier to analyze physical phenomena. In that role they have been hugely successful.

PROBLEMS 1.6

SELF-QUIZ

I. $\mathbf{i}\times\mathbf{k} - \mathbf{k}\times\mathbf{i} =$ ________.
 a. **0** **b.** **j** **c.** 2**j** **d.** −2**j**

II. $\mathbf{i}\cdot(\mathbf{j}\times\mathbf{k}) =$ ________.
 a. 0 **b.** **0**
 c. 1 **d.** $\mathbf{i}-\mathbf{j}+\mathbf{k}$

III. $\mathbf{i}\times\mathbf{j}\times\mathbf{k}$ ________.
 a. = 0 **b.** = **0**
 c. = 1 **d.** is undefined

IV. $(\mathbf{i}+\mathbf{j})\times(\mathbf{j}+\mathbf{k}) =$ ________.
 a. 0 **b.** **0** **c.** 1 **d.** $\mathbf{i}-\mathbf{j}+\mathbf{k}$

V. The sine of the angle between vectors **u** and **w** is ________.
 a. $\dfrac{|\mathbf{u}\times\mathbf{w}|}{|\mathbf{u}||\mathbf{w}|}$ **b.** $\dfrac{|\mathbf{u}\times\mathbf{w}|}{|\mathbf{u}\cdot\mathbf{w}|}$
 c. $\dfrac{|\mathbf{u}\cdot\mathbf{w}|}{|\mathbf{u}\times\mathbf{w}|}$ **d.** $\big||\mathbf{u}\times\mathbf{w}| - |\mathbf{u}\cdot\mathbf{w}|\big|$

VI. $\mathbf{u}\times\mathbf{u} =$ ________.
 a. $|\mathbf{u}|^2$ **b.** 1 **c.** **0** **d.** 0

In Problems 1–20, calculate the cross product $\mathbf{u} \times \mathbf{v}$.

1. $\mathbf{u} = \mathbf{i} - 2\mathbf{j}$; $\mathbf{v} = 3\mathbf{k}$
2. $\mathbf{u} = 3\mathbf{i} - 7\mathbf{j}$; $\mathbf{v} = \mathbf{i} + \mathbf{k}$
3. $\mathbf{u} = \mathbf{i} - \mathbf{j}$; $\mathbf{v} = \mathbf{j} + \mathbf{k}$
4. $\mathbf{u} = -7\mathbf{k}$; $\mathbf{v} = \mathbf{j} + 2\mathbf{k}$
5. $\mathbf{u} = -2\mathbf{i} + 3\mathbf{j}$; $\mathbf{v} = 7\mathbf{i} + 4\mathbf{k}$
6. $\mathbf{u} = a\mathbf{i} + b\mathbf{j}$; $\mathbf{v} = c\mathbf{i} + d\mathbf{j}$
7. $\mathbf{u} = a\mathbf{i} + b\mathbf{k}$; $\mathbf{v} = c\mathbf{i} + d\mathbf{k}$
8. $\mathbf{u} = a\mathbf{j} + b\mathbf{k}$; $\mathbf{v} = c\mathbf{i} + d\mathbf{k}$
9. $\mathbf{u} = 2\mathbf{i} - 3\mathbf{j} + \mathbf{k}$; $\mathbf{v} = \mathbf{i} + 2\mathbf{j} + \mathbf{k}$
10. $\mathbf{u} = 3\mathbf{i} - 4\mathbf{j} + 2\mathbf{k}$; $\mathbf{v} = 6\mathbf{i} - 3\mathbf{j} + 5\mathbf{k}$
11. $\mathbf{u} = -3\mathbf{i} - 2\mathbf{j} + \mathbf{k}$; $\mathbf{v} = 6\mathbf{i} + 4\mathbf{j} - 2\mathbf{k}$
12. $\mathbf{u} = \mathbf{i} + 7\mathbf{j} - 3\mathbf{k}$; $\mathbf{v} = -\mathbf{i} - 7\mathbf{j} + 3\mathbf{k}$
13. $\mathbf{u} = \mathbf{i} - 7\mathbf{j} - 3\mathbf{k}$; $\mathbf{v} = -\mathbf{i} + 7\mathbf{j} - 3\mathbf{k}$
14. $\mathbf{u} = 2\mathbf{i} - 3\mathbf{j} + 5\mathbf{k}$; $\mathbf{v} = 3\mathbf{i} - \mathbf{j} - \mathbf{k}$
15. $\mathbf{u} = 10\mathbf{i} + 7\mathbf{j} - 3\mathbf{k}$; $\mathbf{v} = -3\mathbf{i} + 4\mathbf{j} - 3\mathbf{k}$
16. $\mathbf{u} = 2\mathbf{i} + 4\mathbf{j} - 6\mathbf{k}$; $\mathbf{v} = -\mathbf{i} - \mathbf{j} + 3\mathbf{k}$
17. $\mathbf{u} = 2\mathbf{i} - \mathbf{j} + \mathbf{k}$; $\mathbf{v} = 4\mathbf{i} + 2\mathbf{j} + 2\mathbf{k}$
18. $\mathbf{u} = 3\mathbf{i} - \mathbf{j} + 8\mathbf{k}$; $\mathbf{v} = \mathbf{i} + \mathbf{j} - 4\mathbf{k}$
19. $\mathbf{u} = a\mathbf{i} + a\mathbf{j} + a\mathbf{k}$; $\mathbf{v} = b\mathbf{i} + b\mathbf{j} + b\mathbf{k}$
20. $\mathbf{u} = a\mathbf{i} + b\mathbf{j} + c\mathbf{k}$; $\mathbf{v} = a\mathbf{i} + b\mathbf{j} - c\mathbf{k}$
21. Find two unit vectors orthogonal to both $\mathbf{u} = 2\mathbf{i} - 3\mathbf{j}$ and $\mathbf{v} = 4\mathbf{j} + 3\mathbf{k}$.
22. Find two unit vectors orthogonal to both $\mathbf{u} = \mathbf{i} + \mathbf{j} + \mathbf{k}$ and $\mathbf{v} = \mathbf{i} - \mathbf{j} - \mathbf{k}$.
23. Use the cross product to find the sine of the angle φ between the vectors $\mathbf{u} = 2\mathbf{i} + \mathbf{j} - \mathbf{k}$ and $\mathbf{v} = -3\mathbf{i} - 2\mathbf{j} + 4\mathbf{k}$.
24. Use the dot product to find the cosine of the angle φ between the vectors of the preceding problem; then verify that for the sine and cosine values you have calculated, $\sin^2 \varphi + \cos^2 \varphi = 1$.

In Problems 25–28, find a line whose direction vector is orthogonal to the direction vectors of the two given lines and that passes through the specified point.

25. $\dfrac{x+2}{-3} = \dfrac{y-1}{4} = \dfrac{z}{-5}$; $\dfrac{x-3}{7} = \dfrac{y+2}{-2} = \dfrac{z-8}{3}$; $(1, -3, 2)$
26. $\dfrac{x-2}{-4} = \dfrac{y+3}{-7} = \dfrac{z+1}{3}$; $\dfrac{x+2}{3} = \dfrac{y-5}{-4} = \dfrac{z+3}{-2}$; $(-4, 7, 3)$
27. $x = 3 - 2t$, $y = 4 + 3t$, $z = -7 + 5t$; $x = -2 + 4s$, $y = 3 - 2s$, $z = 3 + s$; $(-2, 3, 4)$
28. $x = 4 + 10t$, $y = -4 - 8t$, $z = 3 + 7t$; $x = -2s$, $y = 1 + 4s$, $z = -7 - 3s$; $(4, 6, 0)$

In Problems 29–34, calculate the area of a parallelogram having the specified points as consecutive vertices.

29. $(1, -2, 3)$; $(2, 0, 1)$; $(0, 4, 0)$
30. $(-2, 1, 1)$; $(2, 2, 3)$; $(-1, -2, 4)$
31. $(-2, 1, 0)$; $(1, 4, 2)$; $(-3, 1, 5)$
32. $(7, -2, -3)$; $(-4, 1, 6)$; $(5, -2, 3)$
33. $(a, 0, 0)$; $(0, b, 0)$; $(0, 0, c)$
34. $(a, b, 0)$; $(a, 0, b)$; $(0, a, b)$

35. Use equation (6) to calculate the area of the triangle with vertices at $(2, 1, -4)$, $(1, 7, 2)$, and $(3, -2, 3)$.
36. Calculate the area of the triangle with vertices at $(3, 1, 7)$, $(2, -3, 4)$, and $(7, -2, 4)$.
37. Sketch the triangle with vertices at $(1, 0, 0)$, $(0, 1, 0)$, and $(0, 0, 1)$. Then calculate its area.

*38. Sketch the triangle with vertices at $(0, 2, 2)$, $(2, 0, 2)$, and $(2, 2, 0)$. Then sketch the triangle whose vertices are the midpoints of the sides of the first triangle. Calculate the area of each triangle.

39. Calculate the volume of the parallelepiped determined by the vectors $\mathbf{u} = 2\mathbf{i} - \mathbf{j} + \mathbf{k}$, $\mathbf{v} = 3\mathbf{i} + 2\mathbf{j} - 2\mathbf{k}$, and $\mathbf{w} = 3\mathbf{i} + 2\mathbf{j}$.
40. Calculate the volume of the parallelepiped determined by the vectors $\mathbf{u} = \mathbf{i} + \mathbf{j}$, $\mathbf{v} = \mathbf{j} - \mathbf{k}$, and $\mathbf{w} = \mathbf{k} + \mathbf{i}$.
41. Calculate the volume of the parallelepiped determined by the vectors $\mathbf{i} - \mathbf{j}$, $3\mathbf{i} + 2\mathbf{k}$, and $-7\mathbf{j} + 3\mathbf{k}$.
42. Calculate the volume of the parallelepiped determined by the vectors $\overrightarrow{PQ}$, $\overrightarrow{PR}$, and $\overrightarrow{PS}$ where $P = (2, 1, -1)$, $Q = (-3, 1, 4)$, $R = (-1, 0, 2)$, and $S = (-3, -1, 5)$.

*43. Calculate the distance between the lines

$$L_1: \frac{x-2}{3} = \frac{y-5}{2} = \frac{z-1}{-1}$$

and

$$L_2: \frac{x-4}{-4} = \frac{y-5}{4} = \frac{z+2}{1}.$$

[*Hint:* The distance is measured along a vector $\mathbf{v}$ that is perpendicular to both L_1 and L_2. Let P be a point on L_1 and Q a point on L_2. Then the length of the projection of $\overrightarrow{PQ}$ on $\mathbf{v}$ is the distance between the lines, measured along a vector that is perpendicular to them both. See Figure 9.]

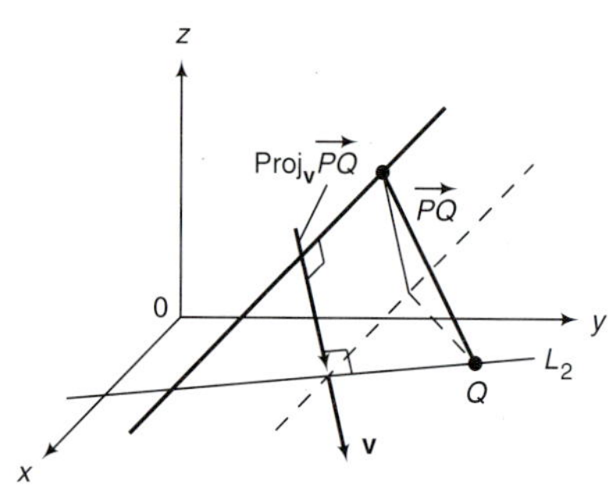

FIGURE 9

*44. Find the distance between the lines

$$L_1: \frac{x+2}{3} = \frac{y-7}{-4} = \frac{z-2}{2}$$

and

$$L_2: \frac{x-1}{-3} = \frac{y+2}{4} = \frac{z+1}{1}.$$

*45. Find the distance between the lines $x = 2 - 3t$, $y = 1 + 2t$, $z = -2 - t$ and $x = 1 + 4s$, $y = -2 - s$, $z = 3 + s$.

*46. Find the distance between the lines $x = -2 - 5t$, $y = -3 - 2t$, $z = 1 + 4t$ and $x = 2 + 3s$, $y = -1 + s$, $z = 3s$.

47. Show that if $\mathbf{a} = (a_1, a_2, a_3)$, $\mathbf{b} = (b_1, b_2, b_3)$, and $\mathbf{c} = (c_1, c_2, c_3)$, then

$$\mathbf{a} \cdot (\mathbf{b} \times \mathbf{c}) = \begin{vmatrix} a_1 & a_2 & a_3 \\ b_1 & b_2 & b_3 \\ c_1 & c_2 & c_3 \end{vmatrix}.$$

48. Show that $|\mathbf{a} \times \mathbf{b}|^2 = |\mathbf{a}|^2|\mathbf{b}|^2 - (\mathbf{a} \cdot \mathbf{b})^2$. [*Hint:* Write out in terms of components.]

*49. Suppose $\mathbf{a}$ and $\mathbf{b}$ are two vectors in $\mathbb{R}^3$; let $\mathbf{c} = \mathbf{a} - \mathbf{b}$. Show that the law of sines can be inferred from the fact that $\mathbf{c} \times \mathbf{c} = \mathbf{0}$ implies $\mathbf{a} \times \mathbf{c} = \mathbf{b} \times \mathbf{c}$.[†]

*50. Suppose $\mathbf{a}$, $\mathbf{b}$, and $\mathbf{c}$ are vectors in $\mathbb{R}^3$. Prove each of the following assertions.
   a. There are scalars β and γ such that $\mathbf{a} \times (\mathbf{b} \times \mathbf{c}) = \beta\mathbf{b} + \gamma\mathbf{c}$.
   b. $\mathbf{a} \times (\mathbf{b} \times \mathbf{c}) = (\mathbf{a} \cdot \mathbf{c})\mathbf{b} - (\mathbf{a} \cdot \mathbf{b})\mathbf{c}$.
   c. **Jacobi identity**: $\mathbf{a} \times (\mathbf{b} \times \mathbf{c}) + \mathbf{b} \times (\mathbf{c} \times \mathbf{a}) + \mathbf{c} \times (\mathbf{a} \times \mathbf{b}) = \mathbf{0}$.

*51. The parallelepiped in Figure 6 has six faces. The diagonals of each face meet at a point. Show that these six "center points" are themselves vertices of a solid. Calculate the volume of this solid. The solid is called an **octahedron**.

ANSWERS TO SELF-QUIZ

I. d **II.** c
III. b = zero vector [*Note:* $\mathbf{i} \times \mathbf{j} \times \mathbf{k}$ *is* defined because $(\mathbf{i} \times \mathbf{j}) \times \mathbf{k} = \mathbf{0} = \mathbf{i} \times (\mathbf{j} \times \mathbf{k})$.]
IV. d **V.** a **VI.** c = zero vector

1.7 PLANES

In Section 1.5, we derived the equation of a line in space by specifying a point on the line and a vector *parallel* to this line. We can derive the equation of a plane in space by specifying a point in the plane and a vector orthogonal to every vector in the plane. This orthogonal vector is called a **normal vector** and is denoted by **N**. (See Figure 1.)

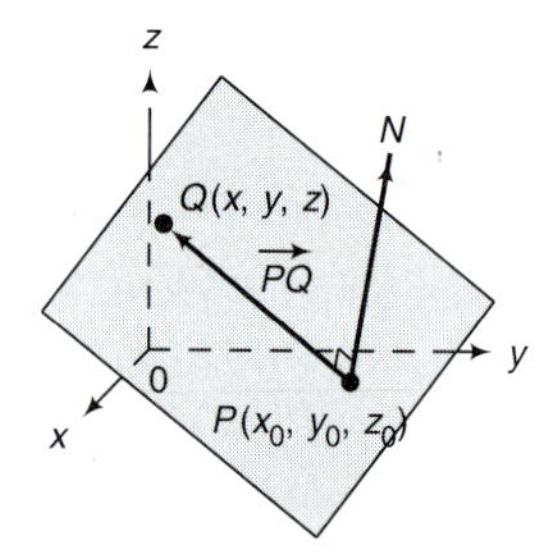

FIGURE 1
A plane is determined by a point and a normal vector.

DEFINITION PLANE

Let P be a point in space and let $\mathbf{N}$ be a given nonzero vector. Then P and the set of all points Q for which $\overrightarrow{PQ}$ and $\mathbf{N}$ are orthogonal constitutes a **plane** in $\mathbb{R}^3$. ■

[†] Analogously, we could interpret the computations in the text's proof of Theorem 2 in Section 1.2 as showing that the law of cosines follows from "multiplying out" the relation $\mathbf{c} \cdot \mathbf{c} = (\mathbf{a} - \mathbf{b}) \cdot (\mathbf{a} - \mathbf{b})$.

NOTATION: We will usually denote a plane by the symbol Π (a capital pi).

Let $P = (x_0, y_0, z_0)$ and $\mathbf{N} = a\mathbf{i} + b\mathbf{j} + c\mathbf{k}$. Then if $Q = (x, y, z)$,

$$\overrightarrow{PQ} = (x - x_0)\mathbf{i} + (y - y_0)\mathbf{j} + (z - z_0)\mathbf{k}.$$

If $\overrightarrow{PQ} \perp \mathbf{N}$, then $\overrightarrow{PQ} \cdot \mathbf{N} = 0$. But this implies that

$$a(x - x_0) + b(y - y_0) + c(z - z_0) = 0. \tag{1}$$

A more common way to write the equation of a plane is easily derived from (1):

EQUATION OF A PLANE

$$ax + by + cz = d \tag{2}$$

where

$$d = ax_0 + by_0 + cz_0 = \overrightarrow{0P} \cdot \mathbf{N}.$$

EXAMPLE 1

FINDING AN EQUATION OF THE PLANE PASSING THROUGH A GIVEN POINT WITH GIVEN NORMAL VECTOR

Find an equation of the plane Π passing through the point $(2, 5, 1)$ and normal to the vector $\mathbf{N} = \mathbf{i} - 2\mathbf{j} + 3\mathbf{k}$.

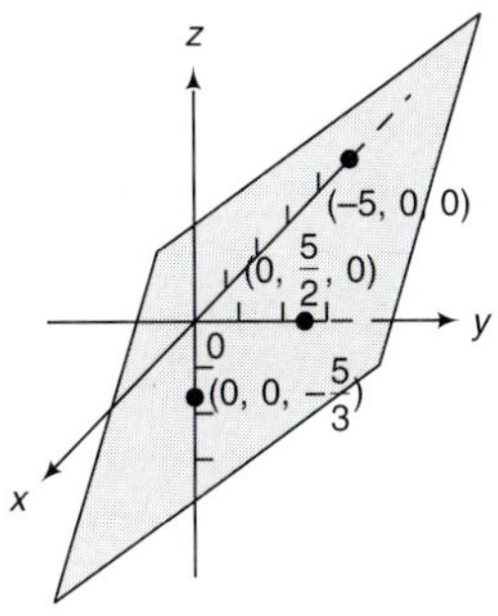

SOLUTION: From (1) we immediately obtain

$$(x - 2) - 2(y - 5) + 3(z - 1) = 0,$$

or

$$x - 2y + 3z = -5. \tag{3}$$

This plane is sketched in Figure 2.

z
y
x
0
(−5, 0, 0)
(0, 5/2, 0)
(0, 0, −5/3)

FIGURE 2
The plane $x - 2y + 3z = -5$

REMARK: We show you how to draw a plane on page 56.

The three coordinate planes are represented as follows:

i. The *xy*-plane passes through the origin $(0, 0, 0)$ and any vector lying along the z-axis is normal to it. The simplest such vector is $\mathbf{k}$. Thus, from (1), we obtain

$$0(x - 0) + 0(y - 0) + 1(z - 0) = 0,$$

which yields

$$z = 0 \tag{4}$$

as the equation of the xy-plane. (This result should not be very surprising.)

REMARK: The equation $z = 0$ is really a shorthand notation for the equation of the xy-plane. The full notation is

$$\text{the } xy\text{-plane} = \{(x, y, z)\colon z = 0\}.$$

The shorthand notation is fine as long as we don't lose sight of the fact that we are in $\mathbb{R}^3$.

ii. The xz-plane has the equation

$$y = 0. \tag{5}$$

iii. The yz-plane has the equation

$$x = 0. \tag{6}$$

Three points that are not collinear determine a plane, since they determine two nonparallel vectors that intersect at a point.

EXAMPLE 2 FINDING AN EQUATION OF THE PLANE PASSING THROUGH THREE GIVEN POINTS

Find an equation of the plane Π passing through the points $P = (1, 2, 1)$, $Q = (-2, 3, -1)$, and $R = (1, 0, 4)$.

SOLUTION: The vectors $\overrightarrow{PQ} = -3\mathbf{i} + \mathbf{j} - 2\mathbf{k}$ and $\overrightarrow{QR} = 3\mathbf{i} - 3\mathbf{j} + 5\mathbf{k}$ lie on the plane and are therefore orthogonal to the normal vector. Thus,

$$\mathbf{N} = \overrightarrow{PQ} \times \overrightarrow{QR} = \begin{vmatrix} \mathbf{i} & \mathbf{j} & \mathbf{k} \\ -3 & 1 & -2 \\ 3 & -3 & 5 \end{vmatrix} = -\mathbf{i} + 9\mathbf{j} + 6\mathbf{k},$$

and we obtain

$$\Pi\colon -(x-1) + 9(y-2) + 6(z-1) = 0,$$

or

$$-x + 9y + 6z = 23.$$

Note that if we choose another point, say Q, we get the equation

$$-(x+2) + 9(y-3) + 6(z+1) = 0,$$

which reduces to

$$-x + 9y + 6z = 23.$$

This plane is sketched in Figure 3.

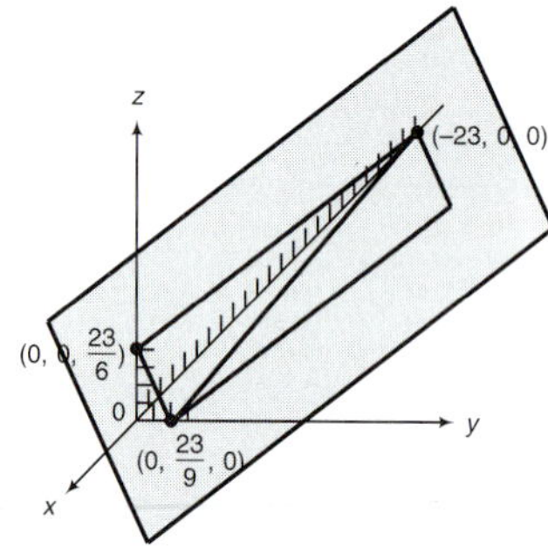

FIGURE 3
The plane $-x + 9y + 6z = 23$

DEFINITION PARALLEL PLANES

Two planes are **parallel** if their normal vectors are parallel.† ■

Two parallel planes are drawn in Figure 4.

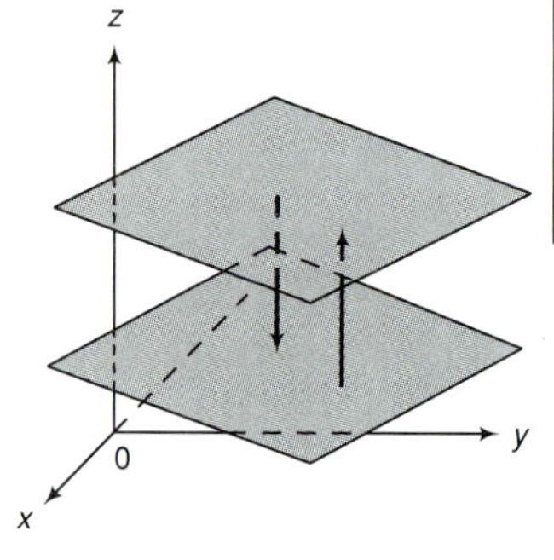

FIGURE 4
Parallel planes

EXAMPLE 3 TWO PARALLEL PLANES

The planes Π_1: $2x + 3y - z = 3$ and Π_2: $-4x - 6y + 2z = 8$ are parallel since $\mathbf{N}_1 = 2\mathbf{i} + 3\mathbf{j} - \mathbf{k}$, $\mathbf{N}_2 = -4\mathbf{i} - 6\mathbf{j} + 2\mathbf{k}$, and $\mathbf{N}_2 = -2\mathbf{N}_1$ (note also that $\mathbf{N}_1 \times \mathbf{N}_2 = \mathbf{0}$).

† In a few books parallel planes are defined as two planes that never intersect. With this definition it is not difficult to prove that if two planes are parallel, then their normal vectors are parallel.

If two distinct planes are not parallel, then they intersect in a straight line.

EXAMPLE 4

FINDING POINTS OF INTERSECTION OF PLANES

Find all points of intersection of the planes $2x - y - z = 3$ and $x + 2y + 3z = 7$.

SOLUTION: When the planes intersect, we have

$$2x - y - z = 3$$

and

$$x + 2y + 3z = 7.$$

Multiplying the first equation by 2 and adding it to the second, we obtain

$$\begin{array}{r} 4x - 2y - 2z = 6 \\ x + 2y + 3z = 7 \\ \hline 5x + \qquad\quad z = 13 \end{array}$$

or $z = -5x + 13$. From the first equation,

$$y = 2x - z - 3 = 2x - (-5x + 13) - 3 = 7x - 16.$$

Then setting $x = t$, we obtain the parametric representation of the line of intersection:

$$x = t, \qquad y = -16 + 7t, \qquad z = 13 - 5t.$$

Note that this line is orthogonal to both normal vectors $\mathbf{N}_1 = 2\mathbf{i} - \mathbf{j} - \mathbf{k}$ and $\mathbf{N}_2 = \mathbf{i} + 2\mathbf{j} + 3\mathbf{k}$.

WARNING: Notice the difference in information required to find the equation of a plane and the equation of a line. For a plane, we need a point on the plane and a vector *normal* to the plane; for a line, we need a point on the line and a vector *parallel* to the line.

ANOTHER GEOMETRIC WAY TO LOOK AT A PLANE

Consider equation (2),

$$ax + by + cz = d,$$

where at least one of the numbers a, b, c is nonzero. Let $\mathbf{N} = (a, b, c)$ and $\mathbf{x} = (x, y, z)$. Then (2) can be written

$$\mathbf{N} \cdot \mathbf{x} = d, \tag{7}$$

or, since $\mathbf{N} \neq \mathbf{0}$, we can divide both sides of (7) by $|\mathbf{N}|$ to obtain

$$\frac{\mathbf{N}}{|\mathbf{N}|} \cdot \mathbf{x} = \frac{d}{|\mathbf{N}|}.$$

Let $\mathbf{v} = \dfrac{\mathbf{N}}{|\mathbf{N}|}$, a unit vector. Then, $\mathbf{v} \cdot \mathbf{x} = \dfrac{\mathbf{N} \cdot \mathbf{x}}{|\mathbf{N}|} = \dfrac{d}{|\mathbf{N}|}$ and, from equation (9) on page 21,

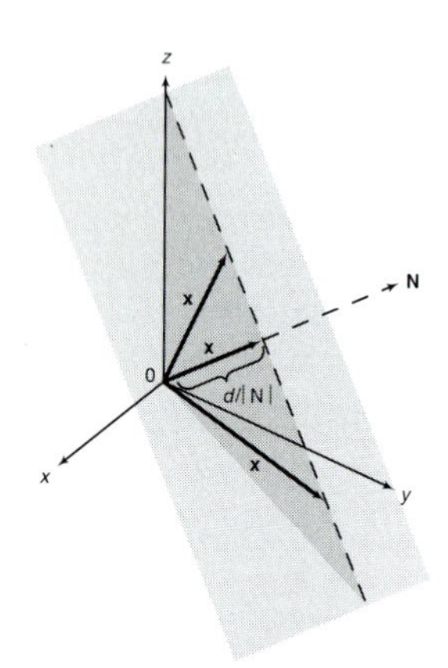

FIGURE 5
The vector **x** is on the plane if the component of **x** in the direction **N** is equal to $\frac{d}{|\mathbf{N}|}$

$$\operatorname{Proj}_{\mathbf{v}} \mathbf{x} = \frac{\mathbf{v} \cdot \mathbf{x}}{\underset{\substack{\uparrow \\ |\mathbf{v}|=1}}{|\mathbf{v}|^2}} \mathbf{v} = \frac{d}{|\mathbf{N}|} \mathbf{v}$$

and the component of **x** in the direction **N** (= the direction of **v**) is $\frac{d}{|\mathbf{N}|}$. That is,

> the plane is the set of vectors **x** such that the component of **x** in the direction **N** is $\frac{d}{|\mathbf{N}|}$.
>
> Moreover, $\frac{|d|}{|\mathbf{N}|}$ is the distance from the origin to the plane.

This is illustrated in Figure 5. Vectors **x** such that $\frac{\mathbf{N}}{|\mathbf{N}|} \cdot \mathbf{x} > \frac{d}{|\mathbf{N}|}$ and $\frac{\mathbf{N}}{|\mathbf{N}|} \cdot \mathbf{x} < \frac{d}{|\mathbf{N}|}$ lie on opposite sides of the plane.

HOW TO DRAW A PLANE

In many places in this and subsequent chapters it will be useful to draw a plane. This is not difficult to do. We consider two cases:

Case 1: The Plane Is Parallel to a Coordinate Plane If the plane is parallel to one of the coordinate planes, then the equation of the plane is one of the following:

$x = a$ (parallel to the yz-plane)

$y = b$ (parallel to the xz-plane)

$z = c$ (parallel to the xy-plane).

We draw each plane as a rectangle with sides parallel to the two other coordinate axes. We sketch three of these planes in Figure 6.

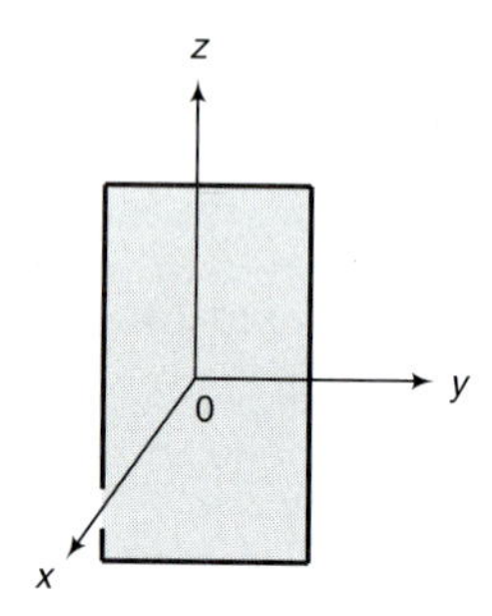

(a) The plane $x = a$ (parallel to the yz-plane).

(b) The plane $y = b$ (parallel to the xy-plane).

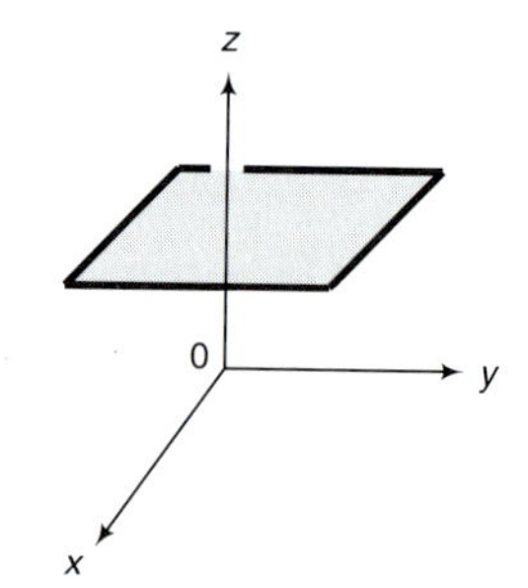

(c) The plane $z = c$ (parallel to the xy-plane).

FIGURE 6
Three planes parallel to a coordinate plane

Case 2: The Plane Intersects Each Coordinate Axis Suppose an equation of the plane is

$$ax + by + cz = d \qquad \text{with } abc \neq 0.$$

The x-intercept is the point $\left(\frac{d}{a}, 0, 0\right)$, the y-intercept is the point $\left(0, \frac{d}{b}, 0\right)$, and the z-intercept is the point $\left(0, 0, \frac{d}{c}\right)$.

Step 1: Plot the three intercepts.
Step 2: Join the three intercepts to form a triangle.
Step 3: By drawing two parallel lines, draw a parallelogram with one diagonal being the third side of the triangle.
Step 4: Expand the parallelogram by drawing four parallel lines.

We illustrate this process by drawing the plane $x + 2y + 3z = 6$ in Figure 7 (a second illustration appears in Figure 3). The intercepts are (6, 0, 0), (0, 3, 0), and (0, 0, 2).

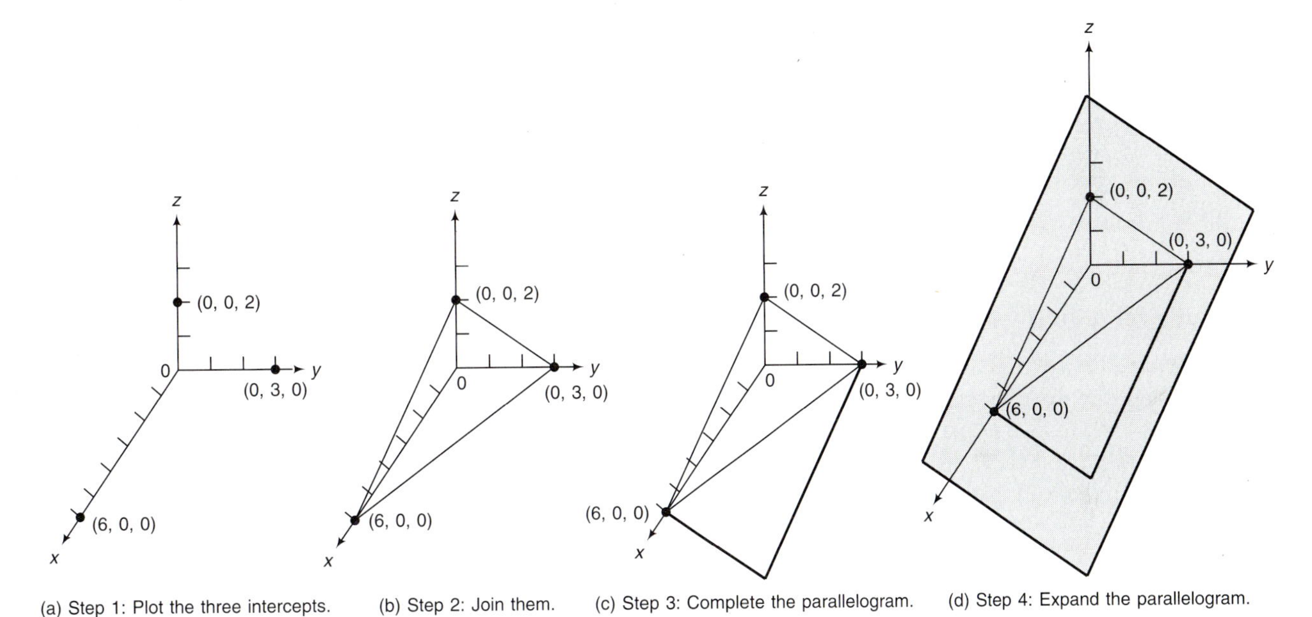

FIGURE 7
Drawing the plane $x + 2y + 3z = 6$ in four steps

PROBLEMS 1.7

SELF-QUIZ

I. The plane passing through $(5, -4, 3)$ that is orthogonal to $\mathbf{j}$ satisfies ________.

a. $y = -4$
b. $(x - 5) + (z - 3) = 0$
c. $x + y + z = 4$
d. $5x - 4y + 3z = -4$

II. The plane passing through $(5, -4, 3)$ that is orthogonal to $\mathbf{i} + \mathbf{j} + \mathbf{k}$ satisfies ________.

a. $y = -4$
b. $(x - 5)/1 = (y + 4)/1 = (z - 3)/1$
c. $x + y + z = 4$
d. $5x - 4y + 3z = -4$

III. The vector ________ is orthogonal to the plane satisfying $2(x-3)-3(y+2)+5(z-5)=0$.

a. $-3\mathbf{i}+2\mathbf{j}-5\mathbf{k}$
b. $2\mathbf{i}-3\mathbf{j}+5\mathbf{k}$
c. $(2-3)\mathbf{i}+(-3+2)\mathbf{j}+(5-5)\mathbf{k}$
d. $(2)(-3)\mathbf{i}+(-3)(2)\mathbf{j}+(5)(-5)\mathbf{k}$

IV. The plane satisfying $6x+18y-12z=17$ is ________ to the plane $-5x-15y+10z=29$.

a. identical
b. parallel
c. orthogonal
d. neither parallel nor orthogonal

In Problems 1–12, find an equation of the plane passing through the given point such that the specified vector **N** is normal to the plane. Sketch the plane.

1. $P=(0,0,0)$; $\mathbf{N}=\mathbf{i}$
2. $P=(0,0,0)$; $\mathbf{N}=\mathbf{j}$
3. $P=(0,0,0)$; $\mathbf{N}=\mathbf{k}$
4. $P=(1,2,3)$; $\mathbf{N}=\mathbf{i}+\mathbf{j}$
5. $P=(1,2,3)$; $\mathbf{N}=\mathbf{i}+\mathbf{k}$
6. $P=(1,2,3)$; $\mathbf{N}=\mathbf{j}+\mathbf{k}$
7. $P=(2,-1,6)$; $\mathbf{N}=3\mathbf{i}-\mathbf{j}+2\mathbf{k}$
8. $P=(-4,-7,5)$; $\mathbf{N}=-3\mathbf{i}-4\mathbf{j}+\mathbf{k}$
9. $P=(-3,11,2)$; $\mathbf{N}=4\mathbf{i}+\mathbf{j}-7\mathbf{k}$
10. $P=(3,-2,5)$; $\mathbf{N}=2\mathbf{i}-7\mathbf{j}-8\mathbf{k}$
11. $P=(4,-7,-3)$; $\mathbf{N}=-\mathbf{i}-\mathbf{j}-\mathbf{k}$
12. $P=(8,1,0)$; $\mathbf{N}=-7\mathbf{i}+\mathbf{j}+2\mathbf{k}$

In Problems 13–16, find an equation for the plane passing through the three given points. Sketch the plane.

13. $(1,2,-4)$, $(2,3,7)$, and $(4,-1,3)$
14. $(-7,1,0)$, $(2,-1,3)$, and $(4,1,6)$
15. $(1,0,0)$, $(0,1,0)$, and $(0,0,1)$
16. $(2,3,-2)$, $(4,-1,-1)$, and $(3,1,2)$

Two planes are said to be **orthogonal** if their normal vectors are orthogonal. In Problems 17–24, determine whether the given planes are parallel, orthogonal, coincident (i.e., the same), or none of these.

17. Π_1: $x+y+z=2$; Π_2: $2x+2y+2z=4$
18. Π_1: $x-y+z=3$; Π_2: $-3x+3y-3z=-9$
19. Π_1: $2x-y+z=3$; Π_2: $x+y-z=7$
20. Π_1: $2x-y+z=3$; Π_2: $x+y+z=3$
21. Π_1: $3x-2y+7z=4$; Π_2: $-2x+4y+2z=16$
22. Π_1: $-4x+4y-6z=7$; Π_2: $2x-2y+3z=-3$
23. Π_1: $-4x+4y-6z=6$; Π_2: $2x-2y+3z=-3$
24. Π_1: $3x-y+z=y-2x$; Π_2: $5x-4y+3z=2z-2y$

In Problems 25–28, find an equation for the set of all points common to the two given planes (i.e., describe the points of intersection).

25. Π_1: $x+y+z=1$; Π_2: $x-y-z=-3$
26. Π_1: $x-y+z=2$; Π_2: $2x-3y+4z=7$
27. Π_1: $3x-y+4z=3$; Π_2: $-4x-2y+7z=8$
28. Π_1: $-2x-y+17z=4$; Π_2: $2x-y-z=-7$

The **angle between two planes** is defined to be the acute angle between their normal vectors. In Problems 29–32, calculate the angle between the two planes given in the specified problem.

29. Problem 25
30. Problem 26
31. Problem 27
32. Problem 28

In Problems 33–36, use the result of Problem 51 to determine whether the three given position vectors (i.e., with one endpoint at the origin) are coplanar. If they are coplanar, then find an equation for the plane containing them.

33. $\mathbf{u}=2\mathbf{i}-3\mathbf{j}+4\mathbf{k}$; $\mathbf{v}=7\mathbf{i}-2\mathbf{j}+3\mathbf{k}$; $\mathbf{w}=9\mathbf{i}-5\mathbf{j}+7\mathbf{k}$
34. $\mathbf{u}=-3\mathbf{i}+\mathbf{j}+8\mathbf{k}$; $\mathbf{v}=-2\mathbf{i}-3\mathbf{j}+5\mathbf{k}$; $\mathbf{w}=2\mathbf{i}+14\mathbf{j}-4\mathbf{k}$
35. $\mathbf{u}=2\mathbf{i}+\mathbf{j}-2\mathbf{k}$; $\mathbf{v}=2\mathbf{i}-\mathbf{j}-2\mathbf{k}$; $\mathbf{w}=2\mathbf{i}-\mathbf{j}+2\mathbf{k}$
36. $\mathbf{u}=3\mathbf{i}-2\mathbf{j}+\mathbf{k}$; $\mathbf{v}=\mathbf{i}+\mathbf{j}-5\mathbf{k}$; $\mathbf{w}=-\mathbf{i}+5\mathbf{j}-16\mathbf{k}$

In Problems 37–40, use the result of Problem 56 to find the distance between the given point and plane.

37. $(2,-1,4)$; $3x-y+7z=2$
38. $(4,0,1)$; $2x-y+8z=3$
39. $(-7,-2,-1)$; $-2x+8z=-5$
40. $(-3,0,2)$; $-3x+y+5z=0$

In Problems 41–50, draw each plane.

41. $x=1$
42. $y=5$
43. $z=-2$
44. $y=-3$
45. $x+y+z=1$
46. $x-y+z=1$
47. $x-y-z=1$
48. $x+2y-4z=4$
49. $-2x+3y+5z=10$
50. $3x-y-2z=6$

51. Three vectors **u**, **v**, and **w** are said to be **coplanar** if and only if they have representative directed line segments that all lie in the same plane Π. Show that if **u**, **v**, and **w** all have an endpoint at the origin, then they are coplanar if and only if their scalar triple product equals zero: $\mathbf{u} \cdot (\mathbf{v} \times \mathbf{w}) = 0$.

***52.** Let **u** and **v** be two nonparallel vectors that lie in a particular plane Π. Show that if **w** is any other vector in Π, then there are scalars α and β such that $\mathbf{w} = \alpha\mathbf{u} + \beta\mathbf{v}$. This expression is called the **parametric representation** of the plane Π. [*Hint:* Draw a parallelogram in which $\alpha\mathbf{u}$ and $\beta\mathbf{v}$ form adjacent sides and the diagonal vector is **w**.]

***53.** Let $P = (x_1, y_1, z_1)$, $Q = (x_2, y_2, z_2)$, and $R = (x_3, y_3, z_3)$ be three points in $\mathbb{R}^3$ that are not collinear. Show that the plane passing through those three points satisfies the equation

$$\begin{vmatrix} x & y & z & 1 \\ x_1 & y_1 & z_1 & 1 \\ x_2 & y_2 & z_2 & 1 \\ x_3 & y_3 & z_3 & 1 \end{vmatrix} = 0.$$

Be sure to show where you used the fact that P, Q, and R are not collinear.

***54.** Let three planes be described by the equations $a_1x + b_1y + c_1z = d_1$, $a_2x + b_2y + c_2z = d_2$, and $a_3x + b_3y + c_3z = d_3$. Show that the planes have a unique point in common if

$$\begin{vmatrix} a_1 & b_1 & c_1 \\ a_2 & b_2 & c_2 \\ a_3 & b_3 & c_3 \end{vmatrix} \neq 0.$$

***55.** Let Π be a plane, P a point on the plane, **N** a vector normal to that plane, and Q a point not on the plane (see Figure 8). Show that the perpendicular distance D from Q to the plane Π is given by

$$D = |\text{Proj}_{\mathbf{N}}\, \overrightarrow{PQ}| = \frac{|\overrightarrow{PQ} \cdot \mathbf{N}|}{|\mathbf{N}|}.$$

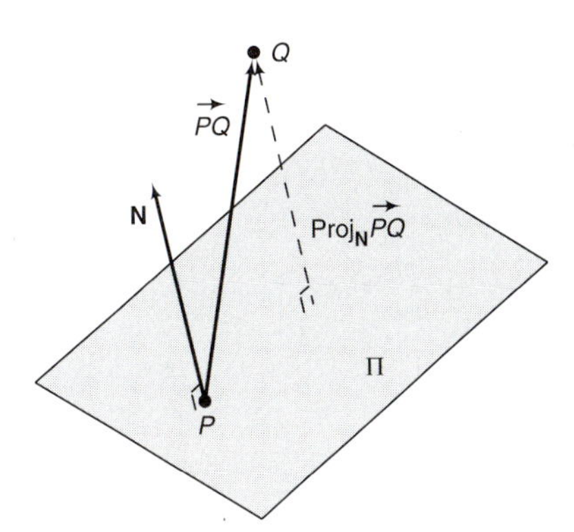

FIGURE 8

***56.** Prove that the distance between the plane $ax + by + cz = d$ and the point (x_0, y_0, z_0) is given by

$$D = \frac{|ax_0 + by_0 + cz_0 - d|}{\sqrt{a^2 + b^2 + c^2}}.$$

ANSWERS TO SELF-QUIZ

I. a **II.** c **III.** b **IV.** b

1.8 QUADRIC SURFACES

A **surface** in space is defined as the set of points in $\mathbb{R}^3$ satisfying the equation $F(x, y, z) = 0$ for F a continuous function. For example, the equation

$$F(x, y, z) = x^2 + y^2 + z^2 - 1 = 0 \tag{1}$$

is the equation of the unit sphere, as we saw in Section 1.3. In this section we will take a brief look at some of the most commonly encountered surfaces in $\mathbb{R}^3$. We will take a more detailed look at general surfaces in $\mathbb{R}^3$ in Chapter 3.

Having already discussed the sphere, we turn our attention to the cylinder.

DEFINITION CYLINDER

Let a plane curve C and a line L not in the plane of C be given. A **cylinder** is the surface generated when a line parallel to L moves around C, remaining parallel to L. The line L is called the **generatrix** of the cylinder, and the curve C is called its **directrix**. ■

EXAMPLE 1 SKETCHING A RIGHT CIRCULAR CYLINDER

Let L be the z-axis and C the circle $x^2 + y^2 = a^2$ in the xy-plane. Sketch the cylinder.

SOLUTION: As we move a line along the circle $x^2 + y^2 = a^2$ and parallel to the z-axis, we obtain the **right circular cylinder** $x^2 + y^2 = a^2$ sketched in Figure 1.

REMARK: We can write the equation $x^2 + y^2 = 4$ as

$$x^2 + y^2 + 0 \cdot z^2 = 4.$$

This illustrates that we are talking about a cylinder in $\mathbb{R}^3$ rather than the circle $x^2 + y^2 = 4$ in $\mathbb{R}^2$.

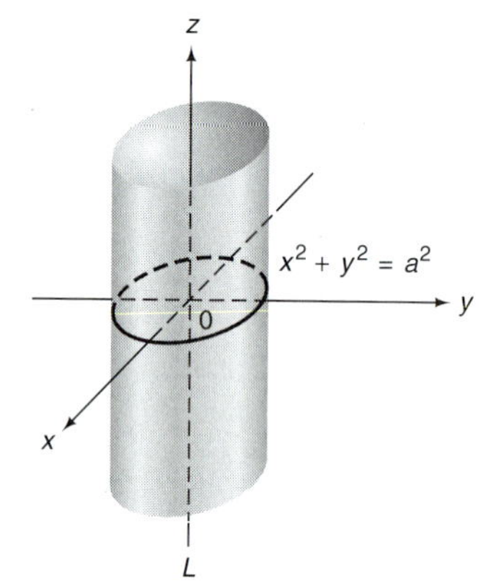

FIGURE 1
Right circular cylinder

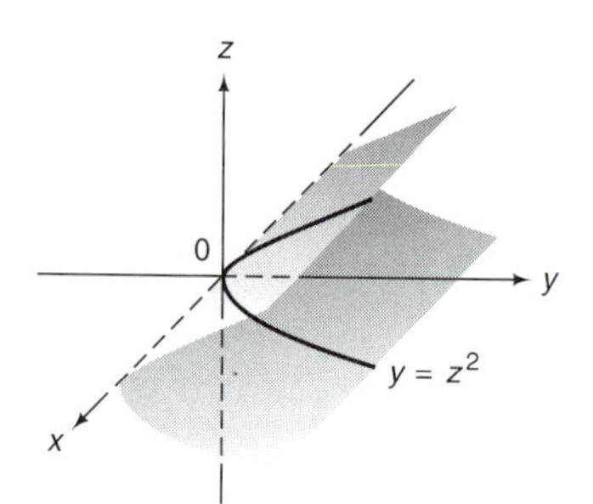

FIGURE 2
Parabolic cylinder

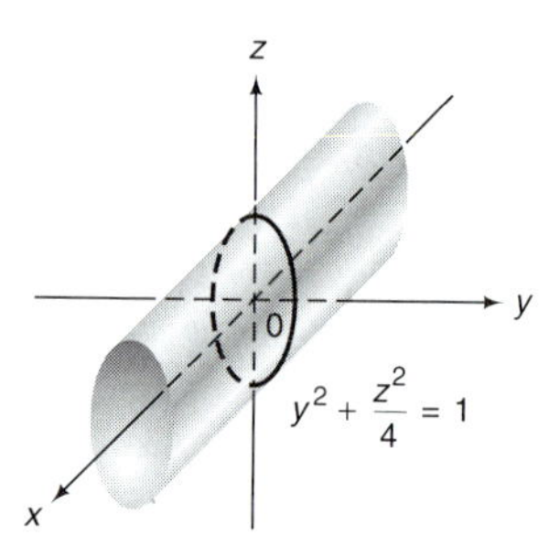

FIGURE 3
Elliptic cylinder

EXAMPLE 2 SKETCHING A PARABOLIC CYLINDER

Suppose L is the x-axis and C is given by $y = z^2$. Sketch the resulting cylinder.

SOLUTION: The curve $y = z^2$ is a parabola in the yz-plane. As we move along it, parallel to the x-axis, we obtain the **parabolic cylinder** sketched in Figure 2.

EXAMPLE 3 SKETCHING AN ELLIPTIC CYLINDER

Suppose L is the x-axis and C is given by $y^2 + (z^2/4) = 1$. Sketch the resulting cylinder.

SOLUTION: $y^2 + (z^2/4) = 1$ is an ellipse in the yz-plane. As we move along it parallel to the x-axis, we obtain the **elliptic cylinder** sketched in Figure 3.

EXAMPLE 4

SKETCHING A HYPERBOLIC CYLINDER

The hyperbolic cylinder: $(y^2/b^2) - (x^2/a^2) = 1$. This is the equation of a hyperbola in the xy-plane. See Figure 4.

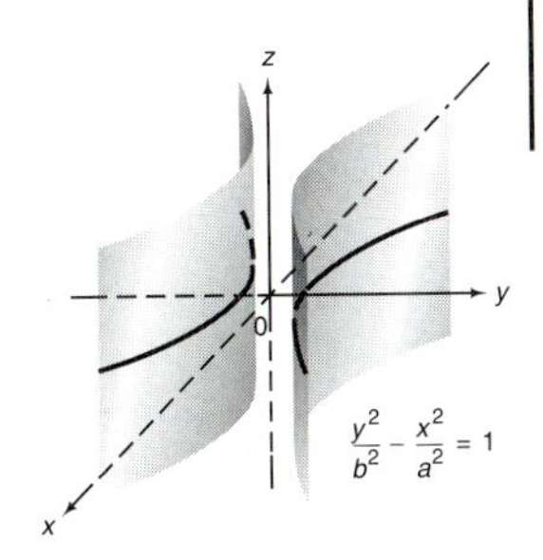

FIGURE 4
Hyperbolic cylinder

The graph of a second-degree equation in the variables x and y is a circle, parabola, ellipse, or hyperbola (or a degenerate form of one of these such as a single point or a straight line or a pair of straight lines) in the plane. In $\mathbb{R}^3$ we have the following definition.

DEFINITION QUADRIC SURFACE

A **quadric surface** in $\mathbb{R}^3$ is the graph of a second-degree equation in the variables x, y, and z. Such an equation takes the form

$$Ax^2 + By^2 + Cz^2 + Dxy + Exz + Fyz + Gx + Hy + Jz + K = 0. \tag{2}$$

■

We have already seen sketches of several quadric surfaces. We list below the eleven types of nondegenerate[†] quadric surfaces. Although we will not prove this result here, any quadric surface can be written in one of these forms by a translation or rotation of the coordinate axes. Here is the list:

1. sphere
2. right circular cylinder
3. parabolic cylinder
4. elliptic cylinder
5. hyperbolic cylinder
6. ellipsoid
7. hyperboloid of one sheet
8. hyperboloid of two sheets
9. elliptic paraboloid
10. hyperbolic paraboloid
11. elliptic cone

We have already seen sketches of surfaces 1, 2, 3, 4, and 5. We can best describe the remaining six surfaces by looking at **cross-sections** parallel to a given coordinate plane. Cross-sections are obtained by setting one of the variables x, y, or z equal to a constant. For example, in the unit sphere $x^2 + y^2 + z^2 = 1$, cross-sections parallel to the xy-plane are circles. To see this, let $z = c$, where $-1 < c < 1$ (this is a plane parallel to the xy-plane). Then,

$$x^2 + y^2 = 1 - c^2,$$

which is the equation of a circle in the xy-plane.

Typical graphs and specific, computer-drawn graphs of the six remaining quadric surfaces appear in Table 1. The equations given in Table 1 are examples of **standard forms**[‡] of the given surfaces. We now describe the six remaining quadric surfaces.

6. **The ellipsoid**: $(x^2/a^2) + (y^2/b^2) + (z^2/c^2) = 1$. Cross-sections parallel to the xy-plane, the xz-plane, and the yz-plane are all ellipses.

NOTE: If the ellipse $(x^2/a^2) + (y^2/b^2) = 1$ is revolved about the x-axis, the resulting surface is the ellipsoid $(x^2/a^2) + (y^2/b^2) + (z^2/b^2) = 1$. What do you get if the ellipse is rotated around the y-axis? (See Problem 37.)

[†] By **degenerate** we mean a surface whose graph contains a finite number of points (or none) or consists of a pair of planes. We saw examples of degenerate spheres (zero or one point) in Section 1.3.

[‡] A standard form is one in which the surface has its center or vertex at the origin and its axes parallel to the coordinate axes.

Name of Surface and Typical Equation	Typical Graph	Computer-Drawn Graph
Ellipsoid $\frac{x^2}{a^2} + \frac{y^2}{b^2} + \frac{z^2}{c^2} = 1$	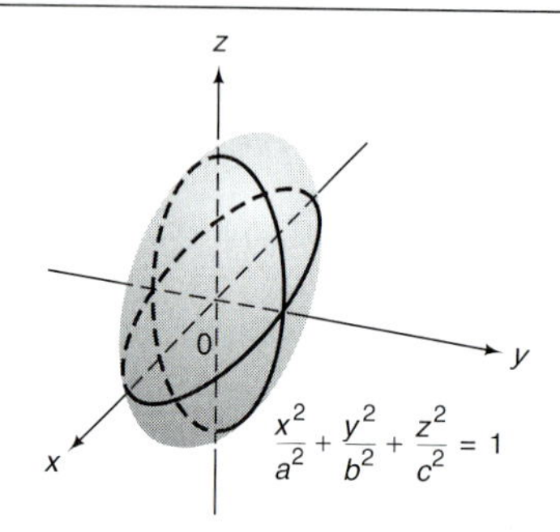	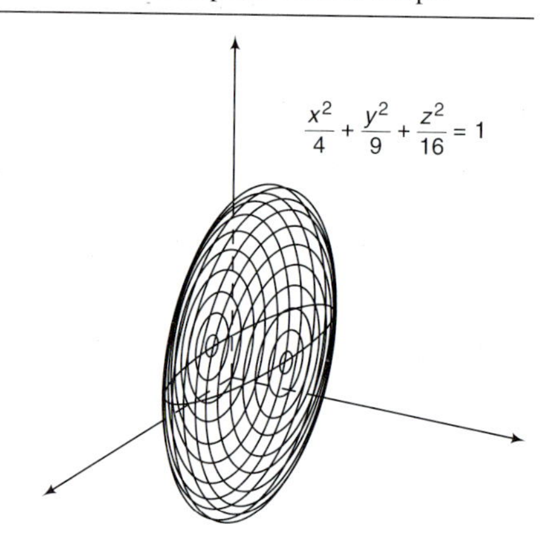
Hyperboloid of One Sheet $\frac{x^2}{a^2} + \frac{y^2}{b^2} - \frac{z^2}{c^2} = 1$	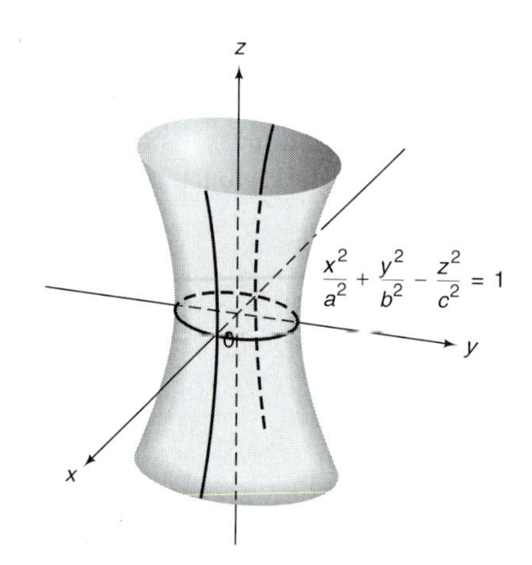	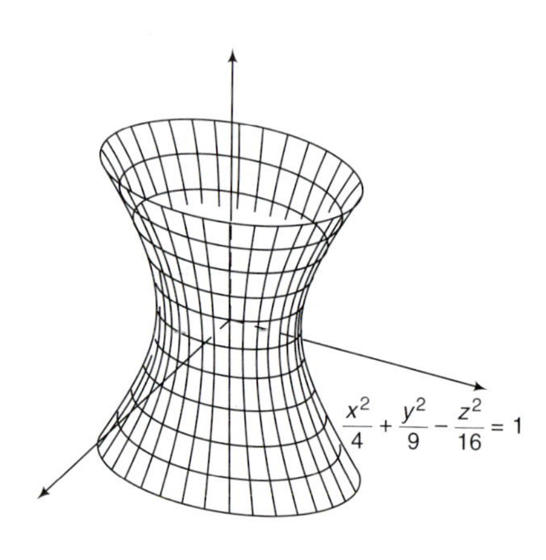
Hyperboloid of Two Sheets $\frac{z^2}{c^2} - \frac{x^2}{a^2} - \frac{y^2}{b^2} = 1$	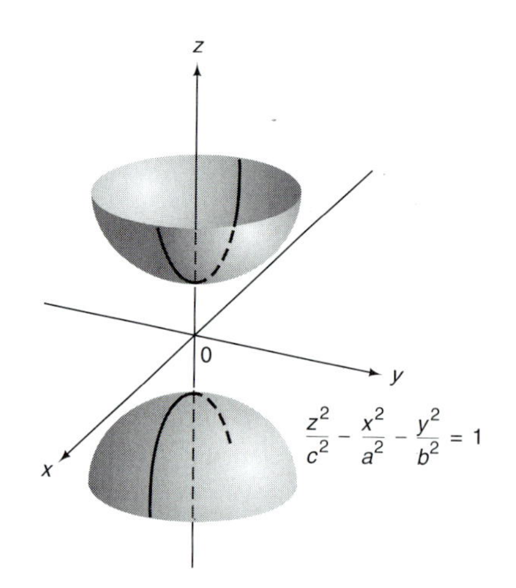	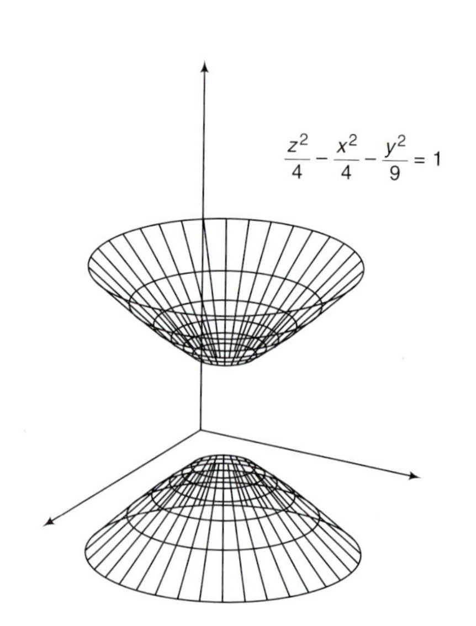

(continued)

TABLE 1 CONTINUED

<table>
<tr><th>Name of Surface and Typical Equation</th><th>Typical Graph</th><th>Computer-Drawn Graph</th></tr>
<tr><td>Elliptic Paraboloid
$z = \frac{x^2}{a^2} + \frac{y^2}{b^2}$</td><td>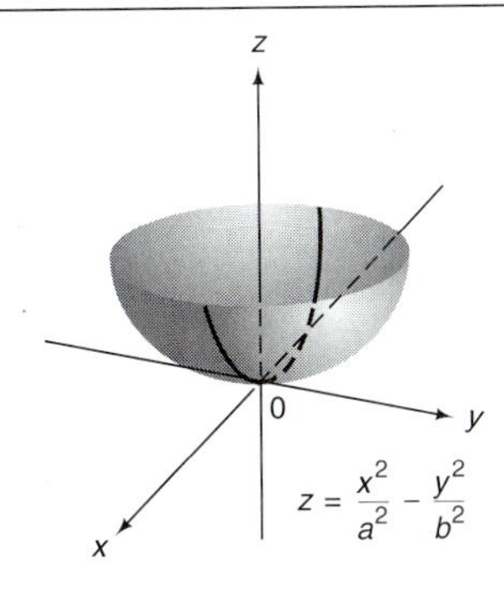
</td><td>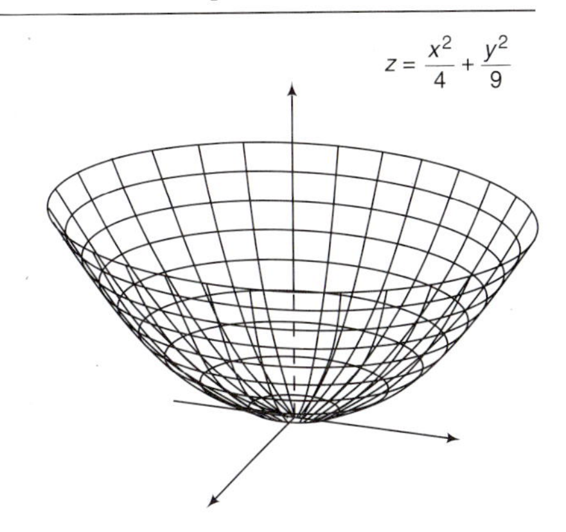
</td></tr>
<tr><td>Hyperbolic Paraboloid
(or Saddle Surface)
$z = \frac{x^2}{a^2} - \frac{y^2}{b^2}$</td><td>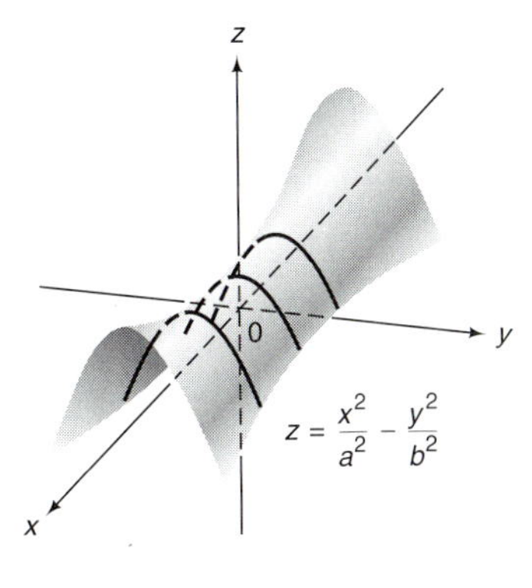
</td><td>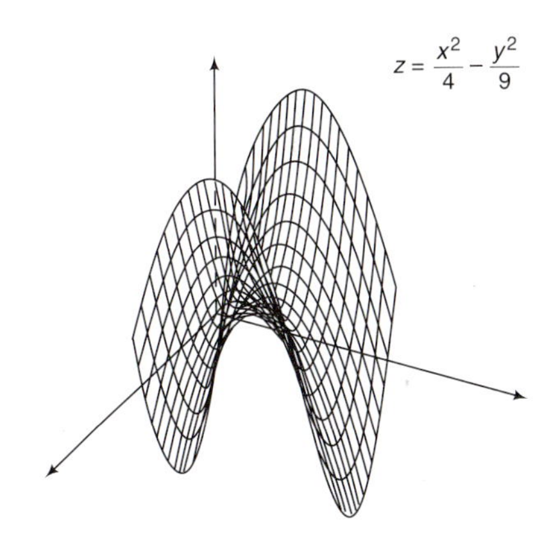
</td></tr>
<tr><td>Elliptic Cone
$\frac{x^2}{a^2} + \frac{y^2}{b^2} = z^2$</td><td>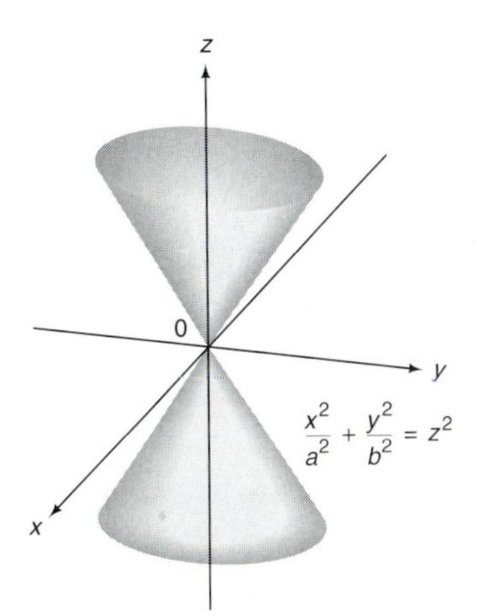
</td><td>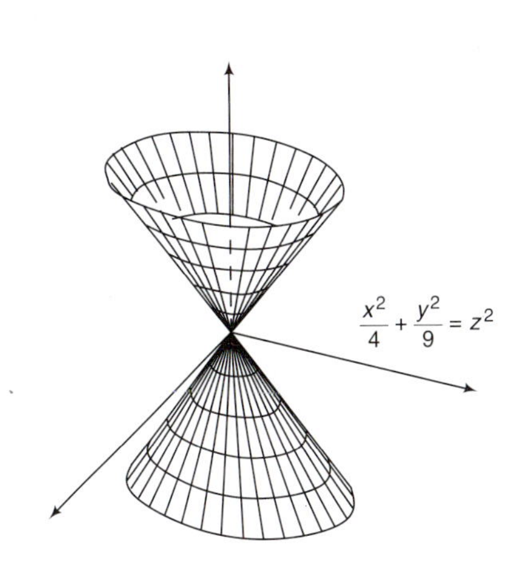
</td></tr>
</table>

7. **The hyperboloid of one sheet**: $(x^2/a^2) + (y^2/b^2) - (z^2/c^2) = 1$. Cross-sections parallel to the xy-plane are ellipses. Cross-sections parallel to the xz-plane and the yz-plane are hyperbolas.
8. **The hyperboloid of two sheets**: $(z^2/c^2) - (x^2/a^2) - (y^2/b^2) = 1$. Cross-sections are the same as those of the hyperboloid of one sheet. Note that the equation implies that $|z| \geq c$ (explain why).
9. **The elliptic paraboloid**: $z = (x^2/a^2) + (y^2/b^2)$. For each positive fixed z, $(x^2/a^2) + (y^2/b^2) = z$ is the equation of an ellipse. Hence, cross-sections parallel to the xy-plane are ellipses. If x or y is fixed, then we obtain parabolas. Hence cross-sections parallel to the xz- or yz-planes are parabolas.
10. **The hyperbolic paraboloid**: $z = (x^2/a^2) - (y^2/b^2)$. For each fixed z we obtain a hyperbola parallel to the xy-plane. Hence, cross-sections parallel to the xy-plane are hyperbolas. If x or y is fixed, we obtain parabolas. Thus cross-sections parallel to the xz- and yz-planes are parabolas. The shape of the graph suggests why the hyperbolic paraboloid is often called a **saddle surface**.
11. **The elliptic cone**: $(x^2/a^2) + (y^2/b^2) = z^2$. We get one **nappe** of the cone for $z > 0$ and another for $z < 0$. Cross-sections cut by planes not passing through the origin are either parabolas, circles, ellipses, or hyperbolas. That is, the cross-sections are the **conic sections.** If $a = b$, we obtain the equation of a **circular cone**.

For all the quadric surfaces discussed above, we can interchange the roles of x, y, and z.

EXAMPLE 5

A HYPERBOLOID OF TWO SHEETS

Describe the surface given by $x^2 - 4y^2 - 9z^2 = 25$.

SOLUTION: Dividing by 25, we obtain

$$\frac{x^2}{25} - \frac{4y^2}{25} - \frac{9z^2}{25} = 1, \quad \text{or} \quad \frac{x^2}{25} - \frac{y^2}{(\frac{5}{2})^2} - \frac{z^2}{(\frac{5}{3})^2} = 1.$$

This is the equation of a hyperboloid of two sheets. Note here that cross-sections parallel to the yz-plane are ellipses. The surface is sketched in Figure 5.

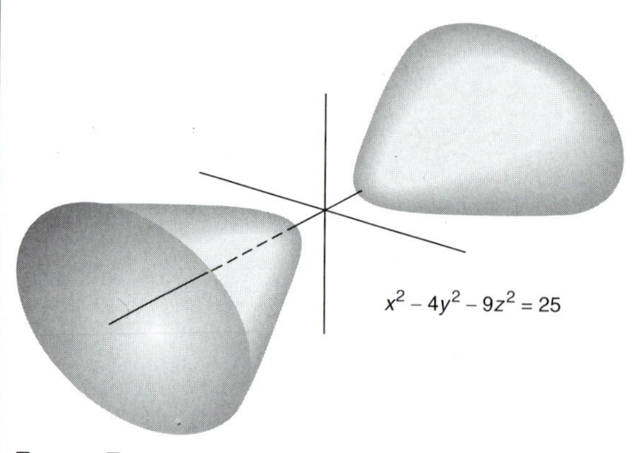

FIGURE 5

The hyperboloid of two sheets $\frac{x^2}{25} - \frac{y^2}{(\frac{5}{2})^2} - \frac{z^2}{(\frac{5}{3})^2} = 1$

All of the surfaces discussed so far have been centered at the origin. We can describe other quadrics as well.

EXAMPLE 6

A HYPERBOLOID OF ONE SHEET

Describe the surface $x^2 + 8x - 2y^2 + 8y + z^2 = 0$.

SOLUTION: Completing the squares, we obtain

$$(x+4)^2 - 2(y-2)^2 + z^2 = 8,$$

or dividing by 8,

$$\frac{(x+4)^2}{8} - \frac{(y-2)^2}{4} + \frac{z^2}{8} = 1.$$

This surface is a hyperboloid of one sheet, which, however, is centered at $(-4, 2, 0)$ instead of the origin. Moreover, cross-sections parallel to the xz-plane are ellipses, while those parallel to the other coordinate planes are hyperbolas. The surface is sketched in Figure 6.

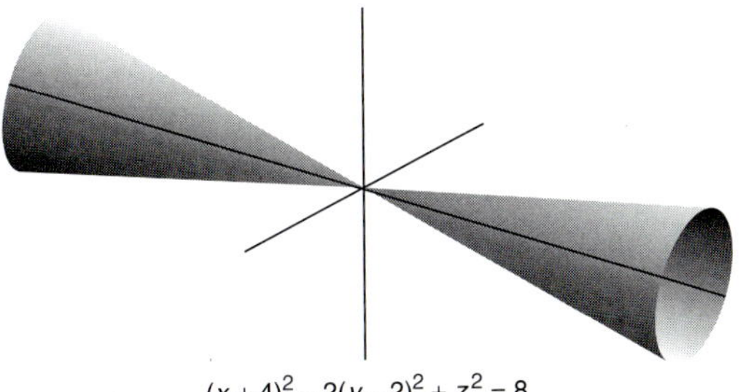

$(x+4)^2 - 2(y-2)^2 + z^2 = 8$

FIGURE 6
The hyperboloid of one sheet, $\frac{(x+4)^2}{8} - \frac{(y-2)^2}{4} + \frac{z^2}{8} = 1$, which is centered at $(-4, 2, 0)$

Generally, to identify a quadric surface with x, y, and z terms, completing the square will resolve the problem. If terms of the form cxy, dyz, or exz are present, also rotate the axes. We will not discuss this technique here.

WARNING: Watch for degenerate surfaces. For example, no points satisfy

$$x^2 + \frac{y^2}{4} + \frac{z^2}{9} + 1 = 0.$$

PROBLEMS 1.8

SELF-QUIZ

I. The graph of ________ is a parabolic cylinder.
a. $z = y^2$ **b.** $y^2 + z^2 = 1$
c. $x^2 - z^2 = 1$ **d.** $(x/3)^2 + z^2 = 1$

II. The graph of ________ is an elliptic cylinder.
a. $x = z^2$ **b.** $x^2 + y^2 = 9$
c. $y^2 - z^2 = 9$ **d.** $(y/5)^2 + (z/2)^2 = 1$

III. The graph of ________ is a right circular cylinder.
a. $z = x^2$ **b.** $x^2 + z^2 = 25$
c. $x^2 - y^2 = 25$ **d.** $(x/2)^2 + (z/7)^2 = 1$

IV. The graph of ________ is a hyperbolic cylinder.
a. $y = z^2$ **b.** $y^2 + z^2 = 36$
c. $z^2 - y^2 = 36$ **d.** $(y/5)^2 + (z/6)^2 = 1$

In Problems 1–8, draw a sketch of the given cylinder. Here the directrix C is given; the generatrix L is the axis of the variable missing in the equation.

1. $y = \sin x$ **2.** $z = \sin y$ **3.** $y = \cos z$
4. $y = \cosh x$ **5.** $z = x^3$ **6.** $z = |y|$
7. $|y| + |z - 5| = 1$
8. $x^2 + y^2 + 2y = 0$
[*Hint:* Complete the square.]

In Problems 9–36, identify the quadric surface and sketch it.

9. $x^2 + y^2 = 4$ **10.** $y^2 + z^2 = 4$
11. $x^2 + z^2 = 4$ **12.** $(x/2)^2 - (y/3)^2 = 1$
13. $x^2 + 4z^2 = 1$ **14.** $y^2 - 2y + 4z^2 = 1$
15. $x^2 - z^2 = 1$ **16.** $3x^2 - 4y^2 = 4$
17. $x^2 + y^2 + z^2 = 1$ **18.** $x^2 + y^2 - z^2 = 1$
19. $y^2 + 2y - z^2 + x^2 = 1$
20. $x^2 + 2y^2 + 3z^2 = 4$
21. $x + 2y^2 + 3z^2 - 4$
22. $x^2 - y^2 - 3z^2 = 4$
23. $x^2 + 4x + y^2 + 6y - z^2 - 8z = 2$
24. $4x - x^2 + y^2 + z^2 = 0$
25. $5x^2 + 7y^2 + 8z^2 = 8z$
26. $x^2 - y^2 - z^2 = 1$
27. $x^2 - 2y^2 - 3z^2 = 4$
28. $z^2 - x^2 - y^2 = 2$
29. $y^2 - 3x^2 - 3z^2 = 27$
30. $x^2 + y^2 + z^2 = 2(x + y + z)$
31. $4x^2 + 4y^2 + 16z^2 = 16$
32. $4y^2 - 4x^2 + 8z^2 = 16$
33. $z + x^2 - y^2 = 0$
34. $-x^2 + 2x + y^2 - 6y = z$
35. $x + y + z = x^2 + z^2$
36. $x - y + z = y^2 - z^2$

37. Identify and sketch the surface generated when the ellipse $(x/a)^2 + (y/b)^2 = 1$ is revolved about the y-axis.

ANSWERS TO SELF-QUIZ

In each of the four quiz problems,

- (a) is a parabolic cylinder,
- (b) is a right circular cylinder,
- (c) is a hyperbolic cylinder, and
- (d) is an elliptic cylinder.

Therefore, the answers are
I. a
II. d, b (a circle is also an ellipse)
III. b
IV. c

1.9 CYLINDRICAL AND SPHERICAL COORDINATES

In this chapter, so far, we have represented points using rectangular (Cartesian) coordinates. However, there are many ways to represent points in space. In this section, we will briefly introduce two common ways to do so. The first is the generalization of the polar coordinate system in the plane.

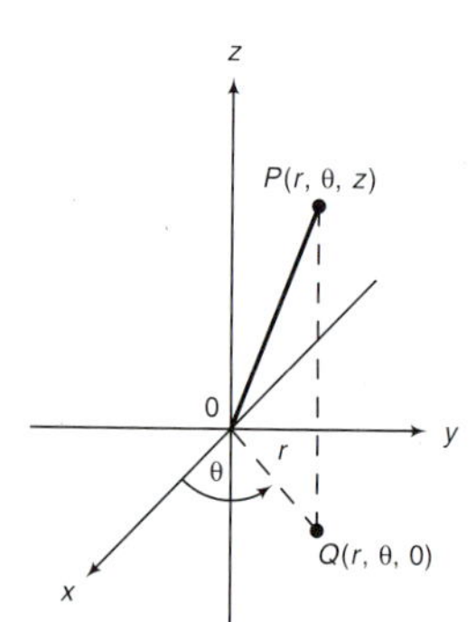

FIGURE 1
Cylindrical coordinates

CYLINDRICAL COORDINATES

In the **cylindrical coordinate system**, a point P is given by

> **A POINT IN CYLINDRICAL COORDINATES**
>
> $$P = (r, \theta, z), \tag{1}$$
>
> where $r \geq 0$, $0 \leq \theta < 2\pi$, r and θ are polar coordinates of the projection of P onto the xy-plane, called the **polar plane**, and z is the directed distance (measured in the positive direction) of this plane from P (see Figure 1). In this figure, $\overrightarrow{0Q}$ is the projection of $\overrightarrow{0P}$ on the xy-plane.

EXAMPLE 1

FOUR STANDARD GRAPHS IN CYLINDRICAL COORDINATES

Discuss the graphs of the equations in cylindrical coordinates:

a. $r = c,\ c > 0$ **b.** $\theta = c$

c. $z = c$ **d.** $r_1 \le r \le r_2,\ \theta_1 \le \theta \le \theta_2,\ z_1 \le z \le z_2$

SOLUTION:

a. If $r = c$, a constant, then θ and z can vary freely, and we obtain a right circular cylinder with radius c [see Figure 2(a)]. This is the analog of the circle whose equation in polar coordinates is $r = c$ and is the reason the system is called the *cylindrical* coordinate system.

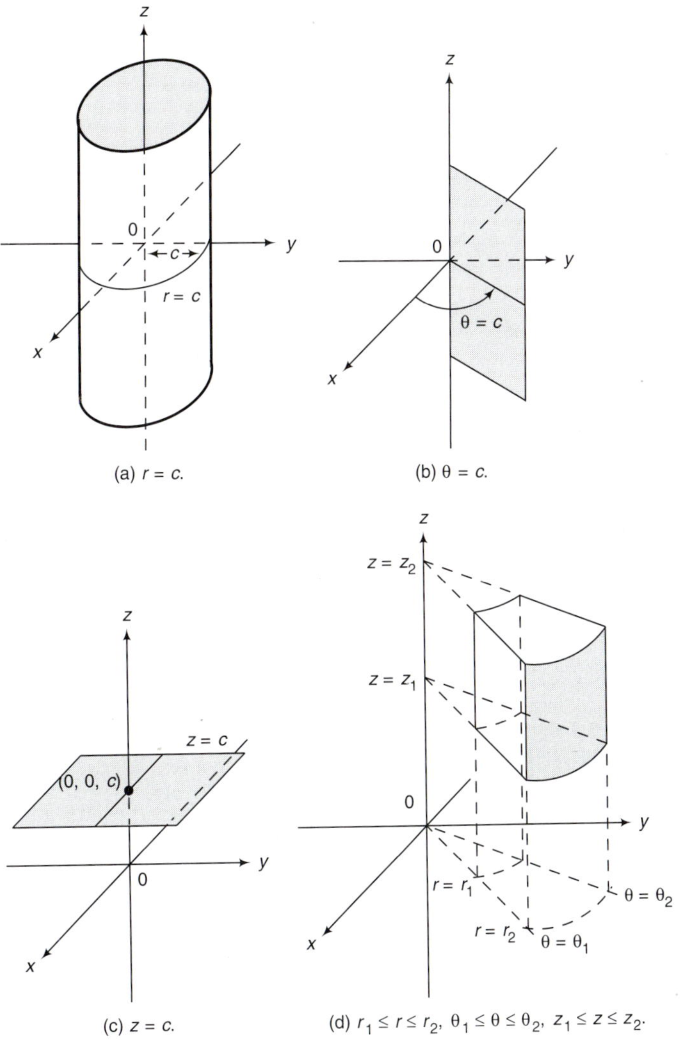

FIGURE 2
Four standard graphs in cylindrical coordinates

b. If $\theta = c$, we obtain a half plane through the z-axis [see Figure 2(b)].

c. If $z = c$, we obtain a plane parallel to the polar plane. This plane is sketched in Figure 2(c).

d. $r_1 \le r \le r_2$ gives the region between the cylinders $r = r_1$ and $r = r_2$. $\theta_1 \le \theta \le \theta_2$ is the wedge-shaped region between the half planes $\theta = \theta_1$ and $\theta = \theta_2$. Finally, $z_1 \le z \le z_2$ is the "slice" of space between the planes $z = z_1$ and $z = z_2$. Putting these results together, we get the rectangular shaped solid in Figure 2(d).

The following formulas need no proof, because they follow from the formulas for converting from Cartesian to polar coordinates—and vice versa.

TO CHANGE FROM CYLINDRICAL TO RECTANGULAR COORDINATES

$$x = r\cos\theta, \qquad y = r\sin\theta, \qquad z = z \tag{2}$$

TO CHANGE FROM RECTANGULAR TO CYLINDRICAL COORDINATES

$$\text{If } x \neq 0, \qquad r = \sqrt{x^2 + y^2}, \qquad \tan\theta = \frac{y}{x}, \qquad z = z. \tag{3}$$

If $x = 0$, then $\theta = \dfrac{\pi}{2}$ if $y > 0$ and $\theta = \dfrac{3\pi}{2}$ if $y < 0$.

EXAMPLE 2 CONVERTING FROM CYLINDRICAL TO RECTANGULAR COORDINATES

Convert $P = (2, \pi/3, -5)$ from cylindrical to rectangular coordinates.

SOLUTION: $x = r\cos\theta = 2\cos(\pi/3) = 1$, $y = r\sin\theta = 2\sin(\pi/3) = \sqrt{3}$, and $z = -5$. Thus, in rectangular coordinates, $P = (1, \sqrt{3}, -5)$.

EXAMPLE 3 CONVERTING FROM RECTANGULAR TO CYLINDRICAL COORDINATES

Convert $(-3, 3, 7)$ from rectangular to cylindrical coordinates.

SOLUTION: $r = \sqrt{(-3)^2 + 3^2} = 3\sqrt{2}$, $\tan\theta = (3/-3) = -1$, so $\theta = 3\pi/4$ [since $(-3, 3)$ is in the second quadrant], and $z = 7$, so $(-3, 3, 7) = (3\sqrt{2}, 3\pi/4, 7)$ in cylindrical coordinates.

SPHERICAL COORDINATES

The second new coordinate system in space is the **spherical coordinate system**. This system is not a generalization of any system in the plane. A typical point P in space is represented as

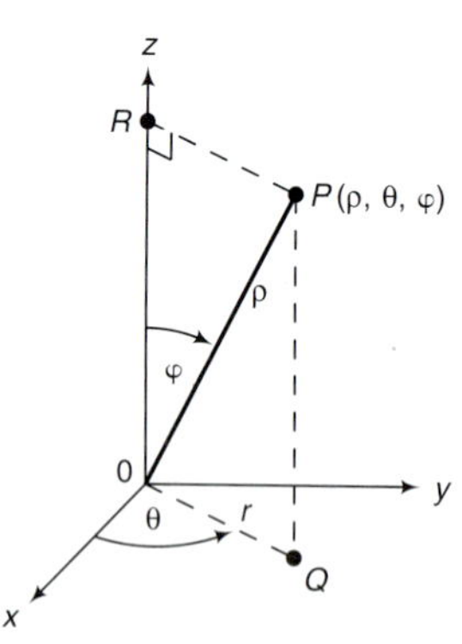

FIGURE 3
Spherical coordinates

A POINT IN SPHERICAL COORDINATES

$$P = (\rho, \theta, \varphi) \tag{4}$$

where $\rho \ge 0$, $0 \le \theta < 2\pi$, $0 \le \varphi \le \pi$, and

ρ is the (positive) distance between the point and the origin,

θ is the same as in cylindrical coordinates, and

φ is the angle between $\overrightarrow{0P}$ and the positive z-axis.

This system is illustrated in Figure 3.

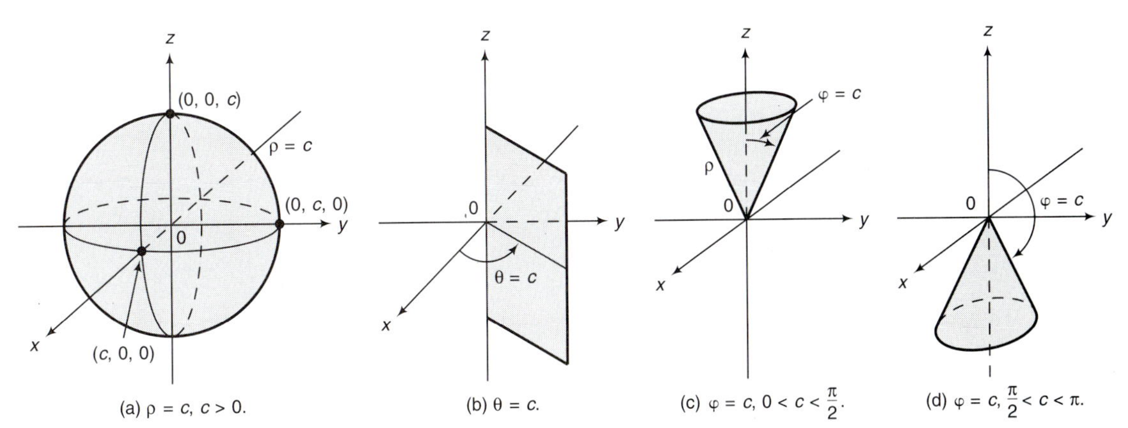

FIGURE 4
Four standard graphs in spherical coordinates

EXAMPLE 4

FOUR STANDARD GRAPHS IN SPHERICAL COORDINATES

Discuss the graphs of the equations in spherical coordinates:

a. $\rho = c,\ c > 0$ **b.** $\theta = c$
c. $\varphi = c$ **d.** $\rho_1 \leq \rho \leq \rho_2,\ \theta_1 \leq \theta \leq \theta_2,\ \varphi_1 \leq \varphi \leq \varphi_2$

SOLUTION:

a. The set of points of which $\rho = c$ is the set of points c units from the origin. This set constitutes a sphere centered at the origin with radius c and is what gives the coordinate system its name. This surface is sketched in Figure 4(a).

b. As with cylindrical coordinates, the graph is a half plane containing the z-axis. See Figure 4(b).

c. If $0 < c < \pi/2$, then the graph of $\varphi = c$ is obtained by rotating the vector $\overrightarrow{0P}$ around the z-axis. This yields the circular cone sketched in Figure 4(c). If $\pi/2 < c < \pi$, then we obtain the cone of Figure 4(d). Finally, if $c = 0$, we obtain the positive z-axis; if $c = \pi$, we obtain the negative z-axis; and if $c = \pi/2$, we obtain the xy-plane.

d. $\rho_1 \leq \rho \leq \rho_2$ gives the region between the sphere $\rho = \rho_1$ and the sphere $\rho = \rho_2$. $\theta_1 \leq \theta \leq \theta_2$ consists, as before, of the wedge-shaped region between the half planes $\theta = \theta_1$ and $\theta = \theta_2$. Finally, $\varphi_1 \leq \varphi \leq \varphi_2$ yields the region between the cones $\varphi = \varphi_1$ and $\varphi = \varphi_2$. Putting these results together, we obtain the solid (called a **spherical wedge**) sketched in Figure 5.

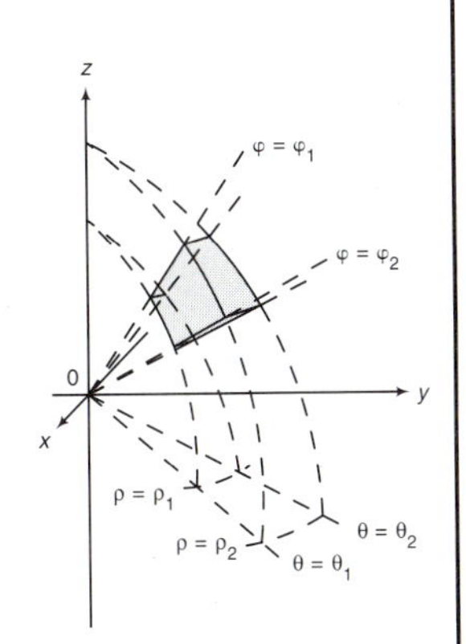

FIGURE 5
Spherical wedge

To convert from spherical to rectangular coordinates, we have, from Figure 3,

$$x = \overline{0Q} \cos \theta \quad \text{and} \quad y = \overline{0Q} \sin \theta.$$

But from Figure 3,

$$\overline{0Q} = \overline{PR} = \rho \sin \varphi, \qquad \text{so } x = \rho \sin \varphi \cos \theta \text{ and } y = \rho \sin \varphi \sin \theta.$$

Finally, from triangle $0PR$ we have $z = \rho \cos \varphi$. In sum, we have the following:

TO CHANGE POINTS FROM SPHERICAL TO RECTANGULAR COORDINATES

$$x = \rho \sin \varphi \cos \theta \qquad (5)$$

$$y = \rho \sin \varphi \sin \theta \qquad (6)$$

$$z = \rho \cos \varphi \qquad (7)$$

TO CHANGE POINTS FROM RECTANGULAR TO SPHERICAL COORDINATES

$$\rho = \sqrt{x^2 + y^2 + z^2} \qquad (8)$$

$$\varphi = \cos^{-1} \frac{z}{\rho}, \qquad 0 \le \varphi \le \pi \qquad (9)$$

$$\cos \theta = \frac{x}{\rho \sin \varphi} \quad \text{or} \quad \sin \theta = \frac{y}{\rho \sin \varphi}, \qquad 0 \le \theta \le 2\pi \qquad (10)$$

Formulas (8), (9), and (10) follow from formulas (5), (6), and (7). First, we observe, from Figure 3, that $\overline{0P} = \sqrt{x^2 + y^2 + z^2} = \rho$, so that, from (7), $\cos \varphi = \frac{z}{\rho} = \frac{z}{\sqrt{x^2 + y^2 + z^2}}$. Knowing ρ and φ, we can then calculate θ from equations (5) and (6).

EXAMPLE 5

CONVERTING FROM SPHERICAL TO RECTANGULAR COORDINATES

Convert $(4, \pi/6, \pi/4)$ from spherical to rectangular coordinates.

SOLUTION:

$$x = \rho \sin \varphi \cos \theta = 4 \sin \frac{\pi}{4} \cos \frac{\pi}{6} = 4 \frac{\sqrt{2}}{2} \frac{\sqrt{3}}{2} = \sqrt{6},$$

$$y = \rho \sin \varphi \sin \theta = 4 \sin \frac{\pi}{4} \sin \frac{\pi}{6} = 4 \frac{\sqrt{2}}{2} \frac{1}{2} = \sqrt{2}, \text{ and}$$

$$z = \rho \cos \varphi = 4 \cos \frac{\pi}{4} = 2\sqrt{2}.$$

Therefore, in rectangular coordinates $(4, \pi/6, \pi/4)$ is $(\sqrt{6}, \sqrt{2}, 2\sqrt{2})$.

EXAMPLE 6

CONVERTING FROM RECTANGULAR TO SPHERICAL COORDINATES

Convert $(1, \sqrt{3}, -2)$ from rectangular to spherical coordinates.

SOLUTION: $\rho = \sqrt{x^2 + y^2 + z^2} = \sqrt{1 + 3 + 4} = \sqrt{8} = 2\sqrt{2}$. Then from (7),

$$\cos \varphi = \frac{z}{\rho} = \frac{-2}{2\sqrt{2}} = -\frac{1}{\sqrt{2}},$$

so that $\varphi = 3\pi/4$. Finally, from (5) and (6),

$$\cos\theta = \frac{x}{\rho\sin\varphi} = \frac{1}{(2\sqrt{2})1/\sqrt{2})} = \frac{1}{2} \text{ and } \sin\theta = \frac{\sqrt{3}}{(2\sqrt{2})(1/\sqrt{2})} = \frac{\sqrt{3}}{2},$$

so that θ is in the first quadrant and $\theta = \pi/3$. Thus $(1, \sqrt{3}, -2)$ is $(2\sqrt{2}, \pi/3, 3\pi/4)$ in spherical coordinates.

We will not discuss the graphs of different kinds of surfaces given in cylindrical and spherical coordinates, because that would take us too far afield. Both coordinate systems are useful in a wide variety of physical applications. For example, points on the earth and its interior are much more easily described by using a spherical coordinate system than a rectangular system: $\{(\rho, \theta, \varphi): \rho \le a\}$, where $a =$ the radius of the earth.

PROBLEMS 1.9

SELF-QUIZ

I. True–False: Cylindrical coordinates are ''just'' polar coordinates (r, θ) with the Cartesian z-coordinate appended.

II. True–False: In spherical coordinates, the angles θ and φ have a constant sum.

III. True–False: In comparing cylindrical and spherical coordinates for a single point, $r^2 = \rho^2$.

IV. The ________ coordinate system is a generalization of polar coordinates.
a. solar **b.** spherical
c. cylindrical **d.** cubical

V. In converting from spherical coordinates to rectangular coordinates, $y =$ ________.
a. $\rho\sin\varphi$ **b.** $\rho\cos\varphi$
c. $\rho\sin\theta$ **d.** $\rho\cos\theta$
e. $\rho\sin\varphi\cos\theta$ **f.** $\rho\sin\varphi\sin\theta$
g. $\rho\cos\varphi\cos\theta$ **h.** $\rho\cos\varphi\sin\theta$

In Problems 1–10, convert from cylindrical to rectangular coordinates.

1. $(2, \pi/3, 5)$ **2.** $(1, 0, -3)$
3. $(8, 2\pi/3, 1)$ **4.** $(4, \pi/2, -7)$
5. $(3, 3\pi/4, 2)$ **6.** $(10, 5\pi/3, 1)$
7. $(10, \pi, -3)$ **8.** $(13, 3\pi/2, 4)$
9. $(7, 5\pi/4, 2)$ **10.** $(5, 11\pi/6, -5)$

In Problems 11–20, convert from rectangular to cylindrical coordinates.

11. $(1, 0, 0)$ **12.** $(0, 1, 0)$
13. $(0, 0, 1)$ **14.** $(1, 1, 2)$
15. $(-1, 1, 4)$ **16.** $(2, 2\sqrt{3}, -5)$
17. $(2\sqrt{3}, -2, 8)$ **18.** $(2, -2\sqrt{3}, 4)$
19. $(-2\sqrt{3}, -2, 1)$ **20.** $(-5, -5, -3\sqrt{2})$

In Problems 21–30, convert from spherical to rectangular coordinates.

21. $(2, 0, \pi/3)$ **22.** $(4, \pi/4, \pi/4)$
23. $(6, \pi/2, \pi/3)$ **24.** $(3, \pi/6, 5\pi/6)$
25. $(7, 7\pi/4, 3\pi/4)$ **26.** $(3, \pi/2, \pi/2)$
27. $(4, \pi/3, 2\pi/3)$ **28.** $(4, 2\pi/3, \pi/3)$
29. $(5, 11\pi/6, 5\pi/6)$ **30.** $(\sqrt{3}, 5\pi/4, \pi/4)$

In Problems 31–40, convert from rectangular to spherical coordinates.

31. $(1, 1, 0)$ **32.** $(1, 1, \sqrt{2})$
33. $(1, -1, \sqrt{2})$ **34.** $(-1, -1, \sqrt{2})$
35. $(1, -\sqrt{3}, 2)$ **36.** $(-\sqrt{3}, -1, 2)$
37. $(2, \sqrt{3}, 4)$ **38.** $(-2, \sqrt{3}, -4)$
39. $(-\sqrt{3}, -2, -4)$
40. $(-1/\sqrt{2}, -1/\sqrt{2}, -1/\sqrt{2})$

In Problems 41–54, convert the given equation into the specified coordinate system. (For instance, Problem 41 asks for a particular sphere to be described by an equation in cylindrical coordinates and by an equation in spherical coordinates.)

41. $x^2 + y^2 + z^2 = 25$, cylindrical and spherical
42. $ax + by + cz = d$, cylindrical and spherical
43. $r = 9\sin\theta$, rectangular
44. $r^2\sin 2\theta = z^3$, rectangular

45. $x^2 + y^2 - z^2 = 1$, cylindrical and spherical
46. $x^2 + 4y^2 + 4z^2 = 1$, cylindrical and spherical
47. $z = r^2$, rectangular
48. $\rho \cos \varphi = 1$, rectangular
49. $\rho^2 \sin \varphi \cos \varphi = 1$, rectangular
50. $\rho = \sin \theta \cos \varphi$, rectangular
51. $z = r^2 \sin 2\theta$, rectangular
52. $\rho = 2 \cot \theta$, rectangular
53. $x^2 + (y - 3)^2 + z^2 = 9$, spherical
54. $r = 6 \cos \theta$, spherical
55. Use the dot product of appropriate vectors to show that $\cos \varphi = z/\rho$.

ANSWERS TO SELF-QUIZ

I. True **II.** False **III.** False **IV.** c **V.** f

1.10 THE SPACE $\mathbb{R}^n$ AND THE SCALAR PRODUCT

In this chapter, we have discussed vectors in $\mathbb{R}^2$ and $\mathbb{R}^3$. Recall that a vector in $\mathbb{R}^2$ is an ordered pair that we can write as $\mathbf{v} = (x, y)$ where x and y are real numbers. Similarly, a vector in $\mathbb{R}^3$ is an ordered triple $\mathbf{v} = (x, y, z)$ where x, y, and z are real numbers.

There are many practical situations where we would not like to be limited to vectors in $\mathbb{R}^2$ or $\mathbb{R}^3$. For example, the buyer for a large department store might have to place orders for varying quantities of 13 different items. He can represent his purchase order by a vector having 13 components. Such a vector might be given as

$$\mathbf{p} = (130, 40, 12, 46, 120, 5, 8, 250, 80, 60, 75, 310, 50).$$

Thus, the buyer would order 130 units of the first item, 40 units of the second item, 12 units of the third item, and so on. Note that the order in which the numbers in **p** are written is important. If the first two numbers were interchanged, for example, the buyer would then be ordering 40 units of the first item and 130 units of the second item—quite a change from the original order.

One important thing to note about the example above is that the vector given cannot be represented geometrically—at least not in terms of anything we can draw on a piece of paper. Nevertheless, such a vector has great practical significance.

DEFINITION *n*-VECTOR

i. An ***n*-component row vector** is an *ordered* set of n numbers written as

$$(x_1, x_2, \ldots, x_n) \tag{1}$$

ii. An ***n*-component column vector** is an ordered set of n numbers written as

$$\begin{pmatrix} x_1 \\ x_2 \\ \vdots \\ x_n \end{pmatrix} \tag{2}$$

In (1) or (2), x_1 is called the **first component** of the vector, x_2 is the **second component**, and so on. In general, x_k is called the ***k*th component** of the vector.

For simplicity, we shall often refer to an n-component row vector as a **row vector** or an ***n*-vector**. Similarly, we shall use the term **column vector** (or ***n*-vector**) to denote an n-component column vector. Any vector whose entries are all zero is called a **zero vector**. ■

EXAMPLE 1 FOUR VECTORS

The following are examples of vectors:

i. (3, 6) is a row vector (or a 2-vector).

ii. $\begin{pmatrix} 2 \\ -1 \\ 5 \end{pmatrix}$ is a column vector (or a 3-vector).

iii. (2, −1, 0, 4) is a row vector (or a 4-vector).

iv. $\begin{pmatrix} 0 \\ 0 \\ 0 \\ 0 \\ 0 \end{pmatrix}$ is a column vector and a zero vector.

WARNING: The word "ordered" in the definition of a vector is essential. Two vectors with the same components written in different orders are *not* the same. Thus, for example, the row vectors (1, 2) and (2, 1) are not equal.

As before, we shall denote vectors with boldface lowercase letters such as **u**, **v**, **a**, **b**, and **c**. A zero vector is denoted by **0**.

It is time to describe some properties of vectors. Since it would be repetitive to do so first for row vectors and then for column vectors, we shall give all definitions in terms of column vectors. Similar definitions hold for row vectors.

$\mathbb{R}^n$

The components of all the vectors in this text are either real or complex numbers.[†] We use the symbol $\mathbb{R}^n$ to denote the set of all n-vectors $\begin{pmatrix} a_1 \\ a_2 \\ \vdots \\ a_n \end{pmatrix}$, where each a_i is a real number. Similarly, we use the symbol $\mathbb{C}^n$ to denote the set of all n-vectors $\begin{pmatrix} c_1 \\ c_2 \\ \vdots \\ c_n \end{pmatrix}$, where each c_i is a complex number. In this chapter we discussed the sets $\mathbb{R}^2$ and $\mathbb{R}^3$. In Chapter 8 we shall examine arbitrary sets of vectors.

DEFINITION EQUAL VECTORS

Two column (or row) vectors **a** and **b** are **equal** if and only if they have the same number of components and their corresponding components are equal. In symbols, the vectors $\mathbf{a} = \begin{pmatrix} a_1 \\ a_2 \\ \vdots \\ a_n \end{pmatrix}$ and $\mathbf{b} = \begin{pmatrix} b_1 \\ b_2 \\ \vdots \\ b_n \end{pmatrix}$ are equal if and only if $a_1 = b_1$, $a_2 = b_2, \ldots, a_n = b_n$. ■

[†] A complex number is a number of the form $a + ib$, where a and b are real numbers and $i = \sqrt{-1}$. A description of complex numbers is given in Appendix 3. We shall not encounter complex vectors again until Chapter 8. Therefore, unless otherwise stated, we assume, for the time being, that all vectors have real components.

DEFINITION ADDITION OF VECTORS

Let $\mathbf{a} = \begin{pmatrix} a_1 \\ a_2 \\ \vdots \\ a_n \end{pmatrix}$ and $\mathbf{b} = \begin{pmatrix} b_1 \\ b_2 \\ \vdots \\ b_n \end{pmatrix}$ be n-vectors. Then the sum of $\mathbf{a}$ and $\mathbf{b}$ is defined by

$$\mathbf{a} + \mathbf{b} = \begin{pmatrix} a_1 + b_1 \\ a_2 + b_2 \\ \vdots \\ a_n + b_n \end{pmatrix}. \tag{3}$$

■

EXAMPLE 2 THE SUM OF TWO VECTORS IN $\mathbb{R}^3$

$$\begin{pmatrix} 1 \\ 2 \\ 4 \end{pmatrix} + \begin{pmatrix} -6 \\ 7 \\ 5 \end{pmatrix} = \begin{pmatrix} -5 \\ 9 \\ 9 \end{pmatrix}.$$

EXAMPLE 3 THE SUM OF TWO VECTORS IN $\mathbb{R}^4$

$$\begin{pmatrix} 2 \\ -1 \\ 3 \\ -4 \end{pmatrix} + \begin{pmatrix} -2 \\ 1 \\ -3 \\ 4 \end{pmatrix} = \begin{pmatrix} 0 \\ 0 \\ 0 \\ 0 \end{pmatrix} = \mathbf{0}.$$

WARNING: It is essential that **a** and **b** have the same number of components. For example, the sum $\begin{pmatrix} 2 \\ 3 \end{pmatrix} + \begin{pmatrix} 1 \\ 2 \\ 3 \end{pmatrix}$ is not defined, since 2-vectors and 3-vectors are different kinds of objects and cannot be added together. Moreover, it is not possible to add a row and a column vector together. For example, the sum $\begin{pmatrix} 1 \\ 2 \end{pmatrix} + (3, 5)$ is *not* defined.

When dealing with vectors, we shall refer to numbers as **scalars** (which may be real or complex, depending on whether the vectors in question are real or complex).

DEFINITION SCALAR MULTIPLICATION OF VECTORS

Let $\mathbf{a} = \begin{pmatrix} a_1 \\ a_2 \\ \vdots \\ a_n \end{pmatrix}$ be a vector and α a scalar. Then the product $\alpha\mathbf{a}$ is given by

$$\alpha\mathbf{a} = \begin{pmatrix} \alpha a_1 \\ \alpha a_2 \\ \vdots \\ \alpha a_n \end{pmatrix}. \quad (4)$$

That is, to multiply a vector by a scalar, we simply multiply each component of the vector by the scalar. ■

EXAMPLE 4

A SCALAR MULTIPLE OF A VECTOR IN $\mathbb{R}^3$

$$3\begin{pmatrix} 2 \\ -1 \\ 4 \end{pmatrix} = \begin{pmatrix} 6 \\ -3 \\ 12 \end{pmatrix}.$$

NOTE: Putting the last two definitions together, we can define the difference of two vectors by

$$\mathbf{a} - \mathbf{b} = \mathbf{a} + (-1)\mathbf{b}. \quad (5)$$

This means that if $\mathbf{a} = \begin{pmatrix} a_1 \\ a_2 \\ \vdots \\ a_n \end{pmatrix}$ and $\mathbf{b} = \begin{pmatrix} b_1 \\ b_2 \\ \vdots \\ b_n \end{pmatrix}$, then $\mathbf{a} - \mathbf{b} = \begin{pmatrix} a_1 - b_1 \\ a_2 - b_2 \\ \vdots \\ a_n - b_n \end{pmatrix}$.

EXAMPLE 5

ADDITION AND SCALAR MULTIPLICATION IN $\mathbb{R}^4$

Let $\mathbf{a} = \begin{pmatrix} 4 \\ 6 \\ 1 \\ 3 \end{pmatrix}$ and $\mathbf{b} = \begin{pmatrix} -2 \\ 4 \\ -3 \\ 0 \end{pmatrix}$. Calculate $2\mathbf{a} - 3\mathbf{b}$.

SOLUTION: $2\mathbf{a} - 3\mathbf{b} = 2\begin{pmatrix} 4 \\ 6 \\ 1 \\ 3 \end{pmatrix} + (-3)\begin{pmatrix} -2 \\ 4 \\ -3 \\ 0 \end{pmatrix} = \begin{pmatrix} 8 \\ 12 \\ 2 \\ 6 \end{pmatrix} + \begin{pmatrix} 6 \\ -12 \\ 9 \\ 0 \end{pmatrix} = \begin{pmatrix} 14 \\ 0 \\ 11 \\ 6 \end{pmatrix}$.

Once we know how to add vectors and multiply them by scalars, we can prove a number of facts relating these operations. Several of these facts are given in Theorem 1. We prove parts **ii** and **iii** and leave the remaining parts as exercises (see Problems 21–23).

THEOREM 1

Let **a**, **b**, and **c** be n-vectors and let α and β be scalars. Then:

i. $\mathbf{a} + \mathbf{0} = \mathbf{a}$.

ii. $0\mathbf{a} = \mathbf{0}$.

iii. $\mathbf{a} + \mathbf{b} = \mathbf{b} + \mathbf{a}$ (commutative law).

iv. $(\mathbf{a} + \mathbf{b}) + \mathbf{c} = \mathbf{a} + (\mathbf{b} + \mathbf{c})$ (associative law).

v. $\alpha(\mathbf{a} + \mathbf{b}) = \alpha\mathbf{a} + \alpha\mathbf{b}$ (distributive law for scalar multiplication).

vi. $(\alpha + \beta)\mathbf{a} = \alpha\mathbf{a} + \beta\mathbf{a}$.

vii. $(\alpha\beta)\mathbf{a} = \alpha(\beta\mathbf{a})$.

Proof of **ii** and **iii**.

ii. If $\mathbf{a} = \begin{pmatrix} a_1 \\ a_2 \\ \vdots \\ a_n \end{pmatrix}$, then $0\mathbf{a} = 0\begin{pmatrix} a_1 \\ a_2 \\ \vdots \\ a_n \end{pmatrix} = \begin{pmatrix} 0 \cdot a_1 \\ 0 \cdot a_2 \\ \vdots \\ 0 \cdot a_n \end{pmatrix} = \begin{pmatrix} 0 \\ 0 \\ \vdots \\ 0 \end{pmatrix} = \mathbf{0}$.

iii. Let $\mathbf{b} = \begin{pmatrix} b_1 \\ b_2 \\ \vdots \\ b_n \end{pmatrix}$. Then $\mathbf{a} + \mathbf{b} = \begin{pmatrix} a_1 + b_1 \\ a_2 + b_2 \\ \vdots \\ a_n + b_n \end{pmatrix} = \begin{pmatrix} b_1 + a_1 \\ b_2 + a_2 \\ \vdots \\ b_n + a_n \end{pmatrix} = \mathbf{b} + \mathbf{a}$.

Here we used the fact that for any two numbers x and y, $x + y = y + x$ and $0 \cdot x = 0$.

NOTE: In **ii**, the zero on the left is the scalar zero (i.e., the real number 0) and the zero on the right is the zero vector. These two things are different. ■

EXAMPLE 6

ILLUSTRATION OF THE ASSOCIATIVE LAW IN $\mathbb{R}^3$

To illustrate the associative law, we note that

$$\left[\begin{pmatrix} 3 \\ 1 \\ 2 \end{pmatrix} + \begin{pmatrix} -2 \\ 4 \\ -1 \end{pmatrix}\right] + \begin{pmatrix} 6 \\ -3 \\ 5 \end{pmatrix} = \begin{pmatrix} 1 \\ 5 \\ 1 \end{pmatrix} + \begin{pmatrix} 6 \\ -3 \\ 5 \end{pmatrix} = \begin{pmatrix} 7 \\ 2 \\ 6 \end{pmatrix}$$

while

$$\begin{pmatrix} 3 \\ 1 \\ 2 \end{pmatrix} + \left[\begin{pmatrix} -2 \\ 4 \\ -1 \end{pmatrix} + \begin{pmatrix} 6 \\ -3 \\ 5 \end{pmatrix}\right] = \begin{pmatrix} 3 \\ 1 \\ 2 \end{pmatrix} + \begin{pmatrix} 4 \\ 1 \\ 4 \end{pmatrix} = \begin{pmatrix} 7 \\ 2 \\ 6 \end{pmatrix}.$$

Example 6 illustrates the importance of the associative law of vector addition, since if we wish to add three or more vectors, we can do so only by adding them two at a time. The associative law tells us that we can do this in two different ways and still come up with the same answer. If this were not the case, the sum of three or more vectors would be more difficult to define since we would have to specify whether we wanted $(\mathbf{a} + \mathbf{b}) + \mathbf{c}$ or $\mathbf{a} + (\mathbf{b} + \mathbf{c})$ to be the definition of the sum $\mathbf{a} + \mathbf{b} + \mathbf{c}$.

In Sections 1.2 and 1.4, we discussed the dot or scalar product of vectors in $\mathbb{R}^2$ and $\mathbb{R}^3$. We now generalize this notion to vectors in $\mathbb{R}^n$. We shall use the term *scalar product* when referring to this operation.

DEFINITION SCALAR PRODUCT

Let $\mathbf{a} = \begin{pmatrix} a_1 \\ a_2 \\ \vdots \\ a_n \end{pmatrix}$ and $\mathbf{b} = \begin{pmatrix} b_1 \\ b_2 \\ \vdots \\ b_n \end{pmatrix}$ be two n-vectors. Then the **scalar product** of **a** and **b**, denoted by $\mathbf{a} \cdot \mathbf{b}$, is given by

$$\mathbf{a} \cdot \mathbf{b} = a_1b_1 + a_2b_2 + \cdots + a_nb_n. \tag{6}$$

WARNING: When taking the scalar product of **a** and **b**, it is necessary that **a** and **b** have the same number of components.

We shall often be taking the scalar product of a row vector and column vector. In this case we have

THE SCALAR PRODUCT OF A ROW VECTOR AND A COLUMN VECTOR

$$(a_1, a_2, \ldots, a_n) \cdot \begin{pmatrix} b_1 \\ b_2 \\ \vdots \\ b_n \end{pmatrix} = a_1b_1 + a_2b_2 + \cdots + a_nb_n. \tag{7}$$

■

EXAMPLE 7 THE SCALAR PRODUCT OF TWO VECTORS IN $\mathbb{R}^3$

Let $\mathbf{a} = \begin{pmatrix} 1 \\ -2 \\ 3 \end{pmatrix}$ and $\mathbf{b} = \begin{pmatrix} 3 \\ -2 \\ 4 \end{pmatrix}$. Calculate $\mathbf{a} \cdot \mathbf{b}$.

SOLUTION: $\mathbf{a} \cdot \mathbf{b} = (1)(3) + (-2)(-2) + (3)(4) = 3 + 4 + 12 = 19.$

EXAMPLE 8 THE SCALAR PRODUCT OF TWO VECTORS IN $\mathbb{R}^4$

Let $\mathbf{a} = (2, -3, 4, -6)$ and $\mathbf{b} = \begin{pmatrix} 1 \\ 2 \\ 0 \\ 3 \end{pmatrix}$. Compute $\mathbf{a} \cdot \mathbf{b}$.

SOLUTION: Here $\mathbf{a} \cdot \mathbf{b} = (2)(1) + (-3)(2) + (4)(0) + (-6)(3) = 2 - 6 + 0 - 18 = -22.$

EXAMPLE 9 THE SCALAR PRODUCT OF A DEMAND VECTOR AND A PRICE VECTOR

Suppose that a manufacturer produces four items. The demand for the items is given by the demand vector $\mathbf{d} = (30, 20, 40, 10)$. The price per unit that he receives for the items is given by the price vector $\mathbf{p} = (20, 15, 18, 40)$. If he meets his demand, how much money will he receive?

SOLUTION: The demand for the first item is 30 and the manufacturer receives \$20 for each of the first item sold. He therefore receives $(30)(20) = \$600$ from the sale of the first item. By continuing this reasoning we see that the total cash received is given by $\mathbf{d} \cdot \mathbf{p}$. Thus income received $= \mathbf{d} \cdot \mathbf{p} = (30)(20) + (20)(15) + (40)(18) + (10)(40) = 600 + 300 + 720 + 400 = \$2020$.

The next result follows directly from the definition of the scalar product (see Problem 48).

THEOREM 2

Let $\mathbf{a}$, $\mathbf{b}$, and $\mathbf{c}$ be n-vectors and let α and β be scalars. Then:

i. $\mathbf{a} \cdot \mathbf{0} = 0$.

ii. $\mathbf{a} \cdot \mathbf{b} = \mathbf{b} \cdot \mathbf{a}$ (commutative law for scalar product).

iii. $\mathbf{a} \cdot (\mathbf{b} + \mathbf{c}) = \mathbf{a} \cdot \mathbf{b} + \mathbf{a} \cdot \mathbf{c}$ (distributive law for scalar product).

iv. $(\alpha\mathbf{a}) \cdot \mathbf{b} = \alpha(\mathbf{a} \cdot \mathbf{b})$. ■

Note that there is *no* associative law for the scalar product. The expression $(\mathbf{a} \cdot \mathbf{b}) \cdot \mathbf{c} = \mathbf{a} \cdot (\mathbf{b} \cdot \mathbf{c})$ does not make sense because neither side of the equation is defined. For the left side, this follows from the fact that $\mathbf{a} \cdot \mathbf{b}$ is a scalar and the scalar product of the scalar $\mathbf{a} \cdot \mathbf{b}$ and the vector $\mathbf{c}$ is not defined.

Recall that in $\mathbb{R}^2$ the vector $\mathbf{v} = (x, y)$ has length $|\mathbf{v}| = \sqrt{x^2 + y^2}$. In $\mathbb{R}^3$ the vector $|\mathbf{v}| = (x, y, z)$ has length $|\mathbf{v}| = \sqrt{x^2 + y^2 + z^2}$. Another term for length is *norm*. We now extend this concept to $\mathbb{R}^n$.

DEFINITION NORM OF A VECTOR

The **norm** of a vector $\mathbf{v}$ in $\mathbb{R}^n$, denoted by $|\mathbf{v}|$,[†] is given by

$$|\mathbf{v}| = \sqrt{\mathbf{v} \cdot \mathbf{v}}. \tag{8}$$

NOTE: If $\mathbf{v} = (a_1, a_2, \ldots, a_n)$, then $\mathbf{v} \cdot \mathbf{v} = a_1{}^2 + a_2{}^2 + \cdots + a_n{}^2$, so that

$$|\mathbf{v}| = \sqrt{a_1{}^2 + a_2{}^2 + \cdots + a_n{}^2}. \tag{9}$$

Previously, the norm of a vector in $\mathbb{R}^2$ and $\mathbb{R}^3$ was called the **magnitude** of that vector. ■

EXAMPLE 10 COMPUTING THE NORM OF A VECTOR IN $\mathbb{R}^5$

Let $\mathbf{v} = (7, -1, 2, 4, 5)$. Compute $|\mathbf{v}|$.

SOLUTION: $|\mathbf{v}| = \sqrt{7^2 + (-1)^2 + 2^2 + 4^2 + 5^2} = \sqrt{95}$.

[†] In some books the norm of $\mathbf{v}$ is denoted by $\|\mathbf{v}\|$.

PROBLEMS 1.10

SELF-QUIZ

I. If $\mathbf{u} = (1, -1, 2, 3)$ and $\mathbf{v} = (2, 0, 1, 4)$ then $\mathbf{u} \cdot \mathbf{v} =$ ________.

a. 12 **b.** 16
c. -12 **d.** $(3, -1, 3, 7)$

II. If $\mathbf{u} = \begin{pmatrix} 1 \\ 2 \end{pmatrix}$, $\mathbf{v} = \begin{pmatrix} -1 \\ 1 \end{pmatrix}$, and $\mathbf{w} = \begin{pmatrix} 4 \\ 3 \end{pmatrix}$, then $\mathbf{u} \cdot \mathbf{v} \cdot \mathbf{w} =$ ________.

a. $-4 + 6 = 2$ **b.** $\begin{pmatrix} -4 \\ 6 \end{pmatrix}$
c. -2 **d.** 1
e. is not defined.

III. If $\mathbf{x} = (2, -1, 4, 3)$, then $|\mathbf{x}| =$ ________.

a. 56 **b.** 30
c. 28 **d.** $\sqrt{30}$
e. $\sqrt{28}$

IV. Which of the following is always true if $\mathbf{u}$ and $\mathbf{v}$ are n-component row vectors and α is a real scalar?

a. $\mathbf{u} + \mathbf{v} = \mathbf{v} + \mathbf{u}$
b. $\mathbf{u} \cdot \mathbf{v} = \mathbf{v} \cdot \mathbf{u}$
c. $\mathbf{u} \cdot (\mathbf{v} + \mathbf{w}) = \mathbf{u} \cdot \mathbf{v} + \mathbf{u} \cdot \mathbf{w}$
d. $\alpha(\mathbf{u} + \mathbf{v}) = \alpha\mathbf{u} + \alpha\mathbf{v}$
e. $(\alpha\mathbf{u}) \cdot \mathbf{v} = \mathbf{u} \cdot (\alpha\mathbf{v})$

In Problems 1–10 perform the indicated computation with $\mathbf{a} = \begin{pmatrix} -3 \\ 1 \\ 4 \end{pmatrix}$, $\mathbf{b} = \begin{pmatrix} 5 \\ -4 \\ 7 \end{pmatrix}$, and $\mathbf{c} = \begin{pmatrix} 2 \\ 0 \\ -2 \end{pmatrix}$.

1. $\mathbf{a} + \mathbf{b}$ **2.** $3\mathbf{b}$
3. $-2\mathbf{c}$ **4.** $\mathbf{b} + 3\mathbf{c}$
5. $2\mathbf{a} - 5\mathbf{b}$ **6.** $-3\mathbf{b} + 2\mathbf{c}$
7. $0\mathbf{c}$ **8.** $\mathbf{a} + \mathbf{b} + \mathbf{c}$
9. $3\mathbf{a} - 2\mathbf{b} + 4\mathbf{c}$ **10.** $3\mathbf{b} - 7\mathbf{c} + 2\mathbf{a}$

In Problems 11–20 perform the indicated computation with $\mathbf{a} = (3, -1, 4, 2)$, $\mathbf{b} = (6, 0, -1, 4)$, and $\mathbf{c} = (-2, 3, 1, 5)$. Of course, it is first necessary to extend the definitions in this section to row vectors.

11. $\mathbf{a} + \mathbf{c}$ **12.** $\mathbf{b} - \mathbf{a}$
13. $4\mathbf{c}$ **14.** $-2\mathbf{b}$
15. $2\mathbf{a} - \mathbf{c}$ **16.** $4\mathbf{b} - 7\mathbf{a}$
17. $\mathbf{a} + \mathbf{b} + \mathbf{c}$ **18.** $\mathbf{c} - \mathbf{b} + 2\mathbf{a}$
19. $3\mathbf{a} - 2\mathbf{b} + 4\mathbf{c}$ **20.** $\alpha\mathbf{a} + \beta\mathbf{b} + \gamma\mathbf{c}$

21. Let $\mathbf{a} = \begin{pmatrix} a_1 \\ a_2 \\ \vdots \\ a_n \end{pmatrix}$ and let $\mathbf{0}$ denote the n-component zero column vector. Use appropriate definitions to show that $\mathbf{a} + \mathbf{0} = \mathbf{a}$ and $0\mathbf{a} = \mathbf{0}$.

22. Let $\mathbf{a} = \begin{pmatrix} a_1 \\ a_2 \\ \vdots \\ a_n \end{pmatrix}$, $\mathbf{b} = \begin{pmatrix} b_1 \\ b_2 \\ \vdots \\ b_n \end{pmatrix}$, and $\mathbf{c} = \begin{pmatrix} c_1 \\ c_2 \\ \vdots \\ c_n \end{pmatrix}$.

Compute $(\mathbf{a} + \mathbf{b}) + \mathbf{c}$ and $\mathbf{a} + (\mathbf{b} + \mathbf{c})$ and show that they are equal.

23. Let $\mathbf{a}$ and $\mathbf{b}$ be as in Problem 22 and let α and β be scalars. Compute $\alpha(\mathbf{a} + \mathbf{b})$ and $\alpha\mathbf{a} + \alpha\mathbf{b}$ and show that they are equal. Similarly, compute $(\alpha + \beta)\mathbf{a}$ and $\alpha\mathbf{a} + \beta\mathbf{a}$ and show that they are equal. Finally, show that $(\alpha\beta)\mathbf{a} = \alpha(\beta\mathbf{a})$.

24. Find numbers α, β, and γ such that $(2, -1, 4) + (\alpha, \beta, \gamma) = \mathbf{0}$.

25. In the manufacture of a certain product, four raw materials are needed. The vector $\mathbf{d} = \begin{pmatrix} d_1 \\ d_2 \\ d_3 \\ d_4 \end{pmatrix}$ represents a given factory's demand for each of the four raw materials to produce one unit of its product. If $\mathbf{d}_1$ is the demand vector for factory 1 and $\mathbf{d}_2$ is the demand vector for factory 2, what is represented by the vectors $\mathbf{d}_1 + \mathbf{d}_2$ and $2\mathbf{d}_1$?

26. Let $\mathbf{a} = \begin{pmatrix} 1 \\ 3 \\ 2 \end{pmatrix}$, $\mathbf{b} = \begin{pmatrix} -2 \\ 4 \\ 1 \end{pmatrix}$, and $\mathbf{c} = \begin{pmatrix} 0 \\ 1 \\ 4 \end{pmatrix}$.

Find a vector $\mathbf{v}$ such that $2\mathbf{a} - \mathbf{b} + 3\mathbf{v} = 4\mathbf{c}$.

In Problems 27–33 calculate the scalar product of the two vectors.

27. $\begin{pmatrix} 2 \\ 3 \\ -5 \end{pmatrix}$; $\begin{pmatrix} 3 \\ 0 \\ 4 \end{pmatrix}$

28. $(1, 2, -1, 0)$; $(3, -7, 4, -2)$

29. $\begin{pmatrix} 5 \\ 7 \end{pmatrix}$; $\begin{pmatrix} 3 \\ -2 \end{pmatrix}$

30. $(8, 3, 1)$; $(7, -4, 3)$

31. (a, b); (c, d)

32. $\begin{pmatrix} x \\ y \\ z \end{pmatrix}; \begin{pmatrix} y \\ z \\ x \end{pmatrix}$

33. $(-1, -3, 4, 5); (-1, -3, 4, 5)$

34. Let $\mathbf{a}$ be an n-vector. Show that $\mathbf{a} \cdot \mathbf{a} \geq 0$.

35. Find conditions on a vector $\mathbf{a}$ such that $\mathbf{a} \cdot \mathbf{a} = 0$.

In Problems 36–40 perform the indicated computation with $\mathbf{a} = \begin{pmatrix} 1 \\ -2 \\ 4 \end{pmatrix}$, $\mathbf{b} = \begin{pmatrix} 0 \\ -3 \\ -7 \end{pmatrix}$, and $\mathbf{c} = \begin{pmatrix} 4 \\ -1 \\ 5 \end{pmatrix}$.

36. $(2\mathbf{a}) \cdot (3\mathbf{b})$

37. $\mathbf{a} \cdot (\mathbf{b} + \mathbf{c})$

38. $\mathbf{c} \cdot (\mathbf{a} - \mathbf{b})$

39. $(2\mathbf{b}) \cdot (3\mathbf{c} - 5\mathbf{a})$

40. $(\mathbf{a} - \mathbf{c}) \cdot (3\mathbf{b} - 4\mathbf{a})$

Two vectors $\mathbf{a}$ and $\mathbf{b}$ are said to be **orthogonal** if $\mathbf{a} \cdot \mathbf{b} = 0$. In Problems 41–45 determine which pairs of vectors are orthogonal.

41. $\begin{pmatrix} 2 \\ -3 \end{pmatrix}; \begin{pmatrix} 3 \\ 2 \end{pmatrix}$

42. $\begin{pmatrix} 1 \\ 4 \\ -7 \end{pmatrix}; \begin{pmatrix} 2 \\ 3 \\ 2 \end{pmatrix}$

43. $\begin{pmatrix} 5 \\ 4 \\ 6 \\ -1 \end{pmatrix}; \begin{pmatrix} 3 \\ -4 \\ 1 \\ 5 \end{pmatrix}$

44. $(1, 0, 1, 0); (0, 1, 0, 1)$

45. $\begin{pmatrix} a \\ 0 \\ b \\ 0 \\ c \end{pmatrix}; \begin{pmatrix} 0 \\ d \\ 0 \\ e \\ 0 \end{pmatrix}$

46. Determine a number α such that $(1, -2, 3, 5)$ is orthogonal to $(-4, \alpha, 6, -1)$.

47. Determine all numbers α and β such that the vectors $\begin{pmatrix} 1 \\ -\alpha \\ 2 \\ 3 \end{pmatrix}$ and $\begin{pmatrix} 4 \\ 5 \\ -2\beta \\ 7 \end{pmatrix}$ are orthogonal.

48. Using the definition of the scalar product, prove Theorem 2.

In Problems 49–53 compute the norm of the given vector.

49. $\begin{pmatrix} 1 \\ 2 \\ 3 \\ 4 \end{pmatrix}$

50. $(0, 0, 1, 0, 0, 0)$

51. (a, b, c, d, e)

52. $\begin{pmatrix} 1 \\ \frac{1}{2} \\ \frac{1}{3} \\ \frac{1}{4} \end{pmatrix}$

53. $(1, 1, 1, 1, 1, 1, 1, 1, 1, 1)$

54. Let α be a scalar and let $\mathbf{v}$ be a vector in $\mathbb{R}^n$. Show that $|\alpha\mathbf{v}| = |\alpha||\mathbf{v}|$.

55. Compute the scalar product of $(1, 3, -1, 2, 4)$ and $(5, 0, 1, 2, -3)$.

56. Find a number α such that $\begin{pmatrix} 2 \\ -1 \\ 4 \\ 6 \end{pmatrix}$ and $\begin{pmatrix} 1 \\ 5 \\ \alpha \\ 4 \end{pmatrix}$ are orthogonal.

ANSWERS TO SELF-QUIZ

I. b **II.** e **III.** d **IV.** all of them

CHAPTER 1 SUMMARY OUTLINE

- The **directed line segment** extending from P to Q in $\mathbb{R}^2$ or $\mathbb{R}^3$ denoted by $\overrightarrow{PQ}$, is the straight line segment that extends from P to Q.
- Two directed line segments in $\mathbb{R}^2$ or $\mathbb{R}^3$ are equivalent if they have the same magnitude (length) and direction.
- **Geometric Definition of a Vector** A **vector** in $\mathbb{R}^2[\mathbb{R}^3]$ is the set of all directed line segments in $\mathbb{R}^2[\mathbb{R}^3]$ equivalent to a given directed line segment. One representation of a vector has its initial point at the origin and is denoted by $\overrightarrow{0P}$.
- **Algebraic Definition of a Vector** A **vector** $\mathbf{v}$ in the xy-

plane ($\mathbb{R}^2$) is an ordered pair of real numbers (a, b). The numbers a and b are called the **components** of the vector $\mathbf{v}$. The **zero vector** is the vector $(0, 0)$. In $\mathbb{R}^3$, a vector $\mathbf{v}$ is an ordered triple of real numbers (a, b, c). The **zero vector** in $\mathbb{R}^3$ is the vector $(0, 0, 0)$.

- The geometric and algebraic definitions of a vector in $\mathbb{R}^2[\mathbb{R}^3]$ are related in the following way: If $\mathbf{v} = (a, b)$ $[(a, b, c)]$, then one representation of $\mathbf{v}$ is $\overrightarrow{0R}$ where $R = (a, b)[R = (a, b, c)]$.

- If $\mathbf{v} = (a, b)$, then the **magnitude of** $\mathbf{v}$, denoted by $|\mathbf{v}|$, is given by $|\mathbf{v}| = \sqrt{a^2 + b^2}$. If $\mathbf{v} = (a, b, c)$, then $|\mathbf{v}| = \sqrt{a^2 + b^2 + c^2}$.

- If $\mathbf{v}$ is a vector in $\mathbb{R}^2$, then the **direction** of $\mathbf{v}$ is the angle in $[0, 2\pi)$ that any representation of $\mathbf{v}$ makes with the positive x-axis.

- **Triangle Inequality** In $\mathbb{R}^2$ or $\mathbb{R}^3$,

$$|\mathbf{u} + \mathbf{v}| \le |\mathbf{u}| + |\mathbf{v}|.$$

- In $\mathbb{R}^2$, let $\mathbf{i} = (1, 0)$ and $\mathbf{j} = (0, 1)$. Then $\mathbf{v} = (a, b)$ can be written

$$\mathbf{v} = a\mathbf{i} + b\mathbf{j}$$

In $\mathbb{R}^3$, let $\mathbf{i} = (1, 0, 0)$, $\mathbf{j} = (0, 1, 0)$, and $\mathbf{k} = (0, 0, 1)$. Then $\mathbf{v} = (a, b, c)$ can be written

$$\mathbf{v} = a\mathbf{i} + b\mathbf{j} + c\mathbf{k}.$$

- A **unit vector** $\mathbf{u}$ in $\mathbb{R}^2$ or $\mathbb{R}^3$ is a vector which satisfies $|\mathbf{u}| = 1$. In $\mathbb{R}^2$, a unit vector can be written

$$\mathbf{u} = (\cos\theta)\mathbf{i} + (\sin\theta)\mathbf{j}$$

where θ is the direction of $\mathbf{u}$.

- Let $\mathbf{u} = (a_1, b_1)$ and $\mathbf{v} = (a_2, b_2)$. Then the **scalar product** or **dot product** of $\mathbf{u}$ and $\mathbf{v}$, written $\mathbf{u} \cdot \mathbf{v}$, is given by

$$\mathbf{u} \cdot \mathbf{v} = a_1a_2 + b_1b_2.$$

If $\mathbf{u} = (a_1, b_1, c_1)$ and $\mathbf{v} = (a_2, b_2, c_2)$, then

$$\mathbf{u} \cdot \mathbf{v} = a_1a_2 + b_1b_2 + c_1c_2.$$

- **Properties of the Dot Product in $\mathbb{R}^2$ or $\mathbb{R}^3$** For any vectors $\mathbf{u}$, $\mathbf{v}$, $\mathbf{w}$, and scalar α,

  i. $\mathbf{u} \cdot \mathbf{v} = \mathbf{v} \cdot \mathbf{u}$
  ii. $(\mathbf{u} + \mathbf{v}) \cdot \mathbf{w} = \mathbf{u} \cdot \mathbf{w} + \mathbf{v} \cdot \mathbf{w}$
  iii. $(\alpha\mathbf{u}) \cdot \mathbf{v} = \alpha(\mathbf{u} \cdot \mathbf{v})$
  iv. $\mathbf{u} \cdot \mathbf{u} \ge 0$ and $\mathbf{u} \cdot \mathbf{u} = 0$ if and only if $\mathbf{u} = \mathbf{0}$
  v. $|\mathbf{u}|^2 = \mathbf{u} \cdot \mathbf{u}$.

- The angle φ between two nonzero vectors $\mathbf{u}$ and $\mathbf{v}$ in $\mathbb{R}^2$ or $\mathbb{R}^3$ is the unique number in $[0, \pi]$ that satisfies

$$\cos\varphi = \frac{\mathbf{u} \cdot \mathbf{v}}{|\mathbf{u}||\mathbf{v}|}$$

- Two vectors in $\mathbb{R}^2$ or $\mathbb{R}^3$ are **parallel** if the angle between them is 0 or π. $\mathbf{u}$ and $\mathbf{v}$ are parallel if and only if $\mathbf{v} = \alpha\mathbf{u}$ for some scalar α.

- Two vectors in $\mathbb{R}^2$ or $\mathbb{R}^3$ are **orthogonal** if the angle between them is $\frac{\pi}{2}$. $\mathbf{u}$ and $\mathbf{v}$ are orthogonal if and only if $\mathbf{u} \cdot \mathbf{v} = 0$.

- Let $\mathbf{u}$ and $\mathbf{v}$ be two nonzero vectors in $\mathbb{R}^2$ or $\mathbb{R}^3$. Then the **projection** of $\mathbf{u}$ on $\mathbf{v}$ is a vector, denoted by $\text{Proj}_\mathbf{v}\,\mathbf{u}$, which is defined by

$$\text{Proj}_\mathbf{v}\,\mathbf{u} = \frac{\mathbf{u} \cdot \mathbf{v}}{|\mathbf{v}|^2}\mathbf{v}.$$

The vector $\frac{\mathbf{u} \cdot \mathbf{v}}{|\mathbf{v}|}$ is called the **component** of $\mathbf{u}$ in the direction $\mathbf{v}$.

- $\text{Proj}_\mathbf{v}\,\mathbf{u}$ is parallel to $\mathbf{v}$ and $\mathbf{u} - \text{Proj}_\mathbf{v}\,\mathbf{u}$ is orthogonal to $\mathbf{v}$.
- The **direction** of a vector $\mathbf{v}$ in $\mathbb{R}^3$ is the unit vector

$$\mathbf{u} = \frac{\mathbf{v}}{|\mathbf{v}|}.$$

- If $\mathbf{v} = (a, b, c)$, then $\cos\alpha = \frac{a}{|\mathbf{v}|}$, $\cos\beta = \frac{b}{|\mathbf{v}|}$, and $\cos\gamma = \frac{c}{|\mathbf{v}|}$ are called the **direction cosines of** $\mathbf{v}$.

- Let $\mathbf{u} = a_1\mathbf{i} + b_1\mathbf{j} + c_1\mathbf{k}$ and $\mathbf{v} = a_2\mathbf{i} + b_2\mathbf{j} + c_2\mathbf{k}$. Then the **cross product** or **vector product** of $\mathbf{u}$ and $\mathbf{v}$, denoted by $\mathbf{u} \times \mathbf{v}$, is given by

$$\mathbf{u} \times \mathbf{v} = \begin{vmatrix} \mathbf{i} & \mathbf{j} & \mathbf{k} \\ a_1 & b_1 & c_1 \\ a_2 & b_2 & c_2 \end{vmatrix}$$

- **Properties of the Cross Product**

  i. $\mathbf{u} \times \mathbf{0} = \mathbf{0} \times \mathbf{u} = \mathbf{0}$
  ii. $\mathbf{u} \times \mathbf{v} = -\mathbf{v} \times \mathbf{u}$
  iii. $(\alpha\mathbf{u}) \times \mathbf{v} = \alpha(\mathbf{u} \times \mathbf{v})$
  iv. $\mathbf{u} \times (\mathbf{v} + \mathbf{w}) = (\mathbf{u} \times \mathbf{v}) + (\mathbf{u} \times \mathbf{w})$
  v. $(\mathbf{u} \times \mathbf{v}) \cdot \mathbf{w} = \mathbf{u} \cdot (\mathbf{v} \times \mathbf{w})$ (the **scalar triple product**)
  vi. $\mathbf{u} \times \mathbf{v}$ is orthogonal to both $\mathbf{u}$ and $\mathbf{v}$
  vii. If $\mathbf{u}$ and $\mathbf{v}$ are parallel, then $\mathbf{u} \times \mathbf{v} = \mathbf{0}$

- If φ is the angle between $\mathbf{u}$ and $\mathbf{v}$, then $|\mathbf{u} \times \mathbf{v}| = |\mathbf{u}||\mathbf{v}| \sin\varphi$ = area of the parallelogram with sides $\mathbf{u}$ and $\mathbf{v}$.
- Let $P = (x_1, y_1, z_1)$ and $Q = (x_2, y_2, z_2)$ be two points on a line L in $\mathbb{R}^3$. Let $\mathbf{v} = (x_2 - x_1)\mathbf{i} + (y_2 - y_1)\mathbf{j} + (z_2 - z_1)\mathbf{k}$ and let $a = x_2 - x_1$, $b = y_2 - y_1$, and $c = z_2 - z_1$. Then

  vector equation of the line: $\overrightarrow{0R} = \overrightarrow{0P} + t\mathbf{v}$

parametric equations of the line: $x = x_1 + at$
$y = y_1 + bt$
$z = z_1 + ct$

symmetric equations of the line:

$$\frac{x - x_1}{a} = \frac{y - y_1}{b} = \frac{z - z_1}{c},$$

if a, b, and c are nonzero.

- Let P be a point in $\mathbb{R}^3$ and let $\mathbf{N}$ be a given nonzero vector. Then the set of all points Q for which $\overrightarrow{PQ} \cdot \mathbf{N} = 0$ constitutes a plane in $\mathbb{R}^3$. The vector $\mathbf{N}$ is called the **normal vector** of the plane.
- If $\mathbf{N} = a\mathbf{i} + b\mathbf{j} + c\mathbf{k}$ and $P = (x_0, y_0, z_0)$, then the equation of the plane can be written

$$ax + by + cz = d,$$

where

$$d = ax_0 + by_0 + cz_0 = \overrightarrow{0P} \cdot \mathbf{N}$$

- The ***xy*-plane** has the equation $z = 0$; the ***xz*-plane** has the equation $y = 0$; the ***yz*-plane** has the equation $x = 0$.
- Two planes are **parallel** if their normal vectors are parallel. If two planes are not parallel, then they intersect in a straight line.
- A **surface** in space is defined as the set of points in $\mathbb{R}^3$ satisfying the equation $F(x, y, z) = 0$ for F a continuous function.
- **Cylinder** Let a plane curve C and a line L not in the plane of C be given. A **cylinder** is the surface generated when a line parallel to L moves around C, remaining parallel to L. The line L is called the **generatrix** of the cylinder, and the curve C is called its **directrix**.
- **Quadric Surface** A **quadric surface** in $\mathbb{R}^3$ is the graph of a second-degree equation in the variables x, y, and z. Such an equation takes the form

$$Ax^2 + By^2 + Cz^2 + Dxy + Exz + Fyz + Gx + Hy + Jz + K = 0.$$

Here is a list of the eleven quadric surfaces discussed in Section 1.8:

1. sphere
2. right circular cylinder
3. parabolic cylinder
4. elliptic cylinder
5. hyperbolic cylinder
6. ellipsoid
7. hyperboloid of one sheet
8. hyperboloid of two sheets
9. elliptic paraboloid
10. hyperbolic paraboloid
11. elliptic cone

Equations and graphs of these surfaces are given in Table 1 on pages 62 and 63.

- **Cylindrical Coordinate System** In the **cylindrical coordinate system,** a point P is given by

$$P = (r, \theta, z),$$

where $r \geq 0$, $0 \leq \theta < 2\pi$, r and θ are polar coordinates of the projection of P onto the xy-plane, called the **polar plane**, and z is the distance (measured in the positive direction) of this plane from P.

- **To Change from Polar to Rectangular Coordinates**

$$x = r\cos\theta, \qquad y = r\sin\theta, \qquad z = z$$

- **To Change from Rectangular to Polar Coordinates**

$$r = \sqrt{x^2 + y^2}, \qquad \tan\theta = \frac{y}{x}, \qquad z = z$$

- **Spherical Coordinate System** In the **spherical coordinate system**, a typical point P in space is represented as

$$P = (\rho, \theta, \varphi)$$

where $\rho \geq 0$, $0 \leq \theta < 2\pi$, $0 \leq \varphi \leq \pi$, and

ρ is the (positive) distance between the point and the origin.

θ is the same as in cylindrical coordinates.

φ is the angle between $\overrightarrow{0P}$ and the positive z-axis.

- **To Change Points from Spherical to Rectangular Coordinates**

$$x = \rho\sin\varphi\cos\theta$$
$$y = \rho\sin\varphi\sin\theta$$
$$z = \rho\cos\varphi$$

- **To Change Points from Rectangular to Spherical Coordinates**

$$\rho = \sqrt{x^2 + y^2 + z^2}$$

$$\varphi = \cos^{-1}\frac{z}{\rho}, \qquad 0 \leq \varphi \leq \pi$$

$$\cos\theta = \frac{x}{\rho\sin\varphi} \quad \text{or} \quad \sin\theta = \frac{y}{\rho\sin\varphi},$$
$$0 \leq \theta < 2\pi$$

CHAPTER 1 REVIEW EXERCISES

1. Let $\mathbf{u} = (2, 1)$ and $\mathbf{v} = (-3, 4)$. Calculate the following:
   a. $5\mathbf{u}$ b. $\mathbf{u} - \mathbf{v}$
   c. $-8\mathbf{u} + 5\mathbf{v}$ d. $3\mathbf{u} + 2\mathbf{v}$
2. Let $\mathbf{u} = -4\mathbf{i} + \mathbf{j}$ and $\mathbf{v} = -3\mathbf{i} - 4\mathbf{j}$. Calculate the following:
   a. $-3\mathbf{v}$ b. $\mathbf{u} + \mathbf{v}$
   c. $4\mathbf{u} + \mathbf{v}$ d. $3\mathbf{u} - 6\mathbf{v}$
3. Let $\mathbf{u} = (1, -3, 5)$ and $\mathbf{v} = (5, 0, 2)$. Calculate the following:
   a. $3\mathbf{v}$ b. $\mathbf{v} - \mathbf{u}$
   c. $2\mathbf{u} - 2\mathbf{v}$ d. $5\mathbf{u} - \mathbf{v}$
4. Let $\mathbf{u} = \mathbf{i} - 2\mathbf{j} + 3\mathbf{k}$ and $\mathbf{v} = -3\mathbf{i} + 2\mathbf{j} + 5\mathbf{k}$. Calculate the following:
   a. $2\mathbf{v}$ b. $\mathbf{u} + \mathbf{v}$
   c. $3\mathbf{u} + 5\mathbf{v}$ d. $5\mathbf{u} - 3\mathbf{v}$

In Problems 5–14, find the magnitude and direction of the given vector.

5. $\mathbf{u} = (3, 3)$
6. $\mathbf{u} = (2, -2\sqrt{3})$
7. $\mathbf{u} = -12\mathbf{i} - 12\mathbf{j}$
8. $\mathbf{u} = \mathbf{i} + 4\mathbf{j}$
9. $\mathbf{w} = (\sqrt{3}, 1, 0)$
10. $\mathbf{w} = (3, -12, -5)$
11. $\mathbf{w} = 7\mathbf{i} - 3\mathbf{j}$ (in $\mathbb{R}^3$)
12. $\mathbf{w} = 2\mathbf{i} - \mathbf{k}$
13. $\mathbf{w} = -3\mathbf{i} + 4\mathbf{j} + 5\mathbf{k}$
14. $\mathbf{w} = \sqrt{6}\mathbf{i} + 3\mathbf{j} - \sqrt{10}\mathbf{k}$

In Problems 15–18, sketch the two given points and then calculate the distance between them.

15. $(3, 1, 2)$; $(-1, -3, -4)$
16. $(4, -1, 7)$; $(-5, 1, 3)$
17. $(-2, 4, -8)$; $(0, 0, 6)$
18. $(2, -7, 0)$; $(0, 5, -8)$

In Problems 19–26, write the vector $\mathbf{u} = a\mathbf{i} + b\mathbf{j}$ or $\mathbf{w} = \alpha\mathbf{i} + \beta\mathbf{j} + \gamma\mathbf{k}$ that is represented by $\overrightarrow{PQ}$.

19. $P = (2, 3)$; $Q = (4, 5)$
20. $P = (1, -2)$; $Q = (7, 12)$
21. $P = (-1, -6)$; $Q = (3, -4)$
22. $P = (-1, 3)$; $Q = (3, -1)$
23. $P = (2, -3, 0)$; $Q = (-2, 3, 0)$
24. $P = (0, 8, 9)$; $Q = (0, -3, 9)$
25. $P = (-3, 5, 12)$; $Q = (-1, 1, 8)$
26. $P = (4, -5, 6)$; $Q = (6, -5, 4)$

In Problems 27–36, find a unit vector whose direction is opposite to that of the given vector $\mathbf{v}$.

27. $\mathbf{v} = \mathbf{i} + \mathbf{j}$
28. $\mathbf{v} = 5\mathbf{i} + 2\mathbf{j}$
29. $\mathbf{v} = 10\mathbf{i} - 7\mathbf{j}$
30. $\mathbf{v} = a\mathbf{i} - a\mathbf{j}$
31. $\mathbf{v} = 3\mathbf{j} + 11\mathbf{k}$
32. $\mathbf{v} = 4\mathbf{i} - 3\mathbf{k}$
33. $\mathbf{v} = \mathbf{i} - 2\mathbf{j} - 3\mathbf{k}$
34. $\mathbf{v} = -4\mathbf{i} + \mathbf{j} + 6\mathbf{k}$
35. $\mathbf{v} = \overrightarrow{PQ}$ where $P = (3, -1)$ and $Q = (-4, 1)$
36. $\mathbf{v} = \overrightarrow{PQ}$ where $P = (1, -3, 0)$ and $Q = (-7, 1, -4)$

In Problems 37–44, find a vector $\mathbf{v}$ having the given magnitude and direction.

37. $|\mathbf{v}| = 2$; $\theta = \pi/3$
38. $|\mathbf{v}| = 1$; $\theta = \pi/2$
39. $|\mathbf{v}| = 4$; $\theta = \pi$
40. $|\mathbf{v}| = 7$; $\theta = 5\pi/6$
41. $|\mathbf{v}| = 5$; direction $= (0.6, 0.8, 0)$
42. $|\mathbf{v}| = 2$; direction $= (0, 1/2, \sqrt{3}/2)$
43. $|\mathbf{v}| = \sqrt{2}$; direction $= (1/2, -1/\sqrt{2}, -1/2)$
44. $|\mathbf{v}| = 12$; direction $= (\cos \pi/3, \cos 4\pi/3, \cos \pi/4)$

In Problems 45–52, calculate the dot product of the two given vectors and then calculate the cosine of the angle between them.

45. $\mathbf{u} = \mathbf{i} - \mathbf{j}$; $\mathbf{v} = \mathbf{i} + 2\mathbf{j}$
46. $\mathbf{u} = -4\mathbf{i}$; $\mathbf{v} = 11\mathbf{j}$
47. $\mathbf{u} = 4\mathbf{i} - 7\mathbf{j}$; $\mathbf{v} = 5\mathbf{i} + 6\mathbf{j}$
48. $\mathbf{u} = -\mathbf{i} - 2\mathbf{j}$; $\mathbf{v} = 4\mathbf{i} + 5\mathbf{j}$
49. $\mathbf{u} = \mathbf{i} + 2\mathbf{k}$; $\mathbf{v} = 2\mathbf{j} - \mathbf{k}$
50. $\mathbf{u} = -3\mathbf{i} + \mathbf{k}$; $\mathbf{v} = \mathbf{i} - \mathbf{j} + 2\mathbf{k}$
51. $\mathbf{u} = \mathbf{i} + \mathbf{j} + \mathbf{k}$; $\mathbf{v} = 2\mathbf{i} - \mathbf{j} - \mathbf{k}$
52. $\mathbf{u} = \mathbf{i} + \mathbf{j} - \mathbf{k}$; $\mathbf{v} = -\mathbf{i} - \mathbf{j} + \mathbf{k}$

In Problems 53–60, determine whether the given vectors are orthogonal, parallel, or neither, then sketch each pair of vectors.

53. $\mathbf{u} = 2\mathbf{i} - 6\mathbf{j}$; $\mathbf{v} = -\mathbf{i} + 3\mathbf{j}$
54. $\mathbf{u} = -7\mathbf{i} - 7\mathbf{j}$; $\mathbf{v} = -\mathbf{i} + \mathbf{j}$
55. $\mathbf{u} = 4\mathbf{i} - 5\mathbf{j}$; $\mathbf{v} = 5\mathbf{i} - 4\mathbf{j}$
56. $\mathbf{u} = 4\mathbf{i} + 5\mathbf{j}$; $\mathbf{v} = -5\mathbf{i} + 4\mathbf{j}$
57. $\mathbf{u} = 3\mathbf{i} - 4\mathbf{j} + 5\mathbf{k}$; $\mathbf{v} = -5\mathbf{i} + 4\mathbf{j} + 3\mathbf{k}$
58. $\mathbf{u} = 3\mathbf{i} - 4\mathbf{j} + 5\mathbf{k}$; $\mathbf{v} = -5\mathbf{i} + 3\mathbf{k}$
59. $\mathbf{u} = 2\mathbf{i} + \mathbf{j} - 3\mathbf{k}$; $\mathbf{v} = 6\mathbf{i} + 3\mathbf{j} + 9\mathbf{k}$
60. $\mathbf{u} = -3\mathbf{i} + 2\mathbf{j} + 5\mathbf{k}$; $\mathbf{v} = 6\mathbf{i} - 4\mathbf{j} - 10\mathbf{k}$

In Problems 61–68, calculate $\text{Proj}_{\mathbf{v}}\,\mathbf{u}$.

61. $\mathbf{u} = 14\mathbf{i}$; $\mathbf{v} = \mathbf{i} + \mathbf{j}$
62. $\mathbf{u} = 14\mathbf{i}$; $\mathbf{v} = \mathbf{i} - \mathbf{j}$
63. $\mathbf{u} = 3\mathbf{i} - 2\mathbf{j}$; $\mathbf{v} = 3\mathbf{i} + 2\mathbf{j}$
64. $\mathbf{u} = 3\mathbf{i} + 2\mathbf{j}$; $\mathbf{v} = \mathbf{i} - 3\mathbf{j}$
65. $\mathbf{u} = 2\mathbf{i} - 5\mathbf{j} + \mathbf{k}$; $\mathbf{v} = -3\mathbf{i} - 7\mathbf{j} + \mathbf{k}$
66. $\mathbf{u} = 4\mathbf{i} - 5\mathbf{j} - \mathbf{k}$; $\mathbf{v} = -3\mathbf{i} - \mathbf{j} + \mathbf{k}$
67. $\mathbf{u} = \mathbf{i} - 2\mathbf{j} + 3\mathbf{k}$; $\mathbf{v} = 2\mathbf{i} - 4\mathbf{j} + \mathbf{k}$
68. $\mathbf{u} = 2\mathbf{i} - 4\mathbf{j} + \mathbf{k}$; $\mathbf{v} = -3\mathbf{i} + 2\mathbf{j} + 5\mathbf{k}$

In Problems 69–72, calculate $\mathbf{u} \times \mathbf{v}$.

69. $\mathbf{u} = 7\mathbf{j}$; $\mathbf{v} = \mathbf{i} - \mathbf{k}$
70. $\mathbf{u} = 3\mathbf{i} - \mathbf{j}$; $\mathbf{v} = 2\mathbf{i} + 4\mathbf{k}$
71. $\mathbf{u} = 4\mathbf{i} - \mathbf{j} + 7\mathbf{k}$; $\mathbf{v} = -7\mathbf{i} + \mathbf{j} - 2\mathbf{k}$
72. $\mathbf{u} = -2\mathbf{i} + 3\mathbf{j} - 4\mathbf{k}$; $\mathbf{v} = -3\mathbf{i} + \mathbf{j} - 10\mathbf{k}$

In Problems 73–76, find a vector equation, parametric equations, and symmetric equations for the specified line.

73. containing $(3, -1, 4)$ and $(-1, 6, 2)$

74. containing $(-4, 1, 0)$ and $(3, 0, 7)$

75. containing $(3, 1, 2)$ and parallel to $3\mathbf{i} - \mathbf{j} - \mathbf{k}$

76. containing $(1, 2, -3)$ and parallel to the line $(x + 1)/5 = (y - 2)/(-3) = (z - 4)/2$.

In Problems 77–80, find an equation for the plane containing the given point and having the given vector as a normal vector.

77. $P = (-1, 1, 1)$; $\mathbf{N} = \mathbf{j} - \mathbf{k}$

78. $P = (1, 3, -2)$; $\mathbf{N} = \mathbf{i} + \mathbf{k}$

79. $P = (1, -4, 6)$; $\mathbf{N} = 2\mathbf{j} - 3\mathbf{k}$

80. $P = (-4, 1, 6)$; $\mathbf{N} = 2\mathbf{i} - 3\mathbf{j} + 5\mathbf{k}$

In Problems 81–88, convert from one coordinate system to another as specified.

81. $(3, \pi/6, -1)$; cylindrical to rectangular

82. $(2, 2\pi/3, 4)$; cylindrical to rectangular

83. $(2, 2, -4)$; rectangular to cylindrical

84. $(-2, 2\sqrt{3}, -4)$; rectangular to cylindrical

85. $(3, \pi/3, \pi/4)$; spherical to rectangular

86. $(2, 7\pi/3, 2\pi/3)$; spherical to rectangular

87. $(-1, 1, -\sqrt{2})$; rectangular to spherical

88. $(2, -\sqrt{3}, 4)$; rectangular to spherical

89. Convert the equation $x^2 + y^2 + z^2 = 25$ into cylindrical and spherical coordinates.

90. Convert the equation $r^2 \cos 2\theta = z^3$ into rectangular coordinates.

91. Convert the equation $x^2 - y^2 + z^2 = 1$ into cylindrical and spherical coordinates.

92. Convert the equation $\rho \sin \varphi = 1$ into rectangular coordinates.

In Problems 93–96, draw a sketch of the specified cylinder. The directrix curve C is given; the generatrix line L is the axis of the variable missing from the equation.

93. $y = 3 - 5x$ **94.** $x = \cos y$

95. $y = z^2$ **96.** $z = \sqrt[3]{x}$

97. Find two unit vectors in $\mathbb{R}^2$ that are orthogonal to $\mathbf{u} = \mathbf{i} - \mathbf{j}$.

98. Find two unit vectors in $\mathbb{R}^3$ that are orthogonal to both $\mathbf{u} = \mathbf{i} - \mathbf{j} + 3\mathbf{k}$ and $\mathbf{v} = -2\mathbf{i} - 3\mathbf{j} + 4\mathbf{k}$.

99. Let $\mathbf{u} = 2\mathbf{i} + 3\mathbf{j}$ and $\mathbf{v} = 4\mathbf{i} + \alpha\mathbf{j}$. Determine α such that

a. $\mathbf{u}$ and $\mathbf{v}$ are parallel,

b. $\mathbf{u}$ and $\mathbf{v}$ are orthogonal,

c. the angle between $\mathbf{u}$ and $\mathbf{v}$ is $\pi/4$,

d. the angle between $\mathbf{u}$ and $\mathbf{v}$ is $\pi/6$.

100. Let $\mathbf{u} = 6\mathbf{i} + \beta\mathbf{j} - 15\mathbf{k}$ and $\mathbf{v} = -2\mathbf{i} + 4\mathbf{j} + 5\mathbf{k}$. Determine β such that

a. $\mathbf{u}$ and $\mathbf{v}$ are parallel,

b. $\mathbf{u}$ and $\mathbf{v}$ are orthogonal,

c. the angle between $\mathbf{u}$ and $\mathbf{v}$ is $\pi/4$,

d. the angle between $\mathbf{u}$ and $\mathbf{v}$ is $\pi/3$.

101. Find an equation for the plane containing the points $(-2, 4, 1)$, $(3, -7, 5)$, and $(-1, -2, -1)$.

102. Find an equation for the plane containing the points $(6, -5, 1)$, $(1, -6, 5)$, and $(5, -1, 6)$.

103. Find all points of intersection of the planes Π_1: $-x + y + z = 3$ and Π_2: $-4x + 2y - 7z = 5$.

104. Find all points of intersection of the planes Π_1: $-4x + 6y + 8z = 12$ and Π_2: $2x - 3y - 4z = 5$.

105. Find the distance from the point $P = (2, 3)$ to the line with equation $(x, y) = (7, -3) + t(1, 2)$.

106. Find the distance from the point $P = (3, -1, 2)$ to the line passing through the points $Q = (-2, -1, 6)$ and $R = (0, 1, -8)$.

107. Find the distance from $(1, -2, 3)$ to the plane $2x - y - z = 6$.

108. Find the distance from the origin to the line passing through the point $(3, 1, 5)$ and having the direction $\mathbf{v} = 2\mathbf{i} - \mathbf{j} + \mathbf{k}$.

109. Find an equation for the line passing through $(-1, 2, 4)$ and orthogonal to both L_1: $(x - 1)/4 = (y + 6)/3 = z/(-2)$ and L_2: $(x + 3)/5 = y - 1 = (z + 3)/4$.

110. Find an equation for the line passing through $(-1, 2, 4)$ and parallel to the intersection of the planes Π_1: $-x + y + z = 3$ and Π_2: $-4x + 2y - 7z = 5$.

111. Do calculations to determine whether or not the points $(1, 3, 0)$, $(3, -1, -2)$, and $(-1, 7, 2)$ are collinear.

112. Calculate the area of the triangle with vertices $(2, 1, 3)$, $(-4, 1, 7)$, and $(-1, -1, 3)$.

113. Calculate the area of a parallelogram with adjacent vertices $(1, 4, -2)$, $(-3, 1, 6)$, and $(1, -2, 3)$.

114. Calculate the volume of the parallelepiped determined by the vectors $\mathbf{i} + \mathbf{j}$, $2\mathbf{i} - 3\mathbf{k}$, and $2\mathbf{j} + 7\mathbf{k}$.

In Problems 115–120, identify the quadric surface satisfying the given equation; then sketch its graph.

115. $x^2 + y^2 = 9$

116. $4x^2 + y^2 + 4z^2 - 8y = 0$

117. $x^2 - \dfrac{y^2}{4} + \dfrac{z^2}{9} = 1$

118. $x^2 - \dfrac{y^2}{4} - \dfrac{z^2}{9} = 1$

119. $-9x^2 + 16y^2 - 9z^2 = 25$

120. $x = \dfrac{y^2}{4} - \dfrac{z^2}{9}$

121. Show that the lines L_1: $x = 3 - 2t$, $y = 4 + t$, $z = -2 + 7t$ and L_2: $x = -3 + s$, $y = 2 - 4s$, $z = 1 + 6s$ have no points of intersection.

122. Show that the position vectors $\mathbf{u} = \mathbf{i} - 2\mathbf{j} + \mathbf{k}$, $\mathbf{v} = 3\mathbf{i} + 2\mathbf{j} - 3\mathbf{k}$, and $\mathbf{w} = 9\mathbf{i} - 2\mathbf{j} - 3\mathbf{k}$ are coplanar. Then find an equation for the plane containing them.

CHAPTER 2

VECTOR FUNCTIONS, VECTOR DIFFERENTIATION, AND PARAMETRIC EQUATIONS

2.1 VECTOR FUNCTIONS AND PARAMETRIC EQUATIONS

In Chapter 1, we considered vectors that could be written as

$$\mathbf{v} = (a, b) = a\mathbf{i} + b\mathbf{j} \text{ in } \mathbb{R}^2 \text{ or}$$

$$\mathbf{v} = (a, b, c) = a\mathbf{i} + b\mathbf{j} + c\mathbf{k} \text{ in } \mathbb{R}^3. \tag{1}$$

In this chapter, we see what happens when the numbers a, b, and c are replaced by functions $f_1(t)$, $f_2(t)$, and $f_3(t)$.

DEFINITION VECTOR FUNCTION IN $\mathbb{R}^2$

Let f_1 and f_2 be functions of the real variable t. Then for all values of t for which $f_1(t)$ and $f_2(t)$ are defined, we define the **vector-valued function f** by

$$\mathbf{f}(t) = (f_1(t), f_2(t)) = f_1(t)\mathbf{i} + f_2(t)\mathbf{j}. \tag{2}$$

■

The **domain** of **f** is the intersection of the domains of f_1 and f_2. It is a set of real numbers. The **range** of **f** is a set of vectors in $\mathbb{R}^2$.

REMARK: For simplicity, we will refer to vector-valued functions as **vector functions.**

EXAMPLE 1

FINDING THE DOMAIN OF A VECTOR FUNCTION

Let $\mathbf{f}(t) = f_1(t)\mathbf{i} + f_2(t)\mathbf{j} = (1/t)\mathbf{i} + \sqrt{t+1}\mathbf{j}$. Find the domain of **f**.

SOLUTION: The domain of **f** is the set of all t for which f_1 and f_2 are defined. Since $f_1(t)$ is defined for $t \neq 0$ and $f_2(t)$ is defined for $t \geq -1$, we see that the domain of **f** is the set $\{t: t \geq -1 \text{ and } t \neq 0\}$.

DEFINITION PLANE CURVES AND PARAMETRIC EQUATIONS

Let **f** be a vector function. Then for each t in the domain of **f**, the endpoint of the vector $f_1(t)\mathbf{i} + f_2(t)\mathbf{j}$ is a point (x, y) in the xy-plane, where

$$x = f_1(t) \quad \text{and} \quad y = f_2(t). \tag{3}$$

Suppose that the interval $[a, b]$ is in the domain of the function **f** and that both f_1 and f_2 are continuous in $[a, b]$. Then the set of points $(f_1(t), f_2(t))$ for $a \leq t \leq b$ is called a **plane curve** C. Equation (2) is called the **vector equation** of C, while equations (3) are called the **parametric equations** or **parametric representation** of C. In this context, the variable t is called a **parameter.** ■

REMARK: Any plane curve can be thought of as the range of a vector function whose domain is restricted to the interval $[a, b]$.

EXAMPLE 2

PARAMETRIC EQUATIONS OF THE UNIT CIRCLE

Describe the curve given by the vector equation

$$\mathbf{f}(t) = (\cos t)\mathbf{i} + (\sin t)\mathbf{j}, \qquad 0 \leq t \leq 2\pi. \tag{4}$$

SOLUTION: We first see that for every t, $|\mathbf{f}(t)| = 1$ since $|\mathbf{f}(t)| = \sqrt{\cos^2 t + \sin^2 t} = 1$. Moreover, if we write the curve in its parametric representation, we find that

$$x = \cos t, \qquad y = \sin t, \tag{5}$$

and since $\cos^2 t + \sin^2 t = 1$, we have

$$x^2 + y^2 = 1,$$

which is the equation of the unit circle. This curve is sketched in Figure 1. Note that in the sketch the parameter t represents both the length of the arc from $(1, 0)$ to the endpoint of the vector and the angle (measured in radians) the vector makes with the positive x-axis. The representation $x^2 + y^2 = 1$ is called the *Cartesian equation* of the curve given by (5).

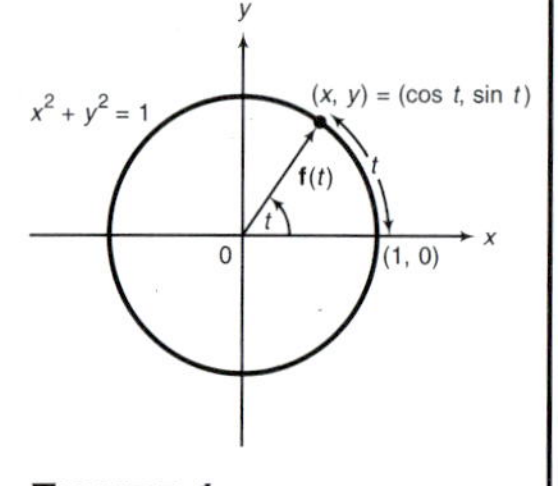

FIGURE 1
The unit circle

REMARK: As t increases from 0 to 2π, we move around the unit circle in the counterclockwise direction. If t were instead restricted to the range $0 \leq t \leq \pi$, then we would not get the entire circle. Rather, we would stop at the point $(\cos \pi, \sin \pi) = (-1, 0)$, which would give us the upper semicircle only.

On the other hand, suppose that $0 \leq t \leq 4\pi$. Starting at $(1, 0)$ when $t = 0$, we get back to $(1, 0)$ when $t = 2\pi$. Then, because $\cos t$ and $\sin t$ are periodic of period 2π, we simply go around the circle again. That is, the curve given by $\mathbf{f}(t) = (\cos t)\mathbf{i} + (\sin t)\mathbf{j}$, $0 \leq t \leq 4\pi$ is the unit circle. It appears as in Figure 1 but, in this case, it is traversed *twice.* Finally, the curve given by $\mathbf{f}(t) = (\cos t)\mathbf{i} + (\sin t)\mathbf{j}$, $0 \leq t \leq 2n\pi$ is the unit circle traversed n times.

DEFINITION CARTESIAN EQUATION OF A PLANE CURVE†

A **Cartesian equation** of the curve $\mathbf{f}(t) = x(t)\mathbf{i} + y(t)\mathbf{j}$ is an equation relating the variables x and y only. ■

EXAMPLE 3 OBTAINING A CARTESIAN EQUATION FROM THE PARAMETRIC EQUATIONS OF A CURVE

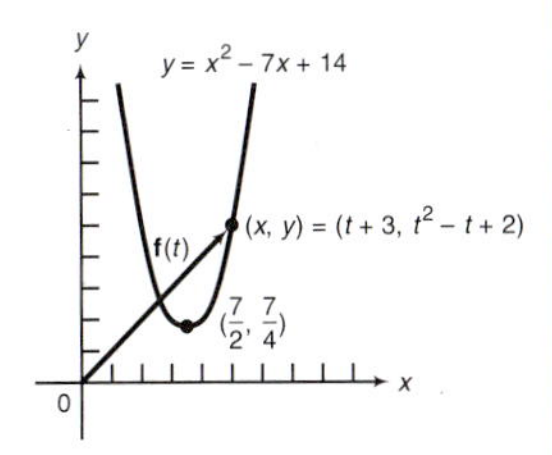

FIGURE 2
Graph of the parabola
$x = t + 3,\ y = t^2 - t + 2$

Describe and sketch the curve given parametrically by $x = t + 3$, $y = t^2 - t + 2$.

SOLUTION: With problems of this type, the easiest thing to do is to write t as a function of x or y, if possible. Since $x = t + 3$, we immediately see that $t = x - 3$ and $y = t^2 - t + 2 = (x - 3)^2 - (x - 3) + 2 = x^2 - 7x + 14$. This is the Cartesian equation of the curve and is the equation of a parabola. It is sketched in Figure 2. Note that in the Cartesian equation of the parabola, the parameter t does not appear.

EXAMPLE 4 OBTAINING A CARTESIAN EQUATION FROM THE PARAMETRIC EQUATIONS OF A CURVE

Describe and sketch the curve given by the vector equation

$$\mathbf{f}(r) = (1 - r^4)\mathbf{i} + r^2\mathbf{j}. \tag{6}$$

SOLUTION: First, we note that the parameter in this problem is r instead of t. This makes absolutely no difference since, as in the case of the variable of integration, the parameter is a ‘‘dummy’’ variable. Now to get a feeling for the shape of this curve, we display in Table 1 values of x and y for various values of r. Plotting some of these points leads to the sketch in Figure 3. To write the Cartesian equation for this curve, we square both sides of the equation $y = r^2$ to obtain $y^2 = r^4$ and $x = 1 - r^4 = 1 - y^2$, which is the equation of the parabola sketched in Figure 3. Note that this curve is *not* the graph of the parabola $x = 1 - y^2$ since the parametric representation $y = r^2$ requires that y be nonnegative. Thus, the curve described by (6) is only the *upper half* of the parabola described by the equation $x = 1 - y^2$.

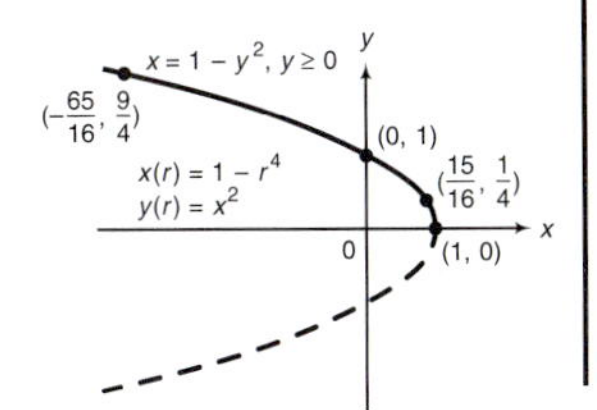

FIGURE 3
Upper half of parabola
$x = 1 - y^2$

TABLE 1

r	0	$\pm\frac{1}{2}$	± 1	$\pm\frac{3}{2}$	± 2
$x = 1 - r^4$	1	$\frac{15}{16}$	0	$-\frac{65}{16}$	-15
$y = r^2$	0	$\frac{1}{4}$	1	$\frac{9}{4}$	4

Having seen how vectors in $\mathbb{R}^2$ generalize to vectors in $\mathbb{R}^3$, we can imagine how properties of vector functions are extended to $\mathbb{R}^3$.

DEFINITION VECTOR FUNCTION IN $\mathbb{R}^3$ AND CURVE IN SPACE

Let f_1, f_2, and f_3 be functions of the real variable t. Then for all values of t for which $f_1(t)$, $f_2(t)$, and $f_3(t)$ are defined, we define the **vector-valued function f of a real**

†Named after the French philosopher and mathematician René Descartes (1596–1650). See the biographical sketch on the next page.

variable t by

$$\mathbf{f}(t) = (f_1(t), f_2(t), f_3(t)) = f_1(t)\mathbf{i} + f_2(t)\mathbf{j} + f_3(t)\mathbf{k}. \tag{7}$$

If f_1, f_2, and f_3 are continuous over an interval I, then as t varies over I, the set of points traced out by the end of the vector $\mathbf{f}$ is called a **curve in space.** ■

EXAMPLE 5

AN ELLIPTICAL HELIX

Sketch the curve $\mathbf{f}(t) = (\cos t)\mathbf{i} + 4(\sin t)\mathbf{j} + t\mathbf{k}$.

SOLUTION: Here $x = \cos t$ and $y = 4 \sin t$, so eliminating t, we obtain $x^2 + (y^2/16) = 1$, which is the equation of an ellipse in the xy-plane. Since $z = t$ increases as t increases, the curve is a spiral that climbs up the side of an elliptical cylinder, as sketched in Figure 4. This curve is called an **elliptical helix.**

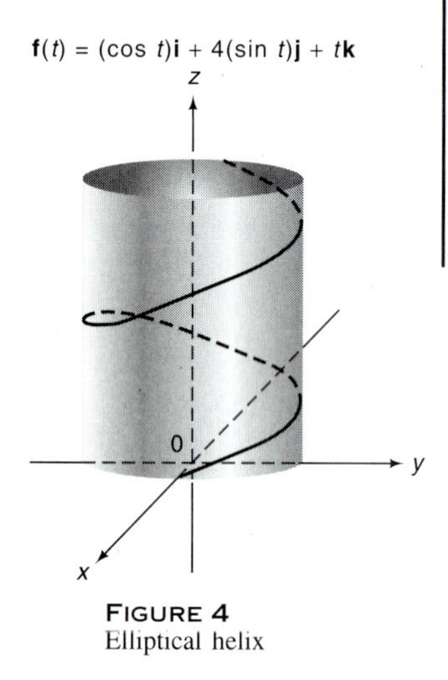

FIGURE 4
Elliptical helix

OF HISTORICAL INTEREST

RENÉ DESCARTES (1596–1650)

RENÉ DESCARTES, 1596–1650

The Cartesian plane is named after the great French mathematician and philosopher René Descartes. Born near the city of Tours in 1596, Descartes received his education first at the Jesuit school at La Flèche and later at Poitier, where he studied law. He had delicate health and, while still in school, developed the habit of spending the greater part of each morning in bed. Later, he considered these morning hours the most productive period of the day.

At the age of 16, Descartes left school and moved to Paris, where he began his study of mathematics. In 1618, he joined the army of Maurice, Prince of Nassau. He also served with Duke Maximillian I of Bavaria and with the French army at the siege of La Rochelle.

Descartes was not a professional soldier, however, and his periods of military service were broken by periods of travel and study in various European cities. After leaving the army for good, he resettled in Paris to continue his mathematical studies and then moved to Holland, where he lived for 24 years.

Much stimulated by the scientists and philosophers he met in France, Holland, and elsewhere, Descartes later became known as the "father of modern philosophy." His statement "Cogito ergo sum" ("I think, therefore I am") played a central role in his philosophical writings.

Descartes's program for philosophical research was enunciated in his famous *Discours de la méthode pour bien conduire sa raison et chercher la vérité dans les sciences* (A Discourse on the

(CONTINUED)

OF HISTORICAL INTEREST

(CONCLUDED)

Method of Rightly Conducting the Reason and Seeking Truth in the Sciences) published in 1637. This work was accompanied by three appendixes: *La dioptrique* (in which the law of refraction—discovered by Snell—was first published), *Les météores* (which contained the first accurate explanation of the rainbow), and *La géométrie. La géométrie,* the third and most famous appendix, took up about a hundred pages of the *Discours.* One of the major achievements of *La géométrie* was that it connected figures of geometry with the equations of algebra. The work established Descartes as the founder of analytic geometry.

In 1649 Descartes was invited to Sweden by Queen Christina. He accepted, reluctantly, but was unable to survive the harsh, Scandinavian winter. He died in Stockholm in early 1650.

PROBLEMS 2.1

SELF-QUIZ

I. The graph of $x = 1 + y$, $y = 3 - 2t$ is the same as the graph of ________.

a. $x + y = 4$
b. $2x + y = 5$
c. $y = -2x + 1$ and $x \geq 0$
d. $|2x| + |y| = 5$

II. The graph of $x = 1 + t^2$, $y = 3 + t^2$ is the same as the graph of ________.

a. $y = x + 2$
b. $x = y + 2$
c. $y = x + 2$ and $x \geq 0$
d. $y = x + 2$ and $x \geq 1$

III. The graph of $x = \cos t$, $y = -\cos t$ is the same as the graph of ________.

a. $x = -y$
b. $x = -y$ and $x \geq 0$
c. $y = -x$ and $x \leq 1$
d. $y = -x$ and $-1 \leq x \leq 1$

IV. The unit circle $x^2 + y^2 = 1$ can be parametrized by ________.

a. $x = \sin t$, $y = \cos t$
b. $x = \cos(t - \pi/2)$, $y = \sin(t - \pi/2)$
c. $x = -\cos t$, $y = \sin t$
d. $x = \cos 2t$, $y = \sin 2t$

In Problems 1–8, find the domain of the given vector-valued function.

1. $\mathbf{f}(t) = \dfrac{1}{t}\mathbf{i} + \dfrac{1}{t-1}\mathbf{j}$

2. $\mathbf{f}(t) = \sqrt{t}\mathbf{i} + \dfrac{1}{t}\mathbf{j}$

3. $\mathbf{f}(s) = \dfrac{1}{s^2 - 1}\mathbf{i} + (s^2 - 1)\mathbf{j}$

4. $\mathbf{f}(s) = e^{1/s}\mathbf{i} + e^{-1/(s+1)}\mathbf{j}$

5. $\mathbf{f}(r) = (\ln r)\mathbf{i} + \ln(1 - r)\mathbf{j} + \ln(1 + r)\mathbf{k}$

6. $\mathbf{f}(r) = (\sin r)\mathbf{i} + r\mathbf{j} - (1 + r^2)\mathbf{k}$

7. $\mathbf{f}(r) = (\sec t)\mathbf{i} + (\csc t)\mathbf{j} + (\cos 2t)\mathbf{k}$

8. $\mathbf{f}(t) = (\tan t)\mathbf{i} + (\tan 2t)\mathbf{j} + (\cot t)\mathbf{k}$

In Problems 9–28, find the Cartesian equation for each curve; then sketch the curve (in $\mathbb{R}^2$ or $\mathbb{R}^3$).

9. $\mathbf{f}(t) = t^2\mathbf{i} + 2t\mathbf{j}$

10. $\mathbf{f}(t) = (2t - 3)\mathbf{i} + t^2\mathbf{j}$

11. $\mathbf{f}(t) = t^2\mathbf{i} + t^3\mathbf{j}$

12. $\mathbf{f}(t) = 3(\sin t)\mathbf{i} + 3(\cos t)\mathbf{j}$

13. $\mathbf{f}(t) = (2t - 1)\mathbf{i} + (4t + 3)\mathbf{j}$

14. $\mathbf{f}(t) = 2(\cosh t)\mathbf{i} + 2(\sinh t)\mathbf{j}$

15. $\mathbf{f}(t) = (t^4 + t^2 + 1)\mathbf{i} + t^2\mathbf{j}$

16. $\mathbf{f}(t) \doteq t^2\mathbf{i} + t^8\mathbf{j}$

17. $\mathbf{f}(t) = t^3\mathbf{i} + (t^9 - 1)\mathbf{j}$

18. $\mathbf{f}(t) = t\mathbf{i} + e^t\mathbf{j}$

19. $\mathbf{f}(t) = e^t\mathbf{i} + t^2\mathbf{j}$

20. $\mathbf{f}(t) = (t^2 + t - 3)\mathbf{i} + \sqrt{t}\mathbf{j}$

***21.** $\mathbf{f}(t) = e^t(\sin t)\mathbf{i} + e^t(\cos t)\mathbf{j}$
[*Hint:* Show that $|\mathbf{f}(t)| = e^t$.]

22. $\mathbf{f}(t) = \mathbf{i} + (\tan t)\mathbf{j}$

23. $\mathbf{f}(t) = e^t\mathbf{i} + e^{2t}\mathbf{j}$

***24.** $\mathbf{f}(t) = \dfrac{6t}{1+t^3}\mathbf{i} + \dfrac{6t^2}{1+t^3}\mathbf{j}$

25. $\mathbf{g}(t) = 2t\mathbf{i} - 3t\mathbf{j} + t\mathbf{k}$

26. $\mathbf{g}(t) = (1-t)\mathbf{i} + (3+5t)\mathbf{j} + 3t\mathbf{k}$

27. $\mathbf{g}(t) = t\mathbf{i} + (\cos t)\mathbf{j} + (\sin t)\mathbf{k}$

28. $\mathbf{g}(t) = (\cos t)\mathbf{i} - (\sin t)\mathbf{j} + (\cos t)\mathbf{k}$

In Problems 29–38, use the result of Problem 49 to find a parametric representation of the straight line that passes through the two given points.

29. $(2, 4); (1, 6)$

30. $(-3, 2); (0, 4)$

31. $(3, 5); (-1, -7)$

32. $(4, 6); (7, 9)$

33. $(-2, 3); (4, 7)$

34. $(-4, 0); (3, -2)$

35. $(1, 3, 5); (2, 4, 6)$

36. $(1, 3, 5); (-2, 4, -6)$

37. $(-3, 0, 7); (-5, 3, 0)$

38. $(7, -3, 0); (5, 0, 3)$

39. Verify that a vector equation for the ellipse $(x^2/a^2) + (y^2/b^2) = 1$ is

$$\mathbf{f}(t) = a(\cos t)\mathbf{i} + b(\sin t)\mathbf{j}.$$

40. Verify that a parametric representation for the hyperbola $(x^2/a^2) - (y^2/b^2) = 1$ is

$$x(\theta) = a\sec\theta \quad \text{and} \quad y(\theta) = b\tan\theta.$$

***41.** Suppose that a wheel of radius r is rolling in a straight line without slipping. Let P be a fixed point on the wheel which is distance s from the center. As the wheel rotates, the point P traces out a curve known as a **trochoid**.† Look at Figure 5. Suppose that P moves through an angle of α radians to reach its position shown in part (b) of the figure.

a. Show that the new position vector is $\overrightarrow{0P} = \overrightarrow{0R} + \overrightarrow{RC} + \overrightarrow{CP}$.

b. Show that $\overrightarrow{0R} = \alpha r\mathbf{i}$.

c. Show that $\overrightarrow{CP} = s[(\cos\theta)\mathbf{i} + (\sin\theta)\mathbf{j}]$.

d. Show that $\overrightarrow{0P} = [\alpha r + s\cos\theta]\mathbf{i} + [r + s\sin\theta]\mathbf{j}$.

e. Show that $\theta = (3\pi/2) - \alpha$.

f. Show that $\cos\theta = -\sin\alpha$ and $\sin\theta = -\cos\alpha$.

g. Show that the trochoid satisfies the parametric equations

$$x = r\alpha - s\sin\alpha \quad \text{and} \quad y = r - s\cos\alpha. \qquad (8)$$

42. A point is located 25 cm from the center of a wheel 1 m in diameter. Find a parametric representation of the trochoid traced by that point as the wheel rolls.

***43.** A trochoid generated by a point P on the circumference of the wheel (i.e., $s = r$) has the special name **cycloid**.‡

a. Find parametric equations for the cycloid.

b. Verify that the portion of the cycloid for $0 \le x \le 2r$ has the Cartesian equation

$$x = r\cos^{-1}\left(\frac{r-y}{r}\right) - \sqrt{2ry - y^2}.$$

44. A point is located on the circumference of a wheel 1 m in diameter. Find a parametric representation for the cycloid traced by that point as the wheel rolls.

***45.** A **hypocycloid**§ is a curve generated by the motion of a point P on the circumference of a circle that rolls internally (without slipping) on a larger circle (see Figure 6). Assume that the radius of the larger circle is a while that of the smaller circle is b. Let θ be the angle indicated in Figure 6. Verify that the hypocycloid has parametric representation

$$x = (a-b)\cos\theta + b\cos\left[\left(\frac{a-b}{b}\right)\theta\right],$$

$$y = (a-b)\sin\theta - b\sin\left[\left(\frac{a-b}{b}\right)\theta\right].$$

[*Hint:* First show that $a\theta = b\alpha$.]

46. If $a = 4b$ in Figure 6, then the curve generated is called a **hypocycloid of four cusps**. Sketch this curve.

† From the Greek word *trochos,* meaning *wheel.*

‡ From the Greek words *kyklos,* meaning *circle.* The cycloid was a source of great controversy in the seventeenth century. Many of its properties were discovered by the French mathematician Gilles Personne de Roberval (1602–1675), although the curve was first discussed by Galileo. Unfortunately, Roberval, for unknown reasons, did not publish discoveries concerning the cycloid, which meant that he lost credit for most of them. The ensuing arguments over who discovered what were so bitter that the cycloid became known as the "Helen of geometers" (after Helen of Troy—the source of the intense jealousy that led to the Trojan War).

§ The *hypo* comes from the Greek *hupo,* meaning *under.*

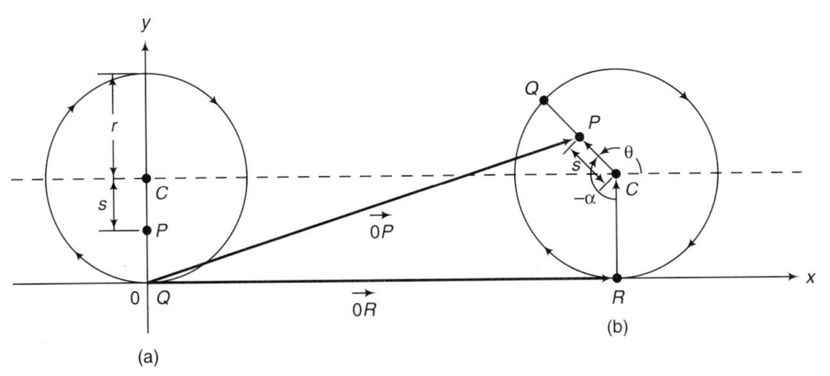

FIGURE 5

FIGURE 6

***47.** Verify that the hypocycloid of four cusps has the Cartesian equation

$$x^{2/3} + y^{2/3} = a^{2/3}.$$

[*Hint:* Use the identities $\cos 3\theta = 4\cos^3\theta - 3\cos\theta$ and $\sin 3\theta = 3\sin\theta - 4\sin^3\theta$ together with the facts that $a - b = 3b$ and $(a - b)/b = 3$.]

48. Calculate the area bounded by the hypocycloid of four cusps.

49. Figure 7 shows three points P, Q, R on a straight line.

a. Show that $\overrightarrow{PR} = t\overrightarrow{PQ}$ for some real number t.

b. Show that $\overrightarrow{0R} = \overrightarrow{0P} + \overrightarrow{PR} = \overrightarrow{0P} + t\overrightarrow{PQ}$.

c. Use the results of parts (a) and (b) to show that the line passing through the points (x_1, y_1) and (x_2, y_2) satisfies the parametric equations.

$$x = x_1 + t(x_2 - x_1)$$

and

$$y = y_1 + t(y_2 - y_1).$$

50. Use the substitution $\tan\theta = \sin t$ to show that the lemniscate $r^2 = \cos 2\theta$ has the Cartesian parametric representation

$$x = \frac{\cos t}{1 + \sin^2 t} \qquad \text{and} \qquad y = \frac{\sin t \cos t}{1 + \sin^2 t}.$$

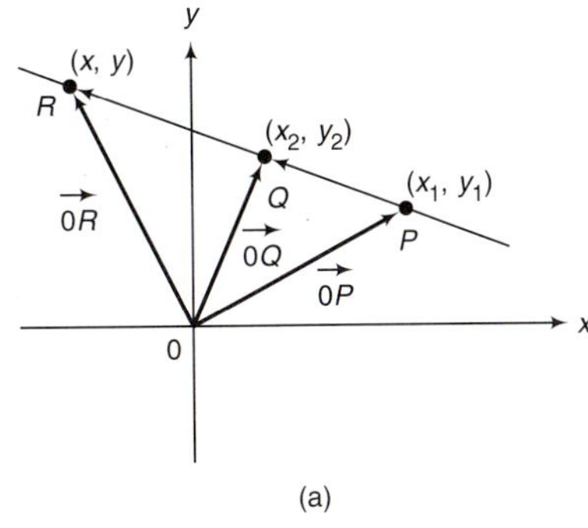

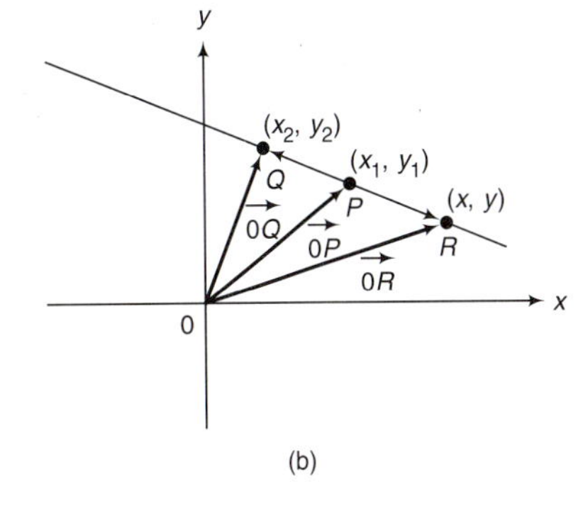

FIGURE 7

ANSWERS TO SELF-QUIZ

I. b **II.** d **III.** d **IV.** a, b, c, d

2.2 THE EQUATION OF THE TANGENT LINE TO A PLANE CURVE AND SMOOTHNESS

Suppose that $x = f_1(t)$ and $y = f_2(t)$ is the parametric representation of a curve C. We would like to be able to calculate the equation of the line tangent to the curve without having to determine the Cartesian equation of the curve. However, there are complications that might occur since the curve could intersect itself. This happens if there are two numbers $t_1 \neq t_2$ such that $f_1(t_1) = f_1(t_2)$ and $f_2(t_1) = f_2(t_2)$. Thus, there are three possibilities. At a given point the curve could have the following.

i. a unique tangent
ii. no tangent
iii. two or more tangents.

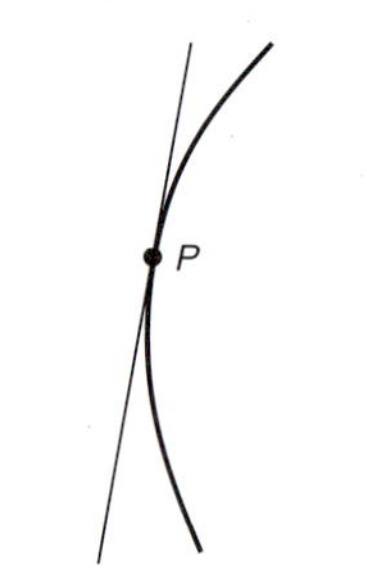

(a) Unique tangent at P.

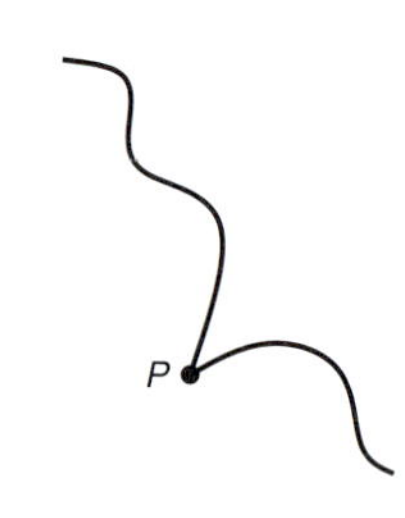

(b) No tangent at P.

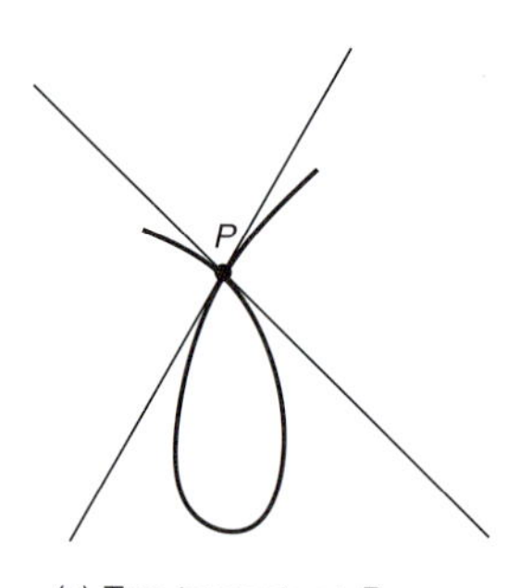

(c) Two tangents at P.

FIGURE 1
Three plane curves

In Figure 1 we sketch three plane curves.

A condition that ensures that there is at least one tangent line at each point is that

$$f_1' \text{ and } f_2' \text{ exist and } [f_1'(t)]^2 + [f_2'(t)]^2 \neq 0. \tag{1}$$

REMARK: Condition (1) simply states that the derivatives of f_1 and f_2 are not zero at the same value of t.

DEFINITION SMOOTH CURVE

The parametric curve $\mathbf{f}(t) = f_1(t)\mathbf{i} + f_2(t)\mathbf{j}$ is said to be **smooth** on an interval $I = (a, b)$ if $f_1'(t)$ and $f_2'(t)$ exist and are continuous on I and, for every t in I, $[f_1'(t)]^2 + [f_2'(t)]^2 \neq 0$. ■

THEOREM 1

Let (x_0, y_0) be on the smooth curve C given by $x = f_1(t)$ and $y = f_2(t)$. If the curve passes through (x_0, y_0) when $t = t_0$,† then the slope m of the line tangent to C at (x_0, y_0) is given by

$$m = \lim_{t \to t_0} \frac{f_2'(t)}{f_1'(t)}, \tag{2}$$

provided that this limit exists.

PROOF: We refer to Figure 2. From the figure we see that

$$m = \lim_{\Delta t \to 0} \frac{f_2(t_0 + \Delta t) - f_2(t_0)}{f_1(t_0 + \Delta t) - f_1(t_0)} \overset{\text{L'Hôpital's rule}}{=} \lim_{\Delta t \to 0} \frac{f_2'(t_0 + \Delta t)}{f_1'(t_0 + \Delta t)} \overset{\text{let } t = t_0 + \Delta t}{=} \lim_{t \to t_0} \frac{f_2'(t)}{f_1'(t)}.$$ ■

In your one-variable calculus course you studied inverse functions. There you saw the following theorem: Suppose that $y = f(x)$ is differentiable in some open interval containing x_0 and, in that interval, $f'(x) \neq 0$. Then f has an inverse function that is differentiable in an open interval containing $y_0 = f(x_0)$. If $x = g(y) = f^{-1}(y)$ is that inverse function,

† The curve may pass through the point (x_0, y_0) for other values of t as well, and therefore it may have other tangent lines at that point, as in Figure 1(c).

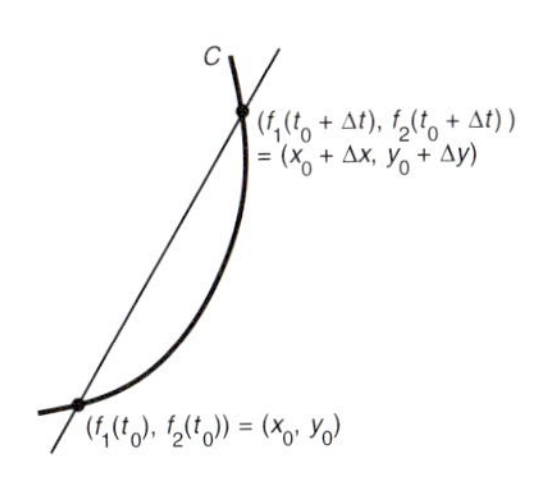

FIGURE 2
The secant line joins (x_0, y_0) and $(x_0 + \Delta x, y_0 + \Delta y)$.

then

$$g'(y_0) = \frac{1}{f'(x_0)} \qquad \text{or equivalently} \qquad \frac{dx}{dy} = \frac{1}{dy/dx}.$$

Suppose now that $f_1'(t_0) \neq 0$. Then, by the result above, f_1 has an inverse that is differentiable in a neighborhood of $x_0 = f_1(t_0)$ and, in this neighborhood, we may write $t = f_1^{-1}(x)$. Then $y = f_2(t) = f_2(f_1^{-1}(x))$ in this neighborhood of x_0. That is, we may write y as a function of x.

COROLLARY TO THEOREM 1

If $f_1'(t_0) \neq 0$, then the smooth curve C has a representation $y = f(x) = f_2(f_1^{-1}(x))$ in a neighborhood of $x_0 = f_1(t_0)$. Moreover, the derivative $f'(x_0)$ exists and is given by

$$\frac{dy}{dx} = f'(x_0) = \frac{f_2'(t_0)}{f_1'(t_0)}. \tag{3}$$

NOTE: This result is often written in the symbolic form

$$\frac{dy}{dx} = \frac{dy/dt}{dx/dt}.$$

PROOF: By the quotient rule for limits and the continuity of f_1' and f_2', we have

$$\lim_{t\to t_0} \frac{f_2'(t)}{f_1'(t)} = \frac{\lim_{t\to t_0} f_2'(t)}{\lim_{t\to t_0} f_1'(t)} = \frac{f_2'(t_0)}{f_1'(t_0)},$$

so, by Theorem 1, $\dfrac{f_2'(t_0)}{f_1'(t_0)}$ is the slope of the line tangent to the curve at $t = t_0$. This result also follows from the chain rule and the theorem cited above:

$$y(x_0) = f_2(f_1^{-1}(x_0))$$

so

$$\left.\frac{dy}{dx}\right|_{x=x_0} \overset{\text{chain rule}}{=} f_2'(f_1^{-1}(x_0))[f_1^{-1}(x_0)]' = \frac{f_2'(t_0)}{f_1'(t_0)}$$

since $f_1^{-1}(x_0) = t_0$. ■

Using the corollary, we find that the tangent line to C at t_0 is given by

$$y - y_0 = \frac{f_2'(t_0)}{f_1'(t_0)}(x - x_0)$$

or after simplification,

EQUATION OF THE TANGENT LINE TO A SMOOTH CURVE AT $(x_0, y_0) = (f_1(t_0), f_2(t_0))$

$$f_2'(t_0)(x - x_0) - f_1'(t_0)(y - y_0) = 0. \tag{4}$$

NOTE: Equation (4) holds even when $f_1'(t_0) = 0$. In that case the tangent line is parallel to the y-axis.

If f is smooth, then $f_1'(t)$ and $f_2'(t)$ are never zero at the same time. If a curve has a derivative of zero at a point, then the tangent is horizontal at that point. If the derivative approaches infinity as we approach a point, then the tangent line is vertical at that point. The following facts follow from the discussion above.

VERTICAL AND HORIZONTAL TANGENTS

i. C has a vertical tangent line at t_0 if

$$\left.\frac{dx}{dt}\right|_{t=t_0} = f_1'(t_0) = 0 \qquad \text{and} \qquad f_2'(t_0) \neq 0.$$

ii. C has a horizontal tangent line at t_0 if

$$\left.\frac{dy}{dt}\right|_{t=t_0} = f_2'(t_0) = 0 \qquad \text{and} \qquad f_1'(t_0) \neq 0.$$

EXAMPLE 1

FINDING THE TANGENT LINE TO A CURVE

Find the equation of the line tangent to the curve

$$x = f_1(t) = 2t^3 - 15t^2 + 24t + 7, \qquad y = f_2(t) = t^2 + t + 1 \qquad \text{at } t = 2.$$

SOLUTION: Here $f_1'(2) = (6t^2 - 30t + 24)|_{t=2} = -12$ and $f_2'(2) = (2t + 1)|_{t=2} = 5$. When $t = 2$, $x = 11$ and $y = 7$. Then using (4), we obtain

$$5(x - 11) + 12(y - 7) = 0, \qquad \text{or} \qquad 5x + 12y = 139.$$

EXAMPLE 2

FINDING HORIZONTAL AND VERTICAL TANGENTS

Find all horizontal and vertical tangents to the curve of Example 1.

SOLUTION: There are vertical tangents when

$$f_1'(t) = 6t^2 - 30t + 24 = 0 = 6(t^2 - 5t + 4) = 6(t - 4)(t - 1),$$

or when $t = 1$ and $t = 4$, since $f_2'(t)$ is nonzero at these points. When $t = 1$, $x = 18$, and when $t = 4$, $x = -9$. Thus, the vertical tangents are the lines $x = 18$ and $x = -9$. There is a horizontal tangent when $2t + 1 = 0$, or $t = -\frac{1}{2}$. When $t = -\frac{1}{2}$, $y = \frac{3}{4}$, and the line $y = \frac{3}{4}$ is a horizontal tangent.

PROBLEMS 2.2

SELF-QUIZ

I. The smooth curve given by $x = 3\cos t$, $y = 4\sin t$ has a horizontal tangent when $t =$ _________.

a. $\pi/4$ **b.** $\pi/2$
c. $3\pi/4$ **d.** π

II. The smooth curve given by $x = 4\cos t$, $y = 3\sin t$ has a vertical tangent when $t =$ _________.

a. $\pi/4$ **b.** $\pi/2$
c. $3\pi/4$ **d.** π

III. The smooth curve given by $x = \cos 2t$, $y = \sin 2t$ has a horizontal tangent when $t =$ ________.

a. $\pi/4$ **b.** $\pi/2$
c. $3\pi/4$ **d.** π

IV. The smooth curve given by $x = 3 \cos 2t$, $y = 4 \sin 2t$ has a vertical tangent when $t =$ ________.

a. $\pi/4$ **b.** $\pi/2$
c. $3\pi/4$ **d.** π

In Problems 1–10, find the slope of the line tangent to the given curve for the specified value of the parameter.

1. $x = t^3;\ y = t^4 - 5;\ t = -1$
2. $x = t^2;\ y = \sqrt{1 - t};\ t = 1/2$
3. $x = \cos \theta;\ y = \sin \theta;\ \theta = \pi/4$
4. $x = \tan \theta;\ y = \sec \theta;\ \theta = \pi/4$
5. $x = \sec \theta;\ y = \tan \theta;\ \theta = \pi/4$
6. $x = \cosh t;\ y = \sinh t;\ t = 0$
7. $x = 8 \cos \theta;\ y = -3 \sin \theta;\ \theta = 2\pi/3$
8. $x = \cos^3 \theta;\ y = \sin^3 \theta;\ \theta = \pi/2$
9. $x = \theta;\ y = 1/\theta;\ \theta = \pi/4$
10. $x = t \cosh t;\ y = (\tanh t)/(1 + t);\ t = 0$

In Problems 11–16, find an equation for the line tangent to the given curve for the specified value of the parameter.

11. $x = t^2 - 2;\ y = 4t;\ t = 3$
12. $x = t + 4;\ y = t^3 - t + 4;\ t = 2$
13. $x = e^{2t};\ y = e^{-2t};\ t = 1$
14. $x = \cos 2\theta;\ y = \sin 2\theta;\ \theta = \pi/4$
15. $x = \sqrt{1 - \sin \theta};\ y = \sqrt{1 + \cos \theta};\ \theta = 0$
16. $x = \cos^3 \theta;\ y = \sin^3 \theta;\ \theta = \pi/6$

In Problems 17–26, find all points [in the form (x, y)] at which the given curve has a vertical or horizontal tangent.

17. $x = t^2 - 1;\ y = 4 - t^2$
18. $x = 1/\sqrt{1 - t^2};\ y = \sqrt{1 - t^2}$
19. $x = \sinh t;\ y = \cosh t$
20. $x = \sin \theta + \cos \theta;\ y = \sin \theta - \cos \theta$
21. $x = 2 \cos \theta;\ y = 3 \sin \theta$
22. $x = 2 + 7 \cos \theta;\ y = 8 + 3 \sin \theta$
****23.** $x = \ln(1 + t^2);\ y = \ln(1 + t^3)$
[*Hint:* Use l'Hôpital's rule.]
24. $x = e^{3t};\ y = e^{-5t}$
***25.** $x = \sin 3\theta;\ y = \cos 5\theta$
***26.** $x = \theta \sin \theta;\ y = \theta \cos \theta$

In Problems 27–34, use the result of Problem 39 to calculate the slope of the tangent line to the given curve for the specified value of θ.

27. $r = 5 \sin \theta;\ \theta = \pi/6$
28. $r = 5 \cos \theta + 5 \sin \theta;\ \theta = \pi/4$
29. $r = -4 + 2 \cos \theta;\ \theta = \pi/3$
30. $r = 4 + 3 \sin \theta;\ \theta = 2\pi/3$
31. $r = 3 \sin 2\theta;\ \theta = \pi/6$
32. $r = 5 \sin 3\theta;\ \theta = \pi/4$
33. $r = e^{\theta/2};\ \theta = 0$
34. $r^2 = \cos 2\theta;\ \theta = \pi/6$

35. Find an equation for each of the two tangents to the curve $x = t^3 - 2t^2 - 3t + 11$ and $y = t^2 - 2t - 5$ at the point $(11, -2)$. [*Hint:* First find the relevant values of the parameter by solving the equation $y = t^2 - 2t - 5 = -2$.]

***36.** Write an equation for each of the three lines that are tangent to the curve $x = t^3 - 2t^2 - t + 3$ and $y = -t^2 + 2t - 2/t$ at the point $(1, -1)$.

37. Calculate the slope of the tangent to the cycloid given in Problem 2.1.43, and then show that the tangent is vertical when α is a multiple of 2π.

38. For what values of α does the trochoid given by equations (2.1.8) (in Problem 2.1.41) have a vertical tangent?

39. Suppose a curve is given in polar coordinates by the equation $r = f(\theta)$. Show that the curve can then be given parametrically by

$$x = f(\theta) \cos \theta \qquad \text{and} \qquad y = f(\theta) \sin \theta.$$

ANSWERS TO SELF-QUIZ

I. b **II.** d **III.** a, c **IV.** b, d

2.3 THE DIFFERENTIATION AND INTEGRATION OF A VECTOR FUNCTION

In Section 2.2, we showed how the derivative dy/dx could be calculated when x and y were given parametrically in terms of t. In this section, we will show how to calculate the derivative of a vector function. The definitions of limit, continuity, and differentiability are virtually the same in $\mathbb{R}^2$ and $\mathbb{R}^3$. For that reason, we give definitions of these basic concepts in terms of vectors in the plane. We begin with the definition of a limit.

DEFINITION LIMIT OF A VECTOR FUNCTION

Let $\mathbf{f}(t) = f_1(t)\mathbf{i} + f_2(t)\mathbf{j}$. Let t_0 be any real number, $+\infty$, or $-\infty$. If $\lim_{t\to t_0} f_1(t)$ and $\lim_{t\to t_0} f_2(t)$ both exist, then we define

$$\lim_{t\to t_0} \mathbf{f}(t) = \left[\lim_{t\to t_0} f_1(t)\right]\mathbf{i} + \left[\lim_{t\to t_0} f_2(t)\right]\mathbf{j}. \tag{1}$$

■

That is, *the limit of a vector function is determined by the limits of its component functions.* Thus, in order to calculate the limit of a vector function, it is only necessary to calculate two ordinary limits.

EXAMPLE 1 COMPUTING A LIMIT

Compute

$$\lim_{t\to 0}\left[\frac{\sin t}{t}\mathbf{i} + 3\cos t\mathbf{j}\right]$$

SOLUTION: We know that $\lim_{t\to 0} \frac{\sin t}{t} = 1$. Also,

$$\lim_{t\to 0} 3\cos t = 3\lim_{t\to 0}\cos t = 3\cdot 1 = 3.$$

Thus

$$\lim_{t\to 0}\left[\frac{\sin t}{t}\mathbf{i} + 3\cos t\mathbf{j}\right] = \left(\lim_{t\to 0}\frac{\sin t}{t}\right)\mathbf{i} + \left(\lim_{t\to 0} 3\cos t\right)\mathbf{j}$$
$$= 1\mathbf{i} + 3\mathbf{j} = \mathbf{i} + 3\mathbf{j}.$$

DEFINITION CONTINUITY OF A VECTOR FUNCTION

$\mathbf{f}$ is **continuous** at t_0 if the component functions f_1 and f_2 are continuous at t_0. Thus, $\mathbf{f}$ is continuous at t_0 if

$$\lim_{t\to t_0} \mathbf{f}(t) = \mathbf{f}(t_0).$$

This means that the following three conditions hold:

i. $\mathbf{f}$ is defined at t_0,

ii. $\lim_{t\to t_0} \mathbf{f}(t)$ exists,

iii. $\lim_{t\to t_0} \mathbf{f}(t) = \mathbf{f}(t_0)$.

■

Definition Derivative of a Vector Function

Let $\mathbf{f}$ be defined at t. Then $\mathbf{f}$ is **differentiable** at t if

$$\lim_{\Delta t \to 0} \frac{\mathbf{f}(t + \Delta t) - \mathbf{f}(t)}{\Delta t} \tag{2}$$

exists and is finite. The vector function $\mathbf{f}'$ defined by

$$\mathbf{f}'(t) = \frac{d\mathbf{f}}{dt} = \lim_{\Delta t \to 0} \frac{\mathbf{f}(t + \Delta t) - \mathbf{f}(t)}{\Delta t} \tag{3}$$

is called the **derivative** of $\mathbf{f}$, and the domain of $\mathbf{f}'$ is the set of all t such that the limit in (2) exists. ■

Definition Differentiability in an Open Interval

The vector function $\mathbf{f}$ is **differentiable** on the open interval I if $\mathbf{f}'(t)$ exists for every t in I. ■

Before giving examples of the calculation of derivatives, we prove a theorem that makes this calculation no more difficult than the calculation of "ordinary" derivatives.

Theorem 1

If $\mathbf{f}(t) = f_1(t)\mathbf{i} + f_2(t)\mathbf{j}$, then at any value t for which $f_1'(t)$ and $f_2'(t)$ exist,

$$\mathbf{f}'(t) = f_1'(t)\mathbf{i} + f_2'(t)\mathbf{j}. \tag{4}$$

That is, the derivative of a vector function is determined by the derivatives of its component functions.

Proof:

$$\begin{aligned}
\mathbf{f}'(t) &= \lim_{\Delta t \to 0} \frac{\mathbf{f}(t + \Delta t) - \mathbf{f}(t)}{\Delta t} \\
&= \lim_{\Delta t \to 0} \frac{[f_1(t + \Delta t)\mathbf{i} + f_2(t + \Delta t)\mathbf{j}] - [f_1(t)\mathbf{i} + f_2(t)\mathbf{j}]}{\Delta t} \\
&= \lim_{\Delta t \to 0} \frac{[f_1(t + \Delta t) - f_1(t)]\mathbf{i} + [f_2(t + \Delta t) - f_2(t)]\mathbf{j}}{\Delta t} \\
&= \lim_{\Delta t \to 0} \left[\frac{f_1(t + \Delta t) - f_1(t)}{\Delta t}\right]\mathbf{i} + \lim_{\Delta t \to 0} \left[\frac{f_2(t + \Delta t) - f_2(t)}{\Delta t}\right]\mathbf{j} \\
&= f_1'(t)\mathbf{i} + f_2'(t)\mathbf{j}.
\end{aligned}$$

■

Example 2 Differentiating a Vector Function

Let $\mathbf{f}(t) = (\cos t)\mathbf{i} + e^{2t}\mathbf{j}$. Calculate $\mathbf{f}'(t)$.

Solution:

$$\mathbf{f}'(t) = \frac{d}{dt}(\cos t)\mathbf{i} + \frac{d}{dt}e^{2t}\mathbf{j} = -(\sin t)\mathbf{i} + 2e^{2t}\mathbf{j}$$

Once we know how to calculate the first derivative of **f**, we can calculate higher derivatives as well.

DEFINITION SECOND DERIVATIVE

If the function $\mathbf{f}'$ is differentiable at t, we define the **second derivative** of **f** to be the derivative of $\mathbf{f}'$. That is,

$$\mathbf{f}'' = (\mathbf{f}')'. \tag{5}$$

■

EXAMPLE 3 FINDING A SECOND DERIVATIVE

Find the second derivative of $\mathbf{f}(t) = (\cos t)\mathbf{i} + (\sin t)\mathbf{j} + t\mathbf{k}$.

SOLUTION: $\mathbf{f}'(t) = -(\sin t)\mathbf{i} + (\cos t)\mathbf{j} + \mathbf{k}$ and $\mathbf{f}''(t) = -(\cos t)\mathbf{i} - (\sin t)\mathbf{j} + 0\mathbf{k} = -\cos t\mathbf{i} - \sin t\mathbf{j}$.

As with real-valued functions, the following are true:

> **VELOCITY AND ACCELERATION VECTOR FUNCTIONS**
>
> the (vector-valued) velocity function is the derivative of the position function;
>
> the (vector-valued) acceleration function is the derivative of the velocity function and so is the second derivative of the position function.

EXAMPLE 4 PARAMETRIC EQUATIONS OF MOTION

A cannonball shot from a cannon has an initial velocity of 600 m/sec. The muzzle of the cannon is inclined at an angle of 30°. Ignoring air resistance and the rotation of the earth, determine the path of the cannonball.

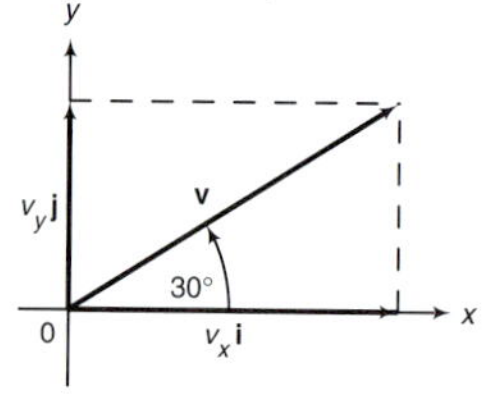

FIGURE 1
Resolution of a velocity vector

SOLUTION: If the earth's rotation is ignored, then all motion takes place in a plane perpendicular to the earth. We place the x- and y-axes so that the mouth of the cannon is at the origin (see Figure 1). The velocity vector **v** can be resolved into its vertical and horizontal components:

$$\mathbf{v} = v_x\mathbf{i} + v_y\mathbf{j}.$$

A unit vector in the direction of **v** is

$$\mathbf{u} = (\cos 30°)\mathbf{i} + (\sin 30°)\mathbf{j} = \frac{\sqrt{3}}{2}\mathbf{i} + \frac{1}{2}\mathbf{j},$$

so initially,

$$\mathbf{v} = |\mathbf{v}|\mathbf{u} = 600\mathbf{u} = 300\sqrt{3}\mathbf{i} + 300\mathbf{j}.$$

The scalar $|\mathbf{v}|$ is called the **speed** of the cannonball. Thus, initially, $v_x = 300\sqrt{3}$ m/sec (the initial speed in the horizontal direction) and $v_y = 300$ m/sec (the initial speed in the vertical direction). Now the vertical acceleration (due to gravity) is

$$a_y = -9.81 \text{ m/sec}^2 \qquad \text{and} \qquad v_y = \int a_y\, dt = -9.81t + C.$$

Since, initially, $v_y(0) = 300$, we have $C = 300$ and

$$v_y = -9.81t + 300.$$

Then,

$$y(t) = \int v_y\, dt = -\frac{9.81t^2}{2} + 300t + C_1,$$

and since $y(0) = 0$ (we start at the origin), we find that

$$y(t) = -\frac{9.81t^2}{2} + 300t.$$

To calculate the x-component of the position vector, we note that, ignoring air resistance, the velocity in the horizontal direction is constant;† that is,

$$v_x = 300\sqrt{3} \text{ m/sec}.$$

Then $x(t) = \int v_x\, dt = 300\sqrt{3}t + C_2$, and since $x(0) = 0$, we obtain

$$x = 300\sqrt{3}t.$$

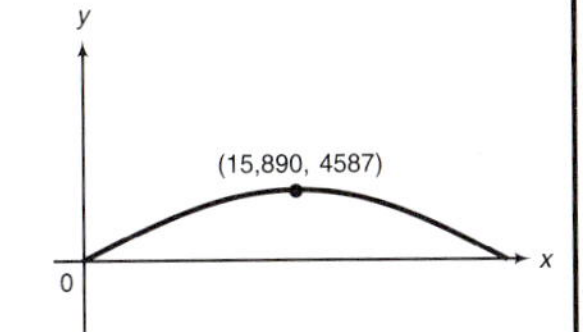

FIGURE 2
Graph of the position function
$y = \dfrac{x}{\sqrt{3}} - \dfrac{9.81}{540{,}000}x^2$

Thus, the **position vector** describing the location of the cannonball is

$$\mathbf{s}(t) = x(t)\mathbf{i} + y(t)\mathbf{j} = 300\sqrt{3}t\mathbf{i} + \left(300t - \frac{9.81t^2}{2}\right)\mathbf{j}.$$

To obtain the Cartesian equation of this curve, we start with

$x = 300\sqrt{3}t$ so $t = x/300\sqrt{3}$ and

$$y = 300\left(\frac{x}{300\sqrt{3}}\right) - \frac{9.81}{2}\frac{x^2}{(300\sqrt{3})^2} = \frac{x}{\sqrt{3}} - \frac{9.81}{540{,}000}x^2.$$

This parabola is sketched in Figure 2.

GEOMETRIC INTERPRETATION OF f′

We now seek a geometric interpretation for $\mathbf{f}'$. As we have seen, the set of vectors $\mathbf{f}(t) = f_1(t)\mathbf{i} + f_2(t)\mathbf{j}$ for t in the domain of $\mathbf{f}$ form a curve C in the plane. We see in Figure 3 that the vector $[\mathbf{f}(t + \Delta t) - \mathbf{f}(t)]/\Delta t$ has the direction of a secant vector whose direction approaches that of the tangent vector as $\Delta t \to 0$. Hence, $\mathbf{f}'(t)$ is tangent to the graph of $\mathbf{f}$ at the point P.

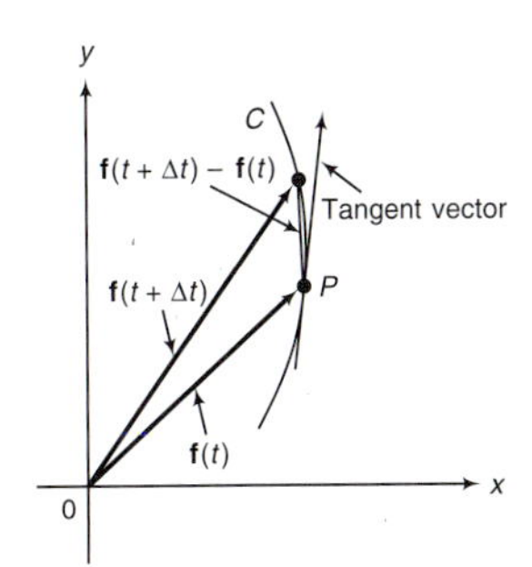

FIGURE 3
The tangent vector to the graph of **f** at P

DEFINITION UNIT TANGENT VECTOR

It is sometimes useful to calculate a **unit tangent vector** to a curve C. This vector is a tangent vector with a magnitude of 1. The unit tangent vector is usually denoted by

† There are no forces acting on the ball in the horizontal direction since the force of gravity acts only in the vertical direction.

T and can be calculated by the formula

$$\mathbf{T}(t) = \frac{\mathbf{f}'(t)}{|\mathbf{f}'(t)|} \tag{6}$$

for any number t, as long as $\mathbf{f}'(t) \neq \mathbf{0}$. This follows since $\mathbf{f}'(t)$ is a tangent vector and $\mathbf{f}'/|\mathbf{f}'|$ is a unit vector. ■

EXAMPLE 5 FINDING A UNIT TANGENT VECTOR

Find the unit tangent vector to the curve $\mathbf{f} = (\ln t)\mathbf{i} + (1/t)\mathbf{j}$ at $t = 1$.

SOLUTION: $\mathbf{f}'(t) = (1/t)\mathbf{i} - (1/t^2)\mathbf{j} = \mathbf{i} - \mathbf{j}$ when $t = 1$. Since $|\mathbf{f}'| = |\mathbf{i} - \mathbf{j}| = \sqrt{2}$, we find that

$$\mathbf{T} = \frac{1}{\sqrt{2}}\mathbf{i} - \frac{1}{\sqrt{2}}\mathbf{j}.$$

EXAMPLE 6 FINDING A UNIT TANGENT VECTOR TO A CIRCULAR HELIX

Find the unit tangent vector to the **circular helix** $\mathbf{f}(t) = \cos t\mathbf{i} + \sin t\mathbf{j} + t\mathbf{k}$ at $t = \pi/3$.

SOLUTION: $\mathbf{f}'(t) = -\sin t\mathbf{i} + \cos t\mathbf{j} + \mathbf{k}$ and $|\mathbf{f}'(t)|^2 = \sin^2 t + \cos^2 t + 1 = 2$ so $|\mathbf{f}'(t)| = \sqrt{2}$. Thus,

$$\mathbf{T}(t) = \frac{1}{\sqrt{2}}(-\sin t\mathbf{i} + \cos t\mathbf{j} + \mathbf{k}) \qquad \text{and}$$

$$\mathbf{T}\left(\frac{\pi}{3}\right) = -\frac{\sqrt{3}}{2\sqrt{2}}\mathbf{i} + \frac{1}{2\sqrt{2}}\mathbf{j} + \frac{1}{\sqrt{2}}\mathbf{k}.$$

We now turn briefly to the integration of vector functions.

DEFINITION INTEGRAL OF A VECTOR FUNCTION

i. Let $\mathbf{f}(t) = f_1(t)\mathbf{i} + f_2(t)\mathbf{j}$ and suppose that the component functions f_1 and f_2 have antiderivatives. Then we define the **indefinite integral** of **f** by

$$\int \mathbf{f}(t)\,dt = \left(\int f_1(t)\,dt\right)\mathbf{i} + \left(\int f_2(t)\,dt\right)\mathbf{j} + \mathbf{C}. \tag{7}$$

where **C** is a constant vector of integration. For each vector **C** the expression (7) is an **antiderivative** of **f**.

ii. If f_1 and f_2 are integrable over the interval $[a, b]$, then we define the **definite integral** of **f** by

$$\int_a^b \mathbf{f}(t)\,dt = \left(\int_a^b f_1(t)\,dt\right)\mathbf{i} + \left(\int_a^b f_2(t)\,dt\right)\mathbf{j}. \tag{8}$$

■

REMARK 1: An antiderivative of $\mathbf{f}$ is a new vector function and is *not unique* since $\int f_1(t)\,dt$ and $\int f_2(t)\,dt$ are not unique.

REMARK 2: The definite integral of $\mathbf{f}$ is a constant vector since $\int_a^b f_1(t)\,dt$ and $\int_a^b f_2(t)\,dt$ are constants.

EXAMPLE 7 INTEGRATING A VECTOR FUNCTION

Let $\mathbf{f}(t) = (\cos t)\mathbf{i} + (\sin t)\mathbf{j}$. Calculate the following:

a. $\int \mathbf{f}(t)\,dt$ **b.** $\int_0^{\pi/2} \mathbf{f}(t)\,dt$

SOLUTION:

a.
$$\int \mathbf{f}(t)\,dt = \left(\int \cos t\,dt\right)\mathbf{i} + \left(\int \sin t\,dt\right)\mathbf{j}$$
$$= (\sin t + C_1)\mathbf{i} + (-\cos t + C_2)\mathbf{j}$$
$$= (\sin t)\mathbf{i} - (\cos t)\mathbf{j} + \mathbf{C},$$

where $\mathbf{C} = C_1\mathbf{i} + C_2\mathbf{j}$ is a constant vector.

b.
$$\int_0^{\pi/2} \mathbf{f}(t)\,dt = \left(\sin t\Big|_0^{\pi/2}\right)\mathbf{i} + \left(-\cos t\Big|_0^{\pi/2}\right)\mathbf{j} = \mathbf{i} + \mathbf{j}$$

PROBLEMS 2.3

SELF-QUIZ

I. If $\mathbf{f}(t) = -3\mathbf{i} + t^2\mathbf{j}$, then $\mathbf{f}'(t) =$ ________.
a. $2t$ **b.** $-3\mathbf{i} + 2t\mathbf{j}$
c. $2t\mathbf{i}$ **d.** $2t\mathbf{j}$

II. If $\mathbf{f}(t) = 15\mathbf{i} - 37\mathbf{j}$, then $\mathbf{f}'(t) =$ ________.
a. 0 **b.** $\mathbf{0}$
c. -22 **d.** $15t\mathbf{i} - 37t\mathbf{j}$

III. If $\mathbf{f}'(t) = \mathbf{j}$ and $\mathbf{f}(0) = 2\mathbf{i} + \mathbf{j}$, then $\mathbf{f}(t) =$ ________.
a. $2\mathbf{i} + t\mathbf{j}$ **b.** $2\mathbf{i} + (1 + t)\mathbf{j}$
c. $(1 + t)\mathbf{j}$ **d.** $t\mathbf{j}$

IV. If $\mathbf{f}(t) = 3\cos t\mathbf{i} + 4\sin s\mathbf{j}$, then the unit tangent vector when $t = \pi/4$ is ________.
a. $0.6\mathbf{i} + 0.8\mathbf{j}$ **b.** $-0.8\mathbf{i} + 0.6\mathbf{j}$
c. $-0.6\mathbf{i} + 0.8\mathbf{j}$ **d.** $(-3/\sqrt{2})\mathbf{i} + (4/\sqrt{2})\mathbf{j}$

In Problems 1–10, calculate the first and second derivatives of the given vector function and then determine their domains.

1. $\mathbf{f}(t) = t\mathbf{i} - t^5\mathbf{j}$

2. $\mathbf{f}(t) = (1 + t^2)\mathbf{i} + \dfrac{2}{t}\mathbf{j}$

3. $\mathbf{f}(t) = (\sin 2t)\mathbf{i} + (\cos 3t)\mathbf{j}$

4. $\mathbf{f}(t) = \dfrac{t}{1 + t}\mathbf{i} + \dfrac{1}{\sqrt{t}}\mathbf{j}$

5. $\mathbf{f}(t) = (\ln t)\mathbf{i} + e^{3t}\mathbf{j}$

6. $\mathbf{f}(t) = e^t(\sin t)\mathbf{i} + e^t(\cos t)\mathbf{j}$

7. $\mathbf{f}(t) = (\tan t)\mathbf{i} + (\sec t)\mathbf{j}$

8. $\mathbf{f}(t) = (\tan^{-1} t)\mathbf{i} + (\sin^{-1} t)\mathbf{j}$

9. $\mathbf{f}(t) = (\ln \cos t)\mathbf{i} + (\ln \sin t)\mathbf{j}$

10. $\mathbf{f}(t) = (\cosh t)\mathbf{i} + (\sinh t)\mathbf{j}$

In Problems 11–26, calculate the unit tangent vector $\mathbf{T}$ for the specified value of t.

11. $\mathbf{f}(t) = t^2\mathbf{i} + t^3\mathbf{j}$; $t = 1$

12. $\mathbf{f}(t) = t\mathbf{i} + \dfrac{1}{t}\mathbf{j}$; $t = 1$

13. $\mathbf{f}(t) = (\cos t)\mathbf{i} + (\sin t)\mathbf{j}$; $t = 0$

14. $\mathbf{f}(t) = (\cos t)\mathbf{i} + (\sin t)\mathbf{j};\ t = \pi/2$
15. $\mathbf{f}(t) = (\cos t)\mathbf{i} + (\sin t)\mathbf{j};\ t = \pi/4$
16. $\mathbf{f}(t) = (\cos t)\mathbf{i} + (\sin t)\mathbf{j};\ t = 3\pi/4$
17. $\mathbf{f}(t) = (\tan t)\mathbf{i} + (\sec t)\mathbf{j};\ t = 0$
18. $\mathbf{f}(t) = (\ln t)\mathbf{i} + e^{2t}\mathbf{j};\ t = 1$
19. $\mathbf{f}(t) = \dfrac{t}{t+1}\mathbf{i} + \dfrac{t+1}{t}\mathbf{j};\ t = 2$
20. $\mathbf{f}(t) = \dfrac{t+1}{t}\mathbf{i} + \dfrac{t}{t+1}\mathbf{j};\ t = 2$
21. $\mathbf{g}(t) = t\mathbf{i} + t^2\mathbf{j} + t^3\mathbf{k};\ t = 1$
22. $\mathbf{g}(t) = t^3\mathbf{i} + t^5\mathbf{j} + t^7\mathbf{k};\ t = 1$
23. $\mathbf{g}(t) = t\mathbf{i} + e^t\mathbf{j} + e^{-t}\mathbf{k};\ t = 0$
24. $\mathbf{g}(t) = t^2\mathbf{i} + t^2\mathbf{j} + t^{5/2}\mathbf{k};\ t = 4$
25. $\mathbf{g}(t) = 4(\cos 2t)\mathbf{i} + 9(\sin 2t)\mathbf{j} + t\mathbf{k};\ t = \pi/4$
26. $\mathbf{g}(t) = (\cosh t)\mathbf{i} + (\sinh t)\mathbf{j} + t^2\mathbf{k};\ t = 0$

In Problems 27–36, calculate the given indefinite or definite integral.

27. $\int [(\sin 2t)\mathbf{i} + e^t\mathbf{j}]\, dt$

28. $\int (t^{-1/2}\mathbf{i} + t^{1/2}\mathbf{j})\, dt$

29. $\int_0^{\pi/4} [(\cos 2t)\mathbf{i} - (\sin 2t)\mathbf{j}]\, dt$

30. $\int_1^e \left[\left(\frac{1}{t}\right)\mathbf{i} - \left(\frac{3}{t}\right)\mathbf{j}\right] dt$

31. $\int [(\ln t)\mathbf{i} + te^t\mathbf{j}]\, dt$

32. $\int [(\tan t)\mathbf{i} + (\sec t)\mathbf{j}]\, dt$

33. $\int_0^2 (t^2\mathbf{i} - t^3\mathbf{j} + t^4\mathbf{k})\, dt$

34. $\int_0^1 [(\sinh t)\mathbf{i} - (\cosh t)\mathbf{j} - (\tanh t)\mathbf{k}]\, dt$

35. $\int 2t[(\sin t^2)\mathbf{i} + (\cos t^2)\mathbf{j} + e^{t^2}\mathbf{k}]\, dt$

36. $\int t[(\cos t)\mathbf{i} + (\sin t)\mathbf{j} + e^t\mathbf{k}]\, dt$

37. Suppose $\mathbf{f}'(t) = t^3\mathbf{i} - t^5\mathbf{j}$ and $\mathbf{f}(0) = 2\mathbf{i} + 5\mathbf{j}$. Find $\mathbf{f}(t)$.

38. Suppose $\mathbf{f}'(t) = (1/\sqrt{t})\mathbf{i} + \sqrt{t}\mathbf{j}$ and $\mathbf{f}(1) = -2\mathbf{i} + \mathbf{j}$. Find $\mathbf{f}(t)$.

39. Suppose $\mathbf{f}'(t) = (\cos t)\mathbf{i} + (\sin t)\mathbf{k}$ and $\mathbf{f}(\pi/2) = \mathbf{i}$. Find $\mathbf{f}(t)$.

40. Suppose $\mathbf{f}'(t) = (\cos t)\mathbf{i} - (\sin t)\mathbf{k}$ and $\mathbf{f}(0) = \mathbf{j} + \mathbf{k}$. Find $\mathbf{f}(t)$.

41. The ellipse $(x/a)^2 + (y/b)^2 = 1$ can be written parametrically as $x = a\cos\theta$, $y = b\sin\theta$. Find a unit tangent vector to this ellipse at $\theta = \pi/4$.

42. Find a unit tangent vector to the cycloid $\mathbf{f}(\alpha) = r(\alpha - \sin\alpha)\mathbf{i} + r(1 - \cos\alpha)\mathbf{j}$ for $\alpha = 0$.

***43.** Find, for $\theta = \pi/6$, a unit tangent vector to the hypocycloid of Problem 2.1.45, assuming that $a = 5$ and $b = 2$.

44. A cannonball is shot upward from ground level at an angle of 45° with an initial speed of 1300 ft/sec. Find a parametric representation of the path of the cannonball; then find a Cartesian equation for this path.

45. How many feet (horizontally) does the cannonball of the preceding problem travel before it hits the ground?

46. An object is thrown down from the top of a 150 m building at an angle of 30° (below the horizontal) with an initial velocity of 100 m/sec. Determine a parametric representation of the path of the object. [*Hint:* Draw a picture.]

47. When the object in the preceding problem hits the ground, how far is it from the base of the building?

ANSWERS TO SELF-QUIZ

I. d **II.** b **III.** b **IV.** c

2.4 SOME DIFFERENTIATION FORMULAS

In your first calculus course, you studied a number of rules for differentiation. Many of these rules carry over with little change to the differentiation of a vector function. The proof of Theorem 1 is a straightforward application of one-variable differentiation rules and is left as an exercise (see Problems 31–36).

THEOREM 1

Let $\mathbf{f}$ and $\mathbf{g}$ be vector functions that are differentiable in an interval I. Let the scalar function h be differentiable in I. Finally, let α be a scalar and let $\mathbf{v}$ be a constant vector. Then, on I, we have

i. $\mathbf{f} + \mathbf{g}$ is differentiable and

$$\frac{d}{dt}(\mathbf{f} + \mathbf{g}) = \frac{d\mathbf{f}}{dt} + \frac{d\mathbf{g}}{dt} = \mathbf{f}' + \mathbf{g}'. \tag{1}$$

ii. $\alpha\mathbf{f}$ is differentiable and

$$\frac{d}{dt}\alpha\mathbf{f} = \alpha\frac{d\mathbf{f}}{dt} = \alpha\mathbf{f}'. \tag{2}$$

iii. $\mathbf{v} \cdot \mathbf{f}$ is differentiable and

$$\frac{d}{dt}\mathbf{v} \cdot \mathbf{f} = \mathbf{v} \cdot \frac{d\mathbf{f}}{dt} = \mathbf{v} \cdot \mathbf{f}'. \tag{3}$$

iv. $h\mathbf{f}$ is differentiable and

$$\frac{d}{dt}h\mathbf{f} = h\frac{d\mathbf{f}}{dt} + \frac{dh}{dt}\mathbf{f} = h\mathbf{f}' + h'\mathbf{f}. \tag{4}$$

v. $\mathbf{f} \cdot \mathbf{g}$ is differentiable and

$$\frac{d}{dt}\mathbf{f} \cdot \mathbf{g} = \mathbf{f} \cdot \frac{d\mathbf{g}}{dt} + \frac{d\mathbf{f}}{dt} \cdot \mathbf{g} = \mathbf{f} \cdot \mathbf{g}' + \mathbf{f}' \cdot \mathbf{g}. \tag{5}$$

vi. If $\mathbf{f}$ and $\mathbf{g}$ are differentiable vector functions in $\mathbb{R}^3$, then $\mathbf{f} \times \mathbf{g}$ is differentiable and

$$(\mathbf{f} \times \mathbf{g})' = (\mathbf{f}' \times \mathbf{g}) + (\mathbf{f} \times \mathbf{g}'). \tag{6}$$

■

EXAMPLE 1 USING THE DIFFERENTIATION FORMULAS

Let $\mathbf{f}(t) = t\mathbf{i} + t^3\mathbf{j}$, $\mathbf{g}(t) = (\cos t)\mathbf{i} + (\sin t)\mathbf{j}$ and $\mathbf{v} = 2\mathbf{i} - 3\mathbf{j}$. Calculate **(a)** $(\mathbf{f} + \mathbf{g})'$, **(b)** $(\mathbf{v} \cdot \mathbf{f})'$, and **(c)** $(\mathbf{f} \cdot \mathbf{g})'$.

SOLUTION:

a. $(\mathbf{f} + \mathbf{g})' = \mathbf{f}' + \mathbf{g}' = (\mathbf{i} + 3t^2\mathbf{j}) + [-(\sin t)\mathbf{i} + (\cos t)\mathbf{j}]$
$= (1 - \sin t)\mathbf{i} + (3t^2 + \cos t)\mathbf{j}$

b. $(\mathbf{v} \cdot \mathbf{f})' = \mathbf{v} \cdot \mathbf{f}' = (2\mathbf{i} - 3\mathbf{j}) \cdot (\mathbf{i} + 3t^2\mathbf{j}) = 2 - 9t^2$

c. $(\mathbf{f} \cdot \mathbf{g})' = \mathbf{f} \cdot \mathbf{g}' + \mathbf{f}' \cdot \mathbf{g}$
$= (t\mathbf{i} + t^3\mathbf{j}) \cdot [-(\sin t)\mathbf{i} + (\cos t)\mathbf{j}] + (\mathbf{i} + 3t^2\mathbf{j}) \cdot [(\cos t)\mathbf{i} + (\sin t)\mathbf{j}]$
$= -t \sin t + t^3 \cos t + \cos t + 3t^2 \sin t$
$= (\cos t)(t^3 + 1) + (\sin t)(3t^2 - t)$

EXAMPLE 2 THE DERIVATIVE OF $\mathbf{f} \cdot \mathbf{f}'$ IS ZERO IF $|\mathbf{f}(t)|$ IS CONSTANT

Let $\mathbf{f}(t) = (\cos t)\mathbf{i} + (\sin t)\mathbf{j}$. Calculate $\mathbf{f} \cdot \mathbf{f}'$.

SOLUTION:

$$\begin{aligned}\mathbf{f} \cdot \mathbf{f}' &= [(\cos t)\mathbf{i} + (\sin t)\mathbf{j}] \cdot [-(\sin t)\mathbf{i} + (\cos t)\mathbf{j}] \\ &= -\cos t \sin t + \sin t \cos t = 0\end{aligned}$$

Example 2 can be generalized to the following interesting result.

THEOREM 2

Let $\mathbf{f}(t) = f_1\mathbf{i} + f_2\mathbf{j}$ be a differentiable vector function such that $|\mathbf{f}(t)| = \sqrt{f_1^2(t) + f_2^2(t)}$ is constant. Then

$$\mathbf{f} \cdot \mathbf{f}' = 0.$$

PROOF: Suppose $|\mathbf{f}(t)| = C$, a constant. Then

$$\mathbf{f} \cdot \mathbf{f} = f_1^2 + f_2^2 = |\mathbf{f}|^2 = C^2,$$

so

$$\frac{d}{dt}(\mathbf{f} \cdot \mathbf{f}) = \frac{d}{dt}C^2 = 0.$$

But

$$\frac{d}{dt}(\mathbf{f} \cdot \mathbf{f}) = \mathbf{f} \cdot \mathbf{f}' + \mathbf{f}' \cdot \mathbf{f} = 2\mathbf{f} \cdot \mathbf{f}' = 0,$$

so

$$\mathbf{f} \cdot \mathbf{f}' = 0. \quad \blacksquare$$

NOTE: In Example 2, $|\mathbf{f}(t)| = \sqrt{\cos^2 t + \sin^2 t} = 1$ for every t.

UNIT NORMAL VECTOR TO A PLANE CURVE

There is an interesting geometric application of Theorem 2. Let $\mathbf{f}(t) = f_1(t)\mathbf{i} + f_2(t)\mathbf{j}$ be a differentiable vector function. For all t for which $\mathbf{f}'(t) \neq \mathbf{0}$, we let $\mathbf{T}(t)$ denote the unit tangent vector to the curve $\mathbf{f}(t)$. Then since $|\mathbf{T}(t)| = 1$ by the definition of a *unit* tangent vector, we have, from Theorem 2 [assuming that $\mathbf{T}'(t)$ exists],

$$\mathbf{T}(t) \cdot \mathbf{T}'(t) = 0. \tag{7}$$

That is, $\mathbf{T}'(t)$ is *orthogonal* to $\mathbf{T}(t)$.

DEFINITION UNIT NORMAL VECTOR

The vector $\mathbf{T}'$ which is orthogonal to the tangent line is called a **normal vector**[†] to the curve $\mathbf{f}$. Whenever $\mathbf{T}'(t) \neq \mathbf{0}$, we can define the **unit normal vector** to the curve $\mathbf{f}$ at

[†] From the Latin *norma,* meaning "square," the carpenter's square. Until the 1830s the English "normal" meant standing at right angles to the ground.

t as

$$\mathbf{n}(t) = \frac{\mathbf{T}'(t)}{|\mathbf{T}'(t)|}. \tag{8}$$

■

NOTE: If $\mathbf{T}'(t) = \mathbf{0}$, then we may define $\mathbf{n}(t)$ to be one of the two unit vectors that are orthogonal to $\mathbf{T}(t)$. For example, if $\mathbf{f}(t) = t\mathbf{i} + 2t\mathbf{j}$, then $\mathbf{f}'(t) = \mathbf{i} + 2\mathbf{j}$ and $\mathbf{T}(t) = \frac{1}{\sqrt{5}}\mathbf{i} + \frac{2}{\sqrt{5}}\mathbf{j}$. Two choices for $\mathbf{n}(t)$ are $\mathbf{n}_1(t) = \frac{2}{\sqrt{5}}\mathbf{i} - \frac{1}{\sqrt{5}}\mathbf{j}$ and $\mathbf{n}_2(t) = -\frac{2}{\sqrt{5}}\mathbf{i} + \frac{1}{\sqrt{5}}\mathbf{j}$.

EXAMPLE 3

FINDING A UNIT NORMAL VECTOR TO THE UNIT CIRCLE

Calculate a unit normal vector to the curve $\mathbf{f}(t) = (\cos t)\mathbf{i} + (\sin t)\mathbf{j}$ at $t = \pi/4$.

SOLUTION: First, we calculate $\mathbf{f}'(t) = -(\sin t)\mathbf{i} + (\cos t)\mathbf{j}$, and since $|\mathbf{f}'(t)| = 1$, we find that

$$\mathbf{T}(t) = \frac{\mathbf{f}'(t)}{|\mathbf{f}'(t)|} = -(\sin t)\mathbf{i} + (\cos t)\mathbf{j}.$$

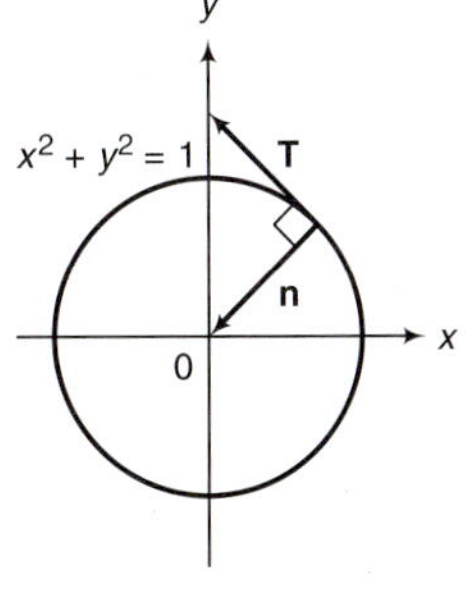

FIGURE 1
Unit tangent and unit normal vectors to the unit circle

Then

$$\mathbf{T}'(t) = -(\cos t)\mathbf{i} - (\sin t)\mathbf{j} = \mathbf{n}(t)$$

since $|\mathbf{T}'(t)| = 1$, so that at $t = \pi/4$,

$$\mathbf{n} = -\frac{1}{\sqrt{2}}\mathbf{i} - \frac{1}{\sqrt{2}}\mathbf{j}.$$

This is sketched in Figure 1 [remember that $\mathbf{f}(t) = (\cos t)\mathbf{i} + (\sin t)\mathbf{j}$ is the parametric equation of the unit circle]. The reason the vector $\mathbf{n}$ points inward is that it is the negative of the position vector at $t = \pi/4$.

EXAMPLE 4

FINDING A UNIT NORMAL VECTOR

Calculate a unit normal vector to the curve $\mathbf{f}(t) = [(t^3/3) - t]\mathbf{i} + t^2\mathbf{j}$ at $t = 3$.

SOLUTION: Here $\mathbf{f}'(t) = (t^2 - 1)\mathbf{i} + 2t\mathbf{j}$ and

$$\begin{aligned}|\mathbf{f}'(t)| &= \sqrt{(t^2-1)^2 + 4t^2} = \sqrt{t^4 - 2t^2 + 1 + 4t^2} \\ &= \sqrt{t^4 + 2t^2 + 1} = t^2 + 1.\end{aligned}$$

Thus

$$\mathbf{T}(t) = \frac{\mathbf{f}'(t)}{|\mathbf{f}'(t)|} = \frac{t^2-1}{t^2+1}\mathbf{i} + \frac{2t}{t^2+1}\mathbf{j}.$$

Then

$$\mathbf{T}'(t) = \frac{d}{dt}\left(\frac{t^2-1}{t^2+1}\right)\mathbf{i} + \frac{d}{dt}\left(\frac{2t}{t^2+1}\right)\mathbf{j} = \frac{4t}{(t^2+1)^2}\mathbf{i} + \frac{2-2t^2}{(t^2+1)^2}\mathbf{j}.$$

Finally,

$$\begin{aligned}|\mathbf{T}'(t)| &= \left\{\left[\frac{4t}{(t^2+1)^2}\right]^2 + \left[\frac{2-2t^2}{(t^2+1)^2}\right]^2\right\}^{1/2} \\ &= \frac{1}{(t^2+1)^2}(16t^2+4-8t^2+4t^4)^{1/2} \\ &= \frac{1}{(t^2+1)^2}\sqrt{4t^4+8t^2+4} = \frac{2}{(t^2+1)^2}\sqrt{t^4+2t^2+1} \\ &= \frac{2(t^2+1)}{(t^2+1)^2} = \frac{2}{t^2+1},\end{aligned}$$

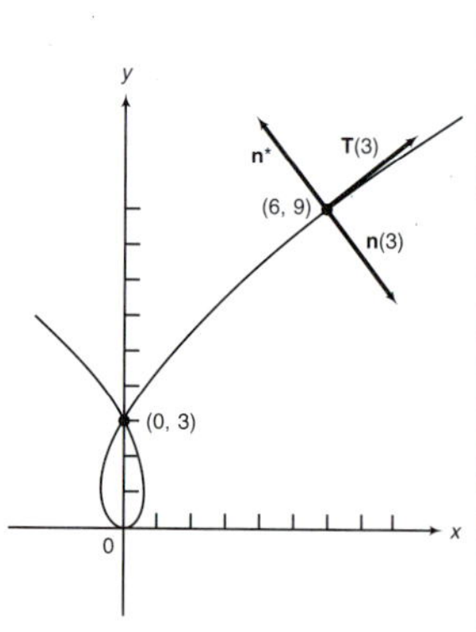

FIGURE 2
Graph of
$\mathbf{f}(t) = \left(\frac{t^3}{3} - t\right)\mathbf{i} + t^2\mathbf{j}$

There are two unit vectors orthogonal to the unit tangent vector when $t = 3$.

so that

$$\begin{aligned}\mathbf{n}(t) &= \frac{\mathbf{T}'(t)}{|\mathbf{T}'(t)|} = \frac{t^2+1}{2}\left[\frac{4t}{(t^2+1)^2}\mathbf{i} + \frac{2-2t^2}{(t^2+1)^2}\mathbf{j}\right] \\ &= \frac{2t}{t^2+1}\mathbf{i} + \frac{1-t^2}{t^2+1}\mathbf{j}.\end{aligned}$$

At $t = 3$, $\mathbf{T}(t) = \frac{4}{5}\mathbf{i} + \frac{3}{5}\mathbf{j}$ and $\mathbf{n} = \frac{3}{5}\mathbf{i} - \frac{4}{5}\mathbf{j}$. Note that $\mathbf{T}(3) \cdot \mathbf{n}(3) = 0$. This is sketched in Figure 2.

THE DIRECTION OF THE NORMAL VECTOR TO A PLANE CURVE

In Figure 2, we see that there are two vectors perpendicular to $\mathbf{T}$ at the point P: one pointing "inward" ($\mathbf{n}$) and one pointing "outward" ($\mathbf{n}^*$). In general, the situation is as shown in Figure 3. How do we know which of the two vectors is $\mathbf{n}$? If φ is the direction of $\mathbf{T}$, then since $\mathbf{T}$ is a unit vector,

$$\mathbf{T} = (\cos\varphi)\mathbf{i} + (\sin\varphi)\mathbf{j},$$

so that

$$\frac{d\mathbf{T}}{dt} = \frac{d\mathbf{T}}{d\varphi}\frac{d\varphi}{dt}$$

and

$$\left|\frac{d\mathbf{T}}{dt}\right| = \left|\frac{d\mathbf{T}}{d\varphi}\right|\left|\frac{d\varphi}{dt}\right| = \left|\frac{d\varphi}{dt}\right| \qquad \left(\text{since } \left|\frac{d\mathbf{T}}{d\varphi}\right| = |-\sin\varphi\mathbf{i} + \cos\varphi\mathbf{j}| = 1\right).$$

Then

$$\mathbf{n}(t) = \frac{d\mathbf{T}/dt}{|d\mathbf{T}/dt|} = \frac{d\mathbf{T}}{d\varphi}\frac{d\varphi/dt}{|d\varphi/dt|} = \frac{d\varphi/dt}{|d\varphi/dt|}[-(\sin\varphi)\mathbf{i} + (\cos\varphi)\mathbf{j}].$$

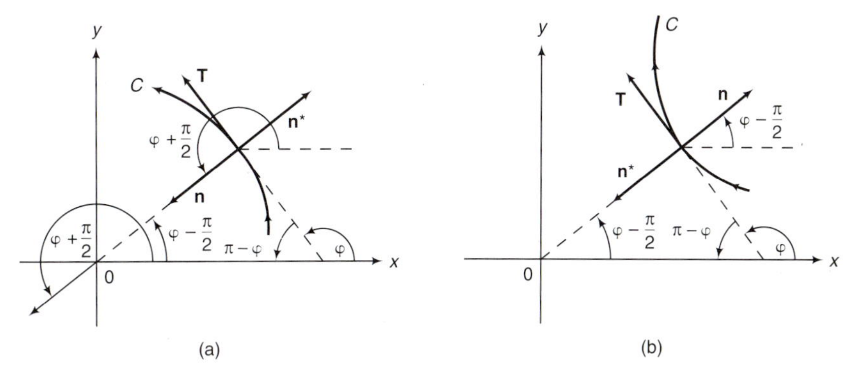

FIGURE 3
The unit normal vector **n** points in the direction of the concave side of the curve

Since $d\varphi/dt$ is a scalar, $(d\varphi/dt)/|d\varphi/dt| = \pm 1$. It is $+1$ in Figure 3(a) since φ increases as t increases so that $d\varphi/dt > 0$, while it is -1 in Figure 3(b) since φ decreases as t increases. In Figure 3(a), then,

$$\mathbf{n}(t) = -(\sin \varphi)\mathbf{i} + (\cos \varphi)\mathbf{j} = \cos[\varphi + (\pi/2)]\mathbf{i} + \sin[\varphi + (\pi/2)]\mathbf{j}.$$

That is, the direction of **n** is $\varphi + (\pi/2)$ and **n** must point inward (on the concave side of C) as in Figure 3(a). In the other case, $(d\varphi/dt)/|d\varphi/dt| = -1$, so

$$\mathbf{n}(t) = (\sin \varphi)\mathbf{i} - (\cos \varphi)\mathbf{j} = \cos[\varphi - (\pi/2)]\mathbf{i} + \sin[\varphi - (\pi/2)]\mathbf{j}.$$

That is, the direction of **n** is $\varphi - (\pi/2)$ and so **n** points as in Figure 3(b). In either case:

> The unit normal vector **n** always points in the direction of the concave side of the curve.

REMARK: It is more difficult to define a unit normal vector in space ($\mathbb{R}^3$) geometrically because there is a whole ''plane-full'' of vectors orthogonal to a given tangent vector **T**. We will define normal vectors in $\mathbb{R}^3$ in Section 2.6, after we have discussed the notion of curvature.

It is easy to extend the results of this section to a function whose range is in $\mathbb{R}^n$.

DEFINITION VECTOR-VALUED FUNCTION

Let $f_1, f_2, \ldots, f_n$ be functions of the real variable t. Then for all values of t for which $f_1(t), f_2(t), \ldots, f_n(t)$ are defined, the **vector-valued function f** is given by

$$\mathbf{f}(t) = (f_1(t), f_2(t), \ldots, f_n(t)). \tag{9}$$

The functions $f_1, f_2, \ldots, f_n$ are called the **component functions** of **f**. ■

NOTATION: We will denote a vector-valued function by

$\mathbf{f}\colon \mathbb{R} \to \mathbb{R}^n$.

DEFINITION CURVE IN $\mathbb{R}^n$

Suppose that the interval $[a, b]$ is in the domain of each of the functions $f_1, f_2, \ldots, f_n$. If $\mathbf{f}\colon \mathbb{R} \to \mathbb{R}^n$ is given by (9), then the set of vectors

$$C = \{f_i(t)\colon a \le t \le b\} \tag{10}$$

is called a **curve in** $\mathbb{R}^n$. ■

DEFINITION DIFFERENTIABILITY

Let $\mathbf{f}(t) = (f_1(t), f_2(t), \ldots, f_n(t))$ where each $f_i(t)$ is differentiable at t_0.

i. Then $\mathbf{f}$ is **differentiable** at t_0 and

$$\mathbf{f}'(t_0) = (f_1'(t_0), f_2'(t_0), \ldots, f_n'(t_0)). \tag{11}$$

ii. If f_i' is continuous at t_0 for $i = 1, 2, \ldots, n$, then f is said to be **continuously differentiable** at t_0. ■

EXAMPLE 5 A DERIVATIVE IN $\mathbb{R}^n$

If $\mathbf{f}(t) = (t, t^2, \ldots, t^n)$, compute $\mathbf{f}'(t)$.

SOLUTION: $\mathbf{f}'(t) = (1, 2t, 3t^2, \ldots, nt^{n-1})$.

EXAMPLE 6 THE DERIVATIVE OF A CONSTANT FUNCTION IN $\mathbb{R}^n$

If $\mathbf{f}(t) = (c_1, c_2, \ldots, c_n)$, a constant vector, then $\mathbf{f}'(t) = (0, 0, \ldots, 0) = \mathbf{0}$.

EXAMPLE 7 THE DERIVATIVE OF A SCALAR FUNCTION MULTIPLE OF $\mathbf{x}$

If $\mathbf{f}(t) = f(t)\mathbf{x}$ where $\mathbf{x} = (x_1, x_2, \ldots, x_n)$ is a constant vector and f is a scalar function, then

$$\mathbf{f}'(t) = \frac{d}{dt}(f(t)x_1, f(t)x_2, \ldots, f(t)x_n) = (f'(t)x_1, f'(t)x_2, \ldots, f'(t)x_n) = f'(t)\mathbf{x}.$$

In particular, if $f(t) = t$, then

$$\frac{d}{dt}(t\mathbf{x}) = \mathbf{x}.$$

The proof of the following theorem is virtually identical to the proof of Theorem 1 and is omitted.

THEOREM 3

Let $\mathbf{f}\colon \mathbb{R} \to \mathbb{R}^n$ and $\mathbf{g}\colon \mathbb{R} \to \mathbb{R}^n$ be differentiable on an open interval (a, b). Let the scalar function h be differentiable on (a, b). Let α be a scalar and let $\mathbf{v}$ be a constant vector in $\mathbb{R}^n$. Then

i. $\mathbf{f} + \mathbf{g}$ is differentiable and

$$\frac{d}{dt}(\mathbf{f} + \mathbf{g}) = \frac{d\mathbf{f}}{dt} + \frac{d\mathbf{g}}{dt} = \mathbf{f}' + \mathbf{g}'. \tag{12}$$

ii. $\alpha\mathbf{f}$ is differentiable and

$$\frac{d}{dt}\alpha\mathbf{f} = \alpha\frac{d\mathbf{f}}{dt} = \alpha\mathbf{f}'. \tag{13}$$

iii. $\mathbf{v} \cdot \mathbf{f}$ is differentiable and

$$\frac{d}{dt}\mathbf{v} \cdot \mathbf{f} = \mathbf{v} \cdot \frac{d\mathbf{f}}{dt} = \mathbf{v} \cdot \mathbf{f}'. \tag{14}$$

iv. $h\mathbf{f}$ is differentiable and

$$\frac{d}{dt}h\mathbf{f} = h\frac{d\mathbf{f}}{dt} + \frac{dh}{dt}\mathbf{f} = h\mathbf{f}' + h'\mathbf{f}. \tag{15}$$

v. $\mathbf{f} \cdot \mathbf{g}$ is differentiable and

$$\frac{d}{dt}\mathbf{f} \cdot \mathbf{g} = \mathbf{f} \cdot \frac{d\mathbf{g}}{dt} + \frac{d\mathbf{f}}{dt} \cdot \mathbf{g} = \mathbf{f} \cdot \mathbf{g}' + \mathbf{f}' \cdot \mathbf{g}. \tag{16}$$

■

THEOREM 4

Let $\mathbf{f}: \mathbb{R} \to \mathbb{R}^n$ be differentiable in $\mathbb{R}$ and suppose that $|\mathbf{f}(t)|$ is constant. Then $\mathbf{f}(t) \cdot \mathbf{f}'(t) = 0$ for every t in $\mathbb{R}$.

PROOF: We have $\mathbf{f} \cdot \mathbf{f} = |\mathbf{f}|^2 = C$, a constant, so that

$$0 = \frac{d}{dt}(\mathbf{f} \cdot \mathbf{f}) = \underbrace{\mathbf{f} \cdot \mathbf{f}' + \mathbf{f}' \cdot \mathbf{f}}_{\text{from (16)}} = 2(\mathbf{f} \cdot \mathbf{f}'),$$

so that $\mathbf{f} \cdot \mathbf{f}' = 0$. ■

Let $\mathbf{f}(t)$ be a differentiable curve in $\mathbb{R}^n$ for $t \in [a, b]$. Then at any point $t \in (a, b)$ with $\mathbf{f}'(t) \neq \mathbf{0}$, we define the **unit tangent vector** $\mathbf{T}(t)$ to the curve by

UNIT TANGENT VECTOR

$$\mathbf{T}(t) = \frac{\mathbf{f}'(t)}{|\mathbf{f}'(t)|}. \tag{17}$$

Since $|\mathbf{T}(t)| = 1$, $|\mathbf{T}(t)|$ is constant and, by Theorem 4,

$$\mathbf{T}(t) \cdot \mathbf{T}'(t) = 0. \tag{18}$$

We see here that many of the ideas in $\mathbb{R}^2$ and $\mathbb{R}^3$ can be extended quite naturally to $\mathbb{R}^n$. The major difference is that, if $n > 3$, we cannot sketch things in $\mathbb{R}^n$.

PROBLEMS 2.4

SELF-QUIZ

I. $\frac{d}{dt}\{[(\cos 3t)\mathbf{i} + (\sin 4t)\mathbf{j}] \cdot [-4\mathbf{i} + 3\mathbf{j}]\} =$ ________.

a. $-3(\sin 3t)\mathbf{i} + 4(\cos 4t)\mathbf{j}$
b. $[-3(\sin 3t)\mathbf{i} + 4(\cos 4t)\mathbf{j}] \cdot [-4\mathbf{i} + 3\mathbf{j}]$
c. $-12 \sin 3t + 12 \cos 4t$
d. $12(\sin 3t + \cos 4t)$

II. If $\mathbf{f}(t) = (\cos 3t)\mathbf{i} + (\sin 3t)\mathbf{j}$, then $\mathbf{f} \cdot \mathbf{f}' =$ ________.

a. 0 **b.** $\mathbf{0}$
c. 1 **d.** $3(\cos 3t)^2 - 3(\sin 3t)^2$

III. If $\mathbf{f}(t) = (\cos 2t)\mathbf{i} + (\sin 2t)\mathbf{j}$, then the unit tangent vector is ________.

a. $-2(\sin 2t)\mathbf{i} + 2(\cos 2t)\mathbf{j}$
b. $-2(\sin t)\mathbf{i} + 2(\cos t)\mathbf{j}$
c. $-(\sin t)\mathbf{i} + (\cos t)\mathbf{j}$
d. $-(\sin 2t)\mathbf{i} + (\cos 2t)\mathbf{j}$

IV. If $\mathbf{f}(t) = (\cos 2t)\mathbf{i} + (\sin 2t)\mathbf{j}$, then the unit normal vector when $t = \pi/6$ is ________.

a. $-\mathbf{i} - \sqrt{3}\mathbf{j}$ **b.** $\frac{1}{2}\mathbf{i} + (\sqrt{3}/2)\mathbf{j}$
c. $-\frac{1}{2}\mathbf{i} - (\sqrt{3}/2)\mathbf{j}$ **d.** $-(\sqrt{3}/2)\mathbf{i} - \frac{1}{2}\mathbf{j}$

In Problems 1–8, calculate the indicated derivative.

1. $\frac{d}{dt}\{[2t\mathbf{i} + (\cos t)\mathbf{j}] + [(\tan t)\mathbf{i} - (\sec t)\mathbf{j}]\}$

2. $\frac{d}{dt}\left[(t^3\mathbf{i} - t^5\mathbf{j}) \cdot \left(\frac{-1}{t^3}\mathbf{i} + \frac{1}{t^5}\mathbf{j}\right)\right]$

3. $\frac{d}{dt}[(t^3\mathbf{i} - t^5\mathbf{j}) + (-t^3\mathbf{i} + t^5\mathbf{j})]$

4. $\frac{d}{dt}\{[(\sinh t)\mathbf{i} + (\cosh t)\mathbf{j}] \cdot [(\sin t)\mathbf{i} + (\cos t)\mathbf{j}]\}$

5. $\frac{d}{dt}\{[\sin^{-1} t\mathbf{i} + \cos^{-1} t\mathbf{j}] + [\cos t\mathbf{i} + \sin t\mathbf{j}]\}$

6. $\frac{d}{dt}\{[(\sin^{-1} t)\mathbf{i} + (\cos^{-1} t)\mathbf{j}] \cdot [2\mathbf{i} - 10\mathbf{j}]\}$

7. $\frac{d}{dt}\{[2\mathbf{i} + t\mathbf{j} - t^2\mathbf{k}] \times [3t\mathbf{i} + 5\mathbf{k}]\}$

8. $\frac{d}{dt}\{[t\mathbf{i} + t^2\mathbf{j} + t^3\mathbf{k}] \times [\mathbf{i} + 2t\mathbf{j} + 3t^2\mathbf{k}]\}$

In Problems 9–16, find the unit tangent $\mathbf{T}(t)$ and the particular vector $\mathbf{T}$ when $t = t_0$.

9. $\mathbf{g}(t) = (\cos t)\mathbf{i} + (\sin t)\mathbf{j} + t\mathbf{k}$; $t = \pi/4$
10. $\mathbf{g}(t) = 2(\cos t)\mathbf{i} + 2(\sin t)\mathbf{j} + t\mathbf{k}$; $t = \pi/6$
11. $\mathbf{g}(t) = 3(\cos t)\mathbf{i} + 4(\sin t)\mathbf{j} + t\mathbf{k}$; $t = 0$
12. $\mathbf{g}(t) = 4(\cos t)\mathbf{i} + 3(\sin t)\mathbf{j} + t\mathbf{k}$; $t = \pi/2$
13. $\mathbf{g}(t) = \mathbf{i} + t\mathbf{j} + t^2\mathbf{k}$; $t = 1$
14. $\mathbf{g}(t) = t\mathbf{i} + t^2\mathbf{j} + t^3\mathbf{k}$; $t = 0$
15. $\mathbf{g}(t) = e^t(\cos 2t)\mathbf{i} + e^t(\sin 2t)\mathbf{j} + e^t\mathbf{k}$; $t = 0$
16. $\mathbf{g}(t) = e^t(\cos 2t)\mathbf{i} + e^t(\sin 2t)\mathbf{j} + e^t\mathbf{k}$; $t = \pi/4$

In Problems 17–30, find unit tangent $\mathbf{T}(t)$, unit normal $\mathbf{n}(t)$, and the particular vectors $\mathbf{T}$ and $\mathbf{n}$ when $t = t_0$. Then sketch the curve near $t = t_0$ and include the vectors $\mathbf{T}$ and $\mathbf{n}$ in your sketch.

17. $\mathbf{f}(t) = (\cos 3t)\mathbf{i} + (\sin 3t)\mathbf{j}$; $t = 0$
18. $\mathbf{f}(t) = (\cos 5t)\mathbf{i} + (\sin 5t)\mathbf{j}$; $t = \pi/2$
19. $\mathbf{f}(t) = 2(\cos 4t)\mathbf{i} + 2(\sin 4t)\mathbf{j}$; $t = \pi/4$
20. $\mathbf{f}(t) = -3(\cos 10t)\mathbf{i} - 3(\sin 10t)\mathbf{j}$; $t = \pi$
21. $\mathbf{f}(t) = 8(\cos t)\mathbf{i} + 8(\sin t)\mathbf{j}$; $t = \pi/4$
22. $\mathbf{f}(t) = 4t\mathbf{i} + 2t^2\mathbf{j}$; $t = 1$
23. $\mathbf{f}(t) = (2 + 3t)\mathbf{i} + (8 - 5t)\mathbf{j}$; $t = 3$
[*Hint:* See the note on page 105.]
24. $\mathbf{f}(t) = (4 - 7t)\mathbf{i} + (-3 + 5t)\mathbf{j}$; $t = -5$
25. $\mathbf{f}(t) = (a + bt)\mathbf{i} + (c + dt)\mathbf{j}$; $t = t_0$
26. $\mathbf{f}(t) = 2t\mathbf{i} + (e^t + e^{-t})\mathbf{j}$; $t = 0$
***27.** $\mathbf{f}(t) = (t - \cos t)\mathbf{i} + (1 - \sin t)\mathbf{j}$; $t = \pi/2$
***28.** $\mathbf{f}(t) = (t - \cos t)\mathbf{i} + (1 - \sin t)\mathbf{j}$; $t = \pi$
***29.** $\mathbf{f}(t) = (t - \cos t)\mathbf{i} + (1 - \sin t)\mathbf{j}$; $t = \pi/4$
***30.** $\mathbf{f}(t) = (\ln \sin t)\mathbf{i} + (\ln \cos t)\mathbf{j}$; $t = \pi/6$

In Problems 31–41, compute the derivative of the given function.

31. $\mathbf{f}(t) = (t, \sin t, \cos t, t^3)$
32. $\mathbf{f}(t) = (\tan t, \ln t, e^t, t^5, \sqrt{t})$
33. $\mathbf{f}(t) = (e^t, e^{t^2}, \ldots, e^{t^n})$
34. $\mathbf{g}(t) = \left(\frac{1}{t}, \frac{1}{t^2}, \frac{1}{t^3}, \ldots, \frac{1}{t^n}\right)$
35. $\mathbf{f}(t) = (\ln t, \ln 2t, \ldots, \ln nt)$
36. $\mathbf{f}(t) = (1, 2, 3, 4, \ldots, n)$
37. $\mathbf{f}(t) = t^2\mathbf{x}$ where $\mathbf{x}$ is a constant vector
38. $\mathbf{h}(t) = \mathbf{f}(t) + \mathbf{g}(t)$ where $\mathbf{f}$ and $\mathbf{g}$ are as in Problems 33 and 34.
39. $\mathbf{h}(t) = 3\mathbf{f}(t)$ where $\mathbf{f}$ is as in Problem 33.
40. $h(t) = \mathbf{g}(t) \cdot (n, n - 1, \ldots, 3, 2, 1)$ where $\mathbf{g}$ is as in Problem 34.

41. $h(t) = \mathbf{f}(t) \cdot \mathbf{g}(t)$ where **f** and **g** are as in Problems 33 and 34.

In Problems 42–46, compute the unit tangent vector to the given curve.

42. $\mathbf{f}(t) = (1, t, t^2, t^3)$

43. $\mathbf{f}(t) = (t, t^2, \ldots, t^n)$

44. $\mathbf{f}(t) = (\sin t, \cos t, \sin t, \cos t, \sin t, \cos t)$

45. $\mathbf{f}(t) = (e^t, e^{2t}, \ldots, e^{nt})$

46. $\mathbf{f}(t) = (1, 2, 3, 4, 5)$

47. Prove part (i) of Theorem 1.

48. Prove part (ii) of Theorem 1.

49. Prove part (iii) of Theorem 1.

50. Prove part (iv) of Theorem 1.

51. Prove part (v) of Theorem 1.

52. Prove part (vi) of Theorem 1.

***53.** Let $\mathbf{f}(t) = a(\cos t)\mathbf{i} + a(\sin t)\mathbf{j} + t\mathbf{k}$ (a circular helix). Show that the angle between the unit tangent vector **T** and the z-axis is constant.

ANSWERS TO SELF-QUIZ

I. b, d **II.** a (scalar zero) **III.** d **IV.** c

2.5 ARC LENGTH REVISITED

One of the standard applications of one-variable integral calculus is to compute arc length. You saw the following definition in your earlier course: Suppose that f and f' are continuous in $[a, b]$. Then the **length of the curve (arc length)** of the graph of $f(x)$ between $x = a$ and $x = b$ is defined by

$$s = \int_a^b \sqrt{1 + [f'(x)]^2}\, dx. \tag{1}$$

We now derive a formula for arc length in a more general setting. Let the curve C be given parametrically by

$$x = f_1(t), \qquad y = f_2(t).$$

We will assume in this section that f_1' and f_2' exist. Let t_0 be a fixed number, which fixes a point $P_0 = (x_0, y_0) = (f_1(t_0), f_2(t_0))$ on the curve (see Figure 1). The arrows in Figure 1 indicate the direction in which a point moves along the curve as t increases. We define the function $s(t)$ by

> $s(t) =$ length along the curve C from $(f_1(t_0), f_2(t_0))$ to $(f_1(t), f_2(t))$.

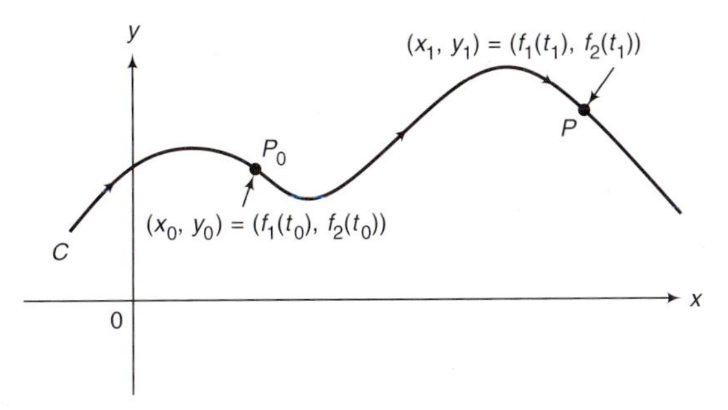

FIGURE 1
$P_0 = (f_1(t_0), f_2(t_0))$ and $P = (f_1(t_1), f_2(t_1))$ are two points on the curve C. As t increases from t_0 to t_1, we move along C from P_0 to P.

The following theorem allows us to calculate the length of a curve given parametrically. In fact, this theorem motivates a reasonable *definition* of arc length.

THEOREM 1

Suppose that f_1' and f_2' are continuous in the interval $[t_0, t_1]$. Then $s(t)$ is a differentiable function of t for $t \in [t_0, t_1]$ and

$$\frac{ds}{dt} = \sqrt{[f_1'(t)]^2 + [f_2'(t)]^2} = \sqrt{\left(\frac{dx}{dt}\right)^2 + \left(\frac{dy}{dt}\right)^2}. \tag{2}$$

SKETCH OF PROOF: The proof of this theorem is quite difficult. However, it is possible to give an intuitive idea of what is going on. Consider an arc of the curve between t and $t + \Delta t$ (see Figure 2).

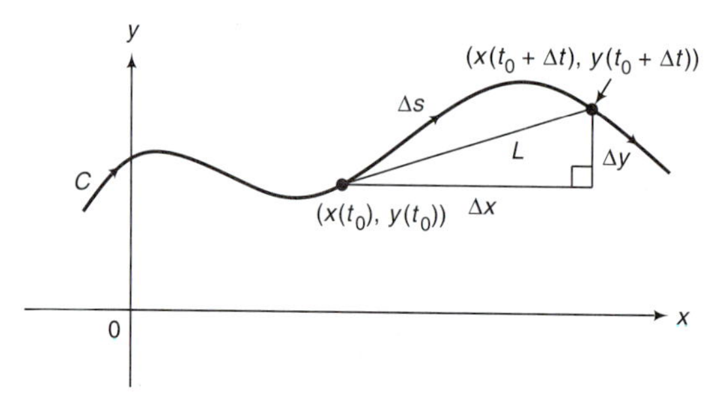

FIGURE 2
The arc length Δs is approximately equal to L. The approximation improves as $\Delta t \to 0$.

First, we note that as $\Delta t \to 0$, the ratio of the length of the secant line L to the length of the arc Δs between the points $(x(t_0), y(t_0))$ and $(x(t_0 + \Delta t), y(t_0 + \Delta t))$ approaches 1. That is,

$$\lim_{\Delta t \to 0} \frac{\Delta s}{L} = 1.^{\dagger} \tag{3}$$

Then since $L = \sqrt{\Delta x^2 + \Delta y^2}$, we have

$$\frac{\Delta s}{\Delta t} = \left(\frac{\Delta s}{L}\right)\left(\frac{L}{\Delta t}\right) = \left(\frac{\Delta s}{L}\right)\left(\frac{\sqrt{\Delta x^2 + \Delta y^2}}{\Delta t}\right)$$

$$= \left(\frac{\Delta s}{L}\right)\sqrt{\left(\frac{\Delta x}{\Delta t}\right)^2 + \left(\frac{\Delta y}{\Delta t}\right)^2}$$

and

$$\frac{ds}{dt} = \lim_{\Delta t \to 0} \frac{\Delta s}{\Delta t} = \lim_{\Delta t \to 0} \frac{\Delta s}{L} \cdot \lim_{\Delta t \to 0} \sqrt{\left(\frac{\Delta x}{\Delta t}\right)^2 + \left(\frac{\Delta y}{\Delta t}\right)^2}$$

$$= \sqrt{\left(\frac{dx}{dt}\right)^2 + \left(\frac{dy}{dt}\right)^2}.$$

† We have not proved this formula. We just indicated by a geometric argument why it ought to be true.

The steps in this proof can be justified under the assumption that dx/dt and dy/dt are continuous.[†] ■

We use equation (2) to define the length of the arc from t_0 to t_1.

DEFINITION ARC LENGTH IN $\mathbb{R}^2$

Suppose **f** has a continuous derivative in the interval $[t_0, t_1]$. Then **f** is said to be **rectifiable** in the interval $[t_0, t_1]$, and the **arc length** $s(t_1)$ of the curve $\mathbf{f}(t)$ in the interval $[t_0, t_1]$ is given by

$$s(t_1) = \text{length of arc from } t_0 \text{ to } t_1$$
$$= \int_{t_0}^{t_1} \left(\frac{ds}{dt}\right) dt = \int_{t_0}^{t_1} \sqrt{\left(\frac{dx}{dt}\right)^2 + \left(\frac{dy}{dt}\right)^2}\, dt. \tag{4}$$

■

REMARK: This formula generalizes equation (1). See Problem 35.

Equation (4) can be rewritten in a slightly different form. Using the chain rule (applied to differentials), we have

$$\frac{ds}{dt} dt = ds, \tag{5}$$

and so (4) becomes

$$s(t_1) = \int_{t_0}^{t_1} ds, \tag{6}$$

where

PARAMETER OF ARC LENGTH

$$ds = \sqrt{\left(\frac{dx}{dt}\right)^2 + \left(\frac{dy}{dt}\right)^2}\, dt. \tag{7}$$

In this context, the variable s is called the **parameter of arc length.** Note that while x measures distance along the horizontal axis and y measures distance along the vertical axis, s measures distance *along the curve* given parametrically by the equations $x = f_1(t)$, $y = f_2(t)$.

EXAMPLE 1 COMPUTING THE LENGTH OF THE UNIT CIRCLE

Calculate the length of the curve $x = \cos t$, $y = \sin t$ in the interval $[0, 2\pi]$.

SOLUTION:

$$ds = \sqrt{\left(\frac{dx}{dt}\right)^2 + \left(\frac{dy}{dt}\right)^2}\, dt = \sqrt{(-\sin t)^2 + (\cos t)^2}\, dt = dt,$$

[†] See, for example, R. C. Buck and E. F. Buck, *Advanced Calculus,* 3rd ed. (New York: McGraw-Hill, 1978), p. 404.

so that

$$s = \int_0^{2\pi} ds = \int_0^{2\pi} dt = 2\pi.$$

Since $x = \cos t$, $y = \sin t$ for t in $[0, 2\pi]$ is a parametric representation of the unit circle, we have verified the accuracy of formula (4) in this instance, since the circumference of the unit circle is 2π.

NOTE: As in the remark on page 86, we can parametrize the unit circle by $\mathbf{f}(t) = (\cos t)\mathbf{i} + (\sin t)\mathbf{j}$, $0 \le t \le 4\pi$. Now the circle is traversed twice and the arc length is $s = \int_0^{4\pi} ds = \int_0^{4\pi} dt = 4\pi$. This does not contradict the result above. Rather, it states the obvious fact that if you go around the circle twice, you cover twice the distance you cover if you go around only once.

EXAMPLE 2

CALCULATING ARC LENGTH

Let the curve C be given by $x = t^2$ and $y = t^3$. Calculate the length of the arc from $t = 0$ to $t = 3$.

SOLUTION: Here,

$$\sqrt{\left(\frac{dx}{dt}\right)^2 + \left(\frac{dy}{dt}\right)^2} = \sqrt{(2t)^2 + (3t^2)^2} = \sqrt{4t^2 + 9t^4} = t\sqrt{4 + 9t^2},$$

so that

$$s = \int_0^3 t\sqrt{4 + 9t^2}\, dt = \frac{1}{27}(4 + 9t^2)^{3/2}\Big|_0^3 = \frac{85^{3/2} - 8}{27}.$$

There is a more concise way to write our formula for arc length by using vector notation. Let the curve C be given by

$$\mathbf{f}(t) = f_1(t)\mathbf{i} + f_2(t)\mathbf{j}.$$

Then

$$\mathbf{f}'(t) = f_1'(t)\mathbf{i} + f_2'(t)\mathbf{j}$$

and

$$|\mathbf{f}'(t)| = \sqrt{[f_1'(t)]^2 + [f_2'(t)]^2} = \frac{ds}{dt}, \tag{8}$$

so that the length of the arc between t_0 and t_1 is given by

THE LENGTH OF AN ARC BETWEEN t_0 AND t_1

$$s = \int_{t_0}^{t_1} |\mathbf{f}'(t)|\, dt. \tag{9}$$

EXAMPLE 3 CALCULATING ARC LENGTH

Calculate the length of the arc of the curve $\mathbf{f}(t) = (2t - t^2)\mathbf{i} + \frac{8}{3}t^{3/2}\mathbf{j}$ between $t = 1$ and $t = 3$.

SOLUTION: $\mathbf{f}'(t) = (2 - 2t)\mathbf{i} + 4\sqrt{t}\mathbf{j}$ and

$$|\mathbf{f}'(t)| = \sqrt{4 - 8t + 4t^2 + 16t} = 2\sqrt{t^2 + 2t + 1} = 2(t + 1),$$

so

$$s = \int_1^3 2(t + 1)\, dt = (t^2 + 2t)\Big|_1^3 = 12.$$

All the results in this section hold, with very little change, in $\mathbb{R}^3$. If

$$\mathbf{f}(t) = x(t)\mathbf{i} + y(t)\mathbf{j} + z(t)\mathbf{k},$$

then

$$\mathbf{f}'(t) = \frac{dx}{dt}\mathbf{i} + \frac{dy}{dt}\mathbf{j} + \frac{dz}{dt}\mathbf{k}$$

and

$$\mathbf{f}'(t) = \sqrt{\left(\frac{dx}{dt}\right)^2 + \left(\frac{dy}{dt}\right)^2 + \left(\frac{dz}{dt}\right)^2}.$$

DEFINITION ARC LENGTH IN $\mathbb{R}^3$

Suppose $\mathbf{f}$ has a continuous derivative in the interval $[t_0, b]$ and suppose that for every t_1 in $[t_0, b]$, $\int_{t_0}^{t_1} |\mathbf{f}'(t)|\, dt$ exists. Then $\mathbf{f}$ is said to be **rectifiable** in the interval $[t_0, b]$, and the **arc length** of the curve $\mathbf{f}(t)$ in the interval $[t_0, t_1]$ is given by

$$s(t_1) = \int_{t_0}^{t_1} |\mathbf{f}'(t)|\, dt$$

where $t_0 \le t_1 \le b$. ■

ONE PHYSICAL INTERPRETATION OF ARC LENGTH

As we stated earlier, if $\mathbf{v}(t)$ is the velocity vector of a moving particle, then $|\mathbf{v}(t)|$ represents its speed. Suppose that $\mathbf{f}(t)$ is a position vector; then the velocity vector is $\mathbf{f}'(t)$ and the speed is $|\mathbf{f}'(t)|$. The equation above states that

arc length is the integral of speed

THEOREM 2

Let $\mathbf{f}$ in $\mathbb{R}^3$ have a continuous derivative and let $s(t)$ denote the length of the arc from t_0 to t, then

$$\frac{ds}{dt} = |\mathbf{f}'(t)|. \tag{10}$$

■

EXAMPLE 4

CALCULATING ARC LENGTH IN $\mathbb{R}^3$

Let $\mathbf{f}(t) = (\cos t)\mathbf{i} + (\sin t)\mathbf{j} + t\mathbf{k}$. Find the length of the arc from $t = 0$ to $t = 4$.

SOLUTION:

$$f'(t) = -\sin t\mathbf{i} + \cos t\mathbf{j} + \mathbf{k}$$

and

$$|\mathbf{f}'(t)| = \sqrt{\sin^2 t + \cos^2 t + 1} = \sqrt{2},$$

so

$$s(4) = \int_0^4 |\mathbf{f}'(t)|\, dt = \int_0^4 \sqrt{2}\, dt = 4\sqrt{2}.$$

ARC LENGTH AS A PARAMETER

In many problems, it is convenient and more natural to use the arc length s as a parameter. We can think of the vector function $\mathbf{f}$ as the *position vector* of a particle moving in the plane or in space. Then, if $P_0 = (x_0, y_0)$ or (x_0, y_0, z_0) is a fixed point on the curve C described by the vector function $\mathbf{f}$, we may write

$$\mathbf{f}(s) = x(s)\mathbf{i} + y(s)\mathbf{j}, \qquad \text{or} \qquad \mathbf{f}(s) = x(s)\mathbf{i} + y(s)\mathbf{j} + z(s)\mathbf{k},$$

where s is the distance along the curve measured from P_0 in the direction of increasing s. In this way we can determine the x-, y-, and z-components of the position vector as we move s units *along the curve,* and we are using a parameter that is intrinsic to the curve.

EXAMPLE 5

THE UNIT CIRCLE PARAMETRIZED BY ARC LENGTH

Write the vector $\mathbf{f}(t) = \cos t\mathbf{i} + \sin t\mathbf{j}$ (which describes the unit circle) with arc length as a parameter. Take $P_0 = (1, 0)$.

SOLUTION: We have $ds/dt = 1$ (from Example 1), so that since $(1, 0)$ is reached when $t = 0$, we find that

$$s = \int_0^t ds = \int_0^t \frac{ds}{du}\, du = \int_0^t du = t.$$

Thus, we may write

$$\mathbf{f}(s) = \cos s\mathbf{i} + \sin s\mathbf{j}.$$

For example, if we begin at the point $(1, 0)$ and move π units along the unit circle (which is half the unit circle), then we move to the point $(\cos \pi, \sin \pi) = (-1, 0)$. This is what we would expect. See Figure 3.

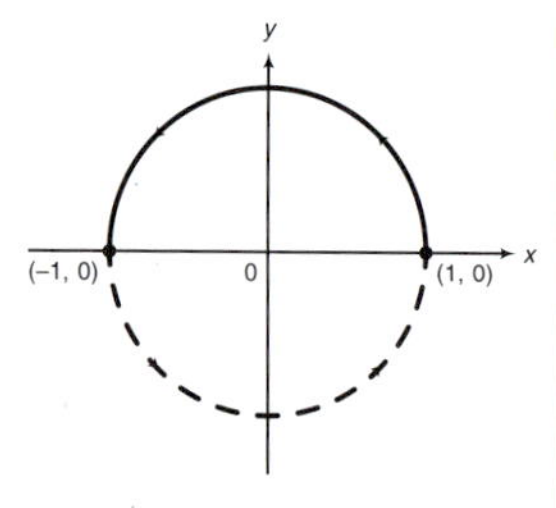

FIGURE 3
Graph of $\mathbf{f}(s) = \cos s\mathbf{i} + \sin s\mathbf{j}$, $0 \le t \le \pi$

EXAMPLE 6

WRITING A CURVE WITH ARC LENGTH AS THE PARAMETER

Let $\mathbf{f}(t) = (2t - t^2)\mathbf{i} + \frac{8}{3}t^{3/2}\mathbf{j}$, $t \geq 0$. Write this curve with arc length as a parameter.

SOLUTION: Suppose that the fixed point is $P_0 = (0, 0)$ when $t = 0$. Then from Example 3 we have

$$\frac{ds}{dt} = 2(t + 1), \qquad \text{and} \qquad s = \int_0^t \frac{ds}{du}\, du = \int_0^t 2(u + 1)\, du = t^2 + 2t.$$

This leads to the equations

$$t^2 + 2t - s = 0 \qquad \text{and} \qquad t = \frac{-2 + \sqrt{4 + 4s}}{2} = \sqrt{1 + s} - 1.$$

We took the positive square root here since it is assumed that t starts at 0 and increases. Then

$$x = 2t - t^2 = 4\sqrt{1 + s} - 4 - s$$
$$y = \tfrac{8}{3}t^{3/2} = \tfrac{8}{3}(\sqrt{1 + s} - 1)^{3/2},$$

and we obtain

$$\mathbf{f}(s) = (4\sqrt{1 + s} - 4 - s)\mathbf{i} + \tfrac{8}{3}(\sqrt{1 + s} - 1)^{3/2}\mathbf{j}.$$

As Example 6 illustrates, writing $\mathbf{f}$ explicitly with arc length as a parameter can be tedious (or, more often, impossible).

EXAMPLE 7

A CIRCULAR HELIX PARAMETRIZED BY ARC LENGTH

Write the equation of a circular helix $\mathbf{f}(t) = \cos t\mathbf{i} + \sin t\mathbf{j} + t\mathbf{k}$ with arc length as a parameter.

SOLUTION: From Example 4, $\dfrac{ds}{dt} = |\mathbf{f}'(t)| = \sqrt{2}$ so $s = \int_0^t \sqrt{2}\, du = \sqrt{2}t$. Then $t = \dfrac{s}{\sqrt{2}}$ and we have

$$\mathbf{f}(s) = \cos\frac{s}{\sqrt{2}}\mathbf{i} + \sin\frac{s}{\sqrt{2}}\mathbf{j} + \frac{s}{\sqrt{2}}\mathbf{k}.$$

There is an interesting and important relationship between position vectors, tangent vectors, and normal vectors that becomes apparent when we use s as a parameter.

THEOREM 3 **THE UNIT TANGENT VECTOR PARAMETRIZED BY ARC LENGTH**

If the curve C is parametrized by $\mathbf{f}(s) = x(s)\mathbf{i} + y(s)\mathbf{j}$ or $x(s)\mathbf{i} + y(s)\mathbf{j} + z(s)\mathbf{k}$, where s is arc length and x, y, and z have continuous derivatives, then the unit tangent vector $\mathbf{T}$ is given by

$$\mathbf{T}(s) = \frac{d\mathbf{f}}{ds}. \tag{11}$$

PROOF: With *any* parametrization of C, the unit tangent vector is given by [see equation (2.3.6) on p. 100]

$$\mathbf{T}(t) = \frac{\mathbf{f}'(t)}{|\mathbf{f}'(t)|}.$$

So choosing $t = s$ yields

$$\mathbf{T}(s) = \frac{d\mathbf{f}/ds}{|d\mathbf{f}/ds|}.$$

But from equation (2) or (10), $|\mathbf{f}'(t)| = ds/dt$, so

$$\left|\frac{d\mathbf{f}}{ds}\right| = \left|\frac{d\mathbf{f}/dt}{ds/dt}\right| = \left|\frac{\mathbf{f}'(t)}{ds/dt}\right| = 1,$$

and the proof is complete. ■

Theorem 3 is quite useful in that it provides a check of our calculation of the parametrization in terms of arc length. For if

$$\mathbf{f}(s) = x(s)\mathbf{i} + y(s)\mathbf{j} \qquad \text{or} \qquad \mathbf{f}(s) = x(s)\mathbf{i} + y(s)\mathbf{j} + z(s)\mathbf{k},$$

then

$$\mathbf{T} = \frac{d\mathbf{f}}{ds} = \frac{dx}{ds}\mathbf{i} + \frac{dy}{ds}\mathbf{j} \qquad \text{or} \qquad \mathbf{T} = \frac{dx}{ds}\mathbf{i} + \frac{dy}{ds}\mathbf{j} + \frac{dz}{ds}\mathbf{k}.$$

But $|\mathbf{T}| = 1$, so that $|\mathbf{T}|^2 = 1$, which implies that

$$\left(\frac{dx}{ds}\right)^2 + \left(\frac{dy}{ds}\right)^2 = 1 \qquad \text{or} \qquad \left(\frac{dx}{ds}\right)^2 + \left(\frac{dy}{ds}\right)^2 + \left(\frac{dz}{ds}\right)^2 = 1. \tag{12}$$

In Problems 37 and 38, you are asked to show that equation (12) holds for the functions computed in Examples 6 and 7.

PROBLEMS 2.5

SELF-QUIZ

I. Let $\mathbf{f}(t) = (3 + t)\mathbf{i} + (2 - t)\mathbf{j} + (1 + 3t)\mathbf{k}$. The arc length between the points where $t = 1$ and $t = 4$ is ________.

a. 3
b. $\sqrt{11}$
c. 99
d. $\sqrt{99}$

II. If we reparametrize $\mathbf{f}(t) = (-1, 7, 2) + t(3, 4, 12)$ in terms of the arc length s from the point where $t = 0$, then $\mathbf{f}(s) =$ ________.

a. $(-1, 7, 2) + s(\frac{3}{13}, \frac{4}{13}, \frac{12}{13})$
b. $s(\frac{3}{13}, \frac{4}{13}, \frac{12}{13})$
c. $s(-1/\sqrt{54}, 7/\sqrt{54}, 2/\sqrt{54})$
d. $(3, 4, 12) + s(-1/\sqrt{54}, 7/\sqrt{54}, 2/\sqrt{54})$

In Problems 1–12, find the length of the given arc over the specified interval or the length of the closed curve.

1. $x = t^3$; $y = t^2$; $-1 \le t \le 1$ [*Hint:* $\sqrt{t^2} = -t$ for $t < 0$.]
2. $x = \cos 2\theta$; $y = \sin 2\theta$; $0 \le \theta \le \pi/2$
3. $x = t^3 + 1$; $y = 3t^2 + 2$; $0 \le t \le 2$
4. $x = 1 + t$; $y = (1 + t)^{3/2}$; $0 \le t \le 1$
5. One arc of the cycloid $x = a(\theta - \sin\theta)$, $y = a(1 - \cos\theta)$
6. The hypocycloid of four cusps $x = a\cos^3\theta$, $y = a\sin^3\theta$ [*Hint:* Calculate the length in the first quadrant and multiply by 4.]

*7. $\mathbf{f}(t) = \dfrac{1}{\sqrt{1+t}}\mathbf{i} + \dfrac{t}{2(1+t)}\mathbf{j}$; $0 \le t \le 4$
[*Hint:* Substitute $u^2 = t + 2$ and integrate by partial fractions.]

8. $\mathbf{g}(t) = \frac{2}{3}t^3\mathbf{i} + (1 + t^{9/2})\mathbf{j} + (1 - t^{9/2})\mathbf{k}$; $0 \le t \le 2$

9. $\mathbf{f}(t) = e^t(\cos t)\mathbf{i} + e^t(\sin t)\mathbf{j}$; $0 \le t \le \pi/2$

10. $\mathbf{g}(t) = e^t(\cos 2t)\mathbf{i} + e^t(\sin 2t)\mathbf{j} + e^t\mathbf{k}$; $1 \le t \le 4$

11. $\mathbf{f}(t) = (\sin^2 t)\mathbf{i} + (\cos^2 t)\mathbf{j}$; $0 \le t \le \pi/2$

12. $\mathbf{g}(t) = 2(\cos 3t)\mathbf{i} + 2(\sin 3t)\mathbf{j} + t^2\mathbf{k}$; $0 \le t \le 10$

In Problems 13–20, use the result of Problem 36 to find the length of the given arc over the specified interval or the length of the given closed curve.

13. $r = a \sin \theta$; $0 \le \theta \le \pi/2$

14. $r = a \cos \theta$; $0 \le \theta \le \pi$

15. $r = \theta^2$; $0 \le \theta \le \pi$

16. $r = 6 \cos^2(\theta/2)$; $0 \le \theta \le \pi/2$

*17. The cardioid $r = a(1 + \sin \theta)$ $(a > 0)$ [*Hint:*

$$\int \sqrt{1 + \sin \theta}\, d\theta = \int \sqrt{1 + \sin \theta} \cdot \frac{\sqrt{1 - \sin \theta}}{\sqrt{1 - \sin \theta}}\, d\theta.$$

Pay attention to signs.]

*18. $r = a\theta$; $0 \le \theta \le 2\pi$; $a > 0$

19. $r = e^\theta$; $0 \le \theta \le 3$

20. $r = \sin^3(\theta/3)$; $0 \le \theta \le \pi/2$
[*Hint:* $\sin^2(\theta/3) = \frac{1}{2}(1 - \cos(2\theta/3))$.]

In Problems 21–32, find parametric equations in terms of the arc length s measured from the point reached when $t = 0$ (suppose all constants a, b, c, d are positive). Verify your solution by using formula (12).

21. $\mathbf{f}(t) = 3t^2\mathbf{i} + 2t^3\mathbf{j}$

22. $\mathbf{f}(t) = t^3\mathbf{i} + t^2\mathbf{j}$

23. $\mathbf{f}(t) = (t^3 + 1)\mathbf{i} + (t^2 - 1)\mathbf{j}$

24. $\mathbf{f}(t) = (3t^2 + a)\mathbf{i} + (2t^3 + b)\mathbf{j}$

25. $\mathbf{f}(t) = 3(\cos t)\mathbf{i} + 3(\sin t)\mathbf{j}$

26. $\mathbf{f}(t) = a(\sin t)\mathbf{i} + a(\cos t)\mathbf{j}$

27. $\mathbf{f}(t) = a(\cos t)\mathbf{i} - a(\sin t)\mathbf{j}$

28. $\mathbf{f}(t) = 3(\cos t + t \sin t)\mathbf{i} + 3(\sin t - t \cos t)\mathbf{j}$

29. $\mathbf{f}(t) = (a + b \cos t)\mathbf{i} + (c + b \sin t)\mathbf{j}$

30. $\mathbf{f}(t) = ae^t(\cos t)\mathbf{i} + ae^t(\sin t)\mathbf{j}$

31. One cusp of the hypocycloid of four cusps $\mathbf{f}(t) = a(\cos^3 t)\mathbf{i} + a(\sin^3 t)\mathbf{j}$; $0 \le t \le \pi/2$.

*32. The cycloid $x = a(\theta - \sin \theta)$, $y = a(1 - \cos \theta)$.

33. The ellipse $(x/a)^2 + (y/b)^2 = 1$ has parametric representation $x = a \cos \theta$, $y = b \sin \theta$. Find an integral that computes the length of the circumference of this ellipse. [*Note:* Do not try to evaluate this **elliptic integral.** It arises in a variety of physical applications, but cannot be integrated in terms of common functions (except numerically) unless $a = b$.]

*34. A tack stuck in the front tire of a bicycle wheel has a diameter of 1 m. What is the total distance traveled by the tack if the bicycle moves forward a total of 30π m?

35. This problem asks you to reconcile the arc length formula presented in equation (4) with that given in equation (1): Suppose $\mathbf{f}(t) = x(t)\mathbf{i} + y(t)\mathbf{j}$ is a smooth function of t for $t_0 \le t \le t_1$. Also suppose y can be expressed as a differentiable monotonic function of x with $x(t_0) = a < x(t_1) = b$. Show that

$$\int_a^b \sqrt{1 + \left(\frac{dy}{dx}\right)^2}\, dx = \int_{t_0}^{t_1} \sqrt{\left(\frac{dx}{dt}\right)^2 + \left(\frac{dy}{dt}\right)^2}\, dt.$$

36. Consider a smooth curve which satisfies the polar coordinate equation $r = f(\theta)$. Show that the length of the arc for $\alpha \le \theta \le \beta$ is

$$s = \int_\alpha^\beta \sqrt{r^2 + \left(\frac{dr}{d\theta}\right)^2}\, d\theta.$$

[*Hint:* Problem 2.2.39 has a useful parametrization for Cartesian coordinates.]

37. In Example 6, show that $(dx/ds)^2 + (dy/ds)^2 = 1$.

38. In Example 7, show that $(dx/ds)^2 + (dy/ds)^2 + (dz/ds)^2 = 1$.

Arc Length in $\mathbb{R}^n$: Let $\mathbf{f}(t)$ be a vector-valued function in $\mathbb{R}^n$. We define the **arc length** $s(t)$ of the curve $\mathbf{f}(t)$ for $t \in [a, b]$ by

$$s(t) = \int_a^t |\mathbf{f}'(u)|\, du.$$

Note that, as in $\mathbb{R}^2$ and $\mathbb{R}^3$, $s(t)$ exists whenever f is continuously differentiable. In Problems 39–41 compute the arc length of the given curve.

39. $\mathbf{f}(t) = (1, t, 2t, 3t, \ldots, nt)$; $t \in [1, 5]$

40. $\mathbf{f}(t) = (t, t^2, t, t^2, t)$; $t \in [0, 1]$

41. $\mathbf{f}(t) = (\sin t, \cos t, \sin t, \cos t, \sin t, \cos t)$; $t \in [0, 2\pi]$

Answers to Self-Quiz

I. d **II.** a

2.6 CURVATURE AND THE ACCELERATION VECTOR (OPTIONAL)

The derivative dy/dx of a curve $y = f(x)$ measures the rate of change of the vertical component of the curve with respect to the horizontal component. As we have seen, the derivative ds/dt represents the change in the length of the arc traced out by the vector $\mathbf{f} = x(t)\mathbf{i} + y(t)\mathbf{j}$ as t increases. Another quantity of interest is the rate of change of the direction of the curve with respect to the length of the curve. That is, how much does the direction change for every one-unit change in the arc length?

We begin our discussion in $\mathbb{R}^2$.

DEFINITION CURVATURE IN $\mathbb{R}^2$ AND RADIUS OF CURVATURE

i. Let the curve C be given by the differentiable vector function $\mathbf{f}(t) = f_1(t)\mathbf{i} + f_2(t)\mathbf{j}$. Let $\varphi(t)$ denote the direction of $\mathbf{f}'(t)$. Then the **curvature** of C, denoted by $\kappa(t)$, is the absolute value of the rate of change of direction with respect to arc length; that is,

$$\kappa(t) = \left|\frac{d\varphi}{ds}\right|. \tag{1}$$

Note that $\kappa(t) \geq 0$.

ii. The **radius of curvature** $\rho(t)$ is defined by

$$\rho(t) = \frac{1}{\kappa(t)} \qquad \text{if } \kappa(t) > 0. \tag{2}$$

■

REMARK 1: The curvature is a measure of *how fast* the curve turns as we move along it.

REMARK 2: If $\kappa(t) = 0$, we say that the radius of curvature is *infinite*. To understand this idea, note that if $\kappa(t) = 0$, then the "curve" does not bend and so is a straight line (see Example 2). A straight line can be thought of as an arc of a circle with *infinite* radius.

In $\mathbb{R}^3$, we will need a definition of curvature that does not depend on the angle φ. We first prove the following theorem in $\mathbb{R}^2$.

THEOREM 1

If $\mathbf{T}(t)$ denotes the unit tangent vector to $\mathbf{f}$, then

$$\kappa(t) = \left|\frac{d\mathbf{T}}{ds}\right|. \tag{3}$$

PROOF: By the chain rule (which applies just as well to vector-valued functions),

$$\frac{d\mathbf{T}}{ds} = \frac{d\mathbf{T}}{d\varphi}\frac{d\varphi}{ds}.$$

Since φ is the direction of $\mathbf{f}'$, and therefore is also the direction of $\mathbf{T}$, we have

$$\mathbf{T} = (\cos\varphi)\mathbf{i} + (\sin\varphi)\mathbf{j},$$

so that

$$\frac{d\mathbf{T}}{d\varphi} = -(\sin\varphi)\mathbf{i} + (\cos\varphi)\mathbf{j} \qquad \text{and}$$

$$\left|\frac{d\mathbf{T}}{d\varphi}\right| = \sqrt{\sin^2\varphi + \cos^2\varphi} = 1.$$

Thus,

$$\left|\frac{d\mathbf{T}}{ds}\right| = \left|\frac{d\mathbf{T}}{d\varphi}\right|\left|\frac{d\varphi}{ds}\right| = 1\left|\frac{d\varphi}{ds}\right| = \kappa(t),$$

and the theorem is proved. ■

We now derive an easier way to calculate $\kappa(t)$.

THEOREM 2

With the curve C given as in the definition of curvature in $\mathbb{R}^2$, the curvature of C is given by the formula

$$\kappa(t) = \frac{|(dx/dt)(d^2y/dt^2) - (dy/dt)(d^2x/dt^2)|}{[(dx/dt)^2 + (dy/dt)^2]^{3/2}}$$

$$= \frac{|x'y'' - y'x''|}{[(x')^2 + (y')^2]^{3/2}} \tag{4}$$

where $x(t) = f_1(t)$ and $y(t) = f_2(t)$.

PROOF: Since x' and y' are not both zero, we may assume that $dx/dt \neq 0$. By the chain rule,

$$\frac{d\varphi}{ds} = \frac{d\varphi}{dt}\frac{dt}{ds} = \frac{d\varphi/dt}{ds/dt}. \tag{5}$$

From equation (2.5.7) on page 113,

$$\frac{ds}{dt} = \sqrt{\left(\frac{dx}{dt}\right)^2 + \left(\frac{dy}{dt}\right)^2}. \tag{6}$$

Let φ denote the direction of $\mathbf{f}'$. Then, from equation (1.1.2) on page 6,

$$\tan\varphi = \frac{dy/dt}{dx/dt}, \tag{7}$$

or

$$\varphi = \tan^{-1}\frac{dy/dt}{dx/dt}. \tag{8}$$

Differentiating both sides of (8) with respect to t, we obtain

$$\frac{d\varphi}{dt} = \frac{1}{1 + \left(\dfrac{dy/dt}{dx/dt}\right)^2}\frac{d}{dt}\left(\frac{dy/dt}{dx/dt}\right)$$

$$= \frac{\left(\frac{dx}{dt}\right)^2}{\left(\frac{dx}{dt}\right)^2 + \left(\frac{dy}{dt}\right)^2} \cdot \frac{\frac{dx}{dt}\frac{d^2y}{dt^2} - \frac{dy}{dt}\frac{d^2x}{dt^2}}{\left(\frac{dx}{dt}\right)^2}. \tag{9}$$

Substitution of (6) and (9) into (5) completes the proof of the theorem. ∎

EXAMPLE 1 THE CURVATURE AND RADIUS OF CURVATURE OF A CIRCLE

We certainly expect that the curvature of a circle is constant and that its radius of curvature is its radius. Show that this is true.

SOLUTION: The circle of radius r centered at the origin is given parametrically by

$$x = r\cos t, \qquad y = r\sin t.$$

Then $dx/dt = -r\sin t$, $d^2x/dt^2 = -r\cos t$, $dy/dt = r\cos t$, and $d^2y/dt^2 = -r\sin t$, and from (4),

$$\kappa(t) = \frac{|r^2\sin^2 t + r^2\cos^2 t|}{[r^2\sin^2 t + r^2\cos^2 t]^{3/2}} = \frac{r^2}{r^3} = \frac{1}{r} \quad \text{and} \quad \rho(t) = \frac{1}{\kappa(t)} = r,$$

as expected.

EXAMPLE 2 THE CURVATURE OF A STRAIGHT LINE IS ZERO

Show that for a straight line $\kappa(t) = 0$.

SOLUTION: A line can be represented parametrically [see Problem 2.1.49] by $x = x_1 + t(x_2 - x_1)$ and $y = y_1 + t(y_2 - y_1)$. Then $dx/dt = x_2 - x_1$, $d^2x/dt^2 = 0$, $dy/dt = y_2 - y_1$, and $d^2y/dt^2 = 0$. Substitution of these values into (4) immediately yields $\kappa(t) = 0$.

EXAMPLE 3 FINDING THE CURVATURE AND RADIUS OF CURVATURE OF A CURVE

Find the curvature and radius of curvature of the curve given parametrically by $x = (t^3/3) - t$ and $y = t^2$ (see Example 2.4.4) at $t = 2\sqrt{2}$.

SOLUTION: We have $dx/dt = t^2 - 1$, $d^2x/dt^2 = 2t$, $dy/dt = 2t$, and $d^2y/dt^2 = 2$, so

$$\kappa(t) = \frac{|2(t^2-1) - 2t(2t)|}{[(t^2-1)^2 + (2t)^2]^{3/2}} = \frac{2(t^2+1)}{(t^4+2t^2+1)^{3/2}} = \frac{2(t^2+1)}{(t^2+1)^3}$$
$$= \frac{2}{(t^2+1)^2}.$$

Then $\kappa(2\sqrt{2}) = 2/81$, and the radius of curvature $\rho(2\sqrt{2}) = 1/\kappa(2\sqrt{2}) = 81/2$. A portion of this curve near $t = 2\sqrt{2}$ is sketched in Figure 1. For reference purposes, the unit tangent vector $\mathbf{T}$ at $t = 2\sqrt{2}$ is included.

NOTE: In Figure 1, the circle with radius of curvature $\rho(t)$ that lies on the concave side of C and is tangent to the curve at a point is called the **circle of curvature**, or **osculating circle**.

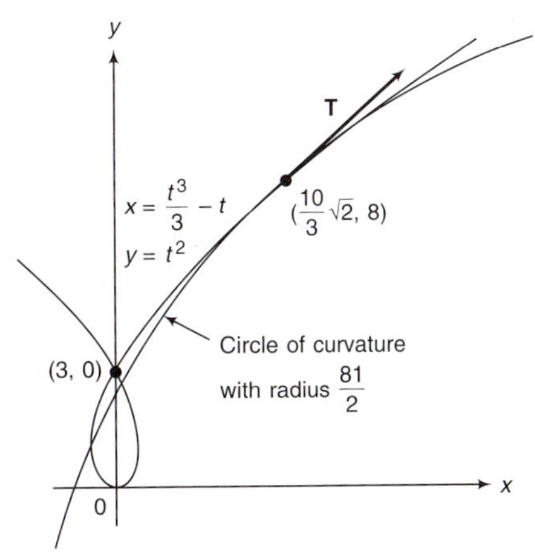

FIGURE 1
The circle of curvature or osculating circle at $t = 2\sqrt{2}$ of the curve $\mathbf{f}(t) = \left(\frac{t^3}{3} - t\right)\mathbf{i} + t^2\mathbf{j}$

CURVATURE AND NORMAL VECTORS IN $\mathbb{R}^3$

The curvature $\kappa(t)$ and the unit normal vector $\mathbf{n}(t)$ are defined differently in $\mathbb{R}^3$. They would have to be. For one thing, there is a whole ''plane-full'' of unit vectors orthogonal to $\mathbf{T}$. Also, curvature cannot be defined as $|d\varphi/ds|$ since there are now three angles that define the direction of C. Therefore, we use an alternative definition of curvature (see Theorem 1).

DEFINITION CURVATURE IN $\mathbb{R}^3$

If $\mathbf{f}$ has a continuous derivative, then the **curvature** of $\mathbf{f}$ is given by

$$\kappa(t) = \left|\frac{d\mathbf{T}}{ds}\right|. \tag{10}$$

■

REMARK: By Theorem 1, equation (10) is equivalent in $\mathbb{R}^2$ to the definition $\kappa(t) = |d\varphi/ds|$, which was our $\mathbb{R}^2$ definition.

DEFINITION PRINCIPAL UNIT NORMAL VECTOR

For any value of t for which $\kappa(t) \neq 0$, the **principal unit normal vector n** is defined by

$$\mathbf{n}(t) = \frac{1}{\kappa(t)}\frac{d\mathbf{T}}{ds}. \tag{11}$$

■

REMARK: Equation (11) also holds in $\mathbb{R}^2$.

Having defined a unit normal vector, we must show that it is orthogonal to the unit tangent vector in order to justify its name.

THEOREM 3

$$\mathbf{n}(t) \perp \mathbf{T}(t).$$

PROOF: Since $1 = \mathbf{T} \cdot \mathbf{T}$, we differentiate both sides with respect to s to obtain

$$0 = \mathbf{T} \cdot \frac{d\mathbf{T}}{ds} + \frac{d\mathbf{T}}{ds} \cdot \mathbf{T} = 2\mathbf{T} \cdot \frac{d\mathbf{T}}{ds} = (2\kappa)\mathbf{T} \cdot \mathbf{n}.$$ ■

EXAMPLE 4

THE CURVATURE AND PRINCIPAL UNIT NORMAL VECTOR FOR A CIRCULAR HELIX

Find the curvature and principal unit normal vector, $\kappa(t)$ and $\mathbf{n}(t)$, for the circular helix $f(\mathbf{t}) = \cos t\mathbf{i} + \sin t\mathbf{j} + t\mathbf{k}$ at $t = \pi/3$.

SOLUTION: From Example 2.5.7, $t = s/\sqrt{2}$. From Example 2.3.6,

$$\mathbf{T}(t) = \frac{1}{\sqrt{2}}(-\sin t\mathbf{i} + \cos t\mathbf{j} + \mathbf{k}).$$

Inserting $t = s/\sqrt{2}$ into this equation yields

$$\mathbf{T}(s) = -\frac{\sin(s/\sqrt{2})}{\sqrt{2}}\mathbf{i} + \frac{\cos(s/\sqrt{2})}{\sqrt{2}}\mathbf{j} + \frac{1}{\sqrt{2}}\mathbf{k}$$

and

$$\frac{d\mathbf{T}}{ds} = -\frac{\cos(s/\sqrt{2})}{2}\mathbf{i} - \frac{\sin(s/\sqrt{2})}{2}\mathbf{j}.$$

Hence,

$$\kappa(t) = \sqrt{\frac{\cos^2(s/\sqrt{2})}{4} + \frac{\sin^2(s/\sqrt{2})}{4}} = \frac{1}{2}$$

and

$$\mathbf{n}(t) = \frac{1}{\kappa(t)}\frac{d\mathbf{T}}{ds} = -(\cos t)\mathbf{i} - (\sin t)\mathbf{j}.$$

At $t = \pi/3$,

$$\mathbf{n}\left(\frac{\pi}{3}\right) = -\frac{1}{2}\mathbf{i} - \frac{\sqrt{3}}{2}\mathbf{j}.$$

REMARK: When it is not convenient to solve for t in terms of s, we can calculate $d\mathbf{T}/ds$ directly by the relation

$$\frac{d\mathbf{T}}{ds} = \frac{d\mathbf{T}/dt}{ds/dt}. \qquad (12)$$

There is a third vector that is often useful in applications.

DEFINITION BINORMAL VECTOR

The **binormal vector B** to the curve **f** in $\mathbb{R}^3$ is defined by

$$\mathbf{B} = \mathbf{T} \times \mathbf{n}. \tag{13}$$

■

From this definition, we see that **B** is orthogonal to both **T** and **n**. Moreover, since $\mathbf{T} \perp \mathbf{n}$, the angle θ between **T** and **n** is $\pi/2$, and

$$|\mathbf{B}| = |\mathbf{T} \times \mathbf{n}| = |\mathbf{T}||\mathbf{n}| \sin \theta = 1,$$

so that **B** is a unit vector. Also, **T**, **n**, and **B** form a right-handed system, just like the vectors **i**, **j**, and **k**. Unlike these last three vectors, **T**, **n**, and **B** are derived from properties of the curve itself.

EXAMPLE 5 CALCULATING A BINORMAL VECTOR

Find the binormal vector to the curve of Example 4 at $t = \pi/3$.

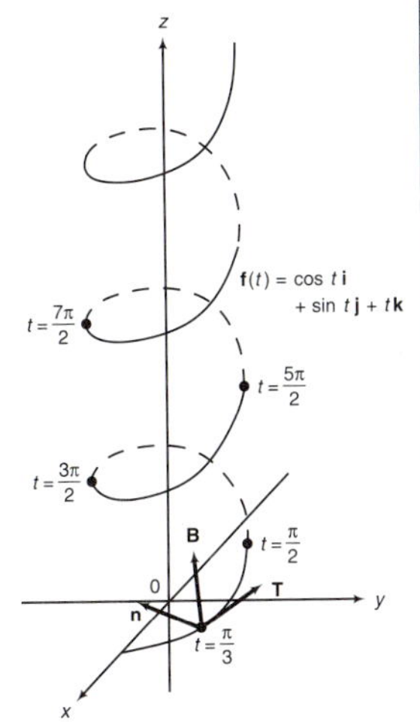

FIGURE 2
Tangent, normal, and binormal vectors for a circular helix

SOLUTION:

$$\mathbf{B} = \mathbf{T} \times \mathbf{n} = \begin{vmatrix} \mathbf{i} & \mathbf{j} & \mathbf{k} \\ -\dfrac{\sin t}{\sqrt{2}} & \dfrac{\cos t}{\sqrt{2}} & \dfrac{1}{\sqrt{2}} \\ -\cos t & -\sin t & 0 \end{vmatrix} = \frac{\sin t}{\sqrt{2}}\mathbf{i} - \frac{\cos t}{\sqrt{2}}\mathbf{j} + \frac{1}{\sqrt{2}}\mathbf{k}$$

When $t = \pi/3$, $\mathbf{B} = (\sqrt{3}/2\sqrt{2})\mathbf{i} - (1/2\sqrt{2})\mathbf{j} + (1/\sqrt{2})\mathbf{k}$. The vectors **T**, **n**, and **B** are sketched in Figure 2.

TANGENTIAL AND NORMAL COMPONENTS OF ACCELERATION

There is an interesting relationship between curvature and acceleration vectors. Suppose that a particle is moving along the curve C with position vector

$$\mathbf{f}(t) = x(t)\mathbf{i} + y(t)\mathbf{j}, \qquad \text{or} \qquad \mathbf{f}(t) = x(t)\mathbf{i} + y(t)\mathbf{j} + z(t)\mathbf{k}.$$

We know that velocity is the derivative of distance and acceleration is the derivative of velocity, so

velocity vector: $\mathbf{v}(t) = x'(t)\mathbf{i} + y'(t)\mathbf{j}$ or

$$\mathbf{v}(t) = x'(t)\mathbf{i} + y'(t)\mathbf{j} + z'(t)\mathbf{k}.$$

acceleration vector: $$\mathbf{a}(t) = \frac{d^2x}{dt^2}\mathbf{i} + \frac{d^2y}{dt^2}\mathbf{j} \quad \text{or} \tag{14}$$

$$\mathbf{a}(t) = \frac{d^2x}{dt^2}\mathbf{i} + \frac{d^2y}{dt^2}\mathbf{j} + \frac{d^2z}{dt^2}\mathbf{k}. \tag{15}$$

The representation (14) resolves **a** into its horizontal and vertical components. However, there is another representation that is often more useful. Imagine yourself driving on the highway. If the car in which you are riding accelerates forward, you are pressed to the back of your seat. If it turns sharply to one side, you are thrown to the other. Both motions are due to acceleration. The second force is related to the rate at which the car turns, which is, of course, related to the curvature of the road. Thus, we would like to express the acceleration vector as a component in the direction of motion and a component somehow related to the curvature of the path. How do we do so? A glance back at Figure 1 on page 123 reveals the answer. The radial line of the circle of curvature is perpendicular to the unit tangent vector **T** since **T** is tangent to the circle of curvature, and in a circle, tangent and radial lines at a point are orthogonal (See Example 2.4.3.) But a vector that is perpendicular to **T** has the direction of the unit normal vector **n**. Thus, the component of acceleration in the direction **n** will be a measure of the acceleration due to turning. We would like to write

$$\mathbf{a} = a_T\mathbf{T} + a_n\mathbf{n} \tag{16}$$

where a_T and a_n are, respectively, the components of **a** in the tangential and normal directions.

THEOREM 4

$$\mathbf{a}(t) = \frac{d^2s}{dt^2}\mathbf{T} + \left(\frac{ds}{dt}\right)^2 \kappa\mathbf{n} \tag{17}$$

PROOF: $\mathbf{v}(t) = (d\mathbf{f}/dt) \overset{\text{Chain rule}}{=} (d\mathbf{f}/ds)(ds/dt) \overset{\text{Theorem 2.5.3}}{=} \mathbf{T}(ds/dt)$. Then

$$\mathbf{a}(t) = \frac{d\mathbf{v}}{dt} \overset{\text{Product rule}}{=} \mathbf{T}\frac{d^2s}{dt^2} + \frac{d\mathbf{T}}{dt}\frac{ds}{dt} = \mathbf{T}\frac{d^2s}{dt^2} + \left(\frac{d\mathbf{T}}{ds}\frac{ds}{dt}\right)\left(\frac{ds}{dt}\right)$$
$$= \frac{d^2s}{dt^2}\mathbf{T} + \kappa\mathbf{n}\left(\frac{ds}{dt}\right)^2.$$

The last step follows from equation (11). ■

NOTE: $v = \dfrac{ds}{dt} = |\mathbf{f}'(t)|$ is called the **speed** of a moving particle. Then, since $\kappa = \dfrac{1}{\rho}$, where ρ is the radius of curvature, we can write (16) as

WRITING a IN TERMS OF T AND n

$$\mathbf{a} = \frac{dv}{dt}\mathbf{T} + v^2\kappa\mathbf{n} = \frac{dv}{dt}\mathbf{T} + \frac{v^2}{\rho}\mathbf{n}. \tag{18}$$

Thus, the tangential component of acceleration is $a_T = dv/dt$, and the normal component of acceleration is $a_n = v^2/\rho$.

Using Theorem 4, we can prove the next theorem.

THEOREM 5

If $\mathbf{f}(t) \in \mathbb{R}^3$ and $\mathbf{f}'$ and $\mathbf{f}''$ exist,

$$\kappa = \frac{|\mathbf{f}' \times \mathbf{f}''|}{|\mathbf{f}'|^3}.$$

PROOF: $\mathbf{f}' \times \mathbf{f}'' = (\mathbf{T}\, ds/dt) \times [(d^2s/dt^2)\mathbf{T} + \kappa\mathbf{n}(ds/dt)^2]$. Now $\mathbf{T} \times \mathbf{T} = \mathbf{0}$, and $\mathbf{T} \times \mathbf{n} = \mathbf{B}$. Thus using Theorem 1.6.2 (iv) on page 44, we have

$$\mathbf{f}' \times \mathbf{f}'' = \left(\frac{ds}{dt}\right)^3 \kappa\mathbf{B}.$$

Then taking absolute values and using the fact that $|ds/dt| = |\mathbf{f}'|$ yields

$$|\mathbf{f}' \times \mathbf{f}''| = \kappa|\mathbf{f}'|^3,$$

from which the result follows. ■

Theorem 5 is useful for calculating curvature when it is not easy to write $\mathbf{f}$ in terms of the arc length parameter s.

EXAMPLE 6

CALCULATING THE CURVATURE OF A CURVE IN $\mathbb{R}^3$

Calculate the curvature of $\mathbf{f} = t^2\mathbf{i} + t^3\mathbf{j} + t^4\mathbf{k}$ at $t = 1$.

SOLUTION: Here,

$$\mathbf{f}' = 2t\mathbf{i} + 3t^2\mathbf{j} + 4t^3\mathbf{k}$$
$$|\mathbf{f}'| = \sqrt{4t^2 + 9t^4 + 16t^6} = t\sqrt{4 + 9t^2 + 16t^4}$$
$$\mathbf{f}'' = 2\mathbf{i} + 6t\mathbf{j} + 12t^2\mathbf{k}$$
$$\mathbf{f}' \times \mathbf{f}'' = \begin{vmatrix} \mathbf{i} & \mathbf{j} & \mathbf{k} \\ 2t & 3t^2 & 4t^3 \\ 2 & 6t & 12t^2 \end{vmatrix} = 12t^4\mathbf{i} - 16t^3\mathbf{j} + 6t^2\mathbf{k}$$
$$|\mathbf{f}' \times \mathbf{f}''| = \sqrt{144t^8 + 256t^6 + 36t^4} = t^2\sqrt{144t^4 + 256t^2 + 36}.$$

Thus,

$$\kappa = \frac{\sqrt{144t^4 + 256t^2 + 36}}{t(4 + 9t^2 + 16t^4)^{3/2}}.$$

At $t = 1$, $\kappa = \sqrt{436}/29^{3/2} \approx 0.1337$.

The result given by (17) or (18) is very important in physics. If an object of constant mass m is traveling along a trajectory, then the force on the object to produce that trajectory is given by

$$\mathbf{F} = m\mathbf{a} = m\frac{dv}{dt}\mathbf{T} + mv^2\kappa\mathbf{n}. \tag{19}$$

The term $mv^2\kappa$ is the magnitude of the force necessary to keep the project from ''moving off'' the trajectory in the direction of $\mathbf{T}$. In the case of automobiles, the force is supplied by tire friction.

EXAMPLE 7 COMPUTING FRICTIONAL FORCE

A 1500-kg race car is driven at a speed of 150 km/hr on an unbanked circular race track of radius 250 m. What frictional force must be exerted by the tires on the road surface to keep the car from skidding?

SOLUTION: The frictional force exerted by the tires must be equal to the component of the force (due to acceleration) normal to the circular race track. That is,

$$F = mv^2\kappa = \frac{mv^2}{\rho} = (1500\text{ kg})\frac{(150{,}000\text{ m})^2}{(3600\text{ sec})^2}\cdot\frac{1}{250\text{ m}}$$
$$= 10{,}416\tfrac{2}{3}(\text{kg})(\text{m})/\text{sec}^2 = 10{,}416\tfrac{2}{3}\text{ N}.$$

EXAMPLE 8 FINDING THE SMALLEST POSSIBLE COEFFICIENT OF FRICTION IN ORDER THAT A CAR REMAIN ON A ROAD

Let the car of Example 7 have the **coefficient of friction** μ. That is, the maximum frictional force that can be exerted by the car on the road surface is μmg, where mg is the **normal force** of the car on the road (the force of the car on the road due to gravity). What is the minimum value μ can take in order that the car not slide off the road?

SOLUTION: We must have $\mu mg \geq 10{,}416\frac{2}{3}$ N. But $\mu mg = \mu(9.81)(1500)$, so we obtain

$$\mu \geq \frac{10{,}416\frac{2}{3}}{(9.81)(1500)} \approx 0.71.$$

PROBLEMS 2.6

SELF-QUIZ

I. True–False:

a. A parabola that opens downward has negative curvature.

b. A circle with radius r has curvature equal to $1/r$.

c. Every straight line has curvature equal to 1.

d. There is exactly one straight line with curvature equal to 1.

e. There is exactly one straight line with curvature equal to 0.

f. $\mathbf{n}(t) \perp \mathbf{T}(t)$ for every smooth curve in $\mathbb{R}^3$.

In Problems 1–28, find the curvature and radius of curvature for each curve in $\mathbb{R}^2$. Sketch the unit tangent vector and the circle of curvature (the osculating circle). [*Note:* For some problems stated in Cartesian coordinates, you may find it convenient to use the result of Problem 55; for those involving polar coordinates, use the result of Problem 56.]

1. $\mathbf{f}(t) = t\mathbf{i} + t^2\mathbf{j}$; $t = 1$

2. $\mathbf{f}(t) = -t\mathbf{i} + t^2\mathbf{j}$; $t = 2$

3. $\mathbf{f}(t) = 2\cos t\mathbf{i} + 2\sin t\mathbf{j}$; $t = \pi/4$

4. $\mathbf{f}(t) = 2\cos t\mathbf{i} + 2\sin t\mathbf{j}$; $t = \pi/2$

5. $\mathbf{f}(t) = 3\sin t\mathbf{i} + 4\cos t\mathbf{j}$; $t = 0$

6. $\mathbf{f}(t) = 3\sin t\mathbf{i} + 4\cos t\mathbf{j}$; $t = \pi/2$

7. $\mathbf{f}(t) = 3\cos t\mathbf{i} + 4\sin t\mathbf{j}$; $t = \pi/4$

8. $\mathbf{f}(t) = 3\cos t\mathbf{i} - 4\sin t\mathbf{j}$; $t = \pi/3$

9. $\mathbf{f}(t) = (t - \sin t)\mathbf{i} + (1 - \cos t)\mathbf{j}$; $t = \pi/3$

10. $\mathbf{f}(t) = (\cos t + t\sin t)\mathbf{i} + (\sin t - t\cos t)\mathbf{j}$; $t = \pi/6$

11. $xy = 1$; $(1, 1)$

12. $y = 3/x$; $(1, 3)$

13. $y = e^x$; $(0, 1)$

14. $y = \ln x$; $(1, 0)$

15. $y = \ln x$; $(e, 1)$

16. $y = e^x$; $(1, e)$

17. $y = \cos x$; $(\pi/3, \frac{1}{2})$
18. $y = \ln \cos x$; $(\pi/4, -\frac{1}{2}\ln 2)$
19. $y = ax^2 + bx + c$; $a \neq 0$; $(0, c)$
20. $y = \sqrt{1 - x^2}$; $(0, 1)$
***21.** $y = \sin^{-1} x$; $(1, \pi/2)$
***22.** $x = \cos y$; $(0, \pi/2)$
23. $r = 2a\cos\theta$; $\theta = \pi/3$
24. $r = a\sin 2\theta$; $\theta = \pi/8$
25. $r = a(1 + \sin\theta)$; $\theta = \pi/2$
26. $r = a(1 - \cos\theta)$; $\theta = \pi$
27. $r = a\theta$; $\theta = 1$
28. $r = e^{a\theta}$; $\theta = 1$

In Problems 29–36, find the unit tangent vector **T**, the curvature κ, the principal unit normal **n**, and the binormal vector **B** at the specified value of t; then verify that $\mathbf{n} \cdot \mathbf{T} = 0$.

29. $\mathbf{g}(t) = \mathbf{i} + t\mathbf{j} + t^2\mathbf{k}$; $t = 1$
30. $\mathbf{g}(t) = t\mathbf{i} + t^2\mathbf{j} + t^3\mathbf{k}$; $t = 0$
31. $\mathbf{g}(t) = a(\sin t)\mathbf{i} + a(\cos t)\mathbf{j} + t\mathbf{k}$; $a > 0$; $t = \pi/4$
32. $\mathbf{g}(t) = a(\sin t)\mathbf{i} + a(\cos t)\mathbf{j} + t\mathbf{k}$; $a > 0$; $t = \pi/6$
33. $\mathbf{g}(t) = a(\cos t)\mathbf{i} + b(\sin t)\mathbf{j} + t\mathbf{k}$; $a > 0$, $b > 0$, $a \neq b$; $t = 0$
34. $\mathbf{g}(t) = a(\cos t)\mathbf{i} + b(\sin t)\mathbf{j} + t\mathbf{k}$; $a > 0$, $b > 0$, $a \neq b$; $t = \pi/2$
35. $\mathbf{g}(t) = e^t(\cos 2t)\mathbf{i} + e^t(\sin 2t)\mathbf{j} + e^t\mathbf{k}$; $t = 0$
36. $\mathbf{g}(t) = e^t(\cos 2t)\mathbf{i} + e^t(\sin 2t)\mathbf{j} + e^t\mathbf{k}$; $t = \pi/4$

In Problems 37–44, find the tangential and normal components of acceleration for the given position vector.

37. $\mathbf{f}(t) = (\cos 2t)\mathbf{i} + (\sin 2t)\mathbf{j}$
38. $\mathbf{f}(t) = 2(\cos t)\mathbf{i} + 3(\sin t)\mathbf{j}$
39. $\mathbf{f}(t) = t\mathbf{i} + t^2\mathbf{j}$
40. $\mathbf{f}(t) = t^2\mathbf{i} + t^3\mathbf{j}$
41. $\mathbf{f}(t) = t\mathbf{i} + (\cos t)\mathbf{j}$
42. $\mathbf{f}(t) = (t^3 - 3t)\mathbf{i} + (t^2 - 1)\mathbf{j}$
43. $\mathbf{f}(t) = e^{-t}\mathbf{i} + e^t\mathbf{j}$
44. $\mathbf{f}(t) = (\sin t^2)\mathbf{i} + (\cos t^2)\mathbf{j}$
45. At what point on the curve $y = \ln x$ is the curvature a maximum?
46. For what value of t in the interval $[0, \pi/2]$ is the curvature of the four-cusp hypocycloid $\mathbf{f}(t) = a(\cos^3 t)\mathbf{i} + a(\sin^3 t)\mathbf{j}$ a minimum? For what value is it a maximum?
47. Suppose that the driver of the race car in Example 7 reduces her speed by a factor of M. The frictional force needed to keep the car from skidding is reduced by what factor?
48. If the race car of Example 7 is placed on a track with half the radius of the original one, how much slower would it have to be driven so as not to increase the normal component of acceleration?
***49.** A truck traveling at 80 km/hr and weighing 10,000 kg is moving on an unbanked curved stretch of track. The equation of the curved section is the parabola $y = x^2 - x$ meters.
  a. What is the frictional force exerted by the wheels of the truck at the "point" $(0, 0)$ on the track?
  b. If the coefficient of friction for the truck is 2.5, what is the maximum speed it can achieve at the point $(0, 0)$ without going off the track?
***50.** A child swings a rope attached to a bucket containing 3 kg of water. The pail rotates in the vertical plane in a circular path with a radius of 1 m. What is the smallest number of revolutions that must be made every minute in order that the water stay in the pail? [*Hint:* Calculate the pressure of the water on the bottom of the pail. This can be determined by first calculating the normal force of the motion. Then the water will stay in the bucket if this normal force exceeds the force due to gravity.]

The local twisting of a curve can be measured by examining $d\mathbf{B}/ds$. Problems 60–63 ask you to think about something called the **torsion** of a curve. Use the computational result of Problem 62 to work Problems 51–54.

51. Calculate the torsion of the straight line given parametrically by $f(t) = t\mathbf{i} + (1 - t)\mathbf{j} + (2 + 3t)\mathbf{k}$.
***52.** Calculate the torsion of the curve $\mathbf{f}(t) = e^t(\cos t)\mathbf{i} + e^t(\sin t)\mathbf{j} + e^t\mathbf{k}$ at $t = 0$.
***53.** Calculate the torsion of the circular helix $\mathbf{f}(t) = (\cos t)\mathbf{i} + (\sin t)\mathbf{j} + t\mathbf{k}$ at $t = \pi/6$.
***54.** Calculate the torsion of the circular helix $\mathbf{f}(t) = a(\cos \omega t)\mathbf{i} + a(\sin \omega t)\mathbf{j} + bt\mathbf{k}$.
***55.** Suppose that a curve satisfies the smooth Cartesian equation $y = f(x)$. Show that its curvature can be computed by

$$\kappa(x) = \left|\frac{d\varphi}{dx}\right| = \frac{\left|\dfrac{d^2y}{dx^2}\right|}{\left[1 + \left(\dfrac{dy}{dx}\right)^2\right]^{3/2}}.$$

***56.** Suppose that a curve satisfies the smooth polar equation $r = g(\theta)$. Show that its curvature can be computed by

$$\kappa(\theta) = \frac{\left| r^2 + 2\left(\frac{dr}{d\theta}\right)^2 - r\frac{d^2r}{d\theta^2} \right|}{\left[r^2 + \left(\frac{dr}{d\theta}\right)^2 \right]^{3/2}}.$$

57. Show that if a particle is moving at a constant speed, then the tangential component of acceleration is zero.

58. Show that the curvature of a straight line in space is zero.

***59.** Let **B** be the binormal vector to a curve **f**, and let s denote arc length. Show that if $d\mathbf{B}/ds \neq \mathbf{0}$, then $d\mathbf{B}/ds \perp \mathbf{B}$ and $d\mathbf{B}/ds \perp \mathbf{T}$. [*Hint:* Differentiate $\mathbf{B} \cdot \mathbf{T} = 0$ and $\mathbf{B} \cdot \mathbf{B} = 1$.]

***60.** $d\mathbf{B}/ds$ must be parallel to **n** because it is orthogonal to **B** and **T** (see the preceding problem). Therefore, there must be some number τ such that

$$\frac{d\mathbf{B}}{ds} = -\tau\mathbf{n}.$$

This number τ is called the **torsion** of the curve. It is a measure of how much the curve twists. Show that

$$\frac{d\mathbf{n}}{ds} = -\kappa\mathbf{T} + \tau\mathbf{B}.$$

****61.** Show that

$$\mathbf{f}'''(t) = \left[\frac{d^3s}{dt^3} - \left(\frac{ds}{dt}\right)^3 \kappa^2\right]\mathbf{T} + \left[3\left(\frac{d^2s}{dt^2}\right)\frac{ds}{dt}\kappa + \left(\frac{ds}{dt}\right)^2\frac{d\kappa}{dt}\right]\mathbf{n} + \left(\frac{ds}{dt}\right)^3 \kappa\tau\mathbf{B}.$$

****62.** Use the result of the preceding problem to show that

$$\tau = \frac{\mathbf{f}''' \cdot (\mathbf{f}' \times \mathbf{f}'')}{\kappa^2\left(\frac{ds}{dt}\right)^6} = \frac{\mathbf{f}''' \cdot (\mathbf{f}' \times \mathbf{f}'')}{|\mathbf{f}' \times \mathbf{f}''|^2}.$$

****63.** Show that any smooth curve with the property that the position vector $\mathbf{r}(t)$ lies in a fixed plane must have torsion equal to zero everywhere.

ANSWERS TO SELF-QUIZ

I. a. False **b.** True **c.** False **d.** False **e.** False **f.** True

CHAPTER 2 SUMMARY OUTLINE

- **Vector Function in $\mathbb{R}^2$** Let f_1 and f_2 be functions of the real variable t. Then for all values of t for which $f_1(t)$ and $f_2(t)$ are defined, we define the **vector-valued function f** by

$$\mathbf{f}(t) = (f_1(t), f_2(t)) = f_1(t)\mathbf{i} + f_2(t)\mathbf{j}. \qquad (*)$$

The **domain** of **f** is the intersection of the domains of f_1 and f_2. It is a set of real numbers. The **range** of **f** is a set of vectors in $\mathbb{R}^2$.

- **Plane Curves and Parametric Equations** Let **f** be a vector function. Then for each t in the domain of **f**, the endpoint of the vector $f_1(t)\mathbf{i} + f_2(t)\mathbf{j}$ is a point (x, y) in the xy-plane, where

$$x = f_1(t) \quad \text{and} \quad y = f_2(t). \qquad (**)$$

Suppose that the interval $[a, b]$ is in the domain of the function **f** and that both f_1 and f_2 are continuous in $[a, b]$. Then the set of points $(f_1(t), f_2(t))$ for $a \leq t \leq b$ is called a **plane curve** C. Equation (*) is called the **vector equation** of C, while equations (**) are called the **parametric equations** or **parametric representation** of C. In this context, the variable t is called a **parameter**.

- **Cartesian Equation of a Plane Curve** A **Cartesian equation** of the curve $f(t) = x(t)\mathbf{i} + y(t)\mathbf{j}$ is an equation relating the variables x and y only

- **Vector Function in $\mathbb{R}^3$** Let f_1, f_2, and f_3 be functions of the real variable t. Then for all values of t for which $f_1(t)$, $f_2(t)$, and $f_3(t)$ are defined, we define the **vector-valued function f of a real variable** t by

$$\mathbf{f}(t) = (f_1(t), f_2(t), f_3(t)) = f_1(t)\mathbf{i} + f_2(t)\mathbf{j} + f_3(t)\mathbf{k}.$$

- **Curve in Space** If f_1, f_2, and f_3 are continuous over an interval I, then as t varies over I, the set of points traced out by the end of the vector **f** is called a **curve** in space.

■ **Smooth Curve** The parametric curve $\mathbf{f}(t) = f_1(t)\mathbf{i} + f_2(t)\mathbf{j}$ is said to be **smooth** on an interval $I = (a, b)$ if $f_1'(t)$ and $f_2'(t)$ exist and are continuous on I and, for every t in I, $[f_1'(t)]^2 + [f_2'(t)]^2 \neq 0$.

■ **The Slope of the Tangent Line to a Smooth Curve** Let (x_0, y_0) be on the smooth curve C given by $x = f_1(t)$ and $y = f_2(t)$. If the curve passes through (x_0, y_0) when $t = t_0$, then the slope m of the line tangent to C at (x_0, y_0) is

$$m = \lim_{t \to t_0} \frac{f_2'(t)}{f_1'(t)}$$

provided that this limit exists.

■ **Vertical and Horizontal Tangents** C has a vertical tangent line at t_0 if

$$\left.\frac{dx}{dt}\right|_{t=t_0} = f_1'(t_0) = 0 \quad \text{and} \quad f_2'(t_0) \neq 0.$$

C has a horizontal tangent line at t_0 if

$$\left.\frac{dy}{dt}\right|_{t=t_0} = f_2'(t_0) = 0 \quad \text{and} \quad f_1'(t_0) \neq 0.$$

■ **Limit of a Vector Function** Let $\mathbf{f}(t) = f_1(t)\mathbf{i} + f_2(t)\mathbf{j}$. Let t_0 be any real number, $+\infty$, or $-\infty$. If $\lim_{t\to t_0} f_1(t)$ and $\lim_{t\to t_0} f_2(t)$ both exist, then

$$\lim_{t\to t_0} \mathbf{f}(t) = \left[\lim_{t\to t_0} f_1(t)\right]\mathbf{i} + \left[\lim_{t\to t_0} f_2(t)\right]\mathbf{j}.$$

■ **Continuity of a Vector Function** $\mathbf{f}$ is **continuous** at t_0 if the component functions f_1 and f_2 are continuous at t_0. Thus, $\mathbf{f}$ is continuous at t_0 if

$$\lim_{t\to t_0} \mathbf{f}(t) = \mathbf{f}(t_0).$$

■ **Derivative of a Vector Function** Let f be defined at t. Then $\mathbf{f}$ is **differentiable** at t if

$$\lim_{\Delta t \to 0} \frac{\mathbf{f}(t + \Delta t) - \mathbf{f}(t)}{\Delta t} \qquad (\surd)$$

exists and is finite. The vector function $\mathbf{f}'$ defined by

$$\mathbf{f}'(t) = \frac{d\mathbf{f}}{dt} = \lim_{\Delta t \to 0} \frac{\mathbf{f}(t + \Delta t) - \mathbf{f}(t)}{\Delta t}$$

is called the **derivative** of $\mathbf{f}$, and the domain of $\mathbf{f}'$ is the set of all t such that the limit in $(\surd)$ exists.

If $\mathbf{f}(t) = f_1(t)\mathbf{i} + f_2(t)\mathbf{j}$, then at any value t for which $f_1'(t)$ and f_2' exist,

$$\mathbf{f}'(t) = f_1'(t)\mathbf{i} + f_2'(t)\mathbf{j}.$$

■ **Differentiability in an Open Interval** The vector function $\mathbf{f}$ is **differentiable** on the open interval I if $\mathbf{f}'(t)$ exists for every t in I.

■ **Unit Tangent Vector** A **unit tangent vector** to a curve C, denoted by $\mathbf{T}$, is given by

$$\mathbf{T}(t) = \frac{\mathbf{f}'(t)}{|\mathbf{f}'(t)|}$$

for any number t, as long as $\mathbf{f}'(t) \neq 0$.

■ **Integral of a Vector Function**

i. Let $\mathbf{f}(t) = f_1(t)\mathbf{i} + f_2(t)\mathbf{j}$ and suppose that the component functions f_1 and f_2 have antiderivatives. Then an **antiderivative**, or the **indefinite integral**, of $\mathbf{f}$ is given by

$$\int \mathbf{f}(t)\,dt = \left(\int f_1(t)\,dt\right)\mathbf{i} + \left(\int f_2(t)\,dt\right)\mathbf{j} + \mathbf{C}$$

where $\mathbf{C}$ is a constant vector of integration.

ii. If f_1 and f_2 are integrable over the interval $[a, b]$, then the **definite integral** of $\mathbf{f}$ is given by

$$\int_a^b \mathbf{f}(t)\,dt = \left(\int_a^b f_1(t)\,dt\right)\mathbf{i} + \left(\int_a^b f_2(t)\,dt\right)\mathbf{j}.$$

■ **Some Differentiation Formulas** Let $\mathbf{f}$ and $\mathbf{g}$ be vector functions that are differentiable in an interval I. Let the scalar function h be differentiable in I. Finally, let α be a scalar and let $\mathbf{v}$ be a constant vector. Then, on I, we have

i. $\mathbf{f} + \mathbf{g}$ is differentiable and

$$\frac{d}{dt}(\mathbf{f} + \mathbf{g}) = \frac{d\mathbf{f}}{dt} + \frac{d\mathbf{g}}{dt} = \mathbf{f}' + \mathbf{g}'.$$

ii. $\alpha\mathbf{f}$ is differentiable and

$$\frac{d}{dt}\alpha\mathbf{f} = \alpha\frac{d\mathbf{f}}{dt} = \alpha\mathbf{f}'.$$

iii. $\mathbf{v} \cdot \mathbf{f}$ is differentiable and

$$\frac{d}{dt}\mathbf{v} \cdot \mathbf{f} = \mathbf{v} \cdot \frac{d\mathbf{f}}{dt} = \mathbf{v} \cdot \mathbf{f}'.$$

iv. $h\mathbf{f}$ is differentiable and

$$\frac{d}{dt}h\mathbf{f} = h\frac{d\mathbf{f}}{dt} + \frac{dh}{dt}\mathbf{f} = h\mathbf{f}' + h'\mathbf{f}.$$

v. $\mathbf{f} \cdot \mathbf{g}$ is differentiable and

$$\frac{d}{dt}\mathbf{f} \cdot \mathbf{g} = \mathbf{f} \cdot \frac{d\mathbf{g}}{dt} + \frac{d\mathbf{f}}{dt} \cdot \mathbf{g} = \mathbf{f} \cdot \mathbf{g}' + \mathbf{f}' \cdot \mathbf{g}.$$

vi. If $\mathbf{f}$ and $\mathbf{g}$ are differentiable vector functions in $\mathbb{R}^3$, then $\mathbf{f} \times \mathbf{g}$ is differentiable and

$$(\mathbf{f} \times \mathbf{g})' = (\mathbf{f}' \times \mathbf{g}) + (\mathbf{f} \times \mathbf{g}')$$

■ **Unit Normal Vector** The vector $\mathbf{T}'$ which is orthogonal to the tangent line is called a **normal vector** to the curve $\mathbf{f}$. Whenever $\mathbf{T}'(t) \neq \mathbf{0}$, we define the **unit normal vector** to the curve $\mathbf{f}$ at t as

$$\mathbf{n}(t) = \frac{\mathbf{T}'(t)}{|\mathbf{T}'(t)|}.$$

The unit normal vector **n** always points in the direction of the concave side of the curve.

■ **The Function $s(t)$** The function $s(t)$ is defined by

$s(t)$ = length along the curve C from $(f_1(t_0), f_2(t_0))$ to $(f_1(t), f_2(t))$.

Suppose that f_1' and f_2' are continuous in the interval $[t_0, t_1]$. Then $s(t)$ is a differentiable function of t for $t \in [t_0, t_1]$ and

$$\frac{ds}{dt} = \sqrt{[f_1'(t)]^2 + [f_2'(t)]^2} = \sqrt{\left(\frac{dx}{dt}\right)^2 + \left(\frac{dy}{dt}\right)^2}.$$

■ **Arc Length in $\mathbb{R}^2$** Suppose **f** has a continuous derivative in the interval $[t_0, b]$. Then **f** is said to be **rectifiable** in the interval $[t_0, b]$, and the **arc length** $s(t_1)$ of the curve $\mathbf{f}(t)$ in the interval $[t_0, t_1]$ (with $t_1 \le b$) is given by

$$s(t_1) = \text{length of arc from } t_0 \text{ to } t_1$$
$$= \int_{t_0}^{t_1} \left(\frac{ds}{dt}\right) dt = \int_{t_0}^{t_1} \sqrt{\left(\frac{dx}{dt}\right)^2 + \left(\frac{dy}{dt}\right)^2}\, dt.$$

■ **Parameter of Arc Length** The parameter s given by

$$ds = \sqrt{\left(\frac{dx}{dt}\right)^2 + \left(\frac{dy}{dt}\right)^2}\, dt$$

is called the **parameter of arc length.**

■ **Arc Length in $\mathbb{R}^3$** Suppose **f** has a continuous derivative in the interval $[t_0, b]$ and suppose that for every t_1 in $[t_0, b]$, $\int_{t_0}^{t_1} |\mathbf{f}'(t)|\, dt$ exists. Then **f** is said to be **rectifiable** in the interval $[t_0, b]$, and the **arc length** of the curve $\mathbf{f}(t)$ in the interval $[t_0, t_1]$ is given by

$$s(t_1) = \int_{t_0}^{t_1} |\mathbf{f}'(t)|\, dt =$$
$$\int_{t_0}^{t_1} \sqrt{\left(\frac{dx}{dt}\right)^2 + \left(\frac{dy}{dt}\right)^2 + \left(\frac{dz}{dt}\right)^2}\, dt.$$

■ **The Unit Tangent Vector Parametrized by Arc Length** If the curve C is parametrized by $\mathbf{f}(s) = x(s)\mathbf{i} + y(s)\mathbf{j}$ or $x(s)\mathbf{i} + y(s)\mathbf{j} + z(s)\mathbf{k}$, where s is arc length and x and y have continuous derivatives, then the unit tangent vector **T** is given by

$$\mathbf{T}(s) = \frac{d\mathbf{f}}{ds}.$$

From this it follows that

$$\left(\frac{dx}{ds}\right)^2 + \left(\frac{dy}{ds}\right)^2 = 1 \quad \text{or}$$
$$\left(\frac{dx}{ds}\right)^2 + \left(\frac{dy}{ds}\right)^2 + \left(\frac{dz}{ds}\right)^2 = 1.$$

■ **Curvature in $\mathbb{R}^2$ and Radius of Curvature**

i. Let the curve C be given by the differentiable vector function $\mathbf{f}(t) = f_1(t)\mathbf{i} + f_2(t)\mathbf{j}$. Let $\varphi(t)$ denote the direction of $\mathbf{f}'(t)$. Then the **curvature** of C, denoted by $\kappa(t)$, is the absolute value of the rate of change of direction with respect to arc length; that is

$$\kappa(t) = \left|\frac{d\varphi}{ds}\right|.$$

ii. The **radius of curvature** $\rho(t)$ is defined by

$$\rho(t) = \frac{1}{\kappa(t)} \qquad \text{if } \kappa(t) > 0.$$

If $\mathbf{T}(t)$ denotes the unit tangent vector to **f**, then

$$\kappa(t) = \left|\frac{d\mathbf{T}}{ds}\right|.$$

Moreover, if $\mathbf{f}(t) = x(t)\mathbf{i} + y(t)\mathbf{j}$, then

$$\kappa(t) = \frac{|(dx/dt)(d^2y/dt^2) - (dy/dt)(d^2x/dt^2)|}{[(dx/dt)^2 + (dy/dt)^2]^{3/2}}$$
$$= \frac{|x'y'' - y'x''|}{[(x')^2 + (y')^2]^{3/2}}.$$

■ **Curvature in $\mathbb{R}^3$** If **f** has a continuous derivative, then the **curvature** of **f** is given by

$$\kappa(t) = \left|\frac{d\mathbf{T}}{ds}\right|.$$

■ **Principal Unit Normal Vector** For any value of t for which $\kappa(t) \neq 0$, the **principal unit normal vector n** is defined by

$$\mathbf{n}(t) = \frac{1}{\kappa(t)} \frac{d\mathbf{T}}{ds}.$$

Moreover,

$$\mathbf{n}(t) \perp \mathbf{T}(t).$$

■ **Binormal Vector** The **binormal vector B** to the curve **f** in $\mathbb{R}^3$ is defined by

$$\mathbf{B} = \mathbf{T} \times \mathbf{n}.$$

■ **Tangential and Normal Components of Acceleration** The acceleration vector

$$\mathbf{a}(t) = \frac{d^2x}{dt^2}\mathbf{i} + \frac{d^2y}{dt^2}\mathbf{j} \qquad \text{or}$$
$$\mathbf{a}(t) = \frac{d^2x}{dt^2}\mathbf{i} + \frac{d^2y}{dt^2}\mathbf{j} + \frac{d^2z}{dt^2}\mathbf{k}$$

can be written

$$\mathbf{a} = a_T\mathbf{T} + a_n\mathbf{n} = \frac{d^2s}{dt^2}\mathbf{T} + \left(\frac{ds}{dt}\right)^2 \kappa\mathbf{n}$$
$$= \frac{dv}{dt}\mathbf{T} + v^2\kappa\mathbf{n} = \frac{dv}{dt}\mathbf{T} + \frac{v^2}{\rho}\mathbf{n}, \text{ where } v = |\mathbf{f}'(t)|.$$

■ **Another Formula for Curvature**

$$\kappa = \frac{|\mathbf{f}' \times \mathbf{f}''|}{|\mathbf{f}'|^3}.$$

CHAPTER 2 REVIEW EXERCISES

In Problems 1–8, find a Cartesian equation for the curve and then sketch the curve in the xy-plane.

1. $\mathbf{f}(t) = t\mathbf{i} + 2t\mathbf{j}$
2. $\mathbf{f}(t) = (2t - 6)\mathbf{i} + t^2\mathbf{j}$
3. $\mathbf{f}(t) = t^2\mathbf{i} + (2t - 6)\mathbf{j}$
4. $\mathbf{f}(t) = t^2\mathbf{i} + t^4\mathbf{j}$
5. $\mathbf{f}(t) = (\cos 4t)\mathbf{i} + (\sin 4t)\mathbf{j}$
6. $\mathbf{f}(t) = 4(\sin t)\mathbf{i} + 9(\sin t)\mathbf{j}$
7. $\mathbf{f}(t) = t^6\mathbf{i} + t^2\mathbf{j}$
8. $\mathbf{f}(t) = e^t(\cos t)\mathbf{i} + e^t(\sin t)\mathbf{j}$

In Problems 9–16, find the slope of the line tangent to the given curve for the specified value of the parameter; then find all points at which the curve has a vertical or horizontal tangent.

9. $x = t^2;\ y = 6t;\ t = 1$
10. $x = t^7;\ y = t^8 - 5;\ t = 2$
11. $x = \sin 5\theta;\ y = \cos 5\theta;\ \theta = \pi/3$
12. $x = \cos^2\theta;\ y = -3\theta;\ \theta = \pi/4$
13. $x = \cosh t;\ y = \sinh t;\ t = 0$
14. $x = 2/\theta;\ y = -3\theta;\ \theta = 10\pi$
15. $x = 3\cos\theta;\ y = 4\sin\theta;\ \theta = \pi/3$
16. $x = 3\cos\theta;\ y = -4\sin\theta;\ \theta = 2\pi/3$

In Problems 17–22, calculate the first and second derivatives of the given vector function.

17. $\mathbf{f}(t) = 2t\mathbf{i} - t^2\mathbf{j}$
18. $\mathbf{f}(t) = (1/t^2)\mathbf{i} + \sqrt[3]{t}\mathbf{j}$
19. $\mathbf{f}(t) = (\cos 5t)\mathbf{i} + 2(\sin t)\mathbf{j}$
20. $\mathbf{f}(t) = (\tan t)\mathbf{i} + (\cot t)\mathbf{j}$
21. $\mathbf{g}(t) = t^3\mathbf{i} - t^2\mathbf{j} + t\mathbf{k}$
22. $\mathbf{g}(t) = (2 - 3t)\mathbf{i} + (e^t - 1 - t)\mathbf{j} + (e^{-t})\mathbf{k}$

In Problems 23–26, calculate the given definite or indefinite integral.

23. $\int_0^3 (t^2\mathbf{i} + t^5\mathbf{j})\,dt$
24. $\int [(\cos 3t)\mathbf{i} + (\sin 3t)\mathbf{j}]\,dt$
25. $\int_0^{\pi/2} [(\cos t)\mathbf{i} + (\sin t)\mathbf{j} + t\mathbf{k}]\,dt$
26. $\int_0^1 [e^t\mathbf{i} + (\cos 3t)\mathbf{j} - 4t\mathbf{k}]\,dt$

In Problems 27–30, calculate the derivative.

27. $\dfrac{d}{dt}[(2t\mathbf{i} + \sqrt{t}\mathbf{j}) \cdot (t\mathbf{i} - 3\mathbf{j})]$
28. $\dfrac{d}{dt}\{(2\mathbf{i} - 11\mathbf{j}) \cdot [-(\tan t)\mathbf{i} + (\sec t)\mathbf{j}]\}$
29. $\dfrac{d}{dt}\{(3\mathbf{i} + 4\mathbf{j}) \times ((\cos t)\mathbf{i} + (\sin t)\mathbf{j} + t\mathbf{k})\}$
30. $\dfrac{d}{dt}\{[(\cos t)\mathbf{i} + (\sin t)\mathbf{j} + t\mathbf{k}] \times (3\mathbf{i} + 4\mathbf{j} + 5\mathbf{k})\}$

In Problems 31–34, find the unit tangent and unit normal vectors to the given curve for the specified value of t.

31. $\mathbf{f}(t) = t^4\mathbf{i} + t^5\mathbf{j};\ t = 1$
32. $\mathbf{f}(t) = (\cos 2t)\mathbf{i} + (\sin 2t)\mathbf{j};\ t = \pi/6$
33. $\mathbf{f}(t) = (\ln t)\mathbf{i} + \sqrt{t}\mathbf{j};\ t = 1$
34. $\mathbf{f}(t) = (\tan t)\mathbf{i} + (\cot t)\mathbf{j};\ t = \pi/3$

In Problems 35–40, calculate the length of the arc over the specified interval or the length of the given closed curve.

35. $x = \cos 4\theta;\ y = \sin 4\theta;\ 0 \le \theta \le \pi/12$
36. $\mathbf{f}(t) = e^t \sin t\mathbf{i} + e^t \cos t\mathbf{j};\ 0 \le t \le \pi/12$
37. $r = 2(1 + \cos\theta)$
38. $r = 5\sin\theta$
39. $x = -2t;\ y = \cos 2t;\ z = \sin 2t;\ 0 \le t \le \pi/6$
40. $\mathbf{g}(t) = 3(\sin 2t)\mathbf{i} + 3(\cos 2t)\mathbf{j} + t^2\mathbf{k};\ 0 \le t \le 5$

In Problems 41–46, the position vector of a moving particle is given. For the specified value of t, calculate the velocity vector, the acceleration vector, the speed, and the acceleration scalar. Then sketch a portion of the trajectory showing the velocity and acceleration vectors.

41. $\mathbf{f}(t) = (\cos 2t)\mathbf{i} + (\sin 2t)\mathbf{j};\ t = \pi/6$
42. $\mathbf{f}(t) = 6t\mathbf{i} + 2t^3\mathbf{j};\ t = 1$
43. $\mathbf{f}(t) = (2^t + e^{-t})\mathbf{i} + 2t\mathbf{j};\ t = 0$
44. $\mathbf{f}(t) = (1 - \cos t)\mathbf{i} + (t - \sin t)\mathbf{j};\ t = \pi/2$
45. $\mathbf{f}(t) = 2(\sinh t)\mathbf{i} + 4t\mathbf{j};\ t = 1$
46. $\mathbf{f}(t) = (3 + 5t)\mathbf{i} + (2 + 8t)\mathbf{j};\ t = 6$

In Problems 47–50, find the tangential and normal components of acceleration for the given position vector.

47. $\mathbf{f}(t) = 2(\sin t)\mathbf{i} + 2(\cos t)\mathbf{j}$
48. $\mathbf{f}(t) = 4(\cos t)\mathbf{i} + 9(\sin t)\mathbf{j}$
49. $\mathbf{f}(t) = 3t^2\mathbf{i} + 2t^3\mathbf{j}$
50. $\mathbf{f}(t) = (\cos t^2)\mathbf{i} + (\sin t^2)\mathbf{j}$

In Problems 51–60, find the curvature and radius of curvature of the given curve at the specified point. Sketch the curve in the neighborhood of that point, the unit tangent vector there, and the circle of curvature.

51. $\mathbf{f}(t) = (\cos 2t)\mathbf{i} + (\sin 2t)\mathbf{j};\ t = \pi/3$
52. $\mathbf{f}(t) = t^2\mathbf{i} + 2t\mathbf{j};\ t = 2$
53. $\mathbf{f}(t) = 4(\cos t)\mathbf{i} + 9(\sin t)\mathbf{j};\ t = \pi/4$
54. $\mathbf{f}(t) = 2t\mathbf{i} + (t^2/2)\mathbf{j};\ t = 0$
55. $y = 2x^2;\ (0, 0)$
56. $xy = 1;\ (2, \frac{1}{2})$
57. $y = e^{-x};\ (1, 1/e)$
58. $y = \sqrt{x};\ (4, 2)$
59. $r = 1 + \sin\theta;\ \theta = \pi/2$
60. $r = 3\theta;\ \theta = \pi$

In Problems 61–64, find parametric equations in terms of the arc length s measured from the point reached when $t = 0$. Verify your answer by using equation (2.5.12).

61. $\mathbf{f}(t) = 3t\mathbf{i} + 4t^{3/2}\mathbf{j}$

62. $\mathbf{f}(t) = \frac{2}{9}t^{9/2}\mathbf{i} + \frac{1}{3}t^3\mathbf{j}$

63. $\mathbf{f}(t) = 2(\cos 3t)\mathbf{i} + 2(\sin 3t)\mathbf{j}$

64. $\mathbf{f}(t) = e^t(\sin t)\mathbf{i} + e^t(\cos t)\mathbf{j}$

In Problems 65–68, find the unit tangent vector $\mathbf{T}$, the curvature κ, the principal unit normal vector $\mathbf{n}$, and the binomial vector $\mathbf{B}$ for the specified value of t.

65. $\mathbf{g}(t) = \mathbf{i} + t\mathbf{j} + \frac{1}{2}t^2\mathbf{k}$; $t = 2$

66. $\mathbf{g}(t) = t^2\mathbf{i} + t^3\mathbf{j} - t\mathbf{k}$; $t = 0$

67. $\mathbf{g}(t) = 2(\cos t)\mathbf{i} + 2(\sin t)\mathbf{j} + t\mathbf{k}$; $t = \pi/6$

68. $\mathbf{g}(t) = e^t[\mathbf{i} + (\cos t)\mathbf{j} + (\sin t)\mathbf{k}]$; $t = 0$

69. Compute the derivative of $\mathbf{f}(t) = (t^3, \cos t, \ln t, e^{2t})$.

70. Compute the derivative of $\mathbf{f}(t) = (\sqrt{t}, t^{5/3}, (1/t), t^5 - 2t + 2)$.

71. Compute the unit tangent vector and the unit normal vector to the curve $\mathbf{f}(t) = (t^2, t^4, t^6, t^8)$ at $t = 1$.

72. Compute the length of the curve $\mathbf{f}(t) = (t^2, t^2, t, t)$ for $t \in [0, 1]$.

CHAPTER 2 COMPUTER EXERCISES

1. This is a marble race. Marbles are started at the origin and allowed to slide down a decreasing path to the point $(\pi, -2)$. The challenge is that each contestant may select the path that his or her marble will follow. (Any point below the x-axis and not on the y-axis would work, but $(\pi, -2)$ turns out to simplify calculations.)

The first contestant chooses the brachistochrone curve given by $(\pi t - \sin \pi t, \cos \pi t - 1)$, $0 \le t \le 1$. The second contestant chooses the straight line $(\pi t, -2t)$, $0 \le t \le 1$, and the third contestant chooses the parabolic arch $(\pi t, 2(t-1)^2 - 2)$, $0 \le t \le 1$. As the fourth contestant, you may select any path you want; just be sure that your parametric curve decreases from the origin to $(\pi, -2)$ as t moves from 0 to 1.

With the aid of some elementary physics, it is not difficult to show that the time required for a marble to slide from the origin along the decreasing parametric curve $(x(t), y(t))$ from $t = 0$ to $t = 1$ is given by the following integral:

$$\text{time} = \frac{1}{\sqrt{g}} \int_0^1 \sqrt{\frac{(x'(t))^2 + (y'(t))^2}{-2y(t)}}\, dt,$$

where g is the constant of gravitational attraction near the surface of the earth.

Calculate the finish times for contestants 1, 2, 3, and for yourself. (You should be able to calculate by hand the time required for contestants 1 and 2. For contestant 3, you will need a computer algebra system to approximate the integral. You are almost certain to need similar help with your own integral.)

Based on the outcome of the race, formulate a conjecture about a physical property of the brachistochrone curve. Can you prove your assertion? (The curve presented here is called a *cycloid* reflected through the x-axis. It is the path taken by a point on the rim of a tire as its rolls down a road.)

Problems 2, 3, and 4 can be completed without electronic aid. However you will find a computer algebra system to be very helpful in your calculations.

2. Show that the curvature of the ellipse $(3 \cos t, \sin t)$ reaches a maximum at $(\pm 3, 0)$ and is a minimum at $(0, \pm 1)$. Calculate the curvature at these points.

3. Calculate the curvature of the elliptical helix $(3 \cos t, \sin t, t)$. Where does the curvature reach a maximum? Where does it reach a minimum?

4. Calculate the curvature of $(\cos at, \sin at, t)$. What happens to the curvature as a approaches 0? What happens to the curvature as a approaches infinity? Give a geometric explanation of your answers above. You may wish to use graphics software to look at a picture of the curve for various values of a.

CHAPTER 3

DIFFERENTIATION OF FUNCTIONS OF TWO OR MORE VARIABLES

For most of the functions we have so far encountered in this book, we have been able to write $y = f(x)$. This means that we could write the variable y explicitly in terms of the single variable x. However, in a great variety of applications it is necessary to write the quantity of interest in terms of two or more variables. We have already encountered this situation. For example, the volume of a right circular cylinder is given by

$$V = \pi r^2 h,$$

where r is the radius of the cylinder and h is its height. That is, V is a function of the *two* variables r and h.

As a second example, the **ideal-gas law**, which relates pressure, volume, and temperature for an ideal gas, can be written as

$$PV = nRT,$$

where P is the pressure of the gas, V is the volume, T is the absolute temperature (i.e., in degrees kelvin), n is the number of moles of the gas, and R is a constant. Solving for P, we find that

$$P = \frac{nRT}{V}.$$

That is, we can write P as a function of the *three* variables n, T, and V.

As a third example, according to **Poiseuille's law**, the resistance R of a blood vessel of length l and radius r is given by

$$R = \frac{\alpha l}{r^4},$$

where α is a constant of proportionality. If l and r are allowed to vary, then R is written as a function of the *two* variables l and r.

As a final example, let the vector $\mathbf{v}$ be given by

$$\mathbf{v} = x\mathbf{i} + y\mathbf{j} + z\mathbf{k}.$$

Then the magnitude of $\mathbf{v}$, $|\mathbf{v}|$, is given by

$$|\mathbf{v}| = \sqrt{x^2 + y^2 + z^2}.$$

That is, the magnitude of $\mathbf{v}$ is written as a function of the *three* variables, x, y, and z.

There are many examples like the four cited above. It is probably fair to say that very few physical, biological, or economic quantities can be properly expressed in terms of one variable alone. Often we write these quantities in terms of one variable simply because functions of only one variable are the easiest functions to handle.

In this chapter, we will see how many of the operations we have studied in our discussion of the "one-variable" calculus can be extended to functions of two and three variables. We will begin by discussing the basic notions of limits and continuity and will go on to differentiation and applications of differentiation. In Chapter 4, we will discuss the integration of functions of two and three variables.

3.1 FUNCTIONS OF TWO OR MORE VARIABLES

In Chapter 2, we discussed the notion of vector-valued functions. For example, if $\mathbf{f}(t) = f_1(t)\mathbf{i} + f_2(t)\mathbf{j}$, then for every t in the domain of $\mathbf{f}$, we obtain a vector $\mathbf{f}(t)$. In defining functions of two or more variables, we obtain a somewhat reversed situation. For example, in the formula for the volume of a right circular cylinder, we have

$$V = \pi r^2 h. \tag{1}$$

Here the volume V is written as a function of the two variables r and h. Put another way, for every ordered pair of positive real numbers (r, h), there is a unique positive number V such that $V = \pi r^2 h$. To indicate the dependence of V on the variables r and h, we write $V(r, h)$. That is, V can be thought of as a function that assigns a positive real number to every ordered pair of positive real numbers. Thus we see what we meant when we said that the situation is "reversed." Instead of having a vector function of one variable (a scalar), we have a scalar function of an ordered pair of variables (which is a vector in $\mathbb{R}^2$). We now give a definition of a function of two variables.

DEFINITION FUNCTION OF TWO VARIABLES

Let D be a subset of $\mathbb{R}^2$. Then a **real-valued function of two variables** f is a rule that assigns to each point (x, y) in D a unique real number that we denote by $f(x, y)$. The set D is called the **domain** of f. The set $\{f(x, y): (x, y) \in D\}$, which is the set of values the function f takes on, is called the **range** of f. ■

EXAMPLE 1 A FUNCTION OF TWO VARIABLES

Let $D = \mathbb{R}^2$. For each point $(x, y) \in \mathbb{R}^2$, we assign the number $f(x, y) = x^2 + y^4$. Since $x^2 + y^4 \geq 0$ for every pair of real numbers (x, y), we see that the range of f is $\mathbb{R}^+$, the set of nonnegative real numbers.

When the domain D is not given, we will take the domain of f to be the largest subset of $\mathbb{R}^2$ for which the expression $f(x, y)$ makes sense.

EXAMPLE 2

FINDING THE DOMAIN AND RANGE OF A FUNCTION OF TWO VARIABLES

Find the domain and range of the function f given by $f(x, y) = \sqrt{4 - x^2 - y^2}$, and find $f(0, 1)$ and $f(-1, 1)$.

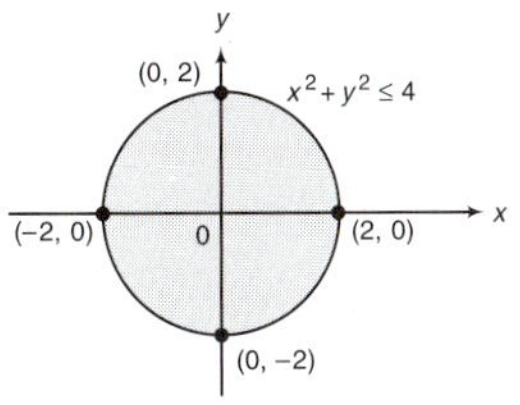

SOLUTION: Clearly, f is defined when the expression under the square-root sign is nonnegative. Thus, $D = \{(x, y): x^2 + y^2 \leq 4\}$. This is the disk† centered at the origin with radius 2. It is sketched in Figure 1. Since x^2 and y^2 are nonnegative, $4 - x^2 - y^2$ is largest when $x = y = 0$. Thus, the largest value of $\sqrt{4 - x^2 - y^2} = \sqrt{4} = 2$. Since $x^2 + y^2 \leq 4$, the smallest value of $4 - x^2 + y^2$ is 0, taken when $x^2 + y^2 = 4$ (at all points on the circle $x^2 + y^2 = 4$). Thus, the range of f is the closed interval $[0, 2]$. Finally,

$$f(0, 1) = \sqrt{4 - 0^2 - 1^2} = \sqrt{3} \quad \text{and}$$
$$f(-1, 1) = \sqrt{4 - (-1)^2 - 1^2} = \sqrt{2}.$$

FIGURE 1
The domain of $f(x, y) = \sqrt{4 - x^2 - y^2}$

REMARK: We emphasize that the domain of f is a subset of $\mathbb{R}^2$, while the range is a subset of $\mathbb{R}$, the real numbers.

CAUTION: In Figure 1, we sketched the *domain* of the function. We did *not* sketch its graph. The graph of a function of two or more variables is more complicated, and will be discussed shortly.

In One Variable Calculus we wrote $y = f(x)$. That is, we used the letter y to denote the value of a function of one variable. Here we can use the letter z (or any other letter for that matter) to denote the value taken by f, which is now a function of two variables. We then write

$$z = f(x, y). \tag{2}$$

EXAMPLE 3

FINDING THE DOMAIN AND RANGE OF A FUNCTION OF TWO VARIABLES

Find the domain and range of $f(x, y) = \tan^{-1}(y/x)$.

SOLUTION: $\tan^{-1}(y/x)$ is defined as long as $x \neq 0$. Hence the domain of $f = \{(x, y): x \neq 0\}$. From the definition of $\tan^{-1} x$, we find that the range of f is the open interval $(-\pi/2, \pi/2)$.

EXAMPLE 4

THE COBB-DOUGLAS PRODUCTION FUNCTION

In a manufacturing process, costs are typically divided between labor costs and capital costs. The total sum spent on production is usually fixed, but often a manufacturer has some choice in allocating money between capital and labor. For example, if part of the process is automated, more money will be spent on capital and less will be spent on labor.

† The **disk** of radius r is the circle of radius r together with all points interior to that circle.

Suppose that L units of labor and K units of capital are used in production. How many units will be produced? Economists have determined that, in some cases, the answer is given by the **Cobb-Douglas production function**:

$$\text{number of units produced} = F(L, K) = cL^aK^{1-a},$$

where c and a are constants that depend on the particular manufacturing process. [Here $c > 0$ and $0 < a < 1$.]

In the manufacture of a certain type of die, the Cobb-Douglas production function is given by

$$F(L, K) = 200L^{2/5}K^{3/5}. \tag{3}$$

a. How many units are produced if 100 units of labor and 300 units of capital are used?

b. If the number of units of labor and capital are both doubled, what is the change in the number of units produced?

SOLUTION:

a. $F(100, 300) = 200(100)^{0.4}(300)^{0.6} \approx 38{,}664$ units

b. We are asked to determine what happens to F if L becomes $2L$ and K becomes $2K$. We have

$$\begin{aligned} F(2L, 2K) &= 200(2L)^{2/5}(2K)^{3/5} = 200 \cdot 2^{2/5}L^{2/5}2^{3/5}K^{3/5} \\ &= \underbrace{2^{2/5}2^{3/5}}_{2^{5/5} = 2^1 = 2}\underbrace{[200L^{2/5}K^{3/5}]}_{F(L, K)} = 2F(L, K). \end{aligned}$$

We have shown that production is doubled if both labor and production costs are doubled.

We now turn to the definition of a function of three variables.

DEFINITION FUNCTION OF THREE VARIABLES

Let D be a subset of $\mathbb{R}^3$. Then a **real-valued function of three variables** f is a rule that assigns to each point (x, y, z) in D a unique real number that we denote by $f(x, y, z)$. The set D is called the **domain** of f, and the set $\{f(x, y, z): (x, y, z) \in D\}$, which is the set of values the function f takes on, is called the **range** of f. ■

NOTE: We will often use the letter w to denote the values that a function of three variables takes. We then write

$$w = f(x, y, z).$$

EXAMPLE 5 FINDING THE DOMAIN AND RANGE OF A FUNCTION OF THREE VARIABLES

Let $w = f(x, y, z) = \sqrt{1 - x^2 - (y^2/4) - (z^2/9)}$. Find the domain and range of f. Calculate $f(0, 1, 1)$.

SOLUTION: $f(x, y, z)$ is defined if $1 - x^2 - (y^2/4) - (z^2/9) \geq 0$, which occurs if $x^2 + (y^2/4) + (z^2/9) \leq 1$. From Section 1.8, we see that the equation $x^2 + (y^2/4) + (z^2/9) = 1$

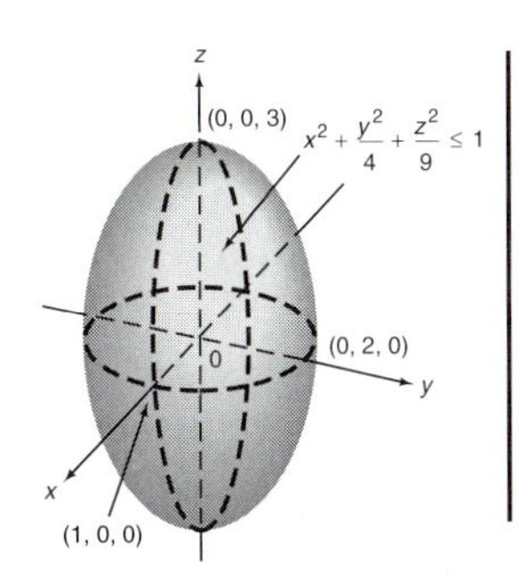

FIGURE 2
Graph of the domain of $f(x, y, z) = \sqrt{1 - x^2 - \frac{y^2}{4} - \frac{z^2}{9}}$

is the equation of an ellipsoid. Thus the domain of f is the set of points (x, y, z) in $\mathbb{R}^3$ that are on and interior to this ellipsoid (sketched in Figure 2). The range of f is the closed interval $[0, 1]$. This follows because $x^2 + (y^2/4) + (z^2/9) \geq 0$, so that $1 - x^2 - (y^2/4) - (z^2/9)$ is maximized when $x = y = z = 0$. Finally,

$$f(0, 1, 1) = \sqrt{1 - 0^2 - \frac{1^2}{4} - \frac{1^2}{9}} = \sqrt{1 - \frac{1}{4} - \frac{1}{9}} = \sqrt{1 - \frac{13}{36}} = \frac{\sqrt{23}}{6}.$$

NOTE: Figure 2 is a sketch of the domain of f, not the graph of f itself.

We now turn to a discussion of the graph of a function. Recall that the graph of a function of one variable f is the set of all points (x, y) in the plane such that $y = f(x)$. Using this definition as our model, we have the following definition.

DEFINITION GRAPH OF A FUNCTION OF TWO VARIABLES AND SURFACE

The **graph** of a function f of two variables x and y is the set of all points (x, y, z) in $\mathbb{R}^3$ such that $z = f(x, y)$. The graph of a continuous† function of two variables is called a **surface** in $\mathbb{R}^3$. ■

EXAMPLE 6

SKETCHING THE GRAPH OF A FUNCTION OF TWO VARIABLES

Sketch the graph of the function

$$z = f(x, y) = \sqrt{1 - x^2 - y^2}. \tag{4}$$

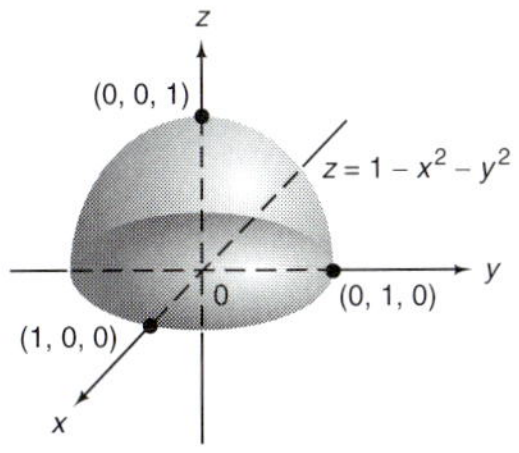

SOLUTION: We first note that $z \geq 0$. Then squaring both sides of (4), we have $z^2 = 1 - x^2 - y^2$, or $x^2 + y^2 + z^2 = 1$. This equation is the equation of the unit sphere. However, since $z \geq 0$, the graph of f is the hemisphere sketched in Figure 3.

FIGURE 3
Graph of the surface $z = \sqrt{1 - x^2 - y^2}$

It is often very difficult to sketch the graph of a function $z = f(x, y)$ since, except for the quadric surfaces we discussed in Section 1.8, we really do not have a vast "catalog" of surfaces to which to refer. Moreover, the techniques of curve plotting in three dimensions are tedious, to say the least, and plotting points in space will, except in the most trivial of cases, not get us very far. The best we can usually do is to describe **cross-sections** of the surface that lie in planes parallel to the coordinate planes. This will give an idea of what the surface looks like. Then if we have access to a computer with graphing capabilities, we can obtain a computer-drawn sketch of the surface. We illustrate this process with two examples.

EXAMPLE 7

SKETCHING A SURFACE AND ITS CROSS-SECTIONS

Obtain a sketch of the surface

$$z = f(x, y) = x^3 + 3x^2 - y^2 - 9x + 2y - 10. \tag{5}$$

† We will discuss continuity in the next section.

SOLUTION: The xz-plane has the equation $y = 0$. Planes parallel to the xz-plane have the equation $y = c$, where c is a constant. Setting $y = c$ in (5), we have

$$\begin{aligned} z &= x^3 + 3x^2 - 9x - 10 + (-c^2 + 2c) \\ &= x^3 + 3x^2 - 9x - 10 + k, \end{aligned}$$

where $k = -c^2 + 2c$ is a constant.

We can obtain the graph of $z = x^3 + 3x^2 - 9x - 10 + k$ by shifting the graph of $z = x^3 + 3x^2 - 9x - 10$ up or down $|k|$ units. For several values of k, cross sections lying in planes parallel to the xz-plane are given in Figure 4(a).

REMARK: The maximum value of the function $g(c) = -c^2 + 2c$ is 1, taken when $c = 1$ (you should verify this). Thus, 1 is the largest value k can assume in the parallel cross-sections, so $-10 + k \leq -9$, and the ''highest'' cross-section is $z = x^3 + 3x^2 - 9x - 9$.

The yz-plane has the equation $x = 0$. Planes parallel to the yz-plane have the equation $x = c$, where c is a constant. Setting $x = c$ in (5), we obtain

$$z = -y^2 + 2y + (c^3 + 3c^2 - 9c - 10) = -y^2 + 2y + k.$$

For several values of k, cross-sections parallel to the yz-plane are drawn in Figure 4(b). Note that this k can take on any real value. In Figure 4(c), we provide a computer-drawn sketch of the surface.

As we have stated, except for the standard quadric surfaces described in Section 1.8, it is extremely difficult to sketch three-dimensional graphs by hand. However, virtually

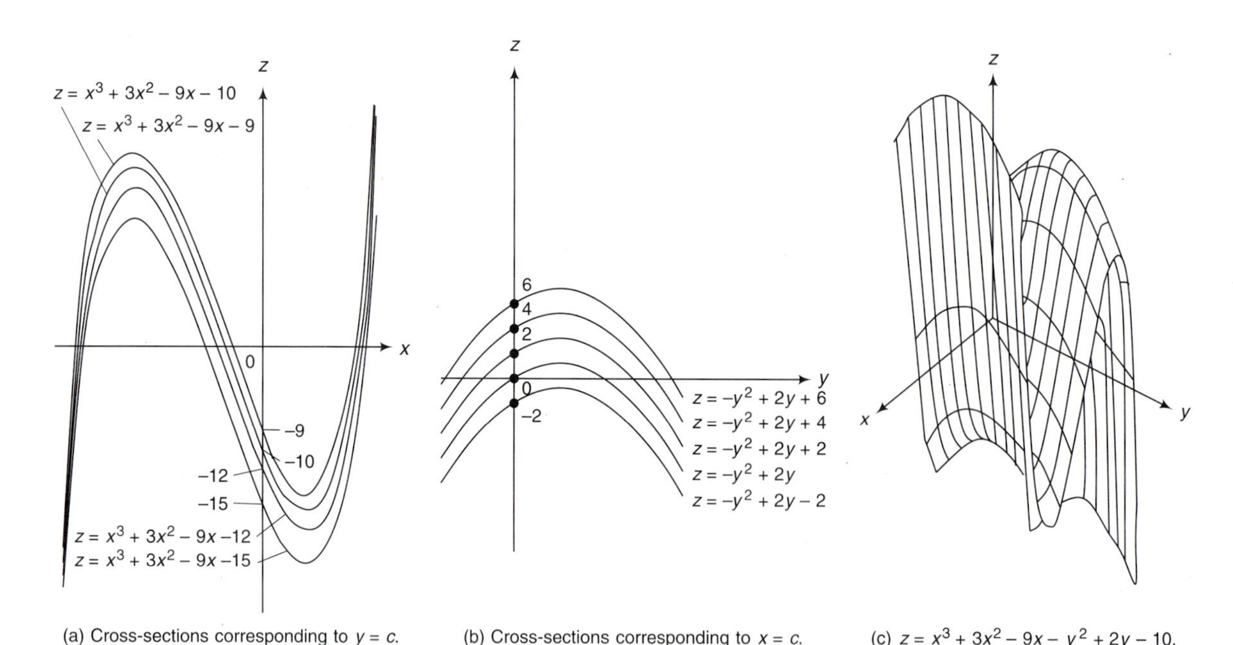

(a) Cross-sections corresponding to $y = c$. (b) Cross-sections corresponding to $x = c$. (c) $z = x^3 + 3x^2 - 9x - y^2 + 2y - 10$.

FIGURE 4
The graph of $z = x^3 + 3x^2 - 9x - y^2 + 2y - 10$

Cross-sections are sketched in (a) and (b) while the surface is sketched in (c).

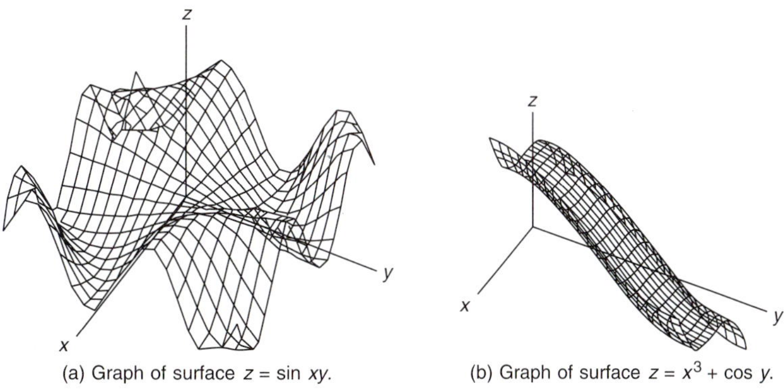

(a) Graph of surface $z = \sin xy$.   (b) Graph of surface $z = x^3 + \cos y$.

FIGURE 5
Computer-drawn graphs of two surfaces

every modern computer (and microcomputer) has graphing capabilities. In Figure 5, we provide computer-drawn sketches of the graphs of two different functions of two variables.

The situation becomes much more complicated when we try to sketch the graph of a function of three variables $w = f(x, y, z)$. We would need *four dimensions* to sketch such a surface. Being human, we are limited to three dimensions, and we see that we have reached the point where our comfortable three-dimensional geometry fails us. We are *not* saying that curves and surfaces in four-dimensional space do not exist—they do—only that we are not able to sketch them.

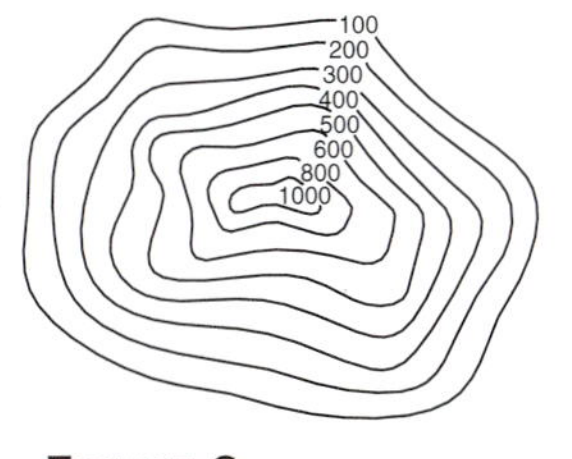

FIGURE 6
Contour curves

We have shown that it is usually very difficult to sketch the graph of a function of two variables. Fortunately, there is a way to represent such a function geometrically in two dimensions. The idea for what we are about to do comes from cartographers (mapmakers). Cartographers have the problem of indicating three-dimensional features (such as mountains and valleys) on a two-dimensional surface. They solve the problem by drawing a **contour map (topographic map)**, which is a map in which points of constant elevation are joined to form curves, called **contour curves**. The closer together these contour curves are drawn, the steeper is the terrain. A portion of a typical contour map is sketched in Figure 6.

LEVEL CURVES

We can use the same idea to depict the function $z = f(x, y)$ geometrically. If z is fixed, then the equation $f(x, y) = z$ is the equation of a curve in the xy-plane, called a **level curve**.

We can think of a level curve as the projection of a cross-section lying in a plane parallel to the xy-plane. Each value of z gives us such a curve. In other words, a level curve is the projection of the intersection of the surface $z = f(x, y)$ with the plane $z = c$. This idea is best illustrated with an example.

EXAMPLE 8

SKETCHING LEVEL CURVES

Sketch some level curves of $z = x^2 + y^2$.

SOLUTION: If $z > 0$, then $z = a^2$ for some positive number $a > 0$. Hence, all level curves are circles of the form $x^2 + y^2 = a^2$. The number a^2 can be thought of as the

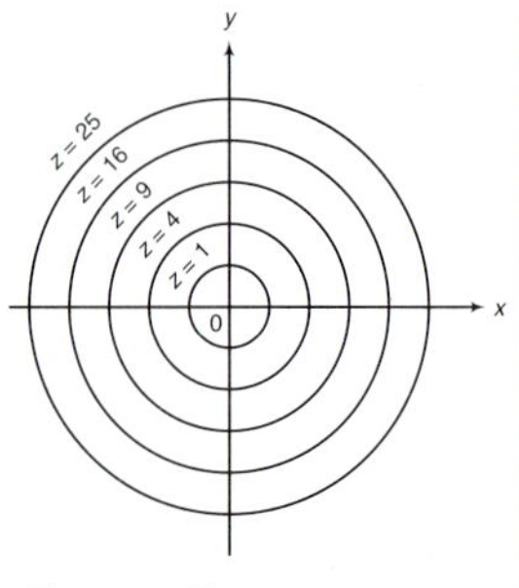

FIGURE 7
Level curves for $z = x^2 + y^2$

"elevation" of points on a level curve. Some of these curves are sketched in Figure 7. Each level curve encloses a projection of a "slice" of the actual graph of the function in three dimensions. In this case, each circle is the projection onto the xy-plane of a part of the surface in space. Actually, this example is especially simple because we can, without much difficulty, sketch the graph in space. From Section 1.8, the equation $z = x^2 + y^2$ is the equation of an elliptic paraboloid (actually, a circular paraboloid). It is sketched in Figure 8. In this easy case, we can see that if we slice this surface parallel to the xy-plane, we obtain the circles whose projections onto the xy-plane are the level curves. In most cases, of course, we will not be able to sketch the graph in space easily, so we will have to rely on our contour map sketch in $\mathbb{R}^2$.

There are some interesting applications of level curves in the sciences and economics. Three are given below.

i. Let $T(x, y)$ denote the temperature at a point (x, y) in the xy-plane. The level curves $T(x, y) = c$ are called **isothermal curves**. All points on such a curve have the same temperature.

ii. Let $V(x, y)$ denote the voltage (or potential) at a point in the xy-plane. The level curves $V(x, y) = c$ are called **equipotential curves**. All points on such a curve have the same voltage.

iii. A manufacturer makes two products. Let x and y denote the number of units of the first and second products produced during a given year. If $P(x, y)$ denotes the profit the manufacturer receives each year, then the level curves $P(x, y) = c$ are **constant profit curves**. All points on such a curve yield the same profit.

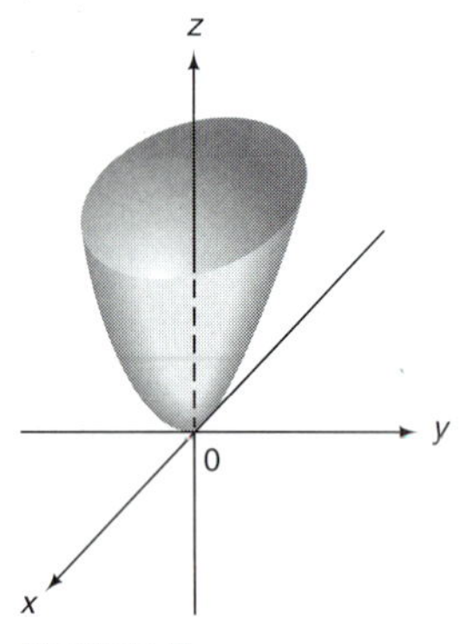

FIGURE 8
Graph of circular paraboloid
$z = x^2 + y^2$

We note that the idea behind level curves can be used to describe functions of three variables of the form $w = f(x, y, z)$. For each fixed w, the equation $f(x, y, z) = w$ is the equation of a surface in space, called a **level surface**. Since most surfaces are so difficult to draw, we will not pursue the notion of level surfaces here.

DEFINITION FUNCTION OF n VARIABLES

Let Ω be a subset of $\mathbb{R}^n$. A (scalar-valued) **function of n variables f** is a rule that assigns to each vector $\mathbf{x} = (x_1, x_2, \ldots, x_n)$ in Ω a unique real number which we denote by $f(\mathbf{x})$ or $f(x_1, x_2, \ldots, x_n)$. The set Ω is called the **domain** of f and is denoted by dom f. The set $\{f(\mathbf{x}): \mathbf{x} \in \Omega\}$, which is the set of values the function f takes on, is called the **range** of f. ■

NOTATION: We will write

$$f: \mathbb{R}^n \to \mathbb{R}$$

to indicate that f is a real-valued function whose domain is a subset of $\mathbb{R}^n$.

EXAMPLE 9 THE RANGE OF A FUNCTION WHOSE DOMAIN IS $\mathbb{R}^5$

Let $\Omega = \mathbb{R}^5$ and let $f(x_1, x_2, x_3, x_4, x_5) = x_1x_2x_3x_4x_5$. Then the range of f is $\mathbb{R}$.

EXAMPLE 10

THE RANGE OF A FUNCTION WHOSE DOMAIN IS $\mathbb{R}^4$

Let $\Omega = \mathbb{R}^4$ and let $f(x_1, x_2, x_3, x_4) = x_1^2 + x_2^2 + x_3^2 + x_4^2$. Then the range of f is $\mathbb{R}^+$.

Usually, as in $\mathbb{R}^2$, when the domain Ω is not given, we take the domain of f to be the largest subset of $\mathbb{R}^n$ for which the expression $f(x)$ makes sense.

EXAMPLE 11

THE RANGE OF A FUNCTION WHOSE DOMAIN IS A SUBSET OF $\mathbb{R}^4$

In $\mathbb{R}^4$, let

$$f(\mathbf{x}) = \frac{1}{x_1^2 + x_2^2 + x_3^2 + x_4^2 - 1}.$$

Then $f(\mathbf{x})$ is defined except when $x_1^2 + x_2^2 + x_3^2 + x_4^2 = 1$. This is the equation of the **unit sphere** in $\mathbb{R}^4$. Thus the domain of $f = \{(x_1, x_2, x_3, x_4): x_1^2 + x_2^2 + x_3^2 + x_4^2 \neq 1\}$. It is a bit more difficult to find the range of f. We note that $x_1^2 + x_2^2 + x_3^2 + x_4^2 - 1$ can be made as close to zero—from either side—as desired. Thus $f(\mathbf{x})$ can take on arbitrarily large values. Since $x_1^2 + x_2^2 + x_3^2 + x_4^2 - 1 \geq -1$, $f(\mathbf{x})$ cannot take on negative values larger than -1. In addition, $f(\mathbf{x}) \neq 0$ for any vector $\mathbf{x}$. Finally, if $|\mathbf{x}|$ is large (see page 78 for the definition of the norm, $|\mathbf{x}|$), then $f(\mathbf{x})$ is small and positive. Thus the range of $f = (-\infty, -1] \cup (0, \infty) = \mathbb{R} - (-1, 0]$.

As in Example 11, we can define spheres and other geometric objects in $\mathbb{R}^n$, where $n > 3$, by analogy with known objects in $\mathbb{R}^2$ and $\mathbb{R}^3$. However, we cannot, of course, draw these objects.

PROBLEMS 3.1

SELF-QUIZ

I. The domain of the function $f(x, y) = \sin x/\cos y$ is ________, where k denotes an integer.
a. $\{(x, y): y \neq k\pi\}$
b. $\{(x, y): y \neq (k + \frac{1}{2})\pi\}$
c. $\{(x, y): x \neq k\pi\}$
d. $\{(x, y): y \neq j\pi,\ y \neq (k + \frac{1}{2})\pi\}$

II. The range of the function $g(x, y) = \cos x + \sin y$ is ________.
a. $[-1, 1]$ **b.** $[-2, 2]$
c. $[-\pi, \pi]$ **d.** $(-\infty, \infty)$

III. The level curves of $z = x^2 + 4y^2$ are ________.
a. circles **b.** ellipses
c. straight lines **d.** diamonds

IV. The level curves of $z = |x| + 2|y|$ are ________.
a. circles **b.** ellipses
c. straight lines **d.** diamonds

V. The level surfaces of $f(x, y, z) = 3x + 2y - 5z$ are ________.
a. spheres **b.** planes
c. straight lines **d.** ellipsoids

In Problems 1–45, find the domain and range of the indicated function.

1. $f(x, y) = \sqrt{x^2 + y^2}$
2. $f(x, y) = \sqrt{1 + x + y}$
3. $f(x, y) = \dfrac{x}{y}$
4. $f(x, y) = \sqrt{1 - x^2 - 4y^2}$
5. $f(x, y) = \sqrt{1 - x^2 + 4y^2}$

6. $f(x, y) = \sin(x + y)$

7. $f(x, y) = e^x + e^y$

8. $f(x, y) = \dfrac{1}{(x^2 - y^2)^{3/2}}$

9. $f(x, y) = \tan(x - y)$

10. $f(x, y) = \sqrt{\dfrac{x + y}{x - y}}$

11. $f(x, y) = \sqrt{\dfrac{x - y}{x + y}}$

12. $f(x, y) = \sin^{-1}(x + y)$

13. $f(x, y) = \cos^{-1}(x - y)$

14. $f(x, y) = \dfrac{y}{|x|}$

15. $f(x, y) = \dfrac{x^2 - y^2}{x + y}$

16. $f(x, y) = \ln(1 + x^2 - y^2)$

***17.** $f(x, y) = \dfrac{x}{2y} + \dfrac{2y}{x}$

18. $f(x, y, z) = x + y + z$

19. $f(x, y, z) = \sqrt{x + y + z}$

20. $f(x, y, z) = \dfrac{1}{\sqrt{x^2 + y^2 + z^2}}$

21. $f(x, y, z) = \dfrac{1}{\sqrt{x^2 - y^2 + z^2}}$

22. $f(x, y, z) = \dfrac{1}{\sqrt{x^2 - y^2 - z^2}}$

23. $f(x, y, z) = \sqrt{-x^2 - y^2 - z^2}$

24. $f(x, y, z) = \ln(x - 2y - 3z + 4)$

25. $f(x, y, z) = \dfrac{xy}{z}$

26. $f(x, y, z) = \sin(x + y - z)$

27. $f(x, y, z) = \sin^{-1}(x + y - z)$

28. $f(x, y, z) = \ln(x + y - z)$

29. $f(x, y, z) = \tan^{-1}\left(\dfrac{x + z}{y}\right)$

30. $f(x, y, z) = e^{xy+z}$

31. $f(x, y, z) = \dfrac{e^x + e^y}{e^z}$

32. $f(x, y, z) = xyz$

33. $f(x, y, z) = \dfrac{1}{xyz}$

34. $f(x, y, z) = \dfrac{x}{y + z}$

35. $f(x, y, z) = \sin x + \cos y + \sin z$

36. $f(x_1, x_2, x_3, x_4) = \sqrt{x_1^2 + x_2^2 + x_3^2 + x_4^2}$

37. $f(x_1, x_2, x_3, x_4) = \sqrt{1 + x_1 + 2x_2 + x_3 - x_4}$

38. $f(x_1, x_2, x_3, x_4) = x_1x_2x_3x_4$

39. $f(x_1, x_2, x_3, x_4) = \dfrac{x_1x_3}{x_2x_4}$

40. $f(x_1, x_2, x_3, x_4) = \dfrac{x_1^2 + x_3^2}{x_2^2 - x_4^2}$

41. $f(x_1, x_2, x_3, x_4) = \dfrac{x_1^2 - x_3^2}{x_2^2 + x_4^2}$

42. $f(x_1, x_2, x_3, x_4, x_5) = \dfrac{x_1 + x_2 + x_4}{x_3 + x_5}$

43. $f(x_1, x_2, x_3, x_4, x_5) = \dfrac{x_1^2 - x_2^2 - x_4^2}{x_3^2 + x_5^2}$

44. $f(x_1, x_2, x_3, x_4, x_5) = \dfrac{x_1 - x_4}{x_2 + x_3 + x_5}$

45. $f(x_1, x_2, x_3, \ldots, x_n) = \sum_{i=1}^{n} x_i^2$

In Problems 46–52, sketch the graph of the given function.

46. $z = 4x^2 + 4y^2$

47. $y = x^2 + 4z^2$

48. $x = 4z^2 - 4y^2$

49. $z = x^2 - 4y^2$

50. $z = \sqrt{x^2 + 4y^2 + 4}$

51. $y = \sqrt{x^2 - 4z^2 + 4}$

52. $x = \sqrt{4 - z^2 - 4y^2}$

In Problems 53–60, describe the level curves of the given function and sketch these curves for the given values of z.

53. $z = \sqrt{1 + x + y}$; $z = 0, 1, 5, 10$

54. $z = \dfrac{x}{y}$; $z = 1, 3, 5, -1, -3$

55. $z = \sqrt{1 - x^2 - 4y^2}$; $z = 0, \frac{1}{4}, \frac{1}{2}, 1$

56. $z = \sqrt{1 + x^2 - y}$; $z = 0, 1, 2, 5$

57. $z = \cos^{-1}(x - y)$; $z = 0, \dfrac{\pi}{6}, \dfrac{\pi}{3}, \dfrac{\pi}{2}$

58. $z = \sqrt{\dfrac{x + y}{x - y}}$; $z = 0, 1, 2, 5$

59. $z = \tan(x + y)$; $z = 0, 1, -1, \sqrt{3}$

60. $z = \tan^{-1}(x - y^2)$; $z = 0, \pi/6, \pi/4$

61. The temperature T at any point on an object in the plane is given by $T(x, y) = 20 + x^2 + 4y^2$. Sketch the isothermal curves for $T = 50$, $T = 60$, and $T = 70$ degrees.

62. The voltage E at a point (x, y) on a metal plate placed in the xy-plane is given by $E(x, y) = \sqrt{1 - 4x^2 - 9y^2}$. Sketch the equipotential curves for $E = 1.0$ V, $E = 0.5$ V, and $E = 0.25$ V.

63. A manufacturer earns $P(x, y) = 100 + 2x^2 + 3y^2$

dollars each year for producing x and y units, respectively, of two products. Sketch the constant profit curves for $P = \$100$, $P = \$200$, and $P = \$1000$.

64. The Cobb-Douglas production function for a given product is

$$F(L, K) = 250L^{0.7}K^{0.3}.$$

a. Compute $F(50, 80)$.

b. Show that if labor and capital costs both triple, then the total output (number of units produced) triples as well.

65. The Cobb-Douglas production function for a given product is

$$F(L, K) = 500L^{1/3}K^{2/3}.$$

a. Compute $F(250, 150)$.

b. If labor costs double while capital costs are halved, what is the change in the total output? (Give this change as a percentage increase or decrease.)

c. Answer the question in part (b) if labor costs are halved while capital costs double.

66. A jeweler sells analog (i.e., with hour and minute hands) and digital watches. Each analog watch costs her \$8, and each digital watch costs her \$25. The demand functions for the two watches are

$$q_1 = 80 - 2.5p_1 + 0.8p_2$$

for the analog watches and

$$q_2 = 120 + p_1 - 1.8p_2$$

for the digital watches.

a. Find the total profit as a function of the prices of the two watches.

b. What is the profit when analog watches are sold for \$24 and digital watches are sold for \$30?

67. In Problem 66, what is the profit if analog and digital watches are sold for \$30 and \$40, respectively?

In Problems 68–79, a function of two variables is given. Match each function with one of the graphs in Figure 9.

68. $z = x^2 + \sin y$ **69.** $z = \ln x + e^{-y}$

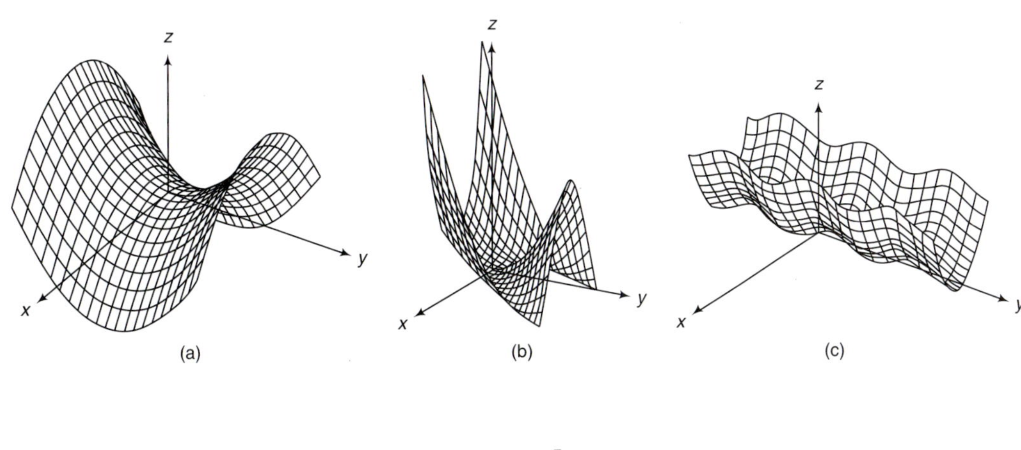

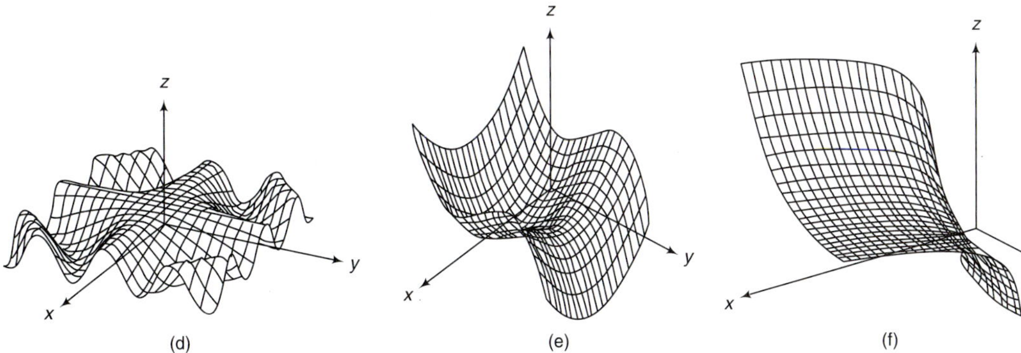

FIGURE 9

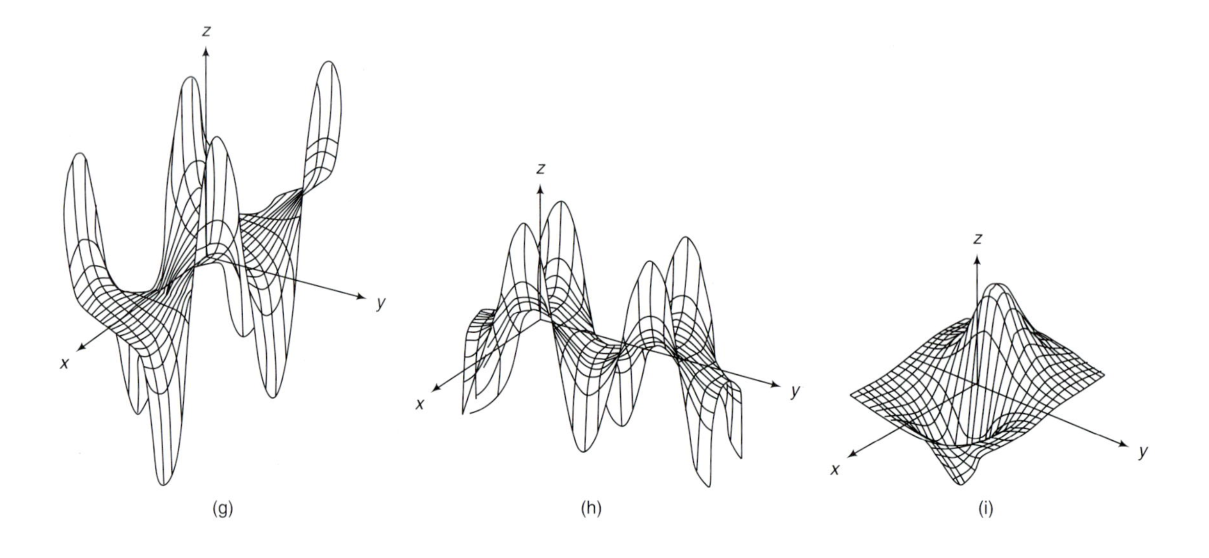

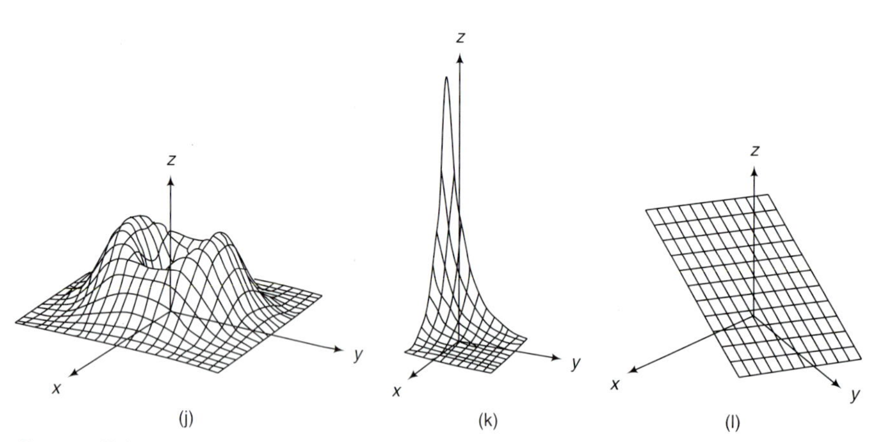

FIGURE 9 (CONTINUED)

70. $z = \cos xy$

71. $z = y^2 - x^2$

72. $z = x^2 - y^3 - x + 2y - 2$

73. $z = (y - 1.5x^2)(y - 0.5x^2)$

74. $z = e^{-(x+y)}$

75. $z = \sin x \tan y, \ -\pi/2 < y < \pi/2$

76. $z = \sec x \cos y, \ -\pi/2 < x < \pi/2$

77. $z = x - 2y + 4$

78. $z = (2x^2 + 3y^2)e^{1-(x^2+y^2)}$

79. $z = \dfrac{1}{(x + 0.15)^2 + y^2 + 0.2} - \dfrac{1}{(x - 0.15)^2 + y^2 + 0.2}$

ANSWERS TO SELF-QUIZ

I. b **II.** b **III.** b **IV.** d **V.** b

3.2 LIMITS AND CONTINUITY

In this section, we discuss the fundamental concepts of limits and continuity for functions of two or more variables. Recall that in the definition of a limit of a function of one variable,[†] the notions of an open and closed interval were fundamental. We could say, for example, that x was close to x_0 if $|x - x_0|$ was sufficiently small or, equivalently, if x was contained in a small open interval (neighborhood) centered at x_0. It is interesting to see how these ideas extend to $\mathbb{R}^2$ or $\mathbb{R}^3$.

Let (x_0, y_0) be a point in $\mathbb{R}^2$. What do we mean by the equation

$$|(x, y) - (x_0, y_0)| = r? \tag{1}$$

This is the definition of the circle of radius r centered at (x_0, y_0). Since (x, y) and (x_0, y_0) are vectors in $\mathbb{R}^2$,

Equation (1.1.1) on page 5

$$|(x, y) - (x_0, y_0)| = |(x - x_0, y - y_0)| = \sqrt{(x - x_0)^2 + (y - y_0)^2}. \tag{2}$$

Inserting (2) in (1) and squaring both sides yields

$$(x - x_0)^2 + (y - y_0)^2 = r^2, \tag{3}$$

which is the equation of the circle of radius r centered at (x_0, y_0). Then the set of points whose coordinates (x, y) satisfy the inequality

$$|(x, y) - (x_0, y_0)| < r \tag{4}$$

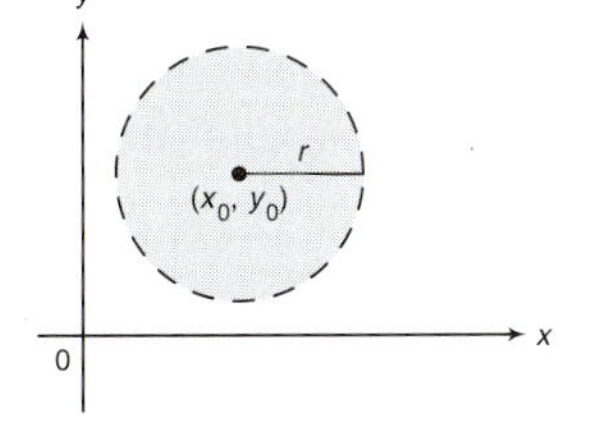

FIGURE 1
An open disk

is the set of all points in $\mathbb{R}^2$ interior to the circle given by (3). This set is sketched in Figure 1. Similarly, the inequality

$$|(x, y) - (x_0, y_0)| \le r \tag{5}$$

describes the set of all points interior to and on the circle given by (3). This is the set of points in Figure 1 plus the points on the circle. Setting $\mathbf{x} = (x, y)$ and $\mathbf{x}_0 = (x_0, y_0)$, we can write (4) and (5) as

$$|\mathbf{x} - \mathbf{x}_0| < r \tag{4'}$$

$$|\mathbf{x} - \mathbf{x}_0| \le r. \tag{5'}$$

This leads to the following definitions:

DEFINITION OPEN AND CLOSED DISKS; NEIGHBORHOODS

i. The **open disk** D_r centered at (x_0, y_0) with radius r is the subset of $\mathbb{R}^2$ given by

$$\{(x, y)\colon |(x, y) - (x_0, y_0)| < r\}.$$

[†] Here is the formal definition of a limit: Suppose that $f(x)$ is defined in a neighborhood (open interval) of the point x_0 (a finite number) except possibly at the point x_0 itself. Then

$$\lim_{x \to x_0} f(x) = L$$

if for every $\epsilon > 0$, there is a $\delta > 0$, such that if

$$0 < |x - x_0| < \delta, \quad \text{then} \quad |f(x) - L| < \epsilon.$$

ii. The **closed disk** centered at (x_0, y_0) with radius r is the subset of $\mathbb{R}^2$ given by

$$\{(x, y): |(x, y) - (x_0, y_0)| \leq r\}.$$

iii. The **boundary** of the open or closed disk defined in (i) or (ii) is the circle

$$\{(x, y): |(x, y) - (x_0, y_0)| = r\}.$$

iv. A **neighborhood** of a point (x_0, y_0) in $\mathbb{R}^2$ is an open disk centered at (x_0, y_0). ■

REMARK: In this definition, the words ''open'' and ''closed'' have meanings very similar to their meanings in the terms ''open interval'' and ''closed interval.'' An open interval does not contain its endpoints. An open disk does not contain any point on its boundary. Similarly, a closed interval contains all its boundary points, as does a closed disk.

With these definitions, it is easy to define a limit of a function of two variables. Intuitively, we say that $f(x, y)$ approaches the limit L as (x, y) approaches (x_0, y_0) if $f(x, y)$ gets arbitrarily ''close'' to L as (x, y) approaches (x_0, y_0) along any **path**.† We define this notion below.

DEFINITION LIMIT

Let $f(x, y)$ be defined in a neighborhood of (x_0, y_0) but not necessarily at (x_0, y_0) itself. Then the **limit** of $f(x, y)$ as (x, y) approaches (x_0, y_0) is the real number L, written

$$\lim_{(x,y)\to(x_0,y_0)} f(x, y) = L, \tag{6}$$

if for every number $\epsilon > 0$, there is a number $\delta > 0$ such that $|f(x, y) - L| < \epsilon$ for every $(x, y) \neq (x_0, y_0)$ in the open disk centered at (x_0, y_0) with radius δ (see Figure 2). ■

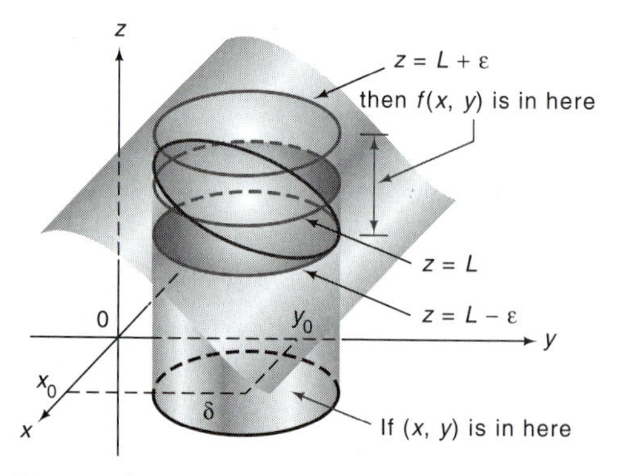

FIGURE 2
Illustration of the definition of a limit in $\mathbb{R}^2$

†A path is another name for a curve joining (x, y) to (x_0, y_0).

EXAMPLE 1 SHOWING FORMALLY THAT A LIMIT EXISTS

Show, using the definition of a limit, that $\lim_{(x,y)\to(1,2)}(3x + 2y) = 7$.

SOLUTION: Let $\epsilon > 0$ be given. We need to choose a $\delta > 0$ such that $|3x + 2y - 7| < \epsilon$ if $0 < \sqrt{(x-1)^2 + (y-2)^2} < \delta$. We start with

$$|3x + 2y - 7| = |3x - 3 + 2y - 4| \underset{\text{Triangle inequality}}{\leq} |3x - 3| + |2y - 4|$$

$$= |3(x-1)| + |2(y-2)| \underset{\text{Explain why}}{=} 3|x-1| + 2|y-2|. \tag{7}$$

Now

$$|x - 1| = \sqrt{(x-1)^2} \leq \sqrt{(x-1)^2 + (y-2)^2}$$

and

$$|y - 2| = \sqrt{(y-2)^2} \leq \sqrt{(x-1)^2 + (y-2)^2}.$$

So from (7),

$$\begin{aligned} |3x + 2y - 7| &\leq 3\sqrt{(x-1)^2 + (y-2)^2} + 2\sqrt{(x-1)^2 + (y-2)^2} \\ &= 5\sqrt{(x-1)^2 + (y-2)^2}. \end{aligned} \tag{8}$$

Now we want $|3x + 2y - 7| < \epsilon$. We choose $\delta = \epsilon/5$. Then from (8),

$$|3x + 2y - 7| \leq 5\sqrt{(x-1)^2 + (y-2)^2} < 5\delta = 5\left(\frac{\epsilon}{5}\right) = \epsilon,$$

and the requested limit is shown.

We can prove the limit much faster if we use the Cauchy-Schwarz inequality (see Problem 1.2.75 on page 23):

$$\sum_{k=1}^{2} a_k b_k \leq \left(\sum_{k=1}^{2} a_k^2\right)^{1/2} \left(\sum_{k=1}^{2} b_k^2\right)^{1/2}.$$

We have

$$|3x + 2y - 7| = |3(x-1) + 2(y-2)|.$$

Set $a_1 = 3$, $b_1 = x - 1$, $a_2 = 2$, $b_2 = y - 2$. Then we have

$$\begin{aligned} |3x + 2y - 7| &= |3(x-1) + 2(y-2)| = |a_1 b_1 + a_2 b_2| \\ &= \left|\sum_{k=1}^{2} a_k b_k\right| \leq \left(\sum_{k=1}^{2} a_k^2\right)^{1/2} \left(\sum_{k=1}^{2} b_k^2\right)^{1/2} \\ &= \sqrt{a_1^2 + a_2^2}\sqrt{b_1^2 + b_2^2} \\ &= \sqrt{3^2 + 2^2}\sqrt{(x-1)^2 + (y-2)^2} \\ &= \sqrt{13}\sqrt{(x-1)^2 + (y-2)^2}. \end{aligned}$$

Now we may choose $\delta = \epsilon/\sqrt{13}$ and the result is proved.

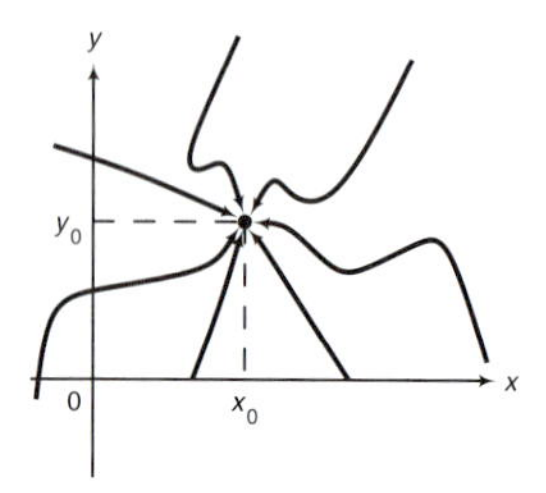

FIGURE 3
Many paths lead to (x_0, y_0)

In the intuitive definition of a limit which preceded the definition of a limit, we used the phrase "if $f(x, y)$ gets arbitrarily close to L as (x, y) approaches (x_0, y_0) *along any path*." Recall that $\lim_{x \to x_0} f(x) = L$ only if $\lim_{x \to x_0^+} f(x) = \lim_{x \to x_0^-} f(x) = L$. That is, the limit exists only if we get the same value from either side of x_0. In $\mathbb{R}^2$ the situation is more complicated because (x, y) can approach (x_0, y_0) not just along two but along an *infinite number* of paths. Some of these are illustrated in Figure 3. Thus, the only way we can verify a limit is by making use of the definition or some appropriate limit theorem that can be proven directly from the definition. In the next two examples we illustrate the kinds of problems we can encounter.

EXAMPLE 2 A LIMIT DOES NOT EXIST AT A POINT IN $\mathbb{R}^2$ IF WE GET DIFFERENT VALUES AS WE APPROACH THE POINT ALONG DIFFERENT PATHS

Let $f(x, y) = (y^2 - x^2)/(y^2 + x^2)$ for $(x, y) \neq (0, 0)$. We will show that $\lim_{(x,y) \to (0,0)} f(x, y)$ does not exist. There are an infinite number of approaches to the origin. For example, if we approach along the x-axis, then $y = 0$ and, if the indicated limit exists,

$$\lim_{(x,y) \to (0,0)} \frac{y^2 - x^2}{y^2 + x^2} = \lim_{(x,y) \to (0,0)} \frac{-x^2}{x^2} = \lim_{(x,y) \to (0,0)} -1 = -1.$$

On the other hand, if we approach along the y-axis, then $x = 0$ and

$$\lim_{(x,y) \to (0,0)} \frac{y^2 - x^2}{y^2 + x^2} = \lim_{(x,y) \to (0,0)} \frac{y^2}{y^2} = \lim_{(x,y) \to (0,0)} 1 = 1.$$

Thus, we get different answers depending on how we approach the origin. To prove that the limit cannot exist, we note that we have shown that in any open disk centered at the origin, there are points at which f takes on the values $+1$ and -1. Hence, f cannot have a limit as $(x, y) \to (0, 0)$.

Example 2 leads to the following general rule:

> RULE FOR NONEXISTENCE OF A LIMIT
>
> If we get two or more different values for $\lim_{(x,y) \to (x_0,y_0)} f(x, y)$ as we approach (x_0, y_0) along different paths, or if the limit fails to exist along some path, then $\lim_{(x,y) \to (x_0,y_0)} f(x, y)$ does not exist.

EXAMPLE 3 A LIMIT THAT DOES NOT EXIST

Let $f(x, y) = xy^2/(x^2 + y^4)$. We will show that $\lim_{(x,y) \to (0,0)} f(x, y)$ does not exist. First, let us approach the origin along a straight line passing through the origin that is not the y-axis. Then $y = mx$ and

$$f(x, y) = \frac{x(m^2x^2)}{x^2 + m^4x^4} = \frac{m^2x}{1 + m^2x^2},$$

which approaches 0 as $x \to 0$. Thus, along every straight line, $f(x, y) \to 0$ as $(x, y) \to (0, 0)$. But if we approach $(0, 0)$ along the parabola $x = y^2$, then

$$f(x, y) = \frac{y^2(y^2)}{y^4 + y^4} = \frac{1}{2},$$

so that along this parabola, $f(x, y) \to \frac{1}{2}$ as $(x, y) \to (0, 0)$, and the limit does not exist.

EXAMPLE 4

A LIMIT THAT DOES EXIST

Prove that $\lim_{(x,y)\to(0,0)}[xy(y^2 - x^2)/(x^2 + y^2)] = 0$.

SOLUTION: It is easy to show that this limit is zero along any straight line passing through the origin. But as we have seen, this is not enough. We must rely on our definition. Let $\epsilon > 0$ be given. Then we must show that there is a $\delta > 0$ such that if $0 < \sqrt{x^2 + y^2} < \delta$, then

$$|f(x, y)| = \left|\frac{xy(y^2 - x^2)}{x^2 + y^2}\right| < \epsilon.$$

But $|x| \le \sqrt{x^2 + y^2}$, $|y| \le \sqrt{x^2 + y^2}$ and $|y^2 - x^2| \le x^2 + y^2$, so

$$\left|\frac{xy(y^2 - x^2)}{x^2 + y^2}\right| \le \frac{\sqrt{x^2 + y^2}\sqrt{x^2 + y^2}(x^2 + y^2)}{x^2 + y^2} = x^2 + y^2 < \delta^2.$$

Hence, if we choose $\delta = \sqrt{\epsilon}$, we will have $|f(x, y)| < \epsilon$ if $\sqrt{x^2 + y^2} < \delta$. A computer-drawn graph of this function is given in Figure 4.

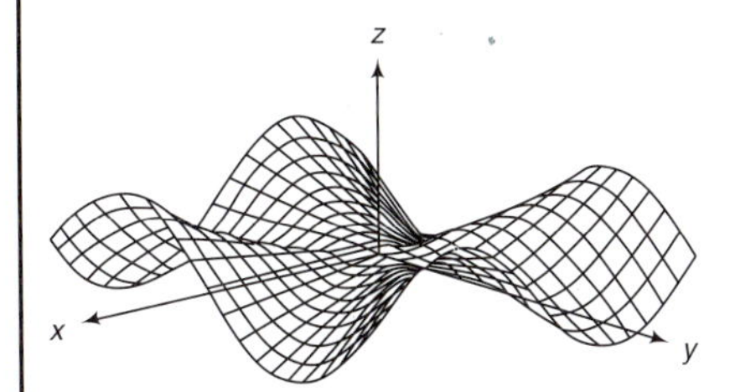

FIGURE 4

Graph of $z = \begin{cases} \dfrac{xy(y^2 - x^2)}{x^2 + y^2} & \text{if } (x, y) \ne (0, 0) \\ 0 & \text{if } (x, y) = (0, 0) \end{cases}$

We can solve this problem more easily by using polar coordinates. Recall that, in polar coordinates,

$$x = r\cos\theta, \qquad y = r\sin\theta, \qquad \text{and} \qquad x^2 + y^2 = r^2.$$

Then $(x, y) \to (0, 0)$ is equivalent to $r \to 0$ and we have

$$\begin{aligned}\lim_{(x,y)\to(0,0)} \frac{xy(y^2 - x^2)}{x^2 + y^2} &= \lim_{r\to 0} \frac{(r\cos\theta)(r\sin\theta)(r^2\sin^2\theta - r^2\cos^2\theta)}{r^2} \\ &= \lim_{r\to 0} \frac{r^4 \sin\theta\cos\theta(\sin^2\theta - \cos^2\theta)}{r^2} \\ &= \lim_{r\to 0} r^2 \sin\theta\cos\theta(\sin^2\theta - \cos^2\theta).\end{aligned}$$

Now $\sin^2\theta - \cos^2\theta = -(\cos^2\theta - \sin^2\theta) = -\cos 2\theta$, so, as $|\sin\theta| \le 1$ and $|\cos\theta| \le 1$

$$|\sin\theta\cos\theta(\sin^2\theta - \cos^2\theta)| = |\sin\theta\cos\theta\cos 2\theta| \le 1.$$

Thus

$$-r^2 \le r^2 \sin\theta\cos\theta(\sin^2\theta - \cos^2\theta) \le r^2$$

and, as both r^2 and $-r^2 \to 0$ as $r \to 0$,

$$r^2 \sin\theta\cos\theta(\sin^2\theta - \cos^2\theta) \to 0 \qquad \text{by the squeezing theorem.}$$

NOTE: $\sin\theta\cos\theta = \frac{1}{2}\sin 2\theta$, so we can replace 1 by $\frac{1}{2}$ in the inequalities above. This, however, makes no difference in the final result.

REMARK: This example illustrates the fact that while the existence of different limits for different paths implies that no limit exists, only the definition or an appropriate limit theorem can be used to prove that a limit does exist.

As Examples 1 and 4 illustrate, it is often tedious to calculate limits from the definition. Fortunately, just as in the case of functions of a single variable, there are a number of theorems that greatly facilitate the calculation of limits. As you will recall, one of our important definitions stated that if f is continuous, then $\lim_{x\to x_0} f(x) = f(x_0)$. We now define continuity of a function of two variables.

DEFINITION CONTINUITY

i. Let $f(x, y)$ be defined at every point (x, y) in a neighborhood of (x_0, y_0). Then f is **continuous** at (x_0, y_0) if all of the following conditions hold:
  a. $f(x_0, y_0)$ exists [i.e., (x_0, y_0) is in the domain of f].
  b. $\lim_{(x,y)\to(x_0,y_0)} f(x, y)$ exists.
  c. $\lim_{(x,y)\to(x_0,y_0)} f(x, y) = f(x_0, y_0)$.

ii. If one or more of these three conditions fail to hold, then f is said to be **discontinuous** at (x_0, y_0).

iii. f is **continuous in a subset** S of $\mathbb{R}^2$ if f is continuous at every point (x, y) in S. ■

REMARK: Condition (c) tells us that if a function f is continuous at (x_0, y_0), then we can calculate $\lim_{(x,y)\to(x_0,y_0)} f(x, y)$ by evaluation of f at (x_0, y_0).

EXAMPLE 5 A FUNCTION THAT IS DISCONTINUOUS AT (0, 0)

Let $f(x, y) = \dfrac{xy^2}{x^2 + y^2}$. Then f is discontinuous at (0, 0) because $f(0, 0)$ is not defined.

EXAMPLE 6 ANOTHER FUNCTION THAT IS DISCONTINUOUS AT (0, 0)

Let

$$f(x, y) = \begin{cases} \dfrac{xy^2}{x^2 + y^4}, & (x, y) \neq (0, 0) \\ 0, & (x, y) = (0, 0). \end{cases}$$

Here f is defined at (0, 0), but it is still discontinuous there because $\lim_{(x,y)\to(0,0)} f(x, y)$ does not exist (see Example 3).

EXAMPLE 7 A FUNCTION THAT IS CONTINUOUS AT (0, 0)

Let

$$f(x, y) = \begin{cases} \dfrac{xy(y^2 - x^2)}{x^2 + y^2}, & (x, y) \neq (0, 0) \\ 0, & (x, y) = (0, 0). \end{cases}$$

Here f is continuous at (0, 0), according to Example 4.

EXAMPLE 8 A FUNCTION THAT IS DISCONTINUOUS AT (0, 0)

Let

$$f(x, y) = \begin{cases} \dfrac{xy(y^2 - x^2)}{x^2 + y^2}, & (x, y) \neq (0, 0) \\ 1, & (x, y) = (0, 0). \end{cases}$$

Then f is discontinuous at (0, 0), because $\lim_{(x,y)\to(0,0)} f(x, y) = 0 \neq f(0, 0) = 1$, so condition (c) is violated.

Naturally, we would like to know what functions are continuous. We can answer this question if we look at the continuous functions of one variable. We start with polynomials.

DEFINITION POLYNOMIAL AND RATIONAL FUNCTION

i. A **polynomial** $p(x, y)$ in the two variables x and y is a finite sum of terms of the form $Ax^m y^n$, where m and n are nonnegative integers and A is a real number.

ii. A **rational function** $r(x, y)$ in the two variables x and y is a function that can be written as the quotient of two polynomials in x and y: $r(x, y) = p(x, y)/q(x, y)$. ■

EXAMPLE 9 A POLYNOMIAL IN x AND y

$p(x, y) = 5x^5y^2 + 12xy^9 - 37x^{82}y^5 + x + 4y - 6$ is a polynomial.

EXAMPLE 10 A RATIONAL FUNCTION IN x AND y

$$r(x, y) = \frac{8x^3y^7 - 7x^2y^4 + xy - 2y}{1 - 3y^3 + 7x^2y^2 + 18yx^7}$$

is a rational function.

The limit theorems for functions of one variable can be extended, with minor modifications, to functions of two variables. We will not state them here. However, by using them it is not difficult to prove the following theorem about continuous functions.

THEOREM 1 FACTS ABOUT CONTINUITY

i. Any polynomial p is continuous at any point in $\mathbb{R}^2$.

ii. Any rational function $r = p/q$ is continuous at any point (x_0, y_0) for which $q(x_0, y_0) \neq 0$. It is discontinuous when $q(x_0, y_0) = 0$ because it is then not defined at (x_0, y_0).

iii. If f and g are continuous at (x_0, y_0), then $f + g$, $f - g$, and $f \cdot g$ are continuous at (x_0, y_0).†

iv. If f and g are continuous at (x_0, y_0) and if $g(x_0, y_0) \neq 0$, then f/g is continuous at (x_0, y_0).

v. If f is continuous at (x_0, y_0) and if h is a function of one variable that is continuous at $f(x_0, y_0)$, then the composite function $h \circ f$, defined by $(h \circ f)(x, y) = h(f(x, y))$, is continuous at (x_0, y_0). ■

EXAMPLE 11 THE LIMIT OF A RATIONAL FUNCTION

Calculate $\lim_{(x,y)\to(4,1)}(x^3y^2 - 4xy)/(x + 6xy^3)$.

SOLUTION: $(x^3y^2 - 4xy)/(x + 6xy^3)$ is continuous at (4, 1) so we can calculate the limit by evaluation. We have

$$\lim_{(x,y)\to(4,1)} \frac{x^3y^2 - 4xy}{x + 6xy^3} = \left.\frac{x^3y^2 - 4xy}{x + 6xy^3}\right|_{(4,\,1)} = \frac{64 \cdot 1 - 4 \cdot 4 \cdot 1}{4 + 6 \cdot 4 \cdot 1} = \frac{48}{28} = \frac{12}{7}.$$

All the ideas in this section can be generalized to functions of three variables. A **neighborhood** of radius r of the point (x_0, y_0, z_0) in $\mathbb{R}^3$ consists of all points in $\mathbb{R}^3$ interior to the sphere

$$(x - x_0)^2 + (y - y_0)^2 + (z - z_0)^2 = r^2.$$

That is, the neighborhood (also called an **open ball**) is described by

$$\{(x, y, z): (x - x_0)^2 + (y - y_0)^2 + (z - z_0)^2 < r^2\}.$$

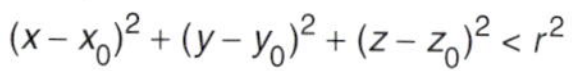

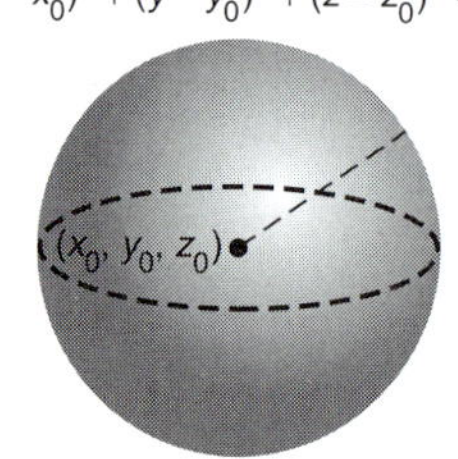

FIGURE 5
An open ball in $\mathbb{R}^3$

A typical neighborhood is sketched in Figure 5. The **closed ball** is described by

$$\{(x, y, z): (x - x_0)^2 + (y - y_0)^2 + (z - z_0)^2 \leq r^2\}.$$

Using an open ball instead of an open disk, the definition of the limit

$$\lim_{(x,y,z)\to(x_0,y_0,z_0)} f(x, y, z) = L$$

is analogous to the definition of $\lim_{(x,y)\to(x_0,y_0)} L$ given on page 148. Rather than discuss $\mathbb{R}^3$ in more detail, we make the leap directly to $\mathbb{R}^n$ for $n > 2$.

† $(f + g)(x, y) = f(x, y) + g(x, y)$, just as in the case of functions of one variable. Similarly, $(f \cdot g)(x, y) = f(x, y)g(x, y)$ and $(f/g)(x, y) = f(x, y)/g(x, y)$.

LIMITS AND CONTINUITY IN $\mathbb{R}^n$

DEFINITION OPEN AND CLOSED BALLS IN $\mathbb{R}^n$

i. The **open ball** $B_r(\mathbf{x}_0)$ in $\mathbb{R}^n$ centered at $\mathbf{x}_0$ with radius r is the subset of $\mathbb{R}^n$ given by

$$B_r(\mathbf{x}_0) = \{\mathbf{x} \in \mathbb{R}^n: |\mathbf{x} - \mathbf{x}_0| < r\}. \tag{9}$$

ii. The **closed ball** $\overline{B}_r(\mathbf{x}_0)$ in $\mathbb{R}^n$ with radius r is the subset of $\mathbb{R}^n$ given by

$$\overline{B}_r(\mathbf{x}_0) = \{\mathbf{x} \in \mathbb{R}^n: |\mathbf{x} - \mathbf{x}_0| \le r\}. \tag{10}$$

iii. The **boundary** of the open or closed ball defined by (1) or (2) is the **sphere** $S_r(\mathbf{x}_0)$ in $\mathbb{R}^n$ given by

$$S_r(\mathbf{x}_0) = \{\mathbf{x} \in \mathbb{R}^n: |\mathbf{x} - \mathbf{x}_0| = r\}. \tag{11}$$

iv. A **neighborhood** of a vector $\mathbf{x}_0$ in $\mathbb{R}^n$ is an open ball centered at $\mathbf{x}_0$.

v. A set Ω in $\mathbb{R}^n$ is **open** if, for every $\mathbf{x}_0 \in \Omega$, there is a neighborhood $B_r(\mathbf{x}_0)$ of $\mathbf{x}_0$ such that $B_r(\mathbf{x}_0) \subset \Omega$.

REMARK 1: The notation in (9), (10), and (11) refers to the norm of a vector defined in Section 1.10 (page 78).

Thus, for example, if

$$\mathbf{x}_0 = (x_1^{(0)}, x_2^{(0)}, \ldots, x_n^{(0)}) \qquad \text{and} \qquad \mathbf{x} = (x_1, x_2, \ldots, x_n),$$

then $|\mathbf{x} - \mathbf{x}_0| < r$ means that

$$\sqrt{(x_1 - x_1^{(0)})^2 + (x_2 - x_2^{(0)})^2 + \cdots + (x_n - x_n^{(0)})^2} < r. \tag{12}$$

REMARK 2: It is easy to visualize an open set in $\mathbb{R}^2$. In Figure 6(a) the set Ω_1 is open because, for every point in the set, it is possible to draw an open disk containing the point that is completely contained in Ω_1. The set Ω_2 in Figure 6(b) is not open because there are points (on the boundary) for which every neighborhood contains points outside Ω_2. ■

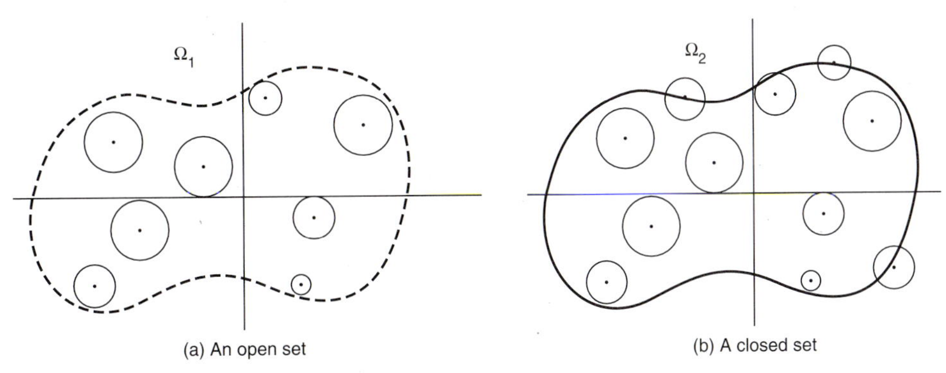

FIGURE 6
An open set and a closed set

DEFINITION LIMIT

Let $f\colon \mathbb{R}^n \to \mathbb{R}$ be defined in a neighborhood of $\mathbf{x}_0$ but not necessarily at $\mathbf{x}_0$ itself. Then the **limit of** $f(\mathbf{x})$ **as x approaches** $\mathbf{x}_0$ **is** L, written

$$\lim_{\mathbf{x}\to\mathbf{x}_0} f(\mathbf{x}) = L,$$

if for every number $\epsilon > 0$ there is a number $\delta > 0$ such that $|f(\mathbf{x}) - L| < \epsilon$ whenever $0 < |\mathbf{x} - \mathbf{x}_0| < \delta$. ■

EXAMPLE 12 THE LIMIT OF A LINEAR FUNCTION DEFINED ON $\mathbb{R}^4$

Show that

$$\lim_{\mathbf{x}\to(-4,1,0,3)} (x_1 + 2x_2 - x_3 + 3x_4) = 7.$$

SOLUTION: Let $\epsilon > 0$ be given. We need to choose a $\delta > 0$ such that

$$|x_1 + 2x_2 - x_3 + 3x_4 - 7| < \epsilon$$

if

$$0 < \sqrt{(x_1 + 4)^2 + (x_2 - 1)^2 + x_3^2 + (x_4 - 3)^2} < \delta.$$

As in Example 1, we start with the inequality $|x_1 + 2x_2 - x_3 + 3x_4 - 7| < \epsilon$ and work backward to find a suitable value for δ. We use the triangle inequality:

$$\begin{aligned} |x_1 + 2x_2 - x_3 + 3x_4 - 7| &= |x_1 + 4 + 2x_2 - 2 - x_3 + 3x_4 - 9| \\ &\le |x_1 + 4| + |2x_1 - 2| + |x_3| + |3x_4 - 9| \\ &= |x_1 + 4| + 2|x - 1| + |x_3 - 0| + 3|x_4 - 3|. \end{aligned} \tag{13}$$

Now each of the quantities $|x_1 + 4|$, $|x - 1|$, $|x_3|$, and $|x_4 - 3|$ is less than or equal to

$$\sqrt{(x_1 + 4)^2 + (x_2 - 1)^2 + x_3^2 + (x_4 - 3)^2}$$

(explain why). Thus, from (13),

$$\begin{aligned} |x_1 + 2x_2 - x_3 + 3x_4 - 7| &\le 7\sqrt{(x_1 + 4)^2 + (x_2 - 1)^2 + x_3^2 + (x_4 - 3)^2} \\ &= 7|(x_1, x_2, x_3, x_4) - (-4, 1, 0, 3)|. \end{aligned}$$

Finally, let $\delta = \epsilon/7$; then if $|\mathbf{x} - (-4, 1, 0, 3)| < \delta$, we have

$$|f(\mathbf{x}) - 7| = |x_1 + 2x_2 - x_3 + 3x_4 - 7| \le 7\delta = 7 \cdot \frac{\epsilon}{7} = \epsilon$$

and, therefore,

$$\lim_{\mathbf{x}\to(-4,1,0,3)} (x_1 + 2x_2 - x_3 + 3x_4) = 7.$$

REMARK: Note the great similarity between the last example and Example 1 (in $\mathbb{R}^2$). This illustrates that, except for our inability to draw things, there is very little difference in the properties of functions of two or three variables and functions of n variables.

EXAMPLE 13 THE LIMIT OF A RATIONAL FUNCTION THAT DOES NOT EXIST

Let

$$f(x_1, x_2, x_3, x_4, x_5) = \frac{x_5^2 - x_4^2 + x_3^2 - x_2^2 + x_1^2}{x_1^2 + x_2^2 + x_3^2 + x_4^2 + x_5^2}$$

for $(x_1, x_2, x_3, x_4, x_5) \neq (0, 0, 0, 0, 0)$. Show that $\lim_{\mathbf{x}\to 0} f(\mathbf{x})$ does not exist.

SOLUTION: We will do this by showing that we get two different answers if we approach the zero vector in two different ways. Let $x_1 = x_2 = x_3 = x_4 = 0$ and let $x_5 \to 0$. That is, we approach zero along the x_5-axis. Then, for $x_5 \neq 0$,

$$f(\mathbf{x}) = \frac{x_5^2}{x_5^2} = 1$$

so that

$$\lim_{(0,0,0,0,x_5)\to\mathbf{0}} f(\mathbf{x}) = 1.$$

Now let $x_1 = x_2 = x_3 = x_5 = 0$ and let $x_4 \to 0$. Then for $x_4 \neq 0$,

$$f(\mathbf{x}) = -\frac{x_4^2}{x_4^2} = -1$$

so that

$$\lim_{(0,0,0,x_4,0)\to\mathbf{0}} f(\mathbf{x}) = -1.$$

Since we get different answers as we approach zero in different ways, we conclude that the limit does not exist.

REMARK: Note the similarity between this example and Example 2.

Example 13 enables us to generalize the rule given earlier in this section.

> **RULE FOR NONEXISTENCE OF A LIMIT**
>
> If we get two or more different values for $\lim_{\mathbf{x}\to\mathbf{x}_0} f(\mathbf{x})$ as we approach $\mathbf{x}_0$ in different ways, then $\lim_{\mathbf{x}\to\mathbf{x}_0} f(\mathbf{x})$ does not exist.

As in $\mathbb{R}^2$ and $\mathbb{R}^3$, we see that the direct computation of limits is quite tedious. Things are simpler once we have defined continuity; for if f is continuous at $\mathbf{x}_0$, then $\lim_{\mathbf{x}\to\mathbf{x}_0} f(\mathbf{x}) = f(\mathbf{x}_0)$.

DEFINITION CONTINUITY

i. Let $f: \mathbb{R}^n \to \mathbb{R}$ be defined at every point $\mathbf{x}$ in a neighborhood of $\mathbf{x}_0$. Then f is **continuous** at $\mathbf{x}_0$ if all of the following conditions hold:
  - **a.** $f(\mathbf{x}_0)$ exists (i.e., $\mathbf{x}_0$ is in the domain of f).
  - **b.** $\lim_{\mathbf{x}\to\mathbf{x}_0} f(\mathbf{x})$ exists.
  - **c.** $\lim_{\mathbf{x}\to\mathbf{x}_0} f(\mathbf{x}) = f(\mathbf{x}_0)$.

ii. If one or more of these three conditions fails to hold, then f is said to be **discontinuous** at $\mathbf{x}_0$.

iii. f is continuous in an open set Ω of $\mathbb{R}^n$ if f is continuous at every point $\mathbf{x}_0$ in Ω. ■

REMARK: Condition (c) tells us that if a function f is continuous at $\mathbf{x}_0$, then we can calculate $\lim_{\mathbf{x}\to\mathbf{x}_0} f(\mathbf{x})$ by evaluation of f at $\mathbf{x}_0$.

EXAMPLE 14

A FUNCTION DEFINED IN A SUBSET OF $\mathbb{R}^5$ THAT IS DISCONTINUOUS AT 0

In $\mathbb{R}^5$,

$$f(x_1, x_2, x_3, x_4, x_5) = \frac{x_5^2 - x_4^2 + x_3^2 - x_2^2 + x_1^2}{x_1^2 + x_2^2 + x_3^2 + x_4^2 + x_5^2}$$

is discontinuous at $\mathbf{0}$ because, as we saw in Example 13, $\lim_{x\to 0}$ does not exist.

As in $\mathbb{R}^2$ and $\mathbb{R}^3$, rational functions are continuous when their denominators are not zero.

DEFINITION POLYNOMIAL AND RATIONAL FUNCTION IN $\mathbb{R}^n$

i. A **polynomial** $p(\mathbf{x}) = p(x_1, x_2, \ldots, x_n)$ in the n variables $x_1, x_2, \ldots, x_n$ is a finite sum of terms of the form $Ax_1^{m_1} x_2^{m_2} \cdots x_n^{m_n}$ where $m_1, m_2, \ldots, m_n$ are nonnegative integers and A is a real number. The **degree** of p is the largest value of the sum $m = m_1 + m_2 + \cdots + m_n$ (i.e., the largest sum among the terms constituting $p(\mathbf{x})$).

ii. A **rational function** $r(\mathbf{x}) = r(x_1, x_2, \ldots, x_n)$ in n variables is a function that can be written as the quotient of two polynomials:

$$r(\mathbf{x}) = \frac{p(\mathbf{x})}{q(\mathbf{x})}.$$ ■

EXAMPLE 15

A POLYNOMIAL IN FOUR VARIABLES OF DEGREE 13

$p(x_1, x_2, x_3, x_4) = 2x_1^3x_2x_3^2x_4^4 + 5x_1x_2^7x_3^2x_4^3 - 11x_1^4x_2^3x_3^4x_4$ is a polynomial of degree 13, since the sums of the exponents of each of the three terms are $3 + 1 + 2 + 4 = 10$, $1 + 7 + 2 + 3 = 13$, and $4 + 3 + 4 + 1 = 12$, respectively, and 13 is the largest sum.

The following theorem, whose proof is omitted (but not difficult), extends the results of Theorem 1. The proof follows directly from an extension to $\mathbb{R}^n$ of the limit theorems of one-variable calculus.

THEOREM 2

i. Any polynomial p is continuous at any point in $\mathbb{R}^n$.

ii. Any rational function $r = p/q$ is continuous at any point $\mathbf{x}_0$ for which

$q(\mathbf{x}_0) \neq 0$. It is discontinuous when $q(\mathbf{x}_0) = 0$ because it is not defined at $\mathbf{x}_0$.

iii. If f and g are continuous at $\mathbf{x}_0$, then $f + g$, $f - g$, and fg are continuous at $\mathbf{x}_0$.

iv. If f and g are continuous at $\mathbf{x}_0$ and if $g(\mathbf{x}_0) \neq 0$, then f/g is continuous at $\mathbf{x}_0$.

v. If f is continuous at $\mathbf{x}_0$ and if h is a function of one variable that is continuous at $f(\mathbf{x}_0)$, then the composite function $h \circ f$, defined by $(h \circ f)(\mathbf{x}_0) = h(f(\mathbf{x}_0))$, is continuous at $\mathbf{x}_0$. ■

EXAMPLE 16 A POLYNOMIAL FUNCTION IS CONTINUOUS

$p(\mathbf{x}) = x_1{}^3x_2{}^5x_5{}^8 - 3x_1x_2{}^4x_4{}^2 + 5x_1{}^2x_2{}^3x_3{}^4x_4{}^2x_5$ is continuous at every $\mathbf{x}$ in $\mathbb{R}^5$.

EXAMPLE 17 A RATIONAL FUNCTION IS CONTINUOUS WHENEVER ITS DENOMINATOR IS NONZERO

$$r(\mathbf{x}) = \frac{x_1{}^2x_5x_4{}^4 + x_1x_2{}^2x_3{}^3x_4{}^5 - x_1{}^6x_2x_5{}^3}{x_1 - 2x_2 + 3x_3 - 4x_4 + 2x_5 - 6}$$

is continuous at every $\mathbf{x}$ in $\mathbb{R}^5$ except at those $\mathbf{x}$ that satisfy

$$x_1 - 2x_2 + 3x_3 - 4x_4 + 2x_5 = 6. \quad (14)$$

REMARK: The set of vectors in $\mathbb{R}^5$ that satisfy (14) is called a **hyperplane**. In general, a hyperplane H in $\mathbb{R}^n$ is defined by

$$H = \{\mathbf{x}: a_1x_1 + a_2x_2 + \cdots + a_nx_n = b\} \quad (15)$$

where $a_1, a_2, \ldots, a_n$ and b are real numbers that are not all equal to zero.

EXAMPLE 18 A CONTINUOUS SINE FUNCTION

The function $\sin(x_1{}^2 + 2x_1x_4 - x_3{}^4x_5{}^5)$ is continuous at every $\mathbf{x}$ in $\mathbb{R}^5$ by part (v) of Theorem 2 since $\sin x$ is continuous and $x_1{}^2 + 2x_1x_4 - x_3{}^4x_5{}^5$ is a polynomial.

PROBLEMS 3.2

SELF-QUIZ

I. Let $f(x, y) = (x^2 - y^2)/(x^2 + y^2)$. Then $\lim_{(x,y)\to(0,0)} f(x, y)$ ________.

a. $= \lim_{x\to 0} (\lim_{y\to 0} f(x, y))$
b. $= \lim_{y\to 0} (\lim_{x\to 0} f(x, y))$
c. $= \lim_{x\to 0} (\lim_{y\to x} f(x, y))$
d. does not exist

II. Let $g(x, y) = x^2 \cdot y/(x^4 + y^2)$. Then $\lim_{(x,y)\to(0,0)} g(x, y)$ ________.

a. $= \lim_{x\to 0} (\lim_{y\to 0} g(x, y))$
b. $= \lim_{y\to 0} (\lim_{x\to 0} g(x, y))$
c. $= \lim_{x\to 0} (\lim_{y\to x} g(x, y))$
d. does not exist

III. Let $h(x, y) = (x^2 - 3y + 5)/(1 + x^2 + y^2)$. Then $\lim_{(x,y)\to(0,0)} h(x, y)$ ________.
a. $= \lim_{x\to 0} (\lim_{y\to 0} h(x, y))$
b. $= \lim_{y\to 0} (\lim_{x\to 0} h(x, y))$
c. $= \lim_{x\to 0} (\lim_{y\to x} h(x, y))$
d. does not exist

IV. Let $F(x, y) = y/x$. Then $\lim_{(x,y)\to(0,0)} F(x, y)$ ________.
a. $= \lim_{x\to 0} (\lim_{y\to 0} F(x, y))$
b. $= \lim_{y\to 0} (\lim_{x\to 0} F(x, y))$
c. $= \lim_{x\to 0} (\lim_{y\to x} F(x, y))$
d. does not exist

In Problems 1–5, sketch the indicated region.

1. The open disk centered at (3, 2) with radius 3.
2. The open disk centered at (3, 0) with radius 2.
3. The open ball centered at (1, 0, 0) with radius 1.
4. The closed disk centered at (−1, 1) with radius 1.
5. The closed ball centered at (0, 1, 1) with radius 2.

In Problems 6–11, use the definition to verify the indicated limit.

6. $\lim_{(x,y)\to(1,2)} (3x + y) = 5$

7. $\lim_{(x,y)\to(3,-1)} (x - 7y) = 10$

8. $\lim_{(x,y)\to(5,-2)} (ax + by) = 5a - 2b$

***9.** $\lim_{(x,y)\to(1,1)} \frac{x}{y} = 1$

10. $\lim_{(x,y)\to(0,0)} \frac{2x^2y}{x^2 + y^2} = 0$

11. $\lim_{(x,y)\to(4,1)} (x^2 + 3y^2) = 19$

In Problems 12–21, show that the given limit does not exist.

12. $\lim_{(x,y)\to(0,0)} \frac{x + y}{x - y}$

13. $\lim_{(x,y)\to(0,0)} \frac{xy}{x^2 - y^2}$

14. $\lim_{(x,y)\to(0,0)} \frac{xy}{x^2 + y^2}$

15. $\lim_{(x,y)\to(0,0)} \frac{xy^3}{x^4 + y^4}$

16. $\lim_{(x,y)\to(0,0)} \frac{xy}{x^3 + y^3}$

17. $\lim_{(x,y)\to(0,0)} \frac{(x^2 + y^2)^2}{x^4 + y^4}$

18. $\lim_{(x,y)\to(0,0)} \frac{x^2 - 2y}{y^2 + 2x}$

19. $\lim_{(x,y)\to(0,0)} \frac{ax^2 + by}{cy^2 + dx}$, $a, b, c, d > 0$

20. $\lim_{(x,y,z)\to(0,0,0)} \frac{xy + 2xz + 3yz}{x^2 + y^2 + z^2}$

21. $\lim_{(x,y,z)\to(0,0,0)} \frac{xyz}{x^3 + y^3 + z^3}$

In Problems 22–25, show that the indicated limit exists and calculate it.

22. $\lim_{(x,y)\to(0,0)} \frac{3xy}{\sqrt{x^2 + y^2}}$

23. $\lim_{(x,y)\to(0,0)} \frac{5x^2y^2}{x^4 + y^2}$

24. $\lim_{(x,y)\to(0,0)} \frac{x^3 + y^3}{x^2 + y^2}$

25. $\lim_{(x,y,z)\to(0,0,0)} \frac{yx^2 + z^3}{x^2 + y^2 + z^2}$

In Problems 26–35, calculate the indicated limit.

26. $\lim_{(x,y)\to(-1,2)} (xy + 4y^2x^3)$

27. $\lim_{(x,y)\to(-1,2)} \frac{4x^3y^2 - 2xy^5 + 7y - 1}{3xy - y^4 + 3x^3}$

28. $\lim_{(x,y)\to(-4,3)} \frac{1 + xy}{1 - xy}$

29. $\lim_{(x,y)\to(\pi,\pi/3)} \ln(x + y)$

30. $\lim_{(x,y)\to(1,2)} \ln(1 + e^{x+y})$

31. $\lim_{(x,y)\to(2,5)} \sinh\left(\frac{x + 1}{y - 2}\right)$

***32.** $\lim_{(x,y)\to(1,1)} \frac{x - y}{x^2 - y^2}$

33. $\lim_{(x,y)\to(2,2)} \frac{x^3 - 2xy + 3x^2 - 2y}{x^2y + 4y^2 - 6x^2 + 24y}$

34. $\lim_{(x,y,z)\to(1,1,3)} \frac{xy^2 - 4xz^2 + 5yz}{3z^2 - 8z^3y^7x^4 + 7x - y + 2}$

35. $\lim_{(x,y,z)\to(4,1,3)} \ln(x - yz + 4x^3y^5z)$

In Problems 36–49, describe the maximum region over which the given function is continuous.

36. $f(x, y) = \sqrt{x - y}$

37. $f(x, y) = \frac{x^3 + 4xy^6 - 7x^4}{x^2 + y^2}$

38. $f(x, y) = \dfrac{x^3 + 4xy^6 - 7x^4}{x^3 - y^3}$

*39. $f(x, y) = \dfrac{xy^3 - 17x^2y^5 + 8x^3y}{xy + 3y - 4x - 12}$

40. $f(x, y) = \ln(3x + 2y + 6)$

41. $f(x, y) = \tan^{-1}(x - y)$

42. $f(x, y) = e^{xy+2}$

43. $f(x, y) = \dfrac{x^3 - 1 + 3y^5x^2}{1 - xy}$

44. $f(x, y) = \dfrac{x}{\sqrt{1 - (x^2/4) - y^2}}$

45. $f(x, y, z) = e^{(xy+yz-\sqrt{x})}$

46. $f(x, y, z) = \dfrac{xyz^2 + yzx^2 - 3x^3yz^5}{x - y + 2z + 4}$

47. $f(x, y, z) = y\ln(xz)$

48. $f(x, y) = \cos^{-1}(x^2 - y)$

49. $f(x, y, z) = \dfrac{1}{\sqrt{1 - x^2 - y^2 - z^2}}$

50. Find a function $g(x)$ such that the function

$$f(x, y) = \begin{cases} \dfrac{x^2 - y^2}{x - y} & x \neq y \\ g(x) & x = y \end{cases}$$

is continuous at every point in $\mathbb{R}^2$.

51. Find a number c such that the function

$$f(x, y) = \begin{cases} \dfrac{3xy}{\sqrt{x^2 + y^2}} & (x, y) \neq (0, 0) \\ c & (x, y) = (0, 0) \end{cases}$$

is continuous at the origin.

52. Find a number c such that

$$f(x, y) = \begin{cases} \dfrac{xy}{|x| + |y|} & \text{if } (x, y) \neq (0, 0) \\ c & \text{if } (x, y) = (0, 0) \end{cases}$$

is continuous at the origin.

*53. Discuss the continuity at the origin of

$$f(x, y, z) = \begin{cases} \dfrac{yz - x^2}{x^2 + y^2 + z^2} & \text{if } (x, y, z) \neq (0, 0, 0) \\ 0 & \text{if } (x, y, z) = (0, 0, 0). \end{cases}$$

In Problems 54–56 use the definition on page 156 to verify the indicated limit.

54. $\lim_{\mathbf{x}\to(1,-1,2,3)} (x_1 - 2x_2 + 3x_3 - x_4) = 6$

55. $\lim_{\mathbf{x}\to(-1,2,4,-1,3)} (2x_1 - 3x_2 + x_3 - 4x_4 + 3x_5) = 9$

56. $\lim_{\mathbf{x}\to(1,-1,0,2)} ({x_1}^2 + {x_2}^2 + {x_3}^2 - {x_4}^2) = -2$

In Problems 57–60 show that the indicated limit does not exist.

57. $\lim_{\mathbf{x}\to\mathbf{0}} \dfrac{x_1 + x_2 + x_3 + x_4}{x_1 - x_2 + x_3 - x_4}$

58. $\lim_{\mathbf{x}\to\mathbf{0}} \dfrac{x_1x_2x_3x_4}{{x_1}^4 - {x_2}^4 + {x_3}^4 - {x_4}^4}$

59. $\lim_{\mathbf{x}\to\mathbf{0}} \dfrac{x_1x_2x_3x_4}{{x_1}^4 + {x_2}^4 + {x_3}^4 + {x_4}^4}$

60. $\lim_{\mathbf{x}\to\mathbf{0}} \dfrac{x_1{x_4}^3 + x_2{x_3}^3 + x_5{x_2}^3}{{x_1}^4 + {x_2}^4 + {x_3}^4 + {x_4}^4 + {x_5}^4}$

In Problems 61–63 show that the given limit exists and calculate it.

61. $\lim_{\mathbf{x}\to\mathbf{0}} \dfrac{2x_1x_2x_3x_4}{\sqrt{{x_1}^2 + {x_2}^2 + {x_3}^2 + {x_4}^2}}$

62. $\lim_{\mathbf{x}\to\mathbf{0}} \dfrac{x_1{x_2}^2 + x_2{x_3}^2 + x_3{x_4}^2 + x_4{x_1}^2}{{x_1}^2 + {x_2}^2 + {x_3}^2 + {x_4}^2}$

63. $\lim_{\mathbf{x}\to\mathbf{0}} \dfrac{2{x_1}^3{x_2}^2 - 3{x_2}^3{x_3}^2 + 5{x_3}^3{x_4}^2}{{x_1}^4 + {x_2}^4 + {x_3}^4 + {x_4}^4}$

In Problems 64–69 calculate the indicated limit.

64. $\lim_{\mathbf{x}\to(1,-1,2,3)} ({x_1}^2 - 4{x_2}^2 + 5x_3x_4)$

65. $\lim_{\mathbf{x}\to(0,1,-2,4)} \dfrac{x_1x_3}{x_2x_4}$

66. $\lim_{\mathbf{x}\to(\pi/2,\pi/4,\pi/3,\pi/6)} \sin(x_1 - 2x_2 + 3x_3 + 5x_4)$

67. $\lim_{\mathbf{x}\to(1,0,-1,2,3)} \ln\left({x_1}^2 + {x_2}^2 + {x_3}^2 + \dfrac{x_4}{x_5}\right)$

68. $\lim_{\mathbf{x}\to(-2,3,1,4,6)} \dfrac{x_1 - x_2{x_3}^3 + {x_4}^5x_5}{{x_1}^2 + x_2x_3x_4x_5}$

69. $\lim_{\mathbf{x}\to(-2,3,1,4,6)} \sqrt{\dfrac{x_1 - x_2{x_3}^3 + {x_4}^5x_5}{{x_1}^2 + x_2x_3x_4x_5}}$

In Problems 70–73 describe the maximum region over which the given function is continuous.

70. $f(x_1, x_2, x_3, x_4) = \sqrt{x_1 + x_2 + x_3 + x_4}$

71. $f(x_1, x_2, x_3, x_4) = \dfrac{x_1x_2}{x_3x_4}$

72. $f(x_1, x_2, x_3, x_4) = \ln({x_1}^2 + {x_2}^2 + {x_3}^2 + {x_4}^2)$

73. $f(x_1, x_2, x_3, x_4, x_5) = \dfrac{1}{\sqrt{1 - {x_1}^2 - {x_2}^2 - {x_3}^2 - {x_4}^2 - {x_5}^2}}$

74. Find a number c such that the following function is continuous at the origin:

$$f(x_1, x_2, x_3, x_4) = \begin{cases} \dfrac{2(x_1x_2 + x_3x_4)}{\sqrt{x_1^2 + x_2^2 + x_3^2 + x_4^2}} & \mathbf{x} \neq \mathbf{0}, \\ c & \mathbf{x} = \mathbf{0}. \end{cases}$$

ANSWERS TO SELF-QUIZ

I. d (a = 1, b = −1, c = 0)

II. d (along the curve $y = x^2$ the limiting value is $\frac{1}{2}$ while along $y = -x^2$ the limiting value is $-\frac{1}{2}$)

III. a = b = c = 5

IV. d (a = 0, b does not exist, c = 1)

3.3 PARTIAL DERIVATIVES

In this section, we show one of the ways a function of several variables can be differentiated. The idea is simple. Let $z = f(x, y)$. If we keep one of the variables, say y, fixed, then f can be treated as a function of x only and we can calculate the derivative (if it exists) of f with respect to x. This new function is called the *partial derivative of f with respect to x* and is denoted by $\partial f/\partial x$.† Before giving a more formal definition, we give an example.

EXAMPLE 1 COMPUTING A PARTIAL DERIVATIVE

Let $z = f(x, y) = x^2y + \sin xy^2$. Calculate $\partial f/\partial x$.

SOLUTION: Treating y as if it were a constant, we have

$$\frac{\partial f}{\partial x} = \frac{\partial}{\partial x}(x^2y + \sin xy^2) = \frac{\partial}{\partial x}x^2y + \frac{\partial}{\partial x}\sin xy^2 = 2xy + y^2\cos xy^2.$$

How should we define a partial derivative of a function f of two variables? Suppose that y is held fixed. Then only x varies. We define

$$g_y(x) = f(x, y).$$

g_y is a function of x only, and by the definition of the derivative of a function of one variable,

$$\frac{dg_y}{dx} = \lim_{\Delta x \to 0} \frac{g_y(x + \Delta x) - g_y(x)}{\Delta x} \overset{g_y(x) = f(x, y)}{\downarrow}{=} \lim_{\Delta x \to 0} \frac{f(x + \Delta x, y) - f(x, y)}{\Delta x}.$$

This suggests the following definitions of partial derivatives.

† See the historical note on page 175.

DEFINITION PARTIAL DERIVATIVES IN $\mathbb{R}^2$

Let $z = f(x, y)$.

i. The **partial derivative of** f **with respect to** x is the function

$$\frac{\partial z}{\partial x} = \frac{\partial f}{\partial x} = \lim_{\Delta x \to 0} \frac{f(x + \Delta x, y) - f(x, y)}{\Delta x}. \tag{1}$$

$\partial f/\partial x$ is defined at every point (x, y) in the domain of f such that the limit (1) exists.

ii. The **partial derivative of** f **with respect to** y is the function

$$\frac{\partial z}{\partial y} = \frac{\partial f}{\partial y} = \lim_{\Delta y \to 0} \frac{f(x, y + \Delta y) - f(x, y)}{\Delta y}. \tag{2}$$

$\partial f/\partial y$ is defined at every point (x, y) in the domain of f such that the limit (2) exists. ■

REMARK 1: This definition allows us to calculate partial derivatives in the same way we calculate ordinary derivatives by allowing only one of the variables to vary while the other one is held fixed. It also allows us to use all the formulas from one-variable calculus.

REMARK 2: The partial derivatives $\partial f/\partial x$ and $\partial f/\partial y$ give us the rate of change of f as each of the variables x and y changes with the other one held fixed. They do *not* tell us how f changes when x and y change simultaneously. We will discuss this different topic in Section 3.5.

REMARK 3: It should be emphasized that while the functions $\partial f/\partial x$ and $\partial f/\partial y$ are computed with one of the variables held constant, each is a function of both variables.

EXAMPLE 2

COMPUTING TWO PARTIAL DERIVATIVES

Let $f(x, y) = \sqrt{x + y^2}$. Calculate $\partial f/\partial x$ and $\partial f/\partial y$ at the point $(2, -3)$.

SOLUTION:

$$\frac{\partial f}{\partial x} = \frac{1}{2\sqrt{x + y^2}} \frac{\partial}{\partial x}(x + y^2) = \frac{1}{2\sqrt{x + y^2}}(1 + 0) = \frac{1}{2\sqrt{x + y^2}}$$

since we are treating y as a constant. At $(2, -3)$,

$$\frac{\partial f}{\partial x} = \frac{1}{2\sqrt{2 + (-3)^2}} = \frac{1}{2\sqrt{11}}$$

$$\frac{\partial f}{\partial y} = \frac{1}{2\sqrt{x + y^2}} \frac{\partial}{\partial y}(x + y^2) = \frac{1}{2\sqrt{x + y^2}}(0 + 2y) = \frac{y}{\sqrt{x + y^2}}$$

since we are treating x as a constant. At $(2, -3)$, $\dfrac{\partial f}{\partial y} = \dfrac{-3}{\sqrt{11}}$.

We now obtain a geometric interpretation of the partial derivative. Let $z = f(x, y)$. As we saw in Section 3.1, this is the equation of a surface in $\mathbb{R}^3$. To obtain $\partial z/\partial x$, we hold y

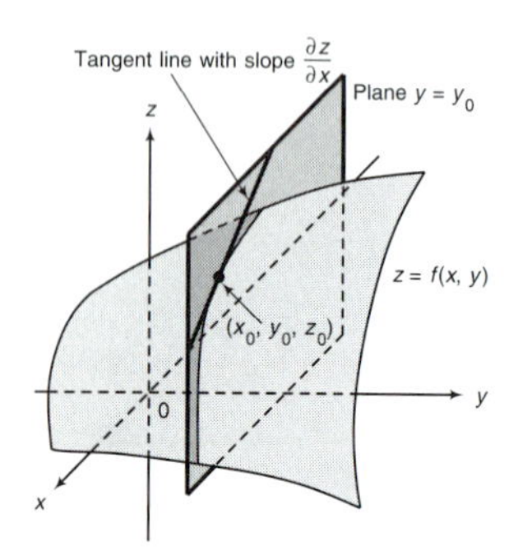

FIGURE 1
Graphical interpretation of $\frac{\partial f}{\partial x}$

fixed at some constant value y_0. The equation $y = y_0$ is a plane in space parallel to the xz-plane (whose equation is $y = 0$). Thus if y is constant, $\partial z/\partial x$ is the rate of change of f with respect to x as x changes along the curve C, which is at the intersection of the surface $z = f(x, y)$ and the plane $y = y_0$. This is illustrated in Figure 1. To be more precise, let (x_0, y_0, z_0) be a point on the surface $z = f(x, y)$. The intersection of this surface and the plane $y = y_0$ is a curve. Then $\partial z/\partial x$ evaluated at (x_0, y_0) is the slope of the tangent line to this curve at the point (x_0, y_0, z_0). Analogously, the intersection of the surface with the plane $x = x_0$ is a different curve and $\partial z/\partial y$ evaluated at (x_0, y_0) is the slope of the line tangent to the new curve at the point (x_0, y_0, z_0). (This is true because x is held fixed in order to calculate $\partial z/\partial y$.) This is illustrated in Figure 2.

There are other ways to denote partial derivatives. We will often write

NOTATION FOR PARTIAL DERIVATIVES

$$f_x = \frac{\partial f}{\partial x} \quad \text{and} \quad f_y = \frac{\partial f}{\partial y}. \tag{3}$$

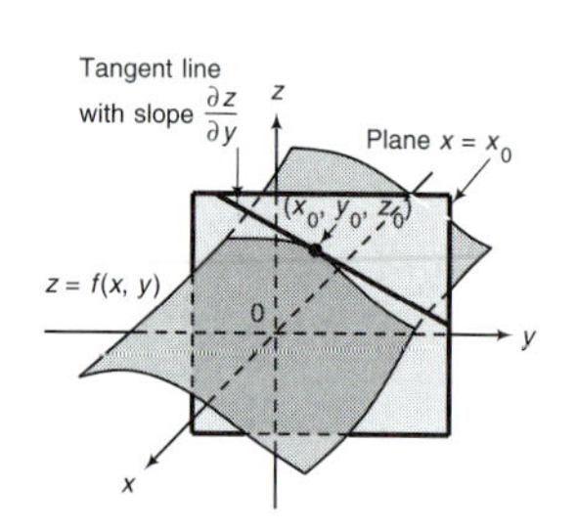

FIGURE 2
Graphical interpretation of $\frac{\partial f}{\partial y}$

As an example of the use of this notation, we may write the result of Example 2 as $f_x(2, -3) = \frac{1}{2\sqrt{11}}$ and $f_y(2, -3) = -\frac{3}{\sqrt{11}}$. If f is a function of other variables, say s and t, then we may write $\partial f/\partial s = f_s$ and $\partial f/\partial t = f_t$.

We now turn to the question of finding partial derivatives of functions of three variables.

DEFINITION PARTIAL DERIVATIVES IN $\mathbb{R}^3$

Let $w = f(x, y, z)$.

i. The **partial derivative of f with respect to x** is the function

$$\frac{\partial w}{\partial x} = \frac{\partial f}{\partial x} = f_x = \lim_{\Delta x \to 0} \frac{f(x + \Delta x, y, z) - f(x, y, z)}{\Delta x}. \tag{4}$$

f_x is defined at every point (x, y, z) in the domain of f at which the limit in (4) exists.

ii. The **partial derivative of f with respect to y** is the function

$$\frac{\partial w}{\partial y} = \frac{\partial f}{\partial y} = f_y = \lim_{\Delta y \to 0} \frac{f(x, y + \Delta y, z) - f(x, y, z)}{\Delta y}. \tag{5}$$

f_y is defined at every point (x, y, z) in the domain of f at which the limit in (5) exists.

iii. The **partial derivative of f with respect to z** is the function

$$\frac{\partial w}{\partial z} = \frac{\partial f}{\partial z} = f_z = \lim_{\Delta z \to 0} \frac{f(x, y, z + \Delta z) - f(x, y, z)}{\Delta z}. \tag{6}$$

f_z is defined at any point (x, y, z) in the domain of f at which the limit in (6) exists. ■

REMARK: As can be seen from this definition, to calculate $\partial f/\partial x$, we simply treat y and z as constants and then calculate an ordinary derivative.

EXAMPLE 3

CALCULATING THREE PARTIAL DERIVATIVES

Let $w = f(x, y, z) = xz + e^{y^2z} + \sqrt{xy^2z^3}$. Calculate $\partial w/\partial x$, $\partial w/\partial y$, and $\partial w/\partial z$.

SOLUTION: To calculate $\partial w/\partial x$, we keep y and z fixed. Then,

$$\begin{aligned}\frac{\partial w}{\partial x} = \frac{\partial f}{\partial x} = f_x &= \frac{\partial}{\partial x}xz + \frac{\partial}{\partial x}e^{y^2z} + \frac{1}{2\sqrt{xy^2z^3}}\frac{\partial}{\partial x}(xy^2z^3)\\ &= z + 0 + \frac{y^2z^3}{2\sqrt{xy^2z^3}} = z + \frac{y^2z^3}{2\sqrt{xy^2z^3}}.\end{aligned}$$

To calculate $\partial w/\partial y$, we keep x and z fixed. Then,

$$\begin{aligned}\frac{\partial w}{\partial y} = \frac{\partial f}{\partial y} = f_y &= \frac{\partial}{\partial y}xz + e^{y^2z}\frac{\partial}{\partial y}(y^2z) + \frac{1}{2\sqrt{xy^2z^3}}\frac{\partial}{\partial y}(xy^2z^3)\\ &= 0 + 2yze^{y^2z} + \frac{2xyz^3}{2\sqrt{xy^2z^3}} = 2yze^{y^2z} + \frac{xyz^3}{\sqrt{xy^2z^3}}.\end{aligned}$$

To calculate $\partial w/\partial z$, we keep x and y fixed. Then,

$$\begin{aligned}\frac{\partial w}{\partial z} = \frac{\partial f}{\partial z} = f_z &= x + e^{y^2z}\frac{\partial}{\partial z}y^2z + \frac{1}{2\sqrt{xy^2z^3}}\frac{\partial}{\partial z}(xy^2z^3)\\ &= x + y^2e^{y^2z} + \frac{3xy^2z^2}{2\sqrt{xy^2z^3}}.\end{aligned}$$

A differentiable function of one variable is continuous. The situation is more complicated for a function of two variables. In fact, there are functions that are discontinuous at a point but for which all partial derivatives exist at that point.

EXAMPLE 4

PARTIAL DERIVATIVES MAY EXIST AT A POINT OF DISCONTINUITY

Let

$$f(x, y) = \begin{cases} \dfrac{xy}{x^2 + y^2}, & (x, y) \neq (0, 0)\\ 0, & (x, y) = (0, 0).\end{cases}$$

Show that f_x and f_y exist at $(0, 0)$ but that f is not continuous there.

SOLUTION: We first show that $\lim_{(x,y)\to(0,0)} f(x, y)$ does not exist, so that f cannot be continuous at $(0, 0)$. To show that, we first let $(x, y) \to (0, 0)$ along the line $y = x$. Then,

$$\frac{xy}{x^2 + y^2} = \frac{x^2}{x^2 + x^2} = \frac{1}{2},$$

so if $\lim_{(x,y)\to(0,0)} f(x, y)$ existed, it would have to equal $\frac{1}{2}$. But if we now let $(x, y) \to$

(0, 0) along the line $y = -x$, we have

$$\frac{xy}{x^2 + y^2} = \frac{-x^2}{x^2 + x^2} = -\frac{1}{2},$$

so along this line the limit is $-\frac{1}{2}$. Hence, the limit does not exist. On the other hand, we have

$$\begin{aligned} f_x(0, 0) &= \lim_{\Delta x \to 0} \frac{f(0 + \Delta x, 0) - f(0, 0)}{\Delta x} = \lim_{\Delta x \to 0} \frac{\dfrac{(0 + \Delta x) \cdot 0}{\Delta x^2 + 0^2}}{\Delta x} \\ &= \lim_{\Delta x \to 0} \frac{0}{\Delta x} = \lim_{\Delta x \to 0} 0 = 0. \end{aligned}$$

Similarly, $f_y(0, 0) = 0$. Hence, both f_x and f_y exist at (0, 0) (and are equal to zero) even though f is not continuous there.

In Section 3.5, we will show a relationship between continuity and partial derivatives and show that a certain kind of differentiability does imply continuity.

We now define partial derivatives in $\mathbb{R}^n$.

DEFINITION PARTIAL DERIVATIVES IN $\mathbb{R}^n$.

Let f: $\mathbb{R}^n \to \mathbb{R}$. Then the **partial derivative of f with respect to x_i** is the function

$$\frac{\partial f}{\partial x_i} = \lim_{\Delta x_i \to 0} \frac{f(x_1, x_2, \ldots, x_{i-1}, x_i + \Delta x_i, x_{i+1}, \ldots, x_n) - f(x_1, x_2, \ldots, x_i, \ldots, x_n)}{\Delta x_i}. \quad (7)$$

The partial derivative $\partial f/\partial x_i$ is defined at every point $\mathbf{x} = (x_1, x_2, \ldots, x_n)$ in the domain of f such that the limit in (7) exists. ■

REMARK 1: Equation (7) defines n functions. That is, $\partial f/\partial x_i$ is a function of $(x_1, x_2, \ldots, x_n)$ for $i = 1, 2, \ldots, n$.

REMARK 2: As in the case in which $n = 2$ or 3, $\partial f/\partial x_i$ is computed by treating all variables except the ith one as if they were fixed.

EXAMPLE 5 THE PARTIAL DERIVATIVES OF A FUNCTION DEFINED ON A SUBSET OF $\mathbb{R}^4$

Let $f(\mathbf{x}) = x_1^2 - x_2^2 + 3x_1x_2x_3 - (x_4/x_1)$. Compute $\partial f/\partial x_i$ for $i = 1, 2, 3, 4$.

SOLUTION:

a. $\dfrac{\partial f}{\partial x_1} = 2x_1 + 3x_2x_3 + \dfrac{x_4}{x_1^2}$.

b. $\dfrac{\partial f}{\partial x_2} = -2x_2 + 3x_1x_3$.

c. $\dfrac{\partial f}{\partial x_3} = 3x_1x_2$.

d. $\dfrac{\partial f}{\partial x_4} = -\dfrac{1}{x_1}$.

NOTATION: We shall write f_i to denote the partial derivative of f with respect to the ith variable. Thus in Example 5 we have

$$f_1 = 2x_1 + 3x_2x_3 + \frac{x_4}{x_1^2}, \qquad f_2 = -2x_3 + 3x_1x_3,$$

and so on.

PROBLEMS 3.3

SELF-QUIZ

I. If $f(x, y) = x^2 - y^3 + x\cos y$, then

$$\frac{\partial f(x, y)}{\partial y} = \_\_\_\_\_\_\_\_.$$

a. $-3y^2 - x\sin y$
b. $-3y^2 + \cos y - x\sin y$
c. $2x - 3y^2 + \cos y - x\sin y$
d. $-\sin y$

II. If $f(x, y, z) = xy - y^2z + xe^z - ye^x$, then $\_\_\_\_\_\_\_\_ = x - 2yz - e^x$.

a. $\frac{\partial f}{\partial x}$ **b.** $\frac{\partial f}{\partial y}$ **c.** $\frac{\partial f}{\partial z}$ **d.** $\frac{dy}{dt}$

III. If $f(x, y) = \_\_\_\_\_\_\_\_$, then $\frac{\partial f}{\partial x} = 0$.

a. $\sin^2 x + \cos^2 y$
b. $e^y + \sin(3x)$
c. $7 + \cos y - \ln|5y + 2|$
d. $(x + y)^2 - (x - y)^2$

IV. If $z = f(x, y)$ and $\frac{\partial z}{\partial x} = 0$, then $\_\_\_\_\_\_\_\_$.

a. $f(x, y) = 7 + \cos y - \ln|5y + 2|$
b. $f(x, y) = e^y - 3y + \tan(2 - y)$
c. $f(x, y)$ is a constant
d. the value of $f(x, y)$ is independent of x

V. True–False: Suppose $f(x, y) = (x + \sin y)(y - \cos x)$ and $g(x) = f(x, -\pi/2)$, then

$$\frac{\partial f}{\partial x}(\pi/3, -\pi/2) = \frac{dg}{dx}(\pi/3).$$

VI. True–False: Suppose $f(x, y) = (x + \sin y)(y - \cos x)$ and $h(y) = f(\pi/3, y)$, then

$$\frac{\partial f}{\partial y}(\pi/3, -\pi/2) = \frac{dh}{dy}(-\pi/2).$$

In Problems 1–16, calculate $\frac{\partial z}{\partial x}$ and $\frac{\partial z}{\partial y}$.

1. $z = x^2y$
2. $z = xy^2$
3. $z = 3e^{xy^3}$
4. $z = \sin(x^2 + y^3)$
5. $z = 4x/y^5$
6. $z = e^y \tan x$
7. $z = \ln(x^3y^5 - 2)$
8. $z = \sqrt{xy + 2y^3}$
9. $z = (x + 5y\sin x)^{4/3}$
10. $z = \sinh(2x - y)$
11. $z = \sin^{-1}(x - y)$
12. $z = \sec xy$
13. $xz + yz^3 = z$
[*Hint:* differentiate implicitly.]
14. $\ln(x + y + z) = yz$
15. $\sin(z - x) = y/z$
16. $2xz + 3xyz^2 = e^{xy^2z^3}$

In Problems 17–24, calculate the value of the given partial derivative at the specified point.

17. $f(x, y) = x^3 - y^4$; $f_x(1, -1)$
18. $f(x, y) = \ln(x^2 + y^4)$; $f_y(3, 1)$
19. $f(x, y) = \sin(x + y)$; $f_x(\pi/6, \pi/3)$
20. $f(x, y) = e^{\sqrt{x^2+y}}$; $f_y(0, 4)$
21. $f(x, y) = \sinh(x - y)$; $f_x(3, 3)$
22. $f(x, y) = \sqrt[3]{x^2y - y^2x^5}$; $f_y(-2, 4)$
23. $f(x, y) = \frac{x^2 - y^2}{x^2 + y^2}$; $f_y(2, -3)$
24. $f(x, y) = \tan^{-1}(y/x)$; $f_x(4, 4)$

In Problems 25–32, calculate $\frac{\partial w}{\partial x}$, $\frac{\partial w}{\partial y}$, and $\frac{\partial w}{\partial z}$.

25. $w = \sqrt{x + y + z}$
26. $w = \sin(xyz)$
27. $w = e^{x+2y+3z}$
28. $w = \cosh\sqrt{x + 2y + 5z}$

29. $w = \dfrac{x + y}{z}$

30. $w = \tan^{-1}(xz/y)$

31. $w = \ln(x^3 + y^2 + z)$

32. $w = \dfrac{x^2 - y^2 + z^2}{x^2 + y^2 + z^2}$

In Problems 33–42, calculate the value of the given partial derivative at the specified point.

33. $f(x, y, z) = xyz$; $f_x(2, 3, 4)$

34. $f(x, y, z) = \sqrt{x + 2y + 3z}$; $f_y(2, -1, 3)$

35. $f(x, y, z) = \dfrac{x - y}{z}$; $f_z(-3, -1, 2)$

36. $f(x, y, z) = \sin(z^2 - y^2 + x)$; $f_y(0, 1, 0)$

37. $f(x, y, z) = \ln(x + 2y + 3z)$; $f_z(2, 2, 5)$

38. $f(x, y, z) = \tan^{-1}\left(\dfrac{xy}{z}\right)$; $f_x(1, 2, -2)$

39. $f(x, y, z) = \dfrac{y^3 - z^5}{x^2y + z}$; $f_y(4, 0, 1)$

40. $f(x, y, z) = \sqrt{\dfrac{x + y - z}{x + y + z}}$; $f_z(1, 1, 1)$

41. $f(x, y, z) = e^{xy}(\cosh z - \sinh z)$; $f_z(2, 3, 0)$

42. $f(x, y, z) = \sqrt{x^2 + y^2 + z^2}$; $f_x(a, b, c)$

43. Find the equation of the line tangent to the surface $z = x^3 - 4y^3$ at the point $(1, -1, 5)$ that
   a. lies in the plane $x = 1$;
   b. lies in the plane $y = -1$.

44. Find the equation of the line tangent to the surface $z = \tan^{-1}(y/x)$ at the point $(\sqrt{3}, 1, \pi/6)$ that
   a. lies in the plane $x = \sqrt{3}$;
   b. lies in the plane $y = 1$.

45. Find the equation of the line tangent to the surface $x^2 + 4y^2 + 4z^2 = 9$ at the point $(1, 1, 1)$ that lies in the plane $y = 1$.

***46.** Find the equation of the line tangent to the surface $x^2 + 4y^2 + 4z^2 = 9$ at the point $(1, 1, 1)$ that lies in the plane $z = 1$.

47. The cost to a manufacturer of producing x units of product A and y units of product B is given (in dollars) by

$$C(x, y) = \frac{50}{2 + x} + \frac{125}{(3 + y)^2}.$$

Calculate the marginal cost of each of the two products.

48. The revenue received by the manufacturer of the preceding problem is given by

$$R(x, y) = \ln(1 + 50x + 75y) + \sqrt{1 + 40x + 125y}.$$

Calculate the marginal revenue from each of the two products.

49. A **partial-differential equation** is an equation involving partial derivatives. Verify that the function $z = f(x, y) = e^{(x+\sqrt{3}y)/4} - 4x - 2y - 4 - 2\sqrt{3}$ satisfies the partial-differential equation

$$\frac{\partial z}{\partial x} + \sqrt{3}\frac{\partial z}{\partial y} - z = 4x + 2y.$$

50. Verify that the function $f(x, y) = z = e^{x+(5/4)y} + \frac{7}{2}(1 - e^{y/2})$ is a solution to the partial-differential equation $3f_x - 4f_y + 2f = 7$ which also satisfies $f(x, 0) = e^x$.

51. The **ideal-gas law** states that $P = nRT/V$. Assume that the number n of moles of an ideal gas and the temperature T of the gas are held constant at the values 10 and 20°C (=293°K), respectively. What is the rate of change of the pressure P as a function of the volume V when the volume of the gas is 2 liters?

52. If the current annual interest rate is i, then the **present value** of an annuity which pays B dollars each year for t years is given by

$$A_0 = \frac{B}{i}\left[1 - \left(\frac{1}{1 + i}\right)^t\right].$$

   a. If time period t is fixed, how does the present value of the annuity change as the rate of interest i changes?
   b. How is the present value changing with respect to i if $B = \$500$ and $i = 6\%$?
   c. If the interest rate i is fixed, how does the present value of the annuity change as the number of years during which the payments are made, t, is increased?

53. Let C denote the oxygen consumption (per unit weight) of a fur-bearing animal. Let T denote its internal body temperature (in deg Celsius). Let t denote the outside temperature of its fur, and let w denote its weight (in kg). It has been experimentally determined that if T is considerably larger than t, then a reasonable model for the oxygen consumption of the animal is given by

$$C = \frac{5(T - t)}{2w^{2/3}}.$$

Consider a particular fur-bearing animal weighing 10 kg which has a constant internal body temperature of 23°C. If the outside temperature is dropping, how is the animal's oxygen consumption changing when the outside temperature of its fur is 5°C?

54. If a particle is falling in a fluid, then according to **Stoke's law**, the velocity of the particle is given by

$$V = \frac{2g}{9}(P - F)\frac{r^2}{\eta},$$

where g is the acceleration due to gravity, P is the density of the particle, F is the density of the fluid, r is the radius of the particle, and η is the absolute viscosity of the liquid. Calculate V_P, V_F, V_r, and V_η.

55. Let $f(x, y) = g(x)h(y)$, where g and h are differentiable functions of a single variable. Show that the partial derivatives $\frac{\partial f}{\partial x}$ and $\frac{\partial f}{\partial y}$ exist and that

$$\frac{\partial f}{\partial x} = g'(x)h(y) \quad \text{and} \quad \frac{\partial f}{\partial y} = g(x)h'(y).$$

56. Suppose g, h, and k are differentiable functions of a single variable. Let $f(x, y, z) = g(x)h(y)k(z)$. Show that $\frac{\partial f}{\partial x}$, $\frac{\partial f}{\partial y}$, and $\frac{\partial f}{\partial z}$ exist and that

$$\frac{\partial f}{\partial x} = g'(x)h(y)k(z), \quad \frac{\partial f}{\partial y} = g(x)h'(y)k(z),$$

$$\frac{\partial f}{\partial z} = g(x)h(y)k'(z).$$

In the following four problems, suppose f and g are functions such that $\frac{\partial f}{\partial x}$ and $\frac{\partial g}{\partial x}$ exist at every point (x, y) in $\mathbb{R}^2$.

57. Prove that $\frac{\partial}{\partial x}(f + g)$ exists and

$$\frac{\partial(f+g)}{\partial x} = \frac{\partial f}{\partial x} + \frac{\partial g}{\partial x}.$$

58. Prove that $\frac{\partial}{\partial x}(af)$ exists for every constant a and

$$\frac{\partial(af)}{\partial x} = a\left(\frac{\partial f}{\partial x}\right).$$

59. Prove that $\frac{\partial}{\partial x}(f \cdot g)$ exists and

$$\frac{\partial(f \cdot g)}{\partial x} = \frac{\partial f}{\partial x} \cdot g + f \cdot \frac{\partial g}{\partial x}.$$

60. Suppose $\frac{\partial g}{\partial x} \neq 0$. Prove that $\frac{\partial}{\partial x}(f/g)$ exists and

$$\frac{\partial(f/g)}{\partial x} = \frac{\frac{\partial f}{\partial x} \cdot g - f \cdot \frac{\partial g}{\partial x}}{g^2}.$$

61. Let

$$f(x, y) = \begin{cases} \dfrac{x+y}{x-y}, & \text{if } (x, y) \neq (0, 0) \\ 0, & \text{if } (x, y) = (0, 0). \end{cases}$$

a. Show that f is not continuous at $(0, 0)$.

b. Do $f_x(0, 0)$ and $f_y(0, 0)$ exist? Explain.

***62.** The ideal-gas law (see Problem 51) allows each of V, T, and P to be expressed as a function of the other two quantities. Suppose that is done so we have $V = V(T, P)$, $T = T(P, V)$, and $P = P(V, T)$. Show that

$$\frac{\partial V}{\partial T} \cdot \frac{\partial T}{\partial P} \cdot \frac{\partial P}{\partial V} = -1.$$

In Problems 63–69 compute all first-order partial derivatives.

63. $f(x_1, x_2, x_3, x_4) = x_1x_2x_3x_4$

64. $f(x_1, x_2, x_3, x_4) = \sqrt{x_1^2 + x_2^2 + x_3^2 + x_4^2}$

65. $f(x_1, x_2, x_3, x_4) = \frac{x_1x_3}{x_2x_4} + e^{x_1x_3/x_2}$

66. $f(x_1, x_2, x_3, x_4, x_5) = x_1^2x_3 - x_2^2x_4 + \sin x_5$

67. $f(x_1, x_2, \ldots, x_n) = \left(\sum_{i=1}^{n} x_i^2\right)^{1/2}$

68. $f(x_1, x_2, \ldots, x_n) = x_1x_2 \cdots x_n$

69. $f(x_1, x_2, \ldots, x_n) = \left(\sum_{i=1}^{n} x_i^2\right)^{-1/2}$

ANSWERS TO SELF-QUIZ

I. a **II.** b **III.** c **IV.** d **V.** True **VI.** True

3.4 HIGHER-ORDER PARTIAL DERIVATIVES

We have seen that if $y = f(x)$, then

$$y' = \frac{df}{dx} \quad \text{and} \quad y'' = \frac{d^2f}{dx^2} = \frac{d}{dx}\left(\frac{df}{dx}\right).$$

That is, the second derivative of f is the derivative of the first derivative of f. Analogously, if $z = f(x, y)$, then we can differentiate each of the two "first" partial derivatives $\partial f/\partial x$ and $\partial f/\partial y$ with respect to both x and y to obtain four **second partial derivatives** as follows:

DEFINITION SECOND PARTIAL DERIVATIVES IN $\mathbb{R}^2$

i. Differentiate twice with respect to x:

$$\frac{\partial^2 z}{\partial x^2} = \frac{\partial^2 f}{\partial x^2} = f_{xx} = \frac{\partial}{\partial x}\left(\frac{\partial f}{\partial x}\right). \tag{1}$$

ii. Differentiate first with respect to x and then with respect to y:

$$\frac{\partial^2 z}{\partial y\,\partial x} = \frac{\partial^2 f}{\partial y\,\partial x} = f_{xy} = \frac{\partial}{\partial y}\left(\frac{\partial f}{\partial x}\right). \tag{2}$$

iii. Differentiate first with respect to y and then with respect to x:

$$\frac{\partial^2 z}{\partial x\,\partial y} = \frac{\partial^2 f}{\partial x\,\partial y} = f_{yx} = \frac{\partial}{\partial x}\left(\frac{\partial f}{\partial y}\right). \tag{3}$$

iv. Differentiate twice with respect to y:

$$\frac{\partial^2 z}{\partial y^2} = \frac{\partial^2 f}{\partial y^2} = f_{yy} = \frac{\partial}{\partial y}\left(\frac{\partial f}{\partial y}\right). \tag{4}$$

The derivatives $\partial^2 f/\partial x\,\partial y$ and $\partial^2 f/\partial y\,\partial x$ are called the **mixed second partials.** ■

REMARK: It is much easier to denote the second partials by f_{xx}, f_{xy}, f_{yx}, and f_{yy}. We will therefore use this notation for the remainder of this section. Note that the symbol f_{xy} indicates that we differentiate first with respect to x and then with respect to y.

EXAMPLE 1 COMPUTING SECOND PARTIAL DERIVATIVES

Let $z = f(x, y) = x^3y^2 - xy^5$. Calculate the four second partial derivatives.

SOLUTION: We have $f_x = 3x^2y^2 - y^5$ and $f_y = 2x^3y - 5xy^4$.

a. $f_{xx} = \dfrac{\partial}{\partial x}(f_x) = 6xy^2$

b. $f_{xy} = \dfrac{\partial}{\partial y}(f_x) = 6x^2y - 5y^4$

c. $f_{yx} = \dfrac{\partial}{\partial x}(f_y) = 6x^2y - 5y^4$

d. $f_{yy} = \dfrac{\partial}{\partial y}(f_y) = 2x^3 - 20xy^3$

In Example 1 we saw that $f_{xy} = f_{yx}$. This result is no accident, as we see by the following theorem.[†]

THEOREM 1 EQUALITY OF MIXED PARTIALS[‡] IN $\mathbb{R}^2$

Suppose that f, f_x, f_y, f_{xy}, and f_{yx} are all continuous at (x_0, y_0). Then

$$f_{xy}(x_0, y_0) = f_{yx}(x_0, y_0). \tag{5}$$

■

The definition of second partial derivatives and the theorem on the equality of mixed partials are easily extended to functions of three variables. If $w = f(x, y, z)$, then we have the nine second partial derivatives (assuming that they exist):

$$\frac{\partial^2 f}{\partial x^2} = f_{xx}, \quad \frac{\partial^2 f}{\partial y\,\partial x} = f_{xy}, \quad \frac{\partial^2 f}{\partial z\,\partial x} = f_{xz},$$

$$\frac{\partial^2 f}{\partial x\,\partial y} = f_{yx}, \quad \frac{\partial^2 f}{\partial y^2} = f_{yy}, \quad \frac{\partial^2 f}{\partial z\,\partial y} = f_{yz},$$

$$\frac{\partial^2 f}{\partial x\,\partial z} = f_{zx}, \quad \frac{\partial^2 f}{\partial y\,\partial z} = f_{zy}, \quad \frac{\partial^2 f}{\partial z^2} = f_{zz}.$$

THEOREM 2 EQUALITY OF MIXED PARTIALS IN $\mathbb{R}^3$

If f, f_x, f_y, f_z, and all six mixed partials are continuous at a point (x_0, y_0, z_0), then at that point

$$f_{xy} = f_{yx}, \quad f_{xz} = f_{zx}, \quad \text{and} \quad f_{yz} = f_{zy}.$$

■

EXAMPLE 2 COMPUTING SECOND PARTIAL DERIVATIVES

Let $f(x, y, z) = xy^3 - zx^5 + x^2yz$. Calculate all nine second partial derivatives and show that all three pairs of mixed partials are equal.

SOLUTION: We have

$$f_x = y^3 - 5zx^4 + 2xyz,$$
$$f_y = 3xy^2 + x^2z,$$

and

$$f_z = -x^5 + x^2y.$$

Then

$$f_{xx} = -20zx^3 + 2yz, \qquad f_{yy} = 6xy, \qquad f_{zz} = 0,$$

$$f_{xy} = \frac{\partial}{\partial y}(y^3 - 5zx^4 + 2xyz) = 3y^2 + 2xz,$$

[†] This theorem is proved on page 173.
[‡] This theorem was first stated by Euler in a 1734 paper devoted to a problem in hydrodynamics. See the biographical sketch of Euler on page 251.

$$f_{yx} = \frac{\partial}{\partial x}(3xy^2 + x^2z) = 3y^2 + 2xz,$$

$$f_{xz} = \frac{\partial}{\partial z}(y^3 - 5zx^4 + 2xyz) = -5x^4 + 2xy,$$

$$f_{zx} = \frac{\partial}{\partial x}(-x^5 + x^2y) = -5x^4 + 2xy,$$

$$f_{yz} = \frac{\partial}{\partial z}(3xy^2 + x^2z) = x^2,$$

$$f_{zy} = \frac{\partial}{\partial y}(-x^5 + x^2y) = x^2.$$

We observe that we can easily define partial derivatives of orders higher than two. For example,

$$f_{zyx} = \frac{\partial^3 f}{\partial x\,\partial y\,\partial z} = \frac{\partial}{\partial x}\left(\frac{\partial^2 f}{\partial y\,\partial z}\right) = \frac{\partial}{\partial x}(f_{zy}).$$

EXAMPLE 3

COMPUTING HIGHER-ORDER PARTIAL DERIVATIVES

Calculate f_{xxx}, f_{xzy}, f_{yxz}, and f_{yxzx} for the function of Example 2.

SOLUTION: We obtain the three third partial derivatives:

$$f_{xxx} = \frac{\partial}{\partial x}(f_{xx}) = \frac{\partial}{\partial x}(-20zx^3 + 2yz) = -60zx^2$$

$$f_{xzy} = \frac{\partial}{\partial y}(f_{xz}) = \frac{\partial}{\partial y}(-5x^4 + 2xy) = 2x$$

$$f_{yxz} = \frac{\partial}{\partial z}(f_{yx}) = \frac{\partial}{\partial z}(3y^2 + 2xz) = 2x.$$

Note that $f_{xzy} = f_{yxz}$. This again is no accident and follows from the generalization of Theorem 2 to mixed third partial derivatives. Finally, the fourth partial derivative f_{yxzx} is given by

$$f_{yxzx} = \frac{\partial}{\partial x}(f_{yxz}) = \frac{\partial}{\partial x}(2x) = 2.$$

Higher-order partial derivatives are defined exactly as in $\mathbb{R}^2$ and $\mathbb{R}^3$.

EXAMPLE 4

SECOND PARTIAL DERIVATIVES IN $\mathbb{R}^4$

Let $f(\mathbf{x}) = x_1{}^2 - x_2{}^2 + 3x_1x_2x_3 - (x_4/x_1)$. Compute f_{13}, f_{31}, and f_{312}.

SOLUTION: From Example 3.3.5, $f_3 = 3x_1x_2$, so that

$$f_{31} = \frac{\partial^2 f}{\partial x_1\,\partial x_3} = \frac{\partial}{\partial x_1}\left(\frac{\partial f}{\partial x_3}\right) = \frac{\partial}{\partial x_1}(3x_1x_2) = 3x_2.$$

Similarly,

$$f_{13} = \left(\frac{\partial f}{\partial x_1}\right)_3 = \left(2x_1 + 3x_2x_3 - \frac{x_4}{x_1{}^2}\right)_3 = 3x_2.$$

Finally,

$$f_{312} = (f_{31})_2 = \frac{\partial}{\partial x_2}(3x_2) = 3.$$

REMARK: In Example 4 we found that

$$\frac{\partial^2 f}{\partial x_1\, \partial x_3} = \frac{\partial^2 f}{\partial x_3\, \partial x_1}.$$

This is not a coincidence.

In Theorem 1 we stated that $f_{xy} = f_{yx}$ if these partial derivatives are continuous. It is also true that if f is a function of n variables, and if all second-order partial derivatives are continuous, then all higher-order partial derivatives involving differentiation with respect to the same variables are equal. That is, the order in which we take successive partial derivatives makes no difference. The proof of the following theorem is difficult and can be omitted at the first reading. Note that this theorem includes Theorems 1 and 2.

THEOREM 3 EQUALITY OF MIXED PARTIALS IN $\mathbb{R}^n$

Suppose that f: $\mathbb{R}^n \to \mathbb{R}$ and, in an open set Ω, f, f_i, f_j, and f_{ij} are continuous. Then f_{ji} exists and is continuous and $f_{ij} = f_{ji}$ at every $\mathbf{x} \in \Omega$.

PROOF: We begin by restating the mean value theorem for a function f of one variable.

> **MEAN-VALUE THEOREM**
>
> Let f be continuous on $[a, b]$ and differentiable on (a, b). Then there is a number c in (a, b) such that
>
> $$f(b) - f(a) = f'(c)(b - a).$$

We first prove the theorem in the case in which $n = 2$. That is, we show that $f_{12}(\mathbf{x}) = f_{21}(\mathbf{x})$ for every $\mathbf{x}$ in Ω.

Let $A(h, k) =$

$$[f(x_0 + h, y_0 + k) - f(x_0 + h, y_0)] - [f(x_0, y_0 + k) - f(x_0, y_0)] \tag{6}$$

In going from one bracket to the next, only the first argument changes. The mean-value theorem gives

$$\begin{aligned} A(h, k) &= h\frac{\partial}{\partial x}[f(x, y_0 + k) - f(x, y_0)]_{x=a} \\ &= h[f_x(a, y_0 + k) - f_x(a, y_0)], \text{ where } x_0 < a < x_0 + h. \end{aligned}$$

In the last equation, only the second argument changes; we apply the mean-value theorem to get

$$A(h, k) = hk\frac{\partial}{\partial y}f_x(a, y)\big|_{y=b} = hkf_{xy}(a, b), \text{ where } y_0 < b < y_0 + k.$$

$$\text{Thus } \frac{A(h, k)}{hk} = f_{xy}(a, b).$$

$$\text{By the continuity of } f_{xy} \text{ we have } \lim_{h\to 0}\lim_{k\to 0}\frac{A(h, k)}{hk} = f_{xy}(x_0, y_0) \tag{7}$$

Rearranging (6) and applying the definition of derivative we have

$$\begin{aligned}
&\lim_{h\to 0}\lim_{k\to 0}\frac{A(h, k)}{hk} \\
&= \lim_{h\to 0}\lim_{k\to 0}\frac{1}{h}\cdot\frac{[f(x_0 + h, y_0 + k) - f(x_0, y_0 + k)] - [f(x_0 + h, y_0) - f(x_0, y_0)]}{k} \\
&= \lim_{h\to 0}\frac{1}{h}\frac{\partial}{\partial y}[f(x_0 + h, y) - f(x_0, y)]_{y=y_0} \\
&= \lim_{h\to 0}\frac{1}{h}[f_y(x_0 + h, y_0) - f_y(x_0, y_0)]
\end{aligned}$$

By equation (7), this limit exists, and by the definition of derivative its value is $f_{yx}(x_0, y_0)$. This proves the result.

We have proved the theorem in the case $n = 2$. If $n > 2$, assume that $i < j$. Let $\mathbf{x}_0 = (x_1^{(0)}, x_2^{(0)}, \ldots, x_n^{(0)})$ and define

$$F(x_i, x_j) = (x_1^{(0)}, x_2^{(0)}, \ldots, x_{i-1}^{(0)}, x_i, x_{i+1}^{(0)}, \ldots, x_{j-1}^{(0)}, x_j, x_{j+1}^{(0)}, \ldots, x_n^{(0)}). \tag{8}$$

That is, only x_i and x_j vary; the other variables stay fixed. But F is a function of two variables only, so that by the part we have already proved,

$$F_{ij}(x_i^{(0)}, x_j^{(0)}) = F_{ji}(x_i^{(0)}, x_j^{(0)}).$$

But

$$F_{ij}(x_i^{(0)}, x_j^{(0)}) = f_{ij}(\mathbf{x}_0) \quad \text{and} \quad F_{ji}(x_i^{(0)}, x_j^{(0)}) = f_{ji}(\mathbf{x}_0).$$

This completes the proof. ■

Let F: $\mathbb{R}^n \to \mathbb{R}$ be a function of n variables. Then the partial derivatives $f_1, f_2, \ldots, f_n$ are called the **first-order partial derivatives** of f. Similarly, the derivatives f_{ij} for $1 \le i, j \le n$ are called the **second-order partial derivatives** of f. Continuing in this manner, we define the **third-order partial derivatives** of f to be derivatives having the form

$$f_{ijk} = \frac{\partial^3 f}{\partial x_k\,\partial x_j\,\partial x_i} \quad \text{for} \quad 1 \le i, j, k \le n.$$

Finally, the **mth-order partial derivatives** are given by

$$f_{i_1 i_2 \ldots i_m} = \frac{\partial^m f}{\partial x_{i_m}\,\partial x_{i_{m-1}} \cdots \partial x_{i_1}}$$

(we emphasize that in this notation $i_1 + i_2 + \cdots + i_m = m$).

OF HISTORICAL INTEREST

DISTINGUISHING BETWEEN ORDINARY AND PARTIAL DERIVATIVES[†]

In the closing years of the seventeenth century it was already becoming evident, as disclosed in the writings of Isaac Newton and Gottfried Wilhelm von Leibniz, that in both the theory and application of mathematics proper attention must be given to the concept of partial derivatives, wherein differentiation of a function of several variables is performed with respect to one variable at a time. In most of the early researches in which partial derivatives appeared, no special notation was used. Thus the symbol du/dx was alternatively interpreted, according to context, to mean the ordinary or total derivative of u with respect to x or the partial derivative of u with respect to x with other independent variables being held fixed, now usually denoted $\partial u/\partial x$. The need for a distinct notation for partial derivatives, however, initiated a conflict in symbols that persisted throughout most of the eighteenth and nineteenth centuries. Of the dozens of proposals made, most died out rather quickly. Under consideration were the symbols d, D, δ, ϑ, and ∂, often in combination with subscripts or superscripts or both. For a period toward the end of the eighteenth century it seemed possible that the letters d and D might be exclusively appropriated by those working in finite differences, and the round ∂ was introduced for ordinary derivatives. The round ∂ was used in 1770 by Marquis de Condorcet (1743–1794) for partial differentials, and in 1776 by Leonhard Euler in the form $\partial^\lambda/p \cdot V$, now written $\partial^\lambda V/\partial p^\lambda$. This symbol was first used in the modern combination $\partial v/\partial x$ in 1786 by Adrien Marie Legendre (1752–1833) and the letter δ was introduced in an identical role in 1824 by William Rowan Hamilton. However, this use of the "round dee" (∂) was not generally adopted until close to the end of the nineteenth century, and Carl Gustav Jacob Jacobi, who outlined the advantages of the symbol ∂ in 1841, is often incorrectly credited with its invention.

A substantial proportion of the application of mathematics, especially to physics and astronomy, rests on partial derivatives and in particular on solutions of partial differential equations. One of the early pioneers in this area was Daniel Bernoulli (1700–1782). In 1747, Jean Le Rond d'Alembert (1717–1783) solved the fundamental problem of the vibrating string by formulating it in terms of the differential equation

$$\frac{\partial^2 y}{\partial t^2} = a^2 \frac{\partial^2 y}{\partial x^2}$$

and expressing the solution in the form (see Problem 27)

$$u = f(x + ct) + \varphi(x - ct).$$

Others whose early work in partial differential equations had far-reaching effects in the history of mathematics and its applications are Euler, Joseph Louis Lagrange (1736–1813), Pierre Simon Laplace (1749–1827), who introduced and studied the equation (see Problem 35)

$$\frac{\partial^2 u}{\partial x^2} + \frac{\partial^2 u}{\partial y^2} + \frac{\partial^2 u}{\partial z^2} = 0,$$

and Jean Joseph Fourier (1768–1830), who exploited his celebrated series (now called *Fourier series*) in the study of the heat equation

$$\frac{\partial T}{\partial t} = \delta \frac{\partial^2 T}{\partial x^2}$$

(see Problem 28). Partial differential equations continue to be a subject of active research activity today.

[†]This note is adapted from an article by John M. H. Olmsted that appears in *Historical Topics for the Mathematics Classroom* published in 1989 by the National Council of Teachers of Mathematics (NCTM). It is used here with the permission of the NCTM.

EXAMPLE 5 A FIFTH-ORDER PARTIAL DERIVATIVE

Let $f(x_1, x_2, x_3, x_4) = x_1{}^2x_2{}^3x_3{}^4x_4{}^5$. Then one fifth-order partial derivative is

$$f_{34312}(\mathbf{x}) = (f_3)_{4312} = (4x_1{}^2x_2{}^3x_3{}^3x_4{}^5)_{4312} = (20x_1{}^2x_2{}^3x_3{}^3x_4{}^4)_{312}$$
$$= (60x_1{}^2x_2{}^3x_3{}^2x_4{}^4)_{12} = (120x_1x_2{}^3x_3{}^2x_4{}^4)_2 = 360x_1x_2{}^2x_3{}^2x_4{}^4.$$

DEFINITION $C^{(m)}$ AND $C^{(\infty)}$ FUNCTIONS

i. The function f: $\mathbb{R}^n \to \mathbb{R}$ is said to be of **class** $C^{(m)}(\Omega)$ if all partial derivatives of f of orders $\le m$ exist and are continuous in Ω.

ii. The function f: $\mathbb{R}^n \to \mathbb{R}$ is said to be of **class** $C^{(\infty)}(\Omega)$ if f is of class $C^{(m)}(\Omega)$ for every integer $m \ge 1$.

We will make use of the ideas of $C^{(m)}(\Omega)$ and $C^{(\infty)}(\Omega)$ functions in Section 9.1. Here we simply note that any polynomial in n variables is a $C^{(\infty)}$ function in $\mathbb{R}^n$. ■

PROBLEMS 3.4

SELF-QUIZ

I. The notation $\partial^2 f/\partial x\, \partial y$ is shorthand for ______.

a. $\dfrac{\partial}{\partial x}\left(\dfrac{\partial}{\partial y}f\right)$ **b.** $\dfrac{\partial}{\partial y}\left(\dfrac{\partial}{\partial x}f\right)$

II. The notation f_{yx} is shorthand for ______.

a. $\dfrac{\partial}{\partial x}\left(\dfrac{\partial}{\partial y}f\right)$ **b.** $\dfrac{\partial}{\partial y}\left(\dfrac{\partial}{\partial x}f\right)$

III. $f_{xy} =$ ______.

a. $\dfrac{\partial^2 f}{\partial y\, \partial x}$ **b.** $\dfrac{\partial^2 f}{\partial x\, \partial y}$

IV. $g_{zxy} =$ ______.

a. $\dfrac{\partial^3 g}{\partial y\, \partial x\, \partial z}$ **b.** $\dfrac{\partial}{\partial y}\left[\dfrac{\partial}{\partial x}\left(\dfrac{\partial g}{\partial z}\right)\right]$

V. True–False:

$$\frac{\partial}{\partial x}\left(\frac{\partial f}{\partial y}\right) \neq \frac{\partial}{\partial y}\left(\frac{\partial f}{\partial x}\right)$$

for all functions f with domain $\mathbb{R}^2$.

VI. True–False:

$$\frac{\partial}{\partial x}\left(\frac{\partial g}{\partial y}\right) = \frac{\partial}{\partial y}\left(\frac{\partial g}{\partial x}\right)$$

for all functions g with domain $\mathbb{R}^2$.

VII. How many third-order partial derivatives are there for a function of

a. two variables? **b.** three variables?

In Problems 1–12, calculate the four second partial derivatives and verify that the two mixed partial derivatives are equal.

1. $f(x, y) = x^2y$
2. $f(x, y) = xy^2$
3. $f(x, y) = 3e^{xy^3}$
4. $f(x, y) = \sin(x^2 + y^3)$
5. $f(x, y) = 4x/y^5$
6. $f(x, y) = e^y \tan x$
7. $f(x, y) = \ln(x^3y^5 - 2)$
8. $f(x, y) = \sqrt{xy + 2y}$
9. $f(x, y) = (x + 5y \sin x)^{4/3}$
10. $f(x, y) = \sinh(2x - y)$
11. $f(x, y) = \sin^{-1}(x - y)$
12. $f(x, y) = \sec(xy)$

In Problems 13–20, calculate the nine second partial derivatives and then verify that the three pairs of mixed partials are equal.

13. $f(x, y, z) = xyz$
14. $f(x, y, z) = \cos(xyz)$
15. $f(x, y, z) = x^2y^3z^4$
16. $f(x, y, z) = \sin(x + 2y + z^2)$
17. $f(x, y, z) = \dfrac{x + y}{z}$

18. $f(x, y, z) = \tan^{-1}\left(\dfrac{xz}{y}\right)$

19. $f(x, y, z) = e^{3xy} \cos z$

20. $f(x, y, z) = \ln(xy + z)$

In Problems 21–26, calculate the indicated partial derivative.

21. $f(x, y) = x^2y^3 + 2y$; f_{xyx}

22. $f(x, y) = \sin(2xy^4)$; f_{xyy}

23. $f(x, y) = \ln(3x - 2y)$; f_{yxy}

24. $f(x, y, z) = \cos(x + 2y + 3z)$; f_{zzx}

25. $f(x, y, z) = x^2y + y^2z - 3\sqrt{xz}$; f_{xyz}

26. $f(x, y, z) = e^{xy} \sin z$; f_{zxyx}

27. Consider a string that is stretched tightly between two fixed points 0 and L on the x-axis. The string is pulled back and released at a time $t = 0$, causing it to vibrate. One position of the string is sketched in Figure 1. Let $y(x, t)$ denote the height of the string at any time $t \geq 0$ and at any point x in the interval $[0, L]$.

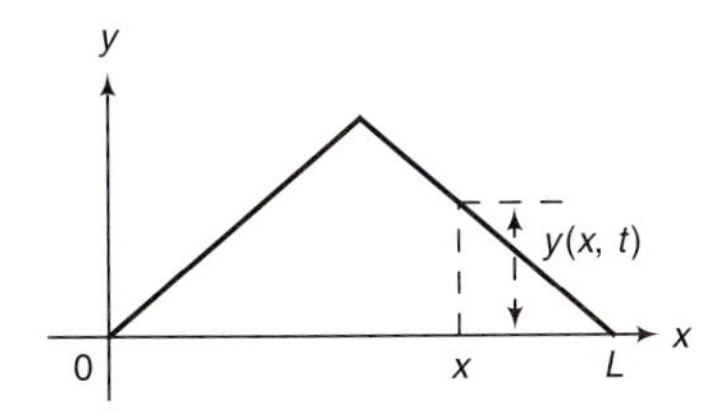

FIGURE 1

It can be shown that $y(x, t)$ satisfies the partial differential equation†

$$\frac{\partial^2 y}{\partial t^2} = c^2 \frac{\partial^2 y}{\partial x^2},$$

where c is a constant. (This equation is called the one-dimensional **wave equation**.) Verify that $y(x, t) = \frac{1}{2}[(x - ct)^2 + (x + ct)^2]$ is a solution to the wave equation.

28. Consider a cylindrical rod composed of a uniform heat-conducting material of length L and radius r. Assume that heat can enter and leave the rod only through its ends. Let $T(x, t)$ denote the absolute temperature (deg Kelvin) at time t at a point x units along the rod (see Figure 2).

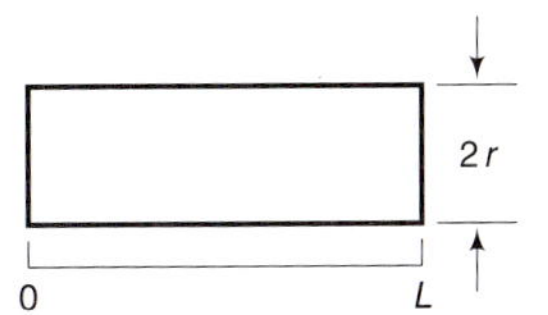

FIGURE 2

Then it can be shown‡ that T satisfies the partial differential equation

$$\frac{\partial T}{\partial t} = \delta \frac{\partial^2 T}{\partial x^2},$$

where δ is a positive constant called the **diffusivity** of the rod. (This equation is called the **heat equation** or **diffusion equation**.) Verify that the function

$$T(x, t) = e^{-\alpha^2 \delta t} \sin(\alpha x)$$

satisfies the heat equation for any constant α.

29. Verify that

$$T(x, t) = \frac{1}{\sqrt{t}} e^{-x^2/(4\delta t)}$$

satisfies the heat equation.

30. Find constants α and β such that $T(x, t) = e^{\alpha x + \beta t}$ satisfies the heat equation.

31. One of the most important partial-differential equations of mathematical physics is **Laplace's equation** in $\mathbb{R}^2$, given by

$$\frac{\partial^2 f}{\partial x^2} + \frac{\partial^2 f}{\partial y^2} = 0.$$

Laplace's equation can be used to model steady-state heat flow in a closed, bounded region in $\mathbb{R}^2$. Verify that the function $f(x, y) = x^2 - y^2$ satisfies Laplace's equation.

32. Verify that $f(x, y) = \tan^{-1}(y/x)$ satisfies Laplace's equation.

33. Verify that $f(x, y) = \ln(x^2 + y^2)$ satisfies Laplace's equation.

34. Verify that $f(x, y) = \sin x \sinh y$ satisfies Laplace's equation.

35. Laplace's equation in $\mathbb{R}^3$ is given by

$$\frac{\partial^2 f}{\partial x^2} + \frac{\partial^2 f}{\partial y^2} + \frac{\partial^2 f}{\partial z^2} = 0.$$

† For a derivation of this wave equation, see W. Derrick and S. Grossman, *Introduction to Differential Equations With Boundary Value Problems,* 3rd ed. (St. Paul: West, 1987), Section 12.5.

‡ See Derrick and Grossman, *Introduction to Differential Equations,* Section 12.7.

Verify that $f(x, y, z) = x^2 + y^2 - 2z^2$ satisfies Laplace's equation in $\mathbb{R}^3$.

***36.** Show that $f_{xy}(0, 0)$ and $f_{yx}(0, 0)$ both exist but are not equal for the following function[†]

$$f(x, y) = \begin{cases} xy\dfrac{x^2 - y^2}{x^2 + y^2} & \text{if } x^2 + y^2 \neq 0, \\ 0 & \text{if } x = 0 \text{ and } y = 0. \end{cases}$$

In Problems 37–42 compute the given higher-order partial derivatives.

37. $f(x_1, x_2, x_3, x_4) = x_1x_2x_3x_4$; f_{13}, f_{11}, f_{134}

38. $f(x_1, x_2, x_3, x_4) = \sqrt{x_1^2 + x_2^2 + x_3^2 + x_4^2}$; f_{12}, f_{33}

39. $f(x_1, x_2, \ldots, x_n) = \left(\sum_{i=1}^{n} x_i^2\right)^{-1/2}$; f_{ij}

40. $f(x_1, x_2) = \sin x_1x_2$; f_{1211}

41. $f(x_1, x_2, x_3) = e^{x_1x_2x_3}$; f_{2211}

42. $f(x_1, x_2, \ldots, x_n) = \sum_{i=1}^{n} x_i^2$; $f_{123\cdots n}$

43. Let f be a function of three variables.

a. How many third-order partial derivatives does f have? (Assume that f_{121} and f_{112}, for example, are different third-order partial derivatives.)

b. How many fourth-order?

c. How many mth-order?

***44.** Determine the number of mth-order partial derivatives of a function $f: \mathbb{R}^n \to \mathbb{R}$.

45. Let $p_k: \mathbb{R}^n \to \mathbb{R}$ be a polynomial of degree k. Show that if $p_k^{(k+1)}$ is a partial derivative of p_k of order $k + 1$, then $p_k^{(k+1)}(\mathbf{x}) = 0$ for every $\mathbf{x} \in \mathbb{R}^n$.

ANSWERS TO SELF-QUIZ

I. a **II.** a **III.** a **IV.** a = b

V. False (Equality holds in certain cases according to Theorem 1.)

VI. False (The hypotheses for Theorem 1 suggest equality does not always hold; see Problem 36 for such an exception.)

VII. a. $2^3 = 8$; for smooth functions only 4 are different.

b. $3^3 = 27$; for smooth functions only 10 are different.

3.5 DIFFERENTIABILITY AND THE GRADIENT

In this section, we discuss the notion of the differentiability of a function of several variables. There are several ways to introduce this subject and the way we have chosen is designed to illustrate the great similarities between differentiation of functions of one variable and differentiation of functions of several variables.

We begin with a function of one variable,

$$y = f(x).$$

If f is differentiable, then

$$f'(x) = \frac{dy}{dx} = \lim_{\Delta x \to 0} \frac{\Delta y}{\Delta x}. \tag{1}$$

If we define the new function $\epsilon(\Delta x)$[‡] by

$$\epsilon(\Delta x) = \frac{\Delta y}{\Delta x} - f'(x), \tag{2}$$

[†] B. R. Gelbaum and J. M. H. Olmstead, *Counterexamples in Analysis* (San Francisco: Holden-Day, 1964), p. 120.

[‡] ϵ is really a function of both x and Δx. We write $\epsilon(\Delta x)$ rather than $\epsilon(x, \Delta x)$ to simplify notation.

we have

$$\lim_{\Delta x \to 0} \epsilon(\Delta x) = \lim_{\Delta x \to 0} \left(\frac{\Delta y}{\Delta x} - f'(x) \right) = \lim_{\Delta x \to 0} \frac{\Delta y}{\Delta x} - f'(x)$$
$$= f'(x) - f'(x) = 0. \tag{3}$$

Multiplying both sides of (2) by Δx and rearranging terms, we obtain

$$\Delta y = f'(x)\,\Delta x + \epsilon(\Delta x)\,\Delta x.$$

Note that here Δy depends on both Δx and x. Finally, since $\Delta y = f(x + \Delta x) - f(x)$, we obtain

$$f(x + \Delta x) + f(x) = f'(x)\,\Delta x + \epsilon(\Delta x)\,\Delta x. \tag{4}$$

Why did we do all this? We do so in order to be able to state the following alternative definition of differentiability of a function f of one variable.

ALTERNATIVE DEFINITION OF DIFFERENTIABILITY OF A FUNCTION OF ONE VARIABLE

Let f be a function of one variable. Then f is **differentiable** at a number x if there is a number $f'(x)$ and a function $g(\Delta x)$ such that

$$f(x + \Delta x) - f(x) = f'(x)\,\Delta x + g(\Delta x), \tag{5}$$

where $\lim_{\Delta x \to 0}[g(\Delta x)/\Delta x] = 0$.

We will soon show how the definition (5) can be extended to a function of two or more variables. First, we give a definition.

DEFINITION DIFFERENTIABILITY OF A FUNCTION OF TWO VARIABLES

Let f be a real-valued function of two variables that is defined in a neighborhood of a point (x, y) and such that $f_x(x, y)$ and $f_y(x, y)$ exist. Then f is **differentiable** at (x, y) if there exist functions $\epsilon_1(\Delta x, \Delta y)$ and $\epsilon_2(\Delta x, \Delta y)$ such that

$$f(x + \Delta x, y + \Delta y) - f(x, y)$$
$$= f_x(x, y)\,\Delta x + f_y(x, y)\,\Delta y + \epsilon_1(\Delta x, \Delta y)\,\Delta x + \epsilon_2(\Delta x, \Delta y)\,\Delta x \tag{6}$$

where

$$\lim_{(\Delta x, \Delta y) \to (0,0)} \epsilon_1(\Delta x, \Delta y) = 0 \qquad \text{and}$$
$$\lim_{(\Delta x, \Delta y) \to (0,0)} \epsilon_2(\Delta x, \Delta y) = 0. \tag{7}$$

■

EXAMPLE 1 SHOWING THAT A FUNCTION IS EVERYWHERE DIFFERENTIABLE IN $\mathbb{R}^2$

Let $f(x, y) = xy$. Show that f is differentiable at every point (x, y) in $\mathbb{R}^2$.

SOLUTION:

$$f(x + \Delta x, y + \Delta y) - f(x, y) = (x + \Delta x)(y + \Delta y) - xy$$
$$= xy + y\,\Delta x + x\,\Delta y + \Delta x\,\Delta y - xy$$
$$= y\,\Delta x + x\,\Delta y + \Delta x\,\Delta y.$$

Now $f_x = y$ and $f_y = x$, so we have

$$f(x+\Delta x, y+\Delta y) - f(x, y) = f_x(x, y)\,\Delta x + f_y(x, y)\,\Delta y + \Delta y\,\Delta x + 0\cdot\Delta y.$$

Setting $\epsilon_1(\Delta x, \Delta y) = \Delta y$ and $\epsilon_2(\Delta x, \Delta y) = 0$, we see that

$$\lim_{(\Delta x,\Delta y)\to(0,0)} \epsilon_1(\Delta x, \Delta y) = \lim_{(\Delta x,\Delta y)\to(0,0)} \epsilon_2(\Delta x, \Delta y) = 0.$$

This shows that $f(x, y) = xy$ is differentiable at every point in $\mathbb{R}^2$.

NOTE: The functions ϵ_1 and ϵ_2 are not unique. For example, we could set $\epsilon_1 = 0$ and $\epsilon_2 = \Delta x$ to show that xy is differentiable.

We now rewrite our definition of differentiability in a more compact form. Since a point (x, y) is a vector in $\mathbb{R}^2$, we write (as we have done before) $\mathbf{x} = (x, y)$. Then if $z = f(x, y)$, we can simply write

$$z = f(\mathbf{x}).$$

Similarly, if $w = f(x, y, z)$, we may write

$$w = f(\mathbf{x}),$$

where $\mathbf{x}$ is the vector (x, y, z). With this notation we may use the symbol $\mathbf{\Delta x}$ to denote the vector $(\Delta x, \Delta y)$ in $\mathbb{R}^2$ or $(\Delta x, \Delta y, \Delta z)$ in $\mathbb{R}^3$.

Next, we write

$$g(\mathbf{\Delta x}) = \epsilon_1(\Delta x, \Delta y)\,\Delta x + \epsilon_2(\Delta x, \Delta y)\,\Delta y. \tag{8}$$

Note that $(\Delta x, \Delta y) \to (0, 0)$ can be written in the compact form $\mathbf{\Delta x} \to \mathbf{0}$. Then if the conditions (7) hold, we see that

$$\lim_{\mathbf{\Delta x}\to\mathbf{0}} \frac{|g(\mathbf{\Delta x})|}{|\mathbf{\Delta x}|} \overset{|\mathbf{\Delta x}| = \sqrt{\Delta x^2+\Delta y^2}}{\le} \lim_{\mathbf{\Delta x}\to\mathbf{0}} |\epsilon_1(\Delta x, \Delta y)| \frac{|\Delta x|}{\sqrt{\Delta x^2+\Delta y^2}} + \lim_{\mathbf{\Delta x}\to\mathbf{0}} |\epsilon_2(\Delta x, \Delta y)| \frac{|\Delta y|}{\sqrt{\Delta x^2+\Delta y^2}}$$

$$\frac{|\Delta x|}{\sqrt{\Delta x^2+\Delta y^2}} \le 1 \text{ and } \frac{|\Delta y|}{\sqrt{\Delta x^2+\Delta y^2}} \le 1$$

$$\le \lim_{\mathbf{\Delta x}\to\mathbf{0}} |\epsilon_1(\Delta x, \Delta y)| + \lim_{\mathbf{\Delta x}\to\mathbf{0}} |\epsilon_2(\Delta x, \Delta y)| = 0 + 0 = 0.$$

Finally, we have the following important definition.

DEFINITION THE GRADIENT IN $\mathbb{R}^2$

Let f be a function of two variables such that f_x and f_y exist at a point $\mathbf{x} = (x, y)$. Then the **gradient** of f at $\mathbf{x}$, denoted by $\nabla f(\mathbf{x})$, is given by

$$\nabla f(\mathbf{x}) = f_x(x, y)\mathbf{i} + f_y(x, y)\mathbf{j}. \tag{9}$$

■

Note that the gradient of f is a **vector function**. That is, for every point $\mathbf{x}$ in $\mathbb{R}^2$ for which $\nabla f(\mathbf{x})$ is defined, we see that $\nabla f(\mathbf{x})$ is a vector in $\mathbb{R}^2$.

HISTORICAL NOTE: The gradient of f is denoted by ∇f, which is read "del" f. This symbol, an inverted Greek delta, was first used in the 1850s, although the name "del" first appeared in print only in 1901. The symbol ∇ is also called *nabla*. This name is used because someone once suggested to the Scottish mathematician Peter Guthrie Tait (1831–1901) that ∇ looks like an Assyrian harp, the Assyrian name of which is nabla.† Tait, incidentally, was one of the mathematicians who helped carry on Hamilton's development of the theory of quaternions and vectors in the nineteenth century.

EXAMPLE 2 COMPUTING A GRADIENT IN $\mathbb{R}^2$

In Example 1, $f(x, y) = xy$, $f_x = y$, and $f_y = x$, so that

$$\nabla f(\mathbf{x}) = \nabla f(x, y) = y\mathbf{i} + x\mathbf{j}.$$

Using this new notation, we observe that

$$\nabla f(\mathbf{x}) \cdot \Delta\mathbf{x} = (f_x\mathbf{i} + f_y\mathbf{j}) \cdot (\Delta x\mathbf{i} + \Delta y\mathbf{j}) = f_x(x, y)\Delta x + f_y(x, y)\Delta y.$$

Also,

$$f(x + \Delta x, y + \Delta y) = f(\mathbf{x} + \Delta\mathbf{x}).$$

Thus we have the following definition, which is equivalent to the definition of differentiability.

DEFINITION DIFFERENTIABILITY IN $\mathbb{R}^2$ (DEFINITION GIVEN IN TERMS OF THE GRADIENT)

Let f be a function of two variables that is defined in a neighborhood of a point $\mathbf{x} = (x, y)$. Let $\Delta\mathbf{x} = (\Delta x, \Delta y)$. If $f_x(x, y)$ and $f_y(x, y)$ exist, then f is **differentiable** at $\mathbf{x}$ if there is a function g such that

$$f(\mathbf{x} + \Delta\mathbf{x}) - f(\mathbf{x}) = \nabla f(\mathbf{x}) \cdot \Delta\mathbf{x} + g(\Delta\mathbf{x}) \tag{10}$$

where

$$\lim_{\Delta\mathbf{x} \to \mathbf{0}} \frac{g(\Delta\mathbf{x})}{|\Delta\mathbf{x}|} = 0. \tag{11}$$

■

REMARK 1: Although formulas (5) and (10) look very similar, there are two fundamental differences. First f' is a scalar, while ∇f is a vector. Second, $f'(x)\,\Delta x$ is an ordinary product of real numbers, while $\nabla f(\mathbf{x}) \cdot \Delta\mathbf{x}$ is a dot product of vectors.

REMARK 2: According to this definition, f is *not* differentiable at $\mathbf{x}$ if one or more of its partial derivatives fails to exist at $\mathbf{x}$.

REMARK 3: We have shown that if f is differentiable at $\mathbf{x} = (x, y)$ according to the definition on page 179, it is also differentiable at $\mathbf{x} = (x, y)$ according to the definition

†Fortunately, most (but certainly not all) of the mathematical terms currently in use have more to do with the objects they describe. We might further point out that ∇ is also called *atled*, which is delta spelled backward.

above. We have not proved this the other way around but the proof is not difficult as it involves writing all vectors in terms of their components. The important point to stress is that if two definitions are given for the same concept, you must make certain that they are equivalent; that is, each one implies the other.

There are two reasons for giving you this new definition. First, it illustrates that the gradient of a function of two variables is the natural extension of the derivative of a function of one variable. Second, the definition (10), (11) can be used to define the notion of differentiability for a function of three or more variables as well. We will say more about this subject shortly.

One important question remains: What functions are differentiable? A partial answer is given in Theorem 1. A proof is given at the end of the section.

THEOREM 1 A FUNCTION WITH CONTINUOUS PARTIAL DERIVATIVES IS DIFFERENTIABLE

Let f, f_x, and f_y be defined and continuous in a neighborhood of $\mathbf{x} = (x, y)$. Then f is differentiable at $\mathbf{x}$. ■

REMARK: We cannot omit the hypothesis that f_x and f_y are continuous in this theorem. In Example 3.3.4 on page 165, we saw an example of a function for which f_x and f_y exist at $(0, 0)$ but f itself is not continuous there. As we will see in Theorem 2, if f is differentiable at a point, then it is continuous there. Thus the function in Example 3.3.4 is not differentiable at $(0, 0)$ even though $f_x(0, 0)$ and $f_y(0, 0)$ exist.

DEFINITION CONTINUOUSLY DIFFERENTIABLE OR SMOOTH FUNCTION

If the hypotheses of Theorem 1 are satisfied, then f is said to be **continuously differentiable** or **smooth** at $\mathbf{x}$. ■

EXAMPLE 3 SHOWING THAT A FUNCTION IS DIFFERENTIABLE AND COMPUTING ITS GRADIENT

Let $z = f(x, y) = \sin xy^2 + e^{x^2y^3}$. Show that f is differentiable and calculate ∇f. Find $\nabla f(1, 1)$.

SOLUTION: $\partial f/\partial x = y^2 \cos xy^2 + 2xy^3e^{x^2y^3}$ and $\partial f/\partial y = 2xy \cos xy^2 + 3x^2y^2e^{x^2y^3}$. Since $\partial f/\partial x$ and $\partial f/\partial y$ are continuous, f is differentiable and

$$\nabla f(x, y) = (y^2 \cos xy^2 + 2xy^3e^{x^2y^3})\mathbf{i} + (2xy \cos xy^2 + 3x^2y^2e^{x^2y^3})\mathbf{j}.$$

At $(1, 1)$, $\nabla f(1, 1) = (\cos 1 + 2e)\mathbf{i} + (2 \cos 1 + 3e)\mathbf{j}$.

In Section 3.3, we showed that the existence of all of its partial derivatives at a point does *not* ensure that a function is continuous at that point. However, differentiability does ensure continuity.

THEOREM 2 A DIFFERENTIABLE FUNCTION IS CONTINUOUS

If f is differentiable at $\mathbf{x}_0 = (x_0, y_0)$, then f is continuous at $\mathbf{x}_0$.

PROOF: We must show that $\lim_{(x,y)\to(x_0,y_0)} f(x, y) = f(x_0, y_0)$, or, equivalently, $\lim_{\mathbf{x}\to\mathbf{x}_0} f(\mathbf{x}) = f(\mathbf{x}_0)$. But if we define $\Delta\mathbf{x}$ by $\Delta\mathbf{x} = \mathbf{x} - \mathbf{x}_0$, this is the same as showing that

$$\lim_{\Delta\mathbf{x}\to\mathbf{0}} f(\mathbf{x}_0 + \Delta\mathbf{x}) = f(\mathbf{x}_0), \tag{12}$$

Since f is differentiable at $\mathbf{x}_0$,

$$f(\mathbf{x}_0 + \Delta\mathbf{x}) - f(\mathbf{x}_0) = \nabla f(\mathbf{x}_0) \cdot \Delta\mathbf{x} + g(\Delta\mathbf{x}). \tag{13}$$

But as $\Delta\mathbf{x} \to \mathbf{0}$, both terms on the right-hand side of (13) approach zero, so

$$\lim_{\Delta\mathbf{x}\to\mathbf{0}} [f(\mathbf{x}_0 + \Delta\mathbf{x}) - f(\mathbf{x}_0)] = 0,$$

which means that (12) holds and the theorem is proved. ■

The converse to this theorem is false, as it is in one-variable calculus. That is, there are functions that are continuous, but not differentiable, at a given point. For example, the function

$$f(x, y) = \sqrt[3]{x} + \sqrt[3]{y}$$

is continuous at any point (x, y) in $\mathbb{R}^2$, but

$$\nabla f(x, y) = \frac{1}{3x^{2/3}}\mathbf{i} + \frac{1}{3y^{2/3}}\mathbf{j},$$

so f is not differentiable at any point (x, y) for which either x or y is zero. Thus, $\nabla f(x, y)$ is not defined on the x- and y-axes; that is, $f_x(0, y)$ and $f_y(x, 0)$ are not defined. Hence f is not differentiable along these axes.

In your first calculus course you saw that

$$(f + g)' = f' + g' \qquad \text{and} \qquad (\alpha f)' = \alpha f'$$

whenever f and g are differentiable functions of one variable. That is, the derivative of the sum of two functions is the sum of the derivatives of the two functions, and the derivative of a scalar multiple of a function is the scalar times the derivative of the function. These results can be extended to the gradient vector. The proof of the following theorem is left as an exercise (see Problems 24 and 25).

THEOREM 3 THE GRADIENT OF A SCALAR MULTIPLE AND SUM OF FUNCTIONS

Let f and g be differentiable in a neighborhood of $\mathbf{x} = (x, y)$. Then for every scalar α, αf and $f + g$ are differentiable at $\mathbf{x}$, and

i. $\nabla(\alpha f) = \alpha\,\nabla f$, and
ii. $\nabla(f + g) = \nabla f + \nabla g$. ■

REMARK: Any rule L that satisfies

$$L(\alpha x) = \alpha L(x) \qquad \text{and} \qquad L(x + y) = L(x) + L(y)$$

is called a **linear mapping** or a **linear operator**. Linear operators play an extremely important role in advanced mathematics. Here we see that ∇ is a linear operator.

All the definitions and theorems in this section hold for functions of three or more variables. We give the equivalent results for functions of three variables below.

DEFINITION THE GRADIENT IN $\mathbb{R}^3$

Let f be a scalar function of three variables such that f_x, f_y, and f_z exist at a point $\mathbf{x} = (x, y, z)$. Then the **gradient** of f at $\mathbf{x}$, denoted by $\nabla f(\mathbf{x})$, is given by the vector

$$\nabla f(\mathbf{x}) = f_x(x, y, z)\mathbf{i} + f_y(x, y, z)\mathbf{j} + f_z(x, y, z)\mathbf{k}. \tag{14}$$

■

DEFINITION DIFFERENTIABILITY IN $\mathbb{R}^3$

Let f be a function of three variables that is defined in a neighborhood of $\mathbf{x} = (x, y, z)$, and let $\mathbf{\Delta x} = (\Delta x, \Delta y, \Delta z)$. If $f_x(x, y, z)$, $f_y(x, y, z)$, and $f_z(x, y, z)$ exist, then f is **differentiable** at $\mathbf{x}$ if there is a function g such that

$$f(\mathbf{x} + \mathbf{\Delta x}) - f(\mathbf{x}) = \nabla f \cdot \mathbf{\Delta x} + g(\mathbf{\Delta x})$$

where

$$\lim_{|\mathbf{\Delta x}| \to \mathbf{0}} \frac{g(\mathbf{\Delta x})}{|\mathbf{\Delta x}|} = 0.$$

Equivalently, we can write

$$f(x + \Delta x, y + \Delta y, z + \Delta z) - f(x, y, z)$$
$$= f_x(x, y, z)\Delta x + f_y(x, y, z)\Delta y + f_z(x, y, z)\Delta z + g(\Delta x, \Delta y, \Delta z)$$

where

$$\lim_{(\Delta x, \Delta y, \Delta z) \to (0,0,0)} \frac{g(\Delta x, \Delta y, \Delta z)}{\sqrt{\Delta x^2 + \Delta y^2 + \Delta z^2}} = 0.$$

■

THEOREM 1′

If f, f_x, f_y, and f_z exist and are continuous in a neighborhood of $\mathbf{x} = (x, y, z)$, then f is differentiable at $\mathbf{x}$.

■

THEOREM 2′

Let f be a function of three variables that is differentiable at $\mathbf{x}_0$. Then f is continuous at $\mathbf{x}_0$.

■

EXAMPLE 4 SHOWING THAT A FUNCTION OF THREE VARIABLES IS DIFFERENTIABLE AND COMPUTING ITS GRADIENT

Let $f(x, y, z) = xy^2z^3$. Show that f is differentiable at any point $\mathbf{x}_0$, calculate ∇f, and find $\nabla f(3, -1, 2)$.

SOLUTION: $\partial f/\partial x = y^2z^3$, $\partial f/\partial y = 2xyz^3$, and $\partial f/\partial z = 3xy^2z^2$. Since f, $\partial f/\partial x$, $\partial f/\partial y$, and

$\partial f/\partial z$ are all continuous, we know that f is differentiable and that

$$\nabla f = y^2z^3\mathbf{i} + 2xyz^3\mathbf{j} + 3xy^2z^2\mathbf{k}$$

and

$$\nabla f(3, -1, 2) = 8\mathbf{i} - 48\mathbf{j} + 36\mathbf{k}.$$

THEOREM 3′

Let f and g be differentiable in a neighborhood of $\mathbf{x} = (x, y, z)$. Then for any scalar α, αf and $f + g$ are differentiable at $\mathbf{x}$, and

i. $\nabla(\alpha f) = \alpha \nabla f$, and
ii. $\nabla(f + g) = \nabla f + \nabla g$. ■

We now give a proof of Theorem 1. The proof of Theorem 1′ is similar.

PROOF OF THEOREM 1: We have assumed that f, f_x, and f_y are all continuous in a neighborhood N of $\mathbf{x} = (x, y)$. Choose $\Delta\mathbf{x}$ so small that $\mathbf{x} + \Delta\mathbf{x}$ is in N. Then

$$\Delta f(\mathbf{x}) = f(x + \Delta x, y + \Delta y) - f(x, y)$$

$$= [f(x + \Delta x, y + \Delta y) - \overbrace{f(x + \Delta x, y)] + [f(x + \Delta x, y)}^{\text{This term was added and subtracted}} - f(x, y)]. \tag{15}$$

If $x + \Delta x$ is fixed, then $f(x + \Delta x, y)$ is a function of y that is continuous and differentiable in the interval $[y, y + \Delta y]$. Hence, by the mean-value theorem of one variable (given on p. 173) there is a number c_2 between y and $y + \Delta y$ such that

$$f(x + \Delta x, y + \Delta y) - f(x + \Delta x, y) = f_y(x + \Delta x, c_2)[(y + \Delta y) - y]$$
$$= f_y(x + \Delta x, c_2)\,\Delta y. \tag{16}$$

Similarly, with y fixed, $f(x, y)$ is a function of x only, and we obtain

$$f(x + \Delta x, y) - f(x, y) = f_x(c_1, y)\,\Delta x, \tag{17}$$

where c_1 is between x and $x + \Delta x$. Thus using (16) and (17) in (15), we have

$$\Delta f(\mathbf{x}) = f_x(c_1, y)\,\Delta x + f_y(x + \Delta x, c_2)\,\Delta y. \tag{18}$$

Now both f_x and f_y are continuous at $\mathbf{x} = (x, y)$, so since c_1 is between x and $x + \Delta x$ and c_2 is between y and $y + \Delta y$, we obtain

$$\lim_{\Delta\mathbf{x}\to\mathbf{0}} f_x(c_1, y) = f_x(x, y) = f_x(\mathbf{x}) \tag{19}$$

and

$$\lim_{\Delta\mathbf{x}\to\mathbf{0}} f_y(x + \Delta x, c_2) = f_y(x, y) = f_y(\mathbf{x}). \tag{20}$$

Let

$$\epsilon_1(\Delta\mathbf{x}) = f_x(c_1, y) - f_x(x, y). \tag{21}$$

From (19) it follows that

$$\lim_{\Delta\mathbf{x}\to\mathbf{0}} \epsilon_1(\Delta\mathbf{x}) = 0. \tag{22}$$

Similarly, if

$$\epsilon_2(\mathbf{\Delta x}) = f_y(x + \Delta x, c_2) - f_y(x, y), \tag{23}$$

then

$$\lim_{\mathbf{\Delta x}\to\mathbf{0}} \epsilon_2(\mathbf{\Delta x}) = 0. \tag{24}$$

Now define

$$g(\mathbf{\Delta x}) = \epsilon_1(\mathbf{\Delta x})\,\Delta x + \epsilon_2(\mathbf{\Delta x})\,\Delta y. \tag{25}$$

From (22) and (24) it follows that

$$\lim_{\mathbf{\Delta x}\to\mathbf{0}} \frac{g(\mathbf{\Delta x})}{|\mathbf{\Delta x}|} = 0. \tag{26}$$

Finally, since

$$f_x(c_1, y) = f_x(x, y) + \epsilon_1(\mathbf{\Delta x}) \qquad \text{From (21)} \tag{27}$$

and

$$f_y(x + \Delta x, c_2) = f_y(x, y) + \epsilon_2(\mathbf{\Delta x}), \qquad \text{From (23)} \tag{28}$$

we may substitute (27) and (28) into (18) to obtain

$$\begin{aligned}\Delta f(\mathbf{x}) &= f(\mathbf{x} + \mathbf{\Delta x}) - f(\mathbf{x})\\ &= [f_x(\mathbf{x}) + \epsilon_1(\mathbf{\Delta x})]\,\Delta x + [f_y(\mathbf{x}) + \epsilon_2(\mathbf{\Delta x})]\,\Delta y\\ &= f_x(\mathbf{x})\,\Delta x + f_y(\mathbf{x})\,\Delta y + g(\mathbf{\Delta x}) = (f_x\mathbf{i} + f_y\mathbf{j})\cdot(\mathbf{\Delta x}) + g(\mathbf{\Delta x}),\end{aligned} \tag{29}$$

where $\lim_{\mathbf{\Delta x}\to\mathbf{0}}[g(\mathbf{\Delta x})/|\mathbf{\Delta x}|] \to 0$, and the proof is (at last) complete. ∎

To define the gradient of a function of n variables, we generalize formula (10).

DEFINITION DIFFERENTIABILITY AND THE GRADIENT IN $\mathbb{R}^n$

Let $f\colon \mathbb{R}^n \to \mathbb{R}$ be defined in a neighborhood of a point $\mathbf{x}$. Let $\mathbf{\Delta x} = (\Delta x_1, \Delta x_2, \ldots, \Delta x_n)$. We say that f is **differentiable** at $\mathbf{x}$ if there exists a vector ∇f and a scalar g such that

$$f(\mathbf{x} + \mathbf{\Delta x}) - f(\mathbf{x}) = \nabla f(\mathbf{x})\cdot\mathbf{\Delta x} + g(\mathbf{\Delta x}) \tag{30}$$

where

$$\lim_{\mathbf{\Delta x}\to\mathbf{0}} \frac{g(\mathbf{\Delta x})}{|\mathbf{\Delta x}|} = 0. \tag{31}$$

If f is differentiable at every point $\mathbf{x}$ in a neighborhood of some vector $\mathbf{x}_0$, then f is said to be **differentiable** in that neighborhood. In that case ∇f is a vector-valued function that depends on the point $\mathbf{x}$ and g is a scalar-valued function. The function ∇f is called the **gradient of** f. ∎

The following theorem has a proof virtually identical to the proof of Theorem 1.

THEOREM 4

Let f and all its first partial derivatives be defined and continuous in a neighborhood of $\mathbf{x} = (x_1, x_2, \ldots, x_n)$. Then f is differentiable at $\mathbf{x}$ and

$$\nabla f(\mathbf{x}) = \left(\frac{\partial f}{\partial x_1}(\mathbf{x}), \frac{\partial f}{\partial x_2}(\mathbf{x}), \ldots, \frac{\partial f}{\partial x_n}(\mathbf{x})\right). \tag{32}$$

In addition, if f is known to be differentiable at $\mathbf{x}$, then $\nabla f(\mathbf{x})$ is given by (32). ■

EXAMPLE 5 THE GRADIENT OF A FUNCTION IN $\mathbb{R}^4$

Let $f(\mathbf{x}) = x_1^2 - x_2^2 + 3x_1x_2x_3 - (x_4/x_1)$. Compute ∇f.

SOLUTION: From Example 3.3.5, we have

$$\begin{aligned}\nabla f(\mathbf{x}) &= (f_1(\mathbf{x}), f_2(\mathbf{x}), f_3(\mathbf{x}), f_4(\mathbf{x})) \\ &= \left(2x_1 + 3x_2x_3 + \frac{x_4}{x_1^2}, -2x_2 + 3x_1x_3, 3x_1x_2, -\frac{1}{x_1}\right).\end{aligned}$$

The following theorems generalize Theorems 2 and 3.

THEOREM 5

Let $f\colon \mathbb{R}^n \to \mathbb{R}$ be differentiable at $\mathbf{x}_0$. Then f is continuous at $\mathbf{x}_0$.

PROOF: We must show that $\lim_{\mathbf{x}\to\mathbf{x}_0} f(\mathbf{x}) = f(\mathbf{x}_0)$. This is the same as showing that $\lim_{\mathbf{x}\to\mathbf{x}_0} [f(\mathbf{x}) - f(\mathbf{x}_0)] = 0$. But since f is differentiable at $\mathbf{x}_0$,

$$\lim_{\mathbf{x}\to\mathbf{x}_0} [f(\mathbf{x}) - f(\mathbf{x}_0)] = \lim_{\mathbf{x}\to\mathbf{x}_0} [\nabla f(\mathbf{x}_0) \cdot (\mathbf{x} - \mathbf{x}_0) + g(\mathbf{x} - \mathbf{x}_0)] = 0$$

since $\nabla f(\mathbf{x}_0)$ is a constant vector (in terms of the limit) and

$$\lim_{\mathbf{x}\to\mathbf{x}_0} g(\mathbf{x} - \mathbf{x}_0) = \lim_{\mathbf{x}\to\mathbf{x}_0} \frac{g(\mathbf{x} - \mathbf{x}_0)}{|\mathbf{x} - \mathbf{x}_0|}|\mathbf{x} - \mathbf{x}_0| = 0$$

by (31). ■

THEOREM 6

Let f and g be differentiable in an open set Ω. Then for every scalar α, αf and $f + g$ are differentiable in Ω and

$$\nabla(f + g) = \nabla f + \nabla g$$

and

$$\nabla(\alpha f) = \alpha\nabla f.$$

■

PROBLEMS 3.5

SELF-QUIZ

I. True–False: If $f(x, y) = \cos y$, then $\nabla f(x, y) = -\sin y$.

II. If $f(x, y) =$ _____, then $\nabla f(x, y) = \mathbf{i} - \sin y\mathbf{j}$.
a. $y + \cos x$ **b.** $x + \cos y$
c. $1 + x\cos y$ **d.** $1 - \sin y$

III. If $f(x, y) = y/x$, then $x^2\nabla f(x, y) =$ _____.
a. $x\mathbf{i} + y\mathbf{j}$ **b.** $x\mathbf{i} - y\mathbf{j}$
c. $y\mathbf{i} - x\mathbf{j}$ **d.** $-y\mathbf{i} + x\mathbf{j}$

IV. If $f(x, y) =$ _____, then $\nabla f(x, y) = y\mathbf{i} + x\mathbf{j}$.
a. $\frac{1}{2}y^2 + \frac{1}{2}x^2$ **b.** x/y
c. xy **d.** y/x

V. If $f(x, y) =$ _____, then $\nabla f(x, y) = x\mathbf{i} + y\mathbf{j}$.
a. $\frac{1}{2}y^2 + \frac{1}{2}x^2$ **b.** x/y
c. xy **d.** y/x

VI. If $g(x, y, z) = xyz$, then $\nabla g(1, 2, 3) =$ _____.
a. $\mathbf{0}$ **b.** $\mathbf{i} + 2\mathbf{j} + 3\mathbf{k}$
c. $yz\mathbf{i} + xz\mathbf{j} + xy\mathbf{k}$ **d.** $6\mathbf{i} + 3\mathbf{j} + 2\mathbf{k}$

In Problems 1–20, calculate the gradient of the given function. If a point is also specified, then evaluate the gradient at that point.

1. $f(x, y) = (x + y)^2$
2. $f(x, y) = \ln(2x - y + 1)$
3. $f(x, y) = e^{\sqrt{xy}}$; $(1, 1)$
4. $f(x, y) = \cos(x - y)$, $(\pi/2, \pi/4)$
5. $f(x, y) = \sqrt{x^2 + y^2}$
6. $f(x, y) = \tan^{-1}(y/x)$; $(3, 3)$
7. $f(x, y) = y\tan(y - x)$
8. $f(x, y) = x^2\sinh y$
9. $f(x, y) = \sec(x + 3y)$; $(0, 1)$
10. $f(x, y) = \dfrac{x - y}{x + y}$; $(3, 1)$
11. $f(x, y) = \dfrac{x^2 - y^2}{x^2 + y^2}$
12. $f(x, y) = \dfrac{e^{x^2} - e^{-y^2}}{3y}$
13. $f(x, y, z) = \sin x\cos y\tan z$; $(\pi/6, \pi/4, \pi/3)$
14. $f(x, y, z) = x\sin y\ln z$; $(1, 0, 1)$
15. $f(x, y, z) = \dfrac{x^2 - y^2 + z^2}{3xy}$; $(1, 2, 0)$
16. $f(x, y, z) = x\ln y - z\ln x$
17. $f(x, y, z) = xy^2 + y^2z^3$; $(2, 3, -1)$
18. $f(x, y, z) = (y - z)e^{x+2y+3z}$; $(-4, -1, 3)$
19. $f(x, y, z) = \dfrac{x - z}{\sqrt{1 - y^2 + x^2}}$; $(0, 0, 1)$
20. $f(x, y, z) = x\cosh z - y\sin x$
21. Let $f(x, y) = x^2 + y^2$. Show, by directly using the definition on page 179, that f is differentiable at any point in $\mathbb{R}^2$. (Do not merely cite Theorem 1.)
22. Let $g(x, y) = x^2y^2$. Show, by using the definition of a limit (don't use Theorem 1), that g is differentiable at any point in $\mathbb{R}^2$.
23. Let $f(x, y)$ be any polynomial function of the two variables x and y. Show that f is differentiable at any point in $\mathbb{R}^2$.
24. Prove Theorem 3 by using the definition of a limit. [*Hint:* (for part (i)): If f and g satisfy equation (10) and if α is constant, then

$$[\alpha f(\mathbf{x} + \Delta\mathbf{x})] - [\alpha f(\mathbf{x})] = [\alpha\nabla f(\mathbf{x})]\cdot\Delta\mathbf{x} + [\alpha g(\Delta\mathbf{x})]$$

and

$$\lim_{\Delta\mathbf{x}\to\mathbf{0}}\frac{\alpha g(\Delta\mathbf{x})}{|\Delta\mathbf{x}|} = \alpha\lim_{\Delta\mathbf{x}\to\mathbf{0}}\frac{g(\Delta\mathbf{x})}{|\Delta\mathbf{x}|}.]$$

25. Show that if f and g are differentiable functions of three variables, then so is $f + g$ and $\nabla(f + g) = \nabla f + \nabla g$.
26. Show that if f and g are differentiable functions of three variables, then fg is also differentiable and $\nabla(fg) = (\nabla f)g + f(\nabla g)$.
***27.** Show that $\nabla f = \mathbf{0}$ at every point in an open set if and only if f is constant over that set.
***28.** Show that if $\nabla f = \nabla g$ at every point in an open set, then there is a constant c for which $f(x, y) = g(x, y) + c$. [*Hint:* Use the result of the preceding problem.]
***29.** Let

$$f(x, y) = \begin{cases} (x^2 + y^2)\sin\left(\dfrac{1}{\sqrt{x^2 + y^2}}\right) & \text{if } (x, y) \neq (0, 0), \\ 0 & \text{if } (x, y) = (0, 0). \end{cases}$$

a. Calculate $f_x(0, 0)$ and $f_y(0, 0)$.

b. Explain why f_x and f_y are *not* continuous at $(0, 0)$.

c. Show that f is differentiable at $(0, 0)$.

***30.** Suppose that f is a differentiable function of one variable and g is a differentiable function of three variables. Show that $f \circ g$ is differentiable and $\nabla(f \circ g) = (f' \circ g)\nabla g$.

***31.** What is the most general function f such that $\nabla f(\mathbf{x}) = \mathbf{x}$ for every $\mathbf{x}$ in $\mathbb{R}^2$?

***32.** Show that the converse of Theorem 1 is not true. [*Hint:* See Problem 29.]

In Problems 33–36 compute the gradient of each function.

33. $f(x_1, x_2, x_3, x_4) = x_1x_2x_3x_4$

34. $f(x_1, x_2, \ldots, x_n) = x_1x_2 \cdots x_n$

35. $f(x_1, x_2, x_3, x_4, x_5) = x_1^2 + x_2^2 + x_3^2 + x_4^2 + x_5^2$

36. $f(x_1, x_2, \ldots, x_n) = \left(\sum_{i=1}^{n} x_i^2\right)^{1/2}$

ANSWERS TO SELF-QUIZ

I. False $(-\sin y\mathbf{j})$ **II.** b **III.** d **IV.** c **V.** a **VI.** d

---

3.6 THE CHAIN RULE

In this section, we derive several chain rules for functions of two or more variables. Let us recall the chain rule for the composition of two functions of one variable:

Let $y = f(u)$ and $u = g(x)$ and assume that f and g are differentiable. Then

$$\frac{dy}{dx} = \frac{dy}{du}\frac{du}{dx} = f'(g(x))g'(x). \tag{1}$$

If $z = f(x, y)$ is a function of two variables, then there are two versions of the chain rule.

THEOREM 1 CHAIN RULE

Let $z = f(x, y)$ be differentiable and suppose that $x = x(t)$ and $y = y(t)$. Assume further that dx/dt and dy/dt exist and are continuous. Then z can be written as a function of the parameter t, and

$$\frac{dz}{dt} = \frac{\partial z}{\partial x}\frac{dx}{dt} + \frac{\partial z}{\partial y}\frac{dy}{dt} = f_x\frac{dx}{dt} + f_y\frac{dy}{dt}. \tag{2}$$

We can also write this result using our gradient notation. If $\mathbf{g}(t) = x(t)\mathbf{i} + y(t)\mathbf{j}$, then $\mathbf{g}'(t) = (dx/dt)\mathbf{i} + (dy/dt)\mathbf{j}$, and (2) can be written as

$$\frac{d}{dt}f(x(t), y(t)) = (f \circ \mathbf{g})'(t) = [f(\mathbf{g}(t))]' = \nabla f \cdot \mathbf{g}'(t). \tag{3}$$

■

THEOREM 2 CHAIN RULE

Let $z = f(x, y)$ be differentiable and suppose that x and y are functions of the two variables r and s. That is, $x = x(r, s)$ and $y = y(r, s)$. Suppose further that $\partial x/\partial r$, $\partial x/\partial s$, $\partial y/\partial r$, and $\partial y/\partial s$ all exist and are continuous. Then z can be written as a function of r and s, and

$$\frac{\partial z}{\partial r} = \frac{\partial z}{\partial x}\frac{\partial x}{\partial r} + \frac{\partial z}{\partial y}\frac{\partial y}{\partial r} \tag{4}$$

$$\frac{\partial z}{\partial s} = \frac{\partial z}{\partial x}\frac{\partial x}{\partial s} + \frac{\partial z}{\partial y}\frac{\partial y}{\partial s}. \tag{5}$$

■

We will leave the proofs of these theorems until the end of this section.

EXAMPLE 1 USING THE CHAIN RULE TO COMPUTE $\frac{dz}{dt}$

Let $z = f(x, y) = xy^2$. Let $x = \cos t$ and $y = \sin t$. Calculate dz/dt.

SOLUTION:

$$\begin{aligned}\frac{dz}{dt} &= \frac{\partial z}{\partial x}\frac{dx}{dt} + \frac{\partial z}{\partial y}\frac{dy}{dt} = y^2(-\sin t) + 2xy(\cos t)\\ &= (\sin^2 t)(-\sin t) + 2(\cos t)(\sin t)(\cos t)\\ &= 2\sin t\cos^2 t - \sin^3 t.\end{aligned}$$

EXAMPLE 2 USING THE CHAIN RULE TO CALCULATE TWO PARTIAL DERIVATIVES

Let $z = f(x, y) = \sin xy^2$. Suppose that $x = r/s$ and $y = e^{r-s}$. Calculate $\partial z/\partial r$ and $\partial z/\partial s$.

SOLUTION:

$$\begin{aligned}\frac{\partial z}{\partial r} &= \frac{\partial z}{\partial x}\frac{\partial x}{\partial r} + \frac{\partial z}{\partial y}\frac{\partial y}{\partial r} = (y^2\cos xy^2)\frac{1}{2} + (2xy\cos xy^2)e^{r-s}\\ &= \frac{e^{2(r-s)}\cos[(r/s)e^{2(r-s)}]}{s} + \frac{2r}{s}\{\cos[(r/s)e^{2(r-s)}]\}e^{2(r-s)}\end{aligned}$$

and

$$\begin{aligned}\frac{\partial z}{\partial s} &= \frac{\partial z}{\partial x}\frac{\partial x}{\partial s} + \frac{\partial z}{\partial y}\frac{\partial y}{\partial s} = (y^2\cos xy^2)\frac{-r}{s^2} + (2xy\cos xy^2)(-e^{r-s})\\ &= \frac{-re^{2(r-s)}\cos[(r/s)e^{2(r-s)}]}{s^2} - \frac{2r}{s}\{\cos[(r/s)e^{2(r-s)}]\}e^{2(r-s)}.\end{aligned}$$

The chain rules given in Theorem 1 and Theorem 2 can be extended to functions of three or more variables.

THEOREM 1 CHAIN RULE

Let $w = f(x, y, z)$ be a differentiable function. If $x = x(t)$, $y = y(t)$, $z = z(t)$, and if dx/dt, dy/dt, and dz/dt exist and are continuous, then w may be considered as a differentiable function of t and

$$\frac{dw}{dt} = \frac{\partial w}{\partial x}\frac{dx}{dt} + \frac{\partial w}{\partial y}\frac{dy}{dt} + \frac{\partial w}{\partial z}\frac{dz}{dt}. \tag{6}$$

■

THEOREM 2 CHAIN RULE

Let $w = f(x, y, z)$ be a differentiable function and let $x = x(r, s, t)$, $y = y(r, s, t)$, and $z = z(r, s, t)$. If all indicated partial derivatives exist and are continuous, then w may be considered as a function of r, s, and t and

$$\begin{aligned} \frac{\partial w}{\partial r} &= \frac{\partial w}{\partial x}\frac{\partial x}{\partial r} + \frac{\partial w}{\partial y}\frac{\partial y}{\partial r} + \frac{\partial w}{\partial z}\frac{\partial z}{\partial r}, \\ \frac{\partial w}{\partial s} &= \frac{\partial w}{\partial x}\frac{\partial x}{\partial s} + \frac{\partial w}{\partial y}\frac{\partial y}{\partial s} + \frac{\partial w}{\partial z}\frac{\partial z}{\partial s}, \\ \frac{\partial w}{\partial t} &= \frac{\partial w}{\partial x}\frac{\partial x}{\partial t} + \frac{\partial w}{\partial y}\frac{\partial y}{\partial t} + \frac{\partial w}{\partial z}\frac{\partial z}{\partial t}. \end{aligned} \tag{7}$$

■

In one variable calculus the chain rule is useful to solve "related rates" types of problems. The chain rules for functions of two or more variables are useful in a similar way.

EXAMPLE 3 APPLYING THE CHAIN RULE TO THE IDEAL-GAS LAW

According to the ideal-gas law, the pressure, volume, and absolute temperature of n moles of an ideal gas are related by

$$PV = nRT,$$

where R is a constant.† Suppose that the volume of an ideal gas is increasing at a rate of $10\ \text{cm}^3/\text{min}$ and the pressure is decreasing at a rate of $0.3\ \text{N/cm}^2/\text{min}$. How is the temperature of the gas changing when the volume of 5 mol of a gas is $100\ \text{cm}^3$ and the pressure is $2\ \text{N/cm}^2$?

SOLUTION: We have $T = PV/nR$, where $n = 5$, $p = 2$, and $dv/dt = -0.3$. Then

$$\begin{aligned} \frac{dT}{dt} &= \frac{\partial T}{\partial P}\frac{dP}{dt} + \frac{\partial T}{\partial V}\frac{dV}{dt} = \frac{V}{nR}(-0.3) + \frac{P}{nR}(10) \\ &= \frac{100}{5R}(-0.3) + \frac{2}{5R}(10) = \frac{-2}{R}\,^\circ\text{K/min}. \end{aligned}$$

We now give a proof of Theorem 2. Theorem 1 follows easily from Theorem 2. (Explain why.)

† $R = 8.316 \times 10^7$ g cm^3/sec^2 deg.

PROOF OF THEOREM 2: We will show that

$$\frac{\partial z}{\partial r} = \frac{\partial z}{\partial x}\frac{\partial x}{\partial r} + \frac{\partial z}{\partial y}\frac{\partial y}{\partial r}.$$

Equation (5) follows in an identical manner. Since $x = x(r, s)$ and $y = y(r, s)$, a change Δr in r will cause a change Δx in x and a change Δy in y. We may therefore write

$$\Delta x = x(r + \Delta r, s) - x(r, s)$$

and

$$\Delta y = y(r + \Delta r, s) - y(r, s).$$

Since z is differentiable, we may write (from equation (3.5.6) on page 179)

$$\Delta z = \frac{\partial z}{\partial x}\Delta x + \frac{\partial z}{\partial y}\Delta y + \epsilon_1(\Delta x, \Delta y)\Delta x + \epsilon_2(\Delta x, \Delta y)\Delta y, \tag{8}$$

where

$$\lim_{(\Delta x,\Delta y)\to(0,0)} \epsilon_1(\Delta x, \Delta y) = \lim_{(\Delta x,\Delta y)\to(0,0)} \epsilon_2(\Delta x, \Delta y) = 0.$$

Then, dividing both sides of (8) by Δr and taking limits, we obtain

$$\lim_{\Delta r\to 0} \frac{\Delta z}{\Delta r} = \lim_{\Delta r\to 0} \left[\frac{\partial z}{\partial x}\frac{\Delta x}{\Delta r} + \frac{\partial z}{\partial y}\frac{\Delta y}{\Delta r} + \epsilon_1(\Delta x, \Delta y)\frac{\Delta x}{\Delta r} + \epsilon_2(\Delta x, \Delta y)\frac{\Delta y}{\Delta r}\right]. \tag{9}$$

Since x and y are continuous functions of r, we have

$$\lim_{\Delta r\to 0} \Delta x = 0 \qquad \text{and} \qquad \lim_{\Delta r\to 0} \Delta y = 0,$$

so that

$$\lim_{\Delta r\to 0} \epsilon_1(\Delta x, \Delta y) = \lim_{(\Delta x,\Delta y)\to(0,0)} \epsilon_1(\Delta x, \Delta y) = 0$$

and

$$\lim_{\Delta r\to 0} \epsilon_2(\Delta x, \Delta y) = \lim_{(\Delta x,\Delta y)\to(0,0)} \epsilon_2(\Delta x, \Delta y) = 0.$$

Thus, the limits in (9) become

$$\frac{\partial z}{\partial r} = \frac{\partial z}{\partial x}\frac{\partial x}{\partial r} + \frac{\partial z}{\partial y}\frac{\partial y}{\partial r} + 0 \cdot \frac{\partial x}{\partial r} + 0 \cdot \frac{\partial y}{\partial r},$$

and the theorem is proved. ■

We now state and prove one extension of the chain rule in $\mathbb{R}^n$.

THEOREM 3 CHAIN RULE

Let $\mathbf{x}(t)$: $\mathbb{R} \to \mathbb{R}^n$ be differentiable at t_0 and let $f(\mathbf{x})$: $\mathbb{R}^n \to \mathbb{R}$ be differentiable at $\mathbf{x}_0 = \mathbf{x}(t_0)$. Then $f(\mathbf{x}(t))$ is differentiable at t_0 and

$$\frac{d}{dt} f(\mathbf{x}(t_0)) = \nabla f(\mathbf{x}_0) \cdot \mathbf{x}'(t_0). \tag{10}$$

PROOF: Since $f(\mathbf{x}(t_0))$ is a function from $\mathbb{R}$ to $\mathbb{R}$, we have

$$\frac{d}{dt}f(\mathbf{x}(t_0)) = \lim_{\Delta t \to 0} \frac{f(\mathbf{x}(t_0 + \Delta t)) - f(\mathbf{x}(t_0))}{\Delta t}. \tag{11}$$

Let $\mathbf{\Delta x}(t_0) = \mathbf{x}(t_0 + \Delta t) - \mathbf{x}(t_0)$. Then, since f is differentiable, we have, using equations (30) and (31) on page 186,

$$f(\mathbf{x}(t_0 + \Delta t) - f(\mathbf{x}(t_0)) = \nabla f(\mathbf{x}(t_0)) \cdot \mathbf{\Delta x}(t_0) + g(\mathbf{\Delta x}(t_0)) \tag{12}$$

where

$$\lim_{|\mathbf{\Delta x}| \to 0} \frac{g(\mathbf{\Delta x})}{|\mathbf{\Delta x}|} = 0.$$

Let $h(\mathbf{\Delta x}) = g(\mathbf{\Delta x})/\mathbf{\Delta x}$. Then (12) can be written as

$$f(\mathbf{x}(t_0 + \Delta t) - f(\mathbf{x}(t_0)) = \nabla f(\mathbf{x}(t_0) \cdot \mathbf{\Delta x}(t_0) + |\mathbf{\Delta x}(t_0)| h(\mathbf{\Delta x}(t_0)) \tag{13}$$

where

$$\lim_{|\mathbf{\Delta x}(t_0)| \to 0} h(\mathbf{\Delta x}(t_0)) = 0. \tag{14}$$

Using (13) in (11), we have

$$\frac{d}{dt}f(\mathbf{x}(t_0)) = \lim_{\Delta t \to 0} \left(\nabla f(\mathbf{x}(t_0)) \cdot \frac{\mathbf{\Delta x}(t_0)}{\Delta t}\right) + \lim_{\Delta t \to 0} \frac{|\mathbf{\Delta x}(t_0)| h(\mathbf{\Delta x}(t_0))}{\Delta t}. \tag{15}$$

Now $\mathbf{\Delta x}(t_0) = \mathbf{x}(t_0 + \Delta t) - \mathbf{x}(t_0)$ and, since each $x_i(t)$ is continuous, $\lim_{\Delta t \to 0} \mathbf{\Delta x}(t) = \mathbf{0}$, so that $\lim_{\Delta t \to 0} |\mathbf{\Delta x}(t)| = 0$. Thus, from (14), $\lim_{\Delta t \to 0} h(\mathbf{\Delta x}(t_0)) = 0$. Also,

$$\lim_{\Delta t \to 0} \frac{x_i(t_0 + \Delta t) - x_i(t_0)}{\Delta t} = x_i'(t_0),$$

so that

$$\lim_{\Delta t \to 0} \frac{|\mathbf{\Delta x}(t_0)|}{|\Delta t|} =$$

$$\lim_{\Delta t \to 0} \sqrt{\left[\frac{x_1(t_0 + \Delta t) - x_1(t_0)}{\Delta t}\right]^2 + \left[\frac{x_2(t_0 + \Delta t) - x_2(t_0)}{\Delta t}\right]^2 + \cdots + \left[\frac{x_n(t_0 + \Delta t) - x_n(t_0)}{\Delta t}\right]^2}$$

$$= \sqrt{[x_1'(t_0)]^2 + [x_2'(t_0)]^2 + \cdots + [x_n'(t_0)]^2}.$$

Therefore the second limit in (15) is 0. Finally, consider the ith term in the scalar product in (15). We have

$$\lim_{\Delta t \to 0} f_i(\mathbf{x}(t_0))\left[\frac{x_i(t_0 + \Delta t) - x_i(t_0)}{\Delta t}\right] = f_i(\mathbf{x}(t_0)) \lim_{\Delta t \to 0} \left[\frac{x_i(t_0 + \Delta t) - x_i(t_0)}{\Delta t}\right] = f_i(\mathbf{x}(t_0))x_i'(t_0),$$

so that, from (15),

$$\frac{d}{dt}f(\mathbf{x}(t_0)) = \sum_{i=1}^{n} [f_i(\mathbf{x}(t))][x_i'(t)] = \nabla f(\mathbf{x}(t_0)) \cdot \mathbf{x}'(t_0).$$

This completes the proof. ■

EXAMPLE 4 USING THE CHAIN RULE IN $\mathbb{R}^4$

Let $f(\mathbf{x}) = x_1^2 + x_2x_3 - x_4^2$ where $x_1 = t^2$, $x_2 = \sin t$, $x_3 = t^3$, and $x_4 = \ln t$. Compute $(d/dt)\,f(\mathbf{x}(t))$.

SOLUTION:

$$\begin{aligned}\frac{d}{dt}f(\mathbf{x}(t)) &= \nabla f \cdot \mathbf{x}'(t)\\ &= (2x_1, x_3, x_2, -2x_4)\cdot\left(2t, \cos t, 3t^2, \frac{1}{t}\right)\\ &= (2t^2, t^3, \sin t, -2\ln t)\cdot\left(2t, \cos t, 3t^2, \frac{1}{t}\right)\\ &= 4t^3 + t^3\cos t + 3t^2\sin t - \frac{2\ln t}{t}.\end{aligned}$$

We will give a more general version of the chain rule in Section 9.4.

We close this section with a discussion of the mean value theorem for functions of n variables. To do so, we first need to define the line segment joining two vectors $\mathbf{x}$ and $\mathbf{y}$ in $\mathbb{R}^n$.

DEFINITION LINE SEGMENT IN $\mathbb{R}^n$

Let $\mathbf{x}$ and $\mathbf{y}$ be two vectors in $\mathbb{R}^n$. Then the **line segment** L joining $\mathbf{x}$ and $\mathbf{y}$ is the set defined by

$$L = \{\mathbf{v}: \mathbf{v} = t\mathbf{x} + (1-t)\mathbf{y} \text{ for } 0 \le t \le 1\}. \tag{16}$$

■

NOTE: We observe that both $\mathbf{x}$ and $\mathbf{y}$ are in L. These vectors are obtained by setting $t = 1$ and $t = 0$ in (16).

THEOREM 4 MEAN VALUE THEOREM

Let f: $\mathbb{R}^n \to \mathbb{R}$ be differentiable at every point in an open set containing the line segment L joining two vectors $\mathbf{x}$ and $\mathbf{y}$ in $\mathbb{R}^n$. Then there is a vector $\mathbf{x}_0$ on L such that

$$f(\mathbf{y}) - f(\mathbf{x}) = \nabla f(\mathbf{x}_0)\cdot(\mathbf{y}-\mathbf{x}). \tag{17}$$

PROOF: Let $g(t) = f(t(\mathbf{y}-\mathbf{x}) + \mathbf{x})$. Note that $g(0) = f(\mathbf{x})$, $g(1) = f(\mathbf{y})$, and for $0 < t < 1$, $t(\mathbf{y}-\mathbf{x}) + \mathbf{x} = t\mathbf{y} + (1-t)\mathbf{x}$ is on L. By the chain rule (Theorem 3),

$$g'(t) = \nabla f(t(\mathbf{y}-\mathbf{x}) + \mathbf{x})\cdot(\mathbf{y}-\mathbf{x}), \tag{18}$$

since

$$\frac{d}{dt}(t(\mathbf{y}-\mathbf{x}) + \mathbf{x}) = \mathbf{y}-\mathbf{x}$$

because $\mathbf{x}$ and $\mathbf{y}$ are constant vectors (see Examples 2.4.6 and 2.4.7 on page 108). Since g is differentiable on $[0, 1]$, the hypotheses of the mean value theorem of one variable are satisfied and there is a number $t_0 \in (0, 1)$ such that

$$g(1) - g(0) = g'(t_0)(1 - 0) = g'(t_0). \tag{19}$$

But, using (18), (19) becomes

$$f(\mathbf{y}) - f(\mathbf{x}) = \nabla f(t_0(\mathbf{y} - \mathbf{x}) + \mathbf{x}) \cdot (\mathbf{y} - \mathbf{x}). \tag{20}$$

Setting $\mathbf{x}_0 = t_0(\mathbf{y} - \mathbf{x}) + \mathbf{x}$ in (20) completes the proof. ■

PROBLEMS 3.6

SELF-QUIZ

I. If $z = f(x, y) = xy$, $x = t$ and $y = -t$, then $\frac{dz}{dt} =$ __________.

a. $-2t\mathbf{k}$ **b.** $-2t$
c. 0 **d.** $0\mathbf{i} + 0\mathbf{j}$

II. If $z = 3x + 4y - 5$, $x = r\cos\theta$ and $y = r\sin\theta$, then $\frac{\partial z}{\partial \theta} =$ __________.

a. $-3y + 4x$
b. $-3r\sin\theta + 4r\cos\theta$
c. $-3r\sin\theta + 4r\cos\theta - 5$
d. $-3\sin\theta + 4\cos\theta$

III. Suppose $w = x^2 + y^2 + z^2$ and $x = \rho\sin\varphi\cos\theta$, $y = \rho\sin\varphi\sin\theta$, $z = \rho\cos\varphi$; then $\frac{\partial w}{\partial \rho} =$ __________.

a. 0 **b.** $(\cos\theta + \sin\theta)\sin\varphi + \cos\varphi$
c. ρ **d.** 2ρ

IV. Suppose $w = x - y + 3z$, $x = \sin z$, and $y = x^2$, then $\frac{\partial w}{\partial z} =$ __________.

a. 3
b. $\cos z - 0 + 3$
c. $\cos z - 2\sin z\cos z + 3$
d. $\cos z - 2\sin z\cos z$

In Problems 1–10, use the chain rule to calculate $\frac{dz}{dt}$ or $\frac{dw}{dt}$. Check your result by writing z or w explicitly as a function of t and then differentiating this expression.

1. $z = xy$, $x = e^t$, $y = e^{2t}$
2. $z = x^2 + y^2$, $x = \cos t$, $y = \sin t$
3. $z = y/x$, $x = t^2$, $y = t^3$
4. $z = e^x \sin y$, $x = \sqrt{t}$, $y = \sqrt[3]{t}$
5. $z = \tan^{-1}(y/x)$, $x = \cos 3t$, $y = \sin 5t$
6. $z = \sinh(x - 2y)$, $x = 2t^4$, $y = t^2 + 1$
7. $w = x^2 + y^2 + z^2$, $x = \cos t$, $y = \sin t$, $z = t$
8. $w = \ln(2x - 3y + 4z)$, $x = e^t$, $y = \ln t$, $z = \cosh t$
9. $w = xy - yz + zx$, $x = e^t$, $y = e^{2t}$, $z = e^{3t}$
10. $w = (x + y)/z$, $x = t$, $y = t^2$, $z = t^3$

In Problems 11–26, use the chain rule to calculate the indicated partial derivatives.

11. $z = xy$, $x = r + s$, $y = r - s$; $\frac{\partial z}{\partial r}$, $\frac{\partial z}{\partial s}$
12. $z = x^2 + y^2$, $x = \cos(r + s)$, $y = \sin(r - s)$; $\frac{\partial z}{\partial r}$, $\frac{\partial z}{\partial s}$
13. $z = y/x$, $x = e^r$, $y = e^s$; $\frac{\partial z}{\partial r}$, $\frac{\partial z}{\partial s}$
14. $z = \sin(y/x)$, $x = r/s$, $y = s/r$; $\frac{\partial z}{\partial r}$, $\frac{\partial z}{\partial s}$
15. $z = \frac{e^{x+y}}{e^{x-y}}$, $x = \ln rs$, $y = \ln(r/s)$; z_r, z_s
16. $z = x^2y^3$, $x = r - s^2$, $y = 2s + r$; z_r, z_s
17. $w = x + y + z$, $x = rs$, $y = r + s$, $z = r - s$; w_r, w_s
18. $w = xy/z$, $x = r$, $y = s$, $z = t$; $\frac{\partial w}{\partial r}$, $\frac{\partial w}{\partial s}$, $\frac{\partial w}{\partial t}$
19. $w = xy/z$, $x = r + s$, $y = t - r$, $z = s + 2t$; w_r, w_s, w_t
20. $w = \sin xyz$, $x = s^2r$, $y = r^2s$, $z = r - s$; w_r, w_s, w_t
21. $w = \sinh(x + 2y + 3z)$, $x = \sqrt{r + s}$, $y = \sqrt[3]{s - t}$, $z = 1/(r + t)$; w_s

22. $w = x^2y + yz^2$, $x = rst$, $y = rs/t$, $z = 1/(rst)$; w_r, w_s, w_t

23. $w = \ln(x + 2y + 3z)$, $x = r^2 + t^2$, $y = s^2 - t^2$, $z = r^2 + s^2$; $\frac{\partial w}{\partial r}, \frac{\partial w}{\partial s}, \frac{\partial w}{\partial t}$

24. $w = e^{xy/z}$, $x = r^2 + t^2$, $y = s^2 - t^2$, $z = r^2 + s^2$; $\frac{\partial w}{\partial r}, \frac{\partial w}{\partial s}, \frac{\partial w}{\partial t}$

***25.** $w = r - \cos\theta + t$, $r = \sqrt{x^2 + y^2}$, $\theta = \tan^{-1}(y/x)$, $t = z$; w_x, w_y, w_z

***26.** $u = xy + w^2 - z^3$, $x = t + r - q$, $y = q^2 + s^2 - t + r$, $z = (qr + st)/r^4$, $w = (r - s)/(t + q)$; u_q, u_r, u_s, u_t

27. The radius of a right circular cone is increasing at a rate of 7 in/min, while its height is decreasing at a rate of 20 in/min. Is the volume of the cone increasing or decreasing when $r =$ 45 in and $h =$ 100 in? How fast is the volume changing then?

28. The radius of a right circular cylinder is decreasing at a rate of 12 cm/sec, while its height is increasing at a rate of 25 cm/sec. How is the volume changing when $r =$ 180 cm and $h =$ 500 cm? Is the volume increasing or decreasing then?

29. The volume of 10 moles of an ideal gas is decreasing at a rate of 25 cm^3/min and its temperature is increasing at a rate of 1°C/min. How fast is the pressure changing when $V =$ 1000 cm^3 and $P =$ 3 N/cm^2? (Leave your answer expressed in terms of R.) Is the pressure increasing or decreasing?

30. The pressure of 8 moles of an ideal gas is decreasing at a rate of 0.4 N/cm^2/min, while the temperature is decreasing at a rate of 0.5°K/min. How fast is the volume of the gas changing when $V =$ 1000 cm^3 and $P =$ 3 N/cm^2? Is the volume increasing or decreasing? [*Hint:* Use the value $R =$ 8.315 J/mol °K.]

31. The angle A of a triangle ABC is increasing at a rate of 3°/sec, the side AB is increasing at a rate of 1 cm/sec, and the side AC is decreasing at a rate of 2 cm/sec. How fast is the side BC changing when $A =$ 30°, $AB =$ 10 cm, and $AC =$ 24 cm? Is the length of BC increasing or decreasing? [*Hint:* Use the law of cosines (after converting degrees to radians).]

32. Suppose $z = f(x, y)$ is differentiable. Write the expression $\left(\frac{\partial z}{\partial x}\right)^2 + \left(\frac{\partial z}{\partial y}\right)^2$ in terms of polar coordinates r and θ.

33. Suppose $w = g(x, y, z)$ is differentiable. Let $x = r\cos\theta$, $y = r\sin\theta$, and $z = t$ (these are cylindrical coordinates). Calculate $\frac{\partial w}{\partial r}$, $\frac{\partial w}{\partial \theta}$, and $\frac{\partial w}{\partial t}$.

34. Suppose $w = g(x, y, z)$ is differentiable. Let $x = \rho\sin\varphi\cos\theta$, $y = \rho\sin\varphi\sin\theta$, and $z = \rho\cos\varphi$ (these are spherical coordinates). Calculate $\frac{\partial w}{\partial \rho}$, $\frac{\partial w}{\partial \theta}$, and $\frac{\partial w}{\partial \varphi}$.

35. The wave equation (see Problem 3.4.27) is the partial-differential equation (PDE) $\frac{\partial^2 y}{\partial t^2} = c^2\frac{\partial^2 y}{\partial x^2}$. Show that if f is any differentiable function, then

$$y(x, t) = \frac{1}{2}[f(x - ct) + f(x + ct)]$$

is a solution to the wave equation.

36. The function $f(x, y)$ is said to be **homogeneous of degree** n if and only if $f(tx, ty) = t^n f(x, y)$ holds for all t, x, y. Show that if f is homogeneous of degree n, then

$$x\frac{\partial f}{\partial x}(x, y) + y\frac{\partial f}{\partial y}(x, y) = nf(x, y).$$

***37.** Suppose $z = F(x, y)$, $x = X(r, s)$, and $y = Y(r, s)$. Show that

$$\frac{\partial^2 z}{\partial r^2} = \frac{\partial}{\partial x}\left(\frac{\partial F}{\partial x}\frac{\partial X}{\partial r} + \frac{\partial F}{\partial y}\frac{\partial Y}{\partial r}\right)\left(\frac{\partial X}{\partial r}\right) + \frac{\partial}{\partial y}\left(\frac{\partial F}{\partial x}\frac{\partial X}{\partial r} + \frac{\partial F}{\partial y}\frac{\partial Y}{\partial r}\right)\left(\frac{\partial Y}{\partial r}\right).$$

****38.** Laplace's equation (see Problem 3.4.31) is the partial-differential equation

$$\frac{\partial^2 f}{\partial x^2} + \frac{\partial^2 f}{\partial y^2} = 0.$$

If we write (x, y) in polar coordinates ($x = r\cos\theta$, $y = r\sin\theta$), show that Laplace's equation becomes

$$\frac{\partial^2 f}{\partial r^2} + \frac{1}{r^2}\frac{\partial^2 f}{\partial \theta^2} + \frac{1}{r}\frac{\partial f}{\partial r} = 0.$$

[*Hint:* Use the result of the preceding problem to write $\partial^2 f/\partial x^2$ and $\partial^2 f/\partial y^2$ in terms of r and θ.]

***39.** Find the general solution to the partial-differential equation (PDE)

$$\frac{\partial z}{\partial y} = \frac{1}{a}\frac{\partial z}{\partial x}$$

where a is some nonzero constant. [*Hint:* Rewrite the given PDE in terms of new independent variables

$$u = y + ax \quad \text{and} \quad v = y - ax.]$$

****40.** Generalize the result of Problem 35 by finding the general smooth solution to the wave equation

$$\frac{\partial^2 y}{\partial t^2} = c^2\frac{\partial^2 y}{\partial x^2} \quad \text{and} \quad c \neq 0.$$

[*Hint:* Rewrite the given PDE in terms of new independent variables

$$u = x + ct \quad \text{and} \quad v = x - ct$$

and assume that mixed partials are equal.]

41. Show that if f: $\mathbb{R}^n \to \mathbb{R}$, then $\nabla f = \mathbf{0}$ at every point in an open set if and only if f is constant on that region.

42. Show that if f and g: $\mathbb{R}^n \to \mathbb{R}$ and $\nabla f = \nabla g$ at every point in that set, then there is a number c such that $f(\mathbf{x}) = g(\mathbf{x}) + c$.

43. Find the most general function f: $\mathbb{R}^n \to \mathbb{R}$ such that $\nabla f(\mathbf{x}) = \mathbf{x}$ for every $\mathbf{x} \in \mathbb{R}^n$.

ANSWERS TO SELF-QUIZ

I. b

II. b = a

III. d $(2\rho\sin^2\varphi(\cos^2\theta + \sin^2\theta) + 2\rho\cos^2\varphi = 2\rho\sin^2\varphi + 2\rho\cos^2\varphi = 2\rho)$

IV. c $\left(\dfrac{dy}{dz} = \dfrac{dy}{dx}\dfrac{dx}{dz}\right)$

3.7 TANGENT PLANES, NORMAL LINES, AND GRADIENTS

Let $z = f(x, y)$ be a continuous function of two variables. As we have seen, the graph of f is a surface in $\mathbb{R}^3$. More generally, the graph of the equation $F(x, y, z) = 0$ is a surface in $\mathbb{R}^3$ if F is continuous. The surface $F(x, y, z) = 0$ is called **continuously differentiable** or **smooth** at a point (x_0, y_0, z_0) if $\partial F/\partial x$, $\partial F/\partial y$, and $\partial F/\partial z$ all exist and are continuous at (x_0, y_0, z_0). In $\mathbb{R}^2$ a differentiable curve has a unique tangent line at each point. A differentiable surface in $\mathbb{R}^3$ has a unique tangent plane at each point at which $\partial F/\partial x$, $\partial F/\partial y$, and $\partial F/\partial z$ are not all zero. We will formally define what we mean by a tangent plane to a surface after a bit, although it should be easy enough to visualize (see Figure 1). We note here that not every surface has a tangent plane at every point. For example, the cone $z = \sqrt{x^2 + y^2}$ has no tangent plane at the origin (see Figure 2).

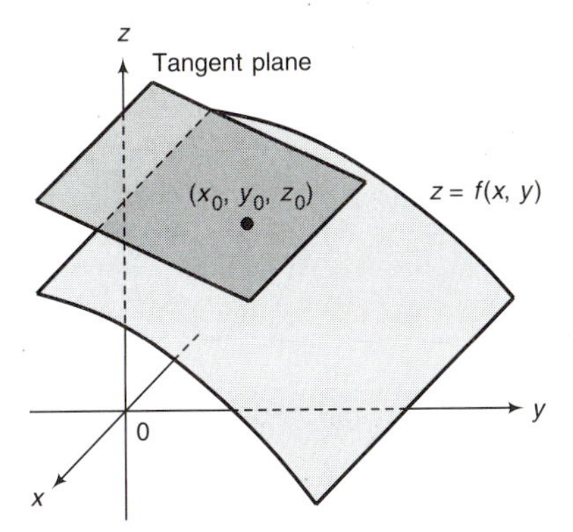

FIGURE 1
A tangent plane to a surface

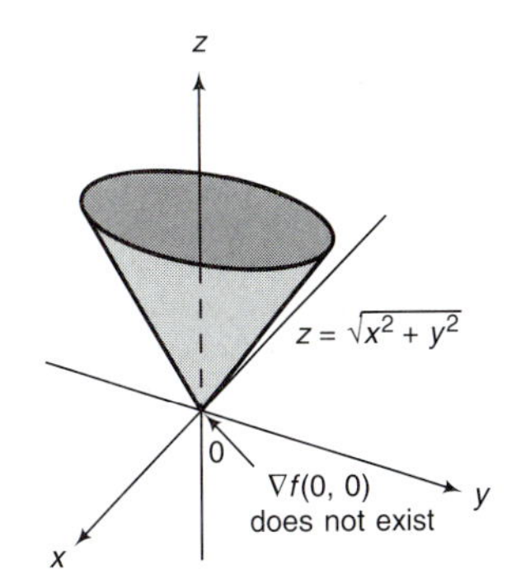

FIGURE 2
There is no tangent plane at the origin.

Assume that the surface S given by $F(x, y, z) = 0$ is smooth. Let C be any curve lying on S. That is, C can be given parametrically by $\mathbf{g}(t) = x(t)\mathbf{i} + y(t)\mathbf{j} + z(t)\mathbf{k}$. (Recall from Section 2.1 (page 88) the definition of a curve in $\mathbb{R}^3$.) Then for points on the curve, $F(x, y, z)$ can be written as a function of t, and from the vector form of the chain rule [equation (3.6.3)] we have

$$F'(t) = \nabla F \cdot \mathbf{g}'(t). \tag{1}$$

But since $F(x(t), y(t), z(t)) = 0$ for all t [since $(x(t), y(t), z(t))$ is on S], we see that $F'(t) = 0$ for all t. But $\mathbf{g}'(t)$ is tangent to the curve C for every number t. Thus (1) implies the following:

> The gradient of F at a point $\mathbf{x}_0 = (x_0, y_0, z_0)$ on S is orthogonal to the tangent vector at $\mathbf{x}_0$ to any curve C remaining on S and passing through $\mathbf{x}_0$.

This statement is illustrated in Figure 3.

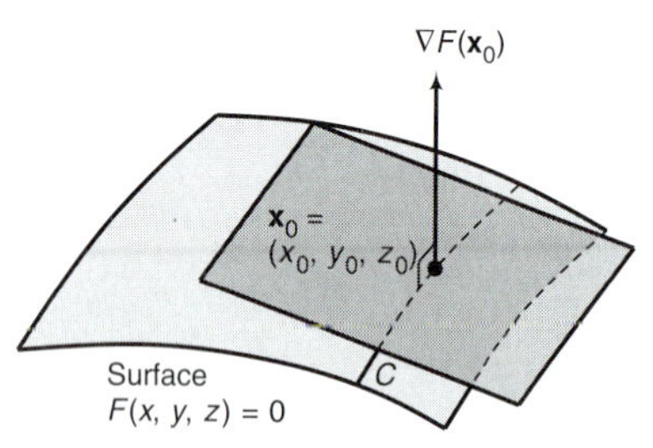

FIGURE 3
The gradient at x_0 is orthogonal to any vector that is tangent to the surface at x_0.

Thus if we think of all the vectors tangent to a surface at a point $\mathbf{x}_0$ as constituting a plane, then $\nabla F(\mathbf{x}_0)$ is a *normal* vector to that plane. This motivates the following definition.

DEFINITION TANGENT PLANE AND NORMAL LINE

Let F be continuously differentiable at $\mathbf{x}_0 = (x_0, y_0, z_0)$ and let the smooth surface S be defined by $F(x, y, z) = 0$.

i. The **tangent plane** to S at (x_0, y_0, z_0) is the plane passing through the point (x_0, y_0, z_0) with normal vector $\nabla F(\mathbf{x}_0)$.

ii. The **normal line** to S at $\mathbf{x}_0$ is the line passing through $\mathbf{x}_0$ having the same direction as $\nabla F(\mathbf{x}_0)$. ■

EXAMPLE 1 FINDING EQUATIONS OF THE TANGENT PLANE AND NORMAL LINE AT A POINT

Find the equation of the tangent plane and symmetric equations of the normal line to the ellipsoid $x^2 + (y^2/4) + (z^2/9) = 3$ at the point $(1, 2, 3)$.

SOLUTION: Since $F(x, y, z) = x^2 + (y^2/4) + (z^2/9) - 3 = 0$, we have

$$\nabla F = \frac{\partial F}{\partial x}\mathbf{i} + \frac{\partial F}{\partial y}\mathbf{j} + \frac{\partial F}{\partial z}\mathbf{k} = 2x\mathbf{i} + \frac{y}{2}\mathbf{j} + \frac{2z}{9}\mathbf{k}.$$

Then $\nabla F(1, 2, 3) = 2\mathbf{i} + \mathbf{j} + \frac{2}{3}\mathbf{k}$, and the equation of the tangent plane is

$$2(x-1) + (y-2) + \frac{2}{3}(z-3) = 0 \text{ or } 2x + y + \frac{2}{3}z = 6.$$

The normal line is given by

$$\frac{x-1}{2} = y - 2 = \frac{3}{2}(z-3).$$

The situation is even simpler if we can write the surface in the form $z = f(x, y)$. That is, the surface is the graph of a function of two variables. Then $F(x, y, z) = z - f(x, y) = 0$, so that

$$F_x = -f_x, \qquad F_y = -f_y, \qquad \text{and} \qquad F_z = 1,$$

and the normal vector $\mathbf{N}$ to the tangent plane is

NORMAL VECTOR TO THE SURFACE $z = f(x, y)$

$$\mathbf{N} = -f_x(x_0, y_0)\mathbf{i} - f_y(x_0, y_0)\mathbf{j} + \mathbf{k}. \tag{2}$$

REMARK: One interesting consequence of this fact is that if $z = f(x, y)$ and if $\nabla f(x_0, y_0) = \mathbf{0}$, then *the tangent plane to the surface at $(x_0, y_0, f(x_0, y_0))$ is parallel to the xy-plane (that is, it is horizontal)*. This occurs because at $(x_0, y_0, f(x_0, y_0))$, $\mathbf{N} = -(\partial f/\partial x)\mathbf{i} - (\partial f/\partial y)\mathbf{j} + \mathbf{k} = -\nabla f + \mathbf{k} = \mathbf{k}$. Thus the z-axis is normal to the tangent plane.

EXAMPLE 2 FINDING EQUATIONS OF THE TANGENT PLANE AND NORMAL LINE AT A POINT

Find the tangent plane and normal line to the surface $z = x^3y^5$ at the point $(2, 1, 8)$.

SOLUTION: $\mathbf{N} = -(\partial z/\partial x)\mathbf{i} - (\partial z/\partial y)\mathbf{j} + \mathbf{k} = -3x^2y^5\mathbf{i} - 5x^3y^4\mathbf{j} + \mathbf{k} = -12\mathbf{i} - 40\mathbf{j} + \mathbf{k}$ at $(2, 1, 8)$. Then the tangent plane is given by

$$-12(x-2) - 40(y-1) + (z-8) = 0,$$

or

$$12x + 40y - z = 56.$$

Symmetric equations of the normal line are

$$\frac{x-2}{12} = \frac{y-1}{40} = \frac{z-8}{-1}.$$

PROBLEMS 3.7

SELF-QUIZ

I. The plane tangent to the surface $x^2 + y^2 + z^2 = 1$ at the point $(1, 0, 0)$ satisfies the equation ______.

a. $x = 1$ **b.** $y + z = 0$
c. $2x + 2y + 2z = 2$ **d.** $2x = 2y + 2z$

II. The line normal to the surface $x^2 + y^2 + z^2 = 1$ at the point $(1, 0, 0)$ satisfies the equation ______.

a. $(x - 1)/2 = (y - 0)/2 = (z - 0)/2$
b. $x\mathbf{i} + y\mathbf{j} + z\mathbf{k} = \mathbf{i} + t(2\mathbf{i})$
c. $y = 0, z = 0$
d. $x = 1, y = z$

III. The plane tangent to the surface $x^2 + y^2 + z^2 = 2$ at the point $(0, 1, 1)$ satisfies the equation ______.

a. $y + z = 0$ **b.** $2y + 2z = 4$
c. $x = 0, y = z$ **d.** $x + y + z = 2$

IV. The plane tangent to the surface $z = 2x + \cos y$ at the point $(1, 0, 3)$ satisfies the equation ______.

a. $2(x - 1) + (-1)(z - 3) = 0$
b. $x\mathbf{i} + y\mathbf{j} + z\mathbf{k} = (\mathbf{i} + 3\mathbf{k}) + t(2\mathbf{i} - \mathbf{k})$
c. $[(x - 1)\mathbf{i} + (y - 0)\mathbf{j} + (z - 3)\mathbf{k}] \cdot (2\mathbf{i} - \mathbf{k}) = 0$
d. $(x - 1)/2 = (z - 3)/(-1), y = 0$
e. $[x\mathbf{i} + y\mathbf{j} + z\mathbf{k}] \cdot (\mathbf{i} + 3\mathbf{k}) = 0$
f. $x\mathbf{i} + y\mathbf{j} + z\mathbf{k} = (2\mathbf{i} - \mathbf{k}) \cdot (\mathbf{i} + 3\mathbf{k})$

V. The line normal to the surface $2x + \cos y = z$ at the point $(1, 0, 3)$ satisfies the equation ______.

a. $2(x - 1) + (-1)(z - 3) = 0$
b. $x\mathbf{i} + y\mathbf{j} + z\mathbf{k} = (\mathbf{i} + 3\mathbf{k}) + t(2\mathbf{i} - \mathbf{k})$
c. $[(x - 1)\mathbf{i} + (y - 0)\mathbf{j} + (z - 3)\mathbf{k}] \cdot (2\mathbf{i} - \mathbf{k}) = 0$
d. $(x - 1)/2 = (z - 3)/(-1), y = 0$
e. $[x\mathbf{i} + y\mathbf{j} + z\mathbf{k}] \cdot (\mathbf{i} + 3\mathbf{k}) = 0$
f. $x\mathbf{i} + y\mathbf{j} + z\mathbf{k} = (2\mathbf{i} - \mathbf{k}) \cdot (\mathbf{i} + 3\mathbf{k})$

In Problems 1–14, a surface is given and a point is specified. Find an equation for the plane tangent to the surface at that point and symmetric (or vector) equations for the normal line.

1. $x^2 + y^2 + z^2 = 1$; $(0, 1, 0)$
2. $x^2 - y^2 + z^2 = 1$; $(1, 1, 1)$
3. $(x/a)^2 + (y/b)^2 + (z/c)^2 = 3$; (a, b, c)
4. $(x/a)^2 + (y/b)^2 + (z/c)^2 = 3$; $(-a, b, -c)$
5. $\sqrt{x} + \sqrt{y} + \sqrt{z} = 6$; $(4, 1, 9)$
6. $ax + by + cz = d$; $(1/a, 1/b, (d - 2)/c)$
7. $xyz = 4$; $(1, 2, 2)$
8. $xy^2 - yz^2 + zx^2 = 1$; $(1, 1, 1)$
9. $4x^2 - y^2 - 5z^2 = 15$; $(3, 1, -2)$
10. $xe^y - ye^z = 1$; $(1, 0, 0)$
11. $\sin xy - 2\cos yz = 0$; $(\pi/2, 1, \pi/3)$
12. $x^2 + y^2 + 4x + 2y + 8z = 7$; $(2, -3, -1)$
13. $e^{xyz} = 5$; $(1, 1, \ln 5)$
14. $\sqrt{\dfrac{x + y}{z - 1}} = 1$; $(1, 1, 3)$

In Problems 15–22, write an equation for the plane tangent to $z = f(\mathbf{x})$ at the point $\mathbf{x}_0$ in the form (discussed in Problem 40 below)

$$z = f(\mathbf{x}_0) + (\mathbf{x} - \mathbf{x}_0) \cdot \nabla f(\mathbf{x}_0)$$

and write a vector equation satisfied by the normal line at that point.

15. $z = xy^2$; $(1, 1, 1)$
16. $z = \ln(x - 2y)$; $(3, 1, 0)$
17. $z = \sin(2x + 5y)$; $(\pi/8, \pi/20, 1)$
18. $z = \sqrt{\dfrac{x + y}{x - y}}$; $(5, 4, 3)$
19. $z = \tan^{-1}(y/x)$; $(-2, 2, -\pi/4)$
20. $z = \sinh xy^2$; $(0, 3, 0)$
21. $z = \sec(x - y)$; $(\pi/2, \pi/6, 2)$
22. $z = e^x \cos y + e^y \cos x$; $(\pi/2, 0, e^{\pi/2})$

In Problems 23–28, use the results of Problem 37 and 38 to find a normal vector and the equation of the line tangent to the curve at the specified point.

23. $xy = 5$; $(1, 5)$
24. $x^2 + xy + y^2 + 3x - 5y = 16$; $(1, -2)$
25. $(x + y)/(x - y) = 7$; $(4, 3)$
26. $xe^{xy} = 1$; $(1, 0)$
27. $(x/2)^2 + (y/4)^2 = 1$; $(\sqrt{2}, 2\sqrt{2})$
28. $\tan(x + y) = 1$; $(\pi/4, 0)$
***29.** Find the two points of intersection of the surface $z = x^2 + y^2$ and the line $(x - 3)/1 = (y + 1)/(-1) = (z + 2)/(-2)$. At each of these intersection points, calculate the cosine of the

angle between the normal line there and the given line.

30. Show that the plane tangent to the surface $z = ax^2 + by^2$ at the point (x_0, y_0, z_0) satisfies the equation

$$\frac{z + z_0}{2} = ax_0x + by_0y.$$

31. Show that the plane tangent to the surface $ax^2 + by^2 + cz^2 = d$ at the point (x_0, y_0, z_0) satisfies the equation $ax_0x + by_0y + cz_0z = d$.

*32. Show that every line normal to the surface of a sphere passes through the center of the sphere.

*33. Show that every line normal to the cone $z^2 = a(x^2 + y^2)$ intersects the z-axis.

*34. Suppose f is a differentiable function of one variable and let $z = yf(y/x)$. Show that all planes tangent to the surface described by this equation have a point in common.

35. The **angle between two surfaces** at a point of intersection is defined to be the acute angle between their normal lines. Show that if two surfaces $F(x, y, z) = 0$ and $G(x, y, z) = 0$ intersect at right angles at a point $\mathbf{x}_0$, then $\nabla F(\mathbf{x}_0) \cdot \nabla G(\mathbf{x}_0) = 0$.

36. Show that the sum of the squares of the x-, y-, and z-intercepts of any plane tangent to the surface $x^{2/3} + y^{2/3} + z^{2/3} = a^{2/3}$ is a constant.

37. Suppose the equation $F(x, y) = 0$ defines a curve in $\mathbb{R}^2$. Show that if F is differentiable, then $\nabla F(x, y)$ is normal to the curve at each point of the curve.

38. Use the result of the preceding problem to find an equation for the line tangent to the curve $F(x, y) = 0$ at a point (x_0, y_0).

39. Show that at any point (x, y), $y \neq 0$, the curve $x/y = a$ is orthogonal to the curve $x^2 + y^2 = r^2$ for any constants a and r.

40. Show that the plane tangent at $\mathbf{x}_0 = (x_0, y_0)$ to the smooth surface $z = F(x, y)$ satisfies the equation

$$z = F(\mathbf{x}_0) + (\mathbf{x} - \mathbf{x}_0) \cdot \nabla F(\mathbf{x}_0).$$

[*Note:* If we focus on a function of a single variable, we can describe the line tangent to the curve $y = f(x)$ at $x = x_0$ by the equation $y = f(x_0) + (x - x_0)f'(x_0)$. This problem asks you to show that the gradient is the "right" generalization of the derivative of a function of one variable.]

ANSWERS TO SELF-QUIZ

I. a **II.** b = c **III.** b **IV.** a = c **V.** b = d

3.8 DIRECTIONAL DERIVATIVES AND THE GRADIENT

Let us take another look at the partial derivatives $\partial f/\partial x$ and $\partial f/\partial y$ of the function $z = f(x, y)$. We have

$$\frac{\partial f}{\partial x}(x_0, y_0) = \lim_{\Delta x \to 0} \frac{f(x_0 + \Delta x, y_0) - f(x_0, y_0)}{\Delta x}. \quad (1)$$

This measures the rate of change of f as we approach the point (x_0, y_0) along a vector parallel to the x-axis [since $(x_0 + \Delta x, y_0) - (x_0, y_0) = (\Delta x, 0) = \Delta x\mathbf{i}$]. Similarly

$$\frac{\partial f}{\partial y}(x_0, y_0) = \lim_{\Delta x \to 0} \frac{f(x_0, y_0 + \Delta y) - f(x_0, y_0)}{\Delta y}. \quad (2)$$

measures the rate of change of f as we approach the point (x_0, y_0) along a vector parallel to the y-axis.

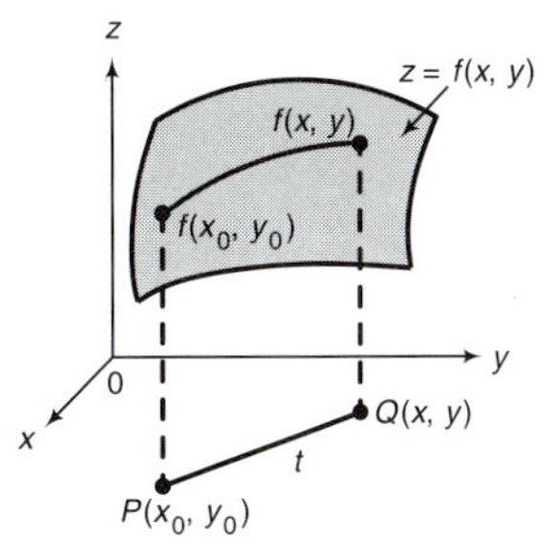

FIGURE 1
(x, y) moves along the vector $\overrightarrow{PQ}$

It is frequently of interest to compute the rate of change of f as we approach (x_0, y_0) along a vector that is not parallel to one of the coordinate axes. The situation is depicted in Figure 1. Suppose that (x, y) approaches the fixed point (x_0, y_0) along the line segment joining them, and let t denote the distance between the two points. We want to determine the relative rate of change in f with respect to a change in t.

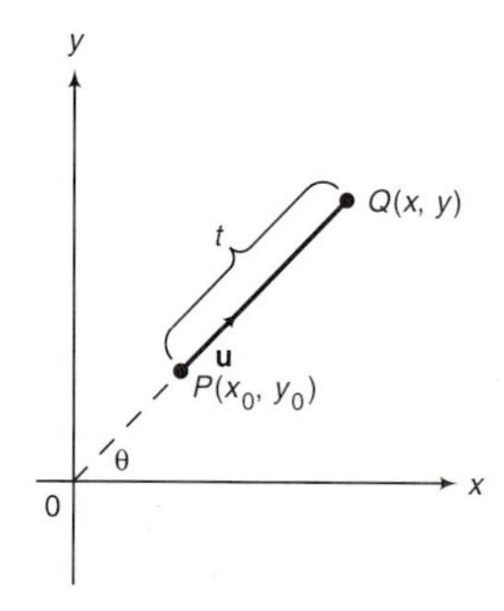

FIGURE 2
u and $\overrightarrow{PQ}$ are parallel

There is a number t such that $\overrightarrow{PQ} = t\mathbf{u}$. Since **u** *is a unit vector, $t = |\overrightarrow{PQ}|$.*

Let **u** denote a unit vector with the initial point at (x_0, y_0) and parallel to $\overrightarrow{PQ}$ (see Figure 2). Since **u** and $\overrightarrow{PQ}$ are parallel, there is, by Theorem 1.2.3, a value of t such that

$$\overrightarrow{PQ} = t\mathbf{u}. \tag{3}$$

Note that $t > 0$ if **u** and $\overrightarrow{PQ}$ have the same direction and $t < 0$ if **u** and $\overrightarrow{PQ}$ have opposite directions. Now

$$\overrightarrow{PQ} = (x - x_0)\mathbf{i} + (y - y_0)\mathbf{j}, \tag{4}$$

and since **u** is a unit vector, we have

$$\mathbf{u} = \cos\theta\mathbf{i} + \sin\theta\mathbf{j}, \tag{5}$$

where θ is the direction of **u**. Thus, inserting (4) and (5) into (3), we have

$$(x - x_0)\mathbf{i} + (y - y_0)\mathbf{j} = t\cos\theta\mathbf{i} + t\sin\theta\mathbf{j},$$

or

$$x = x_0 + t\cos\theta$$
$$y = y_0 + t\sin\theta. \tag{6}$$

The equations (6) are the parametric equations of the line passing through P and Q. Using (6), we have

$$z = f(x, y) = f(x_0 + t\cos\theta, y_0 + t\sin\theta). \tag{7}$$

Remember that θ is fixed—it is the direction of approach. Thus $(x, y) \to (x_0, y_0)$ along $\overrightarrow{PQ}$ is equivalent to $t \to 0$ in (7). Hence, to compute the instantaneous rate of change of f as $(x, y) \to (x_0, y_0)$ along the vector $\overrightarrow{PQ}$, we need to compute dz/dt. But by the chain rule,

$$\frac{dz}{dt} = \frac{\partial f}{\partial x}(x, y)\frac{dx}{dt} + \frac{\partial f}{\partial y}(x, y)\frac{dy}{dt}, \text{ or } \frac{dz}{dt} = f_x(x, y)\cos\theta + f_y(x, y)\sin\theta, \text{ and} \tag{8}$$

$$\frac{dz}{dt} = [f_x(x_0 + t\cos\theta, y_0 + t\sin\theta)]\cos\theta + [f_y(x_0 + t\cos\theta, y_0 + t\sin\theta]\sin\theta. \tag{9}$$

If we set $t = 0$ in (9), we obtain the instantaneous rate of change of f in the direction $\overrightarrow{PQ}$ at the point (x_0, y_0). That is,

$$\left.\frac{dz}{dt}\right|_{t=0} = f_x(x_0, y_0)\cos\theta + f_y(x_0, y_0)\sin\theta. \tag{10}$$

But (10) can be written [using (5)] as

$$\left.\frac{dz}{dt}\right|_{t=0} = \nabla f(x_0, y_0) \cdot \mathbf{u}. \tag{11}$$

This leads to the following definition.

DEFINITION DIRECTIONAL DERIVATIVE

Let f be differentiable at a point $\mathbf{x}_0 = (x_0, y_0)$ in $\mathbb{R}^2$ and let **u** be a unit vector. Then the **directional derivative of f in the direction u**, denoted by $f'_{\mathbf{u}}(\mathbf{x}_0)$, is given by

$$f'_{\mathbf{u}}(\mathbf{x}_0) = \nabla f(\mathbf{x}_0) \cdot \mathbf{u}. \tag{12}$$

■

REMARK 1: Note that if $\mathbf{u} = \mathbf{i}$, then $\nabla f \cdot \mathbf{u} = \partial f/\partial x$ and (12) reduces to the partial derivative $\partial f/\partial x$. Similarly, if $\mathbf{u} = \mathbf{j}$, then (12) reduces to $\partial f/\partial y$.

REMARK 2: This definition makes sense if f is a function of three variables. Then, of course, $\mathbf{u}$ is a unit vector in $\mathbb{R}^3$.

REMARK 3: There is another definition of the directional derivative. It is given by

$$f'_{\mathbf{u}}(\mathbf{x}_0) = \lim_{h\to 0} \frac{f(\mathbf{x}_0 + h\mathbf{u}) - f(x_0)}{h}. \tag{13}$$

It can be shown that if f is differentiable, then the limit in (13) exists, and is equal to $\nabla f(\mathbf{x}_0) \cdot \mathbf{u}$.

EXAMPLE 1

COMPUTING A DIRECTIONAL DERIVATIVE IN $\mathbb{R}^2$

Let $z = f(x, y) = xy^2$. Calculate the directional derivative of f in the direction of the vector $\mathbf{v} = 2\mathbf{i} + 3\mathbf{j}$ at the point $(4, -1)$.

SOLUTION: A unit vector in the direction $\mathbf{v}$ is $\mathbf{u} = (2/\sqrt{13})\mathbf{i} + (3/\sqrt{13})\mathbf{j}$. Also, $\nabla f = y^2\mathbf{i} + 2xy\mathbf{j}$. Thus,

$$f'_{\mathbf{u}}(x, y) = \nabla f(\mathbf{x}) \cdot \mathbf{u} = \frac{2y^2}{\sqrt{13}} + \frac{6xy}{\sqrt{13}} = \frac{2y^2 + 6xy}{\sqrt{13}}.$$

At $(4, -1)$, $f'_{\mathbf{u}}(4, -1) = -22/\sqrt{13}$.

EXAMPLE 2

COMPUTING A DIRECTIONAL DERIVATIVE IN $\mathbb{R}^3$

Let $f(x, y, z) = x \ln y - e^{xz^3}$. Calculate the directional derivative of f in the direction of the vector $\mathbf{v} = \mathbf{i} - \mathbf{j} + 3\mathbf{k}$. Evaluate this derivative at the point $(-5, 1, -2)$.

SOLUTION: A unit vector in the direction of $\mathbf{v}$ is $\mathbf{u} = (1/\sqrt{11})\mathbf{i} - (1/\sqrt{11})\mathbf{i} + (3/\sqrt{11})\mathbf{k}$, and

$$\nabla f = (\ln y - z^3 e^{xz^3})\mathbf{i} + \frac{x}{y}\mathbf{j} - 3xz^2 e^{xz^3}\mathbf{k}.$$

Thus,

$$f'_{\mathbf{u}}(\mathbf{x}) = \nabla f(\mathbf{x}) \cdot \mathbf{u} = \frac{\ln y - z^3 e^{xz^3} - (x/y) - 9xz^2 e^{xz^3}}{\sqrt{11}}, \qquad \text{and}$$

$$f'_{\mathbf{u}}(-5, 1, -2) = \frac{5 + 188e^{40}}{\sqrt{11}}.$$

There is an interesting geometric interpretation of the directional derivative. The projection of ∇f on $\mathbf{u}$ is given by (see page 18)

$$\text{Proj}_{\mathbf{u}} \nabla f = \frac{\nabla f \cdot \mathbf{u}}{|\mathbf{u}|^2}\mathbf{u},$$

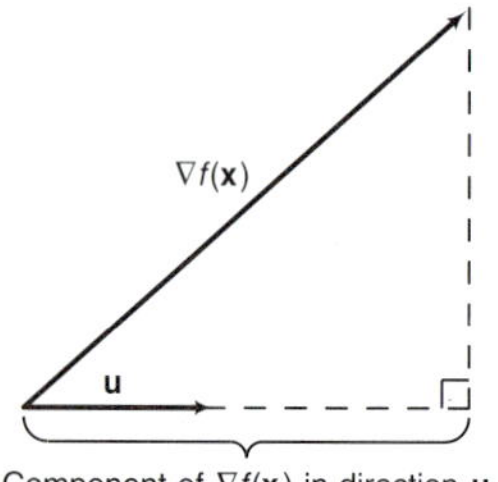

FIGURE 3
The directional derivative of f in the direction **u** is the component of ∇f in the direction **u**.

and since **u** is a unit vector, the component of ∇f in the direction **u** is given by

$$\frac{\nabla f \cdot \mathbf{u}}{|\mathbf{u}|^2} = \nabla f \cdot \mathbf{u}.$$

Thus the *directional derivative of f in the direction* **u** *is the component of the gradient of f in the direction* **u**. This is illustrated in Figure 3.

We now derive another remarkable property of the gradient. Recall that $\mathbf{u} \cdot \mathbf{v} = |\mathbf{u}||\mathbf{v}| \cos\theta$, where θ is the smallest angle between the vectors **u** and **v**. Thus, the directional derivative of f in the direction **u** can be written as

$$f'_{\mathbf{u}}(\mathbf{x}) = \nabla f(\mathbf{x}) \cdot \mathbf{u} = |\nabla f(\mathbf{x})||\mathbf{u}| \cos\theta, \tag{14}$$

or since **u** is a unit vector,

$$f'_{\mathbf{u}}(\mathbf{x}) = |\nabla f(\mathbf{x})| \cos\theta.$$

Now $\cos\theta = 1$ when $\theta = 0$, which occurs when **u** has the direction of ∇f. Similarly, $\cos\theta = -1$ when $\theta = \pi$, which occurs when **u** has the direction of $-\nabla f$. Also, $\cos\theta = 0$ when $\theta = \pi/2$. Thus, since $-1 \le \cos\theta \le 1$, equation (14) implies the following important result.

THEOREM 1 DIRECTION OF MAXIMUM RATE OF INCREASE

Let f be differentiable. Then f increases most rapidly in the direction of its gradient and decreases most rapidly in the direction opposite to that of its gradient. It does not change in a direction perpendicular to its gradient. ■

EXAMPLE 3 FINDING THE PATH OF STEEPEST ASCENT ON A SPHERE

Consider the sphere $x^2 + y^2 + z^2 = 1$. We can write the upper half of this sphere (i.e., the upper hemisphere) as $z = f(x, y) = \sqrt{1 - x^2 - y^2}$. Then

$$\nabla f = \frac{-x}{\sqrt{1 - x^2 - y^2}}\mathbf{i} + \frac{-y}{\sqrt{1 - x^2 - y^2}}\mathbf{j}.$$

Since $\sqrt{1 - x^2 - y^2} > 0$, we see that

$$\text{direction of } \nabla f = \text{direction of } -x\mathbf{i} - y\mathbf{j}.$$

But $-x\mathbf{i} - y\mathbf{j}$ points from (x, y) to $(0, 0)$. Thus, if we start at a point (x, y, z) on the sphere, the path of steepest ascent (increase) is a great circle passing through the point $(0, 0, 1)$ (called the *north pole* of the sphere).

EXAMPLE 4 FINDING THE DIRECTIONS OF MAXIMUM VOLTAGE INCREASE AND DECREASE

The distribution of voltage on a metal plate is given by $V = 50 - x^2 - 4y^2$.

a. At the point $(1, -2)$, in what direction does the voltage increase most rapidly?
b. In what direction does it decrease most rapidly?
c. What is the magnitude of this increase or decrease?
d. In what direction does it change least?

SOLUTION: $\nabla V = V_x\mathbf{i} + V_y\mathbf{j} = -2x\mathbf{i} - 8y\mathbf{j}$. At $(1, -2)$, $\nabla V = -2\mathbf{i} + 16\mathbf{j}$.

a. The voltage increases most rapidly as we move in the direction of $-2\mathbf{i} + 16\mathbf{j}$.

b. It decreases most rapidly in the direction of $2\mathbf{i} - 16\mathbf{j}$.

c. The magnitude of the increase or decrease is $\sqrt{2^2 + 16^2} = \sqrt{260}$.

d. A unit vector perpendicular to ∇V is $(16\mathbf{i} + 2\mathbf{j})/\sqrt{260}$. The voltage does not change in this or the opposite direction.

EXAMPLE 5

FINDING THE PATH OF A PARTICLE THAT MOVES IN THE DIRECTION OF GREATEST VOLTAGE INCREASE

In Example 4, describe the path of a particle that starts at the point $(1, -2)$ and moves in the direction of greatest voltage increase.

SOLUTION: The path of the particle will be that of the gradient. If the particle follows the path $\mathbf{f}(t) = x(t)\mathbf{i} + y(t)\mathbf{j}$, then since the direction of the path is $\mathbf{f}'(t) = x'(t)\mathbf{i} + y'(t)\mathbf{j}$ and since this direction is also given by $\nabla V = -2x\mathbf{i} - 8y\mathbf{j}$, we must have

$$x'(t) = -2x(t) \qquad \text{and} \qquad y'(t) = -8y(t).$$

In your first calculus course you saw that the function $x(t) = ce^{\alpha t}$ satisfies the differential equation $dx/dt = \alpha x(t)$.† You can verify this by differentiation:

$$\frac{d}{dt}ce^{\alpha t} = \alpha ce^{\alpha t} = \alpha x(t) \qquad \text{if} \qquad x(t) = ce^{\alpha t}.$$

Inserting $\alpha = -2$ and $\alpha = -8$ into the two equations above, we obtain

$$x(t) = c_1e^{-2t} \qquad \text{and} \qquad y(t) = c_2e^{-8t},$$

where c_1 and c_2 are arbitrary constants. But $x(0) = 1$ and $y(0) = -2$, so

$$x(t) = e^{-2t} \qquad \text{and} \qquad y(t) = -2e^{-8t}.$$

Then since $e^{-8t} = (e^{-2t})^4$, we see that the particle moves along the path

$$y = -2x^4.$$

REMARK: Technically, a direction is a unit vector, so we should choose the direction $(-2x\mathbf{i} - 8y\mathbf{j})/\sqrt{4x^2 + 64y^2}$ in our computations. But this choice would not change the final answer. A method for obtaining the answer by using unit vectors is suggested in Problem 29.

One other fact about directional derivatives and gradients should be mentioned here. Since the gradient vector is normal to the curve $f(x, y) = C$, for any constant C, we say that the directional derivative of f in the direction of the gradient is the **normal derivative** of f and is denoted by df/dn. We then have, from equation (14),

$$\text{normal derivative} = \frac{df}{dn} = |\nabla f|. \tag{15}$$

† See *Calculus* or *Calculus of One Variable*, page 440.

EXAMPLE 6

CALCULATING A NORMAL DERIVATIVE

Let $f(x, y) = xy^2$. Calculate the normal derivative. Evaluate df/dn at the point $(3, -2)$.

SOLUTION: $\nabla f = y^2\mathbf{i} + 2xy\mathbf{j}$. Then $df/dn = |\nabla f| = \sqrt{y^4 + 4x^2y^2}$. At $(3, -2)$, $df/dn = \sqrt{16 + 144} = \sqrt{160}$.

We now define the directional derivative in $\mathbb{R}^n$.

DEFINITION DIRECTIONAL DERIVATIVE IN $\mathbb{R}^n$

Let $f: \mathbb{R}^n \to \mathbb{R}$ be differentiable at a point $\mathbf{x} = (x_1, x_2, \ldots, x_n)$ in $\mathbb{R}^n$ and let $\mathbf{u}$ be a unit vector in $\mathbb{R}^n$. Then the **directional derivative of f in the direction $\mathbf{u}$**, denoted by $f'_{\mathbf{u}}(\mathbf{x})$, is given by

$$f'_{\mathbf{u}}(\mathbf{x}_0) = \nabla f(\mathbf{x}_0) \cdot \mathbf{u}. \tag{16}$$

■

REMARK: Another, perhaps more common, definition of the directional derivative is

$$f'_{\mathbf{u}}(\mathbf{x}_0) = \lim_{h \to 0} \frac{f(\mathbf{x}_0 + h\mathbf{u}) - f(\mathbf{x}_0)}{h}. \tag{17}$$

Equation (17) illustrates the relationship between the rate of change of f and the direction in which $\mathbf{x} = \mathbf{x}_0 + h\mathbf{u}$ is approaching $\mathbf{x}_0$. For example, if $\mathbf{u} = (0, 0, \ldots, 1, 0, \ldots, 0)$ with a 1 in the ith position, then

$$f(\mathbf{x}_0 + h\mathbf{u}) - f(\mathbf{x}_0) = f(x_1, x_2, \ldots, x_i + h, \ldots, x_n) - f(x_1, x_2, \ldots, x_i, \ldots, x_n)$$

and the limit in (17) defines the partial derivative $\partial f/\partial x_i$. We now show that (17) implies (16).

THEOREM 2

Suppose that the directional derivative $f'_{\mathbf{u}}(\mathbf{x})$ is defined by (17). If f is differentiable at $\mathbf{x}$, then f has a directional derivative in every direction and

$$f'_{\mathbf{u}}(\mathbf{x}) = \nabla f(x) \cdot \mathbf{u}.$$

PROOF: Since f is differentiable at $\mathbf{x}$, we have

$$f(\mathbf{x} + h\mathbf{u}) - f(\mathbf{x}) = \nabla f(\mathbf{x}) \cdot h\mathbf{u} + g(h\mathbf{u}). \tag{18}$$

Now, from the definition of differentiability,

$$\lim_{|h\mathbf{u}| \to 0} \frac{g(h\mathbf{u})}{|h\mathbf{u}|} = 0.$$

But since $\mathbf{u}$ is a unit vector, $|h\mathbf{u}| = |h||\mathbf{u}| = |h|$. Thus we find that

$$\lim_{h \to 0} \frac{g(h\mathbf{u})}{h} = 0.$$

Then, dividing both sides of (18) by h and taking limits, we obtain

$$\lim_{h\to 0} \frac{f(\mathbf{x}+h\mathbf{u})-f(\mathbf{x})}{h} = \lim_{h\to 0} \frac{\nabla f(\mathbf{x})\cdot h\mathbf{u}}{h} + \lim_{h\to 0} \frac{g(h\mathbf{u})}{h}$$

$$= \lim_{h\to 0} \frac{h(\nabla f(\mathbf{x})\cdot \mathbf{u})}{h} + \lim_{h\to 0} \frac{g(h\mathbf{u})}{h} = \nabla f(\mathbf{x})\cdot \mathbf{u}$$

and the proof is complete. ■

EXAMPLE 7

A DIRECTIONAL DERIVATIVE IN $\mathbb{R}^4$

Let $f(\mathbf{x}) = x_1^2 + x_2x_3^4 - x_4^2$. Compute the directional derivative of f in the direction of $(1, 2, 3, 4)$ at the point $(-2, 3, 1, 5)$.

SOLUTION:

$$\nabla f = (2x_1, x_3^4, 4x_2x_3^3, -2x_4)$$
$$= (-4, 1, 12, -10)$$

at $(-2, 3, 1, 5)$. A unit vector in the given direction is $\mathbf{u} = (1/\sqrt{30})(1, 2, 3, 4)$. Thus

$$f'_{\mathbf{u}}(\mathbf{x}) = \nabla f \cdot \mathbf{u} = (-4, 1, 12, 10) \cdot \frac{1}{\sqrt{30}}(1, 2, 3, 4) = \frac{74}{\sqrt{30}}.$$

In Section 1.2 we showed that the cosine of the angle between two vectors $\mathbf{u}$ and $\mathbf{v}$ in $\mathbb{R}^2$ is given by

COSINE OF THE ANGLE BETWEEN TWO VECTORS

$$\cos\theta = \frac{\mathbf{u}\cdot\mathbf{v}}{|\mathbf{u}||\mathbf{v}|}. \tag{19}$$

We can use formula (19) as the *definition*[†] of the angle between two vectors $\mathbf{u}$ and $\mathbf{v}$ in $\mathbb{R}^n$, where θ is always in the interval $[0, \pi]$. Then we have

$$f'_{\mathbf{u}}(\mathbf{x}) = \nabla f(\mathbf{x})\cdot\mathbf{u} = |\nabla f(\mathbf{x})||\mathbf{u}|\cos\theta = |\nabla f(\mathbf{x})|\cos\theta$$

since $\mathbf{u}$ is a unit vector. Thus, as in $\mathbb{R}^2$ and $\mathbb{R}^3$, we see that since $f'_{\mathbf{u}}$ is largest when $\cos\theta = 1$ (so that $\theta = 0$), is smallest when $\cos\theta = -1$, and is 0 when $\cos\theta = 0$ (so that $\theta = \pi/2$). We conclude that

> f increases most rapidly in the direction of its gradient and decreases most rapidly in the direction opposite to that of its gradient. It changes least in a direction orthogonal to its gradient.

[†]It can be shown (as we will do in Problem 8.3.63) that $|\mathbf{u}\cdot\mathbf{v}| \le |\mathbf{u}||\mathbf{v}|$ so that

$$-1 \le \frac{\mathbf{u}\cdot\mathbf{v}}{|\mathbf{u}||\mathbf{v}|} \le 1.$$

This is necessary in order that θ be defined by (19).

We need to say something about the last statement. We say that two vectors in $\mathbb{R}^n$ are *parallel* if one is a scalar multiple of the other. They have the same direction if the scalar is positive and the opposite direction if the scalar is negative. We defined orthogonality in Section 1.10 (page 80). It is important to note here that even though (for $n > 3$) it is impossible to draw parallel and orthogonal vectors, the ideas developed geometrically in $\mathbb{R}^2$ and $\mathbb{R}^3$ are easily extended to higher dimensions.

PROBLEMS 3.8

SELF-QUIZ

I. True–False (assume f is differentiable):
a. If $\mathbf{u} = \mathbf{i}$, then $f'_{\mathbf{u}} = f_x$.
b. If $\mathbf{u} = \mathbf{k}$, then $f'_{\mathbf{u}} = f_z$.
c. $f'_{\mathbf{j}} = f_y$.
d. If $\mathbf{u} = -\mathbf{j}$, then $f'_{\mathbf{u}} = -f_y$.

II. Let $f(x, y) = 7x - 5y$ and $\mathbf{u} = \frac{3}{5}\mathbf{i} + \frac{4}{5}\mathbf{j}$. Then $f'_{\mathbf{u}}(2, 1)$, the directional derivative of f along $\mathbf{u}$ at $(2, 1)$, is ______.
a. $\frac{1}{5}$ **b.** $\frac{1}{5}\mathbf{j}$
c. $-\frac{1}{5}(\mathbf{i} - \mathbf{j})$ **d.** $\frac{22}{5}$

III. Let $g(x, y) = 7x^2 - 5y^2$ and $\mathbf{v} = 3\mathbf{i} + 4\mathbf{j}$. The directional derivative of g at $(2, 1)$ in the direction of $\mathbf{v}$ is ______.
a. $84\mathbf{i} - 40\mathbf{j}$ **b.** 44
c. $\frac{44}{5}$ **d.** $\frac{2}{5}$

IV. Let $f(x, y) = x^3y^4$. At the point $(1, -1)$, f increases most rapidly in the direction of ______.
a. $3x^2y^4\mathbf{i} + 4x^3y^3\mathbf{j}$ **b.** $3\mathbf{i} + 4\mathbf{j}$
c. $3\mathbf{i} - 4\mathbf{j}$ **d.** $\sqrt{3^2 + 4^2}$

V. If $f(x, y) = x^3y^4$, then $|f'_{\mathbf{u}}(1, -1)|$ is minimal for $\mathbf{u} =$ ______.
a. $\frac{3}{5}\mathbf{i} - \frac{4}{5}\mathbf{j}$ **b.** $-\frac{3}{5}\mathbf{i} + \frac{4}{5}\mathbf{j}$
c. $\frac{4}{5}\mathbf{i} + \frac{3}{5}\mathbf{j}$ **d.** $4\mathbf{i} - 3\mathbf{j}$

In Problems 1–14, calculate the directional derivative of the given function at the specified point in the direction of the specified vector $\mathbf{v}$.

1. $f(x, y) = xy$ at $(2, 3)$; $\mathbf{v} = \mathbf{i} + 3\mathbf{j}$
2. $f(x, y) = 2x^2 - 3y^2$ at $(1, -1)$; $\mathbf{v} = -\mathbf{i} + 2\mathbf{j}$
3. $f(x, y) = \ln(x + 3y)$ at $(2, 4)$; $\mathbf{v} = \mathbf{i} + \mathbf{j}$
4. $f(x, y) = ax^2 + by^2$ at (c, d); $\mathbf{v} = \alpha\mathbf{i} + \beta\mathbf{j}$
5. $f(x, y) = \tan^{-1}(y/x)$ at $(2, 2)$; $\mathbf{v} = 3\mathbf{i} - 2\mathbf{j}$
6. $f(x, y) = \dfrac{x - y}{x + y}$ at $(4, 3)$; $\mathbf{v} = -\mathbf{i} - 2\mathbf{j}$
7. $f(x, y) = xe^y + ye^x$ at $(1, 2)$; $\mathbf{v} = \mathbf{i} + \mathbf{j}$
8. $f(x, y) = \sin(2x + 3y)$ at $(\pi/12, \pi/9)$; $\mathbf{v} = -2\mathbf{i} + 3\mathbf{j}$
9. $f(x, y, z) = xy + yz + xz$ at $(1, 1, 1)$; $\mathbf{v} = \mathbf{i} + \mathbf{j} + \mathbf{k}$
10. $f(x, y, z) = xy^3z^5$ at $(-3, -1, 2)$; $\mathbf{v} = -\mathbf{i} - 2\mathbf{j} + \mathbf{k}$
11. $f(x, y, z) = xe^{yz}$ at $(2, 0, -4)$; $\mathbf{v} = -\mathbf{i} + 2\mathbf{j} + 5\mathbf{k}$
12. $f(x, y, z) = x^2y^3 + z\sqrt{x}$ at $(1, -2, 3)$; $\mathbf{v} = 5\mathbf{j} + \mathbf{k}$
13. $f(x, y, z) = e^{-(x^2+y^2+z^2)}$ at $(1, 1, 1)$; $\mathbf{v} = \mathbf{i} + 3\mathbf{j} - 5\mathbf{k}$
14. $f(x, y, z) = 1/\sqrt{x^2 + y^2 + z^2}$ at $(-1, 2, 3)$; $\mathbf{v} = \mathbf{i} - \mathbf{j} + \mathbf{k}$

In Problems 15–18, find the directional derivative of the given function at the specified point in the direction from that point to Q.

15. $f(x, y) = 2x^2y$ at $(-1, 4)$; $Q = (2, -5)$
16. $f(x, y) = \dfrac{x}{y}$ at $(2, 3)$; $Q = (1, 6)$
17. $f(x, y, z) = xy^2z + x^3yz^2$ at $(1, 2, -1)$; $Q = (0, -2, 4)$
18. $f(x, y, z) = \dfrac{x + y}{z}$ at $(3, 2, -1)$; $Q = (4, 2, 5)$

In Problems 19–22, calculate the normal derivative at the given point.

19. $f(x, y) = x + 2y$ at $(1, 4)$
20. $f(x, y) = e^{x+3y}$ at $(1, 0)$
21. $f(x, y) = \tan^{-1}(y/x)$ at $(-1, -1)$
22. $f(x, y) = \sqrt{\dfrac{x - y}{x + y}}$ at $(3, 1)$

23. The voltage (potential) at any point on a metal structure is given by

$$v(x, y, z) = \frac{1}{0.02 + \sqrt{x^2 + y^2 + z^2}}.$$

At the point $(1, -1, 2)$, in what direction does the voltage increase most rapidly?

24. The temperature at any point in a solid metal ball centered at the origin is given by

$$T(x, y, z) = 100e^{-(x^2+y^2+z^2)}.$$

a. Where is the ball hottest?
b. Verify that at any point (x, y, z) on the ball, the direction of greatest increase in temperature is a vector pointing toward the origin.

25. The temperature distribution of a ball centered at the origin is given by

$$T(x, y, z) = \frac{100}{1 + x^2 + y^2 + z^2}.$$

a. Where is the ball hottest?
b. Find the direction of greatest decrease of temperature at the point $(3, -1, 2)$.
c. Find the direction of greatest increase in temperature. Does this vector point toward the origin?

26. The temperature distribution on a plate is given by

$$T(x, y) = 1 - \frac{x^2}{a^2} - \frac{y^2}{b^2}.$$

Find the path of a heat-seeking particle (i.e., a particle that always moves in the direction of greatest increase in temperature) if it starts at the point (a, b).

27. Find the path of the particle in the preceding problem if it starts at the point $(-a, b)$.

28. The height of a mountain is given by $h(x, y) = 3000 - 2x^2 - y^2$, where the y-axis points east, the x-axis points north, and all distances are measured in meters. Suppose a mountain climber is at the point $(30, -20, 800)$.
a. If the climber moves in the southwest direction, will she ascend or descend?
b. In what direction should the climber move so as to ascend most rapidly?

29. We refer to Example 5.
a. Show that a unit vector in the direction of motion is given by

$$\frac{x'}{\sqrt{x'^2 + y'^2}}\mathbf{i} + \frac{y'}{\sqrt{x'^2 + y'^2}}\mathbf{j}.$$

b. Show that a unit vector having the direction of the gradient is

$$\frac{-2x}{\sqrt{4x^2 + 64y^2}}\mathbf{i} + \frac{-8y}{\sqrt{4x^2 + 64y^2}}\mathbf{j}.$$

c. By equating coordinates in parts (a) and (b), show that

$$\frac{y'(t)}{y(t)} = 4\frac{x'(t)}{x(t)}.$$

d. Integrate both sides of the equation in part (c) to show that $\ln|y(t)| = \ln|x^4(t)| + C$.
e. Use part (d) to show that $y(t) = k[x(t)]^4$, where $k = \pm e^C$.
f. Show that $k = -2$ in part (e) by using the point $(1, -2)$ on the curve.

***30.** Prove that if $w = f(x, y, z)$ is differentiable at $\mathbf{x}$, then all the first partials exist at $\mathbf{x}$ and

$$\nabla f(\mathbf{x}) = f_x(\mathbf{x})\mathbf{i} + f_y(\mathbf{x})\mathbf{j} + f_z(\mathbf{x})\mathbf{k}.$$

(This result is a partial converse to Theorem 3.5.1.)

In Problems 31–36 find the directional derivative of the given function in the direction of the given vector $\mathbf{v}$ at the point P.

31. $f(x_1, x_2, x_3, x_4) = x_1 + x_2 + x_3x_4$; $\mathbf{v} = (-1, 1, 0, 1)$; $P = (3, -1, 2, 5)$

32. $f(x_1, x_2, x_3, x_4) = x_1x_2x_3x_4$; $\mathbf{v} = (2, -1, 4, 6)$; $P = (1, 0, -1, 2)$

33. $f(x_1, x_2, x_3, x_4) = x_1^2 + x_2^2 + x_3^2 + x_4^2$; $\mathbf{v} = (-1, 1, 0, 2)$; $P = (1, 1, -2, 3)$

34. $f(x_1, x_2, x_3, x_4) = \sqrt{x_1^2 + x_2^2 + x_3^2 + x_4^2}$; $\mathbf{v} = (-1, 6, 2, 4)$; $P = (-3, 0, 1, 1)$

35. $f(x_1, x_2, \ldots, x_n) = \Sigma x_i^2$; $\mathbf{v} = (1, 1, \ldots, 1)$; $P = (1, 2, 3, \ldots, n)$

36. $f(x_1, x_2, \ldots, x_n) = x_1x_2\cdots x_n$; $\mathbf{v} = (1, 1, \ldots, 1)$; $P = (1, 2, \ldots, n)$

ANSWERS TO SELF-QUIZ

I. a. True **b.** True **c.** True **d.** True
II. a (The directional derivative is always a scalar.)
III. c **IV.** c **V.** c

3.9 THE TOTAL DIFFERENTIAL AND APPROXIMATION

If f is a differentiable function of one variable, then

$$f'(x) = \lim_{\Delta x \to 0} \frac{f(x + \Delta x) - f(x)}{\Delta x}$$

exists. It follows that if Δx is small, then

$$f(x + \Delta x) - f(x) = \Delta y \approx f'(x)\,\Delta x. \tag{1}$$

In your one-variable calculus course the **differential** dy was defined by

$$dy = f'(x)\,dx = f'(x)\,\Delta x \tag{2}$$

(since dx is defined to be equal to Δx). Note that in (2) it is not required that Δx be small.

We now extend these ideas to functions of two or three variables.

DEFINITION INCREMENT AND THE TOTAL DIFFERENTIAL

Let $f = f(\mathbf{x})$ be a function of two or three variables and let $\mathbf{\Delta x} = (\Delta x, \Delta y)$ or $(\Delta x, \Delta y, \Delta z)$.

i. The **increment** of f, denoted by $\mathbf{\Delta f}$, is defined by

$$\Delta f = f(\mathbf{x} + \mathbf{\Delta x}) - f(\mathbf{x}). \tag{3}$$

ii. The **total differential** of f, denoted by df[†], is given by

$$df = \nabla f(\mathbf{x}) \cdot \mathbf{\Delta x}. \tag{4}$$

■

Note that equation (4) is very similar in form to equation (2).

REMARK 1: If f is a function of two variables, then (3) and (4) become

$$\Delta f = f(x + \Delta x, y + \Delta y) - f(x, y), \tag{5}$$

and the total differential is

$$df = f_x(x, y)\,\Delta x + f_y(x, y)\,\Delta y. \tag{6}$$

REMARK 2: If f is a function of three variables, then (3) and (4) become

$$\Delta f = f(x + \Delta x, y + \Delta y, z + \Delta z) - f(x, y, z) \tag{7}$$

and

$$df = f_x(x, y, z)\,\Delta x + f_y(x, y, z)\,\Delta y + f_z(x, y, z)\,\Delta z. \tag{8}$$

REMARK 3: Note that in the definition of the total differential, it is *not* required that $|\mathbf{\Delta x}|$ be small.

From Theorems 3.5.1 and 3.5.1′ and the definition of differentiability, we see that if $|\mathbf{\Delta x}|$ is small and if f is differentiable, then

$$\Delta f \approx df. \tag{9}$$

[†] We sometimes write this as $df(\mathbf{\Delta x})$ to indicate that the differential depends on $\mathbf{\Delta x}$ (as well as on $\mathbf{x}$).

We can use the relation (9) to approximate functions of several variables in much the same way that we used the relation (1) to approximate the values of functions of one variable.

EXAMPLE 1 USING THE TOTAL DIFFERENTIAL TO APPROXIMATE VOLUME

The radius of a cone is measured to be 15 cm and the height of the cone is measured to be 25 cm. There is a maximum error of ± 0.02 cm in the measurement of the radius and ± 0.05 cm in the measurement of the height.

a. What is the approximate volume of the cone?

b. What is the maximum error in the calculation of the volume?

SOLUTION:

a. $V = \frac{1}{3}\pi r^2 h \approx \frac{1}{3}\pi(15)^2 25 = 1875\pi \text{ cm}^3 \approx 5890.5 \text{ cm}^3$.

b. $\nabla V = V_r\mathbf{i} + V_h\mathbf{j} = \frac{2}{3}\pi rh\mathbf{i} + \frac{1}{3}\pi r^2\mathbf{j} = \pi(250\mathbf{i} + 75\mathbf{j})$. Then, choosing $\Delta r = 0.02$ and $\Delta h = 0.05$ to find the maximum error, we have

$$\begin{aligned} \Delta V \approx dV = \nabla V \cdot \mathbf{\Delta x} &= \pi[250(0.02) + 75(0.05)] \\ &= \pi(5 + 3.75) = 8.75\pi \approx 27.5 \text{ cm}^3. \end{aligned}$$

Thus, the maximum error in the calculation is, approximately, 27.5 cm^3, which means that

$$5890.5 - 27.5 < V < 5890.5 + 27.5, \quad \text{or}$$

$$5863 \text{ cm}^3 < V < 5918 \text{ cm}^3.$$

Note that an error of 27.5 cm^3 is only a *relative error* of $27.5/5890.5 \approx 0.0047 < \frac{1}{2}\%$, which is a very small relative error.

PROBLEMS 3.9

SELF-QUIZ

I. Suppose f is a differentiable function of $\mathbf{x} = (x, y)$ and consider the plane tangent to the surface $z = f(\mathbf{x})$ at the point $\mathbf{x} = \mathbf{a}$. True–False:

a. $f(\mathbf{a}) + \nabla f(\mathbf{a}) \cdot \mathbf{\Delta x}$ is the z-value on that tangent plane which corresponds to $\mathbf{x} = \mathbf{a} + \mathbf{\Delta x}$.

b. Approximating $f(\mathbf{a} + \mathbf{\Delta x})$ by $f(\mathbf{a}) + \nabla f(\mathbf{a}) \cdot \mathbf{\Delta x}$ is reasonable when $|\mathbf{\Delta x}|$ is small enough for the tangent plane to be close to the surface.

II. If $f(x, y) = x^2 + \sin y$, then $df(x, y)(\mathbf{\Delta x}) =$ __________.

a. $2x\,\Delta x + \cos y\,\Delta y$

b. $x^2\,\Delta x + \sin y\,\Delta y$

c. $(x^3/3)\Delta x - \cos y\,\Delta y$

d. $2x\mathbf{i} + \cos y\mathbf{j}$

III. We can approximate $f(x, y) = x^2 + \sin y$ in the neighborhood of $\mathbf{a} = (-1, \pi/3)$ by $f(-1, \pi/3) + df(\mathbf{\Delta x}) = 1 + \sqrt{3}/2 + df(\mathbf{\Delta x})$ where $df(\mathbf{\Delta x}) =$ __________.

a. $-2\,\Delta x + \frac{1}{2}\,\Delta y$ **b.** $\Delta x + \frac{1}{2}\,\Delta y$

c. $-\frac{1}{3}\,\Delta x - (\sqrt{3}/2)\Delta y$ **d.** $-2\mathbf{i} + (\sqrt{3}/2)\mathbf{j}$

IV. We can approximate $g(x, y) = y^2 \sin x$ in the neighborhood of $\mathbf{a} = (-\pi/4, 3)$ by $dg(\mathbf{\Delta x}) =$ __________.

a. $\cos x\,\Delta x + 2y\,\Delta y$

b. $(1/\sqrt{2})\Delta x + 6\,\Delta y$

c. $y^2 \cos x\,\Delta x + 2y \sin x\,\Delta y$

d. $(9/\sqrt{2})\Delta x + (-6/\sqrt{2})\Delta y$

In Problems 1–10, calculate the total differential df.

1. $f(x, y) = xy^3$

2. $f(x, y) = \tan^{-1}(y/x)$

3. $f(x, y) = \sqrt{\dfrac{x - y}{x + y}}$

4. $f(x, y) = xe^y$

5. $f(x, y) = \ln(2x + 3y)$

6. $f(x, y) = \sin(x - 4y)$

7. $f(x, y, z) = xy^2z^5$

8. $f(x, y, z) = xy/z$

9. $f(x, y, z) = \cosh(xy - z)$

10. $f(x, y, z) = \dfrac{x - z}{y + 3x}$

11. Let $f(x, y) = xy^2$.
- **a.** Calculate explicitly the difference $\Delta f - df$.
- **b.** Use the result of (a) to find $\Delta f - df$ at the point $(1, 2)$ where $\Delta x = -0.01$ and $\Delta y = 0.03$.

***12.** Repeat the steps of the preceding problem for the function $f(x, y) = x^3y^2$.

13. The radius and height of a cylinder are approximately 10 cm and 20 cm, respectively. The maximum errors in approximation are ± 0.03 cm and ± 0.07 cm.
- **a.** What is the approximate volume of the cylinder?
- **b.** What is the approximate maximum error in your calculation?

14. Two sides of a triangular piece of land were measured to be 50 m and 110 m. There was an error of at most ± 0.3 m in each measurement. The angle between the sides was measured to be $60°$ with an error of at most $\pm 1°$.
- **a.** Using the law of cosines, find the approximate length of the third side of the triangle.
- **b.** What is the approximate maximum error of your measurement? [*Hint:* Convert to radians.]

15. When three resistors are connected in parallel, the total resistance R (measured in ohms, Ω) is given by

$$\frac{1}{R} = \frac{1}{R_1} + \frac{1}{R_2} + \frac{1}{R_3},$$

where R_1, R_2, and R_3 are the three separate resistances. Suppose $R_1 = 6 \pm 0.1\ \Omega$, $R_2 = 8 \pm 0.03\ \Omega$, and $R_3 = 12 \pm 0.15\ \Omega$.
- **a.** Approximate R.
- **b.** Find an approximate value for the maximum error in your approximation of R.

16. The volume of 10 moles of an ideal gas was calculated to be 500 cm^3 at a temperature of 40°C (=313°K). The maximum error in each measurement n, V, and T was $\frac{1}{2}\%$.
- **a.** Calculate the approximate pressure of the gas (in Newtons per square cm).
- **b.** Find the approximate maximum error in your computation. [*Hint:* Recall that according to the ideal-gas law, $PV = nRT$.]

ANSWERS TO SELF-QUIZ

I. a. True   **b.** True   **II.** a   **III.** a   **IV.** d

---

3.10 MAXIMA AND MINIMA FOR A FUNCTION OF TWO VARIABLES

In your first calculus course, you studied methods for obtaining maximum and minimum values for a function of one variable. The first basic fact used was that if f is continuous on a closed bounded interval $[a, b]$, then f takes on a maximum and minimum value in $[a, b]$. A critical point is defined as a number x at which $f'(x) = 0$ or for which $f(x)$ exists but $f'(x)$ does not. Finally, local maxima and minima (which occur at critical points) were defined and you saw conditions on first and second derivatives that ensured that a critical point was a local maximum or minimum.

The theory for functions of two variables is more complicated, but some of the basic ideas are the same. The first thing we need to know is that a function *has* a maximum or minimum. In order to state our basic result, we must say what we mean by a closed, bounded region in $\mathbb{R}^2$.

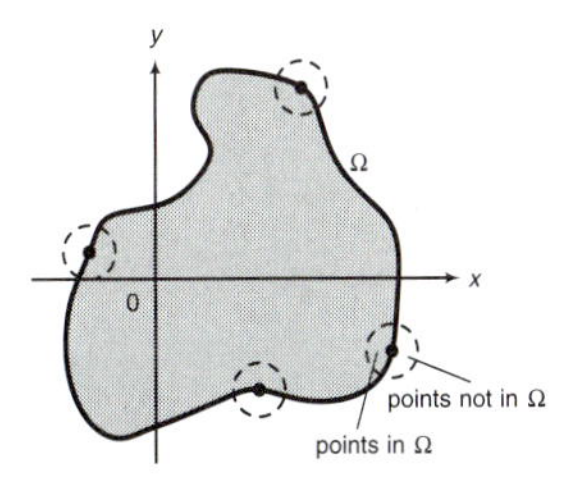

FIGURE 1
A region R showing four of its boundary points

R is closed because it contains all its boundary points.

Let Ω be a region in the plane (we will say a great deal more about regions in $\mathbb{R}^2$ in Section 5.1). A point $\mathbf{P}$ in $\mathbb{R}^2$ is a **boundary point** of Ω if every open disk, D_r (see page 147) centered at $\mathbf{P}$ contains points that are in Ω and points that are not in Ω. The region is **closed** if Ω contains all its boundary points. A closed region is illustrated in Figure 1. Examples of closed regions in $\mathbb{R}^2$ include closed disks (see page 148), triangles and the points inside them, and rectangles and the points inside them (see Figure 2).

A region Ω in $\mathbb{R}^2$ is **bounded** if there is an open disk that completely surrounds it. It doesn't matter how big the disk is, as long as one can be found. The regions in Figures 1 and 2 are all bounded.

A region is called **unbounded** if it is not bounded. The first quadrant, for example, is an unbounded region. The proof of the following theorem can be found in any advanced calculus book.[†]

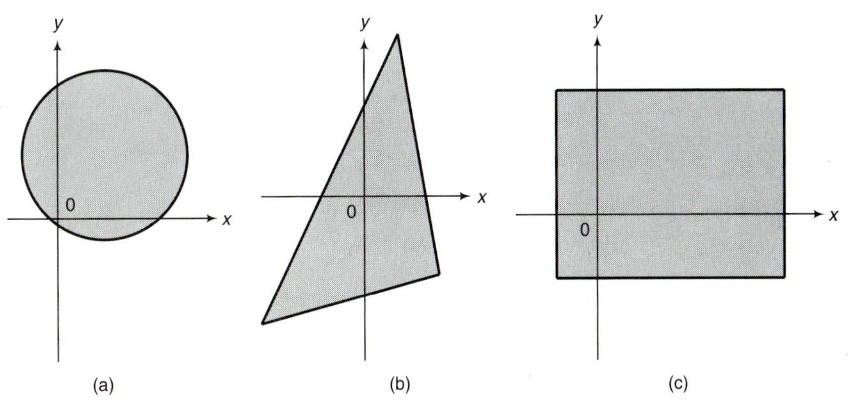

FIGURE 2
Three closed, bounded regions in $\mathbb{R}^2$

THEOREM 1 UPPER AND LOWER BOUND THEOREM IN $\mathbb{R}^2$

Let f be a function of two variables that is continuous on the closed, bounded region Ω. Then there exist points $\mathbf{x}_1$ and $\mathbf{x}_2$ in Ω such that

$$m = f(\mathbf{x}_1) \le f(\mathbf{x}) \le f(\mathbf{x}_2) = M \tag{1}$$

for any other point $\mathbf{x}$ in Ω. The numbers m and M are called, respectively, the **global minimum** and the **global maximum** for the function f over the region Ω.[‡] ■

We now define what we mean by local minima and maxima.

DEFINITION LOCAL MAXIMUM AND LOCAL MINIMUM

Let f be defined in a neighborhood of a point $\mathbf{x}_0 = (x_0, y_0)$.

i. f has a **local** or **relative maximum** at x_0 if there is a neighborhood N_1 of $\mathbf{x}_0$ such that for every point $\mathbf{x}$ in N_1,

$$f(\mathbf{x}) \le f(\mathbf{x}_0). \tag{2}$$

[†] See, e.g., R. C. Buck and E. F. Buck, *Advanced Calculus,* 3rd ed. (New York: McGraw-Hill, 1978), p. 91.
[‡] The first rigorous proof of this theorem was given by the German mathematician Karl Weierstrass (1815–1897). See the biographical sketch on page 220.

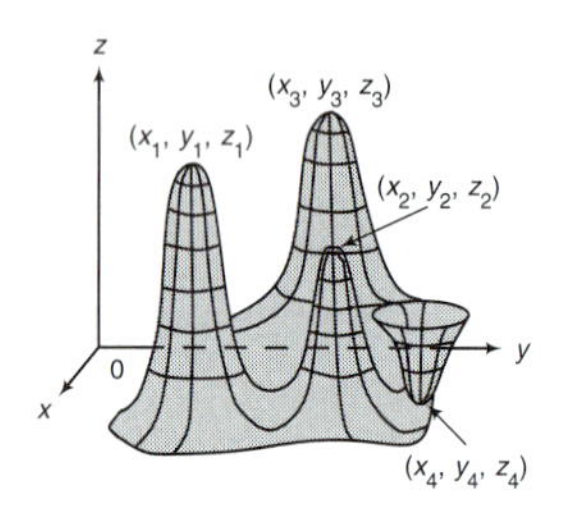

FIGURE 3
A function with three local maxima and one local minimum

ii. f has a **local** or **relative minimum** at $\mathbf{x}_0$ if there is a neighborhood N_2 of $\mathbf{x}_0$ such that for every point $\mathbf{x}$ in N_2,

$$f(\mathbf{x}) \geq f(\mathbf{x}_0). \tag{3}$$

iii. If f has a local maximum or minimum at $\mathbf{x}_0$, then $\mathbf{x}_0$ is called an **extreme point** of f. ■

NOTE: Compare this definition with the definitions in your one-variable calculus book.†

A rough sketch of a function with several local extreme points is given in Figure 3. The following theorem is a natural generalization of the fact that if f has a local maximum or minimum at $\mathbf{x}_0$ and if f is differentiable at $\mathbf{x}_0$, then $f'(\mathbf{x}_0) = 0$.

THEOREM 2 THE GRADIENT IS ZERO AT AN EXTREME POINT

Let f be differentiable at $\mathbf{x}_0$ and suppose that $\mathbf{x}_0$ is an extreme point of f. Then

$$\nabla f(\mathbf{x}_0) = \mathbf{0}. \tag{4}$$

PROOF: We must show that $f_x(x_0, y_0)$ and $f_y(x_0, y_0) = 0$. To prove that $f_x(x_0, y_0) = 0$, define a function h by

$$h(x) = f(x, y_0).$$

Then $h(x_0) = f(x_0, y_0)$ and $h'(x_0) = f_x(x_0, y_0)$. But since f has a local maximum (or minimum) at (x_0, y_0), h has a local maximum (or minimum) at x_0, so that by the one-variable result cited above, $h'(x_0) = 0$. Thus $f_x(x_0, y_0) = 0$. In a very similar way we can show that $f_y(x_0, y_0) = 0$ by defining $h(y) = f(x_0, y)$, and the theorem is proved. ■

DEFINITION CRITICAL POINT

$\mathbf{x}_0$ is a **critical point** of f if either (a) f is differentiable at $\mathbf{x}_0$ and $\nabla f(\mathbf{x}_0) = \mathbf{0}$ or (b) $\mathbf{x}_0$ is in the domain of f but $\nabla f(\mathbf{x}_0)$ does not exist. ■

REMARK 1: The definitions and Theorems 1 and 2 hold for functions of three variables with very little change.

REMARK 2: By the remark on page 199, we see that if $\mathbf{x}_0$ is a critical point of f and if $\nabla f(\mathbf{x}_0)$ exists, then the tangent plane to the surface $z = f(x, y)$ is horizontal (i.e., parallel to the xy-plane).

As in the case of a function of one variable, the fact that $\mathbf{x}_0$ is a critical point of f does not guarantee that $\mathbf{x}_0$ is an extreme point of f, as Example 3 below illustrates.

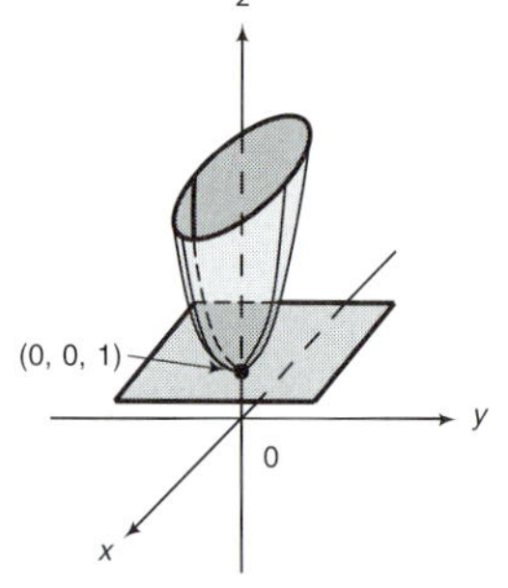

FIGURE 4
The surface
$z = 1 + x^2 + 3y^2$

† See *Calculus* or *Calculus of One-Variable*, page 190.

EXAMPLE 1 A FUNCTION WITH A LOCAL AND GLOBAL MINIMUM AT THE CRITICAL POINT (0, 0)

Let $f(x, y) = 1 + x^2 + 3y^2$. Then $\nabla f = 2x\mathbf{i} + 6y\mathbf{j}$, which is zero only when $(x, y) = (0, 0)$. Thus, $(0, 0)$ is the only critical point, and it is clearly a local (and global) minimum. This is illustrated in Figure 4.

NOTE: Since $x^2 \geq 0$ and $3y^2 \geq 0$ for any real numbers x and y, it is apparent that we do not need calculus to conclude that $1 + x^2 + 3y^2$ has the minimum value of 1 when $x = y = 0$. The point of this example (and the next one) is to illustrate, in a very simple case, the fact that a function may have a local minimum or maximum at a critical point. The point is *not* to solve a problem whose solution is obvious.

EXAMPLE 2 A FUNCTION WITH A LOCAL AND GLOBAL MAXIMUM AT THE CRITICAL POINT (0, 0)

Let $f(x, y) = 1 - x^2 - 3y^2$. Then $\nabla f = -2x\mathbf{i} - 6y\mathbf{j}$, which is zero only at the origin. In this case $(0, 0)$ is a local (and global) maximum for f.

EXAMPLE 3 A FUNCTION FOR WHICH (0, 0) IS A CRITICAL POINT BUT IS NEITHER A LOCAL MAXIMUM NOR A LOCAL MINIMUM

Let $f(x, y) = y^2 - x^2$. Then $\nabla f = -2x\mathbf{i} + 2y\mathbf{j}$, which again is zero only at $(0, 0)$. But $(0, 0)$ is *neither* a local maximum nor a local minimum for f. To see this, we simply note that f can take positive and negative values in any neighborhood of $(0, 0)$ since $f(x, y) > 0$ if $|x| < |y|$ and $f(x, y) < 0$ if $|x| > |y|$. This is illustrated by Figure 5. The figure sketched in Figure 5 is called a **hyperbolic paraboloid**, or **saddle surface**.

We have the following definition.

DEFINITION SADDLE POINT

If $\nabla f(\mathbf{x}_0)$ exists and $\mathbf{x}_0$ is a critical point of f but is not an extreme point of f, then $\mathbf{x}_0$ is called a **saddle point** of f. ■

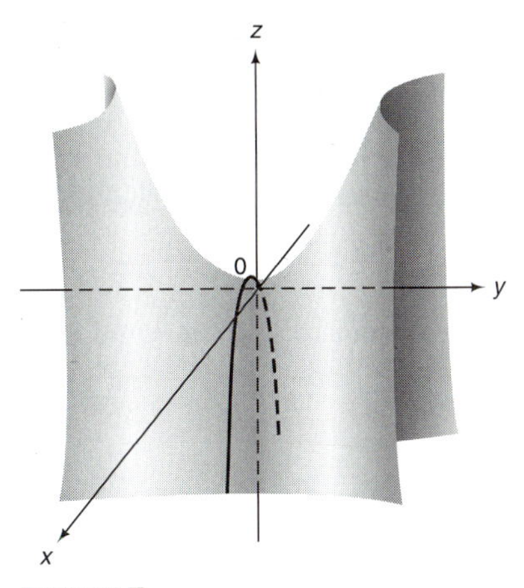

FIGURE 5
Saddle surface $z = y^2 - x^2$

The three examples above illustrate that a critical point of f may be a local maximum, a local minimum, or a saddle point.

EXAMPLE 4 A FUNCTION MAY HAVE A LOCAL MINIMUM AT A POINT WHERE IT IS NOT DIFFERENTIABLE

Let $f(x, y) = \sqrt{x^2 + y^2}$. Then it is obvious that f has a local (and global) minimum at $(0, 0)$. But

$$\nabla f = \frac{x}{\sqrt{x^2 + y^2}}\mathbf{i} + \frac{y}{\sqrt{x^2 + y^2}}\mathbf{j},$$

so that $\nabla f(0, 0)$ *does not exist* even though $(0, 0)$ is a critical point of f (case (b) in the definition of a critical point). We see this clearly in Figure 6. The graph of f (which is a cone) does not have a tangent plane at $(0, 0)$.

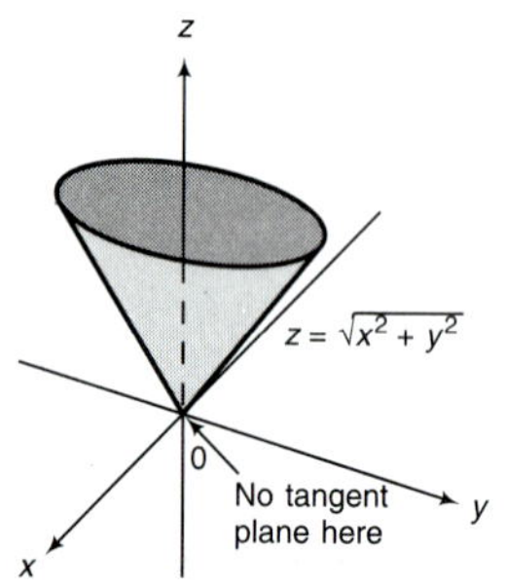

FIGURE 6
The function $z = \sqrt{x^2 + y^2}$ is not differentiable at $(0, 0)$ although it has a local (and global) minimum there.

We generalize the result of Example 4: If $f(x_0, y_0)$ exists but $\nabla f(x_0, y_0)$ does not exist, then f *may* have a local maximum or minimum at (x_0, y_0).

Examples 1, 2, and 3 indicate that more is needed to determine whether a critical point is an extreme point (in most cases it will not be at all obvious). You might suspect that the answer has something to do with the signs of the second partial derivatives of f, as in the case of functions of one variable. The following theorem gives the answer. Its proof is given in Section 9.1 (see page 639).

THEOREM 3 SECOND-DERIVATIVES TEST

Let f and all its first and second partial derivatives be continuous in a neighborhood of (x_0, y_0). Suppose that $f_x(x_0, y_0) = 0$ and $f_y(x_0, y_0) = 0$. That is, (x_0, y_0) is a critical point of f. Let

$$D(x, y) = f_{xx}(x, y)f_{yy}(x, y) - [f_{xy}(x, y)]^2 \tag{5}$$

and let D denote $D(x_0, y_0)$. D is called the **discriminant** of f.

i. If $D > 0$ and $f_{xx}(x_0, y_0) > 0$, then f has a local minimum at (x_0, y_0).
ii. If $D > 0$ and $f_{xx}(x_0, y_0) < 0$, then f has a local maximum at (x_0, y_0).
iii. If $D < 0$, then (x_0, y_0) is a saddle point of f.
iv. If $D = 0$, then any of the preceding is possible. ■

REMARK 1: Using a 2×2 determinant, we can write (5) as

$$D = \begin{vmatrix} f_{xx} & f_{xy} \\ f_{yx} & f_{yy} \end{vmatrix}.$$

REMARK 2: This theorem does not provide the full answer to the questions of what are the extreme points of f in $\mathbb{R}^2$. It is still necessary to check points at which ∇f does not exist (as in Example 4).

EXAMPLE 5

USING THE SECOND-DERIVATIVES TEST TO PROVE THAT A FUNCTION HAS A LOCAL MINIMUM AT (0, 0)

Let $f(x, y) = 1 + x^2 + 3y^2$. Then as we saw in Example 1, (0, 0) is the only critical point of f. But $f_{xx} = 2, f_{yy} = 6$, and $f_{xy} = 0$, so $D(0, 0) = 12$ and $f_{xx} > 0$, which *proves* that f has a local minimum at (0, 0).

EXAMPLE 6

USING THE SECOND-DERIVATIVES TEST TO DETERMINE THE NATURE OF THE CRITICAL POINTS

Let $f(x, y) = -x^2 - y^2 + 2x + 4y + 5$. Determine the nature of the critical points of f.

SOLUTION: $\nabla f = (-2x + 2)\mathbf{i} + (-2y + 4)\mathbf{j}$. We see that $\nabla f = \mathbf{0}$ when

$$-2x + 2 = 0 \qquad \text{and} \qquad -2y + 4 = 0,$$

which occurs only at the point (1, 2). Also $f_{xx} = -2, f_{yy} = -2$, and $f_{xy} = 0$, so $D = 4$ and $f_{xx} < 0$, which implies that there is a local maximum at (1, 2). At (1, 2), $f = 10$.

EXAMPLE 7

USING THE SECOND-DERIVATIVES TEST TO DETERMINE THE NATURE OF THE CRITICAL POINTS

Let $f(x, y) = 2x^3 - 24xy + 16y^3$. Determine the nature of the critical points of f.

SOLUTION: We have $\nabla f(x, y) = (6x^2 - 24y)\mathbf{i} + (-24x + 48y^2)\mathbf{j}$. If $\nabla f = \mathbf{0}$, we have

$$6(x^2 - 4y) = 0 \qquad \text{and} \qquad -24(x - 2y^2) = 0.$$

Clearly, one critical point is (0, 0). To obtain another, we must solve the simultaneous equations

$$x^2 - 4y = 0$$
$$x - 2y^2 = 0.$$

The second equation tells us that $x = 2y^2$. Substituting this expression into the first equation yields

$$4y^4 - 4y = 0,$$

with solutions $y = 0$ and $y = 1$, giving us the critical points (0, 0) and (2, 1). Now $f_{xx} = 12x$, $f_{yy} = 96y$, and $f_{xy} = -24$, so

$$D(x, y) = (12x)(96y) - 24^2 = 1152xy - 576.$$

Then $D(0, 0) = -576$ and $D(2, 1) = 1728$. We find that (0, 0) is a saddle point, and since $f_{xx}(2, 1) = 24 > 0$, (2, 1) is a local minimum.

In Problems 32–34, you are asked to show that any of the three cases can occur when $D = 0$. That is, you will show that the second-derivatives test *fails* when $D = 0$.

EXAMPLE 8

REGRESSION LINES

We can use the theory of this section in an interesting way to derive a result that is very useful for statistical analysis and, in fact, any analysis involving the use of a great deal of data. Suppose n data points $(x_1, y_1), (x_2, y_2), \ldots, (x_n, y_n)$ are collected. For example, the

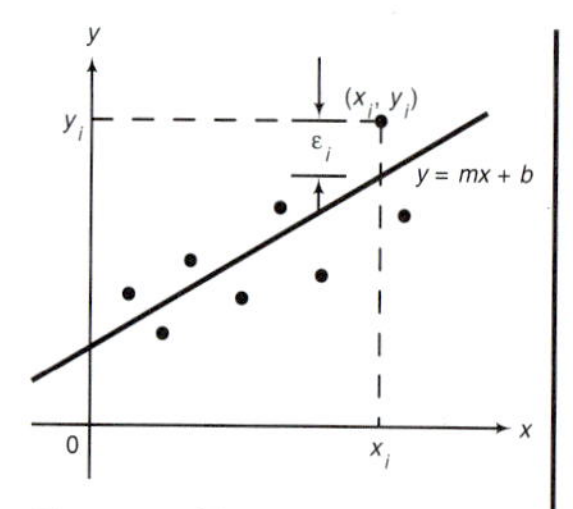

FIGURE 7
The regression line $y = mx + b$

ϵ_i is the difference between the actual y-value corresponding to x_i $(=y_i)$ and the y-value corresponding to x_i that lies on the approximating line $(=mx_i + b)$.

x's may represent average tree growth and the y's average daily temperature in a given year in a certain forest. Or x may represent a week's sales and y a week's profit for a certain business. The question arises as to whether we can "fit" these data points to a straight line. That is, is there a straight line that runs "more or less" through the points? If so, then we can write y as a linear function of x, with obvious computational advantages.

The problem is to find the "best" straight line, $y = mx + b$, passing through or near these points. Look at Figure 7. If (x_i, y_i) is one of our n points, then, on the line $y = mx + b$, corresponding to x_i we obtain $y_i = mx_i + b$. The "error," ϵ_i, between the y-value of our actual point and the "approximating" value on the line is given by

$$\epsilon_i = y - mx_i - b.$$

One way to choose the approximating line is to use the line that minimizes the sum of the squares of the errors. This is called the **least-squares** criterion for choosing the line.

Now we want to choose m and b such that the function

$$f(m, b) = \epsilon_1^2 + \epsilon_2^2 + \cdots + \epsilon_n^2 = \sum \epsilon_i^2 = \sum (y_i - mx_i - b)^2$$

is a minimum. To do this, we calculate

$$\frac{\partial}{\partial m}(y_i - mx_i - b)^2 = -2x_i(y_i - mx_i - b)$$

and

$$\frac{\partial}{\partial b}(y_i - mx_i - b)^2 = -2(y_i - mx_i - b).$$

Hence,

$$\frac{\partial f}{\partial m} = -2 \sum x_i(y_i - mx_i - b)$$

and

$$\frac{\partial f}{\partial b} = -2 \sum (y_i - mx_i - b).$$

Setting $\partial f/\partial m = 0$ and $\partial f/\partial b = 0$ and rearranging terms, we obtain

$$\sum (x_i y_i - mx_i^2 - bx_i) = 0$$
$$\sum (y_i - mx_i - b) = 0.$$

This leads to the system of two equations in the unknowns m and b:

$$(\sum x_i^2)m + (\sum x_i)b = \sum x_i y_i \tag{6}$$

and

$$(\sum x_i)m + nb = \sum y_i. \tag{7}$$

Here we have used the fact that $\sum b = nb$. The system (6) and (7) is not hard to solve for m and b. To do so, we multiply both sides of (6) by n and both sides of (7) by $\sum x_i$ and then subtract to obtain finally

$$m = \frac{n\sum x_i y_i - [\sum x_i][\sum y_i]}{n\sum x_i^2 - [\sum x_i]^2} \tag{8}$$

and

$$b = \frac{[\sum x_i^2][\sum y_i] - [\sum x_i][\sum x_i y_i]}{n\sum x_i^2 - [\sum x_i]^2}. \tag{9}$$

We will leave it to you to check that the numbers m and b given in (8) and (9) do indeed provide a local (and global) minimum. The line $y = mx + b$ given by (8) and (9) is called the **regression line** for the n points.

TABLE 1

i	x_i	y_i	x_i^2	$x_i y_i$
1	1	2	1	2
2	2	4	4	8
3	5	5	25	25
$\sum$	8	11	30	35

REMARK: Equations (8) and (9) make sense only if

$$n\sum x_i^2 - (\sum x_i^2) \neq 0.$$

But in fact,

$$n\sum x_i^2 - (\sum x_i)^2 \geq 0$$

and is equal to zero only when all the x_i's are equal (in which case the regression line is the vertical line $x = x_i$). This fact follows from the Cauchy-Schwarz inequality (see page 23 or 548).

EXAMPLE 9

FINDING A REGRESSION LINE

Find the regression line for the points (1, 2), (2, 4), and (5, 5).

SOLUTION: We tabulate some appropriate values in Table 1. Then, from (8) and (9),

$$m = \frac{3(35) - (8)(11)}{3(30) - 8^2} = \frac{17}{26} \approx 0.654$$

and

$$b = \frac{(30)(11) - 8(35)}{26} = \frac{50}{26} \approx 1.923.$$

Thus the regression line is

$$y = 0.654x + 1.923$$

This is all illustrated in Figure 8.

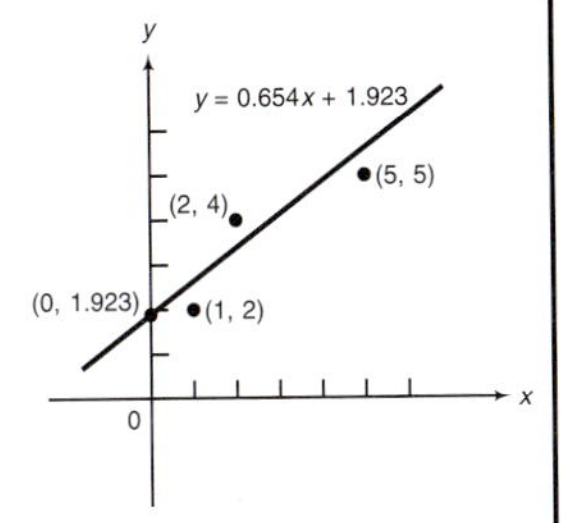

FIGURE 8
The regression line for the points (1, 2), (2, 4), and (5, 5)

OF HISTORICAL INTEREST

KARL WEIERSTRASS (1815–1897)

KARL WEIERSTRASS, 1815–1897

The first rigorous proof of the upper and lower bound theorem in $\mathbb{R}^2$ was given by the German mathematician Karl Weierstrass (1815–1897). Weierstrass began his academic career by studying law and finance and then became a secondary-school teacher. He did not devote himself to mathematics until the age of 40. He was able to work at mathematics full time only after being appointed to a professorship at the University of Berlin in 1864—at the age of 49. Nevertheless, he made significant contributions to many branches of mathematics.

Weierstrass was the first to prove a great number of results in the theory of functions of two or more variables. However his best-known work is his development of the theory of complex functions using power series. Many of the important results in complex function theory today bear his name.

Weierstrass was a very influential teacher. Many of his mathematical discoveries were published by his students, who often got credit for his ideas. One of his most famous students was the Russian mathematician Sonya Kovalevsky (1850–1891). Kovalevsky applied Weierstrass's ideas to solve problems in mathematical physics and other areas. She was one of the first women to gain international prominence as a research mathematician.

Weierstrass developed new standards of mathematical rigor. Using "Weierstrassian analysis," most of what we today term "mathematical analysis" can be logically derived from a set of axioms that characterize the real-number system.

PROBLEMS 3.10

SELF-QUIZ

I. The second-derivatives test for extreme points of a differentiable function $f(x, y)$ includes examination of the discriminant $D =$ ________.

a. $f_{xx}f_{xy} - f_{yy}f_{yx}$ **b.** $f_{xx}{}^2 - f_{yy}{}^2$
c. $f_{xx}f_{yx} - f_{yy}f_{xy}$ **d.** $f_{xx}f_{yy} - (f_{xy})^2$

II. True–False: In searching for extreme points of a function $f(x, y)$, only the various second derivatives of f are examined.

III. Let $f(x, y) = x^2 + y^2 + 4x - 2y$. Then $f_x(-2, 1) = 2(-2) + 4 = 0$, $f_y(-2, 1) = 2(1) - 2 = 0$, $f_{xx}(-2, 1) = 2 > 0$, $f_{xy}(-2, 1) = 0 = f_{yx}(-2, 1)$, $f_{yy}(-2, 1) = 2$, and $D(-2, 1) = (2)(2) - (0)(0) = 4 > 0$. Therefore, $(-2, 1)$ is ________ of the function f.

a. a local minimum
b. a local maximum
c. not a critical point
d. a saddle point

IV. Let $f(x, y) = x^3 + xy^2 + y^4$; then $f_x(1, -1) = 3 + 1 = 4 \neq 0$, $f_y(1, -1) = -2 - 4 = -6 \neq 0$, $f_{xx}(1, -1) = 6 > 0$, $f_{xy}(1, -1) = -2 = f_{yx}(1, -1)$, $f_{yy}(1, -1) = 14$, and $D(1, -1) = 6(14) - (-2)^2 = 80 > 0$. Therefore $(1, -1)$ is ________ of the function f.

a. a local maximum
b. a local minimum
c. not a critical point
d. a saddle point

In Problems 1–18, find the critical points of the given function and determine their nature.

1. $f(x, y) = 7x^2 - 8xy + 3y^2 + 1$
2. $f(x, y) = x^2 + y^3 - 3xy$
3. $f(x, y) = x^2 + 3y^2 + 4x - 6y + 3$
4. $f(x, y) = x^2 + y^2 + 4xy + 6y - 3$
5. $f(x, y) = x^2 + y^2 + 4x - 2y + 3$
6. $f(x, y) = xy^2 + x^2y - 3xy$
7. $f(x, y) = x^3 + 3xy^2 + 3y^2 - 15x + 2$
8. $f(x, y) = x^3 + y^3 - 3xy$
9. $f(x, y) = \dfrac{1}{y} - \dfrac{1}{x} - 4x + y$
10. $f(x, y) = \dfrac{1}{x} + \dfrac{2}{y} + 2x + y + 1$
11. $f(x, y) = x^2 - xy + y^2 + 2x + 2y$
***12.** $f(x, y) = xy + \dfrac{8}{x} + \dfrac{1}{y}$
13. $f(x, y) = (4 - x - y)xy$
***14.** $f(x, y) = \sin x + \sin y + \sin(x + y)$
15. $f(x, y) = 2x^2 + y^2 + \dfrac{2}{x^2y}$
16. $f(x, y) = 4x^2 + 12xy + 9y^2 + 25$
17. $f(x, y) = x^{25} - y^{25}$
18. $f(x, y) = \tan xy$
19. Find three positive numbers whose sum is 50 and such that their product is a maximum.
20. Find three positive numbers whose product is 50 and whose sum is a minimum.
21. Find three positive numbers x, y, and z whose sum is 50 such that the product xy^2z^3 is a maximum.
22. Find three numbers whose sum is 50 and the sum of whose squares is a minimum.
23. Use the methods of this section to find the minimum distance from the point $(1, -1, 2)$ to the plane $x + 2y - z = 4$. [*Hint:* Express the distance between $(1, -1, 2)$ and a point (x, y, z) on the plane in terms of x and y.]
24. A rectangular wooden box with an open top is to contain α cubic centimeters, where α is a given positive number. Ignoring the thickness of the wood, how is the box to be constructed so as to use the smallest amount of wood? [*Hint:* $V = xyz = \alpha$. Find a formula for the total area of the sides and bottom and substitute $z = \alpha/xy$ to write this formula in terms of x and y only.]
25. What is the maximum volume of an open-top rectangular box that can be built from β m^2 of wood? (Assume that any board can be cut without waste.)
***26.** Find the dimensions of the rectangular box of maximum volume that can be inscribed in the ellipsoid

$$\frac{x^2}{a^2} + \frac{y^2}{b^2} + \frac{z^2}{c^2} = 1$$

such that the faces of the box are parallel to the coordinate planes.
27. A company uses two types of raw material, I and II, for its product. If it uses x units of I and y units of II, it can produce U units of the finished product where $U(x, y) = 8xy + 32x + 40y - 4x^2 - 6y^2$. Each unit of I costs \$10 and each unit of II costs \$4; each unit of the product can be sold for \$40. How can the company maximize its profits?
28. A major oil company sells both oil and natural gas. Each unit of oil costs the company \$25, and each unit of gas costs \$15. The revenue (in dollars) received from selling i units of oil and g units of gas is given by

$$R(i, g) = 60i + 50g - 0.02ig - 0.3i^2 - 0.2g^2.$$

a. How many units of each should the company sell in order to maximize profits?
b. What is the maximum profit?

29. A manufacturer is developing a new electronic garage-door opener. She hopes to build the device for \$100 and sell it for \$250. After considerable market research, it is determined that the number of units that can be sold at that price depends on the amount of money spent on advertising (a) and the amount spent on product development (d). The relationship is given by

$$N(a, d) = \text{number of units sold} = \frac{300a}{a + 3} + \frac{160d}{d + 5}.$$

a. Write down a function that gives profit P as a function of a and d.
b. Find the expenditure on advertising and product development that will maximize profit.

30. Find the regression line for the points $(1, 1)$, $(2, 3)$, and $(3, 6)$. Sketch the line and the points.
31. Find the regression line for the points $(-1, 3)$,

$(1, 2)$, $(2, 0)$, and $(4, -2)$. Sketch the line and the points.

32. Let $f(x, y) = 4x^2 - 4xy + y^2 + 5$. Show that there are an infinite number of critical points, that $D = 0$ at each one, and that each is a global (and local) minimum.

33. Let $f(x, y) = -(4x^2 - 4xy + y^2) + 5$. Show that there are an infinite number of critical points, that $D = 0$ at each one, and that each is a global (and local) maximum.

34. Let $f(x, y) = x^3 - y^3$. Show that $D = 0$ at the only critical point and that this point is a saddle point.

REMARK: Problems 32, 33, and 34 illustrate that any of the three cases can occur when $D = 0$.

35. Show that the rectangular box inscribed in a sphere that encloses the greatest volume is a cube.

***36.** Show that the function

$$e^{(x^2+y^2)/y} + \frac{11}{x} + \frac{5}{y} + \sin x^2 y$$

has a global minimum in the first quadrant.

ANSWERS TO SELF-QUIZ

I. d
II. False (First derivatives are examined in search for critical points.)
III. a
IV. c

3.11 CONSTRAINED MAXIMA AND MINIMA—LAGRANGE MULTIPLIERS

In the previous section, we saw how to find the maximum and minimum of a function of two variables by taking gradients and applying a second-derivative test. It often happens that there are side conditions (or **constraints**) attached to a problem. For example, we have been asked to find the shortest distance from a point (x_0, y_0) to a line $y = mx + b$. We could write this problem as follows:

Minimize the function: $z = \sqrt{(x - x_0)^2 + (y - y_0)^2}$

subject to the constraint: $y - mx - b = 0$.

As another example, suppose that a region of space containing a sphere is heated and a function $w = T(x, y, z)$ gives the temperature of every point of the region. Then if the sphere is given by $x^2 + y^2 + z^2 = r^2$, and if we wish to find the hottest point on the sphere, we have the following problem:

Maximize: $w = T(x, y, z)$

subject to the constraint: $x^2 + y^2 + z^2 - r^2 = 0$.

We now generalize these two examples. Let f and g be functions of two variables. Then we can formulate a **constrained maximization** (or **minimization**) problem as follows:

Maximize (or minimize): $z = f(x, y)$ **(1)**

subject to the constraint: $g(x, y) = 0$. **(2)**

If f and g are functions of three variables, we have the following problem:

Maximize (or minimize): $w = f(x, y, z)$ **(3)**

subject to the constraint: $g(x, y, z) = 0$. **(4)**

We now develop a method for dealing with problems of the type (1), (2), or (3), (4). Let C be a curve in $\mathbb{R}^2$ or $\mathbb{R}^3$ given parametrically by the differentiable function $\mathbf{F}(t)$. That is, C is given by

$$\mathbf{F}(t) = x(t)\mathbf{i} + y(t)\mathbf{j} \qquad (\text{in } \mathbb{R}^2) \text{ or } \mathbf{F}(t) = x(t)\mathbf{i} + y(t)\mathbf{j} + z(t)\mathbf{k} \qquad (\text{in } \mathbb{R}^3).$$

Let $f(\mathbf{x})$ denote the function (of two or three variables) that is to be maximized.

THEOREM 1

Suppose that f is differentiable at a point $\mathbf{x}_0$ and that among all points on a differentiable curve C, f takes its maximum (or minimum) value at $\mathbf{x}_0$. Then $\nabla f(\mathbf{x}_0)$ is orthogonal to C at $\mathbf{x}_0$. That is, since $\mathbf{F}'(t)$ is tangent to C, if $\mathbf{x}_0 = \mathbf{F}(t_0)$, then

$$\nabla f(\mathbf{x}_0) \cdot \mathbf{F}'(t_0) = 0. \tag{5}$$

PROOF: For $\mathbf{x}$ on C, $\mathbf{x} = \mathbf{F}(t)$, so that the composite function $f(\mathbf{F}(t))$ has a maximum (or minimum) at t_0. Therefore its derivative at t_0 is 0. By the chain rule,

$$\frac{d}{dt} f(\mathbf{F}(t)) = \nabla f(\mathbf{F}(t)) \cdot \mathbf{F}'(t),$$

and at t_0

$$0 = \nabla f(\mathbf{F}(t_0)) \cdot \mathbf{F}'(t_0) = \nabla f(\mathbf{x}_0) \cdot \mathbf{F}'(t_0),$$

and the theorem is proved. ■

For f as a function of two variables, the result of Theorem 1 is illustrated in Figure 1.

We can use Theorem 1 to make an interesting geometric observation.[†] Suppose that, subject to the constraint $g(x, y) = 0$, f takes its maximum (or minimum) at the point $\mathbf{x}_0 = (x_0, y_0)$. The equation $g(x, y) = 0$ determines a curve C in the xy-plane, and by Theorem 1, $\nabla f(x_0, y_0)$ is orthogonal to C at (x_0, y_0). But from Section 3.7, $\nabla g(x_0, y_0)$ is also orthogonal to C at (x_0, y_0). Thus, we see that

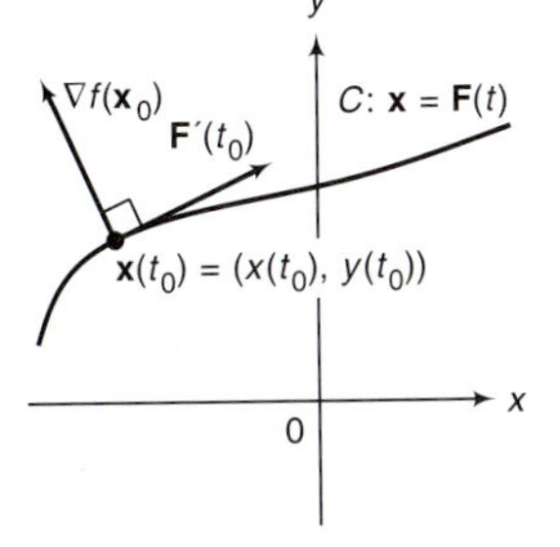

FIGURE 1
The gradient at $\mathbf{x}_0$ is orthogonal to any tangent line at $\mathbf{x}_0$.

LAGRANGE'S OBSERVATION

$\nabla g(x_0, y_0)$ and $\nabla f(x_0, y_0)$ are parallel.

Hence, there is a number λ such that

$$\nabla f(x_0, y_0) = \lambda \nabla g(x_0, y_0).$$

We can extend this observation to the following rule, which applies equally well to functions of three or more variables:

THE BASIC FACT ABOUT LAGRANGE MULTIPLIERS

If, subject to the constraint $g(\mathbf{x}) = 0$, f takes its maximum (or minimum) value at a point $\mathbf{x}_0$, then there is a number λ such that

$$\nabla f(\mathbf{x}_0) = \lambda \nabla g(\mathbf{x}_0). \tag{6}$$

[†] This observation was first made by the French mathematician Joseph-Louis Lagrange (1736–1813). See the biography of Lagrange on page 228.

DEFINITION LAGRANGE MULTIPLIER

The number λ is called a **Lagrange multiplier**. ■

We will illustrate the Lagrange multiplier technique with three examples.

EXAMPLE 1 USING LAGRANGE MULTIPLIERS TO FIND CLOSEST AND FARTHEST POINTS

Find the points on the sphere $x^2 + y^2 + z^2 = 1$ closest to and farthest from the point $(1, 2, 3)$.

SOLUTION: In problems like this, it is almost always easier to work with the square of the distance rather than with the distance itself. The distance from (x, y, z) to $(1, 2, 3)$ is $\sqrt{(x-1)^2 + (y-2)^2 + (z-3)^2}$. The square of the distance is $(x-1)^2 + (y-2)^2 + (z-3)^2$. You should explain why maximum and minimum values of the square of the distance occur at the same points at which the distance is maximized and minimized.

Now, our problem is to minimize and maximize $f(x, y, z) = (x-1)^2 + (y-2)^2 + (z-3)^2$ subject to $g(x, y, z) = x^2 + y^2 + z^2 - 1 = 0$. We have

$$\nabla f(x, y, z) = 2(x-1)\mathbf{i} + 2(y-2)\mathbf{j} + 2(z-3)\mathbf{k},$$

and

$$\nabla g(x, y, z) = 2x\mathbf{i} + 2y\mathbf{j} + 2z\mathbf{k}.$$

Condition (6) implies that if f has a maximum, then, at a maximizing point, $\nabla f = \lambda \nabla g$, so

$$2(x-1) = 2x\lambda$$
$$2(y-2) = 2y\lambda$$
$$2(z-3) = 2z\lambda.$$

If $\lambda \neq 1$, then we find that

$$x - 1 = x\lambda, \qquad \text{or} \qquad x - x\lambda = 1, \qquad x(1-\lambda) = 1, \text{ and } x = \frac{1}{1-\lambda}.$$

Similarly, we obtain

$$y = \frac{2}{1-\lambda} \qquad \text{and} \qquad z = \frac{3}{1-\lambda}.$$

Then

$$1 = x^2 + y^2 + z^2 = \frac{1}{(1-\lambda)^2}(1^2 + 2^2 + 3^2) = \frac{14}{(1-\lambda)^2},$$

so that $(1-\lambda)^2 = 14$, $(1-\lambda) = \pm\sqrt{14}$, and $\lambda = 1 \pm \sqrt{14}$.

If $\lambda = 1 + \sqrt{14}$, then $(x, y, z) = (-1/\sqrt{14}, -2/\sqrt{14}, -3/\sqrt{14})$.

If $\lambda = 1 - \sqrt{14}$, then $(x, y, z) = (1/\sqrt{14}, 2/\sqrt{14}, 3/\sqrt{14})$. Evaluation shows us that f is maximized at $(-1/\sqrt{14}, -2/\sqrt{14}, -3/\sqrt{14})$ and that f is minimized at $(1/\sqrt{14}, 2/\sqrt{14}, 3/\sqrt{14})$. Finally, $f_{\max} = f(-1/\sqrt{14}, -2/\sqrt{14}, -3/\sqrt{14}) \approx 22.48$ and $f_{\min} = f(1/\sqrt{14}, 2/\sqrt{14}, 3/\sqrt{14}) \approx 7.52$.

EXAMPLE 2

MAXIMIZING OUTPUT USING A COBB-DOUGLAS PRODUCTION FUNCTION

In Example 3.1.4 on page 137, we discussed the Cobb-Douglas production formula. In particular, we discussed the function

$$F(L, K) = 200L^{2/5}K^{3/5},$$

which represents the number of units of a certain type of die produced if L units of labor and K units of capital are used in production. Suppose that each unit of labor costs \$400 and each unit of capital costs \$500. If \$50,000 is available for production costs, how many units of labor and capital should be used in order to maximize output? How many units will be produced?

SOLUTION: The total cost is

$$400L + 500K.$$

But the total cost is fixed at \$50,000, so we have

$$400L + 500K = 50{,}000$$

or

$$400L + 500K - 50{,}000 = 0.$$

Thus, our problem is

$$\begin{aligned} &\text{Maximize:} && F(L, K) = 200L^{2/5}K^{3/5} \\ &\text{subject to:} && G(L, K) = 400L + 500K - 50{,}000 = 0. \end{aligned}$$

Now

$$\frac{\partial F}{\partial L} = 80L^{-3/5}K^{3/5}, \qquad \frac{\partial F}{\partial K} = 120L^{2/5}K^{-2/5}, \qquad \frac{\partial G}{\partial L} = 400, \qquad \frac{\partial G}{\partial K} = 500.$$

At a maximizing point,

$$\frac{\partial F}{\partial L} = \lambda\frac{\partial G}{\partial L} \quad \text{and} \quad \frac{\partial F}{\partial K} = \lambda\frac{\partial G}{\partial K},$$

so that

$$80L^{-3/5}K^{3/5} = 400\lambda \quad \text{and} \quad 120L^{2/5}K^{-2/5} = 500\lambda.$$

Thus

$$\lambda = \frac{80L^{-3/5}K^{3/5}}{400} = \frac{120L^{2/5}K^{-2/5}}{500} \text{ and } \tfrac{1}{5}L^{-3/5}K^{3/5} = \tfrac{6}{25}L^{2/5}K^{-2/5}.$$

We multiply both sides by $L^{3/5}K^{2/5}$.

$$\frac{1}{5}(\overbrace{L^{-3/5}L^{3/5}}^{1}\overbrace{K^{3/5}K^{2/5}}^{K}) = \frac{6}{25}(\overbrace{L^{2/5}L^{3/5}}^{L}\overbrace{K^{-2/5}K^{2/5}}^{1}),$$

$$\tfrac{1}{5}K = \tfrac{6}{25}L,$$

$$K = \tfrac{30}{25}L = \tfrac{6}{5}L.$$

But

$$400L + 500K = 50{,}000 \qquad 400L + 500(\tfrac{6}{5})L = 50{,}000$$
$$400L + 600L = 50{,}000 \qquad 1000L = 50{,}000$$
$$L = 50 \qquad K = \tfrac{6}{5}L = 60.$$

Thus 50 units of labor and 60 units of capital should be used, and the maximum number of units produced is

$$F(50, 60) = 200(50)^{2/5}(60)^{3/5} \approx 11{,}156 \text{ units.}$$

We can use the result of the last example to make an interesting observation. The quantity $\partial F/\partial L$ is called the **marginal productivity of labor** and $\partial F/\partial K$ is called the **marginal productivity of capital**. At our maximizing values $L = 50$, $K = 60$, we have

$$\frac{\text{marginal productivity of labor}}{\text{marginal productivity of capital}} = \frac{\partial F/\partial L}{\partial F/\partial K} = \frac{80(50)^{-3/5}(60)^{3/5}}{120(50)^{2/5}(60)^{-2/5}}$$
$$= \frac{2}{3}\left[\frac{(60)^{3/5}(60)^{2/5}}{(50)^{2/5}(50)^{3/5}}\right] = \frac{2}{3}\cdot\frac{60}{50} = \frac{2}{3}\cdot\frac{6}{5} = \frac{4}{5}.$$

In addition,

$$\frac{\text{unit cost of labor}}{\text{unit cost of capital}} = \frac{400}{500} = \frac{4}{5}.$$

This is no coincidence! It is a general law of economics that

> when labor and capital investments are such as to maximize production, the ratio of marginal productivity of labor to marginal productivity of capital is equal to the ratio of the unit cost of labor to the unit cost of capital.

We note that we can use Lagrange multipliers in $\mathbb{R}^3$ if there are two or more constraint equations. Suppose, for example, we wish to maximize (or minimize) $w = f(x, y, z)$ subject to the constraints

$$g(x, y, z) = 0 \tag{7}$$

and

$$h(x, y, z) = 0. \tag{8}$$

Each of the equations (7) and (8) represent a surface in $\mathbb{R}^3$, and their intersection forms a curve in $\mathbb{R}^3$. By an argument very similar to the one we used earlier (but applied in $\mathbb{R}^3$ instead of $\mathbb{R}^2$), we find that if f is maximized (or minimized) at (x_0, y_0, z_0), then $\nabla f(x_0, y_0, z_0)$ is in the plane determined by $\nabla g(x_0, y_0, z_0)$ and $\nabla h(x_0, y_0, z_0)$. Thus, there are numbers λ and μ such that

$$\nabla f(x_0, y_0, z_0) = \lambda\, \nabla g(x_0, y_0, z_0) + \mu\, \nabla h(x_0, y_0, z_0) \tag{9}$$

(see Problem 1.7.52). Formula (9) is the generalization of formula (6) in the case of two constraints.

EXAMPLE 3 USING LAGRANGE MULTIPLIERS IN $\mathbb{R}^3$

Find the maximum value of $w = xyz$ among all points (x, y, z) lying on the line of intersection of planes $x + y + z = 30$ and $x + y - z = 0$.

SOLUTION: Setting $f(x, y, z) = xyz$, $g(x, y, z) = x + y + z - 30$, and $h(x, y, z) = x + y - z$, we obtain

$$\nabla f = yz\mathbf{i} + xz\mathbf{j} + xy\mathbf{k} \qquad \nabla g = \mathbf{i} + \mathbf{j} + \mathbf{k} \qquad \nabla h = \mathbf{i} + \mathbf{j} - \mathbf{k},$$

and using equation (9) to obtain the maximum, we obtain the equations

$$yz = \lambda + \mu \qquad xz = \lambda + \mu \qquad xy = \lambda - \mu.$$

Multiplying the three equations by x, y, and z, respectively, we find that

$$xyz = (\lambda + \mu)x = (\lambda + \mu)y = (\lambda - \mu)z.$$

If $\lambda + \mu = 0$, then $yz = 0$ and $xyz = 0$, which is not a maximum value since xyz can be positive (for example, $x = 8$, $y = 7$, $z = 15$ is in the constraint set and $xyz = 840$). Thus we can divide the first two equations by $\lambda + \mu$ to find that $x = y$.

Since $x + y - z = 0$, we have $2x - z = 0$, or $z = 2x$. But then

$$30 = x + y + z = x + x + 2x = 4x, \qquad \text{or} \qquad x = \frac{15}{2}.$$

Then

$$y = \frac{15}{2}, \qquad z = 15.$$

and the maximum value of xyz occurs at $(\frac{15}{2}, \frac{15}{2}, 15)$ and is equal to $(\frac{15}{2})(\frac{15}{2})15 = 843\frac{3}{4}$.

We conclude this section with two observations. First, while the outlined steps make the method of Lagrange multipliers seem easy, it should be noted that solving three nonlinear equations in three unknowns or four such equations in four unknowns often entails very involved algebraic manipulations. Second, no method is given for determining whether a solution found actually yields a maximum, a minimum, or neither. Fortunately, in many practical applications the existence of a maximum or a minimum can readily be inferred from the nature of the particular problem. For a function of two variables, the following theorem is true. We omit the proof.†

THEOREM 2 TEST TO DETERMINE WHETHER, FOR A CONSTRAINED MAX-MIN PROBLEM, (x_0, y_0) IS A LOCAL MAXIMUM OR A LOCAL MINIMUM

Suppose that f is a function of two variables. Define $\Delta(x_0, y_0)$ by

$$\begin{aligned}\Delta(x_0, y_0) = {} & 2g_x(x_0, y_0)g_y(x_0, y_0)[f_{xy}(x_0, y_0) - \lambda g_{xy}(x_0, y_0)] \\ & - [g_x(x_0, y_0)]^2[f_{yy}(x_0, y_0) - \lambda g_{yy}(x_0, y_0)] \\ & - [g_y(x_0, y_0)]^2[f_{xx}(x_0, y_0) - \lambda g_{xx}(x_0, y_0)].\end{aligned}$$

† For a proof see the paper by David Spring "On The Second Derivative Test For Constrained Local Extrema," *American Mathematical Monthly* 92 (1985), 631–643. The theorem cited here is a special case of Theorem 1 on page 635 in that paper.

If $\Delta(x_0, y_0) > 0$, then f has a local maximum at (x_0, y_0) subject to the constraint $g(x, y) = 0$

If $\Delta(x_0, y_0) < 0$, then f has a local minimum at (x_0, y_0) subject to the constraint $g(x, y) = 0$

If $\Delta(x_0, y_0) = 0$, then the test is inconclusive. ■

OF HISTORICAL INTEREST

JOSEPH LOUIS LAGRANGE, 1736–1813

JOSEPH LOUIS LAGRANGE, 1736–1813

Joseph Louis Lagrange was one of the two greatest mathematicians of the eighteenth century—the other being Leonhard Euler. Born in 1736 in Turin, Italy, Lagrange was the youngest of 11 children of French and Italian parents and the only one to survive to adulthood. Educated in Turin, Joseph Louis became a professor of mathematics in the military academy there when he was still quite young.

Lagrange's early publications established his reputation. When Euler left his post at the court of Frederick the Great in Berlin in 1766, he recommended that Lagrange be appointed his successor. Accepting Euler's advice, Frederick wrote to Lagrange that "the greatest king in Europe" wished to invite to his court "the greatest mathematician in Europe." Lagrange accepted and remained in Berlin for 20 years. Afterward, he accepted a post at the École Polytechnique in France.

Lagrange had a deep influence on nineteenth- and twentieth-century mathematics. He is perhaps best known as the first great mathematician to attempt to make calculus mathematically rigorous. His major work in this area was his 1797 paper "Théorie des fonctions analytiques contenant les principes du calcul différentiel." In this work, Lagrange tried to make calculus more logical—rather than more useful. His key idea was to represent a function $f(x)$ by a Taylor series. For example, we can write $1/(1 - x) = 1 + x + x^2 + \cdots + x^n + \cdots$ (a result that can be obtained by long division). Lagrange multiplied the coefficient of x^n by $n!$ and called the result the nth *derived function* of $1/(1 - x)$ at $x = 0$. This is the origin of the word *derivative*. The notation $f'(x), f''(x), \ldots$ was first used by Lagrange as was the form of the remainder term (3).

Lagrange is known for much else as well. Beginning in the 1750s, he invented the calculus of variations. He made significant contributions to ordinary differential equations, partial differential equations, numerical analysis, number theory, and algebra. In 1787 he published his *Mécanique analytique*, which contained the equations of motion of a dynamical system. Today these equations are known as *Lagrange's equations.*

Lagrange lived in France during the French Revolution. In 1790 he was placed on a committee to reform weights and measures and later became the head of a related committee that, in 1799, recommended the adoption of the system that we know today as the *metric system.* Despite his work for the revolution, however, Lagrange was disgusted by its cruelties. After the great French chemist Lavoisier was guillotined, Lagrange exclaimed, "It took the mob only a moment to remove his head; a century will not suffice to reproduce it."

In his later years, Lagrange was often lonely and depressed. When he was 56, the 17-year-old daughter of his friend the astronomer P. C. Lemonier was so moved by his unhappiness that she proposed to him. The resulting marriage apparently turned out to be ideal for both.

Perhaps the greatest tribute to Lagrange was given by Napoleon Bonaparte, who said, "Lagrange is the lofty pyramid of the mathematical sciences."

PROBLEMS 3.11

SELF-QUIZ

I. True–False: By using the Lagrange multiplier technique, we can look for extreme points of $F(x, y)$ subject to the constraint $3x + 2y = 18$ without needing to solve the constraint for $y = 9 - 1.5x$ and then considering $f(x) = F(x, 9 - 1.5x)$.

II. To find extreme points of xy subject to the constraint $3x + 2y = 18$, we consider the equation ________.

a. $xy = \lambda[3x + 2y - 18]$
b. $y + x = \lambda[3 + 2]$
c. $y\mathbf{i} + x\mathbf{j} = \lambda[3\mathbf{i} + 2\mathbf{j}]$
d. $x\mathbf{i} + y\mathbf{j} = \lambda[3\mathbf{i} + 2\mathbf{j}]$

III. True–False: If we change the constraint of the preceding problem to be $3x + 2y = 42$, the equation for the Lagrange multiplier λ changes also.

IV. To find extreme points of $3x + 2y$ subject to the constraint $xy = 18$, we consider the equation ________.

a. $\lambda[xy - 18] = 3x + 2y$
b. $\lambda[y + x] = 3 + 2$
c. $\lambda[y\mathbf{i} + x\mathbf{j}] = 3\mathbf{i} + 2\mathbf{j}$
d. $\lambda[x\mathbf{i} + y\mathbf{j}] = 3\mathbf{i} + 2\mathbf{j}$

For Problems 1–4, use the technique of Lagrange multipliers to solve Problems 19–22 of the preceding section. Assume that $x \geq 0$, $y \geq 0$, and $z \geq 0$.

1. Maximize xyz subject to $x + y + z = 50$.
2. Minimize $x + y + z$ subject to $xyz = 50$.
3. Maximize xy^2z^3 subject to $x + y + z = 50$.
4. Minimize $x^2 + y^2 + z^2$ subject to $x + y + z = 50$.

In Problems 5–12, use the technique of Lagrange multipliers to find the minimum distance between a point and a line or plane.

5. Point $(1, 2)$; line $2x + 3y = 5$.
6. Point $(3, -2)$; line $y = 2 - x$.
7. Point $(1, -1, 2)$; plane $x + y - z = 3$.
8. Point $(3, 0, 1)$; plane $2x - y + 4z = 5$.
9. Point $(0, 0)$; line $ax + by = d$.
10. Point $(0, 0, 0)$; plane $ax + by + cz = d$.
***11.** Point (x_0, y_0); line $ax + by = d$. (Compare your result here to the formula stated in Problem 1.2.78.)
***12.** Point (x_0, y_0, z_0); plane $ax + by + cz = d$. (Compare your result here to the formula stated in Problem 1.7.56.)
***13.** Find the maximum and minimum values of $x^2 + y^2$ subject to the condition $x^3 + y^3 = 6xy$.
14. Find the maximum and minimum values of $2x^2 + xy + y^2 - 2y$ subject to the condition $y = 2x - 1$.
15. Find the maximum and minimum values of $x^2 + y^2 + z^2$ subject to the condition $z^2 = x^2 - 1$.
16. Find the maximum and minimum values of $x^3 + y^3 + z^3$ if (x, y, z) lies on the sphere $x^2 + y^2 + z^2 = 4$.
17. Find the maximum and minimum values of $x + y + z$ if (x, y, z) lies on the sphere $x^2 + y^2 + z^2 = 1$.
18. Find the maximum and minimum values of xyz if (x, y, z) lies on the ellipsoid $x^2 + (y^2/4) + (z^2/9) = 1$.
19. Solve the preceding problem if (x, y, z) lies on the ellipsoid $(x/a)^2 + (y/b)^2 + (z/c)^2 = 1$.
***20.** Minimize the function $x^2 + y^2 + z^2$ for (x, y, z) on the planes $3x - y + z = 6$ and $x + 2y + 2z = 2$.
21. Find the minimum value of $x^3 + y^3 + z^3$ for (x, y, z) on the planes $x + y + z = 2$ and $x + y - z = 3$.
22. Find the maximum and minimum distances from the origin to a point on the ellipse $(x/a)^2 + (y/b)^2 = 1$.
23. Find the maximum and minimum distances from the origin to a point on the ellipsoid $(x/a)^2 + (y/b)^2 + (z/c)^2 = 1$.
***24.** Find the maximum value of $x_1 + x_2 + x_3 + x_4$ subject to $x_1^2 + x_2^2 + x_3^2 + x_4^2 = 1$. [*Hint:* Use the obvious generalization of Lagrange multipliers to functions of four variables.]
***25.** Find the maximum value of $x_1 + x_2 + \cdots + x_n$ subject to $x_1^2 + x_2^2 + \cdots + x_n^2 = 1$.
26. Find the maximum and minimum values of xyz subject to $x^2 + z^2 = 1$ and $x = y$.
27. Find the volume of the largest rectangular parallelepiped that can be inscribed in the ellipsoid $x^2 + 4y^2 + 9z^2 = 9$.
28. A silo is in the shape of a cylinder topped with a cone. The radius of each is 6 m and the total

surface area (excluding the base) is 200 m^2. What are the heights of the cylinder and cone that maximize the volume enclosed by the silo?

29. The base of an open-top rectangular box costs $\$3/m^2$ to construct, while the sides cost only $\$1/m^2$. Find the dimensions of the box of greatest volume that can be constructed for $36.

30. A manufacturing company has three plants I, II, and III, which produce x, y, and z units, respectively, of a certain product. The annual revenue from this production is given by $R(x, y, z) = 6xyz^2 - 400{,}000(x + y + z)$. If the company is to produce 1000 units annually, how should it allocate production among the three plants so as to maximize its profits?

31. A firm has $250,000 to spend on labor and raw materials. The output of the firm is αxy, where α is a constant and x and y are, respectively, the quantity of labor and raw materials consumed. The unit price of hiring labor is $5000, and the unit price of raw materials is $2500. Find the ratio of x to y that maximizes output.

32. The temperature of a point (x, y, z) on the unit sphere is given by $T(x, y, z) = xy + yz$. What is the hottest point on the sphere?

33. The Cobb-Douglas production function for a certain product is

$$F(L, K) = 250L^{0.7}K^{0.3}.$$

Suppose that each unit of labor costs $200, that each unit of capital costs $350, and that $25,000 is available for production costs.

a. How many units of labor and capital should be used to maximize output?

b. How many units will be produced?

c. Compute the ratio of marginal productivity of labor to marginal productivity of capital at levels of labor and capital costs that maximize output.

34. A product has the Cobb-Douglas production function

$$F(L, K) = 500L^{1/3}K^{2/3}.$$

Answer the questions of Problem 33 assuming that each unit of labor costs $1000, each unit of capital costs $1600, and a total of $250,000 is available for production.

35. A can of dog food is advertised to contain 80 units of protein. Two types of meat are used in making up the food. Each unit of liver costs 30¢, and each unit of horsemeat costs 16¢. If l units of liver and h units of horsemeat are in the can, then the number of units of protein is

$$N(l, h) = 4l^2 + 2.5h^2.$$

a. How many units of each meat should be put in a can of dog food to meet the advertised claim at minimum cost?

b. What is the minimum cost?

36. Bellingham Health Care (BHC) is a nonprofit foundation providing medical treatment to emotionally distressed children. BHC has hired you as a business consultant to aid the foundation in the development of a hiring policy that would be consistent with its overall goal of providing the most meaningful patient service possible given scarce foundation resources. In your initial analysis, you have determined that *service* can be described as a function of medical (M) and social services (S) staff input as follows:

$$\text{service} = M + 0.5S + 0.5MS - S^2.$$

BHC's staff budget for the coming year is $600,000. Annual employment costs total $15,000 for each social service staff member and $30,000 for each medical staff member.

a. Construct the function you might use to determine the optimal (service-maximizing) social service-medical employment combination.

b. Determine the optimal combination of social service and medical staff for BHC.

37. A field representative for a major pharmaceutical firm has just received the following information from a marketing research consultant who has been analyzing his recent sales performance. The consultant estimates that time spent in the two major metropolitan areas that compose his sales territory will result in monthly sales as indicated by the equation

$$\text{sales} = 500A - 20A^2 + 300B - 10B^2.$$

Here A and B represent the number of days spent in each metropolitan area respectively. Assuming that a working month is composed of twenty business days, what is the optimal number of days the salesperson should spend in each city?

38. For a simple lens of focal length f, the object distance d and the image distance i are related by the formula

$$\frac{1}{d} + \frac{1}{i} = \frac{1}{f}.$$

A given lens has a focal length of 50 cm.

a. What is the minimum value of the object-image distance $d + i$?

b. For what values of d and i is this minimum achieved?

39. Show that among all rectangles with the same perimeter, the square encloses the greatest area.

***40.** Show that among all triangles having the same perimeter, the equilateral triangle has the greatest area. [*Hint:* If the sides have lengths a, b, and c, **Heron's formula** gives the area as $\sqrt{s(s-a)(s-b)(s-c)}$ where $s = (a+b+c)/2$.]

41. Prove the **general arithmetic-geometric inequality**: If $p + q + r = 1$, then $x^p y^q z^r \leq px + qy + rz$ with equality if and only if $x = y = z > 0$. Assume that p, q, r, x, y, and z are nonnegative.

42. The task "Find the maximal area of a rectangle inscribed within semicircle of radius r" can be rephrased as "Maximize $A(x, y) = (2x)y$ subject to the constraint that $x^2 + y^2 = r^2$" because there is no loss of generality in fixing the semicircle to be the upper half of a circle centered at the origin. You now know several ways to analyze and solve this maximization task.

a. Think about the following fact:

$$2xy = (x^2 + y^2) - (x - y)^2$$
$$= r^2 - (x - y)^2 \leq r^2.$$

b. Solve the constraint for y; substitute that expression, $y = \sqrt{r^2 - x^2}$, into A; locate and analyze the critical point(s) of this function, $A(x) = (2x)\sqrt{r^2 - x^2}$, of a single variable x.

c. Use the implicit differentiation technique: Assume the constraint determines y as a function of x and differentiate to obtain $2x + 2yy' = 0$. Therefore $y' = -x/y$ and $A' = (2)y + (2x)y' = 2y - 2x^2/y, \ldots.$ (Alternatively, solve the simultaneous pair of equations $2x + 2yy' = 0$ and $2y + 2xy' = 0$.)

d. Use the technique of Lagrange multipliers: analyze the consequences of

$$2y\mathbf{i} + 2x\mathbf{j} = \lambda[2x\mathbf{i} + 2y\mathbf{j}].$$

Do the computations for each of these parts, then write a brief essay comparing the four techniques.

ANSWERS TO SELF-QUIZ

I. True **II.** c **III.** False **IV.** c

3.12 NEWTON'S METHOD FOR FUNCTIONS OF TWO VARIABLES (OPTIONAL)

In one-variable calculus you studied Newton's method for finding solutions of the equation $f(x) = 0$. There are theorems that guarantee that the sequence defined recursively by

$$x_{n+1} = x_n - \frac{f(x_n)}{f'(x_n)} \tag{1}$$

converges quadratically to a solution s of $f(x) = 0$ if we start at any point x_0 in a specified interval $[a, b]$.[†]

In this section we extend Newton's method to find points (x, y) such that

$$f(x, y) = 0 \tag{2}$$

and

$$g(x, y) = 0 \tag{3}$$

for some differentiable functions f and g. We saw a need to solve such a system in Section 3.10 in our calculation of critical points. For example, in Example 7 on page 217, we solved the system

$$\begin{aligned} 6(x^2 - 4y) &= 0 \\ -24(x - 2y^2) &= 0. \end{aligned}$$

We were able to find both solutions, (0, 0) and (2, 1) quite easily, but in most situations the computations are likely to be much more difficult.

The most compact form of Newton's method in two (or more) variables looks very much like equation (1). However, in order to present this form, it is necessary to make use of matrices and their inverses—a topic not discussed until Chapter 6. Instead, we will describe the method without making use of matrix notation. For those of you who have studied matrices, we give the compact form at the end of the section.

NEWTON'S METHOD IN TWO VARIABLES

Let (x_0, y_0) be chosen. Generate the sequence of vectors (x_n, y_n) recursively as follows: Let

$$D(x, y) = f_x(x, y)g_y(x, y) - f_y(x, y)g_x(x, y). \tag{4}$$

Then we define

NEWTON ITERATES IN $\mathbb{R}^2$

$$x_{n+1} = x_n - \frac{f(x_n, y_n)g_y(x_n, y_n) - f_y(x_n, y_n)g(x_n, y_n)}{D(x_n, y_n)} \tag{5}$$

$$y_{n+1} = y_n - \frac{-f(x_n, y_n)g_x(x_n, y_n) + f_x(x_n, y_n)g(x_n, y_n)}{D(x_n, y_n)}. \tag{6}$$

Under suitable conditions, which we do not give here, the sequence $\{(x_n, y_n)\}$ will con-

[†]One such theorem can be found on pages 231–32 of *Calculus* or *Calculus of One-Variable*.

verge to a vector (u, v) which satisfies

$$f(u, v) = g(u, v) = 0.$$

Newton's method is most likely to work when the initial vector (x_0, y_0) is close to a solution to the system (2), (3). It is not necessary that the system have only one solution.

EXAMPLE 1

USING NEWTON'S METHOD FOR A FUNCTION OF TWO VARIABLES

Use Newton's method to find a solution to the system

$$x^2 - 2x - y + \frac{1}{2} = 0$$
$$x^2 + 4y^2 - 4 = 0,$$

starting at the initial vector (2, 0.25). Carry all calculations to 6 decimal places of accuracy.

SOLUTION: We first compute

$$f = x^2 - 2x - y + \frac{1}{2} \qquad f_x = 2x - 2 \qquad \text{and} \qquad f_y = -1$$

$$g = x^2 + 4y^2 - 4 \qquad g_x = 2x \qquad \text{and} \qquad g_y = 8y$$

so

$$\begin{aligned} D(x, y) = f_x g_y - f_y g_x &= (2x - 2)(8y) - (-1)(2x) \\ &= 2x - 16y + 16xy \end{aligned}$$

$$\begin{aligned} fg_y - f_y g &= \left(x^2 - 2x - y + \frac{1}{2}\right)(8y) - (-1)(x^2 + 4y^2 - 4) \\ &= -4 + 4y + x^2 - 4y^2 - 16xy + 8x^2y \end{aligned}$$

$$\begin{aligned} -fg_x + f_x g &= -\left(x^2 - 2x - y + \frac{1}{2}\right)(2x) + (2x - 2)(x^2 + 4y^2 - 4) \\ &= 8 - 9x + 2xy + 2x^2 - 8y^2 + 8xy^2. \end{aligned}$$

Thus,

$$x_{n+1} = x_n - \frac{-4 + 4y_n + x_n^2 - 4y_n^2 - 16x_n y_n + 8x_n^2 y_n}{2x_n - 16y_n + 16x_n y_n}$$

$$y_{n+1} = y_n - \frac{8 - 9x_n + 2x_n y_n + 2x_n^2 - 8y_n^2 + 8x_n y_n^2}{2x_n - 16y_n + 16x_n y_n}.$$

For example, starting at $x_0 = 2$ and $y_0 = 0.25$, we obtain

$$x_1 = 2 - \frac{0.75}{8} = 1.90625$$

and

$$y_1 = 0.25 - \frac{-0.5}{8} = 0.3125.$$

We show the first four iterates in Table 1.

TABLE 1

n	x_n	y_n	x_{n+1}	y_{n+1}
0	2	0.25	1.90625	0.3125
1	1.90625	0.3125	1.900691	0.311213
2	1.900691	0.311213	1.900677	0.311219
3	1.900677	0.311219	1.900677	0.311219

We stop here since $(x_4, y_4) = (x_3, y_3)$ (to 6 decimal places). The vector $(1.900677, 0.311219)$ is correct to 6 decimal places.

CHECK:

$$x^2 - 2x - y + \frac{1}{2} = (1.900677)^2 - 2(1.900677) - 0.311219 + 0.5$$
$$= 0.000000058$$
$$x^2 + 4y^2 - 4 = (1.900677)^2 + 4(0.311219)^2 - 4 = 0.000002112.$$

The errors come from the 6-decimal-place rounding.

We see that even in a relatively simple example, implementing Newton's method in two variables can be very tedious. For that reason, it is best to carry out the computations on a computer or programmable calculator. Of course, you first have to differentiate and write out the rules (5) and (6) to begin the iteration. In Problems 3–9, you are asked to solve systems of equations on a computer.

NEWTON'S METHOD IN MATRIX FORM

We now write Newton's method in matrix form. This material should be read only if you know how to multiply matrices and compute the inverse of a matrix. (These topics are discussed in Chapter 6.)

We write the vector (x, y) as $\mathbf{x}$ and define the vector function $\mathbf{F}$ by

$$\mathbf{F}(\mathbf{x}) = \begin{pmatrix} f(\mathbf{x}) \\ g(\mathbf{x}) \end{pmatrix}.$$

We define the **Jacobian matrix** J by

$$J(\mathbf{x}) = \begin{pmatrix} f_x(\mathbf{x}) & f_y(\mathbf{x}) \\ g_x(\mathbf{x}) & g_y(\mathbf{x}) \end{pmatrix}.$$

Then Newton's method becomes

NEWTON'S ITERATION IN MATRIX NOTATION

$$\mathbf{x}_{n+1} = \mathbf{x}_n - J^{-1}(\mathbf{x}_n)\mathbf{F}(\mathbf{x}_n). \tag{7}$$

The nice thing about this formulation is that it applies equally well to solving systems of n equations in n unknowns, where $n > 2$.

In Problem 12 you are asked to show that (7) reduces to (5) and (6) when J^{-1} is calculated and the matrix multiplication in (7) is carried out.

REMARK: Jacobian matrices are discussed in Section 9.4.

PROBLEMS 3.12

SELF-QUIZ

I. In order to use Newton's method to solve the system (2), (3) it is necessary that
- **a.** f and g be differentiable at (x_0, y_0).
- **b.** f_x, f_y, g_x, and g_y exist at (x_0, y_0).
- **c.** the system (2), (3) have a unique solution.
- **d.** $D(x_0, y_0) \neq 0$.

II. In order to carry out Newton's method in two variables, it is necessary to evaluate ________ functions at every step (not counting additions, subtractions, multiplications, or divisions of functions).
- **a.** 4 **b.** 5 **c.** 6
- **d.** 8 **e.** 10 **f.** 14

1. Consider the system

$$x^2 - y = 0$$
$$y^2 - x = 0.$$

- **a.** Verify that the solutions are (0, 0) and (1, 1).
- **b.** Start with $(x_0, y_0) = (\frac{1}{4}, -\frac{1}{4})$ and use Newton's method to find (x_1, y_1) and (x_2, y_2).
- **c.** Start with $(x_0, y_0) = (0.9, 1.25)$ and use Newton's method to find (x_1, y_1) and (x_2, y_2).

2. Consider the system

$$x^2 + y^2 - 2 = 0$$
$$x^2 - y = 0.$$

- **a.** Verify that the solutions are (1, 1) and (−1, 1).
- **b.** Start with $(x_0, y_0) = (0.8, 0.75)$ and use Newton's method to find (x_1, y_1) and (x_2, y_2).
- **c.** Start with $(x_0, y_0) = (-1.25, 0.75)$ and use Newton's method to find (x_1, y_1) and (x_2, y_2).

In Problems 3–8, a system is given.
- **a.** Start at (x_0, y_0) and use Newton's method to find (x_1, y_1) and (x_2, y_2).
- **b.** Write a computer program to find all the solutions to 8 decimal places of accuracy.

3. $2xy - 3 = 0$
$x^2 - y - 2 = 0$
$(x_0, y_0) = (1.5, 0.9)$

4. $x^2 + 4y^2 - 4 = 0$
$x^2 - 2x - y + 1 = 0$
- **i.** $(x_0, y_0) = (1.5, 0.5)$
- **ii.** $(x_0, y_0) = (-0.25, 1.1)$

5. $3x^2 - 2y^2 - 1 = 0$
$x^2 - 2x + y^2 + 2y - 8 = 0$
- **i.** $(x_0, y_0) = (-1, 1)$
- **ii.** $(x_0, y_0) = (3, -3.4)$

6. $-x + y^2 - 2 = 0$
$x^3 - 3x^2 + 4x - y = 0$
- **i.** $(x_0, y_0) = (0.5, 1.2)$
- **ii.** $(x_0, y_0) = (-0.25, -1.3)$

7. $2x^3 - 12x - y - 1 = 0$
$3y^2 - 6y - x - 3 = 0$
- **i.** $(x_0, y_0) = (2.5, 2.5)$
- **ii.** $(x_0, y_0) = (2.5, -1)$
- **iii.** $(x_0, y_0) = (0, 0)$
- **iv.** $(x_0, y_0) = (-2.5, 2.5)$
- **v.** $(x_0, y_0) = (-2.5, 0)$
- **vi.** $(x_0, y_0) = (0, 2.5)$

8. $3x^2 - 2y^2 - 1 = 0$
$x^2 - 2x + 2y - 8 = 0$
- **i.** $(x_0, y_0) = (2.5, 3)$
- **ii.** $(x_0, y_0) = (-1.6, 1.6)$
- **iii.** $(x_0, y_0) = (5.6, -7)$
- **iv.** $(x_0, y_0) = (-3, -3.6)$

9. Use Newton's method to find all nine solutions to

$$7x^3 - 10x - y - 1 = 0$$
$$8y^3 - 11y + x - 1 = 0.$$

Use the starting points (0, 0), (1, 0), (0, 1), (−1, 0), (0, −1), (1, 1), (−1, 1), (1, −1), and (−1, −1).

***10.** Consider the nonlinear system

$$x^2 + y^2 - 2 = 0$$
$$xy - 1 = 0.$$

a. Verify that the solutions are (1, 1) and (−1, −1).

b. What difficulties might arise if we try to use Newton's method to find the solutions?

(For those who know how to manipulate matrices.)

11. Show that $J^{-1}(\mathbf{x}) = \frac{1}{D}\begin{pmatrix} g_y & -f_y \\ -g_x & f_x \end{pmatrix}$.

12. Show that equation (7) yields equations (5) and (6) when the matrices in (7) are multiplied and added.

Consider the system

$$f(x, y, z) = 0 \qquad (8)$$
$$g(x, y, z) = 0 \qquad (9)$$
$$h(x, y, z) = 0. \qquad (10)$$

Write $\mathbf{x} = (x, y, z)$, $\mathbf{F} = \begin{pmatrix} f \\ g \\ h \end{pmatrix}$, and $J = \begin{pmatrix} f_x & f_y & f_z \\ g_x & g_y & g_z \\ h_x & h_y & h_z \end{pmatrix}$. Then equation (7) defines Newton's method in three variables. In Problems 13 and 14, use Newton's method to find all solutions to the given system.

****13.** $x^2 - x + y^2 + z^2 - 5 = 0$
$x^2 + y^2 - y + z^2 - 4 = 0$
$x^2 + y^2 + z^2 + z - 6 = 0$

a. Start with $(x_0, y_0, z_0) = (-0.8, 0.2, 1.8)$.

b. Start with $(x_0, y_0, z_0) = (1.2, 2.2, -0.2)$.

****14.** $x^2 - x + 2y^2 + yz - 10 = 0$
$5x - 6y + z = 0$
$z - x^2 - y^2 = 0$
$(x_0, y_0, z_0) = (1.1, 1.5, 3.5)$

ANSWERS TO SELF-QUIZ

I. b, d **II.** c

CHAPTER 3 SUMMARY OUTLINE

- **Function of Two Variables** Let D be a subset of $\mathbb{R}^2$. Then a **real-valued function of two variables** f is a rule that assigns to each point (x, y) in D a unique real number that we denote by $f(x, y)$. The set D is called the **domain** of f. The set $\{f(x, y): (x, y) \in D\}$, which is the set of values the function f takes on, is called the **range** of f.
- **Function of Three Variables** Let D be a subset of $\mathbb{R}^3$. Then a **real-valued function of three variables** f is a rule that assigns to each point (x, y, z) in D a unique real number that we denote by $f(x, y, z)$. The set D is called the **domain** of f, and the set $\{f(x, y, z): (x, y, z) \in D\}$, which is the set of values the function f takes on, is called the **range** of f.
- **Graph of a Function of Two Variables** The **graph** of a function f of two variables x and y is the set of all points (x, y, z) in $\mathbb{R}^3$ such that $z = f(x, y)$.
- **Surface** The graph of a continuous function of two variables is called a **surface** in $\mathbb{R}^3$.
- **Level Curve** If z is fixed, then the equation $f(x, y) = z$ is the equation of a curve in the xy-plane, called a **level curve**.
- **Open Disk** The **open disk** D_r centered at (x_0, y_0) with radius r is the subset of $\mathbb{R}^2$ given by

$$\{(x, y): |(x, y) - (x_0, y_0)| < r\}. \qquad \text{(i)}$$

- **Closed Disk** The **closed disk** centered at (x_0, y_0) with radius r is the subset of $\mathbb{R}^2$ given by

$$\{(x, y): |(x, y) - (x_0, y_0)| \le r\}. \qquad \text{(ii)}$$

- **Boundary** The **boundary** of the open or closed disk defined in (i) or (ii) is the circle

$$\{(x, y): |(x, y) - (x_0, y_0)| = r\}.$$

- **Neighborhood** A **neighborhood** of a point (x_0, y_0) in $\mathbb{R}^2$ is an open disk centered at (x_0, y_0).
- **Limit** Let $f(x, y)$ be defined in a neighborhood of (x_0, y_0) but not necessarily at (x_0, y_0) itself. Then the **limit** of $f(x, y)$ as (x, y) approaches (x_0, y_0) is the real number L, written

$$\lim_{(x,y)\to(x_0,y_0)} f(x, y) = L,$$

if for every number $\epsilon > 0$, there is a number $\delta > 0$ such that $|f(x, y) - L| < \epsilon$ for every $(x, y) \neq (x_0, y_0)$ in the open disk centered at (x_0, y_0) with radius δ.

- **Rule for Nonexistence of a Limit** If we get two or more different values for $\lim_{(x,y)\to(x_0,y_0)} f(x, y)$ as we approach (x_0, y_0) along different paths, or if the limit fails to exist along some path, then $\lim_{(x,y)\to(x_0,y_0)} f(x, y)$ does not exist.

- **Continuity**

  i. Let $f(x, y)$ be defined at every point (x, y) in a neighborhood of (x_0, y_0). Then f is **continuous** at (x_0, y_0) if all of the following conditions hold:
  a. $f(x_0, y_0)$ exists [i.e., (x_0, y_0) is in the domain of f].
  b. $\lim_{(x,y)\to(x_0,y_0)} f(x, y)$ exists.
  c. $\lim_{(x,y)\to(x_0,y_0)} f(x, y) = f(x_0, y_0)$.

  ii. If one or more of these three conditions fail to hold, then f is said to be **discontinuous** at (x_0, y_0).

  iii. f is **continuous in a subset** S of $\mathbb{R}^2$ if f is continuous at every point (x, y) in S.

- **Polynomial and Rational Functions**

  i. A **polynomial** $p(x, y)$ in the two variables x and y is a finite sum of terms of the form $Ax^m y^n$, where m and n are nonnegative integers and A is a real number.

  ii. A **rational function** $r(x, y)$ in the two variables x and y is a function that can be written as the quotient of two polynomials in x and y: $r(x, y) = p(x, y)/q(x, y)$.

- **Facts about Continuity**

  i. Any polynomial p is continuous at any point in $\mathbb{R}^2$.

  ii. Any rational function $r = p/q$ is continuous at any point (x_0, y_0) for which $q(x_0, y_0) \neq 0$. It is discontinuous when $q(x_0, y_0) = 0$ because it is then not defined at (x_0, y_0).

  iii. If f and g are continuous at (x_0, y_0), then $f + g$, $f - g$, and $f \cdot g$ are continuous at (x_0, y_0).

  iv. If f and g are continuous at (x_0, y_0) and if $g(x_0, y_0) \neq 0$, then f/g is continuous at (x_0, y_0).

  v. If f is continuous at (x_0, y_0) and if h is a function of one variable that is continuous at $f(x_0, y_0)$, then the composite function $h \circ f$, defined by $(h \circ f)(x, y) = h(f(x, y))$, is continuous at (x_0, y_0).

- **Partial Derivatives in $\mathbb{R}^2$** Let $z = f(x, y)$.

  i. The **partial derivative of f with respect to** x is the function

$$\frac{\partial z}{\partial x} = \frac{\partial f}{\partial x} = \lim_{\Delta x\to 0} \frac{f(x + \Delta x, y) - f(x, y)}{\Delta x}. \tag{1}$$

  $\partial f/\partial x$ is defined at every point (x, y) in the domain of f such that the limit (1) exists.

  ii. The **partial derivative of f with respect to** y is the function

$$\frac{\partial z}{\partial y} = \frac{\partial f}{\partial y} = \lim_{\Delta y\to 0} \frac{f(x, y + \Delta y) - f(x, y)}{\Delta y}. \tag{2}$$

  $\partial f/\partial y$ is defined at every point (x, y) in the domain of f such that the limit (2) exists.

- **Partial Derivatives in $\mathbb{R}^3$** Let $w = f(x, y, z)$.

  i. The **partial derivative of f with respect to** x is the function

$$\frac{\partial w}{\partial x} = \frac{\partial f}{\partial x} = f_x = \lim_{\Delta x\to 0} \frac{f(x + \Delta x, y, z) - f(x, y, z)}{\Delta x}. \tag{3}$$

  f_x is defined at every point (x, y, z) in the domain of f at which the limit in (3) exists.

  ii. The **partial derivative of f with respect to** y is the function

$$\frac{\partial w}{\partial y} = \frac{\partial f}{\partial y} = f_y = \lim_{\Delta y\to 0} \frac{f(x, y + \Delta y, z) - f(x, y, z)}{\Delta y}. \tag{4}$$

  f_y is defined at every point (x, y, z) in the domain of f at which the limit in (4) exists.

  iii. The **partial derivative of f with respect to** z is the function

$$\frac{\partial w}{\partial z} = \frac{\partial f}{\partial z} = f_z = \lim_{\Delta z\to 0} \frac{f(x, y, z + \Delta z) - f(x, y, z)}{\Delta z}. \tag{5}$$

  f_z is defined at any point (x, y, z) in the domain of f at which the limit in (5) exists.

- **Second Partial Derivatives in $\mathbb{R}^2$**

  i. Differentiate twice with respect to x:

$$\frac{\partial^2 z}{\partial x^2} = \frac{\partial^2 f}{\partial x^2} = f_{xx} = \frac{\partial}{\partial x}\left(\frac{\partial f}{\partial x}\right).$$

  ii. Differentiate first with respect to x and then with respect to y:

$$\frac{\partial^2 z}{\partial y\, \partial x} = \frac{\partial^2 f}{\partial y\, \partial x} = f_{xy} = \frac{\partial}{\partial y}\left(\frac{\partial f}{\partial x}\right).$$

  iii. Differentiate first with respect to y and then with respect to x:

$$\frac{\partial^2 z}{\partial x\, \partial y} = \frac{\partial^2 f}{\partial x\, \partial y} = f_{yx} = \frac{\partial}{\partial x}\left(\frac{\partial f}{\partial y}\right).$$

iv. Differentiate twice with respect to y:

$$\frac{\partial^2 z}{\partial y^2} = \frac{\partial^2 f}{\partial y^2} = f_{yy} = \frac{\partial}{\partial y}\left(\frac{\partial f}{\partial y}\right).$$

- **Equality of Mixed Partials in $\mathbb{R}^2$** Suppose that f, f_x, f_y, f_{xy}, and f_{yx} are all continuous at (x_0, y_0). Then

$$f_{xy}(x_0, y_0) = f_{yx}(x_0, y_0)$$

- **Equality of Mixed Partials in $\mathbb{R}^3$** If f, f_x, f_y, f_z, and all six mixed partials are continuous at a point (x_0, y_0, z_0), then at that point

$$f_{xy} = f_{yx}, \qquad f_{xz} = f_{zx}, \qquad f_{yz} = f_{zy}.$$

- **Differentiability of a Function of Two Variables** Let f be a real-valued function of two variables that is defined in a neighborhood of a point (x, y) and such that $f_x(x, y)$ and $f_y(x, y)$ exist. Then f is **differentiable** at (x, y) if there exist functions $\epsilon_1(\Delta x, \Delta y)$ and $\epsilon_2(\Delta x, \Delta y)$ such that

$$f(x + \Delta x, y + \Delta y) - f(x, y)$$
$$= f_x(x, y)\Delta x + f_y(x, y)\Delta y + \epsilon_1(\Delta x, \Delta y)\Delta x + \epsilon_2(\Delta x, \Delta y)\Delta y,$$

where

$$\lim_{(\Delta x,\Delta y)\to(0,0)} \epsilon_1(\Delta x, \Delta y) = 0 \quad \text{and}$$

$$\lim_{(\Delta x,\Delta y)\to(0,0)} \epsilon_2(\Delta x, \Delta y) = 0.$$

- **The Gradient in $\mathbb{R}^2$** Let f be a function of two variables such that f_x and f_y exist at a point $\mathbf{x} = (x, y)$. Then the **gradient** of f at $\mathbf{x}$, denoted by $\nabla f(\mathbf{x})$, is given by

$$\nabla f(\mathbf{x}) = f_x(x, y)\mathbf{i} + f_y(x, y)\mathbf{j}.$$

Note that the gradient of f is a **vector function**. That is, for every point $\mathbf{x}$ in $\mathbb{R}^2$ for which $\nabla f(\mathbf{x})$ is defined, we see that $\nabla f(\mathbf{x})$ is a vector in $\mathbb{R}^2$.

- **Differentiability in $\mathbb{R}^2$** (Gradient Notation) Let f be a function of two variables that is defined in a neighborhood of a point $\mathbf{x} = (x, y)$. Let $\Delta\mathbf{x} = (\Delta x, \Delta y)$. If $f_x(x, y)$ and $f_y(x, y)$ exist, then f is **differentiable** at $\mathbf{x}$ if there is a function g such that

$$f(\mathbf{x} + \Delta\mathbf{x}) - f(\mathbf{x}) = \nabla f(\mathbf{x}) \cdot \Delta\mathbf{x} + g(\Delta\mathbf{x})$$

where

$$\lim_{\Delta\mathbf{x}\to\mathbf{0}} \frac{g(\Delta\mathbf{x})}{|\Delta\mathbf{x}|} = 0.$$

- **Continuously Differentiable (Smooth) Function** A function of two variables is said to be **continuously differentiable** or **smooth** if it has continuous first partial derivatives. A smooth function is differentiable.

- **The Gradient of a Scalar Multiple or Sum of Functions** Let f and g be differentiable in a neighborhood of $\mathbf{x} = (x, y)$. Then for every scalar α, αf and $f + g$ are differentiable at $\mathbf{x}$, and

i. $\nabla(\alpha f) = \alpha\nabla f$, and

ii. $\nabla(f + g) = \nabla f + \nabla g$.

- **The Gradient in $\mathbb{R}^3$** Let f be a scalar function of three variables such that f_x, f_y, and f_z exist at a point $\mathbf{x} = (x, y, z)$. Then the **gradient** of f at $\mathbf{x}$, denoted by $\nabla f(\mathbf{x})$, is given by the vector

$$\nabla f(\mathbf{x}) = f_x(x, y, z)\mathbf{i} + f_y(x, y, z)\mathbf{j} + f_z(x, y, z)\mathbf{k}.$$

- **Differentiability in $\mathbb{R}^3$** Let f be a function of three variables that is defined in a neighborhood of $\mathbf{x} = (x, y, z)$, and let $\Delta\mathbf{x} = (\Delta x, \Delta y, \Delta z)$. If $f_x(x, y, z)$, $f_y(x, y, z)$, and $f_z(x, y, z)$ exist, then f is **differentiable** at $\mathbf{x}$ if there is a function g such that

$$f(\mathbf{x} + \Delta\mathbf{x}) - f(\mathbf{x}) = \nabla f(\mathbf{x}) \cdot \Delta\mathbf{x} + g(\Delta\mathbf{x})$$

where

$$\lim_{\Delta\mathbf{x}\to\mathbf{0}} \frac{g(\Delta\mathbf{x})}{|\Delta\mathbf{x}|} = 0.$$

Equivalently, we can write

$$f(x + \Delta x, y + \Delta y, z + \Delta z) - f(x, y, z)$$
$$= f_x(x, y, z)\Delta x + f_y(x, y, z)\Delta y + f_z(x, y, z)\Delta z + g(\Delta x, \Delta y, \Delta z)$$

where

$$\lim_{(\Delta x,\Delta y,\Delta z)\to(0,0,0)} \frac{g(\Delta x, \Delta y, \Delta z)}{\sqrt{\Delta x^2 + \Delta y^2 + \Delta z^2}} = 0.$$

- **Chain Rules in $\mathbb{R}^2$**

i. Let $z = f(x, y)$ be differentiable and suppose that $x = x(t)$ and $y = y(t)$. Assume further that dx/dt and dy/dt exist and are continuous. Then z can be written as a function of the parameter t, and

$$\frac{dz}{dt} = \frac{\partial z}{\partial x}\frac{dx}{dt} + \frac{\partial z}{\partial y}\frac{dy}{dt} = f_x\frac{dx}{dt} + f_y\frac{dy}{dt}. \qquad (*)$$

We can also write this result using our gradient notation. If $\mathbf{g}(t) = x(t)\mathbf{i} + y(t)\mathbf{j}$, then $\mathbf{g}'(t) = (dx/dt)\mathbf{i} + (dy/dt)\mathbf{j}$, and $(*)$ can be written as

$$\frac{d}{dt}f(x(t), y(t)) = (f \circ \mathbf{g})'(t) = [f(\mathbf{g}(t))]'$$
$$= \nabla f \cdot \mathbf{g}'(t)$$

ii. Let $z = f(x, y)$ be differentiable and suppose that x and y are functions of the two variables r and s. That is, $x = x(r, s)$ and $y = y(r, s)$. Suppose further that $\partial x/\partial r$, $\partial x/\partial s$, $\partial y/\partial r$, and $\partial y/\partial s$ all exist and are continuous. Then z can be written as a differentiable function of r and s and

$$\frac{\partial z}{\partial r} = \frac{\partial z}{\partial x}\frac{\partial x}{\partial r} + \frac{\partial z}{\partial y}\frac{\partial y}{\partial r} \qquad \frac{\partial z}{\partial s} = \frac{\partial z}{\partial x}\frac{\partial x}{\partial s} + \frac{\partial z}{\partial y}\frac{\partial y}{\partial s}.$$

Analogous chain rules hold in $\mathbb{R}^3$.

- **Smooth Surface** The surface $F(x, y, z) = 0$ is called **differentiable** or **smooth** at a point (x_0, y_0, z_0) if $\partial F/\partial x$, $\partial F/\partial y$, and $\partial F/\partial z$ all exist and are continuous at (x_0, y_0, z_0).
- The gradient of F at a point $\mathbf{x}_0 = (x_0, y_0, z_0)$ on S is orthogonal to the tangent vector at $\mathbf{x}_0$ to any curve C remaining on S and passing through $\mathbf{x}_0$.
- **Tangent Plane and Normal Line** Let F be differentiable at $\mathbf{x}_0 = (x_0, y_0, z_0)$ and let the smooth surface S be defined by $F(x, y, z) = 0$.

  i. The **tangent plane** to S at (x_0, y_0, z_0) is the plane passing through the point (x_0, y_0, z_0) with normal vector $\nabla F(\mathbf{x}_0)$.

  ii. The **normal line** to S at $\mathbf{x}_0$ is the line passing through $\mathbf{x}_0$ having the same direction as $\nabla F(\mathbf{x}_0)$.
- **Directional Derivative** Let f be differentiable at a point $\mathbf{x}_0 = (x_0, y_0)$ in $\mathbb{R}^2$ and let $\mathbf{u}$ be a unit vector. Then the **directional derivative of f in the direction u**, denoted by $f'_{\mathbf{u}}(\mathbf{x}_0)$, is given by

$$f'_{\mathbf{u}}(\mathbf{x}_0) = \nabla f(\mathbf{x}_0) \cdot \mathbf{u}.$$

- **Direction of Maximum Rate of Increase** Let f be differentiable. Then f increases most rapidly in the direction of its gradient and decreases most rapidly in the direction opposite to that of its gradient. It does not change in a direction perpendicular to its gradient.
- **Increment and the Total Differential** Let $f = f(\mathbf{x})$ be a function of two or three variables and let $\Delta\mathbf{x} = (\Delta x, \Delta y)$ or $(\Delta x, \Delta y, \Delta z)$.

  i. The **increment of** f, denoted by Δf, is defined by

$$\Delta f = f(\mathbf{x} + \Delta\mathbf{x}) - f(\mathbf{x}).$$

  ii. The **total differential of** f, denoted by df, is given by

$$df = \nabla f(\mathbf{x}) \cdot \Delta\mathbf{x}.$$

- **Boundary Point** A point $\mathbf{P}$ in $\mathbb{R}^2$ is a **boundary point** of a region Ω if every open disk centered at $\mathbf{P}$ contains points that are in Ω and points that are not in Ω.
- **Closed Region** A region in $\mathbb{R}^2$ is closed if it contains all its boundary points.
- **Bounded Region** A region in $\mathbb{R}^2$ is **bounded** if there is an open disk that completely surrounds it.
- **Upper and Lower Bound Theorem in $\mathbb{R}^2$** Let f be a function of two variables that is continuous on the closed, bounded region Ω. Then there exist points $\mathbf{x}_1$ and $\mathbf{x}_2$ in Ω such that

$$m = f(\mathbf{x}_1) \le f(\mathbf{x}) \le f(\mathbf{x}_2) = M$$

  for any other point $\mathbf{x}$ in Ω.
- **Global Minimum and Maximum** The numbers m and M are called, respectively, the **global minimum** and the **global maximum** for the function f over the region Ω.
- **Local Maxima and Minima** Let f be defined in a neighborhood of a point $\mathbf{x}_0 = (x_0, y_0)$.

  i. f has a **local maximum** at $\mathbf{x}_0$ if there is a neighborhood N_1 of $\mathbf{x}_0$ such that for every point $\mathbf{x}$ in N_1,

$$f(\mathbf{x}) \le f(\mathbf{x}_0).$$

  ii. f has a **local minimum** at $\mathbf{x}_0$ if there is a neighborhood N_2 of $\mathbf{x}_0$ such that for every point $\mathbf{x}$ in N_2,

$$f(\mathbf{x}) \ge f(\mathbf{x}_0).$$

  iii. If f has a local maximum or minimum at $\mathbf{x}_0$, then $\mathbf{x}_0$ is called an **extreme point** of f.
- If $\mathbf{x}_0$ is an extreme point for the differentiable function f, then

$$\nabla f(\mathbf{x}_0) = \mathbf{0}.$$

- **Critical Point** $\mathbf{x}_0$ is a **critical point** of f if f is differentiable at $\mathbf{x}_0$ and $\nabla f(\mathbf{x}_0) = \mathbf{0}$ or if $\mathbf{x}_0$ is in the domain of f but $\nabla f(\mathbf{x}_0)$ does not exist.
- **Saddle Point** If $\mathbf{x}_0$ is a critical point of f but is not an extreme point of f, then $\mathbf{x}_0$ is called a **saddle point** of f.
- **Second-Derivatives Test** Let f and all its first and second partial derivatives be continuous in a neighborhood of (x_0, y_0). Suppose that $f_x(x_0, y_0) = 0$ and $f_y(x_0, y_0) = 0$. That is, (x_0, y_0) is a critical point of f. Let

$$D(x, y) = f_{xx}(x, y)f_{yy}(x, y) - [f_{xy}(x, y)]^2$$

  and let D denote $D(x_0, y_0)$. D is called the **discriminant** of f.

  i. If $D > 0$ and $f_{xx}(x_0, y_0) > 0$, then f has a local minimum at (x_0, y_0).

  ii. If $D > 0$ and $f_{xx}(x_0, y_0) < 0$, then f has a local maximum at (x_0, y_0).

  iii. If $D < 0$, then (x_0, y_0) is a saddle point of f.

  iv. If $D = 0$, then any of the preceding is possible.
- **Constrained Maximization or Minimization Problem**

  In $\mathbb{R}^2$: Maximize (or minimize): $z = f(x, y)$

  subject to the constraint: $g(x, y) = 0$.

  In $\mathbb{R}^3$: Maximize (or minimize): $w = f(x, y, z)$

  subject to the constraint: $g(x, y, z) = 0$.
- **Lagrange Multiplier** If, subject to the constraint $g(\mathbf{x}) = 0$, f takes its maximum (or minimum) value at a point $\mathbf{x}_0$, then there is a number λ such that

$$\nabla f(\mathbf{x}_0) = \lambda \nabla g(\mathbf{x}_0).$$

  The number λ is called a **Lagrange multiplier**.
- **Newton Iterates in $\mathbb{R}^2$** In order to find solutions to the system

$$f(x, y) = 0$$
$$g(x, y) = 0,$$

generate a sequence $\{(x_n, y_n)\}$ as follows: Choose a starting vector (x_0, y_0). Then set

$$x_{n+1} = x_n - \frac{f(x_n, y_n)g_y(x_n, y_n) - f_y(x_n, y_n)g(x_n, y_n)}{D(x_n, y_n)}$$

$$y_{n+1} = y_n - \frac{-f(x_n, y_n)g_x(x_n, y_n) + f_x(x_n, y_n)g(x_n, y_n)}{D(x_n, y_n)}$$

where $D = f_x g_y - g_x f_y$. Under suitable conditions, the sequence will converge to a solution of the system. It is helpful if (x_0, y_0) is reasonably close to a solution.

CHAPTER 3 REVIEW EXERCISES

In Problems 1–6, find the domain and range of the given function.

1. $f(x, y) = \sqrt{x^2 - y^2}$
2. $f(x, y) = 1/\sqrt{x^2 + y^2}$
3. $f(x, y) = \cos(x + 3y)$
4. $f(x, y, z) = \sqrt{1 - x^2 - y^2 - z^2}$
5. $f(x, y, z) = 1/\sqrt{x^2 + y^2 + z^2 - 1}$
6. $f(x, y, z) = \ln(x - y + 4z - 3)$

In Problems 7–10, describe the level curves of the given functions; then sketch these curves for the specified values of z.

7. $z = \sqrt{1 - x - y}$; $z = 0, 1, 3, 8$
8. $z = \sqrt{1 - y^2 + x}$; $z = 0, 2, 4, 7$
9. $z = \ln(x - 3y)$; $z = 0, 1, 2, 3$
10. $z = \dfrac{x^2 + y^2}{x^2 - y^2}$; $z = 1, 3, 6$
11. Sketch the open disk centered at $(-1, 2)$ with radius 4.
12. Sketch the closed ball centered at $(1, 2, 3)$ with radius 2.
13. Show that $\displaystyle\lim_{(x,y)\to(0,0)} \frac{xy}{y^2 - x^2}$ does not exist.
14. Show that $\displaystyle\lim_{(x,y)\to(0,0)} \frac{y^2 - 2x}{y^2 + 2x}$ does not exist.
15. Show that $\displaystyle\lim_{(x,y)\to(0,0)} \frac{4xy^3}{x^2 + y^4} = 0.$
16. Show that $\displaystyle\lim_{(x,y,z)\to(0,0,0)} \frac{zy^2 + x^3}{x^2 + y^2 + z^2} = 0.$
17. Calculate $\displaystyle\lim_{(x,y)\to(1,-2)} \frac{1 + x^2y}{2 - y}.$
18. Calculate $\displaystyle\lim_{(x,y,z)\to(2,-1,1)} \frac{x^2 - yz^3}{1 + xyz - 2y^5}.$
19. Find the maximum region over which the function $f(x, y) = \ln(1 - 2x + 3y)$ is continuous.
20. Find the maximum region over which the function $f(x, y, z) = \ln(x - y - z + 4)$ is continuous.
21. Find the maximum region over which the function $f(x, y, z) = 1/\sqrt{1 - x^2 + y^2 - z^2}$ is continuous.
22. Find a number c such that the function

$$f(x, y) = \begin{cases} \dfrac{-2xy}{\sqrt{x^2 + y^2}} & \text{if } (x, y) \neq (0, 0) \\ c & \text{if } (x, y) = (0, 0) \end{cases}$$

is continuous at the origin.

In Problems 23–34, calculate all first partial derivatives.

23. $f(x, y) = y/x$
24. $f(x, y) = \cos(x - 3y)$
25. $f(x, y) = 1/\sqrt{x^2 - y^2}$
26. $f(x, y) = \tan^{-1}\left(\dfrac{y}{1 + x}\right)$
27. $f(x, y, z) = \ln(x - y + 4z)$
28. $f(x, y\, z) = 1/\sqrt{x^2 + y^2 + z^2}$
29. $f(x, y, z) = \cosh(y/x^2)$
30. $f(x, y, z) = \sec\left(\dfrac{x - y}{z}\right)$
31. $f(x, y, z) = (x^2y - y^3z^5 + x\sqrt{z})^{2/3}$
32. $f(x, y, z) = \dfrac{x^2 - y^3}{y^3 + z^4}$
33. $f(x, y, z, w) = \dfrac{x - z + w}{y + 2w - x}$
34. $f(x, y, z, w) = e^{(x-w)/(y+z)}$

In Problems 35–40, calculate all second partial derivatives and show that all pairs of mixed partials are equal.

35. $f(x, y) = xy^3$
36. $f(x, y) = \tan^{-1}(y/x)$
37. $f(x, y) = \sqrt{x^2 - y^2}$
38. $f(x, y) = \dfrac{x + y}{x - y}$
39. $f(x, y, z) = \ln(2 - 3x + 4y - 7z)$
40. $f(x, y, z) = 1/\sqrt{1 - x^2 - y^2 - z^2}$
41. Calculate f_{yzx} if $f(x, y, z) = x^2y^3 - zx^5$.
42. Calculate f_{zxxyz} if $f(x, y, z) = (x - y)/z$.

In Problems 43–50, calculate the gradient of the given function at the specified point.

43. $f(x, y) = x^2 - y^3$; $(1, 2)$

44. $f(x, y) = \tan^{-1}(y/x)$; $(-1, -1)$

45. $f(x, y) = \dfrac{x - y}{x + y}$; $(3, 2)$

46. $f(x, y) = \cos(x - 2y)$; $(\pi/2, \pi/6)$

47. $f(x, y, z) = xy + yz^3$; $(1, 2, -1)$

48. $f(x, y, z) = \dfrac{x - y}{3z}$; $(2, 1, 4)$

49. $f(x, y, z) = 1/\sqrt{x^2 + y^2 + z^2}$; (a, b, c)

50. $f(x, y, z) = e^{-(x^2+y^3+z^4)}$; $(0, -1, 1)$

In Problems 51–56, use the chain rule to calculate the indicated derivative(s).

51. $z = 2xy$, $x = \cos t$, $y = \sin t$; dz/dt

52. $z = y/x$, $x = r - s$, $y = r + s$; $\partial z/\partial s$

53. $z = xy^3$, $x = r/s$, $y = s^2/r$; $\partial z/\partial r$

54. $z = \sin(x - y)$, $x = e^{r+s}$, $y = e^{r-s}$; $\partial z/\partial s$

55. $w = xyz$, $x = rs$, $y = r/s$, $z = s^2r^3$; w_r, w_s

56. $w = x^3y + y^3z$, $x = rst$, $y = rs/t$, $z = rt/s$; w_s, w_t

In Problems 57–62, find an equation for the tangent plane and symmetric equations of the normal line to the given surface at the specified point.

57. $x^2 + y^2 + z^2 = 3$; $(1, 1, 1)$

58. $x^{1/2} + y^{3/2} + z^{1/2} = 3$; $(1, 0, 4)$

59. $3x - y + 5z = 15$; $(-1, 2, 4)$

60. $xy^2 - yz^3 = 0$; $(1, 1, 1)$

61. $xyz = 6$; $(-2, 1, -3)$

62. $\sqrt{\dfrac{x - y}{y + z}} = \dfrac{1}{2}$; $(2, 1, 3)$

In Problems 63–68, calculate the directional derivative of the given function at the specified point in the direction of the specified vector $\mathbf{v}$.

63. $f(x, y) = y/x$ at $(1, 2)$; $\mathbf{v} = \mathbf{i} - \mathbf{j}$

64. $f(x, y) = 3x^2 - 4xy$ at $(3, -1)$; $\mathbf{v} = 2\mathbf{i} + 5\mathbf{j}$

65. $f(x, y) = \tan^{-1}(y/x)$ at $(1, -1)$; $\mathbf{v} = -3\mathbf{i} + 2\mathbf{j}$

66. $f(x, y, z) = xy^2 - zy^3$ at $(1, 2, 3)$; $\mathbf{v} = \mathbf{i} - \mathbf{j} + 2\mathbf{k}$

67. $f(x, y, z) = 1/\sqrt{x^2 + y^2 + z^2}$ at $(1, -1, 2)$; $\mathbf{v} = -2\mathbf{i} + \mathbf{j} - 3\mathbf{k}$

68. $f(x, y, z) = e^{-(x+y^2-xz)}$ at $(1, 0, -1)$; $\mathbf{v} = 2\mathbf{i} + 5\mathbf{j} + \mathbf{k}$

In Problems 69–73, calculate the total differential df.

69. $f(x, y) = x^3y^2$

70. $f(x, y) = \cos^{-1}(y/x)$

71. $f(x, y) = \sqrt{\dfrac{x + 1}{y - 1}}$

72. $f(x, y, z) = xy^5z^3$

73. $f(x, y, z) = \ln(x - y + 4z)$

74. Approximately how much wood is contained in the sides of a rectangular box with sides of inside measurements 1.5 m, 1.3 m, and 2 m if the thickness of the wood making up the sides is 3 cm (=0.03 m)?

In Problems 75–80, determine the nature of the various critical points of the given function.

75. $f(x, y) = 6x^2 + 14y^2 - 16xy + 2$

76. $f(x, y) = x^5 - y^5$

77. $f(x, y) = \dfrac{1}{y} + \dfrac{2}{x} + 2y + x + 4$

78. $f(x, y) = 49 - 16x^2 + 24xy - 9y^2$

79. $f(x, y) = x^2 + y^2 + \dfrac{2}{xy^2}$

80. $f(x, y) = \cot xy$

81. Find the minimum distance from the point $(2, -1, 4)$ to the plane $x - y + 3z = 7$.

82. What is the smallest amount of wood needed to build an open-top rectangular box enclosing a volume of 25 m^3?

83. What is the maximum volume of an open-top rectangular box that can be built from 10 m^2 of wood?

84. What are the dimensions of the rectangular parallelepiped (with faces parallel to the coordinate planes) of maximum volume that can be inscribed in the ellipsoid $x^2 + 9y^2 + 4z^2 = 36$?

85. Minimize the function $x^2 + y^2 + z^2$ for (x, y, z) on the planes $2x + y + z = 2$ and $x - y - 3z = 4$.

***86.** Among all the ellipses $(x/a)^2 + (y/b)^2 = 1$ that pass through the point $(3, 5)$, which one has the smallest area?

87. Show that there is no maximum or minimum value of xy^3z^2 if (x, y, z) lies on the plane $x - y + 2z = 2$.

88. Show that among all rectangles with the same area, the square has the smallest perimeter.

In Exercises 89–92 find the domain and range of the given function.

89. $f(x_1, x_2, x_3, x_4) = \sqrt{x_1^4 + x_2^4 + x_3^4 + x_4^4}$

90. $f(x_1, x_2, x_3, x_4) = \dfrac{x_1x_2}{x_3 - x_4}$

91. $f(x_1, x_2, x_3, x_4) = \dfrac{x_1^2 + x_2^2}{x_3^2 + x_4^2}$

92. $f(x_1, x_2, x_3, x_4) = \sin(x_1 + x_2 - x_3) + \cos x_4$

93. Verify directly from the definition that $\lim_{\mathbf{x}\to(2,-1,1,3)} (2x_1 - 5x_2 + x_3 - 3x_4) = 1$.

94. Show that $\lim_{\mathbf{x}\to\mathbf{0}} \dfrac{x_1x_2 + x_3x_4}{x_1^2 + x_2^2 + x_3^2 + x_4^2}$ does not exist.

95. Compute

$$\lim_{\mathbf{x}\to(1,-1,3,2)} \left(x_1^2 - 4x_2^2 + \frac{3x_3x_4}{x_2}\right).$$

96. Compute

$$\lim_{\mathbf{x}\to(1,0,2,3)} \left[(e^{x_1-2x_2+x_3-x_4}) \cos \frac{\pi x_3}{6x_1} \right].$$

97. Compute all first-order partial derivatives of $f(x_1, x_2, x_3, x_4) = x_1{}^3x_2{}^2x_3 - \ln(x_1 + 2x_2 - x_4) + \dfrac{x_4{}^5}{x_1}$.

98. Do the same as in Problem 97 for $f(x_1, x_2, x_3, x_4) = \dfrac{x_1 - x_3}{x_2 + x_4} + \sin(x_1x_3) - \tan(\sqrt{x_2x_4})$.

99. For the function of Problem 97 compute f_{12}, f_{31}, and f_{124}.

100. Compute the derivative of $\mathbf{f}(t) = (t^3, \cos t, \ln t, e^{2t})$.

101. Compute the derivative of $\mathbf{f}(t) = (\sqrt{t}, t^{5/3}, (1/t), t^5 - 2t + 2)$.

102. Compute the unit tangent vector and the unit normal vector to the curve $\mathbf{f}(t) = (t^2, t^4, t^6, t^8)$ at $t = 1$.

103. Compute the length of the curve $\mathbf{f}(t) = (t^2, t^2, t, t)$ for $t \in [0, 1]$.

104. Compute the gradient of the function in Exercise 89.

105. Compute the gradient of the function in Exercise 90.

106. Find the directional derivative of $f(x_1, x_2, x_3, x_4) = x_1{}^2 + x_2{}^2 + x_3{}^2 + x_4{}^2$ in the direction of $\mathbf{v} = (2, -1, 4, 1)$ at the point $(3, 1, -1, 4)$.

CHAPTER 3 COMPUTER EXERCISES

1. The purpose of this exercise is to demonstrate that f_{xy} and f_{yx} need not always be equal. You may find it helpful to use a computer algebra system that will calculate partial derivatives.

$$\text{Define } f(x, y) = \begin{cases} \dfrac{xy(x^2 - y^2)}{x^2 + y^2} & \text{if } (x, y) \neq (0, 0) \\ 0 & \text{if } (x, y) = (0, 0). \end{cases}$$

a. Calculate
  i. $f_x(x, y)$ and $f_y(x, y)$ for $(x, y) \neq (0, 0)$.
  ii. $f_x(0, 0)$ and $f_y(0, 0)$.

b. Using your answers above, calculate
  i. $f_{xy}(x, y)$ and $f_{yx}(x, y)$ for $(x, y) \neq (0, 0)$.
  ii. $f_{xy}(0, 0)$ and $f_{yx}(0, 0)$.

c. You should be aware of a theorem that says $f_{xy} = f_{yx}$ under certain conditions. What hypothesis of this theorem has been violated in this example? Prove your assertion.

d. Use graphics software to view the graph of this function.

2. Let $f(x, y) = \frac{1}{3}x^3y + x^3 + \frac{1}{2}x^2y^2 + \frac{3}{2}x^2y - xy^2 - 4xy - 3x + y^2 + 6y$

a. Find the critical points of $f(x, y)$.

b. Use the discriminant to determine whether each critical point gives a local maximum, local minimum, or a saddle point.

c. Use graphics software to view the graph of this function.

3. Let $f(x, y) = \dfrac{2x^3}{3} - 2x^2y + \dfrac{x^2}{2} + 2xy - x + y^2 + 2y$

a. Find the critical points of $f(x, y)$.

b. Use the discriminant to test each critical point and determine if it is a local maximum, local minimum, or a saddle point.

c. Use graphics software to view the graph of this function.

4. Find the two points on the graphs of $f(x) = 3x$ and $g(x) = x^2 + 3$ that are nearest.

5. A box with no lid is to be made of special materials. The volume of the box is to be 1 cubic foot. The material for the bottom costs \$1.00 per square foot. The material for the front and back panels costs \$2.00 per square foot, and the material to be used on the left and right hand side panels costs \$3.00 per square foot. The box is to be constructed by gluing the five panels together along their edges. It costs \$1.00 per foot to do the gluing. Find the dimensions of the box of minimal cost.

CHAPTER 4

MULTIPLE INTEGRATION

In your previous calculus course you studied the definite integral of a function of one variable. In this chapter we show how the notion of the definite integral can be extended to functions of several variables. In particular, we will discuss the double integral of a function of two variables and the triple integral of a function of three variables.

Multiple integrals were developed essentially by the great Swiss mathematician Leonhard Euler.[†] Euler had a clear conception of the double integral over a bounded region enclosed by arcs, and in a paper published in 1769,[‡] he gave a procedure for evaluating double integrals by repeated integration. We will discuss Euler's technique in Section 4.2. Multiple integrals were first used by Newton in his work (which appeared in the *Principia*) on the gravitational attraction exerted by spheres on particles. It should be mentioned, however, that Newton had only a geometric interpretation of a multiple integral. The more precise analytical definitions had to await the work of Euler and, later, Lagrange.

We begin our discussion of multiple integration with an introduction to the double integral. Our development will closely parallel the development of the definite integral of a function of one-variable.

4.1 VOLUME UNDER A SURFACE AND THE DOUBLE INTEGRAL

Study of the definite integral usually begins by calculating the area under a curve $y = f(x)$ (and above the x-axis) for x in the interval $[a, b]$. It is initially assumed that, on $[a, b]$, $f(x) \geq 0$. We carry out a similar development here by obtaining an expression which represents a volume in $\mathbb{R}^3$.

We begin by considering an especially simple case. Let R denote the rectangle in $\mathbb{R}^2$ given by

$$R = \{(x, y) \colon a \leq x \leq b \text{ and } c \leq y \leq d\}. \tag{1}$$

[†] See the biographical sketch on page 251.

[‡] *Novi Comm. Acad. Sci. Petrop.*, **14**, 72–103 (1769).

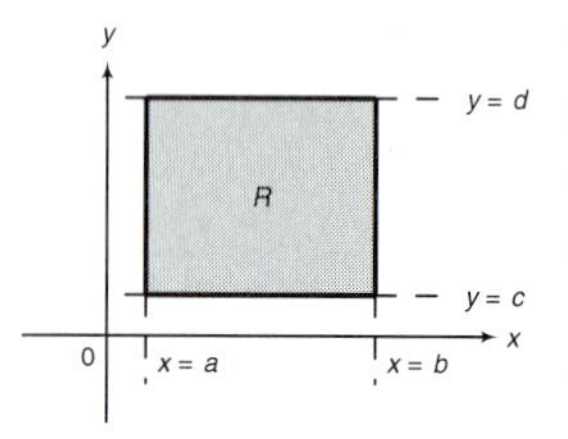

FIGURE 1
A rectangle in $\mathbb{R}^2$

This rectangle is sketched in Figure 1. Let $z = f(x, y)$ be a continuous function that is nonnegative over R. That is, $f(x, y) \geq 0$ for every (x, y) in R. We now ask: What is the volume "under" the surface $z = f(x, y)$ and "over" the rectangle R? The volume requested is sketched in Figure 2.

We will calculate this volume in much the same way the area under a curve was calculated. We begin by "partitioning" the rectangle.

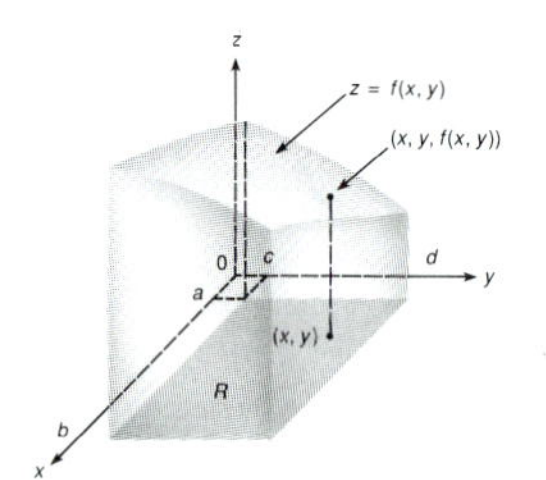

FIGURE 2
The shaded surface lies over the rectangle R.

Step 1: Form a **regular partition** (i.e., all subintervals have the same length) of the intervals $[a, b]$ and $[c, d]$:

$$a = x_0 < x_1 < x_2 < \cdots < x_{n-1} < x_n = b, \tag{2}$$

$$c = y_0 < y_1 < y_2 < \cdots < y_{m-1} < y_m = d. \tag{3}$$

We then define

$$\Delta x = x_i - x_{i-1} = \frac{b-a}{n}, \tag{4}$$

$$\Delta y = y_j - y_{j-1} = \frac{d-c}{m}, \tag{5}$$

and define the subrectangles R_{ij} by

$$R_{ij} = \{(x, y): x_{i-1} \leq x \leq x_i \text{ and } y_{j-1} \leq y \leq y_j\} \tag{6}$$

for $i = 1, 2, \ldots, n$ and $j = 1, 2, \ldots, m$. This is sketched in Figure 3. Note that there are nm subrectangles R_{ij} covering the rectangle R.

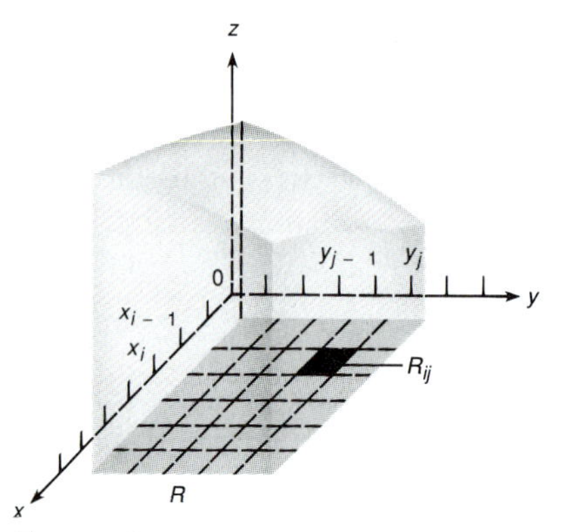

FIGURE 3
R_{ij} is one of the nm subrectangles that partition R.

Step 2: Estimate the volume under the surface and over each subrectangle:

Let (x_i^*, y_j^*) be a point in R_{ij}. Then the volume V_{ij} under the surface and over R_{ij} is approximated by

$$V_{ij} \approx f(x_i^*, y_j^*)\, \Delta x\, \Delta y = f(x_i^*, y_j^*)\, \Delta A, \tag{7}$$

where $\Delta A = \Delta x\, \Delta y$ is the area of R_{ij}. The expression on the right-hand side of (7) is simply the volume of the parallelepiped (three-dimensional box) with base R_{ij} and height $f(x_i^*, y_j^*)$. This volume corresponds to the approximate area $A_i \approx f(x_i^*)\, \Delta x_i$ that is used to approximate an area.†

† See *Calculus* or *Calculus of One Variable*, page 284.

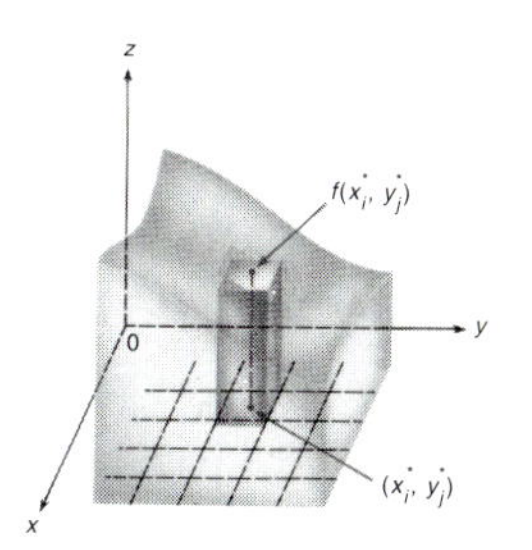

FIGURE 4
The volume of a small rectangular solid approximates the volume under a small part of the surface.

Unless the graph of f is a plane parallel to the xy-plane, the expression $f(x_i^*, y_j^*)\,\Delta A$ will not in general be equal to the volume under the surface S. But if Δx and Δy are small, the approximation will be a good one. The difference between the actual V_{ij} and the approximate volume given in (7) is illustrated in Figure 4.

Step 3: Add up the approximate volumes to obtain an approximation to the total volume sought.

The total volume is

$$V = V_{11} + V_{12} + \cdots + V_{1m} + V_{21} + V_{22} + \cdots + V_{2m} + \cdots + V_{n1} + V_{n2} + \cdots + V_{nm}. \tag{8}$$

To simplify notation, we use the summation sign $\sum$ introduced in your first calculus course. Since we are summing over two variables i and j, we need two such signs:

$$V = \sum_{i=1}^{n} \sum_{j=1}^{m} V_{ij}. \tag{9}$$

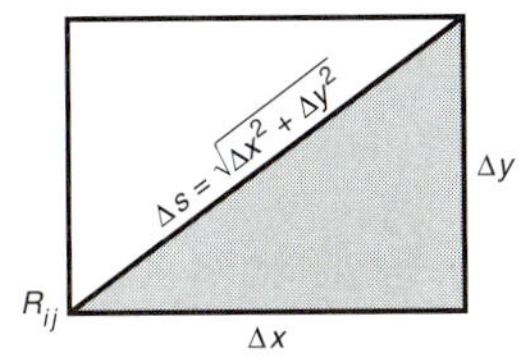

FIGURE 5
Δs is the length of the diagonal of the subrectangle R_{ij}.

The expression in (9) is called a **double sum**. If we "write out"† the expression in (9), we obtain the expression in (8). The combining (7) and (9), we have

$$V \approx \sum_{i=1}^{n} \sum_{j=1}^{m} f(x_i^*, y_j^*)\Delta A. \tag{10}$$

Step 4: Take a limit as both Δx and Δy approach zero.

To indicate that this is happening, we define

$$\Delta s = \sqrt{(\Delta x)^2 + (\Delta y)^2}.$$

Geometrically, Δs is the length of a diagonal of the rectangle R_{ij} whose sides have lengths Δx and Δy (see Figure 5). As $\Delta s \to 0$, the number of subrectangles R_{ij} increases without bound and the area of each R_{ij} approaches zero. This implies that the volume approximation given by (7) is getting closer and closer to the "true" volume over R_{ij}. Thus the approximation (10) gets better and better as $\Delta s \to 0$, which enables us to define

$$V = \lim_{\Delta s \to 0} \sum_{i=1}^{n} \sum_{j=1}^{m} f(x_i^*, y_j^*)\,\Delta A. \tag{11}$$

EXAMPLE 1

COMPUTING THE VOLUME UNDER A PLANE AND OVER A RECTANGLE

Calculate the volume under the plane $z = x + 2y$ and over the rectangle $R = \{(x, y)\colon 1 \le x \le 2 \text{ and } 3 \le y \le 5\}$.

† This writing out is done by summing over j first and then over i. For example,

$$\sum_{i=1}^{3} \sum_{j=1}^{4} a_{ij} = \sum_{i=1}^{3} (a_{i1} + a_{i2} + a_{i3} + a_{i4})$$
$$= a_{11} + a_{12} + a_{13} + a_{14} + a_{21} + a_{22} + a_{23} + a_{24} + a_{31} + a_{32} + a_{33} + a_{34}.$$

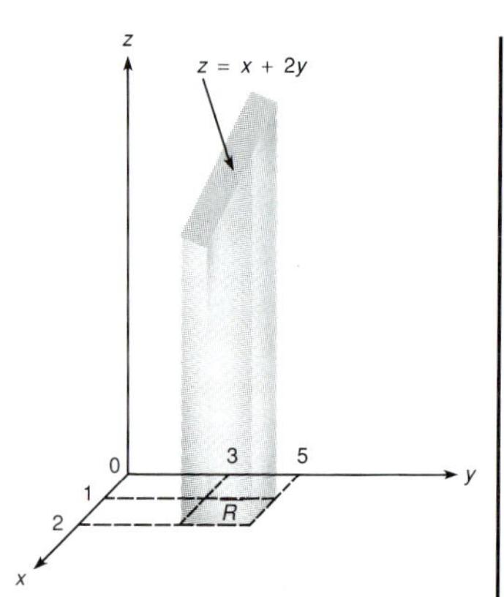

FIGURE 6
The volume of the solid under a plane and over a rectangle

SOLUTION: The solid whose volume we wish to calculate is sketched in Figure 6.

Step 1: For simplicity, we partition each of the intervals [1, 2] and [3, 5] into n subintervals of equal length (i.e., $m = n$):

$$1 = x_0 < x_1 < \cdots < x_n = 2;\ x_i = 1 + \frac{i}{n},\ \Delta x = \frac{1}{n}$$

$$3 = y_0 < y_1 < \cdots < y_n = 5;\ y_j = 3 + \frac{2j}{n},\ \Delta y = \frac{2}{n}.$$

Step 2: Then choosing $x_i{}^* = x_i$ and $y_j{}^* = y_j$, we obtain

$$\begin{aligned} V_{ij} &\approx f(x_i{}^*, y_j{}^*)\,\Delta A = (x_i + 2y_j)\,\Delta x\,\Delta y \\ &= \left[\left(1 + \frac{i}{n}\right) + 2\left(3 + \frac{2j}{n}\right)\right]\frac{1}{n}\cdot\frac{2}{n} \\ &= \left(7 + \frac{i}{n} + \frac{4j}{n}\right)\frac{2}{n^2}. \end{aligned}$$

Step 3:

$$\begin{aligned} V = \sum_{i=1}^{n}\sum_{j=1}^{n} V_{ij} &\approx \sum_{i=1}^{n}\sum_{j=1}^{n}\left(\frac{14}{n^2} + \frac{2i}{n^3} + \frac{8j}{n^3}\right) \\ &= \underbrace{\sum_{i=1}^{n}\sum_{j=1}^{n}\frac{14}{n^2}}_{①} + \underbrace{\sum_{i=1}^{n}\sum_{j=1}^{n}\frac{2i}{n^3}}_{②} + \underbrace{\sum_{i=1}^{n}\sum_{j=1}^{n}\frac{8j}{n^3}}_{③}. \end{aligned}$$

It is not difficult to evaluate each of these double sums. There are n^2 terms in each sum. Since $14/n^2$ does not depend on i or j, we evaluate the sum ① by simply adding up the term $14/n^2$ a total of n^2 times. Thus,

$$\sum_{i=1}^{n}\sum_{j=1}^{n}\frac{14}{n^2} = n^2\left(\frac{14}{n^2}\right) = 14.$$

Next, if we set $i = 1$ in ③, then we have $\sum_{j=1}^{n} 8j/n^3$. Similarly, setting $i = 2, 3, 4, \ldots, n$ in ③ yields $\sum_{j=1}^{n} 8j/n^3$. Thus in ③ we obtain the term $(\sum_{j=1}^{n} 8j/n^3)$ n times. But

$$\sum_{j=1}^{n}\frac{8j}{n^3} = \frac{8}{n^3}\sum_{j=1}^{n} j = \frac{8}{n^3}(1 + 2 + \cdots + n)$$

Example 2 in Appendix 1
$$\downarrow$$
$$= \frac{8}{n^3}\left[\frac{n(n+1)}{2}\right] = \frac{4(n+1)}{n^2}.$$

Thus,

$$\sum_{i=1}^{n}\sum_{j=1}^{n}\frac{8j}{n^3} = n\left[\sum_{j=1}^{n}\frac{8j}{n^3}\right] = n\left[\frac{4(n+1)}{n^2}\right] = \frac{4(n+1)}{n}.$$

Similarly,

$$\sum_{i=1}^{n}\left(\sum_{j=1}^{n}\frac{2i}{n^3}\right)=\sum_{i=1}^{n}\left[n\frac{2i}{n^3}\right]=\frac{2}{n^2}\sum_{i=1}^{n}i=\frac{2}{n^2}\left[\frac{n(n+1)}{2}\right]=\frac{n+1}{n}.$$

Finally, we have

$$\sum_{i=1}^{n}\sum_{j=1}^{n}V_{ij}\approx 14+\frac{4(n+1)}{n}+\frac{n+1}{n}.$$

Step 4: As $\Delta s \to 0$, both Δx and Δy approach 0, so $n = (b-a)/\Delta x \to \infty$. Thus,

$$V=\lim_{\Delta s\to 0}\sum_{i=1}^{n}\sum_{j=1}^{n}f(x_i^*, y_j^*)\,\Delta A=\lim_{n\to\infty}\sum_{i=1}^{n}\sum_{j=1}^{n}f(x_i^*, y_j^*)\,\Delta A$$

$$=\lim_{n\to\infty}\left[14+4\left(\frac{n+1}{n}\right)+\frac{n+1}{n}\right]=14+4+1=19.$$

The calculation we just made was very tedious. Instead of making other calculations like this one, we will define the double integral and, in Section 4.2, show how double integrals can be easily calculated.

DEFINITION THE DOUBLE INTEGRAL

Let $z = f(x, y)$ and let the rectangle R be given by (1). Let $\Delta A = \Delta x\,\Delta y$. Suppose that

$$\lim_{\Delta s\to 0}\sum_{i=1}^{n}\sum_{j=1}^{m}f(x_i^*, y_j^*)\,\Delta A$$

exists and is independent of the way in which the points (x_i^*, y_j^*) are chosen. Then the **double integral of f over R**, written $\iint_R f(x, y)\,dA$, is defined by

$$\iint_R f(x, y)\,dA=\lim_{\Delta s\to 0}\sum_{i=1}^{n}\sum_{j=1}^{m}f(x_i^*, y_j^*)\,\Delta A. \tag{12}$$

If the limit in (12) exists, then the function f is said to be **integrable** over R. ■

We observe that this definition says nothing about volumes (just as the definition of the definite integral says nothing about areas). For example, if $f(x, y)$ takes on negative values in R, then the limit in (12) will not represent the volume under the surface $z = f(x, y)$. However, the limit in (12) may still exist, and in that case f will be integrable over R.

NOTE: $\iint_R f(x, y)\,dA$ is a number, not a function. This is analogous to the fact that the definite integral $\int_a^b f(x)\,dx$ is a number. We will not encounter indefinite double integrals in this book.

As we already stated, we will not calculate any other double integrals in this section but will wait until Section 4.2 to see how these calculations can be made simple. We

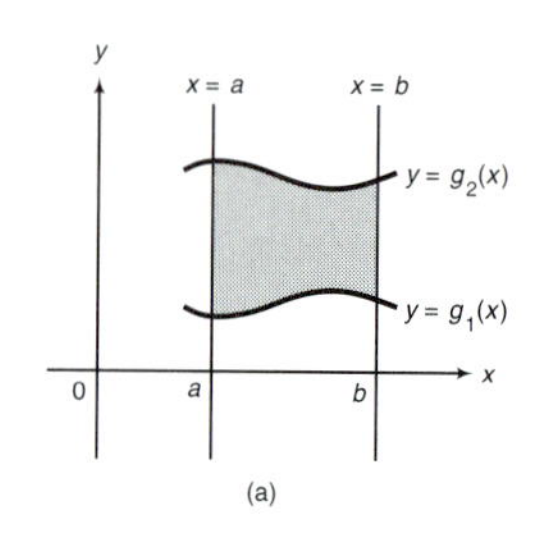

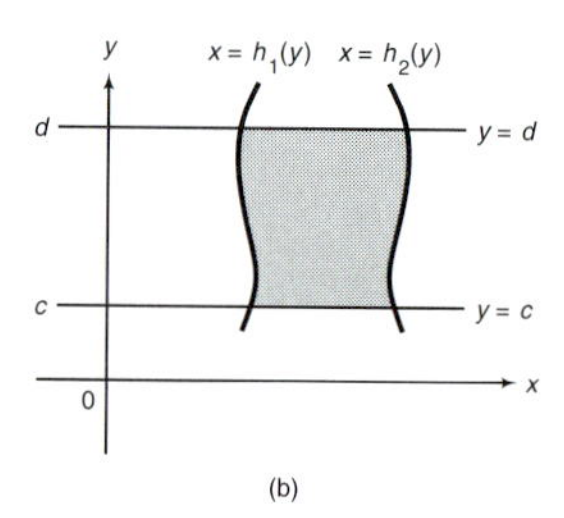

FIGURE 7
Two regions in $\mathbb{R}^2$

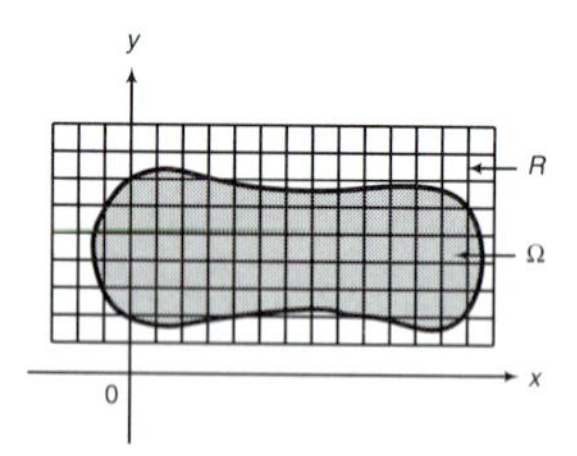

FIGURE 8
Partitioning the region Ω by enclosing Ω in a rectangle R and partitioning R

should note, however, that the result of Example 1 can now be restated as

$$\iint_R (x + 2y)\, dA = 19,$$

where R is the rectangle $\{(x, y)\colon 1 \le x \le 2 \text{ and } 3 \le y \le 5\}$.

What functions are integrable over a rectangle R? The following theorem is the double integral analog of the theorem that states that if f is continuous on $[a, b]$, then f is integrable on $[a, b]$.

THEOREM 1 EXISTENCE OF THE DOUBLE INTEGRAL OVER A RECTANGLE

If f is continuous on R, then f is integrable over R. ■

We will not give proofs of the theorems stated in this section regarding double integrals. The proofs are similar to, but more complicated than, the analogous proofs for theorems on definite integrals of functions of one variable. The proofs of all these theorems can be found in any standard advanced calculus text.†

We now turn to the question of defining double integrals over regions‡ in $\mathbb{R}^2$ that are not rectangular. We will denote a region in $\mathbb{R}^2$ by Ω. The two types of regions in which we will be most interested are illustrated in Figure 7. In this figure g_1, g_2, h_1, and h_2 denote continuous functions. A more general region Ω is sketched in Figure 8. We assume that the region is bounded. This means that there is a number M such that for every (x, y) in Ω, $|(x, y)| = \sqrt{x^2 + y^2} \le M$. Since Ω is bounded, we can draw a rectangle R around it. Let f be defined over Ω. We then define a new function F by

$$F(x, y) = \begin{cases} f(x, y), & \text{for } (x, y) \text{ in } \Omega \\ 0, & \text{for } (x, y) \text{ in } R \text{ but not in } \Omega. \end{cases} \tag{13}$$

DEFINITION INTEGRABILITY OVER A REGION IN $\mathbb{R}^2$

Let f be defined for (x, y) in Ω and let F be defined by (13). Then we write

$$\iint_\Omega f(x, y)\, dA = \iint_R F(x, y)\, dA \tag{14}$$

if the integral on the right exists. In this case, we say that f is **integrable** over Ω. ■

REMARK: If we divide R into nm subrectangles, as in Figure 8, then we can see what is happening. For each subrectangle R_{ij} that lies entirely in Ω, $F = f$, so the volume of the "parallelepiped" above R_{ij} is given by

$$V_{ij} \approx f(x_i^*, y_j^*)\, \Delta x\, \Delta y = F(x_i^*, y_j^*)\, \Delta x\, \Delta y.$$

However, if R_{ij} is in R but not in Ω, then $F = 0$, so

$$V_{ij} \approx F(x_i^*, y_j^*)\, \Delta x\, \Delta y = 0.$$

† See, for example, R. C. Buck and E. F. Buck, *Advanced Calculus*, 3rd ed. (New York: McGraw-Hill, 1978), p. 175.
‡ We will provide a formal definition of a region in Section 5.1. For our purposes in this chapter, a region is a finite union of sets in the plane that takes one of the forms in Figure 7.

Finally, if R_{ij} is partly in Ω and partly outside of Ω, then there is no real problem since, as $\Delta s \to 0$, the sum of the volumes above these rectangles (along the boundary of Ω) will approach zero—unless the boundary of Ω is very complicated indeed. Thus, we see that the limit of the sum of the volumes of the "parallelepipeds" above R is the same as the limit of the sum of the volumes of the "parallelepipeds" above Ω. This should help explain the "reasonableness" of expression (14).

THEOREM 2 EXISTENCE OF THE DOUBLE INTEGRAL OVER A MORE GENERAL REGION

Let Ω be one of the regions depicted in Figure 7 where the functions g_1 and g_2 or h_1 and h_2 are continuous. Let F be defined by (13). If f is continuous over Ω, then f is integrable over Ω and its integral is given by (14). ■

REMARK 1: There are some regions Ω that are so complicated that there are functions continuous but not integrable over Ω. We will not concern ourselves with such regions in this book.

REMARK 2: If f is nonnegative and integrable over Ω, then

$$\iint_{\Omega} f(x, y)\, dA$$

is defined as the volume under the surface $z = f(x, y)$ and over the region Ω.

REMARK 3: If the function $f(x, y) = 1$ is integrable over Ω, then

$$\iint_{\Omega} 1\, dA = \iint_{\Omega} dA \tag{15}$$

is equal to the area of the region Ω. To see this, note that

$$V_{ij} \approx f(x_i^*, y_j^*)\, \Delta A = \Delta A,$$

so the double integral (15) is the limit of the sum of areas of rectangles in Ω.

We close this section by stating five theorems about double integrals. Each one is analogous to a theorem about definite integrals.

THEOREM 3 DOUBLE INTEGRAL OF A CONSTANT TIMES A FUNCTION

If f is integrable over Ω, then for any constant c, cf is integrable over Ω, and

$$\iint_{\Omega} cf(x, y)\, dA = c \iint_{\Omega} f(x, y)\, dA \tag{16}$$

■

THEOREM 4 DOUBLE INTEGRAL OF A SUM

If f and g are integrable over Ω, then $f + g$ is integrable over Ω, and

$$\iint_{\Omega} [f(x, y) + g(x, y)]\, dA = \iint_{\Omega} f(x, y)\, dA + \iint_{\Omega} g(x, y)\, dA \tag{17}$$

■

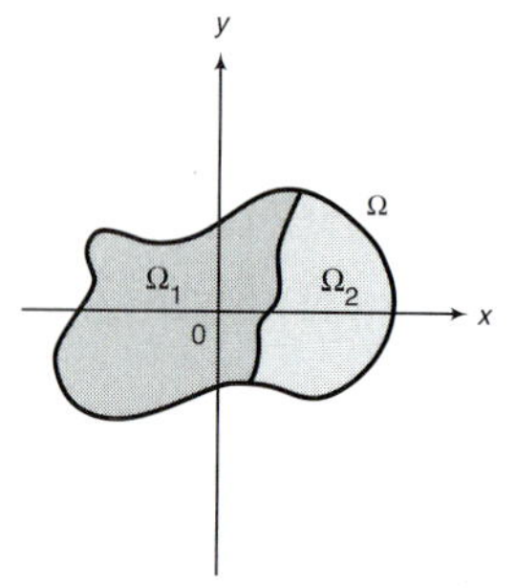

FIGURE 9
$\Omega = \Omega_1 \cup \Omega_2$

THEOREM 5 DOUBLE INTEGRAL OVER THE UNION OF TWO REGIONS

If f is integrable over Ω_1 and Ω_2, where Ω_1 and Ω_2 have no points in common except perhaps those of their common boundary, then f is integrable over $\Omega = \Omega_1 \cup \Omega_2$, and

$$\iint_{\Omega} f(x, y)\, dA = \iint_{\Omega_1} f(x, y)\, dA + \iint_{\Omega_2} f(x, y)\, dA$$

A typical region is depicted in Figure 9. ■

THEOREM 6 COMPARISON THEOREM FOR DOUBLE INTEGRALS

If f and g are integrable over Ω and $f(x, y) \le g(x, y)$ for every (x, y) in Ω, then

$$\iint_{\Omega} f(x, y)\, dA \le \iint_{\Omega} g(x, y)\, dA \tag{18}$$

■

THEOREM 7 UPPER AND LOWER BOUND THEOREM FOR DOUBLE INTEGRALS

Let f be integrable over Ω. Suppose that there exist constants m and M such that

$$m \le f(x, y) \le M \tag{19}$$

for every (x, y) in Ω. If A_Ω denotes the area of Ω, then

$$mA_\Omega \le \iint_{\Omega} f(x, y)\, dA \le MA_\Omega \tag{20}$$

■

Theorem 7 can be useful for estimating double integrals.

EXAMPLE 2 FINDING UPPER AND LOWER BOUNDS FOR A DOUBLE INTEGRAL

Let Ω be the disk $\{(x, y): x^2 + y^2 \le 1\}$. Find upper and lower bounds for

$$\iint_{\Omega} \frac{1}{1 + x^2 + y^2}\, dA.$$

SOLUTION: Since $0 \le x^2 + y^2 \le 1$ in Ω, we easily see that

$$\frac{1}{2} \le \frac{1}{1 + x^2 + y^2} \le 1.$$

Since the area of the disk is π, we have

$$\frac{\pi}{2} \le \iint_{\Omega} \frac{1}{1 + x^2 + y^2}\, dA \le \pi.$$

In fact, it can be shown (see Problem 4.4.24) that the value of the integral is $\pi \ln 2 \approx 0.693\pi$.

OF HISTORICAL INTEREST

LEONHARD EULER, 1707–1783

LEONHARD EULER
(THE GRANGER COLLECTION)

Leonhard Euler (pronounced ''oiler'') was born in Basil, Switzerland. His father was a clergyman who hoped that his son would follow him into the ministry. The father was adept at mathematics, however, and together with Johann Bernoulli, instructed young Leonhard in that subject. Euler also studied theology, astronomy, physics, medicine, and several Eastern languages.

In 1727 Euler applied to and was accepted for a chair of medicine and physiology at the St. Petersburg Academy. The day Euler arrived in Russia, however, Catherine I—founder of the Academy—died, and the Academy was plunged into turmoil. By 1730 Euler was pursuing his mathematical career from the chair of natural philosophy. Accepting an invitation from Frederick the Great, Euler went to Berlin in 1741 to head the Prussian Academy. Twenty-five years later, he returned to St. Petersburg, where he died in 1783 at the age of 76.

The most prolific writer in the history of mathematics, Euler found new results in virtually every branch of pure and applied mathematics. Although German was his native language, he wrote mostly in Latin and occasionally in French. His amazing productivity did not decline even when he became totally blind in 1766. During his lifetime, Euler published 530 books and papers. When he died, he left so many unpublished manuscripts that the St. Petersburg Academy was still publishing his work in its *Proceedings* almost half a century later. Euler's work enriched such diverse areas as hydraulics, celestial mechanics, lunar theory, and the theory of music, as well as mathematics.

Euler had a phenomenal memory. As a young man he memorized the entire *Aeneid* by Virgil (in Latin) and many years later could still recite the entire work. He was able to solve astonishingly complex mathematical problems in his head and is said to have solved, again in his head, problems in astronomy that stymied Newton. The French academician François Arago once commented that Euler could calculate without effort ''just as men breathe, as eagles sustain themselves in the air.''

Euler wrote in a mathematical language that is largely in use today. Among many symbols first used by him are:

$f(x)$	for functional notation
e	for the base of the natural logarithm
Σ	for the summation sign
i	to denote $\sqrt{-1}$

Euler's textbooks were models of clarity. His texts included the *Introductio in analysin infinitorum* (1748), his *Institutiones calculi differentialis* (1755), and the three-volume *Institutiones calculi integralis* (1768–74). This and others of his works served as models for many of today's mathematics textbooks.

It is said that Euler did for mathematical analysis what Euclid did for geometry. It is no wonder that so many later mathematicians expressed their debt to him.

PROBLEMS 4.1

SELF-QUIZ

For each of these Self-quiz problems, let $\Omega = \{(x, y): 1 \le x \le 3, 0 \le y \le 2\}$. Also suppose f is a smooth function of (x, y).

I. If [1, 3] and [0, 2] are each partitioned into two subintervals of length 1, then ________ approximates $\iint_\Omega f(x, y)\, dA$.
a. $f(0, 0) + f(0, 1) + f(1, 0) + f(1, 1)$
b. $f(1, 0) + f(1.5, 0) + f(2, 0) + f(2.5, 0)$
c. $f(0, 0) + f(1, 1) + f(2, 2) + f(3, 3)$
d. $f(1.5, 0.5) + f(2.5, 0.5) + f(1.5, 1.5) + f(2.5, 1.5)$

II. If we are told that the values of $f(x, y)$ lie between 7 and 11 when $(x, y) \in \Omega$, then we also know that $\iint_\Omega f(x, y)\, dA$ must lie between ________.
a. 7 and 11 **b.** 28 and 44
c. 30 and 50 **d.** −4 and 18

III. Suppose $f_x = 0$ and $f_y = 0$ throughout Ω, and also suppose $f(2, 1) = 15$; then $\iint_\Omega f(x, y)\, dA$ ________.
a. does not exist
b. ≤ 57.38
c. $= 60$
d. lies between 30 and 45

In Problems 1–8, let Ω be the rectangle $\{(x, y): 0 \le x \le 3, 1 \le y \le 2\}$. Use the technique employed in Example 1 to calculate the given double integral. Use Theorem 3 or 4 where appropriate.

1. $\iint_\Omega (2x + 3y)\, dA$ **2.** $\iint_\Omega (x - y)\, dA$

3. $\iint_\Omega (y - x)\, dA$ **4.** $\iint_\Omega (ax + by + c)\, dA$

5. $\iint_\Omega (x^2 + y^2)\, dA$
[*Hint:* Use formula (2) in Appendix 1.]

6. $\iint_\Omega (x^2 - y^2)\, dA$

7. $\iint_\Omega (2x^2 + 3y^2)\, dA$

8. $\iint_\Omega (ax^2 + by^2)\, dA$

In Problems 9–14, let Ω be the rectangle $\{(x, y): -1 \le x \le 0, -2 \le y \le 3\}$. Calculate the given integral.

9. $\iint_\Omega (x + y)\, dA$ **10.** $\iint_\Omega (3x - y)\, dA$

11. $\iint_\Omega (y - 2x)\, dA$ **12.** $\iint_\Omega (x^2 + 2y^2)\, dA$

13. $\iint_\Omega (y^2 - x^2)\, dA$ **14.** $\iint_\Omega (3x^2 - 5y^2)\, dA$

In Problems 15–18, suppose that $\iint_\Omega x\, dA = 2$ and $\iint_\Omega y\, dA = 7$. Compute each double integral.

15. $\iint_\Omega (x + y)\, dA$ **16.** $\iint_\Omega (x - 2y)\, dA$

17. $\iint_\Omega (3x + 5y)\, dA$ **18.** $\iint_\Omega (2y - 4x)\, dA$

19. Suppose that Ω_1, Ω_2, and Ω are as in Figure 9 and that $\iint_{\Omega_1} f(x, y)\, dA = 3$ and $\iint_{\Omega_2} f(x, y)\, dA = 8$. Compute $\iint_\Omega f(x, y)\, dA$.

20. In Problem 19 suppose that $\iint_\Omega f(x, y)\, dA = 12$ and $\iint_{\Omega_1} f(x, y)\, dA = 17$. Compute $\iint_{\Omega_2} f(x, y)\, dA$.

In Problems 21–25, use Theorem 7 to obtain lower and upper bounds for the given integral.

21. $\iint_\Omega (x^5y^2 + xy)\, dA$, where Ω is the rectangle $\{(x, y): 0 \le x \le 1, 1 \le y \le 2\}$.

22. $\iint_\Omega e^{-(x^2+y^2)}\, dA$, where Ω is the disk $x^2 + y^2 \le 4$.

***23.** $\displaystyle\iint_\Omega \left[\frac{x - y}{4 - x^2 - y^2}\right] dA$,
where Ω is the disk $x^2 + y^2 \le 1$.

24. $\iint_\Omega \cos(\sqrt{|x|} - \sqrt{|y|})\, dA$, where Ω is the disk $x^2 + y^2 \le 1$.

25. $\iint_\Omega \ln(1 + x + y)\, dA$, where Ω is the region bounded by the lines $y = x$, $y = 1 - x$, and the x-axis.

***26.** Let Ω be one of the regions depicted in Figure 7. Which is greater:

$$\iint_\Omega e^{(x^2+y^2)}\, dA \quad \text{or} \quad \iint_\Omega (x^2 + y^2)\, dA?$$

ANSWERS TO SELF-QUIZ

I. d **II.** b **III.** c

4.2 THE CALCULATION OF DOUBLE INTEGRALS

In this section we derive a method for calculating $\iint_\Omega f(x, y)\, dx\, dy$, where Ω is one of the regions depicted in Figure 4.1.7.

We begin, as in Section 4.1, by considering

$$\iint_R f(x, y)\, dA, \tag{1}$$

where R is the rectangle

$$R = \{(x, y): a \le x \le b \text{ and } c \le y \le d\}. \tag{2}$$

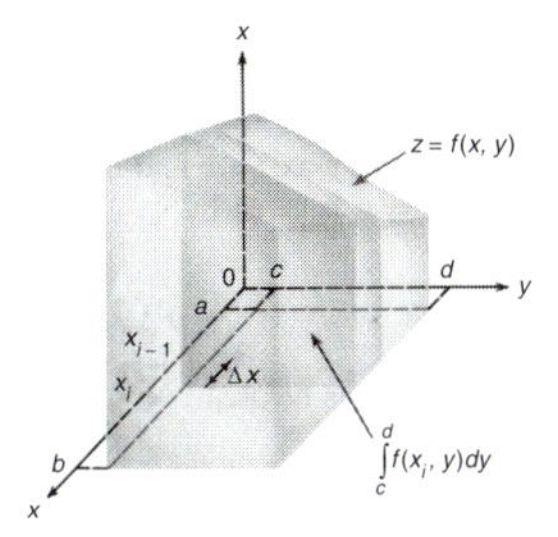

FIGURE 1
$\int_c^d f(x_i, y)\, dy$ is the area of a cross-section parallel to the yz-plane.

If $z = f(x, y) \ge 0$ for (x, y) in R, then the double integral in (1) is the volume under the surface $z = f(x, y)$ and over the rectangle R in the xy-plane. We now calculate this volume by partitioning the x-axis, taking "slices" parallel to the yz-plane. This is illustrated in Figure 1. We can approximate the volume by adding up the volumes of the various "slices." The face of each "slice" lies in the plane $x = x_i$, and the volume of the ith slice is approximately equal to the area of its face times its thickness Δx. What is the area of the face? If x is fixed, then $z = f(x, y)$ can be thought of as a curve lying in the plane $x = x_i$. Thus, the area of the ith face is the area bounded by this curve, the xy-plane, and the planes $y = c$ and $y = d$. This area is sketched in Figure 2. If $f(x_i, y)$ is a continuous function of y, then the area of the ith face, denoted by A_i, is given by

$$A_i = \int_c^d f(x_i, y)\, dy.$$

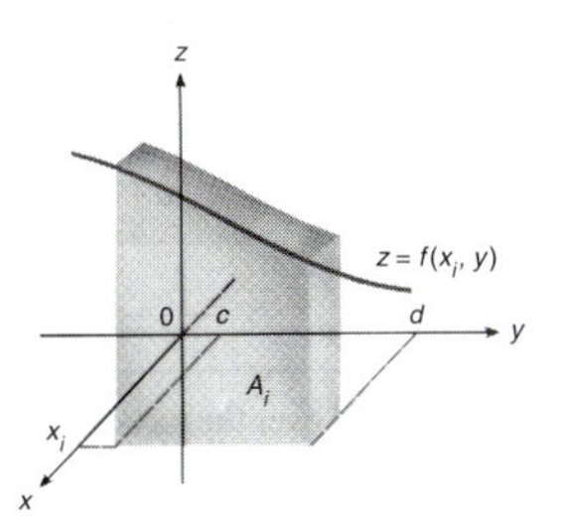

FIGURE 2
The volume of this "slice" is approximated by the area of the face times Δx.

By treating x_i as a constant, we can compute A_i as an ordinary definite integral, where the variable is y. Note, too, that $A(x) = \int_c^d f(x, y)\, dy$ is a function of x only and can therefore be integrated as a definite integral. Then the volume of the ith slice is approximated by

$$V_i \approx \left\{\int_c^d f(x_i, y)\, dy\right\} \Delta x$$

so that, adding up these "subvolumes" and taking the limit as Δx approaches zero, we obtain

$$V = \int_a^b \left\{\int_c^d f(x, y)\, dy\right\} dx = \int_a^b A(x)\, dx. \tag{3}$$

The expression in (3) is called a **repeated integral** or **iterated integral**. Since we also have

$$V = \iint_R f(x, y)\, dA,$$

we obtain

REPEATED INTEGRAL OVER A RECTANGLE: INTEGRATING FIRST WITH RESPECT TO y

$$\iint_R f(x, y)\, dA = \int_a^b \left\{ \int_c^b f(x, y)\, dy \right\} dx. \tag{4}$$

REMARK 1: Usually we will write equation (4) without braces. We then have

$$\iint_R f(x, y)\, dA = \int_a^b \int_c^d f(x, y)\, dy\, dx. \tag{5}$$

REMARK 2: We should emphasize that the first integration in $\int_a^b \int_c^d f(x, y)\, dy\, dx$ is performed by treating x as a constant.

Similarly, if we instead begin by partitioning the y-axis, we find that the area of the face of a "slice" lying in the plane $y = y_i$ is given by

$$A_i = \int_a^b f(x, y_i)\, dx,$$

where now A_i is an integral in the variable x. Thus as before,

$$V = \int_c^d \left\{ \int_a^b f(x, y)\, dx \right\} dy, \tag{6}$$

and

REPEATED INTEGRAL OVER A RECTANGLE: INTEGRATING FIRST WITH RESPECT TO x

$$\iint_R f(x, y)\, dA = \int_c^d \int_a^b f(x, y)\, dx\, dy. \tag{7}$$

EXAMPLE 1 CALCULATING A VOLUME BY EVALUATING A REPEATED INTEGRAL

Calculate the volume under the plane $z = x + 2y$ and over the rectangle

$$R = \{(x, y)\colon 1 \le x \le 2 \text{ and } 3 \le y \le 5\}.$$

SOLUTION: We calculated this volume in Example 4.1.1 (see page 245). We now calculate the volume using a repeated integral. Using equation (5), we have

$$V = \iint_R (x + 2y)\,dA = \int_1^2 \left[\int_3^5 (x + 2y)\,dy \right] dx^\dagger$$

$$= \int_1^2 \left[(xy + y^2)\Big|_{y=3}^{y=5} \right] dx = \int_1^2 [(5x + 25) - (3x + 9)]\,dx$$

$$= \int_1^2 (2x + 16)\,dx = (x^2 + 16x)\Big|_1^2 = 19.$$

Similarly, using equation (7), we have

$$V = \int_3^5 \left[\int_1^2 (x + 2y)\,dx \right] dy = \int_3^5 \left[\left(\frac{x^2}{2} + 2yx \right)\Big|_{x=1}^{x=2} \right] dy$$

$$= \int_3^5 \left[(2 + 4y) - \left(\frac{1}{2} + 2y \right) \right] dy = \int_3^5 \left(2y + \frac{3}{2} \right) dy$$

$$= \left(y^2 + \frac{3}{2}y \right)\Big|_3^5 = 19.$$

EXAMPLE 2

CALCULATING THE VOLUME UNDER A SURFACE AND OVER A RECTANGLE

Calculate the volume of the region beneath the surface $z = xy^2 + y^3$ and over the rectangle

$$R = \{(x, y)\colon 0 \le x \le 2 \text{ and } 1 \le y \le 3\}.$$

SOLUTION: A computer-drawn sketch of this region is given in Figure 3. Using equation (5), we have

$$V = \int_0^2 \int_1^3 (xy^2 + y^3)\,dy\,dx = \int_0^2 \left[\left(\frac{xy^3}{3} + \frac{y^4}{4} \right)\Big|_1^3 \right] dx$$

$$= \int_0^2 \left[\left(9x + \frac{81}{4} \right) - \left(\frac{x}{3} + \frac{1}{4} \right) \right] dx = \int_0^2 \left(\frac{26}{3}x + 20 \right) dx$$

$$= \left(\frac{13x^2}{3} + 20x \right)\Big|_0^2 = \frac{52}{3} + 40 = \frac{172}{3}.$$

You should verify that the same answer is obtained by using equation (7).

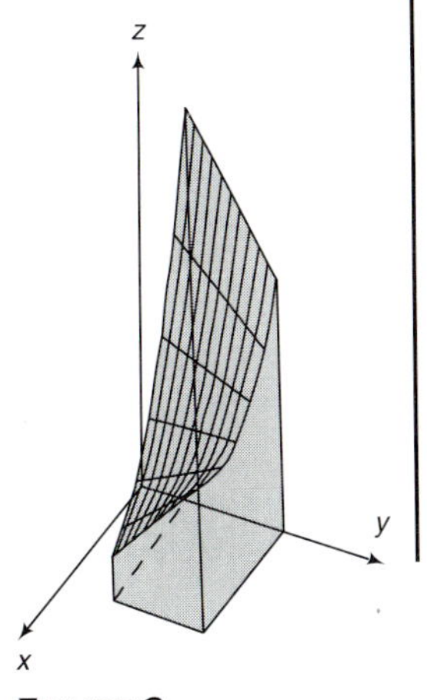

FIGURE 3
$z = xy^2 + y^3$

We now extend our results to more general regions. Let

$$\Omega = \{(x, y)\colon a \le x \le b \text{ and } g_1(x) \le y \le g_2(x)\}. \quad (8)$$

This region is sketched in Figure 4. We assume that for every x in $[a, b]$,

$$g_1(x) \le g_2(x). \quad (9)$$

This means that the *lower* and *upper* boundaries are curves. Such a region will be called a type LU region (for lower-upper). If we partition the x-axis as before, we obtain slices

†Remember, in computing the bracketed integral, we treat x as a constant.

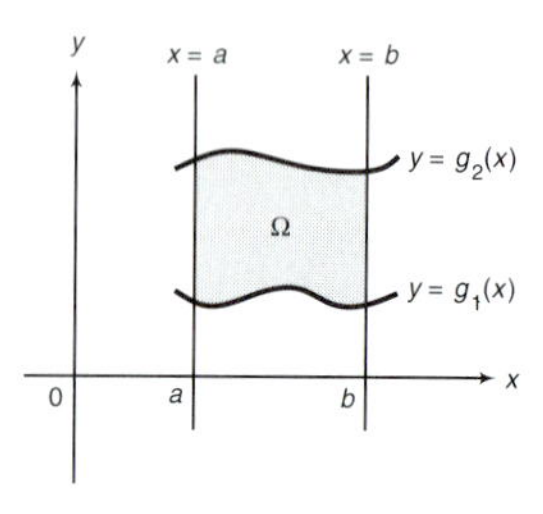

FIGURE 4
A type LU region

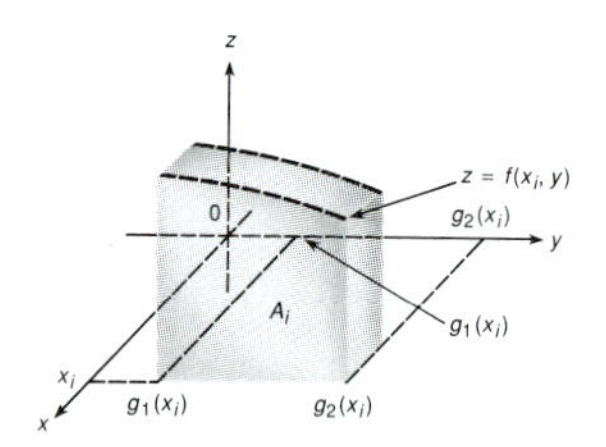

FIGURE 5
A slice of volume over an LU region obtained by partitioning the x-axis

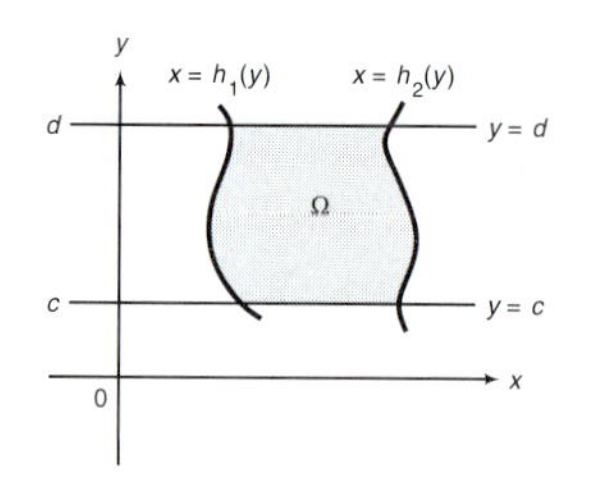

FIGURE 6
A type LR region

lying in the planes $x = x_i$, a typical one of which is sketched in Figure 5. Then

$$A_i = \int_{g_1(x_i)}^{g_2(x_i)} f(x_i, y)\, dy \qquad V_i \approx \left\{ \int_{g_1(x_i)}^{g_2(x_i)} f(x_i, y)\, dy \right\} \Delta x,$$

and

$$V = \iint_\Omega f(x, y)\, dA = \int_a^b \int_{g_1(x)}^{g_2(x)} f(x, y)\, dy\, dx. \tag{10}$$

Similarly, let

$$\Omega = \{(x, y)\text{: } h_1(y) \le x \le h_2(y) \text{ and } c \le y \le d\}. \tag{11}$$

Here the *left* and *right* boundaries are curves (see Figure 6). We call this a type LR region. Then

$$V = \int_c^d \int_{h_1(y)}^{h_2(y)} f(x, y)\, dx\, dy. \tag{12}$$

We summarize these results in the following theorem.

THEOREM 1 DOUBLE INTEGRALS OVER A REGION

Let f be continuous over a region Ω given by equation (8) or (11).

i. If Ω is a type LU region (8), where g_1 and g_2 are continuous, then

$$\iint_\Omega f(x, y)\, dA = \int_a^b \int_{g_1(x)}^{g_2(x)} f(x, y)\, dy\, dx.$$

ii. If Ω is a type LR region (11), where h_1 and h_2 are continuous, then

$$\iint_\Omega f(x, y)\, dA = \int_c^d \int_{h_1(y)}^{h_2(y)} f(x, y)\, dx\, dy.$$

■

REMARK 1: We have not actually proved this theorem here but have merely indicated why it should be so. A rigorous proof can be found in any advanced calculus text.

REMARK 2: Note that this theorem says nothing about volume. It can be used to calculate any double integral if the hypotheses of the theorem are satisfied and if each function being integrated has an antiderivative that can be written in terms of elementary functions.

REMARK 3: Many regions are of the form (8) or (11). In addition, almost all regions that arise in practical applications can be broken into a finite number of regions of the form (8) or (11), and the integration can be carried out using Theorem 4.1.5.

EXAMPLE 3 CALCULATING THE VOLUME UNDER A SURFACE AND OVER A TYPE LU REGION

Find the volume of the solid under the surface $z = x^2 + y^2$ and lying above the region

$$\Omega = \{(x, y)\text{: } 0 \le x \le 1 \text{ and } x^2 \le y \le \sqrt{x}\}.$$

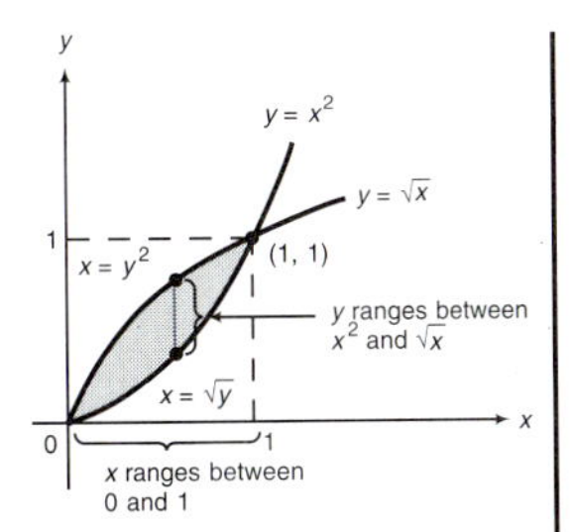

FIGURE 7
The region bounded by $y = x^2$ and $y = \sqrt{x}$ for $0 \le x \le 1$

SOLUTION: Ω is sketched in Figure 7. We see that $0 \le x \le 1$ and $x^2 \le y \le \sqrt{x}$. Then using (10), we have

$$V = \int_0^1 \int_{x^2}^{\sqrt{x}} (x^2 + y^2)\, dy\, dx = \int_0^1 \left\{ \left(x^2 y + \frac{y^3}{3} \right) \Big|_{x^2}^{\sqrt{x}} \right\} dx$$

$$= \int_0^1 \left\{ \left(x^2\sqrt{x} + \frac{(\sqrt{x})^3}{3} \right) - \left(x^2 \cdot x^2 + \frac{(x^2)^3}{3} \right) \right\} dx$$

$$= \int_0^1 \left(x^{5/2} + \frac{x^{3/2}}{3} - x^4 - \frac{x^6}{3} \right) dx$$

$$= \left(\frac{2x^{7/2}}{7} + \frac{2x^{5/2}}{15} - \frac{x^5}{5} - \frac{x^7}{21} \right) \Big|_0^1 = \frac{2}{7} + \frac{2}{15} - \frac{1}{5} - \frac{1}{21} = \frac{18}{105}.$$

We can calculate this integral in another way. We note that x varies between the curves $x = y^2$ and $x = \sqrt{y}$. Then using (12), since $0 \le y \le 1$ and $y^2 \le x \le \sqrt{y}$, we have

$$V = \int_0^1 \int_{y^2}^{\sqrt{y}} (x^2 + y^2)\, dx\, dy,$$

which is easily seen to be equal to $\frac{18}{105}$. The region is sketched in Figure 8.

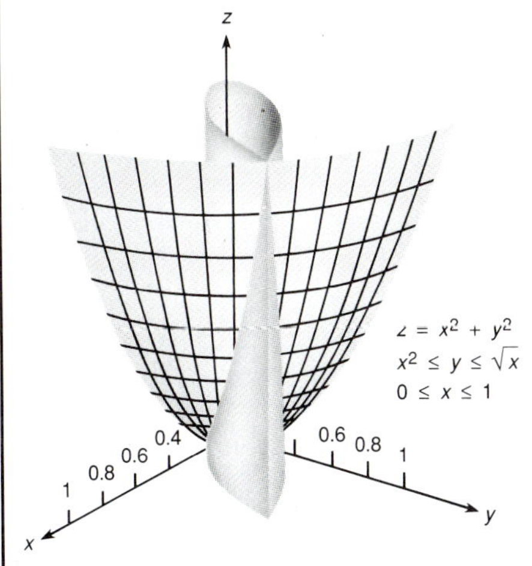

FIGURE 8
Computer-drawn sketch of the region under the circular paraboloid $z = x^2 + y^2$ and over the region $x^2 \le y \le \sqrt{x}$, $0 \le x \le 1$

REVERSING THE ORDER OF INTEGRATION

EXAMPLE 4

EVALUATING A DOUBLE INTEGRAL BY REVERSING THE ORDER OF INTEGRATION

Evaluate $\int_1^2 \int_1^{x^2} (x/y)\, dy\, dx$.

SOLUTION:

$$\int_1^2 \int_1^{x^2} \frac{x}{y}\, dy\, dx = \int_1^2 \left\{ x \ln y \Big|_1^{x^2} \right\} dx = \int_1^2 x \ln x^2\, dx = \int_1^2 2x \ln x\, dx$$

It is necessary to use integration by parts to complete the problem. Setting $u = \ln x$ and $dv = 2x\,dx$, we have $du = (1/x)\,dx$, $v = x^2$, and

$$\int_1^2 2x \ln x\,dx = x^2 \ln x\Big|_1^2 - \int_1^2 x\,dx = 4\ln 2 - \frac{x^2}{2}\Big|_1^2 = 4\ln 2 - \frac{3}{2}.$$

There is an easier way to calculate the repeated integral. We simply **reverse the order of integration**. The region of integration is sketched in Figure 9. If we want to integrate first

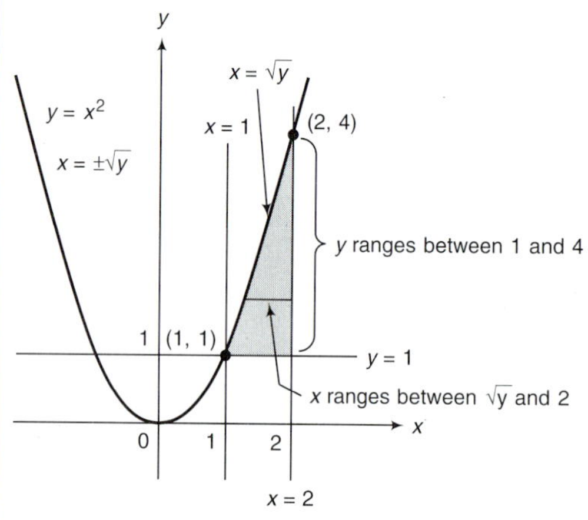

FIGURE 9
The shaded region can be written two ways: $1 \le y \le x^2$, $1 \le x \le 2$ and $\sqrt{y} \le x \le 2$, $1 \le y \le 4$.

with respect to x, we note that we can describe the region by

$$\Omega = \{(x, y)\colon \sqrt{y} \le x \le 2 \text{ and } 1 \le y \le 4\}.$$

Then

$$\int_1^2 \int_1^{x^2} \frac{x}{y}\,dy\,dx = \iint_\Omega \frac{x}{y}\,dA = \int_1^4 \int_{\sqrt{y}}^2 \frac{x}{y}\,dx\,dy = \int_1^4 \left(\frac{x^2}{2y}\Big|_{\sqrt{y}}^2\right) dy$$

$$= \int_1^4 \left(\frac{2}{y} - \frac{1}{2}\right) dy = \left(2\ln y - \frac{y}{2}\right)\Big|_1^4 = 2\ln 4 - \frac{3}{2}$$

$$= 4\ln 2 - \frac{3}{2}.$$

Note that in this case it is easier to integrate first with respect to x.

The technique used in Example 4 suggests the following:

> When changing the order of integration, first sketch the region of integration in the xy-plane.

REMARK: Why is it legitimate to reverse the order of integration? Suppose that the region Ω can be written as

$$\Omega = \{(x, y)\colon a \le x \le b,\ g_1(x) \le y \le g_2(x)\}$$
$$= \{(x, y)\colon c \le y \le d,\ h_1(y) \le x \le h_2(y)\}.$$

Then if f, g_1, g_2, h_1, and h_2 are continuous, we can equate the two parts of Theorem 1 to obtain

$$\iint_{\Omega} f(x, y)\, dA = \int_a^b \int_{g_1(x)}^{g_2(x)} f(x, y)\, dy\, dx = \int_c^d \int_{h_1(y)}^{h_2(y)} f(x, y)\, dx\, dy.$$

EXAMPLE 5

EVALUATING A DOUBLE INTEGRAL BY REVERSING THE ORDER OF INTEGRATION

Compute $\int_0^2 \int_y^2 e^{x^2}\, dx\, dy$.

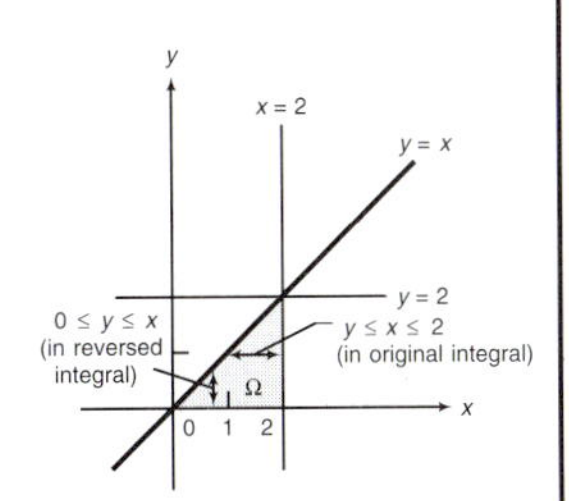

FIGURE 10
The shaded region can be written as $y \le x \le 2$, $0 \le y \le 2$ or $0 \le y \le x$, $0 \le x \le 2$.

SOLUTION: The region of integration is sketched in Figure 10. We first observe that the double integral cannot be evaluated directly since it is impossible to express an antiderivative for e^{x^2} in terms of elementary functions. Instead, we reverse the order of integration. From Figure 10 we see that Ω can be written as $0 \le y \le x$, $0 \le x \le 2$, so

$$\begin{aligned}\int_0^2 \int_y^2 e^{x^2}\, dx\, dy &= \iint_{\Omega} e^{x^2}\, dA = \int_0^2 \int_0^x e^{x^2}\, dy\, dx \\ &= \int_0^2 \left(ye^{x^2} \Big|_{y=0}^{y=x} \right) dx = \int_0^2 xe^{x^2}\, dx \\ &= \frac{1}{2} e^{x^2} \Big|_0^2 = \frac{1}{2}(e^4 - 1).\end{aligned}$$

EXAMPLE 6

REVERSING THE ORDER OF INTEGRATION

Reverse the order of integration in the iterated integral $\int_0^1 \int_{\sqrt{y}}^2 f(x, y)\, dx\, dy$.

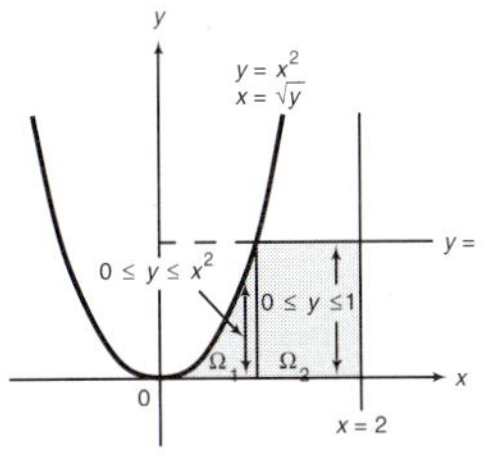

FIGURE 11
The shaded region can be written as $\sqrt{y} \le x \le 2$, $0 \le y \le 1$ or as the union of two regions: $\{(x, y)\colon 0 \le y \le x^2, 0 \le x \le 1\} \cup \{(x, y)\colon 0 \le y \le 1, 1 \le x \le 2\}$.

SOLUTION: The region of integration is sketched in Figure 11. This region is divided into two subregions Ω_1 and Ω_2. What happens if we integrate first with respect to y? In Ω_1, $0 \le y \le x^2$. In Ω_2, $0 \le y \le 1$. Thus,

$$\begin{aligned}\int_0^1 \int_{\sqrt{y}}^2 f(x, y)\, dx\, dy &= \iint_{\Omega} f(x, y)\, dA \overset{\text{Theorem 4.1.5}}{=} \iint_{\Omega_1} f(x, y)\, dA + \iint_{\Omega_2} f(x, y)\, dA \\ &= \int_0^1 \int_0^{x^2} f(x, y)\, dy\, dx + \int_1^2 \int_0^1 f(x, y)\, dy\, dx.\end{aligned}$$

Improper double integrals are defined in much the same way we define improper single integrals in a first calculus course. Rather than providing a formal definition, we give an example.

EXAMPLE 7

COMPUTING THE VOLUME OF A SOLID THAT EXTENDS OVER AN UNBOUNDED REGION LEADS TO AN IMPROPER DOUBLE INTEGRAL

Find the volume in the first octant bounded by the three coordinate planes and the surface $z = 1/(1 + x + 3y)^3$.

SOLUTION: The solid here extends over the unbounded region $\{(x, y): 0 \leq x \leq \infty$ and $0 \leq y \leq \infty\}$. Thus,

$$V = \int_0^{\infty}\int_0^{\infty} \frac{1}{(1 + x + 3y)^3}\, dx\, dy = \int_0^{\infty} \lim_{N\to\infty}\left(-\frac{1}{2(1 + x + 3y)^2}\bigg|_0^N\right) dy$$
$$= \int_0^{\infty} \frac{1}{2(1 + 3y)^2}\, dy = \lim_{N\to\infty}\left(-\frac{1}{6(1 + 3y)}\right)\bigg|_0^N = \frac{1}{6}.$$

Note that improper double integrals can be treated in the same way that we treat improper "single" integrals.

EXAMPLE 8 FINDING THE VOLUME BOUNDED BY A PLANE AND THE THREE COORDINATE PLANES

Find the volume of the solid bounded by the coordinate planes and the plane $2x + y + z = 2$.

SOLUTION: We have $z = 2 - 2x - y$ and this expression must be integrated over the region in the xy-plane bounded by the line $2x + y = 2$ (obtained when $z = 0$) and the x- and y-axes. See Figure 12. We therefore have

$$V = \int_0^1\int_0^{2-2x} (2 - 2x - y)\, dy\, dx = \int_0^1 \left\{\left(2y - 2xy - \frac{y^2}{2}\right)\bigg|_0^{2-2x}\right\} dx$$
$$= \int_0^1 \left\{2(2 - 2x) - 2x(2 - 2x) - \frac{(2 - 2x)^2}{2}\right\} dx$$
$$= \int_0^1 (2x^2 - 4x + 2)\, dx = \left(\frac{2x^3}{3} - 2x^2 + 2x\right)\bigg|_0^1 = \frac{2}{3}.$$

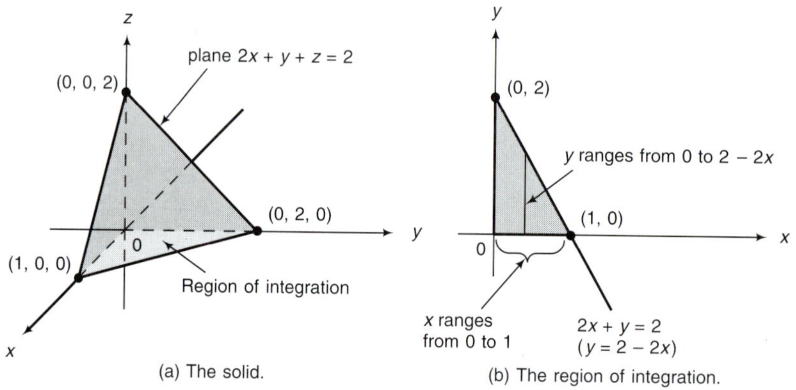

FIGURE 12
To find the volume of the solid in (a), we integrate over the region (b).

COMPUTING DIFFICULT IMPROPER INTEGRALS BY PARAMETRIC INTEGRATION†

There is a very clever technique that can sometimes be used to evaluate an integral that cannot otherwise be integrated using methods from calculus.

† The material here was suggested by the article "Parametric Integration Techniques" by Aurel J. Zajta and Sudkir K. Goel in *Mathematics Magazine*, 62(5), December 1989, pages 318–322.

EXAMPLE 9

EVALUATING AN IMPORTANT IMPROPER INTEGRAL†

Compute $\int_0^\infty \frac{\sin x}{x} dx$.

SOLUTION: Consider $\int_0^\infty e^{-px} \sin x\, dx$. Using integration by parts, it is not difficult to show (see entry 168 in the table of integrals at the back of the book) that

$$\begin{aligned}\int_0^N e^{-px} \sin x\, dx &= \left.\frac{e^{-px}(-p \sin x - \cos x)}{1+p^2}\right|_0^N \\ &= \frac{1}{1+p^2}[1 - e^{-pN}(-p \sin N - \cos N)]\end{aligned}$$

and, for $p > 0$,

$$\int_0^\infty e^{-px} \sin x\, dx = \lim_{N\to\infty} \int_0^N e^{-px} \sin x\, dx = \frac{1}{1+p^2}.$$

Now, let $f(p) = \int_0^\infty e^{-px} \sin x\, dx$. That is, we treat p as a parameter. Then

$$f(p) = \frac{1}{1+p^2},$$

so

$$\begin{aligned}\int_0^\infty f(p)\, dp &= \int_0^\infty \frac{1}{1+p^2} dp = \lim_{N\to\infty} \int_0^N \frac{dp}{1+p^2} = \lim_{N\to\infty} \tan^{-1} p \Big|_0^N \\ &= \lim_{N\to\infty} \tan^{-1} N = \frac{\pi}{2}.\end{aligned}$$

Thus,

$$\frac{\pi}{2} = \int_0^\infty f(p)\, dp = \int_0^\infty \int_0^\infty e^{-px} \sin x\, dx\, dp$$

reverse the order of integration

$$\begin{aligned}&\overset{\downarrow}{=} \int_0^\infty \int_0^\infty e^{-px} \sin x\, dp\, dx^{‡} \\ &= \int_0^\infty \left[\int_0^\infty e^{-px}\, dp\right] \sin x\, dx \\ &= \int_0^\infty \left[\lim_{N\to\infty} \int_0^N e^{-px}\, dp\right] \sin x\, dx \\ &= \int_0^\infty \left[\lim_{N\to\infty} \left.\frac{-e^{-px}}{x}\right|_{p=0}^N\right] \sin x\, dx\end{aligned}$$

† To see why the function $\frac{\sin x}{x}$ is important, read the article "The function $\frac{\sin x}{x}$," by William B. Gearhart and Harris S. Shultz in *The College Mathematics Journal,* 21(2), March 1990, pages 90–99.

‡ We need a theorem to justify reversing the order of integration when the limits are infinite, but we won't worry about that here. The result is correct.

$$= \int_0^\infty \left[\lim_{N\to\infty} \frac{1}{x}(1 - e^{-Nx})\right] \sin x \, dx$$

$$= \int_0^\infty \frac{\sin x}{x} dx.$$

Thus,

$$\int_0^\infty \frac{\sin x}{x} dx = \frac{\pi}{2}.$$

PROBLEMS 4.2

SELF-QUIZ

I. $\int_1^3 \int_2^4 2xy \, dy \, dx =$ ________.

a. $\int_1^3 12x \, dy$ **b.** $\int_1^3 12y \, dx$

c. $\int_1^3 12x \, dx$ **d.** $\int_2^4 8x \, dy$

II. If $\Omega = \{(x, y)\colon 1 \le x \le 3, 2 \le y \le 4\}$ and if f is a smooth function, then $\iint_\Omega f(x, y) \, dA =$ ________.

a. $\int_1^3 A(x) \, dx$ where $A(x) = \int_2^4 f(x, y) \, dy$

b. $\int_1^3 B(y) \, dy$ where $B(y) = \int_2^4 f(x, y) \, dx$

c. $\int_1^3 \int_2^4 f(x, y) \, dx \, dy$

d. $\int_1^3 \int_2^4 f(x, y) \, dy \, dx$

e. $\int_2^4 \int_1^3 f(x, y) \, dy \, dx$

f. $\int_2^4 \int_1^3 f(x, y) \, dx \, dy$

III. The integral of $f(x, y) = x^2 y$ over the region bounded by the x-axis and the semicircle $x^2 + y^2 = 4$, $y \ge 0$ equals ________.

a. $\int_{-2}^2 \int_0^{\sqrt{4-x^2}} x^2 y \, dy \, dx$

b. $\int_{-2}^2 \int_0^{\sqrt{4-x^2}} x^2 y \, dx \, dy$

c. $\int_0^{\sqrt{4-x^2}} \int_{-2}^2 x^2 y \, dx \, dy$

d. $\int_0^2 \int_{-\sqrt{4-y^2}}^{\sqrt{4-y^2}} x^2 y \, dx \, dy$

e. $\int_0^2 \int_{-\sqrt{4-y^2}}^{\sqrt{4-y^2}} x^2 y \, dy \, dx$

f. $\int_{-\sqrt{4-y^2}}^{\sqrt{4-y^2}} \int_0^2 x^2 y \, dy \, dx$

In Problems 1–8, evaluate the given iterated integral.

1. $\int_0^1 \int_0^2 xy^2 \, dx \, dy$

2. $\int_{-1}^3 \int_2^4 (x^2 - y^3) \, dy \, dx$

3. $\int_2^5 \int_0^4 e^{(x-y)} \, dx \, dy$

4. $\int_0^1 \int_{x^2}^x x^3 y \, dy \, dx$

5. $\int_2^4 \int_{1+y}^{2+3y} (x - y^2) \, dx \, dy$

6. $\int_{\pi/4}^{\pi/3} \int_{\sin x}^{\cos x} (x + 2y) \, dy \, dx$

7. $\int_0^3 \int_{-\sqrt{9-y^2}}^{\sqrt{9-y^2}} x^2 y \, dx \, dy$

8. $\int_1^2 \int_{y^5}^{3y^5} \frac{1}{x} \, dx \, dy$

In Problems 9–24, evaluate the given double integral by means of an appropriate iterated integral.

9. $\iint_\Omega (x^2 + y^2) \, dA$, where $\Omega = \{(x, y)\colon 1 \le x \le 2, -1 \le y \le 1\}$.

10. $\iint_\Omega 2xy\, dA$, where $\Omega = \{(x, y): 0 \le x \le 4, 1 \le y \le 3\}$.

11. $\iint_\Omega (x - y)^2\, dA$, where $\Omega = \{(x, y): -2 \le x \le 2, 0 \le y \le 1\}$.

12. $\iint_\Omega \sin(2x + 3y)\, dA$, where $\Omega = \{(x, y): 0 \le x \le \pi/6, 0 \le y \le \pi/18\}$.

13. $\iint_\Omega xe^{(x^2+y)}\, dA$, where Ω is the region of Problem 10.

14. $\iint_\Omega (x - y^2)\, dA$, where Ω is the region in the first quadrant bounded by the x-axis, the y-axis, and the unit circle.

15. $\iint_\Omega (x^2 + y)\, dA$, where Ω is the region of Problem 14.

16. $\iint_\Omega (x^3 - y^3)\, dA$, where Ω is the region of Problem 14.

17. $\iint_\Omega (x + 2y)\, dA$, where Ω is the triangular region bounded by the lines $y = x$, $y = 1 - x$, and the y-axis.

18. $\iint_\Omega e^{(x+2y)}\, dA$, where Ω is the region of Problem 17.

19. $\iint_\Omega (x^2 + y)\, dA$, where Ω is the region in the first quadrant between the parabolas $y = x^2$ and $y = 1 - x^2$.

20. $\iint_\Omega (1/\sqrt{y})\, dA$, where Ω is the region of Problem 19.

***21.** $\iint_\Omega (y/\sqrt{x^2 + y^2})\, dA$, where $\Omega = \{(x, y): 1 \le x \le y, 1 \le y \le 2\}$.

22. $\iint_\Omega [e^{-y}/(1 + x^2)]\, dA$, where Ω is the first quadrant.

23. $\iint_\Omega \sin x \cos y\, dA$ where Ω is the rectangle $\{(x, y): 0 \le x \le \pi/2, 0 \le y \le \pi\}$.

24. $\iint_\Omega \sin^2 y\, dA$ where $\Omega = \{(x, y): 0 \le x \le \cos y, 0 \le y \le \pi/6\}$.

In Problems 25–28, evaluate $\iint_\Omega x^2 y\, dA$ over the given region Ω.

25.

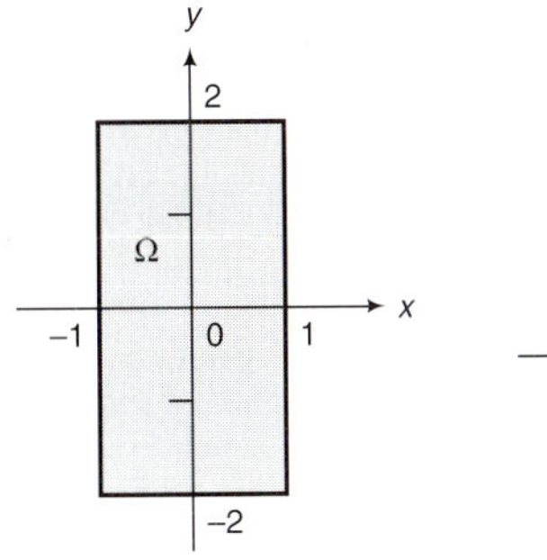

26.

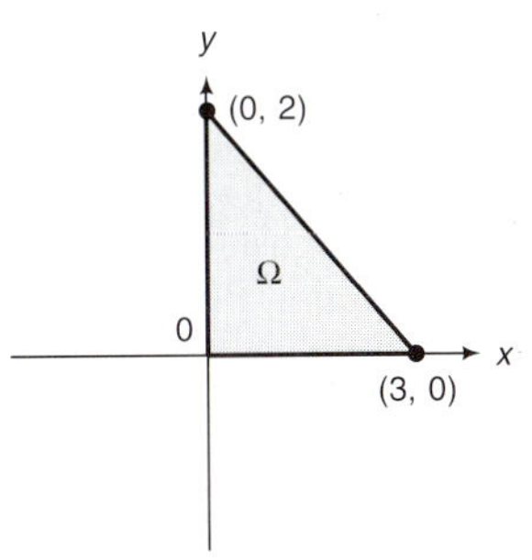

27.

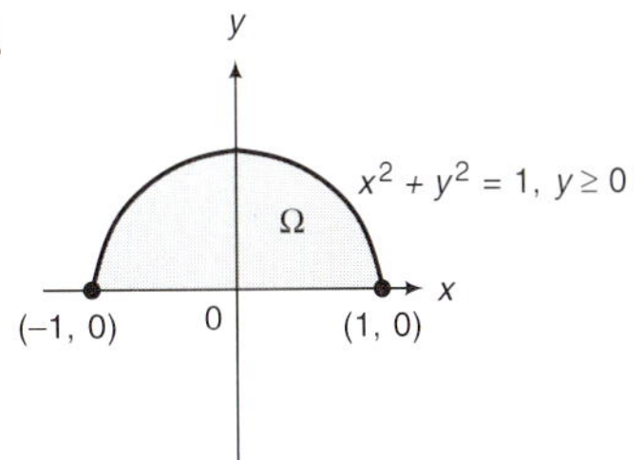

28.

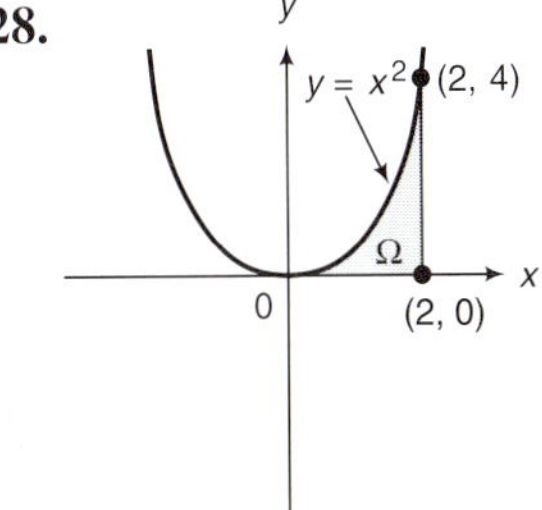

In Problems 29–32, evaluate $\iint_\Omega \dfrac{y}{x^2}\, dA$ over the given region Ω.

29.

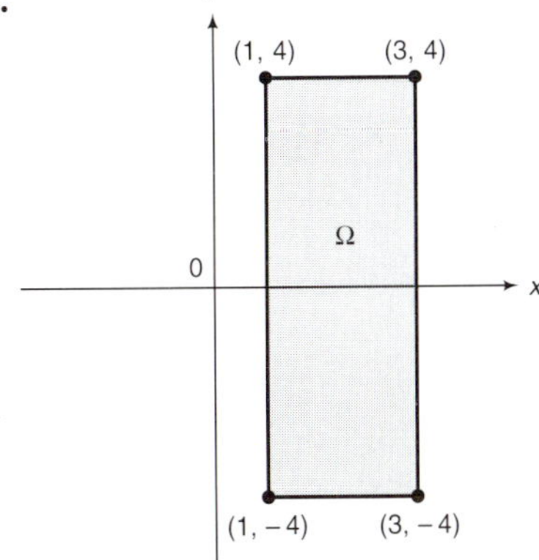

30.

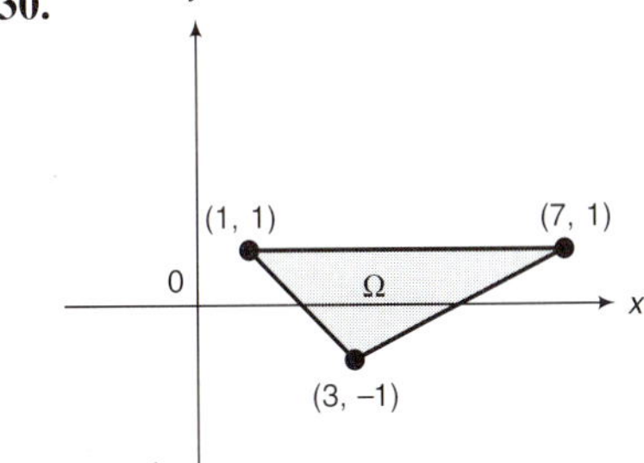

31.

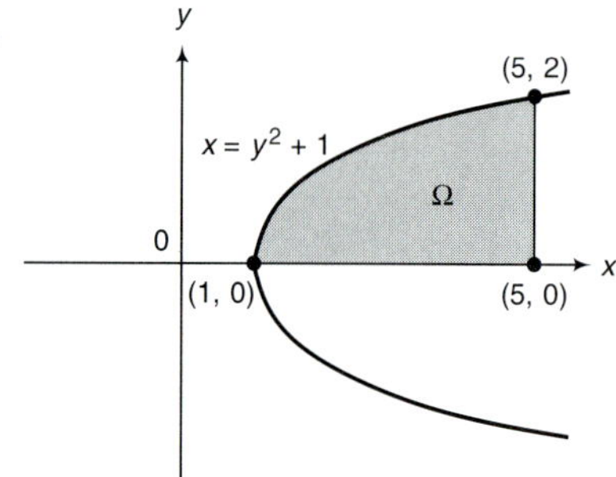

***32.**

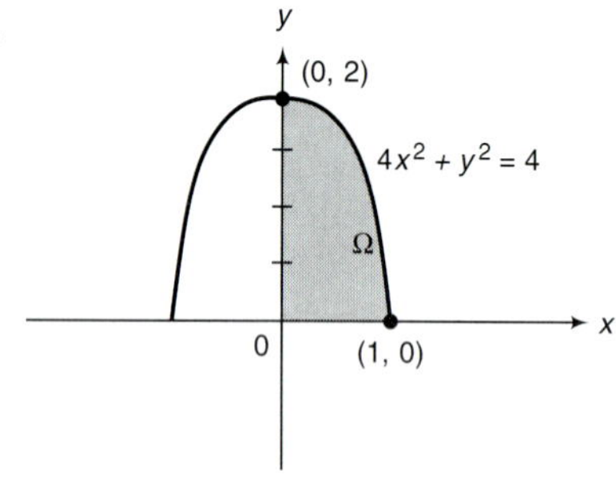

In Problems 33–42, (a) sketch the region over which the integral is taken. Then (b) change the order of integration, and (c) evaluate the given integral.

33. $\int_0^2 \int_{-1}^3 dx\,dy$

34. $\int_0^5 \int_{-5}^8 (x + y)\,dy\,dx$

35. $\int_2^4 \int_1^y (y^3/x^3)\,dx\,dy$

36. $\int_0^1 \int_0^x dy\,dx$

37. $\int_0^1 \int_x^1 dy\,dx$

38. $\int_0^{\pi/2} \int_0^{\cos y} y\,dx\,dy$

39. $\int_0^2 \int_0^{\sqrt{4-y^2}} (4 - x^2)^{3/2}\,dx\,dy$

40. $\int_0^1 \int_{\sqrt{x}}^{\sqrt[3]{x}} (1 + y^6)\,dy\,dx$

41. $\int_0^1 \int_{\sqrt{y}}^1 \sqrt{3 - x^3}\,dx\,dy$

42. $\int_0^\infty \int_x^\infty \frac{1}{(1 + y^2)^{7/5}}\,dy\,dx$

In Problems 43–52, find the volume of the given solid.

43. The solid bounded by the plane $x + y + z = 3$ and the three coordinate planes.

***44.** The solid bounded by the planes $x = 0$, $z = 0$, $x + 2y + z = 6$, and $x - 2y + z = 6$.

45. The solid bounded by the cylinders $x^2 + y^2 = 4$ and $y^2 + z^2 = 4$.

46. The solid bounded by the cylinder $x^2 + z^2 = 1$ and the planes $y = 0$ and $y = 2$.

***47.** The ellipsoid $x^2 + 4y^2 + 9z^2 = 36$.

***48.** The solid bounded above by the sphere $x^2 + y^2 + z^2 = 9$ and below by the plane $z = \sqrt{5}$.

49. The solid bounded by the surface $z = e^{-(x+y)}$ and the three coordinate planes.

***50.** The solid bounded by the parabolic cylinder $x = z^2$ and the planes $y = 1$, $y = 5$, $z = 1$, and $x = 0$.

51. The solid bounded by the circular paraboloid $x = y^2 + z^2$ and the plane $x = 1$.

***52.** The solid bounded by the paraboloid $y = x^2 + z^2$ and the plane $x + y = 3$.

***53.** Use a double integral to find the area of each of the regions bounded by the x-axis and the curves $y = x^3 + 1$ and $y = 3 - x^2$.

54. Use a double integral to find the area in the first quadrant bounded by the curves $y = x^{1/m}$ and $y = x^{1/n}$, where m and n are positive and $n > m$.

55. Sketch the solid whose volume is given by $\int_1^3 \int_0^2 (x + 3y)\,dx\,dy$.

***56.** Sketch the solid whose volume is given by $\int_0^1 \int_{x^2}^{\sqrt{x}} \sqrt{x^2 + y^2}\,dy\,dx$.

57. Show that if both of the following integrals exist, then

$$\int_0^\infty \int_0^x f(x, y)\,dy\,dx = \int_0^\infty \int_y^\infty f(x, y)\,dx\,dy.$$

[*Hint:* Draw a picture.]

58. Suppose $f(x, y) = g(x)h(y)$, where g and h are continuous. Let Ω be the rectangle $\{(x, y)\colon a \le x \le b,\ c \le y \le d\}$. Show that

$$\iint_\Omega f(x, y)\,dA = \left[\int_a^b g(x)\,dx\right] \cdot \left[\int_c^d h(y)\,dy\right].$$

***59.** Evaluate $\int_0^\infty \left(\frac{\sin t}{t}\right)^2 dt$ using the technique of Example 9. [*Hint:* Substitute $t = \frac{x}{p}$ in Example 9 to obtain $\int_0^\infty \frac{\sin pt}{t}\,dt = \frac{\pi}{2}$ and then integrate both sides from 0 to q with respect to p. Then substitute $q = 2$.]

ANSWERS TO SELF-QUIZ

I. c **II.** a = d, f **III.** a, d

4.3 DENSITY, MASS, AND CENTER OF MASS (OPTIONAL)

Let $\rho(x, y)$ denote the density of a plane object (like a thin lamina, for example). Suppose that the object occupies a region Ω in the xy-plane. Then the mass of a small rectangle of sides Δx and Δy centered at the point (x, y) is approximated by

$$\rho(x, y)\,\Delta x\,\Delta y = \rho(x, y)\,\Delta A. \tag{1}$$

MASS

If we sum these approximate masses and take a limit, then we obtain a formula for the total mass:

THE TOTAL MASS μ IS GIVEN BY

$$\mu = \iint_{\Omega} \rho(x, y)\,dA. \tag{2}$$

Compare this formula with the formula for the mass of an object lying along the x-axis with density $\rho(x)$:

$$\mu(x) = \int \rho(x)\,dx.$$

In one-variable calculus we saw how to calculate the first moment and center of mass of an object around the x- and y-axes.[†] For example, we may define

$$M_y = \int_a^b x\rho(x)\,dx \tag{3}$$

to be the first moment about the y-axis when we had a system of masses distributed along the x-axis. Similarly, we may calculate the x-coordinate of the center of mass of the object as

$$\bar{x} = \frac{\displaystyle\int_a^b x\rho(x)\,dx}{\displaystyle\int_a^b \rho(x)\,dx} = \frac{\text{first moment about } y\text{-axis}}{\text{mass}} = \frac{M_y}{\mu} \tag{4}$$

When we calculated the centroid of a plane region, we found that it was necessary to assume that the region had a constant area density ρ. However, by using double integrals, we can get away from this restriction. Consider the plane region whose mass is given by (2). Then we define

FIRST MOMENT AROUND THE y-AXIS

$$M_y = \text{first moment around } y\text{-axis} = \iint_{\Omega} x\rho(x, y)\,dA. \tag{5}$$

[†] See *Calculus* or *Calculus of One Variable,* Section 4.2. See, in particular, equation (4.2.8) on p. 273.

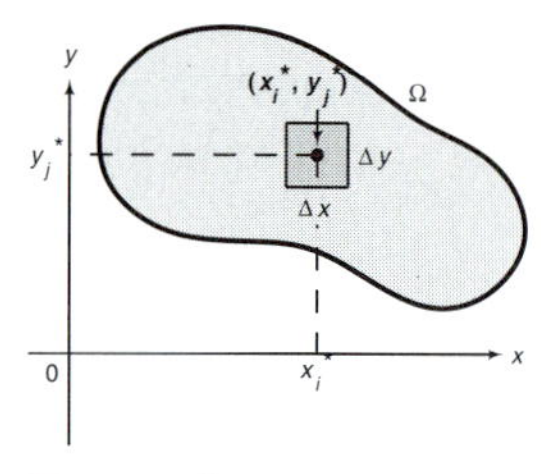

FIGURE 1
The first moment about the y-axis of a small subrectangle centered at (x_i^*, y_j^*) is approximated by $x_i^*\rho(x_i^*, y_j^*)\,\Delta x\,\Delta y$.

Look at Figure 1. The first moment about the y-axis of a small rectangle centered at (x, y) is given approximately by

$$x_i^*\rho(x_i^*, y_j^*)\,\Delta x\,\Delta y,$$

and if we add up these moments for all such "subrectangles" and take a limit, we arrive at equation (5). Finally, we define the **center of mass** of the plane region to be the point $(\bar{x}, \bar{y})$, where

CENTER OF MASS $(\bar{x}, \bar{y})$

$$\bar{x} = \frac{M_y}{\mu} = \frac{\iint_\Omega x\rho(x, y)\,dA}{\iint_\Omega \rho(x, y)\,dA} = \frac{\text{first moment about } y\text{-axis}}{\text{total mass}} \tag{6}$$

and

$$\bar{y} = \frac{M_x}{\mu} = \frac{\iint_\Omega y\rho(x, y)\,dA}{\iint_\Omega \rho(x, y)\,dA} = \frac{\text{first moment about } x\text{-axis}}{\text{total mass}}. \tag{7}$$

If the density of a plane region is constant, then, the center of mass is called the **centroid** of the region.

NOTE: It is not difficult to show that the center of mass of a plane lamina lies on any line that divides the lamina into equal masses.

EXAMPLE 1

COMPUTING THE MASS AND CENTER OF MASS OF A LAMINA

A plane lamina has the shape of the triangle bounded by the lines $y = x$, $y = 2 - x$, and the x-axis. Its density function is given by $\rho(x, y) = 1 + 2x + y$. Distance is measured in meters, and mass is measured in kilograms. Find the mass and center of mass of the lamina.

SOLUTION: The region is sketched in Figure 2. The mass is given by

$$\mu = \iint_\Omega \rho(x, y)\,dA = \int_0^1 \int_y^{2-y} (1 + 2x + y)\,dx\,dy$$

$$= \int_0^1 \left\{(x + x^2 + xy)\Big|_y^{2-y}\right\} dy = \int_0^1 (6 - 4y - 2y^2)\,dy$$

$$= \left(6y - 2y^2 - \frac{2y^3}{3}\right)\Big|_0^1 = \frac{10}{3}\text{ kg.}$$

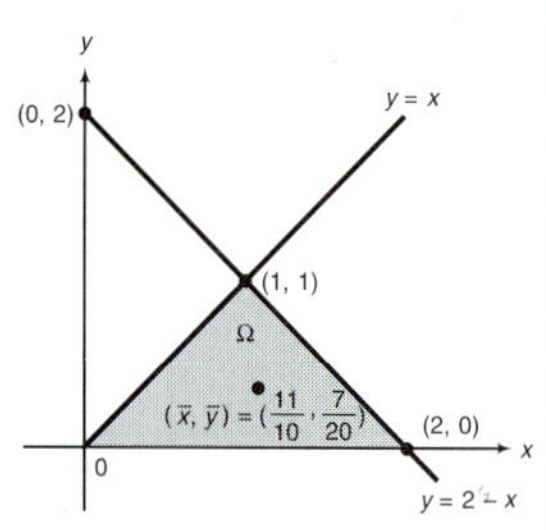

FIGURE 2
The plane lamina bounded by the lines $y = x$, $y = 2 - x$ and the x-axis

Then

$$M_y = \int_0^1 \int_y^{2-y} x(1 + 2x + y)\,dx\,dy$$

$$= \int_0^1 \int_y^{2-y} (x + 2x^2 + xy)\, dx\, dy$$

$$= \int_0^1 \left\{ \left(\frac{x^2}{2} + \frac{2x^3}{3} + \frac{x^2 y}{2} \right) \Bigg|_y^{2-y} \right\} dy$$

$$= \int_0^1 \left(\frac{22}{3} - 8y + 2y^2 - \frac{4y^3}{3} \right) dy$$

$$= \left(\frac{22}{3}y - 4y^2 + \frac{2y^3}{3} - \frac{y^4}{3} \right) \Bigg|_0^1 = \frac{11}{3}\ \text{kg} \cdot \text{m},$$

and

$$M_x = \int_0^1 \int_y^{2-y} y(1 + 2x + y)\, dx\, dy = \int_0^1 \left\{ (xy + x^2 y + xy^2) \Big|_y^{2-y} \right\} dy$$

$$= \int_0^1 (6y - 4y^2 - 2y^3)\, dy = \left(3y^2 - \frac{4y^3}{3} - \frac{y^4}{2} \right) \Bigg|_0^1 = \frac{7}{6}\ \text{kg} \cdot \text{m}.$$

Thus,

$$\bar{x} = \frac{M_y}{\mu} = \frac{\frac{11}{3}}{\frac{10}{3}} = \frac{11}{10}\ \text{m} \qquad \text{and} \qquad \bar{y} = \frac{M_x}{\mu} = \frac{\frac{7}{6}}{\frac{10}{3}} = \frac{7}{20}\ \text{m}.$$

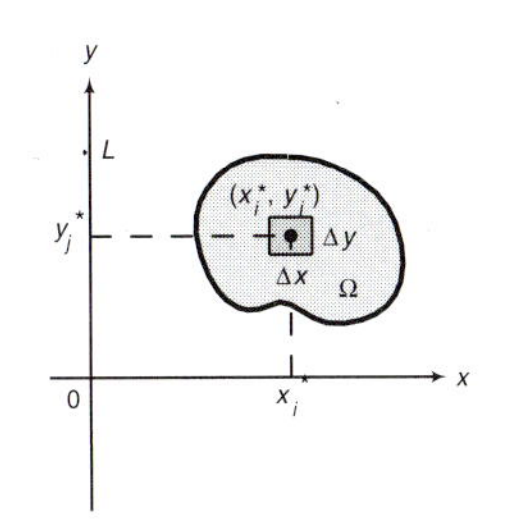

FIGURE 3
(x_i^*, y_j^*) is the center of a small subrectangle.

When the subrectangle is rotated about the y-axis, the resulting ring has a radius approximately equal to x_i^*.

The following result shows how centroids can be used to compute volumes of revolution.

THEOREM 1 FIRST THEOREM OF PAPPUS†

Suppose that the plane region Ω is revolved about a line L in the xy-plane that does not intersect it. Then the volume generated is equal to the product of the area of Ω and the length of the circumference of the circle traced by the centroid of Ω.

PROOF: We construct a coordinate system, placing the y-axis so that it coincides with the line L and Ω is in the first quadrant. The situation is then as depicted in Figure 3. We form a regular partition of the region Ω and choose the point (x_i^*, y_j^*) to be a point in the subrectangle Ω_{ij}. If Ω_{ij} is revolved about L (the y-axis), it forms a ring, as depicted in Figure 4. The volume of the ring is, approximately,

$$V_{ij} \approx (\text{circumference of circle with radius } x_i^*) \times (\text{thickness of ring}) \times (\text{height of ring})$$

$$= 2\pi x_i^*\, \Delta x\, \Delta y = 2\pi x_i^*\, \Delta A.$$

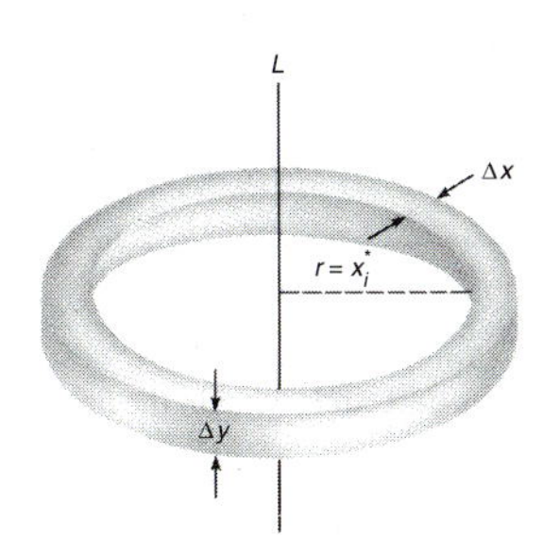

FIGURE 4
The ring obtained by rotating a subrectangle about L

And using a familiar limiting argument, we have

$$V = \lim_{\substack{\Delta x \to 0 \\ \Delta y \to 0}} \sum_{i=1}^{n} \sum_{j=1}^{m} V_{ij} = \iint_\Omega 2\pi x\, dA = 2\pi \iint_\Omega x\, dA. \qquad \textbf{(8)}$$

Now we can think of Ω as a thin lamina with constant density $\rho = 1$. Then from (8) (with $\rho = 1$),

† See the biographical sketch on the next page.

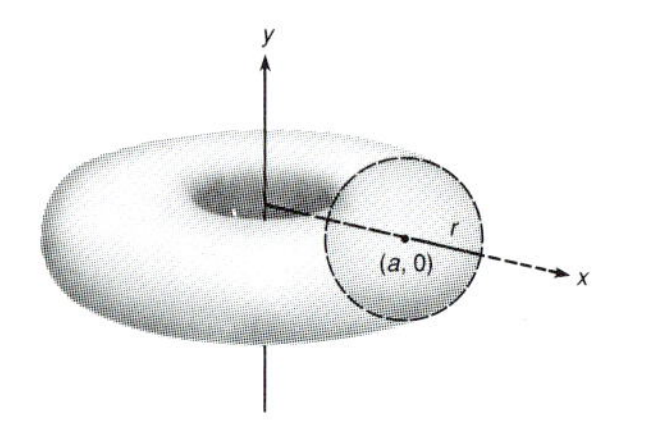

FIGURE 5
A torus

$$\bar{x} = \frac{\iint_\Omega x\,dA}{\iint_\Omega dA} \overset{\text{From (8)}}{=} \frac{V/2\pi}{\iint_\Omega dA} \overset{\text{Formula (4.1.15)}}{=} \frac{V/2\pi}{\text{area of }\Omega},$$

or

$$V = (2\pi\bar{x})(\text{area of }\Omega)$$
$$= \begin{pmatrix}\text{length of the circumference of the circle} \\ \text{traced by the centroid of }\Omega\end{pmatrix} \times (\text{area of }\Omega). \quad \blacksquare$$

EXAMPLE 2 USING THE FIRST THEOREM OF PAPPUS TO COMPUTE THE VOLUME OF A TORUS

Use the first theorem of Pappus to calculate the volume of the torus generated by rotating the circle $(x - a)^2 + y^2 = r^2$ $(r < a)$ about the y-axis.

SOLUTION: The circle and the torus are sketched in Figure 5. The area of the circle is πr^2. The radius of the circle traced by the centroid $(a, 0)$ is a, and the circumference is $2\pi a$. Thus,

$$V = (2\pi a)\pi r^2 = 2\pi^2 a r^2.$$

OF HISTORICAL INTEREST

PAPPUS OF ALEXANDRIA, BORN C. 290 A.D., FL. C. 320 A.D.

A HARBOR VIEW OF ALEXANDRIA, WHERE PAPPUS LIVED IN THE FOURTH CENTURY

Born during the reign of the Roman emperor Diocletian (A.D. 284–305), Pappus of Alexandria lived approximately 600 years after the time of Euclid and Archimedes and devoted much of his life attempting to revitalize interest in the traditional study of Greek geometry.

Pappus's greatest work was his *Mathematical Collection,* written between A.D. 320 and 340. This work is significant for three reasons. First, in it are collected works of more than 30 different mathematicians of antiquity. We owe much of our knowledge of Greek geometry to the *Collection.* Second, it provides alternative proofs of the results of the greatest of the Greeks, including Euclid and Archimedes. Third, the *Collection* contains a variety of discoveries not found in any earlier work.

The *Collection* comprises eight books, each one containing a variety of interesting and sometimes amusing results. In Book V, for example, Pappus showed that if two regular polygons have equal perimeters, then the one with the greater number of sides has the larger area. Pappus used this result to suggest the great wisdom of bees, as bees construct their hives using hexagonal (6-sided) cells, rather than square or triangular ones.

Book VII contains the theorem proved in this section. It is also the book that is the most important to the history of mathematics in that it contains the *Treasury of Analysis.* The *Treasury* is a collection of mathematical facts that, together with Euclid's *Elements,* claimed to contain the material the practicing mathematician in the fourth century needed to know. Although mathematicians wrote in Greek for another thousand or so years, no follower wrote a work of equal significance.

PROBLEMS 4.3

SELF-QUIZ

I. If Ω is a region of constant density, then $(\iint_\Omega x\, dA)/(\iint_\Omega dA)$ computes ______.

a. the x-coordinate of the region's center of mass

b. the first moment of the region about the x-axis

c. the y-coordinate of the region's center of mass

d. the first moment of the region about the y-axis

II. Consider an object filling a region Ω and having density given by $\rho(x, y)$. The first moment of this object about the x-axis equals ______.

a. $\iint_\Omega \rho(x, y)\, dA$

b. $\iint_\Omega y\rho(x, y)\, dA$

c. $\iint_\Omega x\rho(x, y)\, dA$

d. $\left(\iint_\Omega x\rho(x, y)\, dA\right) \Big/ \left(\iint_\Omega \rho(x, y)\, dA\right)$

In Problems 1–12, find the mass and the center of mass of an object that lies in the given region and has the given area density function.

1. $\Omega = \{(x, y)\colon 1 \le x \le 2,\ -1 \le y \le 1\}$; $\rho(x, y) = x^2 + y^2$

2. $\Omega = \{(x, y)\colon 0 \le x \le 4,\ 1 \le y \le 3\}$; $\rho(x, y) = 2xy$

3. $\Omega = \{(x, y)\colon 0 \le x \le \pi/6,\ 0 \le y \le \pi/18\}$; $\rho(x, y) = \sin(2x + 3y)$

4. $\Omega = \{(x, y)\colon -2 \le x \le 2,\ 0 \le y \le 1\}$; $\rho(x, y) = (x - y)^2$

5. Ω is the region of Problem 2; $\rho(x, y) = xe^{x-y}$

6. Ω is the quarter of the unit circle lying in the first quadrant; $\rho(x, y) = x + y^2$

7. Ω is the region of Problem 6; $\rho(x, y) = x^2 + y$

8. Ω is the region of Problem 6; $\rho(x, y) = x^3 + y^3$

9. Ω is the triangular region bounded by the lines $y = x$, $y = 1 - x$, and the x-axis; $\rho(x, y) = x + 2y$

10. Ω is the region of Problem 9; $\rho(x, y) = e^{x+2y}$

11. Ω is the first quadrant; $\rho(x, y) = e^{-y}/(1 + x)^3$

12. Ω is the first quadrant; $\rho(x, y) = (x + y)e^{-(x+y)}$

13. Use the first theorem of Pappus to calculate the volume of the torus generated by rotating the unit circle about the line $y = 4 - x$.

14. Use the first theorem of Pappus to calculate the volume of the "elliptical torus" generated by rotating the ellipse $(x/a)^2 + (y/b)^2 = 1$ about the line $y = 3a$. Assume that $3a > b > 0$.

ANSWERS TO SELF-QUIZ

I. a **II.** b

4.4 DOUBLE INTEGRALS IN POLAR COORDINATES

In this section, we will see how to evaluate double integrals of functions in the form $z = f(r, \theta)$, where r and θ denote the polar coordinates of a point in the plane.

Let $z = f(r, \theta)$ and let Ω denote the "polar rectangle"

$$\theta_1 \le \theta \le \theta_2, \qquad r_1 \le r \le r_2. \tag{1}$$

This region is sketched in Figure 1. We will calculate the volume of the solid between the surface $z = f(r, \theta)$ and the region Ω. We partition Ω by small "polar rectangles" and calculate the volume over such a region. The volume of the part of the solid over the region Ω_{ij} is given, approximately, by

$$f(r_i, \theta_j)A_{ij}, \tag{2}$$

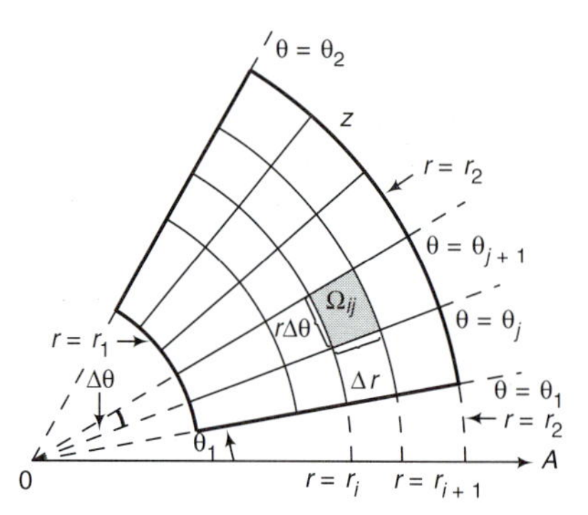

FIGURE 1
The shaded polar rectangle is bounded by the lines $\theta = \theta_j$ and $\theta = \theta_{j+1}$, and the circles $r = r_i$ and $r = r_{i+1}$.

where A_{ij} is the area of Ω_{ij}. Recall that if $r = f(\theta)$, then the area bounded by the lines $\theta = \alpha$, $\theta = \beta$, and the curve $r = f(\theta)$ is given by[†]

$$A = \int_{\alpha}^{\beta} \frac{1}{2}[f(\theta)]^2\, d\theta. \tag{3}$$

Thus,

$$\begin{aligned} A_{ij} &= \frac{1}{2}\int_{\theta_j}^{\theta_{j+1}} (r_{i+1}^2 - r_i^2)\, d\theta = \frac{1}{2}(r_{i+1}^2 - r_i^2)(\theta_{j+1} - \theta_j) \\ &= \frac{1}{2}(r_{i+1} + r_i)(r_{i+1} - r_i)(\theta_{j+1} - \theta_j) = \frac{1}{2}(r_{i+1} + r_i)\,\Delta r\,\Delta\theta. \end{aligned}$$

But if Δr is small, then $r_{i+1} \approx r_i$, and we have

$$A_{ij} \approx \frac{1}{2}(2r_i)\,\Delta r\,\Delta\theta = r_i\,\Delta r\,\Delta\theta.$$

Then

$$V_{ij} \approx f(r_i, \theta_j)A_{ij} \approx f(r_i, \theta_j)r_i\,\Delta r\,\Delta\theta,$$

so that, adding up the individual volumes and taking a limit, we obtain

> **VOLUME IN POLAR COORDINATES**
>
> $$V = \iint_{\Omega} f(r, \theta)r\, dr\, d\theta. \tag{4}$$

NOTE: Do not forget the extra r in the formula above.

EXAMPLE 1 FINDING THE VOLUME ENCLOSED BY A SPHERE

Find the volume enclosed by the sphere $x^2 + y^2 + z^2 = a^2$.

[†] See *Calculus* or *Calculus of One Variable*, p. 591.

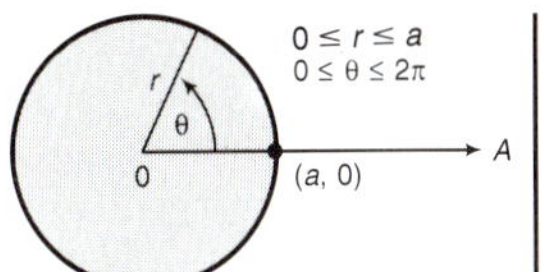

FIGURE 2
A disk of radius r

SOLUTION: We will calculate the volume enclosed by the hemisphere $z = \sqrt{a^2 - x^2 - y^2}$ and then multiply by two. To do so, we first note that this volume is the volume of the solid under the hemisphere and above the disk $x^2 + y^2 \le a^2$. We use polar coordinates, since in polar coordinates $x^2 + y^2 = (r\cos\theta)^2 + (r\sin\theta)^2 = r^2$. On the disk, $0 \le r \le a$ and $0 \le \theta \le 2\pi$ (see Figure 2). Then $\sqrt{a^2 - (x^2 + y^2)} = \sqrt{a^2 - r^2}$, so by (4)

$$V = \iint_{\text{disk}} \sqrt{a^2 - r^2}\, r\, dr\, d\theta = \int_0^{2\pi} \int_0^a \sqrt{a^2 - r^2}\, r\, dr\, d\theta$$

$$= \int_0^{2\pi} \left\{ -\frac{1}{3}(a^2 - r^2)^{3/2} \Big|_0^a \right\} d\theta = \int_0^{2\pi} \frac{1}{3} a^3\, d\theta = \frac{2\pi a^3}{3}.$$

Thus, the volume of the sphere is $2(2\pi a^3/3) = \frac{4}{3}\pi a^3$.

AREA FORMULA

Since area of $\Omega = \iint_\Omega dA$, we can write, using (4),

> **THE AREA OF A POLAR REGION**
>
> $$\iint_\Omega r\, dr\, d\theta = \text{area of } \Omega. \qquad (5)$$

EXAMPLE 2

COMPUTING THE AREA ENCLOSED BY A CARDIOID

Calculate the area enclosed by the cardioid $r = 1 + \sin\theta$.

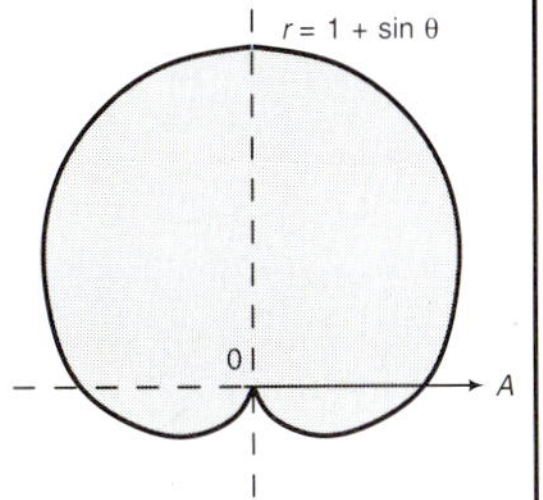

FIGURE 3
The cardioid $r = 1 + \sin\theta$

SOLUTION: The cardioid is sketched in Figure 3. The region is described by the inequalities $0 \le r \le 1 + \sin\theta$, $0 \le \theta \le 2\pi$. Thus,

$$\text{area} = \int_0^{2\pi} \int_0^{1+\sin\theta} r\, dr\, d\theta = \int_0^{2\pi} \left\{ \frac{r^2}{2} \Big|_0^{1+\sin\theta} \right\} d\theta$$

$$= \frac{1}{2} \int_0^{2\pi} (1 + 2\sin\theta + \sin^2\theta)\, d\theta$$

$$= \frac{1}{2} \int_0^{2\pi} \left(1 + 2\sin\theta + \frac{1 - \cos 2\theta}{2} \right) d\theta$$

$$= \frac{1}{2} \left(\frac{3\theta}{2} - 2\cos\theta - \frac{\sin 2\theta}{4} \right) \Big|_0^{2\pi} = \frac{3\pi}{2}.$$

As we will see, it is often very useful to write a double integral in terms of polar coordinates. Let $z = f(x, y)$ be a function defined over a region Ω. Then using polar coordinates, we can write

$$z = f(r\cos\theta, r\sin\theta),$$

and we can also describe Ω in terms of polar coordinates. The volume of the solid under f and over Ω is the same whether we use rectangular or polar coordinates. Thus, writing the volume in both rectangular and polar coordinates, we obtain the useful **change-of-variables formula**:

CONVERTING A DOUBLE INTEGRAL FROM RECTANGULAR TO POLAR COORDINATES

$$\iint_{\Omega} f(x, y)\, dA = \iint_{\Omega} f(r\cos\theta, r\sin\theta) r\, dr\, d\theta. \quad \textbf{(6)}$$

EXAMPLE 3 FINDING THE MASS OF A SEMICIRCULAR PLANE LAMINA USING POLAR COORDINATES

The density at any point on a semicircular plane lamina is proportional to the square of the distance from the point to the center of the circle. Find the mass of the lamina.

SOLUTION: We have $\rho(x, y) = \alpha(x^2 + y^2) = \alpha r^2$ and $\Omega = \{(r, \theta): 0 \le r \le a \text{ and } 0 \le \theta \le \pi\}$, where a is the radius of the circle. Then

$$\mu = \int_0^{\pi}\int_0^{a} (\alpha r^2) r\, dr\, d\theta = \int_0^{\pi} \left\{ \frac{\alpha r^4}{4}\bigg|_0^a \right\} d\theta = \frac{\alpha a^4}{4}\int_0^{\pi} d\theta = \frac{\alpha\pi a^4}{4}.$$

This double integral can be computed without using polar coordinates, but the computation is much more tedious. Try it!

EXAMPLE 4 COMPUTING THE VOLUME OF A SOLID USING POLAR COORDINATES

Find the volume of the solid bounded by the xy-plane, the cylinder $x^2 + y^2 = 4$, and the paraboloid $z = 2(x^2 + y^2)$.

SOLUTION: The volume requested is the volume under the surface $z = 2(x^2 + y^2) = 2r^2$ and above the circle $x^2 + y^2 = 4$. Thus,

$$V = \int_0^{2\pi}\int_0^{2} 2r^2 \cdot r\, dr\, d\theta = \int_0^{2\pi} \left\{ \frac{r^4}{2}\bigg|_0^2 \right\} d\theta = \int_0^{2\pi} 8\, d\theta = 16\pi.$$

The solid is sketched in Figure 4.

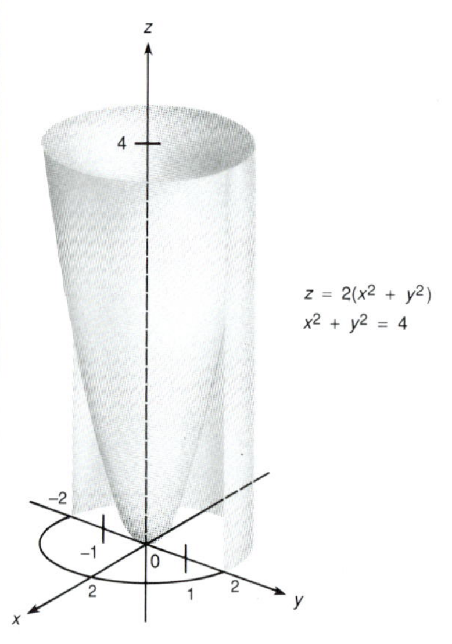

FIGURE 4
The solid above the xy-plane bounded by the cylinder $x^2 + y^2 = 4$ and the paraboloid $z = 2(x^2 + y^2)$

EXAMPLE 5 USING POLAR COORDINATES TO EVALUATE AN INTEGRAL IMPORTANT IN PROBABILITY THEORY

In probability theory, one of the most important integrals is

$$\int_{-\infty}^{\infty} e^{-x^2}\,dx.$$

We now show how a combination of double integrals and polar coordinates can be used to evaluate it. Let

$$I = \int_0^{\infty} e^{-x^2}\,dx.$$

Then by symmetry $\int_{-\infty}^{\infty} e^{-x^2}\,dx = 2I$. Thus, we need only to evaluate I. But since any dummy variable can be used in a definite integral, we also have

$$I = \int_0^{\infty} e^{-y^2}\,dy.$$

Thus,

$$I^2 = \left(\int_0^{\infty} e^{-x^2}\,dx\right)\left(\int_0^{\infty} e^{-y^2}\,dy\right),$$

and from a result similar to that of Problem 4.2.58,†

$$I^2 = \int_0^{\infty}\int_0^{\infty} e^{-x^2}e^{-y^2}\,dx\,dy = \int_0^{\infty}\int_0^{\infty} e^{-(x^2+y^2)}\,dx\,dy = \iint_{\Omega} e^{-(x^2+y^2)}\,dA,$$

where Ω denotes the first quadrant. In polar coordinates the first quadrant can be written as

$$\Omega = \left\{(r,\theta)\colon 0 \le r < \infty \text{ and } 0 \le \theta \le \frac{\pi}{2}\right\}.$$

Thus, since $x^2 + y^2 = r^2$, we obtain

$$\begin{aligned} I^2 &= \int_0^{\pi/2}\int_0^{\infty} e^{-r^2} r\,dr\,d\theta = \int_0^{\pi/2}\left(\lim_{N\to\infty}\int_0^{N} e^{-r^2} r\,dr\right)d\theta \\ &= \int_0^{\pi/2}\left(\lim_{N\to\infty} -\frac{1}{2}e^{-r^2}\Big|_0^N\right)d\theta = \frac{1}{2}\int_0^{\pi/2} d\theta = \frac{\pi}{4}. \end{aligned}$$

Hence, $I^2 = \pi/4$, so $I = \sqrt{\pi}/2$, and

$$\int_{-\infty}^{\infty} e^{-x^2}\,dx = 2I = \sqrt{\pi}.$$

By making the substitution $u = x/\sqrt{2}$, it is straightforward to show that

$$\int_{-\infty}^{\infty} e^{-x^2/2}\,dx = \sqrt{2\pi}.$$

The function $\rho(x) = (1/\sqrt{2\pi})e^{-x^2/2}$ is called the **density function for the unit normal distribution**. We have just shown that $\int_{-\infty}^{\infty} \rho(x)\,dx = 1$.

† Problem 4.2.58 really does not apply to improper integrals such as this one. The answer we obtain is correct, but additional theory is needed to justify this next step. The needed result is best left to a course in advanced calculus.

PROBLEMS 4.4

SELF-QUIZ

I. True–False: The area of region Ω equals the integral of the constant function 1 over the region.

II. The circle of radius A centered at the pole (the origin) has area ________.

a. $\int_0^{\pi}\int_0^{A} r\,dr\,d\theta$ **c.** $\int_0^{2\pi}\int_0^{A} r\,dr\,d\theta$

b. $\int_0^{A}\int_0^{\pi} r\,dr\,d\theta$ **d.** $\int_0^{2\pi}\int_0^{r} A\,dr\,d\theta$

III. The volume under the cone $z = r$ and above the disk $r \le a$, $z = 0$ is ________.

a. $\int_0^{2\pi}\int_0^{a} r^2\,dr\,d\theta$

b. $\int_0^{2\pi}\int_0^{r} ar\,dr\,d\theta$

c. $\int_0^{2\pi}\int_0^{a} (r-a)r\,dr\,d\theta$

d. $\int_{-\pi}^{\pi}\int_0^{a} (r-0)r\,dr\,d\theta$

IV. The volume under the plane $z = x + y$ and above the quarter-circle $x^2 + y^2 \le 25$, $x \ge 0$, $y \ge 0$, $z = 0$ is ________.

a. $\int_0^{\pi}\int_0^{25} (r\cos\theta + r\sin\theta)r\,dr\,d\theta$

b. $\int_0^{\pi/2}\int_0^{5} r(\cos\theta + \sin\theta)\,dr\,d\theta$

c. $\int_0^{\pi/4}\int_{-5}^{5} (r\cos\theta + r\sin\theta)\,dr\,d\theta$

d. $\int_0^{\pi/2}\int_0^{5} (r\cos\theta + r\sin\theta)r\,dr\,d\theta$

In Problems 1–4, calculate the volume under the given surface that lies over the specified region in the plane $z = 0$.

1. $z = r^n$ where n is a positive integer; Ω is the circle of radius a centered at the pole.

2. $z = 3 - r$; Ω is the circle $r = 2\cos\theta$, $-\pi/2 \le \theta \le \pi/2$.

3. $z = r^2$; Ω is the cardioid $r = 4(1 - \cos\theta)$.

4. $z = r^3$; Ω is the region enclosed by the spiral of Archimedes $r = a\theta$ and the polar axis for θ between 0 and 2π.

In Problems 5–14, calculate the area of the region enclosed by the given curve or curves.

5. $r = 1 - \cos\theta$ **6.** $r = 4(1 + \cos\theta)$

7. $r = 1 + 2\cos\theta$ (outer loop)

8. $r = 3 - 2\sin\theta$ **9.** $r^2 = \cos 2\theta$

10. $r^2 = 4\sin 2\theta$ **11.** $r = a + b\sin\theta$, $a > b > 0$

12. $r = \tan\theta$ and the line $\theta = \pi/4$

13. Outside the circle $r = 6$ and inside the cardioid $r = 4(1 + \sin\theta)$.

14. Inside the cardioid $r = 2(1 + \cos\theta)$ but outside the circle $r = 2$.

15. Find the volume of the solid bounded above by the sphere $x^2 + y^2 + z^2 = 4a^2$, below by the xy-plane, and on the sides by the cylinder $x^2 + y^2 = a^2$.

16. Find the area of the region interior to the curve $(x^2 + y^2)^3 = 9y^2$.

***17.** Find the volume of the solid bounded by the cone $x^2 + y^2 = z^2$ and the cylinder $x^2 + y^2 = 4y$.

***18.** Find the volume of the solid bounded by the cone $z^2 = x^2 + y^2$ and the paraboloid $2z = x^2 + y^2$.

***19.** Find the volume of the solid bounded by the cylinder $x^2 + y^2 = 9$ and the hyperboloid $x^2 + y^2 - z^2 = 1$.

20. Find the volume of the solid centered at the origin that is bounded above by the surface $z = e^{-(x^2+y^2)}$ and below by the unit circle.

21. Find the centroid of the region bounded by $r = \cos\theta + 2\sin\theta$.

22. Find the centroid of the region bounded by the limaçon $r = 3 + \sin\theta$.

23. Find the centroid of the region bounded by the limaçon $r = a + b\cos\theta$, $a > b > 0$.

24. Show that if Ω is the unit disk, then

$$\iint_{\Omega} \frac{1}{1 + x^2 + y^2}\,dA = \pi \ln 2.$$

ANSWERS TO SELF-QUIZ

I. True **II.** c **III.** a = d **IV.** d

4.5 THE TRIPLE INTEGRAL

In this section we discuss the idea behind the triple integral of a function of three variables $f(x, y, z)$ over a region S in $\mathbb{R}^3$. This is really a simple extension of the double integral. For that reason we will omit a number of technical details.

We start with a parallelepiped π in $\mathbb{R}^3$, which can be written as

$$\pi = \{(x, y, z): a_1 \le x \le a_2, b_1 \le y \le b_2, c_1 \le z \le c_2\} \tag{1}$$

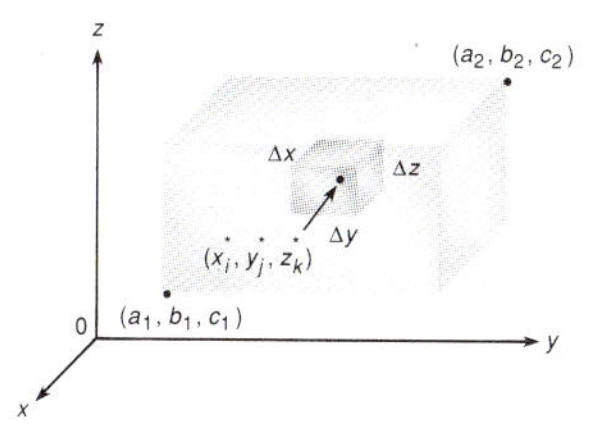

FIGURE 1
The parallelepiped $a_1 \le x \le a_2$, $b_1 \le y \le b_2$, $c_1 \le z \le c_2$

and is sketched in Figure 1. We construct regular partitions of the three intervals $[a_1, a_2]$, $[b_1, b_2]$, and $[c_1, c_2]$:

$$a_1 = x_0 < x_1 < \cdots < x_n = a_2$$
$$b_1 = y_0 < y_1 < \cdots < y_m = b_2$$
$$c_1 = z_0 < z_1 < \cdots < z_p = c_2,$$

to obtain nmp "boxes." The volume of a typical box B_{ijk} is given by

$$\Delta V = \Delta x \, \Delta y \, \Delta z, \tag{2}$$

where $\Delta x = x_i - x_{i-1}$, $\Delta y = y_j - y_{j-1}$, and $\Delta z = z_k - z_{k-1}$. We then form the sum

$$\sum_{i=1}^{n} \sum_{j=1}^{m} \sum_{k=1}^{p} f(x_i^*, y_j^*, z_k^*) \, \Delta V, \tag{3}$$

where (x_i^*, y_j^*, z_k^*) is in B_{ijk}. We now define

$$\Delta u = \sqrt{\Delta x^2 + \Delta y^2 + \Delta z^2}.$$

Geometrically, Δu is the length of a diagonal of a rectangular solid with sides having lengths Δx, Δy, and Δz, respectively. We see that as $\Delta u \to 0$, Δx, Δy, and Δz approach 0, and the volume of each box tends to zero. We then take the limit as $\Delta u \to 0$.

DEFINITION THE TRIPLE INTEGRAL

Let $w = f(x, y, z)$ and let the parallelepiped π be given by (1). Suppose that

$$\lim_{\Delta u \to 0} \sum_{i=1}^{n} \sum_{j=1}^{m} \sum_{k=1}^{p} f(x_i^*, y_j^*, z_k^*) \, \Delta x \, \Delta y \, \Delta z$$

exists and is independent of the way in which the points (x_i^*, y_j^*, z_k^*) are chosen. Then the **triple integral** of f over π, written $\iiint_\pi f(x, y, z) \, dV$, is defined by

$$\iiint_\pi f(x, y, z) \, dV = \lim_{\Delta u \to 0} \sum_{i=1}^{n} \sum_{j=1}^{m} \sum_{k=1}^{p} f(x_i^*, y_j^*, z_k^*) \, \Delta V. \tag{4}$$

■

As with double integrals, we can write triple integrals as iterated (or repeated) integrals. If π is defined by (1), we have

THE TRIPLE INTEGRAL OVER A PARALLELEPIPED

$$\iiint_\pi f(x, y, z) \, dV = \int_{a_1}^{a_2} \int_{b_1}^{b_2} \int_{c_1}^{c_2} f(x, y, z) \, dz \, dy \, dx. \tag{5}$$

EXAMPLE 1 CALCULATING A TRIPLE INTEGRAL OVER A PARALLELEPIPED

Evaluate $\iiint_\pi xy \cos yz \, dV$, where π is the parallelepiped

$$\left\{(x, y, z): 0 \le x \le 1, 0 \le y \le 1, 0 \le z \le \frac{\pi}{2}\right\}.$$

SOLUTION:

$$\begin{aligned}
\iiint_\pi xy \cos yz \, dV &= \int_0^1 \int_0^1 \int_0^{\pi/2} xy \cos yz \, dz \, dy \, dx \\
&= \int_0^1 \int_0^1 \left\{ xy \cdot \frac{1}{y} \sin yz \Big|_0^{\pi/2} \right\} dy \, dx \\
&= \int_0^1 \int_0^1 x \sin\left(\frac{\pi}{2} y\right) dy \, dx \\
&= \int_0^1 \left\{ -\frac{2}{\pi} x \cos \frac{\pi}{2} y \Big|_0^1 \right\} dx = \int_0^1 \frac{2}{\pi} x \, dx = \frac{x^2}{\pi} \Big|_0^1 \\
&= \frac{1}{\pi}.
\end{aligned}$$

THE TRIPLE INTEGRAL OVER A MORE GENERAL REGION

We now define the triple integral over a more general region S. We assume that S is bounded. Then we can enclose S in a parallelepiped π and define a new function F by

$$F(x, y, z) = \begin{cases} f(x, y, z), & \text{if } (x, y, z) \text{ is in } S \\ 0, & \text{if } (x, y, z) \text{ is in } \pi \text{ but not in } S. \end{cases}$$

We then define

$$\iiint_S f(x, y, z) \, dV = \iiint_\pi F(x, y, z) \, dV. \tag{6}$$

REMARK 1: If f is continuous over S and if S is a region of the type we will discuss below, then $\iiint_S f(x, y, z) \, dV$ will exist. The proof of this fact is beyond the scope of this text but can be found in any advanced calculus text.

REMARK 2: If $f \ge 0$ on S, then the triple integral $\iiint_S f(x, y, z) \, dV$ represents the ''volume'' in four-dimensional space $\mathbb{R}^4$ of the region bounded above by f and below by S. We cannot, of course, draw this volume, but otherwise the theory of volumes carries over to four (and more) dimensions.

Now let S take the form

$$S = \{(x, y, z): a_1 \le x \le a_2, g_1(x) \le y \le g_2(x), h_1(x, y) \le z \le h_2(x, y)\}. \tag{7}$$

What does such a solid look like? We first note that the equations $z = h_1(x, y)$ and $z = h_2(x, y)$ are the equations of surfaces in $\mathbb{R}^3$. The equations $y = g_1(x)$ and $y = g_2(x)$ are equations of cylinders in $\mathbb{R}^3$, and the equations $x = a_1$ and $x = a_2$ are equations of planes

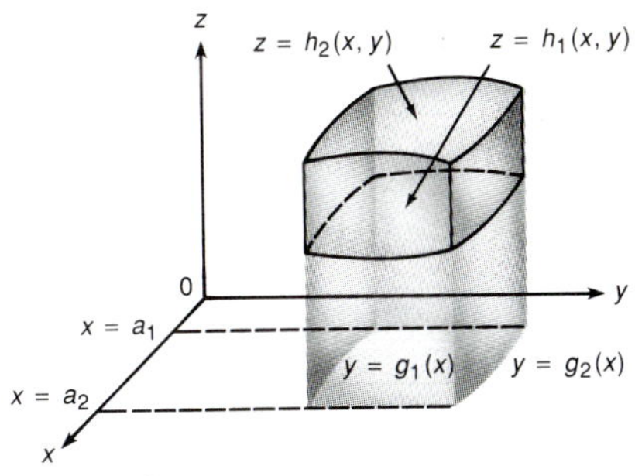

FIGURE 2
A region in $\mathbb{R}^3$ of the form (7)

in $\mathbb{R}^3$. The solid S is sketched in Figure 2. We assume that g_1, g_2, h_1, and h_2 are continuous. If f is continuous, then $\iiint_S f(x, y, z)\, dV$ will exist and

THE TRIPLE INTEGRAL OVER A REGION S IN $\mathbb{R}^3$

$$\iiint_S f(x, y, z)\, dV = \int_{a_1}^{a_2} \int_{g_1(x)}^{g_2(x)} \int_{h_1(x, y)}^{h_2(x, y)} f(x, y, z)\, dz\, dy\, dx. \tag{8}$$

Note the similarity between equation (8) and equation (4.2.10).

EXAMPLE 2

EVALUATING A TRIPLE INTEGRAL OVER A REGION S

Evaluate $\iiint_S 2x^3y^2z\, dV$, where S is the region

$$\{(x, y, z): 0 \le x \le 1, x^2 \le y \le x, x - y \le z \le x + y\}.$$

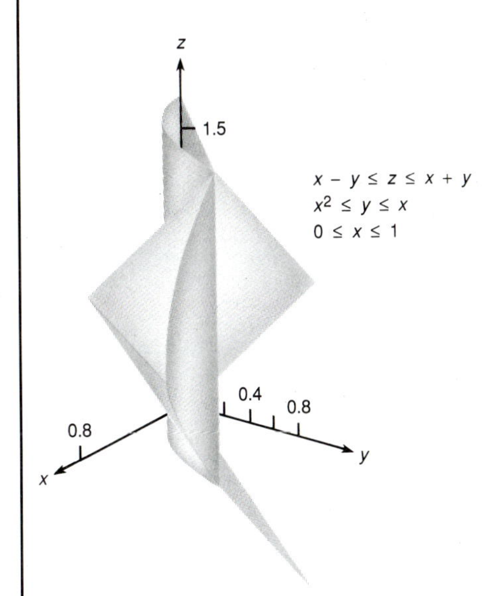

FIGURE 3
Computer-drawn sketch of the region $\{(x, y, z): 0 \le x \le 1, x^2 \le y \le x, x - y \le z \le x + y\}$

SOLUTION: A computer-drawn sketch of the region S is given in Figure 3. We have

$$\begin{aligned}
\iiint_S 2x^3y^2z\,dV &= \int_0^1 \int_{x^2}^x \int_{x-y}^{x+y} 2x^3y^2z\,dz\,dy\,dx \\
&= \int_0^1 \int_{x^2}^x \left\{ x^3y^2z^2 \Big|_{x-y}^{x+y} \right\} dy\,dx \\
&= \int_0^1 \int_{x^2}^x x^3y^2[(x+y)^2 - (x-y)^2]\,dy\,dx \\
&= \int_0^1 \int_{x^2}^x 4x^4y^3\,dy\,dx = \int_0^1 \left\{ x^4y^4 \Big|_{x^2}^x \right\} dx \\
&= \int_0^1 (x^8 - x^{12})\,dx = \frac{1}{9} - \frac{1}{13} = \frac{4}{117}.
\end{aligned}$$

Many of the applications we saw for the double integral can be extended to the triple integral. We present two of them below and one more in the problem sets.

VOLUME

Let the region S be defined by (7). Then, since $\Delta V = \Delta x\,\Delta y\,\Delta z$ represents the volume of a "box" in S, when we add up the volumes of these boxes and take a limit, we obtain the total volume of S. That is,

> **VOLUME AS A TRIPLE INTEGRAL**
>
> $$\text{volume of } S = \iiint_S dV. \qquad (9)$$

EXAMPLE 3

USING A TRIPLE INTEGRAL TO COMPUTE VOLUME

Calculate the volume of the region of Example 2.

SOLUTION:

$$\begin{aligned}
V &= \int_0^1 \int_{x^2}^x \int_{x-y}^{x+y} dz\,dy\,dx = \int_0^1 \int_{x^2}^x \left\{ z \Big|_{x-y}^{x+y} \right\} dy\,dx \\
&= \int_0^1 \int_{x^2}^x 2y\,dy\,dx = \int_0^1 \left\{ y^2 \Big|_{x^2}^x \right\} dx = \int_0^1 (x^2 - x^4)\,dx \\
&= \frac{1}{3} - \frac{1}{5} = \frac{2}{15}.
\end{aligned}$$

EXAMPLE 4

USING A TRIPLE INTEGRAL TO COMPUTE THE VOLUME OF A TETRAHEDRON

Find the volume of the tetrahedron formed by the planes $x = 0$, $y = 0$, $z = 0$, and $x + (y/2) + (z/4) = 1$.

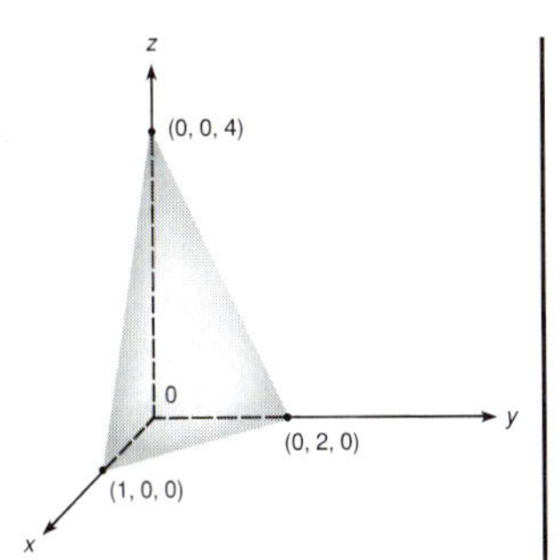

FIGURE 4
The tetrahedron with vertices at (0, 0, 0), (1, 0, 0), (0, 2, 0), and (0, 0, 4)

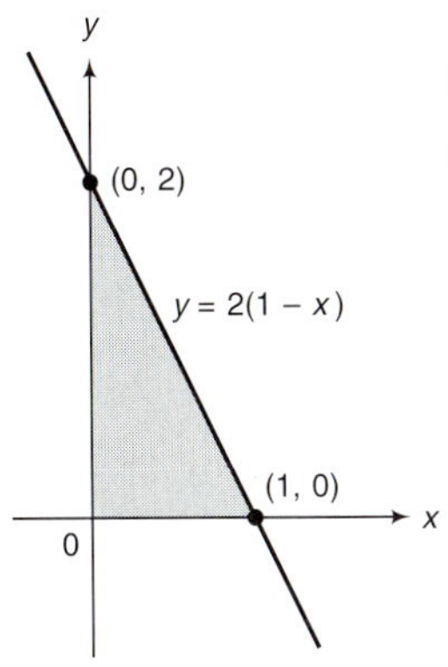

FIGURE 5
The projection of the tetrahedron in Figure 4 onto the xy-plane

SOLUTION: The tetrahedron is sketched in Figure 4. We see that z ranges from 0 to the plane $x + (y/2) + (z/4) = 1$, or $z = 4[1 - x - (y/2)]$. This last plane intersects the xy-plane in a line whose equation (obtained by setting $z = 0$) is given by $0 = 1 - x - (y/2)$ or $y = 2(1 - x)$, so that y ranges from 0 to $2(1 - x)$ (see Figure 5). Finally, this line intersects the x-axis at the point $(1, 0, 0)$, so that x ranges from 0 to 1, and we have

$$\begin{aligned} V &= \int_0^1 \int_0^{2(1-x)} \int_0^{4(1-x-y/2)} dz\, dy\, dx \\ &= \int_0^1 \int_0^{2(1-x)} 4\left(1 - x - \frac{y}{2}\right) dy\, dx = \int_0^1 -4\left(1 - x - \frac{y}{2}\right)^2\Bigg|_0^{2(1-x)} dx \\ &= 4\int_0^1 (1 - x)^2\, dx = -\frac{4}{3}(1 - x)^3\Bigg|_0^1 = \frac{4}{3}. \end{aligned}$$

REMARK: It was not necessary to integrate in the order z, then y, then x. We could write, for example,

$$0 \le x \le 1 - \frac{y}{2} - \frac{z}{4}, \qquad 0 \le y \le 2 - \frac{z}{2}, \qquad 0 \le z \le 4$$

and could continue the problem by integrating in the order x, then y, then z.

DENSITY AND MASS IN $\mathbb{R}^3$

Let the function $\rho(x, y, z)$ denote the density (in kilograms per cubic meter, say) of a solid S in $\mathbb{R}^3$. Then for a "box" of sides Δx, Δy, and Δz, the approximate mass of the box will be equal to $\rho(x_i^*, y_j^*, z_k^*)\,\Delta x\,\Delta y\,\Delta z = \rho(x_i^*, y_j^*, z_k^*)\,\Delta V$ if Δx, Δy, and Δz are small. We then obtain

> **MASS AS A TRIPLE INTEGRAL**
>
> $$\text{total mass of } S = \mu(S) = \iiint_S \rho(x, y, z)\, dV. \qquad \textbf{(10)}$$

EXAMPLE 5 USING A TRIPLE INTEGRAL TO COMPUTE MASS

The density of the solid of Example 2 is given by $\rho(x, y, z) = x + 2y + 4z$ kg/m^3. Calculate the total mass of the solid.

SOLUTION:

$$\begin{aligned} \mu(S) &= \int_0^1 \int_{x^2}^{x} \int_{x-y}^{x+y} (x + 2y + 4z)\, dz\, dy\, dx \\ &= \int_0^1 \int_{x^2}^{x} \left\{[(x + 2y)z + 2z^2]\Bigg|_{x-y}^{x+y}\right\} dy\, dx \\ &= \int_0^1 \int_{x^2}^{x} (10xy + 4y^2)\, dy\, dx = \int_0^1 \left(5xy^2 + \frac{4y^3}{3}\right)\Bigg|_{x^2}^{x} dx \end{aligned}$$

$$= \int_0^1 \left(5x^3 - 5x^5 + \frac{4}{3}x^3 - \frac{4}{3}x^6\right) dx$$

$$= \int_0^1 \left(\frac{19}{3}x^3 - 5x^5 - \frac{4}{3}x^6\right) dx = \frac{19}{12} - \frac{5}{6} - \frac{4}{21} = \frac{47}{84} \text{ kg.}$$

PROBLEMS 4.5

SELF-QUIZ

I. True–False: The volume of a region in $\mathbb{R}^3$ is the triple integral of the constant function 1 over that region.

II. Item ________ below is not equal to the other three items.

a. $\int_0^4 \int_2^5 \int_0^5 f(x, y, z)\, dy\, dz\, dx$

b. $\int_0^5 \int_2^5 \int_0^4 f(x, y, z)\, dx\, dz\, dy$

c. $\int_0^4 \int_0^5 \int_2^5 f(x, y, z)\, dx\, dy\, dz$

d. $\int_0^5 \int_0^4 \int_2^5 f(x, y, z)\, dz\, dx\, dy$

III. If we know that values of f are bounded above by 3 on the region $\Omega = \{(x, y, z): 0 \le x \le 2, 0 \le y \le 5, 1 \le z \le 3\}$, then we can infer that $\iiint_\Omega f(x, y, z)\, dV$ ________.

a. is positive

b. is negative

c. is less than or equal to $3 + 2 + 5 + 3$

d. is less than or equal to $3(2)(5)(3 - 1)$

IV. Suppose f is a function which is smooth throughout $\mathbb{R}^3$. True–False:

$$\int_1^2 \int_4^4 \int_0^5 f(x, y, z)\, dx\, dy\, dz = \int_1^2 \int_0^5 f(x, 4, z)\, dx\, dz.$$

In Problems 1–8, evaluate the repeated triple integral.

1. $\int_0^1 \int_0^y \int_0^x y\, dz\, dx\, dy$

2. $\int_0^1 \int_0^y \int_z^y y\, dx\, dz\, dy$

3. $\int_0^2 \int_{-z}^{z} \int_{y-z}^{y+z} 2xz\, dx\, dy\, dz$

4. $\int_{a_1}^{a_2} \int_{b_1}^{b_2} \int_{c_1}^{c_2} dy\, dx\, dz$

5. $\int_0^{\pi/2} \int_0^{\pi/2} \int_0^z \sin(x/z)\, dx\, dz\, dy$

6. $\int_1^2 \int_{1-y}^{1+y} \int_0^{yz} 6xyz\, dx\, dz\, dy$

7. $\int_0^1 \int_0^{\sqrt{1-x^2}} \int_0^x yz\, dz\, dy\, dx$

***8.** $\int_0^1 \int_{-\sqrt{1-x^2}}^{\sqrt{1-x^2}} \int_{-\sqrt{1-x^2-y^2}}^{\sqrt{1-x^2-y^2}} z^2\, dz\, dy\, dx$
[*Hint:* Use polar coordinates.]

***9.** Change the order of integration in Problem 1 and write the integral in the forms

a. $\int_?^? \int_?^? \int_?^? y\, dx\, dy\, dz$

b. $\int_?^? \int_?^? \int_?^? y\, dy\, dz\, dx.$

[*Hint:* Sketch the region in Problem 1 from the given limits and then find the new limits directly from that sketch.]

***10.** Write the integral of Problem 3 in the forms

a. $\int_?^? \int_?^? \int_?^? 2xz\, dy\, dx\, dz$

b. $\int_?^? \int_?^? \int_?^? 2xz\, dz\, dy\, dx.$

***11.** Write the integral of Problem 7 in the forms

a. $\int_?^? \int_?^? \int_?^? yz\, dy\, dz\, dx$

b. $\int_?^? \int_?^? \int_?^? yz\, dx\, dy\, dz.$

***12.** Write the integral of Problem 8 in the form $\int_?^? \int_?^? \int_?^? z^2\, dx\, dz\, dy.$

13. Find $\iiint_S z\, dV$ where S is the region bounded by $x + y + z = 1$ and the coordinate planes.

14. Find $\iiint_S (x+y)\,dV$ where S is the region of Problem 13.

15. Find $\iiint_S (x+y+z)\,dV$ where S is the region of Problem 13.

16. Find $\iiint_S (x+y-2z)\,dV$ where S is the region of Problem 13.

17. Compute $\iiint_S z^2\,dV$ where S is the region bounded by $x+2y+3z=6$ and the coordinate planes.

18. Compute $\iiint_S (x-y)\,dV$ where S is the region of Problem 17.

In Problems 19–26, find the volume of the given solid.

19. The tetrahedron with vertices at the points $(0,0,0)$, $(1,0,0)$, $(0,1,0)$, and $(0,0,1)$.

20. The tetrahedron with vertices at the points $(0,0,0)$, $(a,0,0)$, $(0,b,0)$, and $(0,0,c)$.

21. The solid in the first octant bounded by the cylinder $x^2+z^2=9$, the plane $x+y=4$, and the three coordinate planes.

***22.** The solid bounded by the planes $x-2y+4z=4$, $-2x+3y-z=6$, $x=0$, and $y=0$.

23. The solid bounded above by the sphere $x^2+y^2+z^2=16$ and below by the plane $z=2$.

24. The solid bounded by the parabolic cylinder $z=5-x^2$ and the planes $z=y$ and $z=2y$ that lies in the half space $y \geq 0$.

25. The solid bounded by the ellipsoid $(x/a)^2+(y/b)^2+(z/c)^2=1$.

26. The solid bounded by the elliptic cylinder $9x^2+y^2=9$ and the planes $z=0$ and $x+y+9z=9$.

In Problems 27–30, find the mass of the given solid having the specified density.

27. The tetrahedron of Problem 19; $\rho(x,y,z)=x$.

28. The tetrahedron of Problem 20; $\rho(x,y,z)=x^2+y^2+z^2$.

29. The solid of Problem 21; $\rho(x,y,z)=z$.

30. The ellipsoid of Problem 25; $\rho(x,y,z)=\alpha x^2+\beta y^2+\gamma z^2$.

In $\mathbb{R}^3$, the first moments of a solid are considered with respect to the various coordinate planes. These moments are defined as weighted integrals of the signed-distances to the designated plane. For instance, the first moment with respect to the yz-plane is

$$M_{yz}=\iiint_{\Omega} x\rho(x,y,z)\,dV.$$

Similarly, the moments with respect to the xz-plane and to the xy-plane are

$$M_{xz}=\iiint_{\Omega} y\rho(x,y,z)\,dV$$

and

$$M_{xy}=\iiint_{\Omega} z\rho(x,y,z)\,dV.$$

The **center of mass** of the region Ω is then given by

$$(\overline{x},\overline{y},\overline{z})=\left(\frac{M_{yz}}{\mu},\frac{M_{xz}}{\mu},\frac{M_{xy}}{\mu}\right),$$

where $\mu=\iiint_\Omega \rho\,dV$ is the total mass. (If the density ρ is constant, then the center of mass is called the **centroid**.)

31. Find the centroid of the tetrahedron of Problem 20.

32. Find the centroid of the ellipsoid of Problem 25.

33. Find the center of mass for the tetrahedron of Problem 27.

34. Find the center of mass for the tetrahedron of Problem 28.

35. Find the center of mass for the solid of Problem 29.

36. Find the center of mass for the ellipsoid of Problem 30.

***37.** The solid S lies in the first octant and is bounded by the planes $z=0$, $y=1$, $x=y$, and the hyperboloid $z=xy$. Its density is given by $\rho(x,y,z)=1+2z$.

a. Find the center of mass of S.

b. Show that the center of mass lies inside S. [This is not obvious.]

ANSWERS TO SELF-QUIZ

I. True **II.** c **III.** d

IV. False (The left-hand side equals zero, the right-hand side need not.)

4.6 THE TRIPLE INTEGRAL IN CYLINDRICAL AND SPHERICAL COORDINATES

In this section, we show how triple integrals can be written by using cylindrical and spherical coordinates.

CYLINDRICAL COORDINATES

Recall from Section 1.9 that the cylindrical coordinates of a point in $\mathbb{R}^3$ are (r, θ, z), where r and θ are the polar coordinates of the projection of the point onto the xy-plane and z is the usual z-coordinate. In order to calculate an integral of a region given in cylindrical coordinates, we go through a procedure very similar to the one we used to write double integrals in polar coordinates in Section 4.4.

Consider the "cylindrical parallelepiped" given by

$$\pi_c = \{(r, \theta, z): r_1 \le r \le r_2;\ \theta_1 \le \theta \le \theta_2;\ z_1 \le z \le z_2\}.$$

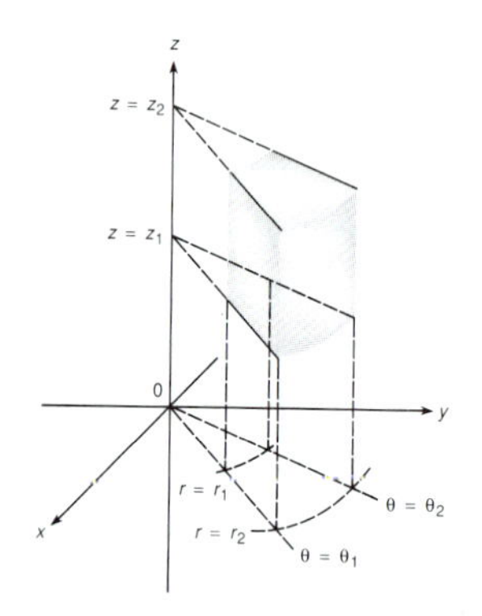

FIGURE 1 The cylindrical parallelepiped bounded by the planes $\theta = \theta_1$, $\theta = \theta_2$, $z = z_1$, $z = z_2$, and the cylinders $r = r_1$, $r = r_2$

This solid is sketched in Figure 1. If we partition the z-axis for z in $[z_1, z_2]$, we obtain "slices." For a fixed z the area of the face of a slice of π_c is, according to equation (4.4.5), given by

$$A_i = \int_{\theta_1}^{\theta_2} \int_{r_1}^{r_2} r\, dr\, d\theta. \tag{1}$$

The volume of a slice is, by (1), given by

$$V_i = A_i\, \Delta z = \left\{ \int_{\theta_1}^{\theta_2} \int_{r_1}^{r_2} r\, dr\, d\theta \right\} \Delta z.$$

Then adding these volumes and taking the limit as before, we obtain

$$\text{volume of } \pi_c = \int_{z_1}^{z_2} \int_{\theta_1}^{\theta_2} \int_{r_1}^{r_2} r\, dr\, d\theta\, dz. \tag{2}$$

In general, let the region S be given in cylindrical coordinates by

$$S = \{(r, \theta, z): \theta_1 \le \theta \le \theta_2,\ 0 \le g_1(\theta) \le r \le g_2(\theta),\ h_1(r, \theta) \le z \le h_2(r, \theta)\}, \tag{3}$$

and let f be a function of r, θ, and z. Then the triple integral of f over S is given by

TRIPLE INTEGRAL IN CYLINDRICAL COORDINATES

$$\iiint_S f = \int_{\theta_1}^{\theta_2} \int_{g_1(\theta)}^{g_2(\theta)} \int_{h_1(r,\theta)}^{h_2(r,\theta)} f(r, \theta, z) r\, dz\, dr\, d\theta. \tag{4}$$

NOTE: We can, of course, integrate in different orders. The order given in (4) is the most common one.

The formula for conversion from rectangular to cylindrical coordinates is virtually identical to formula (4.4.6). Let S be a region in $\mathbb{R}^3$. Since $x = r\cos\theta$, $y = r\sin\theta$, and $z = z$, we have

CONVERTING A TRIPLE INTEGRAL FROM RECTANGULAR TO CYLINDRICAL COORDINATES

$$\underbrace{\iiint\limits_S f(x, y, z)\, dV}_{\text{Given in rectangular coordinates}} = \underbrace{\iiint\limits_S f(r\cos\theta, r\sin\theta, z) r\, dz\, dr\, d\theta.}_{\text{Given in cylindrical coordinates}} \tag{5}$$

REMARK: Again, do not forget the extra r when you convert to cylindrical coordinates.

EXAMPLE 1

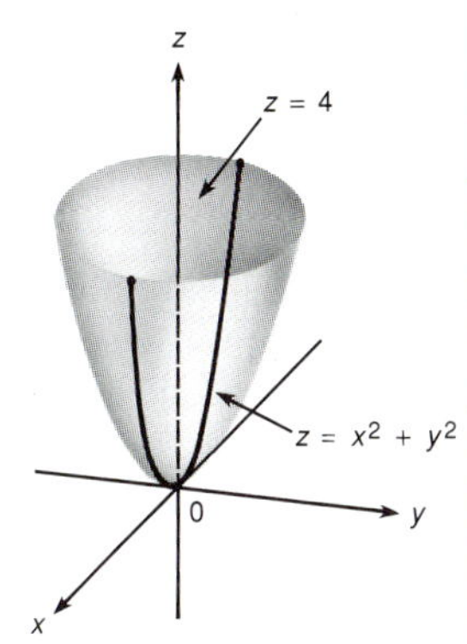

FIGURE 2
The solid bounded by the circular paraboloid $z = x^2 + y^2$ and the plane $z = 4$

COMPUTING MASS BY CONVERTING TO CYLINDRICAL COORDINATES

Find the mass of the solid bounded by the paraboloid $z = x^2 + y^2$ and the plane $z = 4$ if the density at any point is proportional to the distance from the point to the z-axis.

SOLUTION: The solid is sketched in Figure 2. The density is given by $\rho(x, y, z) = \alpha\sqrt{x^2 + y^2}$, where α is a constant of proportionality. Thus since the solid may be written as

$$S = \{(x, y, z)\colon -2 \le x \le 2, -\sqrt{4 - x^2} \le y \le \sqrt{4 - x^2}, x^2 + y^2 \le z \le 4\},$$

we have

$$\mu = \iiint\limits_S \alpha\sqrt{x^2 + y^2}\, dV = \int_{-2}^{2}\int_{-\sqrt{4-x^2}}^{\sqrt{4-x^2}}\int_{x^2+y^2}^{4} \alpha\sqrt{x^2 + y^2}\, dz\, dy\, dx.$$

We write this expression in cylindrical coordinates, using the fact that $\alpha\sqrt{x^2 + y^2} = \alpha r$. We note that the largest value of r is 2, since at the "top" of the solid, $r^2 = x^2 + y^2 = 4$. Then

$$\mu = \int_0^{2\pi}\int_0^2\int_{r^2}^4 (\alpha r) r\, dz\, dr\, d\theta = \int_0^{2\pi}\int_0^2 \alpha r^2(4 - r^2) dr\, d\theta$$
$$= \alpha\int_0^{2\pi}\left\{\left(\frac{4r^3}{3} - \frac{r^5}{5}\right)\Bigg|_0^2\right\} d\theta = \frac{64\alpha}{15}\int_0^{2\pi} d\theta = \frac{128}{15}\pi\alpha.$$

SPHERICAL COORDINATES

Recall from Section 1.9 that a point P in $\mathbb{R}^3$ can be written in the spherical coordinates (ρ, θ, φ), where $\rho \ge 0$, $0 \le \theta < 2\pi$, $0 \le \varphi \le \pi$. Here ρ is the distance between the point and the origin, θ is the same as in cylindrical coordinates, and φ is the angle between $\overrightarrow{OP}$ and the positive z-axis. Consider the "spherical parallelepiped"

$$\pi_S = \{(\rho, \theta, \varphi)\colon \rho_1 \le \rho \le \rho_2, \theta_1 \le \theta \le \theta_2, \varphi_1 \le \varphi \le \varphi_2\}. \tag{6}$$

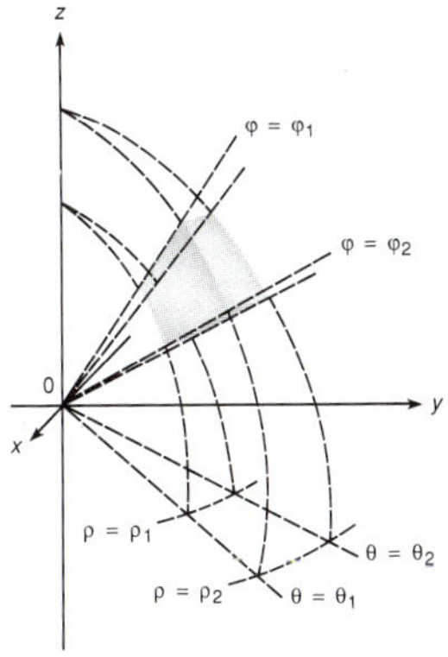

FIGURE 3
The spherical parallelepiped bounded by the planes $\varphi = \varphi_1$, $\varphi = \varphi_2$, $\theta = \theta_1$, $\theta = \theta_2$, and the spheres $\rho = \rho_1$, $\rho = \rho_2$

This solid is sketched in Figure 3. To approximate the volume of π_S, we partition the intervals $[\rho_1, \rho_2]$, $[\theta_1, \theta_2]$, and $[\varphi_1, \varphi_2]$. This partition gives us a number of "spherical boxes," one of which is sketched in Figure 4. The length of an arc of a circle is given by

$$L = r\theta, \tag{7}$$

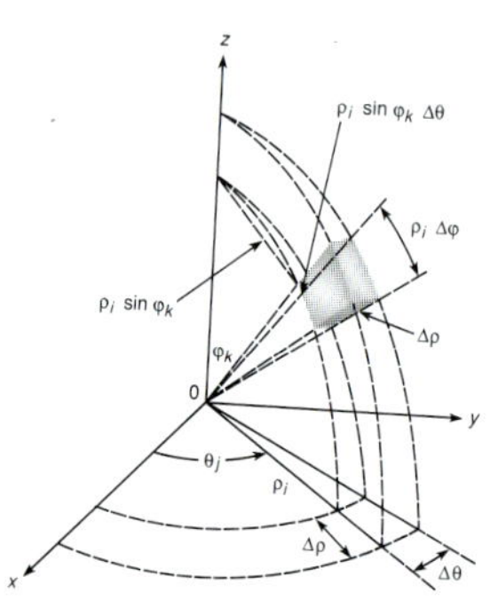

FIGURE 4
A small spherical parallelepiped

where r is the radius of the circle and θ is the angle that "cuts off" the arc. In Figure 4, one side of the spherical box is $\Delta\rho$. Since $\rho = \rho_i$ is the equation of a sphere, we find, from (7), that the length of a second side is $\rho_i\,\Delta\varphi$. Finally, the length of the third side is $\rho_i \sin\varphi_k\,\Delta\theta$ (see Figure 4). Thus, the volume of the box in Figure 4 is given, approximately, by

$$V_i \approx (\Delta\rho)(\rho_i\,\Delta\varphi)(\rho_i \sin\varphi_k\,\Delta\theta),$$

and using a familiar argument, we have

$$\text{volume of } \pi_S = \iiint_S \rho^2 \sin\varphi\,d\rho\,d\varphi\,d\theta = \int_{\theta_1}^{\theta_2}\int_{\varphi_1}^{\varphi_2}\int_{\rho_1}^{\rho_2} \rho^2 \sin\varphi\,d\rho\,d\varphi\,d\theta. \tag{8}$$

If f is a function of the variables ρ, θ, and φ, we have

$$\iiint_{\pi_S} f = \iiint_{\pi_S} f(\rho, \theta, \varphi)\rho^2 \sin\varphi\,d\rho\,d\varphi\,d\theta. \tag{9}$$

More generally, let the region S be defined in spherical coordinates by

$$S = \{(\rho, \theta, \varphi)\colon \theta_1 \le \theta \le \theta_2,\ g_1(\theta) \le \varphi \le g_2(\theta),\ h_1(\theta, \varphi) \le \rho \le h_2(\theta, \varphi)\}. \tag{10}$$

Then

THE TRIPLE INTEGRAL IN SPHERICAL COORDINATES

$$\iiint_S f = \int_{\theta_1}^{\theta_2}\int_{g_1(\theta)}^{g_2(\theta)}\int_{h_1(\theta,\varphi)}^{h_2(\theta,\varphi)} f(\rho, \theta, \varphi)\rho^2 \sin\varphi\,d\rho\,d\varphi\,d\theta. \tag{11}$$

NOTE: As for cylindrical coordinates, we can integrate in different orders; the order given in (11) is the most common one.

Recall that to convert from rectangular to spherical coordinates, we have the formulas [see equations (1.9.5), (1.9.6), and (1.9.7)]

$$x = \rho \sin\varphi \cos\theta, \qquad y = \rho \sin\varphi \sin\theta, \qquad z = \rho\cos\varphi.$$

Thus, to convert from a triple integral in rectangular coordinates to a triple integral in spherical coordinates, we have

CONVERTING A TRIPLE INTEGRAL FROM RECTANGULAR TO SPHERICAL COORDINATES

$$\iiint_S f(x, y, z)\,dV \qquad \text{Given in rectangular coordinates}$$

$$= \iiint_S f(\rho\sin\varphi\cos\theta, \rho\sin\varphi\sin\theta, \rho\cos\varphi)\rho^2 \sin\varphi\,d\rho\,d\varphi\,d\theta. \tag{12}$$

Given in spherical coordinates

EXAMPLE 2 COMPUTING A MASS USING SPHERICAL COORDINATES

Find the mass of the sphere $x^2 + y^2 + z^2 = a^2$ if its density at a point is proportional to the distance from the point to the origin.

SOLUTION: We have density $= \alpha\sqrt{x^2 + y^2 + z^2} = \alpha\rho$. Also, the sphere can be written (in spherical coordinates)

$$S = \{(\rho, \theta, \varphi): 0 \le \rho \le a, 0 \le \theta \le 2\pi, 0 \le \varphi \le \pi\}.$$

Thus,

$$\mu = \int_0^{2\pi}\int_0^{\pi}\int_0^{a} \alpha\rho(\rho^2 \sin\varphi)\, d\rho\, d\varphi\, d\theta = \frac{\alpha a^4}{4}\int_0^{2\pi}\int_0^{\pi} \sin\varphi\, d\varphi\, d\theta$$
$$= \frac{\alpha a^4}{4}(4\pi) = \pi a^4\alpha.$$

PROBLEMS 4.6

SELF-QUIZ

I. True–False: The volume of a region is the integral of r over the region.

II. The volume of the cylinder $\{(r, \theta, z): 0 \le r \le 4, 0 \le \theta \le 2\pi, -4 \le z \le 4\}$ is ______.

a. $\int_0^4\int_0^{2\pi}\int_{-4}^{4} r\, dr\, d\theta\, dz$

b. $\int_{-4}^4\int_0^{2\pi}\int_0^{4} (1)r\, dr\, d\theta\, dz$

c. $\int_{-4}^4\int_0^{2\pi}\int_0^{4} r^2\, dr\, d\theta\, dz$

d. $\int_{-4}^4\int_0^{2\pi}\int_0^{4} r\, dx\, dy\, dz$

III. True–False: The volume of a region is the integral of $\rho^2 \sin\varphi$ over the region.

IV. The volume of the ball $\{(\rho, \theta, \varphi): 0 \le \rho \le 1, 0 \le \theta \le 2\pi, 0 \le \varphi \le \pi\}$ is ______.

a. $\int_0^1\int_0^{2\pi}\int_0^{\pi} \rho^2 \sin\varphi\, d\rho\, d\theta\, d\varphi$

b. $\int_0^1\int_0^{2\pi}\int_0^{\pi} \rho^2 \sin\varphi\, \rho^2 \sin\varphi\, d\rho\, d\theta\, d\varphi$

c. $\int_0^{\pi}\int_0^{2\pi}\int_0^{1} \rho^2 \sin\varphi\, \rho^2 \sin\varphi\, d\rho\, d\theta\, d\varphi$

d. $\int_0^{\pi}\int_0^{2\pi}\int_0^{1} (1)\rho^2 \sin\varphi\, d\rho\, d\theta\, d\varphi$

Solve Problems 1–6 using cylindrical coordinates.

1. Find the volume of the region inside both the sphere $x^2 + y^2 + z^2 = 4$ and the cylinder $(x - 1)^2 + y^2 = 1$.

2. Find the volume of the solid bounded above by the paraboloid $z = 4 - x^2 - y^2$ and below by the xy-plane.

3. Find the volume of the solid bounded by the plane $z = y$ and the paraboloid $z = x^2 + y^2$.

4. Find the volume of the solid bounded by the two cones $z^2 = x^2 + y^2$ and $z^2 = 16x^2 + 16y^2$ between $z = 0$ and $z = 2$.

***5.** Find the volume of the solid bounded by the hyperboloid of two sheets $z^2 - x^2 - y^2 = 1$ and the cone $z^2 = x^2 + y^2$ for $0 \le z \le a$, $a \ge 1$.

6. Evaluate

$$\int_0^1\int_0^{\sqrt{1-x^2}}\int_{\sqrt{x^2+y^2}}^{\sqrt{2-x^2-y^2}} z^3\, dz\, dy\, dx.$$

Solve Problems 7–14 by using spherical coordinates.

7. Evaluate

$$\int_0^1\int_0^{\sqrt{1-x^2}}\int_0^{\sqrt{1-x^2-y^2}} \frac{1}{\sqrt{x^2 + y^2 + z^2}}\, dz\, dy\, dx.$$

8. Find the mass of the unit sphere if the density at any point is proportional to the distance to the boundary of the sphere.

***9.** Find the volume of the solid inside the sphere $x^2 + y^2 + z^2 = 4$ and outside the cone $z^2 = x^2 + y^2$.

10. A solid fills the space between two concentric spheres centered at the origin. The radii of the spheres are a and b, where $0 < a < b$. Find the mass of the solid if the density at each point is inversely proportional to its distance from the origin.

11. Find the volume of one of the smaller wedges cut from the unit sphere by two planes that meet at a diameter with an angle of $\pi/6$.

12. Find the mass of a wedge cut from a sphere of radius a by two planes that meet at a diameter with an angle of b radians if the density at any point on the wedge is proportional to the distance to that diameter.

13. Evaluate

$$\int_{-3}^{3} \int_{-\sqrt{9-x^2}}^{\sqrt{9-x^2}} \int_{-\sqrt{9-x^2-y^2}}^{\sqrt{9-x^2-y^2}} (x^2 + y^2 + z^2)^{3/2}\, dz\, dy\, dx.$$

14. Evaluate

$$\int_{0}^{a} \int_{0}^{\sqrt{a^2-x^2}} \int_{0}^{\sqrt{a^2-x^2-y^2}} \frac{z^3}{\sqrt{x^2 + y^2}}\, dz\, dy\, dx.$$

Problems 15–21 ask you to compute the centroid of a region or the center of mass of a solid object. Both of these were discussed immediately preceding Problem 4.5.31 on page 281; you might want to review that paragraph now before tackling the following problems.

15. Find the centroid of the region of Problem 1.

***16.** Suppose the density of the region of Problem 1 is proportional to the square of the distance to the xy-plane and is measured in kg/m^3. Find the center of mass of the solid region.

17. Find the center of mass for the solid of Problem 2 if the density is proportional to the distance to the xy-plane.

18. Find the center of mass for the solid in Problem 4 if the density at any point is proportional to the distance to the z-axis.

19. Find the centroid of the "ice-cream cone" shaped region which is below the upper half of the sphere $x^2 + y^2 + z^2 = z$ and above the cone $z^2 = x^2 + y^2$.

20. Find the center of mass for the unit sphere if the density at a point is proportional to the distance from the origin.

21. Find the center of mass for the object in Problem 9 if the density at a point is proportional to the square of the distance from the region.

***22.** A "sphere" in $\mathbb{R}^4$ has the equation

$$x^2 + y^2 + z^2 + w^2 = a^2.$$

a. Explain why it is reasonable that the volume of this "sphere" be given by

$$\int_{-a}^{a} \int_{-\sqrt{a^2-x^2}}^{\sqrt{a^2-x^2}} \int_{-\sqrt{a^2-x^2-y^2}}^{\sqrt{a^2-x^2-y^2}} \int_{-\sqrt{a^2-x^2-y^2-z^2}}^{\sqrt{a^2-x^2-y^2-z^2}} dw\, dz\, dy\, dx$$

$$= 16 \int_{0}^{a} \int_{0}^{\sqrt{a^2-x^2}} \int_{0}^{\sqrt{a^2-x^2-y^2}} \int_{0}^{\sqrt{a^2-x^2-y^2-z^2}} dw\, dz\, dy\, dx.$$

b. Using spherical coordinates, evaluate the integral established in part (a). [*Hint:* The first integral is easy and reduces the quadruple integral to a triple integral.]

23. Back in the days when you only knew how to work with functions of one variable, you computed volumes of solids of revolution in several ways; you may have been confused about when to use the "disk method" and when to use the "shell method." This problem offers you the chance to unify those methods.

Show that the "disk method" and the "shell method" only involve different repeated integrals which compute the value of a triple integral over the region in space of the function which has the constant value of 1.

Now that you also know about cylindrical and spherical coordinates and how to express a triple integral as repeated integrals in either system, you may want to include them in your discussion.

ANSWERS TO SELF-QUIZ

I. False (integrate 1 over the region.) **II.** b **III.** False (integrate 1 over the origin.) **IV.** d

CHAPTER 4 SUMMARY OUTLINE

- **The Double Integral over a Rectangle** Let $z = f(x, y)$ and let the rectangle R be given by $a \le x \le b$, $c \le y \le d$. Let $\Delta A = \Delta x\, \Delta y$. Let $\Delta s = \sqrt{\Delta x^2 + \Delta y^2}$. Suppose that

$$\lim_{\Delta s \to 0} \sum_{i=1}^{n} \sum_{j=1}^{m} f(x_i^*, y_j^*)\, \Delta A$$

exists and is independent of the way in which the points (x_i^*, y_j^*) are chosen. Then the **double integral of f over R**, written $\iint_R f(x, y)\, dA$, is defined by

$$\iint_R f(x, y)\, dA = \lim_{\Delta s \to 0} \sum_{i=1}^{n} \sum_{j=1}^{m} f(x_i^*, y_j^*)\, \Delta A. \qquad (*)$$

- **Integrable Function of Two Variables** If the limit in (*) exists, then the function f is said to be **integrable** over R.
- **Existence of the Double Integral over a Rectangle** If f is continuous on R, then f is integrable over R.
- **Integrability over a Region** Let f be defined for (x, y) in Ω and let F be defined by

$$F(x, y) = \begin{cases} f(x, y), & \text{for } (x, y) \text{ in } \Omega \\ 0, & \text{for } (x, y) \text{ in } R \text{ but not in } \Omega, \end{cases}$$

where R is a rectangle that contains Ω. Then

$$\iint_\Omega f(x, y)\, dA = \iint_R F(x, y)\, dA$$

if the integral on the right exists. In this case, we say that f is **integrable** over Ω. If Ω is a region having one of the forms depicted in Figure 7 on page 248 and if f is continuous over Ω, then f is integrable over Ω.
- **Facts about Double Integrals**

i. If f is integrable over Ω, then for any constant c, cf is integrable over Ω, and

$$\iint_\Omega cf(x, y)\, dA = c \iint_\Omega f(x, y)\, dA.$$

ii. If f and g are integrable over Ω, then $f + g$ is integrable over Ω, and

$$\iint_\Omega [f(x, y) + g(x, y)]\, dA = \iint_\Omega f(x, y)\, dA + \iint_\Omega g(x, y)\, dA.$$

iii. If f is integrable over Ω_1 and Ω_2, where Ω_1 and Ω_2 have no points in common except perhaps those of their common boundary, then f is integrable over $\Omega = \Omega_1 \cup \Omega_2$, and

$$\iint_\Omega f(x, y)\, dA = \iint_{\Omega_1} f(x, y)\, dA + \iint_{\Omega_2} f(x, y)\, dA.$$

iv. If f and g are integrable over Ω and $f(x, y) \le g(x, y)$ for every (x, y) in Ω, then

$$\iint_\Omega f(x, y)\, dA \le \iint_\Omega g(x, y)\, dA.$$

v. Let f be integrable over Ω. Suppose that there exist constants m and M such that

$$m \le f(x, y) \le M$$

for every (x, y) in Ω. If A_Ω denotes the area of Ω, then

$$mA_\Omega \le \iint_\Omega f(x, y)\, dA \le MA_\Omega.$$

- **Repeated or Iterated Integral over a Rectangle**

$$\iint_R f(x, y)\, dA = \int_a^b \left\{ \int_c^d f(x, y)\, dy \right\} dx$$

and

$$\iint_R f(x, y)\, dA = \int_c^d \int_a^b f(x, y)\, dx\, dy.$$

- **A Type LU Region** $\Omega = \{(x, y): a \le x \le b \text{ and } g_1(x) \le x \le g_2(x)\}$.
- **A Type LR Region** $\Omega = \{(x, y): h_1(y) \le x \le h_2(y) \text{ and } c \le y \le d\}$.
- **Repeated Integral over a Region** Let f be continuous over a region Ω.

i. If Ω is a type LU region, where g_1 and g_2 are continuous, then

$$\iint_\Omega f(x, y)\, dA = \int_a^b \int_{g_1(x)}^{g_2(x)} f(x, y)\, dy\, dx.$$

ii. If Ω is a type LR region, where h_1 and h_2 are continuous, then

$$\iint_\Omega f(x, y)\, dA = \int_c^d \int_{h_1(y)}^{h_2(y)} f(x, y)\, dx\, dy.$$

- **Mass** Let $\rho(x, y)$ denote the density of a plane object that occupies a region Ω in $\mathbb{R}^2$. Then the **mass**, μ, of the object is given by

$$\mu = \iint_\Omega \rho(x, y)\, dA.$$

■ **Center of Mass** The **center of mass** $(\bar{x}, \bar{y})$ is given by

$$\bar{x} = \frac{M_y}{\mu} = \frac{\displaystyle\iint_\Omega x\rho(x, y)\, dA}{\displaystyle\iint_\Omega \rho(x, y)\, dA} = \frac{\text{first moment about } y\text{-axis}}{\text{total mass}}$$

$$\bar{y} = \frac{M_x}{\mu} = \frac{\displaystyle\iint_\Omega y\rho(x, y)\, dA}{\displaystyle\iint_\Omega \rho(x, y)\, dA} = \frac{\text{first moment about } x\text{-axis}}{\text{total mass}}.$$

■ **First Theorem of Pappus** Suppose that the plane region Ω is revolved about a line L in the xy-plane that does not intersect it. Then the volume generated is equal to the product of the area of Ω and the length of the circumference of the circle traced by the centroid of Ω.

■ **Volume in Polar Coordinates** Let $z = f(r, \theta)$. Then the volume under the surface and over the polar rectangle $\Omega = \{(r, \theta): r_1 \le r \le r_2, \theta_1 \le \theta \le \theta_2\}$ is given by

$$V = \iint_\Omega f(r, \theta) r\, dr\, d\theta.$$

■ **Converting a Double Integral from Rectangular to Polar Coordinates**

$$\iint_\Omega f(x, y)\, dA = \iint_\Omega f(r\cos\theta, r\sin\theta) r\, dr\, d\theta.$$

■ **Parallelepiped in $\mathbb{R}^3$** A **parallelepiped** π in $\mathbb{R}^3$ is a "box" given by

$$\pi = \{(x, y, z): a_1 \le x \le a_2, b_1 \le y \le b_2, c_1 \le z \le c_2\} \quad (*)$$

■ **The Triple Integral** Let $w = f(x, y, z)$ and let the parallelepiped π be given by (*). Let $\Delta u = \sqrt{\Delta x^2 + \Delta y^2 + \Delta z^2}$. Suppose that

$$\lim_{\Delta u \to 0} \sum_{i=1}^{n} \sum_{j=1}^{m} \sum_{k=1}^{p} f(x_i^*, y_j^*, z_k^*)\, \Delta x\, \Delta y\, \Delta z$$

exists and is independent of the way in which the points (x_i^*, y_j^*, z_k^*) are chosen. Then the **triple integral** of f over π, written $\iiint_\pi f(x, y, z)\, dV$, is defined by

$$\iiint_\pi f(x, y, z)\, dV = \lim_{\Delta u \to 0} \sum_{i=1}^{n} \sum_{j=1}^{m} \sum_{k=1}^{p} f(x_i^*, y_j^*, z_k^*)\, \Delta V.$$

■ **Repeated or Iterated Triple Integral over a Parallelepiped**

$$\iiint_\pi f(x, y, z)\, dV = \int_{a_1}^{a_2} \int_{b_1}^{b_2} \int_{c_1}^{c_2} f(x, y, z)\, dz\, dy\, dx.$$

■ **Triple Integral over a Region S** Enclose S in a parallelepiped π and define F by

$$F(x, y, z) = \begin{cases} f(x, y, z), & \text{if } (x, y, z) \text{ is in } S \\ 0 & \text{if } (x, y, z) \text{ is in } \pi \text{ but not in } S. \end{cases}$$

Then

$$\iiint_S f(x, y, z)\, dV = \iiint_\pi F(x, y, z)\, dV.$$

If S takes the form

$$S = \{(x, y, z): a_1 \le x \le a_2, g_1(x) \le y \le g_2(x), h_1(x, y) \le z \le h_2(x, y)\}, \quad (**)$$

then

$$\iiint_S f(x, y, z)\, dV = \int_{a_1}^{a_2} \int_{g_1(x)}^{g_2(x)} \int_{h_1(x, y)}^{h_2(x, y)} f(x, y, z)\, dz\, dy\, dx.$$

■ **Volume of a Region S as a Triple Integral** If S is of the form (**), then

$$\text{volume of } S = \iiint_S dV.$$

■ **Density or Mass in $\mathbb{R}^3$** Let the function $\rho(x, y, z)$ denote the density (in kilograms per cubic meter, say) of a solid S in $\mathbb{R}^3$. Then

$$\text{total mass of } S = \mu(S) = \iiint_S \rho(x, y, z)\, dV.$$

■ **Triple Integral in Cylindrical Coordinates** Let the region S be given in cylindrical coordinates by

$$S = \{(r, \theta, z): \theta_1 \le \theta \le \theta_2, 0 \le g_1(\theta) \le r \le g_2(\theta), h_1(r, \theta) \le z \le h_2(r, \theta)\},$$

and let f be a function of r, θ, and z. Then the triple integral of f over S is given by

$$\iiint_S f = \int_{\theta_1}^{\theta_2} \int_{g_1(\theta)}^{g_2(\theta)} \int_{h_1(r, \theta)}^{h_2(r, \theta)} f(r, \theta, z) r\, dz\, dr\, d\theta.$$

■ **Converting a Triple Integral from Rectangular to Cylindrical Coordinates**

$$\underset{S}{\iiint} f(x, y, z)\, dV = \underset{S}{\iiint} f(r\cos\theta, r\sin\theta, z) r\, dz\, dr\, d\theta.$$

Given in rectangular coordinates — Given in cylindrical coordinates

■ **Triple Integral in Spherical Coordinates** Let the region S be defined in spherical coordinates by

$$S = \{(\rho, \theta, \varphi) \colon \theta_1 \le \theta \le \theta_2, g_1(\theta) \le \varphi \le g_2(\theta), h_1(\theta, \varphi) \le \rho \le h_2(\theta, \varphi)\}.$$

Then

$$\underset{S}{\iiint} f = \int_{\theta_1}^{\theta_2} \int_{g_1(\theta)}^{g_2(\theta)} \int_{h_1(\theta, \varphi)}^{h_2(\theta, \varphi)} f(\rho, \theta, \varphi)\rho^2 \sin\varphi\, d\rho\, d\varphi\, d\theta.$$

■ **Converting a Triple Integral from Rectangular to Spherical Coordinates**

$$\underset{S}{\iiint} f(x, y, z)\, dV$$

Given in rectangular coordinates

$$= \underset{S}{\iiint} f(\rho\sin\varphi\cos\theta, \rho\sin\varphi\sin\theta, \rho\cos\varphi)\rho^2 \sin\varphi\, d\rho\, d\varphi\, d\theta.$$

Given in spherical coordinates

CHAPTER 4 REVIEW EXERCISES

In Problems 1–12, evaluate the integral.

1. $\int_0^1 \int_0^2 x^2 y\, dx\, dy$

2. $\int_0^1 \int_{x^2}^{x} xy^3\, dy\, dx$

3. $\int_2^4 \int_{1+y}^{2+5y} (x - y^2)\, dx\, dy$

4. $\int_0^4 \int_{-\sqrt{16-x^2}}^{\sqrt{16-x^2}} 4y\, dy\, dx$

5. $\int_0^1 \int_0^y \int_0^x x^2\, dz\, dx\, dy$

6. $\int_1^2 \int_{2-x}^{2+x} \int_0^{xz} 12xyz\, dy\, dz\, dx$

7. $\int_{-2}^2 \int_{-\sqrt{4-x^2}}^{\sqrt{4-x^2}} \int_{\sqrt{x^2+y^2}}^{\sqrt{8-x^2-y^2}} z^2\, dz\, dy\, dx$

8. $\int_0^2 \int_0^{\sqrt{4-x^2}} \int_0^{\sqrt{4-x^2-y^2}} \frac{z^2}{\sqrt{x^2+y^2}}\, dz\, dy\, dx$

9. $\iint_\Omega (y - x^2)\, dA$, where $\Omega = \{(x, y) \colon -3 \le x \le 3, 0 \le y \le 2\}$.

10. $\iint_\Omega (y - x^2)\, dA$, where Ω is the region in the first quadrant bounded by the x-axis, the y-axis, and the circle $x^2 + y^2 = 4$.

11. $\iint_\Omega (x + y^2)\, dA$, where Ω is the region in the first quadrant bounded by the x-axis, the y-axis, and the unit circle.

12. $\iint_\Omega (2x + y)e^{-(x+y)}\, dA$, where Ω is the first quadrant.

13. Change the order of integration of $\int_2^5 \int_1^x 3x^2 y\, dy\, dx$ and evaluate.

14. Change the order of integration of $\int_0^3 \int_0^{\sqrt{9-x^2}} (9 - y^2)^{3/2}\, dy\, dx$ and evaluate.

15. Change the order of integration of $\int_0^\infty \int_x^\infty f(x, y)\, dy\, dx$.

16. Change the order of integration in $\int_0^1 \int_0^y \int_0^x x^2\, dz\, dx\, dy$ and write the integral in the following forms:

a. $\int_?^? \int_?^? \int_?^? x^2\, dx\, dy\, dz$

b. $\int_?^? \int_?^? \int_?^? x^2\, dy\, dz\, dx$

17. Evaluate $\iint_\Omega e^{-(x^2+y^2)}\, dA$ where Ω is the circle of radius 3 centered at the origin.

18. Find close lower and upper bounds for $\iint_\Omega e^{-(x^4+y^4)}\, dA$ where Ω is the same region as in the preceding exercise.

19. Find the volume of the solid bounded by the plane $x + 2y + 3z = 6$ and the three coordinate planes.

20. Find the volume of the tetrahedron with vertices at $(0, 0, 0)$, $(2, 0, 0)$, $(0, 1, 0)$, and $(0, 0, 3)$.

21. Find the volume of the solid bounded by the cylinder $y^2 + z^2 = 4$ and the planes $x = 0$ and $x = 3$.

22. Find the volume enclosed by the ellipsoid $4x^2 + y^2 + 25z^2 = 100$.

23. Find the volume of the solid bounded by the paraboloid $x = y^2 + z^2$ and the plane $y = x - 2$.

24. Calculate the volume under the surface $z = r^5$ and over the circle of radius 6 centered at the origin.

25. Calculate the area of the region enclosed by the curve $r = 2(1 + \sin\theta)$.

26. Calculate the area of the region enclosed by the curve $r = 3(1 + \sin\theta)$ and outside the circle $r = 2$.

27. Find the volume of the solid bounded by the cone $z^2 = x^2 + y^2$ and the paraboloid $4z = x^2 + y^2$.

28. Find the volume between the planes $z = x + y$ and $z =$

$3x + 5y$ which lies over the region bounded by the curves $x = y^2$ and $x = y^3$.

29. Find the volume of the solid bounded by the parabolic cylinder $z = 4 - y^2$ and the planes $z = x$ and $z = 2x$ that lies in the half space $z \geq 0$.

30. Find the volume of the region inside both the sphere $x^2 + y^2 + z^2 = a^2$ and the cylinder $x^2 + [y - (a/2)]^2 = (a/2)^2$.

31. Find the volume of the solid bounded by the plane $z = x$ and the paraboloid $z = x^2 + y^2$.

32. Find the volume of the solid inside the sphere $x^2 + y^2 + z^2 = 9$ and outside the cone $z^2 = x^2 + y^2$.

33. Find the volume of a wedge cut from a sphere of radius 2 by two planes that meet at a diameter with an angle of $\pi/3$.

34. Find the mass of the tetrahedron of Problem 20 if the density function is $\rho(x, y, z) = 2x + y^2$.

35. Find the centroid of the region bounded by the right-hand loop of the lemniscate $r^2 = 4 \cos 2\theta$.

36. Find the centroid of the region under the surface $z = \dfrac{y^4}{4} + \dfrac{1}{8y^2}$ and over the rectangle $\{(x, y): 0 \leq x \leq 3, 2 \leq y \leq 4\}$.

37. Find the centroid of the hemisphere $x^2 + y^2 + z^2 = 16$, $z \geq 0$.

38. Find the center of mass of the unit disk if the density at a point is proportional to the distance to the boundary of the disk.

39. Find the center of mass of the region $\Omega = \{(x, y): 0 \leq x \leq 3, 1 \leq y \leq 5\}$ if $\rho(x, y) = 3x^2y$.

40. Find the center of mass of the triangular region bounded by the lines $y = x$, $y = 2 - x$, and the y-axis if $\rho(x, y) = 3xy + y^3$.

41. Find the center of mass of the solid bounded by the ellipsoid $4x^2 + y^2 + 25z^2 = 100$ if the density function is $\rho(x, y, z) = z^2$.

42. Suppose the density of the region of Problem 30 is proportional to the square of the distance to the xy-plane; find the center of mass of the region.

***43.** Find the center of mass of the solid in Problem 32 if the density at a point is proportional to its distance to the origin.

CHAPTER 4 COMPUTER EXERCISES

1. Let D denote the disk of radius 1 with center at the origin. The disk is to be cut along a vertical line so that the center of mass of the right-hand piece lies halfway along a radius. That is, the disk is to be cut along a line $x = a$ so that the center of mass of the resulting piece is located at $(\frac{1}{2}, 0)$.

a. Should a be chosen to be positive or negative? [*Hint:* Calculate the center of mass of the half disk. How does the center compare with $(\frac{1}{2}, 0)$?]

b. Approximate a to two decimal places.

2. What value of a is required if the centroid is to be at $x = \frac{2}{3}$?

3. R is the region between the graphs of $\tan^{-1} x$ and $-\tan^{-1} x$ from $x = 0$ to $x = a$. Find a so that the center of mass of R is located at $(1, 0)$.

CHAPTER 5

INTRODUCTION TO VECTOR ANALYSIS

5.1 VECTOR FIELDS

In Section 3.1, we discussed functions that assign a real number to each vector in a subset, D, of $\mathbb{R}^2$ or $\mathbb{R}^3$. More generally, suppose we assign a vector $\mathbf{F}(\mathbf{p})$ to each point $\mathbf{p}$ in a subset D of a set of vectors. We call this assignment a **vector field** on D and say that $\mathbf{F}(\mathbf{p})$ is a **vector-valued function** (or simply, **vector function**) with domain D. For most purposes, D will be a subset of $\mathbb{R}^2$ or $\mathbb{R}^3$.

Before defining a vector field, we need to define a region in $\mathbb{R}^2$ or $\mathbb{R}^3$.

DEFINITION OPEN SET, CONNECTED SET, AND REGION

A set Ω in $\mathbb{R}^2$ (or $\mathbb{R}^3$) is **open** if for every point $\mathbf{x} \in \Omega$, there is an open disk D (open sphere S) with center at $\mathbf{x}$ that is contained in Ω. An open set Ω in $\mathbb{R}^2$ (or $\mathbb{R}^3$) is **connected** if any two points in Ω can be joined by a piecewise smooth curve lying entirely in Ω. A **region** Ω in $\mathbb{R}^2$ or $\mathbb{R}^3$ is an open, connected set. ■

These definitions are illustrated in Figure 1.

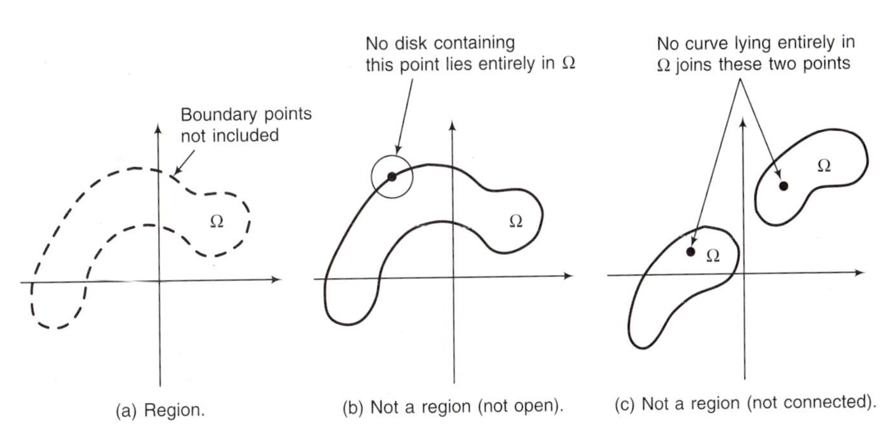

FIGURE 1
Three sets in $\mathbb{R}^2$

DEFINITION VECTOR FIELD IN $\mathbb{R}^2$

Let Ω be a region in $\mathbb{R}^2$. Then $\mathbf{F}$ is a **vector field** in $\mathbb{R}^2$ if $\mathbf{F}$ assigns to every $\mathbf{x}$ in Ω a unique vector $\mathbf{F}(\mathbf{x})$ in $\mathbb{R}^2$. ■

REMARK: Simply put, a vector field in $\mathbb{R}^2$ is a function whose domain is a region in $\mathbb{R}^2$ and whose range is a subset of $\mathbb{R}^2$. One vector field in $\mathbb{R}^2$ and one in $\mathbb{R}^3$ are sketched in Figure 2. The meaning of this sketch is that to every point $\mathbf{x}$ in Ω a unique vector $\mathbf{F}(\mathbf{x})$ is assigned. That is, the function value $\mathbf{F}(\mathbf{x})$ is represented by an arrow with $\mathbf{x}$ at the "tail" of the arrow.

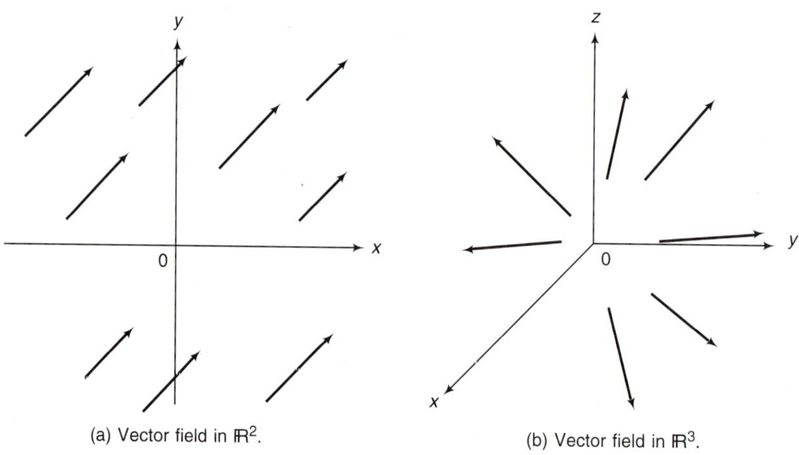

(a) Vector field in $\mathbb{R}^2$. (b) Vector field in $\mathbb{R}^3$.

FIGURE 2
Two vector fields

EXAMPLE 1 GRADIENT FIELD IN $\mathbb{R}^2$

Let $z = f(x, y)$ be a differentiable function. Then $\mathbf{F} = \nabla f = f_x\mathbf{i} + f_y\mathbf{j}$ is a vector field. It is called a **gradient vector field**.

The definition of a vector field can be extended, in an obvious way, to $\mathbb{R}^3$.

DEFINITION VECTOR FIELD IN $\mathbb{R}^3$

Let S be a region in $\mathbb{R}^3$. Then $\mathbf{F}$ is a **vector field** in $\mathbb{R}^3$ if $\mathbf{F}$ assigns to each vector $\mathbf{x}$ in S a unique vector $\mathbf{F}(\mathbf{x})$ in $\mathbb{R}^3$. ■

REMARK: A vector field in $\mathbb{R}^3$ is a function whose domain and range are subsets of $\mathbb{R}^3$.

A vector field in $\mathbb{R}^3$ is sketched in Figure 2(b).

EXAMPLE 2 GRADIENT FIELD IN $\mathbb{R}^3$

If $u = f(x, y, z)$ is differentiable, then $\mathbf{F} = \nabla f = f_x\mathbf{i} + f_y\mathbf{j} + f_z\mathbf{k}$ is a vector field. It is also called a **gradient vector field**.

EXAMPLE 3

GRAVITATIONAL FIELD

Let m_1 represent the mass of a (relatively) fixed object in space and let m_2 denote the mass of an object moving near the fixed object. Then the magnitude of the gravitational force between the objects is given by **Newton's law of universal gravitation**†

$$|\mathbf{F}| = G\frac{m_1 m_2}{r^2}, \tag{1}$$

where r is the distance between the objects and G is a universal constant. If we assume that the first object is at the origin, then we may denote the position of the second object by $\mathbf{x}(t)$, and then, since $r = |\mathbf{x}|$, (1) can be written

$$|\mathbf{F}| = \frac{G m_1 m_2}{|\mathbf{x}|^2}. \tag{2}$$

Also, the force acts toward the origin, that is, in the direction opposite to that of the position vector $\mathbf{x}$. Therefore,

$$\text{direction of } \mathbf{F} = -\frac{\mathbf{x}}{|\mathbf{x}|}, \tag{3}$$

so from (2) and (3),

$$\mathbf{F}(t) = \frac{\alpha \mathbf{x}(t)}{|\mathbf{x}(t)|^3}, \tag{4}$$

where $\alpha = -G m_1 m_2$.

The vector field (4) is called a **gravitational field**. We can sketch this vector field without much difficulty because for every $\mathbf{x} \neq \mathbf{0} \in \mathbb{R}^3$, $\mathbf{F}(\mathbf{x})$ points toward the origin. The sketch appears in Figure 3.

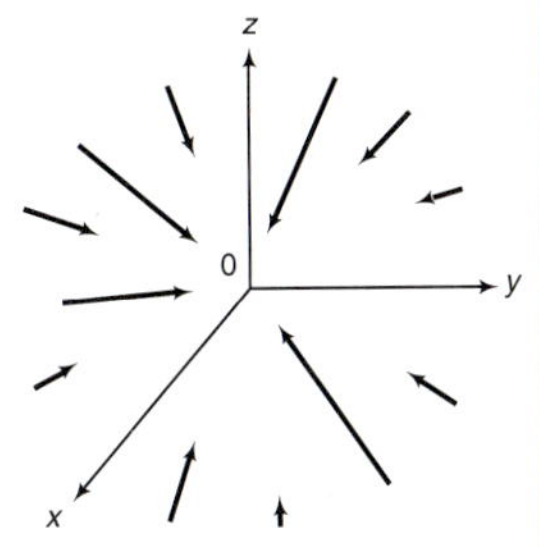

FIGURE 3
A gravitational field

Vector fields arise in a great number of physical applications. There are, for example, mechanical force fields, magnetic fields, electric fields, velocity fields, and direction fields.

Newton's law of universal gravitation was named after Sir Isaac Newton, co-inventor (with Wilhelm Leibniz) of calculus. Because Newton is so important in the history of both mathematics and physics, we provide two somewhat contrasting biographical sketches of his life.

† See *Calculus* or *Calculus of One Variable,* page 143. Also see the biographical sketches of Newton at the end of this section.

OF HISTORICAL INTEREST

SIR ISAAC NEWTON, 1642–1727

SIR ISAAC NEWTON, 1642–1727

Isaac Newton was born in the small English town of Woolsthorpe on Christmas Day 1642, the year of Galileo's death. His father, a farmer, had died before Isaac was born. His mother remarried when he was three and, thereafter, Isaac was raised by his grandmother. As a boy, Newton showed great cleverness and inventiveness—designing a water clock and a toy gristmill, among other things. One of his uncles, a Cambridge graduate, took an interest in the boy's education, and as a result, Newton entered Trinity College, Cambridge, in 1661. His interest at that time was chemistry.

Newton's interest in mathematics began with his discovery of two of the great mathematics books of his day: Euclid's *Elements* and Descartes's *La géométrie.* He also became aware of the work of the great scientists who preceded him (Galileo and Fermat).

By the end of 1664, Newton seems to have mastered all the mathematical knowledge of the time and had begun adding substantially to it. In 1665, he began his study of the rates of change, or *fluxions,* of quantities, such as distances or temperatures that varied continuously. The result of this study was what today we call *differential calculus.*

Newton disliked controversy so much that he delayed the publication of many of his findings for years. An unfortunate result of one of these delays was a conflict with Leibniz over who first discovered calculus. Leibniz made similar discoveries at about the same time as Newton, and to this day there is no universal agreement as to who discovered what first. The conflict stirred up so much ill will that English mathematicians (supporters of Newton) and continental mathematicians (supporters of Leibniz) had virtually no communication for more than a hundred years. English mathematics suffered greatly as a result.

Many of Newton's discoveries governed physics until the discoveries of Einstein early in this century. In 1679, Newton used a new measurement of the radius of the earth, together with an analysis of the earth's motion, to formulate his universal law of gravitational attraction. Although he made many other discoveries at that time, he communicated them to no one for five years. In 1684, Edmund Halley (after whom Halley's comet is named) visited Cambridge to discuss his theories of planetary motion with Newton. The conversations with Halley stimulated Newton's interest in celestial mechanics and led him to work out many of the laws that govern the motion of bodies subject to the forces of gravitation. The result of this work was the 1687 publication of Newton's masterpiece, *Philosophiae naturalis principia mathematica* (known as the *Principia*). It was received with great acclaim throughout Europe.

Newton is considered by many to be the greatest mathematician the world has ever produced. He was the greatest "applied" mathematician, determined by his ability to discover a physical property and analyze it in mathematical terms. Leibniz once said, "Taking mathematics from the beginning of the world to the time when Newton lived, what he did was much the better half." The great English poet Alexander Pope wrote,

> Nature and Nature's law lay hid in night;
> God said, 'Let Newton be,' and all was light.

Newton, by contrast, was modest about his accomplishments. Late in life he wrote, "If I have seen farther than Descartes, it is because I have stood on the shoulders of giants." All who study mathematics today are standing on Isaac Newton's shoulders.

Unlike mathematics, history is often imprecise. In his best selling book *A Brief History of Time* (New York: Bantam Books, 1988), the British mathematician and physicist Stephen Hawking provides an alternative biography of Newton. We reproduce it here.[†]

[†] From A BRIEF HISTORY OF TIME by Stephen W. Hawking. Copyright © 1988 by Stephen W. Hawking. Used by permission of Bantam Books, a division of Bantam Doubleday, Dell Publishing Group, Inc.

OF HISTORICAL INTEREST

ALTERNATIVE BIOGRAPHY OF SIR ISAAC NEWTON, 1642–1727

Isaac Newton was not a pleasant man. His relations with other academics were notorious, with most of his later life spent embroiled in heated disputes. Following publication of *Principia Mathematica*—surely the most influential book ever written in physics—Newton had risen rapidly into public prominence. He was appointed president of the Royal Society and became the first scientist ever to be knighted.

Newton soon clashed with the Astronomer Royal, John Flamsteed, who had earlier provided Newton with much needed data for *Principia,* but was now withholding information that Newton wanted. Newton would not take no for an answer; he had himself appointed to the governing body of the Royal Observatory and then tried to force immediate publication of the data. Eventually he arranged for Flamsteed's work to be seized and prepared for publication by Flamsteed's mortal enemy, Edmond Halley. But Flamsteed took the case to court and, in the nick of time, won a court order preventing distribution of the stolen work. Newton was incensed and sought his revenge by systematically deleting all references to Flamsteed in later editions of *Principia.*

A more serious dispute arose with the German philosopher Gottfried Leibniz. Both Leibniz and Newton had independently developed a branch of mathematics called calculus, which underlies most of modern physics. Although we now know that Newton discovered calculus years before Leibniz, he published his work much later. A major row ensued over who had been first, with scientists vigorously defending both contenders. It is remarkable, however, that most of the articles appearing in defense of Newton were originally written by his own hand—and only published in the name of friends! As the row grew, Leibniz made the mistake of appealing to the Royal Society to resolve the dispute. Newton, as president, appointed an ''impartial'' committee to investigate, coincidentally consisting entirely of Newton's friends! But that was not all: Newton then wrote the committee's report himself and had the Royal Society publish it, officially accusing Leibniz of plagiarism. Still unsatisfied, he then wrote an anonymous review of the report in the Royal Society's own periodical. Following the death of Leibniz, Newton is reported to have declared that he had taken great satisfaction in ''breaking Leibniz's heart.''

During the period of these two disputes, Newton had already left Cambridge and academe. He had been active in anti-Catholic politics at Cambridge, and later in Parliament, and was rewarded eventually with the lucrative post of Warden of the Royal Mint. Here he used his talents for deviousness and vitriol in a more socially acceptable way, successfully conducting a major campaign against counterfeiting, even sending several men to their death on the gallows.

CONSERVATIVE VECTOR FIELDS

Suppose that $\mathbf{F}$ is a vector. If $\mathbf{F} = -\nabla f$ for some function f, then $\mathbf{F}$ is said to be a **conservative vector field** and f is called a **potential function** for $\mathbf{F}$. The reason for this terminology will be made clear shortly. [If $\mathbf{F} = \nabla f$, then $\mathbf{F} = -\nabla(-f)$, so that the introduction of the minus sign does not cause any problem.]

Now let $\mathbf{g}(t) = x(t)\mathbf{i} + y(t)\mathbf{j}$ be a differentiable curve and suppose that a particle of mass m moves along it. Suppose further that the force acting on the particle at any time t is given by $\mathbf{F}(\mathbf{x}(t))$, where $\mathbf{F}$ is assumed to be a conservative vector field. By Newton's second law,

$$\mathbf{F}(\mathbf{x}(t)) = m\mathbf{a}(t) = m\mathbf{x}''(t).$$

But since $\mathbf{F}$ is conservative, $\mathbf{F}(\mathbf{x}) = -\nabla f(\mathbf{x})$ for some differentiable function f. Then we have

$$-\nabla f(\mathbf{x}(t)) = m\mathbf{x}''(t)$$

or

$$m\mathbf{x}'' + \nabla f(\mathbf{x}) = \mathbf{0}. \tag{5}$$

We now take the dot product of both sides of (3) with $\mathbf{x}'$ to obtain

$$m\mathbf{x}' \cdot \mathbf{x}'' + \nabla f(\mathbf{x}) \cdot \mathbf{x}' = 0. \tag{6}$$

But by the product rule,

$$\begin{aligned} \frac{d}{dx}|\mathbf{x}'(t)|^2 &= \frac{d}{dt}(\mathbf{x}'(t) \cdot \mathbf{x}'(t)) = \mathbf{x}'(t) \cdot \mathbf{x}''(t) + \mathbf{x}''(t) \cdot \mathbf{x}'(t) \\ &= 2\mathbf{x}'(t) \cdot \mathbf{x}''(t), \end{aligned} \tag{7}$$

and by equation (3.6.3) on page 189,

$$\frac{d}{dt}f(\mathbf{x}(t)) = \nabla f(\mathbf{x}(t)) \cdot \mathbf{x}'(t). \tag{8}$$

Using (7) and (8) in (6), we obtain

$$\frac{d}{dt}\left[\frac{1}{2}m|\mathbf{x}'|^2 + f(\mathbf{x}(t))\right] = 0,$$

which implies that

$$\frac{1}{2}m|\mathbf{x}'|^2 + f(\mathbf{x}(t)) = C, \tag{9}$$

where C is a constant. This is one of the versions of the **law of conservation of energy**. The term $\frac{1}{2}m|\mathbf{x}'|^2 = \frac{1}{2}m|\mathbf{v}|^2$ is called the **kinetic energy** of the particle, and the term $f(\mathbf{x}(t))$ is called the **potential energy** of the particle. Equation (9) tells us simply that if the force function $\mathbf{F}$ is conservative, then the total energy of the system is constant and, moreover, the potential function f of $\mathbf{F}$ represents the potential energy of the system.

The **principle of the conservation of energy** states that energy may be transformed from one form to another but cannot be created or destroyed; that is, the total energy is constant. Thus it seems reasonable that force fields in classical physics are conservative (although since the work of Einstein, it has been found that energy can be transformed into mass and vice versa, so that there are forces that are not conservative). One example of a conservative force is given by the force of gravitational attraction. Let m_1 represent the mass of a (relatively) fixed object in space and let m_2 denote the mass of an object moving near the fixed object. Then, as in Example 3,

$$\mathbf{F}(t) = \frac{\alpha\mathbf{x}(t)}{|\mathbf{x}(t)|^3}$$

where $\alpha = -Gm_1m_2$. We now prove that $\mathbf{F}$ is conservative. To show this, we must come up with a function f such that $\mathbf{F} = -\nabla f$. Here we will pull f "out of a hat." In Section 5.3 we will show you how to construct such an f (if one exists). Let

$$f(\mathbf{x}) = \frac{\alpha}{|\mathbf{x}|}.$$

Then $f(\mathbf{x}) = \alpha/\sqrt{x^2 + y^2 + z^2}$ and

$$-\nabla f(\mathbf{x}) = -\alpha\left[\frac{-x}{(x^2 + y^2 + z^2)^{3/2}}\mathbf{i} + \frac{-y}{(x^2 + y^2 + z^2)^{3/2}}\mathbf{j} + \frac{-z}{(x^2 + y^2 + z^2)^{3/2}}\mathbf{k}\right]$$

$$= \frac{\alpha}{|\mathbf{x}|^3}(x\mathbf{i} + y\mathbf{j} + z\mathbf{k}) = \frac{\alpha\mathbf{x}}{|\mathbf{x}|^3} = \mathbf{F}.$$

Thus $\mathbf{F}$ is conservative, so that, with respect to gravitational forces, the law of conservation of energy holds. The function f given above is called **gravitational potential**.

PROBLEMS 5.1

SELF-QUIZ

I. ________ is not a vector field in $\mathbb{R}^2$.

a. $\{x\mathbf{i} + y\mathbf{j}: |y| \geq 1\}$ **b.** $3\mathbf{i} + 4\mathbf{j}$

c. $y\mathbf{i} - x\mathbf{j}$ **d.** $xy - \mathbf{i} \cdot \mathbf{j}$

II. Suppose $f(x, y)$ is differentiable, then ________ is not a vector field in $\mathbb{R}^2$.

a. $(f_x + f_y)\mathbf{i} + (f_x - f_y)\mathbf{j}$

b. $f_y\mathbf{i} - f_x\mathbf{j}$

c. $(f_x - f_y)(\mathbf{i} - \mathbf{j})$

d. $(f_x + \mathbf{i}) - (f_y + \mathbf{j})$

In Problems 1–20, compute the gradient vector field of the given function.

1. $f(x, y) = 1/\sqrt{x^2 + y^2}$

2. $f(x, y) = \tan^{-1}(y/x)$

3. $f(x, y) = (x + y)^2$

4. $f(x, y) = e^{\sqrt{xy}}$

5. $f(x, y) = \cos(x - y)$

6. $f(x, y) = \ln(2x - y + 1)$

7. $f(x, y) = y\tan(y - x)$

8. $f(x, y) = x^2 \sinh y$

9. $f(x, y) = \sec(x + 3y)$

10. $f(x, y) = \dfrac{x - y}{x + y}$

11. $f(x, y) = \dfrac{x^2 - y^2}{x^2 + y^2}$

12. $f(x, y) = \dfrac{e^{x^2} - e^{-y^2}}{3y}$

13. $f(x, y, z) = \sqrt{x^2 + y^2 + z^2}$

14. $f(x, y, z) = xyz$

15. $f(x, y, z) = \sin x \cos y \tan z$

16. $f(x, y, z) = \dfrac{x^2 - y^2 + z^2}{3xy}$

17. $f(x, y, z) = x\ln y - z\ln x$

18. $f(x, y, z) = xy^2 + y^2z^3$

19. $f(x, y, z) = (y - z)e^{x+2y+3z}$

20. $f(x, y, z) = \dfrac{x - z}{\sqrt{1 - y^2 + x^2}}$

***21.** Two wires, straight and infinite, pass through the points $(-1, 0)$ and $(1, 0)$. They are perpendicular to the xy-plane. If the wires are uniformly, but oppositely, charged with electricity, then they create an electric field that is the gradient of

$$f(x, y) = \ln\frac{\sqrt{(x - 1)^2 + y^2}}{\sqrt{(x + 1)^2 + y^2}}.$$

Compute and graph this vector field.

22. Verify that

$$\nabla\frac{1}{|\mathbf{x}|} = \frac{-1}{|\mathbf{x}|^2}\mathbf{u}$$

where $\mathbf{u}$ is the unit vector $\mathbf{x}/|\mathbf{x}|$.

23. Show that the force $\mathbf{F}(x, y) = y\mathbf{i} + x\mathbf{j}$ is conservative by finding a potential function for it.

24. Suppose that a moving particle is subjected to a force of constant magnitude that always points towards the origin. Show that this force is conservative. [*Hint:* If the path is given by $\mathbf{x}(t) = x(t)\mathbf{i} + y(t)\mathbf{j}$, find a unit vector that points toward the origin.]

***25.** Show that the force $\mathbf{F}(\mathbf{x}) = -\alpha\mathbf{x}/|\mathbf{x}|^k$ is conservative if k is a positive integer.

26. Show that if $\mathbf{F}$ and $\mathbf{G}$ are conservative, then so is $\alpha\mathbf{F} + \beta\mathbf{G}$ where α and β are arbitrary constants.

27. Show that the force $\mathbf{F}(x, y) = y\mathbf{i} - x\mathbf{j}$ is *not* conservative. [*Hint:* Suppose there were an f such that $\mathbf{F} = -\nabla f$; then $-y = \dfrac{\partial f}{\partial x}$ and $x = \dfrac{\partial f}{\partial y}$; integrate to obtain a contradiction.]

ANSWERS TO SELF-QUIZ

I. a, d [(a): $\{(x, y): |y| \geq 1\}$ is neither open nor connected, (b) is a constant vector field.] **II.** d

5.2 WORK AND LINE INTEGRALS

Recall from Section 1.2 that the work done by a force vector $\mathbf{F}$ in moving an object in the direction $\mathbf{d}$ is given by

$$W = \frac{\mathbf{F}\cdot\mathbf{d}}{|\mathbf{d}|}|\mathbf{d}| = \mathbf{F}\cdot\mathbf{d}. \qquad \text{equation (10) on page 21} \tag{1}$$

The component of $\mathbf{F}$ in the direction of motion is

$$\frac{\mathbf{F}\cdot\mathbf{d}}{|\mathbf{d}|}. \tag{2}$$

The vector $\mathbf{d}$ is called a **displacement vector**. That is, *the work done is the dot product of the force* $\mathbf{F}$ *and the displacement vector* $\mathbf{d}$. Note that if $\mathbf{F}$ acts in the direction $\mathbf{d}$ and if φ denotes the angle (which is zero) between $\mathbf{F}$ and $\mathbf{d}$, then $\mathbf{F}\cdot\mathbf{d} = |\mathbf{F}||\mathbf{d}|\cos\varphi = |\mathbf{F}||\mathbf{d}|\cos 0 = |\mathbf{F}||\mathbf{d}|$, which is the scalar formula

$$W = Fd$$

(see p. 20).

EXAMPLE 1

COMPUTING THE WORK DONE BY A FORCE

A force of 4 N has the direction $\pi/3$. What is the work done in moving an object from the point (1, 2) to the point (5, 4), where distances are measured in meters?

SOLUTION: A unit vector with direction $\pi/3$ is given by $\mathbf{u} = (\cos\pi/3)\mathbf{i} + (\sin\pi/3)\mathbf{j} = (1/2)\mathbf{i} + (\sqrt{3}/2)\mathbf{j}$. Thus $\mathbf{F} = 4\mathbf{u} = 2\mathbf{i} + 2\sqrt{3}\mathbf{j}$. The displacement vector $\mathbf{d}$ is given by $(5 - 1)\mathbf{i} + (4 - 2)\mathbf{j} = 4\mathbf{i} + 2\mathbf{j}$. Thus,

$$\begin{aligned} W = \mathbf{F}\cdot\mathbf{d} &= (2\mathbf{i} + 2\sqrt{3}\mathbf{j})\cdot(4\mathbf{i} + 2\mathbf{j}) = 8 + 4\sqrt{3} \\ &\approx 14.93 \text{ N}\cdot\text{m} = 14.93 \text{ joules (J)}. \end{aligned}$$

The component of $\mathbf{F}$ in the direction of motion is sketched in Figure 1.

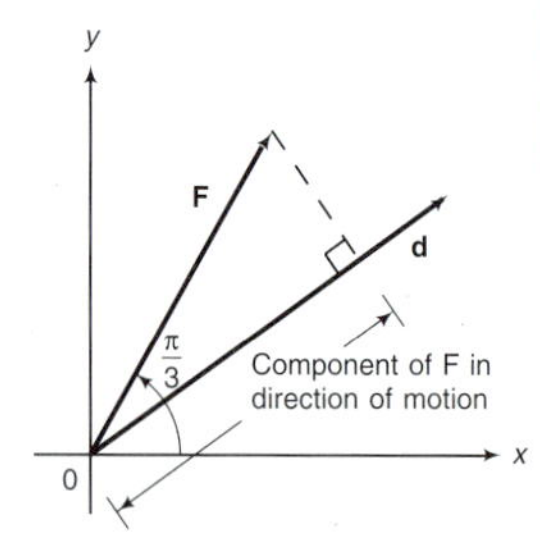

FIGURE 1
The force $\mathbf{F}$ acts in the direction $\mathbf{d}$

We now calculate the work done when a particle moves along a curve C. In doing so, we will define an important concept in applied mathematics—the *line integral.*

Suppose that a curve in the plane is given parametrically by

$$C: \quad \mathbf{x}(t) = f(t)\mathbf{i} + g(t)\mathbf{j}, \qquad \text{or} \qquad \mathbf{x}(t) = x(t)\mathbf{i} + y(t)\mathbf{j}. \tag{3}$$

If a force is applied to a particle moving along C, then such a force will have magnitude and direction, so the force will be a vector function of the vector $\mathbf{x}(t)$. That is, the force will be a vector field. We write

$$\mathbf{F}(\mathbf{x}) = \mathbf{F}(x, y) = P(x, y)\mathbf{i} + Q(x, y)\mathbf{j} \tag{4}$$

where P and Q are scalar-valued functions. The problem is to determine the work done when a particle moves on C from a point $\mathbf{x}(a)$ to a point $\mathbf{x}(b)$ subject to the force $\mathbf{F}$ given by (4). We will assume in our discussion that the curve C is *smooth* or *piecewise smooth.* By that we mean that the functions $f(t)$ and $g(t)$ in (3) are continuously differentiable or that f' and g' exist and are piecewise continuous.†

A typical curve C is sketched in Figure 2. Let

$$W(t) = \text{work done in moving from } \mathbf{x}(a) \text{ to } \mathbf{x}(t).$$

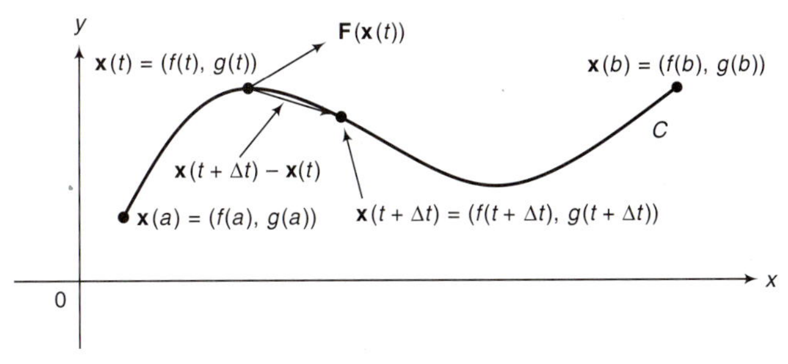

FIGURE 2
Moving a particle along a smooth curve

The work done by the variable force $\mathbf{F}$ *in moving a particle along* C *from* $\mathbf{x}(t)$ *to* $\mathbf{x}(t + \Delta t)$ *is, if* Δt *is small, approximately equal to* $\mathbf{F}(\mathbf{x}(t)) \cdot [(\mathbf{x}(t + \Delta t) - \mathbf{x}(t)]$ *because, if* Δt *is small, the part of the curve from* $\mathbf{x}(t)$ *to* $\mathbf{x}(t + \Delta t)$ *is "almost" a (straight) directed line segment.*

Then the work done in moving from $\mathbf{x}(t)$ to $\mathbf{x}(t + \Delta t)$ is given by

$$W(t + \Delta t) - W(t).$$

Now if Δt is small, then the part of the curve between $\mathbf{x}(t)$ and $\mathbf{x}(t + \Delta t)$ is "close" to a straight line and so can be approximated by the vector

$$\mathbf{x}(t + \Delta t) - \mathbf{x}(t).$$

If Δt is small and if $\mathbf{F}(\mathbf{x})$ is continuous, then the force applied between $\mathbf{x}(t)$ and $\mathbf{x}(t + \Delta t)$ is approximately equal to $\mathbf{F}(\mathbf{x}(t))$. Thus by (1), if Δt is small, then

$$W(t + \Delta t) - W(t) \approx \mathbf{F}(\mathbf{x}(t)) \cdot [\mathbf{x}(t + \Delta t) - \mathbf{x}(t)]. \tag{5}$$

We divide both sides of (7) by Δt and take the limit as $\Delta t \to 0$ to obtain

$$\begin{aligned} W'(t) &= \lim_{\Delta t \to 0} \frac{W(t + \Delta t) - W(t)}{\Delta t} = \lim_{\Delta t \to 0} \left\{ \mathbf{F}(\mathbf{x}(t)) \cdot \frac{[\mathbf{x}(t + \Delta t) - \mathbf{x}(t)]}{\Delta t} \right\} \\ &= \mathbf{F}(\mathbf{x}(t)) \cdot \mathbf{x}'(t). \end{aligned}$$

Also,

$$W(a) = 0 \qquad \text{and} \qquad W(b) = \text{total work done on the particle.} \tag{6}$$

† A function is **piecewise continuous** in the bounded interval (a, b) if f is continuous at all but a finite number of points in that interval and, at each point of discontinuity x_0, f has a **jump discontinuity** at x_0; that is, $\lim_{x \to x_0^+} f(x)$ and $\lim_{x \to x_0^-} f(x)$ both exist, are unequal, and are finite.

Thus,

$$W = \text{total work done} = W(b) - W(a) = \int_a^b W'(t)\,dt, \tag{7}$$

or

WORK DONE BY A FORCE F IN MOVING A PARTICLE ALONG A CURVE x(t) FROM x(a) TO x(b)

$$W = \int_a^b \mathbf{F}(\mathbf{x}(t)) \cdot \mathbf{x}'(t)\,dt. \tag{8}$$

We write equation (8) as

WORK WRITTEN AS A LINE INTEGRAL

$$W = \int_C \mathbf{F}(\mathbf{x}) \cdot \mathbf{dx}. \tag{9}$$

The symbol $\int_C$ is read "the integral along the curve C." The integral in (9) is called a *line integral of* **F** *over C.*

REMARK 1: If C lies along the x-axis, then C is given by $\mathbf{x}(t) = x(t)\mathbf{i} + 0\mathbf{j}$† and $\mathbf{x}'(t) = x'(t)\mathbf{i}$, so that (10) becomes

$$\int_C \mathbf{F}(\mathbf{x}) \cdot \mathbf{dx} = \int_a^b [F(x(t))\mathbf{i} \cdot [x'(t)\mathbf{i}]\,dt = \int_{x(a)}^{x(b)} F(x)\,dx, \tag{10}$$

which is our usual definite integral.

REMARK 2: Since **F** is given by (4), we can write equation (8) in a form more useful in computations:

ALTERNATIVE EQUATION FOR WORK

$$W = \int_a^b [P(x(t), y(t))x'(t) + Q(x(t), y(t))y'(t)]\,dt. \tag{11}$$

EXAMPLE 2 COMPUTING WORK

A particle is moving along the parabola $y = x^2$ subject to a force given by the vector field $2xy\mathbf{i} + (x^2 + y^2)\mathbf{j}$. How much work is done in moving from the point (1, 1) to the point (3, 9) if forces are measured in newtons and distances are measured in meters?

† We have substituted $x(t)$ for $f(t)$ and 0 for $g(t)$ here.

SOLUTION: The curve C is given parametrically by

$$\mathbf{x}(t) = t\mathbf{i} + t^2\mathbf{j} \qquad \text{where } 1 \le t \le 3.$$

We therefore have $f(t) = t$ and $g(t) = t^2$. Then $P = 2xy = 2t^3$ and $Q = x^2 + y^2 = t^2 + t^4$. Also, $f'(t) = 1$ and $g'(t) = 2t$, so by (11),

$$W = \int_1^3 [(2t^3)1 + (t^2 + t^4)(2t)]\, dt = \int_1^3 (4t^3 + 2t^5)\, dt = 322\frac{2}{3}\text{J}.$$

We used the notion of work to motivate the discussion of the line integral. We now give a general definition of the line integral in the plane.

DEFINITION LINE INTEGRAL IN THE PLANE

Let P and Q be continuous on a region S containing the smooth (or piecewise smooth) curve C given by

$$C\colon \quad \mathbf{x}(t) = f(t)\mathbf{i} + g(t)\mathbf{j}, \qquad t \in [a, b].$$

Let the vector field $\mathbf{F}$ be given by

$$\mathbf{F}(x, y) = P(x, y)\mathbf{i} + Q(x, y)\mathbf{j}.$$

The **line integral** of $\mathbf{F}$ over C is given by

$$\int_C \mathbf{F}(\mathbf{x}) \cdot d\mathbf{x} = \int_a^b [P(f(t), g(t))f'(t) + Q(f(t), g(t))g'(t)]\, dt. \tag{12}$$

■

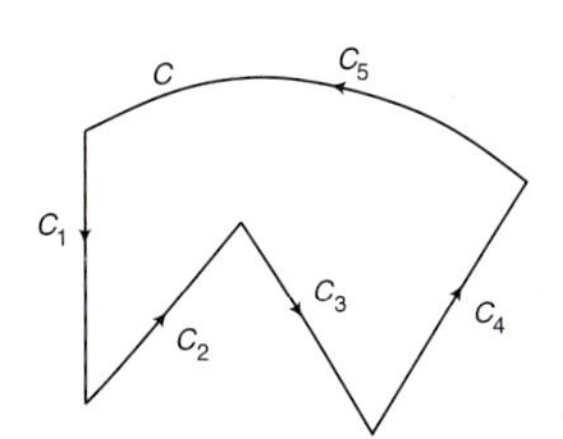

FIGURE 3
A piecewise smooth curve

REMARK: If C is piecewise smooth but not smooth, then C is made up of a number of "sections," each of which is smooth. Since C is continuous, these sections are joined. One typical piecewise smooth curve is sketched in Figure 3. If C is made up of the n smooth curves $C_1, C_2, \ldots, C_n$, then

$$\int_C \mathbf{F}(\mathbf{x}) \cdot d\mathbf{x} = \int_{C_1} \mathbf{F}(\mathbf{x}) \cdot d\mathbf{x} + \int_{C_2} \mathbf{F}(\mathbf{x}) \cdot d\mathbf{x} + \cdots + \int_{C_n} \mathbf{F}(\mathbf{x}) \cdot d\mathbf{x} \tag{13}$$

EXAMPLE 3

COMPUTING A LINE INTEGRAL OVER A RECTANGLE

Calculate $\int_C \mathbf{F}(\mathbf{x}) \cdot d\mathbf{x}$, where $\mathbf{F}(x, y) = xy\mathbf{i} + ye^x\mathbf{j}$ and C is the rectangle joining the points (0, 0), (2, 0), (2, 1), and (0, 1) if C is traversed in the counterclockwise direction.

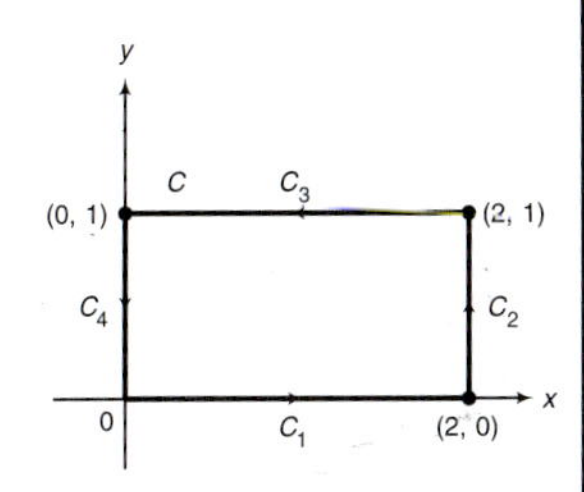

FIGURE 4
A rectangle traversed in the counterclockwise direction

SOLUTION: The rectangle is sketched in Figure 4, and it is made up of four smooth curves (straight lines). We have

C_1: $\mathbf{x}(t) = t\mathbf{i}$, $\quad 0 \le t \le 2 \quad \mathbf{x}'(t) = \mathbf{i}$

C_2: $\mathbf{x}(t) = 2\mathbf{i} + t\mathbf{j}$, $\quad 0 \le t \le 1 \quad \mathbf{x}'(t) = \mathbf{j}$

C_3: $\mathbf{x}(t) = (2 - t)\mathbf{i} + \mathbf{j}$, $\quad 0 \le t \le 1 \quad \mathbf{x}'(t) = -\mathbf{i}$

C_4: $\mathbf{x}(t) = (1 - t)\mathbf{j}$, $\quad 0 \le t \le 1. \quad \mathbf{x}'(t) = -\mathbf{j}$

Note, for example, that on C_3, $t = 0$ corresponds to the point $2\mathbf{i} + \mathbf{j} = (2, 1)$ and $t = 2$ corresponds to $(2 - 2)\mathbf{i} + \mathbf{j} = (0, 1)$. Thus as t increases, we do move along C_3 in the direction indicated by the arrow in Figure 4. This illustrates why our parametrization of the rectangle is correct.

Now

$$\int_C \mathbf{F}(\mathbf{x}) \cdot d\mathbf{x} = \int_{C_1} \mathbf{F}(\mathbf{x}) \cdot d\mathbf{x} + \int_{C_2} \mathbf{F}(\mathbf{x}) \cdot d\mathbf{x} + \int_{C_3} \mathbf{F}(\mathbf{x}) \cdot d\mathbf{x} + \int_{C_4} \mathbf{F}(\mathbf{x}) \cdot d\mathbf{x}.$$

On C_1, $x = t$ and $y = 0$, so that $xy = 0$, $ye^x = 0$, and

$$\int_{C_1} \mathbf{F}(\mathbf{x}) \cdot d\mathbf{x} = 0.$$

On C_2, $x = 2$, $y = t$, $\mathbf{F}(\mathbf{x}) = 2t\mathbf{i} + te^2\mathbf{j}$ and $\mathbf{x}'(t) = \mathbf{j}$, so that

$$\int_{C_2} \mathbf{F}(\mathbf{x}) \cdot d\mathbf{x} = \int_0^1 te^2\, dt = \frac{e^2}{2}.$$

We leave it as an exercise (see Problems 7 and 8) to show that

$$\int_{C_3} \mathbf{F}(\mathbf{x}) \cdot d\mathbf{x} = -2 \qquad \text{and} \qquad \int_{C_4} \mathbf{F}(\mathbf{x}) \cdot d\mathbf{x} = -\frac{1}{2}.$$

Thus

$$\int_C \mathbf{F}(\mathbf{x}) \cdot d\mathbf{x} = 0 + \frac{e^2}{2} - 2 - \frac{1}{2} = \frac{e^2}{2} - \frac{5}{2} \approx 1.2.$$

LINE INTEGRALS IN SPACE

The definition and theory of line integrals in $\mathbb{R}^3$ is very similar to the results in $\mathbb{R}^2$. We summarize some basic facts below.

Suppose that a curve C in space is given parametrically as

$$C: \quad \mathbf{x}(t) = f(t)\mathbf{i} + g(t)\mathbf{j} + h(t)\mathbf{k}, \qquad \text{or} \qquad \mathbf{x}(t) = x(t)\mathbf{i} + y(t)\mathbf{j} + z(t)\mathbf{k}. \tag{14}$$

As defined earlier, a **region** S in space is, as in $\mathbb{R}^2$, an open, connected set.

We now define the line integral in space.

DEFINITION LINE INTEGRAL IN SPACE

Suppose that P, Q, and R are continuous on a region S in $\mathbb{R}^3$ containing the smooth or piecewise smooth curve C given by (14). Let the vector field $\mathbf{F}$ be given by

$$\mathbf{F}(x, y, z) = P(x, y, z)\mathbf{i} + Q(x, y, z)\mathbf{j} + R(x, y, z)\mathbf{k}.$$

Then the **line integral** of $\mathbf{F}$ over C is given by

$$\int_C \mathbf{F}(\mathbf{x}) \cdot d\mathbf{x} = \int_a^b [P(f(t), g(t), h(t))f'(t) + Q(f(t), g(t), h(t))g'(t) + R(f(t), g(t), h(t))h'(t)]\, dt. \tag{15}$$

■

PROBLEMS 5.2

SELF-QUIZ

I. A force of 4 N acts in the direction of the positive x-axis to move a particle from (2, 0) to (5, 0) where the unit distance is meters. The work done is ________.

a. $4(2-5) = -12$ N-m
b. $4(5^2) - 4(2^2) = 84$ N-m
c. $[4\mathbf{i} + 0\mathbf{j}] \cdot [(5-2)\mathbf{i} + 0\mathbf{j}] = 12$ N-m
d. $[0\mathbf{i} + 4\mathbf{j}] \cdot [(5-2)\mathbf{i} + 0\mathbf{j}] = 0$ N-m

II. A force of ________ N acting in the direction of $\mathbf{i} + \mathbf{j}$ does 12 N-m work in moving a particle from (2, 0) to (5, 0) (unit distance is meters).
a. 4 **b.** $4\sqrt{2}$ **c.** $4/\sqrt{2}$ **d.** 8

III. $\int_0^4 [t\mathbf{i} - t\mathbf{j}] \cdot [\mathbf{i} + \mathbf{j}]\, dt = \int_C \mathbf{F}(\mathbf{x}) \cdot \mathbf{dx}$ where $\mathbf{F}(x, y) = x\mathbf{i} - y\mathbf{j}$ and C is ________.

a. the straight line segment from (0, 0) to (4, 4)
b. the straight line segment from (4, 4) to (0, 0)
c. the straight line segment from (0, 0) to (4, 0)
d. the parabola $y = x^2$ from (0, 0) to (4, 16)

IV. If $\mathbf{F}(\mathbf{x}) = \mathbf{x}$ and C is the straight line from (1, 1) to (4, 4), then $\int_C \mathbf{F}(\mathbf{x}) \cdot \mathbf{dx} =$ ________.

a. $\int_1^4 [t\mathbf{i} + t\mathbf{j}] \cdot [\mathbf{i} + \mathbf{j}]\, dt$
b. $\int_0^3 [(1+t)\mathbf{i} + (1+t)\mathbf{j}] \cdot [\mathbf{i} + \mathbf{j}]\, dt$
c. $\int_0^1 [(1+3u)\mathbf{i} + (1+3u)\mathbf{j}] \cdot [3\mathbf{i} + 3\mathbf{j}]\, du$
d. $\int_0^1 [(1+3u)\mathbf{i} + (1+3u)\mathbf{j}] \cdot [\mathbf{i} + \mathbf{j}]\, du$

In Problems 1–6, find the work done when the force with the given magnitude and direction moves an object from P to Q (all forces are in newtons and all distances are in meters).

1. $|\mathbf{F}| = 3$ N, $\theta = 0$, $P = (2, 3)$, $Q = (1, 7)$
2. $|\mathbf{F}| = 2$ N, $\theta = \pi/2$, $P = (5, 7)$, $Q = (1, 1)$
3. $|\mathbf{F}| = 6$ N, $\theta = \pi/4$, $P = (2, 3)$, $Q = (-1, 4)$
4. $|\mathbf{F}| = 4$ N, $\theta = \pi/6$, $P = (-1, 2)$, $Q = (3, 4)$
5. $|\mathbf{F}| = 4$ N, θ is in the direction of $2\mathbf{i} + 3\mathbf{j}$, $P = (2, 0)$, $Q = (-1, 3)$
6. $|\mathbf{F}| = 5$ N, θ is in the direction of $-3\mathbf{i} + 2\mathbf{j}$, $P = (1, 3)$, $Q = (4, -6)$

In Problems 7–26, calculate $\int_C \mathbf{F}(\mathbf{x}) \cdot \mathbf{dx}$.

7. $\mathbf{F}(x, y) = xy\mathbf{i} + ye^x\mathbf{j}$; C is the curve $\mathbf{x}(t) = (2-t)\mathbf{i} + \mathbf{j}$ for $0 \le t \le 2$ [*Note:* This is $\int_{C_3} \mathbf{F}(\mathbf{x}) \cdot \mathbf{dx}$ for Example 3.]
8. $\mathbf{F}(x, y) = xy\mathbf{i} + ye^x\mathbf{j}$; C is the curve $\mathbf{x}(t) = (1-t)\mathbf{j}$ for $0 \le t \le 1$ [*Note:* This is $\int_{C_4} \mathbf{F}(\mathbf{x}) \cdot \mathbf{dx}$ for Example 3.]
9. $\mathbf{F}(x, y) = x^2\mathbf{i} + y^2\mathbf{j}$; C is the straight line segment from (0, 0) to (2, 4)
10. $\mathbf{F}(x, y) = x^2\mathbf{i} + y^2\mathbf{j}$; C is the parabola $y = x^2$ from (0, 0) to (2, 4)
11. $\mathbf{F}(x, y) = xy\mathbf{i} + (y - x)\mathbf{j}$; C is the line $y = 2x - 4$ from $(1, -2)$ to (2, 0)
12. $\mathbf{F}(x, y) = xy\mathbf{i} + (y - x)\mathbf{j}$; C is the curve $y = \sqrt{x}$ from (0, 0) to (1, 1)
13. $\mathbf{F}(x, y) = xy\mathbf{i} + (y - x)\mathbf{j}$; C is the unit circle in the counterclockwise direction
14. $\mathbf{F}(x, y) = xy\mathbf{i} + (y - x)\mathbf{j}$; C is the triangle joining the points (0, 0), (0, 1), and (1, 0) in the counterclockwise direction
15. $\mathbf{F}(x, y) = xy\mathbf{i} + (y - x)\mathbf{j}$; C is the triangle joining the points (0, 0), (1, 0), and (1, 1) in the counterclockwise direction
16. $\mathbf{F}(x, y) = e^x\mathbf{i} + e^y\mathbf{j}$; C is the curve of Problem 12
17. $\mathbf{F}(x, y) = (x^2 + 2y)\mathbf{i} - y^2\mathbf{j}$; C is the part of the ellipse $x^2 + 9y^2 = 9$ joining the points $(0, -1)$ and (0, 1) in the clockwise direction
18. $\mathbf{F}(x, y) = (\cos x)\mathbf{i} - (\sin y)\mathbf{j}$; C is the curve of Problem 17
19. $\mathbf{F}(x, y) = e^{x+y}\mathbf{i} + e^{x-y}\mathbf{j}$; C is the curve of Problem 14
20. $\mathbf{F}(x, y) = e^{x+y}\mathbf{i} + e^{x-y}\mathbf{j}$; C is the curve of Problem 15
21. $\mathbf{F}(x, y) = (y/x^2)\mathbf{i} + (x/y^2)\mathbf{j}$; C is the straight line segment from (2, 1) to (4, 6)
22. $\mathbf{F}(x, y) = (\ln x)\mathbf{i} + (\ln y)\mathbf{j}$; C is the curve $\mathbf{x}(t) = 2t\mathbf{i} + t^3\mathbf{j}$ for $1 \le t \le 4$
23. $\mathbf{F}(x, y, z) = x\mathbf{i} + y\mathbf{j} + z\mathbf{k}$; C is the curve $\mathbf{x}(t) = t\mathbf{i} + t^2\mathbf{j} + t^3\mathbf{k}$ from (0, 0, 0) to (1, 1, 1)
24. $\mathbf{F}(x, y, z) = 2xz\mathbf{i} - xy\mathbf{j} + yz^2\mathbf{k}$; C is the curve of the preceding problem
25. $\mathbf{F}(x, y, z) = x^2\mathbf{i} + y^2\mathbf{j} + z^2\mathbf{k}$; C is the helix $\mathbf{x}(t) = (\cos t)\mathbf{i} + (\sin t)\mathbf{j} + t\mathbf{k}$ from (1, 0, 0) to $(0, 1, \pi/2)$
26. $\mathbf{F}(x, y, z) = yz\mathbf{i} + xz\mathbf{j} + xy\mathbf{k}$; C is the curve of the preceding problem

In Problems 27–34, forces are given in newtons and distances are given in meters.

27. Calculate the work done when a force field $\mathbf{F}(x, y) = x^3\mathbf{i} + xy\mathbf{j}$ moves a particle from the point $(0, 1)$ to the point $(1, e^{\pi/2})$ along the curve $\mathbf{x}(t) = (\sin t)\mathbf{i} + e^t\mathbf{j}$.

28. Calculate the work done by the force field $\mathbf{F}(x, y) = 2xy\mathbf{i} + y^2\mathbf{j}$ when a particle is moved counterclockwise around the triangle with vertices $(0, 0)$, $(1, 0)$, and $(1, 1)$.

29. Calculate the work done when the force field $\mathbf{F}(x, y) = xy\mathbf{i} + (2x^3 - y)\mathbf{j}$ moves a particle around the unit circle in the counterclockwise direction.

30. What is the work done if the particle in the preceding problem is moved in the clockwise direction?

31. Calculate the work done by the force field $\mathbf{F}(x, y) = -xy^2\mathbf{i} + 2x\mathbf{j}$ when a particle is moved around the ellipse $(x/a)^2 + (y/b)^2 = 1$ in the counterclockwise direction.

32. Calculate the work done by the force field $\mathbf{F}(x, y) = 2x\mathbf{i} - xy^2\mathbf{j}$ when a particle is moved around the ellipse $(x/a)^2 + (y/b)^2 = 1$ in the counterclockwise direction.

33. Two electrical charges of like polarity (i.e., both positive or both negative) will repel each other. If a charge of α coulombs is placed at the origin and a charge of 1 coulomb of the same polarity is at the point (x, y), then the force of repulsion is given by

$$\mathbf{F}(x, y) = \frac{\alpha x}{(x^2 + y^2)^{3/2}}\mathbf{i} + \frac{\alpha y}{(x^2 + y^2)^{3/2}}\mathbf{j}.$$

How much work is done by the force on the 1-coulomb charge as that charge moves on the straight line segment from $(1, 0)$ to $(3, -2)$?

34. How much work is done by the force on the charge in the preceding problem if the charge moves in the counterclockwise direction along the semicircle $x^2 + y^2 = a^2$, $y \geq 0$?

35. A particle moves along the elliptical helix $\mathbf{x}(t) = (\cos t)\mathbf{i} + 4(\sin t)\mathbf{j} + t\mathbf{k}$. It is subject to a force given by the vector field $x^2\mathbf{i} + y^2\mathbf{j} + 2xyz\mathbf{k}$. How much work is done in moving from the point $(1, 0, 0)$ to the point $(0, 4, \pi/2)$?

***36.** Show that if $\mathbf{F}$ is continuous and if the curve C is smooth, then the value of $\int_C \mathbf{F}(\mathbf{x}) \cdot d\mathbf{x}$ does not depend on the parametrization of C. Note that the notion of a curve involves both a set of points and a direction. For instance, the straight line segment from $(0, 0)$ to $(1, 1)$ can be parametrized by $t\mathbf{i} + t\mathbf{j}$ for $0 \leq t \leq 1$ and by $\cos\theta\mathbf{i} + \cos\theta\mathbf{j}$ for $-\pi/2 \leq \theta \leq 0$ but $(1 - u^2)\mathbf{i} + (1 - u^2)\mathbf{j}$ for $0 \leq u \leq 1$ goes in the wrong direction. There are many more parametrizations. You will prove that the value of the integral is independent of the one chosen. [Also note that you are *not* dealing with the question of whether a different path connecting the same endpoints, e.g., by way of $(1, 0)$, would yield the same value for the integral.]

ANSWERS TO SELF-QUIZ

I. c **II.** b **III.** a **IV.** a, b, c (See Problem 36.)

5.3 EXACT VECTOR FIELDS AND INDEPENDENCE OF PATH

There are certain conditions under which the calculation of a line integral becomes very easy. We first illustrate what we have in mind.

EXAMPLE 1

COMPUTING A LINE INTEGRAL OVER THREE DIFFERENT PATHS

Let $\mathbf{F}(x, y) = y\mathbf{i} + x\mathbf{j}$. Calculate $\int_C \mathbf{F}(\mathbf{x}) \cdot d\mathbf{x}$, where C is as follows:

a. The straight line from $(0, 0)$ to $(1, 1)$.

b. The parabola $y = x^2$ from $(0, 0)$ to $(1, 1)$.

c. The curve $\mathbf{x}(t) = t^{3/2}\mathbf{i} + t^5\mathbf{j}$ from $(0, 0)$ to $(1, 1)$.

SOLUTION:

a. C is given by $\mathbf{x}(t) = t\mathbf{i} + t\mathbf{j}$. Then $\mathbf{x}'(t) = \mathbf{i} + \mathbf{j}$ and $\mathbf{F}\cdot\mathbf{x}' = (t\mathbf{i} + t\mathbf{j})\cdot(\mathbf{i} + \mathbf{j}) = 2t$, so

$$\int_C \mathbf{F}(\mathbf{x})\cdot d\mathbf{x} = \int_0^1 2t\,dt = 1.$$

b. C is given by $\mathbf{x}(t) = t\mathbf{i} + t^2\mathbf{j}$. Then $\mathbf{x}'(t) = \mathbf{i} + 2t\mathbf{j}$ and $\mathbf{F}\cdot\mathbf{x}' = (t^2\mathbf{i} + t\mathbf{j})\cdot(\mathbf{i} + 2t\mathbf{j}) = 3t^2$, so

$$\int_C \mathbf{F}(\mathbf{x})\cdot d\mathbf{x} = \int_0^1 3t^2\,dt = 1.$$

c. Here $\mathbf{x}'(t) = \frac{3}{2}\sqrt{t}\mathbf{i} + 5t^4\mathbf{j}$ and $\mathbf{F}(x, y) = t^5\mathbf{i} + t^{3/2}\mathbf{j}$, so $\mathbf{F}\cdot\mathbf{x}' = (t^5\mathbf{i} + t^{3/2}\mathbf{j})\cdot(\frac{3}{2}\sqrt{t}\mathbf{i} + 5t^4\mathbf{j}) = \frac{3}{2}t^{11/2} + 5t^{11/2} = \frac{13}{2}t^{11/2}$. Then

$$\int_C \mathbf{F}\cdot\mathbf{x}'\,dt = \int_0^1 \frac{13}{2}t^{11/2}\,dt = 1.$$

INDEPENDENCE OF PATH

In Example 1, we saw that on three very different curves, we obtained the same answer as we moved between the points (0, 0) and (1, 1). In fact, as we will show in a moment, we will get the same answer if we integrate along any piecewise smooth curve C joining these two points. When this happens, we say that the line integral is **independent of the path**. A condition that ensures that a line integral is independent of the path over which it is integrated is given below.

THEOREM 1 A FUNCTION IS A GRADIENT IF AND ONLY IF ITS LINE INTEGRAL IS INDEPENDENT OF THE PATH

Let $\mathbf{F}$ be continuous in a region Ω in $\mathbb{R}^2$. Then $\mathbf{F}$ is the gradient of a differentiable function f if and only if for any piecewise smooth curve C lying in Ω, the value of the line integral $\int_C \mathbf{F}(\mathbf{x})\cdot d\mathbf{x}$ depends only on the endpoints of C.

PROOF: We first assume that $\mathbf{F}(\mathbf{x}) = \nabla f$ for some differentiable function f. Recall the vector form of the chain rule:

$$\frac{d}{dt}f(\mathbf{x}(t)) = \nabla f(\mathbf{x}(t))\cdot\mathbf{x}'(t) \tag{1}$$

[see equation (3.6.3) on page 189]. Now suppose that C is given by $\mathbf{x}(t)$: $a \le t \le b$, $\mathbf{x}(a) = \mathbf{x}_0$, and $\mathbf{x}(b) = \mathbf{x}_1$. We will assume that C is smooth. Otherwise, we could write the line integral in the form (5.2.13) and treat the integral over each smooth curve C_i separately. Using (1), we have

$$\int_C \mathbf{F}(\mathbf{x})\cdot d\mathbf{x} = \int_a^b \mathbf{F}(\mathbf{x}(t))\cdot\mathbf{x}'(t)\,dt \overset{\mathbf{F}=\nabla f}{=} \int_a^b \nabla f(\mathbf{x}(t))\cdot\mathbf{x}'(t)\,dt$$

$$= \int_a^b \frac{d}{dt}f(\mathbf{x}(t))\,dt = f(\mathbf{x}(b)) - f(\mathbf{x}(a))$$

$$= f(\mathbf{x}_1) - f(\mathbf{x}_0).$$

This proves that the line integral is independent of the path since $f(\mathbf{x}_1) - f(\mathbf{x}_0)$ does not depend on the particular curve chosen.

We now assume that $\int_C \mathbf{F}(\mathbf{x}) \cdot \mathbf{dx}$ is independent of the path and prove that $\mathbf{F} = \nabla f$ for some differentiable function f. Let $\mathbf{x}_0$ be a fixed point in Ω and let $\mathbf{x}$ be any other point in Ω. Since Ω is connected, there is at least one piecewise smooth path C joining $\mathbf{x}_0$ and $\mathbf{x}$, with C wholly contained in Ω. We define a function f by

$$f(\mathbf{x}) = \int_C \mathbf{F}(\mathbf{x}) \cdot \mathbf{dx}.$$

This function is well defined because, by hypothesis, $\int_C \mathbf{F}(\mathbf{x}) \cdot \mathbf{dx}$ is the same no matter what path is chosen between $\mathbf{x}_0$ and $\mathbf{x}$. Write $\mathbf{x}_0 = (x_0, y_0)$ and $\mathbf{x} = (x, y)$. Since Ω is open, there is an open disk D centered at (x, y) that is contained in Ω. Choose $\Delta x > 0$ such that $(x + \Delta x, y) \in D$, and let C_1 be the horizontal line segment joining (x, y) to $(x + \Delta x, y)$. The situation is depicted in Figure 1. We see that $C \cup C_1$ is a path joining (x_0, y_0) to $(x + \Delta x, y)$, so that

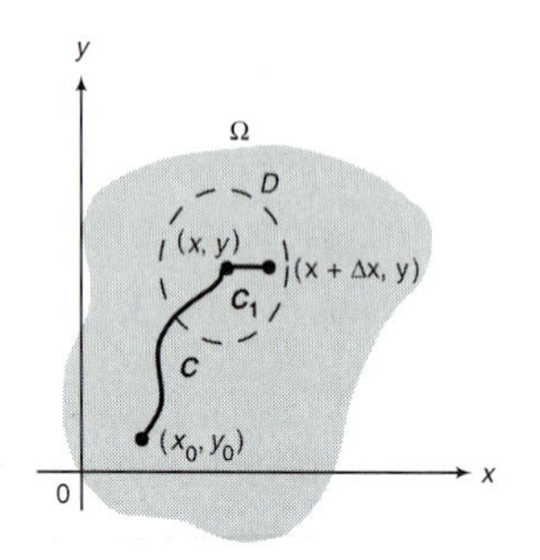

FIGURE 1
The curve C joins (x_0, y_0) to (x, y). The horizontal line segment C_1 joins (x, y) to $(x + \Delta x, y)$.

$$\begin{aligned} f(x + \Delta x, y) &= \int_C \mathbf{F}(\mathbf{x}) \cdot \mathbf{dx} + \int_{C_1} \mathbf{F}(\mathbf{x}) \cdot \mathbf{dx} \\ &= f(x, y) + \int_{C_1} \mathbf{F}(\mathbf{x}) \cdot \mathbf{dx}. \end{aligned} \tag{2}$$

A parametrization for C_1 (which is a horizontal line) is

$$\mathbf{x}(t) = (x + t\,\Delta x, y), \qquad 0 \le t \le 1$$

and

$$\mathbf{x}'(t) = (\Delta x, 0).$$

Suppose that $\mathbf{F}(\mathbf{x}) = P(x, y)\mathbf{i} + Q(x, y)\mathbf{j}$. Then

$$\mathbf{F}(\mathbf{x}) \cdot \mathbf{dx} = P(x, y)\,\Delta x. \tag{3}$$

From (2) and (3) we compute

$$f(x + \Delta x, y) - f(x, y) = \int_{C_1} \mathbf{F}(\mathbf{x}) \cdot \mathbf{dx} = \int_0^1 P(x + t\,\Delta x, y)\,\Delta x\,dt,$$

so that

$$\frac{f(x + \Delta x, y) - f(x, y)}{\Delta x} = \int_0^1 P(x + t\,\Delta x, y)\,dt. \tag{4}$$

In (4), $P(x + t\,\Delta x, y)$ is a function of one variable (t), so we may apply the mean value theorem for integrals[†] to see that there is a number $\bar{x}$ with $x < \bar{x} < x + \Delta x$ such that

$$\int_0^1 P(x + t\,\Delta x, y)\,dt = P(\bar{x}, y) \int_0^1 dt = P(\bar{x}, y).$$

[†] Here is the **mean value theorem for integrals**. Suppose that f and g are continuous on $[a, b]$ and that $g(x)$ is never zero on (a, b). Then there exists a number c in (a, b) such that

$$\int_a^b f(x)g(x)\,dx = f(c) \int_a^b g(x)\,dx.$$

For a proof, see *Calculus* or *Calculus of One Variable,* page 328.

Thus taking the limits as $\Delta x \to 0$ on both sides of (4), we have

$$\frac{\partial f(\mathbf{x})}{\partial x} = \lim_{\Delta x \to 0} \frac{f(x + \Delta x, y) - f(x, y)}{\Delta x}$$

$$= \lim_{\Delta x \to 0} P(\bar{x}, y) \overset{\substack{P \text{ is continuous and } \bar{x} \to x \text{ as } \Delta x \to 0 \\ \downarrow}}{=} P(x, y).$$

In a similar manner, we can show that $(\partial f/\partial y)(\mathbf{x}) = Q(x, y)$. This shows that $\mathbf{F} = \nabla f$, and the proof is complete. ■

In proving this theorem, we also proved the following corollary.

COROLLARY 1

Suppose that $\mathbf{F}$ is continuous in a region Ω in $\mathbb{R}^2$ and $\mathbf{F} = \nabla f$ for some differentiable function f. Then for any piecewise smooth curve C in Ω starting at the point $\mathbf{x}_0$ and ending at the point $\mathbf{x}_1$,

$$\int_C \mathbf{F}(\mathbf{x}) \cdot \mathbf{dx} = f(\mathbf{x}_1) - f(\mathbf{x}_0). \tag{5}$$

That is, the value of the integral depends only on the endpoints of the path. ■

REMARK 1: This corollary is really the line integral analog of the fundamental theorem of calculus. It says that we can evaluate the line integral of a gradient field by evaluating at two points the function for which $\mathbf{F}$ is the gradient.

REMARK 2: In Theorem 1 it is important that $\mathbf{F}$ be continuous on Ω, not only on C. (See Problem 28)

EXAMPLE 2 INTEGRATING A FUNCTION THAT CAN BE WRITTEN AS A GRADIENT

Since $y\mathbf{i} + x\mathbf{j} = \nabla(xy)$, we immediately find that for any curve C starting at (0, 0) and ending at (1, 1)

$$\int_C \mathbf{F}(\mathbf{x}) \cdot \mathbf{dx} = xy\Big|_{(1,1)} - xy\Big|_{(0,0)} = 1 - 0 = 1.$$

We now state a general result that tells us whether $\mathbf{F}$ is a gradient of some function f. Half its proof is difficult and is suggested in Problem 5.4.24. The other half is suggested in Problem 27.

THEOREM 2 CONDITIONS FOR EXACTNESS IN $\mathbb{R}^2$

Let $\mathbf{F}(x, y) = P(x, y)\mathbf{i} + Q(x, y)\mathbf{j}$ and suppose that P, Q, $\partial P/\partial y$, and $\partial Q/\partial x$ are continuous in an open disk D centered at (x, y). Then, in D, $\mathbf{F}$ is the gradient of a function f if and only if

$$\frac{\partial P}{\partial y} = \frac{\partial Q}{\partial x}.$$ ■

DEFINITION EXACT VECTOR FIELD IN $\mathbb{R}^2$

We say that the vector field $P(x, y)\mathbf{i} + Q(x, y)\mathbf{j}$ is **exact** if $\mathbf{F}$ is the gradient of a function f. ■

COROLLARY 2

Let $\mathbf{F}(x, y) = P(x, y)\mathbf{i} + Q(x, y)\mathbf{j}$. Then if P, Q, $\partial P/\partial y$, and $\partial Q/\partial x$ are all continuous on an open disk D containing C and if $\partial P/\partial y = \partial Q/\partial x$, then $\int_C \mathbf{F}(\mathbf{x}) \cdot \mathbf{dx}$ is independent of the path. ■

EXAMPLE 3 INTEGRATING AN EXACT VECTOR FIELD

Let $\mathbf{F}(x, y) = [4x^3y^3 + (1/x)]\mathbf{i} + [3x^4y^2 - (1/y)]\mathbf{j}$. Calculate $\int_C \mathbf{F}(\mathbf{x}) \cdot \mathbf{dx}$ for any smooth curve C, not crossing the x- or y-axis, starting at (1, 1) and ending at (2, 3).

SOLUTION: $P(x, y) = 4x^3y^3 + (1/x)$ and $Q(x, y) = 3x^4y^2 - (1/y)$, so

$$\frac{\partial P}{\partial y} = 12x^3y^2 = \frac{\partial Q}{\partial x}.$$

Thus, $\mathbf{F}$ is exact.

If $\nabla f = \mathbf{F}$, then $\partial f/\partial x = P$, so

$$f(x, y) = \int \left(4x^3y^3 + \frac{1}{x}\right) dx = x^4y^3 + \ln|x| + g(y).$$

Differentiating with respect to y, we have

$$Q = \frac{\partial f}{\partial y} = 3x^4y^2 + g'(y) = 3x^4y^2 - \frac{1}{y}.$$

Thus, $g'(y) = -1/y$, $g(y) = -\ln|y| + C$, and finally,

$$f(x, y) = x^4y^3 + \ln|x| - \ln|y| + C = x^4y^3 + \ln\left|\frac{x}{y}\right| + C.$$

We have shown that $\mathbf{F}$ is exact and that $\mathbf{F} = \nabla f$, where $f(x, y) = x^4y^3 + \ln|x/y|$. Thus $\int_C \mathbf{F}(\mathbf{x}) \cdot \mathbf{dx}$ is independent of the path, and

$$\int_C \mathbf{F}(\mathbf{x}) \cdot \mathbf{dx} = f(2, 3) - f(1, 1) = (16)(27) + \ln\left(\frac{2}{3}\right) - 1$$

$$= 431 + \ln\left(\frac{2}{3}\right).$$

EXAMPLE 4 INTEGRATING A VECTOR FIELD THAT IS NOT EXACT OVER TWO DIFFERENT PATHS

Let $\mathbf{F}(\mathbf{x}) = x\mathbf{i} + (x - y)\mathbf{j}$. Let C_1 be the part of the curve $y = x^2$ that connects (0, 0) to (1, 1), and let C_2 be the part of the curve $y = x^3$ that connects these two points. Compute $\int_{C_1} \mathbf{F}(\mathbf{x}) \cdot \mathbf{dx}$ and $\int_{C_2} \mathbf{F}(\mathbf{x}) \cdot \mathbf{dx}$.

SOLUTION: Along C_1, $\mathbf{x}(t) = t\mathbf{i} + t^2\mathbf{j}$ and $\mathbf{x}'(t) = \mathbf{i} + 2t\mathbf{j}$ for $0 \leq t \leq 1$, so that

$$\int_{C_1} \mathbf{F}(\mathbf{x}) \cdot d\mathbf{x} = \int_0^1 [t \cdot 1 + (t - t^2)2t]\, dt = \int_0^1 (t + 2t^2 - 2t^3)\, dt$$

$$= \left(\frac{t^2}{2} + \frac{2t^3}{3} - \frac{t^4}{2}\right)\Bigg|_0^1 = \frac{2}{3}.$$

Along C_2, $\mathbf{x}(t) = t\mathbf{i} + t^3\mathbf{j}$ and $\mathbf{x}' = \mathbf{i} + 3t^2\mathbf{j}$ for $0 \leq t \leq 1$, so

$$\int_{C_2} \mathbf{F}(x) \cdot d\mathbf{x} = \int_0^1 [t \cdot 1 + (t - t^3)(3t^2)]\, dt = \int_0^1 (t + 3t^3 - 3t^5)\, dt$$

$$= \left(\frac{t^2}{2} + \frac{3}{4}t^4 - \frac{1}{2}t^6\right)\Bigg|_0^1 = \frac{3}{4}.$$

We see that $\int_C \mathbf{F}(\mathbf{x}) \cdot d\mathbf{x}$ is *not* independent of the path. Note that here $\partial P/\partial y = 0$ and $\partial Q/\partial x = 1$, so $\mathbf{F}$ is *not* exact.

There is another important consequence of Theorem 1. Let C be a closed curve (i.e., $\mathbf{x}_0 = \mathbf{x}_1$).

THE LINE INTEGRAL OF AN EXACT VECTOR FIELD OVER A CLOSED CURVE IS ZERO

If $\mathbf{F}$ is continuous in a region Ω and if $\mathbf{F}$ is the gradient of a differentiable function f, then for any closed curve C lying in Ω,

$$\int_C \mathbf{F}(\mathbf{x}) \cdot d\mathbf{x} = f(\mathbf{x}_1) - f(\mathbf{x}_0) = 0. \quad (6)$$

INDEPENDENCE OF PATH IN $\mathbb{R}^3$

As in the plane, the computation of a line integral in space is not difficult if the integral is **independent of the path**. Theorem 3 generalizes Theorem 1. Its proof is essentially identical to the proof of that theorem.

THEOREM 3 CONDITIONS FOR EXACTNESS IN $\mathbb{R}^3$

Let $\mathbf{F}$ be continuous on a region S in $\mathbb{R}^3$. Then $\mathbf{F}$ is the gradient of a function f if and only if the following equivalent conditions hold:

i. For any piecewise smooth curve C lying in S, the line integral

$$\int_C \mathbf{F}(\mathbf{x}) \cdot d\mathbf{x}$$

is independent of the path.

ii. For any piecewise smooth curve C in S starting at the point $\mathbf{x}_0$ and ending at the point $\mathbf{x}_1$,

$$\int_C \mathbf{F}(\mathbf{x}) \cdot d\mathbf{x} = f(\mathbf{x}_1) - f(\mathbf{x}_0).$$

■

In Theorem 2; we saw that $\mathbf{F}(x, y) = P(x, y)\mathbf{i} + Q(x, y)\mathbf{j}$ is the gradient of a function f if $\partial P/\partial y = \partial Q/\partial x$. If $\mathbf{F}(x, y, z) = P(x, y, z)\mathbf{i} + Q(x, y, z)\mathbf{j} + R(x, y, z)\mathbf{k}$, then there is a condition that can be used to check whether there is a differentiable function f such that $\mathbf{F} = \nabla f$. We give this result without proof.†

THEOREM 4 CONDITIONS FOR EXACTNESS IN $\mathbb{R}^3$

Let $\mathbf{F}(x, y, z) = P(x, y, z)\mathbf{i} + Q(x, y, z)\mathbf{j} + R(x, y, z)\mathbf{k}$ and suppose that P, Q, R, $\partial P/\partial y$, $\partial P/\partial z$, $\partial Q/\partial x$, $\partial Q/\partial z$, $\partial R/\partial x$, and $\partial R/\partial y$ are continuous in an open ball centered at (x, y, z).‡ Then $\mathbf{F}$ is the gradient of a function f if and only if

$$\frac{\partial P}{\partial y} = \frac{\partial Q}{\partial x}, \qquad \frac{\partial R}{\partial x} = \frac{\partial P}{\partial z}, \qquad \frac{\partial Q}{\partial z} = \frac{\partial R}{\partial y}. \tag{7}$$

■

DEFINITION EXACT VECTOR FIELD IN $\mathbb{R}^3$

If $\mathbf{F}$ is the gradient of a differentiable function f, then $\mathbf{F}$ is said to be **exact**. ■

EXAMPLE 5 INTEGRATING AN EXACT VECTOR FIELD IN $\mathbb{R}^3$

Let $\mathbf{F}(x, y, z) = yz\mathbf{i} + xz\mathbf{j} + xy\mathbf{k}$. Verify that $\mathbf{F}$ is exact and compute $\int_C \mathbf{F}(\mathbf{x}) \cdot \mathbf{dx}$ where C is a smooth curve joining $(1, 1, 1)$ to $(2, -1, 3)$.

SOLUTION: We find that $\partial P/\partial y = z = \partial Q/\partial x$, $\partial R/\partial x = y = \partial P/\partial z$, and $\partial Q/\partial z = x = \partial R/\partial y$. Thus $\mathbf{F}$ is exact, and there is a differentiable function f such that $\mathbf{F} = \nabla f$, so that

$$f_x = yz, \qquad f_y = xz, \qquad f_z = xy.$$

Then

$$f(x, y, z) = \int f_x\, dx = \int yz\, dx = xyz + g(y, z)$$

for some differentiable function g of y and z only. (This means that g is a constant function *with respect to x*.) Hence,

$$f_y = xz + g_y(y, z) = xz,$$

so $g_y(y, z) = 0$ and $g(y, z) = h(z)$, where h is a differentiable function of z only. Then

$$f(x, y, z) = xyz + h(z)$$

and

$$f_z(x, y, z) = xy + h'(z) = xy,$$

† For a proof, see R. C. Buck and E. F. Buck, *Advanced Calculus,* 3rd ed. (New York: McGraw-Hill, 1978), pp. 497–498.
‡ This theorem and Theorem 2 are also true in a more general region called a **simply connected region**. We define such a region on page 314.

so $h'(z) = 0$ and $h(z) = C$, a constant. Thus,

$$f(x, y, z) = xyz + C,$$

and it is easily verified that $\nabla f = \mathbf{F}$. Finally, we have

$$\int_C \mathbf{F}(\mathbf{x}) \cdot d\mathbf{x} = f(2, -1, 3) - f(1, 1, 1) = (2)(-1)(3) - 1 = -7.$$

There is another way to complete this problem. Since **F** is exact, the integral is independent of the path. The simplest path joining (1, 1, 1) to (2, −1, 3) is the straight line segment joining these two points. One parametrization of this line segment is (see Section 1.5)

$$x = 1 + t \qquad y = 1 - 2t \qquad z = 1 + 2t, \qquad 0 \le t \le 1.$$

Then

$$\begin{aligned} yz &= (1 - 2t)(1 + 2t) = 1 - 4t^2 \\ xz &= (1 + t)(1 + 2t) = 1 + 3t + 2t^2 \\ xy &= (1 + t)(1 - 2t) = 1 - t - 2t^2 \\ d\mathbf{x} &= \mathbf{i} - 2\mathbf{j} + 2\mathbf{k} \end{aligned}$$

and

$$\begin{aligned} \mathbf{F}(\mathbf{x}) \cdot d\mathbf{x} &= 1 - 4t^2 - 2(1 + 3t + 2t^2) + 2(1 - t - 2t^2) \\ &= 1 - 8t - 12t^2. \end{aligned}$$

Thus,

$$\begin{aligned} \int_C \mathbf{F}(\mathbf{x}) \cdot d\mathbf{x} &= \int_0^1 (1 - 8t - 12t^2)\, dt = (t - 4t^2 - 4t^3)\Big|_0^1 = \\ &= 1 - 4 - 4 = -7. \end{aligned}$$

Before leaving this section, we note that line integrals in space are often written in a different way. If $\mathbf{F}(x, y, z) = P(x, y, z)\mathbf{i} + Q(x, y, z)\mathbf{j} + R(x, y, z)\mathbf{k}$, and if we write $d\mathbf{x} = dx\mathbf{i} + dy\mathbf{j} + dz\mathbf{k}$, then

ALTERNATIVE WAY TO WRITE A LINE INTEGRAL IN SPACE

$$\int_C \mathbf{F}(\mathbf{x}) \cdot d\mathbf{x} = \int_C (P\,dx + Q\,dy + R\,dz) \qquad (8)$$

Finally, we remind you of an important definition given in Section 5.1.

DEFINITION CONSERVATIVE VECTOR FIELD AND SCALAR POTENTIAL

Suppose that $\mathbf{F} = -\nabla f$ for some differential function f. Then **F** is said to be **conservative** and f is called a **scalar potential**. ■

PROBLEMS 5.3

SELF-QUIZ

I. True–False: $0.333x\mathbf{i} - 0.667y\mathbf{j}$ is not exact because 0.333 is not exactly equal to 1/3 and -0.667 is not exactly equal to $-2/3$.

II. True–False: $2y\mathbf{i} + 3x\mathbf{j}$ is exact because 2 and 3 are integers.

III. Let C be a smooth path connecting (2, 3) to (8, 7). Then $\int_C \nabla(xy) \cdot d\mathbf{x} =$ ________.

a. $(8)(7) - (2)(3) = 50$
b. $(8 - 2)(7 - 3) = 24$
c. $(8)(2) - (7)(3) = -5$
d. varies with the particular path chosen

IV. Let C be a smooth path connecting (2, 3, 0) to (8, 7, 0). Then $\int_C (y\mathbf{i} + x\mathbf{j} + 0\mathbf{k}) \cdot d\mathbf{x} =$ ________.

a. $(8)(7) - (2)(3) = 50$
b. $(8)(7)(0) - (2)(3)(0) = 0$
c. $(8 - 2)(7 - 3)(0 - 0) = 0$
d. varies with the particular path chosen

V. Suppose C is the square path from (0, 0) to (1, 0) to (1, 1) to (0, 1) and back to (0, 0). Then $\int_C \nabla[\tan^{-1}(x - \ln|e^y + e^x|)] \cdot d\mathbf{x} =$ ________.

a. 0 **b.** $\coth(\ln \sqrt{5}) + (22/7)$
c. $\pi/4 - e^2$ **d.** $\sqrt{e} + \sqrt{\pi}$

VI. Suppose C is the square path from (0, 0) to (1, 0) to (1, 1) to (0, 1) and back to (0, 0). Then $\int_C (y\mathbf{i} - x\mathbf{j}) \cdot d\mathbf{x} =$ ________.

a. 0 **b.** -1
c. -2 **d.** 2

In Problems 1–16, test for exactness. If the given vector field $\mathbf{F}$ is exact, then find all functions f for which $\nabla f = \mathbf{F}$.

1. $\mathbf{F}(x, y) = 2xy\mathbf{i} + (x^2 + 1)\mathbf{j}$
2. $\mathbf{F}(x, y) - (4x^3 - ye^{xy})\mathbf{i} + (\tan y - xe^{xy})\mathbf{j}$
3. $\mathbf{F}(x, y) = (4x^2 - 4y^2)\mathbf{i} + (8xy - \ln y)\mathbf{j}$
4. $\mathbf{F}(x, y) = [x \cos(x + y) + \sin(x + y)]\mathbf{i} + x \cos(x + y)\mathbf{j}$
5. $\mathbf{F}(x, y) = 2x(\cos y)\mathbf{i} + x^2(\sin y)\mathbf{j}$
6. $\mathbf{F}(x, y) = \left[\dfrac{\ln(\ln y)}{x} + \dfrac{2}{3}xy^3\right]\mathbf{i} + \left[\dfrac{\ln x}{y \ln y} + x^2y^2\right]\mathbf{j}$
7. $\mathbf{F}(x, y) = (x - y \cos x)\mathbf{i} - (\sin x)\mathbf{j}$
8. $\mathbf{F}(x, y) = e^{x^2y}\mathbf{i} + e^{x^2y}\mathbf{j}$
9. $\mathbf{F}(x, y) = (3x \ln x + x^5 - y)\mathbf{i} - x\mathbf{j}$
10. $\mathbf{F}(x, y) = \left[\dfrac{1}{x^2} + y^2\right]\mathbf{i} + 2xy\mathbf{j}$
11. $\mathbf{F}(x, y) = (x^2 + y^2 + 1)\mathbf{i} - (xy + y)\mathbf{j}$
12. $\mathbf{F}(x, y) = \left[\dfrac{-1}{x^3} + 4x^3y\right]\mathbf{i} + [\sin y + \sqrt{y} + x^4]\mathbf{j}$
13. $\mathbf{F}(x, y, z) = \mathbf{i} + \mathbf{j} + \mathbf{k}$
14. $\mathbf{F}(x, y, z) = yz\mathbf{i} + xz\mathbf{j} + xy\mathbf{k}$
15. $\mathbf{F}(x, y, z) = [e^{yz} + y]\mathbf{i} + [xze^{yz} - x]\mathbf{j} + [xye^{yz} + 2z]\mathbf{k}$
16. $\mathbf{F}(x, y, z) = [2xy^3 + x + z]\mathbf{i} + [3x^2y^2 - y]\mathbf{j} + [x + \sin z]\mathbf{k}$

In Problems 17–26, verify that $\mathbf{F}$ is exact, then use Corollary 1 to calculate $\int_C \mathbf{F}(\mathbf{x}) \cdot d\mathbf{x}$ where C is a smooth curve starting at $\mathbf{x}_0$ and ending at $\mathbf{x}_1$.

17. $\mathbf{F}(x, y) = 2xy\mathbf{i} + (x^2 + 1)\mathbf{j}$; $\mathbf{x}_0 = (0, 1)$, $\mathbf{x}_1 = (2, 3)$
18. $\mathbf{F}(x, y) = (4x^2 - 4y^2)\mathbf{i} + (\ln y - 8xy)\mathbf{j}$; $\mathbf{x}_0 = (-1, 1)$, $\mathbf{x}_1 = (4, e)$
19. $\mathbf{F}(x, y) = [x \cos(x + y) + \sin(x + y)]\mathbf{i} + [x \cos(x + y)]\mathbf{j}$; $\mathbf{x}_0 = (0, 0)$, $\mathbf{x}_1 = (\pi/6, \pi/3)$
20. $\mathbf{F}(x, y) = \left[\dfrac{1}{x^2} + y^2\right]\mathbf{i} + 2xy\mathbf{j}$; $\mathbf{x}_0 = (1, 4)$, $\mathbf{x}_1 = (3, 2)$
21. $\mathbf{F}(x, y) = =2x(\cos y)\mathbf{i} - x^2(\sin y)\mathbf{j}$; $\mathbf{x}_0 = (0, \pi/2)$, $\mathbf{x}_1 = (\pi/2, 0)$
22. $\mathbf{F}(x, y) = [2xy^3 - 2]\mathbf{i} + [3x^2y^2 + \cos y]\mathbf{j}$; $\mathbf{x}_0 = (1, 0)$, $\mathbf{x}_1 = (0, -\pi)$
23. $\mathbf{F}(x, y) = e^y\mathbf{i} + xe^y\mathbf{j}$; $\mathbf{x}_0 = (0, 0)$, $\mathbf{x}_1 = (5, 7)$
24. $\mathbf{F}(x, y) = (\cosh x)(\cosh y)\mathbf{i} + (\sinh x)(\sinh y)\mathbf{j}$; $\mathbf{x}_0 = (0, 0)$, $\mathbf{x}_1 = (1, 2)$
25. $\mathbf{F}(x, y, z) = y^2z^4\mathbf{i} + 2xyz^4\mathbf{j} + 4xy^2z^3\mathbf{k}$; $\mathbf{x}_0 = (0, 0, 0)$, $\mathbf{x}_1 = (3, 2, 1)$
26. $\mathbf{F}(x, y, z) = [yz + 2]\mathbf{i} + [xz - 3]\mathbf{j} + [xy + 5]\mathbf{k}$; $\mathbf{x}_0 = (2, 1, 2)$, $\mathbf{x}_1 = (-1, 0, 4)$
27. We omitted the proof of Theorem 2, but half of it is not difficult to prove. Show that if each of P, Q, $\dfrac{\partial P}{\partial y}$, and $\dfrac{\partial Q}{\partial x}$ are continuous in an open disk, and there is a smooth function f such that $P(x, y)\mathbf{i} + Q(x, y)\mathbf{j} = \nabla f(x, y)$ in that disk, then

$$\frac{\partial P}{\partial y} = \frac{\partial Q}{\partial x}.$$

[*Hint:* Use the equality of second mixed partials.]

*28. This problem gives an example that shows the warning of Remark 2 on page 307 is necessary. Let the force field **F** be given by

$$\mathbf{F}(x, y) = \frac{y}{x^2 + y^2}\mathbf{i} - \frac{x}{x^2 + y^2}\mathbf{j}.$$

a. Show that **F** is exact.
b. Let C be the unit circle traversed in the counterclockwise direction; calculate $\int_C \mathbf{F}(\mathbf{x}) \cdot \mathbf{dx}$.
c. Explain why the integral computed in the preceding part is not equal to zero. Does your work on this problem contradict Theorem 1 or equation (6)?

29. Show that the work done by a conservative force field as it moves a particle completely around a closed path is zero, provided the path is contained in a region over which the force field is continuous.

*30. If a particle moves around a circle with constant angular speed ω, then the radial force pushing the particle away from the center is called **centrifugal force** and is given by $\mathbf{F}_C = \omega^2|\mathbf{x}|\mathbf{u}$ where $\mathbf{x}$ is the position of the particle and $\mathbf{u}$ is a unit vector pointing away from the center. The **potential**, f_C, of the centrifugal force is defined as the work done in moving the particle from $\mathbf{0}$ to $\mathbf{x}$. Assuming that $f_C(\mathbf{0}) = 0$, show that

$$f_C(\mathbf{x}) = \frac{1}{2}\,\omega^2|\mathbf{x}|^2.$$

ANSWERS TO SELF-QUIZ

I. False ($0.333x\mathbf{i} - 0.667y\mathbf{j}$ is exact.) **II.** False ($2y\mathbf{i} + 3x\mathbf{j}$ is not exact.) **III.** a **IV.** a **V.** a **VI.** c [$y\mathbf{i} - x\mathbf{j}$ is not exact.]

5.4 GREEN'S THEOREM IN THE PLANE

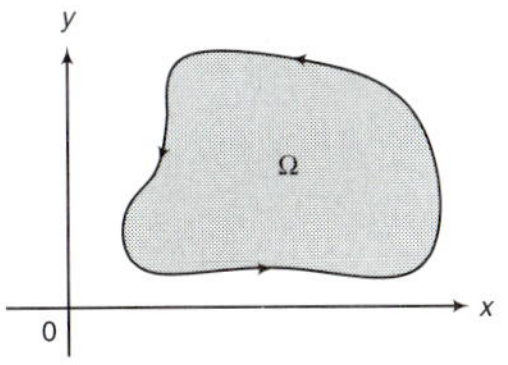

In this section we state a result that gives an important relationship between line integrals and double integrals.

Let Ω be a region in the plane (a typical region is sketched in Figure 1). The curve (indicated by the arrows) that goes around the edge of Ω in the direction that keeps Ω on the left (the *counterclockwise* direction) is called the **boundary of** Ω and is denoted by $\partial\Omega$. Let

$$\mathbf{F}(x, y) = P(x, y)\mathbf{i} + Q(x, y)\mathbf{j}.$$

FIGURE 1
A typical region in $\mathbb{R}^2$

If the curve $\partial\Omega$ is given by

$$\partial\Omega: \quad \mathbf{x}(t) = x(t)\mathbf{i} + y(t)\mathbf{j},$$

we can write

$$\mathbf{F}(\mathbf{x}) \cdot \mathbf{dx} = P\,dx + Q\,dy. \tag{1}$$

We then denote the line integral of **F** around $\partial\Omega$ by

$$\oint_{\partial\Omega} P\,dx + Q\,dy. \tag{2}$$

IMPORTANT NOTATION: The symbol $\oint_{\partial\Omega}$ indicates that $\partial\Omega$ is a closed curve around which we integrate in the counterclockwise direction (the direction of the arrow).

DEFINITION SIMPLE CURVE

A curve C is called **simple** if it does not cross itself. That is, suppose C is given by

$$C: \quad \mathbf{x}(t) = f(t)\mathbf{i} + g(t)\mathbf{j}, \qquad t \in [a, b].$$

Then C is simple if and only if $\mathbf{x}(t_1) \neq \mathbf{x}(t_2)$ whenever $t_1 \neq t_2$ (with the possible exception $t_1 = a$ and $t_2 = b$). ■

This notation is illustrated in Figure 2.

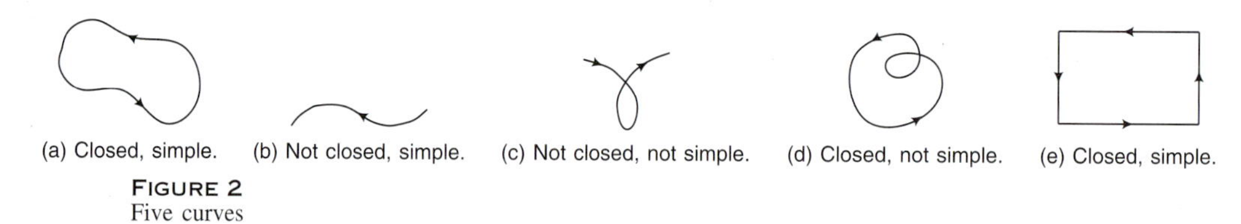

FIGURE 2
Five curves

DEFINITION SIMPLY CONNECTED REGION

A region Ω in the xy plane is called **simply connected** if it has the following property: If C is a simple closed curve contained in Ω, then every point in the region enclosed by C is also in Ω. ■

Intuitively, a region is simply connected if it has no holes. We illustrate this definition in Figure 3.

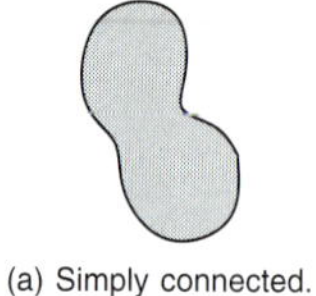
(a) Simply connected.

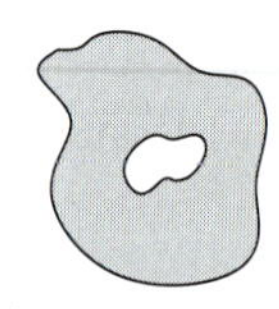
(b) Not simply connected (region has a hole).

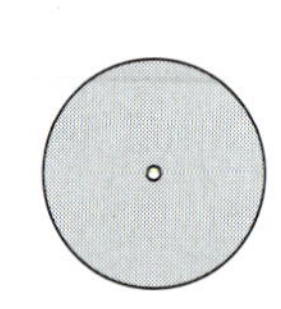
(c) Not simply connected (point is missing in this "punctured disk").

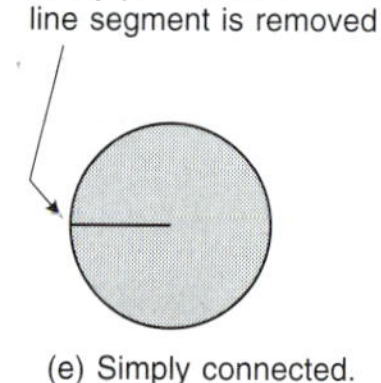

(e) Simply connected.

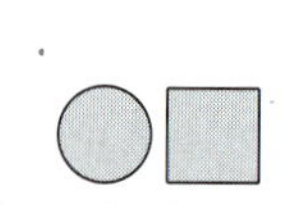
(d) Not a region (not connected).

FIGURE 3
Five sets in the xy-plane

We are now ready to state the principal result of this section.

THEOREM 1 GREEN'S THEOREM IN THE PLANE

Let Ω be a simply connected region in the xy-plane bounded by a piecewise smooth curve $\partial\Omega$. Let P and Q be continuous with continuous first partials in an open disk containing Ω. Then

$$\oint_{\partial\Omega} P\,dx + Q\,dy = \iint_{\Omega} \left(\frac{\partial Q}{\partial x} - \frac{\partial P}{\partial y}\right) dx\,dy. \tag{3}$$
■

REMARK: This theorem shows how the line integral of a function around the boundary of a region is related to a double integral over that region.

PARTIAL PROOF OF GREEN'S THEOREM

We prove Green's theorem in the case in which Ω takes the simple form given in Figure 4. The region Ω can be written

$$\{(x, y)\colon a \le x \le b,\ g_1(x) \le y \le g_2(x)\} \tag{4}$$

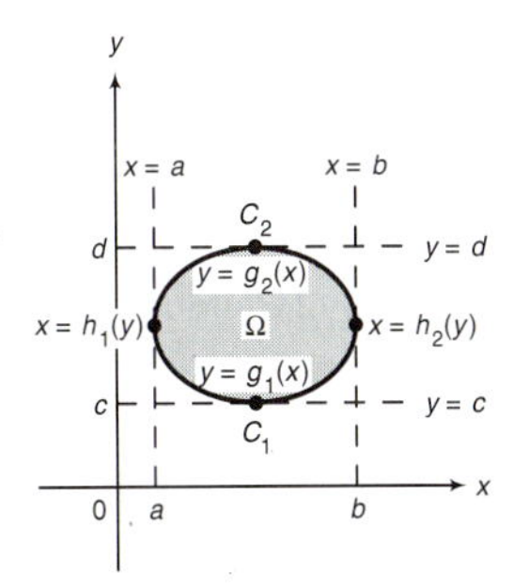

FIGURE 4
The region Ω can be written

$\{(x, y)\colon a \le x \le b,\ g_1(x) \le y \le g_2(x)\}$

and

$\{(x, y)\colon c \le y \le d,\ h_1(y) \le x \le h_2(y)\}$.

and

$$\{(x, y)\colon c \le y \le d,\ h_1(y) \le x \le h_2(y)\}. \tag{5}$$

We first calculate

$$\iint_\Omega \frac{\partial P}{\partial y}\,dx\,dy = \int_a^b \left\{\int_{g_1(x)}^{g_2(x)} \frac{\partial P}{\partial y}\,dy\right\} dx = \int_a^b \left\{P(x, y)\Big|_{y=g_1(x)}^{y=g_2(x)}\right\} dx$$

$$= \int_a^b [P(x, g_2(x)) - P(x, g_1(x))]\,dx. \tag{6}$$

Now

$$\oint_{\partial\Omega} P\,dx = \int_{C_1} P\,dx + \int_{C_2} P\,dx, \tag{7}$$

where C_1 is the graph of $g_1(x)$ from $x = a$ to $x = b$ and C_2 is the graph of $g_2(x)$ from $x = b$ to $x = a$ (note the order). Since on C_1, $y = g_1(x)$, we have

$$\int_{C_1} P(x, y)\,dx = \int_a^b P(x, g_1(x))\,dx. \tag{8}$$

Similarly,

$$\int_{C_2} P(x, y)\,dx = \int_b^a P(x, g_2(x))\,dx = -\int_a^b P(x, g_2(x))\,dx. \tag{9}$$

Thus

$$\oint_{\partial\Omega} P\,dx = \int_a^b P(x, g_1(x))\,dx - \int_a^b P(x, g_2(x))\,dx$$

$$= -\int_a^b [P(x, g_2(x)) - P(x, g_1(x))]\,dx. \tag{10}$$

Comparing (6) and (10), we find that

$$\iint_\Omega -\frac{\partial P}{\partial y}\,dx\,dy = \oint_{\partial\Omega} P\,dx. \tag{11}$$

Similarly, using the representation (5), we have

$$\iint_\Omega \frac{\partial Q}{\partial x}\,dx\,dy = \int_c^d \left\{\int_{h_1(y)}^{h_2(y)} \frac{\partial Q}{\partial x}\,dx\right\} dy$$

$$= \int_c^d [Q(h_2(y), y) - Q(h_1(y), y)]\,dy,$$

which by analogous reasoning yields the equation

$$\iint_\Omega \frac{\partial Q}{\partial x}\,dx\,dy = \oint_{\partial\Omega} Q\,dy. \tag{12}$$

Adding (11) and (12) completes the proof of theorem in the special case that Ω can be written in the form (4) and (5).[†] ■

HISTORICAL NOTE: Green's theorem is named after George Green (1793–1841), a British mathematician and physicist who wrote an essay in 1828 on electricity and magnetism that contained this important theorem. Green was the self-educated son of a baker. His 1828 essay was published for private circulation. It was largely overlooked until it was rediscovered by Lord Kelvin in 1846. The theorem was independently discovered by the Russian mathematician Michel Ostrogradski (1801–1861), and to this day the theorem is known in Russia as *Ostrogradski's theorem.*

EXAMPLE 1

EVALUATING A LINE INTEGRAL USING GREEN'S THEOREM

Evaluate $\oint_{\partial\Omega} xy\,dy + (x-y)\,dy$, where Ω is the rectangle $\{(x, y): 0 \le x \le 1, 1 \le y \le 3\}$.

SOLUTION: $P(x, y) = xy$, $Q(x, y) = x - y$, $\partial Q/\partial x = 1$, and $\partial P/\partial y = x$, so

$$\oint_{\partial\Omega} xy\,dx + (x-y)\,dy = \int_0^1 \int_1^3 (1-x)\,dy\,dx = \int_0^1 \left\{(1-x)y\Big|_1^3\right\} dx$$

$$= \int_0^1 2(1-x)\,dx = 1.$$

EXAMPLE 2

EVALUATING A LINE INTEGRAL USING GREEN'S THEOREM

Evaluate $\oint_C (x^3 + y^3)\,dx + (2y^3 - x^3)\,dy$, where C is the unit circle.

SOLUTION: We first note that $C = \partial\Omega$, where Ω is the unit disk. Next, we have $(\partial Q/\partial x) - (\partial P/\partial y) = -3x^2 - 3y^2 = -3(x^2 + y^2)$. Thus,

$$\oint_C (x^3 + y^3)\,dx + (2y^3 - x^3)\,dy = -3 \iint_{\text{unit disk}} (x^2 + y^2)\,dx\,dy$$

Converting to polar coordinates ↓

$$= -3 \int_0^{2\pi} \int_0^1 r^2 \cdot r\,dr\,d\theta$$

$$= -3 \int_0^{2\pi} \left\{\frac{r^4}{4}\Big|_0^1\right\} d\theta = -\frac{3}{4} \int_0^{2\pi} d\theta$$

$$= -\frac{3\pi}{2}.$$

Green's theorem can be useful for calculating area. Recall that

$$\text{area enclosed by } \Omega = \iint_{\Omega} dA. \tag{13}$$

[†] For a proof of Green's theorem for more general regions, see R. C. Buck and E. F. Buck, *Advanced Calculus*, 3rd ed. (New York: McGraw-Hill, 1978), p. 479.

But by Green's theorem,

FOUR INTEGRALS THAT REPRESENT AREA

$$\iint_{\Omega} dA = \oint_{\partial\Omega} x\,dy = \oint_{\partial\Omega} (-y)\,dx = \frac{1}{2}\oint_{\partial\Omega} [(-y)\,dx + x\,dy] \tag{14}$$

(explain why). Any of the line integrals in (14) can be used to calculate area.

EXAMPLE 3

CALCULATING THE AREA OF AN ELLIPSE USING GREEN'S THEOREM

Use Green's theorem to calculate the area enclosed by the ellipse $(x^2/a^2) + (y^2/b^2) = 1$.

SOLUTION: The ellipse can be written parametrically as

$$\mathbf{x}(t) = a(\cos t)\mathbf{i} + b(\sin t)\mathbf{j}, \qquad 0 \le t \le 2\pi.$$

Then using the first line integral in (14), we obtain

$$\begin{aligned} A = \oint_{\partial\Omega} x\,dy &= \int_0^{2\pi} (a\cos t)\frac{d}{dt}(b\sin t)\,dt \\ &= \int_0^{2\pi} (a\cos t)b\cos t\,dt = ab\int_0^{2\pi} \cos^2 t\,dt \\ &= \frac{ab}{2}\int_0^{2\pi} (1 + \cos 2t)\,dt = \pi ab. \end{aligned}$$

Note how much easier this calculation is than the direct evaluation of $\iint_A dx\,dy$, where A denotes the area enclosed by the ellipse.

PROBLEMS 5.4

SELF-QUIZ

I. Suppose C is the square path from $(0, 0)$ to $(1, 0)$ to $(1, 1)$ to $(0, 1)$ and back to $(0, 0)$; then $\oint_C 3y\,dx - 2x\,dy =$ ________.

a. $\int_0^1\int_0^1 [(-2) - 3]\,dx\,dy$

b. $\int_0^1\int_0^1 [3 - (-2)]\,dx\,dy$

c. $\int_0^3\int_0^{-2} [(-2) - 3]\,dx\,dy$

d. $\int_0^1\int_0^1 [3y - 2x]\,dx\,dy$

II. Suppose D is the unit disk and C is the unit circle (traversed counterclockwise); then $\oint_C 3y^2\,dx - 2x\,dy =$ ________.

a. $\iint_D [(-2x) - 3y^2]\,dA$

b. $\iint_D [6y + (-2)]\,dA$

c. $\iint_D [(-2) - 6y]\,dA$

d. $\iint_D [y^3 - x^2]\,dA$

III. Suppose D is the unit disk and C is the unit circle (traversed counterclockwise); then $\iint_D 0\, dA \neq$ ________.

a. $\oint_C x\, dx + y\, dy$ **b.** $\oint_C y\, dx + x\, dy$

c. $\oint_C x\, dx - y\, dy$ **d.** $\oint_C y\, dx - x\, dy$

IV. Suppose C is the square path from $(0, 0)$ to $(1, 0)$ to $(1, 1)$ to $(0, 1)$ and back to $(0, 0)$; then $\int_0^1 \int_0^1 2x\, dA \neq$ ________.

a. $\oint_C x^2\, dx$

b. $\oint_C x^2\, dy$

c. $\oint_C \sin x\, dx + x^2\, dy$

d. $\oint_C \tan^{-1} x\, dx + (x^2 - \sin y)\, dy$

In Problems 1–15, calculate the value of the line integral by means of Green's theorem.

1. $\oint_{\partial\Omega} 3y\, dx + 5x\, dy$;
$\Omega = \{(x, y)\colon 0 \le x \le 1,\ 0 \le y \le 1\}$

2. $\oint_{\partial\Omega} ay\, dx + bx\, dy$;
$\Omega = \{(x, y)\colon 0 \le x \le A,\ 0 \le y \le B\}$

3. $\oint_{\partial\Omega} e^x \cos y\, dx + e^x \sin y\, dy$; Ω is the region enclosed by the triangle with vertices at $(0, 0)$, $(1, 0)$, and $(0, 1)$.

4. The integral of Problem 3 where Ω is the region enclosed by the triangle with vertices at $(0, 0)$, $(1, 1)$, and $(1, 0)$.

5. The integral of Problem 3 where Ω is the region enclosed by the rectangle with vertices at $(0, 0)$, $(2, 0)$, $(2, 1)$ and $(0, 1)$.

6. $\oint_{\partial\Omega} 2xy\, dx + x^2\, dy$; Ω is the unit disk

7. $\oint_{\partial\Omega} (x^2 + y^2)\, dx - 2xy\, dy$; Ω is the unit disk

8. $\oint_{\partial\Omega} \frac{1}{y}\, dx + \frac{1}{x}\, dy$; Ω is the region bounded by the lines $y = 1$ and $x = 16$ and the curve $y = \sqrt{x}$

9. $\oint_{\partial\Omega} \cos y\, dx + \cos x\, dy$; Ω is the region enclosed by the rectangle $\{(x, y)\colon 0 \le x \le \pi/4,\ 0 \le y \le \pi/3\}$

10. $\oint_{\partial\Omega} x^2 y\, dx - xy^2\, dy$; Ω is the disk $\{(x, y)\colon x^2 + y^2 \le 9\}$

11. $\oint_{\partial\Omega} y \ln x\, dy$; $\Omega = \{(x, y)\colon 1 \le y \le 3, e^y \le x \le e^{y^3}\}$.

12. $\oint_{\partial\Omega} \sqrt{1 + y^2}\, dx$;
$\Omega = \{(x, y)\colon -1 \le y \le 1,\ y^2 \le x \le 1\}$

13. $\oint_{\partial\Omega} ay\, dx + bx\, dy$; Ω is a region satisfying relations of types (4) and (5)

14. $\oint_{\partial\Omega} e^x \sin y\, dx + e^x \cos y\, dy$; Ω is the elliptical region $\{(x, y)\colon (x/a)^2 + (y/b)^2 \le 1\}$

15. $\oint_{\partial\Omega} \frac{-4x}{\sqrt{1 + y^2}}\, dx + \frac{2x^2 y}{(1 + y^2)^{3/2}}\, dy$; Ω is a region satisfying relations of types (4) and (5)

16. Use one of the line integrals in (14) to calculate the area of the circle $\mathbf{x}(t) = a(\cos t)\mathbf{i} + a(\sin t)\mathbf{j}$.

17. Use Green's theorem to calculate the area enclosed by the triangle with vertices at $(0, 0)$, $(5, 2)$, and $(-3, 8)$.

18. Use Green's theorem to calculate the area enclosed by the triangle with vertices at (a_1, b_1), (a_2, b_2), and (a_3, b_3), assuming that the three points are not collinear.

19. Use Green's theorem to calculate the area enclosed by the quadrilateral with vertices at $(0, 0)$, $(2, 1)$, $(-1, 3)$, and $(4, 4)$.

20. Use Green's theorem to calculate the area enclosed by the quadrilateral with vertices at (a_1, b_1), (a_2, b_2), (a_3, b_3), and (a_4, b_4), assuming that no three of these points are collinear and that no point is within the triangle whose vertices are the other three points.

***21.** Let $\partial\Omega$ be the ellipse satisfying $(x/a)^2 + (y/b)^2 = 1$; let $\mathbf{F}(x, y)$ be the vector field $-x\mathbf{i} - y\mathbf{j}$ that, at any point (x, y), points towards the origin. Show that $\oint_{\partial\Omega} \mathbf{F} \cdot \mathbf{T}\, ds = 0$, where $\mathbf{T}$ is the unit tangent vector to $\partial\Omega$ and ds is the differential of arc length (see page 113).

***22.** Let Ω be the disk $\{(x, y): x^2 + y^2 \le a^2\}$. Show that $\oint_{\partial\Omega} \alpha\sqrt{x^2 + y^2}\, dx + \beta\sqrt{x^2 + y^2}\, dy = 0$.

23. Let Ω be the same disk as in the preceding problem; suppose that g is continuously differentiable. Show that

$$\oint_{\partial\Omega} \alpha g(x^2 + y^2)\, dx + \beta g(x^2 + y^2)\, dy = 0.$$

***24.** Use Green's theorem to prove the "if" part of Theorem 5.3.2. Show that if each of P, Q, $\dfrac{\partial P}{\partial y}$, and $\dfrac{\partial Q}{\partial x}$ are continuous in an open disk and $\dfrac{\partial P}{\partial y} = \dfrac{\partial Q}{\partial x}$ then there is a smooth function f such that $P(x, y)\mathbf{i} + Q(x, y)\mathbf{j} = \nabla f(x, y)$ in that disk. [*Hint:* A line integral is independent of path in Ω if and only if the line integral along any closed curve in Ω is zero; use Theorem 5.3.1.]

ANSWERS TO SELF-QUIZ

I. a **II.** c **III.** d **IV.** a

5.5 THE PARAMETRIC REPRESENTATION OF A SURFACE AND SURFACE AREA

In Section 2.1, we saw that a curve in $\mathbb{R}^2$ or $\mathbb{R}^3$ can be represented parametrically in terms of a single variable t. We have

$$\underbrace{\mathbf{f}(t) = x(t)\mathbf{i} + y(t)\mathbf{j},\ a \le t \le b}_{\text{in } \mathbb{R}^2} \qquad \text{or} \qquad \underbrace{\mathbf{f}(t) = x(t)\mathbf{i} + y(t)\mathbf{j} + z(t)\mathbf{k},\ a \le t \le b}_{\text{in } \mathbb{R}^3}$$

where every function of t is continuous. We can think of the function $\mathbf{f}$ in $\mathbb{R}^3$ as a continuous function that takes the interval $I = [a, b]$ and "deforms" it into a curve C in $\mathbb{R}^3$ (see Figure 1).

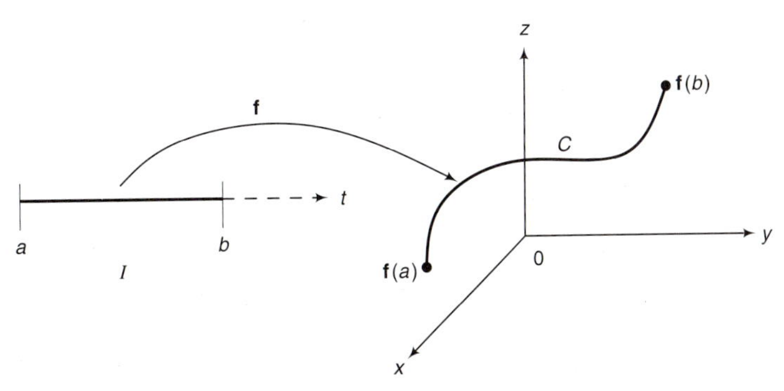

FIGURE 1
As t varies over I, the curve C is traced out in $\mathbb{R}^3$.

A surface in $\mathbb{R}^3$ can be represented parametrically in terms of two parameters.

DEFINITION PARAMETRIC SURFACE

Let $x = x(u, v)$, $y = y(u, v)$, and $z = z(u, v)$ be continuous. Then the set of points S given by

$$\mathbf{r}(u, v) = x(u, v)\mathbf{i} + y(u, v)\mathbf{j} + z(u, v)\mathbf{k} \tag{1}$$

where (u, v) ranges over a region Ω in $\mathbb{R}^2$ is called a **parametric surface**, or simply a **surface**, in $\mathbb{R}^3$. Equation (1) is a **parametric representation** of the surface. ■

The function $\mathbf{r}$ in equation (1) is a function with domain Ω in $\mathbb{R}^2$ and range in $\mathbb{R}^3$. We can think of this function as deforming a set of points Ω in $\mathbb{R}^2$ into a set of points S in $\mathbb{R}^3$. This is illustrated in Figure 2.

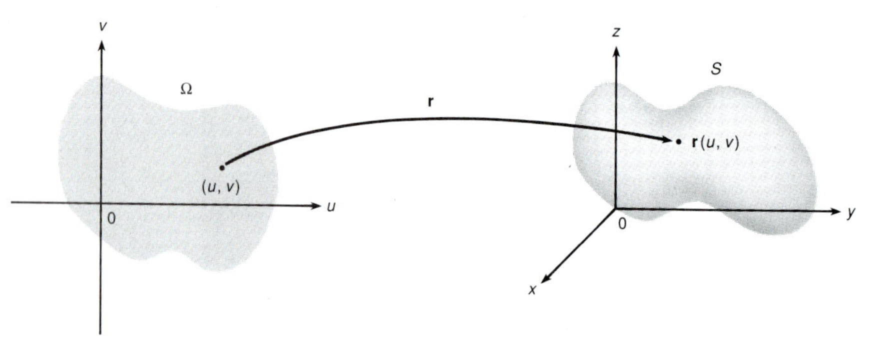

FIGURE 2
As (u, v) varies over Ω, the points in the surface S are obtained.

It may be helpful to think of the function $\mathbf{r}$ as deforming a rectangle R into a "magic carpet," as in Figure 3.

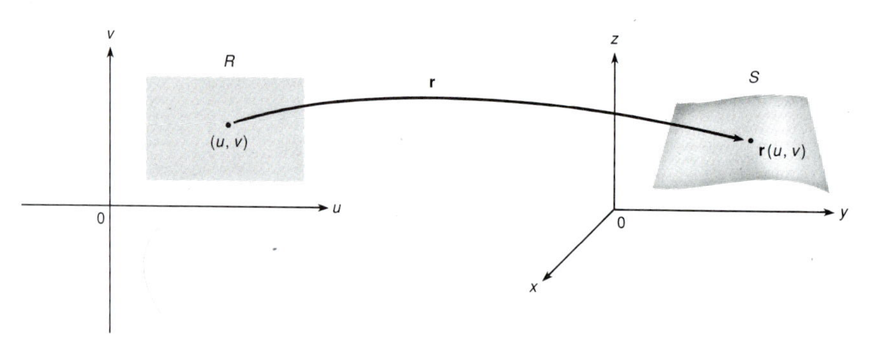

FIGURE 3
As (u, v) varies over the rectangle R, the points in a "magic carpet" are obtained.

EXAMPLE 1

REPRESENTING THE SURFACE $z = f(x, y)$ PARAMETRICALLY

In Section 3.1 (page 139), we said that the graph of the continuous function $z = f(x, y)$ is a surface in $\mathbb{R}^3$. We can write this surface in the form (1) as follows: Let

$$x(u, v) = u \qquad y(u, v) = v \qquad \text{and} \qquad z(u, v) = f(u, v) \tag{2}$$

Then

$$\mathbf{r}(u, v) = u\mathbf{i} + v\mathbf{j} + f(u, v)\mathbf{k}$$

yields the surface which is the graph of $z = f(x, y)$.

EXAMPLE 2 PARAMETRIC REPRESENTATION OF A HYPERBOLIC PARABOLOID

Find a parametric representation of the hyperbolic paraboloid

$$z = \frac{x^2}{4} - \frac{y^2}{9}.$$

SOLUTION: Using (2), we set

$$x = u \qquad y = v \qquad \text{and} \qquad z = \frac{u^2}{4} - \frac{v^2}{9}.$$

Then the surface can be written

$$\mathbf{r}(u, v) = u\mathbf{i} + v\mathbf{j} + \left(\frac{u^2}{4} - \frac{v^2}{9}\right)\mathbf{k} \qquad \text{for } (u, v) \in \mathbb{R}^2.$$

EXAMPLE 3 PARAMETRIC REPRESENTATION OF A PLANE THROUGH THE ORIGIN

Let Π be a plane passing through the origin. If $\mathbf{x}$ and $\mathbf{y}$ are two noncollinear vectors lying on Π, then every other vector $\mathbf{w}$ on Π can be written in the form

$$\mathbf{w} = \mathbf{r}(u, v) = u\mathbf{x} + v\mathbf{y}. \tag{3}$$

The representation (3) is called a **parametric representation of a plane**. You are asked to prove that (3) holds in Problems 40–42.

EXAMPLE 4 PARAMETRIC REPRESENTATION OF THE PLANE $x + 2y + 3z = 0$

Find a parametric representation in the form (3) of the plane

$$\Pi\colon x + 2y + 3z = 0. \tag{4}$$

SOLUTION: From (4), we have

$$z = -\frac{1}{3}(x + 2y).$$

Then every vector on Π has the form

$$\left(x, y, -\frac{1}{3}(x + 2y)\right). \tag{5}$$

Two noncollinear vectors on Π are found by setting $x = 1$, $y = 0$, and $x = 0$, $y = 1$:

$$\mathbf{x} = \left(1, 0, -\frac{1}{3}\right) \qquad \text{and} \qquad \mathbf{y} = \left(0, 1, -\frac{2}{3}\right).$$

If u and v are real numbers, then

$$u\mathbf{x} + v\mathbf{y} = \left(u, v, -\frac{1}{3}(u + 2v)\right),$$

which is of the form (5), so $u\mathbf{x} + v\mathbf{y}$ is on Π. Conversely, if $\mathbf{w} = (\alpha, \beta, \gamma)$ is on Π, then $\gamma = -\frac{1}{3}(\alpha + 2\beta)$, so

$$\mathbf{w} = (\alpha, \beta, \gamma) = \left(\alpha, \beta, -\frac{1}{3}(\alpha + 2\beta)\right)$$
$$= \alpha\left(1, 0, -\frac{1}{3}\right) + \beta\left(0, 1, -\frac{2}{3}\right) = \alpha\mathbf{x} + \beta\mathbf{y}.$$

Thus Π consists precisely of those vectors that can be written in the form $u\mathbf{x} + v\mathbf{y}$.

EXAMPLE 5

PARAMETRIC REPRESENTATION OF A SPHERE

Recall that if $P = (x, y, z)$ is a point or vector in $\mathbb{R}^3$, then the Cartesian coordinates of P can be written in terms of the spherical coordinates of P as follows (see page 68):

$$x = \rho \sin \varphi \cos \theta$$
$$y = \rho \sin \varphi \sin \theta \qquad 0 \le \varphi \le \pi,\ 0 \le \theta \le 2\pi$$
$$z = \rho \cos \varphi$$

If $\rho = a$ is fixed, we have

$$x^2 + y^2 + z^2 = a^2. \qquad \text{Equation (8) on page 70}$$

This is the equation of a sphere centered at the origin with radius a. Thus the parametric representation of this sphere is

$$\mathbf{r}(\varphi, \theta) = a \sin \varphi \cos \theta\mathbf{i} + a \sin \varphi \sin \theta\mathbf{j} + a \cos \varphi\mathbf{k},$$
$$0 \le \varphi \le \pi,\ 0 \le \theta \le 2\pi \tag{6}$$

Note that here the parameters φ and θ lie in the bounded rectangle given by (6).

EXAMPLE 6

PARAMETRIC REPRESENTATION OF A CYLINDER

Consider the right circular cylinder sketched in Figure 4. In order to write it in parametric form, it is convenient to parametrize the side, top, and bottom separately.

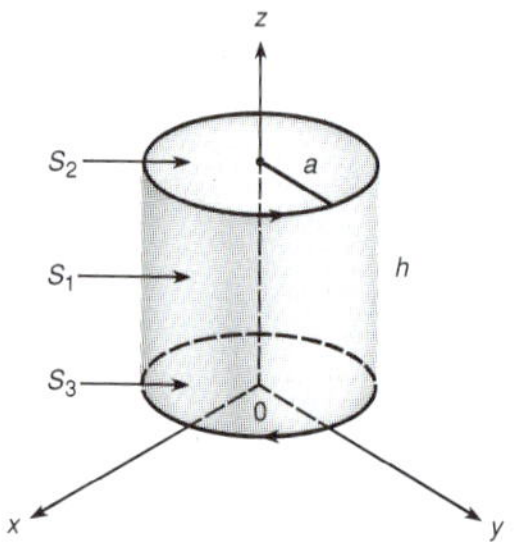

FIGURE 4
Parametrizable subsurfaces of a cylinder

A right circular cylinder with radius a and height h consists of three surfaces: the side S_1, the top S_2, and the bottom S_3.

The Side S_1 We use cylindrical coordinates with $r = a$, a constant (see equation (2) on page 68):

$$x = a \cos \theta, \qquad y = a \sin \theta, \qquad z = z,$$

so

$$\mathbf{r}(\theta, z) = a \cos \theta\mathbf{i} + a \sin \theta\mathbf{j} + z\mathbf{k}, \qquad 0 \le \theta \le 2\pi,\ 0 \le z \le h.$$

The Top S_2 Here $z = h$ but both r and θ vary:

$$\mathbf{r}(r, \theta) = r \cos \theta\mathbf{i} + r \sin \theta\mathbf{j} + h\mathbf{k}, \qquad 0 \le \theta \le 2\pi,\ 0 \le r \le a$$

The Bottom S_3 Now $z = 0$ and

$$\mathbf{r}(r, \theta) = r \cos \theta\mathbf{i} + r \sin \theta\mathbf{j} + 0\mathbf{k}, \qquad 0 \le \theta \le 2\pi,\ 0 \le r \le a.$$

REMARK: If it is convenient, as here, to divide the surface into different ''subsurfaces'' and parametrize each one separately, then each of the parametrizable subsurfaces is called a **surface patch**.

COMPUTING SURFACE AREA

Consider the surface

$$S\colon \mathbf{r}(u, v) = x(u, v)\mathbf{i} + y(u, v)\mathbf{j} + z(u, v)\mathbf{k}. \tag{7}$$

We assume that the six partial derivatives

$$\frac{\partial x}{\partial u}, \frac{\partial x}{\partial v}, \frac{\partial y}{\partial u}, \frac{\partial y}{\partial v}, \frac{\partial z}{\partial u}, \text{ and } \frac{\partial z}{\partial v} \tag{8}$$

exist and are continuous for (u, v) in a region Ω.

We assume further that Ω is a bounded region and, as in Section 4.1, we partition Ω into nm subrectangles, denoted by R_{ij}. A typical subrectangle is sketched in Figure 5(a). The subregion S_{ij} in Figure 5(b) is the image of R_{ij} under the function $\mathbf{r}$. If we know the area of each S_{ij}, we can compute the area of S by adding up the areas of the subsurfaces. If we can only approximate the areas of the subsurfaces, then we can obtain the area of S by adding the approximate values and taking a limit as Δu and Δv tend to zero. This gives us a double integral as in our calculation of volume in Section 4.1.

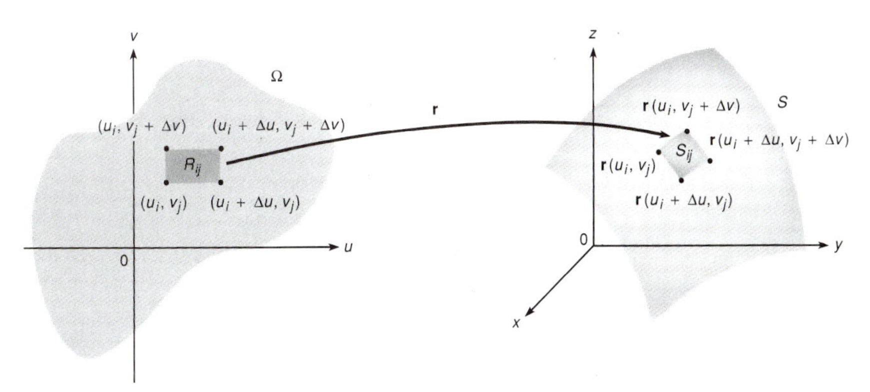

FIGURE 5
The subregion S_{ij} is the image of the rectangle R_{ij} under the function $\mathbf{r}$.

$\mathbf{r}$ takes the region Ω in $\mathbb{R}^2$ into the surface S in $\mathbb{R}^3$. $\mathbf{r}$ takes the subrectangle R_{ij} of Ω into the subsurface S_{ij} in S.

Consider the point $\mathbf{r}(u_i, v_j)$ on the surface S. If v_j is held fixed and u is allowed to vary, then $\mathbf{r}(u, v_j)$ is a curve in $\mathbb{R}^3$ that passes through the point (u_i, v_j). A tangent vector to this curve is, from Chapter 2, given by the derivative of $\mathbf{r}$ with respect to u (since v is fixed). That is,

a tangent vector to the curve $\mathbf{r}(u, v_j)$ at the point (u_i, v_j)

$$= \mathbf{r}_u(u_i, v_j) = \frac{\partial x(u_i, v_j)}{\partial u}\mathbf{i} + \frac{\partial y(u_i, v_j)}{\partial u}\mathbf{j} + \frac{\partial z(u_i, v_j)}{\partial u}\mathbf{k}. \tag{9}$$

Similarly, if we hold $u = u_i$ fixed and we let v vary, we obtain

a tangent vector to the curve $\mathbf{r}(u_i, v)$ at the point (u_i, v_j)

$$= \mathbf{r}_v(u_i, v_j) = \frac{\partial x(u_i, v_j)}{\partial v}\mathbf{i} + \frac{\partial y(u_i, v_j)}{\partial v}\mathbf{j} + \frac{\partial z(u_i, v_j)}{\partial v}\mathbf{k}.$$

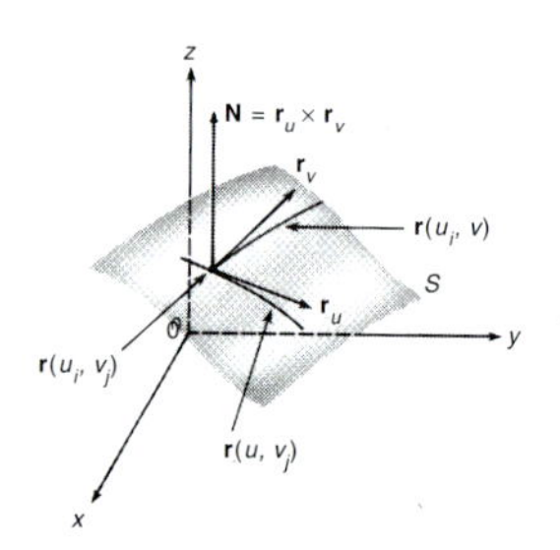

FIGURE 6
Two tangent vectors to the surface S at the point $\mathbf{r}(u_i, v_j)$

$\mathbf{r}_u$ is tangent to the curve $\mathbf{r}(u, v_j)$; $\mathbf{r}_v$ is tangent to the curve $\mathbf{r}(u_i, v)$ and $\mathbf{N} = \mathbf{r}_u \times \mathbf{r}_v$ is orthogonal to both tangent vectors.

We sketch these two tangent vectors in Figure 6. We now assume that

$$\mathbf{r}_u \times \mathbf{r}_v \neq \mathbf{0}.$$

This means that $\mathbf{r}_u$ and $\mathbf{r}_v$ determine a plane, called the **tangent plane** to the surface S at the point $\mathbf{r}(u_i, v_j)$. A normal vector to this plane is given by (see Section 1.7)

$$\mathbf{N} = \mathbf{r}_u(u_i, v_j) \times \mathbf{r}_v(u_i, v_j).$$

Suppose that Δu and Δv are small. We have

$$\frac{\mathbf{r}(u_i + \Delta u, v_j) - \mathbf{r}(u_i, v_j)}{\Delta u} = \frac{x(u_i + \Delta u, v_j) - x(u_i, v_j)}{\Delta u}\mathbf{i} + \frac{y(u_i + \Delta u, v_j) - y(u_i, v_j)}{\Delta u}\mathbf{j} + \frac{z(u_i + \Delta u, v_j) - z(u_i, v_j)}{\Delta u}\mathbf{k}.$$

But by the definition of $\dfrac{\partial x}{\partial u}$, if Δu is small, then

$$\frac{x(u_i + \Delta u, v_j) - x(u_i, v_j)}{\Delta u} \approx \frac{\partial x(u_i, v_j)}{\partial u}$$

or

$$x(u_i + \Delta u, v_j) - x(u_i, v_j) \approx \frac{\partial x(u_i, v_j)}{\partial u}\Delta u.$$

Then, using similar approximations for the two other terms, we obtain

$$\mathbf{r}(u_i + \Delta u, v_j) - \mathbf{r}(u_i, v_j) \approx \mathbf{r}_u(u_i, v_j)\,\Delta u.$$

Thus, one side of the subsurface S_{ij} is approximated by the vector from $\mathbf{r}(u_i, v_j)$ to $\mathbf{r}(u_i + \Delta u, v_j)$, which in turn is approximated by the tangent vector $\mathbf{r}_u(u_i, v_j)\,\Delta u$. Similarly, a second side of S_{ij} is approximated by the tangent vector $\mathbf{r}_v(u_i, v_j)\,\Delta v$. Then, if Δu and Δv are small, we have

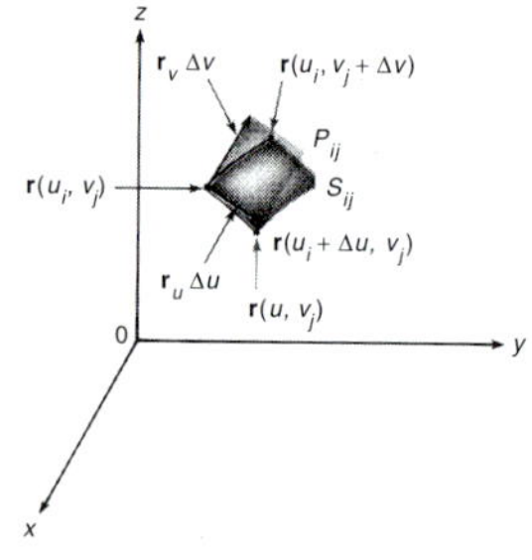

FIGURE 7
The area of S_{ij} is approximated by the area of the parallelogram with sides $\mathbf{r}_u\,\Delta u$ and $\mathbf{r}_v\,\Delta v$.

area of $S_{ij} \approx$ area of parallelogram with sides $\mathbf{r}_u\,\Delta u$ and $\mathbf{r}_v\,\Delta v$

see page 48

$$= |\mathbf{r}_u\,\Delta u \times \mathbf{r}_v\,\Delta v|$$

or

$$\text{area of } S_{ij} \approx |\mathbf{r}_u(u_i, v_j) \times \mathbf{r}_v(u_i, v_j)|\,\Delta u\,\Delta v$$

(see Figure 7). Thus

$$\text{area of } S \approx \sum_{i=1}^{n}\sum_{j=1}^{n} |\mathbf{r}_u(u_i, v_j) \times \mathbf{r}_v(u_i, v_j)|\,\Delta u\,\Delta v$$

and, taking a limit as $\sqrt{\Delta u^2 + \Delta v^2} \to 0$, we obtain

$$\text{area of } S = \iint_\Omega |\mathbf{r}_u(u, v) \times \mathbf{r}_v(u, v)|\, du\, dv. \tag{10}$$

Before continuing, we stress that the computation that led to (10) did not prove anything. But it does make it plausible to use (10) as a definition of surface area. In order to do so, we must make two assumptions. First, we assume that $\mathbf{r}_u \times \mathbf{r}_v \neq \mathbf{0}$.

DEFINITION SMOOTH SURFACE

The surface given by (7) is called **smooth** over the region Ω if

i. all six partial derivatives (8) exist and are continuous for (u, v) in Ω, and

ii. $\mathbf{N}(u, v) = \mathbf{r}_u(u, v) \times \mathbf{r}_v(u, v) \neq \mathbf{0}$ for (u, v) in Ω. ■

We also require that $\mathbf{r}$ be **one-to-one** over Ω (so that we do not count some subsurfaces more than once). That is,

if $(u_1, v_1) \neq (u_2, v_2)$, then $\mathbf{r}(u_1, v_1) \neq \mathbf{r}(u_2, v_2)$.

DEFINITION SURFACE AREA

Let $\mathbf{r}$ be a surface that is both smooth and one-to-one over a region Ω. Then the **surface area** σ of $\mathbf{r}$ over Ω is given by

$$\text{surface area} = \sigma = \iint_\Omega |\mathbf{N}(u, v)|\, du\, dv = \iint_\Omega |\mathbf{r}_u(u, v) \times \mathbf{r}_v(u, v)|\, du\, dv. \tag{11}$$

■

EXAMPLE 7 THE SURFACE AREA OF A SPHERE

Compute the surface area of the sphere

$$x^2 + y^2 + z^2 = a^2.$$

SOLUTION: We write the sphere parametrically, using u and v instead of ϕ and θ (see Example 4):

$$\mathbf{r}(u, v) = a \sin u \cos v\mathbf{i} + a \sin u \sin v\mathbf{j} + a \cos u\mathbf{k},$$
$$0 \le u \le \pi,\ 0 \le v \le 2\pi.$$

Then

$$\mathbf{r}_u = a \cos u \cos v\mathbf{i} + a \cos u \sin v\mathbf{j} - a \sin u\mathbf{k}$$
$$\mathbf{r}_v = -a \sin u \sin v\mathbf{i} + a \sin u \cos v\mathbf{j} + 0\mathbf{k}$$

$$\mathbf{N}(u, v) = \mathbf{r}_u \times \mathbf{r}_v = \begin{vmatrix} \mathbf{i} & \mathbf{j} & \mathbf{k} \\ a \cos u \cos v & a \cos u \sin v & -a \sin u \\ -a \sin u \sin v & a \sin u \cos v & 0 \end{vmatrix}$$

$$= a^2[\sin^2 u \cos v\mathbf{i} + \sin^2 u \sin v\mathbf{j} + \sin u \cos u\overbrace{(\sin^2 v + \cos^2 v)}^{=1}\mathbf{k}]$$

so

$$\begin{aligned}|\mathbf{N}(u, v)| &= |\mathbf{r}_u \times \mathbf{r}_v| \\ &= a^2[\sin^4 u \cos^2 v + \sin^4 u \sin^2 v + \sin^2 u \cos^2 u]^{1/2} \\ &= a^2[\sin^4 u + \sin^2 u \cos^2 u]^{1/2} \\ &= a^2[\sin^2 u(\sin^2 u + \cos^2 u)]^{1/2} \\ &= a^2 \sin u.\end{aligned}$$

Thus,

$$\begin{aligned}\sigma &= \iint_{\Omega} a^2 \sin u\, du\, dv = a^2 \int_0^{2\pi} \int_0^{\pi} \sin u\, du\, dv = a^2 \int_0^{2\pi} \left(-\cos u \Big|_0^{\pi}\right) dv \\ &= 2a^2 \int_0^{2\pi} dv = 2a^2(2\pi) = 4\pi a^2.\end{aligned}$$

EXAMPLE 8

THE LATERAL SURFACE AREA OF A RIGHT CIRCULAR CONE

Compute the lateral surface area of a right circular cone with radius a and height h.

SOLUTION: We place the cone with its vertex at the origin and its axis along the z-axis, as in Figure 8. By the **lateral** surface area, we mean the area of the side, not the circle on the top (which has area πa^2). In Figure 8, we drew the **slant height** $s = \sqrt{h^2 + a^2}$ and the vertex **semiangle** α. Consider the cross-section of the cone having a slant height of v. The radius of the circle is $v \sin \alpha$, so a parametrization of the cross-sectional circle is

$$v \sin \alpha(\cos u\mathbf{i} + \sin u\mathbf{j}), \qquad 0 \le u \le 2\pi.$$

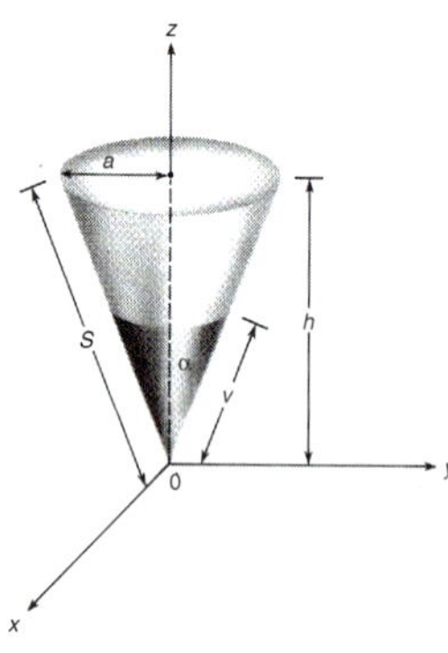

FIGURE 8
A right circular cone with radius a and height h

The z-value corresponding to a slant height v is $v \cos \alpha$. Remember that α is fixed: $\alpha = \tan^{-1} \dfrac{a}{h}$. Then a parametrization of the cone is

$$\begin{aligned}\mathbf{r}(u, v) &= v \cos u \sin \alpha\mathbf{i} + v \sin u \sin \alpha\mathbf{j} + v \cos \alpha\mathbf{k}, \\ &\quad 0 \le u \le 2\pi,\ 0 \le v \le s = \sqrt{a^2 + h^2}.\end{aligned}$$

We compute

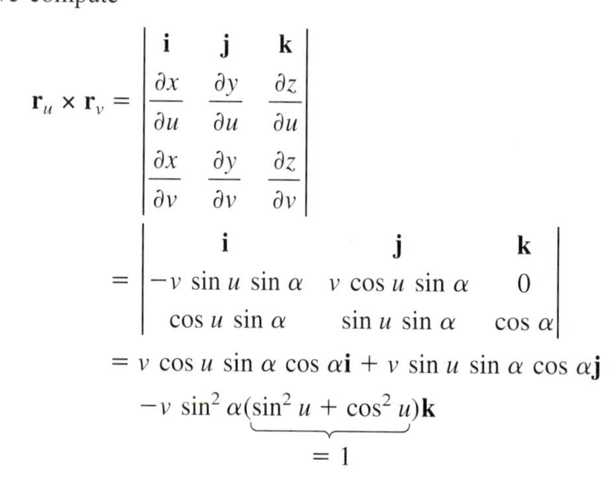

$$\begin{aligned}\mathbf{r}_u \times \mathbf{r}_v &= \begin{vmatrix} \mathbf{i} & \mathbf{j} & \mathbf{k} \\ \dfrac{\partial x}{\partial u} & \dfrac{\partial y}{\partial u} & \dfrac{\partial z}{\partial u} \\ \dfrac{\partial x}{\partial v} & \dfrac{\partial y}{\partial v} & \dfrac{\partial z}{\partial v} \end{vmatrix} \\ &= \begin{vmatrix} \mathbf{i} & \mathbf{j} & \mathbf{k} \\ -v \sin u \sin \alpha & v \cos u \sin \alpha & 0 \\ \cos u \sin \alpha & \sin u \sin \alpha & \cos \alpha \end{vmatrix} \\ &= v \cos u \sin \alpha \cos \alpha\mathbf{i} + v \sin u \sin \alpha \cos \alpha\mathbf{j} \\ &\quad - v \sin^2 \alpha \underbrace{(\sin^2 u + \cos^2 u)}_{=1}\mathbf{k}\end{aligned}$$

so

$$|\mathbf{r}_u \times \mathbf{r}_v| = [v^2 \sin^2 \alpha \cos^2 \alpha \overbrace{(\cos^2 u + \sin^2 u)}^{=1} + v^2 \sin^4 \alpha]^{1/2}$$
$$= [v^2 \sin^2 \alpha(\cos^2 \alpha + \sin^2 \alpha)]^{1/2} = v \sin \alpha.$$

Thus

$$\sigma = \text{lateral surface of the cone} = \int_0^s \int_0^{2\pi} v \sin \alpha \, du \, dv$$
$$= 2\pi \sin \alpha \int_0^s v \, dv = \pi s^2 \sin \alpha.$$

But, $\sin \alpha = \dfrac{a}{s}$. Thus

$$\sigma = \pi s^2\left(\frac{a}{s}\right) = \pi a s = \pi a\sqrt{a^2 + h^2}.$$

SURFACE AREA OF THE SURFACE $z = f(x, y)$

If the surface is given in the form $z = f(x, y)$ for (x, y) in Ω, then, from Example 1, the surface is parametrized by

$$\mathbf{r}(u, v) = u\mathbf{i} + v\mathbf{j} + f(u, v)\mathbf{k}, \qquad (u, v) \in \Omega.$$

We compute

$$\mathbf{r}_u \times \mathbf{r}_v = \begin{vmatrix} \mathbf{i} & \mathbf{j} & \mathbf{k} \\ \dfrac{\partial x}{\partial u} & \dfrac{\partial y}{\partial u} & \dfrac{\partial z}{\partial u} \\ \dfrac{\partial x}{\partial v} & \dfrac{\partial y}{\partial v} & \dfrac{\partial z}{\partial v} \end{vmatrix} = \begin{vmatrix} \mathbf{i} & \mathbf{j} & \mathbf{k} \\ 1 & 0 & \dfrac{\partial f}{\partial u} \\ 0 & 1 & \dfrac{\partial f}{\partial v} \end{vmatrix} = -\frac{\partial f}{\partial u}\mathbf{i} - \frac{\partial f}{\partial v}\mathbf{j} + \mathbf{k}$$

so

$$|\mathbf{N}(u, v)| = |\mathbf{r}_u \times \mathbf{r}_v| = \sqrt{1 + \left(\frac{\partial f}{\partial u}\right)^2 + \left(\frac{\partial f}{\partial v}\right)^2}.$$

Reverting to the original parameters $x = u$ and $y = v$, we have

LATERAL SURFACE AREA OF THE SURFACE $z = f(x, y)$ OVER A REGION Ω

If $\dfrac{\partial f}{\partial x}$ and $\dfrac{\partial f}{\partial y}$ are continuous, then

$$\text{surface area} = \sigma = \iint_\Omega \sqrt{1 + \left(\frac{\partial f}{\partial x}\right)^2 + \left(\frac{\partial f}{\partial y}\right)^2} \, dx \, dy. \tag{12}$$

EXAMPLE 9

COMPUTING THE SURFACE AREA OF PART OF A CIRCULAR PARABOLOID

Find the lateral surface area of the circular paraboloid $z = x^2 + y^2$ between the xy-plane and the plane $z = 9$.

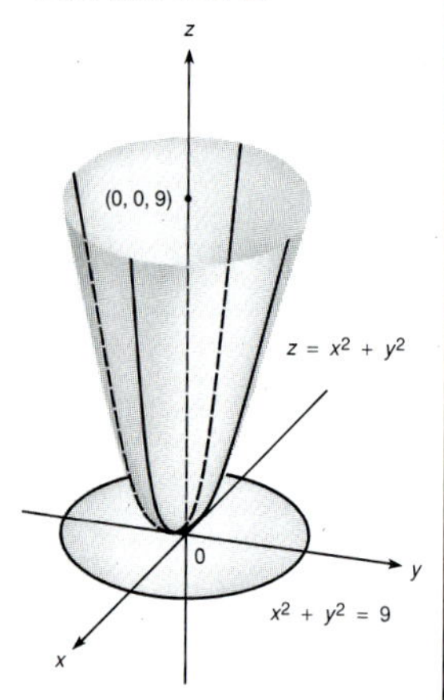

FIGURE 9
The circular paraboloid $z = x^2 + y^2$ for $0 \le z \le 9$

SOLUTION: The surface area requested is sketched in Figure 9. The region Ω is the disk $x^2 + y^2 \le 9$. We have

$$f_x = 2x \quad \text{and} \quad f_y = 2y, \quad \text{so} \quad \sigma = \iint_\Omega \sqrt{1 + 4x^2 + 4y^2}\, dA.$$

Clearly, this problem calls for the use of polar coordinates. We have

$$\sigma = \int_0^{2\pi} \int_0^3 \sqrt{1 + 4r^2}\, r\, dr\, d\theta = \int_0^{2\pi} \left\{ \frac{1}{12}(1 + 4r^2)^{3/2} \bigg|_0^3 \right\} d\theta$$

$$= \frac{\pi}{6}(37^{3/2} - 1) \approx 117.3$$

EXAMPLE 10

OBTAINING A DOUBLE INTEGRAL THAT REPRESENTS SURFACE AREA

Calculate the area of the part of the surface $z = x^3 + y^4$ that lies over the square $\{(x, y)\colon 0 \le x \le 1, 0 \le y \le 1\}$.

SOLUTION: $f_x = 3x^2$ and $f_y = 4y^3$, so $\sigma = \int_0^1 \int_0^1 \sqrt{1 + 9x^4 + 16y^6}\, dx\, dy.$

However, this is as far as we can go unless we resort to numerical techniques to approximate this double integral. As with ordinary definite integrals, many double integrals cannot be integrated in terms of functions that we know. However, there are a great number of techniques for approximating a double integral numerically that parallel the trapezoidal rule and Simpson's rule.† A computer-generated sketch of the solid whose surface area we seek is given in Figure 10.

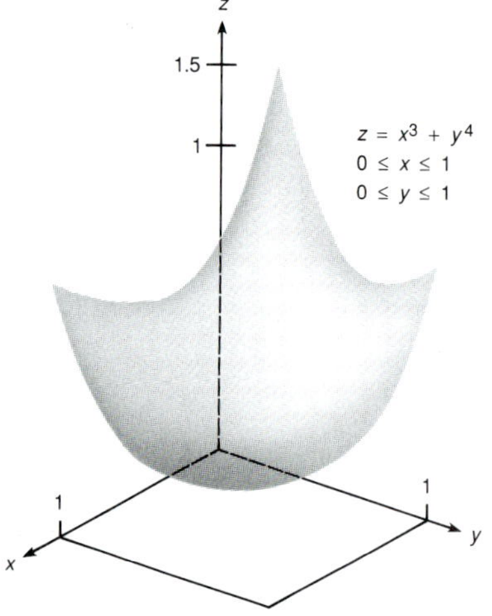

FIGURE 10
The solid below $z = x^3 + y^4$ and over the square $0 \le x \le 1$, $0 \le y \le 1$

† See *Calculus* or *Calculus of One Variable*, Section 7.9.

PROBLEMS 5.5

SELF-QUIZ

I. Which of the following (if any) are parametrizations of the plane $2x - y + 3z = 0$?

a. $\mathbf{r}(u, v) = 2\mathbf{i} - \mathbf{j} + 3\mathbf{k}$
b. $\mathbf{r}(u, v) = 2u\mathbf{i} - v\mathbf{j} + 3(u + v)\mathbf{k}$
c. $\mathbf{r}(u, v) = u\mathbf{i} + v\mathbf{j} + \frac{1}{3}(v - 2u)\mathbf{k}$
d. $\mathbf{r}(u, v) = u\mathbf{i} + (2u + 3v)\mathbf{j} + v\mathbf{k}$
e. $\mathbf{w} = u(0, 3, 1) + v(1, 2, 0)$
f. $\mathbf{w} = u(-1, 1, 1) + v(-2, -1, 1)$

II. The area of that part of the surface $z = 5 - x + y$ which lies over the region $\Omega = \{(x, y): 0 \le x \le 2, 0 \le y \le 3\}$ is computed by ________.

a. $\int_0^3 \int_0^2 \sqrt{1 + (-1)^2 + (1)^2}\, dx\, dy$
b. $\int_0^3 \int_0^2 \sqrt{1 + (-x)^2 + (y)^2}\, dx\, dy$
c. $\int_0^2 \int_0^3 \sqrt{1 + (-x)^2 + (y)^2}\, dx\, dy$
d. $\int_0^2 \int_0^3 \sqrt{1 - (x)^2 + (y)^2}\, dx\, dy$

III. If f is a constant function of (x, y) and if Ω is a simple region in the xy-plane, then the area on the surface $z = f(x, y)$ above Ω is ________ the area of the region Ω.

a. greater than or equal to
b. equal to
c. less than
d. unrelated to (i.e., can be larger than, or smaller than, or the same as)

IV. If f is a smooth function of (x, y) and if Ω is a simple region in the xy-plane, then the area on the surface $z = f(x, y)$ above Ω is ________ the area of the region Ω.

a. greater than or equal to
b. equal to
c. less than
d. unrelated to (i.e., can be larger than, smaller than, or the same as)

In Problems 1–10, parametrize the given surface.

1. the plane $2x + 3y - z = 0$
2. the plane $x - 2y + 5z = 0$
3. the hemisphere $x^2 + y^2 + z^2 = 4$, $z \ge 0$
4. the sphere $(x - 1)^2 + (y + 1)^2 + (z - 2)^2 = 16$
5. the paraboloid $x^2 - y^2 - 4z = 0$; $0 \le x \le 1$, $2 \le y \le 3$
6. the cone $z = \sqrt{x^2 + y^2}$ lying above the disk $x^2 + y^2 \le 9$
7. the part of the plane $z = 4x - y$ that lies inside the cylinder $x^2 + y^2 = 1$
8. the part of the cylinder $x^2 + y^2 = 16$ that lies between the planes $z = 0$ and $z = y + 1$
9. the part of the unit sphere that lies above the cone $z = \sqrt{x^2 + y^2}$
***10.** the **torus** T obtained by revolving a circle in the xz-plane with center at $(b, 0, 0)$ and radius $a < b$ about the z-axis.

In Problems 11–20, find the area of that part of the given surface which lies over the specified region.

11. $z = x + 2y$; $\Omega = \{(x, y): 0 \le x \le y, 0 \le y \le 2\}$
12. $z = 4x + 7y$; Ω is the region between $y = x^2$ and $y = x^5$
13. $z = ax + by$; Ω is upper half of unit circle
14. $z = y^2$; $\Omega = \{(x, y): 0 \le x \le 2, 0 \le y \le 4\}$
***15.** $z = 3 + x^{2/3}$; $\Omega = \{(x, y): -1 \le x \le 1, 0 \le y \le 2\}$
16. $z = (x^4/4) + 1/(8x^2)$; $\Omega = \{(x, y): 1 \le x \le 2, 0 \le y \le 5\}$
17. $z = \frac{1}{3}(y^2 + 2)^{3/2}$; $\Omega = \{(x, y): -4 \le x \le 7, 0 \le y \le 3\}$
18. $z = 2 \ln(1 + y)$; $\Omega = \{(x, y): 0 \le x \le 2, 0 \le y \le 1\}$
***19.** $(z + 1)^2 = 4x^3$; $\Omega = \{(x, y): 0 \le x \le 1, 0 \le y \le 2\}$
***20.** $y^2 + z^2 = 9$; $\Omega = \{(x, y): 0 \le x \le 1, 0 \le y \le 2\}$

In Problems 21–24, compute the surface area of each surface.

21. the surface in Problem 7.
22. the surface in Problem 6.
23. the surface in Problem 9.
***24.** the torus in Problem 10.
***25.** Calculate the lateral surface area of the cylinder $y^{2/3} + z^{2/3} = 1$ for x in the interval $[0, 2]$.
26. Find the surface area of the hemisphere $x^2 + y^2 + z^2 = a^2$, $z \ge 0$.

***27.** Find the surface area of the part of the sphere $x^2 + y^2 + z^2 = a^2$ that is also inside the cylinder $x^2 + y^2 = ay$.

***28.** Find the area of the surface in the first octant cut from the cylinder $x^2 + y^2 = 16$ by the plane $y = z$.

***29.** Find the area of the portion of the sphere $x^2 + y^2 + z^2 = 16z$ lying within the circular paraboloid $z = x^2 + y^2$.

***30.** Find the area of the surface cut from the hyperbolic paraboloid $4z = x^2 - y^2$ by the cylinder $x^2 + y^2 = 16$.

In Problems 31–34, find a repeated integral that represents the area of the given surface over the specified region. Do *not* try to evaluate your integral.

31. $z = x^3 + y^3$; Ω is the unit circle

32. $z = \ln(x + 2y)$;
$\Omega = \{(x, y)\colon 0 \le x \le 1, 0 \le y \le 4\}$

33. $z = \sqrt{1 + x + y}$; Ω is the triangle bounded by $y = x$, $y = 4 - x$, and the y-axis

34. $z = e^{x-y}$; Ω is the ellipse $4x^2 + 9y^2 = 36$.

***35.** Find, but do not evaluate, a repeated integral that represents the surface area of the ellipsoid $(x/a)^2 + (y/b)^2 + (z/c)^2 = 1$.

36. A **spiral ramp** or **helicoid** is given parametrically by

$$\mathbf{r}(u, v) = v\cos u\mathbf{i} + v\sin u\mathbf{j} + u\mathbf{k},\quad 0 \le u \le 2\pi,\ 0 \le v \le 1.$$

a. Sketch this surface.

b. Obtain its surface area.

37. Find a formula for the area of the triangle in $\mathbb{R}^3$ with vertices at $(a, 0, 0)$, $(0, b, 0)$, and $(0, 0, c)$.

***38.** Let $z = f(x, y)$ be the equation of a plane. Show that the area of the portion of this plane lying over the region Ω is

$$\sigma = \iint_\Omega \sec\gamma\, dA,$$

where γ is the angle between the normal vector $\mathbf{N}$ to the plane and the positive z-axis. Further, suppose that $f(x, y) = ax + by + c$; show, using the dot product, that

$$\cos\gamma = \frac{\mathbf{N}\cdot\mathbf{k}}{|\mathbf{N}|} = \frac{1}{\sqrt{1 + a^2 + b^2}}.$$

***39.** Suppose g is a smooth function of one variable. The graph, in the xz-plane, of $z = g(x)$ for $a \le x \le b$ is an arc. By moving this arc parallel to the y-axis from $y = 0$ to $y = c$, we sweep out a patch on a surface in $\mathbb{R}^3$. Show that the surface area of this patch is equal to the arc length of the curve multiplied by the horizontal distance we move the curve parallel to the y-axis.

40. Let Π: $ax + by + cz = 0$ be an equation of a plane passing through the origin. Assume that $c \ne 0$.

a. Show that for (x, y, z) in Π,

$$z = -\frac{1}{c}(ax + by).$$

b. Show that $\mathbf{x} = (c, 0, -a)$ and $\mathbf{y} = (0, c, -b)$ are two noncollinear vectors in Π.

41. In Problem 40, show that $u\mathbf{x} + v\mathbf{y}$ is in Π for any (u, v) in $\mathbb{R}^2$.

42. In Problem 41, suppose that $\mathbf{w} = (\alpha, \beta, \gamma)$ is in Π. Find numbers u and v such that $\mathbf{w} = u\mathbf{x} + v\mathbf{y}$.

43. Suppose that the graph of $y = f(x) > 0$ for $a \le x \le b$ is revolved about the x-axis, as in Figure 11. Obtain a parametrization for the surface in Figure 11(b).

44. Show that the lateral surface area of the surface in Figure 11(b) is given by

$$\sigma = \text{lateral surface area} = 2\pi\int_a^b f(x)\sqrt{1 + [f'(x)]^2}\, dx,$$

assuming that f is continuously differentiable.

45. Use the result of Problem 44 to compute the lateral surface area of a cone with radius a and height h. [*Hint:* Revolve a certain line about the x-axis.]

ANSWERS TO SELF-QUIZ

I. c, d, e, f **II.** a **III.** b (since $\sqrt{1 + f_x^2 + f_y^2} = 1$) **IV.** a (since $\sqrt{1 + f_x^2 + f_y^2} \ge 1$)

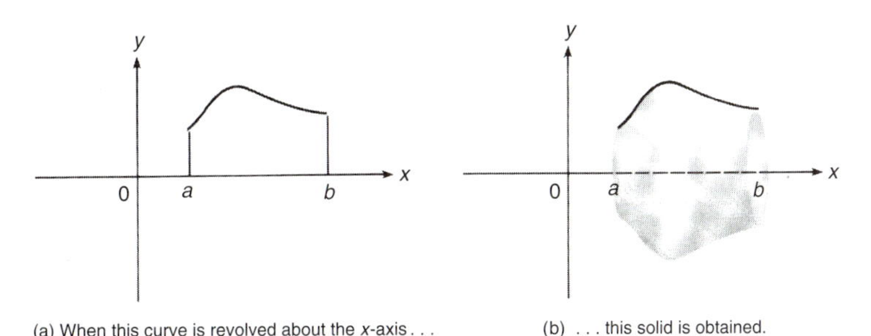

(a) When this curve is revolved about the x-axis . . . (b) . . . this solid is obtained.

FIGURE 11

5.6 SURFACE INTEGRALS

In Section 5.4 we discussed Green's theorem, which gave a relationship between a line integral over a closed curve and a double integral over a region enclosed by that curve. In this section, we discuss the notion of a surface integral. This will enable us, in Section 5.8, to extend Green's theorem to integrals over regions in space.

Recall from Section 1.8 that a surface in space is defined as the set of points satisfying the equation

$$G(x, y, z) = 0, \tag{1}$$

where G is a continuous function defined on a region in $\mathbb{R}^3$. In this section we do not write the surface parametrically. Rather, we will only consider surfaces that can be written in the form

$$z = f(x, y) \tag{2}$$

for some function f. Note that (2) is a special case of (1), as can be seen by defining $G(x, y, z) = z - f(x, y)$ so that $z = f(x, y)$ is equivalent to $G(x, y, z) = 0$. We remind you of a definition given in Section 3.7.

DEFINITION SMOOTH SURFACE

The surface $z = f(x, y)$ is called **smooth** at a point (x_0, y_0, z_0) if $\partial f/\partial x$ and $\partial f/\partial y$ are continuous at (x_0, y_0). If the surface is smooth at all points in the domain of f, we speak of it as a **smooth surface**. That is, if f is continuously differentiable, then the surface is smooth.† ■

A surface integral is very much like a double integral. Suppose that the surface $z = f(x, y)$ is smooth for (x, y) in a bounded region Ω in the xy-plane. Since $f(x, y)$ is continuous on Ω, we find from the definition of a double integral and Theorem 4.1.2, that

$$\iint_\Omega f(x, y)\, dA = \lim_{\Delta s \to 0} \sum_{i=1}^{n} \sum_{j=1}^{m} f(x_i^*, y_j^*)\, \Delta x\, \Delta y \tag{3}$$

exists, where $\Delta s = \sqrt{\Delta x^2 + \Delta y^2}$ and the limit is independent of the way in which the

†This definition does not differ from the definition on page 325 because the added restriction $\mathbf{r}_u \times \mathbf{r}_v \neq \mathbf{0}$ is automatically satisfied in this case: $|\mathbf{r}_u \times \mathbf{r}_v| = \sqrt{1 + \left(\frac{\partial f}{\partial x}\right)^2 + \left(\frac{\partial f}{\partial y}\right)^2} \geq 1.$

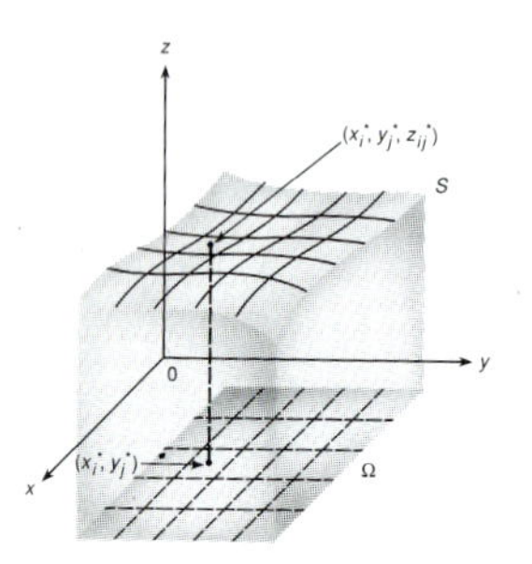

FIGURE 1
A partition of Ω into subrectangles leads to a partition of S into subsurfaces.

points (x_i^*, y_j^*) are chosen in the rectangle R_{ij}. We stress that in (3) the quantity $\Delta x\,\Delta y$ represents the *area* of the rectangle R_{ij}.

The double integral (3) is an integral over a region in the plane. A *surface integral,* which we will soon define, is an integral over a surface in space. Suppose we wish to integrate the function $F(x, y, z)$ over the surface S given by $z = f(x, y)$ where $(x, y) \in \Omega$ and, as before, Ω is a bounded region in the xy-plane. Such a surface is sketched in Figure 1.

We partition Ω into rectangles (and parts of rectangles) as before. This procedure provides a partition of S into mn "subsurfaces" S_{ij}, where

$$S_{ij} = \{(x, y, z)\colon z = f(x, y) \text{ and } (x, y) \in R_{ij}\}. \tag{4}$$

We choose a point in each S_{ij}. Such a point will have the form (x_i^*, y_j^*, z_{ij}^*), where $z_{ij}^* = f(x_i^*, y_j^*)$ and $(x_i^*, y_j^*) \in R_{ij}$. We let $\Delta\sigma_{ij}$ denote the surface area of S_{ij}. This is analogous to the notation $\Delta x\,\Delta y$ as the area of the rectangle R_{ij}. Then we write the double sum

$$\sum_{i=1}^{n}\sum_{j=1}^{m} F(x_i^*, y_j^*, z_{ij}^*)\,\Delta\sigma_{ij}$$

and consider

$$\lim_{\Delta s\to 0}\sum_{i=1}^{n}\sum_{j=1}^{m} F(x_i^*, y_j^*, z_{ij}^*)\,\Delta\sigma_{ij}. \tag{5}$$

DEFINITION INTEGRAL OVER A SURFACE

Suppose the limit in (5) exists and is independent of the way the surface S is partitioned and the way in which the points (x_i^*, y_j^*, z_{ij}^*) are chosen in S_{ij}. Then F is said to be **integrable** over S and the **surface integral** of F over S, denoted by $\iint_S F(x, y, z)\,d\sigma$, is given by

$$\iint_S F(x, y, z)\,d\sigma = \lim_{\Delta s\to 0}\sum_{i=1}^{n}\sum_{j=1}^{m} F(x_i^*, y_j^*, z_{ij}^*)\,\Delta\sigma_{ij}. \tag{6}$$

■

The key to evaluating the limit in (6) is to note that, from formula (12) on page 327,

$$\Delta\sigma_{ij} \approx \sqrt{f_x^2(x_i^*, y_j^*) + f_y^2(x_i^*, y_j^*) + 1}\,\Delta x\,\Delta y. \tag{7}$$

Then inserting (7) into the limit in (6) and noting that $z_{ij}^* = f(x_i^*, y_j^*)$, we have

$$\lim_{\Delta s\to 0}\sum_{i=1}^{n}\sum_{j=1}^{m} F(x_i^*, y_j^*, z_{ij}^*)\,\Delta\sigma_{ij}$$

$$= \lim_{\Delta s\to 0}\sum_{i=1}^{n}\sum_{j=1}^{m} F(x_i^*, y_j^*, f(x_i^*, y_j^*))\sqrt{f_x^2(x_i^*, y_j^*) + f_y^2(x_i^*, y_j^*) + 1}\,\Delta x\,\Delta y. \tag{8}$$

Now if $z = f(x, y)$ is a smooth surface over Ω, then f_x and f_y are continuous over Ω, so that $\sqrt{f_x^2 + f_y^2 + 1}$ is also continuous over Ω. Furthermore, if $F(x, y, z)$ is continuous for

(x, y, z) on S, then by Theorem 4.1.2

$$\iint_\Omega F(x, y, f(x, y))\sqrt{f_x^2(x, y) + f_y^2(x, y) + 1}\, dA$$

exists and is equal to the right-hand limit in (8). We therefore have the following important result.

THEOREM 1 WRITING A SURFACE INTEGRAL AS AN ORDINARY DOUBLE INTEGRAL

Let S: $z = f(x, y)$ be a smooth surface for (x, y) in the bounded region Ω in the xy-plane. Then if F is continuous on S, F is integrable over S, and

$$\iint_S F(x, y, z)\, d\sigma = \iint_\Omega F(x, y, f(x, y))\sqrt{f_x^2(x, y) + f_y^2(x, y) + 1}\, dA. \tag{9}$$

■

REMARK 1: Using (9), we can compute a surface integral over S by transforming it into an ordinary double integral over the region Ω that is the projection of S into the xy-plane.

REMARK 2: If $F(x, y, z) = 1$ in (9), then (9) reduces to

$$\sigma = \iint_S d\sigma = \iint_\Omega \sqrt{f_x^2(x, y) + f_y^2(x, y) + 1}\, dA. \tag{10}$$

This is the formula for surface area given in equation (12) in the last section.

EXAMPLE 1 COMPUTING A SURFACE INTEGRAL

Compute $\iint_S (x^2 + y^2 + 3z^2)\, d\sigma$, where S is the part of the circular paraboloid $z = x^2 + y^2$ with $x^2 + y^2 \le 9$.

SOLUTION: Here $f(x, y) = x^2 + y^2$, so that $f_x = 2x$, $f_y = 2y$, and $d\sigma = \sqrt{1 + 4x^2 + 4y^2}\, dx\, dy$. Thus,

$$\begin{aligned} I &= \iint_S (x^2 + y^2 + 3z^2)\, d\sigma \\ &= \iint_\Omega [x^2 + y^2 + 3(x^2 + y^2)^2]\sqrt{1 + 4x^2 + 4y^2}\, dA, \end{aligned}$$

where Ω is the disk (in the xy-plane) $x^2 + y^2 \le 9$. The problem is greatly simplified by the use of polar coordinates. We have, using $x^2 + y^2 = r^2$,

$$\begin{aligned} I &= \int_0^{2\pi}\int_0^3 [r^2 + 3(r^2)^2]\sqrt{1 + 4r^2}\, r\, dr\, d\theta \\ &= \int_0^{2\pi}\int_0^3 (r^3 + 3r^5)\sqrt{1 + 4r^2}\, dr\, d\theta \\ &= 2\pi\int_0^3 (r^3 + 3r^5)\sqrt{1 + 4r^2}\, dr. \end{aligned}$$

There are several ways to complete the evaluation of this integral. The easiest way is to make the substitution $u^2 = 1 + 4r^2$. The result is

$$I = 2\pi\left\{21(37)^{3/2} + \frac{1}{120}[1 - 55(37)^{5/2}] + \frac{1}{280}(37^{7/2} - 1)\right\} \approx 12{,}629.4.$$

EXAMPLE 2

COMPUTING A SURFACE INTEGRAL

Evaluate $\iint_S (x + y + z)\, d\sigma$, where S is the part of the surface $z = x^3 + y^4$ lying over the square $\{(x, y)\colon 0 \le x \le 1, 0 \le y \le 1\}$.

SOLUTION: We have $f_x = 3x^2$, $f_y = 4y^3$, and $d\sigma = \sqrt{1 + 9x^4 + 16y^6}\, dx\, dy$, so

$$\iint_S (x + y + z)\, d\sigma = \int_0^1 \int_0^1 (x + x^3 + y + y^4)\sqrt{1 + 9x^4 + 16y^6}\, dx\, dy.$$

This is as far as we can go because there is no way to compute this integral directly. The best we can do is to use numerical integration to approximate the answer.

REMARK: Example 2 is typical. Like the computations of arc length and surface area, the direct evaluation of a surface integral will often be impossible since it will involve integrals of functions for which antiderivatives cannot be found.

We can derive another way to represent a surface integral. First we need to define the orientation of a surface.

ORIENTATION OF A SURFACE

Consider the smooth surface $z = f(x, y)$. We can write this surface in two ways:

$$F(x, y, z) = f(x, y) - z = 0 \quad \text{and} \quad G(x, y, z) = z - f(x, y) = 0. \tag{11}$$

From Section 3.7 (page 198), we know that the gradient vectors ∇F and ∇G are normal to the surface at every point on the surface. Then, if ∇F and $\nabla G \ne \mathbf{0}$,

$$\mathbf{n}_1 = \frac{\nabla F}{|\nabla F|} \quad \text{and} \quad \mathbf{n}_2 = \frac{\nabla G}{|\nabla G|} \tag{12}$$

are unit normal vectors. In fact, it is evident from the way F and G are defined that $\mathbf{n}_2 = -\mathbf{n}_1$.

DEFINITION OUTWARD UNIT NORMAL VECTOR

We choose one of these normal vectors, denote it by $\mathbf{n}$, and call it the **outward unit normal vector**. The direction of $\mathbf{n}$ is called the **positive normal direction** to the surface at a point. ■

REMARK: ***Choosing n if S is a closed surface*** If S is a closed surface, such as the surface of a ball, then by convention we choose $\mathbf{n}$ so that it points away from the region bounded by the surface. This explains the use of the term *outward* unit normal vector. Put another way, if S is the boundary of a finite region Ω, then $\mathbf{n}$ points away from Ω.

Choosing n if S does not enclose any region If S does not enclose any part of a region in space (for example, if S is part of a plane), then by convention we choose $\mathbf{n}$ so that its $\mathbf{k}$ component is positive. In this context $\mathbf{n}$ is called the **upper unit normal vector**.

DEFINITION ORIENTABLE SURFACE

Let P_0 be any point on the surface S. Let C be a closed curve on S that passes through P_0. Then S is said to be **orientable** if the outward unit normal vector at P_0 does not change direction as it is displaced continuously around C until it returns to P_0. ■

If a surface is orientable, then the choice of the outward unit normal vector $\mathbf{n}$ determines a positive direction on S by displacing $\mathbf{n}$ along curves lying on S.

All of the surfaces we have considered in this text are orientable. The most famous example of a nonorientable surface is given by the Möbius strip.[†] To construct a Möbius strip, take a long rectangular piece of paper, twist one of the ends once, and paste the ends together (as in Figure 2). In our terminology, the Möbius strip is not orientable because a normal vector moving once around the closed curve indicated by the dotted line will change direction.

Paste ends together

FIGURE 2
A Möbius strip

In this text, we will write the surface $z = f(x, y)$ as $G(x, y, z) = z - f(x, y) = 0$. Then

$$\nabla G = G_x\mathbf{i} + G_y\mathbf{j} + G_z\mathbf{k} = -f_x\mathbf{i} - f_y\mathbf{j} + \mathbf{k}$$

and the outward (and upper) unit normal vector is

$$\mathbf{n} = \frac{\nabla G}{|\nabla G|} = \frac{-f_x\mathbf{i} - f_y\mathbf{j} + \mathbf{k}}{\sqrt{f_x^2 + f_y^2 + 1}}. \qquad (13)$$

The outward unit normal vector to the surface $z = f(x, y)$

This vector determines our orientation (positive direction).

We now define the angle γ to be the acute angle between $\mathbf{n}$ and the positive z-axis. This angle is depicted in Figure 3. Since $\mathbf{k}$ is a unit vector having the direction of the positive z-axis, we have, from Theorem 1.2.2 on page 16,

$|\mathbf{n}| = |\mathbf{k}| = 1$

$$\cos\gamma = \frac{\mathbf{n}\cdot\mathbf{k}}{|\mathbf{n}|\,|\mathbf{k}|} = \frac{1}{\sqrt{f_x^2 + f_y^2 + 1}} \qquad (14)$$

and

$$\sec\gamma = \sqrt{f_x^2 + f_y^2 + 1}. \qquad (15)$$

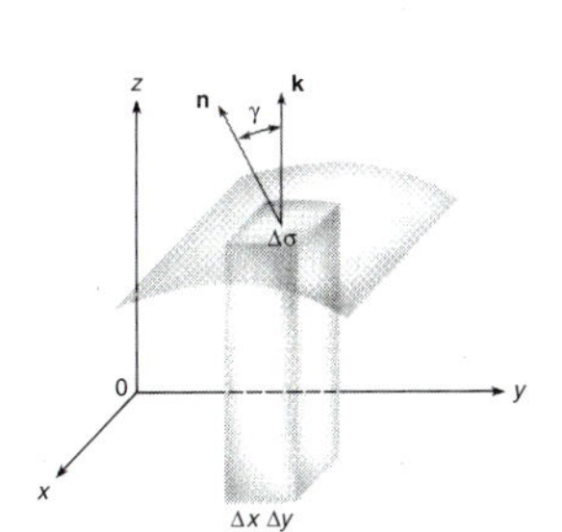

FIGURE 3
γ is the angle between the outward unit normal vector $\mathbf{n}$ and the unit vector $\mathbf{k}$.

Hence, we have

A FORMULA FOR A SURFACE INTEGRAL

$$\iint_S F(x, y, z)\, d\sigma = \iint_\Omega F(x, y, f(x, y)) \sec\gamma(x, y)\, dx\, dy. \qquad (16)$$

[†] Named after the German mathematician August Ferdinand Möbius (1790–1868). Möbius, who was a student of Gauss, did important work in geometry, mechanics, and number theory.

EXAMPLE 3 COMPUTING SURFACE AREA USING EQUATION (16)

Compute $\iint_S (x + 2y + 3z)\, d\sigma$, where S is the part of the plane $2x - y + z = 3$, that lies above the triangular region Ω in the xy-plane bounded by the x- and y-axes and the line $y = 1 - 2x$.

SOLUTION: Here S is the plane $2x - y + z = 3$. From Section 1.7 we know that $\mathbf{N} = 2\mathbf{i} - \mathbf{j} + \mathbf{k}$ is normal to this surface, so $\mathbf{n} = \mathbf{N}/|\mathbf{N}| = (1/\sqrt{6})(2\mathbf{i} - \mathbf{j} + \mathbf{k})$ is the outward unit normal vector. Then $\cos\gamma = \mathbf{n}\cdot\mathbf{k} = 1/\sqrt{6}$ and $\sec\gamma = \sqrt{6}$, so from Figure 4,

$$\iint_S (x + 2y + 3z)\, d\sigma = \iint_\Omega [(x + 2y) + 3(3 - 2x + y)]\sqrt{6}\, dx\, dy,$$

$$= \sqrt{6}\int_0^{1/2}\int_0^{1-2x} (9 - 5x + 5y)\, dy\, dx$$

$$= \sqrt{6}\int_0^{1/2} \left(9y - 5xy + \frac{5y^2}{2}\right)\Bigg|_0^{1-2x} dx$$

$$= \sqrt{6}\int_0^{1/2} \left[9(1 - 2x) - 5x(1 - 2x) + \frac{5}{2}(1 - 2x)^2\right] dx$$

$$= \sqrt{6}\int_0^{1/2} \left(20x^2 - 33x + \frac{23}{2}\right) dx = \frac{59}{24}\sqrt{6}.$$

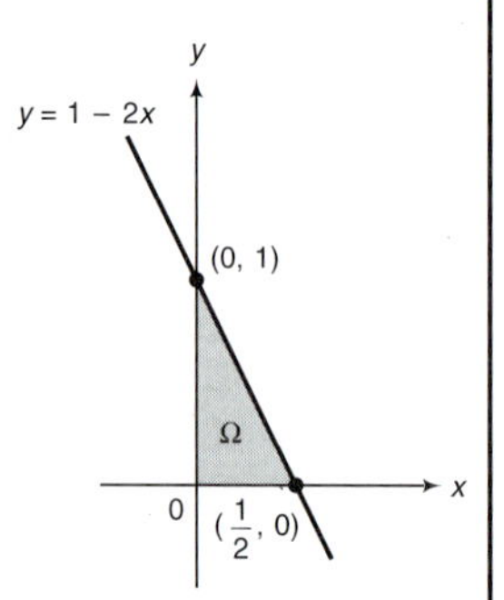

FIGURE 4
The region bounded by the coordinate axes and the line $y = 1 - 2x$

PROBLEMS 5.6

SELF-QUIZ

I. The integral of $F(x, y, z) = 7 + z$ over the surface which is that patch of the plane $z = 3 + x + 5y$ lying above the unit disk $D = \{(x, y): x^2 + y^2 \le 1\}$ is ________.

a. $\iint_D [3 + x + 5y]\, dA$

b. $\iint_D [7 + z]\, dA$

c. $\iint_D [7 + (3 + x + 5y)]\, dA$

d. $\iint_D [7 + (3 + x + 5y)]\sqrt{(1)^2 + (5)^2 + 1}\, dA$

II. Let S be the surface which is that patch of the plane $z = 3 + x + 5y$ lying above the unit disk D. The area of that surface is ________.

a. $\iint_S 1\, d\sigma$

b. $\iint_S \sqrt{(1)^2 + (5)^2 + 1}\, d\sigma$

c. $\iint_S z\, d\sigma$ **d.** $\iint_D 1\, dA$

e. $\iint_D \sqrt{(1)^2 + (5)^2 + 1}\, dA$

f. $\iint_D [3 + x + 5y]\, dA$

III. The integral of $F(x, y, z) = x^2 + y^2 - z$ over the surface S of the preceding problem is ________.

a. $\int_0^{2\pi}\int_0^1 \sqrt{(1)^2 + (5)^2 + 1}\, r\, dr\, d\theta$

b. $\int_0^{2\pi}\int_0^1 [r^2 - 3 - r(\cos\theta + 5\sin\theta)]r\, dr\, d\theta$

c. $\int_0^{2\pi}\int_0^1 [r^2 - 3 - r(\cos\theta + 5\sin\theta)] \times \sqrt{(1)^2 + (5)^2 + 1}\, r\, dr\, d\theta$

d. $\int_0^{2\pi}\int_0^1 [r^2 - 3 - r(\cos\theta + 5\sin\theta)] \times \sqrt{(2r\cos\theta)^2 + (2r\sin\theta)^2 + 1}\, r\, dr\, d\theta$

In Problems 1–16, evaluate the given surface integral over the specified surface S.

1. $\iint_S x\,d\sigma$, where $S = \{(x, y, z)\colon z = x^2,\ 0 \le x \le 1,\ 0 \le y \le 2\}$
2. $\iint_S y\,d\sigma$, where S is as in Problem 1
*3. $\iint_S (x^2 - 2y^2)\,d\sigma$, where S is as in Problem 1
4. $\iint_S \sqrt{1 + 4z}\,d\sigma$, where S is as in Problem 1
*5. $\iint_S x\,d\sigma$, where S is the hemisphere $\{(x, y, z)\colon x^2 + y^2 + z^2 = 4,\ x^2 + y^2 \le 4,\ z \ge 0\}$
*6. $\iint_S xy\,d\sigma$, where S is as in Problem 5
7. $\iint_S (x + y)\,d\sigma$, where S is the planar patch $\{(x, y, z)\colon x + 2y - 3z = 4,\ 0 \le x \le 1,\ 1 \le y \le 2\}$
8. $\iint_S yz\,d\sigma$, where S is as in Problem 7
9. $\iint_S z^2\,d\sigma$, where S is as in Problem 7
*10. $\iint_S (x^2 + y^2 + z^2)\,d\sigma$, where S is the part of the plane $x - y = 4$ that lies inside the cylinder $y^2 + z^2 = 4$
11. $\iint_S \cos z\,d\sigma$, where S is the planar patch $\{(x, y, z)\colon 2x + 3y + z = 1, 0 \le x \le 1, -1 \le y \le 2\}$
*12. $\iint_S z\,d\sigma$, where S is the tetrahedron bounded by the coordinate planes and the plane $4x + 8y + 2z = 16$
13. $\iint_S |x|\,d\sigma$, where S is the hemisphere $\{(x, y, z)\colon x^2 + y^2 + z^2 = 4,\ y^2 + z^2 \le 4,\ x \ge 0\}$
14. $\iint_S z\,d\sigma$, where S is the surface of Problem 13
15. $\iint_S z^2\,d\sigma$, where S is the hemisphere $\{(x, y, z)\colon x^2 + y^2 + z^2 = 9,\ x^2 + z^2 \le 9,\ y \ge 0\}$
16. $\iint_S x^2\,d\sigma$, where S is the surface of Problem 15

MASS: Suppose $z = f(x, y)$ for points (x, y, z) on a smooth thin metallic surface S. Also suppose $\rho(x, y, z)$ is the local density of that metallic surface. Then the **mass** of the surface is

$$\mu = \iint_S \rho(x, y, f(x, y))\,d\sigma.$$

Problem 33 asks you to work through a justification for this formula. For now, use this result to work Problems 17–20.

17. A metallic dome has the shape of a hemisphere centered at the origin with radius 4 m. Its area density at a point (x, y, z) in space is given by $\rho(x, y, z) = 25 - x^2 - y^2$ kg/m^2. Find the total mass of this dome.
18. Find the mass of a metallic sheet in the shape of the hemisphere

$$\{(x, y, z)\colon x^2 + y^2 + z^2 = 9,\ x^2 + y^2 \le 9,\ z \ge 0\}$$

if its density at a point is proportional to the distance from that point to the origin.

19. Find the mass of a triangular metallic sheet with corners at $(1, 0, 0)$, $(0, 1, 0)$, and $(0, 0, 1)$ if its density is a constant α. (Assume the units are meters and kg/m^2.)
20. Find the mass of the sheet of the preceding problem if its density is proportional to x^2.

FLUX: Immerse a porous surface in a moving fluid and think about the rate of flow through the surface. To be a bit more formal, let S be the surface and suppose **F** is the vector field which specifies how the fluid flows. In general, **F** is the product of a scalar density $\rho(x, y, z)$ and a vector velocity $\mathbf{v}(x, y, z)$. If $\mathbf{n} = \mathbf{n}(x, y, z)$ is the unit normal to the surface, then $\mathbf{F}\cdot\mathbf{n}$ is the component of **F** on **n** and it measures the local flow rate per unit of area; if we integrate $\mathbf{F}\cdot\mathbf{n}$ over the surface, then we obtain the total rate of flow through the whole surface:

$$\textbf{flux of F over S} = \iint_S \mathbf{F}\cdot\mathbf{n}\,d\sigma.$$

Problem 34 includes an alternative expression for flux which may be easier to compute. In Problems 21–31, compute the flux $\iint_S \mathbf{F}\cdot\mathbf{n}\,d\sigma$ for the given surface S lying in the given vector field **F**.

21. $S = \{(x, y, z)\colon z = xy,\ 0 \le x \le 1,\ 0 \le y \le 2\}$; $\mathbf{F} = x^2y\mathbf{i} - z\mathbf{j}$
22. $S = \{(x, y, z)\colon z = 4 - x - y,\ x \ge 0,\ y \ge 0,\ z \ge 0\}$; $\mathbf{F} = -3x\mathbf{i} - y\mathbf{j} + 3z\mathbf{k}$
23. S is the unit sphere; $\mathbf{F} = x\mathbf{i} + y\mathbf{j} + z\mathbf{k}$
24. $S = \{(x, y, z)\colon x^2 + y^2 + z^2 = 1,\ z \ge 0\}$; $\mathbf{F} = x\mathbf{i} + y\mathbf{j} + z\mathbf{k}$
25. $S = \{(x, y, z)\colon x^2 + y^2 + z^2 = 1,\ y \ge 0\}$; $\mathbf{F} = x\mathbf{i} + y\mathbf{j} + z\mathbf{k}$
26. $S = \{(x, y, z)\colon x^2 + y^2 + z^2 = 1,\ x \le 0\}$; $\mathbf{F} = x\mathbf{i} + y\mathbf{j} + z\mathbf{k}$
27. $S = \{(x, y, z)\colon z = \sqrt{x^2 + y^2},\ x^2 + y^2 \le 1\}$; $\mathbf{F} = x\mathbf{i} - y\mathbf{j} + xy\mathbf{k}$
28. $S = \{(x, y, z)\colon x = \sqrt{y^2 + z^2},\ y^2 + z^2 \le 1\}$; $\mathbf{F} = y\mathbf{i} - z\mathbf{j} + yz\mathbf{k}$
29. $S = \{(x, y, z)\colon z = \sqrt{x^2 + y^2},\ x^2 + y^2 \le 1,\ x \ge 0,\ y \ge 0\}$; $\mathbf{F} = x^2\mathbf{i} + y^2\mathbf{j} + z\mathbf{k}$
*30. S is region bounded by $y = 1$ and $y = \sqrt{x^2 + z^2}$; $x^2 + z^2 \le 1$; $\mathbf{F} = x\mathbf{i} - z\mathbf{j} + xz\mathbf{k}$
*31. $S = \{(x, y, z)\colon 0 \le x \le 1,\ 0 \le y \le 2,\ 0 \le z \le 3\}$; $\mathbf{F} = x^2y\mathbf{i} - 2yz\mathbf{j} + x^3y^2\mathbf{k}$ [*Hint:* Compute the

flux through each of the six faces separately; then add up those partial results.]

32. Assuming all three of the following integrals exist, show that

$$\iint_S [F(x, y, z) + G(x, y, z)]\, d\sigma = \iint_S F(x, y, z)\, d\sigma + \iint_S G(x, y, z)\, d\sigma.$$

***33.** Suppose $z = f(x, y)$ for points (x, y, z) on a smooth thin metallic surface S. Also suppose $\rho(x, y, z)$ is the local density of that metallic surface. Show that $\iint_S \rho(x, y, f(x, y))\, d\sigma$ is the limit of finite sums which approximate the total mass of the surface and comment on the reasonableness of defining that mass to be the value of the integral expression.

34. Suppose $S = \{(x, y, z): z = f(x, y),\ (x, y) \in \Omega\}$ is a smooth surface with upper unit normal $\mathbf{n} = \mathbf{n}(x, y, z)$. Also suppose $\mathbf{F} = \mathbf{F}(x, y, z) = P(x, y, z)\mathbf{i} + Q(x, y, z)\mathbf{j} + R(x, y, z)\mathbf{k}$ is a vector field which is continuous on S. The paragraph preceding Problem 21 discusses the flux of $\mathbf{F}$ over S and argues that $\iint_S \mathbf{F}\cdot\mathbf{n}\, d\sigma$ computes it.

For this problem, show that

$$\iint_S \mathbf{F}\cdot\mathbf{n}\, d\sigma = \iint_\Omega [-Pf_x - Qf_y + R]\, dx\, dy.$$

35. Suppose S is the sphere $\{(x, y, z): x^2 + y^2 + z^2 = \alpha^2\}$ and $\mathbf{F} = a\mathbf{i} + b\mathbf{j} + c\mathbf{k}$ where a, b, c are constants. Show that $\iint_S \mathbf{F}\cdot\mathbf{n}\, d\sigma = 0$.

36. Let S be the hemisphere $\{(x, y, z): x^2 + y^2 + z^2 = a^2,\ x^2 + y^2 \le a^2,\ z \ge 0\}$.

a. Show that

$$\iint_S (x^2 + y^2)\, d\sigma = \frac{4}{3}\pi a^4.$$

b. Show that

$$\iint_S x^2\, d\sigma = \iint_S y^2\, d\sigma = \iint_S z^2\, d\sigma = \frac{1}{3}\iint_S (x^2 + y^2 + z^2)\, d\sigma.$$

c. Without performing any explicit integrations, explain why the last integral of part (b) equals $\frac{2}{3}\pi a^4$.

d. Use part (c) to explain, without doing an integration, why $\iint_S (x^2 + y^2)\, d\sigma = \frac{4}{3}\pi a^4$.

ANSWERS TO SELF-QUIZ

I. d **II.** a = e = $\sqrt{27\pi}$ **III.** c

5.7 DIVERGENCE AND CURL

Let the function $F(x, y, z)$ be given. Then the gradient of F is given by

$$\boldsymbol{\nabla} F = \frac{\partial F}{\partial x}\mathbf{i} + \frac{\partial F}{\partial y}\mathbf{j} + \frac{\partial F}{\partial z}\mathbf{k}. \tag{1}$$

We can think of the gradient as a function that takes a differentiable function of (x, y, z) into a vector field in $\mathbb{R}^3$. We write this function, symbolically, as

$$\boldsymbol{\nabla} = \frac{\partial}{\partial x}\mathbf{i} + \frac{\partial}{\partial y}\mathbf{j} + \frac{\partial}{\partial z}\mathbf{k}. \tag{2}$$

The operator in (2) provides a useful device for writing things down. For example, (1) can be written as

$$\boldsymbol{\nabla} F = \left(\frac{\partial}{\partial x}\mathbf{i} + \frac{\partial}{\partial y}\mathbf{j} + \frac{\partial}{\partial z}\mathbf{k}\right) F = \frac{\partial F}{\partial x}\mathbf{i} + \frac{\partial F}{\partial y}\mathbf{j} + \frac{\partial F}{\partial z}\mathbf{k}.$$

We now define the divergence and curl of a vector field $\mathbf{F}$ in $\mathbb{R}^3$ given by

$$\mathbf{F}(x, y, z) = P(x, y, z)\mathbf{i} + Q(x, y, z)\mathbf{j} + R(x, y, z)\mathbf{k}. \tag{3}$$

DEFINITION DIVERGENCE AND CURL

Let the vector field **F** be given by (3), where P, Q, and R are differentiable. Then the **divergence** of **F** (div **F**) and **curl** of **F** (curl **F**) are given by

$$\text{div } \mathbf{F} = \frac{\partial P}{\partial x} + \frac{\partial Q}{\partial y} + \frac{\partial R}{\partial z} \tag{4}$$

and

$$\text{curl } \mathbf{F} = \left(\frac{\partial R}{\partial y} - \frac{\partial Q}{\partial z}\right)\mathbf{i} + \left(\frac{\partial P}{\partial z} - \frac{\partial R}{\partial x}\right)\mathbf{j} + \left(\frac{\partial Q}{\partial x} - \frac{\partial P}{\partial y}\right)\mathbf{k}. \tag{5}$$

■

NOTE: div **F** is a scalar function and curl **F** is a vector field.

Before giving examples of divergence and curl, we derive an easy way to remember how to compute them.

THEOREM 1

Let ∇ be given by (2) and let the differentiable vector field **F** be given by (3). Then,

i. $\text{div } \mathbf{F} = \nabla \cdot \mathbf{F}$ (6)

and

ii. $\text{curl } \mathbf{F} = \nabla \times \mathbf{F}$. (7)

PROOF:

i.
$$\nabla \cdot \mathbf{F} = \left(\frac{\partial}{\partial x}\mathbf{i} + \frac{\partial}{\partial y}\mathbf{j} + \frac{\partial}{\partial z}\mathbf{k}\right) \cdot (P\mathbf{i} + Q\mathbf{j} + R\mathbf{k})$$
$$= \frac{\partial P}{\partial x} + \frac{\partial Q}{\partial y} + \frac{\partial R}{\partial z} = \text{div } \mathbf{F}.$$

ii.
$$\nabla \times \mathbf{F} = \begin{vmatrix} \mathbf{i} & \mathbf{j} & \mathbf{k} \\ \frac{\partial}{\partial x} & \frac{\partial}{\partial y} & \frac{\partial}{\partial z} \\ P & Q & R \end{vmatrix}$$
$$= \left(\frac{\partial R}{\partial y} - \frac{\partial Q}{\partial z}\right)\mathbf{i} + \left(\frac{\partial P}{\partial z} - \frac{\partial R}{\partial x}\right)\mathbf{j} + \left(\frac{\partial Q}{\partial x} - \frac{\partial P}{\partial y}\right)\mathbf{k}$$
$$= \text{curl } \mathbf{F}.$$

■

EXAMPLE 1 COMPUTING DIVERGENCE AND CURL

Compute the divergence and curl of $\mathbf{F}(x, y, z) = xy\mathbf{i} + (z^2 - 2y)\mathbf{j} + \cos yz\mathbf{k}$.

SOLUTION:

$$\text{div } \mathbf{F} = \frac{\partial}{\partial x}(xy) + \frac{\partial}{\partial y}(z^2 - 2y) + \frac{\partial}{\partial z}(\cos yz) = y - 2 - y \sin yz$$

and

$$\text{curl } \mathbf{F} = \begin{vmatrix} \mathbf{i} & \mathbf{j} & \mathbf{k} \\ \dfrac{\partial}{\partial x} & \dfrac{\partial}{\partial y} & \dfrac{\partial}{\partial z} \\ xy & z^2 - 2y & \cos yz \end{vmatrix}$$

$$= \left[\frac{\partial}{\partial y} \cos yz - \frac{\partial}{\partial z}(z^2 - 2y)\right]\mathbf{i} + \left(\frac{\partial}{\partial z} xy - \frac{\partial}{\partial x} \cos yz\right)\mathbf{j}$$

$$+ \left[\frac{\partial}{\partial x}(z^2 - 2y) - \frac{\partial}{\partial y} xy\right]\mathbf{k} = (-z \sin yz - 2z)\mathbf{i} - x\mathbf{k}.$$

Recall from Theorem 5.3.4 that the vector field $\mathbf{F} = P\mathbf{i} + Q\mathbf{j} + R\mathbf{k}$ is the gradient of a function f if and only if

$$\frac{\partial P}{\partial y} = \frac{\partial Q}{\partial x}, \qquad \frac{\partial R}{\partial x} = \frac{\partial P}{\partial z}, \qquad \frac{\partial Q}{\partial z} = \frac{\partial R}{\partial y}. \tag{8}$$

Using (8) and (5), we obtain the following interesting result.

THEOREM 2

The differentiable vector field $\mathbf{F}$ defined on a simply connected region S is the gradient of a function f if and only if curl $\mathbf{F} = \mathbf{0}$. ■

As we shall see, the curl of a vector field $\mathbf{F}$ represents the circulation per unit area at the point (x, y, z). If curl $\mathbf{F} = \mathbf{0}$ for every (x, y, z) in some region W in $\mathbb{R}^3$, then the fluid flow $\mathbf{F}$ is called **irrotational**. The divergence of $\mathbf{F}$ at a point (x, y, z) represents the net rate of flow away from (x, y, z). If div $\mathbf{F} = 0$ for every (x, y, z) in W, then the flow $\mathbf{F}$ is called **incompressible** or **solenoidal**. Let us examine these ideas more closely.

PHYSICAL INTERPRETATION OF DIVERGENCE AND CURL

DIVERGENCE: Suppose that a fluid is flowing through the small volume $dx\,dy\,dz$ depicted in Figure 1. Let $\mathbf{v} = v_x\mathbf{i} + v_y\mathbf{j} + v_z\mathbf{k}$ and ρ denote, respectively, the velocity and the density of the fluid at the origin. Fluid is flowing in all directions. We consider the positive x-direction first. The fluid flowing into this volume per unit time through the face $EFGH$ is

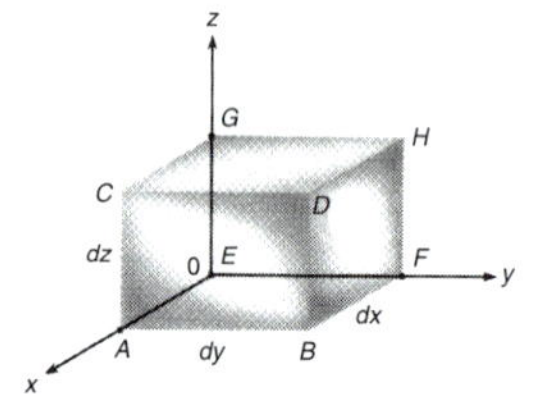

FIGURE 1
A box of volume $dx\,dy\,dz$

$$\text{rate of flow in (face } EFGH) = \rho v_x\Big|_{x=0} dy\,dz. \tag{9}$$

The components ρv_y and ρv_z of the flow are tangential to this face and contribute nothing to the flow through this face. The rate of flow out (in the positive direction) through the face $ABCD$ is

$$\text{rate of flow out (face } ABCD) = \rho v_x\Big|_{x=dx} dy\,dz. \tag{10}$$

By the mean-value theorem, we have

$$\rho v_x\Big|_{x=dx} dy\,dz - \rho v_x\Big|_{x=0} dy\,dz = \left[\frac{\partial}{\partial x}(\rho v_x)\,dx\right] dy\,dz, \tag{11}$$

where the partial derivative is evaluated at some point where x takes a value in $(0, dx)$. Hence,

net flow in positive direction at the point (x, y, z) $= \text{flow out} - \text{flow in} = \dfrac{\partial}{\partial x}(\rho v_x)\, dx\, dy\, dz.$

Similar results hold in the positive y- and z-directions and we have

$$\begin{aligned}\text{net flow per unit time} &= \left[\frac{\partial}{\partial x}(\rho v_x) + \frac{\partial}{\partial y}(\rho v_y) + \frac{\partial}{\partial z}(\rho v_z)\right] dx\, dy\, dz \\ &= \operatorname{div}(\rho v) dx\, dy\, dz.\end{aligned}$$

Therefore, the new flow of the compressible fluid from the volume element $dx\, dy\, dz$ per unit volume per unit time is $\operatorname{div}(\rho v)$. This is why we call it *divergence.*

The divergence appears in a wide variety of physical problems, ranging from a probability current density in quantum mechanics to neutron leakage in a nuclear reactor.

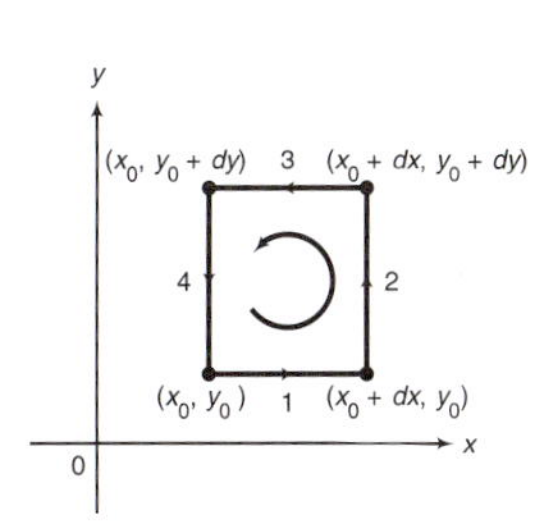

FIGURE 2
Circulation around a differential loop

CURL: Consider the circulation of fluid around the loop in the xy-plane drawn in Figure 2. Let the velocity vector $\mathbf{v} = v_x\mathbf{i} + v_y\mathbf{j}$. Now

$$\text{circulation} = \text{flow along 1} + \text{flow along 2} + \text{flow along 3} + \text{flow along 4}.$$

Recall that directed distance is the integral of velocity. Thus, treating v_x and v_y as constants (since, dx and dy are small), we have

$$\text{flow along 1} = \int_{x_0}^{x_0+dx} v_x(x_0, y_0)\, dx = v_x(x_0, y_0)\, dx$$

$$\text{flow along 2} = \int_{y_0}^{y_0+dy} v_y(x_0 + dx, y_0)\, dy = v_y(x_0 + dx, y_0)\, dy$$

$$\begin{aligned}\text{flow along 3} &= \int_{x_0+dx}^{x_0} v_x(x_0, y_0 + dy)\, dx \\ &= -\int_{x_0}^{x_0+dx} v_x(x_0, y_0 + dy)\, dx = -v_x(x_0, y_0 + dy)\, dx\end{aligned}$$

$$\text{flow along 4} = \int_{y_0+dy}^{y_0} v_y(x_0, y_0)\, dy = -v_y(x_0, y_0)\, dy.$$

Using Taylor expansions around (x_0, y_0), and no longer treating v_x and v_y as constants, we have

$$v_y(x_0 + t, y_0) = v_y(x_0, y_0) + \frac{\partial v_y}{\partial x}(x_0, y_0)t + \text{higher-order terms}$$

and

$$v_y(x_0 + dx, y_0) = v_y(x_0, y_0) + \frac{\partial v_y}{\partial x}(x_0, y_0)dx + \text{higher-order terms},$$

so

$$\text{flow along 2} \approx v_y(x_0, y_0)\, dy + \frac{\partial v_y}{\partial x}(x_0, y_0)\, dx\, dy.$$

Similarly,

$$v_x(x_0, y_0 + dy) = v_x(x_0, y_0) + \frac{\partial v_x}{\partial y}\,dy + \text{higher-order terms}$$

and

$$\text{flow along } 3 = -v_x(x_0, y_0)\,dx - \frac{\partial v_x}{\partial y}(x_0, y_0)\,dx\,dy.$$

Adding the circulations along 1, 2, 3, and 4 and ignoring the higher-order terms, we have

$$\text{circulation}_{1234} = \left(\frac{\partial v_y}{\partial x} - \frac{\partial v_x}{\partial y}\right) dx\,dy,$$

or, dividing by $dx\,dy$,

$$\text{circulation per unit area at } (x_0, y_0) = \text{curl } \mathbf{v}(x_0, y_0).$$

In principle, curl $\mathbf{v}$ at (x_0, y_0) could be determined by inserting a paddle wheel into the moving fluid at the point (x_0, y_0). The rotation of the paddle wheel would be a measure of the curl.

PROBLEMS 5.7

SELF-QUIZ

I. Let $\mathbf{F} = x^2 \cos y\mathbf{i} + 2x \sin y\mathbf{j} + 0\mathbf{k}$; then div $\mathbf{F}$ = ________.

a. 0 **b.** $4x \cos y$
c. $(2 - x^2) \sin y$ **d.** $(2 - x^2) \sin y\mathbf{k}$

II. Let $\mathbf{G} = 2x \sin y\mathbf{i} + x^2 \cos y\mathbf{j} + 0\mathbf{k}$; then curl $\mathbf{G}$ = ________.

a. $0\mathbf{k} = \mathbf{0}$ **b.** $4x \cos y$
c. $(2 - x^2) \sin y\mathbf{k}$ **d.** $4x \cos y\mathbf{k}$

III. True–False: If the smooth vector field $\mathbf{F}(x, y, z)$ is exact; then curl $\mathbf{F} = \mathbf{0}$.

IV. True–False: It is possible to produce an example of a smooth vector field $\mathbf{G}(x, y, z)$ which is conservative, but for which curl $\mathbf{G} \neq \mathbf{0}$.

In Problems 1–10, compute the divergence and curl of the given vector field.

1. $x^2\mathbf{i} + y^2\mathbf{j} + z^2\mathbf{k}$
2. $(\sin y)\mathbf{i} + (\sin z)\mathbf{j} + (\sin x)\mathbf{k}$
3. $a\mathbf{i} + b\mathbf{j} + c\mathbf{k}$; a, b, c are constants
4. $\sqrt{1 + x^2 + y^2}\mathbf{i} + \sqrt{1 + x^2 + y^2}\mathbf{j} + z^4\mathbf{k}$
5. $xy\mathbf{i} + yz\mathbf{j} + xz\mathbf{k}$
6. $(y^2 + z^2)\mathbf{i} + (x^2 + z^2)\mathbf{j} + (x^2 + y^2)\mathbf{k}$
7. $e^{yz}\mathbf{i} + e^{xz}\mathbf{j} + e^{xy}\mathbf{k}$
8. $e^{xy}\mathbf{i} + e^{yz}\mathbf{j} + e^{zx}\mathbf{k}$
9. $\frac{x}{y}\mathbf{i} + \frac{y}{z}\mathbf{j} + \frac{z}{x}\mathbf{k}$
10. $\sqrt{y + z}\mathbf{i} + \sqrt{x + z}\mathbf{j} + \sqrt{x + y}\mathbf{k}$

In Problems 11–18, use the results of Problems 45 and 46 to calculate

a. curl $\mathbf{F}$ **b.** $\oint_{\partial\Omega} \mathbf{F} \cdot \mathbf{T}\,ds$ **c.** div $\mathbf{F}$ **d.** $\oint_{\partial\Omega} \mathbf{F} \cdot \mathbf{n}\,ds$

11. $\mathbf{F}(x, y) = ax\mathbf{i} + by\mathbf{j}$; $\Omega = \{(x, y)\colon 0 \le x \le 1, 0 \le y \le 1\}$
12. $\mathbf{F}(x, y) = Ay\mathbf{i} + Bx\mathbf{j}$; Ω is the region of Problem 11
13. $\mathbf{F}(x, y) = x^2\mathbf{i} + y^2\mathbf{j}$; Ω is the region of Problem 11
14. $\mathbf{F}(x, y) = y^2\mathbf{i} + x^2\mathbf{j}$; Ω is the region of Problem 11
15. $\mathbf{F}(x, y) = x\mathbf{i} + y\mathbf{j}$; Ω is the unit disk
16. $\mathbf{F}(x, y) = y\mathbf{i} - x\mathbf{j}$; Ω is the unit disk
17. $\mathbf{F}(x, y) = y^3\mathbf{i} + x^3\mathbf{j}$; Ω is the unit disk
18. $\mathbf{F}(x, y) = xy\mathbf{i} + (y^2 - x^2)\mathbf{j}$; Ω is the region enclosed by the triangle with vertices at $(0, 0)$, $(1, 0)$, and $(0, 1)$

LAPLACIAN: The **Laplacian** of a twice-differentiable scalar function $f = f(x, y, z)$, denoted by $\nabla^2 f$, is defined by

$$\text{Laplacian of } f = \nabla^2 f = \frac{\partial^2 f}{\partial x^2} + \frac{\partial^2 f}{\partial y^2} + \frac{\partial^2 f}{\partial z^2}.$$

A function that satisfies **Laplace's equation** $\nabla^2 f = 0$ is said to be a **harmonic** function.

In Problems 19–22, compute $\nabla^2 f$ and identify which functions are harmonic.

19. $f(x, y, z) = xyz$ **20.** $f(x, y, z) = x^2 + y^2 + z^2$

21. $f(x, y, z) = 2x^2 + 5y^2 + 3z^2$

22. $f(x, y, z) = 1/\sqrt{x^2 + y^2 + z^2}$

23. It is true, although we will not attempt to prove it, that if div $\mathbf{F} = 0$, then there is some vector field $\mathbf{G}$ such that $\mathbf{F} = \text{curl}\,\mathbf{G}$. Let $\mathbf{F} = 2x\mathbf{i} + y\mathbf{j} - 3z\mathbf{k}$.
 a. Verify that div $\mathbf{F} = 0$.
 b. Find a vector field $\mathbf{G}$ such that $\mathbf{F} = \text{curl}\,\mathbf{G}$.

24. Let $\mathbf{F} = 4xyz\mathbf{i} + 2x^2z\mathbf{j} + 2x^2y\mathbf{k}$.
 a. Verify that curl $\mathbf{F} = \mathbf{0}$.
 b. Find a function f such that $\mathbf{F} = \nabla f$.

25. Let $\mathbf{F}(x, y, z) = x^2\mathbf{i} + y^2\mathbf{j} + z^2\mathbf{k}$; verify that div[curl $\mathbf{F}$] = 0.

26. Let $f(x, y, z) = 1/\sqrt{x^2 + y^2 + z^2}$; verify that curl[grad f] = $\mathbf{0}$.

27. Verify that the vector flow $\mathbf{F}(x, y) = x\mathbf{i} + y\mathbf{j}$ is irrotational.

28. Is the vector flow $\mathbf{G}(x, y, z) = e^x \sin x\,\mathbf{i} + y^{5/2}\mathbf{j} + \tan^{-1} z\,\mathbf{k}$ irrotational?

29. Verify that the vector flow $\mathbf{F}(x, y) = y\sqrt{x^2 + y^2}\mathbf{i} - x\sqrt{x^2 + y^2}\mathbf{j}$ is incompressible.

30. Is the vector flow $\mathbf{G}(x, y, z) = yz^2\mathbf{i} + zx^2\mathbf{j} + xy^2\mathbf{k}$ incompressible?

31. The electrostatic field at $\mathbf{x}$ caused by a point charge q located at the origin is $\mathbf{E}(\mathbf{x}) = \dfrac{q}{4\pi\epsilon_0|\mathbf{x}|^3}\mathbf{x}$. Calculate div $\mathbf{E}$. What happens as $\mathbf{x}$ approaches the origin?

32. Verify that the gravitational force (see example 5.1.3) $\mathbf{F} = \dfrac{-Gm_1m_2}{|\mathbf{x}|^3}\mathbf{x}$ is irrotational.

In Problems 33–41, assume that all given functions are smooth.

33. Show that div($\mathbf{F} + \mathbf{G}$) = div $\mathbf{F}$ + div $\mathbf{G}$.

34. Show that curl($\mathbf{F} + \mathbf{G}$) = curl $\mathbf{F}$ + curl $\mathbf{G}$.

35. If $f = f(x, y, z)$ is a scalar function, show that div($f\mathbf{G}$) = f(div $\mathbf{G}$) + $(\nabla f)\cdot\mathbf{G}$.

36. If f is a scalar function, show that curl($f\mathbf{G}$) = f(curl $\mathbf{G}$) + $(\nabla f)\times\mathbf{G}$.

37. If f is a smooth scalar function, show that curl(∇f) = $\mathbf{0}$.

38. Show that div[curl $\mathbf{F}$] = 0.

39. Show that div($\mathbf{F}\times\mathbf{G}$) = (curl $\mathbf{F}$)$\cdot\mathbf{G} - \mathbf{F}\cdot$(curl $\mathbf{G}$).

40. Show that $\nabla^2 f$ = div(grad f).

41. The **Laplacian** of a vector field $\mathbf{F} = P\mathbf{i} + Q\mathbf{j} + R\mathbf{k}$ is given by $\nabla^2\mathbf{F} = (\nabla^2 P)\mathbf{i} + (\nabla^2 Q)\mathbf{j} + (\nabla^2 R)\mathbf{k}$. Show that curl(curl $\mathbf{F}$) = ∇ div $\mathbf{F} - \nabla^2\mathbf{F}$.

42. Suppose $f(x)$ and $g(y)$ are smooth scalar functions. Show that $\mathbf{F}(x, y) = f(x)\mathbf{i} + g(y)\mathbf{j}$ is irrotational.

43. Suppose g is a smooth scalar function. Show that $\mathbf{F}(x, y) = -yg(x^2 + y^2)\mathbf{i} + xg(x^2 + y^2)\mathbf{j}$ is incompressible.

44. Let f, g, and h be differentiable functions of two variables. Show that the vector field $\mathbf{F}(x, y, z) = f(y, z)\mathbf{i} + g(x, z)\mathbf{j} + h(x, y)\mathbf{k}$ is incompressible.

***45.** Suppose $\mathbf{F}(x, y)$ is smooth on the region Ω and $\mathbf{T}(s)$ is the unit tangent vector to $\partial\Omega$. Show that Green's theorem implies

$$\oint_{\partial\Omega} \mathbf{F}\cdot\mathbf{T}\,ds = \iint_{\Omega} ([\text{curl}\,\mathbf{F}]\cdot\mathbf{k})\,dx\,dy.$$

***46.** Suppose $\mathbf{F}(x, y)$ is smooth on the region Ω and $\mathbf{n}$ is the unit normal vector to $\partial\Omega$. Show that Green's theorem implies

$$\oint_{\partial\Omega} \mathbf{F}\cdot\mathbf{n}\,ds = \iint_{\Omega} \text{div}\,\mathbf{F}\,dx\,dy.$$

47. Let

$$\mathbf{F}(x, y) = \frac{y}{x^2 + y^2}\mathbf{i} - \frac{x}{x^2 + y^2}\mathbf{j}.$$

 a. Show that curl $\mathbf{F} = 0$.
 b. Show that $\oint_C \mathbf{F}\cdot\mathbf{T}\,ds \neq \mathbf{0}$ if C is the unit circle oriented counterclockwise.
 c. Explain why the results of parts (a) and (b) do not contradict Problem 45.

48. The velocity of a two-dimensional flow of liquid is given by $\mathbf{F} = u(x, y)\mathbf{i} + v(x, y)\mathbf{j}$. If the liquid is incompressible and the flow is irrotational, show that

$$\frac{\partial u}{\partial y} = \frac{\partial v}{\partial x} \quad \text{and} \quad \frac{\partial u}{\partial x} = -\frac{\partial v}{\partial y}.$$

(These are known as the **Cauchy-Riemann** equations.)

ANSWERS TO SELF-QUIZ

I. b **II.** a **III.** True **IV.** False (Such an example cannot exist.)

5.8 STOKES'S THEOREM

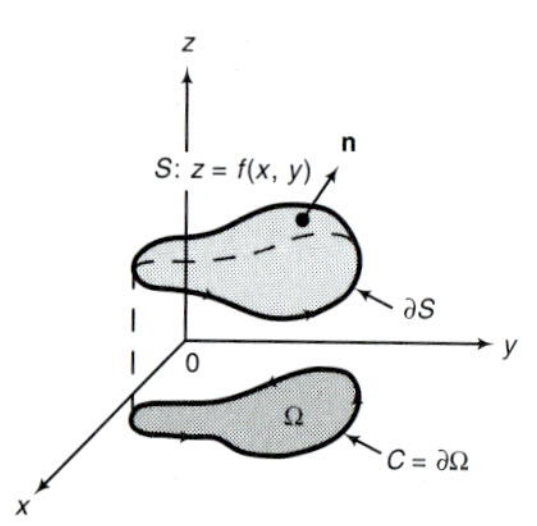

FIGURE 1
The boundary of Ω ($C = \partial\Omega$) is the projection onto the xy-plane of the boundary of S.

Let S: $z = f(x, y)$ be a smooth surface for $(x, y) \in \Omega$, where Ω is bounded. We assume that the boundary of S, denoted by ∂S, is a piecewise smooth, simple closed curve in $\mathbb{R}^3$. The positive direction on ∂S corresponds to the positive direction of $\partial\Omega$, where $\partial\Omega$ is the projection of ∂S into the xy-plane. This orientation is illustrated in Figure 1.

We now state the second major result in vector calculus (the first was Green's theorem).

THEOREM 1 STOKES'S THEOREM[†]

Let $\mathbf{F}(x, y, z) = P(x, y, z)\mathbf{i} + Q(x, y, z)\mathbf{j} + R(x, y, z)\mathbf{k}$ be continuously differentiable on a bounded region W in space that contains the smooth surface S, and let ∂S be a piecewise smooth, simple closed curve traversed in the positive sense. Then

$$\oint_{\partial S} \mathbf{F} \cdot \mathbf{T}\, ds = \iint_S \text{curl}\, \mathbf{F} \cdot \mathbf{n}\, d\sigma \tag{1}$$

where $\mathbf{T}$ is the unit tangent vector to the curve ∂S at a point (x, y, z) and $\mathbf{n}$ is the outward unit normal vector to the surface S at the point (x, y, z). ■

The proof of this theorem is difficult and is given at the end of the section.

REMARK 1: If ∂S is given parametrically by $\mathbf{x}(t) = x(t)\mathbf{i} + y(t)\mathbf{j} + z(t)\mathbf{k}$, then by Theorem 2.5.3,

$$\mathbf{T} = \frac{d\mathbf{x}}{ds}, \quad \text{or} \quad d\mathbf{x} = \mathbf{T}\, ds. \tag{2}$$

Thus Stokes's theorem can be written as

ALTERNATIVE FORM OF STOKES'S THEOREM

$$\oint_{\partial S} \boldsymbol{F} \cdot \boldsymbol{dx} = \iint_S \textit{curl}\, \boldsymbol{F} \cdot \boldsymbol{n}\, d\sigma. \tag{3}$$

REMARK 2: Using the notation given in equation (5.2.15) on page 299 we can also write Stokes's theorem as

SECOND ALTERNATIVE FORM OF STOKES'S THEOREM

$$\oint_{\partial S} P\, dx + Q\, dy + R\, dz = \iint_S \text{curl}\, \mathbf{F} \cdot \mathbf{n}\, d\sigma. \tag{4}$$

[†] Named after the English mathematician and physicist Sir G. G. Stokes (1819–1903). Stokes is also known as one of the first to discuss the notion of uniform convergence (in 1848).

EXAMPLE 1

EVALUATING A LINE INTEGRAL USING STOKES'S THEOREM

Use Stokes's theorem to evaluate $\oint_C \mathbf{F} \cdot \mathbf{dx}$, where $\mathbf{F}(x, y, z) = (z - 2y)\mathbf{i} + (3x - 4y)\mathbf{j} + (z + 3y)\mathbf{k}$ and C is the unit circle in the plane $z = 2$.

SOLUTION: C is given parametrically by $\mathbf{x}(t) = (\cos t)\mathbf{i} + (\sin t)\mathbf{j} + 2\mathbf{k}$. Clearly, C bounds the unit disk $x^2 + y^2 \leq 1$, $z = 2$ in the plane $z = 2$. (It bounds many surfaces—for example, a hemisphere with this circle as its base. But the disk is the simplest one to use.) Then

$$\oint_C \mathbf{F} \cdot \mathbf{dx} = \iint_S \operatorname{curl} \mathbf{F} \cdot \mathbf{n}\, d\sigma.$$

We compute

$$\operatorname{curl} \mathbf{F} = \begin{vmatrix} \mathbf{i} & \mathbf{j} & \mathbf{k} \\ \dfrac{\partial}{\partial x} & \dfrac{\partial}{\partial y} & \dfrac{\partial}{\partial z} \\ z - 2y & 3x - 4y & z + 3y \end{vmatrix} = 3\mathbf{i} + \mathbf{j} + 5\mathbf{k},$$

and $\mathbf{n} = \mathbf{k}$ (since $\mathbf{k}$ is normal to any vector lying in a plane parallel to the xy-plane). Thus

$$\operatorname{curl} \mathbf{F} \cdot \mathbf{n} = 5,$$

so that

$$\oint_C \mathbf{F} \cdot \mathbf{dx} = 5 \iint_S d\sigma.$$

But $\iint_S d\sigma =$ the surface area of $S =$ the area of the unit disk $= \pi$. Thus,

$$\oint_C \mathbf{F} \cdot \mathbf{dx} = 5\pi.$$

EXAMPLE 2

EVALUATING A LINE INTEGRAL AROUND A TRIANGLE USING STOKES'S THEOREM

Compute $\oint_C \mathbf{F} \cdot \mathbf{dx}$, where $\mathbf{F}$ is as in Example 1 and C is the boundary of the triangle joining the points $(1, 0, 0)$, $(0, 1, 0)$, and $(0, 0, 1)$.

SOLUTION: We already have found that $\operatorname{curl} \mathbf{F} = 3\mathbf{i} + \mathbf{j} + 5\mathbf{k}$. The curve C lies in a plane. Two vectors on the plane are $\mathbf{i} - \mathbf{j}$ and $\mathbf{j} - \mathbf{k}$ (explain why). A normal vector to the plane is therefore given by

$$\mathbf{N} = (\mathbf{i} - \mathbf{j}) \times (\mathbf{j} - \mathbf{k}) = \begin{vmatrix} \mathbf{i} & \mathbf{j} & \mathbf{k} \\ 1 & -1 & 0 \\ 0 & 1 & -1 \end{vmatrix} = \mathbf{i} + \mathbf{j} + \mathbf{k},$$

so that $\mathbf{n} = (1/\sqrt{3})(\mathbf{i} + \mathbf{j} + \mathbf{k})$ is the required outward unit normal vector. Thus, $\operatorname{curl} \mathbf{F} \cdot \mathbf{n} = 9/\sqrt{3} = 3\sqrt{3}$, and we have

$$\oint_C \mathbf{F} \cdot \mathbf{dx} = 3\sqrt{3} \iint_S d\sigma = 3\sqrt{3} \times (\text{area of the triangle}).$$

The triangle is an equilateral triangle with the length, ℓ, of one side given by

$$\ell = \text{distance between } (1, 0, 0) \text{ and } (0, 1, 0) = \sqrt{2}$$

Then the area of the triangle is $\sqrt{3}/2$ (since the area of an equilateral triangle with side s is $(\sqrt{3}/4)s^2$). Thus,

$$\oint_C \mathbf{F} \cdot \mathbf{dx} = (3\sqrt{3})\left(\frac{\sqrt{3}}{2}\right) = \frac{9}{2}.$$

REMARK: In the preceding example, tangent vectors to the surface (a plane) are vectors lying on the plane. The vector $\mathbf{j} - \mathbf{k}$ is obtained from the vector $\mathbf{i} - \mathbf{j}$ by rotating $\mathbf{i} - \mathbf{j}$ 120° in the counterclockwise direction when viewed from the side of the plane that does not contain the origin. This procedure leads to $\mathbf{n} = (1/\sqrt{3})(\mathbf{i} + \mathbf{j} + \mathbf{k})$. If we chose the "other" counterclockwise direction, then we would obtain $\mathbf{n} = -(1/\sqrt{3})(\mathbf{i} + \mathbf{j} + \mathbf{k})$ and $\oint_C \mathbf{F} \cdot \mathbf{dx} = -\frac{9}{2}$. Which answer is correct? The first, because we choose $\mathbf{n}$ so that its $\mathbf{k}$-component is positive. If $\mathbf{n} = \dfrac{1}{\sqrt{3}}(\mathbf{i} + \mathbf{j} + \mathbf{k})$, then the $\mathbf{k}$-component is $\dfrac{1}{\sqrt{3}} > 0$. In the other case the $\mathbf{k}$-component is $-\dfrac{1}{\sqrt{3}} < 0$.

From Theorem 5.7.2, in a simply connected region, $\mathbf{F}$ is the gradient of a function f if and only if curl $\mathbf{F} = \mathbf{0}$. This result provides another proof of the fact that

$$\oint_C \mathbf{F} \cdot \mathbf{dx} = 0$$

if $\mathbf{F}$ is the gradient of a function f, since

$$\oint_C \mathbf{F} \cdot \mathbf{dx} = \iint_S \text{curl } \mathbf{F} \cdot \mathbf{n}\, d\sigma = 0,$$

where S is any smooth surface whose boundary is C.

We can combine several results to obtain the following theorem.

THEOREM 2 FIVE EQUIVALENT CONDITIONS

Let $\mathbf{F} = P(x, y, z)\mathbf{i} + Q(x, y, z)\mathbf{j} + R(x, y, z)\mathbf{k}$ be continuously differentiable on a simply connected region S in space. Then the following conditions are equivalent. That is, if one is true, all are true.

i. $\mathbf{F}$ is the gradient of a differentiable function f.

ii. $\dfrac{\partial P}{\partial y} = \dfrac{\partial Q}{\partial x}$, $\dfrac{\partial R}{\partial x} = \dfrac{\partial P}{\partial z}$, and $\dfrac{\partial Q}{\partial z} = \dfrac{\partial R}{\partial y}$.

iii. $\int_C \mathbf{F} \cdot \mathbf{dx} = 0$ for every piecewise smooth, simple closed curve C lying in S.

iv. $\int_C \mathbf{F} \cdot \mathbf{dx}$ is independent of path.

v. Curl $\mathbf{F} = \mathbf{0}$. ■

Many interesting physical results follow from Stokes's theorem, including the following.

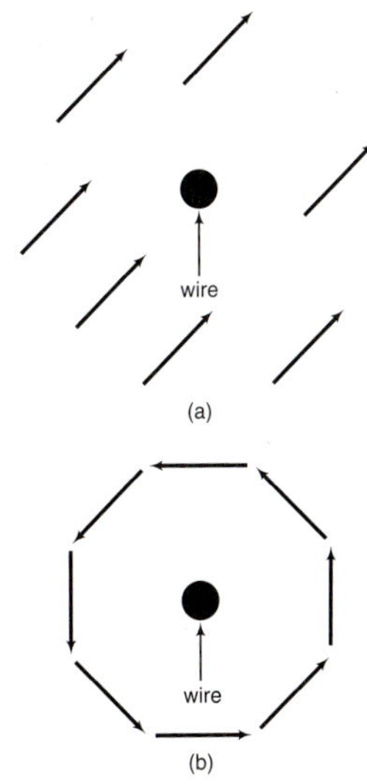

FIGURE 2
In (a) there is no current; in (b) there is a strong current in the wire.

AMPERE'S LAW

Suppose a steady current is flowing in a wire with an electric current density given by the vector field **i**. It is well known in physics that such a current will set up a magnetic field, which is usually denoted by **B**. In Figure 2 we see a collection of compass needles near a wire carrying (a) no current and (b) a very strong current. A special case of one of the famous Maxwell equations† states that

$$\text{curl } \mathbf{B} = \mathbf{i}. \tag{5}$$

From this equation we can deduce an important physical law. Let S be a surface with smooth boundary ∂S. Then $\oint_{\partial S} \mathbf{B} \cdot \mathbf{T}\, ds$ is defined as the **circulation** of the magnetic field around ∂S. By Stokes's theorem

$$\oint_{\partial S} \mathbf{B} \cdot \mathbf{T}\, ds = \iint_S \text{curl } \mathbf{B} \cdot \mathbf{n}\, d\sigma.$$

But by (5), $\iint_S \text{curl } \mathbf{B} \cdot \mathbf{n}\, d\sigma = \iint_S \mathbf{i} \cdot \mathbf{n}\, d\sigma$, so

$$\oint_{\partial S} \mathbf{B} \cdot \mathbf{T}\, ds = \iint_S \mathbf{i} \cdot \mathbf{n}\, d\sigma. \tag{6}$$

In other words, (6) states that *the total current flowing through a surface S is equal to the circulation of the magnetic field induced by* **i** *around the boundary of S*. This important result is known as **Ampère's law**.‡

We close this section with a proof of Stokes's theorem.

PROOF OF STOKES'S THEOREM (OPTIONAL): We have $\mathbf{F} = P\mathbf{i} + Q\mathbf{j} + R\mathbf{k}$ and

$$\text{curl } \mathbf{F} = \left(\frac{\partial R}{\partial y} - \frac{\partial Q}{\partial z}\right)\mathbf{i} + \left(\frac{\partial P}{\partial z} - \frac{\partial R}{\partial x}\right)\mathbf{j} + \left(\frac{\partial Q}{\partial x} - \frac{\partial P}{\partial y}\right)\mathbf{k}. \tag{7}$$

The surface S is given by S: $z = f(x, y)$ for (x, y) in Ω. From equation (13) on page 335

$$\mathbf{n} = \frac{-f_x\mathbf{i} - f_y\mathbf{j} + \mathbf{k}}{\sqrt{f_x^2 + f_y^2 + 1}} \tag{8}$$

and from equation (12) on page 327

$$d\sigma = \sqrt{f_x^2 + f_y^2 + 1}\, dx\, dy \tag{9}$$

Using (7), (8), and (9), we obtain

$$\iint_S \text{curl } \mathbf{F} \cdot \mathbf{n}\, d\sigma =$$

$$\iint_\Omega \left[\left(\frac{\partial R}{\partial y} - \frac{\partial Q}{\partial z}\right)\left(-\frac{\partial f}{\partial x}\right) + \left(\frac{\partial P}{\partial z} - \frac{\partial R}{\partial x}\right)\left(-\frac{\partial f}{\partial y}\right) + \left(\frac{\partial Q}{\partial x} - \frac{\partial P}{\partial y}\right)\right] dx\, dy. \tag{10}$$

† James Maxwell (1831–1879) was a British physicist. He formulated four equations, known as **Maxwell's equations**, which were supposed to explain all electromagnetic phenomena.

‡ Named after André-Marie Ampère (1775–1836), a French physicist, mathematician, chemist, and philosopher. Ampère is credited with founding the science of electromagnetics (which he named electrodynamics). The unit of electric current is named after him.

Suppose that $\mathbf{x}(t) = x(t)\mathbf{i} + y(t)\mathbf{j}$ for $t_0 \le t \le t_1$ parametrizes $\partial\Omega$. Then ∂S is parametrized by

$$\mathbf{x}(t) = x(t)\mathbf{i} + y(t)\mathbf{j} + f(x(t), y(t))\mathbf{k}, \qquad t_0 \le t \le t_1.$$

Thus

$$\oint_{\partial S} \mathbf{F}\cdot\mathbf{T}\, ds = \oint_{\partial S} \mathbf{F}\cdot d\mathbf{x} = \int_{t_0}^{t_1} \left(P\frac{dx}{dt} + Q\frac{dy}{dt} + R\frac{dz}{dt}\right) dt. \tag{11}$$

By the chain rule,

$$\frac{dz}{dt} = \frac{\partial z}{\partial x}\frac{dx}{dt} + \frac{\partial z}{\partial y}\frac{dy}{dt}. \tag{12}$$

Substituting (12) into (11) yields

$$\begin{aligned}\oint_{\partial S} \mathbf{F}\cdot\mathbf{T}\, ds &= \int_{t_0}^{t_1}\left[\left(P + R\frac{\partial z}{\partial x}\right)\frac{dx}{dt} + \left(Q + R\frac{\partial z}{\partial y}\right)\frac{dy}{dt}\right] dt \\ &= \oint_{\partial\Omega} \left(P + R\frac{\partial z}{\partial x}\right) dx + \left(Q + R\frac{\partial z}{\partial y}\right) dy.\end{aligned} \tag{13}$$

We apply Green's theorem to (13) to obtain

$$\oint_{\partial S} \mathbf{F}\cdot\mathbf{T}\, ds = \iint_{\Omega} \left\{\frac{\partial[Q + R(\partial z/\partial y)]}{\partial x} - \frac{\partial[P + R(\partial z/\partial x)]}{\partial y}\right\} dx\, dy. \tag{14}$$

We now apply the chain rule again. This is a bit complicated. For example,

$$\frac{\partial}{\partial x}Q(x, y, z) = \frac{\partial Q}{\partial x} + \frac{\partial Q}{\partial y}\frac{\partial y}{\partial x} + \frac{\partial Q}{\partial z}\frac{\partial z}{\partial x} = \frac{\partial Q}{\partial x} + \frac{\partial Q}{\partial z}\frac{\partial z}{\partial x}$$

since y is not a function of x (so that $\partial y/\partial x = 0$). Also

$$\begin{aligned}\frac{\partial}{\partial x}\left(R\frac{\partial z}{\partial y}\right) &= \left[\frac{\partial}{\partial x}R(x, y, z)\right]\frac{\partial z}{\partial y} + R\frac{\partial}{\partial x}\left\{\frac{\partial[z(x, y)]}{\partial y}\right\} \\ &= \frac{\partial R}{\partial x}\frac{\partial z}{\partial y} + \frac{\partial R}{\partial z}\frac{\partial z}{\partial x}\frac{\partial z}{\partial y} + R\frac{\partial^2 z}{\partial x\,\partial y} + R\frac{\partial^2 z}{\partial y^2}\frac{\partial y}{\partial x}\end{aligned}$$

and the last term is zero because y is not a function of x. Differentiating inside the right-hand integral in (14), we obtain

$$\begin{aligned}\oint_{\partial S} \mathbf{F}\cdot\mathbf{T}\, ds = \iint_{\Omega}\Bigg[&\left(\frac{\partial Q}{\partial x} + \frac{\partial Q}{\partial z}\frac{\partial z}{\partial x} + \frac{\partial R}{\partial x}\frac{\partial z}{\partial y} + \frac{\partial R}{\partial z}\frac{\partial z}{\partial x}\frac{\partial z}{\partial y} + R\frac{\partial^2 z}{\partial x\,\partial y}\right) \\ &- \left(\frac{\partial P}{\partial y} + \frac{\partial P}{\partial z}\frac{\partial z}{\partial y} + \frac{\partial R}{\partial y}\frac{\partial z}{\partial x} + \frac{\partial R}{\partial z}\frac{\partial z}{\partial y}\frac{\partial z}{\partial x} + R\frac{\partial^2 z}{\partial y\,\partial x}\right)\Bigg] dz\, dy \\ \overset{\text{After terms cancel}}{=} \iint_{\Omega}&\left(\frac{\partial Q}{\partial x} + \frac{\partial Q}{\partial z}\frac{\partial z}{\partial x} + \frac{\partial R}{\partial x}\frac{\partial z}{\partial y} - \frac{\partial P}{\partial y} - \frac{\partial P}{\partial z}\frac{\partial z}{\partial y} - \frac{\partial R}{\partial y}\frac{\partial z}{\partial x}\right) dx\, dy.\end{aligned} \tag{15}$$

After rearranging and noting that $\partial f/\partial x = \partial z/\partial x$ and $\partial f/\partial y = \partial z/\partial y$, we see that the integrals in (10) and (15) are identical. Thus

$$\iint_S \text{curl } \mathbf{F} \cdot \mathbf{n} \, d\sigma = \oint_{\partial S} \mathbf{F} \cdot \mathbf{T} \, ds. \qquad \blacksquare$$

NOTE: We have proved this theorem only under the simplifying assumption that the surface can be written in the form $z = f(x, y)$. The proof for more general surfaces is best left to an advanced calculus book.

PROBLEMS 5.8

SELF-QUIZ

I. Let $\mathbf{F}(x, y, z) = x\mathbf{i} + x\mathbf{j} + x\mathbf{k}$ and let C be the square with vertices (0, 0, 0), (1, 0, 0), (1, 1, 0), and (0, 1, 0) traversed counterclockwise. Then $\oint_C \mathbf{F} \cdot \mathbf{dx} =$ ________.

a. $\int_0^1 \int_0^1 -x \, dx \, dy$

b. $\int_0^1 \int_0^1 x \, dx \, dy$

c. $\int_0^1 \int_0^1 -1 \, dx \, dy$

d. $\int_0^1 \int_0^1 1 \, dx \, dy$

II. Let $\mathbf{F}(x, y, z) = y^3\mathbf{i} + y^2\mathbf{j} + y\mathbf{k}$ and let C be the square with vertices (0, 0, 0), (1, 0, 0), (1, 1, 0), and (0, 1, 0) traversed counterclockwise. Then $\oint_C \mathbf{F} \cdot \mathbf{dx} =$ ________.

a. $\int_0^1 \int_0^1 -3y^2 \, dx \, dy$

b. $\int_0^1 \int_0^1 2y \, dx \, dy$

c. $\int_0^1 \int_0^1 -1 \, dx \, dy$

d. $\int_0^1 \int_0^1 0 \, dx \, dy$

III. Let $\mathbf{F}(x, y, z) = z^3\mathbf{i} + z^2\mathbf{j} + z\mathbf{k}$ and let C be the square with vertices (0, 0, 5), (1, 0, 5), (1, 1, 5), and (0, 1, 5) traversed counterclockwise. Then $\oint_C \mathbf{F} \cdot \mathbf{dx} =$ ________.

a. $\int_0^1 \int_0^1 -3(5^2) \, dx \, dy$

b. $\int_0^1 \int_0^1 (2z - 3z^2) \, dx \, dy$

c. $\int_0^1 \int_0^1 -65 \, dx \, dy$

d. $\int_0^1 \int_0^1 0 \, dx \, dy$

IV. Let $\mathbf{F}(x, y, z) = 3\mathbf{i} + 4\mathbf{j} + 5\mathbf{k}$ and let C be the square with vertices (0, 0, 0), (0, 1, 0), (1, 1, 0), and (1, 0, 0) traversed clockwise. Then $\oint_C \mathbf{F} \cdot \mathbf{dx} =$ ________.

a. $\int_0^1 \int_0^1 3 \, dx \, dy$

b. $\int_0^1 \int_0^1 0 \, dx \, dy$

c. $\int_0^1 \int_0^1 -\sqrt{50} \, dx \, dy$

d. $\int_0^1 \int_0^1 (-1/\sqrt{3}) \, dx \, dy$

In Problems 1–8, evaluate the line integral by using Stokes's theorem. In each case S is the plane region enclosed by C.

1. $\oint_C \mathbf{F} \cdot \mathbf{dx}$, where $\mathbf{F}(x, y, z) = (x + y)\mathbf{i} + (z - 2x + y)\mathbf{j} + (y - z)\mathbf{k}$ and C is the unit circle in the plane $z = 5$.

2. $\oint_C \mathbf{F} \cdot \mathbf{dx}$, where $\mathbf{F} = ax\mathbf{i} + by\mathbf{j} + cz\mathbf{k}$ and C is the unit circle in the plane $z = -7$.

3. $\oint_C \mathbf{F} \cdot \mathbf{dx}$, where $\mathbf{F}$ is as in Example 1 but C is the triangle joining the points (2, 0, 0), (0, 2, 0), and (0, 0, 2).

4. $\oint_C \mathbf{F} \cdot \mathbf{dx}$, where $\mathbf{F}$ is as in Example 2 but C is the triangle with vertices $(d, 0, 0)$, $(0, d, 0)$, and $(0, 0, d)$.

***5.** $\oint_C x^3y^2 \, dx + 2xyz^3 \, dy + 3xy^2z^2 \, dz$, where C is given parametrically by $\mathbf{x}(t) = 2 \cos t\mathbf{i} + 3\mathbf{j} + 2 \sin t\mathbf{k}$, $0 \le t \le 2\pi$. [*Hint:* C is a circle.]

6. $\oint_C \mathbf{F} \cdot \mathbf{dx}$, where $\mathbf{F} = 2y(x - z)\mathbf{i} + (x^2 + z^2)\mathbf{j} +$

$y^3\mathbf{k}$ and C is the square $\{(x, y, z): 0 \le x \le 3, 0 \le y \le 3, z = 4\}$.

7. $\oint_C e^x\,dx + x\sin y\,dy + (y^2 - x^2)\,dz$, where C is the equilateral triangle formed by the intersection of the plane $x + y + z = 3$ with the three coordinate planes.

8. $\oint_C \mathbf{F}\cdot\mathbf{T}\,ds$, where $\mathbf{F} = -3y\mathbf{i} + 3x\mathbf{j} + \mathbf{k}$ and C is the circle $\{(x, y, z): x^2 + y^2 = 1, z = 3\}$.

In Problems 9–12, verify Stokes's theorem for the given S and $\mathbf{F}$.

9. $\mathbf{F} = z^2\mathbf{i} + x^2\mathbf{j} + y^2\mathbf{k}$; S is the part of the plane satisfying $x + y + z = 1$ which lies in the first octant (i.e., $x \ge 0$, $y \ge 0$, $z \ge 0$).

10. $\mathbf{F} = y^2\mathbf{i} + z^2\mathbf{j} + x^2\mathbf{k}$ and S is the part of the plane $x + 2y + 3z = 6$ lying in the first octant.

11. $\mathbf{F} = 2y\mathbf{i} + x^2\mathbf{j} + 3x\mathbf{k}$; S is the hemisphere $x^2 + y^2 + z^2 = 16$, $z \ge 0$.

12. $\mathbf{F} = y\mathbf{i} - x\mathbf{j} - z\mathbf{k}$; S is the circle $x^2 + y^2 \le 9$, $z = 2$.

***13.** Let $\mathbf{E}$ be an electric field and let $\mathbf{B}(t)$ be a magnetic field in space induced by $\mathbf{E}$. One of **Maxwell's equations** states that $\text{curl}\,\mathbf{E} = -\dfrac{\partial\mathbf{B}}{\partial t}$.

Let S be a surface in space with smooth boundary C. We define

$$\text{voltage drop around } C = \oint_C \mathbf{E}\cdot\mathbf{dx}.$$

Prove **Faraday's law**, which states that the voltage drop around C is equal to the time rate of decrease of the magnetic flux through S. (Flux of a vector field over a surface is discussed preceding Problem 5.6.21.) [*Warning:* At one point in your work, it will be necessary to interchange the order of differentiation and integration. A proof that such manipulation is "legal" requires techniques from advanced calculus; for now, just assume that it can be done.]

14. Show that Green's theorem is really a special case of Stokes's theorem.

15. Let S be a sphere. Use Stokes's theorem to show that $\iint_S [\text{curl}\,\mathbf{F}]\cdot\mathbf{n}\,d\sigma = 0$ if $\mathbf{F}$ is smooth over a region containing S.

ANSWERS TO SELF-QUIZ

I. d **II.** a **III.** d **IV.** b

5.9 THE DIVERGENCE THEOREM

In this section we discuss the third major result in vector integral calculus: the **divergence theorem**.

Green's theorem gives us a relationship between a line integral in $\mathbb{R}^2$ around a closed curve and a double integral over the region in $\mathbb{R}^2$ enclosed by that curve. Stokes's theorem shows a relationship between a line integral in $\mathbb{R}^3$ around a closed curve and a surface integral over a surface that has the closed curve as a boundary. As we will see, the divergence theorem gives us a relationship between a surface integral over a closed surface and a triple integral over the solid bounded by that surface.

Let S be a surface that forms the complete boundary of a solid W in space. A typical region and its boundary are sketched in Figure 1. We will assume that S is smooth or piecewise smooth. This assumption ensures that $\mathbf{n}(x, y, z)$, the outward unit normal to S, is continuous or piecewise continuous as a function of (x, y, z).

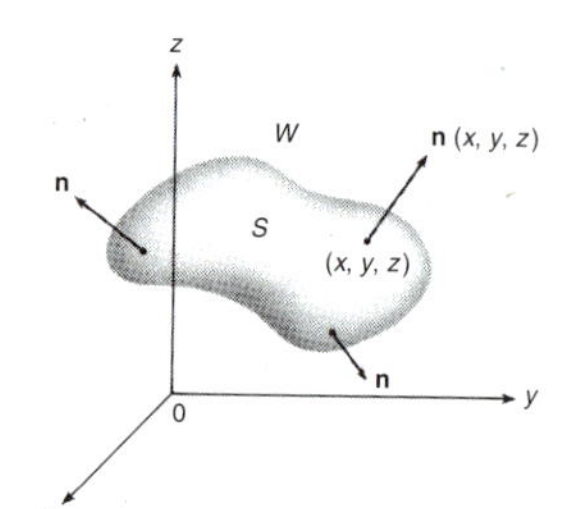

FIGURE 1 The surface S is the boundary of the solid W.

THEOREM 1 THE DIVERGENCE THEOREM†

Let W be a solid in $\mathbb{R}^3$ totally bounded by the smooth or piecewise smooth‡ surface S. Let $\mathbf{F}$ be a smooth vector field on W, and let $\mathbf{n}$ denote the outward unit normal to S.

† This theorem is also known as **Gauss's theorem**, named after the German mathematician Carl Friedrich Gauss (1777–1855). See the biographical sketch on page 353.

‡ A surface is **piecewise smooth** if it can be written as a finite union of smooth surfaces.

Then

$$\iint_S \mathbf{F} \cdot \mathbf{n}\, d\sigma = \iiint_W \operatorname{div} \mathbf{F}\, dV. \tag{1}$$

■

REMARK: Just as Green's theorem transforms a line integral to an ordinary double integral, the divergence theorem transforms a surface integral to an ordinary triple integral.

The proof of the divergence theorem is given at the end of the section.

EXAMPLE 1

EVALUATING A SURFACE INTEGRAL USING THE DIVERGENCE THEOREM

Compute $\iint_S \mathbf{F} \cdot \mathbf{n}\, d\sigma$, where $\mathbf{F} = 2xy\mathbf{i} + 3y\mathbf{j} + 2z\mathbf{k}$ and S is the boundary of the solid bounded by the three coordinate planes and the plane $x + y + z = 1$.

SOLUTION: $\operatorname{div} \mathbf{F} = 2y + 3 + 2 = 2y + 5$, so

$$\begin{aligned}
\iint_S \mathbf{F} \cdot \mathbf{n}\, d\sigma &= \iiint_W (2y + 5)\, dx\, dy\, dz, \quad \int_0^1 \int_0^{1-y} \int_0^{1-x-y} (2y + 5)\, dz\, dx\, dy \\
&= \int_0^1 \int_0^{1-y} (2y + 5)(1 - x - y)\, dx\, dy \\
&= \int_0^1 \left[-(2y + 5) \frac{(1 - x - y)^2}{2} \Bigg|_{x=0}^{x=1-y} \right] dy \\
&= \frac{1}{2} \int_0^1 (2y + 5)(1 - y)^2\, dy = \frac{11}{12}.
\end{aligned}$$

EXAMPLE 2

EVALUATING A SURFACE INTEGRAL USING THE DIVERGENCE THEOREM

Compute $\iint_S \mathbf{F} \cdot \mathbf{n}\, d\sigma$, where $\mathbf{F} = (5x + \sin y \tan z)\mathbf{i} + (y^2 - e^{x-2\cos z^3})\mathbf{j} + (4xy)^{3/5}\mathbf{k}$ and S is the boundary of the solid bounded by the parabolic cylinder $y = 9 - x^2$, the plane $y + z = 1$, and the xy- and xz-planes.

SOLUTION: The solid is sketched in Figure 2. It would be difficult to compute this surface integral directly, but with the divergence theorem it becomes relatively easy. Since $\operatorname{div} \mathbf{F} = 5 + 2y$, we have

$$\begin{aligned}
\iint_S \mathbf{F} \cdot \mathbf{n}\, d\sigma &= \iiint_W (5 + 2y)\, dx\, dy\, dz \\
&= \int_0^1 \int_{-\sqrt{9-y}}^{\sqrt{9-y}} \int_0^{1-y} (5 + 2y)\, dz\, dx\, dy \\
&= 2 \int_0^1 \int_0^{\sqrt{9-y}} (5 + 2y)(1 - y)\, dx\, dy \\
&= 2 \int_0^1 \int_0^{\sqrt{9-y}} (5 - 3y - 2y^2)\, dx\, dy
\end{aligned}$$

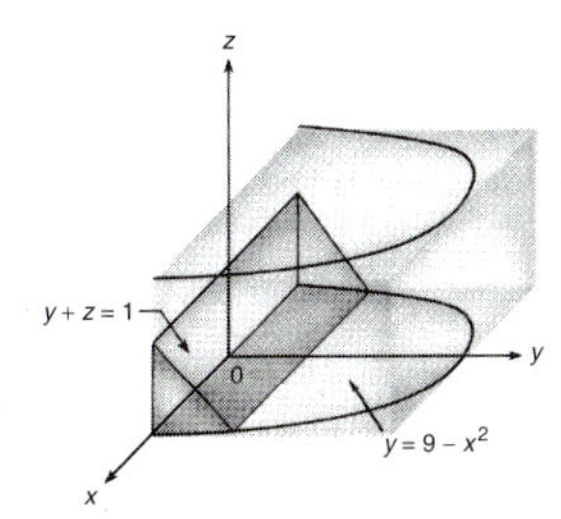

FIGURE 2
The solid bounded by $y = 9 - x^2$, $y + z = 1$, and the xy- and xz-planes

$$= 2\int_0^1 (5 - 3y - 2y^2)\sqrt{9 - y}\, dy$$

$$= \frac{56}{15} 8^{5/2} + \frac{64}{105} 8^{7/2} + 180 - \frac{180{,}792}{105} \approx 16.66.$$

The integral was obtained by integrating by parts.

We now prove the divergence theorem.

PROOF OF THE DIVERGENCE THEOREM: In the proof of Green's theorem we assumed (see Figure 5.4.4) that the region Ω could be enclosed by two curves—an upper curve $y = g_2(x)$ and a lower curve $y = g_1(x)$ or a left curve $x = h_1(y)$ and a right curve $x = h_2(y)$. This is equivalent to saying that any line parallel to the x- or y-axis crosses $\partial\Omega$ in at most two points. In order to prove the divergence theorem, we assume much the same thing. That is, we assume that W can be enclosed by an upper surface and a lower surface so that any line parallel to one of the coordinate axes crosses S in at most two points. This assumption is not necessary; smoothness of S is sufficient. But it allows us to give a reasonably simple proof.

According to equation (14) on page 335,

$$\cos\gamma = \mathbf{n}\cdot\mathbf{k}$$

where γ is the acute angle between $\mathbf{n}$ and the positive z-axis. Similarly, if we define α and β to be the acute angles between $\mathbf{n}$ and the positive x- and y-axes, respectively, then

$$\cos\alpha = \mathbf{n}\cdot\mathbf{i} \qquad \text{and} \qquad \cos\beta = \mathbf{n}\cdot\mathbf{j}.$$

This means we can write $\mathbf{n}$ in terms of its components as

$$\mathbf{n} = (\cos\alpha)\mathbf{i} + (\cos\beta)\mathbf{j} + (\cos\gamma)\mathbf{k}. \tag{2}$$

If $\mathbf{F} = P\mathbf{i} + Q\mathbf{j} + R\mathbf{k}$, then using equation (2), we can write the divergence theorem in the form

$$\iint_S (P\cos\alpha + Q\cos\beta + R\cos\gamma)\, d\sigma = \iiint_W \operatorname{div}\mathbf{F}\, dx\, dy\, dz, \tag{3}$$

where α, β, and γ are the acute angles that $\mathbf{n}$ makes with the positive coordinate axes.

From (3) it is sufficient to prove that

$$\iint_S P\cos\alpha\, d\sigma = \iiint_W \frac{\partial P}{\partial x}\, dx\, dy\, dz, \tag{4}$$

$$\iint_S Q\cos\beta\, d\sigma = \iiint_W \frac{\partial Q}{\partial y}\, dx\, dy\, dz, \tag{5}$$

and

$$\iint_S R\cos\gamma\, d\sigma = \iiint_W \frac{\partial R}{\partial z}\, dx\, dy\, dz. \tag{6}$$

The proofs of (4), (5), and (6) are virtually identical, so we will prove (6) only and leave (4) and (5) to you. By the assumption stated above, we can think of S as "walnut shaped." That is, S consists of two surfaces: an upper surface S_2: $z = f_2(x, y)$

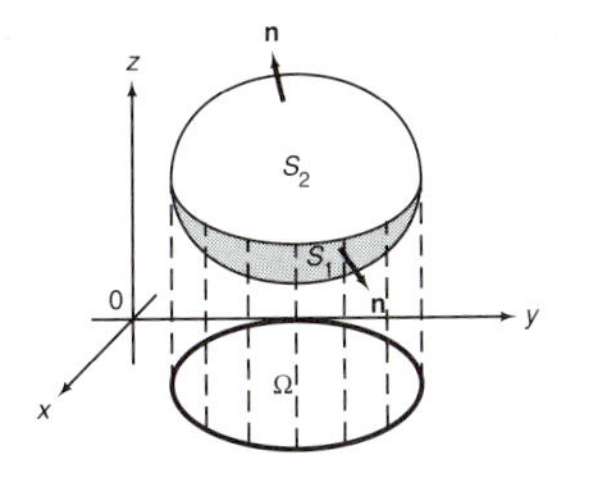

FIGURE 3
An upper surface and a lower surface

and a lower surface S_1: $z = f_1(x, y)$. This is depicted in Figure 3. We then have

$$\iint_S R \cos \gamma \, d\sigma = \iint_{S_1} R \cos \gamma \, d\sigma + \iint_{S_2} R \cos \gamma \, d\sigma.$$

Now from equation (5.6.16),

$$\begin{aligned}\iint_{S_2} R(x, y, z) \cos \gamma \, d\sigma &= \iint_\Omega R(x, y, f_2(x, y)) \cos \gamma \sec \gamma \, dx \, dy \\ &= \iint_\Omega R(x, y, f_2(x, y)) \, dx \, dy,\end{aligned} \tag{7}$$

where Ω is the projection of S_2 into the xy-plane. On S_1 the situation is a bit different because the outer normal **n** on S_1 points downward and we assumed in deriving equation (5.6.16), that **n** points upward (so that the angle γ between **n** and the positive z-axis is an acute angle). We solve this problem by using $-\mathbf{n}$ in (5.6.16) so that on S_1 we have

$$\begin{aligned}\iint_{S_1} R(x, y, z) \cos \gamma \, d\sigma &= -\iint_\Omega R(x, y, f_1(x, y)) \cos \gamma \sec \gamma \, dx \, dy \\ &= -\iint_\Omega R(x, y, f_1(x, y)) \, dx \, dy.\end{aligned}$$

Thus

$$\begin{aligned}\iint_S R \cos \gamma \, ds &= \iint_\Omega [R(x, y, f_2(x, y)) - R(x, y, f_1(x, y)] \, dx \, dy \\ &= \iint_\Omega \left(\int_{f_1(x, y)}^{f_2(x, y)} \frac{\partial R}{\partial z} \, dz \right) dx \, dy = \iiint_W \frac{\partial R}{\partial z} \, dz \, dx \, dy,\end{aligned}$$

and under the restrictions already mentioned, this completes the proof. ■

OF HISTORICAL INTEREST

CARL FRIEDRICH GAUSS, 1777–1855

CARL FRIEDRICH GAUSS, 1777–1855

The greatest mathematician of the nineteenth century, Carl Friedrich Gauss is considered one of the three greatest mathematicians of all time—the others being Archimedes and Newton.

Gauss was born in Brunswick, Germany, in 1777. His father, a hardworking laborer who was exceptionally stubborn and did not believe in formal education, did what he could to keep Gauss from appropriate schooling. Fortunately for Carl (and for mathematics), his mother, while uneducated herself, encouraged her son in his studies and took considerable pride in his achievements until her death at the age of 97.

Gauss was a child prodigy. At the age of three, he found an error in his father's bookkeeping. A famous story tells of Carl, age 10, as a student in the local Brunswick school. The teacher there was known to assign tasks to keep his pupils busy. One day he asked his students to add the numbers from 1 to 100. Almost at

(CONTINUED)

OF HISTORICAL INTEREST (CONCLUDED)

once, Carl placed his slate face down with the words, "There it is." Afterwards, the teacher found that Gauss was the only one with the correct answer, 5050. Gauss had noticed that the numbers could be arranged in 50 pairs, each with the sum 101(1 + 100, 2 + 99, and so on) and $50 \times 101 = 5050$. Later in life, Gauss joked that he could add before he could speak.

When Gauss was 15, the Duke of Brunswick noticed him and became his patron. The Duke helped him enter Brunswick College in 1795 and, three years later, to enter the university at Göttingen. Undecided between careers in mathematics and philosophy, Gauss chose mathematics after two remarkable discoveries. First, he invented the method of least squares a decade before the result was published by Legendre. Second, a month before his 19th birthday, he solved a problem whose solution had been sought for more than two thousand years. Gauss showed how to construct, using compass and ruler, a regular polygon with the number of sides not a multiple of 2, 3, or 5. On March 30, 1796, the day of this discovery, he began a diary, which contained as its first entry rules for construction of a 17-sided regular polygon. The diary, which contains 146 statements of results in only 19 pages, is one of the most important documents in the history of mathematics.

After a short period at Göttingen, Gauss went to the University of Helmstädt and, in 1798 at the age of 20, wrote his now-famous doctoral dissertation. In it he gave the first mathematically rigorous proof of the fundamental theorem of algebra—that every polynomial of degree n, has, counting multiplicities, exactly n roots. Many mathematicians, including Euler, Newton, and Lagrange, had attempted to prove this result.

Gauss made a great number of discoveries in physics as well as in mathematics. For example, in 1801 he used a new procedure to calculate, from very little data, the orbit of the planetoid Ceres. In 1833, he invented the electromagnetic telegraph with his colleague Wilhelm Weber (1804–1891). However, while he did brilliant work in astronomy and electricity, it was Gauss's mathematical output that was astonishing. He made fundamental contributions to algebra and geometry. In 1811, he discovered a result that led to the development of complex variable theory by Cauchy. He is encountered in courses in matrix theory in the Gauss-Jordan method of elimination. Students of numerical analysis study Gaussian quadrature—a technique for numerical integration.

Gauss became a professor of mathematics at Göttingen in 1807 and remained in that post until his death in 1855. Even after his death, his mathematical spirit remained to haunt nineteenth-century mathematicians. Often it turned out that an important new result was discovered earlier by Gauss and could be found in his published notes.

In his mathematical writings, Gauss was a perfectionist and is probably the last mathematician who knew everything in his subject. Claiming that a cathedral was not a cathedral until the last piece of scaffolding was removed, he endeavored to make each of his published works complete, concise, and polished. He used a seal that pictured a tree carrying only a few fruit together with the motto *pauca sed matura* (few, but ripe). But Gauss also believed that mathematics must reflect the real world. At his death, Gauss was honored by a commemorative medal on which was inscribed "George V. King of Hanover to the Prince of Mathematicians."

PROBLEMS 5.9

SELF-QUIZ

I. Let $\mathbf{F}(x, y, z) = x\mathbf{i} + y\mathbf{j} + z\mathbf{k}$ and let S be the surface of the rectangular solid with (0, 0, 0) and (1, 2, 5) at the ends of an interior diagonal. Then $\iint_S \mathbf{F} \cdot \mathbf{n}\, d\sigma =$ __________.

a. $\oint_C \mathbf{F} \cdot \mathbf{dx}$ where C is the rectangle with vertices (0, 0, 0), (1, 0, 5), (1, 2, 5), (0, 2, 5)

b. $\int_0^1 \int_0^2 \int_0^5 [x + y + z]\, dx\, dy\, dz$

c. $\int_0^1 \int_0^1 \int_0^1 \sqrt{3}\, dx\, dy\, dz$

d. $\int_0^1 \int_0^2 \int_0^5 3\, dx\, dy\, dz$

e. $\int_0^1 \int_0^2 \int_0^5 0\, dx\, dy\, dz$

II. Let $\mathbf{F} = \nabla(xyz)$ and let S be the same surface as in the preceding problem. Then $\iint_S \mathbf{F} \cdot \mathbf{n}\, d\sigma =$ ________.

a. $\int_0^1 \int_0^2 \int_0^5 0\, dx\, dy\, dz$

b. $\int_0^1 \int_0^2 \int_0^5 (yz + xz + xy)\, dx\, dy\, dz$

c. $\int_0^1 \int_0^2 \int_0^5 (3xyz)\, dx\, dy\, dz$

e. $\int_0^1 \int_0^2 \int_0^5 3\, dx\, dy\, dz$

III. True–False: If $\mathbf{F}$ is a conservative vector field throughout a region containing the surface S, then $\iint_S \mathbf{F} \cdot \mathbf{n}\, d\sigma = 0$

In Problems 1–15, evaluate the given surface integral by using the divergence theorem.

1. $\iint_S \mathbf{F} \cdot \mathbf{n}\, d\sigma$, where $\mathbf{F}(x, y, z) = x\mathbf{i} + y\mathbf{j} + z\mathbf{k}$ and S is the unit sphere.

2. $\iint_S \mathbf{F} \cdot \mathbf{n}\, d\sigma$, where $\mathbf{F} = x^2\mathbf{i} + y^2\mathbf{j} + z^2\mathbf{k}$ and S is the unit sphere.

3. $\iint_S \mathbf{F} \cdot \mathbf{n}\, d\sigma$, where $\mathbf{F} = x\mathbf{i} + y\mathbf{j} + z\mathbf{k}$ and S is the cylinder $x^2 + y^2 = 4$, $0 \le z \le 3$.

4. $\iint_S \mathbf{F} \cdot \mathbf{n}\, d\sigma$, where $\mathbf{F} = y\mathbf{i} + z\mathbf{j} + x\mathbf{k}$ and S is as in the preceding problem.

5. $\iint_S \mathbf{F} \cdot \mathbf{n}\, d\sigma$, where $\mathbf{F} = (y^2 + z^2)^{3/2}\mathbf{i} + \sin[(x^2 - z^5)^{4/3}]\mathbf{j} + e^{x^2 - y^2}\mathbf{k}$ and S is the ellipsoid $(x/a)^2 + (y/b)^2 + (z/c)^2 = 1$.

6. $\iint_S \mathbf{F} \cdot \mathbf{n}\, d\sigma$, where $\mathbf{F} = x\mathbf{i} + y\mathbf{j} + z\mathbf{k}$ and S is the surface of the unit cube $\{(x, y, z): 0 \le x \le 1, 0 \le y \le 1, 0 \le z \le 1\}$.

7. $\iint_S \mathbf{F} \cdot \mathbf{n}\, d\sigma$, where $\mathbf{F} = x^2\mathbf{i} + y^2\mathbf{j} - xy\mathbf{k}$ and S is as in Problem 6.

8. $\iint_S \mathbf{F} \cdot \mathbf{n}\, d\sigma$, where $\mathbf{F} = xyz\mathbf{i} + yz\mathbf{j} + z\mathbf{k}$ and S is as in Problem 6.

9. $\iint_S \mathbf{F} \cdot \mathbf{n}\, d\sigma$, where $\mathbf{F} = 2x\mathbf{i} + 3y\mathbf{j} + z\mathbf{k}$ and S is the boundary of the hemisphere $\{(x, y, z): x^2 + y^2 + z^2 \le 9, z \ge 0\}$.

10. $\iint_S \mathbf{F} \cdot \mathbf{n}\, d\sigma$, where $\mathbf{F} = x^2\mathbf{i} + y^2\mathbf{j} + z^2\mathbf{k}$ and S is as in the preceding problem.

11. $\iint_S \mathbf{F} \cdot \mathbf{n}\, d\sigma$, where $\mathbf{F} = xy\mathbf{i} + y^2\mathbf{j} + yz\mathbf{k}$ and S is the boundary of the tetrahedron with vertices at (0, 0, 0), (1, 0, 0), (0, 1, 0), and (0, 0, 1).

12. $\iint_S \mathbf{F} \cdot \mathbf{n}\, d\sigma$, where $\mathbf{F} = y^2\mathbf{i} + x^2\mathbf{j} + z^2\mathbf{k}$ and S is as in the preceding problem.

13. $\iint_S \mathbf{F} \cdot \mathbf{n}\, d\sigma$, where $\mathbf{F} = x(1 - \sin y)\mathbf{i} + (y - \cos y)\mathbf{j} + z\mathbf{k}$ and S is as in Problem 11.

14. $\iint_S \mathbf{F} \cdot \mathbf{n}\, d\sigma$, where $\mathbf{F} = x\mathbf{i} + y\mathbf{j} + z\mathbf{k}$ and S is the surface of the region bounded by the parabolic cylinder $z = 1 - y^2$, the plane $x + z = 2$, and the xy- and yz-planes.

15. $\iint_S \mathbf{F} \cdot \mathbf{n}\, d\sigma$, where $\mathbf{F} = (x^2 + e^{y \cos z})\mathbf{i} + (xy - \tan z^{1/3})\mathbf{j} + (x - y^{3/5})^{2/9}\mathbf{k}$ and S is as in the preceding problem.

***16.** One of Maxwell's equations states that an electric field $\mathbf{E}$ in space satisfies div $\mathbf{E} = (1/\epsilon_0)\rho$, where ϵ_0 is a constant and ρ is the charge density. Show that the flux of the displacement $\mathbf{D} = \epsilon_0\mathbf{E}$ across a closed surface is equal to the charge q inside the surface. (This last result is known as **Gauss's law**.)

17. Show that if $\mathbf{F}$ is twice continuously differentiable, then $\iint_S [\text{curl } \mathbf{F}] \cdot \mathbf{n}\, d\sigma = 0$ for any smooth closed surface S.

18. Show that if $\mathbf{F}$ is a constant vector field and if S is a smooth closed surface, then $\iint_S \mathbf{F} \cdot \mathbf{n}\, d\sigma = 0$.

19. Let $\mathbf{F} = x\mathbf{i} + y\mathbf{j} + z\mathbf{k}$ and suppose W is a solid with a smooth closed boundary S. Show that

$$\text{volume of } W = \frac{1}{3} \iint_S \mathbf{F} \cdot \mathbf{n}\, d\sigma.$$

***20.** Suppose that a vector field $\mathbf{F}$ is tangent to a closed smooth surface S, where S is the boundary of a solid W. Show that $\iiint_W [\text{div } \mathbf{F}]\, dx\, dy\, dz = 0$.

ANSWERS TO SELF-QUIZ

I. d **II.** a **III.** False (e.g., let $\mathbf{F} = \nabla[x^2] = 2x\mathbf{i}$.)

5.10 CHANGING VARIABLES IN MULTIPLE INTEGRALS AND THE JACOBIAN

In many places in this and the last chapter we found it useful to evaluate a double integral by first converting to polar coordinates. In Section 4.4 we proved that [see equation (6) on page 272]

$$\iint_{\Omega} f(x, y)\,dx\,dy = \iint_{\Omega} f(\cos\theta,\ r\sin\theta) r\,dr\,d\theta. \tag{1}$$

In Section 4.6 we showed how to convert an ordinary triple integral from an integral in Cartesian coordinates to an integral in cylindrical or spherical coordinates.

In this section we show how, under certain conditions, it is possible to convert from one set of coordinates to another in double and triple integrals. Much of what we do is a generalization of the formula for changing variables in ordinary definite integrals. Let us recall that formula now. If $x = g(u)$ is a differentiable one-to-one function, then

$$\int_a^b f(x)\,dx = \int_c^d f(g(u))g'(u)\,du, \tag{2}$$

where $c = g^{-1}(a)$ and $d = g^{-1}(b)$.

Now suppose we wish to change the variables of integration in the double integral

$$\iint_{\Omega} f(x, y)\,dx\,dy. \tag{3}$$

We assume that the new variables are called u and v and that they are related to the old variables x and y by the relations

$$x = g(u, v) \qquad \text{and} \qquad y = h(u, v). \tag{4}$$

The functional relationship described by (4) is called a **mapping** from the uv-plane into the xy-plane. We will assume that there is a region Σ in the uv-plane that gets mapped onto Ω by the mapping described by (4) and that the mapping is one-to-one. That is:

i. For every $(x, y) \in \Omega$, there is a $(u, v) \in \Sigma$ such that $x = g(u, v)$ and $y = h(u, v)$.

ii. If $g(u_1, v_1) = g(u_2, v_2)$ and $h(u_1, v_1) = h(u_2, v_2)$, then $u_1 = u_2$ and $v_1 = v_2$.

Condition (ii) is the natural extension of the definition of one-to-one for functions of two variables. We illustrate what is going on in Figure 1.

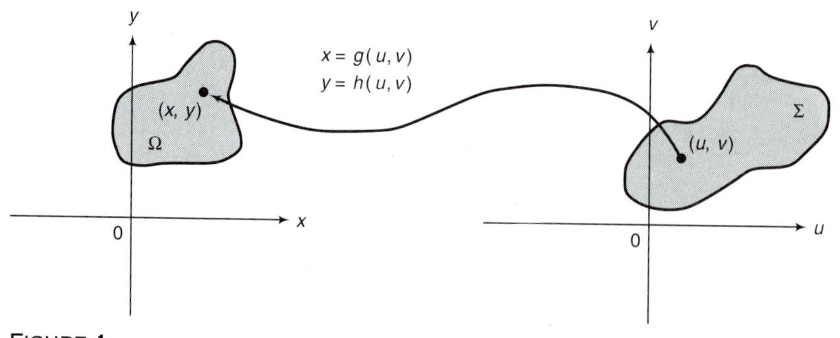

FIGURE 1
The mapping takes a point (u, v) in Σ (in the uv-plane) into a point (x, y) in Ω (in the xy-plane).

EXAMPLE 1

FINDING THE IMAGES OF VERTICAL LINES UNDER A MAPPING

Let $x = u^2 - v^2$ and $y = 3uv$. We will compute the image of the vertical line $u = k$ in the uv-plane under the mapping. Substituting k for u in the equations given above, we have

$$x = k^2 - v^2 \qquad \text{and} \qquad y = 3kv.$$

Thus $v = y/3k$ and

$$x = k^2 - \frac{y^2}{9k^2}. \tag{5}$$

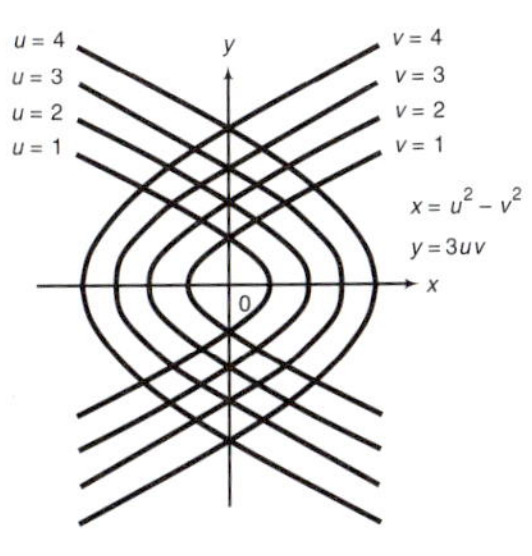

FIGURE 2
The image of $v =$ constant is a parabola opening to the right in the xy-plane. The image of $u =$ constant is a parabola opening to the left.

For every $k \neq 0$, equation (5) is the equation of a parabola in the xy-plane. Moreover, if $v = c$, a constant, then $x = u^2 - c^2$, $y = 3uc$, $u = y/3c$, and

$$x = \frac{y^2}{9c^2} - c^2, \tag{6}$$

which is also a parabola for $c \neq 0$. Thus the functions given above map straight lines into parabolas. Some of these curves are sketched in Figure 2. Note that $u^2 - v^2$ is not one-to-one since $(-1)^2 - (-2)^2 = 1^2 - 2^2 = -3$.

We now give an important definition.

DEFINITION JACOBIAN IN $\mathbb{R}^2$

Let $x = g(u, v)$ and $y = h(u, v)$ be differentiable. Then the **Jacobian**[†] of x and y with respect to u and v, denoted by $\partial(x, y)/\partial(u, v)$, is given by

$$\frac{\partial(x, y)}{\partial(u, v)} = \frac{\partial x}{\partial u}\frac{\partial y}{\partial v} - \frac{\partial x}{\partial v}\frac{\partial y}{\partial u} = \begin{vmatrix} \dfrac{\partial x}{\partial u} & \dfrac{\partial x}{\partial v} \\ \dfrac{\partial y}{\partial u} & \dfrac{\partial y}{\partial v} \end{vmatrix}. \tag{7}$$

■

EXAMPLE 2

COMPUTING A JACOBIAN

If $x = u^2 - v^2$ and $y = 3uv$ as in Example 1, then $\partial x/\partial u = 2u$, $\partial x/\partial v = -2v$, $\partial y/\partial u = 3v$, and $\partial y/\partial v = 3u$, so

$$\begin{vmatrix} \dfrac{\partial x}{\partial u} & \dfrac{\partial x}{\partial v} \\ \dfrac{\partial y}{\partial u} & \dfrac{\partial y}{\partial v} \end{vmatrix} = \begin{vmatrix} 2u & -2v \\ 3v & 3u \end{vmatrix} = 6(u^2 + v^2).$$

EXAMPLE 3

COMPUTING THE JACOBIAN OF THE POLAR COORDINATE MAPPING

Let $x = r \cos \theta$ and $y = r \sin \theta$. Then

$$\frac{\partial x}{\partial r} = \cos \theta, \qquad \frac{\partial x}{\partial \theta} = -r \sin \theta, \qquad \frac{\partial y}{\partial r} = \sin \theta, \qquad \frac{\partial y}{\partial \theta} = r \cos \theta,$$

[†] Named after the German mathematician Carl Gustav Jacob Jacobi (1804–1851). See the biographical sketch on page 363.

so

$$\frac{\partial(x, y)}{\partial(r, \theta)} = \begin{vmatrix} \cos\theta & -r\sin\theta \\ \sin\theta & r\cos\theta \end{vmatrix}$$
$$= r\cos^2\theta + r\sin^2\theta = r(\sin^2\theta + \cos^2\theta) = r.$$

To obtain our main result we need assumptions (i) and (ii) cited earlier. We need also to assume the following:

iii. $C_1 = \partial\Omega$ and $C_2 = \partial\Sigma$ are simple closed curves, and as (u, v) moves once about C_2 in the positive direction, $(x, y) = (g(u, v), h(u, v))$ moves once around C_1 in the positive or negative direction.

iv. All second-order partial derivatives are continuous.

THEOREM 1 CHANGE OF VARIABLES IN A DOUBLE INTEGRAL

If assumptions (i), (ii), (iii), and (iv) hold, then

$$\iint_{\Omega} f(x, y)\, dx\, dy = \pm \iint_{\Sigma} f[g(u, v), h(u, v)] \frac{\partial(x, y)}{\partial(u, v)}\, du\, dv. \tag{8}$$

The plus (minus) sign is taken if as (u, v) moves around C_2 in the positive direction $(x, y) = (g(u, v), h(u, v))$ moves around C_1 in the positive (negative) direction. ■

The proof is given at the end of the section.

REMARK: Conditions (i) and (ii) given on page 356 are often difficult to verify. It can be shown that these conditions hold if all partial derivatives are continuous in Ω and the Jacobian $\partial(x, y)/\partial(u, v)$ is not zero on Ω. See Theorem 9.5.2 on page 666.

EXAMPLE 4 CHANGING TO POLAR COORDINATES IN A DOUBLE INTEGRAL

We can use Theorem 1 to obtain the polar-coordinate formula (1) very easily. For if $x = r\cos\theta$ and $y = r\sin\theta$, then $\partial(x, y)/\partial(r, \theta) = r$ by Example 3, and (1) follows immediately from (8).

EXAMPLE 5 CHANGING VARIABLES IN A DOUBLE INTEGRAL

Let Ω be the region in the upper half of the xy-plane bounded by the parabolas $y^2 = 9 - 9x$ and $y^2 = 9 + 9x$ and by the x-axis. Compute $\iint_\Omega (x + y)\, dx\, dy$ by making the change of variables $x = u^2 - v^2$, $y = 3uv$.

SOLUTION: We saw in Example 1 that this mapping takes straight lines in the uv-plane into parabolas in the xy-plane. For example, if $y^2 = 9 - 9x$, then $y^2 = 9u^2v^2 = 9 - 9x$, or $u^2v^2 = 1 - x = 1 - u^2 + v^2$, or $u^2 + u^2v^2 = 1 + v^2 = u^2(1 + v^2)$, and $u^2 = 1$, so $u = \pm 1$. Similarly, if $y^2 = 9 + 9x$, then $v = \pm 1$. Since, in Ω, $y \geq 0$, we must have $uv \geq 0$, so that u and v have the same sign. We will choose $u = v = 1$ for reasons to be made clear shortly. Note also that if $u = 0$, then $y = 0$ and $x = -v^2 \leq 0$, so that the positive v-axis in

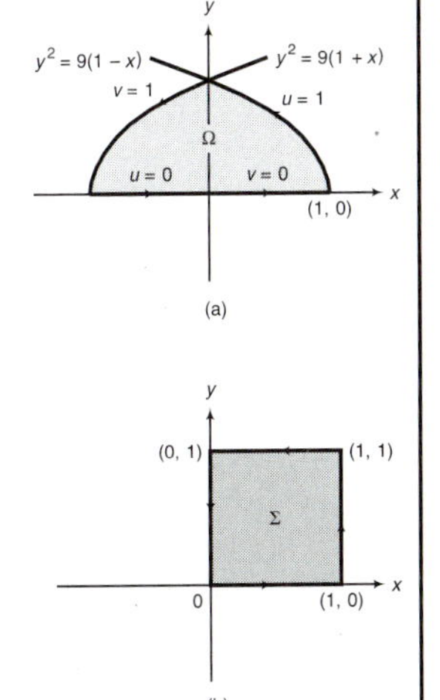

Figure 3
Changing variables in a double integral

Under the mapping $x = u^2 - v^2$, $y = 3uv$, the region bounded by two parabolas and the x-axis in (a) is mapped into the unit square in (b).

the uv-plane is mapped into the negative x-axis in the xy-plane. Similarly, the positive u-axis in the uv-plane is mapped into the positive x-axis in the xy-plane. The situation is sketched in Figure 3. The reason we chose $u = v = 1$ is that moving around Σ in the positive direction corresponds to moving around Ω in the positive direction. [Try it: Take the path $v = 0$ (the u-axis) to $u = 1$ to $v = 1$ to $u = 0$.] Thus the integral around the "parabolic" region given in the problem can be reduced to an integral around a square. Also, as we computed in Example 2,

$$\frac{\partial(x, y)}{\partial(u, v)} = 6(u^2 + v^2).$$

Finally, $x + y = u^2 + 3uv - v^2$, so

$$\iint_{\Omega} (x + y)\,dx\,dy = 6\int_0^1\int_0^1 (u^2 + 3uv - v^2)(u^2 + v^2)\,du\,dv$$
$$= 6\int_0^1\int_0^1 (u^4 + 3u^3v + 3uv^3 - v^4)\,du\,dv = \frac{9}{2}.$$

For triple integrals there is a result analogous to the one we have proven for double integrals. We will state this result without proof.

Let

$$x = g(u, v, w), \quad y = h(u, v, w), \quad z = j(u, v, w).$$

Definition Jacobian in $\mathbb{R}^3$

We define the **Jacobian** of the transformation from a region U in a uvw-space to a region W in xyz-space as

$$\text{Jacobian} = \frac{\partial(x, y, z)}{\partial(u, v, w)} = \begin{vmatrix} \dfrac{\partial x}{\partial u} & \dfrac{\partial x}{\partial v} & \dfrac{\partial x}{\partial w} \\ \dfrac{\partial y}{\partial u} & \dfrac{\partial y}{\partial v} & \dfrac{\partial y}{\partial w} \\ \dfrac{\partial z}{\partial u} & \dfrac{\partial z}{\partial v} & \dfrac{\partial z}{\partial w} \end{vmatrix}.$$ ■

Then under hypotheses similar to the ones made in Theorem 1, we have the following theorem.

Theorem 2 Change of Variables in a Triple Integral

$$\iiint_W F(x, y, z)\,dx\,dy\,dz$$
$$= \pm\iiint_U F(g(u, v, w), h(u, v, w), j(u, v, w))\frac{\partial(x, y, z)}{\partial(u, v, w)}\,du\,dv\,dw.$$ ■

EXAMPLE 6

CHANGING TO CYLINDRICAL COORDINATES IN A TRIPLE INTEGRAL

Let $x = r\cos\theta$, $y = r\sin\theta$, and $z = z$. These are cylindrical coordinates. Then

$$\frac{\partial(x, y, z)}{\partial(r, \theta, z)} = \begin{vmatrix} \frac{\partial x}{\partial r} & \frac{\partial x}{\partial \theta} & \frac{\partial x}{\partial z} \\ \frac{\partial y}{\partial r} & \frac{\partial y}{\partial \theta} & \frac{\partial y}{\partial z} \\ \frac{\partial z}{\partial r} & \frac{\partial z}{\partial \theta} & \frac{\partial z}{\partial z} \end{vmatrix} = \begin{vmatrix} \cos\theta & -r\sin\theta & 0 \\ \sin\theta & r\cos\theta & 0 \\ 0 & 0 & 1 \end{vmatrix}$$

$$= r(\cos^2\theta + \sin^2\theta) = r,$$

so

$$\iiint_W f(x, y, z)\,dx\,dy\,dz = \iiint_U f(r\cos\theta, r\sin\theta, z)r\,dz\,dr\,d\theta.$$

This is equation (4.6.5) on page 283.

EXAMPLE 7

CHANGING TO SPHERICAL COORDINATES IN A TRIPLE INTEGRAL

Let $x = \rho\sin\phi\cos\theta$, $y = \rho\sin\phi\sin\theta$, and $z = \rho\cos\phi$. These are spherical coordinates (see Section 1.9). Then

$$\frac{\partial(x, y, z)}{\partial(\rho, \phi, \theta)} = \begin{vmatrix} \frac{\partial x}{\partial \rho} & \frac{\partial x}{\partial \phi} & \frac{\partial x}{\partial \theta} \\ \frac{\partial y}{\partial \rho} & \frac{\partial y}{\partial \phi} & \frac{\partial y}{\partial \theta} \\ \frac{\partial z}{\partial \rho} & \frac{\partial z}{\partial \phi} & \frac{\partial z}{\partial \theta} \end{vmatrix}$$

$$= \begin{vmatrix} \sin\phi\cos\theta & \rho\cos\phi\cos\theta & -\rho\sin\phi\sin\theta \\ \sin\phi\sin\theta & \rho\cos\phi\sin\theta & \rho\sin\phi\cos\theta \\ \cos\phi & -\rho\sin\phi & 0 \end{vmatrix}$$

Expanding in the last row

$$= \cos\phi \begin{vmatrix} \rho\cos\phi\cos\theta & -\rho\sin\phi\sin\theta \\ \rho\cos\phi\sin\theta & \rho\sin\phi\cos\theta \end{vmatrix}$$

$$+ \rho\sin\theta \begin{vmatrix} \sin\phi\cos\theta & -\rho\sin\phi\sin\theta \\ \sin\phi\sin\theta & \rho\sin\phi\cos\theta \end{vmatrix}$$

$$= \cos\phi(\rho^2\cos\phi\sin\phi\cos^2\theta + \rho^2\sin\phi\cos\phi\sin^2\theta)$$

$$+ \rho\sin\phi(\rho\sin^2\phi\cos^2\theta + \rho\sin^2\phi\sin^2\theta)$$

$$= \cos\phi(\rho^2\sin\phi\cos\phi) + \rho\sin\phi(\rho\sin^2\phi)$$

$$= \rho^2\sin\phi(\cos^2\phi + \sin^2\phi) = \rho^2\sin\phi.$$

Thus

$$\iiint_W f(x, y, z)\, dx\, dy\, dz$$

$$= \iiint_U f(\rho \sin \phi \cos \theta, \rho \sin \phi \sin \theta, \rho \cos \theta)\rho^2 \sin \phi\, d\rho\, d\phi\, d\theta.$$

This is formula (4.6.12) on page 284.

We conclude this section with a proof of Theorem 1.

PROOF OF THEOREM 1: Define $F(x, y) = \int_{x_0}^{x} f(t, y)\, dt$. By the fundamental theorem of calculus, F is continuous and

$$\frac{\partial F}{\partial x} = f(x, y). \tag{9}$$

Then by Green's theorem

$$\iint_\Omega f(x, y)\, dx\, dy = \iint_\Omega \frac{\partial F}{\partial x}\, dx\, dy = \oint_{C_1} F\, dy. \tag{10}$$

We wish to write $\oint_{C_1} F\, dy$ as a line integral in the uv-plane. Let $\mathbf{u} = u(t)\mathbf{i} + v(t)\mathbf{j}$ be a parametric representation of C_2 in the uv-plane for $a \le t \le b$. Then by assumption (iii), $\mathbf{x} = x(t)\mathbf{i} + y(t)\mathbf{j} = g(u(t), v(t))\mathbf{i} + h(u(t), v(t))\mathbf{j}$, $a \le t \le b$, is a parametric representation for C_1. The only difference is that this representation may traverse C_1 in either the positive or the negative direction. Thus

$$\oint_{C_1} F(x, y)\, dy = \int_a^b F(g(u(t), v(t)), h(u(t), v(t))) \frac{dy}{dt}\, dt. \tag{11}$$

Now by the chain rule and the fact that $y = h(u, v)$,

$$\frac{dy}{dt} = \frac{\partial h}{\partial u} u'(t) + \frac{\partial h}{\partial v} v'(t). \tag{12}$$

To simplify notation, let $\overline{F}(u, v) = F(g(u, v), h(u, v))$. Then if we substitute (12) into (11), we obtain

$$\oint_{C_1} F(x, y)\, dy = \int_a^b \overline{F}(u(t), v(t))\left[\frac{\partial h}{\partial u} u'(t) + \frac{\partial h}{\partial v} v'(t)\right] dt$$

$$= \int_a^b \left[\overline{F}\frac{\partial h}{\partial u} u'(t) + \overline{F}\frac{\partial h}{\partial v} v'(t)\right] dt. \tag{13}$$

Using the definition of the line integral, we can write (13) as

$$\oint_{C_1} F(x, y)\, dy = \pm\oint_{C_1} \overline{F}\frac{\partial h}{\partial u}\, du + \overline{F}\frac{\partial h}{\partial v}\, dv, \tag{14}$$

where the $\pm$ depends on whether (x, y) traverses C_1 in the same direction as, or in the direction opposite to, that in which (u, v) traverses C_2 as t goes from a to b.

Let

$$P = \overline{F}\frac{\partial h}{\partial u} \quad \text{and} \quad Q = \overline{F}\frac{\partial h}{\partial v}. \tag{15}$$

Then by Green's theorem

$$\oint_{C_2} P\,du + Q\,dv = \iint_{\Sigma} \left(\frac{\partial Q}{\partial u} - \frac{\partial P}{\partial v}\right) du\,dv. \tag{16}$$

But from the product rule

$$\frac{\partial Q}{\partial u} = \frac{\partial \overline{F}}{\partial u}\frac{\partial h}{\partial v} + \overline{F}\frac{\partial^2 h}{\partial u\,\partial v} \tag{17}$$

and

$$\frac{\partial P}{\partial v} = \frac{\partial \overline{F}}{\partial v}\frac{\partial h}{\partial u} + \overline{F}\frac{\partial^2 h}{\partial v\,\partial u}, \tag{18}$$

so that from (14)–(18),

Equal because mixed partials are continuous

$$\begin{aligned}\oint_{C_1} F\,dy &= \pm\iint_{\Sigma} \left(\frac{\partial \overline{F}}{\partial u}\frac{\partial h}{\partial v} + \overline{F}\frac{\partial^2 h}{\partial u\,\partial v} - \frac{\partial \overline{F}}{\partial v}\frac{\partial h}{\partial u} - \overline{F}\frac{\partial^2 h}{\partial v\,\partial u}\right) du\,dv \\ &= \pm\iint_{\Sigma} \left(\frac{\partial \overline{F}}{\partial u}\frac{\partial h}{\partial v} - \frac{\partial \overline{F}}{\partial v}\frac{\partial h}{\partial u}\right) du\,dv.\end{aligned} \tag{19}$$

But

$$\frac{\partial \overline{F}}{\partial u} = \frac{\partial \overline{F}}{\partial x}\frac{\partial x}{\partial u} + \frac{\partial \overline{F}}{\partial y}\frac{\partial y}{\partial u} = \frac{\partial \overline{F}}{\partial x}\frac{\partial g}{\partial u} + \frac{\partial \overline{F}}{\partial y}\frac{\partial h}{\partial u},$$

and similarly for $\partial\overline{F}/\partial v$. Thus

$$\begin{aligned}\oint_{C_1} F\,dy &= \pm\iint_{\Sigma} \left\{\left(\frac{\partial \overline{F}}{\partial x}\frac{\partial g}{\partial u} + \frac{\partial \overline{F}}{\partial y}\frac{\partial h}{\partial u}\right)\frac{\partial h}{\partial u} - \left(\frac{\partial \overline{F}}{\partial x}\frac{\partial g}{\partial v} + \frac{\partial \overline{F}}{\partial y}\frac{\partial h}{\partial v}\right)\frac{\partial h}{\partial u}\right\} du\,dv \\ &= \pm\iint_{\Sigma} \frac{\partial \overline{F}}{\partial x}\left(\frac{\partial g}{\partial u}\frac{\partial h}{\partial v} - \frac{\partial g}{\partial v}\frac{\partial h}{\partial u}\right) du\,dv.\end{aligned}$$

But $\partial\overline{F}/\partial x = f(g(u, v), h(u, v))$ and $(\partial g/\partial u)(\partial h/\partial v) - (\partial g/\partial v)(\partial h/\partial u) = \partial(x, y)/\partial(u, v)$. Thus

$$\oint_{C_1} F\,dy = \pm\iint_{\Sigma} f(g(u, v),\, h(u, v))\,\frac{\partial(x, y)}{\partial(u, v)}\,du\,dv. \tag{20}$$

Combining (10) with (20) completes the proof. ■

OF HISTORICAL INTEREST

CARL GUSTAV JACOB JACOBI, 1804–1851

CARL GUSTAV JACOB JACOBI, 1804–1851

The son of a prosperous banker, Carl Gustav Jacob Jacobi was born in Potsdam, Germany, in 1804. He was educated at the University of Berlin, where he received his doctorate in 1825. In 1827, he was appointed Extraordinary Professor of Mathematics at the University of Königsberg. Jacobi taught at Königsberg until 1842, when he returned to Berlin under a pension from the Prussian government. He remained in Berlin until his death in 1851.

A prolific writer of mathematical treatises, Jacobi was best known in his time for his results in the theory of elliptic functions. Today, however, he is most remembered for his work on determinants. He was one of the two most creative developers of determinant theory, the other being Cauchy. In 1829, Jacobi published a paper on algebra that contained the notation for the Jacobian that we use today. In 1841 he published an extensive treatise titled *De determinantibus functionalibus,* which was devoted to results about the Jacobian. Jacobi showed the relationship between the Jacobian of functions of several variables and the derivative of a function of one variable. He also showed that n functions of n variables are linearly independent if and only if their Jacobian is not identically zero.

In addition to being a fine mathematician, Jacobi was considered the greatest teacher of mathematics of his generation. He inspired and influenced an astonishing number of students. To dissuade his students from mastering great amounts of mathematics before setting off to do their own research, Jacobi often remarked, "Your father would never have married, and you would not be born, if he had insisted on knowing all the girls in the world before marrying one."

Jacobi believed strongly in research in pure mathematics and frequently defended it against the claim that research should always be applicable to something. He once said, "The real end of science is the honor of the human mind."

PROBLEMS 5.10

SELF-QUIZ

I. The Jacobian of the mapping $x = u + v$, $y = u - v$ is ________.
a. 0 **b.** 1 **c.** 2
d. -1 **e.** -2 **f.** $u^2 - v^2$

II. The Jacobian of the mapping $x = w$, $y = v$, $z = u$ is ________.
a. 1 **b.** -1 **c.** uvw
d. 0 **e.** 3 **f.** -3

In Problems 1–20, compute the Jacobian of the given transformation.

1. $x = u + v$, $y = u - v$
2. $x = u^2 - v^2$, $y = u^2 + v^2$
3. $x = u^2 - v^2$, $y = 2uv$
4. $x = \sin u$, $y = \cos v$
5. $x = u + 3v - 1$, $y = 2u + 4v + 6$
6. $x = v - 2u$, $y = u + 2v$
7. $x = au + bv$, $y = bu - av$
8. $x = e^v$, $y = e^u$
9. $x = ue^v$, $y = ve^u$
***10.** $x = u^v$, $y = v^u$
11. $x = \ln(u + v)$, $y = \ln uv$
12. $x = \tan u$, $y = \sec v$
13. $x = u \sec v$, $y = v \csc u$
14. $x = u \ln v$, $y = v \ln u$

15. $x = u + v + w$, $y = u - v - w$, $z = -u + v + w$

16. $x = au + bv + cw$, $y = au - bv - cw$, $z = -au + bv + cw$

17. $x = u^2 + v^2 + w^2$, $y = u + v + w$, $z = uvw$

18. $x = u \sin v$, $y = v \cos w$, $z = w \sin u$

19. $x = e^u$, $y = e^v$, $z = e^w$

20. $x = u \ln(v + w)$, $y = v \ln(u + w)$, $z = w \ln(u + v)$

In Problems 21–25, transform the integral in (x, y) to an integral in (u, v) by using the given transformation. You need not evaluate the integral.

21. $\int_0^1 \int_y^1 xy\, dx\, dy$; $x = u - v$, $y = u + v$.

22. $\iint_\Omega e^{(x+y)/(x-y)}\, dx\, dy$, where Ω is the region in the first quadrant between the lines $x + y = 1$ and $x + y = 2$; $x = u + v$, $y = u - v$.

23. $\iint_\Omega y\, dx\, dy$, where Ω is the region $7x^2 + 6\sqrt{3}x(y - 1) + 13(y - 1)^2 \leq 16$; use the transformation $x = \sqrt{3}u + (\frac{1}{2})v$, $y = 1 - u + (\sqrt{3}/2)v$.

24. $\int_0^1 \int_0^x (x^2 + y^2)\, dy\, dx$; $x = v$, $y = u$.

***25.** $\iint_\Omega (y - x)\, dy\, dx$, Ω is the region bounded by $y = 2$, $y = x$, and $x = -y^2$; $x = v - u^2$, $y = u + v$.

26. Let Ω be the region in the first quadrant of the xy-plane bounded by the hyperbolas $xy = 1$, $xy = 2$, and the lines $x = y$ and $x = 4y$. Compute $\iint_\Omega x^2y^2\, dx\, dy$ by setting $x = u$ and $y = u/v$.

27. Let W be the solid enclosed by the ellipsoid $(x^2/a^2) + (y^2/b^2) + (z^2/c^2) = 1$. Then

$$\text{volume of } W = \iiint_W dx\, dy\, dz.$$

Compute this volume by making the transformation $x = au$, $y = bv$, $z = cw$. [*Hint:* In uvw-space you'll obtain a sphere.]

28. Compute $\iiint_W (xy + xz + yz)\, dx\, dy\, dz$, where W is the region of Problem 27.

ANSWERS TO SELF-QUIZ

I. e **II.** b

CHAPTER 5 SUMMARY OUTLINE

- **Connected Set** A set Ω in $\mathbb{R}^2$ (or $\mathbb{R}^3$) is **connected** if any two points in Ω can be joined by a piecewise smooth curve lying entirely in Ω.

- **Open Set** A set Ω in $\mathbb{R}^2$ (or $\mathbb{R}^3$) is **open** if for every point $\mathbf{x} \in \Omega$, there is an open disk D (open sphere S) with center at $\mathbf{x}$ that is contained in Ω.

- **Region** A **region** Ω in $\mathbb{R}^2$ or $\mathbb{R}^3$ is an open, connected set.

- **Vector Field in $\mathbb{R}^2$** Let Ω be a region in $\mathbb{R}^2$. Then $\mathbf{F}$ is a **vector field** in $\mathbb{R}^2$ if $\mathbf{F}$ assigns to every $\mathbf{x}$ in Ω a unique vector $\mathbf{F}(\mathbf{x})$ in $\mathbb{R}^2$.

- **Gradient Field in $\mathbb{R}^2$** Let $z = f(x, y)$ be a differentiable function. Then $\mathbf{F} = \nabla f = f_x\mathbf{i} + f_y\mathbf{j}$ is a vector field. It is called a **gradient vector field**.

- **Gradient Field in $\mathbb{R}^3$** If $u = f(x, y, z)$ is differentiable, then $\mathbf{F} = \nabla f = f_x\mathbf{i} + f_y\mathbf{j} + f_z\mathbf{k}$ is a vector field. It is called a **gradient vector field**.

- **Work** done by a force $\mathbf{F}$ moving a particle along a curve $\mathbf{x}(t)$ from $\mathbf{x}(a)$ to $\mathbf{x}(b)$:

$$W = \int_a^b \mathbf{F}(\mathbf{x}(t)) \cdot \mathbf{x}'(t)\, dt.$$

Written as a line integral:

$$W = \int_C \mathbf{F}(\mathbf{x}) \cdot d\mathbf{x}.$$

Alternatively, if $\mathbf{F}(x, y) = P(x, y)\mathbf{i} + Q(x, y)\mathbf{j}$, then

$$W = \int_a^b [P(x(t), y(t))x'(t) + Q(x(t), y(t))y'(t)]\, dt.$$

- **Line Integral in the Plane** Let P and Q be continuous on a region S containing the smooth (or piecewise smooth) curve C given by

$$C: \quad \mathbf{x}(t) = f(t)\mathbf{i} + g(t)\mathbf{j}, \qquad t \in [a, b].$$

Let the vector field $\mathbf{F}$ be given by

$$\mathbf{F}(x, y) = P(x, y)\mathbf{i} + Q(x, y)\mathbf{j}.$$

The **line integral** of $\mathbf{F}$ over C is given by

$$\int_C \mathbf{F}(\mathbf{x}) \cdot d\mathbf{x} = \int_a^b [P(f(t), g(t))f'(t) + Q(f(t), g(t))g'(t)]\, dt.$$

- **Line Integral in Space** Suppose that P, Q, and R are continuous on a region S in $\mathbb{R}^3$ containing the smooth or piecewise smooth curve C given by

$$C:\quad \mathbf{x}(t) = f(t)\mathbf{i} + g(t)\mathbf{j} + h(t)\mathbf{k}, \quad \text{or}$$
$$\mathbf{x}(t) = x(t)\mathbf{i} + y(t)\mathbf{j} + z(t)\mathbf{k}.$$

Let the vector field $\mathbf{F}$ be given by

$$\mathbf{F}(x, y, z) = P(x, y, z)\mathbf{i} + Q(x, y, z)\mathbf{j} + R(x, y, z)\mathbf{k}.$$

Then the **line integral** of $\mathbf{F}$ over C is given by

$$\int_C \mathbf{F}(\mathbf{x})\cdot d\mathbf{x} = \int_a^b [P(f(t), g(t), h(t))f'(t) + Q(f(t), g(t), h(t))g'(t) + R(f(t), g(t), h(t))h'(t)]\, dt.$$

- Let $\mathbf{F}$ be continuous in a region Ω in $\mathbb{R}^2$. Then $\mathbf{F}$ is the gradient of a differentiable function f if and only if for any piecewise smooth curve C lying in Ω, the line integral $\int_C \mathbf{F}(\mathbf{x})\cdot d\mathbf{x}$ is independent of the path.
- Suppose that $\mathbf{F}$ is a continuous in a region Ω in $\mathbb{R}^2$ and $\mathbf{F} = \nabla f$ for some differentiable function f. Then for any piecewise smooth curve C in Ω starting at the point $\mathbf{x}_0$ and ending at the point $\mathbf{x}_1$,

$$\int_C \mathbf{F}(\mathbf{x})\cdot d\mathbf{x} = f(\mathbf{x}_1) - f(\mathbf{x}_0)$$

- **Exact Vector Field in $\mathbb{R}^2$** We say that the vector field $P(x, y)\mathbf{i} + Q(x, y)\mathbf{j}$ is **exact** if $\mathbf{F}$ is the gradient of a function f. A similar definition holds in $\mathbb{R}^3$.
- **Conditions for Exactness in $\mathbb{R}^2$** Let $\mathbf{F}(x, y) = P(x, y)\mathbf{i} + Q(x, y)\mathbf{j}$ and suppose that P, Q, $\partial P/\partial y$, and $\partial Q/\partial x$ are continuous in an open disk D centered at (x, y). Then, in D, $\mathbf{F}$ is the gradient of a function f if and only if

$$\frac{\partial P}{\partial y} = \frac{\partial Q}{\partial x}$$

- **Conditions for Exactness in $\mathbb{R}^3$** Let $\mathbf{F}(x, y, z) = P(x, y, z)\mathbf{i} + Q(x, y, z)\mathbf{j} + R(x, y, z)\mathbf{k}$ and suppose that P, Q, R, $\partial P/\partial y$, $\partial P/\partial z$, $\partial Q/\partial x$, $\partial Q/\partial z$, $\partial R/\partial x$, and $\partial R/\partial y$ are continuous in an open ball centered at (x, y, z). Then $\mathbf{F}$ is the gradient of a differentiable function f if and only if

$$\frac{\partial P}{\partial y} = \frac{\partial Q}{\partial x}, \quad \frac{\partial R}{\partial x} = \frac{\partial P}{\partial z}, \quad \frac{\partial Q}{\partial z} = \frac{\partial R}{\partial y}.$$

- **The Symbol $\oint_{\partial\Omega}$** The symbol $\oint_{\partial\Omega}$ indicates that $\partial\Omega$ is a closed curve (the boundary of the region Ω) around which we integrate in the counterclockwise direction (the direction of the arrow).
- **Simple Curve** A curve C is called **simple** if it does not cross itself. That is, suppose C is given by

$$C:\quad \mathbf{x}(t) = f(t)\mathbf{i} + g(t)\mathbf{j}, \qquad t \in [a, b].$$

Then C is simple if and only if $\mathbf{x}(t_1) \neq \mathbf{x}(t_2)$ whenever $t_1 \neq t_2$ (with the possible exception $t_1 = a$ and $t_2 = b$).

- **Simply Connected Region** A region Ω in the xy-plane is called **simply connected** if it has the following property: If C is a simple closed curve contained in Ω, then every point in the region enclosed by C is also in Ω.

  Intuitively, a region is simply connected if it has no holes.
- **Green's Theorem in the Plane** Let Ω be a simply connected region in the xy-plane bounded by a piecewise smooth curve $\partial\Omega$. Let P and Q be continuous with continuous first partials in an open disk containing Ω. Then

$$\oint_{\partial\Omega} P\,dx + Q\,dy = \iint_\Omega \left(\frac{\partial Q}{\partial x} - \frac{\partial P}{\partial y}\right) dx\,dy.$$

- **Four Integrals that Represent Area**

$$\iint_\Omega dA = \oint_{\partial\Omega} x\,dy = \oint_{\partial\Omega} (-y)\,dx = \frac{1}{2}\oint_{\partial\Omega} [(-y)\,dx + x\,dy]$$

- **Parametric Surface** Let $x = x(u, v)$, $y = y(u, v)$, and $z = z(u, v)$ be continuous. Then the set of points S given by

$$\mathbf{r}(u, v) = x(u, v)\mathbf{i} + y(u, v)\mathbf{j} + z(u, v)\mathbf{k} \qquad (*)$$

where (u, v) ranges over a region Ω in $\mathbb{R}^2$ is called a **parametric surface**, or simply a **surface**, in $\mathbb{R}^3$. Equation (*) is a **parametric representation** of the surface.

- **Two Tangent Vectors to a Parametric Surface and a Normal Vector** The vectors

$$\mathbf{r}_u(u, v) = \frac{\partial x(u, v)}{\partial u}\mathbf{i} + \frac{\partial y(u, v)}{\partial u}\mathbf{j} + \frac{\partial z(u, v)}{\partial u}\mathbf{k}$$

and

$$\mathbf{r}_v(u, v) = \frac{\partial x(u, v)}{\partial v}\mathbf{i} + \frac{\partial y(u, v)}{\partial v}\mathbf{j} + \frac{\partial z(u, v)}{\partial v}\mathbf{k}$$

are tangent to the surface (*) provided that all six partial derivatives exist. The vector

$$\mathbf{N} = \mathbf{r}_u \times \mathbf{r}_v$$

is normal to the surface (*) provided that $\mathbf{r}_u \times \mathbf{r}_v \neq \mathbf{0}$.

- **Smooth Surface** The surface given by (*) is called **smooth** over the region Ω if

  i. all six partial derivatives exist and are continuous for (u, v) in Ω and

  ii. $\mathbf{N}(u, v) = \mathbf{r}_u(u, v) \times \mathbf{r}_v(u, v) \neq \mathbf{0}$ for (u, v) in Ω.
- **Surface Area** Let $\mathbf{r}$ be a surface that is both smooth and one-to-one over a region Ω. Then the **surface area** σ of $\mathbf{r}$ over Ω is given by

$$\text{surface area} = \sigma = \iint_\Omega |\mathbf{N}(u, v)|\, du\, dv = \iint_\Omega |\mathbf{r}_u(u, v) \times \mathbf{r}_v(u, v)|\, du\, dv.$$

■ **Lateral Surface Area** Let f be continuous with continuous partial derivatives in the region Ω in the xy-plane. Then the **lateral surface area** σ of the graph of f over Ω is defined by

$$\sigma = \iint_{\Omega} \sqrt{1 + f_x^2(x, y) + f_y^2(x, y)}\, dA.$$

■ **Integral over a Surface** Suppose that

$$\lim_{\Delta s \to 0} \sum_{i=1}^{n} \sum_{j=1}^{m} F(x_i^*, y_j^*, z_{ij}^*)\, \Delta\sigma_{ij}$$

exists and is independent of the way the surface S is partitioned and the way in which the points (x_i^*, y_j^*, z_{ij}^*) are chosen in S_{ij}. Then F is said to be **integrable** over S and the **surface integral** of F over S, denoted by $\iint_S F(x, y, z)\, d\sigma$, is given by

$$\iint_S F(x, y, z)\, d\sigma = \lim_{\Delta s \to 0} \sum_{i=1}^{n} \sum_{j=1}^{m} F(x_i^*, y_j^*, z_{ij}^*)\, \Delta\sigma_{ij}.$$

■ **Writing a Surface Integral as an Ordinary Double Integral** Let S: $z = f(x, y)$ be a smooth surface for (x, y) in the bounded region Ω in the xy-plane. Then if F is continuous on S, F is integrable over S, and

$$\iint_S F(x, y, z)\, d\sigma = \iint_{\Omega} F(x, y, f(x, y))\sqrt{f_x^2(x, y) + f_y^2(x, y) + 1}\, dA.$$

■ **Outward Unit Normal Vector** The vector $\mathbf{n}_1 = \dfrac{\nabla F}{|\nabla F|}$ is normal to the surface $F(x, y, z) = f(x, y) - z = 0$ while $\mathbf{n}_2 = \dfrac{\nabla G}{|\nabla G|}$ is normal to the surface $G(x, y, z) = z - f(x, y) = 0$.

We choose one of these normal vectors, denote it by $\mathbf{n}$, and call it the **outward unit normal vector**. The direction of $\mathbf{n}$ is called the **positive normal direction** to the surface at a point.

■ **Orientable Surface** Let P_0 be a point on the surface S. Let C be a closed curve on S that passes through P_0. Then S is said to be **orientable** if the outward unit normal vector at P_0 does not change direction as it is displaced continuously around C until it returns to P_0.

■ **Another Representation for a Surface Integral** We define the angle γ to be the acute angle between $\mathbf{n}$ and the positive z-axis. Then

$$\iint_S F(x, y, z)\, d\sigma = \iint_{\Omega} F(x, y, f(x, y)) \sec \gamma(x, y)\, dx\, dy.$$

■ **Divergence and Curl** Let the vector field $\mathbf{F}$ be given by

$$\mathbf{F}(x, y, z) = P(x, y, z)\mathbf{i} + Q(x, y, z)\mathbf{j} + R(x, y, z)\mathbf{k},$$

where P, Q, and R are differentiable. Then the **divergence** of $\mathbf{F}$ (div $\mathbf{F}$) and **curl** of $\mathbf{F}$ (curl $\mathbf{F}$) are given by

$$\text{div}\, \mathbf{F} = \frac{\partial P}{\partial x} + \frac{\partial Q}{\partial y} + \frac{\partial R}{\partial z}$$

and

$$\text{curl}\, \mathbf{F} = \left(\frac{\partial R}{\partial y} - \frac{\partial Q}{\partial z}\right)\mathbf{i} + \left(\frac{\partial P}{\partial z} - \frac{\partial R}{\partial x}\right)\mathbf{j} + \left(\frac{\partial Q}{\partial x} - \frac{\partial P}{\partial y}\right)\mathbf{k}.$$

These can also be written

$$\text{div}\, \mathbf{F} = \nabla \cdot \mathbf{F} \text{ and curl } \mathbf{F} = \nabla \times \mathbf{F}$$

■ **Stokes's Theorem** Let $\mathbf{F}(x, y, z) = P(x, y, z)\mathbf{i} + Q(x, y, z)\mathbf{j} + R(x, y, z)\mathbf{k}$ be continuously differentiable on a bounded region W in space that contains the smooth surface S, and let ∂S be a piecewise smooth, simple closed curve traversed in the positive sense. Then

$$\oint_{\partial S} \mathbf{F} \cdot \mathbf{T}\, ds = \iint_S \text{curl}\, \mathbf{F} \cdot \mathbf{n}\, d\sigma$$

where $\mathbf{T}$ is the unit tangent vector to the curve ∂S at a point (x, y, z) and $\mathbf{n}$ is the outward unit normal vector to the surface S at the point (x, y, z). Alternatively, we may write

$$\oint_{\partial S} \mathbf{F} \cdot d\mathbf{x} = \iint_S \text{curl}\, \mathbf{F} \cdot \mathbf{n}\, d\sigma$$

and

$$\oint_{\partial S} P\, dx + Q\, dy + R\, dz = \iint_S \text{curl}\, \mathbf{F} \cdot \mathbf{n}\, d\sigma.$$

■ **Important Theorem** Let $\mathbf{F} = P(x, y, z)\mathbf{i} + Q(x, y, z)\mathbf{j} + R(x, y, z)\mathbf{k}$ be continuously differentiable on a simply connected region S in space. Then the following conditions are equivalent. That is, if one is true, all are true.

i. $\mathbf{F}$ is the gradient of a differentiable function f.

ii. $\dfrac{\partial P}{\partial y} = \dfrac{\partial Q}{\partial x}$, $\dfrac{\partial R}{\partial x} = \dfrac{\partial P}{\partial z}$, and $\dfrac{\partial Q}{\partial z} = \dfrac{\partial R}{\partial y}$.

iii. $\int_C \mathbf{F} \cdot d\mathbf{x} = 0$ for every piecewise smooth, simple closed curve C lying in S.

iv. $\int_C \mathbf{F} \cdot d\mathbf{x}$ is independent of path.

v. Curl $\mathbf{F} = \mathbf{0}$.

■ **The Divergence Theorem** Let W be a solid in $\mathbb{R}^3$ totally bounded by the smooth or piecewise smooth surface S. Let $\mathbf{F}$ be a smooth vector field on W, and let $\mathbf{n}$ denote the outward unit normal to S. Then

$$\iint_S \mathbf{F}\cdot\mathbf{n}\,d\sigma = \iiint_W \operatorname{div}\mathbf{F}\,dv.$$

■ **Jacobian in $\mathbb{R}^2$** Let $x = g(u, v)$ and $y = h(u, v)$ be differentiable. Then the **Jacobian** of x and y with respect to u and v, denoted by $\partial(x, y)/\partial(u, v)$, is given by

$$\frac{\partial(x, y)}{\partial(u, v)} = \frac{\partial x}{\partial u}\frac{\partial y}{\partial v} - \frac{\partial x}{\partial v}\frac{\partial y}{\partial u} = \begin{vmatrix} \frac{\partial x}{\partial u} & \frac{\partial x}{\partial v} \\ \frac{\partial y}{\partial u} & \frac{\partial y}{\partial v} \end{vmatrix}.$$

■ **Change of Variables in a Double Integral** Suppose that

i. For every $(x, y) \in \Omega$, there is a $(u, v) \in \Sigma$ such that $x = g(u, v)$ and $y = h(u, v)$.

ii. If $g(u_1, v_1) = g(u_2, v_2)$ and $h(u_1, v_1) = h(u_2, v_2)$, then $u_1 = u_2$ and $v_1 = v_2$.

iii. $C_1 = \partial\Omega$ and $C_2 = \partial\Sigma$ are simple closed curves, and as (u, v) moves once about C_2 in the positive direction, $(x, y) = (g(u, v), h(u, v))$ moves once around C_1 in the positive or negative direction.

iv. All second-order partial derivatives are continuous.

Then

$$\iint_\Omega f(x, y)\,dx\,dy = \pm \iint_\Sigma f[g(u, v), h(u, v)]\frac{\partial(x, y)}{\partial(u, v)}\,du\,dv. \quad (**)$$

The plus (minus) sign is taken if as (u, v) moves around C_2 in the positive direction $(x, y) = (g(u, v), h(u, v))$ moves around C_1 in the positive (negative) direction.

■ **Jacobian in $\mathbb{R}^3$** Let $x = g(u, v, w)$, $y = h(u, v, w)$, and $z = j(u, v, w)$, and suppose that all nine partial derivatives exist. We define the **Jacobian** of the transformation from a region U in a uvw-space to a region W in xyz-space as

$$\text{Jacobian} = \frac{\partial(x, y, z)}{\partial(u, v, w)} = \begin{vmatrix} \frac{\partial x}{\partial u} & \frac{\partial x}{\partial v} & \frac{\partial x}{\partial w} \\ \frac{\partial y}{\partial u} & \frac{\partial y}{\partial v} & \frac{\partial y}{\partial w} \\ \frac{\partial z}{\partial u} & \frac{\partial z}{\partial v} & \frac{\partial z}{\partial w} \end{vmatrix}.$$

■ **Change of Variables in a Triple Integral** Under conditions similar to those in (**) (for $\mathbb{R}^2$),

$$\iiint_W F(x, y, z)\,dx\,dy\,dz = \pm \iiint_U F(g(u, v, w), h(u, v, w), j(u, v, w))\frac{\partial(x, y, z)}{\partial(u, v, w)}\,du\,dv\,dw.$$

$\iint_S d\sigma$

CHAPTER 5 REVIEW EXERCISES

In Problems 1–6, compute the gradient vector field of the given function.

1. $f(x, y) = (x + y)^3$

2. $f(x, y) = \sin(x + 2y)$

3. $f(x, y) = \dfrac{x + y}{x - y}$

4. $f(x, y) = \sqrt{x/y}$

5. $f(x, y) = x^2 + y^2 + z^2$

6. $f(x, y) = xyz$

In Problems 7 and 8, compute and sketch the gradient field of the given function.

7. $f(x, y) = xy$

8. $f(x, y) = x^2 - y^2$

In Problems 9–14, calculate $\int_C \mathbf{F}\cdot d\mathbf{x}$.

9. $\mathbf{F}(x, y) = x^2\mathbf{i} + y^2\mathbf{j}$; C is the curve $y = x^{3/2}$ from $(0, 0)$ to $(1, 1)$.

10. $\mathbf{F}(x, y) = x^2y\mathbf{i} - xy^2\mathbf{j}$; C is the unit circle in the counterclockwise direction.

11. $\mathbf{F}(x, y) = 3xy\mathbf{i} - y\mathbf{j}$; C is the triangle joining the points $(0, 0)$, $(1, 1)$, and $(0, 1)$ in the counterclockwise direction.

12. $\mathbf{F}(x, y) = e^y\mathbf{i} + e^x\mathbf{j}$; C is the curve from $(0, 0)$ to $(1, 1)$ and back to $(0, 0)$ which has the parametric description

$$\mathbf{x}(t) = \begin{cases} t\mathbf{i} + \sqrt{t}\mathbf{j} & \text{for } 0 \le t \le 1, \\ (2 - t)\mathbf{i} + (2 - t)^2\mathbf{j} & \text{for } 1 \le t \le 2. \end{cases}$$

13. $\mathbf{F}(x, y, z) = x\mathbf{i} + y\mathbf{j} + z\mathbf{k}$; C is the curve $\mathbf{x}(t) = t^3\mathbf{i} + t^2\mathbf{j} + t\mathbf{k}$ from $(0, 0, 0)$ to $(1, 1, 1)$.

14. $\mathbf{F}(x, y, z) = x^2\mathbf{i} + y^2\mathbf{j} + z^2\mathbf{k}$; C is the helix $\mathbf{x}(t) = (\sin t)\mathbf{i} + (\cos t)\mathbf{j} + 2t\mathbf{k}$ from $(0, 1, 0)$ to $(1, 0, \pi)$.

15. Calculate the work done when the force field $\mathbf{F}(x, y) = x^2y\mathbf{i} + (y^3 + x^3)\mathbf{j}$ moves a particle around the unit circle in the counterclockwise direction.

16. Calculate the work done when the force field $\mathbf{F}(x, y) =$

$3(x-y)\mathbf{i} + x^5\mathbf{j}$ moves a particle around the triangle of Problem 11.

17. Verify that $\mathbf{F}(x, y) = 3x^2y^2\mathbf{i} + 2x^3y\mathbf{j}$ is exact, and calculate $\int_C \mathbf{F}\cdot\mathbf{dx}$, where C starts at $(1, 2)$ and ends at $(3, -1)$.

18. Verify that $\mathbf{F}(x, y) = e^{xy}(1 + xy)\mathbf{i} + x^2e^{xy}\mathbf{j}$ is exact, and calculate $\int_C \mathbf{F}\cdot\mathbf{dx}$, where C starts at $(1, 2)$ and ends at $(3, -1)$.

***19.** Verify that $\mathbf{F}(x, y, z) = -(y/z)\mathbf{i} - (x/z)\mathbf{j} + (xy/z^2)\mathbf{k}$ is exact. Use that fact to evaluate $\int_C \mathbf{F}\cdot\mathbf{dx}$ where C is a piecewise smooth curve joining the points $(1, 1, 1)$ and $(2, -1, 3)$ and not crossing the xy-plane.

20. Evaluate $\oint_{\partial\Omega} 2y\,dx + 4x\,dy$, where $\Omega = \{(x, y): 0 \le x \le 2,\ 0 \le y \le 2\}$.

21. Evaluate $\oint_{\partial\Omega} x^2y\,dx + xy^2\,dy$, where Ω is the region enclosed by the triangle of Problem 11.

22. Evaluate $\oint_{\partial\Omega} (x^3 + y^3)\,dx + (x^2y^2)\,dy$, where Ω is the unit disk.

23. Evaluate $\oint_{\partial\Omega} \sqrt{1 + x^2}\,dy$, where $\Omega = \{(x, y): -1 \le x \le 1,\ x^2 \le y \le 1\}$.

24. Evaluate $$\oint_{\partial\Omega} \frac{2x^2 + y^2 - xy}{\sqrt{x^2 + y^2}}\,dx + \frac{xy - x^2 - 2y^2}{\sqrt{x^2 + y^2}}\,dy$$ where Ω is a region of the type shown in Figure 5.4.4 that does not contain the origin.

25. Let $\mathbf{F}(x, y) = xy^2\mathbf{i} + x^2y\mathbf{j}$ and let Ω denote the disk of radius 2 centered at $(0, 0)$. Calculate

a. curl $\mathbf{F}$ **b.** $\oint_{\partial\Omega} \mathbf{F}\cdot\mathbf{T}\,ds$

c. div $\mathbf{F}$ **d.** $\oint_{\partial\Omega} \mathbf{F}\cdot\mathbf{n}\,ds$

26. Let $\mathbf{F}(x, y) = y^2\mathbf{i} - x^2\mathbf{j}$ and let Ω be the triangle of Problem 11. Calculate the four items specified in the preceding problem.

27. Verify that the vector flow $\mathbf{F}(x, y) = (\cos x^2)\mathbf{i} + e^y\mathbf{j}$ is irrotational.

28. Verify that the vector flow $\mathbf{F}(x, y) = -(x^2y + y^3)\mathbf{i} + (x^3 + xy^2)\mathbf{j}$ is incompressible.

In Problems 29–31, parametrize the given surface.

29. the plane $2x - 3y + z = 0$

30. the hemisphere $x^2 + y^2 + z^2 = 25,\ z \ge 0$

31. the paraboloid $x^2 - y^2 - z = 0;\ 0 \le x \le 2;\ 2 \le y \le 4$

32. Compute the surface area of the surface in Problem 30.

33. Find an integral that gives the surface area of the surface in Problem 31. Do not evaluate it.

34. Compute the lateral surface area of a right circular cone with radius 2 and height 5.

35. Find the surface area of the hemisphere $x^2 + y^2 + z^2 = 16,\ x \le 0$.

36. Find the surface area of the part of the surface $z = \dfrac{y^4}{4} + \dfrac{1}{8y^2}$ over the rectangle $0 \le x \le 3,\ 2 \le y \le 4$.

In Problems 37–42, evaluate the integral of the given function over the specified surface.

37. $\iint_S y\,d\sigma$, where $S = \{(x, y, z): z = y^2,\ 0 \le x \le 2,\ 0 \le y \le 1\}$

38. $\iint_S (x^2 + y^2)\,d\sigma$, where S is as in the preceding problem

39. $\iint_S y^2\,d\sigma$, where S is the hemisphere $\{(x, y, z): x^2 + y^2 + z^2 = 9,\ x^2 + y^2 \le 9,\ z \ge 0\}$

40. $\iint_S xz\,d\sigma$, where S is as in the preceding problem

41. $\iint_S y\,d\sigma$, where S is the surface of the tetrahedron bounded by the coordinate planes and the plane $x + y + z = 1$

42. $\iint_S (x^2 + y^2 + z^2)\,d\sigma$, where S is the part of the plane $y - x = 3$ that lies inside the cylinder $y^2 + z^2 = 9$

43. Find the mass of a triangular metallic sheet with corners at $(1, 0, 0)$, $(0, 1, 0)$, and $(0, 0, 1)$ if its density is proportional to y^2.

44. Find the mass of a metallic sheet in the shape of the hemisphere $x^2 + y^2 + z^2 = 1,\ x^2 + z^2 \le 1,\ y \ge 0$ if its density is proportional to the distance from the y-axis.

In Problems 45–48, compute the flux $\iint_S \mathbf{F}\cdot\mathbf{n}\,d\sigma$ for the given surface immersed in the specified vector field.

45. $S = \{(x, y, z):\ z = 2xy;\ 0 \le x \le 1,\ 0 \le y \le 4\}$; $\mathbf{F} = xy^2\mathbf{i} - 2z\mathbf{j}$

46. $S = \{(x, y, z): x^2 + y^2 + z^2 = 1,\ z \ge 0\}$; $\mathbf{F} = x\mathbf{i} + 2y\mathbf{j} + 3z\mathbf{k}$

47. $S = \{(x, y, z): y = \sqrt{x^2 + z^2},\ x^2 + z^2 \le 4\}$; $\mathbf{F} = x\mathbf{i} - xz\mathbf{j} + 3z\mathbf{k}$

48. S is the unit sphere; $\mathbf{F} = x^2\mathbf{i} + y^2\mathbf{j} + z^2\mathbf{k}$

In Problems 49–54, compute the divergence and curl of the given vector field.

49. $\mathbf{F}(x, y, z) = x\mathbf{i} + y\mathbf{j} + z\mathbf{k}$

50. $\mathbf{F}(x, y, z) = (x - y)\mathbf{i} + (y - z)\mathbf{j} + (z - x)\mathbf{k}$

51. $\mathbf{F}(x, y, z) = yz\mathbf{i} + xz\mathbf{j} + xy\mathbf{k}$

52. $\mathbf{F}(x, y, z) = (\ln x)\mathbf{i} + (\ln y)\mathbf{j} + (\ln z)\mathbf{k}$

53. $\mathbf{F}(x, y, z) = e^{yz}\mathbf{i} + e^{xz}\mathbf{j} + e^{xy}\mathbf{k}$

54. $\mathbf{F}(x, y, z) = (\cos y)\mathbf{i} + (\cos x)\mathbf{j} + (\cos z)\mathbf{k}$

In Problems 55–57, evaluate the line integral by using Stokes's theorem.

55. $\oint_C \mathbf{F}\cdot\mathbf{dx}$, where $\mathbf{F}(x, y, z) = (x + 2y)\mathbf{i} + (y - 3z)\mathbf{j} + (z - x)\mathbf{k}$ and C is the unit circle in the plane $z = 2$.

56. $\oint_C \mathbf{F}\cdot\mathbf{dx}$, where $\mathbf{F}(x, y, z) = x\mathbf{i} + y\mathbf{j} + z\mathbf{k}$ and C is the boundary of the triangle joining the points $(1, 0, 0)$, $(0, 1, 0)$, and $(0, 0, 1)$.

57. $\oint_C \mathbf{F}\cdot\mathbf{dx}$, where $\mathbf{F}(x, y, z) = y^2\mathbf{i} + x^2\mathbf{j} + z^2\mathbf{k}$ and C is the boundary of that part of the plane $x + y + z = 1$ which lies in the first octant.

58. Compute $\iint_S [\text{curl } \mathbf{F}]\cdot\mathbf{n}\,d\sigma$, where S is the unit sphere and $\mathbf{F} = e^{xy}\mathbf{i} + \tan^{-1} z\mathbf{j} + (x + y + z)^{7/3} z^2\mathbf{k}$.

In Problems 59–64, evaluate the surface integral by using the divergence theorem.

59. $\iint_S \mathbf{F}\cdot\mathbf{n}\,d\sigma$, where $\mathbf{F} = x\mathbf{i} + 2y\mathbf{j} + 3z\mathbf{k}$ and S is the unit sphere.

60. $\iint_S \mathbf{F}\cdot\mathbf{n}\,d\sigma$, where $\mathbf{F} = ax\mathbf{i} + by\mathbf{j} + cz\mathbf{k}$ and S is the unit sphere.

61. $\iint_S \mathbf{F}\cdot\mathbf{n}\,d\sigma$, where $\mathbf{F} = x\mathbf{i} + 2y\mathbf{j} + 3z\mathbf{k}$ and S is the cylinder $x^2 + y^2 = 9,\ 0 \le z \le 6$.

62. $\iint_S \mathbf{F}\cdot\mathbf{n}\,d\sigma$, where $\mathbf{F} = (y^3 - z)\mathbf{i} + x^2e^z\mathbf{j} + (\sin xy)\mathbf{k}$ and S is the ellipsoid $(x/2)^2 + (y/4)^2 + (z/5)^2 = 1$.

63. $\iint_S \mathbf{F}\cdot\mathbf{n}\,d\sigma$, where $\mathbf{F} = x\mathbf{i} + 2y\mathbf{j} + 3z\mathbf{k}$ and S is the surface of the unit cube, $0 \le x \le 1$, $0 \le y \le 1$, $0 \le z \le 1$.

64. $\iint_S \mathbf{F}\cdot\mathbf{n}\,d\sigma$, where $\mathbf{F} = xz\mathbf{i} + xy\mathbf{j} + xyz\mathbf{k}$ and S is the unit cube in the first octant.

In Problems 65–72, compute the Jacobian of the given transformation.

65. $x = u + 2v$, $y = 2u - v$

66. $x = u^3 - v^3$, $y = u^3 + v^3$

67. $x = u \ln v$, $y = v \ln u$

68. $x = ve^u$, $y = ue^v$

69. $x = \dfrac{u}{v}$, $y = \dfrac{v}{u}$

70. $x = v \tan u$, $y = \tan uv$

71. $x = u + v + w$, $y = u - 2v + 3w$, $z = -2u + v - 5w$

72. $x = vw$, $y = uw$, $z = uv$

73. Transform the integral $\int_0^1 \int_x^1 xy\,dy\,dx$ by making the transformation $x = u + v$, $y = u - v$.

74. Transform $\iint_\Omega e^{(x-y)/(x+y)}\,dx\,dy$, where Ω is the region in the first quadrant between the lines $x + y = 2$ and $x + y = 3$, by making the transformation $u = x - y$ and $v = x + y$. Then evaluate the integral.

PART II

LINEAR ALGEBRA

CHAPTERS INCLUDE:

CHAPTER 6

SYSTEMS OF LINEAR EQUATIONS AND MATRICES

6.1 INTRODUCTION

This is the first of three chapters on linear algebra. If you look up the word "linear" in a dictionary, you will find something like the following: lin-e-ar (lin′ ē ər), adj. 1. of, consisting of, or using lines.† In mathematics, the word "linear" means a good deal more than that. Nevertheless, much of the theory of elementary linear algebra is in fact a generalization of properties of straight lines. As a review, here are some fundamental facts about straight lines:

i. The **slope** m of a line passing through the points (x_1, y_1) and (x_2, y_2) is given by

$$m = \frac{y_2 - y_1}{x_2 - x_1} = \frac{\Delta y}{\Delta x} \quad \text{if } x_1 \neq x_2$$

ii. If $x_2 - x_1 = 0$ and $y_2 \neq y_1$, then the line is vertical and the slope is said to be **undefined**‡

iii. Any line (except one with undefined slope) can be described by writing its equation in the slope-intercept form $y = mx + b$, where m is the slope of the line and b is the y-intercept of the line (the value of y at the point where the line crosses the y-axis).

iv. Two distinct lines are parallel if and only if they have the same slope.

v. If the equation of a line is written in the form $ax + by = c$ $(b \neq 0)$, then, as is easily computed, $m = -a/b$.

vi. If m_1 is the slope of line L_1, m_2 is the slope of line L_2, and $m_1 \neq 0$, then L_1 and L_2 are perpendicular if and only if $m_2 = -1/m_1$.

---

†Taken from the pocket edition of *The Random House Dictionary*.
‡In some textbooks a vertical line is said to have "an infinite slope."

vii. Lines parallel to the x-axis have a slope of zero.

viii. Lines parallel to the y-axis have an undefined slope.

In the next section we shall illustrate the relationship between solving systems of equations and finding points of intersection of pairs of straight lines.

6.2 TWO LINEAR EQUATIONS IN TWO UNKNOWNS

Consider the following system of two linear equations in the two unknowns x and y:

$$\begin{aligned} a_{11}x + a_{12}y &= b_1 \\ a_{21}x + a_{22}y &= b_2 \end{aligned} \tag{1}$$

where a_{11}, a_{12}, a_{21}, a_{22}, b_1, and b_2 are given numbers. Each of these equations is the equation of a straight line. The slope of the first line is $-a_{11}/a_{12}$; the slope of the second line is $-a_{21}/a_{22}$ (if $a_{12} \neq 0$ and $a_{22} \neq 0$). A **solution** to system (1) is a pair of numbers, denoted by (x, y), that satisfies (1). The questions that naturally arise are whether (1) has any solutions and if so, how many? We answer these questions after looking at some examples. In these examples we make use of two important facts from elementary algebra:

Fact A: If $a = b$ and $c = d$, then $a + c = b + d$.

Fact B: If $a = b$ and c is any real number, then $ca = cb$.

Fact A states that if we add two equations together, we obtain a third, valid equation. Fact B states that if we multiply both sides of an equation by a constant, we obtain a second, valid equation. We shall assume that $c \neq 0$ for although the equation $0 = 0$ is correct, it is not very useful.

EXAMPLE 1 A SYSTEM WITH A UNIQUE SOLUTION

Consider the system

$$\begin{aligned} x - y &= 7 \\ x + y &= 5 \end{aligned} \tag{2}$$

Adding the two equations together gives us, by Fact A, the following equation: $2x = 12$ (or $x = 6$). Then, from the second equation, $y = 5 - x = 5 - 6 = -1$. Thus the pair $(6, -1)$ satisfies system (2) and the way we found the solution shows that it is the only pair of numbers to do so. That is, system (2) has a **unique solution**.

EXAMPLE 2 A SYSTEM WITH AN INFINITE NUMBER OF SOLUTIONS

Consider the system

$$\begin{aligned} x - y &= 7 \\ 2x - 2y &= 14 \end{aligned} \tag{3}$$

It is apparent that these two equations are equivalent. To see this multiply the first by 2. This is permitted by Fact B. Then $x - y = 7$ or $y = x - 7$. Thus the pair $(x, x - 7)$ is a solution to system (3) for any real number x. That is, system (3) has an **infinite number of solutions**. For example, the following pairs are solutions: $(7, 0)$, $(0, -7)$, $(8, 1)$, $(1, -6)$, $(3, -4)$, and $(-2, -9)$.

EXAMPLE 3

A SYSTEM WITH NO SOLUTION

Consider the system

$$\begin{aligned} x - y &= 7 \\ 2x - 2y &= 13 \end{aligned} \tag{4}$$

Multiplying the first equation by 2 (which, again, is permitted by Fact B) gives us $2x - 2y = 14$. This contradicts the second equation. Thus system (4) has **no solution**.

It is easy to explain, geometrically, what is going on in the preceding examples. First we repeat that the equations in system (1) are both equations of straight lines. A solution to (1) is a point (x, y) that lies on both lines. If the two lines are not parallel, then they intersect at a single point. If they are parallel, then they either never intersect (no points in common) or are the same line (infinite number of points in common). In Example 1 the lines have slopes of 1 and -1, respectively. Thus they are not parallel. They have the single point $(6, -1)$ in common. In Example 2 the lines are parallel (slope of 1) and coincident. In Example 3 the lines are parallel and distinct. These relationships are all illustrated in Figure 1.

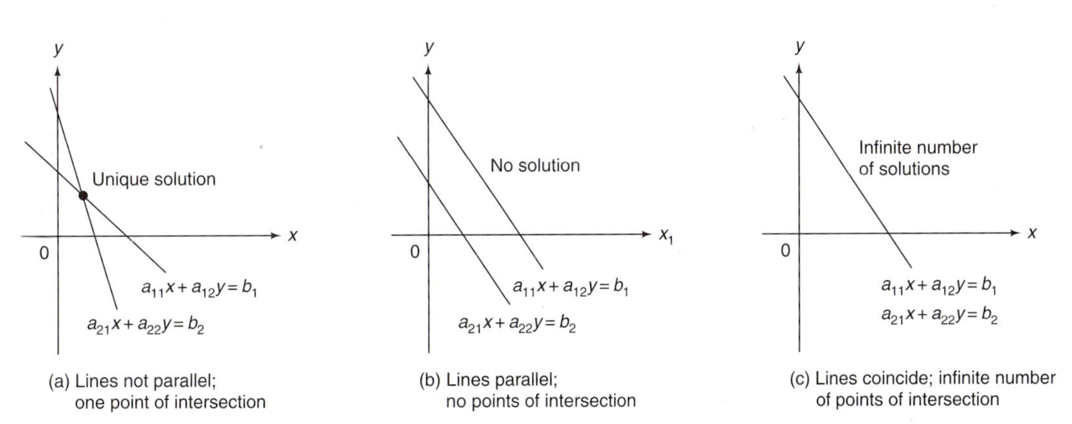

FIGURE 1
Two lines intersect at one point, no points, or (if they coincide) an infinite number of points

Let us now solve system (1) formally. We have

$$\begin{aligned} a_{11}x + a_{12}y &= b_1 \\ a_{21}x + a_{22}y &= b_2 \end{aligned} \tag{1}$$

If $a_{12} = 0$, then $x = \dfrac{b_1}{a_{11}}$ and we can use the second equation to solve for y.

If $a_{22} = 0$, then $x = \dfrac{b_2}{a_{21}}$ and we can use the first equation to solve for y.

If $a_{12} = a_{22} = 0$, then system (1) contains only one unknown x.

Thus we may assume that neither a_{12} nor a_{22} is zero.

Multiplying the first equation by a_{22} and the second by a_{12} yields

$$\begin{aligned} a_{11}a_{22}x + a_{12}a_{22}y &= a_{22}b_1 \\ a_{12}a_{21}x + a_{12}a_{22}y &= a_{12}b_2 \end{aligned} \tag{5}$$

Before continuing we note that system (1) and system (5) are **equivalent**. By that we mean that any solution to system (1) is a solution to system (5) and vice versa. This follows immediately from Fact B assuming that the c in Fact B is not zero. Next we subtract the second equation from the first to obtain

$$(a_{11}a_{22} - a_{12}a_{21})x = a_{22}b_1 - a_{12}b_2 \tag{6}$$

At this point we must pause. If $a_{11}a_{22} - a_{12}a_{21} \neq 0$, then we can divide by it to obtain

$$x = \frac{a_{22}b_1 - a_{12}b_2}{a_{11}a_{22} - a_{12}a_{21}}$$

Then we can substitute this value of x into system (1) to solve for y, and thus we have found the unique solution to the system. We define the **determinant** of system (1) by

$$\text{Determinant of system (1)} = a_{11}a_{22} - a_{12}a_{21} \tag{7}$$

and we have shown the following:

> If the determinant of system (1) $\neq 0$, then the system has a unique solution. **(8)**

How does this statement relate to what we discussed earlier? In system (1) we see that the slope of the first line is $-a_{11}/a_{12}$ and the slope of the second is $-a_{21}/a_{22}$. In Problems 31, 32, and 33 you are asked to show that the determinant of system (1) is zero if and only if the lines are parallel (have the same slope). So, if the determinant is *not* zero, the lines are not parallel and the system has a unique solution.

We now put the facts discussed above together in a theorem. It is a theorem that will be generalized in later sections of this chapter and in subsequent chapters. We shall keep track of our progress by referring to the theorem as our "Summing Up Theorem." When all its parts have been proved, we shall see a remarkable relationship among several important concepts in linear algebra.

THEOREM 1 SUMMING UP THEOREM—VIEW 1

The system

$$\begin{aligned} a_{11}x + a_{12}y &= b_1 \\ a_{21}x + a_{22}y &= b_2 \end{aligned}$$

of two equations in the two unknowns x and y has no solution, a unique solution, or an infinite number of solutions. It has:

i. A unique solution if and only if its determinant is not zero.

ii. No solution or an infinite number of solutions if and only if its determinant is zero. ■

In Section 6.3 we shall discuss systems of m equations in n unknowns and shall see that there is always either no solution, one solution, or an infinite number of solutions. In Chapter 7 we define and calculate determinants for systems of n equations in n unknowns and shall find that our Summing Up Theorem—Theorem 1—is true in this general setting.

PROBLEMS 6.2

SELF-QUIZ

Multiple Choice

I. Which of the following is *not* true about the solution for a system of two linear equations in two unknowns?
- **a.** It is an ordered pair that satisfies both equations.
- **b.** Its graph consists of the point(s) of intersection of the graphs of the equations.
- **c.** Its graph is the x-intercept of the graphs of the equations.
- **d.** If the system is inconsistent, there is no solution.

II. Which of the following is true of an inconsistent system of two linear equations?
- **a.** There is no solution.
- **b.** The graph of the system is on the y-axis.
- **c.** The graph of the solution is one line.
- **d.** The graph of the solution is the point of intersection of two lines.

III. Which of the following is true of the system of equations below?

$$3x - 2y = 8$$
$$4x + y = 7$$

- **a.** The system is inconsistent.
- **b.** The solution is $(-1, 2)$.
- **c.** The solution is on the line $x = 2$.
- **d.** The equations are equivalent.

IV. Which of the following is a second equation for the system whose first equation is $x - 2y = -5$ if there are to be an infinite number of solutions for the system?
- **a.** $6y = 3x + 15$ **b.** $6x - 3y = -15$
- **c.** $y = -\frac{1}{2}x + \frac{5}{2}$ **d.** $\frac{3}{2}x = 3y + \frac{15}{2}$

V. The graph of which of the following systems is a pair of parallel lines?
- **a.** $3x - 2y = 7$, $4y = 6x - 14$ **b.** $x - 2y = 7$, $3x = 4 + 6y$
- **c.** $2x + 3y = 7$, $3x - 2y = 6$ **d.** $5x + y = 1$, $7y = 3x$

In Problems 1–12 find all solutions (if any) to the given systems. In each case calculate the determinant.

1. $x - 3y = 4$, $-4x + 2y = 6$
2. $2x - y = -3$, $5x + 7y = 4$
3. $2x - 8y = 5$, $-3x + 12y = 8$
4. $2x - 8y = 6$, $-3x + 12y = -9$
5. $6x + y = 3$, $-4x - y = 8$
6. $3x + y = 0$, $2x - 3y = 0$
7. $4x - 6y = 0$, $-2x + 3y = 0$
8. $5x + 2y = 3$, $2x + 5y = 3$
9. $2x + 3y = 4$, $3x + 4y = 5$
10. $ax + by = c$, $ax - by = c$
11. $ax + by = c$, $bx + ay = c$
12. $ax - by = c$, $bx + ay = d$
13. Find conditions on a and b such that the system in Problem 10 has a unique solution.
14. Find conditions on a, b, and c such that the system in Problem 11 has an infinite number of solutions.
15. Find conditions on a, b, c, and d such that the system in Problem 12 has no solutions.

In Problems 16–21 find the point of intersection (if there is one) of the two lines.

16. $x - y = 7$; $2x + 3y = 1$
17. $y - 2x = 4$; $4x - 2y = 6$
18. $4x - 6y = 7$; $6x - 9y = 12$
19. $4x - 6y = 10$; $6x - 9y = 15$
20. $3x + y = 4$; $y - 5x = 2$
21. $3x + 4y = 5$; $6x - 7y = 8$

Let L be a line and let $L_{\perp}$ denote the line perpendicular to L that passes through a given point P. The **distance** from L to P is defined to be the distance[†] between P and the point of intersection of L and $L_{\perp}$. In Problems 22–27 find the distance between the given line and point.

22. $x - y = 6$; $(0, 0)$
23. $2x + 3y = -1$; $(0, 0)$
24. $3x + y = 7$; $(1, 2)$
25. $5x - 6y = 3$; $(2, \frac{16}{5})$
26. $2y - 5x = -2$; $(5, -3)$
27. $6y + 3x = 3$; $(8, -1)$
28. Find the distance between the line $2x - y = 6$

[†] Recall that if (x_1, y_1) and (x_2, y_2) are two points in the xy-plane, then the distance d between them is given by $d = \sqrt{(x_1 - x_2)^2 + (y_1 - y_2)^2}$.

and the point of intersection of the lines $2x - 3y = 1$ and $3x + 6y = 12$.

***29.** Prove that the distance between the point (x_1, y_1) and the line $ax + by = c$ is given by

$$d = \frac{|ax_1 + by_1 - c|}{\sqrt{a^2 + b^2}}$$

30. A zoo keeps birds (two-legged) and beasts (four-legged). If the zoo contains 60 heads and 200 feet, how many birds and how many beasts live there?

31. Suppose that the determinant of system (1) is zero. Show that the lines given in (1) are parallel.

32. If there is a unique solution to system (1), show that its determinant is nonzero.

33. If the determinant of system (1) is nonzero, show that the system has a unique solution.

34. The Sunrise Porcelain Company manufactures ceramic cups and saucers. For each cup or saucer a worker measures a fixed amount of material and puts it into a forming machine, from which it is automatically glazed and dried. On the average, a worker needs 3 minutes to get the process started for a cup and 2 minutes for a saucer. The material for a cup costs 25¢ and the material for a saucer costs 20¢. If $44 is allocated daily for production of cups and saucers, how many of each can be manufactured in an 8-hour workday if a worker is working every minute and exactly $44 is spent on materials?

35. Answer the question of Problem 34 if the materials for a cup and saucer cost 15¢ and 10¢, respectively, and $24 is spent in an 8-hour day.

36. Answer the question of Problem 35 if $25 is spent in an 8-hour day.

37. An ice-cream shop sells only ice-cream sodas and milk shakes. It puts 1 ounce of syrup and 4 ounces of ice cream in an ice-cream soda, and 1 ounce of syrup and 3 ounces of ice cream in a milk shake. If the store used 4 gallons of ice cream and 5 quarts of syrup in a day, how many ice-cream sodas and milk shakes did it sell? [*Hint:* 1 quart = 32 ounces; 1 gallon = 128 ounces]

ANSWERS TO SELF-QUIZ

I. c **II.** a **III.** c **IV.** a **V.** b

6.3 *m* EQUATIONS IN *n* UNKNOWNS: GAUSS-JORDAN AND GAUSSIAN ELIMINATION

In this section we describe a method for finding all solutions (if any) to a system of m linear equations in n unknowns. In doing so we shall see that, like the 2×2 case, such a system has no solutions, one solution, or an infinite number of solutions. Before launching into the general method, let us look at some simple examples. As variables, we use x_1, x_2, x_3, and so on instead of $x, y, z, \ldots$ because the subscripted notation is easier to generalize.

EXAMPLE 1 SOLVING A SYSTEM OF THREE EQUATIONS IN THREE UNKNOWNS: UNIQUE SOLUTION

Solve the system

$$\begin{aligned} 2x_1 + 4x_2 + 6x_3 &= 18 \\ 4x_1 + 5x_2 + 6x_3 &= 24 \\ 3x_1 + x_2 - 2x_3 &= 4 \end{aligned} \qquad (1)$$

SOLUTION: Here we seek three numbers x_1, x_2, and x_3 such that the three equations in (1) are satisfied. Our method of solution will be to simplify the equations as we did in Section 6.2 so that solutions can be readily identified. We begin by dividing the first equation by 2. This gives us

$$\begin{aligned} x_1 + 2x_2 + 3x_3 &= 9 \\ 4x_1 + 5x_2 + 6x_3 &= 24 \\ 3x_1 + x_2 - 2x_3 &= 4 \end{aligned} \qquad (2)$$

As we saw in Section 6.2, adding two equations together leads to a third, valid equation. This equation may replace either of the two equations used to obtain it in the system. We begin simplifying system (2) by multiplying both sides of the first equation in (2) by -4 and adding this new equation to the second equation. This gives us

$$\begin{array}{r} -4x_1 - 8x_2 - 12x_3 = -36 \\ 4x_1 + 5x_2 + 6x_3 = 24 \\ \hline -3x_2 - 6x_3 = -12 \end{array}$$

The equation $-3x_2 - 6x_3 = -12$ is our new second equation and the system is now

$$\begin{aligned} x_1 + 2x_2 + 3x_3 &= 9 \\ -3x_2 - 6x_3 &= -12 \\ 3x_1 + x_2 - 2x_3 &= 4 \end{aligned}$$

We then multiply the first equation by -3 and add it to the third equation:

$$\begin{aligned} x_1 + 2x_2 + 3x_3 &= 9 \\ -3x_2 - 6x_3 &= -12 \\ -5x_2 - 11x_3 &= -23 \end{aligned} \tag{3}$$

Note that in system (3) the variable x_1 has been eliminated from the second and third equations. Next we divide the second equation by -3:

$$\begin{aligned} x_1 + 2x_2 + 3x_3 &= 9 \\ x_2 + 2x_3 &= 4 \\ -5x_2 - 11x_3 &= -23 \end{aligned}$$

We multiply the second equation by -2 and add it to the first and then multiply the second equation by 5 and add it to the third:

$$\begin{aligned} x_1 \qquad - x_3 &= 1 \\ x_2 + 2x_3 &= 4 \\ - x_3 &= -3 \end{aligned}$$

We multiply the third equation by -1:

$$\begin{aligned} x_1 \qquad - x_3 &= 1 \\ x_2 + 2x_3 &= 4 \\ x_3 &= 3 \end{aligned}$$

Finally, we add the third equation to the first and then multiply the third equation by -2 and add it to the second to obtain the following system [which is equivalent to system (1)]:

$$\begin{aligned} x_1 \qquad\qquad &= 4 \\ x_2 \qquad &= -2 \\ x_3 &= 3 \end{aligned}$$

This is the unique solution to the system. We write it in the form $(4, -2, 3)$. The method we used here is called **Gauss-Jordan elimination.**†

† Named after the great German mathematician Karl Friedrich Gauss (1777–1855) and the German engineer Wilhelm Jordan (1844–1899). See the biographical sketch of Gauss on page 353.

Before going on to another example, let us summarize what we have done in this example:

i. We divided to make the coefficient of x_1 in the first equation equal to 1.
ii. We "eliminated" the x_1 terms in the second and third equations. That is, we made the coefficients of these terms equal to zero by multiplying the first equation by appropriate numbers and then adding it to the second and third equations, respectively.
iii. We divided to make the coefficient of the x_2 term in the second equation equal to 1 and then proceeded to use the second equation to eliminate the x_2 terms in the first and third equations.
iv. We divided to make the coefficient of the x_3 term in the third equation equal to 1 and then proceeded to use the third equation to eliminate the x_3 terms in the first and second equations.

We emphasize that, at every step, we obtained systems that were equivalent. That is, each system had the same set of solutions as the one that preceded it. This follows from Facts A and B on page 372.

Before solving other systems of equations, we introduce notation that makes it easier to write down each step in our procedure. A **matrix** is a rectangular array of numbers. We shall discuss matrices in great detail in this chapter. For example, the coefficients of the variables x_1, x_2, x_3 in system (1) can be written as the entries of a matrix A, called the **coefficient matrix** of the system:

$$A = \begin{pmatrix} 2 & 4 & 6 \\ 4 & 5 & 6 \\ 3 & 1 & -2 \end{pmatrix} \tag{4}$$

A matrix with m rows and n columns is called an $\boldsymbol{m \times n}$ **matrix**. The symbol $m \times n$ is read "m by n." The study of matrices will take a large part of Chapters 6, 7 and 8. We introduce them here for convenience of notation.

Using matrix notation, we can write system (1) as the **augmented matrix**

$$\left(\begin{array}{ccc|c} 2 & 4 & 6 & 18 \\ 4 & 5 & 6 & 24 \\ 3 & 1 & -2 & 4 \end{array}\right) \tag{5}$$

We now introduce some terminology. We have seen that multiplying (or dividing) the sides of an equation by a nonzero number gives us a new, valid equation. Moreover, adding a multiple of one equation to another equation in a system gives us another valid equation. Finally, if we interchange two equations in a system of equations, we obtain an equivalent system. These three operations, when applied to the rows of the augmented matrix representation of a system of equations, are called **elementary row operations**.

To sum up, the three elementary row operations applied to the augmented matrix representation of a system of equations are:

ELEMENTARY ROW OPERATIONS

i. Multiply (or divide) one row by a nonzero number.
ii. Add a multiple of one row to another row.
iii. Interchange two rows.

The process of applying elementary row operations to simplify an augmented matrix is called **row reduction**.

NOTATION

1. $R_i \to cR_i$ stands for "replace the ith row by the ith row multiplied by c."
2. $R_j \to R_j + cR_i$ stands for "replace the jth row with the sum of the jth row and the ith row multiplied by c."
3. $R_i \rightleftarrows R_j$ stands for "interchange rows i and j."
4. $A \to B$ indicates that the augmented matrices A and B are equivalent; that is, the systems they represent have the same solution.

In Example 1 we saw that by using the elementary row operations (i) and (ii) several times we could obtain a system in which the solutions to the system were given explicitly. We now repeat the steps in Example 1, using the notation just introduced:

$$\left(\begin{array}{ccc|c} 2 & 4 & 6 & 18 \\ 4 & 5 & 6 & 24 \\ 3 & 1 & -2 & 4 \end{array}\right) \xrightarrow{R_1 \to \frac{1}{2}R_1} \left(\begin{array}{ccc|c} 1 & 2 & 3 & 9 \\ 4 & 5 & 6 & 24 \\ 3 & 1 & -2 & 4 \end{array}\right) \xrightarrow[R_3 \to R_3 - 3R_1]{R_2 \to R_2 - 4R_1} \left(\begin{array}{ccc|c} 1 & 2 & 3 & 9 \\ 0 & -3 & -6 & -12 \\ 0 & -5 & -11 & -23 \end{array}\right)$$

$$\xrightarrow{R_2 \to -\frac{1}{3}R_2} \left(\begin{array}{ccc|c} 1 & 2 & 3 & 9 \\ 0 & 1 & 2 & 4 \\ 0 & -5 & -11 & -23 \end{array}\right) \xrightarrow[R_3 \to R_3 + 5R_2]{R_1 \to R_1 - 2R_2} \left(\begin{array}{ccc|c} 1 & 0 & -1 & 1 \\ 0 & 1 & 2 & 4 \\ 0 & 0 & -1 & -3 \end{array}\right)$$

$$\xrightarrow{R_3 \to -R_3} \left(\begin{array}{ccc|c} 1 & 0 & -1 & 1 \\ 0 & 1 & 2 & 4 \\ 0 & 0 & 1 & 3 \end{array}\right) \xrightarrow[R_2 \to R_2 - 2R_3]{R_1 \to R_1 + R_3} \left(\begin{array}{ccc|c} 1 & 0 & 0 & 4 \\ 0 & 1 & 0 & -2 \\ 0 & 0 & 1 & 3 \end{array}\right)$$

Again we can easily "see" the solution $x_1 = 4$, $x_2 = -2$, $x_3 = 3$.

EXAMPLE 2

SOLVING A SYSTEM OF THREE EQUATIONS IN THREE UNKNOWNS: INFINITE NUMBER OF SOLUTIONS

Solve the system

$$\begin{aligned} 2x_1 + 4x_2 + 6x_3 &= 18 \\ 4x_1 + 5x_2 + 6x_3 &= 24 \\ 2x_1 + 7x_2 + 12x_3 &= 30 \end{aligned}$$

SOLUTION: To solve, we proceed as in Example 1, first writing the system as an augmented matrix:

$$\left(\begin{array}{ccc|c} 2 & 4 & 6 & 18 \\ 4 & 5 & 6 & 24 \\ 2 & 7 & 12 & 30 \end{array}\right)$$

We then obtain, successively,

$$\xrightarrow{R_1 \to \frac{1}{2}R_1} \left(\begin{array}{ccc|c} 1 & 2 & 3 & 9 \\ 4 & 5 & 6 & 24 \\ 2 & 7 & 12 & 30 \end{array}\right) \xrightarrow[R_3 \to R_3 - 2R_1]{R_2 \to R_2 - 4R_1} \left(\begin{array}{ccc|c} 1 & 2 & 3 & 9 \\ 0 & -3 & -6 & -12 \\ 0 & 3 & 6 & 12 \end{array}\right)$$

$$\xrightarrow{R_2 \to -\frac{1}{3}R_2} \left(\begin{array}{ccc|c} 1 & 2 & 3 & 9 \\ 0 & 1 & 2 & 4 \\ 0 & 3 & 6 & 12 \end{array}\right) \xrightarrow[R_3 \to R_3 - 3R_2]{R_1 \to R_1 - 2R_2} \left(\begin{array}{ccc|c} 1 & 0 & -1 & 1 \\ 0 & 1 & 2 & 4 \\ 0 & 0 & 0 & 0 \end{array}\right)$$

This is equivalent to the system of equations

$$\begin{aligned} x_1 \qquad - x_3 &= 1 \\ x_2 + 2x_3 &= 4 \end{aligned}$$

This is as far as we can go. There are now only two equations in the three unknowns x_1, x_2, x_3 and there are an infinite number of solutions. To see this let x_3 be chosen. Then $x_2 = 4 - 2x_3$ and $x_1 = 1 + x_3$. This will be a solution for any number x_3. We write these solutions in the form $(1 + x_3, 4 - 2x_3, x_3)$. For example, if $x_3 = 0$, we obtain the solution $(1, 4, 0)$. For $x_3 = 10$ we obtain the solution $(11, -16, 10)$.

EXAMPLE 3

AN INCONSISTENT SYSTEM

Solve the system

$$\begin{aligned} 2x_1 + 4x_2 + 6x_3 &= 18 \\ 4x_1 + 5x_2 + 6x_3 &= 24 \\ 2x_1 + 7x_2 + 12x_3 &= 40 \end{aligned} \tag{6}$$

SOLUTION: We use the augmented-matrix form and proceed exactly as in Example 2 to obtain, successively, the following systems. (Note how in each step we use either elementary row operation (*i*) or (*ii*).)

$$\left(\begin{array}{ccc|c} 2 & 4 & 6 & 18 \\ 4 & 5 & 6 & 24 \\ 2 & 7 & 12 & 40 \end{array}\right) \xrightarrow{R_1 \to \frac{1}{2}R_1} \left(\begin{array}{ccc|c} 1 & 2 & 3 & 9 \\ 4 & 5 & 6 & 24 \\ 2 & 7 & 12 & 40 \end{array}\right)$$

$$\xrightarrow[R_3 \to R_3 - 2R_1]{R_2 \to R_2 - 4R_1} \left(\begin{array}{ccc|c} 1 & 2 & 3 & 9 \\ 0 & -3 & -6 & -12 \\ 0 & 3 & 6 & 22 \end{array}\right) \xrightarrow{R_2 \to -\frac{1}{3}R_2} \left(\begin{array}{ccc|c} 1 & 2 & 3 & 9 \\ 0 & 1 & 2 & 4 \\ 0 & 3 & 6 & 22 \end{array}\right)$$

$$\xrightarrow[R_3 \to R_3 - 3R_2]{R_1 \to R_1 - 2R_2} \left(\begin{array}{ccc|c} 1 & 0 & -1 & 1 \\ 0 & 1 & 2 & 4 \\ 0 & 0 & 0 & 10 \end{array}\right) \xrightarrow{R_3 \to \frac{1}{10}R_3} \left(\begin{array}{ccc|c} 1 & 0 & -1 & 1 \\ 0 & 1 & 2 & 4 \\ 0 & 0 & 0 & 1 \end{array}\right)$$

The last equation now reads $0x_1 + 0x_2 + 0x_3 = 1$, which is impossible since $0 \neq 1$. Thus system (6) has *no* solution. In this case the system is said to be **inconsistent**.

Let us take another look at these three examples. In Example 1 we began with the coefficient matrix

$$A_1 = \begin{pmatrix} 2 & 4 & 6 \\ 4 & 5 & 6 \\ 3 & 1 & -2 \end{pmatrix}$$

In the process of row reduction A_1 was "reduced" to the matrix

$$R_1 = \begin{pmatrix} 1 & 0 & 0 \\ 0 & 1 & 0 \\ 0 & 0 & 1 \end{pmatrix}$$

In Example 2 we started with

$$A_2 = \begin{pmatrix} 2 & 4 & 6 \\ 4 & 5 & 6 \\ 2 & 7 & 12 \end{pmatrix}$$

and ended up with

$$R_2 = \begin{pmatrix} 1 & 0 & -1 \\ 0 & 1 & 2 \\ 0 & 0 & 0 \end{pmatrix}$$

In Example 3 we began with

$$A_3 = \begin{pmatrix} 2 & 4 & 6 \\ 4 & 5 & 6 \\ 2 & 7 & 12 \end{pmatrix}$$

and again ended up with

$$R_3 = \begin{pmatrix} 1 & 0 & -1 \\ 0 & 1 & 2 \\ 0 & 0 & 0 \end{pmatrix}$$

The matrices R_1, R_2, and R_3 are called the *reduced row echelon forms* of the matrices A_1, A_2, and A_3, respectively. In general, we have the following definition.

DEFINITION REDUCED ROW ECHELON FORM

A matrix is in **reduced row echelon form** if the following four conditions hold:

i. All rows (if any) consisting entirely of zeros appear at the bottom of the matrix.

ii. The first nonzero number (starting from the left) in any row not consisting entirely of zeros is 1.

iii. If two successive rows do not consist entirely of zeros, then the first 1 in the lower row occurs farther to the right than the first 1 in the higher row.

iv. Any column containing the first 1 in a row has zeros everywhere else. ■

EXAMPLE 4 FIVE MATRICES IN REDUCED ROW ECHELON FORM

The following matrices are in reduced row echelon form:

i. $\begin{pmatrix} 1 & 0 & 0 \\ 0 & 1 & 0 \\ 0 & 0 & 1 \end{pmatrix}$ **ii.** $\begin{pmatrix} 1 & 0 & 0 & 0 \\ 0 & 1 & 0 & 0 \\ 0 & 0 & 0 & 1 \end{pmatrix}$ **iii.** $\begin{pmatrix} 1 & 0 & 0 & 5 \\ 0 & 0 & 1 & 2 \end{pmatrix}$

iv. $\begin{pmatrix} 1 & 0 \\ 0 & 1 \end{pmatrix}$ **v.** $\begin{pmatrix} 1 & 0 & 2 & 5 \\ 0 & 1 & 3 & 6 \\ 0 & 0 & 0 & 0 \end{pmatrix}$

DEFINITION ROW ECHELON FORM

A matrix is in **row echelon form** if conditions (*i*), (*ii*), and (*iii*) hold in the definition of reduced row echelon form. ■

EXAMPLE 5 FIVE MATRICES IN ROW ECHELON FORM

The following matrices are in row echelon form:

$$\textbf{i.}\ \begin{pmatrix} 1 & 2 & 3 \\ 0 & 1 & 5 \\ 0 & 0 & 1 \end{pmatrix} \qquad \textbf{ii.}\ \begin{pmatrix} 1 & -1 & 6 & 4 \\ 0 & 1 & 2 & -8 \\ 0 & 0 & 0 & 1 \end{pmatrix}$$

$$\textbf{iii.}\ \begin{pmatrix} 1 & 0 & 2 & 5 \\ 0 & 0 & 1 & 2 \end{pmatrix} \qquad \textbf{iv.}\ \begin{pmatrix} 1 & 2 \\ 0 & 1 \end{pmatrix} \qquad \textbf{v.}\ \begin{pmatrix} 1 & 3 & 2 & 5 \\ 0 & 1 & 3 & 6 \\ 0 & 0 & 0 & 0 \end{pmatrix}$$

NOTE: The row echelon form of a matrix might not be unique.

REMARK 1: The difference between these two forms should be clear from the examples. In row echelon form, all the numbers below the first 1 in a row are zero. In reduced row echelon form, all the numbers above and below the first 1 in a row are zero. Thus reduced row echelon form is more exclusive. That is, every matrix in reduced row echelon form is in row echelon form, but not conversely.

REMARK 2: We can always reduce a matrix to reduced row echelon form or row echelon form by performing elementary row operations. We saw this reduction to reduced row echelon form in Examples 1, 2, and 3.

As we saw in Examples 1, 2, and 3, there is a strong connection between the reduced row echelon form of a matrix and the existence of a unique solution to the system. In Example 1 the reduced row echelon form of the coefficient matrix (that is, the first three columns of the augmented matrix) had a 1 in each row and there was a unique solution. In Examples 2 and 3 the reduced row echelon form of the coefficient matrix had a row of zeros and the system had either no solution or an infinite number of solutions. This turns out always to be true in any system with the same number of equations as unknowns. But before turning to the general case, let us discuss the usefulness of the row echelon form of a matrix. It is possible to solve the system in Example 1 by reducing the coefficient matrix to its row echelon form.

EXAMPLE 6 SOLVING A SYSTEM BY GAUSSIAN ELIMINATION

Solve the system of Example 1 by reducing the coefficient matrix to row echelon form.

SOLUTION: We begin as before:

$$\left(\begin{array}{ccc|c} 2 & 4 & 6 & 18 \\ 4 & 5 & 6 & 24 \\ 3 & 1 & -2 & 4 \end{array}\right) \xrightarrow{R_1 \to \frac{1}{2}R_1} \left(\begin{array}{ccc|c} 1 & 2 & 3 & 9 \\ 4 & 5 & 6 & 24 \\ 3 & 1 & -2 & 4 \end{array}\right)$$

$$\xrightarrow[R_3 \to R_3 - 3R_1]{R_2 \to R_2 - 4R_1} \left(\begin{array}{ccc|c} 1 & 2 & 3 & 9 \\ 0 & -3 & -6 & -12 \\ 0 & -5 & -11 & -23 \end{array}\right) \xrightarrow{R_2 \to -\frac{1}{3}R_2} \left(\begin{array}{ccc|c} 1 & 2 & 3 & 9 \\ 0 & 1 & 2 & 4 \\ 0 & -5 & -11 & -23 \end{array}\right)$$

So far, this process is identical to our earlier one. Now, however, we only make zero the number (-5) below the first 1 in the second row:

$$\xrightarrow{R_3 \to R_3 + 5R_2} \left(\begin{array}{ccc|c} 1 & 2 & 3 & 9 \\ 0 & 1 & 2 & 4 \\ 0 & 0 & -1 & -3 \end{array}\right) \xrightarrow{R_3 \to -R_3} \left(\begin{array}{ccc|c} 1 & 2 & 3 & 9 \\ 0 & 1 & 2 & 4 \\ 0 & 0 & 1 & 3 \end{array}\right)$$

The augmented matrix of the system (and the coefficient matrix) are now in row echelon form and we immediately see that $x_3 = 3$. We then use **back substitution** to solve for x_2 and then x_1. The second equation reads $x_2 + 2x_3 = 4$. Thus $x_2 + 2(3) = 4$ and $x_2 = -2$. Similarly, from the first equation we obtain $x_1 + 2(-2) + 3(3) = 9$ or $x_1 = 4$. Thus we again obtain the solution $(4, -2, 3)$. The method of solution just employed is called **Gaussian elimination**.

We therefore have two methods for solving our sample systems of equations:

i. Gauss-Jordan Elimination
Row-reduce the coefficient matrix to reduced row echelon form using the procedure used in Examples 1, 2 and 3.

ii. Gaussian Elimination
Row-reduce the coefficient matrix to row echelon form, solve for the last unknown, and then use back substitution to solve for the other unknowns.

Which method is more useful? It depends. In solving systems of equations on a computer, Gaussian elimination is the preferred method because it involves fewer elementary row operations. We shall discuss the numerical solution of systems of equations in Section 6.10. On the other hand, there are times when it is essential to obtain the reduced row echelon form of a matrix (one of these is discussed in Section 6.8). In these cases Gauss-Jordan elimination is the preferred method.

We now turn to the solution of a general system of m equations in n unknowns. Because of our need to do so in Section 6.8, we shall be solving most of the systems by Gauss-Jordan elimination. Keep in mind, however, that Gaussian elimination is sometimes the preferred approach.

The general $m \times n$ system of m linear equations in n unknowns is given by

$$\begin{array}{ccccccc} a_{11}x_1 & + a_{12}x_2 & + a_{13}x_3 & + \cdots + & a_{1n}x_n & = & b_1 \\ a_{21}x_1 & + a_{22}x_2 & + a_{23}x_3 & + \cdots + & a_{2n}x_n & = & b_2 \\ a_{31}x_1 & + a_{32}x_2 & + a_{33}x_3 & + \cdots + & a_{3n}x_n & = & b_3 \\ \vdots & \vdots & \vdots & \vdots & \vdots & & \vdots \\ a_{m1}x_1 & + a_{m2}x_2 & + a_{m3}x_3 & + \cdots + & a_{mn}x_n & = & b_m \end{array} \quad (7)$$

In system (7) all the a's and b's are given real numbers. The problem is to find all sets of n numbers, denoted by $(x_1, x_2, x_3, \ldots, x_n)$, that satisfy every one of the m equations in (7). The number a_{ij} is the coefficient of the variable x_j in the ith equation.

We solve system (7) by writing the system as an augmented matrix and row-reducing the matrix to its reduced row echelon form. We start by dividing the first row by a_{11}

[elementary row operation (i)]. If $a_{11} = 0$, then we rearrange† the equations so that, with rearrangement, the new $a_{11} \neq 0$. We then use the first equation to eliminate the x_1 term in each of the other equations [using elementary row operation (ii)]. Then the new second equation is divided by the new a_{22} term and the new, new second equation is used to eliminate the x_2 terms in all the other equations. The process is continued until one of three situations occurs:

i. The last nonzero‡ equation reads $x_n = c$ for some constant c. Then there is either a unique solution or an infinite number of solutions to the system.

ii. The last nonzero equation reads $a'_{ij}x_j + a'_{i,j+1}x_{j+1} + \cdots + a'_{i,j+k}x_n = c$ for some constant c where at least two of the a's are nonzero. That is, the last equation is a linear equation in two or more of the variables. Then there are an infinite number of solutions.

iii. The last equation reads $0 = c$, where $c \neq 0$. Then there is no solution. In this case the system is called **inconsistent**. In cases (i) and (ii) the system is called **consistent**.

EXAMPLE 7 SOLVING A SYSTEM OF TWO EQUATIONS IN FOUR UNKNOWNS

Solve the system

$$\begin{aligned} x_1 + 3x_2 - 5x_3 + x_4 &= 4 \\ 2x_1 + 5x_2 - 2x_3 + 4x_4 &= 6 \end{aligned}$$

SOLUTION: We write this system as an augmented matrix and row-reduce:

$$\left(\begin{array}{cccc|c} 1 & 3 & -5 & 1 & 4 \\ 2 & 5 & -2 & 4 & 6 \end{array}\right) \xrightarrow{R_2 \to R_2 - 2R_1} \left(\begin{array}{cccc|c} 1 & 3 & -5 & 1 & 4 \\ 0 & -1 & 8 & 2 & -2 \end{array}\right)$$

$$\xrightarrow{R_2 \to -R_2} \left(\begin{array}{cccc|c} 1 & 3 & -5 & 1 & 4 \\ 0 & 1 & -8 & -2 & 2 \end{array}\right) \xrightarrow{R_1 \to R_1 - 3R_2} \left(\begin{array}{cccc|c} 1 & 0 & 19 & 7 & -2 \\ 0 & 1 & -8 & -2 & 2 \end{array}\right)$$

This is as far as we can go. The coefficient matrix is in reduced row echelon form—case (ii) above. There are evidently an infinite number of solutions. The variables x_3 and x_4 can be chosen arbitrarily. Then $x_2 = 2 + 8x_3 + 2x_4$ and $x_1 = -2 - 19x_3 - 7x_4$. All solutions are, therefore, represented by $(-2 - 19x_3 - 7x_4, 2 + 8x_3 + 2x_4, x_3, x_4)$. For example, if $x_3 = 1$ and $x_4 = 2$, we obtain the solution $(-35, 14, 1, 2)$.

As you will see if you do a lot of system solving, the computations can become very messy. It is a good rule of thumb to use a calculator whenever the fractions become unpleasant. It should be noted, however, that if computations are carried out on a computer or calculator, "round-off" errors can be introduced.

We close this section with three examples illustrating how a system of linear equations can arise in a practical situation.

---

†To rearrange a system of equations we simply write the same equations in a different order. For example, the first equation can become the fourth equation, the third equation can become the second equation, and so on. This is a sequence of elementary row operations (iii).

‡The "zero equation" is the equation $0 = 0$.

EXAMPLE 8

THE LEONTIEF INPUT-OUTPUT MODEL

A model that is often used in economics is the **Leontief input-output model.**† Suppose an economic system has n industries. There are two kinds of demands on each industry. First there is the *external* demand from outside the system. If the system is a country, for example, then the external demand could be from another country. Second there is the demand placed on one industry by another industry in the same system. In the United States, for example, there is a demand on the output of the steel industry by the automobile industry.

Let e_i represent the external demand placed on the ith industry. Let a_{ij} represent the internal demand placed on the ith industry by the jth industry. More precisely, a_{ij} represents the number of units of the output of industry i needed to produce 1 unit of the output of industry j. Let x_i represent the output of industry i. Now we assume that the output of each industry is equal to its demand (that is, there is no overproduction). The total demand is equal to the sum of the internal and external demands. To calculate the internal demand on industry 2, for example, we note that industry 1 needs a_{21} units of the output of industry 2 to produce 1 unit of its output. If the output from industry 1 is x_1, then $a_{21}x_1$ is the total amount industry 1 needs from industry 2. Thus the total internal demand on industry 2 is $a_{21}x_1 + a_{22}x_2 + \cdots + a_{2n}x_n$.

We are led to the following system of equations obtained by equating the total demand with the output of each industry:

$$\begin{aligned} a_{11}x_1 + a_{12}x_2 + \cdots + a_{1n}x_n + e_1 &= x_1 \\ a_{21}x_1 + a_{22}x_2 + \cdots + a_{2n}x_n + e_2 &= x_2 \\ \vdots \qquad \vdots \qquad\qquad \vdots \qquad \vdots \quad &\quad \vdots \\ a_{n1}x_1 + a_{n2}x_2 + \cdots + a_{nn}x_n + e_n &= x_n \end{aligned} \tag{8}$$

Or, rewriting (8) so it looks like system (7), we get

$$\begin{aligned} (1 - a_{11})x_1 - a_{12}x_2 - \cdots - a_{1n}x_n &= e_1 \\ -a_{21}x_1 + (1 - a_{22})x_2 - \cdots - a_{2n}x_n &= e_2 \\ \vdots \qquad\qquad \vdots \qquad\qquad\qquad \vdots \quad &\quad \vdots \\ -a_{n1}x_1 - a_{n2}x_2 - \cdots + (1 - a_{nn})x_n &= e_n \end{aligned} \tag{9}$$

System (9) of n equations in n unknowns is very important in economic analysis.

EXAMPLE 9

THE LEONTIEF MODEL APPLIED TO AN ECONOMIC SYSTEM WITH THREE INDUSTRIES

In an economic system with three industries, suppose that the external demands are, respectively, 10, 25, and 20. Suppose that $a_{11} = 0.2$, $a_{12} = 0.5$, $a_{13} = 0.15$, $a_{21} = 0.4$, $a_{22} = 0.1$, $a_{23} = 0.3$, $a_{31} = 0.25$, $a_{32} = 0.5$, and $a_{33} = 0.15$. Find the output in each industry such that supply exactly equals demand.

SOLUTION: Here $n = 3$, $1 - a_{11} = 0.8$, $1 - a_{22} = 0.9$, and $1 - a_{33} = 0.85$. Then system (9) is

$$\begin{aligned} 0.8x_1 - 0.5x_2 - 0.15x_3 &= 10 \\ -0.4x_1 + 0.9x_2 - 0.3x_3 &= 25 \\ -0.25x_1 - 0.5x_2 + 0.85x_3 &= 20 \end{aligned}$$

†Named after American economist Wassily W. Leontief. This model was used in his pioneering paper "Qualitative Input and Output Relations in the Economic System of the United States" in *Review of Economic Statistics* 18(1936):105–125. An updated version of this model appears in Leontief's book *Input-Output Analysis* (New York: Oxford University Press, 1966). Leontief won the Nobel Prize in economics in 1973 for his development of input-output analysis.

Solving this system by using a calculator, we obtain, successively (using five-decimal-place accuracy and Gauss-Jordan elimination),

$$\left(\begin{array}{ccc|c} 0.8 & -0.5 & -0.15 & 10 \\ -0.4 & 0.9 & -0.3 & 25 \\ -0.25 & -0.5 & 0.85 & 20 \end{array}\right) \xrightarrow{R_1 \to \frac{1}{0.8}R_1} \left(\begin{array}{ccc|c} 1 & -0.625 & -0.1875 & 12.5 \\ -0.4 & 0.9 & -0.3 & 25 \\ -0.25 & -0.5 & 0.85 & 20 \end{array}\right)$$

$$\xrightarrow[R_3 \to R_3 + 0.25R_1]{R_2 \to R_2 + 0.4R_1} \left(\begin{array}{ccc|c} 1 & -0.625 & -0.1875 & 12.5 \\ 0 & 0.65 & -0.375 & 30 \\ 0 & -0.65625 & 0.80313 & 23.125 \end{array}\right)$$

$$\xrightarrow{R_2 \to \frac{1}{0.65}R_2} \left(\begin{array}{ccc|c} 1 & -0.625 & -0.1875 & 12.5 \\ 0 & 1 & -0.57692 & 46.15385 \\ 0 & -0.65625 & 0.80313 & 23.125 \end{array}\right)$$

$$\xrightarrow[R_3 \to R_3 + 0.65625R_2]{R_1 \to R_1 + 0.625R_2} \left(\begin{array}{ccc|c} 1 & 0 & -0.54808 & 41.34616 \\ 0 & 1 & -0.57692 & 46.15385 \\ 0 & 0 & 0.42453 & 53.41346 \end{array}\right)$$

$$\xrightarrow{R_3 \to \frac{1}{0.42453}R_3} \left(\begin{array}{ccc|c} 1 & 0 & -0.54808 & 41.34616 \\ 0 & 1 & -0.57692 & 46.15385 \\ 0 & 0 & 1 & 125.81787 \end{array}\right)$$

$$\xrightarrow[R_2 \to R_2 + 0.57692R_3]{R_1 \to R_1 + 0.54808R_3} \left(\begin{array}{ccc|c} 1 & 0 & 0 & 110.30442 \\ 0 & 1 & 0 & 118.74070 \\ 0 & 0 & 1 & 125.81787 \end{array}\right)$$

We conclude that the outputs needed for supply to equal demand are, approximately, $x_1 \approx 110$, $x_2 \approx 119$, and $x_3 \approx 126$.

EXAMPLE 10

A PROBLEM IN RESOURCE MANAGEMENT

A State Fish and Game Department supplies three types of food to a lake that supports three species of fish. Each fish of Species 1 consumes, each week, an average of 1 unit of Food 1, 1 unit of Food 2, and 2 units of Food 3. Each fish of Species 2 consumes, each week, an average of 3 units of Food 1, 4 units of Food 2, and 5 units of Food 3. For a fish of Species 3, the average weekly consumption is 2 units of Food 1, 1 unit of Food 2, and 5 units of Food 3. Each week 25,000 units of Food 1, 20,000 units of Food 2, and 55,000 units of Food 3 are supplied to the lake. If we assume that all food is eaten, how many fish of each species can coexist in the lake?

SOLUTION: We let x_1, x_2, and x_3 denote the numbers of fish of the three species being supported by the lake environment. Using the information in the problem, we see that x_1 fish of Species 1 consume x_1 units of Food 1, x_2 fish of Species 2 consume $3x_2$ units of Food 1, and x_3 fish of Species 3 consume $2x_3$ units of Food 1. Thus $x_1 + 3x_2 + 2x_3 = 25{,}000 =$ total weekly supply of Food 1. Obtaining a similar equation for each of the other two foods, we are led to the following system:

$$\begin{aligned} x_1 + 3x_2 + 2x_3 &= 25{,}000 \\ x_1 + 4x_2 + x_3 &= 20{,}000 \\ 2x_1 + 5x_2 + 5x_3 &= 55{,}000 \end{aligned}$$

Upon solving, we obtain

$$\left(\begin{array}{ccc|c} 1 & 3 & 2 & 25{,}000 \\ 1 & 4 & 1 & 20{,}000 \\ 2 & 5 & 5 & 55{,}000 \end{array}\right)$$

$$\xrightarrow[R_3 \to R_3 - 2R_1]{R_2 \to R_2 - R_1} \left(\begin{array}{ccc|c} 1 & 3 & 2 & 25{,}000 \\ 0 & 1 & -1 & -5{,}000 \\ 0 & -1 & 1 & 5{,}000 \end{array}\right) \xrightarrow[R_3 \to R_3 + R_2]{R_1 \to R_1 - 3R_2} \left(\begin{array}{ccc|c} 1 & 0 & 5 & 40{,}000 \\ 0 & 1 & -1 & -5{,}000 \\ 0 & 0 & 0 & 0 \end{array}\right)$$

Thus, if x_3 is chosen arbitrarily, we have an infinite number of solutions given by $(40{,}000 - 5x_3,\ x_3 - 5{,}000,\ x_3)$. Of course, we must have $x_1 \geq 0$, $x_2 \geq 0$ and $x_3 \geq 0$. Since $x_2 = x_3 - 5{,}000 \geq 0$, we have $x_3 \geq 5{,}000$. This means that $0 \leq x_1 \leq 40{,}000 - 5(5{,}000) = 15{,}000$. Finally, since $40{,}000 - 5x_3 \geq 0$, we see that $x_3 \leq 8{,}000$. This means that the populations that can be supported by the lake with all food consumed are

$$x_1 = 40{,}000 - 5x_3$$
$$x_2 = x_3 - 5{,}000$$
$$5{,}000 \leq x_3 \leq 8{,}000$$

For example, if $x_3 = 6{,}000$, then $x_1 = 10{,}000$ and $x_2 = 1{,}000$.

NOTE: The system of equations does have an infinite number of solutions. However, the resource management problem has only a finite number of solutions because x_1, x_2, and x_3 must be integers and there are only 3001 integers in the interval [5000, 8000]. (You can't stock 5237.578 fish, for example.)

PROBLEMS 6.3

SELF-QUIZ

Multiple Choice

I. Which of the following systems has the coefficient matrix given at the right?

$$\begin{pmatrix} 3 & 2 & -1 \\ 0 & 1 & 5 \\ 2 & 0 & 1 \end{pmatrix}$$

a. $3x + 2y = -1$
$y = 5$
$2x = 1$

b. $3x + 2z = 10$
$2x + y = 0$
$-x + 5y + z = 5$

c. $3x = 2$
$2x + y = 0$
$-x + 5y = 1$

d. $3x + 2y - z = -3$
$y + 5z = 15$
$2x + z = 3$

II. Which of the following is an elementary row operation?

a. Replace a row with a nonzero multiple of that row.

b. Add a nonzero constant to each entry in a row.

c. Interchange two columns.

d. Replace a row with a sum of the row and a nonzero constant.

III. Which of the following is true about the given matrix?

$$\begin{pmatrix} 1 & 0 & 0 & 3 \\ 0 & 1 & 1 & 2 \\ 0 & 0 & 0 & 3 \\ 0 & 0 & 0 & 0 \end{pmatrix}$$

a. It is in row-echelon form.

b. It is not in row-echelon form because the fourth number in row 1 is not 1.

c. It is not in row-echelon form because the first nonzero entry in row 3 is 3.

d. It is not in row-echelon form because the last column contains a 0.

IV. Which of the following is true about the system given below?

$$\begin{aligned} x + y + z &= 3 \\ 2x + 2y + 2z &= 6 \\ 3x + 3y + 3z &= 10 \end{aligned}$$

a. It has the unique solution $x = 1$, $y = 1$, $z = 1$.
b. It is inconsistent.
c. It has an infinite number of solutions.

In Problems 1–20 use Gauss-Jordan or Gaussian elimination to find all solutions, if any, to the given systems.

1. $$\begin{aligned} x_1 - 2x_2 + 3x_3 &= 11 \\ 4x_1 + x_2 - x_3 &= 4 \\ 2x_1 - x_2 + 3x_3 &= 10 \end{aligned}$$

2. $$\begin{aligned} -2x_1 + x_2 + 6x_3 &= 18 \\ 5x_1 \quad + 8x_3 &= -16 \\ 3x_1 + 2x_2 - 10x_3 &= -3 \end{aligned}$$

3. $$\begin{aligned} 3x_1 + 6x_2 - 6x_3 &= 9 \\ 2x_1 - 5x_2 + 4x_3 &= 6 \\ -x_1 + 16x_2 - 14x_3 &= -3 \end{aligned}$$

4. $$\begin{aligned} 3x_1 + 6x_2 - 6x_3 &= 9 \\ 2x_1 - 5x_2 + 4x_3 &= 6 \\ 5x_1 + 28x_2 - 26x_3 &= -8 \end{aligned}$$

5. $$\begin{aligned} x_1 + x_2 - x_3 &= 7 \\ 4x_1 - x_2 + 5x_3 &= 4 \\ 2x_1 + 2x_2 - 3x_3 &= 0 \end{aligned}$$

6. $$\begin{aligned} x_1 + x_2 - x_3 &= 7 \\ 4x_1 - x_2 + 5x_3 &= 4 \\ 6x_1 + x_2 + 3x_3 &= 18 \end{aligned}$$

7. $$\begin{aligned} x_1 + x_2 - x_3 &= 7 \\ 4x_1 - x_2 + 5x_3 &= 4 \\ 6x_1 + x_2 + 3x_3 &= 20 \end{aligned}$$

8. $$\begin{aligned} x_1 - 2x_2 + 3x_3 &= 0 \\ 4x_1 + x_2 - x_3 &= 0 \\ 2x_1 - x_2 + 3x_3 &= 0 \end{aligned}$$

9. $$\begin{aligned} x_1 + x_2 - x_3 &= 0 \\ 4x_1 - x_2 + 5x_3 &= 0 \\ 6x_1 + x_2 + 3x_3 &= 0 \end{aligned}$$

10. $$\begin{aligned} 2x_2 + 5x_3 &= 6 \\ x_1 \quad - 2x_3 &= 4 \\ 2x_1 + 4x_2 \quad &= -2 \end{aligned}$$

11. $$\begin{aligned} x_1 + 2x_2 - x_3 &= 4 \\ 3x_1 + 4x_2 - 2x_3 &= 7 \end{aligned}$$

12. $$\begin{aligned} x_1 + 2x_2 - 4x_3 &= 4 \\ -2x_1 - 4x_2 + 8x_3 &= -8 \end{aligned}$$

13. $$\begin{aligned} x_1 + 2x_2 - 4x_3 &= 4 \\ -2x_1 - 4x_2 + 8x_3 &= -9 \end{aligned}$$

14. $$\begin{aligned} x_1 + 2x_2 - x_3 + x_4 &= 7 \\ 3x_1 + 6x_2 - 3x_3 + 3x_4 &= 21 \end{aligned}$$

15. $$\begin{aligned} 2x_1 + 6x_2 - 4x_3 + 2x_4 &= 4 \\ x_1 \quad - x_3 + x_4 &= 5 \\ -3x_1 + 2x_2 - 2x_3 \quad &= -2 \end{aligned}$$

16. $$\begin{aligned} x_1 - 2x_2 + x_3 + x_4 &= 2 \\ 3x_1 \quad + 2x_3 - 2x_4 &= -8 \\ 4x_2 - x_3 - x_4 &= 1 \\ -x_1 + 6x_2 - 2x_3 \quad &= 7 \end{aligned}$$

17. $$\begin{aligned} x_1 - 2x_2 + x_3 + x_4 &= 2 \\ 3x_1 \quad + 2x_3 - 2x_4 &= -8 \\ 4x_2 - x_3 - x_4 &= 1 \\ 5x_1 \quad + 3x_3 - x_4 &= -3 \end{aligned}$$

18. $$\begin{aligned} x_1 - 2x_2 + x_3 + x_4 &= 2 \\ 3x_1 \quad + 2x_3 - 2x_4 &= -8 \\ 4x_2 - x_3 - x_4 &= 1 \\ 5x_1 \quad + 3x_3 - x_4 &= 0 \end{aligned}$$

19. $$\begin{aligned} x_1 + x_2 &= 4 \\ 2x_1 - 3x_2 &= 7 \\ 3x_1 + 2x_2 &= 8 \end{aligned}$$

20. $$\begin{aligned} x_1 + x_2 &= 4 \\ 2x_1 - 3x_2 &= 7 \\ 3x_1 - 2x_2 &= 11 \end{aligned}$$

In Problems 21–29 determine whether the given matrix is in row echelon form (but not reduced row echelon form), reduced row echelon form, or neither.

21. $\begin{pmatrix} 1 & 1 & 0 \\ 0 & 1 & 0 \\ 0 & 0 & 1 \end{pmatrix}$

22. $\begin{pmatrix} 2 & 0 & 0 \\ 0 & 1 & 0 \\ 0 & 0 & -1 \end{pmatrix}$

23. $\begin{pmatrix} 1 & 0 & 1 & 0 \\ 0 & 1 & 1 & 0 \\ 0 & 0 & 0 & 0 \end{pmatrix}$

24. $\begin{pmatrix} 1 & 0 & 0 & 0 \\ 0 & 0 & 1 & 0 \\ 0 & 0 & 0 & 1 \end{pmatrix}$

25. $\begin{pmatrix} 0 & 1 & 0 & 0 \\ 1 & 0 & 0 & 0 \\ 0 & 0 & 0 & 0 \end{pmatrix}$

26. $\begin{pmatrix} 1 & 0 & 1 & 2 \\ 0 & 1 & 3 & 4 \end{pmatrix}$

27. $\begin{pmatrix} 1 & 0 \\ 0 & 1 \\ 0 & 0 \end{pmatrix}$

28. $\begin{pmatrix} 1 & 0 & 0 \\ 0 & 0 & 0 \\ 0 & 0 & 1 \end{pmatrix}$

29. $\begin{pmatrix} 1 & 0 & 0 & 4 \\ 0 & 1 & 0 & 5 \\ 0 & 1 & 1 & 6 \end{pmatrix}$

In Problems 30–35 use the elementary row operations to reduce the given matrices to row echelon form and reduced row echelon form.

30. $\begin{pmatrix} 1 & 1 \\ 2 & 3 \end{pmatrix}$

31. $\begin{pmatrix} -1 & 6 \\ 4 & 2 \end{pmatrix}$

32. $\begin{pmatrix} 1 & -1 & 1 \\ 2 & 4 & 3 \\ 5 & 6 & -2 \end{pmatrix}$

33. $\begin{pmatrix} 2 & -4 & 8 \\ 3 & 5 & 8 \\ -6 & 0 & 4 \end{pmatrix}$

34. $\begin{pmatrix} 2 & -4 & -2 \\ 3 & 1 & 6 \end{pmatrix}$

35. $\begin{pmatrix} 2 & -7 \\ 3 & 5 \\ 4 & -3 \end{pmatrix}$

36. In the Leontief input-output model of Example 8 suppose that there are three industries. Suppose further that $e_1 = 10$, $e_2 = 15$, $e_3 = 30$, $a_{11} = \frac{1}{3}$, $a_{12} = \frac{1}{2}$, $a_{13} = \frac{1}{6}$, $a_{21} = \frac{1}{4}$, $a_{22} = \frac{1}{4}$, $a_{23} = \frac{1}{8}$, $a_{31} = \frac{1}{12}$, $a_{32} = \frac{1}{3}$, and $a_{33} = \frac{1}{6}$. Find the output of each industry such that supply exactly equals demand.

37. In Example 10 assume that there are 15,000 units of the first food, 10,000 units of the second, and 35,000 units of the third supplied to the lake each week. Assuming that all three foods are consumed, what populations of the three species can coexist in the lake? Is there a unique solution?

38. A traveler who just returned from Europe spent \$30 a day for housing in England, \$20 a day in France, and \$20 a day in Spain. For food the traveler spent \$20 a day in England, \$30 a day in France, and \$20 a day in Spain. The traveler spent \$10 a day in each country for incidental expenses. The traveler's records of the trip indicate a total of \$340 spent for housing, \$320 for food, and \$140 for incidental expenses while traveling in these countries. Calculate the number of days the traveler spent in each of the countries or show that the records must be incorrect because the amounts spent are incompatible with each other.

39. An investor remarks to a stockbroker that all her stock holdings are in three companies, Eastern Airlines, Hilton Hotels, and McDonald's, and that 2 days ago the value of her stocks went down \$350 but yesterday the value increased by \$600. The broker recalls that 2 days ago the price of Eastern Airlines stock dropped by \$1 a share, Hilton Hotels dropped \$1.50, but the price of McDonald's stock rose by \$0.50. The broker also remembers that yesterday the price of Eastern Airlines stock rose \$1.50, there was a further drop of \$0.50 a share in Hilton Hotels stock, and McDonald's stock rose \$1. Show that the broker does not have enough information to calculate the number of shares the investor owns of each company's stock, but that when the investor says that she owns 200 shares of McDonald's stock, the broker can calculate the number of shares of Eastern Airlines and Hilton Hotels.

40. An intelligence agent knows that 60 aircraft, consisting of fighter planes and bombers, are stationed at a certain secret airfield. The agent wishes to determine how many of the 60 are fighter planes and how many are bombers. There is a type of rocket carried by both sorts of planes; the fighter carries six of these rockets, the bomber only two. The agent learns that 250 rockets are required to arm every plane at this airfield. Furthermore, the agent overhears a remark that there are twice as many fighter planes as bombers at the base (that is, the number of fighter planes minus twice the number of bombers equals zero). Calculate the number of fighter planes and bombers at the airfield or show that the agent's information must be incorrect, because it is inconsistent.

41. Consider the system

$$\begin{aligned} 2x_1 - x_2 + 3x_3 &= a \\ 3x_1 + x_2 - 5x_3 &= b \\ -5x_1 - 5x_2 + 21x_3 &= c \end{aligned}$$

Show that the system is inconsistent if $c \neq 2a - 3b$.

42. Consider the system

$$\begin{aligned} 2x_1 + 3x_2 - x_3 &= a \\ x_1 - x_2 + 3x_3 &= b \\ 3x_1 + 7x_2 - 5x_3 &= c \end{aligned}$$

Find conditions on a, b, and c such that the system is consistent.

***43.** Consider the general system of three linear equations in three unknowns:

$$\begin{aligned} a_{11}x_1 + a_{12}x_2 + a_{13}x_3 &= b_1 \\ a_{21}x_1 + a_{22}x_2 + a_{23}x_3 &= b_2 \\ a_{31}x_1 + a_{32}x_2 + a_{33}x_3 &= b_3 \end{aligned}$$

Find conditions on the coefficients a_{ij} such that the system has a unique solution.

44. Solve the following system using a hand calculator and carrying five decimal places of accuracy:

$$\begin{aligned} 2x_2 - x_3 - 4x_4 &= 2 \\ x_1 - x_2 + 5x_3 + 2x_4 &= -4 \\ 3x_1 + 3x_2 - 7x_3 - x_4 &= 4 \\ -x_1 - 2x_2 + 3x_3 &= -7 \end{aligned}$$

45. Do the same for the system

$$\begin{aligned} 3.8x_1 + 1.6x_2 + 0.9x_3 &= 3.72 \\ -0.7x_1 + 5.4x_2 + 1.6x_3 &= 3.16 \\ 1.5x_1 + 1.1x_2 - 3.2x_3 &= 43.78 \end{aligned}$$

ANSWERS TO SELF-QUIZ

I. d **II.** a **III.** c **IV.** b

6.4 HOMOGENEOUS SYSTEMS OF EQUATIONS

The general $m \times n$ system of linear equations [system (6.3.7), page 383] is called **homogeneous** if all the constants $b_1, b_2, \ldots, b_m$ are zero. That is, the general homogeneous system is given by

$$\begin{aligned} a_{11}x_1 + a_{12}x_2 + \cdots + a_{1n}x_n &= 0 \\ a_{21}x_1 + a_{22}x_2 + \cdots + a_{2n}x_n &= 0 \\ \vdots \qquad \vdots \qquad\quad \vdots \quad &\;\;\vdots \\ a_{m1}x_1 + a_{m2}x_2 + \cdots + a_{mn}x_n &= 0 \end{aligned} \tag{1}$$

Homogeneous systems arise in a variety of ways. In this section we solve some homogeneous systems—again by the method of Gauss-Jordan elimination.

For the general linear system there are three possibilities: no solution, one solution, or an infinite number of solutions. For the general homogeneous system the situation is simpler. Since $x_1 = x_2 = \cdots = x_n = 0$ is always a solution (called the **trivial solution** or **zero solution**), there are only two possibilities: Either the zero solution is the only solution or there are an infinite number of solutions in addition to the zero solution. Solutions other than the zero solution are called **nontrivial solutions**.

EXAMPLE 1

A HOMOGENEOUS SYSTEM WITH ONLY THE ZERO SOLUTION

Solve the homogeneous system

$$\begin{aligned} 2x_1 + 4x_2 + 6x_3 &= 0 \\ 4x_1 + 5x_2 + 6x_3 &= 0 \\ 3x_1 + x_2 - 2x_3 &= 0 \end{aligned}$$

SOLUTION: This is the homogeneous version of the system in Example 6.3.1, page 376. Reducing successively, we obtain (after dividing the first equation by 2)

$$\left(\begin{array}{ccc|c} 1 & 2 & 3 & 0 \\ 4 & 5 & 6 & 0 \\ 3 & 1 & -2 & 0 \end{array}\right) \xrightarrow[R_3 \to R_3 - 3R_1]{R_2 \to R_2 - 4R_1} \left(\begin{array}{ccc|c} 1 & 2 & 3 & 0 \\ 0 & -3 & -6 & 0 \\ 0 & -5 & -11 & 0 \end{array}\right) \xrightarrow{R_2 \to -\frac{1}{3}R_2}$$

$$\left(\begin{array}{rrr|r} 1 & 2 & 3 & 0 \\ 0 & 1 & 2 & 0 \\ 0 & -5 & -11 & 0 \end{array}\right) \xrightarrow[R_3 \to R_3 + 5R_2]{R_1 \to R_1 - 2R_2} \left(\begin{array}{rrr|r} 1 & 0 & -1 & 0 \\ 0 & 1 & 2 & 0 \\ 0 & 0 & -1 & 0 \end{array}\right)$$

$$\xrightarrow{R_3 \to -R_3} \left(\begin{array}{rrr|r} 1 & 0 & -1 & 0 \\ 0 & 1 & 2 & 0 \\ 0 & 0 & 1 & 0 \end{array}\right) \xrightarrow[R_2 \to R_2 - 2R_3]{R_1 \to R_1 + R_3} \left(\begin{array}{rrr|r} 1 & 0 & 0 & 0 \\ 0 & 1 & 0 & 0 \\ 0 & 0 & 1 & 0 \end{array}\right)$$

Thus the system has the unique solution (0, 0, 0). That is, the system has only the trivial solution.

EXAMPLE 2

A HOMOGENEOUS SYSTEM WITH AN INFINITE NUMBER OF SOLUTIONS

Solve the homogeneous system

$$\begin{aligned} x_1 + 2x_2 - x_3 &= 0 \\ 3x_1 - 3x_2 + 2x_3 &= 0 \\ -x_1 - 11x_2 + 6x_3 &= 0 \end{aligned}$$

SOLUTION: Using Gauss-Jordan elimination, we obtain, successively,

$$\left(\begin{array}{rrr|r} 1 & 2 & -1 & 0 \\ 3 & -3 & 2 & 0 \\ -1 & -11 & 6 & 0 \end{array}\right) \xrightarrow[R_3 \to R_3 + R_1]{R_2 \to R_2 - 3R_1} \left(\begin{array}{rrr|r} 1 & 2 & -1 & 0 \\ 0 & -9 & 5 & 0 \\ 0 & -9 & 5 & 0 \end{array}\right)$$

$$\xrightarrow{R_2 \to -\frac{1}{9}R_2} \left(\begin{array}{rrr|r} 1 & 2 & -1 & 0 \\ 0 & 1 & -\frac{5}{9} & 0 \\ 0 & -9 & 5 & 0 \end{array}\right) \xrightarrow[R_3 \to R_3 + 9R_2]{R_1 \to R_1 - 2R_2} \left(\begin{array}{rrr|r} 1 & 0 & \frac{1}{9} & 0 \\ 0 & 1 & -\frac{5}{9} & 0 \\ 0 & 0 & 0 & 0 \end{array}\right)$$

The augmented matrix is now in reduced row echelon form and, evidently, there are an infinite number of solutions given by $(-\frac{1}{9}x_3, \frac{5}{9}x_3, x_3)$. If $x_3 = 0$, for example, we obtain the trivial solution. If $x_3 = 1$ we obtain the solution $(-\frac{1}{9}, \frac{5}{9}, 1)$.

EXAMPLE 3

A HOMOGENEOUS SYSTEM WITH MORE UNKNOWNS THAN EQUATIONS HAS AN INFINITE NUMBER OF SOLUTIONS

Solve the system

$$\begin{aligned} x_1 + x_2 - x_3 &= 0 \\ 4x_1 - 2x_2 + 7x_3 &= 0 \end{aligned} \tag{2}$$

SOLUTION: Row-reducing, we obtain

$$\left(\begin{array}{rrr|r} 1 & 1 & -1 & 0 \\ 4 & -2 & 7 & 0 \end{array}\right) \xrightarrow{R_2 \to R_2 - 4R_1} \left(\begin{array}{rrr|r} 1 & 1 & -1 & 0 \\ 0 & -6 & 11 & 0 \end{array}\right)$$

$$\xrightarrow{R_2 \to -\frac{1}{6}R_2} \left(\begin{array}{rrr|r} 1 & 1 & -1 & 0 \\ 0 & 1 & -\frac{11}{6} & 0 \end{array}\right) \xrightarrow{R_1 \to R_1 - R_2} \left(\begin{array}{rrr|r} 1 & 0 & \frac{5}{6} & 0 \\ 0 & 1 & -\frac{11}{6} & 0 \end{array}\right)$$

Thus there are an infinite number of solutions given by $(-\frac{5}{6}x_3, \frac{11}{6}x_3, x_3)$. This is not surprising since system (2) contains three unknowns and only two equations.

In fact, if there are more unknowns than equations, the homogeneous system (1) will always have an infinite number of solutions. To see this, note that if there were only the trivial solution, then row reduction would lead us to the system

$$\begin{aligned} x_1 \qquad\qquad &= 0 \\ x_2 \qquad &= 0 \\ &\vdots \\ x_n &= 0 \end{aligned}$$

and, possibly, additional equations of the form $0 = 0$. But this system has at least as many equations as unknowns. Since row reduction does not change either the number of equations or the number of unknowns, we have a contradiction of our assumption that there were more unknowns than equations. Thus we have Theorem 1.

THEOREM 1

The homogeneous system (1) has an infinite number of solutions if $n > m$. ■

PROBLEMS 6.4

SELF-QUIZ

I. Which of the following systems *must* have nontrivial solutions?

a. $a_{11}x_1 + a_{12}x_2 = 0$
$a_{21}x_1 + a_{22}x_2 = 0$

b. $a_{11}x_1 + a_{12}x_2 = 0$
$a_{21}x_1 + a_{22}x_2 = 0$
$a_{31}x_1 + a_{32}x_2 = 0$

c. $a_{11}x_1 + a_{12}x_2 + a_{13}x_3 = 0$
$a_{21}x_1 + a_{22}x_2 + a_{23}x_3 = 0$

II. For what value of k will the following system have nontrivial solutions?

$$\begin{aligned} x + y + z &= 0 \\ 2x + 3y + 4z &= 0 \\ 3x + 4y + kz &= 0 \end{aligned}$$

a. 1 **b.** 2 **c.** 3
d. 4 **e.** 5 **f.** 0

In Problems 1–13 find all solutions to the homogeneous systems.

1. $2x_1 - x_2 = 0$
$3x_1 + 4x_2 = 0$

2. $x_1 - 5x_2 = 0$
$-x_1 + 5x_2 = 0$

3. $x_1 + x_2 - x_3 = 0$
$2x_1 - 4x_2 + 3x_3 = 0$
$3x_1 + 7x_2 - x_3 = 0$

4. $x_1 + x_2 - x_3 = 0$
$2x_1 - 4x_2 + 3x_3 = 0$
$-x_1 - 7x_2 + 6x_3 = 0$

5. $x_1 + x_2 - x_3 = 0$
$2x_1 - 4x_2 + 3x_3 = 0$
$-5x_1 + 13x_2 - 10x_3 = 0$

6. $2x_1 + 3x_2 - x_3 = 0$
$6x_1 - 5x_2 + 7x_3 = 0$

7. $4x_1 - x_2 = 0$
$7x_1 + 3x_2 = 0$
$-8x_1 + 6x_2 = 0$

8. $x_1 - x_2 + 7x_3 - x_4 = 0$
$2x_1 + 3x_2 - 8x_3 + x_4 = 0$

9. $x_1 - 2x_2 + x_3 + x_4 = 0$
$3x_1 + 2x_3 - 2x_4 = 0$
$4x_2 - x_3 - x_4 = 0$
$5x_1 + 3x_3 - x_4 = 0$

10. $-2x_1 + 7x_4 = 0$
$x_1 + 2x_2 - x_3 + 4x_4 = 0$
$3x_1 - x_3 + 5x_4 = 0$
$4x_1 + 2x_2 + 3x_3 = 0$

11. $$\begin{aligned} 2x_1 - x_2 &= 0 \\ 3x_1 + 5x_2 &= 0 \\ 7x_1 - 3x_2 &= 0 \\ -2x_1 + 3x_2 &= 0 \end{aligned}$$

12. $$\begin{aligned} x_1 - 3x_2 &= 0 \\ -2x_1 + 6x_2 &= 0 \\ 4x_1 - 12x_2 &= 0 \end{aligned}$$

13. $$\begin{aligned} x_1 + x_2 - x_3 &= 0 \\ 4x_1 - x_2 + 5x_3 &= 0 \\ -2x_1 + x_2 - 2x_3 &= 0 \\ 3x_1 + 2x_2 - 6x_3 &= 0 \end{aligned}$$

14. Show that the homogeneous system

$$\begin{aligned} a_{11}x_1 + a_{12}x_2 &= 0 \\ a_{21}x_1 + a_{22}x_2 &= 0 \end{aligned}$$

has an infinite number of solutions if and only if $a_{11}a_{22} - a_{12}a_{21} = 0$.

15. Consider the system

$$\begin{aligned} 2x_1 - 3x_2 + 5x_3 &= 0 \\ -x_1 + 7x_2 - x_3 &= 0 \\ 4x_1 - 11x_2 + kx_3 &= 0 \end{aligned}$$

For what value of k will the system have nontrivial solutions?

***16.** Consider the 3×3 homogeneous system

$$\begin{aligned} a_{11}x_1 + a_{12}x_2 + a_{13}x_3 &= 0 \\ a_{21}x_1 + a_{22}x_2 + a_{23}x_3 &= 0 \\ a_{31}x_1 + a_{32}x_2 + a_{33}x_3 &= 0 \end{aligned}$$

Find conditions on the coefficients a_{ij} such that the zero solution is the only solution.

ANSWERS TO SELF-QUIZ

I. c **II.** e

6.5 MATRICES

We first discussed the space $\mathbb{R}^n$ in Section 1.10. In this chapter and in Chapters 7 and 8 we shall investigate properties of this space. An important tool for carrying out algebraic operations in $\mathbb{R}^n$ is the *matrix*. Therefore, we begin by defining a matrix.

MATRIX

An $\boldsymbol{m \times n}$ **matrix** A is a rectangular array of mn numbers arranged in m rows and n columns:

$$A = \begin{pmatrix} a_{11} & a_{12} & \cdots & a_{1j} & \cdots & a_{1n} \\ a_{21} & a_{22} & \cdots & a_{2j} & \cdots & a_{2n} \\ \vdots & \vdots & & \vdots & & \vdots \\ a_{i1} & a_{i2} & \cdots & a_{ij} & \cdots & a_{in} \\ \vdots & \vdots & & \vdots & & \vdots \\ a_{m1} & a_{m2} & \cdots & a_{mj} & \cdots & a_{mn} \end{pmatrix} \tag{1}$$

The symbol $m \times n$ is read "m by n."

In Section 1.10 we defined n-component row and column vectors. Vectors can be thought of as special cases of matrices. The row vector $\mathbf{v} = (v_1, v_2, \ldots, v_n)$ is a matrix with one row and n columns: a $1 \times n$ matrix. Similarly, the vector $\begin{pmatrix} v_1 \\ v_2 \\ \vdots \\ v_n \end{pmatrix}$ is an $n \times 1$ matrix. Unless stated otherwise, we shall always assume that the numbers in a matrix or vector are real. We call the row vector $(a_{i1}, a_{i2} \ldots a_{in})$ **row i** and the column vector $\begin{pmatrix} a_{1j} \\ a_{2j} \\ \vdots \\ a_{mj} \end{pmatrix}$ **column j**. The **ijth component** of A, denoted by a_{ij}, is the number appearing in the ith row and jth column of A. We shall sometimes write the matrix A as $A = (a_{ij})$. Usually, matrices will be denoted by capital letters.

If A is an $m \times n$ matrix with $m = n$, then A is called a **square matrix**. An $m \times n$ matrix with all components equal to zero is called the $m \times n$ **zero matrix**.

An $m \times n$ matrix is said to have the **size** $m \times n$.

HISTORICAL NOTE: The term ''matrix'' was first used in 1850 by the British mathematician James Joseph Sylvester (1814–1897) to distinguish matrices from determinants (which we shall discuss in Chapter 7). In fact, the term ''matrix'' was intended to mean ''mother of determinants.''

EXAMPLE 1

FIVE MATRICES

Five matrices of different sizes are given below:

i. $A = \begin{pmatrix} 1 & 3 \\ 4 & 2 \end{pmatrix}$, 2×2 (square)

ii. $A = \begin{pmatrix} -1 & 3 \\ 4 & 0 \\ 1 & -2 \end{pmatrix}$, 3×2

iii. $\begin{pmatrix} -1 & 4 & 1 \\ 3 & 0 & 2 \end{pmatrix}$, 2×3

iv. $\begin{pmatrix} 1 & 6 & -2 \\ 3 & 1 & 4 \\ 2 & -6 & 5 \end{pmatrix}$, 3×3 (square)

v. $\begin{pmatrix} 0 & 0 & 0 & 0 \\ 0 & 0 & 0 & 0 \end{pmatrix}$, 2×4 zero matrix

BRACKET NOTATION: In some books matrices are given in square brackets rather than parentheses. For example, the first two matrices in Example 1 can be written as

i. $A = \begin{bmatrix} 1 & 3 \\ 4 & 2 \end{bmatrix}$

ii. $A = \begin{bmatrix} -1 & 3 \\ 4 & 0 \\ 1 & -2 \end{bmatrix}$

In this text we shall use the parentheses exclusively.

Throughout this book we shall refer to the ith row, the jth column, and the ijth component of a matrix for various numbers i and j. We illustrate these ideas in the next example.

EXAMPLE 2

FINDING COMPONENTS OF A MATRIX

Find the 1,2, the 3,1, and the 2,2 components of

$$A = \begin{pmatrix} 1 & 6 & 4 \\ 2 & -3 & 5 \\ 7 & 4 & 0 \end{pmatrix}$$

SOLUTION: The 1,2 component is the number in the first row and the second column. We have shaded the first row and the second column; the 1,2 component is 6:

2nd column ↓

1st row ⟶ $\begin{pmatrix} 1 & 6 & 4 \\ 2 & -3 & 5 \\ 7 & 4 & 0 \end{pmatrix}$

From the shaded matrices below, we see that the 3,1 component is 7 and the 2,2 component is -3:

1st column ↓

3rd row ⟶ $\begin{pmatrix} 1 & 6 & 4 \\ 2 & -3 & 5 \\ 7 & 4 & 0 \end{pmatrix}$

2nd column ↓

2nd row ⟶ $\begin{pmatrix} 1 & 6 & 4 \\ 2 & -3 & 5 \\ 7 & 4 & 0 \end{pmatrix}$

Two matrices $A = (a_{ij})$ and $B = (b_{ij})$ are **equal** if (1) they have the same size, and (2) corresponding components are equal.

EXAMPLE 3

EQUAL AND UNEQUAL MATRICES

Are the following matrices equal?

i. $\begin{pmatrix} 4 & 1 & 5 \\ 2 & -3 & 0 \end{pmatrix}$ and $\begin{pmatrix} 1+3 & 1 & 2+3 \\ 1+1 & 1-4 & 6-6 \end{pmatrix}$

ii. $\begin{pmatrix} -2 & 0 \\ 1 & 3 \end{pmatrix}$ and $\begin{pmatrix} 0 & -2 \\ 1 & 3 \end{pmatrix}$

iii. $\begin{pmatrix} 1 & 0 \\ 0 & 1 \end{pmatrix}$ and $\begin{pmatrix} 1 & 0 & 0 \\ 0 & 1 & 0 \end{pmatrix}$

SOLUTION:

i. Yes; both matrices are 2×3, and $1 + 3 = 4$, $2 + 3 = 5$, $1 + 1 = 2$, $1 - 4 = -3$, and $6 - 6 = 0$.

ii. No; $-2 \neq 0$, so the matrices are unequal because, for example, the 1,1 components are unequal. This is true even though the two matrices contain the same numbers. *Corresponding* components must be equal. This means that the 1,1 component in A must be equal to the 1,1 component in B, and so on.

iii. No; the first matrix is 2×2 and the second matrix is 2×3, so they do not have the same size.

Matrices, like vectors, arise in a great number of practical situations. For example, the vector $\begin{pmatrix} 10 \\ 30 \\ 15 \\ 60 \end{pmatrix}$ can represent order quantities for four different products used by one manufacturer. Suppose that there were five different plants. Then the 4×5 matrix

$$Q = \begin{pmatrix} 10 & 20 & 15 & 16 & 25 \\ 30 & 10 & 20 & 25 & 22 \\ 15 & 22 & 18 & 20 & 13 \\ 60 & 40 & 50 & 35 & 45 \end{pmatrix}$$

could represent the orders for the four products in each of the five plants. We can see, for example, that plant 4 orders 25 units of the second product while plant 2 orders 40 units of the fourth product.

Matrices, like vectors, can be added and multiplied by scalars.

DEFINITION ADDITION OF MATRICES

Let $A = (a_{ij})$ and $B = (b_{ij})$ be two $m \times n$ matrices. Then the sum of A and B is the $m \times n$ matrix $A + B$ given by

$$A + B = (a_{ij} + b_{ij}) = \begin{pmatrix} a_{11} + b_{11} & a_{12} + b_{12} & \cdots & a_{1n} + b_{1n} \\ a_{21} + b_{21} & a_{22} + b_{22} & \cdots & a_{2n} + b_{2n} \\ \vdots & \vdots & & \vdots \\ a_{m1} + b_{m1} & a_{m2} + b_{m2} & \cdots & a_{mn} + b_{mn} \end{pmatrix} \quad (2)$$

That is, $A + B$ is the $m \times n$ matrix obtained by adding the corresponding components of A and B. ■

WARNING: The sum of two matrices is defined only when both matrices have the same size. Thus, for example, it is not possible to add together the matrices

$$\begin{pmatrix} 1 & 2 & 3 \\ 4 & 5 & 6 \end{pmatrix} \quad \text{and} \quad \begin{pmatrix} -1 & 0 \\ 2 & -5 \\ 4 & 7 \end{pmatrix}.$$

EXAMPLE 4 THE SUM OF TWO MATRICES

$$\begin{pmatrix} 2 & 4 & -6 & 7 \\ 1 & 3 & 2 & 1 \\ -4 & 3 & -5 & 5 \end{pmatrix} + \begin{pmatrix} 0 & 1 & 6 & -2 \\ 2 & 3 & 4 & 3 \\ -2 & 1 & 4 & 4 \end{pmatrix} = \begin{pmatrix} 2 & 5 & 0 & 5 \\ 3 & 6 & 6 & 4 \\ -6 & 4 & -1 & 9 \end{pmatrix}$$

DEFINITION MULTIPLICATION OF A MATRIX BY A SCALAR

If $A = (a_{ij})$ is an $m \times n$ matrix and if α is a scalar, then the $m \times n$ matrix αA is given by

$$\alpha A = (\alpha a_{ij}) = \begin{pmatrix} \alpha a_{11} & \alpha a_{12} & \cdots & \alpha a_{1n} \\ \alpha a_{21} & \alpha a_{22} & \cdots & \alpha a_{2n} \\ \vdots & \vdots & & \vdots \\ \alpha a_{m1} & \alpha a_{m2} & \cdots & \alpha a_{mn} \end{pmatrix} \quad (3)$$

In other words, $\alpha A = (\alpha a_{ij})$ is the matrix obtained by multiplying each component of A by α. If $\alpha A = B = (b_{ij})$, then $b_{ij} = \alpha a_{ij}$ for $i = 1, 2, \ldots, m$ and $j = 1, 2, \ldots, n$. ■

EXAMPLE 5 SCALAR MULTIPLES OF MATRICES

Let $A = \begin{pmatrix} 1 & -3 & 4 & 2 \\ 3 & 1 & 4 & 6 \\ -2 & 3 & 5 & 7 \end{pmatrix}$. Then $2A = \begin{pmatrix} 2 & -6 & 8 & 4 \\ 6 & 2 & 8 & 12 \\ -4 & 6 & 10 & 14 \end{pmatrix}$,

$-3A = \begin{pmatrix} -3 & 9 & -12 & -6 \\ -9 & -3 & -12 & -18 \\ 6 & -9 & -15 & -21 \end{pmatrix}$, and $0A = \begin{pmatrix} 0 & 0 & 0 & 0 \\ 0 & 0 & 0 & 0 \\ 0 & 0 & 0 & 0 \end{pmatrix}$.

EXAMPLE 6 SUM OF SCALAR MULTIPLES OF MATRICES

Let $A = \begin{pmatrix} 1 & 2 & 4 \\ -7 & 3 & -2 \end{pmatrix}$ and $B = \begin{pmatrix} 4 & 0 & 5 \\ 1 & -3 & 6 \end{pmatrix}$. Calculate $-2A + 3B$.

SOLUTION:

$$-2A + 3B = (-2)\begin{pmatrix} 1 & 2 & 4 \\ -7 & 3 & -2 \end{pmatrix} + (3)\begin{pmatrix} 4 & 0 & 5 \\ 1 & -3 & 6 \end{pmatrix}$$
$$= \begin{pmatrix} -2 & -4 & -8 \\ 14 & -6 & 4 \end{pmatrix} + \begin{pmatrix} 12 & 0 & 15 \\ 3 & -9 & 18 \end{pmatrix} = \begin{pmatrix} 10 & -4 & 7 \\ 17 & -15 & 22 \end{pmatrix}$$

The next theorem is not difficult to prove. We prove part (*iii*) and leave the remaining parts of the proof as an exercise (see Problems 21–23).

THEOREM 1

Let A, B, and C be $m \times n$ matrices and let α be a scalar. Then:

i. $A + 0 = A$

ii. $0A = 0$

iii. $A + B = B + A$ **(commutative law for matrix addition)**

iv. $(A + B) + C = A + (B + C)$ **(associative law for matrix addition)**

v. $\alpha(A + B) = \alpha A + \alpha B$ **(distributive law for scalar multiplication)**

vi. $1A = A$

NOTE: The zero in part (i) of the theorem is the $m \times n$ zero matrix. In part (ii) the zero on the left is a scalar while the zero on the right is the $m \times n$ zero matrix.

PROOF OF III.:

$$\text{Let } A = \begin{pmatrix} a_{11} & a_{12} & \cdots & a_{1n} \\ a_{21} & a_{22} & \cdots & a_{2n} \\ \vdots & \vdots & & \vdots \\ a_{m1} & a_{m2} & \cdots & a_{mn} \end{pmatrix} \text{ and } B = \begin{pmatrix} b_{11} & b_{12} & \cdots & b_{1n} \\ b_{21} & b_{22} & \cdots & b_{2n} \\ \vdots & \vdots & & \vdots \\ b_{m1} & b_{m2} & \cdots & b_{mn} \end{pmatrix}.$$

Then

$$A + B = \begin{pmatrix} a_{11} + b_{11} & a_{12} + b_{12} & \cdots & a_{1n} + b_{1n} \\ a_{21} + b_{21} & a_{22} + b_{22} & \cdots & a_{2n} + b_{2n} \\ \vdots & \vdots & & \vdots \\ a_{m1} + b_{m1} & a_{m2} + b_{m2} & \cdots & a_{mn} + b_{mn} \end{pmatrix}$$

$$\underset{\substack{a+b=b+a \text{ for any} \\ \text{real numbers } a \text{ and } b}}{=} \begin{pmatrix} b_{11} + a_{11} & b_{12} + a_{12} & \cdots & b_{1n} + a_{1n} \\ b_{21} + a_{21} & b_{22} + a_{22} & \cdots & b_{2n} + a_{2n} \\ \vdots & \vdots & & \vdots \\ b_{m1} + a_{m1} & b_{m2} + a_{m2} & \cdots & b_{mn} + a_{mn} \end{pmatrix} = B + A$$

■

EXAMPLE 7 ILLUSTRATING THE ASSOCIATIVE LAW OF MATRIX ADDITION

To illustrate the associative law we note that

$$\left[\begin{pmatrix} 1 & 4 & -2 \\ 3 & -1 & 0 \end{pmatrix} + \begin{pmatrix} 2 & -2 & 3 \\ 1 & -1 & 5 \end{pmatrix}\right] + \begin{pmatrix} 3 & -1 & 2 \\ 0 & 1 & 4 \end{pmatrix}$$

$$= \begin{pmatrix} 3 & 2 & 1 \\ 4 & -2 & 5 \end{pmatrix} + \begin{pmatrix} 3 & -1 & 2 \\ 0 & 1 & 4 \end{pmatrix} = \begin{pmatrix} 6 & 1 & 3 \\ 4 & -1 & 9 \end{pmatrix}$$

Similarly,

$$\begin{pmatrix} 1 & 4 & -2 \\ 3 & -1 & 0 \end{pmatrix} + \left[\begin{pmatrix} 2 & -2 & 3 \\ 1 & -1 & 5 \end{pmatrix} + \begin{pmatrix} 3 & -1 & 2 \\ 0 & 1 & 4 \end{pmatrix}\right]$$

$$= \begin{pmatrix} 1 & 4 & -2 \\ 3 & -1 & 0 \end{pmatrix} + \begin{pmatrix} 5 & -3 & 5 \\ 1 & 0 & 9 \end{pmatrix} = \begin{pmatrix} 6 & 1 & 3 \\ 4 & -1 & 9 \end{pmatrix}$$

As with vectors, the associative law for matrix addition enables us to define the sum of three or more matrices.

PROBLEMS 6.5

SELF-QUIZ

I. Which of the following is true of the matrix

$$\begin{pmatrix} 1 & 2 & 3 \\ 7 & -1 & 0 \end{pmatrix}?$$

a. It is a square matrix.
b. If multiplied by the scalar -1, the product is

$$\begin{pmatrix} -1 & -2 & -3 \\ -7 & 1 & 0 \end{pmatrix}.$$

c. It is a 3×2 matrix.
d. It is the sum of $\begin{pmatrix} 3 & 1 & 4 \\ 7 & 2 & 0 \end{pmatrix}$ and $\begin{pmatrix} -2 & 1 & 1 \\ 0 & 1 & 0 \end{pmatrix}$.

II. Which of the following is $2A - 4B$ if $A = (2 \quad 0 \quad 0)$ and $B = (3 \quad 1)$?
a. $(-8 \quad -4)$
b. $(5 \quad 0 \quad 1)$
c. $(16 \quad -4 \quad 0)$
d. This operation cannot be performed.

III. Which of the following is true when finding the difference of two matrices?
a. The matrices must have the same size.
b. The matrices must be square.
c. The matrices must both be row vectors or both be column vectors.
d. One matrix must be a row vector and the other must be a column vector.

IV. Which of the following would be the entries in the second column of matrix B, if

$$\begin{pmatrix} 3 & -4 & 0 \\ 2 & 8 & -1 \end{pmatrix} + B = \begin{pmatrix} 0 & 0 & 0 \\ 0 & 0 & 0 \end{pmatrix}?$$

a. $-2, -8, 1$ **b.** $4, -8$
c. $2, 8, -1$ **d.** $-4, 8$

V. Which of the following must be the second row of matrix B if $3A - B = 2C$ for

$$A = \begin{pmatrix} 1 & -1 & 1 \\ 0 & 0 & 3 \\ 4 & 2 & 0 \end{pmatrix} \text{ and } C = \begin{pmatrix} 1 & 0 & 0 \\ 0 & 1 & 0 \\ 0 & 0 & 1 \end{pmatrix}?$$

a. $-3 \quad 2 \quad 6$ **b.** $0 \quad -2 \quad 9$
c. $3 \quad -2 \quad 6$ **d.** $0 \quad 2 \quad -9$

In Problems 1–12 perform the indicated computation with $A = \begin{pmatrix} 1 & 3 \\ 2 & 5 \\ -1 & 2 \end{pmatrix}$, $B = \begin{pmatrix} -2 & 0 \\ 1 & 4 \\ -7 & 5 \end{pmatrix}$, and $C = \begin{pmatrix} -1 & 1 \\ 4 & 6 \\ -7 & 3 \end{pmatrix}$.

1. $3A$
2. $A + B$
3. $A - C$
4. $2C - 5A$
5. $0B$ (0 is the scalar zero)
6. $-7A + 3B$
7. $A + B + C$
8. $C - A - B$
9. $2A - 3B + 4C$
10. $7C - B + 2A$
11. Find a matrix D such that $2A + B - D$ is the 3×2 zero matrix.
12. Find a matrix E such that $A + 2B - 3C + E$ is the 3×2 zero matrix.

In Problems 13–20 perform the indicated computation with $A = \begin{pmatrix} 1 & -1 & 2 \\ 3 & 4 & 5 \\ 0 & 1 & -1 \end{pmatrix}$, $B = \begin{pmatrix} 0 & 2 & 1 \\ 3 & 0 & 5 \\ 7 & -6 & 0 \end{pmatrix}$, and $C = \begin{pmatrix} 0 & 0 & 2 \\ 3 & 1 & 0 \\ 0 & -2 & 4 \end{pmatrix}$.

13. $A - 2B$
14. $3A - C$
15. $A + B + C$
16. $2A - B + 2C$
17. $C - A - B$
18. $4C - 2B + 3A$
19. Find a matrix D such that $A + B + C + D$ is the 3×3 zero matrix.
20. Find a matrix E such that $3C - 2B + 8A - 4E$ is the 3×3 zero matrix.
21. Let $A = (a_{ij})$ be an $m \times n$ matrix and let $\overline{0}$ denote the $m \times n$ zero matrix. Use appropriate definitions to show that $0A = \overline{0}$ and $\overline{0} + A = A$. Similarly, show that $1A = A$.
22. If $A = (a_{ij})$, $B = (b_{ij})$, and $C = (c_{ij})$ are $m \times n$

matrices, compute $(A + B) + C$ and $A + (B + C)$ and show that they are equal.

23. If α is a scalar and A and B are $m \times n$ matrices, compute $\alpha(A + B)$ and $\alpha A + \alpha B$ and show that they are equal.

24. Consider the "graph" joining the four points in Figure 1. Construct a 4×4 matrix having the property that $a_{ij} = 0$ if point i is not connected (joined by a line) to point j and $a_{ij} = 1$ if point i is connected to point j. Set $a_{ii} = 0$.

25. Do the same (this time constructing a 5×5 matrix) for the graph in Figure 2.

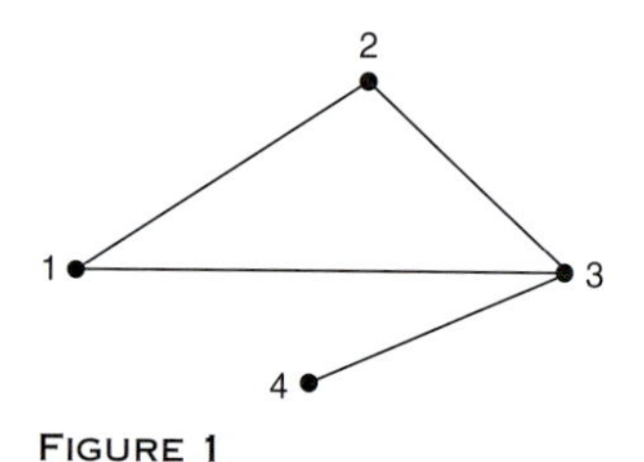

FIGURE 1

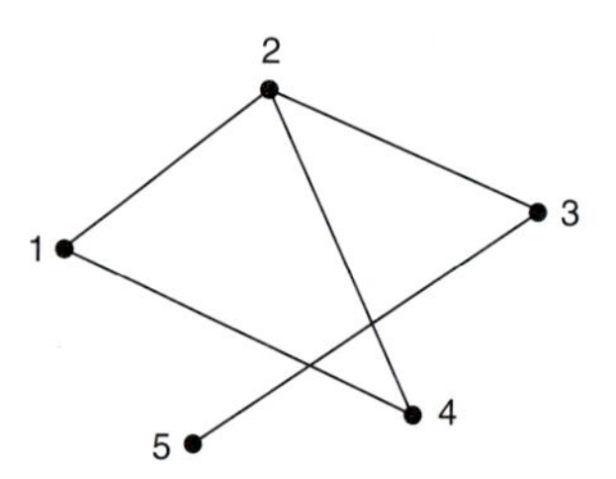

FIGURE 2

ANSWERS TO SELF-QUIZ

I. b **II.** d **III.** a **IV.** b **V.** b

6.6 MATRIX PRODUCTS

In this section we see how two matrices can be multiplied together. Quite obviously, we could define the product of two $m \times n$ matrices $A = (a_{ij})$ and $B = (b_{ij})$ to be the $m \times n$ matrix whose ijth component is $a_{ij}b_{ij}$. However, for just about all the important applications involving matrices, another kind of product is needed. It comes as the generalization of the scalar product (see page 77).

DEFINITION PRODUCT OF TWO MATRICES

Let $A = (a_{ij})$ be an $m \times n$ matrix whose ith row is denoted $\mathbf{a}_i$. Let $B = (b_{ij})$ be an $n \times p$ matrix whose jth column is denoted $\mathbf{b}_j$. Then the product of A and B is an $m \times p$ matrix $C = (c_{ij})$, where

$$c_{ij} = \mathbf{a}_i \cdot \mathbf{b}_j. \tag{1}$$

That is, the ijth element of AB is the dot product of the ith row of A and the jth column of B. If we write this out, we obtain

$$c_{ij} = a_{i1}b_{1j} + a_{i2}b_{2j} + \cdots + a_{in}b_{nj} = \sum_{k=1}^{m} a_{ik}b_{kj} \tag{2}$$

■

WARNING: Two matrices can be multiplied together only if the number of columns of the first matrix is equal to the number of rows of the second. Otherwise the vectors that are the ith row of A and the jth column of B will not have the same number of components, and the dot product in equation (3) will not be defined. To illustrate this, we write the matrices A and B:

$$\text{ith row of } A \rightarrow \begin{pmatrix} a_{11} & a_{12} & \cdots & a_{1n} \\ a_{21} & a_{22} & \cdots & a_{2n} \\ \vdots & \vdots & & \vdots \\ a_{i1} & a_{i2} & \cdots & a_{in} \\ \vdots & \vdots & & \vdots \\ a_{m1} & a_{m2} & \cdots & a_{mn} \end{pmatrix} \begin{pmatrix} b_{11} & b_{12} & \cdots & b_{1j} & \cdots & b_{1p} \\ b_{21} & b_{22} & \cdots & b_{2j} & \cdots & b_{2p} \\ \vdots & \vdots & & \vdots & & \vdots \\ b_{n1} & b_{n2} & \cdots & b_{nj} & \cdots & b_{np} \end{pmatrix}$$

(jth column of B ↓ points to the column $b_{1j}, b_{2j}, \ldots, b_{nj}$)

The shaded row and column vectors must have the same number of components.

EXAMPLE 1

THE PRODUCT OF TWO 2 × 2 MATRICES

If $A = \begin{pmatrix} 1 & 3 \\ -2 & 4 \end{pmatrix}$ and $B = \begin{pmatrix} 3 & -2 \\ 5 & 6 \end{pmatrix}$, calculate AB and BA.

SOLUTION: A is a 2×2 matrix and B is a 2×2 matrix, so $C = AB = (2 \times 2) \times (2 \times 2)$ is also a 2×2 matrix. If $C = (c_{ij})$, what is c_{11}? We know that

$$c_{11} = (\text{1st row of } A) \cdot (\text{1st column of } B)$$

Rewriting the matrices, we have

$$\text{1st row of } A \rightarrow \begin{pmatrix} 1 & 3 \\ -2 & 4 \end{pmatrix} \begin{pmatrix} 3 & -2 \\ 5 & 6 \end{pmatrix}$$

(2nd column of B ↓ points to the first column of B)

Thus

$$c_{11} = (1 \quad 3)\begin{pmatrix} 3 \\ 5 \end{pmatrix} = 3 + 15 = 18$$

Similarly, to compute c_{12} we have

$$\text{1st row of } A \rightarrow \begin{pmatrix} 1 & 3 \\ -2 & 4 \end{pmatrix} \begin{pmatrix} 3 & -2 \\ 5 & 6 \end{pmatrix}$$

(2nd column of B ↓)

and

$$c_{12} = (1 \quad 3)\begin{pmatrix} -2 \\ 6 \end{pmatrix} = -2 + 18 = 16$$

Continuing, we find that

$$c_{21} = (-2 \quad 4)\begin{pmatrix} 3 \\ 5 \end{pmatrix} = -6 + 20 = 14$$

and

$$c_{22} = (-2 \quad 4)\begin{pmatrix} -2 \\ 6 \end{pmatrix} = 4 + 24 = 28$$

Thus

$$C = AB = \begin{pmatrix} 18 & 16 \\ 14 & 28 \end{pmatrix}$$

Similarly, leaving out the intermediate steps, we see that

$$C' = BA = \begin{pmatrix} 3 & -2 \\ 5 & 6 \end{pmatrix}\begin{pmatrix} 1 & 3 \\ -2 & 4 \end{pmatrix} = \begin{pmatrix} 3+4 & 9-8 \\ 5-12 & 15+24 \end{pmatrix} = \begin{pmatrix} 7 & 1 \\ -7 & 39 \end{pmatrix}$$

REMARK: Example 4 illustrates an important fact:

> Matrix products do not, in general, commute.

That is, $AB \neq BA$ in general. It sometimes happens that $AB = BA$, but this will be the exception, not the rule. In fact, as the next example illustrates, it may occur that AB is defined while BA is not. Thus we must be careful of *order* when multiplying two matrices together.

EXAMPLE 2

THE PRODUCT OF A 2 × 3 AND A 3 × 3 MATRIX IS DEFINED WHILE THE PRODUCT OF A 3 × 3 AND A 2 × 3 MATRIX IS NOT DEFINED

Let $A = \begin{pmatrix} 2 & 0 & -3 \\ 4 & 1 & 5 \end{pmatrix}$ and $B = \begin{pmatrix} 7 & -1 & 4 & 7 \\ 2 & 5 & 0 & -4 \\ -3 & 1 & 2 & 3 \end{pmatrix}$. Calculate AB.

SOLUTION: We first note that A is a 2×3 matrix and B is a 3×4 matrix. Hence the number of columns of A equals the number of rows of B. The product AB is therefore defined and is a 2×4 matrix. Let $AB = C = (c_{ij})$. Then

$$c_{11} = (2 \quad 0 \quad -3) \cdot \begin{pmatrix} 7 \\ 2 \\ -3 \end{pmatrix} = 23 \qquad c_{12} = (2 \quad 0 \quad -3) \cdot \begin{pmatrix} -1 \\ 5 \\ 1 \end{pmatrix} = -5$$

$$c_{13} = (2 \quad 0 \quad -3) \cdot \begin{pmatrix} 4 \\ 0 \\ 2 \end{pmatrix} = 2 \qquad c_{14} = (2 \quad 0 \quad -3) \cdot \begin{pmatrix} 7 \\ -4 \\ 3 \end{pmatrix} = 5$$

$$c_{21} = (4 \quad 1 \quad 5) \cdot \begin{pmatrix} 7 \\ 2 \\ -3 \end{pmatrix} = 15 \qquad c_{22} = (4 \quad 1 \quad 5) \cdot \begin{pmatrix} -1 \\ 5 \\ 1 \end{pmatrix} = 6$$

$$c_{23} = (4 \quad 1 \quad 5) \cdot \begin{pmatrix} 4 \\ 0 \\ 2 \end{pmatrix} = 26 \qquad c_{24} = (4 \quad 1 \quad 5) \cdot \begin{pmatrix} 7 \\ -4 \\ 3 \end{pmatrix} = 39$$

Hence $AB = \begin{pmatrix} 23 & -5 & 2 & 5 \\ 15 & 6 & 26 & 39 \end{pmatrix}$. This completes the problem. Note that the product BA is *not* defined since the number of columns of B (four) is not equal to the number of rows of A (two).

EXAMPLE 3

DIRECT AND INDIRECT CONTACT WITH A CONTAGIOUS DISEASE

In this example we show how matrix multiplication can be used to model the spread of a contagious disease. Suppose that four individuals have contracted such a disease. This group has contacts with six people in a second group. We can represent these contacts, called *direct contacts,* by a 4×6 matrix. An example of such a matrix is given below.

DIRECT CONTACT MATRIX: First and second groups

$$A = \begin{pmatrix} 0 & 1 & 0 & 0 & 1 & 0 \\ 1 & 0 & 0 & 1 & 0 & 1 \\ 0 & 0 & 0 & 1 & 1 & 0 \\ 1 & 0 & 0 & 0 & 0 & 1 \end{pmatrix}$$

Here we set $a_{ij} = 1$ if the ith person in the first group has made contact with the jth person in the second group. For example, the 1 in the 2,4 position means that the second person in the first (infected) group has been in contact with the fourth person in the second group. Now suppose that a third group of five people has had a variety of direct contacts with individuals of the second group. We can also represent this by a matrix.

DIRECT CONTACT MATRIX: Second and third groups

$$B = \begin{pmatrix} 0 & 0 & 1 & 0 & 1 \\ 0 & 0 & 0 & 1 & 0 \\ 0 & 1 & 0 & 0 & 0 \\ 1 & 0 & 0 & 0 & 1 \\ 0 & 0 & 0 & 1 & 0 \\ 0 & 0 & 1 & 0 & 0 \end{pmatrix}$$

Note that $b_{64} = 0$, which means that the sixth person in the second group has had no contact with the fourth person in the third group.

The *indirect* or *second-order* contacts between the individuals in the first and third groups is represented by the 4×5 matrix $C = AB$. To see this, observe that a person in group 3 can be infected from someone in group 2, who in turn has been infected by someone in group 1. For example, since $a_{24} = 1$ and $b_{45} = 1$, we see that, indirectly, the fifth person in group 3 has contact (through the fourth person in group 2) with the second person in group 1. The total number of indirect contacts between the second person in group 1 and the fifth person in group 3 is given by

$$\begin{aligned} c_{25} &= a_{21}b_{15} + a_{22}b_{25} + a_{23}b_{35} + a_{24}b_{45} + a_{25}b_{55} + a_{26}b_{65} \\ &= 1 \cdot 1 + 0 \cdot 0 + 0 \cdot 0 + 1 \cdot 1 + 0 \cdot 0 + 1 \cdot 0 = 2 \end{aligned}$$

We now compute.

INDIRECT CONTACT MATRIX: First and third groups

$$C = AB = \begin{pmatrix} 0 & 0 & 0 & 2 & 0 \\ 1 & 0 & 2 & 0 & 2 \\ 1 & 0 & 0 & 1 & 1 \\ 0 & 0 & 2 & 0 & 1 \end{pmatrix}$$

We observe that only the second person in group 3 has no indirect contacts with the disease. The fifth person in this group has $2 + 1 + 1 = 4$ indirect contacts.

We have seen that for matrix multiplication the commutative law does not hold. The next theorem shows that the associative law does hold.

THEOREM 1 ASSOCIATIVE LAW FOR MATRIX MULTIPLICATION

Let $A = (a_{ij})$ be an $n \times m$ matrix, $B = (b_{ij})$ an $m \times p$ matrix, and $C = (c_{ij})$ a $p \times q$ matrix. Then the **associative law**

$$A(BC) = (AB)C \tag{3}$$

holds and ABC, defined by either side of (3), is an $n \times q$ matrix. ■

The proof of this theorem is not difficult, but it is somewhat tedious. It is best given using the summation notation. For that reason let us defer it until the end of the section.

EXAMPLE 4 ILLUSTRATING THE ASSOCIATIVE LAW OF MATRIX MULTIPLICATION

Verify the associative law for $A = \begin{pmatrix} 1 & -3 \\ 0 & 2 \end{pmatrix}$, $B = \begin{pmatrix} 2 & -1 & 4 \\ 3 & 1 & 5 \end{pmatrix}$, and $C = \begin{pmatrix} 0 & -2 & 1 \\ 4 & 3 & 2 \\ -5 & 0 & 6 \end{pmatrix}$.

SOLUTION: We first note that A is 2×2, B is 2×3, and C is 3×3. Hence all products used in the statement of the associative law are defined and the resulting product will be a 2×3 matrix. We then calculate

$$AB = \begin{pmatrix} 1 & -3 \\ 0 & 2 \end{pmatrix}\begin{pmatrix} 2 & -1 & 4 \\ 3 & 1 & 5 \end{pmatrix} = \begin{pmatrix} -7 & -4 & -11 \\ 6 & 2 & 10 \end{pmatrix}$$

$$(AB)C = \begin{pmatrix} -7 & -4 & -11 \\ 6 & 2 & 10 \end{pmatrix}\begin{pmatrix} 0 & -2 & 1 \\ 4 & 3 & 2 \\ -5 & 0 & 6 \end{pmatrix} = \begin{pmatrix} 39 & 2 & -81 \\ -42 & -6 & 70 \end{pmatrix}$$

Similarly,

$$BC = \begin{pmatrix} 2 & -1 & 4 \\ 3 & 1 & 5 \end{pmatrix}\begin{pmatrix} 0 & -2 & 1 \\ 4 & 3 & 2 \\ -5 & 0 & 6 \end{pmatrix} = \begin{pmatrix} -24 & -7 & 24 \\ -21 & -3 & 35 \end{pmatrix}$$

$$A(BC) = \begin{pmatrix} 1 & -3 \\ 0 & 2 \end{pmatrix}\begin{pmatrix} -24 & -7 & 24 \\ -21 & -3 & 35 \end{pmatrix} = \begin{pmatrix} 39 & 2 & -81 \\ -42 & -6 & 70 \end{pmatrix}$$

Thus $(AB)C = A(BC)$.

From now on we shall write the product of three matrices simply as ABC. We can do this because $(AB)C = A(BC)$; thus we get the same answer no matter how the multiplication is carried out (provided that we do not commute any of the matrices).

The associative law can be extended to longer products. For example, suppose that AB, BC, and CD are defined. Then

$$ABCD = A(B(CD)) = ((AB)C)D = A(BC)D = (AB)(CD) \tag{4}$$

There are two distributive laws for matrix multiplication.

THEOREM 2 DISTRIBUTIVE LAWS FOR MATRIX MULTIPLICATION

If all the following sums and products are defined, then

$$A(B + C) = AB + AC \tag{5}$$

and

$$(A + B)C = AC + BC \tag{6}$$

PROOFS OF THEOREMS 1 AND 2: ***Associative Law:*** Since A is $n \times m$ and B is $m \times p$, AB is $n \times p$. Thus $(AB)C = (n \times p) \times (p \times q)$ is an $n \times q$ matrix. Similarly, BC is $m \times q$ and $A(BC)$ is $n \times q$ so that $(AB)C$ and $A(BC)$ are both of the same size. We must show that the ijth component of $(AB)C$ equals the ijth component of $A(BC)$. Define $D = (d_{ij}) = AB$. Then

$$d_{ij} \overset{\text{from (1)}}{=} \sum_{k=1}^{m} a_{ik}b_{kj}$$

The ijth component of $(AB)C = DC$ is

$$\sum_{l=1}^{p} d_{il}c_{lj} = \sum_{l=1}^{p}\left(\sum_{k=1}^{m} a_{ik}b_{kl}\right)c_{lj} = \sum_{k=1}^{m}\sum_{l=1}^{p} a_{ik}b_{kl}c_{lj}$$

Next we define $E = (e_{ij}) = BC$. Then

$$e_{kj} = \sum_{l=1}^{p} b_{kl}c_{lj}$$

and the ijth component of $A(BC) = AE$ is

$$\sum_{k=1}^{m} a_{ik}e_{kj} = \sum_{k=1}^{m}\sum_{l=1}^{p} a_{ik}b_{kl}c_{lj}$$

Thus the ijth component of $(AB)C$ is equal to the ijth component of $A(BC)$. This proves the associative law.

Distributive Laws: We prove the first distributive law [equation (5)]. The proof of the second one [equation (6)] is virtually identical and is therefore omitted. Let A be $n \times m$ and let B and C be $m \times p$. Then the kjth component of $B + C$ is $b_{kj} + c_{kj}$ and the ijth component of $A(B + C)$ is

$$\sum_{k=1}^{m} a_{ik}(b_{kj} + c_{kj}) = \sum_{k=1}^{m} a_{ik}b_{kj} + \sum_{k=1}^{m} a_{ik}c_{kj} = ij\text{th component of } AB \text{ plus}$$

the ijth component of AC and this proves equation (5). ■

OF HISTORICAL INTEREST

ARTHUR CAYLEY AND THE ALGEBRA OF MATRICES

ARTHUR CAYLEY (LIBRARY OF CONGRESS)

The algebra of matrices, that is, the rules by which matrices can be added and multiplied, was developed by the English mathematician Arthur Cayley (1821–1895) in 1857. Cayley was born at Richmond, in Surrey (near London), and was educated at Trinity College, Cambridge, graduating in 1842. In that year he placed first in the very difficult test for the Smith's prize. For a period of several years he studied and practiced law, always being careful not to let his legal practice prevent him from working on mathematics. While a student of the bar he went to Dublin and attended Hamilton's lectures on quaternions. When the Sadlerian professorship was established at Cambridge in 1863, Cayley was offered the chair, which he accepted, thus giving up a lucrative future in the legal profession for the modest provision of an academic life. But then he could devote *all* of his time to mathematics.

Cayley ranks as the third most prolific writer of mathematics in the history of the subject, being surpassed only by Euler and Cauchy. He began publishing while still an undergraduate student at Cambridge, put out between 200 and 300 papers during his years of legal practice, and continued his prolific publication the rest of his long life. The massive *Collected Mathematical Papers* of Cayley contains 966 papers and fills 13 large quarto volumes averaging about 600 pages per volume. There is scarcely an area in pure mathematics that has not been touched and enriched by the genius of Cayley.

Besides developing matrix theory, Cayley made pioneering contributions to analytic geometry, the theory of determinants, higher-dimensional geometry, the theory of curves and surfaces, the study of binary forms, the theory of elliptic functions, and the development of invariant theory.

Cayley's mathematical style reflects his legal training, for his papers are severe, direct, methodical, and clear. He possessed a phenomenal memory and seemed never to forget anything he had once seen or read. He also possessed a singularly serene, even, and gentle temperament. He has been called "the mathematicians' mathematician."

Cayley developed an unusual avidity for novel reading. He read novels while traveling, while waiting for meetings to start, and at any odd moments that presented themselves. During his life he read thousands of novels, not only in English, but also in Greek, French, German, and Italian. He took great delight in painting, especially in water colors, and he exhibited a marked talent as a water colorist. He was also an ardent student of botany and nature in general.

(CONTINUED)

OF HISTORICAL INTEREST

(CONCLUDED)

Cayley was, in the true British tradition, an amateur mountain climber, and he made frequent trips to the Continent for long walks and mountain scaling. A story is told that he claimed the reason he undertook mountain climbing was that, though he found the ascent arduous and tiring, the grand feeling of exhilaration he attained when he conquered the peak was like that he experienced when he solved a difficult mathematics problem or completed an intricate mathematical theory, and it was easier for him to attain the desired feeling by climbing the mountain.

Matrices arose with Cayley in connection with linear transformations of the type

$$\begin{aligned} x' &= ax + by \\ y' &= cx + dy \end{aligned} \tag{15}$$

where a, b, c, d are real numbers, and which may be thought of as functions that take the vector (x, y) into the vector (x', y'). We shall discuss linear transformations in great detail in Chapter 8. Here we observe that the transformation (15) is completely determined by the four coefficients a, b, c, d, and so they can be symbolized by the square array

$$\begin{pmatrix} a & b \\ c & d \end{pmatrix}$$

which we have called a 2×2 matrix. Since two transformations like (15) are identical if and only if they possess the same coefficients, Cayley defined two matrices

$$\begin{pmatrix} a & b \\ c & d \end{pmatrix} \quad \text{and} \quad \begin{pmatrix} e & f \\ g & h \end{pmatrix}$$

to be equal if and only if $a = e$, $b = f$, $c = g$, and $d = h$.

Now suppose that the transformation (15) is followed by a second transformation

$$\begin{aligned} x'' &= ex' + fy' \\ y'' &= gx' + hy' \end{aligned} \tag{16}$$

Then

$$x'' = e(ax + by) + f(cx + dy) = (ea + fc)x + (eb + fd)y$$

and

$$y'' = g(ax + by) + h(cx + dy) = (ga + hc)x + (gb + hd)y$$

This led Cayley to the following definition for the product of two matrices:

$$\begin{pmatrix} e & f \\ g & h \end{pmatrix}\begin{pmatrix} a & b \\ c & d \end{pmatrix} = \begin{pmatrix} ea + fc & eb + fd \\ ga + hc & gb + hd \end{pmatrix}$$

which is, of course, a special case of the general definition of the matrix product we gave on page 400.

It is interesting to observe how, in mathematics, very simple observations can sometimes lead to important, and far reaching, definitions and theorems.

PROBLEMS 6.6

SELF-QUIZ

I. Which of the following is true of matrix multiplication of matrices A and B?

a. It can be performed only if A and B are square matrices.
b. Each entry c_{ij} is the product of a_{ij} and b_{ij}.
c. $AB = BA$.
d. It can be performed only if the number of columns of A is equal to the number of rows of B.

II. Which of the following would be the size of the product matrix AB when a 2×4 matrix A is multiplied by a 4×3 matrix B?

a. 2×3
b. 3×2
c. 4×4
d. This product cannot be found.

III. Which of the following is true of matrices A and B if AB is a column vector?

a. B is a column vector.
b. A is a row vector.
c. A and B are square vectors.
d. The number of rows in A must equal the number of columns in B.

IV. Which of the following is true about a product AB if A is a 4×5 matrix?

a. B must have 4 rows and the result will have 5 columns.
b. B must have 5 columns and the result will be a square matrix.
c. B must have 4 columns and the result will have 5 rows.
d. B must have 5 rows and the result will have 4 rows.

In Problems 1–15 perform the indicated computation.

1. $\begin{pmatrix} 2 & 3 \\ -1 & 2 \end{pmatrix}\begin{pmatrix} 4 & 1 \\ 0 & 6 \end{pmatrix}$

2. $\begin{pmatrix} 3 & -2 \\ 1 & 4 \end{pmatrix}\begin{pmatrix} -5 & 6 \\ 1 & 3 \end{pmatrix}$

3. $\begin{pmatrix} 1 & -1 \\ 1 & 1 \end{pmatrix}\begin{pmatrix} -1 & 0 \\ 2 & 3 \end{pmatrix}$

4. $\begin{pmatrix} -5 & 6 \\ 1 & 3 \end{pmatrix}\begin{pmatrix} 3 & -2 \\ 1 & 4 \end{pmatrix}$

5. $\begin{pmatrix} -4 & 5 & 1 \\ 0 & 4 & 2 \end{pmatrix}\begin{pmatrix} 3 & -1 & 1 \\ 5 & 6 & 4 \\ 0 & 1 & 2 \end{pmatrix}$

6. $\begin{pmatrix} 7 & 1 & 4 \\ 2 & -3 & 5 \end{pmatrix}\begin{pmatrix} 1 & 6 \\ 0 & 4 \\ -2 & 3 \end{pmatrix}$

7. $\begin{pmatrix} 1 & 6 \\ 0 & 4 \\ -2 & 3 \end{pmatrix}\begin{pmatrix} 7 & 1 & 4 \\ 2 & -3 & 5 \end{pmatrix}$

8. $\begin{pmatrix} 1 & 4 & -2 \\ 3 & 0 & 4 \end{pmatrix}\begin{pmatrix} 0 & 1 \\ 2 & 3 \end{pmatrix}$

9. $\begin{pmatrix} 1 & 4 & 6 \\ -2 & 3 & 5 \\ 1 & 0 & 4 \end{pmatrix}\begin{pmatrix} 2 & -3 & 5 \\ 1 & 0 & 6 \\ 2 & 3 & 1 \end{pmatrix}$

10. $\begin{pmatrix} 2 & -3 & 5 \\ 1 & 0 & 6 \\ 2 & 3 & 1 \end{pmatrix}\begin{pmatrix} 1 & 4 & 6 \\ -2 & 3 & 5 \\ 1 & 0 & 4 \end{pmatrix}$

11. $\begin{pmatrix} 1 & 4 & 0 & 2 \end{pmatrix}\begin{pmatrix} 3 & -6 \\ 2 & 4 \\ 1 & 0 \\ -2 & 3 \end{pmatrix}$

12. $\begin{pmatrix} 1 \\ 3 \\ 5 \end{pmatrix}\begin{pmatrix} 2 & -1 & 4 \end{pmatrix}$

13. $\begin{pmatrix} 3 \\ -1 \\ 10 \\ 2 \end{pmatrix}\begin{pmatrix} 1 & 5 & -3 & 8 \end{pmatrix}$

14. $\begin{pmatrix} 1 & 0 & 0 \\ 0 & 1 & 0 \\ 0 & 0 & 1 \end{pmatrix}\begin{pmatrix} 3 & -2 & 1 \\ 4 & 0 & 6 \\ 5 & 1 & 9 \end{pmatrix}$

15. $\begin{pmatrix} a & b & c \\ d & e & f \\ g & h & j \end{pmatrix}\begin{pmatrix} 1 & 0 & 0 \\ 0 & 1 & 0 \\ 0 & 0 & 1 \end{pmatrix}$, where $a, b, c, d, e, f, g, h, j$ are real numbers.

16. Find a matrix $A = \begin{pmatrix} a & b \\ c & d \end{pmatrix}$ such that $A\begin{pmatrix} 2 & 3 \\ 1 & 2 \end{pmatrix} = \begin{pmatrix} 1 & 0 \\ 0 & 1 \end{pmatrix}$.

***17.** Let a_{11}, a_{12}, a_{21}, and a_{22} be given real numbers such that $a_{11}a_{22} - a_{12}a_{21} \neq 0$. Find numbers b_{11}, b_{12}, b_{21}, and b_{22} such that
$$\begin{pmatrix} a_{11} & a_{12} \\ a_{21} & a_{22} \end{pmatrix}\begin{pmatrix} b_{11} & b_{12} \\ b_{21} & b_{22} \end{pmatrix} = \begin{pmatrix} 1 & 0 \\ 0 & 1 \end{pmatrix}.$$

18. Verify the associative law for multiplication for the matrices $A = \begin{pmatrix} 2 & -1 & 4 \\ 1 & 0 & 6 \end{pmatrix}$,
$$B = \begin{pmatrix} 1 & 0 & 1 \\ 2 & -1 & 2 \\ 3 & -2 & 0 \end{pmatrix}, \text{ and } C = \begin{pmatrix} 1 & 6 \\ -2 & 4 \\ 0 & 5 \end{pmatrix}.$$

19. As in Example 3, suppose that a group of people have contracted a contagious disease. These persons have contacts with a second group who in turn have contacts with a third group. Let $A = \begin{pmatrix} 1 & 0 & 1 & 0 \\ 0 & 1 & 1 & 0 \\ 1 & 0 & 0 & 1 \end{pmatrix}$ represent the contacts between the contagious group and the members of group 2, and let
$$B = \begin{pmatrix} 1 & 0 & 1 & 0 & 0 \\ 0 & 0 & 0 & 1 & 0 \\ 1 & 1 & 0 & 0 & 0 \\ 0 & 0 & 1 & 0 & 1 \end{pmatrix}$$
represent the contacts between groups 2 and 3.
(a) How many people are in each group?
(b) Find the matrix of indirect contacts between groups 1 and 3.

20. Answer the questions of Problem 19 for $A = \begin{pmatrix} 1 & 0 & 1 & 1 & 0 \\ 0 & 1 & 0 & 1 & 1 \end{pmatrix}$ and
$$B = \begin{pmatrix} 1 & 0 & 0 & 0 & 0 & 0 & 1 \\ 0 & 1 & 0 & 1 & 0 & 0 & 0 \\ 1 & 1 & 0 & 0 & 1 & 1 & 1 \\ 0 & 0 & 0 & 1 & 1 & 0 & 1 \\ 0 & 1 & 0 & 0 & 0 & 0 & 0 \end{pmatrix}.$$

21. A company pays its executives a salary and gives them shares of its stock as an annual bonus. Last year, the president of the company received \$80,000 and 50 shares of stock, each of the three vice-presidents was paid \$45,000 and 20 shares of stock, and the treasurer was paid \$40,000 and 10 shares of stock.
(a) Express the payments to the executives in money and stock by means of a 2×3 matrix.
(b) Express the number of executives of each rank by means of a column vector.
(c) Use matrix multiplication to calculate the total amount of money and the total number of shares of stock the company paid these executives last year.

22. Sales, unit gross profits, and unit taxes for sales of a large corporation are given in the table below. Find a matrix that shows total profits and taxes in each of the four months.

23. Let A be a square matrix. Then A^2 is defined simply as AA. Calculate $\begin{pmatrix} 2 & -1 \\ 4 & 6 \end{pmatrix}^2$.

24. Calculate A^2, where $A = \begin{pmatrix} 1 & -2 & 4 \\ 2 & 0 & 3 \\ 1 & 1 & 5 \end{pmatrix}$.

25. Calculate A^3, where $A = \begin{pmatrix} -1 & 2 \\ 3 & 4 \end{pmatrix}$.

26. Calculate A^2, A^3, A^4, and A^5, where
$$A = \begin{pmatrix} 0 & 1 & 0 & 0 \\ 0 & 0 & 1 & 0 \\ 0 & 0 & 0 & 1 \\ 0 & 0 & 0 & 0 \end{pmatrix}.$$

TABLE FOR PROBLEM 22

	Product					
	Sales of Item					
Month	i	ii	iii	Item	Unit Profit (in hundreds of dollars)	Unit Taxes (in hundreds of dollars)
January	4	2	20	i	3.5	1.5
February	6	1	9	ii	2.75	2
March	5	3	12	iii	1.5	0.6
April	8	2.5	20			

27. Calculate A^2, A^3, A^4, and A^5, where

$$A = \begin{pmatrix} 0 & 1 & 0 & 0 & 0 \\ 0 & 0 & 1 & 0 & 0 \\ 0 & 0 & 0 & 1 & 0 \\ 0 & 0 & 0 & 0 & 1 \\ 0 & 0 & 0 & 0 & 0 \end{pmatrix}.$$

28. An $n \times n$ matrix A has the property that AB is the zero matrix for any $n \times n$ matrix B. Prove that A is the zero matrix.

29. A **probability matrix** is a square matrix having two properties: (i) every component is nonnegative (≥ 0) and (ii) the sum of the elements in each row is 1. The following are probability matrices:

$$P = \begin{pmatrix} \frac{1}{3} & \frac{1}{3} & \frac{1}{3} \\ \frac{1}{4} & \frac{1}{2} & \frac{1}{4} \\ 0 & 0 & 1 \end{pmatrix} \quad \text{and} \quad Q = \begin{pmatrix} \frac{1}{6} & \frac{1}{6} & \frac{2}{3} \\ 0 & 1 & 0 \\ \frac{1}{5} & \frac{1}{5} & \frac{3}{5} \end{pmatrix}$$

Show that PQ is a probability matrix.

***30.** Let P be a probability matrix. Show that P^2 is a probability matrix.

****31.** Let P and Q be probability matrices of the same size. Prove that PQ is a probability matrix.

32. Prove formula (4) by using the associative law [equation (3)].

***33.** A round robin tennis tournament can be organized in the following way. Each of the n players plays all the others, and the results are recorded in an $n \times n$ matrix R as follows:

$$R_{ij} = \begin{cases} 1 & \text{if the } i\text{th player beats the } j\text{th player} \\ 0 & \text{if the } i\text{th player loses to the } j\text{th player} \\ 0 & \text{if } i = j \end{cases}$$

The ith player is then assigned the score

$$S_i = \sum_{j=1}^{n} R_{ij} + \frac{1}{2} \sum_{j=1}^{n} (R^2)_{ij}{}^{\dagger}$$

(a) In a tournament between four players

$$R = \begin{pmatrix} 0 & 1 & 0 & 0 \\ 0 & 0 & 1 & 1 \\ 1 & 0 & 0 & 0 \\ 1 & 0 & 1 & 0 \end{pmatrix}.$$

Rank the players according to their scores.

(b) Interpret the meaning of the score.

34. Let O be the $m \times n$ zero matrix and let A be an $n \times p$ matrix. Show that $OA = O_1$, where O_1 is the $m \times p$ zero matrix.

35. Verify the distributive law [equation (5)] for the matrices

$$A = \begin{pmatrix} 1 & 2 & 4 \\ 3 & -1 & 0 \end{pmatrix} \quad B = \begin{pmatrix} 2 & 7 \\ -1 & 4 \\ 6 & 0 \end{pmatrix}$$

$$C = \begin{pmatrix} -1 & 2 \\ 3 & 7 \\ 4 & 1 \end{pmatrix}.$$

ANSWERS TO SELF-QUIZ

I. d **II.** a **III.** a **IV.** d

6.7 MATRICES AND LINEAR SYSTEMS OF EQUATIONS

In Section 6.3, page 383, we discussed the following systems of m equations in n unknowns:

$$\begin{aligned} a_{11}x_1 + a_{12}x_2 + \cdots + a_{1n}x_n &= b_1 \\ a_{21}x_1 + a_{22}x_2 + \cdots + a_{2n}x_n &= b_2 \\ \vdots \qquad \vdots \qquad\quad \vdots \quad &\quad \vdots \\ a_{m1}x_1 + a_{m2}x_2 + \cdots + a_{mn}x_n &= b_m \end{aligned} \tag{1}$$

† $(R^2)_{ij}$ is the ijth component of the matrix R^2.

Let

$$A = \begin{pmatrix} a_{11} & a_{12} & \cdots & a_{1n} \\ a_{21} & a_{22} & \cdots & a_{2n} \\ \vdots & \vdots & & \vdots \\ a_{m1} & a_{m2} & \cdots & a_{mn} \end{pmatrix}$$

be the coefficient matrix, **x** the vector $\begin{pmatrix} x_1 \\ x_2 \\ \vdots \\ x_n \end{pmatrix}$, and **b** the vector $\begin{pmatrix} b_1 \\ b_2 \\ \vdots \\ b_m \end{pmatrix}$. Since A is an $m \times n$ matrix and **x** is an $n \times 1$ matrix, the matrix product $A\mathbf{x}$ is defined as an $m \times 1$ matrix. It is not difficult to see that system (1) can be written as

> **MATRIX FORM OF A LINEAR SYSTEM OF EQUATIONS**
>
> $$A\mathbf{x} = \mathbf{b} \tag{2}$$

EXAMPLE 1

WRITING A SYSTEM IN MATRIX FORM

Consider the system

$$\begin{aligned} 2x_1 + 4x_2 + 6x_3 &= 18 \\ 4x_1 + 5x_2 + 6x_3 &= 24 \\ 3x_1 + x_2 - 2x_3 &= 4 \end{aligned} \tag{3}$$

(See Example 6.3.1 on page 376.) This can be written in form $A\mathbf{x} = \mathbf{b}$ with

$$A = \begin{pmatrix} 2 & 4 & 6 \\ 4 & 5 & 6 \\ 3 & 1 & -2 \end{pmatrix}, \quad \mathbf{x} = \begin{pmatrix} x_1 \\ x_2 \\ x_3 \end{pmatrix}, \text{ and } \mathbf{b} = \begin{pmatrix} 18 \\ 24 \\ 4 \end{pmatrix}.$$

It is obviously easier to write out system (1) in the form $A\mathbf{x} = \mathbf{b}$. There are many other advantages, too. In Section 6.8 we shall see how a square system can be solved almost at once if we know a matrix called the *inverse* of A. Even without that, as we saw in Section 6.3, computations are much easier to write down by using an augmented matrix.

If $\mathbf{b} = \begin{pmatrix} 0 \\ 0 \\ \vdots \\ 0 \end{pmatrix}$ is the $m \times 1$ zero vector, then system (1) is homogeneous (see Section 6.4) and can be written

$$A\mathbf{x} = \mathbf{0} \qquad \text{Matrix form of a homogeneous system of equations}$$

There is a fundamental relationship between homogeneous and nonhomogeneous systems. Let A be an $m \times n$ matrix

$$\mathbf{x} = \begin{pmatrix} x_1 \\ x_2 \\ \vdots \\ x_n \end{pmatrix}, \qquad \mathbf{b} = \begin{pmatrix} b_1 \\ b_2 \\ \vdots \\ b_m \end{pmatrix}, \qquad \text{and} \qquad \mathbf{0} = \underbrace{\begin{pmatrix} 0 \\ 0 \\ \vdots \\ 0 \end{pmatrix}}_{m \text{ zeros}}$$

The general nonhomogeneous system can be written as

$$A\mathbf{x} = \mathbf{b} \tag{4}$$

With A and $\mathbf{x}$ as in (4), we define the **associated homogeneous system** by

$$A\mathbf{x} = \mathbf{0} \tag{5}$$

THEOREM 1

Let $\mathbf{x}_1$ and $\mathbf{x}_2$ be solutions of the nonhomogeneous system (4). Then their difference, $\mathbf{x}_1 - \mathbf{x}_2$, is a solution of the related homogeneous system (5).

PROOF:

By the distributive law (5) on page 405

$$A(\mathbf{x}_1 - \mathbf{x}_2) = A\mathbf{x}_1 - A\mathbf{x}_2 = \mathbf{b} - \mathbf{b} = \mathbf{0}.$$ ■

COROLLARY

Let $\mathbf{x}$ be a particular solution to the nonhomogeneous system (4) and let $\mathbf{y}$ be another solution to (4). Then there exists a solution $\mathbf{h}$ to the homogeneous system (5) such that

$$\mathbf{y} = \mathbf{x} + \mathbf{h} \tag{6}$$

PROOF: If $\mathbf{h}$ is defined by $\mathbf{h} = \mathbf{y} - \mathbf{x}$, then $\mathbf{h}$ solves (5) by Theorem 1 and $\mathbf{y} = \mathbf{x} + \mathbf{h}$. ■

Theorem 1 and its corollary are very useful. They tell us that

> In order to find all solutions to the nonhomogeneous system (4), it is sufficient to find *one* solution to (4) and all solutions to the associated homogeneous system (5).

REMARK: A very similar result holds for solutions of homogeneous and nonhomogeneous linear differential equations (See Problems 23 and 24 and Chapter 10). One of the many nice things about mathematics is that seemingly very different topics are closely interrelated.

EXAMPLE 2 WRITING AN INFINITE NUMBER OF SOLUTIONS AS A PARTICULAR SOLUTION TO A NONHOMOGENEOUS SYSTEM PLUS SOLUTIONS TO THE HOMOGENEOUS SYSTEM

Find all solutions to the nonhomogeneous system

$$\begin{aligned} x + 2x_2 - x_3 &= 2 \\ 2x_1 + 3x_2 + 5x_3 &= 5 \\ -x_1 - 3x_2 + 8x_3 &= -1 \end{aligned}$$

by using the result given above.

SOLUTION: First we find one solution by row reduction:

$$\left(\begin{array}{rrr|r} 1 & 2 & -1 & 2 \\ 2 & 3 & 5 & 5 \\ -1 & -3 & 8 & -1 \end{array}\right) \xrightarrow[R_3 \to R_3 + R_1]{R_2 \to R_2 - 2R_1} \left(\begin{array}{rrr|r} 1 & 2 & -1 & 2 \\ 0 & -1 & 7 & 1 \\ 0 & -1 & 7 & 1 \end{array}\right) \xrightarrow[R_3 \to R_3 - R_2]{R_1 \to R_1 + 2R_2} \left(\begin{array}{rrr|r} 1 & 0 & 13 & 4 \\ 0 & -1 & 7 & 1 \\ 0 & 0 & 0 & 0 \end{array}\right)$$

We see that there are an infinite number of solutions. Setting $x_3 = 0$ (any other number would do), we obtain $x_1 = 4$ and $x_2 = -1$. So one particular solution is $\mathbf{x_p} = (4, -1, 0)$.

Row reduction of the associated homogeneous system leads to

$$\left(\begin{array}{rrr|r} 1 & 0 & 13 & 0 \\ 0 & -1 & 7 & 0 \\ 0 & 0 & 0 & 0 \end{array}\right)$$

Therefore all solutions to the homogeneous system satisfy

$$x_1 = -13x_3, \qquad x_2 = 7x_3$$

or

$$\mathbf{x_h} = (x_1, x_2, x_3) = (-13x_3, 7x_3, x_3) = x_3(-13, 7, 1)$$

Thus each solution to (6) can be written

$$\mathbf{x} = \mathbf{x_p} + \mathbf{x_h} = (4, -1, 0) + x_3(-13, 7, 1)$$

for an appropriate value of x_3. For example, $x_3 = 0$ yields the solution $(4, -1, 0)$, whereas $x_3 = 2$ gives the solution $(-22, 13, 2)$.

PROBLEMS 6.7

SELF-QUIZ

I. If the system

$$\begin{aligned} x - \, z &= 2 \\ y + z &= 3 \\ x + 2y &= 4 \end{aligned}$$

is written in the form $A\mathbf{x} = \mathbf{b}$ with $\mathbf{x} = \begin{pmatrix} x \\ y \\ z \end{pmatrix}$ and $\mathbf{b} = \begin{pmatrix} 2 \\ 3 \\ 4 \end{pmatrix}$, then $A =$ ________.

a. $\begin{pmatrix} 1 & 1 & -1 \\ 1 & 1 & 1 \\ 1 & 1 & 2 \end{pmatrix}$ **b.** $\begin{pmatrix} 1 & -1 & 0 \\ 0 & 1 & 1 \\ 1 & 2 & 0 \end{pmatrix}$

c. $\begin{pmatrix} 1 & 0 & -1 \\ 0 & 1 & 1 \\ 1 & 0 & 2 \end{pmatrix}$

d. $\begin{pmatrix} 1 & 0 & -1 \\ 0 & 1 & 1 \\ 1 & 2 & 0 \end{pmatrix}$

In Problems 1–6 write the given system in the form $A\mathbf{x} = \mathbf{b}$.

1. $$\begin{aligned} 2x_1 - x_2 &= 3 \\ 4x_1 + 5x_2 &= 7 \end{aligned}$$

2. $$\begin{aligned} x_1 - x_2 + 3x_3 &= 11 \\ 4x_1 + x_2 - x_3 &= -4 \\ 2x_1 - x_2 + 3x_3 &= 10 \end{aligned}$$

3. $$\begin{aligned} 3x_1 + 6x_2 - 7x_3 &= 0 \\ 2x_1 - x_2 + 3x_3 &= 1 \end{aligned}$$

4. $$\begin{aligned} 4x_1 - x_2 + x_3 - x_4 &= -7 \\ 3x_1 + x_2 - 5x_3 + 6x_4 &= 8 \\ 2x_1 - x_2 + x_3 &= 9 \end{aligned}$$

5. $$\begin{aligned} x_2 - x_3 &= 7 \\ x_1 + x_3 &= 2 \\ 3x_1 + 2x_2 &= -5 \end{aligned}$$

6. $$\begin{aligned} 2x_1 + 3x_2 - x_3 &= 0 \\ -4x_1 + 2x_2 + x_3 &= 0 \\ 7x_1 + 3x_2 - 9x_3 &= 0 \end{aligned}$$

In Problems 7–15 write out the system of equations represented by the given augmented matrix

7. $\left(\begin{array}{ccc|c} 1 & 1 & -1 & 7 \\ 4 & -1 & 5 & 4 \\ 6 & 1 & 3 & 20 \end{array}\right)$

8. $\left(\begin{array}{cc|c} 0 & 1 & 2 \\ 1 & 0 & 3 \end{array}\right)$

9. $\left(\begin{array}{ccc|c} 2 & 0 & 1 & 2 \\ -3 & 4 & 0 & 3 \\ 0 & 5 & 6 & 5 \end{array}\right)$

10. $\left(\begin{array}{ccc|c} 2 & 3 & 1 & 2 \\ 0 & 4 & 1 & 3 \\ 0 & 0 & 0 & 0 \end{array}\right)$

11. $\left(\begin{array}{cccc|c} 1 & 0 & 0 & 0 & 2 \\ 0 & 1 & 0 & 0 & 3 \\ 0 & 0 & 1 & 0 & -5 \\ 0 & 0 & 0 & 1 & 6 \end{array}\right)$

12. $\left(\begin{array}{ccc|c} 2 & 3 & 1 & 0 \\ 4 & -1 & 5 & 0 \\ 3 & 6 & -7 & 0 \end{array}\right)$

13. $\left(\begin{array}{ccc|c} 6 & 2 & 1 & 2 \\ -2 & 3 & 1 & 4 \\ 0 & 0 & 0 & 2 \end{array}\right)$

14. $\left(\begin{array}{ccc|c} 3 & 1 & 5 & 6 \\ 2 & 3 & 2 & 4 \end{array}\right)$

15. $\left(\begin{array}{cc|c} 7 & 2 & 1 \\ 3 & 1 & 2 \\ 6 & 9 & 3 \end{array}\right)$

16. Find a matrix A and vectors $\mathbf{x}$ and $\mathbf{b}$ such that the system represented by the following augmented matrix can be written in the form $A\mathbf{x} = \mathbf{b}$ and solve the system.

$$\left(\begin{array}{ccc|c} 2 & 0 & 0 & 3 \\ 0 & 4 & 0 & 5 \\ 0 & 0 & -5 & 2 \end{array}\right)$$

In Problems 17–22 find all solutions to the given nonhomogeneous system by first finding one solution (if possible) and then finding all solutions to the associated homogeneous system.

17. $$\begin{aligned} x_1 - 3x_2 &= 2 \\ -2x_1 + 6x_2 &= -4 \end{aligned}$$

18. $$\begin{aligned} x_1 - x_2 + x_3 &= 6 \\ 3x_1 - 3x_2 + 3x_3 &= 18 \end{aligned}$$

19. $$\begin{aligned} x_1 - x_2 - x_3 &= 2 \\ 2x_1 + x_2 + 2x_3 &= 4 \\ x_1 - 4x_2 - 5x_3 &= 2 \end{aligned}$$

20. $$\begin{aligned} x_1 - x_2 - x_3 &= 2 \\ 2x_1 + x_2 + 2x_3 &= 4 \\ x_1 - 4x_2 - 5x_3 &= 3 \end{aligned}$$

21. $$\begin{aligned} x_1 + x_2 - x_3 + 2x_4 &= 3 \\ 3x_1 + 2x_2 + x_3 - x_4 &= 5 \end{aligned}$$

22. $$\begin{aligned} x_1 - x_2 + x_3 - x_4 &= -2 \\ -2x_1 + 3x_2 - x_3 + 2x_4 &= 5 \\ 4x_1 - 2x_2 + 2x_3 - 3x_4 &= 6 \end{aligned}$$

23. Consider the linear, homogeneous second-order differential equation

$$y''(x) + a(x)y'(x) + b(x)y(x) = 0 \tag{7}$$

where $a(x)$ and $b(x)$ are continuous and the unknown function y is assumed to have a second derivative. Show that if y_1 and y_2 are solutions to (7), then $c_1y_1 + c_2y_2$ is a solution for any constants c_1 and c_2.

24. Suppose that y_p and y_q are solutions to the nonhomogeneous equation

$$y''(x) + a(x)y'(x) + b(x)y(x) = f(x) \tag{8}$$

Show that $y_p - y_q$ is a solution to (7).

ANSWER TO SELF-QUIZ

I. d

6.8 THE INVERSE OF A SQUARE MATRIX

In this section we define two kinds of matrices that are central to matrix theory. We begin with a simple example. Let $A = \begin{pmatrix} 2 & 5 \\ 1 & 3 \end{pmatrix}$ and $B = \begin{pmatrix} 3 & -5 \\ -1 & 2 \end{pmatrix}$. Then an easy computation shows that $AB = BA = I_2$, where $I_2 = \begin{pmatrix} 1 & 0 \\ 0 & 1 \end{pmatrix}$. The matrix I_2 is called the 2×2 *identity matrix*. The matrix B is called the *inverse* of A and is written A^{-1}.

DEFINITION IDENTITY MATRIX

The $n \times n$ **identity matrix** I_n is the $n \times n$ matrix with 1's down the **main diagonal**† and 0's everywhere else. That is,

$$I_n = (b_{ij}) \quad \text{where} \quad b_{ij} = \begin{cases} 1 & \text{if } i = j \\ 0 & \text{if } i \neq j \end{cases} \tag{1}$$

■

EXAMPLE 1 TWO IDENTITY MATRICES

$$I_3 = \begin{pmatrix} 1 & 0 & 0 \\ 0 & 1 & 0 \\ 0 & 0 & 1 \end{pmatrix} \quad \text{and} \quad I_5 = \begin{pmatrix} 1 & 0 & 0 & 0 & 0 \\ 0 & 1 & 0 & 0 & 0 \\ 0 & 0 & 1 & 0 & 0 \\ 0 & 0 & 0 & 1 & 0 \\ 0 & 0 & 0 & 0 & 1 \end{pmatrix}$$

THEOREM 1

Let A be a square $n \times n$ matrix. Then

$$AI_n = I_nA = A$$

That is, I_n commutes with every $n \times n$ matrix and leaves it unchanged after multiplication on the left or right.

NOTE: I_n functions for $n \times n$ matrices the way the number 1 functions for real numbers (since $1 \cdot a = a \cdot 1 = a$ for every real number a).

PROOF: Let c_{ij} be the ijth element of AI_n and let $B = I_n$. Then

$$c_{ij} = a_{i1}b_{1j} + a_{i2}b_{2j} + \cdots + a_{ij}b_{jj} + \cdots + a_{in}b_{nj}$$

But, from (1), this sum is equal to a_{ij}. Thus $AI_n = A$. In a similar fashion we can show that $I_nA = A$, and this proves the theorem. ■

†The main diagonal of $A = (a_{ij})$ consists of the components a_{11}, a_{22}, a_{33}, and so on. Unless otherwise stated, we shall refer to the main diagonal simply as the **diagonal**.

NOTATION: From now on we shall write the identity matrix simply as I, since if A is $n \times n$, the products IA and AI are defined only if I is also $n \times n$.

DEFINITION THE INVERSE OF A MATRIX

Let A and B be $n \times n$ matrices. Suppose that

$$AB = BA = I$$

Then B is called the **inverse** of A and is written as A^{-1}. We then have

$$AA^{-1} = A^{-1}A = I$$

If A has an inverse, then A is said to be **invertible**. ■

REMARK 1: From this definition it immediately follows that $(A^{-1})^{-1} = A$ if A is invertible.

REMARK 2: This definition does *not* state that every square matrix has an inverse. In fact there are many square matrices that have no inverse. (See, for instance, Example 3 below.)

In the definition above, we defined *the* inverse of a matrix. This statement suggests that inverses are unique. This is indeed the case, as the following theorem shows.

THEOREM 2

If a square matrix A is invertible, then its inverse is unique.

PROOF: Suppose B and C are two inverses for A. We can show that $B = C$. By definition, we have $AB = BA = I$ and $AC = CA = I$. Then $B(AC) = BI = B$ and $(BA)C = IC = C$. But $B(AC) = (BA)C$ by the associative law of matrix multiplication. Hence $B = C$ and the theorem is proved. ■

Another important fact about inverses is given below.

THEOREM 3

Let A and B be invertible $n \times n$ matrices. Then AB is invertible and

$$(AB)^{-1} = B^{-1}A^{-1}$$

PROOF: To prove this result, we refer to the definition of an inverse. That is, $B^{-1}A^{-1} = (AB)^{-1}$ if and only if $B^{-1}A^{-1}(AB) = (AB)(B^{-1}A^{-1}) = I$. But this follows since

$$(B^{-1}A^{-1})(AB) \overset{\text{Equation (4) on page 405}}{=} B^{-1}(A^{-1}A)B = B^{-1}IB = B^{-1}B = I$$

and

$$(AB)(B^{-1}A^{-1}) = A(BB^{-1})A^{-1} = AIA^{-1} = AA^{-1} = I. \qquad \blacksquare$$

Consider the system of n equations in n unknowns

$$A\mathbf{x} = \mathbf{b}$$

and suppose that A is invertible. Then

$$\begin{aligned} A^{-1}A\mathbf{x} &= A^{-1}\mathbf{b} && \text{we multiplied on the left by } A^{-1} \\ I\mathbf{x} &= A^{-1}\mathbf{b} && A^{-1}A = I \\ \mathbf{x} &= A^{-1}\mathbf{b} && I\mathbf{x} = \mathbf{x} \end{aligned}$$

This is a solution to the system because

$$A\mathbf{x} = A(A^{-1}\mathbf{b}) = (AA^{-1})\mathbf{b} = I\mathbf{b} = \mathbf{b}$$

If $\mathbf{y}$ is a vector with $A\mathbf{y} = \mathbf{b}$, then the computation above shows that $\mathbf{y} = A^{-1}\mathbf{b}$. That is, $\mathbf{y} = \mathbf{x}$. We have shown the following:

> If A is invertible, the system $A\mathbf{x} = \mathbf{b}$ has the unique solution $\mathbf{x} = A^{-1}\mathbf{b}$. **(2)**

This is one of the reasons we study matrix inverses.

There are two basic questions that come to mind once we have defined the inverse of a matrix.

Question 1: What matrices have inverses?

Question 2: If a matrix has an inverse, how can we compute it?

We answer both questions in this section. Rather than starting by giving you what seems to be a set of arbitrary rules, we look first at what happens in the 2×2 case.

EXAMPLE 2 FINDING THE INVERSE OF A 2×2 MATRIX

Let $A = \begin{pmatrix} 2 & -3 \\ -4 & 5 \end{pmatrix}$. Compute A^{-1} if it exists.

SOLUTION: Suppose that A^{-1} exists. We write $A^{-1} = \begin{pmatrix} x & y \\ z & w \end{pmatrix}$ and use the fact that $AA^{-1} = I$.

$$AA^{-1} = \begin{pmatrix} 2 & -3 \\ -4 & 5 \end{pmatrix}\begin{pmatrix} x & y \\ z & w \end{pmatrix} = \begin{pmatrix} 2x - 3z & 2y - 3w \\ -4x + 5z & -4y + 5w \end{pmatrix} = \begin{pmatrix} 1 & 0 \\ 0 & 1 \end{pmatrix}$$

The last two matrices can be equal only if each of their corresponding components are equal. This means that

$$\begin{aligned} 2x \qquad - 3z \qquad &= 1 && (3) \\ 2y \qquad - 3w &= 0 && (4) \\ -4x \qquad + 5z \qquad &= 0 && (5) \\ - 4y \qquad + 5w &= 1 && (6) \end{aligned}$$

This is a system of four equations in four unknowns. Note that there are two equations involving x and z only [equations (3) and (5)] and two equations involving y and w only [equations (4) and (6)]. We write these two systems in augmented-matrix form:

$$\left(\begin{array}{rr|r} 2 & -3 & 1 \\ -4 & 5 & 0 \end{array}\right) \tag{7}$$

$$\left(\begin{array}{rr|r} 2 & -3 & 0 \\ -4 & 5 & 1 \end{array}\right) \tag{8}$$

Now, we know from Section 6.3 that if system (7) (in the variables x and z) has a unique solution, then Gauss-Jordan elimination of (7) will result in

$$\left(\begin{array}{rr|r} 1 & 0 & x \\ 0 & 1 & z \end{array}\right)$$

where (x, z) is the unique pair of numbers that satisfies $2x - 3z = 1$ and $-4x + 5z = 0$. Similarly, row reduction of (8) will result in

$$\left(\begin{array}{rr|r} 1 & 0 & y \\ 0 & 1 & w \end{array}\right)$$

where (y, w) is the unique pair of numbers that satisfies $2y - 3w = 0$ and $-4y + 5w = 1$.

Since the coefficient matrices in (7) and (8) are the same, we can perform the row reductions on the two augmented matrices simultaneously by considering the new augmented matrix

$$\left(\begin{array}{rr|rr} 2 & -3 & 1 & 0 \\ -4 & 5 & 0 & 1 \end{array}\right) \tag{9}$$

If A is invertible, then the system defined by (3), (4), (5), and (6) has a unique solution and, by what we said above, row reduction will result in

$$\left(\begin{array}{rr|rr} 1 & 0 & x & y \\ 0 & 1 & z & w \end{array}\right)$$

We now carry out the computation, noting that the matrix on the left in (9) is A and the matrix on the right in (9) is I:

$$\left(\begin{array}{rr|rr} 2 & -3 & 1 & 0 \\ -4 & 5 & 0 & 1 \end{array}\right) \xrightarrow{R_1 \to \frac{1}{2}R_1} \left(\begin{array}{rr|rr} 1 & -\frac{3}{2} & \frac{1}{2} & 0 \\ -4 & 5 & 0 & 1 \end{array}\right)$$

$$\xrightarrow{R_2 \to R_2 + 4R_1} \left(\begin{array}{rr|rr} 1 & -\frac{3}{2} & \frac{1}{2} & 0 \\ 0 & -1 & 2 & 1 \end{array}\right)$$

$$\xrightarrow{R_2 \to -R_2} \left(\begin{array}{rr|rr} 1 & -\frac{3}{2} & \frac{1}{2} & 0 \\ 0 & 1 & -2 & -1 \end{array}\right)$$

$$\xrightarrow{R_1 \to R_1 + \frac{3}{2}R_2} \left(\begin{array}{rr|rr} 1 & 0 & -\frac{5}{2} & -\frac{3}{2} \\ 0 & 1 & -2 & -1 \end{array}\right)$$

Thus $x = -\frac{5}{2}$, $y = -\frac{3}{2}$, $z = -2$, $w = -1$ and $\begin{pmatrix} x & y \\ z & w \end{pmatrix} = \begin{pmatrix} -\frac{5}{2} & -\frac{3}{2} \\ -2 & -1 \end{pmatrix}$. We compute

$$\begin{pmatrix} 2 & -3 \\ -4 & 5 \end{pmatrix}\begin{pmatrix} -\frac{5}{2} & -\frac{3}{2} \\ -2 & -1 \end{pmatrix} = \begin{pmatrix} 1 & 0 \\ 0 & 1 \end{pmatrix}$$

and

$$\begin{pmatrix} -\frac{5}{2} & -\frac{3}{2} \\ -2 & -1 \end{pmatrix}\begin{pmatrix} 2 & -3 \\ -4 & 5 \end{pmatrix} = \begin{pmatrix} 1 & 0 \\ 0 & 1 \end{pmatrix}$$

Thus A is invertible and $A^{-1} = \begin{pmatrix} -\frac{5}{2} & -\frac{3}{2} \\ -2 & -1 \end{pmatrix}$.

Example 3

A 2 × 2 Matrix that Is Not Invertible

Let $A = \begin{pmatrix} 1 & 2 \\ -2 & -4 \end{pmatrix}$. Calculate A^{-1} if it exists.

Solution: If $A^{-1} = \begin{pmatrix} x & y \\ z & w \end{pmatrix}$ exists, then

$$AA^{-1} = \begin{pmatrix} 1 & 2 \\ -2 & -4 \end{pmatrix}\begin{pmatrix} x & y \\ z & w \end{pmatrix} = \begin{pmatrix} x + 2z & y + 2w \\ -2x - 4z & -2y - 4w \end{pmatrix} = \begin{pmatrix} 1 & 0 \\ 0 & 1 \end{pmatrix}$$

This leads to the system

$$\begin{aligned} x \qquad + 2z \qquad &= 1 \\ y \qquad + 2w &= 0 \\ -2x \qquad - 4z \qquad &= 0 \\ - 2y \qquad - 4w &= 1 \end{aligned} \tag{10}$$

Using the same reasoning as in Example 1, we can write this system in the augmented-matrix form $(A|I)$ and row-reduce:

$$\left(\begin{array}{cc|cc} 1 & 2 & 1 & 0 \\ -2 & -4 & 0 & 1 \end{array}\right) \xrightarrow{R_2 \to R_2 + 2R_1} \left(\begin{array}{cc|cc} 1 & 2 & 1 & 0 \\ 0 & 0 & 2 & 1 \end{array}\right)$$

This is as far as we can go. The last line reads $0 = 2$ or $0 = 1$, depending on which of the two systems of equations (in x and z or in y and w) is being solved. Thus system (10) is inconsistent and A is not invertible.

The last two examples illustrate a procedure that always works when you are trying to find the inverse of a matrix.

Procedure for Computing the Inverse of a Square Matrix A

Step 1: Write the augmented matrix $(A|I)$.

Step 2: Use row reduction to reduce the matrix A to its reduced row echelon form.

Step 3: Decide if A is invertible.

- **a.** If A can be reduced to the identity matrix I, then A^{-1} will be the matrix to the right of the vertical bar.
- **b.** If the row reduction of A leads to a row of zeros to the left of the vertical bar, then A is not invertible.

REMARK: We can rephrase (a) and (b) as follows:

> A square matrix A is invertible if and only if its reduced row echelon form is the identity matrix.

Let $A = \begin{pmatrix} a_{11} & a_{12} \\ a_{21} & a_{22} \end{pmatrix}$. Then, as in equation (6.2.7), page 374, we define

$$\text{Determinant of } A = a_{11}a_{22} - a_{12}a_{21} \tag{11}$$

We abbreviate the determinant of A by $\det A$.

THEOREM 4

Let A be a 2×2 matrix. Then:

i. A is invertible if and only if $\det A \neq 0$.
ii. If $\det A \neq 0$, then

$$A^{-1} = \frac{1}{\det A}\begin{pmatrix} a_{22} & -a_{12} \\ -a_{21} & a_{11} \end{pmatrix}^{\dagger} \tag{12}$$

PROOF: First suppose that $\det A \neq 0$ and let $B = (1/\det A)\begin{pmatrix} a_{22} & -a_{12} \\ -a_{21} & a_{11} \end{pmatrix}$. Then

$$\begin{aligned} BA &= \frac{1}{\det A}\begin{pmatrix} a_{22} & -a_{12} \\ -a_{21} & a_{11} \end{pmatrix}\begin{pmatrix} a_{11} & a_{12} \\ a_{21} & a_{22} \end{pmatrix} \\ &= \frac{1}{a_{11}a_{22} - a_{12}a_{21}}\begin{pmatrix} a_{22}a_{11} - a_{12}a_{21} & 0 \\ 0 & -a_{21}a_{12} + a_{11}a_{22} \end{pmatrix} = \begin{pmatrix} 1 & 0 \\ 0 & 1 \end{pmatrix} = I \end{aligned}$$

Similarly, $AB = I$, which shows that A is invertible and that $B = A^{-1}$. We still must show that if A is invertible, then $\det A \neq 0$. To do so we consider the system

$$\begin{aligned} a_{11}x_1 + a_{12}x_2 &= b_1 \\ a_{21}x_1 + a_{22}x_2 &= b_2 \end{aligned} \tag{13}$$

We do this because we know from our Summing Up Theorem (Theorem 6.2.1, page 374) that if this system has a unique solution, then its determinant is nonzero. The system can be written in the form

$$A\mathbf{x} = \mathbf{b} \tag{14}$$

with $\mathbf{x} = \begin{pmatrix} x_1 \\ x_2 \end{pmatrix}$ and $\mathbf{b} = \begin{pmatrix} b_1 \\ b_2 \end{pmatrix}$. Then, since A is invertible, we see from

[†] This formula can be obtained directly by applying our procedure for computing an inverse (see Problem 46).

(2) that system (14) has a unique solution given by

$$\mathbf{x} = A^{-1}\mathbf{b}$$

But by Theorem 6.2.1 the fact that system (13) has a unique solution implies that $\det A \neq 0$. This completes the proof. ■

EXAMPLE 4

CALCULATING THE INVERSE OF A 2 × 2 MATRIX

Let $A = \begin{pmatrix} 2 & -4 \\ 1 & 3 \end{pmatrix}$. Calculate A^{-1} if it exists.

SOLUTION: We find that $\det A = (2)(3) - (-4)(1) = 10$; hence A^{-1} exists. From equation (12) we get

$$A^{-1} = \frac{1}{10}\begin{pmatrix} 3 & 4 \\ -1 & 2 \end{pmatrix} = \begin{pmatrix} \frac{3}{10} & \frac{4}{10} \\ -\frac{1}{10} & \frac{2}{10} \end{pmatrix}$$

Check:

$$A^{-1}A = \frac{1}{10}\begin{pmatrix} 3 & 4 \\ -1 & 2 \end{pmatrix}\begin{pmatrix} 2 & -4 \\ 1 & 3 \end{pmatrix} = \frac{1}{10}\begin{pmatrix} 10 & 0 \\ 0 & 10 \end{pmatrix} = \begin{pmatrix} 1 & 0 \\ 0 & 1 \end{pmatrix}$$

and

$$AA^{-1} = \begin{pmatrix} 2 & -4 \\ 1 & 3 \end{pmatrix}\begin{pmatrix} \frac{3}{10} & \frac{4}{10} \\ -\frac{1}{10} & \frac{2}{10} \end{pmatrix} = \begin{pmatrix} 1 & 0 \\ 0 & 1 \end{pmatrix}$$

EXAMPLE 5

A 2 × 2 MATRIX THAT IS NOT INVERTIBLE

Let $A = \begin{pmatrix} 1 & 2 \\ -2 & -4 \end{pmatrix}$. Calculate A^{-1} if it exists.

SOLUTION: We find that $\det A = (1)(-4) - (2)(-2) = -4 + 4 = 0$, so that A^{-1} does not exist, as we saw in Example 3.

The procedure described above works for $n \times n$ matrices where $n > 2$. We illustrate this with a number of examples.

EXAMPLE 6

CALCULATING THE INVERSE OF A 3 × 3 MATRIX

Let $A = \begin{pmatrix} 2 & 4 & 6 \\ 4 & 5 & 6 \\ 3 & 1 & -2 \end{pmatrix}$ (see Example 6.3.1 on page 376.) Calculate A^{-1} if it exists.

SOLUTION: We first put I next to A in an augmented-matrix form

$$\left(\begin{array}{ccc|ccc} 2 & 4 & 6 & 1 & 0 & 0 \\ 4 & 5 & 6 & 0 & 1 & 0 \\ 3 & 1 & -2 & 0 & 0 & 1 \end{array}\right)$$

and then carry out the row reduction.

$$\xrightarrow{R_1 \to \frac{1}{2}R_1} \left(\begin{array}{ccc|ccc} 1 & 2 & 3 & \frac{1}{2} & 0 & 0 \\ 4 & 5 & 6 & 0 & 1 & 0 \\ 3 & 1 & -2 & 0 & 0 & 1 \end{array}\right) \xrightarrow[R_3 \to R_3 - 3R_1]{R_2 \to R_2 - 4R_1} \left(\begin{array}{ccc|ccc} 1 & 2 & 3 & \frac{1}{2} & 0 & 0 \\ 0 & -3 & -6 & -2 & 1 & 0 \\ 0 & -5 & -11 & -\frac{3}{2} & 0 & 1 \end{array}\right)$$

$$\xrightarrow{R_2 \to -\frac{1}{3}R_2} \left(\begin{array}{ccc|ccc} 1 & 2 & 3 & \frac{1}{2} & 0 & 0 \\ 1 & 1 & 2 & \frac{2}{3} & -\frac{1}{3} & 0 \\ 0 & -5 & -11 & -\frac{3}{2} & 0 & 1 \end{array}\right) \xrightarrow[R_3 \to R_3 + 5R_2]{R_1 \to R_1 - 2R_2} \left(\begin{array}{ccc|ccc} 1 & 0 & -1 & -\frac{5}{6} & \frac{2}{3} & 0 \\ 0 & 1 & 2 & \frac{2}{3} & -\frac{1}{3} & 0 \\ 0 & 0 & -1 & \frac{11}{6} & -\frac{5}{3} & 1 \end{array}\right)$$

$$\xrightarrow{R_3 \to -R_3} \left(\begin{array}{ccc|ccc} 1 & 0 & -1 & -\frac{5}{6} & \frac{2}{3} & 0 \\ 0 & 1 & 2 & \frac{2}{3} & -\frac{1}{3} & 0 \\ 0 & 0 & 1 & -\frac{11}{6} & \frac{5}{3} & -1 \end{array}\right) \xrightarrow[R_2 \to R_2 - 2R_3]{R_1 \to R_1 + R_3} \left(\begin{array}{ccc|ccc} 1 & 0 & 0 & -\frac{8}{3} & \frac{7}{3} & -1 \\ 0 & 1 & 0 & \frac{13}{3} & -\frac{11}{3} & 2 \\ 0 & 0 & 1 & -\frac{11}{6} & \frac{5}{3} & -1 \end{array}\right)$$

Since A has now been reduced to I, we have

$$A^{-1} = \begin{pmatrix} -\frac{8}{3} & \frac{7}{3} & -1 \\ \frac{13}{3} & -\frac{11}{3} & 2 \\ -\frac{11}{6} & \frac{5}{3} & -1 \end{pmatrix} = \frac{1}{6}\begin{pmatrix} -16 & 14 & -6 \\ 26 & -22 & 12 \\ -11 & 10 & -6 \end{pmatrix}$$

We factor out $\frac{1}{6}$ to make computations easier.

Check:

$$A^{-1}A = \frac{1}{6}\begin{pmatrix} -16 & 14 & -6 \\ 26 & -22 & 12 \\ -11 & 10 & -6 \end{pmatrix}\begin{pmatrix} 2 & 4 & 6 \\ 4 & 5 & 6 \\ 3 & 1 & -2 \end{pmatrix} = \frac{1}{6}\begin{pmatrix} 6 & 0 & 0 \\ 0 & 6 & 0 \\ 0 & 0 & 6 \end{pmatrix} = I.$$

We can also verify that $AA^{-1} = I$.

WARNING: It is easy to make numerical errors in computing A^{-1}. Therefore it is important to check the computations by verifying that $A^{-1}A = I$.

EXAMPLE 7

CALCULATING THE INVERSE OF A 3 × 3 MATRIX

Let $A = \begin{pmatrix} 2 & 4 & 3 \\ 0 & 1 & -1 \\ 3 & 5 & 7 \end{pmatrix}$. Calculate A^{-1} if it exists.

SOLUTION: Proceeding as in Example 6, we obtain, successively, the following augmented matrices:

$$\left(\begin{array}{ccc|ccc} 2 & 4 & 3 & 1 & 0 & 0 \\ 0 & 1 & -1 & 0 & 1 & 0 \\ 3 & 5 & 7 & 0 & 0 & 1 \end{array}\right) \xrightarrow{R_1 \to \frac{1}{2}R_1} \left(\begin{array}{ccc|ccc} 1 & 2 & \frac{3}{2} & \frac{1}{2} & 0 & 0 \\ 0 & 1 & -1 & 0 & 1 & 0 \\ 3 & 5 & 7 & 0 & 0 & 1 \end{array}\right)$$

$$\xrightarrow{R_3 \to R_3 - 3R_1} \left(\begin{array}{ccc|ccc} 1 & 2 & \frac{3}{2} & \frac{1}{2} & 0 & 0 \\ 0 & 1 & -1 & 0 & 1 & 0 \\ 0 & -1 & \frac{5}{2} & -\frac{3}{2} & 0 & 1 \end{array}\right)$$

$$\xrightarrow[R_3 \to R_3 + R_2]{R_1 \to R_1 - 2R_2} \left(\begin{array}{ccc|ccc} 1 & 0 & \frac{7}{2} & \frac{1}{2} & -2 & 0 \\ 0 & 1 & -1 & 0 & 1 & 0 \\ 0 & 0 & \frac{3}{2} & -\frac{3}{2} & 1 & 1 \end{array}\right) \xrightarrow{R_3 \to \frac{2}{3}R_3} \left(\begin{array}{ccc|ccc} 1 & 0 & \frac{7}{2} & \frac{1}{2} & -2 & 0 \\ 0 & 1 & -1 & 0 & 1 & 0 \\ 0 & 0 & 1 & -1 & \frac{2}{3} & \frac{2}{3} \end{array}\right)$$

$$\xrightarrow[R_2 \to R_2 + R_3]{R_1 \to R_1 - \frac{7}{2}R_3} \left(\begin{array}{ccc|ccc} 1 & 0 & 0 & 4 & -\frac{13}{3} & -\frac{7}{3} \\ 0 & 1 & 0 & -1 & \frac{5}{3} & \frac{2}{3} \\ 0 & 0 & 1 & -1 & \frac{2}{3} & \frac{2}{3} \end{array}\right)$$

Thus

$$A^{-1} = \begin{pmatrix} 4 & -\frac{13}{3} & -\frac{7}{3} \\ -1 & \frac{5}{3} & \frac{2}{3} \\ -1 & \frac{2}{3} & \frac{2}{3} \end{pmatrix}$$

Check:

$$A^{-1}A = \begin{pmatrix} 4 & -\frac{13}{3} & -\frac{7}{3} \\ -1 & \frac{5}{3} & \frac{2}{3} \\ -1 & \frac{2}{3} & \frac{2}{3} \end{pmatrix}\begin{pmatrix} 2 & 4 & 3 \\ 0 & 1 & -1 \\ 3 & 5 & 7 \end{pmatrix} = \begin{pmatrix} 1 & 0 & 0 \\ 0 & 1 & 0 \\ 0 & 0 & 1 \end{pmatrix}$$

EXAMPLE 8 A 3×3 MATRIX THAT IS NOT INVERTIBLE

Let $A = \begin{pmatrix} 1 & -3 & 4 \\ 2 & -5 & 7 \\ 0 & -1 & 1 \end{pmatrix}$. Calculate A^{-1} if it exists.

SOLUTION: Proceeding as before, we obtain, successively,

$$\left(\begin{array}{ccc|ccc} 1 & -3 & 4 & 1 & 0 & 0 \\ 2 & -5 & 7 & 0 & 1 & 0 \\ 0 & -1 & 1 & 0 & 0 & 1 \end{array}\right) \xrightarrow{R_2 \to R_2 - 2R_1} \left(\begin{array}{ccc|ccc} 1 & -3 & 4 & 1 & 0 & 0 \\ 0 & 1 & -1 & -2 & 1 & 0 \\ 0 & -1 & 1 & 0 & 0 & 1 \end{array}\right)$$

$$\xrightarrow[R_3 \to R_3 + R_2]{R_1 \to R_1 + 3R_2} \left(\begin{array}{ccc|ccc} 1 & 0 & 1 & -5 & 3 & 0 \\ 0 & 1 & -1 & -2 & 1 & 0 \\ 0 & 0 & 0 & -2 & 1 & 1 \end{array}\right)$$

This is as far as we can go. The matrix A *cannot* be reduced to the identity matrix, and we can conclude that A is *not* invertible.

There is another way to see the result of the last example. Let **b** be any 3-vector and consider the system $A\mathbf{x} = \mathbf{b}$. If we tried to solve this by Gaussian elimination, we would end up with an equation that reads $0 = c \neq 0$ as in Example 3, or $0 = 0$. This is case (*ii*) or (*iii*) of Section 6.3 (see page 384). That is, the system either has no solution or it has an infinite number of solutions. The one possibility ruled out is the case in which the system has a unique solution. But if A^{-1} existed, then there would be a unique solution given by $\mathbf{x} = A^{-1}\mathbf{b}$. We are left to conclude that

> If row reduction of A produces a row of zeros, then A is *not* invertible.

DEFINITION ROW EQUIVALENT MATRICES

Suppose that by elementary row operations we can transform the matrix A into the matrix B. Then A and B are said to be **row equivalent**.

The reasoning used above can be used to prove the following theorem (see Problem 47).

THEOREM 5

Let A be an $n \times n$ matrix.

i. A is invertible if and only if A is row equivalent to the identity matrix I_n; that is, the reduced row echelon form of A is I_n.
ii. A is invertible if and only if the system $A\mathbf{x} = \mathbf{b}$ has a unique solution for every n-vector $\mathbf{b}$.
iii. If A is invertible, then this unique solution is given by $\mathbf{x} = A^{-1}\mathbf{b}$. ■

EXAMPLE 9 USING THE INVERSE TO SOLVE A SYSTEM OF EQUATIONS

Solve the system

$$\begin{aligned} 2x_1 + 4x_2 + 3x_3 &= 6 \\ x_2 - x_3 &= -4 \\ 3x_1 + 5x_2 + 7x_3 &= 7 \end{aligned}$$

SOLUTION: This system can be written as $A\mathbf{x} = \mathbf{b}$, where $A = \begin{pmatrix} 2 & 4 & 3 \\ 0 & 1 & -1 \\ 3 & 5 & 7 \end{pmatrix}$ and $\mathbf{b} = \begin{pmatrix} 6 \\ -4 \\ 7 \end{pmatrix}$. In Example 7 we found that A^{-1} exists and

$$A^{-1} = \begin{pmatrix} 4 & -\frac{13}{3} & -\frac{7}{3} \\ -1 & \frac{5}{3} & \frac{2}{3} \\ -1 & \frac{2}{3} & \frac{2}{3} \end{pmatrix}$$

Thus the unique solution is given by

$$\mathbf{x} = \begin{pmatrix} x_1 \\ x_2 \\ x_3 \end{pmatrix} = A^{-1}\mathbf{b} = \begin{pmatrix} 4 & -\frac{13}{3} & -\frac{7}{3} \\ -1 & \frac{5}{3} & \frac{2}{3} \\ -1 & \frac{2}{3} & \frac{2}{3} \end{pmatrix} \begin{pmatrix} 6 \\ -4 \\ 7 \end{pmatrix} = \begin{pmatrix} 25 \\ -8 \\ -4 \end{pmatrix}$$

EXAMPLE 10 THE TECHNOLOGY AND LEONTIEF MATRICES: MODELING THE 1958 AMERICAN ECONOMY

In the Leontief input-output model described in Example 6.3.8 on page 385, we obtained the system

$$\begin{aligned} a_{11}x_1 + a_{12}x_2 + \cdots + a_{1n}x_n + e_1 &= x_1 \\ a_{21}x_1 + a_{22}x_2 + \cdots + a_{2n}x_n + e_2 &= x_2 \\ \vdots \qquad \vdots \qquad\qquad \vdots \qquad \vdots \quad &\quad \vdots \\ a_{n1}x_1 + a_{n2}x_2 + \cdots + a_{nn}x_n + e_n &= x_n \end{aligned} \tag{15}$$

which can be written as

$$A\mathbf{x} + \mathbf{e} = \mathbf{x} = I\mathbf{x}$$

or

$$(I - A)\mathbf{x} = \mathbf{e} \tag{16}$$

The matrix A of internal demands is called the **technology matrix**, and the matrix $I - A$ is called the **Leontief matrix**. If the Leontief matrix is invertible, then systems (15) and (16) have unique solutions.

Leontief used his model to analyze the 1958 U.S. economy.[†] He divided the economy into 81 sectors and grouped them into six families of related sectors. For simplicity, we treat each family of sectors as a single sector so that we can treat the U.S. economy as an economy with six industries. These industries are listed in Table 1.

TABLE 1

Sector	Examples
Final nonmetal (FN)	Furniture, processed food
Final metal (FM)	Household appliances, motor vehicles
Basic metal (BM)	Machine-shop products, mining
Basic nonmetal (BN)	Agriculture, printing
Energy (E)	Petroleum, coal
Services (S)	Amusements, real estate

The input-output table, Table 2, gives internal demands in 1958 based on Leontief's figures. The units in the table are millions of dollars. Thus, for example, the number 0.173 in the 6,5 position means that in order to produce \$1 million worth of energy, it is necessary to provide \$0.173 million = \$173,000 worth of services. Similarly, the 0.037 in the 4,2 position means that in order to produce \$1 million worth of final metal, it is necessary to expend \$0.037 million = \$37,000 on basic nonmetal products.

TABLE 2 INTERNAL DEMANDS IN 1958 U.S. ECONOMY

	FN	FM	BM	BN	E	S
FN	0.170	0.004	0	0.029	0	0.008
FM	0.003	0.295	0.018	0.002	0.004	0.016
BM	0.025	0.173	0.460	0.007	0.011	0.007
BN	0.348	0.037	0.021	0.403	0.011	0.048
E	0.007	0.001	0.039	0.025	0.358	0.025
S	0.120	0.074	0.104	0.123	0.173	0.234

[†] *Scientific American* (April 1965): 26–27.

TABLE 3 EXTERNAL DEMANDS ON 1958 U.S. ECONOMY (MILLIONS OF DOLLARS)

FN	\$99,640
FM	\$75,548
BM	\$14,444
BN	\$33,501
E	\$23,527
S	\$263,985

Finally, Leontief estimated the external demands on the 1958 U.S. economy (in millions of dollars) as listed in Table 3.

In order to run the U.S. economy in 1958 and to meet all external demands, how many units in each of the six sectors had to be produced?

SOLUTION: The technology matrix is given by

$$A = \begin{pmatrix} 0.170 & 0.004 & 0 & 0.029 & 0 & 0.008 \\ 0.003 & 0.295 & 0.018 & 0.002 & 0.004 & 0.016 \\ 0.025 & 0.173 & 0.460 & 0.007 & 0.011 & 0.007 \\ 0.348 & 0.037 & 0.021 & 0.403 & 0.011 & 0.048 \\ 0.007 & 0.001 & 0.039 & 0.025 & 0.358 & 0.025 \\ 0.120 & 0.074 & 0.104 & 0.123 & 0.173 & 0.234 \end{pmatrix}$$

and

$$\mathbf{e} = \begin{pmatrix} 99{,}640 \\ 75{,}548 \\ 14{,}444 \\ 33{,}501 \\ 23{,}527 \\ 263{,}985 \end{pmatrix}$$

To obtain the Leontief matrix, we subtract to obtain

$$I - A = \begin{pmatrix} 1 & 0 & 0 & 0 & 0 & 0 \\ 0 & 1 & 0 & 0 & 0 & 0 \\ 0 & 0 & 1 & 0 & 0 & 0 \\ 0 & 0 & 0 & 1 & 0 & 0 \\ 0 & 0 & 0 & 0 & 1 & 0 \\ 0 & 0 & 0 & 0 & 0 & 1 \end{pmatrix} - \begin{pmatrix} 0.170 & 0.004 & 0 & 0.029 & 0 & 0.008 \\ 0.003 & 0.295 & 0.018 & 0.002 & 0.004 & 0.016 \\ 0.025 & 0.173 & 0.460 & 0.007 & 0.011 & 0.007 \\ 0.348 & 0.037 & 0.021 & 0.403 & 0.011 & 0.048 \\ 0.007 & 0.001 & 0.039 & 0.025 & 0.358 & 0.025 \\ 0.120 & 0.074 & 0.104 & 0.123 & 0.173 & 0.234 \end{pmatrix}$$

$$= \begin{pmatrix} 0.830 & -0.004 & 0 & -0.029 & 0 & -0.008 \\ -0.003 & 0.705 & -0.018 & -0.002 & -0.004 & -0.016 \\ -0.025 & -0.173 & 0.540 & -0.007 & -0.011 & -0.007 \\ -0.348 & -0.037 & -0.021 & 0.597 & -0.011 & -0.048 \\ -0.007 & -0.001 & -0.039 & -0.025 & 0.642 & -0.025 \\ -0.120 & -0.074 & -0.104 & -0.123 & -0.173 & 0.766 \end{pmatrix}$$

The computation of the inverse of a 6×6 matrix is a tedious affair. The following result was obtained using a computer with matrix software. To three decimal places,

$$(I - A)^{-1} \approx \begin{pmatrix} 1.234 & 0.014 & 0.007 & 0.064 & 0.006 & 0.017 \\ 0.017 & 1.436 & 0.056 & 0.014 & 0.019 & 0.032 \\ 0.078 & 0.467 & 1.878 & 0.036 & 0.044 & 0.031 \\ 0.752 & 0.133 & 0.101 & 1.741 & 0.065 & 0.123 \\ 0.061 & 0.045 & 0.130 & 0.083 & 1.578 & 0.059 \\ 0.340 & 0.236 & 0.307 & 0.315 & 0.376 & 1.349 \end{pmatrix}$$

Therefore the "ideal" output vector is given by

$$\mathbf{x} = (I - A)^{-1}\mathbf{e} \approx \begin{pmatrix} 131{,}033.21 \\ 120{,}458.90 \\ 80{,}680.56 \\ 178{,}732.04 \\ 66{,}929.26 \\ 431{,}562.04 \end{pmatrix}$$

This means that it would require approximately 131,033 units ($131,033 million worth) of final nonmetal products, 120,459 units of final metal products, 80,681 units of basic metal products, 178,732 units of basic nonmetal products, 66,929 units of energy, and 431,562 service units to run the U.S. economy and to meet the external demands in 1958.

In Section 6.2 we encountered the first form of our Summing Up Theorem (Theorem 6.2.1, page 374). We are now ready to improve upon it. The next theorem states that several statements involving inverse, uniqueness of solutions, row equivalence, and determinants are equivalent. At this point we can prove the equivalence of parts (*i*), (*ii*), (*iii*), and (*iv*). We shall finish the proof after we have developed some basic theory about determinants (see Theorem 7.3.4 on page 478).

THEOREM 6 SUMMING UP THEOREM—VIEW 2

Let A be an $n \times n$ matrix. Then the following five statements are equivalent. That is, each statement implies the other four (so that if one is true, all are true and if one is false, all are false).

i. A is invertible.

ii. The only solution to the homogeneous system $A\mathbf{x} = \mathbf{0}$ is the trivial solution ($\mathbf{x} = \mathbf{0}$).

iii. The system $A\mathbf{x} = \mathbf{b}$ has a unique solution for every n-vector $\mathbf{b}$.

iv. A is row equivalent to the $n \times n$ identity matrix I_n; that is, the reduced row echelon form of A is I_n.

v. $\det A \neq 0$. (So far, $\det A$ is only defined if A is a 2×2 matrix.)

PROOF: We have already seen that statements (*i*) and (*iii*) are equivalent [Theorem 5, part (*ii*)] and that (*i*) and (*iv*) are equivalent [Theorem 5, part (*i*)]. We shall show that (*ii*) and (*iv*) are equivalent. Suppose that (*ii*) holds. That is, suppose that $A\mathbf{x} = \mathbf{0}$

has only the trivial solution $\mathbf{x} = \mathbf{0}$. If we write out this system, we obtain

$$\begin{aligned} a_{11}x_1 + a_{12}x_2 + \cdots + a_{1n}x_n &= 0 \\ a_{21}x_1 + a_{22}x_2 + \cdots + a_{2n}x_n &= 0 \\ \vdots \qquad \vdots \qquad\qquad \vdots \quad & \\ a_{n1}x_1 + a_{n2}x_2 + \cdots + a_{nn}x_n &= 0 \end{aligned} \tag{17}$$

If A were not equivalent to I_n, then row reduction of the augmented matrix associated with (17) would leave us with a row of zeros. But if, say, the last row is zero, then the last equation reads $0 = 0$. Then the homogeneous system reduces to one with $n - 1$ equations in n unknowns, which by Theorem 6.4.1 on page 392 has an infinite number of solutions. But we assumed that $\mathbf{x} = \mathbf{0}$ was the only solution to system (17). This contradiction shows that A is row equivalent to I_n. Conversely, suppose that (*iv*) holds; that is, suppose that A is row equivalent to I_n. Then by Theorem 5, part (*i*), A is invertible and by Theorem 5, part (*iii*), the unique solution to $A\mathbf{x} = \mathbf{0}$ is $\mathbf{x} = A^{-1}\mathbf{0} = \mathbf{0}$. Thus (*ii*) and (*iv*) are equivalent. In Theorem 6.2.1 we showed that (*i*) and (*v*) are equivalent in the 2×2 case. We shall prove the equivalence of (*i*) and (*v*) in Section 7.3. ■

REMARK: We could add another statement to the theorem. Suppose the system $A\mathbf{x} = \mathbf{b}$ has a unique solution. Let R be a matrix in row echelon form that is row equivalent to A. Then R cannot have a row of zeros because if it did, it could not be reduced to the identity matrix.[†] Thus the row echelon form of A must look like this:

$$\begin{pmatrix} 1 & r_{12} & r_{13} & \cdots & r_{1n} \\ 0 & 1 & r_{23} & \cdots & r_{2n} \\ 0 & 0 & 1 & \cdots & r_{3n} \\ \vdots & \vdots & \vdots & & \vdots \\ 0 & 0 & 0 & \cdots & 1 \end{pmatrix} \tag{18}$$

That is, R is a matrix with 1's down the diagonal and 0's below it. We thus have Theorem 7.

THEOREM 7

If any of the statements in Theorem 6 holds, then the row echelon form of A has the form of matrix (18). ■

We have seen that in order to verify that $B = A^{-1}$, we have to check that $AB = BA = I$. It turns out that only half this work has to be done.

THEOREM 8

Let A and B be $n \times n$ matrices. Then A is invertible and $B = A^{-1}$ if (*i*) $BA = I$ or (*ii*) $AB = I$.

[†] Note that if the ith row of R contains only zeros, then the homogeneous system $R\mathbf{x} = \mathbf{0}$ contains more unknowns than equations (since the ith equation is the zero equation) and the system has an infinite number of solutions. But then $A\mathbf{x} = \mathbf{0}$ has an infinite number of solutions, which is a contradiction of our assumption.

REMARK: This theorem simplifies the work in checking that one matrix is the inverse of another.

PROOF:

i. We assume that $BA = I$. Consider the homogeneous system $A\mathbf{x} = \mathbf{0}$. Multiplying both sides of this equation on the left by B, we obtain

$$BA\mathbf{x} = B\mathbf{0} \tag{19}$$

But $BA = I$ and $B\mathbf{0} = \mathbf{0}$, so (19) becomes $I\mathbf{x} = \mathbf{0}$ or $\mathbf{x} = \mathbf{0}$. This shows that $\mathbf{x} = \mathbf{0}$ is the only solution to $A\mathbf{x} = \mathbf{0}$ and, by Theorem 6, parts (i) and (ii), this means that A is invertible. We still have to show that $B = A^{-1}$. Let $A^{-1} = C$. Then $AC = I$. Thus

$$BAC = B(AC) = BI = B \quad \text{and} \quad BAC = (BA)C = IC = C$$

Hence $B = C$ and part (i) is proved.

ii. Let $AB = I$. Then, from part (i) $A = B^{-1}$. From the definition of an inverse this means that $AB = BA = I$, which proves that A is invertible and that $B = A^{-1}$. This completes the proof. ■

PROBLEMS 6.8

SELF-QUIZ

I. Which of the following is true?

a. Every square matrix has an inverse.
b. A square matrix has an inverse if its row reduction leads to a row of zeros.
c. A square matrix is invertible if it has an inverse.
d. A square matrix B is the inverse of A if $AI = B$.

II. Which of the following is true of a system of equations in matrix form?

a. It is of the form $A^{-1}\mathbf{x} = \mathbf{b}$.
b. If it has a unique solution, the solution will be $\mathbf{x} = A^{-1}\mathbf{b}$.
c. It will have a solution if A is not invertible.
d. It will have a unique solution.

III. Which of the following is invertible?

a. $\begin{pmatrix} 1 & 3 \\ -3 & -9 \end{pmatrix}$ **b.** $\begin{pmatrix} 6 & -1 \\ 1 & -\frac{1}{6} \end{pmatrix}$

c. $\begin{pmatrix} 2 & -3 \\ 1 & -1 \end{pmatrix}$ **d.** $\begin{pmatrix} 1 & 0 \\ 2 & 0 \end{pmatrix}$

IV. Which of the following is true of an invertible matrix A?

a. The product of A and I is A^{-1}.
b. A is a 2×3 matrix.
c. $A = A^{-1}$.
d. A is a square matrix.

V. Which of the following is true of the system

$$4x - 5y = 3$$
$$6x - 7y = 4?$$

a. It has no solution because $\begin{pmatrix} 4 & -5 \\ 6 & -7 \end{pmatrix}$ is not invertible.
b. It has the solution $(-1, -\frac{1}{2})$.
c. If it had a solution, it would be found by solving $\begin{pmatrix} 4 & -5 \\ 6 & -7 \end{pmatrix}\begin{pmatrix} x \\ y \end{pmatrix} = \begin{pmatrix} 3 \\ 4 \end{pmatrix}$.
d. Its solution is $\begin{pmatrix} 4 & -5 \\ 6 & -7 \end{pmatrix}\begin{pmatrix} 3 \\ 4 \end{pmatrix}$.

In Problems 1–15 determine whether the given matrix is invertible. If it is, calculate the inverse.

1. $\begin{pmatrix} 2 & 1 \\ 3 & 2 \end{pmatrix}$

2. $\begin{pmatrix} -1 & 6 \\ 2 & -12 \end{pmatrix}$

3. $\begin{pmatrix} 0 & 1 \\ 1 & 0 \end{pmatrix}$

4. $\begin{pmatrix} 1 & 1 \\ 3 & 3 \end{pmatrix}$

5. $\begin{pmatrix} a & a \\ b & b \end{pmatrix}$

6. $\begin{pmatrix} 1 & 1 & 1 \\ 0 & 2 & 3 \\ 5 & 5 & 1 \end{pmatrix}$

7. $\begin{pmatrix} 3 & 2 & 1 \\ 0 & 2 & 2 \\ 0 & 0 & -1 \end{pmatrix}$

8. $\begin{pmatrix} 1 & 1 & 1 \\ 0 & 1 & 1 \\ 0 & 0 & 1 \end{pmatrix}$

9. $\begin{pmatrix} 1 & 6 & 2 \\ -2 & 3 & 5 \\ 7 & 12 & -4 \end{pmatrix}$

10. $\begin{pmatrix} 3 & 1 & 0 \\ 1 & -1 & 2 \\ 1 & 1 & 1 \end{pmatrix}$

11. $\begin{pmatrix} 2 & -1 & 4 \\ -1 & 0 & 5 \\ 19 & -7 & 3 \end{pmatrix}$

12. $\begin{pmatrix} 1 & 2 & 3 \\ 1 & 1 & 2 \\ 0 & 1 & 2 \end{pmatrix}$

13. $\begin{pmatrix} 1 & 1 & 1 & 1 \\ 1 & 2 & -1 & 2 \\ 1 & -1 & 2 & 1 \\ 1 & 3 & 3 & 2 \end{pmatrix}$

14. $\begin{pmatrix} 1 & 0 & 2 & 3 \\ -1 & 1 & 0 & 4 \\ 2 & 1 & -1 & 3 \\ -1 & 0 & 5 & 7 \end{pmatrix}$

15. $\begin{pmatrix} 1 & -3 & 0 & -2 \\ 3 & -12 & -2 & -6 \\ -2 & 10 & 2 & 5 \\ -1 & 6 & 1 & 3 \end{pmatrix}$

16. Show that if A, B, and C are invertible matrices, then ABC is invertible and $(ABC)^{-1} = C^{-1}B^{-1}A^{-1}$.

17. If $A_1, A_2, \ldots, A_m$ are invertible $n \times n$ matrices, show that $A_1A_2 \ldots A_m$ is invertible and calculate its inverse.

18. Show that the matrix $\begin{pmatrix} 3 & 4 \\ -2 & -3 \end{pmatrix}$ is equal to its own inverse.

19. Show that the matrix $A = \begin{pmatrix} a_{11} & a_{12} \\ a_{21} & a_{22} \end{pmatrix}$ is equal to its own inverse if $A = \pm I$ or if $a_{11} = -a_{22}$ and $a_{21}a_{12} = 1 - a_{11}^2$.

20. Find the output vector $\mathbf{x}$ in the Leontief input-output model if $n = 3$, $\mathbf{e} = \begin{pmatrix} 30 \\ 20 \\ 40 \end{pmatrix}$, and

$$A = \begin{pmatrix} \frac{1}{5} & \frac{1}{5} & 0 \\ \frac{2}{5} & \frac{2}{5} & \frac{3}{5} \\ \frac{1}{5} & \frac{1}{10} & \frac{2}{5} \end{pmatrix}.$$

***21.** Suppose that A is $n \times m$ and B is $m \times n$ so that AB is $n \times n$. Show that AB is not invertible if $n > m$. [*Hint:* Show that there is a nonzero vector $\mathbf{x}$ such that $AB\mathbf{x} = \mathbf{0}$ and then apply Theorem 6.]

***22.** Use the methods of this section to find the inverses of the following matrices with complex entries:

a. $\begin{pmatrix} i & 2 \\ 1 & -i \end{pmatrix}$

b. $\begin{pmatrix} 1-i & 0 \\ 0 & 1+i \end{pmatrix}$

c. $\begin{pmatrix} 1 & i & 0 \\ -i & 0 & 1 \\ 0 & 1+i & 1-i \end{pmatrix}$

23. Show that for every real number θ the matrix $\begin{pmatrix} \sin\theta & \cos\theta & 0 \\ \cos\theta & -\sin\theta & 0 \\ 0 & 0 & 1 \end{pmatrix}$ is invertible and find its inverse.

24. Calculate the inverse of $A = \begin{pmatrix} 2 & 0 & 0 \\ 0 & 3 & 0 \\ 0 & 0 & 4 \end{pmatrix}$.

25. A square matrix $A = (a_{ij})$ is called **diagonal** if all its elements off the main diagonal are zero. That is, $a_{ij} = 0$ if $i \neq j$. (The matrix of Problem 24 is diagonal.) Show that a diagonal matrix is invertible if and only if each of its diagonal components is nonzero.

26. Let

$$A = \begin{pmatrix} a_{11} & 0 & \cdots & 0 \\ 0 & a_{22} & \cdots & 0 \\ \vdots & \vdots & \ddots & \vdots \\ 0 & 0 & \cdots & a_{nn} \end{pmatrix}$$

be a diagonal matrix such that each of its diagonal components is nonzero. Calculate A^{-1}.

27. Calculate the inverse of $A = \begin{pmatrix} 2 & 1 & -1 \\ 0 & 3 & 4 \\ 0 & 0 & 5 \end{pmatrix}$.

28. Show that the matrix $A = \begin{pmatrix} 1 & 0 & 0 \\ -2 & 0 & 0 \\ 4 & 6 & 1 \end{pmatrix}$ is not invertible.

***29.** A square matrix is called **upper (lower) triangular** if all its elements below (above) the main diagonal are zero. (The matrix of Problem 27 is upper triangular and the matrix of Problem 28 is lower triangular.) Show that an upper or lower triangular matrix is invertible if and only if each of its diagonal elements is nonzero.

30. Show that the inverse of an invertible upper triangular matrix is upper triangular. [*Hint:* First prove the result for a 3×3 matrix.]

In Problems 31 and 32 a matrix is given. In each case show that the matrix is not invertible by finding a non-zero vector $\mathbf{x}$ such that $A\mathbf{x} = \mathbf{0}$.

31. $\begin{pmatrix} 2 & -1 \\ -4 & 2 \end{pmatrix}$

32. $\begin{pmatrix} 1 & -1 & 3 \\ 0 & 4 & -2 \\ 2 & -6 & 8 \end{pmatrix}$

33. A factory for the construction of quality furniture has two divisions: a machine shop where the parts of the furniture are fabricated, and an assembly and finishing division where the parts are put together into the finished product. Suppose that there are 12 employees in the machine shop and 20 in the assembly and finishing division and that each employee works an 8-hour day. Suppose further that the factory produces only two products: chairs and tables. A chair requires $\frac{384}{17}$ hours of machine shop time and $\frac{480}{17}$ hours of assembly and finishing time. A table requires $\frac{240}{17}$ hours of machine shop time and $\frac{640}{17}$ hours of assembly and finishing time. Assuming that there is an unlimited demand for these products and that the manufacturer wishes to keep all employees busy, how many chairs and how many tables can this factory produce each day?

34. A witch's magic cupboard contains 10 ounces of ground four-leaf clovers and 14 ounces of powdered mandrake root. The cupboard will replenish itself automatically provided she uses up exactly all her supplies. A batch of love potion requires $3\frac{1}{13}$ ounces of ground four-leaf clovers and $2\frac{2}{13}$ ounces of powdered mandrake root. One recipe of a well-known (to witches) cure for the common cold requires $5\frac{5}{13}$ ounces of four-leaf clovers and $10\frac{10}{13}$ ounces of mandrake root. How much of the love potion and the cold remedy should the witch make in order to use up the supply in the cupboard exactly?

35. A farmer feeds his cattle a mixture of two types of feed. One standard unit of type A feed supplies a steer with 10% of its minimum daily requirement of protein and 15% of its requirement of carbohydrates. Type B feed contains 12% of the requirement of protein and 8% of the requirement of carbohydrates in a standard unit. If the farmer wishes to feed his cattle exactly 100% of their minimum daily requirement of protein and carbohydrates, how many units of each type of feed should he give a steer each day?

36. A much simplified version of an input-output table for the 1958 Israeli economy divides that economy into three sectors—agriculture, manufacturing, and energy—with the following result.[†]

	Agriculture	Manufacturing	Energy
Agriculture	0.293	0	0
Manufacturing	0.014	0.207	0.017
Energy	0.044	0.010	0.216

a. How many units of agricultural production are required to produce one unit of agricultural output?

b. How many units of agricultural production are required to produce 200,000 units of agricultural output?

c. How many units of agricultural product go into the production of 50,000 units of energy?

d. How many units of energy go into the production of 50,000 units of agricultural products?

[†] Wassily Leontief, *Input-Output Economics* (New York: Oxford University Press, 1966), 54–57.

37. Continuing Problem 36, exports (in thousands of Israeli pounds) in 1958 were

Agriculture	13,213
Manufacturing	17,597
Energy	1,786

a. Compute the technology and Leontief matrices.

b. Determine the number of Israeli pounds worth of agricultural products, manufactured goods, and energy required to run this model of the Israeli economy and export the stated value of products.

In Problems 38–45 compute the row echelon form of the given matrix and use it to determine directly whether the given matrix is invertible.

38. The matrix of Problem 3.

39. The matrix of Problem 1.

40. The matrix of Problem 4.

41. The matrix of Problem 7.

42. The matrix of Problem 9.

43. The matrix of Problem 11.

44. The matrix of Problem 13.

45. The matrix of Problem 14.

46. Let $A = \begin{pmatrix} a_{11} & a_{12} \\ a_{21} & a_{22} \end{pmatrix}$ and assume that $a_{11}a_{22} - a_{12}a_{21} \neq 0$. Derive formula (12) by row-reducing the augmented matrix
$\left(\begin{array}{cc|cc} a_{11} & a_{12} & 1 & 0 \\ a_{21} & a_{22} & 0 & 1 \end{array}\right)$.

47. Prove parts (i) and (ii) of Theorem 5.

ANSWERS TO SELF-QUIZ

I. c **II.** b **III.** c **IV.** d **V.** c

6.9 THE TRANSPOSE OF A MATRIX

Corresponding to every matrix is another matrix, which, as we shall see in Chapter 7, has properties very similar to those of the original matrix.

DEFINITION TRANSPOSE

Let $A = (a_{ij})$ be an $m \times n$ matrix. Then the **transpose** of A, written A^t, is the $n \times m$ matrix obtained by interchanging the rows and columns of A. Succinctly, we may write $A^t = (a_{ji})$. In other words,

$$\text{if } A = \begin{pmatrix} a_{11} & a_{12} & \cdots & a_{1n} \\ a_{21} & a_{22} & \cdots & a_{2n} \\ \vdots & \vdots & & \vdots \\ a_{m1} & a_{m2} & \cdots & a_{mn} \end{pmatrix}, \text{ then } A^t = \begin{pmatrix} a_{11} & a_{21} & \cdots & a_{m1} \\ a_{12} & a_{22} & \cdots & a_{m2} \\ \vdots & \vdots & & \vdots \\ a_{1n} & a_{2n} & \cdots & a_{mn} \end{pmatrix} \quad (1)$$

Simply put, the ith row of A is the ith column of A^t and the jth column of A is the jth row of A^t. ■

EXAMPLE 1 FINDING THE TRANSPOSES OF THREE MATRICES

Find the transposes of the matrices

$$A = \begin{pmatrix} 2 & 3 \\ 1 & 4 \end{pmatrix} \quad B = \begin{pmatrix} 2 & 3 & 1 \\ -1 & 4 & 6 \end{pmatrix} \quad C = \begin{pmatrix} 1 & 2 & -6 \\ 2 & -3 & 4 \\ 0 & 1 & 2 \\ 2 & -1 & 5 \end{pmatrix}$$

SOLUTION: Interchanging the rows and columns of each matrix, we obtain

$$A^t = \begin{pmatrix} 2 & 1 \\ 3 & 4 \end{pmatrix} \quad B^t = \begin{pmatrix} 2 & -1 \\ 3 & 4 \\ 1 & 6 \end{pmatrix} \quad C^t = \begin{pmatrix} 1 & 2 & 0 & 2 \\ 2 & -3 & 1 & -1 \\ -6 & 4 & 2 & 5 \end{pmatrix}$$

Note, for example, that 4 is the component in row 2 and column 3 of C while 4 is the component in row 3 and column 2 of C^t. That is, the 23 element of C is the 32 element of C^t.

THEOREM 1

Suppose $A = (a_{ij})$ is an $n \times m$ matrix and $B = (b_{ij})$ is an $m \times p$ matrix. Then:

i. $(A^t)^t = A$. (2)

ii. $(AB)^t = B^tA^t$ (3)

iii. If A and B are $n \times m$, then $(A + B)^t = A^t + B^t$. (4)

iv. If A is invertible, then A^t is invertible and $(A^t)^{-1} = (A^{-1})^t$. (5)

PROOF:

i. This follows directly from the definition of the transpose.

ii. First we note that AB is an $n \times p$ matrix, so $(AB)^t$ is $p \times n$. Also, B^t is $p \times m$ and A^t is $m \times n$, so B^tA^t is $p \times n$. Thus both matrices in equation (3) have the same size. Now the ijth element of AB is $\sum_{k=1}^{m} a_{ik}b_{kj}$ and this is the jith element of $(AB)^t$. Let $C = B^t$ and $D = A^t$. Then the ijth element c_{ij} of C is b_{ji} and the ijth element d_{ij} of D is a_{ji}. Thus the jith element of CD = the jith element of $B^tA^t = \sum_{k=1}^{m} c_{jk}d_{ki} = \sum_{k=1}^{m} b_{kj}a_{ik} = \sum_{k=1}^{m} a_{ik}b_{kj}$ = the jith element of $(AB)^t$. This completes the proof of part (ii).

iii. This part is left as an exercise (see Problem 11).

iv. Let $A^{-1} = B$. Then $AB = BA = I$ so that, from part (ii), $(AB)^t = B^tA^t = I^t = I$ and $(BA)^t = A^tB^t = I$. Thus A^t is invertible and B^t is the inverse of A^t; that is, $(A^t)^{-1} = B^t = (A^{-1})^t$. ■

The transpose plays an important role in matrix theory. Since columns of A^t are rows of A, we shall be able to use facts about the transpose to conclude that just about anything which is true about the rows of a matrix is true about its columns.

We conclude this section with an important definition.

DEFINITION SYMMETRIC MATRIX

The $n \times m$ (square) matrix A is called **symmetric** if $A^t = A$. ■

EXAMPLE 2 FOUR SYMMETRIC MATRICES

The following four matrices are symmetric:

$$I \qquad A = \begin{pmatrix} 1 & 2 \\ 2 & 3 \end{pmatrix} \qquad B = \begin{pmatrix} 1 & -4 & 2 \\ -4 & 7 & 5 \\ 2 & 5 & 0 \end{pmatrix} \qquad C = \begin{pmatrix} -1 & 2 & 4 & 6 \\ 2 & 7 & 3 & 5 \\ 4 & 3 & 8 & 0 \\ 6 & 5 & 0 & -4 \end{pmatrix}$$

We shall see the importance of symmetric matrices in Section 8.12.

ANOTHER WAY TO WRITE THE SCALAR PRODUCT

Let $\mathbf{a} = \begin{pmatrix} a_1 \\ a_2 \\ \vdots \\ a_n \end{pmatrix}$ and $\mathbf{b} = \begin{pmatrix} b_1 \\ b_2 \\ \vdots \\ b_n \end{pmatrix}$ be two n-component column vectors. Then, from equation (6) on page 77

$$\mathbf{a} \cdot \mathbf{b} = a_1b_1 + a_2b_2 + \cdots + a_nb_n$$

Now $\mathbf{a}$ is an $n \times 1$ matrix so $\mathbf{a}^t$ is a $1 \times n$ matrix and

$$\mathbf{a}^t = (a_1 \; a_2 \cdots a_n)$$

Then $\mathbf{a}^t\mathbf{b}$ is a 1×1 matrix (or scalar), and by the definition of matrix multiplication,

$$\mathbf{a}^t\mathbf{b} = (a_1 \; a_2 \cdots a_n)\begin{pmatrix} b_1 \\ b_2 \\ \vdots \\ b_n \end{pmatrix} = a_1b_1 + a_2b_2 + \cdots + a_nb_n$$

Thus, if $\mathbf{a}$ and $\mathbf{b}$ are n-component column vectors, then

$$\mathbf{a} \cdot \mathbf{b} = \mathbf{a}^t\mathbf{b} \qquad (6)$$

Formula (6) will be useful to us later in the book.

PROBLEMS 6.9

SELF-QUIZ

Multiple Choice

I. If A is a 3×4 matrix, then A^t is a ________ matrix.

a. 4×3 **b.** 3×4
c. 3×3 **d.** 4×4

II. The transpose of $\begin{pmatrix} 1 & 2 & 3 \\ -1 & 0 & 0 \end{pmatrix}$ is ________.

a. $\begin{pmatrix} 1 & 0 \\ 2 & 0 \\ 3 & -1 \end{pmatrix}$ **b.** $\begin{pmatrix} 1 & -1 \\ 2 & 0 \\ 3 & 0 \end{pmatrix}$

c. $\begin{pmatrix} 1 & 0 \\ -1 & 3 \\ 2 & 0 \end{pmatrix}$ **d.** $\begin{pmatrix} -1 & -2 & -3 \\ 1 & 0 & 0 \end{pmatrix}$

True-False

III. A^t is defined only if A is a square mtarix.

IV. If A is an $n \times n$ matrix, then the main diagonal of A^t is the same as the main diagonal of A.

V. $[(A^t)^t]^t = A^t$.

In Problems 1–10 find the transpose of the given matrix.

1. $\begin{pmatrix} -1 & 4 \\ 6 & 5 \end{pmatrix}$

2. $\begin{pmatrix} 3 & 0 \\ 1 & 2 \end{pmatrix}$

3. $\begin{pmatrix} 2 & 3 \\ -1 & 2 \\ 1 & 4 \end{pmatrix}$

4. $\begin{pmatrix} 2 & -1 & 0 \\ 1 & 5 & 6 \end{pmatrix}$

5. $\begin{pmatrix} 1 & 2 & 3 \\ -1 & 0 & 4 \\ 1 & 5 & 5 \end{pmatrix}$

6. $\begin{pmatrix} 1 & 2 & 3 \\ 2 & 4 & -5 \\ 3 & -5 & 7 \end{pmatrix}$

7. $\begin{pmatrix} 1 & 0 & 1 & 0 \\ 0 & 1 & 0 & 1 \end{pmatrix}$

8. $\begin{pmatrix} 2 & -1 \\ 2 & 4 \\ 1 & 6 \\ 1 & 5 \end{pmatrix}$

9. $\begin{pmatrix} a & b & c \\ d & e & f \\ g & h & i \end{pmatrix}$

10. $\begin{pmatrix} 0 & 0 & 0 \\ 0 & 0 & 0 \end{pmatrix}$

11. Let A and B be $n \times m$ matrices. Show, using the definition of transpose that $(A + B)^t = A^t + B^t$.

12. Find numbers α and β such that $\begin{pmatrix} 2 & \alpha & 3 \\ 5 & -6 & 2 \\ \beta & 2 & 4 \end{pmatrix}$ is symmetric.

13. If A and B are symmetric $n \times n$ matrices, prove that $A + B$ is symmetric.

14. If A and B are symmetric $n \times n$ matrices, show that $(AB)^t = BA$.

15. Show that for any matrix A, the product matrix AA^t is defined and is a symmetric matrix.

16. Show that every diagonal matrix (see Problem 6.8.25, page 430) is symmetric.

17. Show that the transpose of every upper triangular matrix (see Problem 6.8.29 on page 431) is lower triangular.

18. A square matrix is called **skew-symmetric** if $A^t = -A$ (that is, $a_{ij} = -a_{ji}$). Which of the following matrices are skew-symmetric?

a. $\begin{pmatrix} 1 & -6 \\ 6 & 0 \end{pmatrix}$

b. $\begin{pmatrix} 0 & -6 \\ 6 & 0 \end{pmatrix}$

c. $\begin{pmatrix} 2 & -2 & -2 \\ 2 & 2 & -2 \\ 2 & 2 & 2 \end{pmatrix}$

d. $\begin{pmatrix} 0 & 1 & -1 \\ -1 & 0 & 2 \\ 1 & -2 & 0 \end{pmatrix}$

19. Let A and B be $n \times n$ skew-symmetric matrices. Show that $A + B$ is skew-symmetric.

20. If A is skew-symmetric, show that every component on the main diagonal of A is zero.

21. If A and B are skew-symmetric $n \times n$ matrices, show that $(AB)^t = BA$, so that AB is symmetric if and only if A and B commute.

22. Let A be an $n \times n$ matrix. Show that the matrix $\frac{1}{2}(A + A^t)$ is symmetric.

23. Let A be an $n \times n$ matrix. Show that the matrix $\frac{1}{2}(A - A^t)$ is skew-symmetric.

***24.** Show that any square matrix can be written in a unique way as the sum of a symmetric matrix and a skew-symmetric matrix.

***25.** Let $A = \begin{pmatrix} a_{11} & a_{12} \\ a_{21} & a_{22} \end{pmatrix}$ be a matrix with nonnegative entries having the properties that
i. $a_{11}^2 + a_{12}^2 = 1$ and $a_{21}^2 + a_{22}^2 = 1$ and
ii. $\begin{pmatrix} a_{11} \\ a_{12} \end{pmatrix} \cdot \begin{pmatrix} a_{21} \\ a_{22} \end{pmatrix} = 0$. Show that A is invertible and that $A^{-1} = A^t$.

In Problems 26–29 compute $(A^t)^{-1}$ and $(A^{-1})^t$ and show that they are equal.

26. $A = \begin{pmatrix} 1 & 2 \\ 3 & 4 \end{pmatrix}$

27. $A = \begin{pmatrix} 2 & 1 \\ 3 & 2 \end{pmatrix}$

28. $A = \begin{pmatrix} 3 & 2 & 1 \\ 0 & 2 & 2 \\ 0 & 0 & -1 \end{pmatrix}$

29. $A = \begin{pmatrix} 1 & 1 & 1 \\ 0 & 2 & 3 \\ 5 & 5 & 1 \end{pmatrix}$

ANSWERS TO SELF-QUIZ

I. a **II.** b **III.** False **IV.** True **V.** True

6.10 ELEMENTARY MATRICES AND MATRIX INVERSES

Let A be an $m \times n$ matrix. Then, as we shall soon see, we can perform elementary row operations on A by multiplying A on the left by an appropriate matrix. The elementary row operations are:

i. Multiply the ith row by a nonzero number c. $R_i \to cR_i$
ii. Add a multiple of the ith row to the jth row. $R_j \to R_j + cR_i$
iii. Permute (interchange) the ith and jth rows. $R_i \rightleftarrows R_j$

DEFINITION ELEMENTARY MATRIX

An $n \times n$ (square) matrix E is called an **elementary matrix** if it can be obtained from the $n \times n$ identity matrix I_n by a *single* elementary row operation. ■

NOTATION: We denote an elementary matrix by E or by cR_i, $R_j + cR_i$, or P_{ij} depending on how the matrix is obtained from I. Here P_{ij} is the matrix obtained by permuting the ith and jth rows of I.

EXAMPLE 1

THREE ELEMENTARY MATRICES

We obtain three elementary 3×3 matrices.

i. $$\begin{pmatrix} 1 & 0 & 0 \\ 0 & 1 & 0 \\ 0 & 0 & 1 \end{pmatrix} \xrightarrow{R_2 \to 5R_2} \begin{pmatrix} 1 & 0 & 0 \\ 0 & 5 & 0 \\ 0 & 0 & 1 \end{pmatrix} = 5R_2$$

Matrix obtained by multiplying the second row of I by 5

ii. $$\begin{pmatrix} 1 & 0 & 0 \\ 0 & 1 & 0 \\ 0 & 0 & 1 \end{pmatrix} \xrightarrow{R_3 \to R_3 - 3R_1} \begin{pmatrix} 1 & 0 & 0 \\ 0 & 1 & 0 \\ -3 & 0 & 1 \end{pmatrix} = R_3 - 3R_1$$

Matrix obtained by multiplying the first row of I by -3 and adding it to the third row

iii. $$\begin{pmatrix} 1 & 0 & 0 \\ 0 & 1 & 0 \\ 0 & 0 & 1 \end{pmatrix} \xrightarrow{R_2 \rightleftarrows R_3} \begin{pmatrix} 1 & 0 & 0 \\ 0 & 0 & 1 \\ 0 & 1 & 0 \end{pmatrix} = P_{23}$$

Matrix obtained by permuting the second and third rows of I

The proof of the following theorem is left as an exercise (see Problems 54–56).

THEOREM 1

To perform an elementary row operation on a matrix A, multiply A on the left by the appropriate elementary matrix. ■

EXAMPLE 2

PERFORMING ELEMENTARY ROW OPERATIONS BY MULTIPLYING BY ELEMENTARY MATRICES

Let $A = \begin{pmatrix} 1 & 3 & 2 & 1 \\ 4 & 2 & 3 & -5 \\ 3 & 1 & -2 & 4 \end{pmatrix}$. Perform the following elementary row operations on A by multiplying A on the left by an appropriate elementary matrix.

i. Multiply the second row by 5.
ii. Multiply the first row by -3 and add it to the third row.
iii. Permute the second and third rows.

SOLUTION: Since A is a 3×4 matrix, each elementary matrix E must be 3×3 since E must be square and E is multiplying A on the left. We use the results of Example 1.

i. $$(5R_2)A = \begin{pmatrix} 1 & 0 & 0 \\ 0 & 5 & 0 \\ 0 & 0 & 1 \end{pmatrix}\begin{pmatrix} 1 & 3 & 2 & 1 \\ 4 & 2 & 3 & -5 \\ 3 & 1 & -2 & 4 \end{pmatrix} = \begin{pmatrix} 1 & 3 & 2 & 1 \\ 20 & 10 & 15 & -25 \\ 3 & 1 & -2 & 4 \end{pmatrix}$$

ii. $$(R_3 - 3R_1)A = \begin{pmatrix} 1 & 0 & 0 \\ 0 & 1 & 0 \\ -3 & 0 & 1 \end{pmatrix}\begin{pmatrix} 1 & 3 & 2 & 1 \\ 4 & 2 & 3 & -5 \\ 3 & 1 & -2 & 4 \end{pmatrix} = \begin{pmatrix} 1 & 3 & 2 & 1 \\ 4 & 2 & 3 & -5 \\ 0 & -8 & -8 & 1 \end{pmatrix}$$

iii. $$(P_{23})A = \begin{pmatrix} 1 & 0 & 0 \\ 0 & 0 & 1 \\ 0 & 1 & 0 \end{pmatrix}\begin{pmatrix} 1 & 3 & 2 & 1 \\ 4 & 2 & 3 & -5 \\ 3 & 1 & -2 & 4 \end{pmatrix} = \begin{pmatrix} 1 & 3 & 2 & 1 \\ 3 & 1 & -2 & 4 \\ 4 & 2 & 3 & -5 \end{pmatrix}$$

Consider the following three products, with $c \neq 0$:

$$\begin{pmatrix} 1 & 0 & 0 \\ 0 & c & 0 \\ 0 & 0 & 1 \end{pmatrix}\begin{pmatrix} 1 & 0 & 0 \\ 0 & 1/c & 0 \\ 0 & 0 & 1 \end{pmatrix} = \begin{pmatrix} 1 & 0 & 0 \\ 0 & 1 & 0 \\ 0 & 0 & 1 \end{pmatrix} \tag{1}$$

$$\begin{pmatrix} 1 & 0 & 0 \\ 0 & 1 & 0 \\ c & 0 & 1 \end{pmatrix}\begin{pmatrix} 1 & 0 & 0 \\ 0 & 1 & 0 \\ -c & 0 & 1 \end{pmatrix} = \begin{pmatrix} 1 & 0 & 0 \\ 0 & 1 & 0 \\ 0 & 0 & 1 \end{pmatrix} \tag{2}$$

$$\begin{pmatrix} 1 & 0 & 0 \\ 0 & 0 & 1 \\ 0 & 1 & 0 \end{pmatrix}\begin{pmatrix} 1 & 0 & 0 \\ 0 & 0 & 1 \\ 0 & 1 & 0 \end{pmatrix} = \begin{pmatrix} 1 & 0 & 0 \\ 0 & 1 & 0 \\ 0 & 0 & 1 \end{pmatrix} \tag{3}$$

Equations (1), (2), and (3) suggest that each elementary matrix is invertible and that its inverse is of the same type (Table 1). These facts follow from Theorem 1. Evidently, if the operations $R_j \to R_j + cR_i$ followed by $R_j \to R_j - cR_i$ are performed on the matrix A, the matrix A is unchanged. Also, $R_i \to cR_i$ followed by $R_i \to \frac{1}{c}R_i$, and permuting the same two rows twice leave the matrix A unchanged. We have

$$(cR_i)^{-1} = \frac{1}{c}R_i \tag{4}$$

$$(R_j + cR_i)^{-1} = R_j - cR_i \tag{5}$$

$$(P_{ij})^{-1} = P_{ij} \tag{6}$$

Equation (6) indicates that

> Every elementary permutation matrix is its own inverse.

We summarize our results.

TABLE 1

Elementary Matrix Type E	Effect of Multiplying A on the Left by E	Symbolic Representation of Elementary Row Operation	When Multiplied on the Left, E^{-1} Does the Following	Symbolic Representation of Inverse Operation
Multiplication	Multiplies ith row of A by $c \neq 0$	cR_i	Multiplies ith row of A by $\frac{1}{c}$	$\frac{1}{c}R_i$
Addition	Multiplies ith row of A by c and adds it to jth row	$R_j + cR_i$	Multiplies ith row of A by $-c$ and adds it to jth row	$R_j - cR_i$
Permutation	Permutes the ith and jth rows of A	P_{ij}	Permutes the ith and jth rows of A	P_{ij}

THEOREM 2

Each elementary matrix is invertible. The inverse of an elementary matrix is a matrix of the same type. ■

NOTE: The inverse of an elementary matrix can be found by inspection. No computation is necessary.

THEOREM 3

A square matrix is invertible if and only if it is the product of elementary matrices.

PROOF: Let $A = E_1E_2 \cdots E_m$ where each E_i is an elementary matrix. By Theorem 2, each E_i is invertible. Moreover, by Theorem 6.8.3 on page 416 A is invertible† and

$$A^{-1} = E_m^{-1}E_{m-1}^{-1} \cdots E_2^{-1}E_1^{-1}$$

Conversely, suppose that A is invertible. According to Theorem 6.8.6 (the Summing Up Theorem), A is row equivalent to the identity matrix. This means that A can be reduced to I by a finite number, say, m, of elementary row operations. By Theorem 1 each such operation is accomplished by multiplying A on the left by an elementary matrix. This means that there are elementary matrices $E_1, E_2, \ldots, E_m$ such that

$$E_mE_{m-1} \cdots E_2E_1A = I,$$

Thus from Theorem 6.8.8 on page 428,

$$E_mE_{m-1} \cdots E_2E_1 = A^{-1}$$

† Here we have used the generalization of Theorem 6.8.3 to more than two matrices. See, for example, Problem 6.8.16 on page 430.

and since each E_i is invertible by Theorem 2,

$$A = (A^{-1})^{-1} = (E_m E_{m-1} \cdots E_2 E_1)^{-1} = E_1^{-1} E_2^{-1} \cdots E_{m-1}^{-1} E_m^{-1} \tag{7}$$

Since the inverse of an elementary matrix is an elementary matrix, we have written A as a product of elementary matrices and the proof is complete. ■

EXAMPLE 3

WRITING AN INVERTIBLE MATRIX AS THE PRODUCT OF ELEMENTARY MATRICES

Show that the matrix $A = \begin{pmatrix} 2 & 4 & 6 \\ 4 & 5 & 6 \\ 3 & 1 & -2 \end{pmatrix}$ is invertible and write it as a product of elementary matrices.

SOLUTION: We have encountered this matrix before, first in Example 6.3.1 on page 376. To solve the problem, we reduce A to I and keep track of the elementary row operations. In Example 6.8.6 on page 421 we did reduce A to I by using the following operations:

① $\frac{1}{2}R_1$ ② $R_2 - 4R_1$ ③ $R_3 - 3R_1$ ④ $-\frac{1}{3}R_2$
⑤ $R_1 - 2R_2$ ⑥ $R_3 + 5R_2$ ⑦ $-R_3$ ⑧ $R_1 + R_3$
⑨ $R_2 - 2R_3$

A^{-1} was obtained by starting with I and applying these nine elementary row operations. Thus A^{-1} is the product of nine elementary matrices:

$$A^{-1} = \underset{R_2 - 2R_3}{\begin{pmatrix} 1 & 0 & 0 \\ 0 & 1 & -2 \\ 0 & 0 & 1 \end{pmatrix}} \underset{R_1 + R_3}{\begin{pmatrix} 1 & 0 & 1 \\ 0 & 1 & 0 \\ 0 & 0 & 1 \end{pmatrix}} \underset{-R_3}{\begin{pmatrix} 1 & 0 & 0 \\ 0 & 1 & 0 \\ 0 & 0 & -1 \end{pmatrix}} \underset{R_3 + 5R_2}{\begin{pmatrix} 1 & 0 & 0 \\ 0 & 1 & 0 \\ 0 & 5 & 1 \end{pmatrix}} \underset{R_1 - 2R_2}{\begin{pmatrix} 1 & -2 & 0 \\ 0 & 1 & 0 \\ 0 & 0 & 1 \end{pmatrix}}$$

$$\times \underset{-\frac{1}{3}R_2}{\begin{pmatrix} 1 & 0 & 0 \\ 0 & -\frac{1}{3} & 0 \\ 0 & 0 & 1 \end{pmatrix}} \underset{R_3 - 3R_1}{\begin{pmatrix} 1 & 0 & 0 \\ 0 & 1 & 0 \\ -3 & 0 & 1 \end{pmatrix}} \underset{R_2 - 4R_1}{\begin{pmatrix} 1 & 0 & 0 \\ -4 & 1 & 0 \\ 0 & 0 & 1 \end{pmatrix}} \underset{\frac{1}{2}R_1}{\begin{pmatrix} \frac{1}{2} & 0 & 0 \\ 0 & 1 & 0 \\ 0 & 0 & 1 \end{pmatrix}}$$

Then $A = (A^{-1})^{-1} =$ the product of the inverses of the nine matrices in the opposite order:

$$\begin{pmatrix} 2 & 4 & 6 \\ 4 & 5 & 6 \\ 3 & 1 & -2 \end{pmatrix} = \underset{2R_1}{\begin{pmatrix} 2 & 0 & 0 \\ 0 & 1 & 0 \\ 0 & 0 & 1 \end{pmatrix}} \underset{R_2 + 4R_1}{\begin{pmatrix} 1 & 0 & 0 \\ 4 & 1 & 0 \\ 0 & 0 & 1 \end{pmatrix}} \underset{R_3 + 3R_1}{\begin{pmatrix} 1 & 0 & 0 \\ 0 & 1 & 0 \\ 3 & 0 & 1 \end{pmatrix}} \underset{-3R_2}{\begin{pmatrix} 1 & 0 & 0 \\ 0 & -3 & 0 \\ 0 & 0 & 1 \end{pmatrix}} \underset{R_1 + 2R_2}{\begin{pmatrix} 1 & 2 & 0 \\ 0 & 1 & 0 \\ 0 & 0 & 1 \end{pmatrix}}$$

$$\times \underset{R_3 - 5R_2}{\begin{pmatrix} 1 & 0 & 0 \\ 0 & 1 & 0 \\ 0 & -5 & 1 \end{pmatrix}} \underset{-R_3}{\begin{pmatrix} 1 & 0 & 0 \\ 0 & 1 & 0 \\ 0 & 0 & -1 \end{pmatrix}} \underset{R_1 - R_3}{\begin{pmatrix} 1 & 0 & -1 \\ 0 & 1 & 0 \\ 0 & 0 & 1 \end{pmatrix}} \underset{R_2 + 2R_3}{\begin{pmatrix} 1 & 0 & 0 \\ 0 & 1 & 2 \\ 0 & 0 & 1 \end{pmatrix}}$$

We can use Theorem 3 to extend our Summing Up Theorem, last seen on page 427.

THEOREM 4 SUMMING UP THEOREM—VIEW 3

Let A be an $n \times n$ matrix. Then the following seven statements are equivalent. That is, each one implies the other six (so that if one statement is true, all are true, and if one is false, all are false).

i. A is invertible.

ii. The only solution to the homogeneous system $A\mathbf{x} = \mathbf{0}$ is the trivial solution ($\mathbf{x} = \mathbf{0}$).

iii. The system $A\mathbf{x} = \mathbf{b}$ has a unique solution for every n-vector $\mathbf{b}$.

iv. A is row equivalent to the $n \times n$ identity matrix I_n; that is, the reduced row echelon form of A is I_n.

v. A can be written as a product of elementary matrices.

vi. The row echelon form of A has n pivots.

vii. $\det A \neq 0$ (so far, $\det A$ is defined only if A is a 2×2 matrix). ■

There is one further result that will prove very useful to us. First, we need a definition (given earlier in Problem 6.8.29 on page 431).

DEFINITION UPPER TRIANGULAR MATRIX AND LOWER TRIANGULAR MATRIX

A square matrix is called **upper (lower) triangular** if all its components below (above) the main diagonal are zero. ■

NOTE: a_{ij} is below the main diagonal if $i > j$.

EXAMPLE 4 TWO UPPER TRIANGULAR AND TWO LOWER TRIANGULAR MATRICES

Matrices U and V are upper triangular while matrices L and M are lower triangular:

$$U = \begin{pmatrix} 2 & -3 & 5 \\ 0 & 1 & 6 \\ 0 & 0 & 2 \end{pmatrix} \qquad V = \begin{pmatrix} 1 & 5 \\ 0 & -2 \end{pmatrix}$$

$$L = \begin{pmatrix} 0 & 0 \\ 5 & 1 \end{pmatrix} \qquad M = \begin{pmatrix} 2 & 0 & 0 & 0 \\ -5 & 4 & 0 & 0 \\ 6 & 1 & 2 & 0 \\ 3 & 0 & 1 & 5 \end{pmatrix}$$

THEOREM 5

Let A be a square matrix. Then A can be written as a product of elementary matrices and an upper triangular matrix U. In the product the elementary matrices are on the left and the upper triangular matrix is on the right.

PROOF: Gaussian elimination to solve the system $A\mathbf{x} = \mathbf{b}$ results in an upper triangular matrix. To see this, observe that Gaussian elimination will terminate when the matrix is in row echelon form—and the row echelon form of a square matrix is upper triangular. We denote the row echelon form of A by U. Then A is reduced to U by a sequence of elementary row operations each of which can be obtained by multiplication by an elementary matrix. Thus

$$U = E_m E_{m-1} \cdots E_2 E_1 A$$

and

$$A = E_1^{-1} E_2^{-1} \cdots E_{m-1}^{-1} E_m^{-1} U$$

Since the inverse of an elementary matrix is an elementary matrix, we have written A as the product of elementary matrices and U. ■

EXAMPLE 5

WRITING A MATRIX AS THE PRODUCT OF ELEMENTARY MATRICES AND AN UPPER TRIANGULAR MATRIX

Write the matrix

$$A = \begin{pmatrix} 3 & 6 & 9 \\ 2 & 5 & 1 \\ 1 & 1 & 8 \end{pmatrix}$$

as the product of elementary matrices and an upper triangular matrix.

SOLUTION: We row-reduce A to obtain its row echelon form:

$$\begin{pmatrix} 3 & 6 & 9 \\ 2 & 5 & 1 \\ 1 & 1 & 8 \end{pmatrix} \xrightarrow{R_1 \to \frac{1}{3}R_1} \begin{pmatrix} 1 & 2 & 3 \\ 2 & 5 & 1 \\ 1 & 1 & 8 \end{pmatrix}$$

$$\xrightarrow[R_3 \to R_3 - R_1]{R_2 \to R_2 - 2R_1} \begin{pmatrix} 1 & 2 & 3 \\ 0 & 1 & -5 \\ 0 & -1 & 5 \end{pmatrix} \xrightarrow{R_3 \to R_3 + R_2} \begin{pmatrix} 1 & 2 & 3 \\ 0 & 1 & -5 \\ 0 & 0 & 0 \end{pmatrix} = U$$

Then working backward, we see that

$$U = \begin{pmatrix} 1 & 2 & 3 \\ 0 & 1 & -5 \\ 0 & 0 & 0 \end{pmatrix} = \underset{R_3 + R_2}{\begin{pmatrix} 1 & 0 & 0 \\ 0 & 1 & 0 \\ 0 & 1 & 1 \end{pmatrix}} \underset{R_3 - R_1}{\begin{pmatrix} 1 & 0 & 0 \\ 0 & 1 & 0 \\ -1 & 0 & 1 \end{pmatrix}}$$

$$\times \underset{R_2 - 2R_1}{\begin{pmatrix} 1 & 0 & 0 \\ -2 & 1 & 0 \\ 0 & 0 & 1 \end{pmatrix}} \underset{\frac{1}{3}R_1}{\begin{pmatrix} \frac{1}{3} & 0 & 0 \\ 0 & 1 & 0 \\ 0 & 0 & 1 \end{pmatrix}} \underset{A}{\begin{pmatrix} 3 & 6 & 9 \\ 2 & 5 & 1 \\ 1 & 1 & 8 \end{pmatrix}}$$

and, taking inverses of the four elementary matrices, we obtain

$$A = \begin{pmatrix} 3 & 6 & 9 \\ 2 & 5 & 1 \\ 1 & 1 & 8 \end{pmatrix} = \underset{3R_1}{\begin{pmatrix} 3 & 0 & 0 \\ 0 & 1 & 0 \\ 0 & 0 & 1 \end{pmatrix}} \underset{R_2 + 2R_1}{\begin{pmatrix} 1 & 0 & 0 \\ 2 & 1 & 0 \\ 0 & 0 & 1 \end{pmatrix}}$$

$$\times \underset{R_3 + R_1}{\begin{pmatrix} 1 & 0 & 0 \\ 0 & 1 & 0 \\ 1 & 0 & 1 \end{pmatrix}} \underset{R_3 - R_2}{\begin{pmatrix} 1 & 0 & 0 \\ 0 & 1 & 0 \\ 0 & -1 & 1 \end{pmatrix}} \underset{U}{\begin{pmatrix} 1 & 2 & 3 \\ 0 & 1 & -5 \\ 0 & 0 & 0 \end{pmatrix}}$$

PROBLEMS 6.10

SELF-QUIZ

True-False

I. The product of two elementary matrices is an elementary matrix.

II. The inverse of an elementary matrix is an elementary matrix.

III. Every matrix can be written as the product of elementary matrices.

IV. Every square matrix can be written as the product of elementary matrices.

V. Every invertible matrix can be written as the product of elementary matrices.

VI. Every square matrix can be written as the product of elementary matrices and an upper triangular matrix.

Multiple Choice

VII. The inverse of $\begin{pmatrix} 1 & 0 & 0 \\ 0 & 1 & 0 \\ 0 & 3 & 1 \end{pmatrix}$ is ________.

a. $\begin{pmatrix} 1 & 0 & 0 \\ 0 & 1 & 0 \\ 0 & -3 & 1 \end{pmatrix}$ **b.** $\begin{pmatrix} 1 & 0 & 0 \\ 0 & 1 & 0 \\ 0 & \frac{1}{3} & 1 \end{pmatrix}$

c. $\begin{pmatrix} 1 & -3 & 0 \\ 0 & 1 & 0 \\ 0 & 0 & 1 \end{pmatrix}$ **d.** $\begin{pmatrix} 1 & 0 & 0 \\ 0 & 1 & 0 \\ 0 & 3 & 1 \end{pmatrix}$

VIII. The inverse of $\begin{pmatrix} 1 & 0 & 0 \\ 0 & 1 & 0 \\ 0 & 0 & 4 \end{pmatrix}$ is ________.

a. $\begin{pmatrix} 1 & 0 & 0 \\ 0 & 1 & 0 \\ 0 & 0 & -4 \end{pmatrix}$ **b.** $\begin{pmatrix} 1 & 0 & 0 \\ 0 & 1 & 0 \\ 0 & 0 & \frac{1}{4} \end{pmatrix}$

c. $\begin{pmatrix} \frac{1}{4} & 0 & 0 \\ 0 & 1 & 0 \\ 0 & 0 & 1 \end{pmatrix}$ **d.** $\begin{pmatrix} 1 & 0 & 0 \\ 0 & 1 & 0 \\ 0 & 0 & 4 \end{pmatrix}$

IX. The inverse of $\begin{pmatrix} 0 & 1 & 0 \\ 1 & 0 & 0 \\ 0 & 0 & 1 \end{pmatrix}$ is ________.

a. $\begin{pmatrix} 0 & 1 & 0 \\ 1 & 0 & 0 \\ 0 & 0 & -1 \end{pmatrix}$ **b.** $\begin{pmatrix} 0 & -1 & 0 \\ -1 & 0 & 0 \\ 0 & 0 & -1 \end{pmatrix}$

c. $\begin{pmatrix} 1 & 0 & 0 \\ 0 & 0 & 1 \\ 0 & 1 & 0 \end{pmatrix}$ **d.** $\begin{pmatrix} 0 & 1 & 0 \\ 1 & 0 & 0 \\ 0 & 0 & 1 \end{pmatrix}$

In Problems 1–12 determine which matrices are elementary matrices.

1. $\begin{pmatrix} 0 & 1 \\ 1 & 0 \end{pmatrix}$ **2.** $\begin{pmatrix} 1 & 0 \\ 1 & 1 \end{pmatrix}$

3. $\begin{pmatrix} 0 & 1 \\ 1 & 1 \end{pmatrix}$ **4.** $\begin{pmatrix} 1 & 0 \\ 0 & 2 \end{pmatrix}$

5. $\begin{pmatrix} 3 & 0 \\ 0 & 3 \end{pmatrix}$ **6.** $\begin{pmatrix} 0 & 1 & 0 \\ 1 & 0 & 0 \\ 0 & 0 & 1 \end{pmatrix}$

7. $\begin{pmatrix} 0 & 1 & 0 \\ 0 & 0 & 1 \\ 1 & 0 & 0 \end{pmatrix}$ **8.** $\begin{pmatrix} 1 & 0 & 0 \\ 2 & 1 & 0 \\ 3 & 0 & 1 \end{pmatrix}$

9. $\begin{pmatrix} 1 & 0 & 0 \\ 2 & 1 & 0 \\ 0 & 0 & 1 \end{pmatrix}$ **10.** $\begin{pmatrix} 1 & 0 & 0 & 0 \\ 0 & 1 & 0 & 0 \\ 0 & 0 & 1 & 0 \\ 0 & 1 & 0 & 1 \end{pmatrix}$

11. $\begin{pmatrix} 1 & 0 & 0 & 0 \\ 1 & 1 & 0 & 0 \\ 0 & 0 & 1 & 0 \\ 0 & 0 & 1 & 1 \end{pmatrix}$ **12.** $\begin{pmatrix} 1 & -1 & 0 & 0 \\ 0 & 1 & 0 & 0 \\ 0 & 0 & 1 & 0 \\ 0 & 0 & 0 & 1 \end{pmatrix}$

In Problems 13–20 write the 3×3 elementary matrix that carries out the given row operation on a 3×5 matrix A by left multiplication.

13. $R_2 \to 4R_2$ **14.** $R_2 \to R_2 + 2R_1$

15. $R_1 \to R_1 - 3R_2$ **16.** $R_1 \to R_1 + 4R_3$

17. $R_1 \rightleftarrows R_3$ **18.** $R_2 \rightleftarrows R_3$

19. $R_2 \to R_2 + R_3$ **20.** $R_3 \to -R_3$

In Problems 21–30 find the elementary matrix E such that $EA = B$.

21. $A = \begin{pmatrix} 2 & 3 \\ -1 & 4 \end{pmatrix}$, $B = \begin{pmatrix} 2 & 3 \\ 2 & -8 \end{pmatrix}$

22. $A = \begin{pmatrix} 2 & 3 \\ -1 & 4 \end{pmatrix}$, $B = \begin{pmatrix} 2 & 3 \\ -5 & -2 \end{pmatrix}$

23. $A = \begin{pmatrix} 2 & 3 \\ -1 & 4 \end{pmatrix}$, $B = \begin{pmatrix} 0 & 11 \\ -1 & 4 \end{pmatrix}$

24. $A = \begin{pmatrix} 2 & 3 \\ -1 & 4 \end{pmatrix}$, $B = \begin{pmatrix} -1 & 4 \\ 2 & 3 \end{pmatrix}$

25. $A = \begin{pmatrix} 1 & 2 \\ 3 & 4 \\ 5 & 6 \end{pmatrix}$, $B = \begin{pmatrix} 5 & 6 \\ 3 & 4 \\ 1 & 2 \end{pmatrix}$

26. $A = \begin{pmatrix} 1 & 2 \\ 3 & 4 \\ 5 & 6 \end{pmatrix}$, $B = \begin{pmatrix} 1 & 2 \\ 0 & -2 \\ 5 & 6 \end{pmatrix}$

27. $A = \begin{pmatrix} 1 & 2 \\ 3 & 4 \\ 5 & 6 \end{pmatrix}$, $B = \begin{pmatrix} -1 & -2 \\ 3 & 4 \\ 5 & 6 \end{pmatrix}$

28. $A = \begin{pmatrix} 1 & 2 \\ 3 & 4 \\ 5 & 6 \end{pmatrix}$, $B = \begin{pmatrix} -5 & -6 \\ 3 & 4 \\ 5 & 6 \end{pmatrix}$

29. $A = \begin{pmatrix} 1 & 2 & 5 & 2 \\ 0 & -1 & 3 & 4 \\ 5 & 0 & -2 & 7 \end{pmatrix}$,

$B = \begin{pmatrix} 1 & 2 & 5 & 2 \\ 0 & -1 & 3 & 4 \\ 0 & -10 & -27 & -3 \end{pmatrix}$

30. $A = \begin{pmatrix} 1 & 2 & 5 & 2 \\ 0 & -1 & 3 & 4 \\ 5 & 0 & -2 & 7 \end{pmatrix}$,

$B = \begin{pmatrix} 1 & 0 & 11 & 10 \\ 0 & -1 & 3 & 4 \\ 5 & 0 & -2 & 7 \end{pmatrix}$

In Problems 31–40 find the inverse of the given elementary matrix.

31. $\begin{pmatrix} 0 & 1 \\ 1 & 0 \end{pmatrix}$

32. $\begin{pmatrix} 1 & 3 \\ 0 & 1 \end{pmatrix}$

33. $\begin{pmatrix} 1 & 0 \\ 0 & 4 \end{pmatrix}$

34. $\begin{pmatrix} 0 & 1 & 0 \\ 1 & 0 & 0 \\ 0 & 0 & 1 \end{pmatrix}$

35. $\begin{pmatrix} 1 & -2 & 0 \\ 0 & 1 & 0 \\ 0 & 0 & 1 \end{pmatrix}$

36. $\begin{pmatrix} 1 & 0 & 0 \\ 0 & 1 & 0 \\ -2 & 0 & 1 \end{pmatrix}$

37. $\begin{pmatrix} 1 & 0 & 0 \\ 0 & -\frac{1}{2} & 0 \\ 0 & 0 & 1 \end{pmatrix}$

38. $\begin{pmatrix} 1 & 0 & 1 & 0 \\ 0 & 1 & 0 & 0 \\ 0 & 0 & 1 & 0 \\ 0 & 0 & 0 & 1 \end{pmatrix}$

39. $\begin{pmatrix} 1 & 0 & 0 & 5 \\ 0 & 1 & 0 & 0 \\ 0 & 0 & 1 & 0 \\ 0 & 0 & 0 & 1 \end{pmatrix}$

40. $\begin{pmatrix} 1 & 0 & 0 & 0 \\ 0 & 1 & 0 & 0 \\ 0 & -3 & 1 & 0 \\ 0 & 0 & 0 & 1 \end{pmatrix}$

In Problems 41–48 show that each matrix is invertible and write it as a product of elementary matrices.

41. $\begin{pmatrix} 2 & 1 \\ 3 & 2 \end{pmatrix}$

42. $\begin{pmatrix} 1 & 2 \\ 3 & 4 \end{pmatrix}$

43. $\begin{pmatrix} 1 & 1 & 1 \\ 0 & 2 & 3 \\ 5 & 5 & 1 \end{pmatrix}$

44. $\begin{pmatrix} 3 & 2 & 1 \\ 0 & 2 & 2 \\ 0 & 0 & -1 \end{pmatrix}$

45. $\begin{pmatrix} 0 & -1 & 0 \\ 0 & 1 & -1 \\ 1 & 0 & 1 \end{pmatrix}$

46. $\begin{pmatrix} 2 & 0 & 4 \\ 0 & 1 & 1 \\ 3 & -1 & 1 \end{pmatrix}$

47. $\begin{pmatrix} 2 & 0 & 0 & 0 \\ 0 & 3 & 0 & 0 \\ 0 & 0 & -4 & 0 \\ 0 & 0 & 0 & 5 \end{pmatrix}$

48. $\begin{pmatrix} 2 & 1 & 0 & 0 \\ 0 & 2 & 1 & 0 \\ 0 & 0 & 2 & 1 \\ 0 & 0 & 0 & 2 \end{pmatrix}$

49. Let $A = \begin{pmatrix} a & b \\ 0 & c \end{pmatrix}$ where $ac \neq 0$. Write A as a product of three elementary matrices and conclude that A is invertible.

50. Let $A = \begin{pmatrix} a & b & c \\ 0 & d & e \\ 0 & 0 & f \end{pmatrix}$ where $adf \neq 0$. Write A as a product of six elementary matrices and conclude that A is invertible.

***51.** Let A be an $n \times n$ upper triangular matrix. Prove that if each diagonal component of A is nonzero, then A is invertible [*Hint:* Look at Problems 49 and 50].

***52.** Show that if A is an $n \times n$ upper triangular matrix with nonzero diagonal components, then A^{-1} is upper triangular.

***53.** Use Theorem 6.9.1(*iv*) on page 433 and the result of Problem 52 to show that if A is an $n \times n$ lower triangular matrix with nonzero diagonal components, then A is invertible and A^{-1} is lower triangular.

54. Show that if P_{ij} is the $n \times n$ matrix obtained by permuting the ith and jth rows of I_n, then $P_{ij}A$ is the matrix obtained from A by permuting its ith and jth rows.

55. Let A_{ij} be the matrix with c in the jith position, 1's down the diagonal, and 0's everywhere else.

Show that $A_{ij}A$ is the matrix obtained from A by multiplying the ith row of A by c and adding it to the jth row.

56. Let M_i be the matrix with c in the ii position, 1's in the other diagonal positions, and 0's everywhere else. Show that M_iA is the matrix obtained from A by multiplying the ith row of A by c.

In Problems 57–62 write each square matrix as a product of elementary matrices and an upper triangular matrix.

57. $A = \begin{pmatrix} 1 & 2 \\ 2 & 4 \end{pmatrix}$

58. $A = \begin{pmatrix} 2 & -3 \\ -4 & 6 \end{pmatrix}$

59. $A = \begin{pmatrix} 0 & 0 \\ 1 & 0 \end{pmatrix}$

60. $A = \begin{pmatrix} 1 & -1 & 2 \\ 2 & 1 & 4 \\ 4 & -1 & 8 \end{pmatrix}$

61. $A = \begin{pmatrix} 1 & -3 & 3 \\ 0 & -3 & 1 \\ 1 & 0 & 2 \end{pmatrix}$

62. $A = \begin{pmatrix} 1 & 0 & 0 \\ 2 & 3 & 0 \\ -1 & 4 & 0 \end{pmatrix}$

ANSWERS TO SELF-QUIZ

I. False **II.** True **III.** False **IV.** False **V.** True **VI.** True
VII. a **VIII.** b **IX.** d

CHAPTER 6 SUMMARY OUTLINE

- An **n-component row vector** is an ordered set of n numbers written as $(x_1, x_2, \ldots, x_n)$.
- An **n-component column vector** is an ordered set of n numbers written as

$$\begin{pmatrix} x_1 \\ x_2 \\ \vdots \\ x_n \end{pmatrix}$$

- A vector all of whose components are zero is called a **zero vector**.
- **Vector addition** and **multiplication by scalars** are defined by

$$\mathbf{a} + \mathbf{b} = \begin{pmatrix} a_1 + b_1 \\ a_2 + b_2 \\ \vdots \\ a_n + b_n \end{pmatrix} \quad \text{and} \quad \alpha\mathbf{a} = \begin{pmatrix} \alpha a_1 \\ \alpha a_2 \\ \vdots \\ \alpha a_n \end{pmatrix}$$

- An **$m \times n$ matrix** is a rectangular array of mn numbers arranged in m rows and n columns:

$$A = \begin{pmatrix} a_{11} & a_{12} & \cdots & a_{1n} \\ a_{21} & a_{22} & \cdots & a_{2n} \\ \vdots & \vdots & & \vdots \\ a_{m1} & a_{m2} & \cdots & a_{mn} \end{pmatrix}$$

The matrix A is also written as $A = (a_{ij})$.

- A matrix all of whose components are zero is called a **zero matrix**.
- If A and B are $m \times n$ matrices, then $A + B$ and αA (α a scalar) are $m \times n$ matrices:

  The ijth component of $A + B$ is $a_{ij} + b_{ij}$.

  The ijth component of αA is αa_{ij}.

- The **Scalar product** of two n-component vectors:

$$\mathbf{a} \cdot \mathbf{b} = (a_1\ a_2 \cdots a_n) \cdot \begin{pmatrix} b_1 \\ b_2 \\ \vdots \\ b_n \end{pmatrix}$$

$$= a_1b_1 + a_2b_2 + \cdots + a_nb_n = \sum_{i=1}^{n} a_ib_i$$

- **Matrix Products** Let A be an $m \times n$ matrix and let B be an $n \times p$ matrix. Then AB is an $m \times p$ matrix and

  ijth component of AB

  $= (i\text{th row of } A) \cdot (j\text{th column of } B)$

$$= a_{i1}b_{1j} + a_{i2}b_{2j} + \cdots + a_{in}b_{nj} = \sum_{k=1}^{n} a_{ik}b_{kj}$$

- Matrix products do not in general commute; that is, it is usually the case that $AB \neq BA$.
- **Associative Law for Matrix Multiplication** If A is an $n \times m$ matrix, B is $m \times p$ and C is $p \times q$, then

  $A(BC) = (AB)C$

  and both $A(BC)$ and $(AB)C$ are $n \times q$ matrices.

- **Distributive Laws for Matrix Multiplication** If all the products are defined, then

$$A(B+C)=AB+AC \quad \text{and} \quad (A+B)C=AC+BC$$

- The **coefficient matrix** of the system

$$\begin{aligned} a_{11}x_1 + a_{12}x_2 + \cdots + a_{1n}x_n &= b_1 \\ a_{21}x_1 + a_{22}x_2 + \cdots + a_{2n}x_n &= b_2 \\ \vdots \quad\quad \vdots \quad\quad\quad\quad \vdots \quad &\quad \vdots \\ a_{m1}x_1 + a_{m2}x_2 + \cdots + a_{mn}x_n &= b_m \end{aligned}$$

is the matrix

$$A = \begin{pmatrix} a_{11} & a_{12} & \cdots & a_{1n} \\ a_{21} & a_{22} & \cdots & a_{2n} \\ \vdots & \vdots & & \vdots \\ a_{m1} & a_{m2} & \cdots & a_{mn} \end{pmatrix}$$

- The system above can be written using the **augmented matrix**

$$\left(\begin{array}{cccc|c} a_{11} & a_{12} & \cdots & a_{1n} & b_1 \\ a_{21} & a_{22} & \cdots & a_{2n} & b_2 \\ \vdots & \vdots & & \vdots & \vdots \\ a_{m1} & a_{m2} & \cdots & a_{mn} & b_m \end{array}\right)$$

It can also be written as $A\mathbf{x} = \mathbf{b}$, where

$$\mathbf{x} = \begin{pmatrix} x_1 \\ x_2 \\ \vdots \\ x_n \end{pmatrix} \quad \text{and} \quad \mathbf{b} = \begin{pmatrix} b_1 \\ b_2 \\ \vdots \\ b_m \end{pmatrix}$$

- A matrix is in **reduced row echelon form** if the four conditions on page 381 hold.
- A matrix is in **row echelon form** if the first three conditions on page 381 hold.
- The three **elementary row operations** are:
Multiply the ith row of a matrix by c: $R_i \to cR_i$, where $c \neq 0$.
Multiply the ith row by c and add it to the jth row: $R_j \to R_j + cR_i$, where $c \neq 0$.
Permute the ith and jth rows: $R_i \rightleftarrows R_j$.
- The process of applying elementary row operations to a matrix is called **row reduction**.
- **Gauss-Jordan elimination** is the process of solving a system of equations by row-reducing its augmented matrix to its reduced row echelon form.
- **Gaussian elimination** is the process of solving a system of equations by row-reducing its augmented matrix to row echelon form and using **back substitution**.
- A system that has one or more solutions is called **consistent**.
- A system that has no solution is called **inconsistent**.
- A system having solutions has either a unique solution or an infinite number of solutions.
- A **homogeneous** system of m equations in n unknowns is a system of the form

$$\begin{aligned} a_{11}x_1 + a_{12}x_2 + \cdots + a_{1n}x_n &= 0 \\ a_{21}x_1 + a_{22}x_2 + \cdots + a_{2n}x_n &= 0 \\ \vdots \quad\quad \vdots \quad\quad\quad\quad \vdots \quad &\quad \vdots \\ a_{m1}x_1 + a_{m2}x_2 + \cdots + a_{mn}x_n &= 0 \end{aligned}$$

- A homogeneous system always has the **trivial solution** (or **zero solution**)

$$x_1 = x_2 = \cdots = x_n = 0$$

- Solutions other than the zero solution to a homogeneous system are called **nontrivial solutions**.
- The homogeneous system above has an infinite number of solutions if there are more unknowns than equations ($n > m$).
- The $\boldsymbol{n \times n}$ **identity matrix**, $\boldsymbol{I_n}$, is the $n \times n$ matrix with 1's down the **main diagonal** and 0's everywhere else. I_n is usually denoted by I.
- If A is a square matrix, then $AI = IA = A$.
- The $n \times n$ matrix A is **invertible** if there is an $n \times n$ matrix A^{-1} such that

$$AA^{-1} = A^{-1}A = I$$

In this case the A^{-1} is called the **inverse** of A.
- If A is invertible, the inverse is unique.
- If A and B are invertible $n \times n$ matrices, then AB is invertible and

$$(AB)^{-1} = B^{-1}A^{-1}$$

- To determine whether an $n \times n$ matrix A is invertible:
  - **i.** Write the augmented matrix $(A|I)$.
  - **ii.** Use row reduction to reduce A to its reduced row echelon form.
  - **iii. a.** If the reduced row echelon form of A is I, then A^{-1} will be the matrix to the right of the vertical bar.
    - **b.** If the reduced row echelon form of A contains a row of zeros, then A is not invertible.
- If $A = \begin{pmatrix} a_{11} & a_{12} \\ a_{21} & a_{22} \end{pmatrix}$, then A is invertible if and only if

$$\text{determinant of } A = \det A = a_{11}a_{22} - a_{12}a_{21} \neq 0$$

In that case

$$A^{-1} = \frac{1}{\det A}\begin{pmatrix} a_{22} & -a_{12} \\ -a_{21} & a_{11} \end{pmatrix}$$

- Two matrices A and B are **row equivalent** if A can be transformed into B by row reduction.
- Let A be an $n \times n$ matrix. If $AB = I$ or $BA = I$, then A is invertible and $B = A^{-1}$.

- If $A = (a_{ij})$, then the **transpose of A**, written A^t, is given by $A^t = (a_{ji})$. That is, A^t is obtained by interchanging the rows and columns of A.
- **Facts about Transpose** If all sums and products are defined and if A is invertible, then

$$(A^t)^t = A \qquad (AB)^t = B^tA^t$$
$$(A + B)^t = A^t + B^t \qquad (A^t)^{-1} = (A^{-1})^t$$

- A square matrix A is **symmetric** if $A^t = A$.
- An **elementary matrix** is a square matrix obtained by performing exactly one of the elementary row operations on the identity matrix. The three types of elementary matrices are:

cR_i	multiply the ith row of I by c, $c \neq 0$
$R_j + cR_i$	multiply the ith row of I by c and add it to the jth row, $c \neq 0$
P_{ij}	permute the ith and jth rows

- A square matrix is invertible if and only if it is the product of elementary matrices.
- Any square matrix can be written as the product of elementary matrices and one upper triangular matrix.
- **Summing Up Theorem** Let A be an $n \times n$ matrix. Then the following are equivalent:
  - **i.** A is invertible.
  - **ii.** The only solution to the homogeneous system $A\mathbf{x} = \mathbf{0}$ is the trivial solution ($\mathbf{x} = \mathbf{0}$).
  - **iii.** The system $A\mathbf{x} = \mathbf{b}$ has a unique solution for every n-vector $\mathbf{b}$.
  - **iv.** A is row equivalent to the $n \times n$ identity matrix I_n.
  - **v.** A can be written as a product of elementary matrices.
  - **vi.** $\det A \neq 0$ (so far, $\det A$ is defined only if A is a 2×2 matrix).

CHAPTER 6 REVIEW EXERCISES

In Exercises 1–14 find all solutions (if any) to the given systems.

1. $\begin{aligned} 3x_1 + 6x_2 &= 9 \\ -2x_1 + 3x_2 &= 4 \end{aligned}$

2. $\begin{aligned} 3x_1 + 6x_2 &= 9 \\ 2x_1 + 4x_2 &= 6 \end{aligned}$

3. $\begin{aligned} 3x_1 - 6x_2 &= 9 \\ -2x_1 + 4x_2 &= 6 \end{aligned}$

4. $\begin{aligned} x_1 + x_2 + x_3 &= 2 \\ 2x_1 - x_2 + 2x_3 &= 4 \\ -3x_1 + 2x_2 + 3x_3 &= 8 \end{aligned}$

5. $\begin{aligned} x_1 + x_2 + x_3 &= 0 \\ 2x_1 - x_2 + 2x_3 &= 0 \\ -3x_1 + 2x_2 + 3x_3 &= 0 \end{aligned}$

6. $\begin{aligned} x_1 + x_2 + x_3 &= 2 \\ 2x_1 - x_2 + 2x_3 &= 4 \\ -x_1 + 4x_2 + x_3 &= 2 \end{aligned}$

7. $\begin{aligned} x_1 + x_2 + x_3 &= 2 \\ 2x_1 - x_2 + 2x_3 &= 4 \\ -x_1 + 4x_2 + x_3 &= 3 \end{aligned}$

8. $\begin{aligned} x_1 + x_2 + x_3 &= 0 \\ 2x_1 - x_2 + 2x_3 &= 0 \\ -x_1 + 4x_2 + x_3 &= 0 \end{aligned}$

9. $\begin{aligned} 2x_1 + x_2 - 3x_3 &= 0 \\ 4x_1 - x_2 + x_3 &= 0 \end{aligned}$

10. $\begin{aligned} x_1 + x_2 &= 0 \\ 2x_1 + x_2 &= 0 \\ 3x_1 + x_2 &= 0 \end{aligned}$

11. $\begin{aligned} x_1 + x_2 &= 1 \\ 2x_1 + x_2 &= 3 \\ 3x_1 + x_2 &= 4 \end{aligned}$

12. $\begin{aligned} x_1 + x_2 + x_3 + x_4 &= 4 \\ 2x_1 - 3x_2 - x_3 + 4x_4 &= 7 \\ -2x_1 + 4x_2 + x_3 - 2x_4 &= 1 \\ 5x_1 - x_2 + 2x_3 + x_4 &= -1 \end{aligned}$

13. $\begin{aligned} x_1 + x_2 + x_3 + x_4 &= 0 \\ 2x_1 - 3x_2 - x_3 + 4x_4 &= 0 \\ -2x_1 + 4x_2 + x_3 - 2x_4 &= 0 \\ 5x_1 - x_2 + 2x_3 + x_4 &= 0 \end{aligned}$

14. $\begin{aligned} x_1 + x_2 + x_3 + x_4 &= 0 \\ 2x_1 - 3x_2 - x_3 + 4x_4 &= 0 \\ -2x_1 + 4x_2 + x_3 - 2x_4 &= 0 \end{aligned}$

In Exercises 15–22 perform the indicated computations.

15. $3\begin{pmatrix} -2 & 1 \\ 0 & 4 \\ 2 & 3 \end{pmatrix}$

16. $\begin{pmatrix} 1 & 0 & 3 \\ 2 & -1 & 6 \end{pmatrix} + \begin{pmatrix} 2 & 0 & 4 \\ -2 & 5 & 8 \end{pmatrix}$

17. $5\begin{pmatrix} 2 & 1 & 3 \\ -1 & 2 & 4 \\ -6 & 1 & 5 \end{pmatrix} - 3\begin{pmatrix} -2 & 1 & 4 \\ 5 & 0 & 7 \\ 2 & -1 & 3 \end{pmatrix}$

18. $\begin{pmatrix} 2 & 3 \\ -1 & 4 \end{pmatrix}\begin{pmatrix} 5 & -1 \\ 2 & 7 \end{pmatrix}$

19. $\begin{pmatrix} 2 & 3 & 1 & 5 \\ 0 & 6 & 2 & 4 \end{pmatrix}\begin{pmatrix} 5 & 7 & 1 \\ 2 & 0 & 3 \\ 1 & 0 & 0 \\ 0 & 5 & 6 \end{pmatrix}$

20. $\begin{pmatrix} 2 & 3 & 5 \\ -1 & 6 & 4 \\ 1 & 0 & 6 \end{pmatrix}\begin{pmatrix} 0 & -1 & 2 \\ 3 & 1 & 2 \\ -7 & 3 & 5 \end{pmatrix}$

21. $\begin{pmatrix} 1 & 0 & 3 & -1 & 5 \\ 2 & 1 & 6 & 2 & 5 \end{pmatrix}\begin{pmatrix} 7 & 1 \\ 2 & 3 \\ -1 & 0 \\ 5 & 6 \\ 2 & 3 \end{pmatrix}$

22. $\begin{pmatrix} 1 & -1 & 2 \\ 3 & 5 & 6 \\ 2 & 4 & -1 \end{pmatrix}\begin{pmatrix} 2 \\ 1 \\ 3 \end{pmatrix}$

In Exercises 23–26 determine whether the given matrix is in row echelon form (but not reduced row echelon form), reduced row echelon form, or neither.

23. $\begin{pmatrix} 1 & 0 & 0 & 0 \\ 0 & 1 & 0 & 2 \\ 0 & 0 & 1 & 3 \end{pmatrix}$ **24.** $\begin{pmatrix} 1 & 8 & 1 & 0 \\ 0 & 1 & 5 & -7 \\ 0 & 0 & 1 & 4 \end{pmatrix}$

25. $\begin{pmatrix} 1 & 0 \\ 0 & 3 \\ 0 & 0 \end{pmatrix}$ **26.** $\begin{pmatrix} 1 & 0 & 2 & 0 \\ 0 & 1 & 3 & 0 \end{pmatrix}$

In Exercises 27 and 28 reduce the matrix to row echelon form and reduced row echelon form.

27. $\begin{pmatrix} 2 & 8 & -2 \\ 1 & 0 & -6 \end{pmatrix}$ **28.** $\begin{pmatrix} 1 & -1 & 2 & 4 \\ -1 & 2 & 0 & 3 \\ 2 & 3 & -1 & 1 \end{pmatrix}$

In Exercises 29–33 calculate the row echelon form and the inverse of the given matrix (if the inverse exists).

29. $\begin{pmatrix} 2 & 3 \\ -1 & 4 \end{pmatrix}$ **30.** $\begin{pmatrix} -1 & 2 \\ 2 & -4 \end{pmatrix}$

31. $\begin{pmatrix} 1 & 2 & 0 \\ 2 & 1 & -1 \\ 3 & 1 & 1 \end{pmatrix}$ **32.** $\begin{pmatrix} -1 & 2 & 0 \\ 4 & 1 & -3 \\ 2 & 5 & -3 \end{pmatrix}$

33. $\begin{pmatrix} 2 & 0 & 4 \\ -1 & 3 & 1 \\ 0 & 1 & 2 \end{pmatrix}$

In Exercises 34–36 first write the system in the form $A\mathbf{x} = \mathbf{b}$, then calculate A^{-1}, and, finally, use matrix multiplication to obtain the solution vector.

34. $\begin{aligned} x_1 - 3x_2 &= 4 \\ 2x_1 + 5x_2 &= 7 \end{aligned}$

35. $\begin{aligned} x_1 + 2x_2 \qquad &= 3 \\ 2x_1 + x_2 - x_3 &= -1 \\ 3x_1 + x_2 + x_3 &= 7 \end{aligned}$

36. $\begin{aligned} 2x_1 \qquad + 4x_3 &= 7 \\ -x_1 + 3x_2 + x_3 &= -4 \\ x_2 + 2x_3 &= 5 \end{aligned}$

In Exercises 37–42 calculate the transpose of the given matrix and determine whether the matrix is symmetric or skew-symmetric.†

37. $\begin{pmatrix} 2 & 3 & 1 \\ -1 & 0 & 2 \end{pmatrix}$ **38.** $\begin{pmatrix} 4 & 6 \\ 6 & 4 \end{pmatrix}$

39. $\begin{pmatrix} 2 & 3 & 1 \\ 3 & -6 & -5 \\ 1 & -5 & 9 \end{pmatrix}$ **40.** $\begin{pmatrix} 0 & 5 & 6 \\ -5 & 0 & 4 \\ -6 & -4 & 0 \end{pmatrix}$

41. $\begin{pmatrix} 1 & -1 & 4 & 6 \\ -1 & 2 & 5 & 7 \\ 4 & 5 & 3 & -8 \\ 6 & 7 & -8 & 9 \end{pmatrix}$ **42.** $\begin{pmatrix} 0 & 1 & -1 & 1 \\ -1 & 0 & 1 & -2 \\ 1 & 1 & 0 & 1 \\ 1 & -2 & -1 & 0 \end{pmatrix}$

In Exercises 43–47 find a 3×3 elementary matrix that will carry out the given row operation.

43. $R_2 \to -2R_2$
44. $R_1 \to R_1 + 2R_2$
45. $R_3 \to R_3 - 5R_1$
46. $R_3 \rightleftarrows R_1$
47. $R_2 \to R_2 + \frac{1}{5}R_3$

In Exercises 48–50 find the inverse of the elementary matrix.

48. $\begin{pmatrix} 1 & 3 \\ 0 & 1 \end{pmatrix}$

49. $\begin{pmatrix} 0 & 1 & 0 \\ 1 & 0 & 0 \\ 0 & 0 & 1 \end{pmatrix}$

50. $\begin{pmatrix} 1 & 0 & 0 \\ 0 & 1 & 0 \\ 0 & 0 & -\frac{1}{3} \end{pmatrix}$

In Exercises 51 and 52 write the matrix as the product of elementary matrices.

51. $\begin{pmatrix} 2 & -1 \\ -1 & 1 \end{pmatrix}$

52. $\begin{pmatrix} 1 & 0 & 3 \\ 2 & 1 & -5 \\ 3 & 2 & 4 \end{pmatrix}$

In Exercises 53 and 54 write each matrix as the product of elementary matrices and one upper triangular matrix.

53. $\begin{pmatrix} 2 & -1 \\ -4 & 2 \end{pmatrix}$

54. $\begin{pmatrix} 1 & -2 & 3 \\ 2 & 0 & 4 \\ 1 & 2 & 1 \end{pmatrix}$

† From Problem 6.9.18 on page 435 we have: A is skew-symmetric if $A^t = -A$.

CHAPTER 7

DETERMINANTS

7.1 DEFINITIONS

Let $A = \begin{pmatrix} a_{11} & a_{12} \\ a_{21} & a_{22} \end{pmatrix}$ be a 2×2 matrix. In Section 6.2 on page 374 we defined the determinant of A by

$$\det A = a_{11}a_{22} - a_{12}a_{21} \tag{1}$$

We shall often denote $\det A$ by

$$|A| = \begin{vmatrix} a_{11} & a_{12} \\ a_{21} & a_{22} \end{vmatrix} \tag{2}$$

We showed that A is invertible if and only if $\det A \neq 0$. As we shall see, this important theorem is valid for $n \times n$ matrices.

In this chapter we develop some of the basic properties of determinants and see how they can be used to calculate inverses and solve systems of n linear equations in n unknowns.

We define the determinant of an $n \times n$ matrix *inductively*. In other words, we use our knowledge of a 2×2 determinant to define a 3×3 determinant, use this to define a 4×4 determinant, and so on. We start by defining a 3×3 determinant.[†]

DEFINITION 3×3 DETERMINANT

Let $A = \begin{pmatrix} a_{11} & a_{12} & a_{13} \\ a_{21} & a_{22} & a_{23} \\ a_{31} & a_{32} & a_{33} \end{pmatrix}$. Then

[†]There are several ways to define a determinant and this is one of them. It is important to realize that "det" is a function which assigns a *number* to a *square* matrix.

$$\det A = |A| = a_{11}\begin{vmatrix} a_{22} & a_{23} \\ a_{32} & a_{33} \end{vmatrix} - a_{12}\begin{vmatrix} a_{21} & a_{23} \\ a_{31} & a_{33} \end{vmatrix} + a_{13}\begin{vmatrix} a_{21} & a_{22} \\ a_{31} & a_{32} \end{vmatrix} \quad (3)$$

Note the minus sign before the second term on the right side of (3). ■

EXAMPLE 1

CALCULATING A 3 × 3 DETERMINANT

Let $A = \begin{pmatrix} 3 & 5 & 2 \\ 4 & 2 & 3 \\ -1 & 2 & 4 \end{pmatrix}$. Calculate $|A|$.

SOLUTION:

$$|A| = \begin{vmatrix} 3 & 5 & 2 \\ 4 & 2 & 3 \\ -1 & 2 & 4 \end{vmatrix} = 3\begin{vmatrix} 2 & 3 \\ 2 & 4 \end{vmatrix} - 5\begin{vmatrix} 4 & 3 \\ -1 & 4 \end{vmatrix} + 2\begin{vmatrix} 4 & 2 \\ -1 & 2 \end{vmatrix}$$
$$= 3 \cdot 2 - 5 \cdot 19 + 2 \cdot 10 = -69$$

EXAMPLE 2

CALCULATING A 3 × 3 DETERMINANT

Calculate $\begin{vmatrix} 2 & -3 & 5 \\ 1 & 0 & 4 \\ 3 & -3 & 9 \end{vmatrix}$.

SOLUTION:

$$\begin{vmatrix} 2 & -3 & 5 \\ 1 & 0 & 4 \\ 3 & -3 & 9 \end{vmatrix} = 2\begin{vmatrix} 0 & 4 \\ -3 & 9 \end{vmatrix} - (-3)\begin{vmatrix} 1 & 4 \\ 3 & 9 \end{vmatrix} + 5\begin{vmatrix} 1 & 0 \\ 3 & -3 \end{vmatrix}$$
$$= 2 \cdot 12 + 3(-3) + 5(-3) = 0$$

There is another method for calculating 3×3 determinants. From equation (3) we have

$$\begin{vmatrix} a_{11} & a_{12} & a_{13} \\ a_{21} & a_{22} & a_{23} \\ a_{31} & a_{32} & a_{33} \end{vmatrix} = a_{11}(a_{22}a_{33} - a_{23}a_{32}) - a_{12}(a_{21}a_{33} - a_{23}a_{31}) + a_{13}(a_{21}a_{32} - a_{22}a_{31})$$

or

$$|A| = a_{11}a_{22}a_{33} + a_{12}a_{23}a_{31} + a_{13}a_{21}a_{32} - a_{13}a_{22}a_{31} - a_{12}a_{21}a_{33} - a_{11}a_{32}a_{23} \quad (4)$$

We write A and adjoin its first two columns to it:

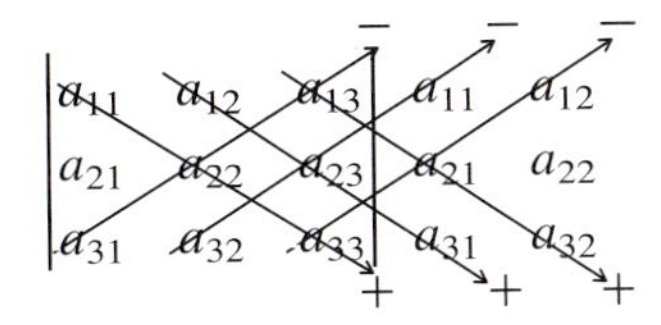

We then calculate the six products, put minus signs before the products with arrows pointing upward, and add. This gives the sum in equation (4).

EXAMPLE 3 CALCULATING A 3 × 3 DETERMINANT BY USING THE NEW METHOD

Calculate $\begin{vmatrix} 3 & 5 & 2 \\ 4 & 2 & 3 \\ -1 & 2 & 4 \end{vmatrix}$ by using this new method.

SOLUTION: Writing $\left|\begin{array}{ccc|cc} 3 & 5 & 2 & 3 & 5 \\ 4 & 2 & 3 & 4 & 2 \\ -1 & 2 & 4 & -1 & 2 \end{array}\right.$ and multiplying as indicated, we obtain

$$\begin{aligned} |A| &= (3)(2)(4) + (5)(3)(-1) + (2)(4)(2) - (-1)(2)(2) - 2(3)(3) - (4)(4)(5) \\ &= 24 - 15 + 16 + 4 - 18 - 80 = -69 \end{aligned}$$

WARNING: The method given above will *not* work for $n \times n$ determinants if $n > 3$. If you try something analogous for 4×4 or higher-order determinants, you will get the wrong answer.

Before defining $n \times n$ determinants, we first note that in equation (3) $\begin{pmatrix} a_{22} & a_{23} \\ a_{32} & a_{33} \end{pmatrix}$ is the matrix obtained by deleting the first row and first column of A; $\begin{pmatrix} a_{21} & a_{23} \\ a_{31} & a_{33} \end{pmatrix}$ is the matrix obtained by deleting the first row and second column of A; and $\begin{pmatrix} a_{21} & a_{22} \\ a_{31} & a_{32} \end{pmatrix}$ is the matrix obtained by deleting the first row and third column of A. If we denote these three matrices by M_{11}, M_{12}, and M_{13}, respectively, and if $A_{11} = \det M_{11}$, $A_{12} = -\det M_{12}$, and $A_{13} = \det M_{13}$, then equation (3) can be written

$$\det A = |A| = a_{11}A_{11} + a_{12}A_{12} + a_{13}A_{13} \quad (5)$$

DEFINITION MINOR

Let A be an $n \times n$ matrix and let M_{ij} be the $(n-1) \times (n-1)$ matrix obtained from A by deleting the ith row and jth column of A. M_{ij} is called the ***ij*th minor** of A. ■

EXAMPLE 4 FINDING TWO MINORS OF A 3 × 3 MATRIX

Let $A = \begin{pmatrix} 2 & -1 & 4 \\ 0 & 1 & 5 \\ 6 & 3 & -4 \end{pmatrix}$. Find M_{13} and M_{32}.

SOLUTION: Deleting the first row and third column of A, we obtain $M_{13} = \begin{pmatrix} 0 & 1 \\ 6 & 3 \end{pmatrix}$. Similarly, by eliminating the third row and second column, we obtain $M_{32} = \begin{pmatrix} 2 & 4 \\ 0 & 5 \end{pmatrix}$.

EXAMPLE 5

FINDING TWO MINORS OF A 4×4 MATRIX

Let $A = \begin{pmatrix} 1 & -3 & 5 & 6 \\ 2 & 4 & 0 & 3 \\ 1 & 5 & 9 & -2 \\ 4 & 0 & 2 & 7 \end{pmatrix}$. Find M_{32} and M_{24}.

SOLUTION: Deleting the third row and second column of A, we find that $M_{32} = \begin{pmatrix} 1 & 5 & 6 \\ 2 & 0 & 3 \\ 4 & 2 & 7 \end{pmatrix}$; similarly, $M_{24} = \begin{pmatrix} 1 & -3 & 5 \\ 1 & 5 & 9 \\ 4 & 0 & 2 \end{pmatrix}$.

DEFINITION COFACTOR

Let A be an $n \times n$ matrix. The ***ij*th cofactor** of A, denoted by A_{ij}, is given by

$$A_{ij} = (-1)^{i+j}|M_{ij}| \qquad (6)$$

That is, the ijth cofactor of A is obtained by taking the determinant of the ijth minor and multiplying it by $(-1)^{i+j}$. Note that

$$(-1)^{i+j} = \begin{cases} 1 & \text{if } i+j \text{ is even} \\ -1 & \text{if } i+j \text{ is odd} \end{cases}$$ ■

REMARK: The definition of a cofactor makes sense because we are going to define an $n \times n$ determinant with the assumption that we already know what an $(n-1) \times (n-1)$ determinant is.

EXAMPLE 6

FINDING TWO COFACTORS OF A 4×4 MATRIX

In Example 5 we have

$$A_{32} = (-1)^{3+2}|M_{32}| = -\begin{vmatrix} 1 & 5 & 6 \\ 2 & 0 & 3 \\ 4 & 2 & 7 \end{vmatrix} = -8$$

$$A_{24} = (-1)^{2+4}\begin{vmatrix} 1 & -3 & 5 \\ 1 & 5 & 9 \\ 4 & 0 & 2 \end{vmatrix} = -192$$

We now consider the general $n \times n$ matrix. Here

$$A = \begin{pmatrix} a_{11} & a_{12} & \cdots & a_{1n} \\ a_{21} & a_{22} & \cdots & a_{2n} \\ \vdots & \vdots & & \vdots \\ a_{n1} & a_{n2} & \cdots & a_{nn} \end{pmatrix} \quad (7)$$

DEFINITION $n \times n$ DETERMINANT

Let A be an $n \times n$ matrix. Then the determinant of A, written $\det A$ or $|A|$, is given by

$$\det A = |A| = a_{11}A_{11} + a_{12}A_{12} + a_{13}A_{13} + \cdots + a_{1n}A_{1n} \quad (8)$$
$$= \sum_{k=1}^{n} a_{1k}A_{1k}$$

The expression on the right side of (8) is called an **expansion of cofactors.** ■

REMARK: In equation (8) we defined the determinant by expanding by cofactors using components of A in the first row. We shall see in the next section (Theorem 7.2.1) that we get the same answer if we expand by cofactors in any row or column.

EXAMPLE 7 CALCULATING THE DETERMINANT OF A 4×4 MATRIX

Calculate $\det A$, where

$$A = \begin{pmatrix} 1 & 3 & 5 & 2 \\ 0 & -1 & 3 & 4 \\ 2 & 1 & 9 & 6 \\ 3 & 2 & 4 & 8 \end{pmatrix}$$

SOLUTION:

$$\begin{vmatrix} 1 & 3 & 5 & 2 \\ 0 & -1 & 3 & 4 \\ 2 & 1 & 9 & 6 \\ 3 & 2 & 4 & 8 \end{vmatrix} = a_{11}A_{11} + a_{12}A_{12} + a_{13}A_{13} + a_{14}A_{14}$$

$$= 1\begin{vmatrix} -1 & 3 & 4 \\ 1 & 9 & 6 \\ 2 & 4 & 8 \end{vmatrix} - 3\begin{vmatrix} 0 & 3 & 4 \\ 2 & 9 & 6 \\ 3 & 4 & 8 \end{vmatrix} + 5\begin{vmatrix} 0 & -1 & 4 \\ 2 & 1 & 6 \\ 3 & 2 & 8 \end{vmatrix} - 2\begin{vmatrix} 0 & -1 & 3 \\ 2 & 1 & 9 \\ 3 & 2 & 4 \end{vmatrix}$$

$$= 1(-92) - 3(-70) + 5(2) - 2(-16) = 160$$

It is clear that calculating the determinant of an $n \times n$ matrix can be tedious. To calculate a 4×4 determinant, we must calculate four 3×3 determinants. To calculate a 5×5 determinant, we must calculate five 4×4 determinants—which is the same as calculating 20 3×3 determinants. Fortunately, techniques exist for greatly simplifying these computations. Some of these methods are discussed in the next section. There are, however, some matrices whose determinants can easily be calculated. We begin by repeating a definition given on page 440.

DEFINITION TRIANGULAR MATRIX

A square matrix is called **upper triangular** if all its components below the diagonal are zero. It is **lower triangular** if all its components above the diagonal are zero. A matrix is called **diagonal** if all its elements not on the diagonal are zero; that is, $A = (a_{ij})$ is upper triangular if $a_{ij} = 0$ for $i > j$, lower triangular if $a_{ij} = 0$ for $i < j$, and diagonal if $a_{ij} = 0$ for $i \neq j$. Note that a diagonal matrix is both upper and lower triangular. ■

EXAMPLE 8 SIX TRIANGULAR MATRICES

The matrices $A = \begin{pmatrix} 2 & 1 & 7 \\ 0 & 2 & -5 \\ 0 & 0 & 1 \end{pmatrix}$ and $B = \begin{pmatrix} -2 & 3 & 0 & 1 \\ 0 & 0 & 2 & 4 \\ 0 & 0 & 1 & 3 \\ 0 & 0 & 0 & -2 \end{pmatrix}$ are upper triangular; $C = \begin{pmatrix} 5 & 0 & 0 \\ 2 & 3 & 0 \\ -1 & 2 & 4 \end{pmatrix}$ and $D = \begin{pmatrix} 0 & 0 \\ 1 & 0 \end{pmatrix}$ are lower triangular; I and $E = \begin{pmatrix} 2 & 0 & 0 \\ 0 & -7 & 0 \\ 0 & 0 & -4 \end{pmatrix}$ are diagonal. Note that the matrix E is both upper and lower triangular.

EXAMPLE 9 THE DETERMINANT OF A LOWER TRIANGULAR MATRIX

The matrix

$$A = \begin{pmatrix} a_{11} & 0 & 0 & 0 \\ a_{21} & a_{22} & 0 & 0 \\ a_{31} & a_{32} & a_{33} & 0 \\ a_{41} & a_{42} & a_{43} & a_{44} \end{pmatrix}$$

is lower triangular. Compute $\det A$.

SOLUTION:

$$\begin{aligned} \det A &= a_{11}A_{11} + 0A_{12} + 0A_{13} + 0A_{14} = a_{11}A_{11} \\ &= a_{11}\begin{vmatrix} a_{22} & 0 & 0 \\ a_{32} & a_{33} & 0 \\ a_{42} & a_{43} & a_{44} \end{vmatrix} \\ &= a_{11}a_{22}\begin{vmatrix} a_{33} & 0 \\ a_{43} & a_{44} \end{vmatrix} \\ &= a_{11}a_{22}a_{33}a_{44} \end{aligned}$$

Example 9 can be generalized to prove the following theorem.

THEOREM 1

Let $A = (a_{ij})$ be an upper[†] or lower triangular $n \times n$ matrix. Then

$$\det A = a_{11}a_{22}a_{33}\cdots a_{nn} \tag{9}$$

That is: **The determinant of a triangular matrix equals the product of its diagonal components.** ■

The following theorem will be useful to us in the next section.

THEOREM 2

Let T be an upper triangular matrix. Then T is invertible if and only if $\det T \neq 0$.

PROOF: Let

$$T = \begin{pmatrix} a_{11} & a_{12} & a_{13} & \cdots & a_{1n} \\ 0 & a_{22} & a_{23} & \cdots & a_{2n} \\ 0 & 0 & a_{33} & \cdots & a_{3n} \\ \vdots & \vdots & \vdots & & \vdots \\ 0 & 0 & 0 & \cdots & a_{nn} \end{pmatrix}$$

From Theorem 1,

$$\det T = a_{11}a_{22}\cdots a_{nn}$$

Thus $\det T \neq 0$ if and only if each of its diagonal components is nonzero.

If $\det T \neq 0$, then T can be row-reduced to I in the following way. For $i = 1, 2, \ldots, n$, divide the ith row of T by $a_{ii} \neq 0$ to obtain

$$\begin{pmatrix} 1 & a'_{12} & \cdots & a'_{1n} \\ 0 & 1 & \cdots & a'_{2n} \\ \vdots & \vdots & & \vdots \\ 0 & 0 & \cdots & 1 \end{pmatrix}$$

Then use the 1's on the diagonal to make all off-diagonal entries equal to 0. Thus the reduced row echelon form of T is I so, by the summing up theorem on page 440, T is invertible.

Suppose that $\det T = 0$. Then at least one of the diagonal components of T is zero. Let a_{ii} be the first such component. Then T can be written

$$T = \begin{pmatrix} a_{11} & a_{12} & \cdots & a_{1,i-1} & a_{1i} & a_{1,i+1} & \cdots & a_{1n} \\ 0 & a_{22} & \cdots & a_{2,i-1} & a_{2i} & a_{2,i+1} & \cdots & a_{2n} \\ \vdots & \vdots & & \vdots & \vdots & \vdots & & \vdots \\ 0 & 0 & \cdots & a_{i-1,i-1} & a_{i-1,i} & a_{i-1,i+1} & \cdots & a_{i-1,n} \\ 0 & 0 & \cdots & 0 & 0 & a_{i,i+1} & \cdots & a_{in} \\ 0 & 0 & \cdots & 0 & 0 & a_{i+1,i+1} & \cdots & a_{i+1,n} \\ \vdots & \vdots & & \vdots & \vdots & \vdots & \ddots & \vdots \\ 0 & 0 & \cdots & 0 & 0 & 0 & \cdots & a_{nn} \end{pmatrix}$$

[†] The proof for the upper triangular case is more difficult at this stage, but it will be just the same once we know that $\det A$ can be evaluated by expanding in any column (Theorem 7.2.1).

Now divide each of the first $i-1$ rows by the number on the diagonal and row reduce further to obtain

$$S=\begin{pmatrix} 1 & 0 & \cdots & 0 & a'_{1i} & a'_{1,i+1} & \cdots & a'_{1n} \\ 0 & 1 & \cdots & 0 & a'_{2i} & a'_{2,i+1} & \cdots & a'_{2n} \\ \vdots & & & & & & & \\ 0 & 0 & \cdots & 1 & a'_{i-1,i} & a'_{i-1,i+1} & \cdots & a'_{i-1,n} \\ 0 & 0 & \cdots & 0 & 0 & a_{i,i+1} & \cdots & a_{in} \\ 0 & 0 & \cdots & 0 & 0 & a_{i+1,i+1} & \cdots & a_{i+1,n} \\ \vdots & \vdots & & \vdots & \vdots & \vdots & \ddots & \vdots \\ 0 & 0 & \cdots & 0 & 0 & 0 & \cdots & a_{nn} \end{pmatrix}$$

Let

$$\mathbf{x}=\begin{pmatrix} -a'_{1i} \\ -a'_{2i} \\ \vdots \\ -a'_{i-1,i} \\ 1 \\ 0 \\ \vdots \\ 0 \end{pmatrix} \leftarrow i\text{th position}$$

Then $\mathbf{x} \neq \mathbf{0}$ and $S\mathbf{x} = \mathbf{0}$. Since T is row equivalent to S, $T\mathbf{x} = \mathbf{0}$ as well. Thus the homogeneous system $T\mathbf{x} = \mathbf{0}$ has nontrivial solutions and we may conclude that T is not invertible. ■

EXAMPLE 10

THE DETERMINANTS OF SIX TRIANGULAR MATRICES

The determinants of the six matrices in Example 8 are $|A| = 2 \cdot 2 \cdot 1 = 4$; $|B| = (-2)(0)(1)(-2) = 0$; $|C| = 5 \cdot 3 \cdot 4 = 60$; $|D| = 0$; $|I| = 1$; $|E| = (2)(-7)(-4) = 56$.

GEOMETRIC INTERPRETATION OF THE 2 × 2 DETERMINANT

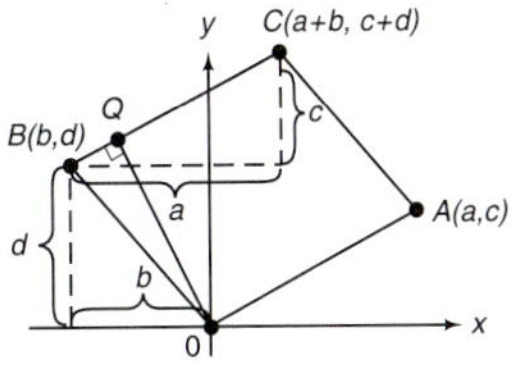

y C(a+b, c+d) Q B(b,d) c a A(a,c) d b 0 x

FIGURE 1

Q is on the line segment BC and is also on a line perpendicular to BC that passes through the origin. The area of the parallelogram is $\overline{0Q} \times \overline{0A}$.

Let $A = \begin{pmatrix} a & b \\ c & d \end{pmatrix}$. In Figure 1 we plot the points (a, c) and (b, d) in the xy-plane and draw lines from $(0, 0)$ to each of these points. We assume that the two lines are not collinear. This is the same as assuming that (b, d) is not a multiple of (a, c).

The area generated by A is defined as the area of the parallelogram with three vertices at $(0, 0)$, (a, c), and (b, d).

THEOREM 3 THE AREA GENERATED BY $A = |\text{DET } A|$

PROOF: We can think of the vectors $\overrightarrow{0B}$ and $\overrightarrow{0A}$ as vectors in $\mathbb{R}^3$ with endpoints $(b, d, 0)$ and $(a, c, 0)$. Let

$$\mathbf{u} = (b, d, 0) \quad \text{and} \quad \mathbf{v} = (a, c, 0)$$

From our discussion of the cross product in Section 1.6 (see equation (8) on page 48)

$$\text{area of the parallelogram} = |\mathbf{u} \times \mathbf{v}|$$

But

$$\mathbf{u} \times \mathbf{v} = \begin{vmatrix} \mathbf{i} & \mathbf{j} & \mathbf{k} \\ b & d & 0 \\ a & c & 0 \end{vmatrix} = (bc - ad)\mathbf{i}$$

and

$$|\mathbf{u} \times \mathbf{v}| = |bc - ad| = |ad - bc| = |\det A|$$ ■

PROBLEMS 7.1

SELF-QUIZ

Multiple Choice

I. Which of the following is the cofactor of 3 in $\begin{vmatrix} 1 & 2 & 3 \\ 2 & -2 & 1 \\ 4 & 0 & 2 \end{vmatrix}$?

a. 8 **b.** −8
c. 3 **d.** 6
e. −10 **f.** 0

II. Which of the following is 0 for all a and b?

a. $\begin{vmatrix} a & b \\ -b & a \end{vmatrix}$ **b.** $\begin{vmatrix} a & -b \\ -a & b \end{vmatrix}$ **c.** $\begin{vmatrix} a & a \\ b & -b \end{vmatrix}$

d. The determinants cannot be determined because values of a and b are not known.

III. The determinant of $A = \begin{pmatrix} 2 & -1 & 5 & 6 \\ 0 & 3 & 2 & 4 \\ 0 & 0 & -2 & 15 \\ 0 & 0 & 0 & 1 \end{pmatrix} =$ ______.

a. 0 **b.** 12 **c.** −12
d. 6 **e.** −6

In Problems 1–10 calculate the determinant.

1. $\begin{vmatrix} 1 & 0 & 3 \\ 0 & 1 & 4 \\ 2 & 1 & 0 \end{vmatrix}$

2. $\begin{vmatrix} -1 & 1 & 0 \\ 2 & 1 & 4 \\ 1 & 5 & 6 \end{vmatrix}$

3. $\begin{vmatrix} 3 & -1 & 4 \\ 6 & 3 & 5 \\ 2 & -1 & 6 \end{vmatrix}$

4. $\begin{vmatrix} -1 & 0 & 6 \\ 0 & 2 & 4 \\ 1 & 2 & -3 \end{vmatrix}$

5. $\begin{vmatrix} -2 & 3 & 1 \\ 4 & 6 & 5 \\ 0 & 2 & 1 \end{vmatrix}$

6. $\begin{vmatrix} 5 & -2 & 1 \\ 6 & 0 & 3 \\ -2 & 1 & 4 \end{vmatrix}$

7. $\begin{vmatrix} 2 & 0 & 3 & 1 \\ 0 & 1 & 4 & 2 \\ 0 & 0 & 1 & 5 \\ 1 & 2 & 3 & 0 \end{vmatrix}$

8. $\begin{vmatrix} -3 & 0 & 0 & 0 \\ -4 & 7 & 0 & 0 \\ 5 & 8 & -1 & 0 \\ 2 & 3 & 0 & 6 \end{vmatrix}$

9. $\begin{vmatrix} -2 & 0 & 0 & 7 \\ 1 & 2 & -1 & 4 \\ 3 & 0 & -1 & 5 \\ 4 & 2 & 3 & 0 \end{vmatrix}$

10. $\begin{vmatrix} 2 & 3 & -1 & 4 & 5 \\ 0 & 1 & 7 & 8 & 2 \\ 0 & 0 & 4 & -1 & 5 \\ 0 & 0 & 0 & -2 & 8 \\ 0 & 0 & 0 & 0 & 6 \end{vmatrix}$

11. Show that if A and B are diagonal $n \times n$ matrices, then $\det AB = \det A \det B$.

***12.** Show that if A and B are lower triangular matrices, then $\det AB = \det A \det B$.

13. Show that, in general, it is not true that $\det(A + B) = \det A + \det B$.

14. Show that if A is triangular, then $\det A \neq 0$ if and only if all the diagonal components of A are nonzero.

15. Prove Theorem 1 for a lower triangular matrix.

****16. More on the geometric interpretation of the determinant**: Let $\mathbf{u}_1$ and $\mathbf{u}_2$ be two 2-vectors and let $\mathbf{v}_1 = A\mathbf{u}_1$ and $\mathbf{v}_2 = A\mathbf{u}_2$. Show that (area generated by $\mathbf{v}_1$ and $\mathbf{v}_2$) = (area generated by $\mathbf{u}_1$ and $\mathbf{u}_2$) $\times$ $|\det A|$

17. Prove Theorem 2 when A has coordinates $(0, c)$ or $(a, 0)$.

ANSWERS TO SELF-QUIZ

I. a **II.** b **III.** c

7.2 PROPERTIES OF DETERMINANTS

Determinants have many properties that can make computations easier. We begin to describe these properties by stating a theorem from which everything else follows. The proof of this theorem is difficult. It is given in Appendix 4 at the back of the book.

THEOREM 1 BASIC THEOREM

Let

$$A = \begin{pmatrix} a_{11} & a_{12} & \cdots & a_{1n} \\ a_{21} & a_{22} & \cdots & a_{2n} \\ \vdots & \vdots & & \vdots \\ a_{n1} & a_{n2} & \cdots & a_{nn} \end{pmatrix}$$

be an $n \times n$ matrix. Then

$$\det A = a_{i1}A_{i1} + a_{i2}A_{i2} + \cdots + a_{in}A_{in} = \sum_{k=1}^{n} a_{ik}A_{ik} \qquad (1)$$

for $i = 1, 2, \ldots, n$. That is, we can calculate $\det A$ by expanding by cofactors in *any* row of A. Furthermore,

$$\det A = a_{1j}A_{1j} + a_{2j}A_{2j} + \cdots + a_{nj}A_{nj} = \sum_{k=1}^{n} a_{kj}A_{kj} \qquad (2)$$

Since the jth column of A is $\begin{pmatrix} a_{1j} \\ a_{2j} \\ \vdots \\ a_{nj} \end{pmatrix}$, equation (2) indicates that we can calculate $\det A$ by expanding by cofactors in any column of A. ■

EXAMPLE 1 OBTAINING THE DETERMINANT BY EXPANDING IN THE SECOND ROW OR THE THIRD COLUMN

For $A = \begin{pmatrix} 3 & 5 & 2 \\ 4 & 2 & 3 \\ -1 & 2 & 4 \end{pmatrix}$, we saw in Example 7.1.1 on page 449 that $\det A =$

-69. Expanding in the second row, we obtain

$$\begin{aligned}\det A &= 4A_{21} + 2A_{22} + 3A_{23}\\ &= 4(-1)^{2+1}\begin{vmatrix}5 & 2\\ 2 & 4\end{vmatrix} + 2(-1)^{2+2}\begin{vmatrix}3 & 2\\ -1 & 4\end{vmatrix} + 3(-1)^{2+3}\begin{vmatrix}3 & 5\\ -1 & 2\end{vmatrix}\\ &= -4(16) + 2(14) - 3(11) = -69\end{aligned}$$

Similarly, if we expand in the third column, say, we obtain

$$\begin{aligned}\det A &= 2A_{13} + 3A_{23} + 4A_{33}\\ &= 2(-1)^{1+3}\begin{vmatrix}4 & 2\\ -1 & 2\end{vmatrix} + 3(-1)^{2+3}\begin{vmatrix}3 & 5\\ -1 & 2\end{vmatrix} + 4(-1)^{3+3}\begin{vmatrix}3 & 5\\ 4 & 2\end{vmatrix}\\ &= 2(10) - 3(11) + 4(-14) = -69\end{aligned}$$

You should verify that we get the same answer if we expand in the third row or the first or second column.

We now list and prove some additional properties of determinants. In each case we assume that A is an $n \times n$ matrix.[†] We shall see that these properties can be used to reduce greatly the work involved in evaluating a determinant.

PROPERTY 1

If any row or column of A is the zero vector, then $\det A = 0$.

PROOF: Suppose the ith row of A contains all zeros. That is, $a_{ij} = 0$ for $j = 1, 2, \ldots, n$. Then $\det A = a_{i1}A_{i1} + a_{i2}A_{i2} + \cdots + a_{in}A_{in} = 0 + 0 + \cdots + 0 = 0$. The same proof works if the jth column is the zero vector. ■

EXAMPLE 2 IF A HAS A ROW OR COLUMN OF ZEROS, THEN $\det A = 0$

It is easy to verify that

$$\begin{vmatrix}2 & 3 & 5\\ 0 & 0 & 0\\ 1 & -2 & 4\end{vmatrix} = 0 \quad \text{and} \quad \begin{vmatrix}-1 & 3 & 0 & 1\\ 4 & 2 & 0 & 5\\ -1 & 6 & 0 & 4\\ 2 & 1 & 0 & 1\end{vmatrix} = 0$$

PROPERTY 2

If the ith row or the jth column of A is multiplied by the constant c, then $\det A$ is multiplied by c. That is, if we call this new matrix B, then

$$|B| = \begin{vmatrix}a_{11} & a_{12} & \cdots & a_{1n}\\ a_{21} & a_{22} & \cdots & a_{2n}\\ \vdots & \vdots & & \vdots\\ ca_{i1} & ca_{i2} & \cdots & ca_{in}\\ \vdots & \vdots & & \vdots\\ a_{n1} & a_{n2} & \cdots & a_{nn}\end{vmatrix} = c\begin{vmatrix}a_{11} & a_{12} & \cdots & a_{1n}\\ a_{21} & a_{22} & \cdots & a_{2n}\\ \vdots & \vdots & & \vdots\\ a_{i1} & a_{i2} & \cdots & a_{in}\\ \vdots & \vdots & & \vdots\\ a_{n1} & a_{n2} & \cdots & a_{nn}\end{vmatrix} = c|A| \tag{3}$$

[†] The proofs of these properties are given in terms of the rows of a matrix. Using Theorem 1 the same properties can be proved for columns.

PROOF: To prove (3) we expand in the ith row of A to obtain

$$\det B = ca_{i1}A_{i1} + ca_{i2}A_{i2} + \cdots + ca_{in}A_{in}$$
$$= c(a_{i1}A_{i1} + a_{i2}A_{i2} + \cdots + a_{in}A_{in}) = c \det A$$

A similar proof works for columns. ■

EXAMPLE 3 ILLUSTRATION OF PROPERTY 2

Let $A = \begin{pmatrix} 1 & -1 & 2 \\ 3 & 1 & 4 \\ 0 & -2 & 5 \end{pmatrix}$. Then $\det A = 16$. If we multiply the second row by 4, we have $B = \begin{pmatrix} 1 & -1 & 2 \\ 12 & 4 & 16 \\ 0 & -2 & 5 \end{pmatrix}$ and $\det B = 64 = 4 \det A$. If the third column is multiplied by -3, we obtain $C = \begin{pmatrix} 1 & -1 & -6 \\ 3 & 1 & -12 \\ 0 & -2 & -15 \end{pmatrix}$ and $\det C = -48 = -3 \det A$.

REMARK: Using Property 2 we can prove (see Problem 28) the following interesting fact: For any scalar α and $n \times n$ matrix A, $\det \alpha A = \alpha^n \det A$.

PROPERTY 3

Let

$$A = \begin{pmatrix} a_{11} & a_{12} & \cdots & a_{1j} & \cdots & a_{1n} \\ a_{21} & a_{22} & \cdots & a_{2j} & \cdots & a_{2n} \\ \vdots & \vdots & & \vdots & & \vdots \\ a_{n1} & a_{n2} & \cdots & a_{nj} & \cdots & a_{nn} \end{pmatrix}, \quad B = \begin{pmatrix} a_{11} & a_{12} & \cdots & \alpha_{1j} & \cdots & a_{1n} \\ a_{21} & a_{22} & \cdots & \alpha_{2j} & \cdots & a_{2n} \\ \vdots & \vdots & & \vdots & & \vdots \\ a_{n1} & a_{n2} & \cdots & \alpha_{nj} & \cdots & a_{nn} \end{pmatrix},$$

$$\text{and} \quad C = \begin{pmatrix} a_{11} & a_{12} & \cdots & a_{1j} + \alpha_{1j} & \cdots & a_{1n} \\ a_{21} & a_{22} & \cdots & a_{2j} + \alpha_{2j} & \cdots & a_{2n} \\ \vdots & \vdots & & \vdots & & \vdots \\ a_{n1} & a_{n2} & \cdots & a_{nj} + \alpha_{nj} & \cdots & a_{nn} \end{pmatrix}$$

Then

$$\det C = \det A + \det B \tag{4}$$

In other words, suppose that A, B, and C are identical except for the jth column and that the jth column of C is the sum of the jth columns of A and B. Then $\det C = \det A + \det B$. The same statement is true for rows.

PROOF: We expand $\det C$ in the jth column to obtain

$$\det C = (a_{1j} + \alpha_{1j})A_{1j} + (a_{2j} + \alpha_{2j})A_{2j} + \cdots + (a_{nj} + \alpha_{nj})A_{nj}$$
$$= (a_{1j}A_{1j} + a_{2j}A_{2j} + \cdots + a_{nj}A_{nj})$$
$$+ (\alpha_{1j}A_{1j} + \alpha_{2j}A_{2j} + \cdots + \alpha_{nj}A_{nj}) = \det A + \det B$$

■

EXAMPLE 4 ILLUSTRATION OF PROPERTY 3

Let $A = \begin{pmatrix} 1 & -1 & 2 \\ 3 & 1 & 4 \\ 0 & -2 & 5 \end{pmatrix}$, $B = \begin{pmatrix} 1 & -6 & 2 \\ 3 & 2 & 4 \\ 0 & 4 & 5 \end{pmatrix}$, and $C = \begin{pmatrix} 1 & -1-6 & 2 \\ 3 & 1+2 & 4 \\ 0 & -2+4 & 5 \end{pmatrix} = \begin{pmatrix} 1 & -7 & 2 \\ 3 & 3 & 4 \\ 0 & 2 & 5 \end{pmatrix}$. Then $\det A = 16$, $\det B = 108$, and $\det C = 124 = \det A + \det B$.

PROPERTY 4

Interchanging any two rows (or columns) of A has the effect of multiplying $\det A$ by -1.

PROOF: We prove the statement for rows and assume first that two adjacent rows are interchanged. That is, we assume that the ith and $(i+1)$st rows are interchanged. Let

$$A = \begin{pmatrix} a_{11} & a_{12} & \cdots & a_{1n} \\ a_{21} & a_{22} & \cdots & a_{2n} \\ \vdots & \vdots & & \vdots \\ a_{i1} & a_{i2} & \cdots & a_{in} \\ a_{i+1,1} & a_{i+1,2} & \cdots & a_{i+1,n} \\ \vdots & \vdots & & \vdots \\ a_{n1} & a_{n2} & & a_{nn} \end{pmatrix}$$

and

$$B = \begin{pmatrix} a_{11} & a_{12} & \cdots & a_{1n} \\ a_{21} & a_{22} & \cdots & a_{2n} \\ \vdots & \vdots & & \vdots \\ a_{i+1,1} & a_{i+1,2} & \cdots & a_{i+1,n} \\ a_{i1} & a_{i2} & \cdots & a_{in} \\ \vdots & \vdots & & \vdots \\ a_{n1} & a_{n2} & \cdots & a_{nn} \end{pmatrix}$$

Then, expanding $\det A$ in its ith row and $\det B$ in its $(i+1)$st row, we obtain

$$\begin{aligned} \det A &= a_{i1}A_{i1} + a_{i2}A_{i2} + \cdots + a_{in}A_{in} \\ \det B &= a_{i1}B_{i+1,1} + a_{i2}B_{i+1,2} + \cdots + a_{in}B_{i+1,n} \end{aligned} \qquad (5)$$

Here $A_{ij} = (-1)^{i+j}|M_{ij}|$, where M_{ij} is obtained by crossing off the ith row and jth column of A. Notice now that if we cross off the $(i+1)$st row and jth column of B, we obtain the same M_{ij}. Thus

$$B_{i+1,j} = (-1)^{i+1+j}|M_{ij}| = -(-1)^{i+j}|M_{ij}| = -A_{ij}$$

so that, from equations (5), $\det B = -\det A$.

Now suppose that $i < j$ and that the ith and jth rows are to be interchanged. We can do this by interchanging adjacent rows several times. It will take $j - i$ interchanges to move row j into the ith row. Then row i will be in the $(i+1)$st row and it will take an additional $j - i - 1$ interchanges to move row i into the jth row. To illustrate, we interchange rows 2 and 6:[†]

$$\begin{array}{cccccccccccccccc} 1 & & 1 & & 1 & & 1 & & 1 & & 1 & & 1 & & 1 \\ 2 & & 2 & & 2 & & 2 & & 6 & & 6 & & 6 & & 6 \\ 3 & & 3 & & 3 & & 6 & & 2 & & 3 & & 3 & & 3 \\ 4 & \to & 4 & \to & 6 & \to & 3 & \to & 3 & \to & 2 & \to & 4 & \to & 4 \\ 5 & & 6 & & 4 & & 4 & & 4 & & 4 & & 2 & & 5 \\ 6 & & 5 & & 5 & & 5 & & 5 & & 5 & & 5 & & 2 \\ 7 & & 7 & & 7 & & 7 & & 7 & & 7 & & 7 & & 7 \end{array}$$

$6 - 2 = 4$ interchanges to move the 6 into the 2 position; $6 - 2 - 1 = 3$ interchanges to get the 2 into the 6 position

Finally, the total number of interchanges of adjacent rows is $(j - i) + (j - i - 1) = 2j - 2i - 1$, which is odd. Thus $\det A$ is multiplied by -1 an odd number of times, which is what we needed to show. ■

EXAMPLE 5

ILLUSTRATION OF PROPERTY 4

Let $A = \begin{pmatrix} 1 & -1 & 2 \\ 3 & 1 & 4 \\ 0 & -2 & 5 \end{pmatrix}$. By interchanging the first and third rows, we obtain $B = \begin{pmatrix} 0 & -2 & 5 \\ 3 & 1 & 4 \\ 1 & -1 & 2 \end{pmatrix}$. By interchanging the first and second columns of A, we obtain $C = \begin{pmatrix} -1 & 1 & 2 \\ 1 & 3 & 4 \\ -2 & 0 & 5 \end{pmatrix}$. Then, by direct calculation, we find that $\det A = 16$ and $\det B = \det C = -16$.

PROPERTY 5

If A has two equal rows or columns, then $\det A = 0$.

PROOF: Suppose the ith and jth rows of A are equal. By interchanging these rows we get a matrix B having the property that $\det B = -\det A$ (from Property 4). But since row i = row j, interchanging them gives us the same matrix. Thus $A = B$ and $\det A = \det B = -\det A$. Thus $2 \det A = 0$, which can happen only if $\det A = 0$. ■

EXAMPLE 6

ILLUSTRATION OF PROPERTY 5

By direct calculation we can verify that for $A = \begin{pmatrix} 1 & -1 & 2 \\ 5 & 7 & 3 \\ 1 & -1 & 2 \end{pmatrix}$ [two equal rows] and $B = \begin{pmatrix} 5 & 2 & 2 \\ 3 & -1 & -1 \\ -2 & 4 & 4 \end{pmatrix}$ [two equal columns], $\det A = \det B = 0$.

[†] Note that all the numbers here refer to rows.

PROPERTY 6

If one row (column) of A is a constant multiple of another row (column), then $\det A = 0$.

PROOF: Let $(a_{j1}, a_{j2}, \ldots, a_{jn}) = c(a_{i1}, a_{i2}, \ldots, a_{in})$. Then, from Property 2,

$$\det A = c \begin{vmatrix} a_{11} & a_{12} & \cdots & a_{1n} \\ a_{21} & a_{22} & \cdots & a_{2n} \\ \vdots & \vdots & & \vdots \\ a_{i1} & a_{i2} & \cdots & a_{in} \\ \vdots & \vdots & & \vdots \\ a_{i1} & a_{i2} & \cdots & a_{in} \\ \vdots & \vdots & & \vdots \\ a_{n1} & a_{n2} & \cdots & a_{nn} \end{vmatrix} = 0 \qquad \text{(from Property 5)}$$

(jth row → marks the second row $a_{i1}\ a_{i2}\ \cdots\ a_{in}$) ■

EXAMPLE 7 ILLUSTRATION OF PROPERTY 6

$$\begin{vmatrix} 2 & -3 & 5 \\ 1 & 7 & 2 \\ -4 & 6 & -10 \end{vmatrix} = 0 \text{ since the third row is } -2 \text{ times the first row.}$$

EXAMPLE 8 ANOTHER ILLUSTRATION OF PROPERTY 6

$$\begin{vmatrix} 2 & 4 & 1 & 12 \\ -1 & 1 & 0 & 3 \\ 0 & -1 & 9 & -3 \\ 7 & 3 & 6 & 9 \end{vmatrix} = 0 \text{ since the fourth column is three times the second column.}$$

PROPERTY 7

If a multiple of one row (column) of A is added to another row (column) of A, then the determinant is unchanged.

PROOF: Let B be the matrix obtained by adding c times the ith row of A to the jth row of A. Then

$$\det B = \begin{vmatrix} a_{11} & a_{12} & \cdots & a_{1n} \\ a_{21} & a_{22} & \cdots & a_{2n} \\ \vdots & \vdots & & \vdots \\ a_{i1} & a_{i2} & \cdots & a_{in} \\ \vdots & \vdots & & \vdots \\ a_{j1} + ca_{i1} & a_{j2} + ca_{i2} & \cdots & a_{jn} + ca_{in} \\ \vdots & \vdots & & \vdots \\ a_{n1} & a_{n2} & \cdots & a_{nn} \end{vmatrix}$$

$$\overset{\text{(from Property 3)}}{=} \begin{vmatrix} a_{11} & a_{12} & \cdots & a_{1n} \\ a_{21} & a_{22} & \cdots & a_{2n} \\ \vdots & \vdots & & \vdots \\ a_{i1} & a_{i2} & \cdots & a_{in} \\ \vdots & \vdots & & \vdots \\ a_{j1} & a_{j2} & \cdots & a_{jn} \\ \vdots & \vdots & & \vdots \\ a_{n1} & a_{n2} & \cdots & a_{nn} \end{vmatrix} + \begin{vmatrix} a_{11} & a_{12} & \cdots & a_{1n} \\ a_{21} & a_{22} & \cdots & a_{2n} \\ \vdots & \vdots & & \vdots \\ a_{i1} & a_{i2} & \cdots & a_{in} \\ \vdots & \vdots & & \vdots \\ ca_{i1} & ca_{i2} & \cdots & ca_{in} \\ \vdots & \vdots & & \vdots \\ a_{n1} & a_{n2} & \cdots & a_{nn} \end{vmatrix}$$

$$= \det A + 0 = \det A \quad \text{(the zero comes from Property 6)}$$ ■

EXAMPLE 9

ILLUSTRATION OF PROPERTY 7

Let $A = \begin{pmatrix} 1 & -1 & 2 \\ 3 & 1 & 4 \\ 0 & -2 & 5 \end{pmatrix}$. Then $\det A = 16$. If we multiply the third row by 4 and add it to the second row, we obtain a new matrix B given by

$$B = \begin{pmatrix} 1 & -1 & 2 \\ 3 + 4(0) & 1 + 4(-2) & 4 + 5(4) \\ 0 & -2 & 5 \end{pmatrix} = \begin{pmatrix} 1 & -1 & 2 \\ 3 & -7 & 24 \\ 0 & -2 & 5 \end{pmatrix}$$

and $\det B = 16 = \det A$.

The properties discussed above make it much easier to evaluate high-order determinants. We simply ''row-reduce'' the determinant, using Property 7, until the determinant is in an easily evaluated form. The most common goal will be to use Property 7 repeatedly until either (1) the new determinant has a row (column) of zeros or one row (column) is a multiple of another row (column)—in which case the determinant is zero—or (2) the new matrix is triangular so that its determinant is the product of its diagonal elements.

EXAMPLE 10

USING THE PROPERTIES OF A DETERMINANT TO CALCULATE A 4 × 4 DETERMINANT

Calculate

$$|A| = \begin{vmatrix} 1 & 3 & 5 & 2 \\ 0 & -1 & 3 & 4 \\ 2 & 1 & 9 & 6 \\ 3 & 2 & 4 & 8 \end{vmatrix}$$

SOLUTION: (See Example 7.1.7, page 452)

There is already a zero in the first column, so it is simplest to reduce other elements in the first column to zero. We then continue to reduce, aiming for a triangular matrix.

Multiply the first row by -2 and add it to the third row and multiply the first row by -3 and add it to the fourth row.

$$|A| = \begin{vmatrix} 1 & 3 & 5 & 2 \\ 0 & -1 & 3 & 4 \\ 0 & -5 & -1 & 2 \\ 0 & -7 & -11 & 2 \end{vmatrix}$$

Multiply the second row by -5 and -7 and add it to the third and fourth rows, respectively.

$$= \begin{vmatrix} 1 & 3 & 5 & 2 \\ 0 & -1 & 3 & 4 \\ 0 & 0 & -16 & -18 \\ 0 & 0 & -32 & -26 \end{vmatrix}$$

Factor out -16 from the third row (using Property 2).

$$= -16\begin{vmatrix} 1 & 3 & 5 & 2 \\ 0 & -1 & 3 & 4 \\ 0 & 0 & 1 & \frac{9}{8} \\ 0 & 0 & -32 & -26 \end{vmatrix}$$

Multiply the third row by 32 and add it to the fourth row.

$$= -16\begin{vmatrix} 1 & 3 & 5 & 2 \\ 0 & -1 & 3 & 4 \\ 0 & 0 & 1 & \frac{9}{8} \\ 0 & 0 & 0 & 10 \end{vmatrix}$$

Now we have an upper triangular matrix and $|A| = -16(1)(-1)(1)(10) = (-16)(-10) = 160$.

EXAMPLE 11 USING THE PROPERTIES TO CALCULATE A 4×4 DETERMINANT

Calculate

$$|A| = \begin{vmatrix} -2 & 1 & 0 & 4 \\ 3 & -1 & 5 & 2 \\ -2 & 7 & 3 & 1 \\ 3 & -7 & 2 & 5 \end{vmatrix}$$

SOLUTION: There are a number of ways to proceed here and it is not apparent which way will get us the answer most quickly. However, since there is already one zero in the first row, we begin our reduction in that row.

Multiply the second column by 2 and -4 and add it to the first and fourth columns, respectively.

$$|A| = \begin{vmatrix} 0 & 1 & 0 & 0 \\ 1 & -1 & 5 & 6 \\ 12 & 7 & 3 & -27 \\ -11 & -7 & 2 & 33 \end{vmatrix}$$

Interchange the first two columns.

$$= -\begin{vmatrix} 1 & 0 & 0 & 0 \\ -1 & 1 & 5 & 6 \\ 7 & 12 & 3 & -27 \\ -7 & -11 & 2 & 33 \end{vmatrix}$$

Multiply the second column by -5 and -6 and add it to the third and fourth columns, respectively.

$$= -\begin{vmatrix} 1 & 0 & 0 & 0 \\ -1 & 1 & 0 & 0 \\ 7 & 12 & -57 & -99 \\ -7 & -11 & 57 & 99 \end{vmatrix}$$

Since the fourth column is now a multiple of the third column (column 4 $= \frac{99}{57} \times$ column 3), we see that $|A| = 0$.

EXAMPLE 12 USING THE PROPERTIES TO CALCULATE A 5 × 5 DETERMINANT

Calculate

$$|A| = \begin{vmatrix} 1 & -2 & 3 & -5 & 7 \\ 2 & 0 & -1 & -5 & 6 \\ 4 & 7 & 3 & -9 & 4 \\ 3 & 1 & -2 & -2 & 3 \\ -5 & -1 & 3 & 7 & -9 \end{vmatrix}$$

SOLUTION: Adding first row 2 and then row 4 to row 5, we obtain

$$|A| = \begin{vmatrix} 1 & -2 & 3 & -5 & 7 \\ 2 & 0 & -1 & -5 & 6 \\ 4 & 7 & 3 & -9 & 4 \\ 3 & 1 & -2 & -2 & 3 \\ 0 & 0 & 0 & 0 & 0 \end{vmatrix} = 0 \qquad \text{(from Property 1)}$$

This example illustrates the fact that a little looking before beginning the computations can simplify matters considerably.

There are three additional facts about determinants that will be very useful to us.

THEOREM 2

Let A be an $n \times n$ matrix. Then

$$a_{i1}A_{j1} + a_{i2}A_{j2} + \cdots + a_{in}A_{jn} = 0 \qquad \text{if } i \neq j \tag{6}$$

NOTE: From Theorem 1 the sum in equation (6) equals $\det A$ if $i = j$.

PROOF: Let

$$B = \begin{pmatrix} a_{11} & a_{12} & \cdots & a_{1n} \\ a_{21} & a_{22} & \cdots & a_{2n} \\ \vdots & \vdots & & \vdots \\ a_{i1} & a_{i2} & \cdots & a_{in} \\ \vdots & \vdots & & \vdots \\ a_{i1} & a_{i2} & \cdots & a_{in} \\ \vdots & \vdots & & \vdots \\ a_{n1} & a_{n2} & \cdots & a_{nn} \end{pmatrix} \begin{matrix} \\ \\ \\ \\ \\ \leftarrow j\text{th row} \\ \\ \\ \end{matrix}$$

Then since two rows of B are equal, $\det B = 0$. But $B = A$ except in the jth row. Thus if we calculate $\det B$ by expanding in the jth row of B, we obtain the sum in (6) and the theorem is proved. Note that when we expand in the jth row, the jth row is deleted in computing the cofactors of B. Thus $B_{jk} = A_{jk}$ for $k = 1, 2, \ldots, n$. ■

THEOREM 3

Let A be an $n \times n$ matrix. Then

$$\det A = \det A^t \tag{7}$$

PROOF: This proof uses mathematical induction. If you are unfamiliar with this important method of proof, refer to Appendix 1. We first prove the theorem in the case $n = 2$. If

$$|A| = \begin{vmatrix} a_{11} & a_{12} \\ a_{21} & a_{22} \end{vmatrix} = a_{11}a_{22} - a_{12}a_{21}$$

then

$$|A^t| = \begin{vmatrix} a_{11} & a_{21} \\ a_{12} & a_{22} \end{vmatrix} = a_{11}a_{22} - a_{21}a_{12} = |A|$$

so the theorem is true for $n = 2$. Next we assume the theorem to be true for $(n-1) \times (n-1)$ matrices and prove it for $n \times n$ matrices. This will prove the theorem. Let $B = A^t$. Then

$$|A| = \begin{vmatrix} a_{11} & a_{12} & \cdots & a_{1n} \\ a_{21} & a_{22} & \cdots & a_{2n} \\ \vdots & \vdots & & \vdots \\ a_{n1} & a_{n2} & \cdots & a_{nn} \end{vmatrix} \quad \text{and} \quad |A^t| = |B| = \begin{vmatrix} a_{11} & a_{21} & \cdots & a_{n1} \\ a_{12} & a_{22} & \cdots & a_{n2} \\ \vdots & \vdots & & \vdots \\ a_{1n} & a_{2n} & \cdots & a_{nn} \end{vmatrix}.$$

We expand $|A|$ in the first row and expand $|B|$ in the first column. This gives us

$$|A| = a_{11}A_{11} + a_{12}A_{12} + \cdots + a_{1n}A_{1n}$$
$$|B| = a_{11}B_{11} + a_{12}B_{21} + \cdots + a_{1n}B_{n1}$$

We need to show that $A_{1k} = B_{k1}$ for $k = 1, 2, \ldots, n$. But $A_{1k} = (-1)^{1+k}|M_{1k}|$ and $B_{k1} = (-1)^{k+1}|N_{k1}|$, where M_{1k} is the $1k$th minor of A and N_{k1} is the $k1$st minor of B. Then

$$|M_{1k}| = \begin{vmatrix} a_{21} & a_{22} & \cdots & a_{2,k-1} & a_{2,k+1} & \cdots & a_{2n} \\ a_{31} & a_{32} & \cdots & a_{3,k-1} & a_{3,k+1} & \cdots & a_{3n} \\ \vdots & \vdots & & \vdots & \vdots & & \vdots \\ a_{n1} & a_{n2} & \cdots & a_{n,k-1} & a_{n,k+1} & \cdots & a_{nn} \end{vmatrix}$$

and

$$|N_{k1}| = \begin{vmatrix} a_{21} & a_{31} & \cdots & a_{n1} \\ a_{22} & a_{32} & \cdots & a_{n2} \\ \vdots & \vdots & & \vdots \\ a_{2,k-1} & a_{3,k-1} & \cdots & a_{n,k-1} \\ a_{2,k+1} & a_{3,k+1} & \cdots & a_{n,k+1} \\ \vdots & \vdots & & \vdots \\ a_{2n} & a_{3n} & \cdots & a_{nn} \end{vmatrix}$$

Clearly $M_{1k} = {N_{k1}}^t$, and since both are $(n-1) \times (n-1)$ matrices, the induction hypothesis tells us that $|M_{1k}| = |N_{k1}|$. Thus $A_{1k} = B_{k1}$ and the proof is complete. ■

EXAMPLE 13

A MATRIX AND ITS TRANSPOSE HAVE THE SAME DETERMINANT

Let $A = \begin{pmatrix} 1 & -1 & 2 \\ 3 & 1 & 4 \\ 0 & -2 & 5 \end{pmatrix}$. Then $A^t = \begin{pmatrix} 1 & 3 & 0 \\ -1 & 1 & -2 \\ 2 & 4 & 5 \end{pmatrix}$ and it is easy to verify that $|A| = |A^t| = 16$.

THEOREM 4

Let A and B be $n \times n$ matrices. Then

$$\det AB = \det A \det B \tag{8}$$

That is: **The determinant of the product is the product of the determinants.** ■

The proof of this important theorem, using elementary matrices, is given at the end of this section.

EXAMPLE 14

ILLUSTRATION OF FACT THAT DET AB = DET A DET B

Verify equation (8) for

$$A = \begin{pmatrix} 1 & -1 & 2 \\ 3 & 1 & 4 \\ 0 & -2 & 5 \end{pmatrix} \quad \text{and} \quad B = \begin{pmatrix} 1 & -2 & 3 \\ 0 & -1 & 4 \\ 2 & 0 & -2 \end{pmatrix}.$$

SOLUTION: $\det A = 16$ and $\det B = -8$. We calculate

$$AB = \begin{pmatrix} 1 & -1 & 2 \\ 3 & 1 & 4 \\ 0 & -2 & 5 \end{pmatrix}\begin{pmatrix} 1 & -2 & 3 \\ 0 & -1 & 4 \\ 2 & 0 & -2 \end{pmatrix} = \begin{pmatrix} 5 & -1 & -5 \\ 11 & -7 & 5 \\ 10 & 2 & -18 \end{pmatrix}$$

and $\det AB = -128 = (16)(-8) = \det A \det B$.

WARNING: The determinant of the sum is *not* equal to the sum of the determinants. That is,

$$\det (A + B) \neq \det A + \det B$$

For example, let $A = \begin{pmatrix} 1 & 2 \\ 3 & 4 \end{pmatrix}$ and $B = \begin{pmatrix} 3 & 0 \\ -2 & 2 \end{pmatrix}$. Then $A + B = \begin{pmatrix} 4 & 2 \\ 1 & 6 \end{pmatrix}$:

$$\det A = -2, \qquad \det B = 6, \qquad \text{and}$$
$$\det (A + B) = 22 \neq \det A + \det B = -2 + 6 = 4$$

In our summing up theorem last seen on page 440 we asserted that A is invertible if and only if $\det A \neq 0$. We prove that fact after giving some preliminary results about elementary matrices (see Section 6.10).

We begin by computing the determinants of elementary matrices.

LEMMA 1

Let E be an elementary matrix:

i. If E is the matrix representing the elementary row operation $R_i \rightleftarrows R_j$, then $\det E = -1$. **(9)**

ii. If E is the matrix representing the elementary row operation $R_j \to R_j + cR_i$, then $\det E = 1$. **(10)**

iii. If E is the matrix representing the elementary row operation $R_i \to cR_i$, then $\det E = c$. **(11)**

PROOF:

i. $\det I = 1$. E is obtained from I by interchanging the ith and jth rows of I. From Property 4 on page 460, $\det E = (-1) \det I = -1$.

ii. E is obtained from I by multiplying the ith row of I by c and adding it to the jth row. Thus by Property 7 on page 462 $\det E = \det I = 1$.

iii. E is obtained from I by multiplying the ith row of I by c. Thus from Property 2 on page 458, $\det E = c \det I = c$. ■

LEMMA 2

Let B be an $n \times n$ matrix and let E be an elementary matrix. Then

$$\det EB = \det E \det B \tag{12}$$

The proof of this lemma follows from Lemma 1 and the results relating elementary row operations to determinants discussed in this section. The steps in the proof are indicated in Problems 41 to 43. ■

The next theorem is a central result in matrix theory.

THEOREM 5

Let A be an $n \times n$ matrix. Then A is invertible if and only if $\det A \neq 0$.

PROOF: From Theorem 6.10.5 on page 440, we know that there are elementary matrices $E_1, E_2, \ldots, E_m$ and an upper triangular matrix T such that

$$A = E_1E_2 \cdots E_mT \tag{13}$$

Using Lemma 2 m times, we see that

$$\begin{aligned}\det A &= \det E_1 \det (E_2E_3 \cdots E_mT)\\ &= \det E_1 \det E_2 \det(E_3 \cdots E_mT)\\ &\vdots\\ &= \det E_1 \det E_2 \cdots \det E_{m-1} \det(E_mT)\end{aligned}$$

or

$$\det A = \det E_1 \det E_2 \cdots \det E_{m-1} \det E_m \det T \tag{14}$$

By Lemma 1, $\det E_i \neq 0$ for $i = 1, 2, \ldots, m$. We conclude that $\det A \neq 0$ if and only if $\det T \neq 0$.

Now suppose that A is invertible. Then by using (13) and the fact that every elementary matrix is invertible, $E_m{}^{-1} \cdots E_1{}^{-1}A$ is the product of invertible matrices. Thus T is invertible, and by Theorem 7.1.2 on page 454, $\det T \neq 0$. Thus $\det A \neq 0$.

If $\det A \neq 0$, then by (14), $\det T \neq 0$, so T is invertible (by Theorem 7.1.2). Then the right side of (14) is the product of invertible matrices, and so A is invertible. This completes the proof. ■

Using Theorem 5, we can prove that $\det AB = \det A \det B$.

PROOF THAT DET AB = DET A DET B:

PROOF: *Case 1:* $\det A = \det B = 0$. Then by Theorem 5, B is not invertible, so by the summing up theorem, there is an n-vector $\mathbf{x} \neq \mathbf{0}$ such that $B\mathbf{x} = \mathbf{0}$. Then $(AB)\mathbf{x} = A(B\mathbf{x}) = A\mathbf{0} = \mathbf{0}$. Therefore, again by the summing up theorem, AB is not invertible. By Theorem 5,

$$\det AB = 0 = 0 \cdot 0 = \det A \det B$$

Case 2: $\det A = 0$ and $\det B \neq 0$. A is not invertible, so there is an n-vector $\mathbf{y} \neq \mathbf{0}$ such that $A\mathbf{y} = \mathbf{0}$. Since $\det B \neq 0$, B is invertible and there is a unique vector $\mathbf{x} \neq \mathbf{0}$ such that $B\mathbf{x} = \mathbf{y}$. Then $AB\mathbf{x} = A(B\mathbf{x}) = A\mathbf{y} = \mathbf{0}$. Thus AB is not invertible, so

$$\det AB = 0 = 0 \det B = \det A \det B$$

Case 3: $\det A \neq 0$. A is invertible and can be written as a product of elementary matrices:

$$A = E_1 E_2 \cdots E_m$$

Then

$$AB = E_1 E_2 \cdots E_m B$$

Using the result of Lemma 2 repeatedly, we see that

$$\begin{aligned}\det AB &= \det(E_1 E_2 \cdots E_m B)\\ &= \det E_1 \det E_2 \cdots \det E_m \det B\\ &= \det(E_1 E_2 \cdots E_m) \det B\\ &= \det A \det B\end{aligned}$$

■

OF HISTORICAL INTEREST

A SHORT HISTORY OF DETERMINANTS

GOTTFRIED WILHELM LEIBNIZ (DAVID EUGENE SMITH COLLECTION, RARE BOOK AND MANUSCRIPT LIBRARY, COLUMBIA UNIVERSITY)

Determinants appeared in mathematical literature over a century before matrices. As pointed out in the note on page 406, the term *matrix* was coined by James Joseph Sylvester and was intended to mean ''mother of determinants.''

Some of the greatest mathematicians of the eighteenth and nineteenth centuries helped to develop properties of determinants. Most historians believe that the theory of determinants originated with the German mathematician Gottfried Wilhelm Leibniz (1646–1716), who, with Newton, was the co-inventor of calculus. Leibniz used determinants in 1693 in reference to systems of simultaneous linear equations. Some believe, however, that a Japanese mathematician, Seki Kōwa, did the same thing about 10 years earlier.

The most prolific contributor to the theory of determinants was the French mathematician Augustin-Louis Cauchy (1789–1857). Cauchy wrote an 84-page memoir in 1812 that contained the first proof of the theorem $\det AB = \det A \det B$. In 1840 Cauchy defined the characteristic equation of the matrix A to be the polynomial equation $\det(A - \lambda I) = 0$. We shall discuss this equation in great detail in Chapter 8.

Cauchy made many other contributions to mathematics. In his 1829 calculus textbook *Leçons sur le calcul différential,* he gave the first reasonably clear definition of a limit.

Cauchy wrote extensively in both pure and applied mathematics. Only Euler wrote more. Cauchy contributed to many areas including real and complex function theory, probability theory, geometry, wave propagation theory, and infinite series.

Cauchy is credited with setting a new standard of rigor in mathematical publication. After Cauchy, it was much more difficult to publish a paper based on intuition; a strict adherence to formal proof was demanded.

The sheer volume of Cauchy's publication was overwhelming. When the French Academy of Sciences began publishing its journal *Comptes Rendus* in 1835, Cauchy sent his work there to be published. Soon the printing bill for Cauchy's work alone became so large that the Academy placed a limit of four pages on each published paper. This rule is still in force today.

Some other mathematicians are worthy of mention here. The expansion of a determinant by cofactors was first used by the French mathematician Pierre-Simon Laplace (1749–1827). Laplace is best known for the Laplace transform studied in applied mathematics courses.

A major contributor to determinant theory (second only to Cauchy) was the German mathematician Carl Gustav Jacobi (1804–1851). It was with him that the word ''determinant'' gained final acceptance. Jacobi first used the determinant applied to functions in the setting of the theory of functions of several variables. This determinant was later named the *Jacobian* by Sylvester. Students today study Jacobians in second year calculus classes.

Finally, no history of determinants would be complete without citing the text book *An Elementary Theory of Determinants,* written in 1867 by Charles Dodgson (1832–1898). In this book Dodgson gives conditions such that systems of equations have nontrivial solutions. These conditions are written in terms of the determinants of the minors of coefficient matrices. Charles Dodgson is better known by his pen name Lewis Carroll. Under this pen name he wrote his much better known book *Alice in Wonderland.*

AUGUSTIN-LOUIS CAUCHY (DAVID EUGENE SMITH COLLECTION, RARE BOOK AND MANUSCRIPT LIBRARY, COLUMBIA UNIVERSITY)

PROBLEMS 7.2

SELF-QUIZ

Multiple Choice

I. Which of the following is 0?

a. $\begin{vmatrix} 1 & 2 & 3 \\ 1 & 2 & 4 \\ 1 & 6 & 4 \end{vmatrix}$ **b.** $\begin{vmatrix} 1 & 2 & 7 \\ 2 & 3 & 8 \\ -1 & -2 & -7 \end{vmatrix}$

c. $\begin{vmatrix} 2 & 1 & 3 \\ -2 & 1 & 3 \\ 0 & 2 & 5 \end{vmatrix}$ **d.** $\begin{vmatrix} 1 & 0 & 0 \\ 0 & -1 & 0 \\ 0 & 0 & 4 \end{vmatrix}$

II. Which of the following is 0?

a. $\begin{vmatrix} 1 & 2 & 3 & 4 \\ -1 & 2 & -3 & 4 \\ 3 & -1 & 5 & 2 \\ 3 & 1 & 5 & 2 \end{vmatrix}$ **b.** $\begin{vmatrix} 1 & 3 & 0 & 1 \\ 0 & 2 & 1 & 4 \\ 3 & 1 & 0 & 2 \\ 0 & 0 & 0 & 5 \end{vmatrix}$

c. $\begin{vmatrix} 1 & 2 & 2 & 1 \\ -1 & 5 & -2 & 0 \\ 2 & 7 & 4 & 2 \\ 3 & 4 & 6 & 5 \end{vmatrix}$ **d.** $\begin{vmatrix} 2 & 1 & -1 & 1 \\ 2 & 1 & 1 & -1 \\ 3 & 0 & 0 & 2 \\ 0 & 3 & 2 & 0 \end{vmatrix}$

III. The determinant of $\begin{pmatrix} 1 & 2 & 3 \\ -1 & 2 & 4 \\ -1 & 2 & 5 \end{pmatrix}$ is ________.

a. 32 **b.** 10 **c.** −10
d. −4 **e.** 4

IV. The determinant of $\begin{pmatrix} 1 & -1 & 2 & 3 \\ -1 & 2 & 5 & 7 \\ 0 & 0 & -1 & 6 \\ -2 & 2 & -4 & -5 \end{pmatrix} =$ ________.

a. 10 **b.** 2 **c.** −8
d. 1 **e.** −1

In Problems 1–20 evaluate the determinant by using the methods of this section.

1. $\begin{vmatrix} 3 & -5 \\ 2 & 6 \end{vmatrix}$

2. $\begin{vmatrix} 4 & 1 \\ 0 & -3 \end{vmatrix}$

3. $\begin{vmatrix} -1 & 0 & 2 \\ 3 & 1 & 4 \\ 2 & 0 & -6 \end{vmatrix}$

4. $\begin{vmatrix} 2 & 1 & -1 \\ 3 & -2 & 0 \\ 5 & 1 & 6 \end{vmatrix}$

5. $\begin{vmatrix} -3 & 2 & 4 \\ 1 & -1 & 2 \\ -1 & 4 & 0 \end{vmatrix}$

6. $\begin{vmatrix} 0 & -2 & 3 \\ 1 & 2 & -3 \\ 4 & 0 & 5 \end{vmatrix}$

7. $\begin{vmatrix} -2 & 3 & 6 \\ 4 & 1 & 8 \\ -2 & 0 & 0 \end{vmatrix}$

8. $\begin{vmatrix} 2 & -1 & 3 \\ 4 & 0 & 6 \\ 5 & -2 & 3 \end{vmatrix}$

9. $\begin{vmatrix} 1 & -1 & 2 & 4 \\ 0 & -3 & 5 & 6 \\ 1 & 4 & 0 & 3 \\ 0 & 5 & -6 & 7 \end{vmatrix}$

10. $\begin{vmatrix} 2 & -3 & 1 & 4 \\ 0 & -2 & 0 & 0 \\ 3 & 7 & -1 & 2 \\ 4 & 1 & -3 & 8 \end{vmatrix}$

11. $\begin{vmatrix} 1 & 1 & -1 & 0 \\ -3 & 4 & 6 & 0 \\ 2 & 5 & -1 & 3 \\ 4 & 0 & 3 & 0 \end{vmatrix}$

12. $\begin{vmatrix} 3 & -1 & 2 & 1 \\ 4 & 3 & 1 & -2 \\ -1 & 0 & 2 & 3 \\ 6 & 2 & 5 & 2 \end{vmatrix}$

13. $\begin{vmatrix} 2 & 0 & 0 & 0 \\ 0 & 0 & 3 & 0 \\ 0 & -1 & 0 & 0 \\ 0 & 0 & 0 & 4 \end{vmatrix}$

14. $\begin{vmatrix} 0 & a & 0 & 0 \\ b & 0 & 0 & 0 \\ 0 & 0 & 0 & c \\ 0 & 0 & d & 0 \end{vmatrix}$

15. $\begin{vmatrix} 1 & 2 & 0 & 0 \\ 3 & -2 & 0 & 0 \\ 0 & 0 & 1 & -5 \\ 0 & 0 & 7 & 2 \end{vmatrix}$

16. $\begin{vmatrix} a & b & 0 & 0 \\ c & d & 0 & 0 \\ 0 & 0 & a & -b \\ 0 & 0 & c & d \end{vmatrix}$

17. $\begin{vmatrix} 2 & -1 & 0 & 4 & 1 \\ 3 & 1 & -1 & 2 & 0 \\ 3 & 2 & -2 & 5 & 1 \\ 0 & 0 & 4 & -1 & 6 \\ 3 & 2 & 1 & -1 & 1 \end{vmatrix}$

18. $\begin{vmatrix} 1 & -1 & 2 & 0 & 0 \\ 3 & 1 & 4 & 0 & 0 \\ 2 & -1 & 5 & 0 & 0 \\ 0 & 0 & 0 & 2 & 3 \\ 0 & 0 & 0 & -1 & 4 \end{vmatrix}$

19. $\begin{vmatrix} a & 0 & 0 & 0 & 0 \\ 0 & 0 & b & 0 & 0 \\ 0 & 0 & 0 & 0 & c \\ 0 & 0 & 0 & d & 0 \\ 0 & e & 0 & 0 & 0 \end{vmatrix}$

20. $\begin{vmatrix} 2 & 5 & -6 & 8 & 0 \\ 0 & 1 & -7 & 6 & 0 \\ 0 & 0 & 0 & 4 & 0 \\ 0 & 2 & 1 & 5 & 1 \\ 4 & -1 & 5 & 3 & 0 \end{vmatrix}$

In Problems 21–27 compute the determinant assuming that

$$\begin{vmatrix} a_{11} & a_{12} & a_{13} \\ a_{21} & a_{22} & a_{23} \\ a_{31} & a_{32} & a_{33} \end{vmatrix} = 8$$

21. $\begin{vmatrix} a_{31} & a_{32} & a_{33} \\ a_{21} & a_{22} & a_{23} \\ a_{11} & a_{12} & a_{13} \end{vmatrix}$

22. $\begin{vmatrix} a_{31} & a_{32} & a_{33} \\ a_{11} & a_{12} & a_{13} \\ a_{21} & a_{22} & a_{23} \end{vmatrix}$

23. $\begin{vmatrix} a_{11} & a_{12} & a_{13} \\ 2a_{21} & 2a_{22} & 2a_{23} \\ a_{31} & a_{32} & a_{33} \end{vmatrix}$

24. $\begin{vmatrix} -3a_{11} & -3a_{12} & -3a_{13} \\ 2a_{21} & 2a_{22} & 2a_{23} \\ 5a_{31} & 5a_{32} & 5a_{33} \end{vmatrix}$

25. $\begin{vmatrix} a_{11} & 2a_{13} & a_{12} \\ a_{21} & 2a_{23} & a_{22} \\ a_{31} & 2a_{33} & a_{32} \end{vmatrix}$

26. $\begin{vmatrix} a_{11} - a_{12} & a_{12} & a_{13} \\ a_{21} - a_{22} & a_{22} & a_{23} \\ a_{31} - a_{32} & a_{32} & a_{33} \end{vmatrix}$

27. $\begin{vmatrix} 2a_{11} - 3a_{21} & 2a_{12} - 3a_{22} & 2a_{13} - 3a_{23} \\ a_{31} & a_{32} & a_{33} \\ a_{21} & a_{22} & a_{23} \end{vmatrix}$

28. Using Property 2, show that if α is a number and A is an $n \times n$ matrix, then

$$\det \alpha A = \alpha^n \det A.$$

***29.** Show that

$$\begin{vmatrix} 1 + x_1 & x_2 & x_3 & \cdots & x_n \\ x_1 & 1 + x_2 & x_3 & \cdots & x_n \\ x_1 & x_2 & 1 + x_3 & \cdots & x_n \\ \vdots & \vdots & \vdots & & \vdots \\ x_1 & x_2 & x_3 & \cdots & 1 + x_n \end{vmatrix} = 1 + x_1 + x_2 + \cdots + x_n$$

***30.** A matrix is **skew-symmetric** if $A^t = -A$. If A is an $n \times n$ skew-symmetric matrix, show that $\det A = (-1)^n \det A$.

31. Using the result of Problem 30, show that if A is a skew-symmetric $n \times n$ matrix and n is odd, then $\det A = 0$.

32. A matrix A is called **orthogonal** if A is invertible and $A^{-1} = A^t$. Show that if A is orthogonal, then $\det A = \pm 1$.

****33.** Let Δ denote the triangle in the plane with vertices at (x_1, y_1), (x_2, y_2), and (x_3, y_3). Show that the area of the triangle is given by

$$\text{Area of } \Delta = \pm\frac{1}{2}\begin{vmatrix} 1 & x_1 & y_1 \\ 1 & x_2 & y_2 \\ 1 & x_3 & y_3 \end{vmatrix}$$

Under what circumstances will this determinant equal zero?

****34.** Three lines, no two of which are parallel, determine a triangle in the plane. Suppose that the lines are given by

$$a_{11}x + a_{12}y + a_{13} = 0$$
$$a_{21}x + a_{22}y + a_{23} = 0$$
$$a_{31}x + a_{32}y + a_{33} = 0$$

Show that the area determined by the lines is

$$\frac{\pm 1}{2A_{13}A_{23}A_{33}}\begin{vmatrix} A_{11} & A_{12} & A_{13} \\ A_{21} & A_{22} & A_{23} \\ A_{31} & A_{32} & A_{33} \end{vmatrix}$$

35. The 3×3 **Vandermonde**[†] **determinant** is given by

$$D_3 = \begin{vmatrix} 1 & 1 & 1 \\ a_1 & a_2 & a_3 \\ a_1^2 & a_2^2 & a_3^2 \end{vmatrix}$$

Show that $D_3 = (a_2 - a_1)(a_3 - a_1)(a_3 - a_2)$.

[†] A. T. Vandermonde (1735–1796) was a French mathematician.

36. $D_4 = \begin{vmatrix} 1 & 1 & 1 & 1 \\ a_1 & a_2 & a_3 & a_4 \\ a_1^2 & a_2^2 & a_3^2 & a_4^2 \\ a_1^3 & a_2^3 & a_3^3 & a_4^3 \end{vmatrix}$ is the 4×4 Vandermonde determinant. Show that $D_4 = (a_2 - a_1)(a_3 - a_1) \times (a_4 - a_1)(a_3 - a_2)(a_4 - a_2)(a_4 - a_3)$.

****37. a.** Define the $n \times n$ Vandermonde determinant D_n.

b. Show that $D_n = \prod_{\substack{i=1 \\ j>i}}^{n} (a_j - a_i)$, where $\prod$ stands for the word "product." Note that the product in Problem 36 can be written $\prod_{\substack{i=1 \\ j>i}}^{4} (a_j - a_i)$.

38. The $n \times n$ matrix A is called **nilpotent** if $A^k = 0$, the $n \times n$ zero matrix, for some integer $k \geq 1$. Show that the following matrices are nilpotent and find the smallest k such that $A^k = 0$.

a. $\begin{pmatrix} 0 & 2 \\ 0 & 0 \end{pmatrix}$ **b.** $\begin{pmatrix} 0 & 1 & 3 \\ 0 & 0 & 4 \\ 0 & 0 & 0 \end{pmatrix}$

39. Show that if A is nilpotent, then $\det A = 0$.

40. The matrix A is called **idempotent** if $A^2 = A$. What are the possible values for $\det A$ if A is idempotent?

41. Let E be the representation of $R_i \rightleftarrows R_j$ and let B be an $n \times n$ matrix. Show that $\det EB = \det E \det B$. [*Hint:* Describe the matrix EB and then use equation (9) and Property 4.]

42. Let E be the representation of $R_j \rightarrow R_j + cR_i$ and let B be an $n \times n$ matrix. Show that $\det EB = \det E \det B$. [*Hint:* Describe the matrix EB and then use equation (10) and Property 7.]

43. Let E be the representation of $R_i \rightarrow cR_i$ and let B be an $n \times n$ matrix. Show that $\det EB = \det E \det B$. [*Hint:* Describe the matrix EB and then use equation (11) and Property 2.]

ANSWERS TO SELF-QUIZ

I. b **II.** c **III.** e **IV.** e

7.3 DETERMINANTS AND INVERSES

In this section we shall see how matrix inverses can be calculated by using determinants. We begin with a simple result.

THEOREM 1

If A is invertible, then $\det A \neq 0$ and

$$\det A^{-1} = \frac{1}{\det A}. \tag{1}$$

PROOF: I is a triangular matrix so $\det I$ = product of diagonal elements = 1. Then, since $\det AB = \det A \det B$ (Theorem 7.2.4), we have

$$1 = \det I = \det AA^{-1} = \det A \det A^{-1} \tag{2}$$

By Theorem 7.2.5 $\det A \neq 0$ so $\det A^{-1} = 1/\det A$. ■

Before using determinants to calculate inverses, we need to define the *adjoint* of a matrix $A = (a_{ij})$. Let $B = (A_{ij})$ be the matrix of cofactors of A. (Remember that a cofactor, defined on page 451, is a number.) Then

$$B = \begin{pmatrix} A_{11} & A_{12} & \cdots & A_{1n} \\ A_{21} & A_{22} & \cdots & A_{2n} \\ \vdots & \vdots & & \vdots \\ A_{n1} & A_{n2} & \cdots & A_{nn} \end{pmatrix}. \tag{3}$$

DEFINITION THE ADJOINT

Let A be an $n \times n$ matrix and let B, given by (3), denote the matrix of its cofactors. Then the **adjoint** of A, written adj A, is the transpose of the $n \times n$ matrix B; that is,

$$\text{adj } A = B^t = \begin{pmatrix} A_{11} & A_{21} & \cdots & A_{n1} \\ A_{12} & A_{22} & \cdots & A_{n2} \\ \vdots & \vdots & & \vdots \\ A_{1n} & A_{2n} & \cdots & A_{nn} \end{pmatrix}. \tag{4}$$

■

NOTE: In some books, the adjoint of a matrix is referred to as the **adjugate** of the matrix.

EXAMPLE 1 COMPUTING THE ADJOINT OF A 3 × 3 MATRIX

Let $A = \begin{pmatrix} 2 & 4 & 3 \\ 0 & 1 & -1 \\ 3 & 5 & 7 \end{pmatrix}$. Compute adj A.

SOLUTION: We have $A_{11} = \begin{vmatrix} 1 & -1 \\ 5 & 7 \end{vmatrix} = 12$, $A_{12} = -\begin{vmatrix} 0 & -1 \\ 3 & 7 \end{vmatrix} = -3$, $A_{13} = -3$,

$A_{21} = -13$, $A_{22} = 5$, $A_{23} = 2$, $A_{31} = -7$, $A_{32} = 2$, and $A_{33} = 2$. Thus

$$B = \begin{pmatrix} 12 & -3 & -3 \\ -13 & 5 & 2 \\ -7 & 2 & 2 \end{pmatrix} \text{ and adj } A = B^t = \begin{pmatrix} 12 & -13 & -7 \\ -3 & 5 & 2 \\ -3 & 2 & 2 \end{pmatrix}.$$

EXAMPLE 2 COMPUTING THE ADJOINT OF A 4 × 4 MATRIX

Let

$$A = \begin{pmatrix} 1 & -3 & 0 & -2 \\ 3 & -12 & -2 & -6 \\ -2 & 10 & 2 & 5 \\ -1 & 6 & 1 & 3 \end{pmatrix}$$

Calculate adj A.

SOLUTION: This is more tedious since we have to compute sixteen 3×3 determinants. For example, we have $A_{12} = -\begin{vmatrix} 3 & -2 & -6 \\ -2 & 2 & 5 \\ -1 & 1 & 3 \end{vmatrix} = -1$, $A_{24} = \begin{vmatrix} 1 & -3 & 0 \\ -2 & 10 & 2 \\ -1 & 6 & 1 \end{vmatrix} = -2$, and $A_{43} = -\begin{vmatrix} 1 & -3 & -2 \\ 3 & -12 & -6 \\ -2 & 10 & 5 \end{vmatrix} = 3$. Completing these calculations, we find that

$$B = \begin{pmatrix} 0 & -1 & 0 & 2 \\ -1 & 1 & -1 & -2 \\ 0 & 2 & -3 & -3 \\ -2 & -2 & 3 & 2 \end{pmatrix}$$

$$\text{adj } A = B^t = \begin{pmatrix} 0 & -1 & 0 & -2 \\ -1 & 1 & 2 & -2 \\ 0 & -1 & -3 & 3 \\ 2 & -2 & -3 & 2 \end{pmatrix}$$

EXAMPLE 3

THE ADJOINT OF A 2×2 MATRIX

Let $A = \begin{pmatrix} a_{11} & a_{12} \\ a_{21} & a_{22} \end{pmatrix}$. Then $\text{adj } A = \begin{pmatrix} A_{11} & A_{21} \\ A_{12} & A_{22} \end{pmatrix} = \begin{pmatrix} a_{22} & -a_{12} \\ -a_{21} & a_{11} \end{pmatrix}$.

WARNING: In taking the adjoint of a matrix, do not forget to transpose the matrix of cofactors.

THEOREM 2

Let A be an $n \times n$ matrix. Then

$$(A)(\text{adj } A) = \begin{pmatrix} \det A & 0 & 0 & \cdots & 0 \\ 0 & \det A & 0 & \cdots & 0 \\ 0 & 0 & \det A & \cdots & 0 \\ \vdots & \vdots & \vdots & & \vdots \\ 0 & 0 & 0 & \cdots & \det A \end{pmatrix} = (\det A)I \tag{5}$$

PROOF: Let $C = (c_{ij}) = (A)(\text{adj } A)$. Then

$$C = \begin{pmatrix} a_{11} & a_{12} & \cdots & a_{1n} \\ a_{21} & a_{22} & \cdots & a_{2n} \\ \vdots & \vdots & & \vdots \\ a_{n1} & a_{n2} & \cdots & a_{nn} \end{pmatrix} \begin{pmatrix} A_{11} & A_{21} & \cdots & A_{n1} \\ A_{12} & A_{22} & \cdots & A_{n2} \\ \vdots & \vdots & & \vdots \\ A_{1n} & A_{2n} & \cdots & A_{nn} \end{pmatrix} \tag{6}$$

We have

$$c_{ij} = (i\text{th row of } A) \cdot (j\text{th column of adj } A)$$

$$= (a_{i1} \quad a_{i2} \cdots a_{in}) \cdot \begin{pmatrix} A_{j1} \\ A_{j2} \\ \vdots \\ A_{jn} \end{pmatrix}$$

Thus

$$c_{ij} = a_{i1}A_{j1} + a_{i2}A_{j2} + \cdots + a_{in}A_{jn} \tag{7}$$

Now if $i = j$, the sum in (7) equals $a_{i1}A_{i1} + a_{i2}A_{i2} + \cdots + a_{in}A_{in}$, which is the expansion of det A in the ith row of A. On the other hand, if $i \neq j$, then from Theorem 7.2.2 on page 465, the sum in (7) equals zero. Thus

$$c_{ij} = \begin{cases} \det A & \text{if } i = j \\ 0 & \text{if } i \neq j \end{cases}$$

This proves the theorem. ■

We can now state the main result.

THEOREM 3

Let A be an $n \times n$ matrix. Then A is invertible if and only if $\det A \neq 0$. If $\det A \neq 0$, then

$$A^{-1} = \frac{1}{\det A} \operatorname{adj} A \tag{8}$$

Note that Theorem 6.8.4 on page 420 for 2×2 matrices is a special case of this theorem.

PROOF: If A is invertible, then $\det A \neq 0$ by Theorem 7.2.5. If $\det A \neq 0$, then from Theorem 2,

$$(A)\left(\frac{1}{\det A} \operatorname{adj} A\right) = \frac{1}{\det A}[A(\operatorname{adj} A)] = \frac{1}{\det A}(\det A)I = I.$$

But, by Theorem 6.8.8 on page 428, if $AB = I$, then $B = A^{-1}$. Thus

$$(1/\det A) \operatorname{adj} A = A^{-1}$$

■

EXAMPLE 4 USING THE DETERMINANT AND THE ADJOINT TO CALCULATE AN INVERSE

Let $A = \begin{pmatrix} 2 & 4 & 3 \\ 0 & 1 & -1 \\ 3 & 5 & 7 \end{pmatrix}$. Determine whether A is invertible and calculate A^{-1} if it is.

SOLUTION: Since $\det A = 3 \neq 0$, we see that A is invertible. From Example 1,

$$\operatorname{adj} A = \begin{pmatrix} 12 & -13 & -7 \\ -3 & 5 & 2 \\ -3 & 2 & 2 \end{pmatrix}$$

Thus

$$A^{-1} = \frac{1}{3}\begin{pmatrix} 12 & -13 & -7 \\ -3 & 5 & 2 \\ -3 & 2 & 2 \end{pmatrix} = \begin{pmatrix} 4 & -\frac{13}{3} & -\frac{7}{3} \\ -1 & \frac{5}{3} & \frac{2}{3} \\ -1 & \frac{2}{3} & \frac{2}{3} \end{pmatrix}$$

Check:

$$A^{-1}A = \frac{1}{3}\begin{pmatrix} 12 & -13 & -7 \\ -3 & 5 & 2 \\ -3 & 2 & 2 \end{pmatrix}\begin{pmatrix} 2 & 4 & 3 \\ 0 & 1 & -1 \\ 3 & 5 & 7 \end{pmatrix} = \frac{1}{3}\begin{pmatrix} 3 & 0 & 0 \\ 0 & 3 & 0 \\ 0 & 0 & 3 \end{pmatrix} = I$$

EXAMPLE 5

CALCULATING THE INVERSE OF A 4×4 MATRIX USING THE DETERMINANT AND THE ADJOINT

Let

$$A = \begin{pmatrix} 1 & -3 & 0 & -2 \\ 3 & -12 & -2 & -6 \\ -2 & 10 & 2 & 5 \\ -1 & 6 & 1 & 3 \end{pmatrix}$$

Determine whether A is invertible and, if so, calculate A^{-1}.

SOLUTION: Using properties of determinants, we compute

$$\begin{vmatrix} 1 & -3 & 0 & -2 \\ 3 & -12 & -2 & -6 \\ -2 & 10 & 2 & 5 \\ -1 & 6 & 1 & 3 \end{vmatrix}$$

Multiply the first column by 3 and 2 and add it to the second and fourth columns, respectively.

$$= \begin{vmatrix} 1 & 0 & 0 & 0 \\ 3 & -3 & -2 & 0 \\ -2 & 4 & 2 & 1 \\ -1 & 3 & 1 & 1 \end{vmatrix}$$

Expand in the first row.

$$= \begin{vmatrix} -3 & -2 & 0 \\ 4 & 2 & 1 \\ 3 & 1 & 1 \end{vmatrix} = -1$$

Thus $\det A = -1 \neq 0$ and A^{-1} exists. By Example 2, we have

$$\operatorname{adj} A = \begin{pmatrix} 0 & -1 & 0 & -2 \\ -1 & 1 & 2 & -2 \\ 0 & -1 & -3 & 3 \\ 2 & -2 & -3 & 2 \end{pmatrix}$$

Thus

$$A^{-1} = \frac{1}{-1}\begin{pmatrix} 0 & -1 & 0 & -2 \\ -1 & 1 & 2 & -2 \\ 0 & -1 & -3 & 3 \\ 2 & -2 & -3 & 2 \end{pmatrix} = \begin{pmatrix} 0 & 1 & 0 & 2 \\ 1 & -1 & -2 & 2 \\ 0 & 1 & 3 & -3 \\ -2 & 2 & 3 & -2 \end{pmatrix}$$

Check:

$$AA^{-1} = \begin{pmatrix} 1 & -3 & 0 & -2 \\ 3 & -12 & -2 & -6 \\ -2 & 10 & 2 & 5 \\ -1 & 6 & 1 & 3 \end{pmatrix}\begin{pmatrix} 0 & 1 & 0 & 2 \\ 1 & -1 & -2 & 2 \\ 0 & 1 & 3 & -3 \\ -2 & 2 & 3 & -2 \end{pmatrix} = \begin{pmatrix} 1 & 0 & 0 & 0 \\ 0 & 1 & 0 & 0 \\ 0 & 0 & 1 & 0 \\ 0 & 0 & 0 & 1 \end{pmatrix}$$

NOTE: As you have noticed, if $n > 3$ it is generally easier to compute A^{-1} by row reduction than by using adj A since, even for the 4×4 case, it is necessary to calculate 17 determinants (16 for the adjoint plus det A). Nevertheless, Theorem 3 is very important since, before you do any row reduction, the calculation of det A (if it can be done easily) will tell you whether or not A^{-1} exists.

We last saw our Summing Up Theorem on page 440. This is the theorem that ties together many of the concepts developed in Chapters 6 and 7. We extend this theorem now.

THEOREM 4 SUMMING UP THEOREM—VIEW 4

Let A be an $n \times n$ matrix. Then the following six statements are equivalent. That is, each one implies the other five (so that, if one is true, all are true):

i. A is invertible.

ii. The only solution to the homogeneous system $A\mathbf{x} = \mathbf{0}$ is the trivial solution ($\mathbf{x} = \mathbf{0}$).

iii. The system $A\mathbf{x} = \mathbf{b}$ has a unique solution for every n-vector $\mathbf{b}$.

iv. A is row equivalent to the $n \times n$ identity matrix I_n.

v. A is the product of elementary matrices.

vi. $\det A \neq 0$.

PROOF: In Theorem 6.8.6 we proved the equivalence of parts (i), (ii), (iii), and (iv). In Theorem 6.10.3 we proved the equivalence of (i) and (v). Theorem 7.2.5 proves the equivalence of (i) and (vi). ■

PROBLEMS 7.3

SELF-QUIZ

Multiple Choice

I. The determinant of $\begin{pmatrix} 1 & 2 & -1 & 4 \\ 2 & 3 & 2 & 4 \\ 5 & 1 & 0 & -3 \\ -4 & 3 & 1 & 6 \end{pmatrix}$ is -149.

The 2,3 component of A^{-1} is given by ______.

a. $-\dfrac{1}{149}\begin{vmatrix} 1 & 2 & 4 \\ 5 & 1 & -3 \\ -4 & 3 & 6 \end{vmatrix}$

b. $\dfrac{1}{149}\begin{vmatrix} 1 & 2 & 4 \\ 5 & 1 & -3 \\ -4 & 3 & 6 \end{vmatrix}$

c. $-\dfrac{1}{149}\begin{vmatrix} 1 & -1 & 4 \\ 2 & 2 & 4 \\ -4 & 1 & 6 \end{vmatrix}$

d. $\dfrac{1}{149}\begin{vmatrix} 1 & -1 & 4 \\ 2 & 2 & 4 \\ -4 & 1 & 6 \end{vmatrix}$

II. The determinant of $\begin{pmatrix} 3 & 7 & 2 \\ -1 & 5 & 8 \\ 6 & -4 & 4 \end{pmatrix}$ is 468.

The 3,1 component of A^{-1} is ______.

a. $-\dfrac{26}{468}$ **b.** $\dfrac{26}{468}$

c. $\dfrac{46}{468}$ **d.** $-\dfrac{46}{468}$

In Problems 1–12 use the methods of this section to determine whether the given matrix is invertible. If so, compute the inverse.

1. $\begin{pmatrix} 3 & 2 \\ 1 & 2 \end{pmatrix}$

2. $\begin{pmatrix} 3 & 6 \\ -4 & -8 \end{pmatrix}$

3. $\begin{pmatrix} 0 & 1 \\ 1 & 0 \end{pmatrix}$

4. $\begin{pmatrix} 1 & 1 & 1 \\ 0 & 2 & 3 \\ 5 & 5 & 1 \end{pmatrix}$

5. $\begin{pmatrix} 3 & 2 & 1 \\ 0 & 2 & 2 \\ 0 & 1 & -1 \end{pmatrix}$

6. $\begin{pmatrix} 1 & 1 & 1 \\ 0 & 1 & 1 \\ 0 & 0 & 1 \end{pmatrix}$

7. $\begin{pmatrix} 1 & 2 & 3 \\ 1 & 1 & 2 \\ 0 & 1 & 2 \end{pmatrix}$

8. $\begin{pmatrix} 3 & 1 & 0 \\ 1 & -1 & 2 \\ 1 & 1 & 1 \end{pmatrix}$

9. $\begin{pmatrix} 2 & -1 & 4 \\ -1 & 0 & 5 \\ 19 & -7 & 3 \end{pmatrix}$

10. $\begin{pmatrix} 1 & 6 & 2 \\ -2 & 3 & 4 \\ 7 & 12 & -4 \end{pmatrix}$

11. $\begin{pmatrix} 1 & 1 & 1 & 1 \\ 1 & 2 & -1 & 2 \\ 1 & -1 & 2 & 1 \\ 1 & 3 & 3 & 2 \end{pmatrix}$

12. $\begin{pmatrix} 1 & -3 & 0 & -2 \\ 3 & -12 & -2 & -6 \\ -2 & 10 & 2 & 5 \\ -1 & 6 & 1 & 3 \end{pmatrix}$

13. Use determinants to show that an $n \times n$ matrix A is invertible if and only if A^t is invertible.

14. For $A = \begin{pmatrix} 1 & 1 \\ 2 & 5 \end{pmatrix}$, verify that $\det A^{-1} = 1/\det A$.

15. For $A = \begin{pmatrix} 1 & -1 & 3 \\ 4 & 1 & 6 \\ 2 & 0 & -2 \end{pmatrix}$, verify that $\det A^{-1} = 1/\det A$.

16. For what values of α is the matrix $\begin{pmatrix} \alpha & -3 \\ 4 & 1-\alpha \end{pmatrix}$ not invertible?

17. For what values of α does the matrix

$$\begin{pmatrix} -\alpha & \alpha - 1 & \alpha + 1 \\ 1 & 2 & 3 \\ 2 - \alpha & \alpha + 3 & \alpha + 7 \end{pmatrix}$$

not have an inverse?

18. Suppose that the $n \times n$ matrix A is not invertible. Show that $(A)(\text{adj } A)$ is the zero matrix.

ANSWERS TO SELF-QUIZ

I. d **II.** a

7.4 CRAMER'S RULE (OPTIONAL)

In this section we examine an old method for solving systems with the same number of unknowns as equations. Consider the system of n equations in n unknowns.

$$\begin{aligned} a_{11}x_1 + a_{12}x_2 + \cdots + a_{1n}x_n &= b_1 \\ a_{21}x_1 + a_{22}x_2 + \cdots + a_{2n}x_n &= b_2 \\ \vdots \qquad \vdots \qquad\quad \vdots \quad &\quad \vdots \\ a_{n1}x_1 + a_{n2}x_2 + \cdots + a_{nn}x_n &= b_n \end{aligned} \tag{1}$$

which can be written in the form

$$A\mathbf{x} = \mathbf{b} \tag{2}$$

We suppose that $\det A \neq 0$. Then system (2) has a unique solution given by $\mathbf{x} = A^{-1}\mathbf{b}$. We can develop a method for finding that solution without row reduction and without computing A^{-1}.

Let $D = \det A$. We define n new matrices:

$$A_1 = \begin{pmatrix} b_1 & a_{12} & \cdots & a_{1n} \\ b_2 & a_{22} & \cdots & a_{2n} \\ \vdots & \vdots & & \vdots \\ b_n & a_{n2} & \cdots & a_{nn} \end{pmatrix}, \quad A_2 = \begin{pmatrix} a_{11} & b_1 & \cdots & a_{1n} \\ a_{21} & b_2 & \cdots & a_{2n} \\ \vdots & \vdots & & \vdots \\ a_{n1} & b_n & \cdots & a_{nn} \end{pmatrix}, \ldots,$$

$$A_n = \begin{pmatrix} a_{11} & a_{12} & \cdots & b_1 \\ a_{21} & a_{22} & \cdots & b_2 \\ \vdots & \vdots & & \vdots \\ a_{n1} & a_{n2} & \cdots & b_n \end{pmatrix}$$

That is, A_i is the matrix obtained by replacing the ith column of A with $\mathbf{b}$. Finally, let $D_1 = \det A_1,\ D_2 = \det A_2, \ldots, D_n = \det A_n$.

THEOREM 1 CRAMER'S RULE

Let A be an $n \times n$ matrix and suppose that $\det A \neq 0$. Then the unique solution to the system $A\mathbf{x} = \mathbf{b}$ is given by

$$x_1 = \frac{D_1}{D},\ x_2 = \frac{D_2}{D}, \ldots, x_i = \frac{D_i}{D}, \ldots, x_n = \frac{D_n}{D} \tag{3}$$

PROOF: The solution to $A\mathbf{x} = \mathbf{b}$ is $\mathbf{x} = A^{-1}\mathbf{b}$. But

$$A^{-1}\mathbf{b} = \frac{1}{D}(\text{adj } A)\mathbf{b} = \frac{1}{D}\begin{pmatrix} A_{11} & A_{21} & \cdots & A_{n1} \\ A_{12} & A_{22} & \cdots & A_{n2} \\ \vdots & \vdots & & \vdots \\ A_{1n} & A_{2n} & \cdots & A_{nn} \end{pmatrix}\begin{pmatrix} b_1 \\ b_2 \\ \vdots \\ b_n \end{pmatrix} \tag{4}$$

Now $(\text{adj } A)\mathbf{b}$ is an n-vector, the jth component of which is

$$(A_{1j}\ A_{2j}\ \ldots\ A_{nj}) \cdot \begin{pmatrix} b_1 \\ b_2 \\ \vdots \\ b_n \end{pmatrix} = b_1A_{1j} + b_2A_{2j} + \cdots + b_nA_{nj} \tag{5}$$

Consider the matrix A_j:

$$A_j = \begin{pmatrix} a_{11} & a_{12} & \cdots & b_1 & \cdots & a_{1n} \\ a_{21} & a_{22} & \cdots & b_2 & \cdots & a_{2n} \\ \vdots & \vdots & & \vdots & & \vdots \\ a_{n1} & a_{n2} & \cdots & b_n & \cdots & a_{nn} \end{pmatrix} \tag{6}$$

↑ jth column

If we expand the determinant of A_j in its jth column, we obtain

$$D_j = b_1 \text{ (cofactor of } b_1) + b_2 \text{ (cofactor of } b_2) + \cdots + b_n \text{ (cofactor of } b_n) \tag{7}$$

But to find the cofactor of b_i, say, we delete the ith row and jth column of A_j (since b_i is in the jth column of A_j). But the jth column of A_j is $\mathbf{b}$ and, with this deleted, we simply have the ij minor, M_{ij}, of A. Thus

$$\text{Cofactor of } b_i \text{ in } A_j = A_{ij}$$

so that (7) becomes

$$D_j = b_1A_{1j} + b_2A_{2j} + \cdots + b_nA_{nj} \tag{8}$$

But this is the same as the right side of (5). Thus the ith component of $(\text{adj } A)\mathbf{b}$ is D_i, and we have

$$\mathbf{x} = \begin{pmatrix} x_1 \\ x_2 \\ \vdots \\ x_n \end{pmatrix} = A^{-1}\mathbf{b} = \frac{1}{D}(\text{adj } A)\mathbf{b} = \frac{1}{D}\begin{pmatrix} D_1 \\ D_2 \\ \vdots \\ D_n \end{pmatrix} = \begin{pmatrix} D_1/D \\ D_2/D \\ \vdots \\ D_n/D \end{pmatrix}$$

and the proof is complete. ■

HISTORICAL NOTE: Cramer's rule is named for the Swiss mathematician Gabriel Cramer (1704–1752). Cramer published the rule in 1750 in his *Introduction to the Analysis of Lines of Algebraic Curves.* Actually, there is much evidence to suggest that the rule was known as early as 1729 to Colin Maclaurin (1698–1746), who was probably the most outstanding British mathematician in the years following the death of Newton. Cramer's rule is one of the most famous results in the history of mathematics. For almost 200 years it was central in the teaching of algebra and the theory of equations. Because of the great number of computations involved, the rule is used today less frequently. However, the result was very important in its time.

EXAMPLE 1

SOLVING A 3 × 3 SYSTEM USING CRAMER'S RULE

Solve, using Cramer's rule, the system

$$\begin{aligned} 2x_1 + 4x_2 + 6x_3 &= 18 \\ 4x_1 + 5x_2 + 6x_3 &= 24 \\ 3x_1 + x_2 - 2x_3 &= 4 \end{aligned} \tag{9}$$

SOLUTION: We have solved this before—using row reduction in Example 6.3.1 on page 376. We could also solve it by calculating A^{-1} (Example 6.8.6, page 421) and then finding $A^{-1}\mathbf{b}$. We now solve it by using Cramer's rule. First we have

$$D = \begin{vmatrix} 2 & 4 & 6 \\ 4 & 5 & 6 \\ 3 & 1 & -2 \end{vmatrix} = 6 \neq 0$$

so that system (9) has a unique solution. Then $D_1 = \begin{vmatrix} 18 & 4 & 6 \\ 24 & 5 & 6 \\ 4 & 1 & -2 \end{vmatrix} = 24$, $D_2 = \begin{vmatrix} 2 & 18 & 6 \\ 4 & 24 & 6 \\ 3 & 4 & -2 \end{vmatrix} = -12$ and $D_3 = \begin{vmatrix} 2 & 4 & 18 \\ 4 & 5 & 24 \\ 3 & 1 & 4 \end{vmatrix} = 18$. Hence $x_1 = \dfrac{D_1}{D} = \dfrac{24}{6} = 4$, $x_2 = \dfrac{D_2}{D} = -\dfrac{12}{6} = -2$ and $x_3 = \dfrac{D_3}{D} = \dfrac{18}{6} = 3$.

EXAMPLE 2 SOLVING A 4 × 4 SYSTEM USING CRAMER'S RULE

Show that the system

$$\begin{aligned} x_1 + 3x_2 + 5x_3 + 2x_4 &= 2 \\ -x_2 + 3x_3 + 4x_4 &= 0 \\ 2x_1 + x_2 + 9x_3 + 6x_4 &= -3 \\ 3x_1 + 2x_2 + 4x_3 + 8x_4 &= -1 \end{aligned} \tag{10}$$

has a unique solution and find it by using Cramer's rule.

SOLUTION: We saw in Example 7.2.10 on page 463 that

$$|A| = \begin{vmatrix} 1 & 3 & 5 & 2 \\ 0 & -1 & 3 & 4 \\ 2 & 1 & 9 & 6 \\ 3 & 2 & 4 & 8 \end{vmatrix} = 160 \neq 0$$

Thus the system has a unique solution. To find it we compute $D_1 = -464$; $D_2 = 280$; $D_3 = -56$; $D_4 = 112$. Thus $x_1 = D_1/D = -464/160$, $x_2 = D_2/D = 280/160$, $x_3 = D_3/D = -56/160$, and $x_4 = D_4/D = 112/160$. These solutions can be verified by direct substitution into system (10).

PROBLEMS 7.4

SELF-QUIZ

I. Consider the systems

$$\begin{aligned} 2x - 3y + 4z &= 7 \\ 3x + 8y - z &= 2 \\ -5x - 12y + 6z &= 11 \end{aligned}$$

If $D = \begin{vmatrix} 2 & -3 & 4 \\ 3 & 8 & -1 \\ -5 & -12 & 6 \end{vmatrix}$, then $y =$ ________.

a. $\frac{1}{D}\begin{vmatrix} 7 & -3 & 4 \\ 2 & 8 & -1 \\ 11 & -12 & 6 \end{vmatrix}$ **b.** $\frac{1}{D}\begin{vmatrix} 2 & -3 & 7 \\ 3 & 8 & 2 \\ -5 & -12 & 11 \end{vmatrix}$

c. $\frac{1}{D}\begin{vmatrix} 2 & 7 & 4 \\ 3 & 2 & -1 \\ -5 & 11 & 6 \end{vmatrix}$ **d.** $\frac{1}{D}\begin{vmatrix} 2 & -7 & 4 \\ 3 & -2 & -1 \\ -5 & -11 & 6 \end{vmatrix}$

In Problems 1–9 solve the given system by using Cramer's rule.

1. $\begin{aligned} 2x_1 + 3x_2 &= -1 \\ -7x_1 + 4x_2 &= 47 \end{aligned}$

2. $\begin{aligned} 3x_1 - x_2 &= 0 \\ 4x_1 + 2x_2 &= 5 \end{aligned}$

3. $\begin{aligned} 2x_1 + x_2 + x_3 &= 6 \\ 3x_1 - 2x_2 - 3x_3 &= 5 \\ 8x_1 + 2x_2 + 5x_3 &= 11 \end{aligned}$

4. $\begin{aligned} x_1 + x_2 + x_3 &= 8 \\ 4x_2 - x_3 &= -2 \\ 3x_1 - x_2 + 2x_3 &= 0 \end{aligned}$

5. $\begin{aligned} 2x_1 + 2x_2 + x_3 &= 7 \\ x_1 + 2x_2 - x_3 &= 0 \\ -x_1 + x_2 + 3x_3 &= 1 \end{aligned}$

6. $\begin{aligned} 2x_1 + 5x_2 - x_3 &= -1 \\ 4x_1 + x_2 + 3x_3 &= 3 \\ -2x_1 + 2x_2 &= 0 \end{aligned}$

7. $\begin{aligned} 2x_1 + x_2 - x_3 &= 4 \\ x_1 + x_3 &= 2 \\ -x_2 + 5x_3 &= 1 \end{aligned}$

8. $\begin{aligned} x_1 + x_2 + x_3 + x_4 &= 6 \\ 2x_1 - x_3 - x_4 &= 4 \\ 3x_3 + 6x_4 &= 3 \\ x_1 - x_4 &= 5 \end{aligned}$

9.
$$\begin{aligned} x_1 \qquad\qquad\quad - x_4 &= 7 \\ 2x_2 + x_3 \qquad &= 2 \\ 4x_1 - x_2 \qquad\qquad &= -3 \\ 3x_3 - 5x_4 &= 2 \end{aligned}$$

***10.** Consider the triangle in Figure 1.

a. Show, using elementary trigonometry, that

$$\begin{aligned} c\cos A \qquad\qquad + a\cos C &= b \\ b\cos A + a\cos B \qquad\qquad &= c \\ c\cos B + b\cos C &= a \end{aligned}$$

b. If the system of part (a) is thought of as a system of three equations in the three unknowns $\cos A$, $\cos B$, and $\cos C$, show that the determinant of the system is nonzero.

c. Use Cramer's rule to solve for $\cos C$.

d. Use part (c) to prove the **law of cosines**; $c^2 = a^2 + b^2 - 2ab\cos C$.

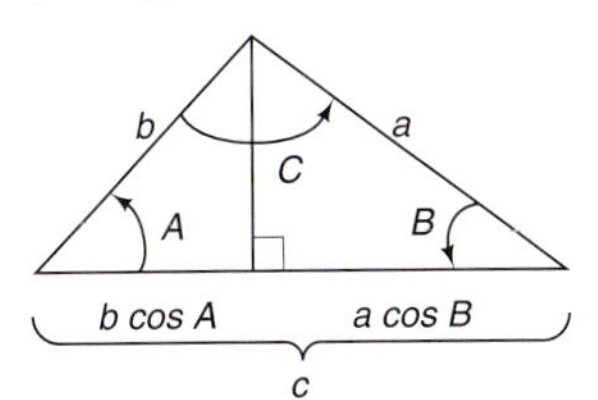

FIGURE 1

ANSWER TO SELF-QUIZ

I. c

CHAPTER 7 SUMMARY OUTLINE

- The **determinant** of a 2×2 matrix $A = \begin{pmatrix} a_{11} & a_{12} \\ a_{21} & a_{22} \end{pmatrix}$ is given by

  Determinant of $A = \det A = |A| = a_{11}a_{22} - a_{12}a_{21}$

- **3 × 3 Determinant**

$$\det\begin{pmatrix} a_{11} & a_{12} & a_{13} \\ a_{21} & a_{22} & a_{23} \\ a_{31} & a_{32} & a_{33} \end{pmatrix} = a_{11}\begin{vmatrix} a_{22} & a_{23} \\ a_{32} & a_{33} \end{vmatrix} - a_{12}\begin{vmatrix} a_{21} & a_{23} \\ a_{31} & a_{33} \end{vmatrix} + a_{13}\begin{vmatrix} a_{21} & a_{22} \\ a_{31} & a_{32} \end{vmatrix}$$

- The ***ij*th minor** of the $n \times n$ matrix A, denoted by M_{ij}, is the $(n-1) \times (n-1)$ matrix obtained by crossing off the ith row and jth column of A.
- The ***ij*th cofactor** of A, denoted by A_{ij}, is given by

  $A_{ij} = (-i)^{i+j} \det M_{ij}$

- ***n* × *n* Determinant** Let A be a $n \times n$ matrix. Then

$$\det A = a_{11}A_{11} + a_{12}A_{12} + \cdots + A_{1n}A_{1n} = \sum_{k=1}^{n} a_{1k}A_{1k}$$

  The sum above is called the **expansion of det *A* by cofactors**.

- If A is an upper triangular, lower triangular, or diagonal $n \times n$ matrix with diagonal components $a_{11}, a_{22}, \ldots, a_{nn}$, then

  $\det A = a_{11}\, a_{22} \cdots a_{nn}$

- **Basic Theorem** If A is an $n \times n$ matrix, then

$$\det A = a_{i1}A_{i1} + a_{i2}A_{i2} + \cdots + a_{in}A_{in} = \sum_{k=1}^{n} a_{ik}A_{ik}$$

  and

$$\det A = a_{1j}A_{1j} + a_{2j}A_{2j} + \cdots + a_{nj}A_{nj} = \sum_{k=1}^{n} a_{kj}A_{kj}$$

  for $i = 1, 2, \ldots, n$ and $j = 1, 2, \ldots, n$.
  That is, the determinant of A can be obtained by expanding in any row or column of A.

- If any row or column A is the zero vector, then $\det A = 0$.
- If any row (column) of A is multiplied by the constant c, then $\det A$ is multiplied by c.
- If A and B are two $n \times n$ matrices that are equal except in the jth column (ith row) and C is the matrix that is identical to A and B except that the jth column (ith row) of C is the sum of the jth column of A and the jth column of B (ith row of A and ith row of B), then $\det C = \det A + \det B$.
- Interchanging any two rows or columns of A has the effect of multiplying $\det A$ by -1.
- If any row (column) of A is multiplied by a constant and added to any other row (column) of A, then $\det A$ is unchanged.
- If one row (column) of A is a multiple of another row (column) of A, then $\det A = 0$.
- $\det A = \det A^t$.

- $\det AB = \det A \det B$
- If A is invertible, then $\det A \neq 0$ and

$$\det A^{-1} = \frac{1}{\det A}$$

- Let A be an $n \times n$ matrix. The **adjoint** of A, denoted by adj A, is the $n \times n$ matrix whose ijth component is A_{ji}, the jith cofactor of A.
- If $\det A \neq 0$, then A is invertible and

$$A^{-1} = \frac{1}{\det A} \operatorname{adj} A$$

- **Summing Up Theorem** Let A be an $n \times n$ matrix. Then the following six statements are equivalent:

  i. A is invertible.

  ii. The only solution to the homogeneous system $A\mathbf{x} = \mathbf{0}$ is the trivial solution ($\mathbf{x} = \mathbf{0}$).

  iii. The system $A\mathbf{x} = \mathbf{b}$ has a unique solution for every n-vector $\mathbf{b}$.

  iv. A is row equivalent to the $n \times n$ identity matrix I_n.

  v. A is the product of elementary matrices.

  vi. $\det A \neq 0$.

- **Cramer's Rule** Let A be an $n \times n$ matrix with $\det A \neq 0$. Then the unique solution to the system $A\mathbf{x} = \mathbf{b}$ is given by

$$x_1 = \frac{D_1}{\det A},\ x_2 = \frac{D_2}{\det A},\ \ldots,\ x_n = \frac{D_n}{\det A}$$

where D_j is the determinant of the matrix obtained by replacing the jth column of A by the column vector $\mathbf{b}$.

CHAPTER 7 REVIEW EXERCISES

In Exercises 1–8 calculate the determinant.

1. $\begin{vmatrix} -1 & 2 \\ 0 & 4 \end{vmatrix}$

2. $\begin{vmatrix} -3 & 5 \\ -7 & 4 \end{vmatrix}$

3. $\begin{vmatrix} 1 & -2 & 3 \\ 0 & 4 & 5 \\ 0 & 0 & 6 \end{vmatrix}$

4. $\begin{vmatrix} 5 & 0 & 0 \\ 6 & 2 & 0 \\ 10 & 100 & 6 \end{vmatrix}$

5. $\begin{vmatrix} 1 & -1 & 2 \\ 3 & 4 & 2 \\ -2 & 3 & 4 \end{vmatrix}$

6. $\begin{vmatrix} 3 & 1 & -2 \\ 4 & 0 & 5 \\ -6 & 1 & 3 \end{vmatrix}$

7. $\begin{vmatrix} 1 & -1 & 2 & 3 \\ 4 & 0 & 2 & 5 \\ -1 & 2 & 3 & 7 \\ 5 & 1 & 0 & 4 \end{vmatrix}$

8. $\begin{vmatrix} 3 & 15 & 17 & 19 \\ 0 & 2 & 21 & 60 \\ 0 & 0 & 1 & 50 \\ 0 & 0 & 0 & -1 \end{vmatrix}$

In Exercises 9–14 use determinants to calculate the inverse (if one exists).

9. $\begin{pmatrix} -3 & 4 \\ 2 & 1 \end{pmatrix}$

10. $\begin{pmatrix} 3 & -5 & 7 \\ 0 & 2 & 4 \\ 0 & 0 & -3 \end{pmatrix}$

11. $\begin{pmatrix} 1 & -1 & 2 \\ 3 & 1 & 4 \\ 5 & -1 & 8 \end{pmatrix}$

12. $\begin{pmatrix} 1 & 1 & 1 \\ 1 & 0 & 1 \\ 0 & 1 & 1 \end{pmatrix}$

13. $\begin{pmatrix} 2 & 1 & 0 & 0 \\ 0 & -1 & 3 & 0 \\ 1 & 0 & 0 & -2 \\ 3 & 0 & -1 & 0 \end{pmatrix}$

14. $\begin{pmatrix} 3 & -1 & 2 & 4 \\ 1 & 1 & 0 & 3 \\ -2 & 4 & 1 & 5 \\ 6 & -4 & 1 & 2 \end{pmatrix}$

In Exercises 15–18 solve the system by using Cramer's rule.

15. $\begin{aligned} 2x_1 - x_2 &= 3 \\ 3x_1 + 2x_2 &= 5 \end{aligned}$

16. $\begin{aligned} x_1 - x_2 + x_3 &= 7 \\ 2x_1 \qquad - 5x_3 &= 4 \\ 3x_2 - x_3 &= 2 \end{aligned}$

17. $\begin{aligned} 2x_1 + 3x_2 - x_3 &= 5 \\ -x_1 + 2x_2 + 3x_3 &= 0 \\ 4x_1 - x_2 + x_3 &= -1 \end{aligned}$

18. $\begin{aligned} x_1 \qquad - x_3 + x_4 &= 7 \\ 2x_2 + 2x_3 - 3x_4 &= -1 \\ 4x_1 - x_2 - x_3 \qquad &= 0 \\ -2x_1 + x_2 - 4x_3 \qquad &= 2 \end{aligned}$

CHAPTER 8

VECTOR SPACES AND LINEAR TRANSFORMATIONS

8.1 VECTOR SPACES

As we saw in Section 1.10, the set $\mathbb{R}^n$ has a number of nice properties. We can add two vectors in $\mathbb{R}^n$ and obtain another vector in $\mathbb{R}^n$. Under addition, vectors in $\mathbb{R}^n$ obey the commutative and associative laws. If $\mathbf{x} \in \mathbb{R}^n$, then $\mathbf{x} + \mathbf{0} = \mathbf{x}$ and $\mathbf{x} + (-\mathbf{x}) = \mathbf{0}$. We can multiply vectors in $\mathbb{R}^n$ by scalars, and we can show that this scalar multiplication satisfies a distributive law.

The set $\mathbb{R}^n$ is called a *vector space*. Loosely speaking, we can say that a vector space is a set of objects that obey the rules given in the preceding paragraph.

In this chapter we make a seemingly great leap from the concrete world of solving equations and dealing with easily visualized vectors to the world of abstract vector spaces. There is a great advantage in doing so. Once we have established a fact about vector spaces in general, we can apply that fact to *every* vector space. Otherwise, we would have to prove that fact again and again, once for each new vector space we encounter (and there is an endless supply of them). But, as you will see, the abstract theorems we shall prove are really no more difficult than the ones already encountered.

DEFINITION REAL VECTOR SPACE

A **real vector space** V is a set of objects, called **vectors**, together with two operations called **addition** and **scalar multiplication** that satisfy the ten axioms listed below. ■

NOTATION: If $\mathbf{x}$ and $\mathbf{y}$ are in V and if α is a real number, then we write $\mathbf{x} + \mathbf{y}$ for the sum of $\mathbf{x}$ and $\mathbf{y}$ and $\alpha\mathbf{x}$ for the scalar product of α and $\mathbf{x}$.

Before we list the properties satisfied by vectors in a vector space, two things should be mentioned. First, while it might be helpful to think of $\mathbb{R}^2$ or $\mathbb{R}^3$ when dealing with a vector space, it often occurs that a vector space may appear to be very different from these comfortable spaces. (We shall see this shortly.) Second, the definition above gives a definition of a *real* vector space. The word ''real'' means that the scalars we use are real numbers. It would be just as easy to define a *complex* vector space by using complex numbers instead of real ones. This book deals primarily with real vector spaces, but generalizations to other sets of scalars present little difficulty.

AXIOMS OF A VECTOR SPACE

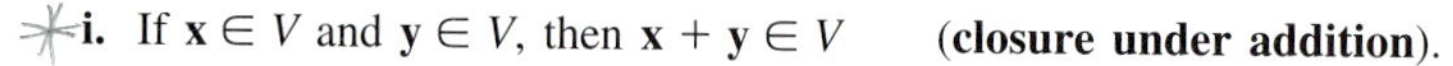

i. If $\mathbf{x} \in V$ and $\mathbf{y} \in V$, then $\mathbf{x} + \mathbf{y} \in V$ (**closure under addition**).

ii. For all $\mathbf{x}$, $\mathbf{y}$, and $\mathbf{z}$ in V, $(\mathbf{x} + \mathbf{y}) + \mathbf{z} = \mathbf{x} + (\mathbf{y} + \mathbf{z})$ (**associative law of vector addition**).

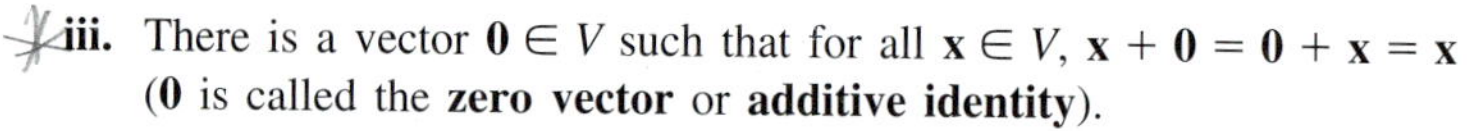

iii. There is a vector $\mathbf{0} \in V$ such that for all $\mathbf{x} \in V$, $\mathbf{x} + \mathbf{0} = \mathbf{0} + \mathbf{x} = \mathbf{x}$ ($\mathbf{0}$ is called the **zero vector** or **additive identity**).

iv. If $\mathbf{x} \in V$, there is a vector $-\mathbf{x}$ in V such that $\mathbf{x} + (-\mathbf{x}) = \mathbf{0}$ ($-\mathbf{x}$ is called the **additive inverse** of $\mathbf{x}$).

v. If $\mathbf{x}$ and $\mathbf{y}$ are in V, then $\mathbf{x} + \mathbf{y} = \mathbf{y} + \mathbf{x}$ (**commutative law of vector addition**).

vi. If $\mathbf{x} \in V$ and α is a scalar, then $\alpha\mathbf{x} \in V$ (**closure under scalar multiplication**).

vii. If $\mathbf{x}$ and $\mathbf{y}$ are in V and α is a scalar, then $\alpha(\mathbf{x} + \mathbf{y}) = \alpha\mathbf{x} + \alpha\mathbf{y}$ (**first distributive law**).

viii. If $\mathbf{x} \in V$ and α and β are scalars, then $(\alpha + \beta)\mathbf{x} = \alpha\mathbf{x} + \beta\mathbf{x}$ (**second distributive law**).

ix. If $\mathbf{x} \in V$ and α and β are scalars, then $\alpha(\beta\mathbf{x}) = (\alpha\beta)\mathbf{x}$ (**associative law of scalar multiplication**).

x. For every vector $\mathbf{x} \in V$, $1\mathbf{x} = \mathbf{x}$ (the scalar 1 is called a **multiplicative identity**).

EXAMPLE 1 THE SPACE $\mathbb{R}^n$

Let $V = \mathbb{R}^n = \{(x_1, x_2, \ldots, x_n): x_i \in \mathbb{R} \text{ for } i = 1, 2, \ldots, n\}$.

From Section 1.10 (see Theorem 1.10.1, page 75) we see that V satisfies all the axioms of a vector space if we take the set of scalars to be $\mathbb{R}$.

EXAMPLE 2 A TRIVIAL VECTOR SPACE

Let $V = \{0\}$. That is, V consists of the single number 0. Since $0 + 0 = 1 \cdot 0 = 0 + (0 + 0) = (0 + 0) + 0 = 0$, we see that V is a vector space. It is often referred to as a **trivial** vector space.

EXAMPLE 3 A SET THAT IS NOT A VECTOR SPACE

Let $V = \{1\}$. That is, V consists of the single number 1. This is *not* a vector space since it violates axiom (i)—the closure axiom. To see this we simply note that $1 + 1 = 2 \notin V$. It also violates other axioms as well. However, all we need to show is that it violates one axiom in order to prove that V is not a vector space.

EXAMPLE 4

THE SET OF POINTS IN $\mathbb{R}^2$ THAT LIE ON A LINE PASSING THROUGH THE ORIGIN CONSTITUTES A VECTOR SPACE

Let

$$V = \{(x, y)\colon y = mx, \text{ where } m \text{ is a fixed real number and } x \text{ is an arbitrary real number}\}.$$

That is, V consists of all points lying on the line $y = mx$ passing through the origin with slope m. To show that V is a vector space, we check each of the axioms.

i. Suppose that $\mathbf{x} = (x_1, y_1)$ and $\mathbf{y} = (x_2, y_2)$ are in V. Then $y_1 = mx_1$, $y_2 = mx_2$, and

$$\begin{aligned}\mathbf{x} + \mathbf{y} = (x_1, y_1) + (x_2, y_2) &= (x_1, mx_1) + (x_2, mx_2)\\ &= (x_1 + x_2, mx_1 + mx_2)\\ &= (x_1 + x_2, m(x_1 + x_2)) \in V\end{aligned}$$

Thus axiom (i) is satisfied.

ii. Let $\mathbf{z} = (x_3, y_3) = (x_3, mx_3)$ be in V. Then

$$\begin{aligned}(\mathbf{x} + \mathbf{y}) + \mathbf{z} &= [(x_1, y_1) + (x_2, y_2)] + (x_3, y_3)\\ &= (x_1 + x_2, mx_1 + mx_2) + (x_3, mx_3)\\ &= (x_1 + x_2 + x_3, mx_1 + mx_2 + mx_3)\\ &= (x_1, mx_1) + (x_2 + x_3, mx_2 + mx_3)\\ &= (x_1, y_1) + [(x_2, y_2) + (x_3, y_3)] = \mathbf{x} + (\mathbf{y} + \mathbf{z})\end{aligned}$$

iii. Let $\mathbf{0} = (0, 0) = (0, m \cdot 0) \in \mathbf{V}$. Then

$$\mathbf{x} + \mathbf{0} = (x_1, mx_1) + (0, 0) = (x_1 + 0, mx_1 + 0) = (x_1, mx_1) = \mathbf{x}$$

iv. If $\mathbf{x} = (x_1, mx_1)$, let $-\mathbf{x} = (-x_1, -mx_1)$. Then

$$\mathbf{x} + (-\mathbf{x}) = (x_1 - x_1, mx_1 - mx_1) = (0, 0) = \mathbf{0}$$

v.
$$\begin{aligned}\mathbf{x} + \mathbf{y} = (x_1, mx_1) + (x_2, mx_2) &= (x_1 + x_2, mx_1 + mx_2)\\ &= (x_2 + x_1, mx_2 + mx_1)\\ &= (x_2, mx_2) + (x_1, mx_1) = \mathbf{y} + \mathbf{x}\end{aligned}$$

vi. $\alpha\mathbf{x} = \alpha(x_1, mx_1) = (\alpha x_1, \alpha mx_1) = (\alpha x_1, m(\alpha x_1)) \in V$

vii.
$$\begin{aligned}\alpha(\mathbf{x} + \mathbf{y}) &= \alpha(x_1 + x_2, mx_1 + mx_2) = (\alpha x_1 + \alpha x_2, \alpha mx_1 + \alpha mx_2)\\ &= (\alpha x_1, \alpha mx_1) + (\alpha x_2, \alpha mx_2)\\ &= \alpha(x_1, mx_1) + \alpha(x_2, mx_2) = \alpha\mathbf{x} + \alpha\mathbf{y}\end{aligned}$$

viii.
$$\begin{aligned}(\alpha + \beta)\mathbf{x} &= (\alpha + \beta)(x_1, mx_1) = ((\alpha + \beta)x_1, (\alpha + \beta)mx_1)\\ &= (\alpha x_1 + \beta x_1, \alpha mx_1 + \beta mx_1) = (\alpha x_1, \alpha mx_1) + (\beta x_1, \beta mx_1)\\ &= \alpha(x_1, mx_1) + \beta(x_1, mx_1) = \alpha\mathbf{x} + \beta\mathbf{x}\end{aligned}$$

ix.
$$\begin{aligned}\alpha(\beta\mathbf{x}) &= \alpha[\beta(x_1, mx_1)] = \alpha(\beta x_1, \beta mx_1) = (\alpha\beta x_1, \alpha\beta mx_1)\\ &= \alpha\beta(x_1, mx_1) = \alpha\beta\mathbf{x}\end{aligned}$$

x. $1\mathbf{x} = 1(x_1, mx_1) = (1 \cdot x_1, 1 \cdot mx_1) = (x_1, mx_1) = \mathbf{x}$

Thus all ten axioms are satisfied, and we see that the set of points in the plane lying on a straight line passing through the origin constitutes a vector space.

NOTE: Checking all ten axioms can be tedious. From now on we shall check only those axioms that are not immediately obvious.

EXAMPLE 5 THE SET OF POINTS IN $\mathbb{R}^2$ LYING ON A LINE NOT PASSING THROUGH THE ORIGIN DOES NOT CONSTITUTE A VECTOR SPACE

Let $V = \{(x, y)\colon y = 2x + 1,\ x \in \mathbb{R}\}$. That is, V is the set of points lying on the line $y = 2x + 1$. V is *not* a vector space because closure is violated, as in Example 3. To see this, let us suppose that (x_1, y_1) and (x_2, y_2) are in V. Then

$$(x_1, y_1) + (x_2, y_2) = (x_1 + x_2, y_1 + y_2)$$

If this last vector were in V, we would have

$$y_1 + y_2 = 2(x_1 + x_2) + 1 = 2x_1 + 2x_2 + 1$$

But $y_1 = 2x_1 + 1$ and $y_2 = 2x_2 + 1$ so that

$$y_1 + y_2 = (2x_1 + 1) + (2x_2 + 1) = 2x_1 + 2x_2 + 2$$

Hence we conclude that

$$(x_1 + x_2, y_1 + y_2) \notin V \qquad \text{if } (x_1, y_1) \in V \qquad \text{and} \qquad (x_2, y_2) \in V.$$

An easier way to see that V is not a vector space is to observe that $\mathbf{0} = (0, 0)$ is not in V because $0 \neq 2 \cdot 0 + 1$.

EXAMPLE 6 THE SET OF POINTS IN $\mathbb{R}^3$ LYING ON A PLANE PASSING THROUGH THE ORIGIN CONSTITUTES A VECTOR SPACE

Let $V = \{(x, y, z);\ ax + by + cz = 0\}$. That is, V is the set of points in $\mathbb{R}^3$ lying on the plane with normal vector (a, b, c) and passing through the origin.

Suppose (x_1, y_1, z_1) and (x_2, y_2, z_2) are in V. Then $(x_1, y_1, z_1) + (x_2, y_2, z_2) = (x_1 + x_2, y_1 + y_2, z_1 + z_2) \in V$ because

$$\begin{aligned} &a(x_1 + x_2) + b(y_1 + y_2) + c(z_1 + z_2) \\ &\qquad = (ax_1 + by_1 + cz_1) + (ax_2 + by_2 + cz_2) = 0 + 0 = 0; \end{aligned}$$

hence axiom (i) is satisfied. The other axioms are easily verified. Thus the set of points lying on a plane in $\mathbb{R}^3$ that passes through the origin constitutes a vector space.

EXAMPLE 7 THE VECTOR SPACE P_n

Let $V = P_n$, the set of polynomials with real coefficients of degree less than or equal to n.[†] If $p \in P_n$, then

$$p(x) = a_n x^n + a_{n-1} x^{n-1} + \cdots + a_1 x + a_0$$

where each a_i is real. The sum $p(x) + q(x)$ is defined in the obvious way: If $q(x) = b_n x^n + b_{n-1} x^{n-1} + \cdots + b_1 x + b_0$, then

$$p(x) + q(x) = (a_n + b_n)x^n + (a_{n-1} + b_{n-1})x^{n-1} + \cdots + (a_1 + b_1)x + (a_0 + b_0)$$

[†] Constant functions (including the function $f(x) = 0$) are said to be polynomials of **degree zero**.

Clearly the sum of two polynomials of degree less than or equal to n is another polynomial with degree less than or equal to n, so axiom (i) is satisfied. Properties (ii) and (v) to (x) are obvious. If we define the zero polynomial $\mathbf{0} = 0x^n + 0x^{n-1} + \cdots + 0x + 0$, then clearly $\mathbf{0} \in P_n$ and axiom (iii) is satisfied. Finally, letting $-p(x) = -a_n x^n - a_{n-1}x^{n-1} - \cdots - a_1 x - a_0$, we see that axiom (iv) holds, so P_n is a real vector space.

EXAMPLE 8

THE VECTOR SPACES $C[0, 1]$ AND $C[a, b]$

Let $V = C[0, 1]$ = the set of real-valued continuous functions defined on the interval $[0, 1]$. We define

$$(f + g)x = f(x) + g(x) \qquad \text{and} \qquad (\alpha f)(x) = \alpha[f(x)].$$

Since the sum of continuous functions is continuous, axiom (i) is satisfied and the other axioms are easily verified with $\mathbf{0}$ = the zero function and $(-f)(x) = -f(x)$. Similarly, $C[a, b]$, the set of real-valued functions defined and continuous on $[a, b]$, constitutes a vector space.

EXAMPLE 9

THE VECTOR SPACE M_{34}

Let $V = M_{34}$ denote the set of 3×4 matrices with real components. Then with the usual sum and scalar multiplication of matrices, it is again easy to verify that M_{34} is a vector space with $\mathbf{0}$ being the 3×4 zero matrix. If $A = (a_{ij})$ is in M_{34}, then $-A = (-a_{ij})$ is also in M_{34}.

EXAMPLE 10

THE VECTOR SPACE M_{mn}

In an identical manner we see that M_{mn}, the set of $m \times n$ matrices with real components, forms a vector space for any positive integers m and n.

EXAMPLE 11

A SET OF INVERTIBLE MATRICES MIGHT NOT FORM A VECTOR SPACE

Let S_3 denote the set of invertible 3×3 matrices. Define the "sum" $A + B$ by $A + B = AB$. If A and B are invertible, then AB is invertible (by Theorem 6.8.3, page 416) so that axiom (i) is satisfied. Axiom (ii) is simply the associative law for matrix multiplication (Theorem 6.6.1, page 404); axioms (iii) and (iv) are satisfied with $\mathbf{0} = I_3$ and $-A = A^{-1}$. Axiom (v) fails, however, since in general $AB \neq BA$ so that S_3 is not a vector space.

EXAMPLE 12

A SET OF POINTS IN A HALF PLANE MIGHT NOT FORM A VECTOR SPACE

Let $V = \{(x, y)\colon y \geq 0\}$. V consists of the points in $\mathbb{R}^2$ in the upper half plane (the first two quadrants). If $y_1 \geq 0$ and $y_2 \geq 0$, then $y_1 + y_2 \geq 0$; hence if $(x_1, y_1) \in V$ and $(x_2, y_2) \in V$, then $(x_1 + x_2, y_1 + y_2) \in V$. V is not a vector space, however, since the vector $(1, 1)$, for example, does not have an inverse in V because $(-1, -1) \notin V$. Moreover, axiom (vi) fails since if $(x, y) \in V$, then $\alpha(x, y) \notin V$ if $\alpha < 0$.

EXAMPLE 13

THE SPACE $\mathbb{C}^n$

Let $V = \mathbb{C}^n = \{(c_1, c_2, \ldots, c_n)$: c_i is a complex number for $i = 1, 2, \ldots, n\}$ and the set of scalars is the set of complex numbers. It is not difficult to verify that $\mathbb{C}^n$, too, is a vector space.

NOTE: See Appendix 3 for an introduction to complex numbers if these are not familiar.

As these examples suggest, there are many different kinds of vector spaces and many kinds of sets that are *not* vector spaces. Before leaving this section, let us prove some elementary results about vector spaces.

THEOREM 1

Let V be a vector space. Then

i. $\alpha\mathbf{0} = \mathbf{0}$ for every real number α.
ii. $0 \cdot \mathbf{x} = \mathbf{0}$ for every $\mathbf{x} \in V$.
iii. If $\alpha\mathbf{x} = \mathbf{0}$, then $\alpha = 0$ or $\mathbf{x} = \mathbf{0}$ (or both).
iv. $(-1)\mathbf{x} = -\mathbf{x}$ for every $\mathbf{x} \in V$.

PROOF:

i. By axiom (*iii*), $\mathbf{0} + \mathbf{0} = \mathbf{0}$; and from axiom (*vii*),

$$\alpha(\mathbf{0} + \mathbf{0}) = \alpha\mathbf{0} + \alpha\mathbf{0} = \alpha\mathbf{0} \tag{1}$$

Adding $-\alpha\mathbf{0}$ to both sides of the last equation in (1) and using the associative law (axiom *ii*), we obtain

$$[\alpha\mathbf{0} + \alpha\mathbf{0}] + (-\alpha\mathbf{0}) = \alpha\mathbf{0} + (-\alpha\mathbf{0})$$
$$\alpha\mathbf{0} + [\alpha\mathbf{0} + (-\alpha\mathbf{0})] = \mathbf{0}$$
$$\alpha\mathbf{0} + \mathbf{0} = \mathbf{0}$$
$$\alpha\mathbf{0} = \mathbf{0}$$

ii. Essentially the same proof as used in part (*i*) works. We start with $0 + 0 = 0$ and use axiom (*viii*) to see that $0\mathbf{x} = (0 + 0)\mathbf{x} = 0\mathbf{x} + 0\mathbf{x}$ or $0\mathbf{x} + (-0\mathbf{x}) = 0\mathbf{x} + [0\mathbf{x} + (-0\mathbf{x})]$ or $\mathbf{0} = 0\mathbf{x} + \mathbf{0} = 0\mathbf{x}$.

iii. Let $\alpha\mathbf{x} = \mathbf{0}$. If $\alpha \neq 0$, we multiply both sides of the equation by $1/\alpha$ to obtain $(1/\alpha)(\alpha\mathbf{x}) = (1/\alpha)\mathbf{0} = \mathbf{0}$ [by part (*i*)]. But $(1/\alpha)(\alpha\mathbf{x}) = 1\mathbf{x} = \mathbf{x}$ (by axiom *ix*), so $\mathbf{x} = \mathbf{0}$.

iv. We start with the fact that $1 + (-1) = 0$. Then, using part (*ii*), we obtain

$$\mathbf{0} = 0\mathbf{x} = [1 + (-1)]\mathbf{x} = 1\mathbf{x} + (-1)\mathbf{x} = \mathbf{x} + (-1)\mathbf{x} \tag{2}$$

We add $-\mathbf{x}$ to both sides of (2) to obtain

$$\mathbf{0} + (-\mathbf{x}) = \mathbf{x} + (-1)\mathbf{x} + (-\mathbf{x}) = \mathbf{x} + (-\mathbf{x}) + (-1)\mathbf{x}$$
$$= \mathbf{0} + (-1)\mathbf{x} = (-1)\mathbf{x}$$

Thus $-\mathbf{x} = (-1)\mathbf{x}$. Note that we were able to reverse the order of addition in the preceding equation by using the commutative law (axiom *v*). ■

REMARK: Part (*iii*) of Theorem 1 is not so obvious as it seems. There are objects which have the property that $xy = 0$ does not imply that either x or y is zero. As an example, we look at the multiplication of 2×2 matrices. If $A = \begin{pmatrix} 0 & 1 \\ 0 & 0 \end{pmatrix}$ and $B = \begin{pmatrix} 0 & -2 \\ 0 & 0 \end{pmatrix}$, then neither A nor B is zero, although, as is easily verified, the product $AB = 0$, the zero matrix.

PROBLEMS 8.1

SELF-QUIZ

True–False

I. The set of vectors $\begin{pmatrix} x \\ y \end{pmatrix}$ in $\mathbb{R}^2$ with $y = -3x$ is a vector space.

II. The set of vectors $\begin{pmatrix} x \\ y \end{pmatrix}$ in $\mathbb{R}^2$ with $y = -3x + 1$ is a vector space.

III. The set of invertible 5×5 matrices forms a vector space under the usual matrix addition and scalar multiplication.

IV. The set of constant multiples of the 2×2 identity matrix is a vector space under the usual matrix addition and scalar multiplication.

V. The set of $n \times n$ identity matrices for $n = 2, 3, 4, \ldots$ is a vector space.

VI. The set of vectors $\begin{pmatrix} x \\ y \\ z \end{pmatrix}$ in $\mathbb{R}^3$ with $2x - y - 12z = 0$ is a vector space.

VII. The set of vectors $\begin{pmatrix} x \\ y \\ z \end{pmatrix}$ in $\mathbb{R}^3$ with $2x - y - 12z = 1$ is a vector space.

VIII. The set of polynomials of degree 3 is a vector space.

In Problems 1–20 determine whether the given set is a vector space. If it is not, list the axioms that do not hold.

1. The set of diagonal $n \times n$ matrices under the usual matrix addition and the usual scalar multiplication.

2. The set of diagonal $n \times n$ matrices under multiplication (that is, $A + B = AB$).

3. $\{(x, y): y \leq 0;\ x, y \text{ real}\}$ with the usual addition and scalar multiplication of vectors.

4. The vectors in the plane lying in the first quadrant.

5. The set of vectors in $\mathbb{R}^3$ in the form (x, x, x).

6. The set of polynomials of degree 4 under the operations of Example 7.

7. The set of $n \times n$ symmetric matrices (see page 433) under the usual addition and scalar multiplication.

8. The set of 2×2 matrices having the form $\begin{pmatrix} 0 & a \\ b & 0 \end{pmatrix}$ under the usual addition and scalar multiplication.

9. The set of matrices of the form $\begin{pmatrix} 1 & \alpha \\ \beta & 1 \end{pmatrix}$ with the matrix operations of addition and scalar multiplication.

10. The set consisting of the single vector $(0, 0)$ under the usual operations in $\mathbb{R}^2$.

11. The set of polynomials of degree $\leq n$ with zero constant term.

12. The set of polynomials of degree $\leq n$ with positive constant term a_0.

13. The set of continuous functions in $[0, 1]$ with $f(0) = 0$ and $f(1) = 0$ under the operations of Example 8.

14. The set of points in $\mathbb{R}^3$ lying on a line passing through the origin.

15. The set of points in $\mathbb{R}^3$ lying on the line $x = t + 1,\ y = 2t,\ z = t - 1$.

16. $\mathbb{R}^2$ with addition defined by $(x_1, y_1) + (x_2, y_2) = (x_1 + x_2 + 1,\ y_1 + y_2 + 1)$ and ordinary scalar multiplication.

17. The set of Problem 16 with scalar multiplication defined by $\alpha(x, y) = (\alpha + \alpha x - 1,\ \alpha + \alpha y - 1)$.

18. The set consisting of one object with addition defined by *object* + *object* = *object* and scalar multiplication defined by $\alpha(object) = object$.

19. The set of differentiable functions defined on [0, 1] with the operations of Example 8.

***20.** The set of real numbers of the form $a + b\sqrt{2}$, where a and b are rational numbers, under the usual addition of real numbers and with scalar multiplication defined only for rational scalars.

21. Show that in a vector space the additive identity element is unique.

22. Show that in a vector space each vector has a unique additive inverse.

23. If $\mathbf{x}$ and $\mathbf{y}$ are vectors in a vector space V, show that there is a unique vector $\mathbf{z} \in V$ such that $\mathbf{x} + \mathbf{z} = \mathbf{y}$.

24. Show that the set of positive real numbers forms a vector space under the operations $x + y = xy$ and $\alpha x = x^\alpha$.

25. Consider the homogeneous second-order differential equation

$$y''(x) + a(x)y'(x) + b(x)y(x) = 0$$

where $a(x)$ and $b(x)$ are continuous functions. Show that the set of solutions to the equation is a vector space under the usual rules for adding functions and multiplying them by real numbers.

ANSWERS TO SELF-QUIZ

I. True **II.** False **III.** False **IV.** True **V.** False **VI.** True **VII.** False **VIII.** False

---

8.2 SUBSPACES

From Example 8.1.1, page 486, we know that $\mathbb{R}^2 = \{(x, y): x \in \mathbb{R} \text{ and } y \in \mathbb{R}\}$ is a vector space. In Example 8.1.4, page 487, we saw that $V = \{(x, y): y = mx\}$ is also a vector space. Moreover, it is clear that $V \subset \mathbb{R}^2$. That is, $\mathbb{R}^2$ has a subset that is also a vector space. In fact, all vector spaces have subsets that are also vector spaces. We examine these important subsets in this section.

DEFINITION SUBSPACE

Let H be a nonempty subset of a vector space V and suppose that H is itself a vector space under the operations of addition and scalar multiplication defined on V. Then H is said to be a **subspace** of V. ■

We can say that the subspace H **inherits** the operations from the "parent" vector space V.

We will encounter many examples of subspaces in this chapter. But first we prove a result that makes it relatively easy to determine whether a subset of V is indeed a subspace of V.

THEOREM 1

A nonempty subset H of the vector space V is a subspace of V if the two closure rules hold:

RULES FOR CHECKING WHETHER A NONEMPTY SUBSET IS A SUBSPACE

i. If $\mathbf{x} \in H$ and $\mathbf{y} \in H$, then $\mathbf{x} + \mathbf{y} \in H$.

ii. If $\mathbf{x} \in H$, then $\alpha\mathbf{x} \in H$ for every scalar α.

PROOF: Evidently, if H is a vector space, then the two closure rules must hold. Conversely, to show that H is a vector space, we must show that axioms (i) to (x) on page 486 hold under the operations of vector addition and scalar multiplication defined in V. The two closure operations [axioms (i) and (vi)] hold by hypothesis. Since vectors in H are also in V, the associative, commutative, distributive, and multiplicative identity laws [axioms (ii), (v), (vii), $(viii)$, (ix), and (x)] hold. Let $\mathbf{x} \in H$. Then $0\mathbf{x} \in H$ by hypothesis (ii). But by Theorem 8.1.1, page 490, (part ii), $0\mathbf{x} = \mathbf{0}$. Thus $\mathbf{0} \in H$ and axiom (iii) holds. Finally, by part (ii), $(-1)\mathbf{x} \in H$ for every $\mathbf{x} \in H$. By Theorem 8.1.1 (part iv), $-\mathbf{x} = (-1)\mathbf{x} \in H$ so that axiom (iv) also holds and the proof is complete. ■

This theorem shows that to test whether H is a subspace of V, it is sufficient to verify that

> $\mathbf{x} + \mathbf{y}$ and $\alpha\mathbf{x}$ are in H when $\mathbf{x}$ and $\mathbf{y}$ are in H and α is a scalar.

The preceding proof contains a fact that is important enough to mention explicitly.

> Every nonempty subspace of a vector space V contains $\mathbf{0}$. (1)

This fact will often make it easy to see that a particular subset of V is *not* a vector space. That is, if a subset does not contain $\mathbf{0}$, then it is not a subspace. Note that the zero vector in H, a subspace of V, is the same as the zero vector in V.

We now give some examples of subspaces.

EXAMPLE 1 THE TRIVIAL SUBSPACE

For any vector space V, the subset $\{\mathbf{0}\}$ consisting of the zero vector alone is a subspace since $\mathbf{0} + \mathbf{0} = \mathbf{0}$ and $\alpha\mathbf{0} = \mathbf{0}$ for every real number α [part (i) of Theorem 8.2.1]. It is called the **trivial subspace**.

EXAMPLE 2 A VECTOR SPACE IS A SUBSPACE OF ITSELF

For every vector space V, V is a subspace of itself.

The first two examples show that every vector space V contains two subspaces $\{\mathbf{0}\}$ and V (which coincide if $V = \{\mathbf{0}\}$). It is more interesting to find other subspaces. Subspaces other than $\{\mathbf{0}\}$ and $\mathbf{V}$ are called **proper subspaces**.

EXAMPLE 3 A PROPER SUBSPACE OF $\mathbb{R}^2$

Let $H = \{(x, y): y = mx\}$ (see Example 8.1.4, page 487). Then as we have already mentioned, H is a subspace of $\mathbb{R}^2$. As we shall see in Section 8.5 (Problem 15 on page 523), if H is a proper subspace of $\mathbb{R}^2$, then H consists of the set of points lying on a straight line through the origin; that is, a set of points lying on a straight line passing through the origin is the only kind of proper subspace of $\mathbb{R}^2$.

EXAMPLE 4

A PROPER SUBSPACE OF $\mathbb{R}^3$

Let $H = \{(x, y, z): x = at, y = bt, \text{ and } z = ct; a, b, c, t \text{ real and at least one of } a, b, c \text{ is nonzero}\}$. Then H consists of the vectors in $\mathbb{R}^3$ lying on a straight line passing through the origin. To see that H is a subspace of $\mathbb{R}^3$, let $\mathbf{x} = (at_1, bt_1, ct_1) \in H$ and $\mathbf{y} = (at_2, bt_2, ct_2) \in H$. Then

$$\mathbf{x} + \mathbf{y} = (a(t_1 + t_2), b(t_1 + t_2), c(t_1 + t_2)) \in H$$

and

$$\alpha\mathbf{x} = (a(\alpha t_1), b(\alpha t_2), c(\alpha t_3)) \in H.$$

Thus H is a subspace of $\mathbb{R}^3$.

EXAMPLE 5

ANOTHER PROPER SUBSPACE OF $\mathbb{R}^3$

Let $\pi = \{(x, y, z): ax + by + cz = 0; a, b, c \text{ real and at least one of } a, b, c \text{ is nonzero}\}$. Then, as we saw in Example 8.1.6, page 488, π is a vector space; thus π is a subspace of $\mathbb{R}^3$.

We shall prove in Section 8.5 that sets of vectors lying on lines and planes through the origin are the only proper subspaces of $\mathbb{R}^3$.

Before studying more examples, we note that *not every vector space has proper subspaces.*

EXAMPLE 6

$\mathbb{R}$ HAS NO PROPER SUBSPACE

Let H be a subspace of $\mathbb{R}$.† If $H \neq \{\mathbf{0}\}$, then H contains a nonzero real number α. Then, by axiom (vi), $1 = (1/\alpha)\alpha \in H$ and $\beta 1 = \beta \in H$ for every real number β. Thus if H is not the trivial subspace then $H = \mathbb{R}$. That is, $\mathbb{R}$ has *no* proper subspace.

EXAMPLE 7

SOME PROPER SUBSPACES OF P_n

If P_n denotes the vector space of polynomials of degree $\leq n$ (Example 8.1.7, page 488), and if $0 \leq m < n$, then P_m is a proper subspace of P_n, as is easily verified.

EXAMPLE 8

A PROPER SUBSPACE OF M_{mn}

Let M_{mn} (Example 8.1.10, page 489) denote the vector space of $m \times n$ matrices with real components and let $H = \{A \in M_{mn}: a_{11} = 0\}$. By the definition of matrix addition and scalar multiplication it is clear that the two closure axioms hold, so that H is a subspace.

† Note that $\mathbb{R}$ is a vector space over itself; that is, $\mathbb{R}$ is a vector space where the scalars are taken to be the reals. This is Example 8.1.1, page 486, with $n = 1$.

EXAMPLE 9

A SUBSET THAT IS NOT A SUBSPACE OF M_{nn}

Let $V = M_{nn}$ (the $n \times n$ matrices) and let $H = \{A \in M_{nn}: A \text{ is invertible}\}$. Then H is not a subspace since the $n \times n$ zero matrix is not in H.

EXAMPLE 10

A PROPER SUBSPACE OF $C[0, 1]$

$P_n[0, 1]^{\dagger} \subset C[0, 1]$ (see Example 8.1.8, page 489) because every polynomial is continuous and P_n is a vector space for every integer n, so that each $P_n[0, 1]$ is a subspace of $C[0, 1]$.

EXAMPLE 11

$C^1[0, 1]$ IS A PROPER SUBSPACE OF $C[0, 1]$

Let $C^1[0, 1]$ denote the set of functions with continuous first derivatives defined on $[0, 1]$. Since every differentiable function is continuous, we have $C^1[0, 1] \subset C[0, 1]$. Since the sum and scalar multiple of two differentiable functions are differentiable, we see that $C^1[0, 1]$ is a subspace of $C[0, 1]$. It is a proper subspace because not every continuous function is differentiable.

EXAMPLE 12

ANOTHER PROPER SUBSPACE OF $C[0, 1]$

If $f \in C[0, 1]$, then $\int_0^1 f(x)\,dx$ exists. Let $H = \{f \in C[0, 1]: \int_0^1 f(x)\,dx = 0\}$. If $f \in H$ and $g \in H$, then $\int_0^1 [f(x) + g(x)]\,dx = \int_0^1 f(x)\,dx + \int_0^1 g(x)\,dx = 0 + 0 = 0$ and $\int_0^1 \alpha f(x)\,dx = \alpha \int_0^1 f(x)\,dx = 0$. Thus $f + g$ and αf are in H for every real number α. This shows that H is a proper subspace of $C[0, 1]$.

As the last three examples illustrate, a vector space can have a great number and variety of proper subspaces. Before leaving this section, we prove an interesting fact about subspaces.

THEOREM 2

Let H_1 and H_2 be subspaces of a vector space V. Then $H_1 \cap H_2$ is a subspace of V.

PROOF: Note that $H_1 \cap H_2$ is nonempty because it contains $\mathbf{0}$. Let $\mathbf{x}_1 \in H_1 \cap H_2$ and $\mathbf{x}_2 \in H_1 \cap H_2$. Then, since H_1 and H_2 are subspaces, $\mathbf{x}_1 + \mathbf{x}_2 \in H_1$ and $\mathbf{x}_1 + \mathbf{x}_2 \in H_2$. This means that $\mathbf{x}_1 + \mathbf{x}_2 \in H_1 \cap H_2$. Similarly, $\alpha\mathbf{x}_1 \in H_1 \cap H_2$. Thus the two closure axioms are satisfied and $H_1 \cap H_2$ is a subspace.[‡] ■

EXAMPLE 13

THE INTERSECTION OF TWO SUBSPACES OF $\mathbb{R}^3$ IS A SUBSPACE

In $\mathbb{R}^3$ let $H_1 = \{(x, y, z): 2x - y - z = 0\}$ and $H_2 = \{(x, y, z): x + 2y + 3z = 0\}$. Then H_1 and H_2 consist of vectors lying on planes through the origin and are, by Example 5,

† $P_n[0, 1]$ denotes the set of polynomials defined on the interval $[0, 1]$ of degree $\leq n$.

‡ Note, in particular, that as $\mathbf{0} \in H_1$ and $\mathbf{0} \in H_2$, we have $\mathbf{0} \in H_1 \cap H_2$

subspaces of $\mathbb{R}^3$. $H_1 \cap H_2$ is the intersection of the two planes which we compute as in Chapter 6:

$$\begin{aligned} x + 2y + 3z &= 0 \\ 2x - y - z &= 0 \end{aligned}$$

or, row-reducing:

$$\left(\begin{array}{ccc|c} 1 & 2 & 3 & 0 \\ 2 & -1 & -1 & 0 \end{array}\right) \xrightarrow{R_2 \to R_2 - 2R_1} \left(\begin{array}{ccc|c} 1 & 2 & 3 & 0 \\ 0 & -5 & -7 & 0 \end{array}\right) \xrightarrow{R_2 \to -\frac{1}{5}R_2}$$

$$\left(\begin{array}{ccc|c} 1 & 2 & 3 & 0 \\ 0 & 1 & \frac{7}{5} & 0 \end{array}\right) \xrightarrow{R_1 \to R_1 - 2R_2} \left(\begin{array}{ccc|c} 1 & 0 & \frac{1}{5} & 0 \\ 0 & 1 & \frac{7}{5} & 0 \end{array}\right)$$

Thus all solutions to the homogeneous system are given by $(-\frac{1}{5}z, -\frac{7}{5}z, z)$. Setting $z = t$, we obtain the parametric equations of a line L in $\mathbb{R}^3$: $x = -\frac{1}{5}t$, $y = -\frac{7}{5}t$, $z = t$. As we saw in Example 4, the set of vectors on L constitutes a subspace of $\mathbb{R}^3$.

REMARK: It is not true that if H_1 and H_2 is a subspace of V, then $H_1 \cup H_2$ is necessarily a subspace of V (it may or may not be). For example, $H_1 = \{(x, y): y = 2x\}$ and $\{(x, y): y = 3x\}$ are subspaces of $\mathbb{R}^2$, but $H_1 \cup H_2$ is not a subspace. To see this, observe that $(1, 2) \in H_1$ and $(1, 3) \in H_2$, so that both $(1, 2)$ and $(1, 3)$ are in $H_1 \cup H_2$. But $(1, 2) + (1, 3) = (2, 5) \notin H_1 \cup H_2$ because $(2, 5) \notin H_1$ and $(2, 5) \notin H_2$. Thus $H_1 \cup H_2$ is not closed under addition, and is therefore not a subspace.

PROBLEMS 8.2

SELF-QUIZ

True–False

I. The set of vectors of the form $\begin{pmatrix} x \\ y \\ 1 \end{pmatrix}$ is a subspace of $\mathbb{R}^3$.

II. The set of vectors of the form $\begin{pmatrix} x \\ 0 \\ z \end{pmatrix}$ is a subspace of $\mathbb{R}^3$.

III. The set of diagonal 3×3 matrices is a subspace of M_{33}.

IV. The set of upper triangular 3×3 matrices is a subspace of M_{33}.

V. The set of triangular 3×3 matrices is a subspace of M_{33}.

VI. Let H be a subspace of M_{22}. Then $\begin{pmatrix} 0 & 0 \\ 0 & 0 \end{pmatrix}$ must be in H.

VII. Let $H = \left\{\begin{pmatrix} x \\ y \\ z \end{pmatrix}: 2x + 3y - z = 0\right\}$ and $K = \left\{\begin{pmatrix} x \\ y \\ z \end{pmatrix}: x - 2y + 5z = 0\right\}$. Then $H \cup K$ is a subspace of $\mathbb{R}^3$.

VIII. If H and K are as in Problem VII, then $H \cap K$ is a subspace of $\mathbb{R}^3$.

IX. The set of polynomials of degree 2 is a subspace of P_3.

In Problems 1–20 determine whether the given subset H of the vector space V is a subspace of V.

1. $V = \mathbb{R}^2$; $H = \{(x, y): y \geq 0\}$

2. $V = \mathbb{R}^2$; $H = \{(x, y): x = y\}$

3. $V = \mathbb{R}^3$; $H =$ the xy-plane

4. $V = \mathbb{R}^2$; $H = \{(x, y): x^2 + y^2 \leq 1\}$

5. $V = M_{nn}$; $H = \{D \in M_{nn}: D \text{ is diagonal}\}$

6. $V = M_{nn}$; $H = \{T \in M_{nn}: T \text{ is upper triangular}\}$

7. $V = M_{nn}$; $H = \{S \in M_{nn}$: S is symmetric$\}$
8. $V = M_{mn}$; $H = \{A \in M_{mn}$: $a_{ij} = 0$ for some i and $j\}$
9. $V = M_{22}$; $H = \left\{A \in M_{22}: A = \begin{pmatrix} a & b \\ -b & c \end{pmatrix}\right\}$
10. $V = M_{22}$; $H = \left\{A \in M_{22}: A = \begin{pmatrix} a & 1+a \\ 0 & 0 \end{pmatrix}\right\}$
11. $V = M_{22}$; $H = \left\{A \in M_{22}: A = \begin{pmatrix} 0 & a \\ b & 0 \end{pmatrix}\right\}$
12. $V = P_4$; $H = \{p \in P_4: \deg p = 4\}$
13. $V = P_4$; $H = \{p \in P_4: p(0) = 0\}$
14. $V = P_n$; $H = \{p \in P_n: p(0) = 0\}$
15. $V = P_n$; $H = \{p \in P_n: p(0) = 1\}$
16. $V = C[0, 1]$; $H = \{f \in C[0, 1]: f(0) = f(1) = 0\}$
17. $V = C[0, 1]$; $H = \{f \in C[0, 1]: f(0) = 2\}$
18. $V = C^1[0, 1]$; $H = \{f \in C^1[0, 1]: f'(0) = 0\}$
19. $V = C[a, b]$, where a and b are real numbers and $a < b$; $H = \{f \in C[a, b]: \int_a^b f(x)\,dx = 0\}$
20. $V = C[a, b]$; $H = \{f \in C[a, b]: \int_a^b f(x)\,dx = 1\}$
21. Let $V = M_{22}$; let $H_1 = \{A \in M_{22}: a_{11} = 0\}$ and $H_2 = \left\{A \in M_{22}: A = \begin{pmatrix} -b & a \\ a & b \end{pmatrix}\right\}$.
    a. Show that H_1 and H_2 are subspaces.
    b. Describe the subset $H = H_1 \cap H_2$ and show that it is a subspace.
22. If $V = C[0, 1]$, let H_1 denote the subspace of Example 10 and H_2 denote the subspace of Example 11. Describe the set $H_1 \cap H_2$ and show that it is a subspace.
23. Let A be an $n \times m$ matrix and let $H = \{\mathbf{x} \in \mathbb{R}^m: A\mathbf{x} = \mathbf{0}\}$. Show that H is a subspace of $\mathbb{R}^m$. H is called the **kernel** of the matrix A.
24. In Problem 23 let $H = \{\mathbf{x} \in \mathbb{R}^m: A\mathbf{x} \neq \mathbf{0}\}$. Show that H is not a subspace of $\mathbb{R}^m$.
25. Let $H = \{(x, y, z, w): ax + by + cz + dw = 0\}$, where a, b, c, and d are real numbers not all zero. Show that H is a proper subspace of $\mathbb{R}^4$. H is called a **hyperplane** in $\mathbb{R}^4$ that contains $\mathbf{0}$.
26. Let $H = \{(x_1, x_2, \ldots, x_n): a_1x_1 + a_2x_2 + \cdots + a_nx_n = 0\}$, where $a_1, a_2, \ldots, a_n$ are real numbers not all zero. Show that H is a proper subspace of $\mathbb{R}^n$. H, as in Problem 25, is called a **hyperplane** in $\mathbb{R}^n$ that contains $\mathbf{0}$.
27. Let H_1 and H_2 be subspaces of a vector space V. Let $H_1 + H_2 = \{\mathbf{v}: \mathbf{v} = \mathbf{v}_1 + \mathbf{v}_2$ with $\mathbf{v}_1 \in H_1$ and $\mathbf{v}_2 \in H_2\}$. Show that $H_1 + H_2$ is a subspace of V.
28. Let $\mathbf{v}_1$ and $\mathbf{v}_2$ be two vectors in $\mathbb{R}^2$. Show that $H = \{\mathbf{v}: \mathbf{v} = a\mathbf{v}_1 + b\mathbf{v}_2;\ a, b \text{ real}\}$ is a subspace of $\mathbb{R}^2$.
*29. In Problem 28 show that if $\mathbf{v}_1$ and $\mathbf{v}_2$ are not collinear, then $H = \mathbb{R}^2$.
30. Let $\mathbf{v}_1, \mathbf{v}_2, \ldots, \mathbf{v}_n$ be arbitrary vectors in a vector space V. Let $H = \{\mathbf{v} \in V: \mathbf{v} = a_1\mathbf{v}_1 + a_2\mathbf{v}_2 + \cdots + a_n\mathbf{v}_n$, where $a_1, a_2, \ldots, a_n$ are scalars$\}$. Show that H is a subspace of V. H is called the subspace *spanned* by the vectors $\mathbf{v}_1, \mathbf{v}_2, \ldots, \mathbf{v}_n$.

ANSWERS TO SELF-QUIZ

I. False **II.** True **III.** True **IV.** True **V.** False **VI.** True
VII. False **VIII.** True **IX.** False

8.3 LINEAR COMBINATION AND SPAN

We have seen that every vector $\mathbf{v} = (a, b, c)$ in $\mathbb{R}^3$ can be written in the form

$$\mathbf{v} = a\mathbf{i} + b\mathbf{j} + c\mathbf{k}$$

In this case we say that $\mathbf{v}$ is a *linear combination* of the three vectors $\mathbf{i}$, $\mathbf{j}$, and $\mathbf{k}$. More generally, we have the following definition.

DEFINITION LINEAR COMBINATION

Let $\mathbf{v}_1, \mathbf{v}_2, \ldots, \mathbf{v}_n$ be vectors in a vector space V. Then any expression of the form

$$a_1\mathbf{v}_1 + a_2\mathbf{v}_2 + \cdots + a_n\mathbf{v}_n \tag{1}$$

where $a_1, a_2, \ldots, a_n$ are scalars is called a **linear combination** of $\mathbf{v}_1, \mathbf{v}_2, \ldots, \mathbf{v}_n$. ■

EXAMPLE 1 A LINEAR COMBINATION IN $\mathbb{R}^3$

In $\mathbb{R}^3$, $\begin{pmatrix} -7 \\ 7 \\ 7 \end{pmatrix}$ is a linear combination of $\begin{pmatrix} -1 \\ 2 \\ 4 \end{pmatrix}$ and $\begin{pmatrix} 5 \\ -3 \\ 1 \end{pmatrix}$ since $\begin{pmatrix} -7 \\ 7 \\ 7 \end{pmatrix} = 2\begin{pmatrix} -1 \\ 2 \\ 4 \end{pmatrix} - \begin{pmatrix} 5 \\ -3 \\ 1 \end{pmatrix}$

EXAMPLE 2 A LINEAR COMBINATION IN M_{23}

In M_{23}, $\begin{pmatrix} -3 & 2 & 8 \\ -1 & 9 & 3 \end{pmatrix} = 3\begin{pmatrix} -1 & 0 & 4 \\ 1 & 1 & 5 \end{pmatrix} + 2\begin{pmatrix} 0 & 1 & -2 \\ -2 & 3 & -6 \end{pmatrix}$, which shows that $\begin{pmatrix} -3 & 2 & 8 \\ -1 & 9 & 3 \end{pmatrix}$ is a linear combination of

$$\begin{pmatrix} -1 & 0 & 4 \\ 1 & 1 & 5 \end{pmatrix} \quad \text{and} \quad \begin{pmatrix} 0 & 1 & -2 \\ -2 & 3 & -6 \end{pmatrix}.$$

EXAMPLE 3 LINEAR COMBINATIONS IN P_n

In P_n every polynomial can be written as a linear combination of the "monomials" $1, x, x^2, \ldots, x^n$.

DEFINITION SPAN

The vectors $\mathbf{v}_1, \mathbf{v}_2, \ldots, \mathbf{v}_n$ in a vector space V are said to **span** V if every vector in V can be written as a linear combination of them. That is, for every $\mathbf{v} \in V$ there are scalars $a_1, a_2, \ldots, a_n$ such that

$$\mathbf{v} = a_1\mathbf{v}_1 + a_2\mathbf{v}_2 + \cdots + a_n\mathbf{v}_n \tag{2}$$

■

EXAMPLE 4 SETS OF VECTORS THAT SPAN $\mathbb{R}^2$ AND $\mathbb{R}^3$

Since $\begin{pmatrix} x \\ y \end{pmatrix} = x\begin{pmatrix} 1 \\ 0 \end{pmatrix} + y\begin{pmatrix} 0 \\ 1 \end{pmatrix}$, we see that the vectors $\mathbf{i} = \begin{pmatrix} 1 \\ 0 \end{pmatrix}$ and $\mathbf{j} = \begin{pmatrix} 0 \\ 1 \end{pmatrix}$

span $\mathbb{R}^2$. In Section 1.4 we saw that $\mathbf{i} = \begin{pmatrix} 1 \\ 0 \\ 0 \end{pmatrix}$, $\mathbf{j} = \begin{pmatrix} 0 \\ 1 \\ 0 \end{pmatrix}$, and $\mathbf{k} = \begin{pmatrix} 0 \\ 0 \\ 1 \end{pmatrix}$ span $\mathbb{R}^3$.

We now look briefly at spanning sets of some other vector spaces.

EXAMPLE 5

$n + 1$ VECTORS THAT SPAN P_n

From Example 3 it follows that the monomials $1, x, x^2, \ldots, x^n$ span P^n.

EXAMPLE 6

FOUR VECTORS THAT SPAN M_{22}

Since $\begin{pmatrix} a & b \\ c & d \end{pmatrix} = a\begin{pmatrix} 1 & 0 \\ 0 & 0 \end{pmatrix} + b\begin{pmatrix} 0 & 1 \\ 0 & 0 \end{pmatrix} + c\begin{pmatrix} 0 & 0 \\ 1 & 0 \end{pmatrix} + d\begin{pmatrix} 0 & 0 \\ 0 & 1 \end{pmatrix}$, we see that $\begin{pmatrix} 1 & 0 \\ 0 & 0 \end{pmatrix}, \begin{pmatrix} 0 & 1 \\ 0 & 0 \end{pmatrix}, \begin{pmatrix} 0 & 0 \\ 1 & 0 \end{pmatrix}$, and $\begin{pmatrix} 0 & 0 \\ 0 & 1 \end{pmatrix}$ span M_{22}.

EXAMPLE 7

NO FINITE SET OF POLYNOMIALS SPANS P

Let P denote the vector space of polynomials. Then no *finite* set of polynomials spans P. To see this suppose that $p_1, p_2, \ldots, p_m$ are polynomials. Let p_k be the polynomial of largest degree in this set and let $N = \deg p_k$. Then the polynomial $p(x) = x^{N+1}$ cannot be written as a linear combination of $p_1, p_2, \ldots, p_m$. For example if $N = 3$, then $x^4 \neq c_0 + c_1x + c_2x^2 + c_3x^3$ for any scalars c_0, c_1, c_2, and c_3.

We now turn to another way of finding subspaces of a vector space V.

DEFINITION SPAN OF A SET OF VECTORS

Let $\mathbf{v}_1, \mathbf{v}_2, \ldots, \mathbf{v}_k$ be k vectors in a vector space V. The **span** of $\{\mathbf{v}_1, \mathbf{v}_2, \ldots, \mathbf{v}_k\}$ is the set of linear combinations of $\mathbf{v}_1, \mathbf{v}_2, \ldots, \mathbf{v}_k$. That is,

$$\text{span}\,\{\mathbf{v}_1, \mathbf{v}_2, \ldots, \mathbf{v}_k\} = \{\mathbf{v}\colon \mathbf{v} = a_1\mathbf{v}_1 + a_2\mathbf{v}_2 + \cdots + a_k\mathbf{v}_k\} \tag{3}$$

where $a_1, a_2, \ldots, a_i$ are scalars. ■

THEOREM 1

Span $\{\mathbf{v}_1, \mathbf{v}_2, \ldots, \mathbf{v}_k\}$ is a subspace of V.

PROOF: The proof is easy and is left as an exercise (see Problem 16). ■

EXAMPLE 8

THE SPAN OF TWO VECTORS IN $\mathbb{R}^3$

Let $\mathbf{v}_1 = (2, -1, 4)$ and $\mathbf{v}_2 = (4, 1, 6)$. Then $H = \text{span}\,\{\mathbf{v}_1, \mathbf{v}_2\} = \{\mathbf{v}\colon \mathbf{v} = a_1(2, -1, 4) + a_2(4, 1, 6)\}$. What does H look like? If $\mathbf{v} = (x, y, z) \in H$, then we have $x = 2a_1 + 4a_2$, $y = -a_1 + a_2$, and $z = 4a_1 + 6a_2$. If we think of (x, y, z) as being fixed, then we can view these equations as a system of three equations in the two unknowns a_1, a_2. We solve this system in the usual way:

$$\left(\begin{array}{cc|c} -1 & 1 & y \\ 2 & 4 & x \\ 4 & 6 & z \end{array}\right) \xrightarrow{R_1 \to -R_1} \left(\begin{array}{cc|c} 1 & -1 & -y \\ 2 & 4 & x \\ 4 & 6 & z \end{array}\right) \xrightarrow[R_3 \to R_3 - 4R_1]{R_2 \to R_2 - 2R_1} \left(\begin{array}{cc|c} 1 & -1 & -y \\ 0 & 6 & x + 2y \\ 0 & 10 & z + 4y \end{array}\right)$$

$$\xrightarrow{R_2 \to \frac{1}{6}R_2} \left(\begin{array}{cc|c} 1 & -1 & -y \\ 0 & 1 & (x + 2y)/6 \\ 0 & 10 & z + 4y \end{array}\right) \xrightarrow[R_3 \to R_3 - 10R_2]{R_1 \to R_1 + R_2} \left(\begin{array}{cc|c} 1 & 0 & x/6 - 2y/3 \\ 0 & 1 & x/6 + y/3 \\ 0 & 0 & -5x/3 + 2y/3 + z \end{array}\right)$$

From Chapter 6 we see that the system has a solution only if $-5x/3 + 2y/3 + z = 0$; or, multiplying through by -3, if

$$5x - 2y - 3z = 0 \tag{4}$$

Equation (4) is the equation of a plane in $\mathbb{R}^3$ passing through the origin.

The last example can be generalized to prove the following interesting fact:

> The span of two nonzero vectors in $\mathbb{R}^3$ that are not parallel is a plane passing through the origin.

For a suggested proof see Problems 19 and 20.

We can give a geometric interpretation of this result. Look at the vectors in Figure 1.

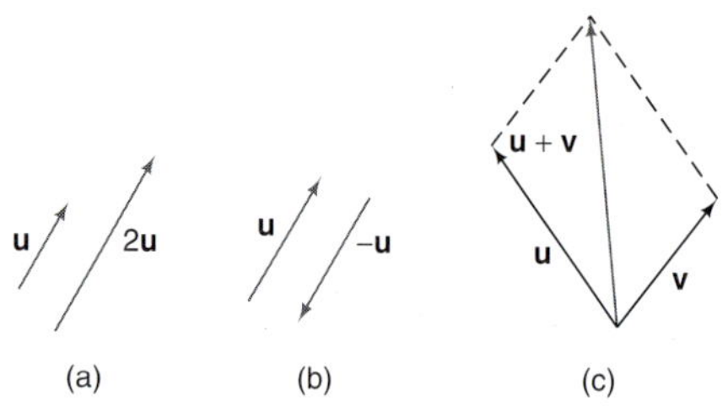

FIGURE 1
$\mathbf{u} + \mathbf{v}$ is obtained from the parallelogram rule

We know (from Section 1.1) the geometric interpretation of the vectors $2\mathbf{u}$, $-\mathbf{u}$, and $\mathbf{u} + \mathbf{v}$, for example. Using these, we see that any other vector in the plane of $\mathbf{u}$ and $\mathbf{v}$ can be obtained as a linear combination of $\mathbf{u}$ and $\mathbf{v}$. Figure 2 shows how in four different situations a third vector $\mathbf{w}$ in the plane of $\mathbf{u}$ and $\mathbf{v}$ can be written as $\alpha\mathbf{u} + \beta\mathbf{v}$ for appropriate choices of the numbers α and β.

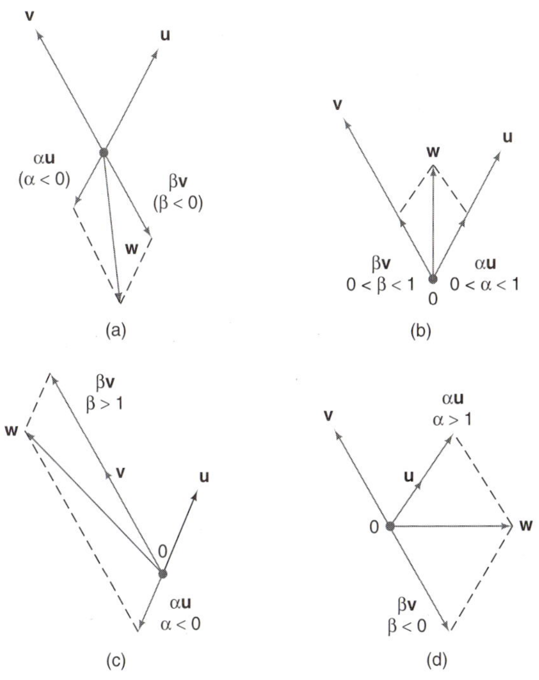

FIGURE 2
In each case $\mathbf{w} = \alpha\mathbf{u} + \beta\mathbf{v}$ for appropriate choices of α and β

REMARK: In the definitions on pages 498 and 499 we used the word ''span'' in two different ways: as a verb and as a noun. We emphasize that

> verb
> ↓
> A set of vectors $\mathbf{v}_1, \mathbf{v}_2, \ldots, \mathbf{v}_n$ *span* V if every vector in V can be written as a linear combination of $\mathbf{v}_1, \mathbf{v}_2, \ldots, \mathbf{v}_n$.

but

> noun
> ↓
> The *span* of the n vectors $\mathbf{v}_1, \mathbf{v}_2, \ldots, \mathbf{v}_k$ is the set of linear combinations of these vectors.

These two concepts are different—even though they use the same word.

We close this section by citing a useful result. Its proof is not difficult and is left as an exercise (see Problem 21).

THEOREM 2

Let $\mathbf{v}_1, \mathbf{v}_2, \ldots, \mathbf{v}_n, \mathbf{v}_{n+1}$ be $n + 1$ vectors that are in a vector space V. If $\mathbf{v}_1, \mathbf{v}_2, \ldots, \mathbf{v}_n$ span V, then $\mathbf{v}_1, \mathbf{v}_2, \ldots, \mathbf{v}_n, \mathbf{v}_{n+1}$ also span V. That is, the addition of one (or more) vectors to a spanning set yields another spanning set. ■

PROBLEMS 8.3

SELF-QUIZ

Multiple Choice

I. Which of the following pairs of vectors could *not* possibly span $\mathbb{R}^2$?

a. $\begin{pmatrix}1\\1\end{pmatrix}, \begin{pmatrix}-3\\-3\end{pmatrix}$ **b.** $\begin{pmatrix}1\\1\end{pmatrix}, \begin{pmatrix}2\\2\end{pmatrix}$

c. $\begin{pmatrix}1\\1\end{pmatrix}, \begin{pmatrix}-1\\1\end{pmatrix}$ **d.** $\begin{pmatrix}1\\3\end{pmatrix}, \begin{pmatrix}0\\0\end{pmatrix}$

e. $\begin{pmatrix}1\\3\end{pmatrix}, \begin{pmatrix}3\\1\end{pmatrix}$

II. Which of the following sets of polynomials span P_2?

a. $1, x^2$ **b.** $3, 2x, -x^2$

c. $1 + x, 2 + 2x, x^2$ **d.** $1, 1 + x, 1 + x^2$

True–False

III. $\begin{pmatrix}3\\5\end{pmatrix}$ is in the span of $\left\{\begin{pmatrix}1\\1\end{pmatrix}, \begin{pmatrix}2\\4\end{pmatrix}\right\}$.

IV. $\begin{pmatrix}1\\2\\3\end{pmatrix}$ is in the span of $\left\{\begin{pmatrix}2\\0\\4\end{pmatrix}, \begin{pmatrix}-1\\0\\3\end{pmatrix}\right\}$.

V. $\{1, x, x^2, x^3, \ldots, x^{10,000}\}$ spans P.

VI. $\left\{\begin{pmatrix}1&0\\0&0\end{pmatrix}, \begin{pmatrix}0&1\\0&0\end{pmatrix}, \begin{pmatrix}0&0\\1&0\end{pmatrix}, \begin{pmatrix}0&0\\0&1\end{pmatrix}\right\}$ spans M_{22}.

VII. Span $\left\{\begin{pmatrix}1\\2\\-1\\3\end{pmatrix}, \begin{pmatrix}7\\1\\0\\4\end{pmatrix}, \begin{pmatrix}-8\\0\\8\\2\end{pmatrix}\right\}$ is a subspace of $\mathbb{R}^3$.

VIII. Span $\left\{\begin{pmatrix}1\\2\\-1\\3\end{pmatrix}, \begin{pmatrix}7\\1\\0\\4\end{pmatrix}, \begin{pmatrix}-8\\0\\8\\2\end{pmatrix}\right\}$ is a subspace of $\mathbb{R}^4$.

IX. If $\left\{\begin{pmatrix}1\\2\end{pmatrix}, \begin{pmatrix}2\\3\end{pmatrix}\right\}$ spans $\mathbb{R}^2$, then $\left\{\begin{pmatrix}1\\2\end{pmatrix}, \begin{pmatrix}2\\3\end{pmatrix}, \begin{pmatrix}-2\\-3\end{pmatrix}\right\}$ also spans $\mathbb{R}^2$.

In Problems 1–13 determine whether the given set of vectors spans the given vector space.

1. In $\mathbb{R}^2$: $\begin{pmatrix}1\\2\end{pmatrix}, \begin{pmatrix}3\\4\end{pmatrix}$

2. In $\mathbb{R}^2$: $\begin{pmatrix}1\\1\end{pmatrix}, \begin{pmatrix}2\\1\end{pmatrix}, \begin{pmatrix}2\\2\end{pmatrix}$

3. In $\mathbb{R}^2$: $\begin{pmatrix}1\\1\end{pmatrix}, \begin{pmatrix}2\\2\end{pmatrix}, \begin{pmatrix}5\\5\end{pmatrix}$

4. In $\mathbb{R}^3$: $\begin{pmatrix}1\\2\\3\end{pmatrix}, \begin{pmatrix}-1\\2\\3\end{pmatrix}, \begin{pmatrix}5\\2\\3\end{pmatrix}$

5. In $\mathbb{R}^3$: $\begin{pmatrix}1\\1\\1\end{pmatrix}, \begin{pmatrix}0\\1\\1\end{pmatrix}, \begin{pmatrix}0\\0\\1\end{pmatrix}$

6. In $\mathbb{R}^3$: $\begin{pmatrix}2\\0\\1\end{pmatrix}, \begin{pmatrix}3\\1\\2\end{pmatrix}, \begin{pmatrix}1\\1\\1\end{pmatrix}, \begin{pmatrix}7\\3\\5\end{pmatrix}$

7. In $\mathbb{R}^3$: $(1, -1, 2), (1, 1, 2), (0, 0, 1)$

8. In $\mathbb{R}^3$: $(1, -1, 2), (-1, 1, 2), (0, 0, 1)$

9. In P_2: $1 - x, 3 - x^2$

10. In P_2: $1 - x, 3 - x^2, x$

11. In M_{22}: $\begin{pmatrix}2&1\\0&0\end{pmatrix}, \begin{pmatrix}0&0\\2&1\end{pmatrix}, \begin{pmatrix}3&-1\\0&0\end{pmatrix}, \begin{pmatrix}0&0\\3&1\end{pmatrix}$

12. In M_{22}: $\begin{pmatrix}1&0\\1&0\end{pmatrix}, \begin{pmatrix}1&2\\0&0\end{pmatrix}, \begin{pmatrix}4&-1\\3&0\end{pmatrix}, \begin{pmatrix}-2&5\\6&0\end{pmatrix}$

13. In M_{23}: $\begin{pmatrix}1&0&0\\0&0&0\end{pmatrix}, \begin{pmatrix}0&1&0\\0&0&0\end{pmatrix}, \begin{pmatrix}0&0&1\\0&0&0\end{pmatrix},$
$\begin{pmatrix}0&0&0\\1&0&0\end{pmatrix}, \begin{pmatrix}0&0&0\\0&1&0\end{pmatrix}, \begin{pmatrix}0&0&0\\0&0&1\end{pmatrix}$

14. Show that two polynomials cannot span P_2.

***15.** If $p_1, p_2, \ldots, p_m$ span P_n, show that $m \geq n + 1$.

16. Show that if $\mathbf{u}$ and $\mathbf{v}$ are in span $\{\mathbf{v}_1, \mathbf{v}_2, \ldots, \mathbf{v}_k\}$, then $\mathbf{u} + \mathbf{v}$ and $\alpha\mathbf{u}$ are in span $\{\mathbf{v}_1, \mathbf{v}_2, \ldots, \mathbf{v}_k\}$ [*Hint:* Using the definition of span write $\mathbf{u} + \mathbf{v}$ and $\alpha\mathbf{u}$ as linear combinations of $\mathbf{v}_1, \mathbf{v}_2, \ldots, \mathbf{v}_k$.]

17. Show that the infinite set $\{1, x, x^2, x^3, \ldots\}$ spans P, the vector space of polynomials.

18. Let H be a subspace of V containing $\mathbf{v}_1, \mathbf{v}_2, \ldots, \mathbf{v}_n$. Show that span $\{\mathbf{v}_1, \mathbf{v}_2, \ldots, \mathbf{v}_n\} \subseteq H$. That is, span $\{\mathbf{v}_1\ \mathbf{v}_2, \ldots, \mathbf{v}_n\}$ is the *smallest* subspace of V containing $\mathbf{v}_1, \mathbf{v}_2, \ldots, \mathbf{v}_n$.

19. Let $\mathbf{v}_1 = (x_1, y_1, z_1)$ and $\mathbf{v}_2 = (x_2, y_2, z_2)$ be in $\mathbb{R}^3$. Show that if $\mathbf{v}_2 = c\mathbf{v}_1$, then span $\{\mathbf{v}_1, \mathbf{v}_2\}$ is a line passing through the origin.

***20.** In Problem 19 assume that $\mathbf{v}_1$ and $\mathbf{v}_2$ are not parallel. Show that $H =$ span $\{\mathbf{v}_1, \mathbf{v}_2\}$ is a plane passing through the origin. What is the equation of that plane? [*Hint:* If $(x, y, z) \in H$, write $\mathbf{v} = a_1\mathbf{v}_1 + a_2\mathbf{v}_2$ and find a condition relating x, y, and z such that the resulting 3×2 system has a solution.]

21. Prove Theorem 2. [*Hint:* If $\mathbf{v} \in V$, write $\mathbf{v}$ as a linear combination of $\mathbf{v}_1, \mathbf{v}_2, \ldots, \mathbf{v}_n, \mathbf{v}_{n+1}$ with the coefficient of $\mathbf{v}_{n+1}$ equal to zero.]

22. Show that M_{22} can be spanned by invertible matrices.

***23.** Let $\{\mathbf{u}_1, \mathbf{u}_2, \ldots, \mathbf{u}_n\}$ and $\{\mathbf{v}_1, \mathbf{v}_2, \ldots, \mathbf{v}_n\}$ be $2n$ vectors in a vector space V. Suppose that

$$\begin{aligned} \mathbf{v}_1 &= a_{11}\mathbf{u}_1 + a_{12}\mathbf{u}_2 + \cdots + a_{1n}\mathbf{u}_n \\ \mathbf{v}_2 &= a_{21}\mathbf{u}_1 + a_{22}\mathbf{u}_2 + \cdots + a_{2n}\mathbf{u}_n \\ &\vdots \\ \mathbf{v}_n &= a_{n1}\mathbf{u}_1 + a_{n2}\mathbf{u}_2 + \cdots + a_{nn}\mathbf{u}_n \end{aligned}$$

Show that if

$$\begin{vmatrix} a_{11} & a_{12} & \cdots & a_{1n} \\ a_{21} & a_{22} & \cdots & a_{2n} \\ \vdots & \vdots & & \vdots \\ a_{n1} & a_{n2} & \cdots & a_{nn} \end{vmatrix} \neq 0$$

then span $\{\mathbf{u}_1, \mathbf{u}_2, \ldots, \mathbf{u}_n\} =$ span $\{\mathbf{v}_1, \mathbf{v}_2, \ldots, \mathbf{v}_n\}$.

ANSWERS TO SELF-QUIZ

I. a, b, d **II.** b, d **III.** True **IV.** False **V.** False **VI.** True
VII. False **VIII.** True **IX.** True

8.4 LINEAR INDEPENDENCE

In the study of linear algebra, one of the central ideas is that of the linear dependence or independence of vectors. In this section we define what we mean by linear independence and show how it is related to the theory of homogeneous systems of equations and determinants.

Is there a special relationship between the vectors $\mathbf{v}_1 = \begin{pmatrix} 1 \\ 2 \end{pmatrix}$ and $\mathbf{v}_2 = \begin{pmatrix} 2 \\ 4 \end{pmatrix}$? Of course, we see that $\mathbf{v}_2 = 2\mathbf{v}_1$ or, writing this equation in another way,

$$2\mathbf{v}_1 - \mathbf{v}_2 = \mathbf{0} \tag{1}$$

That is, the zero vector can be written as a linear combination of $\mathbf{v}_1$ and $\mathbf{v}_2$. What is special about the vectors $\mathbf{v}_1 = \begin{pmatrix} 1 \\ 2 \\ 3 \end{pmatrix}$, $\mathbf{v}_2 = \begin{pmatrix} -4 \\ 1 \\ 5 \end{pmatrix}$, and $\mathbf{v}_3 = \begin{pmatrix} -5 \\ 8 \\ 19 \end{pmatrix}$? This question is more difficult to answer at first glance. It is easy to verify, however, that $\mathbf{v}_3 = 3\mathbf{v}_1 + 2\mathbf{v}_2$, or, rewriting,

$$3\mathbf{v}_1 + 2\mathbf{v}_2 - \mathbf{v}_3 = \mathbf{0} \tag{2}$$

Now we have written the zero vector as a linear combination of $\mathbf{v}_1$, $\mathbf{v}_2$, and $\mathbf{v}_3$. It appears that the two vectors in equation (1) and the three vectors in (2) are more closely related than an arbitrary pair of 2-vectors or an arbitrary triple of 3-vectors. In each case we say that the vectors are *linearly dependent.* In general, we have the following important definition.

DEFINITION LINEAR DEPENDENCE AND INDEPENDENCE

Let $\mathbf{v}_1, \mathbf{v}_2, \ldots, \mathbf{v}_n$ be n vectors in a vector space V. Then the vectors are said to be **linearly dependent** if there exist n scalars $c_1, c_2, \ldots, c_n$ *not all zero* such that

$$c_1\mathbf{v}_1 + c_2\mathbf{v}_2 + \cdots + c_n\mathbf{v}_n = \mathbf{0} \tag{3}$$

If the vectors are not linearly dependent, they are said to be **linearly independent**. ■

Putting this another way, $\mathbf{v}_1, \mathbf{v}_2, \ldots, \mathbf{v}_n$ are linearly independent if the equation $c_1\mathbf{v}_1 + c_2\mathbf{v}_2 + \cdots + c_n\mathbf{v}_n = \mathbf{0}$ holds only for $c_1 = c_2 = \cdots = c_n = 0$. They are linearly dependent if the zero vector in V can be written as a linear combination of $\mathbf{v}_1, \mathbf{v}_2, \ldots, \mathbf{v}_n$ with not all the coefficients equal to zero.

How do we determine whether a set of vectors is linearly dependent or independent? The case for 2-vectors is easy.

THEOREM 1

Two vectors in a vector space V are linearly dependent if and only if one is a scalar multiple of the other.

PROOF: First suppose that $\mathbf{v}_2 = c\mathbf{v}_1$ for some scalar $c \neq 0$. Then $c\mathbf{v}_1 - \mathbf{v}_2 = \mathbf{0}$ and $\mathbf{v}_1$ and $\mathbf{v}_2$ are linearly dependent. On the other hand, suppose that $\mathbf{v}_1$ and $\mathbf{v}_2$ are linearly dependent. Then there are constants c_1 and c_2, not both zero, such that $c_1\mathbf{v}_1 + c_2\mathbf{v}_2 = \mathbf{0}$. If $c_1 \neq 0$, then, dividing by c_1, we obtain $\mathbf{v}_1 + (c_2/c_1)\mathbf{v}_2 = \mathbf{0}$ or

$$\mathbf{v}_1 = \left(-\frac{c_2}{c_1}\right)\mathbf{v}_2$$

That is, $\mathbf{v}_1$ is a scalar multiple of $\mathbf{v}_2$. If $c_1 = 0$, then $c_2 \neq 0$, and hence $\mathbf{v}_2 = \mathbf{0} = 0\mathbf{v}_1$. ■

EXAMPLE 1 TWO LINEARLY DEPENDENT VECTORS IN $\mathbb{R}^4$

The vectors $\mathbf{v}_1 = \begin{pmatrix} 2 \\ -1 \\ 0 \\ 3 \end{pmatrix}$ and $\mathbf{v}_2 = \begin{pmatrix} -6 \\ 3 \\ 0 \\ -9 \end{pmatrix}$ are linearly dependent since $\mathbf{v}_2 = -3\mathbf{v}_1$.

EXAMPLE 2 TWO LINEARLY INDEPENDENT VECTORS IN $\mathbb{R}^3$

The vectors $\begin{pmatrix} 1 \\ 2 \\ 4 \end{pmatrix}$ and $\begin{pmatrix} 2 \\ 5 \\ -3 \end{pmatrix}$ are linearly independent; if they were not, we would have $\begin{pmatrix} 2 \\ 5 \\ -3 \end{pmatrix} = c\begin{pmatrix} 1 \\ 2 \\ 4 \end{pmatrix} = \begin{pmatrix} c \\ 2c \\ 4c \end{pmatrix}$. Then $2 = c$, $5 = 2c$, and $-3 = 4c$, which is clearly impossible for any number c.

EXAMPLE 3 DETERMINING WHETHER THREE VECTORS IN $\mathbb{R}^3$ ARE LINEARLY DEPENDENT OR INDEPENDENT

Determine whether the vectors $\begin{pmatrix}1\\-2\\3\end{pmatrix}$, $\begin{pmatrix}2\\-2\\0\end{pmatrix}$, and $\begin{pmatrix}0\\1\\7\end{pmatrix}$ are linearly dependent or independent.

SOLUTION: Suppose that $c_1\begin{pmatrix}1\\-2\\3\end{pmatrix} + c_2\begin{pmatrix}2\\-2\\0\end{pmatrix} + c_3\begin{pmatrix}0\\1\\7\end{pmatrix} = \mathbf{0} = \begin{pmatrix}0\\0\\0\end{pmatrix}$. Then, multiplying through and adding, we have $\begin{pmatrix}c_1 + 2c_2 \\ -2c_1 - 2c_2 + c_3 \\ 3c_1 + 7c_3\end{pmatrix} = \begin{pmatrix}0\\0\\0\end{pmatrix}$. This yields a homogeneous system of three equations in the three unknowns c_1, c_2, and c_3:

$$\begin{aligned} c_1 + 2c_2 \qquad &= 0 \\ -2c_1 - 2c_2 + c_3 &= 0 \\ 3c_1 \qquad + 7c_3 &= 0 \end{aligned} \tag{4}$$

Thus the vectors will be linearly dependent if and only if system (4) has nontrivial solutions. We write system (4) using an augmented matrix and then row-reduce:

$$\left(\begin{array}{ccc|c} 1 & 2 & 0 & 0 \\ -2 & -2 & 1 & 0 \\ 3 & 0 & 7 & 0 \end{array}\right) \xrightarrow[R_3 \to R_3 - 3R_1]{R_2 \to R_2 + 2R_1} \left(\begin{array}{ccc|c} 1 & 2 & 0 & 0 \\ 0 & 2 & 1 & 0 \\ 0 & -6 & 7 & 0 \end{array}\right)$$

$$\xrightarrow{R_2 \to \frac{1}{2}R_2} \left(\begin{array}{ccc|c} 1 & 2 & 0 & 0 \\ 0 & 1 & \frac{1}{2} & 0 \\ 0 & -6 & 7 & 0 \end{array}\right) \xrightarrow[R_3 \to R_3 + 6R_2]{R_1 \to R_1 - 2R_2} \left(\begin{array}{ccc|c} 1 & 0 & -1 & 0 \\ 0 & 1 & \frac{1}{2} & 0 \\ 0 & 0 & 10 & 0 \end{array}\right)$$

$$\xrightarrow{R_3 \to \frac{1}{10}R_3} \left(\begin{array}{ccc|c} 1 & 0 & -1 & 0 \\ 0 & 1 & \frac{1}{2} & 0 \\ 0 & 0 & 1 & 0 \end{array}\right) \xrightarrow[R_2 \to R_2 - \frac{1}{2}R_3]{R_1 \to R_1 + R_3} \left(\begin{array}{ccc|c} 1 & 0 & 0 & 0 \\ 0 & 1 & 0 & 0 \\ 0 & 0 & 1 & 0 \end{array}\right)$$

The last system of equations reads $c_1 = 0$, $c_2 = 0$, $c_3 = 0$. Hence (4) has no nontrivial solutions and the given vectors are linearly independent.

EXAMPLE 4 DETERMINING WHETHER THREE VECTORS IN $\mathbb{R}^3$ ARE LINEARLY DEPENDENT OR INDEPENDENT

Determine whether the vectors $\begin{pmatrix}1\\-3\\0\end{pmatrix}$, $\begin{pmatrix}3\\0\\4\end{pmatrix}$, and $\begin{pmatrix}11\\-6\\12\end{pmatrix}$ are linearly dependent or independent.

SOLUTION: The equation $c_1\begin{pmatrix}1\\-3\\0\end{pmatrix} + c_2\begin{pmatrix}3\\0\\4\end{pmatrix} + c_3\begin{pmatrix}11\\-6\\12\end{pmatrix} = \begin{pmatrix}0\\0\\0\end{pmatrix}$ leads to the homogeneous system

$$\begin{aligned} c_1 + 3c_2 + 11c_3 &= 0 \\ -3c_1 \qquad - 6c_3 &= 0 \\ 4c_2 + 12c_3 &= 0 \end{aligned} \tag{5}$$

Writing system (5) in augmented-matrix form and row-reducing, we obtain, successively,

$$\left(\begin{array}{ccc|c} 1 & 3 & 11 & 0 \\ -3 & 0 & -6 & 0 \\ 0 & 4 & 12 & 0 \end{array}\right) \xrightarrow{R_2 \to R_2 + 3R_1} \left(\begin{array}{ccc|c} 1 & 3 & 11 & 0 \\ 0 & 9 & 27 & 0 \\ 0 & 4 & 12 & 0 \end{array}\right)$$

$$\xrightarrow{R_2 \to \frac{1}{9}R_2} \left(\begin{array}{ccc|c} 1 & 3 & 11 & 0 \\ 0 & 1 & 3 & 0 \\ 0 & 4 & 12 & 0 \end{array}\right) \xrightarrow[R_3 \to R_3 - 4R_2]{R_1 \to R_1 - 3R_2} \left(\begin{array}{ccc|c} 1 & 0 & 2 & 0 \\ 0 & 1 & 3 & 0 \\ 0 & 0 & 0 & 0 \end{array}\right)$$

We can stop here since the theory of Section 6.4 shows us that system (5) has an infinite number of solutions. For example, the last augmented matrix reads

$$\begin{aligned} c_1 \qquad + 2c_3 &= 0 \\ c_2 + 3c_3 &= 0 \end{aligned}$$

If we choose $c_3 = 1$, we have $c_2 = -3$ and $c_1 = -2$ so that, as is easily verified, $-2\begin{pmatrix} 1 \\ -3 \\ 0 \end{pmatrix} - 3\begin{pmatrix} 3 \\ 0 \\ 4 \end{pmatrix} + \begin{pmatrix} 11 \\ -6 \\ 12 \end{pmatrix} = \begin{pmatrix} 0 \\ 0 \\ 0 \end{pmatrix}$ and the vectors are linearly dependent.

GEOMETRIC INTERPRETATION OF LINEAR DEPENDENCE IN $\mathbb{R}^3$

In Example 3 we found three vectors in $\mathbb{R}^3$ that were linearly independent. In Example 4 we found three vectors that were dependent. What does this mean geometrically?

Suppose that $\mathbf{u}$, $\mathbf{v}$, and $\mathbf{w}$ are three linearly dependent vectors in $\mathbb{R}^3$. Then there are constants c_1, c_2, and c_3, not all zero, such that

$$c_1\mathbf{u} + c_2\mathbf{v} + c_3\mathbf{w} = \mathbf{0} \tag{6}$$

Suppose that $c_3 \neq 0$ (a similar result holds if $c_1 \neq 0$ or $c_2 \neq 0$). Then we may divide both sides of (6) by c_3 and rearrange terms to obtain

$$\mathbf{w} = -\frac{c_1}{c_3}\mathbf{u} - \frac{c_2}{c_3}\mathbf{v} = A\mathbf{u} + B\mathbf{v}$$

where $A = -c_1/c_3$ and $B = -c_2/c_3$. We now show that $\mathbf{u}$, $\mathbf{v}$, and $\mathbf{w}$ are coplanar. We compute

$$\begin{aligned} \mathbf{w} \cdot (\mathbf{u} \times \mathbf{v}) &= (A\mathbf{u} + B\mathbf{v}) \cdot (\mathbf{u} \times \mathbf{v}) = A[\mathbf{u} \cdot (\mathbf{u} \times \mathbf{v})] + B[\mathbf{v} \cdot (\mathbf{u} \times \mathbf{v})] \\ &= A \cdot 0 + B \cdot 0 = 0 \end{aligned}$$

because $\mathbf{u}$ and $\mathbf{v}$ are both orthogonal to $\mathbf{u} \times \mathbf{v}$ (see page 44). Let $\mathbf{n} = \mathbf{u} \times \mathbf{v}$. Then $\mathbf{u}$ and $\mathbf{v}$ lie in the plane consisting of those vectors passing through the origin that are orthogonal to $\mathbf{n}$. But $\mathbf{w}$ is in the same plane because $\mathbf{w} \cdot \mathbf{n} = \mathbf{w} \cdot (\mathbf{u} \times \mathbf{v}) = 0$. This shows that $\mathbf{u}$, $\mathbf{v}$, and $\mathbf{w}$ are coplanar.

In Problem 59 you are asked to show that if $\mathbf{u}$, $\mathbf{v}$, and $\mathbf{w}$ are coplanar, then they are linearly dependent. We conclude that

Three vectors in $\mathbb{R}^3$ are linearly dependent if and only if they are coplanar.

Figure 1 illustrates this fact using the vectors in Examples 3 and 4.

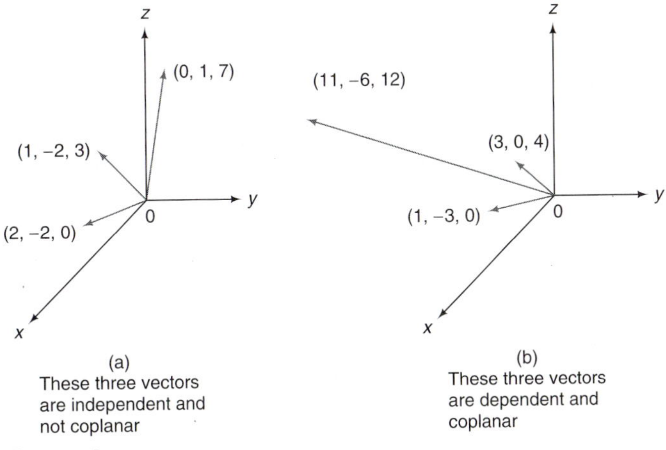

FIGURE 1
Two sets of three vectors

The theory of homogeneous systems can tell us something about the linear dependence or independence of vectors.

THEOREM 2

A set of n vectors in $\mathbb{R}^m$ is always linearly dependent if $n > m$.

PROOF: Let $\mathbf{v}_1, \mathbf{v}_2, \ldots, \mathbf{v}_n$ be n vectors in $\mathbb{R}^m$ and let us try to find constants $c_1, c_2, \ldots, c_n$ not all zero such that

$$c_1\mathbf{v}_1 + c_2\mathbf{v}_2 + \cdots + c_n\mathbf{v}_n = \mathbf{0} \tag{7}$$

Let $\mathbf{v}_1 = \begin{pmatrix} a_{11} \\ a_{21} \\ \vdots \\ a_{m1} \end{pmatrix}$, $\mathbf{v}_2 = \begin{pmatrix} a_{12} \\ a_{22} \\ \vdots \\ a_{m2} \end{pmatrix}, \ldots,$ $\mathbf{v}_n = \begin{pmatrix} a_{1n} \\ a_{2n} \\ \vdots \\ a_{mn} \end{pmatrix}$. Then equation (7) becomes

$$\begin{aligned} a_{11}c_1 + a_{12}c_2 + \cdots + a_{1n}c_n &= 0 \\ a_{21}c_1 + a_{22}c_2 + \cdots + a_{2n}c_n &= 0 \\ \vdots \qquad \vdots \qquad\qquad \vdots \quad &\;\; \vdots \\ a_{m1}c_1 + a_{m2}c_2 + \cdots + a_{mn}c_n &= 0 \end{aligned} \tag{8}$$

But system (8) is system (6.4.1) on page 390 and, according to Theorem 6.4.1, this system has an infinite number of solutions if $n > m$. Thus there are scalars $c_1, c_2, \ldots, c_n$ not all zero that satisfy (8) and the vectors $\mathbf{v}_1, \mathbf{v}_2, \ldots, \mathbf{v}_n$ are therefore linearly dependent. ■

EXAMPLE 5

FOUR VECTORS IN $\mathbb{R}^3$ ARE LINEARLY DEPENDENT

The vectors $\begin{pmatrix} 2 \\ -3 \\ 4 \end{pmatrix}$, $\begin{pmatrix} 4 \\ 7 \\ -6 \end{pmatrix}$, $\begin{pmatrix} 18 \\ -11 \\ 4 \end{pmatrix}$, and $\begin{pmatrix} 2 \\ -7 \\ 3 \end{pmatrix}$ are linearly dependent since they comprise a set of four 3-vectors.

There is a very important (and obvious) corollary to Theorem 2.

COROLLARY

A set of linearly independent vectors in $\mathbb{R}^n$ contains at most n vectors. ■

NOTE: We can rephrase the corollary as follows: If we have n linearly independent n-vectors, then we cannot add any more vectors without making the set linearly dependent.

From system (8) we can make another important observation whose proof is left as an exercise (see Problem 27).

THEOREM 3

Let

$$A = \begin{pmatrix} a_{11} & a_{12} & \cdots & a_{1n} \\ a_{21} & a_{22} & \cdots & a_{2n} \\ \vdots & \vdots & & \vdots \\ a_{m1} & a_{m2} & \cdots & a_{mn} \end{pmatrix}$$

Then the columns of A, considered as vectors, are linearly dependent if and only if system (8), which can be written $A\mathbf{c} = \mathbf{0}$, has nontrivial solutions. Here $\mathbf{c} = \begin{pmatrix} c_1 \\ c_2 \\ \vdots \\ c_n \end{pmatrix}$.

■

EXAMPLE 6

WRITING SOLUTIONS TO A HOMOGENEOUS SYSTEM AS LINEAR COMBINATIONS OF LINEARLY INDEPENDENT SOLUTION VECTORS

Consider the homogeneous system

$$\begin{aligned} x_1 + 2x_2 - x_3 + 2x_4 &= 0 \\ 3x_1 + 7x_2 + x_3 + 4x_4 &= 0 \end{aligned} \tag{9}$$

We solve this by row reduction:

$$\left(\begin{array}{cccc|c} 1 & 2 & -1 & 2 & 0 \\ 3 & 7 & 1 & 4 & 0 \end{array}\right) \xrightarrow{R_2 \to R_2 - 3R_1} \left(\begin{array}{cccc|c} 1 & 2 & -1 & 2 & 0 \\ 0 & 1 & 4 & -2 & 0 \end{array}\right)$$

$$\xrightarrow{R_1 \to R_1 - 2R_2} \left(\begin{array}{cccc|c} 1 & 0 & -9 & 6 & 0 \\ 0 & 1 & 4 & -2 & 0 \end{array}\right)$$

The last system is

$$\begin{aligned} x_1 \qquad - 9x_3 + 6x_4 &= 0 \\ x_2 + 4x_3 - 2x_4 &= 0 \end{aligned}$$

We see that this system has an infinite number of solutions, which we write as a linear combination of column vectors:

$$\begin{pmatrix} x_1 \\ x_2 \\ x_3 \\ x_4 \end{pmatrix} = \begin{pmatrix} 9x_3 - 6x_4 \\ -4x_3 + 2x_4 \\ x_3 \\ x_4 \end{pmatrix} = x_3 \begin{pmatrix} 9 \\ -4 \\ 1 \\ 0 \end{pmatrix} + x_4 \begin{pmatrix} -6 \\ 2 \\ 0 \\ 1 \end{pmatrix} \tag{10}$$

Note that $\begin{pmatrix} 9 \\ -4 \\ 1 \\ 0 \end{pmatrix}$ and $\begin{pmatrix} -6 \\ 2 \\ 0 \\ 1 \end{pmatrix}$ are linearly independent solutions to (9) because neither one is a multiple of the other. (You should verify that they are solutions.) Since x_3 and x_4 are arbitrary real numbers, we see, from (10), that we can express all solutions to the system (9) in terms of two linearly independent solution vectors.

The next two theorems follow directly from Theorem 3.

THEOREM 4

Let $\mathbf{v}_1, \mathbf{v}_2, \ldots, \mathbf{v}_n$ be n vectors in $\mathbb{R}^n$ and let A be the $n \times n$ matrix whose columns are $\mathbf{v}_1, \mathbf{v}_2, \ldots, \mathbf{v}_n$. Then $\mathbf{v}_1, \mathbf{v}_2, \ldots, \mathbf{v}_n$ are linearly independent if and only if the only solution to the homogeneous system $A\mathbf{x} = \mathbf{0}$ is the trivial solution $\mathbf{x} = \mathbf{0}$.

PROOF: This is Theorem 3 in the case $m = n$. ■

THEOREM 5

Let A be an $n \times n$ matrix. Then $\det A \neq 0$ if and only if the columns of A are linearly independent.

PROOF: From Theorem 4 and the Summing Up Theorem (see page 478). Columns of A are linearly independent $\Leftrightarrow$ $\mathbf{0}$ is the only solution to $A\mathbf{x} = \mathbf{0}$ $\Leftrightarrow$ $\det A \neq 0$. Here, $\Leftrightarrow$ stands for the words "if and only if." ■

Theorem 5 enables us to extend our Summing Up Theorem.

THEOREM 6 SUMMING UP THEOREM—VIEW 5

Let A be an $n \times n$ matrix. Then each of the following seven statements are equivalent; that is, each one implies the other six (so that if one is true, all are true).

i. A is invertible.

ii. The only solution to the homogeneous system $A\mathbf{x} = \mathbf{0}$ is the trivial solution ($\mathbf{x} = \mathbf{0}$).

iii. The system $A\mathbf{x} = \mathbf{b}$ has a unique solution for every n-vector $\mathbf{b}$.
iv. A is row equivalent to the $n \times n$ identity matrix I_n.
v. A is the product of elementary matrices.
vi. $\det A \neq 0$.
vii. The columns (and rows) of A are linearly independent.

PROOF: The only part not proved is that the rows of A are linearly independent if $\det A \neq 0$. If the columns are independent, then $\det A \neq 0$. Then $\det A^t = \det A \neq 0$ (see Theorem 7.2.3 on page 466). Thus the columns of A^t are linearly independent. But the columns of A^t are the rows of A, so the rows of A are independent. ■

The following theorem ties together the ideas of linear independence and spanning sets in $\mathbb{R}^n$.

THEOREM 7

Any set of n linearly independent vectors in $\mathbb{R}^n$ spans $\mathbb{R}^n$.

PROOF: Let $\mathbf{v}_1 = \begin{pmatrix} a_{11} \\ a_{21} \\ \vdots \\ a_{n1} \end{pmatrix}$, $\mathbf{v}_2 = \begin{pmatrix} a_{12} \\ a_{22} \\ \vdots \\ a_{n2} \end{pmatrix}, \ldots, \mathbf{v}_n = \begin{pmatrix} a_{1n} \\ a_{2n} \\ \vdots \\ a_{nn} \end{pmatrix}$ be linearly independent and let $\mathbf{v} = \begin{pmatrix} x_1 \\ x_2 \\ \vdots \\ x_n \end{pmatrix}$ be a vector in $\mathbb{R}^n$. We must show that there exist scalars $c_1, c_2, \ldots, c_n$ such that

$$\mathbf{v} = c_1\mathbf{v}_1 + c_2\mathbf{v}_2 + \cdots + c_n\mathbf{v}_n$$

That is,

$$\begin{pmatrix} x_1 \\ x_2 \\ \vdots \\ x_n \end{pmatrix} = c_1\begin{pmatrix} a_{11} \\ a_{21} \\ \vdots \\ a_{n1} \end{pmatrix} + c_2\begin{pmatrix} a_{12} \\ a_{22} \\ \vdots \\ a_{n2} \end{pmatrix} + \cdots + c_n\begin{pmatrix} a_{1n} \\ a_{2n} \\ \vdots \\ a_{nn} \end{pmatrix} \tag{11}$$

In (11) we multiply through, add, and equate components to obtain a system of n equations in the n unknowns $c_1, c_2, \ldots, c_n$:

$$\begin{aligned} a_{11}c_1 + a_{12}c_2 + \cdots + a_{1n}c_n &= x_1 \\ a_{21}c_1 + a_{22}c_2 + \cdots + a_{2n}c_n &= x_2 \\ \vdots \qquad\quad \vdots \qquad\qquad\quad \vdots \quad &\quad \vdots \\ a_{n1}c_1 + a_{n2}c_2 + \cdots + a_{nn}c_n &= x_n \end{aligned} \tag{12}$$

We can write (12) as $A\mathbf{c} = \mathbf{v}$, where

$$A = \begin{pmatrix} a_{11} & a_{12} & \cdots & a_{1n} \\ a_{21} & a_{22} & \cdots & a_{2n} \\ \vdots & \vdots & & \vdots \\ a_{n1} & a_{n2} & \cdots & a_{nn} \end{pmatrix} \quad \text{and} \quad \mathbf{c} = \begin{pmatrix} c_1 \\ c_2 \\ \vdots \\ c_n \end{pmatrix}$$

But det $A \neq 0$ since the columns of A are linearly independent. So system (12) has a unique solution $\mathbf{c}$ by Theorem 6 and the theorem is proved. ■

REMARK: This theorem not only shows that $\mathbf{v}$ can be written as a linear combination of the independent vectors $\mathbf{v}_1, \mathbf{v}_2, \ldots, \mathbf{v}_n$, but also that this can be done in *only one way* (since the solution vector $\mathbf{c}$ is unique).

EXAMPLE 7

THREE VECTORS IN $\mathbb{R}^3$ SPAN $\mathbb{R}^3$ IF THEIR DETERMINANT IS NONZERO

The vectors $(2, -1, 4)$, $(1, 0, 2)$, and $(3, -1, 5)$ span $\mathbb{R}^3$ because $\begin{vmatrix} 2 & 1 & 3 \\ -1 & 0 & -1 \\ 4 & 2 & 5 \end{vmatrix} = -1 \neq 0$, so they are independent.

Every example we have done so far has been in the space $\mathbb{R}^n$. We now look at some other vector spaces.

EXAMPLE 8

THREE LINEARLY INDEPENDENT MATRICES IN M_{23}

In M_{23} let $A_1 = \begin{pmatrix} 1 & 0 & 2 \\ 3 & 1 & -1 \end{pmatrix}$, $A_2 = \begin{pmatrix} -1 & 1 & 4 \\ 2 & 3 & 0 \end{pmatrix}$, and $A_3 = \begin{pmatrix} -1 & 0 & 1 \\ 1 & 2 & 1 \end{pmatrix}$. Determine whether A_1, A_2, and A_3 are linearly dependent or independent.

SOLUTION: Suppose that $c_1A_1 + c_2A_2 + c_3A_3 = 0$. Then

$$\begin{pmatrix} 0 & 0 & 0 \\ 0 & 0 & 0 \end{pmatrix} = c_1\begin{pmatrix} 1 & 0 & 2 \\ 3 & 1 & -1 \end{pmatrix} + c_2\begin{pmatrix} -1 & 1 & 4 \\ 2 & 3 & 0 \end{pmatrix} + c_3\begin{pmatrix} -1 & 0 & 1 \\ 1 & 2 & 1 \end{pmatrix}$$

$$= \begin{pmatrix} c_1 - c_2 - c_3 & c_2 & 2c_1 + 4c_2 + c_3 \\ 3c_1 + 2c_2 + c_3 & c_1 + 3c_2 + 2c_3 & -c_1 + c_3 \end{pmatrix}$$

This gives us a homogeneous system of six equations in the three unknowns c_1, c_2, and c_3, and it is quite easy to verify that the only solution is $c_1 = c_2 = c_3 = 0$. Thus the three matrices are linearly independent.

EXAMPLE 9

FOUR LINEARLY INDEPENDENT POLYNOMIALS IN P_3

In P_3 determine whether the polynomials 1, x, x^2, and x^3 are linearly dependent or independent.

SOLUTION: Suppose that $c_1 + c_2x + c_3x^2 + c_4x^3 = 0$. This must hold for every real number x. In particular, if $x = 0$, we obtain $c_1 = 0$. Then, setting $x = 1, -1, 2$, we obtain, successively,

$$\begin{aligned} c_2 + c_3 + c_4 &= 0 \\ -c_2 + c_3 - c_4 &= 0 \\ 2c_2 + 4c_3 + 8c_4 &= 0 \end{aligned}$$

The determinant of this homogeneous system is

$$\begin{vmatrix} 1 & 1 & 1 \\ -1 & 1 & -1 \\ 2 & 4 & 8 \end{vmatrix} = 12 \neq 0$$

so that the system has the unique solution $c_2 = c_3 = c_4 = 0$ and the four polynomials are linearly independent. We can see this in another way. We know that any polynomial of degree 3 has at most three real roots. But if $c_1 + c_2x + c_3x^2 + c_4x^3 = 0$ for some nonzero constants c_1, c_2, c_3, and c_4 and for every real number x, then we have constructed a cubic polynomial for which every real number is a root. This clearly is impossible.

EXAMPLE 10 THREE LINEARLY DEPENDENT POLYNOMIALS IN P_2

In P_2 determine whether the polynomials $x - 2x^2$, $x^2 - 4x$, and $-7x + 8x^2$ are linearly dependent or independent.

SOLUTION: Let $c_1(x - 2x^2) + c_2(x^2 - 4x) + c_3(-7x + 8x^2) = 0$. Then, rearranging terms, we obtain

$$\begin{aligned} (c_1 - 4c_2 - 7c_3)x &= 0 \\ (-2c_1 + c_2 + 8c_3)x^2 &= 0 \end{aligned}$$

These equations hold for every x if and only if

$$c_1 - 4c_2 - 7c_3 = 0 \quad \text{and} \quad -2c_1 + c_2 + 8c_3 = 0$$

But by Theorem 6.4.1, page 392 this system of two equations in three unknowns has an infinite number of solutions. This shows that the polynomials are linearly dependent.

If we solve this homogeneous system, we obtain, successively,

$$\left(\begin{array}{ccc|c} 1 & -4 & -7 & 0 \\ -2 & 1 & 8 & 0 \end{array}\right) \xrightarrow{R_2 \to R_2 + 2R_1} \left(\begin{array}{ccc|c} 1 & -4 & -7 & 0 \\ 0 & -7 & -6 & 0 \end{array}\right)$$

$$\xrightarrow{R_2 \to -\frac{1}{7}R_2} \left(\begin{array}{ccc|c} 1 & -4 & -7 & 0 \\ 0 & 1 & \frac{6}{7} & 0 \end{array}\right) \xrightarrow{R_1 \to R_1 + 4R_2} \left(\begin{array}{ccc|c} 1 & 0 & -\frac{25}{7} & 0 \\ 0 & 1 & \frac{6}{7} & 0 \end{array}\right)$$

Thus c_3 can be chosen arbitrarily, $c_1 = \frac{25}{7}c_3$ and $c_2 = -\frac{6}{7}c_3$. If $c_3 = 7$, for example, then $c_1 = 25$, $c_2 = -6$, and we have

$$25(x - 2x^2) - 6(x^2 - 4x) + 7(-7x + 8x^2) = 0.$$

PROBLEMS 8.4

SELF-QUIZ

Multiple Choice

I. Which of the following pairs of vectors are linearly independent?

a. $\begin{pmatrix} 1 \\ 1 \end{pmatrix}, \begin{pmatrix} 1 \\ -1 \end{pmatrix}$ **b.** $\begin{pmatrix} 2 \\ 3 \end{pmatrix}, \begin{pmatrix} 3 \\ 2 \end{pmatrix}$

c. $\begin{pmatrix} 11 \\ 0 \end{pmatrix}, \begin{pmatrix} 0 \\ 4 \end{pmatrix}$ **d.** $\begin{pmatrix} -3 \\ -11 \end{pmatrix}, \begin{pmatrix} -6 \\ 11 \end{pmatrix}$

e. $\begin{pmatrix} -2 \\ 4 \end{pmatrix}, \begin{pmatrix} 2 \\ 4 \end{pmatrix}$

II. Which of the following pairs of vectors span $\mathbb{R}^2$?

a. $\begin{pmatrix} 1 \\ 1 \end{pmatrix}, \begin{pmatrix} 1 \\ -1 \end{pmatrix}$ **b.** $\begin{pmatrix} 2 \\ 3 \end{pmatrix}, \begin{pmatrix} 3 \\ 2 \end{pmatrix}$

c. $\begin{pmatrix} 11 \\ 0 \end{pmatrix}, \begin{pmatrix} 0 \\ 4 \end{pmatrix}$ d. $\begin{pmatrix} -3 \\ -11 \end{pmatrix}, \begin{pmatrix} -6 \\ 11 \end{pmatrix}$

e. $\begin{pmatrix} -2 \\ 4 \end{pmatrix}, \begin{pmatrix} 2 \\ 4 \end{pmatrix}$

III. Which of the following sets of vectors *must* be linearly dependent?

a. $\begin{pmatrix} a \\ b \\ c \end{pmatrix}, \begin{pmatrix} d \\ e \\ f \end{pmatrix}$ b. $\begin{pmatrix} a \\ b \end{pmatrix}, \begin{pmatrix} c \\ d \end{pmatrix}, \begin{pmatrix} e \\ f \end{pmatrix}$

c. $\begin{pmatrix} a \\ b \\ c \end{pmatrix}, \begin{pmatrix} d \\ e \\ f \end{pmatrix}, \begin{pmatrix} g \\ h \\ i \end{pmatrix}$ d. $\begin{pmatrix} a \\ b \\ c \end{pmatrix}, \begin{pmatrix} d \\ e \\ f \end{pmatrix}, \begin{pmatrix} g \\ h \\ i \end{pmatrix}, \begin{pmatrix} j \\ k \\ l \end{pmatrix}$

Here $a, b, c, d, e, f, g, h, i, j, k$, and l are real numbers.

True–False

IV. If $\mathbf{v}_1, \mathbf{v}_2, \ldots, \mathbf{v}_n$ are linearly independent, then $\mathbf{v}_1, \mathbf{v}_2, \ldots, \mathbf{v}_n, \mathbf{v}_{n+1}$ are also linearly independent.

V. If $\mathbf{v}_1, \mathbf{v}_2, \ldots, \mathbf{v}_n$ are linearly dependent, then $\mathbf{v}_1, \mathbf{v}_2, \ldots, \mathbf{v}_n, \mathbf{v}_{n+1}$ are linearly dependent.

VI. If A is a 3×3 matrix and $\det A = 0$, then the rows of A are linearly dependent vectors in $\mathbb{R}^3$.

VII. The polynomials $3, 2x, -x^3$, and $3x^4$ are linearly independent in P_4.

VIII. The matrices $\begin{pmatrix} 1 & 0 \\ 0 & 0 \end{pmatrix}, \begin{pmatrix} 0 & 1 \\ 0 & 0 \end{pmatrix}, \begin{pmatrix} 0 & 1 \\ 1 & 0 \end{pmatrix}$, and $\begin{pmatrix} 2 & 3 \\ -5 & 0 \end{pmatrix}$ are linearly independent in M_{22}.

In Problems 1–22 determine whether the given set of vectors is linearly dependent or independent.

1. $\begin{pmatrix} 1 \\ 2 \end{pmatrix}; \begin{pmatrix} -1 \\ -3 \end{pmatrix}$

2. $\begin{pmatrix} 2 \\ -1 \\ 4 \end{pmatrix}; \begin{pmatrix} 4 \\ -2 \\ 7 \end{pmatrix}$

3. $\begin{pmatrix} 2 \\ -1 \\ 4 \end{pmatrix}; \begin{pmatrix} 4 \\ -2 \\ 8 \end{pmatrix}$

4. $\begin{pmatrix} -2 \\ 3 \end{pmatrix}; \begin{pmatrix} 4 \\ 7 \end{pmatrix}$

5. $\begin{pmatrix} -3 \\ 2 \end{pmatrix}; \begin{pmatrix} 1 \\ 10 \end{pmatrix}; \begin{pmatrix} 4 \\ -5 \end{pmatrix}$

6. $\begin{pmatrix} 1 \\ 0 \\ 1 \end{pmatrix}; \begin{pmatrix} 0 \\ 1 \\ 1 \end{pmatrix}; \begin{pmatrix} 1 \\ 1 \\ 0 \end{pmatrix}$

7. $\begin{pmatrix} 1 \\ 0 \\ 0 \end{pmatrix}; \begin{pmatrix} 0 \\ 1 \\ 0 \end{pmatrix}; \begin{pmatrix} 0 \\ 0 \\ 1 \end{pmatrix}$

8. $\begin{pmatrix} -3 \\ 4 \\ 2 \end{pmatrix}; \begin{pmatrix} 7 \\ -1 \\ 3 \end{pmatrix}; \begin{pmatrix} 1 \\ 2 \\ 8 \end{pmatrix}$

9. $\begin{pmatrix} -3 \\ 4 \\ 2 \end{pmatrix}; \begin{pmatrix} 7 \\ -1 \\ 3 \end{pmatrix}; \begin{pmatrix} 1 \\ 1 \\ 8 \end{pmatrix}$

10. $\begin{pmatrix} 1 \\ -2 \\ 1 \\ 1 \end{pmatrix}; \begin{pmatrix} 3 \\ 0 \\ 2 \\ -2 \end{pmatrix}; \begin{pmatrix} 0 \\ 4 \\ -1 \\ -1 \end{pmatrix}; \begin{pmatrix} 5 \\ 0 \\ 3 \\ -1 \end{pmatrix}$

11. $\begin{pmatrix} 1 \\ -2 \\ 1 \\ 1 \end{pmatrix}; \begin{pmatrix} 3 \\ 0 \\ 2 \\ -2 \end{pmatrix}; \begin{pmatrix} 0 \\ 4 \\ -1 \\ 1 \end{pmatrix}; \begin{pmatrix} 5 \\ 0 \\ 3 \\ -1 \end{pmatrix}$

12. $\begin{pmatrix} 1 \\ -1 \\ 2 \end{pmatrix}; \begin{pmatrix} 4 \\ 0 \\ 0 \end{pmatrix}; \begin{pmatrix} -2 \\ 3 \\ 5 \end{pmatrix}; \begin{pmatrix} 7 \\ 1 \\ 2 \end{pmatrix}$

13. In P_2: $1 - x, x$

14. In P_2: $-x, x^2 - 2x, 3x + 5x^2$

15. In P_2: $1 - x, 1 + x, x^2$

16. In P_3: $x, x^2 - x, x^3 - x$

17. In P_3: $2x, x^3 - 3, 1 + x - 4x^3, x^3 + 18x - 9$

18. In M_{22}: $\begin{pmatrix} 2 & -1 \\ 4 & 0 \end{pmatrix}, \begin{pmatrix} 0 & -3 \\ 1 & 5 \end{pmatrix}, \begin{pmatrix} 4 & 1 \\ 7 & -5 \end{pmatrix}$

19. In M_{22}: $\begin{pmatrix} 1 & -1 \\ 0 & 6 \end{pmatrix}, \begin{pmatrix} -1 & 0 \\ 3 & 1 \end{pmatrix}, \begin{pmatrix} 1 & 1 \\ -1 & 2 \end{pmatrix}, \begin{pmatrix} 0 & 1 \\ 1 & 0 \end{pmatrix}$

20. In M_{22}: $\begin{pmatrix} -1 & 0 \\ 1 & 2 \end{pmatrix}, \begin{pmatrix} 2 & 3 \\ 7 & -4 \end{pmatrix}, \begin{pmatrix} 8 & -5 \\ 7 & 6 \end{pmatrix}, \begin{pmatrix} 4 & -1 \\ 2 & 3 \end{pmatrix}, \begin{pmatrix} 2 & 3 \\ -1 & 4 \end{pmatrix}$

***21.** In $C[0, 1]$: $\sin x, \cos x$

***22.** In $C[0, 1]$: $x, \sqrt{x}, \sqrt[3]{x}$

23. Determine a condition on the numbers a, b, c, and d such that the vectors $\begin{pmatrix} a \\ b \end{pmatrix}$ and $\begin{pmatrix} c \\ d \end{pmatrix}$ are linearly dependent.

***24.** Find a condition on the numbers a_{ij} such that the vectors $\begin{pmatrix} a_{11} \\ a_{21} \\ a_{31} \end{pmatrix}$, $\begin{pmatrix} a_{12} \\ a_{22} \\ a_{32} \end{pmatrix}$, and $\begin{pmatrix} a_{13} \\ a_{23} \\ a_{33} \end{pmatrix}$ are linearly dependent.

25. For what value(s) of α will the vectors $\begin{pmatrix} 1 \\ 2 \\ 3 \end{pmatrix}$, $\begin{pmatrix} 2 \\ -1 \\ 4 \end{pmatrix}$, $\begin{pmatrix} 3 \\ \alpha \\ 4 \end{pmatrix}$ be linearly dependent?

26. For what value(s) of α are the vectors $\begin{pmatrix} 2 \\ -3 \\ 1 \end{pmatrix}$, $\begin{pmatrix} -4 \\ 6 \\ -2 \end{pmatrix}$, $\begin{pmatrix} \alpha \\ 1 \\ 2 \end{pmatrix}$ linearly dependent? [*Hint:* Look carefully.]

27. Prove Theorem 3. [*Hint:* Look closely at system (8).]

28. Prove that if the vectors $\mathbf{v}_1, \mathbf{v}_2, \ldots, \mathbf{v}_n$ are linearly dependent vectors in $\mathbb{R}^m$ and if $\mathbf{v}_{n+1}$ is any other vector in $\mathbb{R}^m$, then the set $\mathbf{v}_1, \mathbf{v}_2, \ldots, \mathbf{v}_n, \mathbf{v}_{n+1}$ is linearly dependent.

29. Show that if $\mathbf{v}_1, \mathbf{v}_2, \ldots, \mathbf{v}_n$ $(n \geq 2)$ are linearly independent, then so too are $\mathbf{v}_1, \mathbf{v}_2, \ldots, \mathbf{v}_k$, where $k < n$.

30. Show that if the nonzero vectors $\mathbf{v}_1$ and $\mathbf{v}_2$ in $\mathbb{R}^n$ are orthogonal (see page 80), then the set $\{\mathbf{v}_1, \mathbf{v}_2\}$ is linearly independent.

***31.** Suppose that $\mathbf{v}_1$ is orthogonal to $\mathbf{v}_2$ and $\mathbf{v}_3$ and that $\mathbf{v}_2$ is orthogonal to $\mathbf{v}_3$. If $\mathbf{v}_1$, $\mathbf{v}_2$, and $\mathbf{v}_3$ are nonzero, show that the set $\{\mathbf{v}_1, \mathbf{v}_2, \mathbf{v}_3\}$ is linearly independent.

32. Let A be a square $(n \times n)$ matrix whose columns are the vectors $\mathbf{v}_1, \mathbf{v}_2, \ldots, \mathbf{v}_n$. Show that $\mathbf{v}_1, \mathbf{v}_2, \ldots, \mathbf{v}_n$ are linearly independent if and only if the row echelon form of A does not contain a row of zeros.

In Problems 33–37 write the solutions to the given homogeneous systems in terms of one or more linearly independent vectors.

33. $x_1 + x_2 + x_3 = 0$

34. $\begin{aligned} x_1 - x_2 + 7x_3 - x_4 &= 0 \\ 2x_1 + 3x_2 - 8x_3 + x_4 &= 0 \end{aligned}$

35. $\begin{aligned} x_1 + 2x_2 - x_3 &= 0 \\ 2x_1 + 5x_2 + 4x_3 &= 0 \end{aligned}$

36. $\begin{aligned} x_1 + x_2 + x_3 - x_4 - x_5 &= 0 \\ -2x_1 + 3x_2 + x_3 + 4x_4 - 6x_5 &= 0 \end{aligned}$

37. $x_1 + 2x_2 - 3x_3 + 5x_4 = 0$

38. Let $\mathbf{u} = (1, 2, 3)$.

- **a.** Let $H = \{\mathbf{v} \in \mathbb{R}^3 \colon \mathbf{u} \cdot \mathbf{v} = 0\}$. Show that H is a subspace of $\mathbb{R}^3$.
- **b.** Find two linearly independent vectors in H. Call them $\mathbf{x}$ and $\mathbf{y}$.
- **c.** Compute $\mathbf{w} = \mathbf{x} \times \mathbf{y}$.
- **d.** Show that $\mathbf{u}$ and $\mathbf{w}$ are linearly dependent.
- **e.** Give a geometric interpretation of parts (a) and (c) and explain why (d) must be true.

REMARK: If $V = \{\mathbf{v} \in \mathbb{R}^3 \colon \mathbf{v} = \alpha\mathbf{u}$ for some real number $\alpha\}$, then V is a subspace of $\mathbb{R}^3$ and H is called the **orthogonal complement** of V.

39. Choose a vector $\mathbf{u} \neq \mathbf{0}$ in $\mathbb{R}^3$. Repeat the steps of Problem 38, starting with the vector you have chosen.

40. Show that any four polynomials in P_2 are linearly dependent.

41. Show that two polynomials cannot span P_2.

***42.** Show that any $n + 2$ polynomials in P_n are linearly dependent.

43. Show that any subset of a set of linearly independent vectors is linearly independent. [*Note:* This generalizes Problem 29.]

44. Show that any seven matrices in M_{32} are linearly dependent.

***45.** Prove that any $mn + 1$ matrices in M_{mn} are linearly dependent.

46. Let S_1 and S_2 be two finite, linearly independent sets in a vector space V. Show that $S_1 \cap S_2$ is a linearly independent set.

***47.** Show that in P_n the polynomials $1, x, x^2, \ldots, x^n$ are linearly independent. [*Hint:* This is certainly true if $n = 1$. Assume that $1, x, x^2, \ldots, x^{n-1}$ are linearly independent and show how this implies that $1, x, x^2, \ldots, x^n$ are also independent. This will complete the proof by mathematical induction.]

48. Let $\{\mathbf{v}_1, \mathbf{v}_2, \ldots, \mathbf{v}_n\}$ be a linearly independent set. Show that the vectors $\mathbf{v}_1, \mathbf{v}_1 + \mathbf{v}_2, \mathbf{v}_1 + \mathbf{v}_2 + \mathbf{v}_3, \ldots, \mathbf{v}_1 + \mathbf{v}_2 + \cdots + \mathbf{v}_n$ are linearly independent.

49. Let $S = \{\mathbf{v}_1, \mathbf{v}_2, \ldots, \mathbf{v}_n\}$ be a linearly dependent set of nonzero vectors in a vector space V. Show that at least one of the vectors in S can be written as a linear combination of the vectors that precede it. That is, show that there is an integer $k \leq n$ and scalars $a_1, a_2, \ldots, a_{k-1}$ such that $\mathbf{v}_k = a_1\mathbf{v}_1 + a_2\mathbf{v}_2 + \cdots + a_{k-1}\mathbf{v}_{k-1}$.

50. Let $\{\mathbf{v}_1, \mathbf{v}_2, \ldots, \mathbf{v}_n\}$ be a set of vectors having the property that the set $\{\mathbf{v}_i, \mathbf{v}_j\}$ is linearly dependent when $i \neq j$. Show that each vector in the set is a multiple of a single vector in the set.

51. Let f and g be in $C^1[0, 1]$. Then the **Wronskian**† of f and g is defined by

$$W(f, g)(x) = \begin{vmatrix} f(x) & g(x) \\ f'(x) & g'(x) \end{vmatrix}$$

Show that if f and g are linearly dependent, then $W(f, g)(x) = 0$ for every $x \in [0, 1]$.

52. Determine a suitable definition for the Wronskian of the functions $f_1, f_2, \ldots, f_n \in C^{(n-1)}[0, 1]$.‡

53. Suppose that $\mathbf{u}$, $\mathbf{v}$, and $\mathbf{w}$ are linearly independent. Prove or disprove: $\mathbf{u} + \mathbf{v}$, $\mathbf{u} + \mathbf{w}$, and $\mathbf{v} + \mathbf{w}$ are linearly independent.

54. For what real values of c are the vectors $(1 - c, 1 + c)$ and $(1 + c, 1 - c)$ linearly independent?

55. Show that the vectors $(1, a, a^2)$, $(1, b, b^2)$, and $(1, c, c^2)$ are linearly independent if $a \neq b$, $a \neq c$, and $b \neq c$.

56. Let $\{\mathbf{v}_1, \mathbf{v}_2, \ldots, \mathbf{v}_n\}$ be a linearly independent set and suppose that $\mathbf{v} \notin \text{span}\,\{\mathbf{v}_1, \mathbf{v}_2, \ldots, \mathbf{v}_n\}$. Show that $\{\mathbf{v}_1, \mathbf{v}_2, \ldots, \mathbf{v}_n, \mathbf{v}\}$ is a linearly independent set.

57. Find a set of three linearly independent vectors in $\mathbb{R}^3$ that contains the vectors $\begin{pmatrix} 2 \\ 1 \\ 2 \end{pmatrix}$ and $\begin{pmatrix} -1 \\ 3 \\ 4 \end{pmatrix}$. $\left[\textit{Hint:} \text{ Find a vector } \mathbf{v} \notin \text{span}\left\{\begin{pmatrix} 2 \\ 1 \\ 2 \end{pmatrix}, \begin{pmatrix} -1 \\ 3 \\ 4 \end{pmatrix}\right\}.\right]$

58. Find a set of three linearly independent vectors in P_2 that contains the polynomials $1 - x^2$ and $1 + x^2$.

59. Suppose that $\mathbf{u} = \begin{pmatrix} u_1 \\ u_2 \\ u_3 \end{pmatrix}$, $\mathbf{v} = \begin{pmatrix} v_1 \\ v_2 \\ v_3 \end{pmatrix}$, and $\mathbf{w} = \begin{pmatrix} w_1 \\ w_2 \\ w_3 \end{pmatrix}$ are coplanar.

a. Show that there exist constants a, b, and c not all zero such that

$$au_1 + bu_2 + cu_3 = 0$$
$$av_1 + bv_2 + cv_3 = 0$$
$$aw_1 + bw_2 + cw_3 = 0$$

b. Explain why

$$\det \begin{pmatrix} u_1 & u_2 & u_3 \\ v_1 & v_2 & v_3 \\ w_1 & w_2 & w_3 \end{pmatrix} = 0$$

c. Use Theorem 3 to show that $\mathbf{u}$, $\mathbf{v}$, and $\mathbf{w}$ are linearly dependent.

ANSWERS TO SELF-QUIZ

I. All of them **II.** All of them **III.** b, d **IV.** False **V.** True
VI. True **VII.** True **VIII.** False

8.5 BASIS AND DIMENSION

We have seen that in $\mathbb{R}^2$ it is convenient to write vectors in terms of the vectors $\mathbf{i} = \begin{pmatrix} 1 \\ 0 \end{pmatrix}$ and $\mathbf{j} = \begin{pmatrix} 0 \\ 1 \end{pmatrix}$. In $\mathbb{R}^3$ we wrote vectors in terms of $\begin{pmatrix} 1 \\ 0 \\ 0 \end{pmatrix}$, $\begin{pmatrix} 0 \\ 1 \\ 0 \end{pmatrix}$, and $\begin{pmatrix} 0 \\ 0 \\ 1 \end{pmatrix}$. We now generalize this idea.

---

†Named after the Polish mathematician Józef Maria Hoene-Wroński (1778–1853). Hoene-Wroński spent most of his adult life in France. He worked on the theory of determinants and was also known for his critical writings in the philosophy of mathematics.

‡$C^{(n-1)}[0, 1]$ is the set of functions whose $(n - 1)$st derivatives are defined and continuous on $[0, 1]$.

DEFINITION BASIS

A set of vectors $\{\mathbf{v}_1, \mathbf{v}_2, \ldots, \mathbf{v}_n\}$ is a **basis** for a vector space V if

i. $\{\mathbf{v}_1, \mathbf{v}_2, \ldots, \mathbf{v}_n\}$ is linearly independent.
ii. $\{\mathbf{v}_1, \mathbf{v}_2, \ldots, \mathbf{v}_n\}$ spans V. ■

We have already seen quite a few examples of bases. In Theorem 8.4.7, for instance, we saw that any set of n linearly independent vectors in $\mathbb{R}^n$ spans $\mathbb{R}^n$. Thus

> Every set of n linearly independent vectors in $\mathbb{R}^n$ is a basis in $\mathbb{R}^n$.

In $\mathbb{R}^n$ we define

$$\mathbf{e}_1 = \begin{pmatrix}1\\0\\0\\\vdots\\0\end{pmatrix}, \mathbf{e}_2 = \begin{pmatrix}0\\1\\0\\\vdots\\0\end{pmatrix}, \mathbf{e}_3 = \begin{pmatrix}0\\0\\1\\\vdots\\0\end{pmatrix}, \ldots, \mathbf{e}_n = \begin{pmatrix}0\\0\\0\\\vdots\\1\end{pmatrix}$$

Then since the $\mathbf{e}_i$'s are the columns of the identity matrix (which has determinant 1), $\{\mathbf{e}_1, \mathbf{e}_2, \ldots, \mathbf{e}_n\}$ is linearly independent and therefore constitutes a basis in $\mathbb{R}^n$. This special basis is called the **standard basis** in $\mathbb{R}^n$. We now find bases for some other spaces.

EXAMPLE 1 STANDARD BASIS FOR P_n

By Example 8.4.9, page 511, the polynomials $1, x, x^2, x^3$ are linearly independent in P_3. By Example 8.3.3, page 498, these polynomials span P_3. Thus $\{1, x, x^2, x^3\}$ is a basis for P_3. In general, the monomials $\{1, x, x^2, x^3, \ldots, x^n\}$ constitute a basis for P_n. This is called the **standard basis** for P_n.

EXAMPLE 2 STANDARD BASIS FOR M_{22}

We saw in Example 8.3.6, page 499, that $\begin{pmatrix}1&0\\0&0\end{pmatrix}, \begin{pmatrix}0&1\\0&0\end{pmatrix}, \begin{pmatrix}0&0\\1&0\end{pmatrix}$, and $\begin{pmatrix}0&0\\0&1\end{pmatrix}$ span M_{22}. If $\begin{pmatrix}c_1&c_2\\c_3&c_4\end{pmatrix} = c_1\begin{pmatrix}1&0\\0&0\end{pmatrix} + c_2\begin{pmatrix}0&1\\0&0\end{pmatrix} + c_3\begin{pmatrix}0&0\\1&0\end{pmatrix} + c_4\begin{pmatrix}0&0\\0&1\end{pmatrix} = \begin{pmatrix}0&0\\0&0\end{pmatrix}$, then, obviously, $c_1 = c_2 = c_3 = c_4 = 0$. Thus these four matrices are linearly independent and form a basis for M_{22}. This is called the **standard basis** for M_{22}.

EXAMPLE 3 A BASIS FOR A SUBSPACE OF $\mathbb{R}^3$

Find a basis for the set of vectors lying on the plane

$$\pi = \left\{\begin{pmatrix}x\\y\\z\end{pmatrix}: 2x - y + 3z = 0\right\}$$

SOLUTION: We saw in Example 8.1.6 that π is a vector space. To find a basis, we first note that if x and z are chosen arbitrarily and if $\begin{pmatrix} x \\ y \\ z \end{pmatrix} \in \pi$, then $y = 2x + 3z$. Thus vectors in π have the form $\begin{pmatrix} x \\ 2x + 3z \\ z \end{pmatrix}$. Since x and z are arbitrary, we choose some simple values for them. Choosing $x = 1$, $z = 0$, we obtain $\mathbf{v}_1 = \begin{pmatrix} 1 \\ 2 \\ 0 \end{pmatrix}$; choosing $x = 0$, $z = 1$, we get $\mathbf{v}_2 = \begin{pmatrix} 0 \\ 3 \\ 1 \end{pmatrix}$. Then $\begin{pmatrix} x \\ 2x + 3z \\ z \end{pmatrix} = x\begin{pmatrix} 1 \\ 2 \\ 0 \end{pmatrix} + z\begin{pmatrix} 0 \\ 3 \\ 1 \end{pmatrix}$. Thus $\{\mathbf{v}_1, \mathbf{v}_2\}$ span π, and since they are obviously linearly independent (because one is not a multiple of the other), they form a basis for π.

If $\mathbf{v}_1, \mathbf{v}_2, \ldots, \mathbf{v}_n$ is a basis for V, then any other vector $\mathbf{v} \in V$ can be written $\mathbf{v} = c_1\mathbf{v}_1 + c_2\mathbf{v}_2 + \cdots + c_n\mathbf{v}_n$. Can it be written in another way as a linear combination of the $\mathbf{v}_i$'s? The answer is *no*. (See the remark following the proof of Theorem 8.4.7, page 510, in the case $V = \mathbb{R}^n$.)

THEOREM 1

If $\{\mathbf{v}_1, \mathbf{v}_2, \ldots, \mathbf{v}_n\}$ is a basis for V and if $\mathbf{v} \in V$, then there exists a *unique* set of scalars $c_1, c_2, \ldots, c_n$ such that $\mathbf{v} = c_1\mathbf{v}_1 + c_2\mathbf{v}_2 + \cdots + c_n\mathbf{v}_n$.

PROOF: At least one such set of scalars exists because $\{\mathbf{v}_1, \mathbf{v}_2, \ldots, \mathbf{v}_n\}$ spans V. Suppose then that $\mathbf{v}$ can be written in two ways as a linear combination of the basis vectors. That is, suppose that

$$\mathbf{v} = c_1\mathbf{v}_1 + c_2\mathbf{v}_2 + \cdots + c_n\mathbf{v}_n = d_1\mathbf{v}_1 + d_2\mathbf{v}_2 + \cdots + d_n\mathbf{v}_n$$

Then, subtracting, we obtain the equation

$$(c_1 - d_1)\mathbf{v}_1 + (c_2 - d_2)\mathbf{v}_2 + \cdots + (c_n - d_n)\mathbf{v}_n = \mathbf{0}$$

But since the $\mathbf{v}_i$'s are linearly independent, this equation can hold only if $c_1 - d_1 = c_2 - d_2 = \cdots = c_n - d_n = 0$. Thus $c_1 = d_1, c_2 = d_2, \ldots, c_n = d_n$ and the theorem is proved. ■

We have seen that vector spaces may have many bases. A question naturally arises: Do all bases contain the same number of vectors? In $\mathbb{R}^3$ the answer is certainly yes. To see this we note that any three linearly independent vectors in $\mathbb{R}^3$ form a basis. But fewer than three vectors cannot form a basis since, as we saw in Section 8.3, the span of two linearly independent vectors in $\mathbb{R}^3$ is a plane in $\mathbb{R}^3$—and a plane is not all of $\mathbb{R}^3$. Similarly, a set of four or more vectors in $\mathbb{R}^3$ cannot be linearly independent; for if the first three vectors in the set are linearly independent, then they form a basis, and therefore all other vectors in the set can be written as a linear combination of the first three. Thus all bases in $\mathbb{R}^3$ contain three vectors. The next theorem tells us that the answer to the question posed above is *yes* for all vector spaces.

THEOREM 2

If $\{\mathbf{u}_1, \mathbf{u}_2, \ldots, \mathbf{u}_m\}$ and $\{\mathbf{v}_1, \mathbf{v}_2, \ldots, \mathbf{v}_n\}$ are bases for the vector space V, then $m = n$; that is, any two bases in a vector space V have the same number of vectors.

PROOF:† Let $S_1 = \{\mathbf{u}_1, \ldots, \mathbf{u}_m\}$ and $S_2 = \{\mathbf{v}_1, \ldots, \mathbf{v}_n\}$ be two bases for V. We must show that $m = n$. We prove this by showing that if $m > n$, then S_1 is a linearly dependent set, which contradicts the hypothesis that S_1 is a basis. This will show that $m \le n$. The same proof will then show that $n \le m$, and this will prove the theorem. Hence all we must show is that if $m > n$, then S_1 is dependent. Since S_2 constitutes a basis, we can write each $\mathbf{u}_i$ as a linear combination of the $\mathbf{v}_i$'s. We have

$$\begin{aligned} \mathbf{u}_1 &= a_{11}\mathbf{v}_1 + a_{12}\mathbf{v}_2 + \cdots + a_{1n}\mathbf{v}_n \\ \mathbf{u}_2 &= a_{21}\mathbf{v}_1 + a_{22}\mathbf{v}_2 + \cdots + a_{2n}\mathbf{v}_n \\ &\vdots \\ \mathbf{u}_m &= a_{m1}\mathbf{v}_1 + a_{m2}\mathbf{v}_2 + \cdots + a_{mn}\mathbf{v}_n \end{aligned} \tag{1}$$

To show that S_1 is dependent, we must find scalars $c_1, c_2, \ldots, c_m$, not all zero, such that

$$c_1\mathbf{u}_1 + c_2\mathbf{u}_2 + \cdots + c_m\mathbf{u}_m = \mathbf{0} \tag{2}$$

Inserting (1) into (2), we obtain

$$\begin{aligned} c_1(a_{11}\mathbf{v}_1 + a_{12}\mathbf{v}_2 + \cdots + a_{1n}\mathbf{v}_n) + c_2(a_{21}\mathbf{v}_1 + a_{22}\mathbf{v}_2 + \cdots + a_{2n}\mathbf{v}_n) \\ + \cdots + c_m(a_{m1}\mathbf{v}_1 + a_{m2}\mathbf{v}_2 + \cdots + a_{mn}\mathbf{v}_n) = \mathbf{0} \end{aligned} \tag{3}$$

Equation (3) can be rewritten as

$$\begin{aligned} (a_{11}c_1 + a_{21}c_2 + \cdots + a_{m1}c_m)\mathbf{v}_1 + (a_{12}c_1 + a_{22}c_2 + \cdots + a_{m2}c_m)\mathbf{v}_2 \\ + \cdots + (a_{1n}c_1 + a_{2n}c_2 + \cdots + a_{mn}c_m)\mathbf{v}_n = \mathbf{0} \end{aligned} \tag{4}$$

But since $\mathbf{v}_1, \mathbf{v}_2, \ldots, \mathbf{v}_n$ are linearly independent, we must have

$$\begin{aligned} a_{11}c_1 + a_{21}c_2 + \cdots + a_{m1}c_m &= 0 \\ a_{12}c_1 + a_{22}c_2 + \cdots + a_{m2}c_m &= 0 \\ &\vdots \\ a_{1n}c_1 + a_{2n}c_2 + \cdots + a_{mn}c_m &= 0 \end{aligned} \tag{5}$$

System (5) is a homogeneous system of n equations in the m unknowns $c_1, c_2, \ldots, c_m$, and since $m > n$, Theorem 6.4.1, page 392, tells us that the system has an infinite number of solutions. Thus there are scalars $c_1, c_2, \ldots, c_m$, not all zero, such that (2) is satisfied, and therefore S_1 is a linearly dependent set. This contradiction proves that $m \le n$, and, by exchanging the roles of S_1 and S_2, we can show that $n \le m$ and the proof is complete. ■

With this theorem we can define one of the central concepts in linear algebra.

DEFINITION DIMENSION

If the vector space V has a finite basis, then the **dimension** of V is the number of vectors in every basis and V is called a **finite dimensional vector space**. Otherwise V

†This proof is given for vector spaces with bases containing a finite number of vectors. We also treat the scalars as though they were real numbers. However, the proof works in the complex case as well.

is called an **infinite dimensional vector space**. If $V = \{\mathbf{0}\}$, then V is said to be **zero dimensional**. ■

NOTATION: We write the dimension of V as dim V.

REMARK: We have not proved that every vector space has a basis. This very difficult proof appears in Appendix 6. But we do not need this fact for the definition above to make sense; for *if* V has a finite basis, then V is finite dimensional. Otherwise V is infinite dimensional. Thus, in order to show that V is infinite dimensional, it is only necessary to show that V does not have a finite basis. We can do this by showing that V contains an infinite number of linearly independent vectors (see Example 7 below). It is not necessary to construct an infinite basis for V.

EXAMPLE 4 THE DIMENSION OF $\mathbb{R}^n$

Since n linearly independent vectors in $\mathbb{R}^n$ constitute a basis, we see that

$$\dim \mathbb{R}^n = n$$

EXAMPLE 5 THE DIMENSION OF P_n

By Example 1 and Problem 8.4.47, page 514, the polynomials $\{1, x, x^2, \ldots, x^n\}$ constitute a basis in P_n. Thus $\dim P_n = n + 1$.

EXAMPLE 6 THE DIMENSION OF M_{mn}

In M_{mn} let A_{ij} be the $m \times n$ matrix with a 1 in the ijth position and a zero everywhere else. It is easy to show that the A_{ij} for $i = 1, 2, \ldots, m$ and $j = 1, 2, \ldots, n$ form a basis for M_{mn}. Thus $\dim M_{mn} = mn$.

EXAMPLE 7 P IS INFINITE DIMENSIONAL

In Example 8.3.7, page 499, we saw that no finite set of polynomials spans P. Thus P has no finite basis, and is therefore an infinite dimensional vector space.

There are a number of theorems that tell us something about the dimension of a vector space.

THEOREM 3

Suppose that $\dim V = n$. If $\mathbf{u}_1, \mathbf{u}_2, \ldots, \mathbf{u}_m$ is a set of m linearly independent vectors in V, then $m \le n$.

PROOF: Let $\mathbf{v}_1, \mathbf{v}_2, \ldots, \mathbf{v}_n$ be a basis for V. If $m > n$, then, as in the proof of Theorem 2, we can find constants $c_1, c_2, \ldots, c_m$ not all zero such that equation (2) is satisfied. This would contradict the linear independence of the $\mathbf{u}_i$'s. Thus $m \le n$. ■

THEOREM 4

Let H be a subspace of the finite dimensional vector space V. Then H is finite dimensional and

$$\dim H \leq \dim V \tag{6}$$

PROOF: Let $\dim V = n$. Any set of linearly independent vectors in H is also a linearly independent set in V. By Theorem 3, any linearly independent set in H can contain at most n vectors. Hence H is finite dimensional. Moreover, since any basis in H is a linearly independent set, we see that $\dim H \leq n$. ■

Theorem 4 has some interesting consequences. We give two of them here.

EXAMPLE 8

$C[0, 1]$ AND $C^1[0, 1]$ ARE INFINITE DIMENSIONAL

Let $P[0, 1]$ denote the set of polynomials defined on the interval $[0, 1]$. Then $P[0, 1] \subset C[0, 1]$. If $C[0, 1]$ were finite dimensional, then $P[0, 1]$ would be finite dimensional also. But, by Example 7, this is not the case. Hence $C[0, 1]$ is infinite dimensional. Similarly, since $P[0, 1] \subset C^1[0, 1]$ (since every polynomial is differentiablc), we also see that $C^1[0, 1]$ is infinite dimensional.

In general,

> Any vector space containing an infinite dimensional subspace is infinite dimensional.

EXAMPLE 9

THE SUBSPACES OF $\mathbb{R}^3$

We can use Theorem 4 to find *all* subspaces of $\mathbb{R}^3$. Let H be a subspace of $\mathbb{R}^3$. Then there are four possibilities: $H = \{\mathbf{0}\}$; $\dim H = 1$, $\dim H = 2$, and $\dim H = 3$. If $\dim H = 3$, then H contains a basis of three linearly independent vectors $\mathbf{v}_1, \mathbf{v}_2, \mathbf{v}_3$ in $\mathbb{R}^3$. But then $\mathbf{v}_1, \mathbf{v}_2, \mathbf{v}_3$ also form a basis for $\mathbb{R}^3$. Thus $H = \text{span}\,\{\mathbf{v}_1, \mathbf{v}_2, \mathbf{v}_3\} = \mathbb{R}^3$. Hence the only way to get a *proper* subspace of $\mathbb{R}^3$ is to have $\dim H = 1$ or $\dim H = 2$. If $\dim H = 1$, then H has a basis consisting of the one vector $\mathbf{v} = (a, b, c)$. Let $\mathbf{x}$ be in H. Then $\mathbf{x} = t(a, b, c)$ for some real number t [since (a, b, c) spans H]. If $\mathbf{x} = (x, y, z)$, this means that $x = at$, $y = bt$, $z = ct$. But this is the equation of a line in $\mathbb{R}^3$ passing through the origin with direction vector (a, b, c).

Now suppose $\dim H = 2$ and let $\mathbf{v}_1 = (a_1, b_1, c_1)$ and $\mathbf{v}_2 = (a_2, b_2, c_2)$ be a basis for H. If $\mathbf{x} = (x, y, z) \in H$, then there exist real numbers s and t such that $\mathbf{x} = s\mathbf{v}_1 + t\mathbf{v}_2$ or $(x, y, z) = s(a_1, b_1, c_1) + t(a_2, b_2, c_2)$. Then

$$\begin{aligned} x &= sa_1 + ta_2 \\ y &= sb_1 + tb_2 \\ z &= sc_1 + tc_2 \end{aligned} \tag{7}$$

Let $\mathbf{v}_3 = (\alpha, \beta, \gamma) = \mathbf{v}_1 \times \mathbf{v}_2$. Then, from Theorem 1.6.2 on page 44, part (*vi*), we have $\mathbf{v}_3 \cdot \mathbf{v}_1 = 0$ and $\mathbf{v}_3 \cdot \mathbf{v}_2 = 0$. Now we calculate

$$\begin{aligned}\alpha x + \beta y + \gamma z &= \alpha(sa_1 + ta_2) + \beta(sb_1 + tb_2) + \gamma(sc_1 + tc_2)\\ &= (\alpha a_1 + \beta b_1 + \gamma c_1)s + (\alpha a_2 + \beta b_2 + \gamma c_2)t\\ &= (\mathbf{v}_3 \cdot \mathbf{v}_1)s + (\mathbf{v}_3 \cdot \mathbf{v}_2)t = 0\end{aligned}$$

Thus, if $(x, y, z) \in H$, then $\alpha x + \beta y + \gamma z = 0$, which shows that H is a plane passing through the origin with normal vector $\mathbf{v}_3 = \mathbf{v}_1 \times \mathbf{v}_2$. Therefore we have proved that

> The only proper subspaces of $\mathbb{R}^3$ are sets of vectors lying on a single line or a single plane passing through the origin.

EXAMPLE 10

SOLUTION SPACE AND KERNEL

Let A be an $m \times n$ matrix and let $S = \{\mathbf{x} \in \mathbb{R}^n;\ A\mathbf{x} = \mathbf{0}\}$. Let $\mathbf{x}_1 \in S$ and $\mathbf{x}_2 \in S$; then $A(\mathbf{x}_1 + \mathbf{x}_2) = A\mathbf{x}_1 + A\mathbf{x}_2 = \mathbf{0} + \mathbf{0} = \mathbf{0}$ and $A(\alpha\mathbf{x}_1) = \alpha(A\mathbf{x}_1) = \alpha\mathbf{0} = \mathbf{0}$, so that S is a subspace of $\mathbb{R}^n$ and $\dim S \le n$. S is called the **solution space** of the homogeneous system $A\mathbf{x} = \mathbf{0}$. It is also called the **kernel** of the matrix A.

EXAMPLE 11

FINDING A BASIS FOR THE SOLUTION SPACE OF A HOMOGENEOUS SYSTEM

Find a basis for (and the dimension of) the solution space S of the homogeneous system

$$\begin{aligned} x + 2y - z &= 0\\ 2x - y + 3z &= 0\end{aligned}$$

SOLUTION: Here $A = \begin{pmatrix} 1 & 2 & -1\\ 2 & -1 & 3\end{pmatrix}$. Since A is a 2×3 matrix, S is a subspace of $\mathbb{R}^3$. Row-reducing, we find, successively,

$$\left(\begin{array}{ccc|c} 1 & 2 & -1 & 0\\ 2 & -1 & 3 & 0\end{array}\right) \xrightarrow{R_2 \to R_2 - 2R_1} \left(\begin{array}{ccc|c} 1 & 2 & -1 & 0\\ 0 & -5 & 5 & 0\end{array}\right)$$

$$\xrightarrow{R_2 \to -\frac{1}{5}R_2} \left(\begin{array}{ccc|c} 1 & 2 & -1 & 0\\ 0 & 1 & -1 & 0\end{array}\right) \xrightarrow{R_1 \to R_1 - 2R_2} \left(\begin{array}{ccc|c} 1 & 0 & 1 & 0\\ 0 & 1 & -1 & 0\end{array}\right)$$

Then $y = z$ and $x = -z$, so that all solutions are of the form $\begin{pmatrix} -z\\ z\\ z\end{pmatrix}$. Thus $\begin{pmatrix} -1\\ 1\\ 1\end{pmatrix}$ is a basis for S and $\dim S = 1$. Note that S is the set of vectors lying on the straight line $x = -t$, $y = t$, $z = t$.

EXAMPLE 12

FINDING A BASIS FOR THE SOLUTION SPACE OF A HOMOGENEOUS SYSTEM

Find a basis for the solution space S of the system

$$\begin{aligned} 2x - y + 3z &= 0\\ 4x - 2y + 6z &= 0\\ -6x + 3y - 9z &= 0\end{aligned}$$

SOLUTION: Row-reducing as above, we obtain

$$\left(\begin{array}{rrr|r} 2 & -1 & 3 & 0 \\ 4 & -2 & 6 & 0 \\ -6 & 3 & -9 & 0 \end{array}\right) \xrightarrow[R_3 \to R_3 + 3R_2]{R_2 \to R_2 - 2R_1} \left(\begin{array}{rrr|r} 2 & -1 & 3 & 0 \\ 0 & 0 & 0 & 0 \\ 0 & 0 & 0 & 0 \end{array}\right)$$

giving the single equation $2x - y + 3z = 0$. S is a plane and, by Example 3, a basis is given by $\begin{pmatrix} 1 \\ 2 \\ 0 \end{pmatrix}$ and $\begin{pmatrix} 0 \\ 3 \\ 1 \end{pmatrix}$ and $\dim S = 2$. Note that we have shown that any solution to the homogeneous equation can be written as

$$c_1 \begin{pmatrix} 1 \\ 2 \\ 0 \end{pmatrix} + c_2 \begin{pmatrix} 0 \\ 3 \\ 1 \end{pmatrix}$$

For example, if $c_1 = 2$ and $c_2 = -3$, we obtain the solution

$$\mathbf{x} = 2\begin{pmatrix} 1 \\ 2 \\ 0 \end{pmatrix} - 3\begin{pmatrix} 0 \\ 3 \\ 1 \end{pmatrix} = \begin{pmatrix} 2 \\ 4 \\ 0 \end{pmatrix} + \begin{pmatrix} 0 \\ -9 \\ -3 \end{pmatrix} = \begin{pmatrix} 2 \\ -5 \\ -3 \end{pmatrix}$$

Before leaving this section, we prove a result that is very useful in finding bases in an arbitrary vector space. We have seen that n linearly independent vectors in $\mathbb{R}^n$ constitute a basis for $\mathbb{R}^n$. This fact holds in *any* finite dimensional vector space.

THEOREM 5

Any n linearly independent vectors in a vector space V of dimension n constitute a basis for V.

PROOF: Let $\mathbf{v}_1, \mathbf{v}_2, \ldots, \mathbf{v}_n$ be the n vectors. If they span V, then they constitute a basis. If they do not, then there is a vector $\mathbf{u} \in V$ such that $\mathbf{u} \notin \text{span}\,\{\mathbf{v}_1, \mathbf{v}_2, \ldots, \mathbf{v}_n\}$. This means that the $n + 1$ vectors $\mathbf{v}_1, \mathbf{v}_2, \ldots, \mathbf{v}_n, \mathbf{u}$ are linearly independent. To see this, note that if

$$c_1\mathbf{v}_1 + c_2\mathbf{v}_2 + \cdots + c_n\mathbf{v}_n + c_{n+1}\mathbf{u} = 0 \tag{8}$$

then $c_{n+1} = 0$, for if not we could write $\mathbf{u}$ as a linear combination of $\mathbf{v}_1, \mathbf{v}_2, \ldots, \mathbf{v}_n$ by dividing equation (8) by c_{n+1} and putting all terms except $\mathbf{u}$ on the right-hand side. But if $c_{n+1} = 0$, then (8) reads

$$c_1\mathbf{v}_1 + c_2\mathbf{v}_2 + \cdots + c_n\mathbf{v}_n = 0$$

which means that $c_1 = c_2 = \cdots = c_n = 0$ since the $\mathbf{v}_i$'s are linearly independent. Now let $W = \text{span}\,\{\mathbf{v}_1, \mathbf{v}_2, \ldots, \mathbf{v}_n, \mathbf{u}\}$. Then as all the vectors in braces are in V, W is a subspace of V. Since $\mathbf{v}_1, \mathbf{v}_2, \ldots, \mathbf{v}_n, \mathbf{u}$ are linearly independent, they form a basis for W. Thus $\dim W = n + 1$. But from Theorem 4, $\dim W \leq n$. This contradiction shows that there is *no* vector $\mathbf{u} \in V$ such that $\mathbf{u} \notin \text{span}\,\{\mathbf{v}_1, \mathbf{v}_2, \ldots, \mathbf{v}_n\}$. Thus $\mathbf{v}_1, \mathbf{v}_2, \ldots, \mathbf{v}_n$ span V and therefore constitute a basis for V. ■

PROBLEMS 8.5

SELF-QUIZ

True–False

I. Any three vectors in $\mathbb{R}^3$ form a basis for $\mathbb{R}^3$.

II. Any three linearly independent vectors in $\mathbb{R}^3$ form a basis for $\mathbb{R}^3$.

III. A basis in a vector space is unique.

IV. Let H be a proper subspace of $\mathbb{R}^4$. It is possible to find four linearly independent vectors in H.

V. Let $H = \left\{\begin{pmatrix} x \\ y \\ z \end{pmatrix}: 2x + 11y - 17z = 0\right\}$. Then $\dim H = 2$.

VI. Let $\{\mathbf{v}_1, \mathbf{v}_2, \ldots, \mathbf{v}_n\}$ be a basis for the vector space V. Then it is *not* possible to find a vector $\mathbf{v} \in V$ such that $\mathbf{v} \notin \text{span}\,\{\mathbf{v}_1, \mathbf{v}_2, \ldots, \mathbf{v}_n\}$.

VII. $\left\{\begin{pmatrix} 2 & 0 \\ 0 & 0 \end{pmatrix}, \begin{pmatrix} 0 & 3 \\ 0 & 0 \end{pmatrix}, \begin{pmatrix} 0 & 0 \\ -7 & 0 \end{pmatrix}, \begin{pmatrix} 0 & 0 \\ 0 & 12 \end{pmatrix}\right\}$ is a basis for M_{22}.

In Problems 1–10 determine whether the given set of vectors is a basis for the given vector space.

1. In P_2: $1 - x^2, x$

2. In P_2: $-3x, 1 + x^2, x^2 - 5$

3. In P_2: $x^2 - 1, x^2 - 2, x^2 - 3$

4. In P_3: $1, 1 + x, 1 + x^2, 1 + x^3$

5. In P_3: $3, x^3 - 4x + 6, x^2$

6. In M_{22}: $\begin{pmatrix} 3 & 1 \\ 0 & 0 \end{pmatrix}, \begin{pmatrix} 3 & 2 \\ 0 & 0 \end{pmatrix}, \begin{pmatrix} -5 & 1 \\ 0 & 6 \end{pmatrix}, \begin{pmatrix} 0 & 1 \\ 0 & -7 \end{pmatrix}$

7. In M_{22}: $\begin{pmatrix} a & 0 \\ 0 & 0 \end{pmatrix}, \begin{pmatrix} 0 & b \\ 0 & 0 \end{pmatrix}, \begin{pmatrix} 0 & 0 \\ c & 0 \end{pmatrix}, \begin{pmatrix} 0 & 0 \\ 0 & d \end{pmatrix}$, where $abcd \neq 0$

8. In M_{22}: $\begin{pmatrix} -1 & 0 \\ 3 & 1 \end{pmatrix}, \begin{pmatrix} 2 & 1 \\ 1 & 4 \end{pmatrix}, \begin{pmatrix} -6 & 1 \\ 5 & 8 \end{pmatrix}, \begin{pmatrix} 7 & -2 \\ 1 & 0 \end{pmatrix}, \begin{pmatrix} 0 & 1 \\ 0 & 0 \end{pmatrix}$

9. $H = \{(x, y) \in \mathbb{R}^2: x + y = 0\}$; $(1, -1)$

10. $H = \{(x, y) \in \mathbb{R}^2: x + y = 0\}$; $(1, -1), (-3, 3)$

11. Find a basis in $\mathbb{R}^3$ for the set of vectors in the plane $2x - y - z = 0$.

12. Find a basis in $\mathbb{R}^3$ for the set of vectors in the plane $3x - 2y + 6z = 0$.

13. Find a basis in $\mathbb{R}^3$ for the set of vectors on the line $x/2 = y/3 = z/4$.

14. Find a basis in $\mathbb{R}^3$ for the set of vectors on the line $x = 3t$, $y = -2t$, $z = t$.

15. Show that the only proper subspaces of $\mathbb{R}^2$ are straight lines passing through the origin.

16. In $\mathbb{R}^4$ let $H = \{(x, y, z, w): ax + by + cz + dw = 0\}$, where $abcd \neq 0$.

a. Show that H is a subspace of $\mathbb{R}^4$.

b. Find a basis for H.

c. What is $\dim H$?

***17.** In $\mathbb{R}^n$ a **hyperplane** through $\mathbf{0}$ is a subspace of dimension $n - 1$. If H is a hyperplane through $\mathbf{0}$ in $\mathbb{R}^n$, show that

$$H = \{(x_1, x_2, \ldots, x_n): a_1x_1 + a_2x_2 + \cdots + a_nx_n = 0\}$$

where $a_1, a_2, \ldots, a_n$ are fixed real numbers, not all of which are zero.

18. In $\mathbb{R}^5$ find a basis for the hyperplane

$$H = \{(x_1, x_2, x_3, x_4, x_5): 2x_1 - 3x_2 + x_3 + 4x_4 - x_5 = 0\}$$

In Problems 19–23 find a basis for the solution space of the given homogeneous system.

19. $\begin{aligned} x - y &= 0 \\ -2x + 2y &= 0 \end{aligned}$

20. $\begin{aligned} x - 2y &= 0 \\ 3x + y &= 0 \end{aligned}$

21. $\begin{aligned} x - y - z &= 0 \\ 2x - y + z &= 0 \end{aligned}$

22. $\begin{aligned} x - 3y + z &= 0 \\ -2x + 2y - 3z &= 0 \\ 4x - 8y + 5z &= 0 \end{aligned}$

23. $\begin{aligned} 2x - 6y + 4z &= 0 \\ -x + 3y - 2z &= 0 \\ -3x + 9y - 6z &= 0 \end{aligned}$

24. Find a basis for D_3, the vector space of diagonal 3×3 matrices. What is the dimension of D_3?

25. What is the dimension of D_n, the space of diagonal $n \times n$ matrices?

26. Let S_{nn} denote the set of symmetric $n \times n$ matrices. Show that S_{nn} is a subspace of M_{nn} and that $\dim S_{nn} = [n(n+1)]/2$.

27. Suppose that $\mathbf{v}_1, \mathbf{v}_2, \ldots, \mathbf{v}_m$ are linearly independent vectors in a vector space V of dimension n and $m < n$. Show that $\{\mathbf{v}_1, \mathbf{v}_2, \ldots, \mathbf{v}_m\}$ can be enlarged to a basis for V. That is, there exist vectors $\mathbf{v}_{m+1}, \mathbf{v}_{m+2}, \ldots, \mathbf{v}_n$ such that $\{\mathbf{v}_1, \mathbf{v}_2, \ldots, \mathbf{v}_n\}$ is a basis. [*Hint:* Look at the proof of Theorem 5.]

28. Let $\{\mathbf{v}_1, \mathbf{v}_2, \ldots, \mathbf{v}_n\}$ be a basis for V. Let $\mathbf{u}_1 = \mathbf{v}_1$, $\mathbf{u}_2 = \mathbf{v}_1 + \mathbf{v}_2$, $\mathbf{u}_3 = \mathbf{v}_1 + \mathbf{v}_2 + \mathbf{v}_3, \ldots, \mathbf{u}_n = \mathbf{v}_1 + \mathbf{v}_2 + \cdots + \mathbf{v}_n$. Show that $\{\mathbf{u}_1, \mathbf{u}_2, \ldots, \mathbf{u}_n\}$ is also a basis for V.

29. Show that if $\{\mathbf{v}_1, \mathbf{v}_2, \ldots, \mathbf{v}_n\}$ spans V, then $\dim V \le n$. [*Hint:* Use the result of Problem 8.4.49.]

30. Let H and K be subspaces of V such that $H \subseteq K$ and $\dim H = \dim K < \infty$. Show that $H = K$.

31. Let H and K be subspaces of V and define $H + K = \{\mathbf{h} + \mathbf{k}\text{: } \mathbf{h} \in H \text{ and } \mathbf{k} \in K\}$.
a. Show that $H + K$ is a subspace of V.
b. If $H \cap K = \{\mathbf{0}\}$, show that $\dim (H + K) = \dim H + \dim K$.

***32.** If H is a subspace of the finite dimensional vector space V, show that there exists a unique subspace K of V such that **(a)** $H \cap K = \{\mathbf{0}\}$ and **(b)** $H + K = V$.

33. Show that two vectors $\mathbf{v}_1$ and $\mathbf{v}_2$ in $\mathbb{R}^2$ with endpoints at the origin are collinear if and only if $\dim \text{span } \{\mathbf{v}_1, \mathbf{v}_2\} = 1$.

34. Show that three vectors $\mathbf{v}_1$, $\mathbf{v}_2$, and $\mathbf{v}_3$ in $\mathbb{R}^3$ with endpoints at the origin are coplanar if and only if $\dim \text{span } \{\mathbf{v}_1, \mathbf{v}_2, \mathbf{v}_3\} \le 2$.

35. Show that any n vectors which span an n-dimensional space V form a basis for V. [*Hint:* Show that if the n vectors are not linearly independent, then $\dim V < n$.]

***36.** Show that every subspace of a finite dimensional vector space has a basis.

37. Find two bases for $\mathbb{R}^4$ that contain $(1, 0, 1, 0)$ and $(0, 1, 0, 1)$ and have no other vectors in common.

38. For what values of the real number a do the vectors $(a, 1, 0)$, $(1, 0, a)$, and $(1 + a, 1, a)$ constitute a basis for $\mathbb{R}^3$?

ANSWERS TO SELF-QUIZ

I. False **II.** True **III.** False **IV.** False **V.** True **VI.** True **VII.** True

---

8.6 THE RANK, NULLITY, ROW SPACE, AND COLUMN SPACE OF A MATRIX

In Section 8.4 we introduced the notion of linear independence. We showed that if A is an invertible $n \times n$ matrix, then the columns and rows of A form sets of linearly independent vectors. However, if A is not invertible (so that $\det A = 0$), or if A is not a square matrix, then these results tell us nothing about the number of linearly independent rows or columns of A. In this section we fill in this gap. We also show how a basis for the span of a set of vectors can be obtained by row reduction.

Let A be an $m \times n$ matrix and let

$$N_A = \{\mathbf{x} \in \mathbb{R}^n\text{: } A\mathbf{x} = \mathbf{0}\} \qquad (1)$$

Then, as we saw in Example 8.5.10 on page 521, N_A is a subspace of $\mathbb{R}^n$.

DEFINITION KERNEL AND NULLITY OF A MATRIX

N_A is called the **kernel** of A and $\nu(A) = \dim N_A$ is called the **nullity** of A. If N_A contains only the zero vector, then $\nu(A) = 0$. ∎

EXAMPLE 1 THE KERNEL AND NULLITY OF A 2×3 MATRIX

Let $A = \begin{pmatrix} 1 & 2 & -1 \\ 2 & -1 & 3 \end{pmatrix}$. Then, as we saw in Example 8.5.11 on page 521, N_A is spanned by $\begin{pmatrix} -1 \\ 1 \\ 1 \end{pmatrix}$ and $\nu(A) = 1$.

EXAMPLE 2 THE KERNEL AND NULLITY OF A 3×3 MATRIX

Let $A = \begin{pmatrix} 2 & -1 & 3 \\ 4 & -2 & 6 \\ -6 & 3 & -9 \end{pmatrix}$. Then, by Example 8.5.12 on page 521, $\left\{ \begin{pmatrix} 1 \\ 2 \\ 0 \end{pmatrix}, \begin{pmatrix} 0 \\ 3 \\ 1 \end{pmatrix} \right\}$ is a basis for N_A and $\nu(A) = 2$.

THEOREM 1

Let A be an $n \times n$ matrix. Then A is invertible if and only if $\nu(A) = 0$.

PROOF: By our Summing Up Theorem [Theorem 8.4.6, page 509, parts (i) and (ii)], A is invertible if and only if the homogeneous system $A\mathbf{x} = \mathbf{0}$ has only the trivial solution $\mathbf{x} = \mathbf{0}$. But, from equation (1), this means that A is invertible if and only if $N_A = \{\mathbf{0}\}$. Thus A is invertible if and only if $\nu(A) = \dim N_A = 0$. ■

DEFINITION RANGE OF A MATRIX

Let A be an $m \times n$ matrix. Then the **range** of A, denoted by Range A, is given by

$$\text{Range } A = \{\mathbf{y} \in \mathbb{R}^m : A\mathbf{x} = \mathbf{y} \text{ for some } \mathbf{x} \in \mathbb{R}^n\} \quad (2)$$

■

THEOREM 2

Let A be an $m \times n$ matrix. Then Range A is a subspace of $\mathbb{R}^m$.

PROOF: Suppose that $\mathbf{y}_1$ and $\mathbf{y}_2$ are in Range A. Then there are vectors $\mathbf{x}_1$ and $\mathbf{x}_2$ in $\mathbb{R}^n$ such that $\mathbf{y}_1 = A\mathbf{x}_1$ and $\mathbf{y}_2 = A\mathbf{x}_2$. Therefore

$$A(\alpha\mathbf{x}_1) = \alpha A\mathbf{x}_1 = \alpha\mathbf{y}_1 \qquad \text{and} \qquad A(\mathbf{x}_1 + \mathbf{x}_2) = A\mathbf{x}_1 + A\mathbf{x}_2 = \mathbf{y}_1 + \mathbf{y}_2$$

so $\alpha\mathbf{y}_1$ and $\mathbf{y}_1 + \mathbf{y}_2$ are in Range A. Thus, from Theorem 8.2.1, Range A is a subspace of $\mathbb{R}^m$. ■

DEFINITION RANK OF A MATRIX

Let A be an $m \times n$ matrix. Then the **rank** of A, denoted by $\rho(A)$, is given by

$$\rho(A) = \dim \text{Range } A$$

■

We shall give two definitions and a theorem that make the calculation of rank relatively easy.

DEFINITION ROW AND COLUMN SPACE OF A MATRIX

If A is an $m \times n$ matrix, let $\{\mathbf{r}_1, \mathbf{r}_2, \ldots, \mathbf{r}_m\}$ denote the rows of A and let $\{\mathbf{c}_1, \mathbf{c}_2, \ldots, \mathbf{c}_n\}$ denote the columns of A. Then we define

$$R_A = \textbf{row space} \text{ of } A = \text{span } \{\mathbf{r}_1, \mathbf{r}_2, \ldots, \mathbf{r}_m\} \tag{3}$$

and

$$C_A = \textbf{column space} \text{ of } A = \text{span } \{\mathbf{c}_1, \mathbf{c}_2, \ldots, \mathbf{c}_n\} \tag{4}$$

■

NOTE: R_A is a subspace of $\mathbb{R}^n$ and C_A is a subspace of $\mathbb{R}^m$.

We have introduced a lot of notation in just two pages. Let us stop for a moment to illustrate these ideas with an example.

EXAMPLE 3 FINDING N_A, $\nu(A)$, RANGE A, $\rho(A)$, R_A, AND C_A FOR A 2×3 MATRIX

Let $A = \begin{pmatrix} 1 & 2 & -1 \\ 2 & -1 & 3 \end{pmatrix}$. A is a 2×3 matrix.

i. *The kernel of* $A = N_A = \{\mathbf{x} \in \mathbb{R}^3\colon A\mathbf{x} = \mathbf{0}\}$. As we saw in Example 1,

$$N_A = \text{span}\left\{\begin{pmatrix} -1 \\ 1 \\ 1 \end{pmatrix}\right\}.$$

ii. *The nullity of* $A = \nu(A) = \dim N_A = 1$.

iii. *The range of* A = Range $A = \{\mathbf{y} \in \mathbb{R}^2\colon A\mathbf{x} = \mathbf{y}$ for some $\mathbf{x} \in \mathbb{R}^3\}$. Let $\mathbf{y} = \begin{pmatrix} y_1 \\ y_2 \end{pmatrix}$ be in $\mathbb{R}^2$. Then, if $\mathbf{y} \in$ Range A, there is an $\mathbf{x} \in \mathbb{R}^3$ such that $A\mathbf{x} = \mathbf{y}$.

Writing $\mathbf{x} = \begin{pmatrix} x_1 \\ x_2 \\ x_3 \end{pmatrix}$, we have

$$\begin{pmatrix} 1 & 2 & -1 \\ 2 & -1 & 3 \end{pmatrix}\begin{pmatrix} x_1 \\ x_2 \\ x_3 \end{pmatrix} = \begin{pmatrix} y_1 \\ y_2 \end{pmatrix}$$

or

$$\begin{aligned} x_1 + 2x_2 - x_3 &= y_1 \\ 2x_1 - x_2 + 3x_3 &= y_2. \end{aligned}$$

Row-reducing this system, we have

$$\left(\begin{array}{rrr|c} 1 & 2 & -1 & y_1 \\ 2 & -1 & 3 & y_2 \end{array}\right) \xrightarrow{R_2 \to R_2 - 2R_1} \left(\begin{array}{rrr|c} 1 & 2 & -1 & y_1 \\ 0 & -5 & 5 & y_2 - 2y_1 \end{array}\right)$$

$$\xrightarrow{R_2 \to -\frac{1}{5}R_2} \left(\begin{array}{rrr|c} 1 & 2 & -1 & y_1 \\ 0 & 1 & -1 & \dfrac{2y_1 - y_2}{5} \end{array}\right) \xrightarrow{R_1 \to R_1 - 2R_2} \left(\begin{array}{rrr|c} 1 & 0 & 1 & \dfrac{y_1 + 2y_2}{5} \\ 0 & 1 & -1 & \dfrac{2y_1 - y_2}{5} \end{array}\right)$$

Thus, if x_3 is chosen arbitrarily, we see that

$$x_1 = -x_3 + \frac{y_1 + 2y_2}{5} \qquad \text{and} \qquad x_2 = x_3 + \frac{2y_1 - y_2}{5}$$

That is, for every $\mathbf{y} = \begin{pmatrix} y_1 \\ y_2 \end{pmatrix} \in \mathbb{R}^2$, there are an infinite number of vectors $\mathbf{x} \in \mathbb{R}^3$ such that $A\mathbf{x} = \mathbf{y}$. Thus Range $A = \mathbb{R}^2$. Note, for example, that if $\mathbf{y} = \begin{pmatrix} 2 \\ -3 \end{pmatrix}$, then, choosing $x_3 = 0$ (the simplest choice), we have

$$x_1 = \frac{2 + 2(-3)}{5} = -\frac{4}{5} \qquad x_2 = \frac{2(2) - (-3)}{5} = \frac{7}{5}$$

and

$$A\mathbf{x} = \begin{pmatrix} 1 & 2 & -1 \\ 2 & -1 & 3 \end{pmatrix}\begin{pmatrix} -\frac{4}{5} \\ \frac{7}{5} \\ 0 \end{pmatrix} = \begin{pmatrix} \frac{10}{5} \\ -\frac{15}{5} \end{pmatrix} = \begin{pmatrix} 2 \\ -3 \end{pmatrix} = \mathbf{y}$$

iv. *The rank of* $A = \rho(A) = \dim \text{Range } A = \dim \mathbb{R}^2 = 2$.

v. *The row space of* $A = R_A = \text{span}\,\{(1, 2, -1), (2, -1, 3)\}$. Since these two vectors are linearly independent, we see that R_A is a two-dimensional subspace of $\mathbb{R}^3$. From Example 8.5.9 on page 520, we observe that R_A is a plane passing through the origin.

vi. *The column space of* $A =$

$$C_A = \text{span}\left\{\begin{pmatrix} 1 \\ 2 \end{pmatrix}, \begin{pmatrix} 2 \\ -1 \end{pmatrix}, \begin{pmatrix} -1 \\ 3 \end{pmatrix}\right\} = \mathbb{R}^2$$

since $\begin{pmatrix}1\\2\end{pmatrix}$ and $\begin{pmatrix}2\\-1\end{pmatrix}$, being linearly independent, constitute a basis for $\mathbb{R}^2$.

In Example 3 we may observe that Range $A = C_A = \mathbb{R}^2$ and $\dim R_A = \dim C_A = \dim \text{Range } A = \rho(A) = 2$. This is no coincidence.

THEOREM 3

If A is an $m \times n$ matrix, then:

i. $C_A = \text{Range } A$

ii. $\dim R_A = \dim C_A = \dim \text{Range } A = \rho(A)$ ■

The proof of this theorem is not difficult, but it is quite long. For that reason it is omitted.[†]

EXAMPLE 4

FINDING RANGE A AND $\rho(A)$ FOR A 3×3 MATRIX

Find a basis for Range A and determine the rank of $A = \begin{pmatrix}2 & -1 & 3\\4 & -2 & 6\\-6 & 3 & -9\end{pmatrix}$.

SOLUTION: Since $\mathbf{r}_2 = 2\mathbf{r}_1$ and $\mathbf{r}_3 = -3\mathbf{r}_1$, we see that $\rho(A) = \dim R_A = 1$. Thus any column in C_A is a basis for $C_A = \text{Range } A$. For example, $\begin{pmatrix}2\\4\\-6\end{pmatrix}$ is a basis for Range A.

The following theorem will simplify our computations.

THEOREM 4

If A is row (or column) equivalent to B, then $R_A = R_B$, $\rho(A) = \rho(B)$, and $\nu(A) = \nu(B)$.

PROOF: Recall from the definition on page 424, that A is row equivalent to B if A can be "reduced" to B by elementary row operations. The definition for "column equivalent" is similar. Suppose that C is the matrix obtained by performing an elementary row operation on A. We first show that $R_A = R_C$. Since B is obtained by performing several elementary row operations on A, our first result, applied several times, will imply that $R_A = R_B$.

Case 1: Interchange two rows of A. Then $R_A = R_C$ because the rows of A and C are the same (just written in a different order).

[†] For a proof see *Elementary Linear Algebra, Fifth Edition* by Stanley I. Grossman, Saunders, Philadelphia, 1994, page 351.

Case 2: Multiply the ith row of A by $c \neq 0$. If the rows of A are $\{\mathbf{r}_1, \mathbf{r}_2, \ldots, \mathbf{r}_i, \ldots, \mathbf{r}_m\}$, then the rows of C are $\{\mathbf{r}_1, \mathbf{r}_2, \ldots, c\mathbf{r}_i, \ldots, \mathbf{r}_m\}$. Obviously, $c\mathbf{r}_i = c(\mathbf{r}_i)$ and $\mathbf{r}_i = (1/c)\mathbf{r}_i$. Thus each row of C is a multiple of one row of A and vice versa. This means that each row of C is in the span of the rows of A and vice versa. We have

$$R_A \subseteq R_C \quad \text{and} \quad R_C \subseteq R_A, \quad \text{so } R_C = R_A$$

Case 3: Multiply the ith row of A by $c \neq 0$ and add it to the jth row. Now the rows of C are $\{\mathbf{r}_1, \mathbf{r}_2, \ldots, \mathbf{r}_i, \ldots, c\mathbf{r}_i + \mathbf{r}_j, \ldots, \mathbf{r}_m\}$. Here

$$\mathbf{r}_j = \underbrace{(c\mathbf{r}_i + \mathbf{r}_j)}_{j\text{th row of } C} - \underset{\substack{\uparrow \\ i\text{th row of } C}}{c\mathbf{r}_i}$$

so each row of A can be written as a linear combination of the rows of C and vice versa. Then, as before,

$$R_A \subseteq R_C \quad \text{and} \quad R_C \subseteq R_A, \quad \text{so } R_C = R_A$$

We have shown that $R_A = R_B$. Hence $\rho(R_A) = \rho(R_B)$. Finally, the set of solutions to $A\mathbf{x} = \mathbf{0}$ does not change under elementary row operations. Thus $N_A = N_B$, so $\nu(A) = \nu(B)$. ■

Example 5

Finding $\rho(A)$ and R_A for a 3×3 Matrix

Determine the rank and row space of

$$A = \begin{pmatrix} 1 & -1 & 3 \\ 2 & 0 & 4 \\ -1 & -3 & 1 \end{pmatrix}.$$

Solution: We row-reduce to obtain a simpler matrix:

$$\begin{pmatrix} 1 & -1 & 3 \\ 2 & 0 & 4 \\ -1 & -3 & 1 \end{pmatrix} \xrightarrow[R_3 \to R_3 + R_1]{R_2 \to R_2 - 2R_1} \begin{pmatrix} 1 & -1 & 3 \\ 0 & 2 & -2 \\ 0 & -4 & 4 \end{pmatrix}$$

$$\xrightarrow{R_2 \to \frac{1}{2}R_2} \begin{pmatrix} 1 & -1 & 3 \\ 0 & 1 & -1 \\ 0 & -4 & 4 \end{pmatrix} \xrightarrow{R_3 \to R_3 + 4R_2} \begin{pmatrix} 1 & -1 & 3 \\ 0 & 1 & -1 \\ 0 & 0 & 0 \end{pmatrix} = B$$

Since B has two independent rows, we have $\rho(B) = \rho(A) = 2$ and

$$R_A = \text{span } \{(1, -1, 3), (0, 1, -1)\}$$

Theorem 4 is useful when we want to find a basis for the span of a set of vectors.

Example 6

Finding a Basis for the Span of Four Vectors in $\mathbb{R}^3$

Find a basis for the space spanned by

$$\mathbf{v}_1 = \begin{pmatrix} 1 \\ 2 \\ -3 \end{pmatrix}, \quad \mathbf{v}_2 = \begin{pmatrix} -2 \\ 0 \\ 4 \end{pmatrix}, \quad \mathbf{v}_3 = \begin{pmatrix} 0 \\ 4 \\ -2 \end{pmatrix}, \quad \mathbf{v}_4 = \begin{pmatrix} -2 \\ -4 \\ 6 \end{pmatrix}$$

SOLUTION: We write the vectors as rows of a matrix A and then reduce the matrix to row echelon form. The resulting matrix will have the same row space as A.

$$\begin{pmatrix} 1 & 2 & -3 \\ -2 & 0 & 4 \\ 0 & 4 & -2 \\ -2 & -4 & 6 \end{pmatrix} \xrightarrow[R_4 \to R_4 + 2R_1]{R_2 \to R_2 + 2R_1} \begin{pmatrix} 1 & 2 & -3 \\ 0 & 4 & -2 \\ 0 & 4 & -2 \\ 0 & 0 & 0 \end{pmatrix}$$

$$\xrightarrow{R_3 \to R_3 - R_2} \begin{pmatrix} 1 & 2 & -3 \\ 0 & 4 & -2 \\ 0 & 0 & 0 \\ 0 & 0 & 0 \end{pmatrix} \xrightarrow{R_2 \to \frac{1}{4}R_2} \begin{pmatrix} 1 & 2 & -3 \\ 0 & 1 & -\frac{1}{2} \\ 0 & 0 & 0 \\ 0 & 0 & 0 \end{pmatrix}$$

Thus a basis for span $\{\mathbf{v}_1, \mathbf{v}_2, \mathbf{v}_3, \mathbf{v}_4\}$ is $\left\{ \begin{pmatrix} 1 \\ 2 \\ -3 \end{pmatrix}, \begin{pmatrix} 0 \\ 1 \\ -\frac{1}{2} \end{pmatrix} \right\}$. For example,

$$\begin{pmatrix} -2 \\ 0 \\ 4 \end{pmatrix} = -2\begin{pmatrix} 1 \\ 2 \\ -3 \end{pmatrix} + 4\begin{pmatrix} 0 \\ 1 \\ -\frac{1}{2} \end{pmatrix}.$$

The next theorem gives the relationship between rank and nullity.

THEOREM 5

Let A be an $m \times n$ matrix. Then

$$\rho(A) + \nu(A) = n \tag{5}$$

PROOF: We assume that $k = \rho(A)$ and that the first k columns of A are linearly independent. Let $\mathbf{c}_i$ $(i > k)$ denote any other column of A. Since $\mathbf{c}_1, \mathbf{c}_2, \ldots, \mathbf{c}_k$ form a basis for C_A, we have, for some scalars $a_1, a_2, \ldots, a_k$,

$$\mathbf{c}_i = a_1\mathbf{c}_1 + a_2\mathbf{c}_2 + \cdots + a_k\mathbf{c}_k \tag{6}$$

Thus, by adding $-a_1\mathbf{c}_1, -a_2\mathbf{c}_2, \ldots, -a_k\mathbf{c}_k$ successively to the ith column of A, we obtain a new $m \times n$ matrix B with $\rho(B) = \rho(A)$ and $\nu(B) = \nu(A)$ with the ith column of $B = \mathbf{0}$. We do this to all other columns of A (except the first k) to obtain the matrix

$$D = \begin{pmatrix} a_{11} & a_{12} & \cdots & a_{1k} & 0 & 0 & \cdots & 0 \\ a_{21} & a_{22} & \cdots & a_{2k} & 0 & 0 & \cdots & 0 \\ \vdots & \vdots & & \vdots & \vdots & \vdots & & \vdots \\ a_{m1} & a_{m2} & \cdots & a_{mk} & 0 & 0 & \cdots & 0 \end{pmatrix} \tag{7}$$

where $\rho(D) = \rho(A)$ and $\nu(D) = \nu(A)$. By possibly rearranging the rows of D, we can assume that the first k rows of D are independent. Then we do the same thing to the rows (i.e., add multiples of the first k rows to the last $m - k$ rows) to obtain a new matrix:

$$F = \begin{pmatrix} a_{11} & a_{12} & \cdots & a_{1k} & 0 & \cdots & 0 \\ a_{21} & a_{22} & \cdots & a_{2k} & 0 & \cdots & 0 \\ \vdots & \vdots & & \vdots & \vdots & & \vdots \\ a_{k1} & a_{k2} & \cdots & a_{kk} & 0 & \cdots & 0 \\ 0 & 0 & \cdots & 0 & 0 & \cdots & 0 \\ \vdots & \vdots & & \vdots & \vdots & & \vdots \\ 0 & 0 & \cdots & 0 & 0 & \cdots & 0 \end{pmatrix}$$

where $\rho(F) = \rho(A)$ and $\nu(F) = \nu(A)$. It is now obvious that if $i > k$, then $F\mathbf{e}_i = \mathbf{0}$,[†] so $E_k = \{\mathbf{e}_{k+1}, \mathbf{e}_{k+2}, \ldots, \mathbf{e}_n\}$ is a linearly independent set of $n - k$ vectors in N_F. We now show that E_k spans N_F. Let the vector $\mathbf{x} \in N_F$ have the form

$$\mathbf{x} = \begin{pmatrix} x_1 \\ x_2 \\ \vdots \\ x_k \\ \vdots \\ x_n \end{pmatrix}$$

Then

$$\mathbf{0} = F\mathbf{x} = \begin{pmatrix} a_{11}x_1 + a_{12}x_2 + \cdots + a_{1k}x_k \\ a_{21}x_1 + a_{22}x_2 + \cdots + a_{2k}x_k \\ \vdots \\ a_{k1}x_1 + a_{k2}x_2 + \cdots + a_{kk}x_k \\ 0 \\ \vdots \\ 0 \end{pmatrix} = \begin{pmatrix} 0 \\ 0 \\ \vdots \\ 0 \end{pmatrix}$$

The determinant of the matrix of the $k \times k$ homogeneous system described above is nonzero, since the rows of this matrix are linearly independent. Thus the only solution to the system is $x_1 = x_2 = \cdots = x_k = 0$. Thus $\mathbf{x}$ has the form

$$(0, 0, \ldots, 0, x_{k+1}, x_{k+2}, \ldots, x_n) = x_{k+1}\mathbf{e}_{k+1} + x_{k+2}\mathbf{e}_{k+2} + \cdots + x_n\mathbf{e}_n$$

This means that E_k spans N_F so that $\nu(F) = n - k = n - \rho(F)$. This completes the proof. ■

EXAMPLE 7

ILLUSTRATION THAT $\rho(A) + \nu(A) = n$

For $A = \begin{pmatrix} 1 & 2 & -1 \\ 2 & -1 & 3 \end{pmatrix}$ we calculated (in Examples 1 and 3) that $\rho(A) = 2$ and $\nu(A) = 1$; this illustrates that $\rho(A) + \nu(A) = n$ (=3).

EXAMPLE 8

ILLUSTRATION THAT $\rho(A) + \nu(A) = n$

For $A = \begin{pmatrix} 1 & -1 & 3 \\ 2 & 0 & 4 \\ -1 & -3 & 1 \end{pmatrix}$, calculate $\nu(A)$.

SOLUTION: In Example 5 we found that $\rho(A) = 2$. Thus $\nu(A) = 3 - 2 = 1$.

[†] Recall that $\mathbf{e}_i$ is the vector with a 1 in the ith position and a zero everywhere else.

THEOREM 6

Let A be an $n \times n$ matrix. Then A is invertible if and only if $\rho(A) = n$.

PROOF: By Theorem 1, A is invertible if and only if $\nu(A) = 0$. But, by Theorem 5, $\rho(A) = n - \nu(A)$. Thus A is invertible if and only if $\rho(A) = n - 0 = n$. ■

We next show how the notion of rank can be used to solve linear systems of equations. Again we consider the system of m equations in n unknowns

$$\begin{aligned} a_{11}x_1 + a_{12}x_2 + \cdots + a_{1n}x_n &= b_1 \\ a_{21}x_1 + a_{22}x_2 + \cdots + a_{2n}x_n &= b_2 \\ \vdots \qquad \vdots \qquad\quad \vdots \quad &\quad \vdots \\ a_{m1}x_1 + a_{m2}x_2 + \cdots + a_{mn}x_n &= b_m \end{aligned} \tag{8}$$

which we write as $A\mathbf{x} = \mathbf{b}$. We use the symbol $(A, \mathbf{b})$ to denote the $m \times (n + 1)$ augmented matrix obtained (as in Section 6.3) by adjoining the vector $\mathbf{b}$ to A.

THEOREM 7

The system $A\mathbf{x} = \mathbf{b}$ has at least one solution if and only if $\mathbf{b} \in C_A$. This will occur if and only if A and the augmented matrix $(A, \mathbf{b})$ have the same rank.

PROOF: If $\mathbf{c}_1, \mathbf{c}_2, \ldots, \mathbf{c}_n$ are the columns of A, then we can write system (8) as

$$x_1\mathbf{c}_1 + x_2\mathbf{c}_2 + \cdots + x_n\mathbf{c}_n = \mathbf{b} \tag{9}$$

System (9) will have a solution if and only if $\mathbf{b}$ can be written as a linear combination of the columns of A. That is, to have a solution we must have $\mathbf{b} \in C_A$. If $\mathbf{b} \in C_A$, then $(A, \mathbf{b})$ has the same number of linearly independent columns as A so that A and $(A, \mathbf{b})$ have the same rank. If $\mathbf{b} \notin C_A$, then $\rho(A, \mathbf{b}) = \rho(A) + 1$ and the system has no solutions. This completes the proof. ■

EXAMPLE 9 USING THEOREM 7 TO DETERMINE WHETHER A SYSTEM HAS SOLUTIONS

Determine whether the system

$$\begin{aligned} 2x_1 + 4x_2 + 6x_3 &= 18 \\ 4x_1 + 5x_2 + 6x_3 &= 24 \\ 2x_1 + 7x_2 + 12x_3 &= 40 \end{aligned}$$

has solutions.

SOLUTION: Let $A = \begin{pmatrix} 2 & 4 & 6 \\ 4 & 5 & 6 \\ 2 & 7 & 12 \end{pmatrix}$. Then we row-reduce to obtain, successively,

$$\xrightarrow{R_1 \to \frac{1}{2}R_1} \begin{pmatrix} 1 & 2 & 3 \\ 4 & 5 & 6 \\ 2 & 7 & 12 \end{pmatrix} \xrightarrow[R_3 \to R_3 - 2R_1]{R_2 \to R_2 - 4R_1} \begin{pmatrix} 1 & 2 & 3 \\ 0 & -3 & -6 \\ 0 & 3 & 6 \end{pmatrix}$$

$$\xrightarrow{R_2 \to -\frac{1}{3}R_2} \begin{pmatrix} 1 & 2 & 3 \\ 0 & 1 & 2 \\ 0 & 3 & 6 \end{pmatrix} \xrightarrow[R_3 \to R_3 - 3R_2]{R_1 \to R_1 - 2R_2} \begin{pmatrix} 1 & 0 & -1 \\ 0 & 1 & 2 \\ 0 & 0 & 0 \end{pmatrix}$$

Thus $\rho(A) = 2$. Similarly, we row-reduce $(A, \mathbf{b})$ to obtain

$$\left(\begin{array}{ccc|c} 2 & 4 & 6 & 18 \\ 4 & 5 & 6 & 24 \\ 2 & 7 & 12 & 40 \end{array}\right) \xrightarrow{R_1 \to \frac{1}{2}R_1} \left(\begin{array}{ccc|c} 1 & 2 & 3 & 9 \\ 4 & 5 & 6 & 24 \\ 2 & 7 & 12 & 40 \end{array}\right) \xrightarrow[R_3 \to R_3 - 2R_1]{R_2 \to R_2 - 4R_1} \left(\begin{array}{ccc|c} 1 & 2 & 3 & 9 \\ 0 & -3 & -6 & -12 \\ 0 & 3 & 6 & 22 \end{array}\right)$$

$$\xrightarrow{R_2 \to -\frac{1}{3}R_2} \left(\begin{array}{ccc|c} 1 & 2 & 3 & 9 \\ 0 & 1 & 2 & 4 \\ 0 & 3 & 6 & 22 \end{array}\right) \xrightarrow[R_3 \to R_3 - 3R_2]{R_1 \to R_1 - 2R_2} \left(\begin{array}{ccc|c} 1 & 0 & -1 & 1 \\ 0 & 1 & 2 & 4 \\ 0 & 0 & 0 & 10 \end{array}\right)$$

It is easy to see that the last three columns of the last matrix are linearly independent. Thus $\rho(A, \mathbf{b}) = 3$ and there are no solutions to the system.

EXAMPLE 10 USING THEOREM 7 TO DETERMINE WHETHER A SYSTEM HAS SOLUTIONS

Determine whether the system

$$\begin{aligned} x_1 - x_2 + 2x_3 &= 4 \\ 2x_1 + x_2 - 3x_3 &= -2 \\ 4x_1 - x_2 + x_3 &= 6 \end{aligned}$$

has solutions.

SOLUTION: Let $A = \begin{pmatrix} 1 & -1 & 2 \\ 2 & 1 & -3 \\ 4 & -1 & 1 \end{pmatrix}$. Then $\det A = 0$, so $\rho(A) < 3$. Since the first column is not a multiple of the second, we see that the first two columns are linearly independent; hence $\rho(A) = 2$. To compute $\rho(A, \mathbf{b})$, we row-reduce:

$$\left(\begin{array}{ccc|c} 1 & -1 & 2 & 4 \\ 2 & 1 & -3 & -2 \\ 4 & -1 & 1 & 6 \end{array}\right) \xrightarrow[R_3 \to R_3 - 4R_1]{R_2 \to R_2 - 2R_1} \left(\begin{array}{ccc|c} 1 & -1 & 2 & 4 \\ 0 & 3 & -7 & -10 \\ 0 & 3 & -7 & -10 \end{array}\right)$$

We see that $\rho(A, \mathbf{b}) = 2$ and there are an infinite number of solutions to the system. (If there were a unique solution, we would have $\det A \neq 0$.)

The results of this section allow us to improve on our Summing Up Theorem—last seen in Section 8.4 on page 509.

THEOREM 8 SUMMING UP THEOREM—VIEW 6

Let A be an $n \times n$ matrix. Then the following nine statements are equivalent: that is, each one implies the other eight (so if one is true, all are true.)

i. A is invertible.

ii. The only solution to the homogeneous system $A\mathbf{x} = \mathbf{0}$ is the trivial solution ($\mathbf{x} = \mathbf{0}$).

iii. The system $A\mathbf{x} = \mathbf{b}$ has a unique solution for every n-vector $\mathbf{b}$.

iv. A is row equivalent to the $n \times n$ identity matrix I_n.

v. A is a product of elementary matrices.

vi. The rows (and columns) of A are linearly independent.

vii. $\det A \neq 0$.

viii. $\nu(A) = 0$.

ix. $\rho(A) = n$.

Moreover, if one of the above fails to hold, then for every vector $\mathbf{b} \in \mathbb{R}^n$, the system $A\mathbf{x} = \mathbf{b}$ has either no solution or an infinite number of solutions. It has an infinite number of solutions if and only if $\rho(A) = \rho((A, \mathbf{b}))$. ■

PROBLEMS 8.6

SELF-QUIZ

Multiple Choice

I. The rank of the matrix $\begin{pmatrix} 1 & 2 & 3 & 4 \\ 0 & 2 & -1 & 5 \\ 0 & 0 & 3 & 7 \end{pmatrix}$ is ________.

a. 1 **b.** 2
c. 3 **d.** 4

II. The nullity of the matrix in Problem I is ________.

a. 1 **b.** 2
c. 3 **d.** 4

III. If a 5×7 matrix has nullity 2, then its rank is ________.

a. 5 **b.** 3
c. 2 **d.** 7
e. Cannot be determined without further information.

IV. The rank of the matrix $\begin{pmatrix} 1 & 2 \\ -2 & -4 \\ 3 & 6 \end{pmatrix}$ is ________.

a. 1 **b.** 2 **c.** 3

V. The nullity of the matrix in Problem IV is ________.

a. 0 **b.** 1
c. 2 **d.** 3

VI. If A is a 4×4 matrix and $\det A = 0$, then the maximum possible value for $\rho(A)$ is ________.

a. 1 **b.** 2
c. 3 **d.** 4

VII. In Problem IV $\dim C_A =$ ________.

a. 1 **b.** 2 **c.** 3

VIII. In Problem I $\dim R_A =$ ________.

a. 1 **b.** 2
c. 3 **d.** 4

True–False

IX. In any $m \times n$ matrix, $C_A = R_A$

X. In any $m \times n$ matrix, $C_A = \text{Range } A$

In Problems 1–15 find the rank and nullity of the given matrix.

1. $\begin{pmatrix} 1 & 2 \\ 3 & 4 \end{pmatrix}$

2. $\begin{pmatrix} 1 & -1 & 2 \\ 3 & 1 & 0 \end{pmatrix}$

3. $\begin{pmatrix} -1 & 3 & 2 \\ 2 & -6 & -4 \end{pmatrix}$

4. $\begin{pmatrix} 1 & -1 & 2 \\ 3 & 1 & 4 \\ -1 & 0 & 4 \end{pmatrix}$

5. $\begin{pmatrix} 1 & -1 & 2 \\ 3 & 1 & 4 \\ 5 & -1 & 8 \end{pmatrix}$

6. $\begin{pmatrix} -1 & 2 & 1 \\ 2 & -4 & -2 \\ -3 & 6 & 3 \end{pmatrix}$

7. $\begin{pmatrix} 1 & -1 & 2 & 3 \\ 0 & 1 & 4 & 3 \\ 1 & 0 & 6 & 6 \end{pmatrix}$

8. $\begin{pmatrix} 1 & -1 & 2 & 3 \\ 0 & 1 & 4 & 3 \\ 1 & 0 & 6 & 5 \end{pmatrix}$

9. $\begin{pmatrix} 2 & 3 \\ -1 & 1 \\ 4 & 7 \end{pmatrix}$

10. $\begin{pmatrix} 1 & -1 & 2 & 3 \\ 0 & 1 & 0 & 1 \\ 1 & 0 & 1 & 0 \\ 0 & 0 & 0 & 1 \end{pmatrix}$

11. $\begin{pmatrix} 1 & -1 & 2 & 1 \\ -1 & 0 & 1 & 2 \\ 1 & -2 & 5 & 4 \\ 2 & -1 & 1 & -1 \end{pmatrix}$

12. $\begin{pmatrix} 1 & -1 & 2 & 3 \\ -2 & 2 & -4 & -6 \\ 2 & -2 & 4 & 6 \\ 3 & -3 & 6 & 9 \end{pmatrix}$

13. $\begin{pmatrix} -1 & -1 & 0 & 0 \\ 0 & 0 & 2 & 3 \\ 4 & 0 & -2 & 1 \\ 3 & -1 & 0 & 4 \end{pmatrix}$

14. $\begin{pmatrix} 3 & 0 & 0 \\ 0 & 0 & 0 \\ 0 & 0 & 6 \end{pmatrix}$

15. $\begin{pmatrix} 1 & 2 & 3 \\ 0 & 0 & 4 \\ 0 & 0 & 6 \end{pmatrix}$

In Problems 16–22 find a basis for the range and kernel of the given matrix.

16. The matrix of Problem 2
17. The matrix of Problem 5
18. The matrix of Problem 6
19. The matrix of Problem 8
20. The matrix of Problem 11
21. The matrix of Problem 12
22. The matrix of Problem 13

In Problems 23–26 find a basis for the span of the given set of vectors.

23. $\begin{pmatrix} 1 \\ 4 \\ -2 \end{pmatrix}, \begin{pmatrix} 2 \\ 1 \\ 2 \end{pmatrix}, \begin{pmatrix} -1 \\ 3 \\ -4 \end{pmatrix}$

24. $(1, -2, 3), (2, -1, 4), (3, -3, 3), (2, 1, 0)$

25. $(1, -1, 1, -1), (2, 0, 0, 1), (4, -2, 2, 1), (7, -3, 3, -1)$

26. $\begin{pmatrix} 1 \\ 0 \\ 0 \\ 1 \end{pmatrix}, \begin{pmatrix} 0 \\ 1 \\ 1 \\ 0 \end{pmatrix}, \begin{pmatrix} 1 \\ -2 \\ -2 \\ 1 \end{pmatrix}, \begin{pmatrix} 0 \\ 2 \\ 2 \\ 1 \end{pmatrix}$

In Problems 27–30 use Theorem 7 to determine whether the given system has any solutions.

27.
$$\begin{aligned} x_1 + x_2 - x_3 &= 7 \\ 4x_1 - x_2 + 5x_3 &= 4 \\ 6x_1 + x_2 + 3x_3 &= 20 \end{aligned}$$

28.
$$\begin{aligned} x_1 + x_2 - x_3 &= 7 \\ 4x_1 - x_2 + 5x_3 &= 4 \\ 6x_1 + x_2 + 3x_3 &= 18 \end{aligned}$$

29.
$$\begin{aligned} x_1 - 2x_2 + x_3 + x_4 &= 2 \\ 3x_1 \qquad + 2x_3 - 2x_4 &= -8 \\ 4x_2 - x_3 - x_4 &= 1 \\ 5x_1 \qquad + 3x_3 - x_4 &= -3 \end{aligned}$$

30.
$$\begin{aligned} x_1 - 2x_2 + x_3 + x_4 &= 2 \\ 3x_1 \qquad + 2x_3 - 2x_4 &= -8 \\ 4x_2 - x_3 - x_4 &= 1 \\ 5x_1 \qquad + 3x_3 - x_4 &= 0 \end{aligned}$$

31. Show that the rank of a diagonal matrix is equal to the number of nonzero components on the diagonal.

32. Let A be an upper triangular $n \times n$ matrix with zeros on the diagonal. Show that $\rho(A) < n$.

33. Show that for any matrix A, $\rho(A) = \rho(A^t)$.

34. Show that if A is an $m \times n$ matrix and $m < n$, then **(a)** $\rho(A) \le m$ and **(b)** $\nu(A) \ge n - m$.

35. Let A be an $m \times n$ matrix and let B and C be invertible $m \times m$ and $n \times n$ matrices, respectively. Prove that $\rho(A) = \rho(BA) = \rho(AC)$. That is, multiplying a matrix by an invertible matrix does not change its rank.

36. Let A and B be $m \times n$ and $n \times p$ matrices, respectively. Show that

$$\rho(AB) \le \min\,(\rho(A), \rho(B)).$$

37. Let A be a 5×7 matrix with rank 5. Show that the linear system $A\mathbf{x} = \mathbf{b}$ has at least one solution for every 5-vector $\mathbf{b}$.

***38.** Let A and B be $m \times n$ matrices. Show that if $\rho(A) = \rho(B)$, then there exist invertible matrices C and D such that $B = CAD$.

39. If $B = CAD$, where C and D are invertible, prove that $\rho(A) = \rho(B)$.

40. Suppose that some k rows of A are linearly independent while any $k + 1$ rows of A are linearly dependent. Show that $\rho(A) = k$.

41. If A is an $n \times n$ matrix, show that $\rho(A) < n$ if and only if there is a vector $\mathbf{x} \in \mathbb{R}^n$ such that $\mathbf{x} \neq \mathbf{0}$ and $A\mathbf{x} = \mathbf{0}$.

42. Let A be an $m \times n$ matrix. Suppose that for every $\mathbf{y} \in \mathbb{R}^m$ there is an $\mathbf{x} \in \mathbb{R}^n$ such that $A\mathbf{x} = \mathbf{y}$. Show that $\rho(A) = m$.

ANSWERS TO SELF-QUIZ

I. c **II.** a **III.** a **IV.** a **V.** b **VI.** c **VII.** a **VIII.** c **IX.** False **X.** True

8.7 ORTHONORMAL BASES AND PROJECTIONS IN $\mathbb{R}^n$

In $\mathbb{R}^n$ we saw that n linearly independent vectors constitute a basis. The most commonly used basis is the standard basis $E = \{\mathbf{e}_1, \mathbf{e}_2, \ldots, \mathbf{e}_n\}$. These vectors have two properties:

i. $\mathbf{e}_i \cdot \mathbf{e}_j = 0 \quad$ if $i \neq j$

ii. $\mathbf{e}_i \cdot \mathbf{e}_i = 1$

DEFINITION ORTHONORMAL SET IN $\mathbb{R}^n$

A set of vectors $S = \{\mathbf{u}_1, \mathbf{u}_2, \ldots, \mathbf{u}_k\}$ in $\mathbb{R}^n$ is said to be an **orthonormal set** if

$$\mathbf{u}_i \cdot \mathbf{u}_j = 0 \quad \text{if } i \neq j \tag{1}$$

$$\mathbf{u}_i \cdot \mathbf{u}_i = 1 \tag{2}$$

If only equation (1) is satisfied, the set is called **orthogonal.** ■

Since we shall be working with the scalar product extensively in this section, let us recall some basic facts (see Theorem 1.2.1, page 15). Without mentioning them again explicitly, we shall use these facts often in the rest of this section.

If $\mathbf{u}$, $\mathbf{v}$, and $\mathbf{w}$ are in $\mathbb{R}^n$ and α is a real number, then

$$\mathbf{u} \cdot \mathbf{v} = \mathbf{v} \cdot \mathbf{u} \tag{3}$$

$$(\mathbf{u} + \mathbf{v}) \cdot \mathbf{w} = \mathbf{u} \cdot \mathbf{w} + \mathbf{v} \cdot \mathbf{w} \tag{4}$$

$$\mathbf{u} \cdot (\mathbf{v} + \mathbf{w}) = \mathbf{u} \cdot \mathbf{v} + \mathbf{u} \cdot \mathbf{w} \tag{5}$$

$$(\alpha\mathbf{u}) \cdot \mathbf{v} = \alpha(\mathbf{u} \cdot \mathbf{v}) \tag{6}$$

$$\mathbf{u} \cdot (\alpha\mathbf{v}) = \alpha(\mathbf{u} \cdot \mathbf{v}) \tag{7}$$

DEFINITION LENGTH OR NORM OF A VECTOR

We now give another useful definition. If $\mathbf{v} \in \mathbb{R}^n$, then the **length** or **norm** of $\mathbf{v}$, written $|\mathbf{v}|$, is given by

$$|\mathbf{v}| = \sqrt{\mathbf{v} \cdot \mathbf{v}} \tag{8}$$

■

NOTE: If $\mathbf{v} = (x_1, x_2, \ldots, x_n)$, then $\mathbf{v} \cdot \mathbf{v} = x_1^2 + x_2^2 + \cdots + x_n^2$. This means that

$$\mathbf{v} \cdot \mathbf{v} \geq 0 \quad \text{and} \quad \mathbf{v} \cdot \mathbf{v} = 0 \quad \text{if and only if } \mathbf{v} = \mathbf{0} \tag{9}$$

Thus we can take the square root in (8), and we have

$$|\mathbf{v}| = \sqrt{\mathbf{v} \cdot \mathbf{v}} \geq 0 \quad \text{for every } \mathbf{v} \in \mathbb{R}^n \tag{10}$$

$$|\mathbf{v}| = 0 \quad \text{if and only if } \mathbf{v} = \mathbf{0} \tag{11}$$

EXAMPLE 1 THE NORM OF A VECTOR IN $\mathbb{R}^2$

Let $\mathbf{v} = (x, y) \in \mathbb{R}^2$. Then $|\mathbf{v}| = \sqrt{x^2 + y^2}$ conforms to our usual definition of length of a vector in the plane.

EXAMPLE 2 THE NORM OF A VECTOR IN $\mathbb{R}^3$

If $\mathbf{v} = (x, y, z) \in \mathbb{R}^3$, then $|\mathbf{v}| = \sqrt{x^2 + y^2 + z^2}$ as in Section 1.4.

EXAMPLE 3 THE NORM OF A VECTOR IN $\mathbb{R}^5$

If $\mathbf{v} = (2, -1, 3, 4, -6) \in \mathbb{R}^5$, then $|\mathbf{v}| = \sqrt{4 + 1 + 9 + 16 + 36} = \sqrt{66}$.

We can now restate the definition of an orthonormal set in $\mathbb{R}^n$.

> A set of vectors is orthonormal if any pair of them is orthogonal and each has length 1.

Orthonormal sets of vectors are reasonably easy to work with. Now we prove that any finite orthogonal set of nonzero vectors is linearly independent.

THEOREM 1

If $S = \{\mathbf{v}_1, \mathbf{v}_2, \ldots, \mathbf{v}_k\}$ is an orthogonal set of nonzero vectors, then S is linearly independent.

PROOF: Suppose that $c_1\mathbf{v}_1 + c_2\mathbf{v}_2 + \cdots + c_n\mathbf{v}_k = \mathbf{0}$. Then, for any $i = 1, 2, \ldots, k$,

$$\begin{aligned} 0 = \mathbf{0} \cdot \mathbf{v}_i &= (c_1\mathbf{v}_1 + c_2\mathbf{v}_2 + \cdots + c_i\mathbf{v}_i + \cdots + c_k\mathbf{v}_k) \cdot \mathbf{v}_i \\ &= c_1(\mathbf{v}_1 \cdot \mathbf{v}_i) + c_2(\mathbf{v}_2 \cdot \mathbf{v}_i) + \cdots + c_i(\mathbf{v}_i \cdot \mathbf{v}_i) + \cdots + c_k(\mathbf{v}_k \cdot \mathbf{v}_i) \\ &= c_1 0 + c_2 0 + \cdots + c_i|\mathbf{v}_i|^2 + \cdots + c_k 0 = c_i|\mathbf{v}_i|^2 \end{aligned}$$

Since $\mathbf{v}_i \neq \mathbf{0}$ by hypothesis, $|\mathbf{v}_i|^2 > 0$ and we have $c_i = 0$. This is true for $i = 1, 2, \ldots, k$ and the proof is complete. ■

We now see how *any* basis in $\mathbb{R}^n$ can be "turned into" an orthonormal basis. The method described below is called the **Gram-Schmidt orthonormalization process.**†

THEOREM 2 GRAM-SCHMIDT ORTHONORMALIZATION PROCESS

Let H be an m-dimensional subspace of $\mathbb{R}^n$. Then H has an orthonormal basis.‡

PROOF: Let $S = \{\mathbf{v}_1, \mathbf{v}_2, \ldots, \mathbf{v}_m\}$ be a basis for H. We shall prove the theorem by constructing an orthonormal basis from the vectors in S. Before giving the steps in this construction, we note the simple fact that a linearly independent set of vectors does *not* contain the zero vector (see Problem 21).

Step 1: Choosing the first unit vector:
Let

$$\mathbf{u}_1 = \frac{\mathbf{v}_1}{|\mathbf{v}_1|} \tag{12}$$

Then

$$\mathbf{u}_1 \cdot \mathbf{u}_1 = \left(\frac{\mathbf{v}_1}{|\mathbf{v}_1|}\right) \cdot \left(\frac{\mathbf{v}_1}{|\mathbf{v}_1|}\right) = \left(\frac{1}{|\mathbf{v}_1|^2}\right)(\mathbf{v}_1 \cdot \mathbf{v}_1) = 1$$

so that $|\mathbf{u}_1| = 1$.

Step 2: Choosing a second vector orthogonal to $\mathbf{u}_1$

$$\mathbf{v}_2' = \mathbf{v}_2 - (\mathbf{v}_2 \cdot \mathbf{u}_1)\mathbf{u}_1 \tag{13}$$

Then

$$\mathbf{v}_2' \cdot \mathbf{u}_1 = \mathbf{v}_2 \cdot \mathbf{u}_1 - (\mathbf{v}_2 \cdot \mathbf{u}_1)(\mathbf{u}_1 \cdot \mathbf{u}_1) = \mathbf{v}_2 \cdot \mathbf{u}_1 - \mathbf{v}_2 \cdot \mathbf{u}_1 = 0,$$

so that $\mathbf{v}_2'$ is orthogonal to $\mathbf{u}_1$. Moreover, by Theorem 1, $\mathbf{u}_1$ and $\mathbf{v}_2'$ are linearly independent. $\mathbf{v}_2' \neq \mathbf{0}$ because otherwise $\mathbf{v}_2 = (\mathbf{v}_2 \cdot \mathbf{u}_1)\mathbf{u}_1 = \dfrac{(\mathbf{v}_2 \cdot \mathbf{u}_1)}{|\mathbf{v}_1|}\mathbf{v}_1$, contradicting the independence of $\mathbf{v}_1$ and $\mathbf{v}_2$.

Step 3: Let

$$\mathbf{u}_2 = \frac{\mathbf{v}_2'}{|\mathbf{v}_2'|} \tag{14}$$

Then clearly $\{\mathbf{u}_1, \mathbf{u}_2\}$ is an orthonormal set.

Suppose now that the vectors $\mathbf{u}_1, \mathbf{u}_2, \ldots, \mathbf{u}_k$ $(k < m)$ have been constructed and form an orthonormal set. We show how to construct $\mathbf{u}_{k+1}$.

Step 4: Let

$$\begin{aligned}\mathbf{v}_{k+1}' = \mathbf{v}_{k+1} &- (\mathbf{v}_{k+1} \cdot \mathbf{u}_1)\mathbf{u}_1 \\ &- (\mathbf{v}_{k+1} \cdot \mathbf{u}_2)\mathbf{u}_2 - \cdots - (\mathbf{v}_{k+1} \cdot \mathbf{u}_k)\mathbf{u}_k\end{aligned} \tag{15}$$

† Jörgen Pederson Gram (1850–1916) was a Danish actuary who was very interested in the science of measurement. Erhardt Schmidt (1876–1959) was a German mathematician.

‡ Note that H may be $\mathbb{R}^n$ in this theorem. That is, $\mathbb{R}^n$ itself has an orthonormal basis.

Then, for $i = 1, 2, \ldots, k$,

$$\mathbf{v}'_{k+1} \cdot \mathbf{u}_i = \mathbf{v}_{k+1} \cdot \mathbf{u}_i - (\mathbf{v}_{k+1} \cdot \mathbf{u}_1)(\mathbf{u}_1 \cdot \mathbf{u}_i) - (\mathbf{v}_{k+1} \cdot \mathbf{u}_2)(\mathbf{u}_2 \cdot \mathbf{u}_i) - \cdots - (\mathbf{v}_{k+1} \cdot \mathbf{u}_i)(\mathbf{u}_i \cdot \mathbf{u}_i) - \cdots - (\mathbf{v}_{k+1} \cdot \mathbf{u}_k)(\mathbf{u}_k \cdot \mathbf{u}_i)$$

But $\mathbf{u}_j \cdot \mathbf{u}_i = 0$ if $j \neq i$ and $\mathbf{u}_i \cdot \mathbf{u}_i = 1$. Thus

$$\mathbf{v}'_{k+1} \cdot \mathbf{u}_i = \mathbf{v}_{k+1} \cdot \mathbf{u}_i - \mathbf{v}_{k+1} \cdot \mathbf{u}_i = 0$$

Hence $\{\mathbf{u}_1, \mathbf{u}_2, \ldots, \mathbf{u}_k, \mathbf{v}'_{k+1}\}$ is an orthogonal, linearly independent set and $\mathbf{v}'_{k+1} \neq \mathbf{0}$.

Step 5: Let $\mathbf{u}_{k+1} = \mathbf{v}'_{k+1}/|\mathbf{v}'_{k+1}|$. Then clearly $\{\mathbf{u}_1, \mathbf{u}_2, \ldots, \mathbf{u}_k, \mathbf{u}_{k+1}\}$ is an orthonormal set, and we continue in this manner until $k + 1 = m$ and the proof is complete.

Note that since each $\mathbf{u}_i$ is a linear combination of the $\mathbf{v}_i$'s, span $\{\mathbf{u}_1, \mathbf{u}_2, \ldots, \mathbf{u}_m\}$ is a subspace of span $\{\mathbf{v}_1, \mathbf{v}_2, \ldots, \mathbf{v}_m\}$ and each space has dimension m, the spaces must be equal. ■

EXAMPLE 4

CONSTRUCTING AN ORTHONORMAL BASIS IN $\mathbb{R}^3$

Construct an orthonormal basis in $\mathbb{R}^3$ starting with the basis $\{\mathbf{v}_1, \mathbf{v}_2, \mathbf{v}_3\} = \left\{\begin{pmatrix}1\\1\\0\end{pmatrix}, \begin{pmatrix}0\\1\\1\end{pmatrix}, \begin{pmatrix}1\\0\\1\end{pmatrix}\right\}$.

SOLUTION: We have $|\mathbf{v}_1| = \sqrt{2}$, so $\mathbf{u}_1 = \begin{pmatrix}1/\sqrt{2}\\1/\sqrt{2}\\0\end{pmatrix}$. Then

$$\mathbf{v}'_2 = \mathbf{v}_2 - (\mathbf{v}_2 \cdot \mathbf{u}_1)\mathbf{u}_1 = \begin{pmatrix}0\\1\\1\end{pmatrix} - \frac{1}{\sqrt{2}}\begin{pmatrix}1/\sqrt{2}\\1/\sqrt{2}\\0\end{pmatrix} = \begin{pmatrix}0\\1\\1\end{pmatrix} - \begin{pmatrix}\frac{1}{2}\\\frac{1}{2}\\0\end{pmatrix} = \begin{pmatrix}-\frac{1}{2}\\\frac{1}{2}\\1\end{pmatrix}$$

Since $|\mathbf{v}'_2| = \sqrt{3/2}$, $\mathbf{u}_2 = \sqrt{2/3}\begin{pmatrix}-\frac{1}{2}\\\frac{1}{2}\\1\end{pmatrix} = \begin{pmatrix}-1/\sqrt{6}\\1/\sqrt{6}\\2/\sqrt{6}\end{pmatrix}$. Continuing, we have

$$\begin{aligned}\mathbf{v}'_3 &= \mathbf{v}_3 - (\mathbf{v}_3 \cdot \mathbf{u}_1)\mathbf{u}_1 - (\mathbf{v}_3 \cdot \mathbf{u}_2)\mathbf{u}_2 \\ &= \begin{pmatrix}1\\0\\1\end{pmatrix} - \frac{1}{\sqrt{2}}\begin{pmatrix}1/\sqrt{2}\\1/\sqrt{2}\\0\end{pmatrix} - \frac{1}{\sqrt{6}}\begin{pmatrix}-1/\sqrt{6}\\1/\sqrt{6}\\2/\sqrt{6}\end{pmatrix} = \begin{pmatrix}1\\0\\1\end{pmatrix} - \begin{pmatrix}\frac{1}{2}\\\frac{1}{2}\\0\end{pmatrix} - \begin{pmatrix}-\frac{1}{6}\\\frac{1}{6}\\\frac{2}{6}\end{pmatrix} = \begin{pmatrix}\frac{2}{3}\\-\frac{2}{3}\\\frac{2}{3}\end{pmatrix}\end{aligned}$$

Finally, $|\mathbf{v}'_3| = \sqrt{12/9} = 2/\sqrt{3}$, so that $\mathbf{u}_3 = \frac{\sqrt{3}}{2}\begin{pmatrix}2/3\\-2/3\\2/3\end{pmatrix} = \begin{pmatrix}1/\sqrt{3}\\-1/\sqrt{3}\\1/\sqrt{3}\end{pmatrix}$. Thus an orthonormal basis in $\mathbb{R}^3$ is $\left\{\begin{pmatrix}1/\sqrt{2}\\1/\sqrt{2}\\0\end{pmatrix}, \begin{pmatrix}-1/\sqrt{6}\\1/\sqrt{6}\\2/\sqrt{6}\end{pmatrix}, \begin{pmatrix}1/\sqrt{3}\\-1/\sqrt{3}\\1/\sqrt{3}\end{pmatrix}\right\}$. This result should be checked.

EXAMPLE 5 FINDING AN ORTHONORMAL BASIS FOR A SUBSPACE OF $\mathbb{R}^3$

Find an orthonormal basis for the set of vectors in $\mathbb{R}^3$ lying on the plane

$$\pi = \left\{ \begin{pmatrix} x \\ y \\ z \end{pmatrix} : 2x - y + 3z = 0 \right\}.$$

SOLUTION: As we saw in Example 8.5.3, page 516, a basis for this two-dimensional subspace is $\mathbf{v}_1 = \begin{pmatrix} 1 \\ 2 \\ 0 \end{pmatrix}$ and $\mathbf{v}_2 = \begin{pmatrix} 0 \\ 3 \\ 1 \end{pmatrix}$. Then $|\mathbf{v}_1| = \sqrt{5}$ and $\mathbf{u}_1 = \mathbf{v}_1/|\mathbf{v}_1| = \begin{pmatrix} 1/\sqrt{5} \\ 2/\sqrt{5} \\ 0 \end{pmatrix}$. Continuing, we define

$$\begin{aligned} \mathbf{v}_2' &= \mathbf{v}_2 - (\mathbf{v}_2 \cdot \mathbf{u}_1)\mathbf{u}_1 \\ &= \begin{pmatrix} 0 \\ 3 \\ 1 \end{pmatrix} - (6/\sqrt{5}) \begin{pmatrix} 1/\sqrt{5} \\ 2/\sqrt{5} \\ 0 \end{pmatrix} = \begin{pmatrix} 0 \\ 3 \\ 1 \end{pmatrix} - \begin{pmatrix} \frac{6}{5} \\ \frac{12}{5} \\ 0 \end{pmatrix} = \begin{pmatrix} -\frac{6}{5} \\ \frac{3}{5} \\ 1 \end{pmatrix} \end{aligned}$$

Finally, $|\mathbf{v}_2'| = \sqrt{70/25} = \sqrt{70}/5$, so that $\mathbf{u}_2 = \mathbf{v}_2'/|\mathbf{v}_2'| = (5/\sqrt{70}) \begin{pmatrix} -\frac{6}{5} \\ \frac{3}{5} \\ 1 \end{pmatrix} = \begin{pmatrix} -6/\sqrt{70} \\ 3/\sqrt{70} \\ 5/\sqrt{70} \end{pmatrix}$.

Thus an orthonormal basis is $\left\{ \begin{pmatrix} 1/\sqrt{5} \\ 2/\sqrt{5} \\ 0 \end{pmatrix}, \begin{pmatrix} -6/\sqrt{70} \\ 3/\sqrt{70} \\ 5/\sqrt{70} \end{pmatrix} \right\}$. To check this answer we note that (1) the vectors are orthogonal, (2) each has length 1, and (3) each satisfies $2x - y + 3z = 0$.

In Figure 1*a* we draw the vectors $\mathbf{v}_1$, $\mathbf{v}_2$, and $\mathbf{u}_1$. In Figure 1*b* we draw the vector $-\begin{pmatrix} \frac{6}{5} \\ \frac{12}{5} \\ 0 \end{pmatrix} = \begin{pmatrix} -\frac{6}{5} \\ -\frac{12}{5} \\ 0 \end{pmatrix}$ and add it to $\mathbf{v}_2$ using the parallelogram rule to obtain $\mathbf{v}_2' = \begin{pmatrix} -\frac{6}{5} \\ \frac{3}{5} \\ 0 \end{pmatrix}$. Finally, $\mathbf{u}_2$ lies 1 unit along the vector $\mathbf{v}_2'$.

REMARK: We can see why $\mathbf{u}_1$ and $\mathbf{u}_2$ must be orthogonal. We note that

$$\mathbf{v}_2' = \mathbf{v}_2 - (\mathbf{v}_2 \cdot \mathbf{u}_1)\mathbf{u}_1 = \mathbf{v}_2 - \left(\mathbf{v}_2 \cdot \frac{\mathbf{v}_1}{|\mathbf{v}_1|}\right)\left(\frac{\mathbf{v}_1}{|\mathbf{v}_1|}\right) = \mathbf{v}_2 - \frac{(\mathbf{v}_2 \cdot \mathbf{v}_1)}{|\mathbf{v}_1|^2}\mathbf{v}_1$$

But, from the definitions of projection (on p. 18 for $\mathbb{R}^2$ and p. 34 for $\mathbb{R}^3$), $[(\mathbf{v}_2 \cdot \mathbf{v}_1)/|\mathbf{v}_1|^2]\mathbf{v}_1$ is the projection of $\mathbf{v}_2$ on $\mathbf{v}_1$. Moreover, from Figure 1.2.4, page 18, the vector $\mathbf{v}_2' = \mathbf{v}_2 - \text{proj}_{\mathbf{v}_1} \mathbf{v}_2$ is a vector orthogonal to $\mathbf{v}_1$.

Thus we see that in a certain sense the process we have described here is really a generalization of the notion of projection in $\mathbb{R}^2$ and $\mathbb{R}^3$.

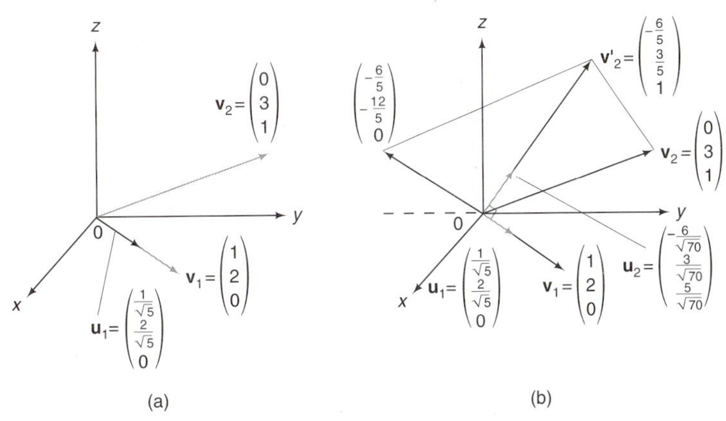

FIGURE 1
The vectors $\mathbf{u}_1$ and $\mathbf{u}_2$ form an orthonormal basis for the plane spanned by the vectors $\mathbf{v}_1$ and $\mathbf{v}_2$

We now define a new kind of matrix that will be very useful in later sections.

DEFINITION ORTHOGONAL MATRIX

The $n \times n$ matrix Q is called **orthogonal** if Q is invertible and

$$Q^{-1} = Q^t \tag{16}$$

■

Note that if $Q^{-1} = Q^t$, then $Q^tQ = I$.

Orthogonal matrices are not difficult to find, according to the next theorem.

THEOREM 3

The $n \times n$ matrix Q is orthogonal if and only if the columns of Q form an orthonormal basis for $\mathbb{R}^n$.

PROOF: Let

$$Q = \begin{pmatrix} a_{11} & a_{12} & \cdots & a_{1n} \\ a_{21} & a_{22} & \cdots & a_{2n} \\ \vdots & \vdots & & \vdots \\ a_{n1} & a_{n2} & \cdots & a_{nn} \end{pmatrix}$$

Then

$$Q^t = \begin{pmatrix} a_{11} & a_{21} & \cdots & a_{n1} \\ a_{12} & a_{22} & \cdots & a_{n2} \\ \vdots & \vdots & & \vdots \\ a_{1n} & a_{2n} & \cdots & a_{nn} \end{pmatrix}$$

Let $B = (b_{ij}) = Q^tQ$. Then

$$b_{ij} = a_{1i}a_{1j} + a_{2i}a_{2j} + \cdots + a_{ni}a_{nj} = \mathbf{c}_i \cdot \mathbf{c}_j \tag{17}$$

where $\mathbf{c}_i$ denotes the ith column of Q. If the columns of Q are orthonormal, then

$$b_{ij} = \begin{cases} 0 & \text{if } i \neq j \\ 1 & \text{if } i = j \end{cases} \tag{18}$$

That is, $B = I$. Conversely, if $Q^t = Q^{-1}$, then $B = I$, so that (18) holds and (17) shows that the columns of Q are orthonormal. This completes the proof. ■

EXAMPLE 6

AN ORTHOGONAL MATRIX

From Example 4, the vectors $\begin{pmatrix} 1/\sqrt{2} \\ 1/\sqrt{2} \\ 0 \end{pmatrix}, \begin{pmatrix} -1/\sqrt{6} \\ 1/\sqrt{6} \\ 2/\sqrt{6} \end{pmatrix}, \begin{pmatrix} 1/\sqrt{3} \\ -1/\sqrt{3} \\ 1/\sqrt{3} \end{pmatrix}$ form an orthonormal basis in $\mathbb{R}^3$. Thus the matrix $Q = \begin{pmatrix} 1/\sqrt{2} & -1/\sqrt{6} & 1/\sqrt{3} \\ 1/\sqrt{2} & 1/\sqrt{6} & -1/\sqrt{3} \\ 0 & 2/\sqrt{6} & 1/\sqrt{3} \end{pmatrix}$ is an orthogonal matrix. To check this we note that

$$Q^tQ = \begin{pmatrix} 1/\sqrt{2} & 1/\sqrt{2} & 0 \\ -1/\sqrt{6} & 1/\sqrt{6} & 2/\sqrt{6} \\ 1/\sqrt{3} & -1/\sqrt{3} & 1/\sqrt{3} \end{pmatrix} \begin{pmatrix} 1/\sqrt{2} & -1/\sqrt{6} & 1/\sqrt{3} \\ 1/\sqrt{2} & 1/\sqrt{6} & -1/\sqrt{3} \\ 0 & 2/\sqrt{6} & 1/\sqrt{3} \end{pmatrix} = \begin{pmatrix} 1 & 0 & 0 \\ 0 & 1 & 0 \\ 0 & 0 & 1 \end{pmatrix}$$

In the proof of Theorem 2 we defined $\mathbf{v}_2' = \mathbf{v}_2 - (\mathbf{v}_2 \cdot \mathbf{u}_1)\mathbf{u}_1$. But, as we have seen, $(\mathbf{v}_2 \cdot \mathbf{u}_1)\mathbf{u}_1 = \text{proj}_{\mathbf{u}_1} \mathbf{v}_2$ (since $|\mathbf{u}_1|^2 = 1$). We now extend this notion from projection onto a vector to projection onto a subspace.

DEFINITION ORTHOGONAL PROJECTION

Let H be a subspace of $\mathbb{R}^n$ with orthonormal basis $\{\mathbf{u}_1, \mathbf{u}_2, \ldots, \mathbf{u}_k\}$. If $\mathbf{v} \in \mathbb{R}^n$, then the **orthogonal projection** of $\mathbf{v}$ onto H, denoted by $\text{proj}_H \mathbf{v}$, is given by

$$\text{proj}_H \mathbf{v} = (\mathbf{v} \cdot \mathbf{u}_1)\mathbf{u}_1 + (\mathbf{v} \cdot \mathbf{u}_2)\mathbf{u}_2 + \cdots + (\mathbf{v} \cdot \mathbf{u}_k)\mathbf{u}_k \tag{19}$$

■

Note that $\text{proj}_H \mathbf{v} \in H$.

EXAMPLE 7

THE ORTHOGONAL PROJECTION OF A VECTOR ONTO A PLANE

Find $\text{proj}_\pi \mathbf{v}$, where π is the plane $\left\{ \begin{pmatrix} x \\ y \\ z \end{pmatrix} : 2x - y + 3z = 0 \right\}$ and $\mathbf{v}$ is the vector $\begin{pmatrix} 3 \\ -2 \\ 4 \end{pmatrix}$.

SOLUTION: From Example 5, an orthonormal basis for π is $\mathbf{u}_1 = \begin{pmatrix} 1/\sqrt{5} \\ 2/\sqrt{5} \\ 0 \end{pmatrix}$ and $\mathbf{u}_2 = \begin{pmatrix} -6/\sqrt{70} \\ 3/\sqrt{70} \\ 5/\sqrt{70} \end{pmatrix}$. Then

$$\text{proj}_\pi \mathbf{v} = \left[\begin{pmatrix} 3 \\ -2 \\ 4 \end{pmatrix} \cdot \begin{pmatrix} 1/\sqrt{5} \\ 2/\sqrt{5} \\ 0 \end{pmatrix}\right]\begin{pmatrix} 1/\sqrt{5} \\ 2/\sqrt{5} \\ 0 \end{pmatrix} + \left[\begin{pmatrix} 3 \\ -2 \\ 4 \end{pmatrix} \cdot \begin{pmatrix} -6/\sqrt{70} \\ 3/\sqrt{70} \\ 5/\sqrt{70} \end{pmatrix}\right]\begin{pmatrix} -6/\sqrt{70} \\ 3/\sqrt{70} \\ 5/\sqrt{70} \end{pmatrix}$$

$$= -\frac{1}{\sqrt{5}}\begin{pmatrix} 1/\sqrt{5} \\ 2/\sqrt{5} \\ 0 \end{pmatrix} - \frac{4}{\sqrt{70}}\begin{pmatrix} -6/\sqrt{70} \\ 3/\sqrt{70} \\ 5/\sqrt{70} \end{pmatrix} = \begin{pmatrix} -\frac{1}{5} \\ -\frac{2}{5} \\ 0 \end{pmatrix} + \begin{pmatrix} \frac{24}{70} \\ -\frac{12}{70} \\ -\frac{20}{70} \end{pmatrix} = \begin{pmatrix} \frac{1}{7} \\ -\frac{4}{7} \\ -\frac{2}{7} \end{pmatrix}$$

The notion of projection gives us a convenient way to write a vector in $\mathbb{R}^n$ in terms of an orthonormal basis.

THEOREM 4

Let $B = \{\mathbf{u}_1, \mathbf{u}_2, \ldots, \mathbf{u}_n\}$ be an orthonormal basis for $\mathbb{R}^n$ and let $\mathbf{v} \in \mathbb{R}^n$. Then

$$\mathbf{v} = (\mathbf{v} \cdot \mathbf{u}_1)\mathbf{u}_1 + (\mathbf{v} \cdot \mathbf{u}_2)\mathbf{u}_2 + \cdots + (\mathbf{v} \cdot \mathbf{u}_n)\mathbf{u}_n \qquad (20)$$

That is, $\mathbf{v} = \text{proj}_{\mathbb{R}^n} \mathbf{v}$.

PROOF: Since B is a basis, we can write $\mathbf{v}$ in a unique way as $\mathbf{v} = c_1\mathbf{u}_1 + c_2\mathbf{u}_2 + \cdots + c_n\mathbf{u}_n$. Then

$$\mathbf{v} \cdot \mathbf{u}_i = c_1(\mathbf{u}_1 \cdot \mathbf{u}_i) + c_2(\mathbf{u}_2 \cdot \mathbf{u}_i) + \cdots + c_i(\mathbf{u}_i \cdot \mathbf{u}_i) + \cdots + c_n(\mathbf{u}_n \cdot \mathbf{u}_i) = c_i$$

since the $\mathbf{u}_i$'s are orthonormal. Since this is true for $i = 1, 2, \ldots, n$, the proof is complete. ■

EXAMPLE 8 WRITING A VECTOR IN TERMS OF AN ORTHONORMAL BASIS

Write the vector $\begin{pmatrix} 2 \\ -1 \\ 3 \end{pmatrix}$ in $\mathbb{R}^3$ in terms of the orthonormal basis $\left\{\begin{pmatrix} 1/\sqrt{2} \\ 1/\sqrt{2} \\ 0 \end{pmatrix}, \begin{pmatrix} -1/\sqrt{6} \\ 1/\sqrt{6} \\ 2/\sqrt{6} \end{pmatrix}, \begin{pmatrix} 1/\sqrt{3} \\ -1/\sqrt{3} \\ 1/\sqrt{3} \end{pmatrix}\right\}$.

SOLUTION:

$$\begin{pmatrix}2\\-1\\3\end{pmatrix} = \left[\begin{pmatrix}2\\-1\\3\end{pmatrix}\cdot\begin{pmatrix}1/\sqrt{2}\\1/\sqrt{2}\\0\end{pmatrix}\right]\begin{pmatrix}1/\sqrt{2}\\1/\sqrt{2}\\0\end{pmatrix} + \left[\begin{pmatrix}2\\-1\\3\end{pmatrix}\cdot\begin{pmatrix}-1/\sqrt{6}\\1/\sqrt{6}\\2/\sqrt{6}\end{pmatrix}\right]\begin{pmatrix}-1/\sqrt{6}\\1/\sqrt{6}\\2/\sqrt{6}\end{pmatrix}$$

$$+\left[\begin{pmatrix}2\\-1\\3\end{pmatrix}\cdot\begin{pmatrix}1/\sqrt{3}\\-1/\sqrt{3}\\1/\sqrt{3}\end{pmatrix}\right]\begin{pmatrix}1/\sqrt{3}\\-1/\sqrt{3}\\1/\sqrt{3}\end{pmatrix}$$

$$= \frac{1}{\sqrt{2}}\begin{pmatrix}1/\sqrt{2}\\1/\sqrt{2}\\0\end{pmatrix} + \frac{3}{\sqrt{6}}\begin{pmatrix}-1/\sqrt{6}\\1/\sqrt{6}\\2/\sqrt{6}\end{pmatrix} + \frac{6}{\sqrt{3}}\begin{pmatrix}1/\sqrt{3}\\-1/\sqrt{3}\\1/\sqrt{3}\end{pmatrix}$$

Before continuing, we need to know that an orthogonal projection is well defined. By this we mean that the definition of $\text{proj}_H \mathbf{v}$ is independent of the orthonormal basis chosen in H. The following theorem takes care of this problem.

THEOREM 5

Let H be a subspace of $\mathbb{R}^n$. Suppose that H has two orthonormal bases, $\{\mathbf{u}_1, \mathbf{u}_2, \ldots, \mathbf{u}_k\}$ and $\{\mathbf{w}_1, \mathbf{w}_2, \ldots, \mathbf{w}_k\}$. Let $\mathbf{v}$ be a vector in $\mathbb{R}^n$. Then

$$(\mathbf{v}\cdot\mathbf{u}_1)\mathbf{u}_1 + (\mathbf{v}\cdot\mathbf{u}_2)\mathbf{u}_2 + \cdots + (\mathbf{v}\cdot\mathbf{u}_k)\mathbf{u}_k = (\mathbf{v}\cdot\mathbf{w}_1)\mathbf{w}_1 + (\mathbf{v}\cdot\mathbf{w}_2)\mathbf{w}_2 + \cdots + (\mathbf{v}\cdot\mathbf{w}_k)\mathbf{w}_k \quad (21)$$

PROOF: Choose vectors $\mathbf{u}_{k+1}, \mathbf{u}_{k+2}, \ldots, \mathbf{u}_n$ such that $B_1 = \{\mathbf{u}_1, \mathbf{u}_2, \ldots, \mathbf{u}_k, \mathbf{u}_{k+1}, \ldots, \mathbf{u}_n\}$ is an orthonormal basis for $\mathbb{R}^n$ (this can be done as in the proof of Theorem 2). Then $B_2 = \{\mathbf{w}_1, \mathbf{w}_2, \ldots, \mathbf{w}_k, \mathbf{u}_{k+1}, \mathbf{u}_{k+2}, \ldots, \mathbf{u}_n\}$ is also an orthonormal basis for $\mathbb{R}^n$. To see this, note first that none of the vectors $\mathbf{u}_{k+1}, \mathbf{u}_{k+2}, \ldots, \mathbf{u}_n$ can be written as a linear combination of $\mathbf{w}_1, \mathbf{w}_2, \ldots, \mathbf{w}_k$ because none of these vectors is in H and $\{\mathbf{w}_1, \mathbf{w}_2, \ldots, \mathbf{w}_k\}$ is a basis for H. Thus B_2 is a basis for $\mathbb{R}^n$ because it contains n linearly independent vectors. The orthonormality of the vectors in B_2 follows from the way these vectors were chosen ($\mathbf{u}_{k+j}$ is orthogonal to every vector in H for $j = 1, 2, \ldots, n-k$). Let $\mathbf{v}$ be a vector in $\mathbb{R}^n$. Then, from Theorem 4 [equation (20)],

$$\begin{aligned}\mathbf{v} &= (\mathbf{v}\cdot\mathbf{u}_1)\mathbf{u}_1 + (\mathbf{v}\cdot\mathbf{u}_2)\mathbf{u}_2 + \cdots + (\mathbf{v}\cdot\mathbf{u}_k)\mathbf{u}_k + (\mathbf{v}\cdot\mathbf{u}_{k+1})\mathbf{u}_{k+1} \\ &\quad + \cdots + (\mathbf{v}\cdot\mathbf{u}_n)\mathbf{u}_n \\ &= (\mathbf{v}\cdot\mathbf{w}_1)\mathbf{w}_1 + (\mathbf{v}\cdot\mathbf{w}_2)\mathbf{w}_2 + \cdots + (\mathbf{v}\cdot\mathbf{w}_k)\mathbf{w}_k + (\mathbf{v}\cdot\mathbf{u}_{k+1})\mathbf{u}_{k+1} \\ &\quad + \cdots + (\mathbf{v}\cdot\mathbf{u}_n)\mathbf{u}_n \end{aligned} \quad (22)$$

Equation (21) now follows from equation (22). ■

DEFINITION ORTHOGONAL COMPLEMENT

Let H be a subspace of $\mathbb{R}^n$. Then the **orthogonal complement** of H, denoted by $H^\perp$, is given by

$$H^\perp = \{\mathbf{x} \in \mathbb{R}^n\colon \mathbf{x}\cdot\mathbf{h} = 0 \quad \text{for every } \mathbf{h} \in H\}$$

■

THEOREM 6

If H is a subspace of $\mathbb{R}^n$, then:

i. $H^\perp$ is a subspace of $\mathbb{R}^n$.
ii. $H \cap H^\perp = \{\mathbf{0}\}$.
iii. $\dim H^\perp = n - \dim H$.

PROOF:

i. If $\mathbf{x}$ and $\mathbf{y}$ are in $H^\perp$ and if $\mathbf{h} \in H$, then $(\mathbf{x} + \mathbf{y}) \cdot \mathbf{h} = \mathbf{x} \cdot \mathbf{h} + \mathbf{y} \cdot \mathbf{h} = 0 + 0 = 0$ and $(\alpha\mathbf{x} \cdot \mathbf{h}) = \alpha(\mathbf{x} \cdot \mathbf{h}) = 0$, so $H^\perp$ is a subspace.

ii. If $\mathbf{x} \in H \cap H^\perp$, then $\mathbf{x} \cdot \mathbf{x} = 0$, so $\mathbf{x} = \mathbf{0}$, which shows that $H \cap H^\perp = \{\mathbf{0}\}$.

iii. Let $\{\mathbf{u}_1, \mathbf{u}_2, \ldots, \mathbf{u}_k\}$ be an orthonormal basis for H. By the result of Problem 8.5.27, page 524, this can be expanded into a basis B for $\mathbb{R}^n$: $B = \{\mathbf{u}_1, \mathbf{u}_2, \ldots, \mathbf{u}_k, \mathbf{v}_{k+1}, \ldots, \mathbf{v}_n\}$. Using the Gram-Schmidt process, we can turn B into an orthonormal basis for $\mathbb{R}^n$. As in the proof of Theorem 2, the already orthonormal $\mathbf{u}_1, \mathbf{u}_2, \ldots, \mathbf{u}_k$ will remain unchanged in the process, and we obtain the orthonormal basis $B_1 = \{\mathbf{u}_1, \mathbf{u}_2, \ldots, \mathbf{u}_k, \mathbf{u}_{k+1}, \ldots, \mathbf{u}_n\}$. To complete the proof, we need only show that $\{\mathbf{u}_{k+1}, \ldots, \mathbf{u}_n\}$ is a basis for $H^\perp$. Since the $\mathbf{u}_i$'s are independent, we must show that they span $H^\perp$. Let $\mathbf{x} \in H^\perp$; then, by Theorem 4,

$$\mathbf{x} = (\mathbf{x} \cdot \mathbf{u}_1)\mathbf{u}_1 + (\mathbf{x} \cdot \mathbf{u}_2)\mathbf{u}_2 + \cdots + (\mathbf{x} \cdot \mathbf{u}_k)\mathbf{u}_k + (\mathbf{x} \cdot \mathbf{u}_{k+1})\mathbf{u}_{k+1} + \cdots + (\mathbf{x} \cdot \mathbf{u}_n)\mathbf{u}_n$$

But $(\mathbf{x} \cdot \mathbf{u}_i) = 0$ for $i = 1, 2, \ldots, k$, since $\mathbf{x} \in H^\perp$ and $\mathbf{u}_i \in H$. Thus $\mathbf{x} = (\mathbf{x} \cdot \mathbf{u}_{k+1})\mathbf{u}_{k+1} + \cdots + (\mathbf{x} \cdot \mathbf{u}_n)\mathbf{u}_n$. This shows that $\{\mathbf{u}_{k+1}, \ldots, \mathbf{u}_n\}$ is a basis for $H^\perp$ which means that $\dim H^\perp = n - k$. ∎

The spaces H and $H^\perp$ allow us to "decompose" any vector in $\mathbb{R}^n$.

THEOREM 7 PROJECTION THEOREM

Let H be a subspace of $\mathbb{R}^n$ and let $\mathbf{v} \in \mathbb{R}^n$. Then there exists a unique pair of vectors $\mathbf{h}$ and $\mathbf{p}$ such that $\mathbf{h} \in H$, $\mathbf{p} \in H^\perp$, and $\mathbf{v} = \mathbf{h} + \mathbf{p}$. In particular, $\mathbf{h} = \text{proj}_H\, \mathbf{v}$ and $\mathbf{p} = \text{proj}_{H^\perp}\, \mathbf{v}$

$$\mathbf{v} = \mathbf{h} + \mathbf{p} = \text{proj}_H\, \mathbf{v} + \text{proj}_{H^\perp}\, \mathbf{v} \tag{23}$$

PROOF: Let $\mathbf{h} = \text{proj}_H\, \mathbf{v}$ and let $\mathbf{p} = \mathbf{v} - \mathbf{h}$. By the definition of orthogonal projection, we have $\mathbf{h} \in H$. We now show that $\mathbf{p} \in H^\perp$. Let $\{\mathbf{u}_1, \mathbf{u}_2, \ldots, \mathbf{u}_k\}$ be a basis for H. Then

$$\mathbf{h} = (\mathbf{v} \cdot \mathbf{u}_1)\mathbf{u}_1 + (\mathbf{v} \cdot \mathbf{u}_2)\mathbf{u}_2 + \cdots + (\mathbf{v} \cdot \mathbf{u}_k)\mathbf{u}_k$$

Let $\mathbf{x}$ be a vector in H. There exist constants $\alpha_1, \alpha_2, \ldots, \alpha_k$ such that

$$\mathbf{x} = \alpha_1\mathbf{u}_1 + \alpha_2\mathbf{u}_2 + \cdots + \alpha_k\mathbf{u}_k$$

Then

$$\begin{aligned}\mathbf{p}\cdot\mathbf{x} &= (\mathbf{v}-\mathbf{h})\cdot\mathbf{x}\\ &= [\mathbf{v}-(\mathbf{v}\cdot\mathbf{u}_1)\mathbf{u}_1-(\mathbf{v}\cdot\mathbf{u}_2)\mathbf{u}_2-\cdots-(\mathbf{v}\cdot\mathbf{u}_k)\mathbf{u}_k]\\ &\quad\cdot[\alpha_1\mathbf{u}_1+\alpha_2\mathbf{u}_2+\cdots+\alpha_k\mathbf{u}_k]\end{aligned} \tag{24}$$

Since $\mathbf{u}_i\cdot\mathbf{u}_j=\begin{cases}0, & i\neq j\\ 1, & i=j\end{cases}$ it is easy to verify that the scalar product in (24) is given by

$$\mathbf{p}\cdot\mathbf{x}=\sum_{i=1}^{k}\alpha_i(\mathbf{v}\cdot\mathbf{u}_i)-\sum_{i=1}^{k}\alpha_i(\mathbf{v}\cdot\mathbf{u}_i)=0$$

Thus $\mathbf{p}\cdot\mathbf{x}=0$ for every $\mathbf{x}\in H$, which means that $\mathbf{p}\in H^\perp$. To show that $\mathbf{p}=\text{proj}_{H^\perp}\mathbf{v}$, we extend $\{\mathbf{u}_1,\mathbf{u}_2,\ldots,\mathbf{u}_k\}$ to an orthonormal basis for $\mathbb{R}^n$: $\{\mathbf{u}_1,\mathbf{u}_2,\ldots,\mathbf{u}_k,\mathbf{u}_{k+1},\ldots,\mathbf{u}_n\}$. Then $\{\mathbf{u}_{k+1},\ldots,\mathbf{u}_n\}$ is a basis for $H^\perp$ and, by Theorem 4,

$$\begin{aligned}\mathbf{v} &= (\mathbf{v}\cdot\mathbf{u}_1)\mathbf{u}_1+(\mathbf{v}\cdot\mathbf{u}_2)\mathbf{u}_2+\cdots+(\mathbf{v}\cdot\mathbf{u}_k)\mathbf{u}_k+(\mathbf{v}\cdot\mathbf{u}_{k+1})\mathbf{u}_{k+1}\\ &\quad+\cdots+(\mathbf{v}\cdot\mathbf{u}_n)\mathbf{u}_n\\ &= \text{proj}_H\mathbf{v}+\text{proj}_{H^\perp}\mathbf{v}\qquad\text{(by the definition of orthogonal projection)}\end{aligned}$$

This proves equation (23). To prove uniqueness, suppose that $\mathbf{v}=\mathbf{h}_1-\mathbf{p}_1=\mathbf{h}_2-\mathbf{p}_2$, where $\mathbf{h}_1,\mathbf{h}_2\in H$ and $\mathbf{p}_1,\mathbf{p}_2\in H^\perp$. Then $\mathbf{h}_1-\mathbf{h}_2=\mathbf{p}_1-\mathbf{p}_2$. But $\mathbf{h}_1-\mathbf{h}_2\in H$ and $\mathbf{p}_1-\mathbf{p}_2\in H^\perp$, so $\mathbf{h}_1-\mathbf{h}_2\in H\cap H^\perp=\{\mathbf{0}\}$. Thus $\mathbf{h}_1-\mathbf{h}_2=\mathbf{0}$ and $\mathbf{p}_1-\mathbf{p}_2=\mathbf{0}$, which completes the proof.

EXAMPLE 9

DECOMPOSING A VECTOR IN $\mathbb{R}^3$

In $\mathbb{R}^3$ let $\pi=\left\{\begin{pmatrix}x\\y\\z\end{pmatrix}: 2x-y+3z=0\right\}$. Write the vector $\begin{pmatrix}3\\-2\\4\end{pmatrix}$ as $\mathbf{h}+\mathbf{p}$, where $\mathbf{h}\in\pi$ and $\mathbf{p}\in\pi^\perp$.

SOLUTION: An orthonormal basis for π is $B_1=\left\{\begin{pmatrix}1/\sqrt{5}\\2/\sqrt{5}\\0\end{pmatrix},\begin{pmatrix}-6/\sqrt{70}\\3/\sqrt{70}\\5/\sqrt{70}\end{pmatrix}\right\}$ and, from Example 7, $\mathbf{h}=\text{proj}_\pi\mathbf{v}=\begin{pmatrix}\frac{1}{7}\\-\frac{4}{7}\\-\frac{2}{7}\end{pmatrix}\in\pi$. Then

$$\mathbf{p}=\mathbf{v}-\mathbf{h}=\begin{pmatrix}3\\-2\\4\end{pmatrix}-\begin{pmatrix}\frac{1}{7}\\-\frac{4}{7}\\-\frac{2}{7}\end{pmatrix}=\begin{pmatrix}\frac{20}{7}\\-\frac{10}{7}\\\frac{30}{7}\end{pmatrix}\in\pi^\perp$$

Note that $\mathbf{p}\cdot\mathbf{h}=0$.

The following theorem is very useful in statistics and other applied areas. We shall provide one application of this theorem in Section 8.8.

THEOREM 8 NORM APPROXIMATION THEOREM

Let H be a subspace of $\mathbb{R}^n$ and let $\mathbf{v}$ be a vector in $\mathbb{R}^n$. Then in H, $\text{proj}_H \mathbf{v}$ is the best approximation to $\mathbf{v}$ in the following sense: If $\mathbf{h}$ is any other vector in H, then

$$|\mathbf{v} - \text{proj}_H \mathbf{v}| < |\mathbf{v} - \mathbf{h}| \tag{25}$$

PROOF: From Theorem 7, $\mathbf{v} - \text{proj}_H \mathbf{v} \in H^\perp$. We write

$$\mathbf{v} - \mathbf{h} = (\mathbf{v} - \text{proj}_H \mathbf{v}) + (\text{proj}_H \mathbf{v} - \mathbf{h})$$

The first term on the right is in $H^\perp$, while the second is in H, so

$$(\mathbf{v} - \text{proj}_H \mathbf{v}) \cdot (\text{proj}_H \mathbf{v} - \mathbf{h}) = 0 \tag{26}$$

Now

$$\begin{aligned}
|\mathbf{v} - \mathbf{h}|^2 &= (\mathbf{v} - \mathbf{h}) \cdot (\mathbf{v} - \mathbf{h}) \\
&= [(\mathbf{v} - \text{proj}_H \mathbf{v}) + (\text{proj}_H \mathbf{v} - \mathbf{h})] \cdot [(\mathbf{v} - \text{proj}_H \mathbf{v}) + (\text{proj}_H \mathbf{v} - \mathbf{h})] \\
&= |\mathbf{v} - \text{proj}_H \mathbf{v}|^2 + 2(\mathbf{v} - \text{proj}_H \mathbf{v}) \cdot (\text{proj}_H \mathbf{v} - \mathbf{h}) + |\text{proj}_H \mathbf{v} - \mathbf{h}|^2 \\
&= |\mathbf{v} - \text{proj}_H \mathbf{v}|^2 + |\text{proj}_H \mathbf{v} - \mathbf{h}|^2
\end{aligned}$$

But $|\text{proj}_H \mathbf{v} - \mathbf{h}|^2 > 0$ because $\mathbf{h} \neq \text{proj}_H \mathbf{v}$. Hence

$$|\mathbf{v} - \mathbf{h}|^2 > |\mathbf{v} - \text{proj}_H \mathbf{v}|^2$$

or

$$|\mathbf{v} - \mathbf{h}| > |\mathbf{v} - \text{proj}_H \mathbf{v}| \qquad \blacksquare$$

ORTHOGONAL BASES IN $\mathbb{R}^3$ WITH INTEGER COEFFICIENTS AND INTEGER NORMS

It is sometimes useful to construct an orthogonal basis of vectors where the coordinates and norm of each vector is an integer. For example,

$$\left\{\begin{pmatrix} 2 \\ 2 \\ -1 \end{pmatrix}, \begin{pmatrix} 2 \\ -1 \\ 2 \end{pmatrix}, \begin{pmatrix} -1 \\ 2 \\ 2 \end{pmatrix}\right\}$$

constitutes an orthogonal basis in $\mathbb{R}^3$ where each vector has norm 3. As another example,

$$\left\{\begin{pmatrix} 12 \\ 4 \\ -3 \end{pmatrix}, \begin{pmatrix} 0 \\ 3 \\ 4 \end{pmatrix}, \begin{pmatrix} -25 \\ 48 \\ -36 \end{pmatrix}\right\}$$

is an orthogonal basis in $\mathbb{R}^3$ whose vectors have norms 13, 5, and 65, respectively. Finding bases like this in $\mathbb{R}^3$ turns out to be not so difficult as you might imagine. A discussion of this topic appears in the interesting paper ''Orthogonal Bases of $\mathbb{R}^3$ with Integer Coordinates and Integer Lengths'' by Anthony Osborne and Hans Liebeck in *The American Mathematical Monthly,* Volume 96, Number 1, January 1989, pp. 49–53.

We close this section with an important theorem.

THEOREM 9 CAUCHY-SCHWARZ INEQUALITY IN $\mathbb{R}^n$

Let $\mathbf{u}$ and $\mathbf{v}$ be vectors in $\mathbb{R}^n$. Then

i. $|\mathbf{u}\cdot\mathbf{v}| \le |\mathbf{u}|\,|\mathbf{v}|$. **(27)**

ii. $|\mathbf{u}\cdot\mathbf{v}| = |\mathbf{u}||\mathbf{v}|$ if and only if $\mathbf{u} = \mathbf{0}$ or $\mathbf{v} = \lambda\mathbf{u}$ for some real number λ.

PROOF:

i. If $\mathbf{u} = \mathbf{0}$ or $\mathbf{v} = \mathbf{0}$ (or both), then (27) holds (both sides are equal to 0). We assume that $\mathbf{u} \ne \mathbf{0}$ and $\mathbf{v} \ne \mathbf{0}$. Then

$$0 \le \left|\frac{\mathbf{u}}{|\mathbf{u}|} - \frac{\mathbf{v}}{|\mathbf{v}|}\right|^2 = \left(\frac{\mathbf{u}}{|\mathbf{u}|} - \frac{\mathbf{v}}{|\mathbf{v}|}\right)\cdot\left(\frac{\mathbf{u}}{|\mathbf{u}|} - \frac{\mathbf{v}}{|\mathbf{v}|}\right)$$

$$= \frac{\mathbf{u}\cdot\mathbf{u}}{|\mathbf{u}|^2} - \frac{2\mathbf{u}\cdot\mathbf{v}}{|\mathbf{u}||\mathbf{v}|} + \frac{\mathbf{v}\cdot\mathbf{v}}{|\mathbf{v}|^2}$$

$$= \frac{|\mathbf{u}|^2}{|\mathbf{u}|^2} - \frac{2\mathbf{u}\cdot\mathbf{v}}{|\mathbf{u}||\mathbf{v}|} + \frac{|\mathbf{v}|^2}{|\mathbf{v}|^2} = 2 - \frac{2\mathbf{u}\cdot\mathbf{v}}{|\mathbf{u}||\mathbf{v}|}$$

Thus $\dfrac{2\mathbf{u}\cdot\mathbf{v}}{|\mathbf{u}||\mathbf{v}|} \le 2$, so $\dfrac{\mathbf{u}\cdot\mathbf{v}}{|\mathbf{u}||\mathbf{v}|} \le 1$ and $\mathbf{u}\cdot\mathbf{v} \le |\mathbf{u}||\mathbf{v}|$. Similarly, starting with $0 \le \left|\dfrac{\mathbf{u}}{|\mathbf{u}|} + \dfrac{\mathbf{v}}{|\mathbf{v}|}\right|^2$, we end up with $\dfrac{\mathbf{u}\cdot\mathbf{v}}{|\mathbf{u}||\mathbf{v}|} \ge -1$ or $\mathbf{u}\cdot\mathbf{v} \ge -|\mathbf{u}||\mathbf{v}|$. Putting these together, we obtain

$$-|\mathbf{u}||\mathbf{v}| \le \mathbf{u}\cdot\mathbf{v} \le |\mathbf{u}||\mathbf{v}| \qquad \text{or} \qquad |\mathbf{u}\cdot\mathbf{v}| \le |\mathbf{u}||\mathbf{v}|$$

ii. If $\mathbf{u} = \lambda\mathbf{v}$, then $|\mathbf{u}\cdot\mathbf{v}| = |\lambda\mathbf{v}\cdot\mathbf{v}| = |\lambda||\mathbf{v}|^2$ and $|\mathbf{u}||\mathbf{v}| = |\lambda\mathbf{v}||\mathbf{v}| = |\lambda||\mathbf{v}||\mathbf{v}| = |\lambda||\mathbf{v}|^2 = |\mathbf{u}\cdot\mathbf{v}|$. Conversely, suppose that $|\mathbf{u}\cdot\mathbf{v}| = |\mathbf{u}||\mathbf{v}|$ with $\mathbf{u} \ne \mathbf{0}$ and $\mathbf{v} \ne \mathbf{0}$. Then $\left|\dfrac{\mathbf{u}\cdot\mathbf{v}}{|\mathbf{u}||\mathbf{v}|}\right| = 1$ so $\dfrac{\mathbf{u}\cdot\mathbf{v}}{|\mathbf{u}||\mathbf{v}|} = \pm 1$

Case 1: $\dfrac{\mathbf{u}\cdot\mathbf{v}}{|\mathbf{u}||\mathbf{v}|} = 1$. Then

$$\left|\frac{\mathbf{u}}{|\mathbf{u}|} - \frac{\mathbf{v}}{|\mathbf{v}|}\right|^2 = \left(\frac{\mathbf{u}}{|\mathbf{u}|} - \frac{\mathbf{v}}{|\mathbf{v}|}\right)\cdot\left(\frac{\mathbf{u}}{|\mathbf{u}|} - \frac{\mathbf{v}}{|\mathbf{v}|}\right) \overset{\text{as in (i)}}{=} 2 - \frac{2\mathbf{u}\cdot\mathbf{v}}{|\mathbf{u}||\mathbf{v}|} = 2 - 2 = 0.$$

Thus

$$\frac{\mathbf{u}}{|\mathbf{u}|} = \frac{\mathbf{v}}{|\mathbf{v}|} \qquad \text{or} \qquad \mathbf{u} = \frac{|\mathbf{u}|}{|\mathbf{v}|}\mathbf{v} = \lambda\mathbf{v}$$

Case 2: $\dfrac{\mathbf{u}\cdot\mathbf{v}}{|\mathbf{u}||\mathbf{v}|} = -1$. Then

$$\left|\frac{\mathbf{u}}{|\mathbf{u}|} + \frac{\mathbf{v}}{|\mathbf{v}|}\right|^2 = 2 + \frac{2\mathbf{u}\cdot\mathbf{v}}{|\mathbf{u}||\mathbf{v}|} = 2 - 2 = 0$$

so

$$\frac{\mathbf{u}}{|\mathbf{u}|} = -\frac{\mathbf{v}}{|\mathbf{v}|} \qquad \text{and} \qquad \mathbf{u} = -\frac{|\mathbf{u}|}{|\mathbf{v}|}\mathbf{v} = \lambda\mathbf{v}$$

■

PROBLEMS 8.7

SELF-QUIZ

True-False

I. The set $\{(1, 1), (1, -1)\}$ is an orthonormal set in $\mathbb{R}^2$.

II. The set $\left\{\begin{pmatrix}\frac{1}{\sqrt{2}}, \frac{1}{\sqrt{2}}\end{pmatrix}, \begin{pmatrix}\frac{1}{\sqrt{2}}, \frac{-1}{\sqrt{2}}\end{pmatrix}\right\}$ is an orthonormal set in $\mathbb{R}^2$.

III. Every basis in $\mathbb{R}^n$ can be turned into an orthonormal basis by using the Gram-Schmidt orthonormalization process.

IV. The matrix $\begin{pmatrix}1 & 1\\ 1 & -1\end{pmatrix}$ is orthogonal.

V. The matrix $\begin{pmatrix}\frac{1}{\sqrt{2}} & \frac{1}{\sqrt{2}}\\ \frac{1}{\sqrt{2}} & \frac{-1}{\sqrt{2}}\end{pmatrix}$ is orthogonal.

Multiple Choice

VI. For which of the following matrices is Q^{-1} equal to Q^t?

a. $\begin{pmatrix}1 & 6\\ 3 & -2\end{pmatrix}$ **b.** $\begin{pmatrix}\frac{1}{\sqrt{10}} & \frac{6}{\sqrt{40}}\\ \frac{3}{\sqrt{10}} & \frac{2}{\sqrt{40}}\end{pmatrix}$

c. $\begin{pmatrix}\frac{1}{\sqrt{10}} & \frac{6}{\sqrt{40}}\\ \frac{3}{\sqrt{10}} & \frac{-2}{\sqrt{40}}\end{pmatrix}$ **d.** $\begin{pmatrix}1 & 6\\ 3 & 2\end{pmatrix}$

In Problems 1–13 construct an orthonormal basis for the given vector space or subspace.

1. In $\mathbb{R}^2$, starting with the basis vectors $\begin{pmatrix}1\\1\end{pmatrix}$, $\begin{pmatrix}-1\\1\end{pmatrix}$

2. $H = \{(x, y) \in \mathbb{R}^2\colon x + y = 0\}$

3. $H = \{(x, y) \in \mathbb{R}^2\colon ax + by = 0\}$

4. In $\mathbb{R}^2$, starting with $\begin{pmatrix}a\\b\end{pmatrix}$, $\begin{pmatrix}c\\d\end{pmatrix}$, where $ad - bc \neq 0$.

5. $\pi = \{(x, y, z)\colon 2x - y - z = 0\}$

6. $\pi = \{(x, y, z)\colon 3x - 2y + 6z = 0\}$

7. $L = \{(x, y, z)\colon x/2 = y/3 = z/4\}$

8. $L = \{(x, y, z)\colon x = 3t,\ y = -2t,\ z = t;\ t \text{ real}\}$

9. $H = \{(x, y, z, w) \in \mathbb{R}^4\colon 2x - y + 3z - w = 0\}$

10. $\pi = \{(x, y, z)\colon ax + by + cz = 0\}$, where $abc \neq 0$

11. $L = \{(x, y, z)\colon x/a = y/b = z/c\}$, where $abc \neq 0$.

12. $H = \{(x_1, x_2, x_3, x_4, x_5) \in \mathbb{R}^5\colon 2x_1 - 3x_2 + x_3 + 4x_4 - x_5 = 0\}$

13. H is the solution space of

$$\begin{aligned} x - 3y + z &= 0\\ -2x + 2y - 3z &= 0\\ 4x - 8y + 5z &= 0 \end{aligned}$$

***14.** Find an orthonormal basis in $\mathbb{R}^4$ that includes the vectors

$$\mathbf{u}_1 = \begin{pmatrix}1/\sqrt{2}\\ 0\\ 1/\sqrt{2}\\ 0\end{pmatrix} \quad\text{and}\quad \mathbf{u}_2 = \begin{pmatrix}-\frac{1}{2}\\ \frac{1}{2}\\ \frac{1}{2}\\ -\frac{1}{2}\end{pmatrix}$$

[*Hint:* First find two vectors $\mathbf{v}_3$ and $\mathbf{v}_4$ to complete the basis.]

15. Show that $Q = \begin{pmatrix}\frac{2}{3} & \frac{1}{3} & \frac{2}{3}\\ \frac{1}{3} & \frac{2}{3} & -\frac{2}{3}\\ -\frac{2}{3} & \frac{2}{3} & \frac{1}{3}\end{pmatrix}$ is an orthogonal matrix.

16. Show that if P and Q are orthogonal $n \times n$ matrices, then PQ is orthogonal.

17. Verify the result of Problem 16 with

$$P = \begin{pmatrix}1/\sqrt{2} & -1/\sqrt{2}\\ 1/\sqrt{2} & 1/\sqrt{2}\end{pmatrix} \quad\text{and}$$

$$Q = \begin{pmatrix}1/3 & -\sqrt{8}/3\\ \sqrt{8}/3 & 1/3\end{pmatrix}$$

18. Show that if Q is a symmetric orthogonal matrix, then $Q^2 = I$.

19. Show that if Q is orthogonal, then $\det Q = \pm 1$.

20. Show that for any real number t, the matrix $A = \begin{pmatrix}\sin t & \cos t\\ \cos t & -\sin t\end{pmatrix}$ is orthogonal.

21. Let $\{\mathbf{v}_1, \mathbf{v}_2, \ldots, \mathbf{v}_k\}$ be a linearly independent set of vectors in $\mathbb{R}^n$. Prove that $\mathbf{v}_i \neq \mathbf{0}$ for $i = 1, 2, \ldots, k$.

In Problems 22–28 a subspace H and a vector $\mathbf{v}$ are given. **(a)** Compute $\text{proj}_H \mathbf{v}$; **(b)** find an orthonormal basis for $H^\perp$; **(c)** write $\mathbf{v}$ as $\mathbf{h} + \mathbf{p}$, where $\mathbf{h} \in H$ and $\mathbf{p} \in H^\perp$.

22. $H = \left\{ \begin{pmatrix} x \\ y \end{pmatrix} \in \mathbb{R}^2\colon x + y = 0 \right\}$; $\mathbf{v} = \begin{pmatrix} -1 \\ 2 \end{pmatrix}$

23. $H = \left\{ \begin{pmatrix} x \\ y \end{pmatrix} \in \mathbb{R}^2\colon ax + by = 0 \right\}$; $\mathbf{v} = \begin{pmatrix} a \\ b \end{pmatrix} \neq \mathbf{0}$

24. $H = \left\{ \begin{pmatrix} x \\ y \\ z \end{pmatrix} \in \mathbb{R}^3\colon ax + by + cz = 0 \right\}$;

$\mathbf{v} = \begin{pmatrix} a \\ b \\ c \end{pmatrix}$, $\mathbf{v} \neq \mathbf{0}$

25. $H = \left\{ \begin{pmatrix} x \\ y \\ z \end{pmatrix} \in \mathbb{R}^3\colon 3x - 2y + 6z = 0 \right\}$; $\mathbf{v} = \begin{pmatrix} -3 \\ 1 \\ 4 \end{pmatrix}$

26. $H = \left\{ \begin{pmatrix} x \\ y \\ z \end{pmatrix} \in \mathbb{R}^3\colon x/2 = y/3 = z/4 \right\}$; $\mathbf{v} = \begin{pmatrix} 1 \\ 1 \\ 1 \end{pmatrix}$

27. $H = \left\{ \begin{pmatrix} x \\ y \\ z \\ w \end{pmatrix} \in \mathbb{R}^4\colon 2x - y + 3z - w = 0 \right\}$;

$\mathbf{v} = \begin{pmatrix} 1 \\ -1 \\ 2 \\ 3 \end{pmatrix}$

28. $H = \left\{ \begin{pmatrix} x \\ y \\ z \\ w \end{pmatrix} \in \mathbb{R}^4\colon x = y \text{ and } w = 3y \right\}$; $\mathbf{v} = \begin{pmatrix} -1 \\ 2 \\ 3 \\ 1 \end{pmatrix}$

29. Let $\mathbf{u}_1$ and $\mathbf{u}_2$ be two orthonormal vectors in $\mathbb{R}^n$. Show that $|\mathbf{u}_1 - \mathbf{u}_2| = \sqrt{2}$.

30. If $\mathbf{u}_1, \mathbf{u}_2, \ldots, \mathbf{u}_n$ are orthonormal, show that

$$|\mathbf{u}_1 + \mathbf{u}_2 + \cdots + \mathbf{u}_n|^2 = |\mathbf{u}_1|^2 + |\mathbf{u}_2|^2 + \cdots + |\mathbf{u}_n|^2 = n$$

31. Find a condition on the numbers a and b such that $\left\{ \begin{pmatrix} a \\ b \end{pmatrix}, \begin{pmatrix} b \\ -a \end{pmatrix} \right\}$ and $\left\{ \begin{pmatrix} a \\ b \end{pmatrix}, \begin{pmatrix} -b \\ a \end{pmatrix} \right\}$ form orthonormal bases in $\mathbb{R}^2$.

32. Show that *any* orthonormal basis in $\mathbb{R}^2$ has one of the forms of the bases in Problem 31.

33. Using the Cauchy-Schwarz inequality, prove that if $|\mathbf{u} + \mathbf{v}| = |\mathbf{u}| + |\mathbf{v}|$, then $\mathbf{u}$ and $\mathbf{v}$ are linearly dependent.

34. Using the Cauchy-Schwarz inequality, prove the **triangle inequality:**

$$|\mathbf{u} + \mathbf{v}| \leq |\mathbf{u}| + |\mathbf{v}|.$$

[*Hint:* Expand $|\mathbf{u} + \mathbf{v}|^2$.]

***35.** Suppose that $\mathbf{x}_1, \mathbf{x}_2, \ldots, \mathbf{x}_k$ are vectors in $\mathbb{R}^n$ (not all zero) and

$$|\mathbf{x}_1 + \mathbf{x}_2 + \cdots + \mathbf{x}_k| = |\mathbf{x}_1| + |\mathbf{x}_2| + \cdots + |\mathbf{x}_k|$$

Show that $\dim \text{span}\,\{\mathbf{x}_1, \mathbf{x}_2, \ldots, \mathbf{x}_k\} = 1$. [*Hint:* Use the results of Problems 33 and 34.]

36. Let $\{\mathbf{u}_1, \mathbf{u}_2, \ldots, \mathbf{u}_n\}$ be an orthonormal basis in $\mathbb{R}^n$ and let $\mathbf{v}$ be a vector in $\mathbb{R}^n$. Prove that $|\mathbf{v}|^2 = |\mathbf{v} \cdot \mathbf{u}_1|^2 + |\mathbf{v} \cdot \mathbf{u}_2|^2 + \cdots + |\mathbf{v} \cdot \mathbf{u}_n|^2$. This equality is called **Parseval's equality** in $\mathbb{R}^n$.

37. Show that for any subspace H of $\mathbb{R}^n$, $(H^\perp)^\perp = H$.

38. Let H_1 and H_2 be two subspaces of $\mathbb{R}^n$ and suppose that $H_1^\perp = H_2^\perp$. Show that $H_1 = H_2$.

39. If H_1 and H_2 are subspaces of $\mathbb{R}^n$, show that if $H_1 \subset H_2$, then $H_2^\perp \subset H_1^\perp$.

40. Prove the **generalized Pythagorean theorem:** Let $\mathbf{u}$ and $\mathbf{v}$ be vectors in $\mathbb{R}^n$ with $\mathbf{u} \perp \mathbf{v}$. Then

$$|\mathbf{u} + \mathbf{v}|^2 = |\mathbf{u}|^2 + |\mathbf{v}|^2$$

ANSWERS TO SELF-QUIZ

I. False **II.** True **III.** True **IV.** False **V.** True **VI.** c

8.8 LEAST SQUARES APPROXIMATION

In many problems in the biological, physical, and social sciences it is useful to describe the relationship among the variables of the problem by means of a mathematical expression. Thus, for example, we may describe the relationship among cost, revenue, and profit by means of the simple formula

$$P = R - C$$

In a different vein, we may represent the relationship among the acceleration due to gravity, the time an object has been falling, and the height of the object by the physical law

$$s = s_0 - v_0 t - \tfrac{1}{2}gt^2$$

where s_0 is the initial height of the object and v_0 is its initial velocity.

Unfortunately, formulas like the ones above do not come easily. It is usually the task of the scientist or economist to sort through large amounts of data in order to find relationships among the variables in the problem. A common way to do this is to fit a curve among the various data points. This curve may be a straight line or a quadratic or a cubic, etc. The object is to find the curve of the given type that "best" fits the given data. In this section we show how to do this when there are two variables in the problem. In every case we assume that there are n data points $(x_1, y_1), (x_2, y_2), \ldots, (x_n, y_n)$.

In Figure 1 we can indicate three of the curves that can be used to fit data.

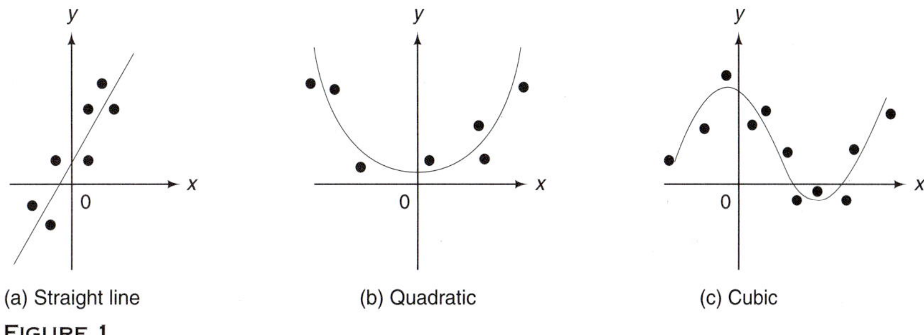

FIGURE 1
Three curves in the xy-plane

STRAIGHT-LINE APPROXIMATION

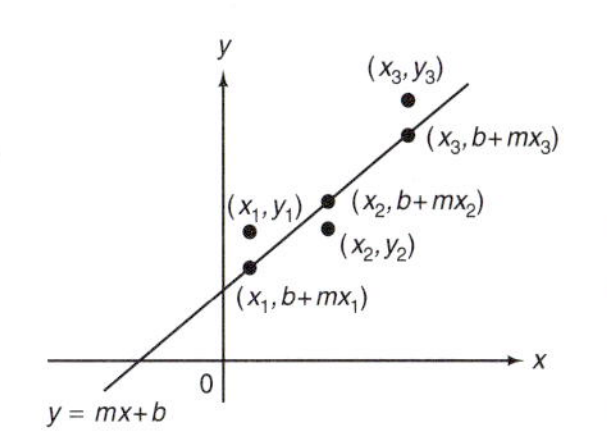

FIGURE 2
Points on the straight line have coordinates $(x, b + mx)$

Before continuing, we must be clear as to what we mean by the "best fit." Suppose we seek a straight line of the form $y = b + mx$ that best represents the n data points (x_1, y_1), $(x_2, y_2), \ldots, (x_n, y_n)$.

Figure 2 illustrates what is going on (using three data points). From the figure we see that if we assume that the x- and y-variables are related by the formula $y = b + mx$, then, for example, for $x = x_1$, the corresponding y-value is $b + mx_1$. This is different from the "true" y-value $y = y_1$.

In $\mathbb{R}^2$ the distance between the points (a_1, b_1) and (a_2, b_2) is given by $d = \sqrt{(a_1 - a_2)^2 + (b_1 - b_2)^2}$. Therefore, in determining how to choose the line $y = b + mx$ that best approximates the given data, it is reasonable to use the criterion of choosing the line that minimizes the sum of the squares of the distances between the points and the line. Note that since the distance between (x_1, y_1) and $(x_1, b + mx_1)$ is $y_1 - (b + mx_1)$, our problem (for n data points) can be stated as follows:

THE LEAST SQUARES PROBLEM FOR A LINE

Find numbers m and b such that the sum

$$[y_1 - (b + mx_1)]^2 + [y_2 - (b + mx_2)]^2 + \cdots + [y_n - (b + mx_n)]^2 \tag{1}$$

is a minimum. For this choice of m and b, the line $y = mx + b$ is called the **least squares straight-line approximation to the data points** $(x_1, y_1), (x_2, y_2), \ldots, (x_n, y_n)$.

Having defined the problem, we now seek a method for finding the least squares approximation. This is most easily done by writing everything in matrix form. If the points $(x_1, y_1), (x_2, y_2), \ldots, (x_n, y_n)$ all lie on the line $y = b + mx$ (that is, if they are collinear), then we have

$$\begin{aligned} y_1 &= b + mx_1 \\ y_2 &= b + mx_2 \\ &\vdots \\ y_n &= b + mx_n \end{aligned}$$

or

$$\mathbf{y} = A\mathbf{u} \tag{2}$$

where

$$\mathbf{y} = \begin{pmatrix} y_1 \\ y_2 \\ \vdots \\ y_n \end{pmatrix}, \quad A = \begin{pmatrix} 1 & x_1 \\ 1 & x_2 \\ \vdots & \vdots \\ 1 & x_n \end{pmatrix}, \quad \text{and} \quad \mathbf{u} = \begin{pmatrix} b \\ m \end{pmatrix} \tag{3}$$

If the points are not collinear, then $\mathbf{y} - A\mathbf{u} \neq \mathbf{0}$ and the problem becomes

VECTOR FORM OF THE LEAST SQUARES PROBLEM

Find a vector $\mathbf{u}$ such that the Euclidean norm

$$|\mathbf{y} - A\mathbf{u}| \tag{4}$$

is a minimum.

Note that in $\mathbb{R}^2$, $|(x, y)| = \sqrt{x^2 + y^2}$; in $\mathbb{R}^3$, $|(x, y, z)| = \sqrt{x^2 + y^2 + z^2}$, etc. Thus minimizing (4) is equivalent to minimizing the sum of the squares in (1).

Finding the minimizing vector $\mathbf{u}$ is not so difficult as it seems. Since A is an $n \times 2$ matrix and $\mathbf{u}$ is a 2×1 matrix, the vector $A\mathbf{u}$ is a vector in $\mathbb{R}^n$ and belongs to the range of A. The range of A is a subspace of $\mathbb{R}^n$ of dimension at most two (since at most two of the columns of A are linearly independent). Thus, by the norm approximation theorem in $\mathbb{R}^n$ (Theorem 8 on page 547), (4) is a minimum when

$$A\mathbf{u} = \text{Proj}_H \mathbf{y}$$

where H is the range of A. We illustrate this graphically in the case $n = 3$.

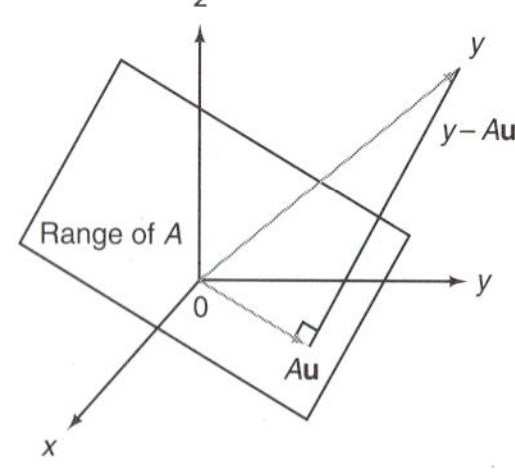

FIGURE 3
$\mathbf{y} - A\mathbf{u}$ is orthogonal to $A\mathbf{u}$

In $\mathbb{R}^3$ the range of A will be a plane or a line passing through the origin (since these are the only subspaces of $\mathbb{R}^3$ of dimension one or two). Look at Figure 3. We denote the minimizing vector $\overline{\mathbf{u}}$. It follows from the figure (and the Pythagorean theorem) that

$|\mathbf{y} - A\mathbf{u}|$ is minimized when $\mathbf{y} - A\mathbf{u}$ is orthogonal to the range of A. That is, if $\overline{\mathbf{u}}$ is the minimizing vector, then, for every vector $\mathbf{u} \in \mathbb{R}^2$,

$$A\mathbf{u} \perp (\mathbf{y} - A\overline{\mathbf{u}}) \tag{5}$$

Using the definition of the scalar product in $\mathbb{R}^n$, we find that (5) becomes

$$A\mathbf{u} \cdot (\mathbf{y} - A\overline{\mathbf{u}}) = 0$$

$$(A\mathbf{u})^t(\mathbf{y} - A\overline{\mathbf{u}}) = 0 \qquad \text{formula (6) on page 434}$$

$$(\mathbf{u}^t A^t)(\mathbf{y} - A\overline{\mathbf{u}}) = 0 \qquad \text{Theorem 1 (ii) on page 433}$$

or

$$\mathbf{u}^t(A^t\mathbf{y} - A^tA\overline{\mathbf{u}}) = 0 \tag{6}$$

Equation (6) can hold for every $\mathbf{u} \in \mathbb{R}^2$ only if

$$A^t\mathbf{y} - A^tA\overline{\mathbf{u}} = 0 \tag{7}$$

Solving (7) for $\overline{\mathbf{u}}$, we obtain

SOLUTION TO THE LEAST SQUARE PROBLEM FOR A STRAIGHT-LINE FIT

If A and $\mathbf{y}$ are as in (3), then the line $y = mx + b$ gives the best straight-line fit (in the least squares sense) to the data points $(x_1, y_1), (x_2, y_2), \ldots, (x_n, y_n)$ when $\begin{pmatrix} b \\ m \end{pmatrix} = \overline{\mathbf{u}}$ and

$$\overline{\mathbf{u}} = (A^tA)^{-1}A^ty \tag{8}$$

Here we have assumed that A^tA is invertible. This is always the case when the n data points are not collinear. The proof of this fact is left to the end of the section.

EXAMPLE 1 FINDING THE BEST STRAIGHT-LINE FIT TO FOUR POINTS

Find the best straight-line fit to the data points (1, 4), (−2, 5), (3, −1), and (4, 1).

SOLUTION: Here

$$A = \begin{pmatrix} 1 & 1 \\ 1 & -2 \\ 1 & 3 \\ 1 & 4 \end{pmatrix}, \qquad A^t = \begin{pmatrix} 1 & 1 & 1 & 1 \\ 1 & -2 & 3 & 4 \end{pmatrix} \qquad \text{and} \qquad \mathbf{y} = \begin{pmatrix} 4 \\ 5 \\ -1 \\ 1 \end{pmatrix}$$

Then

$$A^tA = \begin{pmatrix} 4 & 6 \\ 6 & 30 \end{pmatrix}, \qquad (A^tA)^{-1} = \tfrac{1}{84}\begin{pmatrix} 30 & -6 \\ -6 & 4 \end{pmatrix} \qquad \text{and}$$

$$\overline{\mathbf{u}} = (A^tA)^{-1}A^ty = \tfrac{1}{84}\begin{pmatrix} 30 & -6 \\ -6 & 4 \end{pmatrix}\begin{pmatrix} 1 & 1 & 1 & 1 \\ 1 & -2 & 3 & 4 \end{pmatrix}\begin{pmatrix} 4 \\ 5 \\ -1 \\ 1 \end{pmatrix}$$

$$= \tfrac{1}{84}\begin{pmatrix} 30 & -6 \\ -6 & 4 \end{pmatrix}\begin{pmatrix} 9 \\ -5 \end{pmatrix} = \tfrac{1}{84}\begin{pmatrix} 300 \\ -74 \end{pmatrix} \approx \begin{pmatrix} 3.57 \\ -0.88 \end{pmatrix}$$

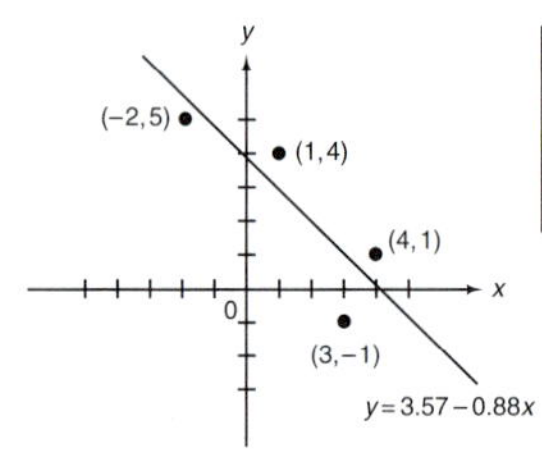

FIGURE 4
The best straight-line fit to the four points is $y = 3.57 - 0.88x$

Therefore, the best straight-line fit is given by

$$y = 3.57 - 0.88x$$

This line and the four data points are sketched in Figure 4.

QUADRATIC APPROXIMATION

Here we wish to fit a quadratic to our n data points. Recall that a quadratic in x is any expression of the form

$$y = a + bx + cx^2 \tag{9}$$

Equation (9) is the equation of a parabola in the plane. If the n data points were on the parabola, we would have

$$\begin{aligned} y_1 &= a + bx_1 + cx_1^2 \\ y_2 &= a + bx_2 + cx_2^2 \\ &\vdots \\ y_n &= a + bx_n + cx_n^2 \end{aligned} \tag{10}$$

For

$$\mathbf{y} = \begin{pmatrix} y_1 \\ y_2 \\ \vdots \\ y_n \end{pmatrix}, \quad A = \begin{pmatrix} 1 & x_1 & x_1^2 \\ 1 & x_2 & x_2^2 \\ \vdots & \vdots & \vdots \\ 1 & x_n & x_n^2 \end{pmatrix} \quad \text{and} \quad \mathbf{u} = \begin{pmatrix} a \\ b \\ c \end{pmatrix} \tag{11}$$

(10) can be rewritten as

$$\mathbf{y} = A\mathbf{u}$$

as before. If the data points do not all lie on the same parabola, then $\mathbf{y} = A\mathbf{u} \neq \mathbf{0}$ for any vector $\mathbf{u}$, and our problem is, again,

> Find a vector $\mathbf{u}$ in $\mathbb{R}^3$ such that $|\mathbf{y} - A\mathbf{u}|$ is a minimum.

Using reasoning similar to that used earlier, we can show that if the data points do not all lie on one parabola, then A^tA is invertible and the minimizing vector $\overline{\mathbf{u}}$ is given by

$$\overline{\mathbf{u}} = (A^tA)^{-1}A^t\mathbf{y} \tag{12}$$

EXAMPLE 2 FINDING THE BEST QUADRATIC FIT TO FOUR POINTS

Find the best quadratic fit to the data points of Example 1.

SOLUTION: Here

$$A = \begin{pmatrix} 1 & 1 & 1 \\ 1 & -2 & 4 \\ 1 & 3 & 9 \\ 1 & 4 & 16 \end{pmatrix}, \quad A^t = \begin{pmatrix} 1 & 1 & 1 & 1 \\ 1 & -2 & 3 & 4 \\ 1 & 4 & 9 & 16 \end{pmatrix} \quad \text{and} \quad \mathbf{y} = \begin{pmatrix} 4 \\ 5 \\ -1 \\ 1 \end{pmatrix}$$

Then

$$A^tA = \begin{pmatrix} 4 & 6 & 30 \\ 6 & 30 & 84 \\ 30 & 84 & 354 \end{pmatrix}, \qquad (A^tA)^{-1} = \tfrac{1}{4752}\begin{pmatrix} 3564 & 396 & -396 \\ 396 & 516 & -156 \\ -396 & -156 & 84 \end{pmatrix}$$

and

$$\bar{\mathbf{u}} = (A^tA)^{-1}A^t\mathbf{y} = \tfrac{1}{4752}\begin{pmatrix} 3565 & 396 & -396 \\ 396 & 516 & -156 \\ -396 & -156 & 84 \end{pmatrix}\begin{pmatrix} 1 & 1 & 1 & 1 \\ 1 & -2 & 3 & 4 \\ 1 & 4 & 9 & 16 \end{pmatrix}\begin{pmatrix} 4 \\ 5 \\ -1 \\ 1 \end{pmatrix}.$$

$$= \tfrac{1}{4752}\begin{pmatrix} 3564 & 396 & -396 \\ 396 & 516 & -156 \\ -396 & -156 & 84 \end{pmatrix}\begin{pmatrix} 9 \\ -5 \\ 31 \end{pmatrix} = \tfrac{1}{4752}\begin{pmatrix} 17820 \\ -3852 \\ -180 \end{pmatrix} \approx \begin{pmatrix} 3.75 \\ -0.81 \\ -0.04 \end{pmatrix}$$

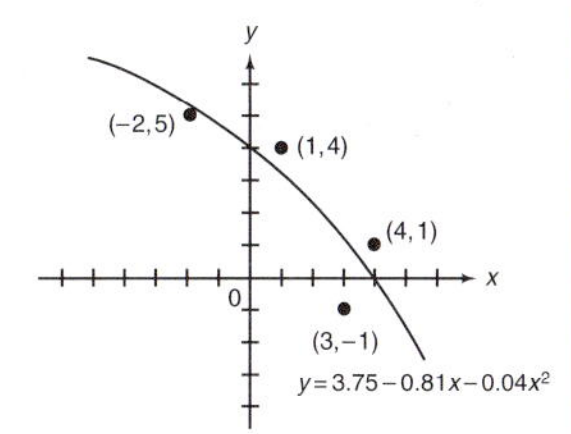

FIGURE 5
The quadratic $y = 3.75 - 0.81x - 0.04x^2$ is the best quadratic fit to the four points

Thus the best quadratic fit to the data is given by the parabola

$$\mathbf{y} = 3.75 - 0.81x - 0.04x^2$$

The parabola and data points are sketched in Figure 5.

EXAMPLE 3

FINDING THE BEST QUADRATIC FIT TO FIVE DATA POINTS CAN PROVIDE AN ESTIMATE FOR g

The method of curve-fitting can be used to measure physical constants. Suppose, for example, that an object is dropped from a height of 200 meters. The following measurements are taken:

Elapsed Time	Height (in meters)
0	200
1	195
2	180
4	120
6	25

If an object at an initial height of 200 meters is dropped from rest, then its height after t seconds is given by

$$s = 200 - \tfrac{1}{2}gt^2.$$

To estimate g, we can fit a quadratic to the five data points given above. The coefficients of the t^2 term will, if our measurements are accurate, be a reasonable approximation to the number $-\frac{1}{2}g$. Using the earlier notation, we have

$$A = \begin{pmatrix} 1 & 0 & 0 \\ 1 & 1 & 1 \\ 1 & 2 & 4 \\ 1 & 4 & 16 \\ 1 & 6 & 36 \end{pmatrix}, \qquad A^t = \begin{pmatrix} 1 & 1 & 1 & 1 & 1 \\ 0 & 1 & 2 & 4 & 6 \\ 0 & 1 & 4 & 16 & 36 \end{pmatrix} \qquad \text{and} \qquad \mathbf{y} = \begin{pmatrix} 200 \\ 195 \\ 180 \\ 120 \\ 25 \end{pmatrix}$$

Then

$$A^tA = \begin{pmatrix} 5 & 13 & 57 \\ 13 & 57 & 289 \\ 57 & 289 & 1569 \end{pmatrix}, \qquad (A^tA)^{-1} = \tfrac{1}{7504}\begin{pmatrix} 5912 & -3924 & 508 \\ -3924 & 4596 & -704 \\ 508 & -704 & 116 \end{pmatrix}$$

and

$$\bar{\mathbf{u}} = \frac{1}{7504}\begin{pmatrix} 5912 & -3924 & 508 \\ -3924 & 4596 & -704 \\ 508 & -704 & 116 \end{pmatrix}\begin{pmatrix} 1 & 1 & 1 & 1 & 1 \\ 0 & 1 & 2 & 4 & 6 \\ 0 & 1 & 4 & 16 & 36 \end{pmatrix}\begin{pmatrix} 200 \\ 195 \\ 180 \\ 120 \\ 25 \end{pmatrix}$$

$$= \frac{1}{7504}\begin{pmatrix} 5912 & -3924 & 508 \\ -3924 & 4596 & -704 \\ 508 & -704 & 116 \end{pmatrix}\begin{pmatrix} 720 \\ 1185 \\ 3735 \end{pmatrix} = \frac{1}{7504}\begin{pmatrix} 1504080 \\ -8460 \\ -35220 \end{pmatrix} \approx \begin{pmatrix} 200.44 \\ -1.13 \\ -4.69 \end{pmatrix}$$

Thus the data points are fitted by the quadratic

$$s(t) = 200.44 - 1.13t - 4.69t^2$$

and we have $\frac{1}{2}g \approx 4.69$ or

$$g \approx 2(4.69) = 9.38 \text{ m/sec}^2$$

This is reasonably close to the correct value 9.81 m/sec^2. To obtain a more accurate approximation for g, we would need to obtain more accurate observations.

We note here that higher-order polynomial approximations are carried out in a virtually identical manner. For details, see Problems 7 and 9.

We conclude this section by proving the result which guarantees that equation (8) will always be valid except when the data points lie on the same vertical line.

THEOREM 1

Let $(x_1, y_1), (x_2, y_2), \ldots, (x_n, y_n)$ be n points in $\mathbb{R}^2$ and suppose that not all the x_i's are equal. Then if A is given as in (3), the matrix A^tA is an invertible 2×2 matrix.

NOTE: If $x_1 = x_2 = x_3 = \cdots = x_n$, then all the data points lie on the vertical line $x = x_1$, and the best linear approximation is, of course, this line.

PROOF: We have

$$A = \begin{pmatrix} 1 & x_1 \\ 1 & x_2 \\ \vdots & \vdots \\ 1 & x_n \end{pmatrix}$$

Since not all the x_i's are equal, the columns of A are linearly independent. Now

$$A^tA = \begin{pmatrix} 1 & 1 & \cdots & 1 \\ x_1 & x_2 & \cdots & x_n \end{pmatrix}\begin{pmatrix} 1 & x_1 \\ 1 & x_2 \\ \vdots & \vdots \\ 1 & x_n \end{pmatrix} = \begin{pmatrix} n & \sum_{i=1}^n x_i \\ \sum_{i=1}^n x_i & \sum_{i=1}^n x_i^2 \end{pmatrix}$$

If A^tA is not invertible, then $\det A^tA = 0$. This means that

$$n\sum_{i=1}^{n} x_i^2 = \left(\sum_{i=1}^{n} x_i\right)^2 \tag{13}$$

Let $\mathbf{u} = \begin{pmatrix} 1 \\ 1 \\ \vdots \\ 1 \end{pmatrix}$ and $\mathbf{x} = \begin{pmatrix} x_1 \\ x_2 \\ \vdots \\ x_n \end{pmatrix}$. Then

$$|\mathbf{u}|^2 = \mathbf{u} \cdot \mathbf{u} = n, \qquad |\mathbf{x}|^2 = \sum_{i=1}^{n} x_i^2, \qquad \text{and} \qquad \mathbf{u} \cdot \mathbf{x} = \sum_{i=1}^{n} x_i$$

so that equation (13) can be restated as

$$|\mathbf{u}|^2|\mathbf{x}|^2 = |\mathbf{u} \cdot \mathbf{x}|^2$$

or, taking square roots, we obtain

$$|\mathbf{u} \cdot \mathbf{x}| = |\mathbf{u}|\,|\mathbf{x}|$$

Now the Cauchy-Schwartz inequality (page 548) states that $|\mathbf{u} \cdot \mathbf{x}| \le |\mathbf{u}|\,|\mathbf{x}|$ with equality if and only if $\mathbf{x}$ is a constant multiple of $\mathbf{u}$. But $\mathbf{u}$ and $\mathbf{x}$ are the columns of A which are linearly independent, by hypothesis. This contradiction proves the theorem. ■

PROBLEMS 8.8

SELF-QUIZ

I. The least squares line for the data points (2, 1), (−1, 2), (3, −5) will minimize ________.

a. $[2 - (b + m)]^2 + [-1 - (b + 2m)]^2 + [3 - (b - 5m)]^2$

b. $[1 - (b + 2m)]^2 + [2 - (b - m)]^2 + [-5 - (b + 3m)]^2$

c. $|1 - (b + 2m)| + |2 - (b - m)| + |-5 - (b + 3m)|$

d. $[1 - (b + 2)]^2 + [2 - (b - 1)]^2 + [-5 - (b + 3)]^2$

In Problems 1–3 find the best straight-line fit to the given data points.

1. (1, 3), (−2, 4), (7, 0)

2. (−3, 7), (4, 9)

3. (1, −3), (4, 6), (−2, 5), (3, −1)

In Problems 4–6 find the best quadratic fit to the given data points.

4. (2, −5), (3, 0), (1, 1), (4, −2)

5. (−7, 3), (2, 8), (1, 5)

6. (1, −1), (3, −6), (5, 2), (−3, 1), (7, 4)

7. The general cubic is given by

$$a + bx + cx^2 + dx^3$$

Show that the best cubic approximation to n data points is given by

$$\mathbf{u} = \begin{pmatrix} a \\ b \\ c \\ d \end{pmatrix} = (A^tA)^{-1}A^t\mathbf{y}$$

where $\mathbf{y}$ is as before, and

$$A = \begin{pmatrix} 1 & x_1 & x_1^2 & x_1^3 \\ 1 & x_2 & x_2^2 & x_2^3 \\ \vdots & \vdots & \vdots & \vdots \\ 1 & x_n & x_n^2 & x_n^3 \end{pmatrix}$$

8. Find the best cubic approximation to the data points $(3, -2)$, $(0, 3)$, $(-1, 4)$, $(2, -2)$, $(1, 2)$

9. The general kth-degree polynomial is given by

$$a_0 + a_1x + a_2x^2 + \cdots + a_kx^k$$

Show that the best kth-degree fit to n data points is given by

$$\bar{\mathbf{u}} = \begin{pmatrix} a_0 \\ a_1 \\ \vdots \\ a_k \end{pmatrix} = (A^tA)^{-1}A^t\mathbf{y}$$

where

$$A = \begin{pmatrix} 1 & x_1 & x_1^2 & \cdots & x_1^k \\ 1 & x_2 & x_2^2 & \cdots & x_2^k \\ \vdots & \vdots & \vdots & \cdots & \vdots \\ 1 & x_n & x_n^2 & \cdots & x_n^k \end{pmatrix}$$

10. The points $(1, 5.52)$, $(-1, 15.52)$, $(3, 11.28)$, and $(-2, 26.43)$ all lie on a parabola.

a. Find the parabola.

b. Show that $|\mathbf{y} - A\bar{\mathbf{u}}| = 0$.

11. A manufacturer buys large quantities of a certain machine replacement part. He finds that his cost depends on the number of cases bought at the same time and that the cost per unit decreases as the number of cases bought increases. He assumes that cost is a quadratic function of volume and, from past invoices, he obtains the following table:

Number of Cases Bought	Total Cost (dollars)
10	150
30	260
50	325
100	500
175	670

Find his total cost function.

12. A person throws a ball straight into the air. Its height is given by $s(t) = s_0 + v_0t + \frac{1}{2}gt^2$. The following measurements are taken:

Elapsed Time (seconds)	Height (feet)
1	57
1.5	67
2.5	68
4	9.5

Using these data, estimate

a. The height at which the ball was released

b. Its initial velocity

c. g (in ft/sec^2)

ANSWER TO SELF-QUIZ

I. b

8.9 LINEAR TRANSFORMATIONS

In this section we discuss a special class of functions, called *linear transformations,* which occur with great frequency in linear algebra and other branches of mathematics. They are also important in a wide variety of applications. Before defining a linear transformation, let us study two simple examples to see what can happen.

EXAMPLE 1 REFLECTION ABOUT THE x-AXIS

In $\mathbb{R}^2$ define a function T by the formula $T\begin{pmatrix} x \\ y \end{pmatrix} = \begin{pmatrix} x \\ -y \end{pmatrix}$. Geometrically, T takes a vector in $\mathbb{R}^2$ and reflects it about the x-axis. This is illustrated in Figure 1. Once we have given our basic definition, we shall see that T is a linear transformation from $\mathbb{R}^2$ into $\mathbb{R}^2$.

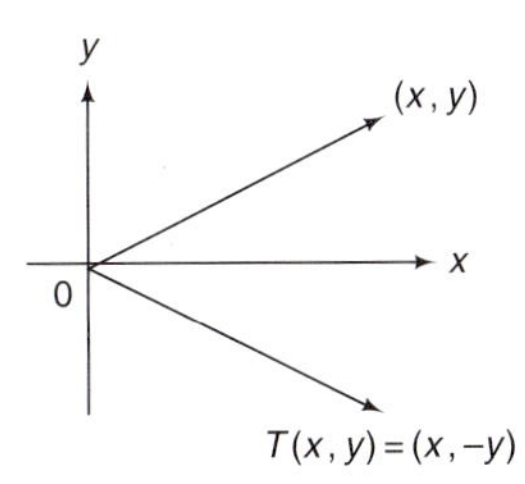

FIGURE 1
The vector $(x, -y)$ is the reflection about the x-axis of the vector (x, y)

EXAMPLE 2

TRANSFORMING A PRODUCTION VECTOR INTO A RAW MATERIAL VECTOR

A manufacturer makes four different products, each of which requires three raw materials. We denote the four products by P_1, P_2, P_3, and P_4 and denote the raw materials by R_1, R_2, and R_3. The accompanying table gives the number of units of each raw material required to manufacture 1 unit of each product.

		Needed to Produce 1 Unit of			
		P_1	P_2	P_3	P_4
Number of Units of Raw Material	R_1	2	1	3	4
	R_2	4	2	2	1
	R_3	3	3	1	2

A natural question arises: If certain numbers of the four products are produced, how many units of each raw material are needed? We let p_1, p_2, p_3, and p_4 denote the number of items of the four products manufactured and let r_1, r_2, and r_3 denote the number of units of the three raw materials needed. Then we define

$$\mathbf{p} = \begin{pmatrix} p_1 \\ p_2 \\ p_3 \\ p_4 \end{pmatrix} \qquad \mathbf{r} = \begin{pmatrix} r_1 \\ r_2 \\ r_3 \end{pmatrix} \qquad A = \begin{pmatrix} 2 & 1 & 3 & 4 \\ 4 & 2 & 2 & 1 \\ 3 & 3 & 1 & 2 \end{pmatrix}$$

For example, suppose that $\mathbf{p} = \begin{pmatrix} 10 \\ 30 \\ 20 \\ 50 \end{pmatrix}$. How many units of R_1 are needed to produce these numbers of units of the four products? From the table, we find that

$$\begin{aligned} r_1 &= p_1 \cdot 2 + p_2 \cdot 1 + p_3 \cdot 3 + p_4 \cdot 4 \\ &= 10 \cdot 2 + 30 \cdot 1 + 20 \cdot 3 + 50 \cdot 4 = 310 \text{ units} \end{aligned}$$

Similarly,

$$r_2 = 10 \cdot 4 + 30 \cdot 2 + 20 \cdot 2 + 50 \cdot 1 = 190 \text{ units}$$

and

$$r_3 = 10 \cdot 3 + 30 \cdot 3 + 20 \cdot 1 + 50 \cdot 2 = 240 \text{ units}$$

In general, we see that

$$\begin{pmatrix} 2 & 1 & 3 & 4 \\ 4 & 2 & 2 & 1 \\ 3 & 3 & 1 & 2 \end{pmatrix} \begin{pmatrix} p_1 \\ p_2 \\ p_3 \\ p_4 \end{pmatrix} = \begin{pmatrix} r_1 \\ r_2 \\ r_3 \end{pmatrix}$$

or

$$\mathbf{r} = A\mathbf{p}$$

We can look at this in another way. If $\mathbf{p}$ is called the **production vector** and $\mathbf{r}$ the **raw material vector**, we define the function T by $\mathbf{r} = T\mathbf{p} = A\mathbf{p}$. That is, T is the function that "transforms" the production vector into the raw material vector. It is defined by ordinary matrix multiplication. As we shall see, this function is also a linear transformation.

Before defining a linear transformation, let us say a bit about functions. In Section 6.7 we wrote a system of equations as

$$A\mathbf{x} = \mathbf{b}$$

where A is an $m \times n$ matrix, $\mathbf{x} \in \mathbb{R}^n$ and $\mathbf{b} \in \mathbb{R}^m$. We were asked to find $\mathbf{x}$ when A and $\mathbf{b}$ were known. However, we can look at this equation in another way: Suppose A is given. Then the equation $A\mathbf{x} = \mathbf{b}$ "says": Give me an $\mathbf{x}$ in $\mathbb{R}^n$ and I'll give you a $\mathbf{b}$ in $\mathbb{R}^m$; that is, A represents a *function* with domain $\mathbb{R}^n$ and range in $\mathbb{R}^m$.

The function defined above has the property that $A(\alpha\mathbf{x}) = \alpha A\mathbf{x}$ if α is a scalar and $A(\mathbf{x} + \mathbf{y}) = A\mathbf{x} + A\mathbf{y}$. This property characterizes linear transformations.

DEFINITION LINEAR TRANSFORMATION

Let V and W be vector spaces. A **linear transformation** T from V into W is a function that assigns to each vector $\mathbf{v} \in V$ a unique vector $T\mathbf{v} \in W$ and that satisfies, for each $\mathbf{u}$ and $\mathbf{v}$ in V and each scalar α,

$$T(\mathbf{u} + \mathbf{v}) = T\mathbf{u} + T\mathbf{v} \quad (1)$$

and

$$T(\alpha\mathbf{v}) = \alpha T\mathbf{v} \quad (2)$$

■

NOTATION: We write $T: V \to W$ to indicate that T takes the vector space V into the vector space W.

TERMINOLOGY: Linear transformations are often called **linear operators**.

EXAMPLE 3 A LINEAR TRANSFORMATION FROM $\mathbb{R}^2$ TO $\mathbb{R}^3$

Let $T: \mathbb{R}^2 \to \mathbb{R}^3$ be defined by $T\begin{pmatrix} x \\ y \end{pmatrix} = \begin{pmatrix} x+y \\ x-y \\ 3y \end{pmatrix}$. For example, $T\begin{pmatrix} 2 \\ -3 \end{pmatrix} = \begin{pmatrix} -1 \\ 5 \\ -9 \end{pmatrix}$. Then

$$T\left[\begin{pmatrix} x_1 \\ y_1 \end{pmatrix} + \begin{pmatrix} x_2 \\ y_2 \end{pmatrix}\right] = T\begin{pmatrix} x_1 + x_2 \\ y_1 + y_2 \end{pmatrix} = \begin{pmatrix} x_1 + x_2 + y_1 + y_2 \\ x_1 + x_2 - y_1 - y_2 \\ 3y_1 + 3y_2 \end{pmatrix} = \begin{pmatrix} x_1 + y_1 \\ x_1 - y_1 \\ 3y_1 \end{pmatrix} + \begin{pmatrix} x_2 + y_2 \\ x_2 - y_2 \\ 3y_2 \end{pmatrix}$$

But

$$\begin{pmatrix} x_1 + y_1 \\ x_1 - y_1 \\ 3y_1 \end{pmatrix} = T\begin{pmatrix} x_1 \\ y_1 \end{pmatrix} \quad \text{and} \quad \begin{pmatrix} x_2 + y_2 \\ x_2 - y_2 \\ 3y_2 \end{pmatrix} = T\begin{pmatrix} x_2 \\ y_2 \end{pmatrix}$$

Thus

$$T\left[\begin{pmatrix} x_1 \\ y_1 \end{pmatrix} + \begin{pmatrix} x_2 \\ y_2 \end{pmatrix}\right] = T\begin{pmatrix} x_1 \\ y_1 \end{pmatrix} + T\begin{pmatrix} x_2 \\ y_2 \end{pmatrix}$$

Similarly,

$$T\left[\alpha\begin{pmatrix} x \\ y \end{pmatrix}\right] = T\begin{pmatrix} \alpha x \\ \alpha y \end{pmatrix} = \begin{pmatrix} \alpha x + \alpha y \\ \alpha x - \alpha y \\ 3\alpha y \end{pmatrix} = \alpha\begin{pmatrix} x + y \\ x - y \\ 3y \end{pmatrix} = \alpha T\begin{pmatrix} x \\ y \end{pmatrix}$$

Thus T is a linear transformation.

EXAMPLE 4 THE ZERO TRANSFORMATION

Let V and W be vector spaces and define $T: V \to W$ by $T\mathbf{v} = \mathbf{0}$ for every $\mathbf{v}$ in V. Then $T(\mathbf{v}_1 + \mathbf{v}_2) = \mathbf{0} = \mathbf{0} + \mathbf{0} = T\mathbf{v}_1 + T\mathbf{v}_2$ and $T(\alpha\mathbf{v}) = \mathbf{0} = \alpha\mathbf{0} = \alpha T\mathbf{v}$. Here T is called the **zero transformation**.

EXAMPLE 5 THE IDENTITY TRANSFORMATION

Let V be a vector space and define $I: V \to V$ by $I\mathbf{v} = \mathbf{v}$ for every $\mathbf{v}$ in V. Here I is obviously a linear transformation. It is called the **identity transformation** or **identity operator**.

EXAMPLE 6 A REFLECTION TRANSFORMATION

Let $T: \mathbb{R}^2 \to \mathbb{R}^2$ be defined by $T\begin{pmatrix} x \\ y \end{pmatrix} = \begin{pmatrix} -x \\ y \end{pmatrix}$. It is easy to verify that T is linear. Geometrically, T takes a vector in $\mathbb{R}^2$ and reflects it about the y-axis (see Figure 2).

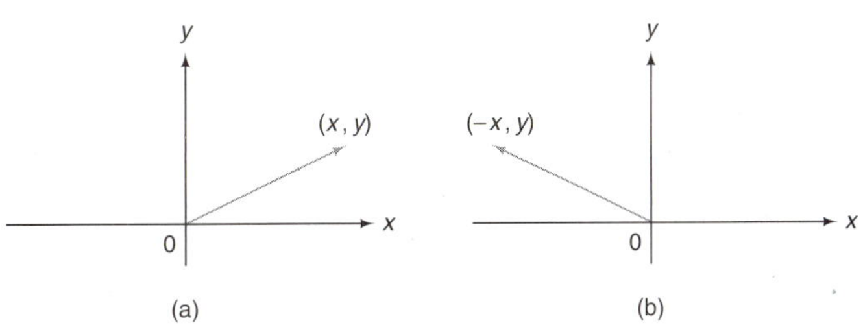

FIGURE 2
The vector $(-x, y)$ is the reflection about the y-axis of the vector (x, y)

EXAMPLE 7

A TRANSFORMATION FROM $\mathbb{R}^n \to \mathbb{R}^m$ GIVEN BY MULTIPLICATION BY AN $m \times n$ MATRIX

Let A be an $m \times n$ matrix and define $T: \mathbb{R}^n \to \mathbb{R}^m$ by $T\mathbf{x} = A\mathbf{x}$. Since $A(\mathbf{x} + \mathbf{y}) = A\mathbf{x} + A\mathbf{y}$ and $A(\alpha\mathbf{x}) = \alpha A\mathbf{x}$ if $\mathbf{x}$ and $\mathbf{y}$ are in $\mathbb{R}^n$, we see that T is a linear transformation. Thus: *Every $m \times n$ matrix A gives rise to a linear transformation from $\mathbb{R}^n$ into $\mathbb{R}^m$.* In Section 8.11 we shall see that a certain converse is true: *Every linear transformation between finite dimensional vector spaces can be represented by a matrix.*

EXAMPLE 8

A ROTATION TRANSFORMATION

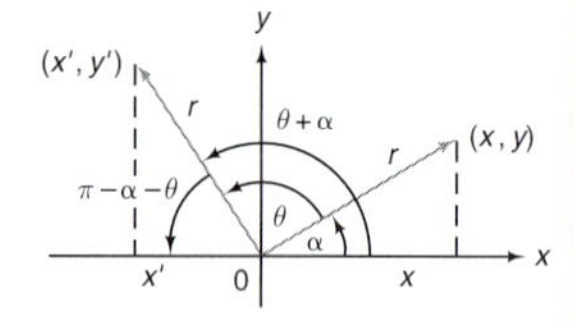

FIGURE 3

(x', y') is obtained by rotating (x, y) through the angle θ

Suppose the vector $\mathbf{v} = \begin{pmatrix} x \\ y \end{pmatrix}$ in the xy-plane is rotated through an angle of θ (measured in degrees or radians) in the counterclockwise direction. Call the new rotated vector $\mathbf{v}' = \begin{pmatrix} x' \\ y' \end{pmatrix}$. Then, as in Figure 3, if r denotes the length of $\mathbf{v}$ (which is unchanged by rotation),

$$x = r\cos\alpha \qquad y = r\sin\alpha$$
$$x' = r\cos(\theta + \alpha) \qquad y' = r\sin(\theta + \alpha)^\dagger$$

But $r\cos(\theta + \alpha) = r\cos\theta\cos\alpha - r\sin\theta\sin\alpha$, so that

$$x' = x\cos\theta - y\sin\theta \tag{3}$$

Similarly, $r\sin(\theta + \alpha) = r\sin\theta\cos\alpha + r\cos\theta\sin\alpha$ or

$$y' = x\sin\theta + y\cos\theta \tag{4}$$

Let

$$A_\theta = \begin{pmatrix} \cos\theta & -\sin\theta \\ \sin\theta & \cos\theta \end{pmatrix} \tag{5}$$

Then, from (3) and (4), we see that $A_\theta \begin{pmatrix} x \\ y \end{pmatrix} = \begin{pmatrix} x' \\ y' \end{pmatrix}$. The linear transformation $T: \mathbb{R}^2 \to \mathbb{R}^2$ defined by $T\mathbf{v} = A_\theta \mathbf{v}$, where A_θ is given by (5), is called a **rotation transformation**.

† These follow from the standard definitions of $\cos\theta$ and $\sin\theta$ as the x- and y-coordinates of a point on the unit circle. If (x, y) is a point on the circle centered at the origin of radius r, then $x = r\cos\varphi$ and $y = r\sin\varphi$, where φ is the angle the vector (x, y) makes with the positive x-axis.

EXAMPLE 9

TWO PROJECTION OPERATORS

Let $T: \mathbb{R}^3 \to \mathbb{R}^3$ be defined by $T\begin{pmatrix} x \\ y \\ z \end{pmatrix} = \begin{pmatrix} x \\ y \\ 0 \end{pmatrix}$. Then T is the projection operator taking a vector in space and projecting it into the xy-plane. Similarly, $T\begin{pmatrix} x \\ y \\ z \end{pmatrix} = \begin{pmatrix} x \\ 0 \\ z \end{pmatrix}$ projects a vector in space into the xz-plane. These two transformations are depicted in Figure 4.

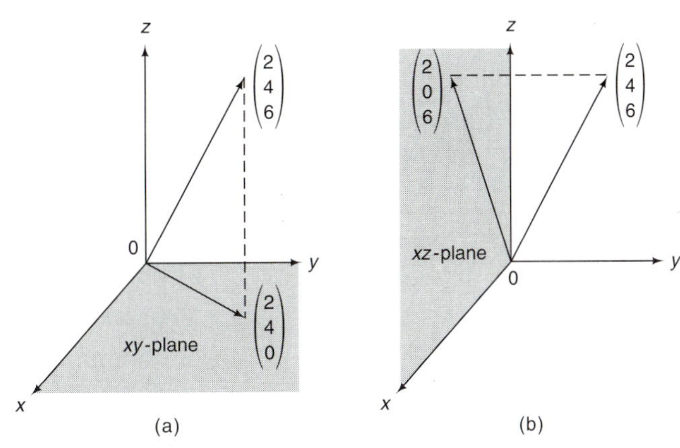

FIGURE 4

(a) Projection onto xy-plane: $T\begin{pmatrix} x \\ y \\ z \end{pmatrix} = \begin{pmatrix} x \\ y \\ 0 \end{pmatrix}$. (b) Projection onto xz-plane: $T\begin{pmatrix} x \\ y \\ z \end{pmatrix} = \begin{pmatrix} x \\ 0 \\ z \end{pmatrix}$

EXAMPLE 10

A TRANSPOSE OPERATOR

Define $T: M_{mn} \to M_{nm}$ by $T(A) = A^t$. Since $(A + B)^t = A^t + B^t$ and $(\alpha A)^t = \alpha A^t$, we see that T, called the **transpose operator**, is a linear transformation.

EXAMPLE 11

AN INTEGRAL OPERATOR

Let J: $C[0, 1] \to \mathbb{R}$ be defined by $Jf = \int_0^1 f(x)\, dx$. Since $\int_0^1 [f(x) + g(x)]\, dx = \int_0^1 f(x)\, dx + \int_0^1 g(x)\, dx$ and $\int_0^1 \alpha f(x)\, dx = \alpha \int_0^1 f(x)\, dx$ if f and g are continuous, we see that J is linear. For example, $J(x^3) = \frac{1}{4}$. J is called an **integral operator**.

EXAMPLE 12

A DIFFERENTIAL OPERATOR

Let D: $C^1[0, 1] \to C[0, 1]$ be defined by $Df = f'$. Since $(f + g)' = f' + g'$ and $(\alpha f)' = \alpha f'$ if f and g are differentiable, we see that D is linear. D is called a **differential operator**.

WARNING: Not every transformation that looks linear actually is linear. For example, define $T: \mathbb{R} \to \mathbb{R}$ by $Tx = 2x + 3$. Then $\{(x, Tx): x \in \mathbb{R}\}$ is a straight line in the xy-plane. But T is not linear since $T(x + y) = 2(x + y) + 3 = 2x + 2y + 3$ and $Tx + Ty = (2x + 3) + (2y + 3) = 2x + 2y + 6$. The only linear transformations from $\mathbb{R}$ to $\mathbb{R}$ are functions of the form $f(x) = mx$ for some real number m. Thus among all straight-line functions, the only ones that are linear are those that pass through the origin. In algebra and calculus, a **linear function** with domain $\mathbb{R}$ is defined as a function having the form $f(x) = mx + b$. Thus we can say that a linear function is a linear transformation from $\mathbb{R}$ to $\mathbb{R}$ if and only if b (the y-intercept) is zero.

EXAMPLE 13 A TRANSFORMATION THAT IS NOT LINEAR

Let $T: C[0, 1] \to \mathbb{R}$ be defined by $Tf = f(0) + 1$. Then T is not linear. To see this, we compute

$$T[f + g] = (f + g)(0) + 1 = f(0) + g(0) + 1$$
$$Tf + Tg = [f(0) + 1] + [g(0) + 1] = f(0) + g(0) + 2$$

This provides another example of a transformation that might look linear but in fact is not.

PROBLEMS 8.9

SELF-QUIZ

True-False

I. If T is a linear transformation, then $T(3\mathbf{x}) = 3T\mathbf{x}$.

II. If T is a linear transformation, then $T(\mathbf{x} + \mathbf{y}) = T\mathbf{x} + T\mathbf{y}$.

III. If T is a linear transformation, then $T(\mathbf{xy}) = T\mathbf{x}\, T\mathbf{y}$.

IV. If A is a 4×5 matrix, then $T\mathbf{x} = A\mathbf{x}$ is a linear transformation from $\mathbb{R}^4$ to $\mathbb{R}^5$.

V. If A is a 4×5 matrix, then $T\mathbf{x} = A\mathbf{x}$ is a linear transformation from $\mathbb{R}^5$ to $\mathbb{R}^4$.

In Problems 1–29 determine whether the given transformation from V to W is linear.

1. $T: \mathbb{R}^2 \to \mathbb{R}^2;\ T\begin{pmatrix} x \\ y \end{pmatrix} = \begin{pmatrix} x \\ 0 \end{pmatrix}$

2. $T: \mathbb{R}^2 \to \mathbb{R}^2;\ T\begin{pmatrix} x \\ y \end{pmatrix} = \begin{pmatrix} 1 \\ y \end{pmatrix}$

3. $T: \mathbb{R}^3 \to \mathbb{R}^2;\ T\begin{pmatrix} x \\ y \\ z \end{pmatrix} = \begin{pmatrix} x \\ y \end{pmatrix}$

4. $T: \mathbb{R}^3 \to \mathbb{R}^2;\ T\begin{pmatrix} x \\ y \\ z \end{pmatrix} = \begin{pmatrix} 0 \\ y \end{pmatrix}$

5. $T: \mathbb{R}^3 \to \mathbb{R}^2;\ T\begin{pmatrix} x \\ y \\ z \end{pmatrix} = \begin{pmatrix} 1 \\ z \end{pmatrix}$

6. $T: \mathbb{R}^2 \to \mathbb{R}^2;\ T\begin{pmatrix} x \\ y \end{pmatrix} = \begin{pmatrix} x^2 \\ y^2 \end{pmatrix}$

7. $T: \mathbb{R}^2 \to \mathbb{R}^2;\ T\begin{pmatrix} x \\ y \end{pmatrix} = \begin{pmatrix} y \\ x \end{pmatrix}$

8. $T: \mathbb{R}^2 \to \mathbb{R}^2;\ T\begin{pmatrix} x \\ y \end{pmatrix} = \begin{pmatrix} x + y \\ x - y \end{pmatrix}$

9. $T: \mathbb{R}^2 \to \mathbb{R};\ T\begin{pmatrix} x \\ y \end{pmatrix} = xy$

10. $T: \mathbb{R}^n \to \mathbb{R};\ T\begin{pmatrix} x_1 \\ x_2 \\ \vdots \\ x_n \end{pmatrix} = x_1 + x_2 + \cdots + x_n$

11. $T: \mathbb{R} \to \mathbb{R}^n;\ T(x) = \begin{pmatrix} x \\ x \\ \vdots \\ x \end{pmatrix}$

12. $T: \mathbb{R}^4 \to \mathbb{R}^2;\ T\begin{pmatrix} x \\ y \\ z \\ w \end{pmatrix} = \begin{pmatrix} x + z \\ y + w \end{pmatrix}$

13. $T: \mathbb{R}^4 \to \mathbb{R}^2;\ T\begin{pmatrix} x \\ y \\ z \\ w \end{pmatrix} = \begin{pmatrix} xz \\ yw \end{pmatrix}$

14. $T: M_{nn} \to M_{nn};\ T(A) = AB$, where B is a fixed $n \times n$ matrix

15. $T: M_{nn} \to M_{nn};\ T(A) = A^tA$

16. $T: M_{mn} \to M_{mp};\ T(A) = AB$, where B is a fixed $n \times p$ matrix

17. $T: D_n \to D_n;\ T(D) = D^2$ (D_n is the set of $n \times n$ diagonal matrices)

18. $T: D_n \to D_n;\ T(D) = I + D$

19. $T: P_2 \to P_1;\ T(a_0 + a_1x + a_2x^2) = a_0 + a_1x$

20. $T: P_2 \to P_1;\ T(a_0 + a_1x + a_2x^2) = a_1 + a_2x$

21. $T: \mathbb{R} \to P_n;\ T(a) = a + ax + ax^2 + \cdots + ax^n$

22. $T: P_2 \to P_4;\ T(p(x)) = [p(x)]^2$

23. $T: C[0, 1] \to C[0, 1];\ Tf(x) = f^2(x)$

24. $T: C[0, 1] \to C[0, 1];\ Tf(x) = f(x) + 1$

25. $T: C[0, 1] \to \mathbb{R};\ Tf = \int_0^1 f(x)g(x)\,dx$, where g is a fixed function in $C[0, 1]$

26. $T: C^1[0, 1] \to C[0, 1];\ Tf = (fg)'$, where g is a fixed function in $C^1[0, 1]$

27. $T: C[0, 1] \to C[1, 2];\ Tf(x) = f(x - 1)$

28. $T: C[0, 1] \to \mathbb{R};\ Tf = f(\frac{1}{2})$

29. $T: M_{nn} \to \mathbb{R};\ T(A) = \det A$

30. Let $T: \mathbb{R}^2 \to \mathbb{R}^2$ be given by $T(x, y) = (-x, -y)$. Describe T geometrically.

31. Let T be a linear transformation from $\mathbb{R}^2 \to \mathbb{R}^3$ such that $T\begin{pmatrix} 1 \\ 0 \end{pmatrix} = \begin{pmatrix} 1 \\ 2 \\ 3 \end{pmatrix}$ and $T\begin{pmatrix} 0 \\ 1 \end{pmatrix} = \begin{pmatrix} -4 \\ 0 \\ 5 \end{pmatrix}$.
Find: **(a)** $T\begin{pmatrix} 2 \\ 4 \end{pmatrix}$ and **(b)** $T\begin{pmatrix} -3 \\ 7 \end{pmatrix}$.

32. In Example 8:
a. Find the rotation matrix A_θ when $\theta = \pi/6$.
b. What happens to the vector $\begin{pmatrix} -3 \\ 4 \end{pmatrix}$ if it is rotated through an angle of $\pi/6$ in the counterclockwise direction?

33. Let $A_\theta = \begin{pmatrix} \cos\theta & -\sin\theta & 0 \\ \sin\theta & \cos\theta & 0 \\ 0 & 0 & 1 \end{pmatrix}$. Describe geometrically the linear transformation $T: \mathbb{R}^3 \to \mathbb{R}^3$ given by $T\mathbf{x} = A_\theta\mathbf{x}$.

34. Answer the questions in Problem 33 for
$$A_\theta = \begin{pmatrix} \cos\theta & 0 & -\sin\theta \\ 0 & 1 & 0 \\ \sin\theta & 0 & \cos\theta \end{pmatrix}.$$

35. Suppose that in a real vector space V, T satisfies $T(\mathbf{x} + \mathbf{y}) = T\mathbf{x} + T\mathbf{y}$ and $T(\alpha\mathbf{x}) = \alpha T\mathbf{x}$ for $\alpha \geq 0$. Show that T is linear.

36. Find a linear transformation $T: M_{33} \to M_{22}$.

37. If T is a linear transformation from V to W, show that $T(\mathbf{x} - \mathbf{y}) = T\mathbf{x} - T\mathbf{y}$.

38. If T is a linear transformation from V to W, show that $T\mathbf{0} = \mathbf{0}$. Are the two zero vectors here the same?

39. Let V and W be vector spaces. Let $L(V, W)$ denote the set of linear transformations from V to W. If T_1 and T_2 are in $L(V, W)$, define αT_1 and $T_1 + T_2$ by $(\alpha T_1)\mathbf{v} = \alpha(T_1\mathbf{v})$ and $(T_1 + T_2)\mathbf{v} = T_1\mathbf{v} + T_2\mathbf{v}$. Prove that $L(V, W)$ is a vector space.

ANSWERS TO SELF-QUIZ

I. True **II.** True **III.** False **IV.** False **V.** True

---

8.10 PROPERTIES OF LINEAR TRANSFORMATIONS: RANGE AND KERNEL

In this section we develop some of the basic properties of linear transformations.

THEOREM 1

Let $T: V \to W$ be a linear transformation. Then, for all vectors $\mathbf{u}, \mathbf{v}, \mathbf{v}_1, \mathbf{v}_2, \ldots, \mathbf{v}_n$ in V and all scalars $\alpha_1, \alpha_2, \ldots, \alpha_n$:

i. $T(\mathbf{0}) = \mathbf{0}$

ii. $T(\mathbf{u} - \mathbf{v}) = T\mathbf{u} - T\mathbf{v}$

iii. $T(\alpha_1\mathbf{v}_1 + \alpha_2\mathbf{v}_2 + \cdots + \alpha_n\mathbf{v}_n) = \alpha_1T\mathbf{v}_1 + \alpha_2T\mathbf{v}_2 + \cdots + \alpha_nT\mathbf{v}_n$

NOTE: In part (i) the $\mathbf{0}$ on the left is the zero vector in V, whereas the $\mathbf{0}$ on the right is the zero vector in W.

PROOF:

i. $T(\mathbf{0}) = T(\mathbf{0} + \mathbf{0}) = T(\mathbf{0}) + T(\mathbf{0})$. Thus

$$\mathbf{0} = T(\mathbf{0}) - T(\mathbf{0}) = T(\mathbf{0}) + T(\mathbf{0}) - T(\mathbf{0}) = T(\mathbf{0}).$$

ii. $T(\mathbf{u} - \mathbf{v}) = T[\mathbf{u} + (-1)\mathbf{v}] = T\mathbf{u} + T[(-1)\mathbf{v}] = T\mathbf{u} + (-1)T\mathbf{v} = T\mathbf{u} - T\mathbf{v}$.

iii. We prove this part by induction (see Appendix 1). For $n = 2$, we get $T(\alpha_1\mathbf{v}_1 + \alpha_2\mathbf{v}_2) = T(\alpha_1\mathbf{v}_1) + T(\alpha_2\mathbf{v}_2) = \alpha_1 T\mathbf{v}_1 + \alpha_2 T\mathbf{v}_2$. Thus the equation holds for $n = 2$. We assume that it holds for $n = k$ and prove it for $n = k + 1$: $T(\alpha_1\mathbf{v}_1 + \alpha_2\mathbf{v}_2 + \cdots + \alpha_k\mathbf{v}_k + \alpha_{k+1}\mathbf{v}_{k+1}) = T(\alpha_1\mathbf{v}_1 + \alpha_2\mathbf{v}_2 + \cdots + \alpha_k\mathbf{v}_k) + T(\alpha_{k+1}\mathbf{v}_{k+1})$, and using the equation in part (iii) for $n = k$, this is equal to $(\alpha_1 T\mathbf{v}_1 + \alpha_2 T\mathbf{v}_2 + \cdots + \alpha_k T\mathbf{v}_k) + \alpha_{k+1}T\mathbf{v}_{k+1}$, which is what we wanted to show. This completes the proof. ■

REMARK: Note that parts (i) and (ii) of Theorem 1 are special cases of part (iii).

An important fact about linear transformations is that they are completely determined by what they do to basis vectors.

THEOREM 2

Let V be a finite dimensional vector space with basis $B = \{\mathbf{v}_1, \mathbf{v}_2, \ldots, \mathbf{v}_n\}$. Let $\mathbf{w}_1, \mathbf{w}_2, \ldots, \mathbf{w}_n$ be n vectors in W. Suppose that T_1 and T_2 are two linear transformations from V to W such that $T_1\mathbf{v}_i = T_2\mathbf{v}_i = \mathbf{w}_i$ for $i = 1, 2, \ldots, n$. Then for any vector $\mathbf{v} \in V$, $T_1\mathbf{v} = T_2\mathbf{v}$. That is, $T_1 = T_2$.

PROOF: Since B is a basis for V, there exists a unique set of scalars $\alpha_1, \alpha_2, \ldots, \alpha_n$ such that $\mathbf{v} = \alpha_1\mathbf{v}_1 + \alpha_2\mathbf{v}_2 + \cdots + \alpha_n\mathbf{v}_n$. Then, from part ($iii$) of Theorem 1,

$$\begin{aligned} T_1\mathbf{v} = T_1(\alpha_1\mathbf{v}_1 + \alpha_2\mathbf{v}_2 + \cdots + \alpha_n\mathbf{v}_n) &= \alpha_1 T_1\mathbf{v}_1 + \alpha_2 T_1\mathbf{v}_2 + \cdots + \alpha_n T_1\mathbf{v}_n \\ &= \alpha_1\mathbf{w}_1 + \alpha_2\mathbf{w}_2 + \cdots + \alpha_n\mathbf{w}_n \end{aligned}$$

Similarly,

$$\begin{aligned} T_2\mathbf{v} = T_2(\alpha_1\mathbf{v}_1 + \alpha_2\mathbf{v}_2 + \cdots + \alpha_n\mathbf{v}_n) &= \alpha_1 T_2\mathbf{v}_1 + \alpha_2 T_2\mathbf{v}_2 + \cdots + \alpha_n T_2\mathbf{v}_n \\ &= \alpha_1\mathbf{w}_1 + \alpha_2\mathbf{w}_2 + \cdots + \alpha_n\mathbf{w}_n \end{aligned}$$

Thus $T_1\mathbf{v} = T_2\mathbf{v}$. ■

Theorem 2 tells us that if $T: V \to W$ and V is finite dimensional, then we only need to know what T does to basis vectors in V. This determines T completely. To see this, let $\mathbf{v}_1, \mathbf{v}_2, \ldots, \mathbf{v}_n$ be a basis in V and let $\mathbf{v}$ be another vector in V. Then, as in the proof of Theorem 2,

$$T\mathbf{v} = \alpha_1 T\mathbf{v}_1 + \alpha_2 T\mathbf{v}_2 + \cdots + \alpha_n T\mathbf{v}_n$$

Thus we can compute $T\mathbf{v}$ for any vector $\mathbf{v} \in V$ if we know $T\mathbf{v}_1, T\mathbf{v}_2, \ldots, T\mathbf{v}_n$.

EXAMPLE 1

IF YOU KNOW WHAT A LINEAR TRANSFORMATION DOES TO BASIS VECTORS, THEN YOU KNOW WHAT IT DOES TO ANY OTHER VECTOR

Let T be a linear transformation from $\mathbb{R}^3$ into $\mathbb{R}^2$ and suppose that $T\begin{pmatrix}1\\0\\0\end{pmatrix} = \begin{pmatrix}2\\3\end{pmatrix}$, $T\begin{pmatrix}0\\1\\0\end{pmatrix} = \begin{pmatrix}-1\\4\end{pmatrix}$, and $T\begin{pmatrix}0\\0\\1\end{pmatrix} = \begin{pmatrix}5\\-3\end{pmatrix}$. Compute $T\begin{pmatrix}3\\-4\\5\end{pmatrix}$.

SOLUTION: We have $\begin{pmatrix}3\\-4\\5\end{pmatrix} = 3\begin{pmatrix}1\\0\\0\end{pmatrix} - 4\begin{pmatrix}0\\1\\0\end{pmatrix} + 5\begin{pmatrix}0\\0\\1\end{pmatrix}$.

Thus

$$\begin{aligned} T\begin{pmatrix}3\\-4\\5\end{pmatrix} &= 3T\begin{pmatrix}1\\0\\0\end{pmatrix} - 4T\begin{pmatrix}0\\1\\0\end{pmatrix} + 5T\begin{pmatrix}0\\0\\1\end{pmatrix} \\ &= 3\begin{pmatrix}2\\3\end{pmatrix} - 4\begin{pmatrix}-1\\4\end{pmatrix} + 5\begin{pmatrix}5\\-3\end{pmatrix} = \begin{pmatrix}6\\9\end{pmatrix} + \begin{pmatrix}4\\-16\end{pmatrix} + \begin{pmatrix}25\\-15\end{pmatrix} = \begin{pmatrix}35\\-22\end{pmatrix} \end{aligned}$$

Another question arises: If $\mathbf{w}_1, \mathbf{w}_2, \ldots, \mathbf{w}_n$ are n vectors in W, does there exist a linear transformation T such that $T\mathbf{v}_i = \mathbf{w}_i$ for $i = 1, 2, \ldots, n$? The answer is yes, as the next theorem shows.

THEOREM 3

Let V be a finite dimensional vector space with basis $B = \{\mathbf{v}_1, \mathbf{v}_2, \ldots, \mathbf{v}_n\}$. Let W be a vector space containing the n vectors $\mathbf{w}_1, \mathbf{w}_2, \ldots, \mathbf{w}_n$. Then there exists a unique linear transformation $T: V \rightarrow W$ such that $T\mathbf{v}_i = \mathbf{w}_i$ for $i = 1, 2, \ldots, n$.

PROOF: Define a function T as follows:

i. $T\mathbf{v}_i = \mathbf{w}_i$

ii. If $\mathbf{v} = \alpha_1\mathbf{v}_1 + \alpha_2\mathbf{v}_2 + \cdots + \alpha_n\mathbf{v}_n$, then

$$T\mathbf{v} = \alpha_1\mathbf{w}_1 + \alpha_2\mathbf{w}_2 + \cdots + \alpha_n\mathbf{w}_n \tag{1}$$

Because B is a basis for V, T is defined for every $\mathbf{v} \in V$; and since W is a vector space, $T\mathbf{v} \in W$. Thus it only remains to show that T is linear. But this follows directly from equation (1). For if $\mathbf{u} = \alpha_1\mathbf{v}_1 + \alpha_2\mathbf{v}_2 + \cdots + \alpha_n\mathbf{v}_n$ and $\mathbf{v} = \beta_1\mathbf{v}_1 + \beta_2\mathbf{v}_2 + \cdots + \beta_n\mathbf{v}_n$, then

$$\begin{aligned} T(\mathbf{u} + \mathbf{v}) &= T[(\alpha_1 + \beta_1)\mathbf{v}_1 + (\alpha_2 + \beta_2)\mathbf{v}_2 + \cdots + (\alpha_n + \beta_n)\mathbf{v}_n] \\ &= (\alpha_1 + \beta_1)\mathbf{w}_1 + (\alpha_2 + \beta_2)\mathbf{w}_2 + \cdots + (\alpha_n + \beta_n)\mathbf{w}_n \\ &= (\alpha_1\mathbf{w}_1 + \alpha_2\mathbf{w}_2 + \cdots + \alpha_n\mathbf{w}_n) + (\beta_1\mathbf{w}_1 + \beta_2\mathbf{w}_2 + \cdots + \beta_n\mathbf{w}_n) \\ &= T\mathbf{u} + T\mathbf{v} \end{aligned}$$

Similarly, $T(\alpha\mathbf{v}) = \alpha T\mathbf{v}$, so T is linear. The uniqueness of T follows from Theorem 2 and the theorem is proved. ■

REMARK: In Theorems 2 and 3 the vectors $\mathbf{w}_1, \mathbf{w}_2, \ldots, \mathbf{w}_n$ need not be independent and, in fact, need not even be distinct. Moreover, we emphasize that the theorems are true if V is any finite dimensional vector space, not just $\mathbb{R}^n$. Note also that W does not have to be finite dimensional.

EXAMPLE 2 FINDING A LINEAR TRANSFORMATION FROM $\mathbb{R}^2$ INTO A SUBSPACE OF $\mathbb{R}^3$

Find a linear transformation from $\mathbb{R}^2$ into the plane

$$W = \left\{ \begin{pmatrix} x \\ y \\ z \end{pmatrix} : 2x - y + 3z = 0 \right\}$$

SOLUTION: From Example 8.5.3, page 516, we know that W is a two-dimensional subspace of $\mathbb{R}^3$ with basis vectors $\mathbf{w}_1 = \begin{pmatrix} 1 \\ 2 \\ 0 \end{pmatrix}$ and $\mathbf{w}_2 = \begin{pmatrix} 0 \\ 3 \\ 1 \end{pmatrix}$. Using the standard basis in $\mathbb{R}^2$, $\mathbf{v}_1 = \begin{pmatrix} 1 \\ 0 \end{pmatrix}$ and $\mathbf{v}_2 = \begin{pmatrix} 0 \\ 1 \end{pmatrix}$, we define the linear transformation T by $T\begin{pmatrix} 1 \\ 0 \end{pmatrix} = \begin{pmatrix} 1 \\ 2 \\ 0 \end{pmatrix}$ and $T\begin{pmatrix} 0 \\ 1 \end{pmatrix} = \begin{pmatrix} 0 \\ 3 \\ 1 \end{pmatrix}$. Then, as the discussion following Theorem 2 shows, T is completely determined. For example,

$$T\begin{pmatrix} 5 \\ -7 \end{pmatrix} = T\left[5\begin{pmatrix} 1 \\ 0 \end{pmatrix} - 7\begin{pmatrix} 0 \\ 1 \end{pmatrix}\right] = 5T\begin{pmatrix} 1 \\ 0 \end{pmatrix} - 7T\begin{pmatrix} 0 \\ 1 \end{pmatrix} = 5\begin{pmatrix} 1 \\ 2 \\ 0 \end{pmatrix} - 7\begin{pmatrix} 0 \\ 3 \\ 1 \end{pmatrix} = \begin{pmatrix} 5 \\ -11 \\ -7 \end{pmatrix}$$

We now turn to two important definitions in the theory of linear transformations.

DEFINITION KERNEL AND RANGE OF A LINEAR TRANSFORMATION

Let V and W be vector spaces and let $T: V \rightarrow W$ be a linear transformation. Then

i. The **kernel** of T, denoted by ker T, is given by

$$\ker T = \{\mathbf{v} \in V: T\mathbf{v} = \mathbf{0}\} \tag{2}$$

ii. The **range** of T, denoted by range T, is given by

$$\text{Range } T = \{\mathbf{w} \in W: \mathbf{w} = T\mathbf{v} \text{ for some } \mathbf{v} \in V\} \tag{3}$$

■

REMARK 1: Note that ker T is nonempty because, by Theorem 1, $T(\mathbf{0}) = \mathbf{0}$ so that $\mathbf{0} \in \ker T$ for any linear transformation T. We shall be interested in finding other vectors in V that get "mapped to zero." Again note that when we write $T(\mathbf{0}) = \mathbf{0}$, the $\mathbf{0}$ on the left is in V and the $\mathbf{0}$ is on the right is in W.

REMARK 2: Range T is simply the set of "images" of vectors in V under the transformation T. In fact, if $\mathbf{w} = T\mathbf{v}$, we say that $\mathbf{w}$ is the **image** of $\mathbf{v}$ under T.

Before giving examples of kernels and ranges, we prove a theorem that will be very useful.

THEOREM 4

If T: $V \to W$ is a linear transformation, then

i. ker T is a subspace of V.
ii. range T is a subspace of W.

PROOF

i. Let $\mathbf{u}$ and $\mathbf{v}$ be in ker T; then $T(\mathbf{u} + \mathbf{v}) = T\mathbf{u} + T\mathbf{v} = \mathbf{0} + \mathbf{0} = \mathbf{0}$ and $T(\alpha\mathbf{u}) = \alpha T\mathbf{u} = \alpha\mathbf{0} = \mathbf{0}$ so that $\mathbf{u} + \mathbf{v}$ and $\alpha\mathbf{u}$ are in ker T.
ii. Let $\mathbf{w}$ and $\mathbf{x}$ be in range T. Then $\mathbf{w} = T\mathbf{u}$ and $\mathbf{x} = T\mathbf{v}$ for two vectors $\mathbf{u}$ and $\mathbf{v}$ in V. This means that $T(\mathbf{u} + \mathbf{v}) = T\mathbf{u} + T\mathbf{v} = \mathbf{w} + \mathbf{x}$ and $T(\alpha\mathbf{u}) = \alpha T\mathbf{u} = \alpha\mathbf{w}$. Thus $\mathbf{w} + \mathbf{x}$ and $\alpha\mathbf{w}$ are in range T. ■

EXAMPLE 3

KERNEL AND RANGE OF THE ZERO TRANSFORMATION

Let $T\mathbf{v} = \mathbf{0}$ for every $\mathbf{v} \in V$. (T is the zero transformation.) Then ker $T = V$ and range $T = \{\mathbf{0}\}$.

EXAMPLE 4

KERNEL AND RANGE OF THE IDENTITY TRANSFORMATION

Let $T\mathbf{v} = \mathbf{v}$ for every $\mathbf{v} \in V$. (T is the identity transformation.) Then ker $T = \{\mathbf{0}\}$ and range $T = V$.

The zero and identity transformations provide two extremes. In the first, everything is in the kernel. In the second, only the zero vector is in the kernel. The cases in between are more interesting.

EXAMPLE 5

KERNEL AND RANGE OF A PROJECTION OPERATOR

Let T: $\mathbb{R}^3 \to \mathbb{R}^3$ be defined by $T\begin{pmatrix} x \\ y \\ z \end{pmatrix} = \begin{pmatrix} x \\ y \\ 0 \end{pmatrix}$. That is (see Example 8.9.9, page 563), T is the projection operator from $\mathbb{R}^3$ into the xy-plane. If $T\begin{pmatrix} x \\ y \\ z \end{pmatrix} = \begin{pmatrix} x \\ y \\ 0 \end{pmatrix} = \mathbf{0} = \begin{pmatrix} 0 \\ 0 \\ 0 \end{pmatrix}$, then $x = y = 0$. Thus ker $T = \{(x, y, z)\colon x = y = 0\}$ = the z-axis, and range $T = \{(x, y, z)\colon z = 0\}$ = the xy-plane. Note that dim ker $T = 1$ and dim range $T = 2$.

DEFINITION NULLITY AND RANK OF A LINEAR TRANSFORMATION

If T is a linear transformation from V to W, then we define

$$\textbf{Nullity of } T = \nu(T) = \dim \ker T \quad (4)$$

$$\textbf{Rank of } T = \rho(T) = \dim \text{range } T \quad (5)$$

■

REMARK: In Section 8.6 we defined the rank and nullity of a matrix. According to Example 8.9.7, every $m \times n$ matrix A gives rise to a linear transformation $T: \mathbb{R}^n \rightarrow \mathbb{R}^m$ defined by $T\mathbf{x} = A\mathbf{x}$. Evidently, $\ker T = N_A$, range T = range $A = C_A$, $\nu(T) = \nu(A)$ and $\rho(T) = \rho(A)$. Thus we see that the definitions of kernel, range, nullity, and rank of a linear transformation are generalizations of the same concepts applied to matrices.

EXAMPLE 6 THE KERNEL AND RANGE OF A TRANSPOSE OPERATOR

Let $V = M_{mn}$ and define $T: M_{mn} \rightarrow M_{nm}$ by $T(A) = A^t$ (see Example 8.9.10, page 564). If $TA = A^t = 0$, then A^t is the $n \times m$ zero matrix so that A is the $m \times n$ zero matrix. Thus $\ker T = \{\mathbf{0}\}$, and clearly range $T = M_{nm}$. This means that $\nu(T) = 0$ and $\rho(T) = nm$.

EXAMPLE 7 THE KERNEL AND RANGE OF A TRANSFORMATION FROM P_3 TO P_2

Let $T: P_3 \rightarrow P_2$ be defined by $T(p) = T(a_0 + a_1x + a_2x^2 + a_3x^3) = a_0 + a_1x + a_2x^2$. Then if $T(p) = 0$, $a_0 + a_1x + a_2x^2 = 0$ for every x, which implies that $a_0 = a_1 = a_2 = 0$. Thus $\ker T = \{p \in P_3: p(x) = a_3x^3\}$ and range $T = P_2$, $\nu(T) = 1$, and $\rho(T) = 3$.

EXAMPLE 8 THE KERNEL AND RANGE OF AN INTEGRAL OPERATOR

Let $V = C[0, 1]$ and define $J: C[0, 1] \rightarrow \mathbb{R}$ by $Jf = \int_0^1 f(x)\,dx$ (see Example 8.8.11, page 564). Then $\ker J = \{f \in C[0, 1]: \int_0^1 f(x)\,dx = 0\}$. Let α be a real number. Then the constant function $f(x) = \alpha$ for $x \in [0, 1]$ is in $C[0, 1]$ and $\int_0^1 \alpha\,dx = \alpha$. Since this is true for every real number α, we have range $J = \mathbb{R}$.

In the next section we shall see how every linear transformation from one finite dimensional vector space to another can be represented by a matrix. This will enable us to compute the kernel and range of any linear transformation between finite dimensional vector spaces by finding the kernel and range of a corresponding matrix.

PROBLEMS 8.10

SELF-QUIZ

True-False

I. Let $T: V \rightarrow W$ be a linear transformation. It is sometimes possible to find three different vectors $\mathbf{v}_1 \in V$, $\mathbf{v}_2 \in V$, and $\mathbf{w} \in W$ such that $T\mathbf{v}_1 = T\mathbf{v}_2 = \mathbf{w}$

II. If $T\mathbf{v}_1 = T\mathbf{v}_2 = \mathbf{w}$ as in Problem I, then $\mathbf{v}_1 - \mathbf{v}_2 \in \ker T$

III. If T is a linear transformation from V into W, then the range of T is W.

IV. Let $\mathbf{v}_1, \mathbf{v}_2, \ldots, \mathbf{v}_n$ be a basis for $\mathbb{R}^n$ and $\mathbf{w}_1, \mathbf{w}_2, \ldots, \mathbf{w}_n$ be a basis for P_{n-1}. Then there may be two linear transformations S and T such that $T\mathbf{v}_i = \mathbf{w}_i$ and $S\mathbf{v}_i = \mathbf{w}_i$ for $i = 1, 2, \ldots, n$.

V. If $T: \mathbb{R}^2 \to \mathbb{R}^2$ is a linear transformation and $T\begin{pmatrix} 0 \\ 0 \end{pmatrix} = \begin{pmatrix} 0 \\ 0 \end{pmatrix}$, then T is the zero transformation.

In Problems 1–10 find the kernel, range, rank, and nullity of the given linear transformation.

1. $T: \mathbb{R}^2 \to \mathbb{R}^2$; $T\begin{pmatrix} x \\ y \end{pmatrix} = \begin{pmatrix} x \\ 0 \end{pmatrix}$

2. $T: \mathbb{R}^3 \to \mathbb{R}^2$; $T\begin{pmatrix} x \\ y \\ z \end{pmatrix} = \begin{pmatrix} z \\ y \end{pmatrix}$

3. $T: \mathbb{R}^2 \to \mathbb{R}$; $T\begin{pmatrix} x \\ y \end{pmatrix} = x + y$

4. $T: \mathbb{R}^4 \to \mathbb{R}^2$; $T\begin{pmatrix} x \\ y \\ z \\ w \end{pmatrix} = \begin{pmatrix} x + z \\ y + w \end{pmatrix}$

5. $T: M_{22} \to M_{22}$: $T(A) = AB$, where $B = \begin{pmatrix} 1 & 2 \\ 0 & 1 \end{pmatrix}$

6. $T: \mathbb{R} \to P_3$: $T(a) = a + ax + ax^2 + ax^3$

***7.** $T: M_{nn} \to M_{nn}$: $T(A) = A^t + A$

8. $T: C^1[0, 1] \to C[0, 1]$; $Tf = f'$

9. $T: C[0, 1] \to \mathbb{R}$; $Tf = f(\frac{1}{2})$

10. $T: \mathbb{R}^2 \to \mathbb{R}^2$: T is a rotation through an angle of $\pi/3$

11. Let $T: V \to W$ be a linear transformation, let $\{\mathbf{v}_1, \mathbf{v}_2, \ldots, \mathbf{v}_n\}$ be a basis for V, and suppose that $T\mathbf{v}_i = \mathbf{0}$ for $i = 1, 2, \ldots, n$. Show that T is the zero transformation.

12. In Problem 11 suppose that $W = V$ and $T\mathbf{v}_i = \mathbf{v}_i$ for $i = 1, 2, \ldots, n$. Show that T is the identity operator.

13. Let $T: V \to \mathbb{R}^3$. Prove that range T is either **(a)** $\{\mathbf{0}\}$, **(b)** a line through the origin, **(c)** a plane through the origin, or **(d)** $\mathbb{R}^3$.

14. Let $T: \mathbb{R}^3 \to V$. Show that $\ker T$ is one of four spaces listed in Problem 13.

15. Find all linear transformations from $\mathbb{R}^2$ into $\mathbb{R}^2$ such that the line $y = 0$ is carried into the line $x = 0$.

16. Find all linear transformations from $\mathbb{R}^2$ into $\mathbb{R}^2$ that carry the line $y = ax$ into the line $y = bx$.

17. Find a linear transformation T from $\mathbb{R}^3 \to \mathbb{R}^3$ such that

$$\ker T = \{(x, y, z): 2x - y + z = 0\}.$$

18. Find a linear transformation T from $\mathbb{R}^3 \to \mathbb{R}^3$ such that

$$\text{range } T = \{(x, y, z): 2x - y + z = 0\}.$$

19. Let $T: M_{nn} \to M_{nn}$ be defined by $TA = A - A^t$. Show that $\ker T = \{\text{symmetric } n \times n \text{ matrices}\}$ and range of $T = \{\text{skew-symmetric } n \times n \text{ matrices}\}$.

***20.** Let $T: C^1[0, 1] \to C[0, 1]$ be defined by $Tf(x) = xf'(x)$. Find the kernel and range of T.

***21.** In Problem 8.9.39 you were asked to show that the set of linear transformations from a vector space V to a vector space W, denoted by $L(V, W)$, is a vector space. Suppose that $\dim V = n < \infty$ and $\dim W = m < \infty$. Find $\dim L(V, W)$.

22. Let H be a subspace of V where $\dim H = k$ and $\dim V = n$. Let U be the subset of $L(V, V)$ having the property that if $T \in U$, then $T\mathbf{h} = \mathbf{0}$ for every $\mathbf{h} \in H$.

a. Prove that U is a subspace of $L(V, V)$.

b. Find $\dim U$.

***23.** Let S and T be in $L(V, V)$ such that ST is the zero transformation. Prove or disprove: TS is the zero transformation.

ANSWERS TO SELF-QUIZ

I. True **II.** True **III.** False **IV.** False **V.** False

8.11 THE MATRIX REPRESENTATION OF A LINEAR TRANSFORMATION

If A is an $m \times n$ matrix and $T: \mathbb{R}^n \to \mathbb{R}^m$ is defined by $T\mathbf{x} = A\mathbf{x}$, then, as we saw in Example 8.9.7 on page 562, T is a linear transformation. We shall now see that for *every* linear transformation from $\mathbb{R}^n$ into $\mathbb{R}^m$, there exists an $m \times n$ matrix A such that $T\mathbf{x} = A\mathbf{x}$ for every $\mathbf{x} \in \mathbb{R}^n$. This fact is extremely useful. As we saw in the remark on page 570, if $T\mathbf{x} = A\mathbf{x}$, then $\ker T = N_A$ and range $T = R_A$. Moreover, $\nu(T) = \dim \ker T = \nu(A)$ and $\rho(T) = \dim \text{range } T = \rho(A)$. Thus we can determine the kernel, range, nullity, and rank of a linear transformation from $\mathbb{R}^n \to \mathbb{R}^m$ by determining the kernel and range space of a corresponding matrix. Moreover, once we know that $T\mathbf{x} = A\mathbf{x}$, we can evaluate $T\mathbf{x}$ for any $\mathbf{x}$ in $\mathbb{R}^n$ by simple matrix multiplication.

But this is not all. As we shall see, any linear transformation between finite dimensional vector spaces can be represented by a matrix.

THEOREM 1

Let $T: \mathbb{R}^n \to \mathbb{R}^m$ be a linear transformation. Then there exists a unique $m \times n$ matrix A_T such that

$$T\mathbf{x} = A_T\mathbf{x} \qquad \text{for every } \mathbf{x} \in \mathbb{R}^n \tag{1}$$

PROOF: Let $\mathbf{w}_1 = T\mathbf{e}_1$, $\mathbf{w}_2 = T\mathbf{e}_2, \ldots, \mathbf{w}_n = T\mathbf{e}_n$. Let A_T be the matrix whose columns are $\mathbf{w}_1, \mathbf{w}_2, \ldots, \mathbf{w}_n$ and let A_T also denote the transformation from $\mathbb{R}^n \to \mathbb{R}^m$ which multiplies a vector in $\mathbb{R}^n$ on the left by A_T. If

$$\mathbf{w}_i = \begin{pmatrix} a_{1i} \\ a_{2i} \\ \vdots \\ a_{mi} \end{pmatrix} \qquad \text{for } i = 1, 2, \ldots, n$$

then

$$A_T\mathbf{e}_i = \begin{pmatrix} a_{11} & a_{12} & \cdots & a_{1i} & \cdots & a_{1n} \\ a_{21} & a_{22} & \cdots & a_{2i} & \cdots & a_{2n} \\ \vdots & \vdots & & \vdots & & \vdots \\ a_{m1} & a_{m2} & \cdots & a_{mi} & \cdots & a_{mn} \end{pmatrix} \begin{pmatrix} 0 \\ 0 \\ \vdots \\ 1 \\ 0 \\ \vdots \\ 0 \end{pmatrix} = \begin{pmatrix} a_{1i} \\ a_{2i} \\ \vdots \\ a_{mi} \end{pmatrix} = \mathbf{w}_i.$$

(1 in the ith position)

Thus $A_T\mathbf{e}_i = \mathbf{w}_i$ for $i = 1, 2, \ldots, n$. By Theorem 8.10.2 on page 566, T and the transformation A_T are the same because they agree on basis vectors.

We can now show that A_T is unique. Suppose that $T\mathbf{x} = A_T\mathbf{x}$ and $T\mathbf{x} = B_T\mathbf{x}$ for every $\mathbf{x} \in \mathbb{R}^n$. Then $A_T\mathbf{x} = B_T\mathbf{x}$ or, setting $C_T = A_T - B_T$, we have $C_T\mathbf{x} = \mathbf{0}$ for every $\mathbf{x} \in \mathbb{R}^n$. In particular, $C_T\mathbf{e}_i = \mathbf{0}$ for $i = 1, 2, \ldots, n$. But, as we see from the proof of the first part of the theorem, $C_T\mathbf{e}_i$ is the ith column of C_T. Thus each of the n columns of C_T is the m-zero vector and $C_T = 0$, the $m \times n$ zero matrix. This shows that $A_T = B_T$ and the theorem is proved. ■

REMARK 1: In this theorem we assumed that every vector in $\mathbb{R}^n$ and $\mathbb{R}^m$ is written in terms of the standard basis vectors in those spaces. If we choose other bases for $\mathbb{R}^n$ and $\mathbb{R}^m$, we shall, of course, get a different matrix A_T. In this book we will limit ourselves to standard bases.

REMARK 2: The proof of the theorem shows us that A_T is easily obtained as the matrix whose columns are the vectors $T\mathbf{e}_i$.

DEFINITION TRANSFORMATION MATRIX

The matrix A_T in Theorem 1 is called the **transformation matrix** corresponding to T. ■

In Section 8.9 we defined the range, rank, kernel, and nullity of a linear transformation. In Section 8.6 we defined the range, rank, kernel, and nullity of a matrix. The proof of the following theorem follows from Theorem 1 and is left as an exercise (see Problem 29).

THEOREM 2

Let A_T be the transformation matrix corresponding to the linear transformation T. Then:

i. range $T = R_{A_T} = C_{A_T}$ **iii.** $\ker T = N_{A_T}$
ii. $\rho(T) = \rho(A_T)$ **iv.** $\nu(T) = \nu(A_T)$ ■

EXAMPLE 1

THE MATRIX REPRESENTATION OF A PROJECTION TRANSFORMATION

Find the transformation matrix A_T corresponding to the projection of a vector in $\mathbb{R}^3$ onto the xy-plane.

SOLUTION: Here $T\begin{pmatrix} x \\ y \\ z \end{pmatrix} = \begin{pmatrix} x \\ y \\ 0 \end{pmatrix}$. In particular, $T\begin{pmatrix} 1 \\ 0 \\ 0 \end{pmatrix} = \begin{pmatrix} 1 \\ 0 \\ 0 \end{pmatrix}$, $T\begin{pmatrix} 0 \\ 1 \\ 0 \end{pmatrix} = \begin{pmatrix} 0 \\ 1 \\ 0 \end{pmatrix}$, and $T\begin{pmatrix} 0 \\ 0 \\ 1 \end{pmatrix} = \begin{pmatrix} 0 \\ 0 \\ 0 \end{pmatrix}$. Thus $A_T = \begin{pmatrix} 1 & 0 & 0 \\ 0 & 1 & 0 \\ 0 & 0 & 0 \end{pmatrix}$. Note that $A_T\begin{pmatrix} x \\ y \\ z \end{pmatrix} = \begin{pmatrix} 1 & 0 & 0 \\ 0 & 1 & 0 \\ 0 & 0 & 0 \end{pmatrix}\begin{pmatrix} x \\ y \\ z \end{pmatrix} = \begin{pmatrix} x \\ y \\ 0 \end{pmatrix}$.

EXAMPLE 2

THE MATRIX REPRESENTATION OF A TRANSFORMATION FROM $\mathbb{R}^3$ TO $\mathbb{R}^4$

Let $T: \mathbb{R}^3 \to \mathbb{R}^4$ be defined by

$$T\begin{pmatrix} x \\ y \\ z \end{pmatrix} = \begin{pmatrix} x - y \\ y + z \\ 2x - y - z \\ -x + y + 2z \end{pmatrix}$$

Find A_T, $\ker T$, range T, $\nu(T)$, and $\rho(T)$.

SOLUTION:

$$T\begin{pmatrix}1\\0\\0\end{pmatrix} = \begin{pmatrix}1\\0\\2\\-1\end{pmatrix},\ T\begin{pmatrix}0\\1\\0\end{pmatrix} = \begin{pmatrix}-1\\1\\-1\\1\end{pmatrix},\ \text{and}\ T\begin{pmatrix}0\\0\\1\end{pmatrix} = \begin{pmatrix}0\\1\\-1\\2\end{pmatrix}.\ \text{Thus}$$

$$A_T = \begin{pmatrix}1 & -1 & 0\\0 & 1 & 1\\2 & -1 & -1\\-1 & 1 & 2\end{pmatrix}$$

Note (as a check) that

$$\begin{pmatrix}1 & -1 & 0\\0 & 1 & 1\\2 & -1 & -1\\-1 & 1 & 2\end{pmatrix}\begin{pmatrix}x\\y\\z\end{pmatrix} = \begin{pmatrix}x - y\\x + y\\2x - y - z\\-x + y + 2z\end{pmatrix}$$

Next we compute the kernel and range of A. Row-reducing, we obtain

$$\begin{pmatrix}1 & -1 & 0\\0 & 1 & 1\\2 & -1 & -1\\-1 & 1 & 2\end{pmatrix} \xrightarrow[R_4 \to R_4 + R_1]{R_3 \to R_3 - 2R_1} \begin{pmatrix}1 & -1 & 0\\0 & 1 & 1\\0 & 1 & -1\\0 & 0 & 2\end{pmatrix}$$

$$\xrightarrow[R_3 \to R_3 - R_2]{R_1 \to R_1 + R_2} \begin{pmatrix}1 & 0 & 1\\0 & 1 & 1\\0 & 0 & -2\\0 & 0 & 2\end{pmatrix} \xrightarrow{R_4 \to R_4 + R_3} \begin{pmatrix}1 & 0 & 1\\0 & 1 & 1\\0 & 0 & -2\\0 & 0 & 0\end{pmatrix}$$

The first three rows of the last matrix are linearly independent (because their determinant is -2) so

$$\rho(A) = 3 \text{ and } \nu(A) \overset{\text{since } \rho(A) + \nu(A) = 3}{=} 3 - 3 = 0.$$

This means that $\ker T = \{\mathbf{0}\}$, range $T = \text{span}\left\{\begin{pmatrix}1\\0\\2\\-1\end{pmatrix}, \begin{pmatrix}-1\\1\\-1\\1\end{pmatrix}, \begin{pmatrix}0\\1\\-1\\2\end{pmatrix}\right\}$, $\nu(T) = 0$, and $\rho(T) = 3$.

EXAMPLE 3

THE MATRIX REPRESENTATION OF A TRANSFORMATION FROM $\mathbb{R}^3$ TO $\mathbb{R}^3$

Let T: $\mathbb{R}^3 \to \mathbb{R}^3$ be defined by $T\begin{pmatrix}x\\y\\z\end{pmatrix} = \begin{pmatrix}2x - y + 3z\\4x - 2y + 6z\\-6x + 3y - 9z\end{pmatrix}$. Find A_T, $\ker T$, range T, $\nu(T)$, and $\rho(T)$.

SOLUTION: Since $T\begin{pmatrix}1\\0\\0\end{pmatrix} = \begin{pmatrix}2\\4\\-6\end{pmatrix}$, $T\begin{pmatrix}0\\1\\0\end{pmatrix} = \begin{pmatrix}-1\\-2\\3\end{pmatrix}$, and $T\begin{pmatrix}0\\0\\1\end{pmatrix} = \begin{pmatrix}3\\6\\-9\end{pmatrix}$, we have

$$A_T = \begin{pmatrix}2 & -1 & 3\\ 4 & -2 & 6\\ -6 & 3 & -9\end{pmatrix}$$

From Example 8.6.4 on page 528 we see that $\rho(A) \overset{\text{Theorem } 2(ii)}{=} \rho(T) = 1$ and range $T = \text{span}\left\{\begin{pmatrix}2\\4\\-6\end{pmatrix}\right\}$. Then $\nu(T) = 2$.

To find $N_A \overset{\text{Theorem } 2(iii)}{=} \ker T$, we row-reduce to solve the system $A\mathbf{x} = \mathbf{0}$:

$$\left(\begin{array}{rrr|r}2 & -1 & 3 & 0\\ 4 & -2 & 6 & 0\\ -6 & 3 & -9 & 0\end{array}\right) \xrightarrow[R_3 \to R_3 + 3R_1]{R_2 \to R_2 - 2R_1} \left(\begin{array}{rrr|r}2 & -1 & 3 & 0\\ 0 & 0 & 0 & 0\\ 0 & 0 & 0 & 0\end{array}\right).$$

This means that $\begin{pmatrix}x\\y\\z\end{pmatrix} \in N_A$ if $2x - y + 3z = 0$ or $y = 2x + 3z$. First setting $x = 1$, $z = 0$ and then $x = 0$, $z = 1$, we obtain a basis for N_A:

$$\ker T = N_A = \text{span}\left\{\begin{pmatrix}1\\2\\0\end{pmatrix}, \begin{pmatrix}0\\3\\1\end{pmatrix}\right\}.$$

EXAMPLE 4 **THE MATRIX REPRESENTATION OF A ZERO AND IDENTITY TRANSFORMATION**

It is easy to verify that if T is the zero transformation from $\mathbb{R}^n \to \mathbb{R}^m$, then A_T is the $m \times n$ zero matrix. Similarly, if T is the identity transformation from $\mathbb{R}^n \to \mathbb{R}^n$, then $A_T = I_n$.

EXAMPLE 5 **THE MATRIX REPRESENTATION OF A ROTATION TRANSFORMATION**

We saw in Example 8.9.8 on page 562 that if T is the function that rotates every vector in $\mathbb{R}^2$ through an angle of θ, then $A_T = \begin{pmatrix}\cos\theta & -\sin\theta\\ \sin\theta & \cos\theta\end{pmatrix}$.

We now generalize the notion of matrix representation to arbitrary finite dimensional vector spaces.

THEOREM 3

Let V be an n-dimensional vector space, W be an m-dimensional vector space, and $T: V \to W$ be a linear transformation. Let $B_1 = \{\mathbf{v}_1, \mathbf{v}_2, \ldots, \mathbf{v}_n\}$ be a basis for V and let $B_2 = \{\mathbf{w}_1, \mathbf{w}_2, \ldots, \mathbf{w}_m\}$ be a basis for W. Then there is a unique $m \times n$ matrix A_T

such that

$$(T\mathbf{x})_{B_2} = A_T(\mathbf{x})_{B_1} \tag{2}$$

REMARK 1: The notation in (2) is interpreted as follows: If $\mathbf{x} \in V = c_1\mathbf{v}_1 + c_2\mathbf{v}_2 + \cdots + c_n\mathbf{v}_n$, then $(\mathbf{x})_{B_1} = \begin{pmatrix} c_1 \\ c_2 \\ \vdots \\ c_n \end{pmatrix}$. Let $\mathbf{c} = \begin{pmatrix} c_1 \\ c_2 \\ \vdots \\ c_n \end{pmatrix}$. Then $A_T\mathbf{c}$ is an m-vector that we denote by $\mathbf{d} = \begin{pmatrix} d_1 \\ d_2 \\ \vdots \\ d_m \end{pmatrix}$. Equation (2) says that $(T\mathbf{x})_{B_2} = \begin{pmatrix} d_1 \\ d_2 \\ \vdots \\ d_m \end{pmatrix}$. That is,

$$T\mathbf{x} = d_1\mathbf{w}_1 + d_2\mathbf{w}_2 + \cdots + d_m\mathbf{w}_m$$

REMARK 2: As in Theorem 1, the uniqueness of A_T is relative to the bases B_1 and B_2. If we change the bases, we change A_T.

PROOF: Let $T\mathbf{v}_1 = \mathbf{y}_1, T\mathbf{v}_2 = \mathbf{y}_2, \ldots, T\mathbf{v}_n = \mathbf{y}_n$. Since $\mathbf{y}_i \in W$, we have, for $i = 1, 2, \ldots, n$,

$$\mathbf{y}_i = a_{1i}\mathbf{w}_1 + a_{2i}\mathbf{w}_2 + \cdots + a_{mi}\mathbf{w}_m$$

for some (unique) set of scalars $a_{1i}, a_{2i}, \ldots, a_{mi}$, and we write

$$(\mathbf{y}_1)_{B_2} = \begin{pmatrix} a_{11} \\ a_{21} \\ \vdots \\ a_{m1} \end{pmatrix}, \ (\mathbf{y}_2)_{B_2} = \begin{pmatrix} a_{12} \\ a_{22} \\ \vdots \\ a_{m2} \end{pmatrix}, \ldots, (\mathbf{y}_n)_{B_2} = \begin{pmatrix} a_{1n} \\ a_{2n} \\ \vdots \\ a_{mn} \end{pmatrix}$$

This means, for example, that $\mathbf{y}_1 = a_{11}\mathbf{w}_1 + a_{21}\mathbf{w}_2 + \cdots + a_{m1}\mathbf{w}_m$. We now define

$$A_T = \begin{pmatrix} a_{11} & a_{12} & \cdots & a_{1n} \\ a_{21} & a_{22} & \cdots & a_{2n} \\ \vdots & \vdots & & \vdots \\ a_{m1} & a_{m2} & \cdots & a_{mn} \end{pmatrix}$$

Since

$$(\mathbf{v}_1)_{B_1} = \begin{pmatrix} 1 \\ 0 \\ \vdots \\ 0 \end{pmatrix}, \ (\mathbf{v}_2)_{B_1} = \begin{pmatrix} 0 \\ 1 \\ 0 \\ \vdots \\ 0 \end{pmatrix}, \ldots, (\mathbf{v}_n)_{B_1} = \begin{pmatrix} 0 \\ 0 \\ \vdots \\ 1 \end{pmatrix}$$

we have, as in the proof of Theorem 1,

$$A_T(\mathbf{v}_i)_{B_1} = \begin{pmatrix} a_{11} & a_{12} & \cdots & a_{1n} \\ a_{21} & a_{22} & \cdots & a_{2n} \\ \vdots & \vdots & \cdots & \vdots \\ a_{i1} & a_{i2} & \cdots & a_{in} \\ \vdots & \vdots & \cdots & \vdots \\ a_{m1} & a_{m2} & & a_{mn} \end{pmatrix} \begin{pmatrix} 0 \\ 0 \\ \vdots \\ 1 \\ 0 \\ \vdots \\ 0 \end{pmatrix} = \begin{pmatrix} a_{1i} \\ a_{2i} \\ \vdots \\ a_{mi} \end{pmatrix} = (\mathbf{y}_i)_{B_2}$$

(ith position)

If $\mathbf{x}$ is V, then

$$\mathbf{x} = c_1\mathbf{v}_1 + c_2\mathbf{v}_2 + \cdots + c_n\mathbf{v}_n$$

$$(\mathbf{x})_{B_1} = \begin{pmatrix} c_1 \\ c_2 \\ \vdots \\ c_n \end{pmatrix}$$

and

$$(A_T(\mathbf{x})_{B_1})_{B_2} = \begin{pmatrix} a_{11} & a_{12} & \cdots & a_{1n} \\ a_{21} & a_{22} & \cdots & a_{2n} \\ \vdots & \vdots & & \vdots \\ a_{m1} & a_{m2} & \cdots & a_{mn} \end{pmatrix} \begin{pmatrix} c_1 \\ c_2 \\ \vdots \\ c_n \end{pmatrix}$$

$$= \begin{pmatrix} a_{11}c_1 + a_{12}c_2 + \cdots + a_{1n}c_n \\ a_{21}c_1 + a_{22}c_2 + \cdots + a_{2n}c_n \\ \vdots \\ a_{m1}c_1 + a_{m2}c_2 + \cdots + a_{mn}c_n \end{pmatrix}$$

$$= c_1\begin{pmatrix} a_{11} \\ a_{21} \\ \vdots \\ a_{m1} \end{pmatrix} + c_2\begin{pmatrix} a_{12} \\ a_{22} \\ \vdots \\ a_{m2} \end{pmatrix} + \cdots + c_n\begin{pmatrix} a_{1n} \\ a_{2n} \\ \vdots \\ a_{mn} \end{pmatrix}$$

$$= c_1(\mathbf{y}_1)_{B_2} + c_2(\mathbf{y}_2)_{B_2} + \cdots + c_n(\mathbf{y}_n)_{B_2}$$

Similarly, $T\mathbf{x} = T(c_1\mathbf{v}_1 + c_2\mathbf{v}_2 + \cdots + c_n\mathbf{v}_n) = c_1T\mathbf{v}_1 + c_2T\mathbf{v}_2 + \cdots + c_nT\mathbf{v}_n = c_1\mathbf{y}_1 + c_2\mathbf{y}_2 + \cdots + c_n\mathbf{y}_n$. Thus $(T\mathbf{x})_{B_2} = (A_T(\mathbf{x})_{B_1})$. The proof of uniqueness is exactly as in the proof of uniqueness in Theorem 1. ■

The following useful result follows immediately from Theorem 8.6.5 on page 530 and generalizes Theorem 2. Its proof is left as an exercise (see Problem 30).

THEOREM 4

Let V and W be finite dimensional vector spaces with dim $V = n$. Let $T: V \rightarrow W$ be a linear transformation and let A_T be a matrix representation of T. Then:

i. $\rho(T) = \rho(A_T)$

ii. $\nu(T) = \nu(A_T)$

iii. $\nu(T) + \rho(T) = n$ ■

EXAMPLE 6 THE MATRIX REPRESENTATION OF A TRANSFORMATION FROM P_2 TO P_3

Let $T: P_2 \rightarrow P_3$ be defined by $(Tp)(x) = xp(x)$. Find A_T and use it to determine the kernel and range of T.

SOLUTION: Using the standard basis $B_1 = \{1, x, x^2\}$ in P_2 and $B_2 = \{1, x, x^2, x^3\}$ in P_3, we have $(T(1))_{B_2} = (x)_{B_2} = \begin{pmatrix} 0 \\ 1 \\ 0 \\ 0 \end{pmatrix}$, $(T(x))_{B_2} = (x^2)_{B_2} = \begin{pmatrix} 0 \\ 0 \\ 1 \\ 0 \end{pmatrix}$, and $(T(x^2))_{B_2} = (x^3)_{B_2} = \begin{pmatrix} 0 \\ 0 \\ 0 \\ 1 \end{pmatrix}$. Thus $A_T = \begin{pmatrix} 0 & 0 & 0 \\ 1 & 0 & 0 \\ 0 & 1 & 0 \\ 0 & 0 & 1 \end{pmatrix}$. Clearly $\rho(A) = 3$ and a basis for R_A is $\left\{ \begin{pmatrix} 0 \\ 1 \\ 0 \\ 0 \end{pmatrix}, \begin{pmatrix} 0 \\ 0 \\ 1 \\ 0 \end{pmatrix}, \begin{pmatrix} 0 \\ 0 \\ 0 \\ 1 \end{pmatrix} \right\}$. Therefore range $T = \text{span}\ \{x, x^2, x^3\}$. Since $\nu(A) = 3 - \rho(A) = 0$, we see that $\ker T = \{0\}$.

EXAMPLE 7 THE MATRIX REPRESENTATION OF A TRANSFORMATION FROM P_3 TO P_2

Let $T: P_3 \rightarrow P_2$ be defined by $T(a_0 + a_1x + a_2x^2 + a_3x^3) = a_1 + a_2x^2$. Compute A_T and use it to find the kernel and range of T.

SOLUTION: Using the standard bases $B_1 = (1, x, x^2, x^3\}$ in P_3 and $B_2 = \{1, x, x^2\}$ in P_2, we immediately see that $(T(1))_{B_2} = \begin{pmatrix} 0 \\ 0 \\ 0 \end{pmatrix}$, $(T(x))_{B_2} = \begin{pmatrix} 1 \\ 0 \\ 0 \end{pmatrix}$, $(T(x^2))_{B_2} = \begin{pmatrix} 0 \\ 0 \\ 1 \end{pmatrix}$, and $(T(x^3))_{B_2} = \begin{pmatrix} 0 \\ 0 \\ 0 \end{pmatrix}$. Thus $A_T = \begin{pmatrix} 0 & 1 & 0 & 0 \\ 0 & 0 & 0 & 0 \\ 0 & 0 & 1 & 0 \end{pmatrix}$. Clearly $\rho(A) = 2$ and a basis for R_A is $\left\{ \begin{pmatrix} 1 \\ 0 \\ 0 \end{pmatrix}, \begin{pmatrix} 0 \\ 0 \\ 1 \end{pmatrix} \right\}$ so that range $T = \text{span}\ \{1, x^2\}$. Then $\nu(A) = 4 - 2 = 2$; and if $A_T \begin{pmatrix} a_0 \\ a_1 \\ a_2 \\ a_3 \end{pmatrix} = \begin{pmatrix} 0 \\ 0 \\ 0 \end{pmatrix}$, then $a_1 = 0$ and $a_2 = 0$. Hence a_0 and a_3 are arbitrary and a basis for N_A is $\left\{ \begin{pmatrix} 1 \\ 0 \\ 0 \\ 0 \end{pmatrix}, \begin{pmatrix} 0 \\ 0 \\ 0 \\ 1 \end{pmatrix} \right\}$, so that a basis for $\ker T$ is $\{1, x^3\}$.

PROBLEMS 8.11

SELF-QUIZ

Multiple Choice

I. If T: $\mathbb{R}^3 \to \mathbb{R}^3$ is the linear transformation $T\begin{pmatrix} x \\ y \\ z \end{pmatrix} = \begin{pmatrix} z \\ -x \\ y \end{pmatrix}$, then $A_T =$ ________.

a. $\begin{pmatrix} 0 & -1 & 0 \\ 0 & 0 & 1 \\ 1 & 0 & 0 \end{pmatrix}$ **b.** $\begin{pmatrix} 0 & 0 & 1 \\ -1 & 0 & 0 \\ 0 & 1 & 0 \end{pmatrix}$

c. $\begin{pmatrix} 1 & 0 & 0 \\ 0 & -1 & 0 \\ 0 & 0 & 1 \end{pmatrix}$ **d.** $\begin{pmatrix} 0 & 0 & 1 \\ 0 & 1 & 0 \\ -1 & 0 & 0 \end{pmatrix}$

II. If T: $P_2 \to P_1$ is the linear transformation $Tp(x) = p'(x)$, then $A_T =$ ________.

a. $\begin{pmatrix} 0 & 1 & 0 \\ 0 & 0 & 2 \end{pmatrix}$ **b.** $\begin{pmatrix} 0 & 1 & 0 \\ 0 & 0 & 1 \end{pmatrix}$

c. $\begin{pmatrix} 0 & 0 \\ 1 & 0 \\ 0 & 2 \end{pmatrix}$ **d.** $\begin{pmatrix} 0 & 0 \\ 1 & 0 \\ 0 & 1 \end{pmatrix}$

True-False

III. There exists a linear transformation T from $\mathbb{R}^5 \to \mathbb{R}^5$ with $\rho(T) = \nu(T)$.

IV. Suppose that T: $M_{22} \to M_{22}$ with $\rho(T) = 4$. Then if $TA = \begin{pmatrix} 0 & 0 \\ 0 & 0 \end{pmatrix}$, then $A = \begin{pmatrix} 0 & 0 \\ 0 & 0 \end{pmatrix}$.

In Problems 1–24 find the matrix representation A_T of the linear transformation T, ker T, range T, $\nu(T)$, and $\rho(T)$. Assume that B_1 and B_2 are standard bases.

1. $\mathbb{R}^2 \to \mathbb{R}^2$; $T\begin{pmatrix} x \\ y \end{pmatrix} = \begin{pmatrix} x - 2y \\ -x + y \end{pmatrix}$

2. T: $\mathbb{R}^2 \to \mathbb{R}^3$; $T\begin{pmatrix} x \\ y \end{pmatrix} = \begin{pmatrix} x + y \\ x - y \\ 2x + 3y \end{pmatrix}$

3. T: $\mathbb{R}^3 \to \mathbb{R}^2$; $T\begin{pmatrix} x \\ y \\ z \end{pmatrix} = \begin{pmatrix} x - y + z \\ -2x + 2y - 2z \end{pmatrix}$

4. T: $\mathbb{R}^2 \to \mathbb{R}^2$; $T\begin{pmatrix} x \\ y \end{pmatrix} = \begin{pmatrix} ax + by \\ cx + dy \end{pmatrix}$, $ad - bc \neq 0$

5. T: $\mathbb{R}^3 \to \mathbb{R}^3$; $T\begin{pmatrix} x \\ y \\ z \end{pmatrix} = \begin{pmatrix} x - y + 2z \\ 3x + y + 4z \\ 5x - y + 8z \end{pmatrix}$

6. T: $\mathbb{R}^3 \to \mathbb{R}^3$; $T\begin{pmatrix} x \\ y \\ z \end{pmatrix} = \begin{pmatrix} -x + 2y + z \\ 2x - 4y - 2z \\ -3x + 6y + 3z \end{pmatrix}$

7. T: $\mathbb{R}^4 \to \mathbb{R}^3$; $T\begin{pmatrix} x \\ y \\ z \\ w \end{pmatrix} = \begin{pmatrix} x - y + 2z + 3w \\ y + 4z + 3w \\ x + 6z + 6w \end{pmatrix}$

8. T: $\mathbb{R}^4 \to \mathbb{R}^4$; $T\begin{pmatrix} x \\ y \\ z \\ w \end{pmatrix} = \begin{pmatrix} x - y + 2z + w \\ -x + z + 2w \\ x - 2y + 5z + 4w \\ 2x - y + z - w \end{pmatrix}$

9. T: $P_2 \to P_3$; $T(a_0 + a_1x + a_2x^2) = a_1 - a_1x + a_0x^3$

10. T: $\mathbb{R} \to P_3$; $T(a) = a + ax + ax^2 + ax^3$

11. T: $P_3 \to \mathbb{R}$; $T(a_0 + a_1x + a_2x^2 + a_3x^3) = a_2$

12. T: $P_3 \to P_1$; $T(a_0 + a_1x + a_2x^2 + a_3x^3) = (a_1 + a_3)x - a_2$

13. T: $P_3 \to P_2$; $T(a_0 + a_1x + a_2x^2 + a_3x^3) = (a_0 - a_1 + 2a_2 + 3a_3) + (a_1 + 4a_2 + 3a_3)x + (a_0 + 6a_2 + 5a_3)x^2$

14. T: $M_{22} \to M_{22}$; $T\begin{pmatrix} a & b \\ c & d \end{pmatrix} = \begin{pmatrix} a - b + 2c + d & -a + 2c + 2d \\ a - 2b + 5c + 4d & 2a - b + c - d \end{pmatrix}$

***15.** D: $P_4 \to P_3$; $Dp(x) = p'(x)$

16. T: $P_4 \to P_4$; $Tp(x) = xp'(x) - p(x)$

***17.** D: $P_n \to P_{n-1}$; $Dp(x) = p'(x)$

18. D: $P_4 \to P_2$; $Dp(x) = p''(x)$

***19.** T: $P_4 \to P_4$; $Tp(x) = p''(x) + xp'(x) + 2p(x)$

***20.** D: $P_n \to P_{n-k}$; $Dp(x) = p^{(k)}(x)$

21. T: $P_n \to P_n$; $Tp(x) = x^n p^{(n)}(x) + x^{n-1}p^{(n-1)}(x) + \cdots + xp'(x) + p(x)$

22. J: $P_n \to \mathbb{R}$; $Jp = \int_0^1 p(x)\,dx$

23. $T\colon \mathbb{R}^3 \to P_2$; $T\begin{pmatrix} a \\ b \\ c \end{pmatrix} = a + bx + cx^2$

24. $T\colon P_3 \to \mathbb{R}^3$; $T(a_0 + a_1x + a_2x^2 + a_3x^3) = \begin{pmatrix} a_3 - a_2 \\ a_1 + a_3 \\ a_2 - a_1 \end{pmatrix}$

25. *Let* $T\colon M_{mn} \to M_{nm}$ be given by $TA = A^t$. Find A_T with respect to the standard bases in M_{mn} and M_{nm}.

***26.** Let $T\colon \mathbb{C}^2 \to \mathbb{C}^2$ be given by $T\begin{pmatrix} x \\ y \end{pmatrix} = \begin{pmatrix} x + iy \\ (1 + i)y - x \end{pmatrix}$. Find A_T.

27. Let $V = \text{span}\ \{1, \sin x, \cos x\}$. Find A_D, where $D\colon V \to V$ is defined by $Df(x) = f'(x)$. Find range D and ker D.

28. Answer the questions of Problems 26 given $V = \text{span}\ \{e^x, xe^x, x^2e^x\}$.

29. Prove Theorem 2.

30. Prove Theorem 4.

ANSWERS TO SELF-QUIZ

I. b **II.** a **III.** False **IV.** True

8.12 ISOMORPHISMS

In this section we introduce some important terminology and then prove a theorem which says that all n-dimensional vector spaces are "essentially" the same.

DEFINITION ONE-TO-ONE TRANSFORMATION

Let $T\colon V \to W$ be a linear transformation. Then T is **one-to-one**, written 1–1, if

$$T\mathbf{v}_1 = T\mathbf{v}_2 \quad \text{implies that} \quad \mathbf{v}_1 = \mathbf{v}_2 \tag{1}$$

That is, T is 1–1 if every vector $\mathbf{w}$ in the range of T is the image of exactly one vector in V. ■

THEOREM 1

Let $T\colon V \to W$ be a linear transformation. Then T is 1–1 if and only if $\ker T = \{\mathbf{0}\}$.

PROOF: Suppose $\ker T = \{\mathbf{0}\}$ and $T\mathbf{v}_1 = T\mathbf{v}_2$. Then $T\mathbf{v}_1 - T\mathbf{v}_2 = T(\mathbf{v}_1 - \mathbf{v}_2) = \mathbf{0}$, which means that $(\mathbf{v}_1 - \mathbf{v}_2) \in \ker T = \{\mathbf{0}\}$. Thus $\mathbf{v}_1 - \mathbf{v}_2 = \mathbf{0}$, so $\mathbf{v}_1 = \mathbf{v}_2$, which shows that T is 1–1. Now suppose that T is 1–1 and $\mathbf{v} \in \ker T$. Then $T\mathbf{v} = \mathbf{0}$. But $T\mathbf{0} = \mathbf{0}$ also. Thus, since T is 1–1, $\mathbf{v} = \mathbf{0}$. This completes the proof. ■

EXAMPLE 1 A 1–1 TRANSFORMATION FROM $\mathbb{R}^2$ TO $\mathbb{R}^2$

Let $T\colon \mathbb{R}^2 \to \mathbb{R}^2$ be defined by $T\begin{pmatrix} x \\ y \end{pmatrix} = \begin{pmatrix} x - y \\ 2x + y \end{pmatrix}$. We easily find $A_T = \begin{pmatrix} 1 & -1 \\ 2 & 1 \end{pmatrix}$ and $\rho(A_T) = 2$; hence $\nu(A_T) = 0$ and $N_{A_T} = \ker T = \{\mathbf{0}\}$. Thus T is 1–1.

EXAMPLE 2

A TRANSFORMATION FROM $\mathbb{R}^2$ TO $\mathbb{R}^2$ THAT IS NOT 1–1

Let $T: \mathbb{R}^2 \to \mathbb{R}^2$ be defined by $T\begin{pmatrix} x \\ y \end{pmatrix} = \begin{pmatrix} x - y \\ 2x - 2y \end{pmatrix}$. Then $A_T = \begin{pmatrix} 1 & -1 \\ 2 & -2 \end{pmatrix}$, $\rho(A_T) = 1$, and $\nu(A_T) = 1$; hence $\nu(T) = 1$ and T is not 1–1. Note, for example, that $T\begin{pmatrix} 1 \\ 1 \end{pmatrix} = \mathbf{0}$ $= T\begin{pmatrix} 0 \\ 0 \end{pmatrix}$.

DEFINITION ONTO TRANSFORMATION

Let $T: V \to W$ be a linear transformation. Then T is said to be **onto** W or simply **onto,** if for every $\mathbf{w} \in W$ there is at least one $\mathbf{v} \in V$ such that $T\mathbf{v} = \mathbf{w}$. That is: *T is onto W if and only if range $T = W$.* ■

EXAMPLE 3

DETERMINING WHETHER A TRANSFORMATION IS ONTO

In Example 1, $\rho(A_T) = 2$; hence range $T = \mathbb{R}^2$ and T is onto. In Example 2, $\rho(A_T) = 1$ and range $T =$ span $\left\{\begin{pmatrix} 1 \\ 2 \end{pmatrix}\right\} \neq \mathbb{R}^2$; hence T is not onto.

THEOREM 2

Let $T: V \to W$ be a linear transformation and suppose that dim $V =$ dim $W = n$.

i. If T is 1–1, then T is onto.
ii. If T is onto, then T is 1–1.

PROOF: Let A_T be the matrix representation of T. Then if T is 1–1, ker $T = \{\mathbf{0}\}$ and $\nu(A_T) = 0$, which means that $\rho(T) = \rho(A_T) = n - 0 = n$ so that range $T = W$. If T is onto, then $\rho(A_T) = n$ so that $\nu(T) = \nu(A_T) = 0$ and T is 1–1. ■

THEOREM 3

Let $T: V \to W$ be a linear transformation. Suppose that dim $V = n$ and dim $W = m$. Then:

i. If $n > m$, T is not 1–1.
ii. If $m > n$, T is not onto.

PROOF:

i. Let $\{\mathbf{v}_1, \mathbf{v}_2, \ldots, \mathbf{v}_n\}$ be a basis for V. Let $\mathbf{w}_i = T\mathbf{v}_i$ for $i = 1, 2, \ldots, n$ and look at the set $S = \{\mathbf{w}_1, \mathbf{w}_2, \ldots, \mathbf{w}_n\}$. Since $m =$ dim $W < n$, the set S is linearly dependent. Thus there exist scalars not all zero such that $c_1\mathbf{w}_1 + c_2\mathbf{w}_2 + \cdots + c_n\mathbf{w}_n = \mathbf{0}$. Let $\mathbf{v} = c_1\mathbf{v}_1 + c_2\mathbf{v}_2 + \cdots + c_n\mathbf{v}_n$. Since the $\mathbf{v}_i$'s are linearly independent and since not all the c_i's are zero, we see that $\mathbf{v} \neq \mathbf{0}$. But $T\mathbf{v} = T(c_1\mathbf{v}_1 + c_2\mathbf{v}_2 + \cdots + c_n\mathbf{v}_n) = c_1T\mathbf{v}_1 + c_2T\mathbf{v}_2 + \cdots + c_nT\mathbf{v}_n = c_1\mathbf{w}_1 + c_2\mathbf{w}_2 + \cdots + c_n\mathbf{w}_n = \mathbf{0}$. Thus $\mathbf{v} \in \ker T$ and $\ker T \neq \{\mathbf{0}\}$.

ii. If $\mathbf{v} \in V$, then $\mathbf{v} = a_1\mathbf{v}_1 + a_2\mathbf{v}_2 + \cdots + a_n\mathbf{v}_n$ for some scalars $a_1, a_2, \ldots, a_n$ and $T\mathbf{v} = a_1T\mathbf{v}_1 + a_2T\mathbf{v}_2 + \cdots + a_nT\mathbf{v}_n = a_1\mathbf{w}_1 + a_2\mathbf{w}_2 + \cdots + a_n\mathbf{w}_n$. Thus $\{\mathbf{w}_1, \mathbf{w}_2, \ldots, \mathbf{w}_n\} = \{T\mathbf{v}_1, T\mathbf{v}_2, \ldots, T\mathbf{v}_n\}$ spans range T. Then, from Problem 8.5.29 on page 524, $\rho(T) = \dim \text{range } T \leq n$. Since $m > n$, this shows that range $T \neq W$. Thus T is not onto. ■

EXAMPLE 4

A TRANSFORMATION FROM $\mathbb{R}^3 \rightarrow \mathbb{R}^2$ IS NOT 1–1

Let T: $\mathbb{R}^3 \rightarrow \mathbb{R}^2$ be given by $T\begin{pmatrix} x \\ y \\ z \end{pmatrix} = \begin{pmatrix} 1 & 2 & 3 \\ 4 & 5 & 6 \end{pmatrix}\begin{pmatrix} x \\ y \\ z \end{pmatrix}$. Here $n = 3$ and $m = 2$, so T is not 1–1. To see this, observe that

$$T\begin{pmatrix} -1 \\ 2 \\ 0 \end{pmatrix} = \begin{pmatrix} 1 & 2 & 3 \\ 4 & 5 & 6 \end{pmatrix}\begin{pmatrix} -1 \\ 2 \\ 0 \end{pmatrix} = \begin{pmatrix} 3 \\ 6 \end{pmatrix} \quad \text{and} \quad T\begin{pmatrix} 2 \\ -4 \\ 3 \end{pmatrix} = \begin{pmatrix} 1 & 2 & 3 \\ 4 & 5 & 6 \end{pmatrix}\begin{pmatrix} 2 \\ -4 \\ 3 \end{pmatrix} = \begin{pmatrix} 3 \\ 6 \end{pmatrix}$$

That is, two different vectors in $\mathbb{R}^3$ have the same image in $\mathbb{R}^2$.

EXAMPLE 5

A TRANSFORMATION FROM $\mathbb{R}^2$ TO $\mathbb{R}^3$ IS NOT ONTO

Let T: $\mathbb{R}^2 \rightarrow \mathbb{R}^3$ be given by $T\begin{pmatrix} x \\ y \end{pmatrix} = \begin{pmatrix} 1 & 2 \\ 3 & 4 \\ 5 & 6 \end{pmatrix}\begin{pmatrix} x \\ y \end{pmatrix}$. Here $n = 2$ and $m = 3$, so T is not onto. To show this, we must find a vector in $\mathbb{R}^3$ which is not in the range of T. One such vector is $\begin{pmatrix} 0 \\ 0 \\ 1 \end{pmatrix}$. That is, there is no vector $\mathbf{x} = \begin{pmatrix} x \\ y \end{pmatrix}$ in $\mathbb{R}^2$ such that $T\mathbf{x} = \begin{pmatrix} 0 \\ 0 \\ 1 \end{pmatrix}$. We prove this by assuming that $T\begin{pmatrix} x \\ y \end{pmatrix} = \begin{pmatrix} 0 \\ 0 \\ 1 \end{pmatrix}$. That is,

$$\begin{pmatrix} 1 & 2 \\ 3 & 4 \\ 5 & 6 \end{pmatrix}\begin{pmatrix} x \\ y \end{pmatrix} = \begin{pmatrix} 0 \\ 0 \\ 1 \end{pmatrix} \quad \text{or} \quad \begin{pmatrix} x + 2y \\ 3x + 4y \\ 5x + 6y \end{pmatrix} = \begin{pmatrix} 0 \\ 0 \\ 1 \end{pmatrix}$$

Row-reducing, we have

$$\left(\begin{array}{cc|c} 1 & 2 & 0 \\ 3 & 4 & 0 \\ 5 & 6 & 1 \end{array}\right) \xrightarrow[R_3 \rightarrow R_3 - 5R_1]{R_2 \rightarrow R_2 - 3R_1} \left(\begin{array}{cc|c} 1 & 2 & 0 \\ 0 & -2 & 0 \\ 0 & -4 & 1 \end{array}\right) \xrightarrow{R_3 \rightarrow R_3 - 2R_2} \left(\begin{array}{cc|c} 1 & 2 & 0 \\ 0 & -2 & 0 \\ 0 & 0 & 1 \end{array}\right)$$

The last line reads $0 \cdot x + 0 \cdot y = 1$, so the system is inconsistent and $\begin{pmatrix} 0 \\ 0 \\ 1 \end{pmatrix}$ is not in the range of T.

DEFINITION ISOMORPHISM

Let $T: V \to W$ be a linear transformation. Then T is an **isomorphism** if T is 1–1 and onto. ■

DEFINITION ISOMORPHIC VECTOR SPACES

The vector spaces V and W are said to be **isomorphic** if there exists an isomorphism T from V onto W. In this case we write $V \cong W$.

REMARK: The word ''isomorphism'' comes from the Greek *isomorphos* meaning ''of equal form'' (*iso* = equal; *morphos* = form). After a few examples we shall see how closely related are the ''forms'' of isomorphic vector spaces.

Let $T: \mathbb{R}^n \to \mathbb{R}^n$ and let A_T be the matrix representation of T. Now T is 1–1 if and only if $\ker T = \{\mathbf{0}\}$, which is true if and only if $\nu(A_T) = 0$ if and only if $\det A_T \neq 0$. Thus we can extend our Summing Up Theorem (last seen on page 533) in another direction.

THEOREM 4 SUMMING UP THEOREM—VIEW 7

Let A be an $n \times n$ matrix. Then the following ten statements are equivalent; that is, each one implies the other nine (so that if one is true, all are true):

i. A is invertible.
ii. The only solution to the homogeneous system $A\mathbf{x} = \mathbf{0}$ is the trivial solution ($\mathbf{x} = \mathbf{0}$).
iii. The system $A\mathbf{x} = \mathbf{b}$ has a unique solution for every n-vector $\mathbf{b}$.
iv. A is row equivalent to the $n \times n$ identity matrix I_n.
v. A is a product of elementary matrices.
vi. The rows (and columns) of A are linearly independent.
vii. $\det A \neq 0$.
viii. $\nu(A) = 0$.
ix. $\rho(A) = n$.
x. The linear transformation T from $\mathbb{R}^n$ to $\mathbb{R}^n$ defined by $T\mathbf{x} = A\mathbf{x}$ is an isomorphism. ■

We now look at some examples of isomorphisms between other pairs of vector spaces.

EXAMPLE 6 AN ISOMORPHISM BETWEEN $\mathbb{R}^3$ AND P_2

Let $T: \mathbb{R}^3 \to P_2$ be defined by $T\begin{pmatrix} a \\ b \\ c \end{pmatrix} = a + bx + cx^2$. It is easy to verify that T is linear. Suppose that $T\begin{pmatrix} a \\ b \\ c \end{pmatrix} = \mathbf{0} = 0 + 0x + 0x^2$. Then $a = b = c = 0$. That is,

$\ker T = \{\mathbf{0}\}$ and T is 1–1. If $p(x) = a_0 + a_1x + a_2x^2$, then $p(x) = T\begin{pmatrix} a_0 \\ a_1 \\ a_2 \end{pmatrix}$. This means that range $T = P_2$ and T is onto. Thus $\mathbb{R}^3 \cong P_2$.

NOTE: $\dim \mathbb{R}^3 = \dim P_2 = 3$. Thus, by Theorem 2, once we know that T is 1–1, we also know that it is onto. We verified that it was onto; but it was unnecessary to do so.

EXAMPLE 7

AN ISOMORPHISM BETWEEN TWO INFINITE DIMENSIONAL VECTOR SPACES

Let $V = \{f \in C^1[0, 1]: f(0) = 0\}$ and $W = C[0, 1]$. Let $D: V \to W$ be given by $Df = f'$. Suppose that $Df = Dg$. Then $f' = g'$ or $(f - g)' = 0$ and $f(x) - g(x) = c$, a constant. But $f(0) = g(0) = 0$, so $c = 0$ and $f = g$. Thus D is 1–1. Let $g \in C[0, 1]$ and let $f(x) = \int_0^x g(t)\, dt$. Then, from the fundamental theorem of calculus, $f \in C^1[0, 1]$ and $f'(x) = g(x)$ for every $x \in [0, 1]$. Moreover, since $\int_0^0 g(t)\, dt = 0$, we have $f(0) = 0$. Thus, for every g in W, there is an $f \in V$ such that $Df = g$. Hence D is onto and we have shown that $V \cong W$.

The following theorem illustrates the similarity of two isomorphic vector spaces.

THEOREM 5

Let $T: V \to W$ be an isomorphism:

i. If $\mathbf{v}_1, \mathbf{v}_2, \ldots, \mathbf{v}_n$ span V, then $T\mathbf{v}_1, T\mathbf{v}_2, \ldots, T\mathbf{v}_n$ span W.

ii. If $\mathbf{v}_1, \mathbf{v}_2, \ldots, \mathbf{v}_n$ are linearly independent in V, then $T\mathbf{v}_1, T\mathbf{v}_2, \ldots, T\mathbf{v}_n$ are linearly independent in W.

iii. If $\{\mathbf{v}_1, \mathbf{v}_2, \ldots, \mathbf{v}_n\}$ is a basis in V, then $\{T\mathbf{v}_1, T\mathbf{v}_2, \ldots, T\mathbf{v}_n\}$ is a basis in W.

iv. If V is finite dimensional, then W is finite dimensional and $\dim V = \dim W$.

PROOF:

i. Let $\mathbf{w} \in W$. Then, since T is onto, there is a $\mathbf{v} \in V$ such that $T\mathbf{v} = \mathbf{w}$. Since the $\mathbf{v}_i$'s span V, we can write $\mathbf{v} = a_1\mathbf{v}_1 + a_2\mathbf{v}_2 + \cdots + a_n\mathbf{v}_n$ so that $\mathbf{w} = T\mathbf{v} = a_1T\mathbf{v}_1 + a_2T\mathbf{v}_2 + \cdots + a_nT\mathbf{v}_n$ and this shows that $\{T\mathbf{v}_1, T\mathbf{v}_2, \ldots, T\mathbf{v}_n\}$ spans W.

ii. Suppose $c_1T\mathbf{v}_1 + c_2T\mathbf{v}_2 + \cdots + c_nT\mathbf{v}_n = \mathbf{0}$. Then $T(c_1\mathbf{v}_1 + c_2\mathbf{v}_2 + \cdots + c_n\mathbf{v}_n) = \mathbf{0}$. Thus, since T is 1–1, $c_1\mathbf{v}_1 + c_2\mathbf{v}_2 + \cdots + c_n\mathbf{v}_n = \mathbf{0}$, which implies that $c_1 = c_2 = \cdots = c_n = 0$ since the $\mathbf{v}_i$'s are independent.

iii. This follows from parts (*i*) and (*ii*).

iv. This follows from part (*iii*). ■

In general, it is difficult to show that two infinite dimensional vector spaces are isomorphic. For finite dimensional spaces, however, it is remarkably easy. Theorem 3 shows that if $\dim V \neq \dim W$, then V and W are not isomorphic. The next theorem shows that if $\dim V = \dim W$, and if V and W are real vector spaces, then V and W are isomorphic. That is,

Two real finite dimensional spaces of the same dimension are isomorphic.

THEOREM 6

Let V and W be two real[†] finite dimensional vector spaces with $\dim V = \dim W$. Then $V \cong W$.

PROOF: Let $\{\mathbf{v}_1, \mathbf{v}_2, \ldots, \mathbf{v}_n\}$ be a basis for V and let $\{\mathbf{w}_1, \mathbf{w}_2, \ldots, \mathbf{w}_n\}$ be a basis for W. Define the linear transformation T by

$$T\mathbf{v}_i = \mathbf{w}_i \qquad \text{for } i = 1, 2, \ldots, n \tag{2}$$

By Theorem 8.10.2 on page 566 there is exactly one linear transformation that satisfies equation (2). Suppose $\mathbf{v} \in V$ and $T\mathbf{v} = \mathbf{0}$. Then if $\mathbf{v} = c_1\mathbf{v}_1 + c_2\mathbf{v}_2 + \cdots + c_n\mathbf{v}_n$, we have $T\mathbf{v} = c_1T\mathbf{v}_1 + \cdots + c_nT\mathbf{v}_n = c_1\mathbf{w}_1 + c_2\mathbf{w}_2 + \cdots + c_n\mathbf{w}_n = \mathbf{0}$. But, since $\mathbf{w}_1, \mathbf{w}_2, \ldots, \mathbf{w}_n$ are linearly independent, $c_1 = c_2 = \cdots = c_n = 0$. Thus $\mathbf{v} = \mathbf{0}$ and T is 1–1. Since V and W are finite dimensional and $\dim V = \dim W$, T is onto by Theorem 2 and the proof is complete. ■

This last result is one of the central results of linear algebra. Loosely speaking, it says that if you know one real n-dimensional vector space, you know all real vector spaces of dimension n. That is, loosely speaking, $\mathbb{R}^n$ is the only n-dimensional vector space over the reals.

PROBLEMS 8.12

SELF-QUIZ

True-False

I. A linear transformation from $\mathbb{R}^n \to \mathbb{R}^m$ with $n \neq m$ cannot be both 1–1 and onto.

II. If $\dim V = 5$ and $\dim W = 7$ it may still be possible to find an isomorphism T from V onto W.

III. If T is 1–1, then $\ker T = \{0\}$.

IV. If T is an isomorphism from a vector space V into $\mathbb{R}^6$, then $\rho(T) = 6$.

V. If A_T is the transformation matrix of an isomorphism from $\mathbb{R}^8$ into $\mathbb{R}^8$, then $\det A_T \neq 0$.

1. Show that $T: M_{mn} \to M_{nm}$ defined by $TA = A^t$ is an isomorphism.
2. Show that $T: \mathbb{R}^n \to \mathbb{R}^n$ is an isomorphism if and only if A_T is invertible.
3. For what values of α is the transformation $T: \mathbb{R}^2 \to \mathbb{R}^2$ given by $T\begin{pmatrix} x \\ y \end{pmatrix} = \begin{pmatrix} 1 & 2 \\ 3 & \alpha \end{pmatrix}\begin{pmatrix} x \\ y \end{pmatrix}$ an isomorphism?

[†] We need the word "real" here because it is important that the sets of scalars in V and W be the same. Otherwise, the condition $T(\alpha\mathbf{v}) = \alpha T\mathbf{v}$ might not hold because $\mathbf{v} \in V$, $T\mathbf{v} \in W$, and either $\alpha\mathbf{v}$ or $\alpha T\mathbf{v}$ might not be defined. Theorem 6 is true if the word "real" is omitted and, instead, we impose the conditions that V and W be defined with the same set of scalars (like $\mathbb{C}$, for example).

4. Find an isomorphism between D_n, the $n \times n$ diagonal matrices with real entries, and $\mathbb{R}^n$. [*Hint:* Look first at the case $n = 2$.]

5. For what value of m is the set of $n \times n$ symmetric matrices isomorphic to $\mathbb{R}^m$?

6. Show that the set of $n \times n$ symmetric matrices is isomorphic to the set of $n \times n$ upper triangular matrices.

7. Let $V = P_4$ and $W = \{p \in P_5: p(0) = 0\}$. Show that $V \cong W$.

* **8.** Define $T: P_n \to P_n$ by $Tp = p + p'$. Show that T is an isomorphism.

9. Find a condition on the numbers m, n, p, q such that $M_{mn} \cong M_{pq}$.

10. Show that $D_n \cong P_{n-1}$.

11. Prove that any two finite dimensional complex vector spaces V and W with $\dim V = \dim W$ are isomorphic.

12. Define $T: C[0, 1] \to C[3, 4]$ by $Tf(x) = f(x - 3)$. Show that T is an isomorphism.

13. Let B be an invertible $n \times n$ matrix. Show that $T: M_{nm} \to M_{nm}$ defined by $TA = AB$ is an isomorphism.

14. Show that the transformation $Tp(x) = xp'(x)$ is not an isomorphism from P_n into P_n.

15. Let H be a subspace of $\mathbb{R}^n$. Show that $T: \mathbb{R}^n \to H$ defined by $T\mathbf{v} = \text{proj}_H \mathbf{v}$ is onto. Under what circumstances will it be 1–1?

16. Show that if $T: V \to W$ is an isomorphism, then there exists an isomorphism $S: W \to V$ such that $S(T\mathbf{v}) = \mathbf{v}$. Here S is called the **inverse transformation** of T and is denoted by T^{-1}.

17. Show that if $T: \mathbb{R}^n \to \mathbb{R}^n$ is defined by $T\mathbf{x} = A\mathbf{x}$ and if T is an isomorphism, then A is invertible and the inverse transformation T^{-1} is given by $T^{-1}\mathbf{x} = A^{-1}\mathbf{x}$.

18. Find T^{-1} for the isomorphism of Problem 7.

***19.** Consider the space $C = \{z = a + ib$, where a and b are real numbers and $i^2 = -1\}$. Show that if the scalars are taken to be the reals, then $C \cong \mathbb{R}^2$.

***20.** Consider the space $\mathbb{C}_{\mathbb{R}}^n = \{(\mathbf{c}_1, \mathbf{c}_2, \ldots, \mathbf{c}_n): \mathbf{c}_i \in C$ and the scalars are the reals$\}$. Show that $\mathbb{C}_{\mathbb{R}}^n \cong \mathbb{R}^{2n}$. [*Hint:* See Problem 19.]

ANSWERS TO SELF-QUIZ

I. True **II.** False **III.** True **IV.** True **V.** True

8.13 EIGENVALUES AND EIGENVECTORS

Let $T: V \to V$ be a linear transformation. In a great variety of applications (one of which is given in the next section) it is useful to find a vector $\mathbf{v}$ in V such that $T\mathbf{v}$ and $\mathbf{v}$ are parallel. That is, we seek a vector $\mathbf{v}$ and a scalar λ such that

$$T\mathbf{v} = \lambda\mathbf{v} \tag{1}$$

If $\mathbf{v} \neq \mathbf{0}$ and λ satisfy (1), then λ is called an *eigenvalue* of T and $\mathbf{v}$ is called an *eigenvector* of T corresponding to the eigenvalue λ. The purpose of this section is to investigate properties of eigenvalues and eigenvectors. If V is finite dimensional, then T can be represented by a matrix A_T. For that reason we shall discuss eigenvalues and eigenvectors of $n \times n$ matrices.

DEFINITION EIGENVALUE AND EIGENVECTOR

Let A be an $n \times n$ matrix with real† components. The number λ (real or complex) is called an **eigenvalue** of A if there is a *nonzero* vector $\mathbf{v}$ in $\mathbb{C}^n$ such that

†This definition is also valid if A has complex components; but as the matrices we shall be dealing with will, for the most part, have real components, the definition is sufficient for our purposes.

$$A\mathbf{v} = \lambda\mathbf{v} \tag{2}$$

The vector $\mathbf{v} \neq \mathbf{0}$ is called an **eigenvector** of A corresponding to the eigenvalue λ. ■

NOTE: The word "eigen" is the German word for "own" or "proper." Eigenvalues are also called **proper values** or **characteristic values** and eigenvectors are called **proper vectors** or **characteristic vectors.**

REMARK: As we shall see (for instance, in Example 6) a matrix with real components can have complex eigenvalues and eigenvectors. That is why, in the definition, we have asserted that $\mathbf{v} \in \mathbb{C}^n$. We shall not be using many facts about complex numbers in this book. For a discussion of those few facts we do need, see Appendix 3.

EXAMPLE 1 EIGENVALUES AND EIGENVECTORS OF A 2 × 2 MATRIX

Let $A = \begin{pmatrix} 10 & -18 \\ 6 & -11 \end{pmatrix}$. Then $A\begin{pmatrix} 2 \\ 1 \end{pmatrix} = \begin{pmatrix} 10 & -18 \\ 6 & -11 \end{pmatrix}\begin{pmatrix} 2 \\ 1 \end{pmatrix} = \begin{pmatrix} 2 \\ 1 \end{pmatrix}$. Thus $\lambda_1 = 1$ is an eigenvalue of A with corresponding eigenvector $\mathbf{v}_1 = \begin{pmatrix} 2 \\ 1 \end{pmatrix}$. Similarly, $A\begin{pmatrix} 3 \\ 2 \end{pmatrix} = \begin{pmatrix} 10 & -18 \\ 6 & -11 \end{pmatrix}\begin{pmatrix} 3 \\ 2 \end{pmatrix} = \begin{pmatrix} -6 \\ -4 \end{pmatrix} = -2\begin{pmatrix} 3 \\ 2 \end{pmatrix}$ so that $\lambda_2 = -2$ is an eigenvalue of A with corresponding eigenvector $\mathbf{v}_2 = \begin{pmatrix} 3 \\ 2 \end{pmatrix}$. As we soon shall see, these are the only eigenvalues of A.

EXAMPLE 2 EIGENVALUES AND EIGENVECTORS OF THE IDENTITY MATRIX

Let $A = I$. Then for any $\mathbf{v} \in \mathbb{C}^n$, $A\mathbf{v} = I\mathbf{v} = \mathbf{v}$. Thus 1 is the only eigenvalue of A and every $\mathbf{v} \neq \mathbf{0} \in \mathbb{C}^n$ is an eigenvector of I.

We shall compute the eigenvalues and eigenvectors of many matrices in this section. But first we need to prove some facts that will simplify our computations.

Suppose that λ is an eigenvalue of A. Then there exists a nonzero vector $\mathbf{v} = \begin{pmatrix} x_1 \\ x_2 \\ \vdots \\ x_n \end{pmatrix} \neq \mathbf{0}$ such that $A\mathbf{v} = \lambda\mathbf{v} = \lambda I\mathbf{v}$. Rewriting this, we have

$$(A - \lambda I)\mathbf{v} = \mathbf{0} \tag{3}$$

If A is an $n \times n$ matrix, equation (3) is a homogeneous system of n equations in the unknowns $x_1, x_2, \ldots, x_n$. Since, by assumption, the system has nontrivial solutions, we conclude that det $(A - \lambda I) = 0$. Conversely, if det $(A - \lambda I) = 0$, then equation (3) has nontrivial solutions and λ is an eigenvalue of A. On the other hand, if det $(A - \lambda I) \neq 0$, then (3) has only the solution $\mathbf{v} = \mathbf{0}$ so that λ is *not* an eigenvalue of A. Summing up these facts, we have the following theorem.

THEOREM 1

Let A be an $n \times n$ matrix. Then λ is an eigenvalue of A if and only if

$$p(\lambda) = \det(A - \lambda I) = 0 \tag{4}$$

■

DEFINITION CHARACTERISTIC EQUATION AND POLYNOMIAL

Equation (4) is called the **characteristic equation** of A; $p(\lambda)$ is called the **characteristic polynomial** of A. ■

As will become apparent in the examples, $p(\lambda)$ is a polynomial of degree n in λ. For example, if $A = \begin{pmatrix} a & b \\ c & d \end{pmatrix}$, then $A - \lambda I = \begin{pmatrix} a & b \\ c & d \end{pmatrix} - \begin{pmatrix} \lambda & 0 \\ 0 & \lambda \end{pmatrix} = \begin{pmatrix} a - \lambda & b \\ c & d - \lambda \end{pmatrix}$ and $p(\lambda) = \det(A - \lambda I) = (a - \lambda)(d - \lambda) - bc = \lambda^2 - (a + d)\lambda + (ad - bc)$.

According to the **fundamental theorem of algebra**, any polynomial of degree n with real or complex coefficients has exactly n roots (counting multiplicities). By this we mean, for example, that the polynomial $(\lambda - 1)^5$ has five roots, all equal to the number 1. Since any eigenvalue of A is a root of the characteristic equation of A, we conclude that

> Counting multiplicities, every $n \times n$ matrix has exactly n eigenvalues.

THEOREM 2

Let λ be an eigenvalue of the $n \times n$ matrix A and let $E_\lambda = \{\mathbf{v}: A\mathbf{v} = \lambda\mathbf{v}\}$. Then E_λ is a subspace of $\mathbb{C}^n$.

PROOF: If $A\mathbf{v} = \lambda\mathbf{v}$, then $(A - \lambda I)\mathbf{v} = \mathbf{0}$. Thus E_λ is the kernel of the matrix $A - \lambda I$, which, by Example 8.5.10 on page 521, is a subspace† of $\mathbb{C}^n$. ■

DEFINITION EIGENSPACE

Let λ be an eigenvalue of A. The subspace E_λ is called the **eigenspace**‡ of A corresponding to the eigenvalue λ. ■

We now prove another useful result.

† In Example 8.5.10 we saw that ker A is a subspace of $\mathbb{R}^n$ if A is a real matrix. The extension of this result to $\mathbb{C}^n$ presents no difficulties.

‡ Note that $\mathbf{0} \in E_\lambda$ since E_λ is a subspace. However, $\mathbf{0}$ is *not* an eigenvector.

THEOREM 3

Let A be an $n \times n$ matrix and let $\lambda_1, \lambda_2, \ldots, \lambda_m$ be distinct eigenvalues of A with corresponding eigenvectors $\mathbf{v}_1, \mathbf{v}_2, \ldots, \mathbf{v}_m$. Then $\mathbf{v}_1, \mathbf{v}_2, \ldots, \mathbf{v}_m$ are linearly independent. That is: *Eigenvectors corresponding to distinct eigenvalues are linearly independent.*

PROOF: We prove this by mathematical induction. We start with $m = 2$. Suppose that

$$c_1\mathbf{v}_1 + c_2\mathbf{v}_2 = \mathbf{0} \tag{5}$$

Then, multiplying both sides of (5) by A, we have

$$\mathbf{0} = A(c_1\mathbf{v}_1 + c_2\mathbf{v}_2) = c_1A\mathbf{v}_1 + c_2A\mathbf{v}_2$$

or (since $A\mathbf{v}_i = \lambda_i\mathbf{v}_i$ for $i = 1, 2$)

$$c_1\lambda_1\mathbf{v}_1 + c_2\lambda_2\mathbf{v}_2 = \mathbf{0} \tag{6}$$

We then multiply (5) by λ_1 and subtract it from (6) to obtain

$$(c_1\lambda_1\mathbf{v}_1 + c_2\lambda_2\mathbf{v}_2) - (c_1\lambda_1\mathbf{v}_1 + c_2\lambda_1\mathbf{v}_2) = \mathbf{0}$$

or

$$c_2(\lambda_2 - \lambda_1)\mathbf{v}_2 = \mathbf{0}$$

Since $\mathbf{v}_2 \neq \mathbf{0}$ (by the definition of an eigenvector) and since $\lambda_2 \neq \lambda_1$, we conclude that $c_2 = 0$. Then inserting $c_2 = 0$ in (5), we see that $c_1 = 0$, which proves the theorem in the case $m = 2$. Now suppose that the theorem is true for $m = k$. That is, we assume that any k eigenvectors corresponding to distinct eigenvalues are linearly independent. We prove the theorem for $m = k + 1$. So we assume that

$$c_1\mathbf{v}_1 + c_2\mathbf{v}_2 + \cdots + c_k\mathbf{v}_k + c_{k+1}\mathbf{v}_{k+1} = \mathbf{0} \tag{7}$$

Then, multiplying both sides of (7) by A and using the fact that $A\mathbf{v}_i = \lambda_i\mathbf{v}_i$, we obtain

$$c_1\lambda_1\mathbf{v}_1 + c_2\lambda_2\mathbf{v}_2 + \cdots + c_k\lambda_k\mathbf{v}_k + c_{k+1}\lambda_{k+1}\mathbf{v}_{k+1} = \mathbf{0} \tag{8}$$

We multiply both sides of (7) by λ_{k+1} and subtract it from (8):

$$c_1(\lambda_1 - \lambda_{k+1})\mathbf{v}_1 + c_2(\lambda_2 - \lambda_{k+1})\mathbf{v}_2 + \cdots + c_k(\lambda_k - \lambda_{k+1})\mathbf{v}_k = \mathbf{0}$$

But, by the induction assumption, $\mathbf{v}_1, \mathbf{v}_2, \ldots, \mathbf{v}_k$ are linearly independent. Thus $c_1(\lambda_1 - \lambda_{k+1}) = c_2(\lambda_2 - \lambda_{k+1}) = \cdots = c_k(\lambda_k - \lambda_{k+1}) = 0$; and, since $\lambda_i \neq \lambda_{k+1}$ for $i = 1, 2, \ldots, k$, we conclude that $c_1 = c_2 = \cdots = c_k = 0$. But, from (7), this means that $c_{k+1} = 0$. Thus the theorem is true for $m = k + 1$ and the proof is complete. ■

If

$$A = \begin{pmatrix} a_{11} & a_{12} & \cdots & a_{1n} \\ a_{21} & a_{22} & \cdots & a_{2n} \\ \vdots & \vdots & & \vdots \\ a_{n1} & a_{n2} & \cdots & a_{nm} \end{pmatrix}$$

then

$$p(\lambda) = \det(A - \lambda I) = \begin{vmatrix} a_{11} - \lambda & a_{12} & \cdots & a_{1n} \\ a_{21} & a_{22} - \lambda & \cdots & a_{2n} \\ \vdots & \vdots & & \vdots \\ a_{n1} & a_{n2} & \cdots & a_{nm} - \lambda \end{vmatrix}$$

and $p(\lambda) = 0$ can be written in the form

$$p(\lambda) = \lambda^n + b_{n-1}\lambda^{n-1} + \cdots + b_1\lambda + b_0 = 0 \qquad (9)$$

Equation (9) has n roots, some of which may be repeated. If $\lambda_1, \lambda_2, \ldots, \lambda_m$ are the distinct roots of (9) with multiplicities $r_1, r_2, \ldots, r_m$, respectively, then (9) may be factored to obtain

$$p(\lambda) = (\lambda - \lambda_1)^{r_1}(\lambda - \lambda_2)^{r_2} \cdots (\lambda - \lambda_m)^{r_m} = 0 \qquad (10)$$

The numbers $r_1, r_2, \ldots, r_m$ are called the **algebraic multiplicities** of the eigenvalues $\lambda_1, \lambda_2, \ldots, \lambda_m$, respectively.

We now calculate eigenvalues and corresponding eigenspaces. We do this in a three-step procedure:

PROCEDURE FOR COMPUTING EIGENVALUES AND EIGENVECTORS

i. Find $p(\lambda) = \det(A - \lambda I)$.

ii. Find the roots $\lambda_1, \lambda_2, \ldots, \lambda_m$ of $p(\lambda) = 0$.

iii. Corresponding to each eigenvalue λ_i, solve the homogeneous system $(A - \lambda_i I)\mathbf{v} = \mathbf{0}$.

REMARK 1: Step (ii) is usually the most difficult one.

REMARK 2: A relatively easy way to find eigenvalues and eigenvectors of 2×2 matrices is suggested in Problems 35 and 36.

EXAMPLE 3

COMPUTING EIGENVALUES AND EIGENVECTORS

Let $A = \begin{pmatrix} 4 & 2 \\ 3 & 3 \end{pmatrix}$. Then $\det(A - \lambda I) = \begin{vmatrix} 4 - \lambda & 2 \\ 3 & 3 - \lambda \end{vmatrix} = (4 - \lambda)(3 - \lambda) - 6 = \lambda^2 - 7\lambda + 6 = (\lambda - 1)(\lambda - 6) = 0$. Thus the eigenvalues of A are $\lambda_1 = 1$ and $\lambda_2 = 6$. For $\lambda_1 = 1$, we solve $(A - I)\mathbf{v} = \mathbf{0}$ or $\begin{pmatrix} 3 & 2 \\ 3 & 2 \end{pmatrix}\begin{pmatrix} x_1 \\ x_2 \end{pmatrix} = \begin{pmatrix} 0 \\ 0 \end{pmatrix}$. Clearly any eigenvector corresponding to $\lambda_1 = 1$ satisfies $3x_1 + 2x_2 = 0$. One such eigenvector is $\mathbf{v}_1 = \begin{pmatrix} 2 \\ -3 \end{pmatrix}$. Thus $E_1 = \text{span}\left\{\begin{pmatrix} 2 \\ -3 \end{pmatrix}\right\}$. Similarly, the equation $(A - 6I)\mathbf{v} = \mathbf{0}$ means that

$\begin{pmatrix} -2 & 2 \\ 3 & -3 \end{pmatrix}\begin{pmatrix} x_1 \\ x_2 \end{pmatrix} = \begin{pmatrix} 0 \\ 0 \end{pmatrix}$ or $x_1 = x_2$. Thus $\mathbf{v}_2 = \begin{pmatrix} 1 \\ 1 \end{pmatrix}$ is an eigenvector corresponding to $\lambda_2 = 6$ and $E_6 = \text{span}\left\{\begin{pmatrix} 1 \\ 1 \end{pmatrix}\right\}$. Note that $\mathbf{v}_1$ and $\mathbf{v}_2$ are linearly independent since one is not a multiple of the other.

EXAMPLE 4

A 3 × 3 MATRIX WITH DISTINCT EIGENVALUES

Let $A = \begin{pmatrix} 1 & -1 & 4 \\ 3 & 2 & -1 \\ 2 & 1 & -1 \end{pmatrix}$. Then $\det(A - \lambda I) = \begin{vmatrix} 1-\lambda & -1 & 4 \\ 3 & 2-\lambda & -1 \\ 2 & 1 & -1-\lambda \end{vmatrix}$

$$= -(\lambda^3 - 2\lambda^2 - 5\lambda + 6) = -(\lambda - 1)(\lambda + 2)(\lambda - 3) = 0$$

Thus the eigenvalues of A are $\lambda_1 = 1$, $\lambda_2 = -2$, and $\lambda_3 = 3$. Corresponding to $\lambda_1 = 1$ we have

$$(A - I)\mathbf{v} = \begin{pmatrix} 0 & -1 & 4 \\ 3 & 1 & -1 \\ 2 & 1 & -2 \end{pmatrix}\begin{pmatrix} x_1 \\ x_2 \\ x_3 \end{pmatrix} = \begin{pmatrix} 0 \\ 0 \\ 0 \end{pmatrix}$$

Solving by row reduction, we obtain, successively,

$$\left(\begin{array}{ccc|c} 0 & -1 & 4 & 0 \\ 3 & 1 & -1 & 0 \\ 2 & 1 & -2 & 0 \end{array}\right) \xrightarrow[R_3 \to R_3 + R_1]{R_2 \to R_2 + R_1} \left(\begin{array}{ccc|c} 0 & -1 & 4 & 0 \\ 3 & 0 & 3 & 0 \\ 2 & 0 & 2 & 0 \end{array}\right)$$

$$\xrightarrow{R_2 \to \frac{1}{3}R_2} \left(\begin{array}{ccc|c} 0 & -1 & 4 & 0 \\ 1 & 0 & 1 & 0 \\ 2 & 0 & 2 & 0 \end{array}\right) \xrightarrow{R_3 \to R_3 - 2R_2} \left(\begin{array}{ccc|c} 0 & -1 & 4 & 0 \\ 1 & 0 & 1 & 0 \\ 0 & 0 & 0 & 0 \end{array}\right)$$

Thus $x_1 = -x_3$, $x_2 = 4x_3$, an eigenvector is $\mathbf{v}_1 = \begin{pmatrix} -1 \\ 4 \\ 1 \end{pmatrix}$, and $E_1 = \text{span}\left\{\begin{pmatrix} -1 \\ 4 \\ 1 \end{pmatrix}\right\}$.

For $\lambda_2 = -2$, we have $[A - (-2I)]\mathbf{v} = (A + 2I)\mathbf{v} = \mathbf{0}$ or

$$\begin{pmatrix} 3 & -1 & 4 \\ 3 & 4 & -1 \\ 2 & 1 & 1 \end{pmatrix}\begin{pmatrix} x_1 \\ x_2 \\ x_3 \end{pmatrix} = \begin{pmatrix} 0 \\ 0 \\ 0 \end{pmatrix}.$$

This leads to

$$\left(\begin{array}{ccc|c} 3 & -1 & 4 & 0 \\ 3 & 4 & -1 & 0 \\ 2 & 1 & -1 & 0 \end{array}\right) \xrightarrow[R_3 \to R_3 + R_1]{R_2 \to R_2 + 4R_1} \left(\begin{array}{ccc|c} 3 & -1 & 4 & 0 \\ 15 & 0 & 15 & 0 \\ 5 & 0 & 5 & 0 \end{array}\right)$$

$$\xrightarrow{R_2 \to \frac{1}{15}R_2} \left(\begin{array}{ccc|c} 3 & -1 & 4 & 0 \\ 1 & 0 & 1 & 0 \\ 5 & 0 & 5 & 0 \end{array}\right) \xrightarrow[R_3 \to R_3 - 5R_2]{R_1 \to R_1 - 4R_2} \left(\begin{array}{ccc|c} -1 & -1 & 0 & 0 \\ 1 & 0 & 1 & 0 \\ 0 & 0 & 0 & 0 \end{array}\right)$$

Then $x_2 = -x_1$, $x_3 = -x_1$, and an eigenvector is $\mathbf{v}_2 = \begin{pmatrix} 1 \\ -1 \\ -1 \end{pmatrix}$. Thus $E_{-2} =$ span $\left\{\begin{pmatrix} 1 \\ -1 \\ -1 \end{pmatrix}\right\}$. Finally, for $\lambda_3 = 3$, we have

$$(A - 3I)\mathbf{v} = \begin{pmatrix} -2 & -1 & 4 \\ 3 & -1 & -1 \\ 2 & 1 & -4 \end{pmatrix}\begin{pmatrix} x_1 \\ x_2 \\ x_3 \end{pmatrix} = \begin{pmatrix} 0 \\ 0 \\ 0 \end{pmatrix}$$

and

$$\left(\begin{array}{rrr|r} -2 & -1 & 4 & 0 \\ 3 & -1 & -1 & 0 \\ 2 & 1 & -4 & 0 \end{array}\right) \xrightarrow[R_3 \to R_3 + R_1]{R_2 \to R_2 - R_1} \left(\begin{array}{rrr|r} -2 & -1 & 4 & 0 \\ 5 & 0 & -5 & 0 \\ 0 & 0 & 0 & 0 \end{array}\right)$$

$$\xrightarrow{R_2 \to \frac{1}{5}R_2} \left(\begin{array}{rrr|r} -2 & -1 & 4 & 0 \\ 1 & 0 & -1 & 0 \\ 0 & 0 & 0 & 0 \end{array}\right) \xrightarrow{R_1 \to R_1 + 4R_2} \left(\begin{array}{rrr|r} 2 & -1 & 0 & 0 \\ 1 & 0 & -1 & 0 \\ 0 & 0 & 0 & 0 \end{array}\right)$$

Hence $x_3 = x_1$, $x_2 = 2x_1$, and $\mathbf{v}_3 = \begin{pmatrix} 1 \\ 2 \\ 1 \end{pmatrix}$ so that $E_3 = \text{span}\left\{\begin{pmatrix} 1 \\ 2 \\ 1 \end{pmatrix}\right\}$.

REMARK: In this and every other example there are always an infinite number of choices for each eigenvector. We arbitrarily choose a simple one by setting one or more of the x_i's equal to a convenient number. Here we have set one of the x_i's equal to 1.

EXAMPLE 5

A 2 × 2 MATRIX WITH ONE OF ITS EIGENVALUES EQUAL TO ZERO

Let $A = \begin{pmatrix} 2 & -1 \\ -4 & 2 \end{pmatrix}$. Then $\det(A - \lambda I) = \begin{vmatrix} 2-\lambda & -1 \\ -4 & 2-\lambda \end{vmatrix} = \lambda^2 - 4\lambda = \lambda(\lambda - 4)$. Thus the eigenvalues are $\lambda_1 = 0$ and $\lambda_2 = 4$. The eigenspace corresponding to zero is simply the kernel of A. We calculate $\begin{pmatrix} 2 & -1 \\ -4 & 2 \end{pmatrix}\begin{pmatrix} x_1 \\ x_2 \end{pmatrix} = \begin{pmatrix} 0 \\ 0 \end{pmatrix}$ or $2x_1 = x_2$ and an eigenvector is $\mathbf{v}_1 = \begin{pmatrix} 1 \\ 2 \end{pmatrix}$. Thus $\ker A = E_0 = \text{span}\left\{\begin{pmatrix} 1 \\ 2 \end{pmatrix}\right\}$. Corresponding to $\lambda_2 = 4$ we have $\begin{pmatrix} -2 & -1 \\ -4 & -2 \end{pmatrix}\begin{pmatrix} x_1 \\ x_2 \end{pmatrix} = \begin{pmatrix} 0 \\ 0 \end{pmatrix}$, so $E_4 = \text{span}\left\{\begin{pmatrix} 1 \\ -2 \end{pmatrix}\right\}$.

EXAMPLE 6 **A 2 × 2 MATRIX WITH COMPLEX CONJUGATE EIGENVALUES**

Let $A = \begin{pmatrix} 3 & -5 \\ 1 & -1 \end{pmatrix}$. Then $\det(A - \lambda I) = \begin{vmatrix} 3-\lambda & -5 \\ 1 & -1-\lambda \end{vmatrix} = \lambda^2 - 2\lambda + 2 = 0$ and

$$\lambda = \frac{-(-2) \pm \sqrt{4 - 4(1)(2)}}{2} = \frac{2 \pm \sqrt{-4}}{2} = \frac{2 \pm 2i}{2} = 1 \pm i$$

Thus $\lambda_1 = 1 + i$ and $\lambda_2 = 1 - i$. We compute

$$[A - (1+i)I]\mathbf{v} = \begin{pmatrix} 2-i & -5 \\ 1 & -1-i \end{pmatrix}^{\dagger} \begin{pmatrix} x_1 \\ x_2 \end{pmatrix} = \begin{pmatrix} 0 \\ 0 \end{pmatrix}$$

and we obtain $(2 - i)x_1 - 5x_2 = 0$ and $x_1 + (-2 - i)x_2 = 0$. Thus $x_1 = (2 + i)x_2$, which yields the eigenvector $\mathbf{v}_1 = \begin{pmatrix} 2+i \\ 1 \end{pmatrix}$ and $E_{1+i} = \text{span}\left\{\begin{pmatrix} 2+i \\ 1 \end{pmatrix}\right\}$. Similarly, $[A - (1-i)I]\mathbf{v} = \begin{pmatrix} 2+i & -5 \\ 1 & -1+i \end{pmatrix}\begin{pmatrix} x_1 \\ x_2 \end{pmatrix} = \begin{pmatrix} 0 \\ 0 \end{pmatrix}$ or $x_1 + (-2 + i)x_2 = 0$, which yields $x_1 = (2 - i)x_2$, $\mathbf{v}_2 = \begin{pmatrix} 2-i \\ 1 \end{pmatrix}$, and $E_{1-i} = \text{span}\left\{\begin{pmatrix} 2-i \\ 1 \end{pmatrix}\right\}$.

REMARK 1: This example illustrates that a real matrix may have complex eigenvalues and eigenvectors. Some texts define eigenvalues of real matrices to be *real* roots of the characteristic equation. With this definition the matrix of the last example has *no* eigenvalues. This might make the computations simpler, but it also greatly reduces the usefulness of the theory of eigenvalues and eigenvectors.

REMARK 2: Note that $\lambda_2 = 1 - i$ is the complex conjugate of $\lambda_1 = 1 + i$. Also, the components of $\mathbf{v}_2$ are complex conjugates of the components of $\mathbf{v}_1$. This is no coincidence. In Problem 33 you are asked to prove that

> The eigenvalues of a real matrix occur in complex conjugate pairs
> and
> corresponding eigenvectors are complex conjugates of one another.

EXAMPLE 7 **A 2 × 2 MATRIX WITH ONE EIGENVALUE AND TWO LINEARLY INDEPENDENT EIGENVECTORS**

Let $A = \begin{pmatrix} 4 & 0 \\ 0 & 4 \end{pmatrix}$. Then det $(A - \lambda I) = \begin{vmatrix} 4-\lambda & 0 \\ 0 & 4-\lambda \end{vmatrix} = (\lambda - 4)^2 = 0$; hence $\lambda = 4$ is an eigenvalue of algebraic multiplicity 2. It is obvious that $A\mathbf{v} = 4\mathbf{v}$ for every vector $\mathbf{v} \in \mathbb{R}^2$ so that $E_4 = \mathbb{R}^2 = \text{span}\left\{\begin{pmatrix} 1 \\ 0 \end{pmatrix}, \begin{pmatrix} 0 \\ 1 \end{pmatrix}\right\}$.

---

† Note that the columns of this matrix are linearly dependent because $\begin{pmatrix} -5 \\ -2-i \end{pmatrix} = (-2-i)\begin{pmatrix} 2-i \\ 1 \end{pmatrix}$.

EXAMPLE 8 A 2×2 MATRIX WITH ONE EIGENVALUE AND ONLY ONE LINEARLY INDEPENDENT EIGENVECTOR

Let $A = \begin{pmatrix} 4 & 1 \\ 0 & 4 \end{pmatrix}$. Then $\det(A - \lambda I) = \begin{vmatrix} 4-\lambda & 1 \\ 0 & 4-\lambda \end{vmatrix} = (\lambda - 4)^2 = 0$; thus $\lambda = 4$ is again an eigenvalue of algebraic multiplicity 2. But this time we have $(A - 4I)\mathbf{v} = \begin{pmatrix} 0 & 1 \\ 0 & 0 \end{pmatrix}\begin{pmatrix} x_1 \\ x_2 \end{pmatrix} = \begin{pmatrix} x_2 \\ 0 \end{pmatrix}$. Thus $x_2 = 0$, $\mathbf{v}_1 = \begin{pmatrix} 1 \\ 0 \end{pmatrix}$ is an eigenvector, and $E_4 = \text{span}\left\{\begin{pmatrix} 1 \\ 0 \end{pmatrix}\right\}$.

EXAMPLE 9 A 3×3 MATRIX WITH TWO DISTINCT EIGENVALUES AND THREE LINEARLY INDEPENDENT EIGENVECTORS

Let $A = \begin{pmatrix} 3 & 2 & 4 \\ 2 & 0 & 2 \\ 4 & 2 & 3 \end{pmatrix}$. Then $\det(A - \lambda I) = \begin{vmatrix} 3-\lambda & 2 & 4 \\ 2 & -\lambda & 2 \\ 4 & 2 & 3-\lambda \end{vmatrix} = -\lambda^3 + 6\lambda^2 + 15\lambda + 8 = -(\lambda + 1)^2(\lambda - 8) = 0$ so that the eigenvalues are $\lambda_1 = 8$ and $\lambda_2 = -1$ (with algebraic multiplicity 2). For $\lambda_1 = 8$, we obtain

$$(A - 8I)\mathbf{v} = \begin{pmatrix} -5 & 2 & 4 \\ 2 & -8 & 2 \\ 4 & 2 & -5 \end{pmatrix}\begin{pmatrix} x_1 \\ x_2 \\ x_3 \end{pmatrix} = \begin{pmatrix} 0 \\ 0 \\ 0 \end{pmatrix}$$

or, row reducing,

$$\left(\begin{array}{ccc|c} -5 & 2 & 4 & 0 \\ 2 & -8 & 2 & 0 \\ 4 & 2 & -5 & 0 \end{array}\right) \xrightarrow[R_3 \to R_3 - R_1]{R_2 \to R_2 + 4R_1} \left(\begin{array}{ccc|c} -5 & 2 & 4 & 0 \\ -18 & 0 & 18 & 0 \\ 9 & 0 & -9 & 0 \end{array}\right)$$

$$\xrightarrow{R_2 \to \frac{1}{18}R_2} \left(\begin{array}{ccc|c} -5 & 2 & 4 & 0 \\ -1 & 0 & 1 & 0 \\ 9 & 0 & -9 & 0 \end{array}\right) \xrightarrow[R_3 \to R_3 + 9R_2]{R_1 \to R_1 - 5R_2} \left(\begin{array}{ccc|c} 0 & 2 & -1 & 0 \\ -1 & 0 & 1 & 0 \\ 0 & 0 & 0 & 0 \end{array}\right)$$

Hence $x_3 = 2x_2$ and $x_1 = x_3$, we obtain the eigenvector $\mathbf{v}_1 = \begin{pmatrix} 2 \\ 1 \\ 2 \end{pmatrix}$, and $E_8 = \text{span}\left\{\begin{pmatrix} 2 \\ 1 \\ 2 \end{pmatrix}\right\}$.

For $\lambda_2 = -1$, we have $(A + I)\mathbf{v} = \begin{pmatrix} 4 & 2 & 4 \\ 2 & 1 & 2 \\ 4 & 2 & 4 \end{pmatrix}\begin{pmatrix} x_1 \\ x_2 \\ x_3 \end{pmatrix} = \begin{pmatrix} 0 \\ 0 \\ 0 \end{pmatrix}$, which gives us the single equation $2x_1 + x_2 + 2x_3 = 0$ or $x_2 = -2x_1 - 2x_3$. If $x_1 = 1$ and $x_3 = 0$, we obtain $\mathbf{v}_2 = \begin{pmatrix} 1 \\ -2 \\ 0 \end{pmatrix}$. If $x_1 = 0$ and $x_3 = 1$, we obtain $\mathbf{v}_3 = \begin{pmatrix} 0 \\ -2 \\ 1 \end{pmatrix}$. Thus $E_{-1} = \text{span}\left\{\begin{pmatrix} 1 \\ -2 \\ 0 \end{pmatrix},\right.$

$\left.\begin{pmatrix} 0 \\ -2 \\ 1 \end{pmatrix}\right\}$. There are other convenient choices for eigenvectors. For example,

$\mathbf{v} = \begin{pmatrix} 1 \\ 0 \\ -1 \end{pmatrix}$ is in E_{-1} since $\mathbf{v} = \mathbf{v}_2 - \mathbf{v}_3$.

EXAMPLE 10

A 3 × 3 MATRIX WITH ONE EIGENVALUE AND ONLY ONE LINEARLY INDEPENDENT EIGENVECTOR

Let $A = \begin{pmatrix} -5 & -5 & -9 \\ 8 & 9 & 18 \\ -2 & -3 & -7 \end{pmatrix}$. Then $\det\ (A - \lambda I) = \begin{vmatrix} -5-\lambda & -5 & -9 \\ 8 & 9-\lambda & 18 \\ -2 & -3 & -7-\lambda \end{vmatrix} =$ $-\lambda^3 - 3\lambda^2 - 3\lambda - 1 = -(\lambda + 1)^3 = 0$. Thus $\lambda = -1$ is an eigenvalue of algebraic multiplicity 3. To compute E_{-1}, we set $(A + I)\mathbf{v} = \begin{pmatrix} -4 & -5 & -9 \\ 8 & 10 & 18 \\ -2 & -3 & -6 \end{pmatrix}\begin{pmatrix} x_1 \\ x_2 \\ x_3 \end{pmatrix} = \begin{pmatrix} 0 \\ 0 \\ 0 \end{pmatrix}$ and row-reduce to obtain, successively,

$$\left(\begin{array}{ccc|c} -4 & -5 & -9 & 0 \\ 8 & 10 & 18 & 0 \\ -2 & -3 & -6 & 0 \end{array}\right) \xrightarrow[R_2 \to R_2 + 4R_3]{R_1 \to R_1 - 2R_3} \left(\begin{array}{ccc|c} 0 & 1 & 3 & 0 \\ 0 & -2 & -6 & 0 \\ -2 & -3 & -6 & 0 \end{array}\right)$$

$$\xrightarrow[R_3 \to R_3 + 3R_1]{R_2 \to R_2 + 2R_1} \left(\begin{array}{ccc|c} 0 & 1 & 3 & 0 \\ 0 & 0 & 0 & 0 \\ -2 & 0 & 3 & 0 \end{array}\right)$$

This yields $x_2 = -3x_3$ and $2x_1 = 3x_3$. Setting $x_3 = 2$, we obtain only one linearly independent eigenvector: $\mathbf{v}_1 = \begin{pmatrix} 3 \\ -6 \\ 2 \end{pmatrix}$. Thus $E_{-1} = \text{span}\left\{\begin{pmatrix} 3 \\ -6 \\ 2 \end{pmatrix}\right\}$.

EXAMPLE 11

A 3 × 3 MATRIX WITH ONE EIGENVALUE AND TWO LINEARLY INDEPENDENT EIGENVECTORS

Let $A = \begin{pmatrix} -1 & -3 & -9 \\ 0 & 5 & 18 \\ 0 & -2 & -7 \end{pmatrix}$. Then $\det\ (A - \lambda I) = \begin{vmatrix} -1-\lambda & -3 & -9 \\ 0 & 5-\lambda & 18 \\ 0 & -2 & -7-\lambda \end{vmatrix} =$ $-(\lambda + 1)^3 = 0$. Thus, as in Example 10, $\lambda = -1$ is an eigenvalue of algebraic multiplicity 3. To find E_{-1}, we compute $(A + I)\mathbf{v} = \begin{pmatrix} 0 & -3 & -9 \\ 0 & 6 & 18 \\ 0 & -2 & -6 \end{pmatrix}\begin{pmatrix} x_1 \\ x_2 \\ x_3 \end{pmatrix} = \begin{pmatrix} 0 \\ 0 \\ 0 \end{pmatrix}$. Thus $-2x_2 - 6x_3 = 0$ or $x_2 = -3x_3$, and x_1 is arbitrary. Setting $x_1 = 0$, $x_3 = 1$, we obtain

$\mathbf{v}_1 = \begin{pmatrix} 0 \\ -3 \\ 1 \end{pmatrix}$. Setting $x_1 = 1$, $x_3 = 1$ yields $\mathbf{v}_2 = \begin{pmatrix} 1 \\ -3 \\ 1 \end{pmatrix}$. Thus

$$E_{-1} = \text{span}\left\{\begin{pmatrix} 0 \\ -3 \\ 1 \end{pmatrix}, \begin{pmatrix} 1 \\ -3 \\ 1 \end{pmatrix}\right\}.$$

In each of the last five examples we found an eigenvalue with an algebraic multiplicity of 2 or more. But, as we saw in Examples 8, 10, and 11, the number of linearly independent eigenvectors is not necessarily equal to the algebraic multiplicity of the eigenvalue (as was the case in Examples 7 and 9). This observation leads to the following definition.

DEFINITION GEOMETRIC MULTIPLICITY

Let λ be an eigenvalue of the matrix A. Then the **geometric multiplicity** of λ is the dimension of the eigenspace corresponding to λ (which is the nullity of the matrix $A - \lambda I$). That is,

$$\text{Geometric multiplicity of } \lambda = \dim E_\lambda = \nu(A - \lambda I)$$

■

In Examples 7 and 9 we saw that for the eigenvalues of algebraic multiplicity 2, the geometric multiplicities were also 2. In Example 8 the geometric multiplicity of $\lambda = 4$ was 1 while the algebraic multiplicity was 2. In Example 10 the algebraic multiplicity was 3 and the geometric multiplicity was 1. In Example 11 the algebraic multiplicity was 3 and the geometric multiplicity was 2. These examples illustrate the fact that if the algebraic multiplicity of λ is greater than 1, then we cannot predict the geometric multiplicity of λ without additional information.

If A is a 2×2 matrix and λ is an eigenvalue with algebraic multiplicity 2, then the geometric multiplicity of λ is ≤ 2 since there can be at most two linearly independent vectors in a two-dimensional space. Let A be a 3×3 matrix having two eigenvalues λ_1 and λ_2 with algebraic multiplicities 1 and 2, respectively. Then the geometric multiplicity of λ_2 is ≤ 2 because otherwise we would have at least four linearly independent vectors in a three-dimensional space. Intuitively, it seems that the geometric multiplicity of an eigenvalue is always less than or equal to its algebraic multiplicity. The proof of the following theorem is not difficult if additional facts about determinants are proved. Since this would take us too far afield, we omit the proof.†

THEOREM 4

Let λ be an eigenvalue of A. Then

$$\text{Geometric multiplicity of } \lambda \leq \text{algebraic multiplicity of } \lambda$$

■

†For a proof see Theorem 11.2.6 in C. R. Wylie's book *Advanced Engineering Mathematics* (New York: McGraw-Hill, 1975).

NOTE: The geometric multiplicity of an eigenvalue is never zero. This follows from the definition of eigenvalue and eigenvector, which states that if λ is an eigenvalue, then there exists a *nonzero* eigenvector corresponding to λ.

An important problem is to determine whether a given $n \times n$ matrix does or does not have n linearly independent eigenvectors. From what we have already discussed in this section, the following theorem is apparent.

THEOREM 5

Let A be an $n \times n$ matrix. Then A has n linearly independent eigenvectors if and only if the geometric multiplicity of every eigenvalue is equal to its algebraic multiplicity. In particular, A has n linearly independent eigenvectors if all the eigenvalues are distinct (since then the algebraic multiplicity of every eigenvalue is 1). ■

In Example 5 we saw a matrix for which zero was an eigenvalue. In fact, from Theorem 1, it is evident that zero is an eigenvalue of A if and only if $\det A = \det (A - 0I) = 0$. This enables us to extend, for the last time, our Summing Up Theorem (see Theorem 8.12.4, page 583).

THEOREM 6 SUMMING UP THEOREM—VIEW 8

Let A be an $n \times n$ matrix. Then the following eleven statements are equivalent; that is, each one implies the other ten (so that if one is true, all are true):

i. A is invertible.

ii. The only solution to the homogeneous system $A\mathbf{x} = \mathbf{0}$ is the trivial solution ($\mathbf{x} = \mathbf{0}$).

iii. The system $A\mathbf{x} = \mathbf{b}$ has a unique solution for every n-vector $\mathbf{b}$.

iv. A is row equivalent to the $n \times n$ identity matrix I_n.

v. A is a product of elementary matrices.

vi. The rows (and columns) of A are linearly independent.

vii. $\det A \neq 0$.

viii. $\nu(A) = 0$.

ix. $\rho(A) = n$.

x. The linear transformation T from $\mathbb{R}^n$ to $\mathbb{R}^n$ defined by $T\mathbf{x} = A\mathbf{x}$ is an isomorphism.

xi. Zero is *not* an eigenvalue of A. ■

PROBLEMS 8.13

SELF-QUIZ

True–False

I. The eigenvalues of a triangular matrix are the numbers on the diagonal of the matrix.

II. If the 3×3 matrix A has three distinct eigenvalues, then the eigenvectors corresponding to those eigenvalues constitute a basis for $\mathbb{R}^3$.

III. If the 3×3 matrix A has two distinct

eigenvalues, then A has at most two linearly independent eigenvectors.

IV. If A has real entries, then A can have exactly one complex eigenvalue (i.e., an eigenvalue $a + ib$ with $b \neq 0$).

Multiple Choice

V. 1 is an eigenvalue for the 3×3 identity matrix. Its geometric multiplicity is ________.

a. 1 **b.** 2 **c.** 3

VI. 1 is the only eigenvalue of $A = \begin{pmatrix} 1 & 2 & 0 \\ 0 & 1 & 0 \\ 0 & 0 & 1 \end{pmatrix}$. Its geometric multiplicity is ________.

a. 1 **b.** 2 **c.** 3

In Problems 1–20 calculate the eigenvalues and eigenspaces of the given matrix. If the algebraic multiplicity of an eigenvalue is greater than 1, calculate its geometric multiplicity.

1. $\begin{pmatrix} -2 & -2 \\ -5 & 1 \end{pmatrix}$

2. $\begin{pmatrix} -12 & 7 \\ -7 & 2 \end{pmatrix}$

3. $\begin{pmatrix} 2 & -1 \\ 5 & -2 \end{pmatrix}$

4. $\begin{pmatrix} -3 & 0 \\ 0 & -3 \end{pmatrix}$

5. $\begin{pmatrix} -3 & 2 \\ 0 & -3 \end{pmatrix}$

6. $\begin{pmatrix} 3 & 2 \\ -5 & 1 \end{pmatrix}$

7. $\begin{pmatrix} 1 & -1 & 0 \\ -1 & 2 & -1 \\ 0 & -1 & 1 \end{pmatrix}$

8. $\begin{pmatrix} 1 & 1 & -2 \\ -1 & 2 & 1 \\ 0 & 1 & -1 \end{pmatrix}$

9. $\begin{pmatrix} 5 & 4 & 2 \\ 4 & 5 & 2 \\ 2 & 2 & 2 \end{pmatrix}$

10. $\begin{pmatrix} 1 & 2 & 2 \\ 0 & 2 & 1 \\ -1 & 2 & 2 \end{pmatrix}$

11. $\begin{pmatrix} 0 & 1 & 0 \\ 0 & 0 & 1 \\ 1 & -3 & 3 \end{pmatrix}$

12. $\begin{pmatrix} 3 & -7 & -5 \\ 2 & 4 & 3 \\ 1 & 2 & 2 \end{pmatrix}$

13. $\begin{pmatrix} 1 & -1 & -1 \\ 1 & -1 & 0 \\ 1 & 0 & -1 \end{pmatrix}$

14. $\begin{pmatrix} 7 & -2 & -4 \\ 3 & 0 & -2 \\ 6 & -2 & -3 \end{pmatrix}$

15. $\begin{pmatrix} 4 & 6 & 6 \\ 1 & 3 & 2 \\ -1 & -5 & -2 \end{pmatrix}$

16. $\begin{pmatrix} 4 & 1 & 0 & 1 \\ 2 & 3 & 0 & 1 \\ -2 & 1 & 2 & -3 \\ 2 & -1 & 0 & 5 \end{pmatrix}$

17. $\begin{pmatrix} a & 0 & 0 & 0 \\ 0 & a & 0 & 0 \\ 0 & 0 & a & 0 \\ 0 & 0 & 0 & a \end{pmatrix}$

18. $\begin{pmatrix} a & b & 0 & 0 \\ 0 & a & 0 & 0 \\ 0 & 0 & a & 0 \\ 0 & 0 & 0 & a \end{pmatrix}$; $b \neq 0$

19. $\begin{pmatrix} a & b & 0 & 0 \\ 0 & a & c & 0 \\ 0 & 0 & a & 0 \\ 0 & 0 & 0 & a \end{pmatrix}$; $bc \neq 0$

20. $\begin{pmatrix} a & b & 0 & 0 \\ 0 & a & c & 0 \\ 0 & 0 & a & d \\ 0 & 0 & 0 & a \end{pmatrix}$; $bcd \neq 0$

21. Show that for any real numbers a and b, the matrix $A = \begin{pmatrix} a & b \\ -b & a \end{pmatrix}$ has the eigenvectors $\begin{pmatrix} 1 \\ i \end{pmatrix}$ and $\begin{pmatrix} 1 \\ -i \end{pmatrix}$.

In Problems 22–28 assume that the matrix A has the eigenvalues $\lambda_1, \lambda_2, \ldots, \lambda_k$.

22. Show that the eigenvalues of A^t are $\lambda_1, \lambda_2, \ldots, \lambda_k$.

23. Show that the eigenvalues of αA are $\alpha\lambda_1, \alpha\lambda_2, \ldots, \alpha\lambda_k$.

24. Show that A^{-1} exists if and only if $\lambda_1\lambda_2 \cdots \lambda_k \neq 0$.

***25.** If A^{-1} exists, show that the eigenvalues of A^{-1} are $1/\lambda_1, 1/\lambda_2, \ldots, 1/\lambda_k$.

26. Show that the matrix $A - \alpha I$ has the eigenvalues $\lambda_1 - \alpha, \lambda_2 - \alpha, \ldots, \lambda_k - \alpha$.

***27.** Show that the eigenvalues of A^2 are $\lambda_1^2, \lambda_2^2, \ldots, \lambda_k^2$.

***28.** Show that the eigenvalues of A^m are $\lambda_1^m, \lambda_2^m, \ldots, \lambda_k^m$ for $m = 1, 2, 3, \ldots$.

29. Let λ be an eigenvalue of A with corresponding eigenvector $\mathbf{v}$. Let $p(\lambda) = a_0 + a_1\lambda + a_2\lambda^2 + \cdots + a_n\lambda^n$. Define the matrix $p(A)$ by $p(A) = a_0 I + a_1 A + a_2 A^2 + \cdots + a_n A^n$. Show that $p(A)\mathbf{v} = p(\lambda)\mathbf{v}$.

30. Using the result of Problem 29, show that if $\lambda_1, \lambda_2, \ldots, \lambda_k$ are eigenvalues of A, then $p(\lambda_1), p(\lambda_2), \ldots, p(\lambda_k)$ are eigenvalues of $p(A)$.

31. Show that if A is an upper triangular matrix, then the eigenvalues of A are the diagonal components of A.

32. Let $A_1 = \begin{pmatrix} 2 & 0 & 0 & 0 \\ 0 & 2 & 0 & 0 \\ 0 & 0 & 2 & 0 \\ 0 & 0 & 0 & 2 \end{pmatrix}$, $A_2 = \begin{pmatrix} 2 & 1 & 0 & 0 \\ 0 & 2 & 0 & 0 \\ 0 & 0 & 2 & 0 \\ 0 & 0 & 0 & 2 \end{pmatrix}$, $A_3 = \begin{pmatrix} 2 & 1 & 0 & 0 \\ 0 & 2 & 1 & 0 \\ 0 & 0 & 2 & 0 \\ 0 & 0 & 0 & 2 \end{pmatrix}$, and $A_4 = \begin{pmatrix} 2 & 1 & 0 & 0 \\ 0 & 2 & 1 & 0 \\ 0 & 0 & 2 & 1 \\ 0 & 0 & 0 & 2 \end{pmatrix}$. Show that, for each matrix, $\lambda = 2$ is an eigenvalue of algebraic multiplicity 4. In each case compute the geometric multiplicity of $\lambda = 2$.

***33.** Let A be a real $n \times n$ matrix. Show that if λ_1 is a complex eigenvalue of A with eigenvector $\mathbf{v}_1$, then $\overline{\lambda}_1$ is an eigenvalue of A with eigenvector $\overline{\mathbf{v}}_1$.

***34.** A **probability matrix** is an $n \times n$ matrix having two properties:

i. $a_{ij} \geq 0$ for every i and j.

ii. The sum of the components in every column is 1.

Prove that 1 is an eigenvalue of every probability matrix.

***35.** Let $A = \begin{pmatrix} a & b \\ c & d \end{pmatrix}$ be a 2×2 matrix. Suppose that $b \neq 0$. Let m be a root (real or complex) of the equation

$$bm^2 + (a - d)m - c = 0$$

Show that $a + bm$ is an eigenvalue of A with corresponding eigenvector $\mathbf{v} = \begin{pmatrix} 1 \\ m \end{pmatrix}$. This gives us an easy way to compute eigenvalues and eigenvectors of 2×2 matrices. [This procedure appeared in the paper "A Simple Algorithm for Finding Eigenvalues and Eigenvectors for 2×2 Matrices" by Tyre A. Newton in The *American Mathematical Monthly,* 97(1), January, 1990, 57–60.]

36. Let $A = \begin{pmatrix} a & 0 \\ c & d \end{pmatrix}$ be a 2×2 matrix. Show that d is an eigenvalue of A with corresponding eigenvector $\begin{pmatrix} 0 \\ 1 \end{pmatrix}$.

ANSWERS TO SELF-QUIZ

I. True **II.** True **III.** False **IV.** False **V.** c **VI.** b

---

8.14 A MODEL OF POPULATION GROWTH (OPTIONAL)

In this section we show how the theory of eigenvalues and eigenvectors can be used to analyze a model of the growth of a bird population.† We begin by discussing a simple model of population growth. We assume that a certain species grows at a constant rate; that is, the population of the species after one time period (which could be an hour, a

†The material in this section is based on a paper by D. Cooke: "A 2×2 Matrix Model of Population Growth," *Mathematical Gazette* 61(416):120–123.

week, a month, a year, etc.) is a constant multiple of the population in the previous time period. One way this could happen, for example, is that each generation is distinct and each organism produces r offspring and then dies. If p_n denotes the population after the nth time period, we would have

$$p_n = rp_{n-1}$$

For example, this model might describe a bacteria population, where, at a given time, an organism splits into two separate organisms. Then $r = 2$. Let p_0 denote the initial population. Then $p_1 = rp_0$, $p_2 = rp_1 = r(rp_0) = r^2p_0$, $p_3 = rp_2 = r(r^2p_0) = r^3p_0$, and so on, so that

$$p_n = r^np_0 \tag{1}$$

From this model we see that the population increases without bound if $r > 1$ and decreases to zero if $r < 1$. If $r = 1$ the population remains at the constant value p_0.

This model is, evidently, very simplistic. One obvious objection is that the number of offspring produced depends, in many cases, on the ages of the adults. For example, in a human population the average female adult over 50 would certainly produce fewer children than the average 21-year-old female. To deal with this difficulty, we introduce a model which allows for age groupings with different fertility rates.

We look at a model of population growth for a species of birds. In this bird population we assume that the number of female birds equals the number of males. Let $p_{j,n-1}$ denote the population of juvenile (immature) females in the $(n - 1)$st year and let $p_{a,n-1}$ denote the number of adult females in the $(n - 1)$st year. Some of the juvenile birds will die during the year. We assume that a certain proportion α of the juvenile birds survive to become adults in the spring of the nth year. Each surviving female bird produces eggs later in the spring, which hatch to produce, on the average, k juvenile female birds in the following spring. Adults also die, and the proportion of adults that survive from one spring to the next is β.

This constant survival rate of birds is not just a simplistic assumption. It appears to be the case with most of the natural bird populations that have been studied. This means that the adult survival rate of many bird species is independent of age. Perhaps few birds in the wild survive long enough to exhibit the effects of old age. Moreover, in many species the number of offspring seems to be uninfluenced by the age of the mother.

In the notation introduced above $p_{j,n}$ and $p_{a,n}$ represent, respectively, the population of juvenile and adult females in the nth year. Putting together all the information given, we arrive at the following 2×2 system:

$$\begin{aligned} p_{j,n} &= \qquad\qquad kp_{a,n-1} \\ p_{a,n} &= \alpha p_{j,n-1} + \beta p_{a,n-1} \end{aligned} \tag{2}$$

or

$$\mathbf{p}_n = A\mathbf{p}_{n-1} \tag{3}$$

where $\mathbf{p}_n = \begin{pmatrix} p_{j,n} \\ p_{a,n} \end{pmatrix}$ and $A = \begin{pmatrix} 0 & k \\ \alpha & \beta \end{pmatrix}$. It is clear, from (3), that $\mathbf{p}_1 = A\mathbf{p}_0$, $\mathbf{p}_2 = A\mathbf{p}_1 = A(A\mathbf{p}_0) = A^2\mathbf{p}_0, \ldots$, and so on. Hence

$$\mathbf{p}_n = A^n\mathbf{p}_0 \tag{4}$$

where $\mathbf{p}_0$ is the vector of initial populations of juvenile and adult females.

Equation (4) is like equation (1), but now we are able to distinguish between the survival rates of juvenile and adult birds.

EXAMPLE 1

AN ILLUSTRATION OF THE MODEL CARRIED THROUGH 20 GENERATIONS

Let $A = \begin{pmatrix} 0 & 2 \\ 0.3 & 0.5 \end{pmatrix}$. This means that each adult female produces two female offspring and, since the number of males is assumed equal to the number of females, at least four eggs—and probably many more, since losses among fledglings are likely to be high. From the model, it is apparent that α and β lie in the interval [0, 1]. Since juvenile birds are not so likely as adults to survive, we must have $\alpha < \beta$.

In Table 1 we assume that, initially, there are 10 female (and 10 male) adults and no juveniles. The computations were done on a computer, but the work would not be too onerous if done on a hand calculator. For example, $\mathbf{p}_1 = \begin{pmatrix} 0 & 2 \\ 0.3 & 0.5 \end{pmatrix}\begin{pmatrix} 0 \\ 10 \end{pmatrix} = \begin{pmatrix} 20 \\ 5 \end{pmatrix}$ so that $p_{j,1} = 20$, $p_{a,1} = 5$, the total female population after 1 year is 25, and the ratio of juvenile to adult females is 4 to 1. In the second year $\mathbf{p}_2 = \begin{pmatrix} 0 & 2 \\ 0.3 & 0.5 \end{pmatrix}\begin{pmatrix} 20 \\ 5 \end{pmatrix} = \begin{pmatrix} 10 \\ 8.5 \end{pmatrix}$, which we round down to $\begin{pmatrix} 10 \\ 8 \end{pmatrix}$ since we cannot have $8\frac{1}{2}$ adult birds. Table 1 tabulates the ratios $p_{j,n}/p_{a,n}$ and the ratios T_n/T_{n-1} of the total number of females in successive years.

TABLE 1

Year n	No. of Juveniles $p_{j,n}$	No. of Adults $p_{a,n}$	Total Female Population T_n in nth Year	$p_{j,n}/p_{a,n}$†	T_n/T_{n-1}†
0	0	10	10	0	—
1	20	5	25	4.00	2.50
2	10	8	18	1.18	0.74
3	17	7	24	2.34	1.31
4	14	8	22	1.66	0.96
5	17	8	25	2.00	1.13
10	22	12	34	1.87	1.06
11	24	12	36	1.88	1.07
12	25	13	38	1.88	1.06
20	42	22	64	1.88	1.06

†The figures in these columns were obtained before the numbers in the previous columns were rounded. Thus, for example, in year 2, $p_{j,2}/p_{a,2} = 10/8.5 \approx 1.176470588 \approx 1.18$.

In Table 1 it seems as if the ratio $p_{j,n}/p_{a,n}$ is approaching the constant 1.88 while the total population seems to be increasing at a constant rate of 6 percent a year. Let us see if we can determine why this is the case.

First, we return to the general case (equation (4)). Suppose that A has the real distinct eigenvalues λ_1 and λ_2 with corresponding eigenvectors $\mathbf{v}_1$ and $\mathbf{v}_2$. Since $\mathbf{v}_1$ and $\mathbf{v}_2$ are linearly independent, they form a basis for $\mathbb{R}^2$ and we can write

$$\mathbf{p}_0 = a_1\mathbf{v}_1 + a_2\mathbf{v}_2 \tag{5}$$

for some real numbers a_1 and a_2. Then (4) becomes

$$\mathbf{p}_n = A^n(a_1\mathbf{v}_1 + a_2\mathbf{v}_2) \tag{6}$$

But $A\mathbf{v}_1 = \lambda_1\mathbf{v}_1$ and $A^2\mathbf{v}_1 = A(A\mathbf{v}_1) = A(\lambda_1\mathbf{v}_1) = \lambda_1 A\mathbf{v}_1 = \lambda_1(\lambda_1\mathbf{v}_1) = \lambda_1^2\mathbf{v}_1$. Thus we can see that $A^n\mathbf{v}_1 = \lambda_1^n\mathbf{v}_1$, $A^n\mathbf{v}_2 = \lambda_2^n\mathbf{v}_2$, and, from (6),

$$\mathbf{p}_n = a_1\lambda_1^n\mathbf{v}_1 + a_2\lambda_2^n\mathbf{v}_2 \tag{7}$$

The characteristic equation of A is $\begin{vmatrix} -\lambda & k \\ \alpha & \beta - \lambda \end{vmatrix} = \lambda^2 - \beta\lambda - k\alpha = 0$ or $\lambda = (\beta \pm \sqrt{\beta^2 + 4\alpha k})/2$. By assumption, $k > 0$, $0 < \alpha < 1$, and $0 < \beta < 1$. Hence $4\alpha k > 0$ and $\beta^2 + 4\alpha k > 0$, which means that the eigenvalues are, indeed, real and distinct and that one eigenvalue λ_1 is positive, one λ_2 is negative, and $|\lambda_1| > |\lambda_2|$. We can write (7) as

$$\mathbf{p}_n = \lambda_1^n\left[a_1\mathbf{v}_1 + \left(\frac{\lambda_2}{\lambda_1}\right)^n a_2\mathbf{v}_2\right] \tag{8}$$

Since $|\lambda_2/\lambda_1| < 1$, it is apparent that $(\lambda_2/\lambda_1)^n$ gets very small as n gets large. Thus, for n large,

$$\mathbf{p}_n \approx a_1\lambda_1^n\mathbf{v}_1 \tag{9}$$

This means that, in the long run, the age distribution stabilizes and is proportional to $\mathbf{v}_1$. Each age group will change by a factor of λ_1 each year. Thus—in the long run—equation (4) acts just like equation (1). In the short term—that is, before ''stability'' is reached—the numbers oscillate. The magnitude of this oscillation depends on the magnitude of λ_2/λ_1 (which is negative, thus explaining the oscillation).

EXAMPLE 1 (CONTINUED) THE EIGENVALUES AND EIGENVECTORS OF *A* DETERMINE THE BEHAVIOR IN FUTURE GENERATIONS

For $A = \begin{pmatrix} 0 & 2 \\ 0.3 & 0.5 \end{pmatrix}$, we have $\lambda^2 - 0.5\lambda - 0.6 = 0$ or $\lambda = (0.5 \pm \sqrt{0.25 + 2.4})/2 = (0.5 \pm \sqrt{2.65})/2$ so that $\lambda_1 \approx 1.06$ and $\lambda_2 \approx -0.56$. This explains the 6 percent increase in population noted in the last column of Table 1. Corresponding to the eigenvalue $\lambda_1 = 1.06$, we compute $(A - 1.06I)\mathbf{v}_1 = \begin{pmatrix} -1.06 & 2 \\ 0.3 & -0.56 \end{pmatrix}\begin{pmatrix} x_1 \\ x_2 \end{pmatrix} = \begin{pmatrix} 0 \\ 0 \end{pmatrix}$ or $1.06x_1 = 2x_2$ so that $\mathbf{v}_1 = \begin{pmatrix} 1 \\ 0.53 \end{pmatrix}$ is an eigenvector. Similarly, $(A + 0.56)\mathbf{v}_2 = \begin{pmatrix} 0.56 & 2 \\ 0.3 & 1.06 \end{pmatrix}\begin{pmatrix} x_1 \\ x_2 \end{pmatrix} = \begin{pmatrix} 0 \\ 0 \end{pmatrix}$ so that $0.56x_1 + 2x_2 = 0$ and $\mathbf{v}_2 = \begin{pmatrix} 1 \\ -0.28 \end{pmatrix}$ is a second eigenvector. Note that in $\mathbf{v}_1$ we have $1/0.53 \approx 1.88$. This explains the ratio $p_{j,n}/p_{a,n}$ in the fifth column of the table.

It is remarkable just how much information is available from a simple computation of eigenvalues. It is of great interest to know whether a population will ultimately increase or decrease. It will increase if $\lambda_1 > 1$, and the condition for that is $(\beta + \sqrt{\beta^2 + 4\alpha k})/2 > 1$ or $\sqrt{\beta^2 + 4\alpha k} > 2 - \beta$ or $\beta^2 + 4\alpha k > (2 - \beta)^2 = 4 - 4\beta + \beta^2$. This leads to $4\alpha k > 4 - 4\beta$ or

$$k > \frac{1 - \beta}{\alpha} \tag{10}$$

In Example 1 we had $\beta = 0.5$ and $\alpha = 0.3$; thus (10) is satisfied if $k > 0.5/0.3 \approx 1.67$.

Before we close this section we indicate two limitations of this model:

i. Birth and death rates often change from year to year and are particularly dependent on the weather. This model assumes a constant environment.

ii. Ecologists have found that for many species birth and death rates vary with the size of the population. In particular, a population cannot grow when it reaches a certain size due to the effects of limited food resources and overcrowding. It is obvious that a population cannot grow indefinitely at a constant rate. Otherwise that population would overrun the earth.

PROBLEMS 8.14

In Problems 1–3 find the numbers of juvenile and adult female birds after 1, 2, 5, 10, 19, and 20 years. Then find the long-term ratios of $p_{j,n}$ to $p_{a,n}$ and T_n to T_{n-1}. [*Hint:* Use equations (7) and (9) and a calculator and round to three decimals.]

1. $\mathbf{p}_0 = \begin{pmatrix} 0 \\ 12 \end{pmatrix}$; $k = 3,\ \alpha = 0.4,\ \beta = 0.6$

2. $\mathbf{p}_0 = \begin{pmatrix} 0 \\ 15 \end{pmatrix}$; $k = 1,\ \alpha = 0.3,\ \beta = 0.4$

3. $\mathbf{p}_0 = \begin{pmatrix} 0 \\ 20 \end{pmatrix}$; $k = 4,\ \alpha = 0.7,\ \beta = 0.8$

4. Show that if $\alpha = \beta$ and $\alpha > \frac{1}{2}$, then the bird population will always increase in the long run if at least one female offspring on the average is produced by each female adult.

5. Show that, in the long run, the ratio $p_{j,n}/p_{a,n}$ approaches the limiting value k/λ_1.

6. Suppose we divide the adult birds into two age groups: those 1–5 years old and those more than 5 years old. Assume that the survival rate for birds in the first group is β, whereas in the second group it is γ (and $\beta > \gamma$). Assume that the birds in the first group are equally divided as to age. (That is, if there are 100 birds in the group, then 20 are 1 year old, 20 are 2 years old, and so on.) Formulate a 3×3 matrix model for this situation.

8.15 SIMILAR MATRICES AND DIAGONALIZATION

In this section we describe an interesting and useful relationship that can hold between two matrices.

DEFINITION SIMILAR MATRICES

Two $n \times n$ matrices A and B are said to be **similar** if there exists an invertible $n \times n$ matrix C such that

$$B = C^{-1}AC \tag{1}$$

■

The function defined by (1) which takes the matrix A into the matrix B is called a **similarity transformation**. We can write this linear transformation as

$$T(A) = C^{-1}AC$$

NOTE: $C^{-1}(A_1 + A_2)C = C^{-1}A_1C + C^{-1}A_2C$ and $C^{-1}(\alpha A)C = \alpha C^{-1}AC$, so that the function defined by (1) is, in fact, a linear transformation. This explains the use of the word "transformation" in the definition of similar matrices.

The purpose of this section is to show that (1) similar matrices have several important properties in common and (2) most matrices are similar to diagonal matrices. (See the remark on the next page.)

EXAMPLE 1 TWO SIMILAR MATRICES

Let $A = \begin{pmatrix} 2 & 1 \\ 0 & -1 \end{pmatrix}$, $B = \begin{pmatrix} 4 & -2 \\ 5 & -3 \end{pmatrix}$, and $C = \begin{pmatrix} 2 & -1 \\ -1 & 1 \end{pmatrix}$. Then $CB = \begin{pmatrix} 2 & -1 \\ -1 & 1 \end{pmatrix}\begin{pmatrix} 4 & -2 \\ 5 & -3 \end{pmatrix} = \begin{pmatrix} 3 & -1 \\ 1 & -1 \end{pmatrix}$ and $AC = \begin{pmatrix} 2 & 1 \\ 0 & -1 \end{pmatrix}\begin{pmatrix} 2 & -1 \\ -1 & 1 \end{pmatrix} = \begin{pmatrix} 3 & -1 \\ 1 & -1 \end{pmatrix}$. Thus $CB = AC$. Since $\det C = 1 \neq 0$, C is invertible; and since $CB = AC$, we have $C^{-1}CB = C^{-1}AC$ or $B = C^{-1}AC$. This shows that A and B are similar.

EXAMPLE 2 A MATRIX THAT IS SIMILAR TO A DIAGONAL MATRIX

Let $D = \begin{pmatrix} 1 & 0 & 0 \\ 0 & -1 & 0 \\ 0 & 0 & 2 \end{pmatrix}$, $A = \begin{pmatrix} -6 & -3 & -25 \\ 2 & 1 & 8 \\ 2 & 2 & 7 \end{pmatrix}$, and $C = \begin{pmatrix} 2 & 4 & 3 \\ 0 & 1 & -1 \\ 3 & 5 & 7 \end{pmatrix}$. C is invertible because $\det C = 3 \neq 0$. We then compute:

$$CA = \begin{pmatrix} 2 & 4 & 3 \\ 0 & 1 & -1 \\ 3 & 5 & 7 \end{pmatrix}\begin{pmatrix} -6 & -3 & -25 \\ 2 & 1 & 8 \\ 2 & 2 & 7 \end{pmatrix} = \begin{pmatrix} 2 & 4 & 3 \\ 0 & -1 & 1 \\ 6 & 10 & 14 \end{pmatrix}$$

$$DC = \begin{pmatrix} 1 & 0 & 0 \\ 0 & -1 & 0 \\ 0 & 0 & 2 \end{pmatrix}\begin{pmatrix} 2 & 4 & 3 \\ 0 & 1 & -1 \\ 3 & 5 & 7 \end{pmatrix} = \begin{pmatrix} 2 & 4 & 3 \\ 0 & -1 & 1 \\ 6 & 10 & 14 \end{pmatrix}$$

Thus $CA = DC$ and $A = C^{-1}DC$, so A and D are similar.

NOTE: In Examples 1 and 2 it was not necessary to compute C^{-1}. It was only necessary to know that C was nonsingular.

THEOREM 1

If A and B are similar $n \times n$ matrices, then A and B have the same characteristic equation, and therefore have the same eigenvalues.

PROOF: Since A and B are similar, $B = C^{-1}AC$ and

$$\begin{aligned} \det(B - \lambda I) &= \det(C^{-1}AC - \lambda I) = \det[C^{-1}AC - C^{-1}(\lambda I)C] \\ &= \det[C^{-1}(A - \lambda I)C] = \det(C^{-1}) \det(A - \lambda I) \det(C) \\ &= \det(C^{-1}) \det(C) \det(A - \lambda I) = \det(C^{-1}C) \det(A - \lambda I) \\ &= \det I \det(A - \lambda I) = \det(A - \lambda I) \end{aligned}$$

This means that A and B have the same characteristic equation, and, since eigenvalues are roots of the characteristic equation, they have the same eigenvalues. ■

EXAMPLE 3

EIGENVALUES OF SIMILAR MATRICES ARE THE SAME

In Example 2 it is obvious that the eigenvalues of $D = \begin{pmatrix} 1 & 0 & 0 \\ 0 & -1 & 0 \\ 0 & 0 & 2 \end{pmatrix}$ are 1, -1, and 2. Thus these are the eigenvalues of $A = \begin{pmatrix} -6 & -3 & -25 \\ 2 & 1 & 8 \\ 2 & 2 & 7 \end{pmatrix}$. Check this by verifying that $\det(A - I) = \det(A + I) = \det(A - 2I) = 0$.

In a variety of applications it is quite useful to "diagonalize" a matrix A—that is, to find a diagonal matrix similar to A.

DEFINITION DIAGONALIZABLE MATRIX

An $n \times n$ matrix A is **diagonalizable** if there is a diagonal matrix D such that A is similar to D. ■

REMARK: If D is a diagonal matrix, then its eigenvalues are its diagonal components. If A is similar to D, then A and D have the same eigenvalues (by Theorem 1). Putting these two facts together, we observe that if A is diagonalizable, then A is similar to a diagonal matrix whose diagonal components are the eigenvalues of A.

The next theorem tells us when a matrix is diagonalizable.

THEOREM 2

An $n \times n$ matrix A is diagonalizable if and only if it has n linearly independent eigenvectors. In that case the diagonal matrix D similar to A is given by

$$D = \begin{pmatrix} \lambda_1 & 0 & 0 & \cdots & 0 \\ 0 & \lambda_2 & 0 & \cdots & 0 \\ 0 & 0 & \lambda_3 & \cdots & 0 \\ \vdots & \vdots & \vdots & & \vdots \\ 0 & 0 & 0 & \cdots & \lambda_n \end{pmatrix} \quad (2)$$

where $\lambda_1, \lambda_2, \ldots, \lambda_n$ are the eigenvalues of A. If C is a matrix whose columns are linearly independent eigenvectors of A, then

$$D = C^{-1}AC \quad (3)$$

PROOF: We first assume that A has n linearly independent eigenvectors $\mathbf{v}_1, \mathbf{v}_2, \ldots, \mathbf{v}_n$ corresponding to the (not necessarily distinct) eigenvalues $\lambda_1, \lambda_2, \ldots, \lambda_n$.

Let

$$\mathbf{v}_1 = \begin{pmatrix} c_{11} \\ c_{21} \\ \vdots \\ c_{n1} \end{pmatrix}, \quad \mathbf{v}_2 = \begin{pmatrix} c_{12} \\ c_{22} \\ \vdots \\ c_{n2} \end{pmatrix}, \ldots, \quad \mathbf{v}_n = \begin{pmatrix} c_{1n} \\ c_{2n} \\ \vdots \\ c_{nn} \end{pmatrix}$$

and let

$$C = \begin{pmatrix} c_{11} & c_{12} & \cdots & c_{1n} \\ c_{21} & c_{22} & \cdots & c_{2n} \\ \vdots & \vdots & & \vdots \\ c_{n1} & c_{n2} & \cdots & c_{nn} \end{pmatrix}$$

Then C is invertible since its columns are linearly independent. Now

$$AC = \begin{pmatrix} a_{11} & a_{12} & \cdots & a_{1n} \\ a_{21} & a_{22} & \cdots & a_{2n} \\ \vdots & \vdots & & \vdots \\ a_{n1} & a_{n2} & \cdots & a_{nn} \end{pmatrix} \begin{pmatrix} c_{11} & c_{12} & \cdots & c_{1n} \\ c_{21} & c_{22} & \cdots & c_{2n} \\ \vdots & \vdots & & \vdots \\ c_{n1} & c_{n2} & \cdots & c_{nn} \end{pmatrix}$$

and we see that the ith column of AC is $A\begin{pmatrix} c_{1i} \\ c_{2i} \\ \vdots \\ c_{ni} \end{pmatrix} = A\mathbf{v}_i = \lambda_i \mathbf{v}_i$. Thus AC is the matrix whose ith column is $\lambda_i \mathbf{v}_i$ and

$$AC = \begin{pmatrix} \lambda_1 c_{11} & \lambda_2 c_{12} & \cdots & \lambda_n c_{1n} \\ \lambda_1 c_{21} & \lambda_2 c_{22} & \cdots & \lambda_n c_{2n} \\ \vdots & \vdots & & \vdots \\ \lambda_1 c_{n1} & \lambda_2 c_{n2} & \cdots & \lambda_n c_{nn} \end{pmatrix}$$

But

$$CD = \begin{pmatrix} c_{11} & c_{12} & \cdots & c_{1n} \\ c_{21} & c_{22} & \cdots & c_{2n} \\ \vdots & \vdots & & \vdots \\ c_{n1} & c_{n2} & \cdots & c_{nn} \end{pmatrix} \begin{pmatrix} \lambda_1 & 0 & \cdots & 0 \\ 0 & \lambda_2 & \cdots & 0 \\ \vdots & \vdots & & \vdots \\ 0 & 0 & \cdots & \lambda_n \end{pmatrix}$$

$$= \begin{pmatrix} \lambda_1 c_{11} & \lambda_2 c_{12} & \cdots & \lambda_n c_{1n} \\ \lambda_1 c_{21} & \lambda_2 c_{22} & \cdots & \lambda_n c_{2n} \\ \vdots & \vdots & & \vdots \\ \lambda_1 c_{n1} & \lambda_2 c_{n2} & \cdots & \lambda_n c_{nn} \end{pmatrix}$$

Thus

$$AC = CD \tag{4}$$

and, since C is invertible, we can multiply both sides of (4) on the left by C^{-1} to obtain

$$D = C^{-1}AC \tag{5}$$

This proves that if A has n linearly independent eigenvectors, then A is diagonalizable. Conversely, suppose that A is diagonalizable. That is, suppose that (5) holds for some invertible matrix C. Let $\mathbf{v}_1, \mathbf{v}_2, \ldots, \mathbf{v}_n$ be the columns of C. Then $AC = CD$, and, reversing the arguments above, we immediately see that $A\mathbf{v}_i = \lambda_i \mathbf{v}_i$ for $i = 1, 2, \ldots, n$. Thus $\mathbf{v}_1, \mathbf{v}_2, \ldots, \mathbf{v}_n$ are eigenvectors of A and are linearly independent because C is invertible.

■

NOTATION: To indicate that D is a diagonal matrix with diagonal components $\lambda_1, \lambda_2, \ldots, \lambda_n$, we write $D = \text{diag}(\lambda_1, \lambda_2, \ldots, \lambda_n)$.

Theorem 2 has a useful corollary that follows immediately from Theorem 8.12.3 on page 589.

COROLLARY

If the $n \times n$ matrix A has n distinct eigenvalues, then A is diagonalizable. ■

REMARK: If the real coefficients of a polynomial of degree n are picked at random, then, with probability 1, the polynomial will have n distinct roots. It is not difficult to see, intuitively, why this is so. If $n = 2$, for example, then the equation $\lambda^2 + a\lambda + b = 0$ has equal roots if and only if $a^2 = 4b$—a highly unlikely event if a and b are chosen at random. We can, of course, write down polynomials having roots of algebraic multiplicity greater than 1, but these polynomials are exceptional. Thus, without attempting to be mathematically precise, it is fair to say that *most* polynomials have distinct roots. Hence *most* matrices have distinct eigenvalues, and, as we stated at the beginning of the section, *most* matrices are diagonalizable.

EXAMPLE 4 **DIAGONALIZING A 2 × 2 MATRIX**

Let $A = \begin{pmatrix} 4 & 2 \\ 3 & 3 \end{pmatrix}$. In Example 8.13.3 on page 590 we found the two linearly independent eigenvectors $\mathbf{v}_1 = \begin{pmatrix} 2 \\ -3 \end{pmatrix}$ and $\mathbf{v}_2 = \begin{pmatrix} 1 \\ 1 \end{pmatrix}$. Then, setting $C = \begin{pmatrix} 2 & 1 \\ -3 & 1 \end{pmatrix}$, we find that

$$C^{-1}AC = \frac{1}{5}\begin{pmatrix} 1 & -1 \\ 3 & 2 \end{pmatrix}\begin{pmatrix} 4 & 2 \\ 3 & 3 \end{pmatrix}\begin{pmatrix} 2 & 1 \\ -3 & 1 \end{pmatrix}$$

$$= \frac{1}{5}\begin{pmatrix} 1 & -1 \\ 3 & 2 \end{pmatrix}\begin{pmatrix} 2 & 6 \\ -3 & 6 \end{pmatrix} = \frac{1}{5}\begin{pmatrix} 5 & 0 \\ 0 & 30 \end{pmatrix} = \begin{pmatrix} 1 & 0 \\ 0 & 6 \end{pmatrix}$$

which is the matrix whose diagonal components are the eigenvalues of A.

EXAMPLE 5 **DIAGONALIZING A 3 × 3 MATRIX WITH THREE DISTINCT EIGENVALUES**

Let $A = \begin{pmatrix} 1 & -1 & 4 \\ 3 & 2 & -1 \\ 2 & 1 & -1 \end{pmatrix}$. In Example 8.13.4 on page 591 we computed the three linearly independent eigenvectors $\mathbf{v}_1 = \begin{pmatrix} -1 \\ 4 \\ 1 \end{pmatrix}$, $\mathbf{v}_2 = \begin{pmatrix} 1 \\ -1 \\ -1 \end{pmatrix}$, and $\mathbf{v}_3 = \begin{pmatrix} 1 \\ 2 \\ 1 \end{pmatrix}$.

Then $C = \begin{pmatrix} -1 & 1 & 1 \\ 4 & -1 & 2 \\ 1 & -1 & 1 \end{pmatrix}$ and

$$C^{-1}AC = -\frac{1}{6}\begin{pmatrix} 1 & -2 & 3 \\ -2 & -2 & 6 \\ -3 & 0 & -3 \end{pmatrix}\begin{pmatrix} 1 & -1 & 4 \\ 3 & 2 & -1 \\ 2 & 1 & -1 \end{pmatrix}\begin{pmatrix} -1 & 1 & 1 \\ 4 & -1 & 2 \\ 1 & -1 & 1 \end{pmatrix}$$

$$= -\frac{1}{6}\begin{pmatrix} 1 & -2 & 3 \\ -2 & -2 & 6 \\ -3 & 0 & -3 \end{pmatrix}\begin{pmatrix} -1 & -2 & 3 \\ 4 & 2 & 6 \\ 1 & 2 & 3 \end{pmatrix}$$

$$= -\frac{1}{6}\begin{pmatrix} -6 & 0 & 0 \\ 0 & 12 & 0 \\ 0 & 0 & -18 \end{pmatrix} = \begin{pmatrix} 1 & 0 & 0 \\ 0 & -2 & 0 \\ 0 & 0 & 3 \end{pmatrix}$$

with eigenvalues 1, −2, and 3.

REMARK: Since there are an infinite number of ways to choose an eigenvector, there are an infinite number of ways to choose the diagonalizing matrix C. The only advice is to choose the eigenvectors and matrix C that are, arithmetically, the easiest to work with. This usually means that you should insert as many 0's and 1's as possible.

EXAMPLE 6

DIAGONALIZING A 3 × 3 MATRIX WITH TWO DISTINCT EIGENVALUES AND THREE LINEARLY INDEPENDENT EIGENVECTORS

Let $A = \begin{pmatrix} 3 & 2 & 4 \\ 2 & 0 & 2 \\ 4 & 2 & 3 \end{pmatrix}$. Then, from Example 8.13.9 on page 594, we have the three linearly independent eigenvectors $\mathbf{v}_1 = \begin{pmatrix} 2 \\ 1 \\ 2 \end{pmatrix}$, $\mathbf{v}_2 = \begin{pmatrix} 1 \\ -2 \\ 0 \end{pmatrix}$, and $\mathbf{v}_3 = \begin{pmatrix} 0 \\ -2 \\ 1 \end{pmatrix}$.

Setting $C = \begin{pmatrix} 2 & 1 & 0 \\ 1 & -2 & -2 \\ 2 & 0 & 1 \end{pmatrix}$, we obtain

$$C^{-1}AC = -\frac{1}{9}\begin{pmatrix} -2 & -1 & -2 \\ -5 & 2 & 4 \\ 4 & 2 & -5 \end{pmatrix}\begin{pmatrix} 3 & 2 & 4 \\ 2 & 0 & 2 \\ 4 & 2 & 3 \end{pmatrix}\begin{pmatrix} 2 & 1 & 0 \\ 1 & -2 & -2 \\ 2 & 0 & 1 \end{pmatrix}$$

$$= -\frac{1}{9}\begin{pmatrix} -2 & -1 & -2 \\ -5 & 2 & 4 \\ 4 & 2 & -5 \end{pmatrix}\begin{pmatrix} 16 & -1 & 0 \\ 8 & 2 & 2 \\ 16 & 0 & -1 \end{pmatrix}$$

$$= -\frac{1}{9}\begin{pmatrix} -72 & 0 & 0 \\ 0 & 9 & 0 \\ 0 & 0 & 9 \end{pmatrix} = \begin{pmatrix} 8 & 0 & 0 \\ 0 & -1 & 0 \\ 0 & 0 & -1 \end{pmatrix}$$

This example illustrates that A is diagonalizable even though its eigenvalues are not distinct.

EXAMPLE 7

A 2 × 2 MATRIX WITH ONLY ONE LINEARLY INDEPENDENT EIGENVECTOR CANNOT BE DIAGONALIZED

Let $A = \begin{pmatrix} 4 & 1 \\ 0 & 4 \end{pmatrix}$. In Example 8.13.8 on page 594 we saw that A did *not* have two linearly independent eigenvectors. Suppose that A were diagonalizable (in contradiction to Theorem 2). Then $D = \begin{pmatrix} 4 & 0 \\ 0 & 4 \end{pmatrix}$ and there would be an invertible matrix C such that $C^{-1}AC = D$. Multiplying this equation on the left by C and on the right by C^{-1}, we find that $A = CDC^{-1} = C\begin{pmatrix} 4 & 0 \\ 0 & 4 \end{pmatrix}C^{-1} = C(4I)C^{-1} = 4CIC^{-1} = 4CC^{-1} = 4I = \begin{pmatrix} 4 & 0 \\ 0 & 4 \end{pmatrix} = D$. But $A \neq D$, so no such C exists.

We have seen that many matrices are similar to diagonal matrices. However, one question remains: What do we do if A is not diagonalizable? The answer to this question involves a discussion of the *Jordan canonical form* of a matrix, a subject not discussed in this text.[†]

PROBLEMS 8.15

SELF-QUIZ

True–False

I. If an $n \times n$ matrix has n distinct eigenvalues, it can always be diagonalized.

II. If the 5×5 matrix A has three distinct eigenvalues, then A cannot be similar to a diagonal matrix.

III. If A is similar to the matrix $\begin{pmatrix} 1 & 2 & 5 \\ 0 & 2 & 4 \\ 0 & 0 & 3 \end{pmatrix}$, then its eigenvalues are 1, 2, and 3.

IV. If $\det A = 0$, then 0 is an eigenvalue of A.

In Problems 1–15 determine whether the given matrix A is diagonalizable. If it is, find a matrix C such that $C^{-1}AC = D$.

1. $\begin{pmatrix} -2 & -2 \\ -5 & 1 \end{pmatrix}$

2. $\begin{pmatrix} 3 & -1 \\ -2 & 4 \end{pmatrix}$

3. $\begin{pmatrix} 2 & -1 \\ 5 & -2 \end{pmatrix}$

4. $\begin{pmatrix} 3 & -5 \\ 1 & -1 \end{pmatrix}$

5. $\begin{pmatrix} 3 & 2 \\ -5 & 1 \end{pmatrix}$

6. $\begin{pmatrix} 1 & -1 & 0 \\ -1 & 2 & -1 \\ 0 & -1 & 1 \end{pmatrix}$

7. $\begin{pmatrix} 1 & 1 & -2 \\ -1 & 2 & 1 \\ 0 & 1 & -1 \end{pmatrix}$

8. $\begin{pmatrix} 2 & 1 & 0 \\ 0 & 0 & 1 \\ 0 & 0 & 0 \end{pmatrix}$

9. $\begin{pmatrix} 3 & 0 & 0 \\ 0 & 0 & 1 \\ 0 & 0 & 2 \end{pmatrix}$

10. $\begin{pmatrix} 3 & -1 & -1 \\ 1 & 1 & -1 \\ 1 & -1 & 1 \end{pmatrix}$

11. $\begin{pmatrix} 7 & -2 & -4 \\ 3 & 0 & -2 \\ 6 & -2 & -3 \end{pmatrix}$

12. $\begin{pmatrix} 4 & 6 & 6 \\ 1 & 3 & 2 \\ -1 & -5 & -2 \end{pmatrix}$

13. $\begin{pmatrix} -3 & -7 & -5 \\ 2 & 4 & 3 \\ 1 & 2 & 2 \end{pmatrix}$

14. $\begin{pmatrix} -2 & -2 & 0 & 0 \\ -5 & 1 & 0 & 0 \\ 0 & 0 & 2 & -1 \\ 0 & 0 & 5 & -2 \end{pmatrix}$

[†] This topic is discussed in Section 6.6 of S. I. Grossman, *Elementary Linear Algebra,* Fifth Edition, Saunders, Philadelphia, 1994.

15. $\begin{pmatrix} 4 & 1 & 0 & 1 \\ 2 & 3 & 0 & 1 \\ -2 & 1 & 2 & -3 \\ 2 & -1 & 0 & 5 \end{pmatrix}$

16. Show that if A is similar to B and B is similar to C, then A is similar to C.

17. If A is similar to B, show that A^n is similar to B^n for any positive integer n.

***18.** If A is similar to B, show that $\rho(A) = \rho(B)$ and $\nu(A) = \nu(B)$. [*Hint:* First prove that if C is invertible, then $\nu(CA) = \nu(A)$ by showing that $\mathbf{x} \in N_A$ if and only if $\mathbf{x} \in N_{CA}$. Next prove that $\rho(AC) = \rho(A)$ by showing that $R_A = R_{AC}$. Conclude that $\rho(AC) = \rho(CA) = \rho(A)$. Finally, use the fact that C^{-1} is invertible to show that $\rho(C^{-1}AC) = \rho(A)$.]

19. Let $D = \begin{pmatrix} 1 & 0 \\ 0 & -1 \end{pmatrix}$. Compute D^{20}.

20. If A is similar to B, show that $\det A = \det B$.

21. Suppose that $C^{-1}AC = D$. Show that for any integer n, $A^n = CD^nC^{-1}$. This gives an easy way to compute powers of a diagonalizable matrix.

22. Let $A = \begin{pmatrix} 3 & -4 \\ 2 & -3 \end{pmatrix}$. Compute A^{20}. [*Hint:* Find a C such that $A = CDC^{-1}$]

***23.** Let A be an $n \times n$ matrix whose characteristic equation is $(\lambda - c)^n = 0$. Show that A is diagonalizable if and only if $A = cI$.

24. Use the result of Problem 21 and Example 6 to compute A^{10}, where $A = \begin{pmatrix} 3 & 2 & 4 \\ 2 & 0 & 2 \\ 4 & 2 & 3 \end{pmatrix}$.

***25.** Let A and B be real $n \times n$ matrices with distinct eigenvalues. Prove that $AB = BA$ if and only if A and B have the same eigenvectors.

26. If A is diagonalizable, show that $\det A = \lambda_1\lambda_2 \cdots \lambda_n$, where $\lambda_1, \lambda_2, \ldots, \lambda_n$ are the eigenvalues of A.

ANSWERS TO SELF-QUIZ

I. True **II.** False **III.** True **IV.** True

8.16 SYMMETRIC MATRICES AND ORTHOGONAL DIAGONALIZATION

In this section we shall see that symmetric matrices[†] have a number of important properties. In particular, we claim that any symmetric matrix has n linearly independent real eigenvectors and therefore, by Theorem 8.14.2, is diagonalizable. We begin by stating that the eigenvalues of a real symmetric matrix are real. The proof of the following theorem makes use of the theory of inner product spaces and so is omitted.[‡]

THEOREM 1

Let A be a real $n \times n$ symmetric matrix. Then the eigenvalues of A are real. ■

We saw in Theorem 8.13.3 on page 589 that eigenvectors corresponding to distinct eigenvalues are linearly independent. For symmetric matrices the result is stronger: *Eigenvectors of a symmetric matrix corresponding to distinct eigenvalues are orthogonal.*

THEOREM 2

Let A be a real symmetric $n \times n$ matrix. If λ_1 and λ_2 are distinct eigenvalues with corresponding real eigenvectors $\mathbf{v}_1$ and $\mathbf{v}_2$, then $\mathbf{v}_1$ and $\mathbf{v}_2$ are orthogonal.

[†] Recall that A is symmetric if and only if $A' = A$.
[‡] A proof can be found on page 576 of *Elementary Linear Algebra, Fifth Edition* by S. I. Grossman.

PROOF: We compute

$$A\mathbf{v}_1 \cdot \mathbf{v}_2 = \lambda_1 \mathbf{v}_1 \cdot \mathbf{v}_2 = \lambda_1(\mathbf{v}_1 \cdot \mathbf{v}_2) \tag{1}$$

and

$$A\mathbf{v}_1 \cdot \mathbf{v}_2 = \mathbf{v}_1 \cdot A^t\mathbf{v}_2{}^{\dagger} = \mathbf{v}_1 \cdot A\mathbf{v}_2 = \mathbf{v}_1 \cdot (\lambda_2\mathbf{v}_2) = \lambda_2(\mathbf{v}_1 \cdot \mathbf{v}_2) \tag{2}$$

Combining (1) and (2), we have $\lambda_1(\mathbf{v}_1 \cdot \mathbf{v}_2) = \lambda_2(\mathbf{v}_1 \cdot \mathbf{v}_2)$ and since $\lambda_1 \neq \lambda_2$, we conclude that $\mathbf{v}_1 \cdot \mathbf{v}_2 = 0$. This is what we wanted to show. ■

We now state the main result of this section. Because of its difficulty, we omit the proof.[‡]

THEOREM 3

Let A be a real symmetric $n \times n$ matrix. Then A has n real orthonormal eigenvectors. ■

REMARK: It follows from this theorem that the geometric multiplicity of each eigenvalue of A is equal to its algebraic multiplicity.

Theorem 3 tells us that if A is symmetric, then $\mathbb{R}^n$ has a basis $B = \{\mathbf{u}_1, \mathbf{u}_2, \ldots, \mathbf{u}_n\}$ consisting of orthonormal eigenvectors of A. Let Q be the matrix whose columns are $\mathbf{u}_1, \mathbf{u}_2, \ldots, \mathbf{u}_n$. Then, by Theorem 8.7.3 on page 541, Q is an orthogonal matrix. This leads to the following definition.

DEFINITION ORTHOGONALLY DIAGONALIZABLE MATRIX

An $n \times n$ matrix A is said to be **orthogonally diagonalizable** if there exists an orthogonal matrix Q such that

$$Q^tAQ = D \tag{3}$$

where $D = \text{diag}\,(\lambda_1, \lambda_2, \ldots, \lambda_n)$ and $\lambda_1, \lambda_2, \ldots, \lambda_n$ are the eigenvalues of A. ■

NOTE: Remember that Q is orthogonal if $Q^t = Q^{-1}$; hence (9) could be written as $Q^{-1}AQ = D$.

THEOREM 4

Let A be a real $n \times n$ matrix. Then A is orthogonally diagonalizable if and only if A is symmetric.

[†] You should verify from the definition of the dot product that if A is an $n \times n$ matrix and if $\mathbf{x}$ and $\mathbf{y}$ are in $\mathbb{R}^n$, then $A\mathbf{x} \cdot \mathbf{y} = \mathbf{x} \cdot A^t\mathbf{y}$.

[‡] A proof is given on page 580 of the linear algebra book cited in the last footnote on page 610.

PROOF: Let A be symmetric. Then, by Theorems 2 and 3, A is orthogonally diagonalizable with Q the matrix whose columns are the orthonormal eigenvectors given in Theorem 3. Conversely, suppose that A is orthogonally diagonalizable. Then there exists an orthogonal matrix Q such that $Q^tAQ = D$. Multiplying this equation on the left by Q and on the right by Q^t and using the fact that $Q^tQ = QQ^t = I$, we obtain

$$A = QDQ^t \tag{4}$$

Then $A^t = (QDQ^t)^t = (Q^t)^tD^tQ^t = QDQ^t = A$. Thus A is symmetric and the theorem is proved. In the last series of equations we used the facts that $(AB)^t = B^tA^t$ [part (*ii*) of Theorem 6.9.1, page 433], $(A^t)^t = A$ [part (*i*) of Theorem 6.9.1], and $D^t = D$ for any diagonal matrix D. ■

Before giving examples, we provide the following three-step procedure for finding the orthogonal matrix Q that diagonalizes the symmetric matrix A.

PROCEDURE FOR FINDING A DIAGONALIZING MATRIX Q

i. Find a basis for each eigenspace of A.

ii. Find an orthonormal basis for each eigenspace of A by using the Gram-Schmidt process.

iii. Write Q as the matrix whose columns are the orthonormal eigenvectors obtained in step (*ii*).

EXAMPLE 1 DIAGONALIZING A 2 × 2 SYMMETRIC MATRIX USING AN ORTHOGONAL MATRIX

Let $A = \begin{pmatrix} 1 & -2 \\ -2 & 3 \end{pmatrix}$. Then the characteristic equation of A is $\det(A - \lambda I) = \begin{vmatrix} 1-\lambda & -2 \\ -2 & 3-\lambda \end{vmatrix} = \lambda^2 - 4\lambda - 1 = 0$, which has the roots $\lambda = (4 \pm \sqrt{20})/2 = (4 \pm 2\sqrt{5})/2 = 2 \pm \sqrt{5}$. For $\lambda_1 = 2 - \sqrt{5}$, we obtain $(A - \lambda I)\mathbf{v} = \begin{pmatrix} -1+\sqrt{5} & -2 \\ -2 & 1+\sqrt{5} \end{pmatrix}\begin{pmatrix} x_1 \\ x_2 \end{pmatrix} = \begin{pmatrix} 0 \\ 0 \end{pmatrix}$. An eigenvector is $\mathbf{v}_1 = \begin{pmatrix} 2 \\ -1+\sqrt{5} \end{pmatrix}$ and $|\mathbf{v}_1| = \sqrt{2^2 + (-1+\sqrt{5})^2} = \sqrt{10 - 2\sqrt{5}}$. Thus

$$\mathbf{u}_1 = \frac{1}{\sqrt{10 - 2\sqrt{5}}}\begin{pmatrix} 2 \\ -1+\sqrt{5} \end{pmatrix}.$$

Next, for $\lambda_2 = 2 + \sqrt{5}$, we compute $(A - \lambda I)\mathbf{v} = \begin{pmatrix} -1-\sqrt{5} & -2 \\ -2 & 1-\sqrt{5} \end{pmatrix}\begin{pmatrix} x_1 \\ x_2 \end{pmatrix} = \begin{pmatrix} 0 \\ 0 \end{pmatrix}$ and $\mathbf{v}_2 = \begin{pmatrix} 1-\sqrt{5} \\ 2 \end{pmatrix}$. Note that $\mathbf{v}_1 \cdot \mathbf{v}_2 = 0$ (which must be true according to Theorem 2). Then $|\mathbf{v}_2| = \sqrt{10 - 2\sqrt{5}}$ so that $\mathbf{u}_2 = \frac{1}{\sqrt{10 - 2\sqrt{5}}}\begin{pmatrix} 1-\sqrt{5} \\ 2 \end{pmatrix}$. Finally,

$$Q = \frac{1}{\sqrt{10 - 2\sqrt{5}}}\begin{pmatrix} 2 & 1-\sqrt{5} \\ -1+\sqrt{5} & 2 \end{pmatrix}$$

$$Q^t = \frac{1}{\sqrt{10-2\sqrt{5}}}\begin{pmatrix} 2 & -1+\sqrt{5} \\ 1-\sqrt{5} & 2 \end{pmatrix}$$

and

$$\begin{aligned} Q^tAQ &= \frac{1}{10-2\sqrt{5}}\begin{pmatrix} 2 & -1+\sqrt{5} \\ 1-\sqrt{5} & 2 \end{pmatrix}\begin{pmatrix} 1 & -2 \\ -2 & 3 \end{pmatrix}\begin{pmatrix} 2 & 1-\sqrt{5} \\ -1+\sqrt{5} & 2 \end{pmatrix} \\ &= \frac{1}{10-2\sqrt{5}}\begin{pmatrix} 2 & -1+\sqrt{5} \\ 1-\sqrt{5} & 2 \end{pmatrix}\begin{pmatrix} 4-2\sqrt{5} & -3-\sqrt{5} \\ -7+3\sqrt{5} & 4+2\sqrt{5} \end{pmatrix} \\ &= \frac{1}{10-2\sqrt{5}}\begin{pmatrix} 30-14\sqrt{5} & 0 \\ 0 & 10+6\sqrt{5} \end{pmatrix} = \begin{pmatrix} 2-\sqrt{5} & 0 \\ 0 & 2+\sqrt{5} \end{pmatrix} \end{aligned}$$

EXAMPLE 2

DIAGONALIZING A 3×3 SYMMETRIC MATRIX USING AN ORTHOGONAL MATRIX

Let $A = \begin{pmatrix} 5 & 4 & 2 \\ 4 & 5 & 2 \\ 2 & 2 & 2 \end{pmatrix}$. Then A is symmetric and $\det(A-\lambda I) = \begin{vmatrix} 5-\lambda & 4 & 2 \\ 4 & 5-\lambda & 2 \\ 2 & 2 & 2-\lambda \end{vmatrix} = -(\lambda-1)^2(\lambda-10)$. Corresponding to $\lambda = 1$ we compute the linearly independent eigenvectors $\mathbf{v}_1 = \begin{pmatrix} -1 \\ 1 \\ 0 \end{pmatrix}$ and $\mathbf{v}_2 = \begin{pmatrix} -1 \\ 0 \\ 2 \end{pmatrix}$. Corresponding to $\lambda = 10$ we find that $\mathbf{v}_3 = \begin{pmatrix} 2 \\ 2 \\ 1 \end{pmatrix}$. To find Q, we apply the Gram-Schmidt process to $\{\mathbf{v}_1, \mathbf{v}_2\}$, a basis for E_1. Since $|\mathbf{v}_1| = \sqrt{2}$, we set $\mathbf{u}_1 = \begin{pmatrix} -1/\sqrt{2} \\ 1/\sqrt{2} \\ 0 \end{pmatrix}$. Next

$$\begin{aligned} \mathbf{v}_2' = \mathbf{v}_2 - (\mathbf{v}_2 \cdot \mathbf{u}_1)\mathbf{u}_1 &= \begin{pmatrix} -1 \\ 0 \\ 2 \end{pmatrix} - \frac{1}{\sqrt{2}}\begin{pmatrix} -1/\sqrt{2} \\ 1/\sqrt{2} \\ 0 \end{pmatrix} \\ &= \begin{pmatrix} -1 \\ 0 \\ 2 \end{pmatrix} - \begin{pmatrix} -1/2 \\ 1/2 \\ 0 \end{pmatrix} = \begin{pmatrix} -1/2 \\ -1/2 \\ 2 \end{pmatrix} \end{aligned}$$

Then $|\mathbf{v}_2| = \sqrt{18/4} = 3\sqrt{2}/2$ and $\mathbf{u}_2 = \dfrac{2}{3\sqrt{2}}\begin{pmatrix} -1/2 \\ -1/2 \\ 2 \end{pmatrix} = \begin{pmatrix} -1/3\sqrt{2} \\ -1/3\sqrt{2} \\ 4/3\sqrt{2} \end{pmatrix}$. We check this by noting that $\mathbf{u}_1 \cdot \mathbf{u}_2 = 0$. Finally, we have $\mathbf{u}_3 = \mathbf{v}_3/|\mathbf{v}_3| = \frac{1}{3}\mathbf{v}_3 = \begin{pmatrix} 2/3 \\ 2/3 \\ 1/3 \end{pmatrix}$. We can check this too by noting that $\mathbf{u}_1 \cdot \mathbf{u}_3 = 0$ and $\mathbf{u}_2 \cdot \mathbf{u}_3 = 0$. Thus

$$Q = \begin{pmatrix} -1/\sqrt{2} & -1/3\sqrt{2} & 2/3 \\ 1/\sqrt{2} & -1/3\sqrt{2} & 2/3 \\ 0 & 4/3\sqrt{2} & 1/3 \end{pmatrix}$$

and

$$Q^tAQ = \begin{pmatrix} -1/\sqrt{2} & 1/\sqrt{2} & 0 \\ -1/3\sqrt{2} & -1/3\sqrt{2} & 4/3\sqrt{2} \\ 2/3 & 2/3 & 1/3 \end{pmatrix} \begin{pmatrix} 5 & 4 & 2 \\ 4 & 5 & 2 \\ 2 & 2 & 2 \end{pmatrix} \begin{pmatrix} -1/\sqrt{2} & -1/3\sqrt{2} & 2/3 \\ 1/\sqrt{2} & -1/3\sqrt{2} & 2/3 \\ 0 & 4/3\sqrt{2} & 1/3 \end{pmatrix}$$

$$= \begin{pmatrix} -1/\sqrt{2} & 1/\sqrt{2} & 0 \\ -1/3\sqrt{2} & -1/3\sqrt{2} & 4/3\sqrt{2} \\ 2/3 & 2/3 & 1/3 \end{pmatrix} \begin{pmatrix} -1/\sqrt{2} & -1/3\sqrt{2} & 20/3 \\ 1/\sqrt{2} & -1/3\sqrt{2} & 20/3 \\ 0 & 4/3\sqrt{2} & 10/3 \end{pmatrix}$$

$$= \begin{pmatrix} 1 & 0 & 0 \\ 0 & 1 & 0 \\ 0 & 0 & 10 \end{pmatrix}$$

In this section we have proved results for real symmetric matrices. If $A = (a_{ij})$ is a complex matrix, then the **conjugate transpose** of A, denoted by A^*, is defined by: the ijth element of $A^* = \overline{a_{ji}}$. The matrix A is called **hermitian** if $A^* = A$. It turns out that Theorems 1, 2, and 3 are also true for hermitian matrices. Moreover, if we define a **unitary** matrix to be a complex matrix U with $U^* = U^{-1}$, then, using the proof of Theorem 4, we can show that a hermitian matrix is unitarily diagonalizable. We leave all these facts as exercises (see Problems 15–17).

PROBLEMS 8.16

SELF-QUIZ

True–False

I. The eigenvalues of a symmetric matrix are real.
II. The eigenvectors of a symmetric matrix are real.
III. Every symmetric matrix is similar to a diagonal matrix.
IV. If A can be diagonalized, then there is an orthogonal matrix Q such that Q^tAQ is diagonal.
V. If A is symmetric, then there is an orthogonal matrix Q such that Q^tAQ is diagonal.

In Problems 1–8 find an orthogonal matrix that diagonalizes the given symmetric matrix.

1. $\begin{pmatrix} 3 & 4 \\ 4 & -3 \end{pmatrix}$

2. $\begin{pmatrix} 2 & 1 \\ 1 & 2 \end{pmatrix}$

3. $\begin{pmatrix} 1 & -1 \\ -1 & 1 \end{pmatrix}$

4. $\begin{pmatrix} 1 & -1 & -1 \\ -1 & 1 & -1 \\ -1 & -1 & 1 \end{pmatrix}$

5. $\begin{pmatrix} -1 & 2 & 2 \\ 2 & -1 & 2 \\ 2 & 2 & 1 \end{pmatrix}$

6. $\begin{pmatrix} 1 & -1 & 0 \\ -1 & 2 & -1 \\ 0 & -1 & 1 \end{pmatrix}$

7. $\begin{pmatrix} 3 & 2 & 2 \\ 2 & 2 & 0 \\ 2 & 0 & 4 \end{pmatrix}$

8. $\begin{pmatrix} 1 & -1 & 0 & 0 \\ -1 & 0 & 0 & 0 \\ 0 & 0 & 0 & 0 \\ 0 & 0 & 0 & 2 \end{pmatrix}$

9. Let Q be a symmetric orthogonal matrix. Show that if λ is an eigenvalue of Q, then $\lambda = \pm 1$.

10. A is **orthogonally similar** to B if there exists an orthogonal matrix Q such that $B = Q^tAQ$. Suppose that A is orthogonally similar to B and that B is orthogonally similar to C. Show that A is orthogonally similar to C.

11. Show that if $Q = \begin{pmatrix} a & b \\ c & d \end{pmatrix}$ is orthogonal, then $b = \pm c$. [*Hint:* Write out the equations that result from the equation $Q^tQ = I$.]

12. Suppose that A is a real symmetric matrix every one of whose eigenvalues is zero. Show that A is the zero matrix.

13. Show that if a real 2×2 matrix A has eigenvectors that are orthogonal, then A is symmetric.

14. Let A be a real skew-symmetric matrix ($A^t = -A$). Prove that every eigenvalue of A is of the form $i\alpha$, where α is a real number. That is, prove that every eigenvalue of A is a **pure imaginary** number.

15. Find a unitary matrix U such that U^*AU is diagonal, where $A = \begin{pmatrix} 2 & 3-3i \\ 3+3i & 5 \end{pmatrix}$.

16. Do the same for $A = \begin{pmatrix} 1 & 1-i \\ 1+i & 0 \end{pmatrix}$.

***17.** Prove that the determinant of a hermitian matrix is real.

ANSWERS TO SELF-QUIZ

I. True **II.** True **III.** True **IV.** False **V.** True

8.17 QUADRATIC FORMS AND CONIC SECTIONS

In this section we use the material of Section 8.16 to discover information about the graphs of quadratic equations. Quadratic equations and quadratic forms, which are defined below, arise in a variety of ways. For example, we can use quadratic forms to obtain information about the conic sections in $\mathbb{R}^2$ (circles, parabolas, ellipses, hyperbolas) and extend this theory to describe certain surfaces, called *quadric surfaces,* in $\mathbb{R}^3$. These topics are discussed later in the section. Although we shall not discuss it in this text, quadratic forms arise in a number of applications ranging from a description of cost functions in economics to an analysis of the control of a rocket traveling in space.

DEFINITION QUADRATIC EQUATION AND QUADRATIC FORM

i. A **quadratic equation in two variables with no linear terms** is an equation of the form

$$ax^2 + bxy + cy^2 = d \tag{1}$$

where $|a| + |b| + |c| \neq 0$. That is, at least one of the numbers a, b, and c is nonzero.

ii. A **quadratic form in two variables** is an expression of the form

$$F(x, y) = ax^2 + bxy + cy^2 \tag{2}$$

where $|a| + |b| + |c| \neq 0$. ■

Obviously, quadratic equations and quadratic forms are closely related. We begin our analysis of quadratic forms with a simple example.

Consider the quadratic form $F(x, y) = x^2 - 4xy + 3y^2$. Let $\mathbf{v} = \begin{pmatrix} x \\ y \end{pmatrix}$ and $A = \begin{pmatrix} 1 & -2 \\ -2 & 3 \end{pmatrix}$. Then

$$A\mathbf{v}\cdot\mathbf{v} = \begin{pmatrix} 1 & -2 \\ -2 & 3 \end{pmatrix}\begin{pmatrix} x \\ y \end{pmatrix}\cdot\begin{pmatrix} x \\ y \end{pmatrix} = \begin{pmatrix} x - 2y \\ -2x + 3y \end{pmatrix}\cdot\begin{pmatrix} x \\ y \end{pmatrix}$$
$$= (x^2 - 2xy) + (-2xy + 3y^2) = x^2 - 4xy + 3y^2 = F(x, y)$$

Thus we have "represented" the quadratic form $F(x, y)$ by the symmetric matrix A in the sense that

$$F(x, y) = A\mathbf{v}\cdot\mathbf{v} \tag{3}$$

Conversely, if A is a symmetric matrix, then equation (3) defines a quadratic form $F(x, y) = A\mathbf{v}\cdot\mathbf{v}$.

We can represent $F(x, y)$ by many matrices but only one symmetric matrix. To see this, let $A = \begin{pmatrix} 1 & a \\ b & 3 \end{pmatrix}$, where $a + b = -4$. Then $A\mathbf{v}\cdot\mathbf{v} = F(x, y)$. If $A = \begin{pmatrix} 1 & 3 \\ -7 & 3 \end{pmatrix}$, for example, then $A\mathbf{v} = \begin{pmatrix} x + 3y \\ -7x + 3y \end{pmatrix}$ and $A\mathbf{v}\cdot\mathbf{v} = x^2 - 4xy + 3y^2$. If, however, we insist that A be symmetric, then we must have $a + b = -4$ and $a = b$. This pair of equations has the unique solution $a = b = -2$.

If $F(x, y) = ax^2 + bxy + cy^2$ is a quadratic form, let

$$A = \begin{pmatrix} a & b/2 \\ b/2 & c \end{pmatrix} \tag{4}$$

Then

$$A\mathbf{v}\cdot\mathbf{v} = \left[\begin{pmatrix} a & b/2 \\ b/2 & c \end{pmatrix}\begin{pmatrix} x \\ y \end{pmatrix}\right]\cdot\begin{pmatrix} x \\ y \end{pmatrix} = \begin{pmatrix} ax + (b/2)y \\ (b/2)x + cy \end{pmatrix}\cdot\begin{pmatrix} x \\ y \end{pmatrix}$$
$$= ax^2 + bxy + cy^2 = F(x, y)$$

Now let us return to the quadratic equation (1). Using (3), we can write (1) as

$$A\mathbf{v}\cdot\mathbf{v} = d \tag{5}$$

where A is symmetric. By Theorem 8.16.4 on page 611, there is an orthogonal matrix Q such that $Q^tAQ = D$, where $D = \text{diag}(\lambda_1, \lambda_2)$ and λ_1 and λ_2 are the eigenvalues of A. Then $A = QDQ^t$ (remember that $Q^t = Q^{-1}$) and (5) can be written

$$(QDQ^t\mathbf{v})\cdot\mathbf{v} = d \tag{6}$$

It is not difficult to prove that $A\mathbf{v}\cdot\mathbf{y} = \mathbf{v}\cdot A^t\mathbf{y}$. Thus

$$Q(DQ^t\mathbf{v})\cdot\mathbf{v} = DQ^t\mathbf{v}\cdot Q^t\mathbf{v} \tag{7}$$

so that (6) reads

$$[DQ^t\mathbf{v}] \cdot Q^t\mathbf{v} = d \tag{8}$$

Let $\mathbf{v}' = Q^t\mathbf{v}$. Then $\mathbf{v}'$ is a 2-vector and (8) becomes

$$D\mathbf{v}' \cdot \mathbf{v}' = d \tag{9}$$

Let us look at (9) more closely. We can write $\mathbf{v}' = \begin{pmatrix} x' \\ y' \end{pmatrix}$. Since a diagonal matrix is symmetric, (9) defines a quadratic form $\overline{F}(x', y')$ in the variables x' and y'. If $D = \begin{pmatrix} a' & 0 \\ 0 & c' \end{pmatrix}$, then $D\mathbf{v}' = \begin{pmatrix} a' & 0 \\ 0 & c' \end{pmatrix}\begin{pmatrix} x' \\ y' \end{pmatrix} = \begin{pmatrix} a'x' \\ c'y' \end{pmatrix}$ and

$$\overline{F}(x', y') = D\mathbf{v}' \cdot \mathbf{v}' = \begin{pmatrix} a'x' \\ c'y' \end{pmatrix} \cdot \begin{pmatrix} x' \\ y' \end{pmatrix} = a'x'^2 + c'y'^2$$

That is: *$\overline{F}(x', y')$ is a quadratic form with the $x'y'$ term missing.* Hence equation (9) is a quadratic equation in the new variables x', y' with the $x'y'$ term missing.

EXAMPLE 1

WRITING A QUADRATIC FORM IN NEW VARIABLES X' AND Y' WITH THE $X'Y'$ TERM MISSING

Consider the quadratic equation $x^2 - 4xy + 3y^2 = 6$. Then, as we have seen, the equation can be written in the form $A\mathbf{x} \cdot \mathbf{x} = 6$, where $A = \begin{pmatrix} 1 & -2 \\ -2 & 3 \end{pmatrix}$. In Example 8.16.1 on page 612 we saw that A can be diagonalized to $D = \begin{pmatrix} 2 - \sqrt{5} & 0 \\ 0 & 2 + \sqrt{5} \end{pmatrix}$ by using the orthogonal matrix

$$Q = \frac{1}{\sqrt{10 - 2\sqrt{5}}}\begin{pmatrix} 2 & 1 - \sqrt{5} \\ -1 + \sqrt{5} & 2 \end{pmatrix}$$

Then

$$\mathbf{x}' = \begin{pmatrix} x' \\ y' \end{pmatrix} = Q^t\mathbf{x} = \frac{1}{\sqrt{10 - 2\sqrt{5}}}\begin{pmatrix} 2 & -1 + \sqrt{5} \\ 1 - \sqrt{5} & 2 \end{pmatrix}\begin{pmatrix} x \\ y \end{pmatrix}$$

$$Q^t\mathbf{x} = \frac{1}{\sqrt{10 - 2\sqrt{5}}}\begin{pmatrix} 2x + (-1 + \sqrt{5})y \\ (1 - \sqrt{5})x + 2y \end{pmatrix}$$

and in the new variables the equation can be written as

$$(2 - \sqrt{5})x'^2 + (2 + \sqrt{5})y'^2 = 6$$

Let us take another look at the matrix Q. Since Q is real and orthogonal, $1 = \det QQ^{-1} = \det QQ^t = \det Q \ \det Q^t = \det Q \ \det Q = (\det Q)^2$. Thus $\det Q = \pm 1$. If $\det Q = -1$, we can interchange the rows of Q to make the determinant of this new Q equal to 1. Then it can be shown (see Problem 36) that $Q = \begin{pmatrix} \cos\theta & -\sin\theta \\ \sin\theta & \cos\theta \end{pmatrix}$ for some number θ with $0 \le \theta < 2\pi$. But, from Example 8.9.8 on page 562, this means that Q is a rotation matrix. We have therefore proved the following theorem.

THEOREM 1 PRINCIPAL AXES THEOREM IN $\mathbb{R}^2$

Let

$$ax^2 + bxy + cy^2 = d \tag{10}$$

be a quadratic equation in the variables x and y. Then there exists a number θ in $[0, 2\pi)$ such that equation (10) can be written in the form

$$a'x'^2 + c'y'^2 = d \tag{11}$$

where x', y' are the axes obtained by rotating the x- and y-axes through an angle of θ in the counterclockwise direction. Moreover, the numbers a' and c' are the eigenvalues of the matrix $A = \begin{pmatrix} a & b/2 \\ b/2 & c \end{pmatrix}$. The x'- and y'-axes are called the **principal axes** of the graph of the quadratic equation (10). ■

We can use Theorem 1 to identify three important conic sections. Recall that the **standard equations** of a circle, ellipse, and hyperbola are

Circle: $$x^2 + y^2 = r^2 \tag{12}$$

Ellipse: $$\frac{x^2}{a^2} + \frac{y^2}{b^2} = 1 \tag{13}$$

Hyperbola: $$\frac{x^2}{a^2} - \frac{y^2}{b^2} = 1 \tag{14}$$

or

$$\frac{y^2}{a^2} - \frac{x^2}{b^2} = 1 \tag{15}$$

EXAMPLE 2 A HYPERBOLA

Identify the conic section whose equation is

$$x^2 - 4xy + 3y^2 = 6 \tag{16}$$

SOLUTION: In Example 1 we found that this can be written as $(2 - \sqrt{5})x'^2 + (2 + \sqrt{5})y'^2 = 6$ or

$$\frac{y'^2}{6/(2 + \sqrt{5})} - \frac{x'^2}{6/(\sqrt{5} - 2)} = 1$$

This is equation (15) with $a = \sqrt{6/(2 + \sqrt{5})} \approx 1.19$ and $b = \sqrt{6/(\sqrt{5} - 2)} \approx 5.04$. Since

$$Q = \frac{1}{\sqrt{10 - 2\sqrt{5}}} \begin{pmatrix} 2 & 1 - \sqrt{5} \\ -1 + \sqrt{5} & 2 \end{pmatrix}$$

and $\det Q = 1$, we have, using Problem 36 and the fact that 2 and $-1 + \sqrt{5}$ are positive,

$$\cos \theta = \frac{2}{\sqrt{10 - 2\sqrt{5}}} \approx 0.85065$$

Thus θ is in the first quadrant and, using a table (or a calculator), we find that $\theta \approx 0.5536$ rad $\approx 31.7°$. Thus (16) is the equation of a standard hyperbola rotated through an angle of 31.7° (see Figure 1).

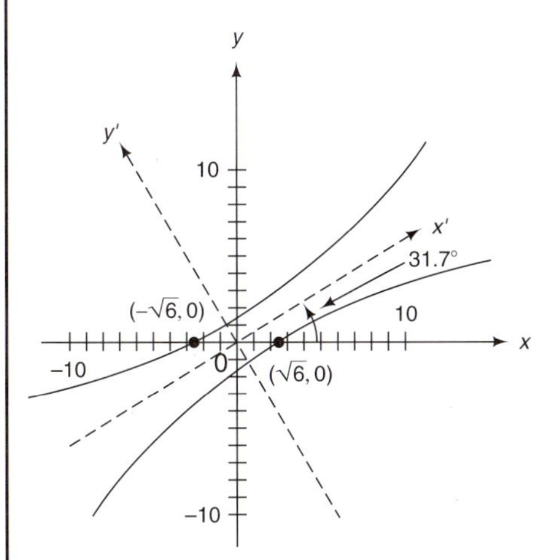

FIGURE 1
The hyperbola $x^2 - 4xy + 3y^2 = 6$

EXAMPLE 3

AN ELLIPSE

Identify the conic section whose equation is

$$5x^2 - 2xy + 5y^2 = 4 \tag{17}$$

SOLUTION: Here $A = \begin{pmatrix} 5 & -1 \\ -1 & 5 \end{pmatrix}$, the eigenvalues of A are $\lambda_1 = 4$ and $\lambda_2 = 6$, and two orthonormal eigenvectors are $\mathbf{v}_1 = \begin{pmatrix} 1/\sqrt{2} \\ 1/\sqrt{2} \end{pmatrix}$ and $\mathbf{v}_2 = \begin{pmatrix} 1/\sqrt{2} \\ -1/\sqrt{2} \end{pmatrix}$. Then $Q = \begin{pmatrix} 1/\sqrt{2} & 1/\sqrt{2} \\ 1/\sqrt{2} & -1/\sqrt{2} \end{pmatrix}$. Before continuing, we note that $\det Q = -1$. For Q to be a rotation matrix, we need $\det Q = 1$. This is easily accomplished by reversing the eigenvectors. Thus we set $\lambda_1 = 6$, $\lambda_2 = 4$, $\mathbf{v}_1 = \begin{pmatrix} 1/\sqrt{2} \\ -1/\sqrt{2} \end{pmatrix}$, $\mathbf{v}_2 = \begin{pmatrix} 1/\sqrt{2} \\ 1/\sqrt{2} \end{pmatrix}$, and $Q = \begin{pmatrix} 1/\sqrt{2} & 1/\sqrt{2} \\ -1/\sqrt{2} & 1/\sqrt{2} \end{pmatrix}$; now $\det Q = 1$. Then $D = \begin{pmatrix} 6 & 0 \\ 0 & 4 \end{pmatrix}$ and (17) can be written as $D\mathbf{v} \cdot \mathbf{v} = 4$ or

$$6x'^2 + 4y'^2 = 4 \tag{18}$$

where

$$\begin{pmatrix} x' \\ y' \end{pmatrix} = Q^t \begin{pmatrix} x \\ y \end{pmatrix} = \begin{pmatrix} 1/\sqrt{2} & -1/\sqrt{2} \\ 1/\sqrt{2} & 1/\sqrt{2} \end{pmatrix} \begin{pmatrix} x \\ y \end{pmatrix} = \begin{pmatrix} 1/\sqrt{2}x - 1/\sqrt{2}y \\ 1/\sqrt{2}x + 1/\sqrt{2}y \end{pmatrix}$$

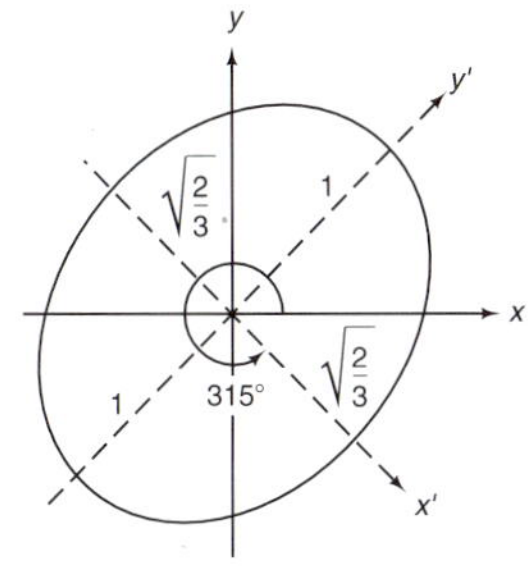

FIGURE 2
The ellipse $5x^2 - 2xy + 5y^2 = 4$

Rewriting (18), we obtain $x'^2/(\frac{4}{6}) + y'^2/1 = 1$, which is equation (13) with $a = \sqrt{\frac{2}{3}}$ and $b = 1$. Moreover, since $1/\sqrt{2} > 0$ and $-1/\sqrt{2} < 0$, we have, from Problem 36, $\theta = 2\pi - \cos^{-1}(1/\sqrt{2}) = 2\pi - \pi/4 = 7\pi/4 = 315°$. Thus (17) is the equation of a standard ellipse rotated through an angle of 315° (or 45° in the clockwise direction). (See Figure 2.)

EXAMPLE 4

A DEGENERATE CONIC SECTION

Identify the conic section whose equation is

$$-5x^2 + 2xy - 5y^2 = 4 \tag{19}$$

SOLUTION: Referring to Example 3, equation (19) can be rewritten as

$$-6x'^2 - 4y'^2 = 4 \tag{20}$$

Since, for any real numbers x' and y', $-6x'^2 - 4y'^2 \leq 0$, we see that there are no real numbers x and y which satisfy (19). The conic section defined by (19) is called a **degenerate conic section**.

There is an easy way to identify the conic section defined by

$$ax^2 + bxy + cy^2 = d \tag{21}$$

If $A = \begin{pmatrix} a & b/2 \\ b/2 & c \end{pmatrix}$, then the characteristic equation of A is

$$\lambda^2 - (a + c)\lambda + (ac - b^2/4) = 0 = (\lambda - \lambda_1)(\lambda - \lambda_2)$$

This means that $\lambda_1\lambda_2 = ac - b^2/4 = \det A$. But equation (21) can, as we have seen, be rewritten as

$$\lambda_1 x'^2 + \lambda_2 y'^2 = d \tag{22}$$

If λ_1 and λ_2 have the same sign, then (21) defines an ellipse (or a circle) or a degenerate conic as in Examples 3 and 4. If λ_1 and λ_2 have opposite signs, then (21) is the equation of a hyperbola (as in Example 2). We can therefore prove the following.

THEOREM 2

If $A = \begin{pmatrix} a & b/2 \\ b/2 & c \end{pmatrix}$, then the quadratic equation (21) with $d \neq 0$ is the equation of:

i. A hyperbola if $\det A < 0$.

ii. An ellipse, circle, or degenerate conic section if $\det A > 0$.

iii. A pair of straight lines or a degenerate conic section if $\det A = 0$.

iv. If $d = 0$, then (21) is the equation of two straight lines if $\det A \neq 0$ and the equation of a single line if $\det A = 0$.

PROOF: We have already shown why (*i*) and (*ii*) are true. To prove part (*iii*), suppose that $\det A = 0$. Then, by our Summing Up Theorem (Theorem 6.1.6), $\lambda = 0$ is an eigenvalue of A and equation (22) reads $\lambda_1 x'^2 = d$ or $\lambda_2 y'^2 = d$. If $\lambda_1 x'^2 = d$ and $d/\lambda_1 > 0$, then $x_1' = \pm\sqrt{d/\lambda_1}$ is the equation of two straight lines in the xy-plane. If $d/\lambda_1 < 0$, then we have $x'^2 < 0$ (which is impossible) and we obtain a degenerate conic. The same facts hold if $\lambda_2 y'^2 = d$. Part (*iv*) is left as an exercise (see Problem 37). ■

NOTE: In Example 2 we had $\det A = ac - b^2/4 = -1$. In Examples 3 and 4 we had $\det A = 24$.

The methods described above can be used to analyze quadratic equations in more than two variables. We give one example below.

EXAMPLE 5 AN ELLIPSOID

Consider the quadratic equation

$$5x^2 + 8xy + 5y^2 + 4xz + 4yz + 2z^2 = 100 \tag{23}$$

If $A = \begin{pmatrix} 5 & 4 & 2 \\ 4 & 5 & 2 \\ 2 & 2 & 2 \end{pmatrix}$ and $\mathbf{v} = \begin{pmatrix} x \\ y \\ z \end{pmatrix}$, then (23) can be written in the form

$$A\mathbf{v} \cdot \mathbf{v} = 100 \tag{24}$$

From Example 8.16.2 on page 613, $Q^tAQ = D = \begin{pmatrix} 1 & 0 & 0 \\ 0 & 1 & 0 \\ 0 & 0 & 10 \end{pmatrix}$, where $Q = \begin{pmatrix} -1/\sqrt{2} & -1/3\sqrt{2} & 2/3 \\ 1/\sqrt{2} & -1/3\sqrt{2} & 2/3 \\ 0 & 4/3\sqrt{2} & 1/3 \end{pmatrix}$
Let

$$\mathbf{v}' = \begin{pmatrix} x' \\ y' \\ z' \end{pmatrix} = Q^t\mathbf{v} = \begin{pmatrix} -1/\sqrt{2} & 1/\sqrt{2} & 0 \\ -1/3\sqrt{2} & -1/3\sqrt{2} & 4/3\sqrt{2} \\ 2/3 & 2/3 & 1/3 \end{pmatrix}\begin{pmatrix} x \\ y \\ z \end{pmatrix}$$

$$= \begin{pmatrix} (-1/\sqrt{2})x + (1/\sqrt{2})y \\ -(1/3\sqrt{2})x - (1/3\sqrt{2})y + (4/3\sqrt{2})z \\ (2/3)x + (2/3)y + (1/3)z \end{pmatrix}$$

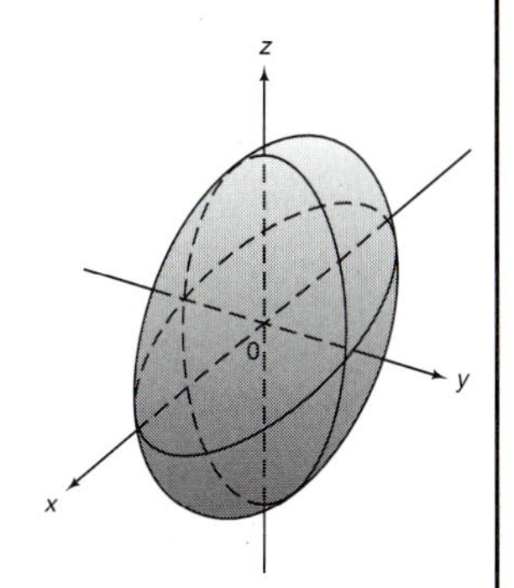

FIGURE 3
The ellipsoid $5x^2 + 8xy + 5y^2 + 4xz + 4yz + 2z^2 = 100$, which can be written in new variables as $x'^2 + y'^2 + 10z'^2 = 100$

Then, as before, $A = QDQ^t$ and $A\mathbf{v} \cdot \mathbf{v} = QDQ^t\mathbf{v} \cdot \mathbf{v} = DQ^t\mathbf{v} \cdot Q^t\mathbf{v} = D\mathbf{v}' \cdot \mathbf{v}'$. Thus (24) can be written in the new variables x', y', z' as $D\mathbf{v}' \cdot \mathbf{v}' = 100$ or

$$x'^2 + y'^2 + 10z'^2 = 100 \tag{25}$$

In $\mathbb{R}^3$ the surface defined by (25) is called an **ellipsoid** (see Figure 3).

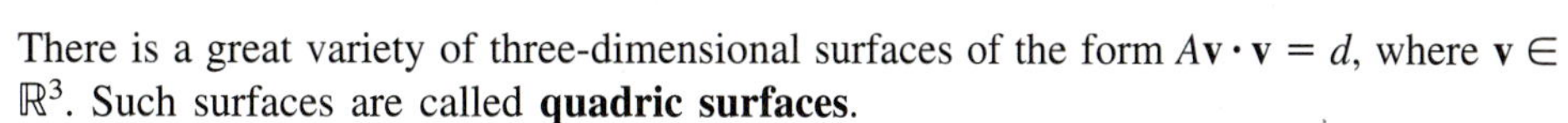

There is a great variety of three-dimensional surfaces of the form $A\mathbf{v} \cdot \mathbf{v} = d$, where $\mathbf{v} \in \mathbb{R}^3$. Such surfaces are called **quadric surfaces**.

We close this section by noting that quadratic forms can be defined in any number of variables.

DEFINITION QUADRATIC FORM

Let $\mathbf{v} = \begin{pmatrix} x_1 \\ x_2 \\ \vdots \\ x_n \end{pmatrix}$ and let A be a symmetric $n \times n$ matrix. Then a **quadratic form** in $x_1, x_2, \ldots, x_n$ is an expression of the form

$$F(x_1, x_2, \ldots, x_n) = A\mathbf{v} \cdot \mathbf{v} \tag{26}$$

■

EXAMPLE 6 A QUADRATIC FORM IN FOUR VARIABLES

Let $A = \begin{pmatrix} 2 & 1 & 2 & -2 \\ 1 & -4 & 6 & 5 \\ 2 & 6 & 7 & -1 \\ -2 & 5 & -1 & 3 \end{pmatrix}$

Then

$$\begin{aligned} A\mathbf{v}\cdot\mathbf{v} &= \left[\begin{pmatrix} 2 & 1 & 2 & -2 \\ 1 & -4 & 6 & 5 \\ 2 & 6 & 7 & -1 \\ -2 & 5 & -1 & 3 \end{pmatrix}\begin{pmatrix} x_1 \\ x_2 \\ x_3 \\ x_4 \end{pmatrix}\right]\cdot\begin{pmatrix} x_1 \\ x_2 \\ x_3 \\ x_4 \end{pmatrix} \\ &= \begin{pmatrix} 2x_1 + x_2 + 2x_3 - 2x_4 \\ x_1 - 4x_2 + 6x_3 + 5x_4 \\ 2x_1 + 6x_2 + 7x_3 - x_4 \\ -2x_1 + 5x_2 - x_3 + 3x_4 \end{pmatrix}\cdot\begin{pmatrix} x_1 \\ x_2 \\ x_3 \\ x_4 \end{pmatrix} \\ &= 2x_1^2 + 2x_1x_2 - 4x_2^2 + 4x_1x_3 + 12x_2x_3 \\ &\quad + 7x_3^2 - 4x_1x_4 + 10x_2x_4 - 2x_3x_4 + 3x_4^2 \end{aligned}$$

(after simplification)

EXAMPLE 7 FINDING A SYMMETRIC MATRIX THAT CORRESPONDS TO A QUADRATIC FORM IN FOUR VARIABLES

Find the symmetric matrix A corresponding to the quadratic form

$$5x_1^2 - 3x_1x_2 + 4x_2^2 + 8x_1x_3 - 9x_2x_3 + 2x_3^2 - x_1x_4 + 7x_2x_4 + 6x_3x_4 + 9x_4^2$$

SOLUTION: If $A = (a_{ij})$, then, by looking at the earlier examples in this section, we see that a_{ii} is the coefficient of the x_i^2 term and $a_{ij} + a_{ji}$ is the coefficient of the x_ix_j term. Since A is symmetric, $a_{ij} = a_{ji}$; hence $a_{ij} = a_{ji} = \frac{1}{2}\cdot$ (coefficient of x_ix_j term). Putting this all together, we obtain

$$A = \begin{pmatrix} 5 & -\frac{3}{2} & 4 & -\frac{1}{2} \\ -\frac{3}{2} & 4 & -\frac{9}{2} & \frac{7}{2} \\ 4 & -\frac{9}{2} & 2 & 3 \\ -\frac{1}{2} & \frac{7}{2} & 3 & 9 \end{pmatrix}$$

PROBLEMS 8.17

SELF-QUIZ

I. If A is a symmetric matrix with positive eigenvalues, then $A\mathbf{v}\cdot\mathbf{v} = d > 0$ is the equation of

a. A parabola
b. An ellipse
c. A hyperbola
d. Two straight lines
e. None of the above

II. If A is a symmetric matrix with one positive and one negative eigenvalue, then $A\mathbf{v}\cdot\mathbf{v} = d > 0$ is the equation of

a. A parabola
b. An ellipse
c. A hyperbola
d. Two straight lines
e. None of the above

III. If A is a symmetric matrix with one positive eigenvalue and one eigenvalue equal to zero, then $A\mathbf{v} \cdot \mathbf{v} = 0$ is the equation of

a. A parabola
b. An ellipse
c. A hyperbola
d. Two straight lines
e. None of the above

IV. If A is a symmetric matrix with two negative eigenvalues, then $A\mathbf{v} \cdot \mathbf{v} = d > 0$ is the equation of

a. A parabola
b. An ellipse
c. A hyperbola
d. Two straight lines
e. None of the above

In Problems 1–13 write the quadratic equation in the form $A\mathbf{v} \cdot \mathbf{v} = d$ (where A is a symmetric matrix) and eliminate the xy-term by rotating the axes through an angle of θ. Write the equation in terms of the new variables and identify the conic section obtained.

1. $3x^2 - 2xy - 5 = 0$
2. $4x^2 + 4xy + y^2 = 9$
3. $4x^2 + 4xy - y^2 = 9$
4. $xy = 1$
5. $xy = a;\ a > 0$
6. $4x^2 + 2xy + 3y^2 + 2 = 0$
7. $xy = a;\ a < 0$
8. $x^2 + 4xy + 4y^2 - 6 = 0$
9. $-x^2 + 2xy - y^2 = 0$
10. $2x^2 + xy + y^2 = 4$
11. $3x^2 - 6xy + 5y^2 = 36$
12. $x^2 - 3xy + 4y^2 = 1$
13. $6x^2 + 5xy - 6y^2 + 7 = 0$
14. What are the possible forms of the graph of $ax^2 + bxy + cy^2 = 0$?

In Problems 15–18 write the quadratic form in new variables x', y', and z' so that no cross-product terms (xy, xz, yz) are present.

15. $x^2 - 2xy + y^2 - 2xz - 2yz + z^2$
16. $-x^2 + 4xy - y^2 + 4xz + 4yz + z^2$
17. $3x^2 + 4xy + 2y^2 + 4xz + 4z^2$
18. $x^2 - 2xy + 2y^2 - 2yz + z^2$

In Problems 19–21 find a symmetric matrix A such that the quadratic form can be written in the form $A\mathbf{x} \cdot \mathbf{x}$.

19. $x_1^2 + 2x_1x_2 + x_2^2 + 4x_1x_3 + 6x_2x_3 + 3x_3^2 + 7x_1x_4 - 2x_2x_4 + x_4^2$
20. $x_1^2 - x_2^2 + x_1x_3 - x_2x_4 + x_4^2$
21. $3x_1^2 - 7x_1x_2 - 2x_2^2 + x_1x_3 - x_2x_3 + 3x_3^2 - 2x_1x_4 + x_2x_4 - 4x_3x_4 - 6x_4^2 + 3x_1x_5 - 5x_3x_5 + x_4x_5 - x_5^2$
22. Suppose that for some nonzero value of d, the graph of $ax^2 + bxy + cy^2 = d$ is a hyperbola. Show that the graph is a hyperbola for any other nonzero value of d.
23. Show that if $a \neq c$, the xy-term in quadratic equation (1) will be eliminated by rotation through an angle θ if θ is given by $\cot 2\theta = (a - c)/b$.
24. Show that if $a = c$ in Problem 23, then the xy-term will be eliminated by a rotation through an angle of either $\pi/4$ or $-\pi/4$.
***25.** Suppose that a rotation converts $ax^2 + bxy + cy^2$ into $a'(x')^2 + b'(x'y') + c'(y')^2$. Show that:
a. $a + c = a' + c'$
b. $b^2 - 4ac = b'^2 - 4a'c'$
***26.** A quadratic form $F(\mathbf{x}) = F(x_1, x_2, \ldots, x_n)$ is said to be **positive definite** if $F(\mathbf{x}) \geq 0$ for every $\mathbf{x} \in \mathbb{R}^n$ and $F(\mathbf{x}) = 0$ if and only if $\mathbf{x} = \mathbf{0}$. Show that F is positive definite if and only if the symmetic matrix A associated with F has positive eigenvalues.
27. A quadratic form $F(\mathbf{x})$ is said to be **positive semidefinite** if $F(\mathbf{x}) \geq 0$ for every $\mathbf{x} \in \mathbb{R}^n$. Show that F is positive semidefinite if and only if the eigenvalues of the symmetric matrix associated with F are all nonnegative.

The definitions of **negative definite** and **negative semidefinite** are the definitions in Problems 26 and 27 with ≤ 0 replacing ≥ 0. A quadratic form is **indefinite** if it is none of the above. In Problems 28–35 determine whether the given quadratic form is positive definite, positive semidefinite, negative definite, negative semidefinite, or indefinite.

28. $3x^2 + 2y^2$
29. $-3x^2 - 3y^2$
30. $3x^2 - 2y^2$
31. $x^2 + 2xy + 2y^2$

32. $x^2 - 2xy + 2y^2$

33. $x^2 - 4xy + 3y^2$

34. $-x^2 + 4xy - 3y^2$

35. $-x^2 + 2xy - 2y^2$

***36.** Let $Q = \begin{pmatrix} a & b \\ c & d \end{pmatrix}$ be a real orthogonal matrix with $\det Q = 1$. Define the number $\theta \in [0, 2\pi)$:

a. If $a \geq 0$ and $c > 0$, then $\theta = \cos^{-1} a$ $(0 < \theta \leq \pi/2)$.

b. If $a \geq 0$ and $c < 0$, then $\theta = 2\pi - \cos^{-1} a$ $(3\pi/2 \leq \theta < 2\pi)$.

c. If $a \leq 0$ and $c > 0$, then $\theta = \cos^{-1} a$ $(\pi/2 \leq \theta < \pi)$.

d. If $a \leq 0$ and $c < 0$, then $\theta = 2\pi - \cos^{-1} a$ $(\pi < \theta \leq 3\pi/2)$.

e. If $a = 1$ and $c = 0$, then $\theta = 0$.

f. If $a = -1$ and $c = 0$, then $\theta = \pi$.

(Here $\cos^{-1} x \in [0, \pi]$ for $x \in [-1, 1]$.) With θ chosen as above, show that

$$Q = \begin{pmatrix} \cos\theta & -\sin\theta \\ \sin\theta & \cos\theta \end{pmatrix}.$$

37. Prove, using formula (22), that equation (21) is the equation of two straight lines in the xy-plane when $d = 0$ and $\det A < 0$. If $\det A = d = 0$, show that equation (21) is the equation of a single line.

38. Let A be the symmetric matrix representation of quadratic equation (1) with $d \neq 0$. Let λ_1 and λ_2 be the eigenvalues of A. Show that (1) is the equation of **(a)** a hyperbola if $\lambda_1\lambda_2 < 0$ and **(b)** a circle, ellipse, or degenerate conic section if $\lambda_1\lambda_2 > 0$.

ANSWERS TO SELF-QUIZ

I. b **II.** c **III.** d **IV.** e

CHAPTER 8 SUMMARY OUTLINE

- A **real vector space** V is a set of objects, called **vectors**, together with two operations called **addition** (denoted by $\mathbf{x} + \mathbf{y}$) and **scalar multiplication** (denoted by $\alpha\mathbf{x}$) that satisfy the following ten axioms:

  i. If $\mathbf{x} \in V$ and $\mathbf{y} \in V$, then $\mathbf{x} + \mathbf{y} \in V$ (closure under addition).

  ii. For all $\mathbf{x}$, $\mathbf{y}$, and $\mathbf{z}$ in V, $(\mathbf{x} + \mathbf{y}) + \mathbf{z} = \mathbf{x} + (\mathbf{y} + \mathbf{z})$ (associative law of vector addition).

  iii. There is a vector $\mathbf{0} \in V$ such that for all $\mathbf{x} \in V$, $\mathbf{x} + \mathbf{0} = \mathbf{0} + \mathbf{x} = \mathbf{x}$ ($\mathbf{0}$ is called the additive identity).

  iv. If $\mathbf{x} \in V$, there is a vector $-\mathbf{x}$ in V such that $\mathbf{x} + (-\mathbf{x}) = \mathbf{0}$ ($-\mathbf{x}$ is called the additive inverse of $\mathbf{x}$).

  v. If $\mathbf{x}$ and $\mathbf{y}$ are in V, then $\mathbf{x} + \mathbf{y} = \mathbf{y} + \mathbf{x}$ (commutative law of vector addition).

  vi. If $\mathbf{x} \in V$ and α is a scalar, then $\alpha\mathbf{x} \in V$ (closure under scalar multiplication).

  vii. If $\mathbf{x}$ and $\mathbf{y}$ are in V and α is a scalar, then $\alpha(\mathbf{x} + \mathbf{y}) = \alpha\mathbf{x} + \alpha\mathbf{y}$ (first distributive law).

  viii. If $\mathbf{x} \in V$ and α and β are scalars, then $(\alpha + \beta)\mathbf{x} = \alpha\mathbf{x} + \beta\mathbf{x}$ (second distributive law).

  ix. If $\mathbf{x} \in V$ and α and β are scalars, then $\alpha(\beta\mathbf{x}) = \alpha\beta\mathbf{x}$ (associative law of scalar multiplication).

  x. For every vector $\mathbf{x} \in V$, $1\mathbf{x} = \mathbf{x}$ (the scalar 1 is called a multiplicative identity).

- **The space** $\mathbb{R}^n = \{(x_1, x_2, \ldots, x_n)\colon\ x_i \in \mathbb{R} \text{ for } i = 1, 2, \ldots, n\}$.
- **The space P_n** = {polynomials of degree less than or equal to n}.
- **The space $C[a, b]$** = {real-valued functions that are continuous on the interval $[a, b]$}.
- **The space M_{mn}** = {$m \times n$ matrices with real coefficients}.
- **The space** $\mathbb{C}_n = \{(c_1, c_2, \ldots, c_n)\colon\ c_i \in \mathbb{C} \text{ for } i = 1, 2, \ldots, n\}$. $\mathbb{C}$ denotes the set of complex numbers.
- A **subspace** H of a vector space V is a subset of V that is itself a vector space.
- A nonempty subset H of a vector space V is a subspace of V if the following two rules hold:

  i. If $\mathbf{x} \in H$ and $\mathbf{y} \in H$, then $\mathbf{x} + \mathbf{y} \in H$.

  ii. If $\mathbf{x} \in H$, then $\alpha\mathbf{x} \in H$ for every scalar α.

- A **proper subspace** of a vector space V is a subspace of V other than $\{\mathbf{0}\}$ or V.

- A **linear combination** of the vectors $\mathbf{v}_1, \mathbf{v}_2, \ldots, \mathbf{v}_n$ in a vector space V is a sum of the form

$$\alpha_1\mathbf{v}_1 + \alpha_2\mathbf{v}_2 + \cdots + \alpha_n\mathbf{v}_n$$

where $\alpha_1, \alpha_2, \ldots, \alpha_n$ are scalars.

- The vectors $\mathbf{v}_1, \mathbf{v}_2, \ldots, \mathbf{v}_n$ in a vector space V are said to **span** V if every vector in V can be written as a linear combination of $\mathbf{v}_1, \mathbf{v}_2, \ldots, \mathbf{v}_n$.
- The **span of a set of vectors** $\mathbf{v}_1, \mathbf{v}_2, \ldots, \mathbf{v}_k$ in a vector space V is the set of linear combinations of $\mathbf{v}_1, \mathbf{v}_2, \ldots, \mathbf{v}_k$.
- span $\{\mathbf{v}_1, \mathbf{v}_2, \ldots, \mathbf{v}_k\}$ is a subspace of V.
- **Linear Dependence and Independence** The vectors $\mathbf{v}_1, \mathbf{v}_2, \ldots, \mathbf{v}_n$ in a vector space V are said to be **linearly dependent** if there exist scalars $c_1, c_2, \ldots, c_n$ not all zero such that

$$c_1\mathbf{v}_1 + c_2\mathbf{v}_2 + \cdots + c_n\mathbf{v}_n = \mathbf{0}$$

If the vectors are not linearly dependent, they are said to be **linearly independent**.

- Two vectors in a vector space V are linearly dependent if and only if one is a scalar multiple of the other.
- Any set of n linearly independent vectors in $\mathbb{R}^n$ spans $\mathbb{R}^n$.
- A set of n vectors in $\mathbb{R}^m$ is linearly dependent if $n > m$.
- **Basis** A set of vectors $\mathbf{v}_1, \mathbf{v}_2, \ldots, \mathbf{v}_n$ is a **basis** for a vector space V if

i. $\{\mathbf{v}_1, \mathbf{v}_2, \ldots, \mathbf{v}_n\}$ is linearly independent

ii. $\{\mathbf{v}_1, \mathbf{v}_2, \ldots, \mathbf{v}_n\}$ spans V

- Every set of n linearly independent vectors in $\mathbb{R}^n$ is a basis in $\mathbb{R}^n$
- The **standard basis** in $\mathbb{R}^n$ consists of the n vectors

$$\mathbf{e}_1 = \begin{pmatrix}1\\0\\0\\\vdots\\0\end{pmatrix}, \mathbf{e}_2 = \begin{pmatrix}0\\1\\0\\\vdots\\0\end{pmatrix}, \mathbf{e}_3 = \begin{pmatrix}0\\0\\1\\\vdots\\0\end{pmatrix}, \ldots, \mathbf{e}_n = \begin{pmatrix}0\\0\\0\\\vdots\\1\end{pmatrix}$$

- **Dimension** If the vector space V has a finite basis, then the **dimension** of V is the number of vectors in every basis and V is called a **finite dimensional vector space**. Otherwise V is called an **infinite dimensional vector space**. If $V = \{\mathbf{0}\}$, then V is said to be **zero dimensional**.

We write the dimension of V as dim V.

- If H is a subspace of the finite dimensional space V, then dim $H \le$ dim V
- The only proper subspaces of $\mathbb{R}^3$ are sets of vectors lying on a single line or a single plane passing through the origin.
- The **kernel** of an $n \times n$ matrix A is the subspace of $\mathbb{R}^n$ given by

$$N_A = \{\mathbf{x} \in \mathbb{R}^n\colon A\mathbf{x} = \mathbf{0}\}$$

- The **nullity** of an $n \times n$ matrix A is the dimension of N_A and is denoted by $\nu(A)$.
- Let A be an $m \times n$ matrix. The **range of A**, denoted by Range A, is the subspace of $\mathbb{R}^m$ given by

$$\text{Range } A = \{\mathbf{y} \in \mathbb{R}^m\colon A\mathbf{x} = \mathbf{y} \text{ for some } \mathbf{x} \in \mathbb{R}^n\}$$

- The **rank of A**, denoted by $\rho(A)$, is the dimension of Range A.
- The **row space of A**, denoted by R_A, is the span of the rows of A and is a subspace of $\mathbb{R}^n$.
- The **column space of A**, denoted by C_A, is the span of the columns of A and is a subspace of $\mathbb{R}^m$
- If A is $m \times n$ matrix, then

$$C_A = \text{Range } A \qquad \text{and} \qquad \dim R_A = \dim C_A = \dim \text{Range } A = \rho(A)$$

- $\rho(A) + \nu(A) = n$
- The system $A\mathbf{x} = \mathbf{b}$ has at least one solution if and only if $\rho(A) = \rho(A, \mathbf{b})$ where $(A, \mathbf{b})$ is the augmented matrix obtained by adjoining the column vector $\mathbf{b}$ to A.
- The vectors $\mathbf{u}_1, \mathbf{u}_2, \ldots, \mathbf{u}_k$ in $\mathbb{R}^n$ form an **orthogonal set** if $\mathbf{u}_i \cdot \mathbf{u}_j = \mathbf{0}$ for $i \ne j$. If, in addition, $\mathbf{u}_i \cdot \mathbf{u}_i = 1$ for $i = 1, 2, \ldots, k$, the set is said to be **orthonormal**.
- $|\mathbf{v}| = |\mathbf{v} \cdot \mathbf{v}|^{1/2}$ is called the **length** or **norm** of $\mathbf{v}$.
- Every subspace of $\mathbb{R}^n$ has an orthonormal basis. **The Gram-Schmidt orthonormalization process** can be used to construct such a basis.
- An **orthogonal matrix** is an invertible $n \times n$ matrix Q such that $Q^{-1} = Q^t$.
- An $n \times n$ matrix is orthogonal if and only if its columns form an orthonormal basis for $\mathbb{R}^n$.
- Let H be a subspace of $\mathbb{R}^n$ with orthonormal basis $\{\mathbf{u}_1, \mathbf{u}_2, \ldots, \mathbf{u}_k\}$. If $\mathbf{v} \in \mathbb{R}^n$, then the **orthogonal projection** of $\mathbf{v}$ onto H, denoted by $\text{proj}_H \mathbf{v}$, is given by

$$\text{proj}_H \mathbf{v} = (\mathbf{v} \cdot \mathbf{u}_1)\mathbf{u}_1 + (\mathbf{v} \cdot \mathbf{u}_2)\mathbf{u}_2 + \cdots + (\mathbf{v} \cdot \mathbf{u}_k)\mathbf{u}_k$$

- Let H be a subspace of $\mathbb{R}^n$. Then the **orthogonal complement** of H, denoted by $H^\perp$, is given by

$$H^\perp = \{\mathbf{x} \in \mathbb{R}^n\colon \mathbf{x} \cdot \mathbf{h} = 0 \quad \text{for every } \mathbf{h} \in H\}$$

- **Projection Theorem** Let H be a subspace of $\mathbb{R}^n$ and let $\mathbf{v} \in \mathbb{R}^n$. Then there exists a unique pair of vectors $\mathbf{h}$ and $\mathbf{p}$ such that $\mathbf{h} \in H$, $\mathbf{p} \in H^\perp$, and

$$\mathbf{v} = \mathbf{h} + \mathbf{p} = \text{proj}_H \mathbf{v} + \text{proj}_{H^\perp} \mathbf{v}$$

- **Norm Approximation Theorem** Let H be a subspace of $\mathbb{R}^n$ and let $\mathbf{v}$ be a vector in $\mathbb{R}^n$. Then, in H, $\text{proj}_H \mathbf{v}$ is the best approximation to $\mathbf{v}$ in the following sense: If $\mathbf{h}$ is any other vector in H, then

$$|\mathbf{v} - \text{proj}_H \mathbf{v}| < |\mathbf{v} - \mathbf{h}|$$

- Let $(x_1, y_1), (x_2, y_2), \ldots, (x_n, y_n)$ be a set of data points. If we wish to represent these data by the straight line $y = mx + b$, then the **least squares problem** is to find the m and b that minimizes the sum of squares

$$[y_1 - (b + mx_1)]^2 + [y_2 - (b + mx_2)]^2 + \cdots + [y_n - (b + mx_n)]^2$$

- The solution to this problem is to set

$$\begin{pmatrix} b \\ m \end{pmatrix} = \mathbf{u} = (A^tA)^{-1}A^t\mathbf{y}$$

where $\mathbf{y} = \begin{pmatrix} y_1 \\ y_2 \\ \vdots \\ y_n \end{pmatrix}$ and $A = \begin{pmatrix} 1 & x_1 \\ 1 & x_2 \\ \vdots & \vdots \\ 1 & x_n \end{pmatrix}$.

- Similar results apply when we attempt to represent the data using a polynomial of degree > 1.
- **Linear Transformation** Let V and W be vector spaces. A **linear transformation** T from V into W is a function that assigns to each vector $\mathbf{v} \in V$ a unique vector $T\mathbf{v} \in W$ and that satisfies, for each $\mathbf{u}$ and $\mathbf{v}$ in V and each scalar α,

$$T(\mathbf{u} + \mathbf{v}) = T\mathbf{u} + T\mathbf{v}$$

and

$$T(\alpha\mathbf{v}) = \alpha T\mathbf{v}$$

- **Basic Properties of Linear Transformations** Let $T: V \rightarrow W$ be a linear transformation. Then for all vectors $\mathbf{u}, \mathbf{v}, \mathbf{v}_1, \mathbf{v}_2, \ldots, \mathbf{v}_n$ in V and all scalars $\alpha_1, \alpha_2, \ldots, \alpha_n$:

  i. $T(\mathbf{0}) = \mathbf{0}$

  ii. $T(\mathbf{u} - \mathbf{v}) = T\mathbf{u} - T\mathbf{v}$

  iii. $T(\alpha_1\mathbf{v}_1 + \alpha_2\mathbf{v}_2 + \cdots + \alpha_n\mathbf{v}_n) = \alpha_1 T\mathbf{v}_1 + \alpha_2 T\mathbf{v}_2 + \cdots + \alpha_n T\mathbf{v}_n$

- **Kernel and Range of a Linear Transformation** Let V and W be vector spaces and let $T: V \rightarrow W$ be a linear transformation. Then the **kernel** of T, denoted by ker T, is given by

$$\ker T = \{\mathbf{v} \in V: T\mathbf{v} = \mathbf{0}\}$$

The **range** of T, denoted by range T, is given by

$$\text{Range } T = \{\mathbf{w} \in W: \mathbf{w} = T\mathbf{v} \text{ for some } \mathbf{v} \in V\}$$

Ker T is a subspace of V and range T is a subspace of W.

- **Nullity and Rank of a Linear Transformation** If T is a linear transformation from V to W, then

  Nullity of $T = \nu(T) = \dim \ker T$

  Rank of $T = \rho(T) = \dim \text{range } T$

- **Transformation Matrix** Let $T: \mathbb{R}^n \rightarrow \mathbb{R}^m$ be a linear transformation. Then there is a unique $m \times n$ matrix A_T such that

$$T\mathbf{x} = A_T\mathbf{x} \qquad \text{for every } \mathbf{x} \in \mathbb{R}^n$$

The matrix A_T is called the **transformation matrix** of T.

- Let A_T be the transformation matrix corresponding to the linear transformation T. Then:

  i. range $T = R_{A_T} = C_{A_T}$

  ii. $\rho(T) = \rho(A_T)$

  iii. $\ker T = N_{A_T}$

  iv. $\nu(T) = \nu(A_T)$

- **Matrix Representation of a Linear Transformation** Let V be a real n-dimensional vector space, W be a real m-dimensional vector space, and $T: V \rightarrow W$ be a linear transformation. Let $B_1 = \{\mathbf{v}_1, \mathbf{v}_2, \ldots, \mathbf{v}_n\}$ be a basis for V and let $B_2 = \{\mathbf{w}_1, \mathbf{w}_2, \ldots, \mathbf{w}_m\}$ be a basis for W. Then there is a unique $m \times n$ matrix A_T such that

$$(T\mathbf{x})_{B_2} = A_T(\mathbf{x})_{B_1}$$

A_T is called the **matrix representation** of T

- Let V and W be finite dimensional vector spaces with $\dim V = n$. Let $T: V \rightarrow W$ be a linear transformation and let A_T be a matrix representation of T. Then:

  i. $\rho(T) = \rho(A_T)$

  ii. $\nu(T) = \nu(A_T)$

  iii. $\nu(T) + \rho(T) = n$

- **One-to-One Transformation** Let $T: V \rightarrow W$ be a linear transformation. Then T is **one-to-one**, written $1-1$, if $T\mathbf{v}_1 = T\mathbf{v}_2$ implies that $\mathbf{v}_1 = \mathbf{v}_2$. That is, T is $1-1$ if every vector $\mathbf{w}$ in the range of T is the image of exactly one vector in V.
- Let $T: V \rightarrow W$ be a linear transformation. Then T is $1-1$ if and only if $\ker T = \{\mathbf{0}\}$.
- **Onto Transformation** Let $T: V \rightarrow W$ be a linear transformation. Then T is said to be **onto** W or simply **onto**, if for every $\mathbf{w} \in W$ there is at least one $\mathbf{v} \in V$ such that $T\mathbf{v} = \mathbf{w}$. That is: *T is onto W if and only if range $T = W$.*
- Let $T: V \rightarrow W$ be a linear transformation and suppose that $\dim V = \dim W = n$:

  i. If T is $1-1$, then T is onto.

  ii. If T is onto, then T is $1-1$.

- Let $T: V \rightarrow W$ be a linear transformation. Suppose that $\dim V = n$ and $\dim W = m$. Then:

  i. If $n > m$, T is not $1-1$.

  ii. If $m > n$, T is not onto.

- **Isomorphism** Let $T: V \rightarrow W$ be a linear transformation. Then T is an **isomorphism** if T is $1-1$ and onto.
- **Isomorphic Vector Spaces** The vector spaces V and W are said to be **isomorphic** if there exists an isomorphism T from V onto W. In this case we write $V \cong W$.
- Any two real finite dimensional vector spaces of the same dimension are isomorphic.

- **Eigenvalue and Eigenvector** Let A be an $n \times n$ matrix with real components. The number λ (real or complex) is called an **eigenvalue** of A if there is a *nonzero* vector $\mathbf{v}$ in $\mathbb{C}^n$ such that

$$A\mathbf{v} = \lambda\mathbf{v}$$

The vector $\mathbf{v} \neq \mathbf{0}$ is called an **eigenvector** of A corresponding to the eigenvalue λ.

- Let A be an $n \times n$ matrix. Then λ is an eigenvalue of A if and only if

$$p(\lambda) = \det(A - \lambda I) = 0$$

The equation $p(\lambda) = 0$ is called the **characteristic equation** of A; $p(\lambda)$ is called the **characteristic polynomial** of A.

- Counting multiplicities, every $n \times n$ matrix has exactly n eigenvalues.
- Eigenvectors corresponding to different eigenvalues are linearly independent.
- **Algebraic Multiplicity** If

$$p(\lambda) = (\lambda - \lambda_1)^{r_1}(\lambda - \lambda_2)^{r_2} \cdots (\lambda - \lambda_m)^{r_m},$$

then r_i is the **algebraic multiplicity** of λ_i.

- The eigenvalues of a real matrix occur in complex conjugate pairs.
- **Eigenspace** If λ is an eigenvalue of the $n \times n$ matrix A, then $E_\lambda = \{\mathbf{v}\colon A\mathbf{v} = \lambda\mathbf{v}\}$ is a subspace of $\mathbb{R}^n$ called the **eigenspace** of A corresponding to λ. It is denoted by E_λ.
- **Geometric Multiplicity** The **geometric multiplicity** of an eigenvalue λ of the matrix A is equal to $\dim E_\lambda = \nu(A - \lambda I)$.
- For any eigenvalue λ, geometric multiplicity $\leq$ algebraic multiplicity.
- Let A be an $n \times n$ matrix. Then A has n linearly independent eigenvectors if and only if the geometric multiplicity of every eigenvalue is equal to its algebraic multiplicity. In particular, A has n linearly independent eigenvectors if all the eigenvalues are distinct (since then the algebraic multiplicity of every eigenvalue is 1).
- **Summing Up Theorem** Let A be an $n \times n$ matrix. Then the following eleven statements are equivalent; that is each one implies the other ten (so that if one is true, all are true):

  i. A is invertible.

  ii. The only solution to the homogeneous system $A\mathbf{x} = \mathbf{0}$ is the trivial solution ($\mathbf{x} = \mathbf{0}$).

  iii. The system $A\mathbf{x} = \mathbf{b}$ has a unique solution for every n-vector $\mathbf{b}$.

  iv. A is row equivalent to the $n \times n$ identity matrix I_n.

  v. A is a product of elementary matrices.

  vi. The rows (and columns) of A are linearly independent.

  vii. $\det A \neq 0$.

  viii. $\nu(A) = 0$.

  ix. $\rho(A) = n$.

  x. The linear transformation T from $\mathbb{R}^n$ to $\mathbb{R}^n$ defined by $T\mathbf{x} = A\mathbf{x}$ is an isomorphism.

  xi. Zero is *not* an eigenvalue of A.

- **Similar Matrices** Two $n \times n$ matrices A and B are said to be **similar** if there exists an invertible $n \times n$ matrix C such that

$$B = C^{-1}AC$$

The function defined above which takes the matrix A into the matrix B is called a **similarity transformation**.

- Similar matrices have the same eigenvalues.
- **Diagonalizable Matrix** An $n \times n$ matrix A is **diagonalizable** if there is a diagonal matrix D such that A is similar to D.
- An $n \times n$ matrix A is diagonalizable if and only if it has n linearly independent eigenvectors. In that case the diagonal matrix D similar to A is given by

$$D = \begin{pmatrix} \lambda_1 & 0 & 0 & \cdots & 0 \\ 0 & \lambda_2 & 0 & \cdots & 0 \\ 0 & 0 & \lambda_3 & \cdots & 0 \\ \vdots & \vdots & \vdots & & \vdots \\ 0 & 0 & 0 & \cdots & \lambda_n \end{pmatrix}$$

where $\lambda_1, \lambda_2, \ldots, \lambda_n$ are the eigenvalues of A. If C is a matrix whose columns are linearly independent eigenvectors of A, then

$$D = C^{-1}AC$$

- If the $n \times n$ matrix A has n distinct eigenvalues, then A is diagonalizable.
- The eigenvalues of a real symmetric matrix are real.
- Eigenvectors of a real symmetric matrix corresponding to distinct eigenvalues are orthogonal.
- A real symmetric $n \times n$ matrix has n real orthonormal eigenvectors.
- **Orthogonally Diagonalizable Matrix** An $n \times n$ matrix A is said to be **orthogonally diagonalizable** if there exists an orthogonal matrix Q such that

$$Q^tAQ = D$$

where $D = \text{diag}\,(\lambda_1, \lambda_2, \ldots, \lambda_n)$ and $\lambda_1, \lambda_2, \ldots, \lambda_n$ are the eigenvalues of A.

- **Procedure for finding an orthogonal diagonalizing matrix Q for a real symmetric matrix A:**

  i. Find a basis for each eigenspace of A.

  ii. Find an orthonormal basis for each eigenspace of A by using the Gram-Schmidt process.

  iii. Write Q as the matrix whose columns are the orthonormal eigenvectors obtained in step (*ii*).

- The **conjugate transpose** of an $m \times n$ matrix $A = (a_{ij})$, denoted by A^*, is the $n \times m$ matrix whose ijth component is $\overline{a}_{ji}$.
- A complex $n \times n$ matrix A is **hermitian** if $A^* = A$.
- A complex $n \times n$ matrix U is **unitary** if $U^* = U^{-1}$.
- **Quadratic Equation and Quadratic Form** A **quadratic equation in two variables with no linear term** is an equation of the form

$$ax^2 + bxy + cy^2 = d$$

where $|a| + |b| + |c| \neq 0$ and a, b, c are real numbers.

A **quadratic form in two variables** is an expression of the form

$$F(x, y) = ax^2 + bxy + cy^2$$

where $|a| + |b| + |c| \neq 0$ and a, b, c are real numbers.

- A quadratic form can be written as

$$F(x, y) = A\mathbf{v} \cdot \mathbf{v}$$

where $A = \begin{pmatrix} a & b/2 \\ b/2 & c \end{pmatrix}$ is a symmetric matrix.

- If the eigenvalues of A are a' and c', then the quadratic form can be written as

$$F(x', y') = a'x'^2 + c'y'^2$$

where $\begin{pmatrix} x' \\ y' \end{pmatrix} = Q^t\begin{pmatrix} x \\ y \end{pmatrix}$ and Q is the orthogonal matrix that diagonalizes A.

- **Principal Axes Theorem in $\mathbb{R}^2$** Let

$$ax^2 + bxy + cy^2 = d \qquad (*)$$

be a quadratic equation in the variables x and y. Then there exists a number θ in $[0, 2\pi)$ such that equation (*) can be written in the form

$$a'x'^2 + c'y'^2 = d$$

where x', y' are the axes obtained by rotating the x- and y-axes through an angle of θ in the counterclockwise direction. Moreover, the numbers a' and c' are the eigenvalues of the matrix $A = \begin{pmatrix} a & b/2 \\ b/2 & c \end{pmatrix}$. The x'- and y'-axes are called the **principal axes** of the graph of the quadratic equation.

CHAPTER 8 REVIEW EXERCISES

In Exercises 1–10 determine whether the given set is a vector space. If so, determine its dimension. If it is finite dimensional, find a basis for it.

1. The vectors (x, y, z) in $\mathbb{R}^3$ satisfying $x + 2y - z = 0$
2. The vectors (x, y, z) in $\mathbb{R}^3$ satisfying $x + 2y - z \leq 0$
3. The vectors (x, y, z, w) in $\mathbb{R}^4$ satisfying $x + y + z + w = 0$
4. The vectors in $\mathbb{R}^3$ satisfying $x - 2 = y + 3 = z - 4$
5. The set of upper triangular $n \times n$ matrices under the operations of matrix addition and scalar multiplication
6. The set of polynomials of degree ≤ 5
7. The set of polynomials of degree 5
8. The set of 3×2 matrices $A = (a_{ij})$, with $a_{12} = 0$, under the operations of matrix addition and scalar multiplication
9. The set in Exercise 8 except that $a_{12} = 1$
10. The set $S = \{f \in C[0, 2]: f(2) = 0\}$

In Exercises 11–19 determine whether the given set of vectors is linearly dependent or independent.

11. $\begin{pmatrix} 2 \\ 3 \end{pmatrix}; \begin{pmatrix} 4 \\ -6 \end{pmatrix}$
12. $\begin{pmatrix} 2 \\ 3 \end{pmatrix}; \begin{pmatrix} 4 \\ 6 \end{pmatrix}$
13. $\begin{pmatrix} 1 \\ -1 \\ 2 \end{pmatrix}; \begin{pmatrix} 3 \\ 0 \\ 1 \end{pmatrix}; \begin{pmatrix} 0 \\ 0 \\ 0 \end{pmatrix}$
14. $\begin{pmatrix} 1 \\ -4 \\ 2 \end{pmatrix}; \begin{pmatrix} 0 \\ 2 \\ -1 \end{pmatrix}; \begin{pmatrix} 2 \\ -10 \\ 5 \end{pmatrix}$
15. $\begin{pmatrix} 1 \\ 0 \\ 0 \\ 0 \end{pmatrix}; \begin{pmatrix} 0 \\ 1 \\ 0 \\ 0 \end{pmatrix}; \begin{pmatrix} 0 \\ 0 \\ 1 \\ 0 \end{pmatrix}; \begin{pmatrix} 0 \\ 0 \\ 0 \\ 1 \end{pmatrix}$
16. In P_3: $1, 2 - x^2, 3 - x, 7x^2 - 8x$
17. In P_3: $1, 2 + x^3, 3 - x, 7x^2 - 8x$
18. In M_{22}: $\begin{pmatrix} 1 & -1 \\ 0 & 0 \end{pmatrix}, \begin{pmatrix} 1 & 1 \\ 0 & 0 \end{pmatrix}, \begin{pmatrix} 0 & 0 \\ 1 & 1 \end{pmatrix}, \begin{pmatrix} 0 & 0 \\ 1 & -1 \end{pmatrix}$
19. In M_{22}: $\begin{pmatrix} 1 & 1 \\ 0 & 0 \end{pmatrix}, \begin{pmatrix} 1 & -1 \\ 0 & 0 \end{pmatrix}, \begin{pmatrix} 0 & 0 \\ 1 & 1 \end{pmatrix}, \begin{pmatrix} 0 & 0 \\ 1 & -1 \end{pmatrix}$
20. Using determinants, determine whether each set of vectors is linearly dependent or independent.
    a. $\begin{pmatrix} 1 \\ 5 \\ 2 \end{pmatrix}; \begin{pmatrix} 3 \\ 0 \\ 4 \end{pmatrix}; \begin{pmatrix} -5 \\ 5 \\ 6 \end{pmatrix}$
    b. $(2, 1, 4); (3, -2, 6); (-1, -4, -2)$

In Exercises 21–26 find a basis for the given vector space and determine its dimension.

21. The vectors in $\mathbb{R}^3$ lying on the plane $2x + 3y - 4z = 0$

22. $H = \{(x, y): 2x - 3y = 0\}$

23. $\{\mathbf{v} \in \mathbb{R}^4; 3x - y - z + w = 0\}$

24. $\{p \in P_3; p(0) = 0\}$

25. The set of diagonal 4×4 matrices

26. M_{23}

In Exercises 27–32 find the kernel, range, nullity, and rank of the given matrix.

27. $A = \begin{pmatrix} 1 & -2 \\ -2 & 4 \end{pmatrix}$ **28.** $A = \begin{pmatrix} 1 & -1 & 3 \\ 2 & 0 & 4 \\ 0 & -2 & 2 \end{pmatrix}$

29. $A = \begin{pmatrix} 1 & -1 & 2 \\ 0 & 1 & 4 \\ 1 & -1 & 0 \end{pmatrix}$ **30.** $A = \begin{pmatrix} 2 & 4 & -2 \\ -1 & -2 & 1 \end{pmatrix}$

31. $A = \begin{pmatrix} 2 & 3 \\ -1 & 2 \\ 4 & 6 \end{pmatrix}$ **32.** $A = \begin{pmatrix} 1 & -1 & 2 & 3 \\ 0 & 1 & -1 & 0 \\ 1 & -2 & 3 & 3 \\ 2 & -3 & 5 & 6 \end{pmatrix}$

33. Find the best straight-line fit to the points $(2, 5)$, $(-1, -3)$, $(1, 0)$, $(3, 11)$

34. Find the best quadratic fit to the points in Exercise 33.

In Exercises 35–40 determine whether the given transformation from V to W is linear.

35. T: $\mathbb{R}^2 \to \mathbb{R}^2$; $T(x, y) = (0, -y)$

36. T: $\mathbb{R}^3 \to \mathbb{R}^3$; $T(x, y, z) = (1, y, z)$

37. T: $\mathbb{R}^2 \to \mathbb{R}$: $T(x, y) = x/y$

38. T: $P_1 \to P_2$: $(Tp)(x) = xp(x)$

39. T: $P_2 \to P_2$; $(Tp)(x) = 1 + p(x)$

40. T: $C[0, 1] \to C[0, 1]$: $Tf(x) = f(1)$

In Exercises 41–46 find the kernel, range, rank, and nullity of the given linear transformation.

41. T: $\mathbb{R}^2 \to \mathbb{R}^2$; $T\begin{pmatrix} x \\ y \end{pmatrix} = \begin{pmatrix} 2 & -1 \\ 4 & 7 \end{pmatrix}\begin{pmatrix} x \\ y \end{pmatrix}$

42. T: $\mathbb{R}^3 \to \mathbb{R}^3$; $T\begin{pmatrix} x \\ y \\ z \end{pmatrix} = \begin{pmatrix} 1 & 2 & -1 \\ 2 & 4 & 3 \\ 1 & 2 & -6 \end{pmatrix}\begin{pmatrix} x \\ y \\ z \end{pmatrix}$

43. T: $\mathbb{R}^3 \to \mathbb{R}^2$; $T\begin{pmatrix} x \\ y \\ z \end{pmatrix} = \begin{pmatrix} y \\ -x \end{pmatrix}$

44. T: $P_2 \to P_4$; $Tp(x) = x^2p(x)$

45. T: $M_{22} \to M_{22}$; $T(A) = AB$, where $B = \begin{pmatrix} 1 & 1 \\ -1 & 1 \end{pmatrix}$

46. T: $C[0, 1] \to \mathbb{R}$; $Tf = f(1)$

In Exercises 47–51 find the matrix representation of the given linear transformation and find the kernel, range, nullity, and rank of the transformation.

47. T: $\mathbb{R}^2 \to \mathbb{R}^2$; $T(x, y) = (0, -y)$

48. T: $\mathbb{R}^3 \to \mathbb{R}^2$; $T(x, y, z) = (y, z)$

49. T: $\mathbb{R}^4 \to \mathbb{R}^2$; $T(x, y, z, w) = (x - 2z, 2y + 3w)$

50. T: $P_3 \to P_4$; $(Tp)(x) = xp(x)$

51. T: $M_{22} \to M_{22}$; $TA = AB$, where $B = \begin{pmatrix} -1 & 0 \\ 1 & 2 \end{pmatrix}$

52. Find an isomorphism T: $P_2 \to \mathbb{R}^3$.

In Exercises 53–56 find an orthonormal basis for the given vector space.

53. $\mathbb{R}^2$ starting with the basis $\left\{\begin{pmatrix} 2 \\ 3 \end{pmatrix}, \begin{pmatrix} -1 \\ 4 \end{pmatrix}\right\}$

54. $\{(x, y, z) \in \mathbb{R}^3: x - y - z = 0\}$

55. $\{(x, y, z) \in \mathbb{R}^3: x = y = z\}$

56. $\{(x, y, z, w) \in \mathbb{R}^4: x = z \text{ and } y = w\}$

In Exercises 57 and 58: (**a**) compute $\text{proj}_H \mathbf{v}$; (**b**) find an orthonormal basis for $H^\perp$; (**c**) write $\mathbf{v}$ as $\mathbf{h} + \mathbf{p}$, where $\mathbf{h} \in H$ and $\mathbf{p} \in H^\perp$.

57. H is the subspace of Problem 54; $\mathbf{v} = \begin{pmatrix} -1 \\ 2 \\ 4 \end{pmatrix}$.

58. H is the subspace of Problem 55; $\mathbf{v} = \begin{pmatrix} 1 \\ 0 \\ -1 \end{pmatrix}$.

In Exercises 59–64 calculate the eigenvalues and eigenspaces of the given matrix.

59. $\begin{pmatrix} -8 & 12 \\ -6 & 10 \end{pmatrix}$ **60.** $\begin{pmatrix} 2 & 5 \\ 0 & 2 \end{pmatrix}$

61. $\begin{pmatrix} 1 & 0 & 0 \\ 3 & 7 & 0 \\ -2 & 4 & -5 \end{pmatrix}$ **62.** $\begin{pmatrix} 1 & -1 & 0 \\ 1 & 2 & 1 \\ -2 & 1 & -1 \end{pmatrix}$

63. $\begin{pmatrix} 5 & -2 & 0 & 0 \\ 4 & -1 & 0 & 0 \\ 0 & 0 & 3 & -1 \\ 0 & 0 & 2 & 3 \end{pmatrix}$ **64.** $\begin{pmatrix} -2 & 1 & 0 \\ 0 & -2 & 1 \\ 0 & 0 & -2 \end{pmatrix}$

In Exercises 65–73 determine whether the given matrix A is diagonalizable. If it is, find a matrix C such that $C^{-1}AC = D$. If A is symmetric, find an orthogonal matrix Q such that $Q^tAQ = D$.

65. $\begin{pmatrix} -18 & -15 \\ 20 & 17 \end{pmatrix}$ **66.** $\begin{pmatrix} \frac{17}{2} & \frac{9}{2} \\ -15 & -8 \end{pmatrix}$

67. $\begin{pmatrix} 1 & 1 & 1 \\ -1 & -1 & 0 \\ -1 & 0 & -1 \end{pmatrix}$ **68.** $\begin{pmatrix} 4 & 2 & 0 \\ 2 & 4 & 0 \\ 0 & 0 & -3 \end{pmatrix}$

69. $\begin{pmatrix} -3 & 2 & 1 \\ -7 & 4 & 2 \\ -5 & 3 & 2 \end{pmatrix}$ **70.** $\begin{pmatrix} 8 & 0 & 12 \\ 0 & -2 & 0 \\ 12 & 0 & -2 \end{pmatrix}$

71. $\begin{pmatrix} 2 & 2 & 0 \\ 2 & 2 & 0 \\ 0 & 0 & -3 \end{pmatrix}$

72. $\begin{pmatrix} 4 & 2 & -2 & 2 \\ 1 & 3 & 1 & -1 \\ 0 & 0 & 2 & 0 \\ 1 & 1 & -3 & 5 \end{pmatrix}$

73. $\begin{pmatrix} 3 & 4 & -4 & 0 \\ 0 & -1 & 0 & 0 \\ 0 & 0 & -1 & 0 \\ 0 & -4 & 4 & 3 \end{pmatrix}$

In Exercises 74–78 identify the conic section and write it in new variables with the xy term absent.

74. $xy = -4$

75. $4x^2 + 2xy + 2y^2 = 8$

76. $4x^2 - 3xy + y^2 = 1$

77. $3y^2 - 2xy - 5 = 0$

78. $x^2 - 4xy + 4y^2 + 1 = 0$

79. Write the quadratic form $2x^2 + 4xy + 2y^2 - 3z^2$ in new variables x', y', and z' so that no cross-product terms are present.

PART III

INTRODUCTION TO INTERMEDIATE CALCULUS

CHAPTERS INCLUDE:

CHAPTER 9

CALCULUS IN $\mathbb{R}^n$

In Chapters 2 and 3 we discussed vector functions and functions of several variables first in $\mathbb{R}^2$ and $\mathbb{R}^3$ and then in $\mathbb{R}^n$. In this chapter we go further and discuss functions with domain in $\mathbb{R}^n$ and range in $\mathbb{R}^m$. We begin by extending Taylor's theorem to n variables.

9.1 TAYLOR'S THEOREM IN n VARIABLES

In your one-variable calculus course you may have studied Taylor's theorem. This important result is stated below. If this topic is unfamiliar, please read the review material in Sections 12.1 and 12.2 before going on in this section.

TAYLOR'S THEOREM FOR A FUNCTION OF ONE VARIABLE

Let the function f and its first $n+1$ derivatives be continuous on the interval $[x_0, x]$. Then

$$f(x) = f(x_0) + \frac{f'(x_0)}{1!}(x - x_0) + \frac{f''(x_0)}{2!}(x - x_0)^2 + \cdots + \frac{f^{(n)}(x_0)}{n!}(x - x_0)^n + R_n(x) \tag{1}$$

where

$$R_n(x) = \frac{f^{(n+1)}(c)}{(n+1)!}(x - x_0)^{n+1} \tag{2}$$

for some number c in (x_0, x). We can rewrite (1) as

$$f(x) = p_n(x) + R_n(x). \tag{3}$$

Before continuing, we list four common Taylor polynomials. These are obtained in Sections 12.1 and 12.2:

$$e^x \approx 1 + x + \frac{x^2}{2!} + \frac{x^3}{3!} + \cdots + \frac{x^n}{n!} \tag{4}$$

$$\sin x \approx x - \frac{x^3}{3!} + \frac{x^5}{5!} - \cdots + \frac{(-1)^n x^{2n+1}}{(2n+1)!} \tag{5}$$

$$\cos x \approx 1 - \frac{x^2}{2!} + \frac{x^4}{4!} - \cdots + \frac{(-1)^n x^{2n}}{(2n)!} \tag{6}$$

$$\ln(1+x) \approx x - \frac{x^2}{2} + \frac{x^3}{3} - \cdots + \frac{(-1)^{n+1} x^n}{n} \tag{7}$$

In this section we shall extend Taylor's theorem to a function of n variables. More precisely, we shall indicate how a function $f(\mathbf{x})$ can be approximated by an nth-degree polynomial $P_n(x)$ with a remainder term similar to (2).

The result we want is obtained by repeated application of the chain rule (Theorem 3.6.3 on page 192). We assume that the function f: $\mathbb{R}^n \to \mathbb{R}$ is defined on an open set Ω and that the line segment joining $\mathbf{x}$ and $\mathbf{x}_0$ in $\mathbb{R}^n$ is in Ω (see the definition on page 194). We define the vector $\mathbf{h}$ by

$$\mathbf{h} = \mathbf{x} - \mathbf{x}_0. \tag{8}$$

Keep in mind that $\mathbf{h} = (h_1, h_2, \ldots, h_n)$ is an n-vector. Let

$$g(t) = f(\mathbf{x}_0 + t\mathbf{h}). \tag{9}$$

Note that g is defined for $t \in [0, 1]$ because $\{\mathbf{x}: \mathbf{x} = \mathbf{x}_0 + t\mathbf{h}, 0 \le t \le 1\}$ is the line segment joining $\mathbf{x}_0$ and $\mathbf{x}$. We assume that f is of class $C^{(m)}(\Omega)$. Then g has m continuous derivatives, which we now compute. We first note that

$$\frac{d}{dt}(\mathbf{x}_0 + t\mathbf{h}) = \mathbf{h}$$

since $\mathbf{h}$ is assumed to be constant (see Example 2.4.7 on page 108). Thus, by the chain rule,

$$g'(t) = \frac{d}{dt} f(\mathbf{x}_0 + t\mathbf{h}) = \nabla f(\mathbf{x}_0 + t\mathbf{h}) \cdot \mathbf{h}. \tag{10}$$

Writing out the terms in the scalar product in (10), we have

$$g'(t) = \sum_{i=1}^{n} f_i(\mathbf{x}_0 + t\mathbf{h}) h_i. \tag{11}$$

The idea now is to compute higher-order derivatives of g and then to apply Taylor's theorem of one variable to g. Consider the function $f_i(\mathbf{x}_0 + t\mathbf{h})$. We can apply the chain rule again to obtain

$$\frac{d}{dt} f_i(\mathbf{x}_0 + t\mathbf{h}) = \nabla f_i(\mathbf{x}_0 + t\mathbf{h}) \cdot \mathbf{h}. \tag{12}$$

But $\nabla f_i = (f_{i1}, f_{i2}, \ldots, f_{in})$, so that, from (12),

$$\frac{d}{dt} f_i(\mathbf{x}_0 + t\mathbf{h}) = \sum_{j=1}^{n} f_{ij}(\mathbf{x}_0 + t\mathbf{h})\mathbf{h}_j. \tag{13}$$

Note that the subscripts on f represent partial derivatives, whereas those on the vector $\mathbf{h}$ represent the components of $\mathbf{h}$. Now, from (11),

$$g''(t) = \sum_{i=1}^{n} \frac{d}{dt} f_i(\mathbf{x}_0 + t\mathbf{h}) h_i. \tag{14}$$

So inserting (13) into (14) yields

$$g''(t) = \sum_{i=1}^{n}\left[\sum_{j=1}^{n} f_{ij}(\mathbf{x}_0 + t\mathbf{h})h_j\right]h_i. \tag{15}$$

This is getting complicated, but if you have come this far, you can probably see the pattern. We have

$$g'''(t) = \sum_{i=1}^{n}\left\{\sum_{j=1}^{n}\left[\sum_{k=1}^{n} f_{ijk}(\mathbf{x}_0 + t\mathbf{h})h_k\right]h_j\right\}h_i \tag{16}$$

which we write in the abbreviated form

$$g'''(t) = \sum_{i,j,k=1}^{n} f_{ijk}(\mathbf{x}_0 + t\mathbf{h})h_i h_j h_k. \tag{17}$$

Finally, we have

$$g^{(m)}(t) = \sum_{i_1,i_2,\ldots,i_m=1}^{n} f_{i_1 i_2 i_3 \cdots i_m}(\mathbf{x}_0 + t\mathbf{h})h_{i_1}h_{i_2}\cdots h_{i_m}. \tag{18}$$

Formula (18) is formidable, but it is useful in deriving Taylor's theorem for n variables.

THEOREM 1 TAYLOR'S THEOREM IN n VARIABLES

Let f: $\mathbb{R}^n \to \mathbb{R}$ be of class $C^{(m+1)}(\Omega)$ and let the line segment joining $\mathbf{x}_0$ and $\mathbf{x}$ be in Ω. Then if $\mathbf{x} = (x_1, x_2, \ldots, x_n)$ and $\mathbf{x}_0 = (x_1^{(0)}, x_2^{(0)}, \ldots, x_n^{(0)})$,

$$\begin{aligned}\mathbf{f}(\mathbf{x}) &= f(\mathbf{x}_0) + \sum_{i=1}^{n} f_i(\mathbf{x}_0)(x_i - x_i^{(0)}) + \frac{1}{2!}\sum_{i=1}^{n}\sum_{j=1}^{n} f_{ij}(\mathbf{x}_0)(x_i - x_i^{(0)})(x_j - x_j^{(0)}) \\ &\quad + \frac{1}{3!}\sum_{i,j,k=1}^{n} f_{ijk}(\mathbf{x}_0)(x_i - x_i^{(0)})(x_j - x_j^{(0)})(x_k - x_k^{(0)}) + \cdots \\ &\quad + \frac{1}{m!}\sum_{i_1,i_2,\ldots,i_m=1}^{n} f_{i_1 i_2 \cdots i_m}(\mathbf{x}_0)(x_{i_1} - x_{i_1}^{(0)})(x_{i_2} - x_{i_2}^{(0)})\cdots(x_{i_m} - x_{i_m}^{(0)}) + R_m(\mathbf{x})\end{aligned} \tag{19}$$

where

$$R_m(\mathbf{x}) = \frac{1}{(m+1)!}\sum_{i_1,i_2,\ldots,i_{m+1}=1}^{n} f_{i_1 i_2 \cdots i_{m+1}}(\mathbf{x}_0 + c\mathbf{h})(x_{i_1} - x_{i_1}^{(0)})(x_{i_2} - x_{i_2}^{(0)})\cdots(x_{i_{m+1}} - x_{i_{m+1}}^{(0)}) \tag{20}$$

for some number c in (0, 1).

PROOF: Since $g(t)$ has $m+1$ continuous derivatives on $[0, 1]$ we have, from Taylor's theorem

$$g(1) = g(0) + g'(0) + \frac{1}{2!}g''(0) + \cdots + \frac{1}{m!}g^{(m)}(0) + \frac{1}{(m+1)!}g^{(m+1)}(c) \quad (21)$$

for some c in $(0, 1)$. But, from (8) and (9),

$$g(1) = f(\mathbf{x}_0 + \mathbf{h}) = f(\mathbf{x}_0 + \mathbf{x} - \mathbf{x}_0) = f(\mathbf{x}) \qquad \text{and} \qquad g(0) = f(\mathbf{x}_0).$$

Also, $h_i = (x_i - x_i^{(0)})$, so inserting (11), (15), (17), and (18) into (21) gives the desired result. ■

NOTATION: We shall denote the polynomial given in (19) by $p_m(\mathbf{x})$.

Taylor's theorem is not terribly difficult to apply, but it does become tedious when computing higher-order terms of a function of n variables. We shall therefore keep our examples simple.

EXAMPLE 1

A THIRD-DEGREE TAYLOR POLYNOMIAL FOR A FUNCTION OF TWO VARIABLES

Compute $p_3(\mathbf{x})$ around $\mathbf{x}_0 = \mathbf{0}$ for the function $f(x, y) = x^2y$.

SOLUTION: $f(\mathbf{x}_0) = f(0, 0) = 0$. The second term in (19) is

$$\sum_{i=1}^{n} f_i(\mathbf{0})(x_i - 0).$$

Here, $x_1 = x$ and $x_2 = y$. Also, $f_1 = 2xy$ and $f_2 = x^2$, so that $f_1(\mathbf{0}) = f_2(\mathbf{0}) = 0$. The third term is

$$\frac{1}{2!}\sum_{i=1}^{2}\sum_{j=1}^{2} f_{ij}(\mathbf{0})x_i x_j. \quad (22)$$

We compute this methodically, starting by setting $i = 1$ in (22). Then we compute $f_{11} = 2y$ and $f_{12} = 2x$, both of which are zero at $(0, 0)$. Setting $i = 2$, we find that $f_{21}(\mathbf{0}) = f_{22}(\mathbf{0}) = 0$, so that the third term is 0. Finally, the fourth term in $p_3(\mathbf{x})$ is

$$\frac{1}{3!}\sum_{i=1}^{2}\sum_{j=1}^{2}\sum_{k=1}^{2} f_{ijk}(\mathbf{0})x_i x_j x_k. \quad (23)$$

Setting $i = 1$, we have

$$\sum_{j=1}^{2}\sum_{k=1}^{2} f_{1jk}(\mathbf{0})x_1 x_j x_k. \quad (24)$$

Setting $j = 1$ in (24) gives us

$$\sum_{k=1}^{2} f_{11k}(\mathbf{0})x^2 x_k. \quad (25)$$

Now $f_{111} = 0$ and $f_{112} = 2$, so that (25) reduces to $2x^2x_2 = 2x^2y$. Next, setting $j = 2$ in (24) yields

$$\sum_{k=1}^{2} f_{12k}(\mathbf{0})xyx_k. \tag{26}$$

But $f_{121} = 2$ and $f_{122} = 0$, so that (26) also reduces to $2x^2y$. Thus for $i = 1$ in (23), we obtain the sum $4x^2y$. Setting $i = 2$ in (23) yields

$$\sum_{j=1}^{2}\sum_{k=1}^{2} f_{2jk}(\mathbf{0})yx_jx_k. \tag{27}$$

Now $f_{211} = 2, f_{212} = 0, f_{221} = 0$, and $f_{222} = 0$, so that (27) reduces to $2x^2y$. Finally, (23) becomes $p_3 = (1/3!)(4x^2y + 2x^2y) = x^2y$. All this work has led to an unsurprising result. The polynomial x^2y is of degree 3 (see the definition on page 158), so that it is equal to its own third-degree Taylor polynomial. Note that all fourth-order derivatives of p_3 are zero so that $R_3 \equiv 0$.

REMARK: It is true that if $f(\mathbf{x})$ is a polynomial of degree m, then $f(\mathbf{x}) = p_m(\mathbf{x})$. We will not attempt to prove this fact. The proof is not conceptually difficult, but it involves computations like those carried out in Example 1 (see Problems 17, 18, and 19).

EXAMPLE 2 A SECOND-DEGREE TAYLOR POLYNOMIAL FOR AN EXPONENTIAL FUNCTION OF n VARIABLES

Compute $p_2(\mathbf{x})$ for $\mathbf{x}_0 = \mathbf{0}$ and $f(\mathbf{x}) = e^{x_1+x_2+\cdots+x_n}$.

SOLUTION: $f(\mathbf{x}_0) = e^0 = 1$. The second term in (19) is

$$\sum_{i=1}^{n} f_i(\mathbf{0})x_i. \tag{28}$$

Now $f_i(x) = e^{x_1+x_2+\cdots+x_n}$, so that $f_i(\mathbf{0}) = 1$ and (28) becomes

$$\sum_{i=1}^{n} x_i = x_1 + x_2 + \cdots + x_n.$$

The third term in (19) is

$$\frac{1}{2!}\sum_{i=1}^{n}\sum_{j=1}^{n} f_{ij}(\mathbf{0})x_ix_j. \tag{29}$$

Again $f_{ij}(\mathbf{0}) = 1$ and (29) becomes

$$\frac{1}{2!}\sum_{i=1}^{n}\sum_{j=1}^{n} x_ix_j = \frac{1}{2!}\sum_{i=1}^{n} x_i \sum_{j=1}^{n} x_j = \frac{1}{2!}\left(\sum_{i=1}^{n} x_i\right)^2 = \frac{1}{2!}(x_1 + x_2 + \cdots + x_n)^2.$$

Thus

$$p_2(\mathbf{x}) = 1 + (x_1 + x_2 + \cdots + x_n) + \frac{1}{2!}(x_1 + x_2 + \cdots + x_n)^2.$$

We can obtain this answer immediately from formula (4), for if $x = x_1 + x_2 + \cdots + x_n$, then

$$e^{x_1+x_2+\cdots+x_n} \approx 1 + (x_1 + x_2 + \cdots + x_n) + \frac{(x_1 + x_2 + \cdots + x_n)^2}{2!} + \cdots + \frac{(x_1 + x_2 + \cdots + x_n)^k}{k!}.$$

TAYLOR POLYNOMIALS AND THE BINOMIAL THEOREM

Before doing any more examples, we illustrate how the computation of Taylor polynomials can be made a bit easier. We first suppose that $n = 2$ so that $\mathbf{x} = (x, y)$ and $\mathbf{x}_0 = (x_0, y_0)$. Let us compute $p_2(x, y)$ with $x_1 = x$, $x_2 = y$, $x_1{}^{(0)} = x_0$ and $x_2{}^{(0)} = y_0$. Also $f_1(\mathbf{x}_0) = f_x(x_0, y_0)$ and $f_2(\mathbf{x}_0) = f_y(x_0, y_0)$. Equation (19) then becomes

$$\begin{aligned} f(x, y) &= f(x_0, y_0) + f_x(x_0, y_0)(x - x_0) + f_y(x_0, y_0)(y - y_0) \\ &\quad + \frac{1}{2!}[f_{xx}(x_0, y_0)(x - x_0)^2 + 2f_{xy}(x_0, y_0)(x - x_0)(y - y_0) + f_{yy}(x, y)(y - y_0)^2] \\ &\quad + R_3(x, y). \end{aligned} \tag{30}$$

Here we have used the fact that $f_{xy} = f_{yx}$.

The linear term is easy to remember. What about the quadratic term? Recall that

$$(a + b)^2 = a^2 + 2ab + b^2. \tag{31}$$

Compare this expression with the quadratic terms in brackets in equation (30)!

What about the cubic term? The binomial theorem states that (see Appendix 2)

$$(a + b)^3 = a^3 + 3a^2b + 3ab^2 + b^3. \tag{32}$$

If we compute the third degree term in (19), again with $n = 2$, we obtain

$$\begin{aligned} &\frac{1}{3!}[f_{xxx}(x_0, y_0)(x - x_0)^3 + 3f_{xxy}(x_0, y_0)(x - x_0)^2(y - y_0) \\ &\qquad + 3f_{xyy}(x_0, y_0)(x - x_0)(y - y_0)^2 + f_{yyy}(y - y_0)^3]. \end{aligned} \tag{33}$$

Compare expressions (32) and (33).

These results for the quadratic and cubic terms suggest the following rule, which we offer without proof:

If $n = 2$, the kth degree term in the Taylor polynomial of f at (x_0, y_0) is obtained as follows:

Step 1: Expand $(a + b)^k$ by the binomial theorem.

Step 2: A typical term in the expansion is ([see equation (2) in Appendix 2])

$$\binom{k}{j} a^{k-j} b^j.$$

Write the term

$$\binom{k}{j} f_{\underbrace{xx\cdots x}_{k-j \text{ times}}\underbrace{yy\cdots y}_{j \text{ times}}}(x - x_0)^{k-j}(y - y_0)^j. \tag{34}$$

There will be $k + 1$ such terms (for $j = 0, 1, 2, \ldots, k$)

Step 3: The kth degree term in the Taylor polynomial is

$$\frac{1}{k!}[\text{sum of the } k + 1 \text{ terms obtained in step (2)}]$$

NOTE: This procedure can be extended to $n > 2$. For example,

$$(a + b + c)^2 = a^2 + 2ab + 2ac + b^2 + 2bc + c^2.$$

Then, as can be verified, the quadratic term in (19) at $\mathbf{x}_0 = (x_0, y_0, z_0)$ for $n = 3$ is

$$\frac{1}{2!}[f_{xx}(x - x_0)^2 + 2f_{xy}(x - x_0)(y - y_0) + 2f_{xz}(x - x_0)(z - z_0) + f_{yy}(y - y_0)^2 + 2f_{yz}(y - y_0)(z - z_0) + f_{zz}(z - z_0)^2]. \quad (35)$$

In (35), f_{xx} denotes $f_{xx}(x_0, y_0, z_0)$, f_{xz} denotes $f_{xz}(x_0, y_0, z_0)$, and so on.

After a few computations, it becomes apparent that computing a Taylor polynomial is not much more difficult than computing expressions of the form $(x_1 + x_2 + \cdots + x_n)^k$.

EXAMPLE 3

A TAYLOR POLYNOMIAL FOR A SINE FUNCTION OF TWO VARIABLES

Compute $p_2(\mathbf{x})$ for $f(\mathbf{x}) = \sin(2x + y)$ at $\mathbf{x}_0 = (\pi/6, \pi/3)$.

SOLUTION: We see that

$$f(x_0, y_0) = \sin\left(\frac{\pi}{3} + \frac{\pi}{3}\right) = \frac{\sqrt{3}}{2}.$$

$f_x(x, y) = 2\cos(2x + y)$ and $f_y(x, y) = \cos(2x + y)$, so that

$$f_x\left(\frac{\pi}{6}, \frac{\pi}{3}\right) = -1, \qquad f_y\left(\frac{\pi}{6}, \frac{\pi}{3}\right) = -\frac{1}{2}$$

Next

$$f_{xx}(x, y) = -4\sin(2x + y) \quad \text{and} \quad f_{xx}\left(\frac{\pi}{6}, \frac{\pi}{3}\right) = -2\sqrt{3}$$

$$f_{xy}(x, y) = -2\sin(2x + y) \quad \text{and} \quad 2f_{xy}\left(\frac{\pi}{6}, \frac{\pi}{3}\right) = -2\sqrt{3}$$

$$f_{yy}(y, y) = -\sin(2x + y) \quad \text{and} \quad f_{yy}\left(\frac{\pi}{6}, \frac{\pi}{3}\right) = -\frac{\sqrt{3}}{2}$$

Then, from (30), we have

$$\begin{aligned} p_2(x, y) &= \frac{\sqrt{3}}{2} - \left(x - \frac{\pi}{6}\right) - \frac{1}{2}\left(y - \frac{\pi}{3}\right) + \frac{1}{2}\left[-2\sqrt{3}\left(x - \frac{\pi}{6}\right)^2 \right. \\ &\quad \left. - 2\sqrt{3}\left(x - \frac{\pi}{6}\right)\left(y - \frac{\pi}{3}\right) - \frac{\sqrt{3}}{2}\left(y - \frac{\pi}{3}\right)^2\right]. \\ &= \frac{\sqrt{3}}{2} - \left(x - \frac{\pi}{6}\right) - \frac{1}{2}\left(y - \frac{\pi}{3}\right) - \sqrt{3}\left(x - \frac{\pi}{6}\right)^2 \\ &\quad - \sqrt{3}\left(x - \frac{\pi}{6}\right)\left(y - \frac{\pi}{3}\right) - \frac{\sqrt{3}}{4}\left(y - \frac{\pi}{3}\right)^2. \end{aligned}$$

EXAMPLE 4

A TAYLOR POLYNOMIAL FOR A LOGARITHMIC FUNCTION OF TWO VARIABLES

Compute $p_2(\mathbf{x})$ for $f(\mathbf{x}) = \ln(1 + 2x + 5y)$ at $\mathbf{x}_0 = (0, 0)$.

SOLUTION: As in Example 2, there are two ways to solve this problem. The easier method is to use formula (7). Substituting $2x + 5y$ for x in (7), we obtain

$$\ln(1 + 2x + 5y) \approx 2x + 5y - \frac{(2x + 5y)^2}{2} + \frac{(2x + 5y)^3}{3} - \cdots + \frac{(-1)^{n+1}(2x + 5y)^n}{n}.$$

Thus

$$p_2(\mathbf{x}) = 2x + 5y - \frac{(2x + 5y)^2}{2} = 2x + 5y - 2x^2 - 10xy - \frac{25}{2}y^2.$$

We next compute $p_2(\mathbf{x})$ the "hard" way to illustrate further the accuracy of formula (30). We have

$$f(x_0, y_0) = f(0, 0) = \ln 1 = 0$$

$$f_x(x, y) = \frac{2}{1 + 2x + 5y} \quad \text{and} \quad f_x(0, 0) = 2$$

$$f_y(x, y) = \frac{5}{1 + 2x + 5y} \quad \text{and} \quad f_y(0, 0) = 5$$

$$f_{xx}(x, y) = -\frac{4}{(1 + 2x + 5y)^2} \quad \text{and} \quad f_{xx}(0, 0) = -4$$

$$f_{xy}(x, y) = -\frac{10}{(1 + 2x + 5y)^2} \quad \text{and} \quad 2f_{xy}(0, 0) = -20$$

$$f_{yy}(x, y) = -\frac{25}{(1 + 2x + 5y)^2} \quad \text{and} \quad f_{yy}(0, 0) = -25.$$

Then, from (30),

$$\begin{aligned} p_2(x, y) &= 2x + 5y + \frac{1}{2!}(-4x^2 - 20xy - 25y^2) \\ &= 2x + 5y - 2x^2 - 10xy - (25/2)y^2. \end{aligned}$$

Once we know how to compute Taylor polynomials of n variables, we can use them to approximate functions as is done Section 12.2. The accuracy of the approximation depends, of course, on the size of the remainder term R_m given by (20). It is evident that if all the $(m + 1)$st-order partial derivatives of f are bounded in some interval, then $p_m(\mathbf{x})$ is a good approximation to $f(\mathbf{x})$ if $\mathbf{x}$ is close to $\mathbf{x}_0$, and a bound on the error $|f(\mathbf{x}) - R_m(\mathbf{x})|$ can be computed. We will not give any examples of this here.

We also note that Taylor's theorem in n variables has many practical applications. For example, it is a useful tool in the study of qualitative properties of systems of ordinary differential equations.

In Section 3.10 we stated the second derivatives test for a function of two variables (see page 216). Using Taylor's theorem in two variables, we can prove it.

THEOREM 2 SECOND DERIVATIVES TEST

Let f and all its first and second partial derivatives be continuous in a neighborhood of the critical point (x_0, y_0). Let

$$D(x, y) = f_{xx}(x, y)f_{yy}(x, y) - [f_{xy}(x, y)]^2 \tag{36}$$

and let D denote $D(x_0, y_0)$.

i. If $D > 0$ and $f_{xx}(x_0, y_0) > 0$, then f has a local minimum at (x_0, y_0).
ii. If $D > 0$ and $f_{xx}(x_0, y_0) < 0$, then f has a local maximum at (x_0, y_0).
iii. If $D < 0$, then (x_0, y_0) is a saddle point of f.
iv. If $D = 0$, then any of the preceding alternatives is possible.

PROOF: Because this is a very long proof we shall give it in steps. We begin by making the simplifying assumption that $f \in C^3(\Omega)$ where Ω is a neighborhood of (x_0, y_0). This assumption is not necessary but it does make things easier. We shall assume, also to make things simpler, that $(x_0, y_0) = (0, 0)$.

Step 1: By Taylor's theorem and formula (30),

$$\begin{aligned} f(x, y) = f(0, 0) &+ f_x(0, 0)x + f_y(0, 0)y \\ &+ \tfrac{1}{2}[f_{xx}(0, 0)x^2 + f_{xy}(0, 0)xy + f_{yx}(0, 0)xy \\ &+ f_{yy}(0, 0)y^2] + R_2(x, y). \end{aligned} \tag{37}$$

Now, since $(0, 0)$ is a critical point, $f_x(0, 0) = f_y(0, 0) = 0$ and, since f_{xy} and f_{yx} are continuous, $f_{xy}(0, 0) = f_{yx}(0, 0)$. Also, by (20), $R_2(x, y)$ contains terms of the form x^3, x^2y, xy^2, and y^3 (since $f \in C^3(\Omega)$). A key step in the proof is to observe that if x and y are very small, then third-order terms in x and y (like x^3, x^2y, xy^2, and y^3) are considerably smaller than the second-order terms x^2, xy, and y^2. For example, if $x = y = 0.1$, then $x^3 = x^2y = xy^2 = y^3 = 0.001$, while $x^2 = xy = y^2 = 0.01$. The difference is a factor of 10. Thus, for x and y small enough, $R_2(x, y)$ is negligible compared to the other terms in (31) and we can write

$$\begin{aligned} f(x, y) \approx\, & f(0, 0) \\ &+ \tfrac{1}{2}[f_{xx}(0, 0)x^2 + 2f_{xy}(0, 0)xy + f_{yy}(0, 0)y^2]. \end{aligned} \tag{38}$$

Let $A = f_{xx}(0, 0)$, $B = f_{xy}(0, 0)$, and $C = f_{yy}(0, 0)$. Then (38) becomes

$$f(x, y) - f(0, 0) \approx \tfrac{1}{2}(Ax^2 + 2Bxy + Cy^2). \tag{39}$$

Step 2: We now show that if $D = AC - B^2 > 0$ and if $A > 0$, then $(Ax^2 + 2Bxy + Cy^2) > 0$ for all vectors $(x, y) \neq (0, 0)$. Using (39), this will show that $f(x, y) > f(0, 0)$ if x and y are sufficiently small, which means that $(0, 0)$ is a local minimum. This will prove part (i). We have

$$Ax^2 + 2Bxy + Cy^2 = A\left(x^2 + \frac{2B}{A}xy + \frac{C}{A}y^2\right)$$

Completing the square

$$\begin{aligned} &= A\left[\left(x + \frac{B}{A}y\right)^2 + \left(\frac{C}{A} - \frac{B^2}{A^2}\right)y^2\right] \\ &= A\left[\left(x + \frac{B}{A}y\right)^2 + \frac{AC - B^2}{A^2}y^2\right]. \end{aligned} \tag{40}$$

If $AC - B^2 > 0$ and if $A > 0$, then the last expression in (40) is nonnegative; it is positive if $(x, y) \neq (0, 0)$. This proves part (i). The proof of part (ii) follows from (40) because if $A < 0$ and $AC - B^2 > 0$, then $Ax^2 + 2Bxy + Cy^2 < 0$ for $(x, y) \neq (0, 0)$ and, from (39), $f(x, y) < f(0, 0)$ if x and y are sufficiently small.

Step 3: We prove part (iii). We assume that $A = f_{xx}(0, 0) > 0$. A similar argument works in the case $A < 0$. We must show that we can find values of x and y as small as we like that make $Ax^2 + 2Bxy + Cy^2$ positive or negative. Choose y small and let $x = (-B/A)y$. Then, from (40),

$$Ax^2 + 2Bxy + Cy^2 = A\left(\frac{AC - B^2}{A^2}y^2\right) < 0 \qquad \text{if } A > 0,\ AC - B^2 < 0.$$

Now, choose $y = 0$ but $x \neq 0$. Then $Ax^2 + 2Bxy + Cy^2 = Ax^2 > 0$. Thus there are vectors (x, y) arbitrarily close to $(0, 0)$ such that $f(x, y) > f(0, 0)$ and $f(x, y) < f(0, 0)$. Hence $(0, 0)$ is a saddle point.

Step 4: We prove part (iv) by means of examples. Let

$$f(x, y) = x^2 + 2xy + y^2 = (x + y)^2.$$

Then, since $f(x, y) \geq 0$, $(0, 0)$ is a local minimum (which is, of course, not unique). But $f_{xx} = f_{xy} = f_{yy} = 2$, so that $D = 2^2 - 2 \cdot 2 = 0$. Similarly, $D = 0$ when $f(x, y) = -x^2 - 2xy - y^2 = -(x + y)^2$ and $(0, 0)$ is a local maximum. Finally, let $f(x, y) = x^3 - y^3$. Then $f_{xx}(0, 0) = f_{xy}(0, 0) = f_{yy}(0, 0) = 0$, so that $D = 0$. However, $(0, 0)$ is a critical point and $x^3 - y^3 > 0$ if $x > y$ and $x^3 - y^3 < 0$ if $x < y$. Thus $(0, 0)$ is a saddle point. These three examples show that if $D = 0$, then a critical point can be a local maximum, a local minimum, or a saddle point. ■

PROBLEMS 9.1

SELF-QUIZ

I. The second-degree Taylor polynomial for $\cos(x + 2y)$ at $(0, 0)$ is ________.

a. $1 + \frac{(x + 2y)^2}{2}$

b. $1 - \frac{(x + 2y)^2}{2}$

c. $1 - \frac{1}{2}(x^2 + 4xy + 4y^2)$

d. $1 - x - 2y$

II. The first-degree Taylor polynomial for $2 - 3x + 5y - xy + x^2 - y^3$ at $(0, 0)$ is ________.

a. $2 - 3x + 5y$

b. $2 - 4x + 4y$

c. $2 + 2x - 3y^2$

d. $2 - 3x + 5y - xy$

III. The third-degree Taylor polynomial for e^{2x-3y} at $(0, 0)$ is ________.

a. $1 + (2x - 3y) + (2x - 3y)^2 + 2(x - 3y)^3$

b. $1 + (2x - 3y) + \frac{(2x - 3y)^2}{2} + \frac{(2x - 3y)^3}{6}$

c. $1 + 2x - 3y + \frac{1}{2}(4x^2 - 12xy + 9y^2) + \frac{1}{6}(8x^3 - 36x^2y + 54xy^2 - 27y^3)$

IV. The first-degree Taylor polynomial for $\ln(x + y + z)$ at $(0, 0, 1)$ is ________.

a. $1 + x + y + z$

b. $1 + x + y + (z - 1)$

c. $x + y + z$

d. $x + y + (z - 1)$

The **linearization** of a function $f: \mathbb{R}^n \to \mathbb{R}$ at $\mathbf{x}_0$ is the first-degree Taylor polynomial that approximates that function at $\mathbf{x}_0$. In Problems 1–7 find the linearization of the given function.

1. $\sin(x+y)$; $\mathbf{x}_0 = \mathbf{0}$
2. $\cos(3x-2y)$; $\mathbf{x}_0 = \mathbf{0}$
3. $e^{x_1-4x_2+x_3}$; $\mathbf{x}_0 = \mathbf{0}$
4. $\ln(1+x_1+x_2+\cdots+x_n)$; $\mathbf{x}_0 = \mathbf{0}$
5. $\sqrt{x+y}$; $\mathbf{x}_0 = (2, 2)$
6. $\sqrt{x_1+x_2+x_3+x_4}$; $\mathbf{x}_0 = (1, 1, 1, 1)$
7. $\dfrac{x}{y}+\dfrac{y}{x}$; $\mathbf{x}_0 = (2, 1)$

In Problems 8–16 find the Taylor polynomial of degree m that approximates the given function at the given point.

8. $f(x, y) = \ln(x+2y)$; $\mathbf{x}_0 = (1, 0)$; $m = 2$
9. $f(x, y) = \sin(x^2+y^2)$; $\mathbf{x}_0 = (0, 0)$; $m = 2$
10. $f(x_1, x_2, x_3, x_4) = \sin(x_1+x_2+x_3+x_4)$;

$$\mathbf{x}_0 = \left(\frac{\pi}{8}, \frac{\pi}{8}, \frac{\pi}{8}, \frac{\pi}{8}\right);\ m = 2$$

11. $f(x, y, z) = \sin xyz$; $\mathbf{x}_0 = \mathbf{0}$; $m = 9$
12. $f(x_1, x_2, \ldots, x_n) = \cos(x_1x_2\cdots x_n)$; $\mathbf{x}_0 = \mathbf{0}$; $m = 4n$
13. $f(x, y) = e^x \sin y$; $\mathbf{x}_0 = \mathbf{0}$; $m = 2$
14. $f(x, y) = \sin x \cos y$; $\mathbf{x}_0 = \mathbf{0}$; $m = 2$
15. $f(x, y) = e^x \sin y$; $\mathbf{x}_0 = \left(2, \dfrac{\pi}{4}\right)$; $m = 2$
16. $f(x, y) = \sin x \cos y$; $\mathbf{x}_0 = \left(\dfrac{\pi}{2}, \pi\right)$; $m = 3$
17. Verify that the second-degree Taylor polynomial of xy around $\mathbf{0}$ is xy.
18. Verify that the fifth-degree Taylor polynomial of x^2yz^2 around $\mathbf{0}$ is x^2yz^2.
***19.** **a.** Show that if $f(x, y)$ is a polynomial of degree m, then its mth-degree Taylor polynomial $p_m(\mathbf{x})$ around $\mathbf{0}$ is equal to f.
b. What can you say about the mth-degree Taylor polynomial around a value $\mathbf{x}_0 \neq \mathbf{0}$?
20. Write out the third-degree Taylor polynomial for a function of three variables. [*Hint:* First compute $(a+b+c)^3$.]
21. Write out the fifth-degree term in the Taylor polynomial for a function of two variables.

ANSWERS TO SELF-QUIZ

I. b = c **II.** a **III.** b = c **IV.** d

9.2 INVERSE FUNCTIONS AND THE IMPLICIT FUNCTION THEOREM: I

In this section we discuss one of the most interesting and important theorems in calculus: the implicit function theorem. We use a special case of the theorem each time we say that $f(x, y) = 0$ defines y as a function of x. The result is discussed here in the setting of functions from $\mathbb{R}^{n+1} \to \mathbb{R}$. We shall return to the implicit function theorem in a more general setting in Section 9.5. We begin by looking at functions from $\mathbb{R}$ to $\mathbb{R}$.

Let $y = f(x)$ be a function of one variable. We know that f has an inverse function defined on its range if f is one-to-one.† But a differentiable f is one-to-one in a neighborhood of a point x_0 if $f'(x_0) \neq 0$. Summarizing, we have

THEOREM 1

Let Ω be an open set in $\mathbb{R}$ and let $f: \mathbb{R} \to \mathbb{R}$ have domain Ω. Suppose that f is differentiable and that $f'(x) \neq 0$ for every $x \in \Omega$. Then there exists a differentiable function $g: \mathbb{R} \to \mathbb{R}$ with domain $f(\Omega)$ such that $(g \circ f)(x) = x$ for every $x \in \Omega$ and $(f \circ g)(y) = y$ for every $y \in f(\Omega)$. In this case g is called the **inverse** of f and we write $g = f^{-1}$. ■

† See Section 6.1 in *Calculus* or *Calculus, Part I*.

We can look at the problem of finding inverse functions in another way. Suppose that $y = f(x)$. Let $F(x, y) = f(x) - y$. Then $y = f(x)$ is equivalent to

$$F(x, y) = 0. \tag{1}$$

More generally, equation (1) gives y *implicitly* as a function of x and x implicitly as a function of y. If F takes the special form $f(x) - y$, then writing x explicitly as a function of y is equivalent to finding an inverse function for f—so that $x = f^{-1}(y)$.

But suppose again that $F(x, y) = y - f(x)$. Now $\partial F/\partial x = -f'(x)$ and if $f'(x) \neq 0$, then f has an inverse function and x can be determined explicitly in terms of y. This leads us to suspect that in (1) we can write x as a function of y if $\partial F/\partial x \neq 0$ and y as a function of x if $\partial F/\partial y \neq 0$. This is the basic idea behind the very deep results we are about to discuss. But first we shall give two examples.

EXAMPLE 1

AN EQUATION THAT DOES NOT DETERMINE A FUNCTION

Let $F(x, y) = x^2 + y^2 - 1 = 0$. This is the equation of the unit circle in $\mathbb{R}^2$. If we make no restrictions on x or y, then neither variable can be written in terms of the other since

$$x = \pm\sqrt{1 - y^2} \qquad \text{and} \qquad y = \pm\sqrt{1 - x^2} \tag{2}$$

and neither expression in (2) is a function. However, if we specify $x > 0$ (so that we have only the right semicircle in Figure 1a), then $x = \sqrt{1 - y^2}$, and we have written x explicitly as a function of y. Note that $(\partial F/\partial x)(x, y) = 2x \neq 0$ if $x > 0$. There are three other regions depicted in Figure 1 in which one of the variables can be written as a function of the other. In each case one of the partial derivatives of F is nonzero over the region.

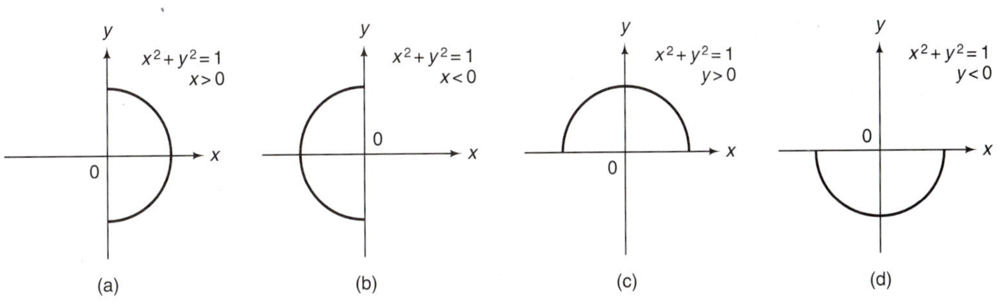

FIGURE 1
In (a) and (b), x can be written as a function of y; in (c) and (d) y can be written as a function of x.

EXAMPLE 2

AN EQUATION FOR WHICH y CAN BE WRITTEN AS A FUNCTION OF x, BUT NOT EXPLICITLY

Let $F(x, y) = x^2y^5 + e^{2x+y} + \sqrt{3x + y} = 0$. Again, we would like to write y as a function of x or x as a function of y. Unfortunately, if you try to do this, you will quickly find it an impossible task. However, we can calculate

$$\frac{\partial F}{\partial x}(x, y) = 2xy^5 + 2e^{2x+y} + \frac{3}{2\sqrt{3x + y}}$$

and

$$\frac{\partial F}{\partial y}(x, y) = 5x^2y^4 + e^{2x+y} + \frac{1}{2\sqrt{3x + y}}.$$

In the region $\Omega = \{(x, y): x > 0, y > 0\}$ both $\partial F/\partial x$ and $\partial F/\partial y$ are nonzero. Then, if our intuition is correct, there are functions f and g such that $y = f(x)$ and $x = g(y)$ in Ω. Note that there is a difference between asserting the existence of the functions f and g and being able to write them explicitly. In Example 1 we could write them; in this example we cannot.

We now generalize the ideas discussed above to higher dimensions. Let $f: \mathbb{R}^n \to \mathbb{R}$ and write

$$x_{n+1} = f(x_1, x_2, \ldots, x_n). \tag{3}$$

In (3) we have written x_{n+1} explicity in terms of the variables $x_1, x_2, \ldots, x_n$. If we define $F: \mathbb{R}^{n+1} \to \mathbb{R}$ by

$$F(x_1, x_2, \ldots, x_n, x_{n+1}) = x_{n+1} - f(x_1, x_2, \ldots, x_n),$$

then (3) can be written

$$F(x_1, x_2, \ldots, x_n, x_{n+1}) = 0. \tag{4}$$

More generally, if $F: \mathbb{R}^{n+1} \to \mathbb{R}$ is a given function, we ask whether the $(n + 1)$st variable x_{n+1} can be written as a function of the other n variables if equation (4) holds.

EXAMPLE 3

EQUATION IN $n + 1$ VARIABLES IN WHICH ONE VARIABLE CANNOT BE WRITTEN AS A FUNCTION OF THE OTHERS

Let $F: \mathbb{R}^{n+1} \to \mathbb{R}$ be given by

$$F(x_1, x_2, \ldots, x_n, x_{n+1}) = x_1^2 + x_2^2 + \cdots + x_n^2 + x_{n+1}^2 - 1.$$

If $F(x_1, x_2, \ldots, x_n, x_{n+1}) = 0$, then, as in Example 1, no one of the variables can be written as a function of the others if we make no restrictions. However, if we specify $x_{n+1} > 0$, say, then we can write

$$x_{n+1} = \sqrt{1 - x_1^2 - x_2^2 - \cdots - x_n^2}.$$

Note that $\partial f/\partial x_{n+1} = 2x_{n+1} \neq 0$ in the open set $\{\mathbf{x} \in \mathbb{R}^{n+1}: x_{n+1} > 0\}$.

We are now ready to state the first form of the implicit function theorem. The proofs of this theorem and its generalization (given in Section 9.5) are far more difficult than any proofs encountered in this text. For that reason they are omitted.†

THEOREM 2 IMPLICIT FUNCTION THEOREM—FIRST FORM

Let $F: \mathbb{R}^{n+1} \to \mathbb{R}$ be defined on an open set Ω and let

$$\mathbf{x}_0 = (x_1^{(0)}, x_2^{(0)}, \ldots, x_n^{(0)}, x_{n+1}^{(0)})$$

be in Ω. Suppose that the partial derivatives $(\partial F/\partial x_i)(\mathbf{x})$ are continuous in a neighbor-

†Reasonably comprehensible proofs can be found in C. H. Edwards, Jr., *Advanced Calculus in Several Variables* (Academic Press, New York, 1973).

hood of $\mathbf{x}_0$ for $i = 1, 2, \ldots, n + 1$. Suppose further that

$$F(\mathbf{x}_0) = 0 \quad \text{and} \quad \frac{\partial F}{\partial x_{n+1}}(\mathbf{x}_0) \neq 0.$$

Then there is an open set $\overline{\Omega}$ in $\mathbb{R}^n$ that contains $\overline{\mathbf{x}}_0 = (x_1^{(0)}, x_2^{(0)}, \ldots, x_n^{(0)})$ and a unique continuous function f: $\mathbb{R}^n \to \mathbb{R}$ such that $x_{n+1}^{(0)} = f(\overline{\mathbf{x}}_0)$ and $F(\underline{\mathbf{x}}, f(\mathbf{x})) = 0$ for all $\mathbf{x}$ in $\overline{\Omega}$. Furthermore, f has continuous first partial derivatives on $\overline{\Omega}$ given by

$$\frac{\partial f}{\partial x_i}(\mathbf{x}) = \frac{-\dfrac{\partial F}{\partial x_i}(\mathbf{x}, f(\mathbf{x}))}{\dfrac{\partial F}{\partial x_{n+1}}(\mathbf{x}, f(\mathbf{x}))} \quad \text{for} \quad 1 \leq i \leq n. \tag{5}$$

■

REMARK 1: In light of this theorem we say that $F(\mathbf{x}, x_{n+1}) = 0$ defines x_{n+1} **implicitly** as a function of $\mathbf{x} = (x_1, x_2, \ldots, x_n)$ near $\overline{\mathbf{x}}_0 = (x_1^{(0)}, x_2^{(0)}, \ldots, x_n^{(0)})$. In this case we write $x_{n+1} = f(\mathbf{x})$.

REMARK 2: It is important to emphasize that the implicit function theorem is a **local** result. That is, we can guarantee that x_{n+1} can be written as a function of $x_1, x_2, \ldots, x_n$ only in some open set in $\mathbb{R}^n$. This open set could be very small or it could be all of $\mathbb{R}^n$. The theorem does not tell you how big $\overline{\Omega}$ is. All we know is that $x_{n+1} = f(\mathbf{x})$ for $\mathbf{x}$ "near" $\mathbf{x}_0$. Nevertheless, even this limited information can be very useful.

REMARK 3: Although the implicit function theorem is, as we have stated, difficult to prove, the condition $(\partial F/\partial x_{n+1})(\mathbf{x}_0) \neq 0$ and equation (5) are very natural. For suppose that $F(x_1, x_2, \ldots, x_n, f(x_1, x_2, \ldots, x_n)) = 0$. Then, using the chain rule to compute $(\partial f/\partial x_i)$, we have

$$\frac{\partial F}{\partial x_i} + \frac{\partial F}{\partial x_{n+1}} \frac{\partial f}{\partial x_i} = 0,$$

which shows that $\partial f/\partial x_i$ is given by (5) if $\partial F/\partial x_{n+1} \neq 0$.

EXAMPLE 4

APPLYING THE IMPLICIT FUNCTION THEOREM

Let $F(x, y, z, w) = x^2 + y^4 + z^6 + w^2 - 1 = 0$. Find a region over which w can be written as a function f of x, y, and z and compute $\partial f/\partial x$, $\partial f/\partial y$, and $\partial f/\partial z$.

SOLUTION: We have $\partial F/\partial w = 2w$, which is nonzero on two different open sets in $\mathbb{R}^4$: $\Omega_1 = \{(x, y, z, w)\text{: } w > 0\}$ and $\Omega_2 = \{(x, y, z, w)\text{: } w < 0\}$. We will choose Ω_1; the analysis for Ω_2 is similar. In Ω_1 we can write

$$w = f(x, y, z) = \sqrt{1 - x^2 - y^4 - z^6},$$

so that

$$\frac{\partial f}{\partial x} = \frac{-x}{\sqrt{1 - x^2 - y^4 - z^6}} \qquad \frac{\partial f}{\partial y} = \frac{-2y^3}{\sqrt{1 - x^2 - y^4 - z^6}}$$

and

$$\frac{\partial f}{\partial z} = \frac{-3z^5}{\sqrt{1 - x^2 - y^4 - z^6}}.$$

Here we have no need for formula (5), since we have found f explicitly. However, we can use our explicit representation to verify formula (5) in this case. We have

$$\frac{\partial f}{\partial x} = \frac{-\partial F/\partial x}{\partial F/\partial w} = \frac{-2x}{2w} = \frac{-x}{w} = \frac{-x}{\sqrt{1 - x^2 - y^4 - z^6}}$$

$$\frac{\partial f}{\partial y} = \frac{-\partial F/\partial y}{\partial F/\partial w} = \frac{-4y^3}{2w} = \frac{-2y^3}{w} = \frac{-2y^3}{\sqrt{1 - x^2 - y^4 - z^6}}$$

and

$$\frac{\partial f}{\partial z} = \frac{-\partial F/\partial z}{\partial F/\partial w} = \frac{-6z^5}{2w} = \frac{-3z^5}{w} = \frac{-3z^5}{\sqrt{1 - x^2 - y^4 - z^6}}.$$

EXAMPLE 5

APPLYING THE IMPLICIT FUNCTION THEOREM

Let $F(x, y, z) = x^2y^3 + 4xyz^8 - 5xyz^4 + z^2 + 1 = 0$. Show that z can be written as a function of x and y in a neighborhood of the point $(2, -1, 1)$ and compute $\partial z/\partial x$ and $\partial z/\partial y$ at that point.

SOLUTION: First we note that $F(2, -1, 1) = -4 - 8 + 10 + 2 = 0$ so that the hypotheses of the implicit function theorem are verified. In this problem it is pretty clear that we cannot easily write z explicitly as a function of x and y. (There is, according to a famous theorem proved by a French mathematician named Galois, no explicit formula for solving all eighth-degree polynomials.) However, we can compute

$$\frac{\partial F}{\partial z} = 32xyz^7 - 20xyz^3 + 2z$$

and, at $(2, -1, 1)$, $\partial F/\partial z = -22 \neq 0$; so that for (x, y) in a neighborhood of $(2, -1)$ there is a unique function f such that $z = f(x, y)$ and $F(x, y, f(x, y)) = 0$. [(Note that $f(2, -1) = 1$.)] Then

$$\frac{\partial z}{\partial x} = \frac{-\partial F/\partial x}{\partial F/\partial z} = \frac{-2xy^3 - 4yz^8 + 5yz^4}{-22} = \frac{3}{-22} = -\frac{3}{22}$$

and

$$\frac{\partial z}{\partial y} = \frac{-\partial F/\partial y}{\partial F/\partial z} = \frac{-3x^2y^2 - 4xz^8 + 5xz^4}{-22} = \frac{-10}{-22} = \frac{5}{11}$$

at the point $(2, -1, 1)$.

REMARK: We emphasize again that the implicit function theorem enables us to compute partial derivatives of the implicitly defined function f even though we may not be able to write f explicitly.

EXAMPLE 6 A FUNCTION MAY BE DEFINED IMPLICITLY EVEN WHEN THE IMPLICIT FUNCTION THEOREM DOES NOT APPLY

Let $F(x, y) = x - y^5 = 0$. Then, clearly, $y = x^{1/5}$ for every $x \in \mathbb{R}$. However, $\partial F/\partial y = -5y^4 = 0$ at $(0, 0)$, so that the hypotheses of the implicit function theorem are not satisfied near $(0, 0)$. This example shows that functions may be implicitly defined in cases where the implicit function theorem cannot be used.

PROBLEMS 9.2

SELF-QUIZ

True–False

I. If $F: \mathbb{R}^{n+1} \to \mathbb{R}$ satisfy $F(\mathbf{x}_0) = 0$ and $\dfrac{\partial F}{\partial x_{n+1}}(\mathbf{x}_0) = 0$. Then x_{n+1} cannot be written as a function of $(x_1, x_2, \ldots, x_n)$ in some neighborhood of $(x_1^{(0)}, x_2^{(0)}, \ldots, x_n^{(0)})$.

II. If $x + y + z + x^2 + y^2 + z^2 = 0$, then z can certainly be described as a function of (x, y) for (x, y) near $(0, 0)$.

III. If $x + 2xy + 3yz^2 + x^4y^3z^5 + 10 = 0$, then z can certainly be described as a function of (x, y) for (x, y) near $(0, 0)$.

Multiple Choice

IV. If $x + 2y + 3z + x^2y^3z^5 - 3(x + y)^2z^{4/3} = 0$, then z can be written as a function of (x, y) near $(0, 0)$. In that case $\partial z/\partial y =$ ________.

a. $\dfrac{1}{3}$ **b.** $-\dfrac{1}{3}$ **c.** $\dfrac{2}{3}$

d. $-\dfrac{2}{3}$ **e.** $\dfrac{3}{2}$ **f.** $-\dfrac{3}{2}$

In Problems 1–9 a function F from $\mathbb{R}^{n+1} \to \mathbb{R}$ is given. Show that $F(x_1, x_2, \ldots, x_n, x_{n+1}) = 0$ and that x_{n+1} can be written as $f(x_1, x_2, \ldots, x_n)$ in a neighborhood of the given point $(x_1, x_2, \ldots, x_{n+1})$. Then compute $\partial f/\partial x_i$ for $i = 1, 2, \ldots, n$ at $(x_1, x_2, \ldots, x_n)$.

1. $F(x, y) = x^2 + xy + y^2 - 3$; $(1, 1)$

2. $F(x, y) = x^3 + x^2y^5 - 2\sqrt{xy}$; $(1, 1)$

3. $F(x, y) = e^{x+y} - \sin\left[\dfrac{\pi}{2}(x - 3y^2)\right] + \dfrac{y}{x} - e + 1$; $(1, 0)$

4. $F(x, y, z) = xy^2 + yz^2 + x^2z - 3$; $(1, 1, 1)$

5. $F(x, y, z) = x + y + z - \sin(xyz)$; $(0, 0, 0)$

6. $F(x, y, z) = e^{x+2y+3z} - 1$; $(0, 0, 0)$

7. $F(x, y, z) = xe^z - ze^y + (xz/y) - 1$; $(1, 3, 0)$

8. $F(x, y, z) = 1 - e^z\cos(y - x)$; $(3\pi/2, 3\pi/2, 0)$

9. $F(x_1, x_2, x_3, \ldots, x_n, x_{n+1}) = \Sigma_{i=1}^{n+1}(x_i)^4 - n - 1$; $(1, 1, 1, \ldots, 1)$

In Problems 10–17 find formulas for $\partial z/\partial x$ and $\partial z/\partial y$, and state where they are valid.

10. $x^2 + y^2 + z^2 - 1 = 0$

11. $x^4 + y^4 + z^4 - 1 = 0$

12. $xy + xz + yz^5 - 2 = 0$

13. $x^3 \sin xyz + y^3 \cos xyz = 0$

14. $z^3e^{xy} - (xy/z) - 3 = 0$

15. $e^{xy}\ln(z/x) + \cos(x^5 - 4y/z) = 0$

16. $\sinh\left(\dfrac{x - 2y}{3z}\right) + \cosh\left(\dfrac{4x}{z^5}\right) = 0$

17. $e^{\sqrt{zx}} - 3x^2y^3z^4 + \sin\left(\dfrac{5z}{x + y}\right) = 0$

18. Let $F: \mathbb{R}^2 \to \mathbb{R}$ be continuously differentiable with $F(x_0, y_0) = 0$ and $(\partial F/\partial y)(x_0, y_0) > 0$. Show that there are intervals (a, b) and (c, d) such that $x_0 \in (a, b)$, $y_0 \in (c, d)$, $F(x, c) < 0$ for all $x \in (a, b)$, $F(x, d) > 0$ for all $x \in (a, b)$, and $(\partial F/\partial x)(x, y) > 0$ for all $x \in (a, b)$ and $y \in (c, d)$. [*Hint:* Use the continuity of F and $\partial F/\partial y$.]

19. Let $\bar{x}$ be in $[a, b]$ and define the function $F_{\bar{x}}(y) = F(\bar{x}, y)$. Show that there is a unique number $\bar{y}$ in $[c, d]$ such that $F_{\bar{x}}(\bar{y}) = F(\bar{x}, \bar{y}) = 0$. [*Hint:* Use the intermediate value theorem.]

20. Use the results of Problems 18 and 19 to prove the following special case of the implicit function theorem: Let $F: \mathbb{R}^2 \to \mathbb{R}$ be continuously differentiable and let $(x_0, y_0) \in \mathbb{R}^2$

be a point at which $F(x_0, y_0) = 0$ and $(\partial F/\partial y)(x_0, y_0) \neq 0$. Then there exists a closed interval $[a, b]$ containing x_0 and a function $f\colon \mathbb{R} \to \mathbb{R}$ with domain $[a, b]$ such that $y = f(x)$ satisfies $F(x, f(x)) = 0$ in a neighborhood of (x_0, y_0).

ANSWERS TO SELF-QUIZ

I. False **II.** True **III.** False **IV.** d

9.3 FUNCTIONS FROM $\mathbb{R}^n$ TO $\mathbb{R}^m$

In Section 3.1 we described functions from $\mathbb{R}^n$ to $\mathbb{R}$ and in Section 2.4 we discussed functions from $\mathbb{R}$ to $\mathbb{R}^n$. In this section we begin our study of functions from $\mathbb{R}^n$ to $\mathbb{R}^m$ where $n \geq 1$ and $m \geq 1$.

DEFINITION FUNCTION FROM $\mathbb{R}^n$ TO $\mathbb{R}^m$

Let Ω be a subset of $\mathbb{R}^n$. Then a **function** or **mapping f** from $\mathbb{R}^n$ to $\mathbb{R}^m$ is a rule that assigns to each $\mathbf{x} = (x_1, x_2, \ldots, x_n)$ in Ω a unique vector $\mathbf{f}(\mathbf{x})$ in $\mathbb{R}^m$. The set Ω is called the **domain** of **f**. The set $\{\mathbf{f}(\mathbf{x})\colon \mathbf{x} \in \Omega\}$ is called the **range** of **f**. ■

NOTATION: We will write $\mathbf{f}\colon \mathbb{R}^n \to \mathbb{R}^m$ and, since $\mathbf{f}(\mathbf{x})$ is a subset of $\mathbb{R}^m$, we will represent it in terms of its **component functions**. That is,

$$\begin{aligned}\mathbf{f}(\mathbf{x}) &= (f_1(\mathbf{x}), f_2(\mathbf{x}), \ldots, f_m(\mathbf{x})) \\ &= (f_1(x_1, x_2, \ldots, x_n), f_2(x_1, x_2, \ldots, x_n), \ldots, f_m(x_1, x_2, \ldots, x_n)).\end{aligned} \tag{1}$$

Note that each of the m component functions is a mapping from $\mathbb{R}^n \to \mathbb{R}$.

EXAMPLE 1 A MAPPING FROM $\mathbb{R}^2$ TO $\mathbb{R}^2$

Let $\mathbf{f}\colon \mathbb{R}^2 \to \mathbb{R}^2$ be defined by

$$\mathbf{f}(x, y) = (x + 2y, -3x + 4y).$$

For example, the image of $(1, 2)$ under this mapping is $(5, 5)$ and the image of $(-4, 2)$ is $(0, 20)$. Note that **f** can be represented by a matrix:

$$\mathbf{f}(x, y) = \begin{pmatrix} 1 & 2 \\ -3 & 4 \end{pmatrix}\begin{pmatrix} x \\ y \end{pmatrix}.$$

That is, **f** is a linear transformation from $\mathbb{R}^2 \to \mathbb{R}^2$. Here the component functions are $f_1(x, y) = x + 2y$ and $f_2(x, y) = -3x + 4y$.

EXAMPLE 2 THE POLAR COORDINATE MAPPING

Let $\mathbf{f}\colon \mathbb{R}^2 \to \mathbb{R}^2$ be defined by $\mathbf{f}(r, \theta) = (r\cos\theta, r\sin\theta)$. This is called the **polar coordinate mapping**. To have an idea how this mapping works, consider the rectangular region $R = \{(r, \theta)\colon 0 \leq r \leq 1, 0 \leq \theta \leq \pi\}$. This rectangular region is sketched in Figure 1a. Keep in mind that r and θ are just numbers here and that the vector (r, θ) is an ordinary vector in $\mathbb{R}^2$. The image of R under the mapping is the semicircular disk S pictured in Figure 1b and given by $S = \{(x, y): x = r\cos\theta, y = r\sin\theta, 0 \le r \le 1, 0 \le \theta \le \pi\}$.

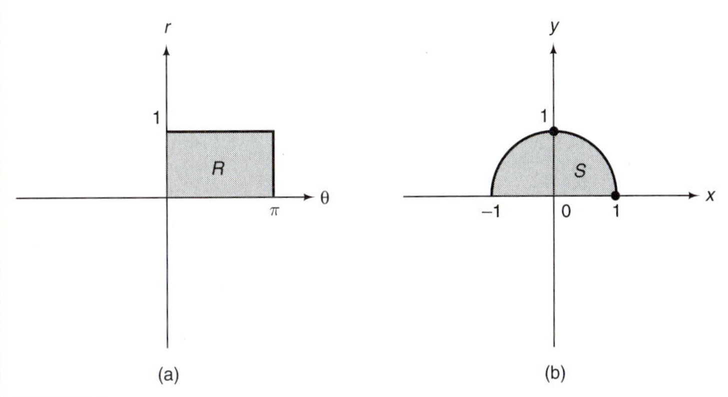

FIGURE 1
The semicircular disk in (b) is the image under the polar coordinate mapping of the rectangular region in (a).

EXAMPLE 3

A MAPPING FROM $\mathbb{R}^3$ TO $\mathbb{R}^4$

Let $\mathbf{f}$: $\mathbb{R}^3 \to \mathbb{R}^4$ be given by $\mathbf{f}(x, y, z) = (xy, xz, y + z, xyz)$. Thus, for example, $\mathbf{f}(1, 2, 3) = (2, 3, 5, 6)$ and $\mathbf{f}(-1, 2, 0) = (-2, 0, 2, 0)$. Here the component functions are $f_1(x, y, z) = xy$, $f_2(x, y, z) = xz$, $f_3(x, y, z) = y + z$, and $f_4(x, y, z) = xyz$.

EXAMPLE 4

LINEAR TRANSFORMATION MAPPINGS

As we saw in Section 8.8, a mapping $\mathbf{f}$: $\mathbb{R}^n \to \mathbb{R}^m$ is called a **linear transformation** if $\mathbf{f}(\mathbf{x} + \mathbf{y}) = \mathbf{f}(\mathbf{x}) + \mathbf{f}(\mathbf{y})$ and $\mathbf{f}(\alpha\mathbf{x}) = \alpha\mathbf{f}(\mathbf{x})$ for any scalar α. If $\mathbf{f}$ is a linear transformation from $\mathbb{R}^n \to \mathbb{R}^m$, then from Section 8.10 there exists an $m \times n$ matrix A such that $\mathbf{f}(\mathbf{x}) = A\mathbf{x}$ for every $\mathbf{x}$ in $\mathbb{R}^n$.

Functions from $\mathbb{R}^n \to \mathbb{R}^m$ have many interesting properties. In Chapter 8 we discussed some of the properties of linear transformations. But most mappings from $\mathbb{R}^n \to \mathbb{R}^m$ are not linear. In this section we shall discuss some basic properties of these functions, including the notions of limits and continuity. In Section 9.4 we shall address the more difficult concept of differentiation.

DEFINITION SUMS AND CONSTANT MULTIPLES OF FUNCTIONS

Let $\mathbf{f}$: $\mathbb{R}^n \to \mathbb{R}^m$ and $\mathbf{g}$: $\mathbb{R}^n \to \mathbb{R}^m$ have domains Ω_1 and Ω_2, respectively. Then we define the functions of $\alpha\mathbf{f}$ and $\mathbf{f} + \mathbf{g}$ by

i. $(\alpha\mathbf{f})(\mathbf{x}) = \alpha\mathbf{f}(\mathbf{x})$ with domain Ω_1, and
ii. $(\mathbf{f} + \mathbf{g})(\mathbf{x}) = f(\mathbf{x}) + g(\mathbf{x})$ with domain $\Omega_1 \cap \Omega_2$. ■

EXAMPLE 5

THE DOMAIN OF A NONLINEAR FUNCTION

Let **f** and **g**: $\mathbb{R}^3 \to \mathbb{R}^2$ be given by

$$\mathbf{f}(x, y, z) = (\sqrt{1 - x^2 - y^2 - z^2}, xy^2z^3) \qquad \text{and} \qquad \mathbf{g}(x, y, z) = \left(\frac{1}{x + y + z}, \sqrt{x}\right).$$

Compute $-4\mathbf{f}$ and $\mathbf{f} + \mathbf{g}$ and determine their respective domains.

SOLUTION: The domain of $\mathbf{f} = \{(x, y, z): x^2 + y^2 + z^2 \leq 1\}$. This is the closed unit ball. Thus, $-4\mathbf{f}(x, y, z) = (-4\sqrt{1 - x^2 - y^2 - z^2}, -4xy^2z^3)$ with the closed unit ball as its domain. In addition,

$$(\mathbf{f} + \mathbf{g})(x, y, z) = \left(\sqrt{1 - x^2 - y^2 - z^2} + \frac{1}{x + y + z}, xy^2z^3 + \sqrt{x}\right).$$

The domain of $\mathbf{g} = \{(x, y, z): x + y + z \neq 0 \text{ and } x \geq 0\}$. This is the set of points in the half-space $x \geq 0$ that are not on the plane $x + y + z = 0$. Hence the domain of $\mathbf{f} + \mathbf{g}$ consists of those points in the closed unit half ball $\{(x, y, z): x^2 + y^2 + z^2 \leq 1, x \geq 0\}$ that do not lie on the plane $x + y + z = 0$.

DEFINITION COMPOSITE FUNCTION

Let **f**: $\mathbb{R}^n \to \mathbb{R}^m$ with domain Ω_1 and **g**: $\mathbb{R}^m \to \mathbb{R}^q$ with domain Ω_2. Then the **composite function** $\mathbf{g} \circ \mathbf{f}$: $\mathbb{R}^n \to \mathbb{R}^q$ is defined by

$$\mathbf{g} \circ \mathbf{f}(\mathbf{x}) = \mathbf{g}(\mathbf{f}(\mathbf{x})). \tag{2}$$

The domain of $\mathbf{g} \circ \mathbf{f}$ is given by

$$\text{dom}(\mathbf{g} \circ \mathbf{f}) = \{\mathbf{x} \in \Omega_1: \mathbf{f}(\mathbf{x}) \in \Omega_2\}. \tag{3}$$

■

EXAMPLE 6

FINDING A COMPOSITE FUNCTION

Let **f**: $\mathbb{R}^2 \to \mathbb{R}^3$ be given by $\mathbf{f}(x, y) = (x^3, xy, y)$ and **g**: $\mathbb{R}^3 \to \mathbb{R}^4$ be given by $\mathbf{g}(x, y, z) = (x^2, y^2, z^2, \sqrt{x})$. Compute $\mathbf{g} \circ \mathbf{f}$ and determine its domain.

SOLUTION:

$$\begin{aligned} \mathbf{g} \circ \mathbf{f}(\mathbf{x}) &= \mathbf{g}(\mathbf{f}(\mathbf{x})) = \mathbf{g}(x^3, xy, y) \\ &= ([x^3]^2, (xy)^2, y^2, \sqrt{x^3}) = (x^6, x^2y^2, y^2, x^{3/2}) \end{aligned}$$

and dom $\mathbf{g} \circ \mathbf{f}$ is $\{(x, y): x \geq 0\}$.

DEFINITION BOUNDARY POINT AND LIMIT

Suppose that $\mathbf{f}$: $\mathbb{R}^n \to \mathbb{R}^m$ is defined on a neighborhood $B_r(\mathbf{x}_0)$ of the vector $\mathbf{x}_0$ except possibly at $\mathbf{x}_0$ itself. Then

$\mathbf{x}_0$ is called a **boundary point** of a set D if every open ball $B_r(\mathbf{x}_0)$ centered at $\mathbf{x}_0$ contains at least one point of D. Suppose that $\mathbf{f}$: $\mathbb{R}^n \to \mathbb{R}^m$ is defined on a set D (= dom $\mathbf{f}$). Suppose further that $\mathbf{x}_0 \in$ D or $\mathbf{x}_0$ is a boundary point of D. Then

$\lim_{\mathbf{x}\to\mathbf{x}_0} \mathbf{f}(\mathbf{x}) = \mathbf{L}$ if for every $\epsilon > 0$, there is a number $\delta > 0$ such that whenever $0 < |\mathbf{x} - \mathbf{x}_0| < \delta$ and $\mathbf{x} \in$ D, we have $|\mathbf{f}(\mathbf{x}) - \mathbf{L}| < \epsilon$. **(4)**

■

The definition of a limit has familiar consequences. It turns out that $\mathbf{f}(\mathbf{x})$ has a limit as $\mathbf{x} \to \mathbf{x}_0$ if and only if each of its component functions has a limit as $\mathbf{x} \to \mathbf{x}_0$.

THEOREM 1

Let $\mathbf{f}$: $\mathbb{R}^n \to \mathbb{R}^m$ with component functions $f_1, f_2, \ldots, f_m$, and let $\mathbf{L} = (L_1, L_2, \ldots, L_m)$. Then

$$\lim_{\mathbf{x}\to\mathbf{x}_0} \mathbf{f}(\mathbf{x}) = \mathbf{L} \text{ if and only if } \lim_{\mathbf{x}\to\mathbf{x}_0} f_i(\mathbf{x}) = L_i \qquad \text{for } i = 1, 2, \ldots, m. \tag{5}$$

PROOF: Suppose that $\lim_{\mathbf{x}\to\mathbf{x}_0} \mathbf{f}(\mathbf{x}) = \mathbf{L}$. Let $\epsilon > 0$ be given and choose δ as in the definition of limit. Then, if $0 < |\mathbf{x} - \mathbf{x}_0| < \delta$,

$$\begin{aligned} |f_i(\mathbf{x}) - L_i| &= \sqrt{(f_i(\mathbf{x}) - L_i)^2} \\ &\le \sqrt{(f_1(\mathbf{x}) - L_1)^2 + (f_2(\mathbf{x}) - L_2)^2 + \cdots + (f_m(\mathbf{x}) - L_m)^2} \\ &= |\mathbf{f}(\mathbf{x}) - \mathbf{L}| < \epsilon. \end{aligned}$$

Thus $\lim_{\mathbf{x}\to\mathbf{x}_0} f_i(\mathbf{x}) = L_i$ for $i = 1, 2, 3, \ldots, m$. Conversely, suppose that $\lim_{\mathbf{x}\to\mathbf{x}_0} f_i(\mathbf{x}) = L_i$ for $i = 1, 2, \ldots, n$. Let $\epsilon > 0$ be given. Then, by the definition on page 156, there is a number $\delta_i > 0$ such that

$$|f_i(\mathbf{x}) - L_i| < \frac{\epsilon}{\sqrt{m}} \qquad \text{if } 0 < |\mathbf{x} - \mathbf{x}_0| < \delta_i.$$

Let δ = minimum of $\delta_1, \delta_2, \ldots, \delta_m$. Then, for $0 < |\mathbf{x} - \mathbf{x}_0| < \delta$,

$$\begin{aligned} |\mathbf{f}(\mathbf{x}) - \mathbf{L}| &= \sqrt{(f_1(\mathbf{x}) - L_1)^2 + (f_2(\mathbf{x}) - L_2)^2 + \cdots + (f_m(\mathbf{x}) - L_m)^2} \\ &\le \sqrt{\underbrace{\frac{\epsilon^2}{m} + \frac{\epsilon^2}{m} + \cdots + \frac{\epsilon^2}{m}}_{m \text{ terms}}} = \sqrt{\epsilon^2} = \epsilon. \end{aligned}$$

■

EXAMPLE 7

FINDING A LIMIT OF A FUNCTION FROM $\mathbb{R}^2$ TO $\mathbb{R}^4$

Find $\lim_{\mathbf{x}\to(1,2)} \mathbf{f}(\mathbf{x})$ if $\mathbf{f}$: $\mathbb{R}^2 \to \mathbb{R}^4$ is given by

$$\mathbf{f}(x, y) = \left(x^2y - y^3, \sqrt{x + y}, \sin \pi\left(\frac{x + y}{12}\right), \frac{16x^4 - y^4}{4x^2 - y^2}\right).$$

SOLUTION: Using Theorem 1, we need only take the limits of the component functions. We can determine the first three limits by evaluation because each component function is continuous at (1, 2). Thus

$$\lim_{(x,y)\to(1,2)} (x^2y - y^3) = -6, \qquad \lim_{(x,y)\to(1,2)} \sqrt{x+y} = \sqrt{3},$$

and

$$\lim_{(x,y)\to(1,2)} \sin \pi\left(\frac{x+y}{12}\right) = \sin\frac{\pi}{4} = \frac{\sqrt{2}}{2}.$$

The last component function is not defined at (1, 2) since the denominator is 0 there. However, we easily calculate

$$\begin{aligned}\lim_{(x,y)\to(1,2)} \frac{16x^4 - y^4}{4x^2 - y^2} &= \lim_{(x,y)\to(1,2)} \frac{(4x^2 - y^2)(4x^2 + y^2)}{4x^2 - y^2} \\ &= \lim_{(x,y)\to(1,2)} (4x^2 + y^2) = 8.\end{aligned}$$

Hence

$$\lim_{\mathbf{x}\to(1,2)} \mathbf{f}(\mathbf{x}) = \left(-6, \sqrt{3}, \frac{\sqrt{2}}{2}, 8\right).$$

NOTE: $\dfrac{16x^4 - y^4}{4x^2 - y^2}$ is defined near (1, 2) except for points on the line $y = 2x$. Thus the point (1, 2) is a boundary point of the domain of this function and the limit exists according to the definition on page 651.

DEFINITION CONTINUITY

Let $\mathbf{f}: \mathbb{R}^n \to \mathbb{R}^m$ be defined in a neighborhood of the vector $\mathbf{x}_0$. Then $\mathbf{f}$ is **continuous** at $\mathbf{x}_0$ if each component function f_i is continuous at $\mathbf{x}_0$.

REMARK: An alternative definition of continuity at $\mathbf{x}_0$ (for $\mathbf{f}$ defined in a neighborhood of $\mathbf{x}_0$) is

$$\lim_{\mathbf{x}\to\mathbf{x}_0} \mathbf{f}(\mathbf{x}) = \mathbf{f}(\mathbf{x}_0). \tag{6}$$

In Problems 28 and 29 you are asked to show that these two definitions are equivalent. ■

The following theorem enables us to compute a large number of limits by evaluation.

THEOREM 2

Let $\mathbf{f}: \mathbb{R}^n \to \mathbb{R}^m$ be continuous at $\mathbf{x}_0$ and let $\mathbf{g}: \mathbb{R}^m \to \mathbb{R}^q$ be continuous at $\mathbf{f}(\mathbf{x}_0)$. Then the composite function $\mathbf{g} \circ \mathbf{f}$ is continuous at $\mathbf{x}_0$.

PROOF: Let $\mathbf{y} = \mathbf{f}(\mathbf{x})$. As $\mathbf{x} \to \mathbf{x}_0$, $\mathbf{y} = \mathbf{f}(\mathbf{x}) \to \mathbf{f}(\mathbf{x}_0) = \mathbf{y}_0$, so that

$$\lim_{\mathbf{x}\to\mathbf{x}_0} (\mathbf{g} \circ \mathbf{f})(\mathbf{x}) = \lim_{\mathbf{x}\to\mathbf{x}_0} \mathbf{g}(\mathbf{f}(\mathbf{x})) = \lim_{\mathbf{y}\to\mathbf{y}_0} \mathbf{g}(\mathbf{y}) = \mathbf{g}(\mathbf{y}_0) = \mathbf{g}(\mathbf{f}(\mathbf{x}_0)) = (\mathbf{g} \circ \mathbf{f})(\mathbf{x}_0).$$ ■

PROBLEMS 9.3

SELF-QUIZ

Multiple Choice

I. If $\mathbf{f}: \mathbb{R}^2 \to \mathbb{R}^3$ is defined by $\mathbf{f}(x, y) = (x + y, x - y, y)$, then $\mathbf{f}(1, -1) =$ ________.
a. 3 **b.** $(0, 2, -1)$ **c.** $(0, 2)$
d. $(2, 0)$ **e.** $(2, 0, -1)$

II. If $\mathbf{f}$ is as in Problem I, then $\lim_{(x,y)\to(0,0)} \mathbf{f}(x, y) =$ ________.
a. 0 **b.** $(0, 0)$
c. $(0, 0, 0)$ **d.** $(0, 0, 0, 0, 0)$

III. If $\mathbf{f}: \mathbb{R}^2 \to \mathbb{R}^2$ is defined by $\mathbf{f}(x, y) = (\sin(x - 2y), e^{x+2y})$, then $\lim_{(x,y)\to(\pi/6,0)} \mathbf{f}(x, y) =$ ________.
a. $\left(\frac{1}{2}, 1\right)$ **b.** $\left(\frac{1}{2}, e^{\pi/6}\right)$ **c.** $\left(\frac{\sqrt{3}}{2}, e^{\pi/6}\right)$
d. $\left(-\frac{1}{2}, 1\right)$ **e.** $\left(-\frac{1}{2}, e^{\pi/6}\right)$

True–False

IV. If A is an $m \times n$ matrix, then the function $\mathbf{f}$ defined by $\mathbf{f}(\mathbf{x}) = A\mathbf{x}$ is a mapping from $\mathbb{R}^m$ to $\mathbb{R}^n$.

V. If A is an $m \times n$ matrix, then the function $\mathbf{f}$ defined by $\mathbf{f}(\mathbf{x}) = A\mathbf{x}$ is a mapping from $\mathbb{R}^n$ to $\mathbb{R}^m$.

In Problems 1–7 find the domain of the given function.

1. $\mathbf{f}(x, y) = (x^2 + y^2, x^2 - y^2, \sqrt{x^2 + y^2})$

2. $f(x, y, z) = \left(\frac{x}{y}, \frac{y}{z}, \frac{z}{x}\right)$

3. $\mathbf{f}(x, y, z) = (e^{x+y}, \ln(x + y + z))$

4. $\mathbf{f}(x, y, z) = (x^2, y^2, z^2, \sqrt{36 - 9x^2 - 4y^2 - z^2})$

5. $\mathbf{f}(x_1, x_2, x_3, x_4, x_5) = (x_1x_2, x_2x_3, x_3x_4, x_4x_5, x_1x_5)$

6. $\mathbf{f}(x_1, x_2, x_3, x_4) =$
$$\left(\frac{1}{x_1 + x_2 + 3x_3 + x_4 - 5}, x_1 + x_2 + x_3 + x_4 - 5\right)$$

7. $\mathbf{f}(x_1, x_2, x_3, x_4) = (\sqrt{1 - x_1^2 - x_2^2 - x_3^2 - x_4^2}, \sqrt[3]{1 - x_1^2 - x_2^2 - x_3^2 - x_4^2})$

8. Let $\mathbf{f}: \mathbb{R}^2 \to \mathbb{R}^2$ be defined by $\mathbf{f}(x, y) = (-x, y)$. Describe the image of the set of points lying on the unit semicircle $x^2 + y^2 = 1$, $x \geq 0$.

9. Do the same as in Problem 8 for $\mathbf{f}(x, y) = (3x, -4y)$.

10. Do the same as in Problem 8 for $\mathbf{f}(x, y) = (ax, by)$.

11. Let $\mathbf{f}: \mathbb{R}^2 \to \mathbb{R}^2$ be given by $\mathbf{f}(x, y) = (e^y \sin x, e^y \cos x)$. Describe the image of the rectangle $0 \leq x \leq 2\pi$, $0 \leq y \leq 1$ under this mapping.

12. Let $\mathbf{f}$ be as in Problem 11. Describe the image of the ellipse $(x^2/4) + (y^2/9) = 1$ under this mapping.

13. Describe the image under the polar coordinate mapping of the rectangle $0 \leq r \leq 2$, $0 \leq \theta \leq \pi/3$.

14. Do the same as in Problem 13 for the rectangle $1 \leq r \leq 3$, $\pi/6 \leq \theta \leq \pi/2$.

In Problems 15–24 find the limit of the given function at the given point (if it exists). Is $\mathbf{f}$ continuous at that point?

15. $\mathbf{f}(x, y) = (x^2 + y^2, x^2 - y^2, xy)$; $(4, 6)$

16. $\mathbf{f}(x, y, z) = \left(\frac{x^2 - y^2}{x - y}, \frac{y^2 - z^2}{y - z}\right)$; $(0, 0, 0)$

17. $\mathbf{f}(x, y, z) = \left(\sin yz, \tan^{-1}\frac{y}{x}, x^3\right)$; $\left(\frac{\pi}{6}, \frac{\pi}{6}, 2\right)$

18. $\mathbf{f}(x_1, x_2, x_3, x_4) =$
$$\left(x_1^2, x_2^2, x_3^2, x_4^2, \frac{\sin(x_1^2 + x_2^2 + x_3^2 + x_4^2)}{x_1^2 + x_2^2 + x_3^2 + x_4^2}\right); \quad (0, 0, 0, 0)$$

19. $\mathbf{f}(x_1, x_2, \ldots, x_n) =$
$$\left(\sum_{i=1}^n x_i, \left[\sum_{i=1}^n x_i\right]^2, \left[\sum_{i=1}^n x_i\right]^3, \ldots, \left[\sum_{i=1}^n x_i\right]^m\right); \quad (1, 1, \ldots, 1)$$

20. $\mathbf{f}(x, y, z) = (\sin xyz, \cos xyz, e^{\sqrt{x^2+y^2+z^2}})$; $(0, 0, 0)$

21. $\mathbf{f}(x, y, z) =$
$$\left(\ln(1 + x + y + z), \frac{x + y + z}{\ln(x + y + z)}, e^{x+y+z}\right); \quad (0, 0, 0)$$

22. $\mathbf{f}(x, y) = \left(x, y, \frac{x^2 - 2y}{y^2 + 2x}\right)$; $(0, 0)$

23. $\mathbf{f}(x_1, x_2, x_3, x_4) = \left(x_1x_2, \sin x_3x_4, \dfrac{x_1 + x_2 + x_3 + x_4}{x_1 - x_2 - x_3 - x_4}\right)$; $(0, 0, 0, 0)$

24. $\mathbf{f}(x_1, x_2, x_3, x_4) = \left(\dfrac{5x_1^2x_2^2 - 3x_1^2x_3^2}{x_1^4 + x_4^2}, \dfrac{x_1x_2^2 - x_3x_4^2 + x_2x_3x_4}{x_1^2 + x_2^2 + x_3^2 + x_4^2}\right)$; $(0, 0, 0, 0)$

In Problems 25–27 write out the composite function $\mathbf{g} \circ \mathbf{f}$. Determine its domain.

25. $\mathbf{f}(x, y) = (xy, x^2, \cos(y/x))$; $\mathbf{g}(x, y, z) = (xyz, e^{x+y-2z})$

26. $\mathbf{f}(x, y, z) = (\sqrt{1 - x^2 - y^2 - z^2}, \ln(1 - x - y + z), xy, yz)$; $\mathbf{g}(x_1, x_2, x_3, x_4) = (x_1 - x_3, x_1{}^3x_3)$

27. $\mathbf{f}(x, y) = (e^{x+y}, \sqrt{1 - x^2 + y^2})$; $\mathbf{g}(x, y) = (\sin(x/y), \sin(y/x), e^{\sqrt{x^2+y^2}})$

28. Suppose that $\mathbf{f}$: $\mathbb{R}^n \to \mathbb{R}^m$ is continuous according to the definition of continuity on page 652. Show that $\lim_{\mathbf{x}\to\mathbf{x}_0} \mathbf{f}(\mathbf{x}) = \mathbf{f}(\mathbf{x}_0)$.

29. Suppose that $\mathbf{f}$ is defined in a neighborhood of $\mathbf{x}_0$ and that $\lim_{\mathbf{x}\to\mathbf{x}_0} \mathbf{f}(\mathbf{x}) = \mathbf{f}(\mathbf{x}_0)$. Prove that $\mathbf{f}$ is continuous at $\mathbf{x}_0$.

30. Suppose that $\mathbf{f}$: $\mathbb{R}^n \to \mathbb{R}^m$ is continuous at $\mathbf{x}_0$. Show that $\alpha\mathbf{f}$ is continuous at $\mathbf{x}_0$ for every scalar α.

31. Suppose that $\mathbf{f}$: $\mathbb{R}^n \to \mathbb{R}^m$ and $\mathbf{g}$: $\mathbb{R}^n \to \mathbb{R}^m$ are continuous at $\mathbf{x}_0$. Show that $\mathbf{f} + \mathbf{g}$ is continuous at $\mathbf{x}_0$.

32. Let $\mathbf{f}$: $\mathbb{R}^n \to \mathbb{R}^m$ be a linear transformation. Show that $\mathbf{f}$ is continuous at every $\mathbf{x}$ in $\mathbb{R}^n$.

33. Let $\mathbf{f}$: $\mathbb{R}^n \to \mathbb{R}^m$ be a linear transformation with matrix representation A_1, and let $\mathbf{g}$: $\mathbb{R}^m \to \mathbb{R}^q$ be a linear transformation with matrix representation A_2. Show that the matrix representation of $\mathbf{g} \circ \mathbf{f}$ is A_2A_1.

ANSWERS TO SELF-QUIZ

I. b **II.** c **III.** b **IV.** False **V.** True

9.4 DERIVATIVES AND THE JACOBIAN MATRIX

In this section we discuss the derivative of a function $\mathbf{f}$ from $\mathbb{R}^n$ to $\mathbb{R}^m$. As you might expect, the notion is closely related to the gradient of a function from $\mathbb{R}^n$ to $\mathbb{R}$. Our definition is similar to the definition of the gradient given in Section 3.5.

DEFINITION DIFFERENTIABILITY

Let f: $\mathbb{R}^n \to \mathbb{R}^m$ be defined on an open set Ω in $\mathbb{R}^n$ and suppose that $\mathbf{x}_0 \in \Omega$. Then $\mathbf{f}$ is **differentiable** at $\mathbf{x}_0$ if there exists a linear transformation $\mathbf{L}$: $\mathbb{R}^n \to \mathbb{R}^m$ and a function $\mathbf{g}$: $\mathbb{R}^n \to \mathbb{R}^m$ such that for every $\mathbf{x}$ in Ω,

$$\mathbf{f}(\mathbf{x}) - \mathbf{f}(\mathbf{x}_0) = \mathbf{L}(\mathbf{x} - \mathbf{x}_0) + \mathbf{g}(\mathbf{x} - \mathbf{x}_0) \quad (1)$$

where

$$\lim_{|\mathbf{x}-\mathbf{x}_0|\to 0} \frac{g(\mathbf{x} - \mathbf{x}_0)}{|\mathbf{x} - \mathbf{x}_0|} = \mathbf{0}. \quad (2)$$

■

REMARK: If we write $\Delta\mathbf{x} = \mathbf{x} - \mathbf{x}_0$, then the definition of differentiability is virtually identical to the definitions on page 186 for a function from $\mathbb{R}^n \to \mathbb{R}$.

We now must answer two questions: First, under what conditions is a given function from $\mathbb{R}^n \to \mathbb{R}^m$ differentiable? Second, how do we determine the linear transformation $\mathbf{L}$? We shall answer the second question first.

If f: $\mathbb{R}^n \to \mathbb{R}$ is differentiable, then

$$\mathbf{L} = \nabla f = \left(\frac{\partial f}{\partial x_1}, \frac{\partial f}{\partial x_2}, \ldots, \frac{\partial f}{\partial x_n}\right).$$

Thus we should not be too surprised if, in this case, the linear transformation $\mathbf{L}$ involves the partial derivatives of the component functions of $\mathbf{f}$. We will soon see that this is the case.

Since $\mathbf{L}$ is a linear transformation from $\mathbb{R}^n \to \mathbb{R}^m$, there is, by Theorem 8.10.1 a unique $m \times n$ matrix, which we denote by $J_{\mathbf{f}}(\mathbf{x}_0)$, such that

$$\mathbf{L}(\mathbf{x}) = J_{\mathbf{f}}(\mathbf{x}_0)\mathbf{x} \quad \text{for every } \mathbf{x} \text{ in } \mathbb{R}^n. \tag{3}$$

We assume that $\mathbf{x}$ is written as a column vector in terms of the standard basis in $\mathbb{R}^n$ and $\mathbf{L}(\mathbf{x})$ is written as a column vector in terms of the standard basis in $\mathbb{R}^m$. The notation $J_{\mathbf{f}}(\mathbf{x}_0)$ is used to emphasize that the matrix J depends on both the function $\mathbf{f}$ and the vector $\mathbf{x}_0$. With this notation, (1) can be written as

$$\mathbf{f}(\mathbf{x}) - \mathbf{f}(\mathbf{x}_0) = J_{\mathbf{f}}(\mathbf{x}_0)(\mathbf{x} - \mathbf{x}_0) + \mathbf{g}(\mathbf{x} - \mathbf{x}_0). \tag{4}$$

REMARK: In equation (4) we assume that all vectors are written as column vectors so that the matrix multiplication makes sense.

DEFINITION JACOBIAN MATRIX

Let $\mathbf{f}$: $\mathbb{R}^n \to \mathbb{R}^m$ be differentiable at $\mathbf{x}_0$. Then the $m \times n$ matrix $J_{\mathbf{f}}(\mathbf{x}_0)$ determined by (4) is called the **Jacobian matrix** of $\mathbf{f}$ at $\mathbf{x}_0$.† ■

REMARK: The matrix $J_{\mathbf{f}}(\mathbf{x}_0)$ is sometimes called the **derivative matrix** or, more simply, the **derivative** of the linear transformation $\mathbf{f}$.

THEOREM 1

Let $\mathbf{f}$: $\mathbb{R}^n \to \mathbb{R}^m$ be differentiable at $\mathbf{x}_0$. Then the Jacobian matrix $J_{\mathbf{f}}(\mathbf{x}_0)$ is given by

$$J_{\mathbf{f}}(\mathbf{x}_0) = \begin{pmatrix} \nabla f_1(\mathbf{x}_0) \\ \nabla f_2(\mathbf{x}_0) \\ \vdots \\ \nabla f_m(\mathbf{x}_0) \end{pmatrix} = \begin{pmatrix} \frac{\partial f_1}{\partial x_1}(\mathbf{x}_0) & \frac{\partial f_1}{\partial x_2}(\mathbf{x}_0) & \frac{\partial f_1}{\partial x_3}(\mathbf{x}_0) & \cdots & \frac{\partial f_1}{\partial x_n}(\mathbf{x}_0) \\ \frac{\partial f_2}{\partial x_1}(\mathbf{x}_0) & \frac{\partial f_2}{\partial x_2}(\mathbf{x}_0) & \frac{\partial f_2}{\partial x_3}(\mathbf{x}_0) & \cdots & \frac{\partial f_2}{\partial x_n}(\mathbf{x}_0) \\ \frac{\partial f_3}{\partial x_1}(\mathbf{x}_0) & \frac{\partial f_3}{\partial x_2}(\mathbf{x}_0) & \frac{\partial f_3}{\partial x_3}(\mathbf{x}_0) & \cdots & \frac{\partial f_3}{\partial x_n}(\mathbf{x}_0) \\ \vdots & \vdots & \vdots & & \vdots \\ \frac{\partial f_m}{\partial x_1}(\mathbf{x}_0) & \frac{\partial f_m}{\partial x_2}(\mathbf{x}_0) & \frac{\partial f_m}{\partial x_3}(\mathbf{x}_0) & \cdots & \frac{\partial f_m}{\partial x_n}(\mathbf{x}_0) \end{pmatrix}. \tag{5}$$

† See the biographical sketch on page 363.

NOTE: The functions $f_1, f_2, \ldots, f_m$ in (5) are the component functions of the function $\mathbf{f}$.

PROOF: Let $\mathbf{J}_1, \mathbf{J}_2, \ldots, \mathbf{J}_m$ denote the rows of $J_\mathbf{f}(\mathbf{x}_0)$. Then, writing (4) in terms of components, we have

$$\begin{pmatrix} f_1(\mathbf{x}) - f_1(\mathbf{x}_0) \\ f_2(\mathbf{x}) - f_2(\mathbf{x}_0) \\ \vdots \\ f_m(\mathbf{x}) - f_m(\mathbf{x}_0) \end{pmatrix} = \begin{pmatrix} \mathbf{J}_1 \cdot (\mathbf{x} - \mathbf{x}_0) \\ \mathbf{J}_2 \cdot (\mathbf{x} - \mathbf{x}_0) \\ \vdots \\ \mathbf{J}_m \cdot (\mathbf{x} - \mathbf{x}_0) \end{pmatrix} + \begin{pmatrix} g_1(\mathbf{x} - \mathbf{x}_0) \\ g_2(\mathbf{x} - \mathbf{x}_0) \\ \vdots \\ g_m(\mathbf{x} - \mathbf{x}_0) \end{pmatrix} \tag{6}$$

where $g_1, g_2, \ldots, g_m$ are the component functions of $\mathbf{g}$. Now

$$\lim_{|\mathbf{x}-\mathbf{x}_0| \to 0} \frac{g(\mathbf{x} - \mathbf{x}_0)}{|\mathbf{x} - \mathbf{x}_0|} = \mathbf{0}.$$

But

$$\frac{\mathbf{g}(\mathbf{x} - \mathbf{x}_0)}{|\mathbf{x} - \mathbf{x}_0|} = \left(\frac{g_1(\mathbf{x} - \mathbf{x}_0)}{|\mathbf{x} - \mathbf{x}_0|}, \frac{g_2(\mathbf{x} - \mathbf{x}_0)}{|\mathbf{x} - \mathbf{x}_0|}, \ldots, \frac{g_m(\mathbf{x} - \mathbf{x}_0)}{|\mathbf{x} - \mathbf{x}_0|} \right).$$

This means that

$$\lim_{|\mathbf{x}-\mathbf{x}_0| \to 0} \frac{g_i(\mathbf{x} - \mathbf{x}_0)}{|\mathbf{x} - \mathbf{x}_0|} = 0 \qquad \text{for } i = 1, 2, \ldots, m. \tag{7}$$

Let's look at the ith row of (6). We have

$$f_i(\mathbf{x}) - f_i(\mathbf{x}_0) = \mathbf{J}_i \cdot (\mathbf{x} - \mathbf{x}_0) + g_i(\mathbf{x} - \mathbf{x}_0). \tag{8}$$

Now f_i: $\mathbb{R}^n \to \mathbb{R}$ and, according to the definition on page 186, f_i is differentiable and $\mathbf{J}_i = \nabla f_i(\mathbf{x})$. This, combined with the fact that

$$\nabla f_i(\mathbf{x}_0) = \left(\frac{\partial f_i}{\partial x_1}(\mathbf{x}_0), \frac{\partial f_i}{\partial x_2}(\mathbf{x}_0), \ldots, \frac{\partial f_i}{\partial x_m}(\mathbf{x}_0) \right),$$

completes the proof. ■

EXAMPLE 1

THE DERIVATIVE OF A MAPPING FROM $\mathbb{R}^3$ TO $\mathbb{R}^2$

Let $\mathbf{f}$: $\mathbb{R}^3 \to \mathbb{R}^2$ be defined by $\mathbf{f}(x, y, z) = (e^{x+y^2+z^3}, xy/z)$. Determine $J_\mathbf{f}(1, -1, 2)$.

SOLUTION:

$$J_\mathbf{f}(\mathbf{x}) = \begin{pmatrix} \frac{\partial f_1}{\partial x}(\mathbf{x}) & \frac{\partial f_1}{\partial y}(\mathbf{x}) & \frac{\partial f_1}{\partial z}(\mathbf{x}) \\ \frac{\partial f_2}{\partial x}(\mathbf{x}) & \frac{\partial f_2}{\partial y}(\mathbf{x}) & \frac{\partial f_2}{\partial z}(\mathbf{x}) \end{pmatrix} = \begin{pmatrix} e^{x+y^2+z^3} & 2ye^{x+y^2+z^3} & 3z^2e^{x+y^2+z^3} \\ \frac{y}{z} & \frac{x}{z} & \frac{-xy}{z^2} \end{pmatrix}$$

Thus

$$J_\mathbf{f}(1, -1, 2) = \begin{pmatrix} e^{10} & -2e^{10} & 12e^{10} \\ -\frac{1}{2} & \frac{1}{2} & \frac{1}{4} \end{pmatrix}.$$

EXAMPLE 2

THE JACOBIAN MATRIX FOR THE POLAR COORDINATE TRANSFORMATION

Determine $J_{\mathbf{f}}(2, \pi/6)$ for the polar coordinate transformation $\mathbf{f}(r, \theta) = (r \cos \theta, r \sin \theta)$.

SOLUTION:

$$J_{\mathbf{f}}(r, \theta) = \begin{pmatrix} \frac{\partial}{\partial r} r \cos \theta & \frac{\partial}{\partial \theta} r \cos \theta \\ \frac{\partial}{\partial r} r \sin \theta & \frac{\partial}{\partial \theta} r \sin \theta \end{pmatrix} = \begin{pmatrix} \cos \theta & -r \sin \theta \\ \sin \theta & r \cos \theta \end{pmatrix},$$

so that

$$J_{\mathbf{f}}\left(2, \frac{\pi}{6}\right) = \begin{pmatrix} \frac{\sqrt{3}}{2} & -1 \\ \frac{1}{2} & \sqrt{3} \end{pmatrix}.$$

DEFINITION TOTAL DERIVATIVE

Let $\mathbf{f}: \mathbb{R}^n \to \mathbb{R}^m$ be differentiable at $\mathbf{x}_0$. Then the Jacobian matrix $J_{\mathbf{f}}(\mathbf{x}_0)$ is called the **total derivative** of $\mathbf{f}$ at $\mathbf{x}_0$.

REMARK: Because of the definition above, $J_{\mathbf{f}}(\mathbf{x})$ is often written as $\mathbf{f}'(\mathbf{x}_0)$. Thus, for example, in Example 1

$$\mathbf{f}'(1, -1, 2) = \begin{pmatrix} e^{10} & -2e^{10} & 12e^{10} \\ -\frac{1}{2} & \frac{1}{2} & \frac{1}{4} \end{pmatrix}$$

and, in Example 2,

$$\mathbf{f}'\left(2, \frac{\pi}{6}\right) = \begin{pmatrix} \frac{\sqrt{3}}{2} & -1 \\ \frac{1}{2} & \sqrt{3} \end{pmatrix}.$$

■

We have seen how to compute the Jacobian matrix or total derivative of $\mathbf{f}$ if $\mathbf{f}$ is differentiable—but how do we know when $\mathbf{f}$ is differentiable? In Theorem 3.5.4 we pointed out that if $\mathbf{f}: \mathbb{R}^n \to \mathbb{R}$ has continuous first-order partial derivatives at $\mathbf{x}$, then f is differentiable at $\mathbf{x}$. In light of this, the answer to the question posed above is easy to obtain.

THEOREM 2

Let $f: \mathbb{R}^n \to \mathbb{R}^m$ with component functions $f_1, f_2, \ldots, f_m$. Suppose that all partial derivatives $\partial f_i/\partial x_j$ for $i = 1, 2, \ldots, m$ and $j = 1, 2, \ldots, n$ exist and are continuous at $\mathbf{x}_0$. Then $\mathbf{f}$ is differentiable at $\mathbf{x}_0$.

PROOF: By Theorem 3.5.4 each component function f_i is differentiable at $\mathbf{x}_0$ and

$$f_i(\mathbf{x}) - f_i(\mathbf{x}_0) = \nabla f_i(\mathbf{x}_0) \cdot (\mathbf{x} - \mathbf{x}_0) + g_i(\mathbf{x} - \mathbf{x}_0) \quad (9)$$

where g_i: $\mathbb{R}^n \to \mathbb{R}$ and

$$\lim_{|\mathbf{x}-\mathbf{x}_0|\to 0} \frac{g_i(\mathbf{x}-\mathbf{x}_0)}{|\mathbf{x}-\mathbf{x}_0|} = 0.$$

Thus, if we define $\mathbf{g}$: $\mathbb{R}^n \to \mathbb{R}^m$ by

$$\mathbf{g}(\mathbf{x}) = \begin{pmatrix} g_1(\mathbf{x}) \\ g_2(\mathbf{x}) \\ \vdots \\ g_m(\mathbf{x}) \end{pmatrix},$$

then

$$\lim_{|\mathbf{x}-\mathbf{x}_0|\to 0} \frac{\mathbf{g}(\mathbf{x}-\mathbf{x}_0)}{|\mathbf{x}-\mathbf{x}_0|} = \mathbf{0} \tag{10}$$

and, using (9), we have

$$\begin{aligned} \mathbf{f}(\mathbf{x}) - \mathbf{f}(\mathbf{x}_0) &= \begin{pmatrix} \nabla f_1(x_0) \cdot (\mathbf{x}-\mathbf{x}_0) \\ \nabla f_2(\mathbf{x}_0) \cdot (\mathbf{x}-\mathbf{x}_0) \\ \vdots \\ \nabla f_m(\mathbf{x}_0) \cdot (\mathbf{x}-\mathbf{x}_0) \end{pmatrix} + \mathbf{g}(\mathbf{x}-\mathbf{x}_0) \\ &= J_{\mathbf{f}}(\mathbf{x}_0)(\mathbf{x}-\mathbf{x}_0) + \mathbf{g}(\mathbf{x}-\mathbf{x}_0), \end{aligned}$$

which, in light of the definition of differentiability and equation (10), completes the proof. ■

We now observe that in yet another case differentiable functions are continuous.

THEOREM 3

Let $\mathbf{f}$: $\mathbb{R}^n \to \mathbb{R}^m$ be differentiable at $\mathbf{x}_0$. Then $\mathbf{f}$ is continuous at $\mathbf{x}_0$.

PROOF: This follows immediately by applying Theorem 3.5.4 to the component functions of $\mathbf{f}$. ■

Since

$$\frac{\partial}{\partial x}(f+g) = \frac{\partial f}{\partial x} + \frac{\partial g}{\partial x} \quad \text{and} \quad \frac{\partial}{\partial x}(\alpha f) = \alpha\frac{\partial}{\partial x}$$

for any scalar α, we can easily prove the following theorem:

THEOREM 4

Let $\mathbf{f}$: $\mathbb{R}^n \to \mathbb{R}^m$ and $\mathbf{g}$: $\mathbb{R}^n \to \mathbb{R}^m$ be differentiable at $\mathbf{x}_0$ and let α be a scalar. Then $\alpha\mathbf{f}$ and $\mathbf{f}+\mathbf{g}$ are differentiable at $\mathbf{x}_0$ and

i. $J_{\alpha\mathbf{f}}(\mathbf{x}_0) = \alpha J_{\mathbf{f}}(\mathbf{x}_0)$ (11)

ii. $J_{\mathbf{f}+\mathbf{g}}(\mathbf{x}_0) = J_{\mathbf{f}}(\mathbf{x}_0) + J_{\mathbf{g}}(\mathbf{x}_0)$ (12)

or, alternatively,

i. $(\alpha \mathbf{f})'(\mathbf{x}_0) = \alpha \mathbf{f}'(\mathbf{x}_0)$ (13)

ii. $(\mathbf{f} + \mathbf{g})'(\mathbf{x}_0) = \mathbf{f}'(\mathbf{x}_0) + \mathbf{g}'(\mathbf{x}_0)$. (14)

■

REMARK: Theorem 4 tells us that the space of differentiable functions from $\mathbb{R}^n$ to $\mathbb{R}^m$ forms a vector space and that the **differentiable operator $\mathbf{f} \rightarrow \mathbf{f}'(\mathbf{x}_0)$** is a linear transformation from this vector space to the mn-dimensional vector space of $m \times n$ matrices.

We close this section by describing the chain rule for functions from $\mathbb{R}^n$ to $\mathbb{R}^m$. This result generalizes Theorem 3.6.3 for functions from $\mathbb{R}^n$ to $\mathbb{R}$, which in turn generalizes the chain rule for scalar functions of one variable.

THEOREM 5 CHAIN RULE

Let $\mathbf{f}$: $\mathbb{R}^n \rightarrow \mathbb{R}^m$ be differentiable at $\mathbf{x}_0$ and let $\mathbf{g}$: $\mathbb{R}^m \rightarrow \mathbb{R}^q$ be differentiable at $\mathbf{f}(\mathbf{x}_0)$. Suppose further that all the component functions of $\mathbf{f}$ and $\mathbf{g}$ have continuous partial derivatives. Then the composite function $\mathbf{g} \circ \mathbf{f}$ is differentiable at $\mathbf{x}_0$ and

$$J_{\mathbf{g}\circ\mathbf{f}}(\mathbf{x}_0) = J_{\mathbf{g}}(\mathbf{f}(\mathbf{x}_0))J_{\mathbf{f}}(\mathbf{x}_0). \tag{15}$$

PROOF: Let

$$\mathbf{h}(\mathbf{x}) = \mathbf{g} \circ \mathbf{f}(\mathbf{x}) = \mathbf{g}(\mathbf{f}(\mathbf{x})) = (g_1(\mathbf{f}(\mathbf{x})), g_2(\mathbf{f}(\mathbf{x})), \ldots, g_q(\mathbf{f}(\mathbf{x}))).$$

Then $\mathbf{h}$: $\mathbb{R}^n \rightarrow \mathbb{R}^q$ and the ith component function h_i: $\mathbb{R}^n \rightarrow \mathbb{R}$ of $\mathbf{h}$ is given by

$$h_i(\mathbf{x}_0) = g_i(\mathbf{f}(\mathbf{x}_0)) = (g_i \circ \mathbf{f})(\mathbf{x}_0). \tag{16}$$

Before continuing, let's take another look at the chain rule proved in Theorem 3.6.3: If f: $\mathbb{R}^n \rightarrow \mathbb{R}$ and $\mathbf{x}$: $\mathbb{R} \rightarrow \mathbb{R}^n$, then

$$\frac{d}{dt} f(\mathbf{x}(t)) = \nabla f(\mathbf{x}) \cdot \mathbf{x}'(t), \tag{17}$$

assuming that f is differentiable at $\mathbf{x}$ and that $\mathbf{x}$ is differentiable at t. Now, in order to compute a partial derivative of a function of n variables, we treat all but one of the variables as if they were constant. Thus we can use the chain rule (17) to compute partial derivatives of the functions h_i given by (16). Letting the jth component x_j of $\mathbf{x}$ play the role of t in (17), and letting $\mathbf{f}(\mathbf{x})$ play the role of $\mathbf{x}$, we have $(\partial \mathbf{f}/\partial x_j)(\mathbf{x})$ instead of $\mathbf{x}'(t)$ and

$$\frac{\partial h_i}{\partial x_j}(\mathbf{x}) = \nabla g_i(\mathbf{f}(\mathbf{x})) \cdot \frac{\partial \mathbf{f}}{\partial x_j}(\mathbf{x}). \tag{18}$$

Note that since $\mathbf{g}$ is differentiable at $\mathbf{f}(\mathbf{x}_0)$, each of its components is differentiable at $\mathbf{f}(\mathbf{x}_0)$. Also

$$\frac{\partial \mathbf{f}}{\partial x_j}(\mathbf{x}_0) = \begin{pmatrix} \frac{\partial f_1}{\partial x_j}(\mathbf{x}_0) \\ \frac{\partial f_2}{\partial x_j}(\mathbf{x}_0) \\ \vdots \\ \frac{\partial f_m}{\partial x_j}(\mathbf{x}_0) \end{pmatrix}$$

exists and is continuous by assumption, so that Theorem 3.6.3 can be applied in this setting.

Now g_i: $\mathbb{R}^m \to \mathbb{R}$ and all the components of the m-vector ∇g_i are the partial derivatives of g_i, and these are continuous at $\mathbf{f}(\mathbf{x}_0)$ by assumption. But since $\mathbf{f}$ is continuous at $\mathbf{x}$, $\nabla g_i(\mathbf{f}(\mathbf{x}_0))$ is continuous at $\mathbf{x}_0$ by Theorem 3.2.2(v). This means that the scalar product in (18) consists of the sum of products of functions that are continuous at $\mathbf{x}_0$, which implies that $\partial h_i/\partial x_j$ is continuous at $\mathbf{x}_0$. Thus, since $\mathbf{h} = \mathbf{g} \circ \mathbf{f}$, we see that all the first-order partial derivatives of the component functions of $\mathbf{g} \circ \mathbf{f}$ are continuous at $\mathbf{x}_0$, so that by Theorem 2, $\mathbf{g} \circ \mathbf{f}$ is differentiable at $\mathbf{x}_0$ and the Jacobian matrix $J_{\mathbf{g}\circ\mathbf{f}}(\mathbf{x}_0)$ exists. Moreover, the ijth component of $J_{\mathbf{g}\circ\mathbf{f}}(\mathbf{x}_0)$ is

$$\nabla g_i(\mathbf{f}(\mathbf{x}_0)) \cdot \frac{\partial \mathbf{f}}{\partial x_j}(\mathbf{x}). \tag{19}$$

Now the ijth component of $J_{\mathbf{g}}(\mathbf{f}(\mathbf{x}_0))J_{\mathbf{f}}(\mathbf{x}_0)$ is the scalar product of the ith row of $J_{\mathbf{g}}(\mathbf{f}(\mathbf{x}_0))$ with the jth column of $J_{\mathbf{f}}(\mathbf{x}_0)$. But the

$$^i\text{th row of } J_{\mathbf{g}}(\mathbf{f}(\mathbf{x}_0)) = \nabla g_i(\mathbf{f}(\mathbf{x}_0))$$

and the

$$j\text{th column of } J_{\mathbf{f}}(\mathbf{x}_0) = \begin{pmatrix} \frac{\partial f_1}{\partial x_j}(\mathbf{x}_0) \\ \frac{\partial f_2}{\partial x_j}(\mathbf{x}_0) \\ \vdots \\ \frac{\partial f_m}{\partial x_j}(\mathbf{x}_0) \end{pmatrix} = \frac{\partial \mathbf{f}}{\partial x_j}(\mathbf{x}_0).$$

Glancing at equation (19), we observe that the ijth component of $J_{\mathbf{g}\circ\mathbf{f}}(\mathbf{x}_0)$ is equal to the ijth component of $J_{\mathbf{g}}(\mathbf{f}(\mathbf{x}_0))J_{\mathbf{f}}(\mathbf{x}_0)$. Since two matrices are equal if and only if their corresponding components are equal, we see that $J_{\mathbf{g}\circ\mathbf{f}}(\mathbf{x}_0) = J_{\mathbf{g}}(\mathbf{f}(\mathbf{x}_0))J_{\mathbf{f}}(\mathbf{x}_0)$ and the proof is complete. ∎

EXAMPLE 3 VERIFYING THE CHAIN RULE

Verify the chain rule at the point $\mathbf{x}_0 = (4, 1)$ for the mappings $\mathbf{f}(x, y) = (xy, x + 2y, x/y)$ and $\mathbf{g}(x, y, z) = (x^2 + z^2, xyz, yz/x, 2x + 3y + 4z)$.

SOLUTION: We have

$$J_{\mathbf{f}}(x, y) = \begin{pmatrix} y & x \\ 1 & 2 \\ \frac{1}{y} & -\frac{x}{y^2} \end{pmatrix} \qquad J_{\mathbf{f}}(4, 1) = \begin{pmatrix} 1 & 4 \\ 1 & 2 \\ 1 & -4 \end{pmatrix}$$

and $\mathbf{f}(4, 1) = (4, 6, 4)$. Similarly,

$$J_{\mathbf{g}}(x, y, z) = \begin{pmatrix} 2x & 0 & 2z \\ yz & xz & xy \\ -\frac{yz}{x^2} & \frac{z}{x} & \frac{y}{x} \\ 2 & 3 & 4 \end{pmatrix} \quad \text{and} \quad J_{\mathbf{g}}(4, 6, 4) = \begin{pmatrix} 8 & 0 & 8 \\ 24 & 16 & 24 \\ -\frac{3}{2} & 1 & \frac{3}{2} \\ 2 & 3 & 4 \end{pmatrix}$$

so that

$$J_{\mathbf{g}}(\mathbf{f}(\mathbf{x}_0))J_{\mathbf{f}}(\mathbf{x}_0) = \begin{pmatrix} 8 & 0 & 8 \\ 24 & 16 & 24 \\ -\frac{3}{2} & 1 & \frac{3}{2} \\ 2 & 3 & 4 \end{pmatrix} \begin{pmatrix} 1 & 4 \\ 1 & 2 \\ 1 & -4 \end{pmatrix} = \begin{pmatrix} 16 & 0 \\ 64 & 32 \\ 1 & -10 \\ 9 & -2 \end{pmatrix}.$$

Now $\mathbf{g} \circ \mathbf{f}: \mathbb{R}^2 \to \mathbb{R}^4$ is given by

$$\begin{aligned} \mathbf{g} \circ \mathbf{f}(x, y) &= \left(x^2y^2 + \frac{x^2}{y^2}, (xy)(x + 2y)\left(\frac{x}{y}\right), \frac{(x + 2y)(x/y)}{xy}, 2xy + 3(x + 2y) + 4\frac{x}{y}\right) \\ &= \left(x^2y^2 + \frac{x^2}{y^2}, x^3 + 2x^2y, \frac{x}{y^2} + \frac{2}{y}, 3x + 6y + 2xy + \frac{4x}{y}\right), \end{aligned}$$

so that

$$J_{\mathbf{g}\circ\mathbf{f}}(x, y) = \begin{pmatrix} 2xy^2 + \frac{2x}{y^2} & 2x^2y - \frac{2x^2}{y^3} \\ 3x^2 + 4xy & 2x^2 \\ \frac{1}{y^2} & -\frac{2x}{y^3} - \frac{2}{y^2} \\ 3 + 2y + \frac{4}{y} & 6 + 2x - \frac{4x}{y^2} \end{pmatrix}$$

and

$$J_{\mathbf{g}\circ\mathbf{f}}(\mathbf{x}_0) = J_{\mathbf{g}\circ\mathbf{f}}(4, 1) = \begin{pmatrix} 16 & 0 \\ 64 & 32 \\ 1 & -10 \\ 9 & -2 \end{pmatrix}.$$

Thus

$$J_{\mathbf{g}\circ\mathbf{f}}(\mathbf{x}_0) = J_{\mathbf{g}}(f(\mathbf{x}_0))J_{\mathbf{f}}(\mathbf{x}_0).$$

Note that in this example, as in most others, it is easier to compute the derivative of the composite function by writing two derivatives and using the chain rule rather than by finding the composite function explicitly and then computing its derivative directly.

PROBLEMS 9.4

SELF-QUIZ

True–False

I. If $\mathbf{f}: \mathbb{R}^n \to \mathbb{R}^m$ is differentiable, then the Jacobian matrix $J_\mathbf{f}(\mathbf{x})$ is an $n \times m$ matrix.

II. If $\mathbf{f}: \mathbb{R}^n \to \mathbb{R}^m$ is differentiable, then the Jacobian matrix is an $m \times n$ matrix.

III. If $\mathbf{f}$ is a mapping from $\mathbb{R}^n$ to $\mathbb{R}^m$ such that $\partial f_i/\partial x_j$ exists at $\mathbf{x}_0$ for $i = 1, 2, \ldots, m$ and $j = 1, 2, \ldots, n$, then $\mathbf{f}$ is differentiable at $\mathbf{x}_0$.

IV. If $\mathbf{f}$ is a mapping from $\mathbb{R}^n$ to $\mathbb{R}^m$ such that $\partial f_i/\partial x_j$ exists and is continuous at $\mathbf{x}_0$ for $i = 1, 2, \ldots, m$ and $j = 1, 2, \ldots, n$, then $\mathbf{f}$ is differentiable at $\mathbf{x}_0$.

Multiple Choice

V. Let $\mathbf{f}: \mathbb{R}^2 \to \mathbb{R}^2$ be given by $\mathbf{f}(x, y) = (x^2 + 3y, e^{x-y})$. Then $J_\mathbf{f}(1, 1) =$ ________.

a. $\begin{pmatrix} 2x & e^{x-y} \\ 3 & -e^{x-y} \end{pmatrix}$ **b.** $\begin{pmatrix} 2x & 3 \\ e^{x-y} & -e^{x-y} \end{pmatrix}$

c. $\begin{pmatrix} 2 & 1 \\ 3 & -1 \end{pmatrix}$ **d.** $\begin{pmatrix} 2 & 3 \\ 1 & -1 \end{pmatrix}$

e. $\begin{pmatrix} 2 & 0 \\ 3 & -1 \end{pmatrix}$

In Problems 1–10 compute the derivative $\mathbf{f}'(\mathbf{x}_0) = J_\mathbf{f}(\mathbf{x}_0)$ of the given function at the point $\mathbf{x}_0$.

1. $\mathbf{f}(x, y) = (x^2, y^2)$; $(1, 3)$

2. $\mathbf{f}(x, y) = (x^2 + y^2, x^2 - y^2, \sqrt{x^2 + y^2})$; $(3, 4)$

3. $\mathbf{f}(x, y, z) = \left(\dfrac{x}{y}, \dfrac{y}{z}, \dfrac{z}{x}\right)$; $(1, 4, 8)$

4. $\mathbf{f}(x, y, z) = (e^{x+y}, \ln(x + y + z))$; $(0, 1, 0)$

5. $\mathbf{f}(x, y, z) = (x^2, y^2, z^2, \sqrt{36 - 9x^2 - y^2 - z^2})$; $(-1, 5, 1)$

6. $f(x_1, x_2, x_3, x_4, x_5) = (x_1x_2, x_2x_3, x_3x_4, x_4x_5, x_1x_5)$; $(1, 2, 3, 4, 5)$

7. $f(x_1, x_2, x_3, x_4) = \left(\dfrac{1}{x_1 + x_2 + 3x_3 + x_4 - 5}, x_1 + x_2 + x_3 + x_4 - 5\right)$; $(1, -1, 0, 2)$

8. $\mathbf{f}(x_1, x_2, x_3, x_4) = \left(\sqrt{x_1{}^2 + x_2{}^2 + x_3{}^2 + x_4{}^2}, \dfrac{2x_1 + x_4}{3x_2x_3}\right)$; $(1, 2, -1, 0)$

9. $\mathbf{f}(x_1x_2, x_3, x_4) = (\sin x_1x_4, \cos x_3x_4, e^{x_1{}^2 - x_2{}^2})$; $\left(1, 2, 3, \dfrac{\pi}{6}\right)$

10. $\mathbf{f}(x_1, x_2, \ldots, x_n) = (x_1x_2 \cdots x_n, x_1 + x_2 + \cdots + x_n)$

In Problems 11–17 find the derivative of the composite function $\mathbf{g} \circ \mathbf{f}$ at the given point $\mathbf{x}_0$ and verify the chain rule at that point.

11. $\mathbf{f}(x, y) = (x^2, y^2)$; $\mathbf{g}(x, y) = (x + y, x - y)$; $(3, 2)$

12. $\mathbf{f}(x, y) = (x + 2y, 3 - 4x)$; $\mathbf{g}(x, y) = \left(\dfrac{x}{y}, \dfrac{y}{x}\right)$; $(-1, 1)$

13. $\mathbf{f}(x, y) = (xy, x^2y, xy^2)$; $\mathbf{g}(x, y, z) = (x^3, y^2)$; $(2, -3)$

14. $\mathbf{f}(x, y, z) = \left(xyz, \dfrac{yx}{z}\right)$; $\mathbf{g}(x, y) = (x, 2x^2y, -3y, xy^2)$; $(1, -1, 2)$

15. $\mathbf{f}(x, y, z) = \left(x + y, x + z, xyz, \dfrac{x^2y}{z^3}\right)$;

$\mathbf{g}(x_1, x_2, x_3, x_4) = \left(\dfrac{x_1x_2}{x_3x_4}, x_1x_2x_3x_4, 3x_1 - 5x_4\right)$; $(2, 3, -1)$

16. $\mathbf{f}(x, y, z) = \left(xy - z^2, 3y - 2z, \dfrac{4z}{x}\right)$;

$\mathbf{g}(x, y, z) = \left(2x - 3y, xy^2, \dfrac{x - y}{y + z}\right)$; $(1, 3, -2)$

17. $\mathbf{f}(x_1, x_2, \ldots, x_n) = (x_n, x_{n-1}, x_{n-2}, \ldots, x_2, x_1)$; $\mathbf{g}(x_1, x_2, \ldots, x_n) = (x_1{}^2, x_2{}^2, \ldots, x_n{}^2)$; $(1, 2, 3, \ldots, n)$

18. Let $\mathbf{f}: \mathbb{R}^n \to \mathbb{R}^m$ be linear. Show that $\mathbf{f}$ is differentiable at every $\mathbf{x}_0$ in $\mathbb{R}^n$.

19. Let $f: \mathbb{R}^n \to \mathbb{R}$ be differentiable. Show that $\nabla f(\mathbf{x}_0) = [J_\mathbf{f}(\mathbf{x}_0)]^t$. (See Section 6.9 for a discussion of the transpose of a matrix.)

20. Let $\mathbf{f}(x, y) = (g(x, y), x + y)$ where

$$g(x, y) = \begin{cases} \dfrac{xy}{y^2 - x^2}, & x \neq \pm y, \\ 0, & x = \pm y. \end{cases}$$

a. Show that $J_{\mathbf{f}}(0, 0)$ exists and compute it.
b. Show that $\mathbf{f}$ is not differentiable at $(0, 0)$.
This problem shows that the existence of the Jacobian matrix is not enough to ensure differentiability.

21. Set $\mathbf{f}(x, y) = (x, g(y))$ where

$$g(y) = \begin{cases} y^2 \sin(1/y), & y \neq 0, \\ 0, & y = 0. \end{cases}$$

a. Show that $\mathbf{f}$ is differentiable at $(0, 0)$.
b. Show that not all the partial derivatives of the component functions of $\mathbf{f}$ are continuous. This problem shows that differentiability of $\mathbf{f}$ does not imply that the first-order partial derivatives of $\mathbf{f}$ are continuous.

22. Suppose that $\mathbf{f}$: $\mathbb{R}^3 \to \mathbb{R}^3$ such that $J_{\mathbf{f}}(\mathbf{x})$ is the zero matrix for every $\mathbf{x} \in \mathbb{R}^3$. Show that $\mathbf{f}(\mathbf{x}) = \mathbf{v}$, a constant vector.

23. Let $\mathbf{f}$: $\mathbb{R}^3 \to \mathbb{R}^3$ and $\mathbf{g}$: $\mathbb{R}^3 \to \mathbb{R}^3$ satisfy $J_{\mathbf{f}}(\mathbf{x}) = J_{\mathbf{g}}(\mathbf{x})$ for every $\mathbf{x}$ in $\mathbb{R}^3$. Show that $\mathbf{f}$ and $\mathbf{g}$ differ by a constant vector.

24. Let $\mathbf{f}$: $\mathbb{R}^n \to \mathbb{R}^n$ satisfy $\mathbf{f}(\mathbf{0}) = \mathbf{0}$ and $J_{\mathbf{f}}(\mathbf{x}) = A$, a constant matrix. Show that $\mathbf{f}$ is a linear mapping.

25. Prove Theorem 4.

ANSWERS TO SELF-QUIZ

I. False **II.** True **III.** False **IV.** True **V.** d

9.5 INVERSE FUNCTIONS AND THE IMPLICIT FUNCTION THEOREM: II

In this section we extend the results of Section 9.2 to mappings from $\mathbb{R}^n$ to $\mathbb{R}^m$. Our first result will be a theorem about the existence of inverse functions. To state such a theorem, we need to have a condition similar to the condition $f'(x_0) \neq 0$ in Theorem 9.2.1.

DEFINITION JACOBIAN

Let $\mathbf{f}$: $\mathbb{R}^n \to \mathbb{R}^n$ be differentiable at a point $\mathbf{x}_0$ in $\mathbb{R}^n$. Then the **Jacobian** of the mapping is the determinant of the Jacobian matrix $J_{\mathbf{f}}(\mathbf{x}_0)$. That is,

$$\text{Jacobian of } \mathbf{f} = \det J_{\mathbf{f}}(\mathbf{x}_0) = \begin{vmatrix} \frac{\partial f_1}{\partial x_1}(\mathbf{x}_0) & \frac{\partial f_1}{\partial x_2}(\mathbf{x}_0) & \cdots & \frac{\partial f_1}{\partial x_n}(\mathbf{x}_0) \\ \frac{\partial f_2}{\partial x_1}(\mathbf{x}_0) & \frac{\partial f_2}{\partial x_2}(\mathbf{x}_0) & \cdots & \frac{\partial f_2}{\partial x_n}(\mathbf{x}_0) \\ \vdots & \vdots & & \vdots \\ \frac{\partial f_n}{\partial x_1}(\mathbf{x}_0) & \frac{\partial f_n}{\partial x_2}(\mathbf{x}_0) & \cdots & \frac{\partial f_n}{\partial x_n}(\mathbf{x}_0) \end{vmatrix} \quad (1)$$

NOTATION: We will denote the Jacobian of $\mathbf{f}$ at $\mathbf{x}_0$ by $|J_{\mathbf{f}}(\mathbf{x}_0)|$.

REMARK 1: The Jacobian of a mapping $\mathbf{f}$: $\mathbb{R}^n \to \mathbb{R}^m$ is defined only when $n = m$, since only square matrices have determinants.

REMARK 2: This definition generalizes the definitions in Section 5.10 (see pages 357 and 359). ■

EXAMPLE 1

THE JACOBIAN OF THE POLAR COORDINATE TRANSFORMATION

Let $\mathbf{f}$: $\mathbb{R}^2 \to \mathbb{R}^2$ be the polar coordinate transformation $f(r, \theta) = (r\cos\theta, r\sin\theta)$. Compute $|J_{\mathbf{f}}(r, \theta)|$.

SOLUTION: From Example 9.4.2 we have

$$J_{\mathbf{f}}(r, \theta) = \begin{pmatrix} \cos\theta & -r\sin\theta \\ \sin\theta & r\cos\theta \end{pmatrix},$$

so

$$|J_{\mathbf{f}}(r, \theta)| = \cos\theta(r\cos\theta) + (r\sin\theta)\sin\theta = r(\cos^2\theta + \sin^2\theta) = r.$$

EXAMPLE 2 THE JACOBIAN OF A MAPPING FROM $\mathbb{R}^3$ TO $\mathbb{R}^3$

Let $\mathbf{f}$: $\mathbb{R}^3 \to \mathbb{R}^3$ be defined by

$$\mathbf{f}(x, y, z) = \left(xy - z, \frac{x^2y}{z}, \frac{xz}{y^2}\right).$$

Compute $|J_{\mathbf{f}}(2, 1, -1)|$.

SOLUTION:

$$J_{\mathbf{f}}(x, y, z) = \begin{pmatrix} y & x & -1 \\ \frac{2xy}{z} & \frac{x^2}{z} & -\frac{x^2y}{z^2} \\ \frac{z}{y^2} & -\frac{2xz}{y^3} & \frac{x}{y^2} \end{pmatrix}$$

So that

$$J_{\mathbf{f}}(2, -1, 1) = \begin{pmatrix} -1 & 2 & -1 \\ -4 & 4 & 4 \\ 1 & 4 & 2 \end{pmatrix}$$

and

$$|J_{\mathbf{f}}(2, -1, 1)| = 52.$$

DEFINITION LOCAL C^1-INVERTIBILITY

Let $\mathbf{f}$: $\mathbb{R}^n \to \mathbb{R}^n$ be in $C^1(\Omega)$ for some open set Ω in $\mathbb{R}^n$. Then $\mathbf{f}$ is said to be **locally C^1-invertible** on Ω if there exists a function $\mathbf{g}$: $\mathbb{R}^n \to \mathbb{R}^n$ which is in $C^1(\mathbf{f}(\Omega))$ such that

$$(\mathbf{g} \circ \mathbf{f})(\mathbf{x}) = \mathbf{x} \quad \text{for every } \mathbf{x} \text{ in } \Omega \tag{2}$$

and

$$(\mathbf{f} \circ \mathbf{g})(\mathbf{y}) = \mathbf{y} \quad \text{for every } \mathbf{y} \text{ in } \mathbf{f}(\Omega). \tag{3}$$

In this case we say that $\mathbf{g}$ is an **inverse function** for $\mathbf{f}$ and write $\mathbf{g} = \mathbf{f}^{-1}$. ■

REMARK: The word "locally" in the definition refers to the region Ω. There are functions that are locally invertible in a neighborhood of every point in $\mathbb{R}^n$ but that are not invertible on all of $\mathbb{R}^n$ (see Example 3 and Problem 27).

EXAMPLE 3

LOCAL BUT NOT GLOBAL INVERTIBILITY

Let $\mathbf{f}$: $\mathbb{R}^2 \to \mathbb{R}^2$ be the polar coordinate mapping $\mathbf{f}(x, y) = (r\cos\theta, r\sin\theta)$. Then $\mathbf{f}$ has continuous first-order partial derivatives of all orders. Let

$$\Omega = \{(r, \theta)\colon 0 < r < 1, 0 < \theta < \pi\}. \tag{4}$$

Then, as we saw in Example 9.3.2 $\mathbf{f}$ maps the open rectangle Ω into the open semicircle $\mathbf{f}(\Omega) = \{(x, y)\colon 0 < x^2 + y^2 < 1, y > 0\}$ (see Figure 9.3.1). We now define

$$\mathbf{g}(x, y) = \left(\sqrt{x^2 + y^2}, \cos^{-1}\frac{x}{\sqrt{x^2 + y^2}}\right).$$

Then, if $(r, \theta) \in \Omega$,

$$(\mathbf{g}\circ\mathbf{f})(r, \theta) = \mathbf{g}(r\cos\theta, r\sin\theta)$$

$$= \left(\sqrt{r^2\cos^2\theta + r^2\sin^2\theta}, \cos^{-1}\frac{r\cos\theta}{\sqrt{r^2\cos^2\theta + r^2\sin^2\theta}}\right) = (r, \theta)^\dagger$$

and if (x, y) is in $\mathbf{f}(\Omega)$, we have

$$(\mathbf{f}\circ\mathbf{g})(x, y) = \mathbf{f}\left(\sqrt{x^2 + y^2}, \cos^{-1}\frac{x}{\sqrt{x^2 + y^2}}\right)$$

$$= \left(\sqrt{x^2 + y^2}\cos\left(\cos^{-1}\frac{x}{\sqrt{x^2 + y^2}}\right), \sqrt{x^2 + y^2}\sin\left(\cos^{-1}\frac{x}{\sqrt{x^2 + y^2}}\right)\right).$$

Now, from the triangle in Figure 1,

$$\sin\left(\cos^{-1}\frac{x}{\sqrt{x^2 + y^2}}\right) = \frac{y}{\sqrt{x^2 + y^2}}.$$

So that

$$(\mathbf{f}\circ\mathbf{g})(x, y) = (x, y)$$

and $\mathbf{g}$ is an inverse function for $\mathbf{f}$. Note that $\mathbf{g}$ is an inverse function for $\mathbf{f}$ on Ω, not on all of $\mathbb{R}^2$. For example,

$$(\mathbf{g}\circ\mathbf{f})\left(1, \frac{5\pi}{2}\right) = \mathbf{g}\left(\mathbf{f}\left(1, \frac{5\pi}{2}\right)\right) = \mathbf{g}(0, 1) = \left(1, \frac{\pi}{2}\right) \neq \left(1, \frac{5\pi}{2}\right),$$

so that $\mathbf{g}$ is *not* an inverse function for $\mathbf{f}$ everywhere in the plane. This illustrates the need for the word "local" in the definition of Jacobian.

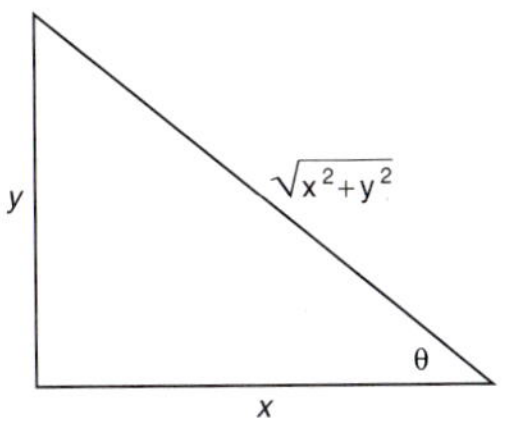

FIGURE 1
In this triangle $\cos\theta = \dfrac{x}{\sqrt{x^2 + y^2}}$

†Note that $\sqrt{r^2} = r$ because $r > 0$ and $\cos^{-1}(\cos\theta) = \theta$ because $0 < \theta < \pi$.

EXAMPLE 4 A LINEAR TRANSFORMATION IS INVERTIBLE IF ITS MATRIX REPRESENTATION IS INVERTIBLE

Let $\mathbf{f}$: $\mathbb{R}^n \to \mathbb{R}^n$ be a linear transformation. Then there is an $n \times n$ matrix A such that

$$\mathbf{f}(\mathbf{x}) = A\mathbf{x} \tag{5}$$

where $\mathbf{x}$ is written as a column vector. Now

$$f(\mathbf{x}) - f(\mathbf{x}_0) = A(\mathbf{x} - \mathbf{x}_0) + 0,$$

so that, by the definition of differentiability, $\mathbf{f}$ is differentiable for every $\mathbf{x}_0$ in $\mathbb{R}^n$ and $\mathbf{J_f}(\mathbf{x}_0)$ is the constant matrix A. Suppose that A is invertible. Then we define a function $\mathbf{g}$: $\mathbb{R}^n \to \mathbb{R}^n$ by

$$\mathbf{g}(\mathbf{y}) = A^{-1}\mathbf{y}.$$

Clearly $\mathbf{g}$ is a linear transformation. Furthermore,

$$(\mathbf{g} \circ \mathbf{f})(\mathbf{x}) = \mathbf{g}(\mathbf{f}(\mathbf{x})) = \mathbf{g}(A\mathbf{x}) = A^{-1}(A\mathbf{x}) = \mathbf{x}$$

and

$$(\mathbf{f} \circ \mathbf{g})(\mathbf{y}) = \mathbf{f}(\mathbf{g}(\mathbf{y})) = \mathbf{f}(A^{-1}\mathbf{y}) = A(A^{-1}\mathbf{y}) = \mathbf{y}.$$

Thus $\mathbf{g} = \mathbf{f}^{-1}$ on all of $\mathbb{R}^n$.

The last result leads to the following theorem.

THEOREM 1 INVERTIBILITY OF A LINEAR TRANSFORMATION

Let $\mathbf{f}$: $\mathbb{R}^n \to \mathbb{R}^n$ be a linear transformation with matrix representation A. Then $\mathbf{f}$ has an inverse on all of $\mathbb{R}^n$ if and only if A is invertible (if and only if $\det A \neq 0$). If A is invertible, then

$$\mathbf{f}^{-1}(\mathbf{x}) = A^{-1}\mathbf{x}. \tag{6}$$

PROOF: We have already proven the theorem in the case in which A is invertible. If A is not invertible, then $\det A = 0$ and the $n \times n$ system of equations $A\mathbf{x} = \mathbf{0}$ has a nontrivial solution $\mathbf{x}_0$. Pick two vectors $\mathbf{y}$ and $\mathbf{z}$ in $\mathbb{R}^n$ such that $\mathbf{y} \neq \mathbf{z}$ and $\mathbf{y} - \mathbf{z} = \mathbf{x}_0$. Then $\mathbf{0} = A\mathbf{x}_0 = A(\mathbf{y} - \mathbf{z}) = A\mathbf{y} - A\mathbf{z}$, so that $A\mathbf{y} = A\mathbf{z}$ and $\mathbf{fy} = \mathbf{fz} = \mathbf{w}$ for some vector $\mathbf{w}$. Thus $\mathbf{f}$ is not one-to-one and $\mathbf{f}^{-1}(\mathbf{w})$ is not defined. This means that $\mathbf{f}$ is not invertible on $\mathbb{R}^n$. ■

As we saw in Theorem 9.2.1, a function f: $\mathbb{R} \to \mathbb{R}$ is invertible in a neighborhood of x_0 if $f'(x_0) \neq 0$. We now generalize that result to nonlinear mappings from $\mathbb{R}^n$ to $\mathbb{R}^n$ (the linear case is dealt with in Theorem 1).

THEOREM 2 INVERSE FUNCTION THEOREM

Let $\mathbf{f}$: $\mathbb{R}^n \to \mathbb{R}^n$ be in $C^1(\Omega)$ on an open set Ω and let $\mathbf{x}_0$ be in Ω. If the Jacobian $|J_\mathbf{f}(\mathbf{x}_0)| \neq 0$, then there is an open set $\Omega_1 \subset \Omega$ such that $\mathbf{x}_0 \in \Omega_1$ and $\mathbf{f}$ is locally C^1-invertible on Ω_1; moreover, if $\mathbf{g} = \mathbf{f}^{-1}$ on $\mathbf{f}(\Omega_1)$, then

$$J_{\mathbf{f}^{-1}}(\mathbf{x}_0) = J_{\mathbf{g}}(\mathbf{f}(\mathbf{x}_0)) = [J_{\mathbf{f}}(\mathbf{x}_0)]^{-1}. \tag{7}$$

PARTIAL PROOF: A proof of the first part of this theorem is best left to an advanced calculus text (see the reference cited in Section 9.2). However, we can justify formula (7); for if $\mathbf{g} = \mathbf{f}^{-1}$, then

$$(\mathbf{g} \circ \mathbf{f})(\mathbf{x}) = \mathbf{x} = I\mathbf{x}$$

where I denotes the identity transformation (see Example 8.9.5 on page 561). Then by the chain rule (Theorem 9.4.5),

$$I = J_I(\mathbf{x}) = J_{\mathbf{g} \circ \mathbf{f}}(\mathbf{x}) = J_{\mathbf{g}}(\mathbf{f}(\mathbf{x}_0))J_{\mathbf{f}}(\mathbf{x}_0). \tag{8}$$

But since $|J_{\mathbf{f}}(\mathbf{x}_0)| \neq 0$, $J_{\mathbf{f}}(\mathbf{x}_0)$ is invertible and (8) shows that $[J_{\mathbf{f}}(\mathbf{x}_0)]^{-1} = J_{\mathbf{g}}(\mathbf{f}(\mathbf{x}_0))$. ■

EXAMPLE 5

INVERTIBILITY OF THE POLAR COORDINATE TRANSFORMATION

Let $\mathbf{f}\colon \mathbb{R}^2 \to \mathbb{R}^2$ be the polar coordinate transformation. Then, as we saw in Example 1, $|J_{\mathbf{f}}(r, \theta)| = r \neq 0$ if $r \neq 0$. Thus, if $\mathbf{g} = \mathbf{f}^{-1}$, we can compute

$$J_{\mathbf{g}}(\mathbf{f}(r, \theta)) = [J_{\mathbf{f}}(r, \theta)]^{-1} = \begin{pmatrix} \cos\theta & -r\sin\theta \\ \sin\theta & r\cos\theta \end{pmatrix}^{-1} = \begin{pmatrix} \cos\theta & \sin\theta \\ -\dfrac{1}{r}\sin\theta & \dfrac{\cos\theta}{r} \end{pmatrix}.$$

We can verify this directly. From Example 3 we have

$$\mathbf{g}(\mathbf{f}(r, \theta)) = \mathbf{g}(r\cos\theta, r\sin\theta) = \mathbf{g}(x, y) = \left(\sqrt{x^2 + y^2}, \cos^{-1}\frac{x}{\sqrt{x^2 + y^2}}\right).$$

Since

$$\frac{d}{du}\cos^{-1} u = \frac{-1}{\sqrt{1 - u^2}},$$

$$J_{\mathbf{g}}(x, y) = \begin{pmatrix} \dfrac{x}{\sqrt{x^2 + y^2}} & \dfrac{y}{\sqrt{x^2 + y^2}} \\ \dfrac{-1}{\sqrt{1 - \dfrac{x^2}{x^2 + y^2}}}\left(\dfrac{\sqrt{x^2 + y^2} - \dfrac{x^2}{\sqrt{x^2 + y^2}}}{x^2 + y^2}\right) & \dfrac{-1}{\sqrt{1 - \dfrac{x^2}{x^2 + y^2}}}\left(\dfrac{-xy}{(x^2 + y^2)^{3/2}}\right) \end{pmatrix}$$

(After some algebra)

$$= \begin{pmatrix} \dfrac{x}{\sqrt{x^2 + y^2}} & \dfrac{y}{\sqrt{x^2 + y^2}} \\ -\dfrac{y}{x^2 + y^2} & \dfrac{x}{x^2 + y^2} \end{pmatrix} = \begin{pmatrix} \dfrac{r\cos\theta}{r} & \dfrac{r\sin\theta}{r} \\ -\dfrac{r\sin\theta}{r^2} & \dfrac{r\cos\theta}{r^2} \end{pmatrix} = \begin{pmatrix} \cos\theta & \sin\theta \\ -\dfrac{1}{r}\sin\theta & \dfrac{1}{r}\cos\theta \end{pmatrix}.$$

EXAMPLE 6 LOCAL INVERTIBILITY OF A MAPPING FROM $\mathbb{R}^3$ TO $\mathbb{R}^3$

Let $\mathbf{f}$: $\mathbb{R}^3 \to \mathbb{R}^3$ be defined by

$$\mathbf{f}(x, y, z) = \left(xy - z, \frac{x^2y}{z}, \frac{xz}{y^2}\right).$$

Show that $\mathbf{f}$ is locally C^1-invertible in a neighborhood of $(2, 1, -1)$, and compute $J_{\mathbf{f}^{-1}}(\mathbf{f}(2, 1, -1))$.

SOLUTION: In Example 2, we found that

$$J_{\mathbf{f}}(2, -1, 1) = \begin{pmatrix} -1 & 2 & -1 \\ -4 & 4 & 4 \\ 1 & 4 & 2 \end{pmatrix}$$

and that $|J_{\mathbf{f}}(2, -1, 1)| = 52$. Thus $\mathbf{f}^{-1}$ exists and is C^1-invertible in a neighborhood of $\mathbf{f}(2, -1, 1) = (-3, -4, 2)$ and

$$J_{\mathbf{f}^{-1}}(-3, -4, 2) = \begin{pmatrix} -1 & 2 & -1 \\ -4 & 4 & 4 \\ 1 & 4 & 2 \end{pmatrix}^{-1} = \frac{1}{52}\begin{pmatrix} -8 & -8 & 12 \\ 12 & -1 & 8 \\ -20 & 6 & 4 \end{pmatrix}.$$

We now state a more general form of the implicit function theorem given in Section 9.2. We first need some notation. Let $\mathbf{x}_0 = (x_1^{(0)}, x_2^{(0)}, \ldots, x_n^{(0)})$ be in $\mathbb{R}^n$ and let $\mathbf{y}_0 = (y_1^{(0)}, y_2^{(0)}, \ldots, y_m^{(0)})$ be in $\mathbb{R}^m$. Then if $\mathbf{F}$: $\mathbb{R}^{n+m} \to \mathbb{R}^m$, the notation $\mathbf{F}(\mathbf{x}_0, \mathbf{y}_0)$ stands for

$$\mathbf{F}(x_1^{(0)}, x_2^{(0)}, \ldots, x_n^{(0)}, y_1^{(0)}, y_2^{(0)}, \ldots, y_m^{(0)}),$$

the $m \times m$ matrix $J_{\mathbf{F}}(\mathbf{y}_0)$ is given by

$$J_{\mathbf{F}}(\mathbf{y}_0) = \begin{pmatrix} \frac{\partial F_1}{\partial y_1}(\mathbf{x}_0, \mathbf{y}_0) & \frac{\partial F_1}{\partial y_2}(\mathbf{x}_0, \mathbf{y}_0) & \cdots & \frac{\partial F_1}{\partial y_m}(\mathbf{x}_0, \mathbf{y}_0) \\ \frac{\partial F_2}{\partial y_1}(\mathbf{x}_0, \mathbf{y}_0) & \frac{\partial F_2}{\partial y_2}(\mathbf{x}_0, \mathbf{y}_0) & \cdots & \frac{\partial F_2}{\partial y_m}(\mathbf{x}_0, \mathbf{y}_0) \\ \vdots & \vdots & & \vdots \\ \frac{\partial F_m}{\partial y_1}(\mathbf{x}_0, \mathbf{y}_0) & \frac{\partial F_m}{\partial y_2}(\mathbf{x}_0, \mathbf{y}_0) & \cdots & \frac{\partial F_m}{\partial y_m}(\mathbf{x}_0, \mathbf{y}_0) \end{pmatrix} \quad (9)$$

and the $m \times n$ matrix $J_{\mathbf{F}}(\mathbf{x}_0)$ is given by

$$J_{\mathbf{F}}(\mathbf{x}_0) = \begin{pmatrix} \frac{\partial F_1}{\partial x_1}(\mathbf{x}_0, \mathbf{y}_0) & \frac{\partial F_1}{\partial x_2}(\mathbf{x}_0, \mathbf{y}_0) & \cdots & \frac{\partial F_1}{\partial x_n}(\mathbf{x}_0, \mathbf{y}_0) \\ \frac{\partial F_2}{\partial x_1}(\mathbf{x}_0, \mathbf{y}_0) & \frac{\partial F_2}{\partial x_2}(\mathbf{x}_0, \mathbf{y}_0) & \cdots & \frac{\partial F_2}{\partial x_n}(\mathbf{x}_0, \mathbf{y}_0) \\ \vdots & \vdots & & \vdots \\ \frac{\partial F_m}{\partial x_1}(\mathbf{x}_0, \mathbf{y}_0) & \frac{\partial F_m}{\partial x_2}(\mathbf{x}_0, \mathbf{y}_0) & \cdots & \frac{\partial F_m}{\partial x_n}(\mathbf{x}_0, \mathbf{y}_0) \end{pmatrix} \quad (10)$$

Note that $J_{\mathbf{F}}(\mathbf{x}_0)$ consists of the first n columns of $J_{\mathbf{F}}(\mathbf{x}_0, \mathbf{y}_0)$ and $J_{\mathbf{F}}(\mathbf{y}_0)$ consists of the last m columns of $J_{\mathbf{F}}(\mathbf{x}_0, \mathbf{y}_0)$.

THEOREM 3 IMPLICIT FUNCTION THEOREM[†]

Let the function $\mathbf{F}: \mathbb{R}^{n+m} \to \mathbb{R}^m$ be C^1 in a neighborhood Ω of the vector $(\mathbf{x}_0, \mathbf{y}_0)$ in $\mathbb{R}^{n+m}$ with $\mathbf{F}(\mathbf{x}_0, \mathbf{y}_0) = \mathbf{0}$. If $|J_\mathbf{F}(\mathbf{y}_0)| \neq 0$, then there exists a neighborhood Ω_1 of $\mathbf{x}_0$ in $\mathbb{R}^n$, a neighborhood Ω_2 of $(\mathbf{x}_0, \mathbf{y}_0)$ in $\mathbb{R}^{n+m}$, and a function $\mathbf{f}: \mathbb{R}^n \to \mathbb{R}^m$ in $C^1(\Omega_1)$ such that $\mathbf{y} = \mathbf{f}(\mathbf{x})$ satisfies $\mathbf{F}(\mathbf{x}, \mathbf{f}(\mathbf{x})) = \mathbf{0}$ in Ω_2. Furthermore,

$$J_\mathbf{f}(\mathbf{x}) = -[J_\mathbf{F}(\mathbf{y})]^{-1} J_\mathbf{F}(\mathbf{x}). \tag{11}$$

■

REMARK 1: If we set $m = 1$ in this theorem, then we obtain Theorem 9.2.2. Note that in this case $f: \mathbb{R}^n \to \mathbb{R}$, so $J_\mathbf{f}(\mathbf{x})$ is an n-component column vector, $\mathbf{y}_0 = x_{n+1}$, and $J_\mathbf{F}(\mathbf{y}_0)$ is a 1×1 matrix, so (11) becomes

$$\begin{pmatrix} \dfrac{\partial f}{\partial x_1}(\mathbf{x}) \\ \dfrac{\partial f}{\partial x_2}(\mathbf{x}) \\ \vdots \\ \dfrac{\partial f}{\partial x_n}(\mathbf{x}) \end{pmatrix} = \left[\frac{\partial F}{\partial x_{n+1}}(\mathbf{x}, x_{n+1}) \right]^{-1} \begin{pmatrix} \dfrac{\partial F}{\partial x_1}(\mathbf{x}, x_{n+1}) \\ \dfrac{\partial F}{\partial x_2}(\mathbf{x}, x_{n+1}) \\ \vdots \\ \dfrac{\partial F}{\partial x_n}(\mathbf{x}, x_{n+1}) \end{pmatrix}$$

which is equation (9.2.5).

REMARK 2: We observe that even though it will generally be impossible to find the implicitly given function **f**, we can use equation (11) to compute its derivative. This will often be all that we need.

REMARK 3: Formula (11) should not be surprising. To see why, let us treat **x** and $\mathbf{y} = f(\mathbf{x})$ as real variables. Then, if we differentiate $F(x, y) = F(x, f(x)) = 0$ and use the chain rule, we obtain

$$F'(x) = \frac{\partial F}{\partial x} + \frac{\partial F}{\partial y} f'(x) = 0,$$

so that $f'(x) = -(\partial F/\partial y)^{-1}(\partial F/\partial x)$. But going back now to the vector situation, we have

$$f'(x) = J_\mathbf{f}(\mathbf{x}), \qquad \frac{\partial F}{\partial y} = J_\mathbf{F}(\mathbf{y}), \qquad \text{and} \qquad \frac{\partial F}{\partial x} = J_\mathbf{F}(\mathbf{x}),$$

and substitution gives us formula (11).

EXAMPLE 7

APPLYING THE IMPLICIT FUNCTION THEOREM TO A LINEAR TRANSFORMATION

Let

$$\mathbf{F}(x_1, x_2, x_3, x_4) = (2x_1 - 3x_2 + 4x_3 - x_4,\ 2x_1 - x_2 + 5x_3 - 2x_4) = (0, 0).$$

[†] A proof of this theorem can be found in C. H. Edwards, Jr., *Advanced Calculus in Several Variables* (Academic Press, New York, 1973).

Here $\mathbf{F}$: $\mathbb{R}^4 \to \mathbb{R}^2$ is a linear transformation and $n = m = 2$. Then with $\mathbf{x} = (x_1, x_2)$ and $\mathbf{y} = (x_3, x_4)$,

$$J_{\mathbf{F}}(\mathbf{y}) = \begin{pmatrix} \dfrac{\partial F_1}{\partial x_3} & \dfrac{\partial F_1}{\partial x_4} \\ \dfrac{\partial F_2}{\partial x_3} & \dfrac{\partial F_2}{\partial x_4} \end{pmatrix} = \begin{pmatrix} 4 & -1 \\ 5 & -2 \end{pmatrix}$$

and $|J_{\mathbf{F}}(\mathbf{y})| = -3 \neq 0$. Since $|J_{\mathbf{F}}(\mathbf{y})|$ is nonzero for every $\mathbf{y}$, we can find a function $\mathbf{f}$: $\mathbb{R}^2 \to \mathbb{R}^2$ such that $(x_3, x_4) = \mathbf{f}(x_1, x_2)$ and

$$J_{\mathbf{f}}(\mathbf{x}) = -[J_{\mathbf{F}}(\mathbf{y})]^{-1} J_{\mathbf{F}}(\mathbf{x}).$$

But

$$J_{\mathbf{F}}(\mathbf{x}) = \begin{pmatrix} \dfrac{\partial F_1}{\partial x_1} & \dfrac{\partial F_1}{\partial x_2} \\ \dfrac{\partial F_2}{\partial x_1} & \dfrac{\partial F_2}{\partial x_2} \end{pmatrix} = \begin{pmatrix} 2 & -3 \\ 2 & -1 \end{pmatrix}.$$

Thus

$$J_{\mathbf{f}}(\mathbf{x}) = \tfrac{1}{3}\begin{pmatrix} -2 & 1 \\ -5 & 4 \end{pmatrix}\begin{pmatrix} 2 & -3 \\ 2 & -1 \end{pmatrix} = \tfrac{1}{3}\begin{pmatrix} -2 & 5 \\ -2 & 11 \end{pmatrix} = \begin{pmatrix} -\frac{2}{3} & \frac{5}{3} \\ -\frac{2}{3} & \frac{11}{3} \end{pmatrix}.$$

Since $\mathbf{F}$ is linear, we can verify this result. The equation $\mathbf{F}(x_1, x_2, x_3, x_4) = (0, 0)$ can be written as a system in augmented matrix form which we can solve by the methods of Section 6.3. Now, however, since we wish to write x_3 and x_4 in terms of x_1 and x_2, we make the coefficients of the x_3 and x_4 terms 1:

$$\left(\begin{array}{cccc|c} 2 & -3 & 4 & -1 & 0 \\ 2 & -1 & 5 & -2 & 0 \end{array}\right) \xrightarrow{R_1 \to \frac{1}{4}R_1} \left(\begin{array}{cccc|c} \frac{1}{2} & -\frac{3}{4} & 1 & -\frac{1}{4} & 0 \\ 2 & -1 & 5 & -2 & 0 \end{array}\right)$$

$$\xrightarrow{R_2 \to R_2 - 5R_1} \left(\begin{array}{cccc|c} \frac{1}{2} & -\frac{3}{4} & 1 & -\frac{1}{4} & 0 \\ -\frac{1}{2} & \frac{11}{4} & 0 & -\frac{3}{4} & 0 \end{array}\right) \xrightarrow{R_2 \to -\frac{4}{3}R_2} \left(\begin{array}{cccc|c} \frac{1}{2} & -\frac{3}{4} & 1 & -\frac{1}{4} & 0 \\ \frac{2}{3} & -\frac{11}{3} & 0 & 1 & 0 \end{array}\right)$$

$$\xrightarrow{R_1 \to R_1 + \frac{1}{4}R_2} \left(\begin{array}{cccc|c} \frac{2}{3} & -\frac{5}{3} & 1 & 0 & 0 \\ \frac{2}{3} & -\frac{11}{3} & 0 & 1 & 0 \end{array}\right).$$

The system represented by the last augmented matrix is

$$\begin{aligned} \tfrac{2}{3}x_1 - \tfrac{5}{3}x_2 + x_3 \qquad &= 0 \\ \tfrac{2}{3}x_1 - \tfrac{11}{3}x_2 \qquad + x_4 &= 0 \end{aligned}$$

which we can write as

$$(x_3, x_4) = \mathbf{f}(x_1, x_2) = (-\tfrac{2}{3}x_1 + \tfrac{5}{3}x_2, -\tfrac{2}{3}x_1 + \tfrac{11}{3}x_2)$$

and

$$J_{\mathbf{f}}(x_1, x_2) = \begin{pmatrix} -\frac{2}{3} & \frac{5}{3} \\ -\frac{2}{3} & \frac{11}{3} \end{pmatrix}.$$

REMARK: Of course, in all but the simplest problems (such as those involving linear transformations), it will be impossible to verify (except by rechecking calculations) that our computed value of $J_{\mathbf{f}}(\mathbf{x})$ is correct. However, it is nice to see that in a verifiable situation the implicit function theorem gives us the correct answer.

EXAMPLE 8 APPLYING THE IMPLICIT FUNCTION THEOREM TO A SYSTEM OF NONLINEAR EQUATIONS

Consider the following system of equations

$$\begin{aligned} x^2 - 2xz + y^2z^3 - 7 &= 0 \\ 2xy^4 - 3y^2 + xz^2 + 5z + 2 &= 0 \end{aligned} \tag{12}$$

We seek to write y and z in terms of x near the point $(2, 1, -1)$. Although a function $(y, z) = \mathbf{f}(\mathbf{x})$ cannot be determined explicitly, we can think of the system (12) as a function $\mathbf{f}: \mathbb{R}^{1+2} \to \mathbb{R}^2$ where $n = 1$ and $m = 2$ and we can use the implicit function theorem. Here $x_0 = 2$ and $\mathbf{y}_0 = (1, -1)$. Then

$$J_{\mathbf{F}}(\mathbf{y}_0) = \begin{pmatrix} \dfrac{\partial F_1}{\partial y} & \dfrac{\partial F_1}{\partial z} \\ \dfrac{\partial F_2}{\partial y} & \dfrac{\partial F_2}{\partial z} \end{pmatrix} = \begin{pmatrix} 2yz^3 & -2x + 3y^2z^2 \\ 8xy^3 - 6y & 2xz + 5 \end{pmatrix} = \begin{pmatrix} -2 & -1 \\ 10 & 1 \end{pmatrix}$$

at $(2, 1, -1)$ and $|J_{\mathbf{F}}(\mathbf{y}_0)| = 8 \neq 0$. Thus we can write

$$(y, z) = f(x)$$

for x in a neighborhood of (an open interval containing) 2. Moreover, $J_{\mathbf{f}}(\mathbf{x}_0) = (J_{\mathbf{F}}(\mathbf{y}_0))^{-1}(J_{\mathbf{F}}(\mathbf{x}_0))$. But

$$J_{\mathbf{F}}(\mathbf{x}_0) = \begin{pmatrix} \dfrac{\partial F_1}{\partial x} \\ \dfrac{\partial F_2}{\partial x} \end{pmatrix} = \begin{pmatrix} 2x - 2z \\ 2y^4 + z^2 \end{pmatrix} = \begin{pmatrix} 6 \\ 3 \end{pmatrix}$$

at $(2, 1, -1)$. Then

$$J_{\mathbf{f}}(2) = -\tfrac{1}{8}\begin{pmatrix} 1 & 1 \\ -10 & -2 \end{pmatrix}\begin{pmatrix} 6 \\ 3 \end{pmatrix} = -\tfrac{1}{8}\begin{pmatrix} 9 \\ -66 \end{pmatrix} = \begin{pmatrix} -\frac{9}{8} \\ \frac{33}{4} \end{pmatrix}.$$

Thus, if $\mathbf{f}(x) = (f_1(x), f_2(x)) = (y, z)$, we have

$$\left.\frac{dy}{dx}\right|_{x=2} = f_1'(2) = -\tfrac{9}{8} \quad \text{and} \quad \left.\frac{dz}{dx}\right|_{x=2} = f_2'(2) = \tfrac{33}{4}.$$

This is a great deal of information considering that we cannot compute $\mathbf{f}$ directly.

EXAMPLE 9 APPLYING THE IMPLICIT FUNCTION THEOREM TO A MAPPING FROM $\mathbb{R}^5$ TO $\mathbb{R}^3$

Let

$$\begin{aligned} &\mathbf{F}(x_1, x_2, x_3, x_4, x_5) \\ &= \left(x_1{}^4 - 2x_3x_4 + x_2x_5{}^3 + 9, x_1x_2x_3x_4x_5 + 4\frac{x_3x_5}{x_2}, \frac{2x_2 + 5x_4}{x_3} + x_1x_3\right) \\ &= (0, 0, 0). \end{aligned}$$

Show that (x_3, x_4, x_5) can be written as a function $\mathbf{f}$ of (x_1, x_2) in a neighborhood of $(1, -2, 3, -1, 2)$ and compute $J_{\mathbf{f}}(1, -2)$.

SOLUTION: Here $\mathbf{F}: \mathbb{R}^{2+3} \to \mathbb{R}^3$, so that $n = 2$, $m = 3$, $\mathbf{x}_0 = (1, -2)$, $\mathbf{y}_0 = (3, -1, 2)$, and

$$J_{\mathbf{F}}(\mathbf{y}) = \begin{pmatrix} \frac{\partial F_1}{\partial x_3} & \frac{\partial F_1}{\partial x_4} & \frac{\partial F_1}{\partial x_5} \\ \frac{\partial F_2}{\partial x_3} & \frac{\partial F_2}{\partial x_4} & \frac{\partial F_2}{\partial x_5} \\ \frac{\partial F_3}{\partial x_3} & \frac{\partial F_3}{\partial x_4} & \frac{\partial F_3}{\partial x_5} \end{pmatrix} = \begin{pmatrix} -2x_4 & -2x_3 & 3x_2x_5^2 \\ x_1x_2x_4x_5 + 4\frac{x_5}{x_2} & x_1x_2x_3x_5 & x_1x_2x_3x_4 + \frac{4x_3}{x_2} \\ \frac{-2x_2 - 5x_4}{x_3^2} + x_1 & \frac{5}{x_3} & 0 \end{pmatrix}$$

$$= \begin{pmatrix} 2 & -6 & -24 \\ 0 & -12 & 0 \\ 2 & \frac{5}{3} & 0 \end{pmatrix}$$

at $(1, -2, 3, -1, 2)$ and $|J_{\mathbf{F}}(\mathbf{y})| = -576 \neq 0$. Thus the required $\mathbf{f}$ exists, and we can write $\mathbf{y} = \mathbf{f}(\mathbf{x})$ in a neighborhood of $(1, -2)$. Finally

$$J_{\mathbf{f}}(1, -2) = -\begin{pmatrix} 2 & -6 & -24 \\ 0 & -12 & 0 \\ 2 & \frac{5}{3} & 0 \end{pmatrix}^{-1} J_{\mathbf{F}}(1, -2).$$

But

$$J_{\mathbf{F}}(\mathbf{x}) = \begin{pmatrix} \frac{\partial F_1}{\partial x_1} & \frac{\partial F_1}{\partial x_2} \\ \frac{\partial F_2}{\partial x_1} & \frac{\partial F_2}{\partial x_2} \\ \frac{\partial F_3}{\partial x_1} & \frac{\partial F_3}{\partial x_2} \end{pmatrix} = \begin{pmatrix} 4x_1^3 & x_5^3 \\ x_2x_3x_4x_5 & x_1x_3x_4x_5 - \frac{4x_3x_5}{x_2^2} \\ x_3 & \frac{2}{x_3} \end{pmatrix} = \begin{pmatrix} 4 & 8 \\ 12 & -12 \\ 3 & \frac{2}{3} \end{pmatrix}$$

at $(\mathbf{x}_0, \mathbf{y}_0)$, so that

$$J_{\mathbf{f}}(1, -2) = \frac{1}{576}\begin{pmatrix} 0 & -40 & -288 \\ 0 & 48 & 0 \\ 24 & -\frac{46}{3} & -24 \end{pmatrix}\begin{pmatrix} 4 & 8 \\ 12 & -12 \\ 3 & \frac{2}{3} \end{pmatrix} = \frac{1}{576}\begin{pmatrix} -1344 & 288 \\ 576 & -576 \\ -160 & 360 \end{pmatrix} = \begin{pmatrix} -\frac{7}{3} & \frac{1}{2} \\ 1 & -1 \\ -\frac{5}{18} & \frac{5}{8} \end{pmatrix}.$$

PROBLEMS 9.5

SELF-QUIZ

Multiple Choice

I. Let $\mathbf{f}: \mathbb{R}^2 \to \mathbb{R}^2$ be defined by $\mathbf{f}(x, y) = (x - 2y, x^2 + 3y^2)$, then the Jacobian of $\mathbf{f}$ is

a. $\begin{pmatrix} 1 & -2 \\ 2x & 6y \end{pmatrix}$ **b.** $\begin{pmatrix} 1 & 2x \\ -2 & 6y \end{pmatrix}$

c. $4x + 6y$ **d.** $(x - 2y)(x^2 + 3y^2)$

True–False

II. The mapping given in Problem I is locally invertible at $(0, 0)$.

III. The mapping given in Problem I is locally invertible at $(2, -3)$.

IV. The mapping given in Problem I is locally invertible at $(3, -2)$.

V. If $\mathbf{f}$ is a linear transformation from $\mathbb{R}^4$ to $\mathbb{R}^4$ and A is its matrix representation, then $\mathbf{f}$ is invertible on all of $\mathbb{R}^4$ if and only if $\det A \neq 0$.

VI. Let $\mathbf{f}(x, y, z, w) = (xy + zw, 2x - 16z + zw^3)$. Then (z, w) can be written as a function of (x, y) in a neighborhood of $(z_0, w_0) = (1, -2)$.

In Problems 1–13 compute the Jacobian of the given mapping at the given point.

1. $\mathbf{f}(x, y) = (x + y, x - y)$; $(1, 2)$

2. $\mathbf{f}(x, y) = (x^2 - y^2, x^2 + y^2)$; $(3, 2)$

3. $\mathbf{f}(x, y) = (\sin x, \cos y)$; $\left(\frac{\pi}{6}, \frac{\pi}{3}\right)$

4. $\mathbf{f}(x, y) = (e^{x+y}, e^{x-y})$; $(1, 1)$

5. $\mathbf{f}(x, y) = (\ln(x + y), \ln xy)$; $(2, 1)$

6. $\mathbf{f}(x, y) = (x \ln y, y \ln x)$; $(4, 2)$

7. $\mathbf{f}(x, y, z) = (x^2, y^2, z^2)$; $(1, 4, 2)$

8. $\mathbf{f}(x, y, z) = \left(xyz, \frac{xy}{z}, x^2 + y^2\right)$; $(2, -1, 1)$

9. $\mathbf{f}(x, y, z) = (x \cos y, y \cos z, z \sin x)$; $\left(\frac{\pi}{6}, \frac{\pi}{3}, \frac{\pi}{2}\right)$

10. $\mathbf{f}(x, y, z) = (x^2 + y^2 + z^2, x + y + z, xyz)$; $(2, -1, 4)$

11. $\mathbf{f}(x, y, z) = (x \ln(y + z), y \ln(x + z), z \ln(x + y))$; $(1, 1, 0)$

12. $\mathbf{f}(x_1, x_2, \ldots, x_n) = (x_1{}^2, x_2{}^2, \ldots, x_n{}^2)$; $(1, 1, \ldots, 1)$

13. $\mathbf{f}(x_1, x_2, \ldots, x_n) = (x_1, x_2{}^2, x_3{}^3, \ldots, x_n{}^n)$; $(1, 1, \ldots, 1)$

14. Let $\mathbf{f}: \mathbb{R}^n \to \mathbb{R}^n$ be differentiable at $\mathbf{x}_0$ and let $\mathbf{g}: \mathbb{R}^n \to \mathbb{R}^n$ be differentiable at $\mathbf{f}(\mathbf{x}_0)$. Show that

$$|J_{\mathbf{g} \circ \mathbf{f}}(\mathbf{x}_0)| = |J_{\mathbf{g}}(\mathbf{f}(\mathbf{x}_0))|\,|J_{\mathbf{f}}(\mathbf{x}_0)|.$$

In Problems 15–26 a mapping $\mathbf{f}$ and a point $\mathbf{x}_0$ are given. Show that the mapping is locally C^1-invertible at the given point and compute $J_{\mathbf{f}^{-1}}(\mathbf{x}_0)$.

15. $\mathbf{f}(x, y) = (2x - 3y, 7x + 2y)$; $(1, 3)$

16. $\mathbf{f}(x, y) = (x - 2y, -3x + y)$; (x_0, y_0)

17. $\mathbf{f}(x, y, z) = (x + y + z, x - y - z, -x + y - z)$; $(2, -1, 5)$

18. $\mathbf{f}(x, y, z) = (2x + 4y + 3z, y - z, 3x + 5y + 7z)$; (x_0, y_0, z_0)

19. $\mathbf{f}(x, y) = (x + y^2 + 1, x^2 + y + 2)$; $(1, 2)$

20. $\mathbf{f}(x, y) = (x^2 + y^2, x^2 - y^2)$; $(3, 5)$

21. $\mathbf{f}(x, y) = (e^{x+y}, e^{x-y})$; $(1, 1)$

22. $\mathbf{f}(x, y, z) = (x^2, y^2, z^2)$; $(2, -1, 3)$

23. $\mathbf{f}(x, y, z) = \left(x^2 y - yz, \frac{2x + y}{z}, -3z^3\right)$; $(3, 0, 1)$

24. $\mathbf{f}(x, y, z) = \left(\sin \frac{\pi}{6}(xy), \ln(x - z), e^{y+z}\right)$; $(2, -1, 1)$

25. $\mathbf{f}(x_1, x_2, \ldots, x_n) = (x_1{}^2, x_2{}^2, \ldots, x_n{}^2)$; $(1, 1, 1, \ldots, 1)$

26. $\mathbf{f}(x_1, x_2, \ldots, x_n) = (x_1, x_1x_2, x_1x_2x_3, \ldots, x_1x_2 \cdots x_n)$; $(1, 2, 3, \ldots, n)$

27. Show that the mapping $\mathbf{f}: \mathbb{R}^2 \to \mathbb{R}^2$ given by $\mathbf{f}(x, y) = (e^y \sin x, e^y \cos x)$ is locally C^1-invertible in a neighborhood of every point in $\mathbb{R}^2$ but is not invertible on all of $\mathbb{R}^2$.

In Problems 28–39 show that the given function $\mathbf{F}: \mathbb{R}^{n+m} \to \mathbb{R}^m$ with $\mathbf{F}(\mathbf{x}, \mathbf{y}) = \mathbf{0}$ determines $\mathbf{y} \in \mathbb{R}^m$ as a function $\mathbf{f}$ of $\mathbf{x} \in \mathbb{R}^n$ near the point $(\mathbf{x}_0, \mathbf{y}_0)$ in $\mathbb{R}^{n+m}$ and compute $J_{\mathbf{f}}(\mathbf{x}_0)$.

28. $\mathbf{F}(x_1, x_2, x_3, x_4) = (x_1 - 3x_2 + x_3 - 5x_4 - 8, 3x_1 - 2x_2 + 2x_3 - 8x_4 - 20) = (0, 0)$; $(\mathbf{x}_0, \mathbf{y}_0) = (2, 1, 4, -1)$

29. $\mathbf{F}(x_1, x_2, x_3, x_4, x_5) = (2x_1 + x_2 + 3x_3 - x_4 + 2x_5 - 2, 5x_1 - 2x_2 + x_3 - 2x_4 + 3x_5 + 1, x_1 + 6x_2 + 6x_3 + x_4 - 4x_5) = (0, 0, 0)$; $(\mathbf{x}_0, \mathbf{y}_0) = (1, 2, -1, 5, 3)$

30. $\mathbf{F}(x_1, x_2, x_3, x_4) = (x_1x_2 - x_3x_4, x_1x_3 - x_2x_4) = (0, 0)$; $(\mathbf{x}_0, \mathbf{y}_0) = (1, 1, 1, 1)$

31. $\mathbf{F}(x_1, x_2, x_3, x_4) = (x_1{}^2 + x_2{}^2 + x_3{}^2 + x_4{}^2 - 4, x_1x_2x_3x_4 - 1) = (0, 0)$; $(\mathbf{x}_0, \mathbf{y}_0) = (1, 1, 1, 1)$

32. $\mathbf{F}(x_1, x_2, x_3, x_4) = \left(2x_1{}^2 - 3x_1x_2 + x_3x_4{}^2 + 1, \frac{-4x_3}{x_2} + \frac{x_2}{x_4} + 8\right) = (0, 0)$; $(\mathbf{x}_0, \mathbf{y}_0) = (1, 2, 3, -1)$

33. $\mathbf{F}(x_1, x_2, x_3, x_4) = (x_1{}^2 + x_2{}^2 - x_3{}^2 - x_4{}^2, 2x_1x_2 + x_2{}^2 + 3x_3{}^2 - 2x_4{}^2 + 8) = (0, 0)$; $(\mathbf{x}_0, \mathbf{y}_0) = (2, -1, 1, 2)$

34. $\mathbf{F}(x_1, x_2, x_3, x_4) = \left(2x_1x_2{}^2x_3x_4{}^3 + 2x_2{}^3x_3{}^2 + 24, \frac{5x_3}{x_1} + \frac{2x_2}{x_4} + 1\right) = (0, 0)$; $(\mathbf{x}_0, \mathbf{y}_0) = (5, -2, 3, 1)$

35. $\mathbf{F}(x_1, x_2, x_3, x_4, x_5) = (x_1{}^2 + x_2{}^2 + x_3{}^2 + x_4{}^2 - 4, x_1x_2x_3x_4x_5 - 1) = (0, 0)$; $(\mathbf{x}_0, \mathbf{y}_0) = (1, 1, 1, 1, 1)$

36. $\mathbf{F}(x_1, x_2, x_3, x_4, x_5) = (x_1{}^2 + x_2{}^2 + x_3{}^2 + x_5{}^2 - 4, x_1x_2x_3x_4x_5 - 1, x_1 + x_2 + x_3 + x_4 - 4) = (0, 0, 0)$; $(\mathbf{x}_0, \mathbf{y}_0) = (1, 1, 1, 1, 1)$

37. $\mathbf{F}(x_1, x_2, x_3, x_4, x_5) = \left(2x_1{}^2 - 3x_2x_3x_4 - 44, 3x_1{}^2 + \frac{5x_4x_5}{x_3} + 28\right) = (0, 0)$; $(\mathbf{x}_0, \mathbf{y}_0) = (2, -3, 1, 4, -2)$

38. $\mathbf{F}(x_1, x_2, x_3, x_4, x_5) = \left(2x_1{}^2 - 3x_2x_3 - 17, 3x_1{}^2 + \frac{5x_4x_5}{x_3} + 28, 2x_3{}^3 - 4x_2x_5 + 22\right) = (0, 0, 0)$; $(\mathbf{x}_0, \mathbf{y}_0) = (2, -3, 1, 4, -2)$

39. $\mathbf{F}(x_1, x_2, x_3, x_4, x_5, x_6) = (e^{x_1}\cos x_2 + e^{x_3}\cos x_4 + e^{x_5}\cos x_6 + 2x_1 - 3, e^{x_1}\sin x_2 + e^{x_3}\sin x_4 + e^{x_5}\cos x_6 - x_2 - 1, e^{x_1}\tan x_2 + e^{x_3}\tan x_4 + e^{x_5}\tan x_6 + x_3) = (0, 0, 0)$; $(\mathbf{x}_0, \mathbf{y}_0) = (0, 0, 0, 0, 0, 0)$

ANSWERS TO SELF-QUIZ

I. c **II.** False **III.** True **IV.** False **V.** True **VI.** False

CHAPTER 9 SUMMARY OUTLINE

■ **Taylor's Theorem in *n* Variables** Let $f\colon \mathbb{R}^n \to \mathbb{R}$ be of class $C^{(m+1)}(\Omega)$ and let the line segment joining $\mathbf{x}_0$ and $\mathbf{x}$ be in Ω. Then if $\mathbf{x} = (x_1, x_2, \ldots, x_n)$ and $\mathbf{x}_0 = (x_1^{(0)}, x_2^{(0)}, \ldots, x_n^{(0)})$,

$$\mathbf{f}(\mathbf{x}) = f(\mathbf{x}_0) + \sum_{i=1}^{n} f_i(\mathbf{x}_0)(x_i - x_i^{(0)}) + \frac{1}{2!}\sum_{i=1}^{n}\sum_{j=1}^{n} f_{ij}(\mathbf{x}_0)(x_i - x_i^{(0)})(x_j - x_j^{(0)})$$

$$+ \frac{1}{3!}\sum_{i,j,k=1}^{n} f_{ijk}(\mathbf{x}_0)(x_i - x_i^{(0)})(x_j - x_j^{(0)})(x_k - x_k^{(0)})$$

$$+ \cdots + \frac{1}{m!}\sum_{i_1,i_2,\cdots,i_m=1}^{n} f_{i_1 i_2 \cdots i_m}(\mathbf{x}_0)(x_{i_1} - x_{i_1}^{(0)})(x_{i_2} - x_{i_2}^{(0)}) \cdots (x_{i_m} - x_{i_m}^{(0)}) + R_m(\mathbf{x})$$

where

$$R_m(\mathbf{x}) = \frac{1}{(m+1)!}\sum_{i_1,i_2,\ldots,i_{m+1}=1}^{n} f_{i_1 i_2 \cdots i_{m+1}}(\mathbf{x}_0 + c\mathbf{h})(x_{i_1} - x_{i_1}^{(0)}) \times (x_{i_2} - x_{i_2}^{(0)}) \cdots (x_{i_{m+1}} - x_{i_{m+1}}^{(0)})$$

for some number c in $(0, 1)$.

■ **Taylor's Theorem in Two Variables**

$$f(x, y) = f(x_0, y_0) + f_x(x_0, y_0)(x - x_0) + f_y(x_0, y_0)(y - y_0) + \frac{1}{2!}[f_{xx}(x_0, y_0)(x - x_0)^2 + 2f_{xy}(x_0, y_0)(x - x_0)(y - y_0) + f_{yy}(x, y)(y - y_0)^2] + R_3(x, y).$$

The next (third-degree) term is

$$\frac{1}{3!}[f_{xxx}(x_0, y_0)(x - x_0)^3 + 3f_{xxy}(x_0, y_0)(x - x_0)^2(y - y_0) + 3f_{xyy}(x_0, y_0)(x - x_0)(y - y_0)^2 + f_{yyy}(y - y_0)^3].$$

■ **Implicit Function Theorem—First Form** Let $F\colon \mathbb{R}^{n+1} \to \mathbb{R}$ be defined on an open set Ω and let

$$\mathbf{x}_0 = (x_1^{(0)}, x_2^{(0)}, \ldots, x_n^{(0)}, x_{n+1}^{(0)})$$

be in Ω. Suppose that the partial derivatives $(\partial F/\partial x_i)(\mathbf{x})$ are continuous in a neighborhood of $\mathbf{x}_0$ for $i = 1, 2, \ldots, n + 1$. Suppose further that

$$F(\mathbf{x}_0) = 0 \quad \text{and} \quad \frac{\partial F}{\partial x_{n+1}}(\mathbf{x}_0) \neq 0.$$

Then there is an open set $\overline{\Omega}$ in $\mathbb{R}^n$ that contains $\overline{\mathbf{x}}_0 = (x_1^{(0)}, x_2^{(0)}, \ldots, x_n^{(0)})$ and a unique continuous function $f\colon \mathbb{R}^n \to \mathbb{R}$ such that $x_{n+1}^{(0)} = f(\overline{\mathbf{x}}_0)$ and $F(\mathbf{x}, f(\mathbf{x})) = 0$ for all $\mathbf{x}$ in $\overline{\Omega}$. Furthermore, f has continuous first partial derivatives on $\overline{\Omega}$ given by

$$\frac{\partial f}{\partial x_i}(\mathbf{x}) = \frac{-\dfrac{\partial F}{\partial x_i}(\mathbf{x}, f(\mathbf{x}))}{\dfrac{\partial F}{\partial x_{n+1}}(\mathbf{x}, f(\mathbf{x}))} \quad \text{for } 1 \le i \le n.$$

■ **Function from $\mathbb{R}^n$ to $\mathbb{R}^m$** Let Ω be a subset of $\mathbb{R}^n$. Then a **function** or **mapping f** from $\mathbb{R}^n$ to $\mathbb{R}^m$ is a rule that assigns to each $\mathbf{x} = (x_1, x_2, \ldots, x_n)$ in Ω a unique vector $\mathbf{f}(\mathbf{x})$ in $\mathbb{R}^m$. The set Ω is called the **domain** of **f**. The set $\{\mathbf{f}(\mathbf{x})\colon \mathbf{x} \in \Omega\}$ is called the **range** of **f**.

NOTATION: We will write $\mathbf{f}\colon \mathbb{R}^n \to \mathbb{R}^m$ and, since $\mathbf{f}(\mathbf{x})$ is a subset of $\mathbb{R}^m$, we represent it in terms of its **component functions**. That is,

$$\mathbf{f}(\mathbf{x}) = (f_1(\mathbf{x}), f_2(\mathbf{x}), \ldots, f_m(\mathbf{x})) = (f_1(x_1, x_2, \ldots, x_n), f_2(x_1, x_2, \ldots, x_n), \ldots, f_m(x_1, x_2, \ldots, x_n)).$$

Note that each of the m component functions is a mapping from $\mathbb{R}^n \to \mathbb{R}$.

■ **Sums and Constant Multiples of Functions from $\mathbb{R}^n \to \mathbb{R}^m$** Let $\mathbf{f}\colon \mathbb{R}^n \to \mathbb{R}^m$ and $\mathbf{g}\colon \mathbb{R}^n \to \mathbb{R}^m$ have domains Ω_1 and Ω_2, respectively. Then we define the functions $\alpha\mathbf{f}$ and

$\mathbf{f} + \mathbf{g}$ by

i. $(\alpha\mathbf{f})(\mathbf{x}) = \alpha\mathbf{f}(\mathbf{x})$ with domain Ω_1, and

ii. $(\mathbf{f} + \mathbf{g})(\mathbf{x}) = f(\mathbf{x}) + g(\mathbf{x})$ with domain $\Omega_1 \cap \Omega_2$.

■ **Composite Function** Let $\mathbf{f}$: $\mathbb{R}^n \to \mathbb{R}^m$ with domain Ω_1 and $\mathbf{g}$: $\mathbb{R}^m \to \mathbb{R}^q$ with domain Ω_2. Then the **composite function** $\mathbf{g} \circ \mathbf{f}$: $\mathbb{R}^n \to \mathbb{R}^q$ is defined by

$$\mathbf{g} \circ \mathbf{f}(\mathbf{x}) = \mathbf{g}(\mathbf{f}(\mathbf{x})).$$

The domain of $\mathbf{g} \circ \mathbf{f}$ is given by

$$\text{dom}(\mathbf{g} \circ \mathbf{f}) = \{\mathbf{x} \in \Omega_1 \colon \mathbf{f}(\mathbf{x}) \in \Omega_2\}.$$

■ **Limit** Suppose that $\mathbf{f}$: $\mathbb{R}^n \to \mathbb{R}^m$ is defined on a neighborhood $B_r(\mathbf{x}_0)$ of the vector $\mathbf{x}_0$ except possibly at $\mathbf{x}_0$ itself. Then

$\lim_{\mathbf{x}\to\mathbf{x}_0} \mathbf{f}(\mathbf{x}) = \mathbf{L}$ if for every $\epsilon > 0$ there is a number $\delta > 0$ such that $|\mathbf{f}(\mathbf{x}) - \mathbf{L}| < \epsilon$ whenever $0 < |\mathbf{x} - \mathbf{x}_0| < \delta$.

■ **An Important Theorem about Limits** Let $\mathbf{f}$: $\mathbb{R}^n \to \mathbb{R}^m$ with component functions $f_1, f_2, \ldots, f_m$, and let $\mathbf{L} = (L_1, L_2, \ldots, L_m)$. Then

$$\lim_{\mathbf{x}\to\mathbf{x}_0} \mathbf{f}(\mathbf{x}) = \mathbf{L} \qquad \text{if and only if}$$

$$\lim_{\mathbf{x}\to\mathbf{x}_0} f_i(\mathbf{x}) = L_i \qquad \text{for } i = 1, 2, \ldots, m.$$

■ **Continuity** Let $\mathbf{f}$: $\mathbb{R}^n \to \mathbb{R}^m$ be defined in a neighborhood of the vector $\mathbf{x}_0$. Then $\mathbf{f}$ is **continuous** at $\mathbf{x}_0$ if each component function f_i is continuous at $\mathbf{x}_0$.

An alternative definition of continuity at $\mathbf{x}_0$ (for $\mathbf{f}$ defined in a neighborhood of $\mathbf{x}_0$) is

$$\lim_{\mathbf{x}\to\mathbf{x}_0} \mathbf{f}(\mathbf{x}) = \mathbf{f}(\mathbf{x}_0).$$

■ **Continuity of a Composite Function** Let $\mathbf{f}$: $\mathbb{R}^n \to \mathbb{R}^m$ be continuous at $\mathbf{x}_0$ and let $\mathbf{g}$: $\mathbb{R}^m \to \mathbb{R}^q$ be continuous at $\mathbf{f}(\mathbf{x}_0)$. Then the composite function $\mathbf{g} \circ \mathbf{f}$ is continuous at $\mathbf{x}_0$.

■ **Differentiability** Let f: $\mathbb{R}^n \to \mathbb{R}^m$ be defined on an open set Ω in $\mathbb{R}^n$ and suppose that $\mathbf{x}_0 \in \Omega$. Then $\mathbf{f}$ is **differentiable** at $\mathbf{x}_0$ if there exists a linear transformation $\mathbf{L}$: $\mathbb{R}^n \to \mathbb{R}^m$ and a function $\mathbf{g}$: $\mathbb{R}^n \to \mathbb{R}^m$ such that for every $\mathbf{x}$ in Ω,

$$\mathbf{f}(\mathbf{x}) - \mathbf{f}(\mathbf{x}_0) = \mathbf{L}(\mathbf{x} - \mathbf{x}_0) + \mathbf{g}(\mathbf{x} - \mathbf{x}_0) \qquad (*)$$

where

$$\lim_{|\mathbf{x}-\mathbf{x}_0|\to 0} \frac{g(\mathbf{x} - \mathbf{x}_0)}{|\mathbf{x} - \mathbf{x}_0|} = \mathbf{0}.$$

Since $\mathbf{L}$ is a linear transformation from $\mathbb{R}^n \to \mathbb{R}^m$, there is a unique $m \times n$ matrix, which we denote by $J_\mathbf{f}(\mathbf{x}_0)$, such that

$$\mathbf{L}(\mathbf{x}) = J_\mathbf{f}(\mathbf{x}_0)\mathbf{x} \qquad \text{for every } \mathbf{x} \text{ in } \mathbb{R}^n.$$

We assume that $\mathbf{x}$ is written as a column vector in terms of the standard basis in $\mathbb{R}^n$ and $\mathbf{L}(\mathbf{x})$ is written as a column vector in terms of the standard basis in $\mathbb{R}^m$. The notation $J_\mathbf{f}(\mathbf{x}_0)$ is used to emphasize that the matrix J depends on both the function $\mathbf{f}$ and the vector $\mathbf{x}_0$. With this notation, (*) can be written as

$$\mathbf{f}(\mathbf{x}) - \mathbf{f}(\mathbf{x}_0) = J_\mathbf{f}(\mathbf{x}_0)(\mathbf{x} - \mathbf{x}_0) + \mathbf{g}(\mathbf{x} - \mathbf{x}_0). \qquad (\surd)$$

■ **Jacobian Matrix** Let $\mathbf{f}$: $\mathbb{R}^n \to \mathbb{R}^m$ be differentiable at $\mathbf{x}_0$. Then the $m \times n$ matrix $J_\mathbf{f}(\mathbf{x}_0)$ determined by ($\surd$) is called the **Jacobian matrix** of $\mathbf{f}$ at $\mathbf{x}_0$.

The matrix $J_\mathbf{f}(\mathbf{x}_0)$ is sometimes called the **derivative matrix** or, more simply, the **derivative** of the linear transformation $\mathbf{f}$.

■ **Form of the Jacobian Matrix** Let $\mathbf{f}$: $\mathbb{R}^n \to \mathbb{R}^m$ be differentiable at $\mathbf{x}_0$. Then the Jacobian matrix $J_\mathbf{f}(\mathbf{x}_0)$ is given by

$$J_\mathbf{f}(\mathbf{x}_0) = \begin{pmatrix} \nabla f_1(\mathbf{x}_0 \\ \nabla f_2(\mathbf{x}_0) \\ \vdots \\ \nabla f_m(\mathbf{x}_0 \end{pmatrix}$$

$$= \begin{pmatrix} \frac{\partial f_1}{\partial x_1}(\mathbf{x}_0) & \frac{\partial f_1}{\partial x_2}(\mathbf{x}_0) & \frac{\partial f_1}{\partial x_3}(\mathbf{x}_0) & \cdots & \frac{\partial f_1}{\partial x_n}(\mathbf{x}_0) \\ \frac{\partial f_2}{\partial x_1}(\mathbf{x}_0) & \frac{\partial f_2}{\partial x_2}(\mathbf{x}_0) & \frac{\partial f_2}{\partial x_3}(\mathbf{x}_0) & \cdots & \frac{\partial f_2}{\partial x_n}(\mathbf{x}_0) \\ \frac{\partial f_3}{\partial x_1}(\mathbf{x}_0) & \frac{\partial f_3}{\partial x_2}(\mathbf{x}_0) & \frac{\partial f_3}{\partial x_3}(\mathbf{x}_0) & \cdots & \frac{\partial f_3}{\partial x_n}(\mathbf{x}_0) \\ \vdots & \vdots & \vdots & & \vdots \\ \frac{\partial f_m}{\partial x_1}(\mathbf{x}_0) & \frac{\partial f_m}{\partial x_2}(\mathbf{x}_0) & \frac{\partial f_m}{\partial x_3}(\mathbf{x}_0) & \cdots & \frac{\partial f_m}{\partial x_n}(\mathbf{x}_0) \end{pmatrix}.$$

■ **Total Derivative** Let $\mathbf{f}$: $\mathbb{R}^n \to \mathbb{R}^m$ be differentiable at $\mathbf{x}_0$. Then the Jacobian matrix $J_\mathbf{f}(\mathbf{x}_0)$ is called the **total derivative** of $\mathbf{f}$ at $\mathbf{x}_0$.

■ **A Theorem about Differentiability** Let $\mathbf{f}$: $\mathbb{R}^n \to \mathbb{R}^m$ with component functions $f_1, f_2, \ldots, f_m$. Suppose that all partial derivatives $\partial f_i/\partial x_j$ for $i = 1, 2, \ldots, m$ and $j = 1, 2, \ldots, n$ exist and are continuous at $\mathbf{x}_0$. Then $\mathbf{f}$ is differentiable at $\mathbf{x}_0$.

■ **Differentiable Functions Are Continuous** Let $\mathbf{f}$: $\mathbb{R}^n \to \mathbb{R}^m$ be differentiable at $\mathbf{x}_0$. Then $\mathbf{f}$ is continuous at $\mathbf{x}_0$.

■ **Sums and Constant Multiples of Differentiable Functions** Let $\mathbf{f}$: $\mathbb{R}^n \to \mathbb{R}^m$ and $\mathbf{g}$: $\mathbb{R}^n \to \mathbb{R}^m$ be differentiable at $\mathbf{x}_0$ and let α be scalar. Then $\alpha\mathbf{f}$ and $\mathbf{f} + \mathbf{g}$ are differentiable at $\mathbf{x}_0$ and

i. $J_{\alpha\mathbf{f}}(\mathbf{x}_0) = \alpha J_\mathbf{f}(\mathbf{x}_0)$

ii. $J_{\mathbf{f}+\mathbf{g}}(\mathbf{x}_0) = J_\mathbf{f}(\mathbf{x}_0) + J_\mathbf{g}(\mathbf{x}_0)$

or, alternatively,

i. $(\alpha\mathbf{f})'(\mathbf{x}_0) = \alpha\mathbf{f}'(\mathbf{x}_0)$

ii. $(\mathbf{f} + \mathbf{g})'(\mathbf{x}_0) = \mathbf{f}'(\mathbf{x}_0) + \mathbf{g}'(\mathbf{x}_0)$.

■ **Chain Rule** Let $\mathbf{f}$: $\mathbb{R}^n \to \mathbb{R}^m$ be differentiable at $\mathbf{x}_0$ and let $\mathbf{g}$: $\mathbb{R}^m \to \mathbb{R}^q$ be differentiable at $\mathbf{f}(\mathbf{x}_0)$. Suppose further that all the component functions of $\mathbf{f}$ and $\mathbf{g}$ have continuous partial derivatives. Then the composite function $\mathbf{g} \circ \mathbf{f}$ is differentiable at $\mathbf{x}_0$ and

$$J_{\mathbf{g}\circ\mathbf{f}}(\mathbf{x}_0) = J_{\mathbf{g}}(\mathbf{f}(\mathbf{x}_0))J_{\mathbf{f}}(\mathbf{x}_0).$$

■ **Jacobian** Let $\mathbf{f}$: $\mathbb{R}^n \to \mathbb{R}^n$ be differentiable at a point $\mathbf{x}_0$ in $\mathbb{R}^n$. Then the **Jacobian** of the mapping is the determinant of the Jacobian matrix $J_{\mathbf{f}}(\mathbf{x}_0)$. That is,

$$\text{Jacobian of } \mathbf{f} = \det J_{\mathbf{f}}(\mathbf{x}_0) = \begin{vmatrix} \frac{\partial f_1}{\partial x_1}(\mathbf{x}_0) & \frac{\partial f_1}{\partial x_2}(\mathbf{x}_0) & \cdots & \frac{\partial f_1}{\partial x_n}(\mathbf{x}_0) \\ \frac{\partial f_2}{\partial x_1}(\mathbf{x}_0) & \frac{\partial f_2}{\partial x_2}(\mathbf{x}_0) & \cdots & \frac{\partial f_2}{\partial x_n}(\mathbf{x}_0) \\ \vdots & \vdots & & \vdots \\ \frac{\partial f_n}{\partial x_1}(\mathbf{x}_0) & \frac{\partial f_n}{\partial x_2}(\mathbf{x}_0) & \cdots & \frac{\partial f_n}{\partial x_n}(\mathbf{x}_0) \end{vmatrix}$$

■ **Local C^1-Invertibility** Let $\mathbf{f}$: $\mathbb{R}^n \to \mathbb{R}^n$ be in $C^1(\Omega)$ for some open set Ω in $\mathbb{R}^n$. Then $\mathbf{f}$ is said to be **locally C^1-invertible** on Ω if there exists a function $\mathbf{g}$: $\mathbb{R}^n \to \mathbb{R}^n$ which is in $C^1(\mathbf{f}(\Omega))$ such that

$$(\mathbf{g} \circ \mathbf{f})(\mathbf{x}) = \mathbf{x} \qquad \text{for every } \mathbf{x} \text{ in } \Omega$$

and

$$(\mathbf{f} \circ \mathbf{g})(\mathbf{y}) = \mathbf{y} \qquad \text{for every } \mathbf{y} \text{ in } \mathbf{f}(\Omega).$$

In this case we say that $\mathbf{g}$ is an **inverse function** for $\mathbf{f}$ and write $\mathbf{g} = \mathbf{f}^{-1}$.

■ **Invertibility of a Linear Transformation** Let $\mathbf{f}$: $\mathbb{R}^n \to \mathbb{R}^n$ be a linear transformation with matrix representation A. Then $\mathbf{f}$ has an inverse on all of $\mathbb{R}^n$ if and only if A is invertible (if and only if $\det A \neq 0$). If A is invertible, then

$$\mathbf{f}^{-1}(\mathbf{x}) = A^{-1}\mathbf{x}.$$

■ **Inverse Function Theorem** Let $\mathbf{f}$: $\mathbb{R}^n \to \mathbb{R}^n$ be in $C^1(\Omega)$ on an open set Ω and let $\mathbf{x}_0$ be in Ω. If the Jacobian $|J_{\mathbf{f}}(\mathbf{x}_0)| \neq 0$, then there is an open set $\Omega_1 \subset \Omega$ such that $\mathbf{x}_0 \in \Omega_1$ and $\mathbf{f}$ is locally C^1-invertible on Ω_1; moreover, if $\mathbf{g} = \mathbf{f}^{-1}$ on $\mathbf{f}(\Omega_1)$, then

$$J_{\mathbf{f}^{-1}}(\mathbf{x}_0) = J_{\mathbf{g}}(\mathbf{f}(\mathbf{x}_0)) = [J_{\mathbf{f}}(\mathbf{x}_0)]^{-1}.$$

■ **Notation for the Implicit Function Theorem** Let $\mathbf{x}_0 = (x_1^{(0)}, x_2^{(0)}, \ldots, x_n^{(0)})$ be in $\mathbb{R}^n$ and let $\mathbf{y}_0 = (y_1^{(0)}, y_2^{(0)}, \ldots, y_m^{(0)})$ be in $\mathbb{R}^m$. Then if $\mathbf{F}$: $\mathbb{R}^{n+m} \to \mathbb{R}^m$, the notation $\mathbf{F}(\mathbf{x}_0, \mathbf{y}_0)$ stands for

$$\mathbf{F}(x_1^{(0)}, x_2^{(0)}, \ldots, x_n^{(0)}, y_1^{(0)}, y_2^{(0)}, \ldots, y_m^{(0)}),$$

the $m \times m$ matrix $J_{\mathbf{F}}(\mathbf{y}_0)$ is given by

$$J_{\mathbf{F}}(\mathbf{y}_0) = \begin{pmatrix} \frac{\partial F_1}{\partial y_1}(\mathbf{x}_0, \mathbf{y}_0) & \frac{\partial F_1}{\partial y_2}(\mathbf{x}_0, \mathbf{y}_0) & \cdots & \frac{\partial F_1}{\partial y_m}(\mathbf{x}_0, \mathbf{y}_0) \\ \frac{\partial F_2}{\partial y_1}(\mathbf{x}_0, \mathbf{y}_0) & \frac{\partial F_2}{\partial y_2}(\mathbf{x}_0, \mathbf{y}_0) & \cdots & \frac{\partial F_2}{\partial y_m}(\mathbf{x}_0, \mathbf{y}_0) \\ \vdots & \vdots & & \vdots \\ \frac{\partial F_m}{\partial y_1}(\mathbf{x}_0, \mathbf{y}_0) & \frac{\partial F_m}{\partial y_2}(\mathbf{x}_0, \mathbf{y}_0) & \cdots & \frac{\partial F_m}{\partial y_m}(\mathbf{x}_0, \mathbf{y}_0) \end{pmatrix}$$

and the $m \times n$ matrix $J_{\mathbf{F}}(\mathbf{x}_0)$ is given by

$$J_{\mathbf{F}}(\mathbf{x}_0) = \begin{pmatrix} \frac{\partial F_1}{\partial x_1}(\mathbf{x}_0, \mathbf{y}_0) & \frac{\partial F_1}{\partial x_2}(\mathbf{x}_0, \mathbf{y}_0) & \cdots & \frac{\partial F_1}{\partial x_n}(\mathbf{x}_0, \mathbf{y}_0) \\ \frac{\partial F_2}{\partial x_1}(\mathbf{x}_0, \mathbf{y}_0) & \frac{\partial F_2}{\partial x_2}(\mathbf{x}_0, \mathbf{y}_0) & \cdots & \frac{\partial F_2}{\partial x_n}(\mathbf{x}_0, \mathbf{y}_0) \\ \vdots & \vdots & & \vdots \\ \frac{\partial F_m}{\partial x_1}(\mathbf{x}_0, \mathbf{y}_0) & \frac{\partial F_m}{\partial x_2}(\mathbf{x}_0, \mathbf{y}_0) & \cdots & \frac{\partial F_m}{\partial x_n}(\mathbf{x}_0, \mathbf{y}_0) \end{pmatrix}$$

Note that $J_{\mathbf{F}}(\mathbf{x}_0)$ consists of the first n columns of $J_{\mathbf{F}}(\mathbf{x}_0, \mathbf{y}_0)$ and $J_{\mathbf{F}}(\mathbf{y}_0)$ consists of the last m columns of $J_{\mathbf{F}}(\mathbf{x}_0, \mathbf{y}_0)$.

■ **Implicit Function Theorem** Let the function $\mathbf{F}$: $\mathbb{R}^{n+m} \to \mathbb{R}^m$ be C^1 in a neighborhood Ω of the vector $(\mathbf{x}_0, \mathbf{y}_0)$ in $\mathbb{R}^{n+m}$ with $\mathbf{F}(\mathbf{x}_0, \mathbf{y}_0) = \mathbf{0}$. If $|J_{\mathbf{F}}(\mathbf{y}_0)| \neq 0$, then there exists a neighborhood Ω_1 of $\mathbf{x}_0$ in $\mathbb{R}^n$, a neighborhood Ω_2 of $(\mathbf{x}_0, \mathbf{y}_0)$ in $\mathbb{R}^{n+m}$, and a function $\mathbf{f}$: $\mathbb{R}^n \to \mathbb{R}^m$ in $C^1(\Omega_1)$ such that $\mathbf{y} = \mathbf{f}(\mathbf{x})$ satisfies $\mathbf{F}(\mathbf{x}, \mathbf{f}(\mathbf{x})) = \mathbf{0}$ in Ω_2. Furthermore,

$$J_{\mathbf{f}}(\mathbf{x}) = -[J_{\mathbf{F}}(\mathbf{y})]^{-1}J_{\mathbf{F}}(\mathbf{x}).$$

CHAPTER 9 REVIEW EXERCISES

1. Find the third-degree Taylor polynomial of $\cos(x + 2y)$ at $\mathbf{x}_0 = \mathbf{0}$.
2. Find the second-degree Taylor polynomial of $\ln(1 + 5x - 4y)$ at $\mathbf{x}_0 = \mathbf{0}$.
3. Let $F(x, y, z) = x^2y + y^2z + z^5x - 3 = 0$. Show that z can be written as a function of x and y in a neighborhood of (1, 1, 1) and compute $\partial z/\partial x$ and $\partial z/\partial y$.
4. Let $F(x, y, z) = e^{2x-y+5z} + xyz - 1 = 0$. Show that z can be written as a function of x and y in a neighborhood of (0, 0, 0) and compute $\partial z/\partial x$ and $\partial z/\partial y$.
5. Let $x^2y^3 - 4xz^2 + 3z^4 - 5y + 4 = 0$. Find formulas for $\partial z/\partial x$ and $\partial z/\partial y$ and state where they are valid.

In Exercises 6–9 find the domain of the given function.

6. $\mathbf{f}(x, y) = (\sqrt{x^2 + y^2}, x^2 - y^3)$
7. $\mathbf{f}(x, y) = \left(\sec \dfrac{y}{x}, \dfrac{1}{x + y}, e^{x^2y}\right)$
8. $\mathbf{f}(x, y, z) = \left(\dfrac{1}{x + y + z}, \dfrac{1}{x + y - z}\right)$

9. $\mathbf{f}(x, y, z) = \left(x^2, y^2, \tan\left(\frac{x+y}{x-y}\right), \sin^{-1}\frac{xy}{z}\right)$

10. Compute $\lim_{x\to(1,2)} \mathbf{f}(\mathbf{x})$ where $\mathbf{f}$ is as in Exercise 6.

11. Compute $\lim_{x\to(1,2,4)} \mathbf{f}(\mathbf{x})$ where $\mathbf{f}$ is as in Exercise 9.

12. Let $\mathbf{f}(x, y) = (xy, x + y, \ln|y|^2)$ and $\mathbf{g}(x, y, z) = (xyz, x + 2y - 3z, e^{xy/z})$. Write out the composite function $\mathbf{g} \circ \mathbf{f}$ and determine its domain.

In Exercises 13–17 compute the derivative $\mathbf{f}'(\mathbf{x}_0) = J_{\mathbf{f}}(\mathbf{x}_0)$ of the given function $\mathbf{f}$ at the given point $\mathbf{x}_0$.

13. $\mathbf{f}(x, y) = (x^3, y^3)$; $(2, 1)$

14. $\mathbf{f}(x, y) = (xy, y/x, x^3y^2)$; $(1, 4)$

15. $\mathbf{f}(x, y, z) = (x^2, 2y^2, 3z^2)$; $(1, 2, 3)$

16. $\mathbf{f}(x, y, z) = (x + 2y - 3z, 2x - 4y + 7z, 4x - y - z, -x - y + 2z)$; $(5, 10, 20)$

17. $\mathbf{f}(x_1, x_2, x_3, x_4) = \left(x_1x_3, x_2x_4, x_1{}^2 + x_2{}^2 + x_3{}^2 + x_4{}^2, \frac{x_1x_4}{x_2x_3}\right)$; $(3, 1, -1, 5)$

18. Let $\mathbf{f}(x, y) = (x^3, y^3)$ and $\mathbf{g}(x, y) = (x + 2y, 2x - y)$. Find the derivative of the composite function $\mathbf{g} \circ \mathbf{f}$ at the point $(2, -1)$ and verify the chain rule at that point.

19. Let $\mathbf{f}(x, y, z) = (x^2y, xz + 3y^2, 3xyz^3)$ and $\mathbf{g}(x, y, z) = (2xy, 3xz, yz, xyz)$. Find the derivative of the composite function $\mathbf{g} \circ \mathbf{f}$ at the point $(2, 1, -1)$ and verify the chain rule at that point.

20. Compute the Jacobian of $\mathbf{f}(x, y) = (x^3, y^3)$ at the point $(2, 1)$.

21. Compute the Jacobian of $\mathbf{f}(x, y, z) = (x^2, 2y^2, 3z^2)$ at the point $(1, 2, 3)$.

22. Compute the Jacobian of $\mathbf{f}(x, y, z) = (x^2 - yz, 3y - 2x^3z^2, 4xyz)$ at the point $(2, -1, 4)$.

In Exercises 23–26 a mapping f and a point $\mathbf{x}_0$ are given. Show that the mapping is locally C^1-invertible at the given point and compute $J_{\mathbf{f}^{-1}}(\mathbf{x}_0)$.

23. $\mathbf{f}(x, y) = (3x + 2y, -x + 5y)$; $(4, -7)$

24. $\mathbf{f}(x, y) = (x^3 + y^3, x^3 - y^3)$; $(2, 1)$

25. $\mathbf{f}(x, y, z) = (x^3, y^3, z^3)$; $(1, -1, 2)$

26. $\mathbf{f}(x, y, z) = (x^2y, xy + 3y^2, 3xyz^3)$; $(-3, 2, 1)$

In Exercises 27–29 show that the given function $\mathbf{F}\colon \mathbb{R}^{n+m} \to \mathbb{R}^m$ with $\mathbf{F}(\mathbf{x}, \mathbf{y}) = \mathbf{0}$ determines $\mathbf{y} \in \mathbb{R}^m$ as a function $\mathbf{f}$ of $\mathbf{x} \in \mathbb{R}^n$ near the point $(\mathbf{x}_0, \mathbf{y}_0)$ in $\mathbb{R}^{n+m}$ and compute $J_{\mathbf{f}}(\mathbf{x}_0)$.

27. $\mathbf{F}(x_1, x_2, x_3, x_4) = (2x_1 - x_2 + x_3 - 4x_4 - 8, 3x_1 + 2x_2 - 2x_3 - 4x_4 + 7) = (0, 0)$; $(\mathbf{x}_0, \mathbf{y}_0) = (3, -2, 4, 1)$

28. $\mathbf{F}(x_1, x_2, x_3, x_4) = (x_1{}^3 + x_2{}^3 + x_3{}^3 + x_4{}^3 - 4, 2x_1x_2x_3x_4 - 2) = (0, 0)$; $(\mathbf{x}_0, \mathbf{y}_0) = (1, 1, 1, 1)$

29. $\mathbf{F}(x_1, x_2, x_3, x_4, x_5) = (x_1{}^3 + x_2{}^3 + x_3{}^3 + x_5{}^3 - 4, x_1x_2x_3x_4x_5 - 1, x_1 + 2x_2 - 5x_3 + 2x_4)$; $(\mathbf{x}_0, \mathbf{y}_0) = (1, 1, 1, 1, 1)$

PART IV

DIFFERENTIAL EQUATIONS

CHAPTERS INCLUDE:

CHAPTER 10

ORDINARY DIFFERENTIAL EQUATIONS

10.1 INTRODUCTION

Many of the basic laws of the physical sciences and, more recently, of the biological and social sciences are formulated in terms of mathematical relations involving certain known and unknown quantities and their derivatives. Such relations are called **differential equations**.

In your study of one-variable calculus you probably encountered the following differential equation:

$$\frac{dy}{dx} = \alpha x \tag{1}$$

where α is a given constant. Equation (1) is called the differential equation of **exponential growth** (if $\alpha > 0$) and of **exponential decay** (if $\alpha < 0$). In this chapter we will discuss this and many other types of differential equations. In Chapter 11 we will discuss systems of differential equations. We begin, in this section, by categorizing the types of differential equations that may be encountered.

The most obvious classification is based on the nature of the derivative (or derivatives) in the equation. A differential equation involving only ordinary derivatives (derivatives of functions of one variable) is called an **ordinary differential equation**. Equation (1) is an ordinary differential equation.

A differential equation containing partial derivatives is called a **partial differential equation**. Examples of partial differential equations are given in Problems 3.3.49, 3.3.50, and 3.4.27–35.

In this chapter we will consider only the solution of certain ordinary differential equations, which we will refer to simply as differential equations, dropping the word "ordinary." Procedures for solving all but the most trivial partial differential equations are beyond the scope of this discussion.

DEFINITION ORDER

The **order** of a differential equation is the order of the highest-order derivative appearing in the equation. ■

EXAMPLE 1 FOUR DIFFERENTIAL EQUATIONS AND THEIR ORDERS

The following are examples of differential equations with indicated orders.

a. $dy/dx = 3x$ (first order).

b. $x''(t) + 4x'(t) - x(t) = \sin t$ (second order).

c. $(dy/dx)^5 - 2e^y = 6\cos x$ (first order).

d. $(y^{(4)})^2 - 2y''' + y' - 6y = \sqrt{x}$ (fourth order).

Consider the following equation having the form of equation (1)

$$\frac{dy}{dx} = 3y \tag{2}$$

together with the **initial condition**

$$y(0) = 2. \tag{3}$$

DEFINITION SOLUTION

A **solution** to (2), (3) is defined to be a function that

i. is differentiable,

ii. satisfies the differential equation (2) on some interval containing 0, and

iii. satisfies the initial condition (3). ■

Let $y = 2e^{3x}$. Then

$$\frac{dy}{dx} = 2\frac{d}{dx}e^{3x} = 2 \cdot 3e^{3x} = 3(2e^{3x}) \overset{y = 2e^{3x}}{\downarrow}= 3y$$

That is, $y = 2e^{3x}$ is a differentiable function that satisfies equation (2). Moreover,

$$y(0) = 2e^{3\cdot 0} = 2e^0 = 2$$

so that y satisfies the initial condition (3). In fact, the function $y = 2e^{3x}$ is the *only* function that satisfies equations (2) and (3).

DEFINITION INITIAL VALUE PROBLEM

The system (2), (3) is called an **initial value problem.** ■

In Appendix 5 we will show that for a certain class of initial value problems, including the problem (2), (3), there is always a unique solution. This makes sense intuitively. For example, suppose that a bacteria population is growing at a rate proportional to itself.

That is, if $P(t)$ denotes the population size at time t, then

$$P'(t) = \alpha P(t) \tag{4}$$

for some constant of proportionality α. Furthermore, we assume that the population at some given time, which we denote by $t = 0$, is 10,000. Then we have

$$P(0) = 10{,}000. \tag{5}$$

It is reasonable to expect that starting with the initial population given by (5) and assuming that population growth is governed by the equation (4), the population in the future will be completely determined. That is, there is one and only one way to write the population $P(t)$; this is a way of saying that the initial value problem (4), (5) has a unique solution.

There are other types of problems in which other than initial conditions are given. For example, let a string be held taut between two points one meter apart, as in Figure 1.

0 1 x

FIGURE 1
A string of length 1 meter

Suppose that the string is then plucked so that it begins to vibrate. Let $\nu(x, t)$ denote the height of the string at a distance x units from the left-hand endpoint and at a time t, where $0 \le x \le 1$ and $t \ge 0$. Then, since the two endpoints are held fixed, we have the conditions

$$\nu(0, t) = \nu(1, t) = 0 \qquad \text{for all} \quad t \ge 0. \tag{6}$$

The conditions (6) are called **boundary conditions** and if we write a differential equation governing the motion of the string (see Problem 3.4.27 on p. 177), then the equation, together with the boundary conditions (6), is called a **boundary value problem**. In general, a boundary value problem is a differential equation together with values given at two or more points, while an initial value problem is a differential equation with values given at only one point.

EXAMPLE 2 A BOUNDARY VALUE PROBLEM

The problem

$$y'' + 2y' + 3y = 0, \qquad y(0) = 1,\ y(1) = 2$$

is a boundary value problem. The problem

$$y'' + 2y' + 3y = 0, \qquad y(0) = 1,\ y'(0) = 2$$

is an initial value problem.

In this chapter we will consider certain kinds of first- and second-order ordinary differential equations and some simple initial value problems. For a more complete discussion see an introductory book on differential equations. In Appendix 5 we shall prove that many first-order initial value problems have unique solutions in some neighborhood of the initial point $(x_0, y(x_0))$.

There is one further classification of differential equations that will be important to us in this book.

DEFINITION LINEAR EQUATION

An nth-order differential equation is **linear** if it can be written in the form

$$\frac{d^n y}{dx^n} + a_{n-1}(x)\frac{d^{n-1}y}{dx^{n-1}} + \cdots + a_1(x)\frac{dy}{dx} + a_0(x)y = f(x). \tag{7}$$

Hence a first-order linear equation has the form

$$\frac{dy}{dx} + a(x)y = f(x), \tag{8}$$

and a second-order linear equation can be written as

$$\frac{d^2y}{dx^2} + a(x)\frac{dy}{dx} + b(x)y = f(x). \tag{9}$$

The notation indicates that $a(x)$, $b(x)$, and $f(x)$ are functions of x alone. ■

A differential equation that is not linear is called **nonlinear**.

EXAMPLE 3 FOUR LINEAR DIFFERENTIAL EQUATIONS

The following are linear differential equations.

a. $\frac{dy}{dx} - 3y = 0$

b. $y'' + 4y = 0$

c. $y'' + 2xy' + e^x y = \sin^2 x$

d. $y^{(4)} + 5y''' - 6y'' + 7y' + 8y = x^5 + x^3 + 10$

EXAMPLE 4 FOUR NONLINEAR DIFFERENTIAL EQUATIONS

The following are nonlinear differential equations.

a. $\frac{dy}{dx} - y^2 = 0$

b. $yy'' = y' + x$

c. $\frac{dy}{dx} + \sqrt{y} = 0$

d. $\sin xy + \cos\left(1 + \frac{dy}{dx}\right) = x^3y^4$

In Sections 10.4, 10.7, and 10.9–12, we will show how solutions to a great number of linear differential equations can be found. Solutions to nonlinear equations can be found only in some special situations. Two of these are given in Sections 10.3 and 10.5.

PROBLEMS 10.1

SELF-QUIZ

Multiple Choice

I. The differential equation $(y')^2 + y^4 + 3x = 0$ is of ________ order.

a. First **b.** Second
c. Third **d.** Fourth

II. The differential equation $\frac{d^2y}{dx^2} + y\frac{dy}{dx} = x^3 \sin x$ is of ________ order.

a. First **b.** Second
c. Third **d.** Fourth

True-False

III. The equation in Problem I is linear.

IV. The equation in Problem II is linear.

In Problems 1–6 state the order of the differential equation.

1. $x'(t) + 2x(t) = t$

2. $(d^2y/dx^2) + 4(dy/dx)^3 - y = x^3$

3. $y''' + y = 0$

4. $(dx/dt)^5 = x^4$

5. $x'' - x^2 = 2x^{(4)}$

6. $(x''(t))^4 - (x'(t))^{3/2} = 2x(t) + e^t$

7. Verify that $y_1 = 2 \sin x$ and $y_2 = -4 \cos x$ are solutions to $y'' + y = 0$.

8. Verify that $y = (x/2)e^x$ is a solution to $y'' - y = e^x$.

9. Verify that $y = 3x/(4x - 3)$ is the solution to the initial value problem

$$x^2\, dy/dx + y^2 = 0, \qquad y(1) = 3.$$

10. Verify that $y = -\frac{1}{2}x^2e^{-x}$ is a solution to the equation $y''' - 3y' - 2y = 3e^{-x}$.

11. Verify that $y = e^{3x}$ satisfies the initial value problem

$$y'' - 6y' + 9y = 0, \qquad y(0) = 1, \qquad y'(0) = 3.$$

[*Hint:* Check that the equation is satisfied, and then verify that the two initial conditions are also satisfied.]

***12.** Verify that $y = e^{-x/2}[\cos(\sqrt{3}/2)x + (7/\sqrt{3})\sin(\sqrt{3}/2)x]$ is the solution to the initial value problem.

$$y'' + y' + y = 0, \qquad y(0) = 1, \qquad y'(0) = 3.$$

In Problems 13–23 determine whether the differential equation is linear or nonlinear.

13. $y'' + 2y' + 3y = 0$

14. $y'' + 2y' + 3y = x^3 + \ln(1 + x^2)$

15. $y'' + x^5y' + 3(\sin x)y = x$

16. $y'' + 2y' + y^2 = 0$

17. $y'' + \sqrt{y} = 0$

18. $\left(\dfrac{dy}{dx}\right)^2 + y = 0$

19. $y\dfrac{dy}{dx} = 2x$

20. $\dfrac{dy}{dx} + 10x^{10}y = \cos x - \dfrac{d^2y}{dx^2}$

21. $\dfrac{dy}{dx} = \dfrac{3}{y}$

22. $\dfrac{dy}{dx} = \dfrac{y}{x}$

23. $x^2\dfrac{d^2y}{dx^2} + y\dfrac{dy}{dx} = 2$

ANSWERS TO SELF-QUIZ

I. a **II.** b **III.** False **IV.** False

10.2 REVIEW OF THE DIFFERENTIAL EQUATION OF EXPONENTIAL GROWTH

In this section we study a differential equation that you encountered in your first calculus course. Before doing so, however, we begin with a differential equation that can be solved directly by integration.

EXAMPLE 1 DETERMINING DISTANCE WHEN THE VELOCITY IS GIVEN

A ball is dropped from rest from a certain height. Its velocity after t seconds due to the earth's gravitational field is given by

$$v(t) = 32.2t \text{ ft/sec} \tag{1}$$

(when air resistance is ignored). How fast has the ball fallen after t seconds?

SOLUTION: Let $s(t)$ denote the distance the ball has fallen after t seconds. Then, as velocity is the derivative of position, we have

$$\frac{ds}{dt} = v(t)$$

or

$$\frac{ds}{dt} = 32.2t \tag{2}$$

Equation (2) is a *differential equation* because it is an equation involving a derivative. We can solve it directly by integration:

$$s = \int v(t)\,dt = \int 32.2t\,dt$$

$$= 32.2\frac{t^2}{2} + C = 16.1t^2 + C$$

But $s(0)$ = distance fallen after 0 seconds = *initial* distance fallen = 0 and

$$s(0) = 16.1 \cdot 0^2 + C = 0 + C = C$$

Thus $C = 0$ and, after t seconds the ball has fallen

$$s(t) = 16.1t^2 \text{ ft}$$

We note that the differential equation $\frac{ds}{dt} = 32.2t$ together with the implicitly given initial condition $s(0) = 0$ is, as we stated in Section 10.1, called an **initial-value problem**. We shall say more about initial-value problems later in this section.

Example 1 is a special case of the equations of motion discussed in calculus and physics courses. In our second example we derive these equations under some simplifying assumptions.

EXAMPLE 2

A MODEL OF FREE FALL

Newton's law of gravitation states that the magnitude of the gravitational force exerted by the earth on a body of mass m is proportional to its mass and inversely proportional to the square of its distance $r = r(t)$ from the center of the earth. Thus, if k denotes the constant of proportionality, then

$$F = \frac{km}{r^2}.$$

By Newton's second law of motion

$$F = ma = m\frac{d^2r}{dt^2},$$

where $a = a(r)$ is the acceleration of the body when the distance to the center of the earth is r. Hence, equating both forces and dividing by m, we get

$$\frac{d^2r}{dt^2} = \frac{k}{r^2}. \tag{3}$$

Equation (3) is a second-order differential equation because it involves a second derivative (and no derivative of order greater than 2).

If the object is falling, the velocity $v = dr/dt$ is negative because the distance to the center of the earth is decreasing. Moreover, if the speed at which the object falls is increasing, the acceleration $a = dv/dt = d^2r/dt^2$ is also negative because v is decreasing, that is, becoming more and more negative. Thus, the constant k in equation (3) is negative.

Let R be the mean radius of the earth ($R \approx 6378$ km ≈ 3963 miles) and denote the acceleration of gravity at the earth's surface by $a(R) = -g$. Then

$$-g = a(R) = \frac{k}{R^2},$$

so that $k = -gR^2$. Hence we obtain the equation of motion

$$\frac{d^2r}{dt^2} = -g\frac{R^2}{r^2}, \tag{4}$$

where g is approximately 9.81 m/sec^2 (=32.2 ft/sec^2). If we substitute $r = R + h$, where $h = h(t)$ is the height of the body from the surface of the earth, then $dr/dt = dh/dt$ and equation (4) becomes

$$\frac{d^2h}{dt^2} = -g\frac{R^2}{(R+h)^2}.$$

When the height h is very small in comparison to the radius of the earth R, the ratio $R/(R + h)$ is very close to 1, so the differential equation is well approximated by the usual equation found in one-variable calculus books:

$$\frac{d^2h}{dt^2} = -g. \tag{5}$$

WARNING: Equation (5) is an equation of constant acceleration, a reasonable model for free fall only if h is very small compared to R. For example, if $h = 1000$ m $= 1$ km, then

$$\frac{h}{R} = \frac{1}{6378} \approx 0.000157 = 0.0157\%$$

Here h is a very small percentage of R. However, if h is, say, 2000 km, then

$$\frac{h}{R} = \frac{2000}{6378} \approx 0.3136 = 31.36\%$$

which is not a small percentage. In this case, the model of constant acceleration would not represent reality very well and, even though we can solve equation (5) without much difficulty, the solution will not be very useful.

Determining the height, h, at which equation (5) ceases to be useful is a very difficult problem. We will not address that problem in this book. It must be left for a more advanced course in differential equations.

We now solve equation (5) by integration. Integrating both sides of equation (5) with respect to t, we obtain

$$h'(t) = -gt + C_1.$$

The constant C_1 can be determined by setting $t = 0$. We obtain $C_1 = h'(0)$, the initial velocity. Thus the velocity of the body at any time t is given by the differential equation

$$h'(t) = -gt + h'(0). \tag{6}$$

Integrating once more with respect to t, we have

$$h(t) = -\frac{gt^2}{2} + h'(0)t + C_2.$$

The constant C_2 can also be found by setting $t = 0$. We obtain $C_2 = h(0)$, the initial height. Thus the height of the body at any time t is

$$h(t) = -\frac{gt^2}{2} + h'(0)t + h(0). \tag{7}$$

Equations (6) and (7) approximate the velocity and altitude of an object subject to free fall. As an example, suppose a ball is dropped at rest from the top of a building 45 meters high. Then the initial velocity is $h'(0) = 0$ and the initial height is $h(0) = 4500$ centimeters. If we wish to find the time it takes for the ball to strike the ground, we substitute these values in equation (7) to obtain

$$0 = h(t) = -\frac{981t^2}{2} + 4500,$$

since the height at impact is zero. Solving for t, we have

$$490.5t^2 = 4500 \qquad \text{or} \qquad t^2 \approx 9.17.$$

Thus $t \approx \pm 3.03$. Since $t = -3.03$ has no physical significance, the answer is $t \approx 3.03$ seconds.

NOTE: The equation

$$\frac{d^2h}{dt^2} = -g$$

together with the given values $h(0)$ and $h'(0)$ is a second-order initial-value problem.

The differential equations in Examples 1 and 2 were solved directly by integration. If it were always possible to do this, then differential equations would be a direct application of integral calculus and there would be no need for separate chapters on differential equations. However, most solvable differential equations can only be solved by other techniques. One type of differential equation that we can solve is often discussed in a beginning calculus class.

Let $y = f(x)$ represent some physical quantity such as the volume of a substance, the population of a certain species, the mass of a decaying radioactive substance, the number of dollars invested in bonds, and so on. Then the growth of $f(x)$ is given by its derivative dy/dx. Thus, if $f(x)$ is growing at a constant rate, then $dy/dx = k$ and $y = kx + C$, that is, $y = f(x)$ is a straight-line function.

It is sometimes more interesting and more appropriate to consider the **relative rate of growth**, defined by

$$\text{relative rate of growth} = \frac{\text{actual rate of growth}}{\text{size of } f(x)} = \frac{f'(x)}{f(x)} = \frac{dy/dx}{y}. \tag{8}$$

The relative rate of growth indicates the percentage increase or decrease in f. For example, an increase of 100 individuals for a species with a population size of 500 would probably have a significant impact, being an increase of 20%. On the other hand, if the population were 1,000,000, then the addition of 100 would hardly be noticed, being an increase of only 0.01%.

In many applications, we are told that the relative rate of growth of the given physical quantity is constant. That is,

$$\frac{dy/dx}{y} = \alpha \tag{9}$$

or

$$\frac{dy}{dx} = \alpha y \tag{10}$$

where α is the constant percentage increase or decrease in the quantity.

Another way to view equation (10) is that it tells us that the function is changing at a rate proportional to itself. If the constant of proportionality α is greater than 0, the quantity is increasing; while if $\alpha < 0$, it is decreasing.

Equation (10) is a first-order differential equation. It is different from equations (2) and (5) because now the unknown function y appears on *both sides* of the equation. If we tried to integrate both sides of equation (10) with respect to x, we would obtain

$$\int \frac{dy}{dx}\,dx = \alpha \int y(x)\,dx + C$$

or

$$y = \alpha \int y(x)\,dx + C$$

But this doesn't help at all because we don't yet know what the function y is.

To solve equation (10), we carry out the following steps:

Step 1: Multiply both sides of equation (10) by $e^{-\alpha x}$ to obtain

$$\frac{dy}{dx}e^{-\alpha x} - \alpha y e^{-\alpha x} = 0. \tag{11}$$

Step 2: Observe that

Product rule

$$\frac{d}{dx}(e^{-\alpha x}y) = e^{-\alpha x}\frac{dy}{dx} + y(-\alpha e^{-\alpha x}). \tag{12}$$

Step 3: Using (12), write (11) as

$$\frac{d}{dx}(e^{-\alpha x}y) = 0.$$

Step 4: Since the derivative of $ye^{-\alpha x}$ is zero,

$$ye^{-\alpha x} = c \qquad \text{for some constant } c. \tag{13}$$

Step 5: Multiply both sides of (13) by $e^{\alpha x}$ to obtain

$$y = ce^{\alpha x}.$$

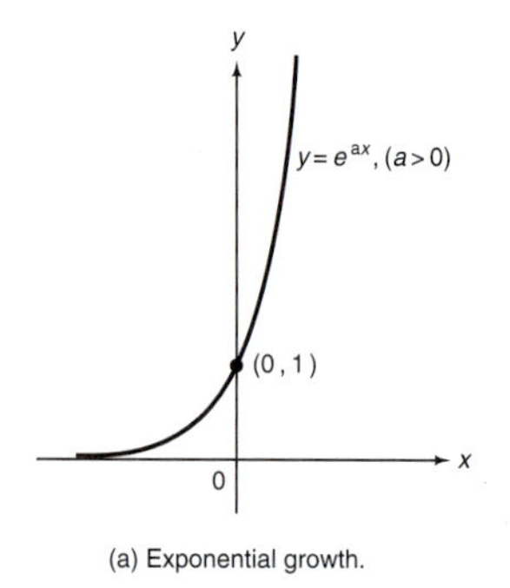

(a) Exponential growth.

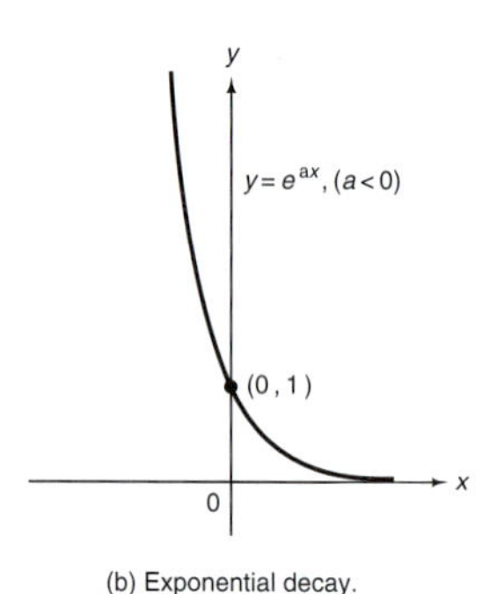

(b) Exponential decay.

FIGURE 1
Curves showing exponential growth and decay

Thus, any solution to (10) can be written as

SOLUTION TO THE DIFFERENTIAL EQUATION OF CONSTANT RELATIVE GROWTH

$$y = ce^{\alpha x} \tag{14}$$

where c can be any real number. To check this, we simply differentiate:

$$\frac{dy}{dx} = \frac{d}{dx}ce^{\alpha x} = c\frac{d}{dx}e^{\alpha x} = c\alpha e^{\alpha x} = \alpha(ce^{\alpha x}) = \alpha y,$$

so that $y = ce^{\alpha x}$ does satisfy (10). We therefore have proven our first theorem.

THEOREM 1

If α is any real number, then there are an infinite number of solutions to the differential equation $y' = \alpha y$. They take the form $y = ce^{\alpha x}$ for any real number c. ■

If $\alpha > 0$, we say that the quantity described by $f(x)$ is **growing exponentially**. If $\alpha < 0$, it is **decaying exponentially** (see Figure 1). Of course, if $\alpha = 0$, then there is no growth and $f(x)$ remains constant.

For a physical problem, we often know more than the rate of growth. We may also know one particular value of x, say $y(x_0) = y_0$. This value is called an **initial condition**, and it will give us a unique solution to the problem. The differential equation (10) together with an initial condition is called an **initial-value problem**. We will find the unique solutions to a number of initial-value problems in this section.

EXAMPLE 3

SOLVING AN INITIAL-VALUE PROBLEM

a. Find all solutions to $dy/dx = 3y$.
b. Find the solution that satisfies the initial condition $y(0) = 2$.

SOLUTION:

a. Since $\alpha = 3$, all solutions are of the form $y = ce^{3x}$.
b. $2 = y(0) = ce^{3\cdot 0} = c \cdot 1 = c$, so $c = 2$ and the unique solution is $y = 2e^{3x}$.

EXAMPLE 4

A MODEL OF POPULATION GROWTH

A bacterial population is growing continuously at a rate equal to 10% of its population each day. Its initial size is 10,000 organisms. How many bacteria are present after 10 days? After 30 days?

SOLUTION: Since the percentage growth of the population is $10\% = 0.1$, we have

$$\frac{dP/dt}{P} = 0.1 \quad \text{or} \quad \frac{dP}{dt} = 0.1P.$$

Here $\alpha = 0.1$, and all solutions have the form

$$P(t) = ce^{0.1t}$$

where t is measured in days. Since $P(0) = 10{,}000$, we have

$$ce^{(0.1)(0)} = c = 10{,}000, \quad \text{and} \quad P(t) = 10{,}000e^{0.1t}.$$

This is the unique solution to the initial-value problem. After 10 days $P(10) = 10{,}000e^{(0.1)(10)} = 10{,}000e \approx 27{,}183$, and after 30 days $P(30) = 10{,}000e^{0.1(30)} = 10{,}000e^3 \approx 200{,}855$ bacteria.

HISTORICAL NOTE: When it applies to human populations, exponential growth is often referred to as **Malthusian growth**. In 1798, the English economist Thomas Robert Malthus (1766–1834) claimed that increases in population tend to exceed increases in the means of subsistence and that therefore sexual restraint should be exercised.† More precisely, he claimed that populations grow exponentially (like the function $e^{\alpha t}$) while food and other necessities grow only geometrically (like the function t^{α}). His theories are still hotly debated.

Before giving further examples, we summarize the principal result of this section:

THE EQUATION OF EXPONENTIAL GROWTH OR DECAY

If $P(t)$ is growing or declining continuously at a constant relative rate (so that $P'(t)/P(t) = \pm\alpha$) with initial size $P(0)$, then $P(t)$ is given by (with $\alpha > 0$)

$$P(t) = P(0)e^{\alpha t} \quad \text{or} \quad P(t) = P(0)e^{-\alpha t}. \tag{15}$$

The term $P(0)e^{\alpha t}$ represents exponential growth and the term $P(0)e^{-\alpha t}$ represents exponential decay.

EXAMPLE 5

POPULATION GROWTH IN INDIA

The population of India was estimated to be 574,220,000 in 1974 and 746,388,000 in 1984. Assume that the relative growth rate remains constant and that growth is continuous.

a. Estimate the population in 1994.

b. When will the population reach 1.5 billion?

SOLUTION:

a. From (15), we have

$$P(t) = P(0)e^{\alpha t}.$$

We have $\alpha > 0$ because the population is increasing.

Treat the year 1974 as year zero. Then 1984 = year 10. We are told that

$$P(0) = 574{,}220{,}000 \quad \text{and} \quad P(10) = 746{,}388{,}000.$$

†T. R. Malthus, "An Essay on the Principle of Population as it Affects the Future Improvement of Society," 1798.

Thus

$$P(t) = 574{,}220{,}000e^{\alpha t}$$

$$P(10) = 746{,}388{,}000 = 574{,}220{,}000e^{10\alpha}$$

$$e^{10\alpha} = \frac{746{,}388{,}000}{574{,}220{,}000} \approx 1.299829334 = \beta$$

$$\ln e^{10\alpha} \approx \ln \beta$$

$$10\alpha \approx \ln \beta \qquad \ln e^{\alpha} = \alpha$$

$$\alpha \approx \frac{\ln \beta}{10} \approx 0.026223297.$$

The year 1994 is year 20. Thus

$$P(20) = 574{,}220{,}000e^{20\alpha} \approx 574{,}220{,}000e^{20(\ln \beta/10)}$$

$$r \ln \alpha = \ln \alpha^r$$

$$= 574{,}220{,}000e^{2 \ln \beta} = 574{,}220{,}000e^{\ln \beta^2}$$

$$e^{\ln \alpha} = \alpha$$

$$P(20) = (574{,}220{,}000)(\beta^2)$$

$$= (574{,}220{,}000)(1.689556297) \approx 970{,}177{,}000.$$

b. We seek a number t such that $P(t) = 1{,}500{,}000{,}000$. That is,

$$P(t) = 1{,}500{,}000{,}000 = 574{,}220{,}000e^{\alpha t}$$

$$e^{\alpha t} = \frac{1{,}500{,}000{,}000}{574{,}220{,}000} \approx 2.612239211$$

$$\ln e^{\alpha t} = \ln 2.612239211$$

$$\alpha t = \ln 2.612239211$$

$$t = \frac{\ln 2.612239211}{\alpha}$$

$$= \frac{\ln 2.612239211}{0.026223297}$$

$$\approx 36.6 \text{ years.}$$

Then

$$\text{Year } 36.6 = 1974 + 36.6 = 2010.6.$$

We conclude that the population of India would reach 1.5 billion sometime in the year 2010 if its population growth rate continued at the same rate as it was between 1974 and 1984.

In this problem we found that

$$\alpha \approx \frac{\ln \beta}{10} \approx 0.0262.$$

This means that the population is growing at a rate of about 2.62% a year. We stress that all the calculations in this problem were made under the assumption that this percentage did not vary and will not vary at all from 1974 until now and into the foreseeable future. This might not be a reasonable assumption. In that case our answers are not reasonable.

In this example, we created a mathematical model to solve a problem based on certain assumptions about population growth. Solving the problem was much easier than checking the validity of our assumptions. You should keep this in mind whenever you use a mathematical model.

EXAMPLE 6

NEWTON'S LAW OF COOLING: A MODEL OF HEATING AND COOLING

Newton's law of cooling states that the rate of change of the temperature difference between an object and its surrounding medium is proportional to the temperature difference. If $D(t)$ denotes this temperature difference at time t and if α denotes the constant of proportionality, then we obtain

$$\frac{dD}{dt} = -\alpha D.$$

The minus sign indicates that this difference decreases. (If the object is cooler than the surrounding medium—usually air—it will warm up; if it is hotter, it will cool.) The solution to this differential equation is

$$D(t) = ce^{-\alpha t}.$$

If we denote the initial ($t = 0$) temperature difference by D_0, then

$$D(t) = D_0 e^{-\alpha t} \tag{16}$$

is the formula for the temperature difference for any $t > 0$. Notice that for t large $e^{-\alpha t}$ is very small, so that, as we have all observed, temperature differences tend to die out rather quickly.

We can rewrite (16) in a form that is often more useful. Let $T(t)$ denote the temperature of the object at time t, let T_s denote the temperature of the surroundings (which is assumed to be constant throughout), and let T_0 denote the initial temperature of the object. Then $D(t) = T(t) - T_s$ and $D_0 = T_0 - T_s$, so (16) becomes

$$T(t) - T_s = (T_0 - T_s)e^{-\alpha t}$$

or

$$T(t) = T_s + (T_0 - T_s)e^{-\alpha t}. \tag{17}$$

We now may ask: In terms of the constant α, how long does it take for the temperature difference to decrease to half its original value?

SOLUTION: The original value is D_0. We are therefore looking for a value of t for which $D(t) = \frac{1}{2}D_0$. That is, $\frac{1}{2}D_0 = D_0 e^{-\alpha t}$, or $e^{-\alpha t} = \frac{1}{2}$. Taking natural logarithms, we obtain

$$-\alpha t = \ln\frac{1}{2} = -\ln 2 \approx -0.6931, \qquad \text{and} \qquad t \approx \frac{0.6931}{\alpha}.$$

Notice that this value of t does *not* depend on the initial temperature difference D_0.

EXAMPLE 7

USING NEWTON'S LAW OF COOLING

With the air temperature equal to 30°C, an object with an initial temperature of 10°C warmed to 14°C in 1 hr.

a. What was its temperature after 2 hr?

b. After how many hours was its temperature 25°C?

SOLUTION: Here $T_s = 30$, $T_0 = 10$ and $T(1) = 14$. From (17), we have

$$T(t) = 30 - 20e^{-\alpha t}.$$

But

$$14 = T(1) = 30 - 20e^{-\alpha}$$
$$20e^{-\alpha} = 16$$
$$e^{-\alpha} = \frac{4}{5} = 0.8.$$

Thus,

$$T(t) = 30 - 20e^{-\alpha t} = 30 - 20(e^{-\alpha})^t = 30 - 20(0.8)^t.$$

We can now answer the two questions.

a. $T(2) = 30 - 20(0.8)^2 = 30 - 20(0.64) = 17.2°\text{C}.$

b. We need to find t such that $T(t) = 25$. That is,

$$25 = 30 - 20(0.8)^t, \quad \text{or} \quad (0.8)^t = \frac{1}{4} \quad \text{or} \quad t\ln(0.8) = -\ln 4,$$

and

$$t = \frac{-\ln 4}{\ln(0.8)} \approx \frac{1.3863}{0.2231} \approx 6.2 \text{ hr} = 6 \text{ hr } 12 \text{ min}.$$

EXAMPLE 8

CARBON DATING: A MODEL FOR ESTIMATING THE AGE OF AN ARTIFACT

Carbon dating is a technique used by archaeologists, geologists, and others who want to estimate the ages of certain artifacts and fossils they uncover. The technique is based on certain properties of the carbon atom. In its natural state the nucleus of the carbon atom ^{12}C has 6 protons and 6 neutrons. The **isotope** carbon-14, ^{14}C, is produced through cosmic-ray bombardment of nitrogen in the atmosphere. Carbon-14 has 6 protons and 8 neutrons and is **radioactive**. It decays by beta emission. That is, when an atom of ^{14}C decays, it gives up an electron to form a stable nitrogen atom ^{14}N. We make the assumption that the ratio of ^{14}C to ^{12}C in the atmosphere is constant. This assumption has been shown experimentally to be approximately valid, for although ^{14}C is being constantly lost through **radioactive decay** (as this process is often termed), new ^{14}C is constantly being produced by the cosmic bombardment of nitrogen in the upper atmosphere. Living plants and animals do not distinguish between ^{12}C and ^{14}C, so at the time of death the ratio of ^{12}C to ^{14}C in an organism is the same as the ratio in the atmosphere. However, this ratio changes after death since ^{14}C is converted to ^{14}N but no further ^{14}C is taken in.

It has been observed that ^{14}C decays at a rate proportional to its mass and that its **half-life** is approximately 5730 years.† That is, if a substance starts with 1 g of ^{14}C, then 5730 years later it will have $\frac{1}{2}$ g of ^{14}C, the other $\frac{1}{2}$ g having been converted to ^{14}N.

We may now pose a question typically asked by an archaeologist. A fossil is unearthed and it is determined that the amount of ^{14}C present is 40% of what it would be for a similarly sized living organism. What is the approximate age of the fossil?

SOLUTION: Let $M(t)$ denote the mass of ^{14}C present in the fossil. Then since ^{14}C decays at a rate proportional to its mass, we have

$$\frac{dM}{dt} = -\alpha M,$$

where α is the constant of proportionality. Then $M(t) = ce^{-\alpha t}$, where $c = M_0$, the initial amount of ^{14}C present. When $t = 0$, $M(0) = M_0$; when $t = 5730$ years, $M(5730) = \frac{1}{2}M_0$, since half the original amount of ^{14}C has been converted to ^{12}C. We can use this fact to solve for α since we have

$$\frac{1}{2}M_0 = M_0 e^{-\alpha \cdot 5730}, \qquad \text{or} \qquad e^{-5730\alpha} = \frac{1}{2}.$$

Thus,

$$(e^{-\alpha})^{5730} = \frac{1}{2}, \qquad \text{or} \qquad e^{-\alpha} = \left(\frac{1}{2}\right)^{1/5730}, \qquad \text{and} \qquad e^{-\alpha t} = \left(\frac{1}{2}\right)^{t/5730},$$

so

$$M(t) = M_0\left(\frac{1}{2}\right)^{t/5730}.$$

Now we are told that after t years (from the death of the fossilized organism to the present) $M(t) = 0.4M_0$, and we are asked to determine t. Then,

$$0.4M_0 = M_0\left(\frac{1}{2}\right)^{t/5730},$$

and taking natural logarithms (after dividing by M_0), we obtain

$$\ln 0.4 = \frac{t}{5730}\ln\left(\frac{1}{2}\right), \qquad \text{or} \qquad t = \frac{5730 \ln(0.4)}{\ln(\frac{1}{2})} \approx 7575 \text{ years.}$$

The carbon-dating method has been used successfully on numerous occasions. It was this technique that established that the Dead Sea scrolls were prepared and buried about two thousand years ago.

---

†This number was first determined in 1949 by the American chemist W. S. Libby, who based his calculations on the wood from sequoia trees, whose ages were determined by rings marking years of growth. Libby's method has come to be regarded as the archaeologist's absolute measuring scale. But in truth, this scale is flawed. Libby used the assumption that the atmosphere had at all times a constant amount of ^{14}C. However, the American chemist C. W. Ferguson of the University of Arizona deduced from his study of tree rings in 4000-year-old American giant trees that before 1500 B.C. the radiocarbon content of the atmosphere was considerably higher than it was later. This result implied that objects from the pre–1500 B.C. era were much older than previously believed, because Libby's "clock" allowed for a smaller amount of ^{14}C than actually was present. For example, a find dated at 1800 B.C. was in fact from 2500 B.C. This has had a considerable impact on the study of prehistoric times. For a fascinating discussion of this subject, see Gerhard Herm, *The Celts* (New York: St. Martin's Press, 1975), pages 90–92.

COMPOUND INTEREST

Suppose P dollars are invested in an enterprise (which may be a bank, or mutual fund) that pays an annual interest rate of r. If the original investment and all amounts earned in interest are left in the account, after one year the account contains the original investment P and the interest Pr, so that it is worth $P + Pr = P(1 + r)$ dollars. At the end of the second year, the interest paid is $rP(1 + r)$, so that now we have

$$P(1 + r) + rP(1 + r) = P(1 + r)(1 + r) = P(1 + r)^2$$

dollars. Continuing in this fashion, we see that after t years, the investment would be worth

$$A(t) = P(1 + r)^t, \tag{18}$$

where $A(t)$ denotes the value of the investment after t years.

EXAMPLE 9

DETERMINING COMPOUND INTEREST

If \$1000 is invested at an annual interest rate of 6%, then $r = 0.06$, and after 5 years the investment is worth

$$A(5) = 1000(1 + 0.06)^5 \approx 1000(1.33823) \approx \$1338.23.$$

The actual interest paid is \$338.23.

In practice interest is compounded more frequently than annually. If it is paid m times per year, then in each interest period the rate of interest is r/m, and in t years there are mt pay periods. Then, following the same procedures that we used to obtain (18), we obtain:

COMPOUND INTEREST FORMULA: COMPOUNDING m TIMES PER YEAR

The value of an investment \$$P$ compounded m times per year with an annual interest rate of r after t years is

$$A(t) = P\left(1 + \frac{r}{m}\right)^{mt}. \tag{19}$$

EXAMPLE 10

QUARTERLY COMPOUNDING

If the interest in Example 9 is compounded quarterly, then after 5 years the investment is worth

$$A(5) = 1000\left(1 + \frac{0.06}{4}\right)^{4(5)} = 1000(1.015)^{20} \approx \$1346.86.$$

The interest paid is now \$346.86, or \$8.63 more than with annual compounding.

It is clear from Examples 9 and 10 that the interest paid increases as the number of pay periods increases. We are naturally led to ask: What is the value of the investment if interest is compounded continuously—that is, if the number of pay periods m approaches infinity?

Suppose each pay period has length Δt, so that there are $m = 1/\Delta t$ pay periods per year. Let $A(t)$ be the amount in the account after t years. Then the amount in the account at the next pay period will be $A(t)$ plus the interest on $A(t)$:

$$A(t+\Delta t) = A(t) + A(t)\frac{r}{1/\Delta t} = A(t) + A(t)r\,\Delta t.$$

Then

$$A(t+\Delta t) - A(t) = A(t)r\,\Delta t,$$

and

$$\frac{A(t+\Delta t) - A(t)}{\Delta t} = A(t)r.$$

Taking the limit as $\Delta t \to 0$ (so that $m \to \infty$), we obtain

$$\frac{dA}{dt} = rA,$$

or, from (14),

$$A(t) = ce^{rt}.$$

But $P = A(0) = c$, and we have proved the following result.

COMPOUND INTEREST FORMULA: CONTINUOUS COMPOUNDING

$$A(t) = Pe^{rt} \tag{20}$$

is the value of an initial investment of $\$P$ after t years at an annual rate of r compounded continuously.

EXAMPLE 11 CONTINUOUS COMPOUNDING

If the investment in Example 9 is compounded continuously, we will have after 5 years

$$A(5) = 1000e^{5(0.06)} = 1000e^{0.3} \approx \$1349.86,$$

and the interest paid is \$349.86.

EXAMPLE 12 DOUBLING TIME

How long would it take for the investment in Example 9 compounded continuously to double its original value?

SOLUTION: We need to determine the value of t so that $A(t) = 2P$. That is

$$2P = Pe^{0.06r} \qquad \text{or} \qquad e^{0.06r} = 2.$$

Taking the natural logarithm, we obtain $0.06r = \ln 2 \approx 0.6931$, so $t \approx 0.6931/.06 \approx 11.55$ years.

OF HISTORICAL INTEREST

COMPOUND INTEREST IN THE SEVENTEENTH CENTURY

JAKOB BERNOULLI
1654–1705

In 1690, the great Swiss mathematician, Jakob Bernoulli (1654–1705), proposed the following problem: "*Quaeritur, si creditor aliquis pecuniam suam foenori exponat, ea lege, ut singulis momentis pars proportionalis usarae annuae sorti annumeretur; quantum ipsi finito anno debeatur?*" ["This question is posed: if a creditor lends his money at interest with the understanding that a proportionate part of the annual interest be added to the principal periodically, how much is due to him at the end of a year?"]

Bernoulli's passage expresses clearly the concept of compound interest on a loan, and reflects a culture in which this practice is new and stands in contrast to the more familiar practice of simple annual interest. Under the older, simple-interest system, the creditor who lent \$100 for a year at 10% would be due \$110 at the end of the year. Under the system described in the question, assuming the *singulis momentis*† to be "quarterly," the principal would increase from \$100 to \$102.50 at the end of three months (\$10 annual interest divided by 4 = \$2.50, added to the principal). During the second quarter then, the annual interest would be reckoned, not on the original \$100, but on \$102.50, and so on.

Bernoulli partially answered his own question by deriving the compound-interest formula (18). But his real question amounted to this: as m, the number of times interest is paid each year, increases without bound, would the depositor acquire an unbounded fortune? In our notation, Bernoulli was asking

$$\text{What happens to } \left(1+\frac{x}{m}\right)^m \text{ as } m \to \infty?$$

We know that the limit is e^x and that the investor would not acquire unlimited wealth. Bernoulli concluded the same thing. But he could not go further than to claim from numerical evidence that the limit for ($x = 1$) was somewhere between 2.7 and 2.8. He did not have the advantage of using the number e because the number e was not introduced by Euler until 1728, 38 years later (and 23 years after Bernoulli's death).

THINKING ABOUT A MATHEMATICAL MODEL

In many of the examples given in this section, we assumed that the relative rate of growth was constant so that

$$y(x) = y(0)e^{kx}$$

Suppose that $k > 0$. Let's give it a value, say $k = 0.1$. This means that something—a population or investment, for example, is growing continuously at a rate of 10% per time period. Did you stop to consider the fact that this rate of growth *cannot possibly continue* for an indefinite period of time?

To illustrate this fact, suppose that a certain insect population grows at 10% a month and starts with 1000 individuals. Then the population after t months is

$$P(t) = 1000e^{0.1t}$$

The population after t months is given in Table 1. We see that after 240 months (20 years) there would be over 26 trillion insects.

TABLE 1

t	$P(t) = 1000e^{0.1t}$
10	2718
20	7389
30	20,086
60	403,429
120	162,754,791
240	2.649×10^{13}
480	7.017×10^{23}

† That is, "periodically;" "from time to time;" "by periodic increments."

After 40 years there would be over 7×10^{23} insects. That is an unimaginably large number. To give you some idea, let us do some calculations.

1. The radius of the earth is approximately 4000 miles.
2. The surface area of a sphere is given by $S = 4\pi r^2$ so surface area of the earth $\approx 4\pi(4000)^2 \approx 200{,}000{,}000$ square miles.
3. There are 5280^2 square feet in a square mile so the surface area of the earth in square feet $\approx (200{,}000{,}000)(5280^2) \approx 5.6 \times 10^{15}$ square feet.
4. We divide:

$$\frac{7 \times 10^{23} \text{ insects}}{5.6 \times 10^{15} \text{ sq ft}} \approx 125{,}000{,}000 \text{ insects per square foot.}$$

That is, 125 million insects occupying every square foot on earth! Where would they fit? What would they eat?

The situation is ridiculous. This should come as no surprise because as $t \to \infty$, $e^{kt} \to \infty$ if $k > 0$, so any population (or anything else) grows without bound if it is growing exponentially.

Since nothing on earth can possibly grow without bound, it seems that the model of this section is useless. This is not true. Many quantities grow exponentially—like continuously compounded interest, for example. But nothing can grow exponentially *indefinitely*. There are always limits to growth. Any population growing exponentially will eventually run out of space or food and suffer other effects from overcrowding. Then nature will force the growth rate to change.

But unlimited growth is not the only problem with this model. Is the assumption that growth continues at a constant relative rate a reasonable one? If a bond pays 7% compounded continuously for 20 years, then you are certain that, at least for 20 years, your money will grow exponentially at the constant rate of 7%.

However, for another kind of problem, the answer is likely to be no. In Example 5 we found that the population of India grew at an average annual rate of 2.6% from 1974 to 1984. Better education in the schools and more prosperity might lead to fewer children born per couple. Better health care could lead to drops in infant mortality and a corresponding increase in the population growth rate. You can undoubtedly think of other factors.

So the assumption of indefinite growth at a constant rate is unrealistic for many different types of problems. Models based on this assumption may be very useful. But they are often useful only for limited periods of time. As with all mathematical models, you must question the validity of the assumptions inherent in the model before you make predictions based on the model.

PROBLEMS 10.2†

SELF-QUIZ

Multiple Choice

I. Suppose y depends smoothly on x (i.e., $dy/dx = y'$ exists). Also suppose that its relative rate of growth is constant. Then y satisfies a differential equation of the form ________ (where α is a nonzero constant).

a. $yy' = \alpha$ **b.** $y'/y = \alpha$

†To complete most of these problems, you will need to use a calculator with ln and e^x function keys (INV ln gives e^x on some calculators).

c. $y' + 1/y = \alpha$ **d.** $y' = \alpha/y$

II. The general solution to $y'/y = 10$ is ______.

a. $\ln|y| = 10x + C$ **b.** $y = K \cdot e^{10x}$

c. $x = K + e^{10y}$ **d.** $y = 5y^2 + K$

III. Suppose y is a smooth function of x such that $y'/y = -3$; then $\lim_{x\to\infty} y$ ______.

a. $= 1$

b. $= -\frac{3}{2}$

c. $= \infty$

d. $= 0$

e. doesn't exist.

f. can't be found from this information alone.

IV. Suppose y is a smooth function of x such that $y'/y = -3$; then $\lim_{x\to 0} y$ ______.

a. $= 1$

b. $= -\frac{3}{2}$

c. $= \infty$

d. $= 0$

e. doesn't exist.

f. can't be determined from this information.

In Problems 1–4, use Newton's law of cooling to determine how long to bake a cake at the given oven temperature, assuming that it takes exactly 30 minutes to change 70°F dough into a 170°F cake in a 350°F oven.

1. 250°F

2. 400°F

3. 300°F

4. 200°F

5. The growth rate of a bacteria population is proportional to its size. Initially the population is 10,000, and after 10 days it is 25,000. What is the population size after 20 days? After 30 days?

6. Suppose that in Problem 5 the population after 10 days is 6000. What is the population after 20 days? After 30 days?

7. The population of a certain city grows 6 percent a year. If the population in 1970 is 250,000, what is the population in 1980? In 2000?

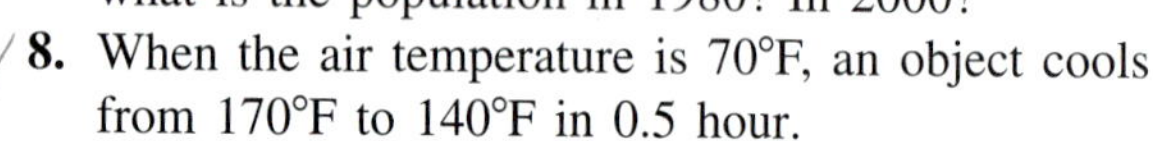

8. When the air temperature is 70°F, an object cools from 170°F to 140°F in 0.5 hour.

a. What is the temperature after 1 hour?

b. When does the temperature reach 90°F?

9. A hot coal (temperature 150°C) is immersed in ice water (temperature 0°C). After 30 seconds the temperature of the coal is 60°C. Assume that the ice water is kept at 0°C.

a. What is the temperature of the coal after 2 minutes?

b. When does the temperature of the coal reach 10°C?

10. A ball is thrown upward with an initial velocity v_0 meters per second from the top of a building h_0 meters high. Find how high the ball travels and determine when it hits the ground for the following choices of v_0 and h_0 (neglect air resistance and let $g = 9.8$ m/s^2).

a. $v_0 = 49$ m/s, $h_0 = 539$ m

b. $v_0 = 14$ m/s, $h_0 = 21$ m

c. $v_0 = 21$ m/s, $h_0 = 175$ m

d. $v_0 = 7$ m/s, $h_0 = 56$ m

e. $v_0 = 7.7$ m/s, $h_0 = 42$ m

11. A fossilized leaf contains 70 percent of a "normal" amount of ^{14}C. How old is the fossil?

12. Forty percent of a radioactive substance disappears in 100 years.

a. What is its half-life?

b. After how many years will 90 percent be gone?

13. Salt decomposes in water into sodium [Na^+] and chloride [Cl^-] ions at a rate proportional to its mass. Suppose there are 25 kilograms of salt initially and 15 kilograms after 10 hours.

a. How much salt is left after one day?

b. After how many hours is there less than 0.5 kilograms of salt left?

14. X rays are absorbed into a uniform, partially opaque body as a function not of time but of penetration distance. The rate of change of the intensity $I(x)$ of the X ray is proportional to the intensity. Here x measures the distance of penetration. The more the X ray penetrates, the lower the intensity is. The constant of proportionality is the density D of the medium being penetrated.

a. Formulate a differential equation describing this phenomenon.

b. Solve for $I(x)$ in terms of x, D, and the initial (surface) intensity $I(0)$.

15. Radioactive beryllium is sometimes used to date fossils found in deep-sea sediment. The decay of beryllium satisfies the equation

$$\frac{dA}{dt} = -\alpha A, \qquad \text{where } \alpha = 1.5 \times 10^{-7},$$

and t is measured in years. What is the half-life of beryllium?

16. In a certain medical treatment a tracer dye is injected into the pancreas to measure its function rate. A normally active pancreas secretes 4 percent of the dye each minute. A physician injects 0.3 gram of the dye, and 30 minutes later 0.1 gram remains. How much dye would remain if the pancreas were functioning normally?†

17. The estimated world population in 1986 was 4,845,000,000. Assume that the population grows at a constant rate of 1.9%. When will the world population reach 8 billion?

18. In Problem 17, at what constant rate would the population grow if it reached 6 billion in the year 2000?

19. In 1900 A.D. the world's population was 1.571 billion while in 1950 it was 2.517 billion. Assuming the Malthusian law of population growth, predict the world's population in the year 2000.

20. The population of the United States was 76,212,168 in 1900 and 92,228,496 in 1910. If population had grown at a constant percentage until 1990, what would the population have been in 1980 and 1990?

21. The population of Australia was approximately 13,400,000 in 1974 and 16,643,000 in 1990. Assume constant relative growth.
   a. Predict the population in the year 2000.
   b. When will the population be 20 million?

22. The population of New York State was 17,558,165 in 1980 and 17,990,455 in 1990. Assume a constant relative growth in population.
   a. Predict the population in 1995.
   b. When will the population reach 18.5 million?

23. The population of Florida was 9,747,197 in 1980 and 12,937,926 in 1990. Assuming that Florida continues its high growth rate, when will the population of Florida exceed the population of New York (see Problem 22)?

24. A sum of $5000 is invested at a return of 7% per year, compounded continuously. What is the investment worth after 8 years?

25. If $10,000 is invested in bonds yielding 9% compounded continuously, what will the bonds be worth in 8 years?

26. An investor buys a bond that pays 12% annual interest compounded continuously. If she invests $10,000 now, what will her investment be worth in (a) 1 year? (b) 4 years? (c) 10 years?

27. A bank offers 5% interest compounded continuously in comparison with its competitor, which offers $5\frac{1}{8}$% compounded annually. Which bank would you choose?

28. A Roman deposited 1¢ in a bank at the beginning of the year A.D. 1. If the bank paid a meager 1% interest, compounded continuously, what would the investment be worth at the beginning of 1995?

29. Atmospheric pressure is a function of altitude above sea level and is given by $dP/da = \beta P$, where β is a constant. The pressure is measured in millibars (mbar). At sea level ($a = 0$), $P(0)$ is 1013.25 mbar which means that the atmosphere at sea level will support a column of mercury 1013.25 millimeters high at a standard temperature of 15°C. At an altitude of $a = 1500$ meters, the pressure is 845.6 mbar.
   a. What is the pressure at $a = 4000$ meters?
   b. What is the pressure at 10 kilometers?
   c. In California the highest and lowest points are Mount Whitney (4418 meters) and Death Valley (86 meters below sea level). What is the difference in their atmospheric pressures?
   d. What is the atmospheric pressure at the top of Mount Everest (elevation 8848 meters)?
   e. At what elevation is the atmospheric pressure equal to 1 mbar?

30. A bacteria population is known to grow exponentially. The following data are collected:

Number of Days	Number of Bacteria
5	936
10	2,190
20	11,993

   a. What is the initial population?
   b. If the present growth rate continues, what is the population after 60 days?

***31.** The president and vice-president sit down for coffee. They are each served a cup of hot black coffee (at the same temperature). The president immediately adds cream to his coffee, stirs it,

†This and similar mathematical models in medicine are discussed by J. S. Rustagi, "Mathematical Models in Medicine," *International Journal of Mathematical Education in Science and Technology* 2(1971): 193–203.

and waits. The vice-president waits 10 minutes and then adds the same amount of cream (which has been kept cool) to her coffee and stirs it in. Then they both drink. Assuming that the temperature of the cream is lower than that of the air, who drinks the hotter coffee? [*Hint:* Use Newton's law of cooling. It is necessary to treat each case separately and to keep track of the volumes of coffee, cream, and the coffee-cream mixture.]†

ANSWERS TO SELF-QUIZ

I. b **II.** a, b (a implies b) **III.** d **IV.** f [The limit is $y(0)$.]

10.3 FIRST-ORDER EQUATIONS—SEPARATION OF VARIABLES

In this section we discuss a technique that can be used to solve a variety of differential equations.

Consider the differential equation

$$\frac{dy}{dx} = f(x, y) \tag{1}$$

and suppose that the function $f(x, y)$ can be factored into a product,

$$f(x, y) = g(x)h(y), \tag{2}$$

where $g(x)$ and $h(y)$ are each functions of only one variable. When this occurs, equation (1) can be solved by the method of **separation of variables**. To solve the equation, we substitute the product (2) into (1) to obtain

$$\frac{dy}{dx} = g(x)h(y),$$

or

$$\frac{1}{h(y)}\frac{dy}{dx} = g(x).$$

Integrating both sides of this equation with respect to x, we have

$$\int \frac{1}{h(y)}\frac{dy}{dx}\,dx = \int g(x)\,dx + C, \tag{3}$$

which can be rewritten as

$$\int \frac{1}{h(y)}\,dy = \int g(x)\,dx + C,^{\ddagger} \tag{4}$$

---

†This is a famous old problem that keeps popping up (with an ever-changing pair of characters) in books on games and puzzles in mathematics. The problem is hard and has stymied many a mathematician. Do not get frustrated if you cannot solve it. The trick is to write everything down and to keep track of all the variables. The fact that the air is warmer than the cream is critical. It should also be noted that guessing the correct answer is fairly easy; proving that your guess is correct is what makes the problem difficult.

‡There is no need to use two constants of integration,

$$\int \frac{dy}{h(y)} + C_1 = \int g(x)\,dx + C_2,$$

since a single arbitrary constant $C = C_2 - C_1$ serves the same purpose.

since the left side of equation (3) is precisely equal to the left side of equation (4) under a change of variables. Observe that this procedure, in effect, allows us to treat the derivative $\frac{dy}{dx}$ as if it were a fraction: we simply separate all the terms involving x on one side and those involving y on the other. This gives the method its name. If both integrals in equation (4) can be evaluated, a solution to the differential equation (1) is obtained. We illustrate this method with several examples.

EXAMPLE 1

SOLVING BY SEPARATING VARIABLES

Solve the differential equation

$$\frac{dy}{dx} = 2xy.$$

SOLUTION: Let $g(x) = 2x$ and $h(y) = y$ and separate variables to obtain

$$\frac{dy}{y} = 2x\,dx \qquad \text{or} \qquad \int \frac{dy}{y} = 2\int x\,dx + C.$$

Integration yields

$$\ln|y| = x^2 + C \qquad \text{or} \qquad |y| = e^{x^2 + C} = e^C e^{x^2},$$

since $e^{\ln a} = a$. The term e^C is an arbitrary *positive* constant. Since we want to solve for y, we need to remove the absolute value. This is done by setting

$$y = \pm e^C e^{x^2}.$$

But $\pm e^C$ can be any nonzero arbitrary real number, so we replace this term by the letter k, yielding

$$y = k\,e^{x^2}. \tag{5}$$

Since k is any nonzero arbitrary real number, it follows that the original differential equation for Example 1 has infinitely many solutions, one for each value of k. Observe that we also have a solution if $k = 0$, since then $y \equiv 0$ and $dy/dx \equiv 0$, satisfying the differential equation. We call equation (5) the *general solution* for this differential equation.

It should not surprise us that a differential equation has infinitely many solutions: this is simply a consequence of the integration process, since each integral involves an arbitrary constant.

In every case where we wish to single out a specific solution, we require additional information. Often that additional information is given as an *initial condition,* such as $y(1) = 2$. Had that condition been given in Example 1, we would set $x = 1$ and $y = 2$ in equation (5) obtaining

$$2 = y(1) = k\,e^{1^2} = k\,e$$

or $k = 2/e$. Then, the *particular solution* for that initial value problem is

$$y(x) = \frac{2}{e} e^{x^2} = 2e^{x^2 - 1}.$$

WARNING: Do not do the following:

There is a common error that students make when they try to separate variables. Consider the following problem, *done incorrectly*:

$$\frac{dy}{dx} = y + x$$

$$dy = y\,dx + x\,dx$$

$$\int dy = \int y\,dx + \int x\,dx$$

$$y = \frac{y^2}{2} + \frac{x^2}{2} + C$$

What is wrong? *It is not correct* that

$$\int y\,dx = \frac{y^2}{2},$$

because y is a function of x. For example, if $y = \cos x$, then

$$\int y\,dx = \int \cos x\,dx = \sin x + C \neq \frac{y^2}{2} + C = \frac{\cos^2 x}{2} + C.$$

Of course

$$\int y\,dy = \frac{y^2}{2} + C.$$

So, be careful. (Using the techniques of Section 10.4, we can show that the solutions to $dy/dx = y + x$ are $y = -x - 1 + Ce^x$ for every real number C.)

EXAMPLE 2 SEPARATION OF VARIABLES

Solve the initial value problem

$$\frac{dy}{dx} = 3x^2 e^{-y}, \qquad y(0) = 2. \tag{6}$$

SOLUTION: Since $e^{-y} = 1/e^y$, when we separate variables we get

$$e^y\,dy = 3x^2\,dx \qquad \text{or} \qquad \int e^y\,dy = \int 3x^2\,dx + C.$$

Integrating, we have

$$e^y = x^3 + C \qquad \text{or} \qquad y = \ln(x^3 + C).$$

To determine C we use the initial condition. Set $x = 0$, then

$$2 = y(0) = \ln(0^3 + C) = \ln C \qquad \text{or} \qquad C = e^2.$$

Thus the particular solution to the initial value problem is $y = \ln(x^3 + e^2)$.

Observe that this solution is defined only in the domain where $x^3 + e^2 > 0$, or $x > -e^{2/3}$, since the natural logarithm is defined only for positive numbers. Thus a solution to a differential equation need not be defined for all values of the independent variable x.

EXAMPLE 3

SEPARATION OF VARIABLES

Solve the differential equation

$$\frac{dy}{dx} = 2x\,(y^2 + 1).$$

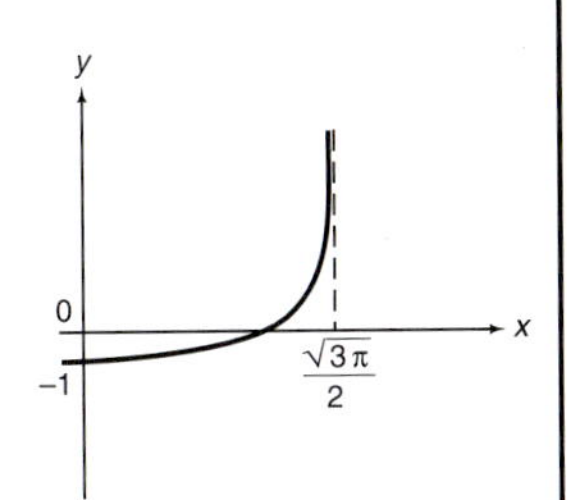

FIGURE 1
Solution is unbounded at $x = \frac{\sqrt{3\pi}}{2}$.

SOLUTION: Separating variables, we have

$$\frac{dy}{1 + y^2} = 2x\,dx \qquad \text{or} \qquad \int \frac{dy}{1 + y^2} = \int 2x\,dx + C,$$

or $\tan^{-1} y = x^2 + C$. Thus, $y = \tan(x^2 + C)$. Note that the solution is only defined for values of x satisfying $|x^2 + C| < \pi/2$, because that is the range of the arc tangent. This suggests that any solution to this differential equation quickly becomes unbounded. In Figure 1 we illustrate the solution that satisfies the initial condition $y(0) = -1$.

EXAMPLE 4

ESCAPE VELOCITY

In Example 10.2.1 we studied the motion of a body falling freely subject to the gravitational force of the earth. In that example we assumed that the distance the body fell was small in comparison to the radius R of the earth.† However, if we wish to study the equation of motion for a communications satellite or an interplanetary vehicle, the distance r of the object from the center of the earth may be considerably larger than R. In such a case the approximation we made in obtaining equation (10.2.3) on page 684 is no longer valid. Returning to equation (10.2.4) on page 685,

$$\frac{d^2r}{dt^2} = -g\frac{R^2}{r^2}, \tag{7}$$

and setting $v = dr/dt$, we see by the chain rule that

$$\frac{d^2r}{dt^2} = \frac{dv}{dt} = \frac{dv}{dr}\frac{dr}{dt} = \frac{v\,dv}{dr}. \tag{8}$$

Hence equation (7) can be rewritten as

$$\frac{v\,dv}{dr} = -g\frac{R^2}{r^2},$$

where g (≈ 9.81 m/s^2) and R are constant. Separating variables and integrating, we have

$$\int v\,dv = -gR^2\int \frac{dr}{r^2} + C,$$

or

$$\tfrac{1}{2}v^2 = \frac{gR^2}{r} + C.$$

†The radius of the earth is approximately 6378 kilometers (3963 miles) at the equator and 6357 kilometers (3950 miles) at the poles.

Assuming that the object is at the surface of the earth when $t = 0$, we get

$$\tfrac{1}{2}v(0)^2 = g\frac{R^2}{R} + C,$$

or

$$C = \tfrac{1}{2}v(0)^2 - gR.$$

Thus

$$v^2 = 2g\frac{R^2}{r} + v(0)^2 - 2gR. \tag{9}$$

For the object to escape the gravitational force of the earth, it is necessary that $v > 0$ for all time t. If we select $v(0) = \sqrt{2gR}$, the last two terms in equation (9) cancel, so that $v^2 > 0$ for all r. Observe that any smaller choice for $v(0)$ allows the right side of equation (9) to be zero for some sufficiently large value of r. Thus $v(0) = \sqrt{2gR} \approx 11.2$ kilometers per second is the initial velocity an object at the surface of the earth needs to escape the gravitational attraction of the earth. This is called the **escape velocity**.

The substitution we used in equation (8) can always be used to reduce an equation involving a second derivative to an equation involving only first derivatives, provided that the independent variable does not appear explicitly in the equation.

EXAMPLE 5

REENTRY INTO ATMOSPHERE

One of the problems facing space vechicles is their reentry into the earth's atmosphere. For simplicity, assume that gravity has a negligible effect in determining the maximum deceleration during reentry of a vehicle heading toward (a flat) earth at a constant angle α and speed V. Let $s(t)$ denote the distance to impact of the reentry vehicle (see Figure 2).

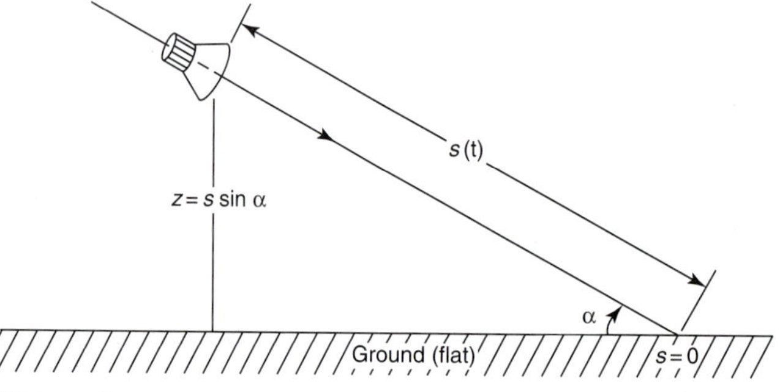

FIGURE 2
Reentry into atmosphere

Then ds/dt is its velocity and d^2s/dt^2 its deceleration. Assume that the density of the earth's atmosphere at height z is

$$\rho(z) = \rho_0 e^{-z/k_0},$$

where ρ_0 and k_0 are constants, and that the **drag force** of the atmosphere is proportional to the product of the air density and the square of the velocity of the vehicle. By Newton's second law of motion,

$$m\frac{d^2s}{dt^2} = \beta\rho_0 e^{-z/k_0}\left(\frac{ds}{dt}\right)^2,$$

where m is the mass of the vehicle and β is the constant of proportionality. Dividing both sides by m and substituting $N = \beta\rho_0/m$ and $k = k_0/\sin\alpha$, we get the equation of motion $[-z/k_0 = -s\sin\alpha/k_0 = -s/(k_0/\sin\alpha) = -s/k]$

$$\frac{d^2s}{dt^2} = Ne^{-s/k}\left(\frac{ds}{dt}\right)^2. \tag{10}$$

Letting $v = ds/dt$ and using the same technique as in equation (8), we have

$$\frac{d^2s}{dt^2} = \frac{dv}{dt} = \frac{dv}{ds}\frac{ds}{dt} = v\frac{dv}{ds},$$

or

$$v\frac{dv}{ds} = Ne^{-s/k}v^2.$$

Separating variables, we get

$$\int\frac{dv}{v} = \int Ne^{-s/k}\,ds + C_1,$$

or

$$\ln|v| = -kNe^{-s/k} + C_1,$$

so

$$v = Ce^{-kNe^{-s/k}}.$$

The arbitrary constant C can now be determined by noting that the vehicle is traveling at velocity V when it is very far from earth. Thus, letting s tend to ∞, we have

$$V = Ce^0 = C,$$

or

$$\frac{ds}{dt} = v(s) = Ve^{-kNe^{-s/k}}. \tag{11}$$

Combining equations (10) and (11), we get

$$\frac{d^2s}{dt^2} = Ne^{-s/k}(Ve^{-kNe^{-s/k}})^2. \tag{12}$$

To find the *maximum* deceleration, we must determine the value of s that maximizes the right side $f(s)$ of equation (12). This can be found by differentiating the right side of equation (12) with respect to s by logarithmic differentiation and using the first derivative test of calculus:

$$f'(s) = f(s)\left(-\frac{1}{k} + 2Ne^{-s/k}\right) = 0.$$

Solving for s, we get $e^{-s/k} = 1/2kN$, or $s = k\ln(2kN)$. This value is easily seen to be a maximum, since $f > 0$ and the term in parentheses is decreasing. Substituting $s_{max} = k\ln(2kN)$ into equation (12), we have

$$\left(\frac{d^2s}{dt^2}\right)\Bigg|_{s_{max}} = Ne^{-\ln 2kN}(Ve^{-kNe^{-\ln 2kN}})^2$$

$$= \frac{N}{2kN}(Ve^{-(kN/2kN)})^2 = \frac{V^2}{2ek} \overset{k = k_0/\sin\alpha}{=} \frac{V^2 \sin\alpha}{2ek_0},$$

which is independent of the drag coefficient $N = \beta\rho_0/m$.

EXAMPLE 6

LOGISTIC GROWTH

The growth rate per individual in a population is the difference between the average birth rate and the average death rate. Suppose that in a given population the average birth rate is a positive constant β, but the average death rate, because of the effects of crowding and increased competition for the available food, is proportional to the size of the population. We call the constant of proportionality δ (which is >0). If $P(t)$ denotes the population at time t, then dP/dt is the growth rate of the population. The growth rate per individual is given by

$$\frac{1}{P}\frac{dP}{dt}.$$

Then, using the conditions described above, we have

$$\frac{1}{P}\frac{dP}{dt} = \beta - \delta P$$

or

$$\frac{dP}{dt} = P(\beta - \delta P). \tag{13}$$

This differential equation, together with the condition

$$P(0) = P_0 \qquad \text{(the initial population)}, \tag{14}$$

is an initial value problem. To solve, we have

$$\frac{dP}{P(\beta - \delta P)} = dt$$

or

$$\int \frac{dP}{P(\beta - \delta P)} = \int dt = t + C. \tag{15}$$

To calculate the integral on the left, we use partial fractions. We have (verify this)

$$\frac{1}{P(\beta - \delta P)} = \frac{1}{\beta P} + \frac{\delta}{\beta(\beta - \delta P)},$$

so that

$$\int \frac{dP}{P(\beta - \delta P)} = \int \frac{1}{\beta}\frac{dP}{P} + \frac{\delta}{\beta}\int \frac{dP}{\beta - \delta P}$$
$$= \frac{1}{\beta}\ln|P| - \frac{1}{\beta}\ln|\beta - \delta P| = t + C$$

or

$$\ln\left|\frac{P}{\beta - \delta P}\right| = \ln|P| - \ln|\beta - \delta P| = \beta t - \beta C = \beta t - C_1$$

and

$$\frac{P}{\beta - \delta P} = e^{\beta t - C_1} = C_2 e^{\beta t}. \tag{16}$$

Here $C_1 = -\beta C$ and $C_2 = \pm e^{-C_1}$.

Using the initial condition (14), we have

$$\frac{P_0}{\beta - \delta P_0} = C_2 e^0 = C_2. \tag{17}$$

Finally, we insert (17) into equation (16) to obtain

$$\frac{P}{\beta - \delta P} = \frac{P_0}{\beta - \delta P_0} e^{\beta t}$$
$$P = \frac{P_0}{\beta - \delta P_0} e^{\beta t}(\beta - \delta P) = \frac{\beta P_0}{\beta - \delta P_0} e^{\beta t} - \frac{\delta P_0 e^{\beta t}}{\beta - \delta P_0} P$$
$$P\left(1 + \frac{\delta P_0}{\beta - \delta P_0} e^{\beta t}\right) = \frac{\beta P_0}{\beta - \delta P_0} e^{\beta t} = P\left[\frac{\beta - \delta P_0 + \delta P_0 e^{\beta t}}{\beta - \delta P_0}\right]$$
$$P(t) = \frac{\beta - \delta P_0}{\beta - \delta P_0 + \delta P_0 e^{\beta t}}\left[\frac{\beta P_0}{\beta - \delta P_0} e^{\beta t}\right] = \frac{\beta P_0 e^{\beta t}}{\beta - \delta P_0 + \delta P_0 e^{\beta t}}$$

multiply and divide by $e^{\beta t}$ / divide top and bottom by P_0

$$\overset{\downarrow}{=} \frac{\beta P_0}{\delta P_0 + (\beta - \delta P_0)e^{-\beta t}} \overset{\downarrow}{=} \frac{\beta}{\delta + \left(\dfrac{\beta}{P_0} - \delta\right)e^{\beta t}}.$$

That is,

$$P(t) = \frac{\beta}{\delta + [(\beta/P_0) - \delta]e^{-\beta t}}. \tag{18}$$

Equation (13) is called the **logistic equation** and we have shown that the solution to the logistic equation with initial population P_0 is given by (18). Sketches of the growth governed by the logistic equation are given in Figure 3.

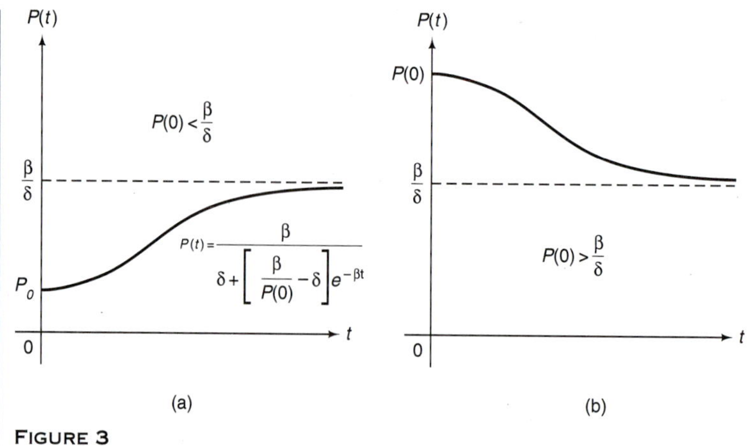

FIGURE 3
Two curves of logistic growth

WHEN IS A DIFFERENTIAL EQUATION SEPARABLE?

In this section we saw how to solve a first-order differential equation when the equation was separable. However, it is not always clear that an equation is separable. For example, it is obvious that $f(x, y) = e^x \cos y$ is separable, but it is not so obvious that $f(x, y) = 2x^2 + y - x^2y + xy - 2x - 2$ is separable.† In this perspective, we give conditions that ensure that an equation is separable.‡

THEOREM 1

Suppose that $f(x, y) = g(x)h(y)$, where both g and h are differentiable. Then

$$f(x, y)f_{xy}(x, y) = f_x(x, y)f_y(x, y). \qquad (19)$$

Here the subscripts denote partial derivatives.

PROOF: Observe that

$$f_x(x, y) = g'(x)h(y)$$
$$f_y(x, y) = g(x)h'(y)$$
$$f_{xy}(x, y) = g'(x)h'(y).$$

Hence

$$\begin{aligned} f(x, y)f_{xy}(x, y) &= g(x)h(y)g'(x)h'(y) = [g'(x)h(y)][g(x)h'(y)] \\ &= f_x(x, y)f_y(x, y). \end{aligned}$$ ■

It turns out that, under further conditions, if equation (19) holds, then $f(x, y)$ is separable. For what follows, D denotes an open disk in the xy-plane; that is, $D = \{(x, y): (x - a)^2 + (y - b)^2 < r^2\}$, where a, b, and r are real numbers and $r > 0$.

† $2x^2 + y - x^2y + xy - 2x - 2 = (1 + x - x^2)(y - 2)$.
‡ The results here are based on the paper "When is an Ordinary Differential Equation Separable?" by David Scott in *American Mathematical Monthly* 92 (1985): 422–423.

THEOREM 2

Suppose that in D, f, f_x, f_y, and f_{xy} exist and are continuous, $f(x, y) \neq 0$, and equation (19) holds. Then there are continuously differentiable functions $g(x)$ and $h(y)$ such that, for every $(x, y) \in D$,

$$f(x, y) = g(x)h(y). \tag{20}$$

PROOF: Since $f(x, y) \neq 0$ and f is continuous on D, f has the same sign on D. Assume $f(x, y) > 0$ for $(x, y) \in D$. A similar proof works if $f(x, y) < 0$ if f is replaced by $-f$. Now, from the quotient rule of differentiation,

$$\frac{\partial}{\partial y}\frac{f_x(x, y)}{f(x, y)} = \frac{f(x, y)f_{xy}(x, y) - f_x(x, y)f_y(x, y)}{f^2(x, y)} \overset{\text{equation (19)}}{=} 0.$$

When the partial derivative with respect to y of a function of x and y is zero, then that function must be a function of x only. Thus there is a function $\alpha(x)$ such that

$$\frac{f_x(x, y)}{f(x, y)} = \alpha(x).$$

Also, since $f(x, y) > 0$, $\ln f(x, y)$ is defined and

$$\frac{\partial}{\partial x}\ln f(x, y) = \frac{f_x(x, y)}{f(x, y)} = \alpha(x).$$

The function $\alpha(x)$ is continuous in D because it is the quotient of continuous functions and the function in the denominator is nonzero. Let $\beta(x) = \int \alpha(x)\,dx$. Then

$$\ln f(x, y) = \int \left[\frac{\partial}{\partial x}\ln f(x, y)\right] dx = \int \alpha(x)\,dx = \beta(x) + \gamma(y),$$

where γ is a function of y only. (The partial derivative with respect to x of a function of y only is zero. Thus $\gamma(y)$ represents the most general constant of integration.) Finally, let $g(x) = e^{\beta(x)}$ and $h(y) = e^{\gamma(y)}$. Then

$$f(x, y) = e^{\ln f(x, y)} = e^{\beta(x)+\gamma(y)} = e^{\beta(x)}e^{\gamma(y)} = g(x)h(y).$$ ■

EXAMPLE 7

A SEPARABLE DIFFERENTIAL FUNCTION

Let $f(x, y) = 2x^2 + y - x^2y + xy - 2x - 2$. Then

$$\begin{aligned} f_x(x, y) &= 4x - 2xy + y - 2 \\ f_y(x, y) &= 1 - x^2 + x \\ f_{xy}(x, y) &= -2x + 1 \\ f(x, y)f_{xy}(x, y) &= (2x^2 + y - x^2y + xy - 2x - 2)(-2x + 1) \\ &= -4x^3 - xy + 2x^3y - 3x^2y + 6x^2 + 2x + y - 2 \end{aligned}$$

and

$$\begin{aligned} f_x(x, y)f_y(x, y) &= (4x - 2xy + y - 2)(1 - x^2 + x) \\ &= 2x - xy + y - 2 - 4x^3 + 2x^3y - 3x^2y + 6x^2. \end{aligned}$$

Since the last two expressions are equal, we conclude by Theorem 2 that $f(x, y)$ is separable.

EXAMPLE 8 A FUNCTION WHICH IS NOT SEPARABLE

Let $f(x, y) = 1 + xy$. Then

$$f_x(x, y) = y$$
$$f_y(x, y) = x$$
$$f_{xy}(x, y) = 1$$
$$f(x, y)f_{xy}(x, y) = 1 + xy$$

and

$$f_x(x, y)f_y(x, y) = xy.$$

Since the last two expressions are unequal, we conclude that $f(x, y)$ is not separable.

PROBLEMS 10.3

SELF-QUIZ

Multiple Choice

I. $f(x) = e^{2x}$ is a solution of the differential equation ________.
a. $y' + y = 0$ **b.** $y' - y = 0$
c. $2y' - y = 0$ **d.** $y' - 2y = 0$

II. $g(x) = x^2$ is a solution of the differential equation ________.
a. $y' - 2y = 0$ **b.** $2y' - y = 0$
c. $xy' - 2y = 0$ **d.** $2xy' - y = 0$

III. ________ is a solution of the differential equation $xy' - y = 0$.
a. $f(x) = 17$ **b.** $g(x) = x$
c. $F(x) = x^2$ **d.** $G(x) = e^{1/x}$

IV. ________ is a solution of the differential equation $xy' - 3y = 0$.
a. $f(x) = e^{3x}$ **b.** $g(x) = 13.807x$
c. $F(x) = e^{2 \ln x}$ **d.** $G(x) = (-5x)^3$

V. ________ is the solution of the initial-value problem $y' + 2y = 0$, $y(0) = 4$.
a. $f(x) = 4e^{2x}$ **b.** $g(x) = 4e^{-2x}$
c. $F(x) = e^{-2x} + 3$ **d.** $G(x) = (e^{-x} + 1)^2$

True–False

VI. $\dfrac{dy}{dx} = x^2 + y$ is separable.

VII. $\dfrac{dy}{dx} = x^2(y + \ln y)$ is separable.

VIII. $\dfrac{dy}{dx} = 3x + xy^2$ is separable.

IX. $\dfrac{dy}{dx} = xy + y - x - 1$ is separable.

X. $\dfrac{dy}{dx} = xy + y - x$ is separable.

In Problems 1–23 find the general solution by separating variables. If a specific condition is given, find the particular solution that satisfies that condition and sketch its graph.

1. $\dfrac{dy}{dx} = x\,e^y$

2. $xy' = 3y$, $y(2) = 5$

3. $\dfrac{dP}{dQ} = 2\sqrt{P}$

4. $\dfrac{dy}{dx} = 2x\sqrt{y}$

5. $\dfrac{dx}{dt} = txe^{t^2}$

6. $\dfrac{dr}{ds} = sre^{s^2}$, $r(0) = 1$

7. $\dfrac{dy}{dt} + 4y = y(e^{-t} + 4)$

8. $\dfrac{dz}{dx} = \dfrac{\sqrt{x(z^2 + 4)}}{z}$, $z(1) = 1$

9. $\dfrac{dy}{dx} = \dfrac{e^y x}{e^y + x^2 e^y}$

10. $\dfrac{dx}{dy} = x \cos y$, $x\left(\dfrac{\pi}{2}\right) = 1$

11. $\dfrac{dz}{dr} = r^2(1 + z^2)$

12. $\dfrac{dy}{dx} + y = y(xe^{x^2} + 1)$, $y(0) = 1$

13. $\dfrac{dP}{dQ} = P(\cos Q + \sin Q)$

14. $\dfrac{dy}{dx} = y^2(1 + x^2)$, $y(0) = 1$

15. $\dfrac{ds}{dt} + 2s = st^2$, $s(0) = 1$

16. $\dfrac{dx}{dt} + (\cos t)e^x = 0$ 17. $\cot x \dfrac{dy}{dx} + y + 3 = 0$

18. $\dfrac{dx}{dt} = x(1 - \sin t)$, $x(0) = 1$

19. $x^2 \dfrac{dy}{dx} + y^2 = 0$, $y(1) = 3$

20. $yy' = e^x$ 21. $y' + y = y(xe^x + 1)$

22. $e^x\left(\dfrac{dx}{dt} + 1\right) = 1$, $x(0) = 1$

23. $\dfrac{dy}{dx} = \dfrac{x}{y} - \dfrac{x}{1 + y}$, $y(0) = 1$

24. The table below shows data for the growth of yeast in a culture. Use equation (17), with $\beta = 0.55$ and $\delta = 8.3 \times 10^{-4}$, to calculate the predicted growth, and find the percentage error between observed and predicted values at $t = 0$, 9, and 18 hours.

Time in Hours	Yeast Biomass	Time in Hours	Yeast Biomass
0	9.6	10	513.3
1	18.3	11	559.7
2	29.0	12	594.8
3	47.2	13	629.4
4	71.1	14	640.8
5	119.1	15	651.1
6	174.6	16	655.9
7	257.3	17	659.6
8	350.7	18	661.8
9	441.0		

Data from R. Pearl, "The Growth of Population," *Quarterly Review of Biology* 2 (1927): 532–548.

25. **Obsidian dating**† is a technique used by archaeologists which allows the dating of certain artifacts well beyond the reliable 40,000-year range of carbon dating. Obsidian, a glassy volcanic rock, absorbs water from the atmosphere. The water forms a hydration layer—a compound of water molecules and obsidian molecules. The depth of the hydration layer $x(t)$ beneath the surface of an obsidian artifact is a function of the time t that has elapsed since the manufacture of that artifact. The velocity at which the hydration layer grows is inversely proportional to its depth. Find the depth of the layer for all time t.

26. A rocket is launched from an initial position (x_0, y_0) with an initial speed v_0 and with an angle θ $(0 \le \theta \le \pi/2)$. Find its horizontal and vertical coordinates $x(t)$ and $y(t)$ as functions of time. Assume that there is no air resistance, and that the force of gravity g is constant.

27. The half-life of a radioactive substance is defined as the time required to decompose 50 percent of the substance. If $r(t)$ denotes the amount of the radioactive substance present after t years, $r(0) = r_0$, and the half-life is H years, what is a differential equation for $r(t)$ taking all side conditions into account?

28. A bacteria population is known to double every 3 hours. If the population initially consists of 1000 bacteria, how long does it take for the population to reach 10,000?

29. The economist Vilfredo Pareto (1848–1923) discovered that the rate of decrease of the number of people y in a stable economy having an income of at least x dollars is directly proportional to the number of such people and inversely proportional to their income. Obtain an expression (**Pareto's law**) for y in terms of x.

30. It is the last lap on a speedway, the finish line is straight ahead 2 miles away, and you lead your nearest opponent by 3 miles. Both of you are going at the same speed, but you just ran out of gasoline. Your car decelerates at a rate proportional to the square of your instantaneous speed. At the end of a mile, your speed is exactly half of what it was when you ran out of gas. Will you win the race?

**31. On a certain day it begins to snow early in the

† See "A New Dating Method Using Obsidian" by I. Friedman and R. L. Smith in *American Antiquity* 25 (1960): 476–522.

morning, and the snow continues to fall at a constant rate. The velocity at which a snowplow is able to clear a road is inversely proportional to the height of the accumulated snow. The snowplow starts at 11 A.M. and clears four miles by 2 P.M. By 5 P.M. it clears another two miles. When did it start snowing?†

**32. (Snowplow chase).‡ If a second snowplow starts at noon along the same path as the snowplow in Problem 31, when does it catch up to the first snowplow?

**33. (Snowplow collision).§ Suppose three identical snowplows start clearing the same road at 10 A.M., 11 A.M., and noon. If all three collide sometime after noon, when did it start snowing?

ANSWERS TO SELF-QUIZ

I. d **II.** c **III.** b **IV.** d **V.** b **VI.** False
VII. True **VIII.** True **IX.** True **X.** False

10.4 LINEAR FIRST-ORDER DIFFERENTIAL EQUATIONS

An nth-order differential equation is **linear** if it can be written in the form

$$\frac{d^n y}{dx^n} + a_{n-1}(x)\frac{d^{n-1}y}{dx^{n-1}} + \cdots + a_1(x)\frac{dy}{dx} + a_0(x)y = f(x).$$

Hence a first-order linear equation has the form

$$\frac{dy}{dx} + a(x)y = f(x), \tag{1}$$

and a second-order linear equation can be written as

$$\frac{d^2y}{dx^2} + a(x)\frac{dy}{dx} + b(x)y = f(x).$$

The notation indicates that $a(x)$, $b(x)$, $f(x)$ and so on are functions of x alone. In most of this book we will assume that they are continuous functions on some interval of the real line.

DEFINITION HOMOGENEOUS AND NONHOMOGENEOUS EQUATIONS

If the function $f(x)$ is the zero function, the linear differential equation is said to be **homogeneous**. Otherwise, we say that the linear differential equation is **nonhomogeneous**. Any differential equation that cannot be written in the form above is said to be **nonlinear**. For example,

$$\frac{dy}{dx} = y^2$$

is nonlinear. ■

---

†Based on problem E275 of the Otto Dunkel Memorial Problem Book, *American Mathematical Monthly* 64 (1957): 54.
‡This problem was first proposed by Fred Wan, *Applied Mathematics Notes,* (January 1975): 6–11.
§This problem was first proposed by M. S. Klamkin, *American Mathematical Monthly* 59 (1952): 42 (problem E963).

Before dealing with the nonhomogeneous first-order linear equation (1), it is important to discuss the solution of the homogeneous equation

$$\frac{dy}{dx} + a(x)y = 0, \tag{2}$$

or

$$\frac{dy}{dx} = -a(x)y.$$

Separating variables, we have

$$\int \frac{dy}{y} = -\int a(x)\,dx + C.$$

Integrating, we have

$$\ln|y| = -\int a(x)\,dx + C$$

and

$$y = C_1 e^{-\int a(x)\,dx}. \tag{3}$$

This is the general solution to equation (2); it indicates that a solution is obtainable whenever the antiderivative can be found. We illustrate this situation with two examples.

EXAMPLE 1

SOLVING A HOMOGENEOUS DIFFERENTIAL EQUATION

Solve the homogeneous differential equation

$$y' + 3y = 0.$$

SOLUTION: Rewriting the equation as

$$y' = -3y$$

and separating variables, we have

$$\int \frac{dy}{y} = -3\int dx + C$$

$$\ln|y| = -3x + C$$

$$y = C_1 e^{-3x}.$$

EXAMPLE 2

SOLVING A HOMOGENEOUS INITIAL VALUE PROBLEM

Solve the equation

$$\frac{dy}{dx} = -2xy, \qquad y(1) = 1 \tag{4}$$

SOLUTION: Separating variables, we have

$$\int \frac{dy}{y} = -2 \int x\,dx + C$$
$$\ln|y| = -x^2 + C$$
$$y = C_1 e^{-x^2}.$$

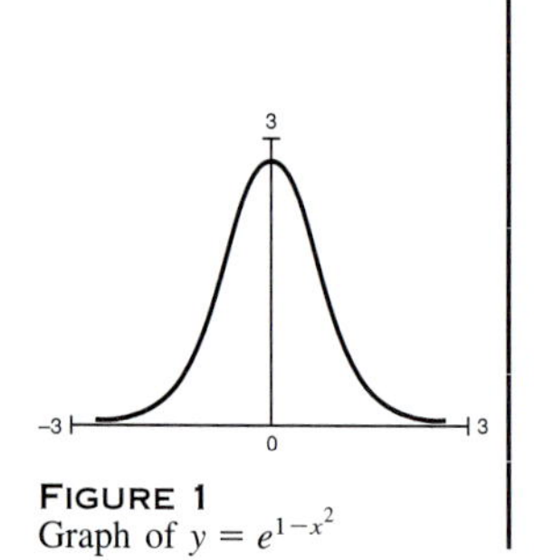

FIGURE 1
Graph of $y = e^{1-x^2}$

Since $y(1) = 1 = C_1 e^{-1}$, it follows that $C_1 = e$; the particular solution to the initial-value problem (4) is therefore

$$y = e^{1-x^2}.$$

A graph of this solution is given in Figure 1.

We see that equation (3) can be written in the form

$$ye^{\int a(x)\,dx} = C \tag{5}$$

by multiplying both sides by $e^{\int a(x)\,dx}$. If we differentiate both sides of equation (5), we obtain

$$y'e^{\int a(x)\,dx} + a(x)ye^{\int a(x)\,dx} = [y' + a(x)y]e^{\int a(x)\,dx} = 0, \tag{6}$$

$$\left(\frac{d}{dx}e^{\int a(x)\,dx} = e^{\int a(x)\,dx}\frac{d}{dx}\int a(x)\,dx = a(x)e^{\int a(x)\,dx}\right)$$

since the derivative of an indefinite integral is the integrand. Notice that the expression in brackets is the left side of the original differential equation (2). We call the exponential

$$e^{\int a(x)\,dx} \tag{7}$$

an **integrating factor** for the linear equation (2), because when we multiply equation (2) by (7), the left side of the result can be integrated.

We can use the integrating factor to obtain the solution of the nonhomogeneous linear equation

$$\frac{dy}{dx} + a(x)y = f(x). \tag{8}$$

We multiply both sides of equation (8) by the integrating factor $e^{\int a(x)\,dx}$ to obtain

$$[y' + a(x)y]e^{\int a(x)\,dx} = f(x)e^{\int a(x)\,dx}. \tag{9}$$

But as we have seen in going from equation (5) to equation (6), the left side of equation (9) is the derivative of $ye^{\int a(x)\,dx}$. Thus

$$\frac{d}{dx}[ye^{\int a(x)\,dx}] = f(x)e^{\int a(x)dx}. \tag{10}$$

Integrating both sides of equation (10) with respect to x, we have

$$\int \frac{d}{dx}[ye^{\int a(x)\,dx}]\,dx = \int f(x)e^{\int a(x)\,dx}\,dx + C,$$

or

$$ye^{\int a(x)\,dx} = \int f(x)e^{\int a(x)\,dx}\,dx + C.$$

THE SOLUTION TO A LINEAR FIRST-ORDER DIFFERENTIAL EQUATION (8)

$$y = \left[\int f(x)e^{\int a(x)\,dx}\,dx + C\right]e^{-\int a(x)\,dx}. \tag{11}$$

Equation (11) provides an expression for the general solution of the first-order nonhomogeneous linear differential equation (8). It is usually better to go through the process of multiplying both sides of (8) by the integrating factor to obtain the solution than to try to memorize equation (11). We illustrate this process with several examples.

EXAMPLE 3

SOLVING A NONHOMOGENEOUS INITIAL-VALUE PROBLEM

Solve the nonhomogeneous linear equation

$$y' = y + x^2, \qquad y(0) = 1. \tag{12}$$

SOLUTION: Rewriting equation (12) in the form

$$y' - y = x^2, \tag{13}$$

we see that $a(x) = -1$; thus the integrating factor is

$$e^{-\int dx} = e^{-x}.$$

Multiplying both sides of equation (13) by e^{-x}, we get

$$e^{-x}(y' - y) = x^2e^{-x},$$

or

$$(ye^{-x})' = x^2e^{-x}.$$

Integrating both sides, we have

$$ye^{-x} = \int x^2e^{-x}\,dx + C$$

$$= C - (x^2 + 2x + 2)e^{-x},$$

↑ Integrate by parts twice

so

$$y = Ce^x - (x^2 + 2x + 2).$$

Finally, setting $x = 0$, we get

$$1 = y(0) = C - 2,$$

so $C = 3$ and the solution of the initial-value problem is

$$y = 3e^x - (x^2 + 2x + 2).$$

Note the solution can be written as the **superposition**, or sum, of two functions: $y = 3e^x$, which is a solution to the homogeneous equation $y' = y$ and $y = -(x^2 + 2x + 2)$ which is a solution to the nonhomogeneous equation $y' = y + x^2$. We will say more about this phenomenon in Section 10.7. A graph of this solution appears in Figure 2.

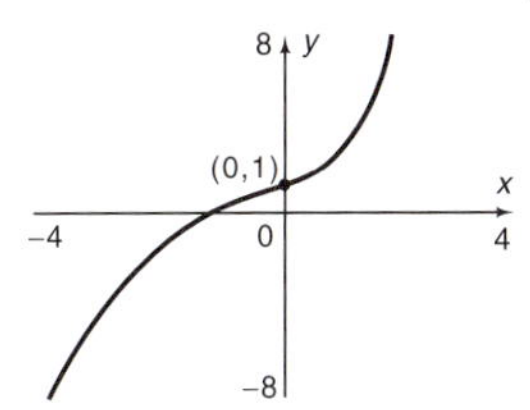

FIGURE 2
Graph of
$y = 3e^x - (x^2 + 2x + 2)$

EXAMPLE 4 SOLVING A NONHOMOGENEOUS INITIAL-VALUE PROBLEM

Consider the equation $dy/dx = x^3 - 2xy$, where $y = 1$ when $x = 1$. Rewriting the equation as $dy/dx + 2xy = x^3$, we see that $a(x) = 2x$ and the integrating factor is $e^{\int a(x)\,dx} = e^{x^2}$. Multiplying both sides by e^{x^2} and integrating, we have

$$e^{x^2}y = \int x^3e^{x^2}\,dx + C,$$

so

$$y = e^{-x^2}\left(\int x^3e^{x^2}\,dx + C\right).$$

We can integrate the integral by parts as follows:

$$\int x^3e^{x^2}\,dx = \int x^2(xe^{x^2})\,dx = \frac{x^2e^{x^2}}{2} - \int xe^{x^2}\,dx = e^{x^2}\left(\frac{x^2-1}{2}\right).$$

Replacing this term for the integral above, we have

$$y = e^{-x^2}\left[e^{x^2}\left(\frac{x^2-1}{2}\right) + C\right] = \frac{x^2-1}{2} + Ce^{-x^2}.$$

Setting $x = 1$ yields

$$1 = y(1) = Ce^{-1}.$$

Thus $C = e$ and the solution to the problem is

$$y = \tfrac{1}{2}(x^2 - 1) + e^{1-x^2}.$$

A graph of the solution is given in Figure 3.

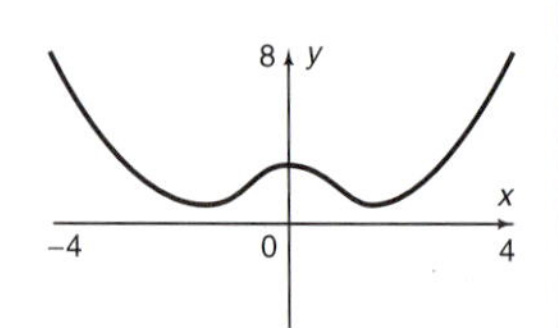

FIGURE 3
Graph of
$y = \frac{1}{2}(x^2 - 1) + e^{1-x^2}$

EXAMPLE 5 DETERMINING TEMPERATURE AS A FUNCTION OF VOLUME[†]

A closed tank containing V gallons of water at temperature T_0 is stirred constantly. It has an inlet pipe which supplies water at temperature T_w. Suppose that water is drained from the tank at the same rate at which it is put in, so that the volume of the water in the tank remains constant at V gallons. Let T denote the temperature of the water in the closed tank after the passage of p gallons through the system. Find T as a function of p.

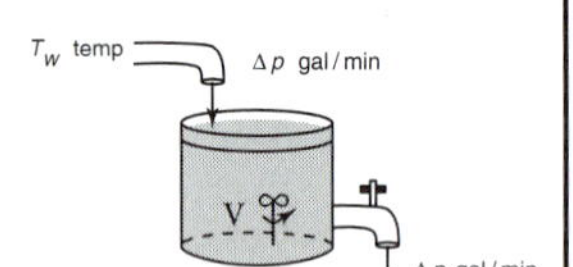

FIGURE 4
The volume V in the tank remains constant

SOLUTION: The situation is depicted in Figure 4.

Let the total heat in the system at time t be $Q(t)$ calories. (One calorie is the amount of heat necessary to raise 1 cubic centimeter of water 1°C.) If the temperature of the water in the tank changes by the amount ΔT, then the change in the total heat in a system of V gallons is

$$\Delta Q = k\,V\,\Delta T,$$

where k is the number of cubic centimeters in a gallon. Similarly, if Δp gallons run through the system at a temperature T, then since water is entering the system at a temper-

[†]This problem is adapted from a problem that appeared in *The American Mathematical Monthly* in March, 1944.

ature T_W, the net change in heat in the system is given by

$$\Delta Q = k(T_W - T)\,\Delta p.$$

Equating the two values of ΔQ yields

$$V\,\Delta T = (T_W - T)\,\Delta p$$

or dividing by $V\,\Delta p$ and rewriting

$$\frac{\Delta T}{\Delta p} = -\frac{T}{V} + \frac{T_W}{V}$$

To obtain the instantaneous rate of change of T with respect to p, we take the limit as $\Delta p \to 0$ to obtain

$$\frac{dT}{dp} = \frac{-T}{V} + \frac{T_W}{V}$$

or

$$\frac{dT}{dp} + \frac{1}{V}T = \frac{T_W}{V}, \tag{14}$$

An integrating factor for (14) is $e^{p/V}$. Thus,

$$\begin{aligned} \frac{d}{dp}\left(Te^{p/V}\right) &= \frac{T_W}{V}e^{p/V} \\ Te^{p/V} &= T_W e^{p/V} + C \\ T(p) &= T_W + Ce^{-p/V} \end{aligned}$$

But, $T(0) = T_0$, the temperature of the water before any water enters or flows out. So

$$T(0) = T_W + C = T_0$$

and

$$C = T_0 - T_W$$

So

$$T(p) = T_W + (T_0 - T_W)e^{-p/V}$$

Note the unsurprising result that as $p \to \infty$, $T(p) \to T_W$, the temperature of the incoming water. Graphs of the solution for $T_0 > T_W$ and $T_0 < T_W$ are given in Figure 5.

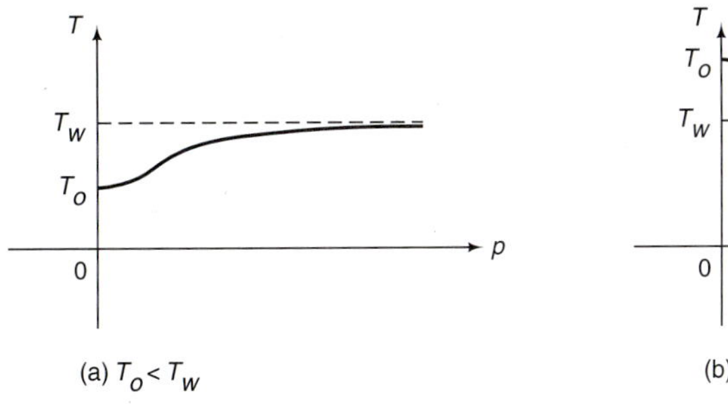

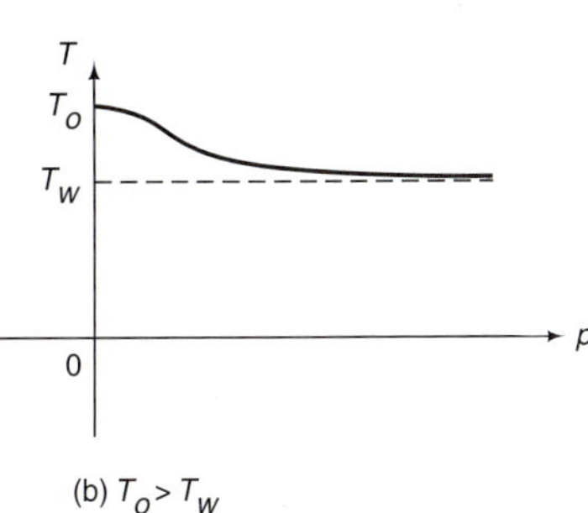

FIGURE 5
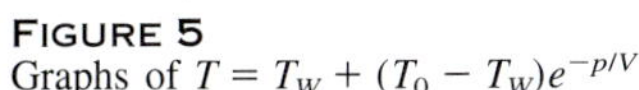
Graphs of $T = T_W + (T_0 - T_W)e^{-p/V}$

BERNOULLI'S EQUATION

Certain nonlinear first-order equations can be reduced to linear equations by a suitable change of variables. The equation

$$\frac{dy}{dx} + a(x)y = f(x)y^n \tag{15}$$

which is known as **Bernoulli's equation**, is of this type. Set $z = y^{1-n}$. Then $z' = (1-n)y^{-n}y'$, so if we multiply both sides of equation (15) by $(1-n)y^{-n}$, we obtain

$$(1-n)y^{-n}y' + (1-n)a(x)y^{1-n} = (1-n)f(x),$$

or

$$\frac{dz}{dx} + (1-n)a(x)z = (1-n)f(x).$$

The equation is now linear and may be solved as before.

EXAMPLE 8

SOLVING BERNOULLI'S EQUATION WHEN $n = 3$

Solve

$$\frac{dy}{dx} - \frac{y}{x} = -\frac{5}{2}x^2y^3. \tag{16}$$

SOLUTION: Here $n = 3$, so we let $z = y^{-2}$. Then $z' = -2y^{-3}y'$, and multiply both sides of equation (16) by $-2y^{-3}$ to obtain

$$-2y^{-3}y' + \frac{2}{x}y^{-2} = 5x^2,$$

or

$$z' + \frac{2z}{x} = 5x^2. \tag{17}$$

An integrating factor for this linear equation is

$$e^{2\int dx/x} = e^{2\ln x} = e^{\ln x^2} = x^2.$$

Multiplying both sides of equation (17) by x^2, we have

$$x^2z' + 2xz = 5x^4$$
$$(x^2z)' = 5x^4.$$

Thus

$$x^2z = \int 5x^4\,dx + C = x^5 + C.$$

Hence

$$y^{-2} = z = x^3 + Cx^{-2}, \qquad \text{or} \qquad y = (x^3 + Cx^{-2})^{-1/2}.$$

A similar procedure can be used to solve

$$\frac{dy}{dx} + a(x)y = f(x)y\ln y. \tag{18}$$

Let $z = \ln y$. Then $z' = y'/y$, so dividing equation (18) by y, we obtain the linear equation

$$\frac{dz}{dx} + a(x) = f(x)z.$$

LINEAR VS NONLINEAR DIFFERENTIAL EQUATIONS

As we have seen in this section, all linear first-order initial value problems can be solved—if not exactly, then to any desired precision by using a numerical integration technique. On the other hand, the only nonlinear equations we can solve precisely are those that are separable or exact (see Section 10.5) or those that have some special form like the Bernoulli equations just discussed.

Later in this chapter and in Chapter 11 we will discuss linear, second and higher-order differential equations and systems. We do not even mention nonlinear equations in those chapters because, except in a few very carefully chosen situations, it is impossible to find closed-form solutions.

For this reason, scientists over the last three centuries have attempted to model physical, biological or economic phenomena by using linear differential equations when a nonlinear equation might give a much better reflection of reality. We have seen this already. On page 696, we showed that, by modeling population growth using the linear, Malthusian equation $\frac{dP}{dt} = \alpha P$, we end up with the absurd result of a population becoming infinitely large. On the other hand, if we use the nonlinear logistic model described on page 706, then we obtain the much more reasonable result of a population approaching a stable equilibrium.

This is not to say that all linear models lead to absurd results. We have seen, and will continue to see, many important phenomena modeled accurately by linear equations. However, when you encounter a linear ''application,'' you should always be at least a little bit skeptical. You should ask, is this linear equation a true representation of reality or is it there because some scientist finds it easy to solve? In the latter case, it is necessary to discard the model and find a nonlinear equation that may be impossible to solve precisely, but which *does* better mirror reality.

PROBLEMS 10.4

SELF-QUIZ

Multiple Choice

I. The general solution to $\frac{dx}{dt} + 3x = 0$ is ________.

a. $y = e^{-3x} + C$ **b.** $y = Ce^{-3x}$
c. $x = e^{-3t} + C$ **d.** $x = Ce^{-3t}$

II. e^{3x} is an integrating factor for ________.

a. $\frac{dy}{dx} - 3y = x^2$ **b.** $\frac{dy}{dx} + y/3 = x - 3$
c. $\frac{dy}{dx} + 3y = x + 4$
d. $3\frac{dy}{dx} + y = x/5$

III. The general solution to $\frac{dy}{dx} + 3y = 9x$ is ________.

a. $y = e^{-3x} + C(3x - 1)$
b. $y = (3x - 1) + Ce^{-3x}$
c. $y = C(e^{-3x} + (3x - 1))$
d. $y = e^{-3x} + (3x - 1) + C$

IV. ________ is an integrating factor for

$$\frac{dy}{dx} + 2xy = -2xe^{-x^2}.$$

a. e^{-x^2} **b.** e^{x^2}
c. $e^{x^2/2}$ **d.** $-2xe^{-x^2}$

V. $\sin x$ is an integrating factor for ________.

a. $\frac{dy}{dx} + (\cot x)y = x$

b. $\frac{dy}{dx} + (\cos x)y = -x$

c. $\frac{dy}{dx} + y = -\cos x$

d. $\frac{dy}{dx} + y/x = \cot x$

True–False

VI. Suppose $f(x)$ is a solution of $\frac{dy}{dx} + y = 3$:

a. $\lim_{x\to\infty} f(x)$ does not exist.
b. $\lim_{x\to\infty} f(x) = 0$.
c. $\lim_{x\to\infty} f(x) = 3$.
d. $\lim_{x\to\infty} f'(x)$ does not exist.
e. $\lim_{x\to\infty} f'(x) = 3$.
f. $\lim_{x\to\infty} f'(x) = 0$.
g. If $f(0) > 3$, then f always decreases.
h. If $f(5) > \pi$, then f always decreases.
i. If $f(0) < \pi$, then $f'(x) > 0$ for all x.
j. If $f(-5) < 3$, then $f'(x) > 0$ for all x.

In Problems 1–21 find the general solution for each equation. When an initial condition is given, find the particular solution that satisfies the condition and sketch its graph.†

1. $\frac{dx}{dt} = 3x$

2. $\frac{dy}{dx} + 22y = 0, \quad y(1) = 2$

3. $\frac{dx}{dt} = x - 1, \quad x(0) = 1$

4. $\frac{dy}{dt} = 2y + 1$

5. $\frac{dy}{dx} - 7y = x$

6. $\frac{dx}{dt} + 3x = t, \quad x(0) = -1$

7. $\frac{dz}{dx} + 5z = x + 1$

8. $\frac{dy}{dx} - 5y = x - 1, \quad y(0) = 1$

9. $\frac{dy}{dx} - 5y = 1 - x, \quad y(0) = -1$

10. $\frac{dy}{dx} + y = x^2$

11. $\frac{dy}{dx} - y = x^2 + 2$

12. $\frac{dy}{dx} - xy = e^{\frac{1}{2}x^2}$

13. $\frac{dx}{dt} = tx + t, \quad x(0) = 1$

14. $\frac{dy}{dx} + y = \sin x, \quad y(0) = 0$

15. $\frac{dx}{dy} - x \ln y = y^y$

16. $\frac{dy}{dx} + y = \frac{1}{1 + e^{2x}}$

17. $\frac{dy}{dx} - \frac{3}{x}y = x^3, \quad y(1) = 4$

18. $\frac{dx}{dt} + x \cot t = 2t \csc t$

19. $x' - 2x = t^2e^{2t}$

20. $y' + \frac{2}{x}y = \frac{\cos x}{x^2}, \quad y(\pi) = 0$

21. $\frac{ds}{du} + s = ue^{-u} + 1$

22. Solve the equation

$$y - x\frac{dy}{dx} = \frac{dy}{dx}y^2e^y$$

---

†Use a graphing calculator if you have access to one.

by reversing the roles of x and y (that is, treat x as the dependent variable).

23. Use the method shown in Problem 22 to solve

$$\frac{dy}{dx} = \frac{1}{e^{-y} - x}.$$

24. Find the solution of $dy/dx = 2(2x - y)$ that passes through the point $(0, -1)$.

25. In a study[†] on the rate at which education is being forgotten or made obsolete, the following linear first-order differential equation was used:

$$x' = 1 - kx,$$

where $x(t)$ denotes the education of an individual at time t and k is a constant given by the rate at which that education is being lost. Obtain an equation for x at time t.

26. Data collected in a botanical experiment[‡] led to the differential equation

$$\frac{dI}{dw} = 0.088(2.4 - I).$$

Find the value of I as $w \to \infty$.

27. Assume that there exists an upper bound B for the size y of a crop in a given field. E. A. Mitscherlich proposed in 1939 the use of the linear differential equation

$$\frac{dy}{dt} = k(B - y)$$

as a model for agricultural growth. Find the general solution of this equation.

28. Suppose a population is growing at a rate proportional to its size and, in addition, individuals are immigrating into the population at a constant rate. (a) Find the linear differential equation governing this situation. (b) Find its general solution.

29. Use the method we have developed in this section for the solution of Bernoulli's equation (15) to solve the logistic equation (see Example 10.3.6)

$$\frac{dP}{dt} = P(\beta - \delta P).$$

***30.** Let $N(t)$ be the biomass of a fish species in a given area of the ocean and suppose the rate of change of the biomass is governed by the logistic equation[§]

$$\frac{dN}{dt} = rN\left(1 - \frac{N}{K}\right),$$

where the net proportional growth rate r is a constant and K is the carrying capacity for that species in that area. Assume that the rate at which fish are caught depends on the biomass. If E is the constant effort expended to harvest that species, then

a. find the resulting growth rate of the biomass;[‖]

b. solve the differential equation in part (a) using the Bernoulli method.

[*Note:* Effort can be measured in person-hours per year, tonnage of the fishing fleet, or volume seined by the fleet's nets.]

***31.** Suppose fish are harvested at a constant rate h independent of their biomass. Answer parts (a) and (b) of Problem 30 for this situation.

32. Find the effort E that maximizes the sustainable yield in Problem 30. [This is the limit of the yield as $t \to \infty$.]

33. Show that if $h > rK/4$, in Problem 31, the species will become extinct regardless of the initial size of the biomass.

34. The differential equation governing the velocity v of an object of mass m subject to air resistance proportional to the instantaneous velocity is

$$m\frac{dv}{dt} = -mg - kv.$$

Solve the equation and determine the limiting velocity of the object as $t \to \infty$.

35. Repeat Problem 34 if air resistance is proportional to the square of the instantaneous velocity.

36. An infectious disease is introduced into a large population. The proportion of people exposed to the disease increases with time. Suppose that $P(t)$ is the proportion of people exposed to the

[†] L. Southwick and S. Zionts, "An Optimal-Control-Theory Approach to the Education Investment Decision." *Operations Research* 22 (1974): 1156–1174.

[‡] R. L. Specht, "Dark Island Heath," *Australian Journal of Botany* 5 (1957): 137–172.

[§] A number of other examples of models of renewable resources are given in C. W. Clark, *Mathematical Bioeconomics* (New York: Wiley-Interscience, 1976).

[‖] This is called the **Schaefer model**, after the biologist M. B. Schaefer.

disease within t years of its introduction. If $P'(t) = [1 - P(t)]/3$ and $P(0) = 0$, after how many years does the proportion increase to 90 percent?

In Problems 37–42 find the general solution for each equation and a particular solution when an initial condition is given.

37. $\dfrac{dy}{dx} = -\dfrac{(6y^2 - x - 1)y}{2x}$

38. $y' = -y^3xe^{-2x} + y$

39. $x\dfrac{dy}{dx} + y = x^4y^3, \qquad y(1) = 1$

40. $tx^2\dfrac{dx}{dt} + x^3 = t\cos t$

41. $\dfrac{dy}{dx} + \dfrac{3}{x}y = x^2y^2, \qquad y(1) = 2$

42. $xyy' - y^2 + x^2 = 0$

43. We have seen that the equation

$$y' + f(x)y = 0$$

has the general solution

$$y = ce^{-\int f(x)\,dx}.$$

This fact prompted J. L. Lagrange (1736–1813) to seek a solution of the equation

$$y' + f(x)y = g(x) \qquad \text{(i)}$$

of the form

$$y = c(x)e^{-\int f(x)\,dx},$$

where $c = c(x)$ is a function of x.†

a. Show that $c'(x) = g(x)e^{\int f(x)\,dx}$;

b. Integrate part (a) to obtain the general solution to equation (i).

44. Use the method in Problem 43 to solve the equation

$$y' + \frac{1}{x}y = e^x.$$

45. Consider the second-order linear equation

$$y'' + 5y' + 6y = 0. \qquad \text{(ii)}$$

a. Let $z = y' + 2y$. Show that equation (ii) reduces to

$$z' + 3z = 0. \qquad \text{(iii)}$$

b. Solve equation (iii), substitute z in the equation $y' + 2y = z$, and use the methods given in this section to obtain the solution to equation (ii).

46. Use the procedure outlined in Problem 45 to find the general solution to

$$y'' + (a + b)y' + aby = 0,$$

where a and b are constants and $a \neq b$. [*Hint:* Let $z = y' + ay$.]

47. Use the method given in Problem 45 to find the general solution to

$$y'' + 2ay' + a^2y = 0,$$

where a is a constant.

In Problems 48 and 49 the function f is given by

$$f(x) = \begin{cases} 1, & \text{if } 0 \le x \le 1, \\ 0, & \text{if } x > 1. \end{cases}$$

48. Solve the initial-value problem

$$y' + y = f(x), \qquad y(0) = 0.$$

49. Solve the initial-value problem

$$y' + f(x)y = 0, \qquad y(0) = 1.$$

[*Hint:* Solve separately on $0 \le x \le 1$ and $x > 1$ and match at $x = 1$.]

****50.** Show that the general solution of the differential equation‡

$$y' - 2xy = x^2$$

is of the form $y = f(x) + ce^{x^2}$, where

$$\left|f(x) + \frac{x}{2}\right| \le \frac{1}{4x}, \qquad \text{for } x \ge 2.$$

****51.** Show that if the family of solutions of§

$$y' + a(x)y = f(x), \qquad a(x)f(x) \neq 0,$$

all cross the line $x = c$, the tangent lines to the solution curves at the points of intersection are **concurrent** (have a point in common).

---

†This technique, called the method of **variation of constants**, or **variation of parameters**, can be extended to equations of higher order (see Section 11.10).

‡R. C. Buck, "On 'Solving' Differential Equations," *American Mathematical Monthly* 63 (1956): 414.

§Problem 3 of the William Lowell Putnam Mathematical Competition, *American Mathematical Monthly* 61 (1954): 545.

**52. Daniel Bernoulli obtained the following differential equation (in 1760),

$$S' = -pS + (S/N)N' + pS^2/mN,$$

in studying the effects of smallpox. Here $N(t)$ is the number of individuals that survive at age t, $S(t)$ is the number that are susceptible to smallpox at age t (the disease imparts lifetime immunity if survived), p is the probability of a susceptible individual getting the disease, and $1/m$ is the proportion of those who die from the disease. Let $y = N/S$ and find a solution of the resulting equation.

ANSWERS TO SELF-QUIZ

I. d	**II.** c	**III.** b	**IV.** b	**V.** a
VI. a. False	**d.** False	**g.** True	**j.** True	
b. False	**e.** False	**h.** True		
c. True	**f.** True	**i.** False		

10.5 EXACT DIFFERENTIAL EQUATIONS (OPTIONAL)

The **total differential** dg of a function of two variables $g(x, y)$ is defined by

$$dg = \frac{\partial g}{\partial x}\,dx + \frac{\partial g}{\partial y}\,dy.$$

EXAMPLE 1

COMPUTING A TOTAL DIFFERENTIAL

Let $g(x, y) = x^2y^3 + e^{4x}\sin y$. Compute the total differential dg.

SOLUTION:

$$\frac{\partial g}{\partial x} = 2xy^3 + 4e^{4x}\sin y,$$

and

$$\frac{\partial g}{\partial y} = 3x^2y^2 + e^{4x}\cos y.$$

Hence

$$dg = (2xy^3 + 4e^{4x}\sin y)\,dx + (3x^2y^2 + e^{4x}\cos y)\,dy.$$

We now use partial derivatives to solve ordinary differential equations. Suppose that we take the total differential of the equation $g(x, y) = c$:

$$dg = \frac{\partial g}{\partial x}\,dx + \frac{\partial g}{\partial y}\,dy = 0. \tag{1}$$

For example, the equation $xy = c$ has the total differential $y\,dx + x\,dy = 0$, which may be rewritten as the differential equation $y' = -y/x$. Reversing the situation, suppose that we start with the differential equation

$$M(x, y)\,dx + N(x, y)\,dy = 0. \tag{2}$$

If we can find a differentiable function $g(x, y)$ such that

$$\frac{\partial g}{\partial x} = M \qquad \text{and} \qquad \frac{\partial g}{\partial y} = N,$$

then equation (2) becomes $dg = 0$, so that $g(x, y) = c$ is the general solution of equation (2). In this case $M\,dx + N\,dy$ is said to be an **exact differential**, and equation (2) is called an **exact differential equation**.

It is very easy to determine whether a differential equation is exact by using the **cross-derivative test** (see page 307):

TEST FOR EXACTNESS

The equation $M(x, y)\,dx + N(x, y)\,dy = 0$ is exact if and only if

$$\frac{\partial M}{\partial y} = \frac{\partial N}{\partial x} \tag{3}$$

If $M\,dx + N\,dy$ is exact, then we can solve the differential equation (2) by finding the function g given above. The procedure for doing this is illustrated in Examples 2 and 3.

EXAMPLE 2 SOLVING AN EXACT EQUATION

Solve the equation

$$(1 - \sin x \tan y)\,dx + (\cos x \sec^2 y)\,dy = 0.$$

SOLUTION: Letting $M(x, y) = 1 - \sin x \tan y$ and $N(x, y) = \cos x \sec^2 y$, we have

$$\frac{\partial M}{\partial y} = -\sin x \sec^2 y = \frac{\partial N}{\partial x},$$

so the equation is exact. We now seek a function g of two variables such that $\partial g/\partial x = M$ and $\partial g/\partial y = N$. But if $\partial g/\partial x = M$, then

$$g(x, y) = \int M\,dx = \int (1 - \sin x \tan y)\,dx$$

$$= x + \cos x \tan y + h(y). \tag{4}$$

The *constant of integration* $h(y)$ occurring in (4) is an arbitrary function of y since we must introduce the most general term that vanishes under partial differentiation with respect to x. But

$$\cos x \sec^2 y = N(x, y) = \frac{\partial g}{\partial y} = \underbrace{\cos x \sec^2 y + h'(y)}_{\substack{\text{differentiating (4)} \\ \text{with respect to } y}}.$$

This means that $h'(y) = 0$, so $h(y) = k$, a constant. Thus the general solution to (4) is

$$g(x, y) = x + \cos x \tan y + k = C \qquad \text{(another constant)}$$

or

$$x + \cos x \tan y = C_1.$$

EXAMPLE 3 SOLVING AN EXACT EQUATION

Find a solution to the equation

$$(y^2 + 2xy + 1)\,dx + (2xy + x^2 + 2)\,dy = 0. \tag{5}$$

SOLUTION: Here $M(x, y) = y^2 + 2xy + 1$ and $N(x, y) = 2xy + x^2 + 2$. Equation (5) is exact because

$$\frac{\partial M}{\partial y} = 2y + 2x = \frac{\partial N}{\partial x},$$

so we calculate

$$g = \int M\,dx = xy^2 + x^2y + x + h(y).$$

To find $h(y)$, we take the partial derivative of g with respect to y:

$$2xy + x^2 + 2 = N = \frac{\partial g}{\partial y} = 2xy + x^2 + h'(y),$$

implying that $h'(y) = 2$. Thus, $h(y) = 2y + C$, and the solution to equation (5) is

$$xy^2 + x^2y + x + 2y = C.$$

MULTIPLYING BY AN INTEGRATING FACTOR TO MAKE AN EQUATION EXACT

It should be apparent that exact equations are comparatively rare, since the condition in equation (3) requires a precise balance of the functions M and N. For example,

$$(3x + 2y)\,dx + x\,dy = 0$$

is not exact. However, if we multiply the equation by x, the new equation

$$(3x^2 + 2xy)\,dx + x^2\,dy = 0$$

is exact.

The problem reduces to finding an **integrating factor** $\mu(x, y)$ so that even if $M\,dx + N\,dy$ is not exact,

$$\mu M\,dx + \mu N\,dy = 0 \tag{6}$$

is exact. Finding integrating factors is generally very difficult. A procedure that is sometimes useful is to note, by (6), that

$$\mu\frac{\partial M}{\partial y} + M\frac{\partial \mu}{\partial y} = \frac{\partial}{\partial y}(\mu M) \overset{(6)}{\downarrow}= \frac{\partial}{\partial x}(\mu N) = \mu\frac{\partial N}{\partial x} + N\frac{\partial \mu}{dx}; \tag{7}$$

therefore

$$\frac{1}{\mu}\left(N\frac{\partial \mu}{\partial x} - M\frac{\partial \mu}{\partial y}\right) = \frac{\partial M}{\partial y} - \frac{\partial N}{\partial x}. \tag{8}$$

In case the integrating factor μ depends only on x, equation (8) becomes

$$\frac{1}{\mu}\frac{d\mu}{dx} = \frac{\partial M/\partial y - \partial N/\partial x}{N} = k(x, y). \tag{9}$$

Since the left-hand side of equation (9) consists only of functions of x, k *must* also be a function of x. If this is indeed true, then μ can be found by separating the variables; $\mu(x) = e^{\int k(x)\,dx}$. A similar result holds if μ is a function of y alone, in which case

$$K = \frac{\partial M/\partial y - \partial N/\partial x}{-M}$$

is also a function of y. In this case, $\mu(y) = e^{\int K(y)\,dy}$ is the integrating factor.

EXAMPLE 4

MAKING A DIFFERENTIAL EQUATION EXACT BY MULTIPLYING BY AN INTEGRATING FACTOR

Solve the equation

$$(3x^2 - y^2)\,dy - 2xy\,dx = 0.$$

SOLUTION: In this problem, $M = -2xy$ and $N = 3x^2 - y^2$, so

$$\frac{\partial M}{\partial y} = -2x \qquad \text{and} \qquad \frac{\partial N}{\partial x} = 6x.$$

Then

$$K = \frac{\partial M/\partial y - \partial N/\partial x}{-M} = \frac{-4}{y}$$

and

$$\mu = e^{-4\int y^{-1}\,dy} = e^{-4\ln y} = y^{-4}.$$

Multiplying the differential equation by y^{-4}, we obtain the exact equation

$$-\frac{2x}{y^3}\,dx + \left(\frac{3x^2 - y^2}{y^4}\right) dy = 0.$$

Then

$$g = \int M\,dx = -\frac{x^2}{y^3} + h(y),$$

so that

$$\frac{3x^2}{y^4} - \frac{1}{y^2} = N = \frac{\partial g}{\partial y} = \frac{3x^2}{y^4} + h'(y),$$

or, $h'(y) = -y^{-2}$. Hence, $h(y) = y^{-1} + C$, and the general solution is

$$g(x, y) = \frac{1}{y} - \frac{x^2}{y^3} = C$$

or

$$Cy^3 - y^2 + x^2 = 0.$$

PROBLEMS 10.5

SELF-QUIZ

True–False

I. $(4y^2 - 4x^2)\,dx + (8xy - \ln y)\,dy$ is exact.
II. $(4x^2 - 4y^2)\,dx + (8xy - \ln y)\,dy$ is exact.
III. $2x\cos y\,dx + x^2 \sin y\,dy$ is exact.
IV. $2x\cos y\,dx - x^2 \sin y\,dy$ is exact.

Multiple Choice

V. Which of the following is an integrating factor for $2y^2\,dx + 3xy\,dy$?
a. x **b.** y **c.** xy **d.** x^2 **e.** y^2

VI. Which of the following is an integrating factor for $2\cos 2x\,dx + 3\sin 2x\,dy$?
a. e^{2y} **b.** e^{3y} **c.** $\sin 3x$
d. $\cos 3y$ **e.** $\dfrac{2}{3}$

In Problems 1–14 verify that each given differential equation is exact and find the general solution. Find a particular solution when an initial condition is given and sketch its graph.

1. $2xy\,dx + (x^2 + 1)\,dy = 0$
2. $3x^2\,dx + (x^3 + 1)\,dy = 0$
3. $y(y + 2x)\,dx + x(2y + x)\,dy = 0$
4. $y\cos(xy)\,dx + x\cos(xy)\,dy = 0$
5. $(x - y\cos x)\,dx - \sin x\,dy = 0,\ y(\pi/2) = 1$
6. $\cosh 2x \cosh 2y\,dx + \sinh 2x \sinh 2y\,dy = 0$
7. $(ye^{xy} + 4y^3)\,dx + (xe^{xy} + 12xy^2 - 2y)\,dy = 0,$ $y(0) = 2$
8. $(3x^2 \ln x + x^2 - y)\,dx - x\,dy = 0,\ y(1) = 5$
9. $(2xy + e^y)\,dx + (x^2 + xe^y)\,dy = 0$
10. $(x^2 + y^2)\,dx + 2xy\,dy = 0,\ y(1) = 1$
11. $\left(\dfrac{1}{x} - \dfrac{y}{x^2 + y^2}\right)dx + \left(\dfrac{x}{x^2 + y^2} - \dfrac{1}{y}\right)dy = 0$
12. $[x\cos(x + y) + \sin(x + y)]\,dx$ $+ x\cos(x + y)\,dy = 0,$ $y(1) = \pi/2 - 1$
13. $\left(4x^3y^3 + \dfrac{1}{x}\right)dx + \left(3x^4y^2 - \dfrac{1}{y}\right)dy = 0,$ $x(e) = 1$
14. $\left(\dfrac{\ln(\ln y)}{x} + \dfrac{2}{3}xy^3\right)dx + \left(\dfrac{\ln x}{y \ln y} + x^2y^2\right)dy = 0$

In Problems 15–20 find an integrating factor for each differential equation and obtain the general solution.

15. $y\,dx + (y - x)\,dy = 0$
16. $(x^2 + y^2 + x)\,dx + y\,dy = 0$
17. $2y^2\,dx + (2x + 3xy)\,dy = 0$
18. $(x^2 + 2y)\,dx - x\,dy = 0$
19. $(x^2 + y^2)\,dx + (3xy)\,dy = 0$
20. $xy\,dx - x^2\,dy = 0$
21. Solve $xy\,dx + (x^2 + 2y^2 + 2)\,dy = 0$.
22. Let $M = yF(xy)$ and $N = xG(xy)$. Show that $1/(xM - yN)$ is an integrating factor for

$$M\,dx + N\,dy = 0.$$

23. Use the result of Problem 26 to solve the equation

$$2x^2y^3\,dx + x^3y^2\,dy = 0.$$

24. Solve $(x^2 + y^2 + 1)\,dx - (xy + y)\,dy = 0$. [*Hint:* Try an integrating factor of the form $\mu(x, y) = (x + 1)^n$.]

ANSWERS TO SELF-QUIZ

I. True **II.** False **III.** False **IV.** True **V.** c **VI.** b

10.6 SIMPLE ELECTRIC CIRCUITS

In this section we consider simple electric circuits containing a resistor and an inductor or capacitor in series with a source of electromotive force (emf). Such circuits are shown in Figure 1. Their action can be understood very easily without any special knowledge of electricity.

a. An electromotive force E (volts), usually a battery or generator, drives an

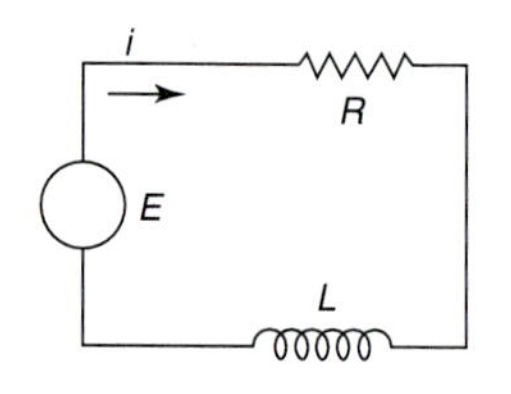

(a) An RL – Circuit

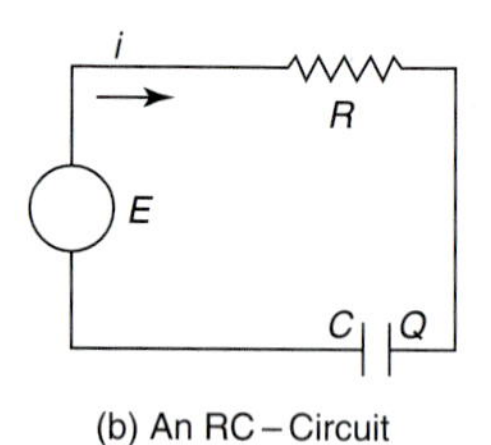

(b) An RC – Circuit

FIGURE 1
Two electric circuits

electric charge Q (coulombs) and produces a current I (amperes). The current is defined as the rate of flow of the charge, and we can write

$$I = \frac{dQ}{dt}. \tag{1}$$

b. A resistor of resistance R (ohms) is a component of the circuit that opposes the current, dissipating the energy in the form of heat. It produces a drop in voltage given by **Ohm's law**:

$$E_R = RI. \tag{2}$$

c. An inductor of inductance L (henrys) opposes any change in current by producing a voltage drop of

$$E_L = L\frac{dI}{dt}. \tag{3}$$

d. A capacitor of capacitance C (farads) stores charge. In so doing, it resists the flow of further charge, causing a drop in the voltage of

$$E_C = \frac{Q}{C}. \tag{4}$$

The quantities R, L, and C are usually constants associated with the particular component in the circuit; E may be a constant or a function of time. The fundamental principle guiding such circuits is given by

> KIRCHHOFF'S VOLTAGE LAW
>
> The algebraic sum of all voltage drops around a closed circuit is zero.

In the circuit of Figure 1(a), the resistor and the inductor cause voltage drops of E_R and E_L, respectively. The emf, however, *provides* a voltage of E (that is, a voltage drop of $-E$). Thus Kirchhoff's voltage law yields

$$E_R + E_L - E = 0.$$

Transposing E to the other side of the equation and using equations (2) and (3) to replace E_R and E_L, we have

$$L\frac{dI}{dt} + RI = E. \tag{5}$$

The following two examples illustrate the use of equation (5) in analyzing the circuit shown in Figure 1(a).

EXAMPLE 1[†]

SOLVING AN RL-CIRCUIT WITH A CONSTANT VOLTAGE

An inductance of 2 henrys and a resistance of 10 ohms are connected in series with an emf of 100 volts. If the current is zero when $t = 0$, what is the current at the end of 0.1 second?

[†]This example is typical in electrical engineering. It is, for example, very similar to Exercise 12a in C. A. Desoer and E. S. Kuh, *Basic Circuit Theory* (New York: McGraw Hill, 1969), p. 169.

SOLUTION: Since $L = 2$, $R = 10$, and $E = 100$, equation (5) and the initial current yield the initial-value problem:

$$2\frac{dI}{dt} + 10I = 100, \qquad I(0) = 0. \tag{6}$$

Dividing both sides of equation (6) by 2, we note that the resulting linear first-order equation has e^{5t} as an integrating factor; that is,

$$\frac{d}{dt}(e^{5t}I) = e^{5t}\left(\frac{dI}{dt} + 5I\right) = 50e^{5t}. \tag{7}$$

Integrating both ends of equation (7), we get

$$e^{5t}I(t) = 10e^{5t} + c,$$

or

$$I(t) = 10 + ce^{-5t}. \tag{8}$$

Setting $t = 0$ in equation (8) and using the initial condition $I(0) = 0$, we have

$$0 = I(0) = 10 + c,$$

which implies that $c = -10$. Substituting this value into equation (8), we obtain an equation for the current at all times t:

$$I(t) = 10(1 - e^{-5t}).$$

Thus, when $t = 0.1$, we have

$$I(0.1) = 10(1 - e^{-0.5}) \approx 3.93 \text{ amperes}.$$

EXAMPLE 2 SOLVING AN RL-CIRCUIT WITH A PERIODIC VOLTAGE

Suppose that the emf $E = 100 \sin 60t$ volts but all other values remain the same as those given in Example 1. In this case equation (5) yields

$$2\frac{dI}{dt} + 10I = 100 \sin 60t, \qquad I(0) = 0. \tag{9}$$

Again dividing by 2 and multiplying both sides by the integrating factor e^{5t}, we have

$$\frac{d}{dt}(e^{5t}I) = e^{5t}\left(\frac{dI}{dt} + 5I\right) = 50e^{5t} \sin 60t. \tag{10}$$

Integrating both ends of equation (10) and using Formula 168 of the Table of Integrals, at the back of the book, we obtain

$$\begin{aligned} I(t) &= e^{-5t}\left[50 \int (\sin 60t)e^{5t}\, dt + c\right] \\ &= e^{-5t}\left[50e^{5t}\left(\frac{5 \sin 60t - 60 \cos 60t}{3625}\right) + c\right] \\ &= \frac{2 \sin 60t - 24 \cos 60t}{29} + ce^{-5t}. \end{aligned}$$

Setting $t = 0$, we find that $c = 24/29$ and

$$I(t) = \frac{2 \sin 60t - 24 \cos 60t}{29} + \frac{24}{29}e^{-5t}.$$

Thus

$$I(0.1) = \frac{2 \sin 6 - 24 \cos 6}{29} + \frac{24}{29}e^{-0.5} \approx -0.31 \text{ amperes.}$$

In the previous example the term $24e^{-5t}/29$ is called the **transient current** because it approaches zero as t increases without bound. The other part of the current, $(2 \sin 60t - 24 \cos 60t)/29$, is called the **steady-state current**.

For the circuit in Figure 1(b) we have $E_R + E_C - E = 0$, or

$$RI + \frac{Q}{C} = E.$$

Using the fact that $I = dQ/dt$, we obtain the linear first-order equation

$$R\frac{dQ}{dt} + \frac{Q}{C} = E. \tag{11}$$

The next example illustrates how to use equation (11).

EXAMPLE 3

SOLVING AN RC-CIRCUIT WITH A CONSTANT VOLTAGE

If a resistance of 2000 ohms and a capacitance of 5×10^{-6} farad are connected in series with an emf of 100 volts, what is the current at $t = 0.1$ second if $I(0) = 0.01$ ampere?

SOLUTION: Setting $R = 2000$, $C = 5 \times 10^{-6}$, and $E = 100$ in equation (11), we have

$$2000\left(\frac{dQ}{dt} + 100Q\right) = 100,$$

or

$$\frac{dQ}{dt} + 100Q = \frac{1}{20}, \tag{12}$$

from which we can determine $Q(0)$ since

$$\frac{1}{20} = Q'(0) + 100Q(0) = I(0) + 100Q(0).$$

Thus

$$Q(0) = \frac{1}{100}\left[\frac{1}{20} - I(0)\right] = \frac{1}{100}\left(\frac{1}{20} - \frac{1}{100}\right)$$

$$= \frac{1}{100}\left(\frac{4}{100}\right) = 4 \times 10^{-4} \text{ coulombs.} \tag{13}$$

Multiplying both sides of equation (12) by the integrating factor e^{100t}, we get

$$\frac{d}{dt}(e^{100t}Q) = \frac{e^{100t}}{20}$$

and integrating this equation yields

$$e^{100t}Q = \frac{e^{100t}}{2000} + c.$$

Dividing both sides by e^{100t} gives us

$$Q(t) = \frac{1}{2000} + ce^{-100t}$$

and setting $t = 0$, we find that

$$c = Q(0) - \frac{1}{2000} = (4 \times 10^{-4}) - (5 \times 10^{-4}) = -10^{-4}.$$

Thus the charge at all times t is

$$Q(t) = (5 - e^{-100t})/10^4$$

and the current is

$$I(t) = Q'(t) = \frac{1}{100}e^{-100t}$$

Thus $I(0.1) = 10^{-2}e^{-10} \approx 4.54 \times 10^{-7}$ amperes.

PROBLEMS 10.6

In Problems 1–5 assume that the RL circuit shown in Figure 1(a) has the given resistance (ohms, Ω), inductance (henrys, H), emf (volts, V), and initial current (amperes, amp). Find an expression for the current at all times t and calculate the current after 1 second.

1. $R = 10\ \Omega$, $L = 1$ H, $E = 12$ V, $I(0) = 0$ amp
2. $R = 8\ \Omega$, $L = 1$ H, $E = 6$ V, $I(0) = 1$ amp
3. $R = 50\ \Omega$, $L = 2$ H, $E = 100$ V, $I(0) = 0$ amp
4. $R = 10\ \Omega$, $L = 5$ H, $E = 10 \sin t$ V, $I(0) = 1$ amp
5. $R = 10\ \Omega$, $L = 10$ H, $E = e^t$ V, $I(0) = 0$ amp

In Problems 6–10 use the given resistance, capacitance (farads, f), emf, and initial charge (coulombs) in the RC circuit shown in Figure 1(b). Find an expression for the charge at all time t.

6. $R = 1\ \Omega$, $C = 1$ f, $E = 12$ V, $Q(0) = 0$ coulomb
7. $R = 10\ \Omega$, $C = 0.001$ f, $E = 10 \cos 60t$ V, $Q(0) = 0$ coulomb
8. $R = 1\ \Omega$, $C = 0.01$ f, $E = \sin 60t$ V, $Q(0) = 0$ coulomb
9. $R = 100\ \Omega$, $C = 10^{-4}$ f, $E = 100$ V, $Q(0) = 1$ coulomb
10. $R = 200\ \Omega$, $C = 5 \times 10^{-5}$ f, $E = 1000$ V, $Q(0) = 1$ coulomb
11. The capacitor C in the circuit illustrated in Figure 2 is charged to 10 volts when the switch is closed. Obtain a differential equation for the capacitor voltage and find the voltage for all times t given that $R = 1000$ ohms and $C = 10^{-6}$ farad.†

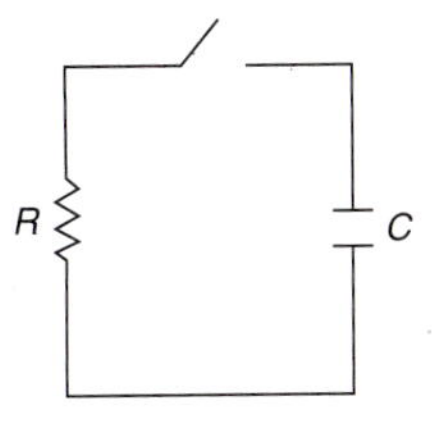

FIGURE 2

12. An inductance of 1 henry and a resistance of 2 ohms are connected in series with a battery of $6e^{-0.0001t}$ volt. No current is flowing initially. When does the current measure 0.5 ampere?
13. A variable resistance $R = 1/(5 + t)$ ohms and a

†This example is Exercise 5.28 in Shearer et al., *System Dynamics* (Reading, Mass.: Addison-Wesley, 1971), p. 141.

capacitance of 5×10^{-6} farad are connected in series with an emf of 100 volts. If $Q(0) = 0$, what is the charge on the capacitor after 1 minute?

14. In the RC circuit [Figure 1(b)] with constant voltage E, how long will it take the current to decrease to one-half its original value?

15. Suppose that the voltage in an RC circuit is $E(t) = E_0 \cos \omega t$, where $2\pi/\omega$ is the period of the cycle. Assuming that the initial charge is zero, what are the charge and current as functions of R, C, ω, and t?

16. Show that the current in Problem 15 consists of two parts: a steady-state term that has a period of $2\pi/\omega$ and a transient term that tends to zero as t increases.

17. Show that if R in Problem 16 is small, then the transient term can be quite large for small values of t. [This is why fuses can blow when a switch is flipped.]

18. Find the steady-state current, given that a resistance of 2000 ohms and a capacitance of 3×10^{-6} farad are connected in series with an alternating emf of $120 \cos 2t$ volts.

19. Find an expression for the current of a series RL circuit, where $R = 100$ ohms, $L = 2$ henrys, $I(0) = 0$ amp, and the emf voltage satisfies

$$E = \begin{cases} 6, & \text{for } 0 \le t \le 10, \\ 7 - e^{10-t}, & \text{for } t \ge 10. \end{cases}$$

20. Repeat Problem 19 with $R = 100/(1 + t)$, all other values remaining the same.

10.7 THEORY OF LINEAR DIFFERENTIAL EQUATIONS

Although there is no procedure for explicitly solving arbitrary differential equations, systematic methods do exist for certain classes of differential equations. In this section we study a class of problems for which there are always unique solutions. In Sections 10.8–10.12 we present some methods for calculating them.

HOMOGENEOUS AND NONHOMOGENEOUS LINEAR DIFFERENTIAL EQUATIONS

Recall from Sections 10.1 and 10.4 that a differential equation is **linear** if it does not involve nonlinear functions (squares, exponentials, etc.) or products of the dependent variable and its derivatives. Thus $y'' + (x^3 \sin x)^5 y' + y = \cos x^3$ is linear, whereas $y'' + (y)^2 + y = 0$ is nonlinear. The most general second-order linear equation can be written

$$y''(x) + a(x)y'(x) + b(x)y(x) = f(x), \tag{1}$$

whereas the most general third-order linear equation can be written

$$y'''(x) + a(x)y''(x) + b(x)y'(x) + c(x)y(x) = f(x), \tag{2}$$

where $a(x)$, $b(x)$, $c(x)$, and $f(x)$ are functions of the independent variable x only. Equations (1) and (2) are special cases of the **general linear nth-order equation**:

$$y^{(n)}(x) + a_{n-1}(x)y^{(n-1)}(x) + \cdots + a_1(x)y'(x) + a_0(x)y(x) = f(x). \tag{3}$$

If the function $f(x)$ is identically zero, we say that equations (1), (2), and (3) are **homogeneous**. Otherwise, they are **nonhomogeneous**.

EXAMPLE 1

HOMOGENEOUS AND NONHOMOGENEOUS LINES DIFFERENTIAL EQUATIONS

a. The equation $y'' + 2xy' + 3y = 0$ is homogeneous.

b. The equation $y'' + 2xy' + 3y = e^x$ is nonhomogeneous.

If the coefficient functions $a(x)$ and $b(x)$ are constants, $a(x) = a$ and $b(x) = b$, the equation is said to have **constant coefficients**. (As we see below, linear differential equations with constant coefficients are the easiest to solve.) If either $a(x)$ or $b(x)$ is not constant, the equation is said to have **variable coefficients**.

EXAMPLE 2 EQUATIONS WITH CONSTANT AND VARIABLE COEFFICIENTS

a. The equation $y'' + 3y' - 10y = 0$ has constant coefficients.
b. The equation $y'' + 3xy' - 10x^2y = 0$ has variable coefficients.

In this section we will answer the questions: when does a second-order linear differential equation have solutions and, if it does, how do we know we have found all of them? Because there is a lot of material to cover, we break the section into four parts.

EXISTENCE AND UNIQUENESS OF SOLUTIONS

In Section 10.4 we discussed the first-order linear differential equation

$$y' + a(x)y = f(x).$$

If $a(x)$ and $f(x)$ are continuous functions in an interval $[x_1, x_2]$, $x_1 < x_2$, this differential equation has a general solution involving an arbitrary constant (see equation 10.4.11) on page 715. The arbitrary constant can be determined if an initial condition $y(x_0) = y_0$ is given at some value x_0 in $[x_1, x_2]$. In this case the initial value problem has a unique solution. We can restate these basic facts as follows:

EXISTENCE-UNIQUENESS THEOREM FOR FIRST-ORDER LINEAR INITIAL VALUE PROBLEMS

If $a(x)$ and $f(x)$ are continuous in the interval $x_1 \le x \le x_2$, then the equation

$$y'(x) + a(x)y(x) = f(x)$$

has one and only one solution that satisfies the initial condition $y(x_0) = y_0$, for x_0 in $[x_1, x_2]$.

This is a very nice result, because it tells us that every linear first-order equation with a given initial condition has a unique solution. We need only set about finding it. It turns out that this special property holds for linear initial-value problems of any order. The only difference is that in order to have a unique solution to a second-order equation, we must specify two initial conditions, for a third-order equation three conditions, and so on. The proof of the following theorem is beyond the scope of this book.†

THEOREM 1 EXISTENCE-UNIQUENESS THEOREM FOR LINEAR INITIAL-VALUE PROBLEMS

Let $a_1(x), a_2(x), \ldots, a_n(x)$, and $f(x)$ be continuous functions on the interval $[x_1, x_2]$, $x_1 < x_2$, and let $c_0, c_1, c_2, \ldots, c_{n-1}$ be n given constants. Then there exists a unique

†A special case of this theorem is proved in Appendix 3 of *Introduction to Differential Equations with Boundary Value Problems, Third Edition* by W. R. Derrick and S. I. Grossman, West, St. Paul, MN, 1988.

function $y(x)$ that satisfies the linear differential equation

$$y^{(n)}(x) + a_{n-1}(x)y^{(n-1)}(x) + a_{n-2}(x)y^{(n-2)}(x) + \cdots + a_0(x)y(x) = f(x)$$

on $[x_1, x_2]$ and the n initial conditions

$$y(x_0) = c_0, \qquad y'(x_0) = c_1, \qquad y''(x_0) = c_2, \ . \ . \ . \ , y^{(n-1)}(x_0) = c_{n-1} \tag{4}$$

at some value x_0 in $[x_1, x_2]$. ■

NOTE: The conditions given in (4) all involve evaluations of the unknown function y and its derivatives at the *same* point x_0. This is a *crucial* requirement for the existence and uniqueness of a solution.

BOUNDARY VALUE PROBLEMS

There is another type of problem involving the differential equation (3) in which conditions at more than one point are given. For example, we might specify $y(x_1) = k_1$, and $y'(x_2) = k_2, \ldots, y^{(n-1)}(x_{n-1}) = k_{n-1}$. Conditions of this sort are called **boundary conditions**, and a differential equation together with a set of boundary conditions is called a **boundary value problem**.

It is important to note that the existence and uniqueness of a solution, guaranteed by Theorem 1 for initial-value problems, does not hold for boundary value problems.

EXAMPLE 3 A BOUNDARY VALUE PROBLEM WITH INFINITELY MANY SOLUTIONS

Observe that the boundary value problem

$$y'' + y = 0, \qquad y(0) = y(\pi) = 0,$$

has infinitely many solutions:

$$y = c \sin x, \qquad \text{for any constant } c.$$

To check, note that $\sin 0 = \sin \pi = 0$, so the boundary conditions are satisfied, and

$$(c \sin x)'' + c \sin x = -c \sin x + c \sin x = 0.$$

It is true (although we shall not prove it here) that the boundary value problem

$$y'' + y = 0, \qquad y(0) = 0, \qquad y(\pi) = 1$$

has *no* solution. On the other hand, the initial-value problem

$$y'' + y = 0, \qquad y(0) = 0, \qquad y'(0) = 1$$

has the unique solution $y = \sin x$; the initial conditions are satisfied since $\sin 0 = 0$, $(\sin x)' = \cos x$, and $\cos 0 = 1$, and

$$(\sin x)'' + \sin x = -\sin x + \sin x = 0.$$

If we apply Theorem 1 to the second-order equation (1), we have the following result.

EXISTENCE-UNIQUENESS THEOREM FOR SECOND-ORDER LINEAR INITIAL-VALUE PROBLEMS

If $a(x)$, $b(x)$, and $f(x)$ are continuous functions on $x_1 \le x \le x_2$, $x_1 < x_2$, and x_0 is in $[x_1, x_2]$, then the equation

$$y''(a) + a(x)y'(x) + b(x)y(x) = f(x) \tag{5}$$

has a unique solution that satisfies the conditions

$$y(x_0) = y_0, \qquad y'(x_0) = y_1$$

for any real numbers y_0 and y_1.

For simplicity we limit most of our discussion in this chapter to second-order linear equations. We emphasize, however, that *every* result we prove can be extended to higher-order linear equations (see Section 10.16).

The requirement that the functions $a(x)$, $b(x)$, and $f(x)$ in (5) be continuous is also an essential requirement for the existence of a unique solution, as the following example demonstrates.

EXAMPLE 4

AN INITIAL-VALUE PROBLEM WITH INFINITELY MANY SOLUTIONS

Verify that the function

$$y = cx^3 + x$$

is a solution (for any constant c) of the initial-value problem

$$x^2y'' - 3xy' + 3y = 0, \qquad y(0) = 0, \qquad y'(0) = 1;$$

that is, the problem does not have a unique solution—it has infinitely many solutions.

SOLUTION: Since $y' = 3cx^2 + 1$, the initial conditions are satisfied. Then $y'' = 6cx$, and

$$x^2(6cx) - 3x(3cx^2 + 1) + 3(cx^3 + x) = 6cx^3 - 9cx^3 - 3x + 3cx^3 + 3x = 0.$$

Theorem 1 does not apply here, because in order to obtain a differential equation of the form (5), we must divide $x^2y'' - 3xy' + 3y = 0$ by x^2:

$$y'' - \frac{3}{x}y' + \frac{3}{x^2}y = 0.$$

Then $a(x) = -3/x$ and $b(x) = 3/x^2$. Theorem 1 requires that $a(x)$ and $b(x)$ be continuous in an interval containing x_0 (here $x_0 = 0$). However, there is *no* interval containing 0 such that $-3/x$ and $3/x^2$ are continuous (neither function is even defined at zero).

There are special techniques for handling some problems of this sort, which we will discuss in Section 12.13.

LINEAR COMBINATIONS AND LINEAR INDEPENDENCE

We now know that the differential equation (5) has solutions if a, b, and f are continuous. What do these solutions look like? How many are there?

Before solving a second-order differential equation, it helps to know what we are seeking. A clue is provided by examining a first-order equation. Consider the equation

$$y' + 2y = 0. \tag{6}$$

In Section 10.2 we saw that one solution to this equation is $y = e^{-2x}$. In fact, $y = ce^{-2x}$ is a solution for any constant c, and every solution to (6) has the form ce^{-2x}. We can summarize this result by noting that once we have found one nonzero solution to (6), we have found all of the solutions, since every other solution is a constant multiple of this one solution.

It turns out that similar results hold for all homogeneous second-order equations:

$$y'' + a(x)y' + b(x)y = 0. \tag{7}$$

The major difference is that now we have to find *two* solutions to (7) where neither solution is a multiple of the other. We now make these ideas more precise.

LINEAR COMBINATION, LINEAR INDEPENDENCE AND LINEAR DEPENDENCE

Let y_1 and y_2 be any two functions. By a **linear combination** of y_1 and y_2 we mean a function $y(x)$ that can be written in the form

$$y(x) = c_1y_1(x) + c_2y_2(x),$$

for some constants c_1 and c_2. Two functions are **linearly independent** on an interval $[x_1, x_2]$, $x_1 < x_2$, whenever the relation

$$c_1y_1(x) + c_2y_2(x) = 0, \tag{8}$$

for all x in $[x_1, x_2]$, implies that $c_1 = c_2 = 0$. Otherwise they are **linearly dependent**.

EXAMPLE 5 TWO LINEARLY INDEPENDENT FUNCTIONS

Verify that the functions $y_1 = 1$ and $y_2 = x$ are linearly independent on the interval [0, 1].

SOLUTION: To determine linear independence or dependence we must consider equation (8):

$$c_1y_1 + c_2y_2 = c_1 \cdot 1 + c_2 \cdot x = 0. \tag{9}$$

This equation must hold for all x in [0, 1]. If $x = 0$, we have

$$c_1 \cdot 1 + c_2 \cdot 0 = c_1 = 0.$$

But then at $x = 1$ we get $c_2 \cdot 1 = 0$. Hence (9) holds for all x in [0, 1] if and only if $c_1 = c_2 = 0$. This proves that $y_1 = 1$ and $y_2 = x$ are linearly independent.

The notions of linear combination, linear independence, and linear dependence extend easily to a collection of n functions $y_1(x), y_2(x), \ldots, y_n(x)$, with $n > 2$. A **linear combination** of these functions is any function of the form

$$y(x) = c_1y_1(x) + c_2y_2(x) + \cdots + c_ny_n(x),$$

where $c_1, c_2, \ldots, c_n$ are constants. We will say that the collection of functions $y_1(x), y_2(x), \ldots, y_n(x)$ are **linearly independent** on an interval $[x_1, x_2]$, $x_1 < x_2$, if the equation

$$c_1y_1(x) + c_2y_2(x) + \cdots + c_ny_n(x) = 0$$

holds for all x in $[x_1, x_2]$ only when $c_1 = c_2 = \cdots = c_n = 0$. Otherwise, we say that the collection of functions is **linearly dependent** on $[x_1, x_2]$.

There is an easy way to see that two functions y_1 and y_2 are linearly dependent. If $c_1y_1(x) + c_2y_2(x) = 0$ (where not both c_1 and c_2 are zero), we may suppose that $c_1 \neq 0$. Dividing the expression above by c_1, we obtain

$$y_1(x) + \frac{c_2}{c_1}y_2(x) = 0,$$

or

$$y_1(x) = -\frac{c_2}{c_1}y_2(x) = cy_2(x). \tag{10}$$

This leads to a useful fact:

DETERMINING WHETHER TWO FUNCTIONS ARE LINEARLY DEPENDENT OR INDEPENDENT

Two functions are linearly dependent on the interval $[x_0, x_1]$ if and only if one of the functions is a constant multiple of the other.

EXAMPLE 5 (REVISITED)

TWO LINEARLY INDEPENDENT FUNCTIONS

It is easy to see that the functions $y_1 = 1$ and $y_2 = x$ are linearly independent on $[0, 1]$, since x is not a *constant* multiple of 1.

THE GENERAL SOLUTION TO A LINEAR, HOMOGENEOUS, SECOND-ORDER DIFFERENTIAL EQUATION

The notions of linear combination and linear independence are central to the theory of linear homogeneous equations, as is illustrated by the results that follow.

THEOREM 2

Every homogeneous linear second-order differential equation

$$y'' + a(x)y' + b(x)y = 0 \tag{11}$$

has two linearly independent solutions.

PROOF: The existence part of Theorem 1 guarantees that we can find a solution $y_1(x)$ to equation (11) satisfying

$$y_1(x_0) = 1 \qquad \text{and} \qquad y_1'(x_0) = 0.$$

Similarly, we can also find a solution $y_2(x)$ to (11) satisfying

$$y_2(x_0) = 0 \qquad \text{and} \qquad y_2'(x_0) = 1.$$

Now consider the equation

$$c_1y_1(x) + c_2y_2(x) = 0 \tag{12}$$

and its derivative

$$c_1 y_1'(x) + c_2 y_2'(x) = 0. \tag{13}$$

Setting $x = x_0$ in (12) yields $c_1 \cdot 1 + c_2 \cdot 0 = 0$, or $c_1 = 0$, whereas substituting $x = x_0$ in (13) gives $c_1 \cdot 0 + c_2 \cdot 1 = c_2 = 0$. Thus (12) holds only when $c_1 = c_2 = 0$, implying that solutions $y_1(x)$ and $y_2(x)$ are linearly independent. ■

EXAMPLE 6 TWO LINEARLY INDEPENDENT SOLUTIONS TO A SECOND-ORDER EQUATION

Verify that $\sin x$ and $\cos x$ are linearly independent solutions to $y'' + y = 0$.

SOLUTION: To check that $\sin x$ and $\cos x$ are solutions, we write

$$(\sin x)'' + \sin x = -\sin x + \sin x = 0,$$
$$(\cos x)'' + \cos x = -\cos x + \cos x = 0.$$

Now consider the equation

$$c_1 \sin x + c_2 \cos x = 0$$

for all values x, and its derivative

$$c_1 \cos x - c_2 \sin x = 0.$$

If $x = 0$, the first equation yields $c_2 = 0$ whereas the second gives $c_1 = 0$. Thus $\sin x$ and $\cos x$ are linearly independent. Alternatively, they are independent according to (10), because

$$\frac{\sin x}{\cos x} = \tan x \neq \text{constant}.$$

Every linear, homogeneous, second-order differential equation has two linearly independent solutions. The next two theorems show us that this is all the information we need; that is, once we have two linearly independent solutions, we can construct them all.

THEOREM 3 PRINCIPLE OF SUPERPOSITION FOR HOMOGENEOUS EQUATIONS

Let $y_1(x)$ and $y_2(x)$ be any two solutions of the homogeneous equation

$$y'' + a(x)y' + b(x)y = 0. \tag{14}$$

Then any linear combination of them is also a solution of (14).

PROOF: Let $y(x) = c_1 y_1(x) + c_2 y_2(x)$. Then

$$\begin{aligned} y'' + ay' + by &= c_1 y_1'' + c_2 y_2'' + c_1 a y_1' + c_2 a y_2' + c_1 b y_1 + c_2 b y_2 \\ &= c_1(y_1'' + a y_1' + b y_1) + c_2(y_2'' + a y_2' + b y_2) \\ &= c_1 \cdot 0 + c_2 \cdot 0 = 0, \end{aligned}$$

since y_1 and y_2 are solutions of the homogeneous equation (14). ■

THEOREM 4 A SECOND-ORDER LINEAR HOMOGENEOUS EQUATION CAN HAVE AT MOST TWO LINEARLY INDEPENDENT SOLUTIONS

Let $y_1(x)$ and $y_2(x)$ be linearly independent solutions to (14) and let $y_3(x)$ be another solution. Then there exist unique constants c_1 and c_2 such that

$$y_3(x) = c_1 y_1(x) + c_2 y_2(x).$$

In other words, any solution of (14) can be written as a linear combination of two given linearly independent solutions of (14). ■

We stress the importance of this theorem. It indicates that once we have found two linearly independent solutions y_1 and y_2 of equation (14), we have essentially found *all* the solutions of (14). We delay the proof of this theorem until the end of the section.

Theorem 4 can be extended to n^{th}-order equations.

AN IMPORTANT RESULT

An nth-order linear, homogeneous differential equation with continuous coefficients can have at most n linearly independent solutions.

DEFINITION GENERAL SOLUTION TO A LINEAR, SECOND-ORDER EQUATION

The **general solution** of (14) is given by the linear combination

$$y(x) = c_1 y_1(x) + c_2 y_2(x)$$

where c_1 and c_2 are arbitrary constants and $y_1(x)$ and $y_2(x)$ are linearly independent solutions of (14). ■

EXAMPLE 6 (REVISITED)

THE GENERAL SOLUTION TO A LINEAR EQUATION

The general solution to $y'' + y = 0$ is

$$y = c_1 \cos x + c_2 \sin x.$$

If we look at the proof of Theorem 2, we see that equations (12) and (13) yield a system of equations

$$c_1 y_1(x) + c_2 y_2(x) = 0 \qquad c_1 y_1'(x) + c_2 y_2'(x) = 0 \tag{15}$$

that can be used to determine whether the solutions $y_1(x)$ and $y_2(x)$ are linearly independent. The procedure involves substituting a value $x = x_0$ in (15), which determines $y_1(x_0)$, $y_2(x_0)$, $y_1'(x_0)$, and $y_2'(x_0)$. We are then left with a linear system of two homogeneous equations in the two unknowns c_1 and c_2. This system[†] has the trivial solution $c_1 = c_2 = 0$

[†] A homogeneous system

$$\begin{aligned} ax + by &= 0 \\ cx + dy &= 0 \end{aligned}$$

has nontrivial solutions for the unknowns x and y if and only if the determinant

$$\begin{vmatrix} a & b \\ c & d \end{vmatrix} = ad - bc$$

equals zero (see page 374).

if and only if the determinant

$$\begin{vmatrix} y_1(x_0) & y_2(x_0) \\ y_1'(x_0) & y_2'(x_0) \end{vmatrix} = y_1(x_0)y_2'(x_0) - y_1'(x_0)y_2(x_0)$$

does not equal zero. This fact leads to the following very useful definition.

WRONSKIAN

Let $y_1(x)$ and $y_2(x)$ be any two solutions to the differential equation

$$y'' + a(x)y' + b(x)y = 0. \tag{16}$$

The **Wronskian**† of y_1 and y_2 is defined as

$$W(y_1, y_2)(x) = \begin{vmatrix} y_1(x) & y_2(x) \\ y_1'(x) & y_2'(x) \end{vmatrix} = y_1(x)y_2'(x) - y_1'(x)y_2(x).$$

The Wronskian is defined at all points x at which $y_1(x)$ and $y_2(x)$ are differentiable.

Since $y_1(x)$ and $y_2(x)$ are solutions to (16), their second derivatives exist. Hence we can differentiate $W(y_1, y_2)(x)$ with respect to x:

$$\begin{aligned} W'(y_1, y_2)(x) &= [y_1(x)y_2'(x) - y_1'(x)y_2(x)]' \\ &= y_1'(x)y_2'(x) + y_1(x)y_2''(x) - y_1''(x)y_2(x) - y_1'(x)y_2'(x) \\ &= y_1(x)y_2''(x) - y_1''(x)y_2(x). \end{aligned}$$

Since y_1 and y_2 are solutions to (16),

$$y_1'' + ay_1' + by_1 = 0 \qquad \text{and} \qquad y_2'' + ay_2' + by_2 = 0.$$

Multiplying the first of these equations by y_2 and the second by y_1 and subtracting, we obtain

$$y_1y_2'' - y_2y_1'' + a(y_1y_2' - y_2y_1') = 0,$$

which is just

$$W' + aW = 0. \tag{17}$$

But (17) is a linear first-order equation similar to equation (10.4.2) (p. 713) with solution [see equation (10.4.3)]

AN EQUATION FOR THE WRONSKIAN

$$W(y_1, y_2)(x) = ce^{-\int a(x)\,dx} \tag{18}$$

for some arbitrary constant c. Equation (18) is known as **Abel's formula**.

Since an exponential is never zero, we see that $W(y_1, y_2)(x)$ is either always zero (when $c = 0$) or never zero (when $c \neq 0$). The importance of this fact is given by the following theorem.

†Named after the Polish philosopher Józef Maria Hoene-Wroński (1778–1853).

THEOREM 5

Two solutions $y_1(x)$ and $y_2(x)$ of equation (16) are linearly independent on $[x_0, x_1]$, $x_0 < x_1$, if and only if $W(y_1, y_2)(x) \neq 0$ at any x in $[x_0, x_1]$. ■

Theorem 5 is useful in at least three ways. First, it provides us with an easy way to determine whether or not two solutions are linearly independent. Second, it greatly simplifies the proof of Theorem 4—as we see later in this section. Third, the Wronskian can be easily extended to third- and higher-order equations with similar results. It is not easy to verify directly that three solutions are linearly independent, but the task is made easy by use of the Wronskian.

EXAMPLE 6 (REVISITED)

USING THE WRONSKIAN TO VERIFY LINEAR INDEPENDENCE

We have seen that $y_1(x) = \sin x$ and $y_2(x) = \cos x$ are linearly independent solutions of $y'' + y = 0$. The linear independence is easily verified using the Wronskian:

$$W(y_1, y_2)(x) = \begin{vmatrix} \sin x & \cos x \\ \cos x & -\sin x \end{vmatrix} = -\sin^2 x - \cos^2 x = -1 \neq 0.$$

EXAMPLE 7

USING THE WRONSKIAN TO VERIFY LINEAR INDEPENDENCE

a. Verify that $y_1(x) = e^{-5x}$ and $y_2(x) = e^{2x}$ are linearly independent solutions of the differential equation

$$y'' + 3y' - 10y = 0. \tag{19}$$

b. Use the general solution obtained from y_1 and y_2 to solve the initial-value problem

$$y'' + 3y' - 10y = 0, \qquad y(0) = 3, \qquad y'(0) = -1. \tag{20}$$

SOLUTION:

a. That $y_1(x)$ and $y_2(x)$ are solutions to the differential equation follows from

$$(e^{-5x})'' + 3(e^{-5x})' - 10(e^{-5x}) = 25e^{-5x} - 15e^{-5x} - 10e^{-5x} = 0,$$
$$(e^{2x})'' + 3(e^{2x})' - 10(e^{2x}) = 4e^{2x} + 6e^{2x} - 10e^{2x} = 0.$$

Also,

$$W(y_1, y_2)(x) = \begin{vmatrix} e^{-5x} & e^{2x} \\ -5e^{-5x} & 2e^{2x} \end{vmatrix} = 2e^{-3x} + 5e^{-3x} = 7e^{-3x} \neq 0,$$

so the solutions are linearly independent. (This is also true since $e^{-5x}/e^{2x} = e^{-7x} \neq$ constant.) Thus the general solution of (19) is

$$y(x) = c_1y_1(x) + c_2y_2(x) = c_1e^{-5x} + c_2e^{2x}, \tag{21}$$

where c_1 and c_2 are arbitrary constants.

b. To find the particular solution of initial-value problem (20), we must determine the constants c_1 and c_2 so that (21) satisfies the initial conditions in

(20). Setting $x = 0$ in (21), we have

$$c_1 + c_2 = y(0) = 3.$$

Differentiating (21), we get

$$y'(x) = -5c_1e^{-5x} + 2c_2e^{2x},$$

and setting $x = 0$ gives

$$-5c_1 + 2c_2 = y'(0) = -1.$$

Hence

$$\begin{aligned} c_1 + c_2 &= 3 \\ -5c_1 + 2c_2 &= -1. \end{aligned}$$

The first equation implies that $c_1 = 3 - c_2$, so

$$-5c_1 + 2c_2 = -5(3 - c_2) + 2c_2 = -1,$$

or

$$\begin{aligned} -15 + 5c_2 + 2c_2 &= -1, \\ 7c_2 &= 14, \\ c_2 &= 2. \end{aligned}$$

Hence $c_1 = 3 - c_2 = 1$, and the particular solution of initial-value problem (20) is

$$y = e^{-5x} + 2e^{2x}.$$

THE GENERAL SOLUTION TO A LINEAR, NONHOMOGENEOUS SECOND-ORDER DIFFERENTIAL EQUATION

We now turn briefly to the nonhomogeneous equation

$$y'' + a(x)y' + b(x)y = f(x). \tag{22}$$

Let y_p be any particular solution to equation (22). If we know the general solution to the associated homogeneous equation

$$y'' + a(x)y' + b(x)y = 0, \tag{23}$$

we can find all solutions to (22).

THEOREM 6 THE DIFFERENCE OF TWO SOLUTIONS TO A NONHOMOGENEOUS EQUATION IS A SOLUTION TO THE ASSOCIATED HOMOGENEOUS EQUATION

Let y_p be a particular solution of (22) and let y^* be any other solution. Then $y^* - y_p$ is a solution of (23); that is,

$$y^*(x) = c_1y_1(x) + c_2y_2(x) + y_p(x)$$

for some constants c_1 and c_2, where y_1, y_2 are two linearly independent solutions of (22).

Thus in order to find all solutions to the nonhomogeneous equation, we need find only one solution to the nonhomogeneous equation and the general solution of the homogeneous equation.

PROOF: We have

$$\begin{aligned}(y^* - y_p)'' + a(y^* - y_p)' + b(y^* - y_p) &= (y^{*''} + ay^{*'} + by^*) \\ &\quad - (y_p'' + ay_p' + by_p) \\ &= f - f = 0.\end{aligned}$$ ■

DEFINITION GENERAL SOLUTION TO A LINEAR, NONHOMOGENEOUS SECOND-ORDER EQUATION

Let $y_p(x)$ be one solution to the nonhomogeneous equation (22) and let $y_1(x)$ and $y_2(x)$ be two linearly independent solutions to the homogeneous equation (23). Then the **general solution** to (22) is given by

$$y(x) = c_1 y_1(x) + c_2 y_2(x) + y_p(x),$$

where c_1 and c_2 are arbitrary constants. ■

EXAMPLE 8 A PARTICULAR SOLUTION TO A NONHOMOGENEOUS EQUATION

It is not difficult to verify that $\frac{1}{2}xe^x$ is a particular solution of $y'' - y = e^x$. Two linearly independent solutions of $y'' - y = 0$ are given by $y_1 = e^x$ and $y_2 = e^{-x}$. The general solution is therefore $y(x) = c_1 y_1 + c_2 y_2 + \frac{1}{2}xe^x = c_1 e^x + c_2 e^{-x} + \frac{1}{2}xe^x$. Note that y_1 and y_2 are linearly independent for all real numbers, since

$$W(y_1, y_2)(x) = e^x(-e^{-x}) - e^{-x}(e^x) = -2 \neq 0.$$

THEOREM 7 PRINCIPLE OF SUPERPOSITION FOR NONHOMOGENEOUS EQUATIONS

Suppose that y_f is a solution to

$$y'' + a(x)y' + b(x)y = f(x) \tag{24}$$

and y_g is a solution to

$$y'' + a(x)y' + b(x)y = g(x) \tag{25}$$

Then $y_f + y_g$ is a solution to

$$y'' + a(x)y' + b(x)y = f(x) + g(x) \tag{26}$$

PROOF: Let $y(x) = y_f(x) + y_g(x)$. Then

$$y'' + a(x)y' + b(x)y = (y_f + y_g)'' + a(x)(y_f + y_g)' + b(x)(y_f + y_g)$$

from (24) and (25)

$$= (y_f'' + a(x)y_f' + b(x)y_f) + (y_g'' + a(x)y_g' + b(x)y_g) \overset{\downarrow}{=} f(x) + g(x)$$ ■

EXAMPLE 9 USING THE PRINCIPLE OF SUPERPOSITION

It is not difficult to verify that one solution to $y'' - y = x^2$ is $-2 - x^2$ and one solution to $y'' - y = 2e^x$ is xe^x. It then follows from the principle of superposition that one solution to $y'' - y = x^2 + 2e^x$ is $-2 - x^2 + xe^x$. To verify this we set

$$y = -2 - x^2 + xe^x$$

Then

$$y' = -2x + (x + 1)e^x$$
$$y'' = -2 + (x + 2)e^x$$

So

$$\begin{aligned} y'' - y &= [-2 + (x + 2)e^x] - [-2 - x^2 + xe^x] \\ &= -2 + xe^x + 2e^x + 2 + x^2 - xe^x = x^2 + 2e^x \end{aligned}$$

In the next three sections we present methods for finding the general solution of homogeneous equations. In Sections 10.11 and 10.12 techniques for obtaining a solution y_p of a nonhomogeneous equation are developed.

PROOFS OF THEOREMS 4 AND 5

THEOREM 5: Two solutions $y_1(x)$ and $y_2(x)$ of

$$y'' + a(x)y' + b(x)y = 0$$

are linearly independent on $[x_0, x_1]$, $x_0 < x_1$ if and only if $W(y_1, y_2)(x) \neq 0$ for all x in $[x_0, x_1]$.

PROOF: We first show that if $W(y_1, y_2)(x) = 0$ for some x_2 in $[x_0, x_1]$, then y_1 and y_2 are linearly dependent. Consider the system of equations

$$\begin{aligned} c_1y_1(x_2) + c_2y_2(x_2) &= 0, \\ c_1y_1'(x_2) + c_2y_2'(x_2) &= 0. \end{aligned} \tag{27}$$

The determinant of this system is

$$y_1(x_2)y_2'(x_2) - y_2(x_2)y_1'(x_2) = W(y_1, y_2)(x_2) = 0.$$

Thus, according to the theory of determinants (see Theorem 7.3.4 on page 478), there exists a solution (c_1, c_2) for (27) where c_1 and c_2 are not both equal to zero. Define $y(x) = c_1y_1(x) + c_2y_2(x)$. By Theorem 3, $y(x)$ is a solution of

$$y'' + a(x)y' + b(x)y = 0.$$

But since c_1 and c_2 satisfy (27),

$$y(x_2) = c_1y_1(x_2) + c_2y_2(x_2) = 0$$

and

$$y'(x_2) = c_1y_1'(x_2) + c_2y_2'(x_2) = 0.$$

Thus $y(x)$ solves the initial-value problem

$$y'' + a(x)y' + b(x)y = 0, \qquad y(x_2) = y'(x_2) = 0.$$

But this initial-value problem also has the solution $y_3(x) \equiv 0$ for all values of x in $x_0 \le x \le x_1$. By Theorem 1, the solution of this initial-value problem is unique, so necessarily $y(x) = y_3(x) \equiv 0$. Thus

$$y(x) = c_1y_1(x) + c_2y_2(x) = 0$$

for all values of x in $x_0 \le x \le x_1$, which proves that y_1 and y_2 are linearly dependent.

We now assume that $W(y_1, y_2)(x) \neq 0$ for all x in $[x_0, x_1]$ and prove that y_1 and y_2 are linearly independent. Consider the equation

$$c_1 y_1(x) + c_2 y_2(x) = 0,$$

for all x in $[x_0, x_1]$. Differentiating with respect to x, we have

$$c_1 y_1'(x) + c_2 y_2'(x) = 0,$$

for all x in $[x_0, x_1]$. Select any x_2 in $[x_0, x_1]$ and consider the homogeneous system

$$c_1 y_1(x_2) + c_2 y_2(x_2) = 0,$$
$$c_1 y_1'(x_2) + c_2 y_2'(x_2) = 0.$$

Its determinant,

$$\begin{vmatrix} y_1(x_2) & y_2(x_2) \\ y_1'(x_2) & y_2'(x_2) \end{vmatrix} = W(y_1, y_2)(x_2),$$

does not equal zero; thus $c_1 = c_2 = 0$. Hence $y_1(x)$ and $y_2(x)$ are linearly independent on $[x_0, x_1]$. ■

THEOREM 4: Let $y_1(x)$ and $y_2(x)$ be linearly independent solutions on $[x_0, x_1]$, $x_0 < x_1$, to

$$y'' + a(x)y' + b(x)y = 0$$

and let $y_3(x)$ be another solution on this interval. Then there exist unique constants c_1 and c_2 such that

$$y_3(x) = c_1 y_1(x) + c_2 y_2(x).$$

PROOF: Let $y_3(x_0) = a$ and $y_3'(x_0) = b$. Consider the linear system of equations in two unknowns c_1 and c_2:

$$y_1(x_0)c_1 + y_2(x_0)c_2 = a$$
$$y_1'(x_0)c_1 + y_2'(x_0)c_2 = b \tag{28}$$

As we saw earlier, the determinant of this system is $W(y_1, y_2)(x_0)$, which is nonzero since the solutions are linearly independent. Thus there is a unique solution (c_1, c_2) to (28) and a solution $y^*(x) = c_1 y_1(x) + c_2 y_2(x)$ that satisfies the conditions $y^*(x_0) = a$ and $y^{*\prime}(x_0) = b$. Since every initial-value problem has a unique solution (by Theorem 1), it must follow that $y_3(x) = y^*(x)$ on the interval $x_0 \le x \le x_1$, and so the proof is complete. ■

PROBLEMS 10.7

SELF-QUIZ

Multiple Choice

I. Which of the following problems is guaranteed by the existence-uniqueness theorem to have a unique solution?

a. $y'' + \frac{1}{x}y' + 3y = x^2$; $y(0) = 0$, $y'(0) = 1$

b. $y'' + \frac{1}{x}y' + 3y = x^2$; $y(1) = 1$, $y'(1) = 3$

c. $y'' + \frac{1}{x}y' + 3y = x^2$; $y(1) = 1$, $y(2) = 3$

d. $xy'' + y' + 4xy = 0$; $y(0) = 1$, $y'(0) = 2$

II. e^x and e^{2x} are two solutions to $y'' - 3y' + 2y = 0$. Which of the following are also solutions to $y'' - 3y' + 2y = 0$?

a. $e^x + e^{2x}$ **b.** $e^x - e^{2x}$
c. $-8e^{2x}$ **d.** $17e^x$
e. $4e^x - 3.8e^{2x}$

III. Which of the following pairs of functions are linearly independent?

a. e^x, e^{2x} **b.** $\frac{-3}{x}, \frac{4}{x}$
c. $\ln x, \ln x^3$ **d.** $\sin x, \tan x$
e. $3, x$

IV. Which of the following is the general solution to $y'' + y = 0$?

a. $c_1 e^x + c_2 e^{-x}$
b. $c_1 \sin x + c_2 \sin 2x$
c. $c_1 \cos x + c_2 \cos(-x)$
d. $c_1 \sin x + c_2 \cos x$
e. $c_1 \cos x - c_2 \sin x$

V. The Wronskian for the functions $y_1(x) = x^2$ and $y_2(x) = x^5$ is

a. $5x^6 - 2x^5$ **b.** x^7
c. $2x^3 - 5x^9$ **d.** $2x - 5x^4$
e. $3x^6$

VI. The general solution to $y'' + 4y = 0$ is $c_1 \cos 2x + c_2 \sin 2x$. One solution to $y'' + 4y = x$ is $y = \frac{1}{4}x$. One solution to $y'' + 4y = e^x$ is $y = e^x/5$. Which of the following are solutions to $y'' + 4y = x + e^x$?

a. $\frac{1}{4}x + e^x/5$
b. $x + 4e^x/5$
c. $\frac{1}{4}x - e^x/5$
d. $\frac{1}{4}x + e^x/5 + 3\cos 2x - 5\sin 2x$
e. $\frac{1}{4}x + \sin 2x$
f. $e^x/5 - 3\cos 2x$
g. $\frac{1}{4}x + e^x/5 + \sin 2x$
h. $\frac{1}{4}x + e^x/5 - \cos 2x$

In Problems 1–10 determine whether the given equation is linear or nonlinear. If it is linear, state whether it is homogeneous or nonhomogeneous with constant or variable coefficients.

1. $y'' + 2x^3y' + y = 0$
2. $y'' + 2y' + y^2 = x$
3. $y'' + 3y' + yy' = 0$
4. $y'' + 3y' + 4y = 0$
5. $y'' + 3y' + 4y = \sin x$
6. $y'' + y(2 + 3y) = e^x$
7. $y'' + 4xy' + 2x^3y = e^{2x}$
8. $y'' + \sin(xe^x)y' + 4xy = 0$
9. $3y'' + 16y' + 2y = 0$
10. $yy'y'' = 1$

In Problems 11–26, determine whether the given set of functions is linearly dependent or independent on $\mathbb{R}$ or the given interval.

11. $y_1 = x, y_2 = x^2$
12. $y_1 = x^3, y_2 = -3x^3$
13. $y_1 = 2e^{3x}, y_2 = 3e^{2x}$
14. $y_1 = \ln(1 + x^2), y_2 = \ln(1 + x^2)^5$
15. $y_1 = x^2 - 1, y_2 = x^2 + 1$
16. $y_1 = 1, y_2 = x, y_3 = x^2$
17. $y_1 = x^2, y_2 = 2x - 5x^2, y_3 = -x$
18. $y = \sin x, y_2 = \cos x, y_3 = \tan x$ on $-\frac{\pi}{4} \leq x \leq \frac{\pi}{4}$.
19. $y_1 = 3, y_2 = \sin^2 x, y_3 = \cos^2 x$
20. $y_1 = e^x, y_2 = e^{-2x}, y_3 = e^{-x}$
21. $y_1 = 1 + x, y_2 = 1 + x^2, y_3 = 3 - x^3$
22. $y_1 = 1 + x, y_2 = 1 - x, y_3 = \frac{1 + x}{1 - x}$ on $|x| < 0.5$
23. $y_1 = 1, y_2 = 1 + x, y_3 = x^2, y_4 = x(1 - x), y_5 = x$
24. $y_1 = \sin^2 x, y_2 = 1, y_3 = \sin x \cos x, y_4 - \cos^2 x, y_5 = \sin 2x$
25. $y_1 = 1 + x, y_2 = 1 - x, y_3 = 1, y_4 = x^2, y_5 = 1 + x^2$
26. $y_1 = 2\cos^2 x - 1, y_2 = \cos^2 x - \sin^2 x, y_3 = 1 - 2\sin^2 x, y_4 = \cos 2x$

In Problems 27–30, use the Wronskian to show that the given functions are linearly independent on the given interval.

27. $\frac{1}{x}, \frac{1}{x^2}$; $[0.1, \infty)$
28. $e^x, \ln x$; $(1, \infty)$
29. $\sin x, \tan x$; $\left[0, \frac{\pi}{3}\right]$

30. $\sin x$, $\sin 2x$; $\mathbb{R}$

In Problems 31–34 test each of the functions 1, x, x^2, and x^3 to see which functions satisfy the given differential equation. Then construct the *general solution* to the equation by writing a linear combination of the linearly independent solutions you have found.

31. $y'' = 0$

32. $y''' = 0$

33. $xy'' - y' = 0$

34. $x^2y'' - 2xy' + 2y = 0$

In Problems 35–42, verify that the two given functions are linearly independent solutions of the given homogeneous equation and then find the general solution.

35. $y'' + 9y = 0$; $\sin 3x$, $\cos 3x$

36. $y'' - y' - 2y = 0$; e^{2x}, e^{-x}

37. $y'' + 2y' - 15y = 0$; e^{3x}, e^{-5x}

38. $y'' + 9y' + 14y = 0$; e^{-2x}, e^{-7x}

39. $y'' - 6y' + 9y = 0$; e^{3x}, xe^{3x}

40. $y'' + 4y' + 4y = 0$; e^{-2x}, xe^{-2x}

41. $y'' + 2y' + 2y = 0$; $e^{-x}\cos x$, $e^{-x}\sin x$

42. $y'' - 6y' + 13y = 0$; $e^{3x}\cos 2x$, $e^{3x}\sin 2x$

In Problems 43–46, a particular solution to a nonhomogeneous equation is given. Using the information found in Problems 35, 36, 37 and 41, find the general solution.

43. $y'' + 9y = 4x$; $y_p = \frac{4}{9}x$

44. $y'' - y' - 2y = e^x$; $y_p = -\frac{1}{2}e^x$

45. $y'' + 2y' - 15y = 15x^2$; $y_p = -x^2 - \frac{4}{15}x - \frac{8}{225}$

46. $y'' + 2y' + 2y = \cos x$; $y_p = \frac{1}{5}\cos x + \frac{2}{5}\sin x$

47. Let $y_1(x)$ be a solution of the homogeneous equation

$$y'' + a(x)y' + b(x)y = 0$$

on the interval $\alpha \le x \le \beta$. Suppose that the curve y_1 is tangent to the x-axis at some point of this interval. Prove that y_1 must be identically zero.

48. Let $y_1(x)$ and $y_2(x)$ be two nontrivial solutions of the homogeneous equation

$$y'' + a(x)y' + b(x)y = 0$$

on the interval $\alpha \le x \le \beta$. Suppose $y_1(x_0) = y_2(x_0) = 0$ for some point $\alpha \le x_0 \le \beta$. Show that y_2 is a constant multiple of y_1.

49. **a.** Show that x and x^3 are linearly independent on $|x| < 1$ even though the Wronskian $W(x, x^3) = 0$ at $x = 0$.

b. Show that x and x^3 are solutions to

$$x^2y'' - 3xy' + 3y = 0.$$

Does this contradict Theorem 5?

50. **a.** Show that $y_1(x) = \sin x^2$ and $y_2(x) = \cos x^2$ are linearly independent solutions of

$$xy'' - y' + 4x^3y = 0.$$

b. Calculate $W(y_1, y_2)(x)$ and show that it is zero when $x = 0$. Does this result contradict Theorem 5? [*Hint:* In Theorem 5, as elsewhere in this section, it is assumed that $a(x)$ and $b(x)$ are continuous on some interval.]

51. Show that

$$y_1(x) = \sin x$$

and

$$y_2(x) = 4\sin x - 2\cos x$$

are linearly independent solutions of $y'' + y = 0$. Write the solution $y_3(x) = \cos x$ as a linear combination of y_1 and y_2.

52. Prove that $e^x \sin x$ and $e^x \cos x$ are linearly independent solutions of the equation

$$y'' - 2y' + 2y = 0.$$

a. Find a solution that satisfies the conditions $y(0) = 1$, $y'(0) = 4$.

b. Find another pair of linearly independent solutions.

53. Assume that some nonzero solution of

$$y'' + a(x)y' + b(x)y = 0, \qquad y(0) = 0$$

vanishes at some point x_1, where $x_1 > 0$. Prove that any other solution vanishes at $x = x_1$.

54. Define the function $s(x)$ to be the unique solution of the initial-value problem

$$y'' + y = 0; \qquad y(0) = 0, \qquad y'(0) = 1,$$

and the function $c(x)$ as the solution of

$$y'' + y = 0; \qquad y(0) = 1, \qquad y'(0) = 0.$$

Without using trigonometry, prove that

a. $\frac{ds}{dx} = c(x)$;

b. $\dfrac{dc}{dx} = -s(x)$; **c.** $s^2 + c^2 = 1$.

55. **a.** Show that $y_1 = \sin \ln x^2$ and $y_2 = \cos \ln x^2$ are linearly independent solutions of

$$y'' + \frac{1}{x}y' + \frac{4}{x^2}y = 0 \qquad (x > 0).$$

b. Calculate $W(y_1, y_2)(x)$.

ANSWERS TO SELF-QUIZ

I. b **II.** All of them **III.** a, d, e **IV.** d or e **V.** e
VI. a, d, g, h

10.8 USING ONE SOLUTION TO FIND ANOTHER: REDUCTION OF ORDER

As we saw in Theorem 10.7.4, it is easy to write down the general solution of the homogeneous equation

$$y'' + a(x)y' + b(x)y = 0 \tag{1}$$

provided we know two linearly independent solutions y_1 and y_2 of equation (1). The general solution is then given by

$$y = c_1y_1 + c_2y_2,$$

where c_1 and c_2 are arbitrary constants. Unfortunately, there is no general procedure for determining y_1 and y_2 unless a and b are constant functions. However, a standard procedure does exist for finding y_2 when y_1 is known. This method is of considerable importance, since it is sometimes possible to find one solution by inspecting the equation or by trial and error.

We assume that y_1 is a nonzero solution of (1) on some interval and seek another solution y_2 such that y_1 and y_2 are linearly independent. Suppose that y_2 can be found. Then, since y_1 and y_2 are linearly independent, $y_2 \neq ky_1$ for any constant k. So

$$\frac{y_2}{y_1} = v(x)$$

must be a nonconstant function of x, and $y_2 = vy_1$ must satisfy (1). Thus

$$(vy_1)'' + a(vy_1)' + b(vy_1) = 0. \tag{2}$$

But

$$(vy_1)' = vy_1' + v'y_1 \tag{3}$$

and

$$\begin{aligned}(vy_1)'' = (vy_1' + v'y_1)' &= vy_1'' + v'y_1' + v'y_1' + v''y_1\\ &= vy_1'' + 2v'y_1' + v''y_1.\end{aligned} \tag{4}$$

Using (3) and (4) in (2), we have

$$(vy_1'' + 2v'y_1' + v''y_1) + a(vy_1' + v'y_1) + bvy_1 = 0,$$

or

$$v(y_1'' + ay_1' + by_1) + v'(2y_1' + ay_1) + v''y_1 = 0. \tag{5}$$

The first term in parentheses in (5) vanishes since y_1 is a solution of (1), so we obtain

$$v''y_1 + v'(2y_1' + ay_1) = 0.$$

Set $z = v'$, so that $z' = v''$, and divide by y_1 to obtain

$$z' + \left(\frac{2y_1'}{y_1} + a(x)\right)z = 0. \tag{6}$$

This is a separable linear first-order equation. Thus we have *reduced the order* of our equation. Multiply by the integrating factor for equation (6)

$$e^{\int (2y_1'/y_1 + a(x))\,dx} = e^{2 \ln y_1 + \int a(x)\,dx} = y_1^2 e^{\int a(x)\,dx},$$

so that equation (6) becomes

$$(y_1^2 e^{\int a(x)\,dx}\, z)' = 0. \tag{7}$$

A solution to equation (7) is

$$z = v' = \frac{1}{{y_1}^2} e^{-\int a(x)\,dx}.$$

Since the exponential is never zero, v is nonconstant. To find v, we perform another integration and obtain

FINDING A SECOND SOLUTION, y_2, WHEN ONE SOLUTION, y_1, IS GIVEN

$$y_2(x) = y_1(x)v(x) = y_1(x) \int \frac{e^{-\int a(x)\,dx}}{{y_1}^2(x)}\,dx. \tag{8}$$

REMARK: It is not advisable to memorize formula (8). It is only necessary to remember the substitution $y_2 = vy_1$ and substitute this into the original differential equation. The following examples illustrate this procedure.

EXAMPLE 1

FINDING A SECOND SOLUTION WHEN ONE IS GIVEN

Note that $y_1 = x$ is a solution of

$$x^2 y'' - xy' + y = 0, \quad x > 0. \tag{9}$$

Setting $y_2 = y_1 v = xv(x)$, it follows that $y_2' = xv' + v$, $y_2'' = xv'' + 2v'$, and (9) becomes

$$x^2(xv'' + 2v') - x(xv' + v) + (xv) = 0,$$

or

$$x^3 v'' + x^2 v' = 0.$$

Setting $z = v'$ and separating variables, we have

$$\frac{dz}{z} = \frac{-dx}{x} \quad \text{or} \quad \ln|z| = -\ln x, \quad \text{(since } x > 0\text{)}$$

from which we obtain by exponentiation

$$v' = z = \frac{1}{x}.$$

Thus $v(x) = \ln x$, so $y_2 = y_1 v = x \ln x$ and the general solution of (9) is

$$y = c_1 x + c_2 x \ln x, \quad x > 0.$$

EXAMPLE 2 FINDING A SECOND SOLUTION TO A LEGENDRE EQUATION

Consider the **Legendre equation of order one**:

$$(1 - x^2)y'' - 2xy' + 2y = 0, \qquad -1 < x < 1. \tag{10}$$

Again, it is not hard to verify that $y_1(x) = x$ is a solution. Setting $y_2 = xv$, we have

$$(1 - x^2)(xv'' + 2v') - 2x(xv' + v) + 2xv = x(1 - x^2)v'' + 2(1 - 2x^2)v' = 0;$$

when we substitute $z = v'$ and divide by $x(1 - x)^2$, we can use partial fractions to obtain

$$\frac{dz}{dx} + \frac{2(1 - 2x^2)z}{x(1 - x^2)} = \frac{dz}{dx} + \left(\frac{2}{x} - \frac{2x}{1 - x^2}\right)z = 0.$$

Separating variables and integrating, we obtain

$$\ln|z| = \int \frac{dz}{z} = \int \left(\frac{2x}{1 - x^2} - \frac{2}{x}\right) dx = -\ln(1 - x^2) - \ln(x^2).$$

Thus, exponentiating and using partial fractions, we get

$$v' = \frac{1}{x^2(1 - x^2)} = \frac{1}{x^2} + \frac{1}{2}\left(\frac{1}{1 + x} + \frac{1}{1 - x}\right).$$

Hence

$$y_2 = y_1 v = x \int \left[\frac{1}{x^2} + \frac{1}{2}\left(\frac{1}{1 + x} + \frac{1}{1 - x}\right)\right] dx = x\left[\frac{-1}{x} + \frac{1}{2}\ln\left(\frac{1 + x}{1 - x}\right)\right],$$

or

$$y_2 = \frac{x}{2}\ln\left(\frac{1 + x}{1 - x}\right) - 1, \qquad |x| < 1.$$

PROBLEMS 10.8

SELF-QUIZ

Multiple Choice

I. Suppose that $y_1(x)$ is a solution to the differential equation $y'' + \frac{1}{x}y' + (\sin x)y = 0$. Then a second solution is given by ________.

a. $y_2(x) = y_1(x) \int \ln \frac{x}{y_1^2}(x)\, dx$

b. $y_2(x) = y_1(x) \int \frac{x}{y_1^2(x)}\, dx$

c. $y_2(x) = y_1(x) \int \frac{1}{xy_1^2(x)}\, dx$

d. $y_2(x) = y_1(x) \int \frac{e^x}{y_1^2(x)}\, dx$

e. $y_2(x) = y_1(x) \int \frac{e^{1/x}}{y_1^2(x)}\, dx$

II. $y_1 = e^{cx}$ is a solution to $y'' - 2cy' + c^2y = 0$ where c is a real number. A second linearly independent solution on $\mathbb{R}$ is ________.

a. $e^{cx} \int dx$

b. $e^{cx} \int \frac{e^{-2cx}}{e^{2cx}}\, dx$

c. $e^{cx} \int ce^{-2cx}\, dx$

d. $e^{cx} \int \frac{e^{-c^2x}}{e^{2cx}}\, dx$

e. $e^{cx} \int \frac{e^{-x^2/2}}{e^{2cx}}\, dx$

In each of Problems 1–25 a second-order differential equation and one solution $y_1(x)$ are given. Verify that $y_1(x)$ is indeed a solution and find a second linearly independent solution.

1. $y'' - 2y' + y = 0,\ y_1(x) = e^x$
2. $y'' + 4y = 0,\ y_1(x) = \sin 2x$
3. $y'' - 4y = 0,\ y_1(x) = e^{2x}$
4. $y'' - 4y' + 4y = 0,\ y_1(x) = xe^{2x}$
5. $y'' + 5y' + 6y = 0,\ y_1(x) = e^{-2x}$
6. $y'' + 2y' + 2y = 0,\ y_1(x) = e^{-x}\cos x$
7. $y'' + y' + 7y = 0,\ y_1(x) = e^{-x/2}\cos\frac{3\sqrt{3}}{2}x$
8. $y'' + 10y' + 25y = 0,\ y_1(x) = xe^{-5x}$
9. $y'' - 30y' + 200y = 0,\ y_1(x) = e^{20x}$
10. $y'' + \left(\frac{3}{x}\right)y' = 0,\ y_1(x) = 1$
11. $y'' + y' + \frac{1}{4}y = 0,\ y_1(x) = e^{-x/2}$
12. $y'' + 0.2y' - 0.03y = 0;\ y_1(x) = e^{x/10}$
13. $x^2y'' + xy' - 4y = 0,\ y_1(x) = x^2$, on $x > 0$
14. $y'' - 2xy' + 2y = 0,\ y_1(x) = x$
15. $x^2y'' - 2xy' + (x^2 + 2)y = 0,\ (x > 0)$, $y_1 = x\sin x$
16. $xy'' + (2x - 1)y' - 2y = 0,\ (x > 0),\ y_1 = e^{-2x}$
17. $xy'' + (x - 1)y' + (3 - 12x)y = 0,\ (x > 0)$, $y_1 = e^{3x}$
18. $xy'' - y' + 4x^3y = 0,\ (x > 0),\ y_1 = \sin(x^2)$
19. $xy'' - (x + n)y' + ny = 0$, (integer $n > 0$), $y_1(x) = e^x,\ x > 0$
20. $x^{1/3}y'' + y' + \left(\frac{1}{4}x^{-1/3} - \frac{1}{6x} - 6x^{-5/3}\right)y = 0$, $y_1 = x^3e^{-3x^{2/3}/4},\ (x > 0)$
21. $(3 + x^2)y'' + 2xy' - 2y = 0;\ y_1(x) = x$
22. $y'' + (\tan x)y' = 0;\ y_1(x) = 1,\ 0 \le x \le \frac{\pi}{4}$
23. $y'' - (2\cot x)y' = 0;\ y_1(x) = 1,\ \frac{\pi}{4} \le x \le \frac{\pi}{2}$
24. $y'' + (2\sec x)y' = 0;\ y_1(x) = 1,\ 0 \le x \le \frac{\pi}{4}$
25. $y'' - (\csc x)y' = 0;\ y_1(x) = 1,\ \frac{\pi}{4} \le x \le \frac{\pi}{2}$

In Problems 26–28, an equation and one solution are given. Find an expression involving an integral that gives a second, linearly independent solution. Do not try to evaluate the integral.

26. $2y'' + xy' - 2y = 0,\ y_1(x) = x^2 - 2$
27. $y'' - 2xy' - 2y = 0;\ y_1(x) = e^{x^2}$
28. $y'' + (\sin x)y' - \frac{\sin x}{x}y = 0;\ y_1(x) = x$
29. The **Bessel differential equation** is given by
$$x^2y'' + xy' + (x^2 - p^2)y = 0.$$
For $p = \frac{1}{2}$, verify that $y_1(x) = (\sin x)/\sqrt{x}$ is a solution for $x > 0$. Find a second linearly independent solution.
30. Letting $p = 0$ in the equation of Problem 29, we obtain the **Bessel differential equation of index zero**. One solution is the **Bessel function of order zero** denoted by $J_0(x)$. In terms of $J_0(x)$, find a second linearly independent solution.

ANSWERS TO SELF-QUIZ

I. c **II.** a

10.9 HOMOGENEOUS EQUATIONS WITH CONSTANT COEFFICIENTS: REAL ROOTS

In this section we present a simple procedure for finding the general solution to the linear homogeneous equation with constant coefficients

$$y'' + ay' + by = 0. \tag{1}$$

Recall that for the comparable first-order equation $y' + ay = 0$ the general solution is $y(x) = ce^{-ax}$. It is then not implausible to "guess" that there may be a solution to (1) of the form $y(x) = e^{\lambda x}$ for some number λ (real or complex). Setting $y(x) = e^{\lambda x}$, we obtain $y' = \lambda e^{\lambda x}$ and $y'' = \lambda^2 e^{\lambda x}$, so that (1) yields

$$\lambda^2 e^{\lambda x} + a\lambda e^{\lambda x} + be^{\lambda x} = 0.$$

Since $e^{\lambda x} \neq 0$, we can divide this equation by $e^{\lambda x}$ to obtain

THE CHARACTERISTIC EQUATION

$$\lambda^2 + a\lambda + b = 0 \tag{2}$$

where a and b are real numbers. Equation (2) is called the **characteristic equation** of the differential equation (1). It is clear that if λ satisfies (2), then $y(x) = e^{\lambda x}$ is a solution to (1). As we saw earlier, we need only obtain two linearly independent solutions. Equation (2) has the roots

$$\lambda_1 = \frac{-a + \sqrt{a^2 - 4b}}{2} \quad \text{and} \quad \lambda_2 = \frac{-a - \sqrt{a^2 - 4b}}{2}. \tag{3}$$

There are three possibilities: $a^2 - 4b > 0$, $a^2 - 4b = 0$, $a^2 - 4b < 0$.

CASE 1—ROOTS REAL AND UNEQUAL: If $a^2 - 4b > 0$, then λ_1 and λ_2 are distinct real numbers, given by (3), and $y_1(x) = e^{\lambda_1 x}$ and $y_2 = e^{\lambda_2 x}$ are distinct solutions.

These two solutions are linearly independent because

$$\frac{y_1}{y_2} = e^{(\lambda_1 - \lambda_2)x}$$

which is clearly not a constant when $\lambda_1 \neq \lambda_2$. Thus we have proved the following theorem:

THEOREM 1 THE GENERAL SOLUTION TO A HOMOGENEOUS EQUATION IN THE CASE OF REAL, UNEQUAL ROOTS

If $a^2 - 4b > 0$, then the roots of the characteristic equation are real and unequal and the general solution to equation (1) is given by

$$y(x) = c_1 e^{\lambda_1 x} + c_2 e^{\lambda_2 x}. \tag{4}$$

where c_1 and c_2 are arbitrary constants and λ_1 and λ_2 are the real roots of (2). ■

EXAMPLE 1 A HOMOGENEOUS EQUATION WITH REAL, DISTINCT ROOTS

Consider the equation

$$y'' + 3y' - 10y = 0.$$

The characteristic equation is $\lambda^2 + 3\lambda - 10 = 0 = (\lambda - 2)(\lambda + 5)$, and the roots are $\lambda_1 = 2$ and $\lambda_2 = -5$ (the order in which the roots are taken is irrelevant). The general solution is

$$y(x) = c_1 e^{2x} + c_2 e^{-5x}.$$

If we specify the initial conditions $y(0) = 1$ and $y'(0) = 3$, for example, then differentiating and substituting $x = 0$, we obtain the simultaneous equations

$$c_1 + c_2 = 1,$$
$$2c_1 - 5c_2 = 3,$$

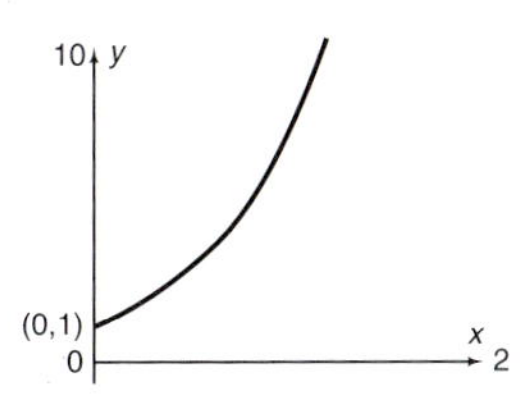

FIGURE 1
Graph of
$y = \frac{1}{7}(8e^{2x} - e^{-5x}),\ x \geq 0$

which have the unique solution $c_1 = \frac{8}{7}$ and $c_2 = -\frac{1}{7}$. The unique solution to the initial-value problem is therefore

$$y(x) = \tfrac{1}{7}(8e^{2x} - e^{-5x}).$$

Note that since $e^{2x} \to \infty$ and $e^{-5x} \to 0$, the solution $y(x) \to \infty$ as $x \to \infty$. It is sketched in Figure 1.

CASE 2—ROOTS REAL AND EQUAL: Suppose $a^2 - 4b = 0$. In this case (2) has the double root $\lambda_1 = \lambda_2 = -a/2$. Thus $y_1(x) = e^{-ax/2}$ is a solution of (1). To find the second solution y_2, we make use of equation (10.8.8), p. 749, since one solution is known:

$$y_2(x) = y_1(x) \int \frac{e^{-ax}}{y_1^{\,2}(x)}\,dx = e^{-ax/2} \int \frac{e^{-ax}}{(e^{-ax/2})^2}\,dx$$

$$= e^{-ax/2} \int \frac{e^{-ax}}{e^{-ax}}\,dx = e^{-ax/2} \int dx = xe^{-ax/2}.$$

Since $y_2/y_1 = x$, it follows that y_1 and y_2 are linearly independent. Hence we have the following result:

THEOREM 2 **THE GENERAL SOLUTION TO A HOMOGENEOUS EQUATION IN THE CASE OF ONE REAL ROOT**

If $a^2 - 4b = 0$, then the roots of the characteristic equation are equal and the general solution to equation (1) is given by

$$y(x) = c_1 e^{-ax/2} + c_2 x e^{-ax/2} \tag{5}$$

where c_1 and c_2 are arbitrary constants. ■

EXAMPLE 2 A HOMOGENEOUS EQUATION WITH ONE REAL ROOT

Consider the equation

$$y'' + 6y' + 9 = 0.$$

The characteristic equation is $\lambda^2 + 6\lambda + 9 = 0 = (\lambda + 3)^2$, yielding the unique double root $\lambda_1 = -a/2 = -3$. The general solution is

$$y(x) = c_1 e^{-3x} + c_2 x e^{-3x}$$

If we use the initial conditions $y(0) = 1$, $y'(0) = 7$, we obtain the simultaneous equations

$$c_1 = 1,$$
$$-3c_1 + c_2 = 7,$$

which yield the unique solution (since $c_2 = 7 + 3c_1 = 10$)

$$y(x) = e^{-3x} + 10xe^{-3x} = e^{-3x}(1 + 10x).$$

Since $\lim_{x\to\infty} e^{-3x} - \lim_{x\to\infty} xe^{-3x} = 0$, the solution $y(x) \to 0$ as $x \to \infty$. It is sketched in Figure 2.

FIGURE 2
Graph of $y(x)$
$= e^{-3x}(1 + 10x)$

We will deal with the more complicated situation of complex roots ($a^2 - 4b < 0$) in Section 10.10.

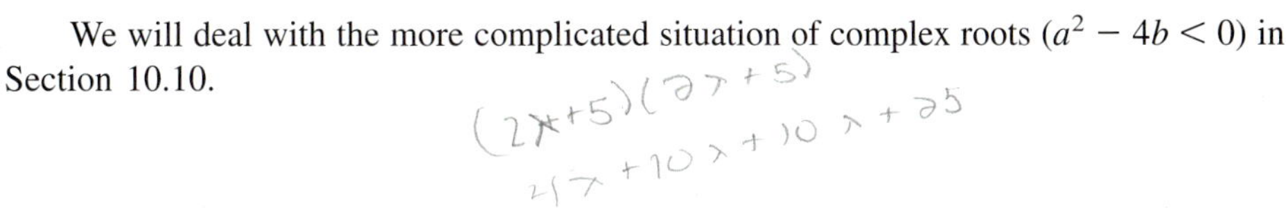

PROBLEMS 10.9

SELF-QUIZ

Multiple Choice

I. Two linearly independent solutions to $y'' + y' - 6y = 0$ are ________.

a. e^{2x}, e^{3x} **b.** e^{-2x}, e^{3x}
c. e^{2x}, e^{-3x} **d.** e^{-2x}, e^{-3x}
e. e^{x}, e^{-6x}

II. Two linearly independent solutions to $y'' + 8y' + 16y = 0$ are ________.

a. e^{8x}, e^{16x} **b.** $e^{4x}, 2e^{4x}$
c. $e^{-4x}, -4e^{-4x}$ **d.** e^{4x}, xe^{4x}
e. e^{-4x}, xe^{-4x}

III. Which condition ensures that the characteristic equation for $y'' + 6y' + ky = 0$ has two real roots?

a. $k \geq 9$ **b.** $k > 9$
c. $k \leq 9$ **d.** $k < 9$
e. k is a real number.

IV. Which condition ensures that the characteristic equation for $y'' + ky' - 5y = 0$ has two real roots?

a. $k^2 < 20$ **b.** $k^2 > 20$
c. $k > \sqrt{5}$ **d.** $0 < k < \sqrt{5}$
e. k is a real number.

In Problems 1–24 find the general solution of each equation. When initial conditions are specified, give the particular solution that satisfies them and sketch a graph.

1. $y'' - 4y = 0$
2. $x'' + x' - 6x = 0,\ x(0) = 0,\ x'(0) = 5$
3. $y'' - 3y' + 2y = 0$
4. $y'' + 5y' + 6y = 0,\ y(0) = 1,\ y'(0) = 2$
5. $4x'' + 20x' + 25x = 0,\ x(0) = 1,\ x'(0) = 2$
6. $y'' + 6y' + 9y = 0$
7. $x'' - x' - 6x = 0,\ x(0) = -1,\ x'(0) = 1$
8. $y'' - 8y' + 16y = 0,\ y(0) = 2,\ y'(0) = -1$
9. $y'' - 5y' = 0$
10. $y'' + 17y' = 0,\ y(0) = 1,\ y'(0) = 0$
11. $y'' + 2\pi y' + \pi^2 y = 0$
12. $y'' - 13y' + 42y = 0$
13. $z'' + 2z' - 15z = 0$
14. $w'' + 8w' + 12w = 0$
15. $y'' - 8y' + 16y = 0,\ y(0) = 1,\ y'(0) = 6$
16. $y'' + 2y' + y = 0,\ y(1) = 2/e,\ y'(1) = -3/e$
17. $y'' - 2y = 0$
18. $y'' + 6y' + 5y = 0$
19. $y'' - 5y = 0,\ y(0) = 3,\ y'(0) = -\sqrt{5}$
20. $y'' - 2y' - 2y = 0,\ y(0) = 1,\ y'(0) = 1 + 3\sqrt{3}$
21. $y'' + 3y' = 0,\ y(0) = 1,\ y'(0) = 1$
22. $y'' - 8y' + 16y = 0$
23. $y'' + 10y' + 16y = 0$
24. $y'' - 6y' - 16y = 0,\ y(1) = y'(1) = 1$

In Problems 25–30 match each initial value problem with the graph of its solution in Figure 3.

25. $y'' - 7y' + 12y = 0,\ y(0) = -1,\ y'(0) = -2$
26. $y'' + y' - 12y = 0,\ y(0) = -1,\ y'(0) = -10$
27. $y'' - y' - 12y = 0,\ y(0) = -1,\ y'(0) = 10$
28. $y'' + 7y' + 12y = 0,\ y(0) = -1,\ y'(0) = 2$
29. $y'' - 7y' + 12y = 0,\ y(0) = -1,\ y'(0) = 2$
30. $y'' + 7y' + 12y = 0,\ y(0) = -1,\ y'(0) = -2$

The Riccati Equation[†]

Linear second-order differential equations may also be used in finding the solution to the **Riccati equation**:

$$y' + y^2 + a(x)y + b(x) = 0. \tag{i}$$

This nonlinear first-order equation frequently occurs in physical applications. To change it into a linear second-order equation, let $y = z'/z$. Then $y' = (z''/z) - (z'/z)^2$, so (i) becomes

$$\frac{z''}{z} - \left(\frac{z'}{z}\right)^2 + \left(\frac{z'}{z}\right)^2 + a(x)\left(\frac{z'}{z}\right) + b(x) = 0.$$

Multiplying by z, we obtain the linear second-order equation

$$z'' + a(x)z' + b(x)z = 0. \tag{ii}$$

If the general solution to (ii) can be found, the quotient $y = z'/z$ is the general solution to (i).

31. Suppose $z = c_1 z_1 + c_2 z_2$ is the general solution

[†] Jacopo Francesco Riccati (1676–1754), an Italian mathematician, physicist, and philosopher, was responsible for bringing much of Newton's work on calculus to the attention of Italian mathematicians.

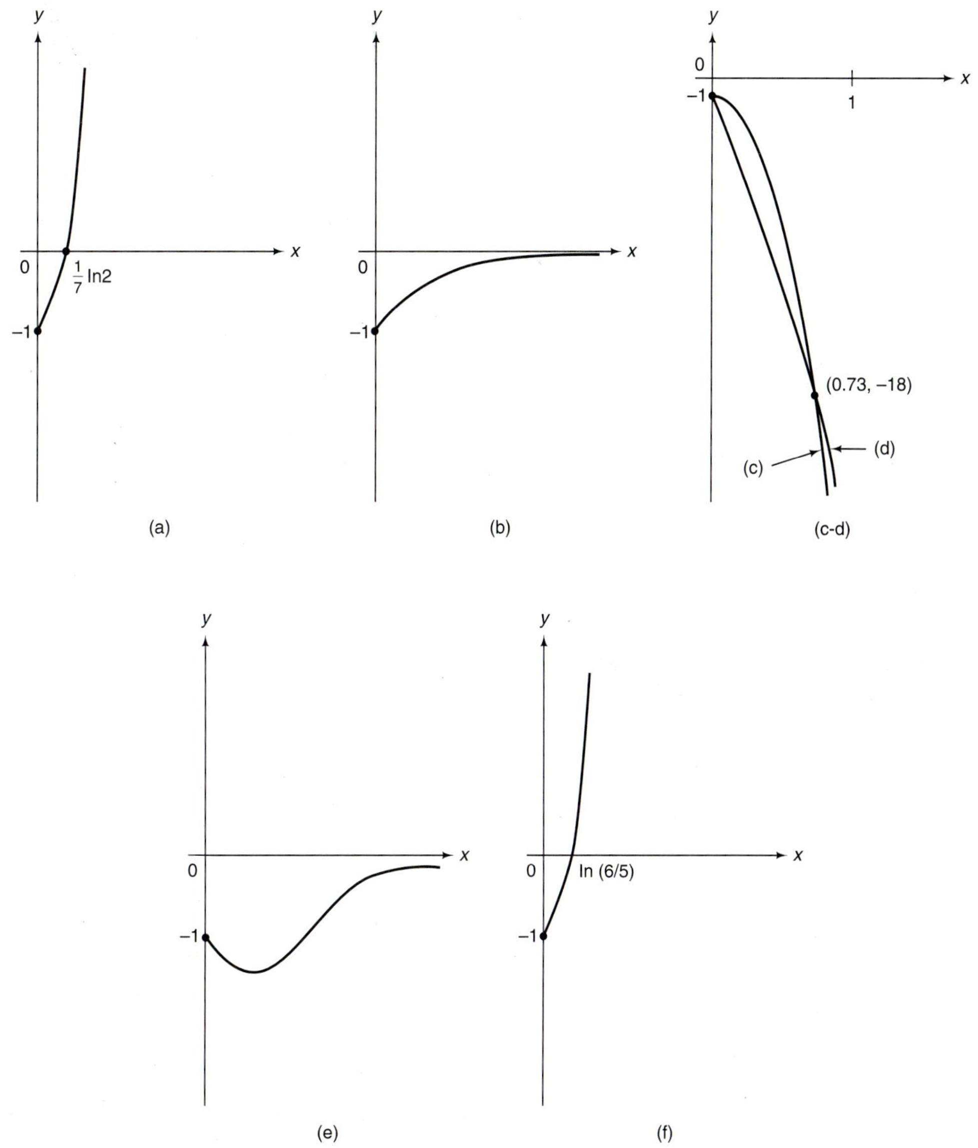

FIGURE 3

to equation (ii). Explain why the quotient z'/z involves only one arbitrary constant. [*Hint:* Divide numerator and denominator by c_1.]

32. For arbitrary constants a, b, and c, find the substitution that changes the nonlinear equation

$$y' + ay^2 + by + c = 0$$

into a linear second-order equation with constant coefficients. What second-order equation is obtained?

Use the method above to find the general solution to the Riccati equations in Problems 33–38. If an initial condition is specified, give the particular solution that satisfies that condition.

33. $y' + y^2 - 1 = 0,\ y(0) = -\frac{1}{3}$

34. $\dfrac{dx}{dt} + x^2 + 1 = 0$

35. $y' + y^2 - 2y + 1 = 0$

36. $y' + y^2 + 3y + 2 = 0,\ y(0) = 1$

37. $y' + y^2 - y - 2 = 0$

38. $y' + y^2 + 2y + 1 = 0,\ y(1) = 0$

39. Suppose that the two roots λ_1 and λ_2 of the characteristic equation (2) are real and satisfy $\lambda_2 = \lambda_1 + h$.

a. Verify that

$$\phi_h(x) = \frac{e^{\lambda_2 x} - e^{\lambda_1 x}}{\lambda_2 - \lambda_1}$$

is a solution of equation (1).

b. Hold λ_1 fixed and evaluate $\phi_h(x)$, as $h \to 0$, with L'Hôpital's rule.

c. Verify that parts (a) and (b) yield Theorem 2's conclusion.

ANSWERS TO SELF-QUIZ

I. c **II.** e **III.** d **IV.** e

10.10 HOMOGENEOUS EQUATIONS WITH CONSTANT COEFFICIENTS: COMPLEX ROOTS

The material in this section requires some familiarity with the basic properties of complex numbers. A review of these properties is given in Appendix 3.

One of the facts we use repeatedly in this section is the **Euler formula**† (see Appendix 3):

> **EULER'S FORMULA**
>
> $e^{i\beta x} = \cos \beta x + i \sin \beta x.$

We return to our examination of the homogeneous equation with constant coefficients

$$y'' + ay' + by = 0, \tag{1}$$

where $a^2 - 4b < 0$.

CASE 3—COMPLEX CONJUGATE ROOTS: Suppose $a^2 - 4b < 0$. The roots of the characteristic equation to equation (1) are

$$\lambda_1 = \alpha + i\beta, \qquad \lambda_2 = \alpha - i\beta, \tag{2}$$

where $\alpha = -a/2$ and $\beta = \sqrt{4b - a^2}/2$. Thus $y_1 = e^{\lambda_1 x}$ and $y_2 = e^{\lambda_2 x}$ are solutions to (1). However, in this case it is useful to recall that any linear combination of solutions is also a solution (see Theorem 10.7.3, p. 738) and to consider instead the solutions

$$y_1^* = \frac{e^{\lambda_1 x} + e^{\lambda_2 x}}{2} \quad \text{and} \quad y_2^* = \frac{e^{\lambda_1 x} - e^{\lambda_2 x}}{2i}.$$

Since $\cos(-\theta) = \cos\theta$ and $\sin(-\theta) = -\sin\theta$, we can write y_1^* as

$$y_1^* = \frac{e^{(\alpha+i\beta)x} + e^{(\alpha-i\beta)x}}{2} = \frac{e^{\alpha x}}{2}(e^{i\beta x} + e^{-i\beta x})$$

Euler formula ↓

$$= \frac{e^{\alpha x}}{2}[\cos\beta x + i\sin\beta x + \cos(-\beta x) + i\sin(-\beta x)]$$

$$= e^{\alpha x}\cos\beta x.$$

---

† See also the biographical sketch of Euler on page 251.

Similarly, $y_2^* = e^{\alpha x}\sin\beta x$, and the linear independence of y_1^* and y_2^* follows easily since

$$\frac{y_1^*}{y_2^*} = \cot\beta x, \qquad \beta \neq 0,$$

which is not a constant. Alternatively, we can compute $W(y_1^*, y_2^*)(x)$:

$$W(y_1^*, y_2^*)(x) = \begin{vmatrix} e^{\alpha x}\cos\beta x & e^{\alpha x}\sin\beta x \\ e^{\alpha x}(\alpha\cos\beta x - \beta\sin\beta x) & e^{\alpha x}(\alpha\sin\beta x + \beta\cos\beta x) \end{vmatrix}$$

$$= e^{2\alpha x}(\alpha\cos\beta x\sin\beta x + \beta\cos^2\beta x - \alpha\cos\beta x\sin\beta x + \beta\sin^2\beta x)$$

$$\overset{\beta(\cos^2\beta x + \sin^2\beta x) = \beta}{\downarrow}$$

$$= \beta e^{2\alpha x} \neq 0.$$

Thus we have proved the following theorem:

THEOREM 7 THE GENERAL SOLUTION TO A HOMOGENEOUS EQUATION IN THE CASE OF COMPLEX CONJUGATE ROOTS

If $a^2 - 4b < 0$, then the characteristic equation has complex conjugate roots, and the general solution to

$$y'' + ay' + by = 0 \tag{3}$$

is given by

$$y(x) = e^{\alpha x}(c_1\cos\beta x + c_2\sin\beta x) \tag{4}$$

where c_1 and c_2 are arbitrary constants and

$$\alpha = -\frac{a}{2}, \qquad \beta = \frac{\sqrt{4b - a^2}}{2}.$$ ■

EXAMPLE 1 HARMONIC MOTION

Solve the differential equation $y'' + y = 0$.

SOLUTION: The characteristic equation is $\lambda^2 + 1 = 0$, and it has the roots $\lambda = \pm i$. By (3) we have $a = 0$ and $b = 1$, so that $\alpha = 0$ and $\beta = 1$. The general solution is

$$y(x) = c_1\cos x + c_2\sin x.$$

Observe that

$$y'(x) = -c_1\sin x + c_2\cos x,$$

so that $y(0) = c_1$ and $y'(0) = c_2$. Clearly, if either c_1 or c_2 is zero, the solution is sinusoidal. We shall see below that this is still the case even if neither c_1 or c_2 is zero.

EXAMPLE 2 AN INITIAL VALUE PROBLEM WITH COMPLEX CONJUGATE ROOTS

Consider the problem

$$y'' + y' + y = 0,\ y(0) = 1,\ y'(0) = 3.$$

We have $\lambda^2 + \lambda + 1 = 0$ with roots $\lambda_1 = (-1 + i\sqrt{3})/2$ and $\lambda_2 = (-1 - i\sqrt{3})/2$.

Then $\alpha = -\frac{1}{2}$ and $\beta = \sqrt{3}/2$, so the general solution is

$$y(x) = e^{-x/2}\left(c_1 \cos \frac{\sqrt{3}}{2}x + c_2 \sin \frac{\sqrt{3}}{2}x\right).$$

To solve the initial-value problem, we differentiate, set $x = 0$, and solve the simultaneous equations

$$c_1 = 1,$$

$$\frac{\sqrt{3}}{2}c_2 - \frac{1}{2}c_1 = 3.$$

Thus $c_1 = 1$, $c_2 = 7/\sqrt{3}$, and

$$y(x) = e^{-x/2}\left(\cos \frac{\sqrt{3}}{2}x + \frac{7}{\sqrt{3}} \sin \frac{\sqrt{3}}{2}x\right). \tag{5}$$

We will often obtain answers in the form (5). It makes things clearer if we can write the answer in terms of a single sine or cosine function. The following formulas will be very useful.

HARMONIC IDENTITIES

$$a \cos \omega t + b \sin \omega t = A \cos(\omega t - \delta) \tag{6}$$

$$a \cos \omega t + b \sin \omega t = A \sin(\omega t + \phi) \tag{7}$$

Here $A = \sqrt{a^2 + b^2}$, δ is the unique number in $[-\pi, \pi]$ such that $\cos \delta = \frac{a}{\sqrt{a^2 + b^2}} = \frac{a}{A}$ and $\sin \delta = \frac{b}{A}$, so $\tan \delta = \frac{b}{a}$. ϕ is the unique number such that $\cos \phi = \frac{b}{A}$ and $\sin \phi = \frac{a}{A}$, so $\tan \phi = \frac{a}{b}$.

REMARK: In (6), A is called the **amplitude** of the function $A \cos(\omega t - \delta)$, $\frac{\omega}{2\pi}$ is called the **frequency** of the function, and $\frac{|\delta|}{\omega}$ is called the **phase shift**.

We prove (6) and leave (7) as an exercise.

We first note that

$$\left(\frac{a}{\sqrt{a^2 + b^2}}\right)^2 + \left(\frac{b}{\sqrt{a^2 + b^2}}\right)^2 = \frac{a^2}{a^2 + b^2} + \frac{b^2}{a^2 + b^2}$$

$$= \frac{a^2 + b^2}{a^2 + b^2} = 1$$

so

$$\left(\frac{a}{\sqrt{a^2 + b^2}}, \frac{b}{\sqrt{a^2 + b^2}}\right)$$

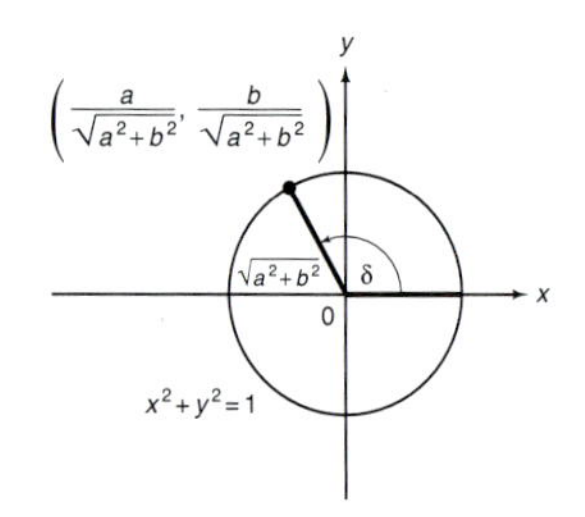

FIGURE 1
In this figure
$\cos \delta = \frac{a}{\sqrt{a^2 + b^2}}$ and
$\sin \delta = \frac{b}{\sqrt{a^2 + b^2}}$

is a point on the unit circle. If δ is chosen as in Figure 1, then

$$\cos \delta = \frac{a}{\sqrt{a^2 + b^2}} = \frac{a}{A} \quad \text{and} \quad \sin \delta = \frac{b}{\sqrt{a^2 + b^2}} = \frac{b}{A}$$

Then

$$\cos(x - y) = \cos x \cos y + \sin x \sin y$$
$$\downarrow$$
$$\begin{aligned} A\cos(\omega t - \delta) &= A[\cos \omega t \cos \delta + \sin \omega t \sin \delta] \\ &= A\cos\delta\cos\omega t + A\sin\delta\sin\omega t \\ &= A\left(\frac{a}{A}\right)\cos\omega t + A\left(\frac{b}{A}\right)\sin\omega t \\ &= a\cos\omega t + b\sin\omega t \end{aligned}$$

■

EXAMPLE 3

REWRITING THE SOLUTION TO EXAMPLE 2 USING THE HARMONIC IDENTITIES

In Example 2 we obtained the solution

$$y(x) = e^{-x/2}\left(\cos\frac{\sqrt{3}}{2}x + \frac{7}{\sqrt{3}}\sin\frac{\sqrt{3}}{2}x\right)$$

In (6), we set $a = 1$ and $b = \dfrac{7}{\sqrt{3}}$ to obtain

$$A = \sqrt{1 + \frac{49}{3}} = \sqrt{\frac{52}{3}} \approx 4.163 \text{ and } \delta = \cos\frac{1}{\sqrt{52/3}} \approx 1.328$$

(δ is in the first quadrant because $\cos\delta > 0$ and $\sin\delta > 0$). Thus

$$y(x) = e^{-x/2}\left(\cos\frac{\sqrt{3}}{2}x + \frac{7}{\sqrt{3}}\sin\frac{\sqrt{3}}{2}x\right) \approx 4.163e^{-x/2}\cos\left(\frac{\sqrt{3}}{2}x - 1.328\right)$$

Using (7), we have $\cos\phi = \dfrac{7/\sqrt{3}}{\sqrt{52/3}} = \dfrac{7}{\sqrt{52}}$ so $\phi \approx 0.2426$ and

$$e^{-x/2}\left(\cos\frac{\sqrt{3}}{2}x + \frac{7}{\sqrt{3}}\sin\frac{\sqrt{3}}{2}x\right) \approx 4.163e^{-x/2}\sin\left(\frac{\sqrt{3}}{2}x + 0.2426\right)$$

A graph of this solution is given in Figure 2.

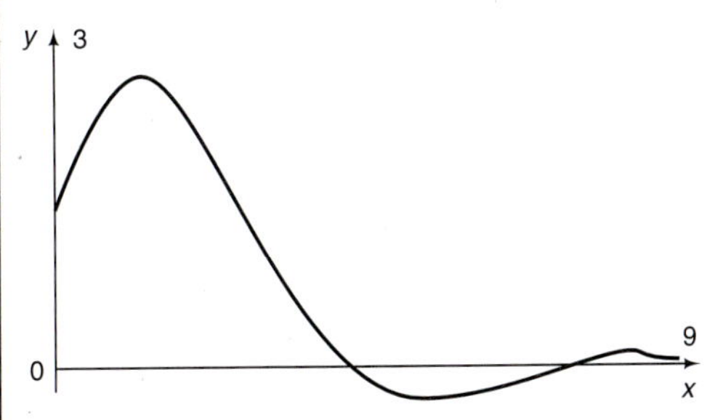

FIGURE 2
Graph of $y = 4.163e^{-x/2}\cos\left(\frac{\sqrt{3}}{2}x - 1.328\right) = 4.163e^{-x/2}\sin\left(\frac{\sqrt{3}}{2}x + 0.2426\right)$, $x \geq 0$

EXAMPLE 4 SIMPLIFYING THE SOLUTION OF AN INITIAL VALUE PROBLEM

Solve the problem

$$y'' - 6y' + 13y = 0, \qquad y(0) = 3, \qquad y'(0) = -5. \tag{8}$$

SOLUTION: The characteristic equation is $\lambda^2 - 6\lambda + 13 = 0$, which has the complex conjugate roots $\lambda = \frac{1}{2}[6 \pm \sqrt{36 - 52}] = 3 \pm 2i$. Thus, the general solution is

$$y(x) = e^{3x}[c_1 \cos 2x + c_2 \sin 2x]. \tag{9}$$

Differentiating with respect to x we have

$$y'(x) = e^{3x}[(3c_1 + 2c_2)\cos 2x + (3c_2 - 2c_1)\sin 2x]. \tag{10}$$

Set $x = 0$ in (9) and (10) and use the initial conditions to obtain

$$3 = y(0) = c_1$$
$$-5 = y'(0) = 3c_1 + 2c_2.$$

Hence $c_2 = -7$, so the particular solution to problem (8) is

$$y(x) = e^{3x}(3 \cos 2x - 7 \sin 2x). \tag{11}$$

If we wish to simplify the term in parentheses in (11) by using the harmonic identity (6), we let $a = 3$, $b = -7$, so that $A = \sqrt{3^2 + (-7)^2} = \sqrt{58}$ and

$$(\cos \delta, \sin \delta) = \left(\frac{3}{\sqrt{58}}, \frac{-7}{\sqrt{58}}\right).$$

Clearly δ is in the fourth quadrant (since x is positive and y is negative), so

$$\delta = \sin^{-1}\left(\frac{-7}{\sqrt{58}}\right) \approx -1.166 \text{ (radians)},$$

and we can rewrite (11) as

$$y(x) = \sqrt{58}e^{3x} \cos(2x + 1.166).$$

A graph of this function is shown in Figure 3.

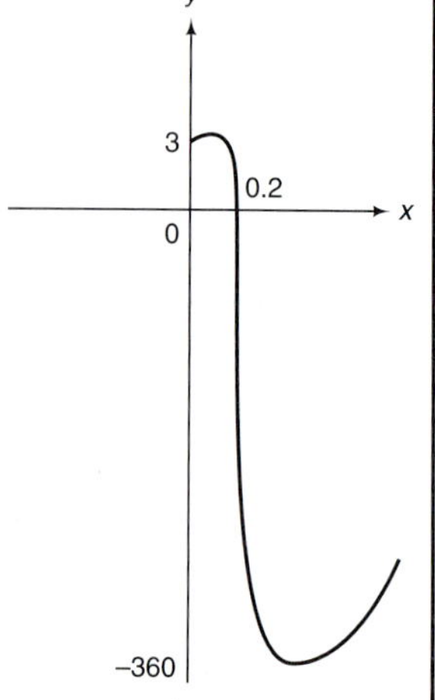

FIGURE 3
Graph of $y = \sqrt{58}e^{3x} \cos(2x - 1.166)$

EXAMPLE 5 WRITING A SUM AS A SINGLE SINE FUNCTION

Write $3 \cos 2t - 7 \sin 2t$ as a sine function.

SOLUTION: Here $a = 3$, $b = -7$, and $A = \sqrt{3^2 + (-7)^2} = \sqrt{58}$. Also, $(\cos \phi, \sin \phi) = \left(\frac{b}{A}, \frac{a}{A}\right) = \left(\frac{-7}{\sqrt{58}}, \frac{3}{\sqrt{58}}\right)$. Then ϕ is in the second quadrant (negative x-coordinate, positive y-coordinate), so

$$\phi = \cos^{-1}\left(\frac{-7}{\sqrt{58}}\right) \approx 2.74 \qquad (\approx 157°)$$

and

$$3 \cos 2t - 7 \sin 2t \approx \sqrt{58} \sin (2t + 2.74)$$

PROBLEMS 10.10

SELF-QUIZ

Multiple Choice

I. What can be said about the solution $y(x)$ to $y'' + 4y = 0$; $y(0) = 1$, $y'(0) = 2$?
a. $y(x) \to 0$ as $x \to \infty$
b. $y(x) \to \infty$ as $x \to \infty$
c. $y(x) \to a$ (nonzero constant) as $x \to \infty$
d. none of the above

II. What can be said about the solution to $y'' + 2y' + 4y = 0$; $y(0) = 1$, $y'(0) = 2$?
a. $y(x) \to 0$ as $x \to \infty$
b. $y(x) \to \infty$ as $x \to \infty$
c. $y(x) \to a$ (nonzero constant) as $x \to \infty$
d. none of the above

III. What can be said about the solution to $y'' - 2y' + 4y = 0$; $y(0) = 1$, $y'(0) = 2$?
a. $y(x) \to 0$ as $x \to \infty$
b. $y(x) \to \infty$ as $x \to \infty$
c. $y(x) \to a$ (nonzero constant) as $x \to \infty$
d. none of the above

IV. If $3 \cos t + 4 \sin t = 5 \cos(t - \delta)$, then $\delta \approx$ ________.
a. 0.6435 **b.** 0.9273
c. 0.7227 **d.** 2.4981

V. If $-3 \cos t + 4 \sin t = 5 \cos(t - \delta)$, then $\delta \approx$ ________.
a. 0.6435 **b.** 0.9273
c. 2.4981 **d.** 2.2143

VI. If $-3 \cos t - 4 \sin t = 5 \cos(t - \delta)$, then $\delta \approx$ ________.
a. −0.6435 **b.** −0.9723
c. −2.2143 **d.** 3.7851

VII. If $3 \cos t - 4 \sin t = 5 \cos(t - \delta)$, then $\delta \approx$ ________.
a. −0.9723 **b.** −0.6435
c. −2.2143 **d.** 3.7851

In Problems 1–12 find the general solution of each equation. When initial conditions are specified, give the particular solution that satisfies them, and use the harmonic identities to rewrite it using equation (6). Sketch the resulting function.

1. $y'' + 2y' + 2y = 0$
2. $8y'' + 4y' + y = 0$, $y(0) = 0$, $y'(0) = 1$
3. $x'' + x' + 7x = 0$
4. $y'' + y' + 2y = 0$
5. $\dfrac{d^2x}{d\theta^2} + 4x = 0$, $x\left(\dfrac{\pi}{4}\right) = 1$, $x'\left(\dfrac{\pi}{4}\right) = 3$
6. $y'' + y = 0$, $y(\pi) = 2$, $y'(\pi) = -1$
7. $y'' + \dfrac{1}{4}y = 0$, $y(\pi) = 1$, $y'(\pi) = -1$
8. $y'' + 6y' + 12y = 0$
9. $y'' + 2y' + 5y = 0$
10. $y'' + 2y' + 5y = 0$, $y(0) = 1$, $y'(0) = -3$
11. $y'' + 2y' + 2y = 0$, $y(\pi) = e^{-\pi}$, $y'(\pi) = -2e^{-\pi}$
12. $y'' + 2y' + 5y = 0$, $y(\pi) = e^{-\pi}$, $y'(\pi) = 3e^{-\pi}$

In Problems 13–24, write each sum as a single cosine function and a single sine function. Use a value of δ or ϕ in the interval $[-\pi, \pi]$.

13. $2 \cos t + 3 \sin t$
14. $-2 \cos t + 3 \sin t$
15. $-2 \cos t - 3 \sin t$
16. $2 \cos t - 3 \sin t$
17. $4 \cos 2\theta + 3 \sin 2\theta$
18. $-3 \cos 2\theta - 4 \sin 2\theta$
19. $-5 \cos \dfrac{\alpha}{2} + 12 \sin \dfrac{\alpha}{2}$
20. $12 \cos \dfrac{\alpha}{2} - 5 \sin \dfrac{\alpha}{2}$
21. $\dfrac{1}{2} \cos 4\beta - \dfrac{1}{4} \sin 4\beta$
22. $3 \cos \pi x + 2 \sin \pi x$
23. $-7 \cos \dfrac{\pi}{2}x - 3 \sin \dfrac{\pi}{2}x$
24. $x_0 \cos \omega_0 t + \dfrac{v_0}{\omega_0} \sin \omega_0 t$

In Problems 25–30, match the given initial value problem with the graph of its solution in Figure 4.

25. $y'' - 2y' + 2y = 0$, $y(0) = 1$, $y'(0) = -2$
26. $y'' + 2y' + 2y = 0$, $y(0) = 1$, $y'(0) = -2$
27. $y'' + y = 0$, $y(0) = 1$, $y'(0) = -2$
28. $y'' + 4y = 0$, $y(0) = 1$, $y'(0) = -2$
29. $y'' - 2y' + 5y = 0$, $y(0) = 1$, $y'(0) = -2$
30. $y'' + 2y' + 5y = 0$, $y(0) = 1$, $y'(0) = -2$
31. Show that the boundary value problem

$$y'' + y = 0, \qquad y(0) = 0, \qquad y(\pi) = 1$$

has no solutions.

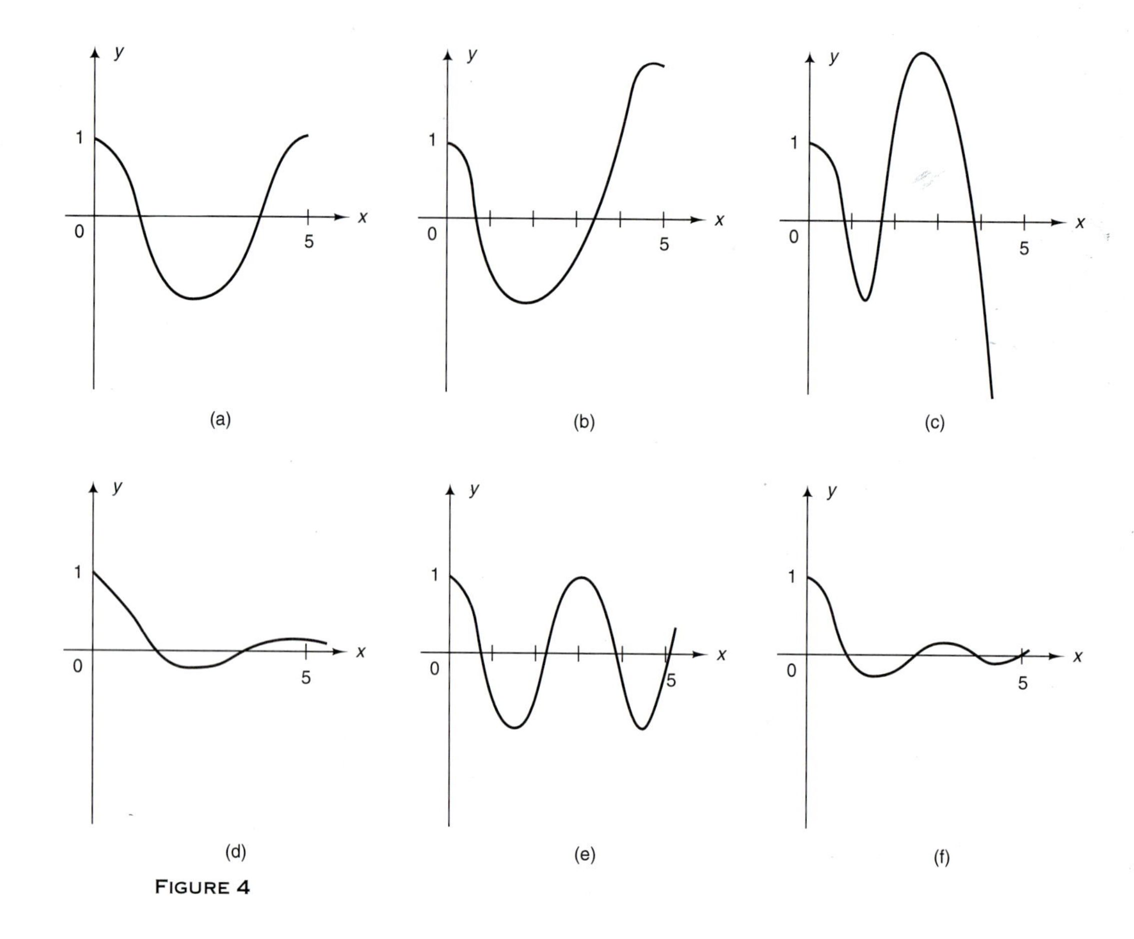

FIGURE 4

ANSWERS TO SELF-QUIZ

I. d **II.** a **III.** d **IV.** b **V.** d **VI.** c **VII.** a

10.11 NONHOMOGENEOUS EQUATIONS: I VARIATION OF PARAMETERS†

In this section we consider a procedure, due to J. L. Lagrange (1736–1813),‡ for finding a particular solution of any nonhomogeneous linear equation

$$y'' + a(x)y' + b(x)y = f(x), \tag{1}$$

where the functions $a(x)$, $b(x)$, and $f(x)$ are continuous. To use this method it is necessary to know the general solution $c_1y_1(x) + c_2y_2(x)$ of the homogeneous equation

$$y'' + a(x)y' + b(x)y = 0. \tag{2}$$

If $a(x)$ and $b(x)$ are constants, the general solution to (2) can always be obtained by the methods of Sections 10.9 and 10.10. If $a(x)$ and $b(x)$ are not both constants, it may be

† This procedure is also called the method of **variation of constants** or **Lagrange's method**.
‡ See the biographical sketch on page 228.

difficult to find this general solution; however, if one solution y_1 of (2) can be found, then the method of reduction of order (see Section 10.8) yields the general solution to (2).

Lagrange noticed that any particular solution y_p of (1) must have the property that y_p/y_1 and y_p/y_2 are not constants, suggesting that we look for a particular solution of (1) of the form

$$y(x) = c_1(x)y_1(x) + c_2(x)y_2(x). \tag{3}$$

This replacement of constants or parameters by variables gives the method its name. Differentiating (3), we obtain

$$y'(x) = c_1(x)y_1'(x) + c_2(x)y_2'(x) + c_1'(x)y_1(x) + c_2'(x)y_2(x).$$

To simplify this expression, it is convenient (but not necessary—see Problem 29) to set

$$c_1'(x)y_1(x) + c_2'(x)y_2(x) = 0. \tag{4}$$

Then

$$y'(x) = c_1(x)y_1'(x) + c_2(x)y_2'(x).$$

Differentiating once again, we obtain

$$y''(x) = c_1(x)y_1''(x) + c_2(x)y_2''(x) + c_1'(x)y_1'(x) + c_2'(x)y_2'(x).$$

Substitution of the expressions for $y(x)$, $y'(x)$, and $y''(x)$ into (1) yields

$$\begin{aligned} y'' + a(x)y' + b(x)y &= c_1(x)(y_1'' + ay_1' + by_1) + c_2(x)(y_2'' + ay_2' + by_2) \\ &\quad + c_1'y_1' + c_2'y_2' \\ &= f(x). \end{aligned}$$

But y_1 and y_2 are solutions to the homogeneous equation, so the equation above reduces to

$$c_1'y_1' + c_2'y_2' = f(x). \tag{5}$$

This gives a second equation relating $c_1'(x)$ and $c_2'(x)$, and we have the simultaneous equations

BASIC EQUATIONS IN THE VARIATION OF PARAMETERS METHOD

$$\begin{aligned} y_1c_1' + y_2c_2' &= 0 \\ y_1'c_1' + y_2'c_2' &= f(x). \end{aligned}$$

The determinant of system (6) is the Wronskian

$$\begin{vmatrix} y_1 & y_2 \\ y_1' & y_2' \end{vmatrix} = W(y_1, y_2)(x) \neq 0. \qquad \text{since } y_1 \text{ and } y_2 \text{ are linearly independent}$$

Thus, for each value of x, $c_1'(x)$ and $c_2'(x)$ are uniquely determined and the problem has essentially been solved. We obtain, from (6),

$$\begin{aligned} y_1y_2'c_1' + y_2y_2'c_2' &= 0, && \text{first equation multiplied by } y_2' \\ y_1'y_2c_1' + y_2y_2'c_2' &= y_2f(x). && \text{second equation multiplied by } y_2 \end{aligned}$$

So

$$(y_1 y_2' - y_1' y_2)c_1' = -y_2 f(x),$$

or

$$c_1' = \frac{-y_2 f(x)}{W(y_1, y_2)(x)}.$$

A similar calculation yields an expression for c_2'. Thus we obtain

SOLUTION OF THE BASIC EQUATIONS

$$c_1'(x) = \frac{-f(x)y_2(x)}{y_1(x)y_2'(x) - y_1'(x)y_2(x)} = \frac{-f(x)y_2(x)}{W(y_1, y_2)(x)}, \tag{7}$$

$$c_2'(x) = \frac{f(x)y_1(x)}{y_1(x)y_2'(x) - y_1'(x)y_2(x)} = \frac{f(x)y_1(x)}{W(y_1, y_2)(x)}. \tag{8}$$

Finally, if we can integrate c_1' and c_2', we can substitute c_1 and c_2 into (3) to obtain a particular solution to the nonhomogeneous equation; that is,

$$y_p(x) = c_1(x)y_1(x) + c_2(x)y_2(x), \tag{9}$$

where

$$c_1(x) = \int c_1'(x)dx, \qquad c_2(x) = \int c_2'(x)dx,$$

with $c_1'(x)$ and $c_2'(x)$ given by (7) and (8).

REMARK: It is not advisable to try to memorize equations (7) and (8), particularly as it is almost always easier to solve system (6) directly. The key concept to this method is system (6). We illustrate with three examples.

EXAMPLE 1 SOLVING A DIFFERENTIAL EQUATION BY THE VARIATION OF PARAMETERS METHOD

Solve $y'' - y = e^{2x}$ by the method of variation of parameters.

SOLUTION: The solutions to the homogeneous equation are $y_1 = e^{-x}$ and $y_2 = e^x$. Using these solutions, system (6) becomes

$$\begin{aligned} e^{-x}c_1' + e^x c_2' &= 0 \\ -e^{-x}c_1' + e^x c_2' &= e^{2x} \end{aligned}$$

Adding these two equations and then subtracting them, we get

$$2e^x c_2' = e^{2x} \qquad \text{and} \qquad 2e^{-x}c_1' = -e^{2x},$$

or

$$c_1' = \frac{-e^{3x}}{2} \qquad \text{and} \qquad c_2' = \frac{e^x}{2}.$$

Integrating these functions, we obtain $c_1(x) = -e^{3x}/6$ and $c_2(x) = e^x/2$. A particular solution is therefore

$$c_1(x)y_1(x) + c_2(x)y_2(x) = \frac{-e^{2x}}{6} + \frac{e^{2x}}{2} = \frac{e^{2x}}{3},$$

and the general solution is

$$y(x) = c_1e^x + c_2e^{-x} + \frac{e^{2x}}{3}.$$

EXAMPLE 2 SOLVING AN INITIAL-VALUE PROBLEM BY THE VARIATION OF PARAMETERS METHOD

Determine the solution of

$$y'' + y = 4 \sin x$$

that satisfies $y(0) = 3$ and $y'(0) = -1$.

SOLUTION: Here $y_1(x) = \cos x$, $y_2(x) = \sin x$, and (6) becomes

$$\cos x \cdot c_1' + \sin x \cdot c_2' = 0$$
$$-\sin x \cdot c_1' + \cos x \cdot c_2' = 4 \sin x$$

Solving these equations simultaneously for c_1' and c_2', we have

$$c_1' = -4 \sin^2 x \qquad \text{and} \qquad c_2' = 4 \sin x \cos x.$$

Since

$$\int \sin^2 x\, dx = \frac{1}{2}(x - \sin x \cos x),$$

we see that

$$c_1 = 2(\sin x \cos x - x), \qquad c_2 = 2 \sin^2 x,$$

and a particular solution is

$$\begin{aligned} y_p(x) &= c_1(x)y_1(x) + c_2(x)y_2(x) \\ &= 2(\sin x \cos x - x)\cos x + 2 \sin^2 x \sin x \\ &= 2 \sin x(\cos^2 x + \sin^2 x) - 2x \cos x = 2 \sin x - 2x \cos x. \end{aligned}$$

Thus the general solution is

$$\begin{aligned} y &= c_1 \cos x + c_2 \sin x + 2 \sin x - 2x \cos x \\ &= c_1 \cos x + c_2^* \sin x - 2x \cos x, \end{aligned}$$

where $c_2^* = c_2 + 2$. To solve the initial-value problem, we differentiate:

$$y' = -c_1 \sin x + c_2^* \cos x + 2x \sin x - 2 \cos x.$$

But

$$3 = y(0) = c_1 \qquad \text{and} \qquad -1 = y'(0) = c_2^* - 2,$$

so $c_2^* = 1$ and the unique solution is

$$y = 3 \cos x + \sin x - 2x \cos x = (3 - 2x)\cos x + \sin x.$$

A graph of this function is illustrated in Figure 1.

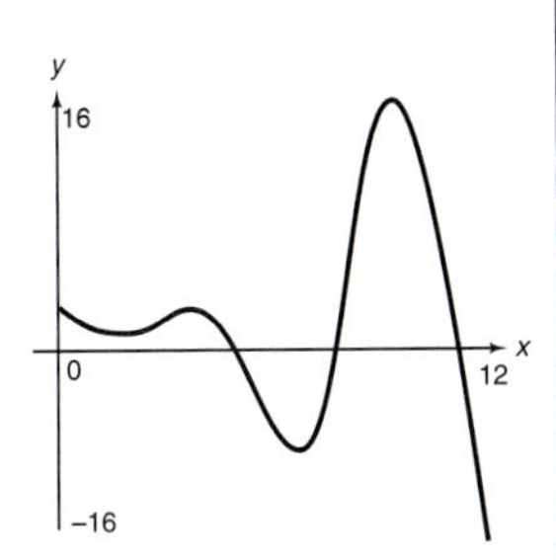

FIGURE 1 Graph of $y = (3 - 2x)\cos x + \sin x$

EXAMPLE 3 USING VARIATION OF PARAMETERS TO SOLVE AN INITIAL VALUE PROBLEM

Solve the problem:

$$y'' + y = \tan x, \qquad y(0) = 1, \qquad y'(0) = 1.$$

SOLUTION: The solutions to the homogeneous equation are $y_1 = \cos x$ and $y_2 = \sin x$. System (6) becomes

$$\cos x \cdot c_1' + \sin x \cdot c_2' = 0,$$
$$-\sin x \cdot c_1' + \cos x \cdot c_2' = \tan x,$$

for which we obtain by elimination

$$c_1'(x) = -\tan x \sin x = -\frac{\sin^2 x}{\cos x} = \frac{\cos^2 x - 1}{\cos x} = \cos x - \sec x,$$
$$c_2'(x) = \tan x \cos x = \sin x.$$

Hence

$$c_1(x) = \sin x - \ln|\sec x + \tan x|$$

and

$$c_2(x) = -\cos x.$$

Thus the particular solution is

$$\begin{aligned} y_p(x) &= c_1(x)y_1(x) + c_2(x)y_2(x) \\ &= \cos x \sin x - \cos x \ln|\sec x + \tan x| - \sin x \cos x \\ &= -\cos x \ln|\sec x + \tan x|, \end{aligned}$$

and the general solution is

$$y(x) = c_1 \cos x + c_2 \sin x - \cos x \ln|\sec x + \tan x|.$$

Differentiating the general solution with respect to x, we get

$$y'(x) = -c_1 \sin x + c_2 \cos x + \sin x \ln|\sec x + \tan x| - 1,$$

and setting $x = 0$ in these last two equations, we have by the initial conditions

$$1 = y(0) = c_1 - \ln|1| = c_1,$$
$$1 = y'(0) = c_2 - 1,$$

so that $c_2 = 2$, and the solution to the initial value problem, shown in Figure 2, is

$$y(x) = \cos x + 2 \sin x - \cos x \ln|\sec x + \tan x|.$$

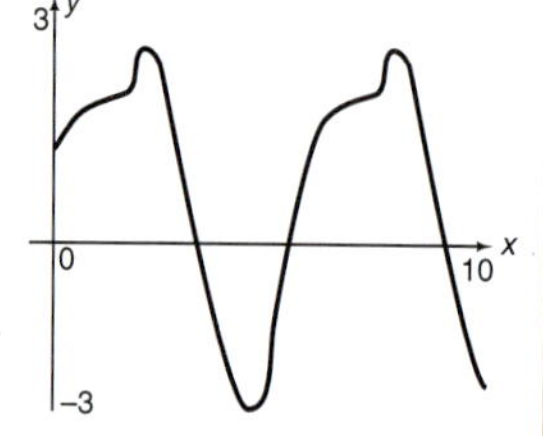

FIGURE 2
Graph of $y(x) = \cos x + 2\sin x - \cos x \ln|\sec x + \tan x|$

PROBLEMS 10.11

SELF-QUIZ

Multiple Choice

I. In using the variation of parameters method to the equation $y'' + y = \sec x$, which system of equations must be solved?

a. $c_1' \sin x + c_2' \cos x = 0$
$c_1' \cos x - c_2' \sin x = \sec x$

b. $c_1' \sin x + c_2' \cos x = 0$
$c_1' \cos x + c_2' \sin x = \sec x$

c. $c_1' \sin x + c_2' \cos x = \sec x$
$c_1' \cos x - c_2' \sin x = 0$

d. $c_1' \sin x + c_2' \cos x = \sec x$
$c_1' \cos x + c_2' \sin x = 0$

II. In using the variation of parameters method applied to the equation $y'' - 4y' + 4y = e^{x^2}$, which system of equations must be solved?

a. $c_1' e^{2x} + c_2' x e^{2x} = e^{x^2}$
$2c_1' e^{2x} + (1 + 2x)c_2' e^{2x} = 0$

b. $c_1' x e^{2x} + c_2' e^{2x} = 0$
$(1 + 2x)c_1' x e^{2x} + 2c_2' e^{2x} = e^{x^2}$

c. $c_1' e^{2x} + c_2' x e^{2x} = 0$
$c_1' x e^{2x} + 2c_2' e^{2x} = e^{x^2}$

d. $c_1' e^{2x} + c_2' x e^{2x} = 0$
$2c_1' e^{2x} + c_2' x e^{2x} = e^{x^2}$

In Problems 1–24 find the general solution to each equation by the method of variation of parameters. If initial conditions are given, find the solution and sketch its graph.

1. $y'' - 3y' + 2y = 10$
2. $y'' - y' - 6y = 20e^{-2x}$
3. $y'' + 4y = 3 \sin x,\ y(0) = 1,\ y'(0) = 2$
4. $y'' + y' = 3x^2,\ y(1) = 0,\ y'(1) = 1$
5. $y'' - 3y' + 2y = 6e^{3x}$
6. $y'' - 4y' + 4y = 6xe^{2x}$
7. $y'' - 2y' + y = -4e^x$
8. $y'' + y = 1 + x + x^2,\ y(0) = 1,\ y'(0) = -1$
9. $y'' - 7y' + 10y = 100x$
10. $y'' + 4y = 16x \sin 2x$
11. $y'' - y' = \sec^2 x - \tan x,\ y(0) = 3,\ y'(0) = 2$
12. $y'' + y = \cot x$
13. $y'' + 4y = \sec 2x$
14. $y'' + 4y = \sec x \tan x,\ y(0) = 1,\ y'(0) = 4$
15. $y'' - 2y' + y = \dfrac{e^x}{(1 - x)^2}$
16. $y'' - y = \sin^2 x$
17. $y'' - y = \dfrac{(2x - 1)e^x}{x^2},\ y(1) = 9,\ y'(1) = 5$
18. $y'' - 3y' - 4y = \dfrac{e^{4x}(5x - 2)}{x^3}$
19. $y'' - 4y' + 4y = \dfrac{e^{2x}}{(1 + x)},\ y(0) = \dfrac{2}{3},\ y'(0) = \dfrac{1}{2}$
20. $y'' + 2y' + y = e^{-x} \ln |x|$
21. Find a particular solution of

$$y'' + \frac{1}{x}y' - \frac{y}{x^2} = \frac{1}{x^2 + x^3}, \qquad (x > 0),$$

given that two solutions of the associated homogeneous equations are $y_1 = x$ and $y_2 = 1/x$.

22. Find a particular solution of

$$y'' - \frac{2}{x}y' + \frac{2}{x^2}y = \frac{\ln |x|}{x}, \qquad (x > 0),$$

given the two homogeneous solutions $y_1 = x$ and $y_2 = x^2$.

23. Verify that

$$y = \frac{1}{\omega}\int_0^x f(t)\sin \omega(x - t)\, dt$$

is a particular solution of $y'' + \omega^2 y = f(x)$. [*Hint:* Use equation (10).]

24. Find a particular solution to the initial-value problem

$$y'' - \omega^2 y = f(x), \qquad y(0) = y'(0) = 0.$$

***25.** This problem shows why there is no loss in generality in equation (4) by setting

$$c_1' y_1 + c_2' y_2 = 0.$$

Suppose that we instead let $c_1' y_1 + c_2' y_2 = z(x)$, with $z(x)$ an undetermined function of x.

a. Show that we then obtain the system

$$c_1' y_1 + c_2' y_2 = z,$$
$$c_1' y_1' + c_2' y_2' = f - z' - az.$$

b. Show that the system in part (a) has the solution

$$c_1' = \frac{-y_2 f}{W(y_1, y_2)} + \frac{(e^{\int a(x)\, dx} z y_2)'}{e^{\int a(x)\, dx} W(y_1, y_2)}$$

$$c_2' = \frac{y_1 f}{W(y_1, y_2)} - \frac{(e^{\int a(x)\, dx} z y_1)'}{e^{\int a(x)\, dx} W(y_1, y_2)}$$

c. Integrate by parts to show that

$$\int \frac{(e^{\int a(x)\, dx} z y_i)'}{e^{\int a(x)\, dx} W(y_1, y_2)}\, dx = \frac{z y_i}{W(y_1, y_2)}, \qquad i = 1, 2.$$

d. Conclude that the particular solution obtained by letting $c_1' y_1 + c_2' y_2 = z$ is identical to that obtained by assuming equation (4).

***26.** Suppose one solution $y_1(x)$ of the homogeneous counterpart of the linear differential equation

$$y'' + a(x)y' + b(x)y = f(x) \tag{i}$$

is known. Use the substitution $y_2 = vy_1$ of Section 10.8 to find the general solution of (i):

$$y = c_1y_1(x) + c_2y_1(x)\int^x \frac{h(t)}{y_1^2(t)}\,dt + y_1(x)\int^x \frac{h(t)}{y_1^2(t)}\left[\int^t \frac{y_1(s)f(s)}{h(s)}\,ds\right]dt,$$

where $h(t) = e^{-\int^t a(u)\,du}$.

ANSWERS TO SELF-QUIZ

I. a **II.** b

10.12 NONHOMOGENEOUS EQUATIONS II: THE METHOD OF UNDETERMINED COEFFICIENTS

In this section we present an alternate method for solving certain linear nonhomogeneous differential equations with constant coefficients. The method is not as general as the method of variation of parameters that we considered in Section 10.11. However, it is sometimes easier to solve specific problems by using it.

Consider the equation

$$y'' + ay' + by = f(x), \tag{1}$$

where a and b are constants, and every term of the function $f(x)$ is a *product* of one or more of the following:

i. a polynomial in x (a constant k is a polynomial of degree zero)
ii. an exponential function $e^{\alpha x}$
iii. $\cos \beta x$, $\sin \beta x$.

The method we shall present applies only when $f(x)$ is a function of this type; if any term in $f(x)$ is not of this type, we cannot use the method of this section. For example, we cannot use the method if $f(x) = \tan x$.

Note that a multitude of cases are of this type:

a. $f(x) = 17$ (f is constant)
b. $f(x) = xe^{2x}$ (f is a polynomial times an exponential)
c. $f(x) = e^{4x}\cos x$ (an exponential times a cosine)
d. $f(x) = x\cos 2x + 3\sin 2x$ (polynomial times a cosine plus a constant times a sine)
e. $f(x) = \frac{1}{2}e^{-3x}\cos 3x - \sqrt{5}(x^2 + 3)e^{2x}\sin x$ (each term is a product of all three types).

Observe that if we differentiate a product of the three types above, we get an expression whose terms are again of this type: for example

$$(P_n(x)e^{\alpha x}\cos \beta x)' = P_n'(x)e^{\alpha x}\cos \beta x + \alpha P_n(x)e^{\alpha x}\cos \beta x - \beta P_n(x)e^{\alpha x}\sin \beta x,$$

for any polynomial $P_n(x)$ in x. This suggests that we may be able to determine a particular solution to equation (1) by substituting a function composed of terms of the same "form" as those in $f(x)$ for the dependent variable y.

The method of undetermined coefficients assumes that each term in the particular solution to equation (1) is of the same "form" as each term in $f(x)$, but has polynomial terms with *undetermined coefficients,* thus giving the method its name. If we substitute the right combination of such terms, the method asserts that when we compare both sides of the resulting equation, it will be possible to "determine" the unknown coefficients. Before attempting to make this procedure precise, we illustrate it with several examples.

EXAMPLE 1

SOLVING A NONHOMOGENEOUS EQUATION WHEN $f(x)$ IS A POLYNOMIAL

Solve the following equation:

$$y'' - y = x^2 \tag{2}$$

SOLUTION: Since $f(x) = x^2$ is a polynomial of degree two, we "guess" that (2) has a solution $y_p(x)$ that is a polynomial of degree two. We try the *most general* polynomial of degree two:

$$y_p(x) = a + bx + cx^2.$$

Then $y_p' = b + 2cx$ and $y_p'' = 2c$, so if we substitute y_p'' and y_p into (2) we obtain

$$2c - (a + bx + cx^2) = x^2. \tag{3}$$

Equating coefficients, in (3) we have

$$\underset{\substack{\text{coefficient of constant}\\ \text{term}}}{\overset{}{2c - a = 0}}, \quad \underset{\substack{\text{coefficient}\\ \text{of } x}}{-b = 0}, \quad \underset{\substack{\text{coefficient}\\ \text{of } x^2}}{-c = 1},$$

which immediately yields $a = -2$, $b = 0$, $c = -1$, and the particular solution

$$y_p(x) = -2 - x^2. \tag{4}$$

This particular solution is easily verified by substitution into (2). Finally, since the general solution of the homogeneous equation $y'' - y = 0$ is given by

$$y = c_1e^x + c_2e^{-x},$$

the general solution of (2) is

$$y = c_1e^x + c_2e^{-x} - 2 - x^2. \tag{5}$$

EXAMPLE 2

SOLVING A NONHOMOGENEOUS EQUATION WHEN $f(x) = e^{\alpha x} \sin \beta x$

Solve

$$y'' - 3y' + 2y = e^x \sin x. \tag{6}$$

SOLUTION: Here $f(x)$ is a product of all three types (with polynomial equal to 1). In collecting the terms for y we must choose not only terms of the form $e^x \sin x$, but also terms of the form $e^x \cos x$, since any derivative of the former will produce a term like the latter. For this reason, we guess that the particular solution has the form

$$y_p(x) = ae^x \sin x + be^x \cos x.$$

Then

$$y_p'(x) = (a - b)e^x \sin x + (a + b)e^x \cos x$$

and

$$y_p''(x) = 2ae^x \cos x - 2be^x \sin x.$$

Substituting these expressions into (6) we have

$$e^x(2a \cos x - 2b \sin x) - 3e^x[(a - b)\sin x + (a + b)\cos x] + 2e^x(a \sin x + b \cos x) = e^x \sin x.$$

Dividing both sides by e^x and equating the coefficients of $\sin x$ and $\cos x$, we have

$$2a - 3(a + b) + 2b = 0$$
$$-2b - 3(a - b) + 2a = 1$$

which yield $a = -\frac{1}{2}$ and $b = \frac{1}{2}$ so that

$$y_p = \frac{e^x}{2}(\cos x - \sin x).$$

Again this result is easily verified by substitution. Finally, the general solution of (6) is

$$y = c_1 e^{2x} + c_2 e^x + \frac{e^x}{2}(\cos x - \sin x).$$

If $f(x)$ had been the function $5e^x \cos x$, we would have used exactly the same guess for the particular solution.

EXAMPLE 3

SOLVING A NONHOMOGENEOUS EQUATION WHEN $f(x)$ IS A POLYNOMIAL TIMES AN EXPONENTIAL

Solve

$$y'' + y = xe^{2x}.$$

SOLUTION: Here $f(x)$ is the product of a first-degree polynomial and an exponential. The *most general* expression of this form is

$$y_p(x) = e^{2x}(a + bx).$$

Then

$$y_p'(x) = e^{2x}(2a + b + 2bx), \qquad y_p''(x) = e^{2x}(4a + 4b + 4bx),$$

and substitution yields

$$e^{2x}(4a + 4b + 4bx) + e^{2x}(a + bx) = xe^{2x}.$$

Dividing both sides by e^{2x} and equating like powers of x, we obtain the equations

$$5a + 4b = 0, \quad 5b = 1.$$

Thus $a = -\frac{4}{25}$, $b = \frac{1}{5}$, and a particular solution is

$$y_p(x) = \frac{e^{2x}}{25}(5x - 4).$$

Therefore the general solution of this example is (since $y'' + y = 0$ is the equation of the harmonic oscillator)

$$y(x) = c_1 \sin x + c_2 \cos x + \frac{e^{2x}}{25}(5x - 4).$$

Difficulties arise in connection with problems of this type whenever any term of the guessed solution is a solution of the homogeneous equation

$$y'' + ay' + by = 0. \tag{7}$$

For example, in the equation

$$y'' + y = (1 + x + x^2)\sin x, \tag{8}$$

the function $f(x)$ is the sum of three functions, one of which ($\sin x$) is a solution to the homogeneous equation $y'' + y = 0$. As another example, in

$$y'' + y = (x + x^2)\sin x \tag{9}$$

the guessed solution is $y_p = (a_0 + a_1x + a_2x^2)\sin x + (b_0 + b_1x + b_2x^2)\cos x$ and $a_0 \sin x + b_0 \cos x$ is a solution to the homogeneous equation $y'' + y = 0$. When this situation occurs, the method of undetermined coefficients must be modified. To see why, consider the following example.

EXAMPLE 4

THE NEED FOR MODIFYING THE METHOD

Find the solution to the equation

$$y'' - y = 2e^x. \tag{10}$$

SOLUTION: The general solution of $y'' - y = 0$ is

$$y(x) = c_1e^x + c_2e^{-x}.$$

Here $f(x) = 2e^x$ is a solution to the homogeneous equation. If we try to find a solution of the form Ae^x we get nowhere, since Ae^x is a solution to the homogeneous equation for every constant A and, therefore, it cannot possibly be a solution to the nonhomogeneous equation.

What do we do? Recall that if λ is a double root of the characteristic equation for a homogeneous differential equation, then two solutions are $e^{\lambda x}$ and $xe^{\lambda x}$. This suggests that we try Axe^x instead of Ae^x as a possible solution to (10). Thus we consider a particular solution of the form

$$y_p = Axe^x.$$

Then $y_p' = Ae^x(x + 1)$, $y_p''(x) = Ae^x(x + 2)$, and

$$y_p'' - y_p = Ae^x(x + 2) - Axe^x = 2Ae^x = 2e^x.$$

Hence $A = 1$ and $y_p = xe^x$. Thus the general solution is

$$y(x) = c_1e^x + c_2e^{-x} + xe^x.$$

The preceding example suggests the following rule:

> **MODIFICATIONS OF THE METHOD**
>
> If any term of the guessed solution $y_p(x)$ is a solution of the homogeneous equation (7), multiply $y_p(x)$ by x repeatedly until no term of the product $x^ky_p(x)$ is a solution of (7). Then use the product $x^ky_p(x)$ to solve equation (1).

EXAMPLE 5

SOLUTION BY THE MODIFIED METHOD

Find the solution to

$$y'' + y = \cos x \tag{11}$$

that satisfies $y(0) = 2$ and $y'(0) = -3$.

SOLUTION: The general solution to $y'' + y = 0$ is $y = c_1 \cos x + c_2 \sin x$. Since $f(x) = \cos x$ is a solution, we must use the modification of the method to find a particular solution to (11). Ordinarily we would guess a solution of the form $y_p = A \cos x + B \sin x$. Instead we multiply by x and try a solution of the form

$$y_p = Ax \cos x + Bx \sin x.$$

Note that now no term of y_p is a solution to $y'' + y = 0$. Then

$$y_p' = A \cos x - Ax \sin x + B \sin x + Bx \cos x$$

and

$$\begin{aligned} \overset{\text{from (11)}}{\downarrow} \\ \cos x = y_p'' + y_p &= (-2A \sin x - Ax \cos x + 2B \cos x - Bx \sin x) \\ &\quad + (Ax \cos x + Bx \sin x) \\ &= -2A \sin x + 2B \cos x. \end{aligned}$$

Therefore

$$-2A = 0, \qquad 2B = 1, \qquad B = \tfrac{1}{2},$$

and

$$y_p = \tfrac{1}{2}x \sin x.$$

Thus the general solution to (11) is

$$y = c_1 \cos x + c_2 \sin x + \tfrac{1}{2}x \sin x.$$

We are not finished yet, as initial conditions were given. We have

$$y' = -c_1 \sin x + c_2 \cos x + \tfrac{1}{2}x \cos x + \tfrac{1}{2} \sin x.$$

Then

$$y(0) = c_1 = 2 \qquad \text{and} \qquad y'(0) = c_2 = -3,$$

which yields the unique solution

$$y(x) = 2 \cos x - 3 \sin x + \tfrac{1}{2}x \sin x.$$

EXAMPLE 6

MODIFYING TWICE

Find the general solution of

$$y'' - 4y' + 4y = e^{2x}.$$

SOLUTION: The homogeneous equation $y'' - 4y' + 4y = 0$ has the independent solutions e^{2x} and xe^{2x}. Thus, multiplying $f(x) = e^{2x}$ by x *twice,* we look for a particular solution of the form $y_p = ax^2e^{2x}$. Then

$$y_p' = ae^{2x}(2x^2 + 2x)$$

and

$$y_p'' = ae^{2x}(4x^2 + 8x + 2),$$

so

$$\begin{aligned} y_p'' - 4y_p' + 4y_p &= ae^{2x}(4x^2 + 8x + 2 - 8x^2 - 8x + 4x^2) \\ &= 2ae^{2x} = e^{2x}, \end{aligned}$$

or $2a = 1$ and $a = \frac{1}{2}$. Thus $y_p = \frac{1}{2}x^2e^{2x}$, and the general solution is

$$y(x) = c_1e^{2x} + c_2xe^{2x} + \tfrac{1}{2}x^2e^{2x} = e^{2x}(c_1 + c_2x + \tfrac{1}{2}x^2).$$

EXAMPLE 7 USING SUPERPOSITION

Consider the equation

$$y'' - y = x^2 + 2e^x.$$

Using the results of Examples 1 and 4 and the principle of superposition (see page 743), we find immediately that a particular solution is given by

$$y_p(x) = -2 - x^2 + xe^x.$$

EXAMPLE 8 SOLVING WHEN $f(x)$ IS A PRODUCT OF A POLYNOMIAL AND A SINE FUNCTION

Find the general solution to

$$y'' + y = x \sin x.$$

SOLUTION: The guessed solution is $y_p = (Ax + B)\cos x + (Cx + D)\sin x$. Since $B \cos x + D \sin x$ solves $y'' + y = 0$, the modification is required. We therefore multiply by x and try a solution of the form

$$y_p = (Ax^2 + Bx)\cos x + (Cx^2 + Dx)\sin x.$$

Then

$$\begin{aligned} y_p' &= [Cx^2 + (2A + D)x + B]\cos x + [-Ax^2 + (2C - B)x + D]\sin x, \\ y_p'' &= [-Ax^2 + (4C - B)x + 2A + 2D]\cos x \\ &\quad + [-Cx^2 - (4A + D)x + 2C - 2B]\sin x, \end{aligned}$$

and

$$y_p'' + y_p = [4Cx + 2A + 2D]\cos x + [-4Ax + 2C - 2B]\sin x \overset{\text{given}}{\downarrow}= x \sin x.$$

This yields $A = -\frac{1}{4}$, $B = 0$, $C = 0$, $D = \frac{1}{4}$, and the particular solution

$$y_p = -\tfrac{1}{4}x^2 \cos x + \tfrac{1}{4}x \sin x.$$

Thus the general solution is

$$y = (c_1 - \tfrac{1}{4}x^2)\cos x + (c_2 + \tfrac{1}{4}x)\sin x.$$

PROBLEMS 10.12

SELF-QUIZ

Multiple Choice

I. In solving $y'' + ay' + by = f(x)$, for which functions f below does the method of undetermined coefficients *not* apply. (There is more than one answer.)

a. $f(x) = x^3e^{4x}$
b. $f(x) = x^{-1}$
c. $f(x) = \sqrt{2}e^{3x}\cos 4x$
d. $f(x) = \sqrt{x}e^x$
e. $f(x) = \cos x/\sin x$
f. $f(x) = x^3e^x - e^{-2x}\cos 4x$

II. To solve $y'' + 2y' + y = e^{-x}$, we must multiply the guessed particular solution $y_p(x) = Ae^{-x}$ by ________.

a. Nothing **b.** x
c. x^2 **d.** x^3

In Problems 1–33 find the general solution of each differential equation. If initial conditions are given, then find the particular solution that satisfies them, and sketch its solution.

1. $y'' + 4y = 3\sin x$
2. $y'' - y' - 6y = 20e^{-2x}$, $y(0) = 0$, $y'(0) = 6$
3. $y'' - 3y' + 2y = 6e^{3x}$
4. $y'' + y' = 3x^2$, $y(0) = 4$, $y'(0) = 0$
5. $y'' - 2y' + y = -4e^x$
6. $y'' - 4y' + 4y = 6xe^{2x}$, $y(0) = 0$, $y'(0) = 3$
7. $y'' - 7y' + 10y = 100x$, $y(0) = 0$, $y'(0) = 5$
8. $y'' + y = 1 + x + x^2$
9. $y'' + y' = x^3 - x^2$
10. $y'' + 4y = 16x\sin 2x$
11. $y'' - 4y' + 5y = 2e^{2x}\cos x$
12. $y'' - y' - 2y = x^2 + \cos x$
13. $y'' + 6y' + 9y = 10e^{-3x}$

Use the principle of superposition to find the general solution of each of the equations in Problems 14–17.

14. $y'' + y = 1 + 2\sin x$
15. $y'' - 2y' - 3y = x - x^2 + e^x$
16. $y'' + 4y = 3\cos 2x - 7x^2$
17. $y'' + 4y' + 4y = xe^x + \sin x$
18. Show by the methods of this section that a particular solution of

$$y'' + 2ay' + b^2y = A\sin\omega x \qquad (a, \omega > 0)$$

is given by

$$y = \frac{A\sin(\omega x - \alpha)}{\sqrt{(b^2 - \omega^2)^2 + 4\omega^2a^2}},$$

where

$$\alpha = \tan^{-1}\frac{2a\omega}{(b^2 - \omega^2)}, \qquad (0 < \alpha < \pi).$$

19. Let $f(x)$ be a polynomial of degree n. Show that, if $b \neq 0$, there is always a solution that is a polynomial of degree n for the equation

$$y'' + ay' + by = f(x).$$

20. Use the method indicated in Problem 19 to find a particular solution of

$$y'' + 3y' + 2y = 9 + 2x - 2x^2.$$

In Problems 21–24 find particular solutions to the given differential equation.

21. $y'' + y = (x + x^2)\sin x$
22. $y'' - y' = x^2$
23. $y'' - 2y' + y = x^2e^x$
24. $y'' - 4y' + 3y = x^3e^{3x}$
25. In many physical problems the nonhomogeneous term $f(x)$ is specified by different formulas in different intervals of x. Find

a. a general solution of the equation

$$y'' + y = \begin{cases} x, & 0 \le x \le 1, \\ 1, & x \ge 1; \end{cases}$$

[*Note:* This "solution" is not differentiable at $x = 1$.]

b. a particular solution of part (a) that satisfies the initial conditions

$$y(0) = 0, \qquad y'(0) = 1.$$

ANSWERS TO SELF-QUIZ

I. b, d, e **II.** c

10.13 EULER EQUATIONS

For most linear second-order equations with variable coefficients it is impossible to write solutions in terms of elementary functions. In most cases it is necessary to use techniques such as the power series method (see Section 12.14) to obtain information about solutions. However, there is one class of such equations that do arise in applications for which solutions in terms of elementary functions can be obtained:

> **EULER EQUATION**
>
> An equation of the form
>
> $$x^2y'' + axy' + by = f(x), \qquad x \neq 0 \tag{1}$$
>
> is called an **Euler equation**.

NOTE: Equation (1) can be written

$$y'' + \frac{a}{x}y' + \frac{b}{x^2}y = \frac{f(x)}{x^2},$$

which is not defined for $x = 0$. This is why we make the restriction that $x \neq 0$. We begin by solving the homogeneous Euler equation

$$x^2y'' + axy' + by = 0, \qquad x \neq 0. \tag{2}$$

If we can find two linearly independent solutions to (2), we can solve (1) by the method of variation of parameters. There are two ways to solve equation (2); each one involves a certain trick. We give one method here and leave the other for the problem set (see Problem 18).

The first method involves guessing an appropriate solution to (2). We note that if $y = x^\lambda$ for some number λ, then $y' = \lambda x^{\lambda-1}$ and $y'' = \lambda(\lambda - 1)x^{\lambda-2}$. This is interesting because x^2y'', xy', and y all can be written as constant multiples of x^λ. Therefore we guess that there is a solution having the form $y = x^\lambda$. Substituting this into equation (2), we obtain

$$\lambda(\lambda - 1)x^\lambda + a\lambda x^\lambda + bx^\lambda = x^\lambda[\lambda(\lambda - 1) + a\lambda + b] = 0.$$

If $x \neq 0$, we can divide by x^λ to obtain the

> **CHARACTERISTIC EQUATION† FOR THE EULER EQUATION**
>
> $$\lambda(\lambda - 1) + a\lambda + b = 0, \tag{3}$$
>
> or
>
> $$\lambda^2 + (a - 1)\lambda + b = 0. \tag{4}$$

As with constant-coefficient equations, there are three cases to consider.

---

†The term "characteristic equation" is generally reserved for linear equations with *constant* coefficients. The only reason we can use this term in this case is that the substitution in Problem 18 converts equation (1) into a second-order constant-coefficient equation with characteristic equation (4). *The method of characteristic equations is generally not applicable to equations with variable coefficients.*

CASE 1: Characteristic equation (4) has two real, distinct roots.

EXAMPLE 1

AN EULER EQUATION WHERE THE ROOTS ARE REAL AND DISTINCT

Find the general solution to

$$x^2y'' + 2xy' - 12y = 0, \qquad x \neq 0.$$

SOLUTION: The characteristic equation is

$$\lambda(\lambda - 1) + 2\lambda - 12 = \lambda^2 + \lambda - 12 = 0 = (\lambda + 4)(\lambda - 3),$$

with roots $\lambda_1 = -4$ and $\lambda_2 = 3$. Thus two solutions (that are linearly independent) are

$$y_1 = x^{-4} = \frac{1}{x^4} \qquad \text{and} \qquad y_2 = x^3,$$

and the general solution is

$$y(x) = \frac{c_1}{x^4} + c_2x^3.$$

In Figure 1, we provide a graph of the solution for $c_1 = c_2 = 1$.

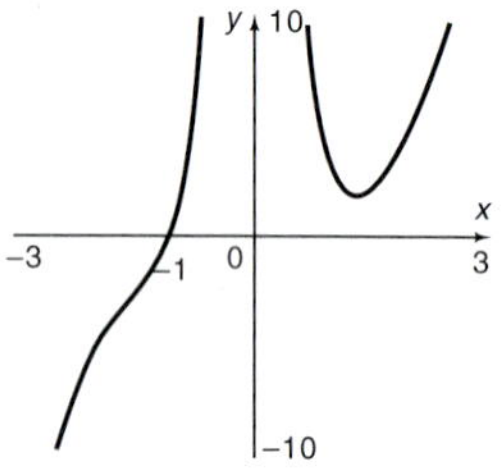

FIGURE 1
Graph of $y = \frac{1}{x^4} + x^3$, $x \neq 0$

THEOREM 1 SOLVING AN EULER EQUATION WHEN THE ROOTS ARE REAL AND UNEQUAL

If λ_1 and λ_2 are real and distinct, then the general solution to equation (2) is

$$y(x) = c_1x^{\lambda_1} + c_2x^{\lambda_2}, \qquad x \neq 0. \tag{5}$$

■

CASE 2: The roots are real and equal ($\lambda_1 = \lambda_2$).

EXAMPLE 2

AN EULER EQUATION WHERE THE ROOTS ARE REAL AND EQUAL

Find the general solution to

$$x^2y'' - 3xy' + 4y = 0, \qquad x > 0. \tag{6}$$

SOLUTION: The characteristic equation is

$$\lambda^2 - 4\lambda + 4 = (\lambda - 2)^2 = 0,$$

with the single root $\lambda = 2$. Thus one solution is $y_1(x) = x^2$. To find a second solution we use the method of reduction of order (see Section 10.8). Let $y_2 = vy_1 = x^2v$. Then $y_2' = x^2v' + 2xv$ and $y_2'' = x^2v'' + 4xv' + 2v$, so (6) becomes

$$x^2(x^2v'' + 4xv' + 2v) - 3x(x^2v' + 2xv) + 4(x^2v) = 0,$$

or, with $z = v'$,

$$x^4v'' + x^3v' = x^4z' + x^3z = 0.$$

Separating variables, we have

$$\ln|z| = \int \frac{dz}{z} = -\int \frac{dx}{x} = -\ln x,$$

so that

$$v' = z = x^{-1}.$$

Then

$$y_2(x) = y_1 v = x^2 \int \frac{dx}{x} = x^2 \ln x.$$

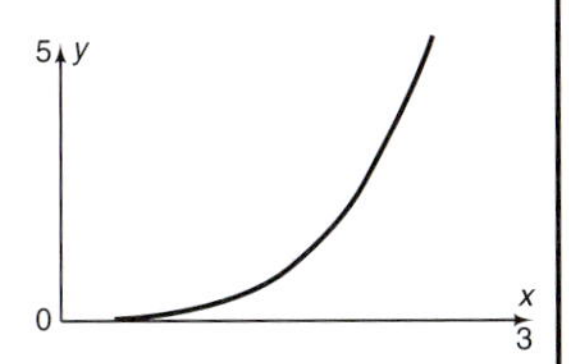

FIGURE 2
Graph of $y = x^2(1 + \ln x)$, $x > 0$.

Thus the general solution to equation (6) is

$$y(x) = c_1 x^2 + c_2 x^2 \ln x = x^2(c_1 + c_2 \ln x).$$

In Figure 2 we provide a graph of the solution for $c_1 = c_2 = 1$.

THEOREM 2 SOLVING AN EULER EQUATION WHEN THE ROOTS ARE REAL AND EQUAL

If λ is the only root of characteristic equation (4), then the general solution to (2) is

$$y(x) = x^\lambda(c_1 + c_2 \ln|x|).$$ ■

CASE 3: The roots are complex conjugates ($\lambda_1 = \alpha + i\beta$, $\lambda_2 = \alpha - i\beta$).

EXAMPLE 3 AN EULER EQUATION WITH COMPLEX CONJUGATE ROOTS

Find the general solution of

$$x^2y'' + 5xy' + 13y = 0, \qquad x > 0. \tag{7}$$

SOLUTION: The characteristic equation is

$$\lambda^2 + 4\lambda + 13 = 0,$$

and

$$\lambda = \frac{-4 \pm \sqrt{16 - 4(13)}}{2} = \frac{-4 \pm \sqrt{-36}}{2} = -2 \pm 3i.$$

Thus two linearly independent solutions are

$$y_1(x) = x^{-2+3i} \qquad \text{and} \qquad y_2(x) = x^{-2-3i}.$$

Using the material in Appendix 3, we can eliminate the imaginary exponents. First we note that

$$x^a = e^{\ln x^a} = e^{a \ln x}.$$

By the Euler formula ($e^{i\beta x} = \cos \beta x + i \sin \beta x$),

$$\begin{aligned} y_1(x) &= (x^{-2})(x^{3i}) = x^{-2}e^{3i \ln x} \\ &= x^{-2}[\cos(3 \ln x) + i \sin(3 \ln x)] \end{aligned}$$

and

$$y_2(x) = (x^{-2})(x^{-3i}) = x^{-2}e^{-3i \ln x}$$
$$= x^{-2}[\cos(3 \ln x) - i \sin(3 \ln x)].$$

We now form two new solutions:

$$y_3(x) = \tfrac{1}{2}[y_1(x) + y_2(x)] = x^{-2} \cos(3 \ln x)$$

and

$$y_4(x) = \frac{1}{2i}[y_1(x) - y_2(x)] = x^{-2} \sin(3 \ln x).$$

These new solutions contain no complex numbers and are easier to work with. The general solution to (7) is

$$y(x) = x^{-2}[c_1 \cos(3 \ln x) + c_2 \sin(3 \ln x)].$$

A sketch of the solution is given in Figure 3 in the case $c_1 = c_2 = 1$.

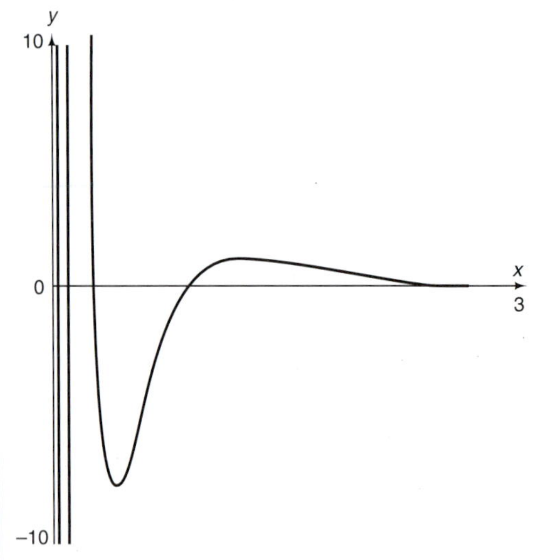

FIGURE 3

Graph of $y = \frac{1}{x^2}[\cos(3 \ln x) + \sin(3 \ln x)]$, $x > 0$.

THEOREM 3 SOLVING AN EULER EQUATION WHEN THE ROOTS ARE COMPLEX CONJUGATES

If $\lambda_1 = \alpha + i\beta$ and $\lambda_2 = \alpha - i\beta$ are complex conjugate roots of characteristic equation (4), then the general solution to (2) is

$$y(x) = x^{\alpha}[c_1 \cos(\beta \ln|x|) + c_2 \sin(\beta \ln|x|)]. \quad (8)$$

■

EXAMPLE 4 SOLVING A NONHOMOGENEOUS EULER EQUATION

Find the general solution to

$$x^2y'' + 2xy' - 12y = \sqrt{x}, \qquad x > 0. \quad (9)$$

SOLUTION: In Example 1 we found the homogeneous solutions

$$y_1 = x^{-4} \qquad \text{and} \qquad y_2 = x^3.$$

Dividing both sides of (9) by x^2, we obtain the standard form

$$y'' + \frac{2}{x}y' - \frac{12}{x^2}y = x^{-3/2}.$$

We now apply the method of variation of parameters, obtaining the system

$$\begin{aligned} x^{-4}c_1' + x^3c_2' &= 0, \\ -4x^{-5}c_1' + 3x^2c_2' &= x^{-3/2}. \end{aligned}$$

Solving these equations simultaneously, we get

$$c_1' = \frac{-x^{7/2}}{7} \qquad \text{and} \qquad c_2' = \frac{x^{-7/2}}{7}.$$

Hence

$$c_1(x) = -\frac{1}{7}\cdot\frac{2}{9}x^{9/2}, \qquad c_2(x) = -\frac{1}{7}\cdot\frac{2}{5}x^{-5/2},$$

so

$$\begin{aligned} y_p(x) = c_1(x)y_1(x) + c_2(x)y_2(x) &= -\frac{1}{7}\left[\frac{2}{9}x^{9/2}\cdot x^{-4} + \frac{2}{5}x^{-5/2}\cdot x^3\right] \\ &= \left(\frac{2}{9}+\frac{2}{5}\right)\frac{-x^{1/2}}{7} = -\frac{4}{45}x^{1/2}. \end{aligned}$$

Thus the general solution is given by

$$y(x) = c_1x^{-4} + c_2x^3 - \frac{4}{45}x^{1/2}.$$

PROBLEMS 10.13

SELF-QUIZ

Multiple Choice

I. Two linearly independent solutions to $x^2y'' + 3xy' + 5y = 0$ are ________.

a. x^{-1}, x^{-5}

b. $\dfrac{\cos(2\ln|x|)}{x}, \dfrac{\sin(2\ln|x|)}{x}$

c. $x^{-3}, x^{-3}\ln|x|$

d. $x^{-3/2}\cos\left(\dfrac{\sqrt{21}\ln|x|}{2}\right), x^{-3/2}\sin\left(\dfrac{\sqrt{21}\ln|x|}{2}\right)$

II. Two linearly independent solutions to $x^2y'' + 5xy' + 3y = 0$ are ________.

a. x^{-1}, x^{-3}

b. $x^{(-5+\sqrt{13})/2}, x^{(-5-\sqrt{13})/2}$

c. $x^{-5/2}\cos\left(\dfrac{\sqrt{13}}{2}\ln|x|\right), x^{-5/2}\sin\left(\dfrac{\sqrt{13}}{2}\ln|x|\right)$

d. $x^{-5/2}, x^{-5/2}\ln|x|$

III. Two linearly independent solutions to $x^2y'' + 7y' + 9y = 0$ are ________.

a. $x^{(-7+\sqrt{13})/2}, x^{(-7-\sqrt{13})/2}$

b. x^3, x^{-3}

c. $x^{-3}\sin(3\ln|x|), x^{-3}\cos(3\ln|x|)$

d. $x^{-3}, x^{-3}\ln|x|$

In Problems 1–17 find the general solution to the given Euler equation for $x > 0$. Find the unique solution when initial conditions are given, and sketch its graph.

1. $x^2y'' + xy' - y = 0$
2. $x^2y'' - 5xy' + 9y = 0$
3. $x^2y'' - xy' + 2y = 0$
4. $x^2y'' - 2y = 0$, $y(1) = 3$, $y'(1) = 1$
5. $4x^2y'' - 4xy' + 3y = 0$, $y(1) = 0$, $y'(1) = 1$
6. $x^2y'' + 3xy' + 2y = 0$
7. $x^2y'' - 3xy' + 3y = 0$
8. $x^2y'' + 5xy' + 4y = 0$, $y(1) = 1$, $y'(1) = 3$
9. $x^2y'' + 5xy' + 5y = 0$
10. $4x^2y'' - 8xy' + 8y = 0$
11. $x^2y'' + 2xy' - 12y = 0$
12. $x^2y'' + xy' + y = 0$
13. $x^2y'' + 3xy' - 15y = \dfrac{1}{x}$
14. $x^2y'' + 3xy' + y = 3x^6$
15. $x^2y'' - 5xy' + 9y = x^3$
16. $x^2y'' + 3xy' - 15y = x^2e^x$
17. $x^2y'' + xy' + y = 10$

***18.** Show that the homogeneous Euler equation (2) can be transformed into the constant-coefficient equation $y'' + (a - 1)y' + by = 0$ by making the substitution $x = e^t$ $(t = \ln x)$. [*Hint:* By the chain rule,

$$\frac{dy}{dt} = \frac{dy}{dx}\frac{dx}{dt} = x\frac{dy}{dx}$$

and

$$\frac{d^2y}{dt^2} = \frac{d}{dx}\left(x\frac{dy}{dx}\right)\frac{dx}{dt} = x^2\frac{d^2y}{dx^2} + x\frac{dy}{dx}.\,]$$

Use the method of Problem 18 to solve Problems 19–22.

19. $x^2y'' + 7xy' + 5y = x$
20. $x^2y'' + 3xy' - 3y = 5x^2$
21. $x^2y'' - 2y = \ln x$, $(x > 0)$
22. $4x^2y'' - 4xy' + 3y = \sin\ln(-x)$ $(x < 0)$
23. The equation $xy'' + 4y' = 0$, $x > 0$, arises in astronomy.† Obtain its general solution.

ANSWERS TO SELF-QUIZ

I. b **II.** a **III.** d

10.14 VIBRATORY MOTION (OPTIONAL)

Differential equations were first studied in attempts to describe the motion of particles. As a simple example, consider a mass m attached to a coiled spring of length l_0, the upper end of which is securely fastened (see Figure 1).

We have denoted by zero the equilibrium position of the mass on the spring, that is, the point where the mass remains at rest. Suppose that the mass is given an initial displacement x_0, and an initial velocity v_0. Can we describe the future movement of the mass? To do so, we make the following assumptions about the force‡ exerted by the spring on the mass:

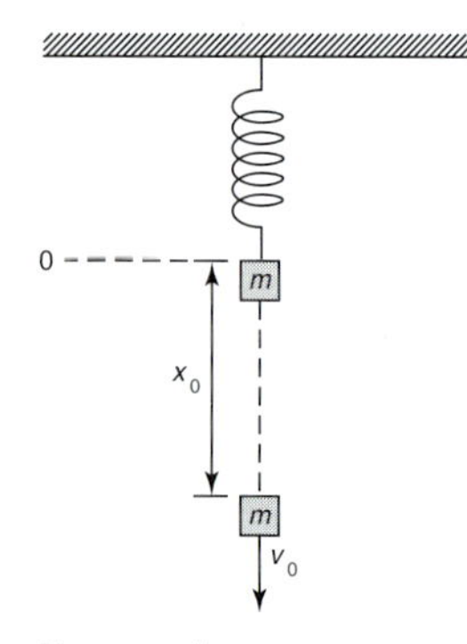

FIGURE 1
A mass attached to a spring

i. The force acts along a vertical line through the center of gravity of the mass (which is then treated as if it were a point mass), and its direction is always from the mass toward the point of equilibrium.

†Z. Kopal, "Stress History of the Moon and of Terrestrial Planets," *Icarus* 2 (1963): 381.
‡The most common systems of units are given in the table below.

System of Units	Force	Length	Mass	Time
International (SI)	Newton (N)	Meter (m)	Kilogram (kg)	Second (sec)
English	Pound (lb)	Foot (ft)	Slug	Second (sec)

$1\text{ N} = 1\text{ kg-m/sec}^2 = 0.22481\text{ lb}$ $1\text{ kg} = 0.06852\text{ slug}$
$1\text{ m} = 3.28084\text{ ft}$ $1\text{ lb} = 1\text{ slug-ft/sec}^2 = 4.4482\text{ N}$

ii. At any time t the magnitude of the force exerted on the mass is proportional to the difference between the length l of the spring and its equilibrium length l_0. The positive constant of proportionality k is called the **spring constant**. This relationship between the force and length of the spring is known as **Hooke's law**.

Newton's second law of motion states that the force F acting on a particle moving with varying velocity v is equal to the time rate of change of the momentum mv. Since the mass is constant,

$$F = \frac{d(mv)}{dt} = ma.$$

Equating the two forces and applying Hooke's law, we have

$$m\frac{d^2x}{dt^2} = -kx, \tag{1}$$

where $x(t)$ denotes the displacement from equilibrium of the spring and is positive when the spring is stretched. The negative sign in equation (1) is present because the force always acts toward the equilibrium position and therefore is in the negative direction when x is positive.

Note that we have assumed that all other forces acting on the spring (such as friction, air resistance, etc.) can be ignored. Equation (1) yields the initial value problem

$$\frac{d^2x}{dt^2} + \frac{k}{m}x = 0, \qquad x(0) = x_0, \qquad x'(0) = v_0. \tag{2}$$

To find the solution of equation (2), the characteristic equation has the complex roots $\pm i\omega_0$, where $\omega_0 = \sqrt{k/m}$. This leads to the general solution

$$x(t) = c_1 \cos \omega_0 t + c_2 \sin \omega_0 t.$$

Using the initial conditions, we find that $c_1 = x_0$ and $c_2 = v_0/\omega_0$, so that the solution of equation (2) is given by

$$x(t) = x_0 \cos \omega_0 t + (v_0/\omega_0)\sin \omega_0 t. \tag{3}$$

It is useful to write $x(t)$ in the form

$$x(t) = A \sin(\omega_0 t + \phi).$$

so that we can graph (and understand) the superposition of the sine and cosine functions in (3). To do so we use the harmonic identity established in Section 10.10 (page 758):

$$a \cos \omega t + b \sin \omega t = A \sin(\omega t + \phi),$$

where $A = \sqrt{a^2 + b^2}$, $\cos \phi = \frac{b}{A}$, $\sin \phi = \frac{a}{A}$. Here, $a = x_0$ and $b = v_0/\omega_0$, so we may rewrite equation (3) as

$$x(t) = A \sin(\omega_0 t + \phi), \tag{4}$$

with A and ϕ determined by

$$A = \sqrt{x_0{}^2 + (v_0/\omega_0)^2}, \qquad \cos \phi = \frac{v_0}{A\omega_0}, \qquad \text{and} \qquad \sin \phi = \frac{x_0}{A}.$$

It is tempting to write

$$\tan\phi = \frac{\sin\phi}{\cos\phi} = \frac{x_0/A}{v_0/(A\omega_0)} = \frac{x_0\omega_0}{v_0},$$

so that $\phi = \tan^{-1}(x_0\omega_0/v_0)$, but this would be correct only if the angle ϕ is in the first or fourth quadrant, because that is where the values of the arctangent are defined.

REMARK: The function $A\sin(\omega_0 t + \phi)$ is easy to graph. We do so in three steps:

i. $\sin\omega_0 t$ is periodic of *period* $2\pi/\omega_0$ since

$$\sin[\omega_0(t + 2\pi/\omega_0)] = \sin(\omega_0 t + 2\pi) = \sin\omega_0 t$$

Thus the graph of $\sin\omega_0 t$ has the same shape as the graph of $\sin t$ except that it repeats every $2\pi/\omega_0$ units (instead of every 2π units).

ii. The graph of $A\sin\omega_0 t$ is the graph of $\sin\omega_0 t$ multiplied by the *amplitude* A. That is, just as $\sin\omega_0 t$ ranges from -1 to 1, $A\sin\omega_0 t$ ranges from $-A$ to A.

iii. The graph of $f(x + c)$, for $c > 0$, is the graph of $f(x)$ shifted c units to the left. For example, the graph of $(x + 2)^2$ is the graph of x^2 shifted two units to the left (see Figure 2).

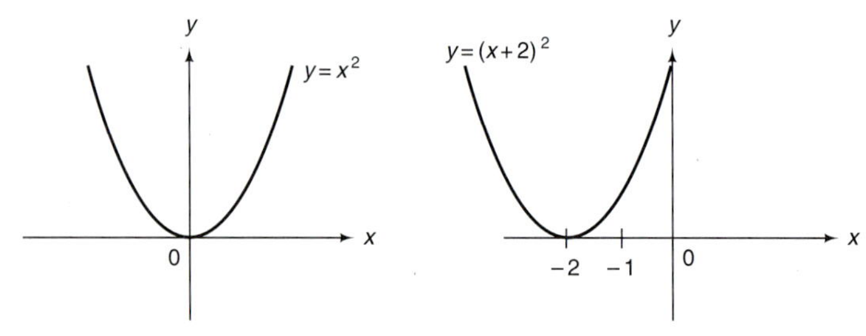

FIGURE 2
The graph of $(x + 2)^2$ is the graph of x^2 shifted two units to the left

Thus the graph of $A\sin(\omega_0 t + \phi) = A\sin\omega_0\left(t + \dfrac{\phi}{\omega_0}\right)$ is the graph of $A\sin\omega_0 t$ shifted ϕ/ω_0 units to the left (if $\phi > 0$).

EXAMPLE 1

SKETCHING A HARMONIC FUNCTION

Sketch the graph of $x = 3(\sin 4t + \pi/6)$

SOLUTION: We do this in three steps, starting with the graph of $\sin x$. Here $\omega_0 = 4$, so the graph is periodic of period $2\pi/4 = \pi/2$. Also, we note that

$$\frac{\phi}{\omega_0} = \frac{\pi/6}{4} = \frac{\pi}{24}.$$

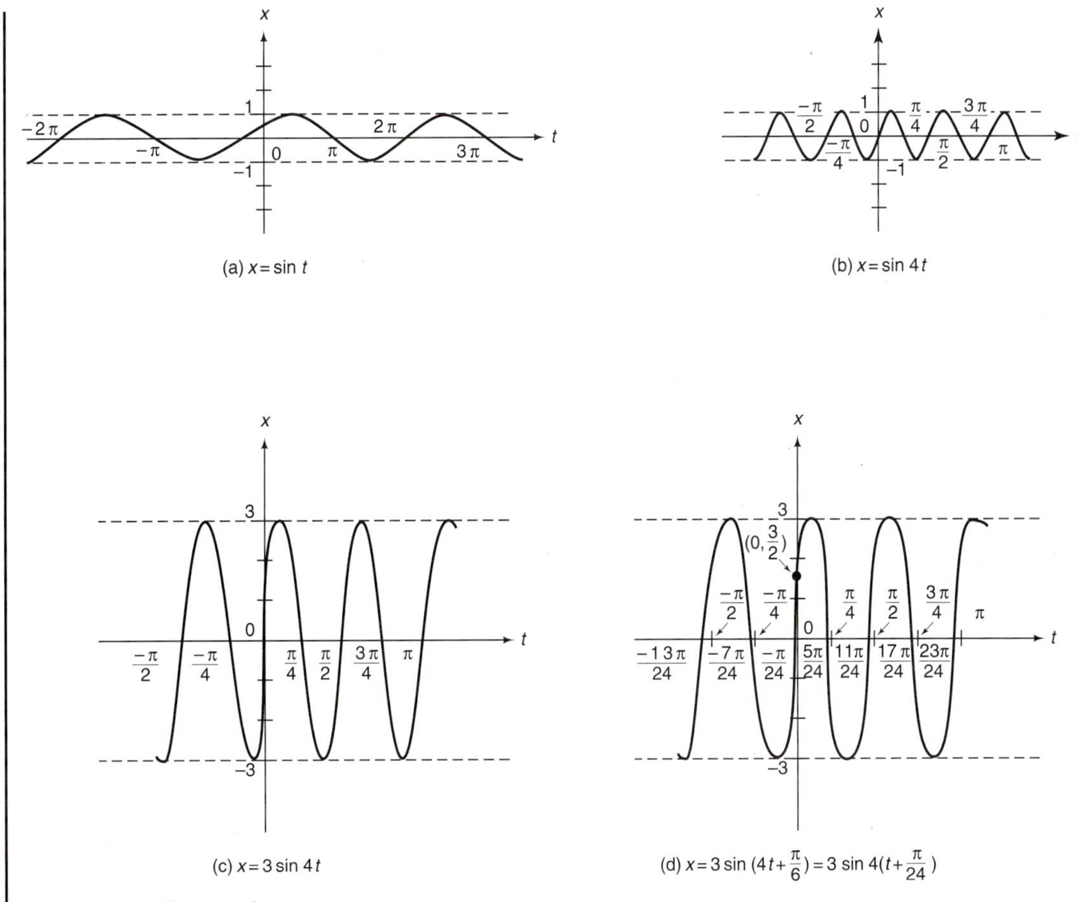

FIGURE 3
Obtaining the graph of $3(\sin 4t + \pi/6)$ from the graph of $\sin t$ in three steps

Because of equation (4), the motion of the mass is called **simple harmonic motion**. That equation indicates that the mass oscillates between the extreme positions $\pm A$; A is called the **amplitude** of the motion. Since the sine term has period $2\pi/\omega_0$, this is the time required for each complete oscillation. The **natural frequency** f of the motion is the number of complete oscillations per unit time:†

$$f = \frac{\omega_0}{2\pi}. \tag{5}$$

Note that although the amplitude depends on the initial conditions, the frequency does not.

EXAMPLE 2

SIMPLE HARMONIC MOTION OF A VIBRATING SPRING

Consider a spring fixed at its upper end and supporting a weight of 10 pounds at its lower end. Suppose the 10-pound weight stretches the spring by 6 inches. Find the equation of

†Cycles/sec = hertz (Hz)

motion of the weight if it is drawn to a position 4 inches below its equilibrium position and released.

SOLUTION: By Hooke's law, since a force of 10 lb stretches the spring by $\frac{1}{2}$ ft, $10 = k(\frac{1}{2})$ or $k = 20$ lb/ft. We are given the initial values $x_0 = \frac{1}{3}$ ft and $v_0 = 0$, so by equation (3) and the identity† $k/m = gk/w = 64 \text{ sec}^{-2}$, we obtain

$$x(t) = \tfrac{1}{3}\cos 8t \text{ ft}.$$

Thus the amplitude is $\frac{1}{3}$ ft (= 4 in.), and the frequency is $f = 4/\pi$ hertz.

DAMPED VIBRATIONS

Throughout the discussion above we made the assumption that there were no external forces acting on the spring. This assumption, however, is not very realistic. To take care of such things as friction in the spring and air resistance, we now assume that there is a *damping* force (that tends to slow things down), which can be thought of as the resultant of all external forces (except gravity) acting on the spring. It is reasonable to assume that the magnitude of the damping force is proportional to the velocity of the particle (for example, the slower the movement, the smaller the air resistance). Therefore, to equation (1) we add the term $c(dx/dt)$, where c is the damping constant that depends on all external factors. This constant could be determined experimentally. The equation of motion then becomes

$$\frac{d^2x}{dt^2} = -\frac{k}{m}x - \frac{c}{m}\frac{dx}{dt}, \qquad x(0) = x_0, \qquad x'(0) = v_0. \tag{6}$$

[Of course, since c depends on external factors, it may very well not be a constant at all but may vary with time and position. In that case, c is really $c(t, x)$, and the equation becomes much harder to analyze than the constant coefficient case.]

To study equation (6), we first find the roots of the characteristic equation $\lambda^2 + \frac{c}{m}\lambda + \frac{k}{m} = 0$:

$$\frac{-c \pm \sqrt{c^2 - 4mk}}{2m}. \tag{7}$$

The nature of the general solution depends on the discriminant $\sqrt{c^2 - 4mk}$. If $c^2 > 4mk$, both roots are negative since $\sqrt{c^2 - 4mk} < c$. In this case

$$x(t) = c_1 \exp\left(\frac{-c + \sqrt{c^2 - 4mk}}{2m}t\right) + c_2 \exp\left(\frac{-c - \sqrt{c^2 - 4mk}}{2m}t\right) \tag{8}$$

becomes small as t becomes large whatever the initial conditions may be. Similarly, in the event the discriminant vanishes, then

$$x(t) = e^{(-c/2m)t}(c_1 + c_2 t), \tag{9}$$

and the solution has a similar behavior. For example, if $c = 5$ (lb-sec/ft) in Example 2, then the discriminant vanishes since $4mk = 4 \cdot \frac{10}{32} \cdot 20 = 25$ and

$$x(t) = e^{-8t}(c_1 + c_2 t).$$

† The identity $w = mg$ may be used to convert weight to mass. Keep in mind that pounds or Newtons are a unit of weight (force) whereas slugs or kilograms are units of mass. The gravitational constant $g = 9.81 \text{ m/sec}^2 = 32.2 \text{ ft/sec}^2$ (approximately).

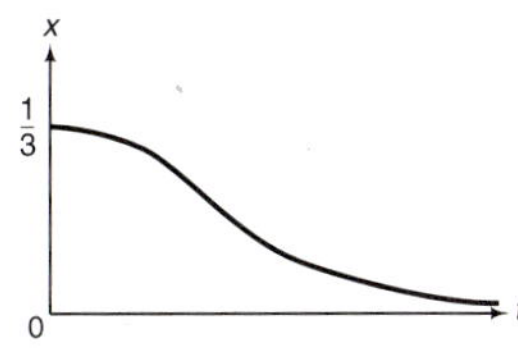

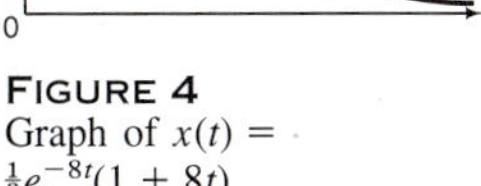
FIGURE 4
Graph of $x(t) = \frac{1}{3}e^{-8t}(1 + 8t)$

Applying the initial conditions yields

$$x(t) = \tfrac{1}{3}e^{-8t}(1 + 8t) \text{ ft},$$

which has the graph shown in Figure 4. We observe that the solution does not oscillate. This type of motion can take place in a highly viscous medium (such as oil or water).

If $c^2 < 4mk$, the general solution is

$$x(t) = e^{(-c/2m)t}\left(c_1 \cos \frac{\sqrt{4mk - c^2}}{2m}t + c_2 \sin \frac{\sqrt{4mk - c^2}}{2m}t\right), \tag{10}$$

which shows an oscillation with frequency

$$f = \frac{\sqrt{4mk - c^2}}{4\pi m}.$$

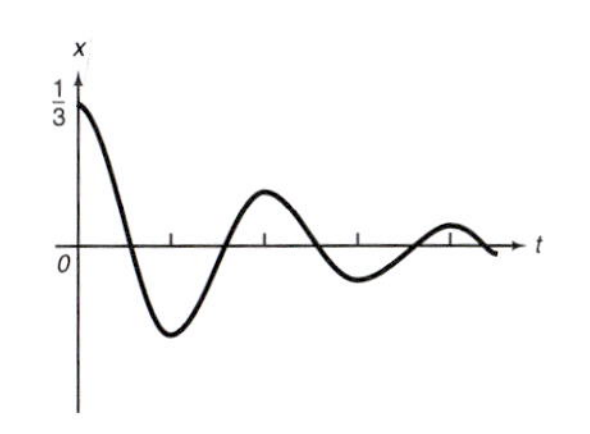

FIGURE 5
Graph of damped harmonic motion

The factor $e^{(-c/2m)t}$ is called the **damping factor**. In Example 2, $F = 10$ so $m = F/a = \frac{10}{32}$. If, in this example, we set $c = 4$ (lb − sec/ft), we are led to the general solution

$$x(t) = e^{-32t/5}(c_1 \cos \tfrac{24}{5}t + c_2 \sin \tfrac{24}{5}t) \text{ ft}.$$

Using the initial values, we find that $c_1 = \frac{1}{3}$, $c_2 = \frac{4}{9}$, and the motion is illustrated in Figure 5.

FORCED VIBRATIONS

The motion of the mass considered in the two cases above is determined by the inherent forces of the spring-weight system and the natural forces acting on the system. Accordingly, the vibrations are called **free** or **natural vibrations**. We will now assume that the mass is also subject to an external periodic force $F_0 \sin \omega t$, due to the motion of the object to which the upper end of the spring is attached. In this case the mass will undergo **forced vibrations**.

Equation (6) may be replaced by the nonhomogeneous second-order differential equation

$$m\frac{d^2x}{dt^2} = -kx - c\frac{dx}{dt} + F_0 \sin \omega t,$$

which we write in the form

$$\frac{d^2x}{dt^2} + \frac{c}{m}\frac{dx}{dt} + \frac{k}{m}x = \frac{F_0}{m} \sin \omega t. \tag{11}$$

By the method of undetermined coefficients, we know that $x(t)$ has a particular solution of the form

$$x_p(t) = b_1 \cos \omega t + b_2 \sin \omega t. \tag{12}$$

Substituting this function into equation (11) yields the simultaneous equations

$$\begin{aligned} (\omega_0{}^2 - \omega^2)b_1 + \frac{c\omega}{m}b_2 &= 0, \\ -\frac{c\omega}{m}b_1 + (\omega_0{}^2 - \omega^2)b_2 &= \frac{F_0}{m}, \end{aligned} \tag{13}$$

where $\omega_0 = \sqrt{k/m}$, from which we obtain

$$b_1 = \frac{-F_0 c\omega}{m^2(\omega_0^2 - \omega^2)^2 + (c\omega)^2},$$

$$b_2 = \frac{F_0 m(\omega_0^2 - \omega^2)}{m^2(\omega_0^2 - \omega^2)^2 + (c\omega)^2}.$$

Using the same method we used to obtain equation (4), we have

$$x_p = A \sin(\omega t + \phi), \tag{14}$$

where

$$A = \frac{F_0/k}{\sqrt{\left[1 - \left(\frac{\omega}{\omega_0}\right)^2\right]^2 + \left(2\frac{c}{c_0}\frac{\omega}{\omega_0}\right)^2}}$$

and

$$\tan \phi = \frac{2\frac{c}{c_0}\frac{\omega}{\omega_0}}{\left(\frac{\omega}{\omega_0}\right)^2 - 1}$$

with $c_0 = 2m\omega_0$. Here A is the amplitude of the motion, ϕ is the **phase angle**, c/c_0 is the **damping ratio**, and ω/ω_0 is the **frequency ratio** of the motion.

The general solution is found by superimposing the periodic function (equation (14)), on the general solution, equation (8), (9), or (10), of the homogeneous equation. Since the solution of the homogeneous equation damps out as t increases, the general solution will be very close to equation (14) for large values of t. Figure 6 illustrates two typical situations.

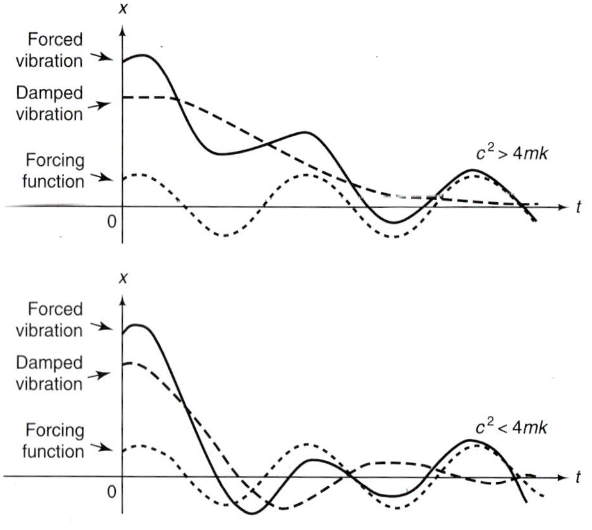

FIGURE 6
Forced and damped vibrations in two cases

It is interesting to see what occurs if the damping constant c vanishes. There are two cases.

CASE 1: If $\omega^2 \neq \omega_0^2$, we superimpose the periodic function of equation (14), on the general solution of the homogeneous equation $x'' + \omega_0^2 x = 0$, obtaining

$$x(t) = c_1 \cos \omega_0 t + c_2 \sin \omega_0 t + \frac{F_0/k}{1 - (\omega/\omega_0)^2} \sin \omega t. \tag{15}$$

Using the initial conditions, we find that

$$c_1 = x_0 \quad \text{and} \quad c_2 = \frac{v_0}{\omega_0} - \frac{(F_0/k)(\omega/\omega_0)}{1 - (\omega/\omega_0)^2}$$

so that

$$x(t) = A \sin(\omega_0 t + \phi) + \frac{F_0/k}{1 - (\omega/\omega_0)^2} \sin \omega t,$$

where

$$A = \sqrt{c_1^2 + c_2^2} \quad \text{and} \quad \tan \phi = c_1/c_2.$$

Hence the motion in this case is simply the sum of two sinusoidal curves as illustrated in Figure 7.

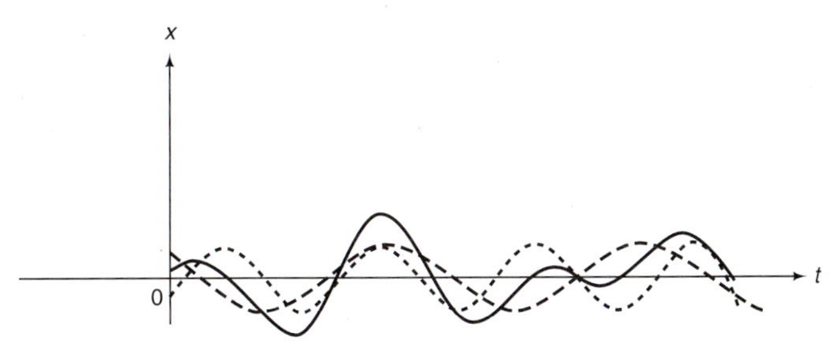

FIGURE 7
The sum of two sine curves

CASE 2: If $\omega^2 = \omega_0^2$, we must seek a particular solution of the form

$$x_p(t) = b_1 t \cos \omega t + b_2 t \sin \omega t, \tag{16}$$

since equation (12) is a solution of the homogeneous equation (2). (See the modification of the method of undetermined coefficients on page 771.) Substituting equation (16) into equation (2), we get

$$b_1 = \frac{-F_0}{2m\omega} \quad \text{and} \quad b_2 = 0,$$

so the general solution has the form

$$x(t) = c_1 \cos \omega t + c_2 \sin \omega t - \frac{F_0}{2m\omega} t \cos \omega t. \tag{17}$$

Note that as t increases the vibrations caused by the last term in equation (17) increase *without bound.* The external force is said to be in **resonance** with the vibrating mass. It is evident that the displacement will become so large that the elastic limit of the spring will be exceeded, leading to fracture or to a permanent distortion in the spring.

Suppose that c is positive but very close to zero while $\omega^2 = \omega_0^2$. Note that equation (13) will yield $b_1 = -F_0/c\omega$ and $b_2 = 0$ when substituted in equation (12). Superimposing $x_p(t) = -(F_0/c\omega)\cos\omega t$ on equation (10) and letting $c_0 = 2m\omega_0$, we obtain

$$x(t) = e^{(-c/c_0)\omega_0 t}\left(c_1 \cos \omega_0 \sqrt{1-\left(\frac{c}{c_0}\right)^2}\,t + c_2 \sin \omega_0 \sqrt{1-\left(\frac{c}{c_0}\right)^2}\,t\right) - \frac{F_0}{c\omega}\cos \omega t. \tag{18}$$

Since c/c_0 is very small, for small values of t we see that equation (18) can be approximated as

$$x(t) \approx c_1 \cos \omega t + c_2 \sin \omega t - \frac{F_0}{2m\omega}\left(\frac{2m}{c}\right)\cos \omega t,$$

which bears a marked resemblance to equation (17) *when equation (17) is evaluated at large values of t* (since $2m/c$ is large). Thus, the *damped* spring problem approaches resonance. This phenomenon is extremely important in engineering since resonance may produce undesirable effects such as metal fatigue and structural fracture, as well as desirable objectives such as sound and light amplification.

PROBLEMS 10.14

In Problems 1–6 determine the equation of motion of a mass m attached to a coiled spring with spring constant k initially displaced a distance x_0 from equilibrium and released with velocity v_0 subject to

a. no damping or external forces,

b. a damping constant c, but no external force,

c. an external force $F_0 \sin \omega t$, but no damping,

d. both a damping constant and external force $F_0 \sin \omega t$.

1. $m = 10$ kg, $k = 1000$ N/m, $x_0 = 1$ m, $v_0 = 0$, $c = 200$ N/(m/s), $F_0 = 1$ N, $\omega = 10$ rad/sec
2. $m = 10$ kg, $k = 10$ N/m, $x_0 = 0$, $v_0 = 1$ m/s, $c = 20$ N/(m/sec), $F_0 = 1$ N, $\omega = 1$ rad/scc
3. $m = 10$ kg, $k = 10$ N/m, $x_0 = 3$ m, $v_0 = 4$ m/s, $c = 10\sqrt{5}$ N/(m/sec), $F_0 = 1$ N, $\omega = 1$ rad/sec
4. $m = 1$ kg, $k = 16$ N/m, $x_0 = 4$ m, $v_0 = 0$, $c = 10$ N/(m/sec), $F_0 = 4$ N, $\omega = 4$ rad/sec
5. $m = 1$ kg, $k = 25$ N/m, $x_0 = 0$ m, $v_0 = 3$ m/sec, $c = 8$ N/(m/sec), $F_0 = 1$ N, $\omega = 3$ rad/sec
6. $m = 9$ kg, $k = 1$ N/m, $x_0 = 4$ m, $v_0 = 1$ m/sec, $c = 10$ N/(m/sec), $F_0 = 2$ N, $\omega = \frac{1}{3}$ rad/sec
7. One end of a rubber band is fixed at a point A. A 1-kg mass, attached to the other end, stretches the rubber band vertically to the point B in such a way that the length AB is 16 cm greater than the natural length of the band. If the mass is further drawn to a position 8 cm below B and released, what will be its velocity (if we neglect resistance) as it passes the position B?
8. If in Problem 7 the mass is released at a position 8 cm above B, what will be its velocity as it passes 1 cm above B?

*9. A cylindrical block of wood of radius and height 1 ft and weighing 12.48 lb floats with its axis vertical in water (62.4 lb/ft^3). If it is depressed so that the surface of the water is tangent to the block, and is then released, what will be its period of vibration and equation of motion? Neglect resistance. [*Hint:* The upward force on the block is equal to the weight of the water displaced by the block.]

*10. A cubical block of wood, 1 ft on a side, is depressed so that its upper face lies along the surface of the water, and is then released. The period of vibration is found to be 1 sec. Neglecting resistance, what is the weight of the block of wood?

11. A 10-kg mass suspended from a spring vibrates freely, the resistance being numerically equal to half the velocity (in m/sec) at any instant. If the period of the motion is 8 sec, what is the spring constant (in kg/sec^2)?
12. A weight w (lb) is suspended from a spring whose constant is 10 lb/ft. The motion of the weight is subject to a resistance (lb) numerically equal to half the velocity (ft/sec). If the motion

is to have a 1-sec period, what are the possible values of w?

13. A 1-g mass is hanging at rest on a spring that is stretched 25 cm by the weight. The upper end of the spring is given the periodic force $0.01 \sin 2t$ N and air resistance has a magnitude 0.0216 (k/sec) times the velocity in meters per second. Find the equation of motion of the mass.

14. An ideal pendulum consists of a weightless rod of length l attached at one end to a frictionless hinge and supporting a body of mass m at the other end. Suppose the pendulum is displaced an angle θ_0 and released (see Figure 8). The tangential acceleration of the ideal pendulum is $l\theta''$, and must be proportional, by Newton's second law of motion, to the tangential component of gravitational force.

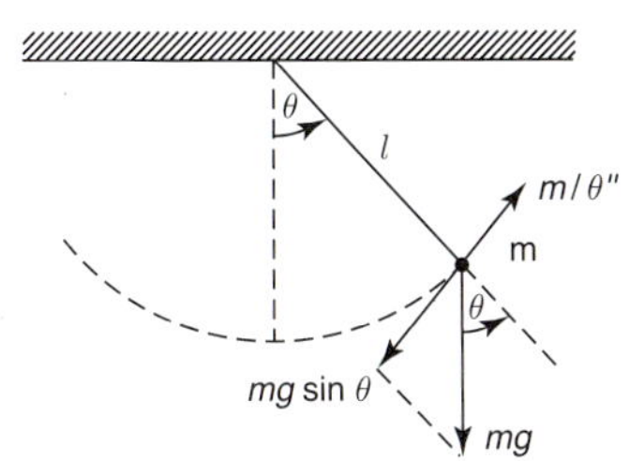

FIGURE 8

a. Neglecting air resistance, show that the ideal pendulum satisfies the nonlinear initial value problem

$$l\frac{d^2\theta}{dt^2} = -g \sin \theta, \quad \theta(0) = \theta_0, \quad \theta'(0) = 0. \tag{19}$$

b. Assuming θ_0 is small, explain why equation (19) may be approximated by the linear initial value problem

$$\frac{d^2\theta}{dt^2} + \frac{g}{l}\theta = 0, \quad \theta(0) = \theta_0, \quad \theta'(0) = 0. \tag{20}$$

c. Solve equation (20) assuming that the rod is 6 inches long and that the initial displacement $\theta_0 = 0.5$ radian. What is the frequency of the pendulum?

15. A grandfather clock has a pendulum that is one meter long. The clock ticks each time the pendulum reaches the rightmost extent of its swing. Neglecting friction and air resistance, and assuming that the motion is small, determine how many times the clock ticks in one minute.

10.15 MORE ON ELECTRIC CIRCUITS (OPTIONAL)

We shall make use of the concepts developed in Section 10.6 and the methods of this chapter to study a simple electric circuit containing a resistor, an inductor, and a capacitor in series with an electromotive force (Figure 1). Suppose that R, L, C, and E are constants. Applying Kirchhoff's law, we obtain

$$L\frac{dI}{dt} + RI + \frac{Q}{C} = E. \tag{1}$$

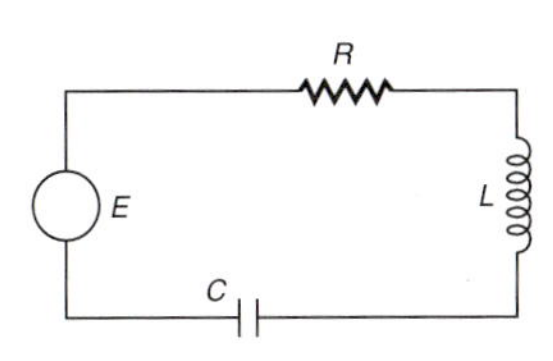

FIGURE 1
An RLC-circuit

Since $dQ/dt = I$, we may differentiate equation (1) to get the second-order homogeneous differential equation

$$L\frac{d^2I}{dt^2} + R\frac{dI}{dt} + \frac{I}{C} = 0. \tag{2}$$

To solve this equation, we note that the characteristic equation

$$\lambda^2 + \frac{R}{L}\lambda + \frac{1}{CL} = 0$$

has the following roots:

$$\lambda_1 = \frac{-R + \sqrt{R^2 - 4L/C}}{2L}, \qquad \lambda_2 = \frac{-R - \sqrt{R^2 - 4L/C}}{2L},$$

or, rewriting the radical, we have

$$\lambda_1 = \frac{R}{2L}\left(-1 + \sqrt{1 - \frac{4L}{CR^2}}\right), \qquad \lambda_2 = \frac{R}{2L}\left(-1 - \sqrt{1 - \frac{4L}{CR^2}}\right). \tag{3}$$

Equation (2) can now be solved using the methods of Sections 10.9 and 10.10.

EXAMPLE 1

SOLVING AN RLC-CIRCUIT

Let $L = 1$ henry (h), $R = 100$ ohms (Ω), $C = 10^{-4}$ farads (f), and $E = 1000$ volts (V) in the circuit shown in Figure 1. Suppose that no charge is present and no current is flowing at time $t = 0$ when E is applied. Here $R^2 - 4L/C = 10{,}000 - 4 \cdot 10^4 = -30{,}000$, $\sqrt{30{,}000} = \sqrt{(10{,}000)(3)} = 100\sqrt{3}$ and $R/2L = 50$ so $\lambda_1 = -50 + 50\sqrt{3}i$ and $\lambda_2 = -50 - 50\sqrt{3}i$. Thus

$$I(t) = e^{-50t}(c_1 \cos 50\sqrt{3}t + c_2 \sin 50\sqrt{3}t).$$

Applying the initial condition $I(0) = 0$, we have $c_1 = 0$. Hence

$$I(t) = c_2 e^{-50t} \sin 50\sqrt{3}t \qquad \text{and}$$
$$I'(t) = c_2 e^{-50t}(50\sqrt{3} \cos 50\sqrt{3}t - 50 \sin 50\sqrt{3}t).$$

To establish the value of c_2, we must make use of equation (1) and the initial condition $Q(0) = 0$. Then

$$Q(t) = C\left(E - L\frac{dI}{dt} - RI\right) = 10^{-4}[1000 - c_2 e^{-50t}(50\sqrt{3} \cos 50\sqrt{3}t - 50 \sin 50\sqrt{3}t + 100 \sin 50\sqrt{3}t)]$$

$50 \cdot 10^{-4} = \frac{1}{200}$ ↓

$$= \frac{1}{10} - \frac{c_2}{200} e^{-50t}(\sin 50\sqrt{3}t + \sqrt{3} \cos 50\sqrt{3}t).$$

Thus $Q(0) = \frac{1}{10} - \frac{c_2}{200}\sqrt{3} = 0$, so $c_2 = 20/\sqrt{3}$, and we have

$$Q(t) = \frac{1}{10} - \frac{1}{10\sqrt{3}} e^{-50t}(\sin 50\sqrt{3}t + \sqrt{3} \cos 50\sqrt{3}t)$$

and

$$I(t) = \frac{20}{\sqrt{3}} e^{-50t} \sin 50\sqrt{3}t.$$

From these equations we observe that the current will rapidly damp out and that the charge will rapidly approach its **steady-state value** of $\frac{1}{10}$ coulomb (coul). Here $I(t)$ is called the **transient current** because it dies out.

EXAMPLE 2

AN RLC-CIRCUIT WITH A PERIODIC EMF

Let the inductance, resistance, and capacitance remain the same as in Example 1, but suppose $E = 962 \sin 60t$. By equation (1) we have

$$\frac{dI}{dt} + 100I + 10^4Q = 962 \sin 60t. \tag{4}$$

Converting equation (4) so that all expressions are in terms of $Q(t)$, we obtain

$$\frac{d^2Q}{dt^2} + 100\frac{dQ}{dt} + 10^4Q = 962 \sin 60t. \tag{5}$$

It is evident that equation (5) has a particular solution of the form

$$Q_p(t) = A_1 \sin 60t + A_2 \cos 60t. \tag{6}$$

To determine the values A_1 and A_2, we substitute equation (6) into equation (4), obtaining the following simultaneous equations:

$$6400A_1 - 6000A_2 = 962$$
$$6000A_1 + 6400A_2 = 0$$

Thus $A_1 = \frac{2}{25}$, $A_2 = -\frac{3}{40}$, and since the general solution of the homogeneous equation is the same as that of equation (2), the general solution of equation (1) is

$$Q(t) = e^{-50t}(c_1 \cos 50\sqrt{3}t + c_2 \sin 50\sqrt{3}t) + \tfrac{2}{25} \sin 60t - \tfrac{3}{40} \cos 60t. \tag{7}$$

Differentiating equation (7), we obtain

$$I(t) = 50e^{-50t}[(\sqrt{3}c_2 - c_1) \cos 50\sqrt{3}t - (c_2 + \sqrt{3}c_1) \sin 50\sqrt{3}t]$$
$$+ \tfrac{24}{5} \cos 60t + \tfrac{9}{2} \sin 60t.$$

Setting $t = 0$ and using the initial conditions, we obtain the simultaneous equations

$$c_1 = \tfrac{3}{40},$$
$$50(\sqrt{3}c_2 - c_1) = -\tfrac{24}{5}.$$

Therefore, $c_1 = \frac{3}{40}$ and $c_2 = -21/1000\sqrt{3}$.

PROBLEMS 10.15

1. In Example 1, let $L = 10$ h, $R = 250\ \Omega$, $C = 10^{-3}$ f, and $E = 900$ V. With the same assumptions, calculate the current and charge for all values of $t \geq 0$.
2. In Problem 1, suppose instead that $E = 50 \cos 30t$. Find $Q(t)$ for $t \geq 0$.

In Problems 3–6, find the steady-state current in the *RLC* circuit of Figure 1 where:

3. $L = 5$ h, $R = 10\ \Omega$, $C = 0.1$ f, $E = 25 \sin t$ V.
4. $L = 10$ h, $R = 40\ \Omega$, $C = 0.025$ f, $E = 100 \cos 5t$ V.
5. $L = 1$ h, $R = 7\ \Omega$, $C = 0.1$ f, $E = 100 \sin 10t$ V.
6. $L = 2.5$ h, $R = 10\ \Omega$, $C = 0.08$ f, $E = 100 \cos 5t$ V.

Find the transient current in the *RLC* circuit of Figure 1 for Problems 7–12.

7. Problem 3.
8. Problem 4.
9. Problem 5.
10. Problem 6.
11. $L = 20$ h, $R = 40\ \Omega$, $C = 10^{-3}$ f, $E = 500 \sin t$ V.
12. $L = 24$ h, $R = 48\ \Omega$, $C = 0.375$ f, $E = 900 \cos 2t$ V.
13. Given that $L = 1$ h, $R = 1200\ \Omega$, $C = 10^{-6}$ f, $I(0) = Q(0) = 0$, and $E = 100 \sin 600t$ V, determine the transient current and the steady-state current.
14. Find the ratio of the current in the circuit of Problem 13 to that which would be flowing if there were no resistance, at $t = 0.001$ sec.
15. Consider the system governed by equation (1) for the case where the resistance is zero and $E =$

$E_0 \sin \omega t$. Show that the solution consists of two parts: a general solution with frequency $1/\sqrt{LC}$ and a particular solution with frequency ω. The frequency $1/\sqrt{LC}$ is called the **natural frequency** of the circuit. Note that if $\omega = 1/\sqrt{LC}$, then the particular solution changes form.

16. To allow for different variations of the voltage, let us assume in equation (1) that $E = E_0 e^{it}$ $(= E_0 \cos t + iE_0 \sin t)$. Assume also, as in Problem 15, that $R = 0$. Finally, for simplicity assume that $E_0 = L = C = 1$. Then $1 = \omega = 1/\sqrt{LC}$.

a. Show that equation (2) becomes

$$\frac{d^2I}{dt^2} + I = e^{it}.$$

b. Determine λ such that $I(t) = \lambda t e^{it}$ is a solution.

c. Calculate the general solution and show that the magnitude of the current increases without bound as t increases. This phenomenon will produce resonance.

17. Let an inductance of L henries, a resistance of R ohms, and a capacitance of C farads be connected in series with an emf of $E_0 \sin \omega t$ volts. Suppose $Q(0) = I(0) = 0$, and $4L > R^2C$.

a. Find the expressions for $Q(t)$ and $I(t)$.

b. What value of ω will produce resonance?

18. Solve Problem 17 for $4L = R^2C$.

19. Solve Problem 17 for $4L < R^2C$.

10.16 HIGHER-ORDER LINEAR DIFFERENTIAL EQUATIONS (OPTIONAL)

In this section we extend the results of the chapter to linear differential equations of order higher than two. There is little theoretical difference between second- and higher-order systems, so we can be relatively brief. We state all theorems without proof.

The *general nonhomogeneous linear nth-order equation* is

$$y^{(n)}(x) + a_{n-1}(x)y^{(n-1)}(x) + \cdots + a_1(x)y'(x) + a_0(x)y(x) = f(x). \tag{1}$$

The *associated homogeneous equation* is

$$y^{(n)}(x) + a_{n-1}(x)y^{(n-1)}(x) + \cdots + a_1(x)y'(x) + a_0(x)y(x) = 0. \tag{2}$$

In Theorem 10.7.1 (p. 733) we stated that equation (1) has a unique solution provided that all the functions in the equation are continuous and n initial conditions are specified. Now we concern ourselves with finding the general solutions to equations (1) and (2). To do so we follow the procedures we have developed for solving second-order equations.

We say that the functions $y_1, y_2, \ldots, y_n$ are **linearly independent** in $[x_0, x_1]$ if the following condition holds:

$$c_1y_1(x) + c_2y_2(x) + \cdots + c_ny_n(x) = 0 \quad \text{for all } x \in [x_0, x_1]$$

implies that $c_1 = c_2 = \cdots = c_n = 0$.

Otherwise the functions are **linearly dependent**. The expression $c_1y_1 + c_2y_2 + \cdots + c_ny_n$ is called a **linear combination** of the functions $y_1, y_2, \ldots, y_n$.

The Wronskian of $y_1, y_2, \ldots, y_n$ is defined by

$$W(y_1, y_2, \ldots, y_n)(x) = \begin{vmatrix} y_1 & y_2 & \cdots & y_n \\ y_1' & y_2' & \cdots & y_n' \\ y_1'' & y_2'' & \cdots & y_n'' \\ \vdots & \vdots & & \vdots \\ y_1^{(n-1)} & y_2^{(n-1)} & \cdots & y_n^{(n-1)} \end{vmatrix}. \tag{3}$$

THEOREM 1 PROPERTIES OF THE WRONSKIAN

Let $a_0, a_1, \ldots, a_{n-1}$ be continuous in $[x_0, x_1]$ and let $y_1, y_2, \ldots, y_n$ be n solutions of equation (2). Then

a. $W(y_1, y_2, \ldots, y_n)(x)$ is zero either for all $x \in [x_0\ x_1]$ or for no $x \in [x_0, x_1]$.

b. $y_1, y_2, \ldots, y_n$ are linearly independent if and only if $W(y_1, y_2, \ldots, y_n)(x) \neq 0$. ■

EXAMPLE 1 THREE LINEARLY INDEPENDENT SOLUTIONS

The functions 1, x, and x^2 are solutions to the equation $y'''(x) = 0$. Determine whether they are linearly independent or dependent for all x.

SOLUTION: The easiest way to test for linear independence is to use the Wronskian:

$$W(y_1, y_2, y_3)(x) = \begin{vmatrix} 1 & x & x^2 \\ 0 & 1 & 2x \\ 0 & 0 & 2 \end{vmatrix} = 2 \neq 0,$$

so the functions are linearly independent. Alternatively, consider

$$c_1 \cdot 1 + c_2 x + c_3 x^2 = 0.$$

Setting $x = 0$, it follows that $c_1 = 0$. Setting $x = \pm 1$ yields the system

$$\begin{aligned} c_2 + c_3 &= 0, \\ -c_2 + c_3 &= 0, \end{aligned}$$

so $c_2 = c_3 = 0$, implying that the functions are linearly independent.

The procedure for solving a linear nth-order equation is as follows:

Step 1: Find n linearly independent solutions, $y_1, y_2, \ldots, y_n$, to the homogeneous equation

$$y^{(n)} + a_{n-1}(x)y^{(n-1)} + \cdots + a_1(x)y' + a_0(x)y = 0.$$

The **general solution** to this equation is then

$$y(x) = c_1 y_1(x) + c_2 y_2(x) + \cdots + c_n y_n(x). \quad (4)$$

Step 2: Find one solution, $y_p(x)$, to the nonhomogeneous equation

$$y^{(n)} + a_{n-1}(x)y^{(n-1)} + \cdots + a_1(x)y' + a_0(x)y = f(x).$$

The **general solution** to this equation is then given by

$$y(x) = c_1 y_1(x) + c_2 y_2(x) + \cdots + c_n y_n(x) + y_p, \quad (5)$$

where $y_1, y_2, \ldots, y_n$ are the n linearly independent solutions of Step 1.

As in the case of second-order equations, we can generally obtain these solutions only when the coefficients $a_k(x)$ are all constants. In this case, equations (1) and (2) are said to have **constant coefficients**. We only deal with the case where these constants are real.

The general nth-order linear, homogeneous constant-coefficient equation is

$$y^{(n)}(x) + a_{n-1}y^{(n-1)}(x) + \cdots + a_1 y'(x) + a_0 y(x) = 0. \tag{6}$$

Note that

$$\frac{d^n}{dx^n} e^{\lambda x} = \lambda^n e^{\lambda x}.$$

If we substitute $y = e^{\lambda x}$ into (6) and then divide by $e^{\lambda x}$, we obtain the **characteristic equation**

$$\lambda^n + a_{n-1}\lambda^{n-1} + \cdots + a_1\lambda + a_0 = 0. \tag{7}$$

Equation (7) has n roots $\lambda_1, \lambda_2, \ldots, \lambda_n$. Some of these roots may be real and distinct, real and equal, distinct complex conjugate pairs, or equal complex conjugate pairs. If a root λ_k (real or complex) occurs m times, we say that it has **multiplicity** m. The following rules tell us how to find the general solution to equation (6).

PROCEDURE FOR SOLVING LINEAR HOMOGENEOUS EQUATIONS WITH CONSTANT COEFFICIENTS

a. Obtain the characteristic equation (7).

b. Find the roots $\lambda_1, \lambda_2, \ldots, \lambda_n$ of (7). (This is usually the most difficult step.)

c. For each real root λ_k of multiplicity 1 (*single root*), one solution to (6) is $y_k = e^{\lambda_k x}$.

d. For each real root λ_k of multiplicity $m > 1$, m solutions to (6) are

$$y_1 = e^{\lambda_k x},\ y_2 = xe^{\lambda_k x}, \ldots, y_m = x^{m-1}e^{\lambda_k x}.$$

e. If $\alpha + i\beta$ and $\alpha - i\beta$ are simple roots, then two solutions to (6) are

$$y_1 = e^{\alpha x}\cos\beta x \quad \text{and} \quad y_2 = e^{\alpha x}\sin\beta x.$$

f. If $\alpha + i\beta$ and $\alpha - i\beta$ are roots of multiplicity $m > 1$, then $2m$ solutions to (6) are

$$y_1 = e^{\alpha x}\cos\beta x,\ y_2 = xe^{\alpha x}\cos\beta x, \ldots, y_m = x^{m-1}e^{\alpha x}\cos\beta x,$$

$$y_{m+1} = e^{\alpha x}\sin\beta x,\ y_{m+2} = xe^{\alpha x}\sin\beta x, \ldots, y_{2m} = x^{m-1}e^{\alpha x}\sin\beta x.$$

g. If $y_1, y_2, \ldots, y_n$ are the n solutions obtained in Steps c–f, then $y_1, y_2, \ldots, y_n$ are linearly independent and the general solution to (6) is given by

$$y(x) = c_1y_1(x) + c_2y_2(x) + \cdots + c_ny_n(x).$$

EXAMPLE 2

SOLVING A THIRD-ORDER EQUATION

Find the general solution of

$$y''' - 3y'' - 10y' + 24y = 0.$$

SOLUTION: The characteristic equation is

$$\lambda^3 - 3\lambda^2 - 10\lambda + 24 = (\lambda - 2)(\lambda + 3)(\lambda - 4) = 0,$$

with roots $\lambda_1 = 2$, $\lambda_2 = -3$, and $\lambda_3 = 4$. Since these roots are real and distinct, three linearly independent solutions are

$$y_1 = e^{2x}, \qquad y_2 = e^{-3x}, \qquad y_3 = e^{4x},$$

and the general solution is

$$y(x) = c_1 e^{2x} + c_2 e^{-3x} + c_3 e^{4x}.$$

EXAMPLE 3 SOLVING A FOURTH-ORDER EQUATION

Find the general solution of

$$y^{(4)} - 4y''' + 6y'' - 4y' + y = 0.$$

SOLUTION: The characteristic equation is

$$\lambda^4 - 4\lambda^3 + 6\lambda^2 - 4\lambda + 1 = (\lambda - 1)^4 = 0,$$

with the single root $\lambda = 1$ of multiplicity 4. Thus four linearly independent solutions are

$$y_1 = e^x, \qquad y_2 = xe^x, \qquad y_3 = x^2 e^x, \qquad y_4 = x^3 e^x,$$

and the general solution is

$$y(x) = e^x(c_1 + c_2 x + c_3 x^2 + c_4 x^3).$$

EXAMPLE 4 SOLVING A FIFTH-ORDER EQUATION

Find the general solution of

$$y^{(5)} - 2y^{(4)} + 8y'' - 12y' + 8y = 0.$$

SOLUTION: The characteristic equation is

$$\lambda^5 - 2\lambda^4 + 8\lambda^2 - 12\lambda + 8 = 0,$$

which can be factored into

$$(\lambda + 2)(\lambda^2 - 2\lambda + 2)^2 = 0.$$

The solutions to $\lambda^2 - 2\lambda + 2 = 0$ are $\lambda = 1 \pm i$. Thus the roots are

$$\lambda_1 = -2 \text{ (simple)}, \qquad \lambda_2 = 1 + i, \qquad \lambda_3 = 1 - i,$$

with the complex roots λ_2 and λ_3 having multiplicity 2. Thus five linearly independent solutions are

$$y_1 = e^{-2x}, \qquad y_2 = e^x \cos x, \qquad y_3 = xe^x \cos x,$$
$$y_4 = e^x \sin x, \qquad y_5 = xe^x \sin x,$$

and the general solution is

$$y(x) = c_1 e^{-2x} + (c_2 + c_3 x)e^x \cos x + (c_4 + c_5 x)e^x \sin x.$$

REMARK: In solving the last three characteristic equations we made the factoring look easy. Finding roots of a polynomial of degree greater than two is, in general, very difficult.

How do we find a particular solution to the nonhomogeneous equation (1)? As with second-order equations, there are two methods: undetermined coefficients and variation of parameters. The method of undetermined coefficients is identical to the technique we used for second-order equations. The method of variation of parameters is discussed in Problems 29 and 30.

Finally, certain equations with variable coefficients can be solved. The higher-order Euler equation is discussed in Problems 31–34.

PROBLEMS 10.16

SELF-QUIZ

Multiple Choice

I. Which of the following are solutions to $y''' - 3y'' + 3y' - y = 0$?

a. e^x **b.** e^{-x} **c.** xe^x
d. xe^{-x} **e.** x^2e^x **f.** x^2e^{-x}
g. x^3e^x **h.** x^3e^{-x}

II. Which of the following are solutions to $y^{(4)} - 4y'' + 4y = 0$?

a. $e^{\sqrt{2}x}$ **b.** $e^{-\sqrt{2}x}$ **c.** $xe^{\sqrt{2}x}$
d. $xe^{-\sqrt{2}x}$ **e.** $\cos\sqrt{2}x$ **f.** $\sin\sqrt{2}x$

III. Which of the following are solutions to a homogeneous differential equation whose characteristic equation is $(\lambda + 1)^3(\lambda^2 - 4)(\lambda^2 + 2\lambda + 2) = 0$?

a. e^{-x} **b.** xe^{-x} **c.** x^2e^{-x}
d. e^{2x} **e.** e^{-2x} **f.** $e^{-x}\cos x$
g. $e^{-x}\sin x$

In Problems 1–16 find the general solution to the given equation. If initial conditions are given, find the particular solutions that satisfy them, and sketch their graphs.

1. $y^{(4)} + 2y'' + y = 0$
2. $y''' - y'' - y' + y = 0$
3. $y''' - 3y'' + 3y' - y = 0$, $y(0) = 1$, $y'(0) = 2$, $y''(0) = 3$
4. $x''' + 5x'' - x' - 5x = 0$
5. $y''' - 9y' = 0$, $y(0) = 3$, $y'(0) = 0$, $y''(0) = 18$
6. $y''' - 6y'' + 3y' + 10y = 0$
7. $y^{(4)} = 0$
8. $y^{(4)} - 9y'' = 0$
9. $y^{(4)} - 5y'' + 4y = 0$
10. $y^{(5)} - 2y''' + y' = 0$
11. $y^{(4)} - 4y'' = 0$, $y(0) = 1$, $y'(0) = 3$, $y''(0) = 0$, $y'''(0) = 16$
12. $y^{(4)} - 4y''' - 7y'' + 22y' + 24y = 0$
13. $y''' - y'' + y' - y = 0$
14. $y''' - 3y'' + 4y' - 2y = 0$, $y(0) = 1$, $y'(0) = 2$, $y''(0) = 3$
15. $y''' - 27y = 0$
16. $y^{(5)} + 2y''' + y' = 0$, $y(\pi/2) = 0$, $y'(\pi/2) = 1$, $y''(\pi/2) = 0$, $y'''(\pi/2) = -3$, $y^{(4)}(\pi/2) = 0$
17. Show that the solutions y_1, y_2, and y_3 of the linear third-order differential equation

$$y''' + a_1(x)y'' + a_2(x)y' + a_3(x)y = 0$$

that satisfy the conditions

$$\begin{aligned} y_1(x_0) &= 1, & y_1'(x_0) &= 0, & y_1''(x_0) &= 0, \\ y_2(x_0) &= 0, & y_2'(x_0) &= 1, & y_2''(x_0) &= 0, \\ y_3(x_0) &= 0, & y_3'(x_0) &= 0, & y_3''(x_0) &= 1, \end{aligned}$$

respectively, are linearly independent.

18. Show that *any* solution of

$$y''' + a_1(x)y'' + a_2(x)y' + a_3(x)y = 0$$

can be expressed as a linear combination of the solutions y_1, y_2, y_3 given in Problem 17. [*Hint:* If $y(x_0) = c_1$, $y'(x_0) = c_2$, and $y''(x_0) = c_3$, consider the linear combination $c_1y_1 + c_2y_2 + c_3y_3$.]

***19.** Consider the third-order equation

$$y''' + a(x)y'' + b(x)y' + c(x)y = 0$$

and let $y_1(x)$ and $y_2(x)$ be two linearly independent solutions. Define $y_3(x) = v(x)y_1(x)$ and assume that $y_3(x)$ is a solution to the equation.

a. Find a second-order differential equation that is satisfied by v'.
b. Show that $(y_2/y_1)'$ is a solution of this equation.
c. Use the result of part (b) to find a second

linearly independent solution of the equation derived in part (a).

***20.** Consider the equation

$$y''' - \left(\frac{3}{x^2}\right)y' + \left(\frac{3}{x^3}\right)y = 0, \qquad (x > 0).$$

a. Show that $y_1(x) = x$ and $y_2(x) = x^3$ are two linearly independent solutions.

b. Use the results of Problem 19 to get a third linearly independent solution.

21. Consider the third-order equation

$$y''' + a(x)y'' + b(x)y' + c(x)y = 0,$$

where a, b, and c are continuous functions of x in some interval I. Prove that if $y_1(x)$, $y_2(x)$, and $y_3(x)$ are solutions to the equation, then so is any linear combination of them.

***22.** In Problem 21, let

$$W(y_1, y_2, y_3)(x) = \begin{vmatrix} y_1 & y_2 & y_3 \\ y_1' & y_2' & y_3' \\ y_1'' & y_2'' & y_3'' \end{vmatrix}.$$

a. Show that W satisfies the differential equation

$$W'(x) = -a(x)W.$$

b. Prove that $W(y_1, y_2, y_3)(x)$ is either always zero or never zero.

***23. a.** Prove that the solutions $y_1(x)$, $y_2(x)$, $y_3(x)$ of the equation in Problem 21 are linearly independent on $[x_0, x_1]$ if and only if $W(y_1, y_2, y_3) \neq 0$.

b. Show that $\sin t$, $\cos t$, and e^t are linearly independent solutions of

$$y''' - y'' + y' - y = 0$$

on any interval (a, b) where $-\infty < a < b < \infty$.

24. Assume that $y_1(x)$ and $y_2(x)$ are two solutions to

$$y''' + a(x)y'' + b(x)y' + c(x)y = f(x).$$

Prove that $y_3(x) = y_1(x) - y_2(x)$ is a solution of the associated homogeneous equation.

In Problems 25–28 use the method of undetermined coefficients to find the general solution of the given equation.

25. $y''' - y'' - y' + y = e^x$

26. $y''' - y'' - y' + y = e^{-x}$

27. $y''' - 3y'' - 10y' + 24y = x + 3$

28. $y^{(4)} + 2y'' + y = 3\cos x$

***29.** Consider the third-order equation

$$y''' + ay'' + by' + cy = f(x).$$

Let $y_1(x)$, $y_2(x)$, and $y_3(x)$ be three linearly independent solutions to the associated homogeneous equation. Assume that there is a solution of this equation of the form $y(x) = c_1(x)y_1(x) + c_2(x)y_2(x) + c_3(x)y_3(x)$.

a. Following the steps used in deriving the method of variation of parameters for second-order equations, derive a method for solving third-order equations.

b. Find a particular solution of the equation

$$y''' - 2y' - 4y = e^{-x}\tan x.$$

30. Use the method derived in Problem 29 to find a particular solution of

$$y''' + 5y'' + 9y' + 5y = 2e^{-2x}\sec x.$$

In Problems 31–33 guess that there is a solution of the form $y = x^\lambda$ to solve the given Euler equation.

31. $x^3y''' + 2x^2y'' - xy' + y = 0$

32. $x^3y''' - 12xy' + 24y = 0$

33. $x^3y''' + 4x^2y'' + 3xy' + y = 0$

34. Show that the substitution $x = e^t$ can be used to solve the third-order Euler equation

$$x^3y''' + x^2y'' - 2xy' + 2y = 0.$$

ANSWERS TO SELF-QUIZ

I. a, c, e **II.** a, b, c, d **III.** All of them

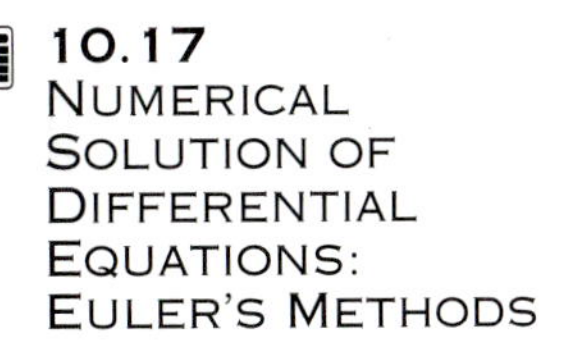

10.17 NUMERICAL SOLUTION OF DIFFERENTIAL EQUATIONS: EULER'S METHODS

In this chapter we have provided a number of methods for solving differential equations. However, as we pointed out earlier, most differential equations cannot be solved by elementary methods. For that reason, a number of numerical techniques have been developed for finding solutions or, more precisely, for finding solutions at particular points. We discuss two of the most elementary numerical techniques for solving first-order equations in this section.

Before presenting these numerical techniques, we should consider *when* numerical methods could or should be employed. Such methods are used primarily when other methods are not applicable. Additionally, even when other methods do apply, there may be an advantage in having a numerical solution, as solutions in terms of more exotic special functions are sometimes difficult to interpret. There may also be computational advantages: the exact solution may be extremely tedious to obtain.

On the other hand, care must always be exercised in using any numerical scheme, as the accuracy of the solution depends not only on the "correctness" of the numerical method being used but also on the precision of the device (hand calculator or computer) used for the computations.

From the general theory we know that, in many cases, the initial value problem

$$\frac{dy}{dx} = f(x, y), \qquad y(x_0) = y_0 \tag{1}$$

has a unique solution $y(x)$. The two techniques we will describe below approximate this solution $y(x)$ only at a finite number of points

$$x_0, \qquad x_1 = x_0 + h, \qquad x_2 = x_0 + 2h, \ldots, \qquad x_n = x_0 + nh,$$

where h is some (nonzero) real number. The methods provide a value y_k that is an approximation to the exact value $y(x_k)$ for $k = 0, 1, \ldots, n$.

EULER'S METHOD†

This procedure is crude but very simple. The idea is to obtain y_1 by assuming that $f(x, y)$ varies so little on the interval $x_0 \leq x \leq x_1$ that only a very small error is made by replacing it by the constant value $f(x_0, y_0)$. Integrating

$$\frac{dy}{dx} = f(x, y)$$

from x_0 to x_1, we obtain

$$y(x_1) - y_0 = y(x_1) - y(x_0) = \int_{x_0}^{x_1} f(x, y)\, dx \approx f(x_0, y_0)(x_1 - x_0), \tag{2}$$

or, since $h = x_1 - x_0$,

$$y_1 = y_0 + hf(x_0, y_0).$$

Repeating the process with (x_1, y_1) to obtain y_2, etc., we obtain the **difference equation**

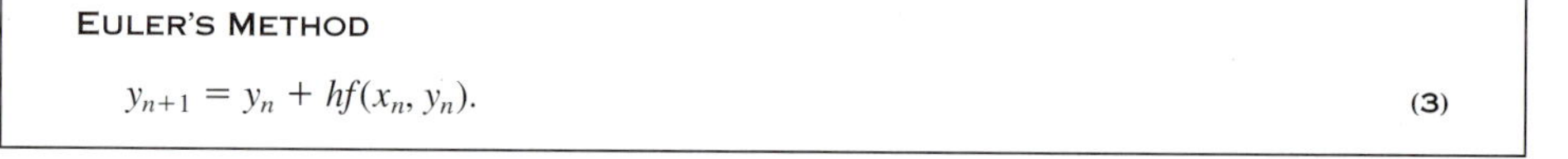

EULER'S METHOD

$$y_{n+1} = y_n + hf(x_n, y_n). \tag{3}$$

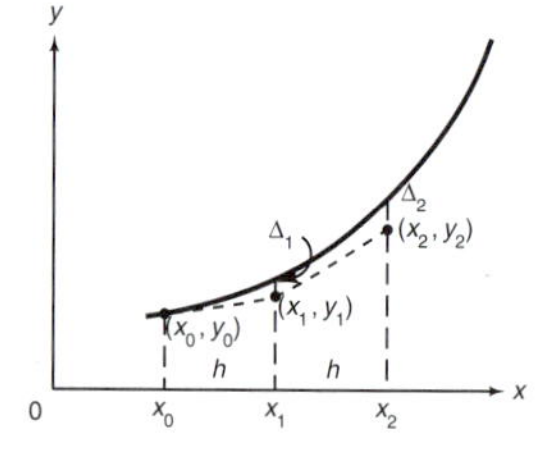

FIGURE 1
The true solution is solid; the approximate solution is dashed

We shall solve equation (3) iteratively—that is, by first finding y_1, then using it to find y_2, and so on.

The geometric meaning of equation (3) is easily seen by considering the solution curve of the differential equation (1): we are simply following the tangent to the solution curve passing through (x_n, y_n) for a small horizontal distance. Looking at Figure 1, where

† See the biographical sketch on page 251.

the smooth curve is the unknown exact solution to the initial value problem (1), we see how equation (3) approximates the exact solution. Since $f(x_0, y_0)$ is the slope of the exact solution at (x_0, y_0), we follow the tangent line to the point (x_1, y_1). Some solution to the differential equation passes through this point. We follow its tangent line at this point to reach (x_2, y_2), and so on. The differences Δ_k are errors at the kth stage in the process.

EXAMPLE 1 USING EULER'S METHOD

Solve

$$\frac{dy}{dx} = y + x^2, \qquad y(0) = 1 \tag{4}$$

at the value $x = 1$.

SOLUTION: We wish to find $y(1)$ by approximating the solution at $x = 0.0, 0.2, 0.4, 0.6, 0.8$, and 1.0. Here $h = 0.2$, $f(x_n, y_n) = y_n + x_n^2$, and Euler's method [equation (3)] yields

$$y_{n+1} = y_n + h \cdot f(x_n, y_n) = y_n + h(y_n + x_n^2).$$

Since $y_0 = y(0) = 1$, we obtain

$$y_1 = y_0 + h \cdot (y_0 + x_0^2) = 1 + 0.2(1 + 0^2) = 1.2,$$
$$y_2 = y_1 + h \cdot (y_1 + x_1^2) = 1.2 + 0.2[1.2 + (0.2)^2] = 1.448 \approx 1.45,$$
$$y_3 = y_2 + h(y_2 + x_2^2) = 1.45 + 0.2[1.45 + (0.4)^2] \approx 1.77,$$
$$y_4 = y_3 + h(y_3 + x_3^2) = 1.77 + 0.2[1.77 + (0.6)^2] \approx 2.20,$$
$$y_5 = y_4 + h(y_4 + x_4^2) = 2.20 + 0.2[2.20 + (0.8)^2] \approx 2.77.$$

We arrange our work as shown in Table 1. The value $y_5 = 2.77$, corresponding to $x_5 = 1.0$, is our approximate value for $y(1)$.

TABLE 1

x_n	y_n	$f(x_n, y_n) = y_n + x_n^2$	$y_{n+1} = y_n + h \cdot f(x_n, y_n)$
0.0	1.00	1.00	1.20
0.2	1.20	1.24	1.45
0.4	1.45	1.61	1.77
0.6	1.77	2.13	2.20
0.8	2.20	2.84	2.77
1.0	2.77		

Equation (4) is a linear equation. We can solve it to obtain the exact solution $y = 3e^x - x^2 - 2x - 2$ (check this), so that $y(1) = 3e - 5 \approx 3.154$. Thus the Euler's method estimate was off by about twelve percent.† This is not surprising because we treated the derivative as a constant over intervals of length of 0.2 unit. The error that arises in this way is called **discretization error**, because the "discrete" function $f(x_n, y_n)$ was substituted

† $y(1) - y_5 \approx 3.15 - 2.77 = 0.38$ and $0.38/3.15 \approx 0.12 = 12\%$.

for the "continuously valued" function $f(x, y)$. It is true that if we reduce the step size h, then we can improve the accuracy of our answer, since, then, the "discretized" function $f(x_n, y_n)$ will be closer to the true value of $f(x, y)$ over the interval $[0, 1]$. This is illustrated in Figure 2 with $h = 0.2$ and $h = 0.1$. Indeed, carrying out similar calculations with $h = 0.1$ yields an approximation of $y(1)$ of 2.94, which is more accurate (an error of about seven percent).†

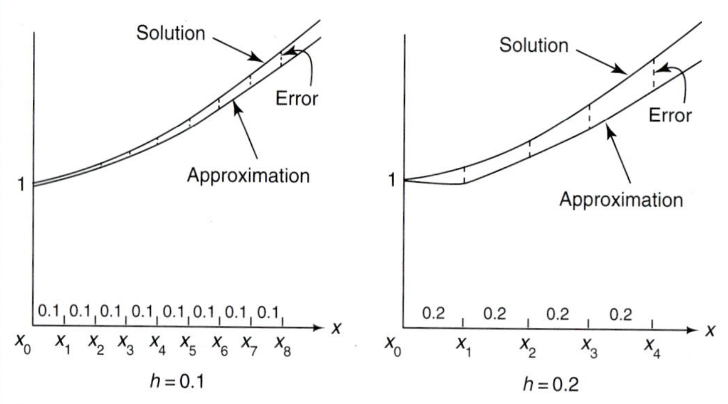

FIGURE 2
The error between the true solution and the Euler approximation

Usually, reducing step size improves accuracy. However, a warning must be attached to this. Reducing the step size obviously also increases the amount of work that must be done. Moreover, at every stage of the computation **round-off errors** are introduced. For example, in our calculations with $h = 0.2$, we rounded off the exact value 1.448 to the value 1.45 (correct to two decimal places). The rounded-off value was then used to calculate further values of y_n. It is not unusual for a computer solution of a more complicated differential equation to take several thousand individual computations, thus having several thousand round-off errors. In some problems the accumulated round-off error can be so large that the resulting computed solution will be sufficiently inaccurate to invalidate the result. Fortunately, this usually does not occur since round-off errors can be positive or negative and tend to cancel one another out. This statement is made under the assumption (usually true) that the average of the round-off errors is zero. In any event, it should be clear that reducing the step size, thereby increasing the number of computations, is a procedure that should be carried out carefully. In general, each problem has an optimal step size, and a smaller than optimal step size will yield a greater error due to accumulated round-off errors.

IMPROVED EULER METHOD

This method has better accuracy than Euler's method and so is more valuable for hand computation. It is based on the fact that an improvement will result if we average the slopes at the left and right endpoints of each interval, thereby reducing the difference between $f(x, y)$ and $f(x_n, y_n)$ in each interval of the form $x_n \leq x < x_{n+1}$ (see Figure 3).

---

† $y(1) - y_{10} \approx 3.15 - 2.94 = 0.21$ and $0.21/3.15 \approx 0.067 \approx 7\%$.

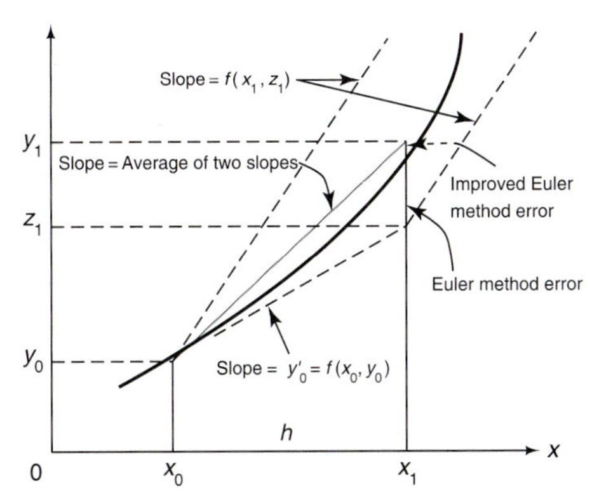

FIGURE 3
The errors in the two methods

This amounts to approximating the integral in equation (2) by the **trapezoidal rule**:

$$\int_{x_0}^{x_1} f(x, y)\, dx \approx \frac{h}{2}\{f(x_0, y_0) + f(x_1, y(x_1))\}.$$

Since $y(x_1)$ is not known, we replace it by the value found by Euler's method, which we call z_1; then equation (2) can be replaced by the system of equations

$$z_1 = y_0 + hf(x_0, y_0),$$

$$y_1 = y_0 + \frac{h}{2}[f(x_0, y_0) + f(x_1, z_1)].$$

This gives us the general procedure

IMPROVED EULER'S METHOD

$$z_{n+1} = y_n + hf(x_n, y_n),$$

$$y_{n+1} = y_n + \frac{h}{2}[f(x_n, y_n) + f(x_{n+1}, z_{n+1})]. \quad (5)$$

Using $x_0 = 0$ and $y_0 = 1$ in Example 1, we obtain, with $h = 0.2$,

$$z_1 = y_0 + hf(x_0, y_0) = 1 + 0.2(1 + 0^2) = 1.2,$$

$$y_1 = y_0 + \frac{h}{2}[f(x_0, y_0) + f(x_1, z_1)] = 1 + 0.1[(1 + 0^2) + 1.2 + 0.2^2]$$

$$= 1 + 0.1[2.24] = 1.224 \approx 1.22,$$

$$z_2 = y_1 + hf(x_1, y_1) = 1.22 + 0.2[1.22 + (0.2)^2] = 1.472 \approx 1.47,$$

$$y_2 = y_1 + \frac{h}{2}[f(x_1, y_1) + f(x_2, z_2)]$$

$$= 1.22 + 0.1[1.22 + (0.2)^2 + 1.47 + (0.4)^2] = 1.509 \approx 1.51,$$

and so on. Table 2 shows the approximating values of the solution of equation (4) used in Example 1. The error this time is less than one percent.

TABLE 2

x_n	y_n	$f(x_n, y_n) = y_n + x_n^2$	z_{n+1}	$f(x_{n+1}, z_{n+1}) = z_{n+1} + x_{n+1}^2$	y_{n+1}
0.0	1.0	1.0	1.20	1.24	1.22
0.2	1.22	1.26	1.47	1.63	1.51
0.4	1.51	1.67	1.84	2.20	1.90
0.6	1.90	2.26	2.35	2.99	2.43
0.8	2.43	3.07	3.04	4.04	3.14
1.0	3.14				

C and Fortran programs for implementing Euler's method are given in the problem set.

PROBLEMS 10.17

SELF-QUIZ

True-False

I. The only error that arises using Euler's method is discretization error.

II. Euler's method always underestimates the exact solution.

Solve Problems 1–10 by using

a. The Euler method with the indicated value of h;

b. The improved Euler method with the given value of h.

1. $\dfrac{dy}{dx} = x + y$, $y(0) = 1$. Find $y(1)$ with $h = 0.2$.

2. $\dfrac{dy}{dx} = x - y$, $y(1) = 2$. Find $y(3)$ with $h = 0.4$.

3. $\dfrac{dy}{dx} = \dfrac{x - y}{x + y}$, $y(2) = 1$. Find $y(1)$ with $h = -0.2$.

4. $\dfrac{dy}{dx} = \dfrac{y}{x} + \left(\dfrac{y}{x}\right)^2$, $y(1) = 1$. Find $y(2)$ with $h = 0.2$.

5. $\dfrac{dy}{dx} = x\sqrt{1 + y^2}$, $y(1) = 0$. Find $y(3)$ with $h = 0.4$.

***6.** $\dfrac{dy}{dx} = x\sqrt{1 - y^2}$, $y(1) = 0$. Find $y(2)$ with $h = 0.125$.

7. $\dfrac{dy}{dx} = \dfrac{y}{x} - \dfrac{5}{2}x^2y^3$, $y(1) = \dfrac{1}{\sqrt{2}}$. Find $y(2)$ with $h = 0.125$.

8. $\dfrac{dy}{dx} = \dfrac{-y}{x} + x^2y^2$, $y(1) = \dfrac{2}{9}$. Find $y(3)$ with $h = \frac{1}{3}$.

9. $\dfrac{dy}{dx} = ye^x$, $y(0) = 2$. Find $y(2)$ with $h = 0.2$.

10. $\dfrac{dy}{dx} = xe^y$, $y(0) = 0$. Find $y(1)$ with $h = 0.1$.

In Problems 11–20 use the improved Euler method to graph approximately the solution of the given initial value problem by plotting the points (x_k, y_k) over the indicated range, where $x_k = x_0 + kh$.

11. $y' = xy^2 + y^3$, $y(0) = 1$, $h = 0.02$, $0 \le x \le 0.1$

12. $y' = x + \sin(\pi y)$, $y(1) = 0$, $h = 0.2$, $1 \le x \le 2$

13. $y' = x + \cos(\pi y)$, $y(0) = 0$, $h = 0.4$, $0 \le x \le 2$

14. $y' = \cos(xy)$, $y(0) = 0$, $h = \pi/4$, $0 \le x \le \pi$

15. $y' = \sin(xy)$, $y(0) = 1$, $h = \pi/4$, $0 \le x \le 2\pi$

16. $y' = \sqrt{x^2 + y^2}$, $y(0) = 1$, $h = 0.5$, $0 \le x \le 5$

17. $y' = \sqrt{y^2 - x^2}$, $y(0) = 1$, $h = 0.1$, $0 \le x \le 1$

18. $y' = \sqrt{x + y^2}$, $y(0) = 1$, $h = 0.2$, $0 \le x \le 1$

19. $y' = \sqrt{x + y^2}$, $y(1) = 2$, $h = -0.2$, $0 \le x \le 1$

20. $y' = \sqrt{x^2 + y^2}$, $y(1) = 5$, $h = -0.2$, $0 \le x \le 1$

COMPUTER EXERCISES

We list below the algorithms in C and FORTRAN for applying the Euler method to the initial value problem in Example 1. The user can modify this program to apply it to other initial value problems.

C

```
/* DEULER.C : Euler Method. */

#include <stdio.h>

main()
{
    float x = 0.0;
    float y = 1.0;
    float xlast = 1.0;
    float h = 0.2;
    float dydx;
    printf( "   x      y/n" );
    while( x < xlast )
    {
        dydx = y + x * x;
        y = y + h * dydx;
        x = x + h;
        printf( "%f %f\n" , x , y );
    }
}
```

FORTRAN

```
      X=0.0
      Y=1.0
      XLAST=1.0
      H=0.2
      N=(XLAST-X)/H
      WRITE(*,'(/7X,A,12X,A)') 'X','Y'
      DO 11 I=0,N
         DYDX=Y+X*X
         X=X+H
         Y=Y+H*DYDX
         WRITE(*,'(1X,F10.4,1X,F14.6)') X,Y
11    CONTINUE
      END
```

In Problems 21–24 modify one of the programs above and solve the given initial value problem using Euler's method.

21. $y' = x + y^2$, $y(1) = -2$, $h = 0.01$. Find $y(2)$.

22. $y' = \sqrt{x^2 + y^2}$, $y(0) = 1$, $h = 0.1$. Find $y(5)$.

23. $y' = \sqrt{x + y^2}$, $y(1) = 2$, $h = 0.02$. Find $y(2)$.

24. $y' = ye^x$, $y(0) = 2$, $h = 0.02$. Find $y(2)$.

ANSWERS TO SELF-QUIZ

I. False **II.** False

CHAPTER 10 SUMMARY OUTLINE

■ **Free Fall** $h(t) = -\frac{g}{2}t^2 + v_0 t + h_0$, $v(t) = -gt + v_2$, $a(t) = -g$.

■ **Relative Rate of Growth** Let $y = f(x)$.

$$\text{relative rate of growth} = \frac{\text{actual rate of growth}}{\text{size of } f(x)} = \frac{f'(x)}{f(x)} = \frac{dy/dx}{y}$$

■ **Exponential Growth and Decay** The differential equation

$$\frac{dy}{dx} = \alpha y \qquad (*)$$

is called the differential equation of exponential growth or decay (growth if $\alpha > 0$ and decay if $\alpha < 0$). All solutions to $(*)$ are given by

$$y = ce^{\alpha x}$$

where c is an arbitrary real number.

■ **The Initial-Value Problem of Exponential Growth or Decay** If $P(t)$ is growing or declining continuously at a constant relative rate (so that $P'(t)/P(t) = \pm\alpha$) with initial size $P(0)$, then $P(t)$ is given by (with $\alpha > 0$)

$$P(t) = P(0)e^{\alpha t} \qquad \text{or} \qquad P(t) = P(0)e^{-\alpha t}$$

- **Newton's Law of Cooling** $T' = \alpha(T - T_s)$ where T is the temperature of an object and T_s is the temperature of the surrounding medium.
- **Compound Interest Formula: Compounding m times per year**

  The value of an investment $\$P$ compounded m times per year with an annual interest rate of r after t years is

$$A(t) = P\left(1 + \frac{r}{m}\right)^{mt}.$$

- **Compound Interest Formula: Continuous Compounding**

$$A(t) = Pe^{rt}$$

  is the value of an initial investment of $\$P$ after t years at an annual rate of r compounded continuously.
- **Order of a Differential Equation** The **order** of a differential equation is defined as the order of the highest derivative appearing in the equation.
- **Solution to an nth-Order Differential Equation** A **solution** to an nth-order differential equation is a function that is n times differentiable and that satisfies the differential equation.
- **Initial-Value Problem** An **initial-value problem** consists of an nth-order differential equation (of any order) together with n initial conditions that must be satisfied by the solution of the differential equation and its derivatives at the initial point.
- **Solution to an Initial-Value Problem** We define a **solution** to an nth-order initial-value problem as a function that is n times differentiable, satisfies the given differential equation, and satisfies the n given initial conditions.
- **Boundary Value Problem** A **boundary value problem** consists of a differential equation and a collection of values that must be satisfied by the solution of the differential equation or its derivatives at no fewer than two different points.
- **Existence-Uniqueness Theorem** Let $y' = f(x, y)$, $y(x_0) = y_0$, and suppose f and $\partial f/\partial y$ are continuous in the region $a < x < b$, $c < y < d$ containing (x_0, y_0). Then, a unique solution exists in some subinterval $|x - x_0| < h$ contained in $a < x < b$.
- **Euler's Method** $y_{n+1} = y_n + hf(x_n, y_n)$, $x_{n+1} = x_n + h$, where $y' = f(x, y)$, $y(0) = y_0$.
- **Improved Euler Method**

$$z_{n+1} = y_n + hf(x_n, y_n),$$

$$y_{n+1} = y_n + \frac{h}{2}[f(x_n, y_n) + f(x_{n+1}, z_{n+1})].$$

- **Separation of Variables** The differential equation $\frac{dy}{dx} = f(x, y)$ can be solved by the method of separation of variables if $f(x, y) = g(x)h(y)$. In this case, we can write

$$\int \frac{dy}{h(y)} = \int g(x)\,dx + C$$

- **Logistic Equation** The equation $\frac{dP}{dt} = P(\beta - \delta P)$ is called the **logistic equation**. The growth shown by this equation is called **logistic growth**.
- The solution of the logistic equation is

$$P(t) = \frac{\beta}{\delta + \left(\dfrac{\beta}{P(0)} - \delta\right)e^{-\beta t}}$$

- **Linear Differential Equation** An nth-order differential equation is **linear** if it can be written in the form

$$\frac{d^ny}{dx^n} + a_{n-1}(x)\frac{d^{n-1}y}{dx^{n-1}} + \cdots + a_1(x)\frac{dy}{dx} + a_0(x)y = f(x).$$

  Hence a first-order linear equation has the form

$$\frac{dy}{dx} + a(x)y = f(x) \tag{2}$$

  and a second-order linear equation can be written as

$$\frac{d^2y}{dx^2} + a(x)\frac{dy}{dx} + b(x)y = f(x).$$

- **Homogeneous and Nonhomogeneous Linear Differential Equation** If the function $f(x)$ is the zero function, the linear differential equation is said to be **homogeneous**. Otherwise, we say that the linear differential equation is **nonhomogeneous**.
- **Nonlinear Differential Equation** Any differential equation that cannot be written in one of the forms above is said to be **nonlinear**.
- The **total differential** dg of a function of two variables $g(x, y)$ is defined by

$$dg = \frac{\partial g}{\partial x}\,dx + \frac{\partial g}{\partial y}\,dy.$$

- **Integrating Factor and Solution to a First-Order Linear Differential Equation** The first-order equation (2) can be solved by multiplying both sides of the equation by the **integrating factor** $e^{\int a(x)\,dx}$. The solutions to (2) are

$$y = \left[\int f(x)e^{\int a(x)\,dx}\,dx + C\right]e^{-\int a(x)\,dx}$$

- **Bernoulli's Equation** The equation

$$\frac{dy}{dx} + a(x)y = f(x)y^n$$

is known as **Bernoulli's equation**. It can be transformed into a linear equation and then solved by making the substitution

$$z = y^{1-n}.$$

- **Exact Differential Equation** The differential equation

$$M(x, y)\, dx + N(x, y)\, dy = 0.$$

is **exact** if we can find a differentiable function $g(x, y)$ such that

$$\frac{\partial g}{\partial x} = M \quad \text{and} \quad \frac{\partial g}{\partial y} = N$$

- **Test for Exactness** The equation $M(x, y)\, dx + N(x, y)\, dy = 0$ is exact if and only if

$$\frac{\partial M}{\partial y} = \frac{\partial N}{\partial x}.$$

- **Some Facts about Simple Electric Circuits**

  a. An electromotive force E (volts), usually a battery or generator, drives an electric charge Q (coulombs) and produces a current I (amperes). The current is defined as the rate of flow of the charge, and we can write

$$I = \frac{dQ}{dt}.$$

  b. A resistor of resistance R (ohms) is a component of the circuit that opposes the current, dissipating the energy in the form of heat. It produces a drop in voltage given by **Ohm's law**:

$$E_R = RI.$$

  c. An inductor of inductance L (henrys) opposes any change in current by producing a voltage drop of

$$E_L = L\frac{dI}{dt}.$$

  d. A capacitor of capacitance C (farads) stores charge. In so doing, it resists the flow of further charge, causing a drop in the voltage of

$$E_C = \frac{Q}{C}.$$

  The quantities R, L, and C are usually constants associated with the particular component in the circuit; E may be a constant or a function of time. The fundamental principle guiding such circuits is given by **Kirchhoff's voltage law**:

  The algebraic sum of all voltage drops around a closed circuit is zero.

- **Homogeneous and Nonhomogeneous Linear Differential Equations** The most general second-order linear equation can be written

$$y''(x) + a(x)y'(x) + b(x)y(x) = f(x), \tag{1}$$

whereas the most general third-order linear equation can be written

$$y'''(x) + a(x)y''(x) + b(x)y'(x) + c(x)y(x) = f(x), \tag{2}$$

where $a(x)$, $b(x)$, $c(x)$, and $f(x)$ are functions of the independent variable x only. Equations (1) and (2) are special cases of the **general linear nth-order equation**:

$$y^{(n)}(x) + a_{n-1}(x)y^{(n-1)}(x) + \cdots + a_1(x)y'(x) + a_0(x)y(x) = f(x). \tag{3}$$

If the function $f(x)$ is identically zero, we say that equations (1), (2), and (3) are **homogeneous**. Otherwise, they are **nonhomogeneous**.

- **Constant and Variable Coefficients** If the coefficient functions $\boldsymbol{a_i(x)}$ in (3) are constants, the equation is said to have **constant coefficients**. Otherwise, it has **variable coefficients**.

- **Existence-Uniqueness Theorem for Linear Initial-Value Problems** Let $a_1(x), a_2(x), \ldots, a_n(x)$, and $f(x)$ be continuous functions on the interval $[x_1, x_2]$, and let $c_0, c_1, c_2, \ldots, c_{n-1}$ be n given constants. Then there exists a unique function $y(x)$ that satisfies the linear differential equation

$$y^{(n)}(x) + a_{n-1}(x)y^{(n-1)}(x) + a_{n-2}(x)y^{(n-2)}(x) + \cdots + a_0(x)y(x) = f(x)$$

on $[x_1, x_2]$ *and* the n initial conditions

$$y(x_0) = c_0, \quad y'(x_0) = c_1, \quad y''(x_0) = c_2, \quad \ldots, \quad y^{(n-1)}(x_0) = c_{n-1}$$

for some value x_0 in $[x_1, x_2]$.

- **Existence-Uniqueness Theorem for Second-Order, Linear Initial-Value Problems** If $a(x)$, $b(x)$, and $f(x)$ are continuous functions, then the equation

$$y''(x) + a(x)y'(x) + b(x)y(x) = f(x)$$

has a unique solution that satisfies the conditions

$$y(x_0) = y_0, \quad y'(x_0) = y_1$$

for any real numbers x_0, y_0, and y_1.

- **Linear Combination, Linear Independence, and Linear Dependence** Let y_1 and y_2 be any two functions. By a **linear combination** of y_1 and y_2 we mean a function $y(x)$ that can be written in the form

$$y(x) = c_1y_1(x) + c_2y_2(x),$$

for some constants c_1 and c_2. Two functions are **linearly independent** on an interval $[x_1, x_2]$ whenever the relation

$$c_1y_1(x) + c_2y_2(x) = 0,$$

for all x in $[x_1, x_2]$, implies that $c_1 = c_2 = 0$. Otherwise they are **linearly dependent.**

- Two functions are linearly dependent on the interval $[x_0, x_1]$ if and only if one of the functions is a constant multiple of the other.
- A **linear combination** of n functions is any function of the form

$$y(x) = c_1 y_1(x) + c_2 y_2(x) + \cdots + c_n y_n(x),$$

where $c_1, c_2, \ldots, c_n$ are constants. We say that the collection of functions $y_1(x), y_2(x), \ldots, y_n(x)$ is **linearly independent** on an interval $[x_1, x_2]$ if the equation

$$c_1 y_1(x) + c_2 y_2(x) + \cdots + c_n y_n(x) = 0 \tag{4}$$

is only true for all x in $[x_1, x_2]$ when $c_1 = c_2 = \cdots = c_n = 0$. Otherwise, we say that the collection of functions is **linearly dependent** on $[x_1, x_2]$.
- Every homogeneous linear second-order differential equation

$$y'' + a(x)y' + b(x)y = 0$$

has two linearly independent solutions.
- **Principle of Superposition for Homogeneous Equations** Let $y_1(x)$ and $y_2(x)$ be any two solutions of the homogeneous equation

$$y'' + a(x)y' + b(x)y = 0.$$

Then any linear combination of them is also a solution.
- An nth-order linear, homogeneous differential equation with continuous coefficients can have at most n linearly independent solutions.
- **General Solution to a Linear, Second-Order Equation** The general solution of (4) is given by the linear combination

$$y(x) = c_1 y_1(x) + c_2 y_2(x)$$

where c_1 and c_2 are arbitrary constants and $y_1(x)$ and $y_2(x)$ are linearly independent solutions of (4).
- **Wronskian** Let $y_1(x)$ and $y_2(x)$ be any two solutions to the differential equation

$$y'' + a(x)y' + b(x)y = 0.$$

The **Wronskian** of y_1 and y_2 is defined as

$$W(y_1, y_2)(x) = \begin{vmatrix} y_1(x) & y_2(x) \\ y_1'(x) & y_2'(x) \end{vmatrix} = y_1(x)y_2'(x) - y_1'(x)y_2(x).$$

- $W(y_1, y_2)(x) = ce^{-\int a(x)\,dx}$
- Two solutions $y_1(x)$ and $y_2(x)$ of equation (4) are linearly independent on $[x_0, x_1]$ if and only if $W(y_1, y_2)(x) \neq 0$.
- Let y_p and y_q be two solutions to the nonhomogeneous equation

$$y'' + a(x)y' + b(x)y = f(x) \tag{5}$$

Then their difference $y_p - y_q$ is a solution to the associated homogeneous equation

$$y'' + a(x)y' + b(x)y = 0 \tag{6}$$

- **General Solution to a Linear, Nonhomogeneous Second-Order Equation** Let y_p be a solution to (5) and let y_1 and y_2 be two linearly independent solutions to (6). Then the general solution to (5) is given by

$$y(x) = c_1 y_1(x) + c_2 y_2(x) + y_p(x)$$

where c_1 and c_2 are arbitrary constants.
- **Principle of Superposition for Nonhomogeneous Equations** Suppose that y_f is a solution to

$$y'' + a(x)y' + b(x)y = f(x)$$

and y_g is a solution to

$$y'' + a(x)y' + b(x)y = g(x)$$

Then $y_f + y_g$ is a solution to

$$y'' + a(x)y' + b(x)y = f(x) + g(x)$$

- **Finding a Second Solution to a Homogeneous Equation when One Is Given** If y_1 is a solution to (6), then a second, linearly independent solution is given by

$$y_2(x) = y_1(x) \int \frac{e^{-\int a(x)dx}}{y_1{}^2(x)}\,dx$$

- The **characteristic equation** of the equation

$$y'' + ay' + by = 0$$

is $\lambda^2 + a\lambda + b = 0$. Here a and b are constants. (7)
- **The General Solution to a Second-Order, Linear Differential Equation with Constant Coefficients**

  i. Let λ_1 and λ_2 be unequal real roots of (7); then the general solution to (6) is

  $$y = c_1 e^{\lambda_1 x} + c_2 e^{\lambda_2 x}$$

  where c_1 and c_2 are arbitrary constants.

  ii. If $\lambda_1 = \lambda_2$, then the general solution to (6) is

  $$y = e^{\lambda x}(c_1 + c_2 x)$$

  iii. If $\lambda_1, \lambda_2 = \alpha \pm i\beta$, then the general solution is

  $$y = e^{\alpha x}(c_1 \cos \beta x + c_2 \sin \beta x)$$

- **Harmonic Identities**

$$a \cos \omega t + b \sin \omega t = A \cos(\omega t - \delta)$$

$$a \cos \omega t + b \sin \omega t = A \sin(\omega t + \phi) \tag{8}$$

Here $A = \sqrt{a^2 + b^2}$, δ is the unique number such that $\cos \delta = \dfrac{a}{\sqrt{a^2 + b^2}} = \dfrac{a}{A}$ and $\sin \delta = \dfrac{b}{A}$, so $\tan \delta = \dfrac{b}{a}$. ϕ

is the unique number such that $\cos\phi = \frac{b}{A}$ and $\sin\phi = \frac{a}{A}$, so $\tan\phi = \frac{a}{b}$.

■ **Variation of Parameters Method** We seek a particular solution to the nonhomogeneous equation

$$y'' + ay' + by = f(x) \tag{9}$$

with the associated homogeneous equation

$$y'' + ay' + by = 0. \tag{10}$$

To find the solution to (9):

Step 1: Write $y(x) = c_1(x)y_1(x) + c_2(x)y_2(x)$ where y_1 and y_2 are linearly independent solutions to (10).

Step 2: Obtain the basic equations

$$y_1c_1' + y_2c_2' = 0$$
$$y_1'c_1' + y_2'c_2' = f(x)$$

Step 3: Solve the basic equations:

$$c_1'(x) = \frac{-f(x)y_2(x)}{y_1(x)y_2'(x) - y_1'(x)y_2(x)} = \frac{-f(x)y_2(x)}{W(y_1, y_2)(x)},$$

$$c_2'(x) = \frac{f(x)y_1(x)}{y_1(x)y_2'(x) - y_1'(x)y_2(x)} = \frac{f(x)y_1(x)}{W(y_1, y_2)(x)}.$$

Step 4: Integrate c_1' and c_2', if possible, to obtain $y = c_1y_1 + c_2y_2$.

■ **Method of Undetermined Coefficients** Let

$$y'' + ay' + by = f(x), \tag{11}$$

where a and b are constants, and every term of the function $f(x)$ is a product of one or more of the following:

i. A polynomial in x, (a constant k is a polynomial of degree zero)

ii. An exponential function $e^{\alpha x}$,

iii. $\cos\beta x$, $\sin\beta x$.

The method of undetermined coefficients assumes that each term in the particular solution to equation (11) is of the same "form" as each term in $f(x)$, but has polynomial terms with undetermined coefficients. Substitute the "guessed" particular solution y_p into (11) to "determine" the unknown coefficients.

■ **Modifications of the Method of Undetermined Coefficients** If any term of the guessed solution $y_p(x)$ is a solution of the homogeneous equation, multiply $y_p(x)$ by x repeatedly until no term of the product $x^k y_p(x)$ is a solution of the homogeneous equation. Then use the product $x^k y_p(x)$ to solve equation (11).

■ **Euler Equation** An equation of the form

$$x^2y'' + axy' + by = f(x), \qquad x \neq 0 \tag{12}$$

is called an **Euler equation**.

■ **Characteristic Equation for an Euler Equation**

$$\lambda^2 + (a-1)\lambda + b = 0 \tag{13}$$

■ **General Solution of an Euler Equation** Let λ_1 and λ_2 be the roots of (13).

Case i: If $\lambda_1 \neq \lambda_2$ are real, the general solution to (12) is

$$y(x) = c_1x^{\lambda_1} + c_2x^{\lambda_2}, \qquad x \neq 0.$$

Case ii: If $\lambda_1 = \lambda_2$, the general solution to (12) is

$$y(x) = x^{\lambda}(c_1 + c_2\ln|x|).$$

Case iii: If $\lambda_1, \lambda_2 = \alpha \pm i\beta$, then the general solution to (12) is

$$y(x) = x^{\alpha}[c_1\cos(\beta\ln|x|) + c_2\sin(\beta\ln|x|)].$$

CHAPTER 10 REVIEW EXERCISES

In Exercises 1–37 find the general solution to the given differential equation. If initial conditions are given, find the unique solution to the initial value problem.

1. $\frac{dy}{dx} = 3x$

2. $\frac{dy}{dx} = e^{x-y}$, $y(0) = 4$

3. $\frac{dx}{dt} = e^x\cos t$, $x(0) = 3$

4. $\frac{dx}{dt} = x^{13}t^{11}$

5. $\frac{dy}{dx} + 3y = \cos x$, $y(0) = 1$

6. $\frac{dx}{dt} + 3x = \frac{1}{1+e^{3t}}$

7. $\frac{dx}{dt} = 3x + t^3e^{3t}$, $x(1) = 2$

8. $\frac{dy}{dx} + y\cot x = \sin x$, $y\left(\frac{\pi}{6}\right) = \frac{1}{2}$

9. $(y - e^y\sec^2 x)\,dx + (x - e^y\tan x)\,dy = 0$

10. $(2x^2y^3 - y^2)\,dx + (x^3y^2 - x)\,dy = 0$

11. $y'' - 5y' + 4y = 0$

12. $y'' - 9y' + 14y = 0,\ y(0) = 2,\ y'(0) = 1$
13. $y'' - 9y = 0$
14. $y'' + 9y = 0$
15. $y'' + 6y' + 9y = 0$
16. $y'' + 8y' + 16y = 0,\ y(0) = -1,\ y'(0) = 3$
17. $y'' - 2y' + 2y = 0,\ y(0) = 0,\ y'(0) = 1$
18. $y'' + 8y' = 0,\ y(0) = 2,\ y'(0) = -3$
19. $y'' + 4y = 2 \sin x$
20. $y'' + y' - 12y = 4e^{2x},\ y(0) = 1,\ y'(0) = -1$
21. $y'' + y' + y = e^{-x/2} \sin(\sqrt{3}/2)x$
22. $y'' + 4y = 6x \cos 2x,\ y(0) = 1,\ y'(0) = 0$
23. $y'' + y = x^3 - x$
24. $y'' - 2y' + y = e^{-x}$
25. $y'' - 8y' + 16y = e^{4x},\ y(0) = 3,\ y'(0) = 1$
26. $y'' + y = x + e^x + \sin x$
27. $y'' - 2y' + 3y = e^x \cos \sqrt{2}x$
28. $y'' + 16y = \cos 4x + x^2 - 3$
29. $y'' + y = 0,\ y(0) = y'(0) = 0$
30. $y'' + y = \sec x,\ 0 < x < \pi/2$
31. $y'' - 2y' + y = 2e^x/x^3$
32. $y'' + 4y' + 4y = e^{-2x}/x^2$
33. $x^2y'' - 2xy' + 3y = 0,\ x > 0$
34. $y''' - y'' + 4y' - 4y = 0$
35. $y''' + 3y'' + 3y' + y = 0$
36. $y^{(4)} - 10y'' + 9y = 0$
37. $y^{(4)} + 18y'' + 81y = 0$
38. Suppose a constant capacitor is connected in series to an emf whose voltage is a sine wave. Show that the current is 90° out of phase with the voltage.
39. Repeat Exercise 38 with the capacitor replaced by an inductor. What can you say in this case?
40. A spring, fixed at its upper end, supports a 10-kg mass that stretches the spring 60 cm. Find the equation of motion of the mass if it is drawn to a position 10 cm below its equilibrium position and released with an initial velocity of 5 cm/sec upward.
41. What are the period, frequency, and amplitude of the motion of the mass in Exercise 40?
42. Find the equation of motion of the mass of Exercise 40 if it is subjected to damping forces having the damping constant $\mu = 3$.
43. In Exercise 40, for what minimum damping constant would the mass fail to oscillate about its equilibrium?
44. For what value of ω will the external force of $10 \sin \omega t$ N produce resonance in the spring of Exercise 43?

In Exercises 45–48 solve the given initial value problem and then obtain an approximate solution at the indicated value of x using the improved Euler method.

45. $\dfrac{dy}{dx} = \dfrac{e^x}{y},\ y(0) = 2$. Find $y(3)$ with $h = \frac{1}{2}$.
46. $\dfrac{dy}{dx} = \dfrac{e^y}{x},\ y(1) = 0$. Find $y(\frac{1}{2})$ with $h = -0.1$.
47. $\dfrac{dy}{dx} - \dfrac{y}{\sqrt{1 + x^2}},\ y(0) = 1$. Find $y(3)$ with $h = \frac{1}{2}$.
48. $xy\dfrac{dy}{dx} = y^2 - x^2,\ y(1) = 2$. Find $y(3)$ with $h = \frac{1}{2}$.

CHAPTER 11

SYSTEMS OF DIFFERENTIAL EQUATIONS

In this chapter we discuss systems of first-order differential equations. We shall show how higher-order differential equations can be written as first-order systems and how all linear first-order systems can be written in matrix notation. We shall see that a great deal of information can be obtained by determining the eigenvalues and eigenvectors of this matrix.

The prerequisites for this chapter are:

For Sections 11.1–11.6: Sections 10.1–10.4, 10.7, 10.9–10.12, 10.16, and 7.1.

For Sections 11.7–11.11: all of the above plus Chapter 6 and Sections 8.1–8.5, 8.12, and 8.14.

11.1 THE METHOD OF ELIMINATION FOR LINEAR SYSTEMS WITH CONSTANT COEFFICIENTS

In Chapter 10 we discussed the problem of finding the solution to a single linear differential equation. In this section we will discuss an elementary method for solving a system of simultaneous first-order linear differential equations by converting the system into a single higher-order linear differential equation that may then be solved by the methods we have already seen. Systems of simultaneous differential equations arise in problems involving more than one unknown function, each of which is a function of a single independent variable (often time). For consistency throughout the remaining sections of this chapter, we denote the independent variable by t and the dependent variables by $x(t)$ and $y(t)$ or by the subscripted letters $x_1(t), x_2(t), \ldots, x_n(t)$.

Although some familiarity with the elementary properties of determinants will be helpful, we will not use matrix methods in solving systems of simultaneous linear differential equations in this section. Here we shall give some examples of how simple systems arise and shall describe an elementary procedure for finding their solution.

EXAMPLE 1

A LINEAR SYSTEM ARISING IN A CHEMICAL FLOW PROBLEM

Suppose that a chemical solution flows from one container into a second container at a rate proportional to the volume of solution in the first vessel. It flows out from the second container at a constant rate. Let $x(t)$ and $y(t)$ denote the volumes of solution in the first and second containers, respectively, at time t. (The containers may be, for example, cells, in which case we are describing a diffusion process across a cell wall.) To establish the necessary equations, we note that the change in volume equals the difference between input and output in each container. The change in volume is the derivative of volume with respect to time. Since no chemical is flowing into the first container, the change in its volume equals the output:

$$\frac{dx}{dt} = -c_1 x$$

where c_1 is a positive constant of proportionality. The amount of solution $c_1 x$ flowing out of the first container is the input of the second container. Let c_2 be the constant output of the second container. Then the change in volume in the second container equals the difference between its input and output:

$$\frac{dy}{dt} = c_1 x - c_2.$$

Thus we can describe the flow of a solution by means of two differential equations. Since more than one differential equation is involved, we say that we have obtained a **system of differential equations**:

$$\begin{aligned} \frac{dx}{dt} &= -c_1 x, \\ \frac{dy}{dt} &= c_1 x - c_2 \end{aligned} \qquad (1)$$

where c_1 and c_2 are positive constants. By a **solution** of the system (1) we shall mean a pair of functions $x(t)$, $y(t)$ that simultaneously satisfy both equations in (1). Since the first equation contains only x and t, it may be solved for x as in Chapter 10. The result is then substituted into the second equation, permitting its solution also. (When more complicated systems arise, this successive solution will usually not be possible.) If we denote the initial volumes in the two containers by $x(0)$ and $y(0)$, respectively, we see that the first equation has the solution

$$x(t) = x(0)e^{-c_1 t}. \qquad (2)$$

Substituting equation (2) into the second equation of (1), we obtain

$$\frac{dy}{dt} = c_1 x(0)e^{-c_1 t} - c_2,$$

which, upon integration from 0 to t, yields the solution

$$y(t) = y(0) + x(0)(1 - e^{-c_1 t}) - c_2 t. \qquad (3)$$

Equations (2) and (3) together constitute the unique solution of system (1) that satisfies the given initial conditions.

EXAMPLE 2

SOLVING A FEEDBACK SYSTEM BY THE METHOD OF ELIMINATION

Let tank X contain 100 gallons of brine in which 100 pounds of salt is dissolved and tank Y contain 100 gallons of water. Suppose water flows into tank X at the rate of 2 gallons per minute, and the mixture flows from tank X into tank Y at 3 gallons per minute. One gallon is pumped from Y back to X (establishing **feedback**) while 2 gallons are flushed away. Find the amount of salt in each tank at any time t (see Figure 1).

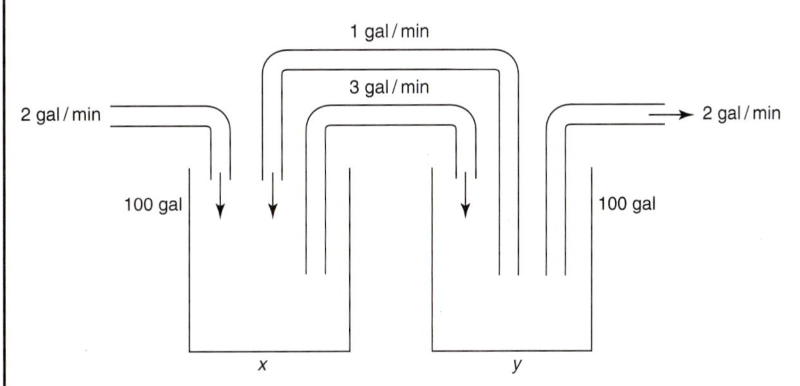

FIGURE 1
Two gallons enter and leave the system each minute

SOLUTION: If we let $x(t)$ and $y(t)$ represent the number of pounds of salt in tanks X and Y at time t and note that the change in weight equals the difference between input and output, we can again derive a system of linear first-order equations. Tanks X and Y initially contain $x(0) = 100$ and $y(0) = 0$ pounds of salt, respectively, at time $t = 0$. The quantities $x/100$ and $y/100$ are, respectively, the amounts of salt contained in each gallon of water taken from tanks X and Y at time t. Three gallons are being removed from tank X and added to tank Y, while only one of the three gallons removed from tank Y is put in tank X. Thus we have the system

$$\begin{aligned} \frac{dx}{dt} &= -3\frac{x}{100} + \frac{y}{100}, \qquad x(0) = 100, \\ \frac{dy}{dt} &= 3\frac{x}{100} - 3\frac{y}{100}, \qquad y(0) = 0. \end{aligned} \tag{4}$$

Since both equations in the system (4) involve *both* dependent variables, we cannot immediately solve for one of the variables, as we did in Example 1. Instead, we shall use differentiation to eliminate one of the dependent variables. Suppose we begin by solving the second equation for x in terms of the dependent variable y and its derivative:

$$x = y + \frac{100}{3}\frac{dy}{dt}. \tag{5}$$

Differentiating equation (5) and equating the right-hand side to the right-hand side of the first equation in system (4), we have

$$\frac{-3x}{100} + \frac{y}{100} = \frac{dx}{dt} = \frac{dy}{dt} + \frac{100}{3}\frac{d^2y}{dt^2}. \tag{6}$$

Replacing the x-term on the left-hand side of equation (6) with equation (5) produces the second-order linear equation

$$\frac{100}{3}\frac{d^2y}{dt^2} + 2\frac{dy}{dt} + \frac{2y}{100} = 0. \tag{7}$$

The initial conditions for equation (7) are obtained directly from system (4), since $y(0) = 0$ and

$$y'(0) = 3\frac{x(0)}{100} - 3\frac{y(0)}{100} = 3. \tag{8}$$

Multiplying both sides of equation (7) by $\frac{3}{100}$, we have the initial value problem

$$y'' + \frac{6}{100}y' + \frac{6}{(100)^2}y = 0, \qquad y(0) = 0, \qquad y'(0) = 3. \tag{9}$$

The characteristic equation for (9) has the roots

$$\lambda_1 = \frac{-3 + \sqrt{3}}{100}, \qquad \lambda_2 = \frac{-3 - \sqrt{3}}{100},$$

so that the general solution is

$$y(t) = c_1 e^{[(-3+\sqrt{3})t]/100} + c_2 e^{[(-3-\sqrt{3})t]/100}.$$

Using the initial conditions, we obtain the simultaneous equations

$$\begin{aligned} c_1 + \qquad\qquad c_2 &= 0, \\ \frac{-3 + \sqrt{3}}{100}c_1 - \frac{3 + \sqrt{3}}{100}c_2 &= 3. \end{aligned}$$

These have the unique solution $c_1 = -c_2 = 50\sqrt{3}$. Hence

$$y(t) = 50\sqrt{3}\,\{e^{[(-3+\sqrt{3})t]/100} - e^{[(-3-\sqrt{3})t]/100}\}$$

and substituting this function into the right-hand side of equation (5), we obtain

$$x(t) = 50\{e^{[(-3+\sqrt{3})t]/100} + e^{[(-3-\sqrt{3})t]/100}\}.$$

As is evident from the problem, the amounts of salt in the two tanks approach zero as time tends to infinity.

The technique we have used in solving Example 2 is called the **method of elimination**, since all but one of the dependent variables are eliminated by repeated differentiation. The method is quite elementary but requires many calculations. In Section 11.3 we will introduce a direct way of obtaining the solution without doing the elimination procedure. Nevertheless, because of its simplicity, elimination can be a very useful tool.

EXAMPLE 3 SOLVING AN INITIAL-VALUE SYSTEM BY ELIMINATION

Solve the system

$$\begin{aligned} x' &= x + y, & x(0) &= 1, \\ y' &= -3x - y, & y(0) &= 0. \end{aligned} \tag{10}$$

SOLUTION: Differentiating the first equation and substituting from the second equation for y', we have

$$x'' = x' + (-3x - y). \tag{11}$$

Solving the first equation of system (10) for y and substituting that expression in equation (11), we obtain

$$x'' + 2x = 0.$$

Thus

$$x(t) = c_1 \cos \sqrt{2}t + c_2 \sin \sqrt{2}t.$$

Then, according to the first equation of system (10),

$$\begin{aligned} y(t) = x'(t) - x(t) &= -\sqrt{2}c_1 \sin \sqrt{2}t + \sqrt{2}c_2 \cos \sqrt{2}t - c_1 \cos \sqrt{2}t \\ &\quad - c_2 \sin \sqrt{2}t \\ &= (c_2 \sqrt{2} - c_1) \cos \sqrt{2}t - (\sqrt{2}c_1 + c_2) \sin \sqrt{2}t. \end{aligned}$$

Using the initial conditions, we find that

$$x(0) = c_1 = 1, \qquad y(0) = c_2\sqrt{2} - c_1 = 0, \qquad \text{or} \qquad c_2 = 1/\sqrt{2}.$$

Therefore, the unique solution of system (10) is given by the pair of functions

$$x(t) = \cos \sqrt{2}t + \frac{1}{\sqrt{2}} \sin \sqrt{2}t,$$

$$y(t) = -\left(\sqrt{2} + \frac{1}{\sqrt{2}}\right) \sin \sqrt{2}t = -\frac{3}{\sqrt{2}} \sin \sqrt{2}t.$$

EXAMPLE 4 SOLVING A NONHOMOGENEOUS SYSTEM

Solve the following system:

$$\begin{aligned} x' &= 4x - y + t^2 \\ y' &= x + 2y + 3t \end{aligned} \tag{12}$$

SOLUTION: Proceeding as before, we obtain

$$\begin{aligned} x'' = 4x' - y' + 2t &= 4x' - \overbrace{(x + 2y + 3t)}^{y'} + 2t \\ &= 4x' - x - 2y - t \end{aligned}$$

From the first equation of (12), $-y = x' - 4x - t^2$, so $-2y = 2x' - 8x - 2t^2$ and

$$x'' = 4x' - x + (2x' - 8x - 2t^2) - t,$$

or, simplifying,

$$x'' - 6x' + 9x = -2t^2 - t. \tag{13}$$

The associated homogeneous equation is

$$x'' - 6x' + 9x = 0$$

with characteristic equation

$$0 = \lambda^2 - 6\lambda + 9 = (\lambda - 3)^2.$$

Thus the solution to the homogeneous equation is $x(t) = c_1e^{3t} + c_2te^{3t}$. Using the method of undetermined coefficients (Section 10.12), we obtain the particular solution

$$-\frac{2}{9}t^2 - \frac{11}{27}t - \frac{2}{9}.$$

Hence the general solution to equation (13) is

$$x(t) = c_1e^{3t} + c_2te^{3t} - \frac{2}{9}t^2 - \frac{11}{27}t - \frac{2}{9}.$$

Also,

$$x'(t) = 3c_1e^{3t} + 3c_2te^{3t} + c_2e^{3t} - \frac{4}{9}t - \frac{11}{27}.$$

From the first equation in (12),

$$y(t) = 4x - x' + t^2,$$

and we obtain

$$y(t) = c_1e^{3t} + c_2te^{3t} - c_2e^{3t} + \frac{1}{9}t^2 - \frac{32}{27}t - \frac{13}{27}.$$

The method illustrated in the last three examples can easily be generalized to apply to linear systems with three or more equations. *A linear system of n first-order equations usually reduces to an nth-order linear differential equation*, because it generally requires one differentiation to eliminate each variable $x_2, \ldots, x_n$ from the system.

EXAMPLE 5 A MASS–SPRING SYSTEM LEADS TO A LINEAR FIRST-ORDER SYSTEM

As a fifth example, we consider the mass–spring system of Figure 2, which is a direct generalization of the system described in Section 10.14. In this example we have two masses suspended by springs in series with spring constants k_1 and k_2 (see page 781). If the vertical displacements from equilibrium of the two masses are denoted by $x_1(t)$ and $x_2(t)$, respectively, then using assumptions (i) and (ii) (Hooke's law) of Section 10.14, we find that the net forces acting on the two masses are given by

$$F_1 = -k_1x_1 + k_2(x_2 - x_1),$$
$$F_2 = -k_2(x_2 - x_1).$$

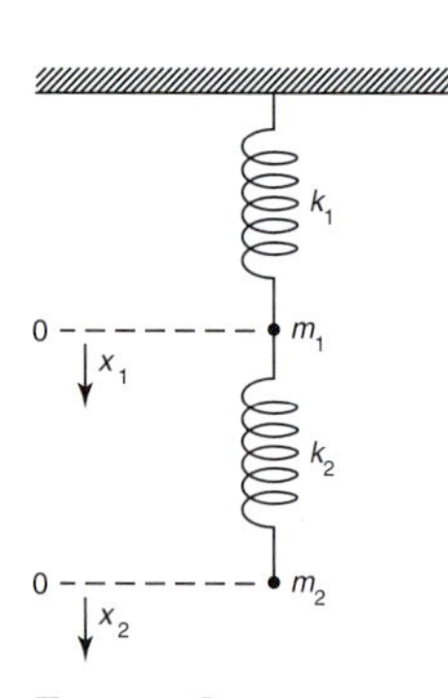

FIGURE 2
Two masses are suspended in series on two springs

Here the positive direction is downward. The first spring is compressed when $x_1 < 0$ and the second spring is compressed with $x_1 > x_2$. The equations of motion are

$$m_1\frac{d^2x_1}{dt^2} = -k_1x_1 + k_2(x_2 - x_1) = -(k_1 + k_2)x_1 + k_2x_2,$$

$$m_2\frac{d^2x_2}{dt^2} = -k_2(x_2 - x_1) = k_2x_1 - k_2x_2, \tag{14}$$

which constitute a system of two second-order linear differential equations with constant coefficients.

We will now show how linear differential equations (and systems) of *any* order can be converted, by the introduction of new variables, into a system of first-order differential equations. This concept is very important, since it means that the study of first-order linear systems provides a unified theory for all linear differential equations and systems. From a practical point of view it means that once we know how to solve first-order linear systems with constant coefficients, we will be able to solve any constant-coefficient linear differential equation or system.

To rewrite system (14) as a first-order system, we define the new variables $x_3 = x_1'$ and $x_4 = x_2'$. Then $x_3' = x_1''$, $x_4' = x_2''$ and (14) can be expressed as the system of four first-order equations

$$\begin{aligned} x_1' &= x_3, \\ x_2' &= x_4, \\ x_3' &= -\left(\frac{k_1 + k_2}{m_1}\right)x_1 + \left(\frac{k_2}{m_1}\right)x_2, \\ x_4' &= \left(\frac{k_2}{m_2}\right)x_1 - \left(\frac{k_2}{m_2}\right)x_2. \end{aligned} \tag{15}$$

If we wish, we can now use the method of elimination to reduce system (15) to a single fourth-order linear differential equation that can be solved by techniques of Section 10.16.

THEOREM 1

The linear nth-order differential equation

$$x^{(n)} + a_1(t)x^{(n-1)} + a_2(t)x^{(n-2)} + \cdots + a_{n-1}(t)x' + a_n(t)x = f(t) \tag{16}$$

can be rewritten as a system of n first-order linear equations.

PROOF: Define $x_1 = x$, $x_2 = x'$, $x_3 = x''$, . . . , $x_n = x^{(n-1)}$. Then we have

$$\begin{aligned} x_1' &= x_2, \\ x_2' &= x_3, \\ &\vdots \\ x_{n-1}' &= x_n, \\ x_n' &= -a_n x_1 - a_{n-1}x_2 - \cdots - a_1 x_n + f. \end{aligned} \tag{17}$$

■

In some cases, Theorem 1 can be extended to nonlinear differential equations (see Problem 28).

Suppose that n initial conditions are specified for the nth-order equation (16):

$$x(t_0) = c_1, \qquad x'(t_0) = c_2, \qquad \ldots, \qquad x^{(n-1)}(t_0) = c_n$$

These initial conditions can be immediately transformed into an initial condition for system (17):

$$x_1(t_0) = c_1, \qquad x_2(t_0) = c_2, \qquad \ldots, \qquad x_n(t_0) = c_n$$

EXAMPLE 6 WRITING A THIRD-ORDER EQUATION AS A FIRST-ORDER SYSTEM

Write the following initial value problem as a first-order system:

$$t^3x''' + 4t^2x'' - 8tx' + 8x = 0, \qquad x(2) = 3, \qquad x'(2) = -6, \qquad x''(2) = 14.$$

SOLUTION: Defining $x_1 = x$, $x_2 = x'$, $x_3 = x''$, we obtain the system

$$x_1' = x_2,$$
$$x_2' = x_3,$$
$$x_3' = \frac{-8}{t^3}x_1 + \frac{8}{t^2}x_2 - \frac{4}{t}x_3,$$

with the initial condition $x_1(2) = 3$, $x_2(2) = -6$, $x_3(2) = 14$.

PROBLEMS 11.1

SELF-QUIZ

Multiple Choice

I. Which of the following systems are linear?

a. $x' = 2x$
$y' = xy$

b. $x' = 2x + 3y + \sin t$
$y' = -3x + 5y - e^t$

c. $x' = (\cos t)\,x + y$
$y' = 3x - ye^t$

d. $x' = e^x + y$
$y' = y - x$

II. If $x'' + 2x' - x = 0$ is transformed into a first-order system, then which system results?

a. $x' = y$
$y' = -x + 2y$

b. $x' = y$
$y' = 2x - y$

c. $x' = y$
$y' = x - 2y$

d. $x' = y$
$y' = -x + 2y$

III. If $x'' - 2tx' + e^t x = \cos t$ is transformed into a first-order system, then which system results?

a. $x' = 2ty$
$y' = e^t x + \cos t$

b. $x' = 2ty - e^t x$
$y' = x$

c. $x' = y$
$y' = e^t x - 2ty + \cos t$

d. $x' = y$
$y' = -e^t x + 2ty + \cos t$

In Problems 1–9 find the general solution of each system of equations. When initial conditions are given, find the unique solution.

1. $x' = x + 2y,$
$y' = 3x + 2y$

2. $x' = x + 2y + t - 1,\ x(0) = 0,$
$y' = 3x + 2y - 5t - 2,\ y(0) = 4$

3. $x' = -4x - y,$
$y' = x - 2y$

4. $x' = x + y,\ x(0) = 1,$
$y' = y,\ y(0) = 0$

5. $x' = 8x - y,$
$y' = 4x + 12y$

6. $x' = 2x + y + 3e^{2t},$
$y' = -4x + 2y + te^{2t}$

7. $x' = 3x + 3y + t,$
$y' = -x - y + 1$

8. $x' = 4x + y,\ x(\pi/4) = 0,$
$y' = -8x + 8y,\ y(\pi/4) = 1$

9. $x' = 12x - 17y,$
$y' = 4x - 4y$

10. By elimination, find a solution to the following nonlinear system:

$$x' = x + \sin x \cos x + 2y,$$
$$y' = (x + \sin x \cos x + 2y) \sin^2 x + x$$

In Problems 11–17 transform each equation into a system of first-order equations.

11. $x'' + 2x' + 3x = 0$

12. $x'' - 6tx' + 3t^3x = \cos t$

13. $x''' - x'' + (x')^2 - x^3 = t$

14. $x^{(4)} - \cos x = t$

15. $x''' + xx'' - x'x^4 = \sin t$

16. $xx'x''x''' = t^5$

17. $x''' - 3x'' + 4x' - x = 0$

18. A mass m moves in xyz-space according to the

following equations of motion:

$$mx'' = f(t, x, y, z),$$
$$my'' = g(t, x, y, z),$$
$$mz'' = h(t, x, y, z).$$

Transform these equations into a system of six first-order equations.

19. Consider the uncoupled system

$$x_1' = x_1, \qquad x_2' = x_2.$$

a. What is the general solution of this system?

b. Show that there is no second-order equation equivalent to this system. [*Hint:* Show that any second-order equation has solutions that are not solutions of this system.] This shows that first-order systems are more general than higher-order equations in the sense that any of the latter can be written as a first-order system, but not vice versa.

Use the method of elimination to solve the systems in Problems 20 and 21.

20. $x_1' = x_1,$
$x_2' = 2x_1 + x_2 - 2x_3,$
$x_3' = 3x_1 + 2x_2 + x_3$

21. $x_1' = x_1 + x_2 + x_3,$
$x_2' = 2x_1 + x_2 - x_3,$
$x_3' = -8x_1 - 5x_2 - 3x_3$

22. In Example 2, when does tank Y contain a maximum amount of salt? How much salt is in tank Y at that time?

23. Suppose in Example 2 that the rate of flow from tank Y to tank X is 2 gallons (gal) per minute (instead of one) and all other facts are unchanged. Find the equations for the amount of salt in each tank at any time t.

24. Tank X contains 500 gal of brine in which 500 lb of salt are dissolved. Tank Y contains 500 gal of water. Water flows into tank X at the rate of 30 gal/min, and the mixture flows into Y at the rate of 40 gal/min. From Y the solution is pumped back into X at the rate of 10 gal/min and into a third tank at the rate of 30 gal/min. Find the maximum amount of salt in Y. When does this concentration occur?

25. Suppose in Problem 24 that tank X contains 1000 gal of brine. Solve the problem, given that all other conditions are unchanged.

26. Consider the mass–spring system illustrated in Figure 3. Here three masses are suspended in series by three springs with spring constants k_1, k_2, and k_3, respectively. Formulate a system of second-order differential equations that describes this system.

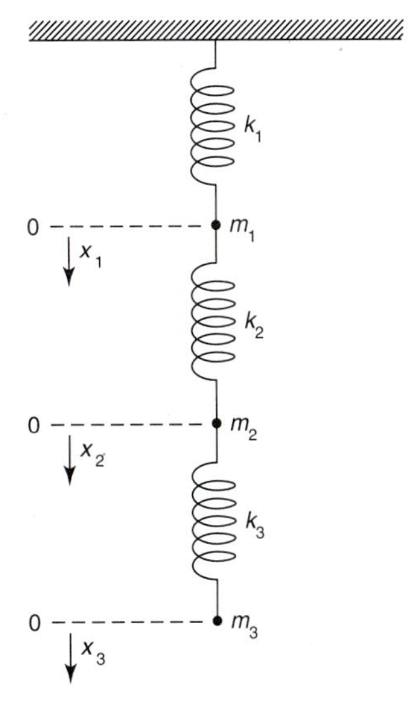

FIGURE 3

27. Find a single fourth-order linear differential equation in terms of the dependent variable x_1 for system (14). Find a solution to the system if $m_1 = 1$ kg, $m_2 = 2$ kg, $k_1 = 5$ N/m and $k_2 =$ 4 N/m.

***28.** Show that the differential equation

$$x^{(n)} = g(t, x, x', \ldots, x^{(n-1)})$$

can be transformed into a system of n first-order equations.

***29.** In a study concerning the distribution of radioactive potassium ^{42}K between red blood cells and the plasma of human blood, C. W. Sheppard and W. R. Martin [*J. Gen. Physiol.* **33**, 703–722 (1950)] added ^{42}K to freshly drawn blood. They discovered that although the total amount of potassium (stable and radioactive) in the red cells and in the plasma remained practically constant during the experiment, the radioactivity was gradually transferred from the plasma to the red cells. Thus the behavior of the radioactivity is that of a linear closed two-compartment system. If 30.1% of the radioactive material is transferred from the plasma to the cells each hour, while 1.7% is transferred back, and the initial radioactivity was 800 counts per minute in the plasma and 25 counts per minute in the red cells, what is the number of counts per minute in the red cells after 300 minutes?

30. Temperature inversions and low wind speeds can trap air pollutants in a mountain valley for an extended period of time. Gaseous sulfur compounds are often a significant air pollution problem, but their study is complicated by their rapid oxidation. Hydrogen sulfide, H_2S, oxidizes into sulfur dioxide, SO_2, which in turn oxidizes into a sulfate. The following model has been proposed for determining the concentrations $x(t)$ and $y(t)$ of H_2S and SO_2, respectively, in a fixed airshed.[†] Let

$$\frac{dx}{dt} = -\alpha x + \gamma, \qquad \frac{dy}{dt} = \alpha x - \beta y + \delta$$

where the constants α and β are the conversion rates of H_2S into SO_2 and SO_2 into sulfate, respectively, and γ and δ are the production rates of H_2S and SO_2, respectively. Solve the equations sequentially and estimate the concentration levels that could be reached under prolonged air pollution.

ANSWERS TO SELF-QUIZ

I. b, c **II.** c **III.** d

11.2 LINEAR SYSTEMS: THEORY

In this section we shall consider the linear system of two first-order equations

$$\begin{aligned} x' &= a_{11}(t)x + a_{12}(t)y + f_1(t), \\ y' &= a_{21}(t)x + a_{22}(t)y + f_2(t), \end{aligned} \tag{1}$$

and the associated homogeneous system (i.e., $f_1 = f_2 = 0$)

$$\begin{aligned} x' &= a_{11}(t)x + a_{12}(t)y, \\ y' &= a_{21}(t)x + a_{22}(t)y. \end{aligned} \tag{2}$$

The point of view here will emphasize the similarities between such systems and the linear second-order equations discussed in Section 10.7. That there is a parallel between the two theories should not be surprising, since we have already shown in Section 11.1 that any linear second-order equation can be transformed into a system of the form of equation (1).

By a **solution** of system (1) (or (2)) we will mean a vector function $(x(t), y(t))$ that is differentiable and that satisfies the given equations. This is a formal statement of what we have been assuming all along. The following theorem is a consequence of the method of elimination and the existence–uniqueness theorem (in Section 10.7.1).

THEOREM 1

If the functions $a_{11}(t)$, $a_{12}(t)$, $a_{21}(t)$, $a_{22}(t)$, $f_1(t)$, and $f_2(t)$, are continuous, then given any numbers t_0, x_0, and y_0, there exists exactly one solution $(x(t), y(t))$ of system (1) that satisfies $x(t_0) = x_0$ and $y(t_0) = y_0$.

[†] R. L. Bohac, "A Mathematical Model for the Conversion of Sulphur Compounds in the Missoula Valley Airshed," *Proceedings of the Montana Academy of Science* (1974).

The vector function $(x_3(t), y_3(t))$ is a **linear combination** of the vectors $(x_1(t), y_1(t))$ and $(x_2(t), y_2(t))$ if there exist constants c_1 and c_2 such that the following two equations hold:

$$\begin{aligned} x_3(t) &= c_1x_1(t) + c_2x_2(t), \\ y_3(t) &= c_1y_1(t) + c_2y_2(t). \end{aligned} \tag{3}$$

The proof of the next theorem is left as an exercise. ■

THEOREM 2

If $(x_1(t), y_1(t))$ and $(x_2(t), y_2(t))$ are solutions of the homogeneous system (2), then any linear combination of them is also a solution of system (2). ■

EXAMPLE 1 A LINEAR COMBINATION OF SOLUTIONS TO A HOMOGENEOUS SYSTEM IS A SOLUTION

Consider the system

$$\begin{aligned} x' &= -x + 6y, \\ y' &= x - 2y. \end{aligned} \tag{4}$$

It is easy to verify that $(-2e^{-4t}, e^{-4t})$ and $(3e^t, e^t)$ are solutions of equations (4). Hence, by Theorem 2, the pair $(-2c_1e^{-4t} + 3c_2e^t, c_1e^{-4t} + c_2e^t)$ is a solution of equations (4) for any constants c_1 and c_2.

We define two vector functions $(x_1(t), y_1(t))$ and $(x_2(t), y_2(t))$ to be **linearly independent** if whenever the equations

$$\begin{aligned} c_1x_1(t) + c_2x_2(t) &= 0, \\ c_1y_1(t) + c_2y_2(t) &= 0 \end{aligned} \tag{5}$$

hold for all values of t, then $c_1 = c_2 = 0$. In Example 1, the two given vector solutions are linearly independent, since $c_1e^{-4t} + c_2e^t$ vanishes for all t only when $c_1 = c_2 = 0$.

Given two solutions $(x_1(t), y_1(t))$ and $(x_2(t), y_2(t))$, we define the **Wronskian**† of the two solutions to be the determinant:

$$W(t) = \begin{vmatrix} x_1(t) & x_2(t) \\ y_1(t) & y_2(t) \end{vmatrix} = x_1(t)y_2(t) - x_2(t)y_1(t). \tag{6}$$

We can then prove the next theorem.

THEOREM 3

If $W(t) \neq 0$ for every t, then equations (3) are the **general solution** of the homogeneous system (2) in the sense that given any solution (x^*, y^*) of system (2), there

†Compare this with the definition of the Wronskian for a second-order equation in Chapter 10. See page 740.

exist constants c_1 and c_2 such that

$$\begin{aligned} x^* &= c_1x_1 + c_2x_2, \\ y^* &= c_1y_1 + c_2y_2. \end{aligned} \tag{7}$$

PROOF: Let t_0 be given and consider the linear system of two equations in the unknown quantities c_1 and c_2:

$$\begin{aligned} c_1x_1(t_0) + c_2x_2(t_0) &= x^*(t_0), \\ c_1y_1(t_0) + c_2y_2(t_0) &= y^*(t_0). \end{aligned} \tag{8}$$

The determinant of this system is $W(t_0)$, which is nonzero by assumption. Thus there is a unique pair of constants (c_1, c_2) satisfying (8). By Theorem 2,

$$(c_1x_1(t) + c_2x_2(t), c_1y_1(t) + c_2y_2(t))$$

is a solution of equations (2). But by equations (8), this solution satisfies the same initial conditions at t_0 as the solution $(x^*(t), y^*(t))$. By the uniqueness part of Theorem 1, these solutions must be identical for all t. ■

EXAMPLE 2

A NONZERO WRONSKIAN

In Example 1 the Wronskian $W(t)$ is

$$W(t) = \begin{vmatrix} -2e^{-4t} & e^{-4t} \\ 3e^t & e^t \end{vmatrix} = -2e^{-3t} - 3e^{-3t} = -5e^{-3t} \neq 0.$$

Hence we need look no further for the general solution of system (4).

In view of the condition required in Theorem 3 that the Wronskian $W(t)$ never vanish, we shall consider the properties of the Wronskian more carefully. Let (x_1, y_1) and (x_2, y_2) be two solutions of the homogeneous system (2). Since $W(t) = x_1y_2 - x_2y_1$, we have

$$\begin{aligned} W'(t) &= x_1y_2' + x_1'y_2 - x_2y_1' - x_2'y_1 \\ &= x_1(a_{21}x_2 + a_{22}y_2) + y_2(a_{11}x_1 + a_{12}y_1) - x_2(a_{21}x_1 + a_{22}y_1) \\ &\quad - y_1(a_{11}x_2 + a_{12}y_1). \end{aligned}$$

Multiplying these expressions through and canceling like terms, we obtain

$$\begin{aligned} W' &= a_{11}x_1y_2 + a_{22}x_1y_2 - a_{11}x_2y_1 - a_{22}x_2y_1 \\ &= (a_{11} + a_{22})(x_1y_2 - x_2y_1) = (a_{11} + a_{22})W. \end{aligned}$$

Thus

$$W(t) = W(t_0) \exp\left\{\int_{t_0}^{t} [a_{11}(u) + a_{22}(u)]\, du\right\}. \tag{9}$$

We have shown the following theorem to be true:

THEOREM 4

Let (x_1, y_1) and (x_2, y_2) be two solutions of the homogeneous system (2). Then the Wronskian $W(t)$ is either always zero or never zero in any interval (since $e^u \neq 0$ for any u). ■

We are now ready to state the theorem that links linear independence with a nonvanishing Wronskian.

THEOREM 5

Two solutions $(x_1(t), y_1(t))$ and $(x_2(t), y_2(t))$ are linearly independent if and only if $W(t) \neq 0$.

PROOF: Let the solutions be linearly independent and suppose $W(t) = 0$. Then $x_1y_2 = x_2y_1$ or $x_1/x_2 = y_1/y_2 = c$ for some constant c. Then $x_1 = cx_2$ and $y_1 = cy_2$, so that the solutions are dependent, in contradiction of the original assumption. Hence $W(t) \neq 0$. Conversely, let $W(t) \neq 0$. If the solutions were dependent, then there would exist constants c_1 and c_2, not both zero, such that

$$c_1x_1 + c_2x_2 = 0,$$
$$c_1y_1 + c_2y_2 = 0.$$

Assuming that $c_1 \neq 0$, we then have $x_1 = cx_2$, $y_1 = cy_2$, where $c = -c_2/c_1$. But then

$$W(t) = x_1y_2 - x_2y_1 = cx_2y_2 - cx_2y_2 = 0.$$

Again this is a contradiction. Therefore the solutions are linearly independent. ■

We may summarize the contents of the previous four theorems in the following statement:

Let (x_1, y_1) and (x_2, y_2) be solutions of the homogeneous linear system

$$x' = a_{11}x + a_{12}y,$$
$$y' = a_{21}x + a_{22}y. \tag{10}$$

Then $(c_1x_1 + c_2x_2, c_1y_1 + c_2y_2)$ will be the general solution of the system (10) provided that $W(t) \neq 0$; that is, provided the solutions (x_1, y_1) and (x_2, y_2) are linearly independent.

Finally, let us consider the nonhomogeneous system (1). The proof of the following theorem is left as an exercise:

THEOREM 6

Let (x^*, y^*) be the general solution of the system (1), and let (x_p, y_p) be any solution of (1). Then $(x^* - x_p, y^* - y_p)$ is the general solution of the homogeneous equation (10). In other words, the general solution of the system (1) can be written as the sum of the general solution of the homogeneous system (10) and any particular solution of the nonhomogeneous system (1). ■

EXAMPLE 3 THE GENERAL SOLUTION OF A NONHOMOGENEOUS SYSTEM

Consider the system

$$x' = 3x + 3y + t,$$
$$y' = -x - y + 1. \tag{11}$$

We could solve this system by the methods given in the previous section. Here we note first that $(1, -1)$ and $(-3e^{2t}, e^{2t})$ are solutions to the homogeneous system

$$x' = 3x + 3y,$$
$$y' = -x - y.$$

A particular solution to the system (11) is $(-\frac{1}{4}(t^2 + 9t + 3), \frac{1}{4}(t^2 + 7t))$. The general solution to (11) is therefore

$$(x(t), y(t)) = (c_1 - 3c_2e^{2t} - \tfrac{1}{4}(t^2 + 9t + 3), -c_1 + c_2e^{2t} + \tfrac{1}{4}(t^2 + 7t)).$$

We close this section by noting that the theorems in this section can readily be generalized to systems of three or more equations.

PROBLEMS 11.2

SELF-QUIZ

I. The Wronskian of $(e^t, 2e^t)$ and $(e^{2t}, -3e^{2t})$ is ________.
a. $5e^{3t}$ **b.** $-5e^{3t}$
c. $5e^{-3t}$ **d.** $-5e^{-3t}$
e. $-e^{3t}$

II. If $(e^{-t}\cos 2t, -e^{-t}\sin 2t)$ and $(4e^{-t}\cos 2t, 3e^{-t}\sin 2t)$ are solutions to a homogeneous system, then which of the following are also solutions?
a. $(5e^{-t}\cos 2t, 2e^{-t}\sin 2t)$
b. $(e^{-t}\cos 2t, 3e^{-t}\sin 2t)$
c. $(4e^{-t}\cos 2t, -e^{-t}\sin 2t)$
d. $(23e^{-t}\cos 2t, 18e^{-t}\sin 2t)$
e. $(-6e^{-t}\cos 2t, 5e^{-t}\sin 2t)$

III. Which of the following pairs of solutions are *not* linearly independent for $t_0 > 0$?
a. $(e^{-t}, 2e^{-t}), (te^{-t}, -2te^{-t})$
b. $(e^{-t}, 4e^{-t}), (-2te^{-t}, -8te^{-t})$
c. $(-e^{-t}, 3e^{-t}), (2te^{-t}, 3te^{-t})$
d. $(-4e^{-t}, 6e^{-t}), (2te^{-t}, -3te^{-t})$

IV. Suppose that $(e^t, 2e^t)$ and $(2e^{2t}, -3e^{2t})$ are two linearly independent solutions of the homogeneous system (2) while $(\sin 3t, \cos 3t)$ is a solution to the nonhomogeneous system (1); which of the following is also a solution to the nonhomogeneous system?
a. $(e^t + \sin 3t, 2e^t + \cos 3t)$
b. $(e^t + 2\sin 3t, 2e^t + 2\cos 3t)$
c. $(2e^{2t} + 5\sin 3t, -3e^{2t} + 5\sin 3t)$
d. $(8e^t + \sin 3t, 16e^t + \cos 3t)$

1. a. Show that

$$(e^{-3t}, -e^{-3t}) \quad \text{and} \quad ((1 - t)e^{-3t}, te^{-3t})$$

are solutions to

$$x' = -4x - y,$$
$$y' = x - 2y.$$

b. Calculate the Wronskian and verify that the solutions are linearly independent.
c. Write the general solution to the system.

2. a. Show that $(e^{2t}\cos 2t, -2e^{2t}\sin 2t)$ and $(e^{2t}\sin 2t, 2e^{2t}\cos 2t)$ are solutions of the system

$$x' = 2x + y,$$
$$y' = -4x + 2y.$$

b. Calculate the Wronskian of these solutions and show that they are linearly independent.
c. Show that $(\frac{1}{4}te^{2t}, -\frac{11}{4}e^{2t})$ is a solution of the nonhomogeneous system

$$x' = 2x + y + 3e^{2t},$$
$$y' = -4x + 2y + te^{2t}.$$

d. Combining (a) and (c), write the general solution of the nonhomogeneous equation in (c).

3. a. Show that

$$(\sin t^2, 2t\cos t^2) \quad \text{and} \quad (\cos t^2, -2t\sin t^2)$$

are solutions of the system

$$x' = y,$$

$$y' = -4t^2x + \frac{1}{t}y.$$

b. Show that the solutions are linearly independent.

c. Show that $W(0) = 0$.

d. Explain the apparent contradiction of Theorem 5.

4. a. Show that $(\sin \ln t^2, (2/t) \cos \ln t^2)$ and $(\cos \ln t^2, -(2/t) \sin \ln t^2)$ are linearly independent solutions of the system

$$x' = y,$$

$$y' = -\frac{4}{t^2}x - \frac{1}{t}y.$$

b. Calculate the Wronskian $W(t)$.

5. Prove Theorem 2.

6. Prove Theorem 6.

ANSWERS TO SELF-QUIZ

I. b **II.** all of them **III.** b, d **IV.** a, d

11.3 THE SOLUTION OF HOMOGENEOUS LINEAR SYSTEMS WITH CONSTANT COEFFICIENTS: THE METHOD OF DETERMINANTS

As we saw in Section 11.1, the method of elimination can be used to solve systems of linear equations with constant coefficients. Since the algebraic manipulations required can be cumbersome, we shall develop in this section a more efficient method of solving homogeneous systems. Nonhomogeneous systems are discussed in Problems 9–14. Consider the homogeneous system

$$\begin{aligned} x' &= a_{11}x + a_{12}y, \\ y' &= a_{21}x + a_{22}y \end{aligned} \tag{1}$$

where the a_{ij} are constants. Our main tool for solving second-order linear homogeneous equations with constant coefficients involved obtaining a characteristic equation by guessing that the solution had the form $y = e^{\lambda x}$.

Parallel to the method of Section 10.9, we guess that there is a solution to system (1) of the form $(\alpha e^{\lambda t}, \beta e^{\lambda t})$, where α, β, and λ are constants yet to be determined. Substituting $x(t) = \alpha e^{\lambda t}$, $y(t) = \beta e^{\lambda t}$, $x'(t) = \alpha\lambda e^{\lambda t}$ and $y'(t) = \beta\lambda e^{\lambda t}$ into equation (1), we obtain

$$\begin{aligned} \alpha\lambda e^{\lambda t} &= a_{11}\alpha e^{\lambda t} + a_{12}\beta e^{\lambda t}, \\ \beta\lambda e^{\lambda t} &= a_{21}\alpha e^{\lambda t} + a_{22}\beta e^{\lambda t}. \end{aligned}$$

After dividing by $e^{\lambda t}$, we obtain the linear system

$$\begin{aligned} (a_{11} - \lambda)\alpha + a_{12}\beta &= 0, \\ a_{21}\alpha + (a_{22} - \lambda)\beta &= 0. \end{aligned} \tag{2}$$

We would like to find values for λ such that the system of equations (2) has a solution (α, β) where α and β are not both zero. According to the theory of determinants, such a solution occurs whenever the determinant of the system

$$\begin{aligned} D &= \begin{vmatrix} a_{11} - \lambda & a_{12} \\ a_{21} & a_{22} - \lambda \end{vmatrix} \\ &= (a_{11} - \lambda)(a_{22} - \lambda) - a_{21}a_{12} \end{aligned} \tag{3}$$

is zero. Setting $D = 0$, we obtain the quadratic equation

$$\boxed{\lambda^2 - (a_{11} + a_{22})\lambda + (a_{11}a_{22} - a_{21}a_{12}) = 0.} \tag{4}$$

We define this equation to be the **characteristic equation** of system (1). That we are using the same term again is no accident, as we shall now demonstrate. Suppose we differentiate the first equation in system (1) and eliminate the function $y(t)$:

$$x'' = a_{11}x' + a_{12}(a_{21}x + a_{22}y).$$

Then

$$x'' - a_{11}x' - a_{12}a_{21}x = a_{22}a_{12}y = a_{22}(x' - a_{11}x),$$

and gathering like terms, we obtain the homogeneous equation

$$x'' - (a_{11} + a_{22})x' + (a_{11}a_{22} - a_{12}a_{21})x = 0. \qquad (5)$$

The characteristic equation for (5), as the term was used in Section 10.9, is the same as equation (4). Hence the algebraic steps needed to obtain (5) can be avoided by setting the determinant $D = 0$.

As in Sections 10.9 and 10.10, there are three cases to consider, depending on whether the two roots λ_1 and λ_2 of the characteristic equation are real and distinct, real and equal, or complex conjugates. We will deal with each case separately.

CASE 1—DISTINCT REAL ROOTS: If λ_1 and λ_2 are distinct real numbers, then corresponding to λ_1 and λ_2 we have the vector solutions to system (1) $(\alpha_1 e^{\lambda_1 t}, \beta_1 e^{\lambda_1 t})$ and $(\alpha_2 e^{\lambda_2 t}, \beta_2 e^{\lambda_2 t})$, respectively. To find the constants α_1 and β_1 (not both zero), replace λ in the system of equations (2) by the value λ_1. The procedure is repeated for α_2, β_2, and λ_2. We note that the constants α_1, β_1, α_2 and β_2 are not unique. In fact, for each number λ_1 or λ_2, there are an infinite number of vectors (α, β) that satisfy system (2). To see this, we observe that if (α, β) is a solution pair, then so is $(c\alpha, c\beta)$ for any real number c. Finally, the vectors given above are linearly independent, since if not, there exists a constant c such that $\alpha_2 e^{\lambda_2 t} = c\alpha_1 e^{\lambda_1 t}$ and $\beta_2 e^{\lambda_2 t} = c\beta_1 e^{\lambda_1 t}$, which is clearly impossible because $\lambda_1 \neq \lambda_2$. We therefore have proved the following theorem:

THEOREM 1

If λ_1 and λ_2 are distinct real roots of equation (4), then two linearly independent solutions of the system (1) are given by

$$(\alpha_1 e^{\lambda_1 t}, \beta_1 e^{\lambda_1 t}), \qquad (\alpha_2 e^{\lambda_2 t}, \beta_2 e^{\lambda_2 t}),$$

where the vectors (α_1, β_1) and (α_1, β_2) are solutions of system (2), with $\lambda = \lambda_1$ and $\lambda = \lambda_2$, respectively. ■

EXAMPLE 1 A SYSTEM WITH DISTINCT REAL ROOTS

Find the general solution of the system

$$x' = -x + 6y,$$
$$y' = x - 2y.$$

SOLUTION: Here $a_{11} = -1$, $a_{12} = 6$, $a_{21} = 1$, $a_{22} = -2$, and equation (3) becomes

$$D = \begin{vmatrix} -1-\lambda & 6 \\ 1 & -2-\lambda \end{vmatrix} = (\lambda+2)(\lambda+1) - 6 = \lambda^2 + 3\lambda - 4 = 0,$$

which has the roots $\lambda_1 = -4$, $\lambda_2 = 1$. For $\lambda_1 = -4$ the system of equations (2) becomes

$$3\alpha_1 + 6\beta_1 = 0,$$
$$\alpha_1 + 2\beta_1 = 0.$$

Ignoring the first equation because it is just a multiple of the second, select β_1 to be 1 to obtain $(-2, 1)$ as a solution of the second equation. Hence a first solution is $(-2e^{-4t}, e^{-4t})$. Similarly, with $\lambda_2 = 1$, we obtain the equations

$$-2\alpha_2 + 6\beta_2 = 0,$$
$$\alpha_2 - 3\beta_2 = 0,$$

which have a solution $\alpha_2 = 3$, $\beta_2 = 1$. Thus a second solution, linearly independent of the first, is given by the pair $(3e^t, e^t)$. By Theorem 11.2.3, the general solution is given by the pair

$$(x(t), y(t)) = (-2c_1e^{-4t} + 3c_2e^t, c_1e^{-4t} + c_2e^t).$$

NOTE: As defined in Section 8.12 the numbers -4 and 1 are eigenvalues of the matrix $\begin{pmatrix} -1 & 6 \\ 1 & -2 \end{pmatrix}$ with corresponding eigenvectors $\begin{pmatrix} -2 \\ 1 \end{pmatrix}$ and $\begin{pmatrix} 3 \\ 1 \end{pmatrix}$.

CASE 2—TWO EQUAL ROOTS: When $\lambda_1 = \lambda_2$, one solution is $(\alpha_1e^{\lambda_1 t}, \beta_1e^{\lambda_1 t})$. The other solution $(\alpha_2e^{\lambda_2 t}, \beta_2e^{\lambda_2 t})$ given by Theorem 1 is not independent of the first unless $a_{11} = a_{22}$ and $a_{12} = a_{21} = 0$. In this case we have the *uncoupled* system of equations

$$x' = a_{11}x, \qquad y' = a_{22}y,$$

with the linearly independent solutions $(\alpha_1e^{\lambda_1 t}, 0)$ and $(0, \beta_2e^{\lambda_1 t})$. (The equations are said to be uncoupled because each involves only one dependent variable.) On the basis of the results of Section 10.9, we would expect that, if the system is not uncoupled, a second linearly independent solution would have the form $(\alpha_2te^{\lambda_1 t}, \beta_2te^{\lambda_1 t})$. This, however, *does not turn out to be the case.* Rather, the second linearly independent solution has the form

$$(x(t), y(t)) = ((\alpha_2 + \alpha_3t)e^{\lambda_1 t}, (\beta_2 + \beta_3t)e^{\lambda_1 t}). \tag{6}$$

To calculate the constants α_2, α_3, β_2, and β_3, it is necessary to substitute back into the original system (1). This is best shown by an example.

EXAMPLE 2 A SYSTEM WITH ONE REAL ROOT

Find the general solution of the system

$$x' = -4x - y,$$
$$y' = x - 2y. \tag{7}$$

SOLUTION: Equation (3) is

$$D = \begin{vmatrix} -4-\lambda & -1 \\ 1 & -2-\lambda \end{vmatrix} = (\lambda+4)(\lambda+2)+1 = \lambda^2+6\lambda+9 = 0,$$

which has the double root $\lambda_1 = \lambda_2 = -3$. From system (2), with $\lambda = -3$, we find that

$$\begin{aligned} -\alpha_1 - \beta_1 &= 0, \\ \alpha_1 + \beta_1 &= 0. \end{aligned}$$

A nontrivial solution is $\alpha_1 = 1$, $\beta_1 = -1$, yielding the solution $(e^{-3t}, -e^{-3t})$.

If we try to find a solution of the form $(\alpha_2 te^{-3t}, \beta_2 te^{-3t})$ we immediately run into trouble, since the derivatives on the left-hand side of system (7) produce terms of the form ce^{-3t} not present on the right-hand side of (7). This explains why we must seek a solution of the form

$$((\alpha_2 + \alpha_3 t)e^{-3t}, (\beta_2 + \beta_3 t)e^{-3t}). \tag{8}$$

Substituting the pair (8) into system (7), we obtain

$$\begin{aligned} e^{-3t}(\alpha_3 - 3\alpha_2 - 3\alpha_3 t) &= -4(\alpha_2 + \alpha_3 t)e^{-3t} - (\beta_2 + \beta_3 t)e^{-3t}, \\ e^{-3t}(\beta_3 - 3\beta_2 - 3\beta_3 t) &= (\alpha_2 + \alpha_3 t)e^{-3t} - 2(\beta_2 + \beta_3 t)e^{-3t}. \end{aligned}$$

Dividing by e^{-3t} and equating constant terms and coefficients of t, we obtain the system of equations

$$\begin{aligned} \alpha_3 - 3\alpha_2 &= -4\alpha_2 - \beta_2, \\ -3\alpha_3 &= -4\alpha_3 - \beta_3, \\ \beta_3 - 3\beta_2 &= \alpha_2 - 2\beta_2, \\ -3\beta_3 &= \alpha_3 - 2\beta_3. \end{aligned}$$

One solution is $\alpha_2 = 1$, $\beta_2 = -2$, $\alpha_3 = 1$, $\beta_3 = -1$. Thus a second solution of system (7) is $((1+t)e^{-3t}, (-2-t)e^{-3t})$. It is easy to verify that the two solutions are linearly independent, since $W(t) = -e^{-6t}$.

We summarize these results by stating the following theorem:

THEOREM 2

Let equation (4) have two equal real roots $\lambda_1 = \lambda_2$. Then there exist constants α_1, α_2, α_3, β_1, β_2, and β_3 such that two linearly independent solutions of system (1) are given by

$$\begin{aligned} (x_1(t), y_1(t)) &= (\alpha_1 e^{\lambda_1 t}, \beta_1 e^{\lambda_1 t}), \\ (x_2(t), y_2(t)) &= ((\alpha_2 + \alpha_3 t)e^{\lambda_1 t}, (\beta_2 + \beta_3 t)e^{\lambda_1 t}). \end{aligned} \tag{9}$$

The constants α_1 and β_1 are found as nontrivial solutions of the homogeneous system of equations (2); the other constants are found by substituting the second equation of (9) back into system (1).

REMARK: In the substitution process, we always obtain a homogeneous system of four equations in the four unknowns α_2, β_2, α_3, and β_3. That this system has nontrivial solutions follows because the determinant of the system is zero. The proof is left as an exercise (see Problem 8). ■

CASE 3—COMPLEX CONJUGATE ROOTS:† Let $\lambda_1 = a + ib$ and $\lambda_2 = a - ib$, where a and b are real and $b \neq 0$. Then the solutions $(\alpha_1 e^{(a+ib)t}, \beta_1 e^{(a-ib)t})$ and $(\alpha_2 e^{(a-ib)t}, \beta_2 e^{(a-ib)t})$ are linearly independent. However, the constants α_1, β_1, α_2, and β_2 obtained from system (2) are complex numbers. To obtain real solution pairs we proceed as follows. Let $\alpha_1 = A_1 + iA_2$, $\beta_1 = B_1 + iB_2$ and apply Euler's formula, $e^{i\theta} = \cos\theta + i\sin\theta$, to the first complex solution, obtaining

$$\begin{aligned} x(t) &= (A_1 + iA_2)e^{at}(\cos bt + i\sin bt), \\ y(t) &= (B_1 + iB_2)e^{at}(\cos bt + i\sin bt). \end{aligned} \tag{10}$$

Multiplying and remembering that $i^2 = -1$, we obtain the equations

$$x(t) = e^{at}[(A_1 \cos bt - A_2 \sin bt) + i(A_1 \sin bt + A_2 \cos bt)],$$
$$y(t) = e^{at}[(B_1 \cos bt - B_2 \sin bt) + i(B_1 \sin bt + B_2 \cos bt)].$$

Now, since the coefficients of system (1) are *real,* the only way (x, y) can be a solution is that all the real terms and, similarly, the imaginary terms, cancel out. Thus the real parts of x and y must form a solution of system (1), as must the imaginary parts:

$$\begin{aligned} (x_1(t), y_1(t)) &= (e^{at}(A_1 \cos bt - A_2 \sin bt), e^{at}(B_1 \cos bt - B_2 \sin bt)), \\ (x_2(t), y_2(t)) &= (e^{at}(A_1 \sin bt + A_2 \cos bt), e^{at}(B_1 \sin bt + B_2 \cos bt)). \end{aligned} \tag{11}$$

The Wronskian of the solutions (11) is

$$W(t) = e^{2at}(A_1B_2 - A_2B_1).$$

We want to show that the solutions (11) are linearly independent. Suppose otherwise. Then $W(t) = 0$, which means that $A_1B_2 = A_2B_1$. This implies that $B_2\alpha_1 = A_2\beta_1$ (according to the definition of α_1 and β_1). Now, neither α_1 nor β_1 vanishes. If either were zero, so would be the other, and the solution (10) would be trivial. Also A_2 cannot vanish, since if it did, so would B_2, and the first equation of system (2) would prevent λ_1 from being complex. Multiplying the first equation of system (2) by A_2, using the identity $B_2\alpha_1 = A_2\beta_1$, and dividing by α_1, we have

$$(a_{11} - \lambda_1)A_2 + a_{12}B_2 = 0.$$

But then λ_1 is not complex. Therefore, it is impossible that $W(t)$ could vanish, and we have proved the following theorem:

THEOREM 3

If equation (4) has the complex roots $\lambda_1 = a + ib$ and $\lambda_2 = a - ib$, then two linearly independent solutions to system (1) are given by (11). ■

†For this discussion we assume you are familiar with the material in Appendix 3.

EXAMPLE 3 A SYSTEM WITH COMPLEX CONJUGATE ROOTS

Find the general solution of the system

$$\begin{aligned} x' &= 4x + y, \\ y' &= -8x + 8y. \end{aligned} \tag{12}$$

SOLUTION: Here

$$D = \begin{vmatrix} 4 - \lambda & 1 \\ -8 & 8 - \lambda \end{vmatrix} = \lambda^2 - 12\lambda + 40 = 0.$$

The roots of the characteristic equation are $\lambda_1 = 6 + 2i$ and $\lambda_2 = 6 - 2i$, so that Theorem 3 yields the linearly independent solutions

$$(x_1(t), y_1(t)) = (e^{6t}(A_1 \cos 2t - A_2 \sin 2t), e^{6t}(B_1 \cos 2t - B_2 \sin 2t)),$$

$$(x_2(t), y_2(t)) = (e^{6t}(A_1 \sin 2t + A_2 \cos 2t), e^{6t}(B_1 \sin 2t - B_2 \cos 2t)).$$

Substituting the first equation into system (12) yields, after a great deal of algebra, the system of equations

$$(2A_1 - 2A_2 - B_1) \cos 2t - (2A_1 + 2A_2 - B_2) \sin 2t = 0,$$

$$(8A_1 - 2B_1 - 2B_2) \cos 2t - (8A_2 + 2B_1 - 2B_2) \sin 2t = 0.$$

Since t is arbitrary and the functions $\sin 2t$ and $\cos 2t$ are linearly independent, the terms in parentheses must all vanish. A choice of values that will accomplish this is $A_1 = 1$, $A_2 = \frac{1}{2}$, $B_1 = 1$, and $B_2 = 3$. Thus two linearly independent solutions to system (12) are $(e^{6t}(\cos 2t - \frac{1}{2}\sin 2t),\ e^{6t}(\cos 2t - 3\sin 2t))$ and $(e^{6t}(\sin 2t + \frac{1}{2}\cos 2t),\ e^{6t}(\sin 2t + 3 \cos 2t))$. The general solution of system (12) is a linear combination of these two solutions.

EXAMPLE 4 A BIOLOGICAL SYSTEM

Most biological systems are controlled by the production of enzymes or hormones that stimulate or inhibit the secretion of some compound. For example, the pancreatic hormone *glucagon* stimulates the release of glucose from the liver to the plasma. A rise in blood glucose inhibits the secretion of glucagon but causes an increase in the production of the hormone insulin. Insulin, in turn, aids in the removal of glucose from the blood and in its conversion to glycogen in the muscle tissue. Let G and I be the deviations of plasma glucose and plasma insulin from the normal (fasting) level, respectively. We then have the system

$$\begin{aligned} \frac{dG}{dt} &= -k_{11}G - k_{12}I, \\ \frac{dI}{dt} &= k_{21}G - k_{22}I \end{aligned} \tag{13}$$

where the positive constants k_{ij} are model parameters, some of which may be determined experimentally. It is known that the system (13) exhibits a strongly damped oscillatory behavior, since direct injection of glucose into the blood will produce a fall of blood glucose to a level below fasting in about one and a half hours, followed by a rise slightly above the fasting level in about three hours. Hence, the characteristic equation of system (13),

$$D = \begin{vmatrix} -k_{11} - \lambda & -k_{12} \\ k_{21} & -k_{22} - \lambda \end{vmatrix} = (k_{11} + \lambda)(k_{22} + \lambda) + k_{12}k_{21}$$
$$= \lambda^2 + (k_{11} + k_{22})\lambda + (k_{11}k_{22} + k_{12}k_{21}) = 0,$$

must have complex conjugate roots $-a \pm ib$, with $a = (k_{11} + k_{22})/2$ and $b = \sqrt{k_{12}k_{21} - (k_{11} - k_{22})^2/4}$, since only complex roots can lead to oscillatory behavior. By Theorem 3, we have the solutions

$$\begin{aligned}(G_1, I_1) &= (e^{-at}(A_1 \cos bt - A_2 \sin bt),\ e^{-at}(B_1 \cos bt - B_2 \sin bt)),\\ (G_2, I_2) &= (e^{-at}(A_1 \sin bt + A_2 \cos bt),\ e^{-at}(B_1 \sin bt + B_2 \cos bt)).\end{aligned} \tag{14}$$

Since the period of the oscillation is approximately three hours, we may set $b = 2\pi/3$ and measure time in hours. Substituting the first equation of system (14) into equations (13) we obtain the following equations:

$$(-aA_1 - bA_2 + k_{11}A_1 + k_{12}B_1) \cos bt + (aA_2 - bA_1 - k_{11}A_2 - k_{12}B_2) \sin bt = 0$$
$$(-aB_1 - bB_2 - k_{21}A_1 + k_{22}B_1) \cos bt + (aB_2 - bB_1 + k_{21}A_2 - k_{22}B_2) \sin bt = 0$$

These equations must hold for all t. Thus all the terms in parentheses must vanish. A choice of values for which this occurs is $A_1 = 1$, $A_2 = 0$, $B_1 = (k_{22} - k_{11})/2k_{12}$, and $B_2 = -b/k_{12}$. Then the general solution of system (13) is given by the vector $(G(t), I(t))$ with

$$G(t) = e^{-at}(c_1 \cos bt + c_2 \sin bt),$$
$$I(t) = e^{-at}\left[\frac{k_{22} - k_{11}}{2k_{12}}(c_1 \cos bt + c_2 \sin bt) + \frac{b}{k_{12}}(c_1 \sin bt - c_2 \cos bt)\right].$$

Assume now that the glucose injection was administered at a time when plasma insulin and glucose were at fasting levels and that the glucose was diffused completely in the blood before the insulin level began to increase ($t = 0$). Also, $G(0) = G_0$ equals the ratio of the volume of glucose administered to blood volume, and $I(0) = 0$. Since $G(t)$ is at a maximum when $t = 0$, it follows that $c_1 = G_0$ and $c_2 = 0$. Hence

$$G(t) = G_0 e^{-at} \cos bt. \tag{15}$$

But

$$0 = I(0) = \frac{k_{22} - k_{11}}{2k_{12}} G_0,$$

so that $k_{11} = k_{22}$, $b = \sqrt{k_{12}k_{21}} = 2\pi/3$, and

$$I(t) = G_0 \frac{b}{k_{12}} e^{-at} \sin bt.$$

If the minimum level $G(\frac{3}{2})$ (<0) is known, then by equation (15),

$$e^{3a/2} = |G(\tfrac{3}{2})|/G_0,$$

so that

$$k_{11} = a = \frac{2}{3} \ln \frac{|G(\frac{3}{2})|}{G_0}.$$

If we determine the plasma insulin at any given time $t_0 > 0$, we can then evaluate the parameters k_{12} and k_{21}.

PROBLEMS 11.3

SELF-QUIZ

Multiple Choice

I. Which are roots of the characteristic equation of the following system?

$$x' = -2x - 2y$$
$$y' = -5x + y$$

a. $-2, 1$ **b.** $4, -3$
c. $-4, 3$ **d.** $4, 3$
e. $-4, -3$

II. Which are the roots of the characteristic equation of the following system?

$$x' = 2x - y$$
$$y' = 5x - 2y$$

a. $1, -1$ **b.** 1
c. -1 **d.** $i, -i$
e. $2, -2$

III. Which are the roots of the characteristic equation of the following system?

$$x' = -3x + 2y$$
$$y' = \quad\quad - 3y$$

a. -3 **b.** $-3, 2$
c. $3, -2$ **d.** $3i, -3i$
e. $1, -5$

True–False

IV. If the roots of the characteristic equation are negative real numbers, then all solutions to the system approach 0 as $t \to \infty$.

V. If one of the roots of the characteristic equations is a positive real number, then *all* solutions to the system approach infinity as $t \to \infty$.

In Problems 1–7 use the method of determinants to find two linearly independent solutions for each given system.

1. $x' = 4x - 3y,$
$y' = 5x - 4y$

2. $x' = 7x + 6y,$
$y' = 2x + 6y$

3. $x' = -x + y,$
$y' = -5x + 3y$

4. $x' = x + y,$
$y' = -x + 3y$

5. $x' = -4x - y,$
$y' = x - 2y$

6. $x' = 4x - 2y,$
$y' = 5x + 2y$

7. $x' = 4x - 3y,$
$y' = 8x - 6y$

8. Substituting the second solution vector (9) into system (1), we obtain the homogeneous system of linear equations

$$\begin{aligned} (\lambda - a_{11})\alpha_2 + \alpha_3 - a_{12}\beta_2 &= 0, \\ (\lambda - a_{11})\alpha_3 - a_{12}\beta_3 &= 0, \\ -a_{21}\alpha_2 + (\lambda - a_{22})\beta_2 + \beta_3 &= 0, \\ -a_{21}\alpha_3 + (\lambda - a_{22})\beta_3 &= 0. \end{aligned} \tag{16}$$

a. Show that since $\lambda_1 = \lambda_2 = (a_{11} + a_{22})/2$, the second and fourth equations of system (16) are identical.

b. Conclude from part (a) that the determinant of system (16) is zero, and from this that (16) has nontrivial solutions.

9. Consider the nonhomogeneous equations

$$\begin{aligned} x' &= a_{11}x + a_{12}y + f_1, \\ y' &= a_{21}x + a_{22}y + f_2. \end{aligned} \tag{17}$$

Let (x_1, y_1) and (x_2, y_2) be two linearly independent solutions of the homogeneous system (1). Show that

$$x_p(t) = v_1(t)x_1(t) + v_2(t)x_2(t),$$
$$y_p(t) = v_1(t)y_1(t) + v_2(t)y_2(t),$$

is a particular solution of system (17) if v_1 and v_2 satisfy the equations

$$v_1'x_1 + v_2'x_2 = f_1,$$
$$v_1'y_1 + v_2'y_2 = f_2.$$

This process for finding a particular solution of the nonhomogeneous system (16) is called the **variation of constants method for systems.**

Note the close parallel between this method and the method given in Section 10.11.

In Problems 10–14 use the variation of constants method to find a particular solution for each given nonhomogeneous system.

10. $x' = 2x + y + 3e^{2t}$,
$y' = -4x + 2y + te^{2t}$

11. $x' = 3x + 3y + t$,
$y' = -x - y + 1$

12. $x' = -2x + y$,
$y' = -3x + 2y + 2 \sin t$

13. $x' = -x + y + \cos t$,
$y' = -5x + 3y$

14. $x' = 3x - 2y + t$,
$y' = 2x - 2y + 3e^t$

***15.** In an experiment on cholesterol turnover in humans, radioactive cholesterol-4–^{14}C was injected intravenously, and the total plasma cholesterol and radioactivity were measured. It was discovered that the turnover of cholesterol behaves like a two-compartment system.† The compartment consisting of the organs and blood has a rapid turnover, while the turnover in the other compartment is much slower. Assume that the body takes in and excretes all cholesterol through the first compartment. Let $x(t)$ and $y(t)$ denote the deviations from normal cholesterol levels in each compartment. Suppose that the daily fractional transfer from compartment x is 0.134, of which 0.036 is the input to compartment y, and that the transfer from compartment y is 0.02.

a. Describe the problem discussed above as a system of homogeneous linear differential equations.
b. Obtain the general solution of the system.

ANSWERS TO SELF-QUIZ

I. c **II.** d **III.** a **IV.** True **V.** False

11.4 ELECTRIC CIRCUITS WITH SEVERAL LOOPS (OPTIONAL)

Here we make use of the concepts developed in Sections 10.6 and 10.15 to study electric networks with two or more coupled closed circuits. The two fundamental principles governing such networks are the two laws of Kirchhoff:

i. The algebraic sum of all voltage drops around any closed circuit is zero.
ii. The algebraic sum of the currents flowing into any junction in the network is zero.

EXAMPLE 1 AN ELECTRIC CIRCUIT WITH TWO LOOPS

Consider the two-loop electric circuit in Figure 1. Suppose we wish to find the current in each loop as a function of time, given that all currents are zero when the switch is closed at $t = 0$. Let I be the current flowing through the inductor L_1 and let I_R and I_L denote the current flowing through the resistor and the inductor L_2, respectively. By Kirchhoff's current law (ii), $I = I_R + I_L$, so when we apply Kirchhoff's voltage law (i) to each loop, we obtain the system

$$L_1\frac{dI}{dt} + RI_R = E, \tag{1}$$

$$L_2\frac{dI_L}{dt} - RI_R = 0. \tag{2}$$

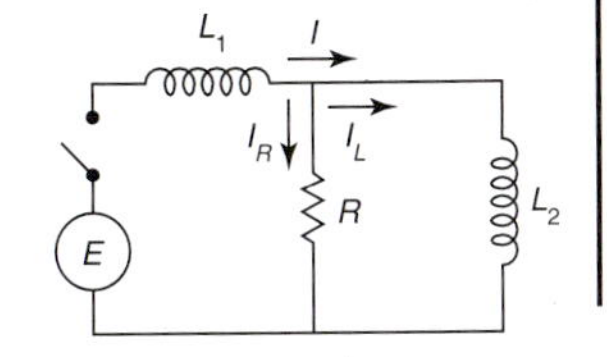

FIGURE 1
An electric circuit with two inductors

†D. S. Goodman and R. P. Noble, "Turnover of Plasma Cholesterol in Man," *J. Clin. Invest.* **47**, 231–241 (1968).

Replacing I by $I_R + I_L$ in (1), we have the system of simultaneous linear differential equations

$$L_1 \frac{dI_R}{dt} + L_1 \frac{dI_L}{dt} + RI_R = E, \tag{3}$$

$$L_2 \frac{dI_L}{dt} - RI_R = 0. \tag{4}$$

If we multiply (4) by L_1/L_2, we obtain

$$L_1 \frac{dI_L}{dt} = \frac{L_1 R}{L_2} I_R,$$

which we can substitute into (3) to eliminate the I_L variable, yielding

$$L_1 \frac{dI_R}{dt} + \left(\frac{L_1}{L_2} + 1\right) RI_R = E. \tag{5}$$

If L_1, L_2, and R are constant, we can solve the linear first-order differential equation by the method in Section 10.4. Assume that $L_1 = 1$ henry, $L_2 = \frac{1}{2}$ henry, $R = 20$ ohms, and $E = 50$ volts. Then

$$\frac{dI_R}{dt} + 60I_R = 50.$$

Multiplying both sides by the integrating factor e^{60t}, we have

$$\frac{d}{dt}(e^{60t}I_R) = 50e^{60t},$$

and an integration yields

$$e^{60t}I_R = \tfrac{5}{6}e^{60t} + k_1.$$

Since $I_R(0) = 0$, it follows that $k_1 = -\frac{5}{6}$ and

$$I_R(t) = \tfrac{5}{6}(1 - e^{-60t}). \tag{6}$$

We can find I_L by substituting (6) into (4) to get

$$\frac{1}{2}\frac{dI_L}{dt} = \frac{100}{6}(1 - e^{-60t}),$$

from which it follows that

$$I_L(t) = \frac{100}{3}\left(t + \frac{e^{-60t}}{60}\right) + k_2.$$

Setting $t = 0$, so that $I_L(0) = 0$, we obtain $k_2 = -\frac{5}{9}$ and

$$I_L(t) = \tfrac{100}{3}t + \tfrac{5}{9}(e^{-60t} - 1). \tag{7}$$

EXAMPLE 2

AN RLC-CIRCUIT WITH TWO LOOPS

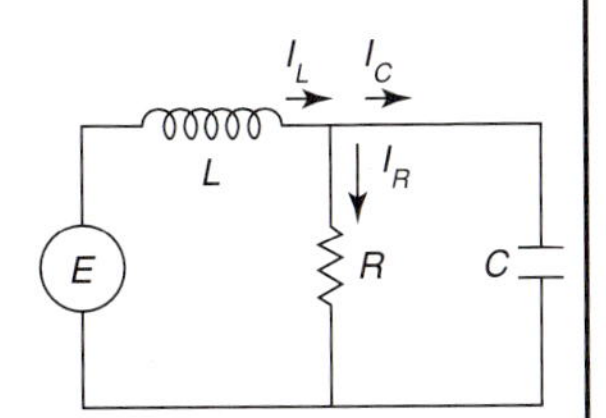

FIGURE 2
An RLC-circuit with two loops

Consider the circuit in Figure 2. There are two loops. By Kirchhoff's voltage law, we obtain

$$L\frac{dI_L}{dt} + RI_R = E, \tag{8}$$

$$\frac{Q_C}{C} - RI_R = 0. \tag{9}$$

Since $I = dQ/dt$, the second equation may be rewritten as

$$\frac{I_C}{C} - R\frac{dI_R}{dt} = 0. \tag{10}$$

By Kirchhoff's current law, we have

$$I_L = I_C + I_R,$$

which, if substituted into (10), yields, together with (8), the nonhomogeneous system of linear first-order differential equations

$$\begin{aligned} \frac{dI_L}{dt} &= -\frac{R}{L}I_R + \frac{E}{L}, \\ \frac{dI_R}{dt} &= \frac{I_L}{RC} - \frac{I_R}{RC}. \end{aligned} \tag{11}$$

The characteristic equation of this system is

$$\begin{aligned} D = \begin{vmatrix} -\lambda & -R/L \\ 1/RC & -\lambda - 1/RC \end{vmatrix} &= \lambda(\lambda + 1/RC) + 1/LC \\ &= \lambda^2 + \lambda/RC + 1/LC = 0. \end{aligned} \tag{12}$$

The roots of (12) are $(-L \pm \sqrt{L^2 - 4R^2LC})/2RLC$.

The value of the discriminant

$$L^2 - 4R^2LC = L(L - 4R^2C)$$

is now important. For simplicity, assume that $R = 100$ ohms, $C = 1.5 \times 10^{-4}$ farad, $E = 100$ volts, and $L = 8$ henrys. Then

$$L - 4R^2C = 8 - 4(100)^2 \cdot 1.5(10^{-4}) = 2,$$

so the roots of (12) are $\lambda_1 = -50$ and $\lambda_2 = -50/3$. Thus, from Section 11.1 or 11.3, the homogeneous system

$$\begin{aligned} \frac{dI_L}{dt} &= -\frac{R}{L}I_R, \\ \frac{dI_R}{dt} &= \frac{I_L}{RC} - \frac{I_R}{RC}. \end{aligned} \tag{13}$$

has a general solution given by

$$I_L(t) = c_1e^{-50t} + c_2e^{-(50/3)t}$$

and

$$I_R(t) = -\frac{L}{R}\frac{dI_L}{dt} = -\frac{8}{100}\left[-50c_1e^{-50t} - \frac{50}{3}c_2e^{-(50/3)t}\right]$$

$$= 4c_1e^{-50t} + \frac{4}{3}c_2e^{-(50/3)t}. \tag{14}$$

Since $E = 100$ volts is constant, when we use the method of undetermined coefficients we assume a particular solution of the form

$$(I_L, I_R)_p = (A, B), \tag{15}$$

where A and B are constants we must determine. Substituting (15) into the nonhomogeneous system (11), we get

$$0 = -\frac{100}{8}B + \frac{100}{8},$$

$$0 = \frac{A - B}{(100)1.5 \times 10^{-4}},$$

so $B = 1 = A$. Finally, we find the constants c_1 and c_2 by using the initial conditions $I_L(0) = 0 = I_R(0)$ in the equations

$$I_L = c_1e^{-50t} + c_2e^{-(50/3)t} + 1,$$
$$I_R = 4c_1e^{-50t} + \tfrac{4}{3}c_2e^{-(50/3)t} + 1.$$

We obtain

$$I_L(0) = c_1 + c_2 + 1 = 0,$$
$$I_R(0) = 4c_1 + \tfrac{4}{3}c_2 + 1 = 0,$$

with solution $c_1 = \frac{1}{8}$ and $c_2 = -\frac{9}{8}$. Thus the unique solution to the initial-value problem is

$$(I_L, I_R) = (\tfrac{1}{8}e^{-50t} - \tfrac{9}{8}e^{(-50/3)t} + 1, \tfrac{1}{2}e^{-50t} - \tfrac{3}{2}e^{(-50/3)t} + 1).$$

EXAMPLE 3 A CIRCUIT WHOSE CHARACTERISTIC EQUATION HAS A DOUBLE ROOT

If $L = 6$ henrys in Example 2 and all other facts remain the same, then $L - 4R^2C = 0$ and (12) has the double root $\lambda_1 = \lambda_2 = -1/2RC = -100/3$. Hence, from page 753 the general solution to the homogeneous system (13) is

$$I_L(t) = (c_1 + c_2t)e^{-(100/3)t}$$

and

$$I_R(t) = -\frac{L}{R}\frac{dI_L}{dt} = -\frac{6}{100}e^{(-100/3)t}\left[\frac{-100}{3}(c_1 + c_2t) + c_2\right]$$

$$= \left[\left(2c_1 - \frac{3}{50}c_2\right) + 2c_2t\right]e^{-(100/3)t}.$$

As in Example 2, we find the particular solutions to (11) to be

$$(I_L, I_R) = (1, 1).$$

Hence the general solution to (11) is

$$I_L(t) = (c_1 + c_2 t)\, e^{(-100/3)t} + 1,$$
$$I_R(t) = [(2c_1 - \tfrac{3}{50}c_2) + 2c_2 t]\, e^{(-100/3)t} + 1.$$

Using the initial conditions, we have

$$I_L(0) = c_1 + 1 = 0,$$
$$I_R(0) = 2c_1 - \tfrac{3}{50}c_2 + 1 = 0,$$

or

$$c_1 = -1 \qquad \text{and} \qquad c_2 = -\tfrac{50}{3}.$$

Thus the unique solution to the initial-value problem is

$$(I_L(t), I_R(t)) = ((-1 - \tfrac{50}{3}t)\, e^{(-100/3)t} + 1, (-1 - \tfrac{100}{3}t)\, e^{(-100/3)t} + 1).$$

EXAMPLE 4

A CIRCUIT WITH OSCILLATORY TRANSIENT CURRENTS

Suppose, in Example 2, that $L = 3$ henrys and all other facts remain unchanged. Then $L - 4R^2C = -3$, so the characteristic equation (12) has the roots $100(-1 \pm i)/3$. Then, from page 757 solutions to the homogeneous system (13) are

$$I_L(t) = e^{-(100/3)t}(c_1\cos\tfrac{100}{3}t + c_2\sin\tfrac{100}{3}t)$$

and

$$I_R(t) = \frac{L}{R}\left(-\frac{dI_L}{dt}\right) = \frac{3}{100}\left(-\frac{dI_L}{dt}\right)$$
$$= \frac{3}{100}e^{(-100/3)t}\left[\frac{100}{3}\left(c_1\cos\frac{100}{3}t + c_2\sin\frac{100}{3}t + c_1\sin\frac{100}{3}t - c_2\cos\frac{100}{3}t\right]\right.$$
$$= e^{(-100/3)t}\left[(c_1 - c_2)\cos\frac{100}{3}t + (c_1 + c_2)\sin\frac{100}{3}t\right].$$

The nonhomogeneous term in system (11) is E/L, a constant. Thus it is reasonable to find a particular solution to (11) of the form $I_L = A$ and $I_R = B$, where A and B are constants. Inserting these values into (11), we obtain, as in Example 3,

$$I_L = I_R = 1.$$

The general solution to (11) is, therefore,

$$I_L(t) = e^{(-100/3)t}(c_1\cos\tfrac{100}{3}t + c_2\sin\tfrac{100}{3}t) + 1$$
$$I_R(t) = e^{(-100/3)t}[(c_1 - c_2)\cos\tfrac{100}{3}t + (c_1 + c_2)\sin\tfrac{100}{3}t] + 1.$$

Finally, setting $I_R(0) = I_L(0) = 0$, we obtain

$$c_1 + 1 = 0,$$
$$c_1 - c_2 + 1 = 0,$$

with solution $c_1 = -1$, $c_2 = 0$.

Thus the unique solution to our initial-value problem is

$$I_L(t) = 1 - e^{(-100/3)t}\cos\tfrac{100}{3}t$$
$$I_R(t) = 1 - e^{(-100/3)t}(\cos\tfrac{100}{3}t + \sin\tfrac{100}{3}t).$$

PROBLEMS 11.4

1. Let $R = 100$ ohms, $L = 4$ henrys, $C = 10^{-4}$ farad, and $E = 100$ volts in the network in Figure 1. Assume that the currents I_R and I_L are both zero at time $t = 0$. Find the currents when $t = 0.001$ second.
2. Let $L = 1$ henry in Problem 1 and assume that all the other facts are unchanged. Find the currents when $t = 0.001$ second and 0.1 second.
3. Let $L = 8$ henrys in Problem 1 and assume that all the other facts are unchanged. Find the currents when $t = 0.001$ second and 0.1 second.
4. Assume that $E = 100e^{-1000t}$ volts and that all the other values are unchanged in Problem 1. Do
   a. Problem 1.
   b. Problem 2.
   c. Problem 3.
5. Repeat Problem 4 for $E = 100 \sin 60\pi t$ volts.
6. Find the current at time t in each loop of the network shown in Figure 3, given that $E = 100$ volts, $R = 10$ ohms, and $L = 10$ henrys.

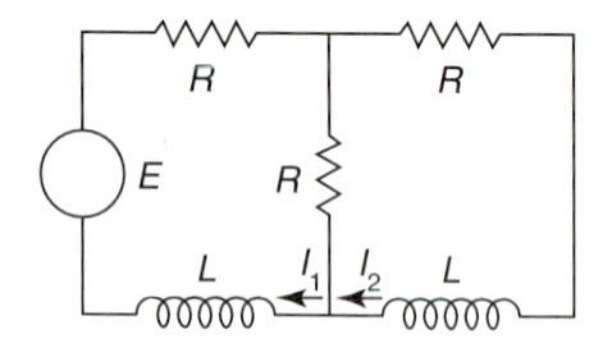

FIGURE 3

7. Repeat Problem 6 for $E = 10 \sin t$ volts.
8. Consider the air-core transformer network shown in Figure 4 with $E = 10 \cos t$ volts, $R = 1$ ohm, $L = 2$ henrys, and mutual inductance $L_* = -1$ (which depends on the relative modes of winding of the two coils involved). Treating the mutual inductance as an inductance for each circuit, find the two circuit currents at all times t assuming they are zero at $t = 0$.

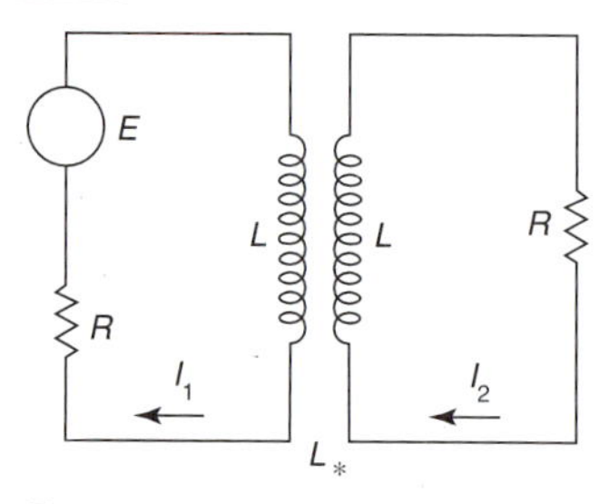

FIGURE 4

9. Consider the circuit shown in Figure 5. Find the current in the top loop if all the resistances are $R = 100$ ohms and the inductances are $L = 1$ henry. Assume that $E = 100$ volts and that there is no current flowing when the switch is closed at time $t = 0$.

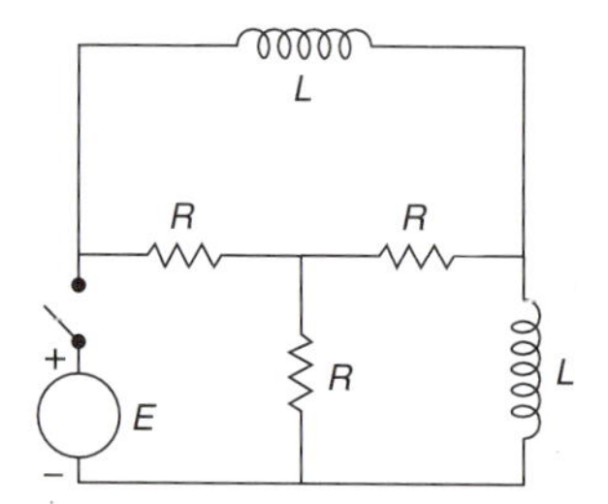

FIGURE 5

10. Repeat Problem 9 for the circuit shown in Figure 6, where $C = 10^{-4}$ farad.

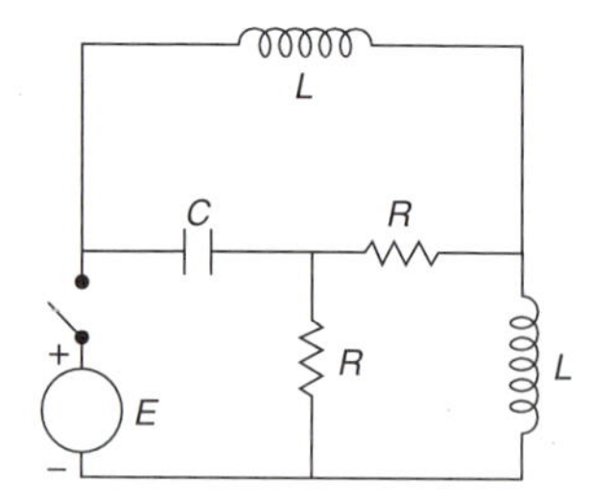

FIGURE 6

11.5 MECHANICAL SYSTEMS

In Section 11.1 we discussed a mass-spring system consisting of two masses and two springs, with one mass and spring pair suspended from another and the latter suspended from a fixed object. In this section we present some additional examples and problems of this type.

EXAMPLE 1 A MASS-SPRING SYSTEM

Consider the mass-spring system in Figure 1, where the point masses m_1 and m_2 are connected in series by springs with spring constants k_1 and k_2. If we denote the vertical

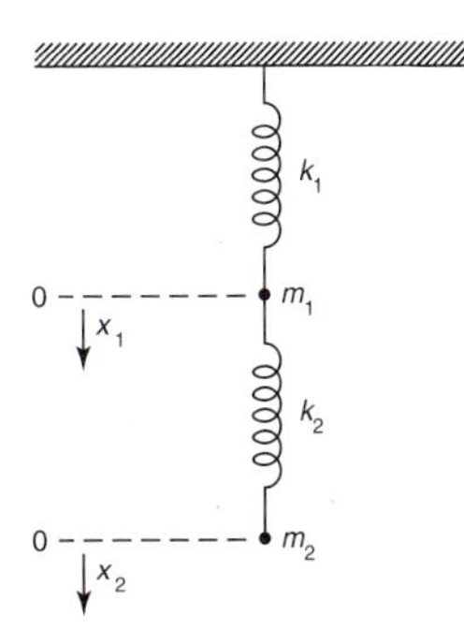

FIGURE 1
Two masses are suspended in series on two springs

displacements from equilibrium of the two masses by $x_1(t)$ and $x_2(t)$, respectively, then using the assumptions of Hooke's law (p. 781), we find that the net forces acting on the two masses are given by

$$F_1 = -k_1x_1 + k_2(x_2 - x_1),$$
$$F_2 = -k_2(x_2 - x_1).$$

Here the positive direction is downward. Note that the first spring is compressed when $x_1 < 0$ and the second spring is compressed when $x_1 > x_2$. The equations of motion are

$$\begin{aligned} m_1\frac{d^2x_1}{dt^2} &= -k_1x_1 + k_2(x_2 - x_1) = -(k_1 + k_2)x_1 + k_2x_2, \\ m_2\frac{d^2x_2}{dt^2} &= -k_2(x_2 - x_1) = k_2x_1 - k_2x_2, \end{aligned} \tag{1}$$

a system of two second-order linear differential equations with constant coefficients.

To rewrite (1) as a system of first-order linear equations we define the variables $x_3 = x_1'$ and $x_4 = x_2'$. Then $x_3' = x_1''$, $x_4' = x_2''$, and (1) can be expressed as the system of four first-order equations

$$\begin{aligned} x_1' &= x_3, \\ x_2' &= x_4, \\ x_1'' = x_3' &= -\left(\frac{k_1 + k_2}{m_1}\right)x_1 + \left(\frac{k_2}{m_1}\right)x_2, \\ x_2'' = x_4' &= \left(\frac{k_2}{m_2}\right)x_1 - \left(\frac{k_2}{m_2}\right)x_2. \end{aligned} \tag{2}$$

The characteristic equation for this system is given by

$$D = \begin{vmatrix} -\lambda & 0 & 1 & 0 \\ 0 & -\lambda & 0 & 1 \\ -\left(\dfrac{k_1 + k_2}{m_1}\right) & \dfrac{k_2}{m_1} & -\lambda & 0 \\ \dfrac{k_2}{m_2} & -\dfrac{k_2}{m_2} & 0 & -\lambda \end{vmatrix} = 0. \tag{3}$$

For simplicity, we assume that $k_1 = 9m_1/2$, $k_2 = 2m_2$, and $7m_1 = 4m_2$. Then (3) becomes

$$0 = \begin{vmatrix} -\lambda & 0 & 1 & 0 \\ 0 & -\lambda & 0 & 1 \\ -8 & \frac{7}{2} & -\lambda & 0 \\ 2 & -2 & 0 & -\lambda \end{vmatrix} = \lambda^4 + 10\lambda^2 + 9 = (\lambda^2 + 9)(\lambda^2 + 1). \tag{4}$$

The four roots are $\lambda_1 = 3i$, $\lambda_2 = -3i$, $\lambda_3 = i$, $\lambda_4 = -i$. Hence, from the theory in Section 11.3, we know that there are constants a_1, a_2, a_3, a_4, b_1, b_2, b_3, and b_4 such that

$$\begin{aligned} x_1 &= a_1\cos 3t + a_2\sin 3t + a_3\cos t + a_4\sin t, \\ x_2 &= b_1\cos 3t + b_2\sin 3t + b_3\cos t + b_4\sin t. \end{aligned} \tag{5}$$

Substituting (5) into (1), we get

$$a_1 = -\tfrac{7}{2}b_1, \qquad a_2 = -\tfrac{7}{2}b_2, \qquad a_3 = \tfrac{1}{2}b_3, \qquad a_4 = \tfrac{1}{2}b_4. \tag{6}$$

If initial conditions are given

$$x_1(0) = x_{10}, \qquad x_1'(0) = x_3(0) = x_{30},$$
$$x_2(0) = x_{20}, \qquad x_2'(0) = x_4(0) = x_{40},$$

we obtain the system of equations

$$\begin{aligned} a_1 \quad + a_3 \quad &= x_{10}, \\ 3a_2 \quad + a_4 &= x_{30}, \\ b_1 \quad + b_3 \quad &= x_{20}, \\ 3b_2 \quad + b_4 &= x_{40}. \end{aligned}$$

Together with (6), this system determines all the coefficients in (5):

$$b_1 = \frac{x_{20} - 2x_{10}}{8}, \qquad b_2 = \frac{x_{40} - 2x_{30}}{24},$$

$$b_3 = \frac{7x_{20} + 2x_{10}}{8}, \qquad b_4 = \frac{7x_{40} + 2x_{30}}{8}.$$

PROBLEMS 11.5

1. Show that the oscillations exhibited by the masses m_1 and m_2 in Example 1 can be represented by the superposition of *two* cosines.
2. Solve Example 1 with $k_1 = 5m_1/2$, $k_2 = 2m_2$, and $3m_1 = 4m_2$.
3. Solve Example 1 with $k_1 = 3m_1/2$, $k_2 = 2m_2$, and $m_1 = 4m_2$.
4. When a water wave travels past any point, the water rises as the wave approaches the point and recedes as it moves past the point. If a ship is in the trough of a wave, its angular acceleration from the vertical is given by

$$\frac{d^2\psi}{dt^2} = -h^2\psi + A \sin \omega t. \qquad \text{(i)}$$

   a. Rewrite (i) as a system of first-order differential equations.
   b. Solve the resulting system by the method of determinants.

*5. A rotating, straight slender shaft may be dynamically unstable at high speeds. When the period ω of rotation is nearly equal to one of its periods of lateral vibration, the shaft is then said to **whirl**. A whirling shaft satisfies the differential equation

$$EI\frac{d^4y}{dx^4} = m\omega^2 y, \qquad \text{(ii)}$$

where $y = y(x)$ is the distance of the shaft from its geometric axis, E denotes Young's modulus of elasticity, I is the moment of inertia of the shaft, and m is the mass. Assume that the shaft is hinged at both ends, has length L, and satisfies

$$y(0) = y(L) = y''(0) = y''(L) = 0.$$

   a. Express (ii) as a system of first-order differential equations.
   b. Show that nontrivial solutions exist if and only if $\omega = n^2\pi^2\sqrt{EI/m}/L^2$.
   c. Find the rotational speed necessary to produce whirling of a steel shaft $1\frac{1}{2}$ inches in diameter and 8 feet long. [*Hint:* $E = 4.32 \times 10^9$, $I = \pi r^4/4$, $m = 6/32.2$.]

6. Consider the coupled mechanical system shown in Figure 2 where two masses m_1 and m_2 rest on a frictionless plane and are attached to fixed walls by two springs with spring constants k_1 and k_3. The masses m_1 and m_2 are connected by a spring with spring constant k_2.

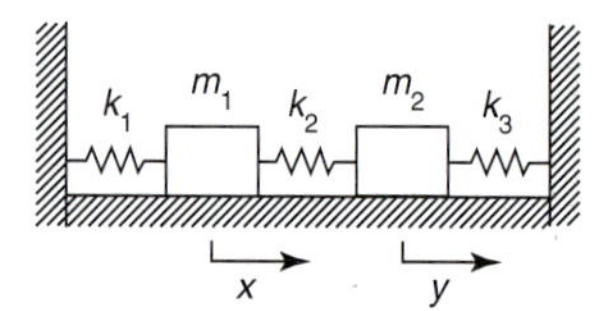

FIGURE 2

a. Obtain the system of differential equations describing the motion of this mechanical system.

b. Solve the system in part (a), and show that the motion of each mass is a superposition of two simple harmonic motions.

***7.** Determine the equations of motion of a double pendulum consisting of two simple pendulums of masses m_1 and m_2 and lengths l_1 and l_2, respectively, shown in Figure 3. You may assume that the angular displacements are so small that $\sin\theta \approx \theta$ and $\sin\phi \approx \phi$.

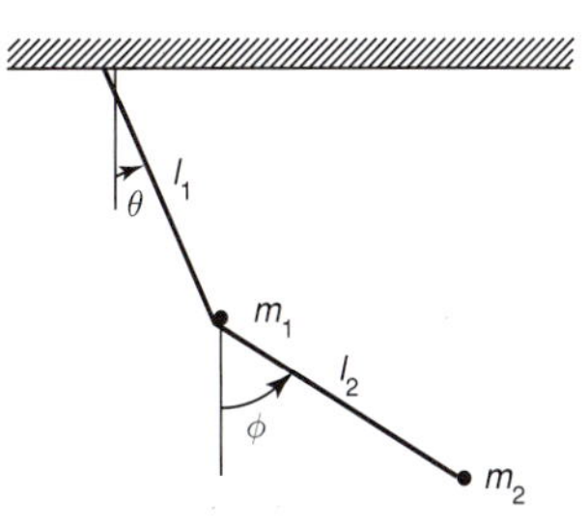

FIGURE 3

8. Using the system in Problem 7, neglecting the terms containing $(\theta')^2$ and $(\phi')^2$, and replacing $\cos(\theta - \phi)$ by 1, show that one obtains the system

$$\begin{aligned} l_1(m_1 + m_2)\theta'' + m_2 l_2\phi'' + g(m_1 + m_2)\theta &= 0, \\ l_1\theta'' + l_2\phi'' + g\phi &= 0. \end{aligned}$$

9. Find the general solution of the system in Problem 8.

11.6 A MODEL FOR EPIDEMICS (OPTIONAL)

In recent years many mathematicians and biologists have attempted to find reasonable mathematical models to describe the growth of an epidemic in a population. One such model has been used by the Center for Disease Control in Atlanta, Georgia, to help form a public testing policy to limit the spread of gonorrhea.† In this section we provide a simple model to describe what may happen in an epidemic.

An **epidemic** is the spread of an infectious disease through a community that affects a significant proportion of the population. The epidemic may begin when a certain number of infected individuals enter the community. This could result, for example, from new people moving to the community or old residents returning from a trip. We make the following assumptions:

a. Everyone in the community is initially susceptible to the disease; that is, no one is immune.

b. The disease is spread only by direct contact between a susceptible person (hereafter called a **susceptible**) and an infected person (called an **infective**).

c. Everyone who has had the disease and has **recovered** is immune. The people who die from the disease are also considered to be in the recovered class in this model.

d. After the disease has been introduced into the community, the total population N of the community remains fixed.

e. The infectives are introduced into the community (that is, the epidemic "starts") at time $t = 0$.

In order to model the spread of the disease, we define three variables:

$x(t)$ is the number of susceptibles at time t

$y(t)$ is the number of infectives at time t

$z(t)$ is the number of recovered persons at time t

† J. A. Yorke, H. W. Hethcote, and A. Nold, "Dynamics and Control of the Transmission of Gonorrhea," *Sexually Transmitted Diseases* 5(2) (1978):51–56.

Then, by assumption (d), we have

$$x(t) + y(t) + z(t) = N. \tag{1}$$

Also, we have

$$y(0) = \text{number of initial infectives}, \tag{2}$$

$$x(0) = N - y(0), \tag{3}$$

$$z(0) = 0. \tag{4}$$

Equation (4) states the obvious fact that no one has yet recovered or died from the disease at the time the epidemic begins.

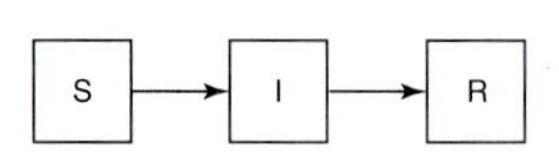

FIGURE 1
An SIR model

To get a better idea of what is going on, look at Figure 1. Here we see that a susceptible can become infective and an infective can recover. These are the only possibilities. For reasons that are obvious, the model we are describing is, in the literature, usually referred to as an **SIR model**.

What mechanism regulates the rate at which the disease spreads? Most likely the disease spreads more rapidly (that is, susceptibles become infectives) if the number of infectives or the number of susceptibles increases, because chances of contact between infectives and susceptibles increase. Thus it is reasonable to assume that the rate of change of the number of susceptibles is proportional both to the number of susceptibles and to the number of infectives. In mathematical terms, we have

$$x'(t) = -\alpha x(t)y(t), \tag{5}$$

where α is a constant of proportionality and the minus sign indicates that the number of susceptibles is decreasing. Equation (5) is called the **law of mass action**.

On the other hand, it is reasonable to assume that the rate at which people recover or die from the disease is proportional to the number of infectives. This gives us the equation

$$z'(t) = \beta y(t). \tag{6}$$

Equations (5) and (6) constitute the system of equations defining our epidemic model. The constant α is often called the **infection rate** and β is called the **removal rate**.

We begin our analysis by dividing equation (5) by equation (6):

$$\frac{x'(t)}{z'(t)} = -\frac{\alpha x(t)y(t)}{\beta y(t)} = -\frac{\alpha}{\beta}x(t). \tag{7}$$

Rearranging the terms in (7) yields

$$\frac{x'(t)}{x(t)} = -\frac{\alpha}{\beta}z'(t),$$

and after integrating both sides we have

$$\int \frac{x'(t)}{x(t)}\,dt = -\frac{\alpha}{\beta}\int z'(t)\,dt + C,$$

or

$$\ln x(t) = -\frac{\alpha}{\beta}z(t) + C. \tag{8}$$

Setting $t = 0$ in (8) and noting that $z(0) = 0$, we have

$$\ln x(0) = 0 + C \qquad \text{or} \qquad C = \ln x(0).$$

Thus

$$\ln x(t) = -\frac{\alpha}{\beta} z(t) + \ln x(0),$$

so

$$\ln \frac{x(t)}{x(0)} = -\frac{\alpha}{\beta} z(t) \qquad \text{or} \qquad \frac{x(t)}{x(0)} = e^{(-\alpha/\beta)z(t)},$$

and, finally,

$$x(t) = x(0)e^{(-\alpha/\beta)z(t)}. \tag{9}$$

A major question about any epidemic is, Can it be controlled? It is clear that the quantity $y'(t)$ is a measure of how bad the epidemic is. The bigger $y'(t)$ is, the greater the number of people becoming infected. We can say that the epidemic has been **controlled** if, at some point, $y'(t) \le 0$. Certainly, at some point $y'(t)$ must become negative. This follows from the fact that the population size N is fixed. Then, if everyone becomes infected, $y'(t)$ will be negative as infectives die or recover, and there cannot possibly be any new infectives (there is no one left to become infected). Thus the epidemic is controlled if $y'(t) \le 0$ for every $t > t_0$.

To analyze this situation, from (1) we have

$$y(t) = N - x(t) - z(t),$$

so

$$y'(t) = -x'(t) - z'(t) = \alpha x(t)y(t) - \beta y(t), \tag{10}$$

or

$$y'(t) = \alpha y(t)\left[x(t) - \frac{\beta}{\alpha}\right]. \tag{11}$$

Thus $y'(t) \le 0$ whenever $x(t) \le \beta/\alpha$. The term β/α is called the **relative removal rate**. We can now answer our question. Since $x' \le 0$ by equation (5), if $x(0) \le \beta/\alpha$, then $x(t) \le \beta/\alpha$ for all t; that is

> The epidemic will be controlled at the start if the condition $x(0) \le \beta/\alpha$ holds.

Biologically, this means that

> An epidemic will not ensue if the initial number of susceptibles (N minus the number of initial infectives) does not exceed the relative removal rate.

EXAMPLE 1 CONDITIONS LEADING TO AN EPIDEMIC

Let $N = 1000$, $y(0) = 50$, $\alpha = 0.0001$, and $\beta = 0.01$. Then $x(0) = 950$ and $\beta/\alpha = 100$, so $x(0) > \beta/\alpha$ and an epidemic will ensue.

EXAMPLE 2 CONDITIONS THAT DO NOT LEAD TO AN EPIDEMIC

Let $N = 1000$, $y(0) = 50$, $\alpha = 0.0001$, and $\beta = 0.1$. Then $x(0) = 950$, $\beta/\alpha = 1000$, and an epidemic will not ensue.

We now ask another question. Suppose that $x(0) > \beta/\alpha$ and an epidemic does ensue. How many people eventually become infected? Equivalently, after the epidemic has run its course, how many susceptibles remain? Since $z(t) \le N$, equation (9) provides a *positive* lower bound for the number of susceptibles at every time t:

$$x(t) = x(0)e^{(-\alpha/\beta)z(t)} \ge x(0)e^{-(\alpha/\beta)N} > 0. \tag{12}$$

Denote the number of susceptibles that never get infected by x^*. The fact that x^* is positive [and greater than or equal to the right side of (12)] has some interesting consequences. By (5) note that $x'(t) < 0$ whenever $x(t)$ and $y(t)$ are positive. Since $x(t)$ does not decrease any lower than x^*, this implies that $x' \to 0$ as $t \to \infty$. But $x \ge x^*$, hence $y \to 0$ as $t \to \infty$. Call $W = N - x^*$ the **extent** of the epidemic: it is the total number of individuals infected. We can obtain a formula for W as follows: Divide equation (10) by equation (5) to obtain

$$\frac{y'(t)}{x'(t)} = \frac{\alpha x(t)y(t) - \beta y(t)}{-\alpha x(t)y(t)} = \frac{\beta}{\alpha x(t)} - 1.$$

Rearranging and integrating both sides, we have

$$\int y'(t)\,dt = \frac{\beta}{\alpha}\int \frac{x'(t)}{x(t)}\,dx - \int x'(t)\,dt + C \quad \text{or} \quad y(t) = \frac{\beta}{\alpha}\ln x(t) - x(t) + C,$$

so that

$$x(t) + y(t) - \frac{\beta}{\alpha}\ln x(t) = C.$$

Setting $t = 0$ and recalling that $N = x(0) + y(0)$, we get

$$C = N - \frac{\beta}{\alpha}\ln x(0), \quad \text{or} \quad x(t) + y(t) - N = \frac{\beta}{\alpha}\ln\frac{x(t)}{x(0)}. \tag{13}$$

Letting $t \to \infty$, we have $x(t) \to x^*$ and $y(t) \to 0$, so equation (13) becomes

$$x^* - N = \frac{\beta}{\alpha}\ln\frac{x^*}{x(0)}, \quad \text{or} \quad -W = \frac{\beta}{\alpha}\ln\frac{N - W}{x(0)}. \tag{14}$$

Multiplying both sides of (14) by α/β and exponentiating, we have

$$e^{(-\alpha/\beta)W} = \frac{N - W}{x(0)},$$

from which we obtain

DETERMINING THE EXTENT, W, OF AN EPIDEMIC

$$N - x(0)e^{(-\alpha/\beta)W} - W = 0. \tag{15}$$

If α and β are > 0, it is impossible to find an explicit solution for W in terms of α, β, and N. The best we can do is solve it numerically. This is a reasonable thing to do using Newton's method from calculus, as we show in the next example.

EXAMPLE 3

USING NEWTON'S METHOD TO COMPUTE THE EXTENT OF THE EPIDEMIC

Let $N = 1000$, $y(0) = 50$, $\alpha = 0.0001$, and $\beta = 0.09$. Compute the extent of the epidemic.

SOLUTION: We first note that $x(0) = 950$ and $\beta/\alpha = 900$, so $x(0) > \beta/\alpha$ and an epidemic will ensue. Since $\alpha/\beta = 1/900$, we must find a root of the equation

$$f(W) = 1000 - 950e^{-W/900} - W = 0. \tag{16}$$

Then

$$f'(W) = \tfrac{950}{900}e^{-W/900} - 1,$$

and by Newton's method we obtain the iterates

$$W_{n+1} = W_n - \frac{f(W_n)}{f'(W_n)},$$

or

$$W_{n+1} = W_n - \frac{1000 - 950e^{-W_n/900} - W_n}{(950/900)e^{-W_n/900} - 1}. \tag{17}$$

We have no idea, initially, what W is (although $W > y(0) = 50$), so we start with a guess: $W_0 = 100$. We then obtain the iterates given in Table 1. After six iterations we obtain the value $W_6 = 370.7732035$, which is correct to ten significant figures, as can be verified by substituting it into equation (16). Thus $W \approx 371$, which means that by the time the epidemic has run its course, 371 individuals have been infected. Of course, we cannot say how many of these remain infected, recover, or die.

TABLE 1

n	W_n	$e^{-W_n/900}$	(a) $1000 - 950e^{-W_n/900} - W_n$	(b) $\frac{950}{900}e^{-W_n/900} - 1$	$\frac{(a)}{(b)}$	$W_{n+1} = W_n - \frac{(a)}{(b)}$
0	100.0000000	0.8948393168	−49.9026490400	−0.0554473878	−900.0000000000	1000.0000000
1	1000.0000000	0.3291929878	−312.7333384000	−0.6525185129	479.2712119000	520.7287882
2	520.7287882	0.5606897580	−53.3840582900	−0.4081608110	130.7917293000	389.9370589
3	389.9370589	0.6483896843	−5.9072589340	−0.3155886666	18.7182226700	371.2188362
4	371.2188362	0.6620161196	−0.1341497754	−0.3012052071	0.4453766808	370.7734595
5	370.7734595	0.6623438079	−0.0000770206	−0.3008593139	0.0002560020	370.7732035
6	370.7732035	0.6623439963	−0.0000000004	−0.3008591150	−0.0000000013	370.7732035

Before leaving this section, we mention some of the limitations of this model. The constants α and β often vary with time. Also, the recovered individuals may, after a time, lose their immunity (or never acquire it) and reenter the susceptible state. This could give rise to periodic epidemics in which individuals get the disease many times. This lack of immunity holds for many diseases, notably gonorrhea and certain types of influenza. Finally, there may be many factors other than relative population sizes that control the sizes of the three classes. Such factors might include weather, the available food supply, living conditions, and the presence of other diseases in the community. However, even a simple model like this one can give us the kind of insight needed to study more complicated situations. In a few cases this has been done with great success.†

PROBLEMS 11.6

In Problems 1–5 values for N, $y(0)$, α, and β are given. Determine whether an epidemic will occur and, if so, find the extent of the epidemic.

1. $N = 1000$, $y(0) = 100$, $\alpha = 0.001$, $\beta = 0.4$.
2. $N = 1000$, $y(0) = 100$, $\alpha = 0.0001$, $\beta = 0.1$.
3. $N = 10{,}000$, $y(0) = 1500$, $\alpha = 10^{-5}$, $\beta = 0.2$.
4. $N = 10{,}000$, $y(0) = 1500$, $\alpha = 10^{-5}$, $\beta = 0.02$.
5. $N = 25{,}000$, $y(0) = 5000$, $\alpha = 10^{-5}$, $\beta = 0.1$.
6. Let $f(W)$ be given by (15). Show that there is exactly one positive value of W for which $f(W) = 0$. [*Hint:* Show that $f(0) > 0$ and that $f''(W) < 0$ for all W and that $f(W) < 0$ if W is sufficiently large.]
7. Prove that the maximum number of infectives, $y_{\max}$, is given by

$$y_{\max} = N + \frac{\beta}{\alpha}\left[\ln\left(\frac{\beta}{\alpha x(0)}\right) - 1\right] \geq N - \frac{\beta}{\alpha}\left[1 + \ln\left(\frac{\alpha N}{\beta}\right)\right] > 0,$$

whenever $x(0) > \beta/\alpha$.

11.7 MATRICES AND SYSTEMS OF LINEAR FIRST-ORDER EQUATIONS

In this section we begin to use the powerful tools of matrix theory to describe the behavior of solutions to systems of differential equations.

We shall assume, from here on, that you are familiar with the elementary properties of vectors and matrices, including vector and matrix addition and scalar multiplication, matrix multiplication, the notion of linear dependence and independence of vectors, and the calculation of the inverse of an invertible matrix.

Before discussing the relationship between matrices and systems of equations, let us consider the notion of a vector and matrix function.

An n-component **vector function**

$$\mathbf{v}(t) = \begin{pmatrix} v_1(t) \\ v_2(t) \\ \vdots \\ v_n(t) \end{pmatrix} \tag{1}$$

† If you are interested in learning more about epidemic models, consult the following: Paul Waltman, *Deterministic Threshold Models in the Theory of Epidemics. Lecture Notes in Biomathematics,* vol. 1. (New York: Springer-Verlag, 1974); Klaus Dietz, "Epidemics and Rumours: A Survey," *Journal of the Royal Statistical Society* Series A., 130 (1967):505–528.

is an n-vector, each of whose components is a function (usually assumed to be continuous). An $n \times n$ **matrix function** $A(t)$ is an $n \times n$ matrix

$$A(t) = \begin{pmatrix} a_{11}(t) & a_{12}(t) & \cdots & a_{1n}(t) \\ a_{21}(t) & a_{22}(t) & \cdots & a_{2n}(t) \\ \vdots & \vdots & & \vdots \\ a_{n1}(t) & a_{n2}(t) & \cdots & a_{nn}(t) \end{pmatrix}, \tag{2}$$

each of whose n^2 components is a function. We may add and multiply vector and matrix functions in the same way that we add and multiply constant vectors and matrices. Thus if, for example,

$$A(t) = \begin{pmatrix} a_{11}(t) & a_{12}(t) \\ a_{21}(t) & a_{22}(t) \end{pmatrix} \quad \text{and} \quad B(t) = \begin{pmatrix} b_{11}(t) & b_{12}(t) \\ b_{21}(t) & b_{22}(t) \end{pmatrix},$$

then

$$A(t)B(t) = \begin{pmatrix} a_{11}(t)b_{11}(t) + a_{12}(t)b_{21}(t) & a_{11}(t)b_{12}(t) + a_{12}(t)b_{22}(t) \\ a_{21}(t)b_{11}(t) + a_{22}(t)b_{21}(t) & a_{21}(t)b_{12}(t) + a_{22}(t)b_{22}(t) \end{pmatrix}.$$

We may also differentiate and integrate vector and matrix functions, component by component. Thus if $\mathbf{v}(t)$ is given by equation (1), then we define

$$\mathbf{v}'(t) = \begin{pmatrix} v_1'(t) \\ v_2'(t) \\ \vdots \\ v_n'(t) \end{pmatrix} \quad \text{and} \quad \int_{t_0}^{t} \mathbf{v}(s)\, ds = \begin{pmatrix} \int_{t_0}^{t} v_1(s)\, ds \\ \int_{t_0}^{t} v_2(s)\, ds \\ \vdots \\ \int_{t_0}^{t} v_n(s)\, ds \end{pmatrix}.$$

Similarly, if $A(t)$ is given by equation (2), then we define

$$A'(t) = \begin{pmatrix} a_{11}'(t) & a_{12}'(t) & \cdots & a_{1n}'(t) \\ a_{21}'(t) & a_{22}'(t) & \cdots & a_{2n}'(t) \\ \vdots & \vdots & \cdots & \vdots \\ a_{n1}'(t) & a_{n2}'(t) & \cdots & a_{nn}'(t) \end{pmatrix}$$

and

$$\int_{t_0}^{t} A(s)\,ds = \begin{pmatrix} \int_{t_0}^{t} a_{11}(s)\,ds & \int_{t_0}^{t} a_{12}(s)\,ds & \cdots & \int_{t_0}^{t} a_{1n}(s)\,ds \\ \int_{t_0}^{t} a_{21}(s)\,ds & \int_{t_0}^{t} a_{22}(s)\,ds & \cdots & \int_{t_0}^{t} a_{2n}(s)\,ds \\ \vdots & \vdots & & \vdots \\ \int_{t_0}^{t} a_{n1}(s)\,ds & \int_{t_0}^{t} a_{n2}(s)\,ds & \cdots & \int_{t_0}^{t} a_{nn}(s)\,ds \end{pmatrix}.$$

EXAMPLE 1

THE DERIVATIVE AND INTEGRAL OF A VECTOR FUNCTION

Let

$$\mathbf{v}(t) = \begin{pmatrix} t \\ t^2 \\ \sin t \\ e^t \end{pmatrix}.$$

Then

$$\mathbf{v}'(t) = \begin{pmatrix} 1 \\ 2t \\ \cos t \\ e^t \end{pmatrix} \quad \text{and} \quad \int_0^t \mathbf{v}(s)\,ds = \begin{pmatrix} t^2/2 \\ t^3/3 \\ 1 - \cos t \\ e^t - 1 \end{pmatrix}.$$

Consider the system of n first-order equations

$$\begin{aligned} x_1' &= a_{11}(t)x_1 + a_{12}(t)x_2 + \cdots + a_{1n}(t)x_n + f_1(t), \\ x_2' &= a_{21}(t)x_1 + a_{22}(t)x_2 + \cdots + a_{2n}(t)x_n + f_2(t), \\ &\vdots \\ x_n' &= a_{n1}(t)x_1 + a_{n2}(t)x_2 + \cdots + a_{nn}(t)x_n + f_n(t), \end{aligned} \tag{3}$$

which is nonhomogeneous if at least one of the functions $f_i(t)$, $i = 1, 2, \ldots, n$, is not the zero function. Consider also the **associated homogeneous system**

$$\begin{aligned} x_1' &= a_{11}(t)x_1 + a_{12}(t)x_2 + \cdots + a_{1n}(t)x_n, \\ x_2' &= a_{21}(t)x_1 + a_{22}(t)x_2 + \cdots + a_{2n}(t)x_n, \\ &\vdots \\ x_n' &= a_{n1}(t)x_1 + a_{n2}(t)x_2 + \cdots + a_{nn}(t)x_n. \end{aligned} \tag{4}$$

As was shown in Section 11.1, systems containing higher-order equations can always be reduced to systems of first-order equations. Hence we shall restrict our discussion to the systems of first-order equations (3) and (4).

We now define the vector function $\mathbf{x}(t)$, the matrix function $A(t)$, and the vector function $\mathbf{f}(t)$ as follows:

$$\mathbf{x}(t) = \begin{pmatrix} x_1(t) \\ x_2(t) \\ \vdots \\ x_n(t) \end{pmatrix}, \qquad A(t) = \begin{pmatrix} a_{11}(t) & a_{12}(t) & \cdots & a_{1n}(t) \\ a_{21}(t) & a_{22}(t) & \cdots & a_{2n}(t) \\ \vdots & \vdots & \cdots & \vdots \\ a_{n1}(t) & a_{n2}(t) & \cdots & a_{nn}(t) \end{pmatrix}, \qquad \mathbf{f}(t) = \begin{pmatrix} f_1(t) \\ f_2(t) \\ \vdots \\ f_n(t) \end{pmatrix}. \tag{5}$$

Then, using equations (5), we can rewrite system (3) as the **vector differential equation**

$$\mathbf{x}'(t) = A(t)\mathbf{x}(t) + \mathbf{f}(t). \tag{6}$$

System (4) becomes

$$\mathbf{x}'(t) = A(t)\mathbf{x}(t). \tag{7}$$

From what has already been said, any linear differential equation or system can be written in the form (7) if it is homogeneous and in the form (6) if it is nonhomogeneous. The reason for writing a system in these forms is that, besides the obvious advantage of compactness of notation, equations (6) and (7) behave very much like first-order linear differential equations, as we shall see. It will be very easy to work with systems of equations in this way once we get used to the notation.

EXAMPLE 2

WRITING A SYSTEM AS A VECTOR DIFFERENTIAL EQUATION

Consider the system

$$\begin{aligned} x_1' &= (2t)x_1 + (\sin t)x_2 - e^t x_3 - e^t, \\ x_2' &= -t^3 x_1 + e^{\sin t} x_2 - (\ln t)x_3 + \cos t, \\ x_3' &= 2x_1 - 5tx_2 + 2tx_3 + \tan t. \end{aligned}$$

It can be rewritten as

$$\begin{pmatrix} x_1 \\ x_2 \\ x_3 \end{pmatrix}' = \begin{pmatrix} 2t & \sin t & -e^t \\ -t^3 & e^{\sin t} & -\ln t \\ 2 & -5t & 2t \end{pmatrix} \begin{pmatrix} x_1 \\ x_2 \\ x_3 \end{pmatrix} + \begin{pmatrix} -e^t \\ \cos t \\ \tan t \end{pmatrix}.$$

EXAMPLE 3

WRITING A SECOND-ORDER EQUATION IN VECTOR FORM

Consider the general second-order linear differential equation

$$x'' + a(t)x' + b(t)x = f(t).$$

Defining $x_1 = x$ and $x_2 = x'$, we obtain the equivalent system

$$\begin{aligned} x_1' &= x_2, \\ x_2' &= -b(t)x_1 - a(t)x_2 + f(t), \end{aligned}$$

which can be rewritten as

$$\mathbf{x}' = A(t)\mathbf{x} + \mathbf{f}(t)$$

where

$$\mathbf{x} = \begin{pmatrix} x_1 \\ x_2 \end{pmatrix}, \qquad A(t) = \begin{pmatrix} 0 & 1 \\ -b(t) & -a(t) \end{pmatrix}, \qquad \mathbf{f}(t) = \begin{pmatrix} 0 \\ f(t) \end{pmatrix}.$$

EXAMPLE 4 WRITING A THIRD-ORDER EQUATION IN VECTOR FORM

Consider the third-order equation with constant coefficients

$$x''' - 6x'' + 11x' - 6x = 0. \tag{8}$$

Defining $x_1 = x$, $x_2 = x'$, and $x_3 = x''$, we obtain the system

$$\begin{aligned} x_1' &= x_2, \\ x_2' &= x_3, \\ x_3' &= 6x_1 - 11x_2 + 6x_3, \end{aligned}$$

or

$$\mathbf{x}' = A\mathbf{x} \tag{9}$$

where

$$\mathbf{x} = \begin{pmatrix} x_1 \\ x_2 \\ x_3 \end{pmatrix} \quad \text{and} \quad A = \begin{pmatrix} 0 & 1 & 0 \\ 0 & 0 & 1 \\ 6 & -11 & 6 \end{pmatrix}.$$

Let us now consider the initial value problem

$$\mathbf{x}'(t) = A(t)\mathbf{x}(t) + \mathbf{f}(t), \qquad \mathbf{x}(t_0) = \mathbf{x}_0, \tag{10}$$

where

$$\mathbf{x}(t) = \begin{pmatrix} x_1(t) \\ x_2(t) \\ \vdots \\ x_n(t) \end{pmatrix} \quad \text{and} \quad \mathbf{x}_0 = \begin{pmatrix} x_{10} \\ x_{20} \\ \vdots \\ x_{n0} \end{pmatrix}.$$

We say that a vector function

$$\boldsymbol{\varphi}(t) = \begin{pmatrix} \varphi_1(t) \\ \vdots \\ \varphi_n(t) \end{pmatrix}$$

is a **solution** to the system (10) if $\boldsymbol{\varphi}$ is differentiable and satisfies the differential equation and the given initial condition. The following theorem is proved in many differential equation texts:†

† See, for example, W. Derrick and S. Grossman, *Elementary Differential Equations with Boundary Value Problems, Third Edition,* West, St. Paul, 1988, Section 10.2

THEOREM 1

Let $A(t)$ and $\mathbf{f}(t)$ be continuous matrix and vector functions, respectively, on some interval $[a, b]$ (i.e., the component functions of both $A(t)$ and $\mathbf{f}(t)$ are continuous). Then there exists a unique vector function $\boldsymbol{\varphi}(t)$ that is a solution to the initial value problem (10) on the entire interval $[a, b]$. ■

EXAMPLE 5 THE UNIQUE VECTOR SOLUTION TO AN INITIAL-VALUE PROBLEM

Referring to Example 4, we find by the methods of Section 10.16 that e^t, e^{2t}, and e^{3t} are solutions of equation (8). Since a solution vector for this problem is

$$\boldsymbol{\varphi}(t) = \begin{pmatrix} x(t) \\ x'(t) \\ x''(t) \end{pmatrix},$$

three vector solutions of equation (9) are

$$\boldsymbol{\varphi}_1(t) = \begin{pmatrix} e^t \\ e^t \\ e^t \end{pmatrix} = e^t\begin{pmatrix} 1 \\ 1 \\ 1 \end{pmatrix}, \qquad \boldsymbol{\varphi}_2(t) = \begin{pmatrix} e^{2t} \\ 2e^{2t} \\ 4e^{2t} \end{pmatrix} = e^{2t}\begin{pmatrix} 1 \\ 2 \\ 4 \end{pmatrix},$$

and

$$\boldsymbol{\varphi}_3(t) = \begin{pmatrix} e^{3t} \\ 3e^{3t} \\ 9e^{3t} \end{pmatrix} = e^{3t}\begin{pmatrix} 1 \\ 3 \\ 9 \end{pmatrix}.$$

If we specify the initial condition

$$\mathbf{x}(0) = \begin{pmatrix} 2 \\ -3 \\ 5 \end{pmatrix},$$

then the unique solution vector is (check it)

$$\boldsymbol{\varphi} = 16\boldsymbol{\varphi}_1 - 23\boldsymbol{\varphi}_2 + 9\boldsymbol{\varphi}_3,$$

or

$$\begin{pmatrix} 16e^t - 23e^{2t} + 9e^{3t} \\ 16e^t - 46e^{2t} + 27e^{3t} \\ 16e^t - 92e^{2t} + 81e^{3t} \end{pmatrix}.$$

EXAMPLE 6 THE UNIQUE VECTOR SOLUTION TO AN INITIAL-VALUE PROBLEM

The system

$$\begin{aligned} x_1' &= -4x_1 - x_2, \qquad & x_1(0) = 1, \\ x_2' &= x_1 - 2x_2, \qquad & x_2(0) = 2, \end{aligned}$$

can be written

$$\begin{pmatrix} x_1 \\ x_2 \end{pmatrix}' = \begin{pmatrix} -4 & -1 \\ 1 & -2 \end{pmatrix}\begin{pmatrix} x_1 \\ x_2 \end{pmatrix}, \qquad \begin{pmatrix} x_1(0) \\ x_2(0) \end{pmatrix} = \begin{pmatrix} 1 \\ 2 \end{pmatrix}.$$

It can be verified that

$$\varphi_1(t) = \begin{pmatrix} e^{-3t} \\ -e^{-3t} \end{pmatrix} \quad \text{and} \quad \varphi_2(t) = \begin{pmatrix} (1-t)e^{-3t} \\ te^{-3t} \end{pmatrix}$$

are solution vectors. We can also show that the unique solution vector that satisfies the given initial conditions is

$$\varphi(t) = \begin{pmatrix} (1-3t)e^{-3t} \\ (2+3t)e^{-3t} \end{pmatrix}.$$

The central problem of the remainder of this chapter is to derive properties of vector solutions and, where possible, to calculate them. In the next section and in Section 11.9 we will show how all solutions to the homogeneous system $\mathbf{x}' = A\mathbf{x}$ can be represented in a convenient form, and in Section 11.10 we will show how information about the solutions to this homogeneous system can be used to find a particular solution to the nonhomogeneous system $\mathbf{x}' = A\mathbf{x} + \mathbf{f}$.

PROBLEMS 11.7

SELF-QUIZ

Multiple Choice

I. The equation $x'' - 2x' + 3x = e^t$ can be written in the form $\mathbf{x}' = A\mathbf{x} + \mathbf{f}(t)$ where $A =$ ________.

a. $\begin{pmatrix} 1 & 0 \\ 3 & -2 \end{pmatrix}$ b. $\begin{pmatrix} 1 & 0 \\ -3 & 2 \end{pmatrix}$

c. $\begin{pmatrix} 0 & 1 \\ 3 & -2 \end{pmatrix}$ d. $\begin{pmatrix} 0 & 1 \\ -3 & 2 \end{pmatrix}$

II. In Problem I, $\mathbf{f}(t) =$ ________.

a. $\begin{pmatrix} e^t \\ e^t \end{pmatrix}$ b. $\begin{pmatrix} e^t \\ 0 \end{pmatrix}$ c. $\begin{pmatrix} 0 \\ e^t \end{pmatrix}$ d. $\begin{pmatrix} 0 \\ 0 \end{pmatrix}$

III. The equation $x''' - t^2x'' + tx' + 3x = 0$ can be written as $\mathbf{x}' = A(t)\mathbf{x}$ where $A(t) =$ ________.

a. $\begin{pmatrix} 0 & 1 & 0 \\ 0 & 0 & 1 \\ -3 & -t & t^2 \end{pmatrix}$ b. $\begin{pmatrix} 0 & 1 & t^2 \\ 0 & -t & 1 \\ -3 & 0 & 0 \end{pmatrix}$

c. $\begin{pmatrix} 0 & 1 & 0 \\ 0 & 0 & 1 \\ 3 & t & -t^2 \end{pmatrix}$ d. $\begin{pmatrix} -3 & -t & t^2 \\ 0 & 1 & 0 \\ 0 & 0 & 1 \end{pmatrix}$

In Problems 1–6 write each equation or system in the matrix-vector form (6), (7), or (10).

1. $x_1' = 2x_1 + 3x_2,$
$x_2' = 4x_1 - 6x_2$

2. $x_1' = (\cos t)x_1 - (\sin t)x_2 + e^{t^2},$
$x_2' = e^t x_1 + 2tx_2 - \ln t,$
$x_1(2) = 3,\ x_2(2) = 7$

3. $x''' - 2x'' + 4tx' - x = \sin t$

4. $x^{(4)} + 2x''' - 3x'' + 4x' - 7x = 0,$
$x(0) = 1,\ x'(0) = 2,\ x''(0) = 3,\ x'''(0) = 4$

5. $x_1' = 2tx_1 - 3t^2x_2 + (\sin t)x_3,$
$x_2' = 2x_1 - 4x_3 - \sin t,$
$x_3' = 17x_2 + 4tx_3 + e^t$

6. $x''' + a(t)x'' + b(t)x' + c(t)x = f(t),$
$x(t_0) = d_1,\ x'(t_0) = d_2,\ x''(t_0) = d_3$

In Problems 7–14, verify that each vector function is a solution to the given system.

7. $\mathbf{x}' = \begin{pmatrix} 1 & 1 \\ -3 & -1 \end{pmatrix}\mathbf{x},$

$$\varphi(t) = \begin{pmatrix} \cos \sqrt{2}t \\ -\sqrt{2} \sin \sqrt{2}t - \cos \sqrt{2}t \end{pmatrix}$$

8. $\mathbf{x}' = \begin{pmatrix} 2 & 1 \\ 1 & 2 \end{pmatrix}\mathbf{x} + \begin{pmatrix} t \\ t^2 \end{pmatrix}$,

$$\varphi(t) = \begin{pmatrix} e^t + \frac{1}{3}t^2 + \frac{2}{9}t + \frac{11}{27} \\ -e^t - \frac{2}{3}t^2 - \frac{7}{9}t - \frac{16}{27} \end{pmatrix}$$

9. $\mathbf{x}' = \begin{pmatrix} -1 & 6 \\ 1 & -2 \end{pmatrix}\mathbf{x}$, $\varphi(t) = \begin{pmatrix} 3e^t \\ e^t \end{pmatrix}$

10. $\mathbf{x}' = \begin{pmatrix} -4 & -1 \\ 1 & -2 \end{pmatrix}\mathbf{x}$, $\varphi(t) = \begin{pmatrix} (1+t)e^{-3t} \\ (-2-t)e^{-3t} \end{pmatrix}$

11. $\mathbf{x}' = \begin{pmatrix} 4 & 1 \\ -8 & 8 \end{pmatrix}\mathbf{x}$, $\varphi(t) = \begin{pmatrix} e^{6t}(\cos 2t - \frac{1}{2}\sin 2t) \\ e^{6t}(\cos 2t - 3\sin 2t) \end{pmatrix}$

12. $\mathbf{x}' = \begin{pmatrix} 1 & 1 & 1 \\ -1 & -1 & 0 \\ -1 & 0 & -1 \end{pmatrix}\mathbf{x}$, $\varphi(t) = \begin{pmatrix} \sin t - \cos t \\ \cos t \\ \cos t \end{pmatrix}$

13. $\mathbf{x}' = \begin{pmatrix} 1 & -1 & 1 & -1 \\ 0 & -1 & 2 & -2 \\ 0 & 0 & 2 & -3 \\ 0 & 0 & 0 & -2 \end{pmatrix}\mathbf{x}$, $\varphi(t) = \begin{pmatrix} e^{-2t} \\ 2e^{-2t} \\ 3e^{-2t} \\ 4e^{-2t} \end{pmatrix}$

14. $\mathbf{x}' = \begin{pmatrix} 3 & 2 & 1 \\ -1 & 0 & -1 \\ 1 & 1 & 2 \end{pmatrix}\mathbf{x}$, $\varphi(t) = \begin{pmatrix} e^{2t} + te^{2t} \\ -te^{2t} \\ te^{2t} \end{pmatrix}$

15. Let $\varphi_1(t)$ and $\varphi_2(t)$ be any two vector solutions of the homogeneous system $\mathbf{x}' = A(t)\mathbf{x}$. Show that $\varphi(t) = c_1\varphi_1(t) + c_2\varphi_2(t)$ is also a solution.

16. Let $\varphi_1(t)$ and $\varphi_2(t)$ be vector solutions of the nonhomogeneous system (6). Show that their difference,

$$\varphi(t) = \varphi_1(t) - \varphi_2(t),$$

is a solution to the homogeneous system (7).

17. Find the derivative and an antiderivative of each of the following vector and matrix functions:

a. $\mathbf{x}(t) = (t, \sin t)$;

b. $\mathbf{y}(t) = \begin{pmatrix} e^t \\ \cos t \\ \tan t \end{pmatrix}$;

c. $A(t) = \begin{pmatrix} \sqrt{t} & t^2 \\ e^{2t} & \sin 2t \end{pmatrix}$;

d. $B(t) = \begin{pmatrix} \ln t & e^t \sin t & e^t \cos t \\ t^{5/2} & -\cos t & -\sin t \\ 1/t & te^{t^2} & t^2e^{t^3} \end{pmatrix}$.

ANSWERS TO SELF-QUIZ

I. d **II.** c **III.** a

11.8 FUNDAMENTAL SETS AND FUNDAMENTAL MATRIX SOLUTIONS OF A HOMOGENEOUS SYSTEM OF DIFFERENTIAL EQUATIONS

In this section we shall discuss properties of the homogeneous system

$$\mathbf{x}' = A(t)\mathbf{x} \tag{1}$$

where $\mathbf{x}(t)$ is an n-vector and $A(t)$ is an $n \times n$ matrix.

Let $\varphi_1(t), \varphi_2(t), \ldots, \varphi_m(t)$ be m vector solutions of the system (1). Recall from Section 8.4 that they are **linearly independent** if the equation

$$c_1\varphi_1(t) + c_2\varphi_2(t) + \cdots + c_m\varphi_m(t) = \mathbf{0}$$

holds only for $c_1 = c_2 = \cdots = c_m = 0$. Since system (1) is equivalent to an nth-order equation, it is natural for us to seek n linearly independent solutions to the system.

DEFINITION FUNDAMENTAL SET OF SOLUTIONS

Any set of n linearly independent solutions to system (1) is called a **fundamental set of solutions**. ■

In Sections 11.2 and 11.3 we saw how a fundamental set of solutions (i.e., two linearly independent solutions) could be found in the case in which A was a 2×2 constant matrix. By Theorem 11.2.5 the vectors

$$\boldsymbol{\varphi}_1(t) = \begin{pmatrix} x_1(t) \\ y_1(t) \end{pmatrix} \quad \text{and} \quad \boldsymbol{\varphi}_2(t) = \begin{pmatrix} x_2(t) \\ y_2(t) \end{pmatrix}$$

are a fundamental set of solutions if and only if the Wronskian $W(t)$, defined by equation (11.2.6), is nonzero.

EXAMPLE 1

A FUNDAMENTAL SET OF SOLUTIONS WHERE THE CHARACTERISTIC EQUATION HAS TWO REAL ROOTS

Let $\mathbf{x}' = A\mathbf{x}$ where

$$A = \begin{pmatrix} -1 & 6 \\ 1 & -2 \end{pmatrix}.$$

In Example 11.3.1 we verified that

$$\boldsymbol{\varphi}_1(t) = \begin{pmatrix} -2e^{-4t} \\ e^{-4t} \end{pmatrix} \quad \text{and} \quad \boldsymbol{\varphi}_2(t) = \begin{pmatrix} 3e^{t} \\ e^{t} \end{pmatrix}$$

are solution vectors. To show that they are fundamental (as we did in Example 11.3.1), we compute the Wronskian:

$$W(\boldsymbol{\varphi}_1, \boldsymbol{\varphi}_2)(t) = \begin{vmatrix} -2e^{-4t} & 3e^{t} \\ e^{-4t} & e^{t} \end{vmatrix} = -5e^{-3t} \neq 0.$$

EXAMPLE 2

A FUNDAMENTAL SET OF SOLUTIONS WHERE THE CHARACTERISTIC EQUATION HAS ONE REAL ROOT

Let $\mathbf{x}' = A\mathbf{x}$ where

$$A = \begin{pmatrix} -4 & -1 \\ 1 & -2 \end{pmatrix}.$$

In Example 11.3.2 we obtained the two solution vectors

$$\boldsymbol{\varphi}_1(t) = \begin{pmatrix} e^{-3t} \\ -e^{-3t} \end{pmatrix} \quad \text{and} \quad \boldsymbol{\varphi}_2(t) = \begin{pmatrix} (1+t)e^{-3t} \\ (-2-t)e^{-3t} \end{pmatrix}.$$

Then

$$W(\boldsymbol{\varphi}_1, \boldsymbol{\varphi}_2)(t) = \begin{vmatrix} e^{-3t} & (1+t)e^{-3t} \\ -e^{-3t} & (-2-t)e^{-3t} \end{vmatrix} = e^{-6t}\begin{vmatrix} 1 & 1+t \\ -1 & -2-t \end{vmatrix} = -e^{-6t} \neq 0.$$

Thus $\boldsymbol{\varphi}_1(t)$ and $\boldsymbol{\varphi}_2(t)$ form a fundamental set of solutions.

EXAMPLE 3

A FUNDAMENTAL SET OF SOLUTIONS WHERE THE CHARACTERISTIC EQUATION HAS COMPLEX CONJUGATE ROOTS

Consider the system $\mathbf{x}' = A\mathbf{x}$ where

$$A = \begin{pmatrix} 4 & 1 \\ -8 & 8 \end{pmatrix}$$

(see Example 11.3.3). Two solution vectors are

$$\boldsymbol{\varphi}_1(t) = \begin{pmatrix} e^{6t}(\cos 2t - \frac{1}{2}\sin 2t) \\ e^{6t}(\cos 2t - 3\sin 2t) \end{pmatrix}, \qquad \boldsymbol{\varphi}_2(t) = \begin{pmatrix} e^{6t}(\sin 2t + \frac{1}{2}\cos 2t) \\ e^{6t}(\sin 2t + 3\cos 2t) \end{pmatrix}.$$

To show that they are linearly independent, let $c_1\boldsymbol{\varphi}_1(t) + c_2\boldsymbol{\varphi}_2(t) = \mathbf{0}$ for every t. Then for $t = 0$, we have

$$\begin{pmatrix} c_1 \\ c_2 \end{pmatrix} + \begin{pmatrix} \frac{1}{2}c_2 \\ 3c_2 \end{pmatrix} = \mathbf{0}$$

or

$$\begin{aligned} c_1 + \tfrac{1}{2}c_2 &= 0, \\ c_1 + 3c_2 &= 0. \end{aligned}$$

The only solution of this system is $c_1 = c_2 = 0$, and so $\boldsymbol{\varphi}_1$ and $\boldsymbol{\varphi}_2$ are, indeed, linearly independent. We can also show this by computing the Wronskian. It is (check it)

$$W(\boldsymbol{\varphi}_1, \boldsymbol{\varphi}_2)(t) = e^{12t}\left(\frac{5}{2}\cos^2 2t + \frac{5}{2}\sin^2 2t\right) = \frac{5}{2}e^{12t} \neq 0.$$

EXAMPLE 4 A FUNDAMENTAL SET OF SOLUTIONS FOR A SYSTEM WITH THREE UNKNOWN FUNCTIONS

Let $\mathbf{x}' = A\mathbf{x}$ where

$$A = \begin{pmatrix} 1 & 1 & -2 \\ -1 & 2 & 1 \\ 0 & 1 & -1 \end{pmatrix}.$$

Then

$$\boldsymbol{\varphi}_1(t) = \begin{pmatrix} e^{-t} \\ 0 \\ e^{-t} \end{pmatrix} = e^{-t}\begin{pmatrix} 1 \\ 0 \\ 1 \end{pmatrix}, \qquad \boldsymbol{\varphi}_2(t) = \begin{pmatrix} 3e^{t} \\ 2e^{t} \\ e^{t} \end{pmatrix} = e^{t}\begin{pmatrix} 3 \\ 2 \\ 1 \end{pmatrix},$$

$$\boldsymbol{\varphi}_3(t) = \begin{pmatrix} e^{2t} \\ 3e^{2t} \\ e^{2t} \end{pmatrix} = e^{2t}\begin{pmatrix} 1 \\ 3 \\ 1 \end{pmatrix}$$

are a fundamental set of solutions. Verify this.

DEFINITION FUNDAMENTAL MATRIX SOLUTION

Let $\boldsymbol{\varphi}_1, \boldsymbol{\varphi}_2, \ldots, \boldsymbol{\varphi}_n$ be n-vector solutions of $\mathbf{x}' = A(t)\mathbf{x}$. Let $\Phi(t)$ be the matrix whose columns are the vectors $\boldsymbol{\varphi}_1, \boldsymbol{\varphi}_2, \ldots, \boldsymbol{\varphi}_n$. That is,

$$\Phi(t) = (\boldsymbol{\varphi}_1(t), \ldots, \boldsymbol{\varphi}_n(t)) = \begin{pmatrix} \varphi_{11}(t) & \varphi_{12}(t) & \cdots & \varphi_{1n}(t) \\ \varphi_{21}(t) & \varphi_{22}(t) & \cdots & \varphi_{2n}(t) \\ \vdots & \vdots & & \vdots \\ \varphi_{n1}(t) & \varphi_{n2}(t) & \cdots & \varphi_{nn}(t) \end{pmatrix}. \tag{2}$$

Such a matrix is called a **matrix solution** of the system $\mathbf{x}' = A\mathbf{x}$. Equivalently, *an $n \times n$ matrix function $\Phi(t)$ is a matrix solution of $\mathbf{x}' = A\mathbf{x}$ if and only if each of its columns is a solution vector of $\mathbf{x}' = A\mathbf{x}$.* If the vectors $\boldsymbol{\varphi}_1, \boldsymbol{\varphi}_2, \ldots, \boldsymbol{\varphi}_n$ form a fundamental set of solutions (i.e., if they are linearly independent), then $\Phi(t)$ is called a **fundamental matrix solution.** ■

In what follows, we shall show that fundamental matrix solutions play a central role in the theory of linear systems of differential equations.

EXAMPLE 5 FOUR FUNDAMENTAL MATRIX SOLUTIONS

In the four previous examples of this section, fundamental matrix solutions were found to be, respectively:

Example 1: $\begin{pmatrix} -2e^{-4t} & 3e^{t} \\ e^{-4t} & e^{t} \end{pmatrix}$;

Example 2: $\begin{pmatrix} e^{-3t} & (1+t)e^{-3t} \\ -e^{-3t} & (-2-t)e^{-3t} \end{pmatrix}$;

Example 3: $\begin{pmatrix} e^{6t}(\cos 2t - \frac{1}{2}\sin 2t) & e^{6t}(\sin 2t + \frac{1}{2}\cos 2t) \\ e^{6t}(\cos 2t - 3\sin 2t) & e^{6t}(\sin 2t + 3\cos 2t) \end{pmatrix}$;

Example 4: $\begin{pmatrix} e^{-t} & 3e^{t} & e^{2t} \\ 0 & 2e^{t} & 3e^{2t} \\ e^{-t} & e^{t} & e^{2t} \end{pmatrix}$.

PRINCIPAL MATRIX SOLUTION

Fundamental matrix solutions are not unique, because a solution vector may be multiplied by any constant and still remain a solution. In addition, any linear combination of solutions is again a solution (see Problem 6). However, we have uniqueness for the **principal matrix solution** $\Psi(t)$, which is defined as that fundamental matrix solution that satisfies the condition

$$\Psi(t_0) = I \tag{3}$$

where I is the $n \times n$ identity matrix and t_0 is the initial value of the independent variable t.

We will show later in this section that if $A(t)$ is continuous, then $\mathbf{x}' = A\mathbf{x}$ always has a unique principal matrix solution. But first we will demonstrate an easy way to determine whether or not a given matrix solution is a fundamental matrix solution.

Let $\Phi(t)$ be a matrix solution of $\mathbf{x}' = A(t)\mathbf{x}$. We define the **Wronskian** of $\Phi(t)$, written $W(t)$, by

$$W(t) = \det \Phi(t). \tag{4}$$

We remind you of two results from matrix theory: First, a matrix A is invertible if and only if $\det A \neq 0$. Second, $\det A \neq 0$ if and only if the columns of A are linearly independent.

From these facts it follows that $\Phi(t)$ will be a fundamental matrix solution if and only if $W(t)$ is nonzero for some t. We will see in the next theorem that $W(t)$ is either always zero or never zero, so that we can calculate $W(t)$ for some especially simple value of t, say $t = 0$, to determine whether Φ is a fundamental matrix solution. Note that many of these theorems are similar to those proven in Section 11.2. In particular, you should compare the present definition (4) of the Wronskian with the definition of the Wronskian in Section 11.2.

THEOREM 1 ABEL'S FORMULA

Let $W(t)$ be the Wronskian of the matrix solution $\Phi(t)$ of the system $\mathbf{x}' = A(t)\mathbf{x}$. Then

$$W(t) = W(t_0)\exp\left(\int_{t_0}^{t} \operatorname{tr} A(s)\,ds\right), \tag{5}$$

where the **trace** of A, written $\operatorname{tr} A(t)$, is the sum of the diagonal elements of the matrix $A(t)$:

$$\operatorname{tr} A(t) = a_{11}(t) + a_{22}(t) + \cdots + a_{nn}(t). \tag{6}$$

PROOF: We prove this theorem for the case of $A(t)$ a 2×2 matrix (see Theorem 11.2.4). The proof for the $n \times n$ case is similar (but more complicated) and is left as an exercise. In the 2×2 case, the system $\mathbf{x}' = A\mathbf{x}$ and the matrix solution may be written as

$$\begin{pmatrix} x_1 \\ x_2 \end{pmatrix}' = \begin{pmatrix} a_{11}(t) & a_{12}(t) \\ a_{21}(t) & a_{22}(t) \end{pmatrix}\begin{pmatrix} x_1 \\ x_2 \end{pmatrix}, \qquad \Phi(t) = (\boldsymbol{\varphi}_1, \boldsymbol{\varphi}_2) = \begin{pmatrix} \varphi_{11} & \varphi_{12} \\ \varphi_{21} & \varphi_{22} \end{pmatrix}.$$

Since $W(t) = \varphi_{11}\varphi_{22} - \varphi_{12}\varphi_{21}$, the derivative

$$W' = \varphi_{11}\varphi_{22}' + \varphi_{11}'\varphi_{22} - \varphi_{12}\varphi_{21}' - \varphi_{12}'\varphi_{21}$$

may be written in determinant form as

$$W' = \begin{vmatrix} \varphi_{11} & \varphi_{12} \\ \varphi_{21}' & \varphi_{22}' \end{vmatrix} + \begin{vmatrix} \varphi_{11}' & \varphi_{12}' \\ \varphi_{21} & \varphi_{22} \end{vmatrix}. \tag{7}$$

But $\varphi_{11}' = a_{11}\varphi_{11} + a_{12}\varphi_{21}$, since $\boldsymbol{\varphi}_1$ is a vector solution, and similarly for φ_{12}', φ_{21}', and φ_{22}'. Replacing these derivatives in (7), we obtain

$$\begin{aligned} W' &= \begin{vmatrix} \varphi_{11} & \varphi_{12} \\ a_{21}\varphi_{11} + a_{22}\varphi_{21} & a_{21}\varphi_{12} + a_{22}\varphi_{22} \end{vmatrix} \\ &\quad + \begin{vmatrix} a_{11}\varphi_{11} + a_{12}\varphi_{21} & a_{11}\varphi_{12} + a_{12}\varphi_{22} \\ \varphi_{21} & \varphi_{22} \end{vmatrix} \\ &= D_1 + D_2. \end{aligned} \tag{8}$$

But according to the theory of determinants (see page 462), a determinant is unchanged when a multiple of one row is added to another row. Also multiplication of every element in one row by a given constant is equivalent to multiplying the determinant by that constant. Hence we may multiply the first row of D_1 by $-a_{21}$ and add it to the second row. Then

$$D_1 = \begin{vmatrix} \varphi_{11} & \varphi_{12} \\ a_{22}\varphi_{21} & a_{22}\varphi_{22} \end{vmatrix} = a_{22}\begin{vmatrix} \varphi_{11} & \varphi_{12} \\ \varphi_{21} & \varphi_{22} \end{vmatrix} = a_{22}W.$$

Similarly, $D_2 = a_{11}W$. Thus

$$W'(t) = [a_{11}(t) + a_{22}(t)]W(t) = [\operatorname{tr} A(t)]W(t). \tag{9}$$

Equation (9) is a first-order (scalar) differential equation that has the solution

$$W(t) = W(t_0)\exp\left(\int_{t_0}^{t} \operatorname{tr} A(s)\, ds\right).$$

This completes the proof for the 2×2 case. This proof has in it the same ideas as the proof of Theorem 11.2.4, but it has to be written in a format that can be extended to the $n \times n$ case (see Problems 13 and 14). ■

EXAMPLE 6

PROOF THAT MATRIX SOLUTIONS ARE FUNDAMENTAL

We consider the four matrix solutions of Example 5. Evaluating each at $t = 0$, we obtain

i. $W(0) = \begin{vmatrix} -2 & 3 \\ 1 & 1 \end{vmatrix} = -5;$

ii. $W(0) = \begin{vmatrix} 1 & 1 \\ -1 & -2 \end{vmatrix} = -1;$

iii. $W(0) = \begin{vmatrix} 1 & \frac{1}{2} \\ 1 & 3 \end{vmatrix} = \frac{5}{2};$

iv. $W(0) = \begin{vmatrix} 1 & 3 & 1 \\ 0 & 2 & 3 \\ 1 & 1 & 1 \end{vmatrix} = 6.$

Therefore, without direct verification of linear independence, we can see that since all four determinants are nonzero, all four matrix solutions are fundamental matrix solutions.

We are now ready to prove the theorem mentioned earlier, namely, that principal matrix solutions exist and are unique. Since principal matrix solutions are fundamental matrix solutions, this theorem also proves the existence of fundamental matrix solutions.

THEOREM 2

Let $A(t)$ be continuous on some interval $[a, b]$. Then for any t_0, $a \le t_0 \le b$, there exists a unique fundamental matrix solution $\Psi(t)$ of the system $\mathbf{x}' = A(t)\mathbf{x}$ satisfying the condition $\Psi(t_0) = I$.

PROOF: Let $\boldsymbol{\delta}_i$, $i = 1, 2, \ldots, n$, denote the n-column vector that has a one in the

ith position (row) and a zero everywhere else:

$$\boldsymbol{\delta}_1 = \begin{pmatrix} 1 \\ 0 \\ 0 \\ \vdots \\ 0 \end{pmatrix}, \quad \boldsymbol{\delta}_2 = \begin{pmatrix} 0 \\ 1 \\ 0 \\ \vdots \\ 0 \end{pmatrix}, \quad \ldots, \quad \boldsymbol{\delta}_n = \begin{pmatrix} 0 \\ 0 \\ \vdots \\ 0 \\ 1 \end{pmatrix}.$$

By the basic existence–uniqueness Theorem 11.7.1 with $\mathbf{f}(t) \equiv \mathbf{0}$, there exists a unique vector solution $\varphi_i(t)$ of $\mathbf{x}' = A\mathbf{x}$ that satisfies $\varphi_i(t_0) = \delta_i$, $i = 1, 2, \ldots, n$. Define the matrix function

$$\Psi(t) = [\varphi_1(t), \varphi_2(t), \ldots, \varphi_n(t)]. \tag{10}$$

Then $\Psi(t)$ is the matrix whose columns are the vector solutions φ_i, $i = 1, 2, \ldots, n$. Clearly $\Psi(t)$ is a matrix solution and $\Psi(t_0) = (\varphi_1(t_0), \ldots, \varphi_n(t_0)) = (\boldsymbol{\delta}_1, \boldsymbol{\delta}_2, \ldots, \boldsymbol{\delta}_n) = I$. It remains to be shown that $\Psi(t)$ is a fundamental matrix solution. This is easy to do. We simply note that $\det \Psi(t_0) = \det I = 1 \neq 0$. ■

ASSOCIATED MATRIX EQUATION

The calculation of a fundamental or principal matrix solution is generally impossible if $A(t)$ is nonconstant. In Section 11.9 we shall show how to compute principal matrix solutions for most constant matrices A and, in Section 11.12, we will show how to compute the principal matrix solution for every constant matrix A. In the remainder of this section we will show how all solutions of $\mathbf{x}' = A(t)\mathbf{x}$ can be expressed in terms of a single fundamental matrix solution.

DEFINITION ASSOCIATED MATRIX EQUATION

Let $X(t)$ denote an $n \times n$ matrix function. Then, we define the **associated matrix equation** to the system $\mathbf{x}' = A(t)\mathbf{x}$:

$$X'(t) = A(t)X(t). \tag{11}$$

We now seek a matrix (instead of a vector) solution of equation (11). The following result is left as an exercise: *$X(t)$ is a solution of the associated matrix equation* (11) *if and only if every column of $X(t)$ is a solution of the system $\mathbf{x}' = A(t)\mathbf{x}$.* ■

THEOREM 3

Let Φ be a matrix solution of $\mathbf{x}' = A(t)\mathbf{x}$ and let C be a constant matrix. Then $\Phi_1 = \Phi C$ is also a matrix solution of $\mathbf{x}' = A(t)\mathbf{x}$.

PROOF: Since a matrix solution of $\mathbf{x}' = A(t)\mathbf{x}$ is also a solution of equation (11), and since Φ is a solution of (11), we must show that Φ_1 is also a solution of (11). If X and Y are differentiable $n \times n$ matrix functions, then it follows from the product rule of differentiation (see Problem 25) that

$$(XY)' = X'Y + XY' \tag{12}$$

Thus, $\Phi_1' = (\Phi C)' = \Phi' C + \Phi C' = \Phi' C$, since $C' = 0$, C being constant. Finally, since Φ is a solution,

$$\Phi_1' = \Phi' C = A\Phi C = A\Phi_1.$$

■

EXAMPLE 7

OBTAINING NEW MATRIX SOLUTIONS BY MATRIX MULTIPLICATION

Consider Example 1, with the fundamental matrix solution

$$\Phi(t) = \begin{pmatrix} -2e^{-4t} & 3e^{t} \\ e^{-4t} & e^{t} \end{pmatrix}.$$

Let

$$C_1 = \begin{pmatrix} 1 & 2 \\ 3 & 4 \end{pmatrix} \quad \text{and} \quad C_2 = \begin{pmatrix} 1 & 2 \\ 2 & 4 \end{pmatrix}.$$

Then

$$\Phi_1 = \Phi C_1 = \begin{pmatrix} -2e^{-4t} + 9e^{t} & -4e^{-4t} + 12e^{t} \\ e^{-4t} + 3e^{t} & 2e^{-4t} + 4e^{t} \end{pmatrix}$$

and

$$\Phi_2 = \Phi C_2 = \begin{pmatrix} -2e^{-4t} + 6e^{t} & -4e^{-4t} + 12e^{t} \\ e^{-4t} + 2e^{t} & 2e^{-4t} + 4e^{t} \end{pmatrix}.$$

It is not difficult to verify that both Φ_1 and Φ_2 are matrix solutions. Note that although Φ_1 is another fundamental matrix solution, Φ_2 is not, since $\det \Phi_2(0) = 0$. Can you explain this (see Problem 10)?

Theorem 3 gives us a way of finding a principal matrix solution when a fundamental matrix solution is known. To see this, let $\Phi(t)$ be a fundamental matrix solution. Since $\det \Phi(t_0) \neq 0$, $\Phi(t_0)$ is invertible, and we define $C = \Phi^{-1}(t_0)$ and $\Psi(t) = \Phi(t)C$. By Theorem 3, $\Psi(t)$ is a matrix solution and $\Psi(t_0) = \Phi(t_0)C = \Phi(t_0)\Phi^{-1}(t_0) = I$, so that $\Psi(t)$ is a principal matrix solution. Thus, if $\Phi(t)$ is a fundamental matrix solution, we can always obtain a principal matrix solution by multiplying $\Phi(t)$ on the right by $\Phi^{-1}(t_0)$.

EXAMPLE 8

FINDING THE PRINCIPAL MATRIX SOLUTION FOR A SYSTEM

A fundamental matrix solution of the system of Example 1 is (see part 1 of Example 5):

$$\Phi(t) = \begin{pmatrix} -2e^{-4t} & 3e^{t} \\ 3e^{-4t} & e^{t} \end{pmatrix} \quad \text{with} \quad \Phi(0) = \begin{pmatrix} -2 & 3 \\ 1 & 1 \end{pmatrix}.$$

Then

$$C = \Phi^{-1}(0) = \begin{pmatrix} -\frac{1}{5} & \frac{3}{5} \\ \frac{1}{5} & \frac{2}{5} \end{pmatrix} = \frac{1}{5}\begin{pmatrix} -1 & 3 \\ 1 & 2 \end{pmatrix},$$

so that

$$\Psi(t) = \Phi(t)C = \tfrac{1}{5}\begin{pmatrix} 2e^{-4t} + 3e^t & -6e^{-4t} + 6e^t \\ -e^{-4t} + e^t & 3e^{-4t} + 2e^t \end{pmatrix}$$

is a principal matrix solution.

THEOREM 4 ANY MATRIX SOLUTION IS A CONSTANT (MATRIX) MULTIPLE OF A FUNDAMENTAL MATRIX SOLUTION

Let $\Phi(t)$ be a fundamental matrix solution and let $X(t)$ be any other matrix solution of the system $\mathbf{x}' = A(t)\mathbf{x}$. Then there exists a constant matrix C such that $X(t) = \Phi(t)C$. That is, *any solution vector of* $\mathbf{x}' = A(t)\mathbf{x}$ *can be written as a linear combination of vectors in a fundamental set.*

Before giving the proof, we remind you that it is important to state on which side of Φ the constant matrix C appears, since matrix multiplication may not commute.

PROOF: Since $\Phi(t)$ is a fundamental matrix solution, $\det \Phi(t) \neq 0$ and $\Phi^{-1}(t)$ exists for every t. We will show that

$$\frac{d}{dt}[\Phi^{-1}(t)X(t)] = 0.$$

This will imply that $\Phi^{-1}(t)X(t)$ is a constant matrix C, and the theorem will be proved. First, we calculate

$$\frac{d}{dt}[\Phi^{-1}(t)].$$

Using the product rule of differentiation, equation (12), we have

$$0 = \frac{dI}{dt} = \frac{d}{dt}(\Phi\Phi^{-1}) = \frac{d\Phi}{dt}\Phi^{-1} + \Phi\frac{d\Phi^{-1}}{dt}. \tag{13}$$

After multiplying both sides of equation (13) on the left by Φ^{-1} and solving for $d\Phi^{-1}/dt$, we obtain

$$\frac{d\Phi^{-1}}{dt} = -\Phi^{-1}\frac{d\Phi}{dt}\Phi^{-1}. \tag{14}$$

Note the analogy between equation (14) and the identity

$$\frac{d}{dt}\left(\frac{1}{f(t)}\right) = -\frac{f'(t)}{(f(t))^2}.$$

Now, by the product formula of derivatives,

$$\frac{d}{dt}(\Phi^{-1}X) = \left(\frac{d}{dt}\Phi^{-1}\right)X + \Phi^{-1}\frac{dX}{dt}, \tag{15}$$

and since both Φ and X are solutions of equation (11), equation (15) becomes

$$\begin{aligned}\frac{d}{dt}(\Phi^{-1}X) &= \left(-\Phi^{-1}\frac{d\Phi}{dt}\Phi^{-1}\right)X + \Phi^{-1}(AX) \\ &= -\Phi^{-1}A\Phi\Phi^{-1}X + \Phi^{-1}AX \\ &= -\Phi^{-1}AX + \Phi^{-1}AX = 0.\end{aligned}$$

■

EXAMPLE 9 ILLUSTRATING THEOREM 4

Consider the system $\mathbf{x}' = A\mathbf{x}$ where

$$A = \begin{pmatrix} 1 & -2 \\ 2 & -3 \end{pmatrix}.$$

It is not difficult to verify that

$$\Phi_1(t) = \begin{pmatrix} e^{-t} & (2t+2)e^{-t} \\ e^{-t} & (2t+1)e^{-t} \end{pmatrix} = e^{-t}\begin{pmatrix} 1 & 2t+2 \\ 1 & 2t+1 \end{pmatrix}$$

is a fundamental matrix solution. Another matrix solution is

$$\Phi_2(t) = e^{-t}\begin{pmatrix} 4t+7 & 8t+1 \\ 4t+5 & 8t-3 \end{pmatrix}.$$

There is a matrix C such that $\Phi_2 = \Phi_1 C$. But $\Phi_1{}^{-1}(t)\Phi_2(t) = C$ for every t, in particular for $t = 0$. Thus

$$C = \Phi_1{}^{-1}(0)\Phi_2(0) = \begin{pmatrix} -1 & 2 \\ 1 & -1 \end{pmatrix}\begin{pmatrix} 7 & 1 \\ 5 & -3 \end{pmatrix} = \begin{pmatrix} 3 & -7 \\ 2 & 4 \end{pmatrix}.$$

EXAMPLE 10 ANOTHER ILLUSTRATION OF THEOREM 4

Consider the system $\mathbf{x}' = A\mathbf{x}$ where

$$A = \begin{pmatrix} 3 & -1 & 1 \\ -1 & 5 & -1 \\ 1 & -1 & 3 \end{pmatrix}.$$

A fundamental matrix solution is

$$\Phi(t) = \begin{pmatrix} e^{2t} & e^{3t} & e^{6t} \\ 0 & e^{3t} & -2e^{6t} \\ -e^{2t} & e^{3t} & e^{6t} \end{pmatrix}.$$

Another matrix solution is

$$X(t) = \begin{pmatrix} e^{2t} + 2e^{3t} + 3e^{6t} & e^{2t} - 3e^{3t} - 2e^{6t} & 2e^{2t} + 5e^{3t} + 7e^{6t} \\ 2e^{3t} - 6e^{6t} & -3e^{3t} + 4e^{6t} & 5e^{3t} - 14e^{6t} \\ -e^{2t} + 2e^{2t} + 3e^{6t} & -e^{2t} - 3e^{3t} - 2e^{6t} & -2e^{2t} + 5e^{3t} + 7e^{6t} \end{pmatrix}.$$

As in the previous example, a matrix C such that $X(t) = \Phi(t)C$ is given by

$$C = \Phi^{-1}(0)X(0) = \tfrac{1}{6}\begin{pmatrix} 3 & 0 & -3 \\ 2 & 2 & 2 \\ 1 & -2 & 1 \end{pmatrix}\begin{pmatrix} 6 & -4 & 14 \\ -4 & 1 & -9 \\ 4 & -6 & 10 \end{pmatrix}$$

$$= \tfrac{1}{6}\begin{pmatrix} 6 & 6 & 12 \\ 12 & -18 & 30 \\ 18 & -12 & 42 \end{pmatrix} = \begin{pmatrix} 1 & 1 & 2 \\ 2 & -3 & 5 \\ 3 & -2 & 7 \end{pmatrix}.$$

THEOREM 5

Let $\Phi(t)$ be a fundamental matrix solution and let $\mathbf{x}(t)$ be any solution of $\mathbf{x}' = A(t)\mathbf{x}$. Then there exists a constant vector $\mathbf{c}$ such that

$$\mathbf{x}(t) = \Phi(t)\mathbf{c}. \tag{16}$$

PROOF: This theorem is an immediate consequence of Theorem 4 if we form the matrix solution $X(t) = (\mathbf{x}(t), \mathbf{x}(t), \ldots, \mathbf{x}(t))$ whose n columns are each the vector solution $\mathbf{x}(t)$. Then a matrix C exists such that $X(t) = \Phi(t)C$. Every column of C is a vector $\mathbf{c}$. ■

EXAMPLE 11 ILLUSTRATING THEOREM 5

In Example 3 we found the fundamental matrix solution

$$\Phi(t) = e^{6t}\begin{pmatrix} \cos 2t - \frac{1}{2}\sin t & \sin 2t + \frac{1}{2}\cos 2t \\ \cos 2t - 3\sin 2t & \sin 2t + 3\cos 2t \end{pmatrix}.$$

The vector

$$\mathbf{x}(t) = e^{6t}\begin{pmatrix} -5\sin 2t \\ -10\cos 2t - 10\sin 2t \end{pmatrix}$$

is also a solution. From equation (16) we obtain

$$\begin{aligned}\mathbf{c} = \Phi^{-1}(t)\mathbf{x}(t) = \Phi^{-1}(0)\mathbf{x}(0) &= \tfrac{2}{5}\begin{pmatrix} 3 & -\frac{1}{2} \\ -1 & 1 \end{pmatrix}\begin{pmatrix} 0 \\ -10 \end{pmatrix} \\ &= \tfrac{2}{5}\begin{pmatrix} 5 \\ -10 \end{pmatrix} = \begin{pmatrix} 2 \\ -4 \end{pmatrix}.\end{aligned}$$

EXAMPLE 12 THE VECTOR SOLUTION TO AN INITIAL VALUE SYSTEM WITH VARIABLE COEFFICIENTS

Consider the system

$$x_1' = \frac{-t}{1-t^2}x_1 + \frac{1}{1-t^2}x_2, \qquad t \neq \pm 1$$

$$x_2' = \frac{1}{1-t^2}x_1 - \frac{t}{1-t^2}x_2, \qquad t \neq \pm 1$$

or

$$\mathbf{x}' = \begin{pmatrix} \dfrac{-t}{1-t^2} & \dfrac{1}{1-t^2} \\ \dfrac{1}{1-t^2} & \dfrac{-t}{1-t^2} \end{pmatrix}\mathbf{x}, \qquad t \neq \pm 1 \tag{17}$$

Here

$$\Phi_1(t) = \begin{pmatrix} t \\ 1 \end{pmatrix} \quad \text{and} \quad \Phi_2(t) = \begin{pmatrix} 1 \\ t \end{pmatrix}$$

are linearly independent solutions, so

$$\Phi(t) = \begin{pmatrix} t & 1 \\ 1 & t \end{pmatrix}$$

is a fundamental matrix solution. First, note that although $W(0) = -1 \neq 0$, we also have $W(1) = 0$. A cursory inspection will help to explain this apparent contradiction of Theorem 1. The matrix $A(t)$ is undefined at $t = 1$; thus there cannot be a solution to equation (17) at $t = 1$. The theorem about Wronskians is, of course, only valid in an interval over which the solution is defined. In this example a suitable interval is $(-1, 1)$, or $(1, \infty)$, or any other interval that does not contain 1 or -1.

Continuing with the example, let us find a solution $\mathbf{x}(t)$ that satisfies the initial conditions

$$\mathbf{x}(0) = \begin{pmatrix} x_1(0) \\ x_2(0) \end{pmatrix} = \begin{pmatrix} 2 \\ -3 \end{pmatrix}.$$

Then

$$\mathbf{c} = \Phi^{-1}(0)\mathbf{x}(0) = \begin{pmatrix} 0 & 1 \\ 1 & 0 \end{pmatrix}\begin{pmatrix} 2 \\ -3 \end{pmatrix} = \begin{pmatrix} -3 \\ 2 \end{pmatrix}.$$

(Note that $\Phi(0) = \begin{pmatrix} 0 & 1 \\ 1 & 0 \end{pmatrix}$ is a matrix that is its own inverse.) Thus

$$\mathbf{x}(t) = \Phi(t)\mathbf{c} = \begin{pmatrix} t & 1 \\ 1 & t \end{pmatrix}\begin{pmatrix} -3 \\ 2 \end{pmatrix} = \begin{pmatrix} 2 - 3t \\ -3 + 2t \end{pmatrix}$$

is a solution vector of equation (17) that satisfies the given initial conditions.

In this section we defined both fundamental and principal matrix solutions (relative to a point t_0) to $\mathbf{x}'(t) = A(t)\mathbf{x}(t)$. Why go to the extra trouble of computing the principal matrix solution? The answer is that in Section 11.10 when we discuss nonhomogeneous systems, we will need to compute the inverse of a fundamental matrix solution. In general, this involves a great deal of algebra. However, the computation of the inverse of a principal matrix solution is sometimes very easy. The following theorem is proved in Section 11.9 (Theorem 11.9.6).

THEOREM 6 THE INVERSE OF THE PRINCIPAL MATRIX SOLUTION

Let $\Psi(t)$ be the principal matrix solution for the system $\mathbf{x}'(t) = A\mathbf{x}(t)$, relative to the point t_0, where A is a constant matrix. Then

$$\Psi^{-1}(t) = \Psi(-t). \tag{18}$$

■

EXAMPLE 13 COMPUTING THE INVERSE OF A PRINCIPAL MATRIX SOLUTION

In Example 8 we found that

$$\Psi(t) = \frac{1}{5}\begin{pmatrix} 2e^{-4t} + 3e^{t} & -6e^{-4t} + 6e^{t} \\ -e^{-4t} + e^{t} & 3e^{-4t} + 2e^{t} \end{pmatrix} \tag{19}$$

is the principal matrix solution relative to $t_0 = 0$ for the system

$$\mathbf{x}' = \begin{pmatrix} -1 & 6 \\ 1 & -2 \end{pmatrix}\mathbf{x}.$$

Then, from (18), we obtain

$$\Psi^{-1}(t) \overset{\text{replace } t \text{ by } -t \text{ in (19)}}{=} \frac{1}{5}\begin{pmatrix} 2e^{4t} + 3e^{-t} & -6e^{4t} + 6e^{-t} \\ -e^{4t} + e^{-t} & 3e^{4t} + 2e^{-t} \end{pmatrix}. \tag{20}$$

It is not difficult to verify that (20) is the inverse matrix for (19).

PROBLEMS 11.8

SELF-QUIZ

Multiple Choice

I. Which of the following could *not* be a fundamental matrix solution for a system of differential equations?

a. $\begin{pmatrix} e^t & e^{2t} \\ 2e^t & e^{2t} \end{pmatrix}$ **b.** $\begin{pmatrix} e^t & e^{2t} \\ 2e^t & 2e^{2t} \end{pmatrix}$

c. $\begin{pmatrix} 4e^t & -2e^{2t} \\ -6e^t & 3e^{2t} \end{pmatrix}$ **d.** $\begin{pmatrix} e^t & -e^{2t} \\ -e^t & -e^{2t} \end{pmatrix}$

II. For $t_0 = 0$, which of the following *could* be the principal matrix solution for a system of differential equations?

a. $\begin{pmatrix} e^{3t}(\cos t + \sin t) & e^{3t}(-\cos t + \sin t) \\ e^{3t}\cos t & e^{3t}\sin t \end{pmatrix}$

b. $\begin{pmatrix} e^{3t}(\cos t + \sin t) & -4e^{3t}\sin t \\ 12e^{3t}\sin t & -e^{3t}(\sin t - \cos t) \end{pmatrix}$

c. $\begin{pmatrix} e^{3t}\sin t & e^{3t}\cos t \\ e^{3t}\cos t & e^{3t}\sin t \end{pmatrix}$

d. $\begin{pmatrix} e^{3t}(\cos t + 5\sin t) & -e^{3t}\sin t \\ -e^{3t}\sin t & \frac{1}{2}e^{3t}(2\cos t - 17\sin t) \end{pmatrix}$

III. Suppose that $\Phi(t)$ is a fundamental matrix solution for a system of differential equations. Then $\Phi(t)C$ is another fundamental matrix solution for which of the following matrices?

a. $C = \begin{pmatrix} 6 & 4 \\ 3 & -2 \end{pmatrix}$ **b.** $C = \begin{pmatrix} 6 & 4 \\ -3 & 2 \end{pmatrix}$

c. $C = \begin{pmatrix} 6 & 4 \\ 3 & 2 \end{pmatrix}$ **d.** $C = \begin{pmatrix} -6 & -4 \\ -3 & 2 \end{pmatrix}$

IV. If $\Phi(t)$ is a fundamental matrix solution for a system, then the principal matrix solution at t_0 is given by

a. $\Phi(t)\Phi^{-1}(t_0)$ **b.** $\Phi^{-1}(t_0)\Phi(t)$
c. $\Phi(t)\Phi^{-1}(t)$ **d.** $\Phi^{-1}(t)\Phi(t)$

V. Suppose that

$$\Psi(t) = \begin{pmatrix} e^{2t}(\cos t + \sin t) & -4e^{2t}\sin t \\ 5e^{2t}\sin t & e^{2t}(\cos t + 2\sin t) \end{pmatrix}$$

is the principal matrix solution for a system of differential equations. Then $\Psi^{-1}(t) =$ ________.

a. $\begin{pmatrix} -e^{-2t}(\cos t + \sin t) & 4e^{-2t}\sin t \\ -5e^{-2t}\sin t & -e^{-2t}(\cos t + 2\sin t) \end{pmatrix}$

b. $\begin{pmatrix} e^{-2t}(\sin t - \cos t) & 4e^{-2t}\sin t \\ 5e^{-2t}\sin t & e^{-2t}(2\sin t - \cos t) \end{pmatrix}$

c. $\begin{pmatrix} e^{-2t}(\cos t - \sin t) & 4e^{-2t}\sin t \\ -5e^{-2t}\sin t & e^{-2t}(\cos t - 2\sin t) \end{pmatrix}$

In Problems 1–5 decide whether each set of solution vectors constitutes a fundamental set of the given system by **(a)** determining whether the vectors are linearly independent, and **(b)** using the Wronskian to determine whether or not $W(t)$ is zero.

1. $\mathbf{x}' = \begin{pmatrix} 2 & 5 \\ 0 & 2 \end{pmatrix}\mathbf{x}$, $\boldsymbol{\varphi}_1(t) = \begin{pmatrix} e^{2t}(1 + 10t) \\ 2e^{2t} \end{pmatrix}$,

$\boldsymbol{\varphi}_2(t) = \begin{pmatrix} e^{2t}(-3 + 20t) \\ 4e^{2t} \end{pmatrix}$

2. $\mathbf{x}' = \begin{pmatrix} 4 & -13 \\ 2 & -6 \end{pmatrix}\mathbf{x}$,

$\varphi_1(t) = \begin{pmatrix} e^{-t}(13\cos t - 26\sin t) \\ e^{-t}(7\cos t - 9\sin t) \end{pmatrix}$,

$\varphi_2(t) = \begin{pmatrix} e^{-t}(26\cos t - 52\sin t) \\ e^{-t}(14\cos t - 18\sin t) \end{pmatrix}$

3. $\mathbf{x}' = \begin{pmatrix} 1 & 1 \\ 4 & 1 \end{pmatrix}\mathbf{x}$, $\varphi_1(t) = \begin{pmatrix} e^{3t} - e^{-t} \\ 2e^{3t} + 2e^{-t} \end{pmatrix}$,

$\varphi_2(t) = \begin{pmatrix} 2e^{3t} \\ 4e^{3t} \end{pmatrix}$

4. $\mathbf{x}' = \begin{pmatrix} 1 & -1 & 4 \\ 3 & 2 & -1 \\ 2 & 1 & -1 \end{pmatrix}\mathbf{x}$,

$\varphi_1(t) = \begin{pmatrix} e^t + 2e^{-2t} + 3e^{3t} \\ -4e^t - 2e^{-2t} + 6e^{3t} \\ -e^t - 2e^{-2t} + 3e^{3t} \end{pmatrix}$,

$\varphi_2(t) = \begin{pmatrix} -2e^t + 2e^{-2t} \\ 8e^t - 2e^{-2t} \\ 2e^t - 2e^{-2t} \end{pmatrix}$,

$\varphi_3(t) = \begin{pmatrix} 3e^t - 6e^{-2t} + 3e^{3t} \\ -12e^t + 6e^{-2t} + 6e^{3t} \\ -3e^t + 6e^{-2t} + 3e^{3t} \end{pmatrix}$

5. $\mathbf{x}' = \begin{pmatrix} 3 & 2 & 1 \\ -1 & 0 & -1 \\ 1 & 1 & 2 \end{pmatrix}\mathbf{x}$, $\varphi_1(t) = \begin{pmatrix} -e^t + te^{2t} + e^{2t} \\ e^t - te^{2t} \\ te^{2t} \end{pmatrix}$,

$\varphi_2(t) = \begin{pmatrix} 2e^{2t} + te^{2t} \\ -e^{2t} - te^{2t} \\ e^{2t} + te^{2t} \end{pmatrix}$,

$\varphi_3(t) = \begin{pmatrix} -e^t + 3e^{2t} + 2te^{2t} \\ e^t - e^{2t} - 2te^{2t} \\ e^{2t} + 2te^{2t} \end{pmatrix}$

6. Let $\varphi_1(t), \varphi_2(t), \ldots, \varphi_m(t)$ be m solutions of the homogeneous system (1). Show that

$$\varphi(t) = c_1\varphi_1(t) + c_2\varphi_2(t) + \cdots + c_m\varphi_m(t)$$

is also a solution.

In each of Problems 7–9 two matrix functions Φ_1 and Φ_2 are given. Find a matrix C such that $\Phi_2(t) = \Phi_1(t)C$.

7. $\Phi_1(t) = e^{6t}\begin{pmatrix} \cos 2t - \frac{1}{2}\sin 2t & \sin 2t + \frac{1}{2}\cos 2t \\ \cos 2t - 3\sin 2t & \sin 2t + 3\cos 2t \end{pmatrix}$,

$\Phi_2(t) = e^{6t}\begin{pmatrix} \frac{1}{2}\cos 2t - \frac{3}{2}\sin 2t & -\frac{3}{2}\cos 2t + 2\sin 2t \\ -2\cos 2t - 4\sin 2t & \cos 2t + 7\sin 2t \end{pmatrix}$

8. $\Phi_1(t) = \begin{pmatrix} \sin e^t & \cos e^t \\ e^t\cos e^t & -e^t\sin e^t \end{pmatrix}$,

$\Phi_2(t) = \begin{pmatrix} \cos e^t & 3\sin e^t + 2\cos e^t \\ -e^t\sin e^t & 3e^t\cos e^t - 2e^t\sin e^t \end{pmatrix}$

9. $\Phi_1(t) = \begin{pmatrix} e^{-t} & 3e^t & e^{2t} \\ 0 & 2e^t & 3e^{2t} \\ e^{-t} & e^t & e^{2t} \end{pmatrix}$,

$\Phi_2(t) = \begin{pmatrix} e^{-t} + e^{2t} & -e^{-t} + 3e^t & e^{-t} + 3e^t \\ 3e^{2t} & 2e^t & 2e^t \\ e^{-t} + e^{2t} & -e^{-t} + e^t & e^{-t} + e^t \end{pmatrix}$

10. Let $\Phi_1(t)$ be a fundamental matrix solution of $\mathbf{x}' = A\mathbf{x}$. Then $\Phi_2 = \Phi_1 C$ is a matrix solution for any constant matrix C. Show that Φ_2 is a fundamental matrix solution if and only if C is nonsingular.

11. Let

$$Y(t) = \begin{vmatrix} a_{11} & a_{12} & a_{13} \\ a_{21} & a_{22} & a_{23} \\ a_{31} & a_{32} & a_{33} \end{vmatrix}$$

where a_{ij} is a differentiable function for $i, j = 1, 2, 3$. Show that

$$Y'(t) = \begin{vmatrix} a'_{11} & a'_{12} & a'_{13} \\ a_{21} & a_{22} & a_{23} \\ a_{31} & a_{32} & a_{33} \end{vmatrix} + \begin{vmatrix} a_{11} & a_{12} & a_{13} \\ a'_{21} & a'_{22} & a'_{23} \\ a_{31} & a_{32} & a_{33} \end{vmatrix} + \begin{vmatrix} a_{11} & a_{12} & a_{13} \\ a_{21} & a_{22} & a_{23} \\ a'_{31} & a'_{32} & a'_{33} \end{vmatrix}$$

***12.** Let

$$Y(t) = \begin{vmatrix} a_{11} & a_{12} & \cdots & a_{1n} \\ a_{21} & a_{22} & \cdots & a_{2n} \\ \vdots & \vdots & & \vdots \\ a_{n1} & a_{n2} & \cdots & a_{nn} \end{vmatrix}$$

where each a_{ij} is a differentiable function for $i, j = 1, 2, \ldots, n$. Use mathematical induction (see Appendix 1) to prove that

$$Y'(t) = \begin{vmatrix} a'_{11} & a'_{12} & \cdots & a'_{1n} \\ a_{21} & a_{22} & \cdots & a_{2n} \\ \vdots & \vdots & & \vdots \\ a_{n1} & a_{n2} & \cdots & a_{nn} \end{vmatrix} + \begin{vmatrix} a_{11} & a_{12} & \cdots & a_{1n} \\ a'_{21} & a'_{22} & \cdots & a'_{2n} \\ \vdots & \vdots & & \vdots \\ a_{n1} & a_{n2} & \cdots & a_{nn} \end{vmatrix}$$

$$+\cdots+\begin{vmatrix} a_{11} & a_{12} & \cdots & a_{1n} \\ a_{21} & a_{22} & \cdots & a_{2n} \\ \vdots & \vdots & & \vdots \\ a'_{n1} & a'_{n2} & \cdots & a'_{nn} \end{vmatrix}$$

13. Let $\Phi(t)$ be a matrix solution of the system $\mathbf{x}' = A(t)\mathbf{x}$ where $A(t)$ is a 3×3 matrix. Prove that

$$W(t) = W(t_0) \exp\left[\int_{t_0}^{t} [a_{11}(s) + a_{22}(s) + a_{33}(s)]\, ds\right].$$

[*Hint:* Use the result of Problem 11.]

***14.** Using the result of Problem 12, prove Theorem 1 for the case in which $A(t)$ is an $n \times n$ matrix.

In Problems 15–18 find the principal matrix solution $\Psi(t)$ corresponding to each fundamental matrix solution $\Phi(t)$. Assume that $t_0 = 0$.

15. $\Phi(t) = e^{6t}\begin{pmatrix} \cos 2t - \frac{1}{2}\sin 2t & \sin 2t + \frac{1}{2}\cos 2t \\ \cos 2t - 3\sin 2t & \sin 2t + 3\cos 2t \end{pmatrix}$

16. $\Phi(t) = \begin{pmatrix} \sin e^t & \cos e^t \\ e^t \cos e^t & -e^t \sin e^t \end{pmatrix}$

17. $\Phi(t) = \begin{pmatrix} e^{-t} & 3e^t & e^{2t} \\ 0 & 2e^t & 3e^{2t} \\ e^{-t} & e^t & e^{2t} \end{pmatrix}$

18. $\Phi(t) = \begin{pmatrix} 2e^t + te^t & te^t + e^t & e^t \\ 0 & e^t & e^t \\ -3e^t - te^t & -te^t - 2e^t & -e^t \end{pmatrix}$

19. Consider the system

$$\mathbf{x}' = \begin{pmatrix} 3 & -2 \\ 2 & -1 \end{pmatrix}\mathbf{x}.$$

Verify that

$$\Phi(t) = \begin{pmatrix} 2te^t + e^t & 2te^t \\ 2te^t & -e^t + 2te^t \end{pmatrix}$$

is a fundamental matrix solution. Then find a solution that satisfies each of the following initial conditions:

a. $\mathbf{x}(0) = \begin{pmatrix} 1 \\ 2 \end{pmatrix}$; **b.** $\mathbf{x}(0) = \begin{pmatrix} -2 \\ 3 \end{pmatrix}$;

c. $\mathbf{x}(1) = \begin{pmatrix} 0 \\ 1 \end{pmatrix}$; **d.** $\mathbf{x}(-1) = \begin{pmatrix} 2 \\ 1 \end{pmatrix}$;

e. $\mathbf{x}(3) = \begin{pmatrix} 3 \\ 3 \end{pmatrix}$; **f.** $\mathbf{x}(a) = \begin{pmatrix} b \\ c \end{pmatrix}$.

20. In Example 5 we saw that a fundamental matrix solution to the system

$$\mathbf{x}' = \begin{pmatrix} 1 & 1 & -2 \\ -1 & 2 & 1 \\ 0 & 1 & -1 \end{pmatrix}\mathbf{x}$$

of Example 4 was

$$\Phi(t) = \begin{pmatrix} e^{-t} & 3e^t & e^{2t} \\ 0 & 2e^t & 3e^{2t} \\ e^{-t} & e^t & e^{2t} \end{pmatrix}.$$

Find a particular solution that satisfies each of the following conditions:

a. $\mathbf{x}(0) = \begin{pmatrix} 1 \\ -1 \\ 2 \end{pmatrix}$; **b.** $\mathbf{x}(0) = \begin{pmatrix} 3 \\ 1 \\ 2 \end{pmatrix}$;

c. $\mathbf{x}(1) = \begin{pmatrix} 1 \\ 0 \\ 1 \end{pmatrix}$; **d.** $\mathbf{x}(-1) = \begin{pmatrix} 2 \\ -3 \\ 5 \end{pmatrix}$.

21. Consider the second-order equation

$$x'' + a(t)x' + b(t)x = 0. \tag{21}$$

a. Write equation (18) in the form $\mathbf{x}' = A(t)\mathbf{x}$.

b. Given that

$$\Phi(t) = \begin{pmatrix} \varphi_1 & \varphi_2 \\ \varphi'_1 & \varphi'_2 \end{pmatrix}$$

is a fundamental matrix solution, show that

$$\det \Phi(t) = \det \Phi(t_0) \exp\left[-\int_{t_0}^{t} a(s)\, ds\right].$$

c. Show that the formula in part (b) can be rearranged as

$$\varphi'_2 - \frac{\varphi'_1}{\varphi_1}\varphi_2 = \frac{\det \Phi(t_0)}{\varphi_1}\exp\left[-\int_{t}^{t} a(s)\, ds\right].$$

Therefore, if one solution $\varphi_1(t)$ of (21) is known, another solution can be calculated by solving this equation.

22. Given that $\varphi_1(t) = \sin(\ln t)$, find a second linearly independent solution of

$$x'' + \frac{1}{t}x' + \frac{1}{t^2}x = 0.$$

23. Given that $\varphi_1(t) = e^{t^2}$ is a solution of

$$x'' - 2tx' - 2x = 0,$$

find a second linearly independent solution.

24. Given that $\varphi_1(t) = \sin t^2$ is a solution of

$$tx'' - x' + 4t^3x = 0,$$

find a second linearly independent solution.

***25.** Let $X(t)$ and $Y(t)$ be differentiable $n \times n$ matrix functions. Show that XY is differentiable and that

$$(XY)' = X'Y + XY'$$

ANSWERS TO SELF-QUIZ

I. b, c **II.** b, d **III.** a, b, d **IV.** a (both c. and d. $= I$) **V.** c

---

11.9 THE COMPUTATION OF THE PRINCIPAL MATRIX SOLUTION TO A HOMOGENEOUS SYSTEM OF EQUATIONS WITH CONSTANT COEFFICIENTS

Consider the homogeneous system of differential equations

$$\mathbf{x}' = A\mathbf{x}(t) \tag{1}$$

where A is a constant $n \times n$ matrix. In this section we show how to use matrix theory directly to compute the principal matrix solution of a system whenever A is diagonalizable. We give a more general procedure for doing this in Section 11.12. The material in this section requires familiarity with eigenvalues, eigenvectors and diagonalization. These topics were discussed in Sections 8.13 and 8.15.

First, let us consider the scalar differential equation

$$x'(t) = ax(t). \tag{2}$$

Note that equation (1) is almost identical to equation (2). The only difference is that now we have a vector function and a matrix whereas before we had a ''scalar'' function and a number (1×1 matrix).

To solve equation (1), we might guess that a solution would have the form e^{At}. But what does e^{At} mean? We shall answer this question in a moment. First, let us recall the series expansion of the function e^t:

$$e^t = 1 + t + \frac{t^2}{2!} + \frac{t^3}{3!} + \frac{t^4}{4!} + \cdots .^{\dagger} \tag{3}$$

This series converges for every real number t. Then, for any real number a,

$$e^{at} = 1 + at + \frac{(at)^2}{2!} + \frac{(at)^3}{3!} + \frac{(at)^4}{4!} + \cdots . \tag{4}$$

DEFINITION THE MATRIX e^A

Let A be an $n \times n$ matrix with real (or complex) entries. Then e^A is an $n \times n$ matrix defined by

$$e^A = I + A + \frac{A^2}{2!} + \frac{A^3}{3!} + \frac{A^4}{4!} + \cdots . \tag{5}$$

■

[†] This series is derived in Section 12.13 of this book.

REMARK: It is possible to define convergence of a series of matrices and to prove that the series of matrices in equation (5) converges for every matrix A, but to do so would take us too far afield. We can, however, give an indication of why it is so. We first define $|A|_i$ to be the sum of the absolute values of the components in the ith row of A. We then define the norm† of A, denoted by $|A|$, by

$$|A| = \max_{1 \leq i \leq n} |A|_i. \tag{6}$$

It can be shown that

$$|AB| \leq |A||B| \tag{7}$$

and

$$|A + B| \leq |A| + |B|. \tag{8}$$

Then, using (7) and (8) in (5), we obtain

$$|e^A| \leq 1 + |A| + \frac{|A|^2}{2!} + \frac{|A|^3}{3!} + \frac{|A|^4}{4!} + \cdots = e^{|A|}.$$

Since $|A|$ is a real number, $e^{|A|}$ is finite. This shows that the series in (5) converges for any matrix A.

We shall now see the usefulness of the series in equation (5).

THEOREM 1

For any constant vector $\mathbf{c}$, $\mathbf{x}(t) = e^{At}\mathbf{c}$ is a solution of (1). Moreover, the solution of (1) given by $\mathbf{x}(t) = e^{At}\mathbf{x}_0$ satisfies $\mathbf{x}(0) = \mathbf{x}_0$.

PROOF: We compute, using (5),

$$\mathbf{x}(t) = e^{At}\mathbf{c} = \left(I + At + A^2\frac{t^2}{2!} + A^3\frac{t^2}{3!} + \cdots\right)\mathbf{c}. \tag{9}$$

Since A is a constant matrix, we have

$$\begin{aligned}\frac{d}{dt}A^k\frac{t^k}{k!} &= \frac{d}{dt}\frac{t^k}{k!}A^k = \frac{kt^{k-1}}{k!}A^k \\ &= \frac{A^k t^{k-1}}{(k-1)!} = A\left[A^{k-1}\frac{t^{k-1}}{(k-1)!}\right].\end{aligned} \tag{10}$$

Then, combining (9) and (10), and using the fact that $\mathbf{c}$ is a constant vector, we obtain

$$\mathbf{x}'(t) = \frac{d}{dt}e^{At}\mathbf{c} = A\left(I + At + A^2\frac{t^2}{2!} + A^3\frac{t^2}{3!} + \cdots\right)\mathbf{c} = Ae^{At}\mathbf{c} = A\mathbf{x}(t).$$

Finally, since $e^{A \cdot 0} = e^0 = I$, we have

$$\mathbf{x}(0) = e^{A\cdot 0}\mathbf{x}_0 = I\mathbf{x}_0 = \mathbf{x}_0.$$

■

REMARK: From Theorem 1 we can show that e^{At} is a solution matrix to (1). Furthermore, since $e^{A0} = I$, e^{At} **is the principal matrix solution.**

---

†This is called the **max-row sum norm** of A.

A major difficulty remains: How do we compute e^{At} without using a series? We begin with an example.

EXAMPLE 1

COMPUTING e^{A} WHEN A IS A DIAGONAL MATRIX

Let $A = \begin{pmatrix} 1 & 0 & 0 \\ 0 & 2 & 0 \\ 0 & 0 & 3 \end{pmatrix}$. Then

$$A^2 = \begin{pmatrix} 1 & 0 & 0 \\ 0 & 2^2 & 0 \\ 0 & 0 & 3^2 \end{pmatrix}, \quad A^3 = \begin{pmatrix} 1 & 0 & 0 \\ 0 & 2^3 & 0 \\ 0 & 0 & 3^3 \end{pmatrix}, \quad \ldots, \quad A^m = \begin{pmatrix} 1 & 0 & 0 \\ 0 & 2^m & 0 \\ 0 & 0 & 3^m \end{pmatrix}$$

and

$$e^{At} = I + At + \frac{A^2t^2}{2!} + \frac{A^3t^3}{3!} + \cdots$$

$$= \begin{pmatrix} 1 & 0 & 0 \\ 0 & 1 & 0 \\ 0 & 0 & 1 \end{pmatrix} + \begin{pmatrix} t & 0 & 0 \\ 0 & 2t & 0 \\ 0 & 0 & 3t \end{pmatrix} + \begin{pmatrix} \frac{t^2}{2!} & 0 & 0 \\ 0 & \frac{2^2t^2}{2!} & 0 \\ 0 & 0 & \frac{3^2t^2}{2!} \end{pmatrix} + \begin{pmatrix} \frac{t^3}{3!} & 0 & 0 \\ 0 & \frac{2^3t^3}{3!} & 0 \\ 0 & 0 & \frac{3^3t^3}{3!} \end{pmatrix} + \cdots$$

$$= \begin{pmatrix} 1 + t + \frac{t^2}{2!} + \frac{t^3}{3!} + \cdots & 0 & 0 \\ 0 & 1 + (2t) + \frac{(2t)^2}{2!} + \frac{(2t)^3}{3!} + \cdots & 0 \\ 0 & 0 & 1 + (3t) + \frac{(3t)^2}{2!} + \frac{(3t)^3}{3!} + \cdots \end{pmatrix}$$

$$= \begin{pmatrix} e^t & 0 & 0 \\ 0 & e^{2t} & 0 \\ 0 & 0 & e^{3t} \end{pmatrix}.$$

As Example 1 illustrates, if $D = \text{diag}(\lambda_1, \lambda_2, \ldots, \lambda_n)$, then $e^{Dt} = \text{diag}(e^{\lambda_1 t}, e^{\lambda_2 t}, \ldots, e^{\lambda_n t})$. It turns out that if A is diagonalizable, this is really all we need to do, as the next theorem suggests.

THEOREM 2

Let A be diagonalizable and suppose that $D = C^{-1}AC$. Then $A = CDC^{-1}$ and

$$e^{At} = Ce^{Dt}C^{-1}. \tag{11}$$

PROOF: We first note that

$$\begin{aligned} A^n &= (CDC^{-1})^n = \overbrace{(CDC^{-1})(CDC^{-1})\cdots(CDC^{-1})}^{n \text{ times}} \\ &= CD(C^{-1}C)D(C^{-1}C)D(C^{-1}C)\cdots(C^{-1}C)DC^{-1} \\ &= CD^nC^{-1}. \end{aligned}$$

It follows that

$$(At)^n = C(Dt)^nC^{-1}. \tag{12}$$

Thus

$$\begin{aligned} e^{At} &= I + (At) + \frac{(At)^2}{2!} + \cdots = CIC^{-1} + C(Dt)C^{-1} + C\frac{(Dt)^2}{2!}C^{-1} + \cdots \\ &= C\left[I + (Dt) + \frac{(Dt)^2}{2!} + \cdots\right]C^{-1} = Ce^{Dt}C^{-1}. \end{aligned}$$

■

There is a way to compute e^{At} for every matrix A—whether or not A is diagonalizable. This method uses an important result in matrix theory—the Cayley-Hamilton theorem. We present this method in Section 11.12.

We now apply our theory to a simple biological model of population growth. Suppose that in an ecosystem there are two interacting species S_1 and S_2. We denote the populations of the species at time t by $x_1(t)$ and $x_2(t)$. One system governing the relative growth of the two species is

$$\begin{aligned} x_1'(t) &= ax_1(t) + bx_2(t), \\ x_2'(t) &= cx_1(t) + dx_2(t). \end{aligned} \tag{13}$$

We can interpret the constants a, b, c, and d as follows. If the species are competing, it is reasonable to have $b < 0$ and $c < 0$. This is true because increases in the population of one species will slow the growth of the other. A second model is a **predator–prey** relationship. If S_1 is the prey and S_2 is the predator (S_2 eats S_1), then it is reasonable to have $b < 0$ and $c > 0$, since an increase in the predator species will cause a decrease in the prey species, while an increase in the prey species will cause an increase in the predator species (since it will have more food). Finally, in a **symbiotic** relationship (each species lives off the other), we would likely have $b > 0$ and $c > 0$. Of course, the constants a, b, c, and d depend on a wide variety of factors including available food, time of year, climate, limits due to overcrowding, other competing species, and so on. We shall analyze three different models by using the material in this section. We assume that t is measured in years.

EXAMPLE 2 A COMPETITIVE MODEL

Consider the system

$$\begin{aligned} x_1'(t) &= 3x_1(t) - x_2(t), \\ x_2'(t) &= -2x_1(t) + 2x_2(t). \end{aligned}$$

Here, an increase in the population of one species causes a decline in the growth rate of another. Suppose that the initial populations are $x_1(0) = 90$ and $x_2(0) = 150$. Find the populations of both species for $t > 0$.

SOLUTION: We have $A = \begin{pmatrix} 3 & -1 \\ -2 & 2 \end{pmatrix}$. The eigenvalues of A are $\lambda_1 = 1$ and $\lambda_2 = 4$ with corresponding eigenvectors $\mathbf{v}_1 = \begin{pmatrix} 1 \\ 2 \end{pmatrix}$ and $\mathbf{v}_2 = \begin{pmatrix} 1 \\ -1 \end{pmatrix}$. Then

$$C = \begin{pmatrix} 1 & 1 \\ 2 & -1 \end{pmatrix} \qquad C^{-1} = -\frac{1}{3}\begin{pmatrix} -1 & -1 \\ -2 & 1 \end{pmatrix} \qquad D = \begin{pmatrix} 1 & 0 \\ 0 & 4 \end{pmatrix} \qquad e^{Dt} = \begin{pmatrix} e^t & 0 \\ 0 & e^{4t} \end{pmatrix}$$

$$\begin{aligned} e^{At} = Ce^{Dt}C^{-1} &= -\frac{1}{3}\begin{pmatrix} 1 & 1 \\ 2 & -1 \end{pmatrix}\begin{pmatrix} e^t & 0 \\ 0 & e^{4t} \end{pmatrix}\begin{pmatrix} -1 & -1 \\ -2 & 1 \end{pmatrix} \\ &= -\frac{1}{3}\begin{pmatrix} 1 & 1 \\ 2 & -1 \end{pmatrix}\begin{pmatrix} -e^t & -e^t \\ -2e^{4t} & e^{4t} \end{pmatrix} \\ &= -\frac{1}{3}\begin{pmatrix} -e^t - 2e^{4t} & -e^t + e^{4t} \\ -2e^t + 2e^{4t} & -2e^t - e^{4t} \end{pmatrix}. \end{aligned}$$

Finally, the solution to the system is given by

$$\begin{aligned} \mathbf{x}(t) = \begin{pmatrix} x_1(t) \\ x_2(t) \end{pmatrix} = e^{At}\mathbf{x}_0 &= -\frac{1}{3}\begin{pmatrix} -e^t - 2e^{4t} & -e^t + e^{4t} \\ -2e^t + 2e^{4t} & -2e^t - e^{4t} \end{pmatrix}\begin{pmatrix} 90 \\ 150 \end{pmatrix} \\ &= -\frac{1}{3}\begin{pmatrix} -240e^t - 30e^{4t} \\ -480e^t + 30e^{4t} \end{pmatrix} = \begin{pmatrix} 80e^t + 10e^{4t} \\ 160e^t - 10e^{4t} \end{pmatrix}. \end{aligned}$$

For example, after six months ($t = \frac{1}{2}$ year), $x_1(t) = 80e^{1/2} + 10e^2 \approx 206$ individuals, while $x_2(t) = 160e^{1/2} - 10e^2 \approx 190$ individuals. More significantly, $160e^t - 10e^{4t} = 0$ when $16e^t = e^{4t}$ or $16 = e^{3t}$ or $3t = \ln 16$ and $t = (\ln 16)/3 \approx 2.77/3 \approx 0.92$ years ≈ 11 months. Thus the second species will be eliminated after only 11 months even though it started with a larger population. In Problems 10 and 11 you are asked to show that neither population will be eliminated if $x_2(0) = 2x_1(0)$ and that the first population will be eliminated if $x_2(0) > 2x_1(0)$. Thus, as was well known to Darwin, survival in this very simple model depends on the relative sizes of the competing species when competition begins.

EXAMPLE 3

A PREDATOR–PREY MODEL

Consider the predator–prey model governed by the system

$$\begin{aligned} x_1'(t) &= x_1(t) + x_2(t), \\ x_2'(t) &= -x_1(t) + x_2(t). \end{aligned}$$

If the initial populations are $x_1(0) = x_2(0) = 1000$, determine the populations of the two species for $t > 0$.

SOLUTION: Here $A = \begin{pmatrix} 1 & 1 \\ -1 & 1 \end{pmatrix}$ with characteristic equation $\lambda^2 - 2\lambda + 2 = 0$, complex roots $\lambda_1 = 1 + i$ and $\lambda_2 = 1 - i$, and eigenvectors $\mathbf{v}_1 = \begin{pmatrix} 1 \\ i \end{pmatrix}$ and $\mathbf{v}_2 = \begin{pmatrix} 1 \\ -i \end{pmatrix}$.[†]

[†] Note that $\lambda_2 = \overline{\lambda_1}$ and $\mathbf{v}_2 = \overline{\mathbf{v}_1}$. This should be no surprise, because, according to the result of Problem 8.13.33, eigenvalues of real matrices occur in complex conjugate pairs and their corresponding eigenvectors are complex conjugates.

Then

$$C = \begin{pmatrix} 1 & 1 \\ i & -i \end{pmatrix} \quad C^{-1} = -\frac{1}{2i}\begin{pmatrix} -i & -1 \\ -i & 1 \end{pmatrix} = \frac{1}{2}\begin{pmatrix} 1 & -i \\ 1 & i \end{pmatrix} \quad D = \begin{pmatrix} 1+i & 0 \\ 0 & 1-i \end{pmatrix}$$

and

$$e^{Dt} = \begin{pmatrix} e^{(1+i)t} & 0 \\ 0 & e^{(1-i)t} \end{pmatrix}.$$

Now, by Euler's formula (see Appendix 3), $e^{it} = \cos t + i \sin t$. Thus $e^{(1+i)t} = e^t e^{it} = e^t(\cos t + i \sin t)$. Similarly, $e^{(1-i)t} = e^t e^{-it} = e^t(\cos t - i \sin t)$. Thus

$$e^{Dt} = e^t\begin{pmatrix} \cos t + i \sin t & 0 \\ 0 & \cos t - i \sin t \end{pmatrix}$$

and

$$\begin{aligned} e^{At} = Ce^{Dt}C^{-1} &= \frac{e^t}{2}\begin{pmatrix} 1 & 1 \\ i & -i \end{pmatrix}\begin{pmatrix} \cos t + i \sin t & 0 \\ 0 & \cos t - i \sin t \end{pmatrix}\begin{pmatrix} 1 & -i \\ 1 & i \end{pmatrix} \\ &= \frac{e^t}{2}\begin{pmatrix} 1 & 1 \\ i & -i \end{pmatrix}\begin{pmatrix} \cos t + i \sin t & -i \cos t + \sin t \\ \cos t - i \sin t & i \cos t + \sin t \end{pmatrix} \\ &= \frac{e^t}{2}\begin{pmatrix} 2 \cos t & 2 \sin t \\ -2 \sin t & 2 \cos t \end{pmatrix} = e^t\begin{pmatrix} \cos t & \sin t \\ -\sin t & \cos t \end{pmatrix}. \end{aligned}$$

Finally,

$$\mathbf{x}(t) = e^{At}\mathbf{x}(0) = e^t\begin{pmatrix} \cos t & \sin t \\ -\sin t & \cos t \end{pmatrix}\begin{pmatrix} 1000 \\ 1000 \end{pmatrix} = \begin{pmatrix} 1000e^t(\cos t + \sin t) \\ 1000e^t(\cos t - \sin t) \end{pmatrix}.$$

The prey species is eliminated when $1000e^t(\cos t - \sin t) = 0$ or when $\sin t = \cos t$. The first positive solution of this last equation is $t = \pi/4 \approx 0.7854$ year ≈ 9.4 months.

EXAMPLE 4

A MODEL OF SPECIES COOPERATION—SYMBIOSIS

Consider the symbiotic model governed by the system

$$\begin{aligned} x_1'(t) &= -\tfrac{1}{2}x_1(t) + x_2(t), \\ x_2'(t) &= \tfrac{1}{4}x_1(t) - \tfrac{1}{2}x_2(t). \end{aligned}$$

In this model the population of each species increases proportionally to the population of the other and decreases proportionally to its own population. Suppose that $x_1(0) = 200$ and $x_2(0) = 500$. Determine the population of each species for $t > 0$.

SOLUTION: Here $A = \begin{pmatrix} -\frac{1}{2} & 1 \\ \frac{1}{4} & -\frac{1}{2} \end{pmatrix}$ with eigenvalues $\lambda_1 = 0$ and $\lambda_2 = -1$ and corresponding eigenvectors $\mathbf{v}_1 = \begin{pmatrix} 2 \\ 1 \end{pmatrix}$ and $\mathbf{v}_2 = \begin{pmatrix} 2 \\ -1 \end{pmatrix}$. Then

$$C = \begin{pmatrix} 2 & 2 \\ 1 & -1 \end{pmatrix}, \quad C^{-1} = -\tfrac{1}{4}\begin{pmatrix} -1 & -2 \\ -1 & 2 \end{pmatrix}, \quad D = \begin{pmatrix} 0 & 0 \\ 0 & -1 \end{pmatrix},$$

and $e^{Dt} = \begin{pmatrix} e^{0t} & 0 \\ 0 & e^{-t} \end{pmatrix} = \begin{pmatrix} 1 & 0 \\ 0 & e^{-t} \end{pmatrix}$. Thus

$$\begin{aligned} e^{At} &= -\frac{1}{4}\begin{pmatrix} 2 & 2 \\ 1 & -1 \end{pmatrix}\begin{pmatrix} 1 & 0 \\ 0 & e^{-t} \end{pmatrix}\begin{pmatrix} -1 & -2 \\ -1 & 2 \end{pmatrix} \\ &= -\frac{1}{4}\begin{pmatrix} 2 & 2 \\ 1 & -1 \end{pmatrix}\begin{pmatrix} -1 & -2 \\ -e^{-t} & 2e^{-t} \end{pmatrix} \\ &= -\frac{1}{4}\begin{pmatrix} -2-2e^{-t} & -4+4e^{-t} \\ -1+e^{-t} & -2-2e^{-t} \end{pmatrix} \end{aligned}$$

and

$$\begin{aligned} \mathbf{x}(t) = e^{At}\mathbf{x}(0) &= -\frac{1}{4}\begin{pmatrix} -2-2e^{-t} & -4+4e^{-t} \\ -1+e^{-t} & -2-2e^{-t} \end{pmatrix}\begin{pmatrix} 200 \\ 500 \end{pmatrix} \\ &= -\frac{1}{4}\begin{pmatrix} -2400+1600e^{-t} \\ -1200-800e^{-t} \end{pmatrix} \\ &= \begin{pmatrix} 600-400e^{-t} \\ 300+200e^{-t} \end{pmatrix}. \end{aligned}$$

We have $e^{-t} \to 0$ as $t \to \infty$. This means that as time goes on, the two cooperating species approach the **equilibrium** populations 600 and 300, respectively. Neither population is eliminated.

We now prove a theorem stated in the last section (Theorem 11.8.6).

THEOREM 6: Let $\Psi(t)$ be the principal matrix solution for the system $\mathbf{x}'(t) = A\mathbf{x}(t)$, $\mathbf{x}(t_0) = \mathbf{x}_0$, where A is a constant matrix. Then

$$\Psi^{-1}(t) = \Psi(-t).$$

PROOF: We prove the theorem in the case $t_0 = 0$. If $t_0 = 0$, we need to show that

$$(e^{At})^{-1} = e^{-At}.$$

But

$$\begin{aligned} e^{At}e^{-At} &= \left(I + At + \frac{A^2t^2}{2!} + \frac{A^3t^3}{3!} + \cdots\right)\left(I - At + \frac{A^2t^2}{2!} - \frac{A^3t^3}{3!} + \cdots\right) \\ &= \left[I + (A-A)t + \left(\frac{A^2}{2!} + \frac{A^2}{2!} - A^2\right)t^2 + \cdots\right] = I, \end{aligned}$$

since all terms except I cancel. Similarly, $e^{-At}e^{At} = I$, which shows that $e^{-At} = (e^{At})^{-1}$.

PROBLEMS 11.9

SELF-QUIZ

Multiple Choice

I. If $A = \begin{pmatrix} 1 & 0 \\ 0 & -1 \end{pmatrix}$, then the principal matrix solution for $\mathbf{x}' = A\mathbf{x}$ is

a. $\begin{pmatrix} e & 1 \\ 1 & e^{-1} \end{pmatrix}$ **b.** $\begin{pmatrix} e & 1 \\ 1 & -e \end{pmatrix}$

c. $\begin{pmatrix} e^t & 1 \\ 1 & e^{-t} \end{pmatrix}$ **d.** $\begin{pmatrix} e^t & 0 \\ 0 & e^{-t} \end{pmatrix}$

II. Suppose that $A = \begin{pmatrix} a & b \\ c & d \end{pmatrix}$ has two real, distinct eigenvalues λ_1 and λ_2. Let C be a matrix whose columns are two linearly independent eigenvectors of A. Then $e^{At} =$ ________.

a. $C\begin{pmatrix} e^{\lambda_1} & 0 \\ 0 & e^{\lambda_2} \end{pmatrix}C^{-1}$ **b.** $C\begin{pmatrix} e^{\lambda_1 t} & 0 \\ 0 & e^{\lambda_2 t} \end{pmatrix}C^{-1}$

c. $C^{-1}\begin{pmatrix} e^{\lambda_1} & 0 \\ 0 & e^{\lambda_2} \end{pmatrix}C$ **d.** $C^{-1}\begin{pmatrix} e^{\lambda_1 t} & 0 \\ 0 & e^{\lambda_2 t} \end{pmatrix}C$

In Problems 1–9 find the principal matrix solution e^{At} of the system $\mathbf{x}'(t) = A\mathbf{x}(t)$.

1. $A = \begin{pmatrix} -2 & -2 \\ -5 & 1 \end{pmatrix}$ **2.** $A = \begin{pmatrix} 3 & -1 \\ -2 & 4 \end{pmatrix}$

3. $A = \begin{pmatrix} 2 & -1 \\ 5 & -2 \end{pmatrix}$ **4.** $A = \begin{pmatrix} 3 & -5 \\ 1 & -1 \end{pmatrix}$

5. $A = \begin{pmatrix} 3 & 2 \\ -5 & 1 \end{pmatrix}$ **6.** $A = \begin{pmatrix} -2 & 1 \\ 5 & 2 \end{pmatrix}$

7. $A = \begin{pmatrix} 7 & -2 & -4 \\ 3 & 0 & -2 \\ 6 & -2 & -3 \end{pmatrix}$

8. $A = \begin{pmatrix} 1 & 1 & -2 \\ -1 & 2 & 1 \\ 0 & 1 & -1 \end{pmatrix}$

9. $A = \begin{pmatrix} 3 & 2 & 4 \\ 2 & 0 & 2 \\ 4 & 2 & 3 \end{pmatrix}$

10. In Example 2, show that if the initial vector $\mathbf{x}(0) = \begin{pmatrix} a \\ 2a \end{pmatrix}$ where a is a constant, then both populations grow at a rate proportional to e^t.

11. In Example 2, show that if $x_2(0) > 2x_1(0)$, then the first population will be eliminated.

***12.** In a water desalinization plant there are two tanks of water. Suppose that tank 1 contains 1000 liters of brine in which 1000 kg of salt is dissolved and tank 2 contains 1000 liters of pure water. Suppose that water flows into tank 1 at the rate of 20 liters per minute and the mixture flows from tank 1 into tank 2 at a rate of 30 liters per minute. From tank 2, 10 liters is pumped back to tank 1 (establishing **feedback**) while 20 liters is flushed away. Find the amount of salt in both tanks at any time t. [*Hint:* Write the information as a 2×2 system and let $x_1(t)$ and $x_2(t)$ denote the amount of salt in each tank.]

13. Consider the second-order differential equation $x''(t) + ax'(t) + bx(t) = 0$.

a. Letting $x_1(t) = x(t)$ and $x_2(t) = x'(t)$, write the preceding equation as a first-order system in the form of equation (1), where A is a 2×2 matrix.

b. Show that the characteristic equation of A is $\lambda^2 + a\lambda + b = 0$.

In Problems 14–17 use the result of Problem 13 to solve the given equation.

14. $x'' + 3x' + 2x = 0$; $x(0) = 1$, $x'(0) = 0$

15. $x'' + 4x' + 4x = 0$; $x(0) = 1$, $x'(0) = 2$

16. $x'' + x = 0$; $x(0) = 0$, $x'(0) = 1$

17. $x'' - 3x' - 10x = 0$; $x(0) = 3$, $x'(0) = 2$

ANSWERS TO SELF-QUIZ

I. d **II.** b

11.10 NONHOMOGENEOUS SYSTEMS

We shall now present a method for solving the nonhomogeneous system

$$\mathbf{x}' = A(t)\mathbf{x} + \mathbf{f}(t), \tag{1}$$

given that a fundamental matrix solution $\Phi(t)$ for the homogeneous system

$$\mathbf{x}' = A(t)\mathbf{x} \tag{2}$$

is known. A solution to (2) can always be found if $A(t)$ is a constant matrix (by the methods of Section 11.3 or Section 11.9).

THEOREM 1

Let $\varphi_p(t)$ and $\varphi_q(t)$ be two solutions of the system (1). Then their difference,

$$\varphi(t) = \varphi_p(t) - \varphi_q(t),$$

is a solution of equation (2).

PROOF: $\varphi' = (\varphi_p - \varphi_q)' = (A\varphi_p + \mathbf{f}) - (A\varphi_q + \mathbf{f}) = A(\varphi_p - \varphi_q) = A\varphi$. Thus, as in the case of linear scalar equations (Section 11.4), it is necessary only to find one particular solution of equation (1). ■

If $\varphi_p(t)$ is such a solution, then the **general solution** to the nonhomogeneous system (1) is of the form

$$\varphi(t) = \Phi(t)\mathbf{c} + \varphi_p(t) \tag{3}$$

where $\mathbf{c}$ is a vector of arbitrary constants and $\Phi(t)$ is a fundamental matrix solution of the homogeneous equation (2). That equation (3) is a solution can be verified as follows:

$$\begin{aligned}\varphi'(t) &= \Phi'(t)\mathbf{c} + \varphi_p'(t)\\ &= [A(t)\Phi(t)\mathbf{c}] + [A(t)\varphi_p(t) + \mathbf{f}(t)],\end{aligned}$$

since Φ is a solution of the associated homogeneous equation and φ_p is a particular solution of equation (2). Combining terms, we have

$$\varphi'(t) = A(t)[\Phi(t)\mathbf{c} + \varphi_p(t)] + \mathbf{f}(t) = A(t)\varphi(t) + \mathbf{f}(t),$$

and φ is a solution of equation (2).

We now derive a **variation-of-constants** formula for the nonhomogeneous system

$$\mathbf{x}' = A(t)\mathbf{x} + \mathbf{f}(t). \tag{4}$$

All variation-of-constants formulas begin by assuming that a solution to the homogeneous equation $\mathbf{x}' = A(t)\mathbf{x}$ is known. Assuming that Φ is a fundamental matrix solution of the homogeneous equation, we seek a particular solution to equation (4) of the form

$$\varphi_p(t) = \Phi(t)\mathbf{k}(t) \tag{5}$$

where $\mathbf{k}(t)$ is a vector function in t. Differentiating both sides of equation (5) with respect to t, we have

$$\varphi_p' = \Phi'\mathbf{k} + \Phi\mathbf{k}' = A\Phi\mathbf{k} + \Phi\mathbf{k}' = A\varphi_p + \Phi\mathbf{k}'.$$

Since φ_p is a particular solution of equation (4), $\Phi\mathbf{k}' = \mathbf{f}$. But every fundamental matrix solution has an inverse, so we can integrate $\mathbf{k}' = \Phi^{-1}\mathbf{f}$, obtaining

$$\varphi_p(t) = \Phi(t)\mathbf{k}(t) = \Phi(t)\int \Phi^{-1}(t)\mathbf{f}(t)\,dt. \tag{6}$$

This is the **variation-of-constants** formula for a particular solution to the nonhomogeneous system (4). Thus, the general solution to system (4) has the form

$$\boldsymbol{\varphi}(t) = \Phi(t)\mathbf{c} + \Phi(t) \int \Phi^{-1}(t)\mathbf{f}(t)\,dt \tag{7}$$

where $\mathbf{c}$ is an arbitrary *constant* vector.

For the initial value problem

$$\mathbf{x}' = A(t)\mathbf{x} + \mathbf{f}(t), \qquad \mathbf{x}(t_0) = \mathbf{x}_0, \tag{8}$$

it is convenient to choose a particular solution $\boldsymbol{\varphi}_p(t)$ that vanishes at t_0. This can be done by selecting the limits of integration in equation (6) to be from t_0 to t, so that

$$\boldsymbol{\varphi}(t) = \Phi(t)\mathbf{c} + \Phi(t) \int_{t_0}^{t} \Phi^{-1}(s)\mathbf{f}(s)\,ds.$$

Substituting $t = t_0$ in this equation, we obtain

$$\mathbf{x}_0 = \boldsymbol{\varphi}(t_0) = \Phi(t_0)\mathbf{c},$$

which implies that $\mathbf{c} = \Phi^{-1}(t_0)\mathbf{x}_0$. Hence, the solution of the initial value problem (8) is

$$\begin{aligned} \boldsymbol{\varphi}(t) &= \Phi(t)\Phi^{-1}(t_0)\mathbf{x}_0 + \Phi(t) \int_{t_0}^{t} \Phi^{-1}(s)\mathbf{f}(s)\,ds \\ &= \boldsymbol{\varphi}_h(t) + \boldsymbol{\varphi}_p(t) \end{aligned} \tag{9}$$

where $\boldsymbol{\varphi}_h$ and $\boldsymbol{\varphi}_p$ are the homogeneous and particular solutions, respectively. Note that if $\Psi(t)$ is the principal matrix solution of $\mathbf{x}' = A\mathbf{x}$, then $\Psi(t_0) = \Psi^{-1}(t_0) = I$. Thus, equation (9) takes the simpler form

$$\boldsymbol{\varphi}(t) = \Psi(t)\mathbf{x}_0 + \Psi(t) \int_{t_0}^{t} \Psi^{-1}(s)\mathbf{f}(s)\,ds \tag{10}$$

when the principal matrix solution $\Psi(t)$ of $\mathbf{x}' = A(t)\mathbf{x}$ is used. We summarize these results in the following theorem:

THEOREM 2 THE VARIATION OF CONSTANTS FORMULA FOR A SYSTEM

Let $\Phi(t)$ be a fundamental matrix solution of the homogeneous system

$$\mathbf{x}' = A(t)\mathbf{x}. \tag{11}$$

Then the solution to the initial value problem

$$\mathbf{x}' = A(t)\mathbf{x} + \mathbf{f}(t), \qquad \mathbf{x}(t_0) = \mathbf{x}_0, \tag{12}$$

is given by

$$\boldsymbol{\varphi}(t) = \Phi(t)\Phi^{-1}(t_0)\mathbf{x}_0 + \boldsymbol{\varphi}_p(t)$$

where

$$\varphi_p(t) = \Phi(t) \int_{t_0}^{t} \Phi^{-1}(s)\mathbf{f}(s)\, ds \tag{13}$$

is a particular solution of the nonhomogeneous system

$$\mathbf{x}' = A(t)\mathbf{x} + \mathbf{f}(t).$$

If $\Psi(t)$ is the principal matrix solution of the homogeneous system (11), then the solution of the initial value problem (12) is

$$\varphi(t) = \Psi(t)\mathbf{x}_0 + \Psi(t) \int_{t_0}^{t} \Psi^{-1}(s)\mathbf{f}(s)\, ds. \tag{14}$$

As we have already seen, the situation is simplest when $A(t)$ is a constant matrix. We now give some examples. ■

EXAMPLE 1

SOLVING A NONHOMOGENEOUS INITIAL-VALUE SYSTEM

Find the unique solution to the system

$$\mathbf{x}' = \begin{pmatrix} x_1 \\ x_2 \end{pmatrix}' = \begin{pmatrix} 4 & 2 \\ 3 & 3 \end{pmatrix}\begin{pmatrix} x_1 \\ x_2 \end{pmatrix} + \begin{pmatrix} e^t \\ e^{2t} \end{pmatrix} = A\mathbf{x} + \mathbf{f}(t), \qquad \mathbf{x}(0) = \begin{pmatrix} 1 \\ 2 \end{pmatrix}.$$

SOLUTION: A fundamental matrix solution for the homogeneous system is

$$\Phi(t) = \begin{pmatrix} 2e^t & e^{6t} \\ -3e^t & e^{6t} \end{pmatrix}.$$

(You should verify this.) Then

$$\Phi^{-1}(t) = \tfrac{1}{5}\begin{pmatrix} e^{-t} & -e^{-t} \\ 3e^{-6t} & 2e^{-6t} \end{pmatrix},$$

and, by equation (12), we have the particular solution

$$\begin{aligned} \varphi_p &= \tfrac{1}{5}\begin{pmatrix} 2e^t & e^{6t} \\ -3e^t & e^{6t} \end{pmatrix} \int_0^t \begin{pmatrix} e^{-s} & -e^{-s} \\ 3e^{-6s} & 2e^{-6s} \end{pmatrix}\begin{pmatrix} e^s \\ e^{2s} \end{pmatrix} ds \\ &= \tfrac{1}{5}\begin{pmatrix} 2e^t & e^{6t} \\ -3e^t & e^{6t} \end{pmatrix} \int_0^t \begin{pmatrix} 1 - e^s \\ 3e^{-5s} + 2e^{-4s} \end{pmatrix} ds. \end{aligned}$$

Since the integral of a vector function is the vector of integrals,

$$\begin{aligned} \varphi_p(t) &= \tfrac{1}{5}\begin{pmatrix} 2e^t & e^{6t} \\ -3e^t & e^{6t} \end{pmatrix}\begin{pmatrix} t - e^t + 1 \\ \dfrac{11}{10} - \dfrac{3}{5}e^{-5t} - \dfrac{e^{-4t}}{2} \end{pmatrix} \\ &= \begin{pmatrix} \dfrac{2}{5}te^t - \dfrac{1}{2}e^{2t} + \dfrac{7}{25}e^t + \dfrac{11}{50}e^{6t} \\ \dfrac{-3}{5}te^t + \dfrac{1}{2}e^{2t} - \dfrac{18}{25}e^t + \dfrac{11}{50}e^{6t} \end{pmatrix}. \end{aligned}$$

Note that $\varphi_p(0) = \mathbf{0}$, which must be the case from the way in which we found φ_p. Next, from equation (9), we see that the unique solution is

$$\begin{aligned}
\boldsymbol{\varphi}(t) &= \Phi(t)\Phi^{-1}(0)\mathbf{x}_0 + \boldsymbol{\varphi}_p(t) \\
&= \tfrac{1}{5}\begin{pmatrix} 2e^t & e^{6t} \\ -3e^t & e^{6t} \end{pmatrix}\begin{pmatrix} 1 & -1 \\ 3 & 2 \end{pmatrix}\begin{pmatrix} 1 \\ 2 \end{pmatrix} + \boldsymbol{\varphi}_p \\
&= \tfrac{1}{5}\begin{pmatrix} 2e^t & e^{6t} \\ -3e^t & e^{6t} \end{pmatrix}\begin{pmatrix} -1 \\ 7 \end{pmatrix} + \boldsymbol{\varphi}_p \\
&= \begin{pmatrix} -\dfrac{2}{5}e^t + \dfrac{7}{5}e^{6t} \\ \dfrac{3}{5}e^t + \dfrac{7}{5}e^{6t} \end{pmatrix} + \begin{pmatrix} \dfrac{2}{5}te^t - \dfrac{1}{2}e^{2t} + \dfrac{7}{25}e^t + \dfrac{11}{50}e^{6t} \\ \dfrac{-3}{5}te^t + \dfrac{1}{2}e^{2t} - \dfrac{18}{25}e^t + \dfrac{11}{50}e^{6t} \end{pmatrix} \\
&= \begin{pmatrix} \dfrac{2}{5}te^t - \dfrac{1}{2}e^{2t} - \dfrac{3}{25}e^t + \dfrac{81}{50}e^{6t} \\ \dfrac{-3}{5}te^t + \dfrac{1}{2}e^{2t} - \dfrac{3}{25}e^t + \dfrac{81}{50}e^{6t} \end{pmatrix}.
\end{aligned}$$

Note, as a check, that $\boldsymbol{\varphi}(0) = \begin{pmatrix} 1 \\ 2 \end{pmatrix}$.

EXAMPLE 2

SOLVING A NONHOMOGENEOUS INITIAL-VALUE PROBLEM

Solve the initial value problem

$$x'' + x = 8 \sin t, \qquad x(0) = 3, \qquad x'(0) = 1.$$

SOLUTION: Using the substitution $x_1 = x$, $x_2 = x'$, we can write this in matrix form:

$$\begin{pmatrix} x_1 \\ x_2 \end{pmatrix}' = \begin{pmatrix} 0 & 1 \\ -1 & 0 \end{pmatrix}\begin{pmatrix} x_1 \\ x_2 \end{pmatrix} + \begin{pmatrix} 0 \\ 8 \sin t \end{pmatrix}.$$

The associated homogeneous system has the principal matrix solution

$$\Psi(t) = \begin{pmatrix} \cos t & \sin t \\ -\sin t & \cos t \end{pmatrix}.$$

Then, from Theorem 11.9.6,

$$\Psi^{-1}(t) = \Psi(-t) = \begin{pmatrix} \cos(-t) & \sin(-t) \\ -\sin(-t) & \cos(-t) \end{pmatrix} = \begin{pmatrix} \cos t & -\sin t \\ \sin t & \cos t \end{pmatrix}$$

and

$$\begin{aligned}
\int_0^t \Psi^{-1}(s)\mathbf{f}(s)\, ds &= \int_0^t \begin{pmatrix} \cos s & -\sin s \\ \sin s & \cos s \end{pmatrix}\begin{pmatrix} 0 \\ 8 \sin s \end{pmatrix} ds \\
&= \int_0^t \begin{pmatrix} -8 \sin^2 s \\ 8 \sin s \cos s \end{pmatrix} ds = \begin{pmatrix} -4t + 4 \sin t \cos t \\ 4 \sin^2 t \end{pmatrix}.
\end{aligned}$$

Since the solution $\boldsymbol{\varphi}(t)$ satisfies the initial conditions

$$\boldsymbol{\varphi}(0) = \begin{pmatrix} 3 \\ 1 \end{pmatrix},$$

we obtain from equation (14)

$$\boldsymbol{\varphi}(t) = \begin{pmatrix} \cos t & \sin t \\ -\sin t & \cos t \end{pmatrix}\begin{pmatrix} 3 \\ 1 \end{pmatrix} + \begin{pmatrix} \cos t & \sin t \\ -\sin t & \cos t \end{pmatrix}\begin{pmatrix} -4t + 4\sin t \cos t \\ 4\sin^2 t \end{pmatrix}$$
$$= \begin{pmatrix} -4t\cos t + 5\sin t + 3\cos t \\ 4t\sin t - 3\sin t + \cos t \end{pmatrix}.$$

Here we have used the identity

$$\sin^3 t + \cos^2 t \sin t = \sin t\,(\sin^2 t + \cos^2 t) = \sin t.$$

Thus, $x(t) = -4t\cos t + 5\sin t + 3\cos t$ is the solution to the problem.

EXAMPLE 3

SOLVING A NONHOMOGENEOUS INITIAL-VALUE PROBLEM WITH VARIABLE COEFFICIENTS

Consider the system

$$\begin{pmatrix} x_1 \\ x_2 \\ x_3 \end{pmatrix}' = \begin{pmatrix} 0 & 1 & 0 \\ 0 & 0 & 1 \\ -2/t^3 & 2/t^2 & -1/t \end{pmatrix}\begin{pmatrix} x_1 \\ x_2 \\ x_3 \end{pmatrix} + \begin{pmatrix} 2t^2 \\ -t^3 \\ t^5 \end{pmatrix} \tag{15}$$

with initial conditions $x_1(1) = 2$, $x_2(1) = 0$, and $x_3(1) = -1$. A fundamental matrix solution (check!) is

$$\Phi(t) = \begin{pmatrix} t & 1/t & t^2 \\ 1 & 1/t^2 & 2t \\ 0 & 2/t^3 & 2 \end{pmatrix}.$$

Note that this solution is valid only for $t > 0$ since $\Phi(t)$ is not defined at $t = 0$. Then

$$\Phi^{-1}(t) = \tfrac{1}{6}\begin{pmatrix} 6/t & 0 & -3t \\ 2t & -2t^2 & t^3 \\ -2/t^2 & 2/t & 2 \end{pmatrix}.$$

Hence, by equation (9), the solution to the initial value problem (15) is

$$\boldsymbol{\varphi}(t) = \Phi(t)\Phi^{-1}(1)\begin{pmatrix} 2 \\ 0 \\ -1 \end{pmatrix} + \Phi(t)\int_1^t \Phi^{-1}(s)\mathbf{f}(s)\,ds = \Phi(t)\mathbf{c} + \boldsymbol{\varphi}_p(t).$$

Setting $t = 1$, we have

$$\Phi^{-1}(1) = \tfrac{1}{6}\begin{pmatrix} 6 & 0 & -3 \\ 2 & -2 & 1 \\ -2 & 2 & 2 \end{pmatrix},$$

so that

$$\Phi(t)\Phi^{-1}(1)\begin{pmatrix} 2 \\ 0 \\ -1 \end{pmatrix} = \tfrac{1}{6}\begin{pmatrix} 15t + 3/t - 6t^2 \\ 15 - 3/t^2 - 12t \\ 6/t^3 - 12 \end{pmatrix}.$$

After a great deal of arithmetic, we arrive at

$$\boldsymbol{\varphi}(t) = \tfrac{1}{6}\begin{pmatrix} 15t + 3/t - 6t^2 \\ 15 - 3/t^2 - 12t \\ 6/t^3 - 12 \end{pmatrix} + \tfrac{1}{6}\begin{pmatrix} t & 1/t & t^2 \\ 1 & -1/t^2 & 2t \\ 0 & 2/t^3 & 2 \end{pmatrix}\int_1^t \begin{pmatrix} 12s - 3s^6 \\ 4s^3 + 2s^5 + s^8 \\ -4 - 2s^2 + 2s^5 \end{pmatrix} ds$$

$$= \begin{pmatrix} \frac{2}{27}t^8 - \frac{1}{14}t^7 - \frac{1}{18}t^5 - \frac{1}{2}t^3 + \frac{13}{18}t^2 + \frac{5}{2}t - \frac{13}{14} + \frac{7}{27t} \\ -\frac{17}{189}t^7 + \frac{1}{9}t^6 - \frac{5}{18}t^4 - \frac{1}{2}t^2 - \frac{5}{9}t + \frac{11}{7} - \frac{7}{27t^2} \\ \frac{4}{27}t^6 - \frac{1}{9}t^3 - t - \frac{5}{9} + \frac{14}{27t^3} \end{pmatrix}.$$

PROBLEMS 11.10

SELF-QUIZ

True–False

I. Let $\Psi(t)$ be the principal matrix solution for the system $\mathbf{x}' = A(t)\mathbf{x}$. Then the solution to the initial-value problem $\mathbf{x}' = A(t)\mathbf{x} + \mathbf{f}(t)$, $\mathbf{x}(t_0) = \mathbf{x}_0$ is

$$\Phi(t) = \Psi(t)\mathbf{x}_0 + \Psi(t)\int_{t_0}^{t} \Psi(-s)\mathbf{f}(s)\,ds$$

II. Let $\Phi(t)$ be a fundamental matrix solution for the system $\mathbf{x}' = A(t)\mathbf{x}$. Then the solution to the initial-value problem $\mathbf{x}' = A(t)\mathbf{x} + \mathbf{f}(t)$, $\mathbf{x}(t_0) = \mathbf{x}_0$ is

$$\boldsymbol{\varphi}(t) = \Phi(t)\Phi^{-1}(t_0)\mathbf{x}_0 + \Phi(t)\int_{t_0}^{t} \Phi^{-1}(s)\mathbf{f}(s)\,ds$$

In Problems 1–9 calculate a fundamental matrix solution for the associated homogeneous system and then use equation (13) and the variation-of-constants formula (9) to obtain a particular solution to the given nonhomogeneous system. Where initial conditions are given, find the unique solution that satisfies them.

1. $\mathbf{x}' = \begin{pmatrix} -2 & -2 \\ -5 & 1 \end{pmatrix}\mathbf{x} + \begin{pmatrix} e^t \\ e^{2t} \end{pmatrix}$

2. $\mathbf{x}' = \begin{pmatrix} 1 & -2 \\ 2 & -3 \end{pmatrix}\mathbf{x} + \begin{pmatrix} t \\ 2 \end{pmatrix}$, $x_1(0) = 1$, $x_2(0) = 0$

3. $\mathbf{x}' = \begin{pmatrix} 2 & -1 \\ 5 & -2 \end{pmatrix}\mathbf{x} + \begin{pmatrix} \sin t \\ \cos t \end{pmatrix}$, $x_1(0) = 0$, $x_2(0) = 1$

4. $\mathbf{x}' = \begin{pmatrix} 3 & 2 \\ -5 & 1 \end{pmatrix}\mathbf{x} + \begin{pmatrix} 2\sin 3t \\ \cos 3t \end{pmatrix}$

5. $\mathbf{x}' = \begin{pmatrix} 1 & 1 & -2 \\ -1 & 2 & 1 \\ 0 & 1 & -1 \end{pmatrix}\mathbf{x} + \begin{pmatrix} e^t \\ e^{2t} \\ e^{2t} \end{pmatrix}$, $x_1(0) = 0$, $x_2(0) = 1$, $x_3(0) = -1$

6. $\mathbf{x}' = \begin{pmatrix} 3 & -1 & 1 \\ -1 & 5 & -1 \\ 1 & -1 & 3 \end{pmatrix}\mathbf{x} + \begin{pmatrix} t^2 \\ 0 \\ 1 \end{pmatrix}$

7. $\mathbf{x}' = \begin{pmatrix} 7 & -2 & -4 \\ 3 & 0 & -2 \\ 6 & -2 & -3 \end{pmatrix}\mathbf{x} + \begin{pmatrix} 0 \\ 1 \\ t \end{pmatrix}$

8. $\mathbf{x}' = \begin{pmatrix} 1 & -1 & -1 \\ 1 & -1 & 0 \\ 1 & 0 & -1 \end{pmatrix}\mathbf{x} + \begin{pmatrix} 2 \\ e^{-t} \\ e^{-t} \end{pmatrix}$, $x_1(0) = 1$, $x_2(0) = -1$, $x_3(0) = 0$

9. $\mathbf{x}' = \begin{pmatrix} 2 & -5 \\ 1 & -2 \end{pmatrix}\mathbf{x} + \begin{pmatrix} 0 \\ \cot t \end{pmatrix}$, $0 < t < \pi$

In Problems 10–12, one homogeneous solution to a given system is given. Use the method of Problem 11.8.21 to obtain a fundamental matrix solution. Then use this solution to find the general solution of the nonhomogeneous system.

10. $\mathbf{x}' = \begin{pmatrix} 0 & 1 \\ -1/4t^2 & 0 \end{pmatrix}\mathbf{x} + \begin{pmatrix} \sqrt{t} \\ 2 \end{pmatrix}$,

$\boldsymbol{\varphi}_1(t) = \begin{pmatrix} \sqrt{t} \\ 1/2\sqrt{t} \end{pmatrix}$, $t > 0$

11. $\mathbf{x}' = \begin{pmatrix} 0 & 1 \\ -1/t^2 & -3/t \end{pmatrix}\mathbf{x} + \begin{pmatrix} t \\ e^t \end{pmatrix}$,

$\boldsymbol{\varphi}_1(t) = \begin{pmatrix} 1/t \\ -1/t^2 \end{pmatrix}$, $t > 0$

12. $\mathbf{x}' = \begin{pmatrix} \dfrac{-t}{1-t^2} & \dfrac{1}{1-t^2} \\ \dfrac{1}{1-t^2} & \dfrac{-t}{1-t^2} \end{pmatrix}\mathbf{x} + \begin{pmatrix} t^2 \\ t^3 \end{pmatrix}$,

$\boldsymbol{\varphi}_1(t) = \begin{pmatrix} 1 \\ t \end{pmatrix}$, $-1 < t < 1$

13. Let $\varphi_1(t)$ be a solution to $\mathbf{x}'(t) = A\mathbf{x}(t) + \mathbf{b}_1(t)$, $\varphi_2(t)$ a solution to $\mathbf{x}'(t) = A\mathbf{x}(t) + \mathbf{b}_2(t), \ldots,$ and $\varphi_n(t)$ a solution to $\mathbf{x}'(t) = A\mathbf{x}(t) + \mathbf{b}_n(t)$. Prove that $\varphi_1(t) + \varphi_2(t) + \cdots + \varphi_n(t)$ is a solution to

$$\mathbf{x}'(t) = A\mathbf{x}(t) + \mathbf{b}_1(t) + \mathbf{b}_2(t) + \cdots + \mathbf{b}_n(t).$$

This again is called the **principle of superposition**.

ANSWERS TO SELF-QUIZ

I. False (true only if $A(t)$ is a constant matrix). **II.** True.

11.11 AN APPLICATION OF NONHOMOGENEOUS SYSTEMS: FORCED OSCILLATIONS (OPTIONAL)

Consider the nonhomogeneous system with constant coefficients

$$\mathbf{x}'(t) = A\mathbf{x}(t) + \mathbf{b}(t). \tag{1}$$

The system (1) can be considered in the following way. If A is fixed, then given an **input vector** $\mathbf{b}(t)$, we can obtain an **output vector** $\mathbf{x}(t)$. That is, the output of the system (the solution) is determined by the input to the system. In this context, vector $\mathbf{b}(t)$ is called the **forcing vector** of the system. With this terminology, system (1) is called an **input–output system**.

It often occurs in practice that the forcing term is periodic (for example, electrical circuits forced by an alternating current). An important question may now be asked: If the forcing term is periodic, does there exist a periodic solution to equation (1) with the same period? An affirmative answer to this question would tell us that an oscillatory input function (an alternating current, for example) will lead to oscillatory behavior of the system governed by the differential equation.

Accordingly, we suppose that

$$\mathbf{b}(t) = e^{i\beta t}\mathbf{c} = (\cos\beta t + i\sin\beta t)\mathbf{c} \tag{2}$$

where $\mathbf{c}$ is a constant vector and $e^{i\beta t}$ is periodic with period $2\pi/\beta$. Suppose that there is a solution $\mathbf{x}(t)$ to equation (1) that is periodic with period $2\pi/\beta$. Then we can write

$$\mathbf{x}(t) = e^{i\beta t}\mathbf{d} \tag{3}$$

where $\mathbf{d}$ is a constant vector. Substituting equations (2) and (3) into (1), we obtain

$$i\beta e^{i\beta t}\mathbf{d} = (e^{i\beta t}\mathbf{d})' = e^{i\beta t}A\mathbf{d} + e^{i\beta t}\mathbf{c}. \tag{4}$$

Dividing equations (4) by $e^{i\beta t}$ and rearranging terms, we obtain

$$(A - i\beta I)\mathbf{d} = -\mathbf{c}. \tag{5}$$

Equation (5) is a nonhomogeneous system of n equations in n unknowns that has a unique solution if and only if

$$\det(A - i\beta I) \neq 0.$$

In other words, there is a unique solution if and only if $i\beta$ is *not* an eigenvalue of the matrix A. We therefore have the following theorem.

THEOREM 1

In the system $\mathbf{x}' = A\mathbf{x} + \mathbf{b}$, let the forcing term $\mathbf{b}(t)$ be periodic with the form $\mathbf{b}(t) = e^{i\beta t}\mathbf{c}$. Then, if $i\beta$ is not an eigenvalue of the matrix A, there exists a unique periodic solution $\mathbf{x}(t) = e^{i\beta t}\mathbf{d}$ of $\mathbf{x}' = A\mathbf{x} + \mathbf{b}$ such that $\mathbf{d} = -(A - i\beta I)^{-1}\mathbf{c}$. If $i\beta$ is an eigenvalue of A, then there are either no periodic solutions or an infinite number of them. ■

NOTE: The differential equation will always have a solution. This theorem tells us something about the nature of the solutions.

REMARK: If $i\beta$ is not an eigenvalue of A, then a periodic (or oscillatory) input will give rise to a periodic output. This phenomenon is called **forced oscillations**.

EXAMPLE 1 A NONHOMOGENEOUS SYSTEM WITH A PERIODIC FORCING TERM

Consider the system

$$\mathbf{x}'(t) = \begin{pmatrix} x_1 \\ x_2 \end{pmatrix}' = \begin{pmatrix} 2 & 1 \\ 1 & 0 \end{pmatrix}\begin{pmatrix} x_1 \\ x_2 \end{pmatrix} + e^{2it}\begin{pmatrix} 2 \\ 1 \end{pmatrix} = A\mathbf{x}(t) + \mathbf{b}(t).$$

Here the forcing term is periodic with period π. Since $2i$ is not an eigenvalue of A, we can use equation (5) to obtain a period solution of period π:

$$\mathbf{x}(t) = e^{2it}\mathbf{d}$$

where

$$\mathbf{d} = -(A - 2iI)^{-1}\begin{pmatrix} 2 \\ 1 \end{pmatrix} = -\begin{pmatrix} 2-2i & 1 \\ 1 & -2i \end{pmatrix}\begin{pmatrix} 2 \\ 1 \end{pmatrix}$$
$$= \frac{-1}{5+4i}\begin{pmatrix} 2i & 1 \\ 1 & -2+2i \end{pmatrix}\begin{pmatrix} 2 \\ 1 \end{pmatrix}.$$

Multiplying the numerator and denominator by $5 - 4i$, we have

$$\mathbf{d} = \frac{-5+4i}{41}\begin{pmatrix} 1+4i \\ 2i \end{pmatrix} = \frac{-1}{41}\begin{pmatrix} 21+16i \\ 8+10i \end{pmatrix}.$$

We would like to write this solution in a more illuminating form. We recall (see Appendix 3) that

$$A + Bi = \sqrt{A^2+B^2}\left(\frac{A}{\sqrt{A^2+B^2}} + i\frac{B}{\sqrt{A^2+B^2}}\right) = \sqrt{A^2+B^2}e^{i\theta}$$

where $\tan\theta = B/A$ (and $\cos\theta = A/\sqrt{A^2+B^2}$, $\sin\theta = B/\sqrt{A^2+B^2}$). Therefore, we have

$$\mathbf{d} = \frac{-1}{41}\begin{pmatrix} \sqrt{697}e^{i\theta_1} \\ \sqrt{164}e^{i\theta_2} \end{pmatrix}$$

where $\tan\theta_1 = 16/21$ and $\tan\theta_2 = 5/4$. Hence

$$\mathbf{x}(t) = \frac{-1}{41}\begin{pmatrix} \sqrt{697}e^{i(\theta_1+2t)} \\ \sqrt{164}e^{i(\theta_2+2t)} \end{pmatrix}.$$

Although $\mathbf{x}(t)$ has the same period, π, as the forcing term, the coordinate functions have been shifted by θ_1 and θ_2. Such a phenomenon is called a **phase shift**.

Returning to equation (5), we answer the question, "What happens if $i\beta$ is an eigenvalue of A?" According to the discussion in Section 11.9, the principal matrix solution e^{At}

of the homogeneous system $\mathbf{x}' = A\mathbf{x}$ contains terms of the form $e^{i\beta t}$, so that $\Phi^{-1}(t)$ contains terms of the form $e^{-i\beta t}$ (since $\Phi\Phi^{-1} = I$). Using equation 11.10.13, we find a particular solution of equation (1):

$$\boldsymbol{\varphi}_p(t) = \Phi(t)\int_{t_0}^{t} \Phi^{-1}(s)\mathbf{c}e^{i\beta s}\,ds.$$

But the product of $\Phi^{-1}(s)$ and $e^{i\beta s}$ contains constant terms (since $e^{-i\beta s}e^{i\beta s} = 1$), and the integral of these constant terms will be of the form ct, which becomes unbounded as t tends to $\pm\infty$. Such a phenomenon is called **resonance**. If $i\beta$ is an eigenvalue of A, then β is called a **natural frequency** of the system. We can summarize the discussion above by stating that **in general, resonance will occur when the frequency of the input vector is a natural frequency of the system**.

EXAMPLE 2 SOLVING AN LC-CIRCUIT

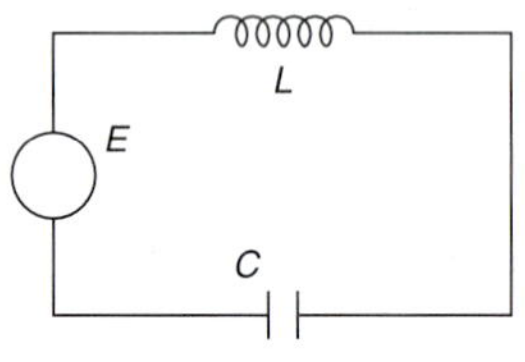

FIGURE 1
An LC-circuit

Consider the circuit shown in Figure 1. Applying Kirchhoff's voltage law (see page 728), we obtain the second-order nonhomogeneous equation

$$L\frac{d^2I}{dt^2} + \frac{I}{C} = E'(t). \tag{6}$$

Letting $x_1 = I$ and $x_2 = I'$, we have

$$x_2' = I'' = \frac{E'}{L} - \frac{I}{LC} = \frac{E'}{L} - \frac{x_1}{LC},$$

so we can rewrite equation (6) in the form

$$\begin{pmatrix} x_1 \\ x_2 \end{pmatrix}' = \begin{pmatrix} 0 & 1 \\ -1/LC & 0 \end{pmatrix}\begin{pmatrix} x_1 \\ x_2 \end{pmatrix} + \begin{pmatrix} 0 \\ E'/L \end{pmatrix}. \tag{7}$$

Suppose that $E = E_0e^{i\omega t}$. Then equation (7) may be treated by the method of this section, and the solution $\mathbf{x}(t)$ is of the form $e^{i\omega t}\mathbf{x}$ where

$$\mathbf{x} = -(A - i\omega I)^{-1}\mathbf{b} \qquad \text{and} \qquad \mathbf{b} = \begin{pmatrix} 0 \\ i\omega E_0/L \end{pmatrix}.$$

But

$$\det(A - i\omega I) = \det\begin{pmatrix} -i\omega & 1 \\ -1/LC & -i\omega \end{pmatrix} = \frac{1}{LC} - \omega^2,$$

which is nonzero if $\omega \neq 1/\sqrt{LC}$. Therefore,

$$\mathbf{x} = \frac{-1}{(1/LC) - \omega^2}\begin{pmatrix} -i\omega & -1 \\ 1/LC & -i\omega \end{pmatrix}\begin{pmatrix} 0 \\ i\omega E_0/L \end{pmatrix} = \frac{\omega CE_0}{LC\omega^2 - 1}\begin{pmatrix} -i \\ \omega \end{pmatrix}$$

and

$$\mathbf{x}(t) = \frac{\omega CE_0}{LC\omega^2 - 1}\begin{pmatrix} -i \\ \omega \end{pmatrix}e^{i\omega t}.$$

Thus the current is

$$x_1(t) = I(t) = -i\omega CE_0e^{i\omega t}$$

or

$$I(t) = \omega C E_0(\sin \omega t - i \cos \omega t).$$

The complex term here simply tells us that the real and imaginary parts of the solution are the responses to the real and imaginary parts of the forcing function.

If $\omega = 1/\sqrt{LC}$ (the natural frequency), then the circuit is in resonance. Since $A = \begin{pmatrix} 0 & 1 \\ -1/LC & 0 \end{pmatrix}$, $A^2 = (-1/LC)I$, and it follows after a short computation that the principal matrix solution e^{At} is

$$\Psi(t) = \begin{pmatrix} \cos(t/\sqrt{LC}) & \sqrt{LC}\sin(t/\sqrt{LC}) \\ (-1/\sqrt{LC})\sin(t/\sqrt{LC}) & \cos(t/\sqrt{LC}) \end{pmatrix},$$

and by equation (11.10.14),

$$\mathbf{x}(t) = \Psi(t)\mathbf{x}(0) + \Psi(t)\int_0^t \Psi^{-1}(s)\mathbf{b}(s)\,ds. \tag{8}$$

To calculate the integral, we observe that since $\cos(-s) = \cos s$ and $\sin(-s) = -\sin s$, we have, from Theorem 11.9.6,

$$\Psi^{-1}(s)\mathbf{b}(s) = \Psi(-s)\mathbf{b}(s) = \begin{pmatrix} \cos\dfrac{s}{\sqrt{LC}} & -\sqrt{LC}\sin\dfrac{s}{\sqrt{LC}} \\ \dfrac{1}{\sqrt{LC}}\sin\dfrac{s}{\sqrt{LC}} & \cos\dfrac{s}{\sqrt{LC}} \end{pmatrix}\begin{pmatrix} 0 \\ \dfrac{iE_0 e^{is/\sqrt{LC}}}{L\sqrt{LC}} \end{pmatrix}$$

$$= \begin{pmatrix} \dfrac{E_0}{L}\sin\dfrac{s}{\sqrt{LC}}\left(\sin\dfrac{s}{\sqrt{LC}} - i\cos\dfrac{s}{\sqrt{LC}}\right) \\ \dfrac{E_0}{L\sqrt{LC}}\cos\dfrac{s}{\sqrt{LC}}\left(-\sin\dfrac{s}{\sqrt{LC}} + i\cos\dfrac{s}{\sqrt{LC}}\right) \end{pmatrix}.$$

After we perform the integration in (8), we obtain

$$\int_0^t \Psi^{-1}(s)\mathbf{b}(s)\,ds = \frac{E_0}{2L}\begin{pmatrix} t - \sqrt{LC}\sin\dfrac{t}{\sqrt{LC}}e^{it/\sqrt{LC}} \\ i\left(\dfrac{t}{\sqrt{LC}} + \sin\dfrac{t}{\sqrt{LC}}e^{it/\sqrt{LC}}\right) \end{pmatrix}. \tag{9}$$

Adding this vector to $\mathbf{x}(0)$ and multiplying $\Psi(t)$ by the result will give the solution to system (7). Note that the vector (9) has components that become arbitrarily large as t approaches $+\infty$. This is an effect of resonance on the system.

PROBLEMS 11.11

In Problems 1–4 for each system find (if possible) a solution $\mathbf{x}(t)$ that has the same period of oscillation as the forcing term. If the natural and forcing frequencies coincide, show that a particular solution is unbounded as t approaches ∞ (i.e., resonance occurs).

1. $\mathbf{x}' = \begin{pmatrix} 2 & -5 \\ 0 & 3 \end{pmatrix}\mathbf{x} + \begin{pmatrix} 1 \\ 4 \end{pmatrix}e^{it}$

2. $\mathbf{x}' = \begin{pmatrix} 1 & -2 \\ \frac{5}{2} & -1 \end{pmatrix}\mathbf{x} + \begin{pmatrix} 2 \\ 3 \end{pmatrix}e^{2it}$

3. $\mathbf{x}' = \begin{pmatrix} 3 & 4 \\ -1 & -2 \end{pmatrix}\mathbf{x} + \begin{pmatrix} -1 \\ 4 \end{pmatrix}e^{-5it}$

4. $\mathbf{x}' = \begin{pmatrix} -1 & -7 \\ 5 & 1 \end{pmatrix}\mathbf{x} + \begin{pmatrix} -2 \\ 7 \end{pmatrix}e^{4it}$

5. Suppose that $\mathbf{b}(t) = \Sigma_{k=1}^{m} \mathbf{b}_k e^{i\beta_k t}$ (a finite sum of periodic inputs). Assume that $i\beta_k$ is not a root of the characteristic equation of A for $k = 1, 2, \ldots, m$. Use the principle of superposition to find a solution $\mathbf{x}(t)$ to equation (1) that can be written in the form

$$\mathbf{x}(t) = \sum_{k=1}^{m} \mathbf{x}_k e^{i\beta_k t}.$$

6. Use Euler's formula to show that

$$2 \cos \beta t = e^{i\beta t} + e^{-i\beta t}$$

and

$$2i \sin \beta t = e^{i\beta t} - e^{-i\beta t}.$$

Then use the results of Problem 5 to obtain (if possible) periodic solutions to the following equations:
   a. $x'' - x' - 2x = \sin t$;
   b. $x'' + 4x = \cos t$;
   c. $x'' + 4x = \cos 2t$;
   d. $x'' + 2x' - 15x = \sin 4t$.
7. Verify the form of the principal matrix solution $\Psi(t)$ if ω equals the natural frequency in Example 2.
8. Verify equation (9).
9. **a.** Express the forced vibration equation (10.14.11) for a spring–mass system with $c = 0$ as a 2×2 system of differential equations.
   b. Obtain a solution by the method of this section, given that $\omega \neq \sqrt{k/m}$.
   c. What happens if $\omega = \sqrt{k/m}$? Justify your answer.

11.12 COMPUTING e^{At}: AN APPLICATION OF THE CAYLEY-HAMILTON THEOREM (OPTIONAL)

In Section 11.9 we showed how to compute the principal matrix solution to $\mathbf{x}' = A\mathbf{x}$ when A was a diagonalizable constant matrix. This solution had the form e^{At}. In this section we use an important theorem in matrix theory to develop an algorithm for computing e^{At} when A is any constant matrix. We first discuss this important theorem.

Let $p(x) = x^n + a_{n-1}x^{n-1} + \cdots + a_1 x + a_0$ be a polynomial and let A be an $n \times n$ matrix. Then powers of A are defined and we define

$$p(A) = A^n + a_{n-1}A^{n-1} + \cdots + a_1 A + a_0 I \tag{1}$$

EXAMPLE 1 EVALUATING $P(A)$

Let $A = \begin{pmatrix} -1 & 4 \\ 3 & 7 \end{pmatrix}$ and $p(x) = x^2 - 5x + 3$. Then

$$P(A) = A^2 - 5A + 3I = \begin{pmatrix} 13 & 24 \\ 18 & 61 \end{pmatrix} + \begin{pmatrix} 5 & -20 \\ -15 & -35 \end{pmatrix} + \begin{pmatrix} 3 & 0 \\ 0 & 3 \end{pmatrix} = \begin{pmatrix} 21 & 4 \\ 3 & 29 \end{pmatrix}$$

Expression (1) is a polynomial with scalar coefficients defined for a matrix variable. We can also define a polynomial with *square matrix* coefficients by

$$Q(\lambda) = B_0 + B_1\lambda + B_2\lambda^2 + \cdots + B_n\lambda^n \tag{2}$$

If A is a matrix of the same size as B, then we define

$$Q(A) = B_0 + B_1 A + B_2 A^2 + \cdots + B_n A^n \tag{3}$$

We must be careful in (3) since matrices do not commute under multiplication.

THEOREM 1

If $P(\lambda)$ and $Q(\lambda)$ are polynomials in the scalar variable λ with square matrix coefficients and if $P(\lambda) = Q(\lambda)(A - \lambda I)$, then $P(A) = 0$.

PROOF: If $Q(\lambda)$ is given by equation (2), then

$$\begin{aligned} P(\lambda) &= (B_0 + B_1\lambda + B_2\lambda^2 + \cdots + B_n\lambda^n)(A - \lambda I) \\ &= B_0A + B_1A\lambda + B_2A\lambda^2 + \cdots + B_nA\lambda^n \\ &\quad - B_0\lambda - B_1\lambda^2 - B_2\lambda^3 - \cdots - B_n\lambda^{n+1} \end{aligned} \tag{4}$$

Then, substituting A for λ in (4), we obtain

$$\begin{aligned} P(A) = B_0A + B_1A^2 + B_2A^3 + \cdots + B_nA^{n+1} \\ - B_0A - B_1A^2 - B_2A^3 - \cdots - B_nA^{n+1} = 0 \end{aligned}$$ ■

NOTE: We cannot prove this theorem by substituting $\lambda = A$ to obtain $P(A) = Q(A)(A - A) = 0$. This is because it is possible to find polynomials $P(\lambda)$ and $Q(\lambda)$ with matrix coefficients such that $F(\lambda) = P(\lambda)Q(\lambda)$ but $F(A) \neq P(A)Q(A)$. (See Problem 23.)

We can now state the main theorem.

THEOREM 2 THE CAYLEY-HAMILTON THEOREM†

Every square matrix satisfies its own characteristic equation. That is, if $p(\lambda) = 0$ is the characteristic equation of A, then $p(A) = 0$.

PROOF: We have

$$p(\lambda) = \det(A - \lambda I) = \begin{vmatrix} a_{11} - \lambda & a_{12} & \cdots & a_{1n} \\ a_{21} & a_{22} - \lambda & \cdots & a_{2n} \\ \vdots & \vdots & & \vdots \\ a_{n1} & a_{n2} & \cdots & a_{nn} - \lambda \end{vmatrix}$$

Clearly any cofactor of $(A - \lambda I)$ is a polynomial in λ. Thus the adjoint of $A - \lambda I$ (see the definition on page 474) is an $n \times n$ matrix each of whose components is a polynomial in λ. That is,

$$\text{adj}\,(A - \lambda I) = \begin{pmatrix} p_{11}(\lambda) & p_{12}(\lambda) & \cdots & p_{1n}(\lambda) \\ p_{21}(\lambda) & p_{22}(\lambda) & \cdots & p_{2n}(\lambda) \\ \vdots & \vdots & & \vdots \\ p_{n1}(\lambda) & p_{22}(\lambda) & \cdots & p_{nn}(\lambda) \end{pmatrix}$$

This means that we can think of adj $(A - \lambda I)$ as a polynomial, $Q(\lambda)$, in λ with $n \times n$ matrix coefficients. To see this, look at the following:

$$\begin{pmatrix} -\lambda^2 - 2\lambda + 1 & 2\lambda^2 - 7\lambda - 4 \\ 4\lambda^2 + 5\lambda - 2 & -3\lambda^2 - \lambda + 3 \end{pmatrix} = \begin{pmatrix} -1 & 2 \\ 4 & -3 \end{pmatrix}\lambda^2 + \begin{pmatrix} -2 & -7 \\ 5 & -1 \end{pmatrix}\lambda + \begin{pmatrix} 1 & -4 \\ -2 & 3 \end{pmatrix}$$

Now, from Theorem 7.3.2 on page 475,

$$\det(A - \lambda I)I = [\text{adj}\,(A - \lambda I)][A - \lambda I] = Q(\lambda)(A - \lambda I) \tag{5}$$

†Named after Sir William Rowan Hamilton (see page 12) and Arthur Cayley (see page 406). Cayley published the first discussion of this famous theorem in 1858. Independently, Hamilton discovered the result in his work on quaternions.

But $\det(A - \lambda I)I = p(\lambda)I$. If

$$p(\lambda) = \lambda^n + a_{n-1}\lambda^{n-1} + \cdots + a_1\lambda + a_0$$

then we define

$$P(\lambda) = p(\lambda)I = \lambda^n I + a_{n-1}\lambda^{n-1}I + \cdots + a_1\lambda I + a_0 I$$

Thus, from (5), we have $P(\lambda) = Q(\lambda)(A - \lambda I)$. Finally, from Theorem 1, $P(A) = 0$. This completes the proof. ■

EXAMPLE 2

ILLUSTRATION OF THE CAYLEY-HAMILTON THEOREM

Let $A = \begin{pmatrix} 1 & -1 & 4 \\ 3 & 2 & -1 \\ 2 & 1 & -1 \end{pmatrix}$. In Example 8.13.4 on page 591 we computed the characteristic equation $\lambda^3 - 2\lambda^2 - 5\lambda + 6 = 0$. Now we compute

$$A^2 = \begin{pmatrix} 6 & 1 & 1 \\ 7 & 0 & 11 \\ 3 & -1 & 8 \end{pmatrix}, \qquad A^3 = \begin{pmatrix} 11 & -3 & 22 \\ 29 & 4 & 17 \\ 16 & 3 & 5 \end{pmatrix}$$

and

$$A^3 - 2A^2 - 5A + 6I = \begin{pmatrix} 11 & -3 & 22 \\ 29 & 4 & 17 \\ 16 & 3 & 5 \end{pmatrix} + \begin{pmatrix} -12 & -2 & -2 \\ -14 & 0 & -22 \\ -6 & 2 & -16 \end{pmatrix}$$
$$+ \begin{pmatrix} -5 & 5 & -20 \\ -15 & -10 & 5 \\ -10 & -5 & 5 \end{pmatrix} + \begin{pmatrix} 6 & 0 & 0 \\ 0 & 6 & 0 \\ 0 & 0 & 6 \end{pmatrix}$$
$$= \begin{pmatrix} 0 & 0 & 0 \\ 0 & 0 & 0 \\ 0 & 0 & 0 \end{pmatrix}$$

We now use the Cayley–Hamilton theorem to calculate a fundamental matrix solution for any system

$$\mathbf{x}' = A\mathbf{x},$$

when the matrix A is constant.

Before explaining the procedure, we will define the notation we plan to use. Let $p(\lambda) = \det(A - \lambda I)$ be the characteristic polynomial of the $n \times n$ matrix A, and suppose that

$$p(\lambda) = (\lambda - \lambda_1)^{r_1} \cdot (\lambda - \lambda_2)^{r_2}, \cdots, (\lambda - \lambda_i)^{r_k}, \tag{6}$$

where $\lambda_1, \lambda_2, \ldots, \lambda_k$ are the eigenvalues of A of algebraic multiplicities $r_1, r_2, \ldots, r_k$, respectively. Using partial fractions, we can write

$$\frac{1}{p(\lambda)} = \frac{a_1(\lambda)}{(\lambda - \lambda_1)^{r_1}} + \frac{a_2(\lambda)}{(\lambda - \lambda_2)^{r_2}} + \cdots + \frac{a_k(\lambda)}{(\lambda - \lambda_k)^{r_k}}, \tag{7}$$

where for each polynomial $a_i(\lambda)$

$$\deg a_i(\lambda) \le r_i - 1. \tag{8}$$

Multiplying both sides of equation (7) by $p(\lambda)$, we get

$$1 = a_1(\lambda)q_1(\lambda) + a_2(\lambda)q_2(\lambda) + \cdots + a_k(\lambda)q_k(\lambda), \tag{9}$$

where $q_i(\lambda)$ is the polynomial consisting of all but the $(\lambda - \lambda_i)^{r_i}$ term of $p(\lambda)$:

$$p(\lambda) = q_i(\lambda)(\lambda - \lambda_i)^{r_i}. \tag{10}$$

We are now ready to begin the procedure for calculating e^{At} for any $n \times n$ matrix A. We shall illustrate each step in the procedure by applying it to the matrix

$$A = \begin{pmatrix} 1 & 2 & 3 \\ 0 & 0 & 4 \\ 0 & 0 & 0 \end{pmatrix}.$$

STEP 1: Find the characteristic polynomial $p(\lambda)$ and use it to determine the polynomials $a_i(\lambda)$ and $q_i(\lambda)$ in equations (9) and (10).

In this case we have

$$p(\lambda) = \det(A - \lambda I) = \det\begin{pmatrix} 1-\lambda & 2 & 3 \\ 0 & -\lambda & 4 \\ 0 & 0 & -\lambda \end{pmatrix} = -\lambda^2(\lambda - 1),$$

so that $\lambda_1 = 0$ and $\lambda_1 = 1$ with $r_1 = 2$ and $r_2 = 1$, respectively. By equation (10) we easily obtain

$$q_1(\lambda) = 1 - \lambda \qquad \text{and} \qquad q_2(\lambda) = -\lambda^2. \tag{11}$$

Using equation (8) we note that $\deg a_1(\lambda) \le 1$ and $\deg a_2(\lambda) \le 0$, so that the general forms of the polynomials a_1 and a_2 are

$$a_1(\lambda) = a\lambda + b \qquad \text{and} \qquad a_2(\lambda) = c.$$

Substituting these polynomials into equation (9) and equating like powers of λ, we obtain

$$a_1(\lambda) = \lambda + 1 \qquad \text{and} \qquad a_2(\lambda) = -1. \tag{12}$$

At this point we need to develop some additional facts. Look at the Cayley–Hamilton theorem. Since A satisfies its characteristic equation, equation (10) becomes

$$0 = p(A) = q_i(A)(A - \lambda_i I)^{r_i}. \tag{13}$$

Equation (9) is also satisfied by the matrix A, since it was derived from the characteristic equation. Thus,

$$I = a_1(A)q_1(A) + a_2(A)q_2(A) + \cdots + a_k(A)q_k(A). \tag{14}$$

We can easily check equation (14) for our particular example:

$$I = (A + I)(I - A) + IA^2.$$

Using equation (5), we observe that $e^{\lambda_i t}I = e^{\lambda_i t}I$ so, since $e^{A+B} = e^A e^B$, we have

$$e^{At} = e^{\lambda_i t I}e^{(A-\lambda_i I)t} = e^{\lambda_i t}\sum_{j=0}^{\infty}\frac{(A - \lambda_i I)^j t^j}{j!}. \tag{15}$$

Multiplying both ends of equation (15) on the left by $q_i(A)$, we obtain

$$q_i(A)e^{At} = e^{\lambda_i t} \sum_{j=0}^{\infty} \frac{q_i(A)(A - \lambda_i I)^j t^j}{j!}. \tag{16}$$

By equation (13), all the terms in the series for $j \geq r_i$ are equal to the zero matrix so equation (16) reduces to

$$q_i(A)e^{At} = e^{\lambda_i t} \sum_{j=0}^{r_i - 1} \frac{q_i(A)(A - \lambda_i I)^j t^j}{j!}. \tag{17}$$

Multiplying each side of equation (17) on the left by $a_i(A)$ yields

$$a_i(A)q_i(A)e^{At} = e^{\lambda_i t} \sum_{j=0}^{r_i - 1} \frac{a_i(A)q_i(A)(A - \lambda_i I)^j t^j}{j!}. \tag{18}$$

Finally, summing equation (18) over all the indices i and using equation (14), we conclude that

$$e^{At} = Ie^{At} = \sum_{i=1}^{k} a_i(A)q_i(A)e^{At}$$

or

$$e^{At} = \sum_{i=1}^{k} \left\{ e^{\lambda_i t} a_i(A)q_i(A) \sum_{j=0}^{r_i - 1} \frac{(A - \lambda_i I)^j t^j}{j!} \right\}.$$

Although this equation looks formidable, we shall see that for most applications it is very easy to use. Thus, the final step in computing e^{At} is:

STEP 2: Using the polynomials $a_i(\lambda)$, $q_i(\lambda)$, and the eigenvalues λ_i with multiplicities r_i that were obtained in Step 1, compute:

$$e^{At} = \sum_{i=1}^{k} \left\{ e^{\lambda_i t} a_i(A)q_i(A) \sum_{j=0}^{r_i - 1} \frac{(A - \lambda_i I)^j t^j}{j!} \right\}. \tag{19}$$

For the particular matrix A we are considering, we have

$$\lambda_1 = 0, \qquad r_1 = 2, \qquad \lambda_2 = 1, \qquad r_2 = 1,$$

and

$$q_1(A) = I - A, \qquad q_2(A) = -A^2, \qquad a_1(A) = A + I, \qquad a_2(A) = -I.$$

Substituting these values in equation (19), we get

$$\begin{aligned} e^{At} &= \left\{ e^{0t}(A + I)(I - A) \sum_{j=0}^{1} \frac{A^j t^j}{j!} \right\} + \left\{ e^t I A^2 \sum_{j=0}^{0} \frac{(A - I)^j t^j}{j!} \right\} \\ &= (A + I)(I - A)(I + At) + e^t A^2 \\ &= (I - A^2)(I + At) + e^t A^2. \end{aligned}$$

Since

$$A^2 = \begin{pmatrix} 1 & 2 & 3 \\ 0 & 0 & 4 \\ 0 & 0 & 0 \end{pmatrix}\begin{pmatrix} 1 & 2 & 3 \\ 0 & 0 & 4 \\ 0 & 0 & 0 \end{pmatrix} = \begin{pmatrix} 1 & 2 & 11 \\ 0 & 0 & 0 \\ 0 & 0 & 0 \end{pmatrix},$$

we have

$$\begin{aligned} e^{At} &= \begin{pmatrix} 0 & -2 & -11 \\ 0 & 1 & 0 \\ 0 & 0 & 1 \end{pmatrix}\begin{pmatrix} 1+t & 2t & 3t \\ 0 & 1 & 4t \\ 0 & 0 & 1 \end{pmatrix} + e^t\begin{pmatrix} 1 & 2 & 11 \\ 0 & 0 & 0 \\ 0 & 0 & 0 \end{pmatrix} \\ &= \begin{pmatrix} 0 & -2 & -8t-11 \\ 0 & 1 & 4t \\ 0 & 0 & 1 \end{pmatrix} + \begin{pmatrix} e^t & 2e^t & 11e^t \\ 0 & 0 & 0 \\ 0 & 0 & 0 \end{pmatrix} \\ &= \begin{pmatrix} e^t & 2(e^t-1) & 11e^t - 8t - 11 \\ 0 & 1 & 4t \\ 0 & 0 & 1 \end{pmatrix}. \end{aligned}$$

Summarizing, we find that the procedure reduces to two steps:

1. Find the characteristic polynomial for the matrix A

$$\begin{aligned} p(\lambda) &= \det(A - \lambda I) \\ &= (\lambda - \lambda_1)^{r_1}(\lambda - \lambda_2)^{r_2}\cdots(\lambda - \lambda_k)^{r_k}, \end{aligned}$$

and use it first to determine the polynomials

$$q_i(\lambda) = \frac{p(\lambda)}{(\lambda - \lambda_i)^{r_i}},$$

and then the polynomials $a_i(\lambda)$ that satisfy

$$a_1(\lambda)q_1(\lambda) + a_2(\lambda)q_2(\lambda) + \cdots + a_k(\lambda)q_k(\lambda) = 1.$$

2. Replace each λ^m by A^m in the expressions for $a_i(\lambda)$ and $q_i(\lambda)$ and compute

$$e^{At} = \sum_{i=1}^{k}\left\{e^{\lambda_i t}a_i(A)q_i(A)\sum_{j=0}^{r_i-1}\frac{(A-\lambda_i I)^j t^j}{j!}\right\}.$$

We illustrate the procedure with several examples.

EXAMPLE 3

COMPUTING e^{At}: TWO DISTINCT EIGENVALUES

Find the principal matrix solution of the system

$$\begin{pmatrix} x_1 \\ x_2 \end{pmatrix}' = \begin{pmatrix} 4 & 2 \\ 3 & 3 \end{pmatrix}\begin{pmatrix} x_1 \\ x_2 \end{pmatrix}.$$

By Example 3 in Section 8.13, we know that the characteristic polynomial is

$$p(\lambda) = (\lambda - 1)(\lambda - 6) = 0.$$

Thus $\lambda_1 = 1$, $r_1 = 1$, $\lambda_2 = 6$, $r_2 = 1$, $q_1(\lambda) = \lambda - 6$, $q_2(\lambda) = \lambda - 1$, and $\deg a_1(\lambda) = \deg a_2(\lambda) \le 0$, implying that a_1 and a_2 are constants:

$$1 = a_1(\lambda - 6) + a_2(\lambda - 1) = (a_1 + a_2)\lambda - (6a_1 + a_2).$$

Solving the system

$$\begin{aligned} a_1 + a_2 &= 0, \\ -6a_1 - a_2 &= 1, \end{aligned}$$

we get $a_1 = -\frac{1}{5}$ and $a_2 = \frac{1}{5}$. Hence, the principal matrix solution is

$$\begin{aligned} \Psi(t) = e^{At} &= \left\{e^t\left(-\frac{1}{5}\right)I(A - 6I)\right\} + \left\{e^{6t}\left(\frac{1}{5}\right)I(A - I)\right\} \\ &= \frac{-e^t}{5}(A - 6I) + \frac{e^{6t}}{5}(A - I) = \frac{-e^t}{5}\begin{pmatrix} -2 & 2 \\ 3 & -3 \end{pmatrix} + \frac{e^{6t}}{5}\begin{pmatrix} 3 & 2 \\ 3 & 2 \end{pmatrix} \\ &= \frac{1}{5}\begin{pmatrix} 3e^{6t} + 2e^t & 2e^{6t} - 2e^t \\ 3e^{6t} - 3e^t & 2e^{6t} + 3e^t \end{pmatrix}. \end{aligned}$$

EXAMPLE 4

COMPUTING e^{At}: ONE REPEATED EIGENVALUE

Find the principal matrix solution for

$$\mathbf{x}' = \begin{pmatrix} 4 & 1 \\ 0 & 4 \end{pmatrix}\mathbf{x}.$$

This is the matrix of Example 8 in Section 8.13 with characteristic polynomial $p(\lambda) = (\lambda - 4)^2$ and eigenspace spanned by the single vector $\binom{1}{0}$. There is only one eigenvalue $\lambda_1 = 4$ with $r_1 = 2$, so $q_1(\lambda) = 1 = a_1(\lambda)$. Thus

$$\begin{aligned} \Psi(t) = e^{At} &= e^{4t}I \cdot I \sum_{j=0}^{1} \frac{(A - 4I)^j t^j}{j!} \\ &= e^{4t}[I + (A - 4I)t] = e^{4t}\left[\begin{pmatrix} 1 & 0 \\ 0 & 1 \end{pmatrix} + \begin{pmatrix} 0 & 1 \\ 0 & 0 \end{pmatrix}t\right] \\ &= \begin{pmatrix} e^{4t} & te^{4t} \\ 0 & e^{4t} \end{pmatrix}. \end{aligned}$$

EXAMPLE 5

COMPUTING e^{At}: COMPLEX CONJUGATE EIGENVALUES

Find the principal matrix solution of

$$\mathbf{x}' = \begin{pmatrix} 3 & -5 \\ 1 & -1 \end{pmatrix}\mathbf{x}.$$

In Example 6 of Section 8.13 we found the characteristic polynomial $p(\lambda) = \lambda^2 - 2\lambda + 2$ with eigenvalues $\lambda_1 = 1 + i$ and $\lambda_2 = 1 - i$ of algebraic multiplicity $r_1 = r_2 = 1$. Thus $p(\lambda) = (\lambda - 1 - i)(\lambda - 1 + i)$, $q_1(\lambda) = \lambda - 1 + i$, $q_2(\lambda) = \lambda - 1 - i$, and a_1 and a_2 are both constants satisfying

$$1 = a_1(\lambda - 1 + i) + a_2(\lambda - 1 - i).$$

Solving, we get $a_1 = 1/2i$, $a_2 = -1/2i$, and

$$\Psi(t) = e^{At} = \left\{e^{(1+i)t}\left(\frac{1}{2i}\right)(A - (1-i)I)\right\} + \left\{e^{(1-i)t}\left(\frac{-1}{2i}\right)(A - (1+i)I)\right\}$$

$$= e^t\begin{pmatrix} 2\left(\frac{e^{it} - e^{-it}}{2i}\right) + \left(\frac{e^{it} + e^{-it}}{2}\right) & -5\left(\frac{e^{it} - e^{-it}}{2i}\right) \\ \frac{e^{it} - e^{-it}}{2i} & -2\left(\frac{e^{it} - e^{-it}}{2i}\right) + \left(\frac{e^{it} + e^{-it}}{2}\right) \end{pmatrix}$$

$$= e^t\begin{pmatrix} 2\sin t + \cos t & -5\sin t \\ \sin t & -2\sin t + \cos t \end{pmatrix}.$$

EXAMPLE 6

COMPUTING e^{At} FOR A 3×3 SYSTEM

Find the principal matrix solution to the system

$$\mathbf{x}' = \begin{pmatrix} 3 & 2 & 4 \\ 2 & 0 & 2 \\ 4 & 2 & 3 \end{pmatrix}\mathbf{x}.$$

The characteristic polynomial is

$$p(\lambda) = -\lambda^3 + 6\lambda^2 + 15\lambda + 8 = (\lambda + 1)^2(8 - \lambda) = 0$$

so the eigenvalues are $\lambda_1 = -1$ and $\lambda_2 = 8$ with multiplicities $r_1 = 2$ and $r_2 = 1$, respectively. Thus $q_1(\lambda) = 8 - \lambda$, $q_2(\lambda) = (\lambda + 1)^2$, $\deg a_1(\lambda) \le 1$, and $\deg a_2(\lambda) \le 0$. Setting $a_1(\lambda) = a\lambda + b$ and $a_2(\lambda) = c$, we have

$$1 = (a\lambda + b)(8 - \lambda) + c(\lambda + 1)^2,$$

which has the solution $a = c = \frac{1}{81}$ and $b = \frac{10}{81}$. Thus

$$\Psi(t) = e^{At} = \left\{e^{-t}\left(\frac{1}{81}\right)(A + 10I)(8I - A)\sum_{j=0}^{1}\frac{(A+I)^j t^j}{j!}\right\} + \left\{\frac{e^{8t}}{81}(A + I)^2 I\right\}$$

$$= \frac{e^{-t}}{81}(80I - 2A - A^2)(I + (A + I)t) + \frac{e^{8t}}{81}(A^2 + 2A + I)$$

$$= \frac{e^{-t}}{81}\begin{pmatrix} 45 & -18 & -36 \\ -18 & 72 & -18 \\ -36 & -18 & 45 \end{pmatrix}\begin{pmatrix} 1 + 4t & 2t & 4t \\ 2t & 1 + t & 2t \\ 4t & 2t & 1 + 4t \end{pmatrix}$$

$$+ \frac{e^{8t}}{81}\begin{pmatrix} 36 & 18 & 36 \\ 18 & 9 & 18 \\ 36 & 18 & 36 \end{pmatrix}$$

$$= \frac{1}{9}\begin{pmatrix} 4e^{8t} + 5e^{-t} & 2(e^{8t} - e^{-t}) & 4(e^{8t} - e^{-t}) \\ 2(e^{8t} - e^{-t}) & e^{8t} + 8e^{-t} & 2(e^{8t} - e^{-t}) \\ 4(e^{8t} - e^{-t}) & 2(e^{8t} - e^{-t}) & 4e^{8t} + 5e^{-t} \end{pmatrix}.$$

EXAMPLE 7 SOLVING AN INITIAL VALUE PROBLEM

Solve the initial value problem

$$\mathbf{x}' = \begin{pmatrix} 2 & -1 \\ -4 & 2 \end{pmatrix}\mathbf{x}, \qquad \mathbf{x}(0) = \begin{pmatrix} 0 \\ 4 \end{pmatrix}.$$

The characteristic polynomial $p(\lambda) = \lambda(\lambda - 4)$ has the eigenvalues $\lambda_1 = 0$ and $\lambda_2 = 4$, both with algebraic multiplicity $r_1 = r_2 = 1$. The polynomials $q_1(\lambda) = \lambda - 4$, $q_2(\lambda) = \lambda$, and $\deg a_1 = \deg a_2 \leq 0$ imply that a_1 and a_2 are constants satisfying

$$1 = a_1 \cdot (\lambda - 4) + a_2\lambda,$$

so that $a_1 = -a_2 = -\frac{1}{4}$. Hence, the principal matrix solution is

$$\Psi(t) = e^{At} = \left\{-\frac{1}{4}(A - 4I)\right\} + \left\{\frac{e^{4t}}{4}A\right\} = I + \left(\frac{e^{4t} - 1}{4}\right)A$$

$$= \begin{pmatrix} \dfrac{1 + e^{4t}}{2} & \dfrac{1 - e^{4t}}{4} \\ 1 - e^{4t} & \dfrac{1 + e^{4t}}{2} \end{pmatrix}.$$

To solve the initial value problem we simply multiply the principal matrix solution by $\mathbf{x}(0)$, which yields the solution

$$\mathbf{x}(t) = \begin{pmatrix} 1 - e^{4t} \\ 2 + 2e^{4t} \end{pmatrix}.$$

PROBLEMS 11.12

SELF-QUIZ

Multiple Choice

I. Which equation is satisfied by $A = \begin{pmatrix} 1 & 3 \\ 0 & 2 \end{pmatrix}$?

a. $A^2 + 3A + 2I = 0$
b. $A^2 - 2A = 0$
c. $A^2 + 2A - 3I = 0$
d. $A^2 - 3A + 2I = 0$

True–False

II. If A cannot be diagonalized, then e^{At} cannot be computed by the method of this section.

III. If λ is an eigenvalue of A of algebraic multiplicity 2 and geometric multiplicity 1, then the terms $e^{\lambda t}$ and $te^{\lambda t}$ appear in the principal matrix solution of $\mathbf{x}' = A\mathbf{x}$.

Use the methods of this section to calculate the principal matrix solution of the system $\mathbf{x}' = A\mathbf{x}$, where A is the given constant matrix. If initial conditions are given, find the solution of the initial-value problem.

1. $\begin{pmatrix} -2 & -2 \\ -5 & 1 \end{pmatrix}$

2. $\begin{pmatrix} -12 & 7 \\ -7 & 2 \end{pmatrix}$

3. $\begin{pmatrix} 2 & -1 \\ 5 & -2 \end{pmatrix}$

4. $\begin{pmatrix} 3 & 2 \\ -5 & 1 \end{pmatrix}$

5. $\begin{pmatrix} 3 & -2 \\ 8 & -5 \end{pmatrix}$

6. $\begin{pmatrix} 1 & 1 \\ 1 & -1 \end{pmatrix}$

7. $\begin{pmatrix} 3 & -2 \\ 8 & -5 \end{pmatrix}$, $\mathbf{x}(0) = \begin{pmatrix} \frac{3}{4} \\ 1 \end{pmatrix}$

8. $\begin{pmatrix} 3 & -2 \\ 8 & -5 \end{pmatrix}$, $\mathbf{x}(0) = \begin{pmatrix} 1 \\ 2 \end{pmatrix}$

9. $\begin{pmatrix} 3 & -2 \\ 8 & -5 \end{pmatrix}$, $\mathbf{x}(0) = \begin{pmatrix} 2 \\ 5 \end{pmatrix}$

10. $\begin{pmatrix} 4 & 1 \\ -8 & 8 \end{pmatrix}$, $\mathbf{x}(0) = \begin{pmatrix} 1 \\ 0 \end{pmatrix}$

11. $\begin{pmatrix} 1 & -1 & 0 \\ -1 & 2 & -1 \\ 0 & -1 & 1 \end{pmatrix}$ **12.** $\begin{pmatrix} 1 & 1 & -2 \\ -1 & 2 & 1 \\ 0 & 1 & -1 \end{pmatrix}$

13. $\begin{pmatrix} 4 & 6 & 6 \\ 1 & 3 & 2 \\ -1 & -5 & -2 \end{pmatrix}$ **14.** $\begin{pmatrix} 7 & -2 & -4 \\ 3 & 0 & -2 \\ 6 & -2 & -3 \end{pmatrix}$

15. $\begin{pmatrix} 5 & 4 & 2 \\ 4 & 5 & 2 \\ 2 & 2 & 2 \end{pmatrix}$ **16.** $\begin{pmatrix} -3 & 0 & 2 \\ 1 & -1 & 0 \\ -2 & -1 & 0 \end{pmatrix}$

17. $\begin{pmatrix} 4 & 1 & 0 & 1 \\ 2 & 3 & 0 & 1 \\ -2 & 1 & 2 & -3 \\ 2 & -1 & 0 & 5 \end{pmatrix}$

18. $\begin{pmatrix} 0 & -1 & -2 \\ 1 & 0 & 1 \\ 2 & -1 & 0 \end{pmatrix}$

19. $\begin{pmatrix} 1 & -1 & 0 \\ -1 & 2 & -1 \\ 0 & -1 & 1 \end{pmatrix}$, $\mathbf{x}(0) = \begin{pmatrix} 1 \\ 0 \\ 1 \end{pmatrix}$

20. $\begin{pmatrix} 5 & 4 & 2 \\ 4 & 5 & 2 \\ 2 & 2 & 2 \end{pmatrix}$, $\mathbf{x}(0) = \begin{pmatrix} 1 \\ 2 \\ 3 \end{pmatrix}$

21. $\begin{pmatrix} 4 & 6 & 6 \\ 1 & 3 & 2 \\ -1 & -5 & -2 \end{pmatrix}$, $\mathbf{x}(0) = \begin{pmatrix} -1 \\ 0 \\ 2 \end{pmatrix}$

22. $\begin{pmatrix} 4 & 1 & 0 & 1 \\ 2 & 3 & 0 & 1 \\ -2 & 1 & 2 & -3 \\ 2 & -1 & 0 & 5 \end{pmatrix}$, $\mathbf{x}(0) = \begin{pmatrix} 4 \\ 5 \\ 6 \\ 2 \end{pmatrix}$

23. Let $P(\lambda) = B_0 + B_1\lambda$ and $Q(\lambda) = C_0 + C_1\lambda$, where B_0, B_1, C_0, and C_1 are $n \times n$ matrices.
   a. Compute $F(\lambda) = P(\lambda)Q(\lambda)$.
   b. Let A be an $n \times n$ matrix. Show that $F(A) = P(A)Q(A)$ if and only if A commutes with both C_0 and C_1.

ANSWERS TO SELF-QUIZ

I. d **II.** False **III.** True

CHAPTER 11 SUMMARY OUTLINE

■ **A Typical System of Differential Equations**

$$\begin{aligned} x_1' &= f_1(t, x_1, x_2, \ldots, x_n), \\ x_2' &= f_2(t, x_1, x_2, \ldots, x_n), \\ &\vdots \\ x_n' &= f_n(t, x_1, x_2, \ldots, x_n), \end{aligned} \tag{1}$$

where each f_i, $i = 1, 2, \ldots, n$, is a function of $n + 1$ variables.

■ **Solution to a System** Any collection of n differentiable functions $x_1(t), x_2(t), \ldots, x_n(t)$ that satisfy the system (1) for all values t in some interval $\alpha < t < \beta$ is called a **solution** of the system (1) on that interval.

■ **First-Order System** System (1) is called a **first-order** system because it involves only first derivatives of the functions $x_1, x_2, \ldots, x_n$.

■ **Linear, First-Order System** A first-order system is said to be **linear** if it can be written in the form

$$\begin{aligned} \frac{dx_1}{dt} &= a_{11}(t)x_1 + a_{12}(t)x_2 + \cdots + a_{1n}(t)x_n + f_1(t), \\ \frac{dx_2}{dt} &= a_{21}(t)x_1 + a_{22}(t)x_2 + \cdots + a_{2n}(t)x_n + f_2(t), \\ &\vdots \\ \frac{dx_n}{dt} &= a_{n1}(t)x_1 + a_{n2}(t)x_2 + \cdots + a_{nn}(t)x_n + f_n(t), \end{aligned} \tag{2}$$

where a_{ij} and f_i are functions that are continuous on some common interval $\alpha < t < \beta$. Any system for which this cannot be done is said to be **nonlinear**. The functions a_{ij} are called the **coefficients** of the linear system (2). If every a_{ij} is constant, system (2) is said to have **constant coefficients**. If every function f_i satisfies $f_i(t) \equiv 0$ for $\alpha < t < \beta$, $i = 1, 2, \ldots, n$, then system (2) is said to be **homogeneous**; otherwise it is called a **nonhomogeneous** system.

■ **Initial-Value Problem** In addition to the given system of differential equations (1) or (2), initial conditions of the form

$$x_1(t_0) = x_{10}, x_2(t_0) = x_{20}, \ldots, x_n(t_0) = x_{n0} \tag{3}$$

may also be given for some t_0 satisfying $\alpha < t_0 < \beta$. We call system (1) or (2) together with the initial condition (3) an **initial-value problem**.

- **Writing a Higher-Order Equation as a First-Order System** The equation

$$y^{(n)} = F(t, y, y', y'', \ldots, y^{(n-1)}),$$

can be written as a system of first-order equations of form (1). The procedure is as follows: we define n new variables $x_1, x_2, \ldots, x_n$ by setting

$$x_1 = y, x_2 = y', x_3 = y'', \ldots, x_n = y^{(n-1)}.$$

Then, since $y' = x_1' = x_2$, $y'' = x_2' = x_3, \ldots, y^{(n-1)} = x_{n-1}' = x_n$ and $y^{(n)} = x_n'$, we have the following first-order system

$$\begin{aligned} x_1' &= x_2, \\ x_2' &= x_3, \\ &\vdots \\ x_{n-1}' &= x_n, \\ x_n' &= F(t, x_1, x_2, \ldots, x_n), \end{aligned}$$

which is a special case of system (1).

- **Linear, Homogeneous Systems in Two Unknown Functions** These are systems that can be written in the form

$$\begin{aligned} x' &= a_{11}(t)x + a_{12}(t)y \\ y' &= a_{21}(t)x + a_{22}(t)y \end{aligned} \qquad (4)$$

- **Linear, Nonhomogeneous Systems in Two Unknown Functions**

$$\begin{aligned} x' &= a_{11}(t)x + a_{12}(t)y + f_1(t) \\ y' &= a_{21}(t)x + a_{22}(t)y + f_2(t) \end{aligned} \qquad (5)$$

- **Existence–Uniqueness Theorem** If the functions $a_{11}(t)$, $a_{12}(t)$, $a_{21}(t)$, $a_{22}(t)$, $f_1(t)$, and $f_2(t)$ are continuous on I, then given any numbers t_0, x_0, and y_0, with t_0 in I, there exists exactly one solution $(x(t), y(t))$ of system (1) that satisfies $x(t_0) = x_0$ and $y(t_0) = y_0$.

- **Solution to the System (4) or (5)** By a **solution** of system (4) [or (5)] we mean a *pair* of functions $(x(t), y(t))$ that possess first derivatives and that satisfy the given equations.

- **Linear Combination** The pair of functions $(x_3(t), y_3(t))$ is a **linear combination** of the pairs $(x_1(t), y_1(t))$ and $(x_2(t), y_2(t))$ if there exist constants c_1 and c_2 such that the following two equations hold:

$$\begin{aligned} x_3(t) &= c_1x_1(t) + c_2x_2(t) \\ y_3(t) &= c_1y_1(t) + c_2y_2(t) \end{aligned}$$

If the pairs $(x_1(t), y_1(t))$ and $(x_2(t), y_2(t))$ are solutions of the homogeneous system (5), then any linear combination of them is also a solution of system (5).

- **Linear Independence** We define two pairs of functions $(x_1(t), y_1(t))$ and $(x_2(t), y_2(t))$ to be **linearly independent** on an interval I if, whenever the equations

$$\begin{aligned} c_1x_1(t) + c_2x_2(t) &= 0, \\ c_1y_1(t) + c_2y_2(t) &= 0 \end{aligned}$$

hold for all values of t in I, then $c_1 = c_2 = 0$.

- **Wronskian** Given two solutions $(x_1(t), y_1(t))$ and $(x_2(t), y_2(t))$ of system (4), we define the **Wronskian** of the two solutions by the following determinant:

$$W(t) = \begin{vmatrix} x_1(t) & y_1(t) \\ x_2(t) & y_2(t) \end{vmatrix} = x_1(t)y_2(t) - x_2(t)y_1(t).$$

Moreover,

$$W(t) = W(t_0)\exp\left(\int_{t_0}^{t} [a_{11}(u) + a_{22}(u)]\,du\right).$$

Two solutions $(x_1(t), y_1(t))$ and $(x_2(t), y_2(t))$ of the homogeneous system (4) are linearly independent on an interval I if and only if $W(t) \neq 0$ in I.

- **General Solution to a Homogeneous System** Let (x_1, y_1) and (x_2, y_2) be solutions of the homogeneous linear system (4).

  Then $(c_1x_1 + c_2x_2, c_1y_1 + c_2y_2)$ is the **general solution** of system (4) provided that $W(t) \neq 0$; that is, provided that the solutions (x_1, y_1) and (x_2, y_2) are linearly independent.

- **General Solution to a Nonhomogeneous System** Let (x^*, y^*) be the general solution of system (4), and let (x_p, y_p) be any solution of (5). Then $(x^* - x_p, y^* - y_p)$ is the general solution of the homogeneous equation (4). In other words, the general solution of system (5) can be written as the sum of the general solution of the homogeneous system (4) and any particular solution of the nonhomogeneous system (5).

- **Characteristic Equation** Consider the homogeneous system with constant coefficients

$$\begin{aligned} x' &= a_{11}x + a_{12}y \\ y' &= a_{21}x + a_{22}y \end{aligned} \qquad (6)$$

The characteristic equation of (6) is given by

$$\begin{vmatrix} a_{11} - \lambda & a_{12} \\ a_{21} & a_{22} - \lambda \end{vmatrix} = 0$$

or

$$\lambda^2 - (a_{11} + a_{22})\lambda + (a_{11}a_{22} - a_{21}a_{12}) = 0$$

- **Method of Determinants for Solving a Linear, Homogeneous System with Constant Coefficients** To find the general solution of the system

$$\begin{aligned} x' &= a_{11}x + a_{12}y, \\ y' &= a_{21}x + a_{22}y, \end{aligned}$$

in the case $a_{12} \neq 0$, first find two numbers λ_1 and λ_2 that satisfy the characteristic equation

$$\begin{vmatrix} a_{11} - \lambda & a_{12} \\ a_{21} & a_{22} - \lambda \end{vmatrix} = 0.$$

CASE 1: If $\lambda_1 \neq \lambda_2$ are real, then

$$x(t) = c_1 e^{\lambda_1 t} + c_2 e^{\lambda_2 t},$$

and $y(t)$ can be obtained from the equation

$$y = \frac{1}{a_{12}}(x' - a_{11}x). \tag{7}$$

CASE 2: If $\lambda_1 = \lambda_2$, then

$$x(t) = e^{\lambda_1 t}(c_1 + c_2 t),$$

and $y(t)$ can be obtained from (7).

CASE 3: If $\lambda_1 = a + ib$ and $\lambda_2 = a - ib$, then

$$x(t) = e^{at}(c_1 \cos bt + c_2 \sin bt),$$

and again $y(t)$ can be obtained from (7).

- **Vector Function** An n-component **vector function**

$$\mathbf{v}(t) = \begin{pmatrix} v_1(t) \\ v_2(t) \\ \vdots \\ v_n(t) \end{pmatrix} \tag{8}$$

is an n-vector each of whose components is a function (usually assumed to be continuous).

- **Matrix Function** An $n \times n$ **matrix function** $A(t)$ is an $n \times n$ matrix

$$A(t) = \begin{pmatrix} a_{11}(t) & a_{12}(t) & \cdots & a_{1n}(t) \\ a_{21}(t) & a_{22}(t) & \cdots & a_{2n}(t) \\ \vdots & \vdots & & \vdots \\ a_{n1}(t) & a_{n2}(t) & \cdots & a_{nn}(t) \end{pmatrix} \tag{9}$$

- **Derivatives and Integrals** If $\mathbf{v}(t)$ is given by (8), then

$$\mathbf{v}'(t) = \begin{pmatrix} v_1'(t) \\ v_2'(t) \\ \vdots \\ v_n'(t) \end{pmatrix} \quad \text{and} \quad \int_{t_0}^{t} \mathbf{v}(s)\, ds = \begin{pmatrix} \int_{t_0}^{t} v_1(s)\, ds \\ \int_{t_0}^{t} v_2(s)\, ds \\ \vdots \\ \int_{t_0}^{t} v_n(s)\, ds \end{pmatrix};$$

if $A(t)$ is given by (9), then

$$A'(t) = \begin{pmatrix} a_{11}'(t) & a_{12}'(t) & \cdots & a_{1n}'(t) \\ a_{21}'(t) & a_{22}'(t) & \cdots & a_{2n}'(t) \\ \vdots & \vdots & & \vdots \\ a_{n1}'(t) & a_{n2}'(t) & \cdots & a_{nn}'(t) \end{pmatrix}$$

and

$$\int_{t_0}^{t} A(s)\, ds = \begin{pmatrix} \int_{t_0}^{t} a_{11}(s)\, ds & \int_{t_0}^{t} a_{12}(s)\, ds & \cdots & \int_{t_0}^{t} a_{1n}(s)\, ds \\ \int_{t_0}^{t} a_{21}(s)\, ds & \int_{t_0}^{t} a_{22}(s)\, ds & \cdots & \int_{t_0}^{t} a_{2n}(s)\, ds \\ \vdots & \vdots & & \vdots \\ \int_{t_0}^{t} a_{n1}(s)\, ds & \int_{t_0}^{t} a_{n2}(s)\, ds & \cdots & \int_{t_0}^{t} a_{nn}(s)\, ds \end{pmatrix}.$$

- **Vector Differential Equation** Consider the following linear system of n first-order equations

$$\begin{aligned} x_1' &= a_{11}(t)x_1 + a_{12}(t)x_2 + \cdots + a_{1n}(t)x_n + f_1(t), \\ x_2' &= a_{21}(t)x_1 + a_{22}(t)x_2 + \cdots + a_{2n}(t)x_n + f_2(t), \\ &\vdots \\ x_n' &= a_{n1}(t)x_1 + a_{n2}(t)x_2 + \cdots + a_{nn}(t)x_n + f_n(t) \end{aligned} \tag{10}$$

which is **nonhomogeneous** if at least one of the functions $f_i(t)$, $i = 1, 2, \ldots, n$ is not the zero function, and the **associated homogeneous system**

$$\begin{aligned} x_1' &= a_{11}(t)x_1 + a_{12}(t)x_2 + \cdots + a_{1n}(t)x_n, \\ x_2' &= a_{21}(t)x_1 + a_{22}(t)x_2 + \cdots + a_{2n}(t)x_n, \\ &\vdots \\ x_n' &= a_{n1}(t)x_1 + a_{n2}(t)x_2 + \cdots + a_{nn}(t)x_n. \end{aligned} \tag{11}$$

Define the vector function $\mathbf{x}(t)$, the matrix function $A(t)$, and the vector function $\mathbf{f}(t)$ as follows:

$$\mathbf{x}(t) = \begin{pmatrix} x_1(t) \\ x_2(t) \\ \vdots \\ x_n(t) \end{pmatrix}, \qquad A(t) = \begin{pmatrix} a_{11}(t) & a_{12}(t) & \cdots & a_{1n}(t) \\ a_{21}(t) & a_{22}(t) & \cdots & a_{2n}(t) \\ \vdots & \vdots & & \vdots \\ a_{n1}(t) & a_{n2}(t) & \cdots & a_{nn}(t) \end{pmatrix},$$

$$\mathbf{f}(t) = \begin{pmatrix} f_1(t) \\ f_2(t) \\ \vdots \\ f_n(t) \end{pmatrix}.$$

Then system (10) can be written

$$\mathbf{x}'(t) = A(t)\mathbf{x}(t) + \mathbf{f}(t) \tag{12}$$

System (11) becomes

$$\mathbf{x}'(t) = A(t)\mathbf{x}(t) \tag{13}$$

- **Vector Initial-Value Problem**

$$\mathbf{x}'(t) = A(t)\mathbf{x}(t) + \mathbf{f}(t), \qquad \mathbf{x}(t_0) = \mathbf{x}_0 \tag{14}$$

- **Existence–Uniqueness Theorem** Let $A(t)$ and $\mathbf{f}(t)$ be continuous matrix and vector functions, respectively, on some interval $[a, b]$ containing t_0 (that is, the component functions of both $A(t)$ and $\mathbf{f}(t)$ are continuous). Then there exists a unique vector function $\boldsymbol{\varphi}(t)$ that is a solution to the initial-value problem (14) on the entire interval $[a, b]$.

- **Linearly Independent and Fundamental Sets of Solutions** Let $\boldsymbol{\varphi}_1(t), \boldsymbol{\varphi}_2(t), \ldots, \boldsymbol{\varphi}_m(t)$ be m vector solutions of system (13). We say that they are **linearly independent** on an interval if the equation (for all t in the common interval)

$$c_1\boldsymbol{\varphi}_1(t) + c_2\boldsymbol{\varphi}_2(t) + \cdots + c_m\boldsymbol{\varphi}_m(t) = \mathbf{0}$$

holds only for $c_1 = c_2 = \cdots = c_m = 0$. Since system (1) is equivalent to an nth-order equation, it is natural for us to seek n linearly independent solutions to the system. Any set of n linearly independent solutions of (13) is called a **fundamental set of solutions**.

- **Matrix Solution and its Wronskian** Let $\boldsymbol{\varphi}_1, \boldsymbol{\varphi}_2, \ldots, \boldsymbol{\varphi}_n$ be n-vector solutions of $\mathbf{x}' = A(t)\mathbf{x}$. Let $\Phi(t)$ be the matrix whose columns are the vectors $\boldsymbol{\varphi}_1, \boldsymbol{\varphi}_2, \ldots, \boldsymbol{\varphi}_n$; that is,

$$\Phi(t) = (\boldsymbol{\varphi}_1(t), \ldots, \boldsymbol{\varphi}_n(t)) = \begin{pmatrix} \phi_{11}(t) & \phi_{12}(t) & \cdots & \phi_{1n}(t) \\ \phi_{21}(t) & \phi_{22}(t) & \cdots & \phi_{2n}(t) \\ \vdots & \vdots & & \vdots \\ \phi_{n1}(t) & \phi_{n2}(t) & \cdots & \phi_{nn}(t) \end{pmatrix}.$$

Such a matrix is called a **matrix solution** of the system $\mathbf{x}' = A\mathbf{x}$. Equivalently, an $n \times n$ matrix function $\Phi(t)$ is a matrix solution of $\mathbf{x}' = A\mathbf{x}$ if and only if each of its columns is a solution vector of $\mathbf{x}' = A\mathbf{x}$.

We define the **Wronskian** of a matrix solution $\Phi(t)$ of the system $\mathbf{x}' = A(t)\mathbf{x}$ by

$$W(t) = \det \Phi(t).$$

- **Fundamental Matrix Solution** If $W(t) \neq 0$, then the vectors $\boldsymbol{\varphi}_1, \boldsymbol{\varphi}_2, \ldots, \boldsymbol{\varphi}_n$ form a fundamental set of solutions (that is, they are linearly independent), and $\Phi(t)$ is called a **fundamental matrix solution**.

- **Principal Matrix Solution** The fundamental matrix solution $\Psi(t)$ for the system

$$\mathbf{x}' = A(t)\mathbf{x}$$

that satisfies $\Psi(t_0) = I$, where I is the identity matrix, is called the **principal matrix solution**.

Let $A(t)$ be continuous on some interval $[a, b]$. Then for any t_0, $a \leq t_0 \leq b$, there exists a unique fundamental matrix solution $\Psi(t)$ of the system $\mathbf{x}' = A(t)\mathbf{x}$ satisfying the condition $\Psi(t_0) = I$.

- **Obtaining the Principal Matrix Solution from a Fundamental Matrix Solution** If $\Phi(t)$ is a fundamental matrix solution for $\mathbf{x}' = A(t)\mathbf{x}$, then

$$\Psi(t) = \Phi(t)\Phi^{-1}(t_0)$$

is the principal matrix solution (associated with the point t_0).

- Let $\Phi(t)$ be a fundamental matrix solution and let $\mathbf{x}(t)$ be any solution of $\mathbf{x}' = A(t)\mathbf{x}$. Then there exists a constant vector $\mathbf{c}$ such that

$$\mathbf{x}(t) = \Phi(t)\mathbf{c}.$$

- **General Solution** The **general solution** of the system $\mathbf{x}' = A(t)\mathbf{x}$ is obtained by multiplying any fundamental matrix solution $\Phi(t)$ by an arbitrary vector of constants $\mathbf{c}$; that is,

$$\Phi(t)\mathbf{c} = (\boldsymbol{\varphi}_1(t), \boldsymbol{\varphi}_2(t), \ldots, \boldsymbol{\varphi}_n(t))\begin{pmatrix} c_1 \\ c_2 \\ \vdots \\ c_n \end{pmatrix}$$
$$= c_1\boldsymbol{\varphi}_1(t) + c_2\boldsymbol{\varphi}_2(t) + \cdots + c_n\boldsymbol{\varphi}_n(t).$$

- **Solving an Initial-Value Problem** The unique solution to the initial-value problem

$$\mathbf{x}' = A\mathbf{x}, \qquad \mathbf{x}(t_0) = \mathbf{x}_0$$

is given by

$$\boldsymbol{\varphi}(t) = \Psi(t)\mathbf{x}_0,$$

where $\Psi(t)$ is the principal matrix solution relative to the initial point t_0.

- **Principal Matrix Solution** For $\mathbf{x}'(t) = A\mathbf{x}(t)$, if A is a constant matrix and $t_0 = 0$, then the principal matrix solution for the system $\mathbf{x}'(t) = A\mathbf{x}(t)$ is given by

$$\Psi(t) = e^{At} = I + At + \frac{A^2t^2}{2!} + \cdots + \frac{A^nt^n}{n!} + \cdots$$

- If $\Psi(t) = e^{At}$, then $\Psi^{-1}(t) = \Psi(-t) = e^{-At}$

- If A is a diagonal matrix with diagonal components $\lambda_1, \lambda_2, \ldots, \lambda_n$, then e^{At} is a diagonal matrix with diagonal components $e^{\lambda_1 t}, e^{\lambda_2 t}, \ldots, e^{\lambda_n t}$.

- **Finding e^{At} when A is diagonalizable** Let A be diagonalizable and suppose that $D = C^{-1}AC$. Then $A = CDC^{-1}$ and

$$e^{At} = Ce^{Dt}C^{-1}.$$

- **General Solution to a Nonhomogeneous System** Let $\Phi(t)$ be a fundamental matrix solution of (13). Then the general solution to $\mathbf{x}' = A(t)\mathbf{x} + \mathbf{f}(t)$ is given by

$$\boldsymbol{\varphi}(t) = \Phi(t)\mathbf{c} + \Phi(t)\int \Phi^{-1}(t)\mathbf{f}(t)\,dt$$

- **Solution to a Nonhomogeneous Initial Value Problem** Let $\Psi(t)$ be the principal matrix solution of the homogeneous system

$$\mathbf{x}' = A(t)\mathbf{x}, \qquad \mathbf{x}(t_0) = \mathbf{x}_0.$$

Then the solution to the initial-value problem

$$\mathbf{x}' = A(t)\mathbf{x} + \mathbf{f}(t), \qquad \mathbf{x}(t_0) = \mathbf{x}_0,$$

is given by

$$\boldsymbol{\varphi}(t) = \Psi(t)\mathbf{x}_0 + \Psi(t)\int_{t_0}^{t} \Psi^{-1}(s)\mathbf{f}(s)\,ds.$$

If $A(t)$ is constant, then

$$\boldsymbol{\varphi}(t) = \Psi(t)\mathbf{x}_0 + \Psi(t)\int_{t_0}^{t} \Psi(-s)\mathbf{f}(s)\,ds.$$

- **The Cayley-Hamilton Theorem** Every square matrix satisfies its own characteristic equation. That is, if $p(\lambda) = 0$ is the characteristic equation of A, then $p(A) = 0$.

CHAPTER 11
REVIEW EXERCISES

In Exercises 1–3 transform the equation into a first-order system.

1. $x''' - 6x'' + 2x' - 5x = 0$

2. $x'' - 3x' + 4t^2x = \sin t$

3. $xx'' + x'x''' = \ln t$

In Exercises 4–8 find the general solution for each system.

4. $x' = x + y,$
$y' = 9x + y$

5. $x' = x + 2y,$
$y' = 4x + 3y$

6. $x' = 4x - y,$
$y' = x + 2y$

7. $x' = 3x + 2y,$
$y' = -5x + y$

8. $x' = x - 4y,$
$y' = x + y$

9. Find the general solution to

$$x' = -x - 3e^{-2t},$$
$$y' = -2x - y - 6e^{-2t}.$$

10. Find the unique solution to

$$x' = -4x - 6y + 9e^{-3t}, \qquad x(0) = -9,$$
$$y' = x + y - 5e^{-3t}, \qquad y(0) = 4.$$

In Exercises 11–14 write the given system of equations in vector–matrix form.

11. $x_1' = 3x_1 - 4x_2,$
$x_2' = -2x_1 + 7x_2$

12. $x_1' = (\sin t)x_1 + e^t x_2,$
$x_2' = -x_1 + (\tan t)x_2$

13. $x_1' = x_1 + x_2 + e^t,$
$x_2' = -3x_1 + 2x_2 + e^{2t}$

14. $x_1' = -tx_1 + t^2x_2 + t^3,$
$x_2' = -\sqrt{t}x_1 + \sqrt[3]{t}x_2 + t^{3/5}$

15. Consider the system

$$\mathbf{x}' = \begin{pmatrix} 4 & 2 \\ 3 & 3 \end{pmatrix}\mathbf{x}.$$

A fundamental matrix solution is

$$\Phi(t) = \begin{pmatrix} 2e^t & e^{6t} \\ -3e^t & e^{6t} \end{pmatrix}.$$

Find a solution that satisfies each of the following initial conditions.

a. $\mathbf{x}(0) = \begin{pmatrix} 2 \\ 3 \end{pmatrix}$; **b.** $\mathbf{x}(0) = \begin{pmatrix} -1 \\ 0 \end{pmatrix}$;

c. $\mathbf{x}(0) = \begin{pmatrix} 0 \\ 0 \end{pmatrix}$; **d.** $\mathbf{x}(0) = \begin{pmatrix} 7 \\ -2 \end{pmatrix}$;

e. $\mathbf{x}(0) = \begin{pmatrix} a \\ b \end{pmatrix}$.

In Exercises 16–19 find the principal matrix solution of the given system.

16. $\mathbf{x}' = \begin{pmatrix} -18 & -15 \\ 20 & 17 \end{pmatrix}\mathbf{x}$

17. $\mathbf{x}' = \begin{pmatrix} -3 & 4 \\ -2 & 3 \end{pmatrix}\mathbf{x}$

18. $\mathbf{x}' = \begin{pmatrix} 1 & 1 & 1 \\ -1 & -1 & 0 \\ -1 & 0 & -1 \end{pmatrix}\mathbf{x}$

19. $\mathbf{x}' = \begin{pmatrix} 4 & 2 & 0 \\ 2 & 4 & 0 \\ 0 & 0 & -3 \end{pmatrix}\mathbf{x}$

20. Solve the system

$$\mathbf{x}' = \begin{pmatrix} 2 & 1 & 0 \\ -2 & -1 & 2 \\ 1 & 1 & 1 \end{pmatrix}\mathbf{x} + \begin{pmatrix} 0 \\ 1 \\ e^t \end{pmatrix}, \qquad \mathbf{x}(0) = \begin{pmatrix} 1 \\ 2 \\ 3 \end{pmatrix}.$$

21. Solve the system

$$\mathbf{x}' = \begin{pmatrix} 2 & 1 \\ -4 & 2 \end{pmatrix}\mathbf{x} + \begin{pmatrix} 3 \\ t \end{pmatrix}e^{2t}, \qquad \mathbf{x}(0) = \begin{pmatrix} 3 \\ 2 \end{pmatrix}.$$

PART V

SEQUENCES AND SERIES

CHAPTERS INCLUDE:

CHAPTER 12

TAYLOR POLYNOMIALS, SEQUENCES, AND SERIES

Many functions arising in applications are difficult to deal with. A continuous function, for example, may take a complicated form, or it may take a simple form that, nevertheless, cannot be integrated.

For this reason mathematicians and physicists have developed methods for approximating certain functions by other functions that are much easier to handle. Some of the easiest functions to deal with are the polynomials, since in addition to having other useful properties, they can be differentiated and integrated any number of times and still remain polynomials. In the first three sections of this chapter we will show how certain continuous functions can be approximated by polynomials.

12.1 TAYLOR'S THEOREM AND TAYLOR POLYNOMIALS

In this section we show how a function can be approximated as closely as desired by a polynomial, provided that the function possesses a sufficient number of derivatives.

We begin by reminding you of the factorial notation defined for each positive integer n:

$$n! = n(n-1)(n-2)\cdots 3 \cdot 2 \cdot 1.$$

That is, $n!$ is the product of the first n positive integers. For example, $3! = 3 \cdot 2 \cdot 1 = 6$ and $5! = 5 \cdot 4 \cdot 3 \cdot 2 \cdot 1 = 120$. By convention, we define $0!$ to be equal to 1.

DEFINITION TAYLOR† POLYNOMIAL

Let the function f and its first n derivatives exist on the closed interval $[x_0, x_1]$. Then, for $a \in (x_0, x_1)$ and $x \in (x_0, x_1)$ the nth-degree **Taylor polynomial** of f at a is the

†Named after the English mathematician Brook Taylor (1685–1731), who published what we now call *Taylor's formula* in *Methodus Incrementorum* in 1715. There was a considerable controversy over whether Taylor's discovery was, in fact, a plagiarism of an earlier result of the Swiss mathematician Johann Bernoulli (1667–1748). We discussed Johann Bernoulli on page 696.

nth-degree polynomial $P_n(x)$, given by

$$P_n(x) = f(a) + \frac{f'(a)}{1!}(x-a) + \frac{f''(a)}{2!}(x-a)^2 + \frac{f'''(a)}{3!}(x-a)^3 + \cdots + \frac{f^{(n)}(a)}{n!}(x-a)^n$$

$$= \sum_{k=0}^{n} \frac{f^{(k)}(a)(x-a)^k}{k!}.^{\dagger} \qquad (1)$$

■

EXAMPLE 1 A TAYLOR POLYNOMIAL FOR $\sin x$ AT 0

Calculate the fifth-degree Taylor polynomial of $f(x) = \sin x$ at 0.

SOLUTION: We have $f(x) = \sin x$, $f'(x) = \cos x$, $f''(x) = -\sin x$, $f'''(x) = -\cos x$, $f^{(4)}(x) = \sin x$, and $f^{(5)}(x) = \cos x$. Then $f(0) = 0$, $f'(0) = 1$, $f''(0) = 0$, $f'''(0) = -1$, $f^{(4)}(0) = 0$, $f^{(5)}(0) = 1$, and we obtain

$$P_5(x) = f(0) + \frac{f'(0)}{1!}x + \frac{f''(0)}{2!}x^2 + \frac{f'''(0)}{3!}x^3 + \frac{f^{(4)}(0)}{4!}x^4 + \frac{f^{(5)}(0)}{5!}x^5$$

$$= x - \frac{x^3}{3!} + \frac{x^5}{5!} = x - \frac{x^3}{6} + \frac{x^5}{120}.$$

DEFINITION REMAINDER TERM

Let $P_n(x)$ be the nth-degree Taylor polynomial of the function f. Then the **remainder term**, denoted by $R_n(x)$, is given by

$$R_n(x) = f(x) - P_n(x). \qquad (2)$$

■

Why do we study Taylor polynomials? Because of the following remarkable result that tells us that a Taylor polynomial provides a good approximation to a function f. The proof is given at the end of this section.

THEOREM 1 TAYLOR'S THEOREM (TAYLOR'S FORMULA WITH REMAINDER)

Suppose that $f^{(n+1)}(x)$ exists on the closed interval $[x_0, x_1]$. Let a be in (x_0, x_1) and let x be any number in $[x_0, x_1]$. Then there is a number $c^{\ddagger}$ in (a, x) or (x, a) such that

$$R_n(x) = \frac{f^{(n+1)}(c)}{(n+1)!}(x-a)^{n+1}. \qquad (3)$$

■

† In this notation we have $f^{(0)}(a) = f(a)$.

‡ c depends on x.

The expression in (3) is called **Lagrange's form of the remainder.**[†] Using (3), we can write Taylor's formula as

$$f(x) = f(a) + \frac{f'(a)}{1!}(x-a) + \frac{f''(a)}{2!}(x-a)^2 + \cdots + \frac{f^{(n)}(a)}{n!}(x-a)^n + \frac{f^{(n+1)}(c)}{(n+1)!}(x-a)^{n+1}. \quad (4)$$

EXAMPLE 2

A TAYLOR POLYNOMIAL FOR $\sin x$ AT $\pi/6$

Calculate the fifth-degree Taylor polynomial of $f(x) = \sin x$ at $\pi/6$.

SOLUTION: Using the derivatives found in Example 1, we have $f(\pi/6) = \frac{1}{2}$, $f'(\pi/6) = \sqrt{3}/2$, $f''(\pi/6) = -\frac{1}{2}$, $f'''(\pi/6) = -\sqrt{3}/2$, $f^{(4)}(\pi/6) = \frac{1}{2}$, and $f^{(5)}(\pi/6) = \sqrt{3}/2$, so that in this case

$$\begin{aligned} P_5(x) &= \frac{1}{2} + \frac{\sqrt{3}}{2}\left(x - \frac{\pi}{6}\right) - \frac{1}{2}\frac{[x-(\pi/6)]^2}{2!} - \frac{\sqrt{3}}{2}\frac{[x-(\pi/6)]^3}{3!} \\ &\quad + \frac{1}{2}\frac{[x-(\pi/6]^4}{4!} + \frac{\sqrt{3}}{2}\frac{[x-(\pi/6]^5}{5!} \\ &= \frac{1}{2} + \frac{\sqrt{3}}{2}\left(x-\frac{\pi}{6}\right) - \frac{1}{4}\left(x-\frac{\pi}{6}\right)^2 - \frac{\sqrt{3}}{12}\left(x-\frac{\pi}{6}\right)^3 \\ &\quad + \frac{1}{48}\left(x-\frac{\pi}{6}\right)^4 + \frac{\sqrt{3}}{240}\left(x-\frac{\pi}{6}\right)^5. \end{aligned}$$

Examples 1 and 2 illustrate that in many cases it is easiest to calculate the Taylor polynomial at 0. In this situation, we have

TAYLOR POLYNOMIAL AT 0

$$P_n(x) = f(0) + f'(0)x + \frac{f''(0)}{2!}x^2 + \cdots + \frac{f^{(n)}(0)}{n!}x^n. \quad (5)$$

EXAMPLE 3

A TAYLOR POLYNOMIAL FOR e^x AT 0

Find the eighth-degree Taylor polynomial of $f(x) = e^x$ at 0.

SOLUTION: Here $f(x) = f'(x) = f''(x) = \cdots = f^{(8)}(x) = e^x$, and $e^0 = 1$, so that

$$P_8(x) = 1 + x + \frac{x^2}{2!} + \frac{x^3}{3!} + \frac{x^4}{4!} + \frac{x^5}{5!} + \frac{x^6}{6!} + \frac{x^7}{7!} + \frac{x^8}{8!} = \sum_{k=0}^{8} \frac{x^k}{k!}.$$

EXAMPLE 4

A TAYLOR POLYNOMIAL FOR $1/(1-x)$ AT 0

Find the fifth-degree Taylor polynomial of $f(x) = 1/(1-x)$ at 0.

[†] See the biographical sketch of Lagrange on page 228.

SOLUTION: Here,

$$f(x) = \frac{1}{1-x}, \qquad f'(x) = \frac{1}{(1-x)^2}, \qquad f''(x) = \frac{2}{(1-x)^3},$$

$$f'''(x) = \frac{6}{(1-x)^4}, \qquad f^{(4)}(x) = \frac{24}{(1-x)^5}, \qquad f^{(5)}(x) = \frac{120}{(1-x)^6},$$

Thus, $f(0) = 1$, $f'(0) = 1$, $f''(0) = 2$, $f'''(0) = 6$, $f^{(4)}(0) = 24$, $f^{(5)}(x) = 120$, and

$$\begin{aligned} P_5(x) &= 1 + x + \frac{2x^2}{2!} + \frac{6x^3}{3!} + \frac{24x^4}{4!} + \frac{120x^5}{5!} \\ &= 1 + x + x^2 + x^3 + x^4 + x^5 = \sum_{k=0}^{5} x^k. \end{aligned}$$

Note that in Examples 1, 2, and 3, the given function had continuous derivatives of all orders defined for all real numbers. In Example 4, $f(x)$ is defined over intervals of the form $[-b, b]$, where $b < 1$. Thus Taylor's theorem does *not* apply, for example, in any interval containing 1.

> It is always necessary to check whether the hypothesis of Taylor's theorem holds over a given interval.

Before leaving this section, we observe that **the nth-degree Taylor polynomial at a of a function is the polynomial that agrees with the function and each of its first n derivatives at a**. This follows immediately from (1):

$$P_n(a) = f(a)$$

$$P_n'(a) = \left[f'(a) + f''(a)(x-a) + f'''(a)\frac{(x-a)^2}{2!} + \cdots + \frac{f^{(n)}(a)}{(n-1)!}(x-a)^{n-1} \right]\Bigg|_{x=a} = f'(a),$$

$$P_n'' = \left[f''(a) + f'''(a)(x-a) + \cdots + \frac{f^{(n)}(a)}{(n-2)!}(x-a)^{n-2} \right]\Bigg|_{x=a} = f''(a),$$

and so on. In particular, since the $(n+1)$st derivative of an nth-degree polynomial is zero, we find that **if $Q(x)$ is a polynomial of degree n, then $P_n(x) = Q(x)$**. This follows immediately from the fact that the remainder term given by (3) will be zero since $Q^{(n+1)}(c) = 0$.

EXAMPLE 5

A TAYLOR POLYNOMIAL FOR A POLYNOMIAL

Let $Q(x) = 3x^4 + 2x^3 - 4x^2 + 5x - 8$. Compute $P_4(x)$ at 0.

SOLUTION: We have

$$Q(0) = -8,$$

$$Q'(0) = (12x^3 + 6x^2 - 8x + 5)\Big|_{x=0} = 5,$$

$$Q''(0) = (36x^2 + 12x - 8)\Big|_{x=0} = -8,$$

$$Q'''(0) = (72x + 12)\Big|_{x=0} = 12,$$

$$Q^{(4)}(0) = 72.$$

Therefore,

$$\begin{aligned} P_4(x) &= -8 + 5x - \frac{8x^2}{2!} + \frac{12x^3}{3!} + \frac{72x^4}{4!} \\ &= -8 + 5x - 4x^2 + 2x^3 + 3x^4 = Q(x), \end{aligned}$$

as expected.

PROOF OF TAYLOR'S THEOREM

We show that if $f^{(n+1)}(x)$ exists in $[a, b]$, then for any number $x \in [a, b]$, there is a number c in $[a, x]$ such that

$$f(x) = f(a) + \frac{f'(a)}{1!}(x - a) + \frac{f''(a)}{2!}(x - a)^2 + \cdots + \frac{f^{(n)}(a)}{n!}(x - a)^n + R_n(x),$$

where

$$f(x) - P_n(x) = \underbrace{R_n(x)}_{\text{definition of } R_n(x)} = \frac{f^{(n+1)}(c)}{(n + 1)!}(x - a)^{n+1}.$$

Let $x \in [a, b]$ be fixed. We define the new function $h(t)$ by

$$\begin{aligned} h(t) = f(x) - f(t) - f'(t)(x - t) - \frac{f''(t)}{2!}(x - t)^2 - \cdots \\ - \frac{f^{(n)}(t)}{n!}(x - t)^n - \frac{R_n(x)(x - t)^{n+1}}{(x - a)^{n+1}}. \end{aligned} \qquad (6)$$

Then, $h(x) = 0$ and

$$h(a) = f(x) - P_n(x) - R_n(x) = R_n(x) - R_n(x) = 0.$$

Since $f^{(n+1)}$ exists, $f^{(n)}$ is differentiable so that h, being a sum of products of differentiable functions, is also differentiable for t in (a, x). Remember that x is fixed so h is a function of t only.

Recall Rolle's theorem,† which states that if h is continuous on $[a, b]$ and differentiable on (a, b), and if $h(a) = h(b) = 0$, then there is at least one number c in (a, b), such that $h'(c) = 0$. We see that the conditions of Rolle's theorem hold in the interval $[a, x]$ so that there is a number c in (a, x) with $h'(c) = 0$. In Problems 30 and 31, you are asked to show

†See page 239 of *Calculus* or *Calculus of One Variable*.

that

$$h'(t) = \frac{-f^{(n+1)}(t)(x-t)^n}{n!} + \frac{(n+1)R_n(x)(x-t)^n}{(x-a)^{n+1}}.$$

Then, setting $t = c$, we obtain

$$0 = h'(c) = \frac{-f^{(n+1)}(c)(x-c)^n}{n!} + \frac{(n+1)R_n(x)(x-c)^n}{(x-a)^{n+1}}.$$

Finally, dividing the equations above through by $(x-c)^n$ and solving for $R_n(x)$, we obtain

$$R_n(x) = \frac{f^{(n+1)}(c)(x-a)^n}{(n+1)n!} = \frac{f^{(n+1)}(c)(x-a)^{n+1}}{(n+1)!}.$$

This is what we wanted to prove. ■

REMARK: If you go over the proof, you may observe that we didn't need to assume that $x > a$; if we replace the interval $[a, x]$ with the interval $[x, a]$, then all results are still valid.

PROBLEMS 12.1

SELF-QUIZ

Multiple Choice

I. The zero-degree Taylor polynomial of $f(x) = \cos x$ at 0 is ______.

a. $0 \cdot x$ **b.** $f(0)$ **c.** 1 **d.** 0

II. The zero-degree Taylor polynomial of $f(x) = \cos x$ at π is ______.

a. $0 \cdot x$ **b.** $1 \cdot (x - \pi)$ **c.** $f(0)$
d. $f(\pi)$ **e.** π **f.** -1

III. The first-degree Taylor polynomial of $f(x) = \cos x$ at π is ______.

a. $0 \cdot (x - \pi)$ **b.** $-1 + 1 \cdot x$
c. $-1 + 0 \cdot x$ **d.** $-1 + 0 \cdot (x - \pi)$

IV. The second-degree Taylor polynomial of $f(x) = \cos x$ at π is ______.

a. $-1 + 0 \cdot (x - \pi)$
b. $-1 + 0 \cdot (x - \pi) + 1 \cdot (x - \pi)^2$
c. $-1 + 0 \cdot x + \frac{1}{2} \cdot x^2$
d. $-1 + 0 \cdot (x - \pi) + \frac{1}{2} \cdot (x - \pi)^2$

V. The second-degree Taylor polynomial of $f(x) = 2 - 3x$ at -5 is ______.

a. $17 - 3 \cdot (x + 5)$
b. $f(0) + f(1)(x - (-5)) + f(2)(x - (-5))^2/2$
c. $17 + (-\frac{3}{1}) \cdot (x - 5) + \frac{0}{2} \cdot (x - 5)^2$
d. $2 - 3x$
e. $2 - 3 \cdot (x + 5) + 4 \cdot (x + 5)^2$
f. $2 - 3 \cdot (x - 5) + 4 \cdot (x - 5)^2$

VI. The second-degree Taylor polynomial of $f(x) = 2 - 3x^2$ at -5 is ______.

a. $-73 + (-\frac{6}{2})(x + 5)^2$
b. $-73 + 30(x + 5) - 3(x + 5)^2$
c. $-73 + 30x - 3x^2$
d. $-73 + \frac{30}{1}(x - 5) + (-\frac{6}{2})(x - 5)^2$

In Problems 1–26, a function f, a positive integer n, and a real number a are specified. Find the nth-degree Taylor polynomial of f at a. [*Note:* In some nth-degree Taylor polynomials, the coefficient of x^n is zero.]

1. $f(x) = \cos x$; $a = \pi/4$; $n = 6$
2. $f(x) = \sqrt{x}$; $a = 1$; $n = 4$
3. $f(x) = \ln x$; $a = e$; $n = 5$
4. $f(x) = \ln(1 + x)$; $a = 0$; $n = 5$
5. $f(x) = \dfrac{1}{x}$; $a = 1$; $n = 4$
6. $f(x) = \dfrac{1}{1 + x}$; $a = 0$; $n = 5$
7. $f(x) = \tan x$; $a = 0$; $n = 4$
8. $f(x) = \tan^{-1} x$; $a = 0$; $n = 6$
9. $f(x) = \tan x$; $a = \pi$; $n = 4$
10. $f(x) = \tan^{-1} x$; $a = 1$; $n = 6$
11. $f(x) = \dfrac{1}{1 + x^2}$; $a = 0$; $n = 4$
12. $f(x) = \dfrac{1}{\sqrt{x}}$; $a = 4$; $n = 3$

13. $f(x) = \sinh x$; $a = 0$; $n = 4$
14. $f(x) = \cosh x$; $a = 0$; $n = 4$
15. $f(x) = \ln \sin x$; $a = \pi/2$; $n = 3$
16. $f(x) = \ln \cos x$; $a = 0$; $n = 3$
17. $f(x) = \dfrac{1}{\sqrt{4-x}}$; $a = 0$; $n = 4$
18. $f(x) = \dfrac{1}{\sqrt{4-x}}$; $a = 3$; $n = 4$
19. $f(x) = e^{\beta x}$, β real; $a = 0$; $n = 6$
20. $f(x) = \sin(\beta x)$, β real; $a = 0$; $n = 6$
21. $f(x) = \sin^{-1} x$; $a = 0$; $n = 3$
22. $f(x) = 1 + x + x^2$; $a = 0$; $n = 10$
23. $f(x) = a_0 + a_1 x + a_2 x^2 + a_3 x^3$; $a = 1$; $n = 10$
24. $f(x) = e^{x^2}$; $a = 0$; $n = 4$
25. $f(x) = \sin(x^2)$; $a = 0$; $n = 4$
26. $f(x) = \cos(x^2)$; $a = 0$; $n = 4$

27. Show that the nth-degree Taylor polynomial of $f(x) = 1/(1-x)$ at 0 is

$$1 + x + x^2 + \cdots + x^n = \sum_{k=0}^{n} x^k.$$

28. Show that the nth-degree Taylor polynomial of $f(x) = \ln(1+x)$ at 0 is

$$x - \frac{x^2}{2} + \frac{x^3}{3} - \frac{x^4}{4} + \cdots + (-1)^{n-1}\frac{x^n}{n} = \sum_{k=1}^{n} (-1)^{k-1}\frac{x^k}{k}.$$

29. Show that the $(2n+1)$st and $(2n+2)$nd degree Taylor polynomials for $f(x) = \sin x$ at $a = 0$ are equal and are given by

$$x - \frac{x^3}{3!} + \frac{x^5}{5!} - \cdots + \frac{(-1)^n x^{2n+1}}{(2n+1)!} = \sum_{k=0}^{n} \frac{(-1)^k x^{2k+1}}{(2k+1)!}$$

30. Show that for $1 \le k \le n$,

$$\frac{d}{dt}\left[\frac{f^{(k)}(t)}{k!} \cdot (x-t)^k\right] = \frac{-f^{(k)}(t)}{(k-1)!} \cdot (x-t)^{k-1} + \frac{f^{(k+1)}(t)}{k!} \cdot (x-t)^k.$$

31. Use the results of the preceding problem to show that the function h defined by equation (6) has the following derivative:

$$h'(t) = \frac{-f^{(n+1)}(t)}{n!} \cdot (x-t)^n + \frac{(n+1) \cdot R_n(x) \cdot (x-t)^n}{(x-a)^{n+1}}.$$

***32.** Suppose the function f and its first $n-1$ derivatives exist throughout an open interval containing a. Also suppose that $f^{(n)}(a)$ exists. Prove that $f(x) = P_n(x) + R(x)$, where P_n is the nth-degree Taylor polynomial of f centered at a and

$$\lim_{x \to a} \frac{R(x)}{(x-a)^n} = 0.$$

Suppose that f and its first $n+1$ derivatives are continuous on the closed interval $[a, b]$. (Note that this assumption is a bit different from and slightly stronger than the one used in the preceding problem.) The following three problems constitute a proof that Taylor polynomials are unique: P_n is the only nth-degree polynomial whose remainder term satisfies (7).

33. Show that

$$\lim_{x \to a^+} \frac{R_n(x)}{(x-a)^n} = 0. \qquad (7)$$

34. Suppose that $S_n(x)$ satisfies (7); show that

$$\lim_{x \to a^+} \frac{S_k(x)}{(x-a)^k} = 0 \qquad \text{for } k = 0, 1, 2, \ldots, n. \qquad (8)$$

***35. Uniqueness Theorem for Taylor Polynomials** Suppose that

$$f(x) = P_n(x) + R_n(x) = Q_n(x) + S_n(x)$$

where $P_n(x)$ is the nth-degree Taylor polynomial, $Q_n(x)$ is another nth-degree polynomial, and $S_n(x)$ satisfies (7). Show that $P_n(x) = Q_n(x)$ for all $x \in [a, b]$. [*Hints:*

a. Let $D(x) = P_n(x) - Q_n(x) = d_0 + d_1(x-a) + \cdots + d_n(x-a)^n$.

b. Use the preceding problem to show that

$$\lim_{x \to a^+} \frac{D_k(x)}{(x-a)^k} = 0 \quad \text{for } k = 0, 1, 2, \ldots, n.$$

c. Conclude that $d_0 = d_1 = \cdots = d_n = 0$.]

ANSWERS TO SELF-QUIZ

I. b = c **II.** d = f **III.** d (c gives same answer, but for the wrong reason)
IV. d **V.** a = d **VI.** b

12.2 APPROXIMATION USING TAYLOR POLYNOMIALS

In this section we show how Taylor's formula can be used as a tool for making approximations. In many of the examples that follow, results have been obtained by making use of a hand calculator.

For convenience, we summarize the results of the preceding section.

Let $f, f', f'', \ldots, f^{(n+1)}$ exist in $[x_0, x_1]$. Then if $a \in (x_0, x_1)$ and $x \in [x_0, x_1]$,

$$f(x) = f(a) + f'(a)(x-a) + f''(a)\frac{(x-a)^2}{2!} + \cdots + f^{(n)}(a)\frac{(x-a)^n}{n!} + R_n(x)$$

$$= \sum_{k=0}^{n} \frac{f^{(k)}(a)(x-a)^k}{k!} + R_n(x), \quad (1)$$

where

$$R_n(x) = f^{(n+1)}(c)\frac{(x-a)^{n+1}}{(n+1)!}, \quad (2)$$

for some number c (which depends on x) in the interval (a, x) or (x, a).

THEOREM 1 THE MAXIMUM ERROR OF THE TAYLOR POLYNOMIAL APPROXIMATION

If f has $n+1$ continuous derivatives on $[a, b]$, there exists a positive number M_n such that

$$|R_n(x)| \le M_n \frac{|x-a|^{n+1}}{(n+1)!} \quad (3)$$

for all x in $[a, b]$. Here, M_n is an upper bound for the $(n+1)$st derivative of f in the interval $[a, b]$.

PROOF: Since $f^{(n+1)}$ is continuous on $[a, b]$, it is bounded above and below on that interval.† That is, there is a number M_n such that $|f^{(n+1)}(x)| \le M_n$ for every x in $[a, b]$. Since

$$R_n(x) = f^{(n+1)}(c)\frac{(x-a)^{n+1}}{(n+1)!},$$

with c in (a, x), we see that

$$|R_n(x)| = |f^{(n+1)}(c)|\frac{|x-a|^{n+1}}{(n+1)!} \le M_n \frac{|x-a|^{n+1}}{(n+1)!},$$

and the theorem is proved. ■

We stress that (3) provides an upper bound for the error. In many cases the actual error [the difference $|f(x) - P_n(x)|$] will be considerably less than $M_n(x-a)^{n+1}/(n+1)!$.

† See page 117 of *Calculus* or *Calculus of One Variable*.

EXAMPLE 1 THE ERROR IN APPROXIMATING $\sin x$ BY A TAYLOR POLYNOMIAL

In Example 1 in Section 12.1 we found that the fifth-degree Taylor polynomial of $f(x) = \sin x$ at 0 is $P_5(x) = x - (x^3/3!) + (x^5/5!)$. We then have

$$\sin x = x - \frac{x^3}{3!} + \frac{x^5}{5!} + R_5(x), \tag{4}$$

where

$$R_5(x) = \frac{f^{(6)}(c)(x-0)^6}{6!}.$$

But $f^{(6)}(c) = \sin^{(6)}(c) = -\sin c$ and $|-\sin c| \le 1$. Thus for x in $[0, 1]$,

$$|R_5(x)| \le \frac{1(x-0)^6}{6!} = \frac{x^6}{720}.$$

For example, suppose we wish to calculate $\sin(\pi/10)$. From (4)

$$\sin\frac{\pi}{10} = \frac{\pi}{10} - \frac{\pi^3}{3!10^3} + \frac{\pi^5}{5!10^5} + R_5\left(\frac{\pi}{10}\right)$$

with

$$\left|R_5\left(\frac{\pi}{10}\right)\right| \le \frac{(\pi/10)^6}{720} \approx 0.000001335.$$

We find that

$$\sin\frac{\pi}{10} \approx \frac{\pi}{10} - \frac{1}{3!}\frac{\pi^3}{10^3} + \frac{1}{5!}\frac{\pi^5}{10^5} \approx 0.3090170542.$$

The actual value of $\sin \pi/10 = \sin 18° = 0.3090169944$, correct to 10 decimal places, so our actual error is 0.0000000598, which is quite a bit less than 0.000001335. This illustrates the fact that the actual error (the value of the remainder term) is often quite a bit smaller than the theoretical upper bound on the error given by formula (3).

In this case we can explain the overly large error bound. Since $\sin^{(6)}(x) = -\sin x$ and $\sin^6(0) = -\sin 0 = 0$, the sixth-degree Taylor polynomial for $\sin x$ at 0 is

$$P_6(x) = x - \frac{x^3}{3!} + \frac{x^5}{5!} + 0 \cdot \frac{x^6}{6!} = x - \frac{x^3}{3!} + \frac{x^5}{5!}$$

so $P_5(x) = P_6(x)$ and we can use the error estimate for P_6. Since $|\sin^{(7)}(c)| = |-\cos c| \le 1$, we have

$$R_6(x) \le \frac{x^7}{7!} = \frac{x^7}{5040}.$$

If $x = \pi/10$, we obtain

$$\left|R_6\left(\frac{\pi}{10}\right)\right| \le \frac{(\pi/10)^7}{5040} \approx 0.0000000599.$$

Now we see that our estimate on the remainder term is really quite accurate.

EXAMPLE 2 APPROXIMATING AN INTEGRAL USING A TAYLOR POLYNOMIAL

Approximate $\int_0^{1/4} e^{x^2}\,dx$ with an error less than 0.001.

SOLUTION: We solve this problem by approximating e^{x^2} by its nth-degree Taylor polynomial at 0 and then integrating this polynomial. If we do this, we find that

$$\begin{aligned}
\text{error} &= \left|\int_0^{1/4} e^{x^2}\,dx - \int_0^{1/4} P_n(x)\,dx\right| = \left|\int_0^{1/4} [e^{x^2} - P_n(x)]\,dx\right| \\
&\le \int_0^{1/4} |R_n(x)|\,dx \le \int_0^{1/4} (\text{maximum value of } |R_n|)\,dx \\
&= \frac{1}{4}[\text{maximum value of } |R_n(x)|] \\
&= \frac{1}{4}\left[\frac{M_n \cdot \left(\frac{1}{4}\right)^{n+1}}{(n+1)!}\right] \\
&= \frac{M_n}{(n+1)!}\left(\frac{1}{4}\right)^{n+2} \qquad \text{where } M_n = \text{upper bound for } f^{(n+1)} \text{ on } [0, \tfrac{1}{4}].
\end{aligned}$$

We need $\dfrac{M_n}{(n+1)!}\left(\dfrac{1}{4}\right)^{n+2} \le 0.001$. Since $f(x) = e^{x^2}$, we have

$$f'(x) = 2xe^{x^2}, \qquad f''(x) = e^{x^2}(4x^2 + 2), \qquad f'''(x) = e^{x^2}(8x^3 + 12x)$$

and

$$f^{(4)}(x) = e^{x^2}(16x^4 + 48x^2 + 12).$$

On $[0, \frac{1}{4}]$,

$$|f'''(x)| \le e^{(1/4)^2}\left[8\left(\frac{1}{4}\right)^3 + 12\left(\frac{1}{4}\right)\right] \approx 3.3 = M_2 \qquad \text{and}$$

$$\frac{M_2}{3!}\left(\frac{1}{4}\right)^4 \approx 0.002,$$

which is too big. On $[0, \frac{1}{4}]$,

$$|f^{(4)}(x)| \le e^{(1/4)^2}\left[16\left(\frac{1}{4}\right)^4 + 48\left(\frac{1}{4}\right)^2 + 12\right] \approx 16.03 = M_3$$

and

$$\frac{M_3}{4!}\left(\frac{1}{4}\right)^5 \approx 0.00065.$$

Thus, $\int_0^{1/4} P_3(x)\,dx$ will approximate $\int_0^{1/4} e^{x^2}\,dx$ with an error no greater than 0.00065. Now $f(0) = 1$, $f'(0) = 0$, $f''(0) = 2$, and $f'''(0) = 0$, so $P_3(x) = P_2(x) = 1 + x^2$ and

$$\begin{aligned}
\int_0^{1/4} e^{x^2}\,dx &\approx \int_0^{1/4} (1 + x^2)\,dx = \left(x + \frac{x^3}{3}\right)\Bigg|_0^{1/4} = \frac{1}{4} + \left(\frac{1}{4}\right)^3 \cdot \frac{1}{3} \\
&\approx 0.255208.
\end{aligned}$$

This answer is within 0.00065 of the actual answer, so adding and subtracting 0.00065, we obtain the bounds

$$0.254558 < \int_0^{1/4} e^{x^2}\,dx < 0.255858.$$

We now consider a more general example.

THE LOGARITHM [ln(1 + x)]

The following product can be obtained by direct multiplication:

$$(1-u)(1+u+u^2+\cdots+u^n) = 1-u^{n+1}.$$

Thus, if $u \neq 1$, we have

$$1+u+u^2+\cdots+u^n = \frac{1-u^{n+1}}{1-u} = \frac{1}{1-u} - \frac{u^{n+1}}{1-u}. \tag{5}$$

The expression (5) is called the sum of a **geometric progression** (see also Section 12.5). From equation (5) we obtain the following:

$$\frac{1}{1-u} = 1+u+u^2+\cdots+u^n+\frac{u^{n+1}}{1-u}. \tag{6}$$

Setting $u = -t$ in (6) gives us

$$\frac{1}{1+t} = 1-t+t^2-t^3+\cdots+(-1)^n t^n + (-1)^{n+1}\frac{t^{n+1}}{1+t}. \tag{7}$$

Integration of both sides of (7) from 0 to x yields

$$\ln(1+x) = x - \frac{x^2}{2} + \frac{x^3}{3} - \cdots + (-1)^n\frac{x^{n+1}}{n+1} + (-1)^{n+1}\int_0^x \frac{t^{n+1}}{t+1}\,dt, \tag{8}$$

which is valid if $x > -1$ [since $\ln(1+x)$ is not defined for $x \leq -1$]. Now let

$$R_{n+1}(x) = (-1)^{n+1}\int_0^x \frac{t^{n+1}}{t+1}\,dt.$$

In Problems 31–33, you are asked to prove two things. First, for $-1 < x \leq 1$,

$$\lim_{n\to\infty} R_{n+1}(x) = 0.$$

This will ensure that the polynomial given in (8) provides an increasingly good approximation to $\ln(1+x)$ as n increases. Second, for every $n \geq 1$,

$$\lim_{x\to 0} \frac{R_{n+1}(x)}{x^{n+1}} = 0.$$

Then, according to the uniqueness theorem (Problem 35 in Section 12.1), we know that the polynomial in (8) is *the* $(n+1)$st-degree Taylor polynomial for $\ln(1+x)$ in the interval $-1 < x \leq 1$. We conclude that the Taylor polynomial

$$x - \frac{x^2}{2} + \frac{x^3}{3} - \cdots + (-1)^n\frac{x^{n+1}}{n+1}$$

may be used to approximate $\ln(1 + x)$ for sufficiently large n, provided that $-1 < x \le 1$.

If $0 \le x \le 1$, then (see Problem 31)

$$R_{n+1}(x) \le \frac{x^{n+2}}{n+2}. \tag{9}$$

EXAMPLE 3 APPROXIMATING ln 1.4 USING A TAYLOR POLYNOMIAL

Calculate ln 1.4 with an error of less than 0.001.

SOLUTION: Here $x = 0.4$, and from (9) we need to find an n such that $(0.4)^{n+2}/(n + 2) < 0.001$. We have $(0.4)^5/5 \approx 0.00205$ and $(0.4)^6/6 \approx 0.00068$, so choosing $n = 4$ $(n + 2 = 6)$, we obtain the Taylor polynomial [from (8)]

$$P_5(x) = x - \frac{x^2}{2} + \frac{x^3}{3} - \frac{x^4}{4} + \frac{x^5}{5}$$

[remember, from (8), the last term in $P_{n+1}(x)$ is $(-1)^n x^{n+1}/(n + 1)$], and

$$\ln 1.4 \approx P_5(0.4) = 0.4 - \frac{(0.4)^2}{2} + \frac{(0.4)^3}{3} - \frac{(0.4)^4}{4} + \frac{(0.4)^5}{5}$$

$$\approx 0.4 - 0.08 + 0.02133 - 0.0064 + 0.00205 = 0.33698.$$

The actual value of ln 1.4 is 0.33647, correct to five decimal places, so that the error is $0.33698 - 0.33647 = 0.00051$. This error is slightly less than the maximum possible error of 0.00068, and so in this case our error bound is fairly sharp. Note that the error formula (9) used in this problem is *not* the same as the error formula given by (3). To obtain the bound (3), we would have to compute the sixth derivative of $\ln(1 + x)$. Surprisingly, if you use formula (3), you will obtain the same bound $[(0.4)^6/6]$. Try it.

THE ARC TANGENT ($\tan^{-1} x$)

If in equation (6) we substitute $u = -t^2$, then for any number t we obtain

$$\frac{1}{1+t^2} = 1 - t^2 + t^4 - t^6 + \cdots + (-1)^n t^{2n} + (-1)^{n+1}\frac{t^{2(n+1)}}{1+t^2}. \tag{10}$$

Integration of both sides of (10) from 0 to x yields

$$\tan^{-1} x = x - \frac{x^3}{3} + \frac{x^5}{5} - \frac{x^7}{5} + \cdots + \frac{(-1)^n x^{2n+1}}{2n+1} + (-1)^{n+1}\int_0^x \frac{t^{2(n+1)}}{1+t^2}\,dt. \tag{11}$$

This equation is valid for any real number $x \ge 0$.

In Problem 34, you are asked to show that

$$\lim_{x\to 0} \frac{R_{2n+1}(x)}{x^{2n+1}} = 0.$$

This ensures, according to Problem 35 in Section 12.1, that the polynomial in (11) is *the* Taylor polynomial for $\tan^{-1} x$ if $0 \le x \le 1$. In Problem 34, you are also asked to show that $R_{2n+1}(x) \to 0$ as $n \to \infty$ so the Taylor polynomial provides an increasingly good approximation to $\tan^{-1} x$ as n increases.

Equation (11) gives us a formula for calculating π.† Setting $x = 1$, we have

$$\frac{\pi}{4} = \tan^{-1} 1 \approx 1 - \frac{1}{3} + \frac{1}{5} - \frac{1}{7} + \frac{1}{9} - \frac{1}{11} + \cdots + \frac{(-1)^n}{2n+1}.$$

The error here (see Problem 34(b)) is bounded by $1/(2n + 3)$, which approaches zero very slowly. To get an error less than 0.001, we would need $1/(2n + 3) < \frac{1}{1000}$, or $2n + 3 > 1000$, and $n \geq 499$. For example, $1 - \frac{1}{3} + \frac{1}{5} - \frac{1}{7} + \frac{1}{9} - \frac{1}{11} + \frac{1}{13} - \frac{1}{15} = 0.75427$, while $\pi/4$ is 0.78540. A better way to approximate π is suggested in Problems 27 and 28.

Under certain conditions, it is easy to show that the remainder term $R_n(x)$ approaches 0 as $n \to \infty$. The proof of the following theorem is left as an exercise (see Problems 35 and 36).

THEOREM 2

Suppose that $f^{(n)}(x)$ exists for $x \in [a, b]$ for $n = 0, 1, 2, \ldots$ and that there exists a number $M > 0$ such that $|f^{(n)}(x)| \leq M$ for $a \leq x \leq b$ and all nonnegative integers n. Then,

$$\lim_{n\to\infty} |R_n(x)| = 0. \qquad \blacksquare$$

REMARK: Theorem 2 can be restated as follows: If all derivatives of f are uniformly bounded, then the remainder term approaches zero as $n \to \infty$.

PROBLEMS 12.2

SELF-QUIZ

Multiple Choice

I. $\sin x \approx$ ________ on the interval $[-0.25, 0.25]$.
a. x **b.** $-x$
c. $1 + x$ **d.** $1/2 - x^2$

II. $\cos x \approx$ ________ on the interval $[\pi/2 - 0.25, \pi/2 + 0.25]$.
a. $x - \pi/2$ **b.** $-(x - \pi/2)$
c. $1 - x^2/2$ **d.** $(x - \pi/2) + x^2/2$

True–False

III. Let f be a function whose fifth derivative is continuous on $[0, 6]$. Let P_4 be the fourth-degree Taylor polynomial for f at 1. Suppose $P_4(1.7)$ is a good approximation to $f(1.7)$. Then is it True or False that $P_4(3.8)$ is just as good an approximation to $f(3.8)$?

Multiple Choice

IV. Let $f(x) = 2 - 7x + 5x^2$. The first-degree Taylor polynomial for f at 2 is $P_1(x) = f(2) + f'(2)(x - 2) = 8 + 13(x - 2)$. Also notice that f'' is constant: $f''(x) = 10$ for all x. Which of the following statements about the remainder are true?
a. $f(x) - P_1(x)$ is constant
b. $f(x) - P_1(x) = 10/2!$
c. $|f(x) - P_1(x)| = 10|x - 2|^2$
d. $|f(x) - P_1(x)| = 10|x - 2|^2/2!$
e. $|f(x) - P_1(x)| \leq 10|x - 2|^2/2!$
f. $f(x) - P_1(x) = 10|x - 2|^2/2!$

V. Let $f(x) = \cos x$, then $P_3(x) = -(x - \pi/2) + (x - \pi/2)^3/3!$ is the third-degree Taylor polynomial for f at $\pi/2$. Because all derivatives of f are bounded by 1, we can make the

†This formula was discovered by the Scottish mathematician James Gregory (1638–1675) and was first published in 1712.

following statement about error when approximating $f(x)$ by $P_3(x)$:

a. $|f(x) - P_3(x)| \le (x - \pi/2)^3$
b. $|f(x) - P_3(x)| = (x - \pi/2)^4$
c. $|f(x) - P_3(x)| \le (x - \pi/2)^4/4!$
d. $|f(x) - P_3(x)| \le (x - \pi/2)^4/5!$

In Problems 1–10, a function f, point a, degree n, and interval are given. Find a close bound for $|R_n(x)|$ where $R_n = f - P_n$ is the remainder from the nth-degree Taylor polynomial P_n.

1. $f(x) = \sin x;\ a = \pi/4;\ n = 6;\ [0, \pi/2]$
2. $f(x) = 1/x;\ a = 1;\ n = 4;\ [\frac{1}{2}, 2]$
3. $f(x) = \sqrt{x};\ a = 1;\ n = 4;\ [\frac{1}{4}, 4]$
4. $f(x) = 1/\sqrt{x};\ a = 5;\ n = 5;\ [\frac{19}{4}, \frac{21}{4}]$
5. $f(x) = e^{\beta x};\ a = 0;\ n = 4;\ [0, 1]$
6. $f(x) = \sinh x;\ a = 0;\ n = 4;\ [0, 1]$
7. $f(x) = \tan x;\ a = 0;\ n = 4;\ [0, \pi/4]$
8. $f(x) = \ln \cos x;\ a = 0;\ n = 3;\ [0, \pi/6]$
9. $f(x) = e^{x^2};\ a = 0;\ n = 4;\ [0, \frac{1}{3}]$
10. $f(x) = \sin(x^2);\ a = 0;\ n = 4;\ [0, \pi/4]$

In Problems 11–16, use a Taylor polynomial of appropriate degree to approximate the given function value with the specified accuracy.

11. $\sin\left(\frac{\pi}{6} + 0.2\right);\ |\text{error}| < 0.001$

12. $\tan\left(\frac{\pi}{4} + 0.1\right);\ |\text{error}| < 0.01$

13. e; $|\text{error}| < 0.0001$ [*Note:* You may use $2 < e < 3$.]

14. e^{-1}; $|\text{error}| < 0.001$

15. $\sin 33°$; $|\text{error}| < 0.001$ [*Hint:* Convert 33° to radians.]

16. $\tan^{-1} 0.5$; $|\text{error}| < 0.001$

17. Use the result of Problem 7 to approximate

$$\int_0^{\pi/4} \tan x\, dx.$$

What is the maximum error of your approximation?

18. Use the result of Problem 8 to approximate

$$\int_0^{\pi/6} \ln \cos x\, dx.$$

What is the maximum error of your approximation?

19. Use the result of Problem 9 to approximate

$$\int_0^{1/3} e^{x^2}\, dx.$$

What is the maximum error of your approximation?

20. Use the result of Problem 10 to approximate

$$\int_0^{\pi/4} \sin(x^2)\, dx.$$

What is the maximum error of your approximation?

In Problems 21–24, use Taylor polynomials to approximate each integral with the indicated maximum error.

21. $\int_0^{1/5} e^{x^3}\, dx;\ 0.00001$

22. $\int_0^{1/2} \sin(x^2)\, dx;\ 0.0001$

23. $\int_0^{\pi/6} \cos(x^2)\, dx;\ 0.0001$

***24.** $\int_0^{1/4} x^2 e^{x^2}\, dx;\ 0.001$

25. Use the binomial theorem (Problem 29) to calculate

a. 1.03^3 **b.** 0.97^4
c. 1.2^4 **d.** 0.8^5

26. Use the general binomial expansion (Problem 30) to approximate the following numbers with four decimal place accuracy (i.e., $|\text{error}| \le 0.5 \cdot 10^{-4}$).

a. $\sqrt{1.2}$ **b.** $0.9^{3/4}$ **c.** $1.8^{1/4}$
d. $1.01^{-1/3}$ **e.** $2^{5/3}$ **f.** $0.4^{1/6}$

***27.** Use the result of Problem 28 to approximate π to five decimal places.

***28.** Use the formula

$$\tan(A \pm B) = \frac{\tan A \pm \tan B}{1 \mp \tan A \cdot \tan B}$$

to prove that

$$\tan\left(4 \tan^{-1}\frac{1}{5} - \tan^{-1}\frac{1}{239}\right) = 1.$$

[*Hint:* First calculate $\tan[2 \tan^{-1}(\frac{1}{5})]$ and $\tan[4 \tan^{-1}(\frac{1}{5})]$. Then observe that $4 \tan^{-1}(\frac{1}{5}) - \tan^{-1}(\frac{1}{239}) = \pi/4$.]

***29.** Let $f(x) = (1 + x)^n$, where n is a positive integer.

a. Show that $f^{(n+1)}(x) = 0$ for all x.

b. Show that

$$f(x) = 1 + nx + \frac{n(n-1)}{2!}x^2 + \frac{n(n-1)(n-2)}{3!}x^3 + \cdots + \frac{n!}{n!}x^n.$$

[*Hint:* Obtain the nth-degree Taylor polynomial for f at 0 and then examine its remainder.] This result is called the **Binomial Theorem.**

30. Let $f(x) = (1+x)^{\alpha}$, where α is any real number.

a. Consider the nth-degree Taylor polynomial for f at 0 and show that

$$f(x) = 1 + \alpha x + \frac{\alpha(\alpha-1)}{2!}x^2 + \frac{\alpha(\alpha-1)(\alpha-2)}{3!}x^3 + \cdots + \frac{\alpha(\alpha-1)(\alpha-2)\cdots(\alpha-n+1)}{n!}x^n + R_n(x).$$

b. Show that if $|x| < 1$, then $R_n(x) \to 0$ as $n \to \infty$. This result is called the **general binomial theorem.**

31. Consider the expression (see equation (8)) which represents the error in the Taylor polynomial approximation for $\ln(1+x)$:

$$R_n(x) = (-1)^n \int_0^x \frac{t^n}{1+t}\,dt.$$

Show that if $0 \le x \le 1$, then

$$|R_n(x)| \le \int_0^x t^n\,dt = \frac{x^{n+1}}{n+1} \le \frac{1}{n+1}.$$

[*Hint:* $1/(1+t) \le 1$ if $0 \le t \le 1$.]

32. In the preceding problem, suppose that $-1 < x < 0$.

a. Show that

$$R_n(x) = (-1)^{n-1} \int_x^0 \frac{t^n}{1+t}\,dt.$$

b. Show that $1 \le \frac{1}{1+t} \le \frac{1}{1+x}$ if $-1 < x \le t \le 0$.

c. Show that

$$|R_n(x)| \le \int_x^0 \frac{|t|^n}{|1+t|}\,dt \le \int_x^0 \frac{(-t)^n}{1+x}\,dt \le \frac{(-x)^{n+1}}{(n+1)(1+x)}.$$

33. Using the results of the preceding two problems, show that if $-1 < x \le 1$,

a. $|R_n(x)| \to 0$ as $n \to \infty$.

b. $\lim_{x\to 0} \frac{|R_n(x)|}{x^n} = 0 n > 0.$

34. Let (see equation (11))

$$R_{2n+1}(x) = (-1)^{n+1} \int_0^x \frac{t^{2(n+1)}}{1+t^2}\,dt.$$

a. Show that $1/(1+t^2) \le 1$ for all t.

b. Show that $|R_{2n+1}(x)| \le x^{2n+3}/(2n+3)$ for all x.

c. Show that

$$\lim_{x\to 0} \frac{|R_{2n+1}(x)|}{x^{2n+1}} = 0 \quad \text{for all integers } n \ge 0.$$

d. Show that

$$\lim_{n\to\infty} |R_{2n+1}(x)| = 0 \quad \text{if } |x| \le 1.$$

***35.** Show that for any real number x,

$$\lim_{n\to\infty} \frac{x^n}{n!} = 0.$$

36. Prove Theorem 2. [*Hint:* Use the result of the preceding problem.]

ANSWERS TO SELF-QUIZ

I. a **II.** b **III.** False **IV.** d, e, f

V. c ($P_3 = P_4$; thus (d) is true too.)

12.3 SEQUENCES OF REAL NUMBERS

According to a popular dictionary,[†] a *sequence* is "the following of one thing after another." In mathematics we could define a sequence intuitively as a succession of numbers that never terminates. The numbers in the sequence are called the *terms* of the sequence. In a sequence there is one term for each positive integer.

EXAMPLE 1

A SEQUENCE

Consider the sequence

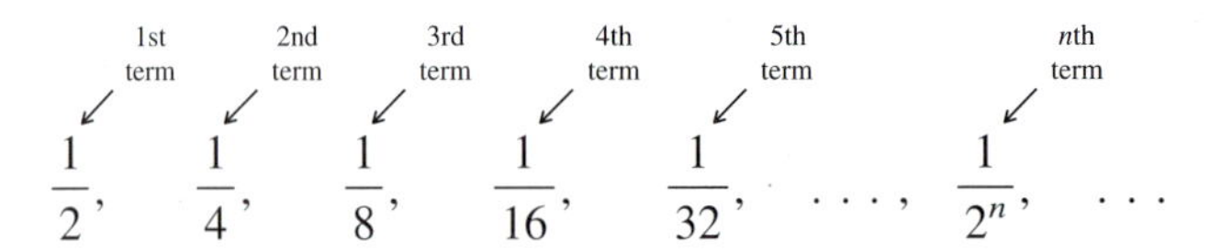

$$\frac{1}{2}, \quad \frac{1}{4}, \quad \frac{1}{8}, \quad \frac{1}{16}, \quad \frac{1}{32}, \quad \ldots, \quad \frac{1}{2^n}, \quad \ldots$$

We see that there is one term for each positive integer. The terms in this sequence form an infinite set of real numbers, which we write as

$$A = \frac{1}{2}, \frac{1}{4}, \frac{1}{8}, \ldots, \frac{1}{2^n}, \ldots. \tag{1}$$

That is, the set A consists of all numbers of the form $1/2^n$, where n is a positive integer. There is another way to describe this set. We define the function f by the rule $f(n) = 1/2^n$, where the domain of f is the set of positive integers. Then the set A is precisely the set of values taken by the function f.

We have the following formal definition.

DEFINITION SEQUENCE

A **sequence** of real numbers is a function whose domain is the set of positive integers. The values taken by the function are called **terms** of the sequence. ■

NOTATION: We will often denote the terms of a sequence by a_n. Thus if the function given in the definition above is f, then $a_n = f(n)$. With this notation, *we can denote the set of values taken by the sequence by* $\{a_n\}$. Also, we will use n, m, and so on as integer variables and x, y, and so on as real variables.

EXAMPLE 2

SIX SEQUENCES AND THE VALUES THEY TAKE WRITTEN OUT

Sequence	The Terms of the Sequence Written Out
a. $\{a_n\} = \left\{\frac{1}{n}\right\}$	$1, \frac{1}{2}, \frac{1}{3}, \frac{1}{4}, \ldots, \frac{1}{n}, \ldots$
b. $\{a_n\} = \{\sqrt{n}\}$	$1, \sqrt{2}, \sqrt{3}, \sqrt{4}, \ldots, \sqrt{n}, \ldots$
c. $\{a_n\} = \left\{\frac{1}{n!}\right\}$	$1, \frac{1}{2}, \frac{1}{6}, \frac{1}{24}, \ldots, \frac{1}{n!}, \ldots$

[†] *Webster's Encyclopedic Unabridged Dictionary of the English Language* (New York: Portland House, 1989).

d. $\{a_n\} = \{\sin n\}$ $\qquad \sin 1, \sin 2, \sin 3, \sin 4, \ldots, \sin n, \ldots$

e. $\{a_n\} = \left\{\dfrac{e^n}{n!}\right\}$ $\qquad e, \dfrac{e^2}{2}, \dfrac{e^3}{6}, \dfrac{e^4}{24}, \ldots, \dfrac{e^n}{n!}, \ldots$

f. $\{a_n\} = \left\{\dfrac{n-1}{n}\right\}$ $\qquad 0, \dfrac{1}{2}, \dfrac{2}{3}, \dfrac{3}{4}, \ldots, \dfrac{n-1}{n}, \ldots$

REMARK: In Example 2 the first term in each sequence corresponds to $n = 1$. However it is sometimes more convenient to start at $n = 0$, $n = 2$, or some other integer. All the definitions and theorems in this and the next section are unchanged if we start at $n = k$ where $k \neq 1$.

Because a sequence is a function, it has a graph. In Figure 1 we draw part of the graph of the first sequence in Example 2.

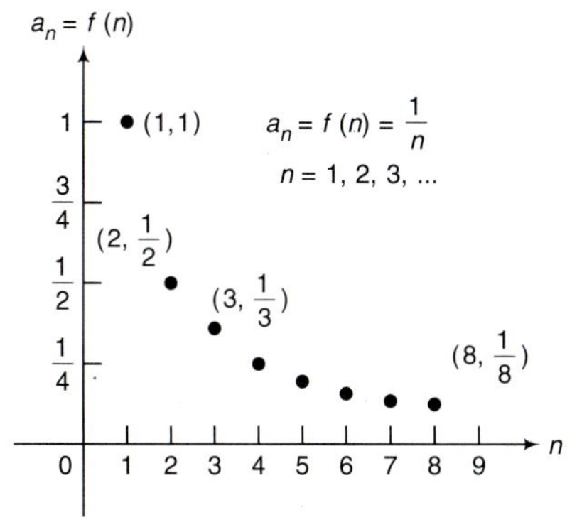

FIGURE 1
Graph of $\{a_n\} = \left\{\dfrac{1}{n}\right\}$

Observe that $f(n)$ is defined only when n is a positive integer so there is one point on the graph for each positive integer.

EXAMPLE 3

FINDING THE GENERAL TERM OF A SEQUENCE

Find the general term a_n of the sequence

$$-1, 1, -1, 1, -1, 1, -1, \ldots.$$

SOLUTION: We see that $a_1 = -1$, $a_2 = 1$, $a_3 = -1$, $a_4 = 1, \ldots$. Hence

$$a_n = \begin{cases} -1, & \text{if } n \text{ is odd} \\ 1, & \text{if } n \text{ is even.} \end{cases}$$

A more concise way to write this term is

$$a_n = (-1)^n.$$

It is evident that as n gets large, the numbers $1/n$ get small. We can write

$$\lim_{n\to\infty} \frac{1}{n} = 0.$$

This is also suggested by the graph in Figure 1. Similarly, it is not hard to show that as n

gets large, $(n-1)/n$ gets close to 1. We write

$$\lim_{n\to\infty} \frac{n-1}{n} = 1.$$

On the other hand, it is clear that $a_n = (-1)^n$ does not get close to any one number as n increases. It simply oscillates back and forth between the numbers $+1$ and -1.

For the remainder of this section we will be concerned with calculating the limit of a sequence as $n \to \infty$. Since a sequence is a special type of function, our formal definition of the limit of a sequence is going to be very similar to the definition of $\lim_{x\to\infty} f(x)$.

DEFINITION FINITE LIMIT OF A SEQUENCE

A sequence $\{a_n\}$[†] has the **limit** L if for every $\epsilon > 0$ there exists an integer $N > 0$ such that if $n \geq N$, then $|a_n - L| < \epsilon$. We write

$$\lim_{n\to\infty} a_n = L. \tag{2}$$

■

Intuitively, this definition states that $a_n \to L$ if as n increases without bound, a_n gets arbitrarily close to L. We illustrate this definition in Figure 2.

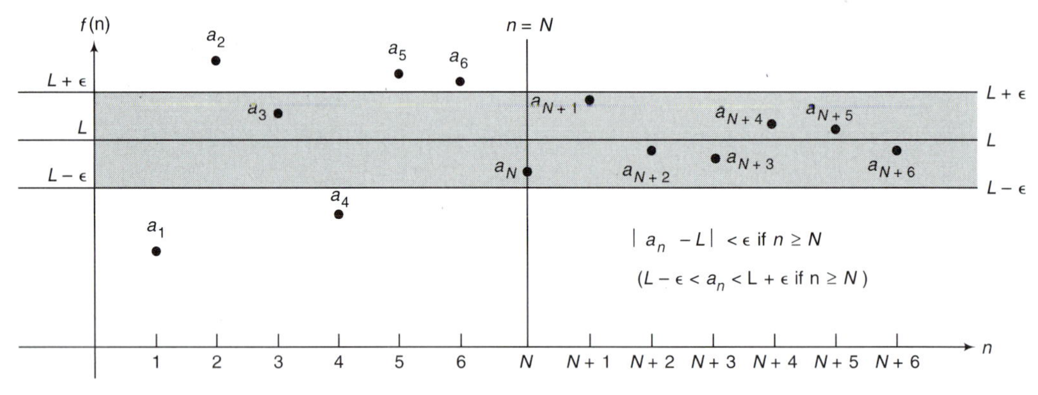

FIGURE 2
Illustration of the definition of a limit

For $n > N$, the values taken by a_n lie between $L - \epsilon$ and $L + \epsilon$ (i.e. in the shaded strip).

DEFINITION INFINITE LIMIT OF A SEQUENCE

The sequence $\{a_n\}$ has the limit ∞ if for every positive number M there is an integer $N > 0$ such that if $n > N$, then $a_n > M$. In this case we write

$$\lim_{n\to\infty} a_n = \infty.$$

■

Intuitively, $\lim_{n\to\infty} a_n = \infty$ means that as n increases without bound, a_n also increases without bound. The expression $\lim_{n\to\infty} a_n = -\infty$ is defined in a similar way. The following theorem gives us a very useful result.

[†]To be precise, $\{a_n\}$ denotes the set of values taken by the sequence. There is a difference between the sequence, which is a function f, and the values $a_n = f(n)$ taken by this function. However, because it is more convenient to write down the values the sequence takes, we will, from now on, use the notation $\{a_n\}$ to denote a sequence.

THEOREM 1 LIMIT OF A GEOMETRIC SEQUENCE

Let r be a real number. Then,

$$\lim_{n\to\infty} r^n = 0 \qquad \text{if } |r| < 1$$

and

$$\lim_{n\to\infty} |r^n| = \infty \qquad \text{if } |r| > 1.$$

NOTE: The sequence $\{a_n\} = \{r^n\}$ for $r \neq 1$ is called a **geometric sequence**.

PROOF:

Case 1: $r = 0$. Then $r^n = 0$, and the sequence has the limit 0.

Case 2: $0 < |r| < 1$. For a given $\epsilon > 0$, choose N such that

$$N > \frac{\ln \epsilon}{\ln|r|}.$$

Note that since $|r| < 1$, $\ln|r| < 0$. Now if $n > N$,

$$n > \frac{\ln \epsilon}{\ln|r|} \qquad \text{and} \qquad n \ln|r| < \ln \epsilon.$$

The second inequality follows from the fact that $\ln|r|$ is negative, and multiplying both sides of an inequality by a negative number reverses the inequality. Thus,

$$\ln|r^n - 0| = \ln|r^n| = \ln|r|^n = n \ln|r| < \ln \epsilon.$$

Since $\ln|r^n - 0| < \ln \epsilon$ and $\ln x$ is an increasing function, we conclude that $|r^n - 0| < \epsilon$. Thus according to the definition of a finite limit of a sequence,

$$\lim_{n\to\infty} r^n = 0.$$

Case 3: $|r| > 1$. Let $M > 0$ be given. Choose $N > \ln M/\ln|r|$. Then if $n > N$,

$$\ln|r^n| = n \ln|r| \overset{n>N}{>} N \ln|r| > \left(\frac{\ln M}{\ln|r|}\right)(\ln|r|) = \ln M,$$

so that

$$|r^n| > M \qquad \text{if } n > N.$$

From the definition of an infinite limit of a sequence, we see that

$$\lim_{n\to\infty} |r^n| = \infty. \qquad \blacksquare$$

In your study of one variable calculus you saw a number of limit theorems. All these results can be applied when n, rather than x, approaches infinity. The only difference is that as n grows, it takes on integer values. For convenience, we state without proof the major limit theorems we need for sequences.

THEOREM 2

Suppose that $\lim_{n\to\infty} a_n$ and $\lim_{n\to\infty} b_n$ both exist and are finite.

i. $\lim_{n\to\infty} \alpha a_n = \alpha \lim_{n\to\infty} a_n$ for any real number α. **(3)**

ii. $\lim_{n\to\infty}(a_n + b_n) = \lim_{n\to\infty} a_n + \lim_{n\to\infty} b_n$. **(4)**

iii. $\lim_{n\to\infty} a_n b_n = (\lim_{n\to\infty} a_n)(\lim_{n\to\infty} b_n)$. **(5)**

iv. If $\lim_{n\to\infty} b_n \neq 0$, then

$$\lim_{n\to\infty} \frac{a_n}{b_n} = \frac{\lim\limits_{n\to\infty} a_n}{\lim\limits_{n\to\infty} b_n}. \tag{6}$$

■

THEOREM 3 CONTINUITY THEOREM

Suppose that L is finite and $\lim_{n\to\infty} a_n = L$. If f is continuous in an open interval containing L, then

$$\lim_{n\to\infty} f(a_n) = f\left(\lim_{n\to\infty} a_n\right) = f(L). \tag{7}$$

■

THEOREM 4 SQUEEZING THEOREM

Suppose that $\lim_{n\to\infty} a_n = \lim_{n\to\infty} b_n = L$ and that $\{c_n\}$ is a sequence having the property that for $n > N$ (a positive integer), $a_n \le c_n \le b_n$. Then,

$$\lim_{n\to\infty} c_n = L. \tag{8}$$

■

We now give a central definition in the theory of sequences.

DEFINITION CONVERGENCE AND DIVERGENCE OF A SEQUENCE

If the limit in (2) exists and if L is finite, we say that the sequence **converges** or is **convergent**. Otherwise, we say that the sequence **diverges** or is **divergent**. ■

EXAMPLE 4 A CONVERGENT GEOMETRIC SEQUENCE

The geometric sequence $\{1/2^n\} = \{(\frac{1}{2})^n\}$ is convergent since, by Theorem 1, $\lim_{n\to\infty} 1/2^n = \lim_{n\to\infty}(1/2)^n = 0$.

EXAMPLE 5 A DIVERGENT GEOMETRIC SEQUENCE

The geometric sequence $\{r^n\}$ is divergent for $r > 1$ since $\lim_{n\to\infty} r^n = \infty$ if $r > 1$.

EXAMPLE 6 A SEQUENCE MAY DIVERGE EVEN THOUGH ITS VALUES DO NOT APPROACH INFINITY

The sequence $\{(-1)^n\}$ is divergent since the values a_n alternate between -1 and $+1$ but do not stay close to any fixed number as n becomes large.

Since we have a large body of theory and experience behind us in the calculation of ordinary limits, we would like to make use of that experience to calculate limits of sequences. The following theorem, whose proof is left as a problem (see Problem 31), is extremely useful.

THEOREM 5

Suppose that $\lim_{x\to\infty} f(x) = L$, a finite number, ∞, or $-\infty$. If f is defined for every positive integer, then the limit of the sequence $\{a_n\} = \{f(n)\}$ is also equal to L. That is, $\lim_{x\to\infty} f(x) = \lim_{n\to\infty} a_n = L$. ■

EXAMPLE 7

A CONVERGENT SEQUENCE

Calculate $\lim_{n\to\infty} 1/n^2$.

SOLUTION: Since $\lim_{x\to\infty} 1/x^2 = 0$, we have $\lim_{n\to\infty} 1/n^2 = 0$ (by Theorem 5).

EXAMPLE 8

DETERMINING CONVERGENCE OR DIVERGENCE BY USING L'HÔPITAL'S RULE

Does the sequence $\{e^n/n\}$ converge or diverge?

SOLUTION: Since $\lim_{x\to\infty} e^x/x = \lim_{x\to\infty} e^x/1$ (by l'Hôpital's rule) $= \infty$, we find that the sequence diverges.

REMARK: It should be emphasized that Theorem 5 does *not* say that if $\lim_{x\to\infty} f(x)$ does not exist, then $\{a_n\} = \{f(n)\}$ diverges. For example, let

$$f(x) = \sin \pi x.$$

Then $\lim_{x\to\infty} f(x)$ does not exist, but $\lim_{n\to\infty} f(n) = \lim_{n\to\infty} \sin \pi n = 0$ since $\sin \pi n = 0$ for every integer n.

EXAMPLE 9

A SEQUENCE THAT CONVERGES TO e

Let $\{a_n\} = \{[1 + (1/n)]^n\}$. Does this sequence converge or diverge?

SOLUTION: Since $\lim_{x\to\infty}[1 + (1/x)]^x = e$, we see that a_n converges to the limit e.

EXAMPLE 10

USING L'HÔPITAL'S RULE

Determine the convergence or divergence of the sequence $\{(\ln n)/n\}$.

SOLUTION: $\lim_{x\to\infty}[(\ln x)/x] = \lim_{x\to\infty}[(1/x)/1] = 0$ by l'Hôpital's rule, so that the sequence converges to 0.

EXAMPLE 11 LIMIT OF A SEQUENCE DETERMINED BY A RATIONAL FUNCTION

Let $p(x) = c_0 + c_1x + \cdots + c_mx^m$ and $q(x) = d_0 + d_1x + \cdots + d_rx^r$. In one variable calculus you saw that if the rational function $r(x) = p(x)/q(x)$ and if $c_md_r \neq 0$, then

$$\lim_{x\to\infty} \frac{p(x)}{q(x)} = \begin{cases} 0, & \text{if } m < r \\ \pm\infty, & \text{if } m > r \\ \dfrac{c_m}{d_r}, & \text{if } m = r. \end{cases}$$

Thus the sequence $\{p(n)/q(n)\}$ converges to 0 if $m < r$, converges to c_m/d_r if $m = r$, and diverges if $m > r$.

EXAMPLE 12 USING THE SQUEEZING THEOREM TO PROVE CONVERGENCE OF A SEQUENCE

Determine the convergence or divergence of the sequence $\{\sin \alpha n/n^\beta\}$, where α is a real number and $\beta > 0$.

SOLUTION: Since $-1 \le \sin \alpha x \le 1$, we see that

$$-\frac{1}{x^\beta} \le \frac{\sin \alpha x}{x^\beta} \le \frac{1}{x^\beta} \quad \text{for any } x > 0.$$

But $\pm\lim_{x\to\infty} 1/x^\beta = 0$, and so by the squeezing theorem for limits, $\lim_{x\to\infty}[(\sin \alpha x)/x^\beta] = 0$. Therefore, the sequence $\{(\sin \alpha n)/n^\beta\}$ converges to 0.

As in Example 12, the squeezing theorem can often be used to calculate the limit of a sequence.

PROBLEMS 12.3

SELF-QUIZ

Multiple Choice

I. The first three terms of the sequence $\{n/(n + 1)\}$, starting with $n = 1$, are ________.
- **a.** $a/(a + 1)$, $b/(b + 1)$, $c/(c + 1)$
- **b.** $\frac{0}{1}, \frac{1}{2}, \frac{2}{3}$
- **c.** $\frac{1}{2}, \frac{2}{3}, \frac{3}{4}$
- **d.** $\frac{2}{1}, \frac{3}{2}, \frac{4}{3}$

II. The first four terms of the sequence $\{(-n)^n\}$, starting with $n = 1$, are ________.
- **a.** $-1, 2, -3, 4$
- **b.** $1, 4, 27, 256$
- **c.** $1, -4, 27, -256$
- **d.** $-1, 4, -27, 256$

III. The first five terms of the sequence $\{\cos(n\pi)\}$, starting at $n = 1$, are ________.
- **a.** $-1, 0, 1, 0, -1$
- **b.** $1, 0, -1, 0, 1$
- **c.** $0, -1, 0, 1, 0$
- **d.** $1, -1, 1, -1, 1$
- **e.** $-1, 1, -1, 1, -1$
- **f.** $-1, 1, -1, 1, -1, \ldots$

IV. The first five terms of the sequence $\{\sin(n\pi)\}$, starting at $n = 1$, are ________.
- **a.** $0, 0, 0, 0, 0, \ldots$
- **b.** $0, 0, 0, 0, 0$
- **c.** $-1, 0, 1, 0, -1$
- **d.** $0, -1, 0, 1, 0$

V. The general term, a_n, of the sequence $1, -2, 3, -4, 5, -6, \ldots$ is ________.
- **a.** $(-n)^n$
- **b.** n^{-n}
- **c.** $(-1)^n n$
- **d.** $n(-1)^{n-1}$

VI. The general term, a_n, of the sequence $\frac{1}{2}, \frac{3}{4}, \frac{7}{8}, \frac{15}{16}, \frac{31}{32}, \ldots$ is ________.

a. $\frac{2n-1}{2n}$ **b.** $1 - \frac{1}{2^n}$

c. $\frac{n}{2^n}$ **d.** $\frac{2^{n-1}}{2^n}$

VII. $\lim_{x\to\infty}(e - \pi)^n =$ ________.

a. ∞ **b.** $-\infty$

c. 0 **d.** Does not converge

In Problems 1–6, find the first five terms of the given sequence.

1. $\left\{\frac{1}{3^n}\right\}$ **2.** $\left\{\frac{n+1}{n}\right\}$

3. $\left\{1 - \frac{1}{4^n}\right\}$ **4.** $\{n \cos n\}$

5. $\left\{\sin \frac{n\pi}{2}\right\}$ **6.** $\left\{\cos \frac{8\pi}{2^n}\right\}$

In Problems 7–10, find the general term, a_n, of the given sequence

7. $\frac{1}{2}, \frac{2}{3}, \frac{3}{4}, \frac{4}{5}, \frac{5}{6}, \ldots$

8. $1, -\frac{1}{3}, \frac{1}{9}, -\frac{1}{27}, \ldots$

9. $1, 2 \cdot 5, 3 \cdot 5^2, 4 \cdot 5^3, 5 \cdot 5^4, \ldots$

10. $\frac{1}{3}, \frac{2}{5}, \frac{3}{7}, \frac{4}{9}, \frac{5}{11}, \ldots$

In Problems 11–28, determine whether the given sequence is convergent or divergent. If it is convergent, find its limit.

11. $\left\{\frac{17}{\sqrt{n}}\right\}$ **12.** $\left\{\frac{(-1)^n}{\sqrt{n}}\right\}$

***13.** $\{\sin n\}$ **14.** $\{\sin n\pi\}$

15. $\left\{\frac{3}{5n}\right\}$ **16.** $\left\{\frac{(-1)^n n^3}{n^3+1}\right\}$

17. $\left\{\frac{n^5 + 3n^2 + 1}{n^6 + 4n}\right\}$ **18.** $\left\{\frac{4n^5 - 3}{7n^5 + n^2 + 2}\right\}$

19. $\left\{\left(1 + \frac{4}{n}\right)^n\right\}$ **20.** $\left\{\left(1 + \frac{1}{4n}\right)^n\right\}$

21. $\left\{\frac{\sqrt{n}}{\ln n}\right\}$ **22.** $\{\sqrt{n+3} - \sqrt{n}\}$

[*Hint:* Multiply and divide by $\sqrt{n+3} + \sqrt{n}$.]

23. $\left\{\frac{2^n}{n!}\right\}$

24. $\left\{\frac{\beta^n}{n!}\right\}$, β is a fixed real number

25. $\left\{\frac{n+1}{n^{5/2}}\right\}$ **26.** $\left\{\frac{4}{\sqrt{n^2 - n + 3}}\right\}$

26. $\{(-1)^n \cos n\pi\}$ **28.** $\left\{\cos\left(n + \frac{\pi}{2}\right)\right\}$

***29.** Suppose that $\{a_n\}$ is a sequence such that $a_{n+1} = a_n - a_n^2$ for $n \geq 0$. Find all values of a_0 such that $\lim_{n\to\infty} a_n = 0$.

****30.** Suppose that $a_{n+1} = 1/(2 + a_n)$ for $n \geq 0$. For what choices of a_0 does the sequence $\{a_n\}$ diverge?

***31.** Prove Theorem 5. [*Hint:* Use the definition of limit at infinity.]

32. Prove that if $|r| < 1$, then the sequence $\{nr^n\}$ converges to 0.

33. Suppose that $\{a_n\}$ and $\{b_n\}$ are two sequences such that $|a_n| \leq |b_n|$ for each n. Show that if $|b_n|$ converges to 0, then $|a_n|$ also converges to 0. [*Hint:* Use the squeezing theorem.]

34. Use the result of the preceding problem to show that the sequence

$$\left\{\frac{a \sin bn + c \cos dn}{n^{p^2}}\right\}$$

converges to 0 for any real numbers a, b, c, d, and $p \neq 0$.

35. Show that $a_n = [1 + (\beta/n)]^n$ converges to e^β as $n \to \infty$.

***36.** Show that if $\{a_n\}$ converges, then its limit is unique. [*Hint:* Suppose that $\lim_{n\to\infty} a_n = L$, $\lim_{n\to\infty} a_n = M$, and $L \neq M$. Then choose $\epsilon = \frac{1}{2}|L - M|$ to show that the definition of a finite limit is violated.]

***37.** Prove or disprove: If $\lim_{n\to\infty} n^p a_n = 0$, then there exists an $\epsilon > 0$, such that $\lim_{n\to\infty} n^{p+\epsilon} a_n$ also exists (i.e., is finite).

ANSWERS TO SELF-QUIZ

I. c **II.** d **III.** e (but not f) **IV.** b (but not a)

V. d **VI.** b **VII.** c ($e - \pi \approx -0.42331$ satisfies $|r| < 1$.)

12.4 BOUNDED AND MONOTONIC SEQUENCES

There are certain kinds of sequences that have special properties worthy of mention.

DEFINITION BOUNDEDNESS

i. The sequence $\{a_n\}$ is **bounded above** if there is a number M_1 such that

$$a_n \le M_1 \text{ for every positive integer } n. \tag{1}$$

ii. It is **bounded below** if there is a number M_2 such that

$$M_2 \le a_n \text{ for every positive integer } n. \tag{2}$$

iii. It is **bounded** if there is a number $M > 0$ such that

$$|a_n| \le M \text{ for every positive integer } n.$$

The numbers M_1, M_2, and M are called, respectively, an **upper bound**, a **lower bound**, and a **bound** for $\{a_n\}$.

iv. If the sequence is not bounded, it is called **unbounded**. ■

REMARK: If $\{a_n\}$ is bounded above and below, then it is bounded. Simply set $M = \max\{|M_1|, |M_2|\}$.

EXAMPLE 1 A BOUNDED SEQUENCE

The sequence $\{\sin n\}$ has the upper bound of 1, the lower bound of -1, and the bound of 1 since $-1 \le \sin n \le 1$ for every n. Of course, any number greater than 1 is also a bound.

EXAMPLE 2 A BOUNDED SEQUENCE

The sequence $\{(-1)^n\}$ has the upper bound 1, the lower bound -1, and the bound 1.

EXAMPLE 3 AN UNBOUNDED SEQUENCE THAT IS BOUNDED BELOW

The sequence $\{2^n\}$ is bounded below by 2 for $n \ge 1$ but has no upper bound and so is unbounded.

EXAMPLE 4 AN UNBOUNDED SEQUENCE THAT IS NEITHER BOUNDED ABOVE NOR BOUNDED BELOW

The sequence $\{(-1)^n 2^n\}$ is bounded neither below nor above.

It turns out that the following statement is true:

Every convergent sequence is bounded.

THEOREM 1

If the sequence $\{a_n\}$ is convergent, then it is bounded.

PROOF: Before giving the technical details, we remark that the idea behind the proof is easy. For if $\lim_{n\to\infty} a_n = L$, then a_n is close to the finite number L if n is large. Thus, for example, $|a_n| \le |L| + 1$ if n is large enough. Since a_n is a real number for every n, the first few terms of the sequence are bounded, and these two facts give us a bound for the entire sequence.

Now to the details: Let $\epsilon = 1$. Then there is an $N > 0$ such that (according to the definition of finite limit of a sequence, in Section 12.3)

$$|a_n - L| < 1 \qquad \text{if } n \ge N. \tag{3}$$

Let

$$K = \max\{|a_1|, |a_2|, \ldots, |a_N|\}. \tag{4}$$

Since each a_n is finite, K, being the maximum of a finite number of terms, is also finite. Now let

$$M = \max\{|L| + 1, K\}. \tag{5}$$

It follows from (4) that if $n \le N$, then $|a_n| \le K$. If $n \ge N$, then from (3), $|a_n| < |L| + 1$; so in either case $|a_n| \le M$, and the theorem is proved. ■

Theorem 1 is useful in another way. Since every convergent sequence is bounded, it follows that:

> Every unbounded sequence is divergent.

EXAMPLE 5 THREE DIVERGENT SEQUENCES

The following sequences are divergent since they are clearly unbounded:

a. $\{\ln \ln n\}$ (starting at $n = 2$)

b. $\left\{n \sin \frac{n\pi}{2}\right\} = \{1, 0, -3, 0, 5, 0, -7, 0, 9, \ldots\}$

c. $\{(-\sqrt{2})^n\}$

The converse of Theorem 1 is *not* true. That is, it is not true that every bounded sequence is convergent. For example, the sequences $\{(-1)^n\}$ and $\left\{\sin \frac{n\pi}{2}\right\}$ are both bounded *and* divergent. Since boundedness alone does not ensure convergence, we need some other property. We investigate this idea now.

DEFINITION MONOTONICITY

i. The sequence $\{a_n\}$ is **monotone increasing** if $a_{n+1} \ge a_n$ for every $n \ge 1$.

ii. The sequence $\{a_n\}$ is **monotone decreasing** if $a_{n+1} \le a_n$ for every $n \ge 1$.

iii. The sequence $\{a_n\}$ is **monotonic** if it is either monotone increasing or monotone decreasing. ■

DEFINITION STRICT MONOTONICITY

i. The sequence $\{a_n\}$ is **strictly increasing** if $a_{n+1} > a_n$ for every $n \geq 1$.
ii. The sequence $\{a_n\}$ is **strictly decreasing** if $a_{n+1} < a_n$ for every $n \geq 1$.
iii. The sequence $\{a_n\}$ is **strictly monotonic** if it is either strictly increasing or strictly decreasing. ■

EXAMPLE 6 A STRICTLY DECREASING SEQUENCE

The sequence $\{1/2^n\}$ is strictly decreasing since $1/2^n > 1/2^{n+1}$ for every n.

EXAMPLE 7 DETERMINING MONOTONICITY

Determine whether the sequence $\{2n/(3n + 2)\}$ is increasing, decreasing, or not monotonic.

SOLUTION: If we write out the first few terms of the sequence, we find that $\{2n/(3n + 2)\} = \{\frac{2}{5}, \frac{4}{8}, \frac{6}{11}, \frac{8}{14}, \frac{10}{17}, \frac{12}{20}, \ldots\}$. Since these terms are strictly increasing, we suspect that $\{2n/(3n + 2)\}$ is an increasing sequence. To check this, we note that $a_n = \dfrac{2}{3 + (2/n)}$ is increasing because its denominator is decreasing.

In all the examples we have given, a divergent sequence diverges for one of two reasons: It goes to infinity (it is unbounded) or it oscillates [like $(-1)^n$, which oscillates between -1 and 1]. But if a sequence is bounded, it does not go to infinity. And if it is monotonic, then it does not oscillate. Thus, the following theorem should not be surprising.

THEOREM 2 A BOUNDED MONOTONIC SEQUENCE IS CONVERGENT

PROOF: We will prove this theorem for the case in which the sequence $\{a_n\}$ is increasing. The proof of the other case is similar. Since $\{a_n\}$ is bounded, there is a number M such that $a_n \leq M$ for every n. Let L be the least upper bound.[†] Now let $\epsilon > 0$ be given. Then there is a number $N > 0$ such that $a_N > L - \epsilon$. If this were not true, then we would have $a_n < L - \epsilon$ for all $n \geq 1$. Then $L - \epsilon$ would be an upper bound for $\{a_n\}$, and since $L - \epsilon < L$, this would contradict the choice of L as the *least* upper bound. Since $\{a_n\}$ is increasing, we have, for $n \geq N$,

$$L - \epsilon < a_N \leq a_n \leq L < L + \epsilon. \tag{6}$$

But the inequalities in (6) imply that $|a_n - L| < \epsilon$ for $n \geq N$, which proves, according to the definition of convergence, that $\lim_{n\to\infty} a_n = L$. ■

We have actually proved a stronger result. Namely, that ***if the sequence $\{a_n\}$ is bounded above and increasing, then it converges to its least upper bound. Similarly, if $\{a_n\}$ is bounded below and decreasing, then it converges to its greatest lower bound.***

[†] According to the **completeness axiom** of the real numbers, every set of real numbers that is bounded above has a **least upper bound** and every set that is bounded below has a **greatest lower bound.**

EXAMPLE 8 A BOUNDED MONOTONIC SEQUENCE IS CONVERGENT

In Example 7, we saw that the sequence $\{2n/(3n+2)\}$ is strictly increasing. Also, since $2n/(3n+2) < 3n/(3n+2) < 3n/3n = 1$, we see that $\{a_n\}$ is also bounded, so that by Theorem 2, $\{a_n\}$ is convergent. We find that $\lim_{n\to\infty} 2n/(3n+2) = \frac{2}{3}$.

SEQUENCES GENERATED BY FIXED-POINT ITERATION

When you studied Newton's method in your one variable calculus course, you saw a number of sequences that converged to zeros of certain functions. We recall that if f is a differentiable function, then we may define the sequence $\{x_n\}$ by

$$x_{n+1} = x_n - \frac{f(x_n)}{f'(x_n)}. \tag{7}$$

If we start at some specified initial value x_0, then we can find the terms of the sequence **recursively**, that is, one after the other. Several theorems give conditions under which the sequence of iterates generated by (7) will converge to the unique solution s of $f(x) = 0$ in an interval $[a, b]$ for any initial value x_0 in $[a, b]$.†

Suppose now that F is a function and we seek a solution to the equation

$$F(x) = x. \tag{8}$$

Any number that satisfies (8) is called a **fixed point** of F. Note that if s is a zero of f, then

$$f(s) = 0 \quad \text{and} \quad f(s) + s = s$$

so that s is a fixed point of the function $F(x) = f(x) + x$.

EXAMPLE 9 FIXED POINTS OF THREE FUNCTIONS

a. 0 is a fixed point of $\sin x$ because $\sin 0 = 0$.

b. 2 and 3 are fixed points of $x^2 - 4x + 6$ because

$$2^2 - 4\cdot 2 + 6 = 2 \quad \text{and} \quad 3^2 - 4\cdot 3 + 6 = 3.$$

c. e is a fixed point of $f(x) = \dfrac{x}{\ln x}$ because $\dfrac{e}{\ln e} = \dfrac{e}{1} = e$.

In order to find fixed points for a function F, we generate a sequence $\{x_n\}$ according to the following algorithm: Choose x_1. Then define

$$x_{n+1} = F(x_n). \tag{9}$$

THEOREM 3 FIXED-POINT CONVERGENCE THEOREM

Suppose that the function F has a fixed point u and that there is a number c such that

i. F is continuous in $[u - c, u + c]$,

ii. F is differentiable in $(u - c, u + c)$,

iii. $|F'(x)| \le M < 1$ for every x in $(u - c, u + c)$.

† See *Calculus* or *Calculus of One Variable*, Sections 3.6 and 3.9.

Then u is the only fixed point of F in $(u - c, u + c)$ and the sequence $\{x_n\}$ generated by the fixed-point iteration (9) converges to u for every choice of x_1 in $(u - c, u + c)$. ■

An outline of the proof of this theorem is given in Problems 45–49. Note that if $f(x) = F(x) - x$, then a fixed point of F is a zero of f. By the intermediate value theorem, if $f(u - c)f(u + c) < 0$, then f has a zero in $(u - c, u + c)$ so F has a fixed point in $(u - c, u + c)$.

The convergence of the sequence generated by (9) is illustrated in Figure 1.

EXAMPLE 10

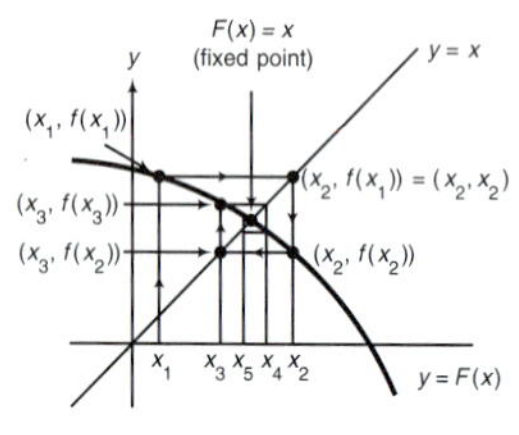

FIGURE 1
Fixed-point convergence

To obtain the next value (x_{n+1}) in the sequence, start at the point $(x_n, f(x_n))$ and move horizontally until you hit the line $y = x$. The x-coordinate (or the equal y-coordinate) of the point that you hit is the value of x_{n+1}.

FINDING A FIXED POINT BY ITERATION

Find a fixed point of $F(x) = \dfrac{x}{\ln x}$ in the interval $(2, 10)$.

SOLUTION: We observe that $F'(x) = \dfrac{\ln x - 1}{\ln^2 x} = \dfrac{1}{\ln x}\left(\dfrac{1}{\ln x} - 1\right)$. Since $\ln x$ is an increasing function, F' is decreasing. We have

$$F'(2) = \frac{1}{\ln 2}\left(\frac{1}{\ln 2} - 1\right) \approx 0.64 \qquad \text{and}$$

$$F'(10) = \frac{1}{\ln 10}\left(\frac{1}{\ln 10} - 1\right) = -0.25,$$

so, on $(2, 10)$, $|F'(x)| < 0.65 < 1$. Starting at $x_1 = 5$, we obtain the sequence in Table 1.

TABLE 1 SEQUENCE OF FIXED-POINT ITERATES OF $f(x) = \dfrac{x}{\ln x}$

n	x_n	$x_{n+1} = \dfrac{x_n}{\ln x_n}$
1	5	3.106674673
2	3.106674673	2.740652532
3	2.740652532	2.718372635
4	2.718372635	2.71828183
5	2.71828183	2.718281828

We see that $x_5 = 2.718281828 \approx e$ (to 9 decimal places). This is no surprise since $\dfrac{e}{\ln e} = \dfrac{e}{1} = e$.

NOTE: This example is a special case of the following problem, which is given on page 442 of the November 1989 issue of the *College Mathematics Journal*:

Let $x_{n+1} = \dfrac{x_n}{\ln x_n}$ where $1 < x_1 < \infty$. Evaluate $\lim_{n\to\infty} x_n$.

Since $\frac{x}{\ln x} = x$ only when $x = e$, e is the only fixed point of $\frac{x}{\ln x}$. Then, using the fixed-point convergence theorem, it is not difficult to show that $\lim_{n\to\infty} x_n = e$ for every choice of x_1 in $(1, \infty)$.

EXAMPLE 11

FINDING A FIXED POINT OF A SINE FUNCTION

Find, by iteration, the unique fixed point of $F(x) = 2 \sin x$ in the interval $\left[\frac{\pi}{2}, \pi\right]$.

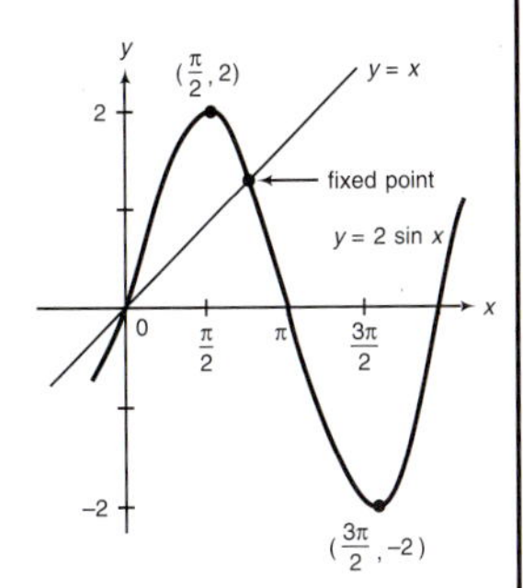

FIGURE 2
The function $2 \sin x$ has a unique positive fixed point

SOLUTION: In Figure 2 we have sketched the graphs of $y = 2 \sin x$ and $y = x$. It is clear that there is a unique positive fixed point and that it lies in the interval $\left(\frac{\pi}{2}, \pi\right)$. (0 is also a fixed point, but it is not very interesting.) Note that $F'(x) = 2 \cos x$ and $F'(\pi) = -2$, so the condition $|F'(x)| < 1$ in $\left[\frac{\pi}{2}, \pi\right]$ does not hold. Nevertheless we perform the iteration and see what happens. The results are given in Table 2. We start with $x_1 = 2$ and stop after 10 iterations.

TABLE 2 SEQUENCE OF FIXED-POINT ITERATES OF $F(x) = 2 \sin x$

n	x_n	$x_{n+1} = 2 \sin x_n$
1	2	1.818594854
2	1.818594854	1.938909453
3	1.938909453	1.866016016
4	1.866016016	1.913476493
5	1.913476493	1.883714959
6	1.883714959	1.902878322
7	1.902878322	1.890731275
8	1.890731275	1.898511758
9	1.898511758	1.893560345
10	1.893560345	1.896724651

Using Newton's method one can find that the unique fixed point is 1.895494267, correct to 9 decimal places.

NOTE: Let $g(x) = 2 \sin x - x$. Then

$$g\left(\frac{\pi}{3}\right) = 2 \sin \frac{\pi}{3} - \frac{\pi}{3} = 2\frac{\sqrt{3}}{2} - \frac{\pi}{3}$$

$$= \sqrt{3} - \frac{\pi}{3} \approx 0.68 > 0$$

and

$$g\left(\frac{2\pi}{3}\right) = 2 \sin \frac{2\pi}{3} - \frac{2\pi}{3} = \sqrt{3} - \frac{2\pi}{3} \approx -0.36 < 0.$$

Thus g has a zero in the interval $\left(\frac{\pi}{3}, \frac{2\pi}{3}\right)$, which means that $2 \sin x$ has a fixed point in $\left(\frac{\pi}{3}, \frac{2\pi}{3}\right)$. Moreover, $2 \cos \frac{\pi}{3} = 1$ and $2 \cos \frac{2\pi}{3} = -1$, so $F'(x)$ decreases from 1 to -1 and $|F'(x)| < 1$ in $\left[\frac{\pi}{3} + \epsilon, \frac{2\pi}{3} - \epsilon\right]$ where ϵ is a small number. This shows that the fixed-point iteration will converge if we choose x_0 in $\left(\frac{\pi}{3}, \frac{2\pi}{3}\right)$. The number 2 is in this interval $\left(\frac{2\pi}{3} \approx 2.094\right)$, so, as we have seen, the iteration does converge if we start at $x_0 = 2$.

PROBLEMS 12.4

SELF-QUIZ

Multiple Choice

I. The sequence $\left\{n + \left(\frac{1}{n}\right)\right\}$ is ________.

- **a.** Unbounded
- **b.** Bounded but not convergent
- **c.** Decreasing
- **d.** Increasing and bounded

II. The sequence $\{1 + (-1)^n\}$ is ________.

- **a.** Convergent but unbounded
- **b.** Bounded
- **c.** Decreasing
- **d.** Increasing

III. The sequence $\left\{5 + \frac{(-1)^n}{n}\right\}$ is ________.

- **a.** Unbounded and divergent
- **b.** Convergent but unbounded
- **c.** Bounded and convergent
- **d.** Bounded but not convergent

IV. The sequence $\left\{\pi^2 + \frac{1}{n}\right\}$ is ________.

- **a.** Unbounded but convergent
- **b.** Increasing but not convergent
- **c.** Decreasing and unbounded
- **d.** Decreasing, bounded, and convergent

V. For each positive integer n, consider a regular n-sided polygon inscribed within a circle of radius 1. (Mark n equally spaced points on that circle, then join consecutive points with a line segment.) Let A_n be the area of this n-gon. The sequence $\{A_n\}$ is ________.

- **a.** Bounded, increasing, and convergent
- **b.** Increasing and bounded but not convergent
- **c.** Decreasing, bounded, and convergent
- **d.** Increasing and convergent but unbounded

In Problems 1–12, determine whether the given sequence is bounded or unbounded. If it is bounded, find the least upper bound for $|a_n|$.

1. $\left\{\frac{1}{n+1}\right\}$ **2.** $\{\sin n\pi\}$

3. $\{\cos n\pi\}$ **4.** $\{\sqrt{n} \sin n\}$

5. $\left\{\frac{2^n}{1 + 2^n}\right\}$ **6.** $\left\{\frac{2^n + 1}{2^n}\right\}$

7. $\left\{\frac{1}{n!}\right\}$ **8.** $\left\{\frac{3^n}{n!}\right\}$

9. $\left\{\frac{n^2}{n!}\right\}$ **10.** $\left\{\frac{2n}{2^n}\right\}$

11. $\left\{\frac{\ln n}{n}\right\}$ ***12.** $\{ne^{-n}\}$

In Problems 13–28, determine whether the given sequence is monotone increasing, strictly increasing, monotone decreasing, strictly decreasing, or not monotonic.

13. $\{\sin n\pi\}$ **14.** $\left\{\frac{3^n}{2 + 3^n}\right\}$

15. $\left\{\left(\frac{n}{25}\right)^{1/3}\right\}$ **16.** $\{n + (-1)^n\sqrt{n}\}$

17. $\left\{\frac{\sqrt{n+1}}{n}\right\}$ **18.** $\left\{\frac{n!}{n^n}\right\}$

***19.** $\left\{\frac{n^n}{n!}\right\}$ ***20.** $\left\{\frac{2(n!)}{1 \cdot 3 \cdot 5 \cdot 7 \cdot \cdots \cdot (2n-1)}\right\}$

***21.** $\{n + \cos n\}$ **22.** $\left\{\frac{2^{2n}}{n!}\right\}$

23. $\left\{\frac{\sqrt{n}-1}{n}\right\}$ **24.** $\left\{\frac{n-1}{n+1}\right\}$

25. $\left\{\ln\left(\frac{3n}{n+1}\right)\right\}$ **26.** $\{\ln n - \ln(n+2)\}$

***27.** $\left\{\left(1 - \frac{3}{n}\right)^n\right\}$ ***28.** $\left\{\left(1 + \frac{3}{n}\right)^{1/n}\right\}$

29. **a.** Show graphically that e^{-x} has a unique fixed point in (0, 1).
b. Find the fixed point, to 5 decimal places, by iterating starting with $x_1 = 1$.

30. **a.** Show graphically that $\cos x$ has a unique fixed point for $x > 0$.
b. Find the fixed point, to 8 decimal places, by iterating starting with $x_1 = 1$. [This is easy to do; just push the cos key repeatedly, making sure that the calculator is set to radians.]

31. Find, to 5 significant figures, a fixed point for $F(x) = 5 \ln x$.

32. Let $F(x) = x^2 - 3x + 3$.
a. Show that 3 is a fixed point for F.
b. Starting with $x_1 = 3.01$, compute the iterates x_2, x_3, x_4, x_5, x_6, x_7, and x_8 and conclude that the sequence of iterates diverges even though we started within 0.01 of the fixed point. [This behavior is explained in Problem 51.]

33. Let $F(x) = x^2 + x - 2$.
a. Compute the two fixed points of F.
b. Show that, starting with $x_1 = 0$, the iterates oscillate between 0 and -2 so that the sequences of iterates diverges.

34. Let $F(x) = \frac{1}{5}x^3 + \frac{1}{4}x^2 - 1$.
a. Find an interval (a, b) over which $|F'(x)| < 1$.
b. Find a fixed point for F in that interval (correct to 5 decimal places).

35. **a.** Show that $\{5^n/n!\}$ is bounded. [*Hint:* See Problem 12.2.35.]
b. Find the smallest integer M such that $5^n/n! \le M$ for all positive integers n.

***36.** **a.** Show that for $n > 2^{10}$, $n^{10}/n! > (n+1)^{10}/(n+1)!$. Use this result to infer that $\{n^{10}/n!\}$ is bounded.
b. Find a relatively small integer M such that $n^{10}/n! \le M$ for all positive integers n.

***37.** Show that the sequence $\{(2^n + 3^n)^{1/n}\}$ is convergent.

***38.** Show that $\{(a^n + b^n)^{1/n}\}$ is convergent for any positive real numbers a and b. [*Hint:* First do the preceding problem, then treat the cases $a = b$ and $a \ne b$ separately.]

***39.** Show that the sequence $\{n!/n^n\}$ is bounded. [*Hint:* Show that $n!/n^n > (n+1)!/(n+1)^{n+1}$ for sufficiently large n.]

40. Prove that the sequence $\{n!/n^n\}$ converges. [*Hint:* Use the result of the preceding problem.]

41. **a.** Show that the sequence $\{(\ln n)/n\}$, $n \ge 1$ is not monotonic.
b. Show that the sequence $\{(\ln n)/n\}$, $n > 2$ is monotonic.

42. Use Theorem 2 to show that $\{\ln n - \ln(n+4)\}$ converges.

43. Show that the sequence of Problem 20 is convergent.

44. Show that the sequence $\{2 \cdot 5 \cdot 8 \cdot 11 \cdot \cdots \cdot (3n-1)/3^n n!\}$ is convergent.

In Problems 45–49, assume that the hypotheses of Theorem 3 (the fixed-point convergence theorem) hold. In particular, assume that u is a fixed point for F.

45. Show that if v is a fixed point for F in $(u - c, u + c)$, then $v = u$. [*Hint:* Use the fact that $F(u) - F(v) = u - v$ and apply the mean-value theorem to the expression $F(u) - F(v)$.]

46. Let E_n, the error term, be defined by $E_n = x_n - u$. Show that $|E_{n+1}| \le M|E_n|$ for $n = 1, 2, 3, \ldots$.

47. Use the result of Problem 46 to show that if x_1 is in $(u - c, u + c)$, then $x_n \in (u - c, u + c)$ for $n = 2, 3, 4, \ldots$. This shows that the iterates are all defined.

48. Use the result of Problem 46 to show that

$$|E_n| \le M^{n-1}|E_1|.$$

49. Show that $E_n \to 0$ as $n \to \infty$ and conclude that $\{x_n\}$ converges to u.

***50.** Assume that F is continuous and that the sequence of iterates given by (9) converges to a number p. Prove that p must be a fixed point of F. [*Hint:* Use the continuity theorem on page 918.]

***51.** **Fixed-Point Divergence Theorem** Suppose that the hypotheses of Theorem 3 hold except that $|F'(x)| \ge k > 1$ for every x in $(u - c, u + c)$.

Show that if $x_1 \neq u$, then the sequence of iterates $\{x_n\}$ given by (9) does not converge to u (it may converge to a different fixed point of F in a different interval—if there is one). [*Hint:* Show that if $x_n \in (u - c, u + c)$, then $|E_{n+1}| > |E_n|$ where E_n is defined as in Problem 46.]

ANSWERS TO SELF-QUIZ

I. a **II.** b **III.** c **IV.** d
V. a (It converges to π, the area of the circle.)

12.5 GEOMETRIC SERIES

Consider the sum

$$S_7 = 1 + 2 + 4 + 8 + 16 + 32 + 64 + 128.$$

This can be written as

$$S_7 = 1 + 2 + 2^2 + 2^3 + 2^4 + 2^5 + 2^6 + 2^7 = \sum_{k=0}^{7} 2^k.$$

GEOMETRIC PROGRESSION

In general, the sum of a **geometric progression** is a sum of the form

$$S_n = 1 + r + r^2 + r^3 + \cdots + r^{n-1} + r^n = \sum_{k=0}^{n} r^k, \qquad (1)$$

where r is a real number and n is a fixed positive integer.

The following result is proved in a more elementary book:[†]

THEOREM 1

If $r \neq 1$, the sum of a geometric progression (1) is given by

$$S_n = \frac{1 - r^{n+1}}{1 - r}. \qquad (2)$$

■

NOTE: If $r = 1$, we obtain

$$S_n = \overbrace{1 + 1 + \cdots + 1}^{n+1 \text{ terms}} = n + 1.$$

EXAMPLE 1 COMPUTING THE SUM OF A GEOMETRIC PROGRESSION

Calculate $S_7 = 1 + 2 + 4 + 8 + 16 + 32 + 64 + 128$, using formula (2).

SOLUTION: Here $r = 2$ and $n = 7$, so that

$$S_7 = \frac{1 - 2^8}{1 - 2} = 2^8 - 1 = 256 - 1 = 255.$$

[†] See *Calculus* or *Calculus of One Variable*, page 629.

EXAMPLE 2

COMPUTING THE SUM OF A GEOMETRIC PROGRESSION

Calculate $\Sigma_{k=0}^{10} (\frac{1}{2})^k$.

SOLUTION: Here $r = \frac{1}{2}$ and $n = 10$, so that

$$S_{10} = \frac{1 - (\frac{1}{2})^{11}}{1 - \frac{1}{2}} = \frac{1 - \frac{1}{2048}}{\frac{1}{2}} = 2\left(\frac{2047}{2048}\right) = \frac{2047}{1024}.$$

EXAMPLE 3

COMPUTING THE SUM OF A GEOMETRIC PROGRESSION WITH A NEGATIVE *r*

Calculate

$$S_6 = 1 - \frac{2}{3} + \left(\frac{2}{3}\right)^2 - \left(\frac{2}{3}\right)^3 + \left(\frac{2}{3}\right)^4 - \left(\frac{2}{3}\right)^5 + \left(\frac{2}{3}\right)^6 = \sum_{k=0}^{6} \left(-\frac{2}{3}\right)^k.$$

SOLUTION: Here $r = -\frac{2}{3}$ and $n = 6$, so that

$$S_6 = \frac{1 - (-\frac{2}{3})^7}{1 - (-\frac{2}{3})} = \frac{1 + \frac{128}{2187}}{\frac{5}{3}} = \frac{3}{5}\left(\frac{2315}{2187}\right) = \frac{463}{729}.$$

The sum of a geometric progression is the sum of a finite number of terms. We now see what happens if the number of terms is infinite. Consider the sum

$$S = 1 + \frac{1}{2} + \frac{1}{4} + \frac{1}{8} + \frac{1}{16} + \cdots = \sum_{k=0}^{\infty} \left(\frac{1}{2}\right)^k.$$

What can such a sum mean? We will give a formal definition in a moment. For now, let us show why it is reasonable to say that $S = 2$. Let $S_n = \Sigma_{k=0}^{n} (\frac{1}{2})^k = 1 + \frac{1}{2} + \frac{1}{4} + \cdots + (\frac{1}{2})^n$. Then,

$$S_n = \frac{1 - (\frac{1}{2})^{n+1}}{1 - \frac{1}{2}} = 2\left[1 - \left(\frac{1}{2}\right)^{n+1}\right] = 2 - \left(\frac{1}{2}\right)^n.$$

Thus, for any n (no matter how large), $1 \le S_n < 2$. Hence, the numbers S_n are bounded. Also, since $S_{n+1} = S_n + (\frac{1}{2})^{n+1} > S_n$, the numbers S_n are monotone increasing. Thus, the sequence $\{S_n\}$ converges. But

$$S = \lim_{n\to\infty} S_n.$$

Thus, S has a finite sum. To compute it, we note that

$$S = \lim_{n\to\infty} S_n = \lim_{n\to\infty} 2[1 - (\tfrac{1}{2})^{n+1}] = 2 \lim_{n\to\infty}[1 - (\tfrac{1}{2})^{n+1}] = 2,$$

since $\lim_{n\to\infty}(\frac{1}{2})^{n+1} = 0$.

GEOMETRIC SERIES

The infinite sum $\Sigma_{k=0}^{\infty} (\frac{1}{2})^k$ is called a *geometric series*. In general, a **geometric series** is an infinite sum of the form

$$S = \sum_{k=0}^{\infty} r^k = 1 + r + r^2 + r^3 + \cdots. \quad (3)$$

DEFINITION CONVERGENCE AND DIVERGENCE OF A GEOMETRIC SERIES

Let $S_n = \Sigma_{k=0}^{n} r^k$. Then we say that the geometric series **converges** if $\lim_{n\to\infty} S_n$ exists and is finite. Otherwise, the series is said to **diverge**. ■

EXAMPLE 4 THE GEOMETRIC SERIES DIVERGES WHEN $r = 1$

Let $r = 1$. Then

$$S_n = \sum_{k=0}^{n} 1^k = \sum_{k=0}^{n} 1 = \underbrace{1 + 1 + \cdots + 1}_{n+1 \text{ times}} = n + 1.$$

Since $\lim_{n\to\infty}(n + 1) = \infty$, the series $\Sigma_{k=0}^{\infty} 1^k$ diverges.

THEOREM 2

Let $S = \Sigma_{k=0}^{\infty} r^k = 1 + r + r^2 + \cdots$ be a geometric series.

i. The series converges to

$$\frac{1}{1-r} \quad \text{if } |r| < 1.$$

ii. The series diverges if $|r| \geq 1$.

PROOF:

i. If $|r| < 1$, then $\lim_{n\to\infty} r^{n+1} = 0$. Thus,

$$S = \lim_{n\to\infty} S_n = \lim_{n\to\infty} \frac{1 - r^{n+1}}{1 - r} = \frac{1}{1 - r} \lim_{n\to\infty} (1 - r^{n+1})$$

$$= \frac{1}{1 - r}(1 - 0) = \frac{1}{1 - r}.$$

ii. If $|r| > 1$, then $\lim_{n\to\infty}|r|^{n+1} = \infty$. Thus, $1 - r^{n+1}$ does not have a finite limit and the series diverges. Finally, if $r = 1$, then the series diverges, by Example 4, and if $r = -1$, then S_n alternates between the numbers 0 and 1, so that the series diverges. ■

EXAMPLE 5 THE GEOMETRIC SERIES FOR $r = -\frac{2}{3}$

$$1 - \frac{2}{3} + \left(\frac{2}{3}\right)^2 - \cdots = \sum_{k=0}^{\infty} \left(-\frac{2}{3}\right)^k = \frac{1}{1 - (-\frac{2}{3})} = \frac{1}{\frac{5}{3}} = \frac{3}{5}.$$

EXAMPLE 6

THE GEOMETRIC SERIES FOR $r = \pi/4$

$$1 + \frac{\pi}{4} + \left(\frac{\pi}{4}\right)^2 + \left(\frac{\pi}{4}\right)^3 + \cdots = \sum_{k=0}^{\infty} \left(\frac{\pi}{4}\right)^k = \frac{1}{1 - (\pi/4)}$$

$$= \frac{4}{4 - \pi} \approx 4.66.$$

EXAMPLE 7

A GEOMETRIC SERIES THAT DOES NOT START WITH THE TERM 1

Compute $\left(\frac{4}{5}\right)^2 + \left(\frac{4}{5}\right)^3 + \left(\frac{4}{5}\right)^4 + \cdots = \sum_{k=2}^{\infty} \left(\frac{4}{5}\right)^k.$

SOLUTION: There are at least two ways to compute this sum.

Method 1—Factor: We factor out the first term:

$$\left(\frac{4}{5}\right)^2 + \left(\frac{4}{5}\right)^3 + \left(\frac{4}{5}\right)^4 + \cdots = \left(\frac{4}{5}\right)^2\left[1 + \frac{4}{5} + \left(\frac{4}{5}\right)^2 + \cdots\right]$$

$$= \left(\frac{4}{5}\right)^2 \sum_{k=0}^{\infty} \left(\frac{4}{5}\right)^k = \left(\frac{4}{5}\right)^2\left(\frac{1}{1 - \frac{4}{5}}\right)$$

$$= \left(\frac{4}{5}\right)^2\left(\frac{1}{\frac{1}{5}}\right) = \left(\frac{4}{5}\right)^2 \cdot 5 = \frac{16}{5}.$$

Method 2: Add and subtract the two missing terms:

$$\left(\frac{4}{5}\right)^2 + \left(\frac{4}{5}\right)^3 + \left(\frac{4}{5}\right)^4 + \cdots = \overbrace{1 + \frac{4}{5}}^{\text{these terms added}} + \left(\frac{4}{5}\right)^2 + \left(\frac{4}{5}\right)^3 + \cdots \overbrace{- 1 - \frac{4}{5}}^{\text{the same two terms subtracted}}$$

$$= \frac{1}{1 - \frac{4}{5}} - 1 - \frac{4}{5} = 5 - \frac{9}{5} = \frac{16}{5}.$$

PROBLEMS 12.5

SELF-QUIZ

Multiple Choice

I. $\sum_{k=0}^{5} \left(\frac{1}{2}\right)^k =$ ________.

a. $\frac{1}{2^6}$ **b.** $\frac{1 - 2^6}{1 - 2}$

c. $\frac{1 - (\frac{1}{2})^5}{1 - (\frac{1}{2})}$ **d.** $\frac{1 - (\frac{1}{2})^6}{1 - (\frac{1}{2})}$

e. $\frac{1 - (\frac{1}{2})}{1 - (\frac{1}{2})^6}$ **f.** $\frac{1 - 2}{1 - 2^5}$

II. $\sum_{k=0}^{4} 2^k =$ ________.

a. $\frac{1}{2^5}$ **b.** $\frac{1 - 2^5}{1 - 2}$

c. $\dfrac{1-(\frac{1}{2})^5}{1-(\frac{1}{2})}$ **d.** $\dfrac{1-(\frac{1}{2})^4}{1-(\frac{1}{2})}$

e. $\dfrac{1-(\frac{1}{2})}{1-(\frac{1}{2})^4}$ **f.** $\dfrac{1-2}{1-2^5}$

III. $2-\sum_{k=0}^{7}\dfrac{1}{2^k}=$ ________.

a. $\dfrac{1}{2}$ **b.** $\dfrac{1}{2^7}$

c. $\dfrac{1}{2^8}$ **d.** $\dfrac{-1}{2^8}$

IV. $\sum_{k=0}^{\infty}\left(\dfrac{1}{10}\right)^k=$ ________.

a. $\frac{10}{9}$ **b.** $\frac{9}{10}$

c. $\frac{10}{11}$ **d.** 1.23456 . . .

V. $\sum_{k=0}^{\infty}\left(\dfrac{-1}{10}\right)^k=$ ________.

a. $\frac{10}{9}$ **b.** $\frac{9}{10}$

c. $\frac{10}{11}$ **d.** 1.23456 . . .

VI. $0.9+0.09+0.009+\cdots=\sum_{k=1}^{\infty}9\cdot\left(\dfrac{1}{10}\right)^k=$ ________.

a. $1.0-\left(\dfrac{1}{10}\right)^k$ **b.** $1.0-\left(\dfrac{1}{10}\right)^{k+1}$

c. 0.99999 . . . **d.** 1.0

In Problems 1–12, calculate the sum of the given geometric progression.

1. $1+3+9+27+81+243$

2. $1+\dfrac{1}{4}+\dfrac{1}{16}+\cdots+\dfrac{1}{4^8}$

3. $1-5+25-125+625-3125$

4. $0.2+0.2^2+0.2^3+\cdots+0.2^9$

5. $0.3^2-0.3^3+0.3^4-0.3^5+0.3^6-0.3^7+0.3^8$

6. $1+b^3+b^6+b^9+b^{12}+b^{15}+b^{18}+b^{21}$

7. $1-\dfrac{1}{b^2}+\dfrac{1}{b^4}-\dfrac{1}{b^6}+\dfrac{1}{b^8}-\dfrac{1}{b^{10}}+\dfrac{1}{b^{12}}-\dfrac{1}{b^{14}}$

8. $\pi-\pi^3+\pi^5-\pi^7+\pi^9-\pi^{11}+\pi^{13}$

9. $1+\sqrt{2}+2+2^{3/2}+4+2^{5/2}+8+2^{7/2}+16$

10. $1-\dfrac{1}{\sqrt{3}}+\dfrac{1}{3}-\dfrac{1}{3\sqrt{3}}+\dfrac{1}{9}-\dfrac{1}{9\sqrt{3}}+\dfrac{1}{27}$

11. $-16+64-256+1024-4096$

12. $\frac{3}{4}-\frac{9}{8}+\frac{27}{16}-\frac{81}{32}+\frac{243}{64}-\frac{729}{128}$

In Problems 13–20, calculate the sum of the given geometric series.

13. $1+\dfrac{1}{4}+\dfrac{1}{4^2}+\dfrac{1}{4^3}+\cdots$

14. $1-\dfrac{1}{2}+\dfrac{1}{4}-\dfrac{1}{8}+\dfrac{1}{16}-\cdots$

15. $1-\dfrac{1}{3}+\dfrac{1}{9}-\dfrac{1}{27}+\dfrac{1}{81}-\cdots$

16. $1+\dfrac{1}{\pi}+\dfrac{1}{\pi^2}+\dfrac{1}{\pi^3}+\cdots$

17. $1+0.7+0.7^2+0.7^3+\cdots$

18. $1-0.62+0.62^2-0.62^3+0.62^4-\cdots$

19. $\dfrac{1}{4}+\dfrac{1}{16}+\dfrac{1}{64}+\cdots$

[*Hint:* Factor out the term $\frac{1}{4}$.]

20. $-\dfrac{3}{5}+\dfrac{3}{25}-\dfrac{3}{125}+\cdots$

21. How large must n be in order that $(\frac{1}{2})^n<0.01$?

22. How large must n be in order that $0.8^n<0.01$?

23. How large must n be in order that $0.99^n<0.01$?

24. A bacteria population initially contains 1000 organisms; each bacterium produces two live bacteria every 2 hrs. If none of the bacteria dies during a 12-hr growth period, then how many organisms will be alive at its end?

25. Show that if $x>1$, then

$$1+\frac{1}{x}+\frac{1}{x^2}+\frac{1}{x^3}+\cdots=\frac{x}{x-1}.$$

[*Note:* You need to do two tasks here: Show that the series converges; then show that the right-hand-side formula does equal the limit of the finite sums.]

***26.** **Assertion:** *The legendary Greek hero Achilles could never overtake a turtle in a footrace.*

Argument: Suppose that Achilles can run 100 times faster than the turtle; also suppose the turtle starts with a 1000 m lead. While Achilles runs his first 1000 m, the turtle runs 10 m; when Achilles has run 10 m, the turtle has run $\frac{1}{10}$ m further. By the time Achilles reaches the turtle's current position, the turtle will have moved on—thus maintaining its lead.†

---

†The Greek mathematician and philosopher Zeno of Elea (ca. 495–430 B.C.) posed this paradox.

Your problem: Resolve this paradox using tools and insights from your current studies. (If you feel more detail is required, you may assume that the turtle runs at 1 km/hr and that Achilles runs 100 km/hr.)

ANSWERS TO SELF-QUIZ

I. d **II.** b **III.** b **IV.** a **V.** c **VI.** c = d

12.6 INFINITE SERIES

In Section 12.5 we defined the geometric series $\Sigma_{k=0}^{\infty} r^k$ and showed that if $|r| < 1$, the series converges to $1/(1 - r)$. Let us again look at what we did. If S_n denotes the sum of the first $n + 1$ terms of the geometric series, then

$$S_n = 1 + r + r^2 + \cdots + r^n = \frac{1 - r^{n+1}}{1 - r}, \qquad r \neq 1. \tag{1}$$

For each n we obtain the number S_n, and therefore we can define a new sequence $\{S_n\}$ to be the sequence of **partial sums** of the geometric series. If $|r| < 1$, then

$$\lim_{n\to\infty} S_n = \lim_{n\to\infty} \frac{1 - r^{n+1}}{1 - r} = \frac{1}{1 - r}.$$

That is, the convergence of the geometric series is implied by the convergence of the sequence of partial sums $\{S_n\}$.

We now give a more general definition of these concepts.

DEFINITION INFINITE SERIES

Let $\{a_n\}$ be a sequence. Then the infinite sum

$$\sum_{k=1}^{\infty} a_k = a_1 + a_2 + a_3 + \cdots + a_n + \cdots \tag{2}$$

is called an **infinite series** (or simply, **series**). Each a_k in (2) is called a **term** of the series. The **partial sums** of the series are given by

$$S_n = \sum_{k=1}^{n} a_k.$$

The sum S_n is called the ***n*th partial sum** of the series. If the sequence of partial sums $\{S_n\}$ converges to L, then we say that the infinite series $\Sigma_{k=1}^{\infty} a_k$ **converges** to L, and we write

$$\sum_{k=1}^{\infty} a_k = L.$$

Otherwise, we say that the series $\Sigma_{k=1}^{\infty} a_k$ **diverges**. ■

REMARK: Occasionally a series will be written with the first term other than a_1. For example, $\Sigma_{k=0}^{\infty} (\frac{1}{2})^k$ and $\Sigma_{k=2}^{\infty} 1/(\ln k)$ are both examples of infinite series. In the second case we must start with $k = 2$ since $1/(\ln 1)$ is not defined.

EXAMPLE 1

EXPRESSING A REPEATING DECIMAL AS A RATIONAL NUMBER BY USING A GEOMETRIC SERIES

Express the **repeating decimal** 0.123123123 . . . as a rational number (the quotient of two integers).

SOLUTION:

$$\begin{aligned} 0.123123123\ldots &= 0.123 + 0.000123 + 0.000000123 + \cdots \\ &= \frac{123}{10^3} + \frac{123}{10^6} + \frac{123}{10^9} + \cdots \\ &= \frac{123}{10^3}\left[1 + \frac{1}{10^3} + \frac{1}{(10^3)^2} + \cdots\right] \\ &= \frac{123}{1000}\sum_{k=0}^{\infty}\left(\frac{1}{1000}\right)^k = \frac{123}{1000}\left[\frac{1}{1-\left(\frac{1}{1000}\right)}\right] \\ &= \frac{123}{1000}\cdot\frac{1}{\frac{999}{1000}} \\ &= \frac{123}{1000}\cdot\frac{1000}{999} = \frac{123}{999} = \frac{41}{333}. \end{aligned}$$

As a matter of fact, any decimal number x can be thought of as a convergent infinite series, for if $x = 0.\,a_1a_2a_3\ldots a_n\ldots$, then

$$x = \frac{a_1}{10} + \frac{a_2}{100} + \frac{a_3}{1000} + \cdots + \frac{a_n}{10^n} + \cdots = \sum_{k=1}^{\infty}\frac{a_k}{10^k}.^\dagger$$

In general, we can use the geometric series to write any repeating decimal in the form of a fraction by using the technique of Example 1. In fact,

> the rational numbers are exactly those real numbers that can be written as repeating decimals.

Repeating decimals include numbers like $3 = 3.00000\ldots$, $\frac{1}{4} = 0.25 = 0.25000000\ldots$, and $\frac{1}{3} = 0.3333\ldots$.

EXAMPLE 2

A TELESCOPING SERIES

Consider the infinite series $\sum_{k=1}^{\infty} 1/k(k+1)$. We write the first three partial sums:

$$S_1 = \sum_{k=1}^{1}\frac{1}{k(k+1)} = \frac{1}{1\cdot 2} = \frac{1}{2} = 1 - \frac{1}{2},$$

---

† Since $0 \le a_k < 10$,

$$\sum_{k=1}^{\infty}\frac{a_k}{10^k} < \sum_{k=1}^{\infty}\frac{10}{10^k} = \sum_{k=1}^{\infty}\frac{1}{10^{k-1}} = 1 + \frac{1}{10} + \left(\frac{1}{10}\right)^2 + \cdots = \frac{1}{1-\frac{1}{10}} = \frac{10}{9}.$$

In Section 12.7 we will prove the comparison test. Once we have this test, the inequality given above implies that $\sum_{k=1}^{\infty}(a_k/10^k)$ converges.

$$S_2 = \sum_{k=1}^{2} \frac{1}{k(k+1)} = \frac{1}{1 \cdot 2} + \frac{1}{2 \cdot 3} = \frac{1}{2} + \frac{1}{6} = \frac{2}{3} = 1 - \frac{1}{3},$$

$$S_3 = \sum_{k=1}^{3} \frac{1}{k(k+1)} = \frac{1}{1 \cdot 2} + \frac{1}{2 \cdot 3} + \frac{1}{3 \cdot 4} = \frac{1}{2} + \frac{1}{6} + \frac{1}{12}$$

$$= \frac{3}{4} = 1 - \frac{1}{4}.$$

We can use partial fractions to rewrite the general term as

$$a_k = \frac{1}{k(k+1)} = \frac{1}{k} - \frac{1}{k+1},$$

from which we can get a better view of the nth partial sum:

$$S_n = \left(\frac{1}{1} - \frac{1}{2}\right) + \left(\frac{1}{2} - \frac{1}{3}\right) + \left(\frac{1}{3} - \frac{1}{4}\right) + \cdots$$

$$+ \left(\frac{1}{n-1} - \frac{1}{n}\right) + \left(\frac{1}{n} - \frac{1}{n+1}\right) = 1 - \frac{1}{n+1}$$

because all other terms cancel. Since

$$\lim_{n\to\infty} S_n = \lim_{n\to\infty} \left(1 - \frac{1}{n-1}\right) = 1,$$

we see that

$$\sum_{k=1}^{\infty} \frac{1}{k(k+1)} = 1.$$

When, as here, alternate terms cancel, we say that the series is a **telescoping series**.

REMARK: Often, it is not possible to calculate the exact sum of an infinite series, even if it can be shown that the series converges.

EXAMPLE 3

THE HARMONIC SERIES: AN EXAMPLE OF A DIVERGENT SERIES

Consider the series

$$\sum_{k=1}^{\infty} \frac{1}{k} = 1 + \frac{1}{2} + \frac{1}{3} + \frac{1}{4} + \cdots + \frac{1}{n} + \cdots. \tag{3}$$

This series is called the **harmonic series**. Although $a_n = 1/n \to 0$ as $n \to \infty$, it is not difficult to show that the harmonic series diverges. To see this, we write

$$\sum_{k=1}^{\infty} \frac{1}{k} = 1 + \frac{1}{2} + \underbrace{\overbrace{\left(\frac{1}{3} + \frac{1}{4}\right)}^{\text{2 terms}}}_{>\frac{1}{2}} + \underbrace{\overbrace{\left(\frac{1}{5} + \frac{1}{6} + \frac{1}{7} + \frac{1}{8}\right)}^{\text{4 terms}}}_{>\frac{1}{2}} + \underbrace{\overbrace{\left(\frac{1}{9} + \cdots + \frac{1}{16}\right)}^{\text{8 terms}}}_{>\frac{1}{2}} + \cdots.$$

Here we have written the terms in groups containing 2^n numbers. Note that $\frac{1}{3} + \frac{1}{4} > \frac{2}{4} = \frac{1}{2}$, $\frac{1}{5} + \frac{1}{6} + \frac{1}{7} + \frac{1}{8} > \frac{1}{8} + \frac{1}{8} + \frac{1}{8} + \frac{1}{8} = \frac{1}{2}$, and so on. Thus $\sum_{k=1}^{\infty} 1/k > 1 + \frac{1}{2} + \frac{1}{2} + \cdots$, and the series diverges.

WARNING: Example 3 clearly shows that even though the sequence $\{a_n\}$ converges to 0, the series Σa_n may, in fact, diverge. That is, if $a_n \to 0$, then $\Sigma_{k=1}^{\infty} a_k$ may or may not converge. Some additional test is needed to determine convergence or divergence.

It is often difficult to determine whether a series converges or diverges. For that reason, a number of techniques have been developed to make it easier to do so. We will present some easy facts here, and then we will develop additional techniques in the three sections that follow.

THEOREM 1

Let c be a constant. Suppose that $\Sigma_{k=1}^{\infty} a_k$ and $\Sigma_{k=1}^{\infty} b_k$ both converge. Then $\Sigma_{k=1}^{\infty} (a_k + b_k)$ and $\Sigma_{k=1}^{\infty} ca_k$ converge, and

i. $$\sum_{k=1}^{\infty} (a_k + b_k) = \sum_{k=1}^{\infty} a_k + \sum_{k=1}^{\infty} b_k, \tag{4}$$

ii. $$\sum_{k=1}^{\infty} ca_k = c \sum_{k=1}^{\infty} a_k. \tag{5}$$

■

This theorem should not be surprising. Since the sum in a series is the limit of a sequence (the sequence of partial sums), the first part, for example, simply restates the fact that the limit of the sum is the sum of the limits. This is Theorem 12.3.2 (ii).

PROOF OF THEOREM 1:

i. Let $S = \Sigma_{k=1}^{\infty} a_k$ and $T = \Sigma_{k=1}^{\infty} b_k$. The partial sums are given by $S_n = \Sigma_{k=1}^{n} a_k$ and $T_n = \Sigma_{k=1}^{n} b_k$. Then,

$$\begin{aligned}\sum_{k=1}^{\infty} (a_k + b_k) &= \lim_{n\to\infty} \sum_{k=1}^{n} (a_k + b_k) = \lim_{n\to\infty} \left(\sum_{k=1}^{n} a_k + \sum_{k=1}^{n} b_k\right) \\ &= \lim_{n\to\infty} (S_n + T_n) = \lim_{n\to\infty} S_n + \lim_{n\to\infty} T_n \\ &= S + T = \sum_{k=1}^{\infty} a_k + \sum_{k=1}^{\infty} b_k.\end{aligned}$$

ii. $$\sum_{k=1}^{\infty} ca_k = \lim_{n\to\infty} \sum_{k=1}^{n} ca_k = \lim_{n\to\infty} c \sum_{k=1}^{n} a_k = \lim_{n\to\infty} cS_n = c \lim_{n\to\infty} S_n = cS = c \sum_{k=1}^{\infty} a_k.$$

EXAMPLE 4 A CONSTANT MULTIPLE OF THE HARMONIC SERIES DIVERGES

Does $\Sigma_{k=1}^{\infty} 1/50k$ converge or diverge?

SOLUTION: We show that the series diverges by assuming that it converges to obtain a contradiction. If $\Sigma_{k=1}^{\infty} 1/50k$ did converge, then $50\, \Sigma_{k=1}^{\infty} 1/50k$ would also converge by Theorem 1. But then $50\, \Sigma_{k=1}^{\infty} 1/50k = \Sigma_{k=1}^{\infty} 50 \cdot 1/50k = \Sigma_{k=1}^{\infty} 1/k$, and this series is the harmonic series, which we know diverges. Hence $\Sigma_{k=1}^{\infty} 1/50k$ diverges.

Another useful test is given by the following theorem and corollary.

THEOREM 2

If $\Sigma_{k=1}^{\infty} a_k$ converges, then $\lim_{n\to\infty} a_n = 0$.

PROOF: Let $S = \Sigma_{k=1}^{\infty} a_k$. Then the partial sums S_n and S_{n-1} are given by

$$S_n = \sum_{k=1}^{n} a_k = a_1 + a_2 + \cdots + a_{n-1} + a_n$$

and

$$S_{n-1} = \sum_{k=1}^{n-1} a_k = a_1 + a_2 + \cdots + a_{n-1},$$

so that

$$S_n - S_{n-1} = a_n.$$

Then

$$\begin{aligned}\lim_{n\to\infty} a_n &= \lim_{n\to\infty}(S_n - S_{n-1}) = \lim_{n\to\infty} S_n - \lim_{n\to\infty} S_{n-1}\\ &= S - S = 0.\end{aligned}$$ ■

We have already seen that the converse of this theorem is false. The convergence of $\{a_n\}$ to 0 does *not* imply that $\Sigma_{k=1}^{\infty} a_k$ converges. For example, the harmonic series does not converge, but the sequence $\{1/n\}$ does converge to zero.

COROLLARY A TEST FOR DIVERGENCE

If $\{a_n\}$ does not converge to 0, then $\Sigma_{k=1}^{\infty} a_k$ diverges. ■

EXAMPLE 5 A SERIES THAT DIVERGES BECAUSE THE PARTIAL SUMS OSCILLATE

$\Sigma_{k=1}^{\infty} (-1)^k$ diverges since the sequence $\{(-1)^k\}$ does not converge to zero. To see this more clearly, look at the partial sums:

$$S_1 = (-1)^1 = -1.$$
$$S_2 = (-1)^1 + (-1)^2 = -1 + 1 = 0.$$
$$S_3 = (-1)^1 + (-1)^2 + (-1)^3 = -1 + 1 - 1 = -1.$$
$$S_4 = (-1)^1 + (-1)^2 + (-1)^3 + (-1)^4 = -1 + 1 - 1 + 1 = 0.$$

We see that

$$S_n = \begin{cases} -1 & \text{if } n \text{ is odd,} \\ 0 & \text{if } n \text{ is even.} \end{cases}$$

Thus the partial sums oscillate between -1 and 0 and do not converge.

EXAMPLE 6 A DIVERGENT SERIES

$\Sigma_{k=1}^{\infty} k/(k + 100)$ diverges since $\lim_{n\to\infty} a_n = \lim_{n\to\infty} n/(n + 100) = 1 \neq 0$.

OF HISTORICAL INTEREST

THE STRUGGLE TO UNDERSTAND INFINITE SUMS

The basic idea in the study of infinite series is that an infinite number of numbers can have a finite sum. This concept may seem natural now, but it took mathematicians over two thousand years to come to grips with it. More generally, the notion of the "infinite" was poorly understood and this lack of understanding led to some very confusing results.

Some of the early work on series was motivated by unresolved questions that had been posed by Greek mathematicians. For example, the fifth-century B.C. philosopher and mathematician Zeno (ca. 495–435 B.C.) posed four problems which came to be known as *Zeno's paradoxes.* In the second of these, Zeno argued that the legendary Greek hero Achilles could never overtake a tortoise. Suppose that the tortoise starts 100 yards ahead and that Achilles can run 10 times as fast as the tortoise. Then when Achilles has run 100 yards, the tortoise has run 10 yards, and when Achilles has covered 10 yards, the tortoise is still a yard ahead, and so on. It seems that the tortoise will stay ahead!

We can view this seeming paradox in another way which is equally contradictory of common sense. Suppose that a woman is standing a certain distance, say 10 feet, from a door (see Figure 1). Using Zeno's reasoning, we may claim that it is impossible for the woman to walk to the door. In order to reach the door, she must walk half the distance (5 feet) to the door. She then reaches point 1 on Figure 1. From point 1, 5 feet from the door, she must again walk halfway ($2\frac{1}{2}$

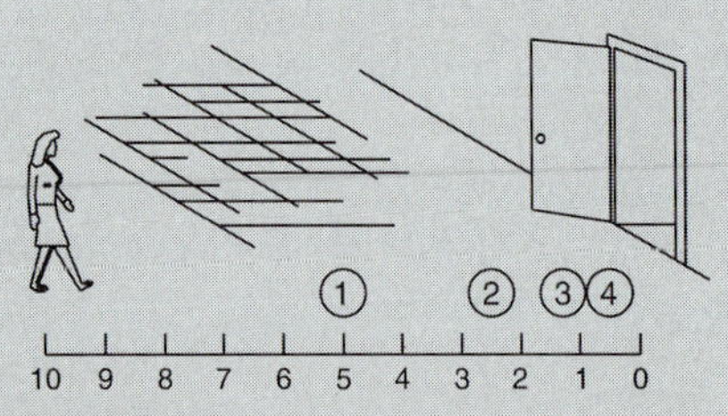

FIGURE 1
A woman walking toward a door must cover an infinite number of distances, each half as long as the preceding one

feet) to the door, to point 2. Continuing in this manner, no matter how close she comes to the door, she must walk halfway to the door and halfway from there and halfway from there . . . and so on. Thus, no matter how close the woman gets to the door, she still has half of some remaining distance to cover. It seems that the woman will never actually reach the door. Of course, this contradicts our common sense. But where is the flaw in Zeno's reasoning?

With our knowledge of geometric series, we can reason as follows: Suppose the woman walks at a speed of 5 feet/second ($\approx$3.4 miles/hour). Then

$$t_1 = \text{time required to walk first 5 feet} = \frac{\text{distance}}{\text{velocity}} = \frac{5}{5} = 1 \text{ second}$$

$$t_2 = \text{time required to walk next } \frac{5}{2} \text{ feet} = \frac{\frac{5}{2}}{5} = \frac{1}{2} \text{ second}$$

$$t_3 = \text{time required to walk next } \frac{5}{4} \text{ feet} = \frac{\frac{5}{4}}{5} = \frac{1}{4} \text{ second}$$

$$t_4 = \text{time required to walk next } \frac{5}{8} \text{ feet} = \frac{\frac{5}{8}}{5} = \frac{1}{8} \text{ second}$$

and so on. We see that

$$\text{total time to walk 10 feet} = 1 + \tfrac{1}{2} + \tfrac{1}{4} + \tfrac{1}{8} + \cdots = 2 \text{ seconds.}$$

(CONTINUED)

OF HISTORICAL INTEREST

(CONTINUED)

This is not surprising as it tells us that it takes 2 seconds to walk 10 feet if she walks at a rate of 5 feet/second. This result seems hardly worth the bother, yet the paradox troubled mathematicians and philosophers for over 2000 years.

The great Archimedes used the idea of an infinite series (sort of) in order to compute area. In his *Quadrature of the Parabola,* Archimedes used the following procedure to obtain the area of a parabolic segment: He began by drawing the triangle T with the same base as the parabolic segment and its third vertex at the vertex of the parabola (see Figure 2). He chose points C and D and inscribed two triangles (denoted by T_1 and T_2 in the figure). He then inscribed four more triangles, using the sides of T_1 and T_2 as bases. One such triangle, T_3, is drawn in Figure 2.

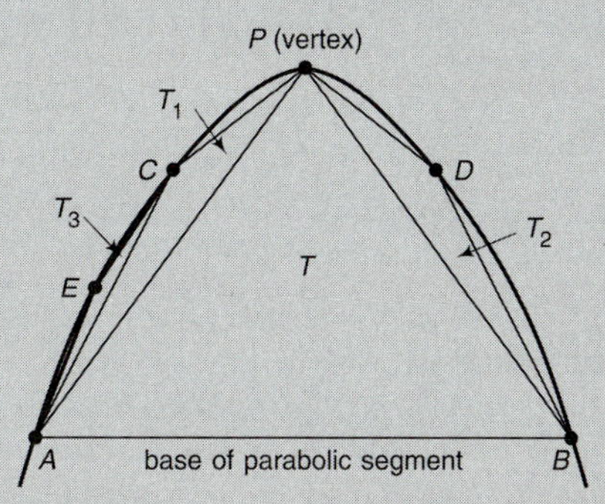

FIGURE 2
The area of the parabolic segment is approximated by the sum of the areas of inscribed triangles

Continuing in this manner, Archimedes obtained a sequence of inscribed polygons which gave better and better approximations to the area of the parabolic segment. Using geometric arguments, Archimedes showed that

Area of nth inscribed polygon

$$= (\text{area of } T) \times \left(1 + \frac{1}{4} + \frac{1}{4^2} + \cdots + \frac{1}{4^{n-1}}\right).$$

Since $1 + \frac{1}{4} + \frac{1}{4^2} + \cdots = \frac{1}{1 - \frac{1}{4}} = \frac{4}{3}$, it follows that

$$\text{area of parabolic segment} = \frac{4}{3} \times \text{area of triangle } T.$$

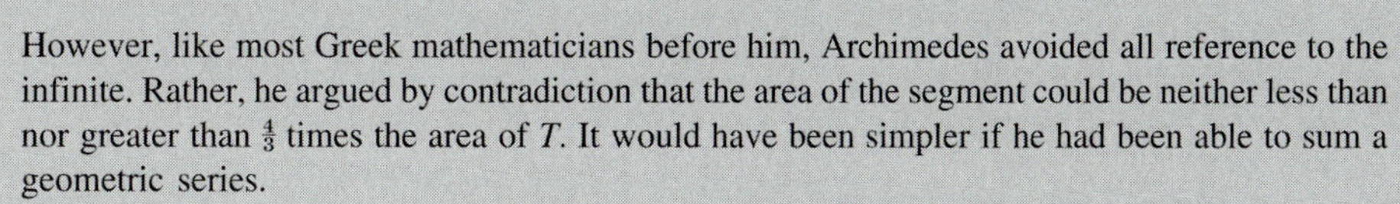

However, like most Greek mathematicians before him, Archimedes avoided all reference to the infinite. Rather, he argued by contradiction that the area of the segment could be neither less than nor greater than $\frac{4}{3}$ times the area of T. It would have been simpler if he had been able to sum a geometric series.

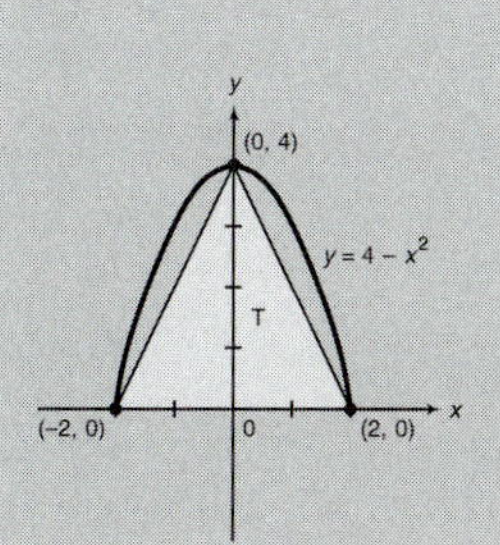

FIGURE 3
The area under the curve $y = 4 - x^2$ equals $\frac{4}{3} \times$ area of T

To see that Archimedes' method works, consider the area, A, above the x-axis and under the curve $y = 4 - x^2$ (Figure 3). T is the triangle with vertices at $(-2, 0)$, $(2, 0)$, and $(0, 4)$. We have

$$\text{area of } T = \frac{1}{2}(\text{base} \times \text{height}) = \frac{1}{2}(4 \times 4) = 8.$$

Then

$$A = \frac{4}{3} \cdot 8 = \frac{32}{3}.$$

(CONTINUED)

OF HISTORICAL INTEREST

(CONTINUED)

We can verify this using an integral:

$$A = \int_{-2}^{2} (4 - x^2)\, dx = \left(4x - \frac{x^3}{3}\right)\Big|_{-2}^{2} = \frac{32}{3}.$$

In thinking about infinite sums, Aristotle distinguished between two infinities: the *possible* and the *actual*. Put another way, he claimed that there was a difference between thinking about an infinite sum and actually having one. The concept of an infinite sum was so difficult that he categorically denied that such a sum could exist.

This state of ignorance continued well into the middle ages. In the thirteenth century, Petrus Hisparus, who became Pope John XXI, maintained the Aristotelian distinction between the two kinds of infinity by describing what he called the **categorematic infinity** (the realizable infinity) and the **syncategorematic infinity** (the merely conceivable infinity).

One of the most important and interesting philosophers of the fourteenth century was William of Occam (or Ockham) (*c.* 1285–*c.* 1349). Occam studied and taught at Oxford University from 1310 until 1324. In 1324 he was summoned to the papal court in Avignon, France, to answer charges of heresy. His principal crime was to deny the existence of universal truths except in language. This directly contradicted the teachings of the church. Occam is best remembered for the principle known as "Occam's razor," which can be stated as "what can be done with fewer [assumptions] is done in vain with more." A modern statement of Occam's razor, highly simplified, is "less is more." Occam's principle of economy led him into agreement with Aristotle that the categorematic infinity could never be realized. And, no, Occam was not executed for his heresy. He stayed in Avignon until 1328 when, it seemed, Pope John XXII was about to condemn his writings. At the last moment he fled to the protection of the Holy Roman Emperor Louis IV. It is believed that Occam died in the Black Death that swept through Europe in the 1340s.

During the Middle Ages, the question of the convergence of an infinite sum was largely philosophical. This changed in the seventeenth century. Galileo attempted to resolve Zeno's paradox by vague arguments that the "end of motion" (the time just before Achilles catches the tortoise) is a time of infinite speed just as the time just before an object that is thrown straight up begins its descent is a time of infinite slowness. Newton made similar arguments and is the first to have realized, but only in a vague sense, that some sort of limiting argument was necessary. In 1669 Newton wrote a treatise *De Analysi per Aequationes Numero Terminorum Infinitas* (On analysis by means of equations with an infinite number of terms) published in 1711. In attempting to compute a certain area, Newton was led to the integral of $\frac{1}{1+x^2}$. He divided to obtain

$$\frac{1}{1+x^2} = 1 - x^2 + x^4 - x^6 + \cdots$$

This is a geometric series.

so

$$\int \frac{dx}{1+x^2} = x - \frac{x^3}{3} + \frac{x^5}{5} - \frac{x^7}{7} + \cdots.$$

He then argued that this formula could be used when x was small enough. On the other hand, if x is large, he argued, then the following expansion works:

$$\frac{1}{1+x^2} = \frac{1}{x^2+1} = \frac{1}{x^2\left(1+\frac{1}{x^2}\right)} = \frac{1}{x^2}\left(1 - \frac{1}{x^2} + \frac{1}{x^4} - \frac{1}{x^6} + \cdots\right)$$

$$= \frac{1}{x^2} - \frac{1}{x^4} + \frac{1}{x^6} - \frac{1}{x^8} + \cdots$$

(CONTINUED)

OF HISTORICAL INTEREST

(CONTINUED)

and

$$\int \frac{dx}{1+x^2} = -\frac{1}{x} + \frac{1}{3x^3} - \frac{1}{5x^5} + \frac{1}{7x^7} - \cdots.$$

We know that the first series converges if $|x| < 1$ and the second converges if $|x| > 1$. However Newton had no sense of convergence. In his treatise, he wrote

> And whatever the common Analysis performs by Means of Equations of a finite Number of Terms (provided that can be done) this can always perform the same by Means of infinite Equations so that I have not made any question of giving this the name of Analysis likewise. For the reasonings in this are no less certain than in the other; nor the equations less exact; albeit we Mortals whose reasoning powers are confined within narrow limits, can neither express, nor so conceive all the Terms of these Equations, as to know exactly from thence the quantities we want.

In this rather vague statement Newton seems to suggest that term-by-term integration of finite sums can be extended to infinite sums without any loss of precision but that human beings can only conceive of the finite. This seems a reversal of Aristotle's claim that the infinite could be conceived, but not realized.

Throughout the seventeenth and eighteenth centuries, mathematicians worked with infinite series but had no clear sense of convergence and divergence. This sometimes led to some absurd problems. For example, seventeenth-century mathematicians were perplexed by the series

$$1 - 1 + 1 - 1 + 1 - \cdots.$$

In Example 5, we saw that this series diverges because the partial sums oscillate. But without the concept of divergence, one is led to problems by the simple process of placing parentheses:

$$\begin{aligned} 1 - 1 + 1 - 1 + 1 - 1 + \cdots &= (1-1) + (1-1) + (1-1) + \cdots \\ &= 0 + 0 + 0 + \cdots = 0, \end{aligned}$$

but

$$\begin{aligned} 1 - 1 + 1 - 1 + 1 - 1 + \cdots &= 1 - (1-1) - (1-1) - \cdots \\ &= 1 - 0 - 0 - 0 - \cdots = 1. \end{aligned}$$

Thus $0 = 1$. This paradoxical result confused mathematicians for about two centuries.

Leonhard Euler (1707–1783) found the following absurd series:

$$\begin{aligned} 0 &= \frac{x}{1-x} - \frac{x}{1-x} = \frac{x}{1-x} + \frac{x}{x-1} = x\left(\frac{1}{1-x}\right) + \frac{1}{1-\dfrac{1}{x}} \\ &= x(1 + x + x^2 + x^3 + \cdots) + 1 + \frac{1}{x} + \frac{1}{x^2} + \cdots \end{aligned} \tag{6}$$

which results in

$$\sum_{k=-\infty}^{\infty} x^k = \cdots + \frac{1}{x^3} + \frac{1}{x^2} + \frac{1}{x} + 1 + x + x^2 + x^3 + \cdots = 0.$$

This is false for every real number x (look at $x = 1$, for instance). The problem is that the geometric series $1 + x + x^2 + \cdots$ converges only when $|x| < 1$, while the geometric series $1 + \frac{1}{x} + \frac{1}{x^2} + \cdots$ converges only when $\left|\frac{1}{x}\right| < 1$ or $|x| > 1$. Thus the series (6) converges only

(CONTINUED)

OF HISTORICAL INTEREST (CONCLUDED)

when $|x| < 1$ and $|x| > 1$, an impossibility. This is clear to us but it was a mystery to Euler, who had no notion of convergence.

The term "convergent series" was first used by the Scottish mathematician James Gregory (1638–1675) in 1668 and the term "divergent series" was coined by Nicolaus Bernoulli (1687–1759) in 1721, but these concepts were not really understood until much later. From the middle of the eighteenth century, progress was made on developing the idea of convergence. This culminated in a formal definition of convergence given by Cauchy (see page 470) in 1821 in his monograph *Analyse Algébrique.* Cauchy was the first to give a rigorous definition of a limit, and it was this definition that he applied to the limit of partial sums. Before Cauchy, some mathematicians gave criteria for convergence of certain series. Notable among these was the great German mathematician Carl Friedrich Gauss (1775–1855) (see page 353), who discussed the hypergeometric series important in probability theory.

For more details about the history of criteria for determining convergence of series, see *History of Mathematics, 3rd Edition,* by Florian Cajori (New York: Chelsea, 1980), pages 373–375.

PROBLEMS 12.6

SELF-QUIZ

Multiple Choice

I. The difference between 1.0 and the repeating decimal 0.999 . . . is ________.

a. 0.0 **b.** 0.001
c. 0.0001 **d.** 0.00 . . . 01

II. $\sum_{k=1}^{\infty} \frac{1}{k}$ ________.

a. Converges to 2.75
b. Converges to some number between 5 and 12.3
c. Diverges to 13.47
d. Diverges

III. $\frac{2}{3} + \frac{2}{8} + \frac{2}{15} + \frac{2}{24} + \frac{2}{35} + \cdots$

$= \frac{2}{1 \cdot 3} + \frac{2}{2 \cdot 4} + \frac{2}{3 \cdot 5} + \frac{2}{4 \cdot 6} + \frac{2}{5 \cdot 7} + \cdots$

$= \sum_{k=2}^{\infty} \frac{2}{(k-1) \cdot (k+1)}$

$= \sum_{k=2}^{\infty} \left(\frac{1}{k-1} - \frac{1}{k+1}\right)$ ________.

a. $= 1 + \frac{1}{2} = \frac{3}{2}$ **b.** $= 1 - \frac{1}{2} = \frac{1}{2}$
c. $= 1 - \frac{1}{3} = \frac{2}{3}$ **d.** Diverges

In Problems 1–10, write the repeating decimal as a rational number.

1. 0.333 . . . **2.** 0.6666 . . .
3. 0.353535 . . . **4.** 0.282828 . . .
5. 0.717171 . . . **6.** 0.214214214 . . .
7. 0.501501501 . . . **8.** 0.124242424 . . .
9. 0.11362362362 . . .
10. 0.0513651365136 . . .

In Problems 11–26, a convergent infinite series is given; find its sum.

11. $\sum_{k=0}^{\infty} \frac{9}{10^k}$ **12.** $\sum_{k=0}^{\infty} \frac{1}{4^k}$

13. $\sum_{k=2}^{\infty} \frac{1}{2^k}$ **14.** $\sum_{k=1}^{\infty} \frac{3}{2^{k-1}}$

15. $\sum_{k=0}^{\infty} \left(-\frac{2}{3}\right)^k$ **16.** $\sum_{k=3}^{\infty} \left(\frac{2}{3}\right)^k$

17. $\sum_{k=0}^{\infty} \frac{100}{5^k}$ **18.** $\sum_{k=0}^{\infty} \frac{5}{100^k}$

19. $\sum_{k=2}^{\infty} \frac{1}{k(k+1)}$

20. $\sum_{k=3}^{\infty} \frac{1}{k(k-1)}$

21. $\sum_{k=0}^{\infty} \frac{1}{(k+1)(k+2)}$

22. $\sum_{k=-1}^{\infty} \frac{1}{(k+3)(k+4)}$

23. $\sum_{k=2}^{\infty} \frac{2^{k+3}}{3^k}$

24. $\sum_{k=2}^{\infty} \frac{2^{k+4}}{3^{k-1}}$

25. $\sum_{k=4}^{\infty} \frac{5^{k-2}}{6^{k+1}}$

26. $\sum_{k=-3}^{\infty} \frac{\sqrt{2}^k}{2^{k+3}}$

In Problems 27–32, use Theorem 1 to calculate the sum of the given convergent series.

27. $\sum_{k=0}^{\infty} \left[\left(\frac{1}{3}\right)^k + \left(\frac{2}{3}\right)^k\right]$

28. $\sum_{k=0}^{\infty} \left[\frac{1}{2^k} + \frac{1}{5^k}\right]$

29. $\sum_{k=0}^{\infty} \left[\frac{3}{5^k} - \frac{7}{4^k}\right]$

30. $\sum_{k=3}^{\infty} \left[\frac{12 \cdot 2^{k+1}}{3^{k-2}} - \frac{15 \cdot 3^{k+1}}{4^{k+2}}\right]$

31. $\sum_{k=1}^{\infty} \left[\frac{8}{5^k} - \frac{7}{(k+3)(k+4)}\right]$

32. $\sum_{k=1}^{\infty} \left[\frac{1}{k(k+1)} + \frac{1}{(k+1)(k+2)}\right]$

*33. At what time between 1 P.M. and 2 P.M. is the minute hand of a clock exactly over the hour hand? [*Hint:* The minute hand moves 12 times as fast as the hour hand. Start at 1:00 P.M. When the minute hand has reached 1, the hour hand points to $1 + \frac{1}{12}$; when the minute hand has reached $1 + \frac{1}{12}$, the hour hand has reached $1 + \frac{1}{12} + \frac{1}{12} \cdot \frac{1}{12}$; etc. Now add up the geometric series.]

*34. At what time between 7 A.M. and 8 A.M. is the minute hand exactly over the hour hand?

35. A ball is dropped from a height of 8 m. Each time it hits the ground, it rebounds to a height of two-thirds the height from which it fell. Find the total distance traveled by the ball until it comes to rest (i.e., until it stops bouncing).

36. All banks in the state of Mondaho are required to maintain cash reserves of 20% for their accounts—only 80% of each deposit is available for loan to other customers. Suppose one bank receives an out-of-state deposit for $25,000 and immediately loans the maximum permissible amount to a customer; suppose that customer then deposits that amount in a second bank which, in its turn, then loans its maximum permissible amount to a third customer who deposits it in another bank. . . . What is the least upper bound on the increase in capitalization caused by this process triggered by the initial $25,000?

37. Use the result of Problem 42 to show that $\sum_{k=1}^{\infty} 1/(k+6)$ diverges.

38. Use the result of Problem 43 to show that $\sum_{k=1}^{\infty} [(3/2^k) + 2 \cdot 5^k]$ diverges.

39. Use the geometric series to show that

$$\frac{1}{1+x} = \sum_{k=0}^{\infty} (-1)^k x^k$$

for any real number x with $|x| < 1$.

40. Prove or disprove: If $|x| < 1$, then

$$\frac{1}{1+x^2} = \sum_{k=0}^{\infty} (-1)^k x^{2k}.$$

41. Prove or disprove: For any nonzero real numbers a and b, $\sum_{k=1}^{\infty} a/(bk)$ diverges.

42. Show that if the sequences $\{a_k\}$ and $\{b_k\}$ differ only for a finite number of terms, then either both $\sum_{k=1}^{\infty} a_k$ and $\sum_{k=1}^{\infty} b_k$ converge or both diverge.

*43. Show that if $\sum_{k=1}^{\infty} a_k$ converges and $\sum_{k=1}^{\infty} b_k$ diverges, then $\sum_{k=1}^{\infty} (a_k + b_k)$ diverges. [*Hint:* Show that if $\sum_{k=1}^{\infty} (a_k + b_k)$ converges, then we get a contradiction of Theorem 1.]

44. Give an example in which $\sum_{k=1}^{\infty} a_k$ and $\sum_{k=1}^{\infty} b_k$ both diverge, but $\sum_{k=1}^{\infty} (a_k + b_k)$ converges.

*45. Pick a_0 and a_1. For $n \geq 2$, compute a_n recursively so that

$$n(n-1)a_n = (n-1)(n-2)a_{n-1} - (n-3)a_{n-2}.$$

Evaluate $\sum_{n=0}^{\infty} a_n$.

Answers to Self-Quiz

I. a **II.** d **III.** a

12.7 SERIES WITH NONNEGATIVE TERMS I: TWO COMPARISON TESTS AND THE INTEGRAL TEST

In this section and the next one, we consider series of the form $\sum_{k=1}^{\infty} a_k$, where each a_k is nonnegative. Such series are often easier to handle than others. One fact is easy to prove. The sequence $\{S_n\}$ of partial sums is a monotone increasing sequence since $S_{n+1} = S_n + a_{n+1}$ and $a_{n+1} \geq 0$ for every n. Then if $\{S_n\}$ is bounded, it is convergent by Theorem 12.4.2, and we have the following theorem:

THEOREM 1

An infinite series of nonnegative terms is convergent if and only if its sequence of partial sums is bounded. ■

EXAMPLE 1 A CONVERGENT SERIES

Show that $\sum_{k=1}^{\infty} 1/k^2$ is convergent.

SOLUTION: We group the terms as follows:

$$\sum_{k=1}^{\infty} \frac{1}{k^2} = \frac{1}{1^2} + \overbrace{\frac{1}{2^2} + \frac{1}{3^2}}^{2 \text{ terms}} + \overbrace{\frac{1}{4^2} + \frac{1}{5^2} + \frac{1}{6^2} + \frac{1}{7^2}}^{4 \text{ terms}} + \overbrace{\frac{1}{8^2} + \cdots + \frac{1}{15^2}}^{8 \text{ terms}} + \cdots$$

$$\leq 1 + \overbrace{\frac{1}{2^2} + \frac{1}{2^2}}^{2 \text{ terms}} + \overbrace{\frac{1}{4^2} + \frac{1}{4^2} + \frac{1}{4^2} + \frac{1}{4^2}}^{4 \text{ terms}} + \overbrace{\frac{1}{8^2} + \cdots + \frac{1}{8^2}}^{8 \text{ terms}} + \cdots$$

$$= 1 + \frac{2}{2^2} + \frac{4}{4^2} + \frac{8}{8^2} + \cdots = 1 + \frac{1}{2} + \frac{1}{4} + \frac{1}{8} + \cdots$$

$$= \sum_{k=0}^{\infty} \frac{1}{2^k} = 2.$$

Thus the sequence of partial sums is bounded by 2 and is therefore convergent. If we draw a picture, it is easy to see that the series is convergent. Figure 1 is self-explanatory.[†]

Since we know that the series converges, we can approximate its value by computing some partial sums on a computer.

$$S_{10} = \sum_{k=1}^{10} \frac{1}{k^2} = 1.5497677312$$

$$S_{50} = \sum_{k=1}^{50} \frac{1}{k^2} = 1.6251327336$$

$$S_{500} = \sum_{k=1}^{500} \frac{1}{k^2} = 1.6429360655$$

[†] Figure 1 was adapted from the interesting article "Convergence with pictures" by P. J. Rippon in *American Mathematical Monthly,* 93(1986): 476.

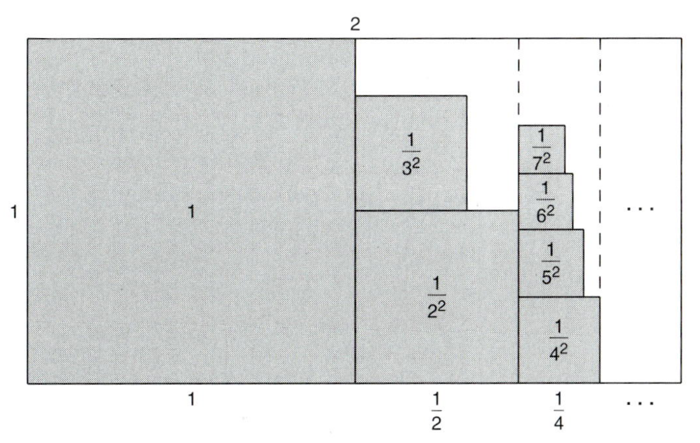

FIGURE 1
Illustration of fact that
$1+\frac{1}{2^2}+\frac{1}{3^2}+\frac{1}{4^2}+\cdots<1+\frac{1}{2}+\frac{1}{4}+\cdots=2$

$$S_{10,000}=\sum_{k=1}^{10,000}\frac{1}{k^2}=1.6448340718$$

It can be shown that $\sum_{k=1}^{\infty}\frac{1}{k^2}=\frac{\pi^2}{6}\approx 1.644934067.$†

WARNING: You cannot use a computer to determine whether a series converges or diverges. You can, as in Example 1, get approximate values for the sum of a series once you *know* that it converges, but you cannot use a calculator to prove that it does converge. If you have trouble with this idea, try to use a calculator to determine convergence or divergence of the harmonic series (see Problem 43).

With Theorem 1 the convergence or divergence of a series of nonnegative terms depends on whether or not its partial sums are bounded. There are several tests that can be used to determine whether or not the sequence of partial sums of a series is bounded. We will deal with these one at a time.

THEOREM 2 COMPARISON TEST

Let $\Sigma_{k=1}^{\infty} a_k$ be a series with $a_k \ge 0$ for every k.

i. If there exists a convergent series $\Sigma_{k=1}^{\infty} b_k$ and a number N such that $a_k \le b_k$ for every $k \ge N$, then $\Sigma_{k=1}^{\infty} a_k$ converges.

ii. If there exists a divergent series $\Sigma_{k=1}^{\infty} c_k$ and a number N such that $a_k \ge c_k \ge 0$ for every $k \ge N$, then $\Sigma_{k=1}^{\infty} a_k$ diverges.

PROOF: In either case the sum of the first N terms is finite, so we need only consider the series $\Sigma_{k=N+1}^{\infty} a_k$ since if this is convergent or divergent, then the addition of a finite number of terms does not affect the convergence or divergence.

† See "Six Ways to Sum a Series" in *The College Mathematics Journal,* November 1993, 402–421.

i. $\sum_{k=N+1}^{\infty} b_k$ is a nonnegative series (since $b_k \geq a_k \geq 0$ for $k > N$) and is convergent. Thus the partial sums $T_n = \sum_{k=N+1}^{\infty} b_k$ are bounded. If $S_n = \sum_{k=N+1}^{n} a_k$, then $S_n \leq T_n$, and so the partial sums of $\sum_{k=N+1}^{\infty} a_k$ are also bounded, implying that $\sum_{k=N+1}^{\infty} a_k$ is convergent.

ii. Let $U_n = \sum_{k=N+1}^{n} c_k$. By Theorem 1 these partial sums are unbounded since $\sum_{k=N+1}^{\infty} c_k$ diverges. Since in this case $S_n \geq U_n$, the partial sums of $\sum_{k=N+1}^{\infty} a_k$ are also unbounded, and the series $\sum_{k=N+1}^{\infty} a_k$ diverges. ■

REMARK: One fact mentioned in the proof of (i) is important enough to state again:

> If for some positive integer N, $\sum_{k=N+1}^{\infty} a_k$ converges, then $\sum_{k=1}^{\infty} a_k$ also converges. If $\sum_{k=N+1}^{\infty} a_k$ diverges, then $\sum_{k=1}^{\infty} a_k$ diverges. That is, the addition of a finite number of terms does not affect convergence or divergence.

EXAMPLE 2 A SERIES THAT DIVERGES BECAUSE EACH OF ITS TERMS IS GREATER THAN OR EQUAL TO THE CORRESPONDING TERM OF THE HARMONIC SERIES

Determine whether $\sum_{k=1}^{\infty} 1/\sqrt{k}$ converges or diverges.

SOLUTION: Since $1/\sqrt{k} \geq 1/k$ for $k \geq 1$, and since $\sum_{k=1}^{\infty} 1/k$ diverges, we see that by the comparison test, $\sum_{k=1}^{\infty} 1/\sqrt{k}$ diverges.

EXAMPLE 3 A CONVERGENT SERIES

Determine whether $\sum_{k=1}^{\infty} 1/k!$ converges or diverges.

SOLUTION: If $k \geq 4$, $k! \geq 2^k$. To see this, note that $4! = 24$ and $2^4 = 16$. Then $5! = 5 \cdot 24$ and $2^5 = 2 \cdot 16$ and since $5 > 2$, $5! > 2^5$, and so on. Since $\sum_{k=1}^{\infty} 1/2^k$ converges, we see that $\sum_{k=1}^{\infty} 1/k!$ converges. In fact, as we will show in Section 12.11, it converges to $e - 1$. That is,

$$e = 1 + 1 + \frac{1}{2!} + \frac{1}{3!} + \frac{1}{4!} + \cdots. \tag{1}$$

This series converges very rapidly because $\frac{1}{n!}$ approaches 0 very rapidly. For example,

$$10! = 3{,}628{,}800 \quad \text{and} \quad \frac{1}{10!} = 0.000000275573 \approx 2.76 \times 10^{-7};$$

$$20! \approx 2.43 \times 10^{18} \quad \text{and} \quad \frac{1}{20!} = 4.11 \times 10^{-19}.$$

We have (remembering that $0! = 1$)

$$\sum_{k=0}^{10} \frac{1}{k!} = 2.718281801 \quad \text{and} \quad \sum_{k=0}^{15} \frac{1}{k!} = 2.718281828.$$

THEOREM 3 THE INTEGRAL TEST

Let f be a function that is continuous, positive, and decreasing for all $x \geq 1$. Then the series

$$\sum_{k=1}^{\infty} f(k) = f(1) + f(2) + f(3) + \cdots + f(n) + \cdots \tag{2}$$

converges if and only if $\int_1^\infty f(x)\,dx$ converges, and diverges if and only if $\int_1^n f(x)\,dx \to \infty$ as $n \to \infty$.

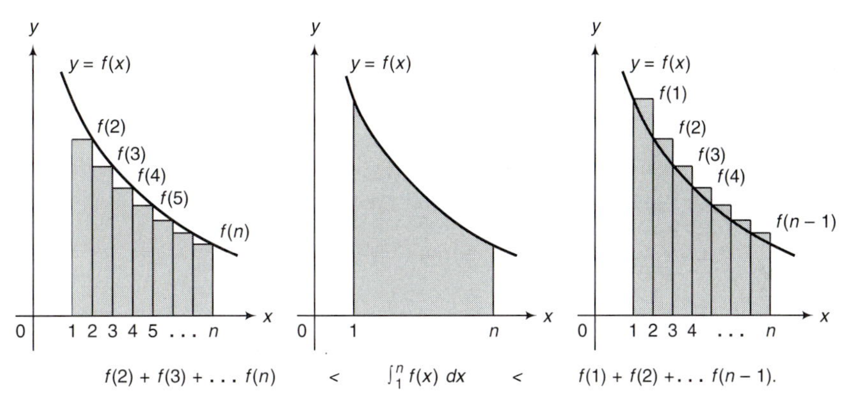

FIGURE 2
Illustration of the proof of the integral test

PROOF: The idea behind this proof is fairly easy. Take a look at Figure 2. Comparing areas, we immediately see that

$$f(2) + f(3) + \cdots + f(n) \leq \int_1^n f(x)\,dx \leq f(1) + f(2) + \cdots + f(n-1).$$

If $\lim\limits_{n\to\infty} \int_1^n f(x)\,dx$ is finite, then the partial sums $[f(2) + f(3) + \cdots + f(n)]$ are bounded and the series converges. On the other hand, if $\lim\limits_{n\to\infty} \int_1^n f(x)\,dx = \infty$, then the partial sums $[f(1) + f(2) + \cdots + f(n-1)]$ are unbounded and the series diverges.

Conversely, if the series diverges, then $f(2) + f(3) + \cdots + f(n) \to \infty$, so $\int_1^n f(x)\,dx \to \infty$. If it converges, then $f(1) + f(2) + \cdots + f(n-1)$ is bounded, so $\int_1^n f(x)\,dx$ converges. ∎

EXAMPLE 4 THE p SERIES

Consider the series $\Sigma_{k=1}^{\infty} 1/k^p$ with $p > 0$. We have already seen that this series diverges for $p = 1$ (the harmonic series) and converges for $p = 2$ (Example 1). Now let $f(x) = 1/x^p$. From one variable calculus we know that the improper integral $\int_1^\infty (1/x^p)\,dx$ diverges if $0 < p \leq 1$ and converges to $1/(p-1)$ if $p > 1$. Thus, by the integral test,

$$\sum_{k=1}^{\infty} \frac{1}{k^p} \quad \begin{cases} \text{diverges if } 0 < p \leq 1, \\ \text{converges if } p > 1. \end{cases}$$

The series $\Sigma_{k=1}^{\infty} 1/k^p$ is called the ***p* series**.

EXAMPLE 5

USING BOTH THE COMPARISON AND INTEGRAL TESTS TO PROVE CONVERGENCE

Determine whether $\Sigma_{k=1}^{\infty} (\ln k)/k^2$ converges or diverges.

SOLUTION: We see, using l'Hôpital's rule, that

$$\lim_{x\to\infty} \frac{\ln x}{\sqrt{x}} = \lim_{x\to\infty} \frac{1/x}{1/2\sqrt{x}} = 2 \lim_{x\to\infty} \frac{\sqrt{x}}{x} = 2 \lim_{x\to\infty} \frac{1}{\sqrt{x}} = 0,$$

so that for k sufficiently large, $\ln k \le \sqrt{k}$. Thus,

$$\frac{\ln k}{k^2} \le \frac{\sqrt{k}}{k^2} = \frac{1}{k^{3/2}}.$$

But $\Sigma_{k=1}^{\infty} 1/k^{3/2}$ converges by the result of Example 4, and therefore, by the comparison test, $\Sigma_{k=1}^{\infty} (\ln k)/k^2$ also converges.

NOTE: The integral test can also be used directly here since $\int_1^{\infty} (\ln x/x^2)\, dx$ can be integrated by parts with $u = \ln x$.

We now give another test that is an extension of the comparison test.

THEOREM 4 THE LIMIT COMPARISON TEST

Let $\Sigma_{k=1}^{\infty} a_k$ and $\Sigma_{k=1}^{\infty} b_k$ be series with positive terms. If there is a number $c > 0$ such that

$$\lim_{k\to\infty} \frac{a_k}{b_k} = c, \tag{3}$$

then either both series converge or both series diverge.

PROOF: We have $\lim_{k\to\infty} (a_k/b_k) = c > 0$. In the definition of a limit on page 916, let $\epsilon = c/2$. Then there is a number $N > 0$ such that

$$\left|\frac{a_k}{b_k} - c\right| < \frac{c}{2} \qquad \text{if } k \ge N. \tag{4}$$

Equation (4) is equivalent to

$$-\frac{c}{2} < \frac{a_k}{b_k} - c < \frac{c}{2}, \qquad \text{or} \qquad \frac{c}{2} < \frac{a_k}{b_k} < \frac{3c}{2}. \tag{5}$$

From the last inequality in (5), we obtain

$$a_k < \frac{3c}{2} b_k. \tag{6}$$

If Σb_k is convergent, then so is $(3c/2)\Sigma b_k = \Sigma(3c/2)b_k$. Thus from (6) and the comparison test, Σa_k is convergent. From the next-to-the-last inequality in (5), we have

$$a_k > \frac{c}{2} b_k. \tag{7}$$

If Σb_k is divergent, then so is $(c/2)\Sigma b_k = \Sigma(c/2)b_k$. Then using (7) and the comparison test, we find that Σa_k is also divergent. Thus if Σb_k is convergent, then Σa_k is convergent; and if Σb_k is divergent, then Σa_k is divergent. This proves the theorem. ■

EXAMPLE 6 USING THE LIMIT COMPARISON TEST TO PROVE CONVERGENCE

Show that $\Sigma_{k=1}^{\infty} 1/(ak^2 + bk + c)$ is convergent, where a, b, and c are positive real numbers.

SOLUTION: We know from Example 1 that $\Sigma_{k=1}^{\infty} 1/k^2$ is convergent. If

$$a_k = \frac{1}{k^2} \quad \text{and} \quad b_k = \frac{1}{ak^2 + bk + c},$$

then

$$\frac{a_k}{b_k} = \frac{1/k^2}{1/(ak^2 + bk + c)} = \frac{ak^2 + bk + c}{k^2} = a + \frac{b}{k} + \frac{c}{k^2},$$

so that $\lim_{k\to\infty} (a_k/b_k) = a > 0$. Thus, by the limit comparison test,

$$\sum_{k=1}^{\infty} \frac{1}{ak^2 + bk + c}$$

is convergent.

PROBLEMS 12.7

SELF-QUIZ

True–False

I. a. $\Sigma_{k=1}^{\infty} (5/k)$ diverges because we can compare each term with corresponding, but smaller, terms of the harmonic series $\Sigma_{k=1}^{\infty} (1/k)$ which is already known to diverge.

b. If $k \geq 1$, then $k^2 \geq k$ and $1/k \geq 1/k^2$. Therefore, by comparing terms of $\Sigma_{k=1}^{\infty} (1/k^2)$ with the divergent harmonic series $\Sigma_{k=1}^{\infty} (1/k)$, we infer that $\Sigma_{k=1}^{\infty} (1/k^2)$ also diverges.

II. a. If $2 \leq k$, then $0 < k^2 - 1 < k^2$ and $1/k^2 < 1/(k^2 - 1)$. Because $\Sigma_{k=2}^{\infty} (1/k^2)$ is known to converge (see Example 1 and the remark following Theorem 2), our term-by-term comparison allows us to conclude that $\Sigma_{k=2}^{\infty} (1/[k^2 - 1])$ must also converge.

b. Since each term of $\Sigma_{k=1}^{\infty} (-3/k)$ is less than a corresponding term of the convergent series $\Sigma_{k=2}^{\infty} (1/k^2)$ ($-3/k < 0 < 1/k^2$ for all $k \geq 2$), Theorem 2 implies that $\Sigma_{k=1}^{\infty} (-3/k)$ converges.

III. a. $\Sigma_{k=2}^{\infty} (1/[k^2 - 1])$ converges because $\int_2^{\infty} (1/[x^2 - 1])\, dx$ converges (to $\frac{1}{2} \ln 3$).

b. $\Sigma_{k=1}^{\infty} (1/[7k^2 + 3 \sin k + 13])$ converges because

$$\lim_{k\to\infty} \frac{1/[7k^2 + 3 \sin k + 13]}{1/k^2} = \frac{1}{7}$$

and $\Sigma_{k=1}^{\infty} (1/k^2)$ converges.

In Problems 1–34, determine the convergence or divergence of the given series.

1. $\sum_{k=1}^{\infty} (-1)^k$

2. $\sum_{k=1}^{\infty} \frac{k}{k+100}$

3. $\sum_{k=4}^{\infty} \frac{1}{5k+50}$

4. $\sum_{k=1}^{\infty} \frac{k+1}{(k+3)(k+5)}$

5. $\sum_{k=1}^{\infty} \frac{1}{k^2+1}$

6. $\sum_{k=10}^{\infty} \frac{1}{k(k-3)}$

7. $\sum_{k=1}^{\infty} \frac{1}{\sqrt{k^2+2k}}$

8. $\sum_{k=1}^{\infty} \frac{1}{\sqrt{k^3+1}}$

9. $\sum_{k=0}^{\infty} ke^{-k}$

10. $\sum_{k=3}^{\infty} k^3 e^{-k^4}$

11. $\sum_{k=2}^{\infty} \frac{4}{k \ln k}$

12. $\sum_{k=1}^{\infty} \frac{1}{k \ln(k+5)}$

13. $\sum_{k=5}^{\infty} \frac{1}{k(\ln k)^3}$

14. $\sum_{k=4}^{\infty} \frac{1}{k^2\sqrt{\ln k}}$

15. $\sum_{k-1}^{\infty} \frac{1}{(3k-1)^{3/2}}$

16. $\sum_{k=1}^{\infty} \frac{1}{\sqrt{k^2+3}}$

17. $\sum_{k=2}^{\infty} \frac{1}{k\sqrt{\ln k}}$

18. $\sum_{k=1}^{\infty} \frac{1}{50+\sqrt{k}}$

19. $\sum_{k=3}^{\infty} \left(\frac{k}{k+1}\right)^k$

20. $\sum_{k=1}^{\infty} \left(\frac{k}{k+1}\right)^{1/k}$

21. $\sum_{k=4}^{\infty} \frac{1}{k \ln \ln k}$

22. $\sum_{k=1}^{\infty} \sin \frac{1}{k}$

23. $\sum_{k=1}^{\infty} \frac{1}{(k+2)\sqrt{\ln(k+1)}}$

24. $\sum_{k=1}^{\infty} \frac{e^{1/k}}{k^2}$

25. $\sum_{k=1}^{\infty} \tan^{-1} k$

26. $\sum_{k=1}^{\infty} \frac{\tan^{-1} k}{1+k^2}$

27. $\sum_{k=1}^{\infty} \frac{\ln k}{k^3}$

28. $\sum_{k=10}^{\infty} \frac{1}{k(\ln k)(\ln \ln k)}$

29. $\sum_{k=1}^{\infty} \operatorname{sech} k$

30. $\sum_{k=1}^{\infty} \frac{1}{\cosh^2 k}$

***31.** $\sum_{k=2}^{\infty} \frac{1}{\sqrt{k} \ln^{10} k}$

32. $\sum_{k=1}^{\infty} \frac{\sqrt{k}}{3k^2+2k+20}$

33. $\sum_{k=1}^{\infty} \frac{(k+1)^{7/8}}{k^3+k^2+3}$

34. $\sum_{k=1}^{\infty} \frac{k^5+2k^4+3k+7}{k^6+3k^4+2k^2+1}$

35. Suppose that $\Sigma_{k=0}^{\infty} a_k$ is a convergent series of positive terms; prove that $\Sigma_{k=0}^{\infty} a_k^2$ also converges.

36. Suppose that $\Sigma_{k=0}^{\infty} a_k$ and $\Sigma_{k=0}^{\infty} b_k$ are both convergent series of positive terms; prove that $\Sigma_{k=0}^{\infty} a_k \cdot b_k$ also converges.
[*Hint:* $(a+b)^2 - a^2 - b^2 = 2ab$.]

***37.** Let $p(x)$ be a polynomial of degree n with positive coefficients and let $q(x)$ be a polynomial of degree $\le n+1$ with positive coefficients. Show that $\Sigma_{k=1}^{\infty} p(k)/q(k)$ diverges.

***38.** With $p(x)$ as in the preceding problem and with $r(x)$ a polynomial of degree $\ge n+2$ having positive coefficients, show that $\Sigma_{k=1}^{\infty} p(k)/r(k)$ converges.

***39.** Determine whether $\sum_{n=1}^{\infty} \left(\frac{1}{n}\right)^{1+(1/n)}$ converges or diverges.

40. Let $S_n = \ln(n!) = \ln 2 + \ln 3 + \cdots + \ln n$. By calculating $\int_1^n \ln x \, dx$ and comparing areas, as in the proof of the integral test, show that for $n \ge 2$

$$\ln[(n-1)!] < n \ln n - n + 1 < \ln[n!]$$

and thus

$$(n-1)! < n^n e^{-(n-1)} < n!.$$

***41.** Let $S_n = \Sigma_{k=1}^{n} 1/k$. Show that

$$\ln(n+1) < S_n < \ln n + 1.$$

[*Hint:* Use the inequality $1/(k+1) \le 1/x \le 1/k$ when $0 < k \le x \le k+1$, integrate and add (as in the proof of the integral test).]

42. Let $S_n = \Sigma_{k=1}^{n} 1/k$.

a. Show that the sequence $\{S_n - \ln(n+1)\}$ is increasing.

***b.** Show that this sequence is bounded by 1.

[*Hint:* $S_n - \ln(n+1) =$ the sum of the areas of "triangular" shaped regions which touch the graph of $y = 1/x$. Move those "triangles" over so that they all fit into a single rectangle with area 1.]

c. Show that $\lim_{n\to\infty} [S_n - \ln(n+1)]$ exists. This limit is denoted by γ; it is called the

Euler–Mascheroni constant:

$$\gamma = \lim_{n\to\infty}\left[1+\frac{1}{2}+\frac{1}{3}+\cdots+\frac{1}{n}-\ln(n+1)\right]$$
$$= \lim_{n\to\infty}\left[1+\frac{1}{2}+\frac{1}{3}+\cdots+\frac{1}{n}-\ln n\right].$$

This number arises in physical applications. To seven decimal places, $\gamma = 0.5772157$.[†]

*43. Consider a computer programmed to add terms of the harmonic series. Suppose it can add one million terms per second. Also suppose it works with perfect accuracy (i.e., no round-off or truncation errors).

a. How many hours will it need to work before the divergent sum exceeds 25?
b. How many days will it need to work before the divergent sum exceeds 30?
c. How many years will it need to work before the divergent sum exceeds 35?[‡]

[*Hint:* Use the preceding problem and the value given there for the Euler–Mascheroni constant.]

ANSWERS TO SELF-QUIZ

I. a. True **b.** False
II. a. False (It does converge, but not for the reason given.) **b.** False
III. a. True **b.** True

12.8 SERIES WITH NONNEGATIVE TERMS II: THE RATIO AND ROOT TESTS

In this section we discuss two more tests that can be used to determine whether an infinite series converges or diverges. The first of these, the ratio test, is useful in a wide variety of applications.

THEOREM 1 THE RATIO TEST

Let $\Sigma_{k=1}^{\infty} a_k$ be a series with $a_k > 0$ for every k, and suppose that

$$\lim_{a_n\to\infty}\frac{a_{n+1}}{a_n} = L. \qquad (1)$$

i. If $L < 1$, $\Sigma_{k=1}^{\infty} a_k$ converges.
ii. If $L > 1$, $\Sigma_{k=1}^{\infty} a_k$ diverges.
iii. If $L = 1$, $\Sigma_{k=1}^{\infty} a_k$ may converge or diverge and the ratio test is inconclusive; some other test must be used.

PROOF:

i. Pick $\epsilon > 0$ such that $L + \epsilon < 1$. By the definition of the limit in (1), there is a number $N > 0$ such that if $n \ge N$, we have

$$\frac{a_{n+1}}{a_n} < L + \epsilon.$$

Then

$$a_{n+1} < a_n(L+\epsilon), \qquad a_{n+2} < a_{n+1}(L+\epsilon) < a_n(L+\epsilon)^2,$$

[†] It is an interesting fact that as of September, 1994, no one has determined whether or not γ is rational.
[‡] There are many interesting ways to use computers to help us understand calculus better; doing such a summation is not one of them.

and

$$a_{n+k} < a_n(L + \epsilon)^k \tag{2}$$

for each $k \geq 1$ and each $n \geq N$. In particular, for $k \geq N$ we use (2) to obtain

$$a_k = a_{(k-N)+N} \leq a_N(L + \epsilon)^{k-N}.$$

Then

$$S_n = \sum_{k=N}^{n} a_k \leq \sum_{k=N}^{n} a_N(L + \epsilon)^{k-N} = \frac{a_N}{(L + \epsilon)^N} \sum_{k=N}^{n} (L + \epsilon)^k.$$

But since $L + \epsilon < 1$, $\Sigma_{k=0}^{\infty} (L + \epsilon)^k = 1/[1 - (L + \epsilon)]$ (since this last sum is the sum of a geometric series). Thus,

$$S_n \leq \frac{a_N}{(L + \epsilon)^N} \cdot \frac{1}{1 - (L + \epsilon)},$$

and so the partial sums of $\Sigma_{k=N}^{\infty} a_k$ are bounded, implying that $\Sigma_{k=N}^{\infty} a_k$ converges. Thus, $\Sigma_{k=1}^{\infty} a_k = \Sigma_{k=1}^{N-1} a_k + \Sigma_{k=N}^{\infty} a_k$ also converges.

ii. If $1 < L < \infty$, pick ϵ such that $L - \epsilon > 1$. Then for $n \geq N$, the same proof as before (with the inequalities reversed) shows that

$$\frac{a_{n+1}}{a_n} > L - \epsilon > 1.$$

Then $\{a_n\}$ is an increasing sequence and so does not have the limit 0. Thus $\Sigma_{k=1}^{\infty} a_k$ diverges. The proof in the case $L = \infty$ is suggested in Problem 35.

iii. To illustrate (iii), we show that $L = 1$ can occur for a converging or diverging series.

a. The harmonic series $\Sigma_{k=1}^{\infty} 1/k$ diverges. But

$$\lim_{n\to\infty} \frac{a_{n+1}}{a_n} = \lim_{n\to\infty} \frac{1/(n + 1)}{1/n} = \lim_{n\to\infty} \frac{n}{n + 1} = 1.$$

b. The series $\Sigma_{k=1}^{\infty} 1/k^2$ converges. Here

$$\lim_{n\to\infty} \frac{a_{n+1}}{a_n} = \lim_{n\to\infty} \frac{1/(n + 1)^2}{1/n^2} = \lim_{n\to\infty} \left(\frac{n}{n + 1}\right)^2 = 1. \quad \blacksquare$$

REMARK 1: The ratio test is very useful. But in those cases where $L = 1$, we must try another test to determine whether the series converges or diverges.

REMARK 2: If the terms of the series $\Sigma_{k=1}^{\infty} a_k$ are nonzero, then we can apply the ratio test to the series $\Sigma_{k=1}^{\infty} |a_k|$, which is a series of positive terms. We then have

$$\sum_{k=1}^{\infty} |a_k| \text{ converges if } \lim_{n\to\infty} \frac{|a_{n+1}|}{|a_n|} < 1$$

and

$$\sum_{k=1}^{\infty} |a_k| \text{ diverges if } \lim_{n\to\infty} \frac{|a_{n+1}|}{|a_n|} > 1.$$

COROLLARY

i. If $0 < \dfrac{a_{n+1}}{a_n} \le L < 1$ for $n \ge 1$, then $\sum\limits_{k=1}^{\infty} a_k$ converges.

ii. If $\dfrac{a_{n+1}}{a_n} \ge L > 1$ for $n \ge 1$, then $\sum\limits_{k=1}^{\infty} a_k$ diverges. ■

The proof of this result is almost identical to the proof of Theorem 1. Note that, using the corollary, we can prove convergence or divergence in cases where $\lim\limits_{n\to\infty} \dfrac{a_{n+1}}{a_n}$ does not exist.

EXAMPLE 1

USING THE RATIO TEST TO PROVE CONVERGENCE

We have used the comparison test to show that $\Sigma_{k=1}^{\infty} 1/k!$ converges. Using the ratio test, we find that

$$\lim_{n\to\infty} \frac{a_{n+1}}{a_n} = \lim_{n\to\infty} \frac{1/(n+1)!}{1/n!} = \lim_{n\to\infty} \frac{n!}{(n+1)!} = \lim_{n\to\infty} \frac{1}{n+1} = 0 < 1,$$

so that the series converges.

EXAMPLE 2

USING THE RATIO TEST TO PROVE CONVERGENCE

Determine whether the series $\Sigma_{k=0}^{\infty} (100)^k/k!$ converges or diverges.

SOLUTION: Here

$$\lim_{n\to\infty} \frac{a_{n+1}}{a_n} = \lim_{n\to\infty} \frac{(100)^{n+1}/(n+1)!}{(100)^n/n!} = \lim_{n\to\infty} \frac{100}{n+1} = 0,$$

so that the series converges.

EXAMPLE 3

USING THE RATIO TEST TO PROVE DIVERGENCE

Determine whether the series $\Sigma_{k=1}^{\infty} k^k/k!$ converges or diverges.

SOLUTION:

$$\lim_{n\to\infty} \frac{a_{n+1}}{a_n} = \lim_{n\to\infty} \frac{[(n+1)^{n+1}/(n+1)!]}{n^n/n!} = \lim_{n\to\infty} \frac{(n+1)^{n+1}}{(n+1)n^n}$$

$$= \lim_{n\to\infty} \left(\frac{n+1}{n}\right)^n = \lim_{n\to\infty} \left(1 + \frac{1}{n}\right)^n = e > 1,$$

so that the series diverges.

EXAMPLE 4

A DIVERGENT SERIES FOR WHICH THE RATIO TEST DOES NOT WORK

Determine whether the series $\Sigma_{k=1}^{\infty} (k + 1)/[k(k + 2)]$ converges or diverges.

SOLUTION: Here,

$$\lim_{n\to\infty} \frac{a_{n+1}}{a_n} = \lim_{n\to\infty} \frac{(n + 2)/(n + 1)(n + 3)}{(n + 1)/n(n + 2)} = \lim_{n\to\infty} \frac{n(n + 2)^2}{(n + 1)^2(n + 3)}$$
$$= \lim_{n\to\infty} \frac{n^3 + 4n^2 + 4n}{n^3 + 5n^2 + 7n + 3} = 1.$$

Thus the ratio test fails. However, $\lim_{k\to\infty} [(k + 1)/k(k + 2)]/(1/k) = 1$, so that $\Sigma_{k=1}^{\infty} (k + 1)/[k(k + 2)]$ diverges by the limit comparison test.

THEOREM 2 THE ROOT TEST

Let $\Sigma_{k=1}^{\infty} a_k$ be a series with $a_k > 0$ and suppose that $\lim_{n\to\infty} (a_n)^{1/n} = R$.

i. If $R < 1$, $\Sigma_{k=1}^{\infty} a_k$ converges.
ii. If $R > 1$, $\Sigma_{k=1}^{\infty} a_k$ diverges.
iii. If $R = 1$, the series either converges or diverges, and no conclusions can be drawn from this test.

The proof of this theorem is similar to the proof of the ratio test and is left as an exercise (see Problems 31–33). ■

EXAMPLE 5

USING THE ROOT TEST TO PROVE CONVERGENCE

Determine whether $\Sigma_{k=2}^{\infty} 1/(\ln k)^k$ converges or diverges.

SOLUTION: Note first that we start at $k = 2$ since $1/(\ln 1)^1$ is not defined.

$$\lim_{n\to\infty} \left[\frac{1}{(\ln n)^n}\right]^{1/n} = \lim_{n\to\infty} \frac{1}{\ln n} = 0,$$

so that the series converges.

EXAMPLE 6

USING THE ROOT TEST TO PROVE DIVERGENCE

Determine whether the series $\Sigma_{k=1}^{\infty} (k^k/3^{4k+5})$ converges or diverges.

SOLUTION: $\lim_{n\to\infty} (n^n/3^{4n+5})^{1/n} = \lim_{n\to\infty} (n/3^{4+5/n}) = \infty$, since $\lim_{n\to\infty} 3^{4+5/n} = 3^4 = 81$. Thus, the series diverges.

PROBLEMS 12.8

SELF-QUIZ

Multiple Choice

I. The ratio test is inconclusive for which of the following series?

a. $\sum_{k=1}^{\infty} \frac{1}{k^3}$ **b.** $\sum_{k=1}^{\infty} \left(\frac{5}{13}\right)^k$

c. $\sum_{k=1}^{\infty} \frac{k}{k+7}$ **d.** $\sum_{k=1}^{\infty} \frac{5^k}{k!}$

e. $\sum_{k=1}^{\infty} \frac{1 \cdot 3 \cdot 5 \cdot \cdots \cdot (2k-1)}{4 \cdot 7 \cdot 10 \cdot \cdots \cdot (3k+1)}$

f. $\sum_{k=1}^{\infty} 2^{-n+(-1)^n}$

II. The root test is inconclusive for which of the following series?

a. $\sum_{k=1}^{\infty} \frac{1}{k^4}$ **b.** $\sum_{k=1}^{\infty} \left(\frac{11}{13}\right)^k$

c. $\sum_{k=1}^{\infty} \frac{k-27}{k}$ **d.** $\sum_{k=1}^{\infty} \frac{3^k}{k!}$

e. $\sum_{k=1}^{\infty} 2^{-n-(-1)^n}$

In Problems 1–26, determine whether the given series converges or diverges.

1. $\sum_{k=1}^{\infty} \frac{2^k}{k^2}$ **2.** $\sum_{k=1}^{\infty} \frac{5^k}{k^5}$

3. $\sum_{k=1}^{\infty} \frac{r^k}{k^r}$, $0 < r < 1$ **4.** $\sum_{k=1}^{\infty} \frac{r^k}{k^r}$, $r > 1$

5. $\sum_{k=2}^{\infty} \frac{k!}{k^k}$ **6.** $\sum_{k=1}^{\infty} \frac{k^k}{(2k)!}$

7. $\sum_{k=1}^{\infty} \frac{e^k}{k^5}$ **8.** $\sum_{k=1}^{\infty} \frac{e^k}{k!}$

9. $\sum_{k=1}^{\infty} \frac{k^{2/3}}{10^k}$ **10.** $\sum_{k=1}^{\infty} \frac{3^k + k}{k! + 2}$

11. $\sum_{k=2}^{\infty} \frac{k}{(\ln k)^k}$ **12.** $\sum_{k=1}^{\infty} \frac{4^k}{k^3}$

13. $\sum_{k=2}^{\infty} \left(1 + \frac{1}{k}\right)^k$ **14.** $\sum_{k=1}^{\infty} \frac{\sqrt{k} \ln k}{k^3 + 1}$

15. $\sum_{k=1}^{\infty} \frac{3^{4k+5}}{k^k}$ **16.** $\sum_{k=1}^{\infty} \frac{a^{mk+b}}{k^k}$, $a > 1$, $m > 0$

17. $\sum_{k=1}^{\infty} \frac{k^k}{a^{mk+b}}$, $a > 1$, $m > 0$ **18.** $\sum_{k=1}^{\infty} \frac{k^6 5^k}{(k+1)!}$

19. $\sum_{k=1}^{\infty} \frac{k^2 k!}{(2k)!}$ **20.** $\sum_{k=1}^{\infty} \frac{(2k)!}{k^2 k!}$

21. $\sum_{k=1}^{\infty} \left(\frac{k!}{k^k}\right)^k$ **22.** $\sum_{k=1}^{\infty} \left(\frac{k^k}{k!}\right)^k$

23. $\sum_{k=2}^{\infty} \frac{e^k}{(\ln k)^k}$ **24.** $\sum_{k=1}^{\infty} \frac{(\ln k)^k}{k^2}$

25. $\sum_{k=1}^{\infty} \left(\frac{k}{3k+2}\right)^k$ **26.** $\sum_{k=1}^{\infty} \left(\frac{1}{2} + \frac{1}{k}\right)^k$

27. Let

$$a_k = \begin{cases} 3/k^2 & \text{if } k \text{ is even,} \\ 1/k^2 & \text{if } k \text{ is odd.} \end{cases}$$

Verify that $\lim_{n\to\infty} (a_{n+1}/a_n)$ does not exist, but that $\Sigma_{k=1}^{\infty} a_k$ converges.

28. Construct a series of positive terms for which $\lim_{n\to\infty} (a_{n+1}/a_n)$ does not exist but for which $\Sigma_{k=1}^{\infty} a_k$ diverges.

29. Show that $\Sigma_{k=0}^{\infty} x^k/k!$ converges for every real number x.

30. **a.** Show that for any real number a,

$$\lim_{k\to\infty} (k^a e^{-k})^{1/k} = \frac{1}{e}.$$

[*Hint:* Use l'Hôpital's rule.]

b. Show that $\Sigma_{k=1}^{\infty} k^a e^{-k}$ converges.

c. Prove that $\lim_{k\to\infty} k^a e^{-k} = 0$ for any real number a.

***31.** Prove part (i) of the root test (Theorem 2). [*Hint:* If $R < 1$, choose $\epsilon > 0$ so that $R + \epsilon < 1$.

Show that there is an N such that if $n \geq N$, then $a_n < (R + \epsilon)^n$. Then complete the proof by comparing Σa_k with the sum of a geometric series.]

32. Prove part (ii) of the root test. [*Hint:* Parallel the steps suggested for the preceding problem.]

33. Show that if $\sqrt[n]{a_n} \to 1$, then $\Sigma_{k=1}^{\infty} a_k$ may converge or diverge. [*Hint:* Consider $\Sigma 1/k$ and $\Sigma 1/k^2$.]

***34.** Prove that $k!/k^k \to 0$ as $k \to \infty$.

35. Prove that if $a_n > 0$ and $\lim_{n\to\infty} (a_{n+1}/a_n) = \infty$, then $\Sigma_{k=1}^{\infty} a_k$ diverges. [*Hint:* Show that for sufficiently large N, $a_k \geq 2^{k-N} \cdot a_N$.]

36. Prove both parts of the corollary to Theorem 1.

ANSWERS TO SELF-QUIZ

I. a (converges), c (diverges), f (converges)

II. a (converges), c (diverges)

12.9 ABSOLUTE AND CONDITIONAL CONVERGENCE: ALTERNATING SERIES

In Sections 12.7 and 12.8 all the series we dealt with had positive terms. In this section we consider special types of series that have positive and negative terms.

DEFINITION ABSOLUTE CONVERGENCE

The series $\Sigma_{k=1}^{\infty} a_k$ is said to **converge absolutely** if the series $\Sigma_{k=1}^{\infty} |a_k|$ converges. ■

Absolute convergence is important for the following reason: The series $\Sigma|a_k|$ is a series with nonnegative terms. Therefore, all the tests we used in the last two sections can be applied to determine whether or not the series $\Sigma|a_k|$ converges; that is, whether or not Σa_k converges absolutely. But, as Theorem 1 below tells us, if $\Sigma|a_k|$ converges, then Σa_k converges also. That is, we can use the ratio, root, comparison, and integral tests to prove that a series converges even when some of its terms are negative.

EXAMPLE 1 AN ABSOLUTELY CONVERGENT SERIES

The series

$$\sum_{k=1}^{\infty} \frac{(-1)^{k+1}}{k^2} = \frac{1}{1^2} - \frac{1}{2^2} + \frac{1}{3^2} - \frac{1}{4^2} + \cdots$$

converges absolutely because $\Sigma_{k=1}^{\infty} |(-1)^{k+1}/k^2| = \Sigma_{k=1}^{\infty} 1/k^2$ converges.

EXAMPLE 2 A SERIES THAT DOES NOT CONVERGE ABSOLUTELY

The series

$$\sum_{k=1}^{\infty} \frac{(-1)^{k+1}}{k} = \frac{1}{1} - \frac{1}{2} + \frac{1}{3} - \frac{1}{4} + \frac{1}{5} + \cdots$$

does not converge absolutely because $\Sigma_{k=1}^{\infty} 1/k$ diverges.

The importance of absolute convergence is given in the theorem below.

THEOREM 1

If $\Sigma_{k=1}^{\infty} |a_k|$ converges, then $\Sigma_{k=1}^{\infty} a_k$ also converges. That is, absolute convergence implies convergence. ■

REMARK: The converse of this theorem is false. That is, there are series that are convergent but not absolutely convergent. We will see examples of this phenomenon shortly.

PROOF: Since $a_k \leq |a_k|$, we have $0 \leq a_k + |a_k| \leq 2|a_k|$.

Since $\Sigma_{k=1}^{\infty} |a_k|$ converges, we see that $\Sigma_{k=1}^{\infty} (a_k + |a_k|)$ converges by the comparison test. Then since $a_k = (a_k + |a_k|) - |a_k|$, $\Sigma_{k=1}^{\infty} a_k$ converges because it is the sum of two convergent series.

EXAMPLE 3 ABSOLUTE CONVERGENCE IMPLIES CONVERGENCE

The series $\Sigma_{k=1}^{\infty} (-1)^{k+1}/k^2$ considered in Example 1 converges since it converges absolutely.

DEFINITION ALTERNATING SERIES

A series in which successive terms have opposite signs is called an **alternating series.** ■

EXAMPLE 4 AN ALTERNATING SERIES

The series

$$\sum_{k=1}^{\infty} \frac{(-1)^{k+1}}{k} = 1 - \frac{1}{2} + \frac{1}{3} - \frac{1}{4} + \frac{1}{5} - \frac{1}{6} + \cdots$$

is an alternating series.

EXAMPLE 5 A SERIES WITH BOTH POSITIVE AND NEGATIVE TERMS THAT IS NOT AN ALTERNATING SERIES

The series $1 + \frac{1}{2} - \frac{1}{3} - \frac{1}{4} + \frac{1}{5} + \frac{1}{6} - \cdots$ is not an alternating series because two successive terms have the same sign.

Let us consider the series of Example 4:

$$S = 1 - \frac{1}{2} + \frac{1}{3} - \frac{1}{4} + \frac{1}{5} - \frac{1}{6} + \cdots.$$

Calculating successive partial sums, we find that

$$S_1 = 1, \quad S_2 = \frac{1}{2}, \quad S_3 = \frac{5}{6}, \quad S_4 = \frac{7}{12}, \quad S_5 = \frac{47}{60}, \quad \ldots .$$

It seems that this series is not diverging to infinity (indeed, $\frac{1}{2} \le S_n \le 1$) and that the partial sums are getting "narrowed down." At this point it is reasonable to suspect that the series converges. But it does *not* converge absolutely (since the series of absolute values is the harmonic series), and we cannot use any of the tests of the previous section since the terms are not nonnegative. The result we need is given in the theorem below.

THEOREM 2 ALTERNATING SERIES TEST

Let $\{a_k\}$ be a decreasing sequence of positive numbers such that $\lim_{k\to\infty} a_k = 0$. Then the alternating series $\Sigma_{k=1}^{\infty} (-1)^{k+1} a_k = a_1 - a_2 + a_3 - a_4 + \cdots$ converges. ■

PROOF: Looking at the odd-numbered partial sums of this series, we find that

$$S_{2n+1} = (a_1 - a_2) + (a_3 - a_4) + (a_5 - a_6) + \cdots + (a_{2n-1} - a_{2n}) + a_{2n+1}.$$

Since $\{a_k\}$ is decreasing, all the terms in parentheses are nonnegative, so that $S_{2n+1} \ge 0$ for every n. Moreover,

$$S_{2n+3} = S_{2n+1} - a_{2n+2} + a_{2n+3} = S_{2n+1} - (a_{2n+2} - a_{2n+3}),$$

and since $a_{2n+2} - a_{2n+3} \ge 0$, we have

$$S_{2n+3} \le S_{2n+1}.$$

Hence, the sequence of odd-numbered partial sums is bounded below by 0 and is decreasing and is therefore convergent by Theorem 2 in Section 12.4. Thus S_{2n+1} converges to some limit L. Now let us consider the sequence of even-numbered partial sums. We find that $S_{2n+2} = S_{2n+1} - a_{2n+2}$ and since $a_{2n+2} \to 0$,

$$\lim_{n\to\infty} S_{2n+2} = \lim_{n\to\infty} S_{2n+1} - \lim_{n\to\infty} a_{2n+2} = L - 0 = L,$$

so that the even partial sums also converge to L. Since both the odd and even sums converge to L, we see that the partial sums converge to L, and the proof is complete.

EXAMPLE 6 FOUR CONVERGENT ALTERNATING SERIES

The following alternating series are convergent by the alternating series test:

a. $1 - \frac{1}{2} + \frac{1}{3} - \frac{1}{4} + \frac{1}{5} - \frac{1}{6} + \cdots$

b. $1 - \frac{1}{\sqrt{2}} + \frac{1}{\sqrt{3}} - \frac{1}{\sqrt{4}} + \frac{1}{\sqrt{5}} - \frac{1}{\sqrt{6}} + \frac{1}{\sqrt{7}} - \cdots$

c. $\frac{1}{\ln 2} - \frac{1}{\ln 3} + \frac{1}{\ln 4} - \frac{1}{\ln 5} + \frac{1}{\ln 6} - \cdots$

d. $1 - \frac{1}{2} + \frac{1}{2^2} - \frac{1}{2^3} + \frac{1}{2^4} - \frac{1}{2^5} + \frac{1}{2^6} - \frac{1}{2^7} + \cdots$

DEFINITION CONDITIONAL CONVERGENCE

An alternating series is said to be **conditionally convergent** if it is convergent but not absolutely convergent. ■

In Example 6 all the series are conditionally convergent except the last one, which is absolutely convergent.

It is not difficult to estimate the sum of a convergent alternating series. We again consider the series

$$S = 1 - \frac{1}{2} + \frac{1}{3} - \frac{1}{4} + \frac{1}{5} - \cdots .$$

Suppose we wish to approximate S by its nth partial sum S_n. Then,

$$S - S_n = \pm\left(\frac{1}{n+1} - \frac{1}{n+2} + \frac{1}{n+3} - \frac{1}{n+4} + \cdots\right) = R_n.$$

But we can estimate the remainder term R_n:

$$|R_n| = \left|\left[\frac{1}{n+1} - \left(\frac{1}{n+2} - \frac{1}{n+3}\right) - \left(\frac{1}{n+4} - \frac{1}{n+5}\right) - \cdots\right]\right| \leq \frac{1}{n+1}.$$

That is, the error is less than the first term that we left out! For example, $|S - S_{20}| \leq \frac{1}{21} \approx 0.0476$.

In general, we have the following result, whose proof is left as an exercise (see Problem 43).

THEOREM 3

If $S = \sum_{k=1}^{\infty} (-1)^{k+1}a_k$ is a convergent alternating series with monotone decreasing terms, then for any n,

$$|S - S_n| \leq |a_{n+1}|. \tag{1}$$

■

EXAMPLE 7

APPROXIMATING A CONVERGENT ALTERNATING SERIES

The series

$$\sum_{k=1}^{\infty} \frac{(-1)^{k+1}}{\ln(k+1)} = \frac{1}{\ln 2} - \frac{1}{\ln 3} + \frac{1}{\ln 4} - \frac{1}{\ln 5} + \cdots$$

can be approximated by S_n with an error of less than $1/\ln(n+2)$. For example, with $n = 10$, $1/\ln(n+2) = 1/\ln 12 \approx 0.4$. Hence, the sum

$$\sum_{k=1}^{\infty} \frac{(-1)^{k+1}}{\ln(k+1)} = \frac{1}{\ln 2} - \frac{1}{\ln 3} + \cdots$$

can be approximated by

$$S_{10} = \frac{1}{\ln 2} - \frac{1}{\ln 3} + \frac{1}{\ln 4} - \frac{1}{\ln 5} + \frac{1}{\ln 6} - \frac{1}{\ln 7} + \frac{1}{\ln 8} - \frac{1}{\ln 9} + \frac{1}{\ln 10} - \frac{1}{\ln 11} \approx 0.7197,$$

with an error of less than 0.4.

By modifying Theorem 3, we can significantly improve on the last result.

THEOREM 4

Suppose that the hypotheses of Theorem 3 hold and that, in addition, the sequence $\{|a_n - a_{n+1}|\}$ is monotone decreasing. Let $T_n = S_{n-1} - (-1)^n \frac{1}{2} a_n$. Then,

$$|S - T_n| \le \tfrac{1}{2}|a_n - a_{n+1}|. \tag{2}$$

■

This result follows from Theorem 3 and is also left as an exercise (see Problem 44).

EXAMPLE 8 IMPROVING THE APPROXIMATION OF A CONVERGENT ALTERNATING SERIES

We can improve the estimate in Example 7. We may approximate $\Sigma_{k=1}^{\infty} (-1)^{k+1}/\ln(k+1)$ by

$$\frac{1}{\ln 2} - \frac{1}{\ln 3} + \frac{1}{\ln 4} - \frac{1}{\ln 5} + \frac{1}{\ln 6} - \frac{1}{\ln 7} + \frac{1}{\ln 8} - \frac{1}{\ln 9} + \frac{1}{\ln 10} - \frac{1}{2 \ln 11} \approx 0.9282.$$

With $n = 10$ (so that $n + 1 = 11$),

$$T_{10} = S_9 - \frac{1}{2}\left(\frac{1}{\ln 11}\right),$$

which is precisely the sum given above. Thus,

$$|S - T_{10}| < \frac{1}{2}|a_{10} - a_{11}| = \frac{1}{2}\left(\frac{1}{\ln 11} - \frac{1}{\ln 12}\right) \approx 0.0073.$$

This result is a considerable improvement.

Note that in order to justify this result, we must verify that $|a_n - a_{n+1}|$ is monotone decreasing. This fact is left as an exercise (see Problem 50).

There is one fascinating fact about an alternating series that is conditionally convergent:

> By reordering the terms of a conditionally convergent alternating series, the new series of rearranged terms can be made to converge to any real number.

Let us illustrate this fact with the series

$$S = 1 - \frac{1}{2} + \frac{1}{3} - \frac{1}{4} + \frac{1}{5} - \frac{1}{6} + \cdots.$$

The odd-numbered terms sum to a divergent series:

$$1 + \frac{1}{3} + \frac{1}{5} + \frac{1}{7} + \cdots. \tag{3}$$

The even-numbered terms are likewise a divergent series:

$$-\frac{1}{2} - \frac{1}{4} - \frac{1}{6} - \cdots. \tag{4}$$

If either of these series converged, then the other one would too (by Theorem 1(i) in Section 12.6) and then the entire series would be absolutely convergent (which we know to be false). Now choose any real number, say 1.5. Then,

i. Choose enough terms from the series (3) so that the sum exceeds 1.5. We can do so since the series diverges.

$$1 + \frac{1}{3} + \frac{1}{5} = 1.53333. \ . \ . \ .$$

ii. Add enough negative terms from (4) so that the sum is now just under 1.5.

$$1 + \frac{1}{3} + \frac{1}{5} - \frac{1}{2} = 1.0333. \ . \ . \ .$$

iii. Add more terms from (3) until 1.5 is exceeded.

$$1 + \frac{1}{3} + \frac{1}{5} - \frac{1}{2} + \frac{1}{7} + \frac{1}{9} + \frac{1}{11} + \frac{1}{13} + \frac{1}{15} \approx 1.5218.$$

iv. Again subtract terms from (4) until the sum is under 1.5.

$$1 + \frac{1}{3} + \frac{1}{5} - \frac{1}{2} + \frac{1}{7} + \frac{1}{9} + \frac{1}{11} + \frac{1}{13} + \frac{1}{15} - \frac{1}{4} \approx 1.2718.$$

We continue the process to "converge" to 1.5. Since the terms in each series are decreasing to 0, the amount above or below 1.5 will approach 0 and the partial sums converge.

We will indicate in Section 12.12 that without rearranging, we have

$$\sum_{k=1}^{\infty} \frac{(-1)^{k+1}}{k} = 1 - \frac{1}{2} + \frac{1}{3} - \frac{1}{4} + \frac{1}{5} - \frac{1}{6} + \cdots = \ln 2 \approx 0.693147. \tag{5}$$

REMARK: *Any* rearrangement of the terms of an *absolutely converging* series converges to the same number.

EXAMPLE 9

AN ALTERNATING SERIES MAY DIVERGE IF $\{a_n\}$ IS NOT A MONOTONE DECREASING SEQUENCE

The following series was given by the great English mathematician G. H. Hardy (1877–

1947)[†] to illustrate the necessity of the "decreasing" condition in Theorem 2:

$$\frac{1}{\sqrt{2}+1}-\frac{1}{\sqrt{3}-1}+\frac{1}{\sqrt{4}+1}-\frac{1}{\sqrt{5}-1}+\cdots=\sum_{k=2}^{\infty}\frac{(-1)^k}{\sqrt{k}+(-1)^k}. \tag{6}$$

For $k \geq 2$, $\sqrt{k} \pm 1 > 0$ and $\frac{1}{\sqrt{k}\pm 1} \to 0$ as $k \to \infty$ so the series (6) is an alternating series with terms approaching zero. However, the terms are not decreasing. For example, since $\sqrt{5}-1<\sqrt{4}+1$, $a_5=\frac{1}{\sqrt{5}-1}>\frac{1}{\sqrt{4}+1}=a_4\left(\frac{1}{\sqrt{5}-1}\approx 0.81 > \frac{1}{\sqrt{4}+1}\approx 0.33\right)$. We can show, using a bit of algebra, that the series (6) diverges. We compute

$$\begin{aligned}\frac{(-1)^k}{\sqrt{k}}-\frac{1}{k+(-1)^k\sqrt{k}} &= \frac{(-1)^k[k+(-1)^k\sqrt{k}]-\sqrt{k}}{\sqrt{k}[k+(-1)^k\sqrt{k}]}\\ &= \frac{(-1)^k k+(-1)^{2k}\sqrt{k}-\sqrt{k}}{\sqrt{k}[k+(-1)^k\sqrt{k}]}\\ &\overset{(-1)^{2k}=1}{=} \frac{(-1)^k k}{\sqrt{k}[k+(-1)^k\sqrt{k}]} \overset{\text{divide numerator and denominator by } k}{=} \frac{(-1)^k}{\sqrt{k}+(-1)^k}.\end{aligned}$$

Thus

$$\sum_{k=2}^{\infty}\frac{(-1)^k}{\sqrt{k}+(-1)^k}=\sum_{k=2}^{\infty}\frac{(-1)^k}{\sqrt{k}}-\sum_{k=2}^{\infty}\frac{1}{k+(-1)^k\sqrt{k}}.$$

The first series on the right converges by the alternating series test. The second series diverges. To see this, note that $k-\sqrt{k}>0$ if $k \geq 2$, so the terms are positive. Moreover, $k \pm \sqrt{k} < 2k$, so $\frac{1}{k+(-1)^k}>\frac{1}{2k}$ and $\sum_{k=2}^{\infty}\frac{1}{2k}$ diverges. This implies that $\sum_{k=2}^{\infty}\frac{1}{k+(-1)^k\sqrt{k}}$ diverges by the comparison test. Thus the series (6) diverges.

EXAMPLE 10 THE COMPARISON TEST CANNOT BE USED FOR AN ALTERNATING SERIES

We again consider the series (6):

$$\sum_{k=2}^{\infty}\frac{(-1)^k}{\sqrt{k}+(-1)^k}.$$

[†] See G. H. Hardy, *Pure Mathematics* (Cambridge University Press, 1967), pages 377–378.

Since $\sqrt{k}+1 > \sqrt{k}-1$, we then have $\dfrac{1}{\sqrt{k}+1} < \dfrac{1}{\sqrt{k}-1}$. The series $\sum_{k=2}^{\infty} \dfrac{(-1)^k}{\sqrt{k}-1}$ converges by the alternating series test. Then, because $\dfrac{(-1)^k}{\sqrt{k}+(-1)^k} \le \dfrac{(-1)^k}{\sqrt{k}-1}$, the series (6) converges by the comparison test. But, as we saw in Example 9, this series diverges. We conclude that

> the comparison test cannot be used to determine convergence or divergence of an alternating series.

Of course, the comparison test *can* be used to determine absolute convergence of an alternating series, because $\Sigma|a_k|$ is a series of nonnegative terms.

NOTE: Examples 9 and 10 are slight modifications of examples found in the article "Counterexamples to a Comparison Test for Alternating Series" by J. Richard Morris in *The College Mathematics Journal* 17(2), March, 1986, pages 165–166.

We close this section by providing in Table 1 a summary of the convergence tests we have discussed.

TABLE 1 TESTS OF CONVERGENCE

Test	First Discussed on Page	Description	Examples and Comments
Convergence test for a geometric series	932	$\Sigma_{k=0}^{\infty} r^k$ converges to $1/(1-r)$ if $\lvert r\rvert < 1$ and diverges if $\lvert r\rvert > 1$	$\Sigma_{k=0}^{\infty} (\frac{1}{2})^k$ converges to 2; $\Sigma_{k=0}^{\infty} 2^k$ diverges
Look at the terms of the series—the limit test	939	If $\lvert a_k\rvert$ does not converge to 0, then Σa_n diverges	If $a_k \to 0$, then $\Sigma_0^{\infty} a_k$ may converge ($\Sigma_{k=0}^{\infty} 1/k^2$) or it may not (the harmonic series $\Sigma_{k=0}^{\infty} 1/k$)
Comparison test	947	If $0 \le a_k \le b_k$ and Σb_k converges, then Σa_k converges If $a_k \ge b_k \ge 0$ and Σb_k diverges, then Σa_k diverges	It is not necessary that $a_k \le b_k$ or $a_k \ge b_k$ for *all* k, only for $k \ge N$ for some integer N; convergence or divergence of a series is not affected by the values of the first few terms
Integral test	949	If $a_k = f(k) \ge 0$ and f is decreasing, then $\Sigma_{k=1}^{\infty} a_k$ converges if $\int_1^{\infty} f(x)\,dx$ converges and $\Sigma_{k=1}^{\infty} a_k$ diverges if $\int_1^{\infty} f(x)\,dx$ diverges	Use this test whenever $f(x)$ can easily be integrated
$\Sigma_{k=1}^{\infty} 1/k^p$	949	$\Sigma_{k=1}^{\infty} 1/k^p$ diverges if $0 \le p \le 1$ and converges if $p > 1$	

(continued)

TABLE 1 (CONTINUED)

Test	First Discussed on Page	Description	Examples and Comments
Limit comparison test	950	If $a_k > 0$, $b_k > 0$ and there is a number $c > 0$ such that $\lim_{k\to\infty} a_k/b_k = c$, then either both series converge or both series diverge	Use the limit comparison test when a series Σb_k can be found such that (a) it is known whether Σb_k converges or diverges and (b) it appears that a_k/b_k has an easily computed limit; (b) will be true, for instance, when $a_k = 1/p(k)$ and $b_k = 1/q(k)$ where $p(k)$ and $q(k)$ are polynomials.
Ratio test	953	If $a_k > 0$ and $\lim_{n\to\infty} a_{n+1}/a_n = L$, then $\Sigma_{k=1}^{\infty} a_k$ converges if $L < 1$ and diverges when $L > 1$	This is often the easiest test to apply; note that if $L = 1$, then the series may either converge ($\Sigma 1/k^2$) or diverge ($\Sigma 1/k$)
Root test	956	If $a_k > 0$ and $\lim_{n\to\infty}(a_n)^{1/n} = R$, then $\Sigma_{k=1}^{\infty} a_k$ converges if $R < 1$ and diverges if $R > 1$	If $R = 1$, the series may either converge ($\Sigma 1/k^2$) or diverge ($\Sigma 1/k$); the root test is the hardest test to apply; it is most useful when a_k is something raised to the kth power [$\Sigma 1/(\ln k)^k$, for example]
Alternating series test	960	$\Sigma(-1)^k a_k$ with $a_k \geq 0$ converges if (i) $a_k \to 0$ as $k \to \infty$ and (ii) $\{a_k\}$ is a decreasing sequence; also, $\Sigma(-1)^k a_k$ diverges if $\lim_{k\to\infty} a_k \neq 0$	This test can only be applied when the terms are alternately positive and negative; if there are two or more positive (or negative) terms in a row, then try another test
Absolute convergence test for a series with both positive and negative terms	959	Σa_k converges absolutely if $\Sigma \lvert a_k\rvert$ converges	To determine whether $\Sigma \lvert a_k\rvert$ converges, try any of the tests that apply to series with nonnegative terms

PROBLEMS 12.9

SELF-QUIZ

True–False

I. $1 + \frac{1}{2} - \frac{1}{3} - \frac{1}{4} + \frac{1}{5} + \frac{1}{6} - \frac{1}{7} + \frac{1}{8} - \frac{1}{9} + \frac{1}{10} - \frac{1}{11} + \cdots + \frac{(-1)^n}{n} + \cdots$ does not converge because it is not a strictly alternating series.

Multiple Choice

II. Suppose $\Sigma_{k=0}^{\infty} a_k$ is absolutely convergent. Which of the following must converge?

a. $\sum_{k=0}^{\infty} (-1)^k \cdot a_k$ **b.** $\sum_{k=0}^{\infty} \frac{a_k}{2^k}$

c. $\sum_{k=0}^{\infty} 2^k \cdot a_k$ **d.** $\sum_{k=0}^{\infty} e^{a_k}$

e. $\sum_{k=0}^{\infty} \frac{1}{a_k}$ **f.** $\sum_{k=0}^{\infty} a_k^2$

g. $\sum_{k=0}^{\infty} \lvert a_k\rvert$ **h.** $\sum_{k=0}^{\infty} \sqrt{\lvert a_k\rvert}$

III. Suppose that the alternating series $\Sigma_{k=0}^{\infty} b_k$ is conditionally convergent. Which of the following must diverge?

a. $\sum_{k=0}^{\infty} (-1)^k \cdot b_k$ **b.** $\sum_{k=0}^{\infty} \frac{b_k}{2^k}$ **e.** $\sum_{k=0}^{\infty} \frac{1}{b_k}$ **f.** $\sum_{k=0}^{\infty} b_k^{\,2}$

c. $\sum_{k=0}^{\infty} \frac{1-b_k}{1+|b_k|}$ **d.** $\sum_{k=0}^{\infty} e^{b_k}$ **g.** $\sum_{k=0}^{\infty} |b_k|$ **h.** $\sum_{k=0}^{\infty} \sqrt{|b_k|}$

In Problems 1–30, determine whether the given series is absolutely convergent, conditionally convergent, or divergent.

1. $\sum_{k=1}^{\infty} (-1)^k$ **2.** $\sum_{k=1}^{\infty} \frac{(-1)^{k+1}}{2k}$

3. $\sum_{k=0}^{\infty} \cos\frac{k\pi}{2}$ **4.** $\sum_{k=0}^{\infty} \sin\frac{k\pi}{2}$

5. $\sum_{k=1}^{\infty} \frac{(-1)^k\sqrt{k}}{k+3}$ **6.** $\sum_{k=1}^{\infty} \frac{(-1)^k}{k^{3/2}}$

7. $\sum_{k=2}^{\infty} \frac{(-1)^k}{k\ln k}$ **8.** $\sum_{k=2}^{\infty} \frac{(-1)^k k}{\ln k}$

9. $\sum_{k=2}^{\infty} \frac{(-1)^k}{k\sqrt{\ln k}}$ **10.** $\sum_{k=2}^{\infty} \frac{(-1)^k}{\sqrt[3]{\ln k}}$

11. $\sum_{k=1}^{\infty} \frac{(-1)^{k+1}}{5k-4}$ **12.** $\sum_{k=1}^{\infty} \frac{(-1)^k \ln k}{k}$

13. $\sum_{k=1}^{\infty} \frac{k!}{(-3)^k}$ **14.** $\sum_{k=1}^{\infty} \frac{k^2}{(-2)^k}$

15. $\sum_{k=1}^{\infty} \frac{(-3)^k}{k^2}$ **16.** $\sum_{k=1}^{\infty} \frac{(-2)^k}{k!}$

17. $\sum_{k=2}^{\infty} \frac{(-1)^k k^2}{k^3+1}$ **18.** $\sum_{k=2}^{\infty} \frac{(-1)^{k+1}}{\sqrt{k(k-1)}}$

19. $\sum_{k=3}^{\infty} \frac{\sin(k\pi/7)}{k^3}$ **20.** $\sum_{k=1}^{\infty} \frac{\cos(k\pi/6)}{k^2}$

21. $\sum_{k=1}^{\infty} \frac{(-1)^k(k+2)}{k(k+1)}$ **22.** $\sum_{k=2}^{\infty} \frac{(-1)^k k(k+1)}{(k+2)^3}$

23. $\sum_{k=2}^{\infty} \frac{(-1)^k k(k+1)}{(k+2)^4}$ **24.** $\sum_{k=1}^{\infty} \frac{(-1)^k k^k}{k!}$

25. $\sum_{k=1}^{\infty} \frac{(-1)^k 2^k}{k}$ **26.** $\sum_{k=1}^{\infty} \frac{(-1)^{k+1}}{k!}$

27. $\sum_{k=1}^{\infty} \frac{(-1)^k k^2}{4+k^2}$ **28.** $\sum_{k=1}^{\infty} (-1)^k\left(1+\frac{1}{k}\right)^k$

29. $\sum_{k=2}^{\infty} \frac{(-1)^k(k^2+3)}{k^3+4}$ **30.** $\sum_{k=2}^{\infty} \frac{(-1)^k k^3}{k^3+2k^2+k-1}$

In Problems 31–36, use the result of Theorem 3 or of Theorem 4 to approximate the given sum within the specified accuracy.

31. $\sum_{k=1}^{\infty} \frac{(-1)^{k+1}}{k!}$; $|\text{error}| < 0.001$

32. $\sum_{k=1}^{\infty} \frac{(-1)^{k+1}}{k^2}$; $|\text{error}| < 0.01$

33. $\sum_{k=1}^{\infty} \frac{(-1)^{k+1}}{k^4}$; $|\text{error}| < 0.0001$

34. $\sum_{k=2}^{\infty} \frac{(-1)^{k+1}}{k\ln k}$; $|\text{error}| < 0.05$

35. $\sum_{k=1}^{\infty} \frac{(-1)^{k+1}}{k^k}$; $|\text{error}| < 0.0001$

36. $\sum_{k=1}^{\infty} \frac{(-1)^{k+1}}{\sqrt{k}}$; $|\text{error}| < 0.1$

37. Find the first 10 terms of a rearrangement of the series $\sum_{k=1}^{\infty} (-1)^{k+1}/k$ that converges to 0. [As on page 963, start with 1, then add just enough negative terms in the order $-\frac{1}{2}, -\frac{1}{4}, \ldots$ to obtain a negative sum. Then add positive terms, starting with $\frac{1}{3}$, and so on.]

38. Find the first 10 terms of a rearrangement of the series $\sum_{k=1}^{\infty} (-1)^{k+1}/k$ that converges to 0.3. [Use the same method as in Problem 37.]

39. Give an example of a sequence $\{a_n\}$ such that $\sum_{k=1}^{\infty} a_k^{\,2}$ converges but $\sum_{k=1}^{\infty} a_k$ diverges.

***40.** Give an example of a sequence $\{a_n\}$ such that $\sum_{k=1}^{\infty} a_k$ converges but $\sum_{k=1}^{\infty} a_k^{\,3}$ diverges.

41. Show that if $\sum_{k=1}^{\infty} a_k$ is absolutely convergent, then $\sum_{k=1}^{\infty} a_k^{\,p}$ is convergent for any integer $p \geq 1$.

42. Prove that if $\sum_{k=1}^{\infty} a_k$ is a convergent series of nonzero terms, then $\sum_{k=1}^{\infty} 1/|a_k|$ diverges.

***43.** Prove Theorem 3. [*Hint:* Assume that the odd-numbered terms are positive. Show that the sequence $\{S_{2n}\}$ is increasing and that $S_{2n} < S_{2n+2} < S$ for all $n \geq 1$. Then show that the sequence of odd-numbered partial sums is decreasing and that $S < S_{2n+1} < S_{2n-1}$ for all

$n \geq 1$. Conclude that for all $n \geq 1$,
a. $0 < S - S_{2n} < a_{2n+1}$ and
b. $0 < S_{2n-1} - S < -a_{2n}$.
Use inequalities (a) and (b) to prove the theorem.]

***44.** Prove Theorem 4. [*Hint:* Think about writing the original series as

$$S = \frac{1}{2}a_1 + \frac{1}{2}(a_1 - a_2) - \frac{1}{2}(a_2 - a_3) + \frac{1}{2}(a_3 - a_4) - \cdots$$

$$= \frac{1}{2}a_1 + \sum_{k=1}^{\infty} (-1)^{k+1}\frac{a_k - a_{k+1}}{2}.$$

Then apply Theorem 3.]

45. Explain why there is no rearrangement of the series $\Sigma_{k=1}^{\infty} (-1)^k/k^2$ that converges to -1.

***46.** Consider the following rearrangement of the alternating harmonic series:

$$S^* = 1 - \frac{1}{2} - \frac{1}{4} + \frac{1}{3} - \frac{1}{6} - \frac{1}{8} + \frac{1}{5} - \frac{1}{10} - \frac{1}{12} + \cdots. \quad (7)$$

By inserting parentheses in (6), show that $S^* = \frac{1}{2}\ln 2$.† $\left[\textit{Hint: } \sum_{k=1}^{\infty} \frac{(-1)^{n+1}}{n} = \ln 2.\right]$

***47.** Show that

$$1 + \frac{1}{3} - \frac{1}{2} + \frac{1}{5} + \frac{1}{7} - \frac{1}{4} + \frac{1}{9} + \frac{1}{11} - \frac{1}{6} + \cdots = \frac{3}{2}\ln 2.$$

****48.** Show that

$$1 - \frac{1}{2} - \frac{1}{4} - \frac{1}{6} + \frac{1}{3} - \frac{1}{8} - \frac{1}{10} - \frac{1}{12} + \frac{1}{5} - \cdots = \ln 2 - \frac{1}{2}\ln 3.$$

***49.** Suppose $\Sigma_{k=0}^{\infty} a_k$ is conditionally convergent. Prove or disprove: $\Sigma_{k=0}^{\infty} 2^k a_k$ must diverge.

50. For this problem, you are asked to fill in some details missing from Example 8.
a. Let

$$f(x) = \frac{1}{\ln(x+1)} - \frac{1}{\ln(x+2)}.$$

Show that $f'(x) < 0$ for all $x \geq 0$.

b. Let $a_n = 1/\ln(n+1)$. Use part (a) to show that $|a_n - a_{n+1}|$ is monotone decreasing.

51. Show that $\sum_{k=1}^{\infty} \frac{(-1)^k}{k[2 + (-1)^k]}$ diverges. This provides another example of a divergent alternating series with nondecreasing terms that approach zero.

52. Show that $\frac{(-1)^k}{k[2 + (-1)^k]} \leq \frac{(-1)^k}{k}$. Since $\sum_{k=1}^{\infty} \frac{(-1)^k}{k}$ converges, this provides another example of the failure of the comparison test when applied to alternating series.

ANSWERS TO SELF-QUIZ

I. False: The series does converge, even though it is not alternating as its start.
II. a, b, f, and g converge; d and e must diverge.
III. a, c, d, e, g, and h diverge; b must converge [c diverges because the nth term $\to 1$].

†The sums in Problems 46–48 are discussed in the paper "Rearranging the Alternating Harmonic Series" by C. C. Cowen, K. R. Davidson, and R. P. Kaufman, *American Mathematical Monthly, 87* (December 1980): 817–819.

12.10 POWER SERIES

In previous sections, we discussed infinite series of real numbers. Here we discuss series of functions.

DEFINITION POWER SERIES

i. A **power series in *x*** is a series of the form

$$\sum_{k=0}^{\infty} a_k x^k = a_0 + a_1 x + a_2 x^2 + \cdots + a_k x^k + \cdots \tag{1}$$

ii. A **power series in $(x - x_0)$** is a series of the form

$$\sum_{k=0}^{\infty} a_k(x - x_0)^k = a_0 + a_1(x - x_0) + a_2(x - x_0)^2 + \cdots + a_k(x - x_0)^k + \cdots \tag{2}$$

where x_0 is a real number. ■

A power series in $(x - x_0)$ can be converted to a power series in u by the change of variables $u = x - x_0$. Then $\Sigma_{k=0}^{\infty} a_k(x - x_0)^k = \Sigma_{k=0}^{\infty} a_k u^k$. For example,

$$\sum_{k=0}^{\infty} \frac{(x-3)^k}{k!} = \sum_{k=0}^{\infty} \frac{u^k}{k!} \qquad \text{if } u = x - 3.$$

DEFINITION CONVERGENCE AND DIVERGENCE OF A POWER SERIES

i. A power series is said to **converge** at x_0 if the series of real numbers $\Sigma_{k=0}^{\infty} a_k x_0^k$ converges. Otherwise, it is said to **diverge** at x_0.

ii. A power series is said to converge in a set D of real numbers if it converges for every real number x in D. ■

EXAMPLE 1 DETERMINING THE SET OF VALUES AT WHICH A POWER SERIES CONVERGES

For what real numbers does the power series

$$\sum_{k=0}^{\infty} \frac{x^k}{3^k} = 1 + \frac{x}{3} + \frac{x^2}{3^2} + \frac{x^3}{3^3} + \cdots$$

converge?

SOLUTION: The nth term in this series is $x^n/3^n$. Using the ratio test, we find that

$$\lim_{n\to\infty} \frac{|a_{n+1}|}{|a_n|} = \lim_{n\to\infty} \frac{|x^{n+1}/3^{n+1}|}{|x^n/3^n|} = \lim_{n\to\infty} \left|\frac{x}{3}\right| = \left|\frac{x}{3}\right|.$$

We put in the absolute value bars since the ratio test only applies to a series of *positive* terms. However, as this example shows, we can use the ratio test to test for the absolute convergence of any series of nonzero terms by inserting absolute value bars, thereby making all the terms positive. (See Remark 2 on page 954.)

Thus, the power series converges absolutely if $|x/3| < 1$ or $|x| < 3$ and diverges if $|x| > 3$. The case $|x| = 3$ has to be treated separately. For $x = 3$,

$$\sum_{k=0}^{\infty} \frac{x^k}{3^k} = \sum_{k=0}^{\infty} \frac{3^k}{3^k} = \sum_{k=0}^{\infty} 1^k,$$

which diverges. For $x = -3$,

$$\sum_{k=0}^{\infty} \frac{x^k}{3^k} = \sum_{k=0}^{\infty} (-1)^k,$$

which also diverges. Thus, the series converges in the open interval $(-3, 3)$. We will show in Theorem 1 that since the series diverges for $|x| = 3$, it diverges for $|x| > 3$, so that conditional convergence at any x for which $|x| > 3$ is ruled out.

EXAMPLE 2 DETERMINING THE SET OF VALUES AT WHICH A POWER SERIES CONVERGES

For what values of x does the series $\sum_{k=0}^{\infty} x^k/(k+1)$ converge?

SOLUTION: Here $a_n = x^n/(n+1)$ so that

$$\lim_{n\to\infty} \frac{n+1}{n+2} = 1$$

$$\lim_{n\to\infty} \frac{|a_{n+1}|}{|a_n|} = \lim_{n\to\infty} \left| \frac{x^{n+1}/(n+2)}{x^n/(n+1)} \right| = |x| \lim_{n\to\infty} \frac{n+1}{n+2} = |x|.$$

Thus the series converges absolutely for $|x| < 1$ and diverges for $|x| > 1$. If $x = 1$, then

$$\sum_{k=0}^{\infty} \frac{x^k}{k+1} = \sum_{k=0}^{\infty} \frac{1}{k+1},$$

which diverges since this series is the harmonic series. If $x = -1$, then

$$\sum_{k=0}^{\infty} \frac{x^k}{k+1} = \sum_{k=0}^{\infty} \frac{(-1)^k}{k+1} = 1 - \frac{1}{2} + \frac{1}{3} - \frac{1}{4} + \cdots,$$

which converges conditionally by the alternating series test (see Example 6(a) in Section 12.9). Hence, the power series $\sum_{k=0}^{\infty} x^k/(k+1)$ converges in the half-open interval $[-1, 1)$.

The following theorem is of great importance in determining the range of values over which a power series converges.

THEOREM 1

i. If $\sum_{k=0}^{\infty} a_k x^k$ converges at x_0, $x_0 \neq 0$, then it converges absolutely at all x such that $|x| < |x_0|$.

ii. If $\sum_{k=0}^{\infty} a_k x^k$ diverges at x_0, then it diverges at all x such that $|x| > |x_0|$.

PROOF:

i. Since $\Sigma_{k=0}^{\infty} a_k x_0^k$ converges, $a_k x_0^k \to 0$ as $k \to \infty$ by Theorem 2 in Section 12.6. This implies that for all k sufficiently large, $|a_k x_0^k| < 1$. Then if $|x| < |x_0|$ and if k is sufficiently large,

$$|a_k x^k| = \left| a_k \frac{x_0^k x^k}{x_0^k} \right| = |a_k x_0^k| \left| \frac{x}{x_0} \right|^k < \left| \frac{x}{x_0} \right|^k.$$

Since $|x| < |x_0|$, $|x/x_0| < 1$, and the geometric series $\Sigma_{k=0}^{\infty} |x/x_0|^k$ converges. Thus, $\Sigma_{k=0}^{\infty} |a_k x^k|$ converges by the comparison test.

ii. Suppose $|x| > |x_0|$ and $\Sigma_{k=0}^{\infty} a_k x_0^k$ diverges. If $\Sigma_{k=0}^{\infty} a_k x^k$ did converge, then by part (i), $\Sigma_{k=0}^{\infty} a_k x_0^k$ would also converge. This contradiction completes the proof of the theorem. ■

Theorem 1 is very useful, for it enables us to place all power series in one of three categories:

DEFINITION RADIUS OF CONVERGENCE

Category 1: $\Sigma_{k=0}^{\infty} a_k x^k$ converges only at 0.

Category 2: $\Sigma_{k=0}^{\infty} a_k x^k$ converges for all real numbers

Category 3: There exists a positive real number R, called the **radius of convergence** of the power series, such that $\Sigma_{k=0}^{\infty} a_k x^k$ converges if $|x| < R$ and diverges if $|x| > R$. At $x = R$ and at $x = -R$, the series may converge or diverge. ■

We can extend the notion of radius of convergence to Categories 1 and 2:

1. In Category 1, we say that the radius of convergence is 0.
2. In Category 2, we say that the radius of convergence is ∞.

NOTE: The series in Examples 1 and 2 both fall into Category 3. In Example 1, $R = 3$; and in Example 2, $R = 1$.

EXAMPLE 3 A POWER SERIES THAT CONVERGES ONLY AT 0

For what values of x does the series $\Sigma_{k=0}^{\infty} k!x^k$ converge?

SOLUTION: Here,

$$\lim_{n\to\infty} \left| \frac{a_{n+1}}{a_n} \right| = \lim_{n\to\infty} \left| \frac{(n+1)!x^{n+1}}{n!x^n} \right| = |x| \lim_{n\to\infty} (n+1) = \infty,$$

so that if $x \neq 0$, the series diverges. Thus, $R = 0$ and the series falls into Category 1.

EXAMPLE 4 A POWER SERIES THAT CONVERGES EVERYWHERE

For what values of x does the series $\Sigma_{k=0}^{\infty} x^k/k!$ converge?

SOLUTION: Here,

$$\lim_{n\to\infty}\left|\frac{a_{n+1}}{a_n}\right| = \lim_{n\to\infty}\left|\frac{x^{n+1}/(n+1)!}{x^n/n!}\right| = |x| \lim_{n\to\infty}\frac{1}{n+1} = 0,$$

so that the series converges for every real number x. Here, $R = \infty$ and the series falls into Category 2.

In going through these examples, we find that there is an easy way to calculate the radius of convergence. The proof of the following theorem is left as an exercise (see Problems 37 and 38).

THEOREM 2 CALCULATING THE RADIUS OF CONVERGENCE

Consider the power series $\Sigma_{k=0}^{\infty} a_k x^k$ and then suppose that $\lim_{n\to\infty}|a_{n+1}/a_n|$ exists and is equal to L or that $\lim_{n\to\infty}|a_n|^{1/n}$ exists and is equal to L.

i. If $L = \infty$, then $R = 0$ and the series falls into Category 1.
ii. If $L = 0$, then $R = \infty$ and the series falls into Category 2.
iii. If $0 < L < \infty$, then $R = 1/L$ and the series falls into Category 3. ■

DEFINITION INTERVAL OF CONVERGENCE

The **interval of convergence** of a power series is the interval over which the power series converges. ■

Using Theorem 2, we can calculate the interval of convergence of a power series in one or two steps:

i. Calculate R. If $R = 0$, the series converges only at 0; and if $R = \infty$, the interval of convergence is $(-\infty, \infty)$.
ii. If $0 < R < \infty$, check the values $x = -R$ and $x = R$. Then the interval of convergence is $(-R, R)$, $[-R, R)$, $(-R, R]$, or $[-R, R]$, depending on the convergence or divergence of the series at $x = R$ and $x = -R$.

NOTE: In Example 1 the interval of convergence is $(-3, 3)$ and in Example 2 the interval of convergence is $[-1, 1)$.

EXAMPLE 5 DETERMINING THE RADIUS AND INTERVAL OF CONVERGENCE

Find the radius of convergence and interval of convergence of the power series $\Sigma_{k=0}^{\infty} 2^k(x-4)^k/\ln(k+2)$.

SOLUTION: Let $u = x - 4$. Here $a_n = 2^n/\ln(n+2)$ and

$$L = \lim_{n\to\infty}\left|\frac{a_{n+1}}{a_n}\right| = \lim_{n\to\infty}\left|\frac{2^{n+1}/\ln(n+3)}{2^n/\ln(n+2)}\right| = 2\lim_{n\to\infty}\frac{\ln(n+2)}{\ln(n+3)} = 2.$$

Thus $R = 1/L = \frac{1}{2}$. For $u = \frac{1}{2}$,

$$\sum_{k=0}^{\infty} \frac{2^k u^k}{\ln(k+2)} = \sum_{k=0}^{\infty} \frac{1}{\ln(k+2)},$$

which diverges by comparison with the harmonic series since $1/\ln(k+2) > 1/(k+2)$ if $k \geq 1$. If $u = -\frac{1}{2}$, then

$$\sum_{k=0}^{\infty} \frac{2^k u^k}{\ln(k+2)} = \sum_{k=0}^{\infty} \frac{(-1)^k}{\ln(k+2)},$$

which converges by the alternating series test. Thus, the interval of convergence for $\Sigma_{k=0}^{\infty} 2^k u^k/\ln(k+2)$ is $[-\frac{1}{2}, \frac{1}{2})$. But $x = u + 4$ so the interval of convergence for the original series is $[-\frac{1}{2} + 4, \frac{1}{2} + 4) = [\frac{7}{2}, \frac{9}{2})$.

EXAMPLE 6

DETERMINING THE RADIUS AND INTERVAL OF CONVERGENCE

Find the radius of convergence and interval of convergence of the power series $\Sigma_{k=0}^{\infty} x^{2k} = 1 + x^2 + x^4 + \cdots$.

SOLUTION:

$$\sum_{k=0}^{\infty} x^{2k} = 1 + 0 \cdot x + 1 \cdot x^2 + 0 \cdot x^3 + 1 \cdot x^4 + 0 \cdot x^5 + 1 \cdot x^6 + \cdots.$$

This example illustrates the pitfalls of blindly applying formulas. We have $a_0 = 1, a_1 = 0, a_2 = 1, a_3 = 0, \ldots$. Thus, the ratio a_{n+1}/a_n is 0 if n is even and is undefined if n is odd. The simplest thing to do here is to apply the ratio test directly. The ratio of consecutive terms is $x^{2k+2}/x^{2k} = x^2$. Thus, the series converges if $|x| < 1$, diverges if $|x| > 1$, and the radius of convergence is 1. If $x = \pm 1$, then $x^2 = 1$ and the series diverges. Finally, the interval of convergence is $(-1, 1)$.

PROBLEMS 12.10

SELF-QUIZ

Multiple Choice

I. The interval of convergence for $\Sigma_{k=0}^{\infty} (x/2)^k$ is ________.

a. $(-3, 1)$ **b.** $[-4, 8]$
c. $(-2, 2)$ **d.** $[-2, 2]$
e. $[-2, 2)$ **f.** $(-2, 2]$

II. The interval of convergence for $\Sigma_{k=0}^{\infty} (x-3)^k$ is ________.

a. $(2, 4]$ **b.** $[1, 5)$
c. $(2, 4)$ **d.** $(0, 6)$

III. The interval of convergence for $\displaystyle\sum_{k=0}^{\infty} \frac{(x+1)^k}{k \cdot 3^k}$ is ________.

a. $[-4, 2)$ **b.** $[-2, 4)$
c. $(-4/3, -2/3]$ **d.** $(-4, 2]$

IV. If $\Sigma_{k=0}^{\infty} a_k$ converges, then the interval of convergence for $\Sigma_{k=0}^{\infty} a_k x^k$ must include ________.

a. $[0, 1]$ **b.** -1 **c.** $[-1, 1]$ **d.** $(-1, 1]$

V. Suppose $\sum_{k=0}^{\infty} a_k x^k$ converges for $x = 3$. At which of the following values of x must it also converge?
a. 2 **b.** -2
c. -3 **d.** 5

VI. Suppose $\sum_{k=0}^{\infty} b_k(x-5)^k$ converges for $x = -3$. At which of the following values of x must it also converge?
a. 3 **b.** 12
c. -11 **d.** 9

In Problems 1–34, find the radius of convergence and the interval of convergence for the given power series.

1. $\sum_{k=0}^{\infty} \frac{x^k}{6^k}$ **2.** $\sum_{k=0}^{\infty} \frac{(-1)^k x^k}{8^k}$

3. $\sum_{k=0}^{\infty} \frac{(x+1)^k}{3^k}$ **4.** $\sum_{k=0}^{\infty} \frac{(-1)^k (x-3)^k}{4^k}$

5. $\sum_{k=0}^{\infty} (3x)^k$ **6.** $\sum_{k=0}^{\infty} \frac{x^k}{k^2+1}$

7. $\sum_{k=0}^{\infty} \frac{(x-1)^k}{k^3+3}$ **8.** $\sum_{k=2}^{\infty} \frac{x^k}{(\ln k)^2}$

9. $\sum_{k=0}^{\infty} \frac{(x+17)^k}{k!}$ **10.** $\sum_{k=2}^{\infty} \frac{x^k}{k \ln k}$

11. $\sum_{k=0}^{\infty} x^{2k}$ **12.** $\sum_{k=1}^{\infty} \frac{x^{2k}}{k}$

13. $\sum_{k=1}^{\infty} \frac{(-1)^k x^{2k}}{k^k}$ **14.** $\sum_{k=1}^{\infty} \frac{k x^k}{\ln(k+1)}$

15. $\sum_{k=0}^{\infty} \frac{(-1)^k k x^k}{\sqrt{k+1}}$ **16.** $\sum_{k=1}^{\infty} \frac{x^k}{k^k}$

***17.** $\sum_{k=2}^{\infty} \frac{x^k}{(\ln k)^k}$ [*Hint:* Use the root test.]

***18.** $\sum_{k=1}^{\infty} \frac{3^k x^k}{k^5}$ **19.** $\sum_{k=1}^{\infty} \frac{(-2x)^k}{k^4}$

20. $\sum_{k=0}^{\infty} \frac{(2x+3)^k}{k!}$ **21.** $\sum_{k=0}^{\infty} \frac{(2x+3)^k}{5^k}$

22. $\sum_{k=0}^{\infty} \frac{(3x-5)^k}{3^{2k}}$ **23.** $\sum_{k=0}^{\infty} \left(\frac{k}{15}\right)^k x^k$

24. $\sum_{k=0}^{\infty} (-1)^k x^{2k}$ **25.** $\sum_{k=0}^{\infty} (-1)^k x^{2k+1}$

26. $\sum_{k=1}^{\infty} \frac{(\ln k)(x+3)^k}{k+1}$ ***27.** $\sum_{k=1}^{\infty} k^k (x+1)^k$

***28.** $\sum_{k=1}^{\infty} \frac{k^k}{k!} x^k$

[*Hint:* See Example 12.8.3, and use Stirling's formula: $\lim_{n\to\infty} \frac{n!}{\sqrt{2\pi n}(n/e)^n} = 1$.]

29. $\sum_{k=0}^{\infty} \frac{(x+10)^k}{(k+1)3^k}$ ***30.** $\sum_{k=1}^{\infty} \frac{k!}{k^k} x^k$

31. $\sum_{k=0}^{\infty} [1+(-1)^k]\, x^k$ **32.** $\sum_{k=0}^{\infty} \frac{[1+(-1)^k]}{k!} x^k$

33. $\sum_{k=1}^{\infty} \frac{[1+(-1)^k]}{k} x^k$ **34.** $\sum_{k=0}^{\infty} \frac{(-1)^k (x-3)^k}{(k+1)^2}$

35. Show that the interval of convergence of the power series $\sum_{k=0}^{\infty} (ax-b)^k/c^k$ with $a > 0$ and $c > 0$ is $((b-c)/a, (b+c)/a)$.

36. Prove that if the interval of convergence of a power series is $(a, b]$, then the power series is conditionally convergent at b.

37. Prove the ratio limit part of Theorem 2. [*Hint:* Show that if $|x| < 1/L$, then the series converges absolutely by applying the ratio test. Then show that if $|x| > 1/L$, the series diverges.]

38. Show that if $\lim_{n\to\infty} |a_n|^{1/n} = L$, then the radius of convergence of $\sum_{k=0}^{\infty} a_k x^k$ is $1/L$.

39. Show that if the radius of convergence of $\sum_{k=0}^{\infty} a_k x^k$ is R, and if $m > 0$ is a positive integer, then the radius of convergence of the power series $\sum_{k=0}^{\infty} a_k x^{mk}$ is $R^{1/m}$.

ANSWERS TO SELF-QUIZ

I. c **II.** c **III.** a **IV.** a, d (note that $[0, 1]$ is contained in $(-1, 1]$)
V. a, b **VI.** a, b, d

12.11 DIFFERENTIATION AND INTEGRATION OF POWER SERIES

Consider the power series

$$\sum_{k=0}^{\infty} a_k(x-x_0)^k = a_0 + a_1(x-x_0) + a_2(x-x_0)^2 + \cdots \tag{1}$$

with a given interval of convergence I. For each x in I we may define a function f by

$$f(x) = \sum_{k=0}^{\infty} a_k(x-x_0)^k. \tag{2}$$

As we will see in this section and in Section 12.12, many familiar functions can be written as power series. In this section we will discuss some properties of a function given in the form of equation (2).

EXAMPLE 1 WRITING $\frac{1}{1-x}$ AS A POWER SERIES (A GEOMETRIC SERIES)

We know that

$$\sum_{k=0}^{\infty} x^k = 1 + x + x^2 + \cdots = \frac{1}{1-x} \qquad \text{if } |x| < 1. \tag{3}$$

Thus the function $1/(1-x)$ for $|x| < 1$ can be defined by the geometric series

$$f(x) = \frac{1}{1-x} = \sum_{k=0}^{\infty} x^k \qquad \text{if } |x| < 1.$$

EXAMPLE 2 WRITING $\frac{1}{1-x^4}$ AS A POWER SERIES

Substituting x^4 for x in (3) leads to the equality

$$g(x) = \frac{1}{1-x^4} = \sum_{k=0}^{\infty} x^{4k} = 1 + x^4 + x^8 + x^{12} + \cdots \qquad \text{if } |x| < 1.$$

EXAMPLE 3 WRITING $\frac{1}{1+x}$ AS A POWER SERIES

Substituting $-x$ for x in (3) leads to the equality

$$h(x) = \frac{1}{1-(-x)} = \frac{1}{1+x} = \sum_{k=0}^{\infty} (-1)^k x^k$$
$$= 1 - x + x^2 - x^3 + x^4 - \cdots \qquad \text{if } |x| < 1.$$

Once we see that certain functions can be written as power series, the next question that naturally arises is whether such functions can be differentiated and integrated. The remarkable theorem given next ensures that every function represented as a power series can be differentiated and integrated at any x such that $|x| < R$, the radius of convergence. Moreover, we see how the derivative and integral can be calculated. The proof of this theorem is long (but not conceptually difficult) and so is omitted.[†]

THEOREM 1 DIFFERENTIATING AND INTEGRATING A POWER SERIES

Let the power series $\sum_{k=0}^{\infty} a_k x^k$ have the radius of convergence $R > 0$. Let

$$f(x) = \sum_{k=0}^{\infty} a_k x^k = a_0 + a_1 x + a_2 x^2 + \cdots \qquad \text{for } |x| < R.$$

Then for $|x| < R$ we have the following:

i. $f(x)$ is continuous.

ii. The derivative $f'(x)$ exists, and

$$f'(x) = \frac{d}{dx} a_0 + \frac{d}{dx} a_1 x + \frac{d}{dx} a_2 x^2 + \cdots$$

$$= a_1 + 2a_2 x + 3a_3 x^2 + \cdots = \sum_{k=1}^{\infty} k a_k x^{k-1}.$$

iii. The indefinite integral $\int f(x)\, dx$ exists and

$$\int f(x)\, dx = \int a_0\, dx + \int a_1 x\, dx + \int a_2 x^2\, dx + \cdots$$

$$= a_0 x + a_1 \frac{x^2}{2} + a_2 \frac{x^3}{3} + \cdots + C$$

$$= \sum_{k=0}^{\infty} a_k \frac{x^{k+1}}{k+1} + C.$$

Moreover, the two series $\sum_{k=1}^{\infty} k a_k x^{k-1}$ and $\sum_{k=0}^{\infty} a_k x^{k+1}/(k+1)$ both have the radius of convergence R. ■

Simply put, this theorem says that the derivative of a converging power series is the series of derivatives of its terms and that the integral of a converging power series is the series of integrals of its terms.

A more concise statement of Theorem 1 is:

> A power series may be differentiated and integrated term-by-term within its radius of convergence.

EXAMPLE 4 FINDING A SERIES FOR $\ln u$ BY INTEGRATING A KNOWN POWER SERIES

From Example 3, we have

$$\frac{1}{1+x} = 1 - x + x^2 - x^3 + \cdots = \sum_{k=0}^{\infty} (-1)^k x^k \qquad \textbf{(4)}$$

[†] See R. C. Buck and E. F. Buck, *Advanced Calculus*, 3rd ed. (New York: McGraw-Hill, 1978), p. 278.

for $|x| < 1$. Substituting $u = x + 1$, we have $x = u - 1$ and $1/(1 + x) = 1/u$. If $-1 < x < 1$, then $0 < u < 2$, and we obtain

$$\frac{1}{u} = 1 - (u-1) + (u-1)^2 - (u-1)^3 + \cdots = \sum_{k=0}^{\infty} (-1)^k (u-1)^k$$

for $0 < u < 2$. Integration then yields

$$\ln u = \int \frac{du}{u} = u - \frac{(u-1)^2}{2} + \frac{(u-1)^3}{3} - \cdots + C.$$

Since $\ln 1 = 0$, we immediately find that $C = -1$, so that

$$\ln u = \sum_{k=0}^{\infty} (-1)^k \frac{(u-1)^{k+1}}{k+1} \tag{5}$$

for $0 < u < 2$. Here we have expressed the logarithmic function defined on the interval $(0, 2)$ as a power series.

EXAMPLE 5

WRITING e^x AS A POWER SERIES

The series

$$f(x) = 1 + x + \frac{x^2}{2!} + \frac{x^3}{3!} + \cdots = \sum_{k=0}^{\infty} \frac{x^k}{k!} \tag{6}$$

converges for every real number x (i.e., $R = \infty$; see Example 4 in Section 12.10). But

$$f'(x) = \frac{d}{dx}1 + \frac{d}{dx}x + \frac{d}{dx}\frac{x^2}{2!} + \cdots = 1 + x + \frac{x^2}{2!} + \cdots = f(x).$$

Thus f satisfies the differential equation

$$f' = f,$$

and so from the discussion in Section 10.2, we find that

$$f(x) = ce^x \tag{7}$$

for some constant c. Then substituting $x = 0$ into equations (6) and (7) yields

$$f(0) = 1 = ce^0 = c,$$

so that $f(x) = e^x$. We have obtained the important expansion that is valid for any real number x:

$$e^x = 1 + x + \frac{x^2}{2!} + \frac{x^3}{3!} + \cdots = \sum_{k=0}^{\infty} \frac{x^k}{k!}. \tag{8}$$

For example, if we substitute the value $x = 1$ into (8), we obtain partial sum approximations for $e = 1 + 1 + 1/2! + 1/3! + \cdots$ (see Table 1). The last value $(\Sigma_{k=0}^{8} 1/k!)$ is correct to five decimal places.

TABLE 1

n	0	1	2	3	4	5	6	7	8
$S_n = \sum_{k=0}^{n} \frac{1}{k!}$	1	2	2.5	2.66667	2.70833	2.71667	2.71806	2.71825	2.71828

EXAMPLE 6 OBTAINING THE SERIES FOR e^{-x}

Substituting $x = -x$ in (8), we obtain

$$e^{-x} = 1 - x + \frac{x^2}{2!} - \frac{x^3}{3!} + \cdots = \sum_{k=0}^{\infty} (-1)^k \frac{x^k}{k!}. \qquad (9)$$

Since this is an alternating series if $x > 0$, Theorem 3 in Section 12.9 tells us that the error $|S - S_n|$ in approximating e^{-x} for $x > 0$ is bounded by $|a_{n+1}| = x^{n+1}/(n+1)!$.† For example, to calculate e^{-1} with an error of less than 0.0001, we must have $|S - S_n| \le 1/(n+1)! < 0.0001 = 1/10{,}000$, or $(n+1)! > 10{,}000$. If $n = 7$, $(n+1)! = 8! = 40{,}320$, so that $\Sigma_{k=0}^{7} (-1)^k/k!$ will approximate e^{-1} correct to four decimal places. We obtain

$$e^{-1} \approx 1 - 1 + \frac{1}{2!} - \frac{1}{3!} + \frac{1}{4!} - \frac{1}{5!} + \frac{1}{6!} - \frac{1}{7!} \approx 0.36786.$$

Note that $e^{-1} \approx 0.36788$, correct to five decimal places.

EXAMPLE 7 THE UNIT NORMAL DISTRIBUTION

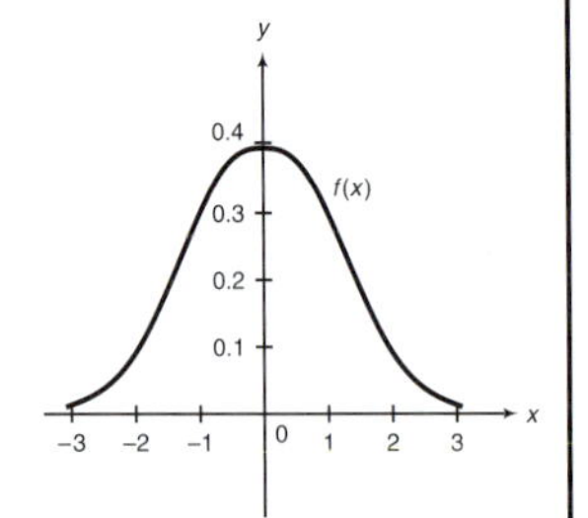

FIGURE 1
Graph of the function $f(x) = \frac{1}{\sqrt{2\pi}} e^{-t^2/2}$

The function $f(x) = \frac{1}{\sqrt{2\pi}} e^{-x^2/2}$ is very important in probability theory. Its graph is the famous "bell-shaped" curve sketched in Figure 1.

The probabilistic interpretation of this function is as follows: If a random variable is normally distributed with mean 0 and standard deviation 1, and if $P_a^{\,b}$ denotes the probability that the random variable takes a value between a and b, then

$$P_a^{\,b} = \frac{1}{\sqrt{2\pi}} \int_a^b e^{-t^2/2}\, dt.$$

In this context $f(x) = \frac{1}{\sqrt{2\pi}} e^{-x^2/2}$ is called the **density function of the unit normal distribution**. In particular,

$$P_0^{\,x} = \begin{matrix}\text{probability that the random variable} \\ \text{takes a value between 0 and } x\end{matrix} = \frac{1}{\sqrt{2\pi}} \int_0^x e^{-t^2/2}\, dt.$$

Estimate $P_0^{\,1}$ to 5 decimal places.

SOLUTION: Substituting $-\frac{t^2}{2}$ for x in (8), we obtain

$$\begin{aligned} e^{-t^2/2} &= 1 - \frac{t^2}{2} + \frac{(t^2/2)^2}{2!} - \frac{(t^2/2)^3}{3!} + \frac{(t^2/2)^4}{4!} - \cdots \\ &= 1 - \frac{t^2}{2} + \frac{t^4}{2^2 2!} - \frac{t^6}{2^3 3!} + \frac{t^8}{2^4 4!} - \cdots = \sum_{k=0}^{\infty} \frac{(-1)^k t^{2k}}{2^k k!}. \end{aligned}$$

† We can apply Theorem 3 in Section 12.9 here because the terms in the sequence $\{x^n/n!\}$ are monotone decreasing for $n > x - 1$.

This series converges for every real number t and we can integrate it term-by-term to obtain

$$\begin{aligned}\int_0^x e^{-t^2/2}\,dt &= \int_0^x \left(1 - \frac{t^2}{2} + \frac{t^4}{2^2 2!} - \frac{t^6}{2^3 3!} + \frac{t^8}{2^4 4!} - \cdots\right) dt \\ &= \left(t - \frac{t^3}{3\cdot 2} + \frac{t^5}{5\cdot 2^2\cdot 2!} - \frac{t^7}{7\cdot 2^3\cdot 3!} + \frac{t^9}{9\cdot 2^4\cdot 4!} - \cdots\right)\Bigg|_0^x \\ &= x - \frac{x^3}{3\cdot 2} + \frac{x^5}{5\cdot 2^2\cdot 2!} - \frac{x^7}{7\cdot 2^3\cdot 3!} + \frac{x^9}{9\cdot 2^4\cdot 4!} - \cdots \\ &= \sum_{k=0}^{\infty} \frac{(-1)^k x^{2k+1}}{(2k+1)2^k k!}\end{aligned}$$

and

$$\frac{1}{\sqrt{2\pi}}\int_0^x e^{-t^2/2}\,dt = \frac{1}{\sqrt{2\pi}}\sum_{k=0}^{\infty}\frac{(-1)^k x^{2k+1}}{(2k+1)2^k k!}.$$

In particular,

$$\frac{1}{\sqrt{2\pi}}\int_0^1 e^{-t^2/2}\,dt \overset{\text{set } x=1}{=} \frac{1}{\sqrt{2\pi}}\sum_{k=0}^{\infty}\frac{(-1)^k}{(2k+1)2^k k!}.$$

If we approximate this integral by taking a partial sum of this alternating series, then the error is, by Theorem 3 in Section 12.9,[†] less than the absolute value of the first term omitted. In Table 2 we give the values of $\dfrac{1}{\sqrt{2\pi}(2k+1)2^k k!}$ for $k = 1, 2, 3, 4, 5,$ and 6.

We see that if we compute $\displaystyle\sum_{k=0}^{5}\frac{(-1)^k}{\sqrt{2\pi}(2k+1)2^k k!}$, then our error will be less than 0.000000665 and our answer will be correct to 5 decimal places. [Explain why stopping at $k = 4$ might not work.] We compute

$$\begin{aligned}\sum_{k=0}^{5}\frac{(-1)^k}{\sqrt{2\pi}(2k+1)2^k k!} &= \frac{1}{\sqrt{2\pi}}\left(1 - \frac{1}{6} + \frac{1}{40} - \frac{1}{336} + \frac{1}{3456} - \frac{1}{42{,}240}\right) \\ &= (0.39894228)(0.85562282) \approx 0.34134.\end{aligned}$$

Tables of values of $\displaystyle\frac{1}{\sqrt{2\pi}}\int_0^x e^{-t^2/2}\,dt \left(\text{or } \frac{1}{\sqrt{2\pi}}\int_{-\infty}^x e^{-t^2/2}\,dt\right)$ can be found in virtually every book on statistics, although most of these books do not mention how these values are obtained.

[†] As in Example 6, we can apply this theorem as long as the terms are decreasing. Here $|a_{n+1}| < |a_n|$ when $2(2n+3)(n+1)/(2n+1) > x^2$. This holds for every n when $x = 1$.

TABLE 2 VALUES OF $\frac{1}{\sqrt{2\pi}(2k+1)2^k k!}$

k	$(2k+1)2^k k!$	$\frac{1}{\sqrt{2\pi}(2k+1)2^k k!}$
1	6	0.06649038
2	40	0.009973557
3	336	0.001187328
4	3456	0.000115434
5	42,240	0.000009444
6	599,040	0.000000665

EXAMPLE 8

A SERIES REPRESENTATION OF COS x

Consider the series

$$f(x) = 1 - \frac{x^2}{2!} + \frac{x^4}{4!} - \frac{x^6}{6!} + \cdots = \sum_{k=0}^{\infty} (-1)^k \frac{x^{2k}}{(2k)!}. \tag{10}$$

It is easy to see that $R = \infty$ since the series is absolutely convergent for every x by comparison with the series (8) for e^x. [The series (8) is larger than the series (10) since it contains the terms $x^n/n!$ for n both even and odd, not just for n even, as in (10).] Differentiating, we obtain

$$f'(x) = -x + \frac{x^3}{3!} - \frac{x^5}{5!} + \frac{x^7}{7!} - \cdots = \sum_{k=0}^{\infty} (-1)^{k+1} \frac{x^{2k+1}}{(2k+1)!}. \tag{11}$$

Since this series has a radius of convergences $R = \infty$, we can differentiate once more to obtain

$$f''(x) = -1 + \frac{x^2}{2!} - \frac{x^4}{4!} + \frac{x^6}{6!} - \cdots = \sum_{k=0}^{\infty} \frac{(-1)^{k+1} x^{2k}}{(2k)!} = -f(x).$$

Thus, we see that f satisfies the differential equation

$$f'' + f = 0. \tag{12}$$

Moreover, from equations (10) and (11), we see that

$$f(0) = 1 \quad \text{and} \quad f'(0) = 0. \tag{13}$$

It is easily seen that the function $f(x) = \cos x$ satisfies equation (12) together with the conditions (13). In fact, although we do not prove it here, it is the only function that does so.† Thus, we have

$$\cos x = 1 - \frac{x^2}{2!} + \frac{x^4}{4!} - \frac{x^6}{6!} + \cdots = \sum_{k=0}^{\infty} (-1)^k \frac{x^{2k}}{(2k)!}. \tag{14}$$

†This follows from a basic existence-uniqueness result for differential equations given on page 733.

Since

$$\frac{d}{dx}\cos x = -\sin x,$$

we obtain, from (14), the series (after multiplying both sides by -1)

$$\sin x = x - \frac{x^3}{3!} + \frac{x^5}{5!} - \frac{x^7}{7!} + \cdots = \sum_{k=0}^{\infty} (-1)^k \frac{x^{2k+1}}{(2k+1)!}. \quad (15)$$

EXAMPLE 9 APPROXIMATING A COSINE INTEGRAL

Approximate $\int_0^{\pi/4} \cos t^2\,dt$ with an error $<10^{-10}$.

SOLUTION: Substituting t^2 for x in (14) yields

$$\cos t^2 = 1 - \frac{(t^2)^2}{2!} + \frac{(t^2)^4}{4!} - \frac{(t^2)^6}{6!} + \cdots = \sum_{k=0}^{\infty} \frac{(-1)^k t^{4k}}{(2k)!},$$

so

$$\int_0^x \cos t^2\,dt = \int_0^x \left(1 - \frac{t^4}{2!} + \frac{t^8}{4!} - \frac{t^{12}}{6!} + \cdots\right) dt$$

$$= \left(t - \frac{t^5}{5\cdot 2!} + \frac{t^9}{9\cdot 4!} - \frac{t^{13}}{13\cdot 6!} + \cdots\right)\Bigg|_0^x$$

$$= x - \frac{x^5}{5\cdot 2!} + \frac{x^9}{9\cdot 4!} - \frac{x^{13}}{13\cdot 6!} + \cdots = \sum_{k=0}^{\infty} \frac{(-1)^k x^{4k+1}}{(4k+1)(2k)!}$$

and, in particular,

$$\int_0^{\pi/4} \cos t^2\,dt = \frac{\pi}{4} - \frac{(\pi/4)^5}{5\cdot 2!} + \frac{(\pi/4)^9}{9\cdot 4!} - \frac{(\pi/4)^{13}}{13\cdot 6!} + \cdots.$$

The error in stopping after the $(k-1)$st term is less than $\dfrac{(\pi/4)^{4k+1}}{(4k+1)(2k)!}$. We compute

$k = 3$ $\quad \dfrac{(\pi/4)^{13}}{13\cdot 6!} = 0.000004622.$

$k = 4$ $\quad \dfrac{(\pi/4)^{17}}{17\cdot 8!} = 0.000000024.$

$k = 5$ $\quad \dfrac{(\pi/4)^{21}}{21\cdot 10!} \approx 8.22 \times 10^{-11} < 10^{-10}.$

Thus the desired accuracy is obtained by stopping at $k = 4$:

$$\int_0^{\pi/4} \cos t^2\,dt \approx \frac{\pi}{4} - \frac{(\pi/4)^5}{5\cdot 2!} + \frac{(\pi/4)^9}{9\cdot 4!} - \frac{(\pi/4)^{13}}{13\cdot 6!} + \frac{(\pi/4)^{17}}{17\cdot 8!}$$

$$= 0.75603527754.$$

The trapezoidal rule and Simpson's rule are examples of some more general ways to calculate definite integrals that do not require the existence of a series expansion of the function being integrated. However, as shown in Examples 7 and 9, a power series provides an easy method of numerical integration when the power series representation of a function is readily obtainable.

PROBLEMS 12.11

SELF-QUIZ

Multiple Choice

I. By differentiating the power series

$$1 + x + x^2 + x^3 + \cdots = \sum_{k=0}^{\infty} x^k,$$

we obtain

$$0 + 1 + 2x + 3x^2 + \cdots = \sum_{k=0}^{\infty} k \cdot x^{k-1},$$

which, for $|x| < 1$, converges to ______.

a. $-\ln|1 - x|$ **b.** $\dfrac{1}{1 - x}$

c. $\dfrac{-1}{(1 - x)^2}$ **d.** $\dfrac{1}{(1 - x)^2}$

II. If $|x| < 1$, then

$$x + \frac{x^2}{2} + \frac{x^3}{3} + \frac{x^4}{4} + \cdots$$

$$= \sum_{j=1}^{\infty} \frac{x^j}{j}$$

$$= \sum_{k=0}^{\infty} \frac{x^{k+1}}{k + 1} = \sum_{k=0}^{\infty} \left(\int_0^x t^k \, dt \right)$$

$$= \int_0^x \left(\sum_{k=0}^{\infty} t^k \right) dt = \_\_\_\_\_\_.$$

a. $-\ln|1 - x|$ **b.** $\ln|1 - x|$

c. $\dfrac{1}{1 - x}$ **d.** $\dfrac{1}{(1 - x)^2}$

III. If $|x| < 1$, then

$$x - \frac{x^2}{2} + \frac{x^3}{3} - \frac{x^4}{4} + \cdots$$

$$= \sum_{j=1}^{\infty} (-1)^{j-1} \frac{x^j}{j}$$

$$= \int_0^x \left(\sum_{k=0}^{\infty} (-1)^k t^k \right) dt = \_\_\_\_\_\_.$$

a. $\ln|1 + x|$ **b.** $-\ln|1 + x|$

c. $\dfrac{1}{1 + x}$ **d.** $\dfrac{-1}{(1 + x)^2}$

IV. If $|x| < 1$, then

$$x + \frac{x^3}{3} + \frac{x^5}{5} + \frac{x^7}{7} + \cdots$$

$$= \sum_{k=0}^{\infty} \frac{x^{2k+1}}{2k + 1}$$

$$= \frac{1}{2} \left(\sum_{j=1}^{\infty} \frac{x^j}{j} + \sum_{j=1}^{\infty} (-1)^{j-1} \frac{x^j}{j} \right)$$

$$= \_\_\_\_\_\_.$$

a. $\dfrac{1}{2} \ln|1 - x^2|$ **b.** $\dfrac{1}{2} \ln\left|\dfrac{1 - x}{1 + x}\right|$

c. $\dfrac{1}{2} \ln\left|\dfrac{1 + x}{1 - x}\right|$ **d.** $\dfrac{1}{2(1 - x^2)}$

1. By substituting x^2 for x in equation (4), find a series expansion for $1/(1 + x^2)$ that is valid for $|x| < 1$.

2. Find a series expansion for xe^x that is valid for all real values of x.

3. Integrate the series obtained in Problem 1 to obtain a series expansion for $\tan^{-1} x$.

4. Use the result of Problem 2 to find a series expansion for $\int_0^x te^t\,dt$.

5. Expand $1/x$ as a power series of the form $\sum_{k=0}^{\infty} a_k(x-1)^k$. What is the interval of convergence of this series?

6. Find a power series expansion for

$$\int_0^x \frac{\ln(1+t)}{t}\,dt.$$

7. Use the result of Problem 3 to obtain an approximation of π that is correct to two decimal places (i.e., error bounded by 0.5×10^{-2}). [*Hint:* The series for $\pi/4 = \tan^{-1} 1$ converges very slowly. Instead, use one of the following facts: $\pi/6 = \tan^{-1}(1/\sqrt{3})$, $\pi/12 = \tan^{-1}(2 - \sqrt{3})$.]

8. Use the series expansion for $\ln x$ (see Example 4) to calculate the following to two decimal places of accuracy:
a. $\ln 0.5$ **b.** $\ln 1.6$

In Problems 9–16, approximate the given integral within the specified accuracy.

9. $\int_0^1 e^{-t^2}\,dt$; $|\text{error}| < 0.01$

10. $\int_0^1 e^{-t^3}\,dt$; $|\text{error}| < 0.001$

11. $\int_0^{1/2} \cos t^2\,dt$; $|\text{error}| < 0.001$

12. $\int_0^{1/2} \sin t^2\,dt$; $|\text{error}| < 0.0001$

13. $\int_0^1 \cos\sqrt{t}\,dt$; $|\text{error}| < 0.01$

14. $\int_0^1 t\sin\sqrt{t}\,dt$; $|\text{error}| < 0.001$

15. $\int_0^1 t^2e^{-t^2}\,dt$; $|\text{error}| < 0.01$

[*Hint:* The series expansion of $t^2e^{-t^2}$ can be obtained by multiplying each term of the series expansion for e^{-t^2} by t^2.]

16. $\int_0^{1/2} \frac{dt}{1+t^8}$; $|\text{error}| < 0.0001$

In Problems 17–20, approximate $\frac{1}{\sqrt{2\pi}}\int_0^x e^{-t^2/2}\,dt$ to 5 decimal places.

17. $x = 0.25$ **18.** $x = 0.5$

19. $x = 2$ ***20.** $x = 50$

21. Use the result of Problem 4 to show that

$$\sum_{k=0}^{\infty} \frac{1}{(k+2)k!} = 1.$$

***22.** Define the function J_0 by

$$J_0(x) = \sum_{k=0}^{\infty} \frac{(-1)^k}{(k!)^2}\left(\frac{x}{2}\right)^{2k}.$$

a. What is the interval of convergence of this series?

b. Show that J_0 satisfies the differential equation

$$x^2J''(x) + xJ'(x) + x^2J(x) = 0.$$

The function J_0 is called a **Bessel function of order zero.**†

23. Use the power series for $\ln|(1 + x)/(1 - x)|$ which was obtained in the Self-quiz (II–IV) to approximate ln 1.5, ln 0.5, and ln 2 within four decimal places. Consider that series and those for $\ln|1 \pm x|$ as computational tools. Which seems to be the best? Why?

ANSWERS TO SELF-QUIZ

I. d **II.** a **III.** a **IV.** c

† Named after the German physicist and mathematician Friedrich Wilhelm Bessel (1784–1846), who used the function in his study of planetary motion. The Bessel functions of various orders arise in many applications in modern physics and engineering.

12.12 TAYLOR AND MACLAURIN SERIES

In the last two sections, we used the fact that within its interval of convergence, the function

$$f(x) = \sum_{k=0}^{\infty} a_k(x - x_0)^k$$

is differentiable and integrable. In this section, we look more closely at the coefficients a_k and show that they can be represented in terms of derivatives of the function f.

We begin with the case $x_0 = 0$ and assume that $R > 0$, so that the theorem on power series differentiation applies. We have

$$f(x) = \sum_{k=0}^{\infty} a_k x^k = a_0 + a_1 x + a_2 x^2 + \cdots + a_n x^n + \cdots, \tag{1}$$

and clearly,

$$f(0) = a_0 + 0 + 0 + \cdots + 0 + \cdots = a_0. \tag{2}$$

If we differentiate (1), we obtain

$$f'(x) = \sum_{k=1}^{\infty} k a_k x^{k-1} = a_1 + 2a_2 x + 3a_3 x^2 + \cdots + n a_n x^{n-1} + \cdots \tag{3}$$

and

$$f'(0) = a_1. \tag{4}$$

Continuing to differentiate, we obtain

$$\begin{aligned} f''(x) &= \sum_{k=2}^{\infty} k(k-1)a_k x^{k-2} \\ &= 2a_2 + 3 \cdot 2a_3 x + 4 \cdot 3a_4 x^2 + \cdots + n(n-1)a_n x^{n-2} + \cdots \end{aligned}$$

and

$$f''(0) = 2a_2, \qquad \text{or} \qquad a_2 = \frac{f''(0)}{2} = \frac{f''(0)}{2!}. \tag{5}$$

Similarly,

$$f'''(x) = \sum_{k=3}^{\infty} k(k-1)(k-2)a_k x^{k-3},$$

so

$$f'''(0) = 3 \cdot 2a_3 \qquad \text{and} \qquad a_3 = \frac{f'''(0)}{3 \cdot 2} = \frac{f'''(0)}{3!}. \tag{6}$$

It is not difficult to see that this pattern continues and that for every positive integer n,

THE nTH COEFFICIENT OF THE POWER SERIES FOR f

$$a_n = \frac{f^{(n)}(0)}{n!}. \tag{7}$$

For $n = 0$, we use the convention $0! = 1$ and $f^{(0)}(x) = f(x)$. Then formula (7) holds for every n, and we have the following:

MACLAURIN SERIES

If $f(x) = \sum_{k=0}^{\infty} a_k x^k$, then

$$f(x) = \sum_{k=0}^{\infty} \frac{f^{(k)}(0)}{k!} x^k$$

$$= f(0) + f'(0)x + f''(0)\frac{x^2}{2!} + \cdots + f^{(n)}(0)\frac{x^n}{n!} + \cdots \tag{8}$$

for every x in the interval of convergence.

In the general case, if

$$f(x) = \sum_{k=0}^{\infty} a_k(x - x_0)^k$$

$$= a_0 + a_1(x - x_0) + a_2(x - x_0)^2 + \cdots + a_n(x - x_0)^n + \cdots, \tag{9}$$

then $f(x_0) = a_0$, and differentiating as before, we find that

$$a_n = \frac{f^{(n)}(x_0)}{n!}. \tag{10}$$

Thus, we have the following:

TAYLOR SERIES

If $f(x) = \sum_{k=0}^{\infty} a_k(x - x_0)^k$, then

$$f(x) = \sum_{k=0}^{\infty} \frac{f^{(k)}(x_0)}{k!}(x - x_0)^k$$

$$= f(x_0) + f'(x_0)(x - x_0) + f''(x_0)\frac{(x - x_0)^2}{2!} + \cdots$$

$$+ f^{(n)}(x_0)\frac{(x - x_0)^n}{n!} + \cdots \tag{11}$$

for every x in the interval of convergence.

DEFINITION TAYLOR AND MACLAURIN SERIES

The series in (11) is called the **Taylor series**[†] of the function f at x_0. The special case

[†] The history of the Taylor series is somewhat muddied. It has been claimed that the basis for its development was found in India before 1550. (Taylor published the result in 1713.) For an interesting discussion of this controversy, see the paper by C. T. Rajagopal and T. V. Vedamurthi, "On the Hindu proof of Gregory's series," *Stripta Mathematics,* 17 (1951): 65–74.

$x_0 = 0$ in (8) is called a **Maclaurin series.**† We see that the first n terms of the Taylor series of a function are simply the Taylor polynomial described in Section 12.1. ■

WARNING: We have shown here that *if* $f(x) = \sum_{k=0}^{\infty} a_k(x - x_0)^k$, then f is infinitely differentiable (i.e., f has derivatives of all orders) and that the series for f is the Taylor series (or Maclaurin series if $x_0 = 0$) of f. What we have *not* shown is that if f is infinitely differentiable at x_0, then f has a Taylor series expansion at x_0. In general, this last statement is false, as we will see in Example 3.

EXAMPLE 1

THE MACLAURIN SERIES FOR e^x

Find the Maclaurin series for e^x.

SOLUTION: If $f(x) = e^x$, then $f(0) = f'(0) = \cdots = f^{(k)}(0) = 1$, and

$$e^x = \sum_{k=0}^{\infty} \frac{x^k}{k!} = 1 + x + \frac{x^2}{2!} + \frac{x^3}{3!} + \cdots + \frac{x^n}{n!} + \cdots. \tag{12}$$

This series is the series we obtained in Example 5 in Section 12.11. It is important to note here that this example shows that *if* e^x has a Maclaurin series expansion, then the series must be the series (12). It does *not* show that e^x actually does have such a series expansion. To prove that the series in (12) is really equal to e^x, we differentiate, as in Example 5 in Section 12.11, and use the fact that the only continuous function that satisfies

$$f'(x) = f(x), \qquad f(0) = 1,$$

is the function e^x.

EXAMPLE 2

THE MACLAURIN SERIES FOR $\cos x$

Assuming that the function $f(x) = \cos x$ can be written as a Maclaurin series, find that series.

SOLUTION: If $f(x) = \cos x$, then $f(0) = 1, f'(0) = 0, f''(0) = -1, f'''(0) = 0, f^{(4)}(0) = 1$, and so on, so that if

$$\cos x = \sum_{k=0}^{\infty} a_k x^k,$$

then

$$\cos x = f(0) = f'(0) + \frac{f''(0)x^2}{2!} + \frac{f'''(0)x^3}{3!} + \frac{f^{(4)}(0)x^4}{4!} + \cdots,$$

† See the biographical sketch of Maclaurin on page 992.

or

$$\cos x = 1 - \frac{x^2}{2!} + \frac{x^4}{4!} - \frac{x^6}{6!} + \cdots = \sum_{k=0}^{\infty} \frac{(-1)^k x^{2k}}{(2k)!} \tag{13}$$

This series is the series found in Example 8 in Section 12.11.

NOTE: Again, this does not prove that the equality in (13) is correct. It only shows that *if* $\cos x$ has a Maclaurin expansion, then the expansion must be given by (13). We will show that $\cos x$ has a Maclaurin series in Example 5.

EXAMPLE 3 AN INFINITELY DIFFERENTIABLE FUNCTION THAT CANNOT BE REPRESENTED BY A MACLAURIN SERIES

Let

$$f(x) = \begin{cases} e^{-1/x^2}, & \text{if } x \neq 0 \\ 0, & \text{if } x = 0. \end{cases}$$

Find a Maclaurin expansion for f if one exists.

SOLUTION: First, we note that since $\lim_{x \to 0} e^{-1/x^2} = 0$, f is continuous. Now recall from your one-variable calculus course (or from using L'Hôpital's rule) that $\lim_{x \to \infty} x^a e^{-bx} = 0$ if $b > 0$. Let $y = 1/x^2$. Then, as $x \to 0$, $y \to \infty$. Also, $1/x^n = (1/x^2)^{n/2}$, so that $\lim_{x \to 0}(e^{-1/x^2}/x^n) = \lim_{y \to \infty} y^{n/2}e^{-y} = 0$.

Now for $x \neq 0$, $f'(x) = (2/x^3)e^{-1/x^2} \to 0$ as $x \to 0$, so that f' is continuous at 0. Similarly, $f''(x) = [(4/x^6) - (6/x^4)]e^{-1/x^2}$, which also approaches 0 as $x \to 0$ by the limit result above. In fact, *every* derivative of f is continuous and $f^{(n)}(0) = 0$ for every n. Thus, f is infinitely differentiable, and *if* it had a Maclaurin series that represented the function, then we would have

$$f(x) = f(0) + f'(0)x + f''(0)\frac{x^2}{2!} + \cdots.$$

But $f(0) = f'(0) = f''(0) = \cdots = 0$, so that the Maclaurin series would be the zero series. But since f is obviously not the zero function, we can only conclude that there is *no* Maclaurin series that represents f at any point other than 0.

Example 3 illustrates that infinite differentiability is not sufficient to guarantee that a given function can be represented by its Taylor series. Something more is needed.

DEFINITION ANALYTIC FUNCTION

We say that a function f is **analytic** at x_0 if f can be represented by a Taylor series in some neighborhood of x_0. ■

We see that the functions e^x and $\cos x$ are analytic at 0, while the function

$$f(x) = \begin{cases} e^{-1/x^2}, & x \neq 0 \\ 0, & x = 0 \end{cases}$$

is not. A condition that guarantees analyticity of an infinitely differentiable function is given next.

THEOREM 1

Suppose that the function f has continuous derivatives of all orders in a neighborhood $N(x_0)$ of the number x_0.

Then f is analytic at x_0 if and only if

$$\lim_{n\to\infty} R_n(x) = \lim_{n\to\infty} \frac{f^{(n+1)}(c_n)}{(n+1)!}(x - x_0)^{n+1} = 0 \tag{14}$$

for every x in $N(x_0)$ where c_n is between x_0 and x. ■

REMARK: The expression between the equal signs in (14) is simply the remainder term given by Taylor's theorem (see page 900).

PROOF: The hypotheses of Taylor's theorem apply, so that we can write, for any n,

$$f(x) = P_n(x) + R_n(x), \tag{15}$$

where $P_n(x)$ is the nth-degree Taylor polynomial for f. To show that f is analytic, we must show that

$$\lim_{n\to\infty} P_n(x) = f(x) \tag{16}$$

for every x in $N(x_0)$. But if x is in $N(x_0)$, we obtain, from (14) and (15),

$$\lim_{n\to\infty} P_n(x) = \lim_{n\to\infty} [f(x) - R_n(x)] = f(x) - \lim_{n\to\infty} R_n(x)$$

$$= f(x) - 0 = f(x).$$

Conversely, if f is analytic, then $f(x) = \lim_{n\to\infty} P_n(x)$ so $R_n(x) \to 0$ as $n \to \infty$. ■

EXAMPLE 4 SHOWING THAT e^x IS ANALYTIC

If $f(x) = e^x$, then $f^{(n)}(x) = e^x$, and

$$\lim_{n\to\infty} \left| \frac{f^{(n+1)}(c_n)}{(n+1)!}(x - 0)^{n+1} \right| = \lim_{n\to\infty} \frac{e^{c_n}|x|^{n+1}}{(n+1)!} \overset{0 < c_n < |x|}{\leq} e^{|x|} \lim_{n\to\infty} \frac{|x|^{n+1}}{(n+1)!} \to 0,$$

since $|x|^{n+1}/(n+1)!$ is the $(n+2)$nd term in the converging power series $\Sigma_{k=0}^{\infty} |x|^k/k!$ and the terms in a converging power series $\to 0$ by Theorem 2 in Section 12.6. Since this result is true for any $x \in \mathbb{R}$, we may take $N = (-\infty, \infty)$ to conclude that the series (12) is valid for every real number x.

EXAMPLE 5 SHOWING THAT $\cos x$ IS ANALYTIC

Let $f(x) = \cos x$. Since all derivatives of $\cos x$ are equal to $\pm\sin x$ or $\pm\cos x$, we see that $|f^{(n+1)}(c_n)| \leq 1$. Then for $x_0 = 0$,

$$|R_n(x)| \leq |x|^{n+1}/(n+1)!$$

which $\to 0$ as $n \to \infty$, so that the series (13) is also valid for every real number x.

EXAMPLE 6

A FUNCTION THAT IS NOT ANALYTIC

It is evident that for the function in Example 3, $R_n(x) \not\to 0$ if $x \neq 0$. This follows from the fact that $R_n(x) = f(x) - P_n(x) = e^{-1/x^2} - 0 = e^{-1/x^2} \neq 0$ if $x \neq 0$.

In the rest of this section we will not prove that remainder terms go to zero. However, you should be aware that unless this is done, there is no guarantee that the series you obtain by using formula (8) or (11) will be valid.

EXAMPLE 7

THE TAYLOR SERIES FOR $\ln x$ AT $x = 1$

Find the Taylor expansion for $f(x) = \ln x$ at $x = 1$.

SOLUTION: Since $f'(x) = 1/x$, $f''(x) = -1/x^2$, $f'''(x) = 2/x^3$, $f^{(4)}(x) = -6/x^4$, . . . , $f^{(n)}(x) = (-1)^{n+1}(n-1)!/x^n$, we find that $f(1) = 0$, $f'(1) = 1$, $f''(1) = -1$, $f'''(1) = 2$, $f^{(4)}(1) = -6$, . . . , $f^{(n)}(1) = (-1)^{n+1}(n-1)!$. Then wherever valid,

$$\ln x = \sum_{k=0}^{\infty} f^{(k)}(1)\frac{(x-1)^k}{k!}$$

$$= 0 + (x-1) - \frac{(x-1)^2}{2} + \frac{2(x-1)^3}{3!} - \frac{3!(x-1)^4}{4!} + \frac{4!(x-1)^5}{5!} + \cdots,$$

or

$$\ln x = (x-1) - \frac{(x-1)^2}{2} + \frac{(x-1)^3}{3} - \frac{(x-1)^4}{4} + \cdots$$

$$= \sum_{k=1}^{\infty} \frac{(-1)^{k+1}(x-1)^k}{k}.$$

The radius of convergence of this power series is 1. Thus, the series converges to $\ln x$ for $-1 < x - 1 < 1$ or $0 < x < 2$. When $x = 2$, we obtain the series $\sum \frac{(-1)^{k+1}}{k}$, which converges by the alternating series test. This implies that the series (17) converges to $\ln(x)$ for $0 < x \leq 2$. When $x = 2$, we obtain, from (17),

$$\ln 2 = 1 - \frac{1}{2} + \frac{1}{3} - \frac{1}{4} + \frac{1}{5} - \cdots = \sum_{k=1}^{\infty} \frac{(-1)^{k+1}}{k}. \tag{18}$$

EXAMPLE 8

THE TAYLOR SERIES FOR $\sin x$ AT $x = \pi/3$

Find a Taylor series for $f(x) = \sin x$ at $x = \pi/3$.

SOLUTION: Here we have $f(\pi/3) = \sqrt{3}/2, f'(\pi/3) = \frac{1}{2}, f''(\pi/3) = -\sqrt{3}/2, f'''(\pi/3) = -\frac{1}{2}$, and so on, so that

$$\sin x = \frac{\sqrt{3}}{2} + \frac{1}{2}\left(x - \frac{\pi}{3}\right) - \frac{\sqrt{3}}{2}\frac{[x - (\pi/3)]^2}{2!} - \frac{1}{2}\frac{[x - (\pi/3)]^3}{3!}$$

$$+ \frac{\sqrt{3}}{2}\frac{[x - (\pi/3)]^4}{4!} + \cdots.$$

The proof that this series is valid for every real number x is similar to the proof in Example 5.

We provide here a list of useful Maclaurin series:

SOME USEFUL MACLAURIN SERIES

$$e^x = \sum_{k=0}^{\infty} \frac{x^k}{k!} = 1 + x + \frac{x^2}{2!} + \frac{x^3}{3!} + \cdots \tag{19}$$

$$\cos x = \sum_{k=0}^{\infty} \frac{(-1)^k x^{2k}}{(2k)!} = 1 - \frac{x^2}{2!} + \frac{x^4}{4!} - \frac{x^6}{6!} + \cdots \tag{20}$$

$$\sin x = \sum_{k=0}^{\infty} \frac{(-1)^k x^{2k+1}}{(2k+1)!} = x - \frac{x^3}{3!} + \frac{x^5}{5!} - \frac{x^7}{7!} + \cdots \tag{21}$$

$$\cosh x = \sum_{k=0}^{\infty} \frac{x^{2k}}{(2k)!} = 1 + \frac{x^2}{2!} + \frac{x^4}{4!} + \frac{x^6}{6!} + \cdots \tag{22}$$

$$\sinh x = \sum_{k=0}^{\infty} \frac{x^{2k+1}}{(2k+1)!} = x + \frac{x^3}{3!} + \frac{x^5}{5!} + \frac{x^7}{7!} + \cdots \tag{23}$$

$$\frac{1}{1-x} = \sum_{k=0}^{\infty} x^k = 1 + x + x^2 + \cdots, \qquad |x| < 1 \tag{24}$$

$$\ln(1+x) = \sum_{k=0}^{\infty} \frac{(-1)^k x^{k+1}}{k+1}, \qquad -1 < x \le 1 \tag{25}$$

You are asked to prove, in Problems 27 and 28, that the series (21), (22), and (23) are valid for every real number x.

BINOMIAL SERIES

We close this section by deriving another series that is quite useful. Let $f(x) = (1+x)^r$, where r is a real number not equal to an integer. We have

$$\begin{aligned}
f'(x) &= r(1+x)^{r-1}, \\
f''(x) &= r(r-1)(1+x)^{r-2}, \\
f'''(x) &= r(r-1)(r-2)(1+x)^{r-3}, \\
&\vdots \\
f^{(n)}(x) &= r(r-1)(r-2)\cdots(r-n+1)(1+x)^{r-n}.
\end{aligned}$$

Note that since r is not an integer, $r - n$ is never equal to 0, and all derivatives exist and are nonzero as long as $x \neq -1$. Then,

$$\begin{aligned}
f(0) &= 1, \\
f'(0) &= r, \\
f''(0) &= r(r-1), \\
&\vdots \\
f^{(n)}(0) &= r(r-1)\cdots(r-n+1),
\end{aligned}$$

and we can write

$$(1+x)^r = 1 + rx + \frac{r(r-1)}{2!}x^2 + \frac{r(r-1)(r-2)}{3!}x^3 + \cdots$$
$$+ \frac{r(r-1)\cdots(r-n+1)}{n!}x^n + \cdots$$
$$= 1 + \sum_{k=1}^{\infty} \frac{r(r-1)\cdots(r-k+1)}{k!}x^k, \qquad |x| < 1 \tag{26}$$

The series (26) is called the **binomial series**.

EXAMPLE 9

USING THE BINOMIAL SERIES TO APPROXIMATE AN INTEGRAL

Approximate $\int_0^{0.6} \sqrt{1+t^5}\, dt$ with an error <0.00001 using an appropriate binomial series.

SOLUTION: Substituting $r = \frac{1}{2}$ and $x = t^5$ in (26) yields

$$(1+t^5)^{1/2} = 1 + \frac{t^5}{2} + \frac{\frac{1}{2}(\frac{1}{2}-1)}{2!}t^{10} + \frac{\frac{1}{2}(\frac{1}{2}-1)(\frac{1}{2}-2)}{3!}t^{15} + \cdots$$
$$= 1 + \frac{t^5}{2} - \frac{1}{4\cdot 2!}t^{10} + \frac{3}{8\cdot 3!}t^{15} - \frac{15}{16\cdot 4!}t^{20} + \cdots.$$

Then

$$\int_0^x (1+t^5)^{1/2}\, dt = x + \frac{x^6}{6\cdot 2} - \frac{x^{11}}{11\cdot 4\cdot 2!} + \frac{3x^{16}}{16\cdot 8\cdot 3!} - \frac{15x^{21}}{21\cdot 16\cdot 4!} + \cdots$$

and

$$\int_0^{0.6} (1+t^5)^{1/2}\, dt = 0.6 + \frac{0.6^6}{12} - \frac{0.6^{11}}{88} + \frac{3(0.6)^{16}}{768} - \frac{15(0.6)^{21}}{8064} + \cdots.$$

After the first term, the series above is alternating. Since $\frac{0.6^{11}}{88} \approx 0.000041$ and $\frac{3(0.6)^{16}}{768} \approx 0.0000011$, we can obtain the desired accuracy by stopping after the term $-\frac{0.6^{11}}{88}$. We have

$$\int_0^{0.6} (1+t^5)^{1/2}\, dt \approx 0.6 + \frac{0.6^6}{12} - \frac{0.6^{11}}{88} \approx 0.603847.$$

OF HISTORICAL INTEREST

COLIN MACLAURIN, 1698–1746

COLIN MACLAURIN
1698–1746

Considered the finest British mathematician of the generation after Newton, Colin Maclaurin was certainly one of the best mathematicians of the eighteenth century.

Born in Scotland, Maclaurin was a mathematical prodigy and entered Glasgow University at the age of 11. By the age of 19, he was a professor of mathematics in Aberdeen and later obtained a post at the University of Edinburgh.

Maclaurin is best known for the term *Maclaurin series,* which is the Taylor series in the case $x_0 = 0$. He used this series in his 1742 work, *Treatise of Fluxions.* (Maclaurin acknowledged that the series had first been used by Taylor in 1715.) The *Treatise of Fluxions* was most significant in that it presented the first logical description of Newton's method of fluxions. This work was written to defend Newton from the attacks of the powerful Bishop George Berkeley (1685–1753). Berkeley was troubled (as are many of today's calculus students) by the idea of a quotient that takes the form $\frac{0}{0}$. This, of course, is what we obtain when we take a derivative. Berkeley wrote:

> And what are these fluxions? The velocities of evanescent increments. And what are these same evanescent increments? They are neither finite quantities nor quantities infinitely small nor yet nothing. May we not call them ghosts of departed quantities?

Maclaurin answered Berkeley using geometric arguments. Later, Newton's calculus was put on an even firmer footing by the work of Lagrange in 1797 (see page 228).

Maclaurin made many other contributions to mathematics—especially in the areas of geometry and algebra. He published his *Geometria organica* when only 21 years old. His posthumous work *Treatise of Algebra,* published in 1748, contained many important results, including the well-known *Cramer's rule* for solving a system of equations (Cramer published the result in 1750).

In 1745, when "Bonnie Prince Charlie" marched against Edinburgh, Maclaurin helped defend the city. When the city fell, Maclaurin escaped, fleeing to York, where he died in 1746 at the age of 48.

PROBLEMS 12.12

SELF-QUIZ

Each of the following power series is the Maclaurin series for a well-known function. Identify each one.

I. $1 + x + x^2 + x^3 + \cdots + x^k + \cdots$

II. $1 - x + x^2 - x^3 + \cdots + (-1)^k x^k + \cdots$

III. $1 + x + \dfrac{x^2}{2!} + \dfrac{x^3}{3!} + \cdots + \dfrac{x^k}{k!} + \cdots$

IV. $1 - x + \dfrac{x^2}{2!} - \dfrac{x^3}{3!} + \cdots + \dfrac{(-x)^k}{k!} + \cdots$

V. $1 - \dfrac{x^2}{2!} + \dfrac{x^4}{4!} - \dfrac{x^6}{6!} + \cdots + \dfrac{(-1)^k x^{2k}}{(2k)!} + \cdots$

VI. $x - \dfrac{x^3}{3!} + \dfrac{x^5}{5!} - \dfrac{x^7}{7!} + \cdots + \dfrac{(-1)^k x^{2k+1}}{(2k+1)!} + \cdots$

VII. $x + \dfrac{x^3}{3!} + \dfrac{x^5}{5!} + \dfrac{x^7}{7!} + \cdots + \dfrac{x^{2k+1}}{(2k+1)!} + \cdots$

VIII. $1 + \dfrac{x^2}{2!} + \dfrac{x^4}{4!} + \dfrac{x^6}{6!} + \cdots + \dfrac{x^{2k}}{(2k)!} + \cdots$

IX. $1 - x^2 + x^4 - x^6 + \cdots + (-1)^k (x^2)^k + \cdots$

X. $x - \dfrac{x^3}{3} + \dfrac{x^5}{5} - \dfrac{x^7}{7} + \cdots + \dfrac{(-1)^k x^{2k+1}}{(2k+1)} + \cdots$

1. Find the Taylor series for e^x at 1.
2. Find the Maclaurin series for e^{-x}.
3. Find the Taylor series for $\cos x$ at $\pi/4$.
4. Find the Taylor series for $\sinh x$ at $\ln 2$.
5. Find the Maclaurin series for $e^{\beta x}$, β real.
6. Find the Taylor series for e^x at $x = -1$.
7. Find the Maclaurin series for xe^x.
8. Find the Maclaurin series for $x^2 e^{-x^2}$.
9. Find the Maclaurin series for $(\sin x)/x$.
10. Find the Maclaurin series for $\sin^2 x$. [*Hint:* $\sin^2 x = (1 - \cos 2x)/2$.]
11. Find the Taylor series for $(x - 1) \ln x$ at 1. Over what interval is this representation valid?
12. Find the first three nonzero terms of the Maclaurin series for $\tan x$. What is the interval of convergence for that Maclaurin series?
13. Find the first four terms of the Taylor series for $\csc x$ at $\pi/2$. What is the interval of convergence for that Taylor series?
14. Find the first three terms of the Maclaurin series for $\ln|\cos x|$. What is its interval of convergence? [*Hint:* $\int \tan x\, dx = -\ln|\cos x|$.]
15. Find the Taylor series of $\sqrt{x}$ at $x = 4$. What is its radius of convergence?

*16. Find the Maclaurin series of $\sin^{-1} x$. What is its radius of convergence? [*Hint:* Find the Maclaurin series for $1/\sqrt{1 - x}$; then find the Maclaurin series for $1/\sqrt{1 - x^2}$; then integrate.]

17. Use the Maclaurin series for $\sin x$ to obtain the Maclaurin series for $\sin x^2$.
18. Find the Maclaurin series for $\cos x^2$.
19. Find the Maclaurin series for $\tan^{-1} x$. What is its radius of convergence? [*Hint:* Integrate the Maclaurin series for $1/(1 + x^2)$.]
20. Find the Maclaurin series for $\ln|(1 + x)/(1 - x)|$. What is its radius of convergence? [*Hint:* Integrate the Maclaurin series for $1/(1 - x^2)$.]
21. Use the binomial series (equation (26)) to find a power series representation for $\sqrt[4]{1 + x}$.
22. Use the result of the preceding problem to find a power series representation for $\sqrt[4]{1 + x^3}$.
23. Use the result of Problem 22 to approximate $\int_0^{0.5} \sqrt[4]{1 + x^3}\, dx$ to four significant figures.
24. Use the technique used for Problems 21–23 or Example 9 to approximate $\int_0^{1/4} (1 + \sqrt{x})^{3/5}\, dx$ to four significant figures.
25. The **error function** (which arises in mathematical statistics) is defined by

$$\operatorname{erf}(x) = \frac{2}{\sqrt{\pi}} \int_0^x e^{-t^2}\, dt.$$

a. Find a Maclaurin series for $\operatorname{erf}(x)$ by integrating the Maclaurin series for e^{-x^2}.

b. Use the series obtained in part (a) to approximate $\operatorname{erf}(1)$ and $\operatorname{erf}(0.5)$, each with an error less than 0.0001.

26. The **complementary error function** is defined by

$$\operatorname{erfc}(x) = 1 - \operatorname{erf}(x) = 1 - \frac{2}{\sqrt{\pi}} \int_0^x e^{-t^2}\, dt$$

$$= \frac{2}{\sqrt{\pi}} \int_x^\infty e^{-t^2}\, dt.$$

Find the Maclaurin series for $\operatorname{erfc}(x)$; use it to approximate $\operatorname{erfc}(1)$ and $\operatorname{erfc}(0.5)$ with a maximum error of 0.0001. (Note that for large values of x, $\operatorname{erfc}(x)$ can be approximated by integrating the last integral by parts.)

27. Prove that the series (21) represents $\sin x$ for all real x.
28. **a.** Prove that the series (22) represents $\cosh x$ for all real x.

b. Use the fact that $\sinh x = \dfrac{d}{dx} \cosh x$ to derive the series in (23).

29. Differentiate the Maclaurin series for $\sin x$ and show that it is equal to the Maclaurin series for $\cos x$.
30. Differentiate the Maclaurin series for $\sinh x$ and show that it is equal to the Maclaurin series for $\cosh x$.
31. Using the fact that if f has a Taylor series at x_0, then the Taylor series is given by (11), show that $1 + x + x^2 + x^3 + \cdots$ is the Taylor series for $1/(1 - x)$ at $x_0 = 0$ with interval of convergence $(-1, 1)$.
32. **a.** Show that the Maclaurin series for f has only even powers of x (i.e., the odd powers have coefficient zero) if and only if f is an even function (i.e., $f(-x) = f(x)$ for all x).

b. The Maclaurin series for f has only odd powers if and only if f is an odd function (i.e., $f(-x) = -f(x)$ for all x). Prove this.

33. Show that the binomial series (equation (26)) converges if $|x| < 1$.
34. Show that for any real number r

$$1 + \frac{r}{2} + \frac{r(r-1)}{2^2 2!} + \cdots$$

$$+ \frac{r(r-1)\cdots(r-k+1)}{2^k k!} + \cdots = \left(\frac{3}{2}\right)^r.$$

***35.** The **sine integral** is defined by

$$\mathrm{Si}(x) = \int_0^x \frac{\sin t}{t}\,dt.$$

a. Show that $\mathrm{Si}(x)$ is defined and continuous for all real x.

b. Show that $\lim_{x\to\infty} \mathrm{Si}(x)$ exists and is finite.

c. Find a Maclaurin series expansion for $\mathrm{Si}(x)$.

d. Approximate $\mathrm{Si}(1)$ and $\mathrm{Si}(0.5)$ with a maximum error of 0.0001.

***36.** Suppose f is a well-behaved function with values tabulated at $a + k \cdot \Delta x$ where $\Delta x = 0.05$. Compare

$$\frac{f(a + \Delta x) - f(a)}{\Delta x} \quad \text{and} \quad \frac{f(a + \Delta x) - f(a - \Delta x)}{2\,\Delta x}$$

as numerical approximations to $f'(a)$.

***37.** Suppose f, f', and f'' are continuous in a neighborhood of a; also suppose $f'(a) = 0$. Prove the second-derivative test for maxima-minima.

***38.** Prove that e is irrational. [*Hint:* If $e = p/q$, then consider the qth-degree Taylor polynomial for e^1 and its remainder.]

****39.** Suppose that $|f(x)| \le 1$ and $|f''(x)| \le 1$ on $[-1, 1]$. Prove that, for every x in $[-1, 1]$, $|f'(x)| \le \sqrt{2}$.

ANSWERS TO SELF-QUIZ

I. $1/(1 - x)$ for $|x| < 1$ **II.** $1/(1 + x)$ for $|x| < 1$ **III.** e^x for all x
IV. e^{-x} for all x **V.** $\cos x$ for all x **VI.** $\sin x$ for all x **VII.** $\sinh x$ for all x
VIII. $\cosh x$ for all x **IX.** $1/(1 + x^2)$ for $|x| < 1$ **X.** $\tan^{-1} x$ for $|x| < 1$

12.13 USING POWER SERIES TO SOLVE ORDINARY DIFFERENTIAL EQUATIONS (OPTIONAL)

Consider the second-order linear differential equation

$$y'' + a(x)y' + b(x)y = f(x).$$

If the functions a and b are not constant functions, then there is, in general, no way to obtain a closed form solution to the equation even in the homogeneous case ($f = 0$). In this section we show how power series can be used to obtain series solutions to the equation above in some cases. The examples we present are merely illustrative of a technique that *sometimes* works. For a more complete discussion you should consult a book on differential equations.

The fundamental assumption used in solving a differential equation by the power series method is that the solution of the differential equation can be expressed in the form of a power series, say,

$$y = \sum_{n=0}^{\infty} c_n x^n. \tag{1}$$

Once this assumption has been made, power series expansions for y', y'', . . . can be obtained by differentiating equation (1) term by term:

$$y' = \sum_{n=1}^{\infty} nc_n x^{n-1}, \tag{2}$$

$$y'' = \sum_{n=2}^{\infty} n(n-1)c_n x^{n-2}, \text{ etc.} \tag{3}$$

These can then be substituted into the given differential equation. After all the indicated operations have been carried out, and like powers of x have been collected, we obtain an expression of the form

$$k_0 + k_1 x + k_2 x^2 + \cdots = \sum_{n=0}^{\infty} k_n x^n = 0 \tag{4}$$

where the coefficients $k_0, k_1, k_2, \ldots$ are expressions involving the unknown coefficients $c_0, c_1, c_2, \ldots$. Since equation (4) must hold for all values of x in some interval, all the coefficients $k_0, k_1, k_2, \ldots$ must vanish. From the equations

$$k_0 = 0, \qquad k_1 = 0, \qquad k_2 = 0, \qquad \ldots$$

it is then possible to determine successively the coefficients $c_0, c_1, c_2, \ldots$. In this section we illustrate this procedure by means of several examples, without concerning ourselves with questions of the convergence of the power series under consideration or the inherent limitations of the method. We shall see that power series provide a powerful method for solving certain linear differential equations with variable coefficients. First, however, in order to check that the power series method does provide the required solution, we shall solve three problems that could be solved more easily by other methods.

EXAMPLE 1

USING A POWER SERIES TO SOLVE A FIRST-ORDER DIFFERENTIAL EQUATION

Consider the initial value problem

$$y' = y + x^2, \qquad y(0) = 1. \tag{5}$$

Inserting equations (1) and (2) into the equation, we have

$$c_1 + 2c_2x + 3c_3x^2 + 4c_4x^3 + \cdots = (c_0 + c_1x + c_2x^2 + c_3x^3 + \cdots) + x^2.$$

Collecting like powers of x, we obtain

$$(c_1 - c_0) + (2c_2 - c_1)x + (3c_3 - c_2 - 1)x^2 + (4c_4 - c_3)x^3 + \cdots = 0.$$

Equating each of the coefficients to zero, we obtain the identities

$$c_1 - c_0 = 0, \qquad 2c_2 - c_1 = 0, \qquad 3c_3 - c_2 - 1 = 0, \qquad 4c_4 - c_3 = 0, \qquad \cdots,$$

from which we find that

$$c_1 = c_0, \quad c_2 = \frac{c_1}{2} = \frac{c_0}{2!}, \quad c_3 = \frac{c_2 + 1}{3} = \frac{c_0 + 2}{3!}, \quad c_4 = \frac{c_3}{4} = \frac{c_0 + 2}{4!}, \quad \ldots.$$

With these values, equation (1) becomes

$$\begin{aligned} y &= c_0 + c_0x + \frac{c_0}{2!}x^2 + \frac{c_0 + 2}{3!}x^3 + \frac{c_0 + 2}{4!}x^4 + \frac{c_0 + 2}{5!}x^5 + \cdots \\ &= (c_0 + 2)\left(1 + x + \frac{x^2}{2!} + \frac{x^3}{3!} + \frac{x^4}{4!} + \cdots\right) - 2\left(1 + x + \frac{x^2}{2!}\right). \end{aligned}$$

Looking carefully at the series in parentheses, we recognize the expansion for e^x, so we have the general solution

$$y = (c_0 + 2)e^x - x^2 - 2x - 2.$$

To solve the initial value problem, we set $x = 0$ to obtain

$$1 = y(0) = c_0 + 2 - 2 = c_0.$$

Thus the solution of the initial value problem (5) is given by the equation

$$y = 3e^x - x^2 - 2x - 2.$$

EXAMPLE 2 USING A POWER SERIES TO SOLVE A SECOND-ORDER DIFFERENTIAL EQUATION

Solve.

$$y'' + y = 0. \tag{6}$$

Using equations (1) and (3), we have

$$(2c_2 + 3 \cdot 2c_3x + 4 \cdot 3c_4x^2 + \cdots) + (c_0 + c_1x + c_2x^2 + \cdots) = 0.$$

Gathering like powers of x yields

$$(2c_2 + c_0) + (3 \cdot 2c_3 + c_1)x + (4 \cdot 3c_4 + c_2)x^2 + \cdots = 0.$$

Setting each of the coefficients to zero, we obtain

$$2c_2 + c_0 = 0, \quad 3 \cdot 2c_3 + c_1 = 0, \quad 4 \cdot 3c_4 + c_2 = 0, \quad 5 \cdot 4c_5 + c_3 = 0, \quad \ldots,$$

and

$$c_2 = -\frac{c_0}{2!}, \quad c_3 = -\frac{c_1}{3!}, \quad c_4 = -\frac{c_2}{4 \cdot 3} = \frac{c_0}{4!}, \quad c_5 = -\frac{c_3}{5 \cdot 4} = \frac{c_1}{5!}, \quad \ldots.$$

Substituting these values into the power series (1) for y yields

$$y = c_0 + c_1x - \frac{c_0}{2!}x^2 - \frac{c_1}{3!}x^3 + \frac{c_0}{4!}x^4 + \frac{c_1}{5!}x^5 + \cdots.$$

Splitting this series into two parts, we have

$$y = c_0\left(1 - \frac{x^2}{2!} + \frac{x^4}{4!} - \cdots\right) + c_1\left(x - \frac{x^3}{3!} + \frac{x^5}{5!} - \cdots\right).$$

Using equations (12.12.20) and (12.12.21) reveals the familiar general solution

$$y = c_0 \cos x + c_1 \sin x.$$

We observe that in this case the power series method produces two arbitrary constants c_0, c_1, and yields the general solution for equation (6).

So far we have considered only linear equations with constant coefficients. We turn now to linear equations with variable coefficients.

EXAMPLE 3 USING A POWER SERIES TO SOLVE A DIFFERENTIAL EQUATION WITH VARIABLE COEFFICIENTS

Consider the initial value problem

$$(1 + x^2)y' = 2pxy, \qquad y(0) = 1, \tag{7}$$

where p is a constant. Applying equations (1) and (2), we have

$$(1 + x^2) \sum_{n=1}^{\infty} nc_nx^{n-1} = 2px \sum_{n=0}^{\infty} c_nx^n.$$

Equation (7) can be rewritten in the form

$$\sum_{n=1}^{\infty} nc_nx^{n-1} + \sum_{n=1}^{\infty} nc_nx^{n+1} = \sum_{n=0}^{\infty} 2pc_nx^{n+1}. \tag{8}$$

We would like to rewrite each of the sums in equation (8) so that each general term contains the same power of x. This can be done by assuming that each general term contains the term x^k. For the first sum, this amounts to substituting $k = n - 1$. Since n ranges from 1 to ∞, $k = n - 1$ will range from 0 to ∞. Substituting $k = n + 1$ with k ranging from 2 to ∞ into the second sum, and $k = n + 1$ with k ranging from 1 to ∞ into the third sum, allows us to rewrite these sums so that the general term will involve the power x^k. We then obtain

$$\sum_{k=0}^{\infty} (k + 1)c_{k+1}x^k + \sum_{k=2}^{\infty} (k - 1)c_{k-1}x^k = \sum_{k=1}^{\infty} 2pc_{k-1}x^k.$$

Now we can gather like terms in x. We take out the $k = 0$ and $k = 1$ terms first:

$$c_1 + (2c_2 - 2pc_0)x + \sum_{k=2}^{\infty} \{(k + 1)c_{k+1} + [(k - 1) - 2p]c_{k-1}\}x^k = 0.$$

Equating each coefficient to zero yields

$$c_1 = 0, \qquad 2c_2 - 2pc_0 = 0, \qquad 3c_3 + (1 - 2p)c_1 = 0,$$

and in general

$$(k + 1)c_{k+1} + [(k - 1) - 2p]c_{k-1} = 0, \qquad k \geq 1. \tag{9}$$

We note that equation (9) is a difference equation with variable coefficients. This equation is called a **recursion formula** and can be used to evaluate the constants $c_0, c_1, c_2, \ldots$ successively. We see that

$$c_1 = 0, \qquad c_2 = pc_0, \qquad c_3 = 0,$$

and by equation (9), in general

$$c_{k+1} = \frac{2p - k + 1}{k + 1} c_{k-1}.$$

Thus

$$c_4 = \frac{2p - 2}{4} c_2 = \frac{p(p - 1)}{1 \cdot 2} c_0, \qquad c_5 = 0,$$

$$c_6 = \frac{2p - 4}{6} c_4 = \frac{p(p - 1)(p - 2)}{1 \cdot 2 \cdot 3} c_0, \qquad c_7 = 0, \ldots,$$

since $c_3 = 0$. Thus the coefficients with odd-numbered subscripts vanish and the power series for y is given by

$$\begin{aligned} y &= c_0 + \frac{p}{1} c_0 x^2 + \frac{p(p - 1)}{1 \cdot 2} c_0 x^4 + \frac{p(p - 1)(p - 2)}{1 \cdot 2 \cdot 3} c_0 x^6 + \cdots \\ &= c_0 \left(1 + \frac{p}{1} x^2 + \frac{p(p - 1)}{1 \cdot 2} x^4 + \frac{p(p - 1)(p - 2)}{1 \cdot 2 \cdot 3} x^6 + \cdots \right). \end{aligned}$$

The binomial series (see equation (12.12.26), states that

$$(1 + x)^p = 1 + \frac{p}{1} x + \frac{p(p - 1)}{1 \cdot 2} x^2 + \frac{p(p - 1)(p - 2)}{1 \cdot 2 \cdot 3} x^3 + \cdots.$$

Replacing x by x^2 in this equation yields the general solution of the differential equation:

$$y = c_0(1 + x^2)^p.$$

Since $y(0) = 1$, it follows that $c_0 = 1$ and $y = (1 + x^2)^p$.

EXAMPLE 4

USING A POWER SERIES TO OBTAIN ONE OF TWO SOLUTIONS TO A DIFFERENTIAL EQUATION

Consider the differential equation

$$y'' + xy' + y = 0. \tag{10}$$

Using equations (1), (2), and (3), we obtain the equation

$$\sum_{n=2}^{\infty} n(n-1)c_n x^{n-2} + x\sum_{n=1}^{\infty} nc_n x^{n-1} + \sum_{n=0}^{\infty} c_n x^n = 0.$$

Reindexing to obtain equal powers of x, we have

$$\sum_{k=0}^{\infty} (k+2)(k+1)c_{k+2}x^k + \sum_{k=1}^{\infty} kc_k x^k + \sum_{k=0}^{\infty} c_k x^k = 0.$$

Note that the second sum can also be allowed to range from 0 to ∞. Gathering like terms in x produces the equation

$$\sum_{k=0}^{\infty} [(k+2)(k+1)c_{k+2} + (k+1)c_k]x^k = 0.$$

Setting the coefficients equal to zero, we obtain the general recursion formula

$$(k+2)(k+1)c_{k+2} + (k+1)c_k = 0.$$

Therefore $(k+2)c_{k+2} = -c_k$, and

$$c_2 = -\frac{c_0}{2}, \qquad c_3 = -\frac{c_1}{3}, \qquad c_4 = -\frac{c_2}{4} = \frac{c_0}{2\cdot 4},$$

$$c_5 = -\frac{c_3}{5} = \frac{c_1}{3\cdot 5}, \qquad c_6 = -\frac{c_4}{6} = -\frac{c_0}{2\cdot 4\cdot 6}, \qquad \text{etc.}$$

Hence the power series for y can be written in the form

$$\begin{aligned} y &= c_0 + c_1 x - \frac{c_0}{2}x^2 - \frac{c_1}{3}x^3 + \frac{c_0}{2\cdot 4}x^4 + \frac{c_1}{3\cdot 5}x^5 - \cdots \\ &= c_0\left(1 - \frac{x^2}{2} + \frac{x^4}{2\cdot 4} - \frac{x^6}{2\cdot 4\cdot 6} + \cdots\right) + c_1\left(x - \frac{x^3}{3} + \frac{x^5}{3\cdot 5} - \frac{x^7}{3\cdot 5\cdot 7} + \cdots\right) \end{aligned} \tag{11}$$

by separating the terms that involve c_0 and c_1. At this point we try to see whether we recognize the two series that have been obtained by the power series method. Very frequently this is an unproductive task, but in this instance we are fortunate:

$$\begin{aligned} 1 - \frac{x^2}{2} + \frac{x^4}{2\cdot 4} - \frac{x^6}{2\cdot 4\cdot 6} + \cdots &= 1 + \left(-\frac{x^2}{2}\right) + \frac{1}{2!}\left(-\frac{x^2}{2}\right)^2 + \frac{1}{3!}\left(-\frac{x^2}{2}\right)^3 + \cdots \\ &= e^{-x^2/2}. \end{aligned}$$

The second series is not a familiar one, so we use the method given in Section 10.8 of finding one solution when another is known. By equation (10.8.8), we have

$$\begin{aligned} y_2 &= y_1 \int \frac{e^{-\int x\,dx}}{y_1{}^2}\,dx = e^{-x^2/2} \int \frac{e^{-x^2/2}}{(e^{-x^2/2})^2}\,dx \\ &= e^{-x^2/2} \int e^{x^2/2}\,dx. \end{aligned} \tag{12}$$

The integral in equation (12) does not have a closed form solution. That this is indeed the second series in equation (11) can be verified by integrating the series for $e^{x^2/2}$ term by term and multiplying the result by the series for $e^{-x^2/2}$. Hence the general solution of equation (10) is given by

$$y = c_0 e^{-x^2/2} + c_1 e^{-x^2/2} \int e^{x^2/2}\,dx.$$

EXAMPLE 5

THE BESSEL FUNCTION OF INDEX ZERO

Solve the equation

$$xy'' + y' + xy = 0. \tag{13}$$

SOLUTION: Using the power series (1), (2), and (3) for equation (13) yields the equation

$$\sum_{n=2}^{\infty} n(n-1)c_n x^{n-1} + \sum_{n=1}^{\infty} nc_n x^{n-1} + \sum_{n=0}^{\infty} c_n x^{n+1} = 0.$$

Reindexing the series to obtain like powers of x, we have

$$\sum_{k=1}^{\infty} (k+1)kc_{k+1}x^k + \sum_{k=0}^{\infty} (k+1)c_{k+1}x^k + \sum_{k=1}^{\infty} c_{k-1}x^k = 0.$$

Condensing the three series in one yields, after some algebra,

$$c_1 + \sum_{k=1}^{\infty} [(k+1)^2 c_{k+1} + c_{k-1}]x^k = 0.$$

Setting the coefficients equal to zero, we have $c_1 = 0$ and

$$(k+1)^2 c_{k+1} = -c_{k-1}, \qquad k = 1, 2, 3, \ldots. \tag{14}$$

The recursion formula (14) together with $c_1 = 0$ implies that all coefficients with odd-numbered subscripts vanish, and

$$c_2 = -\frac{c_0}{2^2}, \qquad c_4 = -\frac{c_2}{4^2} = \frac{c_0}{2^2 4^2}, \qquad c_6 = -\frac{c^4}{6^2} = -\frac{c_0}{2^2 4^2 6^2}, \qquad \ldots.$$

Hence

$$\begin{aligned} y &= c_0 - \frac{c_0}{2^2}x^2 + \frac{c_0}{2^2 4^2}x^4 - \frac{c_0}{2^2 4^2 6^2}x^6 + \cdots \\ &= c_0 \sum_{n=0}^{\infty} \frac{1}{(n!)^2}\left(-\frac{x^2}{4}\right)^n. \end{aligned} \tag{15}$$

It is unlikely that you are familiar with the series in equation (15). This series is often used in applied mathematics and is known as the **Bessel function of index zero**, $J_0(x)$. Note also that the power series method has produced only *one* of the solutions of equation (13). To find the other solution, we can again proceed as in Example 4. Thus

$$y_2(x) = J_0(x) \int \frac{dx}{x J_0^{\,2}(x)}.$$

Finally, the general solution of equation (13) is given by

$$y(x) = AJ_0(x) + BJ_0(x) \int \frac{dx}{x J_0^{\,2}(x)}.$$

In our next example we meet a situation in which the power series method fails to yield any solution.

EXAMPLE 6

AN EXAMPLE WHERE THE POWER SERIES METHOD FAILS

Solve the Euler equation

$$x^2 y'' + xy' + y = 0. \tag{16}$$

SOLUTION: Making use of series (1), (2), and (3), and multiplying by the appropriate powers of x, we have

$$\sum_{n=2}^{\infty} n(n-1)c_n x^n + \sum_{n=1}^{\infty} nc_n x^n + \sum_{n=0}^{\infty} c_n x^n = 0$$

or

$$\sum_{n=0}^{\infty} (n^2 + 1)c_n x^n = 0. \tag{17}$$

Clearly, if we equate each of the coefficients of equation (17) to zero, all the coefficients c_n vanish and $y \equiv 0$. Thus in this case the power series method fails completely in helping us find the general solution

$$y = A \cos(\ln |x|) + B \sin(\ln |x|)$$

of equation (16) (check!).

When initial conditions are given, there is another method based on the Taylor series that can also be used.

EXAMPLE 1 (REVISITED)

USING TAYLOR'S METHOD TO SOLVE AN INITIAL-VALUE PROBLEM

Consider again the initial value problem

$$y' = y + x^2, \qquad y(0) = 1. \tag{18}$$

Differentiating both sides of the differential equation repeatedly and evaluating each derivative at the initial value of $x = 0$, we have

$$y'(0) = y + x^2|_{x=0} = y(0) + (0)^2 = 1,$$
$$y''(0) = y' + 2x|_{x=0} = y'(0) + 2(0) = 1,$$
$$y'''(0) = y'' + 2|_{x=0} = y''(0) + 2 = 3,$$
$$y^{(4)}(0) = y'''|_{x=0} = y'''(0) = 3, \ldots .$$

Substituting these derivatives in the Taylor series

$$y(x) = \sum_{n=0}^{\infty} \frac{y^{(n)}(x_0)}{n!}(x - x_0)^n \tag{19}$$

with $x_0 = 0$, we have

$$y(x) = 1 + x + \frac{x^2}{2!} + \frac{3x^3}{3!} + \frac{3x^4}{4!} + \cdots = 3e^x - 2 - 2x - x^2,$$

which is the result that we obtained before.

Taylor's method is easily adapted to higher-order initial value problems by rewriting the differential equation so that the highest-order derivative is expressed in terms of the other derivatives and the independent variable. Successive differentiations again yields the values $y^{(n)}(x_0)$ for substitution into the Taylor series.

PROBLEMS 12.13

SELF-QUIZ

Multiple Choice

I. If $\sum_{n=0}^{\infty} c_n x^n$ is the solution to $y'' + xy' + y = 0$, $y(0) = 1$, $y'(0) = 1$, then $c_0 = 1$, $c_1 = 1$ and $c_2 =$ ________.

a. 1 **b.** -1 **c.** $\frac{1}{2}$

d. $-\frac{1}{2}$ **e.** $\frac{2}{3}$ **f.** $-\frac{2}{3}$

True–False

II. If $a(x)$, $b(x)$ and $c(x)$ are analytic, then the power series method will *always* yield at least one solution to the differential equation $a(x)y''(x) + b(x)y'(x) + c(x)y(x) = 0$.

In Problems 1–16 find the general solution of each equation by the power series method. When initial conditions are specified, give the solution that satisfies them.

1. $y' = y - x$, $y(0) = 2$
2. $y' = x^3 - 2xy$, $y(0) = 1$
3. $y'' + y = x$
4. $y'' + 4y = 0$, $y(0) = 1$, $y'(0) = 0$
5. $(1 + x^2)y'' + 2xy' - 2y = 0$
6. $xy'' - xy' + y = e^x$, $y(0) = 1$, $y'(0) = 2$
7. $xy'' - x^2y' + (x^2 - 2)y = 0$, $y(0) = 0$, $y'(0) = 1$
8. $(1 - x)y'' - y' + xy = 0$, $y(0) = y'(0) = 1$
9. $y'' - 2xy' + 4y = 0$, $y(0) = 1$, $y'(0) = 0$
10. $(1 - x^2)y'' - xy' + y = 0$, $y(0) = 0$, $y'(0) = 1$
11. $y'' - xy' + y = -x \cos x$, $y(0) = 0$, $y'(0) = 2$
12. $y'' - xy' + xy = 0$, $y(0) = 2$, $y'(0) = 1$
13. $(1 - x)^2y'' - (1 - x)y' - y = 0$, $y(0) = y'(0) = 1$
14. $y'' - 2xy' + 2y = 0$
15. $y'' - 2xy' - 2y = x$, $y(0) = 1$, $y'(0) = -\frac{1}{4}$
16. $y'' - x^2y = 0$
17. **Airy's equation**
$$y'' - xy = 0$$

has applications in the theory of diffraction. Find the general solution of this equation.

18. **Hermite's equation**

$$y'' - 2xy' + 2py = 0,$$

where p is constant, arises in quantum mechanics in connection with the Schrödinger equation for a harmonic oscillator. Show that if p is a positive integer, one of the two linearly independent solutions of Hermite's equation is a polynomial, called the **Hermite polynomial** $H_p(x)$.

19. Use the Taylor series method to solve Airy's equation

$$y'' - xy = 0, \qquad y(1) = 1, \qquad y'(1) = 0.$$

20. Use the Taylor series method to solve

$$y'' - xy' - y = 0, \qquad y(0) = 1, \qquad y'(0) = 0.$$

21. Does the power series method yield a solution to the equation

a. $x^2y' = y$?

b. $x^3y' = y$?

***22.** Solve $y' = y\sqrt{y^2 - 1}$ by squaring the power series for y.

23. Show that the power series method fails for

$$x^2y'' + x^2y' + y = 0.$$

ANSWERS TO SELF-QUIZ

I. d **II.** False

CHAPTER 12 SUMMARY OUTLINE

- **Factorial Notation** $n! = n(n-1)(n-2)\cdots 3\cdot 2\cdot 1.$

- **Taylor Polynomial** Let the function f and its first n derivatives exist on the closed interval $[x_0, x_1]$. Then, for $a \in (x_0, x_1)$ and $x \in (x_0, x_1)$, the nth-degree **Taylor polynomial of** f at a is the nth-degree polynomial $P_n(x)$, given by

$$P_n(x) = f(a) + \frac{f'(a)}{1!}(x-a) + \frac{f''(a)}{2!}(x-a)^2 + \frac{f'''(a)}{3!}(x-a)^3 + \cdots + \frac{f^{(n)}(a)}{n!}(x-a)^n$$

$$= \sum_{k=0}^{n} \frac{f^{(k)}(a)(x-a)^k}{k!}.$$

- **Remainder Term** Let $P_n(x)$ be the nth-degree Taylor polynomial of the function f. Then the **remainder term**, denoted by $R_n(x)$, is given by

$$R_n(x) = f(x) - P_n(x).$$

- **Taylor's Theorem (Taylor's Formula with Remainder)** Suppose that $f^{n+1}(x)$ exists on the closed interval $[x_0, x_1]$. Let a be in (x_0, x_1) and let x be any number in $[x_0, x_1]$. Then there is a number c in (a, x) or (x, a) such that

$$R_n(x) = \frac{f^{(n+1)}(c)}{(n+1)!}(x-a)^{n+1}. \qquad (*)$$

The expression in (*) is called **Lagrange's form of the remainder**.

- **Taylor Polynomial at 0**

$$P_n(x) = f(0) + f'(0)x + \frac{f''(0)}{2!}x^2 + \cdots + \frac{f^{(n)}(0)}{n!}x^n.$$

- If Q is a polynomial of degree n and P_n is its nth-degree Taylor polynomial at a, then $P_n(x) = Q(x)$ for every real number x and every real number a.

- **The Maximum Error of the Taylor Polynomial Approximation** If f has $n+1$ continuous derivatives on $[a, b]$ and if M_n is the maximum value of $|f^{(n)}(x)|$ on $[a, b]$, then for every x in $[a, b]$

$$|R_n(x)| \le M_n \frac{|x-a|^{n+1}}{(n+1)!}.$$

- **Sequence** A **sequence** of real numbers is a function whose domain is the set of positive integers. The values taken by the function are called **terms** of the sequence.

- **Finite Limit of a Sequence** A sequence $\{a_n\}$ has the **limit** L if for every $\epsilon > 0$ there exists an integer $N > 0$ such that if $n \ge N$, then $|a_n - L| < \epsilon$. We write

$$\lim_{n\to\infty} a_n = L. \qquad (**)$$

- **Infinite Limit of a Sequence** The sequence $\{a_n\}$ has the limit ∞ if for every positive number M there is an integer $N > 0$ such that if $n > N$, then $a_n > M$. In this case we write

$$\lim_{n\to\infty} a_n = \infty.$$

■ **Limit of a Geometric Sequence**

$\lim_{n\to\infty} r^n = 0$ if $|r| < 1$ and $\lim_{n\to\infty} |r^n| = \infty$ if $|r| > 1$.

■ **Some Limit Theorems for Sequences** Suppose that $\lim_{n\to\infty} a_n$ and $\lim_{n\to\infty} b_n$ both exist and are finite.

i. $\lim_{n\to\infty} \alpha a_n = \alpha \lim_{n\to\infty} a_n$ for any real number α.

ii. $\lim_{n\to\infty}(a_n + b_n) = \lim_{n\to\infty} a_n + \lim_{n\to\infty} b_n$.

iii. $\lim_{n\to\infty} a_n b_n = (\lim_{n\to\infty} a_n)(\lim_{n\to\infty} b_n)$.

iv. If $\lim_{n\to\infty} b_n \neq 0$, then

$$\lim_{n\to\infty} \frac{a_n}{b_n} = \frac{\lim_{n\to\infty} a_n}{\lim_{n\to\infty} b_n}.$$

■ **Continuity Theorem** Suppose that L is finite and $\lim_{n\to\infty} a_n = L$. If f is continuous in an open interval containing L, then

$$\lim_{n\to\infty} f(a_n) = f(\lim_{n\to\infty} a_n) = f(L).$$

■ **Squeezing Theorem** Suppose that $\lim_{n\to\infty} a_n = \lim_{n\to\infty} b_n = L$ and that $\{c_n\}$ is a sequence having the property that for $n > N$ (a positive integer), $a_n \leq c_n \leq b_n$. Then,

$$\lim_{n\to\infty} c_n = L.$$

■ **Convergence and Divergence of a Sequence** If the limit in (**) exists and if L is finite, we say that the sequence **converges** or is **convergent**. Otherwise, we say that the sequence **diverges** or is **divergent**.

■ **Boundedness**

i. The sequence $\{a_n\}$ is **bounded above** if there is a number M_1 such that

$$a_n \leq M_1 \text{ for every positive integer } n.$$

ii. It is **bounded below** if there is a number M_2 such that

$$M_2 \leq a_n \text{ for every positive integer } n.$$

iii. It is **bounded** if there is a number $M > 0$ such that

$$|a_n| \leq M \text{ for every positive integer } n.$$

The numbers M_1, M_2, and M are called, respectively, an **upper bound**, a **lower bound**, and a **bound** for $\{a_n\}$.

iv. If the sequence is not bounded, it is called **unbounded**.

■ **Every convergent sequence is bounded.**

■ **Every unbounded sequence is divergent.**

■ **Monotonicity**

i. The sequence $\{a_n\}$ is **monotone increasing** if $a_n \leq a_{n+1}$ for every $n \geq 1$.

ii. The sequence $\{a_n\}$ is **monotone decreasing** if $a_n \geq a_{n+1}$ for every $n \geq 1$.

iii. The sequence $\{a_n\}$ is **monotonic** if it is either monotone increasing or monotone decreasing.

■ **Strict Monotonicity**

i. The sequence $\{a_n\}$ is **strictly increasing** if $a_n < a_{n+1}$ for every $n \geq 1$.

ii. The sequence $\{a_n\}$ is **strictly decreasing** if $a_n > a_{n+1}$ for every $n \geq 1$.

iii. The sequence $\{a_n\}$ is **strictly monotonic** if it is either strictly increasing or strictly decreasing.

■ **A bounded monotonic sequence is convergent.**

■ **Fixed-Point Convergence Theorem** Suppose that the function F has a fixed point u (i.e., $F(u) = u$) and that there is a number c such that

i. F is continuous in $[u - c, u + c]$.

ii. F is differentiable in $(u - c, u + c)$.

iii. $|F'(x)| \leq M < 1$ for every x in $(u - c, u + c)$.

Then u is the only fixed point of F in $(u - c, u + c)$ and the sequence $\{x_n\}$ generated by the fixed-point iteration $x_{n+1} = F(x_n)$ converges to u for every choice of x_1 in $(u - c, u + c)$.

■ **Sum of a Geometric Progression** The sum of a **geometric progression** is a sum of the form

$$S_n = 1 + r + r^2 + r^3 + \cdots + r^{n-1} + r^n = \sum_{k=0}^{n} r^k,$$

where r is a real number and n is a fixed positive integer. If $r \neq 1$, then

$$S_n = \frac{1 - r^{n+1}}{1 - r}.$$

■ **Geometric Series** A **geometric series** is an infinite sum of the form

$$S = \sum_{k=0}^{\infty} r^k = 1 + r + r^2 + r^3 + \cdots.$$

■ **Convergence and Divergence of a Geometric Series** Let $S_n = \sum_{k=0}^{n} r^k$. Then we say that the geometric series **converges** if $\lim_{n\to\infty} S_n$ exists and is finite. Otherwise, the series is said to **diverge**.

i. The geometric series converges to

$$\frac{1}{1 - r} \quad \text{if } |r| < 1.$$

ii. The geometric series diverges if $|r| \geq 1$.

■ **Infinite Series** Let $\{a_n\}$ be a sequence. Then the infinite sum

$$\sum_{k=1}^{\infty} a_k = a_1 + a_2 + a_3 + \cdots + a_n + \cdots \qquad (\surd)$$

is called an **infinite series** (or simply, **series**).

- **Terms of the Series** Each a_k in ($\surd$) is called a **term** of the series.
- **Partial Sums** The **partial sums** of the series are given by

$$S_n = \sum_{k=1}^{n} a_k.$$

- **Convergent Series** The sum S_n is called the ***n*th partial sum** of the series. If the sequence of partial sums $\{S_n\}$ converges to L, a finite number, then we say that the infinite series $\sum_{k=1}^{\infty} a_k$ **converges** to L, and we write

$$\sum_{k=1}^{\infty} a_k = L.$$

- **Divergent Series** Otherwise, we say that the series $\sum_{k=1}^{\infty} a_k$ **diverges**.
- **Harmonic Series** The series

$$\sum_{k=1}^{\infty} \frac{1}{k} = 1 + \frac{1}{2} + \frac{1}{3} + \frac{1}{4} + \cdots + \frac{1}{n} + \cdots$$

is called the **harmonic series**. The harmonic series diverges.

- **Adding Series and Multiplying a Series by a Constant** Let c be a constant. Suppose that $\sum_{k=1}^{\infty} a_k$ and $\sum_{k=1}^{\infty} b_k$ both converge. Then $\sum_{k=1}^{\infty} (a_k + b_k)$ and $\sum_{k=1}^{\infty} ca_k$ converge, and

i. $\displaystyle\sum_{k=1}^{\infty} (a_k + b_k) = \sum_{k=1}^{\infty} a_k + \sum_{k=1}^{\infty} b_k$

ii. $\displaystyle\sum_{k=1}^{\infty} ca_k = c\sum_{k=1}^{\infty} a_k.$

- **A Criterion for Convergence** If $\sum_{k=1}^{\infty} a_k$ converges, then $\lim_{n\to\infty} a_n = 0$ (The converse is not true; see the harmonic series)
- **Comparison Test** Let $\sum_{k=1}^{\infty} a_k$ be a series with $a_k \geq 0$ for every k.

  i. If there exists a convergent series $\sum_{k=1}^{\infty} b_k$ and a number N such that $a_k \leq b_k$ for every $k \geq N$, then $\sum_{k=1}^{\infty} a_k$ converges.

  ii. If there exists a divergent series $\sum_{k=1}^{\infty} c_k$ and a number N such that $a_k \geq c_k \geq 0$ for every $k \geq N$, then $\sum_{k=1}^{\infty} a_k$ diverges.

- **The first few terms of a series do not affect convergence or divergence** If for some positive integer N, $\sum_{k=N+1}^{\infty} a_k$ converges, then $\sum_{k=1}^{\infty} a_k$ also converges. If $\sum_{k=N+1}^{\infty} a_k$ diverges, then $\sum_{k=1}^{\infty} a_k$ diverges.
- **The Integral Test** Let f be a function that is continuous, positive, and decreasing for all $x \geq 1$. Then the series

$$\sum_{k=1}^{\infty} f(k) = f(1) + f(2) + f(3) + \cdots + f(n) + \cdots$$

converges if and only if $\int_1^{\infty} f(x)\,dx$ converges, and diverges if and only if $\int_1^{n} f(x)\,dx \to \infty$ as $n \to \infty$.

- **The *p*-series**

$$\sum_{k=1}^{\infty} \frac{1}{k^p} \quad \begin{cases} \text{diverges if } 0 < p \leq 1, \\ \text{converges if } p > 1. \end{cases}$$

- **The Limit Comparison Test** Let $\sum_{k=1}^{\infty} a_k$ and $\sum_{k=1}^{\infty} b_k$ be series with positive terms. If there is a number $c > 0$ such that

$$\lim_{k\to\infty} \frac{a_k}{b_k} = c,$$

then either both series converge or both series diverge.

- **The Ratio Test** Let $\sum_{k=1}^{\infty} a_k$ be a series with $a_k > 0$ for every k, and suppose that

$$\lim_{n\to\infty} \frac{a_{n+1}}{a_n} = L.$$

  i. If $L < 1$, $\sum_{k=1}^{\infty} a_k$ converges.

  ii. If $L > 1$, $\sum_{k=1}^{\infty} a_k$ diverges.

  iii. If $L = 1$, $\sum_{k=1}^{\infty} a_k$ may converge or diverge and the ratio test is inconclusive; some other test must be used.

- **The Root Test** Let $\sum_{k=1}^{\infty} a_k$ be a series with $a_k > 0$ and suppose that $\lim_{n\to\infty}(a_n)^{1/n} = R$.

  i. If $R < 1$, $\sum_{k=1}^{\infty} a_k$ converges.

  ii. If $R > 1$, $\sum_{k=1}^{\infty} a_k$ diverges.

  iii. If $R = 1$, the series either converges or diverges, and no conclusions can be drawn from this test.

- **Absolute Convergence** The series $\sum_{k=1}^{\infty} a_k$ is said to **converge absolutely** if the series $\sum_{k=1}^{\infty} |a_k|$ converges.
- **Absolute convergence implies convergence** If $\sum_{k=1}^{\infty} |a_k|$ converges, then $\sum_{k=1}^{\infty} a_k$ also converges.
- **Alternating Series** A series in which successive terms have opposite signs is called an **alternating series**.
- **Alternating Series Test** Let $\{a_k\}$ be a decreasing sequence of positive numbers such that $\lim_{k\to\infty} a_k = 0$. Then the alternating series $\sum_{k=1}^{\infty} (-1)^{k+1} a_k = a_1 - a_2 + a_3 - a_4 + \cdots$ converges.
- **Conditional Convergence** An alternating series is said to be **conditionally convergent** if it is convergent but not absolutely convergent.
- **An Error Bound on the Partial Sums of a Converging Alternating Series** If $S = \sum_{k=1}^{\infty} (-1)^{k+1} a_k$ is a convergent alternating series with monotone decreasing terms, then for any n,

$$|S - S_n| \leq |a_{n+1}|$$

- **A Better Error Bound** Suppose $S = \sum_{k=1}^{\infty} (-1)^{k+1} a_k$ is a converging alternating series with monotone decreasing terms and that, in addition, the sequence $\{|a_n - a_{n+1}|\}$ is monotone decreasing. Let $T_n = S_{n-1} - (-1)^n \frac{1}{2} a_n$. Then,

$$|S - T_n| \leq \tfrac{1}{2}|a_n - a_{n+1}|.$$

- **Reordering a Conditionally Convergent Series** By reordering the terms of a conditionally convergent alternating series, the new series of rearranged terms can be made to converge to any real number.

- **Power Series**

  i. A **power series** in x is a series of the form

$$\sum_{k=0}^{\infty} a_k x^k = a_0 + a_1 x + a_2 x^2 + \cdots + a_k x^k + \cdots.$$

  ii. A **power series** in $(x - x_0)$ is a series of the form

$$\sum_{k=0}^{\infty} a_k (x - x_0)^k = a_0 + a_1(x - x_0) + a_2(x - x_0)^2 + \cdots + a_k(x - x_0)^k + \cdots$$

  where x_0 is a real number.

- **Convergence and Divergence of a Power Series**

  i. A power series is said to **converge** at x_0 if the series of real numbers $\Sigma_{k=0}^{\infty} a_k x_0^k$ converges. Otherwise, it is said to **diverge** at x_0.

  ii. A power series is said to converge in a set D of real numbers if it converges for every real number x in D.

- **Radius of Convergence** A power series may be placed in one of three categories:

  Category 1: $\Sigma_{k=0}^{\infty} a_k x^k$ converges only at 0.

  Category 2: $\Sigma_{k=0}^{\infty} a_k x^k$ converges for all real numbers.

  Category 3: There exists a positive real number R, called the **radius of convergence** of the power series, such that $\Sigma_{k=0}^{\infty} a_k x^k$ converges if $|x| < R$ and diverges if $|x| > R$. At $x = R$ and at $x = -R$, the series may converge or diverge.

  In Category 1, $R = 0$. In Category 2, $R = \infty$.

- **Interval of Convergence** The **interval of convergence** of a power series is the interval over which the power series converges.

- **Differentiating and Integrating a Power Series** Let the power series $\Sigma_{k=0}^{\infty} a_k x^k$ have the radius of convergence $R > 0$, and suppose that

$$f(x) = \sum_{k=0}^{\infty} a_k x^k = a_0 + a_1 x + a_2 x^2 + \cdots \quad \text{for } |x| < R.$$

  Then for $|x| < R$ we have the following:

  i. $f(x)$ is continuous.

  ii. The derivative $f'(x)$ exists, and

$$f'(x) = \frac{d}{dx} a_0 + \frac{d}{dx} a_1 x + \frac{d}{dx} a_2 x^2 + \cdots$$

$$= a_1 + 2a_2 x + 3a_3 x^2 + \cdots = \sum_{k=1}^{\infty} k a_k x^{k-1}.$$

  iii. The indefinite integral $\int f(x)\,dx$ exists and

$$\int f(x)\,dx = \int a_0\,dx + \int a_1 x\,dx + \int a_2 x^2\,dx + \cdots$$

$$= a_0 x + a_1 \frac{x^2}{2} + a_2 \frac{x^3}{3} + \cdots + C$$

$$= \sum_{k=0}^{\infty} a_k \frac{x^{k+1}}{k+1} + C.$$

  Moreover, the two series $\Sigma_{k=1}^{\infty} k a_k x^{k-1}$ and $\Sigma_{k=0}^{\infty} a_k x^{k+1}/(k+1)$ both have the radius of convergence R.

- **Taylor Series and Maclaurin Series** Suppose that

$$f(x) = \sum_{k=0}^{\infty} a_k (x - x_0)^k$$

  for every x in the interval of convergence and that $R > 0$. Then

$$f(x) = \sum_{k=0}^{\infty} \frac{f^{(k)}(x_0)}{k!}(x - x_0)^k$$

$$= f(x_0) + f'(x_0)(x - x_0) + f''(x_0)\frac{(x - x_0)^2}{2!} + \cdots + f^{(n)}(x_0)\frac{(x - x_0)^n}{n!} + \cdots$$

  for every x in the interval of convergence.

  This last series is called the **Taylor series** for f at x_0. If $x_0 = 0$, then the series is called the **Maclaurin series** for f.

- **Analytic Function** We say that a function f is **analytic** at x_0 if f can be represented by a Taylor series in some neighborhood of x_0.

- **A Theorem about Analytic Functions** Suppose that the function f has continuous derivatives of all orders in a neighborhood $N(x_0)$ of the number x_0. Then f is analytic at x_0 if and only if

$$\lim_{n\to\infty} R_n(x) = \lim_{n\to\infty} \frac{f^{(n+1)}(c_n)}{(n+1)!}(x - x_0)^{n+1} = 0$$

  for every x in $N(x_0)$ where c_n is between x_0 and x.

- **Some Useful Maclaurin Series**

$$e^x = \sum_{k=0}^{\infty} \frac{x^k}{k!} = 1 + x + \frac{x^2}{2!} + \frac{x^3}{3!} + \cdots$$

$$\cos x = \sum_{k=0}^{\infty} \frac{(-1)^k x^{2k}}{(2k)!} = 1 - \frac{x^2}{2!} + \frac{x^4}{4!} - \frac{x^6}{6!} + \cdots$$

$$\sin x = \sum_{k=0}^{\infty} \frac{(-1)^k x^{2k+1}}{(2k+1)!} = x - \frac{x^3}{3!} + \frac{x^5}{5!} - \frac{x^7}{7!} + \cdots$$

$$\cosh x = \sum_{k=0}^{\infty} \frac{x^{2k}}{(2k)!} = 1 + \frac{x^2}{2!} + \frac{x^4}{4!} + \frac{x^6}{6!} + \cdots$$

$$\sinh x = \sum_{k=0}^{\infty} \frac{x^{2k+1}}{(2k+1)!} = x + \frac{x^3}{3!} + \frac{x^5}{5!} + \frac{x^7}{7!} + \cdots$$

$$\frac{1}{1-x} = \sum_{k=0}^{\infty} x^k = 1 + x + x^2 + \cdots \qquad |x| < 1 \qquad \text{geometric series}$$

$$\ln(1+x) = \sum_{k=0}^{\infty} \frac{(-1)^k x^{k+1}}{k+1}, \qquad |x| < 1$$

■ The Binomial Series

$$(1+x)^r = 1 + rx + \frac{r(r-1)}{2!}x^2 + \frac{r(r-1)(r-2)}{3!}x^3 + \cdots + \frac{r(r-1)\cdots(r-n+1)}{n!}x^n + \cdots$$
$$= 1 + \sum_{k=1}^{\infty} \frac{r(r-1)\cdots(r-k+1)}{k!}x^k, \qquad |x| < 1.$$

CHAPTER 12 REVIEW EXERCISES

In Problems 1–8, find the nth-degree Taylor polynomial for f at a.

1. $f(x) = e^x$; $a = 0$; $n = 3$
2. $f(x) = \ln x$; $a = 1$; $n = 4$
3. $f(x) = \sin x$; $a = \pi/6$; $n = 3$
4. $f(x) = \cos x$; $a = \pi/2$; $n = 5$
5. $f(x) = \cot x$; $a = \pi/2$; $n = 4$
6. $f(x) = \sinh x$; $a = 0$; $n = 3$
7. $f(x) = x^3 - x^2 + 2x + 3$; $a = 0$; $n = 8$
8. $f(x) = e^{-x^2}$; $a = 0$; $n = 5$

In Problems 9–12, find a close bound on $|R_n(x)|$ for the specified interval.

9. $f(x) = \cos x$; $a = \pi/6$; $n = 5$; $[0, \pi/2]$
10. $f(x) = \sqrt[3]{x}$; $a = 1$; $n = 4$; $[\frac{7}{8}, \frac{9}{8}]$
11. $f(x) = e^x$; $a = 0$; $n = 6$; $[-\ln e, \ln e]$
12. $f(x) = \cot x$; $a = \pi/2$; $n = 2$; $[\pi/4, 3\pi/4]$
13. Use a Taylor polynomial of degree 4 to approximate

$$\int_0^{1/2} \cos x^2\, dx.$$

Find the maximum error of the approximation.

***14.** Use a Taylor polynomial of degree 4 to approximate

$$\int_{-\pi}^{\pi} e^{\cos x}\, dx.$$

Find the maximum error of the approximation.

15. Find the first five terms of the sequence $\{(n-2)/n\}$.
16. Find the first seven terms of the sequence $\{n^2 \sin n\}$.
17. Find the general term of the sequence $\{\frac{1}{8}, \frac{3}{16}, \frac{5}{32}, \frac{7}{64}, \ldots\}$.
18. Find the general term of the sequence $\{1, -\frac{1}{5}, \frac{1}{25}, -\frac{1}{125}, \frac{1}{625}, \ldots\}$.

In Problems 18–24, determine whether the given sequence is convergent or divergent. If it is convergent, find its limit.

19. $\left\{\frac{-7}{n}\right\}$ **20.** $\{\cos \pi n\}$
21. $\left\{\frac{\ln n}{\sqrt{n}}\right\}$ **22.** $\left\{\frac{7^n}{n!}\right\}$
23. $\left\{\left(1 - \frac{2}{n}\right)^n\right\}$ **24.** $\left\{\frac{3}{\sqrt{n^2+8} - n}\right\}$

In Problems 25–32, determine whether the given sequence is bounded or unbounded and whether it is increasing, decreasing, or not monotonic. Start with $n = 1$.

25. $\{\sqrt{n} \cos n\}$ **26.** $\left\{\frac{3}{n+2}\right\}$
27. $\left\{\frac{2^n}{1+2^n}\right\}$ **28.** $\left\{\frac{n!}{n^n}\right\}$
29. $\left\{\frac{\sqrt{n+1}}{n}\right\}$ **30.** $\left\{\left(1 - \frac{1}{n}\right)^{1/n}\right\}$
31. $\left\{\frac{n-7}{n+4}\right\}$ **32.** $\{(3^n + 5^n)^{1/n}\}$

In Problems 33–36, evaluate the given sum.

33. $\sum_{k=2}^{10} 4^k$ **34.** $\sum_{k=1}^{\infty} \frac{1}{3^k}$

35. $\sum_{k=3}^{\infty}\left[\left(\frac{3}{4}\right)^k-\left(\frac{2}{5}\right)^k\right]$ **36.** $\sum_{k=2}^{\infty}\frac{1}{k(k-1)}$

37. Write 0.797979 . . . as a rational number.

38. Write 14.2314231423 . . . as a rational number.

In Problems 39–50, determine whether the given series converges or diverges.

39. $\sum_{k=1}^{\infty}\frac{1}{k^3-5}$ **40.** $\sum_{k=5}^{\infty}\frac{1}{k(k+6)}$

41. $\sum_{k=1}^{\infty}\frac{1}{\sqrt{k^3+4}}$ **42.** $\sum_{k=2}^{\infty}\frac{3}{\ln k}$

43. $\sum_{k=4}^{\infty}\frac{1}{\sqrt[3]{k^3+50}}$ **44.** $\sum_{k=1}^{\infty}\frac{r^k}{k^r}$, $0<r<1$

45. $\sum_{k=2}^{\infty}\frac{10^k}{k^5}$ **46.** $\sum_{k=1}^{\infty}\frac{k^{6/5}}{8^k}$

47. $\sum_{k=1}^{\infty}\frac{\sqrt{k}\ln(k+3)}{k^2+2}$ **48.** $\sum_{k=1}^{\infty}\operatorname{csch} k$

49. $\sum_{k=2}^{\infty}\frac{e^{1/k}}{k^{3/2}}$

50. $\sum_{k=1}^{\infty}\frac{k(k+6)}{(k+1)(k+3)(k+5)}$

In Problems 51–60, determine whether the given alternating series is absolutely convergent, conditionally convergent, or divergent.

51. $\sum_{k=1}^{\infty}\frac{(-1)^{k+1}}{50k}$

52. $\sum_{k=2}^{\infty}\frac{(-1)^k\sqrt{k}}{\ln k}$

53. $\sum_{k=2}^{\infty}\frac{(-1)^{k+1}}{\sqrt{k(k-1)}}$

54. $\sum_{k=2}^{\infty}\frac{(-1)^k k^2}{k^3+1}$

55. $\sum_{k=2}^{\infty}\frac{(-1)^k k^2}{k^4+1}$

56. $\sum_{k=2}^{\infty}\frac{(-1)^k k^3}{k^3+1}$

57. $\sum_{k=3}^{\infty}\frac{(-1)^k(k+2)(k+3)}{(k+1)^3}$

58. $\sum_{k=2}^{\infty}\frac{(-1)^k 3^k}{3^{k+1}}$

59. $\sum_{k=1}^{\infty}\frac{(-1)^k k^k}{k!}$

60. $\sum_{k=1}^{\infty}\frac{(-1)^k k^4}{k^4+20k^3+17k+2}$

61. $\sum_{k=1}^{\infty}(-1)^k\left(1+\frac{1}{k}\right)^k$

62. $\sum_{k=1}^{\infty}\frac{(-1)^k k!}{k^k}$

63. Approximate $\sum_{k=1}^{\infty}(-1)^{k+1}/k^3$ with an error less than 0.001.

64. Approximate $\sum_{k=0}^{\infty}(-1)^k/k!$ with an error less than 0.0001.

65. At what time between 9 P.M. and 10 P.M. is the minute hand of a clock exactly over the hour hand?

66. Find the first 10 terms of a rearrangement of the conditionally convergent series $\sum_{k=1}^{\infty}(-1)^{k+1}/k$ that converges to 0.5.

In Problems 67–76, find the radius of convergence and the interval of convergence of the given power series.

67. $\sum_{k=0}^{\infty}\frac{x^k}{3^k}$ **68.** $\sum_{k=0}^{\infty}\frac{(-1)^k x^k}{3^k}$

69. $\sum_{k=0}^{\infty}\frac{x^k}{k^2+2}$ **70.** $\sum_{k=1}^{\infty}\frac{x^k}{k!}$

71. $\sum_{k=2}^{\infty}\frac{x^k}{(2\ln k)^k}$ **72.** $\sum_{k=0}^{\infty}\frac{(3x+5)^k}{k!}$

73. $\sum_{k=0}^{\infty}\frac{(3x-5)^k}{3^k}$ **74.** $\sum_{k=0}^{\infty}\left(\frac{k}{6}\right)^k x^k$

75. $\sum_{k=0}^{\infty}(-1)^k x^{3k}$ **76.** $\sum_{k=1}^{\infty}\frac{(\ln k)(x-1)^k}{k+2}$

77. Approximate $\int_0^{1/2} e^{-t^2}\,dt$ with an error less than 0.00001.

78. Approximate $\int_0^{1/2}\sin t^2\,dt$ with an error less than 0.0001.

79. Approximate $\int_0^{1/2} t^3 e^{-t^3}\,dt$ with an error less than 0.001.

80. Approximate $\int_0^{1/2}[1/(t^4+1)]\,dt$ with an error less than 0.00001.

81. Find the Maclaurin series for x^2e^x.

82. Find the Taylor series for e^x at ln 3.

83. Find the Maclaurin series for $\cos^2 x$. [*Hint:* $\cos x = (1+\cos 2x)/2$.]

84. Find the Maclaurin series for $\sin\beta x$, β real.

In Exercises 85–87 use the power series method to obtain the solution to the initial value problem.

85. $\frac{dy}{dx}=3y;\ y(0)=2$

86. $y''+9y=0;\ y(0)=1,\ y'(0)=0$

87. $y''-2y'+y=\cos x;\ y(0)=0,\ y'(0)=3$

88. Use the power series method to obtain the general solution to $y''-xy'-y=0$.

CHAPTER 12
COMPUTER EXERCISES

1. a. Show that $\sum_{n=1}^{\infty} \frac{1}{n^2}$ is a convergent series.

b. Use a computer algebra system to evaluate the following partial sums.

i. $\sum_{n=1}^{10} \frac{1}{n^2}$ **ii.** $\sum_{n=1}^{100} \frac{1}{n^2}$ **iii.** $\sum_{n=1}^{500} \frac{1}{n^2}$

c. How does the integral $\int_{500}^{\infty} \frac{1}{x^2}\,dx$ compare with the sum $\sum_{n=501}^{\infty} \frac{1}{n^2}$? (See Figure 1.)

d. If $\sum_{n=1}^{500} \frac{1}{n^2}$ is used to approximate $\sum_{n=1}^{\infty} \frac{1}{n^2}$, find a bound on the error. Express your answer as an integral and evaluate it.

2. In the eighteenth century, Leonhard Euler showed that $\sum_{n=1}^{\infty} \frac{1}{n^2} = \frac{\pi^2}{6}$. While Euler was in fact able to evaluate $\sum_{n=1}^{\infty} \frac{1}{n^k}$ for every positive even integer k, he was unable to obtain an "exact value" for $\sum_{n=1}^{\infty} \frac{1}{n^3}$. This problem has resisted the efforts of the strongest mathematicians for two centuries and remains unsolved today. The series was only recently shown to converge to an irrational number. Find, accurate to 10 decimal places,

a. $\sum_{n=1}^{100} \frac{1}{n^3}$ **b.** $\sum_{n=1}^{1000} \frac{1}{n^3}$ **c.** $\sum_{n=1}^{5000} \frac{1}{n^3}$

3. Euler's constant is defined to be $\lim_{n\to\infty} \gamma_n = \gamma$ where $\gamma_n = \left(1 + \frac{1}{2} + \cdots + \frac{1}{n}\right) - \ln(n+1)$. (Notice that $\lim_{n\to\infty}\left(1 + \frac{1}{2} + \cdots + \frac{1}{n}\right)$ is the harmonic series and so diverges and that $\lim_{n\to\infty} \ln(n+1) = \infty$. One of the goals of this exercise is to show that the limit of the difference exists. Does this square with theorems you know concerning sums and differences of limits?)

a. Show that Euler's constant is in fact the sum of the shaded areas in Figure 2.

b. Use Figure 2 to show that the sequence $\{\gamma_n\}$ is increasing and bounded above. (Therefore the limit exists.)

c. Calculate γ_n for $n = 100$, 500, and 1,000. Use your answers to obtain an approximation of γ.

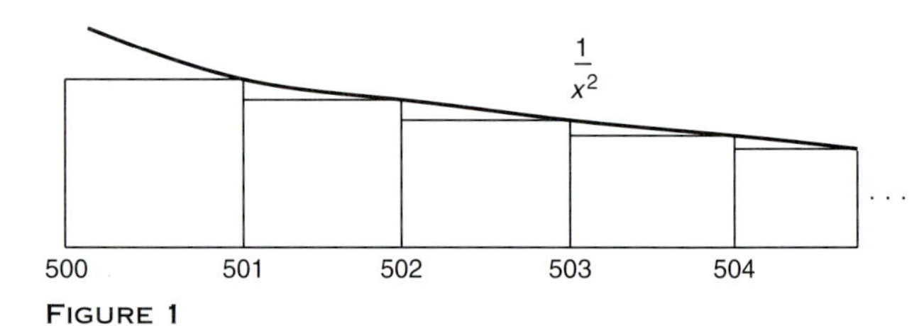

FIGURE 1

d. Write down and evaluate an integral that gives a bound on the error in using $\gamma_{1,000}$ to approximate γ. (Look back at Figure 2.)

e. Use the technique you used to obtain an error estimate in (d) to find the value of n that is needed if γ_n is to approximate γ with error $\leq 10^{-10}$.

4. If $f(x)$ is a function on an open interval about the origin, the partial sums of its Maclaurin series are polynomials which can be graphed along with f. If you compare the graphs, you can see what the geometric meaning of convergence is when f is analytic. Use a computer algebra system to do the calculations and the graphing in each of the following:

a. Graph $f(x) = e^x$ along with the first three distinct partial sums of its Maclaurin series on the same axes.

b. By zooming into the relevant spots on the graph, estimate the largest value of a such that

i. the first-degree Maclaurin polynomial is a "good" approximation to $f(x)$ on the interval $[0, a)$. (By "good" we mean an error of about 0.05.)

ii. the second-degree Maclaurin polynomial is a "good" approximation to $f(x)$ on the interval $[0, a)$.

iii. the third-degree Maclaurin polynomial is a "good" approximation to $f(x)$ on the interval $[0, a)$.

c. Explain how you can use your computer algebra system to approximate these three numbers without resorting to a graph. What results do you get to two decimal places?

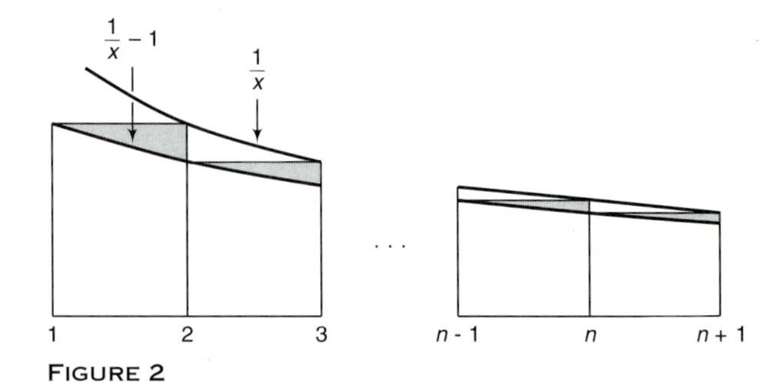

FIGURE 2

5. **a.** Graph $f(x) = \tan x$ along with the first three distinct partial sums of its Maclaurin series on the same axes.

b. By zooming into the relevant spots on the graph, estimate the largest value of a such that

i. the first-degree Maclaurin polynomial is a ''good'' approximation to $f(x)$ on the interval $[0, a)$. (By ''good'' we mean an error of about 0.05.)

ii. the third-degree Maclaurin polynomial is a ''good'' approximation to $f(x)$ on the interval $[0, a)$.

iii. the fifth-degree Maclaurin polynomial is a ''good'' approximation to $f(x)$ on the interval $[0, a)$.

c. Explain how you can use your computer algebra system to approximate these three numbers without resorting to a graph. What results do you get to two decimal places?

APPENDIX 1

MATHEMATICAL INDUCTION

Mathematical induction† is the name given to an elementary logical principle that can be used to prove a certain type of mathematical statement. Typically, we use mathematical induction to prove that a certain statement or equation holds for every positive integer. For example, we may need to prove that $2^n > n$ for all integers $n \geq 1$.

To do so, we proceed in two steps:

i. We prove that the statement is true for some integer N (usually $N = 1$).

ii. We *assume* that the statement is true for an integer k and then *prove* that it is true for the integer $k + 1$.

If we can complete these two steps, then we will have demonstrated the validity of the statement for *all* positive integers greater than or equal to N. To convince you of this fact, we reason as follows: Since the statement is true for N [by step (i)] it is true for the integer $N + 1$ [by step (ii)]. Then it is also true for the integer $(N + 1) + 1 = N + 2$ [again by step (ii)], and so on. We now demonstrate the procedure with some examples.

EXAMPLE 1

USING MATHEMATICAL INDUCTION TO PROVE AN INEQUALITY

Show that $2^n > n$ for all integers $n \geq 1$.

SOLUTION:

i. If $n = 1$, then $2^n = 2^1 = 2 > 1 = n$, so $2^n > n$ for $n = 1$.

ii. Assume that $2^k > k$, where $k \geq 1$ is an integer. Then

$$2^{k+1} = 2 \cdot 2^k = 2^k + 2^k \overset{\text{Since } 2^k > k}{\overset{\downarrow}{>}} k + k \geq k + 1.$$

This completes the proof since we have shown that $2^1 > 1$, which implies, by step (ii), that $2^2 > 2$, so that, again by step (ii), $2^3 > 3$, so that $2^4 > 4$, and so on.

---

†This technique was first used in a mathematical proof by the great French mathematician Pierre de Fermat (1601–1665).

EXAMPLE 2

A FORMULA FOR A SUM OF INTEGERS

Use mathematical induction to prove the formula for the sum of the first n positive integers:

$$1 + 2 + 3 + \cdots + n = \frac{n(n+1)}{2}. \tag{1}$$

SOLUTION:

i. If $n = 1$, then the sum of the first one integer is 1. But $(1)(1 + 1)/2 = 1$, so equation (1) holds in the case in which $n = 1$.

ii. Assume that (1) holds for $n = k$; that is,

$$1 + 2 + 3 + \cdots + k = \frac{k(k+1)}{2}.$$

We must now show that it holds for $n = k + 1$. That is, we must show that

$$1 + 2 + 3 + \cdots + k + (k+1) = \frac{(k+1)(k+2)}{2}.$$

But

$$\begin{aligned} 1 + 2 + 3 + \cdots + k + (k+1) &= (1 + 2 + 3 + \cdots + k) + (k+1) \\ &= \frac{k(k+1)}{2} + (k+1) \\ &= \frac{k(k+1) + 2(k+1)}{2} = \frac{(k+1)(k+2)}{2}, \end{aligned}$$

and the proof is complete.

You may wish to try a few examples to illustrate that formula (1) really works. For example,

$$1 + 2 + 3 + 4 + 5 + 6 + 7 + 8 + 9 + 10 = \frac{10(11)}{2} = 55.$$

EXAMPLE 3

A FORMULA FOR A SUM OF SQUARES

Use mathematical induction to prove the formula for the sum of the squares of the first n positive integers:

$$1^2 + 2^2 + 3^2 + \cdots + n^2 = \frac{n(n+1)(2n+1)}{6}. \tag{2}$$

SOLUTION:

i. Since $1(1 + 1)(2 \cdot 1 + 1)/6 = 1 = 1^2$, equation (2) is valid for $n = 1$.

ii. Suppose that equation (2) holds for $n = k$; that is, suppose that

$$1^2 + 2^2 + 3^2 + \cdots + k^2 = \frac{k(k+1)(2k+1)}{6}.$$

Then to prove that (2) is true for $n = k + 1$, we have

$$\begin{aligned} 1^2 + 2^2 + 3^2 + \cdots + k^2 + (k+1)^2 &= \frac{k(k+1)(2k+1)}{6} + (k+1)^2 \\ &= \frac{k(k+1)(2k+1) + 6(k+1)^2}{6} \\ &= \frac{k+1}{6}[k(2k+1) + 6(k+1)] \\ &= \frac{k+1}{6}(2k^2 + 7k + 6) \\ &= \frac{k+1}{6}[(k+2)(2k+3)] \\ &= \frac{(k+1)(k+2)[2(k+1)+1]}{6}, \end{aligned}$$

which is equation (2) for $n = k + 1$, and the proof is complete.

Again you may wish to experiment with this formula. For example,

$$\begin{aligned} 1^2 + 2^2 + 3^2 + 4^2 + 5^2 + 6^2 + 7^2 &= \frac{7(7+1)(2 \cdot 7 + 1)}{6} \\ &= \frac{7 \cdot 8 \cdot 15}{6} = 140. \end{aligned}$$

EXAMPLE 4

THE SUM OF A GEOMETRIC PROGRESSION

For $a \neq 1$, use mathematical induction to prove the formula for the sum of a geometric progression:

$$1 + a + a^2 + \cdots + a^n = \frac{1 - a^{n+1}}{1 - a}. \tag{3}$$

SOLUTION:

i. If $n = 0$, then

$$\frac{1 - a^{0+1}}{1 - a} = \frac{1 - a}{1 - a} = 1.$$

Thus equation (3) holds for $n = 0$. (We use $n = 0$ instead of $n = 1$ since $a^0 = 1$ is the first term.)

ii. Assume that (3) holds for $n = k$; that is,

$$1 + a + a^2 + \cdots + a^k = \frac{1 - a^{k+1}}{1 - a}.$$

Then

$$1 + a + a^2 + \cdots + a^k + a^{k+1} = \frac{1 - a^{k+1}}{1 - a} + a^{k+1}$$
$$= \frac{1 - a^{k+1} + (1 - a)a^{k+1}}{1 - a} = \frac{1 - a^{k+2}}{1 - a},$$

so that equation (3) also holds for $n = k + 1$, and the proof is complete.

EXAMPLE 5

THE DERIVATIVE OF A SUM

Let $f_1, f_2, \ldots, f_n$ be differentiable functions. Use mathematical induction to prove that

$$\frac{d}{dx}(f_1 + f_2 + \cdots + f_n) = \frac{df_1}{dx} + \frac{df_2}{dx} + \cdots + \frac{df_n}{dx}. \quad (4)$$

SOLUTION:

i. For $n = 2$, equation (4) is a standard result in one-variable calculus.†

ii. Assume that equation (4) is valid for $n = k$; that is,

$$\frac{d}{dx}(f_1 + f_2 + \cdots + f_k) = \frac{df_1}{dx} + \frac{df_2}{dx} + \cdots + \frac{df_k}{dx}.$$

Let $g(x) = f_1(x) + f_2(x) + \cdots + f_k(x)$. Then

(by the case $n = 2$)

$$\frac{d}{dx}(f_1 + f_2 + \cdots + f_k + f_{k+1}) = \frac{d}{dx}(g + f_{k+1}) \overset{\downarrow}{=} \frac{dg}{dx} + \frac{df_{k+1}}{dx}$$
$$= \frac{d}{dx}(f_1 + f_2 + \cdots + f_k) + \frac{df_{k+1}}{dx} = \frac{df_1}{dx} + \frac{df_2}{dx} + \cdots + \frac{df_k}{dx} + \frac{df_{k+1}}{dx},$$

which is equation (4) in the case $n = k + 1$, and the theorem is proved.

PROBLEMS

1. Use mathematical induction to prove that the sum of the cubes of the first n positive integers is given by

$$1^3 + 2^3 + 3^3 + \cdots + n^3 = \frac{n^2(n + 1)^2}{4}. \quad (5)$$

2. Let the functions $f_1, f_2, \ldots, f_n$ be integrable on $[0, 1]$. Show that $f_1 + f_2 + \cdots + f_n$ is integrable on $[0, 1]$ and that

$$\int_0^1 [f_1(x) + f_2(x) + \cdots + f_n(x)]\, dx$$
$$= \int_0^1 f_1(x)\, dx + \int_0^1 f_2(x)\, dx + \cdots + \int_0^1 f_n(x)\, dx.$$

† $\frac{d}{dx}(f + g) = \frac{df}{dx} + \frac{dg}{dx}$

3. Use mathematical induction to prove that the nth derivative of the nth-order polynomial

$$P_n(x) = x^n + a_{n-1}x^{n-1} + a_{n-2}x^{n-2} + \cdots + a_1x^1 + a_0$$

is equal to $n!$ $[n! = n(n-1)(n-2)\cdots 3 \cdot 2 \cdot 1]$.

4. Show that if $a \neq 1$,

$$1 + 2a + 3a^2 + \cdots + na^{n-1} = \frac{1-(n+1)a^n + na^{n+1}}{(1-a)^2}.$$

***5.** Prove, using mathematical induction, that there are exactly 2^n subsets of a set containing n elements.

6. Use mathematical induction to prove that

$$\ln(a_1a_2a_3\cdots a_n) = \ln a_1 + \ln a_2 + \cdots + \ln a_n,$$

if $a_k > 0$ for $k = 1, 2, \ldots, n$.

7. Let $\mathbf{u}, \mathbf{v}_1, \mathbf{v}_2, \ldots, \mathbf{v}_n$ be $n+1$ vectors in $\mathbb{R}^2$. Prove that (see Section 1.2)

$$\mathbf{u}\cdot(\mathbf{v}_1 + \mathbf{v}_2 + \cdots + \mathbf{v}_n) = \mathbf{u}\cdot\mathbf{v}_1 + \mathbf{u}\cdot\mathbf{v}_2 + \cdots + \mathbf{u}\cdot\mathbf{v}_n.$$

APPENDIX 2

THE BINOMIAL THEOREM

The binomial theorem provides a useful device for evaluating expressions of the form $(x + y)^n$, where n is a positive integer. You are familiar with the expression

$$(x + y)^2 = x^2 + 2xy + y^2$$

In addition, it is not difficult to show that

$$(x + y)^3 = x^3 + 3x^2y + 3xy^2 + y^3$$
$$(x + y)^4 = x^4 + 4x^3y + 6x^2y^2 + 4xy^3 + y^4$$
$$(x + y)^5 = x^5 + 5x^4y + 10x^3y^2 + 10x^2y^3 + 5xy^4 + y^5$$
$$(x + y)^6 = x^6 + 6x^5y + 15x^4y^2 + 20x^3y^3 + 15x^2y^4 + 6xy^5 + y^6$$

We add to these expressions

$$(x + y)^0 = 1 \qquad \text{and} \qquad (x + y)^1 = x + y$$

The goal of this section is to obtain an expansion for $(x + y)^n$. To help you see a pattern, we arrange the coefficients of $(x + y)^n$ for $n = 0, 1, 2, 3, 4, 5$, and 6 in a triangular form:

$$\binom{0}{0}$$
$$\binom{1}{0} \quad \binom{1}{1}$$
$$\binom{2}{0} \quad \binom{2}{1} \quad \binom{2}{2}$$
$$\binom{3}{0} \quad \binom{3}{1} \quad \binom{3}{2} \quad \binom{3}{3}$$
$$\binom{4}{0} \quad \binom{4}{1} \quad \binom{4}{2} \quad \binom{4}{3} \quad \binom{4}{4}$$
$$\binom{5}{0} \quad \binom{5}{1} \quad \binom{5}{2} \quad \binom{5}{3} \quad \binom{5}{4} \quad \binom{5}{5}$$
$$\binom{6}{0} \quad \binom{6}{1} \quad \binom{6}{2} \quad \binom{6}{3} \quad \binom{6}{4} \quad \binom{6}{5} \quad \binom{6}{6}$$

FIGURE 1
Pascal's triangle

This triangle is called **Pascal's triangle**, named for the great French mathematician Blaise Pascal (1623–1662). In his 1662 work *Traité du Triangle Arithmétique,* Pascal used mathematical induction to prove that every number in the triangle, except the 1's at the end of each row, is the sum of the two numbers diagonally above it. For example, in the shaded minitriangle in Figure 1, we see that $15 = 10 + 5$.

Before we go on to a statement of the binomial theorem, we introduce a new notation.

THE BINOMIAL COEFFICIENT

Let n and k denote nonnegative integers. Then

$$\binom{n}{k} = \frac{n(n-1)(n-2)\cdots(n-k+1)}{k(k-1)\cdots 3\cdot 2\cdot 1} = \frac{n!}{k!(n-k)!} \tag{1}$$

where $n! = n(n-1)(n-2)\cdots 3\cdot 2\cdot 1$, and, by convention, $0! = 1$.

The number $\binom{n}{k}$ is called a **binomial coefficient**.

EXAMPLE 1

EVALUATING TWO BINOMIAL COEFFICIENTS

Compute

a. $\binom{6}{0}$

b. $\binom{8}{8}$.

SOLUTION:

a. $\binom{6}{0} = \frac{6!}{0!6!} = \frac{6!}{1(6!)} = 1$

b. $\binom{8}{8} = \frac{8!}{8!0!} = \frac{8!}{(8!)1} = 1$

Generalizing Example 1, we see that

$$\binom{n}{0} = 1 = \binom{n}{n} \quad \text{for any positive integer } n$$

We now write the binomial coefficients (up to $n = 6$) in a triangular form (see Figure 2). It turns out that the numbers in Figure 2 are the same as the numbers in Figure 1. For example, in the shaded row,

$$\binom{5}{0} = 1 \quad \binom{5}{1} = 5 \quad \binom{5}{2} = 10 \quad \binom{5}{3} = 10 \quad \binom{5}{4} = 5 \quad \binom{5}{5} = 1$$

This is the sixth row in Figure 1. Thus, we see that the triangular table of binomial coefficients is the Pascal triangle.

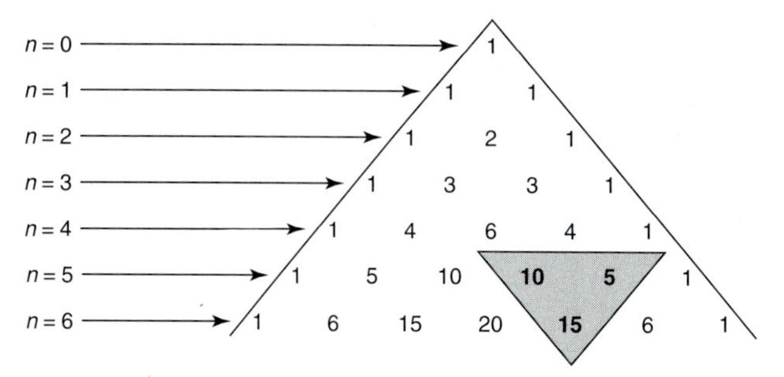

FIGURE 2
Pascal's triangle using binomial coefficients

There is a formula that is useful for writing Pascal's triangle in terms of the binomial coefficients. We have

$$\binom{n}{k-1}+\binom{n}{k}=\frac{n!}{[n-(k-1)]!(k-1)!}+\frac{n!}{(n-k!)k!}$$

$$=\frac{n!}{(n+1-k)!(k-1)!}+\frac{n!}{(n-k)!k!}$$

But

$$\frac{n!}{(n+1-k)!(k-1)!}\overset{\substack{\text{Multiply and}\\\text{divide by } k\\\downarrow}}{=}\frac{n!k}{(n+1-k)!k(k-1)!}\overset{\substack{k(k-1)!=k!\\\downarrow}}{=}k\left[\frac{n!}{(n+1-k)!k!}\right]$$

and

$$\frac{n!}{(n-k)!k!}\overset{\substack{\text{Multiply and}\\\text{divide by } n+1-k\\\downarrow}}{=}\frac{n!(n+1-k)}{(n+1-k)(n-k)!k!}\overset{\substack{(n+1-k)(n-k)!=(n+1-k)!\\\downarrow}}{=}(n+1-k)\left[\frac{n!}{(n+1-k)!k!}\right]$$

Thus

$$\binom{n}{k-1}+\binom{n}{k}=\frac{n!}{(n+1-k)!k!}[k+(n+1-k)]=\frac{n!(n+1)}{(n+1-k)!k!}$$

$$\overset{\substack{(n+1)n!=(n+1)!\\\downarrow}}{=}\frac{(n+1)!}{(n+1-k)!k!}=\binom{n+1}{k}$$

That is,

$$\binom{n}{k-1}+\binom{n}{k}=\binom{n+1}{k} \qquad \textbf{(2)}$$

EXAMPLE 2

THE SUM OF SUCCESSIVE BINOMIAL COEFFICIENTS

From (2) we have, for example,

$$\binom{5}{2} + \binom{5}{3} = \binom{6}{3} \quad \text{and} \quad \binom{6}{1} + \binom{6}{2} = \binom{7}{2}$$

This shows how to get successive rows in Pascal's triangle.

The binomial theorem enables us to compute $(x + y)^n$ without writing down the rows of the Pascal triangle. To see what we should get, observe that the sixth row of the triangle in Figure 2 is equal to

1 5 10 10 5 1

But, as we have seen, these numbers are the coefficients of the expansion of $(x + y)^5$. Thus, we have

$$(x + y)^5 = \overset{=1}{\overset{\downarrow}{\binom{5}{0}}}x^5 + \binom{5}{1}x^4y + \binom{5}{2}x^3y^2 + \binom{5}{3}x^2y^3 + \binom{5}{4}xy^4 + \overset{=1}{\overset{\downarrow}{\binom{5}{5}}}y^5$$

The binomial theorem generalizes this result.

THE BINOMIAL THEOREM

Let n be a positive integer. Then

$$(x + y)^n = x^n + \binom{n}{1}x^{n-1}y + \binom{n}{2}x^{n-2}y^2 + \cdots + \binom{n}{j}x^{n-j}y^j + \cdots + \binom{n}{n-1}xy^{n-1} + y^n \tag{3}$$

The proof of this theorem will be given after we have done some examples. Using the summation notation, we may write the binomial theorem in the following form:

$$(x + y)^n = \sum_{j=0}^{n} \binom{n}{j} x^{n-j}y^j \tag{3'}$$

Note that in (3′) the first term ($j = 0$) is $\binom{n}{0}x^{n-0}y^0 = x^n$ because $\binom{n}{0} = 1$ and $y^0 = 1$. Also, the last term ($j = n$) is $\binom{n}{n}x^{n-n}y^n = x^0y^n = y^n$ because $\binom{n}{n} = 1$.

EXAMPLE 3

USING THE BINOMIAL THEOREM

Compute $(x + y)^7$.

SOLUTION:

$$\begin{aligned}(x+y)^7 &= x^7 + \binom{7}{1}x^6y + \binom{7}{2}x^5y^2 + \binom{7}{3}x^4y^3 + \binom{7}{4}x^3y^4 \\ &\quad + \binom{7}{5}x^2y^5 + \binom{7}{6}xy^6 + y^7 \\ &= x^7 + 7x^6y + 21x^5y^2 + 35x^4y^3 + 35x^3y^4 + 21x^2y^5 + 7xy^6 + y^7\end{aligned}$$

EXAMPLE 4

FINDING A COEFFICIENT USING THE BINOMIAL THEOREM

Find the coefficient of the term containing x^3y^6 in the expansion of $(x + y)^9$.

SOLUTION: In (3), we obtain the term x^3y^6 by setting $j = 6$ (so that $9 - j = 3$). The coefficient is

$$\binom{9}{6} = \frac{9!}{6!3!} = \frac{9 \cdot 8 \cdot 7}{3!} = \frac{9 \cdot 8 \cdot 7}{3 \cdot 2} = 84$$

EXAMPLE 5

USING THE BINOMIAL THEOREM

Calculate $(2x - 3y)^4$.

SOLUTION:

$$\begin{aligned}(2x-3y)^4 &= (2x)^4 + \binom{4}{1}(2x)^3(-3y)^1 + \binom{4}{2}(2x)^2(-3y)^2 \\ &\quad + \binom{4}{3}(2x)^1(-3y)^3 + (-3y)^4 \\ &= 16x^4 + 4(8x^3)(-3y) + 6(4x^2)(9y^2) + 4(2x)(-27y^3) + 81y^4 \\ &= 16x^4 - 96x^3y + 216x^2y^2 - 216xy^3 + 81y^4\end{aligned}$$

NOTE: Binomial coefficients can be obtained on most hand-held calculators. The most common notation for $\binom{n}{r}$ on a calculator is nC_r.

EXAMPLE 6

FINDING A COEFFICIENT USING THE BINOMIAL THEOREM

Find the coefficient of the term containing y^6 in the expansion $(2x + y^2)^9$.

SOLUTION: The term containing y^6 will be of the form $(2x)^6(y^2)^3$ since $6 + 3 = 9$. This term is then

$$\binom{9}{3}(2x)^6(y^2)^3 = 84 \cdot 2^6x^6y^6 = 84 \cdot 64x^6y^6$$

and the coefficient is $84 \cdot 64 = 5376$.

PROOF OF THE BINOMIAL THEOREM: We prove the theorem by mathematical induction (Appendix 1).

Step 1: $n = 1$. Then

$$(x + y)^1 = x^1 + y^1 = x^1 + \binom{1}{1}x^{1-1}y$$

which is formula (3) in the case $n = 1$.

Step 2: Assume that (3) holds for $n = k$. That is,

$$(x + y)^k = x^k + \binom{k}{1}x^{k-1}y + \binom{k}{2}x^{k-2}y^2 + \cdots + \binom{k}{k-1}xy^{k-1} + y^k \quad \textbf{(4)}$$

$$= \sum_{j=0}^{k}\binom{k}{j}x^{k-j}y^j$$

Then

$$(x + y)^{k+1} = (x + y)^k(x + y) = x(x + y)^k + y(x + y)^k$$

$$= x\sum_{j=0}^{k}\binom{k}{j}x^{k-j}y^j + y\sum_{j=0}^{k}\binom{k}{j}x^{k-j}y^j$$

$$= \sum_{j=0}^{k}\binom{k}{j}x^{k+1-j}y^j + \sum_{j=0}^{k}\binom{k}{j}x^{k-j}y^{j+1}$$

Let $m = j + 1$ in the second sum above. Then $j = m - 1$ and $k - j = k + 1 - m$ so we have

$$(x + y)^{k+1} = \sum_{j=0}^{k}\binom{k}{j}x^{k+1-j}y^j + \sum_{m=1}^{k+1}\binom{k}{m-1}x^{k+1-m}y^m$$

setting $j = m$ ↓ ; $j = 0$ in first sum ↓ ; $m = k + 1$ in second sum ↓

$$= x^{k+1} + \sum_{j=1}^{k}\left[\binom{k}{j} + \binom{k}{j-1}\right]x^{k+1-j}y^j + y^{k+1}$$

formula (2)

$$\downarrow$$

$$= x^{k+1} + \sum_{j=1}^{k} \binom{k+1}{j} x^{k+1-j} y^j + y^{k+1}$$

$$= \sum_{j=0}^{k+1} \binom{k+1}{j} x^{k+1-j} y^j$$

which is formula (3) in the case $n = k + 1$. This completes the proof. ■

PROBLEMS

In Problems 1–8 calculate the binomial coefficients.

1. $\binom{5}{3}$ **2.** $\binom{7}{4}$

3. $\binom{7}{3}$ **4.** $\binom{10}{5}$

5. $\binom{11}{3}$ **6.** $\binom{35}{35}$

7. $\binom{8}{0}$ **8.** $\binom{12}{7}$

In Problems 9–43 carry out the indicated binomial expansion.

9. $(x - y)^5$ **10.** $(x - 2y)^3$

11. $(x - 2y)^4$ **12.** $(4x + 5y)^3$

13. $(a + b)^8$ **14.** $(u - w)^4$

15. $(2a - 3b)^5$ **16.** $\left(\frac{u}{2} + \frac{v}{3}\right)^3$

17. $\left(\frac{u}{3} - \frac{v}{4}\right)^3$ **18.** $\left(\frac{v}{2} + u\right)^5$

19. $(x^2 + 2y)^4$ **20.** $(d^2 + d^4)^3$

21. $(a^2 + b^3)^4$ **22.** $(2a^3 - 3b^2)^4$

23. $(\sqrt{x} + \sqrt{y})^6$ **24.** $(\sqrt{x} - \sqrt{y})^6$

25. $(3\sqrt{x} + 3\sqrt{y})^4$ **26.** $(xy^2 + z)^5$

27. $(ab - cd)^3$ **28.** $\left(\frac{u}{v} + w\right)^5$

29. $\left(w - \frac{u}{v}\right)^4$ **30.** $(1 + x)^8$

31. $(1 - a)^{10}$ **32.** $(x + z)^3$

33. $(1 + \sqrt{x})^3$ **34.** $(\sqrt{2} - y)^4$

35. $(\sqrt{5} + \sqrt{7})^4$ **36.** $(1 - c^2)^4$

37. $(1 + z^6)^4$

***38.** $(a + b + 1)^4 = ((a + b) + 1)^4$

***39.** $(u + v - 2)^4$

***40.** $(x + y + z)^3$

***41.** $(x + y + z)^4$

***42.** $(x + y + z)^6$

43. $(x^n + y^n)^5$

44. Find the coefficient of x^5y^7 in the expansion of $(x + y)^{12}$.

45. Find the coefficient of a^7b^3 in the expansion of $(a - b)^{10}$.

46. Find the coefficient of u^4v^2 in the expansion of $(2u - 3v)^6$.

47. Find the coefficient of x^8y^{12} in the expansion of $(x^2 + y^3)^8$.

48. Show that in the expansion of $(x + y)^n$ the coefficient of x^ky^{n-k} is equal to the coefficient of $x^{n-k}y^k$.

49. Show that $\binom{n}{k} = \binom{n}{n-k}$ and explain why this answers the question in Problem 48.

50. Show that $\binom{n}{1} = \binom{n}{n-1} = n$.

51. Show that for any integer n,

$$\binom{n}{0} + \binom{n}{1} + \binom{n}{2} + \cdots + \binom{n}{n} = 2^n$$

[*Hint:* Expand $(1 + 1)^n$.]

52. Show that for any positive integer n,

$$\binom{n}{0} - \binom{n}{1} + \binom{n}{2} - \binom{n}{3} + \cdots + (-1)^n\binom{n}{n} = 0$$

[*Hint:* Expand $(1 - 1)^n$.]

53. According to **Stirling's formula**, when n is large,

$$n! \approx \sqrt{2\pi n}\left(\frac{n}{e}\right)^n$$

Use Stirling's formula to estimate (a) 100! (b) 200! [*Hint:* Use common logarithms.]

***54.** Use the result of Problem 51 to prove that each set containing n elements has precisely 2^n subsets.

55. Find the eighth row of Pascal's triangle from the seventh row in Figure 1.

56. Find the ninth and tenth rows of Pascal's triangle.

APPENDIX 3

COMPLEX NUMBERS

In algebra we encounter the problem of finding the roots of the polynomial

$$\lambda^2 + a\lambda + b = 0. \tag{1}$$

To find the roots, we use the quadratic formula to obtain

$$\lambda = \frac{-a \pm \sqrt{a^2 - 4b}}{2}. \tag{2}$$

If $a^2 - 4b > 0$, there are two real roots. If $a^2 - 4b = 0$, we obtain the single root (of multiplicity 2) $\lambda = -a/2$. To deal with the case $a^2 - 4b < 0$, we introduce the **imaginary number**[†]

$$i = \sqrt{-1}. \tag{3}$$

Then for $a^2 - 4b < 0$,

$$\sqrt{a^2 - 4b} = \sqrt{(4b - a^2)(-1)} = \sqrt{4b - a^2}\sqrt{-1} = \sqrt{4b - a^2}\, i,$$

and the two roots of (1) are given by

$$\lambda_1 = -\frac{a}{2} + \frac{\sqrt{4b - a^2}}{2} i \qquad \text{and} \qquad \lambda_2 = -\frac{a}{2} - \frac{\sqrt{4b - a^2}}{2} i.$$

[†] You should not be troubled by the term "imaginary." It's just a name. The British mathematician Alfred North Whitehead, in the chapter on imaginary numbers in his *Introduction to Mathematics,* wrote:

> At this point it may be useful to observe that a certain type of intellect is always worrying itself and others by discussion as to the applicability of technical terms. Are the incommensurable numbers properly called numbers? Are the positive and negative numbers really numbers? Are the imaginary numbers imaginary, and are they numbers?—are types of such futile questions. Now, it cannot be too clearly understood that, in science, technical terms are names arbitrarily assigned, like Christian names to children. There can be no question of the names being right or wrong. They may be judicious or injudicious; for they can sometimes be so arranged as to be easy to remember, or so as to suggest relevant and important ideas. But the essential principle involved was quite clearly enunciated in Wonderland to Alice by Humpty Dumpty, when he told her, apropos of his use of words, 'I pay them extra and make them mean what I like'. So we will not bother as to whether imaginary numbers are imaginary, or as to whether they are numbers, but will take the phrase as the arbitrary name of a certain mathematical idea, which we will now endeavour to make plain.

EXAMPLE 1

SOLVING A QUADRATIC EQUATION WITH COMPLEX ROOTS

Find the roots of the quadratic equation $\lambda^2 + 2\lambda + 5 = 0$.

SOLUTION: We have $a = 2$, $b = 5$, and $a^2 - 4b = -16$. Thus $\sqrt{a^2 - 4b} = \sqrt{-16} = \sqrt{16}\sqrt{-1} = 4i$, and the roots are

$$\lambda_1 = \frac{-2 + 4i}{2} = -1 + 2i \quad \text{and} \quad \lambda_2 = -1 - 2i.$$

DEFINITION COMPLEX NUMBER

A **complex number** is a number of the form $z = \alpha + i\beta$, **(4)**

where α and β are real numbers. α is called the **real part** of z and is denoted by Re z. β is called the **imaginary part** of z and is denoted by Im z. Representation (4) is sometimes called the **Cartesian form** of the complex number z. ■

REMARK: If $\beta = 0$ in equation (4), then $z = \alpha$ is a real number. In this context we can regard the set of real numbers as a subset of the set of complex numbers.

EXAMPLE 2

THE REAL AND IMAGINARY PARTS OF A COMPLEX NUMBER

In Example 1, Re $\lambda_1 = -1$ and Im $\lambda_1 = 2$.

We can add and multiply complex numbers by using the standard rules of algebra.

EXAMPLE 3

ADDING AND MULTIPLYING COMPLEX NUMBERS

Let $z = 2 + 3i$ and $w = 5 - 4i$. Calculate **a.** $z + w$, **b.** $3w - 5z$, and **c.** zw.

SOLUTION:

a. $z + w = (2 + 3i) + (5 - 4i) = (2 + 5) + (3 - 4)i = 7 - i$.

b. $3w = 3(5 - 4i) = 15 - 12i$, $5z = 10 + 15i$, and $3w - 5z = (15 - 12i) - (10 + 15i) = (15 - 10) + i(-12 - 15) = 5 - 27i$.

c. $zw = (2 + 3i)(5 - 4i) = (2)(5) + 2(-4i) + (3i)(5) + (3i)(-4i) = 10 - 8i + 15i - 12i^2 = 10 + 7i + 12 = 22 + 7i$. Here we used the fact that $i^2 = -1$.

We can plot a complex number z in the xy-plane by plotting Re z along the x-axis and Im z along the y-axis. Thus each complex number can be thought of as a point in the xy-plane. With this representation the xy-plane is called the **complex plane**. Some representative points are plotted in Figure 1.

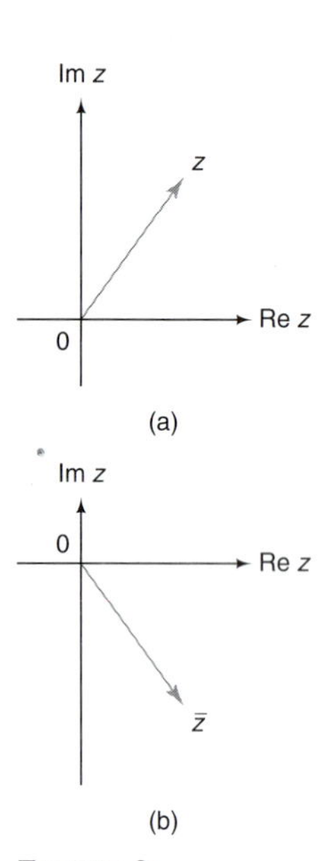

FIGURE 2
z and its conjugate $\bar{z}$

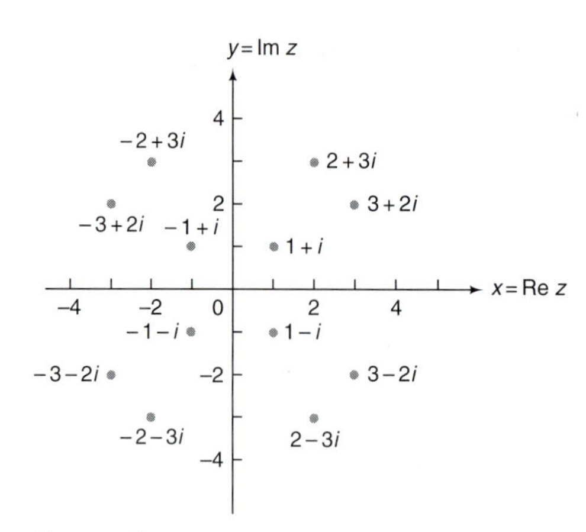

FIGURE 1
Twelve complex numbers

If $z = \alpha + i\beta$, then we define the **conjugate** of z, denoted by $\bar{z}$, by

$$\bar{z} = \alpha - i\beta. \tag{5}$$

Figure 2 depicts a representative value of z and $\bar{z}$.

EXAMPLE 4

COMPUTING FOUR COMPLEX CONJUGATES

Compute the complex conjugates of **a.** $1 + i$, **b.** $3 - 4i$, **c.** $-7 + 5i$, and **d.** -3.

SOLUTION:

a. $\overline{1 + i} = 1 - i.$
b. $\overline{3 - 4i} = 3 + 4i.$
c. $\overline{-7 + 5i} = -7 - 5i.$
d. $\overline{-3} = -3.$

It is not difficult to show (see Problem 35) that

$$\bar{z} = z \qquad \text{if and only if } z \text{ is real.} \tag{6}$$

If $z = \beta i$ with β real, then z is said to be **pure imaginary**. We can then show (see Problem 36) that

$$\bar{z} = -z \qquad \text{if and only if } z \text{ is pure imaginary.} \tag{7}$$

Let $p_n(x) = a_0 + a_1x + a_2x^2 + \cdots + a_nx^n$ be a polynomial with real coefficients. Then it can be shown (see Problem 41) that the complex roots of the equation $p_n(x) = 0$ occur in complex conjugate pairs. That is, if z is a root of $p_n(x) = 0$, then so is $\bar{z}$. We saw this fact illustrated in Example 1 in the case in which $n = 2$.

For $z = \alpha + i\beta$ we define the **magnitude** of z, denoted by $|z|$, by

$$|z| = \sqrt{\alpha^2 + \beta^2}, \tag{8}$$

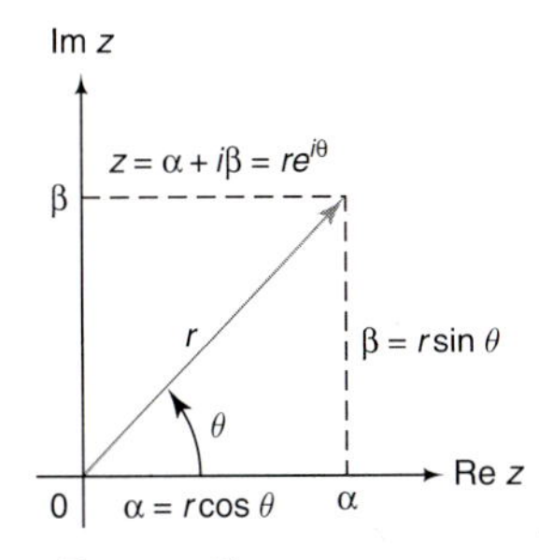

FIGURE 3
$r = |z| = \sqrt{\alpha^2 + \beta^2}$ and $\tan\theta = \dfrac{\beta}{\alpha}$.

and we define the **argument** of z, denoted by arg z, as the angle θ between the line $0z$ and the positive x-axis. From Figure 3 we see that $r = |z|$ is the distance from z to the origin, and

$$\theta = \arg z = \tan^{-1}\frac{\beta}{\alpha}. \tag{9}$$

By convention, we always choose a value of $\tan^{-1}\beta/\alpha$ that lies in the interval

$$-\pi < \theta \le \pi. \tag{10}$$

From Figure 4 we see that

$$|\bar{z}| = |z| \tag{11}$$

and

$$\arg \bar{z} = -\arg z. \tag{12}$$

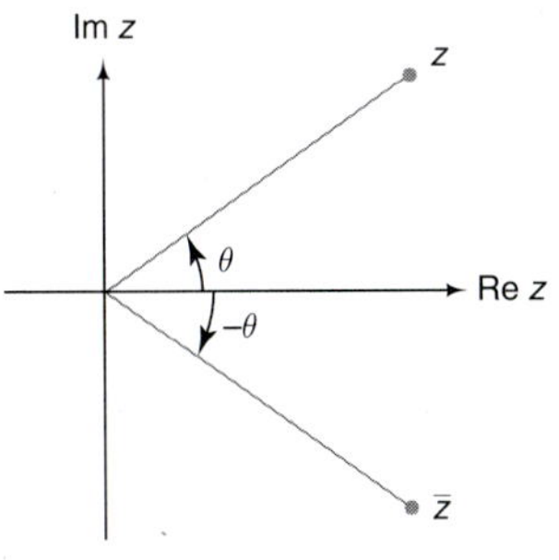

FIGURE 4
$|z| = |\bar{z}|$ and $\arg\bar{z} = -\arg z$.

We can use $|z|$ and arg z to describe what is often a more convenient way to represent complex numbers.[†] From Figure 3 it is evident that if $z = \alpha + i\beta$, $r = |z|$, and $\theta = \arg z$, then

$$\alpha = r\cos\theta \qquad \text{and} \qquad \beta = r\sin\theta. \tag{13}$$

We will see at the end of this appendix that

$$e^{i\theta} = \cos\theta + i\sin\theta. \tag{14}$$

Since $\cos(-\theta) = \cos\theta$ and $\sin(-\theta) = -\sin\theta$, we also have

$$e^{-i\theta} = \cos(-\theta) + i\sin(-\theta) = \cos\theta - i\sin\theta. \tag{14'}$$

Formula (14) is called **Euler's formula**.[‡] Using Euler's formula and equation (13), we have

$$z = \alpha + i\beta = r\cos\theta + ir\sin\theta = r(\cos\theta + i\sin\theta),$$

or

$$z = re^{i\theta}. \tag{15}$$

Representation (15) is called the **polar form** of the complex number z.

EXAMPLE 5 THE POLAR FORMS OF SIX COMPLEX NUMBERS

Determine the polar forms of the following complex numbers:

a. 1 **b.** -1 **c.** i
d. $1 + i$ **e.** $-1 - \sqrt{3}$ **f.** $-2 + 7i$

[†] If you have studied polar coordinates, you will find this new representation very familiar.
[‡] Named for the Swiss mathematician Leonhard Euler (1707–1783). See page 251.

SOLUTION: The six points are plotted in Figure 5.

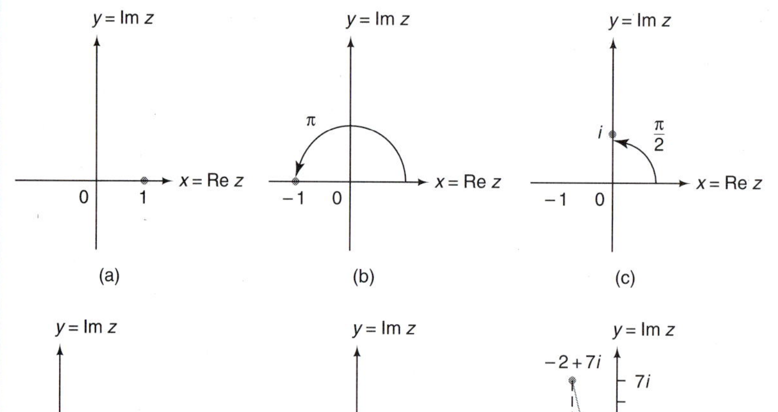

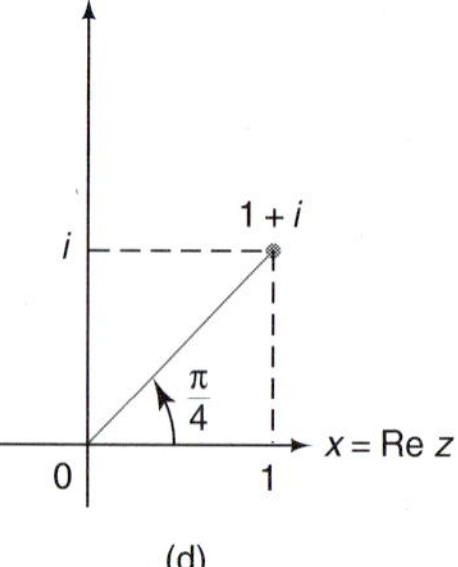

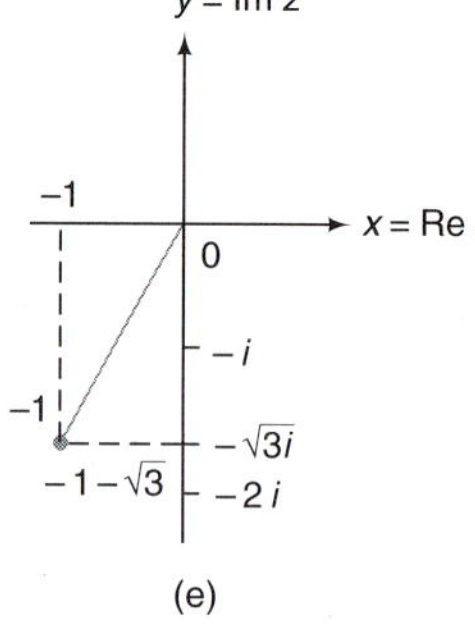

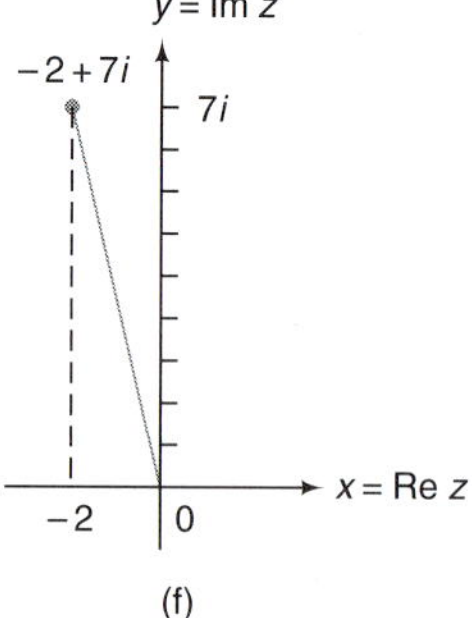

FIGURE 5
Six complex numbers

a. From Figure 5(a) it is clear that arg $1 = 0$. Since Re $1 = 1$, we see that, in polar form,

$$1 = 1e^{i0} = 1e^0 = 1.$$

b. Since $\arg(-1) = \pi$ (Figure 5(b)) and $|-1| = 1$, we have

$$-1 = 1e^{\pi i} = e^{i\pi}.$$

c. From Figure 5(c) we see that $\arg i = \pi/2$. Since $|i| = \sqrt{0^2 + 1^2} = 1$, it follows that

$$i = e^{i\pi/2}.$$

d. $\arg(1 + i) = \tan^{-1}\frac{1}{1} = \pi/4$, and $|1 + i| = \sqrt{1^2 + 1^2} = \sqrt{2}$, so that

$$1 + i = \sqrt{2}e^{i\pi/4}.$$

e. Here $\tan^{-1}\beta/\alpha = \tan^{-1}\sqrt{3} = \pi/3$. However, $\arg z$ is in the third quadrant, so $\theta = (\pi/3) - \pi = -2\pi/3$. Also, $|-1 - \sqrt{3}i| = \sqrt{1^2 + (\sqrt{3})^2} = \sqrt{1 + 3} = 2$, so that $-1 - \sqrt{3} = 2e^{-2\pi i/3}$.

f. To compute this complex number, we need a calculator. A calculator indicates that

$$\arg z = \tan^{-1}(-\tfrac{7}{2}) = \tan^{-1}(-3.5) \approx -1.2925.$$

But $\tan^{-1} x$ is defined as a number in the interval $(-\pi/2, \pi/2)$. Since from Figure 5(f) θ is in the second quadrant, we see that $\arg z = \tan^{-1}(-3.5) + \pi \approx 1.8491$. Next, we see that

$$|-2 + 7i| = \sqrt{(-2)^2 + 7^2} = \sqrt{53}.$$

Hence

$$-2 + 7i \approx \sqrt{53}e^{1.8491i}.$$

EXAMPLE 6

CONVERTING FROM POLAR TO CARTESIAN FORM

Convert the following complex numbers from polar to Cartesian form:

a. $2e^{i\pi/3}$
b. $4e^{3\pi i/2}$

SOLUTION:

a. $e^{i\pi/3} = \cos \pi/3 + i \sin \pi/3 = \frac{1}{2} + (\sqrt{3}/2)i$. Thus $2e^{i\pi/3} = 1 + \sqrt{3}i$.
b. $e^{3\pi i/2} = \cos 3\pi/2 + i \sin 3\pi/2 = 0 + i(-1) = -i$. Thus $4e^{3\pi i/2} = -4i$.

If $\theta = \arg z$, then by equation (12), $\arg \bar{z} = -\theta$. Thus since $|\bar{z}| = |z|$, we have the following:

$$\text{If } z = re^{i\theta}, \text{ then } \bar{z} = re^{-i\theta}. \tag{16}$$

Suppose we write a complex number in its polar form $z = re^{i\theta}$. Then

$$z^n = (re^{i\theta})^n = r^n(e^{i\theta})^n = r^n e^{in\theta} = r^n(\cos n\theta + i \sin n\theta). \tag{17}$$

Formula (17) is useful for a variety of computations. In particular, when $r = |z| = 1$, we obtain the **De Moivre formula**:†

$$(\cos \theta + i \sin \theta)^n = \cos n\theta + i \sin n\theta. \tag{18}$$

EXAMPLE 7

USING THE DE MOIVRE FORMULA

Compute $(1 + i)^5$.

SOLUTION: In Example 5(d) we showed that $1 + i = \sqrt{2}e^{\pi i/4}$. Then

$$\begin{aligned}(1 + i)^5 &= (\sqrt{2}e^{\pi i/4})^5 = (\sqrt{2})^5 e^{5\pi i/4} = 4\sqrt{2}\left(\cos \frac{5\pi}{4} + i \sin \frac{5\pi}{4}\right)\\ &= 4\sqrt{2}\left(-\frac{1}{\sqrt{2}} - \frac{1}{\sqrt{2}}i\right) = -4 - 4i.\end{aligned}$$

This can be checked by direct calculation. If the direct calculation seems no more difficult, then try to compute $(1 + i)^{20}$ directly. Proceeding as above, we obtain

$$\begin{aligned}(1 + i)^{20} &= (\sqrt{2})^{20}e^{20\pi i/4} = 2^{10}(\cos 5\pi + i \sin 5\pi)\\ &= 2^{10}(-1 + 0) = -1024.\end{aligned}$$

PROOF OF EULER'S FORMULA: We will show that

$$e^{i\theta} = \cos \theta + i \sin \theta \tag{19}$$

by using power series. We have

$$e^x = 1 + x + \frac{x^2}{2!} + \frac{x^3}{3!} + \cdots,^{\ddagger} \tag{20}$$

† Abraham De Moivre (1667–1754) was a French mathematician well known for his work in probability theory, infinite series, and trigonometry. He was so highly reg led that Newton often told those who came to him with questions on mathematics, ''Go to M. De Moivre; he knows these things better than I do.''

‡ Although we will not prove it here, these series expansions are also valid when x is a complex number.

$$\sin x = x - \frac{x^3}{3!} + \frac{x^5}{5!} - \cdots, \tag{21}$$

$$\cos x = 1 - \frac{x^2}{2!} + \frac{x^4}{4!} - \cdots. \tag{22}$$

Then

$$e^{i\theta} = 1 + (i\theta) + \frac{(i\theta)^2}{2!} + \frac{(i\theta)^3}{3!} + \frac{(i\theta)^4}{4!} + \frac{(i\theta)^5}{5!} + \cdots. \tag{23}$$

Now $i^2 = -1$, $i^3 = -i$, $i^4 = 1$, $i^5 = i$, and so on. Thus (23) can be written

$$\begin{aligned} e^{i\theta} &= 1 + i\theta - \frac{\theta^2}{2!} - \frac{i\theta^3}{3!} + \frac{\theta^4}{4!} + \frac{i\theta^5}{5!} - \cdots \\ &= \left(1 - \frac{\theta^2}{2!} + \frac{\theta^4}{4!} - \cdots\right) + i\left(\theta - \frac{\theta^3}{3!} + \frac{\theta^5}{5!} - \cdots\right) \\ &= \cos\theta + i\sin\theta. \end{aligned}$$

This completes the proof. ■

PROBLEMS

In Problems 1–5, perform the indicated operation.

1. $(2 - 3i) + (7 - 4i)$
2. $3(4 + i) - 5(-3 + 6i)$
3. $(1 + i)(1 - i)$
4. $(2 - 3i)(4 + 7i)$
5. $(-3 + 2i)(7 + 3i)$

In Problems 6–15, convert the complex number to its polar form.

6. $5i$
7. $5 + 5i$
8. $-2 - 2i$
9. $3 - 3i$
10. $2 + 2\sqrt{3}i$
11. $3\sqrt{3} + 3i$
12. $1 - \sqrt{3}i$
13. $4\sqrt{3} - 4i$
14. $-6\sqrt{3} - 6i$
15. $-1 - \sqrt{3}i$

In Problems 16–25, convert from polar to Cartesian form.

16. $e^{3\pi i}$
17. $2e^{-7\pi i}$
18. $\frac{1}{2}e^{3\pi i/4}$
19. $\frac{1}{2}e^{-3\pi i/4}$
20. $6e^{\pi i/6}$
21. $4e^{5\pi i/6}$
22. $4e^{-5\pi i/6}$
23. $3e^{-2\pi i/3}$
24. $\sqrt{3}e^{23\pi i/4}$
25. e^{i}

In Problems 26–34, compute the conjugate of the given number.

26. $3 - 4i$
27. $4 + 6i$
28. $-3 + 8i$
29. $-7i$
30. 16
31. $2e^{\pi i/7}$
32. $4e^{3\pi i/5}$
33. $3e^{-4\pi i/11}$
34. $e^{0.012i}$

35. Show that $z = \alpha + i\beta$ is real if and only if $z = \bar{z}$. [*Hint:* if $z = \bar{z}$, show that $\beta = 0$.]

36. Show that $z = \alpha + i\beta$ is pure imaginary if and only if $z = -\bar{z}$. [*Hint:* if $z = -\bar{z}$, show that $\alpha = 0$.]

37. For any complex number z, show that $z\bar{z} = |z|^2$.

38. Show that the circle of radius 1 centered at the origin (the *unit circle*) is the set of points in the complex plane that satisfy $|z| = 1$.

39. For any complex number z_0 and real number a, describe $\{z: |z - z_0| = a\}$.

40. Describe $\{z: |z - z_0| \le a\}$, where z_0 and a are as in Problem 39.

*41. Let $p(\lambda) = \lambda^n + a_{n-1}\lambda^{n-1} + a_{n-2}\lambda^{n-2} + \cdots + a_1\lambda + a_0$ with $a_0, a_1, \ldots, a_{n-1}$ real numbers. Show that if $p(z) = 0$, then $p(\bar{z}) = 0$. That is, *the roots of polynomials with real coefficients occur in complex conjugate pairs.*

42. Derive expressions for $\cos 4\theta$ and $\sin 4\theta$ by comparing the De Moivre formula and the expansion of $(\cos\theta + i\sin\theta)^4$.

*43. Prove De Moivre's formula for n a positive integer by mathematical induction. [*Hint:* Recall the trigonometric identities $\cos(x + y) = \cos x\cos y - \sin x\sin y$ and $\sin(x + y) = \sin x\cos y + \cos x\sin y$.]

APPENDIX 4

PROOF OF THE BASIC THEOREM ABOUT DETERMINANTS

THEOREM 1 BASIC THEOREM

Let $A = (a_{ij})$ be an $n \times n$ matrix. Then

$$\begin{aligned} \det A &= a_{11}A_{11} + a_{12}A_{12} + \cdots + a_{1n}A_{1n} \\ &= a_{i1}A_{i1} + a_{i2}A_{i2} + \cdots + a_{in}A_{in} \quad (1) \\ &= a_{1j}A_{1j} + a_{2j}A_{2j} + \cdots + a_{nj}A_{nj} \quad (2) \end{aligned}$$

for $i = 1, 2, \ldots, n$ and $j = 1, 2, \ldots, n$. ■

NOTE: The first equality is the definition of the determinant (on page 452) by cofactor expansion in the first row; the second equality says that the expansion by cofactors in any other row yields the determinant; the third equality says that expansion by cofactors in any column gives the determinant.

PROOF: We prove equality (1) by mathematical induction. For the 2×2 matrix $A = \begin{pmatrix} a_{11} & a_{12} \\ a_{21} & a_{22} \end{pmatrix}$, we first expand the first row by cofactors:

$$\det A = a_{11}A_{11} + a_{12}A_{12} = a_{11}(a_{22}) + a_{12}(-a_{21}) = a_{11}a_{22} - a_{12}a_{21}.$$

Similarly, expanding in the second row, we obtain

$$a_{21}A_{21} + a_{22}A_{22} = a_{21}(-a_{12}) + a_{22}(a_{11}) = a_{11}a_{22} - a_{12}a_{21}.$$

Thus we get the same result by expanding in any row of a 2×2 matrix, and this proves equality (1) in the 2×2 case.

We now assume that equality (1) holds for all $(n-1) \times (n-1)$ matrices. We must show that it holds for $n \times n$ matrices. Our procedure will be to expand by cofactors in the first and ith rows and show that the expansions are identical. If we expand in the first row, then a typical term in the cofactor expansion is

$$a_{1k}A_{1k} = (-1)^{1+k}a_{1k}|M_{1k}|. \tag{3}$$

Note that this is the only place in the expansion of $|A|$ that the term a_{1k} occurs, since another typical term is $a_{1m}A_{1m} = (-1)^{1+m}|M_{1m}|$, $k \neq m$, and M_{1m} is obtained by deleting the first row and mth column of A (and a_{1k} is in the first row of A). Since M_{1k} is an $(n-1) \times (n-1)$ matrix, we can, by the induction hypothesis, calculate $|M_{1k}|$ by expanding in the ith row of A (which is the $(i-1)$st row of M_{1k}). A typical term in this expansion is

$$a_{il} \text{ (cofactor of } a_{il} \text{ in } M_{1k}) \qquad (k \neq l). \tag{4}$$

For the reasons outlined above, this is the only term in the expansion of $|M_{1k}|$ in the ith row of A that contains the term a_{il}. Substituting (4) into (3), we find that

$$(-1)^{1+k}a_{1k}a_{il} \text{ (cofactor of } a_{il} \text{ in } M_{1k}) \qquad (k \neq l) \tag{5}$$

is the only occurrence of the term $a_{1k}a_{il}$ in the cofactor expansion of det A in the first row.

Now if we expand by cofactors in the ith row of A (where $i \neq 1$), a typical term is

$$(-1)^{i+l}a_{il}|M_{il}| \tag{6}$$

and a typical term in the expansion of $|M_{il}|$ in the first row of M_{il} is

$$a_{1k} \text{ (cofactor of } a_{1k} \text{ in } M_{il}) \qquad (k \neq l) \tag{7}$$

and inserting (7) in (6), we find that the only occurrence of the term $a_{il}a_{1k}$ in the expansion of det A along its ith row is

$$(-1)^{i+l}a_{1k}a_{il} \text{ (cofactor of } a_{1k} \text{ in } M_{il}) \qquad (k \neq l). \tag{8}$$

If we can show that the expressions in (5) and (8) are the same, then (1) will be proved, for the term in (5) is the only occurrence of $a_{1k}a_{il}$ in the first row expansion, the term in (8) is the only occurrence of $a_{1k}a_{il}$ in the ith row expansion, and k, i, and l are arbitrary. This will show that the sums of the terms in the first and ith row expansions are the same.

Now let $M_{1i,kl}$ denote the $(n-2) \times (n-2)$ matrix obtained by deleting the first and ith rows and kth and lth columns of A. (This is called a **second-order minor** of A.) We first suppose that $k < l$. Then

$$M_{1k} = \begin{pmatrix} a_{21} & \cdots & a_{2,k-1} & a_{2,k+1} & \cdots & a_{2l} & \cdots & a_{2n} \\ \vdots & & \vdots & \vdots & & \vdots & & \vdots \\ a_{i1} & \cdots & a_{i,k-1} & a_{i,k+1} & \cdots & a_{il} & \cdots & a_{in} \\ \vdots & & \vdots & \vdots & & \vdots & & \vdots \\ a_{n1} & \cdots & a_{n,k-1} & a_{n,k+1} & \cdots & a_{nl} & \cdots & a_{nn} \end{pmatrix}, \tag{9}$$

$$M_{il} = \begin{pmatrix} a_{11} & \cdots & a_{1k} & \cdots & a_{1,l-1} & a_{1,l+1} & \cdots & a_{1n} \\ \vdots & & \vdots & & \vdots & \vdots & & \vdots \\ a_{i-1,1} & \cdots & a_{i-1,k} & \cdots & a_{i-1,l-1} & a_{i-1,l+1} & \cdots & a_{i-1,n} \\ a_{i+1,1} & \cdots & a_{i+1,k} & \cdots & a_{i+1,l-1} & a_{i+1,l+1} & \cdots & a_{i+1,n} \\ \vdots & & \vdots & & \vdots & \vdots & & \vdots \\ a_{n1} & \cdots & a_{nk} & \cdots & a_{n,l-1} & a_{n,l+1} & \cdots & a_{nn} \end{pmatrix}. \tag{10}$$

From (9) and (10), we see that the

$$\text{cofactor of } a_{il} \text{ in } M_{1k} = (-1)^{(i-1)+(l-1)}|M_{1i,kl}|, \tag{11}$$

$$\text{cofactor of } a_{1k} \text{ in } M_{il} = (-1)^{1+k}|M_{1i,kl}|. \tag{12}$$

Thus (5) becomes

$$(-1)^{1+k}a_{1k}a_{il}(-1)^{(i-1)+(l-1)}|M_{1i,kl}| = (-1)^{i+k+l-1}a_{1k}a_{il}|M_{1i,kl}| \quad (13)$$

and (8) becomes

$$(-1)^{i+l}a_{1k}a_{il}(-1)^{1+k}|M_{1i,kl}| = (-1)^{i+k+l+1}a_{1k}a_{il}|M_{1i,kl}|. \quad (14)$$

But $(-1)^{i+k+l-1} = (-1)^{i+k+l+1}$, so the right-hand sides of equations (13) and (14) are equal. Hence expressions (5) and (8) are equal and (1) is proved in the case in which $k < l$. If $k > l$, then by similar reasoning we find that the

$$\text{cofactor of } a_{il} \text{ in } M_{1k} = (-1)^{(i-1)+l}|M_{1i,kl}|$$
$$\text{cofactor of } a_{1k} \text{ in } M_{il} = (-1)^{1+(k-1)}|M_{1i,kl}|$$

so that (5) becomes

$$(-1)^{1+k}a_{1k}a_{il}(-1)^{(i-1)+l}|M_{1i,kl}| = (-1)^{i+k+l}a_{1k}a_{il}|M_{1i,kl}|$$

and (8) becomes

$$(-1)^{i+l}a_{1k}a_{il}(-1)^{k}|M_{1i,kl}| = (-1)^{i+k+l}a_{1k}a_{il}|M_{1i,kl}|.$$

This completes the proof of equation (1).

To prove equation (2) we go through a similar process. If we expand in the kth and lth columns, we find that the only occurrences of the term $a_{1k}a_{il}$ will be given by (5) and (8). (See Problems 1 and 2.) This shows that the expansion by cofactors in any two columns is the same and that each is equal to the expansion along any row. This completes the proof. ■

PROBLEMS

1. Show that if A is expanded along its kth column, then the only occurrence of the term $a_{1k}a_{il}$ is given by equation (5).

2. Show that if A is expanded along its lth column, then the only occurrence of the term $a_{1k}a_{il}$ is given by equation (8).

3. Show that if A is expanded along its kth column, then the only occurrence of the term $a_{ik}a_{jl}$ is $(-1)^{i+k}a_{ik}a_{jl}$ (cofactor of a_{jl} in M_{ik}) for $l \neq k$.

4. Let $A = \begin{pmatrix} 1 & 5 & 7 \\ 2 & -1 & 3 \\ 4 & 5 & -2 \end{pmatrix}$. Compute det A by expanding in each of the rows and columns.

5. Do the same for the matrix

$$A = \begin{pmatrix} 1 & -1 & 4 \\ 0 & 1 & 5 \\ -3 & 7 & 2 \end{pmatrix}.$$

APPENDIX 5

EXISTENCE AND UNIQUENESS FOR FIRST-ORDER INITIAL VALUE PROBLEMS

In Chapters 10 and 11 we sought techniques for finding solutions to differential equations, assuming the *existence* of *unique* solutions for ordinary differential equations with specified initial conditions. The aim of this appendix is to prove some basic existence and uniqueness theorems for solutions of initial value problems and to show that the solution depends continuously on the initial conditions. At this point three questions suggest themselves:

i. Why do we need to prove the existence of a solution, particularly if we know that the differential equation arises from a physical problem that has a solution?
ii. Why must we worry about uniqueness?
iii. Why is continuous dependence on initial conditions important?

To answer the first question, it is important to remember that a differential equation is only a model of a physical problem. It is possible that the differential equation is a very bad model, so bad in fact that it has no solution. Countless hours could be spent using the techniques we have developed looking for a solution that may not even exist. Thus existence theorems are not only of theoretical value in telling us which equations have solutions, but are also of value in developing mathematical models of physical problems.

Similarly, uniqueness theorems have both theoretical and practical implications. If we know that a problem has a unique solution, then once we have found one solution, we are done. If the solution is not unique, then we cannot talk about ''the'' solution, but must, instead, worry about *which* solution is being discussed. In practice, if the physical problem has a unique solution, so should any mathematical model of the problem.

Finally, continuous dependence on the initial conditions is very important, since some inaccuracy is always present in practical situations. We need to know that if the initial conditions are slightly changed, the solution of the differential equation will change only slightly. Otherwise, slight inaccuracies could yield very different solutions. This property will be discussed in Problem 18.

Before continuing with our discussion, we point out that in this appendix we shall use theoretical tools from calculus that have not been widely used earlier in the text. In particular, we shall need the following facts about continuity and convergence of functions. They are discussed in most intermediate and advanced calculus texts:†

i. Let $f(t, x)$ be a continuous function of the two variables t and x and let the closed, bounded region D be defined by

$$D = \{(t, x) \colon a \le t \le b,\ c \le x \le d\}$$

where a, b, c, and d are finite real numbers. Then $f(t, x)$ is bounded for (t, x) in D. That is, there is a number $M > 0$ such that $|f(t, x)| \le M$ for every pair (t, x) in D.

ii. Let $f(x)$ be continuous on the closed interval $a \le x \le b$ and differentiable on the open interval $a < x < b$. Then the *mean value theorem* of differential calculus states that there is a number ξ between a and b $(a < \xi < b)$ such that

$$f(b) - f(a) = f'(\xi)(b - a).$$

The equation can be written as

$$\frac{f(b) - f(a)}{b - a} = f'(\xi),$$

which says, geometrically, that the slope of the tangent to the curve $y = f(x)$ at the point ξ between a and b is equal to the slope of the secant line passing through the points $(a, f(a))$ and $(b, f(b))$ (see Figure 1).

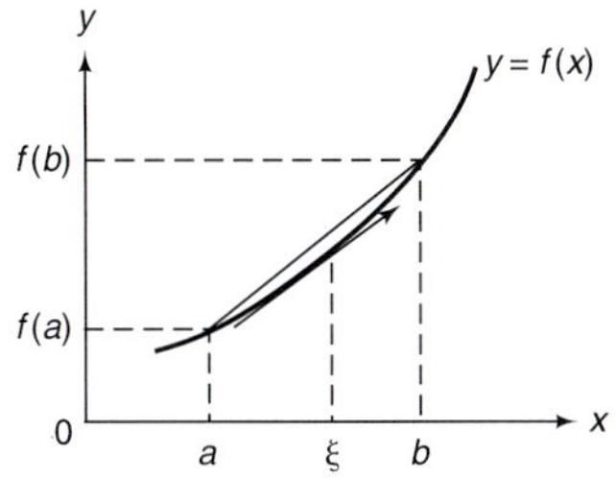

FIGURE 1
Illustration of the mean value theorem

iii. Let $\{x_n(t)\}$ be a sequence of functions. Then $x_n(t)$ is said to **converge uniformly** to a (limit) function $x(t)$ on the interval $a \le t \le b$ if for every real number $\epsilon > 0$ there exists an integer $N > 0$ such that whenever $n \ge N$, we have

$$|x_n(t) - x(t)| < \epsilon$$

for every t, $a \le t \le b$.

iv. If the functions $\{x_n(t)\}$ of statement (iii) are continuous on the interval $a \le t \le b$, then the limit function $x(t)$ is also continuous there. This fact is often stated as "the uniform limit of continuous functions is continuous."

v. Let $f(t, x)$ be a continuous function in the variable x and suppose that $x_n(t)$ converges to $x(t)$ uniformly as $n \to \infty$. Then

$$\lim_{n \to \infty} f(t, x_n(t)) = f(t, x(t)).$$

† See, for example, R. C. Buck, *Advanced Calculus* (McGraw-Hill, New York, 1978).

vi. Let $f(t)$ be an integrable function on the interval $a \le t \le b$. Then

$$\left| \int_a^b f(t)\, dt \right| \le \int_a^b |f(t)|\, dt,$$

and if $|f(t)| \le M$, then

$$\int_a^b |f(t)|\, dt \le M \int_a^b dt = M(b - a).$$

vii. Let $\{x_n(t)\}$ be a sequence of functions with $|x_n(t)| \le M_n$ for $a \le t \le b$. Then, if $\Sigma_{n=0}^{\infty} |M_n| < \infty$ (i.e., if $\Sigma_{n=0}^{\infty} M_n$ converges absolutely), then $\Sigma_{n=0}^{\infty} x_n(t)$ converges uniformly on the interval $a \le t \le b$ to a unique limit function $x(t)$. This is often called the **Weierstrass *M*-test** for uniform convergence.

viii. Let $\{x_n(t)\}$ converge uniformly to $x(t)$ on the interval $a \le t \le b$ and let $f(t, x)$ be a continuous function of t and x in the region D defined in statement (i). Then

$$\lim_{n \to \infty} \int_a^b f(s, x_n(s))\, ds = \int_a^b \lim_{n \to \infty} f(s, x_n(s))\, ds = \int_a^b f(s, x(s))\, ds.$$

Using the facts stated above, we shall prove a general theorem about the existence and uniqueness of solutions of the first-order initial value problem

$$x'(t) = f(t, x(t)), \qquad x(t_0) = x_0, \tag{1}$$

where t_0 and x_0 are real numbers. Equation (1) includes all the first-order equations we have discussed in this book. For example, for the linear nonhomogeneous equation $x' + a(t)x = b(t)$,

$$f(t, x) = -a(t)x + b(t).$$

We shall show that if $f(t, x)$ and $(\partial f/\partial x)(t, x)$ are continuous in some region containing the point (t_0, x_0), then there is an interval (containing t_0) on which a unique solution of equation (1) exists. First, we need some preliminary results.

THEOREM 1

Let $f(t, x)$ be continuous for all values t and x. Then the initial value problem (1) is equivalent to the integral equation

$$x(t) = x_0 + \int_{t_0}^{t} f(s, x(s))\, ds \tag{2}$$

in the sense that $x(t)$ is a solution of equation (1) if and only if $x(t)$ is a solution of equation (2).

PROOF: If $x(t)$ satisfies equation (1), then

$$\int_{t_0}^{t} f(s, x(s))\, ds = \int_{t_0}^{t} x'(s)\, ds = x(s)\Big|_{t_0}^{t} = x(t) - x_0,$$

which shows that $x(t)$ satisfies equation (2). Conversely, if $x(t)$ satisfies equation (2), then differentiating equation (2), we have

$$x'(t) = \frac{d}{dt}\int_{t_0}^{t} f(s, x(s))\, ds = f(t, x(t))$$

and

$$x(t_0) = x_0 + \int_{t_0}^{t_0} f(s, x(s))\, ds = x_0.$$

Hence $x(t)$ also satisfies equation (1). ■

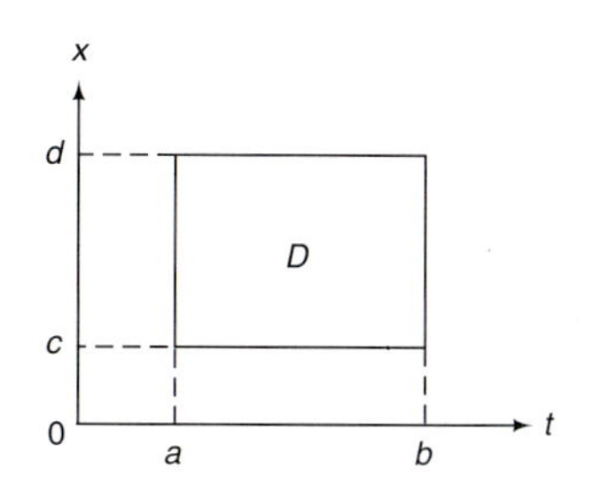

FIGURE 2
A rectangular region in the tx-plane

Let D denote the rectangular region in the tx-plane defined by

$$D: a \le t \le b,\ c \le x \le d, \tag{3}$$

where $-\infty < a < b < +\infty$ and $-\infty < c < d < +\infty$. (See Figure 2.) We say that the function $f(t, x)$ is **Lipschitz continuous** *in x over D* if there exists a constant k, $0 < k < \infty$, such that

$$|f(t, x_1) - f(t, x_2)| \le k|x_1 - x_2| \tag{4}$$

whenever (t, x_1) and (t, x_2) belong to D. The constant k is called a **Lipschitz constant**. Clearly, according to equation (4), every Lipschitz continuous function is continuous in x for each fixed t. However, *not every continuous function is Lipschitz continuous.*

EXAMPLE 1

A FUNCTION THAT DOES NOT SATISFY A LIPSCHITZ CONDITION

Let $f(t, x) = \sqrt{x}$ on the set $0 \le t \le 1$, $0 \le x \le 1$. Then $f(t, x)$ is certainly continuous on this region. But

$$|f(t, x) - f(t, 0)| = |\sqrt{x} - 0| = \frac{1}{\sqrt{x}}|x - 0|$$

for all $0 < x < 1$, and $x^{-1/2}$ tends to ∞ as x approaches zero. Thus no finite Lipschitz constant can be found to satisfy equation (4).

However, Lipschitz continuity is not a rare occurrence, as shown by the following theorem.

THEOREM 2

Let $f(t, x)$ and $(\partial f/\partial x)(t, x)$ be continuous on D. Then $f(t, x)$ is Lipschitz continuous in x over D.

PROOF: Let (t, x_1) and (t, x_2) be points in D. For fixed t, $(\partial f/\partial x)(t, x)$ is a function of x, and so we may apply the mean value theorem of differential calculus [statement (ii)] to obtain

$$|f(t, x_1) - f(t, x_2)| = \left|\frac{\partial f}{\partial x}(t, \xi)\right| |x_1 - x_2|$$

where $x_1 < \xi < x_2$. But since $\partial f/\partial x$ is continuous in D, it is bounded there [according to statement (i)]. Hence there is a constant k, $0 < k < \infty$, such that

$$\left|\frac{\partial f}{\partial x}(t, x)\right| \leq k$$

for all (t, x) in D. ■

EXAMPLE 2

A FUNCTION THAT DOES SATISFY A LIPSCHITZ CONDITION

If $f(t, x) = tx^2$ on $0 \leq t \leq 1$, $0 \leq x \leq 1$, then

$$\left|\frac{\partial f}{\partial x}\right| = |2tx| \leq 2,$$

so that

$$|f(t, x_1) - f(t, x_2)| \leq 2|x_1 - x_2|.$$

We now define a sequence of functions $\{x_n(t)\}$, called **Picard**† **iterations**, by the successive formulas

PICARD ITERATIONS

$$x_0(t) = x_0,$$

$$x_1(t) = x_0 + \int_{t_0}^{t} f(s, x_0(s))\, ds,$$

$$x_2(t) = x_0 + \int_{t_0}^{t} f(s, x_1(s))\, ds,$$

$$\vdots$$

$$x_n(t) = x_0 + \int_{t_0}^{t} f(s, x_{n-1}(s))\, ds. \tag{5}$$

We shall show that under certain conditions the Picard iterations defined by (5) converge uniformly to a solution of equation (2). First we illustrate the process of this iteration by a simple example.

EXAMPLE 3

SOLVING AN INITIAL VALUE PROBLEM BY PICARD ITERATION

Consider the initial value problem

$$x'(t) = x(t), \qquad x(0) = 1. \tag{6}$$

As we know, equation (6) has the unique solution $x(t) = e^t$. In this case, the function $f(t, x)$ in equation (1) is given by $f(t, x(t)) = x(t)$, so that the Picard iterations defined by

†Emile Picard (1856–1941), one of the most eminent French mathematicians of the past century, made several outstanding contributions to mathematical analysis.

(5) yield successively

$$x_0(t) = x_0 = 1,$$

$$x_1(t) = 1 + \int_0^t (1)\, ds = 1 + t,$$

$$x_2(t) = 1 + \int_0^t (1 + s)\, ds = 1 + t + \frac{t^2}{2},$$

$$x_3(t) = 1 + \int_0^t \left(1 + s + \frac{s^2}{2}\right) ds = 1 + t + \frac{t^2}{2!} + \frac{t^3}{3!},$$

and clearly,

$$x_n(t) = 1 + t + \frac{t^2}{2!} + \cdots + \frac{t^n}{n!} = \sum_{k=0}^{n} \frac{t^k}{k!}.$$

Hence

$$\lim_{n\to\infty} x_n(t) = \sum_{k=0}^{\infty} \frac{t^k}{k!} = e^t$$

by equation (12.12.19) on page 990.

We now state and prove the main result of this appendix.

THEOREM 3 EXISTENCE THEOREM

Let $f(t, x)$ be Lipschitz continuous in x with the Lipschitz constant k on the region D of all points (t, x) satisfying the inequalities

$$|t - t_0| \le a, \qquad |x - x_0| \le b.$$

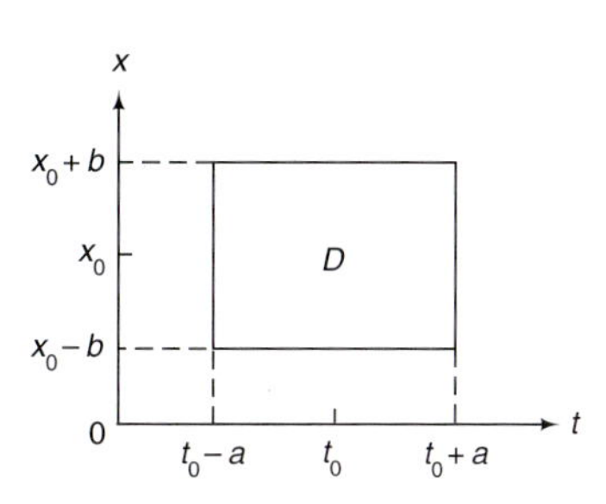

FIGURE 3
A region D in the tx-plane

(See Figure 3.) Then there exists a number $\delta > 0$ with the property that the initial value problem

$$x' = f(t, x) \qquad x(t_0) = x_0,$$

has a solution $x = x(t)$ on the interval $|t - t_0| \le \delta$.

PROOF: The proof of this theorem is complicated and will be done in several stages. However, the basic idea is simple: We need only justify that the Picard iterations converge uniformly and yield, in the limit, the solution of the integral equation (2).

Since f is continuous on D, it is bounded there [statement (i)] and we may begin by letting M be a finite upper bound for $|f(t, x)|$ on D. We then define

$$\delta = \min\{a, b/M\}. \tag{7}$$

1. We first show that the iterations $\{x_n(t)\}$ are continuous and satisfy the inequality

$$|x_n(t) - x_0| \le b. \tag{8}$$

Inequality (8) is necessary in order that $f(t, x_n(t))$ be defined for $n = 0, 1, 2, \ldots$. To show the continuity of $x_n(t)$, we first note that $x_0(t) = x_0$ is continuous (a constant function is always continuous). Then

$$x_1(t) = x_0 + \int_{t_0}^{t} f(t, x_0(s))\, ds.$$

But $f(t, x_0)$ is continuous (since $f(t, x)$ is continuous in t and x), and the integral of a continuous function is continuous. Thus $x_1(t)$ is continuous. In a similar fashion, we can show that

$$x_2(t) = x_1(t) + \int_{t_0}^{t} f(t, x_1(s))\, ds$$

is continuous and so on for $n = 3, 4, \ldots$.

Obviously the inequality (8) holds when $n = 0$, because $x_0(t) = x_0$. For $n \neq 0$, we use the definition (5) and equation (7) to obtain

$$\begin{aligned} |x_n(t) - x_0| &= \left| \int_{t_0}^{t} f(s, x_{n-1}(s))\, ds \right| \\ &\le \left| \int_{t_0}^{t} | f(s, x_{n-1}(s))|\, ds \right| \le M \left| \int_{t_0}^{t} ds \right| \\ &= M|t - t_0| \le M\delta \le b. \end{aligned}$$

These inequalities follow from statement (vi). Note that the last inequality helps explain the choice of δ in equation (7).

2. Next, we show by mathematical induction that

$$|x_n(t) - x_{n-1}(t)| \le Mk^{n-1} \frac{|t - t_0|^n}{n!} \le \frac{Mk^{n-1}\delta^n}{n!}. \tag{9}$$

If $n = 1$, we obtain

$$\begin{aligned} |x_1(t) - x_0(t)| &\le \left| \int_{t_0}^{t} f(s, x_0(s))\, ds \right| \le M \left| \int_{t_0}^{t} ds \right| \\ &= M|t - t_0| \le M\delta. \end{aligned}$$

Thus the result is true for $n = 1$.

We assume that the result is true for $n = m$ and prove that it holds for $n = m + 1$. That is, we assume that

$$|x_m(t) - x_{m-1}(t)| \le \frac{Mk^{m-1}|t - t_0|^m}{m!} \le \frac{Mk^{m-1}\delta^m}{m!}.$$

Then, since $f(t, x)$ is Lipschitz continuous in x over D,

$$\begin{aligned} |x_{m+1}(t) - x_m(t)| &= \left| \int_{t_0}^{t} f(s, x_m(s))\, ds - \int_{t_0}^{t} f(s, x_{m-1}(s))\, ds \right| \\ &\le \left| \int_{t_0}^{t} | f(s, x_m(s)) - f(s, x_{m-1}(s))|\, ds \right| \\ &\le k \left| \int_{t_0}^{t} |x_m(s) - x_{m-1}(s)|\, ds \right| \end{aligned}$$

$$\leq \frac{Mk^m}{m!}\left|\int_{t_0}^{t}(s-t_0)^m\,ds\right|^{\dagger} = \frac{Mk^m|t-t_0|^{m+1}}{(m+1)!}$$
$$\leq \frac{Mk^m\delta^{m+1}}{(m+1)!},$$

which is what we wanted to show.

3. We will now show that $x_n(t)$ converges uniformly to a limit function $x(t)$ on the interval $|t-t_0| \leq \delta$. By statement (iv), this will show that $x(t)$ is continuous.

We first note that

$$x_n(t) - x_0(t) = x_n(t) - x_{n-1}(t) + x_{n-1}(t) - x_{n-2}(t) + \cdots + x_1(t) - x_0(t)$$
$$= \sum_{k=0}^{n} [x_m(t) - x_{m-1}(t)]. \tag{10}$$

But by inequality (9),

$$|x_m(t) - x_{m-1}(t)| \leq \frac{Mk^{m-1}\delta^m}{m!} = \frac{M}{k}\frac{k^m\delta^m}{m!},$$

so that

$$\sum_{m=1}^{\infty} |x_m(t) - x_{m-1}(t)| \leq \frac{M}{k}\sum_{m=1}^{\infty}\frac{(k\delta)^m}{m!} = \frac{M}{k}(e^{k\delta}-1),$$

since

$$e^{k\delta} = \sum_{m=0}^{\infty}\frac{(k\delta)^m}{m!} = 1 + \sum_{m=1}^{\infty}\frac{(k\delta)^m}{m!}.$$

By the Weierstrass M-test [(statement (vii))], we conclude that the series

$$\sum_{m=1}^{\infty} [x_m(t) - x_{m-1}(t)]$$

converges absolutely and uniformly on $|t-t_0| \leq \delta$ to a unique limit function $y(t)$. But

$$y(t) = \lim_{n\to\infty} \sum_{m=1}^{\infty} [x_m(t) - x_{m-1}(t)]$$
$$= \lim_{n\to\infty} [x_n(t) - x_0(t)] = \lim_{n\to\infty} x_n(t) - x_0(t)$$

or

$$\lim_{n\to\infty} x_n(t) = y(t) + x_0(t).$$

† This inequality follows from the induction assumption that the inequality (9) holds for $n = m$.

We denote the right-hand side of this equation by $x(t)$. Thus the limit of the Picard iterations $x_n(t)$ exists and the convergence $x_n(t) \to x(t)$ is uniform for all t in the interval $|t - t_0| \le \delta$.

4. It remains to be shown that $x(t)$ is a solution to equation (2) for $|t - t_0| < \delta$. Since $f(t, x)$ is a continuous function of x and $x_n(t) \to x(t)$ as $n \to \infty$, we have, by statement (v),

$$\lim_{n\to\infty} f(t, x_n(t)) = f(t, x(t)).$$

Hence, by equation (5),

$$x(t) = \lim_{n\to\infty} x_{n+1}(t) = x_0 + \lim_{n\to\infty} \int_{t_0}^{t} f(s, x_n(s))\, ds$$

$$= x_0 + \int_{t_0}^{t} \lim_{n\to\infty} f(s, x_n(s))\, ds = x_0 + \int_{t_0}^{t} f(s, x(s))\, ds.$$

The step in which we interchange the limit and integral is justified by statement (viii). Thus $x(t)$ solves equation (2), and therefore it solves the initial value problem (1). ■

It turns out that the solution obtained in Theorem 3 is unique. Before proving this, however, we shall derive a simple version of a very useful result known as **Gronwall's inequality**.

THEOREM 4 GRONWALL'S INEQUALITY

Let $x(t)$ be a continuous nonnegative function and suppose that

$$x(t) \le A + B\left|\int_{t_0}^{t} x(s)\, ds\right|, \tag{11}$$

where A and B are positive constants, for all values of t such that $|t - t_0| \le \delta$. Then

$$x(t) \le Ae^{B|t-t_0|} \tag{12}$$

for all t in the interval $|t - t_0| \le \delta$.

PROOF: We shall prove this result for $t_0 \le t \le t_0 + \delta$. The proof for $t_0 - \delta \le t \le t_0$ is similar (see Problem 14). We define

$$y(t) = B\int_{t_0}^{t} x(s)\, ds.$$

Then

$$y'(t) = Bx(t) \le B\left[A + B\int_{t_0}^{t} x(s)\, ds\right] = AB + By$$

or

$$y'(t) - By(t) \le AB. \tag{13}$$

We note that

$$\frac{d}{dt}[y(t)e^{-B(t-t_0)}] = e^{-B(t-t_0)}[y'(t) - By(t)].$$

Therefore, multiplying both sides of equation (13) by the integrating factor $e^{-B(t-t_0)}$ (which is greater than zero), we have

$$\frac{d}{dt}[y(t)e^{-B(t-t_0)}] \leq ABe^{-B(t-t_0)}.$$

An integration of both sides of the inequality from t_0 to t yields

$$y(s)e^{-B(s-t_0)}\Big|_{t_0}^{t} \leq AB\int_{t_0}^{t} e^{-B(s-t_0)}\,ds = -Ae^{-B(s-t_0)}\Big|_{t_0}^{t}.$$

But $y(t_0) = 0$, so that

$$y(t)e^{-B(t-t_0)} \leq A(1 - e^{-B(t-t_0)}),$$

from which, after multiplying both sides by $e^{B(t-t_0)}$, we obtain

$$y(t) \leq A[e^{B(t-t_0)} - 1].$$

Then by equation (11),

$$x(t) \leq A + y(t) \leq Ae^{B(t-t_0)}.$$ ■

THEOREM 5 UNIQUENESS THEOREM

Let the conditions of Theorem 3 (existence theorem) hold. Then $x(t) = \lim_{n\to\infty} x_n(t)$ is the only continuous solution of the initial value problem (1) in $|t - t_0| \leq \delta$.

PROOF: Let $x(t)$ and $y(t)$ be two continuous solutions of equation (2) in the interval $|t - t_0| \leq \delta$ and suppose that $(t, y(t))$ belongs to the region D for all t in that interval.† Define $v(t) = |x(t) - y(t)|$. Then $v(t) \geq 0$ and $v(t)$ is continuous. Since $f(t, x)$ is Lipschitz continuous in x over D,

$$\begin{aligned} v(t) &= \left|\left[x_0 + \int_{t_0}^{t} f(s, x(s)\,ds\right] - \left[x_0 + \int_{t_0}^{t} f(s, y(s))\,ds\right]\right| \\ &\leq k\left|\int_{t_0}^{t} |x(s) - y(s)|\,ds\right| = k\left|\int_{t_0}^{t} v(s)\,ds\right| \\ &\leq \epsilon + k\left|\int_{t_0}^{t} v(s)\,ds\right| \end{aligned}$$

for every $\epsilon > 0$. By Gronwall's inequality, we have

$$v(t) \leq \epsilon\, e^{k|t-t_0|}.$$

But $\epsilon > 0$ can be chosen arbitrarily close to zero, so that $v(t) \leq 0$. Since $v(t) \geq 0$, it follows that $v(t) \equiv 0$, implying that $x(t)$ and $y(t)$ are identical. Hence the limit of the Picard iterations is the only continuous solution. ■

†Note that without this assumption, the function $f(t, y(t))$ may not even be defined at points where $(t, y(t))$ is not in D.

THEOREM 6

Let $f(t, x)$ and $\partial f/\partial x$ be continuous on D. Then there exists a constant $\delta > 0$ such that the Picard iteration $\{x_n(t)\}$ converges to a unique continuous solution of the initial value problem (1) on $|t - t_0| \leq \delta$.

PROOF: This theorem follows directly from Theorem 2 and the existence and uniqueness theorems. ■

We note that Theorems 3, 5, and 6 are *local* results. By this we mean that unique solutions are guaranteed to exist only ''near'' the initial point (t_0, x_0).

EXAMPLE 4 AN INITIAL VALUE PROBLEM WITH A UNIQUE SOLUTION

$$x'(t) = x^2(t), \qquad x(1) = 2.$$

Without solving this equation, we can show that there is a unique solution in some interval $|t - t_0| = |t - 1| \leq \delta$. Let $a = b = 1$. Then $|f(t, x)| = x^2 \leq 9\ (=M)$ for all $|x - x_0| \leq 1$, $x_0 = x(1) = 2$. Therefore, $\delta = \min\{a, b/M\} = \frac{1}{9}$, and Theorem 6 guarantees the existence of a unique solution on the interval $|t - 1| \leq \frac{1}{9}$. The solution of this initial value problem is easily found by a separation of variables to be $x(t) = 2/(3 - 2t)$. This solution exists so long as $t \neq \frac{3}{2}$. Starting at $t_0 = 1$, we see that the maximum interval of existences is $|t - t_0| < \frac{1}{2}$. Hence the value $\delta = \frac{1}{9}$ is not the best possible.

EXAMPLE 5 AN INITIAL VALUE PROBLEM WITH TWO SOLUTIONS

Consider the initial value problem

$$x' = \sqrt{x}, \qquad x(0) = 0.$$

As we saw in Example 1, $f(t, x) = \sqrt{x}$ does *not* satisfy a Lipschitz condition in any region containing the point (0, 0). By a separation of variables, it is easy to calculate the solution

$$x(t) = \left(\frac{t}{2}\right)^2.$$

However, $y(t) = 0$ is also a solution. Hence without a Lipschitz condition, the solution to an initial value problem (if one exists) may fail to be unique.

The last two examples illustrate the local nature of our existence–uniqueness theorem. We ask you to show (in Problem 17) that it is possible to derive *global* existence–uniqueness results for certain linear differential equations. That is, we can show that a unique solution exists for every real number t.

PROBLEMS

For each initial value problem of Problems 1–10, determine whether a unique solution can be guaranteed. If so, let $a = b = 1$ if possible, and find the number δ as given by equation (7). When possible, solve the equation and find a better value for δ, as in Example 4.

1. $x' = x^3$, $x(2) = 5$

2. $x' = x^3$, $x(1) = 2$

3. $x' = \dfrac{x}{t - x}$, $x(0) = 1$

4. $x' = x^{1/3}$, $x(1) = 0$

5. $x' = \sin x$, $x(1) = \pi/2$
6. $x' = \sqrt{x(x-1)}$, $x(1) = 2$
7. $x' = \ln|\sin x|$, $x(\pi/2) = 1$
8. $x' = \sqrt{x(x-1)}$, $x(2) = 3$
9. $x' = |x|$, $x(0) = 1$
10. $x' = tx$, $x(5) = 10$
11. Compute a Lipschitz constant for each of the following functions on the indicated region D:
  a. $f(t, x) = te^{-2x/t}$, $t > 0$, $x > 0$
  b. $f(t, x) = \sin tx$, $|t| \le 1$, $|x| \le 2$
  c. $f(t, x) = e^{-t^2} x^2 \sin(1/x)$, for all t, $-1 \le x < 1$, $x \ne 0$
  (Part (c) shows that a Lipschitz constant may exist even when $\partial f/\partial x$ is not bounded in D.)
  d. $f(t, x) = (t^2x^3)^{3/2}$, $|t| \le 2$, $|x| \le 3$
12. Consider the initial value problem

$$x' = x^2, \qquad x(1) = 3.$$

Show that the Picard iterations converge to the unique solution of this problem.
13. Construct the sequence $\{x_n(t)\}$ of Picard iterations for the initial value problem

$$x' = -x, \qquad x(0) = 3,$$

and show that it converges to the unique solution $x(t) = 3e^{-t}$.
14. Prove Gronwall's inequality (Theorem 4) for $t_0 - \delta \le t \le t_0$. [*Hint:* For $t < t_0$, $y(t) \le 0$.]
15. Let $v(t)$ be a positive function that satisfies the inequality

$$v(t) \le A + \int_{t_0}^{t} r(s)v(s)\, ds \tag{14}$$

where $A \ge 0$, $r(t)$ is a continuous positive function, and $t \ge t_0$. Prove that

$$v(t) \le A \exp\left[\int_{t_0}^{t} r(s)\, ds\right] \tag{15}$$

for all $t \ge t_0$. What kind of result holds for $t \le t_0$? This is a general form of Gronwall's inequality. [*Hint:* Define $y(t) \equiv \int_{t_0}^{t} r(s)v(s)\, ds$ and show that $y'(t) = r(t)v(t) \le r(t)[A + y(t)]$. Finish the proof by following the steps of the proof of Theorem 4, using the integrating factor $\exp[-\int_{t_0}^{t} r(s)\, ds]$.]
16. Consider the initial value problem

$$x'' = f(t, x), \qquad x(0) = x_0, x'(0) = x_1, \tag{16}$$

where f is defined on the rectangle D: $|t| \le a$, $|x - x_0| \le b$. Prove under appropriate hypotheses that if a solution exists, then it must be unique. [*Hint:* Let $x(t)$ and $y(t)$ be continuous solutions of (16). Verify by differentiation that

$$x(t) = x_0 + x_1 t - \int_0^t (t - s)f(s, x(s))\, ds,$$

$$y(t) = x_0 + x_1 t - \int_0^t (t - s)f(s, y(s))\, ds$$

in some interval $|t| \le \delta$, $\delta > 0$. Then subtract these two expressions, use an appropriate Lipschitz condition, and apply the Gronwall inequality of Problem 15.]
17. Consider the first-order linear problem

$$x'(t) = a(t)x + b(t), \qquad x(t_0) = x_0, \tag{17}$$

where $a(t)$ and $b(t)$ are continuous in the interval and $a \le t_0 \le b$.
  a. Using $f(t, x) = a(t)x + b(t)$, show that the Picard iterations $\{x_n(t)\}$ exist and are continuous for all t in the interval $a \le t \le b$.
  b. Show that $f(t, x)$ defined in part (a) is Lipschitz continuous in x over the region D of points (t, x) satisfying the conditions $a \le t \le b$, $|x| < \infty$.
  c. Modify the proofs of Theorems 3 and 5 to show that (17) has a unique solution in the *entire* interval $[a, b]$.
  d. Show that if $a(t)$ and $b(t)$ are continuous for all values t, then (17) has a unique solution that is defined for *all* values $-\infty < t < \infty$. (This is a **global** existence–uniqueness theorem.)
18. Consider the initial value problem

$$x' = f(t, x), \qquad x(t_0) = x_0. \tag{18}$$

Show that if $f(t, x)$ satisfies the conditions of Theorem 3, then the solution $x(t) \equiv x(t, x_0)$ depends continuously on the value of x_0. Thus a small change in x_0 produces a small change in the solution over the interval $|t - t_0| \le \delta$. This result is often termed **continuous dependence on initial conditions**. [*Hint:* Define $x_0(t)$ to be the unique solution of (18) and $x_1(t)$ to be the unique solution of

$$x' = f(t, x), \qquad x(t_0) = x_1. \tag{19}$$

Rewrite the initial value problems (18) and (19) as integrals (2) and subtract one from the other to obtain

$$|x_1(t) - x_0(t)| \le |x_1 - x_0| + \left| \int_{t_0}^{t} | f(s, x_1(s)) - f(s, x_0(s))| \, ds \right|.$$

Then apply the Lipschitz condition to show that

$$|x_1(t) - x_0(t)| \le |x_1 - x_0| + k \left| \int_{t_0}^{t} |x_1(s) - x_0(s)| \, ds \right|.$$

Finally, apply Gronwall's inequality (Theorem 4) to show that

$$|x_1(t) - x_0(t)| \le |x_1 - x_0| e^{k|t-t_0|} \le |x_1 - x_0| e^{k\delta}$$

for $|t - t_0| \le \delta$, from which the desired result is immediately obtained.]

APPENDIX 6

THE FOUNDATIONS OF VECTOR SPACE THEORY: THE EXISTENCE OF A BASIS

In this appendix we prove one of the most important results in linear algebra: **Every vector space has a basis**. The proof is more difficult than any other proof in this book; it involves concepts that are part of the foundation of mathematics. It will take some hard work to go through the details of this proof. However, after you have done so, you should have a deeper appreciation of a fundamental mathematical idea.

We begin with some definitions.

DEFINITION PARTIAL ORDERING

Let S be a set. A **partial ordering** on S is a relation, denoted by $\leq$, which is defined for some of the ordered pairs of elements of S and satisfies three conditions:

i.	$x \leq x$ for all $x \in S$	**reflexive law**
ii.	If $x \leq y$ and $y \leq x$, then $x = y$	**antisymmetric law**
iii.	If $x \leq y$ and $y \leq z$, then $x \leq z$	**transitive law**

It may be the case that there are elements x and y in S such that neither $x \leq y$ nor $y \leq x$. However, if for every $x, y \in S$, either $x \leq y$ or $y \leq x$, then the ordering is said to be a **total ordering**. If $x \leq y$ or $y \leq x$, then x and y are said to be **comparable**. ■

NOTATION: $x < y$ means $x \leq y$ and $x \neq y$.

EXAMPLE 1 A PARTIAL ORDERING ON $\mathbb{R}$

The real numbers are partially ordered by $\leq$ where $\leq$ stands for ''less than or equal to.'' The ordering here is a total ordering.

EXAMPLE 2

A PARTIAL ORDER ON A SET OF SUBSETS

Let S be a set and let $P(S)$, called the **power set** of S, denote the set of all subsets of S.

We say that $A \leq B$ if $A \subseteq B$. The inclusion relation is a partial ordering on $P(S)$. This is easy to prove. We have

i. $A \subseteq A$ for every set A.
ii. $A \subseteq B$ and $B \subseteq A$ if and only if $A = B$.
iii. Suppose $A \subseteq B$ and $B \subseteq C$. If $x \in A$, then $x \in B$, so $x \in C$. This means that $A \subseteq C$.

Except in unusual circumstances (for example, if S contains only one element), the ordering will not be a total ordering. This is illustrated in Figure 1.

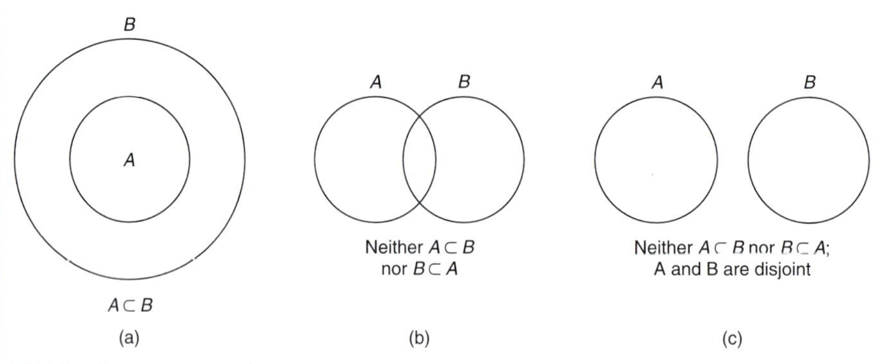

FIGURE 1
Three possibilities for set inclusion

DEFINITION CHAIN, UPPER BOUND, AND MAXIMAL ELEMENT

Let S be a set partially ordered by $\leq$.

i. A subset T of S is called a **chain** if it is totally ordered; that is, if x and y are distinct elements of T, then $x \leq y$ or $y \leq x$.
ii. Let C be a subset of S. An element $u \in S$ is an **upper bound** for C if $c \leq u$ for every element $c \in C$.
iii. The element $m \in S$ is a **maximal element** for S if there is no $s \in S$ with $m < s$.

REMARK 1: In (ii) the upper bound for C must be comparable with every element in C but it need not be in C (although it must be in S). For example, the number 1 is an upper bound for the set $(0, 1)$ but is not in $(0, 1)$. Any number greater than 1 is an upper bound. However, there is no number in $(0, 1)$ that is an upper bound for $(0, 1)$.

REMARK 2: If m is a maximal element for S, it is not necessarily the case that $s \leq m$ for every $s \in S$. In fact, m may be comparable to very few elements of S. The only condition for maximality is that there is no element of S "larger than" m. ■

EXAMPLE 3 A CHAIN OF SUBSETS OF $\mathbb{R}^2$

Let $S = \mathbb{R}^2$. Then $P(S)$ consists of subsets of the xy-plane. Let $D_r = \{(x, y): x^2 + y^2 < r^2\}$; that is, D_r is an open disk of radius r—the interior of the circle of radius r centered at the origin. Let

$$T = \{D_r: r > 0\}$$

Clearly T is a chain. For if D_{r_1} and D_{r_2} are in T, then

$$D_{r_1} \subseteq D_{r_2} \quad \text{if} \quad r_1 \le r_2 \quad \text{and} \quad D_{r_2} \subseteq D_{r_1} \quad \text{if} \quad r_2 \le r_1$$

Before going further, we need some new notation. Let V be a vector space. We have seen that a linear combination of vectors in V is a finite sum $\sum_{i=1}^{n} \alpha_i \mathbf{v}_i = \alpha_1\mathbf{v}_1 + \alpha_2\mathbf{v}_2 + \cdots + \alpha_n\mathbf{v}_n$. In your study of power series, you saw infinite sums of the form $\sum_{n=0}^{\infty} a_n x^n$. For example,

$$e^x = \sum_{n=0}^{\infty} \frac{x^n}{n!} = 1 + x + \frac{x^2}{2!} + \frac{x^3}{3!} + \cdots$$

Here we need a different kind of sum. Let C be a set of vectors in V.† For each $\mathbf{v} \in C$, let $\alpha_{\mathbf{v}}$ denote a scalar (the set of scalars is given in the definition of V). Then when we write

$$\mathbf{x} = \sum_{\mathbf{v} \in C} \alpha_{\mathbf{v}} \mathbf{v} \tag{1}$$

it will be understood that only a finite number of the scalars $\alpha_{\mathbf{v}}$ are nonzero and that all the terms with $\alpha_{\mathbf{v}} = 0$ are to be left out of the summation. We can describe the sum in (1) as follows:

> For each $\mathbf{v} \in C$, assign a scalar $\alpha_{\mathbf{v}}$ and form the product $\alpha_{\mathbf{v}} \mathbf{v}$. Then $\mathbf{x}$ is the sum of the finite subset of the vectors $\alpha_{\mathbf{v}} \mathbf{v}$ for which $\alpha_{\mathbf{v}} \neq 0$.

DEFINITION LINEAR COMBINATION, SPANNING SET, LINEAR INDEPENDENCE, AND BASIS

i. Let C be a subset of a vector space V. Then any vector that can be written in form (1) is called a **linear combination** of vectors in C. The set of linear combinations of vectors in C is denoted by $\mathrm{L}(C)$.

ii. The set C is said to **span** the vector space V if $V \subseteq \mathrm{L}(C)$.

iii. A subset C of a vector space V is said to be **linearly independent** if

$$\sum_{\mathbf{v} \in C} \alpha_{\mathbf{v}} \mathbf{v} = \mathbf{0}$$

holds only when $\alpha_{\mathbf{v}} = 0$ for every $\mathbf{v} \in C$.

iv. The subset B of a vector space V is a **basis** for V if it spans V and is linearly independent.

† C is not necessarily a subspace of V.

REMARK: If C contains only a finite number of vectors, then these definitions are precisely the ones we gave in Chapter 8. ■

THEOREM 1

Let B be a linearly independent subset of a vector space V. Then B is a basis if and only if it is maximal; that is, if $B \subsetneqq D$, then D is linearly dependent.

PROOF: Suppose that B is a basis and that $B \subsetneqq D$. Choose $\mathbf{x}$ such that $\mathbf{x} \in D$ but $\mathbf{x} \notin B$. Since B is a basis, $\mathbf{x}$ can be written as a linear combination of vectors in B:

$$\mathbf{x} = \sum_{\mathbf{v} \in B} \alpha_{\mathbf{v}} \mathbf{v}$$

If $\alpha_{\mathbf{v}} = 0$ for every $\mathbf{v}$, then $\mathbf{x} = \mathbf{0}$ and D is dependent. Otherwise $\alpha_{\mathbf{v}} \neq 0$ for some $\mathbf{v}$ and so the sum

$$\mathbf{x} - \sum_{\mathbf{v} \in B} \alpha_{\mathbf{v}} \mathbf{v} = \mathbf{0}$$

shows that D is dependent. Thus B is maximal.

Conversely, suppose that B is maximal. Let $\mathbf{x}$ be a vector in V that is not in B. Let $D = B \cup \{\mathbf{x}\}$. Then D is dependent (since B is maximal) and there is an equation

$$\sum_{\mathbf{v} \in B} \alpha_{\mathbf{v}} \mathbf{v} + \beta \mathbf{x} = \mathbf{0}$$

in which not every coefficient is zero. But $\beta \neq 0$ since otherwise we would obtain a contradiction of the linear independence of B. Thus we can write

$$\mathbf{x} = -\beta^{-1} \sum_{\mathbf{v} \in B} \alpha_{\mathbf{v}} \mathbf{v}^{\dagger}$$

Thus B is a spanning set and is therefore a basis for V. ■

Where is this all leading? Perhaps you can see the general direction. We have defined ordering on sets and maximal elements. We have shown that a linearly independent set is a basis if it is maximal. We now are lacking only a result that can help us prove the existence of a maximal element. That result is one of the basic assumptions of mathematics.

Many of you studied Euclidean geometry in high school. There, you had perhaps your first contact with mathematical proof. To prove things, Euclid made certain assumptions, which he called *axioms*. For example, he assumed that the shortest distance between two points is a straight line. Starting with these axioms, he, and students of geometry, were able to prove a number of theorems.

In all branches of mathematics it is necessary to have some axioms. If we assume nothing, we can prove nothing. To complete our proof we need the following axiom:

---

† If the scalars are real or complex numbers, then $\beta^{-1} = 1/\beta$.

AXIOM ZORN'S LEMMA[†]

If S is a nonempty, partially ordered set such that every nonempty chain has an upper bound, then S has a maximal element. ■

REMARK: The **axiom of choice** says, roughly, that given a number (finite or infinite) of nonempty sets, there is a function that chooses one element from each set. This axiom is equivalent to Zorn's lemma; that is, if you assume the axiom of choice, then you can prove Zorn's lemma and vice versa. For a proof of this equivalence and other interesting results, see the excellent book *Naive Set Theory* by Paul R. Halmos (New York: Van Nostrand, 1960), especially page 63.

We can now, finally, state and prove our main result.

THEOREM 2

Every vector space V has a basis.

PROOF: We show that V has a maximal linearly independent subset. We do this in steps.

i. Let S be the collection of all linearly independent subsets partially ordered by inclusion.

ii. A chain in S is a subset T of S such that if A and B are in T, either $A \subseteq B$ or $B \subseteq A$.

iii. Let T be a chain. Define

$$M(T) = \bigcup_{A \in T} A$$

Clearly $M(T)$ is a subset of V and $A \subseteq M(T)$ for every $A \in T$. We want to show that $M(T)$ is an upper bound for T. Since $A \subseteq M(T)$ for every $A \in T$, we need only show that $M(T) \in S$; that is, we must show that $M(T)$ is linearly independent.

iv. Suppose $\sum_{\mathbf{v} \in M(T)} \alpha_{\mathbf{v}} \mathbf{v} = \mathbf{0}$ where only a finite number of the $\alpha_{\mathbf{v}}$'s are nonzero. Denote these scalars by $\alpha_1, \alpha_2, \ldots, \alpha_n$ and the corresponding vectors by $\mathbf{v}_1, \mathbf{v}_2, \ldots, \mathbf{v}_n$. For each i, $i = 1, 2, \ldots, n$, there is a set $A_i \in T$ such that $\mathbf{v}_i \in A_i$ (because each $\mathbf{v}_i$ is in $M(T)$ and $M(T)$ is the union of sets in T). But T is totally ordered, so one of the A_i's contains all the others (see Problem 3); call it A_k. (We can only draw this conclusion because $\{A_1, A_2, \ldots, A_n\}$ is finite.) Thus $A_i \subseteq A_k$ for $i = 1, 2, \ldots, n$ and $\mathbf{v}_1, \mathbf{v}_2, \ldots, \mathbf{v}_n \in A_k$. Since A_k is linearly independent and $\sum_{i=1}^{n} \alpha_i \mathbf{v}_i = \mathbf{0}$, it follows that $\alpha_1 = \alpha_2 = \cdots = \alpha_n = 0$. Thus $M(T)$ is linearly independent.

[†] Max A. Zorn (1906–1993) spent a number of years at Indiana University where he was a Professor Emeritus until his death on March 9, 1993. He published his famous result in 1935 ("A Remark on Method in Transfinite Algebra," *Bulletin of the American Mathematical Society* 41 (1935): 667–670).

v. S is nonempty because $\varnothing \in S$ ($\varnothing$ denotes the empty set). We have shown that every chain T in S has an upper bound $M(T)$ which is in S. By Zorn's lemma, S has a maximal element. But S consists of all linearly independent subsets of V. The maximal element $B \in S$ is therefore a maximal linearly independent subset of V. Thus, by Theorem 1, B is a basis for V. ■

PROBLEMS

1. Show that every linearly independent set in a vector space V can be expanded to a basis.

2. Show that every spanning set in a vector space V has a subset that is a basis.

3. Let $A_1, A_2, \ldots, A_n$ be n sets in a chain T. Show that one of the sets contains all the others. [*Hint:* Because T is a chain, either $A_1 \subseteq A_2$ or $A_2 \subseteq A_1$. Thus the result is true if $n = 2$. Complete the proof by mathematical induction.]

TABLE OF INTEGRALS

All angles are measured in radians.

STANDARD FORMS

1. $\int a\,dx = ax + C$

2. $\int af(x)\,dx = a\int f(x)\,dx + C$

3. $\int u\,dv = uv - \int v\,du$ (integration by parts)

4. $\int u^n\,du = \dfrac{u^{n+1}}{n+1} + C, \quad n \neq -1$

5. $\int \dfrac{du}{u} = \ln u$ if $u > 0$ or $\ln(-u)$ if $u < 0$
$= \ln|u| + C$

6. $\int e^u\,du = e^u + C$

7. $\int a^u\,du = \int e^{u\ln a}\,du$
$= \dfrac{e^{u\ln a}}{\ln a} = \dfrac{a^u}{\ln a} + C, \quad a > 0, a \neq 1$

8. $\int \sin u\,du = -\cos u + C$

9. $\int \cos u\,du = \sin u + C$

10. $\int \tan u\,du = \ln|\sec u| = -\ln|\cos u| + C$

11. $\int \cot u\,du = \ln|\sin u| + C$

12. $\int \sec u\,du = \ln|\sec u + \tan u|$
$= \ln \tan\left|\dfrac{u}{2} + \dfrac{\pi}{4}\right| + C$

13. $\int \csc u\,du = \ln|\csc u - \cot u| = \ln\left|\tan\dfrac{u}{2}\right| + C$

14. $\int \sec^2 u\,du = \tan u + C$

15. $\int \csc^2 u\,du = -\cot u + C$

16. $\int \sec u \tan u\,du = \sec u + C$

17. $\int \csc u \cot u\,du = -\csc u + C$

18. $\int \dfrac{du}{u^2 + a^2} = \dfrac{1}{a}\tan^{-1}\dfrac{u}{a} + C$

19. $\int \dfrac{du}{u^2 - a^2} = \dfrac{1}{2a}\ln\left|\dfrac{u-a}{u+a}\right| + C$
$= -\dfrac{1}{a}\coth^{-1}\dfrac{u}{a} + C, \quad u^2 > a^2$

20. $\int \dfrac{du}{a^2 - u^2} = \dfrac{1}{2a}\ln\left|\dfrac{a+u}{a-u}\right| + C$
$= \dfrac{1}{a}\tanh^{-1}\dfrac{u}{a} + C, \quad u^2 < a^2$

21. $\int \dfrac{du}{\sqrt{a^2 - u^2}} = \sin^{-1}\dfrac{u}{|a|} + C$

22. $\int \dfrac{du}{\sqrt{u^2 + a^2}} = \ln(u + \sqrt{u^2 + a^2}) \quad C$

23. $\int \dfrac{du}{\sqrt{u^2 - a^2}} = \ln|u + \sqrt{u^2 - a^2}| + C$

24. $\int \dfrac{du}{u\sqrt{u^2 - a^2}} = \dfrac{1}{|a|}\sec^{-1}\left|\dfrac{u}{a}\right| + C$

25. $\int \dfrac{du}{u\sqrt{u^2 + a^2}} = -\dfrac{1}{a}\ln\left|\dfrac{a + \sqrt{u^2 + a^2}}{u}\right| + C$

26. $\int \dfrac{du}{u\sqrt{a^2 - u^2}} = -\dfrac{1}{a}\ln\left|\dfrac{a + \sqrt{a^2 - u^2}}{u}\right| + C$

INTEGRALS INVOLVING *au* + *b*

27. $\int \dfrac{du}{au + b} = \dfrac{1}{a}\ln|au + b| + C$

28. $\int \dfrac{u\,du}{au + b} = \dfrac{u}{a} - \dfrac{b}{a^2}\ln|au + b| + C$

29. $\int \dfrac{u^2\,du}{au + b} = \dfrac{(au + b)^2}{2a^3} - \dfrac{2b(au + b)}{a^3} + \dfrac{b^2}{a^3}\ln|au + b| + C$

30. $\int \dfrac{du}{u(au + b)} = \dfrac{1}{b}\ln\left|\dfrac{u}{au + b}\right| + C$

31. $\int \dfrac{du}{u^2(au + b)} = -\dfrac{1}{bu} + \dfrac{a}{b^2}\ln\left|\dfrac{au + b}{u}\right| + C$

32. $\int \dfrac{du}{(au + b)^2} = \dfrac{-1}{a(au + b)} + C$

33. $\int \dfrac{u\,du}{(au + b)^2} = \dfrac{b}{a^2(au + b)} + \dfrac{1}{a^2}\ln|au + b| + C$

34. $\int \dfrac{du}{u(au + b)^2} = \dfrac{1}{b(au + b)} + \dfrac{1}{b^2}\ln\left|\dfrac{u}{au + b}\right| + C$

35. $\int (au + b)^n\,du = \dfrac{(au + b)^{n+1}}{(n + 1)a} + C, \quad n \neq -1$

36. $\int u(au+b)^n\,du = \frac{(au+b)^{n+2}}{(n+2)a^2} - \frac{b(au+b)^{n+1}}{(n+1)a^2} + C, \quad n \neq -1, -2$

37. $\int u^m(au+b)^n\,du = \begin{cases} \frac{u^{m+1}(au+b)^n}{m+n+1} + \frac{nb}{m+n+1}\int u^m(au+b)^{n-1}\,du \\ \frac{u^m(au+b)^{n+1}}{(m+n+1)a} - \frac{mb}{(m+n+1)a}\int u^{m-1}(au+b)^n\,du \\ \frac{-u^{m+1}(au+b)^{n+1}}{(n+1)b} + \frac{m+n+2}{(n+1)b}\int u^m(au+b)^{n+1}\,du \end{cases}$

INTEGRALS INVOLVING $\sqrt{au+b}$

38. $\int \frac{du}{\sqrt{au+b}} = \frac{2\sqrt{au+b}}{a} + C$

39. $\int \frac{u\,du}{\sqrt{au+b}} = \frac{2(au-2b)}{3a^2}\sqrt{au+b} + C$

40. $\int \frac{du}{u\sqrt{au+b}} = \begin{cases} \frac{1}{\sqrt{b}}\ln\left|\frac{\sqrt{au+b}-\sqrt{b}}{\sqrt{au+b}+\sqrt{b}}\right| + C, & b > 0 \\ \frac{2}{\sqrt{-b}}\tan^{-1}\sqrt{\frac{au+b}{-b}} + C, & b < 0 \end{cases}$

41. $\int \sqrt{au+b}\,du = \frac{2\sqrt{(au+b)^3}}{3a} + C$

42. $\int u\sqrt{au+b}\,du = \frac{2(3au-2b)}{15a^2}\sqrt{(au+b)^3} + C$

43. $\int \frac{\sqrt{au+b}}{u}\,du = 2\sqrt{au+b} + b\int \frac{du}{u\sqrt{au+b}}$ (See 40.)

INTEGRALS INVOLVING $u^2 + a^2$

44. $\int \frac{du}{u^2+a^2} = \frac{1}{a}\tan^{-1}\frac{u}{a} + C$

45. $\int \frac{u\,du}{u^2+a^2} = \frac{1}{2}\ln(u^2+a^2) + C$

46. $\int \frac{u^2\,du}{u^2+a^2} = u - a\tan^{-1}\frac{u}{a} + C$

47. $\int \frac{du}{u(u^2+a^2)} = \frac{1}{2a^2}\ln\left(\frac{u^2}{u^2+a^2}\right) + C$

48. $\int \frac{du}{u^2(u^2+a^2)} = -\frac{1}{a^2u} - \frac{1}{a^3}\tan^{-1}\frac{u}{a} + C$

49. $\int \frac{du}{(u^2+a^2)^n} = \frac{u}{2(n-1)a^2(u^2+a^2)^{n-1}} + \frac{2n-3}{(2n-2)a^2}\int \frac{du}{(u^2+a^2)^{n-1}}$

50. $\int \frac{u\,du}{(u^2+a^2)^n} = \frac{-1}{2(n-1)(u^2+a^2)^{n-1}} + C, \quad n \neq 1$

51. $\int \frac{du}{u(u^2+a^2)^n} = \frac{1}{2(n-1)a^2(u^2+a^2)^{n-1}} + \frac{1}{a^2}\int \frac{du}{u(u^2+a^2)^{n-1}}, \quad n \neq 1$

INTEGRALS INVOLVING $u^2 - a^2$, $u^2 > a^2$

52. $\int \frac{du}{u^2-a^2} = \frac{1}{2a}\ln\left|\frac{u-a}{u+a}\right| + C$

53. $\int \frac{u\,du}{u^2-a^2} = \frac{1}{2}\ln(u^2-a^2) + C$

54. $\int \frac{u^2\,du}{u^2-a^2} = u + \frac{a}{2}\ln\left|\frac{u-a}{u+a}\right| + C$

55. $\int \frac{du}{u(u^2-a^2)} = \frac{1}{2a^2}\ln\left|\frac{u^2-a^2}{u^2}\right| + C$

56. $\int \frac{du}{u^2(u^2-a^2)} = \frac{1}{a^2u} + \frac{1}{2a^3}\ln\left|\frac{u-a}{u+a}\right| + C$

57. $\int \frac{du}{(u^2-a^2)^2} = \frac{-u}{2a^2(u^2-a^2)} - \frac{1}{4a^3}\ln\left|\frac{u-a}{u+a}\right| + C$

58. $\int \frac{du}{(u^2-a^2)^n} = \frac{-u}{2(n-1)a^2(u^2-a^2)^{n-1}} - \frac{2n-3}{(2n-2)a^2}\int \frac{du}{(u^2-a^2)^{n-1}}$

59. $\int \frac{u\,du}{(u^2-a^2)^n} = \frac{-1}{2(n-1)(u^2-a^2)^{n-1}} + C$

60. $\int \frac{du}{u(u^2-a^2)^n} = \frac{-1}{2(n-1)a^2(u^2-a^2)^{n-1}} - \frac{1}{a^2}\int \frac{du}{u(u^2-a^2)^{n-1}}$

INTEGRALS INVOLVING $a^2 - u^2$, $< a^2$

61. $\int \frac{du}{a^2 - u^2} = \frac{1}{2a} \ln\left|\frac{a+u}{a-u}\right| + C = \frac{1}{a}\tanh^{-1}\frac{u}{a} + C$

62. $\int \frac{u\,du}{a^2 - u^2} = -\frac{1}{2}\ln|a^2 - u^2| + C$

63. $\int \frac{u^2 du}{a^2 - u^2} = -u + \frac{a}{2}\ln\left|\frac{a+u}{a-u}\right| + C$

64. $\int \frac{du}{u(a^2 - u^2)} = \frac{1}{2a^2}\ln\left|\frac{u^2}{a^2 - u^2}\right| + C$

65. $\int \frac{du}{(a^2 - u^2)^2} = \frac{u}{2a^2(a^2 - u^2)} + \frac{1}{4a^3}\ln\left|\frac{a+u}{a-u}\right| + C$

66. $\int \frac{u\,du}{(a^2 - u^2)^2} = \frac{1}{2(a^2 - u^2)} + C$

INTEGRALS INVOLVING $\sqrt{u^2 + a^2}$

67. $\int \frac{du}{\sqrt{u^2 + a^2}} = \ln(u + \sqrt{u^2 + a^2}) + C = \sinh^{-1}\frac{u}{|a|} + C$

68. $\int \frac{u\,du}{\sqrt{u^2 + a^2}} = \sqrt{u^2 + a^2} + C$

69. $\int \frac{u^2 du}{\sqrt{u^2 + a^2}} = \frac{u\sqrt{u^2 + a^2}}{2} - \frac{a^2}{2}\ln(u + \sqrt{u^2 + a^2}) + C$

70. $\int \frac{du}{u\sqrt{u^2 + a^2}} = -\frac{1}{a}\ln\left|\frac{a + \sqrt{u^2 + a^2}}{u}\right| + C$

71. $\int \sqrt{u^2 + a^2}\,du = \frac{u\sqrt{u^2 + a^2}}{2} + \frac{a^2}{2}\ln(u + \sqrt{u^2 + a^2}) + C$

72. $\int u\sqrt{u^2 + a^2}\,du = \frac{(u^2 + a^2)^{3/2}}{3} + C$

73. $\int u^2\sqrt{u^2 + a^2}\,du = \frac{u(u^2 + a^2)^{3/2}}{4} - \frac{a^2 u\sqrt{u^2 + a^2}}{8} - \frac{a^4}{8}\ln(u + \sqrt{u^2 + a^2}) + C$

74. $\int \frac{\sqrt{u^2 + a^2}}{u}\,du = \sqrt{u^2 + a^2} - a\ln\left|\frac{a + \sqrt{u^2 + a^2}}{u}\right| + C$

75. $\int \frac{\sqrt{u^2 + a^2}}{u^2}\,du = -\frac{\sqrt{u^2 + a^2}}{u} + \ln(u + \sqrt{u^2 + a^2}) + C$

INTEGRALS INVOLVING $\sqrt{u^2 - a^2}$

76. $\int \frac{du}{\sqrt{u^2 - a^2}} = \ln|u + \sqrt{u^2 - a^2}| + C$

77. $\int \frac{u\,du}{\sqrt{u^2 - a^2}} = \sqrt{u^2 - a^2} + C$

78. $\int \frac{u^2\,du}{\sqrt{u^2 - a^2}} = \frac{u\sqrt{u^2 - a^2}}{2} + \frac{a^2}{2}\ln|u + \sqrt{u^2 - a^2}| + C$

79. $\int \frac{du}{u\sqrt{u^2 - a^2}} = \frac{1}{|a|}\sec^{-1}\left|\frac{u}{a}\right| + C$

80. $\int \sqrt{u^2 - a^2}\,du = \frac{u\sqrt{u^2 - a^2}}{2} - \frac{a^2}{2}\ln|u + \sqrt{u^2 - a^2}| + C$

81. $\int u\sqrt{u^2 - a^2}\,du = \frac{(u^2 - a^2)^{3/2}}{3} + C$

82. $\int u^2\sqrt{u^2 - a^2}\,du = \frac{u(u^2 - a^2)^{3/2}}{4} + \frac{a^2 u\sqrt{u^2 - a^2}}{8} - \frac{a^4}{8}\ln|u + \sqrt{u^2 - a^2}| + C$

83. $\int \frac{\sqrt{u^2 - a^2}}{u}\,du = \sqrt{u^2 - a^2} - |a|\sec^{-1}\left|\frac{u}{a}\right| + C$

84. $\int \frac{\sqrt{u^2 - a^2}}{u^2}\,du = -\frac{\sqrt{u^2 - a^2}}{u} + \ln|u + \sqrt{u^2 - a^2}| + C$

85. $\int \frac{du}{(u^2 - a^2)^{3/2}} = -\frac{u}{a^2\sqrt{u^2 - a^2}} + C$

INTEGRALS INVOLVING $\sqrt{a^2 - u^2}$

86. $\int \frac{du}{\sqrt{a^2 - u^2}} = \sin^{-1}\frac{u}{|a|} + C$

87. $\int \frac{u\,du}{\sqrt{a^2 - u^2}} = -\sqrt{a^2 - u^2} + C$

88. $\int \frac{u^2\,du}{\sqrt{a^2 - u^2}} = -\frac{u\sqrt{a^2 - u^2}}{2} + \frac{a^2}{2}\sin^{-1}\frac{u}{|a|} + C$

89. $\int \frac{du}{u\sqrt{a^2 - u^2}} = -\frac{1}{a}\ln\left|\frac{a + \sqrt{a^2 - u^2}}{u}\right| + C$

90. $\int \frac{du}{u^2\sqrt{a^2 - u^2}} = -\frac{\sqrt{a^2 - u^2}}{a^2 u} + C$

91. $\int \sqrt{a^2 - u^2}\,du = \frac{u\sqrt{a^2 - u^2}}{2} + \frac{a^2}{2}\sin^{-1}\frac{u}{|a|} + C$

92. $\int u\sqrt{a^2 - u^2}\,du = -\frac{(a^2 - u^2)^{3/2}}{3} + C$

93. $\int u^2\sqrt{a^2-u^2}\,du = -\frac{u(a^2-u^2)^{3/2}}{4} + \frac{a^2u\sqrt{a^2-u^2}}{8} + \frac{a^4}{8}\sin^{-1}\frac{u}{|a|} + C$

94. $\int \frac{\sqrt{a^2-u^2}}{u}\,du = \sqrt{a^2-u^2} - a\ln\left|\frac{a+\sqrt{a^2-u^2}}{u}\right| + C$

95. $\int \frac{\sqrt{a^2-u^2}}{u^2}\,du = -\frac{\sqrt{a^2-u^2}}{u} - \sin^{-1}\frac{u}{|a|} + C$

INTEGRALS INVOLVING THE TRIGONOMETRIC FUNCTIONS

96. $\int \sin au\,du = -\frac{\cos au}{a} + C$

97. $\int u\sin au\,du = \frac{\sin au}{a^2} - \frac{u\cos au}{a} + C$

98. $\int u^2\sin au\,du = \frac{2u}{a^2}\sin au + \left(\frac{2}{a^3} - \frac{u^2}{a}\right)\cos au + C$

99. $\int \frac{du}{\sin au} = \frac{1}{a}\ln(\csc au - \cot au) = \frac{1}{a}\ln\left|\tan\frac{au}{2}\right| + C$

100. $\int \sin^2 au\,du = \frac{u}{2} - \frac{\sin 2au}{4a} + C$

101. $\int u\sin^2 au\,du = \frac{u^2}{4} - \frac{u\sin 2au}{4a} - \frac{\cos 2au}{8a^2} + C$

102. $\int \frac{du}{\sin^2 au} = -\frac{1}{a}\cot au + C$

103. $\int \sin pu\sin qu\,du = \frac{\sin(p-q)u}{2(p-q)} - \frac{\sin(p+q)u}{2(p+q)} + C, \quad p \neq \pm q$

104. $\int \frac{du}{1-\sin au} = \frac{1}{a}\tan\left(\frac{\pi}{4} + \frac{au}{2}\right) + C$

105. $\int \frac{u\,du}{1-\sin au} = \frac{u}{a}\tan\left(\frac{\pi}{4} + \frac{au}{2}\right) + \frac{2}{a^2}\ln\left|\sin\left(\frac{\pi}{4} - \frac{au}{2}\right)\right| + C$

106. $\int \frac{du}{1+\sin au} = -\frac{1}{a}\tan\left(\frac{\pi}{4} - \frac{au}{2}\right) + C$

107. $$\int \frac{du}{p+q\sin au} = \begin{cases} \dfrac{2}{a\sqrt{p^2-q^2}}\tan^{-1}\dfrac{p\tan\frac{1}{2}au + q}{\sqrt{p^2-q^2}} + C, & |p| > |q| \\ \dfrac{1}{a\sqrt{q^2-p^2}}\ln\left|\dfrac{p\tan\frac{1}{2}au + q - \sqrt{q^2-p^2}}{p\tan\frac{1}{2}au + q + \sqrt{q^2-p^2}}\right| + C, & |p| < |q| \end{cases}$$

108. $\int u^m\sin au\,du = -\frac{u^m\cos au}{a} + \frac{mu^{m-1}\sin au}{a^2} - \frac{m(m-1)}{a^2}\int u^{m-2}\sin au\,du$

109. $\int \sin^n au\,du = -\frac{\sin^{n-1} au\cos au}{an} + \frac{n-1}{n}\int \sin^{n-2} au\,du$

110. $\int \frac{du}{\sin^n au} = \frac{-\cos au}{a(n-1)\sin^{n-1} au} + \frac{n-2}{n-1}\int \frac{du}{\sin^{n-2} au}, \quad n \neq 1$

111. $\int \cos au\,du = \frac{\sin au}{a} + C$

112. $\int u\cos au\,du = \frac{\cos au}{a^2} + \frac{u\sin au}{a} + C$

113. $\int u^2\cos au\,du = \frac{2u}{a^2}\cos au + \left(\frac{u^2}{a} - \frac{2}{a^3}\right)\sin au + C$

114. $\int \frac{du}{\cos au} = \frac{1}{a}\ln|\sec au + \tan au| = \frac{1}{a}\ln\left|\tan\left(\frac{\pi}{4} + \frac{au}{2}\right)\right| + C$

115. $\int \cos^2 au\,du = \frac{u}{2} + \frac{\sin 2au}{4a} + C$

116. $\int u\cos^2 au\,du = \frac{u^2}{4} + \frac{u\sin 2au}{4a} + \frac{\cos 2au}{8a^2} + C$

117. $\int \frac{du}{\cos^2 au} = \frac{\tan au}{a} + C$

118. $\int \cos qu\cos pu\,du = \frac{\sin(q-p)u}{2(q-p)} + \frac{\sin(q+p)u}{2(q+p)} + C, \quad q \neq \pm p$

119. $\displaystyle\int \frac{du}{p + q\cos au} = \begin{cases} \dfrac{2}{a\sqrt{p^2 - q^2}} \tan^{-1}\left[\sqrt{(p - q)/(p + q)} \tan \tfrac{1}{2}au\right] + C, & |p| > |q| \\ \dfrac{1}{a\sqrt{q^2 - p^2}} \ln\left[\dfrac{\tan \frac{1}{2}au + \sqrt{(q + p)/(q - p)}}{\tan \frac{1}{2}au - \sqrt{(q + p)/(q - p)}}\right] + C, & |p| < |q| \end{cases}$

120. $\displaystyle\int u^m \cos au\, du = \frac{u^m \sin au}{a} + \frac{mu^{m-1}}{a^2}\cos au - \frac{m(m - 1)}{a^2}\int u^{m-2}\cos au\, du$

121. $\displaystyle\int \cos^n au\, du = \frac{\sin au \cos^{n-1} au}{an} + \frac{n - 1}{n}\int \cos^{n-2} au\, du$

122. $\displaystyle\int \frac{du}{\cos^n au} = \frac{\sin au}{a(n - 1)\cos^{n-1} au} + \frac{n - 2}{n - 1}\int \frac{du}{\cos^{n-2} au}$

123. $\displaystyle\int \sin au \cos au\, du = \frac{\sin^2 au}{2a} + C$

124. $\displaystyle\int \sin pu \cos qu\, du = -\frac{\cos(p - q)u}{2(p - q)} - \frac{\cos(p + q)u}{2(p + q)} + C, \quad p \neq \pm q$

125. $\displaystyle\int \sin^n au \cos au\, du = \frac{\sin^{n+1} au}{(n + 1)a} + C, \quad n \neq -1$

126. $\displaystyle\int \cos^n au \sin au\, du = -\frac{\cos^{n+1} au}{(n + 1)a} + C, \quad n \neq -1$

127. $\displaystyle\int \sin^2 au \cos^2 au\, du = \frac{u}{8} - \frac{\sin 4au}{32a} + C$

128. $\displaystyle\int \frac{du}{\sin au \cos au} = \frac{1}{a}\ln|\tan au| + C$

129. $\displaystyle\int \frac{du}{\cos au(1 \pm \sin au)} = \mp\frac{1}{2a(1 \pm \sin au)} + \frac{1}{2a}\ln\left|\tan\left(\frac{au}{2} + \frac{\pi}{4}\right)\right| + C$

130. $\displaystyle\int \frac{du}{\sin au(1 \pm \cos au)} = \pm\frac{1}{2a(1 \pm \cos au)} + \frac{1}{2a}\ln\left|\tan\frac{au}{2}\right| + C$

131. $\displaystyle\int \frac{du}{\sin au \pm \cos au} = \frac{1}{a\sqrt{2}}\ln\left|\tan\left(\frac{au}{2} \pm \frac{\pi}{8}\right)\right| + C$

132. $\displaystyle\int \frac{\sin au\, du}{\sin au \pm \cos au} = \frac{u}{2} \mp \frac{1}{2a}\ln|\sin au \pm \cos au| + C$

133. $\displaystyle\int \frac{\cos au\, du}{\sin au \pm \cos au} = \pm\left[\frac{u}{2} + \frac{1}{2a}\ln|\sin au \pm \cos au|\right] + C$

134. $\displaystyle\int \frac{\sin au\, du}{p + q\cos au} = -\frac{1}{aq}\ln|p + q\cos au| + C$

135. $\displaystyle\int \frac{\cos au\, du}{p + q\sin au} = \frac{1}{aq}\ln|p + q\sin au| + C$

136. $\displaystyle\int \sin^m au \cos^n au\, du = \begin{cases} -\dfrac{\sin^{m-1} au \cos^{n+1} au}{a(n + m)} + \dfrac{m - 1}{m + n}\displaystyle\int \sin^{m-2} au \cos^n au\, du, & m \neq -n \\ \dfrac{\sin^{m+1} au \cos^{n-1} au}{a(m + n)} + \dfrac{n - 1}{m + n}\displaystyle\int \sin^m au \cos^{n-2} au\, du, & m \neq -n \end{cases}$

137. $\displaystyle\int \tan au\, du = -\frac{1}{a}\ln|\cos au| = \frac{1}{a}\ln|\sec au| + C$

138. $\displaystyle\int \tan^2 au\, du = \frac{\tan au}{a} - u + C$

139. $\displaystyle\int \tan^n au \sec^2 au\, du = \frac{\tan^{n+1} au}{(n + 1)a} + C, \quad n \neq -1$

140. $\displaystyle\int \tan^n au\, du = \frac{\tan^{n-1} au}{(n - 1)a} - \int \tan^{n-2} au\, du + C, \quad n \neq 1$

141. $\displaystyle\int \cot au\, du = \frac{1}{a}\ln|\sin au| + C$

142. $\displaystyle\int \cot^2 au\, du = -\frac{\cot au}{a} - u + C$

143. $\displaystyle\int \cot^n au \csc^2 au\, du = -\frac{\cot^{n+1} au}{(n + 1)a} + C, \quad n \neq -1$

144. $\displaystyle\int \cot^n au\, du = -\frac{\cot^{n-1} au}{(n - 1)a} - \int \cot^{n-2} au\, du, \quad n \neq 1$

145. $\displaystyle\int \sec au\, du = \frac{1}{a}\ln|\sec au + \tan au| = \frac{1}{a}\ln\left|\tan\left(\frac{au}{2} + \frac{\pi}{4}\right)\right| + C$

146. $\displaystyle\int \sec^2 au\, du = \frac{\tan au}{a} + C$

147. $\int \sec^3 au\, du = \dfrac{\sec au \tan au}{2a} + \dfrac{1}{2a} \ln|\sec au + \tan au| + C$

148. $\int \sec^n au \tan au\, du = \dfrac{\sec^n au}{na} + C$

149. $\int \sec^n au\, du = \dfrac{\sec^{n-2} au \tan au}{a(n-1)} + \dfrac{n-2}{n-1} \int \sec^{n-2} au\, du, \quad n \neq 1$

150. $\int \csc au\, du = \dfrac{1}{a} \ln|\csc au - \cot au| = \dfrac{1}{a} \ln\left|\tan \dfrac{au}{2}\right| + C$

151. $\int \csc^2 au\, du = -\dfrac{\cot au}{a} + C$

152. $\int \csc^n au \cot au\, du = -\dfrac{\csc^n au}{na} + C$

153. $\int \csc^n au\, du = -\dfrac{\csc^{n-2} au \cot au}{a(n-1)} + \dfrac{n-2}{n-1} \int \csc^{n-2} au\, du, \quad n \neq 1$

INTEGRALS INVOLVING INVERSE TRIGONOMETRIC FUNCTIONS

154. $\int \sin^{-1} \dfrac{u}{a}\, du = u \sin^{-1} \dfrac{u}{a} + \sqrt{a^2 - u^2} + C$

155. $\int u \sin^{-1} \dfrac{u}{a}\, du = \left(\dfrac{u^2}{2} - \dfrac{a^2}{4}\right) \sin^{-1} \dfrac{u}{a} + \dfrac{u\sqrt{a^2 - u^2}}{4} + C$

156. $\int \cos^{-1} \dfrac{u}{a}\, du = u \cos^{-1} \dfrac{u}{a} - \sqrt{a^2 - u^2} + C$

157. $\int u \cos^{-1} \dfrac{u}{a}\, du = \left(\dfrac{u^2}{2} - \dfrac{a^2}{4}\right) \cos^{-1} \dfrac{u}{a} - \dfrac{u\sqrt{a^2 - u^2}}{4} + C$

158. $\int \tan^{-1} \dfrac{u}{a}\, du = u \tan^{-1} \dfrac{u}{a} - \dfrac{a}{2} \ln(u^2 + a^2) + C$

159. $\int u \tan^{-1} \dfrac{u}{a}\, du = \dfrac{1}{2}(u^2 + a^2) \tan^{-1} \dfrac{u}{a} - \dfrac{au}{2} + C$

160. $\int u^m \sin^{-1} \dfrac{u}{a}\, du = \dfrac{u^{m+1}}{m+1} \sin^{-1} \dfrac{u}{a} - \dfrac{1}{m+1} \int \dfrac{u^{m+1}}{\sqrt{a^2 - u^2}}\, du$

161. $\int u^m \cos^{-1} \dfrac{u}{a}\, du = \dfrac{u^{m+1}}{m+1} \cos^{-1} \dfrac{u}{a} + \dfrac{1}{m+1} \int \dfrac{u^{m+1}}{\sqrt{a^2 - u^2}}\, du$

162. $\int u^m \tan^{-1} \dfrac{u}{a}\, du = \dfrac{u^{m+1}}{m+1} \tan^{-1} \dfrac{u}{a} - \dfrac{a}{m+1} \int \dfrac{u^{m+1}}{u^2 + a^2}\, du$

INTEGRALS INVOLVING e^{au}

163. $\int e^{au}\, du = \dfrac{e^{au}}{a} + C$

164. $\int ue^{au}\, du = \dfrac{e^{au}}{a}\left(u - \dfrac{1}{a}\right) + C$

165. $\int u^2 e^{au}\, du = \dfrac{e^{au}}{a}\left(u^2 - \dfrac{2u}{a} + \dfrac{2}{a^2}\right) + C$

166. $\int u^n e^{au}\, du = \dfrac{u^n e^{au}}{a} - \dfrac{n}{a} \int u^{n-1} e^{au}\, du$

$= \dfrac{e^{au}}{a}\left[u^n - \dfrac{nu^{n-1}}{a} + \dfrac{n(n-1)u^{n-2}}{a^2} - \cdots + \dfrac{(-1)^n n!}{a^n}\right]$ if n is a positive integer

167. $\int \dfrac{du}{p + qe^{au}} = \dfrac{u}{p} - \dfrac{1}{ap} \ln|p + qe^{au}| + C$

168. $\int e^{au} \sin bu\, du = \dfrac{e^{au}(a \sin bu - b \cos bu)}{a^2 + b^2} + C$

169. $\int e^{au} \cos bu\, du = \dfrac{e^{au}(a \cos bu + b \sin bu)}{a^2 + b^2} + C$

170. $\int ue^{au} \sin bu\, du = \dfrac{ue^{au}(a \sin bu - b \cos bu)}{a^2 + b^2} - \dfrac{e^{au}[(a^2 - b^2) \sin bu - 2ab \cos bu]}{(a^2 + b^2)^2} + C$

171. $\int ue^{au} \cos bu\, du = \dfrac{ue^{au}(a \cos bu + b \sin bu)}{a^2 + b^2} - \dfrac{e^{au}[(a^2 - b^2) \cos bu + 2ab \sin bu]}{(a^2 + b^2)^2} + C$

172. $\int e^{au} \sin^n bu\,du = \dfrac{e^{au} \sin^{n-1} bu}{a^2 + n^2b^2}(a \sin bu - nb \cos bu) + \dfrac{n(n-1)b^2}{a^2 + n^2b^2} \int e^{au} \sin^{n-2} bu\,du$

173. $\int e^{au} \cos^n bu\,du = \dfrac{e^{au} \cos^{n-1} bu}{a^2 + n^2b^2}(a \cos bu + nb \sin bu) + \dfrac{n(n-1)b^2}{a^2 + n^2b^2} \int e^{au} \cos^{n-2} bu\,du$

INTEGRALS INVOLVING ln *u*

174. $\int \ln u\,du = u \ln u - u + C$

175. $\int u \ln u\,du = \dfrac{u^2}{2}\left(\ln u - \dfrac{1}{2}\right) + C$

176. $\int u^m \ln u\,du = \dfrac{u^{m+1}}{m+1}\left(\ln u - \dfrac{1}{m+1}\right), \quad m \neq -1$

177. $\int \dfrac{\ln u}{u}\,du = \dfrac{1}{2}\ln^2 u + C$

178. $\int \dfrac{\ln^n u\,du}{u} = \dfrac{\ln^{n+1} u}{n+1} + C \quad \text{if } n \neq -1$

179. $\int \dfrac{du}{u \ln u} = \ln|\ln u| + C$

180. $\int \ln^n u\,du = u \ln^n u - n \int \ln^{n-1} u\,du + C, \quad n \neq -1$

181. $\int u^m \ln^n u\,du = \dfrac{u^{m+1} \ln^n u}{m+1} - \dfrac{n}{m+1} \int u^m \ln^{n-1} u\,du + C, \quad m, n \neq -1$

182. $\int \ln(u^2 + a^2)\,du = u \ln(u^2 + a^2) - 2u + 2a \tan^{-1} \dfrac{u}{a} + C$

183. $\int \ln|u^2 - a^2|\,du = u \ln|u^2 - a^2| - 2u + a \ln\left|\dfrac{u+a}{u-a}\right| + C$

INTEGRALS INVOLVING HYPERBOLIC FUNCTIONS

184. $\int \sinh au\,du = \dfrac{\cosh au}{a} + C$

185. $\int u \sinh au\,du = \dfrac{u \cosh au}{a} - \dfrac{\sinh au}{a^2} + C$

186. $\int \cosh au\,du = \dfrac{\sinh au}{a} + C$

187. $\int u \cosh au\,du = \dfrac{u \sinh au}{a} - \dfrac{\cosh au}{a^2} + C$

188. $\int \cosh^2 au\,du = \dfrac{u}{2} + \dfrac{\sinh au \cosh au}{2a} + C$

189. $\int \sinh^2 au\,du = \dfrac{\sinh au \cosh au}{2a} - \dfrac{u}{2} + C$

190. $\int \sinh^n au\,du = \dfrac{\sinh^{n-1} au \cosh au}{an} - \dfrac{n-1}{n} \int \sinh^{n-2} au\,du$

191. $\int \cosh^n au\,du = \dfrac{\cosh^{n-1} au \sinh au}{an} + \dfrac{n-1}{n} \int \cosh^{n-2} au\,du$

192. $\int \sinh au \cosh au\,du = \dfrac{\sinh^2 au}{2a} + C$

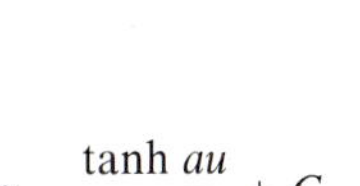

193. $\int \sinh pu \cosh qu\,du = \dfrac{\cosh(p+q)u}{2(p+q)} + \dfrac{\cosh(p-q)u}{2(p-q)} + C$

194. $\int \tanh au\,du = \dfrac{1}{a} \ln \cosh au + C$

195. $\int \tanh^2 au\,du = u - \dfrac{\tanh au}{a} + C$

196. $\int \tanh^n au\,du = \dfrac{-\tanh^{n-1} au}{a(n-1)} + \int \tanh^{n-2} au\,du$

197. $\int \coth au\,du = \dfrac{1}{a} \ln|\sinh au| + C$

198. $\int \coth^2 au\,du = u - \dfrac{\coth au}{a} + C$

199. $\int \operatorname{sech} au\,du = \dfrac{2}{a} \tan^{-1} e^{au} + C$

200. $\int \operatorname{sech}^2 au\,du = \dfrac{\tanh au}{a} + C$

201. $\int \operatorname{sech}^n au\,du = \dfrac{\operatorname{sech}^{n-2} au \tanh au}{a(n-1)} + \dfrac{n-2}{n-1} \int \operatorname{sech}^{n-2} au\,du$

202. $\int \operatorname{csch} au\, du = \frac{1}{a} \ln \left| \tanh \frac{au}{2} \right| + C$

203. $\int \operatorname{csch}^2 au\, du = -\frac{\coth au}{a} + C$

204. $\int \operatorname{sech} u \tanh u\, du = -\operatorname{sech} u + C$

205. $\int \operatorname{csch} u \coth u\, du = -\operatorname{csch} u + C$

SOME DEFINITE INTEGRALS

Unless otherwise stated, all letters stand for positive numbers.

206. $\int_0^\infty \frac{dx}{x^2 + a^2} = \frac{\pi}{2a}$

207. $\int_0^\infty \frac{x^{p-1}}{1+x}\, dx = \frac{\pi}{\sin p\pi}$

208. $\int_0^a \frac{dx}{\sqrt{a^2 - x^2}} = \frac{\pi}{2}$

209. $\int_0^a \sqrt{a^2 - x^2}\, dx = \frac{\pi a^2}{4}$

210. $\int_0^\pi \sin mx \sin nx\, dx = \begin{cases} 0, & \text{if } m, n \text{ integers and } m \neq n \\ \frac{\pi}{2}, & \text{if } m, n \text{ integers and } m = n \end{cases}$

211. $\int_0^\pi \cos mx \cos nx\, dx = \begin{cases} 0, & \text{if } m, n \text{ integers and } m \neq n \\ \frac{\pi}{2}, & \text{if } m, n \text{ integers and } m = n \end{cases}$

212. $\int_0^\pi \sin mx \cos nx\, dx = \begin{cases} 0, & \text{if } m, n \text{ integers and } m + n \text{ is even} \\ \frac{2m}{(m^2 - n^2)}, & \text{if } m, n \text{ integers and } m + n \text{ is odd} \end{cases}$

213. $\int_0^{\pi/2} \sin^2 x\, dx = \int_0^{\pi/2} \cos^2 x\, dx = \frac{\pi}{4}$

214. $\int_0^\infty e^{-ax} \cos bx\, dx = \frac{a}{a^2 + b^2}$

215. $\int_0^\infty e^{-ax} \sin bx\, dx = \frac{b}{a^2 + b^2}$

216. $\int_0^\infty e^{-a^2x^2}\, dx = \frac{\sqrt{\pi}}{2a}$

217. $\int_0^{\pi/2} \sin^{2m} x\, dx = \int_0^{\pi/2} \cos^{2m} x\, dx = \frac{1 \cdot 3 \cdot 5 \cdot \cdots \cdot (2m-1)}{2 \cdot 4 \cdot 6 \cdot \cdots \cdot 2m} \frac{\pi}{2}, \quad m = 1, 2, 3, \ldots$

218. $\int_0^{\pi/2} \sin^{2m+1} x\, dx = \int_0^{\pi/2} \cos^{2m+1} x\, dx = \frac{2 \cdot 4 \cdot 6 \cdot \cdots \cdot 2m}{1 \cdot 3 \cdot 5 \cdot \cdots \cdot (2m+1)}, \quad m = 1, 2, 3, \ldots$

219. $\int_0^\infty \frac{e^{-x}}{\sqrt{x}}\, dx = \sqrt{\pi}$

220. $\int_0^1 x^m (\ln x)^n\, dx = \frac{(-1)^n n!}{(m+1)^{n+1}}, \quad m \neq -1$

ANSWERS TO ODD-NUMBERED PROBLEMS AND REVIEW EXERCISES

CHAPTER 1

Problems 1.1, page 13

1. $Q = (3, 3)$ **3.** $Q = (4, 4)$ **5.** $Q = (-2, -5)$ **7.** $|\mathbf{v}| = 4\sqrt{2}$, $\theta = \pi/4$ **9.** $|\mathbf{v}| = 4\sqrt{2}$, $\theta = 7\pi/4$
11. $|\mathbf{v}| = 2$, $\theta = \pi/6$ **13.** $|\mathbf{v}| = 2$, $\theta = 2\pi/3$ **15.** $|\mathbf{v}| = 2$, $\theta = 4\pi/3$
17. $|\mathbf{v}| = \sqrt{89}$, $\theta = \pi + \tan^{-1}(-\frac{8}{5}) \approx 2.13$ (in second quadrant) **19.** $\mathbf{v} = \mathbf{j}$ **21.** $\mathbf{v} = -6\mathbf{i} + \mathbf{j}$
23. $\mathbf{v} = -9\mathbf{i} + 5\mathbf{j}$ **25.** $\mathbf{v} = -5\mathbf{i} + 5\mathbf{j}$ **27.** **a.** $(6, 9)$ **b.** $(-3, 7)$ **c.** $(-7, 1)$ **d.** $(39, -22)$
29. $(2/\sqrt{13})\mathbf{i} + (3/\sqrt{13})\mathbf{j}$ **31.** $\frac{3}{5}\mathbf{i} + \frac{4}{5}\mathbf{j}$ **33.** $-\frac{3}{5}\mathbf{i} + \frac{4}{5}\mathbf{j}$ **35.** $\sin\theta = -3/\sqrt{13}$, $\cos\theta = 2/\sqrt{13}$
37. $-(1/\sqrt{2})\mathbf{i} - (1/\sqrt{2})\mathbf{j}$ **39.** $(\frac{3}{5}, -\frac{4}{5})$ **41.** $\frac{3}{5}\mathbf{i} + \frac{4}{5}\mathbf{j}$
43. **a.** $(1/\sqrt{2})\mathbf{i} - (1/\sqrt{2})\mathbf{j}$ **b.** $(3/\sqrt{34})\mathbf{i} - (5/\sqrt{34})\mathbf{j}$ **c.** $(7/\sqrt{193})\mathbf{i} - (12/\sqrt{193})\mathbf{j}$ **d.** $-(2/\sqrt{53})\mathbf{i} + (7/\sqrt{53})\mathbf{j}$
45. $(3\sqrt{3}/2)\mathbf{i} + (\frac{3}{2})\mathbf{j}$ **47.** $-7\mathbf{i}$ **49.** $(1/\sqrt{2})\mathbf{i} + (1/\sqrt{2})\mathbf{j}$ **51.** $-8\mathbf{j}$

Problems 1.2, page 21

1. 0; 0 **3.** 0; 0 **5.** 20; $\frac{20}{29}$ **7.** -22; $-22/(5\sqrt{53})$ **9.** 100; $20/\sqrt{481}$ **11.** parallel
13. neither **15.** orthogonal **17.** parallel **19.** $\frac{3}{2}\mathbf{i} + \frac{3}{2}\mathbf{j}$ **21.** $\mathbf{0}$ **23.** $-\frac{2}{13}\mathbf{i} + \frac{3}{13}\mathbf{j}$ **25.** $\frac{14}{5}\mathbf{i} + \frac{28}{5}\mathbf{j}$
27. $-\frac{14}{5}\mathbf{i} + \frac{28}{5}\mathbf{j}$ **29.** $\dfrac{\alpha + \beta}{2}\mathbf{i} + \dfrac{\alpha + \beta}{2}\mathbf{j}$ **31.** $\dfrac{\alpha - \beta}{2}\mathbf{i} + \dfrac{\alpha - \beta}{2}\mathbf{j}$
33. **a.** $-\frac{3}{4}$ **b.** $\frac{4}{3}$ **c.** $(-96 - \sqrt{7500})/78 \approx -2.34$ **d.** $(-96 + \sqrt{7500})/78 \approx -0.12$
35. $52/(5/\sqrt{113}) \approx 0.9783$; $61/(\sqrt{34}\sqrt{113}) \approx 0.9841$; $-27/(5\sqrt{34}) \approx -0.9261$
37. $\text{Proj}_{\overrightarrow{PQ}}\overrightarrow{RS} = \dfrac{21}{13}\mathbf{i} + \dfrac{14}{13}\mathbf{j}$; $\text{Proj}_{\overrightarrow{RS}}\overrightarrow{PQ} = -\dfrac{7}{26}\mathbf{i} + \dfrac{35}{26}\mathbf{j}$ **39.** $\sqrt{5}$ **41.** $-2\mathbf{i} - 5\mathbf{j}$ N; $2\mathbf{i} + 5\mathbf{j}$ N
43. $-2\mathbf{i} - 3\mathbf{j}$ N; $2\mathbf{i} + 3\mathbf{j}$ N **45.** $-2\sqrt{3}\mathbf{i} - 7\mathbf{j}$ lb; $2\sqrt{3}\mathbf{i} + 7\mathbf{j}$ lb
47. $(3/\sqrt{2})\mathbf{i} - [(3/\sqrt{2}) + 2]\mathbf{j}$ N; $-(3/\sqrt{2})\mathbf{i} + [(3/\sqrt{2}) + 2]\mathbf{j}$ N
49. $(7\sqrt{2} - (7\sqrt{3}/2) - \frac{7}{2})(\mathbf{i} + \mathbf{j})$ N; $(-7\sqrt{2} + (7\sqrt{3}/2) + \frac{7}{2})(\mathbf{i} + \mathbf{j})$ N **51.** -12 **J** **53.** $(8\sqrt{3} + 4)$ **J**
55. $3\sqrt{2}$ **J** **57.** $12/\sqrt{13}$ **J** **59.** $500(\sin 20°/\sin 30°) \approx 342$ N
61. tugboat 1: $(500)(\cos 20°)(750) \approx 352{,}385$ **J**; tugboat 2: $(342)(\cos 30°)(750) \approx 221{,}149$ **J** **63.** *Hint:* $\mathbf{u} \cdot \mathbf{v} = 0$.
65. *Hint:* Disprove. **67.** $a_1a_2 + b_1b_2 > 0$

Problems 1.3, page 27

1.

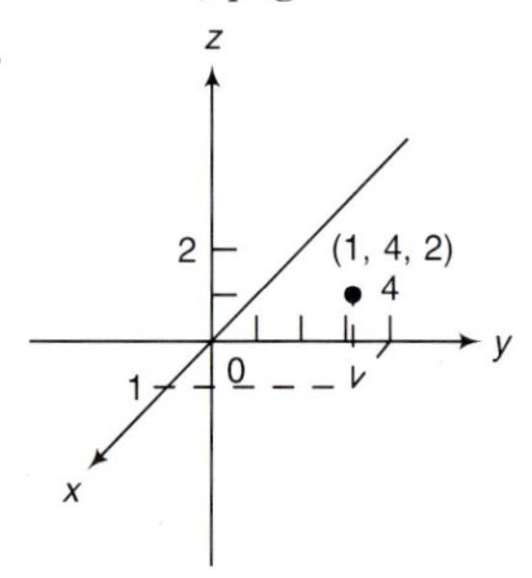

3.

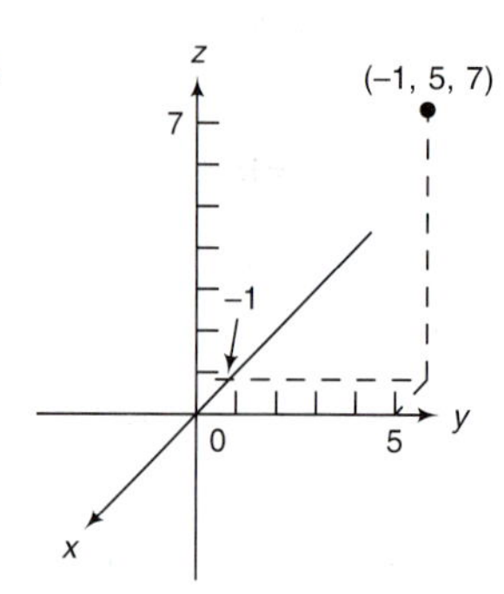

5.

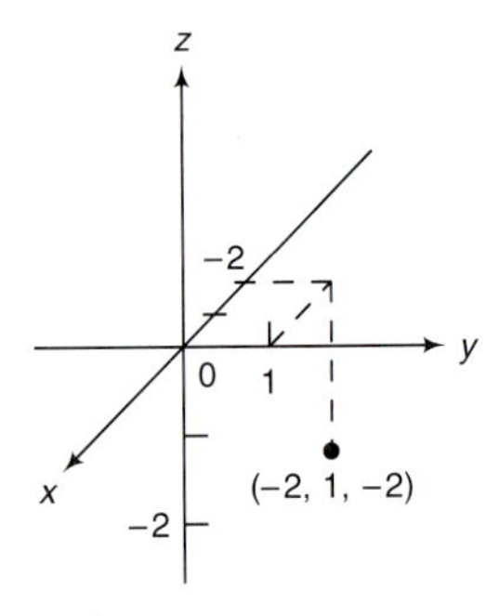

7.

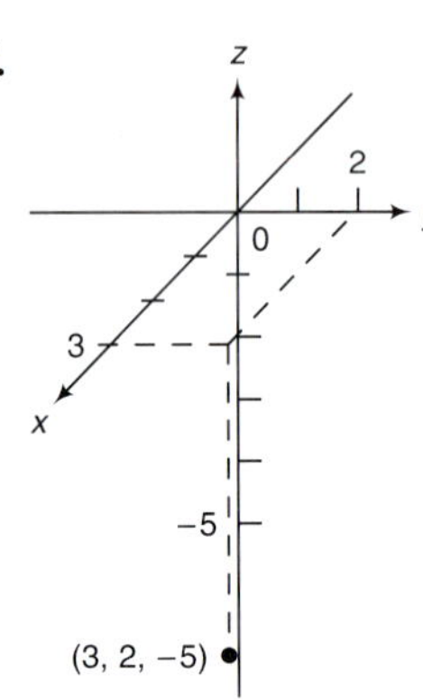

9.

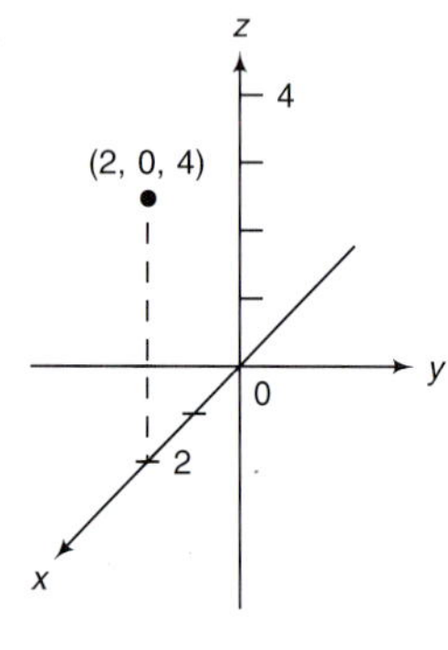

11.

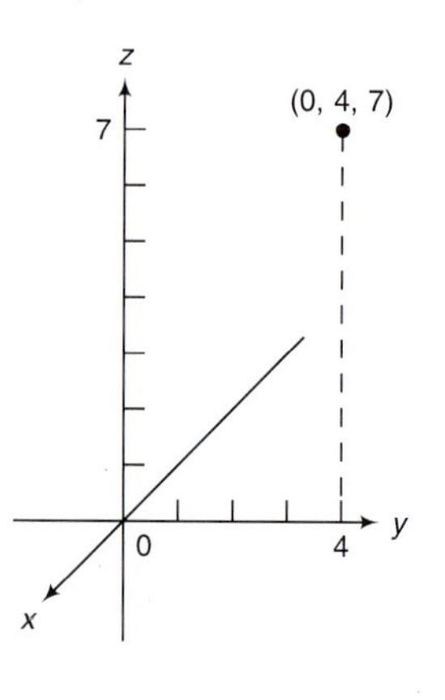

13.

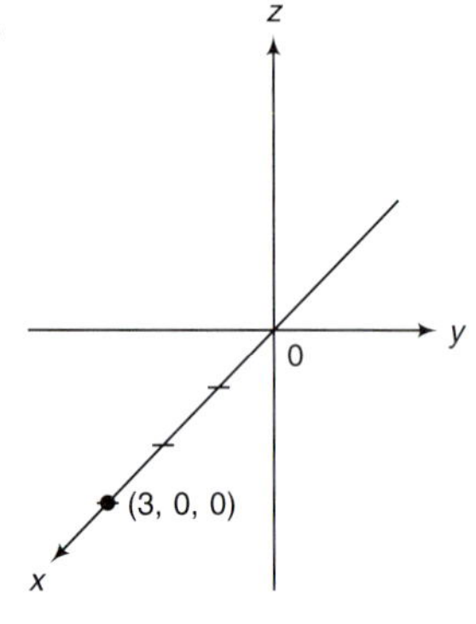

15.

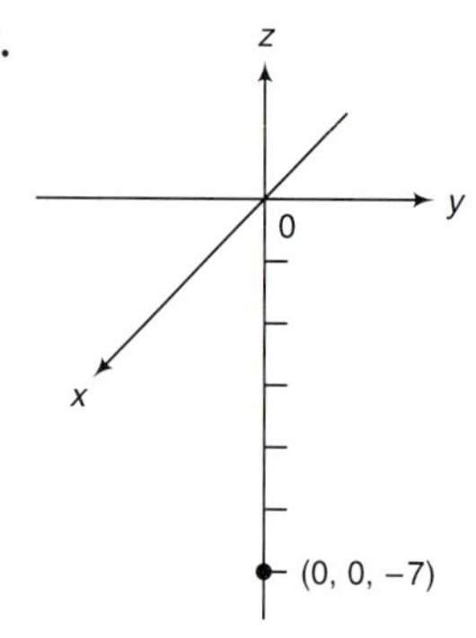

17. 2 **19.** 2 **21.** $6\sqrt{2}$ **23.** $\sqrt{6}$

25. $\sqrt{118}$ **27.** $(x-2)^2+(y+1)^2+(z-4)^2=4$

29. *Hint:* (3, 0, 1) is the midpoint of the segment between (0, −4, 0) and (6, 4, 2). **31.** Center: (2, 2, −4); radius: 4.

33. $(x-1)^2+(y-2)^2+(z+\frac{1}{2})^2=1$ **35.** $\frac{93}{4}$

Problems 1.4, page 35

1. $|\mathbf{v}|=3$; 0, 1, 0 **3.** $|\mathbf{v}|=14$; 0, 0, 1 **5.** $|\mathbf{v}|=\sqrt{17}$; $4/\sqrt{17}$, $-1/\sqrt{17}$, 0

7. $|\mathbf{v}|=\sqrt{13}$, $-2/\sqrt{13}$, $3/\sqrt{13}$, 0 **9.** $|\mathbf{v}|=\sqrt{3}$; $1/\sqrt{3}$, $-1/\sqrt{3}$, $1/\sqrt{3}$ **11.** $|\mathbf{v}|=\sqrt{3}$; $-1/\sqrt{3}$, $1/\sqrt{3}$, $1/\sqrt{3}$

13. $|\mathbf{v}|=\sqrt{3}$; $-1/\sqrt{3}$, $1/\sqrt{3}$, $-1/\sqrt{3}$ **15.** $|\mathbf{v}|=\sqrt{3}$; $-1/\sqrt{3}$, $-1/\sqrt{3}$, $-1/\sqrt{3}$

17. $|\mathbf{v}|=\sqrt{222}$; $-7/\sqrt{222}$, $2/\sqrt{222}$, $-13/\sqrt{222}$ **19.** $|\mathbf{v}|=\sqrt{82}$; $-3/\sqrt{82}$, $-3/\sqrt{82}$, $8/\sqrt{82}$ **21.** $-6\mathbf{j}+9\mathbf{k}$

23. $-36\mathbf{i}+54\mathbf{j}-72\mathbf{k}$ **25.** $8\mathbf{i}-14\mathbf{j}+9\mathbf{k}$ **27.** $16\mathbf{i}+29\mathbf{j}+42\mathbf{k}$ **29.** $\sqrt{59}$

31. $\cos^{-1}(35/(\sqrt{29}\sqrt{59}))\approx 0.562\approx 32°$ **33.** $\cos^{-1}(7/(\sqrt{38}\sqrt{50}))\approx 1.410\approx 81°$ **35.** $\frac{25}{38}\mathbf{v}=-\frac{25}{19}\mathbf{i}-\frac{75}{38}\mathbf{j}+\frac{125}{38}\mathbf{k}$

37. $-\frac{1}{5}\mathbf{t}=-\frac{3}{5}\mathbf{i}-\frac{4}{5}\mathbf{j}-\mathbf{k}$ **39.** $\frac{35}{59}\mathbf{w}=\frac{35}{59}\mathbf{i}-\frac{245}{59}\mathbf{j}+\frac{105}{59}\mathbf{k}$ **41.** $\frac{1}{\sqrt{3}}\mathbf{i}+\frac{1}{\sqrt{3}}\mathbf{j}+\frac{1}{\sqrt{3}}\mathbf{k}$

43. $\dfrac{1}{\sqrt{26}}\mathbf{i} - \dfrac{3}{\sqrt{26}}\mathbf{j} + \dfrac{4}{\sqrt{26}}\mathbf{k}$ **45.** *Hint:* $\cos^2(\pi/6) + \cos^2(\pi/3) + \cos^2(\pi/4) = 1.5$.
47. $\{x, y, z: x + 3 = 0\}$. This is a plane (see Section 1.7).
49. $\sqrt{5610}/51 = \sqrt{\frac{110}{51}}$ (Note that the distance is given by $|\overrightarrow{QP} - \text{Proj}_{\overrightarrow{QR}}\overrightarrow{QP}|$; draw a picture.)
51. *Hint:* $\overrightarrow{PQ} \cdot \overrightarrow{PR} = 0$. **53.** $\left(\dfrac{\pm 1}{\sqrt{419}}\right)(7\mathbf{i} - 17\mathbf{j} + 9\mathbf{k})$

Problems 1.5, page 40

1. $x\mathbf{i} + y\mathbf{j} + z\mathbf{k} = (2\mathbf{i} + \mathbf{j} + 3\mathbf{k}) + t(-\mathbf{i} + \mathbf{j} - 4\mathbf{k})$; $x = 2 - t$, $y = 1 + t$, $z = 3 - 4t$; $(x - 2)/(-1) = y - 1 = (z - 3)/(-4)$
3. $x\mathbf{i} + y\mathbf{j} + z\mathbf{k} = (\mathbf{i} + 3\mathbf{j} + 2\mathbf{k}) + t(\mathbf{i} + \mathbf{j} - 4\mathbf{k})$; $x = 1 + t$, $y = 3 + t$, $z = 2 - 4t$; $x - 1 = y - 3 = (z - 2)/(-4)$
5. $x\mathbf{i} + y\mathbf{j} + z\mathbf{k} = (-4\mathbf{i} + \mathbf{j} + 3\mathbf{k}) + t(-\mathbf{j} - 2\mathbf{k})$; $x = -4$, $y = 1 - t$, $z = 3 - 2t$; $x = -4$ and $(y - 1)/(-1) = (z - 3)/(-2)$
7. $x\mathbf{i} + y\mathbf{j} + z\mathbf{k} = (\mathbf{i} + 2\mathbf{j} + 3\mathbf{k}) + t(2\mathbf{i} - 2\mathbf{k})$; $x = 1 + 2t$, $y = 2$, $z = 3 - 2t$; $y = 2$ and $(x - 1)/2 = (z - 3)/(-2)$
9. $x\mathbf{i} + y\mathbf{j} + z\mathbf{k} = (\mathbf{i} + 2\mathbf{j} + 4\mathbf{k}) + t(3\mathbf{k})$; $x = 1$, $y = 2$, $z = 4 + 3t$; $x = 1$ and $y = 2$
11. $x\mathbf{i} + y\mathbf{j} + z\mathbf{k} = (2\mathbf{i} + 2\mathbf{j} + \mathbf{k}) + t(2\mathbf{i} - \mathbf{j} - \mathbf{k})$; $x = 2 + 2t$, $y = 2 - t$, $z = 1 - t$; $(x - 2)/2 = (y - 2)/(-1) = (z - 1)/(-1)$
13. $x\mathbf{i} + y\mathbf{j} + z\mathbf{k} = (\mathbf{i} + 3\mathbf{k}) + t(\mathbf{i} - \mathbf{j})$; $x = 1 + t$, $y = -t$, $z = 3$; $x - 1 = -y$ or $x + y = 1$ and $z = 3$
15. $x\mathbf{i} + y\mathbf{j} + z\mathbf{k} = (-\mathbf{i} - 2\mathbf{j} + 5\mathbf{k}) + t(-3\mathbf{j} + 4\mathbf{k})$; $x = -1$, $y = -2 - 3t$, $z = 5 + 4t$; $x = -1$ and $\dfrac{y + 2}{-3} = \dfrac{z - 5}{4}$ or $4y + 3z = 7$
17. $x\mathbf{i} + y\mathbf{j} + z\mathbf{k} = (-\mathbf{i} - 3\mathbf{j} + \mathbf{k}) + t(-7\mathbf{j})$; $x = -1$, $y = -3 - 7t$, $z = 1$; $x = -1$ and $z = 1$
19. $x\mathbf{i} + y\mathbf{j} + z\mathbf{k} = (a\mathbf{i} + b\mathbf{j} + c\mathbf{k}) + t(d\mathbf{i} + e\mathbf{j})$; $x = a + dt$, $y = b + et$, $z = c$; $z = c$ and $(x - a)/d = (y - b)/e$
21. $x\mathbf{i} + y\mathbf{j} + z\mathbf{k} = (4\mathbf{i} + \mathbf{j} - 6\mathbf{k}) + t(3\mathbf{i} + 6\mathbf{j} + 2\mathbf{k})$; $x = 4 + 3t$, $y = 1 + 6t$, $z = -6 + 2t$; $(x - 4)/3 = (y - 1)/6 = (z + 6)/2$
23. The lines meet at $(2, -1, -3)$. **25.** The lines meet at $(4, 3, 11)$. **27.** no intersection
29. no intersection **33.** $t = -(\overrightarrow{OP} \cdot \mathbf{v})/(\mathbf{v} \cdot \mathbf{v}) = -(\overrightarrow{OP} \cdot \mathbf{v})/|\mathbf{v}|^2$
35. $x = 2 + t$, $y = -3 + t$, $z = 1$; $x = 2 + 3t$, $y = -3$, $z = 1 - 4t$ or any linear combination of these lines (i.e. a plane)

Problems 1.6, page 50

1. $-6\mathbf{i} - 3\mathbf{j}$ **3.** $-\mathbf{i} - \mathbf{j} + \mathbf{k}$ **5.** $12\mathbf{i} + 8\mathbf{j} - 21\mathbf{k}$ **7.** $(bc - ad)\mathbf{j}$ **9.** $-5\mathbf{i} - \mathbf{j} + 7\mathbf{k}$ **11.** $\mathbf{0}$
13. $42\mathbf{i} + 6\mathbf{j}$ **15.** $-9\mathbf{i} + 39\mathbf{j} + 61\mathbf{k}$ **17.** $-4\mathbf{i} + 8\mathbf{k}$ **19.** $\mathbf{0}$ **21.** $\left(\dfrac{\pm 1}{\sqrt{181}}\right)(-9\mathbf{i} - 6\mathbf{j} + 8\mathbf{k})$
23. $\sqrt{30}/(\sqrt{6}\sqrt{29}) \approx 0.415$ **25.** $(x - 1)/2 = (y + 3)/(-26) = (z - 2)/(-22)$
27. $x = -2 + 13u$, $y = 3 + 22u$, $z = 4 - 8u$ **29.** $5\sqrt{5}$ **31.** $\sqrt{523}$ **33.** $\sqrt{a^2b^2 + b^2c^2 + c^2a^2}$
35. $\frac{1}{2}\sqrt{3778}$ **37.** $\frac{1}{2}\sqrt{3}$ **39.** 14 **41.** 23 **43.** $48/\sqrt{437}$ **45.** $23/\sqrt{27}$
51. $\frac{1}{6}$ × volume of the parallelepiped

Problems 1.7, page 57

1. $x = 0$ (yz-plane) **3.** $z = 0$ (xy-plane) **5.** $x + z = 4$ **7.** $3x - y + 2z = 19$ **9.** $4x + y - 7z = -15$
11. $x + y + z = -6$ **13.** $20x + 13y - 3z = 58$ **15.** $x + y + z = 1$ **17.** coincident **19.** orthogonal
21. orthogonal **23.** coincident **25.** $x = -1$ and $y + z = 2$ or $x = -1$, $y = t$, $z = 2 - t$
27. $x = t$, $y = -11 - 37t$, $z = -2 - 10t$ **29.** $\cos^{-1}(\frac{1}{3}) \approx 1.23 \approx 70.5°$ **31.** $\cos^{-1}(18/(\sqrt{26}\sqrt{69})) \approx 1.132 \approx 64.9°$
33. coplanar: $x - 22y - 17z = 0$ **35.** not coplanar **37.** $33/\sqrt{59}$ **39.** $11/\sqrt{68}$

41.

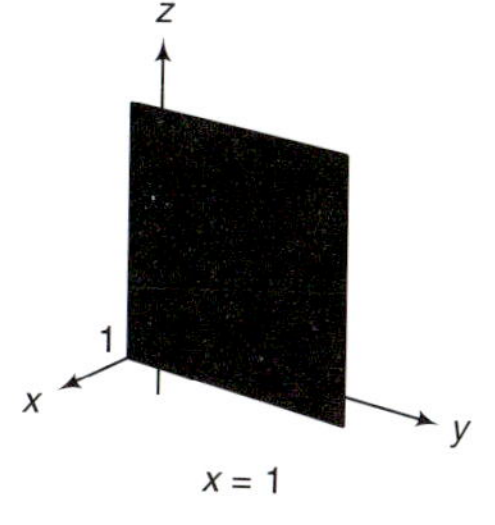

x = 1

43.

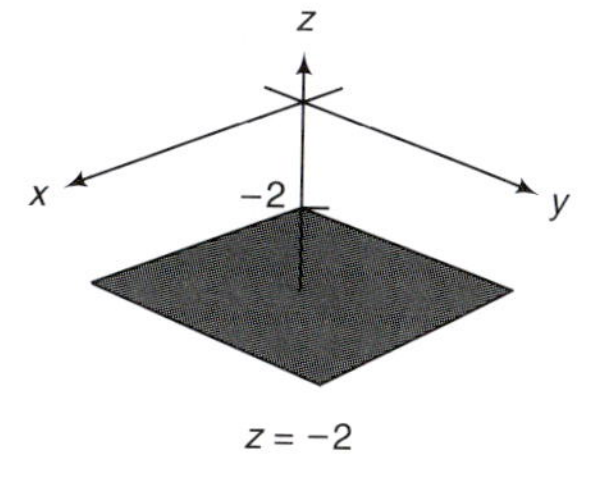

z = −2

45.

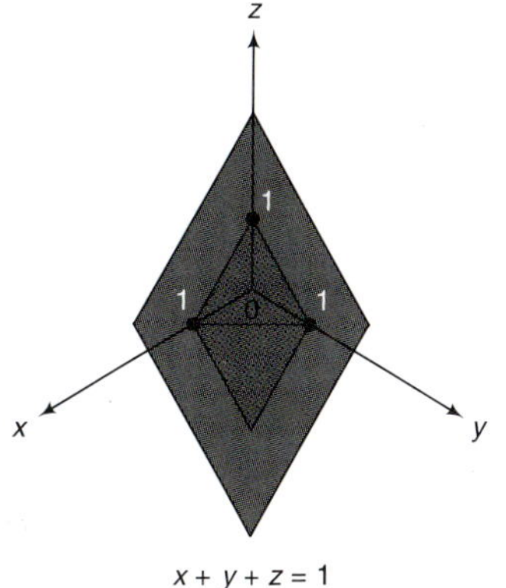

x + y + z = 1

47.

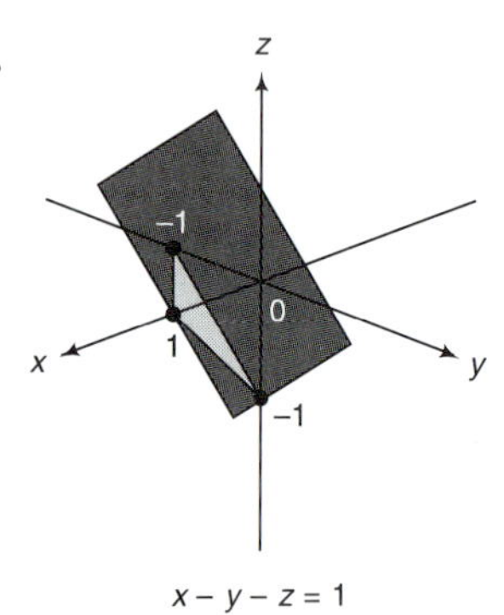

$x - y - z = 1$

49.

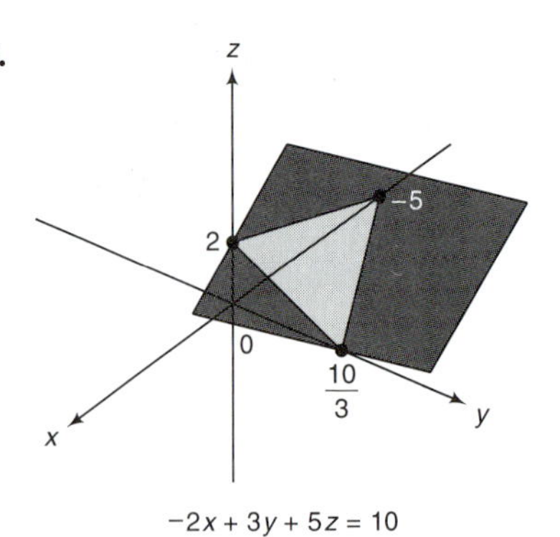

$-2x + 3y + 5z = 10$

Problems 1.8, page 65

1. $y = \sin x$

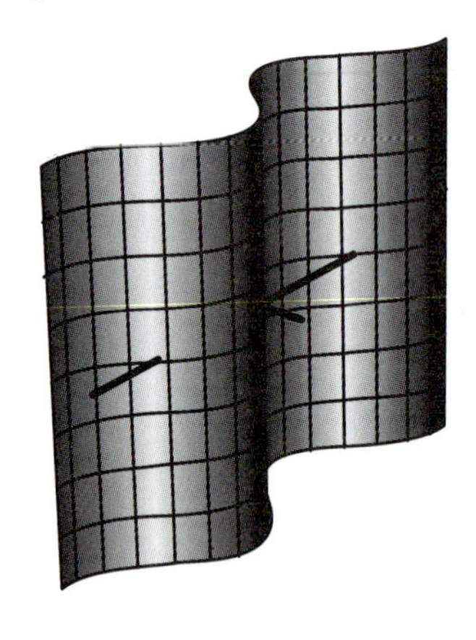

3. $y = \cos x$

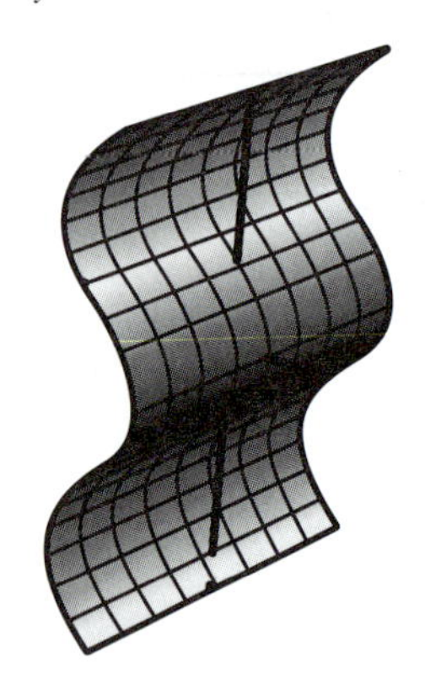

5. $z = x^3$

7. $|y| + |z - 5| = 1$

9. right circular cylinder, radius 2, centered on z-axis

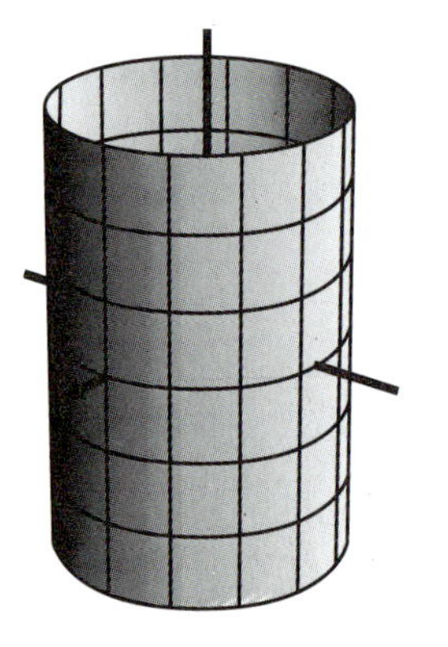

11. right circular cylinder, radius 2, centered on y-axis

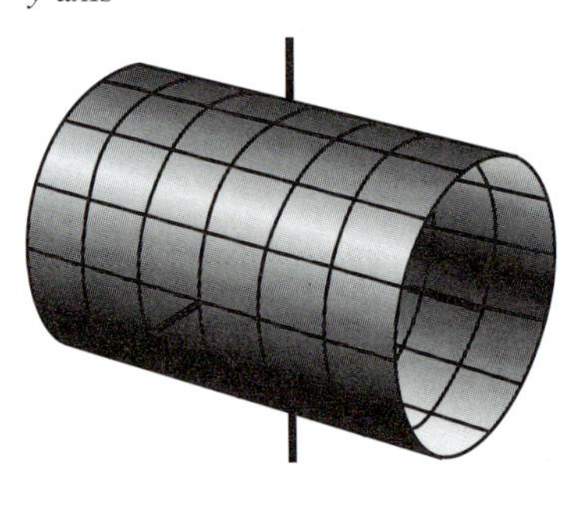

13. elliptic cylinder

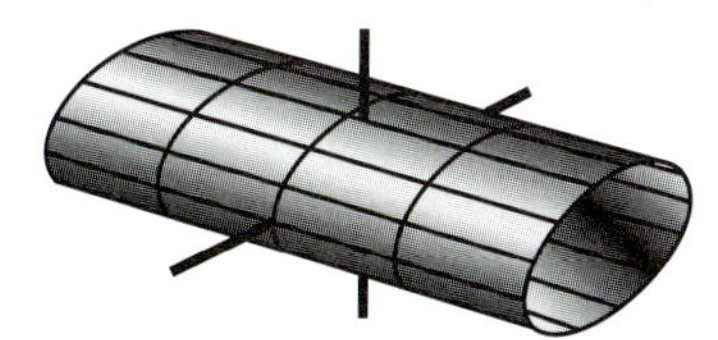

15. hyperbolic cylinder

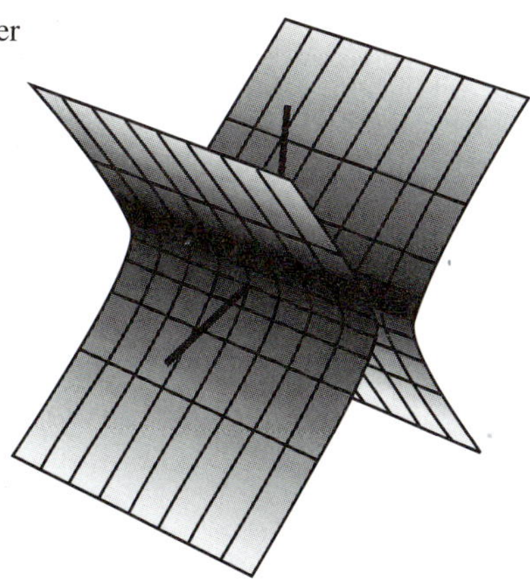

17. sphere of radius 1 centered at origin

19. hyperboloid of one sheet centered at $(0, -1, 0)$; cross-sections parallel to the xy-plane are ellipses

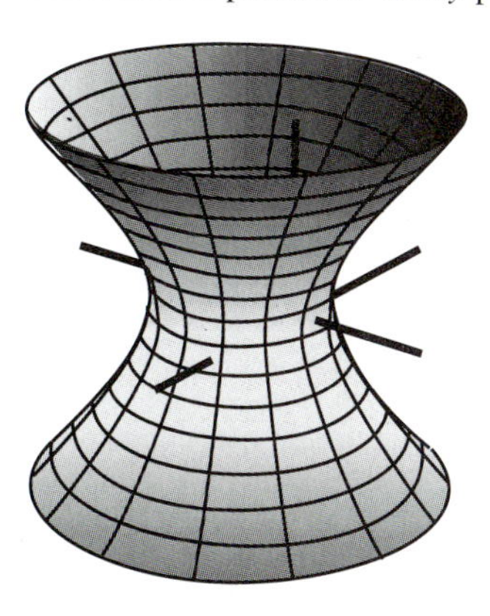

21. elliptic paraboloid; cross-sections parallel to the yz-plane are ellipses; defined for $x \leq 4$

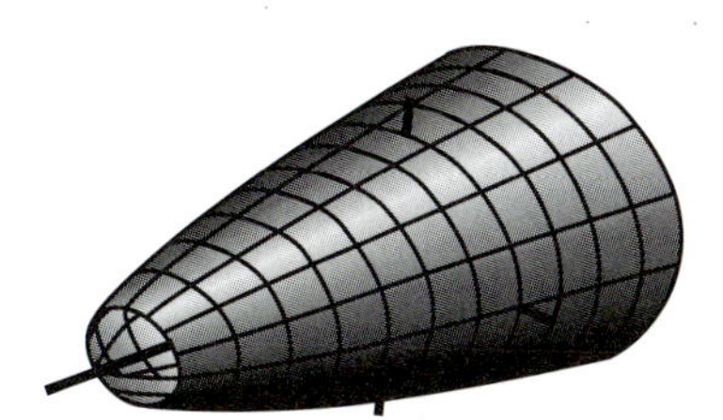

23. hyperboloid of two sheets centered at $(-2, -3, -4)$; cross-sections parallel to the xy-plane are ellipses; defined for $|z + 4| \geq 1$

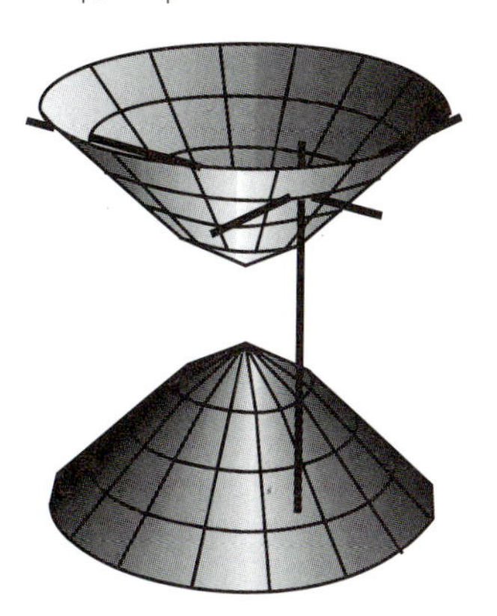

25. ellipsoid centered at $(0, 0, \frac{1}{2})$

27. hyperboloid of two sheets; cross-sections parallel to the yz-plane are ellipses; defined for $|x| \geq 2$

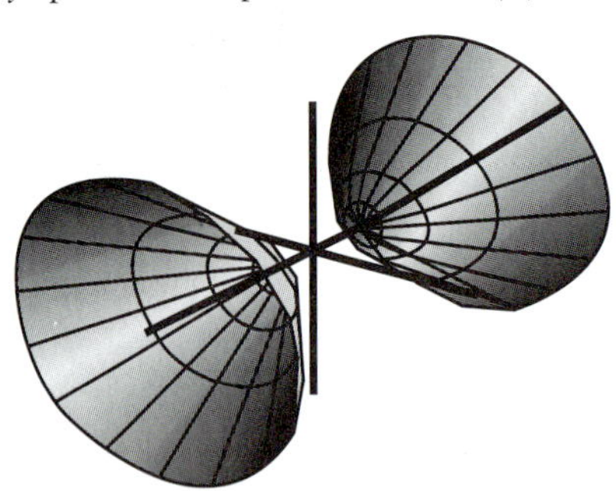

29. hyperboloid of two sheets; cross-sections parallel to the xz-plane are ellipses; defined for $|y| \geq 1$

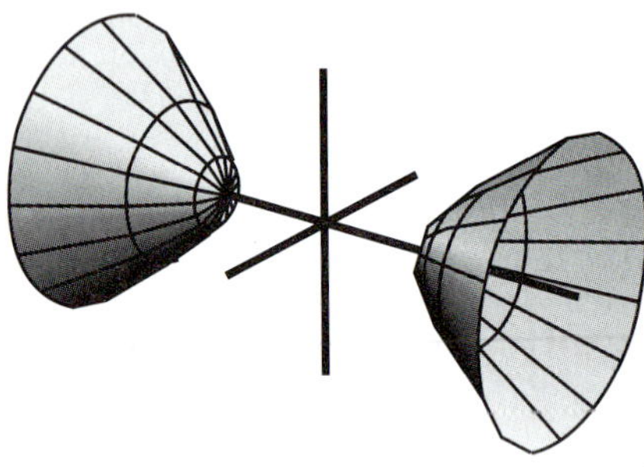

31. ellipsoid

33. hyperbolic paraboloid; cross-sections parallel to the xy-plane are hyperbolas

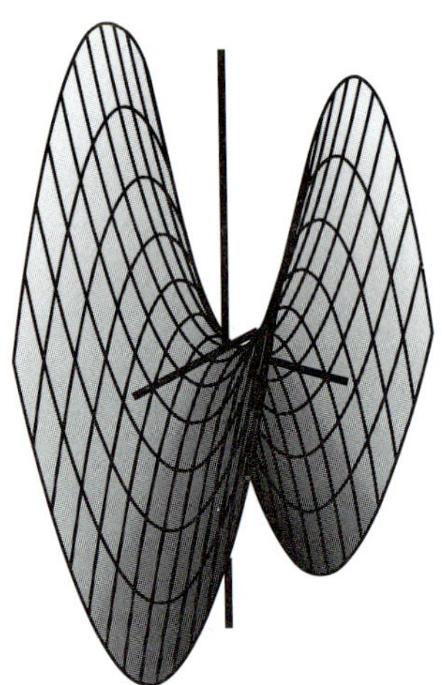

35. elliptic (actually circular) paraboloid centered at $(\frac{1}{2}, -\frac{1}{2}, \frac{1}{2})$, cross-sections parallel to the xz-plane are ellipses

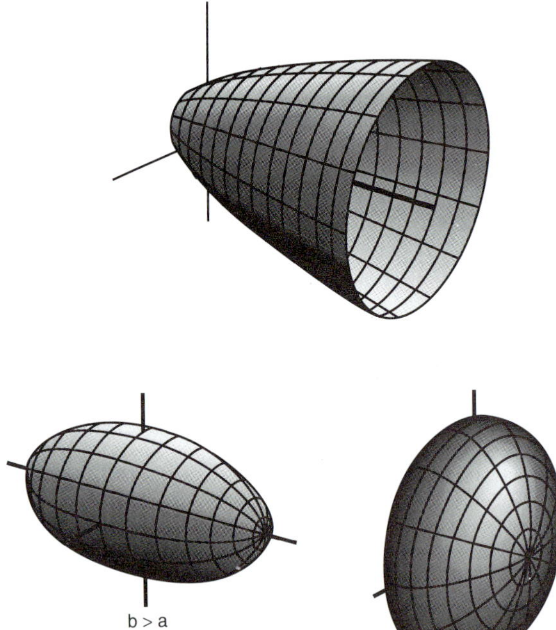

37. ellipsoid with equation $\dfrac{x^2 + z^2}{a^2} + \dfrac{y^2}{b^2} = 1$

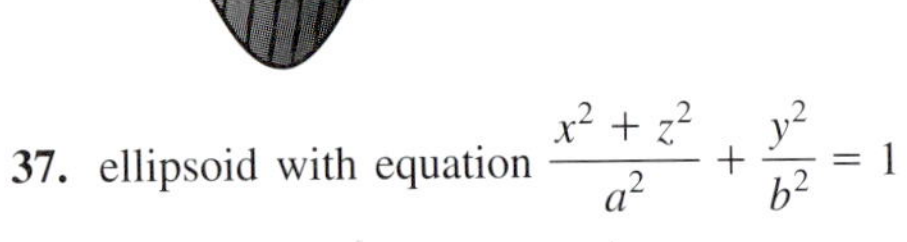

b > a

a > b

Problems 1.9, page 71

1. $(1, \sqrt{3}, 5)$ **3.** $(-4, 4\sqrt{3}, 1)$ **5.** $(-3/\sqrt{2}, 3/\sqrt{2}, 2)$ **7.** $(-10, 0, -3)$ **9.** $(-7/\sqrt{2}, -7/\sqrt{2}, 2)$
11. $(1, 0, 0)$ **13.** $(0, \theta, 1)$ where θ is arbitrary **15.** $(\sqrt{2}, 3\pi/4, 4)$ **17.** $(4, 11\pi/6, 8)$ **19.** $(4, 7\pi/6, 1)$
21. $(\sqrt{3}, 0, 1)$ **23.** $(0, 3\sqrt{3}, 3)$ **25.** $(\frac{7}{2}, -\frac{7}{2}, -7/\sqrt{2})$ **27.** $(\sqrt{3}, 3, -2)$ **29.** $(5\sqrt{3}/4, -\frac{5}{4}, -5\sqrt{3}/2)$
31. $(\sqrt{2}, \pi/4, \pi/2)$ **33.** $(2, 7\pi/4, \pi/4)$ **35.** $(2\sqrt{2}, 5\pi/3, \pi/4)$
37. $(\sqrt{23}, \tan^{-1}(\sqrt{3}/2), \cos^{-1}(4/\sqrt{23})) \approx (4.796, 0.714, 0.584)[\approx(4.796, 41°, 33°)]$
39. $(\sqrt{23}, \pi + \tan^{-1}(2/\sqrt{3}), \cos^{-1}(-4/\sqrt{23})) \approx (4.796, 3.999, 2.557)[\approx(4.796, 229°, 147°)]$ [*Note:* $\cos\theta < 0$ and $\sin\theta < 0$ so θ is in the third quadrant.]
41. cylindrical: $r^2 + z^2 = 25$; spherical: $\rho = 5$ **43.** $x^2 + y^2 - 9y = 0$
45. cylindrical: $r^2 - z^2 = 1$; spherical: $\rho^2(\sin^2\varphi - \cos^2\varphi) = 1$ **47.** $z = x^2 + y^2$ **49.** $z\sqrt{x^2 + y^2} = 1$
51. $z = 2xy$ **53.** $\rho = 6\sin\phi\sin\theta$

Problems 1.10, page 79

1. $\begin{pmatrix} 2 \\ -3 \\ 11 \end{pmatrix}$ **3.** $\begin{pmatrix} -4 \\ 0 \\ 4 \end{pmatrix}$ **5.** $\begin{pmatrix} -31 \\ 22 \\ -27 \end{pmatrix}$ **7.** $\begin{pmatrix} 0 \\ 0 \\ 0 \end{pmatrix}$ **9.** $\begin{pmatrix} -11 \\ 11 \\ -10 \end{pmatrix}$ **11.** $(1, 2, 5, 7)$
13. $(-8, 12, 4, 20)$ **15.** $(8, -5, 7, -1)$ **17.** $(7, 2, 4, 11)$ **19.** $(-11, 9, 18, 18)$
25. $\mathbf{d}_1 + \mathbf{d}_2$ represents the combined demand of the two factories for each of the four raw materials needed to produce one unit of each factory's product; $2\mathbf{d}_1$ represents the demand of factory 1 for each of the four raw materials needed to produce two units of its product.
27. -14 **29.** 1 **31.** $ac + bd$ **33.** 51 **35.** $\mathbf{a} = \mathbf{0}$ **37.** 4 **39.** 28 **41.** orthogonal
43. orthogonal **45.** orthogonal **47.** all α and β that satisfy $5\alpha + 4\beta = 25(\beta = (25 - 5\alpha)/4$, α arbitrary)
49. $\sqrt{30}$ **51.** $\sqrt{a^2 + b^2 + c^2 + d^2 + e^2}$ **53.** $\sqrt{10}$ **55.** -4

Chapter 1—Review Exercises, page 83

1. a. $(10, 5)$ **b.** $(5, -3)$ **c.** $(-31, 12)$ **d.** $(0, 11)$
3. a. $(15, 0, 6)$ **b.** $(4, 3, -3)$ **c.** $(-8, -6, 6)$ **d.** $(0, -15, 23)$ **5.** $|\mathbf{v}| = 3\sqrt{2}$, $\theta = \pi/4$
7. $|\mathbf{u}| = 12\sqrt{2}$, $\theta = 5\pi/4$ **9.** $|\mathbf{w}| = 2$, $(\sqrt{3}/2, 1/2, 0)$ **11.** $|\mathbf{w}| = \sqrt{58}$, $(7/\sqrt{58}, -3/\sqrt{58}, 0)$
13. $|\mathbf{w}| = 5\sqrt{2}$, $(-3/5\sqrt{2}, 4/5\sqrt{2}, 1/\sqrt{2})$ **15.** $\sqrt{68}$ **17.** $\sqrt{216}$ **19.** $2\mathbf{i} + 2\mathbf{j}$ **21.** $4\mathbf{i} + 2\mathbf{j}$
23. $-4\mathbf{i} + 6\mathbf{j}$ **25.** $2\mathbf{i} - 4\mathbf{j} - 4\mathbf{k}$ **27.** $-(1/\sqrt{2})\mathbf{i} - (1/\sqrt{2})\mathbf{j}$ **29.** $-(10/\sqrt{149})\mathbf{i} + (7/\sqrt{149})\mathbf{j}$
31. $-(3/\sqrt{130})\mathbf{j} - (11/\sqrt{130})\mathbf{k}$ **33.** $-(1/\sqrt{14})\mathbf{i} + (2/\sqrt{14})\mathbf{j} + (3/\sqrt{14})\mathbf{k}$ **35.** $(7/\sqrt{53})\mathbf{i} - (2/\sqrt{53})\mathbf{j}$
37. $(1, \sqrt{3})$ **39.** $(-4, 0)$ **41.** $(3, 4, 0)$ **43.** $(1/\sqrt{2}, -1, -1/\sqrt{2})$ **45.** -1; $-1/\sqrt{10}$
47. -22; $-22/\sqrt{3965}$ **49.** -2; $-\frac{2}{5}$ **51.** 0; 0 **53.** parallel **55.** neither **57.** neither
59. neither **61.** $7\mathbf{i} + 7\mathbf{j}$ **63.** $\frac{15}{13}\mathbf{i} + \frac{10}{13}\mathbf{j}$ **65.** $-\frac{90}{59}\mathbf{i} - \frac{210}{59}\mathbf{j} + \frac{30}{59}\mathbf{k}$ **67.** $\frac{26}{21}\mathbf{i} - \frac{52}{21}\mathbf{j} + \frac{13}{21}\mathbf{k}$
69. $-7\mathbf{i} - 7\mathbf{k}$ **71.** $-5\mathbf{i} - 41\mathbf{j} - 3\mathbf{k}$
73. $x\mathbf{i} + y\mathbf{j} + z\mathbf{k} = (3\mathbf{i} - \mathbf{j} + 4\mathbf{k}) + t(4\mathbf{i} - 7\mathbf{j} + 2\mathbf{k})$; $x = 3 + 4t$, $y = -1 - 7t$, $z = 4 + 2t$; $(x - 3)/4 = (y + 1)/(-7) = (z - 4)/2$
75. $x\mathbf{i} + y\mathbf{j} + z\mathbf{k} = (3\mathbf{i} + \mathbf{j} + 2\mathbf{k}) = t(3\mathbf{i} - \mathbf{j} - \mathbf{k})$; $x = 3 + 3t$, $y = 1 - t$, $z = 2 - t$; $(x - 3)/3 = (y - 1)/(-1) = (z - 2)/(-1)$
77. $y - z = 0$ **79.** $2y - 3z = -26$ **81.** $(3\sqrt{3}/2, \frac{3}{2}, -1)$ **83.** $(2\sqrt{2}, \pi/4, -4)$
85. $(3\sqrt{2}/4, 3\sqrt{6}/4, 3/\sqrt{2})$ **87.** $(2, 3\pi/4, 3\pi/4)$ **89.** cylindrical: $r^2 + z^2 = 5^2$; spherical: $\rho = 5$
91. cylindrical: $r^2(\cos^2\theta - \sin^2\theta) + z^2 = 1$; spherical: $\rho^2(\sin^2\varphi(\cos^2\theta - \sin^2\theta) + \cos^2\varphi) = 1$
93. plane parallel to z-axis which intersects xy-plane in the line $y = 3 - 5x$

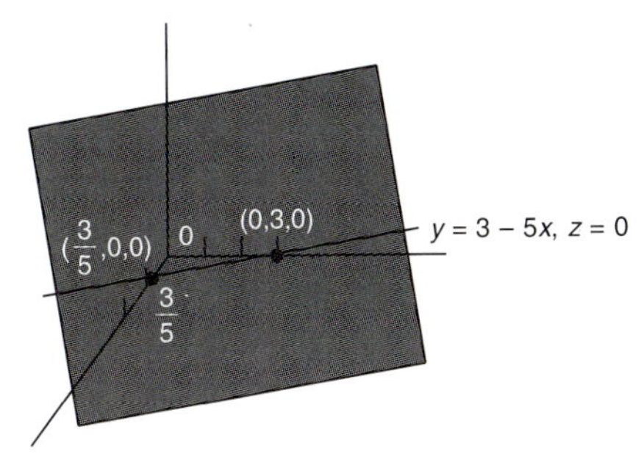

95. parabolic cylinder

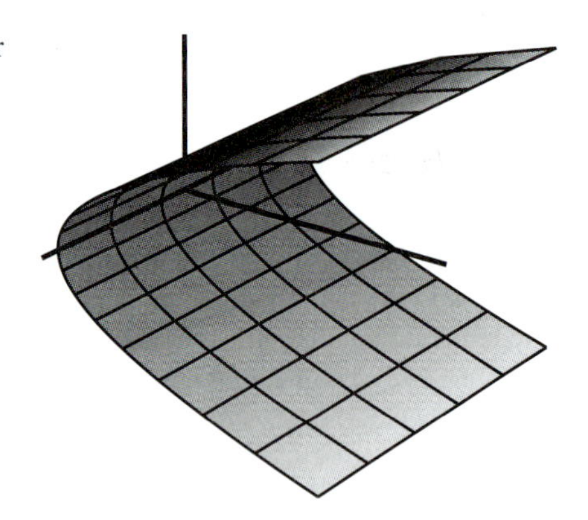

97. $\left(\frac{\pm 1}{\sqrt{2}}\right)(\mathbf{i} + \mathbf{j})$ **99.** **a.** 6 **b.** $-\frac{8}{3}$ **c.** $\frac{4}{5}$ **d.** $(192 \pm \sqrt{32448})/6$ **101.** $46x + 14y - 19z = -55$

103. $x = \frac{1}{2} - \frac{9}{2}t,\ y = \frac{7}{2} - \frac{11}{2}t,\ z = t$ **105.** $16/\sqrt{5}$ **107.** $5/\sqrt{6}$

109. $(x + 1)/14 = (y - 2)/(-26) = (z - 4)/(-11)$ **111.** the points are collinear **113.** $\sqrt{2065}$

115. right circular cylinder **117.** hyperboloid of one sheet; cross-sections parallel to xz-plane are ellipses

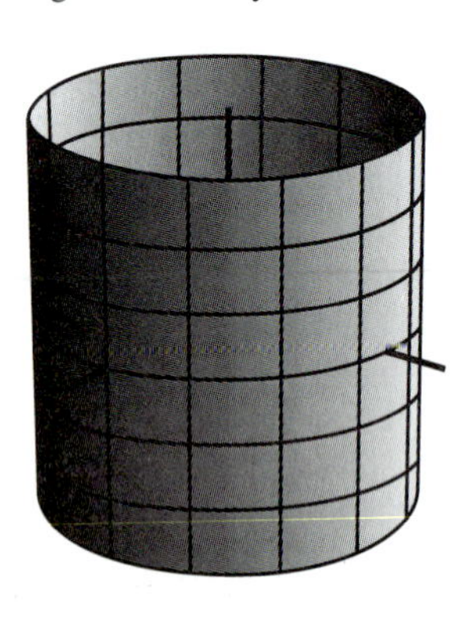

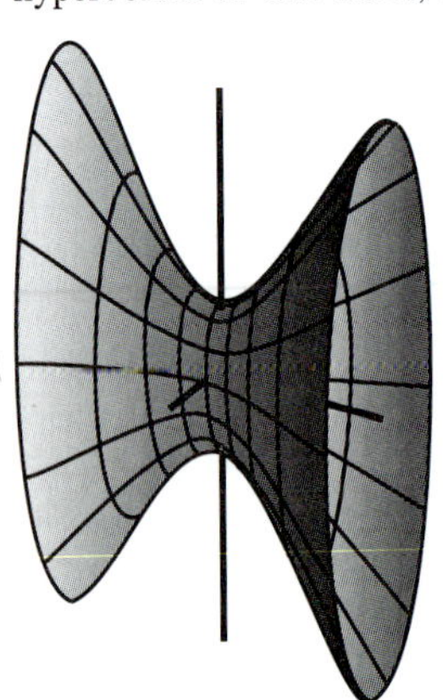

119. hyperboloid of two sheets; cross-sections parallel to xz-plane are ellipses; surface defined only for $|y| \geq \frac{5}{4}$

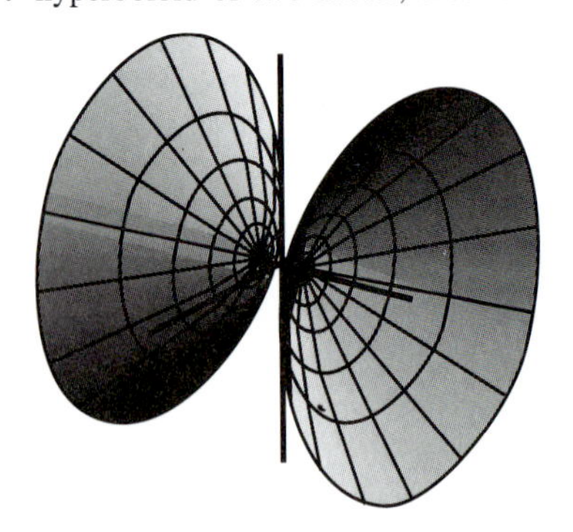

CHAPTER 2

Problems 2.1, page 89

1. $\mathbb{R} - \{0, 1\}$ **3.** $\mathbb{R} - \{-1, 1\}$ **5.** $(0, 1)$ **7.** $\mathbb{R} - \{k\pi/2\colon k \text{ is an integer}\}$

9. $x = y^2/4$

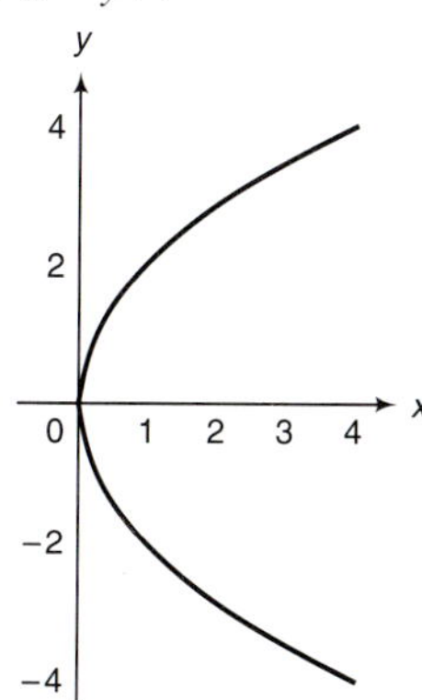

11. $x = y^{2/3}$ or $y = \pm x^{3/2}$

13. $y = 2x + 5$

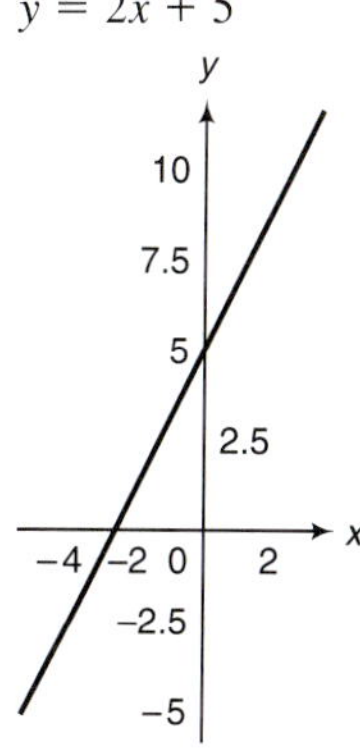

15. $x = y^2 + y + 1,\ y \geq 0$

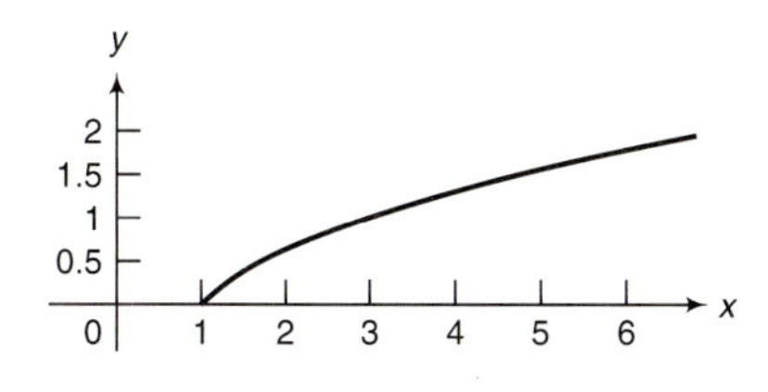

17. $y = x^3 - 1$

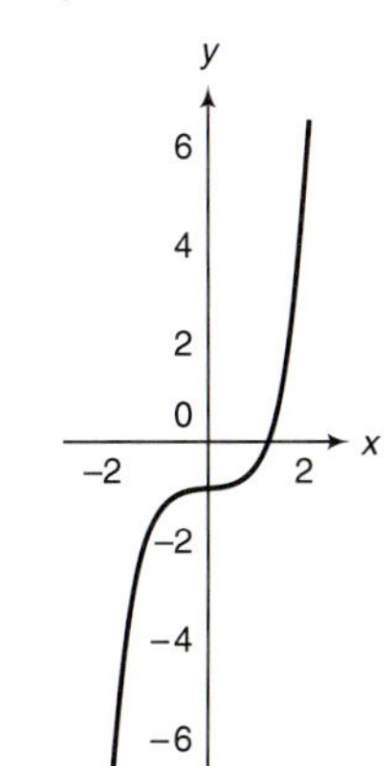

19. $y = (\ln x)^2,\ x > 0$

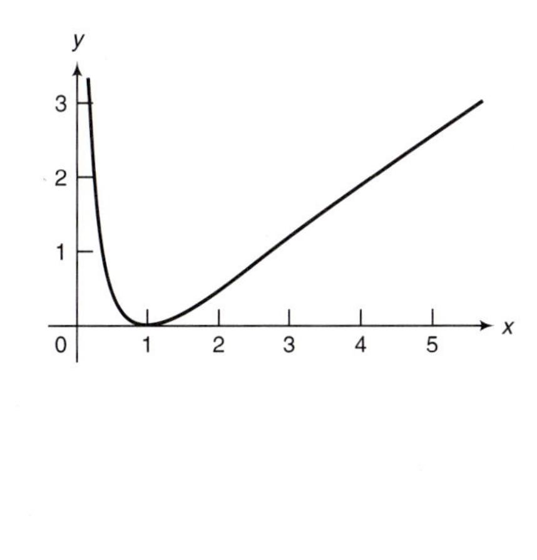

21. $x/y = \tan[(1/2)\ln(x^2 + y^2)]$. This is a logarithmic spiral. In polar coordinates, $r = e^{-\theta}$.

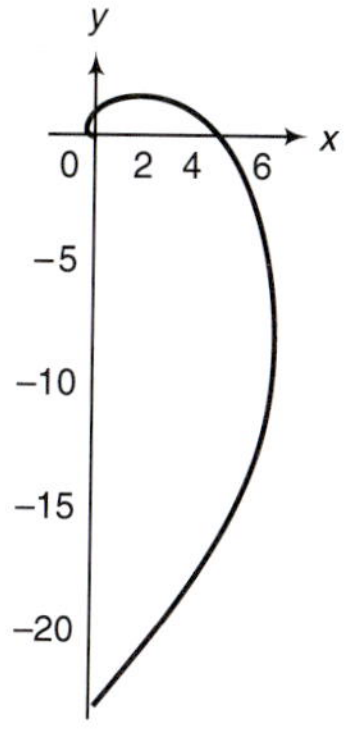

23. $y = x^2,\ x > 0$

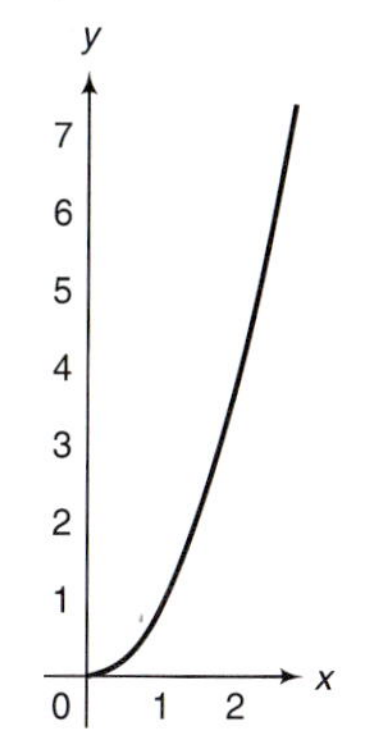

25. $z = x/2 = y/(-3)$

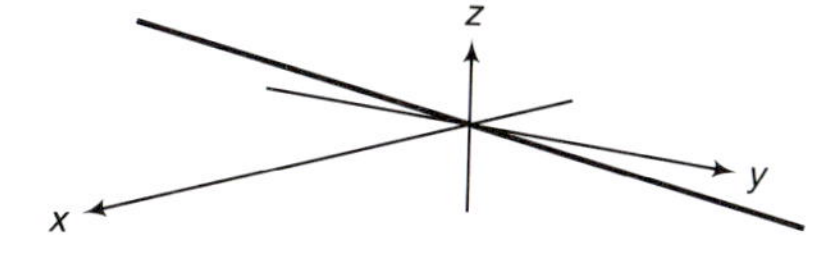

27. $y = \cos x$ and $z = \sin x$ (a circular helix) **29.** $x = 2 - t,\ y = 4 + 2t$
31. $x = 3 - 4t,\ y = 5 - 12t$ **33.** $x = -2 + 6t,\ y = 3 + 4t$ **35.** $x = 1 + t,\ y = 3 + t,\ z = 5 + t$
37. $x = -3 - 2t,\ y = 3t,\ z = 7 - 7t$ **43. a.** $x = r(\alpha - \sin\alpha)$ and $y = r(1 - \cos\alpha)$

Problems 2.2, page 94

1. $-\frac{4}{3}$ **3.** -1 **5.** $\sqrt{2}$ **7.** $-\sqrt{3}/8$ **9.** $-16/\pi^2$ **11.** $y = \frac{2}{3}x + \frac{22}{3}$ **13.** $y = -e^{-4}x + 2e^{-2}$
15. $y = \sqrt{2}$ **17.** V: none; H: none **19.** V: none; H: $(0, 1)$ **21.** V: $(\pm 2, 0)$; H: $(0, \pm 3)$
23. V: none $\left(\text{but } \dfrac{dy}{dx} \to \infty \text{ as } t \to -1^+\right)$; H: $(0, 0)$
25. V: $((-1)^k, \cos(\frac{5}{3}(k + \frac{1}{2})\pi))$ for $k = 0, 1, 2, 3, 4, 5$; H: $(\sin(\frac{3}{5}k\pi), (-1)^k)$ for $k = 0, 1, 2, 3, 4, 5, 6, 7, 8, 9$.
27. $\sqrt{3}$ **29.** $-\sqrt{3}$ **31.** $5/\sqrt{3}$ **33.** 2 **35.** $y = -x + 9;\ y = \frac{1}{3}x - \frac{17}{3}$
37. $m = \sin\alpha/(1 - \cos\alpha)$

Problems 2.3, page 101

1. $\mathbf{i} - 5t^4\mathbf{j};\ -20t^3\mathbf{j};\ \mathbb{R}$ **3.** $(2\cos 2t)\mathbf{i} - (3\sin 3t)\mathbf{j};\ (-4\sin 2t)\mathbf{i} - (9\cos 3t)\mathbf{j};\ \mathbb{R}$
5. $t^{-1}\mathbf{i} + 3e^{3t}\mathbf{j};\ -t^{-2}\mathbf{i} + 9e^{3t}\mathbf{j};\ t > 0$
7. $(\sec^2 t)\mathbf{i} + (\sec t \tan t)\mathbf{j};\ (2\sec^2 t \tan t)\mathbf{i} + (\sec^3 t + \sec t \tan^2 t)\mathbf{j};\ t \neq \pi/2 + k\pi,\ k = 0, \pm 1, \pm 2, \ldots$
9. $-(\tan t)\mathbf{i} + (\cot t)\mathbf{j};\ -(\sec^2 t)\mathbf{i} - (\csc^2 t)\mathbf{j};\ \{t: \cos t > 0, \sin t > 0\} = \{t: 0 < t - 2k\pi < \pi/2\},\ k = 0, \pm 1, \pm 2, \ldots$
11. $(1/\sqrt{13})(2\mathbf{i} + 3\mathbf{j})$ **13.** $\mathbf{j}$ **15.** $(1/\sqrt{2})(-\mathbf{i} + \mathbf{j})$ **17.** $\mathbf{i}$ **19.** $(1/\sqrt{97})(4\mathbf{i} - 9\mathbf{j})$
21. $(1/\sqrt{14})(\mathbf{i} + 2\mathbf{j} + 3\mathbf{k})$ **23.** $(1/\sqrt{3})(\mathbf{i} + \mathbf{j} - \mathbf{k})$ **25.** $(1/\sqrt{65})(-8\mathbf{i} + \mathbf{k})$
27. $(-\frac{1}{2}\cos 2t + C_1)\mathbf{i} + (e^t + C_2)\mathbf{j}$ **29.** $\frac{1}{2}(\mathbf{i} - \mathbf{j})$ **31.** $(t \ln t - t + C_1)\mathbf{i} + (te^t - e^t + C_2)\mathbf{j}$
33. $\frac{8}{3}\mathbf{i} - 4\mathbf{j} + \frac{32}{5}\mathbf{k}$ **35.** $(-\cos t^2 + C_1)\mathbf{i} + (\sin t^2 + C_2)\mathbf{j} + (e^{t^2} + C_3)\mathbf{k}$
37. $[(t^4/4) + 2]\mathbf{i} + [5 - (t^6/6)]\mathbf{j}$ **39.** $(\sin t)\mathbf{i} - (\cos t)\mathbf{k}$ **41.** $(1/\sqrt{a^2 + b^2})(-a\mathbf{i} + b\mathbf{j})$
43. $[1/(8 + 2\sqrt{2} - 2\sqrt{6})^{1/2}][(-1 - \sqrt{2})\mathbf{i} + (\sqrt{3} - \sqrt{2})\mathbf{j}] =$
$\{1/[18 + (9/\sqrt{2})(1 - \sqrt{3})]^{1/2}\}[(-\frac{3}{2} - 3/\sqrt{2})\mathbf{i} + (3\sqrt{3}/2 - 3/\sqrt{2})\mathbf{j}]$
45. $\dfrac{845{,}000}{16.1} \approx 52{,}484.5$ ft ≈ 9.94 miles **47.** ≈ 209.9 m after ≈ 2.42 sec)

Problems 2.4, page 110

1. $(2 + \sec^2 t)\mathbf{i} - [\sin t + \sec t \tan t]\mathbf{j}$ **3.** $\mathbf{0}$ **5.** $\left[\dfrac{1}{\sqrt{1 - t^2}} - \sin t\right]\mathbf{i} + \left[\cos t - \dfrac{1}{\sqrt{1 - t^2}}\right]\mathbf{j}$
7. $5\mathbf{i} - 9t^2\mathbf{j} - 6t\mathbf{k}$ **9.** $(1/\sqrt{2})[(-\sin t)\mathbf{i} + (\cos t)\mathbf{j} + \mathbf{k}];\ -\frac{1}{2}\mathbf{i} + \frac{1}{2}\mathbf{j} + (1/\sqrt{2})\mathbf{k}$
11. $\dfrac{1}{\sqrt{10 + 7\cos^2 t}}[(-3\sin t)\mathbf{i} + (4\cos t)\mathbf{j} + \mathbf{k}];\ (1/\sqrt{17})(4\mathbf{j} + \mathbf{k})$ **13.** $(1/\sqrt{1 + 4t^2})(\mathbf{j} + 2t\mathbf{k});\ (1/\sqrt{5})(\mathbf{j} + 2\mathbf{k})$
15. $(1/\sqrt{6})[\cos 2t - 2\sin 2t)\mathbf{i} + (\sin 2t + 2\cos 2t)\mathbf{j} + \mathbf{k}];\ (1/\sqrt{6})(\mathbf{i} + 2\mathbf{j} + \mathbf{k})$
17. $\mathbf{T}(t) = (-\sin 3t)\mathbf{i} + (\cos 3t)\mathbf{j};\ \mathbf{n}(t) = (-\cos 3t)\mathbf{i} + (-\sin 3t)\mathbf{j};\ \mathbf{T}(0) = \mathbf{j};\ \mathbf{n}(0) = -\mathbf{i}$

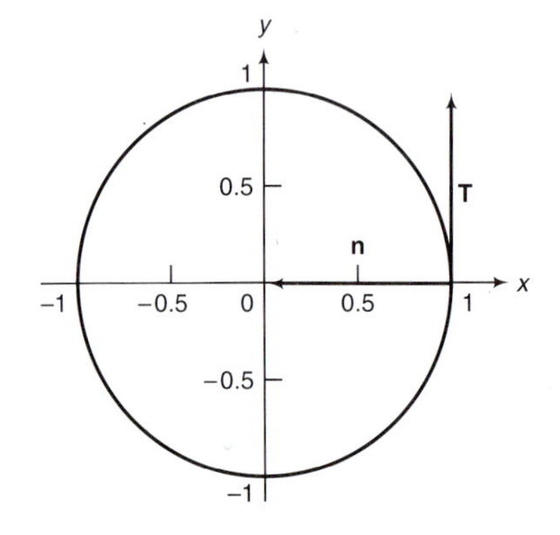

19. $\mathbf{T}(t) = (-\sin 4t)\mathbf{i} + (\cos 4t)\mathbf{j}$; $\mathbf{n}(t) = (-\cos 4t)\mathbf{i} + (-\sin 4t)\mathbf{j}$; $\mathbf{T}(\pi/4) = -\mathbf{j}$; $\mathbf{n}(\pi/4) = \mathbf{i}$

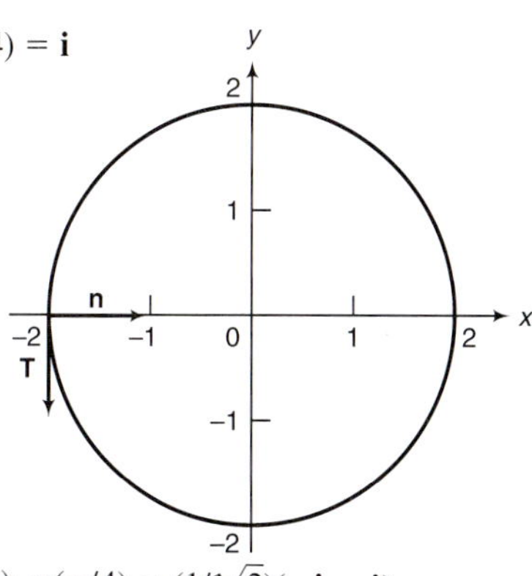

21. $\mathbf{T}(t) = (-\sin t)\mathbf{i} + (\cos t)\mathbf{j}$; $\mathbf{n}(t) = (-\cos t)\mathbf{i} + (-\sin t)\mathbf{j}$; $\mathbf{T}(\pi/4) = (1/\sqrt{2})(-\mathbf{i} + \mathbf{j})$; $\mathbf{n}(\pi/4) = (1/\sqrt{2})(-\mathbf{i} - \mathbf{j})$

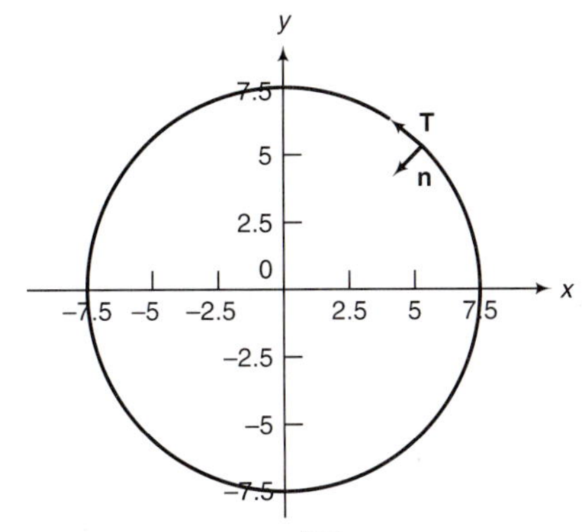

23. $\mathbf{T}(t) = \mathbf{T}(3) = (1/\sqrt{34})(3\mathbf{i} - 5\mathbf{j})$; $\mathbf{n}(t) = \mathbf{n}(3) = \pm(1/\sqrt{34})(5\mathbf{i} + 3\mathbf{j})$

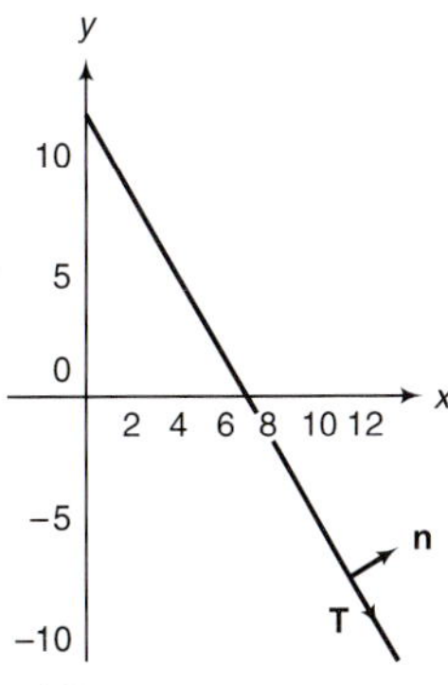

25. $\mathbf{T}(t) = \mathbf{T}(t_0) = (1/\sqrt{b^2 + d^2})(b\mathbf{i} + d\mathbf{j})$; $\mathbf{n}(t) = \mathbf{n}(t_0) = \pm(1/\sqrt{b^2 + d^2})(d\mathbf{i} - b\mathbf{j})$

27. $\mathbf{T}(t) = \sqrt{\dfrac{1 + \sin t}{2}}\mathbf{i} - \dfrac{\cos t}{\sqrt{2 + 2\sin t}}\mathbf{j}$; $\mathbf{n}(t) = \dfrac{\cos t}{\sqrt{2 + 2\sin t}}\mathbf{i} + \sqrt{\dfrac{1 + \sin t}{2}}\mathbf{j}$; $\mathbf{T}(\pi/2) = \mathbf{i}$; $\mathbf{n}(\pi/2) = \mathbf{j}$

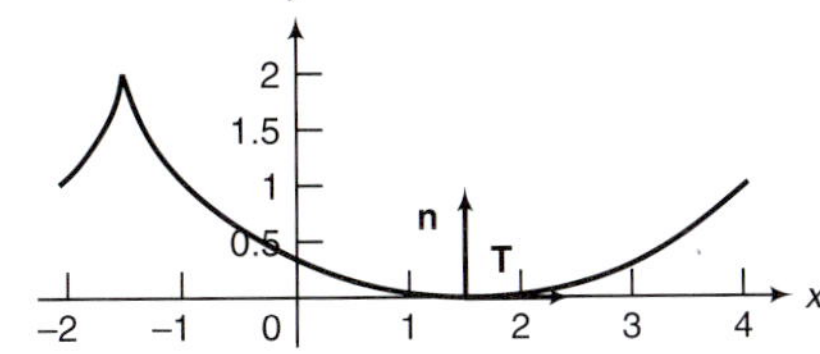

29. $\mathbf{T}(t)$ and $\mathbf{n}(t)$ are the same as for Problem 27;

$$\mathbf{T}\left(\frac{\pi}{4}\right) = \sqrt{\frac{\sqrt{2} + 1}{2\sqrt{2}}}\mathbf{i} - \frac{1}{\sqrt{4 + 2\sqrt{2}}}\mathbf{j} \approx 0.9239\mathbf{i} - 0.3827\mathbf{j}$$

$$\mathbf{n}\left(\frac{\pi}{4}\right) = \frac{1}{\sqrt{4 + 2\sqrt{2}}}\mathbf{i} + \sqrt{\frac{\sqrt{2} + 1}{2\sqrt{2}}}\mathbf{j} \approx 0.3827\mathbf{i} + 0.9239\mathbf{j}$$

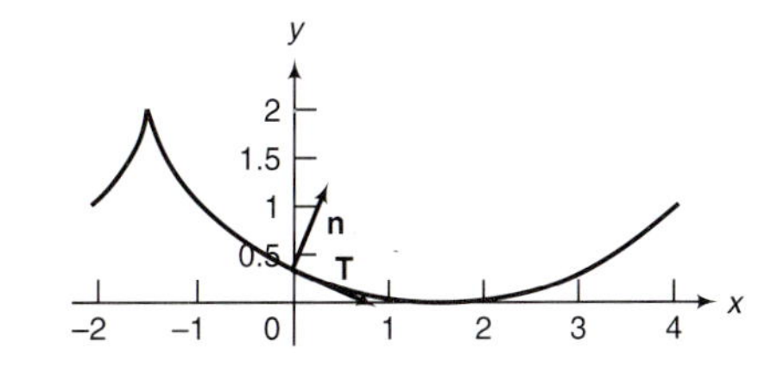

31. $\mathbf{f}'(t) = (1, \cos t, -\sin t, 3t^2)$ **33.** $\mathbf{f}'(t) = (e^t, 2te^{t^2}, \ldots, nt^{n-1}e^{t^n})$ **35.** $\mathbf{f}'(t) = (1/t, 1/t, \ldots, 1/t), (t > 0)$

37. $\mathbf{f}'(t) = 2t\mathbf{x}$ **39.** $\mathbf{h}'(t) = (3e^t, 6te^{t^2}, \ldots, 3nt^{n-1}e^{t^n})$ **41.** $h'(t) = \sum_{k=1}^{n} \frac{ke^{t^n}(t^k - 1)}{t^{k+1}}$

43. $\mathbf{T}(t) = \left(\sum_{k=1}^{n} k^2t^{2k-2}\right)^{-1/2}(1, 2t, \ldots, nt^{n-1})$ **45.** $\mathbf{T}(t) = \left(\sum_{k=1}^{n} k^2e^{2kt}\right)^{-1/2}(e^t, 2e^{2t}, \ldots, ne^{nt})$

Problems 2.5, page 118

1. $\frac{2}{27}[13^{3/2} - 8]$ **3.** $8^{3/2} - 8$ **5.** $8|a|$ **7.** $\frac{1}{2}\left[\sqrt{2} - \left(\frac{\sqrt{6}}{5}\right) + \ln\left(\frac{\sqrt{5} + \sqrt{10}}{1 + \sqrt{6}}\right)\right] \approx 0.6861$
9. $\sqrt{2}(e^{\pi/2} - 1)$ **11.** $\sqrt{2}$ **13.** $|a|\pi/2$ **15.** $\frac{1}{3}[(4 + \pi^2)^{3/2} - 8]$ **17.** $8a$ **19.** $\sqrt{2}(e^3 - 1)$
21. $\mathbf{f} = 3\{[(s + 2)/2]^{3/2} - 1\}\mathbf{i} + 2\{[(s + 2)/2]^{2/3} - 1\}^{3/2}\mathbf{j}$
23. $\mathbf{f} = \frac{1}{27}\{[(27s + 8)^{2/3} - 4]^{3/2} + 27\}\mathbf{i} + \frac{1}{9}\{27s + 8)^{2/3} - 13\}\mathbf{j}$ **25.** $\mathbf{f} = 3\cos(s/3)\mathbf{i} + 3\sin(s/3)\mathbf{j}$
27. $\mathbf{f} = a\cos(s/a)\mathbf{i} - a\sin(s/a)\mathbf{j}$ **29.** $\mathbf{f} = [a + b\cos(s/b)]\mathbf{i} + [c + b\sin(s/b)]\mathbf{j}$
31. $\mathbf{f} = a[1 - (2s/3a)]^{3/2}\mathbf{i} + a(2s/3a)^{3/2}\mathbf{j}$ **33.** $4\int_0^{\pi/2} \sqrt{a^2\sin^2\theta + b^2\cos^2\theta}\, d\theta$ **39.** $4\sqrt{n(n + 1)(2n + 1)/6}$
41. $2\sqrt{3}\,\pi$

Problems 2.6, page 128

1. $\kappa = 2/5^{3/2}$; $\rho = 5^{3/2}/2$ **3.** $\kappa = \frac{1}{2}$; $\rho = 2$ **5.** $\kappa = \frac{4}{9}$; $\rho = \frac{9}{4}$ **7.** $\kappa = 24\sqrt{2}/125$; $\rho = 125/24\sqrt{2}$
9. $\kappa = \frac{1}{2}$; $\rho = 2$ **11.** $\kappa = 1/\sqrt{2}$; $\rho = \sqrt{2}$ **13.** $\kappa = 1/2\sqrt{2}$; $\rho = 2\sqrt{2}$
15. $\kappa = e/(e^2 + 1)^{3/2}$; $\rho = (e^2 + 1)^{3/2}/e$ **17.** $\kappa = 4/7^{3/2}$; $\rho = 7^{3/2}/4$
19. $\kappa = 2|a|/(1 + b^2)^{3/2}$; $\rho = (1 + b^2)^{3/2}/2|a|$ **21.** $\kappa = \rho = 1$ (compute limit as $x \to 1^-$) **23.** $\kappa = 1/|a|$; $\rho = |a|$
25. $\kappa = 3/4|a|$; $\rho = 4|a|/3$ **27.** $\kappa = 3/2\sqrt{2}|a|$; $\rho = 2\sqrt{2}|a|/3$
29. $\mathbf{T} = (1/\sqrt{5})\mathbf{j} + (2/\sqrt{5})\mathbf{k}$; $\kappa = 2/5^{3/2}$; $\mathbf{n} = -(2\sqrt{5})\mathbf{j} + (1/\sqrt{5})\mathbf{k}$; $\mathbf{B} = \mathbf{i}$
31. $\mathbf{T} = (1/\sqrt{a^2 + 1})[(a/\sqrt{2})\mathbf{i} - (a/\sqrt{2})\mathbf{j} + \mathbf{k}]$; $\kappa = a/(a^2 + 1)$; $\mathbf{n} = -(1/\sqrt{2})(\mathbf{i} + \mathbf{j})$;
$\mathbf{B} = (1/\sqrt{a^2 + 1}[(1/\sqrt{2})\mathbf{i} - (1/\sqrt{2})\mathbf{j} - a\mathbf{k}]$
33. $\mathbf{T} = (1/\sqrt{b^2 + 1})(b\mathbf{j} + \mathbf{k})$; $\kappa = a/(b^2 + 1)$; $\mathbf{n} = -\mathbf{i}$; $\mathbf{B} = (1/\sqrt{b^2 + 1})(-\mathbf{j} + b\mathbf{k})$
35. $\mathbf{T} = (1/\sqrt{6})(\mathbf{i} + 2\mathbf{j} + \mathbf{k})$; $\kappa = \sqrt{5}/3$; $\mathbf{n} = (1/\sqrt{5})(-2\mathbf{i} + \mathbf{j}$; $\mathbf{B} = (1/\sqrt{30})(-\mathbf{i} - 2\mathbf{j} + 5\mathbf{k})$
37. $a_T = 0$, $a_n = 4$ **39.** $a_T = 4t/\sqrt{1 + 4t^2}$, $a_n = 2/\sqrt{1 + 4t^2}$
41. $a_T = \sin t\cos t/\sqrt{1 + \sin^2 t}$, $a_n = |\cos t|/\sqrt{1 + \sin^2 t}$
43. $a_T = (-e^{-2t} + e^{2t})/\sqrt{e^{-2t} + e^{2t}}$, $a_n = 2/\sqrt{e^{-2t} + e^{2t}}$ **45.** $\kappa_{max} = 2/(3\sqrt{3})$ at $x = 1/\sqrt{2}$ **47.** M^2
49. a. $[(10{,}000)(80{,}000)^2/(3600)^2] \cdot (1/\sqrt{2}) \approx 3{,}491{,}885.3$ N
b. $\sqrt{2.5\sqrt{2}(3600)^2(9.81)} \approx 21{,}201.4$ m/hr ≈ 21.2 km/hr $= 5.89$ m/sec **51.** 0 **53.** $\frac{1}{2}$

Chapter 2—Review Exercises, page 133

1. $y = 2x$ **3.** $x = (3 + y/2)^2$ **5.** $x^2 + y^2 = 1$ **7.** $x = y^3, x \geq 0$ **9.** 3; V: (0, 0), H: none
11. $\sqrt{3}$; V: $(\pm1, 0)$, H: $(0, \pm1)$ **13.** undefined; V: (1, 0), H: none **15.** $-4/(3\sqrt{3})$; H: $(0, \pm4)$, V: $(\pm3, 0)$
17. $2\mathbf{i} - 2t\mathbf{j}$; $-2\mathbf{j}$ **19.** $-5(\sin 5t)\mathbf{i} + 2(\cos t)\mathbf{j}$; $-25(\cos 5t)\mathbf{i} - 2(\sin t)\mathbf{j}$ **21.** $3t^2\mathbf{i} - 2t\mathbf{j} + \mathbf{k}$; $6t\mathbf{i} - 2\mathbf{j}$
23. $9\mathbf{i} + \frac{243}{2}\,\mathbf{j}$ **25.** $\mathbf{i} + \mathbf{j} + (\pi^2/8)\mathbf{k}$ **27.** $4t - 3/2\sqrt{t}$ **29.** $4\mathbf{i} - 3\mathbf{j} + (3\cos t + 4\sin t)\mathbf{k}$
31. $(1/\sqrt{41})(4\mathbf{i} + 5\mathbf{j})$; $(1/\sqrt{41})(-5\mathbf{i} + 4\mathbf{j})$ **33.** $(1/\sqrt{5})(2\mathbf{i} + \mathbf{j})$; $(1/\sqrt{5})(-\mathbf{i} + 2\mathbf{j})$ **35.** $\pi/3$ **37.** 16
39. $\sqrt{2}\pi/3$ **41.** $\mathbf{v} = -\sqrt{3}\mathbf{i} + \mathbf{j}$, $|\mathbf{v}| = 2$; $\mathbf{a} = -2\mathbf{i} - 2\sqrt{3}\mathbf{j}$, $|\mathbf{a}| = 4$
43. $\mathbf{v} = (\ln 2 - 1)\mathbf{i} + 2\mathbf{j}$, $|\mathbf{v}| = \sqrt{(\ln 2 - 1)^2 + 4}$; $\mathbf{a} = (\ln^2 2 + 1)\mathbf{i}$, $|\mathbf{a}| = \ln^2 2 + 1$
45. $\mathbf{v} = 2(\cosh 1)\mathbf{i} + 4\mathbf{j}$, $|\mathbf{v}| = \sqrt{4\cosh^2 1 + 16}$; $\mathbf{a} = 2(\sinh 1)\mathbf{i}$, $|\mathbf{a}| = 2\sinh 1$ **47.** $a_T = 0$, $a_n = 2$
49. $a_T = (6t + 12t^3)/\sqrt{t^2 + t^4}$, $a_n = 6t^2/\sqrt{t^2 + t^4}$ **51.** $\kappa = \rho = 1$ **53.** $\kappa = 36/(\frac{97}{2})^{3/2}$, $\rho = (\frac{97}{2})^{3/2}/36$
55. $\kappa = 4$, $\rho = 1/4$ **57.** $\kappa = e^{-1}/(1 + e^{-2})^{3/2}$, $\rho = e(1 + e^{-2})^{3/2}$ **59.** $\kappa = \frac{3}{4}$, $\rho = \frac{4}{3}$
61. $\mathbf{f} = \frac{3}{4}[2s + 1)^{2/3} - 1]\mathbf{i} + [(2s + 1)^{2/3} - 1]^{3/2}\mathbf{j}$ **63.** $\mathbf{f} = 2\cos(s/2)\mathbf{i} + 2\sin(s/2)\mathbf{j}$
65. $\mathbf{T}(2) = (1/\sqrt{5})(\mathbf{j} + 2\mathbf{k})$; $\kappa(2) = 5^{-3/2}$; $\mathbf{n}(2) = (1/\sqrt{5})(-2\mathbf{j} + \mathbf{k})$; $\mathbf{B}(2) = \mathbf{i}$
67. $\mathbf{T}(\pi/6) = (1/\sqrt{5})(-\mathbf{i} + \sqrt{3}\mathbf{j} + \mathbf{k})$; $\kappa(\pi/6) = 2/5$; $\mathbf{n}(\pi/6) = \frac{1}{2}(-\sqrt{3}\mathbf{i} - \mathbf{j})$; $\mathbf{B}(\pi/6) = (1/2\sqrt{5})(\mathbf{i} - \sqrt{3}\mathbf{j} + 4\mathbf{k})$
69. $\mathbf{f}'(t) = (3t^2 - \sin t, 1/t, 2e^{2t})$ **71.** $\mathbf{T}(1) = \frac{1}{\sqrt{120}}(2, 4, 6, 8) = \frac{1}{\sqrt{30}}(1, 2, 3, 4)$; $\mathbf{n}(1) = \frac{1}{\sqrt{186}}(-7, -8, -3, 8)$

Computer Exercises, page 134

1. Contestant 1: $\pi/\sqrt{g} \approx 3.14/\sqrt{g}$; Contestant 2: $\approx 3.72/\sqrt{g}$; Contestant 3: $\approx 3.28/\sqrt{g}$ (The pass of quickest descent is the brachistochrome.)

3. The curvature is $\dfrac{\sqrt{-32\cos^4 t - 4\cos^2 t(8\sin^2 t - 9) + 5}}{2(5 - 4\cos^2 t)^{3/2}}$. The critical points of the curvature occur where $\sin t \cos t = 0$.

CHAPTER 3

Problems 3.1, page 143

1. $\mathbb{R}^2$; $[0, \infty)$ **3.** $\{(x, y): y \neq 0\}$; $\mathbb{R}$ **5.** $\{(x, y): x^2 - 4y^2 \leq 1\}$; $[0, \infty)$ **7.** $\mathbb{R}^2$; $(0, \infty)$
9. $\{(x, y): x - y \neq (n + \frac{1}{2})\pi, n = 0, \pm 1, \pm 2, \ldots\}$; $\mathbb{R}$ **11.** $\{(x, y): |x| \geq |y| \text{ and } x \neq -y\}$; $[0, \infty)$
13. $\{(x, y): |x - y| \leq 1\}$: $[0, \pi]$ **15.** $\{(x, y): x \neq -y\}$; $\mathbb{R}$
17. $\{(x, y): x \neq 0 \text{ and } y \neq 0\}$; $(-\infty, -2] \cup [2, \infty)$ (The range is obtained from $f(x, y) = \pm\sqrt{2 + x^2/4y^2 + 4y^2/x^2}$
19. $\{(x, y, z): x + y + z \geq 0\}$—this is the half-space "in front of" the plane $x + y + z = 0$; $[0, \infty)$
21. $\{(x, y, z): y^2 < x^2 + z^2\}$—this is the region "outside" the cone $y^2 = x^2 + z^2$; $(0, \infty)$ **23.** $\{(0, 0, 0)\}$; $\{0\}$
25. $\{(x, y, z): z \neq 0\}$; $\mathbb{R}$
27. $\{(x, y, z): |x + y - z| \leq 1\}$—this is the part of $\mathbb{R}^3$ between the planes $x + y - z = -1$ and $x + y - z = 1$; $[-\pi/2, \pi/2]$
29. $\{(x, y, z): y \neq 0\}$; $(-\pi/2, \pi/2)$ **31.** $\mathbb{R}^3$; $(0, \infty)$ **33.** $\{(x, y, z): x \neq 0, y \neq 0 \text{ and } z \neq 0\}$; $\mathbb{R} - \{0\}$
35. $\mathbb{R}^3$; $[-3, 3]$ **37.** $\{(x_1, x_2, x_3, x_4): x_1 + 2x_2 + x_3 - x_4 + 1 \geq 0\}$; $[0, \infty)$ **39.** $\{(x_1, x_2, x_3, x_4): x_2x_4 \neq 0\}$; $\mathbb{R}$
41. $\{(x_1, x_2, x_3, x_4): x_2^2 + x_4^2 \neq 0\}$; $\mathbb{R}$ **43.** $\{(x_1, x_2, x_3, x_4, x_5): x_3^2 + x_5^2 \neq 0\}$; $\mathbb{R}$ **45.** $\mathbb{R}^n$; $[0, \infty)$
47. $y = x^2 + 4z^2$, elliptic paraboloid **49.** $z = x^2 - 4y^2$, hyperbolic paraboloid

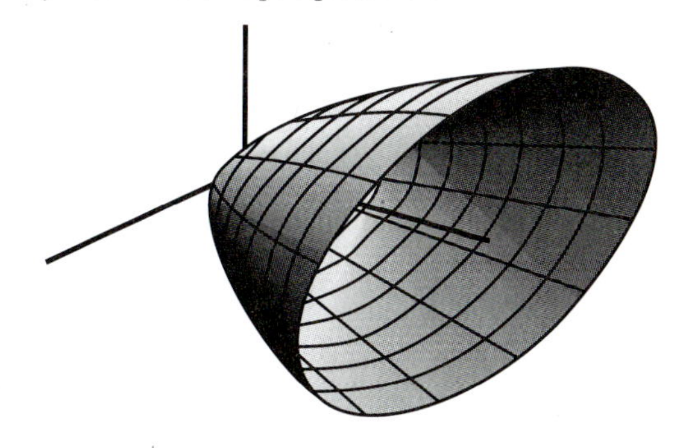

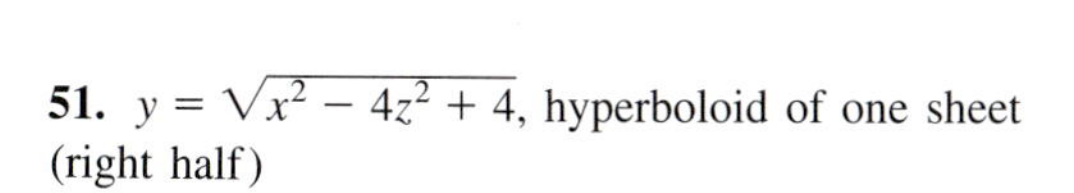

51. $y = \sqrt{x^2 - 4z^2 + 4}$, hyperboloid of one sheet (right half)

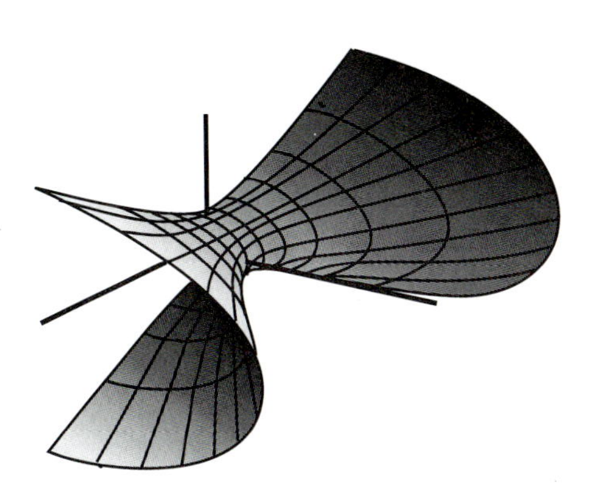

53. parallel straight lines (with slopes of -1)
$y = -x + (z^2 - 1)$

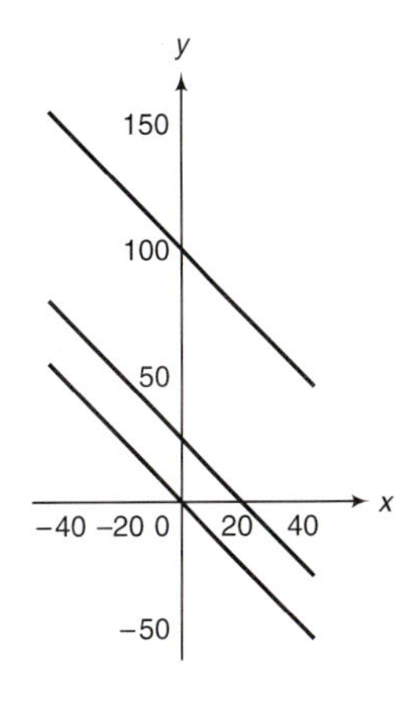

55. concentric ellipses (centered at the origin) with equations $x^2 + 4y^2 = 1 - z^2$ if $z < 1$; if $z = 1$, we obtain the single point $(0, 0, 0)$

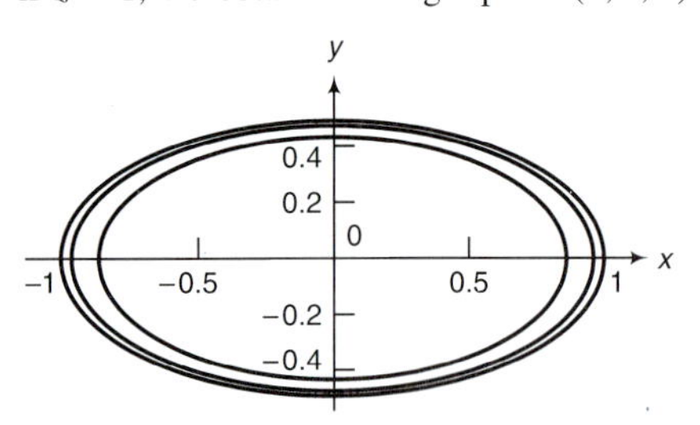

57. parallel straight lines: $y = x - \cos z$

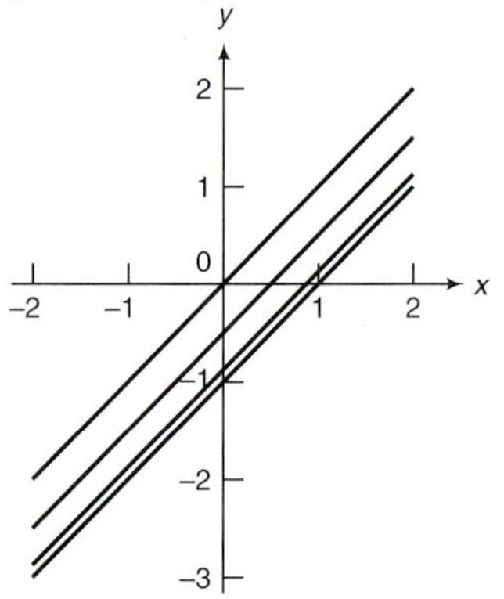

59. for each value of z we get a family of straight lines all of which have a slope of -1: for $z = 0$ we obtain $y = -x + n\pi$ where n is an integer; for $z = 1$ we obtain $y = -x + (n + \frac{1}{4})\pi$; for $z = -1$ we obtain $y = -x + (n - \frac{1}{4})\pi$; for $z = \sqrt{3}$ we obtain $y = -x + (n + \frac{1}{3})\pi$

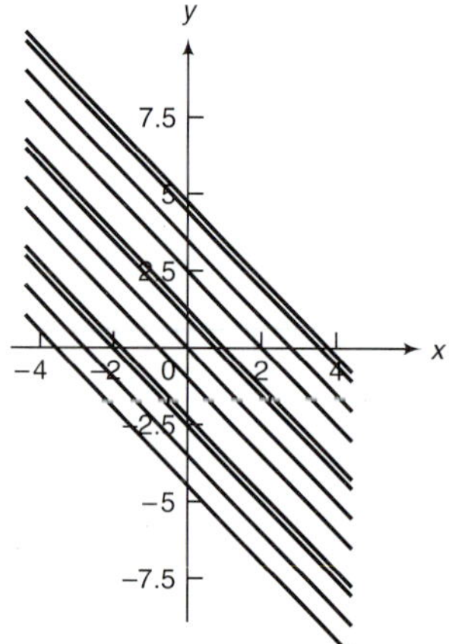

61. concentric ellipses $x^2/a^2 + y^2/b^2 = 1$ (centered at the origin) with $a = \sqrt{T - 20}$ and $b = \frac{1}{2}a$ for $T > 20$

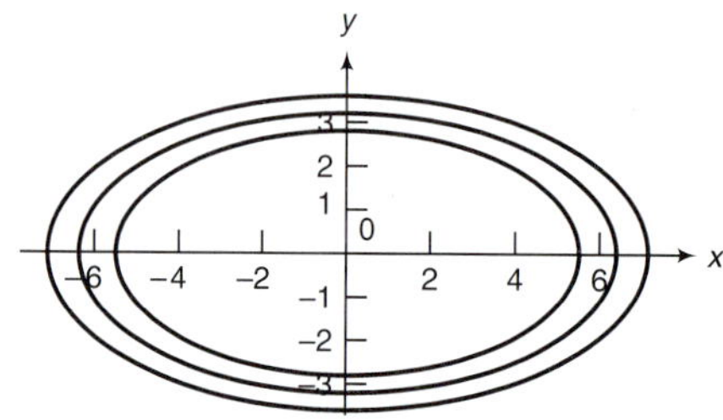

63. concentric ellipses (centered at the origin) with $a = \sqrt{(P - 100)/2}$ and $b = \sqrt{(P - 100)/3}$ *for* $P > 100$; for $P = 100$ we obtain the single point $(0, 0, 0)$

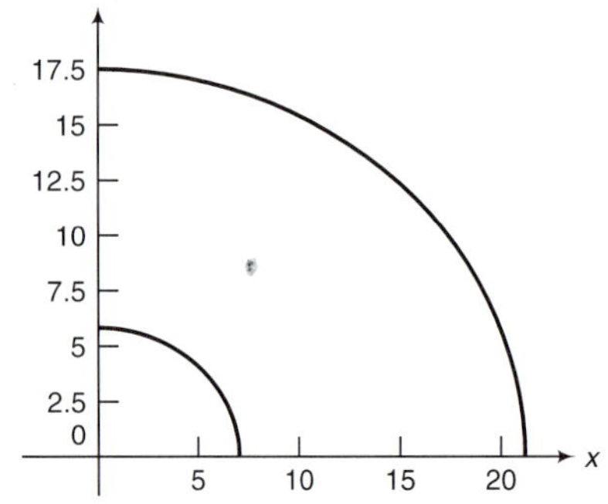

65. **a.** $25{,}000 \cdot 5^{1/3} \cdot 3^{2/3} \approx 88{,}922.3$ **b.** $F(2L, K/2) = 2^{-1/3}F(L, K) \approx 0.79\, F(L, K)$ (i.e., output decreases about 21%) **c.** $F(L/2, 2K) = 2^{1/3}F(L, K) \approx 1.26F(L, K)$ (i.e., output increases about 26%)
67. \$1984 **69.** f **71.** a **73.** b **75.** g **77.** l **79.** i

Problems 3.2, page 159

1. (the circle itself is not included in this region)

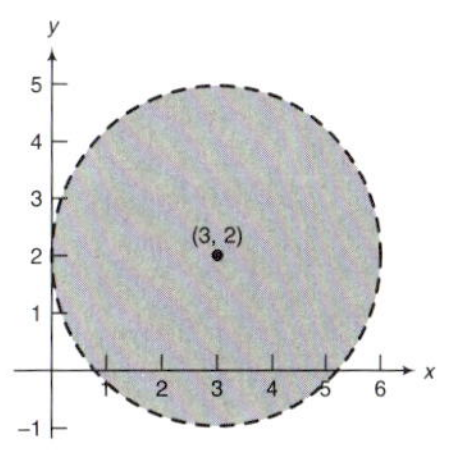

3. (the sphere itself is not included in this region)

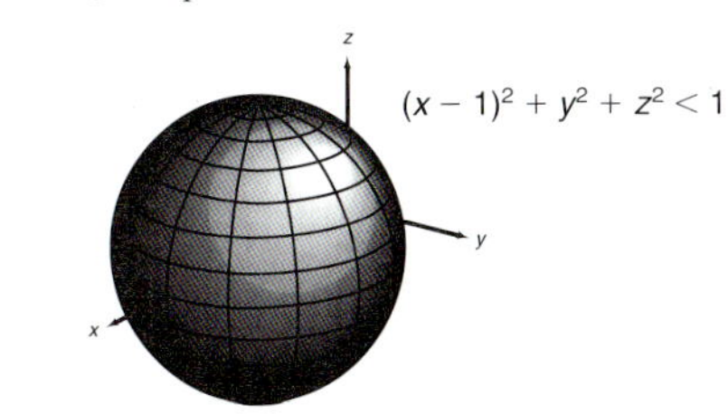

5. (the sphere itself *is* included in this region)

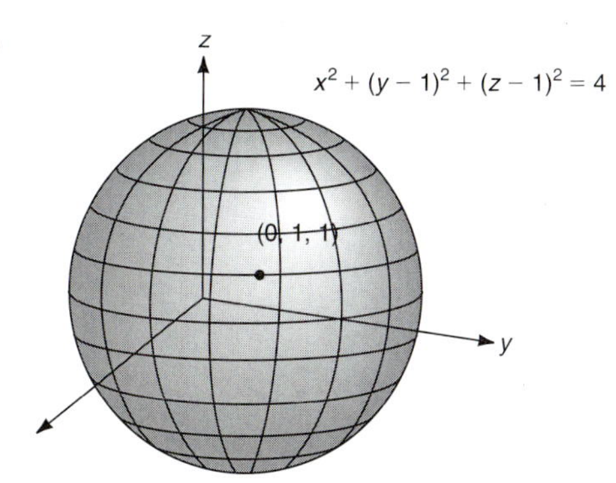

13. along the line $x = at$, $y = bt$, $xy/(x^2 - y^2) = ab/(a^2 - b^2)$; the limit depends on the choice of a and b
15. along the line $x = at$, $y = bt$, $xy^3/(x^4 + y^4) = ab^3/(a^4 + b^4)$
17. along the line $ax + by = 0$, $(x^2 + y^2)^2/(x^4 + y^4) = (a^2 + b^2)^2/(a^4 + b^4)$
19. along the line $y = kx$, $(ax^2 + by)/(cy^2 + dx) = (ax + bk)/(ck^2x + d) \to bk/d$ as $(x, y) \to (0, 0)$
21. along the line $y = kx = z$, $xyz/(x^3 + y^3 + z^3) = k^2/(1 + 2k^3)$
23. 0; since $0 \le 5x^2y^2/(x^4 + y^2) = 5x^2y^2/[(x^2 - |y|)^2 + 2x^2|y|] \le 5x^2y^2/2x^2|y| = \frac{5}{2}|y|$
25. 0; since $0 < |(yx^2 + z^3)/(x^2 + y^2 + z^2)| \le (|y|x^2 + |z|z^2)/(x^2 + y^2 + z^2) \le |y| + |z| \to 0$.
27. $-\frac{61}{25}$ **29.** $\ln(4\pi/3)$ **31.** sinh 1 **33.** $\frac{1}{6}$ **35.** ln 769 **37.** $\mathbb{R}^2 - \{(0, 0)\}$
39. $\{(x, y)\colon y \ne 4 \text{ and } x \ne -3\}$ **41.** $\mathbb{R}^2$ **43.** $\{(x, y)\colon xy \ne 1\}$ **45.** $\{(x, y, z)\colon x > 0)$
47. $\{(x, y, z)\colon xz > 0\}$ **49.** $\{(x, y, z)\colon x^2 + y^2 + z^2 < 1\}$ = the "interior" of the unit sphere **51.** $c = 0$
53. Along the line $y = kx = z$, $(yz - x^2)/(x^2 + y^2 + z^2) = (k^2 - 1)/(1 + 2k^2)$ which shows that $\lim_{(x, y, z)\to(0, 0, 0)} (yz - x^2)/(x^2 + y^2 + z^2)$ does not exist. The function is, therefore, not continuous at the origin.
61. 0 **63.** 0 **65.** 0 **67.** $\ln\left(\frac{8}{3}\right)$ **69.** $\sqrt{6139/76}$ **71.** $\{(x_1, x_2, x_3, x_4)\colon x_3x_4 \ne 0\}$
73. $\{(x_1, x_2, x_3, x_4, x_5)\colon {x_1}^2 + {x_2}^2 + {x_3}^2 + {x_4}^2 + {x_5}^2 < 1\}$. This is the interior (inside) of the unit sphere in $\mathbb{R}^5$ (also called a **hypersphere**).

Problems 3.3, page 167

1. $2xy$; x^2 **3.** $3y^3e^{xy^3}$; $9xy^2e^{xy^3}$ **5.** $4/y^5$; $-20x/y^6$ **7.** $3x^2y^5/(x^3y^5 - 2)$; $5x^3y^4/(x^3y^5 - 2)$
9. $\frac{4}{3}(x + 5y \sin x)^{1/3}(1 + 5y \cos x)$; $\frac{4}{3}(x + 5y \sin x)^{1/3}(5 \sin x)$ **11.** $1/\sqrt{1 - (x - y)^2}$; $-1/\sqrt{1 - (x - y)^2}$
13. $z/(1 - x - 3yz^2)$; $z^3/(1 - x - 3yz^2)$ **15.** $\cos(z - x)/[\cos(z - x) + y/z^2]$; $1/z[\cos(z - x) + y/z^2]$ **17.** 3
19. 0 **21.** 1 **23.** $\frac{48}{169}$ **25.** $w_x = w_y = w_z = \frac{1}{2}(x + y + z)^{-1/2}$ **27.** $e^{x+2y+3z}$; $2e^{x+2y+3z}$; $3e^{x+2y+3z}$
29. $1/z$; $1/z$; $-(x + y)/z^2$ **31.** $3x^2/(x^3 + y^2 + z)$; $2y/(x^3 + y^2 + z)$; $1/(x^3 + y^2 + z)$ **33.** 12 **35.** $\frac{1}{2}$
37. $\frac{1}{7}$ **39.** 16 **41.** $-e^6$ **43.** **a.** $x = 1$, $(y + 1)/1 = (z - 5)/(-12)$ **b.** $y = -1$, $(x - 1)/1 = (z - 5)/3$

45. $y = 1,\ (x-1)/(-4) = z - 1$ **47.** $C_x = -50(2+x)^{-2},\ C_y = -250(3+y)^{-3}$
51. $P_V = -10R293/2^2 = -732.5R$ [T is given in °K] where $R \approx 8315$ joules per kg-mole-degree
53. $C_t = -5/(2 \cdot 10^{2/3}) \approx -0.5386$ **61. b.** no **63.** $f_1 = x_2x_3x_4,\ f_2 = x_1x_3x_4,\ f_3 = x_1x_2x_4,\ f_4 = x_1x_2x_3$
65. $f_1 = \dfrac{x_3}{x_2x_4} + \dfrac{x_3}{x_2}e^{x_1x_4/x_2},\ f_2 = -\dfrac{x_1x_3}{x_2^2x_4} - \dfrac{x_1x_3}{x_2^2}e^{x_1x_4/x_2},\ f_3 = \dfrac{x_1}{x_2x_4} + \dfrac{x_1}{x_2}e^{x_1x_4/x_2},\ f_4 = -\dfrac{x_1x_3}{x_2x_4^2}$
67. $f_j = x_j\left(\sum_{i=1}^{n} x_i^2\right)^{-1/2}$ for $j = 1, 2, \ldots, n$ **69.** $f_j = -x_j\left(\sum_{i=1}^{n} x_i^2\right)^{-3/2}$ for $j = 1, 2, \ldots, n$

Problems 3.4, page 176

1. $f_{xx} = 2y;\ f_{xy} = f_{yx} = 2x;\ f_{yy} = 0$ **3.** $f_{xx} = 3y^6e^{xy^3};\ f_{xy} = f_{yx} = 9y^2(1 + xy^3)e^{xy^3};\ f_{yy} = 9xy(2 + 3xy^3)e^{xy^3}$
5. $f_{xx} = 0;\ f_{xy} = f_{yx} = -20/y^6;\ f_{yy} = 120x/y^7$
7. $f_{xx} = \dfrac{-3xy^5(4 + x^3y^5)}{(x^3y^5 - 2)^2};\ f_{xy} = f_{yx} = \dfrac{-30x^2y^4}{(x^3y^5 - 2)^2};\ f_{yy} = \dfrac{-5x^3y^3(8 + x^3y^5)}{(x^3y^5 - 2)^2}$
9. $f_{xx} = \frac{4}{9}(1 + 5y\cos x)^2(x + 5y\sin x)^{-2/3} - \frac{20}{3}y\sin x(x + 5y\sin x)^{1/3};$
$f_{xy} = f_{yx} = \frac{20}{9}\sin x(1 + 5y\cos x)(x + 5y\sin x)^{-2/3} + \frac{20}{3}\cos x(x + 5y\sin x)^{1/3};\ f_{yy} = \frac{100}{9}\sin^2 x(x + 5y\sin x)^{-2/3}$
11. $f_{xx} = (x-y)[1-(x-y)^2]^{-3/2};\ f_{xy} = f_{yx} = -(x-y)[1-(x-y)^2]^{-3/2};\ f_{yy} = (x-y)[1-(x-y)^2]^{-3/2}$
13. $f_{xx} = 0;\ f_{yy} = 0;\ f_{zz} = 0;\ f_{yz} = f_{zy} = x;\ f_{xz} = f_{zx} = y;\ f_{xy} = f_{yx} = z$
15. $f_{xx} = 2y^3z^4;\ f_{yy} = 6x^2yz^4;\ f_{zz} = 12x^2y^3z^2;\ f_{yz} = f_{zy} = 12x^2y^2z^3;\ f_{xz} = f_{zx} = 8xy^3z^3;\ f_{xy} = f_{yx} = 6xy^2z^4$
17. $f_{xx} = 0;\ f_{yy} = 0;\ f_{zz} = 2(x+y)/z^3;\ f_{yz} = f_{zy} = -1/z^2,\ f_{xz} = f_{zx} = -1/z^2;\ f_{xy} = f_{yx} = 0$
19. $f_{xx} = 9y^2e^{3xy}\cos z;\ f_{yy} = 9x^2e^{3xy}\cos z;\ f_{zz} = -e^{3xy}\cos z;\ f_{yz} = f_{zy} = -3xe^{3xy}\sin z;\ f_{xz} = f_{zx} = -3ye^{3xy}\sin z;$
$f_{xy} = f_{yx} = 3(1 + 3xy)e^{3xy}\cos z$
21. $6y^2$ **23.** $24/(3x - 2y)^3$ **25.** 0 **37.** $f_{13} = x_2x_4,\ f_{11} = 0,\ f_{134} = x_2$
39. $f_{jk} = 3x_jx_k\left(\sum_{i=1}^{n} x_i^2\right)^{-5/2}$ if $j \neq k$: $f_{ii} = -\left(\sum_{i=1}^{n} x_i^2\right)^{-3/2} + 3x_i^2\left(\sum_{i=1}^{n} x_i^2\right)^{-5/2}$
41. $f_{2211} = e^{x_1x_2x_3}(2x_3^2 + 4x_1x_2x_3^3 + x_1^2x_2^2x_3^4)$ **43. a.** $3^3 = 27$ **b.** $3^4 = 81$ **c.** 3^m

Problems 3.5, page 188

1. $2(x+y)\mathbf{i} + 2(x+y)\mathbf{j}$ **3.** $(e/2)\mathbf{i} + (e/2)\mathbf{j}$ **5.** $(x/\sqrt{x^2+y^2})\mathbf{i} + (y/\sqrt{x^2+y^2})\mathbf{j}$
7. $-y\sec^2(y-x)\mathbf{i} + [\tan(y-x) + y\sec^2(y-x)]\mathbf{j}$ **9.** $\sec 3\tan 3\mathbf{i} + 3\sec 3\tan 3\mathbf{j}$
11. $\dfrac{4xy^2}{(x^2+y^2)^2}\mathbf{i} - \dfrac{4x^2y}{(x^2+y^2)^2}\mathbf{j}$ **13.** $(3\sqrt{2}/4)\mathbf{i} - (\sqrt{6}/4)\mathbf{j} + \sqrt{2}\mathbf{k}$ **15.** $\frac{5}{6}\mathbf{i} - \frac{5}{12}\mathbf{j}$ **17.** $9\mathbf{i} + 6\mathbf{j} + 27\mathbf{k}$
19. $\mathbf{i} - \mathbf{k}$ **29. a.** $f_x(0,0) = 0 = f_y(0,0)$ **31.** $f(x,y) = \frac{1}{2}(x^2+y^2) + C$
33. $\nabla f = (x_2x_3x_4, x_1x_3x_4, x_1x_2x_4, x_1x_2x_3)$ **35.** $\nabla f = (2x_1, 2x_2, 2x_3, 2x_4, 2x_5) = 2\mathbf{x}$

Problems 3.6, page 195

1. $3e^{3t}$ **3.** 1 **5.** $\dfrac{5(\cos 5t)(\cos 3t) + 3(\sin 5t)(\sin 3t)}{\sin^2 5t + \cos^2 3t}$ **7.** $2t$ **9.** $3e^{3t} + 4e^{4t} - 5e^{5t}$
11. $z_r = 2r,\ z_s = -2s$ **13.** $z_r = -e^{s-r},\ z_s = e^{s-r}$ **15.** $z_r = 2r/s^2;\ z_s = -2r^2/s^3$ **17.** $w_r = s + 2,\ w_s = r$
19. $w_r = (-2r - s + t)/(s + 2t),\ w_s = (t - r)(2t - r)/(s + 2t)^2,\ w_t = (r + s)(2r + s)/(s + 2t)^2$
21. $w_s = \cosh(\sqrt{r+s} + 2\sqrt[3]{s-t} + 3/(r+t))[1/2\sqrt{r+s} + 2/3(s-t)^{2/3}]$
23. $w_r = 8r/(4r^2 + 5s^2 - t^2),\ w_s = 10s/(4r^2 + 5s^2 - t^2),\ w_t = -2t/(4r^2 + 5s^2 - t^2)$
25. $w_x = x/\sqrt{x^2+y^2} - y^2/(x^2+y^2)^{3/2},\ w_y = y/\sqrt{x^2+y^2} + xy/(x^2+y^2)^{3/2},\ w_z = 1$
27. $V_t = 7500\pi$ in^3/min; increasing **29.** $P_t = (10R + 75)/1000$ N/cm^2/min; increasing
31. $(2\pi - 38 - 2\sqrt{3})/\sqrt{676 - 240\sqrt{3}} \approx -2.18$ cm/sec; decreasing
33. $w_r = g_x\cos\theta + g_y\sin\theta,\ w_\theta = -g_x r\sin\theta + g_y r\cos\theta,\ w_t = g_z$
39. $z = g(y + ax)$ where g is an arbitrary smooth function of one variable **43.** $f(\mathbf{x}) = \dfrac{1}{2}\sum_{i=1}^{n} x_i^2 + C$

Problems 3.7, page 200

1. T: $y = 1$; N: $x = z = 0$ **3.** T: $(x - a)/a + (y - b)/b + (z - c)/c = 0$; N: $a(x - a) = b(y - b) = c(z - c)$
5. T: $(x - 4)/4 + (y - 1)/2 + (z - 9)/6 = 0$, or $3x + 6y + 2z = 36$; N: $4(x - 4) = 2(y - 1) = 6(z - 9)$
7. T: $2(x - 1) + (y - 2) + (z - 2) = 0$, or $2x + y + z = 6$; N: $(x - 1)/2 = y - 2 = z - 2$
9. T: $24(x - 3) - 2(y - 1) + 20(z + 2) = 0$, or $12x - y + 10z = 15$; N: $(x - 3)/24 = (y - 1)/(-2) = (z + 2)/20$
11. T: $(\pi/\sqrt{3})(y - 1) + \sqrt{3}(z - \pi/3) = 0$, or $\pi y + 3z = 2\pi$; N: $(\sqrt{3}/\pi)(y - 1) = (1/\sqrt{3})(z - \pi/3)$, $x = \pi/2$
13. T: $(\ln 5)(x - 1) + (\ln 5)(y - 1) + (z - \ln 5) = 0$, or $\ln 5(x + y) + z = 3 \ln 5$; N: $(x - 1)/(\ln 5) =$ $(y - 1)/(\ln 5) = z - \ln 5$
15. T: $z = x + 2y - 2$; N: $x\mathbf{i} + y\mathbf{j} + z\mathbf{k} = (\mathbf{i} + \mathbf{j} + \mathbf{k}) + t(-\mathbf{i} - 2\mathbf{j} + \mathbf{k})$
17. T: $z = 1$; N: $x\mathbf{i} + y\mathbf{j} + z\mathbf{k} = (\pi/8)\mathbf{i} + (\pi/20)\mathbf{j} + (1 + t)\mathbf{k}$
19. T: $x + y + 4z = -\pi$; N: $x\mathbf{i} + y\mathbf{j} + z\mathbf{k} = (-2\mathbf{i} + 2\mathbf{j} + (-\pi/4)\mathbf{k}) + t(\frac{1}{4}\mathbf{i} + \frac{1}{4}\mathbf{j} + \mathbf{k})$
21. T: $z = 2\sqrt{3}x - 2\sqrt{3}y + 2 - (2\sqrt{3}/3)\pi$; N: $x\mathbf{i} + y\mathbf{j} + z\mathbf{k} = ((\pi/2)\mathbf{i} + (\pi/6)\mathbf{j} + 2\mathbf{k}) + t(-2\sqrt{3}\mathbf{i} + 2\sqrt{3}\mathbf{j} + \mathbf{k})$
23. $5\mathbf{i} + \mathbf{j}$; $((x - 1)\mathbf{i} + (y - 5)\mathbf{j}) \cdot (5\mathbf{i} + \mathbf{j}) = 0$, or $5x + y = 10$
25. $-6\mathbf{i} + 8\mathbf{j}$; $((x - 4)\mathbf{i} + (y - 3)\mathbf{j}) \cdot (-6\mathbf{i} + 8\mathbf{j}) = 0$, or $-6x + 8y = 0$
27. $(\sqrt{2}/2)\mathbf{i} + (\sqrt{2}/4)\mathbf{j}$; $((x - \sqrt{2})\mathbf{i} + (y - 2\sqrt{2})\mathbf{j}) \cdot ((\sqrt{2}/2)\mathbf{i} + (\sqrt{2}/4)\mathbf{j}) = 0$, or $(\sqrt{2}/2)x + (\sqrt{2}/4)y = 2$
29. at $(0, 2, 4)$, $\cos\theta = 2/\sqrt{102}$; at $(1, 1, 2)$, $\cos\theta = 2/(3\sqrt{6})$

Problems 3.8, page 208

1. $9/\sqrt{10}$ **3.** $\sqrt{2}/7$ **5.** $-5/(4\sqrt{13})$ **7.** $(2e^2 + 3e)/\sqrt{2}$ **9.** $2\sqrt{3}$ **11.** $-17/\sqrt{30}$
13. $2e^{-3}/\sqrt{35}$ **15.** $-22/\sqrt{10}$ **17.** $10/\sqrt{42}$ **19.** $\sqrt{5}$ **21.** $1/\sqrt{2}$ **23.** $(1/\sqrt{6})(-\mathbf{i} + \mathbf{j} - 2\mathbf{k})$
25. a. $(0, 0, 0)$ **b.** $(1/\sqrt{14})(3\mathbf{i} - \mathbf{j} + 2\mathbf{k})$ **c.** $-(1/\sqrt{14})(3\mathbf{i} - \mathbf{j} + 2\mathbf{k})$, yes **27.** $\left|\frac{x}{a}\right|^{a^2} = \left|\frac{y}{b}\right|^{b^2}$ **31.** $2/\sqrt{3}$
33. $12/\sqrt{6} = 2\sqrt{6}$ **35.** $\frac{2}{\sqrt{n}} \cdot \frac{n(n + 1)}{2} = \sqrt{n(n + 1)}$

Problems 3.9, page 211

1. $y^3\,\Delta x + 3xy^2\,\Delta y$ **3.** $[(x - y)(x + y)^3]^{-1/2}(y\,\Delta x - x\,\Delta y)$ **5.** $(2x + 3y)^{-1}(2\,\Delta x + 3\,\Delta y)$
7. $y^2z^5\,\Delta x + 2xyz^5\,\Delta y + 5xy^2z^4\,\Delta z$ **9.** $\sinh(xy - z)(y\,\Delta x + x\,\Delta y - \Delta z)$
11. a. $x(\Delta y)^2 + 2y(\Delta x)(\Delta y) + (\Delta x)(\Delta y)^2$ **b.** -0.000309 **13. a.** $2000\pi \approx 6283.2\text{ cm}^3$ **b.** $19\pi \approx 59.7\text{ cm}^3$
15. a. $\frac{8}{3}\,\Omega$ **b.** $dR \approx 0.0305\,\Omega$

Problems 3.10, page 220

1. local minimum at $(0, 0)$ **3.** local minimum at $(-2, 1)$ **5.** local minimum at $(-2, 1)$
7. local minimum at $(\sqrt{5}, 0)$; local maximum at $(-\sqrt{5}, 0)$; $(-1, \pm 2)$ are saddle points
9. $(\frac{1}{2}, 1)$ and $(-\frac{1}{2}, -1)$ are saddle points; local minimum at $(-\frac{1}{2}, 1)$; local maximum at $(\frac{1}{2}, -1)$
11. local minimum at $(-2, -2)$ **13.** local maximum at $(\frac{4}{3}, \frac{4}{3})$; $(0, 0)$, $(0, 4)$ and $(4, 0)$ are saddle points
15. local minima at $(\pm 1, 1)$
17. $(0, 0)$ is a saddle point. (Note that $D = 0$ at the origin, but f can take positive and negative values in any neighborhood of the origin.)
19. $x = y = z = 50/3$; the maximum is $(50/3)^3$ **21.** Maximum product $\approx$36,168,981 at $(\frac{25}{3}, \frac{50}{3}, 25)$.
23. $7/\sqrt{6}$ at $(\frac{13}{6}, \frac{4}{3}, \frac{5}{6})$ **25.** $\sqrt{\beta^3/108}$ when $x = y = \sqrt{\beta/3}$, $z = \frac{1}{2}\sqrt{\beta/3}$
27. Profit $P(x, y) = 40(8xy + 32x + 40y + 4x^2 + 6y^2) - 10x - 4y$ is maximized when $x = \frac{3501}{160} \approx 21.88 \approx 22$ and $y =$ $1433/80 \approx 17.91 \approx 18$. Maximum profit $\approx$ \$28,188.77.
29. a. $P(a, b) = 150[300a/(a + 3) + 160d/(d + 5)] - a - d$ **b.** $a = -3 + \sqrt{135{,}000} \approx \$364.42$; $d = -5 + \sqrt{120{,}000} \approx \$341.40$ **31.** $y = \frac{30}{13} - \frac{27}{26}x$

Problems 3.11, page 229

1. $x = y = z = \frac{50}{3}$; the maximum is $(50/3)^3$ **3.** $x = \frac{25}{3}$, $y = \frac{50}{3}$, $z = 25$; the maximum $\approx$36,168,981
5. $3/\sqrt{13}$ to the point $(\frac{7}{13}, \frac{17}{13})$ **7.** $5/\sqrt{3}$ to the point $(\frac{8}{3}, \frac{2}{3}, \frac{1}{3})$
9. $|d|/\sqrt{a^2 + b^2}$ to the point $(ad/\sqrt{a^2 + b^2}, bd/\sqrt{a^2 + b^2})$ **11.** $|ax_0 + by_0 - d|/\sqrt{a^2 + b^2}$
13. $0 = 0^2 + 0^2$ is the minimum value; $x^2 + y^2$ is unbounded even when restricted to $x^3 + y^3 = 6xy$. [x or y can be negative]
15. $1 = (\pm 1)^2 + 0^2 + 0^2$ is the minimum value; $x^2 + y^2 + z^2$ is unbounded since y is unconstrained.

17. $\sqrt{3}$ is the maximum value and $-\sqrt{3}$ is the minimum value (at the points $(\pm 1)(1/\sqrt{3}, 1/\sqrt{3}, 1/\sqrt{3})$, respectively)
19. maximum is $\frac{|abc|}{3\sqrt{3}}$, at $\left(\frac{|a|}{\sqrt{3}}, \frac{|b|}{\sqrt{3}}, \frac{|c|}{\sqrt{3}}\right)$ and three others points, minimum is $\frac{-|abc|}{3\sqrt{3}}$ at $\left(-\frac{|a|}{\sqrt{3}}, -\frac{|b|}{\sqrt{3}}, -\frac{|c|}{\sqrt{3}}\right)$ and three other points
21. $\frac{121}{32}$ at $(\frac{5}{4}, \frac{5}{4}, -\frac{1}{2})$ **23.** $\min\{|a|, |b|, |c|\}$ is the minimum and $\max\{|a|, |b|, |c|\}$ is the maximum **25.** $\sqrt{n}$
27. $4\sqrt{3}$ **29.** base is 2 m by 2 m; height is 3 m **31.** $x/y = \frac{1}{2}$
33. a. $L = \frac{175}{2}$, $K = \frac{150}{7}$ **b.** $250(\frac{175}{3})^{0.7}(\frac{150}{7})^{0.3} \approx 14{,}343$ **c.** marginal productivity of labor/marginal productivity of capital $= \frac{4}{7}$
35. a. $\ell = 0$, $h = \sqrt{32} \approx 5.60$ **b.** $16\sqrt{32} \approx 90.5¢$ **37.** 10 days in each city

Problems 3.12, page 235

1. b. $(x_1, y_1) = (-0.025, -0.075)$; $(x_2, y_2) = (-0.005573048, -0.0003463476)$
c. $(x_1, y_1) = (1.025, 1035)$; $(x_2, y_2) = (1.00077655, 1.00096693)$
3. a. $(x_1, y_1) = (1.70833333, 0.875)$; $(x_2, y_2) = (1.69806217, 0.88330962)$ **b.** $(1.69804806, 0.88336722)$
5. i. a. $(x_1, y_1) = (-1.2, 1.3)$; $(x_2, y_2) = (-1.192857143, 1.278571423)$ **b.** $(-1.19287310, 1.27844411)$
ii. a. $(x_1, y_1) = (2.925, -3.5125)$; $(x_2, y_2) = (2.923492752, -3.510016747)$ **b.** $(2.92349211, -3.51001556)$
7. i. a. $(x_1, y_1) = (2.596280088, 2.705142232)$; $(x_2, y_2) = (2.590876452, 2.692273937)$
b. $(2.59085858, 2.69222325)$
ii. a. $(x_1, y_1) = (2.462540717, -0.7052117264)$;
$(x_2, y_2) = (2.46272607, -0.6797490992)$ **b.** $(2.4627339306, -0.6795568791)$
iii. a. $(x_1, y_1) = (-0.0422535211, -0.492957746)$;
$(x_2, y_2) = (-0.0491181274, -0.410806881)$ **b.** $(-0.04932091, -0.40838904)$
iv. a. $(x_1, y_1) = (-2.327133479, 2.15809628)$; $(x_2, y_2) = (-2.308094033, 2.110366365)$ **b.** $(-2.30789104, 2.10937054)$
v. a. $(x_1, y_1) = (-2.415584416, -0.097402597)$;
$(x_2, y_2) = (-2.410834885, -0.093801326)$ **b.** $(-2.41082044, -0.09379760)$
vi. a. $(x_1, y_1) = (-0.2821100917, 2.385321101)$;
$(x_2, y_2) = (-0.2855592166, 2.380159497)$ **b.** $(-0.2855601249, 2.3801497352)$
9. $(-0.090533054, -0.0998636709)$; $(1.2433857533, 0.0221338638)$; $(-0.2311447876, 1.225000791)$; $(-1.1531141693, -0.2017059939)$; $(0.012485153, -1.1248379068)$; $(1.2912933376, 1.159132058)$; $(-1.0604369649, 1.2569429491)$; $(1.1860751213, -1.180972109)$; $(-1.1980103894, -1.0558299813)$
13. a. $(-0.8471270884, 0.1528729116, 1.8471270884)$ **b.** $(1.1804604217, 2.1804604217, -0.1804604217)$

Chapter 3—Review Exercises, page 240

1. $\{(x, y): |x| \geq |y|\}$; $[0, \infty)$ **3.** $\mathbb{R}^2$; $[-1, 1]$ **5.** $\{(x, y, z): x^2 + y^2 + z^2 > 1\}$; $(0, \infty)$
7. parallel straight lines with slopes of -1; $y = -x + (1 - z^2)$

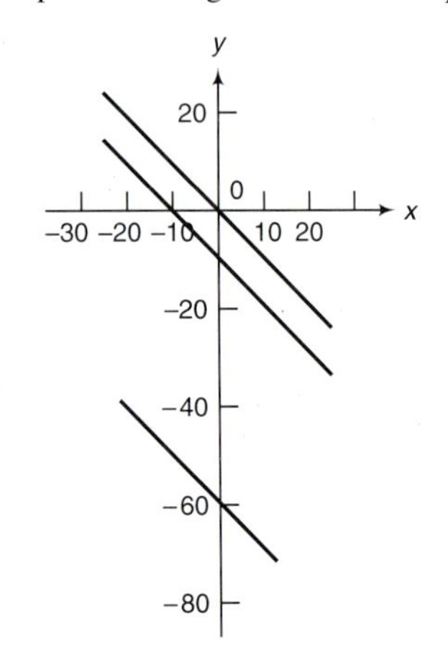

9. parallel straight lines with slopes of $\frac{1}{3}$; $x - 3y = e^x$

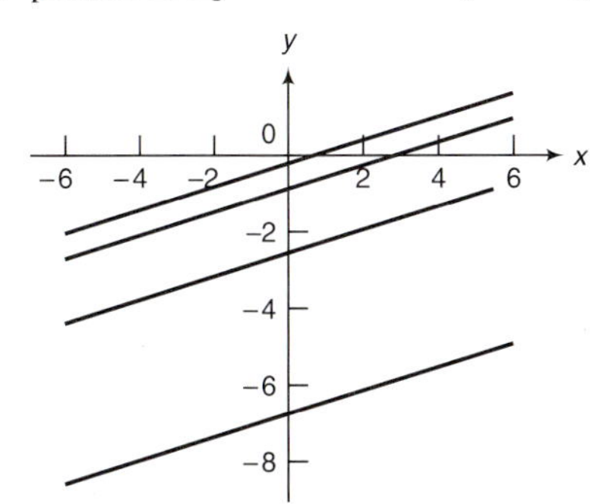

11.

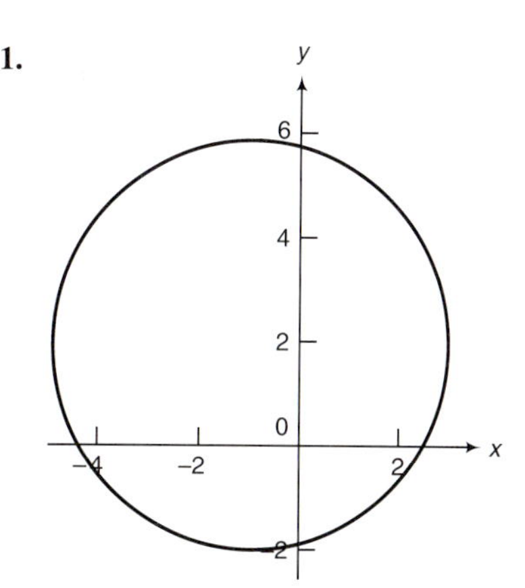

13. along $y = kx$ limit is $k/(k^2 - 1)$
15. since $0 \le (x - y^2)^2 = x^2 - 2xy^2 + y^4$, $|2xy^2| \le x^2 + y^4$ so that $|4xy^3/(x^2 + y^4)| \le |4xy^3/2xy^2| = 2|y|$
17. $-\frac{1}{4}$ **19.** $\{(x, y)\colon 2x - 3y < 1\}$ **21.** $\{(x, y, z)\colon x^2 - y^2 + z^2 < 1\}$ **23.** $f_x = -y/x^2$; $f_y = 1/x$
25. $f_x = -x/(x^2 - y^2)^{3/2}$; $f_y = y/(x^2 - y^2)^{3/2}$ **27.** $f_x = 1/(x - y + 4z)$; $f_y = -1/(x - y + 4z)$; $f_z = 4/(x - y + 4z)$
29. $f_x = -2(y/x^3)\sinh(y/x^2)$; $f_y = (1/x^2)\sinh(y/x^2)$; $f_z = 0$
31. Let A denote $\frac{2}{3}(x^2y - y^3z^5 + x\sqrt{z})^{-1/3}$; then $f_x = A \cdot (2xy + \sqrt{z})$, $f_y = A \cdot (x^2 - 3y^2z^5)$, and $f_z = A \cdot (-5y^3z^4 + x/2\sqrt{z})$
33. $f_x = (y - z + 3w)/(y + 2w - x)^2$; $f_y = (-x + z - w)/(y + 2w - x)^2$; $f_z = -1/(y + 2w - x)$; $f_w = (-3x + y + 2z)/(y + 2w - x)^2$ **35.** $f_{xx} = 0$; $f_{xy} = f_{yx} = 3y^2$; $f_{yy} = 6xy$
37. $f_{xx} = -y^2(x^2 - y^2)^{-3/2}$; $f_{xy} = f_{yx} = xy(x^2 - y^2)^{-3/2}$; $f_{yy} = -x^2(x^2 - y^2)^{-3/2}$
39. Let $A = (2 - 3x + 4y - 7z)^{-2}$. Then $f_{xx} = -9A$, $f_{xy} = f_{yx} = 12A$, $f_{yy} = -16A$, $f_{xz} = f_{zx} = -21A$, $f_{zz} = -49A$, $f_{yz} = f_{zy} = 28A$
41. 0 **43.** $2\mathbf{i} - 12\mathbf{j}$ **45.** $\frac{4}{25}\mathbf{i} - \frac{6}{25}\mathbf{j}$ **47.** $2\mathbf{i} + 6\mathbf{k}$ **49.** $-(a^2 + b^2 + c^2)^{-3/2}(a\mathbf{i} + b\mathbf{j} + c\mathbf{k})$
51. $2\cos 2t$ **53.** $-2s^5/r^3$ **55.** $w_r = 5r^4s^2$, $w_s = 2r^5s$ **57.** T: $x + y + z = 3$; N: $x = y = z$
59. T: $3x - y + 5z = 15$; N: $(x + 1)/3 = (y - 2)/(-1) = (z - 4)/5$
61. T: $-3x + 6y - 2z = 18$; N: $(x + 2)/(-3) = (y - 1)/6 = (z + 3)/(-2)$ **63.** $-3/\sqrt{2}$ **65.** $-1/(2\sqrt{13})$
67. $(9/\sqrt{14})6^{-3/2}$ **69.** $3x^2y^2\,\Delta x + 2x^3y\,\Delta y$ **71.** $\frac{1}{2}[(x + 1)(y - 1)]^{-1/2}\,\Delta x - \frac{1}{2}\sqrt{x + 1}\,(y - 1)^{-3/2}\,\Delta y$
73. $(x - y + 4z)^{-1}(\Delta x - \Delta y + 4\,\Delta z)$ **75.** local minimum at $(0, 0)$
77. local minimum at $(\sqrt{2}, 1/\sqrt{2})$; $(\sqrt{2}, -1/\sqrt{2})$ and $(-\sqrt{2}, 1/\sqrt{2})$ are saddle points; local maximum at $(-\sqrt{2}, -1/\sqrt{2})$
79. local minima at $(2^{-1/5}, \pm 2^{3/10})$ **81.** $8/\sqrt{11}$ at $(\frac{14}{11}, -\frac{3}{11}, \frac{20}{11})$ **83.** $\frac{1}{2}(\frac{10}{3})^{3/2}$ [dimensions are $\sqrt{\frac{10}{3}} \times \sqrt{\frac{10}{3}} \times \frac{1}{2}\sqrt{\frac{10}{3}}$]
85. $86/31$ at $(\frac{44}{31}, \frac{1}{31}, -\frac{27}{31})$ **89.** $\operatorname{dom} f = \mathbb{R}^4$, range $f = [0, \infty)$
91. $\operatorname{dom} f = \{(x_1, x_2, x_3, x_4)\colon {x_3}^2 + {x_4}^2 \ne 0\}$, range $f = [0, \infty)$ **95.** -21
97. $\dfrac{\partial f}{\partial x_1} = 3{x_1}^2{x_2}^2x_3 - \dfrac{1}{x_1 + 2x_2 - x_4} - \dfrac{{x_4}^5}{{x_1}^2}$; $\dfrac{\partial f}{\partial x_2} = 2{x_1}^3x_2x_3 - \dfrac{2}{x_1 + 2x_2 - x_4}$; $\dfrac{\partial f}{\partial x_3} = {x_1}^3{x_2}^2$; $\dfrac{\partial f}{\partial x_4} = \dfrac{1}{x_1 + 2x_2 - x_4} + \dfrac{5{x_4}^4}{x_1}$
99. $f_{12} = 6{x_1}^2x_2x_3 + 2(x_1 + 2x_2 - x_4)^{-2}$; $f_{31} = 3{x_1}^2{x_2}^2$; $f_{124} = 4(x_1 + 2x_2 - x_4)^{-3}$
101. $\mathbf{f}'(t) = \left(\dfrac{1}{2\sqrt{t}}, \dfrac{5}{3}t^{2/3}, -\dfrac{1}{t^2}, 5t^4 - 2\right)$ **103.** $\displaystyle\int_0^1 \sqrt{2 + 8t^2}\,dt = \frac{\sqrt{10}}{2} + \frac{1}{\sqrt{8}}\ln(2 + \sqrt{5})$
105. $\nabla f = \left(\dfrac{x_2}{x_3 - x_4}, \dfrac{x_1}{x_3 - x_4}, -\dfrac{x_1x_2}{(x_3 - x_4)^2}, \dfrac{x_1x_2}{(x_3 - x_4)^2}\right)$

Computer Exercises, page 242

1. a. (i) $f_x = \dfrac{y(x^4 + 4x^2y^2 - y^4)}{(x^2 + y^2)^2}$, $f_y = \dfrac{x(x^4 - 4x^2y^2 - y^4)}{(x^2 + y^2)^2}$, $(x, y) \ne (0, 0)$ (ii) $f_x = f_y = 0$ at $(0, 0)$
b. (i) $f_{xy} = f_{yx} = \dfrac{x^6 + 9x^4y^2 - 9x^2y^4 - y^6}{(x^2 + y^2)^3}$, $(x, y) \ne (0, 0)$ (ii) $f_{xy}(0, 0) = -1$, $f_{yx}(0, 0) = 1$
c. f_{xy} is not continuous at $(0, 0)$; for example $\lim_{x\to 0} f_{xy}(x, 0) = 1$ while $\lim_{y\to 0} f_{xy}(0, y) = -1$

d.

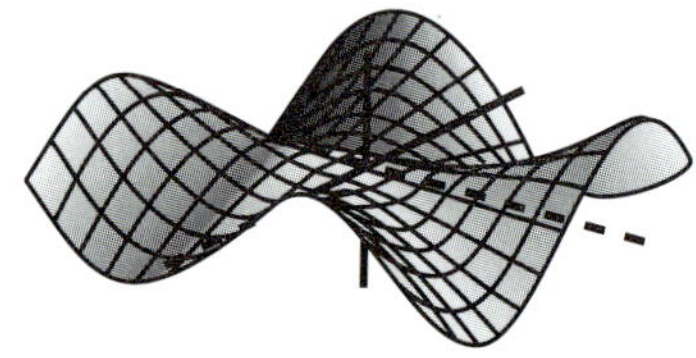

3. **a.** and **b.** local minimum at $(\frac{1}{2}, -\frac{5}{4})$; saddle points at $\left(\frac{3-\sqrt{33}}{4}, \frac{7-\sqrt{33}}{8}\right)$ and $\left(\frac{3+\sqrt{33}}{4}, \frac{7+\sqrt{33}}{8}\right)$

c.

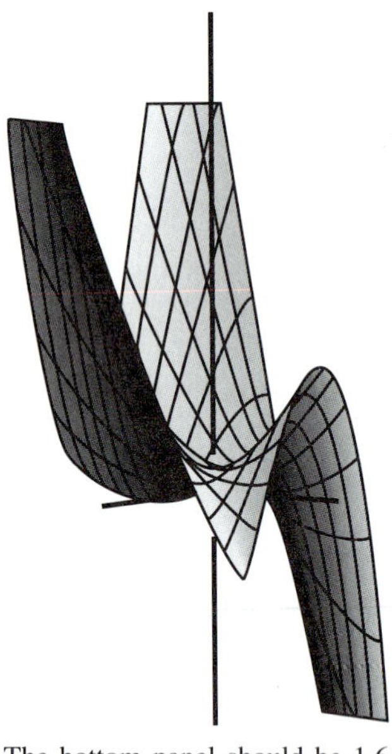

5. The bottom panel should be 1.64550 by 1.32817. The height is 0.457560. (The shorter of the dimensions of the bottom panel is the length of the left and right side panels.) The total cost of the box is \$16.62.

CHAPTER 4

Problems 4.1, page 252

1. $\frac{45}{2}$ **3.** 0 **5.** 16 **7.** 39 **9.** 0 **11.** $\frac{15}{2}$ **13.** 10 **15.** 9 **17.** 41 **19.** 11
21. $0 \le \iint \le 6$ **23.** $-\sqrt{2}\pi/3 \le \iint \le \sqrt{2}\pi/3$ (These are crude bounds obtained from $|x-y| \le \sqrt{2}$ and $1/(4-x^2-y^2) \le \frac{1}{3}$ on the unit disk; a symmetry argument shows that $\iint = 0$.)
25. $0 \le \iint \le \frac{1}{4}\ln 2$ [The area of the triangle is $\frac{1}{4}$ and $x + y \le 1$ over the triangle.]

Problems 4.2, page 262

1. $\frac{2}{3}$ **3.** $e^{-5} - e^{-1} - e^{-2} + e^2$ **5.** -31 **7.** $\frac{162}{5}$ **9.** $\frac{16}{3}$ **11.** $\frac{20}{3}$ **13.** $\frac{1}{2}(e^{19} - e^{17} - e^3 + e)$
15. $\frac{1}{3} + (\pi/16)$ **17.** $\int_0^{1/2}\int_x^{1-x} (x+2y)\,dy\,dx = \frac{7}{24}$ **19.** $\int_0^{1/\sqrt{2}}\int_{x^2}^{1-x^2} (x^2+y)\,dy\,dx = \frac{\sqrt{2}}{5}$
21. $\int_1^2\int_x^2 \frac{y}{\sqrt{x^2+y^2}}\,dy\,dx = \frac{1}{\sqrt{2}} - \frac{\sqrt{5}}{2} + 2\ln\left(\frac{2+2\sqrt{2}}{1+\sqrt{5}}\right)$ **23.** 0 **25.** 0 **27.** $\frac{2}{15}$ **29.** 0
31. $-\frac{2}{5} + \frac{1}{2}\ln 5$ **33.** $\int_{-1}^3\int_0^2 dy\,dx = 8$

35. $\int_1^2 \int_2^4 x^{-3}y^3\, dy\, dx + \int_2^4 \int_x^4 x^{-3}y^3\, dy\, dx = 27$

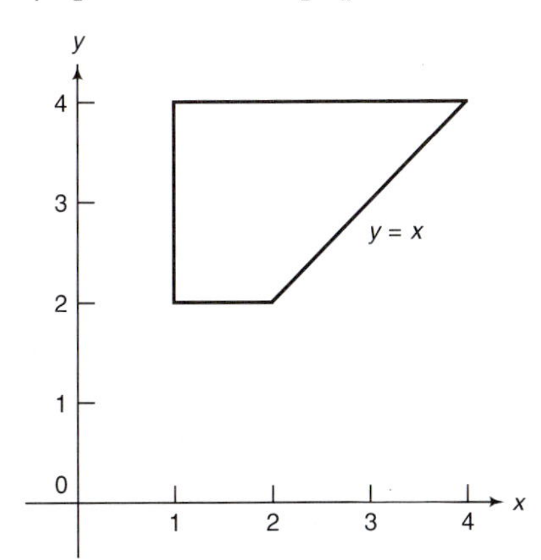

37. $\int_0^1 \int_0^y dx\, dy = \frac{1}{2}$

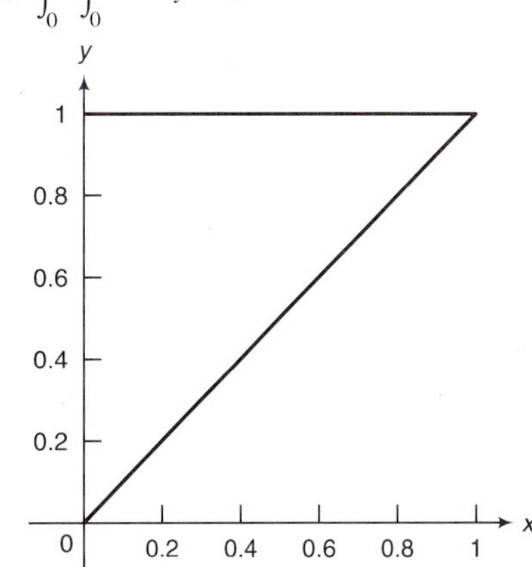

39. $\int_0^2 \int_0^{\sqrt{4-x^2}} (4 - x^2)^{3/2}\, dy\, dx = \dfrac{256}{15}$

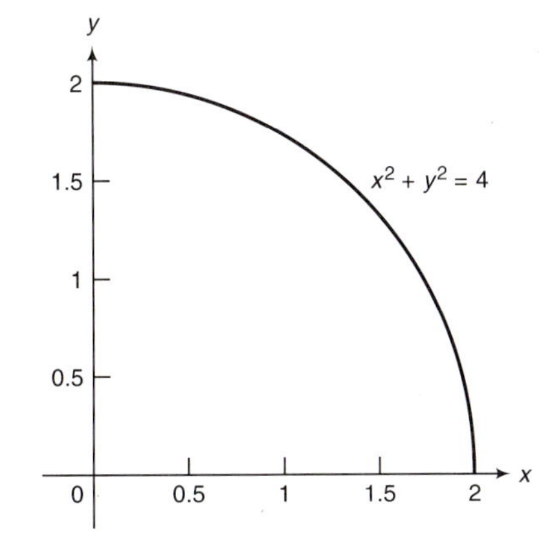

41. $\int_0^1 \int_0^{x^2} \sqrt{3 - x^3}\, dy\, dx = \frac{2}{9}(3^{3/2} - 2^{3/2})$

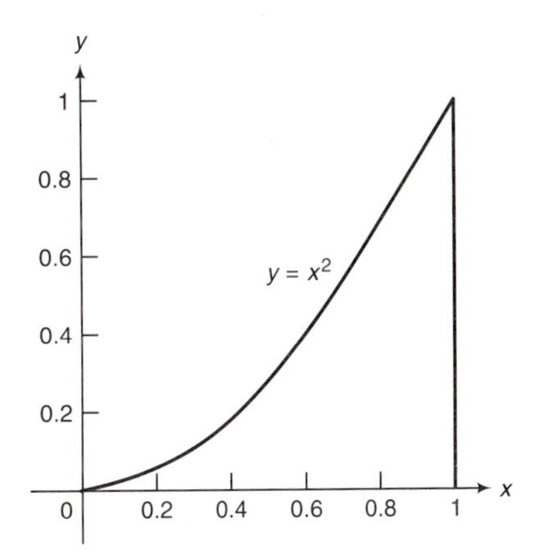

43. $\frac{9}{2}$ **45.** $\frac{128}{3}$ **47.** 48π **49.** 1 **51.** $\dfrac{\pi}{2}$

53. $\int_{-\sqrt{3}}^{-1} \int_0^{3-x^2} dy\, dx + \int_{-1}^{1} \int_{x^3+1}^{3-x^2} dy\, dx = 2\sqrt{3} + \frac{2}{3}$; also $\int_{-1}^{1} \int_0^{x^3+1} dy\, dx + \int_1^{\sqrt{3}} \int_0^{3-x^2} dy\, dx = 2\sqrt{3} - \frac{2}{3}$

55.

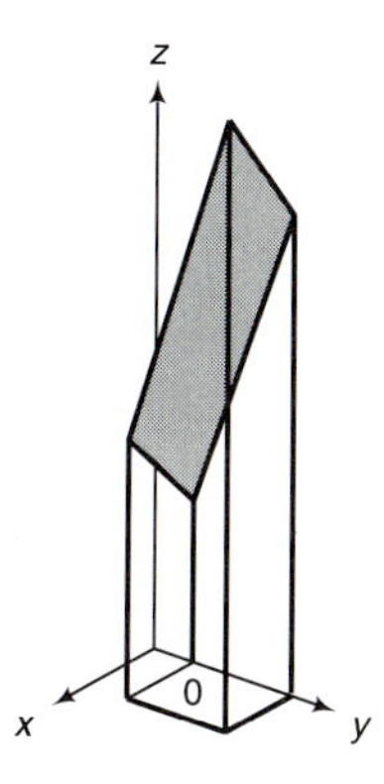

59. 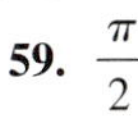 $\dfrac{\pi}{2}$

Problems 4.3, page 269

1. $\frac{16}{3}$; $(\frac{51}{32}, 0)$ **3.** $(\sqrt{3} - 1)/12$; $\left(\frac{1}{2} + \frac{\pi}{6} - \frac{\pi}{6(\sqrt{3} - 1)}, \frac{1}{3} - \frac{\pi}{18(\sqrt{3} - 1)}\right)$

5. $3e^3 - 3e + e^{-1} - e^{-3}$; $\left(\frac{10e^3 - 10e - 2e^{-1} + 2e^{-3}}{3e^3 - 3e + e^{-1} - e^{-3}}, \frac{6e^3 - 12e + 2e^{-1} - 4e^{-3}}{3e^3 - 3e + e^{-1} - e^{-3}}\right)$

7. $(\pi/16) + \frac{1}{3}$; $\left(\frac{62}{5(16 + 3\pi)}, \frac{16 + 15\pi}{5(16 + 3\pi)}\right)$ **9.** $\frac{5}{24}$; $(\frac{11}{20}, \frac{1}{5})$ **11.** $\frac{1}{2}$; $(1, 1)$ **13.** $4\sqrt{2}\pi^2$

Problems 4.4, page 274

1. $2\pi a^{n+2}/(n + 2)$ **3.** 560π **5.** $3\pi/2$ **7.** $2\pi + 3\sqrt{3}/2$ **9.** 1 **11.** $(a^2 + \frac{1}{2}b^2)\pi$

13. $18\sqrt{3} - 4\pi$ **15.** $\frac{2}{3}\pi a^3(8 - 3\sqrt{3})$ **17.** $\frac{512}{9}$ **19.** $\frac{4}{3}\pi 8^{3/2}$ **21.** $(\frac{1}{2}, 1)$ **23.** $\left(\frac{b(b^2 + 4a^2)}{2(b^2 + 2a^2)}, 0\right)$

Problems 4.5, page 280

1. $\frac{1}{8}$ **3.** 0 **5.** $(\pi^3/16)(1 - \cos 1)$ **7.** $\frac{1}{30}$ **9. a.** $\int_0^1 \int_z^1 \int_z^y y\,dx\,dy\,dz$ **b.** $\int_0^1 \int_0^x \int_x^1 y\,dy\,dz\,dx$

11. a. $\int_0^1 \int_0^x \int_0^{\sqrt{1-x^2}} yz\,dy\,dz\,dx$ **b.** $\int_0^1 \int_0^{\sqrt{1-z^2}} \int_z^{\sqrt{1-y^2}} yz\,dx\,dy\,dz$ **13.** $\frac{1}{24}$ **15.** $\frac{1}{8}$ **17.** $\frac{12}{5}$ **19.** $\frac{1}{6}$

21. $9(\pi - 1)$ **23.** $40\pi/3$ **25.** $\frac{4}{3}\pi abc$ **27.** $\frac{1}{24}$ **29.** $\frac{207}{8}$ **31.** $(a/4, b/4, c/4)$ **33.** $(\frac{2}{5}, \frac{1}{5}, \frac{1}{5})$

35. $(\frac{108}{115}, \frac{176}{115}, 18\pi/23 - \frac{72}{115})$ **37. a.** $(\frac{258}{455}, \frac{372}{455}, \frac{7}{26})$

Problems 4.6, page 285

1. $\frac{16}{3}(\pi - \frac{4}{3})$ **3.** $\pi/32$ **5.** $\frac{\pi}{3} + (a - 1)\pi = (a - \frac{2}{3})\pi$ **7.** $\pi/4$ **9.** $16\sqrt{2}\,\pi/3$ **11.** $\pi/9$

13. 486π **15.** $(24/5(3\pi - 4), 0, 0)$ **17.** $(0, 0, 2)$ **19.** $(0, 0, \frac{7}{12})$ **21.** $(0, 0, 0)$

Chapter 4—Review Exercises, page 289

1. $\frac{4}{3}$ **3.** $\frac{67}{3}$ **5.** $\frac{1}{20}$ **7.** $128\pi(2\sqrt{2} - 1)/15$ **9.** -24 **11.** $\frac{1}{3} + \pi/16$ **13.** $\left(\int_1^2 \int_2^5 + \int_2^5 \int_y^5\right) 3x^2y\,dx\,dy = \frac{434}{5}$

15. $\int_0^\infty \int_0^y f(x, y)\,dx\,dy$ **17.** $(1 - e^{-9})\pi$ **19.** 6 **21.** 12π **23.** $81\pi/32$

25. 6π **27.** $32\pi/3$ **29.** $\frac{128}{15}$ **31.** $\pi/32$ **33.** $16\pi/9$ **35.** $(\pi/2\sqrt{2}, 0)$ **37.** $(0, 0, \frac{3}{2})$

39. $(\frac{9}{4}, \frac{31}{9})$ **41.** $(0, 0, 0)$ **43.** $\left(\frac{3\sqrt{3}}{5}, \frac{3}{5}, 0\right)$

CHAPTER 5

Problems 5.1, page 297

1. $(x^2 + y^2)^{-3/2}(-x\mathbf{i} - y\mathbf{j})$ **3.** $2(x + y)(\mathbf{i} + \mathbf{j})$ **5.** $\sin(x - y)(-\mathbf{i} + \mathbf{j})$

7. $-y\sec^2(y - x)\mathbf{i} + [\tan(y - x) + y\sec^2(y - x)]\mathbf{j}$ **9.** $\sec(x + 3y)\tan(x + 3y)(\mathbf{i} + 3\mathbf{j})$

11. $4xy(x^2 + y^2)^{-2}(y\mathbf{i} - x\mathbf{j})$ **13.** $(x^2 + y^2 + z^2)^{-1/2}(x\mathbf{i} + y\mathbf{j} + z\mathbf{k})$

15. $\cos x \cos y \tan z\mathbf{i} - \sin x \sin y \tan z\mathbf{j} + \sin x \cos y \sec^2 z\mathbf{k}$ **17.** $(\ln y - z/x)\mathbf{i} + (x/y)\mathbf{j} - \ln x\mathbf{k}$

19. $e^{x+2y+3z}[(y - z)\mathbf{i} + (1 + 2y - 2z)\mathbf{j} + (-1 + 3y - 3z)\mathbf{k}]$

21. $\nabla F(x, y) = \dfrac{2(x^2 - y^2 - 1)\mathbf{i} + 4xy\mathbf{j}}{[(x+1)^2 + y^2][(x-1)^2 + y^2]}$ **23.** $f(x, y) = -xy$

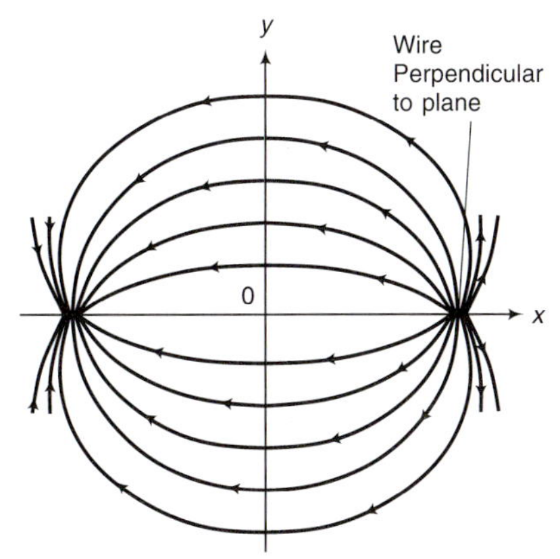

25. If $k = 1$, let $f(\mathbf{x}) = \alpha|\mathbf{x}|$; if $k = 2$, $f(\mathbf{x}) = \alpha \ln|\mathbf{x}|$; if $k > 2$, then $f(\mathbf{x}) = [\alpha/(k-2)]|\mathbf{x}|^{2-k}$ is a potential function.

Problems 5.2, page 303

1. -3J **3.** $-6\sqrt{2}$J **5.** $12/\sqrt{13}$J **7.** -2 **9.** 24 **11.** $-\frac{19}{3}$ **13.** $-\pi$ **15.** $-\frac{5}{6}$
17. $3\pi - \frac{2}{3}$ **19.** $\frac{1}{2}e + \frac{1}{2}e^{-1} - 2 = \cosh 1 - 2$ **21.** $\frac{5}{2}\ln 2 + \frac{2}{5}\ln 6 + \frac{1}{3}$ **23.** $\frac{3}{2}$ **25.** $\pi^3/24$
27. $\frac{9}{20} + (2e^{\pi}/5) \approx 9.706$ J **29.** $\frac{3}{2}\pi$ J **31.** $2ab\pi$ J **33.** $\alpha(1 - 1/\sqrt{13})$ J **35.** $(21 + \pi)$ J

Problems 5.3, page 312

1. $x^2y + y + C$ **3.** not exact **5.** not exact **7.** $\frac{1}{2}x^2 - y\sin x + C$ **9.** $\frac{3}{2}x^2\ln x - \frac{3}{4}x^2 + \frac{1}{6}x^6 - xy + C$

11. not exact **13.** $x + y + z + C$ **15.** not exact **17.** $\mathbf{F} = \nabla(x^2y + y)$; $\int_C \mathbf{F(x)}\cdot d\mathbf{x} = 14$

19. $\mathbf{F} = \nabla[x\sin(x+y)]$; $\int_C \mathbf{F(x)}\cdot d\mathbf{x} = \pi/6$ **21.** $\mathbf{F} = \nabla(x^2\cos y)$; $\int_C \mathbf{F(x)}\cdot d\mathbf{x} = \pi^2/4$

23. $\mathbf{F} = \nabla(xe^y)$; $\int_C \mathbf{F(x)}\cdot d\mathbf{x} = 5e^7$ **25.** $\mathbf{F} = \nabla(xy^2z^4)$; $\int_C \mathbf{F(x)}\cdot d\mathbf{x} = 12$

Problems 5.4, page 317

1. 2 **3.** $\cos 1 - \sin 1 + e - 2$ **5.** $2(e^2 - 1)(1 - \cos 1)$ **7.** 0 **9.** $\frac{\pi}{24}(4\sqrt{2} - 5)$ **11.** $\frac{596}{15}$

13. $(b - a)$(areas of Ω) **15.** 0 **17.** 23 **19.** 10

Problems 5.5, page 329

1. $\mathbf{r}(u, v) = u\mathbf{i} + v\mathbf{j} + (2u + 3v)\mathbf{k}$
3. $\mathbf{r}(\theta, \varphi) = 2\cos\theta\sin\varphi\mathbf{i} + 2\sin\theta\sin\varphi\mathbf{j} + 2\cos\varphi\mathbf{k}$, $0 \le \theta \le 2\pi$, $0 \le \varphi \le \pi/2$
5. $\mathbf{r}(u, v) = u\mathbf{i} + v\mathbf{j} + \dfrac{u^2 - v^2}{4}\mathbf{k}$, $0 \le u \le 1$, $2 \le v \le 3$
7. $\mathbf{r}(r, \theta) = r\cos\theta\mathbf{i} + r\sin\theta\mathbf{j} + r(4\cos\theta - \sin\theta)\mathbf{k}$, $0 \le \theta \le 2\pi$, $0 \le r \le 1$
9. $\mathbf{r}(\theta, \varphi) = \cos\theta\sin\varphi\mathbf{i} + \sin\theta\sin\varphi\mathbf{j} + \cos\varphi\mathbf{k}$, $0 \le \theta \le 2\pi$, $0 \le \varphi \le \pi/4$ **11.** $2\sqrt{6}$ **13.** $(\pi/2)\sqrt{1 + a^2 + b^2}$
15. $2\int_0^1\int_0^2 \sqrt{1 + \frac{4}{9}x^{-2/3}}\,dy\,dx = \frac{4}{27}(13^{3/2} - 8)$ $\Big[$Note that $\int_{-1}^1\int_0^2 \sqrt{1 + \frac{4}{9}x^{-2/3}}\,dy\,dx$ is *wrong* since the integrand is not defined at $x = 0$; evaluate the area as the sum of two improper integrals: $\int_{-1}^0\int_0^2 + \int_0^1\int_0^2.\Big]$

17. 132 **19.** $\int_0^1\int_0^2 \sqrt{1 + 9x}\,dy\,dx = \frac{4}{27}(10^{3/2} - 1)$ (This is the surface area for $z \ge -1$.) **21.** $3\sqrt{2}\pi$

23. $(2-\sqrt{2})\pi$ **25.** 12 **27.** $4a^2(\pi/2-1)$ **29.** 16π **31.** $\int_{-1}^{1}\int_{-\sqrt{1-y^2}}^{\sqrt{1-y^2}} \sqrt{1+9x^4+9y^4}\,dx\,dy$

33. $\int_0^2\int_x^{4-x} \sqrt{1+\dfrac{1}{2(1+x+y)}}\,dy\,dx = \left(\int_0^2\int_0^y + \int_2^4\int_0^{4-y}\right)\sqrt{1+\dfrac{1}{2(1+x+y)}}\,dx\,dy$

35. $2\int_{-a}^{a}\int_{-b\sqrt{1-x^2/a^2}}^{b\sqrt{1-x^2/a^2}} \sqrt{\dfrac{1+\left(\dfrac{c^2}{a^2}-1\right)\dfrac{x^2}{a^2}+\left(\dfrac{c^2}{b^2}-1\right)\dfrac{y^2}{b^2}}{1-\dfrac{x^2}{a^2}-\dfrac{y^2}{b^2}}}\,dy\,dx$ **37.** $\frac{1}{2}\sqrt{b^2c^2+c^2a^2+a^2b^2}$

43. $\mathbf{r}(u,v) = u\mathbf{i} + f(u)\cos v\mathbf{j} + f(u)\sin v\mathbf{k},\ a \le u \le b,\ 0 \le v \le 2\pi$ **45.** $\pi a\sqrt{a^2+h^2}$

Problems 5.6, page 336

1. $(5^{3/2}-1)/6$ **3.** $-\frac{101}{48}\sqrt{5} - \frac{131}{96}\ln(2+\sqrt{5})$ **5.** 0 **7.** $2\sqrt{14}/3$ **9.** $2\sqrt{14}/81$

11. $(\sqrt{14}/6)(-\cos 7 + \cos 5 + \cos 2 - \cos 4)$ **13.** 8π **15.** 54π **17.** $\dfrac{1376}{3}\pi$ **19.** $(\sqrt{3}/2)\alpha$ kg

21. $-\frac{2}{9}$ **23.** 4π **25.** 2π **27.** 0 **29.** $(\pi-2)/6$ **31.** -12

Problems 5.7, page 342

1. div $\mathbf{F} = 2(x+y+z)$; curl $\mathbf{F} = \mathbf{0}$ **3.** div $\mathbf{F} = 0$; curl $\mathbf{F} = \mathbf{0}$ **5.** div $\mathbf{F} = x+y+z$; curl $\mathbf{F} = -y\mathbf{i} - z\mathbf{j} - x\mathbf{k}$

7. div $\mathbf{F} = 0$; curl $\mathbf{F} = x(e^{xy}-e^{xz})\mathbf{i} + y(e^{yz}-e^{yz})\mathbf{j} + z(e^{zx}-e^{zy})\mathbf{k}$

9. div $\mathbf{F} = \dfrac{1}{y}+\dfrac{1}{z}+\dfrac{1}{x}$; curl $\mathbf{F} = \left(\dfrac{y}{z^2}\right)\mathbf{i} + \left(\dfrac{z}{x^2}\right)\mathbf{j} + \left(\dfrac{x}{y^2}\right)\mathbf{k}$ **11. a.** $\mathbf{0}$ **b.** 0 **c.** $a+b$ **d.** $a+b$

13. a. $\mathbf{0}$ **b.** 0 **c.** $2x+2y$ **d.** 2 **15. a.** $\mathbf{0}$ **b.** 0 **c.** 2 **d.** 2π **17. a.** $3(x^2-y^2)\mathbf{k}$ **b.** 0 **c.** 0 **d.** 0

19. 0 (harmonic) **21.** 20 (not harmonic)

23. $G = Cyz + (C-3)xz + (C-1)xy$ where C is an arbitrary constant

31. div $\mathbf{E} = 0$ if $|\mathbf{x}| \neq 0$ so the limit is 0

47. c. $\mathbf{F}$ is undefined at the origin so it is not smooth throughout the unit disk

Problems 5.8, page 349

1. -3π **3.** 18 **5.** -108π **7.** $21 - \sin 3$ **9.** both integrals equal 1 **11.** both integrals equal -32π

Problems 5.9, page 354

1. 4π **3.** 36π **5.** 0 **7.** 2 **9.** 108π **11.** $\frac{1}{6}$ **13.** $\frac{1}{2}$ **15.** $\frac{184}{35}$

Problems 5.10, page 363

1. $[\partial(x,y)/\partial(u,v)] = -2$ **3.** $[\partial(x,y)/\partial(u,v) = 4(u^2+v^2)$

5. $[\partial(x,y)/\partial(u,v)] = -2$ **7.** $[\partial(x,y)/\partial(u,v)] = -(a^2+b^2)$

9. $[\partial(x,y)/\partial(u,v)] = (1-uv)e^{u+v}$ **11.** $[\partial(x,y)/\partial(u,v)] = [1/(u+v)][(1/v)-(1/u)]$

13. $[\partial(x,y)/\partial(u,v)] = \sec v \csc u(1 + uv\cot u \tan v)$ **15.** $[\partial(x,y,z)/\partial(u,v,w)] = 0$

17. $[\partial(x,y,z)/\partial(u,v,w)] = 2(u^2v - uv^2 + v^2w - vw^2 + w^2u - wu^2) = 2(u-v)(v-w)(u-w)$

19. $[\partial(x,y,z)/\partial(u,v,w)] = e^{u+v+w}$ **21.** $\int_0^1\int_y^1 xy\,dx\,dy = \int_{-1/2}^{0}\int_{-v}^{1+v}(u^2-v^2)(2)\,du\,dv = \frac{1}{8}$

23. 2π **25.** $\frac{128}{15}$ **27.** $4\pi abc/3$

Chapter 5—Review Exercises, page 367

1. $3(x+y)^2(\mathbf{i}+\mathbf{j})$ **3.** $(x-y)^{-2}(-2y\mathbf{i}+2x\mathbf{j})$ **5.** $2x\mathbf{i}+2y\mathbf{j}+2z\mathbf{k}$

7. $y\mathbf{i} + x\mathbf{j}$

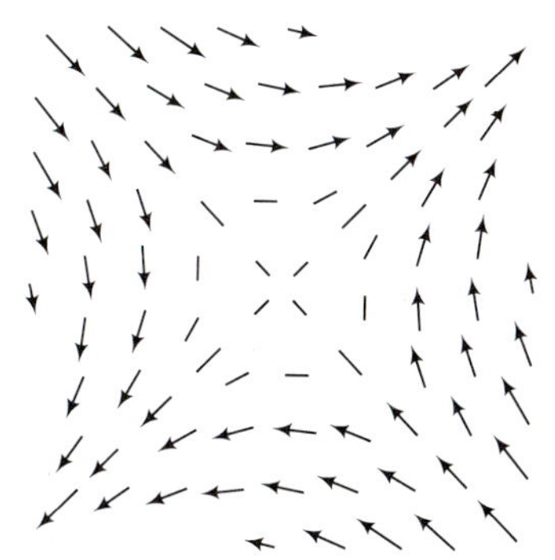

9. $\frac{2}{3}$ **11.** $-\frac{1}{2}$ **13.** $\frac{3}{2}$ **15.** $\pi/2$ **17.** 23 **19.** $\frac{5}{3}$

21. $\frac{1}{6}$ **23.** 0 **25. a.** $\mathbf{0}$ **b.** 0 **c.** $y^2 + x^2$ **d.** 8π **29.** $\mathbf{r}(u, v) = u\mathbf{i} + v\mathbf{j} + (3v - 2u)\mathbf{k}$

31. $\mathbf{r}(u, v) = u\mathbf{i} + v\mathbf{j} + (u^2 - v^2)\mathbf{k}$, $0 \le u \le 2$, $2 \le v \le 4$ **33.** $\int_0^2 \int_2^4 \sqrt{1 + 4x^2 + 4y^2}\, dy\, dx$ **35.** 32π

37. $(5^{3/2} - 1)/6$ **39.** 54π **41.** $(\sqrt{3} - 2)/6$ **43.** $c\sqrt{3}/12$ where c is the constant of proportionality

45. $-\frac{128}{3}$ **47.** $-32\pi/3$ **49.** div $\mathbf{F} = 3$; curl $\mathbf{F} = \mathbf{0}$ **51.** 0; $\mathbf{0}$

53. 0; curl $\mathbf{F} = x(e^{xy} - e^{xz})\mathbf{i} + y(e^{yz} - e^{yx})\mathbf{j} + z(e^{zx} - e^{zy})\mathbf{k}$ **55.** -2π **57.** 0 **59.** 8π **61.** 324π

63. 6 **65.** -5 **67.** $(\ln u)(\ln v) - 1$ **69.** 0 **71.** 3 **73.** $\int_{-1/2}^{0} \int_{-v}^{1+v} 2(u^2 - v^2)\, du\, dv = \dfrac{1}{8}$

CHAPTER 6

Problems 6.2, page 375

1. $x = -\frac{13}{5}$, $y = -\frac{11}{5}$; $a_{11}a_{22} - a_{12}a_{21} = -10$ **3.** no solutions; $a_{11}a_{22} - a_{12}a_{21} = 0$
5. $x = \frac{11}{2}$, $y = -30$; $a_{11}a_{22} - a_{12}a_{21} = -2$
7. infinite number of solutions; $y = \frac{2}{3}x$, where x is arbitrary; $a_{11}a_{22} - a_{12}a_{21} = 0$
9. $x = -1$, $y = 2$; $a_{11}a_{22} - a_{12}a_{21} = -1$
11. $a_{11}a_{22} - a_{12}a_{21} = a^2 - b^2$; if $a^2 - b^2 \ne 0$ (i.e., if $a \ne \pm b$), then $x = y = c/(a + b)$. If $a^2 - b^2 = 0$, then $a = \pm b$. If $a \ne 0$ and $a = b$, then there is an infinite number of solutions given by $y = c/a - x$. If $a \ne 0$ and $a = -b$, then there are no solutions (unless $c = 0$ in which case $x = y$ is a solution).
13. $a_{11}a_{22} - a_{12}a_{21} = -2ab$; so unique solution if both a and b are nonzero **15.** $a = b = 0$ and $c \ne 0$ or $d \ne 0$
17. no point of intersection
19. The lines are coincident. Any point of the form $(x, (4x - 10)/6)$ is a point of intersection.
21. $(\frac{67}{45}, \frac{2}{15})$ **23.** $\sqrt{13}/13$ **25.** $\sqrt{61}/5$ **27.** $\sqrt{5}$
33. The unique solution can be found to be $x = \dfrac{a_{22}b_1 - a_{12}b_2}{a_{11}a_{22} - a_{12}a_{21}}$ and $y = \dfrac{a_{11}b_2 - a_{21}b_1}{a_{11}a_{22} - a_{12}a_{21}}$.
35. Let x = no. of cups; y = no. of saucers; solutions are $(x, 240 - \frac{3}{2}x)$. [There is a finite number of solutions because x and y must be positive integers.] **37.** 32 sodas, 128 milk shakes

Problems 6.3, page 387

Note: Where there are an infinite number of solutions, we wrote the solutions with the last variable chosen arbitrarily. The solutions can be written in other ways as well.

1. $(2, -3, 1)$ **3.** $(3 + \frac{2}{9}x_3, \frac{8}{9}x_3, x_3)$, x_3 arbitrary **5.** $(-9, 30, 14)$ **7.** no solution
9. $(-\frac{4}{5}x_3, \frac{9}{5}x_3, x_3)$, x_3 arbitrary **11.** $(-1, \frac{5}{2} + \frac{1}{2}x_3, x_3)$, x_3 arbitrary **13.** no solution
15. $(\frac{20}{13} - \frac{4}{13}x_4, -\frac{28}{13} + \frac{3}{13}x_4, -\frac{45}{13} + \frac{9}{13}x_4, x_4)$, x_4 arbitrary **17.** $(18 - 4x_4, -\frac{15}{2} + 2x_4, -31 + 7x_4, x_4)$, x_4 arbitrary
19. no solution **21.** row echelon form **23.** reduced row echelon form **25.** neither
27. reduced row echelon form **29.** neither
31. row echelon form: $\begin{pmatrix} 1 & -6 \\ 0 & 1 \end{pmatrix}$; reduced row echelon form: $\begin{pmatrix} 1 & 0 \\ 0 & 1 \end{pmatrix}$

33. row echelon form: $\begin{pmatrix} 1 & -2 & 4 \\ 0 & 1 & -\frac{4}{11} \\ 0 & 0 & 1 \end{pmatrix}$ reduced row echelon form: $\begin{pmatrix} 1 & 0 & 0 \\ 0 & 1 & 0 \\ 0 & 0 & 1 \end{pmatrix}$

35. row echelon form: $\begin{pmatrix} 1 & -\frac{7}{2} \\ 0 & 1 \\ 0 & 0 \end{pmatrix}$ reduced row echelon form: $\begin{pmatrix} 1 & 0 \\ 0 & 1 \\ 0 & 0 \end{pmatrix}$

37. $x_1 = 30{,}000 - 5x_3$; $x_2 = x_3 - 5000$; $5000 \le x_3 \le 6000$; no

39. no unique solution (2 equations in 3 unknowns); if 200 shares of McDonald's, then 100 shares of Hilton and 300 shares of Delta

41. The row echelon form of the augmented matrix representing this system is $\left(\begin{array}{ccc|c} 1 & -\frac{1}{2} & \frac{3}{2} & a/2 \\ 0 & 1 & -\frac{19}{5} & \frac{2}{5}(b - \frac{3}{2}a) \\ 0 & 0 & 0 & -2a + 3b + c \end{array}\right)$ which is inconsistent if $-2a + 3b + c \ne 0$ or $c \ne 2a - 3b$.

43. $a_{11}a_{22}a_{33} + a_{12}a_{23}a_{31} + a_{13}a_{32}a_{21} - a_{13}a_{22}a_{31} - a_{12}a_{21}a_{33} - a_{11}a_{32}a_{23} \ne 0$

Problems 6.4, page 392

1. (0, 0) **3.** (0, 0, 0) **5.** $(\frac{1}{6}x_3, \frac{5}{6}x_3, x_3)$, x_3 arbitrary **7.** (0, 0) **9.** $(-4x_4, 2x_4, 7x_4, x_4)$, x_4 arbitrary
11. (0, 0) **13.** (0, 0, 0) **15.** $k = \frac{95}{11}$

Problems 6.5, page 399

1. $\begin{pmatrix} 3 & 9 \\ 6 & 15 \\ -3 & 6 \end{pmatrix}$ **3.** $\begin{pmatrix} 2 & 2 \\ -2 & -1 \\ 6 & -1 \end{pmatrix}$ **5.** $\begin{pmatrix} 0 & 0 \\ 0 & 0 \\ 0 & 0 \end{pmatrix}$ **7.** $\begin{pmatrix} -2 & 4 \\ 7 & 15 \\ -15 & 10 \end{pmatrix}$ **9.** $\begin{pmatrix} 4 & 10 \\ 17 & 22 \\ -9 & 1 \end{pmatrix}$

11. $\begin{pmatrix} 0 & 6 \\ 5 & 14 \\ -9 & 9 \end{pmatrix}$ **13.** $\begin{pmatrix} 1 & -5 & 0 \\ -3 & 4 & -5 \\ -14 & 13 & -1 \end{pmatrix}$ **15.** $\begin{pmatrix} 1 & 1 & 5 \\ 9 & 5 & 10 \\ 7 & -7 & 3 \end{pmatrix}$ **17.** $\begin{pmatrix} -1 & -1 & -1 \\ -3 & -3 & -10 \\ -7 & 3 & 5 \end{pmatrix}$

19. $\begin{pmatrix} -1 & -1 & -5 \\ -9 & -5 & -10 \\ -7 & 7 & -3 \end{pmatrix}$ **25.** $\begin{pmatrix} 0 & 1 & 0 & 1 & 0 \\ 1 & 0 & 1 & 1 & 0 \\ 0 & 1 & 0 & 0 & 1 \\ 1 & 1 & 0 & 0 & 0 \\ 0 & 0 & 1 & 0 & 0 \end{pmatrix}$

Problems 6.6, page 408

1. $\begin{pmatrix} 8 & 20 \\ -4 & 11 \end{pmatrix}$ **3.** $\begin{pmatrix} -3 & -3 \\ 1 & 3 \end{pmatrix}$ **5.** $\begin{pmatrix} 13 & 35 & 18 \\ 20 & 26 & 20 \end{pmatrix}$ **7.** $\begin{pmatrix} 19 & -17 & 34 \\ 8 & -12 & 20 \\ -8 & -11 & 7 \end{pmatrix}$

9. $\begin{pmatrix} 18 & 15 & 35 \\ 9 & 21 & 13 \\ 10 & 9 & 9 \end{pmatrix}$ **11.** (7 16) **13.** $\begin{pmatrix} 3 & 15 & -9 & 24 \\ -1 & -5 & 3 & -8 \\ 10 & 50 & -30 & 80 \\ 2 & 10 & -6 & 16 \end{pmatrix}$ **15.** $\begin{pmatrix} a & b & c \\ d & e & f \\ g & h & j \end{pmatrix}$

17. If $D = a_{11}a_{22} - a_{12}a_{21}$, then $\begin{pmatrix} b_{11} & b_{12} \\ b_{21} & b_{22} \end{pmatrix} = \begin{pmatrix} a_{22}/D & -a_{12}/D \\ -a_{21}/D & a_{11}/D \end{pmatrix}$

19. a. 3 in group 1, 4 in group 2, 5 in group 3 **b.** $\begin{pmatrix} 2 & 1 & 1 & 0 & 0 \\ 1 & 1 & 0 & 1 & 0 \\ 1 & 0 & 2 & 0 & 1 \end{pmatrix}$

21. **a.** $\begin{pmatrix} 80{,}000 & 45{,}000 & 40{,}000 \\ 50 & 20 & 10 \end{pmatrix}$ **b.** $\begin{pmatrix} 1 \\ 3 \\ 1 \end{pmatrix}$ **c.** money: 255,000; shares: 120 **23.** $\begin{pmatrix} 0 & -8 \\ 32 & 32 \end{pmatrix}$ **25.** $\begin{pmatrix} 11 & 38 \\ 57 & 106 \end{pmatrix}$

27. $A^2 = \begin{pmatrix} 0&0&1&0&0 \\ 0&0&0&1&0 \\ 0&0&0&0&1 \\ 0&0&0&0&0 \\ 0&0&0&0&0 \end{pmatrix}$; $A^3 = \begin{pmatrix} 0&0&0&1&0 \\ 0&0&0&0&1 \\ 0&0&0&0&0 \\ 0&0&0&0&0 \\ 0&0&0&0&0 \end{pmatrix}$; $A^4 = \begin{pmatrix} 0&0&0&0&1 \\ 0&0&0&0&0 \\ 0&0&0&0&0 \\ 0&0&0&0&0 \\ 0&0&0&0&0 \end{pmatrix}$; $A^5 = \begin{pmatrix} 0&0&0&0&0 \\ 0&0&0&0&0 \\ 0&0&0&0&0 \\ 0&0&0&0&0 \\ 0&0&0&0&0 \end{pmatrix}$

29. $PQ = \begin{pmatrix} \frac{11}{90} & \frac{41}{90} & \frac{19}{45} \\ \frac{11}{120} & \frac{71}{120} & \frac{19}{60} \\ \frac{1}{5} & \frac{1}{5} & \frac{3}{5} \end{pmatrix}$; all entries are nonnegative and $\frac{11}{90} + \frac{41}{90} + \frac{19}{45} = \frac{11}{120} + \frac{71}{120} + \frac{19}{60} = \frac{1}{5} + \frac{1}{5} + \frac{3}{5} = 1$.

33. **a.** player 2 > player 4 > player 1 > player 3 **b.** score = number of games won plus one-half the number of games that were won by each player that this given player beat

Problems 6.7, page 413

1. $\begin{pmatrix} 2 & -1 \\ 4 & 5 \end{pmatrix}\begin{pmatrix} x_1 \\ x_2 \end{pmatrix} = \begin{pmatrix} 3 \\ 7 \end{pmatrix}$ **3.** $\begin{pmatrix} 3 & 6 & -7 \\ 2 & -1 & 3 \end{pmatrix}\begin{pmatrix} x_1 \\ x_2 \\ x_3 \end{pmatrix} = \begin{pmatrix} 0 \\ 1 \end{pmatrix}$ **5.** $\begin{pmatrix} 0 & 1 & -1 \\ 1 & 0 & 1 \\ 3 & 2 & 0 \end{pmatrix}\begin{pmatrix} x_1 \\ x_2 \\ x_3 \end{pmatrix} = \begin{pmatrix} 7 \\ 2 \\ -5 \end{pmatrix}$

7. $\begin{aligned} x_1 + x_2 - x_3 &= 7 \\ 4x_1 - x_2 + 5x_3 &= 4 \\ 6x_1 + x_2 + 3x_3 &= 20 \end{aligned}$ **9.** $\begin{aligned} 2x_1 + x_3 &= 2 \\ -3x_1 + 4x_2 &= 3 \\ 5x_2 + 6x_3 &= 5 \end{aligned}$ **11.** $\begin{aligned} x_1 &= 2 \\ x_2 &= 3 \\ x_3 &= -5 \\ x_4 &= 6 \end{aligned}$ **13.** $\begin{aligned} 6x_1 + 2x_2 + x_3 &= 2 \\ -2x_1 + 3x_2 + x_3 &= 4 \\ 0x_1 + 0x_2 + 0x_3 &= 2 \end{aligned}$ **15.** $\begin{aligned} 7x_1 + 2x_2 &= 1 \\ 3x_1 + x_2 &= 2 \\ 6x_1 + 9x_2 &= 3 \end{aligned}$

17. The simplest solution to the nonhomogeneous equation is obtained by setting $x_2 = 0$. Then the general solution is $(2, 0) + x_2(3, 1)$; x_2 arbitrary.
19. If $x_3 = 0$, one nonhomogeneous solution is $(2, 0, 0)$ and the general solution is $(2, 0, 0) + x_3(-\frac{1}{3}, -\frac{4}{3}, 1)$; x_3 arbitrary.
21. If $x_3 = x_4 = 0$, one nonhomogeneous solution is $(-1, 4, 0, 0)$ and the general solution is $(-1, 4, 0, 0) + x_3(-3, 4, 1, 0) + x_4(5, -7, 0, 1)$.
23. $(c_1y_1 + c_2y_2)'' + a(x)(c_1y_1 + c_2y_2)' + b(x)(c_1y_1 + c_2y_2) = c_1y_1'' + c_2y_2'' + a(x)c_1y_1' + a(x)c_2y_2' + b(x)c_1y_1 + b(x)c_2y_2 = c_1(y_1'' + a(x)y_1' + b(x)y_1) + c_2(y_2'' + a(x)y_2' + b(x)y_2) = c_1 \cdot 0 + c_2 \cdot 0 = 0$, since y_1 and y_2 solve (7).

Problems 6.8, page 429

1. $\begin{pmatrix} 2 & -1 \\ -3 & 2 \end{pmatrix}$ **3.** $\begin{pmatrix} 0 & 1 \\ 1 & 0 \end{pmatrix}$ **5.** not invertible **7.** $\begin{pmatrix} \frac{1}{3} & -\frac{1}{3} & -\frac{1}{3} \\ 0 & \frac{1}{2} & 1 \\ 0 & 0 & -1 \end{pmatrix}$ **9.** not invertible

11. not invertible **13.** $\begin{pmatrix} \frac{7}{3} & -\frac{1}{3} & -\frac{1}{3} & -\frac{2}{3} \\ \frac{4}{9} & -\frac{1}{9} & -\frac{4}{9} & \frac{1}{9} \\ -\frac{1}{9} & -\frac{2}{9} & \frac{1}{9} & \frac{2}{9} \\ -\frac{5}{3} & \frac{2}{3} & \frac{2}{3} & \frac{1}{3} \end{pmatrix}$ **15.** $\begin{pmatrix} 0 & 1 & 0 & 2 \\ 1 & -1 & -2 & 2 \\ 0 & 1 & 3 & -3 \\ -2 & 2 & 3 & -2 \end{pmatrix}$

23. $\begin{pmatrix} \sin\theta & \cos\theta & 0 \\ \cos\theta & -\sin\theta & 0 \\ 0 & 0 & 1 \end{pmatrix}$ is its own inverse (since $\sin^2\theta + \cos^2\theta = 1$). **27.** $\begin{pmatrix} \frac{1}{2} & -\frac{1}{6} & \frac{7}{30} \\ 0 & \frac{1}{3} & -\frac{4}{15} \\ 0 & 0 & \frac{1}{5} \end{pmatrix}$

31. any nonzero multiple of $\begin{pmatrix} 1 \\ 2 \end{pmatrix}$ **33.** 3 chairs and 2 tables **35.** 4 units of A and 5 units of B

37. **a.** $A = \begin{pmatrix} 0.293 & 0 & 0 \\ 0.014 & 0.207 & 0.017 \\ 0.044 & 0.010 & 0.216 \end{pmatrix}$; $I - A = \begin{pmatrix} 0.707 & 0 & 0 \\ -0.014 & 0.793 & -0.017 \\ -0.044 & -0.010 & 0.784 \end{pmatrix}$ **b.** $\begin{pmatrix} 18{,}689 \\ 22{,}598 \\ 3{,}615 \end{pmatrix}$

39. $\begin{pmatrix} 1 & \frac{1}{2} \\ 0 & 1 \end{pmatrix}$; yes **41.** $\begin{pmatrix} 1 & \frac{2}{3} & \frac{1}{3} \\ 0 & 1 & 1 \\ 0 & 0 & 1 \end{pmatrix}$; yes **43.** $\begin{pmatrix} 1 & -\frac{1}{2} & 2 \\ 0 & 1 & -14 \\ 0 & 0 & 0 \end{pmatrix}$; no

45. $\begin{pmatrix} 1 & 0 & 2 & 3 \\ 0 & 1 & 2 & 7 \\ 0 & 0 & 1 & \frac{10}{7} \\ 0 & 0 & 0 & 0 \end{pmatrix}$; no

Problems 6.9, page 434

1. $\begin{pmatrix} -1 & 6 \\ 4 & 5 \end{pmatrix}$ **3.** $\begin{pmatrix} 2 & -1 & 1 \\ 3 & 2 & 4 \end{pmatrix}$ **5.** $\begin{pmatrix} 1 & -1 & 1 \\ 2 & 0 & 5 \\ 3 & 4 & 5 \end{pmatrix}$ **7.** $\begin{pmatrix} 1 & 0 \\ 0 & 1 \\ 1 & 0 \\ 0 & 1 \end{pmatrix}$ **9.** $\begin{pmatrix} a & d & g \\ b & e & h \\ c & f & i \end{pmatrix}$

27. $\begin{pmatrix} 2 & -3 \\ -1 & 2 \end{pmatrix}$ **29.** $\begin{pmatrix} \frac{13}{8} & -\frac{15}{8} & \frac{5}{4} \\ -\frac{1}{2} & \frac{1}{2} & 0 \\ -\frac{1}{8} & \frac{3}{8} & -\frac{1}{4} \end{pmatrix}$

Problems 6.10, page 442

1. yes, $R_1 \rightleftarrows R_2$ **3.** no [two operations are used: $R_1 \rightleftarrows R_2$ followed by $R_2 \to R_2 + R_1$]
5. no [two operations are used: $R_1 \to 3R_1$ and $R_2 \to 3R_2$]
7. no [two operations are used: $R_1 \rightleftarrows R_3$ followed by $R_1 \rightleftarrows R_2$]

9. yes, $R_2 \to R_2 + 2R_1$ **11.** no [two operations are used: $R_2 \to R_2 + R_1$ and $R_4 \to R_4 + R_3$] **13.** $\begin{pmatrix} 1 & 0 & 0 \\ 0 & 4 & 0 \\ 0 & 0 & 1 \end{pmatrix}$

15. $\begin{pmatrix} 1 & -3 & 0 \\ 0 & 1 & 0 \\ 0 & 0 & 1 \end{pmatrix}$ **17.** $\begin{pmatrix} 0 & 0 & 1 \\ 0 & 1 & 0 \\ 1 & 0 & 0 \end{pmatrix}$ **19.** $\begin{pmatrix} 1 & 0 & 0 \\ 0 & 1 & 1 \\ 0 & 0 & 1 \end{pmatrix}$ **21.** $\begin{pmatrix} 1 & 0 \\ 0 & -2 \end{pmatrix}$ **23.** $\begin{pmatrix} 1 & 2 \\ 0 & 1 \end{pmatrix}$

25. $\begin{pmatrix} 0 & 0 & 1 \\ 0 & 1 & 0 \\ 1 & 0 & 0 \end{pmatrix}$ **27.** $\begin{pmatrix} -1 & 0 & 0 \\ 0 & 1 & 0 \\ 0 & 0 & 1 \end{pmatrix}$ **29.** $\begin{pmatrix} 1 & 0 & 0 \\ 0 & 1 & 0 \\ -5 & 0 & 1 \end{pmatrix}$ **31.** $\begin{pmatrix} 0 & 1 \\ 1 & 0 \end{pmatrix}$ **33.** $\begin{pmatrix} 1 & 0 \\ 0 & \frac{1}{4} \end{pmatrix}$

35. $\begin{pmatrix} 1 & 2 & 0 \\ 0 & 1 & 0 \\ 0 & 0 & 1 \end{pmatrix}$ **37.** $\begin{pmatrix} 1 & 0 & 0 \\ 0 & -2 & 0 \\ 0 & 0 & 1 \end{pmatrix}$ **39.** $\begin{pmatrix} 1 & 0 & 0 & -5 \\ 0 & 1 & 0 & 0 \\ 0 & 0 & 1 & 0 \\ 0 & 0 & 0 & 1 \end{pmatrix}$ **41.** $\begin{pmatrix} 2 & 0 \\ 0 & 1 \end{pmatrix}\begin{pmatrix} 1 & 0 \\ 3 & 1 \end{pmatrix}\begin{pmatrix} 1 & 0 \\ 0 & \frac{1}{2} \end{pmatrix}\begin{pmatrix} 1 & \frac{1}{2} \\ 0 & 1 \end{pmatrix}$

43. $\begin{pmatrix} 1 & 0 & 0 \\ 0 & 1 & 0 \\ 5 & 0 & 1 \end{pmatrix}\begin{pmatrix} 1 & 0 & 0 \\ 0 & 2 & 0 \\ 0 & 0 & 1 \end{pmatrix}\begin{pmatrix} 1 & 1 & 0 \\ 0 & 1 & 0 \\ 0 & 0 & 1 \end{pmatrix}\begin{pmatrix} 1 & 0 & 0 \\ 0 & 1 & 0 \\ 0 & 0 & -4 \end{pmatrix}\begin{pmatrix} 1 & 0 & -\frac{1}{2} \\ 0 & 1 & 0 \\ 0 & 0 & 1 \end{pmatrix}\begin{pmatrix} 1 & 0 & 0 \\ 0 & 1 & \frac{3}{2} \\ 0 & 0 & 1 \end{pmatrix}$

45. $\begin{pmatrix} 0 & 0 & 1 \\ 0 & 1 & 0 \\ 1 & 0 & 0 \end{pmatrix}\begin{pmatrix} 1 & 0 & 0 \\ 0 & 1 & 0 \\ 0 & -1 & 1 \end{pmatrix}\begin{pmatrix} 1 & 0 & 0 \\ 0 & 1 & 0 \\ 0 & 0 & -1 \end{pmatrix}\begin{pmatrix} 1 & 0 & 1 \\ 0 & 1 & 0 \\ 0 & 0 & 1 \end{pmatrix}\begin{pmatrix} 1 & 0 & 0 \\ 0 & 1 & -1 \\ 0 & 0 & 1 \end{pmatrix}$

47. $\begin{pmatrix} 2 & 0 & 0 & 0 \\ 0 & 1 & 0 & 0 \\ 0 & 0 & 1 & 0 \\ 0 & 0 & 0 & 1 \end{pmatrix}\begin{pmatrix} 1 & 0 & 0 & 0 \\ 0 & 3 & 0 & 0 \\ 0 & 0 & 1 & 0 \\ 0 & 0 & 0 & 1 \end{pmatrix}\begin{pmatrix} 1 & 0 & 0 & 0 \\ 0 & 1 & 0 & 0 \\ 0 & 0 & -4 & 0 \\ 0 & 0 & 0 & 1 \end{pmatrix}\begin{pmatrix} 1 & 0 & 0 & 0 \\ 0 & 1 & 0 & 0 \\ 0 & 0 & 1 & 0 \\ 0 & 0 & 0 & 5 \end{pmatrix}$

49. $\begin{pmatrix} a & 0 \\ 0 & 1 \end{pmatrix}\begin{pmatrix} 1 & 0 \\ 0 & c \end{pmatrix}\begin{pmatrix} 1 & b/a \\ 0 & 1 \end{pmatrix}$; the first two matrices are elementary because $a \neq 0$ and $c \neq 0$

57. $\begin{pmatrix} 1 & 0 \\ 2 & 1 \end{pmatrix}\begin{pmatrix} 1 & 2 \\ 0 & 0 \end{pmatrix}$ **59.** $\begin{pmatrix} 0 & 1 \\ 1 & 0 \end{pmatrix}\begin{pmatrix} 1 & 0 \\ 0 & 0 \end{pmatrix}$ **61.** $\begin{pmatrix} 1 & 0 & 0 \\ 0 & 1 & 0 \\ 1 & 0 & 1 \end{pmatrix}\begin{pmatrix} 1 & 0 & 0 \\ 0 & -3 & 0 \\ 0 & 0 & 1 \end{pmatrix}\begin{pmatrix} 1 & 0 & 0 \\ 0 & 1 & 0 \\ 0 & 3 & 1 \end{pmatrix}\begin{pmatrix} 1 & -3 & 3 \\ 0 & 1 & -\frac{1}{3} \\ 0 & 0 & 0 \end{pmatrix}$

Chapter 6—Review Exercises, page 446

1. $(\frac{1}{7}, \frac{10}{7})$ **3.** no solution **5.** $(0, 0, 0)$ **7.** $(-\frac{1}{2}, 0, \frac{5}{2})$ **9.** $(\frac{1}{3}x_3, \frac{7}{3}x_3, x_3)$, x_3 arbitrary

11. no solution **13.** $(0, 0, 0, 0)$ **15.** $\begin{pmatrix} -6 & 3 \\ 0 & 12 \\ 6 & 9 \end{pmatrix}$ **17.** $\begin{pmatrix} 16 & 2 & 3 \\ -20 & 10 & -1 \\ -36 & 8 & 16 \end{pmatrix}$ **19.** $\begin{pmatrix} 17 & 39 & 41 \\ 14 & 20 & 42 \end{pmatrix}$

21. $\begin{pmatrix} 9 & 10 \\ 30 & 32 \end{pmatrix}$ **23.** reduced row echelon form **25.** neither

27. row echelon form: $\begin{pmatrix} 1 & 4 & -1 \\ 0 & 1 & \frac{5}{4} \end{pmatrix}$; reduced row echelon form: $\begin{pmatrix} 1 & 0 & -6 \\ 0 & 1 & \frac{5}{4} \end{pmatrix}$

29. $\begin{pmatrix} 1 & \frac{3}{2} \\ 0 & 1 \end{pmatrix}$; inverse is $\begin{pmatrix} \frac{4}{11} & -\frac{3}{11} \\ \frac{1}{11} & \frac{2}{11} \end{pmatrix}$

31. $\begin{pmatrix} 1 & 2 & 0 \\ 0 & 1 & \frac{1}{3} \\ 0 & 0 & 1 \end{pmatrix}$; inverse is $\begin{pmatrix} -\frac{1}{4} & \frac{1}{4} & \frac{1}{4} \\ \frac{5}{8} & -\frac{1}{8} & -\frac{1}{8} \\ \frac{1}{8} & -\frac{5}{8} & \frac{3}{8} \end{pmatrix}$ **33.** $\begin{pmatrix} 1 & 0 & 2 \\ 0 & 1 & 1 \\ 0 & 0 & 1 \end{pmatrix}$; inverse is $\begin{pmatrix} \frac{5}{6} & \frac{2}{3} & -2 \\ \frac{1}{3} & \frac{2}{3} & -1 \\ -\frac{1}{6} & -\frac{1}{3} & 1 \end{pmatrix}$

35. $\begin{pmatrix} 1 & 2 & 0 \\ 2 & 1 & -1 \\ 3 & 1 & 1 \end{pmatrix}\begin{pmatrix} x_1 \\ x_2 \\ x_3 \end{pmatrix} = \begin{pmatrix} 3 \\ -1 \\ 7 \end{pmatrix}$; A^{-1} is given in Exercise 31: $x_1 = \frac{3}{4}$, $x_2 = \frac{9}{8}$, $x_3 = \frac{29}{8}$

37. $\begin{pmatrix} 2 & -1 \\ 3 & 0 \\ 1 & 2 \end{pmatrix}$; neither **39.** $\begin{pmatrix} 2 & 3 & 1 \\ 3 & -6 & -5 \\ 1 & -5 & 9 \end{pmatrix}$; symmetric **41.** $\begin{pmatrix} 1 & -1 & 4 & 6 \\ -1 & 2 & 5 & 7 \\ 4 & 5 & 3 & -8 \\ 6 & 7 & -8 & 9 \end{pmatrix}$; symmetric

43. $\begin{pmatrix} 1 & 0 & 0 \\ 0 & -2 & 0 \\ 0 & 0 & 1 \end{pmatrix}$ **45.** $\begin{pmatrix} 1 & 0 & 0 \\ 0 & 1 & 0 \\ -5 & 0 & 1 \end{pmatrix}$ **47.** $\begin{pmatrix} 1 & 0 & 0 \\ 0 & 1 & \frac{1}{5} \\ 0 & 0 & 1 \end{pmatrix}$ **49.** $\begin{pmatrix} 0 & 1 & 0 \\ 1 & 0 & 0 \\ 0 & 0 & 1 \end{pmatrix}$

51. $\begin{pmatrix} 2 & 0 \\ 0 & 1 \end{pmatrix}\begin{pmatrix} 1 & 0 \\ -1 & 1 \end{pmatrix}\begin{pmatrix} 1 & 0 \\ 0 & \frac{1}{2} \end{pmatrix}\begin{pmatrix} 1 & -\frac{1}{2} \\ 0 & 1 \end{pmatrix}$ **53.** $\begin{pmatrix} 2 & 0 \\ 0 & 1 \end{pmatrix}\begin{pmatrix} 1 & 0 \\ -4 & 1 \end{pmatrix}\begin{pmatrix} 1 & -\frac{1}{2} \\ 0 & 0 \end{pmatrix}$

CHAPTER 7

Problems 7.1, page 456

1. -10 **3.** 47 **5.** 4 **7.** 56 **9.** 274

Problems 7.2, page 471

1. 28 **3.** 2 **5.** 32 **7.** -36 **9.** -260 **11.** -183 **13.** 24 **15.** -296 **17.** 138

19. *abcde* **21.** -8 **23.** 16 **25.** -16 **27.** -16 **37. a.** $D_n = \begin{vmatrix} 1 & 1 & \cdots & 1 \\ a_1 & a_2 & \cdots & a_n \\ a_1^2 & a_2^2 & \cdots & a_n^2 \\ \vdots & \vdots & & \vdots \\ a_1^{n-1} & a_2^{n-1} & \cdots & a_n^{n-1} \end{vmatrix}$

Problems 7.3, page 478

1. $\begin{pmatrix} \frac{1}{2} & -\frac{1}{2} \\ -\frac{1}{4} & \frac{3}{4} \end{pmatrix}$ **3.** $\begin{pmatrix} 0 & 1 \\ 1 & 0 \end{pmatrix}$ **5.** $\begin{pmatrix} \frac{1}{3} & -\frac{1}{4} & -\frac{1}{6} \\ 0 & \frac{1}{4} & \frac{1}{2} \\ 0 & \frac{1}{4} & -\frac{1}{2} \end{pmatrix}$ **7.** $\begin{pmatrix} 0 & 1 & -1 \\ 2 & -2 & -1 \\ -1 & 1 & 1 \end{pmatrix}$ **9.** not invertible

11. $\begin{pmatrix} \frac{7}{3} & -\frac{1}{3} & -\frac{1}{3} & -\frac{2}{3} \\ \frac{4}{9} & -\frac{1}{9} & -\frac{4}{9} & \frac{1}{9} \\ -\frac{1}{9} & -\frac{2}{9} & \frac{1}{9} & \frac{2}{9} \\ -\frac{5}{3} & \frac{2}{3} & \frac{2}{3} & \frac{1}{3} \end{pmatrix}$ **15.** $A^{-1} = \begin{pmatrix} \frac{1}{14} & \frac{1}{14} & \frac{9}{28} \\ -\frac{5}{7} & \frac{2}{7} & -\frac{3}{14} \\ \frac{1}{14} & \frac{1}{14} & -\frac{5}{28} \end{pmatrix}$, $\det A = -28$, $\det A^{-1} = -\frac{1}{28}$

17. no inverse if α is any real number

Problems 7.4, page 482

1. $x_1 = -5, x_2 = 3$ **3.** $x_1 = 2, x_2 = 5, x_3 = -3$ **5.** $x_1 = \frac{45}{13}, x_2 = -\frac{11}{13}, x_3 = \frac{23}{13}$ **7.** $x_1 = \frac{3}{2}, x_2 = \frac{3}{2}, x_3 = \frac{1}{2}$
9. $x_1 = \frac{21}{29}, x_2 = \frac{171}{29}, x_3 = -\frac{284}{29}, x_4 = -\frac{182}{29}$

Chapter 7—Review Exercises, page 484

1. -4 **3.** 24 **5.** 60 **7.** 34 **9.** $\begin{pmatrix} -\frac{1}{11} & \frac{4}{11} \\ \frac{2}{11} & \frac{3}{11} \end{pmatrix}$ **11.** not invertible

13. $\begin{pmatrix} \frac{1}{11} & \frac{1}{11} & 0 & \frac{3}{11} \\ \frac{9}{11} & -\frac{2}{11} & 0 & -\frac{6}{11} \\ \frac{3}{11} & \frac{3}{11} & 0 & -\frac{2}{11} \\ \frac{1}{22} & \frac{1}{22} & -\frac{1}{2} & \frac{3}{22} \end{pmatrix}$ **15.** $x_1 = \frac{11}{7}, x_2 = \frac{1}{7}$ **17.** $x_1 = \frac{1}{4}, x_2 = \frac{5}{4}, x_3 = -\frac{3}{4}$

CHAPTER 8

Problems 8.1, page 491

1. yes **3.** no; (iv); also (vi) does not hold if $\alpha < 0$ **5.** yes **7.** yes
9. no; (i), (iii), (iv), (vi) do not hold **11.** yes **13.** yes **15.** no; (i), (iii), (iv), (vi) do not hold
17. yes **19.** yes

Problems 8.2, page 496

1. no; because $\alpha(x, y) \notin H$ if $\alpha < 0$ **3.** yes **5.** yes **7.** yes **9.** yes **11.** yes **13.** yes
15. no; the zero polynomial $\notin H$ **17.** no; the function $f(x) \equiv 0 \notin V$ **19.** yes

21. b. $H = H_1 \cap H_2 = \left\{ A \in M_{22};\ A = \begin{pmatrix} 0 & a \\ a & 0 \end{pmatrix} \text{ for some scalar } a \right\}$.

Problems 8.3, page 502

1. yes **3.** no; for example, $\begin{pmatrix} 1 \\ 2 \end{pmatrix} \notin$ span of the three vectors $\left[\text{each is a multiple of } \begin{pmatrix} 1 \\ 1 \end{pmatrix} \right]$.

5. yes **7.** yes **9.** no; for example $x \notin \text{span}\ \{1 - x, 3 - x^2\}$ **11.** yes **13.** yes

Problems 8.4, page 512

1. independent **3.** dependent; $-2\begin{pmatrix} 2 \\ -1 \\ 4 \end{pmatrix} + \begin{pmatrix} 4 \\ -2 \\ 8 \end{pmatrix} = \begin{pmatrix} 0 \\ 0 \\ 0 \end{pmatrix}$ **5.** dependent (from Theorem 2)

7. independent **9.** independent **11.** independent **13.** independent **15.** independent
17. dependent **19.** independent **21.** independent **23.** $ad - bc = 0$ **25.** $\alpha = -\frac{13}{2}$

33. $x_2\begin{pmatrix}-1\\1\\0\end{pmatrix}+x_3\begin{pmatrix}-1\\0\\1\end{pmatrix}$ **35.** $x_3\begin{pmatrix}13\\-6\\1\end{pmatrix}$ **37.** $x_2\begin{pmatrix}-2\\1\\0\\0\end{pmatrix}+x_3\begin{pmatrix}3\\0\\1\\0\end{pmatrix}+x_4\begin{pmatrix}-5\\0\\0\\1\end{pmatrix}$

39. Any nonzero **u** will lead to a similar result. For example, let $\mathbf{u}=(1,0,0)$. Then $\mathbf{x}=(0,1,0)$ and $\mathbf{y}=(0,0,1)$ are in H. $\mathbf{w}=\mathbf{x}\times\mathbf{y}=(1,0,0)=\mathbf{u}$. **e.** H is a plane orthogonal to **u** and **w** is orthogonal to this plane and so it must be parallel to **u**.

57. $\begin{pmatrix}2\\1\\2\end{pmatrix},\begin{pmatrix}-1\\3\\4\end{pmatrix},\begin{pmatrix}1\\2\\2\end{pmatrix}$. (There are many choices for the third vector.)

59. b. Think of the system in (a) as a homogeneous system of three equations in the three unknowns a, b, and c. Since the system has nontrivial solutions, its determinant is 0.

Problems 8.5, page 523

1. no; does not span **3.** no; dependent **5.** no; does not span **7.** yes **9.** yes

11. $\left\{\begin{pmatrix}0\\1\\-1\end{pmatrix},\begin{pmatrix}1\\0\\2\end{pmatrix}\right\}$ **13.** $\left\{\begin{pmatrix}2\\3\\4\end{pmatrix}\right\}$ **19.** $\left\{\begin{pmatrix}1\\1\end{pmatrix}\right\}$ **21.** $\left\{\begin{pmatrix}-2\\-3\\1\end{pmatrix}\right\}$ **23.** $\left\{\begin{pmatrix}3\\1\\0\end{pmatrix},\begin{pmatrix}-2\\0\\1\end{pmatrix}\right\}$

25. n

37. $B_1=\left\{\begin{pmatrix}1\\0\\1\\0\end{pmatrix},\begin{pmatrix}0\\1\\0\\1\end{pmatrix},\begin{pmatrix}1\\0\\0\\0\end{pmatrix},\begin{pmatrix}0\\1\\0\\0\end{pmatrix}\right\}$; $B_2=\left\{\begin{pmatrix}1\\0\\1\\0\end{pmatrix},\begin{pmatrix}0\\1\\0\\1\end{pmatrix},\begin{pmatrix}0\\0\\1\\0\end{pmatrix},\begin{pmatrix}0\\0\\0\\1\end{pmatrix}\right\}$ There are infinitely many other choices.

Problems 8.6, page 534

1. $\rho=2,\ \nu=0$ **3.** $\rho=1,\ \nu=2$ **5.** $\rho=2,\ \nu=1$ **7.** $\rho=2,\ \nu=2$ **9.** $\rho=2,\ \nu=0$

11. $\rho=2,\ \nu=2$ **13.** $\rho=3,\ \nu=1$ **15.** $\rho=2,\ \nu=1$

17. range basis $=\left\{\begin{pmatrix}1\\3\\5\end{pmatrix},\begin{pmatrix}-1\\1\\-1\end{pmatrix}\right\}$; these are the first two columns of A. kernel basis $=\left\{\begin{pmatrix}-\frac{3}{2}\\\frac{1}{2}\\1\end{pmatrix}\right\}$

19. range basis $=\left\{\begin{pmatrix}1\\0\\1\end{pmatrix},\begin{pmatrix}-1\\1\\0\end{pmatrix},\begin{pmatrix}3\\3\\5\end{pmatrix}\right\}$; these are the first three (linearly independent) columns of A corresponding to pivots in the row echelon form of the matrix. kernel basis $=\left\{\begin{pmatrix}-6\\-4\\1\\0\end{pmatrix}\right\}$

21. range basis $=\left\{\begin{pmatrix}1\\-2\\2\\3\end{pmatrix}\right\}$; kernel basis $=\left\{\begin{pmatrix}1\\1\\0\\0\end{pmatrix},\begin{pmatrix}-2\\0\\1\\0\end{pmatrix},\begin{pmatrix}-3\\0\\0\\1\end{pmatrix}\right\}$

23. $\left\{\begin{pmatrix}1\\4\\-2\end{pmatrix},\begin{pmatrix}0\\1\\-\frac{6}{7}\end{pmatrix}\right\}$ **25.** $\{(1,0,0,\frac{1}{2}),(0,1,-1,\frac{3}{2}),(0,0,0,1)\}$ **27.** no **29.** yes

Problems 8.7, page 549

1. $\begin{pmatrix} 1/\sqrt{2} \\ 1/\sqrt{2} \end{pmatrix}, \begin{pmatrix} -1/\sqrt{2} \\ 1/\sqrt{2} \end{pmatrix}$

3. i. If $a = b = 0$, $\{(1, 0), (0, 1)\}$ **ii.** If $a = 0, b \neq 0$, $\{(1, 0)\}$ **iii.** If $a \neq 0, b = 0$, $\{(0, 1)\}$
iv. If $a \neq 0, b \neq 0$, $\{(b/\sqrt{a^2 + b^2}, -a/\sqrt{a^2 + b^2})\}$

5. $\{(1/\sqrt{5}, 0, 2/\sqrt{5}), (2/\sqrt{30}, 5/\sqrt{30}, -1/\sqrt{30})\}$ **7.** $\{(2/\sqrt{29}, 3/\sqrt{29}, 4/\sqrt{29})\}$

9. $\{(1/\sqrt{5}, 0, 0, 2/\sqrt{5}), (2/\sqrt{30}, 5/\sqrt{30}, 0, -1/\sqrt{30}), (-2/\sqrt{10}, 1/\sqrt{10}, 2/\sqrt{10}, 1/\sqrt{10})\}$

11. $\{(a/\sqrt{a^2 + b^2 + c^2}, b/\sqrt{a^2 + b^2 + c^2}, c/\sqrt{a^2 + b^2 + c^2})\}$ **13.** $\{(-7/\sqrt{66}, -1/\sqrt{66}, 4/\sqrt{66})\}$

23. a. $\mathbf{0}$ **b.** $\dfrac{1}{\sqrt{a^2 + b^2}}\begin{pmatrix} a \\ b \end{pmatrix}$ **c.** $\mathbf{v} = \begin{pmatrix} 0 \\ 0 \end{pmatrix} + \begin{pmatrix} a \\ b \end{pmatrix}$

25. a. $\dfrac{1}{49}\begin{pmatrix} -186 \\ 75 \\ 118 \end{pmatrix}$ **b.** $\dfrac{1}{7}\begin{pmatrix} 3 \\ -2 \\ 6 \end{pmatrix}$ **c.** $\mathbf{v} = \dfrac{1}{49}\begin{pmatrix} -186 \\ 75 \\ 118 \end{pmatrix} + \dfrac{13}{49}\begin{pmatrix} 3 \\ -2 \\ 6 \end{pmatrix}$

27. a. $\dfrac{1}{5}\begin{pmatrix} 1 \\ -3 \\ 4 \\ 17 \end{pmatrix}$ **b.** $\dfrac{1}{\sqrt{15}}\begin{pmatrix} 2 \\ -1 \\ 3 \\ -1 \end{pmatrix}$ **c.** $\dfrac{1}{5}\begin{pmatrix} 1 \\ -3 \\ 4 \\ 17 \end{pmatrix} + \dfrac{2}{5}\begin{pmatrix} 2 \\ -1 \\ 3 \\ -1 \end{pmatrix}$ **31.** $a^2 + b^2 = 1$

Problems 8.8, page 557

1. $y = \frac{408}{126} - \frac{57}{126}x \approx 3.24 - 0.45x$ **3.** $y = \frac{162}{84} - \frac{10}{84}x \approx 1.93 - 0.12x$

5. $y = \frac{13536}{5184} + \frac{10800}{5184}x + \frac{1584}{5184}x^2 \approx 2.61 + 2.08x + 0.31x^2$. This is the equation of the parabola passing through the three points.

11. $y \approx 108.71 + 4.906x - 0.00973x^2$

Problems 8.9, page 564

1. linear **3.** linear **5.** not linear, since $T\left(\alpha\begin{pmatrix} x \\ y \\ z \end{pmatrix}\right) = T\begin{pmatrix} \alpha x \\ \alpha y \\ \alpha z \end{pmatrix} = \begin{pmatrix} 1 \\ \alpha z \end{pmatrix}$ while $\alpha T\begin{pmatrix} x \\ y \\ z \end{pmatrix} = \alpha\begin{pmatrix} 1 \\ z \end{pmatrix} = \begin{pmatrix} \alpha \\ \alpha z \end{pmatrix}$

7. linear **9.** not linear, since $T\left(\alpha\begin{pmatrix} x \\ y \end{pmatrix}\right) = T\begin{pmatrix} \alpha x \\ \alpha y \end{pmatrix} = (\alpha x)(\alpha y) = \alpha^2 xy$ while $\alpha T\begin{pmatrix} x \\ y \end{pmatrix} = \alpha xy$ **11.** linear

13. not linear, since $T\left(\alpha\begin{pmatrix} x \\ y \\ z \\ w \end{pmatrix}\right) = \alpha^2 T\begin{pmatrix} x \\ y \\ z \\ w \end{pmatrix} \neq \alpha T\begin{pmatrix} x \\ y \\ z \\ w \end{pmatrix}$ if $\alpha \neq 1$ or 0

15. not linear, since $T(A + B) = (A + B)^t(A + B) = (A^t + B^t)(A + B) = A^tA + A^tB + B^tA + B^tB$ but $T(A) + T(B) = A^tA + B^tB \neq T(A + B)$ unless $A^tB + B^tA = 0$.

17. not linear, since $T(\alpha D) = (\alpha D)^2 = \alpha^2 D^2 \neq \alpha T(D) = \alpha D^2$ unless $\alpha = 1$ or 0. **19.** linear **21.** linear

23. not linear, since $T(f + g) = (f + g)^2 \neq f^2 + g^2 = T(f) + T(g)$ **25.** linear **27.** linear

29. not linear, since $T(\alpha A) = \det(\alpha A) = \alpha^n \det A \neq \alpha \det A = \alpha T(A)$ unless $\alpha = 0$ or 1. [$\det \alpha A = \alpha^n \det A$ by Problem 7.2.28 on page 472.] Also, in general $\det(A + B) \neq \det A + \det B$

31. a. $\begin{pmatrix} -14 \\ 4 \\ 26 \end{pmatrix}$ **b.** $\begin{pmatrix} -31 \\ -6 \\ 26 \end{pmatrix}$

33. It rotates a vector counterclockwise around the z-axis through an angle of θ in a plane parallel to the xy-plane.

Problems 8.10, page 570

1. kernel $= \{(0, y): y \in \mathbb{R}\}$, that is, the y-axis; range $= \{(x, 0): x \in \mathbb{R}\}$, that is, the x-axis; $\rho(T) = \nu(T) = 1$

3. kernel $= \{(x, -x): x \in \mathbb{R}\}$—this is the line $x + y = 0$; range $= \mathbb{R}$; $\rho(T) = \nu(T) = 1$.

5. kernel $= \left\{\begin{pmatrix} 0 & 0 \\ 0 & 0 \end{pmatrix}\right\}$; range $= M_{22}$; $\rho(T) = 4$, $\nu(T) = 0$

7. kernel $= \{A: A^t = -A\} = \{A: A$ is skew-symmetric$\}$; range $= \{A: A$ is symmetric$\}$; $\rho(T) = (n^2 + n)/2$; $\nu(T) = (n^2 - n)/2$

9. kernel $= \{f \in C[0, 1]: f(\frac{1}{2}) = 0\}$; range $= \mathbb{R}$, $\rho(T) = 1$; the kernel is an infinite dimensional space so that $\nu(T) = \infty$. For example, the linearly independent functions $x - \frac{1}{2}, (x - \frac{1}{2})^2, (x - \frac{1}{2})^3, (x - \frac{1}{2})^4, \ldots, (x - \frac{1}{2})^n, \ldots$ all satisfy $f(\frac{1}{2}) = 0$.

15. $T\mathbf{x} = A\mathbf{x}$, where $A = \begin{pmatrix} 0 & a \\ b & c \end{pmatrix}$, a, b, c real

17. $T\mathbf{x} = A\mathbf{x}$, where $A = \begin{pmatrix} 2 & -1 & 1 \\ 2 & -1 & 1 \\ 2 & -1 & 1 \end{pmatrix}$

21. Let $T_{ij}(\mathbf{u}_i) = \mathbf{w}_j$ and $T_{ij}(\mathbf{u}_k) = \mathbf{0}$ if $k \neq i$. These form a basis for $L(V, W)$, so $\dim L(V, W) = nm$.

23. False. Let S and T: $\mathbb{R}^2 \to \mathbb{R}^2$ be given by $S(\mathbf{x}) = A\mathbf{x}$ and $T(\mathbf{x}) = B\mathbf{x}$, where $A = \begin{pmatrix} 0 & 1 \\ 0 & 0 \end{pmatrix}$ and $B = \begin{pmatrix} 1 & 0 \\ 0 & 0 \end{pmatrix}$. Then

$ST(\mathbf{x}) = AB\mathbf{x} = \begin{pmatrix} 0 & 0 \\ 0 & 0 \end{pmatrix}\mathbf{x} = \mathbf{0}$. However, $TS(\mathbf{x})$ is not the zero transformation because

$BA = \begin{pmatrix} 0 & 1 \\ 0 & 0 \end{pmatrix} \neq$ the zero matrix.

Problems 8.11, page 579

1. $\begin{pmatrix} 1 & -2 \\ -1 & 1 \end{pmatrix}$; $\ker T = \{\mathbf{0}\}$; range $T = \mathbb{R}^2$; $\nu(T) = 0$, $\rho(T) = 2$

3. $\begin{pmatrix} 1 & -1 & 1 \\ -2 & 2 & -2 \end{pmatrix}$; range $T = \text{span}\left\{\begin{pmatrix} 1 \\ -2 \end{pmatrix}\right\}$; $\ker T = \text{span}\left\{\begin{pmatrix} 1 \\ 1 \\ 0 \end{pmatrix}, \begin{pmatrix} 0 \\ 1 \\ 1 \end{pmatrix}\right\}$; $\rho(T) = 1$, $\nu(T) = 2$

5. $\begin{pmatrix} 1 & -1 & 2 \\ 3 & 1 & 4 \\ 5 & -1 & 8 \end{pmatrix}$; range $T = \text{span}\left\{\begin{pmatrix} 1 \\ 3 \\ 5 \end{pmatrix}, \begin{pmatrix} -1 \\ 1 \\ -1 \end{pmatrix}\right\}$; $\ker T = \text{span}\left\{\begin{pmatrix} -3 \\ 1 \\ 2 \end{pmatrix}\right\}$; $\rho(T) = 2$, $\nu(T) = 1$

7. $\begin{pmatrix} 1 & -1 & 2 & 3 \\ 0 & 1 & 4 & 3 \\ 1 & 0 & 6 & 6 \end{pmatrix}$; range $T = \text{span}\left\{\begin{pmatrix} 1 \\ 0 \\ 1 \end{pmatrix}, \begin{pmatrix} -1 \\ 1 \\ 0 \end{pmatrix}\right\}$; $\ker T = \text{span}\left\{\begin{pmatrix} -6 \\ -4 \\ 1 \\ 0 \end{pmatrix}, \begin{pmatrix} -6 \\ -3 \\ 0 \\ 1 \end{pmatrix}\right\}$; $\rho(T) = 2$, $\nu(T) = 2$

9. $\begin{pmatrix} 0 & 1 & 0 \\ 0 & -1 & 0 \\ 0 & 0 & 0 \\ 1 & 0 & 0 \end{pmatrix}$; range $T = \text{span}\ \{1 - x, x^3\}$; $\ker T = \text{span}\ \{x^2\}$; $\rho(T) = 2$, $\nu(T) = 1$

11. $(0, 0, 1, 0)$; range $T = \mathbb{R}$; $\ker T = \text{span}\ \{1, x, x^3\}$; $\rho(T) = 1$, $\nu(T) = 3$

13. $\begin{pmatrix} 1 & -1 & 2 & 3 \\ 0 & 1 & 4 & 3 \\ 1 & 0 & 6 & 5 \end{pmatrix}$; range $T = \text{span}\ \{1 + x^2, -1 + x, 3 + 3x + 5x^2\} = P_2$; $\ker T = \text{span}\ \{x^2 - 4x - 6\}$; $\rho(T) = 3$, $\nu(T) = 1$

15. $\begin{pmatrix} 0 & 1 & 0 & 0 & 0 \\ 0 & 0 & 2 & 0 & 0 \\ 0 & 0 & 0 & 3 & 0 \\ 0 & 0 & 0 & 0 & 4 \end{pmatrix}$; range $D = P_3$; $\ker D = \mathbb{R}$; $\rho(D) = 4$, $\nu(D) = 1$

17. $\begin{pmatrix} 0 & 1 & 0 & 0 & \cdots & 0 \\ 0 & 0 & 2 & 0 & \cdots & 0 \\ 0 & 0 & 0 & 3 & \cdots & 0 \\ \vdots & \vdots & \vdots & \vdots & & \vdots \\ 0 & 0 & 0 & 0 & \cdots & n \end{pmatrix}$; range $D = P_{n-1}$; ker $D = \mathbb{R}$; $\rho(D) = n$, $\nu(D) = 1$

19. $\begin{pmatrix} 2 & 0 & 2 & 0 & 0 \\ 0 & 3 & 0 & 6 & 0 \\ 0 & 0 & 4 & 0 & 12 \\ 0 & 0 & 0 & 5 & 0 \\ 0 & 0 & 0 & 0 & 6 \end{pmatrix}$; range $T = P_4$; ker $T = \{\mathbf{0}\}$; $\rho(T) = 5$, $\nu(T) = 0$

21. $A_T = \text{diag}\,(b_0, b_1, b_2, \ldots, b_n)$, where $b_j = \sum_{i=1}^{j+1} \frac{j!}{(j+1-i)!}$; range $T = P_n$; ker $T = \{\mathbf{0}\}$; $\rho(T) = n + 1$, $\nu(T) = 0$

23. $\begin{pmatrix} 1 & 0 & 0 \\ 0 & 1 & 0 \\ 0 & 0 & 1 \end{pmatrix}$; range $T = P_2$; ker $T = \{\mathbf{0}\}$, $\rho(T) = 3$, $\nu(T) = 0$

25. For example, in M_{34},

$$A_T = \begin{pmatrix} 1 & 0 & 0 & 0 & 0 & 0 & 0 & 0 & 0 & 0 & 0 & 0 \\ 0 & 0 & 0 & 0 & 1 & 0 & 0 & 0 & 0 & 0 & 0 & 0 \\ 0 & 0 & 0 & 0 & 0 & 0 & 0 & 0 & 1 & 0 & 0 & 0 \\ 0 & 1 & 0 & 0 & 0 & 0 & 0 & 0 & 0 & 0 & 0 & 0 \\ 0 & 0 & 0 & 0 & 0 & 1 & 0 & 0 & 0 & 0 & 0 & 0 \\ 0 & 0 & 0 & 0 & 0 & 0 & 0 & 0 & 0 & 1 & 0 & 0 \\ 0 & 0 & 1 & 0 & 0 & 0 & 0 & 0 & 0 & 0 & 0 & 0 \\ 0 & 0 & 0 & 0 & 0 & 0 & 1 & 0 & 0 & 0 & 0 & 0 \\ 0 & 0 & 0 & 0 & 0 & 0 & 0 & 0 & 0 & 0 & 1 & 0 \\ 0 & 0 & 0 & 1 & 0 & 0 & 0 & 0 & 0 & 0 & 0 & 0 \\ 0 & 0 & 0 & 0 & 0 & 0 & 0 & 1 & 0 & 0 & 0 & 0 \\ 0 & 0 & 0 & 0 & 0 & 0 & 0 & 0 & 0 & 0 & 0 & 1 \end{pmatrix}$$

In general, $A_T = (a_{ij})$, where

$$a_{ij} = \begin{cases} 1, & \text{if } i = km + l, \\ & \text{and } j = (l-1)n + k + 1 \\ & \text{for } k = 1, 2, \ldots, n-1 \\ & \text{and } l = 1, 2, \ldots, m \\ 0, & \text{otherwise} \end{cases}$$

33. $\begin{pmatrix} 0 & 0 & 0 \\ 0 & 0 & -1 \\ 0 & 1 & 0 \end{pmatrix}$; range $D = \text{span}\,\{\sin x, \cos x\}$; ker $D = \mathbb{R}$; $\rho(D) = 2$, $\nu(D) = 1$

Problems 8.12, page 585

3. $\alpha \neq 6$ **5.** $m = [n(n+1)]/2 = \dim\{A\colon A \text{ is } n \times n \text{ and symmetric}\}$. **9.** $mn = pq$

15. If $H = \mathbb{R}^n$, then T is also $1 - 1$.

Problems 8.13, page 597

1. $-4, 3$; $E_{-4} = \text{span}\left\{\begin{pmatrix} 1 \\ 1 \end{pmatrix}\right\}$; $E_3 = \text{span}\left\{\begin{pmatrix} 2 \\ -5 \end{pmatrix}\right\}$

3. $i, -i$; $E_i = \text{span}\left\{\begin{pmatrix} 2+i \\ 5 \end{pmatrix}\right\}$; $E_{-i} = \text{span}\left\{\begin{pmatrix} 2-i \\ 5 \end{pmatrix}\right\}$

5. $-3, -3$; $E_{-3} = \text{span}\left\{\begin{pmatrix} 1 \\ 0 \end{pmatrix}\right\}$; geom. mult. is 1

7. $0, 1, 3$; $E_0 = \text{span}\left\{\begin{pmatrix} 1 \\ 1 \\ 1 \end{pmatrix}\right\}$; $E_1 = \text{span}\left\{\begin{pmatrix} -1 \\ 0 \\ 1 \end{pmatrix}\right\}$; $E_3 = \text{span}\left\{\begin{pmatrix} 1 \\ -2 \\ 1 \end{pmatrix}\right\}$

9. $1, 1, 10$; $E_1 = \text{span}\left\{\begin{pmatrix} 1 \\ 0 \\ -2 \end{pmatrix}, \begin{pmatrix} 0 \\ 1 \\ -2 \end{pmatrix}\right\}$; $E_{10} = \text{span}\left\{\begin{pmatrix} 2 \\ 2 \\ 1 \end{pmatrix}\right\}$; geom. mult. of 1 is 2

11. $1, 1, 1$; $E_1 = \text{span}\left\{\begin{pmatrix} 1 \\ 1 \\ 1 \end{pmatrix}\right\}$; geom. mult. is 1 (alg. mult. is 3)

13. $-1, i, -i$; $E_{-1} = \text{span}\left\{\begin{pmatrix} 0 \\ -1 \\ 1 \end{pmatrix}\right\}$; $E_i = \text{span}\left\{\begin{pmatrix} 1+i \\ 1 \\ 1 \end{pmatrix}\right\}$; $E_{-i} = \text{span}\left\{\begin{pmatrix} 1-i \\ 1 \\ 1 \end{pmatrix}\right\}$

15. $1, 2, 2$; $E_1 = \text{span}\left\{\begin{pmatrix} 4 \\ 1 \\ -3 \end{pmatrix}\right\}$; $E_2 = \text{span}\left\{\begin{pmatrix} 3 \\ 1 \\ -2 \end{pmatrix}\right\}$; geom. mult. of 2 is 1

17. a, a, a, a; $E_a = \mathbb{R}^4$; geom. mult. of a = alg. mult. of $a = 4$

19. a, a, a, a; $E_a = \text{span}\left\{\begin{pmatrix} 1 \\ 0 \\ 0 \\ 0 \end{pmatrix}, \begin{pmatrix} 0 \\ 0 \\ 0 \\ 1 \end{pmatrix}\right\}$; alg. mult. of $a = 4$; geom. mult. of $a = 2$.

Problems 8.14, page 603

1.

n	$p_{j,n}$	$p_{a,n}$	T_n	$p_{j,n}/p_{a,n}$	T_n/T_{n-1}
0	0	12	12	0	—
1	36	7	43	5.14	3.58
2	21	19	40	1.11	0.930
5	104	45	149	2.31	—
10	600	291	891	2.06	—
19	16,090	7737	23827	2.08	—
20	23,170	11140	34310	2.08	1.44

Note that the eigenvalues are 1.44 and −0.836. The corresponding eigenvectors are $\begin{pmatrix} 2.09 \\ 1 \end{pmatrix}$ and $\begin{pmatrix} -3.57 \\ 1 \end{pmatrix}$.

3.

n	$p_{j,n}$	$p_{a,n}$	T_n	$p_{j,n}/p_{a,n}$	T_n/T_{n-1}
0	0	20	20	0	—
1	80	16	96	5	4.8
2	64	69	133	0.928	1.39
5	1092	498	1590	2.19	—
10	42,412	22,807	65,219	1.86	—
19	3.69×10^7	1.95×10^7	5.64×10^7	1.89	—
20	7.82×10^7	4.14×10^7	11.96×10^7	1.89	2.12

The eigenvalues are 2.12 and -1.32 with corresponding eigenvectors $\begin{pmatrix} 1.89 \\ 1 \end{pmatrix}$ and $\begin{pmatrix} -3.03 \\ 1 \end{pmatrix}$.

5. From equation (9), $p_n \approx a_1\lambda_1^n \mathbf{v}_i$ for n large. If $\mathbf{v}_1 = \begin{pmatrix} x \\ y \end{pmatrix}$, then $\dfrac{p_{j,n}}{p_{a,n}} \approx \dfrac{a_1\lambda_1^n x}{a_1\lambda_1^n y} = \dfrac{x}{y}$; but

$$\begin{pmatrix} -\lambda_1 & k \\ \alpha & \beta - \lambda_1 \end{pmatrix}\begin{pmatrix} x \\ y \end{pmatrix} = \begin{pmatrix} 0 \\ 0 \end{pmatrix}$$

so that $-\lambda_1 x + ky = 0$ and $\dfrac{x}{y} = \dfrac{k}{\lambda_1}$. Thus $\dfrac{p_{j,n}}{p_{a,n}} \approx \dfrac{x}{y} = \dfrac{k}{\lambda_1}$ for n large.

Problems 8.15, page 609

1. yes; $C = \begin{pmatrix} 1 & 2 \\ 1 & -5 \end{pmatrix}$, $C^{-1}AC = \begin{pmatrix} -4 & 0 \\ 0 & 3 \end{pmatrix}$ **3.** yes; $C = \begin{pmatrix} 1 & 1 \\ 2-i & 2+i \end{pmatrix}$; $C^{-1}AC = \begin{pmatrix} i & 0 \\ 0 & -i \end{pmatrix}$

5. yes; $C = \begin{pmatrix} 2 & 2 \\ -1+3i & -1-3i \end{pmatrix}$; $C^{-1}AC = \begin{pmatrix} 2+3i & 0 \\ 0 & 2-3i \end{pmatrix}$

7. yes; $C = \begin{pmatrix} 3 & 1 & 1 \\ 2 & 3 & 0 \\ 1 & 1 & 1 \end{pmatrix}$; $C^{-1}AC = \begin{pmatrix} 1 & 0 & 0 \\ 0 & 2 & 0 \\ 0 & 0 & -1 \end{pmatrix}$

9. yes; $C = \begin{pmatrix} 0 & 0 & 1 \\ 1 & 1 & 0 \\ 0 & 2 & 0 \end{pmatrix}$; $C^{-1}AC = \begin{pmatrix} 0 & 0 & 0 \\ 0 & 2 & 0 \\ 0 & 0 & 3 \end{pmatrix}$ **11.** $C = \begin{pmatrix} 1 & 0 & 2 \\ 3 & -2 & 1 \\ 0 & 1 & 2 \end{pmatrix}$; $C^{-1}AC = \begin{pmatrix} 1 & 0 & 0 \\ 0 & 1 & 0 \\ 0 & 0 & 2 \end{pmatrix}$

13. no, since 1 is an eigenvalue of algebraic multiplicity 3 and geometric multiplicity 1

15. yes; $\begin{pmatrix} 0 & -1 & 1 & 1 \\ 0 & 1 & 1 & 1 \\ 1 & 0 & 1 & -1 \\ 0 & 1 & -1 & 1 \end{pmatrix}$; $C^{-1}AC = \begin{pmatrix} 2 & 0 & 0 & 0 \\ 0 & 2 & 0 & 0 \\ 0 & 0 & 4 & 0 \\ 0 & 0 & 0 & 6 \end{pmatrix}$ **19.** $\begin{pmatrix} 1 & 0 \\ 0 & 1 \end{pmatrix}$

Problems 8.16, page 614

1. $Q = \begin{pmatrix} 2/\sqrt{5} & 1/\sqrt{5} \\ 1/\sqrt{5} & -2/\sqrt{5} \end{pmatrix}$, $D = \begin{pmatrix} 5 & 0 \\ 0 & -5 \end{pmatrix}$ **3.** $Q = \begin{pmatrix} 1/\sqrt{2} & 1/\sqrt{2} \\ 1/\sqrt{2} & -1/\sqrt{2} \end{pmatrix}$, $D = \begin{pmatrix} 0 & 0 \\ 0 & 2 \end{pmatrix}$

5. $Q = \begin{pmatrix} 1/\sqrt{2} & \frac{1}{2} & \frac{1}{2} \\ -1/\sqrt{2} & \frac{1}{2} & \frac{1}{2} \\ 0 & 1/\sqrt{2} & -1/\sqrt{2} \end{pmatrix}$. $D = \begin{pmatrix} -3 & 0 & 0 \\ 0 & 1+2\sqrt{2} & 0 \\ 0 & 0 & 1 - 2\sqrt{2} \end{pmatrix}$

7. $Q = \begin{pmatrix} -\frac{2}{3} & \frac{1}{3} & \frac{2}{3} \\ \frac{2}{3} & \frac{2}{3} & \frac{1}{3} \\ \frac{1}{3} & -\frac{2}{3} & \frac{2}{3} \end{pmatrix}$, $D = \begin{pmatrix} 0 & 0 & 0 \\ 0 & 3 & 0 \\ 0 & 0 & 6 \end{pmatrix}$

15. $U = \dfrac{1}{\sqrt{3}}\begin{pmatrix} -1+i & 1 \\ 1 & 1+i \end{pmatrix}$; $U^*AU = \begin{pmatrix} -1 & 0 \\ 0 & 8 \end{pmatrix}$

Problems 8.17, page 622

1. $\begin{pmatrix} 3 & -1 \\ -1 & 0 \end{pmatrix}\begin{pmatrix} x \\ y \end{pmatrix} \cdot \begin{pmatrix} x \\ y \end{pmatrix} = 5$; $Q = \begin{pmatrix} \dfrac{2}{\sqrt{26 - 6\sqrt{13}}} & \dfrac{2}{\sqrt{26 + 6\sqrt{13}}} \\ \dfrac{3 - \sqrt{13}}{\sqrt{26 - 6\sqrt{13}}} & \dfrac{3 + \sqrt{13}}{\sqrt{26 + 6\sqrt{13}}} \end{pmatrix} \approx \begin{pmatrix} 0.9571 & 0.2898 \\ -0.2898 & 0.9571 \end{pmatrix}$;

$$\frac{x'^2}{\left(\dfrac{10}{\sqrt{13}+3}\right)} - \frac{y'^2}{\left(\dfrac{10}{\sqrt{13}-3}\right)} = 1;\ \text{hyperbola};\ \theta \approx 5.989 \approx 343^\circ$$

3. $\begin{pmatrix} 4 & 2 \\ 2 & -1 \end{pmatrix}\begin{pmatrix} x \\ y \end{pmatrix} \cdot \begin{pmatrix} x \\ y \end{pmatrix} = 9$; $Q = \begin{pmatrix} \dfrac{5+\sqrt{41}}{\sqrt{82+10\sqrt{41}}} & \dfrac{5-\sqrt{41}}{\sqrt{82-10\sqrt{41}}} \\ \dfrac{4}{\sqrt{82+10\sqrt{41}}} & \dfrac{4}{\sqrt{82-10\sqrt{41}}} \end{pmatrix} \approx \begin{pmatrix} 0.9436 & -0.3310 \\ 0.3310 & 0.9436 \end{pmatrix}$;
$\dfrac{x'^2}{\left(\dfrac{18}{\sqrt{41}+3}\right)} - \dfrac{y'^2}{\left(\dfrac{18}{\sqrt{41}-3}\right)} = 1$; hyperbola; $\theta \approx 0.3374 \approx 19.33°$

5. $\begin{pmatrix} 0 & \frac{1}{2} \\ \frac{1}{2} & 0 \end{pmatrix}\begin{pmatrix} x \\ y \end{pmatrix} \cdot \begin{pmatrix} x \\ y \end{pmatrix} = a > 0$; $Q = \begin{pmatrix} 1/\sqrt{2} & 1/\sqrt{2} \\ -1/\sqrt{2} & 1/\sqrt{2} \end{pmatrix}$; $\dfrac{x'^2}{2a} - \dfrac{y'^2}{2a} = 1$; hyperbola; $\theta = 7\pi/4 = 315°$.

7. Same as Problem 5 except that now we have a hyperbola with the roles of x' and y' reversed; since $a < 0$, we have $\dfrac{y'^2}{(-2a)} - \dfrac{x'^2}{(-2a)} = 1$

9. $\begin{pmatrix} -1 & 1 \\ 1 & -1 \end{pmatrix}\begin{pmatrix} x \\ y \end{pmatrix} \cdot \begin{pmatrix} x \\ y \end{pmatrix} = 0$; $Q = \begin{pmatrix} 1/\sqrt{2} & -1/\sqrt{2} \\ 1/\sqrt{2} & 1/\sqrt{2} \end{pmatrix}$; $y'^2 = 0$, which is the equation of a straight line through the origin; $\theta = \pi/4 = 45°$.

11. $\begin{pmatrix} 3 & -3 \\ -3 & 5 \end{pmatrix}\begin{pmatrix} x \\ y \end{pmatrix} \cdot \begin{pmatrix} x \\ y \end{pmatrix} = 36$; $Q = \begin{pmatrix} \dfrac{1+\sqrt{10}}{\sqrt{20+2\sqrt{10}}} & \dfrac{1-\sqrt{10}}{\sqrt{20-2\sqrt{10}}} \\ \dfrac{3}{\sqrt{20+2\sqrt{10}}} & \dfrac{3}{\sqrt{20-2\sqrt{10}}} \end{pmatrix} \approx \begin{pmatrix} 0.8112 & -0.5847 \\ 0.5847 & 0.8112 \end{pmatrix}$;
$\dfrac{x'^2}{\left(\dfrac{36}{4-\sqrt{10}}\right)} + \dfrac{y'^2}{\left(\dfrac{36}{4+\sqrt{10}}\right)} = 1$; ellipse; $\theta \approx 0.6245 \approx 35.78°$

13. $\begin{pmatrix} 6 & \frac{5}{2} \\ \frac{5}{2} & -6 \end{pmatrix}\begin{pmatrix} x \\ y \end{pmatrix} \cdot \begin{pmatrix} x \\ y \end{pmatrix} = -7$; $Q = \begin{pmatrix} 5/\sqrt{26} & -1/\sqrt{26} \\ 1/\sqrt{26} & 5/\sqrt{26} \end{pmatrix}$; $\dfrac{y'^2}{(14/13)} - \dfrac{x'^2}{(14/13)} = 1$; hyperbola; $\theta \approx 0.197 \approx 11.31°$

15. $\begin{pmatrix} 1 & -1 & -1 \\ -1 & 1 & -1 \\ -1 & -1 & 1 \end{pmatrix}\begin{pmatrix} x \\ y \\ z \end{pmatrix} \cdot \begin{pmatrix} x \\ y \\ z \end{pmatrix}$; $Q = \begin{pmatrix} 1/\sqrt{3} & 1/\sqrt{2} & 1/\sqrt{6} \\ 1/\sqrt{3} & -1/\sqrt{2} & 1/\sqrt{6} \\ 1/\sqrt{3} & 0 & -2/\sqrt{6} \end{pmatrix}$; $-x'^2 + 2y'^2 + 2z'^2$

17. $\begin{pmatrix} 3 & 2 & 2 \\ 2 & 2 & 0 \\ 2 & 0 & 4 \end{pmatrix}\begin{pmatrix} x \\ y \\ z \end{pmatrix} \cdot \begin{pmatrix} x \\ y \\ z \end{pmatrix}$; $Q = \begin{pmatrix} -\frac{2}{3} & \frac{1}{3} & \frac{2}{3} \\ \frac{2}{3} & \frac{2}{3} & \frac{1}{3} \\ \frac{1}{3} & -\frac{2}{3} & \frac{2}{3} \end{pmatrix}$; $3y'^2 + 6z'^2$

19. $\begin{pmatrix} 1 & 1 & 2 & \frac{7}{2} \\ 1 & 1 & 3 & -1 \\ 2 & 3 & 3 & 0 \\ \frac{7}{2} & -1 & 0 & 1 \end{pmatrix}$

21. $\begin{pmatrix} 3 & -\frac{7}{2} & \frac{1}{2} & -1 & \frac{3}{2} \\ -\frac{7}{2} & -2 & -\frac{1}{2} & \frac{1}{2} & 0 \\ \frac{1}{2} & -\frac{1}{2} & 3 & -2 & -\frac{5}{2} \\ -1 & \frac{1}{2} & -2 & -6 & \frac{1}{2} \\ \frac{3}{2} & 0 & -\frac{5}{2} & \frac{1}{2} & -1 \end{pmatrix}$

29. negative definite **31.** positive definite **33.** indefinite

35. negative definite

Chapter 8—Review Exercises, page 628

1. yes; dimesnion 2; basis $\{(1, 0, 1), (0, 1, 2)\}$ **3.** yes; dimension 3; basis $\{(1, 0, 0, -1), (0, 1, 0, -1), (0, 0, 1, -1)\}$
5. yes; dimension $[n(n+1)]/2$; basis $\{(E_{ij}: j \geq i\}$, where E_{ij} is the matrix with 1 in the i, j position and 0 everywhere else
7. no; for example, $(x^5 - 2x) + (-x^5 + x^2) = x^2 - 2x$, which is not a polynomial of degree 5, so the set is not closed under addition.

9. no; for example, $\begin{pmatrix} 1 & 1 \\ 0 & 2 \\ 3 & 1 \end{pmatrix} + \begin{pmatrix} 2 & 1 \\ -1 & 2 \\ 1 & 0 \end{pmatrix} = \begin{pmatrix} 3 & 2 \\ -1 & 4 \\ 4 & 1 \end{pmatrix}$, which does not satisfy $a_{12} = 1$.

11. independent **13.** dependent **15.** independent **17.** independent **19.** independent

21. dimension 2; basis $\{(2, 0, 1), (0, 4, 3)\}$ **23.** dimension 3; basis $\{(1, 0, 3, 0), (0, 1, -1, 0), (0, 0, 1, 1)\}$

25. dimension 4; basis $\{D_1, D_2, D_3, D_4\}$, where D_i is the matrix with a 1 in the i, i position and 0 everywhere else

27. range $A = \text{span}\left\{\begin{pmatrix} 1 \\ -2 \end{pmatrix}\right\}$; kernel $= \text{span}\left\{\begin{pmatrix} 2 \\ 1 \end{pmatrix}\right\}$; $\rho(A) = \nu(A) = 1$

29. range $A = \mathbb{R}^3$; kernel $= \{\mathbf{0}\}$; $\rho(A) = 3,\ \nu(A) = 0$

31. range $A = \text{span}\left\{\begin{pmatrix} 2 \\ -1 \\ 4 \end{pmatrix}, \begin{pmatrix} 3 \\ 2 \\ 6 \end{pmatrix}\right\}$; kernel $= \{\mathbf{0}\}$; $\rho(T) = 2,\ \nu(T) = 0$ **33.** $y = -1 + 3.4x$ **35.** linear

37. not linear, since $T(\alpha(x, y)) = T(\alpha x, \alpha y) = \alpha x/\alpha y = x/y = T(x, y) \neq \alpha T(x, y)$ unless $\alpha = 1$.

39. not linear, since $T(p_1 + p_2) = 1 + p_1 + p_2$, but $Tp_1 + Tp_2 = (1 + p_1) + (1 + p_2) = 2 + p_1 + p_2$.

41. $\ker T = \left\{\begin{pmatrix} 0 \\ 0 \end{pmatrix}\right\}$; range $T = \mathbb{R}^2$; $\rho(T) = 2$; $\nu(T) = 0$

43. $\ker T = \text{span}\left\{\begin{pmatrix} 0 \\ 0 \\ 1 \end{pmatrix}\right\}$; range $T = \mathbb{R}^2$; $\rho(T) = 2$; $\nu(T) = 1$

45. $\ker T = \left\{\begin{pmatrix} 0 & 0 \\ 0 & 0 \end{pmatrix}\right\}$; range $T = M_{22}$; $\rho(T) = 4$; $\nu(T) = 0$

47. $\begin{pmatrix} 0 & 0 \\ 0 & -1 \end{pmatrix}$; range $T = \text{span}\left\{\begin{pmatrix} 0 \\ 1 \end{pmatrix}\right\}$; $\ker T = \text{span}\left\{\begin{pmatrix} 1 \\ 0 \end{pmatrix}\right\}$; $\rho(T) = \nu(T) = 1$

49. $\begin{pmatrix} 1 & 0 & -2 & 0 \\ 0 & 2 & 0 & 3 \end{pmatrix}$; range $T = \mathbb{R}^2$; $\ker T = \text{span}\left\{\begin{pmatrix} 2 \\ 0 \\ 1 \\ 0 \end{pmatrix}, \begin{pmatrix} 0 \\ -3 \\ 0 \\ 2 \end{pmatrix}\right\}$; $\rho(A) = \nu(A) = 2$

51. $\begin{pmatrix} -1 & 1 & 0 & 0 \\ 0 & 2 & 0 & 0 \\ 0 & 0 & -1 & 1 \\ 0 & 0 & 0 & 2 \end{pmatrix}$; range $T = M_{22}$; $\ker T = \{\mathbf{0}\}$; $\rho(T) = 4,\ \nu(T) = 0$

53. $\left\{\frac{1}{\sqrt{13}}\begin{pmatrix} 2 \\ 3 \end{pmatrix}, \frac{1}{\sqrt{13}}\begin{pmatrix} -3 \\ 2 \end{pmatrix}\right\}$

55. $\begin{pmatrix} 1/\sqrt{3} \\ 1/\sqrt{3} \\ 1/\sqrt{3} \end{pmatrix}$ **57. a.** $\begin{pmatrix} \frac{4}{3} \\ -\frac{1}{3} \\ \frac{5}{3} \end{pmatrix}$ **b.** $\begin{pmatrix} -1/\sqrt{3} \\ 1/\sqrt{3} \\ 1/\sqrt{3} \end{pmatrix}$ **c.** $\begin{pmatrix} \frac{4}{3} \\ -\frac{1}{3} \\ \frac{5}{3} \end{pmatrix} + \begin{pmatrix} -\frac{7}{3} \\ \frac{7}{3} \\ \frac{7}{3} \end{pmatrix}$

59. $4\ -2$; $E_4 = \text{span}\left\{\begin{pmatrix} 1 \\ 1 \end{pmatrix}\right\}$; $E_{-2} = \text{span}\left\{\begin{pmatrix} 2 \\ 1 \end{pmatrix}\right\}$

61. $1, 7, -5$; $E_1 = \text{span}\left\{\begin{pmatrix} -6 \\ 3 \\ 4 \end{pmatrix}\right\}$; $E_7 = \text{span}\left\{\begin{pmatrix} 0 \\ 3 \\ 1 \end{pmatrix}\right\}$; $E_{-5} = \text{span}\left\{\begin{pmatrix} 0 \\ 0 \\ 1 \end{pmatrix}\right\}$

63. $1, 3, 3+\sqrt{2}i, 3-\sqrt{2}i$; $E_1 = \text{span}\left\{\begin{pmatrix}1\\2\\0\\0\end{pmatrix}\right\}$; $E_3 = \text{span}\left\{\begin{pmatrix}1\\1\\0\\0\end{pmatrix}\right\}$; $E_{3+\sqrt{2}i} = \text{span}\left\{\begin{pmatrix}0\\0\\-1\\\sqrt{2}i\end{pmatrix}\right\}$;

$E_{3-\sqrt{2}i} = \text{span}\left\{\begin{pmatrix}0\\0\\1\\\sqrt{2}i\end{pmatrix}\right\}$

65. $C = \begin{pmatrix}-3 & 1\\4 & -1\end{pmatrix}$; $C^{-1}AC = \begin{pmatrix}2 & 0\\0 & -3\end{pmatrix}$ **67.** $C = \begin{pmatrix}0 & -1-i & -1+i\\1 & 1 & 1\\-1 & 1 & 1\end{pmatrix}$; $C^{-1}AC = \begin{pmatrix}-1 & 0 & 0\\0 & i & 0\\0 & 0 & -i\end{pmatrix}$

69. not diagonalizable **71.** $Q = \begin{pmatrix}1/\sqrt{2} & 0 & 1/\sqrt{2}\\1/\sqrt{2} & 0 & -1/\sqrt{2}\\0 & 1 & 0\end{pmatrix}$; $Q^tAQ = \begin{pmatrix}4 & 0 & 0\\0 & -3 & 0\\0 & 0 & 0\end{pmatrix}$

73. $C = \begin{pmatrix}1 & 1 & 1 & -1\\-1 & 0 & 0 & 0\\0 & 1 & 0 & 0\\-1 & -1 & -1 & 2\end{pmatrix}$; $C^{-1}AC = \begin{pmatrix}-1 & 0 & 0 & 0\\0 & -1 & 0 & 0\\0 & 0 & 3 & 0\\0 & 0 & 0 & 3\end{pmatrix}$ **75.** $\dfrac{x'2}{8/(3+\sqrt{2})} + \dfrac{y'^2}{8/(3-\sqrt{2})} = 1$: ellipse

77. $\dfrac{y'2}{10/(\sqrt{13}+3)} - \dfrac{x'^2}{10/(\sqrt{13}-3)} = 1$: hyperbola **79.** $4x'^2 - 3y'^2$

CHAPTER 9

Problems 9.1, page 641

1. $x+y$ **3.** $1 + x_1 - 4x_2 + x_3$ **5.** $2 + \frac{1}{4}(x-2) + \frac{1}{4}(y-2)$ **7.** $\frac{5}{2} + \frac{3}{4}(x-2) - \frac{3}{2}(y-1)$
9. $x^2 + y^2$ **11.** $xyz - (x^3y^3z^3/3!)$ **13.** $y + xy$
15. $\dfrac{e^2\sqrt{2}}{2} + \dfrac{e^2\sqrt{2}}{2}(x-2) + \dfrac{e^2\sqrt{2}}{2}\left(y - \dfrac{\pi}{4}\right) + \dfrac{e^2\sqrt{2}}{4}(x-2)^2 + \dfrac{e^2\sqrt{2}}{2}(x-2)\left(y-\dfrac{\pi}{4}\right) - \dfrac{e^2\sqrt{2}}{4}\left(y-\dfrac{\pi}{4}\right)^2$
19. b. It is still equal to f. This may be hard to see, however. For example, the second-degree Taylor polynomial of $f(x) = x^2$ around $x = 3$ is $9 + 6(x-3) + (x-3)^2 = x^2$.
21. $\dfrac{1}{5!}[f_{xxxxx}(x_0, y_0)(x-x_0)^5 + 5f_{xxxxy}(x_0, y_0)(x-x_0)^4(y-y_0) + 10f_{xxxyy}(x_0, y_0)(x-x_0)^3(y-y_0)^2 +$
$10f_{xxyyy}(x_0, y_0)(x-x_0)^2(y-y_0)^3 + 5f_{xyyyy}(x_0, y_0)(x-x_0)(y-y_0)^4 + f_{yyyyy}(x_0, y_0)(y-y_0)^5]$

Problems 9.2, page 647

1. $\dfrac{\partial y}{\partial x} = -1$ at $(1, 1)$ **3.** $\dfrac{\partial y}{\partial x} = -\dfrac{e}{e+1}$ at $(1, 0)$ **5.** $\dfrac{\partial z}{\partial x} = \dfrac{\partial z}{\partial y} = -1$ at $(0, 0, 0)$
7. $\dfrac{\partial z}{\partial x} = -\dfrac{1}{\frac{4}{3} - e^3} = \dfrac{3}{3e^3 - 4}$ and $\dfrac{\partial z}{\partial y} = 0$ at $(1, 3, 0)$ **9.** $\dfrac{\partial x_{n+1}}{\partial x_i} = -1$ at $(1, 1, 1, \ldots, 1)$ for $i = 1, 2, \ldots, n$
11. $\dfrac{\partial z}{\partial x} = -\dfrac{x^3}{z^3}, \dfrac{\partial z}{\partial y} = -\dfrac{y^3}{z^3}$; valid for $z \neq 0$
13. $\dfrac{\partial z}{\partial x} = -\dfrac{(3x^2 - y^4z)\sin xyz + x^3yz\cos xyz}{x^4y\cos xyz - xy^4\sin xyz}, \dfrac{\partial z}{\partial y} = -\dfrac{(x^4z + 3y^2)\cos xyz - xy^3z\sin xyz}{x^4y\cos xyz - xy^4\sin xyz}$;
valid over $\{(x, y, z): x^4y\cos xyz - xy^4\sin xyz \neq 0\}$
15. $\dfrac{\partial z}{\partial x} = -\dfrac{z^2ye^{xy}\ln(z/x) - (z^2/x)e^{xy} - 5x^4z^2\sin[x^5 - (4y/z)]}{ze^{xy} - 4y\sin(x^5 - (4y/z))}, \dfrac{\partial z}{\partial y} = -\dfrac{xz^2e^{xy}\ln(z/x) + 4z\sin[x^5 - (4y/z)]}{ze^{xy} - 4y\sin(x^5 - (4y/z))}$; valid over $\{(x, y, z): ze^{xy} \neq 4y\sin[x^5 - (4y/z)]\}$ Note that the function is not defined when $z = 0$ or $x = 0$.

17. $\dfrac{\partial z}{\partial x} = -\dfrac{\frac{1}{2}\sqrt{z/x}e^{\sqrt{zx}} - 6xy^3z^4 - (5z/(x+y)^2)\cos[5z/(x+y)]}{\frac{1}{2}\sqrt{x/z}e^{\sqrt{zx}} - 12x^2y^3z^3 + [5/(x+y)]\cos[5z/(x+y)]}$,
$\dfrac{\partial z}{\partial y} = -\dfrac{-9x^2y^2z^4 - [5z/(x+y)^2]\cos[5z/(x+y)]}{\frac{1}{2}\sqrt{x/z}e^{\sqrt{zx}} - 12x^2y^3z^3 + [5/(x+y)]\cos[5z/(x+y)]}$; valid over
$\left\{(x, y, z): \dfrac{1}{2}\sqrt{\dfrac{x}{z}}e^{\sqrt{zx}} - 12x^2y^3z^3 + \dfrac{5}{x+y}\cos\left(\dfrac{5z}{x+y}\right) \neq 0 \text{ and } zx \neq 0\right\}$ (*Note:* The function is defined only when $zx \geq 0$ and $x \neq -y$.)

Problems 9.3, page 653

1. $\text{dom } \mathbf{f} = \mathbb{R}^2$ **3.** $\text{dom } \mathbf{f} = \{(x, y, z): x + y + z > 0\}$. This is a half-space. **5.** $\mathbb{R}^5$
7. $\text{dom } \mathbf{f} = \{(x_1, x_2, x_3, x_4): x_1^2 + x_2^2 + x_3^2 + x_4^2 \leq 1\}$. This is the set of points on and "inside" the "unit sphere" in $\mathbb{R}^4$.
9. Let $u = 3x$, $v = -4y$. Then $(v/3)^2 + (-v/4)^2 = 1$, since $x \geq 0$, $u \geq 0$. Thus, the image is half of an ellipse in the uv-plane.
11. Note that if $u = e^y \sin x$ and $v = e^y \cos x$, then $u^2 + v^2 = e^{2y}$. Therefore, each value of y in the interval $[0, 1]$ gives rise to a circle of radius e^y in the uv-plane. Thus the image of the rectangle in the xy-plane is an annulus. That is, it is a ring-shaped region in the uv-plane including and between the circles $u^2 + v^2 = 1$ and $u^2 + v^2 = e^2$.

13.

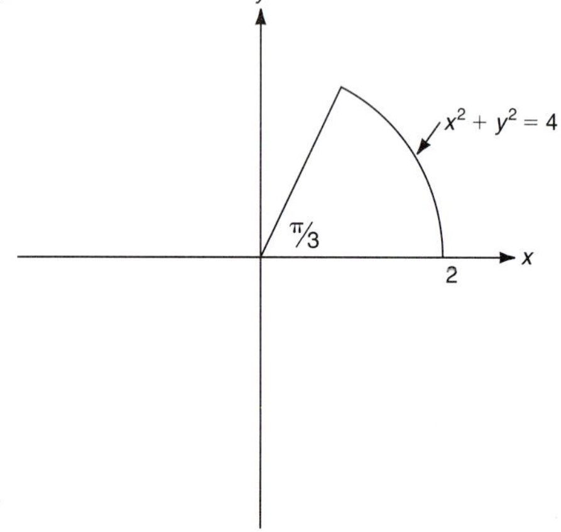

15. $(52, -20, 24)$; yes

17. $(\sqrt{3}/2, \pi/4, \pi^3/216)$; yes **19.** $(n, n^2, n^3, \ldots, n^m)$; yes
21. limit does not exist since f is not defined in a deleted neighborhood of $(0, 0, 0)$. Note that $\lim\limits_{t\to 0^+} \dfrac{t}{\ln t} = 0$; no
23. limit does not exist; no **25.** $\mathbf{g} \circ \mathbf{f}(x, y) = (x^3y\cos(y/x), e^{xy + x^2 - \cos(y/x)})$; $\text{dom}(\mathbf{g} \circ \mathbf{f}) = \{(x, y); x \neq 0\}$
27. $\mathbf{g} \circ \mathbf{f}(x, y) = \left(\sin\dfrac{e^{x+y}}{\sqrt{1 - x^2 + y^2}}, \sin\dfrac{\sqrt{1 - x^2 - y^2}}{e^{x+y}}, e^{\sqrt{e^{2x+y} + (1 - x^2 + y^2)}}\right)$; $\text{dom}(\mathbf{g} \circ \mathbf{f}) = \{(x, y): x^2 - y^2 < 1\}$

Problems 9.4, page 662

1. $\begin{pmatrix} 2 & 0 \\ 0 & 6 \end{pmatrix}$ **3.** $\begin{pmatrix} \frac{1}{4} & -\frac{1}{16} & 0 \\ 0 & \frac{1}{8} & -\frac{1}{16} \\ -8 & 0 & 1 \end{pmatrix}$ **5.** $\begin{pmatrix} -2 & 0 & 0 \\ 0 & 10 & 0 \\ 0 & 0 & 2 \\ 9 & -5 & -1 \end{pmatrix}$ **7.** $\begin{pmatrix} -\frac{1}{9} & -\frac{1}{9} & -\frac{1}{3} & -\frac{1}{9} \\ 1 & 1 & 1 & 1 \end{pmatrix}$

9. $\begin{pmatrix} \pi\sqrt{3}/12 & 0 & 0 & \sqrt{3}/2 \\ 0 & 0 & -\pi/6 & -3 \\ 2e^{-3} & -4e^{-3} & 0 & 0 \end{pmatrix}$ **11.** $\begin{pmatrix} 6 & 4 \\ 6 & -4 \end{pmatrix} = \begin{pmatrix} 1 & 1 \\ 1 & -1 \end{pmatrix}\begin{pmatrix} 6 & 0 \\ 0 & 4 \end{pmatrix}$

13. $\begin{pmatrix} -324 & 216 \\ 288 & -96 \end{pmatrix} = \begin{pmatrix} 108 & 0 & 0 \\ 0 & -24 & 0 \end{pmatrix}\begin{pmatrix} -3 & 2 \\ -12 & 4 \\ 9 & -12 \end{pmatrix}$

15. $\begin{pmatrix} \frac{3}{16} & \frac{13}{216} & \frac{5}{24} \\ 972 & 312 & 1080 \\ -57 & -17 & -180 \end{pmatrix} = \begin{pmatrix} \frac{1}{72} & \frac{5}{72} & -\frac{5}{432} & -\frac{5}{864} \\ 72 & 360 & -60 & -30 \\ 3 & 0 & 0 & 5 \end{pmatrix} \begin{pmatrix} 1 & 1 & 0 \\ 1 & 0 & 1 \\ -3 & -2 & 6 \\ -12 & -4 & -36 \end{pmatrix}$

17. $\begin{pmatrix} 0 & 0 & \cdots & 0 & 2n \\ 0 & 0 & \cdots & (2n-1) & 0 \\ \vdots & \vdots & & \vdots & \vdots \\ 0 & 2 & \cdots & 0 & 0 \\ 1 & 0 & \cdots & 0 & 0 \end{pmatrix} = \begin{pmatrix} 2n & 0 & \cdots & 0 & 0 \\ 0 & 2(n-1) & \cdots & 0 & 0 \\ \vdots & \vdots & & \vdots & \vdots \\ 0 & 0 & \cdots & 2 & 0 \\ 0 & 0 & \cdots & 0 & 1 \end{pmatrix} \begin{pmatrix} 0 & 0 & \cdots & 0 & 1 \\ 0 & 0 & \cdots & 1 & 0 \\ \vdots & \vdots & & \vdots & \vdots \\ 0 & 1 & \cdots & 0 & 0 \\ 1 & 0 & \cdots & 0 & 0 \end{pmatrix}$

Problems 9.5, page 672

1. -2 **3.** $-\frac{3}{4}$ **5.** $\frac{1}{6}$ **7.** 64 **9.** $\pi^3/48$ **11.** $-\ln 2$ **13.** $n!$ **15.** $\frac{1}{25}\begin{pmatrix} 2 & 3 \\ -7 & 2 \end{pmatrix}$

17. $\begin{pmatrix} \frac{1}{2} & \frac{1}{2} & 0 \\ \frac{1}{2} & 0 & \frac{1}{2} \\ 0 & -\frac{1}{2} & -\frac{1}{2} \end{pmatrix}$ **19.** $-\frac{1}{7}\begin{pmatrix} 1 & -4 \\ -2 & 1 \end{pmatrix}$ **21.** $-\frac{1}{2e^2}\begin{pmatrix} -1 & -e^2 \\ -1 & e^2 \end{pmatrix}$ **23.** $\frac{1}{144}\begin{pmatrix} -9 & 72 & -48 \\ 18 & 0 & 0 \\ 0 & 0 & -16 \end{pmatrix}$

25. $\frac{1}{2}I$ **29.** $\mathbf{x}_0 = (1, 2)$, $\mathbf{y}_0 = (-1, 5, 3)$, $J_f(\mathbf{x}_0) = -\frac{1}{19}\begin{pmatrix} 5 & -2 & 1 \\ 22 & -24 & -7 \\ 13 & -9 & -5 \end{pmatrix}\begin{pmatrix} 2 & 1 \\ 5 & -2 \\ 1 & 6 \end{pmatrix} = -\frac{1}{19}\begin{pmatrix} 1 & 15 \\ -83 & 28 \\ -24 & 1 \end{pmatrix}$

31. $\mathbf{x}_0 = (1, 1)$, $\mathbf{y}_0 = (1, 1)$, $J_f(\mathbf{x}_0) = -I$

33. $\mathbf{x}_0 = (2, -1)$, $\mathbf{y}_0 = (1, 2)$, $J_f(\mathbf{x}_0) = -\frac{1}{56}\begin{pmatrix} -16 & 4 \\ -6 & -2 \end{pmatrix}\begin{pmatrix} 4 & -2 \\ 2 & 2 \end{pmatrix} = \begin{pmatrix} 1 & -\frac{5}{7} \\ \frac{1}{2} & -\frac{1}{7} \end{pmatrix}$

35. $\mathbf{x}_0 = (1, 1, 1)$, $\mathbf{y}_0 = (1, 1)$, $J_f(\mathbf{x}_0) = -\frac{1}{2}\begin{pmatrix} 1 & 0 \\ -1 & 2 \end{pmatrix}\begin{pmatrix} 2 & 2 & 2 \\ 1 & 1 & 1 \end{pmatrix} = \begin{pmatrix} -1 & -1 & -1 \\ 0 & 0 & 0 \end{pmatrix}$

37. $\mathbf{x}_0 = (2, -3, 1)$, $\mathbf{y}_0 = (4, -2)$, $J_f(\mathbf{x}_0) = -\frac{1}{180}\begin{pmatrix} 20 & 0 \\ 10 & 9 \end{pmatrix}\begin{pmatrix} 8 & -12 & 36 \\ 12 & 0 & 40 \end{pmatrix} = -\frac{1}{180}\begin{pmatrix} 160 & -240 & 720 \\ 188 & -120 & 720 \end{pmatrix}$

39. $\mathbf{x}_0 = (0, 0, 0)$, $\mathbf{y}_0 = (0, 0, 0)$, $J_f(\mathbf{x}_0) = \begin{pmatrix} 1 & -1 & 0 \\ -1 & 0 & 0 \\ -1 & 1 & -1 \end{pmatrix}\begin{pmatrix} 3 & 0 & 1 \\ 0 & 0 & 0 \\ 0 & 1 & 1 \end{pmatrix} = \begin{pmatrix} 3 & 0 & 1 \\ -3 & 0 & -1 \\ -3 & -1 & -2 \end{pmatrix}$

Chapter 9—Review Exercises, page 676

1. $1 - \frac{(x+2y)^2}{2!}$ **3.** $\frac{\partial z}{\partial x} = \frac{\partial z}{\partial y} = -\frac{1}{2}$

5. $\frac{\partial z}{\partial x} = -\frac{2xy^3 - 4z^2}{-8xz + 12z^3}$; $\frac{\partial z}{\partial y} = -\frac{3x^2y^2 - 5}{-8xz + 12z^3}$; valid over $\{(x, y, z): 8xz \neq 12z^3\}$

7. dom $\mathbf{f} = \{(x, y, z): x \neq 0, x \neq -y, \text{ and } y/x \neq \pi/2 + n\pi\}$

9. dom $\mathbf{f} = \{(x, y, z): x \neq y, z \neq 0, (x+y)/(x-y) \neq \pi/2 + n\pi, \text{ and } -1 \leq xy/2 \leq 1\}$ **11.** $(1, 4, -\tan 3, \pi/6)$

13. $\begin{pmatrix} 12 & 0 \\ 0 & 3 \end{pmatrix}$ **15.** $\begin{pmatrix} 2 & 0 & 0 \\ 0 & 8 & 0 \\ 0 & 0 & 18 \end{pmatrix}$ **17.** $\begin{pmatrix} -1 & 0 & 3 & 0 \\ 0 & 5 & 0 & 1 \\ 6 & 2 & -2 & 10 \\ -5 & 15 & -15 & -3 \end{pmatrix}$

19. $\begin{pmatrix} 0 & 56 & 16 \\ -108 & -144 & 216 \\ 3 & -42 & 6 \\ -12 & -192 & 24 \end{pmatrix} = \begin{pmatrix} 2 & 8 & 0 \\ -18 & 0 & 12 \\ 0 & -6 & 1 \\ -6 & -24 & 4 \end{pmatrix}\begin{pmatrix} 4 & 4 & 0 \\ -1 & 6 & 2 \\ -3 & -6 & 18 \end{pmatrix}$ **21.** 288 **23.** $\frac{1}{17}\begin{pmatrix} 5 & -2 \\ 1 & 3 \end{pmatrix}$

25. $\begin{pmatrix} \frac{1}{3} & 0 & 0 \\ 0 & \frac{1}{3} & 0 \\ 0 & 0 & \frac{1}{12} \end{pmatrix}$ **27.** $\mathbf{x}_0 = (3, -2)$, $\mathbf{y}_0 = (4, 1)$, $J_f(\mathbf{x}_0) = \frac{1}{12}\begin{pmatrix} -4 & 4 \\ 2 & 1 \end{pmatrix}\begin{pmatrix} 2 & -1 \\ 3 & 2 \end{pmatrix} = \frac{1}{12}\begin{pmatrix} 4 & 12 \\ 7 & 0 \end{pmatrix}$

29. $\mathbf{x}_0 = (1, 1)$, $\mathbf{y}_0 = (1, 1, 1)$, $J_f(\mathbf{x}_0) = -\frac{1}{15}\begin{pmatrix} -2 & 6 & -3 \\ -5 & 15 & 0 \\ 7 & -6 & 3 \end{pmatrix}\begin{pmatrix} 3 & 3 \\ 1 & 1 \\ 1 & 2 \end{pmatrix} = -\frac{1}{15}\begin{pmatrix} -3 & -6 \\ 0 & 0 \\ 18 & 21 \end{pmatrix}$

CHAPTER 10

Problems 10.1, page 682

1. first **3.** third **5.** fourth **13.** linear **15.** linear **17.** nonlinear **19.** nonlinear
21. nonlinear **23.** nonlinear

Problems 10.2, page 697

1. 55.1 minutes **3.** 38.7 minutes **5.** 62,500 after 20 days; 156,250 after 30 days
7. $250{,}000(1.06)^{10} \approx 447{,}712$ in 1980; $250{,}000(1.06)^{30} \approx 1{,}435{,}873$ in 2000
9. a. $150(0.16)^2 \approx 3.84$°C **b.** $-\ln 15/\ln 0.16 \approx 1.48$ minutes **11.** $5580 \ln 0.7/\ln 0.5 \approx 2871$ years
13. a. $25(0.6)^{2.4} \approx 7.34$ kg **b.** $10 \ln 0.02/\ln 0.6 \approx 76.6$ hours **15.** $\ln 0.5/(-\alpha) \approx 4.62 \times 10^6$ years
17. in 2012 (23.69 years later) **19.** 4.0326 billion **21. a.** 19,057,278 **b.** in 2004 (29.56 years after 1974)
23. in 2003 (22.74 years after 1980) **25.** $20,544.33
27. 5% interest compounded continuously yields approximately 5.127% annually, so choose the first bank.
29. $\beta = (1/1500) \ln (845.6/1013.25) \approx -1.205812 \times 10^{-4}$ **a.** 625.526 mbar **b.** 303.416 mbar
c. 429.033 mbar **d.** 348.632 mbar **e.** $-(1/\beta) \ln 1013.25 \approx 57.4$ km
31. the president

Problems 10.3, page 710

1. $y = -\ln(c - x^2/2)$ **3.** $P = (Q + c)^2$ **5.** $x = ce^{\frac{1}{2}e^{t^2}}$ **7.** $y = ce^{-e^{-t}}$ **9.** $y = \frac{1}{2} \ln (x^2 + 1) + c$
11. $z = \tan\left(\frac{r^3}{3} + c\right)$ **13.** $P = ce^{\sin Q - \cos Q}$ **15.** $s = e^{(t^3/3) - 2t}$ **17.** $y = c \cos x - 3$ **19.** $y = \frac{3x}{4x - 3}$
21. $y = ce^{(x-1)e^x}$ **23.** $2y^3 + 3y^2 = 3x^2 + 5$ **25.** $x = \sqrt{2kt}$ where k is the constant of (inverse) proportionality
27. $\frac{dr}{dt} = -kr$; $r(0) = r_0$, $r(H) = \frac{1}{2}r_0$ **29.** $y = cx^{-k}$ where k is the constant of proportionality
31. at about 9:08:45 A.M. ($\frac{3}{2}(\sqrt{5} - 1)$ hours before 11 A.M.) **33.** see the reference

Problems 10.4, page 719

1. $x = ce^{3t}$ **3.** $x = 1$ **5.** $y = -\frac{1}{49} - \frac{x}{7} + ce^{7x}$ **7.** $z = ce^{-5x} + \frac{1}{5}x + \frac{4}{25}$
9. $y = -\frac{21}{25}e^{5x} + \frac{x}{5} - \frac{4}{25}$ **11.** $y = ce^x - x^2 - 2x - 4$ **13.** $x = -1 + 2e^{t^2/2}$ **15.** $x = y^y(1 + ce^{-y})$
17. $y = x^4 + 3x^3$ **19.** $x = ce^{2t} + \frac{t^3e^{2t}}{3}$ **21.** $s = \left(c + \frac{u^2}{2}\right)e^{-u} + 1$ **23.** $x = e^{-y}(c + y)$
25. $x = \frac{1}{k} + ce^{-kt}$ **27.** $y = B + ce^{-kt}$ **29.** $P = \frac{\beta}{\delta + (\beta/P_0 - \delta)e^{-\beta t}}$
31. a. $\frac{dN}{dt} = rN\left(1 - \frac{N}{K}\right) - h = -\frac{r}{K}\left(N^2 - KN + \frac{hK}{r}\right)$ **b.** $N = \frac{1}{2}K + \frac{1}{2}D \tanh\left[\frac{1}{2}D\frac{r}{K}t + c\right]$ where $D = \sqrt{K^2 - \frac{4Kh}{r}}$ and $c = \tanh^{-1}[2N_0 - K/D]$
33. Graph $y = rN(1 - N/K)$ and observe maximum $rK/4$ occurs at $N = K/2$
35. The object satisfies $m\dot{v} = -mg + kv^2$. Thus $\lim_{t\to\infty} v = -\sqrt{mg/k}$ **37.** $y = \pm\sqrt{\frac{x}{6 + ce^{-x}}}$

39. $y = \dfrac{1}{x\sqrt{2 - x^2}}$, $0 < x < \sqrt{2}$ **41.** $y = \dfrac{1}{x^3(\frac{1}{2} - \ln x)}$, $0 < x < \sqrt{e}$ **43.** $y = \left[\int g(x)e^{\int f(x)\,dx}\,dx + k\right]e^{-\int f(x)\,dx}$

45. b. $y = -ce^{-3x} + ke^{-2x}$ **47.** $y = (cx + k)e^{-ax}$ **49.** $y = \begin{cases} e^{-x}, & \text{if } 0 \le x \le 1 \\ e^{-1}, & \text{if } x > 1 \end{cases}$ **51.** See the reference.

Problems 10.5, page 727

1. $x^2y + y = c$ **3.** $xy^2 + x^2y = c$ **5.** $x^2 - 2y\sin x = \dfrac{\pi^2}{4} - 2$ **7.** $e^{xy} + 4xy^3 - y^2 + 3 = 0$

9. $x^2y + xe^y = c$ **11.** $\tan^{-1}\dfrac{y}{x} + \ln\left|\dfrac{x}{y}\right| = c$ **13.** $x^4y^3 + \ln\left|\dfrac{x}{y}\right| = e^3 - 1$

15. $\dfrac{x}{y} + \ln|y| = c$; $y \equiv 0$ is also a solution **17.** $2xy^{3/2}e^{-1/y} = c$; other solutions are $x \equiv 0$ and $y \equiv 0$.

19. $\frac{3}{8}x^{8/3} + \frac{3}{2}x^{2/3}y^2 = c$ **21.** $y^2(x^2 + y^2 + 2) = c$ **23.** $x^2y = c$

Problems 10.6, page 731

1. $I = \frac{6}{5}(1 - e^{-10t})$ **3.** $I = 2(1 - e^{-25t})$ **5.** $I = \frac{1}{20}(e^t - e^{-t})$

7. $Q = \frac{1}{680}[3\sin 60t + 5\cos 60t - 5e^{-100t}]$ **9.** $Q = \frac{1}{100}(1 + 99e^{-100t})$

11. $1000\dfrac{dQ}{dt} + 10^6Q = 0$; $Q(0) = 10$, $Q = 10e^{-1000t}$ **13.** $Q(60) = \frac{1}{2000}[1 - e^{-10^5(600+3600)}]$

15. $I(t) = \dfrac{E_0C}{1 + (\omega RC)^2}\left[RC\omega^2\cos\omega t - \omega\sin\omega t + \dfrac{1}{RC}e^{-t/RC}\right]$; $Q(t) = \dfrac{E_0C}{1 + (\omega RC)^2}[(\cos\omega t - e^{-t/RC}) + \omega RC\sin\omega t]$

17. $I_{\text{transient}}(0) = \dfrac{E_0}{R[1 + (\omega RC)^2]} \approx \dfrac{E_0}{R}$ for very small R **19.** $I(t) = \begin{cases} \dfrac{3}{50}(1 - e^{-50t}), & 0 \le t \le 10 \\ \left[\dfrac{e^{500}}{(70)^2} - \dfrac{3}{50}\right]e^{-50t} + \dfrac{7}{100} - \dfrac{e^{10-t}}{98}, & t \ge 10 \end{cases}$

Problems 10.7, page 745

1. linear, homogeneous, variable coefficients **3.** nonlinear **5.** linear, nonhomogeneous, constant coefficients
7. linear, nonhomogeneous, variable coefficients **9.** linear, homogeneous, constant coefficients
11. independent **13.** independent **15.** independent **17.** dependent **19.** dependent
21. independent **23.** dependent **25.** dependent **31.** $y = c_1 + c_2x$ **33.** $y = c_1 + c_2x^2$
35. $c_1\sin 3x + c_2\cos 3x$ **37.** $c_1e^{3x} + c_2e^{-5x}$ **39.** $(c_1 + c_2x)e^{3x}$ **41.** $e^{-x}(c_1\cos x + c_2\sin x)$
43. $c_1\sin 3x + c_2\cos 3x + \frac{4}{9}x$ **45.** $c_1e^{3x} + c_2e^{-5x} - x^2 - \frac{4}{15}x - \frac{8}{225}$
47. At that point $y_1(x_1) = y_1'(x_1) = 0$. But $y \equiv 0$ also satisfies that initial condition.
49. b. No; $a(x) = -3/x$ is not continuous on $|x| < 1$ **51.** $y_3 = 2y_1 - \frac{1}{2}y_2$ **55. b.** $W = -2/x$

Problems 10.8, page 750

1. $y_2(x) = xe^x$ **3.** $y_2(x) = e^{-2x}$ **5.** $y_2(x) = e^{-3x}$ **7.** $y_2(x) = e^{-x/2}\sin\dfrac{3\sqrt{3}}{2}x$ **9.** $y_2(x) = e^{10x}$

11. $y_2(x) = xe^{-x/2}$ **13.** $y_2(x) = x^{-2}$ **15.** $y_2(x) = x\cos x$ **17.** $y_2(x) = (7x + 1)e^{-4x}$

19. $y_2(x) = e^x\int x^ne^{-x}\,dx = -(x^n + nx^{n-1} + n(n-1)x^{n-2} + \cdots + n!)$ **21.** $y_2 = 1 + \dfrac{x}{\sqrt{3}}\tan^{-1}\dfrac{x}{\sqrt{3}}$

23. $y_2 = x - \frac{1}{2}\sin 2x$ **25.** $y_2 = \ln|\csc x - \cot x| - \ln|\sin x| = \ln\left|\dfrac{1 - \cos x}{\sin^2 x}\right|$ **27.** $y_2 = e^{x^2}\int e^{-x^2}\,dx$

29. $y_2(x) = \cos x/\sqrt{x}$

Problems 10.9, page 754

1. $y = c_1e^{2x} + c_2e^{-2x}$ **3.** $y = c_1e^x + c_2e^{2x}$ **5.** $x = (1 + \frac{9}{2}t)e^{-5t/2}$ **7.** $x = -\frac{1}{5}(e^{3t} + 4e^{-2t})$
9. $y = c_1 + c_2e^{5x}$ **11.** $y = (c_1 + c_2x)e^{-\pi x}$ **13.** $z = c_1e^{-5x} + c_2e^{3x}$ **15.** $y = (1 + 2x)e^{4x}$

17. $y = c_1 e^{\sqrt{2}x} + c_2 e^{-\sqrt{2}x}$ **19.** $y = e^{\sqrt{5}x} + 2e^{-\sqrt{5}x}$ **21.** $y = \frac{4}{3} - \frac{1}{3}e^{-3x}$ **23.** $y = c_1 e^{-2x} + c_2 e^{-8x}$

25. d **27.** a **29.** f **33.** $y = 1 - \dfrac{4}{e^{2x} + 2}$ **35.** $y = 1 + \dfrac{1}{c + x}$ **37.** $y = 2 - \dfrac{3c}{e^{3x} + c}$

Problems 10.10, page 761

1. $y = e^{-x}(c_1 \cos x + c_2 \sin x)$ **3.** $x = e^{-t/2}\left(c_1 \cos\dfrac{3\sqrt{3}}{2}t + c_2 \sin\dfrac{3\sqrt{3}}{2}t\right)$ **5.** $x = -\frac{3}{2}\cos 2\theta + \sin 2\theta$

7. $y = \sin\dfrac{x}{2} + 2\cos\dfrac{x}{2}$ **9.** $y = e^{-x}(c_1 \cos 2x + c_2 \sin 2x)$ **11.** $y = e^{-x}(\sin x - \cos x)$

13. $\sqrt{13}\cos(t - 0.9828) = \sqrt{13}\sin(t + 0.5880)$ $\left[\delta = \cos^{-1}\dfrac{2}{\sqrt{13}} \text{ and } \varphi = \sin^{-1}\dfrac{2}{\sqrt{13}}\right]$

15. $\sqrt{13}\cos(t + 2.1588) = \sqrt{13}\sin(t - 2.5536)$ $\left[\delta = \cos^{-1}\dfrac{2}{\sqrt{13}} - \pi \text{ and } \varphi = \sin^{-1}\dfrac{2}{\sqrt{13}} - \pi\right]$

17. $5\cos(2\theta - 0.6435) = 5\sin(2\theta + 0.9273)$ $[\delta = \cos^{-1}\frac{4}{5} \text{ and } \varphi = \sin^{-1}\frac{4}{5}]$

19. $13\cos\left(\dfrac{\alpha}{2} - 1.9656\right) = 13\sin\left(\dfrac{\alpha}{2} + 2.7468\right)$ $[\delta = \pi - \cos^{-1}\frac{5}{13} \text{ and } \varphi = \pi - \sin^{-1}\frac{5}{13}]$

21. $\dfrac{\sqrt{5}}{4}\cos(4\beta + 0.4636) = \dfrac{\sqrt{5}}{4}\sin(4\beta - 1.1071)$ $\left[\delta = -\cos^{-1}\dfrac{2}{\sqrt{5}} \text{ and } \varphi = -\sin^{-1}\dfrac{2}{\sqrt{5}}\right]$

23. $\sqrt{58}\cos\left(\dfrac{\pi}{2}x + 2.7367\right) = \sqrt{58}\sin\left(\dfrac{\pi}{2}x - 1.9757\right)$ $\left[\delta = \cos^{-1}\dfrac{7}{\sqrt{58}} - \pi \text{ and } \varphi = \sin^{-1}\dfrac{7}{\sqrt{58}} - \pi\right]$ **25.** b

27. a **29.** c **31.** $y = c_1 \sin x$ is zero at $x = \pi$.

Problems 10.11, page 766

1. $y = c_1 e^x + c_2 e^{2x} + 5$ **3.** $y = \cos 2x + \frac{1}{2}\sin 2x + \sin x$ **5.** $y = c_1 e^x + c_2 e^{2x} + 3e^{3x}$

7. $y = (c_1 + c_2 x - 2x^2)e^x$ **9.** $y = c_1 e^{2x} + c_2 e^{5x} + 10x + 7$ **11.** $y = 1 + 2e^x - \ln|\cos x|$

13. $y = c_1 \cos 2x + c_2 \sin 2x + \frac{1}{2}x \sin 2x + \frac{1}{4}\cos 2x \ln|\cos 2x|$ **15.** $y = e^x[c_1 + c_2 x - \ln|1 - x|]$

17. $y = \left(\dfrac{7}{e} - \dfrac{1}{2}\right)e^x + \left(2e + \dfrac{e^2}{2}\right)e^{-x} + e^x \ln x$ **19.** $y = e^{2x}(\frac{2}{3} - \frac{11}{6}x + (x + 1)\ln|x + 1|)$

21. $y_p = \frac{1}{2}\left[-1 - x\ln x + \left(x - \dfrac{1}{x}\right)\ln(1 + x)\right]$

Problems 10.12, page 774

1. $y = c_1 \sin 2x + c_2 \cos 2x + \sin x$ **3.** $y = c_1 e^x + c_2 e^{2x} + 3e^{3x}$ **5.** $y = e^x(c_1 + c_2 x - 2x^2)$

7. $y = 3e^{5x} - 10e^{2x} + 10x + 7$ **9.** $y = c_1 + c_2 e^{-x} + \dfrac{x^4}{4} - \dfrac{4}{3}x^3 + 4x^2 - 8x$

11. $y = e^{2x}(c_1 \cos x + c_2 \sin x) + xe^{2x}\sin x$ **13.** $y = e^{-3x}(c_1 + c_2 x + 5x^2)$

15. $y = c_1 e^{3x} + c_2 e^{-x} + \frac{20}{27} - \frac{7}{9}x + \frac{1}{3}x^2 - \frac{1}{4}e^x$ **17.** $y = (c_1 + c_2 x)e^{-2x} + (\frac{1}{9}x - \frac{2}{27})e^x + \frac{3}{25}\sin x - \frac{4}{25}\cos x$

21. $y = (-\frac{1}{6}x^3 - \frac{1}{4}x^2 + \frac{1}{4}x)\cos x + (\frac{1}{4}x^2 + \frac{1}{4}x)\sin x$ **23.** $y = (\frac{1}{12}x^4 + c_1 x + c_2)e^x$

25. a. $y = \begin{cases} a\sin x + b\cos x + x, & 0 \le x \le 1 \\ a\sin x + b\cos x + 1, & x \ge 1 \end{cases}$ **b.** $y = \begin{cases} x, & 0 \le x \le 1 \\ 1, & x \ge 1 \end{cases}$

Problems 10.13, page 779

1. $y = c_1 x + c_2 x^{-1}$ **3.** $y = x[c_1 \cos(\ln x) + c_2 \sin(\ln x)]$ **5.** $y = x^{3/2} - x^{1/2}$ **7.** $y = c_1 x + c_2 x^3$

9. $y = x^{-2}[c_1 \cos(\ln x) + c_2 \sin(\ln x)]$ **11.** $y = c_1 x^3 + c_2 x^{-4}$ **13.** $y = c_1 x^3 + c_2 x^{-5} - \dfrac{1}{16x}$

15. $y = x^3(c_1 + c_2 \ln x + \frac{1}{2}\ln^2 x)$ **17.** $y = c_1 \cos(\ln x) + c_2 \sin(\ln x) + 10$ **19.** $y = c_1 x^{-5} + c_2 x^{-1} + x/12$

21. $y = c_1 x^2 + c_2 x^{-1} + \frac{1}{4} - \frac{1}{2}\ln x$ **23.** $y = c_1 + c_2 x^{-3}$

Problems 10.14, page 788

1. a. $x = \cos 10t$; **b.** $x = (1 + 10t)e^{-10t}$; **c.** $x = \cos 10t + \dfrac{1}{2000}\sin 10t - \dfrac{t}{200}\cos 10t$;
d. $x = \left(\dfrac{201}{200} + \dfrac{201t}{20}\right)e^{-10t} - \dfrac{1}{200}\cos 10t$

3. a. $x = 5\sin(t + \alpha)$, $\tan\alpha = 3/4$; **b.** $x = \left(\dfrac{3\sqrt{5} + 11}{2}\right)e^{[(1-\sqrt{5})t/2]} - \left(\dfrac{3\sqrt{5} + 5}{2}\right)e^{[(-1-\sqrt{5})t/2]}$;
c. $x = 3\cos t + \dfrac{81}{20}\sin t - \dfrac{t}{20}\cos t$; **d.** $x = \frac{1}{100}[(555 + 151\sqrt{5})e^{(1-\sqrt{5})t/2} - (255 + 149\sqrt{5})e^{(-1-\sqrt{5})t/2} - 2\sqrt{5}\cos t]$

5. a. $x = \frac{3}{5}\sin 5t$; **b.** $x = e^{-4t}\sin 3t$; **c.** $x = \frac{9}{16}\sin 5t + \frac{1}{16}\sin 3t$; **d.** $x = \frac{1}{104}[e^{-4t}(106\sin 3t + 3\cos 3t) - 3\cos 3t + 2\sin 3t]$

7. $\pm(\sqrt{g}/5)$ m/s **9.** $\dfrac{2\pi}{\sqrt{5\pi g}}$ s; $x = \left(\dfrac{1}{5\pi} - 1\right)\cos\sqrt{5\pi g}\,t$ m **11.** $k = \dfrac{100\pi^2 + 1}{160}$ N/m = kg/s^2

13. $x = \dfrac{25}{19{,}408}\left(\dfrac{1213}{11}e^{-2t} - \dfrac{400}{11}e^{-19.6t} + 88\sin 2t - 108\cos 2t\right)$ **15.** $\dfrac{7\sqrt{5}}{10\pi}$ Hz $= \dfrac{42\sqrt{5}}{\pi} \approx 29.89$ ticks/min

Problems 10.15, page 791

1. $I(t) = 6(e^{-5t} - e^{-20t})$; $Q(t) = \frac{1}{10}(9 - 12e^{-5t} + 3e^{-20t})$ **3.** $I_{\text{steady state}} = \cos t + 2\sin t$
5. $I_{\text{steady state}} = \frac{1}{13}(70\sin 10t - 90\cos 10t)$ **7.** $I_{\text{transient}} = e^{-t}(c_1\cos t + c_2\sin t)$ **9.** $I_{\text{transient}} = c_1e^{-2t} + c_2e^{-5t}$
11. $I_{\text{transient}} = e^{-t}(c_1\cos 7t + c_2\sin 7t)$

13. $I_{\text{transient}} = \dfrac{e^{-600t}}{2320}(297\sin 800t - 96\cos 800t)$; $I_{\text{steady state}} = \frac{1}{580}(24\cos 600t - 27\sin 600t)$

15. $$Q = \begin{cases} c_1\cos\dfrac{t}{\sqrt{LC}} + c_2\sin\dfrac{t}{\sqrt{LC}} + (CE_0 + LC\omega^2)\sin\omega t, & \omega \neq \dfrac{1}{\sqrt{LC}}, \\ \left(c_1 - \dfrac{E_0t}{2\omega L}\right)\cos\omega t + c_2\sin\omega t, & \omega = \dfrac{1}{\sqrt{LC}} \end{cases}$$

17. b. $\omega = \dfrac{\sqrt{4L/C - R^2}}{2L}$ **19. b.** No ω produces resonance.

Problems 10.16, page 796

1. $y = (c_1 + c_2x)\cos x + (c_3 + c_4x)\sin x$ **3.** $y = (1 + x)e^x$ **5.** $y = 1 + e^{3x} + e^{-3x}$
7. $y = c_1 + c_2x + c_3x^2 + c_4x^3$ **9.** $y = c_1e^x + c_2e^{-x} + c_3e^{2x} + c_4e^{-2x}$ **11.** $y = 1 - x + e^{2x} - e^{-2x}$
13. $y = c_1e^x + c_2\cos x + c_3\sin x$ **15.** $y = c_1e^{3x} + e^{-3x/2}\left(c_2\cos\dfrac{3\sqrt{3}}{2}x + c_3\sin\dfrac{3\sqrt{3}}{2}x\right)$
17. If $c_1y_1 + c_2y_2 + c_3y_3 \equiv 0$, then we also must have $c_1y_1' + c_2y_2' + c_3y_3' \equiv 0$, $c_1y_1'' + c_2y_2'' + c_3y_3'' \equiv 0$. In order for all three of these identities to hold at x_0, it must be that $c_1 = c_2 = c_3 = 0$.
19. a. $y_1(v')'' + (3y_1' + ay_1)(v')' + (3y_1'' + 2ay_1' + by_1)v' = 0$ **c.** $v' = \dfrac{W(y_1, y_2)}{y_1^2}\displaystyle\int \dfrac{y_1e^{-\int a(x)\,dx}}{W^2(y_1, y_2)}\,dx$
25. $y = c_1e^x + c_2xe^x + c_3e^{-x} + \frac{1}{4}x^2e^x$ **27.** $y = c_1e^{2x} + c_2e^{-3x} + c_3e^{4x} + \frac{1}{24}x + \frac{41}{288}$
29. b. $y_p = \frac{1}{10}e^{2x}\int e^{-3x}\tan x\,dx + \frac{1}{10}e^{-x}[(3\cos x - \sin x)\ln|\sec x + \tan x| + 1]$ **31.** $y = c_1x^{-1} + (c_2 + c_3\ln|x|)x$
33. $y = c_1x^{-1} + c_2\cos(\ln|x|) + c_3\sin(\ln|x|)$

Problems 10.17, page 802

1. $y = 2e^x - x - 1$, $y(1) = 2(e - 1) \approx 3.4366$; **a.** $y_E = 2.98$; **b.** $y_{IE} = 3.405$
3. $y = \sqrt{2x^2 + 1} - x$; $y(1) = \sqrt{3} - 1 \approx 0.73205$ **a.** $y_E = 0.71$; **b.** $y_{IE} = 0.73207$
5. $y = \sinh(\frac{1}{2}x^2 - \frac{1}{2})$; $y(3) \approx 27.2899$ **a.** $y_E = 8.31$; **b.** $y_{IE} = 21.671$
7. $y = (x^3 + x^{-2})^{-1/2}$, $y(2) = 2/\sqrt{33} \approx 0.3481553$; **a.** $y_E = 0.343$ **b.** $y_{IE} = 0.34939$
9. $y = 2e^{e^x - 1}$, $y(2) = 2e^{e^2 - 1} \approx 1190.59$; **a.** $y_E = 156$; **b.** $y_{IE} = 781.56$ **11.** $y_1 = 1.02$, $y_2 = 1.04$, $y_3 = 1.07$, $y_4 = 1.09$, $y_5 = 1.12$ **13.** $y_1 = 0.34$, $y_2 = 0.56$, $y_3 = 0.76$, $y_4 = 0.98$, $y_5 = 1.34$
15. $y_1 = 1.28$, $y_2 = 1.65$, $y_3 = 1.46$, $y_4 = 1.09$, $y_5 = 0.87$, $y_6 = 0.79$, $y_7 = 0.93$, $y_8 = 0.95$
17. $y_1 = 1.10$, $y_2 = 1.22$, $y_3 = 1.35$, $y_4 = 1.48$, $y_5 = 1.63$, $y_6 = 1.79$, $y_7 = 1.97$, $y_8 = 2.16$, $y_9 = 2.37$, $y_{10} = 2.60$
19. $y_1 = 1.60$, $y_2 = 1.27$, $y_3 = 1.00$, $y_4 = 0.80$, $y_5 = 0.64$ **21.** 0.326994 **23.** 5.747786

Chapter 10—Review Exercises, page 807

1. $y = \frac{3}{2}x^2 + C$ **3.** $x = -\ln(e^{-3} - \sin t)$ **5.** $y = 0.7e^{-3x} + 0.3\cos x + 0.1\sin x$
7. $x = (2e^{-3} - \frac{1}{4} + \frac{1}{4}t^4)e^{3t}$ **9.** $xy + c = e^y \tan x$ **11.** $y = c_1e^x + c_2e^{4x}$ **13.** $y = c_1e^{3x} + c_2e^{-3x}$
15. $y = (c_1 + c_2x)e^{-3x}$ **17.** $y = e^x \sin x$ **19.** $y = \frac{2}{3}\sin x + c_1\cos 2x + c_2 \sin 2x$
21. $y = c_1e^{-x/2}\sin(\sqrt{3}/2)x + c_2e^{-x/2}\cos(\sqrt{3}/2)x - (x/\sqrt{3})[\cos(\sqrt{3}/2)x]e^{-x/2}$ **23.** $y = x^3 - 7x + c_1\cos x + c_2 \sin x$
25. $y = (3 - 11x + x^2/2)e^{4x}$ **27.** $y = c_1e^x \cos\sqrt{2}x + c_2e^x\sin\sqrt{2}x + (1/2\sqrt{2})xe^x\cos\sqrt{2}x$
29. $y \equiv 0$ (This is a harmonic oscillator that never gets going.) **31.** $y = c_1e^x + c_2xe^x + (e^x/x)$
33. $y = c_1x^{3/2}\cos\left(\frac{\sqrt{3}}{2}\ln x\right) + c_2x^{3/2}\sin\left(\frac{\sqrt{3}}{2}\ln x\right)$ **35.** $y = e^{-x}(c_1 + c_2x + c_3x^2)$
37. $y = (c_1 + c_2x)\cos 3x + (c_3 + c_4x)\sin 3x$
39. $I = I(0) - (1/L)\cos t$, so the current is 90° out of phase with the voltage of $I(0) = 0$.
41. period $= 2\pi\sqrt{60/g}$; frequency $= (\sqrt{g/60}/2\pi$; amplitude $= \sqrt{10^2 + (-5)^2 10/(g/6)} \approx 10.076$ cm
43. $\mu = \sqrt{4(10)(g/6)} \approx 80.87$ **45.** $y^2 = 2(e^x + 1)$, $y(3) = \sqrt{2(e^3 + 1)} \approx 6.4939$; $y_{IE} = 6.56$
47. $y = x + \sqrt{1 + x^2}$, $y(3) = 3 + \sqrt{10} \approx 6.1623$; $y_{IE} = 5.96$

CHAPTER 11

Problems 11.1, page 816

1. $x = c_1e^{-t} + c_2e^{4t}$, $y = -c_1e^{-t} + \frac{3}{2}c_2e^{4t}$ **3.** $x = c_1e^{-3t} + c_2te^{-3t}$, $y = -(c_1 + c_2)e^{-3t} + c_2te^{-3t}$
5. $x = c_1e^{10t} + c_2te^{10t}$, $y = -(2c_1 + c_2)e^{10t} - 2c_2te^{10t}$
7. $x = -\frac{1}{4}(t^2 + 9t + 3) + c_1 - 3c_2e^{2t}$, $y = \frac{1}{4}(t^2 + 7t) - c_1 + c_2e^{2t}$
9. $x = e^{4t}(17c_1\cos 2t + 17c_2\sin 2t)$, $y = e^{4t}[(8c_1 - 2c_2)\cos 2t + (8c_2 + 2c_1)\sin 2t]$
11. $x_1' = x_2$, $x_2' = -3x_1 - 2x_2$ **13.** $x_1' = x_2$, $x_2' = x_3$, $x_3' = x_1{}^3 - x_2{}^2 + x_3 + t$
15. $x_1' = x_2$, $x_2' = x_3$, $x_3' = x_1{}^4x_2 - x_1x_3 + \sin t$ **17.** $x_1' = x_2$, $x_2' = x_3$, $x_3' = x_1 - 4x_2 + 3x_3$
19. a. $x_1 = c_1e^t$, $x_2 = c_2e^t$; c_1, c_2 constant
21. $x_1 = 4c_1e^{-2t} + 3c_2e^{-t}$, $x_2 = -5c_1e^{-2t} - 4c_2e^{-t} + c_3e^{2t}$, $x_3 = -7c_1e^{-2t} - 2c_2e^{-t} - c_3e^{2t}$
23. $x' = \frac{-3x}{100 + t} + \frac{2y}{100 - t}$, $y' = \frac{3x}{100 + t} - \frac{4y}{100 - t}$
25. $y_{\max} = \frac{500}{\sqrt{3}}[(2 + \sqrt{3})^{(1-\sqrt{3})/2} - (2 + \sqrt{3})^{(-1-\sqrt{3})/2}]$ lb, $t_{\max} = \frac{25}{\sqrt{3}}\ln(2 + \sqrt{3})$ min
27. $m_1m_2x_1{}^{(4)} + [k_1m_2 + k_2(m_2 + m_1)]x_1'' + k_1k_2x_1 = 0$, $x_1 = c_1\cos\sqrt{10}t + c_2\sin\sqrt{10}t + c_3\cos t + c_4 \sin t$,
$x_2 = -(c_1/4)\cos\sqrt{10}t - (c_2/4)\sin\sqrt{10}t + 2c_3\cos t + 2c_4 \sin t$
29. 780.9 counts per minute

Problems 11.2, page 822

1. b. $W = e^{-6t}$; **c.** $\{c_1e^{-3t} + c_2(1 - t)e^{-3t}, -c_1e^{-3t} + c_2te^{-3t}\}$ **3. d.** $\frac{1}{t}$ is not continuous at 0.

Problems 11.3, page 830

1. $\{e^t, e^t\}$, $\{3e^{-t}, 5e^{-t}\}$ **3.** $\{e^t\cos t, e^t(2\cos t - \sin t)\}$, $\{e^t \sin t, e^t(2\sin t + \cos t)\}$
5. $\{e^{-3t}, -e^{-3t}\}$, $\{(t - 1)e^{-3t}, -te^{-3t}\}$ **7.** $\{3, 4\}$, $\{e^{-2t}, 2e^{-2t}\}$ **11.** $\{-\frac{1}{4}(t^2 + 9t + 3), \frac{1}{4}(t^2 + 7t)\}$
13. $\{\sin t - \cos t, 2\sin t - \cos t\}$
15. a. $x' = -0.134x + 0.02y$, $y' = 0.036x - 0.02y$ **b.** $10c_1e^{-0.14t} + c_2e^{-0.014t}, -3c_1e^{-0.14t} + 6c_2e^{-0.014t}\}$

Problem 11.4, page 836

1. $(I_L, I_R) = (1 - 1.025e^{-0.05}, 1 - 1.05e^{-0.05})$
3. $(100(k_1e^{\lambda_1t} + k_2e^{\lambda_2t}) + 1, -8(k_1\lambda_1e^{\lambda_1t} + k_2\lambda_2e^{\lambda_2t}) + 1)$ where $\left.\begin{matrix}k_1\\k_2\end{matrix}\right\} = \frac{\mp 3\sqrt{2} - 4}{800}$, $\left.\begin{matrix}\lambda_1\\\lambda_2\end{matrix}\right\} = -50 \pm 25\sqrt{2}$

5. The general solution $(I_L, I_R) = (I_L, I_R)_h + (I_L, I_R)_p$ is given by $(I_L, I_R)_p = (A \sin \omega t + B \cos \omega t, C \sin \omega t + D \cos \omega t)$, where $\omega = 60\pi$, $\Delta = \omega^2 + 2500$, $A = (2500/\Delta)^2$, $B = -25\omega(\omega^2 + 7500)/\Delta^2$, $C = -2500(\omega^2 - 2500)/\Delta^2$, $D = -250{,}000\ \omega/\Delta^2$, and **a.** $(I_L, I_R)_h = \left(\frac{e^{-50t}}{2}\left[k_1 + k_2\left(t + \frac{1}{25}\right)\right], e^{-50t}\left[k_1 + k_2\left(t + \frac{1}{25}\right)\right]\right)$
b. $(I_L, I_R)_h = (e^{-50t}[k_1 \cos 50\sqrt{3}t + k_2 \sin 50\sqrt{3}t], e^{-50t}[k_3 \cos 50\sqrt{3}t + k_4 \sin 50\sqrt{3}t])$
c. $(I_L, I_R)_h = (e^{-50t}(k_1 e^{25\sqrt{2}t} + k_2 e^{-25\sqrt{2}t}), e^{-50t}[(2 - \sqrt{2})k_1 e^{25\sqrt{2}t} + (2 + \sqrt{2})k_2 e^{-25\sqrt{2}t}])$
7. $I_1 = \frac{1}{20}(8 \sin t - 6 \cos t + 5e^{-t} + e^{-3t})$; $I_2 = \frac{1}{20}(2 \sin t - 4 \cos t + 5e^{-t} - e^{-3t})$ **9.** $I = 50t$

Problems 11.5, page 838

1. If $\tan \theta = B/A$, then $A \cos \omega t + B \sin \omega t = \sqrt{A^2 + B^2} \cos (\omega t - \theta)$.
3. $x_1 = (\frac{1}{2}x_{10} + \frac{1}{4}x_{20}) \cos t + (\frac{1}{2}x_{30} + \frac{1}{4}x_{40}) \sin t + (\frac{1}{2}x_{10} - \frac{1}{4}x_{20}) \cos \sqrt{3}t + (\frac{1}{2}x_{30} - \frac{1}{4}x_{40}) \sin \sqrt{3}t/\sqrt{3}$
$x_2 = (x_{10} + \frac{1}{2}x_{20}) \cos t + (x_{30} + \frac{1}{2}x_{40}) \sin t - (x_{10} - \frac{1}{2}x_{20}) \cos \sqrt{3}t - (x_{30} - \frac{1}{2}x_{40}) \sin \sqrt{3}t/\sqrt{3}$
5. $\omega_n = \frac{n^2\pi^2}{64}\sqrt{\frac{4.32 \times 10^9}{24(8)^4/\pi g}} \approx 325n^2$ rev/sec. The lowest frequency is 325 rev/sec.
7. $l_1(m_1 + m_2)\theta'' + m_2 l_2 \phi'' \cos (\theta - \phi) + m_2 l_2 \phi^2 \sin (\theta - \phi) + g(m_1 + m_2) \sin \theta = 0$; $l_1\theta'' \cos (\theta - \phi) + l_2\phi'' - l_1\theta'^2 \sin (\theta - \phi) + g \sin \phi = 0$
9. The characteristic equation is $m_1 l_1 l_2 \lambda^4 + (m_1 + m_2)(l_1 + l_2)g\lambda^2 + (m_1 + m_2)g^2 = 0$. The discriminant as a quadratic in λ^2 is positive, so the roots are pure imaginary, and the solution is a superposition of two simple harmonics.

Problems 11.6, page 844

1. yes; $W = 907$ **3.** no epidemic **5.** yes; $W = 22{,}994$
7. from (11) $x(t) = \beta/\alpha$ at $y_{\max}$. Substitute this into (13). Then use $N \geq x(0)$ and a power series for the logarithm.

Problems 11.7, page 850

1. $\mathbf{x}'(t) = A\mathbf{x}(t)$; $A = \begin{pmatrix} 2 & 3 \\ 4 & -6 \end{pmatrix}$ **3.** $\mathbf{x}'(t) = A(t)\mathbf{x}(t) + \mathbf{f}(t)$; $A(t) = \begin{pmatrix} 0 & 1 & 0 \\ 0 & 0 & 1 \\ 1 & 4t & -2 \end{pmatrix}$; $\mathbf{f}(t) = \begin{pmatrix} 0 \\ 0 \\ \sin t \end{pmatrix}$

5. $\mathbf{x}'(t) = A(t)\mathbf{x}(t) + \mathbf{f}(t)$; $A(t) = \begin{pmatrix} 2t & -3t^2 & \sin t \\ 2 & 0 & -4 \\ 0 & 17 & 4t \end{pmatrix}$; $\mathbf{f}(t) = \begin{pmatrix} 0 \\ -\sin t \\ e^t \end{pmatrix}$

17. a. $\mathbf{x}'(t) = (1, \cos t)$; $\int \mathbf{x}(t)\, dt = \left(\frac{t^2}{2}, -\cos t\right)$ **b.** $\mathbf{y}'(t) = \begin{pmatrix} e^t \\ -\sin t \\ \sec^2 t \end{pmatrix}$, $\int \mathbf{y}(t)\, dt = \begin{pmatrix} e^t \\ \sin t \\ -\ln |\cos t| \end{pmatrix}$

c. $A'(t) = \begin{pmatrix} 1/2\sqrt{t} & 2t \\ 2e^{2t} & 2 \cos 2t \end{pmatrix}$; $\int A(t)\, dt = \begin{pmatrix} \frac{2}{3}t^{3/2} & t^3/3 \\ e^{2t}/2 & -(\cos 2t)/2 \end{pmatrix}$

d. $B'(t) = \begin{pmatrix} 1/t & e^t(\sin t + \cos t) & e^t(-\sin t + \cos t) \\ \frac{5}{2}t^{3/2} & \sin t & -\cos t \\ -(1/t^2) & e^{t^2}(1 + 2t^2) & e^{t^3}(2t + 3t^4) \end{pmatrix}$; $\int B(t)\, dt = \begin{pmatrix} t \ln|t| - t & \frac{1}{2}e^t(\sin t - \cos t) & \frac{1}{2}e^t(\sin t + \cos t) \\ \frac{2}{7}t^{7/2} & -\sin t & \cos t \\ \ln |t| & e^{t^2}/2 & e^{t^3}/3 \end{pmatrix}$

Problems 11.8, page 863

1. fundamental **3.** fundamental **5.** not fundamental **7.** $C = \begin{pmatrix} 1 & -2 \\ -1 & 1 \end{pmatrix}$

9. $C = \begin{pmatrix} 1 & -1 & 1 \\ 0 & 1 & 1 \\ 1 & 0 & 0 \end{pmatrix}$ **15.** $\Psi(t) = e^{6t}\begin{pmatrix} \cos 2t - \sin t & \frac{1}{2} \sin t \\ -4 \sin t & \cos 2t + \sin 2t \end{pmatrix}$

17. $\Psi(t) = \frac{1}{6}\begin{pmatrix} -e^{-t} + 9e^t - 2e^{2t} & -2e^{-t} + 2e^{2t} & 7e^{-t} - 9e^t + 2e^{2t} \\ 6e^t - 6e^{2t} & 6e^{2t} & -6e^t + 6e^{2t} \\ -e^{-t} + 3e^t - 2e^{2t} & -2e^{-t} + 2e^{2t} & 7e^{-t} - 3e^t + 2e^{2t} \end{pmatrix}$

19. a. $e^t\begin{pmatrix}1-2t\\2-2t\end{pmatrix}$ **b.** $e^t\begin{pmatrix}-10t-2\\-10t+3\end{pmatrix}$ **c.** $e^{t-1}\begin{pmatrix}2-2t\\3-2t\end{pmatrix}$ **d.** $e^{t+1}\begin{pmatrix}4+2t\\3+2t\end{pmatrix}$ **e.** $e^{t-3}\begin{pmatrix}3\\3\end{pmatrix}$ **f.** $e^{t-a}\begin{pmatrix}2(c-b)(a-t)+b\\2(c-b)(a-t)+c\end{pmatrix}$

21. a. $\mathbf{x}' = A(t)\mathbf{x}$, where $A(t) = \begin{pmatrix}0 & 1\\-b(t) & -a(t)\end{pmatrix}$ **23.** $\varphi_2(t) = e^{t^2}\int e^{-t^2}\,dt$

Problems 11.9, page 872

1. $\frac{1}{7}\begin{pmatrix}5e^{-4t}+2e^{3t} & 2e^{-4t}-2e^{3t}\\5e^{-4t}-5e^{3t} & 2e^{-4t}+5e^{3t}\end{pmatrix}$ **3.** $\begin{pmatrix}2\sin t+\cos t & -\sin t\\5\sin t & -2\sin t+\cos t\end{pmatrix}$

5. $\frac{2}{3}e^{2t}\begin{pmatrix}\frac{3}{2}\cos 3t+\frac{1}{2}\sin 3t & \sin 3t\\-\frac{5}{2}\sin 3t & -\frac{1}{2}\sin 3t+\frac{3}{2}\cos 3t\end{pmatrix}$ **7.** $\begin{pmatrix}-5e^t+6e^{2t} & 2e^t-2e^{2t} & 4e^t-4e^{2t}\\-3e^t+3e^{2t} & 2e^t-e^{2t} & 2e^t-2e^{2t}\\-6e^t+6e^{2t} & 2e^t-2e^{2t} & 5e^t-4e^{2t}\end{pmatrix}$

9. $-\frac{1}{9}\begin{pmatrix}-4e^{8t}-5e^{-t} & -2e^{8t}+2e^{-t} & -4e^{8t}+4e^{-t}\\-2e^{8t}+2e^{-t} & -e^{8t}-8e^{-t} & -2e^{8t}+2e^{-t}\\-4e^{8t}+4e^{-t} & -2e^{8t}+2e^{-t} & -4e^{8t}-5e^{-t}\end{pmatrix}$ **13. a.** $A = \begin{pmatrix}0 & 1\\-b & -a\end{pmatrix}$ **15.** $(1+4t)e^{-2t}$

17. $\frac{8}{7}e^{5t}+\frac{13}{7}e^{-2t}$

Problems 11.10, page 879

1. $\varphi_p(t) = \begin{pmatrix}\frac{1}{3}e^{2t}\\\frac{1}{2}e^t-\frac{2}{3}e^{2t}\end{pmatrix}$ **3.** $\varphi(t) = \begin{pmatrix}-t\cos t\\-t\sin t-2t\cos t+\cos t+\sin t\end{pmatrix}$

5. $\varphi(t) = \begin{pmatrix}\frac{3}{2}te^t+\frac{1}{3}te^{2t}+\frac{5}{2}e^t-\frac{115}{72}e^{-t}-\frac{7}{9}e^{2t}-\frac{1}{8}e^{3t}\\te^t+te^{2t}+\frac{5}{2}e^t-2e^{2t}+\frac{1}{2}e^{3t}\\\frac{1}{2}te^t+\frac{1}{3}te^{2t}+e^t-\frac{115}{72}e^{-t}-\frac{7}{9}e^{2t}+\frac{3}{8}e^{3t}\end{pmatrix}$ **7.** $\varphi_p(t) = \begin{pmatrix}6e^t-2e^{2t}-2t-4\\4e^t-e^{2t}-t-3\\7e^t-2e^{2t}-3t-5\end{pmatrix}$

9. $\varphi_p(t) = \begin{pmatrix}5\sin t\ln|\csc t+\cot t|\\1+(2\sin t-\cos t)\ln|\csc t+\cot t|\end{pmatrix}$ **11.** $\Phi(t) = \begin{pmatrix}1/t & (\ln t)/t\\-1/t^2 & (1-\ln t)/t^2\end{pmatrix}$

Problems 11.11, page 883

1. $\mathbf{x}(t) = -\frac{1}{5}\begin{pmatrix}12+11i\\6+2i\end{pmatrix}e^{it}$ **3.** $\mathbf{x}(t) = -\frac{1}{754}\begin{pmatrix}353+205i\\-197-595i\end{pmatrix}e^{-5it}$

9. a. Let $x_1 = x$, $x_2 = x'$; then $\begin{pmatrix}x_1\\x_2\end{pmatrix}' = \begin{pmatrix}0 & 1\\-k/m & 0\end{pmatrix}\begin{pmatrix}x_1\\x_2\end{pmatrix} + \begin{pmatrix}0\\F_0/m\end{pmatrix}e^{i\omega t}$

b. $\mathbf{x}(t) = \frac{F_0}{k-\omega^2 m}\begin{pmatrix}1\\i\omega\end{pmatrix}e^{i\omega t}$ **c.** $\Psi(t) = \begin{pmatrix}\cos\sqrt{k/m}t & \sqrt{m/k}\sin\sqrt{k/m}t\\-\sqrt{k/m}\sin\sqrt{k/m}t & \cos\sqrt{k/m}t\end{pmatrix}$. Resonance.

Problems 11.12, page 892

1. $\frac{1}{7}\begin{pmatrix}5e^{-4t}+2e^{3t} & 2e^{-4t}-2e^{3t}\\5e^{-4t}-5e^{3t} & 2e^{-4t}+5e^{3t}\end{pmatrix}$ **3.** $\begin{pmatrix}2\sin t+\cos t & -\sin t\\5\sin t & -2\sin t+\cos t\end{pmatrix}$ **5.** $e^{-t}\begin{pmatrix}1+4t & -2t\\8t & -4t+1\end{pmatrix}$

7. $e^{-t}\begin{pmatrix}3/4+t\\1+2t\end{pmatrix}$ **9.** $e^{-t}\begin{pmatrix}2-2t\\5-4t\end{pmatrix}$ **11.** $\frac{1}{6}\begin{pmatrix}2+3e^t+e^{3t} & 2-2e^{3t} & 2-3e^t+e^{3t}\\2-2e^{3t} & 2+4e^{3t} & 2-2e^{3t}\\2-3e^t+e^{3t} & 2-2e^{3t} & 2+3e^t+e^{3t}\end{pmatrix}$

13. $\begin{pmatrix}4e^t-3e^{2t}+6te^{2t} & -12e^t+12e^{2t}-6te^{2t} & 6te^{2t}\\e^t-e^{2t}+2te^{2t} & -3e^t+4e^{2t}-2te^{2t} & 2te^{2t}\\-3e^t+3e^{2t}-4te^{2t} & 9e^t-9e^{2t}+4te^{2t} & -4te^{2t}+e^{2t}\end{pmatrix}$

15. $\frac{1}{9}\begin{pmatrix}5e^t+4e^{10t} & -4e^t+4e^{10t} & -2e^t+2e^{10t}\\-4e^t+4e^{10t} & 5e^t+4e^{10t} & -2e^t+2e^{10t}\\-2e^t+2e^{10t} & -2e^t+2e^{10t} & 8e^t+e^{10t}\end{pmatrix}$ **17.** $\frac{1}{2}\begin{pmatrix}e^{2t}+e^{6t} & -e^{2t}+e^{4t} & 0 & -e^{4t}+e^{6t}\\-e^{2t}+e^{6t} & e^{2t}+e^{4t} & 0 & -e^{4t}+e^{6t}\\e^{2t}-e^{6t} & -e^{2t}+e^{4t} & 2e^{2t} & 2e^{2t}-e^{4t}-e^{6t}\\-e^{2t}+e^{6t} & e^{2t}-e^{4t} & 0 & e^{4t}+e^{6t}\end{pmatrix}$

19. $\frac{1}{6}\begin{pmatrix} 4+2e^{3t} \\ 4-4e^{3t} \\ 4+2e^{3t} \end{pmatrix}$ **21.** $\begin{pmatrix} -4e^t + 3e^{2t} + 6te^{2t} \\ -e^t + e^{2t} + 2te^{2t} \\ 3e^t - e^{2t} - 4te^{2t} \end{pmatrix}$

Chapter 11—Review Exercises, page 897

1. $x_1' = x_2, x_2' = x_3, x_3' = 6x_3 - 2x_2 + 5x_1$ **3.** $x_1' = x_2, x_2' = x_3, x_3' = (\ln t - x_1x_3)/x_2$

5. $x = c_1e^{5t} + c_2e^{-t}, y = 2c_1e^{5t} - c_2e^{-t}$

7. $x = e^{2t}(c_1 \cos 3t + c_2 \sin 3t), y = \frac{1}{2}e^{2t}((3c_2 - c_1)\cos 3t - (c_2 + 3c_1)\sin 3t)$

9. $x = c_1e^{-t} + 3e^{-2t}, y = (c_2 - 2c_1t)e^{-t} + 12e^{-2t}$ **11.** $\begin{pmatrix} x_1 \\ x_2 \end{pmatrix}' = \begin{pmatrix} 3 & -4 \\ -2 & 7 \end{pmatrix}\begin{pmatrix} x_1 \\ x_2 \end{pmatrix}$

13. $\begin{pmatrix} x_1 \\ x_2 \end{pmatrix}' = \begin{pmatrix} 1 & 1 \\ -3 & 2 \end{pmatrix}\begin{pmatrix} x_1 \\ x_2 \end{pmatrix} + \begin{pmatrix} e^t \\ e^{2t} \end{pmatrix}$

15. a. $\frac{1}{5}\begin{pmatrix} -2e^t + 12e^{6t} \\ 3e^t + 12e^{6t} \end{pmatrix}$ **b.** $\frac{1}{5}\begin{pmatrix} -2e^t - 3e^{6t} \\ 3e^t - 3e^{6t} \end{pmatrix}$ **c.** $\begin{pmatrix} 0 \\ 0 \end{pmatrix}$ **d.** $\frac{1}{5}\begin{pmatrix} 18e^t + 17e^{6t} \\ -27e^t + 17e^{6t} \end{pmatrix}$ **e.** $\frac{1}{5}\begin{pmatrix} 2(a-b)e^t + (3a+2b)e^{6t} \\ -3(a-b)e^t + (3a+2b)e^{6t} \end{pmatrix}$

17. $\begin{pmatrix} 2e^{-t} - e^t & -2e^{-t} + 2e^t \\ e^{-t} - e^t & -e^{-t} + 2e^t \end{pmatrix}$ **19.** $\begin{pmatrix} \frac{1}{2}e^{2t} + \frac{1}{2}e^{6t} & -\frac{1}{2}e^{2t} + \frac{1}{2}e^{6t} & 0 \\ -\frac{1}{2}e^{2t} + \frac{1}{2}e^{6t} & \frac{1}{2}e^{2t} + \frac{1}{2}e^{6t} & 0 \\ 0 & 0 & e^{-3t} \end{pmatrix}$

21. $\varphi(t) = e^{2t}\begin{pmatrix} 3\cos 2t + \frac{19}{8}\sin 2t + \frac{1}{4}t \\ -6\sin 2t + \frac{19}{4}\cos 2t - \frac{11}{4} \end{pmatrix}$

CHAPTER 12

Problems 12.1, page 904

1. $\left(\frac{1}{\sqrt{2}}\right)\left[1 - \left(x - \frac{\pi}{4}\right) - \frac{\left(x-\frac{\pi}{4}\right)^2}{2!} + \frac{\left(x-\frac{\pi}{4}\right)^3}{3!} + \frac{\left(x-\frac{\pi}{4}\right)^4}{4!} - \frac{\left(x-\frac{\pi}{4}\right)^5}{5!} - \frac{\left(x-\frac{\pi}{4}\right)^6}{6!}\right]$

3. $1 + \frac{(x-e)}{e} - \frac{(x-e)^2}{2e^2} + \frac{(x-e)^3}{3e^3} - \frac{(x-e)^4}{4e^4} + \frac{(x-e)^5}{5e^5}$ **5.** $1 - (x-1) + (x-1)^2 - (x-1)^3 + (x-1)^4$

7. $x + \frac{1}{3}x^3$ **9.** $(x-\pi) + \frac{1}{3}(x-\pi)^3$ **11.** $1 - x^2 + x^4$ **13.** $x + x^3/3!$ **15.** $-\frac{1}{2}[x - \pi/2]^2$

17. $\frac{1}{2} + \frac{1}{2^4}x + \frac{3}{2^7 \cdot 2!}x^2 + \frac{3 \cdot 5}{2^{10} \cdot 3!}x^3 + \frac{3 \cdot 5 \cdot 7}{2^{13} \cdot 4!}x^4 = \frac{1}{2} + \frac{1}{16}x + \frac{3}{256}x^2 + \frac{5}{2{,}048}x^3 + \frac{35}{65{,}536}x^4$

19. $1 + \beta x + \left(\frac{\beta^2}{2!}\right)x^2 + \left(\frac{\beta^3}{3!}\right)x^3 + \left(\frac{\beta^4}{4!}\right)x^4 + \left(\frac{\beta^5}{5!}\right)x^5 + \left(\frac{\beta^6}{6!}\right)x^6$ **21.** $x + x^3/3!$

23. $(a_0 + a_1 + a_2 + a_3) + (a_1 + 2a_2 + 3a_3)(x-1) + (a_2 + 3a_3)(x-1)^2 + a_3(x-1)^3 = a_0 + a_1x + a_2x^2 + a_3x^3$

25. x^2

Problems 12.2, page 911

1. $(\pi/4)^7/7! \approx 3.658 \times 10^{-5}$

3. $(\frac{105}{32})(\frac{1}{4})^{-9/2}(4-1)^5/5! = 3402$ (This error bound can be improved dramatically, to about 6.64, by considering subintervals $[\frac{1}{4}, 1]$ and $[1, 4]$ separately.)

5. $|\beta|^5 \cdot \max\{e^0, e^\beta\}/5!$ **7.** $[16 \cdot 8 + 88 \cdot 4 + 16 \cdot 2] \cdot (\pi/4)^5/5! \approx 1.275$

9. $e^{1/9}\left(\frac{120}{3} + \frac{160}{27} + \frac{32}{243}\right)\left(\frac{(\frac{1}{3})^5}{5!}\right) \approx 0.001765$

11. $\frac{1}{2} + \frac{\sqrt{3}}{2}(0.2) - \frac{1}{2}\left(\frac{(0.2)^2}{2!}\right) - \frac{\sqrt{3}}{2}\left(\frac{(0.2)^3}{3!}\right) \approx 0.66205$ **13.** $\sum_{j=0}^{7}(1/j!) \approx 2.71825$

15. $\sin 33° = \sin\left(\frac{\pi}{6} + \frac{\pi}{60}\right); \frac{1}{2} + \left(\frac{\sqrt{3}}{2}\right)\left(\frac{\pi}{60}\right) - \frac{1}{2}\left(\frac{\pi}{60}\right)^2 \Big/ 2 \approx 0.54466\left[\text{Actually } P_1\left(\frac{\pi}{60}\right) \text{ gives the desired accuracy}\right]$

17. $\int_0^{\pi/4}\left(x+\frac{x^3}{3}\right)dx = \left(\frac{\pi}{4}\right)^2\Big/2 + \frac{(\pi/4)^4}{12} \approx 0.34013$; $|\text{max error}| \le 512(\pi/4)^6/5! \approx 1.00145$ (The precise answer is $\frac{1}{2}\ln 2 \approx 0.34657$.)
19. $\int_0^{1/3}(1+x^2+x^4/2)\,dx = \frac{1}{3}+\frac{1}{3}(\frac{1}{3})^3+\frac{1}{10}(\frac{1}{3})^5 \approx 0.34609$; $|\text{error}| \le \frac{1}{3}\cdot 0.001765 \approx 0.000588$
21. $\int_0^{1/5}(1+x^3)\,dx = 0.2+(0.2)^4/4 = 0.20040$ **23.** $\int_0^{\pi/6}\left(1-\frac{x^4}{2}\right)dx = \frac{\pi}{6}-\frac{1}{10}\left(\frac{\pi}{6}\right)^5 \approx 0.519663$
25. a. $(1+0.03)^3 = 1+3\cdot 0.03+3\cdot 0.03^2+0.03^3 = 1+0.09+0.0027+0.000027 = 1.092727$
b. $(1-0.03)^4 = 1+4\cdot(-0.03)+6\cdot(-0.03)^2+4\cdot(-0.03)^3+(-0.03)^4 =$
$1-0.12+0.0054-0.000108+0.00000081 = 0.88529281$ **c.** $(1+0.2)^4 = 1+4\cdot 0.2+6\cdot 0.2^2+4\cdot 0.2^3+0.2^4 =$
$1+0.8+0.24+0.032+0.0016 = 2.0736$ **d.** $(1-0.2)^5 = 1+5\cdot(-0.2)+10\cdot(-0.2)^2+10\cdot(-0.2)^3+$
$5\cdot(-0.2)^4+(-0.2)^5 = 1-1.0+0.4-0.08+0.008-0.00032 = 0.32768$
27. $16\tan^{-1}(\frac{1}{5}) \approx 16[(\frac{1}{5})-(\frac{1}{5})^3/3+(\frac{1}{5})^5/5-(\frac{1}{5})^7/7] \approx 3.158328076$ with an error bounded by $16(\frac{1}{5})^9/9 \approx 9.1\times 10^{-7}$;
$4\tan^{-1}(\frac{1}{239}) \approx 4\cdot\frac{1}{239} \approx 0.01673640167$ with an error bounded by $4(\frac{1}{239})^3/3 \approx 9.8\times 10^{-8}$. Thus,
$\pi \approx 3.158328076 - 0.01673640167 \approx 3.141591675$ with error bounded by 1.008×10^{-6}.

Problems 12.3, page 920

1. $\frac{1}{3}, \frac{1}{9}, \frac{1}{27}, \frac{1}{81}, \frac{1}{243}$ **3.** $\frac{3}{4}, \frac{15}{16}, \frac{63}{64}, \frac{255}{256}, \frac{1023}{1024}$ **5.** 1, 0, −1, 0, 1 **7.** $a_n = \frac{n}{n+1}$ **9.** $a_n = n\cdot 5^{n-1}$
11. 0 **13.** divergent **15.** 0 **17.** 0 **19.** e^4 **21.** diverges to ∞ **23.** 0 **25.** 0
27. 1 **29.** $a_0 \in [0, 1]$ **37.** Disprove

Problems 12.4, page 928

1. $\frac{1}{2}$ **3.** 1 **5.** 1 **7.** 1 **9.** 2 **11.** $(\ln 3)/3 \approx 0.3662$
13. $\sin n\pi = 0$; the sequence is constant (the sequence is both monotone increasing and monotone decreasing)
15. strictly increasing **17.** strictly decreasing **19.** strictly increasing **21.** strictly increasing
23. not monotonic **25.** strictly increasing **27.** not monotonic **29. b.** 0.56714
31. 12.713 [*Note:* $x \approx 1.2959$ is another fixed point but fixed point iteration will not find it because $f'(x) > 1$ in $(0, 5)$]
33. a. $x = \pm\sqrt{2}$ **36. b.** 27

Problems 12.5, page 933

1. 364 **3.** $[1-(-5)^6]/[1-(-5)] = -2604$ **5.** $(0.3)^2[1-(-0.3)^7]/1.3 \approx 0.0692$
7. $(b^{16}-1)/(b^{16}+b^{14})$ **9.** $(1-16\sqrt{2})/(1-\sqrt{2}) = 31+15\sqrt{2}$ **11.** $(-16)(\frac{1025}{5}) = -3280$ **13.** $\frac{4}{3}$
15. $\frac{3}{4}$ **17.** $\frac{10}{3}$ **19.** $\frac{1}{3}$ **21.** $n \ge 7$ **23.** $n \ge 459$

Problems 12.6, page 944

1. $\frac{1}{3}$ **3.** $\frac{35}{99}$ **5.** $\frac{71}{99}$ **7.** $\frac{501}{999} = \frac{167}{333}$ **9.** $\frac{11351}{99900}$ **11.** 10 **13.** $2-1-\frac{1}{2} = \frac{1}{2}$ **15.** $\frac{3}{5}$
17. 125 **19.** $\frac{1}{2}$ **21.** 1 **23.** $2^3\cdot(3-1-\frac{2}{3}) = \frac{32}{3}$ **25.** $(5^{-2}/6^1)(\frac{5}{6})^4[1/(1-5/6)] = 5^2/6^4 = \frac{25}{1296}$
27. $\frac{3}{2}+3 = \frac{9}{2}$ **29.** $\frac{15}{4}-\frac{28}{3} = -\frac{67}{12}$ **31.** $2-\frac{7}{4} = \frac{1}{4}$ **33.** $1\frac{1}{11}$ hr = 1:05$\frac{5}{11}$ P.M. ≈ 1.05:27 P.M.
35. $8+8\cdot 2\cdot\frac{2}{3}+8\cdot 2\cdot(\frac{2}{3})^2+8\cdot 2\cdot(\frac{2}{3})^3+\cdots = 40$ meters **41.** Prove
45. $\sum_{n=0}^{\infty} a_n = a_1 + a_0\sum_{k=1}^{\infty}(1/k!) = a_1 + a_0(e-1)$; see Section 12.11

Problems 12.7, page 951

1. diverges **3.** diverges **5.** converges **7.** diverges **9.** converges **11.** diverges
13. converges **15.** converges **17.** diverges **19.** diverges **21.** diverges **23.** diverges
25. diverges **27.** converges **29.** converges **31.** diverges **33.** converges **39.** diverges
43. a. ≈11.23 hours **b.** ≈69.44 days **c.** ≈28.22 years (using 1 year = 365.25 days)

Problems 12.8, page 957

1. diverges **3.** converges **5.** converges **7.** diverges **9.** converges **11.** converges
13. diverges **15.** converges **17.** diverges **19.** converges **21.** converges **23.** converges
25. converges

Problems 12.9, page 966

1. divergent **3.** divergent **5.** conditionally convergent **7.** conditionally convergent **9.** conditionally convergent **11.** conditionally convergent **13.** divergent **15.** divergent **17.** conditionally convergent **19.** absolutely convergent **21.** conditionally convergent **23.** absolutely convergent **25.** divergent **27.** divergent **29.** conditionally convergent

31. $\sum_{k=1}^{6} \frac{(-1)^{k+1}}{k!} \approx 0.63194$ **33.** $\sum_{k=1}^{9} \frac{(-1)^{k+1}}{k^4} \approx 0.947093$ **35.** $\sum_{k=1}^{5} \frac{(-1)^{k+1}}{k^k} \approx 0.783451$

37. $1 - \frac{1}{2} - \frac{1}{4} - \frac{1}{6} - \frac{1}{8} + \frac{1}{3} - \frac{1}{10} - \frac{1}{12} - \frac{1}{14} - \frac{1}{16}$ **39.** The most obvious example is to let $a_n = 1/n$.

45. The series is absolutely convergent; it, and any rearrangement, converges to $-\pi^2/12 \approx -0.822$.

49. *Hint:* If $\sum_{k=0}^{\infty} 2^k a_k$ converges, then $\sum_{k=0}^{\infty} a_k$ is absolutely convergent.

Problems 12.10, page 973

1. 6; $(-6, 6)$ **3.** 3; $(-4, 2)$ **5.** $\frac{1}{3}$, $(-\frac{1}{3}, \frac{1}{3})$ **7.** 1; $[0, 2]$ **9.** ∞; $(-\infty, \infty)$ **11.** 1; $(-1, 1)$ **13.** ∞; $(-\infty, \infty)$ **15.** 1; $(-1, 1)$ **17.** ∞; $(-\infty, \infty)$ **19.** $\frac{1}{2}$; $[-\frac{1}{2}, \frac{1}{2}]$ **21.** $\frac{5}{2}$; $(-4, 1)$ **23.** 0; $x = 0$ **25.** 1; $(-1, 1)$ **27.** 0; $x = -1$ **29.** 3; $[-13, -7)$ **31.** 1; $(-1, 1)$ **33.** 1; $(-1, 1)$

Problems 12.11, page 982

1. $\sum_{k=0}^{\infty} (-1)^k x^{2k}$ **3.** $\sum_{k=0}^{\infty} \frac{(-1)^k x^{2k+1}}{2k+1}$ **5.** $\sum_{k=0}^{\infty} (-1)^k (x-1)^k$ for $0 < x < 2$

7. $T_{14} \approx 3.139220$ with an error of $|\pi - T_{14}| \leq 0.00445$. Note that by Theorem 12.9.4 on page 962,

$|\pi - T_n| \leq \frac{4}{(2n+1)(2n+3)}$. Alternatively, $\pi = 6 \tan^{-1} \frac{1}{\sqrt{3}} = S = 6 \sum_{k=0}^{\infty} \frac{(-1)^k}{(2k+1)} \left(\frac{1}{\sqrt{3}}\right)^{2k+1}$; then

$S_5 = 6\left(\frac{1}{\sqrt{3}} - \frac{1}{3(\sqrt{3})^3} + \frac{1}{5(\sqrt{3})^5} - \frac{1}{7(\sqrt{3})^7} + \frac{1}{9(\sqrt{3})^9} - \frac{1}{11(\sqrt{3})^{11}}\right) \approx 3.141309$; error < 0.00037.

9. $S_3 \approx 0.743$ with error ≤ 0.0046; $T_3 \approx 0.755$ with error ≤ 0.0096

11. $S_2 \approx 0.496884$ with error ≤ 0.000009; $T_3 \approx 0.496795$ with error ≤ 0.000005

13. $S_2 \approx 0.763542$ with error ≤ 0.000347; $T_2 \approx 0.7569$ with error ≤ 0.0068

15. $S_4 \approx 0.190031$ with error ≤ 0.0038; $T_3 \approx 0.1955$ with error ≤ 0.0074 **17.** 0.09871 **19.** 0.47725

23. $\ln 1.5 \approx 0.4055$; $\ln 0.5 \approx -0.6931$; $\ln 2 \approx 0.6931$

Problems 12.12, page 992

1. $\sum_{k=0}^{\infty} e(x-1)^k/k!$ **3.** $\frac{1}{\sqrt{2}}\left[1 - \left(x - \frac{\pi}{4}\right) - \frac{(x-\pi/4)^2}{2!} + \frac{(x-\pi/4)^3}{3!} + \frac{(x-\pi/4)^4}{4!} - \cdots\right]$

5. $\sum_{k=0}^{\infty} (\beta x)^k/k!$ **7.** $\sum_{k=0}^{\infty} x^{k+1}/k!$ **9.** $\sum_{k=0}^{\infty} (-1)^k x^{2k}/(2k+1)!$ **11.** $\sum_{k=1}^{\infty} (-1)^{k+1}(x-1)^{k+1}/k$; $(0, 2]$

13. $1 + 0 + (x - \pi/2)^2/2! + 0$; $(0, \pi)$ **15.** $2 + \frac{1}{4}(x-4) + \sum_{k=2}^{\infty} \frac{(-1)^{k+1} 1 \cdot 3 \cdot \cdots \cdot (2k-3)}{2^{3k-1} k!}(x-4)^k$; $R = 4$

17. $\sum_{k=0}^{\infty} \frac{(-1)^k x^{4k+2}}{(2k+1)!}$ **19.** $\sum_{k=0}^{\infty} (-1)^k x^{2k+1}/(2k+1)$; $R = 1$

21. $1 + \frac{1}{4}x - \frac{1}{4} \cdot \frac{3}{4} \frac{x^2}{2!} + \frac{1}{4} \cdot \frac{3}{4} \cdot \frac{7}{4} \frac{x^3}{3!} - \frac{1}{4} \cdot \frac{3}{4} \cdot \frac{7}{4} \cdot \frac{11}{4} \frac{x^4}{4!} + \cdots$

23. Using the first four terms we obtain the approximation 0.503807.

25. a. $\frac{2}{\sqrt{\pi}} \sum_{k=0}^{\infty} \frac{(-1)^k x^{2k+1}}{k!(2k+1)}$ **b.** $\text{erf}(1) \approx 0.842699$; $\text{erf}(0.5) \approx 0.520500$

35. c. $\text{Si}(x) = \sum_{k=0}^{\infty} \frac{(-1)^k x^{2k+1}}{(2k+1)!(2k+1)}$ **d.** $\text{Si}(1) \approx 0.946083$; $\text{Si}(0.5) \approx 0.493108$

Problems 12.13, page 1001

1. $y = e^x + x + 1$ **3.** $y = c_0 \cos x + (c_1 - 1)\sin x + x$ **5.** $y = c_0(1 + x\tan^{-1}x) + c_1 x$ **7.** $y = xe^x$
9. $y = 1 - 2x^2$ **11.** $y = x + \sin x$ **13.** $y = 1/(1 - x)$ **15.** $y = e^{x^2} - (x/4)$

17. $y = c_0 \sum_{n=0}^{\infty} \frac{1 \cdot 4 \cdot \cdots \cdot (3n-2)}{(3n)!} x^{3n} + c_1 \sum_{n=1}^{\infty} \frac{2 \cdot 5 \cdot \cdots \cdot (3n-1)}{(3n+1)!} x^{3n+1}$

19. $y = 1 + \frac{(x-1)^2}{2!} + \frac{(x-1)^3}{3!} + \frac{(x-1)^4}{4!} + \frac{4(x-1)^5}{5!} + \cdots$ **21.** **a.** no; **b.** no

Chapter 12—Review Exercises, page 1006

1. $1 + x + x^2/2! + x^3/3!$ **3.** $\frac{1}{2} + \frac{\sqrt{3}}{2}\left(x - \frac{\pi}{6}\right) - \frac{1}{4}\left(x - \frac{\pi}{6}\right)^2 - \frac{\sqrt{3}}{12}\left(x - \frac{\pi}{6}\right)^3$
5. $-(x - \pi/2) - \frac{1}{3}(x - \pi/2)^3$ **7.** $3 + 2x - x^2 + x^3$ **9.** $(\pi/3)^6/6! \approx 0.00183$
11. $e/7! < 3/7! \approx 0.000595$
13. $\int_0^{1/2} (1 - x^4/2)\,dx = \frac{159}{320} = 0.496875$ with error bounded by $0.5^9/4! \approx 8.14 \times 10^{-5}$ **15.** $-1, 0, \frac{1}{3}, \frac{1}{2}, \frac{3}{5}$
17. $a_n = (2n - 1)/2^{n+2}$ **19.** converges to 0 **21.** converges to 0 **23.** converges to e^{-2}
25. unbounded, not monotonic **27.** bounded below by $\frac{2}{3}$ and above by 1, strictly increasing
29. bounded above by 2 and below by 0, strictly decreasing
31. bounded below by $\frac{-6}{5}$ and above by 1, strictly increasing
33. $(4^{11} - 1)/3 - 5 = 1{,}398{,}096$ **35.** $\frac{27}{16} - \frac{8}{75} = \frac{1897}{1200} \approx 1.58$ **37.** $\frac{79}{99}$ **39.** converges **41.** converges
43. diverges **45.** diverges **47.** converges **49.** converges **51.** conditionally convergent
53. conditionally convergent **55.** absolutely convergent **57.** conditionally convergent **59.** divergent
61. divergent **63.** $T_6 \approx 0.9021$ with error ≤ 0.00086; $S_{10} \approx 0.901116$ with error ≤ 0.000751
65. $9\frac{9}{11}$ hr $\approx$ 9:49:05 P.M. **67.** 3; $(-3, 3)$ **69.** 1; $[-1, 1]$ **71.** ∞; $(-\infty, \infty)$ **73.** 1; $(\frac{2}{3}, \frac{8}{3})$
75. 1; $(-1, 1)$ **77.** 0.461281 with error $\le 9.04 \times 10^{-6}$ **79.** 0.014509 with error ≤ 0.0000488

81. $\sum_{k=0}^{\infty} \frac{x^{k+2}}{k!}$ **83.** $\frac{1}{2} + \sum_{k=0}^{\infty} \frac{(-1)^k x^{2k} 2^{2k-1}}{(2k)!} = \frac{1}{2}\left(1 + \sum_{k=0}^{\infty} \frac{(-1)^k 4^k x^{2k}}{(2k)!}\right)$ **85.** $y(x) = 2\sum_{k=0}^{\infty} \frac{3^k x^k}{k!}$

87. $y(x) = \frac{7}{2}\sum_{k=0}^{\infty} \frac{x^{k+1}}{k!} - \frac{1}{2}\sum_{k=0}^{\infty} \frac{(-1)^k x^{2k+1}}{(2k+1)!}$

Computer Exercises, page 1008

1. **b.** $\sum_{n=1}^{10} \frac{1}{n^2} = \frac{1968329}{1270080} \approx 1.54976\,77311\,66540\,69035\,02142.$ $\sum_{n=1}^{100} \frac{1}{n^2} \approx 1.63498\,39001\,84892\,86507\,71695.$
$\sum_{n=1}^{500} \frac{1}{n^2} \approx 1.64293\,60655\,14894\,16980\,27009.$ **c.** $\sum_{n=501}^{\infty} \frac{1}{n^2} < \int_{500}^{\infty} \frac{dx}{x^2} < \sum_{n=500}^{\infty} \frac{1}{n^2}.$

d. $\left|\sum_{n=1}^{\infty} \frac{1}{n^2} - \sum_{n=1}^{500} \frac{1}{n^2}\right| = \sum_{n=501}^{\infty} \frac{1}{n^2} < \int_{500}^{\infty} \frac{dx}{x^2} = \frac{1}{500} = 0.002.$

3. **b.** $\gamma_n \le 1 - \ln 2 + \int_2^{\infty} \left(\frac{1}{x-1} - \frac{1}{x}\right) dx = 1$ **c.** 0.572256, 0.572617, 0.576716

d. error $\le \int_{1001}^{\infty} \left(\frac{1}{x-1} - \frac{1}{x}\right) dx = -\ln \frac{1000}{1001} \approx 0.000995$

e. error $\le \int_{n+1}^{\infty} \left(\frac{1}{x-1} - \frac{1}{x}\right) dx = \ln \frac{n+1}{n} \le 10^{-10}$; therefore $n \ge \frac{1}{e^{10^{-10}} - 1} \approx 10^{10}$

5. **b.** i. $a \approx 0.5$ ii. $a \approx 0.77$ iii. $a \approx 0.93$

APPENDIX 1, PAGE A-4

1. *Hint:* Verify the identity $[n^2(n+1)^2/4] + (n+1)^3 = (n+1)^2(n+2)^2/4$.

3. *Hint:* Apply the product rule to $xP_n(x) + b$.

5. *Hint:* Pick one element from the set and color it orange. Now count how many subsets contain this special orange element; count how many subsets do not contain it. Can there be any other subsets?

7. *Hint:* See Theorem 1.2.1(ii).

APPENDIX 2, PAGE A-12

1. 10 **3.** 35 **5.** 165 **7.** 1 **9.** $x^5 - 5x^4y + 10x^3y^2 - 10x^2y^3 + 5xy^4 - y^5$

11. $x^4 - 8x^3y + 24x^2y^2 - 32xy^3 + 16y^4$

13. $a^8 + 8a^7b + 28a^6b^2 + 56a^5b^3 + 70a^4b^4 + 56a^3b^5 + 28a^2b^6 + 8ab^7 + b^8$

15. $32a^5 - 240a^4b + 720a^3b^2 - 1080a^2b^3 + 810ab^4 - 243b^5$ **17.** $\frac{u^3}{27} - \frac{u^2v}{12} + \frac{uv^2}{16} - \frac{v^3}{64}$

19. $x^8 + 8x^6y + 24x^4y^2 + 32x^2y^3 + 16y^4$ **21.** $a^8 + 4a^6b^3 + 6a^4b^6 + 4a^2b^9 + b^{12}$

23. $x^3 + 6x^{5/2}y^{1/2} + 15x^2y + 20x^{3/2}y^{3/2} + 15xy^2 + 6x^{1/2}y^{5/2} + y^3$

25. $81x^2 + 324x^{3/2}y^{1/2} + 486xy + 324x^{1/2}y^{3/2} + 81y^2$ **27.** $a^3b^3 - 3a^2b^2cd + 3abc^2d^2 - c^3d^3$

29. $w^4 - 4w^3u/v + 6w^2u^2/v^2 - 4wu^3/v^3 + u^4/v^4$

31. $1 - 10a + 45a^2 - 120a^3 + 210a^4 - 252a^5 + 210a^6 - 120a^7 + 45a^8 - 10a^9 + a^{10}$ **33.** $1 + 3\sqrt{x} + 3x + x^{3/2}$

35. $25 + 20\sqrt{35} + 210 + 28\sqrt{35} + 49 = 284 + 48\sqrt{35}$ **37.** $1 + 4z^6 + 6z^{12} + 4z^{18} + z^{24}$

39. $u^4 + 4u^3v + 6u^2v^2 + 4uv^3 + v^4 - 8u^3 - 24u^2v - 24uv^2 - 8v^3 + 24u^2 + 48uv + 24v^2 - 32u - 32v + 16$

41. $x^4 + 4x^3y + 6x^2y^2 + 4xy^3 + y^4 + 4x^3z + 12x^2yz + 12xy^2z + 4y^3z + 6x^2z^2 + 12xyz^2 + 6y^2z^2 + 4xz^3 + 4yz^3 + z^4$

43. $x^{5n} + 5x^{4n}y^n + 10x^{3n}y^{2n} + 10x^{2n}y^{3n} + 5x^ny^{4n} + y^{5n}$ **45.** -120 **47.** 70

53. a. $100! \approx 9.3248476 \times 10^{157}$ **b.** $200! \approx 7.883293288 \times 10^{374}$ **55.** 1 7 21 35 35 21 7 1

APPENDIX 3, PAGE A-20

1. $9 - 7i$ **3.** 2 **5.** $-27 + 5i$ **7.** $5\sqrt{2}e^{\pi i/4}$ **9.** $3\sqrt{2}e^{-\pi i/4}$ **11.** $6e^{\pi i/6}$ **13.** $8e^{-\pi i/6}$

15. $2e^{-2\pi i/3}$ **17.** -2 **19.** $-1/(2\sqrt{2}) - (1/2\sqrt{2})i$ **21.** $-2\sqrt{3} + 2i$ **23.** $-\frac{3}{2} - (3\sqrt{3}/2)i$

25. $\cos 1 + i \sin 1 \approx 0.54 + 0.84i$ **27.** $4 - 6i$ **29.** $7i$ **31.** $2e^{-\pi i/7}$ **33.** $3e^{4\pi i/11}$

39. circle of radius a centered at z_0 if $a > 0$, single point z_0 if $a = 0$, empty set if $a < 0$

APPENDIX 4, PAGE A-23

5. -6

APPENDIX 5, PAGE A-34

1. yes; $\delta = \frac{1}{216}$ (better: $\delta = \frac{1}{50}$)

3. Yes, but $a = b = 1$ is not possible. You need $a + b < 1$. If $a = b = \frac{1}{3}$, then $\delta = \frac{1}{4}$.

5. yes; $\delta = 1$ **7.** Yes; but you need $b < 1$. If $b = \frac{1}{2}$, then $\delta = 1$.

9. yes; $\delta = \frac{1}{2}$ (better: $\delta = +\infty$; there is a unique solution defined for $-\infty < t < \infty$)

11. a. $k = 1$; **b.** $k = 1$; **c.** $k = 3$ $\left(\text{a smaller constant is } \sup_{|x|\le 1}\left|2x \sin\frac{1}{x} - \cos\frac{1}{x}\right|\right)$; **d.** $972\sqrt{3} \approx 1683$.

13. $x_n(t) = 3\sum_{k=0}^{n}\frac{(-1)^k x^k}{k!}$

APPENDIX 6, PAGE A-42

1. Let L be a linearly independent set in V. Let S be the collection of all linearly independent subsets of V, partially ordered by inclusion, such that every set in S contains L. The proof then follows as in the proof of Theorem 2.

3. The result is true for $n = 2$. Assume it is true for $n = k$. Consider the $k + 1$ sets $A_1, A_2, \ldots, A_k, A_{k+1}$ in a chain. The first k sets form a chain, and, by the induction assumption, one of them contains the other $k - 1$ sets. Call this set A_i. Then either $A_i \subseteq A_{k+1}$ or $A_{k+1} \subseteq A_i$. In either case we have found a set that contains the other k sets and the result is true for $n = k + 1$. This completes the induction proof.

INDEX